NIST Handbook of Mathematical Functions

Modern developments in theoretical and applied science depend on knowledge of the properties of mathematical functions, from elementary trigonometric functions to the multitude of special functions. These functions appear whenever natural phenomena are studied, engineering problems are formulated, and numerical simulations are performed. They also crop up in statistics, financial models, and economic analysis. Using them effectively requires practitioners to have ready access to a reliable collection of their properties.

This handbook results from a 10-year project conducted by the National Institute of Standards and Technology with an international group of expert authors and validators. Printed in full color, it is destined to replace its predecessor, the classic but long-outdated *Handbook of Mathematical Functions*, edited by Abramowitz and Stegun. Included with every copy of the book is a CD with a searchable PDF.

Frank W. J. Olver is Professor Emeritus in the Institute for Physical Science and Technology and the Department of Mathematics at the University of Maryland. From 1961 to 1986 he was a Mathematician at the National Bureau of Standards in Washington, D.C. Professor Olver has published 76 papers in refereed and leading mathematics journals, and he is the author of *Asymptotics and Special Functions* (1974). He has served as editor of *SIAM Journal on Numerical Analysis*, *SIAM Journal on Mathematical Analysis*, *Mathematics of Computation*, *Methods and Applications of Analysis*, and the *NBS Journal of Research*.

Daniel W. Lozier leads the Mathematical Software Group in the Mathematical and Computational Sciences Division of NIST. He received his Ph.D. in applied mathematics from the University of Maryland in 1979 and has been at NIST since 1970. He is an active member of the SIAM Activity Group on Orthogonal Polynomials and Special Functions, having served two terms as chair and one as vice-chair, and currently is serving as secretary. He has been an editor of *Mathematics of Computation* and the *NIST Journal of Research*.

Ronald F. Boisvert leads the Mathematical and Computational Sciences Division of the Information Technology Laboratory at NIST. He received his Ph.D. in computer science from Purdue University in 1979 and has been at NIST since then. He has served as editor-in-chief of the *ACM Transactions on Mathematical Software*. He is currently co-chair of the Publications Board of the Association for Computing Machinery (ACM) and chair of the International Federation for Information Processing (IFIP) Working Group 2.5 (Numerical Software).

Charles W. Clark received his Ph.D. in physics from the University of Chicago in 1979. He is a member of the U.S. Senior Executive Service and Chief of the Electron and Optical Physics Division and acting Group Leader of the NIST Synchrotron Ultraviolet Radiation Facility (SURF III). Clark serves as Program Manager for Atomic and Molecular Physics at the U.S. Office of Naval Research and is a Fellow of the Joint Quantum Institute of NIST and the University of Maryland at College Park and a Visiting Professor at the National University of Singapore.

Rainbow over Woolsthorpe Manor

From the frontispiece of the *Notes and Records of the Royal Society of London*, v. 36 (1981–82), with permission. Photograph by Dr. Roy L. Bishop, Physics Department, Acadia University, Nova Scotia, Canada, with permission.

Commentary

The faint line below the main colored arc is a *supernumerary rainbow*, produced by the interference of different sun-rays traversing a raindrop and emerging in the same direction. For each color, the intensity profile across the rainbow is an Airy function. Airy invented his function in 1838 precisely to describe this phenomenon more accurately than Young had done in 1800 when pointing out that supernumerary rainbows require the wave theory of light and are impossible to explain with Newton's picture of light as a stream of independent corpuscles. The house in the picture is Newton's birthplace.

Sir Michael V. Berry
H. H. Wills Physics Laboratory
Bristol, United Kingdom

NIST Handbook of Mathematical Functions

Frank W. J. Olver
Editor-in-Chief and Mathematics Editor

Daniel W. Lozier
General Editor

Ronald F. Boisvert
Information Technology Editor

Charles W. Clark
Physical Sciences Editor

National Institute of
Standards and Technology
U.S. Department of Commerce

and

CAMBRIDGE UNIVERSITY PRESS
Cambridge, New York, Melbourne, Madrid, Cape Town, Singapore,
São Paulo, Delhi, Dubai, Tokyo

Cambridge University Press
32 Avenue of the Americas, New York, NY 10013-2473, USA

www.cambridge.org
Information on this title: www.cambridge.org/9780521140638

© National Institute of Standards and Technology 2010

Pursuant to Title 17 USC 105, the National Institute of Standards and Technology (NIST), United States Department of Commerce, is authorized to receive and hold copyrights transferred to it by assignment or otherwise. Authors of the work appearing in this publication have assigned copyright to the work to NIST, United States Department of Commerce, as represented by the Secretary of Commerce. These works are owned by NIST.

Limited copying and internal distribution of the content of this publication is permitted for research and teaching. Reproduction, copying, or distribution for any commercial purpose is strictly prohibited. Bulk copying, reproduction, or redistribution in any form is not permitted. Questions regarding this copyright policy should be directed to NIST.

While NIST has made every effort to ensure the accuracy and reliability of the information in this publication, it is expressly provided "as is." NIST and Cambridge University Press together and separately make no warranty of any type, including warranties of merchantability or fitness for a particular purpose. NIST and Cambridge University Press together and separately make no warranties or representations as to the correctness, accuracy, or reliability of the information. As a condition of using it, you explicitly release NIST and Cambridge University Press from any and all liability for any damage of any type that may result from errors or omissions.

Certain products, commercial and otherwise, are mentioned in this publication. These mentions are for informational purposes only, and do not imply recommendation or endorsement by NIST.
All rights reserved.

This publication is in copyright. Subject to statutory exception
and to the provisions of relevant collective licensing agreements,
no reproduction of any part may take place without the written
permission of Cambridge University Press.

First published 2010

Printed in the United States of America

A catalog record for this publication is available from the British Library.

ISBN 978-0-521-19225-5 Hardback
ISBN 978-0-521-14063-8 Paperback

Additional resources for this publication at http://dlmf.nist.gov/.

Cambridge University Press and the National Institute of Standards
and Technology have no responsibility for the persistence or
accuracy of URLs for external or third-party Internet Web sites
referred to in this publication and do not guarantee that any
content on such Web sites is, or will remain, accurate or
appropriate.

Contents

Foreword vii
Preface ix
Mathematical Introduction xiii

1. **Algebraic and Analytic Methods**
   R. Roy, F. W. J. Olver, R. A. Askey, R. Wong . . . 1
2. **Asymptotic Approximations**
   F. W. J. Olver, R. Wong 41
3. **Numerical Methods**
   N. M. Temme 71
4. **Elementary Functions**
   R. Roy, F. W. J. Olver 103
5. **Gamma Function**
   R. A. Askey, R. Roy 135
6. **Exponential, Logarithmic, Sine, and Cosine Integrals**
   N. M. Temme 149
7. **Error Functions, Dawson's and Fresnel Integrals**
   N. M. Temme 159
8. **Incomplete Gamma and Related Functions**
   R. B. Paris 173
9. **Airy and Related Functions**
   F. W. J. Olver 193
10. **Bessel Functions**
    F. W. J. Olver, L. C. Maximon 215
11. **Struve and Related Functions**
    R. B. Paris 287
12. **Parabolic Cylinder Functions**
    N. M. Temme 303
13. **Confluent Hypergeometric Functions**
    A. B. Olde Daalhuis 321
14. **Legendre and Related Functions**
    T. M. Dunster 351
15. **Hypergeometric Function**
    A. B. Olde Daalhuis 383
16. **Generalized Hypergeometric Functions and Meijer G-Function**
    R. A. Askey, A. B. Olde Daalhuis 403
17. **q-Hypergeometric and Related Functions**
    G. E. Andrews 419
18. **Orthogonal Polynomials**
    T. H. Koornwinder, R. Wong, R. Koekoek, R. F. Swarttouw 435
19. **Elliptic Integrals**
    B. C. Carlson 485
20. **Theta Functions**
    W. P. Reinhardt, P. L. Walker 523
21. **Multidimensional Theta Functions**
    B. Deconinck 537
22. **Jacobian Elliptic Functions**
    W. P. Reinhardt, P. L. Walker 549
23. **Weierstrass Elliptic and Modular Functions**
    W. P. Reinhardt, P. L. Walker 569
24. **Bernoulli and Euler Polynomials**
    K. Dilcher 587
25. **Zeta and Related Functions**
    T. M. Apostol 601
26. **Combinatorial Analysis**
    D. M. Bressoud 617
27. **Functions of Number Theory**
    T. M. Apostol 637
28. **Mathieu Functions and Hill's Equation**
    G. Wolf 651
29. **Lamé Functions**
    H. Volkmer 683
30. **Spheroidal Wave Functions**
    H. Volkmer 697
31. **Heun Functions**
    B. D. Sleeman, V. B. Kuznetsov 709
32. **Painlevé Transcendents**
    P. A. Clarkson 723
33. **Coulomb Functions**
    I. J. Thompson 741
34. **$3j, 6j, 9j$ Symbols**
    L. C. Maximon 757
35. **Functions of Matrix Argument**
    D. St. P. Richards 767
36. **Integrals with Coalescing Saddles**
    M. V. Berry, C. J. Howls 775

Bibliography 795
Notations 873
Index 887

Foreword

In 1964 the National Institute of Standards and Technology[1] published the *Handbook of Mathematical Functions with Formulas, Graphs, and Mathematical Tables*, edited by Milton Abramowitz and Irene A. Stegun. That 1046-page tome proved to be an invaluable reference for the many scientists and engineers who use the special functions of applied mathematics in their day-to-day work, so much so that it became the most widely distributed and most highly cited NIST publication in the first 100 years of the institution's existence.[2] The success of the original handbook, widely referred to as "Abramowitz and Stegun" ("A&S"), derived not only from the fact that it provided critically useful scientific data in a highly accessible format, but also because it served to standardize definitions and notations for special functions. The provision of standard reference data of this type is a core function of NIST.

Much has changed in the years since A&S was published. Certainly, advances in applied mathematics have continued unabated. However, we have also seen the birth of a new age of computing technology, which has not only changed how we utilize special functions, but also how we communicate technical information. The document you are now holding, or the Web page you are now reading, represents an effort to extend the legacy of A&S well into the 21st century. The new printed volume, the *NIST Handbook of Mathematical Functions*, serves a similar function as the original A&S, though it is heavily updated and extended. The online version, the *NIST Digital Library of Mathematical Functions (DLMF)*, presents the same technical information along with extensions and innovative interactive features consistent with the new medium. The DLMF may well serve as a model for the effective presentation of highly mathematical reference material on the Web.

The production of these new resources has been a very complex undertaking some 10 years in the making. This could not have been done without the cooperation of many mathematicians, information technologists, and physical scientists both within NIST and externally. Their unfailing dedication is acknowledged deeply and gratefully. Particular attention is called to the generous support of the National Science Foundation, which made possible the participation of experts from academia and research institutes worldwide.

Dr. Patrick D. Gallagher
Director, NIST
November 20, 2009
Gaithersburg, Maryland

---

[1] Then known as the National Bureau of Standards.
[2] D. R. Lide (ed.), *A Century of Excellence in Measurement, Standards, and Technology*, CRC Press, 2001.

Preface

The *NIST Handbook of Mathematical Functions*, together with its Web counterpart, the *NIST Digital Library of Mathematical Functions (DLMF)*, is the culmination of a project that was conceived in 1996 at the National Institute of Standards and Technology (NIST). The project had two equally important goals: to develop an authoritative replacement for the highly successful *Handbook of Mathematical Functions with Formulas, Graphs, and Mathematical Tables*, published in 1964 by the National Bureau of Standards (M. Abramowitz and I. A. Stegun, editors); and to disseminate essentially the same information from a public Web site operated by NIST. The new Handbook and DLMF are the work of many hands: editors, associate editors, authors, validators, and numerous technical experts. A summary of the responsibilities of these groups may help in understanding the structure and results of this project.

Executive responsibility was vested in the editors: Frank W. J. Olver (University of Maryland, College Park, and NIST), Daniel W. Lozier (NIST), Ronald F. Boisvert (NIST), and Charles W. Clark (NIST). Olver was responsible for organizing and editing the mathematical content after receiving it from the authors; for communicating with the associate editors, authors, validators, and other technical experts; and for assembling the **Notations** section and the **Index**. In addition, Olver was author or co-author of five chapters. Lozier directed the NIST research, technical, and support staff associated with the project, administered grants and contracts, together with Boisvert compiled the **Software** sections for the Web version of the chapters, conducted editorial and staff meetings, represented the project within NIST and at professional meetings in the United States and abroad, and together with Olver carried out the day-to-day development of the project. Boisvert and Clark were responsible for advising and assisting in matters related to the use of information technology and applications of special functions in the physical sciences (and elsewhere); they also participated in the resolution of major administrative problems when they arose.

The associate editors are eminent domain experts who were recruited to advise the project on strategy, execution, subject content, format, and presentation, and to help identify and recruit suitable candidate authors and validators. The associate editors were:

Richard A. Askey
University of Wisconsin, Madison

Michael V. Berry
University of Bristol

Walter Gautschi (resigned 2002)
Purdue University

Leonard C. Maximon
George Washington University

Morris Newman
University of California, Santa Barbara

Ingram Olkin
Stanford University

Peter Paule
Johannes Kepler University

William P. Reinhardt
University of Washington

Nico M. Temme
Centrum voor Wiskunde en Informatica

Jet Wimp (resigned 2001)
Drexel University

The technical information provided in the Handbook and DLMF was prepared by subject experts from around the world. They are identified on the title pages of the chapters for which they served as authors and in the table of Contents.

The validators played a critical role in the project, one that was absent in its 1964 counterpart: to provide critical, independent reviews during the development of each chapter, with attention to accuracy and appropriateness of subject coverage. These reviews have contributed greatly to the quality of the product. The validators were:

T. M. Apostol
California Institute of Technology

A. R. Barnett
University of Waikato, New Zealand

A. I. Bobenko
Technische Universität, Berlin

B. B. L. Braaksma
University of Groningen

D. M. Bressoud
Macalester College

B. C. Carlson
Iowa State University

B. Deconinck
University of Washington

T. M. Dunster
University of California, San Diego

A. Gil
Universidad de Cantabria

A. R. Its
Indiana University–Purdue University, Indianapolis

B. R. Judd
Johns Hopkins University

R. Koekoek
Delft University of Technology

T. H. Koornwinder
University of Amsterdam

R. J. Muirhead
Pfizer Global R&D

E. Neuman
University of Illinois, Carbondale

A. B. Olde Daalhuis
University of Edinburgh

R. B. Paris
University of Abertay Dundee

R. Roy
Beloit College

S. N. M. Ruijsenaars
University of Leeds

J. Segura
Universidad de Cantabria

R. F. Swarttouw
Vrije Universiteit Amsterdam

N. M. Temme
Centrum voor Wiskunde en Informatica

H. Volkmer
University of Wisconsin, Milwaukee

G. Wolf
Universität Duisberg-Essen

R. Wong
City University of Hong Kong

All of the mathematical information contained in the Handbook is also contained in the DLMF, along with additional features such as more graphics, expanded tables, and higher members of some families of formulas; in consequence, in the Handbook there are occasional gaps in the numbering sequences of equations, tables, and figures. The Web address where additional DLMF content can be found is printed in blue at appropriate places in the Handbook. The home page of the DLMF is accessible at http://dlmf.nist.gov/.

The DLMF has been constructed specifically for effective Web usage and contains features unique to Web presentation. The Web pages contain many active links, for example, to the definitions of symbols within the DLMF, and to external sources of reviews, full texts of articles, and items of mathematical software. Advanced capabilities have been developed at NIST for the DLMF, and also as part of a larger research effort intended to promote the use of the Web as a tool for doing mathematics. Among these capabilities are: a facility to allow users to download LaTeX and MathML encodings of every formula into document processors and software packages (eventually, a fully semantic downloading capability may be possible); a search engine that allows users to locate formulas based on queries expressed in mathematical notation; and user-manipulable 3-dimensional color graphics.

Production of the Handbook and DLMF was a mammoth undertaking, made possible by the dedicated leadership of Bruce R. Miller (NIST), Bonita V. Saunders (NIST), and Abdou S. Youssef (George Washington University and NIST). Miller was responsible for information architecture, specializing LaTeX for the needs of the project, translation from LaTeX to MathML, and the search interface. Saunders was responsible for mesh generation for curves and surfaces, data computation and validation, graphics production, and interactive Web visualization. Youssef was responsible for mathematics search indexing and query processing. They were assisted by the following NIST staff: Marjorie A. McClain (LaTeX, bibliography), Joyce E. Conlon (bibliography), Gloria Wiersma (LaTeX), Qiming Wang (graphics generation, graphics viewers), and Brian Antonishek (graphics viewers).

The editors acknowledge the many other individuals who contributed to the project in a variety of ways. Among the research, technical, and support staff at NIST these are B. K. Alpert, T. M. G. Arrington, R. Bickel, B. Blaser, P. T. Boggs, S. Burley, G. Chu, A. Dienstfrey, M. J. Donahue, K. R. Eberhardt, B. R. Fabijonas, M. Fancher, S. Fletcher, J. Fowler, S. P. Frechette, C. M. Furlani, K. B. Gebbie, C. R. Hagwood, A. N. Heckert, M. Huber, P. K. Janert, R. N. Kacker, R. F. Kayser, P. M. Ketcham, E. Kim, M. J. Lieberman, R. R. Lipman, M. S. Madsen, E. A. P. Mai, W. Mehuron, P. J. Mohr, S. Olver, D. R. Penn, S. Phoha, A. Possolo, S. P. Ressler, M. Rubin, J. Rumble, C. A. Schanzle, B. I. Schneider, N. Sedransk, E. L. Shirley, G. W. Stewart, C. P. Sturrock, G. Thakur, S. Wakid, and S. F. Zevin. Individuals from outside NIST are S. S. Antman, A. M. Ashton, C. M. Bender, J. J. Benedetto, R. L. Bishop, J. M. Borwein, H. W. Braden, C. Brezinski, F. Chyzak, J. N. L. Connor, R. Cools, A. Cuyt, I. Daubechies, P. J. Davis, C. F. Dunkl, J. P. Goedbloed, B. Gordon, J. W. Jenkins, L. H. Kellogg, C. D. Kemp, K. S. Kölbig, S. G. Krantz, M. D. Kruskal, W. Lay, D. A. Lutz, E. L. Mansfield, G. Marsaglia, B. M. McCoy, W. Miller, Jr., M. E. Muldoon, S. P. Novikov, P. J. Olver, W. C. Parke, M. Petkovsek, W. H. Reid, B. Salvy, C. Schneider, M. J. Seaton, N. C. Severo, I. A. Stegun, F. Stenger, M. Steuerwalt, W. G. Strang, P. R. Turner, J. Van Deun, M. Vuorinen, E. J. Weniger, H. Wiersma, R. C. Winther, D. B. Zagier, and M. Zelen. Undoubtedly, the editors have overlooked some individuals who contributed, as is inevitable in a large long-lasting project. Any oversight is unintentional, and the editors apologize in advance.

The project was funded in part by NSF Award 9980036, administered by the NSF's Knowledge and Distributed Intelligence Program. Within NIST financial resources and staff were committed by the Information Technology Laboratory, Physics Laboratory, Systems Integration for Manufacturing Applications Program of the Manufacturing Engineering Laboratory, Standard Reference Data Program, and Advanced Technology Program.

Notwithstanding the great care that has been exercised by the editors, authors, validators, and the NIST staff, it is almost inevitable that in a work of the magnitude and scope of the NIST Handbook and DLMF errors will still be present. Users need to be aware that none of these individuals nor the National Institute of Standards and Technology can assume responsibility for any possible consequences of such errors.

Lastly, the editors appreciate the skill, and long experience, that was brought to bear by the publisher, Cambridge University Press, on the production and publication of the new Handbook.

Frank W. J. Olver
Editor-in-Chief and Mathematics Editor

Daniel W. Lozier
General Editor

Ronald F. Boisvert
Information Technology Editor

Charles W. Clark
Physical Sciences Editor

Mathematical Introduction

Organization and Objective

The mathematical content of the *NIST Handbook of Mathematical Functions* has been produced over a ten-year period. This part of the project has been carried out by a team comprising the mathematics editor, authors, validators, and the NIST professional staff. Also, valuable initial advice on all aspects of the project was provided by ten external associate editors.

The NIST Handbook has essentially the same objective as the *Handbook of Mathematical Functions* that was issued in 1964 by the National Bureau of Standards as Number 55 in the NBS Applied Mathematics Series (AMS). This objective is to provide a reference tool for researchers and other users in applied mathematics, the physical sciences, engineering, and elsewhere who encounter special functions in the course of their everyday work.

The mathematical project team has endeavored to take into account the hundreds of research papers and numerous books on special functions that have appeared since 1964. As a consequence, in addition to providing more information about the special functions that were covered in AMS 55, the NIST Handbook includes several special functions that have appeared in the interim in applied mathematics, the physical sciences, and engineering, as well as in other areas. See, for example, Chapters 16, 17, 18, 19, 21, 27, 29, 31, 32, 34, 35, and 36.

Two other ways in which this Handbook differs from AMS 55, and other handbooks, are as follows.

First, the editors instituted a validation process for the whole technical content of each chapter. This process greatly extended normal editorial checking procedures. All chapters went through several drafts (nine in some cases) before the authors, validators, and editors were fully satisfied.

Secondly, as described in the **Preface**, a Web version (the NIST DLMF) is also available.

Methodology

The first three chapters of the NIST Handbook and DLMF are methodology chapters that provide detailed coverage of, and references for, mathematical topics that are especially important in the theory, computation, and application of special functions. (These chapters can also serve as background material for university graduate courses in complex variables, classical analysis, and numerical analysis.)

Particular care is taken with topics that are not dealt with sufficiently thoroughly from the standpoint of this Handbook in the available literature. These include, for example, multivalued functions of complex variables, for which new definitions of branch points and principal values are supplied (§§1.10(vi), 4.2(i)); the Dirac delta (or delta function), which is introduced in a more readily comprehensible way for mathematicians (§1.17); numerically satisfactory solutions of differential and difference equations (§§2.7(iv), 2.9(i)); and numerical analysis for complex variables (Chapter 3).

In addition, there is a comprehensive account of the great variety of analytical methods that are used for deriving and applying the extremely important asymptotic properties of the special functions, including double asymptotic properties (Chapter 2 and §§10.41(iv), 10.41(v)).

Notation for the Special Functions

The first section in each of the special function chapters (Chapters 5–36) lists notation that has been adopted for the functions in that chapter. This section may also include important alternative notations that have appeared in the literature. With a few exceptions the adopted notations are the same as those in standard applied mathematics and physics literature.

The exceptions are ones for which the existing notations have drawbacks. For example, for the hypergeometric function we often use the notation $\mathbf{F}(a,b;c;z)$ (§15.2(i)) in place of the more conventional $_2F_1(a,b;c;z)$ or $F(a,b;c;z)$. This is because $\mathbf{F}$ is akin to the notation used for Bessel functions (§10.2(ii)), inasmuch as $\mathbf{F}$ is an entire function of each of its parameters a, b, and c: this results in fewer restrictions and simpler equations. Similarly in the case of confluent hypergeometric functions (§13.2(i)).

Other examples are: (a) the notation for the Ferrers functions—also known as associated Legendre functions on the cut—for which existing notations can easily be confused with those for other associated Legendre functions (§14.1); (b) the spherical Bessel functions for which existing notations are unsymmetric and inelegant (§§10.47(i) and 10.47(ii)); and (c) elliptic integrals for which both Legendre's forms and the more recent symmetric forms are treated fully (Chapter 19).

The **Notations** section beginning on p. 873 includes all the notations for the special functions adopted in this Handbook. In the corresponding section for the DLMF some of the alternative notations that appear in the first section of the special function chapters are also included.

Common Notations and Definitions

$\mathbb{C}$	complex plane (excluding infinity).
D	decimal places.
det	determinant.
$\delta_{j,k}$ or δ_{jk}	Kronecker delta: 0 if $j \neq k$; 1 if $j = k$.
Δ (or Δ_x)	forward difference operator: $\Delta f(x) = f(x+1) - f(x)$.
∇ (or ∇_x)	backward difference operator: $\nabla f(x) = f(x) - f(x-1)$. (See also del operator in the **Notations** section.)
empty sums	zero.
empty products	unity.
$\in$	element of.
$\notin$	not an element of.
$\forall$	for every.
$\implies$	implies.
$\iff$	is equivalent to.
$n!$	factorial: $1 \cdot 2 \cdot 3 \cdots n$ if $n = 1, 2, 3, \ldots$; 1 if $n = 0$.
$n!!$	double factorial: $2 \cdot 4 \cdot 6 \cdots n$ if $n = 2, 4, 6, \ldots$; $1 \cdot 3 \cdot 5 \cdots n$ if $n = 1, 3, 5, \ldots$; 1 if $n = 0, -1$.
$\lfloor x \rfloor$	floor or integer part: the integer such that $x - 1 < \lfloor x \rfloor \leq x$, with x real.
$\lceil x \rceil$	ceiling: the integer such that $x \leq \lceil x \rceil < x + 1$, with x real.
$f(z)\|_C = 0$	$f(z)$ is continuous at all points of a simple closed contour C in $\mathbb{C}$.
$< \infty$	is finite, or converges.
$\gg$	much greater than.
$\Im$	imaginary part.
iff	if and only if.
inf	greatest lower bound (infimum).
sup	least upper bound (supremum).
$\cap$	intersection.
$\cup$	union.
(a, b)	open interval in $\mathbb{R}$, or open straight-line segment joining a and b in $\mathbb{C}$.
$[a, b]$	closed interval in $\mathbb{R}$, or closed straight-line segment joining a and b in $\mathbb{C}$.
$(a, b]$ or $[a, b)$	half-closed intervals.
$\subset$	is contained in.
$\subseteq$	is, or is contained in.
lim inf	least limit point.
$[a_{j,k}]$ or $[a_{jk}]$	matrix with (j, k)th element $a_{j,k}$ or a_{jk}.
$\mathbf{A}^{-1}$	inverse of matrix $\mathbf{A}$.
tr $\mathbf{A}$	trace of matrix $\mathbf{A}$.
$\mathbf{A}^\mathrm{T}$	transpose of matrix $\mathbf{A}$.
$\mathbf{I}$	unit matrix.
mod or modulo	$m \equiv n \pmod{p}$ means p divides $m - n$, where m, n, and p are positive integers with $m > n$.
$\mathbb{N}$	set of all positive integers.
$(\alpha)_n$	Pochhammer's symbol: $\alpha(\alpha+1)(\alpha+2)\cdots(\alpha+n-1)$ if $n = 1, 2, 3, \ldots$; 1 if $n = 0$.
$\mathbb{Q}$	set of all rational numbers.
$\mathbb{R}$	real line (excluding infinity).
$\Re$	real part.
res	residue.
S	significant figures.
sign x	-1 if $x < 0$; 0 if $x = 0$; 1 if $x > 0$.
$\setminus$	set subtraction.
$\mathbb{Z}$	set of all integers.
$n\mathbb{Z}$	set of all integer multiples of n.

Graphics

Special functions with one real variable are depicted graphically with conventional two-dimensional (2D) line graphs. See, for example, Figures 10.3.1–10.3.4.

With two real variables, special functions are depicted as 3D surfaces, with vertical height corresponding to the value of the function, and coloring added to emphasize the 3D nature. See Figures 10.3.5–10.3.8 for examples.

Special functions with a complex variable are depicted as colored 3D surfaces in a similar way to functions of two real variables, but with the vertical height corresponding to the modulus (absolute value) of the function. See, for example, Figures 5.3.4–5.3.6. However, in many cases the coloring of the surface is chosen instead to indicate the quadrant of the plane to which the phase of the function belongs, thereby achieving a 4D effect. In these cases the phase colors that correspond to the 1st, 2nd, 3rd, and 4th quadrants are arranged in alphabetical order: blue, green, red, and yellow, respectively, and a "Quadrant Colors" icon appears alongside the figure. See, for example, Figures 10.3.9–10.3.16.

Lastly, users may notice some lack of smoothness in the color boundaries of some of the 4D-type surfaces; see, for example, Figure 10.3.9. This nonsmoothness arises because the mesh that was used to generate the figure was optimized only for smoothness of the surface, and not for smoothness of the color boundaries.

Applications

All of the special function chapters include sections devoted to mathematical, physical, and sometimes other applications of the main functions in the chapter. The purpose of these sections is simply to illustrate the importance of the functions in other disciplines; no attempt is made to provide exhaustive coverage.

Computation

All of the special function chapters contain sections that describe available methods for computing the main functions in the chapter, and most also provide references to numerical tables of, and approximations for, these functions. In addition, the DLMF provides references to research papers in which software is developed, together with links to sites where the software can be obtained.

In referring to the numerical tables and approximations we use notation typified by $x = 0(.05)1$, 8D or 8S. This means that the variable x ranges from 0 to 1 in intervals of 0.05, and the corresponding function values are tabulated to 8 decimal places or 8 significant figures.

Another numerical convention is that decimals followed by dots are unrounded; without the dots they are rounded. For example, to 4D π is 3.1415... (unrounded) and 3.1416 (rounded).

Verification

For all equations and other technical information this Handbook and the DLMF either provide references to the literature for proof or describe steps that can be followed to construct a proof. In the Handbook this information is grouped at the section level and appears under the heading **Sources** in the **References** section. In the DLMF this information is provided in pop-up windows at the subsection level.

For equations or other technical information that appeared previously in AMS 55, the DLMF usually includes the corresponding AMS 55 equation number, or other form of reference, together with corrections, if needed. However, none of these citations are to be regarded as supplying proofs.

Special Acknowledgment

I pay tribute to my friend and predecessor Milton Abramowitz. His genius in the creation of the *National Bureau of Standards Handbook of Mathematical Functions* paid enormous dividends to the world's scientific, mathematical, and engineering communities, and paved the way for the development of the *NIST Handbook of Mathematical Functions* and *NIST Digital Library of Mathematical Functions*.

Frank W. J. Olver, *Mathematics Editor*

Chapter 1
Algebraic and Analytic Methods

R. Roy[1], F. W. J. Olver[2], R. A. Askey[3] and R. Wong[4]

Notation   2
- 1.1   Special Notation   2

Areas   2
- 1.2   Elementary Algebra   2
- 1.3   Determinants   3
- 1.4   Calculus of One Variable   4
- 1.5   Calculus of Two or More Variables   7
- 1.6   Vectors and Vector-Valued Functions   9
- 1.7   Inequalities   12
- 1.8   Fourier Series   13
- 1.9   Calculus of a Complex Variable   14
- 1.10   Functions of a Complex Variable   18
- 1.11   Zeros of Polynomials   22
- 1.12   Continued Fractions   24
- 1.13   Differential Equations   25
- 1.14   Integral Transforms   27
- 1.15   Summability Methods   33
- 1.16   Distributions   35
- 1.17   Integral and Series Representations of the Dirac Delta   37

References   39

[1] Department of Mathematics and Computer Science, Beloit College, Beloit, Wisconsin.
[2] Institute for Physical Science and Technology and Department of Mathematics, University of Maryland, College Park, Maryland.
[3] Department of Mathematics, University of Wisconsin, Madison, Wisconsin.
[4] Liu Bie Ju Centre for Mathematical Sciences, City University of Hong Kong, Kowloon, Hong Kong.
  Acknowledgments: The authors thank Leonard Maximon and William Parke for their assistance with the writing of §1.17.
  Copyright © 2009 National Institute of Standards and Technology. All rights reserved.

Notation

1.1 Special Notation

(For other notation see pp. xiv and 873.)

x, y	real variables.
z	real variable in §§1.5–1.6.
z, w	complex variables in §§1.9–1.11.
j, k, ℓ	integers.
m, n	nonnegative integers, unless specified otherwise.
$\langle f, g \rangle$	distribution.
deg	degree.
primes	derivatives with respect to the variable, except where indicated otherwise.

Areas

1.2 Elementary Algebra

1.2(i) Binomial Coefficients

In (1.2.1)–(1.2.5) k and n are nonnegative integers and $k \leq n$.

1.2.1
$$\binom{n}{k} = \frac{n!}{(n-k)!k!} = \binom{n}{n-k}.$$

Binomial Theorem

1.2.2
$$(a+b)^n = a^n + \binom{n}{1}a^{n-1}b + \binom{n}{2}a^{n-2}b^2 + \cdots + \binom{n}{n-1}ab^{n-1} + b^n.$$

1.2.3
$$\binom{n}{0} + \binom{n}{1} + \cdots + \binom{n}{n} = 2^n.$$

1.2.4
$$\binom{n}{0} - \binom{n}{1} + \cdots + (-1)^n\binom{n}{n} = 0.$$

1.2.5
$$\binom{n}{0} + \binom{n}{2} + \binom{n}{4} + \cdots + \binom{n}{k} = 2^{n-1},$$

where k is n or $n-1$ according as n is even or odd.

In (1.2.6)–(1.2.9) k and m are nonnegative integers and n is unrestricted.

1.2.6
$$\binom{n}{k} = \frac{n(n-1)\cdots(n-k+1)}{k!} = \frac{(-1)^k(-n)_k}{k!} = (-1)^k\binom{k-n-1}{k}.$$

1.2.7
$$\binom{n+1}{k} = \binom{n}{k} + \binom{n}{k-1}.$$

1.2.8
$$\sum_{k=0}^{m}\binom{n+k}{k} = \binom{n+m+1}{m}.$$

1.2.9
$$\binom{n}{0} - \binom{n}{1} + \cdots + (-1)^m\binom{n}{m} = (-1)^m\binom{n-1}{m}.$$

1.2(ii) Finite Series

Arithmetic Progression

1.2.10
$$a + (a+d) + (a+2d) + \cdots + (a+(n-1)d) = na + \tfrac{1}{2}n(n-1)d = \tfrac{1}{2}n(a+\ell),$$

where ℓ = last term of the series = $a + (n-1)d$.

Geometric Progression

1.2.11
$$a + ax + ax^2 + \cdots + ax^{n-1} = \frac{a(1-x^n)}{1-x}, \qquad x \neq 1.$$

1.2(iii) Partial Fractions

Let $\alpha_1, \alpha_2, \ldots, \alpha_n$ be distinct constants, and $f(x)$ be a polynomial of degree less than n. Then

1.2.12
$$\frac{f(x)}{(x-\alpha_1)(x-\alpha_2)\cdots(x-\alpha_n)} = \frac{A_1}{x-\alpha_1} + \frac{A_2}{x-\alpha_2} + \cdots + \frac{A_n}{x-\alpha_n},$$

where

1.2.13
$$A_j = \frac{f(\alpha_j)}{\prod_{k \neq j}(\alpha_j - \alpha_k)}.$$

Also,

1.2.14
$$\frac{f(x)}{(x-\alpha_1)^n} = \frac{B_1}{x-\alpha_1} + \frac{B_2}{(x-\alpha_1)^2} + \cdots + \frac{B_n}{(x-\alpha_1)^n},$$

where

1.2.15
$$B_j = \frac{f^{(n-j)}(\alpha_1)}{(n-j)!},$$

and $f^{(k)}$ is the k-th derivative of f (§1.4(iii)).

If $m_1, m_2, \ldots, m_n$ are positive integers and $\deg f < \sum_{j=1}^{n} m_j$, then there exist polynomials $f_j(x)$, $\deg f_j < m_j$, such that

1.2.16
$$\frac{f(x)}{(x-\alpha_1)^{m_1}(x-\alpha_2)^{m_2}\cdots(x-\alpha_n)^{m_n}} = \frac{f_1(x)}{(x-\alpha_1)^{m_1}} + \frac{f_2(x)}{(x-\alpha_2)^{m_2}} + \cdots + \frac{f_n(x)}{(x-\alpha_n)^{m_n}}.$$

To find the polynomials $f_j(x)$, $j = 1, 2, \ldots, n$, multiply both sides by the denominator of the left-hand side and equate coefficients. See Chrystal (1959, pp. 151–159).

1.2(iv) Means

The *arithmetic mean* of n numbers $a_1, a_2, \ldots, a_n$ is

1.2.17 $\qquad A = \dfrac{a_1 + a_2 + \cdots + a_n}{n}.$

The *geometric mean* G and *harmonic mean* H of n positive numbers $a_1, a_2, \ldots, a_n$ are given by

1.2.18 $\qquad G = (a_1 a_2 \cdots a_n)^{1/n},$

1.2.19 $\qquad \dfrac{1}{H} = \dfrac{1}{n}\left(\dfrac{1}{a_1} + \dfrac{1}{a_2} + \cdots + \dfrac{1}{a_n}\right).$

If r is a nonzero real number, then the *weighted mean* $M(r)$ of n nonnegative numbers $a_1, a_2, \ldots, a_n$, and n positive numbers $p_1, p_2, \ldots, p_n$ with

1.2.20 $\qquad p_1 + p_2 + \cdots + p_n = 1,$

is defined by

1.2.21 $\qquad M(r) = (p_1 a_1^r + p_2 a_2^r + \cdots + p_n a_n^r)^{1/r},$

with the exception

1.2.22 $\qquad M(r) = 0, \quad r < 0 \text{ and } a_1 a_2 \ldots a_n = 0.$

1.2.23 $\qquad \lim_{r \to \infty} M(r) = \max(a_1, a_2, \ldots, a_n),$

1.2.24 $\qquad \lim_{r \to -\infty} M(r) = \min(a_1, a_2, \ldots, a_n).$

For $p_j = 1/n$, $j = 1, 2, \ldots, n$,

1.2.25 $\qquad M(1) = A, \quad M(-1) = H,$

and

1.2.26 $\qquad \lim_{r \to 0} M(r) = G.$

The last two equations require $a_j > 0$ for all j.

1.3 Determinants

1.3(i) Definitions and Elementary Properties

1.3.1 $\qquad \det[a_{jk}] = \begin{vmatrix} a_{11} & a_{12} \\ a_{21} & a_{22} \end{vmatrix} = a_{11}a_{22} - a_{12}a_{21}.$

1.3.2
$$\det[a_{jk}] = \begin{vmatrix} a_{11} & a_{12} & a_{13} \\ a_{21} & a_{22} & a_{23} \\ a_{31} & a_{32} & a_{33} \end{vmatrix}$$
$$= a_{11}\begin{vmatrix} a_{22} & a_{23} \\ a_{32} & a_{33} \end{vmatrix} - a_{12}\begin{vmatrix} a_{21} & a_{23} \\ a_{31} & a_{33} \end{vmatrix} + a_{13}\begin{vmatrix} a_{21} & a_{22} \\ a_{31} & a_{32} \end{vmatrix}$$
$$= a_{11}a_{22}a_{33} - a_{11}a_{23}a_{32} - a_{12}a_{21}a_{33}$$
$$\quad + a_{12}a_{23}a_{31} + a_{13}a_{21}a_{32} - a_{13}a_{22}a_{31}.$$

Higher-order determinants are natural generalizations. The *minor* M_{jk} of the entry a_{jk} in the nth-order determinant $\det[a_{jk}]$ is the $(n-1)$th-order determinant derived from $\det[a_{jk}]$ by deleting the jth row and the kth column. The *cofactor* A_{jk} of a_{jk} is

1.3.3 $\qquad A_{jk} = (-1)^{j+k} M_{jk}.$

An nth-order determinant expanded by its jth row is given by

1.3.4 $\qquad \det[a_{jk}] = \sum_{\ell=1}^{n} a_{j\ell} A_{j\ell}.$

If two rows (or columns) of a determinant are interchanged, then the determinant changes sign. If two rows (columns) of a determinant are identical, then the determinant is zero. If all the elements of a row (column) of a determinant are multiplied by an arbitrary factor μ, then the result is a determinant which is μ times the original. If μ times a row (column) of a determinant is added to another row (column), then the value of the determinant is unchanged.

1.3.5 $\qquad \det[a_{jk}]^{\mathrm{T}} = \det[a_{jk}],$

1.3.6 $\qquad \det[a_{jk}]^{-1} = \dfrac{1}{\det[a_{jk}]},$

1.3.7 $\qquad \det([a_{jk}][b_{jk}]) = (\det[a_{jk}])(\det[b_{jk}]).$

Hadamard's Inequality

For real-valued a_{jk},

1.3.8 $\qquad \begin{vmatrix} a_{11} & a_{12} \\ a_{21} & a_{22} \end{vmatrix}^2 \leq (a_{11}^2 + a_{12}^2)(a_{21}^2 + a_{22}^2),$

1.3.9 $\quad \det[a_{jk}]^2 \leq \left(\sum_{k=1}^{n} a_{1k}^2\right)\left(\sum_{k=1}^{n} a_{2k}^2\right)\cdots\left(\sum_{k=1}^{n} a_{nk}^2\right).$

Compare also (1.3.7) for the left-hand side. Equality holds iff

1.3.10 $\qquad a_{j1}a_{k1} + a_{j2}a_{k2} + \cdots + a_{jn}a_{kn} = 0$

for every distinct pair of j, k, or when one of the factors $\sum_{k=1}^{n} a_{jk}^2$ vanishes.

1.3(ii) Special Determinants

An *alternant* is a determinant function of n variables which changes sign when two of the variables are interchanged. Examples:

1.3.11 $\qquad \det[f_k(x_j)], \qquad j = 1, \ldots, n;\ k = 1, \ldots, n,$

1.3.12 $\qquad \det[f(x_j, y_k)], \qquad j = 1, \ldots, n;\ k = 1, \ldots, n.$

Vandermonde Determinant or Vandermondian

1.3.13 $\qquad \begin{vmatrix} 1 & x_1 & x_1^2 & \cdots & x_1^{n-1} \\ 1 & x_2 & x_2^2 & \cdots & x_2^{n-1} \\ \vdots & \vdots & \vdots & \ddots & \vdots \\ 1 & x_n & x_n^2 & \cdots & x_n^{n-1} \end{vmatrix} = \prod_{1 \leq j < k \leq n} (x_k - x_j).$

Cauchy Determinant

1.3.14
$$\det\left[\frac{1}{a_j - b_k}\right]$$
$$= (-1)^{n(n-1)/2}$$
$$\times \prod_{1 \leq j < k \leq n}(a_k - a_j)(b_k - b_j) \Big/ \prod_{j,k=1}^{n}(a_j - b_k).$$

Circulant

1.3.15
$$\begin{vmatrix} a_1 & a_2 & \cdots & a_n \\ a_n & a_1 & \cdots & a_{n-1} \\ \vdots & \vdots & \ddots & \vdots \\ a_2 & a_3 & \cdots & a_1 \end{vmatrix}$$
$$= \prod_{k=1}^{n}(a_1 + a_2\omega_k + a_3\omega_k^2 + \cdots + a_n\omega_k^{n-1}),$$

where $\omega_1, \omega_2, \ldots, \omega_n$ are the nth roots of unity (1.11.21).

Krattenthaler's Formula

For

1.3.16
$$t_{jk} = (x_j + a_n)(x_j + a_{n-1})\cdots(x_j + a_{k+1})$$
$$\times (x_j + b_k)(x_j + b_{k-1})\cdots(x_j + b_2),$$

1.3.17 $\det[t_{jk}] = \prod_{1 \leq j < k \leq n}(x_j - x_k) \prod_{2 \leq j \leq k \leq n}(b_j - a_k).$

1.3(iii) Infinite Determinants

Let $a_{j,k}$ be defined for all integer values of j and k, and $D_n[a_{j,k}]$ denote the $(2n+1) \times (2n+1)$ determinant

1.3.18
$$D_n[a_{j,k}] = \begin{vmatrix} a_{-n,-n} & a_{-n,-n+1} & \cdots & a_{-n,n} \\ a_{-n+1,-n} & a_{-n+1,-n+1} & \cdots & a_{-n+1,n} \\ \vdots & \vdots & \ddots & \vdots \\ a_{n,-n} & a_{n,-n+1} & \cdots & a_{n,n} \end{vmatrix}.$$

If $D_n[a_{j,k}]$ tends to a limit L as $n \to \infty$, then we say that the *infinite determinant* $D_\infty[a_{j,k}]$ *converges* and $D_\infty[a_{j,k}] = L$.

Of importance for special functions are infinite determinants of *Hill's type*. These have the property that the double series

1.3.19
$$\sum_{j,k=-\infty}^{\infty}|a_{j,k} - \delta_{j,k}|$$

converges (§1.9(vii)). Here $\delta_{j,k}$ is the Kronecker delta. Hill-type determinants always converge.

For further information see Whittaker and Watson (1927, pp. 36–40) and Magnus and Winkler (1966, §2.3).

1.4 Calculus of One Variable

1.4(i) Monotonicity

If $f(x_1) \leq f(x_2)$ for every pair x_1, x_2 in an interval I such that $x_1 < x_2$, then $f(x)$ is *nondecreasing* on I. If the $\leq$ sign is replaced by $<$, then $f(x)$ is *increasing* (also called *strictly increasing*) on I. Similarly for *nonincreasing* and *decreasing* (*strictly decreasing*) functions. Each of the preceding four cases is classified as *monotonic*; sometimes *strictly monotonic* is used for the strictly increasing or strictly decreasing cases.

1.4(ii) Continuity

A function $f(x)$ is *continuous on the right* (or *from above*) at $x = c$ if

1.4.1
$$f(c+) \equiv \lim_{x \to c+} f(x) = f(c),$$

that is, for every arbitrarily small positive constant ϵ there exists $\delta \ (> 0)$ such that

1.4.2
$$|f(c+\alpha) - f(c)| < \epsilon,$$

for all α such that $0 \leq \alpha < \delta$. Similarly, it is *continuous on the left* (or *from below*) at $x = c$ if

1.4.3
$$f(c-) \equiv \lim_{x \to c-} f(x) = f(c).$$

And $f(x)$ is *continuous at* c when both (1.4.1) and (1.4.3) apply.

If $f(x)$ is continuous at each point $c \in (a,b)$, then $f(x)$ is *continuous on the interval* (a,b) and we write $f \in C(a,b)$. If also $f(x)$ is continuous on the right at $x = a$, and continuous on the left at $x = b$, then $f(x)$ is *continuous on the interval* $[a,b]$, and we write $f(x) \in C[a,b]$.

A *removable singularity* of $f(x)$ at $x = c$ occurs when $f(c+) = f(c-)$ but $f(c)$ is undefined. For example, $f(x) = (\sin x)/x$ with $c = 0$.

A *simple discontinuity* of $f(x)$ at $x = c$ occurs when $f(c+)$ and $f(c-)$ exist, but $f(c+) \neq f(c-)$. If $f(x)$ is continuous on an interval I save for a finite number of simple discontinuities, then $f(x)$ is *piecewise* (or *sectionally*) continuous on I. For an example, see Figure 1.4.1

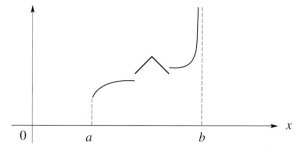

Figure 1.4.1: Piecewise continuous function on $[a, b]$.

1.4(iii) Derivatives

The *derivative* $f'(x)$ of $f(x)$ is defined by

1.4.4 $\quad f'(x) = \dfrac{df}{dx} = \lim_{h \to 0} \dfrac{f(x+h) - f(x)}{h}.$

When this limit exists f is *differentiable* at x.

1.4.5 $\quad (f+g)'(x) = f'(x) + g'(x),$

1.4.6 $\quad (fg)'(x) = f'(x)g(x) + f(x)g'(x),$

1.4.7 $\quad \left(\dfrac{f}{g}\right)'(x) = \dfrac{f'(x)g(x) - f(x)g'(x)}{(g(x))^2}.$

Higher Derivatives

1.4.8 $\quad f^{(2)}(x) = \dfrac{d^2 f}{dx^2} = \dfrac{d}{dx}\left(\dfrac{df}{dx}\right),$

1.4.9 $\quad f^{(n)} = f^{(n)}(x) = \dfrac{d}{dx}f^{(n-1)}(x).$

If $f^{(n)}$ exists and is continuous on an interval I, then we write $f \in C^n(I)$. When $n \geq 1$, f is *continuously differentiable* on I. When n is unbounded, f is *infinitely differentiable* on I and we write $f \in C^\infty(I)$.

Chain Rule

For $h(x) = f(g(x))$,

1.4.10 $\quad h'(x) = f'(g(x))g'(x).$

Maxima and Minima

A necessary condition that a differentiable function $f(x)$ has a *local maximum (minimum)* at $x = c$, that is, $f(x) \leq f(c)$, $(f(x) \geq f(c))$ in a *neighborhood* $c - \delta \leq x \leq c + \delta$ $(\delta > 0)$ of c, is $f'(c) = 0$.

Mean Value Theorem

If $f(x)$ is continuous on $[a,b]$ and differentiable on (a,b), then there exists a point $c \in (a,b)$ such that

1.4.11 $\quad f(b) - f(a) = (b-a)f'(c).$

If $f'(x) \geq 0$ (≤ 0) $(= 0)$ for all $x \in (a,b)$, then f is nondecreasing (nonincreasing) (constant) on (a,b).

Leibniz's Formula

1.4.12 $\quad (fg)^{(n)} = f^{(n)}g + \binom{n}{1}f^{(n-1)}g' + \cdots + \binom{n}{k}f^{(n-k)}g^{(k)} + \cdots + fg^{(n)}.$

Faà Di Bruno's Formula

1.4.13 $\quad \dfrac{d^n}{dx^n}f(g(x)) = \sum \left(\dfrac{n!}{m_1! m_2! \cdots m_n!}\right) f^{(k)}(g(x)) \times \left(\dfrac{g'(x)}{1!}\right)^{m_1}\left(\dfrac{g''(x)}{2!}\right)^{m_2} \cdots \left(\dfrac{g^{(n)}(x)}{n!}\right)^{m_n},$

where the sum is over all nonnegative integers $m_1, m_2, \ldots, m_n$ that satisfy $m_1 + 2m_2 + \cdots + nm_n = n$, and $k = m_1 + m_2 + \cdots + m_n$.

L'Hôpital's Rule

If

1.4.14 $\quad \lim_{x \to a} f(x) = \lim_{x \to a} g(x) = 0$ (or ∞),

then

1.4.15 $\quad \lim_{x \to a} \dfrac{f(x)}{g(x)} = \lim_{x \to a} \dfrac{f'(x)}{g'(x)},$

when the last limit exists.

1.4(iv) Indefinite Integrals

If $F'(x) = f(x)$, then $\int f \, dx = F(x) + C$, where C is a constant.

Integration by Parts

1.4.16 $\quad \int fg \, dx = \left(\int f \, dx\right) g - \int \left(\int f \, dx\right) \dfrac{dg}{dx} \, dx.$

1.4.17 $\quad \int x^n \, dx = \begin{cases} \dfrac{x^{n+1}}{n+1} + C, & n \neq -1, \\ \ln|x| + C, & n = -1. \end{cases}$

For the function ln see §4.2(i).

See §§4.10, 4.26(ii), 4.26(iv), 4.40(ii), and 4.40(iv) for indefinite integrals involving the elementary functions.

For extensive tables of integrals, see Apelblat (1983), Bierens de Haan (1867), Gradshteyn and Ryzhik (2000), Gröbner and Hofreiter (1949, 1950), and Prudnikov et al. (1986a,b, 1990, 1992a,b).

1.4(v) Definite Integrals

Suppose $f(x)$ is defined on $[a,b]$. Let $a = x_0 < x_1 < \cdots < x_n = b$, and ξ_j denote any point in $[x_j, x_{j+1}]$, $j = 0, 1, \ldots, n-1$. Then

1.4.18 $\quad \displaystyle\int_a^b f(x) \, dx = \lim \sum_{j=0}^{n-1} f(\xi_j)(x_{j+1} - x_j)$

as $\max(x_{j+1} - x_j) \to 0$. Continuity, or piecewise continuity, of $f(x)$ on $[a,b]$ is sufficient for the limit to exist.

1.4.19
$$\int_a^b (cf(x) + dg(x)) \, dx = c \int_a^b f(x) \, dx + d \int_a^b g(x) \, dx,$$

c and d constants.

1.4.20 $\quad \displaystyle\int_a^b f(x) \, dx = -\int_b^a f(x) \, dx.$

1.4.21 $\quad \displaystyle\int_a^b f(x) \, dx = \int_a^c f(x) \, dx + \int_c^b f(x) \, dx.$

Infinite Integrals

1.4.22 $$\int_a^\infty f(x)\,dx = \lim_{b\to\infty} \int_a^b f(x)\,dx.$$

Similarly for $\int_{-\infty}^a$. Next, if $f(b) = \pm\infty$, then

1.4.23 $$\int_a^b f(x)\,dx = \lim_{c\to b-} \int_a^c f(x)\,dx.$$

Similarly when $f(a) = \pm\infty$.

When the limits in (1.4.22) and (1.4.23) exist, the integrals are said to be *convergent*. If the limits exist with $f(x)$ replaced by $|f(x)|$, then the integrals are *absolutely convergent*. Absolute convergence also implies convergence.

Cauchy Principal Values

Let $c \in (a,b)$ and assume that $\int_a^{c-\epsilon} f(x)\,dx$ and $\int_{c+\epsilon}^b f(x)\,dx$ exist when $0 < \epsilon < \min(c-a, b-c)$, but not necessarily when $\epsilon = 0$. Then we define

1.4.24
$$\fint_a^b f(x)\,dx = P\int_a^b f(x)\,dx$$
$$= \lim_{\epsilon \to 0+} \left(\int_a^{c-\epsilon} f(x)\,dx + \int_{c+\epsilon}^b f(x)\,dx \right),$$

when this limit exists.

Similarly, assume that $\int_{-b}^b f(x)\,dx$ exists for all finite values of $b \;(> 0)$, but not necessarily when $b = \infty$. Then we define

1.4.25
$$\fint_{-\infty}^\infty f(x)\,dx = P\int_{-\infty}^\infty f(x)\,dx = \lim_{b\to\infty} \int_{-b}^b f(x)\,dx,$$

when this limit exists.

Fundamental Theorem of Calculus

For $F'(x) = f(x)$ with $f(x)$ continuous,

1.4.26 $$\int_a^b f(x)\,dx = F(b) - F(a),$$

1.4.27 $$\frac{d}{dx}\int_a^x f(t)\,dt = f(x).$$

Change of Variables

If $\phi'(x)$ is continuous or piecewise continuous, then

1.4.28 $$\int_a^b f(\phi(x))\phi'(x)\,dx = \int_{\phi(a)}^{\phi(b)} f(t)\,dt.$$

First Mean Value Theorem

For $f(x)$ continuous and $\phi(x) \geq 0$ and integrable on $[a,b]$, there exists $c \in [a,b]$, such that

1.4.29 $$\int_a^b f(x)\phi(x)\,dx = f(c)\int_a^b \phi(x)\,dx.$$

Second Mean Value Theorem

For $f(x)$ monotonic and $\phi(x)$ integrable on $[a,b]$, there exists $c \in [a,b]$, such that

1.4.30 $$\int_a^b f(x)\phi(x)\,dx = f(a)\int_a^c \phi(x)\,dx + f(b)\int_c^b \phi(x)\,dx.$$

Repeated Integrals

If $f(x)$ is continuous or piecewise continuous on $[a,b]$, then

1.4.31
$$\int_a^b dx_n \int_a^{x_n} dx_{n-1} \cdots \int_a^{x_2} dx_1 \int_a^{x_1} f(x)\,dx$$
$$= \frac{1}{n!}\int_a^b (b-x)^n f(x)\,dx.$$

Square-Integrable Functions

A function $f(x)$ is *square-integrable* if

1.4.32 $$\|f\|_2^2 \equiv \int_a^b |f(x)|^2\,dx < \infty.$$

Functions of Bounded Variation

With $a < b$, the *total variation* of $f(x)$ on a finite or infinite interval (a,b) is

1.4.33 $$\mathcal{V}_{a,b}(f) = \sup \sum_{j=1}^n |f(x_j) - f(x_{j-1})|,$$

where the supremum is over all sets of points $x_0 < x_1 < \cdots < x_n$ in the *closure* of (a,b), that is, (a,b) with a, b added when they are finite. If $\mathcal{V}_{a,b}(f) < \infty$, then $f(x)$ is of *bounded variation* on (a,b). In this case, $g(x) = \mathcal{V}_{a,x}(f)$ and $h(x) = \mathcal{V}_{a,x}(f) - f(x)$ are nondecreasing bounded functions and $f(x) = g(x) - h(x)$.

If $f(x)$ is continuous on the closure of (a,b) and $f'(x)$ is continuous on (a,b), then

1.4.34 $$\mathcal{V}_{a,b}(f) = \int_a^b |f'(x)\,dx|,$$

whenever this integral exists.

Lastly, whether or not the real numbers a and b satisfy $a < b$, and whether or not they are finite, we *define* $\mathcal{V}_{a,b}(f)$ by (1.4.34) whenever this integral exists. This definition also applies when $f(x)$ is a complex function of the real variable x. For further information on total variation see Olver (1997b, pp. 27–29).

1.4(vi) Taylor's Theorem for Real Variables

If $f(x) \in C^{n+1}[a,b]$, then

1.4.35 $$f(x) = \sum_{k=0}^n \frac{f^{(k)}(a)}{k!}(x-a)^k + R_n,$$

1.4.36 $$R_n = \frac{f^{(n+1)}(c)}{(n+1)!}(x-a)^{n+1}, \quad a < c < x,$$

and

1.4.37 $$R_n = \frac{1}{n!}\int_a^x (x-t)^n f^{(n+1)}(t)\,dt.$$

1.4(vii) Maxima and Minima

If $f(x)$ is twice-differentiable, and if also $f'(x_0) = 0$ and $f''(x_0) < 0$ (> 0), then $x = x_0$ is a local maximum (minimum) (§1.4(iii)) of $f(x)$. The overall maximum (minimum) of $f(x)$ on $[a,b]$ will either be at a local maximum (minimum) or at one of the end points a or b.

1.4(viii) Convex Functions

A function $f(x)$ is *convex* on (a,b) if

1.4.38 $\quad f((1-t)c + td) \leq (1-t)f(c) + tf(d)$

for any $c,d \in (a,b)$, and $t \in [0,1]$. See Figure 1.4.2. A similar definition applies to closed intervals $[a,b]$.

If $f(x)$ is twice differentiable, then $f(x)$ is convex iff $f''(x) \geq 0$ on (a,b). A continuously differentiable function is convex iff the curve does not lie below its tangent at any point.

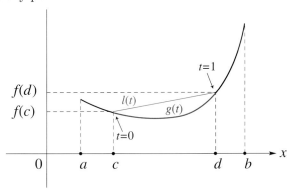

Figure 1.4.2: Convex function $f(x)$. $g(t) = f((1-t)c + td)$, $l(t) = (1-t)f(c) + tf(d)$, $c,d \in (a,b)$, $0 \leq t \leq 1$.

1.5 Calculus of Two or More Variables

1.5(i) Partial Derivatives

A function $f(x,y)$ is *continuous at a point* (a,b) if

1.5.1 $\quad \lim_{(x,y) \to (a,b)} f(x,y) = f(a,b),$

that is, for every arbitrarily small positive constant ϵ there exists δ (> 0) such that

1.5.2 $\quad |f(a+\alpha, b+\beta) - f(a,b)| < \epsilon,$

for all α and β that satisfy $|\alpha|, |\beta| < \delta$.

A function is *continuous on a point set* D if it is continuous at all points of D. A function $f(x,y)$ is *piecewise continuous* on $I_1 \times I_2$, where I_1 and I_2 are intervals, if it is piecewise continuous in x for each $y \in I_2$ and piecewise continuous in y for each $x \in I_1$.

1.5.3 $\quad \dfrac{\partial f}{\partial x} = D_x f = f_x = \lim_{h \to 0} \dfrac{f(x+h,y) - f(x,y)}{h},$

1.5.4 $\quad \dfrac{\partial f}{\partial y} = D_y f = f_y = \lim_{h \to 0} \dfrac{f(x,y+h) - f(x,y)}{h}.$

1.5.5 $\quad \dfrac{\partial^2 f}{\partial x \, \partial y} = \dfrac{\partial}{\partial x}\left(\dfrac{\partial f}{\partial y}\right), \quad \dfrac{\partial^2 f}{\partial y \, \partial x} = \dfrac{\partial}{\partial y}\left(\dfrac{\partial f}{\partial x}\right).$

The function $f(x,y)$ is *continuously differentiable* if f, $\partial f / \partial x$, and $\partial f / \partial y$ are continuous, *and twice-continuously differentiable* if also $\partial^2 f / \partial x^2$, $\partial^2 f / \partial y^2$, $\partial^2 f / \partial x \, \partial y$, and $\partial^2 f / \partial y \, \partial x$ are continuous. In the latter event

1.5.6 $\quad \dfrac{\partial^2 f}{\partial x \, \partial y} = \dfrac{\partial^2 f}{\partial y \, \partial x}.$

Chain Rule

1.5.7 $\quad \dfrac{d}{dt} f(x(t), y(t)) = \dfrac{\partial f}{\partial x}\dfrac{dx}{dt} + \dfrac{\partial f}{\partial y}\dfrac{dy}{dt},$

1.5.8 $\quad \dfrac{\partial}{\partial u} f(x(u,v), y(u,v)) = \dfrac{\partial f}{\partial x}\dfrac{\partial x}{\partial u} + \dfrac{\partial f}{\partial y}\dfrac{\partial y}{\partial u},$

1.5.9
$$\dfrac{\partial}{\partial v} f(x(u,v), y(u,v), z(u,v))$$
$$= \dfrac{\partial f}{\partial x}\dfrac{\partial x}{\partial v} + \dfrac{\partial f}{\partial y}\dfrac{\partial y}{\partial v} + \dfrac{\partial f}{\partial z}\dfrac{\partial z}{\partial v}.$$

Implicit Function Theorem

If $F(x,y)$ is continuously differentiable, $F(a,b) = 0$, and $\partial F/\partial y \neq 0$ at (a,b), then in a *neighborhood* of (a,b), that is, an open disk centered at a,b, the equation $F(x,y) = 0$ defines a continuously differentiable function $y = g(x)$ such that $F(x, g(x)) = 0$, $b = g(a)$, and $g'(x) = -F_x/F_y$.

1.5(ii) Coordinate Systems

Polar Coordinates

With $0 \leq r < \infty$, $0 \leq \phi \leq 2\pi$,

1.5.10 $\quad x = r\cos\phi, \quad y = r\sin\phi,$

1.5.11 $\quad \dfrac{\partial}{\partial x} = \cos\phi \dfrac{\partial}{\partial r} - \dfrac{\sin\phi}{r}\dfrac{\partial}{\partial \phi},$

1.5.12 $\quad \dfrac{\partial}{\partial y} = \sin\phi \dfrac{\partial}{\partial r} + \dfrac{\cos\phi}{r}\dfrac{\partial}{\partial \phi}.$

The *Laplacian* is given by

1.5.13 $\quad \nabla^2 f = \dfrac{\partial^2 f}{\partial x^2} + \dfrac{\partial^2 f}{\partial y^2} = \dfrac{\partial^2 f}{\partial r^2} + \dfrac{1}{r}\dfrac{\partial f}{\partial r} + \dfrac{1}{r^2}\dfrac{\partial^2 f}{\partial \phi^2}.$

Cylindrical Coordinates

With $0 \leq r < \infty$, $0 \leq \phi \leq 2\pi$, $-\infty < z < \infty$,

1.5.14 $\quad x = r\cos\phi, \quad y = r\sin\phi, \quad z = z.$

Equations (1.5.11) and (1.5.12) still apply, but

1.5.15
$$\nabla^2 f = \dfrac{\partial^2 f}{\partial x^2} + \dfrac{\partial^2 f}{\partial y^2} + \dfrac{\partial^2 f}{\partial z^2} = \dfrac{\partial^2 f}{\partial r^2} + \dfrac{1}{r}\dfrac{\partial f}{\partial r} + \dfrac{1}{r^2}\dfrac{\partial^2 f}{\partial \phi^2} + \dfrac{\partial^2 f}{\partial z^2}.$$

Spherical Coordinates

With $0 \le \rho < \infty$, $0 \le \phi \le 2\pi$, $0 \le \theta \le \pi$,

1.5.16 $\quad x = \rho \sin\theta \cos\phi, \quad y = \rho \sin\theta \sin\phi, \quad z = \rho \cos\theta.$

The Laplacian is given by

1.5.17
$$\begin{aligned}\nabla^2 f &= \frac{\partial^2 f}{\partial x^2} + \frac{\partial^2 f}{\partial y^2} + \frac{\partial^2 f}{\partial z^2} \\ &= \frac{1}{\rho^2}\frac{\partial}{\partial \rho}\left(\rho^2 \frac{\partial f}{\partial \rho}\right) + \frac{1}{\rho^2 \sin^2\theta}\frac{\partial^2 f}{\partial \phi^2} \\ &\quad + \frac{1}{\rho^2 \sin\theta}\frac{\partial}{\partial \theta}\left(\sin\theta \frac{\partial f}{\partial \theta}\right).\end{aligned}$$

For applications and other coordinate systems see §§12.17, 14.19(i), 14.30(iv), 28.32, 29.18, 30.13, 30.14. See also Morse and Feshbach (1953a, pp. 655-666).

1.5(iii) Taylor's Theorem; Maxima and Minima

If f is $n+1$ times continuously differentiable, then

1.5.18
$$\begin{aligned}f(a+\lambda, b+\mu) &= f + \left(\lambda\frac{\partial}{\partial x} + \mu\frac{\partial}{\partial y}\right)f + \cdots \\ &\quad + \frac{1}{n!}\left(\lambda\frac{\partial}{\partial x} + \mu\frac{\partial}{\partial y}\right)^n f + R_n,\end{aligned}$$

where f and its partial derivatives on the right-hand side are evaluated at (a,b), and $R_n/(\lambda^2 + \mu^2)^{n/2} \to 0$ as $(\lambda, \mu) \to (0,0)$.

$f(x,y)$ has a *local minimum* (*maximum*) at (a,b) if

1.5.19 $\quad \dfrac{\partial f}{\partial x} = \dfrac{\partial f}{\partial y} = 0 \quad$ at (a,b),

and the second-order term in (1.5.18) is *positive definite* (*negative definite*), that is,

1.5.20 $\quad \dfrac{\partial^2 f}{\partial x^2} > 0 \quad (<0) \quad$ at (a,b),

and

1.5.21 $\quad \dfrac{\partial^2 f}{\partial x^2}\dfrac{\partial^2 f}{\partial y^2} - \left(\dfrac{\partial^2 f}{\partial x \partial y}\right)^2 > 0 \quad$ at (a,b).

1.5(iv) Leibniz's Theorem for Differentiation of Integrals

Finite Integrals

1.5.22
$$\begin{aligned}\frac{d}{dx}\int_{\alpha(x)}^{\beta(x)} f(x,y)\,dy &= f(x,\beta(x))\beta'(x) - f(x,\alpha(x))\alpha'(x) \\ &\quad + \int_{\alpha(x)}^{\beta(x)} \frac{\partial f}{\partial x}\,dy.\end{aligned}$$

Sufficient conditions for validity are: (a) f and $\partial f/\partial x$ are continuous on a rectangle $a \le x \le b$, $c \le y \le d$; (b) when $x \in [a,b]$ both $\alpha(x)$ and $\beta(x)$ are continuously differentiable and lie in $[c,d]$.

Infinite Integrals

Suppose that a, b, c are finite, d is finite or $+\infty$, and $f(x,y)$, $\partial f/\partial x$ are continuous on the partly-closed rectangle or infinite strip $[a,b] \times [c,d)$. Suppose also that $\int_c^d f(x,y)\,dy$ converges and $\int_c^d (\partial f/\partial x)\,dy$ converges *uniformly* on $a \le x \le b$, that is, given any positive number ϵ, however small, we can find a number $c_0 \in [c,d)$ that is independent of x and is such that

1.5.23 $\quad \left|\int_{c_1}^d (\partial f/\partial x)\,dy\right| < \epsilon,$

for all $c_1 \in [c_0, d)$ and all $x \in [a,b]$. Then

1.5.24 $\quad \dfrac{d}{dx}\int_c^d f(x,y)\,dy = \int_c^d \dfrac{\partial f}{\partial x}\,dy, \quad a < x < b.$

1.5(v) Multiple Integrals

Double Integrals

Let $f(x,y)$ be defined on a closed rectangle $R = [a,b] \times [c,d]$. For

1.5.25 $\quad a = x_0 < x_1 < \cdots < x_n = b,$

1.5.26 $\quad c = y_0 < y_1 < \cdots < y_m = d,$

let (ξ_j, η_k) denote any point in the rectangle $[x_j, x_{j+1}] \times [y_k, y_{k+1}]$, $j = 0,\ldots,n-1$, $k = 0,\ldots,m-1$. Then the *double integral* of $f(x,y)$ over R is defined by

1.5.27
$$\iint_R f(x,y)\,dA = \lim \sum_{j,k} f(\xi_j, \eta_k)(x_{j+1}-x_j)(y_{k+1}-y_k)$$

as $\max((x_{j+1}-x_j) + (y_{k+1}-y_k)) \to 0$. Sufficient conditions for the limit to exist are that $f(x,y)$ is continuous, or piecewise continuous, on R.

For $f(x,y)$ defined on a point set D contained in a rectangle R, let

1.5.28 $\quad f^*(x,y) = \begin{cases} f(x,y), & \text{if } (x,y) \in D, \\ 0, & \text{if } (x,y) \in R \setminus D.\end{cases}$

Then

1.5.29 $\quad \displaystyle\iint_D f(x,y)\,dA = \iint_R f^*(x,y)\,dA,$

provided the latter integral exists.

If $f(x,y)$ is continuous, and D is the set

1.5.30 $\quad a \le x \le b, \quad \phi_1(x) \le y \le \phi_2(x),$

with $\phi_1(x)$ and $\phi_2(x)$ continuous, then

1.5.31 $\quad \displaystyle\iint_D f(x,y)\,dA = \int_a^b \int_{\phi_1(x)}^{\phi_2(x)} f(x,y)\,dy\,dx,$

where the right-hand side is interpreted as the repeated integral

1.5.32 $\quad \displaystyle\int_a^b \left(\int_{\phi_1(x)}^{\phi_2(x)} f(x,y)\,dy\right) dx.$

In particular, $\phi_1(x)$ and $\phi_2(x)$ can be constants.

Similarly, if D is the set

1.5.33 $\qquad c \leq y \leq d, \quad \psi_1(y) \leq x \leq \psi_2(y),$

with $\psi_1(y)$ and $\psi_2(y)$ continuous, then

1.5.34 $\qquad \iint_D f(x,y)\,dA = \int_c^d \int_{\psi_1(y)}^{\psi_2(y)} f(x,y)\,dx\,dy.$

Change of Order of Integration

If D can be represented in both forms (1.5.30) and (1.5.33), and $f(x,y)$ is continuous on D, then

1.5.35 $\qquad \int_a^b \int_{\phi_1(x)}^{\phi_2(x)} f(x,y)\,dy\,dx = \int_c^d \int_{\psi_1(y)}^{\psi_2(y)} f(x,y)\,dx\,dy.$

Infinite Double Integrals

Infinite double integrals occur when $f(x,y)$ becomes infinite at points in D or when D is unbounded. In the cases (1.5.30) and (1.5.33) they are defined by taking limits in the repeated integrals (1.5.32) and (1.5.34) in an analogous manner to (1.4.22)–(1.4.23).

Moreover, if a, b, c, d are finite or infinite constants and $f(x,y)$ is piecewise continuous on the set $(a,b) \times (c,d)$, then

1.5.36 $\qquad \int_a^b \int_c^d f(x,y)\,dy\,dx = \int_c^d \int_a^b f(x,y)\,dx\,dy,$

whenever both repeated integrals exist and at least one is absolutely convergent.

Triple Integrals

Finite and infinite integrals can be defined in a similar way. Often the (x,y,z) sets are of the form

1.5.37 $\qquad \begin{aligned} a \leq x \leq b, \quad \phi_1(x) \leq y \leq \phi_2(x), \\ \psi_1(x,y) \leq z \leq \psi_2(x,y). \end{aligned}$

1.5(vi) Jacobians and Change of Variables

Jacobian

1.5.38 $\qquad \dfrac{\partial(f,g)}{\partial(x,y)} = \begin{vmatrix} \partial f/\partial x & \partial f/\partial y \\ \partial g/\partial x & \partial g/\partial y \end{vmatrix},$

1.5.39 $\qquad \dfrac{\partial(x,y)}{\partial(r,\phi)} = r \quad$ (polar coordinates).

1.5.40 $\qquad \dfrac{\partial(f,g,h)}{\partial(x,y,z)} = \begin{vmatrix} \partial f/\partial x & \partial f/\partial y & \partial f/\partial z \\ \partial g/\partial x & \partial g/\partial y & \partial g/\partial z \\ \partial h/\partial x & \partial h/\partial y & \partial h/\partial z \end{vmatrix},$

1.5.41 $\qquad \dfrac{\partial(x,y,z)}{\partial(\rho,\theta,\phi)} = \rho^2 \sin\theta \quad$ (spherical coordinates).

Change of Variables

1.5.42 $\qquad \begin{aligned} &\iint_D f(x,y)\,dx\,dy \\ &= \iint_{D^*} f(x(u,v), y(u,v)) \left| \dfrac{\partial(x,y)}{\partial(u,v)} \right| du\,dv, \end{aligned}$

where D is the image of D^* under a mapping $(u,v) \to (x(u,v), y(u,v))$ which is one-to-one except perhaps for a set of points of area zero.

1.5.43 $\qquad \begin{aligned} &\iiint_D f(x,y,z)\,dx\,dy\,dz \\ &= \iiint_{D^*} f(x(u,v,w), y(u,v,w), z(u,v,w)) \\ &\quad \times \left| \dfrac{\partial(x,y,z)}{\partial(u,v,w)} \right| du\,dv\,dw. \end{aligned}$

Again the mapping is one-to-one except perhaps for a set of points of volume zero.

1.6 Vectors and Vector-Valued Functions

1.6(i) Vectors

1.6.1 $\qquad \mathbf{a} = (a_1, a_2, a_3), \quad \mathbf{b} = (b_1, b_2, b_3).$

Dot Product (or Scalar Product)

1.6.2 $\qquad \mathbf{a} \cdot \mathbf{b} = a_1 b_1 + a_2 b_2 + a_3 b_3.$

Magnitude and Angle of Vector a

1.6.3 $\qquad \|\mathbf{a}\| = \sqrt{\mathbf{a} \cdot \mathbf{a}},$

1.6.4 $\qquad \cos\theta = \dfrac{\mathbf{a} \cdot \mathbf{b}}{\|\mathbf{a}\|\,\|\mathbf{b}\|};$

θ is the angle between $\mathbf{a}$ and $\mathbf{b}$.

Unit Vectors

1.6.5 $\quad \mathbf{i} = (1,0,0), \quad \mathbf{j} = (0,1,0), \quad \mathbf{k} = (0,0,1),$

1.6.6 $\qquad \mathbf{a} = a_1 \mathbf{i} + a_2 \mathbf{j} + a_3 \mathbf{k}.$

Cross Product (or Vector Product)

1.6.7 $\qquad \mathbf{i} \times \mathbf{j} = \mathbf{k}, \quad \mathbf{j} \times \mathbf{k} = \mathbf{i}, \quad \mathbf{k} \times \mathbf{i} = \mathbf{j},$

1.6.8 $\qquad \mathbf{j} \times \mathbf{i} = -\mathbf{k}, \quad \mathbf{k} \times \mathbf{j} = -\mathbf{i}, \quad \mathbf{i} \times \mathbf{k} = -\mathbf{j}.$

1.6.9
$$\begin{aligned} \mathbf{a} \times \mathbf{b} &= \begin{vmatrix} \mathbf{i} & \mathbf{j} & \mathbf{k} \\ a_1 & a_2 & a_3 \\ b_1 & b_2 & b_3 \end{vmatrix} \\ &= (a_2 b_3 - a_3 b_2)\mathbf{i} + (a_3 b_1 - a_1 b_3)\mathbf{j} + (a_1 b_2 - a_2 b_1)\mathbf{k} \\ &= \|\mathbf{a}\|\|\mathbf{b}\|(\sin\theta)\mathbf{n}, \end{aligned}$$

where $\mathbf{n}$ is the unit vector normal to $\mathbf{a}$ and $\mathbf{b}$ whose direction is determined by the right-hand rule; see Figure 1.6.1.

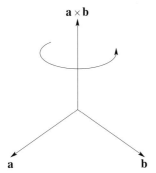

Figure 1.6.1: Vector notation. Right-hand rule for cross products.

Area of parallelogram with vectors **a** and **b** as sides = $\|\mathbf{a} \times \mathbf{b}\|$.

Volume of a parallelepiped with vectors **a**, **b**, and **c** as edges = $|\mathbf{a} \cdot (\mathbf{b} \times \mathbf{c})|$.

1.6.10 $\quad \mathbf{a} \times (\mathbf{b} \times \mathbf{c}) = \mathbf{b}(\mathbf{a} \cdot \mathbf{c}) - \mathbf{c}(\mathbf{a} \cdot \mathbf{b}),$

1.6.11 $\quad (\mathbf{a} \times \mathbf{b}) \times \mathbf{c} = \mathbf{b}(\mathbf{a} \cdot \mathbf{c}) - \mathbf{a}(\mathbf{b} \cdot \mathbf{c}).$

1.6(ii) Vectors: Alternative Notations

The following notations are often used in the physics literature; see for example Lorentz et al. (1923, pp. 122–123).

Einstein Summation Convention

Much vector algebra involves summation over suffices of products of vector components. In almost all cases of repeated suffices, we can suppress the summation notation entirely, if it is understood that an implicit sum is to be taken over any repeated suffix. Thus pairs of indefinite suffices in an expression are resolved by being summed over (or "traced" over).

Example

1.6.12 $\quad a_j b_j = \sum_{j=1}^{3} a_j b_j = \mathbf{a} \cdot \mathbf{b}.$

Next,

1.6.13 $\quad \mathbf{e_1} = (1,0,0), \quad \mathbf{e_2} = (0,1,0), \quad \mathbf{e_3} = (0,0,1);$
compare (1.6.5). Thus $a_j \mathbf{e_j} = \mathbf{a}$.

Levi-Civita Symbol

1.6.14
$$\epsilon_{jk\ell} = \begin{cases} +1, & \text{if } j,k,\ell \text{ is even permutation of } 1,2,3, \\ -1, & \text{if } j,k,\ell \text{ is odd permutation of } 1,2,3, \\ 0, & \text{otherwise.} \end{cases}$$

Examples

1.6.15 $\quad \epsilon_{123} = \epsilon_{312} = 1, \quad \epsilon_{213} = \epsilon_{321} = -1, \quad \epsilon_{221} = 0.$

1.6.16 $\quad \epsilon_{jk\ell}\epsilon_{\ell m n} = \delta_{j,m}\delta_{k,n} - \delta_{j,n}\delta_{k,m},$
where $\delta_{j,k}$ is the Kronecker delta.

1.6.17 $\quad \mathbf{e_j} \times \mathbf{e_k} = \epsilon_{jk\ell}\mathbf{e_\ell};$
compare (1.6.8).

1.6.18 $\quad a_j\mathbf{e_j} \times b_k\mathbf{e_k} = \epsilon_{jk\ell}a_jb_k\mathbf{e_\ell};$
compare (1.6.7)–(1.6.8).

Lastly, the volume of a parallelepiped with vectors **a**, **b**, and **c** as edges is $|\epsilon_{jk\ell}a_jb_kc_\ell|$.

1.6(iii) Vector-Valued Functions

Del Operator

1.6.19 $\quad \nabla = \mathbf{i}\dfrac{\partial}{\partial x} + \mathbf{j}\dfrac{\partial}{\partial y} + \mathbf{k}\dfrac{\partial}{\partial z}.$

The *gradient* of a differentiable scalar function $f(x,y,z)$ is

1.6.20 $\quad \operatorname{grad} f = \nabla f = \dfrac{\partial f}{\partial x}\mathbf{i} + \dfrac{\partial f}{\partial y}\mathbf{j} + \dfrac{\partial f}{\partial z}\mathbf{k}.$

The *divergence* of a differentiable vector-valued function $\mathbf{F} = F_1\mathbf{i} + F_2\mathbf{j} + F_3\mathbf{k}$ is

1.6.21 $\quad \operatorname{div}\mathbf{F} = \nabla \cdot \mathbf{F} = \dfrac{\partial F_1}{\partial x} + \dfrac{\partial F_2}{\partial y} + \dfrac{\partial F_3}{\partial z}.$

The *curl* of **F** is

1.6.22
$$\operatorname{curl}\mathbf{F} = \nabla \times \mathbf{F} = \begin{vmatrix} \mathbf{i} & \mathbf{j} & \mathbf{k} \\ \dfrac{\partial}{\partial x} & \dfrac{\partial}{\partial y} & \dfrac{\partial}{\partial z} \\ F_1 & F_2 & F_3 \end{vmatrix}$$
$$= \left(\dfrac{\partial F_3}{\partial y} - \dfrac{\partial F_2}{\partial z}\right)\mathbf{i} + \left(\dfrac{\partial F_1}{\partial z} - \dfrac{\partial F_3}{\partial x}\right)\mathbf{j}$$
$$+ \left(\dfrac{\partial F_2}{\partial x} - \dfrac{\partial F_1}{\partial y}\right)\mathbf{k}.$$

1.6.23 $\quad \nabla(fg) = f\nabla g + g\nabla f,$

1.6.24 $\quad \nabla(f/g) = (g\nabla f - f\nabla g)/g^2,$

1.6.25 $\quad \nabla \cdot (f\mathbf{F}) = f(\nabla \cdot \mathbf{F}) + \mathbf{F} \cdot \nabla f,$

1.6.26 $\quad \nabla \cdot (\mathbf{F} \times \mathbf{G}) = \mathbf{G} \cdot (\nabla \times \mathbf{F}) - \mathbf{F} \cdot (\nabla \times \mathbf{G}),$

1.6.27 $\quad \nabla \cdot (\nabla \times \mathbf{F}) = \operatorname{div}\operatorname{curl}\mathbf{F} = 0,$

1.6.28 $\quad \nabla \times (f\mathbf{F}) = f(\nabla \times \mathbf{F}) + (\nabla f) \times \mathbf{F},$

1.6.29 $\quad \nabla \times (\nabla f) = \operatorname{curl}\operatorname{grad} f = 0,$

1.6.30 $\quad \nabla^2 f = \nabla \cdot (\nabla f),$

1.6.31 $\quad \nabla^2(fg) = f\nabla^2 g + g\nabla^2 f + 2(\nabla f \cdot \nabla g),$

1.6.32 $\quad \nabla \cdot (\nabla f \times \nabla g) = 0,$

1.6.33 $\quad \nabla \cdot (f\nabla g - g\nabla f) = f\nabla^2 g - g\nabla^2 f,$

1.6.34 $\quad \nabla \times (\nabla \times \mathbf{F}) = \operatorname{curl}\operatorname{curl}\mathbf{F} = \nabla(\nabla \cdot \mathbf{F}) - \nabla^2\mathbf{F}.$

1.6(iv) Path and Line Integrals

Note: The terminology *open* and *closed sets* and *boundary points* in the (x, y) plane that is used in this subsection and §1.6(v) is analogous to that introduced for the complex plane in §1.9(ii).

$\mathbf{c}(t) = (x(t), y(t), z(t))$, with t ranging over an interval and $x(t), y(t), z(t)$ differentiable, defines a *path*.

1.6.35 $$\mathbf{c}'(t) = (x'(t), y'(t), z'(t)).$$

The *length* of a path for $a \leq t \leq b$ is

1.6.36 $$\int_a^b \|\mathbf{c}'(t)\| \, dt.$$

The *path integral* of a continuous function $f(x, y, z)$ is

1.6.37 $$\int_{\mathbf{c}} f \, ds = \int_a^b f(x(t), y(t), z(t)) \|\mathbf{c}'(t)\| \, dt.$$

The *line integral* of a vector-valued function $\mathbf{F} = F_1 \mathbf{i} + F_2 \mathbf{j} + F_3 \mathbf{k}$ along $\mathbf{c}$ is given by

1.6.38 $$\int_{\mathbf{c}} \mathbf{F} \cdot d\mathbf{s} = \int_a^b \mathbf{F}(\mathbf{c}(t)) \cdot \mathbf{c}'(t) \, dt$$
$$= \int_a^b \left(F_1 \frac{dx}{dt} + F_2 \frac{dy}{dt} + F_3 \frac{dz}{dt} \right) dt$$
$$= \int_{\mathbf{c}} F_1 \, dx + F_2 \, dy + F_3 \, dz.$$

A path $\mathbf{c}_1(t)$, $t \in [a, b]$, is a *reparametrization* of $\mathbf{c}(t')$, $t' \in [a', b']$, if $\mathbf{c}_1(t) = \mathbf{c}(t')$ and $t' = h(t)$ with $h(t)$ differentiable and monotonic. If $h(a) = a'$ and $h(b) = b'$, then the reparametrization is called *orientation-preserving*, and

1.6.39 $$\int_{\mathbf{c}} \mathbf{F} \cdot d\mathbf{s} = \int_{\mathbf{c}_1} \mathbf{F} \cdot d\mathbf{s}.$$

If $h(a) = b'$ and $h(b) = a'$, then the reparametrization is *orientation-reversing* and

1.6.40 $$\int_{\mathbf{c}} \mathbf{F} \cdot d\mathbf{s} = -\int_{\mathbf{c}_1} \mathbf{F} \cdot d\mathbf{s}.$$

In either case

1.6.41 $$\int_{\mathbf{c}} f \, ds = \int_{\mathbf{c}_1} f \, ds,$$

when f is continuous, and

1.6.42 $$\int_{\mathbf{c}} \nabla f \cdot d\mathbf{s} = f(\mathbf{c}(b)) - f(\mathbf{c}(a)),$$

when f is continuously differentiable.

The geometrical image C of a path $\mathbf{c}$ is called a *simple closed curve* if $\mathbf{c}$ is one-to-one, with the exception $\mathbf{c}(a) = \mathbf{c}(b)$. The curve C is *piecewise differentiable* if $\mathbf{c}$ is piecewise differentiable. Note that C can be given an orientation by means of $\mathbf{c}$.

Green's Theorem

Let

1.6.43 $$\mathbf{F}(x, y) = F_1(x, y) \mathbf{i} + F_2(x, y) \mathbf{j}$$

and S be the closed and bounded point set in the (x, y) plane having a simple closed curve C as boundary. If C is oriented in the positive (anticlockwise) sense, then

1.6.44 $$\iint_S \left(\frac{\partial F_2}{\partial x} - \frac{\partial F_1}{\partial y} \right) dA = \int_C \mathbf{F} \cdot d\mathbf{s} = \int_C F_1 \, dx + F_2 \, dy.$$

Sufficient conditions for this result to hold are that $F_1(x, y)$ and $F_2(x, y)$ are continuously differentiable on S, and C is piecewise differentiable.

The area of S can be found from (1.6.44) by taking $\mathbf{F}(x, y) = -y\mathbf{i}$, $x\mathbf{j}$, or $-\frac{1}{2} y \mathbf{i} + \frac{1}{2} x \mathbf{j}$.

1.6(v) Surfaces and Integrals over Surfaces

A *parametrized surface* S is defined by

1.6.45 $$\mathbf{\Phi}(u, v) = (x(u, v), y(u, v), z(u, v))$$

with $(u, v) \in D$, an open set in the plane.

For x, y, and z continuously differentiable, the vectors

1.6.46 $$\mathbf{T}_u = \frac{\partial x}{\partial u}(u_0, v_0) \mathbf{i} + \frac{\partial y}{\partial u}(u_0, v_0) \mathbf{j} + \frac{\partial z}{\partial u}(u_0, v_0) \mathbf{k}$$

and

1.6.47 $$\mathbf{T}_v = \frac{\partial x}{\partial v}(u_0, v_0) \mathbf{i} + \frac{\partial y}{\partial v}(u_0, v_0) \mathbf{j} + \frac{\partial z}{\partial v}(u_0, v_0) \mathbf{k}$$

are tangent to the surface at $\mathbf{\Phi}(u_0, v_0)$. The surface is *smooth* at this point if $\mathbf{T}_u \times \mathbf{T}_v \neq 0$. A surface is *smooth* if it is smooth at every point. The vector $\mathbf{T}_u \times \mathbf{T}_v$ at (u_0, v_0) is normal to the surface at $\mathbf{\Phi}(u_0, v_0)$.

The *area* $A(S)$ of a parametrized smooth surface is given by

1.6.48 $$A(S) = \iint_D \|\mathbf{T}_u \times \mathbf{T}_v\| \, du \, dv,$$

and

1.6.49 $$\|\mathbf{T}_u \times \mathbf{T}_v\| = \sqrt{\left(\frac{\partial(x, y)}{\partial(u, v)} \right)^2 + \left(\frac{\partial(y, z)}{\partial(u, v)} \right)^2 + \left(\frac{\partial(x, z)}{\partial(u, v)} \right)^2}.$$

The area is independent of the parametrizations.

For a sphere $x = \rho \sin \theta \cos \phi$, $y = \rho \sin \theta \sin \phi$, $z = \rho \cos \theta$,

1.6.50 $$\|\mathbf{T}_\theta \times \mathbf{T}_\phi\| = \rho^2 |\sin \theta|.$$

For a surface $z = f(x, y)$,

1.6.51 $$A(S) = \iint_D \sqrt{1 + \left(\frac{\partial f}{\partial x} \right)^2 + \left(\frac{\partial f}{\partial y} \right)^2} \, dA.$$

For a surface of revolution, $y = f(x)$, $x \in [a,b]$, about the x-axis,

1.6.52 $\quad A(S) = 2\pi \int_a^b |f(x)|\sqrt{1 + (f'(x))^2}\, dx,$

and about the y-axis,

1.6.53 $\quad A(S) = 2\pi \int_a^b |x|\sqrt{1 + (f'(x))^2}\, dx.$

The integral of a continuous function $f(x,y,z)$ over a surface S is

1.6.54
$$\iint_S f(x,y,z)\, dS = \iint_D f(\boldsymbol{\Phi}(u,v))\|\mathbf{T}_u \times \mathbf{T}_v\|\, du\, dv.$$

For a vector-valued function $\mathbf{F}$,

1.6.55 $\quad \iint_S \mathbf{F} \cdot d\mathbf{S} = \iint_D \mathbf{F} \cdot (\mathbf{T}_u \times \mathbf{T}_v)\, du\, dv,$

where $d\mathbf{S}$ is the surface element with an attached normal direction $\mathbf{T}_u \times \mathbf{T}_v$.

A surface is *orientable* if a continuously varying normal can be defined at all points of the surface. An orientable surface is *oriented* if suitable normals have been chosen. A parametrization $\boldsymbol{\Phi}(u,v)$ of an oriented surface S is *orientation preserving* if $\mathbf{T}_u \times \mathbf{T}_v$ has the same direction as the chosen normal at each point of S, otherwise it is *orientation reversing*.

If $\boldsymbol{\Phi}_1$ and $\boldsymbol{\Phi}_2$ are both orientation preserving or both orientation reversing parametrizations of S defined on open sets D_1 and D_2 respectively, then

1.6.56 $\quad \iint_{\boldsymbol{\Phi}_1(D_1)} \mathbf{F} \cdot d\mathbf{S} = \iint_{\boldsymbol{\Phi}_2(D_2)} \mathbf{F} \cdot d\mathbf{S};$

otherwise, one is the negative of the other.

Stokes's Theorem

Suppose S is an oriented surface with boundary ∂S which is oriented so that its direction is clockwise relative to the normals of S. Then

1.6.57 $\quad \iint_S (\nabla \times \mathbf{F}) \cdot d\mathbf{S} = \int_{\partial S} \mathbf{F} \cdot d\mathbf{s},$

when $\mathbf{F}$ is a continuously differentiable vector-valued function.

Gauss's (or Divergence) Theorem

Suppose S is a piecewise smooth surface which forms the complete boundary of a bounded closed point set V, and S is oriented by its normal being outwards from V. Then

1.6.58 $\quad \iiint_V (\nabla \cdot \mathbf{F})\, dV = \iint_S \mathbf{F} \cdot d\mathbf{S},$

when $\mathbf{F}$ is a continuously differentiable vector-valued function.

Green's Theorem (for Volume)

For f and g twice-continuously differentiable functions

1.6.59 $\quad \iiint_V (f\nabla^2 g + \nabla f \cdot \nabla g)\, dV = \iint_S f\dfrac{\partial g}{\partial n}\, dA,$

and

1.6.60
$$\iiint_V (f\nabla^2 g - g\nabla^2 f)\, dV = \iint_S \left(f\frac{\partial g}{\partial n} - g\frac{\partial f}{\partial n}\right) dA,$$

where $\partial g/\partial n = \nabla g \cdot \mathbf{n}$ is the derivative of g normal to the surface outwards from V and $\mathbf{n}$ is the unit outer normal vector.

1.7 Inequalities

1.7(i) Finite Sums

In this subsection A and B are positive constants.

Cauchy–Schwarz Inequality

1.7.1 $\quad \left(\displaystyle\sum_{j=1}^n a_j b_j\right)^2 \leq \left(\displaystyle\sum_{j=1}^n a_j^2\right)\left(\displaystyle\sum_{j=1}^n b_j^2\right).$

Equality holds iff $a_j = c b_j$, $\forall j$; $c =$ constant.

Conversely, if $\left(\sum_{j=1}^n a_j b_j\right)^2 \leq AB$ for all b_j such that $\sum_{j=1}^n b_j^2 \leq B$, then $\sum_{j=1}^n a_j^2 \leq A$.

Hölder's Inequality

For $p > 1$, $\dfrac{1}{p} + \dfrac{1}{q} = 1$, $a_j \geq 0$, $b_j \geq 0$,

1.7.2 $\quad \displaystyle\sum_{j=1}^n a_j b_j \leq \left(\sum_{j=1}^n a_j^p\right)^{1/p} \left(\sum_{j=1}^n b_j^q\right)^{1/q}.$

Equality holds iff $a_j^p = c b_j^q$, $\forall j$; $c =$ constant.

Conversely, if $\sum_{j=1}^n a_j b_j \leq A^{1/p} B^{1/q}$ for all b_j such that $\sum_{j=1}^n b_j^q \leq B$, then $\sum_{j=1}^n a_j^p \leq A$.

Minkowski's Inequality

For $p > 1$, $a_j \geq 0$, $b_j \geq 0$,

1.7.3 $\quad \left(\displaystyle\sum_{j=1}^n (a_j + b_j)^p\right)^{1/p} \leq \left(\sum_{j=1}^n a_j^p\right)^{1/p} + \left(\sum_{j=1}^n b_j^p\right)^{1/p}.$

The direction of the inequality is reversed, that is, $\geq$, when $0 < p < 1$. Equality holds iff $a_j = c b_j$, $\forall j$; $c =$ constant.

1.7(ii) Integrals

In this subsection a and b $(> a)$ are real constants that can be $\mp\infty$, provided that the corresponding integrals converge. Also A and B are constants that are not simultaneously zero.

Cauchy–Schwarz Inequality

1.7.4
$$\left(\int_a^b f(x)g(x)\,dx\right)^2 \leq \int_a^b (f(x))^2\,dx \int_a^b (g(x))^2\,dx.$$

Equality holds iff $Af(x) = Bg(x)$ for all x.

Hölder's Inequality

For $p > 1$, $\dfrac{1}{p} + \dfrac{1}{q} = 1$, $f(x) \geq 0$, $g(x) \geq 0$,

1.7.5
$$\int_a^b f(x)g(x)\,dx \leq \left(\int_a^b (f(x))^p\,dx\right)^{1/p} \left(\int_a^b (g(x))^q\,dx\right)^{1/q}.$$

Equality holds iff $A(f(x))^p = B(g(x))^q$ for all x.

Minkowski's Inequality

For $p > 1$, $f(x) \geq 0$, $g(x) \geq 0$,

1.7.6
$$\left(\int_a^b (f(x)+g(x))^p\,dx\right)^{1/p} \leq \left(\int_a^b (f(x))^p\,dx\right)^{1/p} + \left(\int_a^b (g(x))^p\,dx\right)^{1/p}.$$

The direction of the inequality is reversed, that is, $\geq$, when $0 < p < 1$. Equality holds iff $Af(x) = Bg(x)$ for all x.

1.7(iii) Means

For the notation, see §1.2(iv).

1.7.7 $\qquad H \leq G \leq A,$

with equality iff $a_1 = a_2 = \cdots = a_n$.

1.7.8 $\min(a_1, a_2, \ldots, a_n) \leq M(r) \leq \max(a_1, a_2, \ldots, a_n),$

with equality iff $a_1 = a_2 = \cdots = a_n$, or $r < 0$ and some $a_j = 0$.

1.7.9 $\qquad M(r) \leq M(s), \qquad r < s,$

with equality iff $a_1 = a_2 = \cdots = a_n$, or $s \leq 0$ and some $a_j = 0$.

1.7(iv) Jensen's Inequality

For f integrable on $[0,1]$, $a < f(x) < b$, and ϕ convex on (a,b) (§1.4(viii)),

1.7.10 $\qquad \phi\left(\int_0^1 f(x)\,dx\right) \leq \int_0^1 \phi(f(x))\,dx,$

1.7.11 $\qquad \exp\left(\int_0^1 \ln(f(x))\,dx\right) < \int_0^1 f(x)\,dx.$

For exp and ln see §4.2.

1.8 Fourier Series

1.8(i) Definitions and Elementary Properties

Formally,

1.8.1 $\quad f(x) = \tfrac{1}{2}a_0 + \sum_{n=1}^{\infty}(a_n \cos(nx) + b_n \sin(nx)),$

1.8.2
$$a_n = \frac{1}{\pi}\int_{-\pi}^{\pi} f(x)\cos(nx)\,dx, \qquad n = 0, 1, 2, \ldots,$$
$$b_n = \frac{1}{\pi}\int_{-\pi}^{\pi} f(x)\sin(nx)\,dx, \qquad n = 1, 2, \ldots.$$

The series (1.8.1) is called the *Fourier series* of $f(x)$, and a_n, b_n are the *Fourier coefficients* of $f(x)$.

If $f(-x) = f(x)$, then $b_n = 0$ for all n.
If $f(-x) = -f(x)$, then $a_n = 0$ for all n.

Alternative Form

1.8.3 $\qquad f(x) = \sum_{n=-\infty}^{\infty} c_n e^{inx},$

1.8.4 $\qquad c_n = \dfrac{1}{2\pi}\int_{-\pi}^{\pi} f(x) e^{-inx}\,dx.$

Bessel's Inequality

1.8.5 $\quad \tfrac{1}{2}a_0^2 + \sum_{n=1}^{\infty}(a_n^2 + b_n^2) \leq \dfrac{1}{\pi}\int_{-\pi}^{\pi}(f(x))^2\,dx.$

1.8.6 $\qquad \sum_{n=-\infty}^{\infty} |c_n|^2 \leq \dfrac{1}{2\pi}\int_{-\pi}^{\pi} |f(x)|^2\,dx.$

Asymptotic Estimates of Coefficients

If $f(x)$ is of period 2π, and $f^{(m)}(x)$ is piecewise continuous, then

1.8.7 $\qquad a_n, b_n, c_n = o(n^{-m}), \qquad n \to \infty.$

Uniqueness of Fourier Series

If $f(x)$ and $g(x)$ are continuous, have the same period and same Fourier coefficients, then $f(x) = g(x)$ for all x.

Lebesgue Constants

1.8.8 $\qquad L_n = \dfrac{1}{\pi}\int_0^{\pi} \dfrac{|\sin(n+\tfrac{1}{2})t|}{\sin(\tfrac{1}{2}t)}\,dt, \quad n = 0, 1, \ldots.$

As $n \to \infty$

1.8.9 $\qquad L_n \sim (4/\pi^2)\ln n;$

see Frenzen and Wong (1986).

Riemann–Lebesgue Lemma

For $f(x)$ piecewise continuous on $[a,b]$ and real λ,

1.8.10
$$\int_a^b f(x)e^{i\lambda x}\,dx \to 0, \qquad \text{as } \lambda \to \infty.$$

(1.8.10) continues to apply if either a or b or both are infinite and/or $f(x)$ has finitely many singularities in (a,b), provided that the integral converges uniformly (§1.5(iv)) at a,b, and the singularities for all sufficiently large λ.

1.8(ii) Convergence

Let $f(x)$ be an absolutely integrable function of period 2π, and continuous except at a finite number of points in any bounded interval. Then the series (1.8.1) converges to the sum

1.8.11
$$\tfrac{1}{2}f(x-) + \tfrac{1}{2}f(x+)$$

at every point at which $f(x)$ has both a left-hand derivative (that is, (1.4.4) applies when $h \to 0-$) and a right-hand derivative (that is, (1.4.4) applies when $h \to 0+$). The convergence is non-uniform, however, at points where $f(x-) \neq f(x+)$; see §6.16(i).

For other tests for convergence see Titchmarsh (1962, pp. 405–410).

1.8(iii) Integration and Differentiation

If a_n and b_n are the Fourier coefficients of a piecewise continuous function $f(x)$ on $[0, 2\pi]$, then

1.8.12
$$\int_0^x (f(t) - \tfrac{1}{2}a_0)\,dt = \sum_{n=1}^{\infty} \frac{a_n \sin(nx) + b_n(1 - \cos(nx))}{n},$$
$$0 \le x \le 2\pi.$$

If a function $f(x) \in C^2[0, 2\pi]$ is periodic, with period 2π, then the series obtained by differentiating the Fourier series for $f(x)$ term by term converges at every point to $f'(x)$.

1.8(iv) Transformations

Parseval's Formula

1.8.13
$$\frac{1}{\pi}\int_{-\pi}^{\pi} f(x)g(x)\,dx = \tfrac{1}{2}a_0 a_0' + \sum_{n=1}^{\infty}(a_n a_n' + b_n b_n'),$$

when $f(x)$ and $g(x)$ are square-integrable and a_n, b_n and a_n', b_n' are their respective Fourier coefficients.

Poisson's Summation Formula

Suppose that $f(x)$ is twice continuously differentiable and $f(x)$ and $|f''(x)|$ are integrable over $(-\infty, \infty)$. Then

1.8.14
$$\sum_{n=-\infty}^{\infty} f(x+n) = \sum_{n=-\infty}^{\infty} e^{2\pi inx}\int_{-\infty}^{\infty} f(t)e^{-2\pi int}\,dt.$$

An alternative formulation is as follows. Suppose that $f(x)$ is continuous and of bounded variation on $[0, \infty)$. Suppose also that $f(x)$ is integrable on $[0, \infty)$ and $f(x) \to 0$ as $x \to \infty$. Then

1.8.15
$$\tfrac{1}{2}f(0) + \sum_{n=1}^{\infty} f(n) = \int_0^{\infty} f(x)\,dx$$
$$+ 2\sum_{n=1}^{\infty}\int_0^{\infty} f(x)\cos(2\pi nx)\,dx.$$

As a special case

1.8.16
$$\sum_{n=-\infty}^{\infty} e^{-(n+x)^2\omega}$$
$$= \sqrt{\frac{\pi}{\omega}}\left(1 + 2\sum_{n=1}^{\infty} e^{-n^2\pi^2/\omega}\cos(2n\pi x)\right),$$
$$\Re\omega > 0.$$

1.8(v) Examples

For collections of Fourier-series expansions see Prudnikov et al. (1986a, v. 1, pp. 725–740), Gradshteyn and Ryzhik (2000, pp. 45–49), and Oberhettinger (1973).

1.9 Calculus of a Complex Variable

1.9(i) Complex Numbers

1.9.1
$$z = x + iy, \qquad x, y \in \mathbb{R}.$$

Real and Imaginary Parts

1.9.2
$$\Re z = x, \quad \Im z = y.$$

Polar Representation

1.9.3
$$x = r\cos\theta, \quad y = r\sin\theta,$$

where

1.9.4
$$r = (x^2 + y^2)^{1/2},$$

and when $z \neq 0$,

1.9.5
$$\theta = \omega, \ \pi - \omega, \ -\pi + \omega, \text{ or } -\omega,$$

according as z lies in the 1st, 2nd, 3rd, or 4th quadrants. Here

1.9.6
$$\omega = \arctan(|y/x|) \in \left[0, \tfrac{1}{2}\pi\right].$$

1.9 Calculus of a Complex Variable

Modulus and Phase

1.9.7 $$|z| = r, \quad \operatorname{ph} z = \theta + 2n\pi, \quad n \in \mathbb{Z}.$$

The *principal value* of $\operatorname{ph} z$ corresponds to $n = 0$, that is, $-\pi \leq \operatorname{ph} z \leq \pi$. It is single-valued on $\mathbb{C} \setminus \{0\}$, except on the interval $(-\infty, 0)$ where it is discontinuous and two-valued. *Unless indicated otherwise*, these principal values are assumed throughout this Handbook. (However, if we require a principal value to be single-valued, then we can restrict $-\pi < \operatorname{ph} z \leq \pi$.)

1.9.8 $$|\Re z| \leq |z|, \quad |\Im z| \leq |z|,$$

1.9.9 $$z = re^{i\theta},$$

where

1.9.10 $$e^{i\theta} = \cos\theta + i\sin\theta;$$

see §4.14.

Complex Conjugate

1.9.11 $$\overline{z} = x - iy,$$

1.9.12 $$|\overline{z}| = |z|,$$

1.9.13 $$\operatorname{ph}\overline{z} = -\operatorname{ph} z.$$

Arithmetic Operations

If $z_1 = x_1 + iy_1$, $z_2 = x_2 + iy_2$, then

1.9.14 $$z_1 \pm z_2 = x_1 \pm x_2 + i(y_1 \pm y_2),$$

1.9.15 $$z_1 z_2 = x_1 x_2 - y_1 y_2 + i(x_1 y_2 + x_2 y_1),$$

1.9.16 $$\frac{z_1}{z_2} = \frac{z_1 \overline{z_2}}{|z_2|^2} = \frac{x_1 x_2 + y_1 y_2 + i(x_2 y_1 - x_1 y_2)}{x_2^2 + y_2^2},$$

provided that $z_2 \neq 0$. Also,

1.9.17 $$|z_1 z_2| = |z_1| \, |z_2|,$$

1.9.18 $$\operatorname{ph}(z_1 z_2) = \operatorname{ph} z_1 + \operatorname{ph} z_2,$$

1.9.19 $$\left|\frac{z_1}{z_2}\right| = \frac{|z_1|}{|z_2|},$$

1.9.20 $$\operatorname{ph}\frac{z_1}{z_2} = \operatorname{ph} z_1 - \operatorname{ph} z_2.$$

Equations (1.9.18) and (1.9.20) hold for general values of the phases, but not necessarily for the principal values.

Powers

1.9.21 $$z^n = \left(x^n - \binom{n}{2}x^{n-2}y^2 + \binom{n}{4}x^{n-4}y^4 - \cdots\right) + i\left(\binom{n}{1}x^{n-1}y - \binom{n}{3}x^{n-3}y^3 + \cdots\right),$$
$$n = 1, 2, \ldots.$$

DeMoivre's Theorem

1.9.22 $$\cos n\theta + i\sin n\theta = (\cos\theta + i\sin\theta)^n, \quad n \in \mathbb{Z}.$$

Triangle Inequality

1.9.23 $$||z_1| - |z_2|| \leq |z_1 + z_2| \leq |z_1| + |z_2|.$$

1.9(ii) Continuity, Point Sets, and Differentiation

Continuity

A function $f(z)$ is *continuous* at a point z_0 if $\lim_{z \to z_0} f(z) = f(z_0)$. That is, given any positive number ϵ, however small, we can find a positive number δ such that $|f(z) - f(z_0)| < \epsilon$ for all z in the open disk $|z - z_0| < \delta$.

A function of two complex variables $f(z, w)$ is *continuous* at (z_0, w_0) if $\lim_{(z,w) \to (z_0,w_0)} f(z, w) = f(z_0, w_0)$; compare (1.5.1) and (1.5.2).

Point Sets in $\mathbb{C}$

A *neighborhood of a point* z_0 is a disk $|z - z_0| < \delta$. An *open set* in $\mathbb{C}$ is one in which each point has a neighborhood that is contained in the set.

A point z_0 is a *limit point* (*limiting point* or *accumulation point*) of a set of points S in $\mathbb{C}$ (or $\mathbb{C} \cup \infty$) if every neighborhood of z_0 contains a point of S distinct from z_0. (z_0 may or may not belong to S.) As a consequence, every neighborhood of a limit point of S contains an infinite number of points of S. Also, the union of S and its limit points is the *closure* of S.

A *domain* D, say, is an open set in $\mathbb{C}$ that is *connected*, that is, any two points can be joined by a polygonal arc (a finite chain of straight-line segments) lying in the set. Any point whose neighborhoods always contain members and nonmembers of D is a *boundary point* of D. When its boundary points are added the domain is said to be *closed*, but unless specified otherwise a domain is assumed to be open.

A *region* is an open domain together with none, some, or all of its boundary points. Points of a region that are not boundary points are called *interior points*.

A function $f(z)$ is *continuous on a region* R if for each point z_0 in R and any given number $\epsilon \; (> 0)$ we can find a neighborhood of z_0 such that $|f(z) - f(z_0)| < \epsilon$ for all points z in the intersection of the neighborhood with R.

Differentiation

A function $f(z)$ is *differentiable* at a point z if the following limit exists:

1.9.24 $$f'(z) = \frac{df}{dz} = \lim_{h \to 0} \frac{f(z+h) - f(z)}{h}.$$

Differentiability automatically implies continuity.

Cauchy–Riemann Equations

If $f'(z)$ exists at $z = x+iy$ and $f(z) = u(x,y)+iv(x,y)$, then

1.9.25
$$\frac{\partial u}{\partial x} = \frac{\partial v}{\partial y}, \quad \frac{\partial u}{\partial y} = -\frac{\partial v}{\partial x}$$

at (x,y).

Conversely, if at a given point (x,y) the partial derivatives $\partial u/\partial x$, $\partial u/\partial y$, $\partial v/\partial x$, and $\partial v/\partial y$ exist, are continuous, and satisfy (1.9.25), then $f(z)$ is differentiable at $z = x+iy$.

Analyticity

A function $f(z)$ is said to be *analytic* (*holomorphic*) at $z = z_0$ if it is differentiable in a neighborhood of z_0.

A function $f(z)$ is *analytic in a domain* D if it is analytic at each point of D. A function analytic at every point of $\mathbb{C}$ is said to be *entire*.

If $f(z)$ is analytic in an open domain D, then each of its derivatives $f'(z)$, $f''(z)$, ... exists and is analytic in D.

Harmonic Functions

If $f(z) = u(x,y)+iv(x,y)$ is analytic in an open domain D, then u and v are *harmonic* in D, that is,

1.9.26
$$\frac{\partial^2 u}{\partial x^2} + \frac{\partial^2 u}{\partial y^2} = \frac{\partial^2 v}{\partial x^2} + \frac{\partial^2 v}{\partial y^2} = 0,$$

or in polar form ((1.9.3)) u and v satisfy

1.9.27
$$\frac{\partial^2 u}{\partial r^2} + \frac{1}{r}\frac{\partial u}{\partial r} + \frac{1}{r^2}\frac{\partial^2 u}{\partial \theta^2} = 0$$

at all points of D.

1.9(iii) Integration

An *arc* C is given by $z(t) = x(t) + iy(t)$, $a \le t \le b$, where x and y are continuously differentiable. If $x(t)$ and $y(t)$ are continuous and $x'(t)$ and $y'(t)$ are piecewise continuous, then $z(t)$ defines a *contour*.

A contour is *simple* if it contains no multiple points, that is, for every pair of distinct values t_1, t_2 of t, $z(t_1) \neq z(t_2)$. A *simple closed contour* is a simple contour, except that $z(a) = z(b)$.

Next,

1.9.28
$$\int_C f(z)\,dz = \int_a^b f(z(t))(x'(t)+iy'(t))\,dt,$$

for a contour C and $f(z(t))$ continuous, $a \le t \le b$. If $f(z(t_0)) = \infty$, $a \le t_0 \le b$, then the integral is defined analogously to the infinite integrals in §1.4(v). Similarly when $a = -\infty$ or $b = +\infty$.

Jordan Curve Theorem

Any simple closed contour C divides $\mathbb{C}$ into two open domains that have C as common boundary. One of these domains is bounded and is called the *interior domain* of C; the other is unbounded and is called the *exterior domain* of C.

Cauchy's Theorem

If $f(z)$ is continuous within and on a simple closed contour C and analytic within C, then

1.9.29
$$\int_C f(z)\,dz = 0.$$

Cauchy's Integral Formula

If $f(z)$ is continuous within and on a simple closed contour C and analytic within C, and if z_0 is a point within C, then

1.9.30
$$f(z_0) = \frac{1}{2\pi i}\int_C \frac{f(z)}{z-z_0}\,dz,$$

and

1.9.31
$$f^{(n)}(z_0) = \frac{n!}{2\pi i}\int_C \frac{f(z)}{(z-z_0)^{n+1}}\,dz, \quad n = 1, 2, 3, \ldots,$$

provided that in both cases C is described in the positive rotational (anticlockwise) sense.

Liouville's Theorem

Any bounded entire function is a constant.

Winding Number

If C is a closed contour, and $z_0 \notin C$, then

1.9.32
$$\frac{1}{2\pi i}\int_C \frac{1}{z-z_0}\,dz = \mathcal{N}(C, z_0),$$

where $\mathcal{N}(C, z_0)$ is an integer called the *winding number of C with respect to z_0*. If C is simple and oriented in the positive rotational sense, then $\mathcal{N}(C, z_0)$ is 1 or 0 depending whether z_0 is inside or outside C.

Mean Value Property

For $u(z)$ harmonic,

1.9.33
$$u(z) = \frac{1}{2\pi}\int_0^{2\pi} u(z + re^{i\phi})\,d\phi.$$

Poisson Integral

If $h(w)$ is continuous on $|w| = R$, then with $z = re^{i\theta}$

1.9.34
$$u(re^{i\theta}) = \frac{1}{2\pi}\int_0^{2\pi}\frac{(R^2-r^2)h(Re^{i\phi})\,d\phi}{R^2 - 2Rr\cos(\phi-\theta) + r^2}$$

is harmonic in $|z| < R$. Also with $|w| = R$, $\lim_{z \to w} u(z) = h(w)$ as $z \to w$ within $|z| < R$.

1.9(iv) Conformal Mapping

The *extended complex plane*, $\mathbb{C} \cup \{\infty\}$, consists of the points of the complex plane $\mathbb{C}$ together with an ideal point ∞ called the *point at infinity*. A system of *open disks around infinity* is given by

1.9.35
$$S_r = \{z \mid |z| > 1/r\} \cup \{\infty\}, \quad 0 < r < \infty.$$

Each S_r is a *neighborhood* of ∞. Also,

1.9.36
$$\infty \pm z = z \pm \infty = \infty,$$

1.9 CALCULUS OF A COMPLEX VARIABLE

1.9.37 $$\infty \cdot z = z \cdot \infty = \infty, \qquad z \neq 0,$$

1.9.38 $$z/\infty = 0,$$

1.9.39 $$z/0 = \infty, \qquad z \neq 0.$$

A function $f(z)$ is *analytic at* ∞ if $g(z) = f(1/z)$ is analytic at $z = 0$, and we set $f'(\infty) = g'(0)$.

Conformal Transformation

Suppose $f(z)$ is analytic in a domain D and C_1, C_2 are two arcs in D passing through z_0. Let C_1', C_2' be the images of C_1 and C_2 under the mapping $w = f(z)$. The *angle between* C_1 *and* C_2 *at* z_0 is the angle between the tangents to the two arcs at z_0, that is, the difference of the signed angles that the tangents make with the positive direction of the real axis. If $f'(z_0) \neq 0$, then the angle between C_1 and C_2 equals the angle between C_1' and C_2' both in magnitude and sense. We then say that the mapping $w = f(z)$ is *conformal* (angle-preserving) at z_0.

The *linear transformation* $f(z) = az + b$, $a \neq 0$, has $f'(z) = a$ and $w = f(z)$ maps $\mathbb{C}$ conformally onto $\mathbb{C}$.

Bilinear Transformation

1.9.40 $$w = f(z) = \frac{az + b}{cz + d}, \quad ad - bc \neq 0, \ c \neq 0.$$

1.9.41 $$f(-d/c) = \infty, \quad f(\infty) = a/c.$$

1.9.42 $$f'(z) = \frac{ad - bc}{(cz + d)^2}, \qquad z \neq -d/c.$$

1.9.43 $$f'(\infty) = \frac{bc - ad}{c^2}.$$

1.9.44 $$z = \frac{dw - b}{-cw + a}.$$

The transformation (1.9.40) is a one-to-one conformal mapping of $\mathbb{C} \cup \{\infty\}$ onto itself.

The *cross ratio* of $z_1, z_2, z_3, z_4 \in \mathbb{C} \cup \{\infty\}$ is defined by

1.9.45 $$\frac{(z_1 - z_2)(z_3 - z_4)}{(z_1 - z_4)(z_3 - z_2)},$$

or its limiting form, and is invariant under bilinear transformations.

Other names for the bilinear transformation are *fractional linear transformation*, *homographic transformation*, and *Möbius transformation*.

1.9(v) Infinite Sequences and Series

A sequence $\{z_n\}$ *converges* to z if $\lim_{n\to\infty} z_n = z$. For $z_n = x_n + iy_n$, the sequence $\{z_n\}$ converges iff the sequences $\{x_n\}$ and $\{y_n\}$ separately converge. A series $\sum_{n=0}^{\infty} z_n$ *converges* if the sequence $s_n = \sum_{k=0}^{n} z_k$ converges. The series is *divergent* if s_n does not converge. The series converges *absolutely* if $\sum_{n=0}^{\infty} |z_n|$ converges.

A series $\sum_{n=0}^{\infty} z_n$ converges (diverges) absolutely when $\lim_{n\to\infty} |z_n|^{1/n} < 1$ (> 1), or when $\lim_{n\to\infty} |z_{n+1}/z_n| < 1$ (> 1). Absolutely convergent series are also convergent.

Let $\{f_n(z)\}$ be a sequence of functions defined on a set S. This sequence *converges pointwise* to a function $f(z)$ if

1.9.46 $$f(z) = \lim_{n\to\infty} f_n(z)$$

for each $z \in S$. The sequence *converges uniformly* on S, if for every $\epsilon > 0$ there exists an integer N, independent of z, such that

1.9.47 $$|f_n(z) - f(z)| < \epsilon$$

for all $z \in S$ and $n \geq N$.

A series $\sum_{n=0}^{\infty} f_n(z)$ *converges uniformly* on S, if the sequence $s_n(z) = \sum_{k=0}^{n} f_k(z)$ converges uniformly on S.

Weierstrass M-test

Suppose $\{M_n\}$ is a sequence of real numbers such that $\sum_{n=0}^{\infty} M_n$ converges and $|f_n(z)| \leq M_n$ for all $z \in S$ and all $n \geq 0$. Then the series $\sum_{n=0}^{\infty} f_n(z)$ converges uniformly on S.

A doubly-infinite series $\sum_{n=-\infty}^{\infty} f_n(z)$ converges (uniformly) on S iff each of the series $\sum_{n=0}^{\infty} f_n(z)$ and $\sum_{n=1}^{\infty} f_{-n}(z)$ converges (uniformly) on S.

1.9(vi) Power Series

For a series $\sum_{n=0}^{\infty} a_n(z - z_0)^n$ there is a number R, $0 \leq R \leq \infty$, such that the series converges for all z in $|z - z_0| < R$ and diverges for z in $|z - z_0| > R$. The circle $|z - z_0| = R$ is called the *circle of convergence* of the series, and R is the *radius of convergence*. Inside the circle the sum of the series is an analytic function $f(z)$. For z in $|z - z_0| \leq \rho$ ($< R$), the convergence is absolute and uniform. Moreover,

1.9.48 $$a_n = \frac{f^{(n)}(z_0)}{n!},$$

and

1.9.49 $$R = \liminf_{n\to\infty} |a_n|^{-1/n}.$$

For the converse of this result see §1.10(i).

Operations

When $\sum a_n z^n$ and $\sum b_n z^n$ both converge

1.9.50 $$\sum_{n=0}^{\infty}(a_n \pm b_n)z^n = \sum_{n=0}^{\infty} a_n z^n \pm \sum_{n=0}^{\infty} b_n z^n,$$

and

1.9.51 $$\left(\sum_{n=0}^{\infty} a_n z^n\right)\left(\sum_{n=0}^{\infty} b_n z^n\right) = \sum_{n=0}^{\infty} c_n z^n,$$

where

1.9.52 $$c_n = \sum_{k=0}^{n} a_k b_{n-k}.$$

Next, let

1.9.53 $$f(z) = a_0 + a_1 z + a_2 z^2 + \cdots, \qquad a_0 \neq 0.$$

Then the expansions (1.9.54), (1.9.57), and (1.9.60) hold for all sufficiently small $|z|$.

1.9.54 $$\frac{1}{f(z)} = b_0 + b_1 z + b_2 z^2 + \cdots,$$

where

1.9.55 $b_0 = 1/a_0, \quad b_1 = -a_1/a_0^2, \quad b_2 = (a_1^2 - a_0 a_2)/a_0^3,$

1.9.56
$$b_n = -(a_1 b_{n-1} + a_2 b_{n-2} + \cdots + a_n b_0)/a_0, \quad n \geq 1.$$

With $a_0 = 1$,

1.9.57 $\quad \ln f(z) = q_1 z + q_2 z^2 + q_3 z^3 + \cdots,$

(principal value), where

1.9.58
$$q_1 = a_1, \quad q_2 = (2a_2 - a_1^2)/2,$$
$$q_3 = (3a_3 - 3a_1 a_2 + a_1^3)/3,$$

and

1.9.59
$$q_n = (na_n - (n-1)a_1 q_{n-1} - (n-2)a_2 q_{n-2} - \cdots$$
$$- a_{n-1} q_1)/n,$$
$$n \geq 2.$$

Also,

1.9.60 $\quad (f(z))^\nu = p_0 + p_1 z + p_2 z^2 + \cdots,$

(principal value), where $\nu \in \mathbb{C}$,

1.9.61 $\quad p_0 = 1, \quad p_1 = \nu a_1, \quad p_2 = \nu((\nu-1)a_1^2 + 2a_2)/2,$

and

1.9.62
$$p_n = ((\nu - n + 1)a_1 p_{n-1} + (2\nu - n + 2)a_2 p_{n-2} + \cdots$$
$$+ ((n-1)\nu - 1)a_{n-1} p_1 + n\nu a_n)/n,$$
$$n \geq 1.$$

For the definitions of the principal values of $\ln f(z)$ and $(f(z))^\nu$ see §§4.2(i) and 4.2(iv).

Lastly, a power series can be differentiated any number of times within its circle of convergence:

1.9.63
$$f^{(m)}(z) = \sum_{n=0}^{\infty} (n+1)_m a_{n+m}(z-z_0)^n,$$
$$|z - z_0| < R, \, m = 0, 1, 2, \ldots.$$

1.9(vii) Inversion of Limits

Double Sequences and Series

A set of complex numbers $\{z_{m,n}\}$ where m and n take all positive integer values is called a *double sequence*. It *converges to* z if for every $\epsilon > 0$, there is an integer N such that

1.9.64 $\quad\quad\quad |z_{m,n} - z| < \epsilon$

for all $m, n \geq N$. Suppose $\{z_{m,n}\}$ converges to z and the repeated limits

1.9.65 $\quad \lim_{m \to \infty} \left(\lim_{n \to \infty} z_{m,n} \right), \quad \lim_{n \to \infty} \left(\lim_{m \to \infty} z_{m,n} \right)$

exist. Then both repeated limits equal z.

A *double series* is the limit of the double sequence

1.9.66 $$z_{p,q} = \sum_{m=0}^{p} \sum_{n=0}^{q} \zeta_{m,n}.$$

If the limit exists, then the double series is *convergent*; otherwise it is *divergent*. The double series is *absolutely convergent* if it is convergent when $\zeta_{m,n}$ is replaced by $|\zeta_{m,n}|$.

If a double series is absolutely convergent, then it is also convergent and its sum is given by either of the repeated sums

1.9.67 $$\sum_{m=0}^{\infty} \left(\sum_{n=0}^{\infty} \zeta_{m,n} \right), \quad \sum_{n=0}^{\infty} \left(\sum_{m=0}^{\infty} \zeta_{m,n} \right).$$

Term-by-Term Integration

Suppose the series $\sum_{n=0}^{\infty} f_n(z)$, where $f_n(z)$ is continuous, converges uniformly on every *compact set* of a domain D, that is, every closed and bounded set in D. Then

1.9.68 $$\int_C \sum_{n=0}^{\infty} f_n(z)\, dz = \sum_{n=0}^{\infty} \int_C f_n(z)\, dz$$

for any finite contour C in D.

Dominated Convergence Theorem

Let (a, b) be a finite or infinite interval, and $f_0(t), f_1(t), \ldots$ be real or complex continuous functions, $t \in (a, b)$. Suppose $\sum_{n=0}^{\infty} f_n(t)$ converges uniformly in any compact interval in (a, b), and at least one of the following two conditions is satisfied:

1.9.69 $$\int_a^b \sum_{n=0}^{\infty} |f_n(t)|\, dt < \infty,$$

1.9.70 $$\sum_{n=0}^{\infty} \int_a^b |f_n(t)|\, dt < \infty.$$

Then

1.9.71 $$\int_a^b \sum_{n=0}^{\infty} f_n(t)\, dt = \sum_{n=0}^{\infty} \int_a^b f_n(t)\, dt.$$

1.10 Functions of a Complex Variable

1.10(i) Taylor's Theorem for Complex Variables

Let $f(z)$ be analytic on the disk $|z - z_0| < R$. Then

1.10.1 $$f(z) = \sum_{n=0}^{\infty} \frac{f^{(n)}(z_0)}{n!} (z - z_0)^n.$$

The right-hand side is the *Taylor series for* $f(z)$ *at* $z = z_0$, and its radius of convergence is at least R.

Examples

1.10.2 $$e^z = 1 + \frac{z}{1!} + \frac{z^2}{2!} + \cdots, \qquad |z| < \infty,$$

1.10.3 $$\ln(1+z) = z - \frac{z^2}{2} + \frac{z^3}{3} - \cdots, \qquad |z| < 1,$$

1.10.4 $$(1-z)^{-\alpha} = 1 + \alpha z + \frac{\alpha(\alpha+1)}{2!}z^2 + \frac{\alpha(\alpha+1)(\alpha+2)}{3!}z^3 + \cdots, \qquad |z| < 1.$$

Again, in these examples $\ln(1+z)$ and $(1-z)^{-\alpha}$ have their principal values; see §§4.2(i) and 4.2(iv).

Zeros

An analytic function $f(z)$ has a *zero of order* (or *multiplicity*) m (≥ 1) at z_0 if the first nonzero coefficient in its Taylor series at z_0 is that of $(z-z_0)^m$. When $m=1$ the zero is *simple*.

1.10(ii) Analytic Continuation

Let $f_1(z)$ be analytic in a domain D_1. If $f_2(z)$, analytic in D_2, equals $f_1(z)$ on an arc in $D = D_1 \cap D_2$, or on just an infinite number of points with a limit point in D, then they are equal throughout D and $f_2(z)$ is called an *analytic continuation* of $f_1(z)$. We write (f_1, D_1), (f_2, D_2) to signify this continuation.

Suppose $z(t) = x(t) + iy(t)$, $a \leq t \leq b$, is an arc and $a = t_0 < t_1 < \cdots < t_n = b$. Suppose the subarc $z(t)$, $t \in [t_{j-1}, t_j]$ is contained in a domain D_j, $j = 1, \ldots, n$. The function $f_1(z)$ on D_1 is said to be *analytically continued along the path* $z(t)$, $a \leq t \leq b$, if there is a chain $(f_1, D_1), (f_2, D_2), \ldots, (f_n, D_n)$.

Analytic continuation is a powerful aid in establishing transformations or functional equations for complex variables, because it enables the problem to be reduced to: (a) deriving the transformation (or functional equation) with real variables; followed by (b) finding the domain on which the transformed function is analytic.

Schwarz Reflection Principle

Let C be a simple closed contour consisting of a segment AB of the real axis and a contour in the upper half-plane joining the ends of AB. Also, let $f(z)$ be analytic within C, continuous within and on C, and real on AB. Then $f(z)$ can be continued analytically across AB by *reflection*, that is,

1.10.5 $$f(\overline{z}) = \overline{f(z)}.$$

1.10(iii) Laurent Series

Suppose $f(z)$ is analytic in the *annulus* $r_1 < |z - z_0| < r_2$, $0 \leq r_1 < r_2 \leq \infty$, and $r \in (r_1, r_2)$. Then

1.10.6 $$f(z) = \sum_{n=-\infty}^{\infty} a_n (z - z_0)^n,$$

where

1.10.7 $$a_n = \frac{1}{2\pi i} \int_{|z-z_0|=r} \frac{f(z)}{(z-z_0)^{n+1}} \, dz,$$

and the integration contour is described once in the positive sense. The series (1.10.6) converges uniformly and absolutely on compact sets in the annulus.

Let $r_1 = 0$, so that the annulus becomes the *punctured neighborhood* N: $0 < |z - z_0| < r_2$, and assume that $f(z)$ is analytic in N, but not at z_0. Then $z = z_0$ is an *isolated singularity* of $f(z)$. This singularity is *removable* if $a_n = 0$ for all $n < 0$, and in this case the Laurent series becomes the Taylor series. Next, z_0 is a *pole* if $a_n \neq 0$ for at least one, but only finitely many, negative n. If $-n$ is the first negative integer (counting from $-\infty$) with $a_{-n} \neq 0$, then z_0 is a *pole of order* (or *multiplicity*) n. Lastly, if $a_n \neq 0$ for infinitely many negative n, then z_0 is an *isolated essential singularity*.

The singularities of $f(z)$ at infinity are classified in the same way as the singularities of $f(1/z)$ at $z = 0$.

An isolated singularity z_0 is always removable when $\lim_{z \to z_0} f(z)$ exists, for example $(\sin z)/z$ at $z = 0$.

The coefficient a_{-1} of $(z-z_0)^{-1}$ in the Laurent series for $f(z)$ is called the *residue* of $f(z)$ at z_0, and denoted by $\operatorname{res}_{z=z_0}[f(z)]$, $\operatorname*{res}_{z=z_0}[f(z)]$, or (when there is no ambiguity) $\operatorname{res}[f(z)]$.

A function whose only singularities, other than the point at infinity, are poles is called a *meromorphic function*. If the poles are infinite in number, then the point at infinity is called an *essential singularity*: it is the limit point of the poles.

Picard's Theorem

In any neighborhood of an isolated essential singularity, however small, an analytic function assumes every value in $\mathbb{C}$ with at most one exception.

1.10(iv) Residue Theorem

If $f(z)$ is analytic within a simple closed contour C, and continuous within and on C—except in both instances for a finite number of singularities within C—then

1.10.8 $$\frac{1}{2\pi i} \int_C f(z) \, dz = \text{sum of the residues of } f(z) \text{ within } C.$$

Here and elsewhere in this subsection the path C is described in the positive sense.

Phase (or Argument) Principle

If the singularities within C are poles and $f(z)$ is analytic and nonvanishing on C, then

1.10.9 $\quad N - P = \dfrac{1}{2\pi i} \displaystyle\int_C \dfrac{f'(z)}{f(z)} \, dz = \dfrac{1}{2\pi} \Delta_C(\operatorname{ph} f(z)),$

where N and P are respectively the numbers of zeros and poles, counting multiplicity, of f within C, and $\Delta_C(\operatorname{ph} f(z))$ is the change in any continuous branch of $\operatorname{ph}(f(z))$ as z passes once around C in the positive sense. For examples of applications see Olver (1997b, pp. 252–254).

In addition,

1.10.10
$$\dfrac{1}{2\pi i} \int_C \dfrac{z f'(z)}{f(z)} \, dz = \text{(sum of locations of zeros)}$$
$$- \text{(sum of locations of poles)},$$

each location again being counted with multiplicity equal to that of the corresponding zero or pole.

Rouché's Theorem

If $f(z)$ and $g(z)$ are analytic on and inside a simple closed contour C, and $|g(z)| < |f(z)|$ on C, then $f(z)$ and $f(z) + g(z)$ have the same number of zeros inside C.

1.10(v) Maximum-Modulus Principle

Analytic Functions

If $f(z)$ is analytic in a domain D, $z_0 \in D$ and $|f(z)| \leq |f(z_0)|$ for all $z \in D$, then $f(z)$ is a constant in D.

Let D be a bounded domain with boundary ∂D and let $\overline{D} = D \cup \partial D$. If $f(z)$ is continuous on $\overline{D}$ and analytic in D, then $|f(z)|$ attains its maximum on ∂D.

Harmonic Functions

If $u(z)$ is harmonic in D, $z_0 \in D$, and $u(z) \leq u(z_0)$ for all $z \in D$, then $u(z)$ is constant in D. Moreover, if D is bounded and $u(z)$ is continuous on $\overline{D}$ and harmonic in D, then $u(z)$ is maximum at some point on ∂D.

Schwarz's Lemma

In $|z| < R$, if $f(z)$ is analytic, $|f(z)| \leq M$, and $f(0) = 0$, then

1.10.11 $\quad |f(z)| \leq \dfrac{M|z|}{R} \quad \text{and} \quad |f'(0)| \leq \dfrac{M}{R}.$

Equalities hold iff $f(z) = Az$, where A is a constant such that $|A| = M/R$.

1.10(vi) Multivalued Functions

Functions which have more than one value at a given point z are called *multivalued* (or *many-valued*) functions. Let $F(z)$ be a multivalued function and D be a domain. If we can assign a unique value $f(z)$ to $F(z)$ at each point of D, and $f(z)$ is analytic on D, then $f(z)$ is a *branch* of $F(z)$.

Example

$F(z) = \sqrt{z}$ is two-valued for $z \neq 0$. If $D = \mathbb{C} \setminus (-\infty, 0]$ and $z = re^{i\theta}$, then one branch is $\sqrt{r}e^{i\theta/2}$, the other branch is $-\sqrt{r}e^{i\theta/2}$, with $-\pi < \theta < \pi$ in both cases. Similarly if $D = \mathbb{C} \setminus [0, \infty)$, then one branch is $\sqrt{r}e^{i\theta/2}$, the other branch is $-\sqrt{r}e^{i\theta/2}$, with $0 < \theta < 2\pi$ in both cases.

A *cut domain* is one from which the points on finitely many nonintersecting simple contours (§1.9(iii)) have been removed. Each contour is called a *cut*. A *cut neighborhood* is formed by deleting a ray emanating from the center. (Or more generally, a simple contour that starts at the center and terminates on the boundary.)

Suppose $F(z)$ is multivalued and a is a point such that there exists a branch of $F(z)$ in a cut neighborhood of a, but there does not exist a branch of $F(z)$ in any punctured neighborhood of a. Then a is a *branch point* of $F(z)$. For example, $z = 0$ is a branch point of $\sqrt{z}$.

Branches can be constructed in two ways:

(a) By introducing appropriate cuts from the branch points and restricting $F(z)$ to be single-valued in the cut plane (or domain).

(b) By specifying the value of $F(z)$ at a point z_0 (not a branch point), and requiring $F(z)$ to be continuous on any path that begins at z_0 and does not pass through any branch points or other singularities of $F(z)$.

If the path circles a branch point at $z = a$ k times in the positive sense, and returns to z_0 without encircling any other branch point, then its value is denoted conventionally as $F((z_0 - a)e^{2k\pi i} + a)$.

Example

Let α and β be real or complex numbers that are not integers. The function $F(z) = (1-z)^\alpha (1+z)^\beta$ is many-valued with branch points at ± 1. Branches of $F(z)$ can be defined, for example, in the cut plane D obtained from $\mathbb{C}$ by removing the real axis from 1 to ∞ and from -1 to $-\infty$; see Figure 1.10.1. One such branch is obtained by assigning $(1-z)^\alpha$ and $(1+z)^\beta$ their principal values (§4.2(iv)).

Figure 1.10.1: Domain D.

Alternatively, take z_0 to be any point in D and set $F(z_0) = e^{\alpha \ln(1-z_0)} e^{\beta \ln(1+z_0)}$ where the logarithms assume their principal values. (Thus if z_0 is in the interval $(-1, 1)$, then the logarithms are real.) Then the value of $F(z)$ at any other point is obtained by analytic continuation.

1.10(vii) Inverse Functions

Thus if $F(z)$ is continued along a path that circles $z = 1$ m times in the positive sense and returns to z_0 without circling $z = -1$, then $F((z_0 - 1)e^{2m\pi i} + 1) = e^{\alpha \ln(1-z_0)} e^{\beta \ln(1+z_0)} e^{2\pi i m\alpha}$. If the path also circles $z = -1$ n times in the clockwise or negative sense before returning to z_0, then the value of $F(z_0)$ becomes $e^{\alpha \ln(1-z_0)} e^{\beta \ln(1+z_0)} e^{2\pi i m\alpha} e^{-2\pi i n\beta}$.

1.10(vii) Inverse Functions

Lagrange Inversion Theorem

Suppose $f(z)$ is analytic at $z = z_0$, $f'(z_0) \neq 0$, and $f(z_0) = w_0$. Then the equation

1.10.12 $$f(z) = w$$

has a unique solution $z = F(w)$ analytic at $w = w_0$, and

1.10.13 $$F(w) = z_0 + \sum_{n=1}^{\infty} F_n (w - w_0)^n$$

in a neighborhood of w_0, where nF_n is the residue of $1/(f(z) - f(z_0))^n$ at $z = z_0$. (In other words nF_n is the coefficient of $(z-z_0)^{-1}$ in the Laurent expansion of $1/(f(z) - f(z_0))^n$ in powers of $(z - z_0)$; compare §1.10(iii).)

Furthermore, if $g(z)$ is analytic at z_0, then

1.10.14 $$g(F(w)) = g(z_0) + \sum_{n=1}^{\infty} G_n (w - w_0)^n,$$

where nG_n is the residue of $g'(z)/(f(z) - f(z_0))^n$ at $z = z_0$.

Extended Inversion Theorem

Suppose that

1.10.15 $$f(z) = f(z_0) + \sum_{n=0}^{\infty} f_n (z - z_0)^{\mu+n},$$

where $\mu > 0$, $f_0 \neq 0$, and the series converges in a neighborhood of z_0. (For example, when μ is an integer $f(z) - f(z_0)$ has a zero of order μ at z_0.) Let $w_0 = f(z_0)$. Then (1.10.12) has a solution $z = F(w)$, where

1.10.16 $$F(w) = z_0 + \sum_{n=1}^{\infty} F_n (w - w_0)^{n/\mu}$$

in a neighborhood of w_0, nF_n being the residue of $1/(f(z) - f(z_0))^{n/\mu}$ at $z = z_0$.

It should be noted that different branches of $(w - w_0)^{1/\mu}$ used in forming $(w - w_0)^{n/\mu}$ in (1.10.16) give rise to different solutions of (1.10.12). Also, if in addition $g(z)$ is analytic at z_0, then

1.10.17 $$g(F(w)) = g(z_0) + \sum_{n=1}^{\infty} G_n (w - w_0)^{n/\mu},$$

where nG_n is the residue of $g'(z)/(f(z) - f(z_0))^{n/\mu}$ at $z = z_0$.

1.10(viii) Functions Defined by Contour Integrals

Let D be a domain and $[a, b]$ be a closed finite segment of the real axis. Assume that for each $t \in [a, b]$, $f(z, t)$ is an analytic function of z in D, and also that $f(z, t)$ is a continuous function of both variables. Then

1.10.18 $$F(z) = \int_a^b f(z, t)\, dt$$

is analytic in D and its derivatives of all orders can be found by differentiating under the sign of integration.

This result is also true when $b = \infty$, or when $f(z, t)$ has a singularity at $t = b$, with the following conditions. For each $t \in [a, b)$, $f(z, t)$ is analytic in D; $f(z, t)$ is a continuous function of both variables when $z \in D$ and $t \in [a, b)$; the integral (1.10.18) converges at b, and this convergence is uniform with respect to z in every compact subset S of D.

The last condition means that given ϵ (> 0) there exists a number $a_0 \in [a, b)$ that is independent of z and is such that

1.10.19 $$\left| \int_{a_1}^b f(z, t)\, dt \right| < \epsilon,$$

for all $a_1 \in [a_0, b)$ and all $z \in S$; compare §1.5(iv).

M-test

If $|f(z, t)| \leq M(t)$ for $z \in S$ and $\int_a^b M(t)\, dt$ converges, then the integral (1.10.18) converges uniformly and absolutely in S.

1.10(ix) Infinite Products

Let $p_{k,m} = \prod_{n=k}^m (1 + a_n)$. If for some $k \geq 1$, $p_{k,m} \to p_k \neq 0$ as $m \to \infty$, then we say that the infinite product $\prod_{n=1}^\infty (1 + a_n)$ *converges*. (The integer k may be greater than one to allow for a finite number of zero factors.) The convergence of the product is *absolute* if $\prod_{n=1}^\infty (1 + |a_n|)$ converges. The product $\prod_{n=1}^\infty (1 + a_n)$, with $a_n \neq -1$ for all n, converges iff $\sum_{n=1}^\infty \ln(1 + a_n)$ converges; and it converges absolutely iff $\sum_{n=1}^\infty |a_n|$ converges.

Suppose $a_n = a_n(z)$, $z \in D$, a domain. The convergence of the infinite product is *uniform* if the sequence of partial products converges uniformly.

M-test

Suppose that $a_n(z)$ are analytic functions in D. If there is an N, independent of $z \in D$, such that

1.10.20 $$|\ln(1 + a_n(z))| \leq M_n, \qquad n \geq N,$$

and

1.10.21 $$\sum_{n=1}^\infty M_n < \infty,$$

then the product $\prod_{n=1}^\infty (1 + a_n(z))$ converges uniformly to an analytic function $p(z)$ in D, and $p(z) = 0$ only when at least one of the factors $1+a_n(z)$ is zero in D. This conclusion remains true if, in place of (1.10.20), $|a_n(z)| \leq M_n$ for all n, and again $\sum_{n=1}^{\infty} M_n < \infty$.

Weierstrass Product

If $\{z_n\}$ is a sequence such that $\sum_{n=1}^{\infty} |z_n^{-2}|$ is convergent, then

$$1.10.22 \qquad P(z) = \prod_{n=1}^{\infty} \left(1 - \frac{z}{z_n}\right) e^{z/z_n}$$

is an entire function with zeros at z_n.

1.10(x) Infinite Partial Fractions

Suppose D is a domain, and

$$1.10.23 \qquad F(z) = \prod_{n=1}^{\infty} a_n(z), \qquad z \in D,$$

where $a_n(z)$ is analytic for all $n \geq 1$, and the convergence of the product is uniform in any compact subset of D. Then $F(z)$ is analytic in D.

If, also, $a_n(z) \neq 0$ when $n \geq 1$ and $z \in D$, then $F(z) \neq 0$ on D and

$$1.10.24 \qquad \frac{F'(z)}{F(z)} = \sum_{n=1}^{\infty} \frac{a_n'(z)}{a_n(z)}.$$

Mittag-Leffler's Expansion

If $\{a_n\}$ and $\{z_n\}$ are sequences such that $z_m \neq z_n$ ($m \neq n$) and $\sum_{n=1}^{\infty} |a_n z_n^{-2}|$ is convergent, then

$$1.10.25 \qquad f(z) = \sum_{n=1}^{\infty} a_n \left(\frac{1}{z - z_n} + \frac{1}{z_n}\right)$$

is analytic in $\mathbb{C}$, except for simple poles at $z = z_n$ of residue a_n.

1.11 Zeros of Polynomials

1.11(i) Division Algorithm

Horner's Scheme

Let

$$1.11.1 \qquad f(z) = a_n z^n + a_{n-1} z^{n-1} + \cdots + a_0.$$

Then

$$1.11.2 \qquad f(z) = (z-\alpha)(b_n z^{n-1} + b_{n-1} z^{n-2} + \cdots + b_1) + b_0,$$

where $b_n = a_n$,

$$1.11.3 \qquad b_k = \alpha b_{k+1} + a_k, \quad k = n-1, n-2, \ldots, 0,$$

$$1.11.4 \qquad f(\alpha) = b_0.$$

Extended Horner Scheme

With b_k as in (1.11.1)–(1.11.3) let $c_n = a_n$ and

$$1.11.5 \qquad c_k = \alpha c_{k+1} + b_k, \quad k = n-1, n-2, \ldots, 1.$$

Then

$$1.11.6 \qquad f'(\alpha) = c_1.$$

More generally, for polynomials $f(z)$ and $g(z)$, there are polynomials $q(z)$ and $r(z)$, found by equating coefficients, such that

$$1.11.7 \qquad f(z) = g(z)q(z) + r(z),$$

where $0 \leq \deg r(z) < \deg g(z)$.

1.11(ii) Elementary Properties

A polynomial of degree n with real or complex coefficients has exactly n real or complex zeros counting multiplicity. Every *monic* (coefficient of highest power is one) polynomial of odd degree with real coefficients has at least one real zero with sign opposite to that of the constant term. A monic polynomial of even degree with real coefficients has at least two zeros of opposite signs when the constant term is negative.

Descartes' Rule of Signs

The number of positive zeros of a polynomial with real coefficients cannot exceed the number of times the coefficients change sign, and the two numbers have same parity. A similar relation holds for the changes in sign of the coefficients of $f(-z)$, and hence for the number of negative zeros of $f(z)$.

Example

$$1.11.8 \qquad \begin{aligned} f(z) &= z^8 + 10z^3 + z - 4, \\ f(-z) &= z^8 - 10z^3 - z - 4. \end{aligned}$$

Both polynomials have one change of sign; hence for each polynomial there is one positive zero, one negative zero, and six complex zeros.

Next, let $f(z) = a_n z^n + a_{n-1} z^{n-1} + \cdots + a_0$. The zeros of $z^n f(1/z) = a_0 z^n + a_1 z^{n-1} + \cdots + a_n$ are reciprocals of the zeros of $f(z)$.

The *discriminant* of $f(z)$ is defined by

$$1.11.9 \qquad D = a_n^{2n-2} \prod_{j<k} (z_j - z_k)^2,$$

where $z_1, z_2, \ldots, z_n$ are the zeros of $f(z)$. The *elementary symmetric functions* of the zeros are (with $a_n \neq 0$)

$$1.11.10 \qquad \begin{aligned} z_1 + z_2 + \cdots + z_n &= -a_{n-1}/a_n, \\ \sum_{1 \leq j < k \leq n} z_j z_k &= a_{n-2}/a_n, \\ &\vdots \\ z_1 z_2 \cdots z_n &= (-1)^n a_0/a_n. \end{aligned}$$

1.11(iii) Polynomials of Degrees Two, Three, and Four

Quadratic Equations

The roots of $az^2 + bz + c = 0$ are

1.11.11 $\quad \dfrac{-b \pm \sqrt{D}}{2a}, \quad D = b^2 - 4ac.$

The sum and product of the roots are respectively $-b/a$ and c/a.

Cubic Equations

Set $z = w - \frac{1}{3}a$ to reduce $f(z) = z^3 + az^2 + bz + c$ to $g(w) = w^3 + pw + q$, with $p = (3b - a^2)/3$, $q = (2a^3 - 9ab + 27c)/27$. The discriminant of $g(w)$ is

1.11.12 $\quad D = -4p^3 - 27q^2.$

Let

1.11.13 $\quad A = \sqrt[3]{-\frac{27}{2}q + \frac{3}{2}\sqrt{-3D}}, \quad B = -3p/A.$

The roots of $g(w) = 0$ are

1.11.14 $\quad \frac{1}{3}(A + B), \quad \frac{1}{3}(\rho A + \rho^2 B), \quad \frac{1}{3}(\rho^2 A + \rho B),$

with

1.11.15 $\quad \rho = -\frac{1}{2} + \frac{1}{2}\sqrt{-3} = e^{2\pi i/3}, \quad \rho^2 = e^{-2\pi i/3}.$

Addition of $-\frac{1}{3}a$ to each of these roots gives the roots of $f(z) = 0$.

Example

$f(z) = z^3 - 6z^2 + 6z - 2$, $g(w) = w^3 - 6w - 6$, $A = 3\sqrt[3]{4}$, $B = 3\sqrt[3]{2}$. Roots of $f(z) = 0$ are $2 + \sqrt[3]{4} + \sqrt[3]{2}$, $2 + \sqrt[3]{4}\rho + \sqrt[3]{2}\rho^2$, $2 + \sqrt[3]{4}\rho^2 + \sqrt[3]{2}\rho$.

For another method see §4.43.

Quartic Equations

Set $z = w - \frac{1}{4}a$ to reduce $f(z) = z^4 + az^3 + bz^2 + cz + d$ to

$$g(w) = w^4 + pw^2 + qw + r,$$

1.11.16 $\quad p = (-3a^2 + 8b)/8, \quad q = (a^3 - 4ab + 8c)/8,$
$r = (-3a^4 + 16a^2 b - 64ac + 256d)/256.$

The discriminant of $g(w)$ is

1.11.17 $\quad D = 16p^4 r - 4p^3 q^2 - 128p^2 r^2 + 144pq^2 r - 27q^4 + 256r^3.$

For the roots $\alpha_1, \alpha_2, \alpha_3, \alpha_4$ of $g(w) = 0$ and the roots $\theta_1, \theta_2, \theta_3$ of the *resolvent cubic equation*

1.11.18 $\quad z^3 - 2pz^2 + (p^2 - 4r)z + q^2 = 0,$

we have

1.11.19
$\begin{aligned}
2\alpha_1 &= \sqrt{-\theta_1} + \sqrt{-\theta_2} + \sqrt{-\theta_3}, \\
2\alpha_2 &= \sqrt{-\theta_1} - \sqrt{-\theta_2} - \sqrt{-\theta_3}, \\
2\alpha_3 &= -\sqrt{-\theta_1} + \sqrt{-\theta_2} - \sqrt{-\theta_3}, \\
2\alpha_4 &= -\sqrt{-\theta_1} - \sqrt{-\theta_2} + \sqrt{-\theta_3}.
\end{aligned}$

The square roots are chosen so that

1.11.20 $\quad \sqrt{-\theta_1}\sqrt{-\theta_2}\sqrt{-\theta_3} = -q.$

Add $-\frac{1}{4}a$ to the roots of $g(w) = 0$ to get those of $f(z) = 0$.

Example

$f(z) = z^4 - 4z^3 + 5z + 2$, $g(w) = w^4 - 6w^2 - 3w + 4$. Resolvent cubic is $z^3 + 12z^2 + 20z + 9 = 0$ with roots $\theta_1 = -1$, $\theta_2 = -\frac{1}{2}(11 + \sqrt{85})$, $\theta_3 = -\frac{1}{2}(11 - \sqrt{85})$, and $\sqrt{-\theta_1} = 1$, $\sqrt{-\theta_2} = \frac{1}{2}(\sqrt{17} + \sqrt{5})$, $\sqrt{-\theta_3} = \frac{1}{2}(\sqrt{17} - \sqrt{5})$. So $2\alpha_1 = 1 + \sqrt{17}$, $2\alpha_2 = 1 - \sqrt{17}$, $2\alpha_3 = -1 + \sqrt{5}$, $2\alpha_4 = -1 - \sqrt{5}$, and the roots of $f(z) = 0$ are $\frac{1}{2}(3 \pm \sqrt{17})$, $\frac{1}{2}(1 \pm \sqrt{5})$.

1.11(iv) Roots of Unity and of Other Constants

The roots of

1.11.21 $\quad z^n - 1 = (z - 1)(z^{n-1} + z^{n-2} + \cdots + z + 1) = 0$

are $1, e^{2\pi i/n}, e^{4\pi i/n}, \ldots, e^{(2n-2)\pi i/n}$, and of $z^n + 1 = 0$ they are $e^{\pi i/n}, e^{3\pi i/n}, \ldots, e^{(2n-1)\pi i/n}$.

The roots of

1.11.22 $\quad z^n = a + ib, \quad a, b \text{ real},$

are

1.11.23 $\quad \sqrt[n]{R}\left(\cos\left(\dfrac{\alpha + 2k\pi}{n}\right) + i\sin\left(\dfrac{\alpha + 2k\pi}{n}\right)\right),$

where $R = (a^2 + b^2)^{1/2}$, $\alpha = \mathrm{ph}(a + ib)$, with the principal value of phase (§1.9(i)), and $k = 0, 1, \ldots, n - 1$.

1.11(v) Stable Polynomials

1.11.24 $\quad f(z) = a_0 + a_1 z + \cdots + a_n z^n,$

with real coefficients, is called *stable* if the real parts of all the zeros are strictly negative.

Hurwitz Criterion

Let

1.11.25

$D_1 = a_1, \quad D_2 = \begin{vmatrix} a_1 & a_3 \\ a_0 & a_2 \end{vmatrix}, \quad D_3 = \begin{vmatrix} a_1 & a_3 & a_5 \\ a_0 & a_2 & a_4 \\ 0 & a_1 & a_3 \end{vmatrix},$

and

1.11.26 $\quad D_k = \det[h_k^{(1)}, h_k^{(3)}, \ldots, h_k^{(2k-1)}],$

where the column vector $h_k^{(m)}$ consists of the first k members of the sequence $a_m, a_{m-1}, a_{m-2}, \ldots$ with $a_j = 0$ if $j < 0$ or $j > n$.

Then $f(z)$, with $a_n \neq 0$, is stable iff $a_0 \neq 0$; $D_{2k} > 0$, $k = 1, \ldots, \lfloor\frac{1}{2}n\rfloor$; $\operatorname{sign} D_{2k+1} = \operatorname{sign} a_0$, $k = 0, 1, \ldots, \lfloor\frac{1}{2}n - \frac{1}{2}\rfloor$.

1.12 Continued Fractions

1.12(i) Notation

The notation used throughout this Handbook for the continued fraction

1.12.1
$$b_0 + \cfrac{a_1}{b_1 + \cfrac{a_2}{b_2 + \cdots}}$$

is

1.12.2
$$b_0 + \frac{a_1}{b_1 +} \frac{a_2}{b_2+} \cdots.$$

1.12(ii) Convergents

1.12.3 $\qquad C = b_0 + \dfrac{a_1}{b_1 +} \dfrac{a_2}{b_2 +} \cdots, \qquad a_n \neq 0,$

1.12.4 $\qquad C_n = b_0 + \dfrac{a_1}{b_1 +} \dfrac{a_2}{b_2 +} \cdots \dfrac{a_n}{b_n} = \dfrac{A_n}{B_n}.$

C_n is called the nth *approximant* or *convergent* to C. A_n and B_n are called the nth *(canonical) numerator* and *denominator* respectively.

Recurrence Relations

1.12.5 $\quad A_k = b_k A_{k-1} + a_k A_{k-2}, \quad B_k = b_k B_{k-1} + a_k B_{k-2},$
$\qquad\qquad k = 1, 2, 3, \ldots,$

1.12.6 $\quad A_{-1} = 1, \quad A_0 = b_0, \quad B_{-1} = 0, \quad B_0 = 1.$

Determinant Formula

1.12.7
$$A_n B_{n-1} - B_n A_{n-1} = (-1)^{n-1} \prod_{k=1}^{n} a_k, \quad n = 0, 1, 2, \ldots.$$

1.12.8 $\quad C_n - C_{n-1} = \dfrac{(-1)^{n-1} \prod_{k=1}^{n} a_k}{B_{n-1} B_n}, \quad n = 1, 2, 3, \ldots,$

1.12.9 $\quad C_n = b_0 + \dfrac{a_1}{B_0 B_1} - \cdots + (-1)^{n-1} \dfrac{\prod_{k=1}^{n} a_k}{B_{n-1} B_n}.$

1.12.10 $\quad a_n = \dfrac{A_{n-1} B_n - A_n B_{n-1}}{A_{n-1} B_{n-2} - A_{n-2} B_{n-1}}, \quad n = 1, 2, 3, \ldots,$

1.12.11 $\quad a_n = \dfrac{B_n}{B_{n-2}} \dfrac{C_{n-1} - C_n}{C_{n-1} - C_{n-2}}, \quad n = 2, 3, 4, \ldots,$

1.12.12 $\quad b_n = \dfrac{A_n B_{n-2} - A_{n-2} B_n}{A_{n-1} B_{n-2} - A_{n-2} B_{n-1}}, \quad n = 1, 2, 3, \ldots,$

1.12.13 $\quad b_n = \dfrac{B_n}{B_{n-1}} \dfrac{C_n - C_{n-2}}{C_{n-1} - C_{n-2}}, \quad n = 2, 3, 4, \ldots,$

1.12.14 $\quad b_0 = A_0 = C_0, \quad b_1 = B_1, \quad a_1 = A_1 - A_0 B_1.$

Equivalence

Two continued fractions are *equivalent* if they have the same convergents.

$b_0 + \dfrac{a_1}{b_1 +} \dfrac{a_2}{b_2 +} \cdots$ is equivalent to $b'_0 + \dfrac{a'_1}{b'_1 +} \dfrac{a'_2}{b'_2 +} \cdots$ if there is a sequence $\{d_n\}_{n=0}^{\infty}$, $d_0 = 1$, $d_n \neq 0$, such that

1.12.15 $\qquad a'_n = d_n d_{n-1} a_n, \quad n = 1, 2, 3, \ldots,$

and

1.12.16 $\qquad b'_n = d_n b_n, \quad n = 0, 1, 2, \ldots.$

Formally,

1.12.17
$$b_0 + \frac{a_1}{b_1+} \frac{a_2}{b_2+} \frac{a_3}{b_3+} \cdots = b_0 + \frac{a_1/b_1}{1+} \frac{a_2/(b_1 b_2)}{1+} \frac{a_3/(b_2 b_3)}{1+} \cdots \frac{a_n/(b_{n-1} b_n)}{1+} \cdots$$
$$= b_0 + \frac{1}{(1/a_1)b_1+} \frac{1}{(a_1/a_2)b_2+} \frac{1}{(a_2/(a_1 a_3))b_3+} \frac{1}{(a_1 a_3/(a_2 a_4))b_4+} \cdots.$$

Series

1.12.18
$$p_0 + \sum_{k=1}^{n} p_1 p_2 \cdots p_k = p_0 + \frac{p_1}{1-} \frac{p_2}{1+p_2-} \frac{p_3}{1+p_3-} \cdots \frac{p_n}{1+p_n}, \qquad n = 0, 1, 2, \ldots,$$

when $p_k \neq 0$, $k = 1, 2, 3, \ldots$.

1.12.19
$$\sum_{k=0}^{n} c_k x^k = c_0 + \frac{c_1 x}{1-} \frac{(c_2/c_1)x}{1+(c_2/c_1)x-} \frac{(c_3/c_2)x}{1+(c_3/c_2)x-} \cdots \frac{(c_n/c_{n-1})x}{1+(c_n/c_{n-1})x}, \qquad n = 0, 1, 2, \ldots,$$

when $c_k \neq 0$, $k = 1, 2, 3, \ldots$.

Fractional Transformations

Define

1.12.20
$$C_n(w) = b_0 + \frac{a_1}{b_1 +} \frac{a_2}{b_2 +} \cdots \frac{a_n}{b_n + w}.$$

Then

1.12.21
$$C_n(w) = \frac{A_n + A_{n-1}w}{B_n + B_{n-1}w}, \quad C_n(0) = C_n, \quad C_n(\infty) = C_{n-1} = \frac{A_{n-1}}{B_{n-1}}.$$

1.12(iii) Existence of Convergents

A sequence $\{C_n\}$ in the extended complex plane, $\mathbb{C} \cup \{\infty\}$, can be a sequence of convergents of the continued fraction (1.12.3) iff

1.12.22
$$C_0 \neq \infty, \quad C_n \neq C_{n-1}, \quad\quad n = 1, 2, 3, \ldots.$$

1.12(iv) Contraction and Extension

A *contraction* of a continued fraction C is a continued fraction C' whose convergents $\{C'_n\}$ form a subsequence of the convergents $\{C_n\}$ of C. Conversely, C is called an *extension* of C'. If $C'_n = C_{2n}$, $n = 0, 1, 2, \ldots$, then C' is called the *even part* of C. The even part of C exists iff $b_{2k} \neq 0$, $k = 1, 2, \ldots$, and up to equivalence is given by

1.12.23
$$b_0 + \frac{a_1 b_2}{a_2 + b_1 b_2 -} \frac{a_2 a_3 b_4}{a_3 b_4 + b_2(a_4 + b_3 b_4) -} \frac{a_4 a_5 b_2 b_6}{a_5 b_6 + b_4(a_6 + b_5 b_6) -} \frac{a_6 a_7 b_4 b_8}{a_7 b_8 + b_6(a_8 + b_7 b_8) -} \cdots.$$

If $C'_n = C_{2n+1}$, $n = 0, 1, 2, \ldots$, then C' is called the *odd part* of C. The odd part of C exists iff $b_{2k+1} \neq 0$, $k = 0, 1, 2, \ldots$, and up to equivalence is given by

1.12.24
$$\frac{a_1 + b_0 b_1}{b_1} - \frac{a_1 a_2 b_3 / b_1}{a_2 b_3 + b_1(a_3 + b_2 b_3) -} \frac{a_3 a_4 b_1 b_5}{a_4 b_5 + b_3(a_5 + b_4 b_5) -} \frac{a_5 a_6 b_3 b_7}{a_6 b_7 + b_5(a_7 + b_6 b_7) -} \cdots.$$

1.12(v) Convergence

A continued fraction *converges* if the convergents C_n tend to a finite limit as $n \to \infty$.

Pringsheim's Theorem

The continued fraction $\dfrac{a_1}{b_1 +} \dfrac{a_2}{b_2 +} \cdots$ converges when

1.12.25
$$|b_n| \geq |a_n| + 1, \quad n = 1, 2, 3, \ldots.$$

With these conditions the convergents C_n satisfy $|C_n| < 1$ and $C_n \to C$ with $|C| \leq 1$.

Van Vleck's Theorem

Let the elements of the continued fraction $\dfrac{1}{b_1 +} \dfrac{1}{b_2 +} \cdots$ satisfy

1.12.26
$$-\tfrac{1}{2}\pi + \delta < \operatorname{ph} b_n < \tfrac{1}{2}\pi - \delta, \quad n = 1, 2, 3, \ldots,$$

where δ is an arbitrary small positive constant. Then the convergents C_n satisfy

1.12.27
$$-\tfrac{1}{2}\pi + \delta < \operatorname{ph} C_n < \tfrac{1}{2}\pi - \delta, \quad n = 1, 2, 3, \ldots,$$

and the even and odd parts of the continued fraction converge to finite values. The continued fraction converges iff, in addition,

1.12.28
$$\sum_{n=1}^{\infty} |b_n| = \infty.$$

In this case $|\operatorname{ph} C| \leq \tfrac{1}{2}\pi$.

1.12(vi) Applications

For analytical and numerical applictions of continued fractions to special functions see §3.10.

1.13 Differential Equations

1.13(i) Existence of Solutions

A domain in the complex plane is *simply-connected* if it has no "holes"; more precisely, if its complement in the extended plane $\mathbb{C} \cup \{\infty\}$ is connected.

The equation

1.13.1
$$\frac{d^2w}{dz^2} + f(z)\frac{dw}{dz} + g(z)w = 0,$$

where $z \in D$, a simply-connected domain, and $f(z)$, $g(z)$ are analytic in D, has an infinite number of analytic solutions in D. A solution becomes unique, for example, when w and dw/dz are prescribed at a point in D.

Fundamental Pair

Two solutions $w_1(z)$ and $w_2(z)$ are called a *fundamental pair* if any other solution $w(z)$ is expressible as

1.13.2 $\qquad w(z) = Aw_1(z) + Bw_2(z),$

where A and B are constants. A fundamental pair can be obtained, for example, by taking any $z_0 \in D$ and requiring that

1.13.3
$$w_1(z_0) = 1, \quad w_1'(z_0) = 0, \quad w_2(z_0) = 0, \quad w_2'(z_0) = 1.$$

Wronskian

The *Wronskian* of $w_1(z)$ and $w_2(z)$ is defined by

1.13.4 $\quad \mathscr{W}\{w_1(z), w_2(z)\} = w_1(z)w_2'(z) - w_2(z)w_1'(z).$

Then

1.13.5 $\qquad \mathscr{W}\{w_1(z), w_2(z)\} = c e^{-\int f(z)\,dz},$

where c is independent of z. If $f(z) = 0$, then the Wronskian is constant.

The following three statements are equivalent: $w_1(z)$ and $w_2(z)$ comprise a fundamental pair in D; $\mathscr{W}\{w_1(z), w_2(z)\}$ does not vanish in D; $w_1(z)$ and $w_2(z)$ are *linearly independent*, that is, the only constants A and B such that

1.13.6 $\qquad Aw_1(z) + Bw_2(z) = 0, \qquad \forall z \in D,$

are $A = B = 0$.

1.13(ii) Equations with a Parameter

Assume that in the equation

1.13.7 $\qquad \dfrac{d^2w}{dz^2} + f(u,z)\dfrac{dw}{dz} + g(u,z)w = 0,$

u and z belong to domains U and D respectively, the coefficients $f(u,z)$ and $g(u,z)$ are continuous functions of both variables, and for each fixed u (fixed z) the two functions are analytic in z (in u). Suppose also that at (a fixed) $z_0 \in D$, w and $\partial w/\partial z$ are analytic functions of u. Then at each $z \in D$, w, $\partial w/\partial z$ and $\partial^2 w/\partial z^2$ are analytic functions of u.

1.13(iii) Inhomogeneous Equations

The *inhomogeneous* (or *nonhomogeneous*) equation

1.13.8 $\qquad \dfrac{d^2w}{dz^2} + f(z)\dfrac{dw}{dz} + g(z)w = r(z)$

with $f(z)$, $g(z)$, and $r(z)$ analytic in D has infinitely many analytic solutions in D. If $w_0(z)$ is any one solution, and $w_1(z), w_2(z)$ are a fundamental pair of solutions of the corresponding homogeneous equation (1.13.1), then every solution of (1.13.8) can be expressed as

1.13.9 $\qquad w(z) = w_0(z) + Aw_1(z) + Bw_2(z),$

where A and B are constants.

Variation of Parameters

With the notation of (1.13.8) and (1.13.9)

1.13.10
$$w_0(z) = w_2(z) \int \frac{w_1(z) r(z)}{\mathscr{W}\{w_1(z), w_2(z)\}}\,dz \\ - w_1(z) \int \frac{w_2(z) r(z)}{\mathscr{W}\{w_1(z), w_2(z)\}}\,dz.$$

1.13(iv) Change of Variables

Transformation of the Point at Infinity

The substitution $\xi = 1/z$ in (1.13.1) gives

1.13.11 $\qquad \dfrac{d^2 W}{d\xi^2} + F(\xi)\dfrac{dW}{d\xi} + G(\xi) W = 0,$

where

1.13.12
$$W(\xi) = w\left(\frac{1}{\xi}\right), \\ F(\xi) = \frac{2}{\xi} - \frac{1}{\xi^2} f\left(\frac{1}{\xi}\right), \\ G(\xi) = \frac{1}{\xi^4} g\left(\frac{1}{\xi}\right).$$

Elimination of First Derivative by Change of Dependent Variable

The substitution

1.13.13 $\qquad w(z) = W(z) \exp\left(-\tfrac{1}{2} \int f(z)\,dz\right)$

in (1.13.1) gives

1.13.14 $\qquad \dfrac{d^2 W}{dz^2} - H(z) W = 0,$

where

1.13.15 $\qquad H(z) = \tfrac{1}{4} f^2(z) + \tfrac{1}{2} f'(z) - g(z).$

Elimination of First Derivative by Change of Independent Variable

In (1.13.1) substitute

1.13.16 $\qquad \eta = \int \exp\left(-\int f(z)\,dz\right) dz.$

Then

1.13.17 $\qquad \dfrac{d^2 w}{d\eta^2} + g(z) \exp\left(2 \int f(z)\,dz\right) w = 0.$

Liouville Transformation

Let $W(z)$ satisfy (1.13.14), $\zeta(z)$ be any thrice-differentiable function of z, and

1.13.18 $\qquad U(z) = (\zeta'(z))^{1/2} W(z).$

Then

1.13.19 $\qquad \dfrac{d^2 U}{d\zeta^2} = \left(\dot{z}^2 H(z) - \tfrac{1}{2}\{z, \zeta\}\right) U.$

Here dots denote differentiations with respect to ζ, and $\{z,\zeta\}$ is the *Schwarzian derivative*:

1.13.20 $\quad \{z,\zeta\} = -2\dot{z}^{1/2}\dfrac{d^2}{d\zeta^2}(\dot{z}^{-1/2}) = \dfrac{\dddot{z}}{\dot{z}} - \dfrac{3}{2}\left(\dfrac{\ddot{z}}{\dot{z}}\right)^2.$

Cayley's Identity

For arbitrary ξ and ζ,

1.13.21 $\quad \{z,\zeta\} = (d\xi/d\zeta)^2\{z,\xi\} + \{\xi,\zeta\}.$

1.13.22 $\quad \{z,\zeta\} = -(dz/d\zeta)^2\{\zeta,z\}.$

1.13(v) Products of Solutions

The product of any two solutions of (1.13.1) satisfies

1.13.23
$$\dfrac{d^3w}{dz^3} + 3f\dfrac{d^2w}{dz^2} + (2f^2 + f' + 4g)\dfrac{dw}{dz} + (4fg + 2g')w = 0.$$

If $U(z)$ and $V(z)$ are respectively solutions of

1.13.24 $\quad \dfrac{d^2U}{dz^2} + IU = 0, \quad \dfrac{d^2V}{dz^2} + JV = 0,$

then $W = UV$ is a solution of

1.13.25
$$\dfrac{d}{dz}\left(\dfrac{W''' + 2(I+J)W' + (I'+J')W}{I-J}\right) = -(I-J)W.$$

1.13(vi) Singularities

For classification of singularities of (1.13.1) and expansions of solutions in the neighborhoods of singularities, see §2.7.

1.13(vii) Closed-Form Solutions

For an extensive collection of solutions of differential equations of the first, second, and higher orders see Kamke (1977).

1.14 Integral Transforms

1.14(i) Fourier Transform

The *Fourier transform* of a real- or complex-valued function $f(t)$ is defined by

1.14.1 $\quad F(x) = \dfrac{1}{\sqrt{2\pi}}\displaystyle\int_{-\infty}^{\infty} f(t)e^{ixt}\,dt.$

(Some references replace ixt by $-ixt$.)

If $f(t)$ is absolutely integrable on $(-\infty,\infty)$, then $F(x)$ is continuous, $F(x) \to 0$ as $x \to \pm\infty$, and

1.14.2 $\quad |F(x)| \le \dfrac{1}{\sqrt{2\pi}}\displaystyle\int_{-\infty}^{\infty} |f(t)|\,dt.$

Inversion

Suppose that $f(t)$ is absolutely integrable on $(-\infty,\infty)$ and of bounded variation in a neighborhood of $t = u$ (§1.4(v)). Then

1.14.3 $\quad \tfrac{1}{2}(f(u+) + f(u-)) = \dfrac{1}{\sqrt{2\pi}}\displaystyle\fint_{-\infty}^{\infty} F(x)e^{-ixu}\,dx,$

where the last integral denotes the Cauchy principal value (1.4.25).

In many applications $f(t)$ is absolutely integrable and $f'(t)$ is continuous on $(-\infty,\infty)$. Then

1.14.4 $\quad f(t) = \dfrac{1}{\sqrt{2\pi}}\displaystyle\int_{-\infty}^{\infty} F(x)e^{-ixt}\,dx.$

Convolution

For Fourier transforms, the *convolution* $(f*g)(t)$ of two functions $f(t)$ and $g(t)$ defined on $(-\infty,\infty)$ is given by

1.14.5 $\quad (f*g)(t) = \dfrac{1}{\sqrt{2\pi}}\displaystyle\int_{-\infty}^{\infty} f(t-s)g(s)\,ds.$

If $f(t)$ and $g(t)$ are absolutely integrable on $(-\infty,\infty)$, then so is $(f*g)(t)$, and its Fourier transform is $F(x)G(x)$, where $G(x)$ is the Fourier transform of $g(t)$.

Parseval's Formula

Suppose $f(t)$ and $g(t)$ are absolutely integrable on $(-\infty,\infty)$, and $F(x)$ and $G(x)$ are their respective Fourier transforms. Then

1.14.6 $\quad (f*g)(t) = \dfrac{1}{\sqrt{2\pi}}\displaystyle\int_{-\infty}^{\infty} F(x)G(x)e^{-itx}\,dx,$

1.14.7 $\quad \displaystyle\int_{-\infty}^{\infty} F(x)G(x)\,dx = \int_{-\infty}^{\infty} f(t)g(-t)\,dt,$

1.14.8 $\quad \displaystyle\int_{-\infty}^{\infty} |F(x)|^2\,dx = \int_{-\infty}^{\infty} |f(t)|^2\,dt.$

(1.14.8) is *Parseval's formula*.

Uniqueness

If $f(t)$ and $g(t)$ are continuous and absolutely integrable on $(-\infty,\infty)$, and $F(x) = G(x)$ for all x, then $f(t) = g(t)$ for all t.

1.14(ii) Fourier Cosine and Sine Transforms

These are defined respectively by

1.14.9 $\quad F_c(x) = \sqrt{\dfrac{2}{\pi}}\displaystyle\int_0^{\infty} f(t)\cos(xt)\,dt,$

1.14.10 $\quad F_s(x) = \sqrt{\dfrac{2}{\pi}}\displaystyle\int_0^{\infty} f(t)\sin(xt)\,dt.$

Inversion

If $f(t)$ is absolutely integrable on $[0,\infty)$ and of bounded variation (§1.4(v)) in a neighborhood of $t = u$, then

1.14.11 $\quad \frac{1}{2}(f(u+) + f(u-)) = \sqrt{\frac{2}{\pi}} \int_0^\infty F_c(x) \cos(ux)\, dx,$

1.14.12 $\quad \frac{1}{2}(f(u+) + f(u-)) = \sqrt{\frac{2}{\pi}} \int_0^\infty F_s(x) \sin(ux)\, dx.$

Parseval's Formula

If $\int_0^\infty |f(t)|\, dt < \infty$, $g(t)$ is of bounded variation on $(0,\infty)$ and $g(t) \to 0$ as $t \to \infty$, then

1.14.13 $\quad \int_0^\infty F_c(x) G_c(x)\, dx = \int_0^\infty f(t) g(t)\, dt,$

1.14.14 $\quad \int_0^\infty F_s(x) G_s(x)\, dx = \int_0^\infty f(t) g(t)\, dt,$

1.14.15 $\quad \int_0^\infty (F_c(x))^2\, dx = \int_0^\infty (f(t))^2\, dt,$

1.14.16 $\quad \int_0^\infty (F_s(x))^2\, dx = \int_0^\infty (f(t))^2\, dt,$

where $G_c(x)$ and $G_s(x)$ are respectively the cosine and sine transforms of $g(t)$.

1.14(iii) Laplace Transform

Suppose $f(t)$ is a real- or complex-valued function and s is a real or complex parameter. The *Laplace transform* of f is defined by

1.14.17 $\quad \mathscr{L}(f(t); s) = \int_0^\infty e^{-st} f(t)\, dt.$

Alternative notations are $\mathscr{L}(f(t))$, $\mathscr{L}(f; s)$, or even $\mathscr{L}(f)$, when it is not important to display all the variables.

Convergence and Analyticity

Assume that on $[0,\infty)$ $f(t)$ is piecewise continuous and of *exponential growth*, that is, constants M and α exist such that

1.14.18 $\quad\quad\quad |f(t)| \le M e^{\alpha t}, \quad\quad 0 \le t < \infty.$

Then $\mathscr{L}(f(t); s)$ is an analytic function of s for $\Re s > \alpha$. Moreover,

1.14.19 $\quad\quad\quad \mathscr{L}(f(t); s) \to 0, \quad\quad \Re s \to \infty.$

Throughout the remainder of this subsection we assume (1.14.18) is satisfied and $\Re s > \alpha$.

Inversion

If $f(t)$ is continuous and $f'(t)$ is piecewise continuous on $[0,\infty)$, then

1.14.20 $\quad f(t) = \frac{1}{2\pi i} \lim_{T \to \infty} \int_{\sigma - iT}^{\sigma + iT} e^{ts} \mathscr{L}(f(t); s)\, ds, \quad \sigma > \alpha.$

Moreover, if $\mathscr{L}(f(t); s) = O(s^{-K})$ in some half-plane $\Re s \ge \gamma$ and $K > 1$, then (1.14.20) holds for $\sigma > \gamma$.

Translation

If $\Re s > \max(\Re(a + \alpha), \alpha)$, then

1.14.21 $\quad \mathscr{L}(f(t); s - a) = \mathscr{L}(e^{at} f(t); s).$

Also, if $a \ge 0$ then

1.14.22 $\quad \mathscr{L}(H(t - a) f(t - a); s) = e^{-as} \mathscr{L}(f(t); s),$

where H is the Heaviside function; see (1.16.13).

Differentiation and Integration

If $f(t)$ is piecewise continuous, then

1.14.23
$$\frac{d^n}{ds^n} \mathscr{L}(f(t); s) = \mathscr{L}((-t)^n f(t); s), \quad n = 1, 2, 3, \ldots.$$

If also $\lim_{t \to 0+} f(t)/t$ exists, then

1.14.24 $\quad \int_s^\infty \mathscr{L}(f(t); u)\, du = \mathscr{L}\left(\frac{f(t)}{t}; s\right).$

Periodic Functions

If $a > 0$ and $f(t + a) = f(t)$ for $t > 0$, then

1.14.25 $\quad \mathscr{L}(f(t); s) = \frac{1}{1 - e^{-as}} \int_0^a e^{-st} f(t)\, dt.$

Alternatively if $f(t + a) = -f(t)$ for $t > 0$, then

1.14.26 $\quad \mathscr{L}(f(t); s) = \frac{1}{1 + e^{-as}} \int_0^a e^{-st} f(t)\, dt.$

Derivatives

If $f(t)$ is continuous on $[0,\infty)$ and $f'(t)$ is piecewise continuous on $(0,\infty)$, then

1.14.27 $\quad \mathscr{L}(f'(t); s) = s \mathscr{L}(f(t); s) - f(0+).$

If $f(t)$ and $f'(t)$ are piecewise continuous on $[0,\infty)$ with discontinuities at $(0 =) t_0 < t_1 < \cdots < t_n$, then

1.14.28
$$\mathscr{L}(f'(t); s) = s \mathscr{L}(f(t); s) - f(0+) \\ - \sum_{k=1}^n e^{-s t_k}(f(t_k+) - f(t_k-)).$$

Next, assume $f(t), f'(t), \ldots, f^{(n-1)}(t)$ are continuous and each satisfies (1.14.18). Also assume that $f^{(n)}(t)$ is piecewise continuous on $[0,\infty)$. Then

1.14.29
$$\mathscr{L}\left(f^{(n)}(t); s\right) = s^n \mathscr{L}(f(t); s) - s^{n-1} f(0+) \\ - s^{n-2} f'(0+) - \cdots - f^{(n-1)}(0+).$$

Convolution

For Laplace transforms, the *convolution* of two functions $f(t)$ and $g(t)$, defined on $[0,\infty)$, is

1.14.30 $\quad (f * g)(t) = \int_0^t f(u) g(t - u)\, du.$

If $f(t)$ and $g(t)$ are piecewise continuous, then

1.14.31 $\quad\quad\quad \mathscr{L}(f * g) = \mathscr{L}(f) \mathscr{L}(g).$

1.14(iv) Mellin Transform

Uniqueness

If $f(t)$ and $g(t)$ are continuous and $\mathscr{L}(f) = \mathscr{L}(g)$, then $f(t) = g(t)$.

1.14(iv) Mellin Transform

The *Mellin transform* of a real- or complex-valued function $f(x)$ is defined by

$$1.14.32 \quad \mathscr{M}(f;s) = \int_0^\infty x^{s-1} f(x)\, dx.$$

Alternative notations for $\mathscr{M}(f;s)$ are $\mathscr{M}(f(x);s)$ and $\mathscr{M}(f)$.

If $x^{\sigma-1} f(x)$ is integrable on $(0,\infty)$ for all σ in $a < \sigma < b$, then the integral (1.14.32) converges and $\mathscr{M}(f;s)$ is an analytic function of s in the vertical strip $a < \Re s < b$. Moreover, for $a < \sigma < b$,

$$1.14.33 \quad \lim_{t\to\pm\infty} \mathscr{M}(f;\sigma+it) = 0.$$

Note: If $f(x)$ is continuous and α and β are real numbers such that $f(x) = O(x^\alpha)$ as $x \to 0+$ and $f(x) = O(x^\beta)$ as $x \to \infty$, then $x^{\sigma-1} f(x)$ is integrable on $(0,\infty)$ for all $\sigma \in (-\alpha, -\beta)$.

Inversion

Suppose the integral (1.14.32) is absolutely convergent on the line $\Re s = \sigma$ and $f(x)$ is of bounded variation in a neighborhood of $x = u$. Then

1.14.34
$$\tfrac{1}{2}(f(u+) + f(u-)) = \frac{1}{2\pi i} \lim_{T\to\infty} \int_{\sigma-iT}^{\sigma+iT} u^{-s} \mathscr{M}(f;s)\, ds.$$

If $f(x)$ is continuous on $(0,\infty)$ and $\mathscr{M}(f;\sigma+it)$ is integrable on $(-\infty,\infty)$, then

$$1.14.35 \quad f(x) = \frac{1}{2\pi i} \int_{\sigma-i\infty}^{\sigma+i\infty} x^{-s} \mathscr{M}(f;s)\, ds.$$

Parseval-type Formulas

Suppose $x^{-\sigma} f(x)$ and $x^{\sigma-1} g(x)$ are absolutely integrable on $(0,\infty)$ and either $\mathscr{M}(g;\sigma+it)$ or $\mathscr{M}(f;1-\sigma-it)$ is absolutely integrable on $(-\infty,\infty)$. Then for $y > 0$,

1.14.36
$$\int_0^\infty f(x) g(yx)\, dx = \frac{1}{2\pi i} \int_{\sigma-i\infty}^{\sigma+i\infty} y^{-s} \mathscr{M}(f; 1-s) \mathscr{M}(g;s)\, ds,$$

1.14.37
$$\int_0^\infty f(x) g(x)\, dx = \frac{1}{2\pi i} \int_{\sigma-i\infty}^{\sigma+i\infty} \mathscr{M}(f; 1-s) \mathscr{M}(g;s)\, ds.$$

When f is real and $\sigma = \tfrac{1}{2}$,

$$1.14.38 \quad \int_0^\infty (f(x))^2\, dx = \frac{1}{2\pi} \int_{-\infty}^\infty \left|\mathscr{M}\left(f;\tfrac{1}{2}+it\right)\right|^2 dt.$$

Convolution

Let

$$1.14.39 \quad (f * g)(x) = \int_0^\infty f(y) g\!\left(\frac{x}{y}\right) \frac{dy}{y}.$$

If $x^{\sigma-1} f(x)$ and $x^{\sigma-1} g(x)$ are absolutely integrable on $(0,\infty)$, then for $s = \sigma + it$,

$$1.14.40 \quad \int_0^\infty x^{s-1} (f * g)(x)\, dx = \mathscr{M}(f;s) \mathscr{M}(g;s).$$

1.14(v) Hilbert Transform

The *Hilbert transform* of a real-valued function $f(t)$ is defined in the following equivalent ways:

$$1.14.41 \quad \mathcal{H}(f;x) = \mathcal{H}(f(t);x) = \mathcal{H}(f) = \frac{1}{\pi}\!\!\fint_{-\infty}^{\infty} \frac{f(t)}{t-x}\, dt,$$

$$1.14.42 \quad \mathcal{H}(f;x) = \lim_{y\to 0+} \frac{1}{\pi} \int_{-\infty}^\infty \frac{t-x}{(t-x)^2 + y^2} f(t)\, dt,$$

$$1.14.43 \quad \mathcal{H}(f;x) = \lim_{\epsilon\to 0+} \frac{1}{\pi} \int_{\epsilon}^\infty \frac{f(x+t) - f(x-t)}{t}\, dt.$$

Inversion

Suppose $f(t)$ is continuously differentiable on $(-\infty,\infty)$ and vanishes outside a bounded interval. Then

$$1.14.44 \quad f(x) = -\frac{1}{\pi}\!\!\fint_{-\infty}^\infty \frac{\mathcal{H}(f;u)}{u-x}\, du.$$

Inequalities

If $|f(t)|^p$, $p > 1$, is integrable on $(-\infty,\infty)$, then so is $|\mathcal{H}(f;x)|^p$ and

$$1.14.45 \quad \int_{-\infty}^\infty |\mathcal{H}(f;x)|^p\, dx \le A_p \int_{-\infty}^\infty |f(t)|^p\, dt,$$

where $A_p = \tan(\tfrac{1}{2}\pi/p)$ when $1 < p \le 2$, or $\cot(\tfrac{1}{2}\pi/p)$ when $p \ge 2$. These bounds are sharp, and equality holds when $p = 2$.

Fourier Transform

When $f(t)$ satisfies the same conditions as those for (1.14.44),

$$1.14.46 \quad \frac{1}{\sqrt{2\pi}} \int_{-\infty}^\infty \mathcal{H}(f;t) e^{ixt}\, dt = -i(\operatorname{sign} x) F(x),$$

where $F(x)$ is given by (1.14.1).

1.14(vi) Stieltjes Transform

The *Stieltjes transform* of a real-valued function $f(t)$ is defined by

$$1.14.47 \quad \mathcal{S}(f;s) = \mathcal{S}(f(t);s) = \mathcal{S}(f) = \int_0^\infty \frac{f(t)}{s+t}\, dt.$$

Sufficient conditions for the integral to converge are that s is a positive real number, and $f(t) = O(t^{-\delta})$ as $t \to \infty$, where $\delta > 0$.

If the integral converges, then it converges uniformly in any compact domain in the complex s-plane not containing any point of the interval $(-\infty, 0]$. In this case, $\mathcal{S}(f;s)$ represents an analytic function in the s-plane cut along the negative real axis, and

1.14.48 $\quad \dfrac{d^m}{ds^m}\mathcal{S}(f;s) = (-1)^m m! \displaystyle\int_0^\infty \dfrac{f(t)\,dt}{(s+t)^{m+1}},$
$$m = 0, 1, 2, \ldots.$$

Inversion

If $f(t)$ is absolutely integrable on $[0, R]$ for every finite R, and the integral (1.14.47) converges, then

1.14.49 $\quad \displaystyle\lim_{t \to 0+} \dfrac{\mathcal{S}(f;-\sigma-it) - \mathcal{S}(f;-\sigma+it)}{2\pi i}$
$= \tfrac{1}{2}(f(\sigma+) + f(\sigma-)),$

for all values of the positive constant σ for which the right-hand side exists.

Laplace Transform

If $f(t)$ is piecewise continuous on $[0, \infty)$ and the integral (1.14.47) converges, then

1.14.50 $\quad \mathcal{S}(f) = \mathscr{L}(\mathscr{L}(f)).$

1.14(vii) Tables

Table 1.14.1: Fourier transforms.

$f(t)$	$\dfrac{1}{\sqrt{2\pi}}\displaystyle\int_{-\infty}^\infty f(t)e^{ixt}\,dt$	
$\begin{cases} 1, & \|t\| < a, \\ 0, & \text{otherwise} \end{cases}$	$\sqrt{\dfrac{2}{\pi}}\dfrac{\sin(ax)}{x}$	
$e^{-a\|t\|}$	$\sqrt{\dfrac{2}{\pi}}\dfrac{a}{a^2+x^2},$	$a > 0$
$te^{-a\|t\|}$	$\sqrt{\dfrac{2}{\pi}}\dfrac{2iax}{(a^2+x^2)^2},$	$a > 0$
$\|t\|e^{-a\|t\|}$	$\sqrt{\dfrac{2}{\pi}}\dfrac{a^2-x^2}{(a^2+x^2)^2},$	$a > 0$
$\dfrac{e^{-a\|t\|}}{\|t\|^{1/2}}$	$\dfrac{(a+(a^2+x^2)^{1/2})^{1/2}}{(a^2+x^2)^{1/2}},$	$a > 0$
$\dfrac{\sinh(at)}{\sinh(\pi t)}$	$\dfrac{1}{\sqrt{2\pi}}\dfrac{\sin a}{\cosh x + \cos a},$	$-\pi < a < \pi$
$\dfrac{\cosh(at)}{\cosh(\pi t)}$	$\sqrt{\dfrac{2}{\pi}}\dfrac{\cos(\tfrac{1}{2}a)\cosh(\tfrac{1}{2}x)}{\cosh x + \cos a},$	$-\pi < a < \pi$
e^{-at^2}	$\dfrac{1}{\sqrt{2a}}e^{-x^2/(4a)},$	$a > 0$
$\sin(at^2)$	$-\dfrac{1}{\sqrt{2a}}\sin\left(\dfrac{x^2}{4a}-\dfrac{\pi}{4}\right),$	$a > 0$
$\cos(at^2)$	$\dfrac{1}{\sqrt{2a}}\cos\left(\dfrac{x^2}{4a}-\dfrac{\pi}{4}\right),$	$a > 0$

1.14 Integral Transforms

Table 1.14.2: Fourier cosine transforms.

$f(t)$	$\sqrt{\dfrac{2}{\pi}}\displaystyle\int_0^\infty f(t)\cos(xt)\,dt,\quad x>0$	
$\begin{cases} 1, & 0<t\le a, \\ 0, & \text{otherwise} \end{cases}$	$\sqrt{\dfrac{2}{\pi}}\dfrac{\sin(ax)}{x}$	
$\dfrac{1}{a^2+t^2}$	$\sqrt{\dfrac{\pi}{2}}\dfrac{e^{-ax}}{a},$	$\Re a>0$
$\dfrac{1}{(a^2+t^2)^2}$	$\sqrt{\dfrac{\pi}{2}}\dfrac{(1+ax)e^{-ax}}{2a^3},$	$\Re a>0$
$\dfrac{4a^3}{4a^4+t^4}$	$\sqrt{\pi}\,e^{-ax}\sin\!\left(ax+\tfrac{1}{4}\pi\right),$	$\Re a>0$
e^{-at}	$\sqrt{\dfrac{2}{\pi}}\dfrac{a}{a^2+x^2},$	$\Re a>0$
e^{-at^2}	$\dfrac{1}{\sqrt{2a}}e^{-x^2/(4a)},$	$\Re a>0$
$\sin(at^2)$	$-\dfrac{1}{\sqrt{2a}}\sin\!\left(\dfrac{x^2}{4a}-\dfrac{\pi}{4}\right),$	$a>0$
$\cos(at^2)$	$\dfrac{1}{\sqrt{2a}}\cos\!\left(\dfrac{x^2}{4a}-\dfrac{\pi}{4}\right),$	$a>0$
$\ln\!\left(1+\dfrac{a^2}{t^2}\right)$	$\sqrt{2\pi}\,\dfrac{1-e^{-ax}}{x},$	$\Re a>0$
$\ln\!\left(\dfrac{a^2+t^2}{b^2+t^2}\right)$	$\sqrt{2\pi}\,\dfrac{e^{-bx}-e^{-ax}}{x},$	$\Re a>0,$ $\Re b>0$

Table 1.14.3: Fourier sine transforms.

$f(t)$	$\sqrt{\dfrac{2}{\pi}}\displaystyle\int_0^\infty f(t)\sin(xt)\,dt,\quad x>0$	
t^{-1}	$\sqrt{\dfrac{\pi}{2}}$	
$t^{-1/2}$	$x^{-1/2}$	
$t^{-3/2}$	$2x^{1/2}$	
$\dfrac{t}{a^2+t^2}$	$\sqrt{\dfrac{\pi}{2}}\,e^{-ax},$	$\Re a>0$
$\dfrac{t}{(a^2+t^2)^2}$	$\sqrt{\dfrac{\pi}{8}}\dfrac{x}{a}e^{-ax},$	$\Re a>0$
$\dfrac{1}{t(a^2+t^2)}$	$\sqrt{\dfrac{\pi}{2}}\dfrac{1-e^{-ax}}{a^2},$	$\Re a>0$
$\dfrac{e^{-at}}{t}$	$\sqrt{\dfrac{2}{\pi}}\arctan\!\left(\dfrac{x}{a}\right),$	$\Re a>0$
e^{-at}	$\sqrt{\dfrac{2}{\pi}}\dfrac{x}{a^2+x^2},$	$\Re a>0$
te^{-at}	$\sqrt{\dfrac{2}{\pi}}\dfrac{2ax}{(a^2+x^2)^2},$	$\Re a>0$
te^{-at^2}	$(2a)^{-3/2}xe^{-x^2/(4a)},$	$\lvert\operatorname{ph}a\rvert<\tfrac{1}{2}\pi$
$\dfrac{\sin(at)}{t}$	$\dfrac{1}{\sqrt{2\pi}}\ln\!\left\lvert\dfrac{x+a}{x-a}\right\rvert,$	$a>0$
$\arctan\!\left(\dfrac{t}{a}\right)$	$\sqrt{\dfrac{\pi}{2}}\dfrac{e^{-ax}}{x},$	$a>0$
$\ln\left\lvert\dfrac{t+a}{t-a}\right\rvert$	$\sqrt{2\pi}\,\dfrac{\sin(ax)}{x},$	$a>0$

Table 1.14.4: Laplace transforms.

$f(t)$	$\int_0^\infty e^{-st} f(t)\, dt$	
1	$\dfrac{1}{s},$	$\Re s > 0$
$\dfrac{t^n}{n!}$	$\dfrac{1}{s^{n+1}},$	$\Re s > 0$
$\dfrac{1}{\sqrt{\pi t}}$	$\dfrac{1}{\sqrt{s}},$	$\Re s > 0$
e^{-at}	$\dfrac{1}{s+a},$	$\Re(s+a) > 0$
$\dfrac{t^n e^{-at}}{n!}$	$\dfrac{1}{(s+a)^{n+1}},$	$\Re(s+a) > 0$
$\dfrac{e^{-at} - e^{-bt}}{b-a}$	$\dfrac{1}{(s+a)(s+b)},$	$a \neq b,$ $\Re s > -\Re a,$ $\Re s > -\Re b$
$\sin(at)$	$\dfrac{a}{s^2+a^2},$	$\Re s > |\Im a|$
$\cos(at)$	$\dfrac{s}{s^2+a^2},$	$\Re s > |\Im a|$
$\sinh(at)$	$\dfrac{a}{s^2-a^2},$	$\Re s > |\Re a|$
$\cosh(at)$	$\dfrac{s}{s^2-a^2},$	$\Re s > |\Re a|$
$t\sin(at)$	$\dfrac{2as}{(s^2+a^2)^2},$	$\Re s > |\Im a|$
$t\cos(at)$	$\dfrac{s^2-a^2}{(s^2+a^2)^2},$	$\Re s > |\Im a|$
$\dfrac{e^{-bt}-e^{-at}}{t}$	$\ln\left(\dfrac{s+a}{s+b}\right),$	$\Re s > -\Re a,$ $\Re s > -\Re b$
$\dfrac{2(1-\cosh(at))}{t}$	$\ln\left(1 - \dfrac{a^2}{s^2}\right),$	$\Re(s+a) > 0$
$\dfrac{2(1-\cos(at))}{t}$	$\ln\left(1 + \dfrac{a^2}{s^2}\right),$	$\Re s > 0$
$\dfrac{\sin(at)}{t}$	$\arctan\left(\dfrac{a}{s}\right),$	$\Re s > 0$

Table 1.14.5: Mellin transforms.

$f(x)$	$\int_0^\infty x^{s-1} f(x)\, dx$	
$\begin{cases} 1, & x < a, \\ 0, & x \geq a \end{cases}$	$\dfrac{a^s}{s},$	$a \geq 0, \Re s > 0$
$\begin{cases} \ln(a/x), & x < a, \\ 0, & x \geq a \end{cases}$	$\dfrac{a^s}{s^2},$	$a \geq 0, \Re s > 1$
$\dfrac{1}{1-x}$	$\pi \cot(s\pi),$	$0 < \Re s < 1,$ (Cauchy p. v.)
$\dfrac{1}{1+x}$	$\pi \csc(s\pi),$	$0 < \Re s < 1$
$\ln(1+ax)$	$\dfrac{\pi \csc(s\pi)}{sa^s},$	$|\operatorname{ph} a| < \pi,$ $-1 < \Re s < 0$
$\ln\left|\dfrac{1+x}{1-x}\right|$	$\dfrac{\pi \tan(\frac{1}{2}s\pi)}{s},$	$-1 < \Re s < 1$
$\dfrac{\ln(1+x)}{x}$	$\dfrac{\pi \csc(s\pi)}{1-s},$	$0 < \Re s < 1$
$\arctan x$	$-\dfrac{\pi \sec(\frac{1}{2}s\pi)}{2s},$	$-1 < \Re s < 0$
$\operatorname{arccot} x$	$\dfrac{\pi \sec(\frac{1}{2}s\pi)}{2s},$	$0 < \Re s < 1$
$\dfrac{1+x\cos\theta}{1+2x\cos\theta+x^2}$	$\dfrac{\pi \cos(s\theta)}{\sin(s\pi)},$	$-\pi < \theta < \pi,$ $0 < \Re s < 1$
$\dfrac{x\sin\theta}{1+2x\cos\theta+x^2}$	$\dfrac{\pi \sin(s\theta)}{\sin(s\pi)},$	$-\pi < \theta < \pi,$ $0 < \Re s < 1$

1.14(viii) Compendia

For more extensive tables of the integral transforms of this section and tables of other integral transforms, see Erdélyi et al. (1954a,b), Gradshteyn and Ryzhik (2000), Marichev (1983), Oberhettinger (1972, 1974, 1990), Oberhettinger and Badii (1973), Oberhettinger and Higgins (1961), Prudnikov et al. (1986a,b, 1990, 1992a,b).

1.15 Summability Methods

1.15(i) Definitions for Series

1.15.1
$$s_n = \sum_{k=0}^{n} a_k.$$

Abel Summability

1.15.2
$$\sum_{n=0}^{\infty} a_n = s \quad (A),$$

if

1.15.3
$$\lim_{x \to 1-} \sum_{n=0}^{\infty} a_n x^n = s.$$

Cesàro Summability

1.15.4
$$\sum_{n=0}^{\infty} a_n = s \quad (C,1),$$

if

1.15.5
$$\lim_{n \to \infty} \frac{s_0 + s_1 + \cdots + s_n}{n+1} = s.$$

General Cesàro Summability

For $\alpha > -1$,

1.15.6
$$\sum_{n=0}^{\infty} a_n = s \quad (C,\alpha),$$

if

1.15.7
$$\lim_{n \to \infty} \frac{n!}{(\alpha+1)_n} \sum_{k=0}^{n} \frac{(\alpha+1)_k}{k!} a_{n-k} = s.$$

Borel Summability

1.15.8
$$\sum_{n=0}^{\infty} a_n = s \quad (B),$$

if

1.15.9
$$\lim_{t \to \infty} e^{-t} \sum_{n=0}^{\infty} \frac{s_n}{n!} t^n = s.$$

1.15(ii) Regularity

Methods of summation are *regular* if they are consistent with conventional summation. All of the methods described in §1.15(i) are regular. For example if

1.15.10
$$\sum_{n=0}^{\infty} a_n = s,$$

then

1.15.11
$$\sum_{n=0}^{\infty} a_n = s \quad (A).$$

1.15(iii) Summability of Fourier Series

Poisson Kernel

1.15.12
$$P(r,\theta) = \frac{1-r^2}{1-2r\cos\theta+r^2} = \sum_{n=-\infty}^{\infty} r^{|n|} e^{in\theta}, \quad 0 \le r < 1,$$

1.15.13
$$\frac{1}{2\pi} \int_0^{2\pi} P(r,\theta)\, d\theta = 1.$$

As $r \to 1-$

1.15.14
$$P(r,\theta) \to 0,$$

uniformly for $\theta \in [\delta, 2\pi - \delta]$. (Here and elsewhere in this subsection δ is a constant such that $0 < \delta < \pi$.)

Fejér Kernel

For $n = 0, 1, 2, \ldots$,

1.15.15
$$K_n(\theta) = \frac{1}{n+1}\left(\frac{\sin\left(\frac{1}{2}(n+1)\theta\right)}{\sin\left(\frac{1}{2}\theta\right)}\right)^2,$$

1.15.16
$$\frac{1}{2\pi} \int_0^{2\pi} K_n(\theta)\, d\theta = 1.$$

As $n \to \infty$

1.15.17
$$K_n(\theta) \to 0,$$

uniformly for $\theta \in [\delta, 2\pi - \delta]$.

Abel Means

1.15.18
$$A(r,\theta) = \sum_{n=-\infty}^{\infty} r^{|n|} F(n) e^{in\theta},$$

where

1.15.19
$$F(n) = \frac{1}{2\pi} \int_0^{2\pi} f(t) e^{-int}\, dt.$$

$A(r,\theta)$ is a harmonic function in polar coordinates ((1.9.27)), and

1.15.20
$$A(r,\theta) = \frac{1}{2\pi} \int_0^{2\pi} P(r, \theta - t) f(t)\, dt.$$

Cesàro (or (C,1)) Means

Let

1.15.21
$$\sigma_n(\theta) = \frac{s_0(\theta) + s_1(\theta) + \cdots + s_n(\theta)}{n+1},$$

$n = 0, 1, 2, \ldots$, where

1.15.22
$$s_n(\theta) = \sum_{k=-n}^{n} F(k) e^{ik\theta}.$$

Then

1.15.23
$$\sigma_n(\theta) = \frac{1}{2\pi} \int_0^{2\pi} K_n(\theta - t) f(t)\, dt.$$

Convergence

If $f(\theta)$ is periodic and integrable on $[0, 2\pi]$, then as $n \to \infty$ the Abel means $A(r,\theta)$ and the *(C,1)* means $\sigma_n(\theta)$ converge to

1.15.24 $\quad \tfrac{1}{2}(f(\theta+) + f(\theta-))$

at every point θ where both limits exist. If $f(\theta)$ is also continuous, then the convergence is uniform for all θ.

For real-valued $f(\theta)$, if

1.15.25 $\quad \sum_{n=-\infty}^{\infty} F(n)e^{in\theta}$

is the Fourier series of $f(\theta)$, then the series

1.15.26 $\quad F(0) + 2\sum_{n=1}^{\infty} F(n)e^{in\theta}$

can be extended to the interior of the unit circle as an analytic function

1.15.27
$$G(z) = G(x + iy) = u(x,y) + iv(x,y)$$
$$= F(0) + 2\sum_{n=1}^{\infty} F(n)z^n.$$

Here $u(x,y) = A(r,\theta)$ is the *Abel* (or *Poisson*) *sum* of $f(\theta)$, and $v(x,y)$ has the series representation

1.15.28 $\quad -\sum_{n=-\infty}^{\infty} i(\operatorname{sign} n) F(n) r^{|n|} e^{in\theta};$

compare §1.15(v).

1.15(iv) Definitions for Integrals

Abel Summability

$\int_{-\infty}^{\infty} f(t)\,dt$ is *Abel summable* to L, or

1.15.29 $\quad \int_{-\infty}^{\infty} f(t)\,dt = L \quad (A),$

when

1.15.30 $\quad \lim_{\epsilon \to 0+} \int_{-\infty}^{\infty} e^{-\epsilon|t|} f(t)\,dt = L.$

Cesàro Summability

$\int_{-\infty}^{\infty} f(t)\,dt$ is *(C,1) summable* to L, or

1.15.31 $\quad \int_{-\infty}^{\infty} f(t)\,dt = L \quad (C,1),$

when

1.15.32 $\quad \lim_{R \to \infty} \int_{-R}^{R} \left(1 - \frac{|t|}{R}\right) f(t)\,dt = L.$

If $\int_{-\infty}^{\infty} f(t)\,dt$ converges and equals L, then the integral is Abel and Cesàro summable to L.

1.15(v) Summability of Fourier Integrals

Poisson Kernel

1.15.33 $\quad P(x,y) = \dfrac{2y}{x^2 + y^2}, \quad y > 0, \; -\infty < x < \infty.$

1.15.34 $\quad \dfrac{1}{2\pi} \int_{-\infty}^{\infty} P(x,y)\,dx = 1.$

For each $\delta > 0$,

1.15.35 $\quad \int_{|x| \geq \delta} P(x,y)\,dx \to 0, \qquad \text{as } y \to 0.$

Let

1.15.36 $\quad h(x,y) = \dfrac{1}{\sqrt{2\pi}} \int_{-\infty}^{\infty} e^{-y|t|} e^{-ixt} F(t)\,dt,$

where $F(t)$ is the Fourier transform of $f(x)$ (§1.14(i)). Then

1.15.37 $\quad h(x,y) = \dfrac{1}{2\pi} \int_{-\infty}^{\infty} f(t) P(x - t, y)\,dt$

is the *Poisson integral* of $f(t)$.

If $f(x)$ is integrable on $(-\infty, \infty)$, then

1.15.38 $\quad \lim_{y \to 0+} \int_{-\infty}^{\infty} |h(x,y) - f(x)|\,dx = 0.$

Suppose now $f(x)$ is real-valued and integrable on $(-\infty, \infty)$. Let

1.15.39 $\quad \Phi(z) = \Phi(x + iy) = \dfrac{i}{\pi} \int_{-\infty}^{\infty} f(t) \dfrac{1}{(x - t) + iy}\,dt,$

where $y > 0$ and $-\infty < x < \infty$. Then $\Phi(z)$ is an analytic function in the upper half-plane and its real part is the Poisson integral $h(x, y)$; compare (1.9.34). The imaginary part

1.15.40 $\quad \Im\Phi(x + iy) = \dfrac{1}{\pi} \int_{-\infty}^{\infty} f(t) \dfrac{x - t}{(x - t)^2 + y^2}\,dt$

is the *conjugate Poisson integral* of $f(x)$. Moreover, $\lim_{y \to 0+} \Im\Phi(x + iy)$ is the Hilbert transform of $f(x)$ (§1.14(v)).

Fejér Kernel

1.15.41 $\quad K_R(s) = \dfrac{1}{\pi R} \dfrac{1 - \cos(Rs)}{s^2},$

1.15.42 $\quad \int_{-\infty}^{\infty} K_R(s)\,ds = 1.$

For each $\delta > 0$,

1.15.43 $\quad \int_{|s| \geq \delta} K_R(s)\,ds \to 0, \qquad \text{as } R \to \infty.$

Let

1.15.44 $\quad \sigma_R(\theta) = \dfrac{1}{\sqrt{2\pi}} \int_{-R}^{R} \left(1 - \dfrac{|t|}{R}\right) e^{-i\theta t} F(t)\,dt,$

then

1.15.45 $\quad \sigma_R(\theta) = \int_{-\infty}^{\infty} f(t) K_R(\theta - t)\,dt.$

If $f(\theta)$ is integrable on $(-\infty, \infty)$, then

$$\text{1.15.46} \qquad \lim_{R \to \infty} \int_{-\infty}^{\infty} |\sigma_R(\theta) - f(\theta)| \, d\theta = 0.$$

1.15(vi) Fractional Integrals

For $\Re\alpha > 0$, the *fractional integral operator of order* α is defined by

$$\text{1.15.47} \qquad I^\alpha f(x) = \frac{1}{\Gamma(\alpha)} \int_0^x (x-t)^{\alpha-1} f(t) \, dt.$$

For $\Gamma(\alpha)$ see §5.2, and compare (1.4.31) in the case when α is a positive integer.

$$\text{1.15.48} \qquad I^\alpha I^\beta = I^{\alpha+\beta}, \qquad \Re\alpha > 0, \Re\beta > 0.$$

For extensions of (1.15.48) see Love (1972b).

If

$$\text{1.15.49} \qquad f(x) = \sum_{k=0}^{\infty} a_k x^k,$$

then

$$\text{1.15.50} \qquad I^\alpha f(x) = \sum_{k=0}^{\infty} \frac{k!}{\Gamma(k+\alpha+1)} a_k x^{k+\alpha}.$$

1.15(vii) Fractional Derivatives

For $0 < \Re\alpha < n$, n an integer,

$$\text{1.15.51} \qquad D^\alpha f(x) = \frac{d^n}{dx^n} I^{n-\alpha} f(x),$$

$$\text{1.15.52} \qquad D^k I^\alpha = D^n I^{\alpha+n-k}, \quad k = 1, 2, \ldots, n.$$

When none of α, β, and $\alpha + \beta$ is an integer

$$\text{1.15.53} \qquad D^\alpha D^\beta = D^{\alpha+\beta}.$$

Note that $D^{1/2} D \neq D^{3/2}$. See also Love (1972b).

1.15(viii) Tauberian Theorems

If

$$\text{1.15.54} \qquad \sum_{n=0}^{\infty} a_n = s \ (A), \quad a_n > -\frac{K}{n}, \ n > 0, K > 0,$$

then

$$\text{1.15.55} \qquad \sum_{n=0}^{\infty} a_n = s.$$

If

$$\text{1.15.56} \qquad \lim_{x \to 1-} (1-x) \sum_{n=0}^{\infty} a_n x^n = s,$$

and either $|a_n| \leq K$ or $a_n \geq 0$, then

$$\text{1.15.57} \qquad \lim_{n \to \infty} \frac{a_0 + a_1 + \cdots + a_n}{n+1} = s.$$

1.16 Distributions

1.16(i) Test Functions

Let ϕ be a function defined on an open interval $I = (a, b)$, which can be infinite. The closure of the set of points where $\phi \neq 0$ is called the *support* of ϕ. If the support of ϕ is a compact set (§1.9(vii)), then ϕ is called a *function of compact support*. A *test function* is an infinitely differentiable function of compact support.

A sequence $\{\phi_n\}$ of test functions *converges* to a test function ϕ if the support of every ϕ_n is contained in a fixed compact set K and as $n \to \infty$ the sequence $\{\phi_n^{(k)}\}$ converges uniformly on K to $\phi^{(k)}$ for $k = 0, 1, 2, \ldots$.

The linear space of all test functions with the above definition of convergence is called *a test function space*. We denote it by $\mathcal{D}(I)$.

A mapping Λ on $\mathcal{D}(I)$ is a *linear functional* if it takes complex values and

$$\text{1.16.1} \qquad \Lambda(\alpha_1 \phi_1 + \alpha_2 \phi_2) = \alpha_1 \Lambda(\phi_1) + \alpha_2 \Lambda(\phi_2),$$

where α_1 and α_2 are real or complex constants. $\Lambda : \mathcal{D}(I) \to \mathbb{C}$ is called a *distribution* if it is a continuous linear functional on $\mathcal{D}(I)$, that is, it is a linear functional and for every $\phi_n \to \phi$ in $\mathcal{D}(I)$,

$$\text{1.16.2} \qquad \lim_{n \to \infty} \Lambda(\phi_n) = \Lambda(\phi).$$

From here on we write $\langle \Lambda, \phi \rangle$ for $\Lambda(\phi)$. The space of all distributions will be denoted by $\mathcal{D}^*(I)$. A distribution Λ is called *regular* if there is a function f on I, which is absolutely integrable on every compact subset of I, such that

$$\text{1.16.3} \qquad \langle \Lambda, \phi \rangle = \int_I f(x) \phi(x) \, dx.$$

We denote a regular distribution by Λ_f, or simply f, where f is the function giving rise to the distribution. (If a distribution is not regular, it is called *singular*.)

Define

$$\text{1.16.4} \qquad \langle \Lambda_1 + \Lambda_2, \phi \rangle = \langle \Lambda_1, \phi \rangle + \langle \Lambda_2, \phi \rangle,$$

$$\text{1.16.5} \qquad \langle c\Lambda, \phi \rangle = c \langle \Lambda, \phi \rangle = \langle \Lambda, c\phi \rangle,$$

where c is a constant. More generally, if $\alpha(x)$ is an infinitely differentiable function, then

$$\text{1.16.6} \qquad \langle \alpha \Lambda, \phi \rangle = \langle \Lambda, \alpha \phi \rangle.$$

We say that a sequence of distributions $\{\Lambda_n\}$ *converges* to a distribution Λ in $\mathcal{D}^*$ if

$$\text{1.16.7} \qquad \lim_{n \to \infty} \langle \Lambda_n, \phi \rangle = \langle \Lambda, \phi \rangle$$

for all $\phi \in \mathcal{D}(I)$.

1.16(ii) Derivatives of a Distribution

The *derivative* Λ' of a distribution is defined by

$$\text{1.16.8} \qquad \langle \Lambda', \phi \rangle = -\langle \Lambda, \phi' \rangle, \qquad \phi \in \mathcal{D}(I).$$

Similarly

1.16.9 $\quad \langle \Lambda^{(k)}, \phi \rangle = (-1)^k \langle \Lambda, \phi^{(k)} \rangle, \quad k = 1, 2, \ldots.$

For any locally integrable function f, its *distributional derivative* is $Df = \Lambda'_f$.

1.16(iii) Dirac Delta Distribution

1.16.10 $\quad \langle \delta, \phi \rangle = \phi(0), \qquad \phi \in \mathcal{D}(I),$

1.16.11 $\quad \langle \delta_{x_0}, \phi \rangle = \phi(x_0), \qquad \phi \in \mathcal{D}(I),$

1.16.12 $\quad \langle \delta_{x_0}^{(n)}, \phi \rangle = (-1)^n \phi^{(n)}(x_0), \qquad \phi \in \mathcal{D}(I).$

The Dirac delta distribution is singular.

1.16(iv) Heaviside Function

1.16.13 $\quad H(x) = \begin{cases} 1, & x > 0, \\ 0, & x \leq 0. \end{cases}$

1.16.14 $\quad H(x - x_0) = \begin{cases} 1, & x > x_0, \\ 0, & x \leq x_0. \end{cases}$

1.16.15 $\quad DH = \delta,$

1.16.16 $\quad DH(x - x_0) = \delta_{x_0}.$

Suppose $f(x)$ is infinitely differentiable except at x_0, where left and right derivatives of all orders exist, and

1.16.17 $\quad \sigma_n = f^{(n)}(x_0+) - f^{(n)}(x_0-).$

Then

1.16.18 $\quad \begin{aligned} D^m f &= f^{(m)} + \sigma_0 \delta_{x_0}^{(m-1)} + \sigma_1 \delta_{x_0}^{(m-2)} + \cdots \\ &\quad + \sigma_{m-1} \delta_{x_0}, \qquad m = 1, 2, \ldots. \end{aligned}$

For $\alpha > -1$,

1.16.19 $\quad x_+^\alpha = x^\alpha H(x) = \begin{cases} x^\alpha, & x > 0, \\ 0, & x \leq 0. \end{cases}$

For $\alpha > 0$,

1.16.20 $\quad D x_+^\alpha = \alpha x_+^{\alpha - 1}.$

For $\alpha < -1$ and α not an integer, define

1.16.21 $\quad x_+^\alpha = \frac{1}{(\alpha+1)_n} D^n x_+^{\alpha+n},$

where n is an integer such that $\alpha + n > -1$. Similarly, we write

1.16.22 $\quad \ln_+ x = H(x) \ln x = \begin{cases} \ln x, & x > 0, \\ 0, & x \leq 0, \end{cases}$

and define

1.16.23 $\quad (-1)^n n! x_+^{-1-n} = D^{(n+1)} \ln_+ x, \quad n = 0, 1, 2, \ldots.$

1.16(v) Tempered Distributions

The space $\mathcal{T}(\mathbb{R})$ of test functions for tempered distributions consists of all infinitely-differentiable functions such that the function and all its derivatives are $O(|x|^{-N})$ as $|x| \to \infty$ for all N.

A sequence $\{\phi_n\}$ of functions in $\mathcal{T}$ is said to *converge* to a function $\phi \in \mathcal{T}$ as $n \to \infty$ if the sequence $\{\phi_n^{(k)}\}$ converges uniformly to $\phi^{(k)}$ on every finite interval and if the constants $c_{k,N}$ in the inequalities

1.16.24 $\quad |x^N \phi_n^{(k)}| \leq c_{k,N}$

do not depend on n.

A *tempered distribution* is a continuous linear functional Λ on $\mathcal{T}$. (See the definition of a distribution in §1.16(i).) The set of tempered distributions is denoted by $\mathcal{T}^*$.

A sequence of tempered distributions Λ_n *converges* to Λ in $\mathcal{T}^*$ if

1.16.25 $\quad \lim_{n \to \infty} \langle \Lambda_n, \phi \rangle = \langle \Lambda, \phi \rangle,$

for all $\phi \in \mathcal{T}$.

The derivatives of tempered distributions are defined in the same way as derivatives of distributions.

For a detailed discussion of tempered distributions see Lighthill (1958).

1.16(vi) Distributions of Several Variables

Let $\mathcal{D}(\mathbb{R}^n) = \mathcal{D}_n$ be the set of all infinitely differentiable functions in n variables, $\phi(x_1, x_2, \ldots, x_n)$, with compact support in $\mathbb{R}^n$. If $k = (k_1, \ldots, k_n)$ is a multi-index and $x = (x_1, \ldots, x_n) \in \mathbb{R}^n$, then we write $x^k = x_1^{k_1} \cdots x_n^{k_n}$ and $\phi^{(k)}(x) = \partial^k \phi / (\partial x_1^{k_1} \cdots \partial x_n^{k_n})$. A sequence $\{\phi_m\}$ of functions in $\mathcal{D}_n$ *converges* to a function $\phi \in \mathcal{D}_n$ if the supports of ϕ_m lie in a fixed compact subset K of $\mathbb{R}^n$ and $\phi_m^{(k)}$ converges uniformly to $\phi^{(k)}$ in K for every multi-index $k = (k_1, k_2, \ldots, k_n)$. A *distribution* in $\mathbb{R}^n$ is a continuous linear functional on $\mathcal{D}_n$.

The partial derivatives of distributions in $\mathbb{R}^n$ can be defined as in §1.16(ii). A locally integrable function $f(x) = f(x_1, x_2, \ldots, x_n)$ gives rise to a distribution Λ_f defined by

1.16.26 $\quad \langle \Lambda_f, \phi \rangle = \int_{\mathbb{R}^n} f(x) \phi(x) \, dx, \qquad \phi \in \mathcal{D}_n.$

The *distributional derivative* $D^k f$ of f is defined by

1.16.27 $\quad \langle D^k f, \phi \rangle = (-1)^{|k|} \int_{\mathbb{R}^n} f(x) \phi^{(k)}(x) \, dx, \quad \phi \in \mathcal{D}_n,$

where k is a multi-index and $|k| = k_1 + k_2 + \cdots + k_n$.

For tempered distributions the space of test functions $\mathcal{T}_n$ is the set of all infinitely-differentiable functions ϕ of n variables that satisfy

1.16.28 $\quad |x^m \phi^{(k)}(x)| \leq c_{m,k}, \qquad x \in \mathbb{R}^n.$

Here $m = (m_1, m_2, \ldots, m_n)$ and $k = (k_1, k_2, \ldots, k_n)$ are multi-indices, and $c_{m,k}$ are constants. Tempered distributions are continuous linear functionals on this space of test functions. The space of tempered distributions is denoted by $\mathcal{T}_n^*$.

1.16(vii) Fourier Transforms of Distributions

Suppose ϕ is a test function in $\mathcal{T}_n$. Then its *Fourier transform* is

1.16.29 $\quad F(\mathbf{x}) = F = \dfrac{1}{(2\pi)^{n/2}} \displaystyle\int_{\mathbb{R}^n} \phi(\mathbf{t}) e^{i\mathbf{x}\cdot\mathbf{t}} \, d\mathbf{t},$

where $\mathbf{x} = (x_1, x_2, \ldots, x_n)$ and $\mathbf{x} \cdot \mathbf{t} = x_1 t_1 + \cdots + x_n t_n$. $F(\mathbf{x})$ is also in $\mathcal{T}_n$. For a multi-index $\boldsymbol{\alpha} = (\alpha_1, \alpha_2, \ldots, \alpha_n)$, set $|\alpha| = \alpha_1 + \alpha_2 + \cdots + \alpha_n$ and

1.16.30 $\quad D_{\boldsymbol{\alpha}} = i^{-|\alpha|} D^{\boldsymbol{\alpha}} = \left(\dfrac{1}{i}\dfrac{\partial}{\partial x_1}\right)^{\alpha_1} \cdots \left(\dfrac{1}{i}\dfrac{\partial}{\partial x_n}\right)^{\alpha_n},$

1.16.31 $\quad P(\mathbf{x}) = P = \sum c_{\boldsymbol{\alpha}} \mathbf{x}^{\boldsymbol{\alpha}} = \sum c_{\boldsymbol{\alpha}} x_1^{\alpha_1} \cdots x_n^{\alpha_n},$

and

1.16.32 $\quad P(D) = \sum c_{\boldsymbol{\alpha}} D_{\boldsymbol{\alpha}}.$

Then

1.16.33 $\quad \dfrac{1}{(2\pi)^{n/2}} \displaystyle\int_{\mathbb{R}^n} (P(D)\phi)(\mathbf{t}) e^{i\mathbf{x}\cdot\mathbf{t}} \, d\mathbf{t} = P(-\mathbf{x}) F(\mathbf{x}),$

and

1.16.34 $\quad \dfrac{1}{(2\pi)^{n/2}} \displaystyle\int_{\mathbb{R}^n} P(\mathbf{t}) \phi(\mathbf{t}) e^{i\mathbf{x}\cdot\mathbf{t}} \, d\mathbf{t} = P(D) F(\mathbf{x}).$

If $u \in \mathcal{T}_n^*$ is a tempered distribution, then its *Fourier transform* $\mathcal{F}(u)$ is defined by

1.16.35 $\quad \langle \mathcal{F}(u), \phi \rangle = \langle u, F \rangle, \qquad \phi \in \mathcal{T}_n,$

where F is given by (1.16.29). The *Fourier transform* $\mathcal{F}(u)$ of a tempered distribution is again a tempered distribution, and

1.16.36 $\quad \mathcal{F}(P(D)u) = P(-\mathbf{x}) \mathcal{F}(u),$

1.16.37 $\quad \mathcal{F}(Pu) = P(D) \mathcal{F}(u).$

In (1.16.36) and (1.16.37) the derivatives in $P(D)$ are understood to be in the sense of distributions.

1.17 Integral and Series Representations of the Dirac Delta

1.17(i) Delta Sequences

In applications in physics and engineering, the Dirac delta distribution (§1.16(iii)) is historically and customarily replaced by the *Dirac delta* (or *Dirac delta function*) $\delta(x)$. This is an operator with the properties:

1.17.1 $\quad \delta(x) = 0, \qquad x \in \mathbb{R}, \, x \neq 0,$

and

1.17.2 $\quad \displaystyle\int_{-\infty}^{\infty} \delta(x-a) \phi(x) \, dx = \phi(a), \qquad a \in \mathbb{R},$

subject to certain conditions on the function $\phi(x)$. From the mathematical standpoint the left-hand side of (1.17.2) can be interpreted as a generalized integral in the sense that

1.17.3 $\quad \displaystyle\lim_{n \to \infty} \int_{-\infty}^{\infty} \delta_n(x-a) \phi(x) \, dx = \phi(a),$

for a suitably chosen sequence of functions $\delta_n(x)$, $n = 1, 2, \ldots$. Such a sequence is called a *delta sequence* and we write, symbolically,

1.17.4 $\quad \displaystyle\lim_{n \to \infty} \delta_n(x) = \delta(x), \qquad x \in \mathbb{R}.$

An example of a delta sequence is provided by

1.17.5 $\quad \delta_n(x-a) = \sqrt{\dfrac{n}{\pi}} e^{-n(x-a)^2}.$

In this case

1.17.6 $\quad \displaystyle\lim_{n \to \infty} \sqrt{\dfrac{n}{\pi}} \int_{-\infty}^{\infty} e^{-n(x-a)^2} \phi(x) \, dx = \phi(a),$

for all functions $\phi(x)$ that are continuous when $x \in (-\infty, \infty)$, and for each a, $\int_{-\infty}^{\infty} e^{-n(x-a)^2} \phi(x) \, dx$ converges absolutely for all sufficiently large values of n. The last condition is satisfied, for example, when $\phi(x) = O\left(e^{\alpha x^2}\right)$ as $x \to \pm\infty$, where α is a real constant.

More generally, assume $\phi(x)$ is piecewise continuous (§1.4(ii)) when $x \in [-c, c]$ for any finite positive real value of c, and for each a, $\int_{-\infty}^{\infty} e^{-n(x-a)^2} \phi(x) \, dx$ converges absolutely for all sufficiently large values of n. Then

1.17.7
$$\lim_{n \to \infty} \sqrt{\dfrac{n}{\pi}} \int_{-\infty}^{\infty} e^{-n(x-a)^2} \phi(x) \, dx = \tfrac{1}{2}\phi(a-) + \tfrac{1}{2}\phi(a+).$$

1.17(ii) Integral Representations

Formal interchange of the order of integration in the Fourier integral formula ((1.14.1) and (1.14.4)):

1.17.8 $\quad \dfrac{1}{2\pi} \displaystyle\int_{-\infty}^{\infty} e^{-iat} \left(\int_{-\infty}^{\infty} \phi(x) e^{itx} \, dx\right) dt = \phi(a)$

yields

1.17.9 $\quad \displaystyle\int_{-\infty}^{\infty} \left(\dfrac{1}{2\pi} \int_{-\infty}^{\infty} e^{i(x-a)t} \, dt\right) \phi(x) \, dx = \phi(a).$

The inner integral does not converge. However, for $n = 1, 2, \ldots,$

1.17.10 $\quad \dfrac{1}{2\pi} \displaystyle\int_{-\infty}^{\infty} e^{-t^2/(4n)} e^{i(x-a)t} \, dt = \sqrt{\dfrac{n}{\pi}} e^{-n(x-a)^2}.$

Hence comparison with (1.17.5) shows that (1.17.9) can be interpreted as a generalized integral (1.17.3) with

1.17.11 $\quad \delta_n(x-a) = \dfrac{1}{2\pi} \displaystyle\int_{-\infty}^{\infty} e^{-t^2/(4n)} e^{i(x-a)t} \, dt,$

provided that $\phi(x)$ is continuous when $x \in (-\infty, \infty)$, and for each a, $\int_{-\infty}^{\infty} e^{-n(x-a)^2} \phi(x)\,dx$ converges absolutely for all sufficiently large values of n (as in the case of (1.17.6)). Then comparison of (1.17.2) and (1.17.9) yields the formal integral representation

1.17.12 $\quad \delta(x-a) = \dfrac{1}{2\pi} \int_{-\infty}^{\infty} e^{i(x-a)t}\,dt.$

Other similar integral representations of the Dirac delta that appear in the physics literature include the following:

Bessel Functions and Spherical Bessel Functions (§§10.2(ii), 10.47(ii))

1.17.13 $\quad \delta(x-a) = x \int_0^{\infty} t\, J_\nu(xt)\, J_\nu(at)\,dt,$
$\Re \nu > -1,\ x > 0,\ a > 0,$

1.17.14
$$\delta(x-a) = \dfrac{2xa}{\pi} \int_0^{\infty} t^2\, \mathsf{j}_\ell(xt)\, \mathsf{j}_\ell(at)\,dt,\quad x>0,\ a>0.$$

See Arfken and Weber (2005, Eq. (11.59)) and Konopinski (1981, p. 242). For a generalization of (1.17.14) see Maximon (1991).

Coulomb Functions (§33.14(iv))

1.17.15
$$\delta(x-a) = \int_0^{\infty} s(x, \ell; r)\, s(a, \ell; r)\,dr,\quad a>0,\ x>0.$$

See Seaton (2002).

Airy Functions (§9.2)

1.17.16 $\quad \delta(x-a) = \int_{-\infty}^{\infty} \mathrm{Ai}(t-x)\, \mathrm{Ai}(t-a)\,dt.$

See Vallée and Soares (2004, §3.5.3).

1.17(iii) Series Representations

Formal interchange of the order of summation and integration in the Fourier summation formula ((1.8.3) and (1.8.4)):

1.17.17 $\quad \dfrac{1}{2\pi} \sum_{k=-\infty}^{\infty} e^{-ika} \left(\int_{-\pi}^{\pi} \phi(x) e^{ikx}\,dx \right) = \phi(a),$

yields

1.17.18 $\quad \int_{-\pi}^{\pi} \phi(x) \left(\dfrac{1}{2\pi} \sum_{k=-\infty}^{\infty} e^{ik(x-a)} \right) dx = \phi(a).$

The sum $\sum_{k=-\infty}^{\infty} e^{ik(x-a)}$ does not converge, but (1.17.18) can be interpreted as a generalized integral in the sense that

1.17.19 $\quad \lim_{n \to \infty} \int_{-\pi}^{\pi} \delta_n(x-a) \phi(x)\,dx = \phi(a),$

where

1.17.20
$$\delta_n(x-a) = \dfrac{1}{2\pi} \sum_{k=-n}^{n} e^{ik(x-a)} \left(= \dfrac{\sin\bigl((n+\tfrac{1}{2})(x-a)\bigr)}{2\pi \sin\bigl(\tfrac{1}{2}(x-a)\bigr)} \right),$$

provided that $\phi(x)$ is continuous and of period 2π; see §1.8(ii).

By analogy with §1.17(ii) we have the formal series representation

1.17.21 $\quad \delta(x-a) = \dfrac{1}{2\pi} \sum_{k=-\infty}^{\infty} e^{ik(x-a)}.$

Other similar series representations of the Dirac delta that appear in the physics literature include the following:

Legendre Polynomials (§§14.7(i) and 18.3)

1.17.22 $\quad \delta(x-a) = \sum_{k=0}^{\infty} (k+\tfrac{1}{2})\, P_k(x)\, P_k(a).$

Laguerre Polynomials (§18.3)

1.17.23 $\quad \delta(x-a) = e^{-(x+a)/2} \sum_{k=0}^{\infty} L_k(x)\, L_k(a).$

Hermite Polynomials (§18.3)

1.17.24 $\quad \delta(x-a) = \dfrac{e^{-(x^2+a^2)/2}}{\sqrt{\pi}} \sum_{k=0}^{\infty} \dfrac{H_k(x)\, H_k(a)}{2^k k!}.$

Spherical Harmonics (§14.30)

1.17.25
$$\delta(\cos\theta_1 - \cos\theta_2)\,\delta(\phi_1 - \phi_2) = \sum_{\ell=0}^{\infty} \sum_{m=-\ell}^{\ell} Y_{\ell,m}(\theta_1, \phi_1)\, Y_{\ell,m}^{*}(\theta_2, \phi_2).$$

(1.17.22)–(1.17.24) are special cases of Morse and Feshbach (1953a, Eq. (6.3.11)). For (1.17.25) see Arfken and Weber (2005, p. 792).

1.17(iv) Mathematical Definitions

The references given in §§1.17(ii)–1.17(iii) are from the physics literature. For mathematical interpretations of (1.17.13), (1.17.15), (1.17.16) and (1.17.22)–(1.17.25) that resemble those given in §§1.17(ii) and 1.17(iii) for (1.17.12) and (1.17.21), see Li and Wong (2008). For (1.17.14) combine (1.17.13) and (10.47.3).

References

Sources

The following list gives the references or other indications of proofs that were used in constructing the various sections of this chapter. These sources supplement the references that are quoted in the text.

§1.2 Chrystal (1959, pp. 62–70, 482–483, 489), Hardy *et al.* (1967, pp. 12–15).

§1.3 Vein and Dale (1999, pp. 3–12, 33–34, 51–52, 57, 79–81), For (1.3.17) see Bressoud (1999, p. 67).

§1.4 Hardy (1952, Chapters 5–7, and pp. 234–235, 247–248, 258, 285–292, 327–328), Olver (1997b, pp. 28, 73), Rudin (1976, Chapter 5), Hardy *et al.* (1967, pp. 70–77). For (1.4.13) see Riordan (1958, pp. 35–36) and Knuth (1968, p. 50). For (1.4.31) integrate by parts.

§1.5 Marsden and Tromba (1996, Chapters 2, 3, 5, 6, and pp. 358–371), Davis and Snider (1987, Chapter 5), Protter and Morrey (1991, pp. 288, 298) For (1.5.36) see Love (1970, 1972a).

§1.6 Marsden and Tromba (1996, Chapter 1 and pp. 144–147, 273–283, 396–417, 421–459, 470, 485, 506). For (1.6.9) see Hubbard and Hubbard (2002, pp. 82–84).

§1.7 Hardy *et al.* (1967, pp. 1–32, 130–147, 151).

§1.8 Protter and Morrey (1991, Chapter 10), Tolstov (1962, Chapter 1 and p. 77), Titchmarsh (1962, Chapter 13 and pp. 419, 421). For the Riemann–Lebesgue lemma see Olver (1997b, p. 73). For Poisson's summation formula see Rademacher (1973, pp. 71–75), Titchmarsh (1986a, p. 61). For (1.8.16) set $f(x) = e^{-\omega x^2}$ in (1.8.14).

§1.9 Copson (1935, Chapters 1–3 and pp. 56–69, 92–98), Levinson and Redheffer (1970, Chapters 1–3, and pp. 259–277, 349–351, 360), Markushevich (1983, pp. 14–18, 41–46, 131–135), Markushevich (1985, vol. 1, §34), Ahlfors (1966, pp. 168–169). For a proof of the Jordan Curve Theorem see, for example, Dienes (1931, pp. 177–197). The theorem is valid with less restrictive conditions than those assumed here. For the operations on series, see Henrici (1974, Chapter 1) or Olver (1997b, pp. 19–22). For (1.9.69)–(1.9.71), see Titchmarsh (1962, §1.77).

§1.10 Copson (1935, pp. 72–81, 106–113, 117–120, 192–193, 438–440), Levinson and Redheffer (1970, pp. 64–77, 140–143, 162–170, 392–395, 398–402), Markushevich (1983, pp. 106–121, 234–245), Titchmarsh (1962, pp. 13–19, 165–169, 246–250). For (1.10.13) and (1.10.14) see Copson (1935, §6.23). See also Andrews *et al.* (1999, pp. 629–631) and Henrici (1974, pp. 57–59). The Extended Inversion Theorem is proved in a similar way.

§1.11 Burnside and Panton (1960, Chapter 2 and pp. 80–81), Dummit and Foote (1999, pp. 300–301, 591–595, 611–616), Henrici (1977, vol. 2, pp. 555–559). For the Horner scheme, see Burnside and Panton (1960, pp. 8–9). The double Horner scheme is derived similarly.

§1.12 Jones and Thron (1980, pp. 20, 31–37, 42–43, 88, 92), Lorentzen and Waadeland (1992, pp. 8–9, 30, 32, 84–85).

§1.13 Olver (1997b, pp. 141–142, 145–147, 190–191), Temme (1996a, pp. 84, 103), Watson (1944, pp. 145–146). For (1.13.10) see Simmons (1972, pp. 90–92).

§1.14 Titchmarsh (1986a, pp. 3–15, 42, 50–60, 119–132, and 176–210), Schiff (1999, pp. 12–57, 91–93, 151–157, and 209–218), Paris and Kaminski (2001, pp. 79–89), Wong (1989, pp. 147–152 and 192–194), Henrici (1986, vol. 3, pp. 197–202), Widder (1941, pp. 325–328, 340–341), Davies (1984, pp. 11–13, 103–108, 152–153, 209–211), Pinkus and Zafrany (1997, pp. 147–149). For (1.14.46) see Sneddon (1972, p. 234).

§1.15 Hardy (1949, pp. 10, 154–155), Weiss (1965, pp. 131–135, 143–148), Andrews *et al.* (1999, pp. 111–114, 602–607), Wong (1989, pp. 197–198), Widder (1941, Chapter 5). For (1.15.24) see Körner (1989, Chapters 2, 27).

§1.16 Wong (1989, pp. 241–254, 261–279).

§1.17 (1.17.6) is a special case of Theorem 7.1 of Olver (1997b, Chapter 3) when $\phi(a) \neq 0$. This theorem also extends straightforwardly to cover $\phi(a) = 0$. (1.17.7) is proved in a similar manner. For (1.17.10) complete the square in the total power of e, make the change of variable $\tau = (t/(2\sqrt{n}) - i(x-a)\sqrt{n}$, and use $\int_{-\infty}^{\infty} e^{-\tau^2} d\tau = \sqrt{\pi}$.

Chapter 2
Asymptotic Approximations

F. W. J. Olver[1] and R. Wong[2]

Areas   **42**
- 2.1 Definitions and Elementary Properties . . 42
- 2.2 Transcendental Equations 43
- 2.3 Integrals of a Real Variable 43
- 2.4 Contour Integrals 46
- 2.5 Mellin Transform Methods 48
- 2.6 Distributional Methods 51
- 2.7 Differential Equations 55
- 2.8 Differential Equations with a Parameter . 58
- 2.9 Difference Equations 61
- 2.10 Sums and Sequences 63
- 2.11 Remainder Terms; Stokes Phenomenon . 66

References   **69**

---

[1] Institute for Physical Science and Technology and Department of Mathematics, University of Maryland, College Park, Maryland.
[2] Liu Bie Ju Centre for Mathematical Sciences, City University of Hong Kong, Kowloon, Hong Kong.
Copyright © 2009 National Institute of Standards and Technology. All rights reserved.

Areas

2.1 Definitions and Elementary Properties

2.1(i) Asymptotic and Order Symbols

Let $\mathbf{X}$ be a point set with a limit point c. As $x \to c$ in $\mathbf{X}$

2.1.1 $\quad f(x) \sim \phi(x) \iff f(x)/\phi(x) \to 1.$

2.1.2 $\quad f(x) = o(\phi(x)) \iff f(x)/\phi(x) \to 0.$

2.1.3 $\quad f(x) = O(\phi(x)) \iff |f(x)/\phi(x)|$ is bounded.

The symbol O can also apply to the whole set $\mathbf{X}$, and not just as $x \to c$.

Examples

2.1.4 $\qquad \tanh x \sim x, \qquad x \to 0$ in $\mathbb{C}$.

2.1.5 $\qquad e^{-x} = o(1), \qquad x \to +\infty$ in $\mathbb{R}$.

2.1.6 $\qquad \sin(\pi x + x^{-1}) = O(x^{-1}), \quad x \to \pm\infty$ in $\mathbb{Z}$.

2.1.7 $\qquad e^{ix} = O(1), \qquad x \in \mathbb{R}$.

In (2.1.5) $\mathbb{R}$ can be replaced by any fixed ray in the sector $|\operatorname{ph} x| < \frac{1}{2}\pi$, or by the whole of the sector $|\operatorname{ph} x| \leq \frac{1}{2}\pi - \delta$. (Here and elsewhere in this chapter δ is an arbitrary small positive constant.) But (2.1.5) does not hold as $x \to \infty$ in $|\operatorname{ph} x| < \frac{1}{2}\pi$ (for example, set $x = 1 + it$ and let $t \to \pm\infty$.)

If $\sum_{s=0}^{\infty} a_s z^s$ converges for all sufficiently small $|z|$, then for each nonnegative integer n

2.1.8 $\qquad \sum_{s=n}^{\infty} a_s z^s = O(z^n), \qquad z \to 0$ in $\mathbb{C}$.

Example

2.1.9 $\qquad e^z = 1 + z + O(z^2), \qquad z \to 0$ in $\mathbb{C}$.

The symbols o and O can be used generically. For example,

2.1.10 $\qquad o(\phi) = O(\phi), \quad o(\phi) + o(\phi) = o(\phi),$

it being understood that these equalities are not reversible. (In other words = here really means $\subseteq$.)

2.1(ii) Integration and Differentiation

Integration of asymptotic and order relations is permissible, subject to obvious convergence conditions. For example, suppose $f(x)$ is continuous and $f(x) \sim x^\nu$ as $x \to +\infty$ in $\mathbb{R}$, where $\nu \; (\in \mathbb{C})$ is a constant. Then

2.1.11 $\qquad \int_x^\infty f(t)\,dt \sim -\dfrac{x^{\nu+1}}{\nu+1}, \qquad \Re\nu < -1,$

2.1.12 $\qquad \int f(x)\,dx \sim \begin{cases} \text{a constant}, & \Re\nu < -1, \\ \ln x, & \nu = -1, \\ x^{\nu+1}/(\nu+1), & \Re\nu > -1. \end{cases}$

Differentiation requires extra conditions. For example, if $f(z)$ is analytic for all sufficiently large $|z|$ in a sector $\mathbf{S}$ and $f(z) = O(z^\nu)$ as $z \to \infty$ in $\mathbf{S}$, ν being real, then $f'(z) = O(z^{\nu-1})$ as $z \to \infty$ in any closed sector properly interior to $\mathbf{S}$ and with the same vertex (*Ritt's theorem*). This result also holds with both O's replaced by o's.

2.1(iii) Asymptotic Expansions

Let $\sum a_s x^{-s}$ be a formal power series (convergent or divergent) and for each positive integer n,

2.1.13 $\qquad f(x) = \sum_{s=0}^{n-1} a_s x^{-s} + O(x^{-n})$

as $x \to \infty$ in an unbounded set $\mathbf{X}$ in $\mathbb{R}$ or $\mathbb{C}$. Then $\sum a_s x^{-s}$ is a *Poincaré asymptotic expansion*, or simply *asymptotic expansion*, of $f(x)$ as $x \to \infty$ in $\mathbf{X}$. Symbolically,

2.1.14 $\quad f(x) \sim a_0 + a_1 x^{-1} + a_2 x^{-2} + \cdots, \quad x \to \infty$ in $\mathbf{X}$.

Condition (2.1.13) is equivalent to

2.1.15 $\qquad x^n \left(f(x) - \sum_{s=0}^{n-1} a_s x^{-s} \right) \to a_n, \quad x \to \infty \text{ in } \mathbf{X},$

for each $n = 0, 1, 2, \ldots$. If $\sum a_s x^{-s}$ converges for all sufficiently large $|x|$, then it is automatically the asymptotic expansion of its sum as $x \to \infty$ in $\mathbb{C}$.

If c is a finite limit point of $\mathbf{X}$, then

2.1.16
$f(x) \sim a_0 + a_1(x-c) + a_2(x-c)^2 + \cdots, \quad x \to c \text{ in } \mathbf{X},$

means that for each n, the difference between $f(x)$ and the nth partial sum on the right-hand side is $O((x-c)^n)$ as $x \to c$ in $\mathbf{X}$.

Most operations on asymptotic expansions can be carried out in exactly the same manner as for convergent power series. These include addition, subtraction, multiplication, and division. Substitution, logarithms, and powers are also permissible; compare Olver (1997b, pp. 19–22). Differentiation, however, requires the kind of extra conditions needed for the O symbol (§2.1(ii)). For reversion see §2.2.

Asymptotic expansions of the forms (2.1.14), (2.1.16) are unique. But for any given set of coefficients $a_0, a_1, a_2, \ldots$, and suitably restricted $\mathbf{X}$ there is an infinity of analytic functions $f(x)$ such that (2.1.14) and (2.1.16) apply. For (2.1.14) $\mathbf{X}$ can be the positive real axis or any unbounded sector in $\mathbb{C}$ of finite angle. As an example, in the sector $|\operatorname{ph} z| \leq \frac{1}{2}\pi - \delta \; (< \frac{1}{2}\pi)$ each of the functions $0, e^{-z}$, and $e^{-\sqrt{z}}$ (principal value) has the null asymptotic expansion

2.1.17 $\qquad 0 + 0 \cdot z^{-1} + 0 \cdot z^{-2} + \cdots, \qquad z \to \infty.$

2.1(iv) Uniform Asymptotic Expansions

If the set $\mathbf{X}$ in §2.1(iii) is a closed sector $\alpha \leq \operatorname{ph} x \leq \beta$, then by definition the asymptotic property (2.1.13) holds uniformly with respect to $\operatorname{ph} x \in [\alpha, \beta]$ as $|x| \to \infty$. The asymptotic property may also hold uniformly with respect to parameters. Suppose u is a parameter (or set of parameters) ranging over a point set (or sets) $\mathbf{U}$, and for each nonnegative integer n

$$\left| x^n \left(f(u,x) - \sum_{s=0}^{n-1} a_s(u) x^{-s} \right) \right|$$

is bounded as $x \to \infty$ in $\mathbf{X}$, uniformly for $u \in \mathbf{U}$. (The coefficients $a_s(u)$ may now depend on u.) Then

2.1.18 $$f(u,x) \sim \sum_{s=0}^{\infty} a_s(u) x^{-s}$$

as $x \to \infty$ in $\mathbf{X}$, uniformly with respect to $u \in \mathbf{U}$.

Similarly for finite limit point c in place of ∞.

2.1(v) Generalized Asymptotic Expansions

Let $\phi_s(x)$, $s = 0, 1, 2, \ldots$, be a sequence of functions defined in $\mathbf{X}$ such that for each s

2.1.19 $$\phi_{s+1}(x) = o(\phi_s(x)), \qquad x \to c \text{ in } \mathbf{X},$$

where c is a finite, or infinite, limit point of $\mathbf{X}$. Then $\{\phi_s(x)\}$ is an *asymptotic sequence* or *scale*. Suppose also that $f(x)$ and $f_s(x)$ satisfy

2.1.20 $$f(x) = \sum_{s=0}^{n-1} f_s(x) + O(\phi_n(x)), \quad x \to c \text{ in } \mathbf{X},$$

for $n = 0, 1, 2, \ldots$. Then $\sum f_s(x)$ is a *generalized asymptotic expansion* of $f(x)$ with respect to the scale $\{\phi_s(x)\}$. Symbolically,

2.1.21 $$f(x) \sim \sum_{s=0}^{\infty} f_s(x); \; \{\phi_s(x)\}, \quad x \to c \text{ in } \mathbf{X}.$$

As in §2.1(iv), generalized asymptotic expansions can also have uniformity properties with respect to parameters. For an example see §14.15(i).

Care is needed in understanding and manipulating generalized asymptotic expansions. Many properties enjoyed by Poincaré expansions (for example, multiplication) do not always carry over. It can even happen that a generalized asymptotic expansion converges, but its sum is not the function being represented asymptotically; for an example see §18.15(iii).

2.2 Transcendental Equations

Let $f(x)$ be continuous and strictly increasing when $a < x < \infty$ and

2.2.1 $$f(x) \sim x, \qquad x \to \infty.$$

Then for $y > f(a)$ the equation $f(x) = y$ has a unique root $x = x(y)$ in (a, ∞), and

2.2.2 $$x(y) \sim y, \qquad y \to \infty.$$

Example

2.2.3 $$t^2 - \ln t = y.$$

With $x = t^2$, $f(x) = x - \frac{1}{2} \ln x$. We may take $a = \frac{1}{2}$. From (2.2.2)

2.2.4 $$t = y^{\frac{1}{2}} (1 + o(1)), \qquad y \to \infty.$$

Higher approximations are obtainable by successive re-substitutions. For example

2.2.5 $$t^2 = y + \ln t = y + \tfrac{1}{2} \ln y + o(1),$$

and hence

2.2.6 $$t = y^{\frac{1}{2}} \left(1 + \tfrac{1}{4} y^{-1} \ln y + o\left(y^{-1}\right) \right), \qquad y \to \infty.$$

An important case is the reversion of asymptotic expansions for zeros of special functions. In place of (2.2.1) assume that

2.2.7 $$f(x) \sim x + f_0 + f_1 x^{-1} + f_2 x^{-2} + \cdots, \quad x \to \infty.$$

Then

2.2.8 $$x \sim y - F_0 - F_1 y^{-1} - F_2 y^{-2} - \cdots, \quad y \to \infty,$$

where $F_0 = f_0$ and sF_s ($s \geq 1$) is the coefficient of x^{-1} in the asymptotic expansion of $(f(x))^s$ (*Lagrange's formula for the reversion of series*). Conditions for the validity of the reversion process in $\mathbb{C}$ are derived in Olver (1997b, pp. 14–16). Applications to real and complex zeros of Airy functions are given in Fabijonas and Olver (1999). For other examples see de Bruijn (1961, Chapter 2).

2.3 Integrals of a Real Variable

2.3(i) Integration by Parts

Assume that the Laplace transform

2.3.1 $$\int_0^{\infty} e^{-xt} q(t) \, dt$$

converges for all sufficiently large x, and $q(t)$ is infinitely differentiable in a neighborhood of the origin. Then

2.3.2 $$\int_0^{\infty} e^{-xt} q(t) \, dt \sim \sum_{s=0}^{\infty} \frac{q^{(s)}(0)}{x^{s+1}}, \qquad x \to +\infty.$$

If, in addition, $q(t)$ is infinitely differentiable on $[0, \infty)$ and

2.3.3 $$\sigma_n = \sup_{(0,\infty)} \left(t^{-1} \ln |q^{(n)}(t)/q^{(n)}(0)| \right)$$

is finite and bounded for $n = 0, 1, 2, \ldots$, then the nth *error term* (that is, the difference between the integral and nth partial sum in (2.3.2)) is bounded in absolute value by $|q^{(n)}(0)/(x^n(x - \sigma_n))|$ when x exceeds both 0 and σ_n.

For the Fourier integral
$$\int_a^b e^{ixt} q(t)\, dt$$
assume a and b are finite, and $q(t)$ is infinitely differentiable on $[a,b]$. Then

2.3.4
$$\int_a^b e^{ixt} q(t)\, dt \sim e^{iax} \sum_{s=0}^\infty q^{(s)}(a) \left(\frac{i}{x}\right)^{s+1} - e^{ibx} \sum_{s=0}^\infty q^{(s)}(b) \left(\frac{i}{x}\right)^{s+1},$$
$$x \to +\infty.$$

Alternatively, assume $b = \infty$, $q(t)$ is infinitely differentiable on $[a, \infty)$, and each of the integrals $\int e^{ixt} q^{(s)}(t)\, dt$, $s = 0, 1, 2, \ldots$, converges as $t \to \infty$ uniformly for all sufficiently large x. Then

2.3.5
$$\int_a^\infty e^{ixt} q(t)\, dt \sim e^{iax} \sum_{s=0}^\infty q^{(s)}(a) \left(\frac{i}{x}\right)^{s+1}, \quad x \to +\infty.$$

In both cases the nth error term is bounded in absolute value by $x^{-n} \mathcal{V}_{a,b}\bigl(q^{(n-1)}(t)\bigr)$, where the *variational operator* $\mathcal{V}_{a,b}$ is defined by

2.3.6
$$\mathcal{V}_{a,b}(f(t)) = \int_a^b |f'(t)|\, dt;$$

see §1.4(v). For other examples, see Wong (1989, Chapter 1).

2.3(ii) Watson's Lemma

Assume again that the integral (2.3.1) converges for all sufficiently large x, but now

2.3.7
$$q(t) \sim \sum_{s=0}^\infty a_s t^{(s+\lambda-\mu)/\mu}, \quad t \to 0+,$$

where λ and μ are positive constants. Then the series obtained by substituting (2.3.7) into (2.3.1) and integrating formally term by term yields an asymptotic expansion:

2.3.8
$$\int_0^\infty e^{-xt} q(t)\, dt \sim \sum_{s=0}^\infty \Gamma\left(\frac{s+\lambda}{\mu}\right) \frac{a_s}{x^{(s+\lambda)/\mu}}, \quad x \to +\infty.$$

For the function Γ see §5.2(i).

This result is probably the most frequently used method for deriving asymptotic expansions of special functions. Since $q(t)$ need not be continuous (as long as the integral converges), the case of a finite integration range is included.

Other types of singular behavior in the integrand can be treated in an analogous manner. For example,

2.3.9
$$\int_0^\infty e^{-xt} q(t) \ln t\, dt \sim \sum_{s=0}^\infty \Gamma'\left(\frac{s+\lambda}{\mu}\right) \frac{a_s}{x^{(s+\lambda)/\mu}}$$
$$- (\ln x) \sum_{s=0}^\infty \Gamma\left(\frac{s+\lambda}{\mu}\right) \frac{a_s}{x^{(s+\lambda)/\mu}},$$

provided that the integral on the left-hand side of (2.3.9) converges for all sufficiently large values of x. (In other words, differentiation of (2.3.8) with respect to the parameter λ (or μ) is legitimate.)

Another extension is to more general factors than the exponential function. In addition to (2.3.7) assume that $f(t)$ and $q(t)$ are piecewise continuous (§1.4(ii)) on $(0, \infty)$, and

2.3.10 $\quad |f(t)| \leq A \exp(-at^\kappa), \quad 0 \leq t < \infty,$

2.3.11 $\quad q(t) = O(\exp(bt^\kappa)), \quad t \to +\infty,$

where A, a, b, κ are positive constants. Then

2.3.12
$$\int_0^\infty f(xt) q(t)\, dt \sim \sum_{s=0}^\infty \mathscr{M}\left(f; \frac{s+\lambda}{\mu}\right) \frac{a_s}{x^{(s+\lambda)/\mu}},$$
$$x \to +\infty,$$

where $\mathscr{M}(f; \alpha)$ is the *Mellin transform* of $f(t)$ (§2.5(i)).

For a more detailed treatment of the integral (2.3.12) see §§2.5, 2.6.

2.3(iii) Laplace's Method

When $p(t)$ is real and x is a large positive parameter, the main contribution to the integral

2.3.13
$$I(x) = \int_a^b e^{-xp(t)} q(t)\, dt$$

derives from the neighborhood of the minimum of $p(t)$ in the integration range. Without loss of generality, we assume that this minimum is at the left endpoint a. Furthermore:

(a) $p'(t)$ and $q(t)$ are continuous in a neighborhood of a, save possibly at a, and the minimum of $p(t)$ in $[a,b)$ is approached only at a.

(b) As $t \to a+$

2.3.14
$$p(t) \sim p(a) + \sum_{s=0}^\infty p_s (t-a)^{s+\mu},$$
$$q(t) \sim \sum_{s=0}^\infty q_s (t-a)^{s+\lambda-1},$$

and the expansion for $p(t)$ is differentiable. Again λ and μ are positive constants. Also $p_0 > 0$ (consistent with (a)).

(c) The integral (2.3.13) converges absolutely for all sufficiently large x.

2.3 Integrals of a Real Variable

Then

$$
\text{2.3.15} \qquad \int_a^b e^{-xp(t)} q(t)\, dt \sim e^{-xp(a)} \sum_{s=0}^{\infty} \Gamma\!\left(\frac{s+\lambda}{\mu}\right) \frac{b_s}{x^{(s+\lambda)/\mu}}, \qquad x \to +\infty,
$$

where the coefficients b_s are defined by the expansion

$$
\text{2.3.16} \qquad \frac{q(t)}{p'(t)} \sim \sum_{s=0}^{\infty} b_s v^{(s+\lambda-\mu)/\mu}, \qquad v \to 0+,
$$

in which $v = p(t) - p(a)$. For example,

$$
\text{2.3.17} \qquad
\begin{aligned}
b_0 &= \frac{q_0}{\mu p_0^{\lambda/\mu}}, \\
b_1 &= \left(\frac{q_1}{\mu} - \frac{(\lambda+1)p_1 q_0}{\mu^2 p_0}\right)\frac{1}{p_0^{(\lambda+1)/\mu}}, \\
b_2 &= \left(\frac{q_2}{\mu} - \frac{(\lambda+2)(p_1 q_1 + p_2 q_0)}{\mu^2 p_0} \right. \\
&\qquad \left. + \frac{(\lambda+2)(\lambda+\mu+2)p_1^2 q_0}{2\mu^3 p_0^2}\right)\frac{1}{p_0^{(\lambda+2)/\mu}}.
\end{aligned}
$$

In general

$$
\text{2.3.18} \qquad b_s = \frac{1}{\mu}\operatorname*{res}_{t=a}\left[\frac{q(t)}{(p(t)-p(a))^{(\lambda+s)/\mu}}\right], \quad s = 0, 1, 2, \ldots.
$$

Watson's lemma can be regarded as a special case of this result.

For error bounds for Watson's lemma and Laplace's method see Boyd (1993) and Olver (1997b, Chapter 3). These references and Wong (1989, Chapter 2) also contain examples.

2.3(iv) Method of Stationary Phase

When the parameter x is large the contributions from the real and imaginary parts of the integrand in

$$
\text{2.3.19} \qquad I(x) = \int_a^b e^{ixp(t)} q(t)\, dt
$$

oscillate rapidly and cancel themselves over most of the range. However, cancellation does not take place near the endpoints, owing to lack of symmetry, nor in the neighborhoods of zeros of $p'(t)$ because $p(t)$ changes relatively slowly at these stationary points.

The first result is the analog of Watson's lemma (§2.3(ii)). Assume that $q(t)$ again has the expansion (2.3.7) and this expansion is infinitely differentiable, $q(t)$ is infinitely differentiable on $(0,\infty)$, and each of the integrals $\int e^{ixt} q^{(s)}(t)\, dt$, $s = 0, 1, 2, \ldots$, converges at $t = \infty$, uniformly for all sufficiently large x. Then

$$
\text{2.3.20} \qquad \int_0^{\infty} e^{ixt} q(t)\, dt \sim \sum_{s=0}^{\infty} \exp\!\left(\frac{(s+\lambda)\pi i}{2\mu}\right) \Gamma\!\left(\frac{s+\lambda}{\mu}\right) \frac{a_s}{x^{(s+\lambda)/\mu}}, \qquad x \to +\infty,
$$

where the coefficients a_s are given by (2.3.7).

For the more general integral (2.3.19) we assume, without loss of generality, that the stationary point (if any) is at the left endpoint. Furthermore:

(a) On (a,b), $p(t)$ and $q(t)$ are infinitely differentiable and $p'(t) > 0$.

(b) As $t \to a+$ the asymptotic expansions (2.3.14) apply, and each is infinitely differentiable. Again λ, μ, and p_0 are positive.

(c) If the limit $p(b)$ of $p(t)$ as $t \to b-$ is finite, then each of the functions

$$
\text{2.3.21} \qquad P_s(t) = \left(\frac{1}{p'(t)}\frac{d}{dt}\right)^s \frac{q(t)}{p'(t)}, \quad s = 0, 1, 2, \ldots,
$$

tends to a finite limit $P_s(b)$.

(d) If $p(b) = \infty$, then $P_0(b) = 0$ and each of the integrals

$$
\text{2.3.22} \qquad \int e^{ixp(t)} P_s(t) p'(t)\, dt, \quad s = 0, 1, 2, \ldots,
$$

converges at $t = b$ uniformly for all sufficiently large x.

If $p(b)$ is finite, then both endpoints contribute:

$$
\text{2.3.23} \qquad
\begin{aligned}
&\int_a^b e^{ixp(t)} q(t)\, dt \\
&\sim e^{ixp(a)} \sum_{s=0}^{\infty} \exp\!\left(\frac{(s+\lambda)\pi i}{2\mu}\right) \Gamma\!\left(\frac{s+\lambda}{\mu}\right) \frac{b_s}{x^{(s+\lambda)/\mu}} \\
&\quad - e^{ixp(b)} \sum_{s=0}^{\infty} P_s(b)\left(\frac{i}{x}\right)^{s+1}, \qquad x \to +\infty.
\end{aligned}
$$

But if (d) applies, then the second sum is absent. The coefficients b_s are defined as in §2.3(iii).

For proofs of the results of this subsection, error bounds, and an example, see Olver (1974). For other estimates of the error term see Lyness (1971). For extensions to oscillatory integrals with logarithmic singularities see Wong and Lin (1978).

2.3(v) Coalescing Peak and Endpoint: Bleistein's Method

In the integral

$$
\text{2.3.24} \qquad I(\alpha, x) = \int_0^k e^{-xp(\alpha,t)} q(\alpha,t) t^{\lambda-1}\, dt
$$

$k\ (\leq \infty)$ and λ are positive constants, α is a variable parameter in an interval $\alpha_1 \leq \alpha \leq \alpha_2$ with $\alpha_1 \leq 0$ and $0 < \alpha_2 \leq k$, and x is a large positive parameter. Assume also that $\partial^2 p(\alpha,t)/\partial t^2$ and $q(\alpha,t)$ are continuous in α and t, and for each α the minimum value of $p(\alpha,t)$

in $[0, k)$ is at $t = \alpha$, at which point $\partial p(\alpha, t)/\partial t$ vanishes, but both $\partial^2 p(\alpha, t)/\partial t^2$ and $q(\alpha, t)$ are nonzero. When $x \to +\infty$ Laplace's method (§2.3(iii)) applies, but the form of the resulting approximation is discontinuous at $\alpha = 0$. In consequence, the approximation is nonuniform with respect to α and deteriorates severely as $\alpha \to 0$.

A uniform approximation can be constructed by quadratic change of integration variable:

2.3.25 $\quad p(\alpha, t) = \tfrac{1}{2} w^2 - aw + b,$

where a and b are functions of α chosen in such a way that $t = 0$ corresponds to $w = 0$, and the stationary points $t = \alpha$ and $w = a$ correspond. Thus

2.3.26 $\quad a = (2p(\alpha, 0) - 2p(\alpha, \alpha))^{1/2}, \quad b = p(\alpha, 0),$

2.3.27
$$w = (2p(\alpha, 0) - 2p(\alpha, \alpha))^{1/2} \pm (2p(\alpha, t) - 2p(\alpha, \alpha))^{1/2},$$
the upper or lower sign being taken according as $t \gtrless \alpha$. The relationship between t and w is one-to-one, and because

2.3.28 $\quad \dfrac{dw}{dt} = \pm \dfrac{1}{(2p(\alpha, t) - 2p(\alpha, \alpha))^{1/2}} \dfrac{\partial p(\alpha, t)}{\partial t}$

it is free from singularity at $t = \alpha$.

The integral (2.3.24) transforms into

2.3.29
$$I(\alpha, x) = e^{-xp(\alpha, 0)}$$
$$\times \int_0^\kappa \exp\!\left(-x\left(\tfrac{1}{2}w^2 - aw\right)\right) f(\alpha, w) w^{\lambda - 1}\, dw,$$
where

2.3.30 $\quad f(\alpha, w) = q(\alpha, t) \left(\dfrac{t}{w}\right)^{\lambda - 1} \dfrac{dt}{dw},$

$\kappa = \kappa(\alpha)$ being the value of w at $t = k$. We now expand $f(\alpha, w)$ in a Taylor series centered at the peak value $w = a$ of the exponential factor in the integrand:

2.3.31 $\quad f(\alpha, w) = \displaystyle\sum_{s=0}^\infty \phi_s(\alpha) (w - a)^s,$

with the coefficients $\phi_s(\alpha)$ continuous at $\alpha = 0$. The desired uniform expansion is then obtained formally as in Watson's lemma and Laplace's method. We replace the limit κ by ∞ and integrate term-by-term:

2.3.32 $\quad I(\alpha, x) \sim \dfrac{e^{-xp(\alpha, 0)}}{x^{\lambda/2}} \displaystyle\sum_{s=0}^\infty \phi_s(\alpha) \dfrac{F_s(a\sqrt{x})}{x^{s/2}}, \quad x \to \infty,$

where

2.3.33 $\quad F_s(y) = \displaystyle\int_0^\infty \exp\!\left(-\tfrac{1}{2}\tau^2 + y\tau\right) (\tau - y)^s \tau^{\lambda - 1}\, d\tau.$

For examples and proofs see Olver (1997b, Chapter 9), Bleistein (1966), Bleistein and Handelsman (1975, Chapter 9), and Wong (1989, Chapter 7).

2.4 Contour Integrals

2.4(i) Watson's Lemma

The result in §2.3(ii) carries over to a complex parameter z. Except that λ is now permitted to be complex, with $\Re \lambda > 0$, we assume the same conditions on $q(t)$ and also that the Laplace transform in (2.3.8) converges for all sufficiently large values of $\Re z$. Then

2.4.1 $\quad \displaystyle\int_0^\infty e^{-zt} q(t)\, dt \sim \sum_{s=0}^\infty \Gamma\!\left(\dfrac{s + \lambda}{\mu}\right) \dfrac{a_s}{z^{(s+\lambda)/\mu}}$

as $z \to \infty$ in the sector $|\operatorname{ph} z| \leq \tfrac{1}{2}\pi - \delta$ ($< \tfrac{1}{2}\pi$), with $z^{(s+\lambda)/\mu}$ assigned its principal value.

If $q(t)$ is analytic in a sector $\alpha_1 < \operatorname{ph} t < \alpha_2$ containing $\operatorname{ph} t = 0$, then the region of validity may be increased by rotation of the integration paths. We assume that in any closed sector with vertex $t = 0$ and properly interior to $\alpha_1 < \operatorname{ph} t < \alpha_2$, the expansion (2.3.7) holds as $t \to 0$, and $q(t) = O(e^{\sigma|t|})$ as $t \to \infty$, where σ is a constant. Then (2.4.1) is valid in any closed sector with vertex $z = 0$ and properly interior to $-\alpha_2 - \tfrac{1}{2}\pi < \operatorname{ph} z < -\alpha_1 + \tfrac{1}{2}\pi$. (The branches of $t^{(s+\lambda - \mu)/\mu}$ and $z^{(s+\lambda)/\mu}$ are extended by continuity.)

For examples and extensions (including uniformity and loop integrals) see Olver (1997b, Chapter 4), Wong (1989, Chapter 1), and Temme (1985).

2.4(ii) Inverse Laplace Transforms

On the interval $0 < t < \infty$ let $q(t)$ be differentiable and $e^{-ct} q(t)$ be absolutely integrable, where c is a real constant. Then the Laplace transform

2.4.2 $\quad Q(z) = \displaystyle\int_0^\infty e^{-zt} q(t)\, dt$

is continuous in $\Re z \geq c$ and analytic in $\Re z > c$, and by inversion (§1.14(iii))

2.4.3 $\quad q(t) = \dfrac{1}{2\pi i} \displaystyle\lim_{\eta \to \infty} \int_{\sigma - i\eta}^{\sigma + i\eta} e^{tz} Q(z)\, dz, \quad 0 < t < \infty,$

where $\sigma\ (\geq c)$ is a constant.

Now assume that $c > 0$ and we are given a function $Q(z)$ that is both analytic and has the expansion

2.4.4 $\quad Q(z) \sim \displaystyle\sum_{s=0}^\infty \Gamma\!\left(\dfrac{s + \lambda}{\mu}\right) \dfrac{a_s}{z^{(s+\lambda)/\mu}}, \quad z \to \infty,$

in the half-plane $\Re z \geq c$. Here $\Re \lambda > 0$, $\mu > 0$, and $z^{(s+\lambda)/\mu}$ has its principal value. Assume also (2.4.4) is differentiable. Then by integration by parts the integral

2.4.5 $\quad q(t) = \dfrac{1}{2\pi i} \displaystyle\int_{\sigma - i\infty}^{\sigma + i\infty} e^{tz} Q(z)\, dz, \quad 0 < t < \infty,$

is seen to converge absolutely at each limit, and be independent of $\sigma \in [c, \infty)$. Furthermore, as $t \to 0+$, $q(t)$ has the expansion (2.3.7).

For large t, the asymptotic expansion of $q(t)$ may be obtained from (2.4.3) by *Haar's method*. This depends on the availability of a comparison function $F(z)$ for $Q(z)$ that has an inverse transform

2.4.6 $$f(t) = \frac{1}{2\pi i} \lim_{\eta \to \infty} \int_{\sigma-i\eta}^{\sigma+i\eta} e^{tz} F(z)\, dz$$

with known asymptotic behavior as $t \to +\infty$. By subtraction from (2.4.3)

2.4.7
$$q(t) - f(t) = \frac{e^{\sigma t}}{2\pi} \lim_{\eta \to \infty} \int_{-\eta}^{\eta} e^{it\tau} (Q(\sigma+i\tau) - F(\sigma+i\tau))\, d\tau.$$

If this integral converges uniformly at each limit for all sufficiently large t, then by the Riemann–Lebesgue lemma (§1.8(i))

2.4.8 $$q(t) = f(t) + o(e^{ct}), \qquad t \to +\infty.$$

If, in addition, the corresponding integrals with Q and F replaced by their derivatives $Q^{(j)}$ and $F^{(j)}$, $j = 1, 2, \ldots, m$, converge uniformly, then by repeated integrations by parts

2.4.9 $$q(t) = f(t) + o(t^{-m} e^{ct}), \qquad t \to +\infty.$$

The most successful results are obtained on moving the integration contour as far to the left as possible. For examples see Olver (1997b, pp. 315–320).

2.4(iii) Laplace's Method

Let $\mathscr{P}$ denote the path for the contour integral

2.4.10 $$I(z) = \int_a^b e^{-zp(t)} q(t)\, dt,$$

in which a is finite, b is finite or infinite, and ω is the angle of slope of $\mathscr{P}$ at a, that is, $\lim(\mathrm{ph}(t-a))$ as $t \to a$ along $\mathscr{P}$. Assume that $p(t)$ and $q(t)$ are analytic on an open domain $\mathbf{T}$ that contains $\mathscr{P}$, with the possible exceptions of $t = a$ and $t = b$. Other assumptions are:

(a) In a neighborhood of a

2.4.11
$$p(t) = p(a) + \sum_{s=0}^{\infty} p_s (t-a)^{s+\mu},$$
$$q(t) = \sum_{s=0}^{\infty} q_s (t-a)^{s+\lambda-1},$$

with $\Re\lambda > 0$, $\mu > 0$, $p_0 \neq 0$, and the branches of $(t-a)^\lambda$ and $(t-a)^\mu$ continuous and constructed with $\mathrm{ph}(t-a) \to \omega$ as $t \to a$ along $\mathscr{P}$.

(b) z ranges along a ray or over an annular sector $\theta_1 \leq \theta \leq \theta_2$, $|z| \geq Z$, where $\theta = \mathrm{ph}\, z$, $\theta_2 - \theta_1 < \pi$, and $Z > 0$. $I(z)$ converges at b absolutely and uniformly with respect to z.

(c) Excluding $t = a$, $\Re(e^{i\theta} p(t) - e^{i\theta} p(a))$ is positive when $t \in \mathscr{P}$, and is bounded away from zero uniformly with respect to $\theta \in [\theta_1, \theta_2]$ as $t \to b$ along $\mathscr{P}$.

Then

2.4.12 $$I(z) \sim e^{-zp(a)} \sum_{s=0}^{\infty} \Gamma\left(\frac{s+\lambda}{\mu}\right) \frac{b_s}{z^{(s+\lambda)/\mu}}$$

as $z \to \infty$ in the sector $\theta_1 \leq \mathrm{ph}\, z \leq \theta_2$. The coefficients b_s are determined as in §2.3(iii), the branch of $\mathrm{ph}\, p_0$ being chosen to satisfy

2.4.13 $$|\theta + \mu\omega + \mathrm{ph}\, p_0| \leq \tfrac{1}{2}\pi.$$

For examples see Olver (1997b, Chapter 4). For error bounds see Boyd (1993).

2.4(iv) Saddle Points

Now suppose that in (2.4.10) the minimum of $\Re(zp(t))$ on $\mathscr{P}$ occurs at an interior point t_0. Temporarily assume that $\theta\ (=\mathrm{ph}\, z)$ is fixed, so that t_0 is independent of z. We may subdivide

2.4.14 $$I(z) = \int_{t_0}^b e^{-zp(t)} q(t)\, dt - \int_{t_0}^a e^{-zp(t)} q(t)\, dt,$$

and apply the result of §2.4(iii) to each integral on the right-hand side, the role of the series (2.4.11) being played by the Taylor series of $p(t)$ and $q(t)$ at $t = t_0$. If $p'(t_0) \neq 0$, then $\mu = 1$, λ is a positive integer, and the two resulting asymptotic expansions are identical. Thus the right-hand side of (2.4.14) reduces to the error terms. However, if $p'(t_0) = 0$, then $\mu \geq 2$ and different branches of some of the fractional powers of p_0 are used for the coefficients b_s; again see §2.3(iii). In consequence, the asymptotic expansion obtained from (2.4.14) is no longer null.

Zeros of $p'(t)$ are called *saddle points* (or *cols*) owing to the shape of the surface $|p(t)|$, $t \in \mathbb{C}$, in their vicinity. Cases in which $p'(t_0) \neq 0$ are usually handled by deforming the integration path in such a way that the minimum of $\Re(zp(t))$ is attained at a saddle point or at an endpoint. Additionally, it may be advantageous to arrange that $\Im(zp(t))$ is constant on the path: this will usually lead to greater regions of validity and sharper error bounds. Paths on which $\Im(zp(t))$ is constant are also the ones on which $|\exp(-zp(t))|$ decreases most rapidly. For this reason the name *method of steepest descents* is often used. However, for the purpose of simply deriving the asymptotic expansions the use of steepest descent paths is not essential.

In the commonest case the interior minimum t_0 of $\Re(zp(t))$ is a simple zero of $p'(t)$. The final expansion then has the form

2.4.15
$$\int_a^b e^{-zp(t)}q(t)\,dt \sim 2e^{-zp(t_0)}\sum_{s=0}^{\infty}\Gamma\!\left(s+\tfrac{1}{2}\right)\frac{b_{2s}}{z^{s+(1/2)}},$$

in which

2.4.16
$$b_0 = \frac{q}{(2p'')^{1/2}},$$
$$b_2 = \left(2q'' - \frac{2p'''q'}{p''} + \left(\frac{5(p''')^2}{6(p'')^2} - \frac{p^{\mathrm{iv}}}{2p''}\right)q\right)\frac{1}{(2p'')^{3/2}},$$

with p, q and their derivatives evaluated at t_0. The branch of $\omega_0 = \mathrm{ph}(p''(t_0))$ is the one satisfying $|\theta + 2\omega + \omega_0| \leq \tfrac{1}{2}\pi$, where ω is the limiting value of $\mathrm{ph}(t - t_0)$ as $t \to t_0$ from b.

Higher coefficients b_{2s} in (2.4.15) can be found from (2.3.18) with $\lambda = 1$, $\mu = 2$, and s replaced by $2s$. For integral representations of the b_{2s} and their asymptotic behavior as $s \to \infty$ see Boyd (1995). The last reference also includes examples, as do Olver (1997b, Chapter 4), Wong (1989, Chapter 2), and Bleistein and Handelsman (1975, Chapter 7).

2.4(v) Coalescing Saddle Points: Chester, Friedman, and Ursell's Method

Consider the integral

2.4.17
$$I(\alpha, z) = \int_{\mathscr{P}} e^{-zp(\alpha,t)} q(\alpha, t)\,dt$$

in which z is a large real or complex parameter, $p(\alpha, t)$ and $q(\alpha, t)$ are analytic functions of t and continuous in t and a second parameter α. Suppose that on the integration path $\mathscr{P}$ there are two simple zeros of $\partial p(\alpha, t)/\partial t$ that coincide for a certain value $\widehat{\alpha}$ of α. The problem of obtaining an asymptotic approximation to $I(\alpha, z)$ that is uniform with respect to α in a region containing $\widehat{\alpha}$ is similar to the problem of a coalescing endpoint and saddle point outlined in §2.3(v).

The change of integration variable is given by

2.4.18 $\quad p(\alpha, t) = \tfrac{1}{3}w^3 + aw^2 + bw + c,$

with a and b chosen so that the zeros of $\partial p(\alpha, t)/\partial t$ correspond to the zeros $w_1(\alpha), w_2(\alpha)$, say, of the quadratic $w^2 + 2aw + b$. Then

2.4.19
$$I(\alpha, z) = e^{-cz}\int_{\mathscr{Q}} \exp\!\left(-z\left(\tfrac{1}{3}w^3 + aw^2 + bw\right)\right)f(\alpha, w)\,dw,$$

where $\mathscr{Q}$ is the w-map of $\mathscr{P}$, and

2.4.20 $\quad f(\alpha, w) = q(\alpha, t)\dfrac{dt}{dw} = q(\alpha, t)\dfrac{w^2 + 2aw + b}{\partial p(\alpha, t)/\partial t}.$

The function $f(\alpha, w)$ is analytic at $w = w_1(\alpha)$ and $w = w_2(\alpha)$ when $\alpha \neq \widehat{\alpha}$, and at the confluence of these points when $\alpha = \widehat{\alpha}$. For large $|z|$, $I(\alpha, z)$ is approximated uniformly by the integral that corresponds to (2.4.19) when $f(\alpha, w)$ is replaced by a constant. By making a further change of variable

2.4.21 $\quad w = z^{-1/3}v - a,$

and assigning an appropriate value to c to modify the contour, the approximating integral is reducible to an Airy function or a Scorer function (§§9.2, 9.12).

For examples, proofs, and extensions see Olver (1997b, Chapter 9), Wong (1989, Chapter 7), Olde Daalhuis and Temme (1994), Chester et al. (1957), and Bleistein and Handelsman (1975, Chapter 9).

For a symbolic method for evaluating the coefficients in the asymptotic expansions see Vidūnas and Temme (2002).

2.4(vi) Other Coalescing Critical Points

The problems sketched in §§2.3(v) and 2.4(v) involve only two of many possibilities for the coalescence of endpoints, saddle points, and singularities in integrals associated with the special functions. For a coalescing saddle point and a pole see Wong (1989, Chapter 7) and van der Waerden (1951); in this case the uniform approximants are complementary error functions. For a coalescing saddle point and endpoint see Olver (1997b, Chapter 9) and Wong (1989, Chapter 7); if the endpoint is an algebraic singularity then the uniform approximants are parabolic cylinder functions with fixed parameter, and if the endpoint is not a singularity then the uniform approximants are complementary error functions.

For two coalescing saddle points and an endpoint see Leubner and Ritsch (1986). For two coalescing saddle points and an algebraic singularity see Temme (1986), Jin and Wong (1998). For a coalescing saddle point, a pole, and a branch point see Ciarkowski (1989). For many coalescing saddle points see §36.12. For double integrals with two coalescing stationary points see Qiu and Wong (2000).

2.5 Mellin Transform Methods

2.5(i) Introduction

Let $f(t)$ be a *locally integrable* function on $(0, \infty)$, that is, $\int_\rho^T f(t)\,dt$ exists for all ρ and T satisfying $0 < \rho < T < \infty$. The *Mellin transform* of $f(t)$ is defined by

2.5.1
$$\mathscr{M}(f; z) = \int_0^\infty t^{z-1} f(t)\,dt,$$

when this integral converges. The domain of analyticity of $\mathscr{M}(f; z)$ is usually an infinite strip $a < \Re z < b$ parallel to the imaginary axis. The inversion formula is given by

2.5.2
$$f(t) = \frac{1}{2\pi i}\int_{c-i\infty}^{c+i\infty} t^{-z}\mathscr{M}(f; z)\,dz,$$

2.5 MELLIN TRANSFORM METHODS

with $a < c < b$.

One of the two convolution integrals associated with the Mellin transform is of the form

2.5.3 $$I(x) = \int_0^\infty f(t)\,h(xt)\,dt, \qquad x > 0,$$

and

2.5.4 $$\mathscr{M}(I;z) = \mathscr{M}(f;1-z)\,\mathscr{M}(h;z).$$

If $\mathscr{M}(f;1-z)$ and $\mathscr{M}(h;z)$ have a common strip of analyticity $a < \Re z < b$, then

2.5.5 $$I(x) = \frac{1}{2\pi i}\int_{c-i\infty}^{c+i\infty} x^{-z}\,\mathscr{M}(f;1-z)\,\mathscr{M}(h;z)\,dz,$$

where $a < c < b$. When $x = 1$, this identity is a Parseval-type formula; compare §1.14(iv).

If $\mathscr{M}(f;1-z)$ and $\mathscr{M}(h;z)$ can be continued analytically to meromorphic functions in a left half-plane, and if the contour $\Re z = c$ can be translated to $\Re z = d$ with $d < c$, then

2.5.6 $$I(x) = \sum_{d < \Re z < c} \operatorname{res}\left[x^{-z}\,\mathscr{M}(f;1-z)\,\mathscr{M}(h;z)\right] + E(x),$$

where

2.5.7 $$E(x) = \frac{1}{2\pi i}\int_{d-i\infty}^{d+i\infty} x^{-z}\,\mathscr{M}(f;1-z)\,\mathscr{M}(h;z)\,dz.$$

The sum in (2.5.6) is taken over all poles of $x^{-z}\,\mathscr{M}(f;1-z)\,\mathscr{M}(h;z)$ in the strip $d < \Re z < c$, and it provides the asymptotic expansion of $I(x)$ for small values of x. Similarly, if $\mathscr{M}(f;1-z)$ and $\mathscr{M}(h;z)$ can be continued analytically to meromorphic functions in a right half-plane, and if the vertical line of integration can be translated to the right, then we obtain an asymptotic expansion for $I(x)$ for large values of x.

Example

2.5.8 $$I(x) = \int_0^\infty \frac{J_\nu^2(xt)}{1+t}\,dt, \qquad \nu > -\tfrac{1}{2},$$

where J_ν denotes the Bessel function (§10.2(ii)), and x is a large positive parameter. Let $h(t) = J_\nu^2(t)$ and $f(t) = 1/(1+t)$. Then from Table 1.14.5 and Watson (1944, p. 403)

2.5.9 $$\mathscr{M}(f;1-z) = \frac{\pi}{\sin(\pi z)}, \qquad 0 < \Re z < 1,$$

2.5.10 $$\mathscr{M}(h;z) = \frac{2^{z-1}\,\Gamma\!\left(\nu + \tfrac{1}{2}z\right)}{\Gamma^2\!\left(1 - \tfrac{1}{2}z\right)\Gamma\!\left(1+\nu - \tfrac{1}{2}z\right)\Gamma(z)}\frac{\pi}{\sin(\pi z)},$$
$$-2\nu < \Re z < 1.$$

In the half-plane $\Re z > \max(0, -2\nu)$, the product $\mathscr{M}(f;1-z)\,\mathscr{M}(h;z)$ has a pole of order two at each positive integer, and

2.5.11 $$\operatorname*{res}_{z=n}\left[x^{-z}\,\mathscr{M}(f;1-z)\,\mathscr{M}(h;z)\right] = (a_n \ln x + b_n)x^{-n},$$

where

2.5.12 $$a_n = \frac{2^{n-1}\,\Gamma\!\left(\nu + \tfrac{1}{2}n\right)}{\Gamma^2\!\left(1 - \tfrac{1}{2}n\right)\Gamma\!\left(1+\nu - \tfrac{1}{2}n\right)\Gamma(n)},$$

2.5.13 $$\begin{aligned}b_n = -a_n\big(&\ln 2 + \tfrac{1}{2}\psi(\nu + \tfrac{1}{2}n) + \psi(1 - \tfrac{1}{2}n)\\ &+ \tfrac{1}{2}\psi(1+\nu - \tfrac{1}{2}n) - \psi(n)\big),\end{aligned}$$

and ψ is the logarithmic derivative of the gamma function (§5.2(i)).

We now apply (2.5.5) with $\max(0, -2\nu) < c < 1$, and then translate the integration contour to the right. This is allowable in view of the asymptotic formula

2.5.14 $$|\Gamma(x+iy)| = \sqrt{2\pi}\,e^{-\pi|y|/2}|y|^{x-(1/2)}(1+o(1)),$$

as $y \to \pm\infty$, uniformly for bounded $|x|$; see (5.11.9). Then as in (2.5.6) and (2.5.7), with $d = 2n+1-\epsilon$ ($0 < \epsilon < 1$), we obtain

2.5.15 $$I(x) = -\sum_{s=0}^{2n}(a_s \ln x + b_s)x^{-s} + O\!\left(x^{-2n-1+\epsilon}\right),$$
$$n = 0, 1, 2, \ldots.$$

From (2.5.12) and (2.5.13), it is seen that $a_s = b_s = 0$ when s is even. Hence

2.5.16 $$I(x) = \sum_{s=0}^{n-1}(c_s \ln x + d_s)x^{-2s-1} + O\!\left(x^{-2n-1+\epsilon}\right),$$

where $c_s = -a_{2s+1}$, $d_s = -b_{2s+1}$.

2.5(ii) Extensions

Let $f(t)$ and $h(t)$ be locally integrable on $(0,\infty)$ and

2.5.17 $$f(t) \sim \sum_{s=0}^\infty a_s t^{\alpha_s}, \qquad t \to 0+,$$

where $\Re\alpha_s > \Re\alpha_{s'}$ for $s > s'$, and $\Re\alpha_s \to +\infty$ as $s \to \infty$. Also, let

2.5.18 $$h(t) \sim \exp(i\kappa t^p)\sum_{s=0}^\infty b_s t^{-\beta_s}, \qquad t \to +\infty,$$

where κ is real, $p > 0$, $\Re\beta_s > \Re\beta_{s'}$ for $s > s'$, and $\Re\beta_s \to +\infty$ as $s \to \infty$. To ensure that the integral (2.5.3) converges we assume that

2.5.19 $$f(t) = O(t^{-b}), \qquad t \to +\infty,$$

with $b + \Re\beta_0 > 1$, and

2.5.20 $$h(t) = O(t^c), \qquad t \to 0+,$$

with $c + \Re\alpha_0 > -1$. To apply the Mellin transform method outlined in §2.5(i), we require the transforms $\mathscr{M}(f;1-z)$ and $\mathscr{M}(h;z)$ to have a common strip of analyticity. This, in turn, requires $-b < \Re\alpha_0$, $-c < \Re\beta_0$, and either $-c < \Re\alpha_0 + 1$ or $1 - b < \Re\beta_0$. Following Handelsman and Lew (1970, 1971) we now give an extension of this method in which none of these conditions is required.

First, we introduce the truncated functions $f_1(t)$ and $f_2(t)$ defined by

2.5.21 $\quad f_1(t) = \begin{cases} f(t), & 0 < t \leq 1, \\ 0, & 1 < t < \infty, \end{cases}$

2.5.22 $\quad f_2(t) = f(t) - f_1(t).$

Similarly,

2.5.23 $\quad h_1(t) = \begin{cases} h(t), & 0 < t \leq 1, \\ 0, & 1 < t < \infty, \end{cases}$

2.5.24 $\quad h_2(t) = h(t) - h_1(t).$

With these definitions and the conditions (2.5.17)–(2.5.20) the Mellin transforms converge absolutely and define analytic functions in the half-planes shown in Table 2.5.1.

Table 2.5.1: Domains of convergence for Mellin transforms.

Transform	Domain of Convergence
$\mathscr{M}(f_1; z)$	$\Re z > -\Re \alpha_0$
$\mathscr{M}(f_2; z)$	$\Re z < b$
$\mathscr{M}(h_1; z)$	$\Re z > -c$
$\mathscr{M}(h_2; z)$	$\Re z < \Re \beta_0$

Furthermore, $\mathscr{M}(f_1; z)$ can be continued analytically to a meromorphic function on the entire z-plane, whose singularities are simple poles at $-\alpha_s$, $s = 0, 1, 2, \ldots$, with principal part

2.5.25 $\quad a_s/(z + \alpha_s).$

By Table 2.5.1, $\mathscr{M}(f_2; z)$ is an analytic function in the half-plane $\Re z < b$. Hence we can extend the definition of the Mellin transform of f by setting

2.5.26 $\quad \mathscr{M}(f; z) = \mathscr{M}(f_1; z) + \mathscr{M}(f_2; z)$

for $\Re z < b$. The extended transform $\mathscr{M}(f; z)$ has the same properties as $\mathscr{M}(f_1; z)$ in the half-plane $\Re z < b$.

Similarly, if $\kappa = 0$ in (2.5.18), then $\mathscr{M}(h_2; z)$ can be continued analytically to a meromorphic function on the entire z-plane with simple poles at β_s, $s = 0, 1, 2, \ldots$, with principal part

2.5.27 $\quad -b_s/(z - \beta_s).$

Alternatively, if $\kappa \neq 0$ in (2.5.18), then $\mathscr{M}(h_2; z)$ can be continued analytically to an entire function.

Since $\mathscr{M}(h_1; z)$ is analytic for $\Re z > -c$ by Table 2.5.1, the analytically-continued $\mathscr{M}(h_2; z)$ allows us to extend the Mellin transform of h via

2.5.28 $\quad \mathscr{M}(h; z) = \mathscr{M}(h_1; z) + \mathscr{M}(h_2; z)$

in the same half-plane. From (2.5.26) and (2.5.28), it follows that both $\mathscr{M}(f; 1 - z)$ and $\mathscr{M}(h; z)$ are defined in the half-plane $\Re z > \max(1 - b, -c)$.

We are now ready to derive the asymptotic expansion of the integral $I(x)$ in (2.5.3) as $x \to \infty$. First we note that

2.5.29 $\quad I(x) = \sum_{j,k=1}^{2} I_{jk}(x),$

where

2.5.30 $\quad I_{jk}(x) = \int_0^\infty f_j(t) h_k(xt)\,dt.$

By direct computation

2.5.31 $\quad I_{21}(x) = 0, \qquad$ for $x \geq 1$.

Next from Table 2.5.1 we observe that the integrals for the transform pair $\mathscr{M}(f_j; 1 - z)$ and $\mathscr{M}(h_k; z)$ are absolutely convergent in the domain D_{jk} specified in Table 2.5.2, and these domains are nonempty as a consequence of (2.5.19) and (2.5.20).

Table 2.5.2: Domains of analyticity for Mellin transforms.

Transform Pair	Domain D_{jk}
$\mathscr{M}(f_1; 1-z)$, $\mathscr{M}(h_1; z)$	$-c < \Re z < 1 + \Re\alpha_0$
$\mathscr{M}(f_1; 1-z)$, $\mathscr{M}(h_2; z)$	$\Re z < \min(1 + \Re\alpha_0, \Re\beta_0)$
$\mathscr{M}(f_2; 1-z)$, $\mathscr{M}(h_1; z)$	$\max(-c, 1 - b) < \Re z$
$\mathscr{M}(f_2; 1-z)$, $\mathscr{M}(h_2; z)$	$1 - b < \Re z < \Re\beta_0$

For simplicity, write

2.5.32 $\quad G_{jk}(z) = \mathscr{M}(f_j; 1 - z)\,\mathscr{M}(h_k; z).$

From Table 2.5.2, we see that each $G_{jk}(z)$ is analytic in the domain D_{jk}. Furthermore, each $G_{jk}(z)$ has an analytic or meromorphic extension to a half-plane containing D_{jk}. Now suppose that there is a real number p_{jk} in D_{jk} such that the Parseval formula (2.5.5) applies and

2.5.33 $\quad I_{jk}(x) = \dfrac{1}{2\pi i} \int_{p_{jk}-i\infty}^{p_{jk}+i\infty} x^{-z} G_{jk}(z)\,dz.$

If, in addition, there exists a number $q_{jk} > p_{jk}$ such that

2.5.34 $\quad \sup_{p_{jk} \leq x \leq q_{jk}} |G_{jk}(x + iy)| \to 0, \qquad y \to \pm\infty,$

then

2.5.35 $\quad I_{jk}(x) = \sum_{p_{jk} < \Re z < q_{jk}} \mathrm{res}\left[-x^{-z} G_{jk}(z)\right] + E_{jk}(x),$

where

2.5.36 $\quad E_{jk}(x) = \dfrac{1}{2\pi i} \int_{q_{jk}-i\infty}^{q_{jk}+i\infty} x^{-z} G_{jk}(z)\,dz = o\!\left(x^{-q_{jk}}\right)$

as $x \to +\infty$. (The last order estimate follows from the Riemann–Lebesgue lemma, §1.8(i).) The asymptotic expansion of $I(x)$ is then obtained from (2.5.29).

For further discussion of this method and examples, see Wong (1989, Chapter 3), Paris and Kaminski (2001, Chapter 5), and Bleistein and Handelsman (1975, Chapters 4 and 6). The first reference also contains explicit expressions for the error terms, as do Soni (1980) and Carlson and Gustafson (1985).

The Mellin transform method can also be extended to derive asymptotic expansions of multidimensional integrals having algebraic or logarithmic singularities, or both; see Wong (1989, Chapter 3), Paris and Kaminski (2001, Chapter 7), and McClure and Wong (1987). See also Brüning (1984) for a different approach.

2.5(iii) Laplace Transforms with Small Parameters

Let $h(t)$ satisfy (2.5.18) and (2.5.20) with $c > -1$, and consider the Laplace transform

2.5.37 $$\mathscr{L}(h;\zeta) = \int_0^\infty h(t) e^{-\zeta t}\, dt.$$

Put $x = 1/\zeta$ and break the integration range at $t = 1$, as in (2.5.23) and (2.5.24). Then

2.5.38 $$\zeta \mathscr{L}(h;\zeta) = I_1(x) + I_2(x),$$

where

2.5.39 $$I_j(x) = \int_0^\infty e^{-t} h_j(xt)\, dt, \qquad j = 1, 2.$$

Since $\mathscr{M}(e^{-t}; z) = \Gamma(z)$, by the Parseval formula (2.5.5), there are real numbers p_1 and p_2 such that $-c < p_1 < 1$, $p_2 < \min(1, \Re \beta_0)$, and

2.5.40 $$I_j(x) = \frac{1}{2\pi i} \int_{p_j - i\infty}^{p_j + i\infty} x^{-z} \Gamma(1-z) \mathscr{M}(h_j; z)\, dz, \; j = 1, 2.$$

Since $\mathscr{M}(h; z)$ is analytic for $\Re z > -c$, by (2.5.14),

2.5.41 $$I_1(x) = \mathscr{M}(h_1; 1) x^{-1} \\ + \frac{1}{2\pi i} \int_{\rho - i\infty}^{\rho + i\infty} x^{-z} \Gamma(1-z) \mathscr{M}(h_1; z)\, dz,$$

for any ρ satisfying $1 < \rho < 2$. Similarly, since $\mathscr{M}(h_2; z)$ can be continued analytically to a meromorphic function (when $\kappa = 0$) or to an entire function (when $\kappa \neq 0$), we can choose ρ so that $\mathscr{M}(h_2; z)$ has no poles in $1 < \Re z \le \rho < 2$. Thus

2.5.42 $$I_2(x) = \sum_{\Re \beta_0 \le \Re z \le 1} \operatorname{res}\left[-x^{-z} \Gamma(1-z) \mathscr{M}(h_2; z)\right] \\ + \frac{1}{2\pi i} \int_{\rho - i\infty}^{\rho + i\infty} x^{-z} \Gamma(1-z) \mathscr{M}(h_2; z)\, dz.$$

On substituting (2.5.41) and (2.5.42) into (2.5.38), we obtain

2.5.43 $$\mathscr{L}(h;\zeta) = \mathscr{M}(h_1; 1) \\ + \sum_{\Re \beta_0 \le \Re z \le 1} \operatorname{res}\left[-\zeta^{z-1} \Gamma(1-z) \mathscr{M}(h_2; z)\right] \\ + \sum_{1 < \Re z < l} \operatorname{res}\left[-\zeta^{z-1} \Gamma(1-z) \mathscr{M}(h; z)\right] \\ + \frac{1}{2\pi i} \int_{l-\delta-i\infty}^{l-\delta+i\infty} \zeta^{z-1} \Gamma(1-z) \mathscr{M}(h; z)\, dz,$$

where $l \;(\ge 2)$ is an arbitrary integer and δ is an arbitrary small positive constant. The last term is clearly $O(\zeta^{l-\delta-1})$ as $\zeta \to 0+$.

If $\kappa = 0$ in (2.5.18) and $c > -1$ in (2.5.20), and if none of the exponents in (2.5.18) are positive integers, then the expansion (2.5.43) gives the following useful result:

2.5.44 $$\mathscr{L}(h;\zeta) \sim \sum_{n=0}^\infty b_n \Gamma(1 - \beta_n) \zeta^{\beta_n - 1} \\ + \sum_{n=0}^\infty \frac{(-\zeta)^n}{n!} \mathscr{M}(h; n+1), \quad \zeta \to 0+.$$

Example

2.5.45 $$\mathscr{L}(h;\zeta) = \int_0^\infty \frac{e^{-\zeta t}}{1+t}\, dt, \qquad \Re \zeta > 0.$$

With $h(t) = 1/(1+t)$, we have $\mathscr{M}(h; z) = \pi \csc(\pi z)$ for $0 < \Re z < 1$. In the notation of (2.5.18) and (2.5.20), $\kappa = 0$, $\beta_s = s+1$, and $c = 0$. Straightforward calculation gives

2.5.46 $$\operatorname*{res}_{z=k}\left[-\zeta^{z-1} \Gamma(1-z) \pi \csc(\pi z)\right] \\ = (-\ln \zeta + \psi(k)) \frac{\zeta^{k-1}}{(k-1)!},$$

where $\psi(z) = \Gamma'(z)/\Gamma(z)$. From (2.5.28)

2.5.47 $$\operatorname*{res}_{z=1}\left[-\zeta^{z-1} \Gamma(1-z) \mathscr{M}(h_2; z)\right] \\ = (-\ln \zeta - \gamma) - \mathscr{M}(h_1; 1),$$

where γ is Euler's constant (§5.2(ii)). Insertion of these results into (2.5.43) yields

2.5.48 $$\mathscr{L}(h;\zeta) \sim (-\ln \zeta) \sum_{k=0}^\infty \frac{\zeta^k}{k!} + \sum_{k=0}^\infty \psi(k+1) \frac{\zeta^k}{k!}, \quad \zeta \to 0+.$$

To verify (2.5.48) we may use

2.5.49 $$\mathscr{L}(h;\zeta) = e^\zeta E_1(\zeta);$$

compare (6.2.2) and (6.6.3).

For examples in which the integral defining the Mellin transform $\mathscr{M}(h; z)$ does not exist for any value of z, see Wong (1989, Chapter 3), Bleistein and Handelsman (1975, Chapter 4), and Handelsman and Lew (1970).

2.6 Distributional Methods

2.6(i) Divergent Integrals

Consider the integral

2.6.1 $$S(x) = \int_0^\infty \frac{1}{(1+t)^{1/3}(x+t)}\, dt,$$

where $x > 0$. For $t > 1$,

2.6.2 $$(1+t)^{-1/3} = \sum_{s=0}^\infty \binom{-\frac{1}{3}}{s} t^{-s-(1/3)}.$$

Motivated by Watson's lemma (§2.3(ii)), we substitute (2.6.2) in (2.6.1), and integrate term by term. This leads to integrals of the form

2.6.3 $$\int_0^\infty \frac{t^{-s-(1/3)}}{x+t}\,dt, \qquad s=1,2,3,\dots.$$

Although divergent, these integrals may be interpreted in a generalized sense. For instance, we have

2.6.4 $$\int_0^\infty \frac{t^{\alpha-1}}{(x+t)^{\alpha+\beta}}\,dt = \frac{\Gamma(\alpha)\,\Gamma(\beta)}{\Gamma(\alpha+\beta)}\frac{1}{x^\beta}, \qquad \Re\alpha>0,\ \Re\beta>0.$$

But the right-hand side is meaningful for all values of α and β, other than nonpositive integers. We may therefore define the integral on the left-hand side of (2.6.4) by the value on the right-hand side, except when $\alpha, \beta = 0, -1, -2, \dots$. With this interpretation

2.6.5 $$\int_0^\infty \frac{t^{-s-(1/3)}}{x+t}\,dt = \frac{2\pi}{\sqrt{3}}\frac{(-1)^s}{x^{s+(1/3)}}, \qquad s=0,1,2,\dots.$$

Inserting (2.6.2) into (2.6.1) and integrating formally term-by-term, we obtain

2.6.6 $$S(x) \sim \frac{2\pi}{\sqrt{3}}\sum_{s=0}^\infty (-1)^s \binom{-\frac{1}{3}}{s} x^{-s-(1/3)}, \quad x\to\infty.$$

However this result is incorrect. The correct result is given by

2.6.7 $$S(x) \sim \frac{2\pi}{\sqrt{3}}\sum_{s=0}^\infty (-1)^s \binom{-\frac{1}{3}}{s} x^{-s-(1/3)}$$
$$-\sum_{s=1}^\infty \frac{3^s(s-1)!}{2\cdot 5\cdots(3s-1)}x^{-s};$$

see §2.6(ii).

The fact that expansion (2.6.6) misses all the terms in the second series in (2.6.7) raises the question: what went wrong with our process of reaching (2.6.6)? In the following subsections, we use some elementary facts of distribution theory (§1.16) to study the proper use of divergent integrals. An important asset of the distribution method is that it gives explicit expressions for the remainder terms associated with the resulting asymptotic expansions.

For an introduction to distribution theory, see Wong (1989, Chapter 5). For more advanced discussions, see Gel'fand and Shilov (1964) and Rudin (1973).

2.6(ii) Stieltjes Transform

Let $f(t)$ be locally integrable on $[0,\infty)$. The *Stieltjes transform* of $f(t)$ is defined by

2.6.8 $$\mathcal{S}(f;z) = \int_0^\infty \frac{f(t)}{t+z}\,dt.$$

To derive an asymptotic expansion of $\mathcal{S}(f;z)$ for large values of $|z|$, with $|\operatorname{ph} z|<\pi$, we assume that $f(t)$ possesses an asymptotic expansion of the form

2.6.9 $$f(t) \sim \sum_{s=0}^\infty a_s t^{-s-\alpha}, \qquad t\to+\infty,$$

with $0<\alpha\le 1$. For each $n=1,2,3,\dots$, set

2.6.10 $$f(t) = \sum_{s=0}^{n-1} a_s t^{-s-\alpha} + f_n(t).$$

To each function in this equation, we shall assign a *tempered distribution* (i.e., a continuous linear functional) on the space $\mathcal{T}$ of rapidly decreasing functions on $\mathbb{R}$. Since $f(t)$ is locally integrable on $[0,\infty)$, it defines a distribution by

2.6.11 $$\langle f, \phi \rangle = \int_0^\infty f(t)\phi(t)\,dt, \qquad \phi\in\mathcal{T}.$$

In particular,

2.6.12 $$\langle t^{-\alpha},\phi\rangle = \int_0^\infty t^{-\alpha}\phi(t)\,dt, \qquad \phi\in\mathcal{T},$$

when $0<\alpha<1$. Since the functions $t^{-s-\alpha}$, $s=1,2,\dots$, are not locally integrable on $[0,\infty)$, we cannot assign distributions to them in a similar manner. However, they are multiples of the derivatives of $t^{-\alpha}$. Motivated by the definition of distributional derivatives, we can assign them the distributions defined by

2.6.13 $$\langle t^{-s-\alpha},\phi\rangle = \frac{1}{(\alpha)_s}\int_0^\infty t^{-\alpha}\phi^{(s)}(t)\,dt, \quad \phi\in\mathcal{T},$$

where $(\alpha)_s = \alpha(\alpha+1)\cdots(\alpha+s-1)$. Similarly, in the case $\alpha=1$, we define

2.6.14 $$\langle t^{-s-1},\phi\rangle = -\frac{1}{s!}\int_0^\infty (\ln t)\phi^{(s+1)}(t)\,dt, \quad \phi\in\mathcal{T}.$$

To assign a distribution to the function $f_n(t)$, we first let $f_{n,n}(t)$ denote the nth repeated integral (§1.4(v)) of f_n:

2.6.15 $$f_{n,n}(t) = \frac{(-1)^n}{(n-1)!}\int_t^\infty (\tau-t)^{n-1} f_n(\tau)\,d\tau.$$

For $0<\alpha<1$, it is easily seen that $f_{n,n}(t)$ is bounded on $[0,R]$ for any positive constant R, and is $O(t^{-\alpha})$ as $t\to\infty$. For $\alpha=1$, we have $f_{n,n}(t) = O(t^{-1})$ as $t\to\infty$ and $f_{n,n}(t) = O(\ln t)$ as $t\to 0+$. In either case, we define the distribution associated with $f_n(t)$ by

2.6.16 $$\langle f_n, \phi\rangle = (-1)^n \int_0^\infty f_{n,n}(t)\phi^{(n)}(t)\,dt, \quad \phi\in\mathcal{T},$$

since the nth derivative of $f_{n,n}$ is f_n.

We have now assigned a distribution to each function in (2.6.10). A natural question is: what is the exact relation between these distributions? The answer is provided by the identities (2.6.17) and (2.6.20) given below.

For $0 < \alpha < 1$ and $n \geq 1$, we have

2.6.17
$$\langle f, \phi \rangle = \sum_{s=0}^{n-1} a_s \langle t^{-s-\alpha}, \phi \rangle - \sum_{s=1}^{n} c_s \langle \delta^{(s-1)}, \phi \rangle + \langle f_n, \phi \rangle$$

for any $\phi \in \mathcal{T}$, where

2.6.18
$$c_s = \frac{(-1)^s}{(s-1)!} \mathscr{M}(f; s),$$

$\mathscr{M}(f; z)$ being the Mellin transform of $f(t)$ or its analytic continuation (§2.5(ii)). The Dirac delta distribution in (2.6.17) is given by

2.6.19
$$\langle \delta^{(s)}, \phi \rangle = (-1)^s \phi^{(s)}(0), \quad s = 0, 1, 2, \ldots;$$

compare §1.16(iii).

For $\alpha = 1$

2.6.20
$$\langle f, \phi \rangle = \sum_{s=0}^{n-1} a_s \langle t^{-s-1}, \phi \rangle - \sum_{s=1}^{n} d_s \langle \delta^{(s-1)}, \phi \rangle + \langle f_n, \phi \rangle$$

for any $\phi \in \mathcal{T}$, where

2.6.21
$$(-1)^{s+1} d_{s+1} = \frac{a_s}{s!} \sum_{k=1}^{s} \frac{1}{k} + \frac{1}{s!} \lim_{z \to s+1} \left(\mathscr{M}(f; z) + \frac{a_s}{z - s - 1} \right),$$

for $s = 0, 1, 2, \ldots$.

To apply the results (2.6.17) and (2.6.20) to the Stieltjes transform (2.6.8), we take a specific function $\phi \in \mathcal{T}$. Let ε be a positive number, and

2.6.22
$$\phi_\varepsilon(t) = \frac{e^{-\varepsilon t}}{t + z}, \quad t \in (0, \infty).$$

From (2.6.13) and (2.6.14)

2.6.23
$$\lim_{\varepsilon \to 0} \langle t^{-s-\alpha}, \phi_\varepsilon \rangle = \frac{\pi}{\sin(\pi \alpha)} \frac{(-1)^s}{z^{s+\alpha}},$$

2.6.24
$$\lim_{\varepsilon \to 0} \langle t^{-s-1}, \phi_\varepsilon \rangle = \frac{(-1)^{s+1}}{z^{s+1}} \sum_{k=1}^{s} \frac{1}{k} + \frac{(-1)^s}{z^{s+1}} \ln z,$$

with $s = 0, 1, 2, \ldots$. From (2.6.11) and (2.6.16), we also have

2.6.25
$$\lim_{\varepsilon \to 0} \langle f, \phi_\varepsilon \rangle = \mathcal{S}(f; z),$$

2.6.26
$$\lim_{\varepsilon \to 0} \langle f_n, \phi_\varepsilon \rangle = n! \int_0^\infty \frac{f_{n,n}(t)}{(t+z)^{n+1}} dt.$$

On substituting (2.6.15) into (2.6.26) and interchanging the order of integration, the right-hand side of (2.6.26) becomes

$$\frac{(-1)^n}{z^n} \int_0^\infty \frac{\tau^n f_n(\tau)}{\tau + z} d\tau.$$

To summarize,

2.6.27
$$\mathcal{S}(f; z) = \frac{\pi}{\sin(\pi \alpha)} \sum_{s=0}^{n-1} (-1)^s \frac{a_s}{z^{s+\alpha}} - \sum_{s=1}^{n} (s-1)! \frac{c_s}{z^s} + R_n(z),$$

if $\alpha \in (0, 1)$ in (2.6.9), or

2.6.28
$$\mathcal{S}(f; z) = \ln z \sum_{s=0}^{n-1} (-1)^s \frac{a_s}{z^{s+1}} + \sum_{s=0}^{n-1} (-1)^s \frac{\widetilde{d}_s}{z^{s+1}} + R_n(z),$$

if $\alpha = 1$ in (2.6.9). Here c_s is given by (2.6.18),

2.6.29
$$\widetilde{d}_s = \lim_{z \to s+1} \left(\mathscr{M}(f; z) + \frac{a_s}{z - s - 1} \right),$$

and

2.6.30
$$R_n(z) = \frac{(-1)^n}{z^n} \int_0^\infty \frac{\tau^n f_n(\tau)}{\tau + z} d\tau.$$

The expansion (2.6.7) follows immediately from (2.6.27) with $z = x$ and $f(t) = (1+t)^{-(1/3)}$; its region of validity is $|\operatorname{ph} x| \leq \pi - \delta \; (< \pi)$. The distribution method outlined here can be extended readily to functions $f(t)$ having an asymptotic expansion of the form

2.6.31
$$f(t) \sim e^{ict} \sum_{s=0}^{\infty} a_s t^{-s-\alpha}, \quad t \to +\infty,$$

where $c \; (\neq 0)$ is real, and $0 < \alpha \leq 1$. For a more detailed discussion of the derivation of asymptotic expansions of Stieltjes transforms by the distribution method, see McClure and Wong (1978) and Wong (1989, Chapter 6). Corresponding results for the *generalized Stieltjes transform*

2.6.32
$$\int_0^\infty \frac{f(t)}{(t+z)^\rho} dt, \quad \rho > 0,$$

can be found in Wong (1979). An application has been given by López (2000) to derive asymptotic expansions of standard symmetric elliptic integrals, complete with error bounds; see §19.27(vi).

2.6(iii) Fractional Integrals

The Riemann–Liouville *fractional integral* of order μ is defined by

2.6.33
$$I^\mu f(x) = \frac{1}{\Gamma(\mu)} \int_0^x (x-t)^{\mu-1} f(t) dt, \quad \mu > 0;$$

see §1.15(vi). We again assume $f(t)$ is locally integrable on $[0, \infty)$ and satisfies (2.6.9). We now derive an asymptotic expansion of $I^\mu f(x)$ for large positive values of x.

In terms of the *convolution product*

2.6.34
$$(f * g)(x) = \int_0^x f(x-t) g(t) dt$$

of two locally integrable functions on $[0, \infty)$, (2.6.33) can be written

2.6.35
$$I^\mu f(x) = \frac{1}{\Gamma(\mu)} (t^{\mu-1} * f)(x).$$

The replacement of $f(t)$ by its asymptotic expansion (2.6.9), followed by term-by-term integration leads to convolution integrals of the form

2.6.36 $$(t^{\mu-1} * t^{-s-\alpha})(x) = \int_0^x (x-t)^{\mu-1} t^{-s-\alpha}\, dt,$$
$$s = 0, 1, 2, \ldots.$$

Of course, except when $s = 0$ and $0 < \alpha < 1$, none of these integrals exists in the usual sense. However, the left-hand side can be considered as the convolution of the two distributions associated with the functions $t^{\mu-1}$ and $t^{-s-\alpha}$, given by (2.6.12) and (2.6.13).

To define convolutions of distributions, we first introduce the space K^+ of all distributions of the form $D^n f$, where n is a nonnegative integer, f is a locally integrable function on $\mathbb{R}$ which vanishes on $(-\infty, 0]$, and $D^n f$ denotes the nth derivative of the distribution associated with f. For $F = D^n f$ and $G = D^m g$ in K^+, we define

2.6.37 $$F * G = D^{n+m}(f * g).$$

It is easily seen that K^+ forms a commutative, associative linear algebra. Furthermore, K^+ contains the distributions H, δ, and t^λ, $t > 0$, for any real (or complex) number λ, where H is the distribution associated with the Heaviside function $H(t)$ (§1.16(iv)), and t^λ is the distribution defined by (2.6.12)–(2.6.14), depending on the value of λ. Since $\delta = DH$, it follows that for $\mu \neq 1, 2, \ldots$,

2.6.38 $$t^{\mu-1} * \delta^{(s-1)} = \frac{\Gamma(\mu)}{\Gamma(\mu+1-s)} t^{\mu-s}, \qquad t > 0.$$

Using (5.12.1), we can also show that when $\mu \neq 1, 2, \ldots$ and $\mu - \alpha$ is not a nonnegative integer,

2.6.39
$$t^{\mu-1} * t^{-s-\alpha} = \frac{\Gamma(\mu)\Gamma(1-s-\alpha)}{\Gamma(\mu+1-s-\alpha)} t^{\mu-s-\alpha}, \quad t > 0,$$

and

2.6.40
$$t^{\mu-1} * t^{-s-1} = \frac{(-1)^s}{\mu \cdot s!} D^{s+1}\left(t^\mu(\ln t - \gamma - \psi(\mu+1))\right),$$
$$t > 0,$$

where γ is Euler's constant (§5.2(ii)).

To derive the asymptotic expansion of $I^\mu f(x)$, we recall equations (2.6.17) and (2.6.20). In the sense of distributions, they can be written

2.6.41 $$f = \sum_{s=0}^{n-1} a_s t^{-s-\alpha} - \sum_{s=1}^{n} c_s \delta^{(s-1)} + f_n,$$

and

2.6.42 $$f = \sum_{s=0}^{n-1} a_s t^{-s-1} - \sum_{s=1}^{n} d_s \delta^{(s-1)} + f_n.$$

Substituting into (2.6.35) and using (2.6.38)–(2.6.40), we obtain

2.6.43
$$t^{\mu-1} * f = \sum_{s=0}^{n-1} a_s \frac{\Gamma(\mu)\Gamma(1-s-\alpha)}{\Gamma(\mu+1-s-\alpha)} t^{\mu-s-\alpha}$$
$$- \sum_{s=1}^{n} c_s \frac{\Gamma(\mu)}{\Gamma(\mu-s+1)} t^{\mu-s} + t^{\mu-1} * f_n$$

when $0 < \alpha < 1$, or

2.6.44
$$t^{\mu-1} * f = \sum_{s=0}^{n-1} \frac{(-1)^s a_s}{\mu \cdot s!} D^{s+1}\left(t^\mu (\ln t - \gamma - \psi(\mu+1))\right)$$
$$- \sum_{s=1}^{n} d_s \frac{\Gamma(\mu)}{\Gamma(\mu-s+1)} t^{\mu-s} + t^{\mu-1} * f_n$$

when $\alpha = 1$. These equations again hold only in the sense of distributions. Since the function $t^\mu(\ln t - \gamma - \psi(\mu+1))$ and all its derivatives are locally absolutely continuous in $(0, \infty)$, the distributional derivatives in the first sum in (2.6.44) can be replaced by the corresponding ordinary derivatives. Furthermore, since $f_{n,n}^{(n)}(t) = f_n(t)$, it follows from (2.6.37) that the remainder terms $t^{\mu-1} * f_n$ in the last two equations can be associated with a locally integrable function in $(0, \infty)$. On replacing the distributions by their corresponding functions, (2.6.43) and (2.6.44) give

2.6.45
$$I^\mu f(x) = \sum_{s=0}^{n-1} a_s \frac{\Gamma(1-s-\alpha)}{\Gamma(\mu+1-s-\alpha)} x^{\mu-s-\alpha}$$
$$- \sum_{s=1}^{n} \frac{c_s}{\Gamma(\mu+1-s)} x^{\mu-s} + \frac{1}{x^n} \delta_n(x),$$

when $0 < \alpha < 1$, or

2.6.46
$$I^\mu f(x)$$
$$= \sum_{s=0}^{n-1} \frac{(-1)^s a_s}{s!\, \Gamma(\mu+1)} \frac{d^{s+1}}{dx^{s+1}}\left(x^\mu (\ln x - \gamma - \psi(\mu+1))\right)$$
$$- \sum_{s=1}^{n} \frac{d_s}{\Gamma(\mu-s+1)} x^{\mu-s} + \frac{1}{x^n} \delta_n(x),$$

when $\alpha = 1$, where

2.6.47 $$\delta_n(x) = \sum_{j=0}^{n} \binom{n}{j} \frac{\Gamma(\mu+1)}{\Gamma(\mu+1-j)} I^\mu\left(t^{n-j} f_{n,j}\right)(x),$$

$f_{n,j}(t)$ being the jth repeated integral of f_n; compare (2.6.15).

Example

Let $f(t) = t^{1-\alpha}/(1+t)$, $0 < \alpha < 1$. Then

2.6.48 $$I^\mu f(x) = \frac{1}{\Gamma(\mu)} \int_0^x (x-t)^{\mu-1} t^{1-\alpha}(1+t)^{-1}\, dt,$$

where $\mu > 0$. For $0 < t < \infty$

2.6.49 $$f(t) = \sum_{s=0}^{n-1}(-1)^s t^{-s-\alpha} + (-1)^n \frac{t^{1-n-\alpha}}{1+t}.$$

In the notation of (2.6.10), $a_s = (-1)^s$ and

2.6.50 $$f_n(t) = (-1)^n \frac{t^{1-n-\alpha}}{1+t}.$$

Since

2.6.51 $$\mathscr{M}(f;s) = (-1)^s \pi/\sin(\pi\alpha),$$

from (2.6.45) it follows that

2.6.52
$$I^\mu f(x) = \sum_{s=0}^{n-1}(-1)^s \frac{\Gamma(1-s-\alpha)}{\Gamma(\mu+1-s-\alpha)} x^{\mu-s-\alpha}$$
$$- \frac{\pi}{\sin(\pi\alpha)} \sum_{s=1}^{n} \frac{1}{\Gamma(\mu+1-s)} \frac{x^{\mu-s}}{(s-1)!}$$
$$+ \frac{1}{x^n} \delta_n(x).$$

Moreover,

2.6.53
$$|\delta_n(x)| \leq \frac{\Gamma(\mu+1)\Gamma(1-\alpha)}{\Gamma(\mu+1-\alpha)\Gamma(n+\alpha)}$$
$$\times \sum_{j=0}^{n}\binom{n}{j}\frac{\Gamma(n+\alpha-j)}{|\Gamma(\mu+1-j)|}x^{\mu-\alpha}$$

for $x > 0$.

It may be noted that the integral (2.6.48) can be expressed in terms of the hypergeometric function $_2F_1(1, 2-\alpha; 2-\alpha+\mu; -x)$; see §15.2(i).

For proofs and other examples, see McClure and Wong (1979) and Wong (1989, Chapter 6). If both f and g in (2.6.34) have asymptotic expansions of the form (2.6.9), then the distribution method can also be used to derive an asymptotic expansion of the convolution $f * g$; see Li and Wong (1994).

2.6(iv) Regularization

The method of distributions can be further extended to derive asymptotic expansions for convolution integrals:

2.6.54 $$I(x) = \int_0^\infty f(t) h(xt)\, dt.$$

We assume that for each $n = 1, 2, 3, \ldots$,

2.6.55 $$f(t) = \sum_{s=0}^{n-1} a_s t^{s+\alpha-1} + f_n(t),$$

where $0 < \alpha \leq 1$ and $f_n(t) = O(t^{n+\alpha-1})$ as $t \to 0+$. Also,

2.6.56 $$h(t) = \sum_{s=0}^{n-1} b_s t^{-s-\beta} + h_n(t),$$

where $0 < \beta \leq 1$, and $h_n(t) = O(t^{-n-\beta})$ as $t \to \infty$. Multiplication of these expansions leads to

2.6.57
$$f(t)h(xt) = \sum_{j=0}^{n-1}\sum_{k=0}^{n-1} a_j b_k t^{j+\alpha-1-k-\beta} x^{-k-\beta}$$
$$+ \sum_{j=0}^{n-1} a_j t^{j+\alpha-1} h_n(xt)$$
$$+ \sum_{k=0}^{n-1} b_k x^{-k-\beta} t^{-k-\beta} f_n(t) + f_n(t) h_n(xt).$$

On inserting this identity into (2.6.54), we immediately encounter divergent integrals of the form

2.6.58 $$\int_0^\infty t^\lambda\, dt, \qquad \lambda \in \mathbb{R}.$$

However, in the theory of *generalized functions* (distributions), there is a method, known as "regularization", by which these integrals can be interpreted in a meaningful manner. In this sense

2.6.59 $$\int_0^\infty t^\lambda\, dt = 0, \qquad \lambda \in \mathbb{C}.$$

From (2.6.55) and (2.6.59)

2.6.60 $$\mathscr{M}(f;z) = \mathscr{M}(f_n;z),$$

where $\mathscr{M}(f;z)$ is the Mellin transform of f or its analytic continuation. Also, when $\alpha \neq \beta$,

2.6.61 $$\mathscr{M}(h_x; j+\alpha) = x^{-j-\alpha}\mathscr{M}(h; j+\alpha),$$

where $h_x(t) = h(xt)$. Inserting (2.6.57) into (2.6.54), we obtain from (2.6.59)–(2.6.61)

2.6.62
$$I(x) = \sum_{j=0}^{n-1} a_j \mathscr{M}(h; j+\alpha) x^{-j-\alpha}$$
$$+ \sum_{k=0}^{n-1} b_k \mathscr{M}(f; 1-k-\beta) x^{-k-\beta} + \delta_n(x)$$

when $\alpha \neq \beta$, where

$$\delta_n(x) = \int_0^\infty f_n(t) h_n(xt)\, dt.$$

There is a similar expansion, involving logarithmic terms, when $\alpha = \beta$. For rigorous derivations of these results and also order estimates for $\delta_n(x)$, see Wong (1979) and Wong (1989, Chapter 6).

2.7 Differential Equations

2.7(i) Regular Singularities: Fuchs–Frobenius Theory

An *ordinary point* of the differential equation

2.7.1 $$\frac{d^2w}{dz^2} + f(z)\frac{dw}{dz} + g(z)w = 0$$

is one at which the coefficients $f(z)$ and $g(z)$ are analytic. All solutions are analytic at an ordinary point, and their Taylor-series expansions are found by equating coefficients.

Other points z_0 are *singularities* of the differential equation. If both $(z-z_0)f(z)$ and $(z-z_0)^2 g(z)$ are analytic at z_0, then z_0 is a *regular singularity* (or *singularity of the first kind*). All other singularities are classified as *irregular*.

In a punctured neighborhood **N** of a regular singularity z_0

2.7.2
$$f(z) = \sum_{s=0}^{\infty} f_s(z-z_0)^{s-1}, \quad g(z) = \sum_{s=0}^{\infty} g_s(z-z_0)^{s-2},$$

with at least one of the coefficients f_0, g_0, g_1 nonzero. Let α_1, α_2 denote the *indices* or *exponents*, that is, the roots of the *indicial equation*

2.7.3 $\quad Q(\alpha) \equiv \alpha(\alpha-1) + f_0\alpha + g_0 = 0.$

Provided that $\alpha_1 - \alpha_2$ is not zero or an integer, equation (2.7.1) has independent solutions $w_j(z)$, $j = 1, 2$, such that

2.7.4 $\quad w_j(z) = (z-z_0)^{\alpha_j} \sum_{s=0}^{\infty} a_{s,j}(z-z_0)^s, \quad z \in \mathbf{N},$

with $a_{0,j} = 1$, and

2.7.5 $\quad Q(\alpha_j + s)a_{s,j} = -\sum_{r=0}^{s-1}((\alpha_j + r)f_{s-r} + g_{s-r})a_{r,j},$

when $s = 1, 2, 3, \ldots$.

If $\alpha_1 - \alpha_2 = 0, 1, 2, \ldots$, then (2.7.4) applies only in the case $j = 1$. But there is an independent solution

2.7.6
$$w_2(z) = (z-z_0)^{\alpha_2} \sum_{\substack{s=0 \\ s \neq \alpha_1 - \alpha_2}}^{\infty} b_s(z-z_0)^s$$
$$+ cw_1(z)\ln(z-z_0), \quad z \in \mathbf{N}.$$

The coefficients b_s and constant c are again determined by equating coefficients in the differential equation, beginning with $c = 1$ when $\alpha_1 - \alpha_2 = 0$, or with $b_0 = 1$ when $\alpha_1 - \alpha_2 = 1, 2, 3, \ldots$.

The radii of convergence of the series (2.7.4), (2.7.6) are not less than the distance of the next nearest singularity of the differential equation from z_0.

To include the point at infinity in the foregoing classification scheme, we transform it into the origin by replacing z in (2.7.1) with $1/z$; see Olver (1997b, pp. 153–154). For corresponding definitions, together with examples, for linear differential equations of arbitrary order see §§16.8(i)–16.8(ii).

2.7(ii) Irregular Singularities of Rank 1

If the singularities of $f(z)$ and $g(z)$ at z_0 are no worse than poles, then z_0 has *rank* $\ell - 1$, where ℓ is the least integer such that $(z-z_0)^\ell f(z)$ and $(z-z_0)^{2\ell} g(z)$ are analytic at z_0. Thus a regular singularity has rank 0. The most common type of irregular singularity for special functions has rank 1 and is located at infinity. Then

2.7.7 $\quad f(z) = \sum_{s=0}^{\infty} \frac{f_s}{z^s}, \quad g(z) = \sum_{s=0}^{\infty} \frac{g_s}{z^s},$

these series converging in an annulus $|z| > a$, with at least one of f_0, g_0, g_1 nonzero.

Formal solutions are

2.7.8 $\quad e^{\lambda_j z} z^{\mu_j} \sum_{s=0}^{\infty} \frac{a_{s,j}}{z^s}, \quad j = 1, 2,$

where λ_1, λ_2 are the roots of the *characteristic equation*

2.7.9 $\quad \lambda^2 + f_0 \lambda + g_0 = 0,$

2.7.10 $\quad \mu_j = -(f_1 \lambda_j + g_1)/(f_0 + 2\lambda_j),$

$a_{0,j} = 1$, and

2.7.11
$$(f_0 + 2\lambda_j)s a_{s,j} = (s - \mu_j)(s - 1 - \mu_j)a_{s-1,j}$$
$$+ \sum_{r=1}^{s} (\lambda_j f_{r+1} + g_{r+1}$$
$$- (s - r - \mu_j)f_r)a_{s-r,j},$$

when $s = 1, 2, \ldots$. The construction fails iff $\lambda_1 = \lambda_2$, that is, when $f_0^2 = 4g_0$: this case is treated below.

For large s,

2.7.12
$$a_{s,1} \sim \frac{\Lambda_1}{(\lambda_1 - \lambda_2)^s}$$
$$\times \sum_{j=0}^{\infty} a_{j,2}(\lambda_1 - \lambda_2)^j \, \Gamma(s + \mu_2 - \mu_1 - j),$$

2.7.13
$$a_{s,2} \sim \frac{\Lambda_2}{(\lambda_2 - \lambda_1)^s}$$
$$\times \sum_{j=0}^{\infty} a_{j,1}(\lambda_2 - \lambda_1)^j \, \Gamma(s + \mu_1 - \mu_2 - j),$$

where Λ_1 and Λ_2 are constants, and the Jth remainder terms in the sums are $O(\Gamma(s + \mu_2 - \mu_1 - J))$ and $O(\Gamma(s + \mu_1 - \mu_2 - J))$, respectively (Olver (1994a)). Hence unless the series (2.7.8) terminate (in which case the corresponding Λ_j is zero) they diverge. However, there are unique and linearly independent solutions $w_j(z)$, $j = 1, 2$, such that

2.7.14 $\quad w_j(z) \sim e^{\lambda_j z}((\lambda_2 - \lambda_1)z)^{\mu_j} \sum_{s=0}^{\infty} \frac{a_{s,j}}{z^s}$

as $z \to \infty$ in the sectors

2.7.15 $\quad -\tfrac{3}{2}\pi + \delta \leq \mathrm{ph}((\lambda_2 - \lambda_1)z) \leq \tfrac{3}{2}\pi - \delta, \quad j = 1,$

2.7.16 $\quad -\tfrac{1}{2}\pi + \delta \leq \mathrm{ph}((\lambda_2 - \lambda_1)z) \leq \tfrac{5}{2}\pi - \delta, \quad j = 2,$

δ being an arbitrary small positive constant.

Although the expansions (2.7.14) apply only in the sectors (2.7.15) and (2.7.16), each solution $w_j(z)$ can

2.7 Differential Equations

be continued analytically into any other sector. Typical connection formulas are

2.7.17
$$w_1(z) = e^{2\pi i \mu_1} w_1(ze^{-2\pi i}) + C_1 w_2(z),$$
$$w_2(z) = e^{-2\pi i \mu_2} w_2(ze^{2\pi i}) + C_2 w_1(z),$$

in which C_1, C_2 are constants, the so-called *Stokes multipliers*. In combination with (2.7.14) these formulas yield asymptotic expansions for $w_1(z)$ in $\frac{1}{2}\pi + \delta \leq \mathrm{ph}((\lambda_2 - \lambda_1)z) \leq \frac{5}{2}\pi - \delta$, and $w_2(z)$ in $-\frac{3}{2}\pi + \delta \leq \mathrm{ph}((\lambda_2 - \lambda_1)z) \leq \frac{1}{2}\pi - \delta$. Furthermore,

2.7.18 $\Lambda_1 = -ie^{(\mu_2 - \mu_1)\pi i}C_1/(2\pi), \quad \Lambda_2 = iC_2/(2\pi).$

Note that the coefficients in the expansions (2.7.12), (2.7.13) for the "late" coefficients, that is, $a_{s,1}, a_{s,2}$ with s large, are the "early" coefficients $a_{j,2}, a_{j,1}$ with j small. This phenomenon is an example of *resurgence*, a classification due to Écalle (1981a,b). See §2.11(v) for other examples.

The exceptional case $f_0^2 = 4g_0$ is handled by *Fabry's transformation*:

2.7.19 $w = e^{-f_0 z/2} W, \quad t = z^{1/2}.$

The transformed differential equation either has a regular singularity at $t = \infty$, or its characteristic equation has unequal roots.

For error bounds for (2.7.14) see Olver (1997b, Chapter 7). For the calculation of Stokes multipliers see Olde Daalhuis and Olver (1995b). For extensions to singularities of higher rank see Olver and Stenger (1965). For extensions to higher-order differential equations see Stenger (1966a,b), Olver (1997a, 1999), and Olde Daalhuis and Olver (1998).

2.7(iii) Liouville–Green (WKBJ) Approximation

For irregular singularities of nonclassifiable rank, a powerful tool for finding the asymptotic behavior of solutions, complete with error bounds, is as follows:

Liouville–Green Approximation Theorem

In a finite or infinite interval (a_1, a_2) let $f(x)$ be real, positive, and twice-continuously differentiable, and $g(x)$ be continuous. Then in (a_1, a_2) the differential equation

2.7.20
$$\frac{d^2 w}{dx^2} = (f(x) + g(x))w$$

has twice-continuously differentiable solutions

2.7.21
$$w_1(x) = f^{-1/4}(x) \exp\left(\int f^{1/2}(x)\,dx\right)(1 + \epsilon_1(x)),$$

2.7.22
$$w_2(x) = f^{-1/4}(x) \exp\left(-\int f^{1/2}(x)\,dx\right)(1 + \epsilon_2(x)),$$

such that

2.7.23 $|\epsilon_j(x)|, \ \tfrac{1}{2}f^{-1/2}(x)|\epsilon'_j(x)| \leq \exp\left(\tfrac{1}{2}\mathcal{V}_{a_j,x}(F)\right) - 1,$
$$j = 1, 2,$$

provided that $\mathcal{V}_{a_j,x}(F) < \infty$. Here $F(x)$ is the *error-control function*

2.7.24
$$F(x) = \int \left(\frac{1}{f^{1/4}}\frac{d^2}{dx^2}\left(\frac{1}{f^{1/4}}\right) - \frac{g}{f^{1/2}}\right)dx,$$

and $\mathcal{V}$ denotes the variational operator (§2.3(i)). Thus

2.7.25
$$\mathcal{V}_{a_j,x}(F) = \int_{a_j}^{x}\left|\left(\frac{1}{f^{1/4}(t)}\frac{d^2}{dt^2}\left(\frac{1}{f^{1/4}(t)}\right) - \frac{g(t)}{f^{1/2}(t)}\right)dt\right|.$$

Assuming also $\mathcal{V}_{a_1,a_2}(F) < \infty$, we have

2.7.26
$$w_1(x) \sim f^{-1/4}(x) \exp\left(\int f^{1/2}(x)\,dx\right), \quad x \to a_1+,$$

2.7.27
$$w_2(x) \sim f^{-1/4}(x) \exp\left(-\int f^{1/2}(x)\,dx\right), \quad x \to a_2-.$$

Suppose in addition $|\int f^{1/2}(x)\,dx|$ is unbounded as $x \to a_1+$ and $x \to a_2-$. Then there are solutions $w_3(x)$, $w_4(x)$, such that

2.7.28
$$w_3(x) \sim f^{-1/4}(x) \exp\left(\int f^{1/2}(x)\,dx\right), \quad x \to a_2-,$$

2.7.29
$$w_4(x) \sim f^{-1/4}(x) \exp\left(-\int f^{1/2}(x)\,dx\right), \quad x \to a_1+.$$

The solutions with the properties (2.7.26), (2.7.27) are unique, but not those with the properties (2.7.28), (2.7.29). In fact, since

2.7.30 $w_1(x)/w_4(x) \to 0, \quad x \to a_1+,$

$w_1(x)$ is a *recessive* (or *subdominant*) solution as $x \to a_1+$, and $w_4(x)$ is a *dominant* solution as $x \to a_1+$. Similarly for $w_2(x)$ and $w_3(x)$ as $x \to a_2-$.

Example

2.7.31
$$\frac{d^2 w}{dx^2} = (x + \ln x)w, \qquad 0 < x < \infty.$$

We cannot take $f = x$ and $g = \ln x$ because $\int g f^{-1/2}\,dx$ would diverge as $x \to +\infty$. Instead set $f = x + \ln x$, $g = 0$. By approximating

2.7.32 $f^{1/2} = x^{1/2} + \tfrac{1}{2}x^{-1/2}\ln x + O\!\left(x^{-3/2}(\ln x)^2\right),$

we arrive at

2.7.33 $w_2(x) \sim x^{-(1/4)-\sqrt{x}} \exp\!\left(2x^{1/2} - \tfrac{2}{3}x^{3/2}\right),$

2.7.34 $w_3(x) \sim x^{-(1/4)+\sqrt{x}} \exp\!\left(\tfrac{2}{3}x^{3/2} - 2x^{1/2}\right),$

as $x \to +\infty$, $w_2(x)$ being recessive and $w_3(x)$ dominant.

For other examples, and also the corresponding results when $f(x)$ is negative, see Olver (1997b, Chapter 6), Olver (1980a), Taylor (1978, 1982), and Smith (1986). The first of these references includes extensions to complex variables and reversions for zeros.

2.7(iv) Numerically Satisfactory Solutions

One pair of independent solutions of the equation

2.7.35 $$d^2w/dz^2 = w$$

is $w_1(z) = e^z$, $w_2(z) = e^{-z}$. Another is $w_3(z) = \cosh z$, $w_4(z) = \sinh z$. In theory either pair may be used to construct any other solution

2.7.36 $$w(z) = Aw_1(z) + Bw_2(z),$$

or

2.7.37 $$w(z) = Cw_3(z) + Dw_4(z),$$

where A, B, C, D are constants. From the numerical standpoint, however, the pair $w_3(z)$ and $w_4(z)$ has the drawback that severe numerical cancellation can occur with certain combinations of C and D, for example if C and D are equal, or nearly equal, and z, or $\Re z$, is large and negative. This kind of cancellation cannot take place with $w_1(z)$ and $w_2(z)$, and for this reason, and following Miller (1950), we call $w_1(z)$ and $w_2(z)$ a *numerically satisfactory pair* of solutions.

The solutions $w_1(z)$ and $w_2(z)$ are respectively recessive and dominant as $\Re z \to -\infty$, and *vice versa* as $\Re z \to +\infty$. This is characteristic of numerically satisfactory pairs. In a neighborhood, or sectorial neighborhood of a singularity, one member has to be recessive. In consequence, if a differential equation has more than one singularity in the extended plane, then usually more than two standard solutions need to be chosen in order to have numerically satisfactory representations everywhere.

In oscillatory intervals, and again following Miller (1950), we call a pair of solutions numerically satisfactory if asymptotically they have the same amplitude and are $\tfrac{1}{2}\pi$ out of phase.

2.8 Differential Equations with a Parameter

2.8(i) Classification of Cases

Many special functions satisfy an equation of the form

2.8.1 $$d^2w/dz^2 = \left(u^2 f(z) + g(z)\right) w,$$

in which u is a real or complex parameter, and asymptotic solutions are needed for large $|u|$ that are uniform with respect to z in a point set $\mathbf{D}$ in $\mathbb{R}$ or $\mathbb{C}$. For example, u can be the order of a Bessel function or degree of an orthogonal polynomial. The form of the asymptotic expansion depends on the nature of the *transition points* in $\mathbf{D}$, that is, points at which $f(z)$ has a zero or singularity. Zeros of $f(z)$ are also called *turning points*.

There are three main cases. In Case I there are no transition points in $\mathbf{D}$ and $g(z)$ is analytic. In Case II $f(z)$ has a simple zero at z_0 and $g(z)$ is analytic at z_0. In Case III $f(z)$ has a simple pole at z_0 and $(z-z_0)^2 g(z)$ is analytic at z_0.

The same approach is used in all three cases. First we apply the *Liouville transformation* (§1.13(iv)) to (2.8.1). This introduces new variables W and ξ, related by

2.8.2 $$W = \dot{z}^{-1/2} w,$$

dots denoting differentiations with respect to ξ. Then

2.8.3 $$\frac{d^2 W}{d\xi^2} = \left(u^2 \dot{z}^2 f(z) + \psi(\xi)\right) W,$$

where

2.8.4 $$\psi(\xi) = \dot{z}^2 g(z) + \dot{z}^{1/2} \frac{d^2}{d\xi^2}(\dot{z}^{-1/2}).$$

The transformation is now specialized in such a way that: (a) ξ and z are analytic functions of each other at the transition point (if any); (b) the approximating differential equation obtained by neglecting $\psi(\xi)$ (or part of $\psi(\xi)$) has solutions that are functions of a single variable. The actual choices are as follows:

2.8.5 $$\dot{z}^2 f(z) = 1, \quad \xi = \int f^{1/2}(z)\, dz,$$

for Case I,

2.8.6 $$\dot{z}^2 f(z) = \xi, \quad \tfrac{2}{3}\xi^{3/2} = \int_{z_0}^{z} f^{1/2}(t)\, dt,$$

for Case II,

2.8.7 $$\dot{z}^2 f(z) = 1/\xi, \quad 2\xi^{1/2} = \int_{z_0}^{z} f^{1/2}(t)\, dt,$$

for Case III.

The transformed equation has the form

2.8.8 $$d^2 W/d\xi^2 = \left(u^2 \xi^m + \psi(\xi)\right) W,$$

with $m = 0$ (Case I), $m = 1$ (Case II), $m = -1$ (Case III). In Cases I and II the asymptotic solutions are in terms of the functions that satisfy (2.8.8) with $\psi(\xi) = 0$. These are elementary functions in Case I, and Airy functions (§9.2) in Case II. In Case III the approximating equation is

2.8.9 $$\frac{d^2 W}{d\xi^2} = \left(\frac{u^2}{\xi} + \frac{\rho}{\xi^2}\right) W,$$

where $\rho = \lim(\xi^2 \psi(\xi))$ as $\xi \to 0$. Solutions are Bessel functions, or modified Bessel functions, of order $\pm(1 + 4\rho)^{1/2}$ (§§10.2, 10.25).

For another approach to these problems based on convergent inverse factorial series expansions see Dunster et al. (1993) and Dunster (2001a, 2004).

2.8(ii) Case I: No Transition Points

The transformed differential equation is

2.8.10 $$d^2W/d\xi^2 = (u^2 + \psi(\xi))W,$$

in which ξ ranges over a bounded or unbounded interval or domain $\boldsymbol{\Delta}$, and $\psi(\xi)$ is C^∞ or analytic on $\boldsymbol{\Delta}$. The parameter u is assumed to be real and positive. Corresponding to each positive integer n there are solutions $W_{n,j}(u,\xi)$, $j=1,2$, that depend on arbitrarily chosen reference points α_j, are C^∞ or analytic on $\boldsymbol{\Delta}$, and as $u \to \infty$

2.8.11
$$W_{n,1}(u,\xi) = e^{u\xi}\left(\sum_{s=0}^{n-1}\frac{A_s(\xi)}{u^s} + O\left(\frac{1}{u^n}\right)\right), \quad \xi \in \boldsymbol{\Delta}_1(\alpha_1),$$

2.8.12
$$W_{n,2}(u,\xi) = e^{-u\xi}\left(\sum_{s=0}^{n-1}(-1)^s\frac{A_s(\xi)}{u^s} + O\left(\frac{1}{u^n}\right)\right), \quad \xi \in \boldsymbol{\Delta}_2(\alpha_2),$$

with $A_0(\xi) = 1$ and

2.8.13
$$A_{s+1}(\xi) = -\tfrac{1}{2}A_s'(\xi) + \tfrac{1}{2}\int \psi(\xi)A_s(\xi)\,d\xi, \quad s=0,1,2,\ldots,$$

(the constants of integration being arbitrary). The expansions (2.8.11) and (2.8.12) are both uniform and differentiable with respect to ξ. The regions of validity $\boldsymbol{\Delta}_j(\alpha_j)$ comprise those points ξ that can be joined to α_j in $\boldsymbol{\Delta}$ by a path $\mathscr{Q}_j$ along which $\Re v$ is nondecreasing ($j=1$) or nonincreasing ($j=2$) as v passes from α_j to ξ. In addition, $\mathcal{V}_{\mathscr{Q}_j}(A_1)$ and $\mathcal{V}_{\mathscr{Q}_j}(A_n)$ must be bounded on $\boldsymbol{\Delta}_j(\alpha_j)$.

For error bounds, extensions to pure imaginary or complex u, an extension to inhomogeneous differential equations, and examples, see Olver (1997b, Chapter 10). This reference also supplies sufficient conditions to ensure that the solutions $W_{n,1}(u,\xi)$ and $W_{n,2}(u,\xi)$ having the properties (2.8.11) and (2.8.12) are independent of n.

2.8(iii) Case II: Simple Turning Point

The transformed differential equation is

2.8.14 $$d^2W/d\xi^2 = (u^2\xi + \psi(\xi))W,$$

and for simplicity ξ is assumed to range over a finite or infinite interval (α_1,α_2) with $\alpha_1 < 0$, $\alpha_2 > 0$. Again, $u > 0$ and $\psi(\xi)$ is C^∞ on (α_1,α_2). Corresponding to each positive integer n there are solutions $W_{n,j}(u,\xi)$, $j=1,2$, that are C^∞ on (α_1,α_2), and as $u \to \infty$

2.8.15 $$W_{n,1}(u,\xi) = \operatorname{Ai}\!\left(u^{2/3}\xi\right)\left(\sum_{s=0}^{n-1}\frac{A_s(\xi)}{u^{2s}} + O\!\left(\frac{1}{u^{2n-1}}\right)\right) + \operatorname{Ai}'\!\left(u^{2/3}\xi\right)\left(\sum_{s=0}^{n-2}\frac{B_s(\xi)}{u^{2s+(4/3)}} + O\!\left(\frac{1}{u^{2n-1}}\right)\right),$$

2.8.16 $$W_{n,2}(u,\xi) = \operatorname{Bi}\!\left(u^{2/3}\xi\right)\left(\sum_{s=0}^{n-1}\frac{A_s(\xi)}{u^{2s}} + O\!\left(\frac{1}{u^{2n-1}}\right)\right) + \operatorname{Bi}'\!\left(u^{2/3}\xi\right)\left(\sum_{s=0}^{n-2}\frac{B_s(\xi)}{u^{2s+(4/3)}} + O\!\left(\frac{1}{u^{2n-1}}\right)\right).$$

Here $A_0(\xi) = 1$,

2.8.17 $$B_s(\xi) = \begin{cases} \dfrac{1}{2\xi^{1/2}}\displaystyle\int_0^\xi (\psi(v)A_s(v) - A_s''(v))\dfrac{dv}{v^{1/2}}, & \xi > 0, \\ \dfrac{1}{2(-\xi)^{1/2}}\displaystyle\int_\xi^0 (\psi(v)A_s(v) - A_s''(v))\dfrac{dv}{(-v)^{1/2}}, & \xi < 0, \end{cases}$$

and

2.8.18 $$A_{s+1}(\xi) = -\tfrac{1}{2}B_s'(\xi) + \tfrac{1}{2}\int \psi(\xi)B_s(\xi)\,d\xi,$$

when $s=0,1,2,\ldots$. For Ai and Bi see §9.2. The expansions (2.8.15) and (2.8.16) are both uniform and differentiable with respect to ξ. These results are valid when $\mathcal{V}_{\alpha_1,\alpha_2}(|\xi|^{1/2}B_0)$ and $\mathcal{V}_{\alpha_1,\alpha_2}(|\xi|^{1/2}B_{n-1})$ are finite.

An alternative way of representing the error terms in (2.8.15) and (2.8.16) is as follows. Let $c = -0.36604\ldots$ be the real root of the equation

2.8.19 $$\operatorname{Ai}(x) = \operatorname{Bi}(x)$$

of smallest absolute value, and define the *envelopes* of $\operatorname{Ai}(x)$ and $\operatorname{Bi}(x)$ by

2.8.20 $$\operatorname{env}\operatorname{Ai}(x) = \operatorname{env}\operatorname{Bi}(x) = \left(\operatorname{Ai}^2(x) + \operatorname{Bi}^2(x)\right)^{1/2}, \quad -\infty < x \le c,$$

2.8.21 $$\operatorname{env}\operatorname{Ai}(x) = \sqrt{2}\operatorname{Ai}(x), \quad \operatorname{env}\operatorname{Bi}(x) = \sqrt{2}\operatorname{Bi}(x), \quad c \le x < \infty.$$

These envelopes are continuous functions of x, and as $u \to \infty$

$$W_{n,1}(u,\xi) = \operatorname{Ai}\left(u^{2/3}\xi\right) \sum_{s=0}^{n-1} \frac{A_s(\xi)}{u^{2s}}$$

2.8.22
$$+ \operatorname{Ai}'\left(u^{2/3}\xi\right) \sum_{s=0}^{n-2} \frac{B_s(\xi)}{u^{2s+(4/3)}}$$

$$+ \operatorname{env}\operatorname{Ai}\left(u^{2/3}\xi\right) O\left(\frac{1}{u^{2n-1}}\right),$$

$$W_{n,2}(u,\xi) = \operatorname{Bi}\left(u^{2/3}\xi\right) \sum_{s=0}^{n-1} \frac{A_s(\xi)}{u^{2s}}$$

2.8.23
$$+ \operatorname{Bi}'\left(u^{2/3}\xi\right) \sum_{s=0}^{n-2} \frac{B_s(\xi)}{u^{2s+(4/3)}}$$

$$+ \operatorname{env}\operatorname{Bi}\left(u^{2/3}\xi\right) O\left(\frac{1}{u^{2n-1}}\right),$$

uniformly with respect to $\xi \in (\alpha_1, \alpha_2)$.

For error bounds, more delicate error estimates, extensions to complex ξ and u, zeros, connection formulas, extensions to inhomogeneous equations, and examples, see Olver (1997b, Chapters 11, 13), Olver (1964b), Reid (1974a,b), Boyd (1987), and Baldwin (1991).

For other examples of uniform asymptotic approximations and expansions of special functions in terms of Airy functions see especially §10.20 and §§12.10(vii), 12.10(viii); also §§12.14(ix), 13.20(v), 13.21(iii), 13.21(iv), 15.12(iii), 18.15(iv), 30.9(i), 30.9(ii), 32.11(ii), 32.11(iii), 33.12(i), 33.12(ii), 33.20(iv), 36.12(ii), 36.13.

2.8(iv) Case III: Simple Pole

The transformed equation (2.8.8) is renormalized as

2.8.24
$$\frac{d^2 W}{d\xi^2} = \left(\frac{u^2}{4\xi} + \frac{\nu^2 - 1}{4\xi^2} + \frac{\psi(\xi)}{\xi}\right) W.$$

We again assume $\xi \in (\alpha_1, \alpha_2)$ with $-\infty \leq \alpha_1 < 0$, $0 < \alpha_2 \leq \infty$. Also, $\psi(\xi)$ is C^∞ on (α_1, α_2), and $u > 0$. The constant $\nu \, (= \sqrt{1+4\rho})$ is real and nonnegative.

There are two cases: $\xi \in (0, \alpha_2)$ and $\xi \in (\alpha_1, 0)$. In the former, corresponding to any positive integer n there are solutions $W_{n,j}(u, \xi)$, $j = 1, 2$, that are C^∞ on $(0, \alpha_2)$, and as $u \to \infty$

2.8.25 $\quad W_{n,1}(u,\xi) = \xi^{1/2} I_\nu\left(u\xi^{1/2}\right) \sum_{s=0}^{n-1} \frac{A_s(\xi)}{u^{2s}} + \xi I_{\nu+1}\left(u\xi^{1/2}\right) \sum_{s=0}^{n-2} \frac{B_s(\xi)}{u^{2s+1}} + \xi^{1/2} I_\nu\left(u\xi^{1/2}\right) O\left(\frac{1}{u^{2n-1}}\right),$

2.8.26 $\quad W_{n,2}(u,\xi) = \xi^{1/2} K_\nu\left(u\xi^{1/2}\right) \sum_{s=0}^{n-1} \frac{A_s(\xi)}{u^{2s}} - \xi K_{\nu+1}\left(u\xi^{1/2}\right) \sum_{s=0}^{n-2} \frac{B_s(\xi)}{u^{2s+1}} + \xi^{1/2} K_\nu\left(u\xi^{1/2}\right) O\left(\frac{1}{u^{2n-1}}\right).$

Here $A_0(\xi) = 1$,

2.8.27 $\quad\quad\quad\quad B_s(\xi) = -A'_s(\xi) + \frac{1}{\xi^{1/2}} \int_0^\xi \left(\psi(v) A_s(v) - \left(\nu + \tfrac{1}{2}\right) A'_s(v)\right) \frac{dv}{v^{1/2}},$

2.8.28 $\quad\quad\quad\quad A_{s+1}(\xi) = \nu B_s(\xi) - \xi B'_s(\xi) + \int \psi(\xi) B_s(\xi)\, d\xi,$

$s = 0, 1, 2, \ldots$. For I_ν and K_ν see §10.25(ii). The expansions (2.8.25) and (2.8.26) are both uniform and differentiable with respect to ξ. These results are valid when $\mathcal{V}_{0,\alpha_2}\left(\xi^{1/2} B_0\right)$ and $\mathcal{V}_{0,\alpha_2}\left(\xi^{1/2} B_{n-1}\right)$ are finite.

If $\xi \in (\alpha_1, 0)$, then there are solutions $W_{n,j}(u, \xi)$, $j = 3, 4$, that are C^∞ on $(\alpha_1, 0)$, and as $u \to \infty$

2.8.29 $\quad W_{n,3}(u,\xi) = |\xi|^{1/2} J_\nu\left(u|\xi|^{1/2}\right) \left(\sum_{s=0}^{n-1} \frac{A_s(\xi)}{u^{2s}} + O\left(\frac{1}{u^{2n-1}}\right)\right) - |\xi| J_{\nu+1}\left(u|\xi|^{1/2}\right) \left(\sum_{s=0}^{n-2} \frac{B_s(\xi)}{u^{2s+1}} + O\left(\frac{1}{u^{2n-2}}\right)\right),$

2.8.30 $\quad W_{n,4}(u,\xi) = |\xi|^{1/2} Y_\nu\left(u|\xi|^{1/2}\right) \left(\sum_{s=0}^{n-1} \frac{A_s(\xi)}{u^{2s}} + O\left(\frac{1}{u^{2n-1}}\right)\right) - |\xi| Y_{\nu+1}\left(u|\xi|^{1/2}\right) \left(\sum_{s=0}^{n-2} \frac{B_s(\xi)}{u^{2s+1}} + O\left(\frac{1}{u^{2n-2}}\right)\right).$

Here $A_0(\xi) = 1$,

2.8.31 $\quad\quad\quad\quad B_s(\xi) = -A'_s(\xi) + \frac{1}{|\xi|^{1/2}} \int_\xi^0 \left(\psi(v) A_s(v) - \left(\nu + \tfrac{1}{2}\right) A'_s(v)\right) \frac{dv}{|v|^{1/2}},$

$s = 0, 1, 2, \ldots$, and (2.8.28) again applies. For J_ν and Y_ν see §10.2(ii). The expansions (2.8.29) and (2.8.30) are both uniform and differentiable with respect to ξ. These results are valid when $\mathcal{V}_{\alpha_1,0}\left(|\xi|^{1/2} B_0\right)$ and $\mathcal{V}_{\alpha_1,0}\left(|\xi|^{1/2} B_{n-1}\right)$ are finite.

Again, an alternative way of representing the error terms in (2.8.29) and (2.8.30) is by means of envelope functions. Let $x = X_\nu$ be the smallest positive root of the equation

2.8.32 $\quad\quad\quad\quad\quad\quad\quad\quad\quad\quad J_\nu(x) + Y_\nu(x) = 0.$

2.9 Difference Equations

Define

2.8.33 $\qquad \operatorname{env} J_\nu(x) = \sqrt{2}\, J_\nu(x), \quad \operatorname{env} Y_\nu(x) = \sqrt{2}\,|Y_\nu(x)|, \qquad 0 < x \le X_\nu,$

2.8.34 $\qquad \operatorname{env} J_\nu(x) = \operatorname{env} Y_\nu(x) = \left(J_\nu^2(x) + Y_\nu^2(x)\right)^{1/2}, \qquad X_\nu \le x < \infty.$

Then as $u \to \infty$

2.8.35 $\quad W_{n,3}(u,\xi) = |\xi|^{1/2} J_\nu\!\left(u|\xi|^{1/2}\right) \sum_{s=0}^{n-1} \dfrac{A_s(\xi)}{u^{2s}} - |\xi|\, J_{\nu+1}\!\left(u|\xi|^{1/2}\right) \sum_{s=0}^{n-2} \dfrac{B_s(\xi)}{u^{2s+1}} + |\xi|^{1/2} \operatorname{env} J_\nu\!\left(u|\xi|^{1/2}\right) O\!\left(\dfrac{1}{u^{2n-1}}\right),$

2.8.36 $\quad W_{n,4}(u,\xi) = |\xi|^{1/2} Y_\nu\!\left(u|\xi|^{1/2}\right) \sum_{s=0}^{n-1} \dfrac{A_s(\xi)}{u^{2s}} - |\xi|\, Y_{\nu+1}\!\left(u|\xi|^{1/2}\right) \sum_{s=0}^{n-2} \dfrac{B_s(\xi)}{u^{2s+1}} + |\xi|^{1/2} \operatorname{env} Y_\nu\!\left(u|\xi|^{1/2}\right) O\!\left(\dfrac{1}{u^{2n-1}}\right),$

uniformly with respect to $\xi \in (\alpha_1, 0)$.

For error bounds, more delicate error estimates, extensions to complex ξ, ν, and u, zeros, and examples see Olver (1997b, Chapter 12), Boyd (1990a), and Dunster (1990a).

For other examples of uniform asymptotic approximations and expansions of special functions in terms of Bessel functions or modified Bessel functions of fixed order see §§13.8(iii), 13.21(i), 13.21(iv), 14.15(i), 14.15(iii), 14.20(vii), 15.12(iii), 18.15(i), 18.15(iv), 18.24, 33.20(iv).

2.8(v) Multiple and Fractional Turning Points

The approach used in preceding subsections for equation (2.8.1) also succeeds when z_0 is a *multiple* or *fractional turning point*. For the former $f(z)$ has a zero of multiplicity $\lambda = 2, 3, 4, \ldots$ and $g(z)$ is analytic. For the latter $(z - z_0)^{-\lambda} f(z)$ and $g(z)$ are both analytic at z_0, $\lambda\ (> -2)$ being a real constant. In both cases uniform asymptotic approximations are obtained in terms of Bessel functions of order $1/(\lambda + 2)$. More generally, $g(z)$ can have a simple or double pole at z_0. (In the case of the double pole the order of the approximating Bessel functions is fixed but no longer $1/(\lambda+2)$.) However, in all cases with $\lambda > -2$ and $\lambda \ne 0$ or ± 1, only uniform asymptotic approximations are available, not uniform asymptotic expansions. For results, including error bounds, see Olver (1977c).

For connection formulas for Liouville–Green approximations across these transition points see Olver (1977b,a, 1978).

2.8(vi) Coalescing Transition Points

Corresponding to the problems for integrals outlined in §§2.3(v), 2.4(v), and 2.4(vi), there are analogous problems for differential equations.

For two coalescing turning points see Olver (1975a, 1976) and Dunster (1996a); in this case the uniform approximants are parabolic cylinder functions. (For envelope functions for parabolic cylinder functions see §14.15(v)).

For a coalescing turning point and double pole see Boyd and Dunster (1986) and Dunster (1990b); in this case the uniform approximants are Bessel functions of variable order.

For a coalescing turning point and simple pole see Nestor (1984) and Dunster (1994b); in this case the uniform approximants are Whittaker functions (§13.14(i)) with a fixed value of the second parameter.

For further examples of uniform asymptotic approximations in terms of parabolic cylinder functions see §§13.20(iii), 13.20(iv), 14.15(v), 15.12(iii), 18.24.

For further examples of uniform asymptotic approximations in terms of Bessel functions or modified Bessel functions of variable order see §§13.21(ii), 14.15(ii), 14.15(iv), 14.20(viii), 30.9(i), 30.9(ii).

For examples of uniform asymptotic approximations in terms of Whittaker functions with fixed second parameter see §18.15(i) and §28.8(iv).

Lastly, for an example of a fourth-order differential equation, see Wong and Zhang (2007).

2.9 Difference Equations

2.9(i) Distinct Characteristic Values

Many special functions that depend on parameters satisfy a three-term linear recurrence relation

2.9.1 $\quad w(n+2) + f(n) w(n+1) + g(n) w(n) = 0, \qquad n = 0, 1, 2, \ldots,$

or equivalently the second-order homogeneous linear difference equation

2.9.2 $\quad \Delta^2 w(n) + (2 + f(n)) \Delta w(n) + (1 + f(n) + g(n)) w(n) = 0, \quad n = 0, 1, 2, \ldots,$

in which Δ is the forward difference operator (§3.6(i)).

Often $f(n)$ and $g(n)$ can be expanded in series

2.9.3
$$f(n) \sim \sum_{s=0}^{\infty} \frac{f_s}{n^s}, \quad g(n) \sim \sum_{s=0}^{\infty} \frac{g_s}{n^s}, \quad n \to \infty,$$

with $g_0 \neq 0$. (For the case $g_0 = 0$ see the final paragraph of §2.9(ii) with Q negative.) This situation is analogous to second-order homogeneous linear differential equations with an irregular singularity of rank 1 at infinity (§2.7(ii)). Formal solutions are

2.9.4
$$\rho_j^n n^{\alpha_j} \sum_{s=0}^{\infty} \frac{a_{s,j}}{n^s}, \quad j = 1, 2,$$

where ρ_1, ρ_2 are the roots of the *characteristic equation*

2.9.5
$$\rho^2 + f_0 \rho + g_0 = 0,$$

2.9.6
$$\alpha_j = (f_1 \rho_j + g_1)/(f_0 \rho_j + 2g_0),$$

$a_{0,j} = 1$, and

2.9.7
$$\rho_j(f_0 + 2\rho_j) s a_{s,j} = \sum_{r=1}^{s} \left(\rho_j^2 2^{r+1} \binom{\alpha_j + r - s}{r+1} + \rho_j \sum_{q=0}^{r+1} \binom{\alpha_j + r - s}{r+1-q} f_q + g_{r+1} \right) a_{s-r,j},$$

$s = 1, 2, 3, \ldots$. The construction fails iff $\rho_1 = \rho_2$, that is, when $f_0^2 = 4g_0$.

When $f_0^2 \neq 4g_0$, there are linearly independent solutions $w_j(n)$, $j = 1, 2$, such that

2.9.8
$$w_j(n) \sim \rho_j^n n^{\alpha_j} \sum_{s=0}^{\infty} \frac{a_{s,j}}{n^s}, \quad n \to \infty.$$

If $|\rho_2| > |\rho_1|$, or if $|\rho_2| = |\rho_1|$ and $\Re\alpha_2 > \Re\alpha_1$, then $w_1(n)$ is recessive and $w_2(n)$ is dominant as $n \to \infty$. As in the case of differential equations (§§2.7(iii), 2.7(iv)) recessive solutions are unique and dominant solutions are not; furthermore, one member of a numerically satisfactory pair has to be recessive. When $|\rho_2| = |\rho_1|$ and $\Re\alpha_2 = \Re\alpha_1$ neither solution is dominant and both are unique.

For proofs see Wong and Li (1992a). For error bounds see Zhang et al. (1996). See also Olver (1967b).

For asymptotic expansions in inverse factorial series see Olde Daalhuis (2004a).

2.9(ii) Coincident Characteristic Values

When the roots of (2.9.5) are equal we denote them both by ρ. Assume first $2g_1 \neq f_0 f_1$. Then (2.9.1) has independent solutions $w_j(n)$, $j = 1, 2$, such that

2.9.9
$$w_j(n) \sim \rho^n \exp\big((-1)^j \kappa \sqrt{n}\big) n^\alpha \sum_{s=0}^{\infty} (-1)^{js} \frac{c_s}{n^{s/2}},$$

where

2.9.10
$$\sqrt{g_0}\kappa = \sqrt{2 f_0 f_1 - 4 g_1}, \quad 4 g_0 \alpha = g_0 + 2 g_1,$$

$c_0 = 1$, and higher coefficients are determined by formal substitution.

Alternatively, suppose that $2g_1 = f_0 f_1$. Then the indices α_1, α_2 are the roots of

2.9.11 $\quad 2 g_0 \alpha^2 - (f_0 f_1 + 2 g_0) \alpha + 2 g_2 - f_0 f_2 = 0.$

Provided that $\alpha_2 - \alpha_1$ is not zero or an integer, (2.9.1) has independent solutions $w_j(n)$, $j = 1, 2$, of the form

2.9.12
$$w_j(n) \sim \rho^n n^{\alpha_j} \sum_{s=0}^{\infty} \frac{a_{s,j}}{n^s}, \quad n \to \infty,$$

with $a_{0,j} = 1$ and higher coefficients given by (2.9.7) (in the present case the coefficients of $a_{s,j}$ and $a_{s-1,j}$ are zero).

If $\alpha_2 - \alpha_1 = 0, 1, 2, \ldots$, then (2.9.12) applies only in the case $j = 1$. But there is an independent solution

2.9.13
$$w_2(n) \sim \rho^n n^{\alpha_2} \sum_{\substack{s=0 \\ s \neq \alpha_2 - \alpha_1}}^{\infty} \frac{b_s}{n^s} + c w_1(n) \ln n, \quad n \to \infty.$$

The coefficients b_s and constant c are again determined by formal substitution, beginning with $c = 1$ when $\alpha_2 - \alpha_1 = 0$, or with $b_0 = 1$ when $\alpha_2 - \alpha_1 = 1, 2, 3, \ldots$. (Compare (2.7.6).)

For proofs and examples, see Wong and Li (1992a). For error bounds see Zhang et al. (1996).

For analogous results for difference equations of the form

2.9.14 $\quad w(n+2) + n^P f(n) w(n+1) + n^Q g(n) w(n) = 0,$

in which P and Q are any integers see Wong and Li (1992b).

2.9(iii) Other Approximations

For asymptotic approximations to solutions of second-order difference equations analogous to the Liouville–Green (WKBJ) approximation for differential equations (§2.7(iii)) see Spigler and Vianello (1992, 1997) and Spigler et al. (1999). Error bounds and applications are included.

2.10 Sums and Sequences

For discussions of turning points, transition points, and uniform asymptotic expansions for solutions of linear difference equations of the second order see Wang and Wong (2003, 2005).

For an introduction to, and references for, the general asymptotic theory of linear difference equations of arbitrary order, see Wimp (1984, Appendix B).

2.10(i) Euler–Maclaurin Formula

As in §24.2, let B_n and $B_n(x)$ denote the nth Bernoulli number and polynomial, respectively, and $\widetilde{B}_n(x)$ the nth Bernoulli periodic function $B_n(x - \lfloor x \rfloor)$.

Assume that a, m, and n are integers such that $n > a$, $m > 0$, and $f^{(2m)}(x)$ is absolutely integrable over $[a, n]$. Then

2.10.1
$$\sum_{j=a}^{n} f(j) = \int_a^n f(x)\,dx + \tfrac{1}{2}f(a) + \tfrac{1}{2}f(n)$$
$$+ \sum_{s=1}^{m-1} \frac{B_{2s}}{(2s)!}\left(f^{(2s-1)}(n) - f^{(2s-1)}(a)\right)$$
$$+ \int_a^n \frac{B_{2m} - \widetilde{B}_{2m}(x)}{(2m)!} f^{(2m)}(x)\,dx.$$

This is the *Euler–Maclaurin formula*. Another version is the *Abel–Plana formula*:

2.10.2
$$\sum_{j=a}^{n} f(j) = \int_a^n f(x)\,dx + \tfrac{1}{2}f(a) + \tfrac{1}{2}f(n)$$
$$- 2\int_0^\infty \frac{\Im(f(a+iy))}{e^{2\pi y} - 1}\,dy$$
$$+ \sum_{s=1}^{m} \frac{B_{2s}}{(2s)!} f^{(2s-1)}(n)$$
$$+ 2\frac{(-1)^m}{(2m)!}\int_0^\infty \Im(f^{(2m)}(n + i\vartheta_n y))\frac{y^{2m}\,dy}{e^{2\pi y}-1},$$

ϑ_n being some number in the interval $(0,1)$. Sufficient conditions for the validity of this second result are:

(a) On the strip $a \leq \Re z \leq n$, $f(z)$ is analytic in its interior, $f^{(2m)}(z)$ is continuous on its closure, and $f(z) = o(e^{2\pi|\Im z|})$ as $\Im z \to \pm\infty$, uniformly with respect to $\Re z \in [a, n]$.

(b) $f(z)$ is real when $a \leq z \leq n$.

(c) The first infinite integral in (2.10.2) converges.

Example

2.10.3
$$S(n) = \sum_{j=1}^{n} j \ln j$$

for large n. From (2.10.1)

2.10.4
$$S(n) = \tfrac{1}{2}n^2 \ln n - \tfrac{1}{4}n^2 + \tfrac{1}{2}n\ln n + \tfrac{1}{12}\ln n + C$$
$$+ \sum_{s=2}^{m-1} \frac{(-B_{2s})}{2s(2s-1)(2s-2)} \frac{1}{n^{2s-2}} + R_m(n),$$

where $m\ (\geq 2)$ is arbitrary, C is a constant, and

2.10.5
$$R_m(n) = \int_n^\infty \frac{\widetilde{B}_{2m}(x) - B_{2m}}{2m(2m-1)x^{2m-1}}\,dx.$$

From §24.12(i), (24.2.2), and (24.4.27), $\widetilde{B}_{2m}(x) - B_{2m}$ is of constant sign $(-1)^m$. Thus $R_m(n)$ and $R_{m+1}(n)$ are of opposite signs, and since their difference is the term corresponding to $s = m$ in (2.10.4), $R_m(n)$ is bounded in absolute value by this term and has the same sign.

Formula (2.10.2) is useful for evaluating the constant term in expansions obtained from (2.10.1). In the present example it leads to

2.10.6
$$C = \frac{\gamma + \ln(2\pi)}{12} - \frac{\zeta'(2)}{2\pi^2} = \frac{1}{12} - \zeta'(-1),$$

where γ is Euler's constant (§5.2(ii)) and ζ' is the derivative of the Riemann zeta function (§25.2(i)). e^C is sometimes called *Glaisher's constant*. For further information on C see §5.17.

Other examples that can be verified in a similar way are:

2.10.7
$$\sum_{j=1}^{n-1} j^\alpha \sim \zeta(-\alpha) + \frac{n^{\alpha+1}}{\alpha+1}\sum_{s=0}^{\infty} \binom{\alpha+1}{s}\frac{B_s}{n^s}, \quad n \to \infty,$$

where $\alpha\ (\neq -1)$ is a real constant, and

2.10.8
$$\sum_{j=1}^{n-1} \frac{1}{j} \sim \ln n + \gamma - \frac{1}{2n} - \sum_{s=1}^{\infty} \frac{B_{2s}}{2s}\frac{1}{n^{2s}}, \quad n \to \infty.$$

In both expansions the remainder term is bounded in absolute value by the first neglected term in the sum, and has the same sign, provided that in the case of (2.10.7), truncation takes place at $s = 2m - 1$, where m is any positive integer satisfying $m \geq \tfrac{1}{2}(\alpha + 1)$.

For extensions of the Euler–Maclaurin formula to functions $f(x)$ with singularities at $x = a$ or $x = n$ (or both) see Sidi (2004). See also Weniger (2007).

For an extension to integrals with Cauchy principal values see Elliott (1998).

2.10(ii) Summation by Parts

The formula for summation by parts is

2.10.9
$$\sum_{j=1}^{n-1} u_j v_j = U_{n-1} v_n + \sum_{j=1}^{n-1} U_j (v_j - v_{j+1}),$$

where

2.10.10 $$U_j = u_1 + u_2 + \cdots + u_j.$$

This identity can be used to find asymptotic approximations for large n when the factor v_j changes slowly with j, and u_j is oscillatory; compare the approximation of Fourier integrals by integration by parts in §2.3(i).

Example

2.10.11 $$S(\alpha, \beta, n) = \sum_{j=1}^{n-1} e^{ij\beta} j^\alpha,$$

where α and β are real constants with $e^{i\beta} \neq 1$.

As a first estimate for large n

2.10.12 $$|S(\alpha, \beta, n)| \leq \sum_{j=1}^{n-1} j^\alpha = O(1), \, O(\ln n), \text{ or } O(n^{\alpha+1}),$$

according as $\alpha < -1$, $\alpha = -1$, or $\alpha > -1$; see (2.10.7), (2.10.8). With $u_j = e^{ij\beta}$, $v_j = j^\alpha$,

2.10.13 $$U_j = e^{i\beta}(e^{ij\beta} - 1)/(e^{i\beta} - 1),$$

and

2.10.14 $$S(\alpha, \beta, n) = \frac{e^{i\beta}}{e^{i\beta} - 1} \left(e^{i(n-1)\beta} n^\alpha - 1 + \sum_{j=1}^{n-1} e^{ij\beta} (j^\alpha - (j+1)^\alpha) \right).$$

Since

2.10.15 $j^\alpha - (j+1)^\alpha = -\alpha j^{\alpha-1} + \alpha(\alpha - 1) O(j^{\alpha-2})$

for any real constant α and the set of all positive integers j, we derive

2.10.16 $$S(\alpha, \beta, n) = \frac{e^{i\beta}}{e^{i\beta} - 1} \left(e^{i(n-1)\beta} n^\alpha - \alpha S(\alpha - 1, \beta, n) + O(n^{\alpha-1}) + O(1) \right).$$

From this result and (2.10.12)

2.10.17 $S(\alpha, \beta, n) = O(n^\alpha) + O(1).$

Then replacing α by $\alpha - 1$ and resubstituting in (2.10.16), we have

2.10.18 $$S(\alpha, \beta, n) = \frac{e^{in\beta}}{e^{i\beta} - 1} n^\alpha + O(n^{\alpha-1}) + O(1), \quad n \to \infty,$$

which is a useful approximation when $\alpha > 0$.

For extensions to $\alpha \leq 0$, higher terms, and other examples, see Olver (1997b, Chapter 8).

2.10(iii) Asymptotic Expansions of Entire Functions

The asymptotic behavior of entire functions defined by Maclaurin series can be approached by converting the sum into a contour integral by use of the residue theorem and applying the methods of §§2.4 and 2.5.

Example

From §§16.2(i)–16.2(ii)

2.10.19 $$_0F_2(-; 1, 1; x) = \sum_{j=0}^{\infty} \frac{x^j}{(j!)^3}.$$

We seek the behavior as $x \to +\infty$. From (1.10.8)

2.10.20 $$\sum_{j=0}^{n-1} \frac{x^j}{(j!)^3} = \frac{1}{2i} \int_{\mathscr{C}} \frac{x^t}{(\Gamma(t+1))^3} \cot(\pi t) \, dt,$$

where $\mathscr{C}$ comprises the two semicircles and two parts of the imaginary axis depicted in Figure 2.10.1.

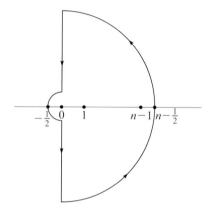

Figure 2.10.1: t-plane. Contour $\mathscr{C}$.

From the identities

2.10.21 $$\frac{\cot(\pi t)}{2i} = -\frac{1}{2} - \frac{1}{e^{-2\pi it} - 1} = \frac{1}{2} + \frac{1}{e^{2\pi it} - 1},$$

and Cauchy's theorem, we have

2.10.22 $$\sum_{j=0}^{n-1} \frac{x^j}{(j!)^3} = \int_{-1/2}^{n-(1/2)} \frac{x^t}{(\Gamma(t+1))^3} \, dt$$
$$- \int_{\mathscr{C}_1} \frac{x^t}{(\Gamma(t+1))^3} \frac{dt}{e^{-2\pi it} - 1}$$
$$+ \int_{\mathscr{C}_2} \frac{x^t}{(\Gamma(t+1))^3} \frac{dt}{e^{2\pi it} - 1},$$

where $\mathscr{C}_1, \mathscr{C}_2$ denote respectively the upper and lower halves of $\mathscr{C}$. (5.11.7) shows that the integrals around the large quarter circles vanish as $n \to \infty$. Hence

2.10.23 $$_0F_2(-; 1, 1; x) = \int_{-1/2}^{\infty} \frac{x^t}{(\Gamma(t+1))^3} \, dt$$
$$+ 2\Re \int_{-1/2}^{i\infty} \frac{x^t}{(\Gamma(t+1))^3} \frac{dt}{e^{-2\pi it} - 1}$$
$$= \int_0^{\infty} \frac{x^t}{(\Gamma(t+1))^3} \, dt + O(1),$$
$$x \to +\infty,$$

the last step following from $|x^t| \leq 1$ when t is on the interval $[-\frac{1}{2}, 0]$, the imaginary axis, or the small semicircle. By application of Laplace's method (§2.3(iii)) and use again of (5.11.7), we obtain

2.10.24 $\quad {}_0F_2(-;1,1;x) \sim \dfrac{\exp(3x^{1/3})}{2\pi 3^{1/2} x^{1/3}}, \quad x \to +\infty.$

For generalizations and other examples see Olver (1997b, Chapter 8), Ford (1960), and Berndt and Evans (1984). See also Paris and Kaminski (2001, Chapter 5) and §§16.11(i)–16.11(ii).

2.10(iv) Taylor and Laurent Coefficients: Darboux's Method

Let $f(z)$ be analytic on the annulus $0 < |z| < r$, with Laurent expansion

2.10.25 $\quad f(z) = \displaystyle\sum_{n=-\infty}^{\infty} f_n z^n, \quad 0 < |z| < r.$

What is the asymptotic behavior of f_n as $n \to \infty$ or $n \to -\infty$? More specially, what is the behavior of the higher coefficients in a Taylor-series expansion?

These problems can be brought within the scope of §2.4 by means of Cauchy's integral formula

2.10.26 $\quad f_n = \dfrac{1}{2\pi i} \displaystyle\int_{\mathscr{C}} \dfrac{f(z)}{z^{n+1}} \, dz,$

where $\mathscr{C}$ is a simple closed contour in the annulus that encloses $z = 0$. For examples see Olver (1997b, Chapters 8, 9).

However, if r is finite and $f(z)$ has algebraic or logarithmic singularities on $|z| = r$, then *Darboux's method* is usually easier to apply. We need a "comparison function" $g(z)$ with the properties:

(a) $g(z)$ is analytic on $0 < |z| < r$.

(b) $f(z) - g(z)$ is continuous on $0 < |z| \leq r$.

(c) The coefficients in the Laurent expansion

2.10.27 $\quad g(z) = \displaystyle\sum_{n=-\infty}^{\infty} g_n z^n, \quad 0 < |z| < r,$

have known asymptotic behavior as $n \to \pm\infty$.

By allowing the contour in Cauchy's formula to expand, we find that

2.10.28
$$f_n - g_n = \dfrac{1}{2\pi i} \int_{|z|=r} \dfrac{f(z) - g(z)}{z^{n+1}} \, dz$$
$$= \dfrac{1}{2\pi r^n} \int_0^{2\pi} \left(f(re^{i\theta}) - g(re^{i\theta}) \right) e^{-ni\theta} \, d\theta.$$

Hence by the Riemann–Lebesgue lemma (§1.8(i))

2.10.29 $\quad f_n = g_n + o(r^{-n}), \quad n \to \pm\infty.$

This result is refinable in two important ways. First, the conditions can be weakened. It is unnecessary for $f(z) - g(z)$ to be continuous on $|z| = r$: it suffices that the integrals in (2.10.28) converge uniformly. For example, Condition (b) can be replaced by:

(b′) On the circle $|z| = r$, the function $f(z) - g(z)$ has a finite number of singularities, and at each singularity z_j, say,

2.10.30 $\quad f(z) - g(z) = O\!\left((z - z_j)^{\sigma_j - 1}\right), \quad z \to z_j,$

where σ_j is a positive constant.

Secondly, when $f(z) - g(z)$ is m times continuously differentiable on $|z| = r$ the result (2.10.29) can be strengthened. In these circumstances the integrals in (2.10.28) are integrable by parts m times, yielding

2.10.31 $\quad f_n = g_n + o\!\left(r^{-n}|n|^{-m}\right), \quad n \to \pm\infty.$

Furthermore, (2.10.31) remains valid with the weaker condition

2.10.32 $\quad f^{(m)}(z) - g^{(m)}(z) = O\!\left((z - z_j)^{\sigma_j - 1}\right),$

in the neighborhood of each singularity z_j, again with $\sigma_j > 0$.

Example

Let α be a constant in $(0, 2\pi)$ and P_n denote the Legendre polynomial of degree n. From §14.7(iv)

2.10.33
$$f(z) \equiv \dfrac{1}{(1 - 2z\cos\alpha + z^2)^{1/2}}$$
$$= \sum_{n=0}^{\infty} P_n(\cos\alpha) z^n, \quad |z| < 1.$$

The singularities of $f(z)$ on the unit circle are branch points at $z = e^{\pm i\alpha}$. To match the limiting behavior of $f(z)$ at these points we set

2.10.34
$$g(z) = e^{-\pi i/4}(2\sin\alpha)^{-1/2}\left(e^{-i\alpha} - z\right)^{-1/2}$$
$$+ e^{\pi i/4}(2\sin\alpha)^{-1/2}\left(e^{i\alpha} - z\right)^{-1/2}.$$

Here the branch of $\left(e^{-i\alpha} - z\right)^{-1/2}$ is continuous in the z-plane cut along the outward-drawn ray through $z = e^{-i\alpha}$ and equals $e^{i\alpha/2}$ at $z = 0$. Similarly for $\left(e^{i\alpha} - z\right)^{-1/2}$. In Condition (c) we have

2.10.35 $\quad g_n = \left(\dfrac{2}{\pi \sin\alpha}\right)^{1/2} \dfrac{\Gamma\!\left(n + \tfrac{1}{2}\right)}{n!} \cos\!\left(n\alpha + \tfrac{1}{2}\alpha - \tfrac{1}{4}\pi\right),$

and in the supplementary conditions we may set $m = 1$. Then from (2.10.31) and (5.11.7)

2.10.36
$$P_n(\cos\alpha) = \left(\dfrac{2}{\pi n \sin\alpha}\right)^{1/2} \cos\!\left(n\alpha + \tfrac{1}{2}\alpha - \tfrac{1}{4}\pi\right) + o(n^{-1}).$$

For higher terms see §18.15(iii).

For uniform expansions when two singularities coalesce on the circle of convergence see Wong and Zhao (2005).

For other examples and extensions see Olver (1997b, Chapter 8), Olver (1970), Wong (1989, Chapter 2), and Wong and Wyman (1974). See also Flajolet and Odlyzko (1990).

2.11 Remainder Terms; Stokes Phenomenon

2.11(i) Numerical Use of Asymptotic Expansions

When a rigorous bound or reliable estimate for the remainder term is unavailable, it is unsafe to judge the accuracy of an asymptotic expansion merely from the numerical rate of decrease of the terms at the point of truncation. Even when the series converges this is unwise: the tail needs to be majorized rigorously before the result can be guaranteed. For divergent expansions the situation is even more difficult. First, it is impossible to bound the tail by majorizing its terms. Secondly, the asymptotic series represents an infinite class of functions, and the remainder depends on which member we have in mind.

As an example consider

2.11.1 $$I(m) = \int_0^\pi \frac{\cos(mt)}{t^2+1}\,dt,$$

with m a large integer. By integration by parts (§2.3(i))

2.11.2 $$I(m) \sim (-1)^m \sum_{s=1}^\infty \frac{q_s(\pi)}{m^{2s}}, \qquad m \to \infty,$$

with

2.11.3 $$q_1(t) = -\frac{2t}{(t^2+1)^2}, \quad q_2(t) = \frac{24(t^3-t)}{(t^2+1)^4},$$
$$q_3(t) = -\frac{240(3t^5 - 10t^3 + 3t)}{(t^2+1)^6}.$$

On rounding to 5D, we have $q_1(\pi) = -0.05318$, $q_2(\pi) = 0.04791$, $q_3(\pi) = -0.08985$. Hence

2.11.4 $$I(10) \sim -0.00053\,18 + 0.00000\,48 - 0.00000\,01$$
$$= -0.0052\,71.$$

But this answer is incorrect: to 7D $I(10) = -0.00045\,58$. The error term is, in fact, approximately 700 times the last term obtained in (2.11.4). The explanation is that (2.11.2) is a more accurate expansion for the function $I(m) - \frac{1}{2}\pi e^{-m}$ than it is for $I(m)$; see Olver (1997b, pp. 76–78).

In order to guard against this kind of error remaining undetected, the wanted function may need to be computed by another method (preferably nonasymptotic) for the smallest value of the (large) asymptotic variable x that is intended to be used. If the results agree within S significant figures, then it is likely—*but not certain*—that the truncated asymptotic series will yield at least S correct significant figures for larger values of x. For further discussion see Bosley (1996).

In $\mathbb{C}$ both the modulus and phase of the asymptotic variable z need to be taken into account. Suppose an asymptotic expansion holds as $z \to \infty$ in any closed sector within $\alpha < \mathrm{ph}\,z < \beta$, say, but not in $\alpha \le \mathrm{ph}\,z \le \beta$. Then numerical accuracy will disintegrate as the boundary rays $\mathrm{ph}\,z = \alpha$, $\mathrm{ph}\,z = \beta$ are approached. In consequence, practical application needs to be confined to a sector $\alpha' \le \mathrm{ph}\,z \le \beta'$ well within the sector of validity, and independent evaluations carried out on the boundaries for the smallest value of $|z|$ intended to be used. The choice of α' and β' is facilitated by a knowledge of the relevant Stokes lines; see §2.11(iv) below.

However, regardless whether we can bound the remainder, the accuracy achievable by direct numerical summation of a divergent asymptotic series is always limited. The rest of this section is devoted to general methods for increasing this accuracy.

2.11(ii) Connection Formulas

From §8.19(i) the generalized exponential integral is given by

2.11.5 $$E_p(z) = \frac{e^{-z}z^{p-1}}{\Gamma(p)} \int_0^\infty \frac{e^{-zt}t^{p-1}}{1+t}\,dt$$

when $\Re p > 0$ and $|\mathrm{ph}\,z| < \frac{1}{2}\pi$, and by analytic continuation for other values of p and z. Application of Watson's lemma (§2.4(i)) yields

2.11.6 $$E_p(z) \sim \frac{e^{-z}}{z} \sum_{s=0}^\infty (-1)^s \frac{(p)_s}{z^s}$$

when p is fixed and $z \to \infty$ in any closed sector within $|\mathrm{ph}\,z| < \frac{3}{2}\pi$. As noted in §2.11(i), poor accuracy is yielded by this expansion as $\mathrm{ph}\,z$ approaches $\frac{3}{2}\pi$ or $-\frac{3}{2}\pi$. However, on combining (2.11.6) with the connection formula (8.19.18), with $m = 1$, we derive

2.11.7 $$E_p(z) \sim \frac{2\pi i e^{-p\pi i}}{\Gamma(p)} z^{p-1} + \frac{e^{-z}}{z}\sum_{s=0}^\infty (-1)^s \frac{(p)_s}{z^s},$$

valid as $z \to \infty$ in any closed sector within $\frac{1}{2}\pi < \mathrm{ph}\,z < \frac{7}{2}\pi$; compare (8.20.3). Since the ray $\mathrm{ph}\,z = \frac{3}{2}\pi$ is well away from the new boundaries, the compound expansion (2.11.7) yields much more accurate results when $\mathrm{ph}\,z \to \frac{3}{2}\pi$. In effect, (2.11.7) "corrects" (2.11.6) by introducing a term that is relatively exponentially small in the neighborhood of $\mathrm{ph}\,z = \pi$, is increasingly significant as $\mathrm{ph}\,z$ passes from π to $\frac{3}{2}\pi$, and becomes the dominant contribution after $\mathrm{ph}\,z$ passes $\frac{3}{2}\pi$. See also §2.11(iv).

2.11(iii) Exponentially-Improved Expansions

The procedure followed in §2.11(ii) enabled $E_p(z)$ to be computed with as much accuracy in the sector $\pi \leq \mathrm{ph}\, z \leq 3\pi$ as the original expansion (2.11.6) in $|\mathrm{ph}\, z| \leq \pi$. We now increase substantially the accuracy of (2.11.6) in $|\mathrm{ph}\, z| \leq \pi$ by re-expanding the remainder term.

Optimum truncation in (2.11.6) takes place at $s = n - 1$, with $|p + n - 1| = |z|$, approximately. Thus

2.11.8 $$n = \rho - p + \alpha,$$

where $z = \rho e^{i\theta}$, and $|\alpha|$ is bounded as $n \to \infty$. From (2.11.5) and the identity

2.11.9 $$\frac{1}{1+t} = \sum_{s=0}^{n-1}(-1)^s t^s + (-1)^n \frac{t^n}{1+t}, \quad t \neq -1,$$

we have

2.11.10
$$E_p(z) = \frac{e^{-z}}{z}\sum_{s=0}^{n-1}(-1)^s \frac{(p)_s}{z^s} + (-1)^n \frac{2\pi}{\Gamma(p)} z^{p-1} F_{n+p}(z),$$

where

2.11.11
$$F_{n+p}(z) = \frac{e^{-z}}{2\pi}\int_0^\infty \frac{e^{-zt}t^{n+p-1}}{1+t}\,dt = \frac{\Gamma(n+p)}{2\pi}\frac{E_{n+p}(z)}{z^{n+p-1}}.$$

With n given by (2.11.8), we have

2.11.12
$$F_{n+p}(z) = \frac{e^{-z}}{2\pi}\int_0^\infty \exp\bigl(-\rho\,(te^{i\theta} - \ln t)\bigr)\frac{t^{\alpha-1}}{1+t}\,dt.$$

For large ρ the integrand has a saddle point at $t = e^{-i\theta}$. Following §2.4(iv), we rotate the integration path through an angle $-\theta$, which is valid by analytic continuation when $-\pi < \theta < \pi$. Then by application of Laplace's method (§§2.4(iii) and 2.4(iv)), we have

2.11.13
$$F_{n+p}(z) \sim \frac{e^{-i(\rho+\alpha)\theta}}{1+e^{-i\theta}}\frac{e^{-\rho-z}}{(2\pi\rho)^{1/2}}\sum_{s=0}^{\infty}\frac{a_{2s}(\theta,\alpha)}{\rho^s}, \quad \rho \to \infty,$$

uniformly when $\theta \in [-\pi+\delta, \pi-\delta]$ ($\delta > 0$) and $|\alpha|$ is bounded. The coefficients are rational functions of α and $1 + e^{i\theta}$, for example, $a_0(\theta, \alpha) = 1$, and

2.11.14
$$a_2(\theta, \alpha) = \frac{1}{12}(6\alpha^2 - 6\alpha + 1) - \frac{\alpha}{1+e^{i\theta}} + \frac{1}{(1+e^{i\theta})^2}.$$

Owing to the factor $e^{-\rho}$, that is, $e^{-|z|}$ in (2.11.13), $F_{n+p}(z)$ is uniformly exponentially small compared with $E_p(z)$. For this reason the expansion of $E_p(z)$ in $|\mathrm{ph}\,z| \leq \pi - \delta$ supplied by (2.11.8), (2.11.10), and (2.11.13) is said to be *exponentially improved*.

If we permit the use of nonelementary functions as approximants, then even more powerful re-expansions become available. One is uniformly valid for $-\pi + \delta \leq \mathrm{ph}\,z \leq 3\pi - \delta$ with bounded $|\alpha|$, and achieves uniform exponential improvement throughout $0 \leq \mathrm{ph}\,z \leq \pi$:

2.11.15
$$F_{n+p}(z) \sim (-1)^n i e^{-p\pi i}\Biggl(\tfrac{1}{2}\mathrm{erfc}\bigl(\sqrt{\tfrac{1}{2}\rho}\,c(\theta)\bigr) - i\frac{e^{i\rho(\pi-\theta)}e^{-\rho-z}}{(2\pi\rho)^{1/2}}\sum_{s=0}^{\infty}\frac{h_{2s}(\theta,\alpha)}{\rho^s}\Biggr).$$

Here erfc is the complementary error function (§7.2(i)), and

2.11.16 $$c(\theta) = \sqrt{2(1+e^{i\theta}+i(\theta-\pi))},$$

the branch being continuous with $c(\theta) \sim \pi - \theta$ as $\theta \to \pi$. Also,

2.11.17
$$h_{2s}(\theta,\alpha) = \frac{e^{i\alpha(\pi-\theta)}}{1+e^{-i\theta}}a_{2s}(\theta,\alpha) + (-1)^{s-1}i\frac{1\cdot 3\cdot 5\cdots(2s-1)}{(c(\theta))^{2s+1}},$$

with $a_{2s}(\theta,\alpha)$ as in (2.11.13), (2.11.14). In particular,

2.11.18 $$h_0(\theta,\alpha) = \frac{e^{i\alpha(\pi-\theta)}}{1+e^{-i\theta}} - \frac{i}{c(\theta)}.$$

For the sector $-3\pi + \delta \leq \mathrm{ph}\,z \leq \pi - \delta$ the conjugate result applies.

Further details for this example are supplied in Olver (1991a, 1994b). See also Paris and Kaminski (2001, Chapter 6), and Dunster (1996b, 1997).

2.11(iv) Stokes Phenomenon

Two different asymptotic expansions in terms of elementary functions, (2.11.6) and (2.11.7), are available for the generalized exponential integral in the sector $\tfrac{1}{2}\pi < \mathrm{ph}\,z < \tfrac{3}{2}\pi$. That the change in their forms is discontinuous, even though the function being approximated is analytic, is an example of the *Stokes phenomenon*. Where should the change-over take place? Can it be accomplished smoothly?

Satisfactory answers to these questions were found by Berry (1989); see also the survey by Paris and Wood (1995). These answers are linked to the terms involving the complementary error function in the more powerful expansions typified by the combination of (2.11.10) and (2.11.15). Thus if $0 \leq \theta \leq \pi - \delta$ ($< \pi$), then $c(\theta)$ lies in the right half-plane. Hence from §7.12(i) $\mathrm{erfc}\bigl(\sqrt{\tfrac{1}{2}\rho}\,c(\theta)\bigr)$ is of the same exponentially-small order of magnitude as the contribution from the other terms in (2.11.15) when ρ is large. On the other hand, when $\pi + \delta \leq \theta \leq 3\pi - \delta$, $c(\theta)$ is in the left half-plane and $\mathrm{erfc}\bigl(\sqrt{\tfrac{1}{2}\rho}\,c(\theta)\bigr)$ differs from 2 by an exponentially-small quantity. In the transition through $\theta = \pi$, $\mathrm{erfc}\bigl(\sqrt{\tfrac{1}{2}\rho}\,c(\theta)\bigr)$ changes very rapidly, but smoothly, from one form to the other; compare the graph of its modulus in Figure 2.11.1 in the case $\rho = 100$.

Figure 2.11.1: Graph of $|\operatorname{erfc}(\sqrt{50}\,c(\theta))|$.

In particular, on the ray $\theta = \pi$ greatest accuracy is achieved by (a) taking the average of the expansions (2.11.6) and (2.11.7), followed by (b) taking account of the exponentially-small contributions arising from the terms involving $h_{2s}(\theta,\alpha)$ in (2.11.15).

Rays (or curves) on which one contribution in a compound asymptotic expansion achieves maximum dominance over another are called *Stokes lines* ($\theta = \pi$ in the present example). As these lines are crossed exponentially-small contributions, such as that in (2.11.7), are "switched on" smoothly, in the manner of the graph in Figure 2.11.1.

For higher-order Stokes phenomena see Olde Daalhuis (2004b) and Howls et al. (2004).

2.11(v) Exponentially-Improved Expansions (continued)

Expansions similar to (2.11.15) can be constructed for many other special functions. However, to enjoy the resurgence property (§2.7(ii)) we often seek instead expansions in terms of the F-functions introduced in §2.11(iii), leaving the connection of the error-function type behavior as an implicit consequence of this property of the F-functions. In this context the F-functions are called *terminants*, a name introduced by Dingle (1973).

For illustration, we give re-expansions of the remainder terms in the expansions (2.7.8) arising in differential-equation theory. For notational convenience assume that the original differential equation (2.7.1) is normalized so that $\lambda_2 - \lambda_1 = 1$. (This means that, if necessary, z is replaced by $z/(\lambda_2 - \lambda_1)$.) From (2.7.12), (2.7.13) it is then seen that the optimum number of terms, n, in (2.7.14) is approximately $|z|$. We set

2.11.19 $$w_j(z) = e^{\lambda_j z} z^{\mu_j} \sum_{s=0}^{n-1} \frac{a_{s,j}}{z^s} + R_n^{(j)}(z), \quad j=1,2,$$

and expand

2.11.20 $$R_n^{(1)}(z) = (-1)^{n-1} i e^{(\mu_2-\mu_1)\pi i} e^{\lambda_2 z} z^{\mu_2} \left(C_1 \sum_{s=0}^{m-1} (-1)^s a_{s,2} \frac{F_{n+\mu_2-\mu_1-s}(z)}{z^s} + R_{m,n}^{(1)}(z) \right),$$

2.11.21 $$R_n^{(2)}(z) = (-1)^n i e^{(\mu_2-\mu_1)\pi i} e^{\lambda_1 z} z^{\mu_1} \left(C_2 \sum_{s=0}^{m-1} (-1)^s a_{s,1} \frac{F_{n+\mu_1-\mu_2-s}(ze^{-\pi i})}{z^s} + R_{m,n}^{(2)}(z) \right),$$

with $m = 0, 1, 2, \ldots$, and C_1, C_2 as in (2.7.17). Then as $z \to \infty$, with $|n - |z||$ bounded and m fixed,

2.11.22 $$R_{m,n}^{(1)}(z) = \begin{cases} O(e^{-|z|-z} z^{-m}), & |\operatorname{ph} z| \leq \pi, \\ O(z^{-m}), & \pi \leq |\operatorname{ph} z| \leq \tfrac{5}{2}\pi - \delta, \end{cases}$$

2.11.23 $$R_{m,n}^{(2)}(z) = \begin{cases} O(e^{-|z|+z} z^{-m}), & 0 \leq \operatorname{ph} z \leq 2\pi, \\ O(z^{-m}), & -\tfrac{3}{2}\pi + \delta \leq \operatorname{ph} z \leq 0 \text{ and } 2\pi \leq \operatorname{ph} z \leq \tfrac{7}{2}\pi - \delta, \end{cases}$$

uniformly with respect to $\operatorname{ph} z$ in each case.

The relevant Stokes lines are $\operatorname{ph} z = \pm \pi$ for $w_1(z)$, and $\operatorname{ph} z = 0, 2\pi$ for $w_2(z)$. In addition to achieving uniform exponential improvement, particularly in $|\operatorname{ph} z| \leq \pi$ for $w_1(z)$, and $0 \leq \operatorname{ph} z \leq 2\pi$ for $w_2(z)$, the re-expansions (2.11.20), (2.11.21) are resurgent.

For further details see Olde Daalhuis and Olver (1994). For error bounds see Dunster (1996c). For other examples see Boyd (1990b), Paris (1992a,b), and Wong and Zhao (2002b).

Often the process of re-expansion can be repeated any number of times. In this way we arrive at *hyperasymptotic expansions*. For integrals, see Berry and Howls (1991), Howls (1992), and Paris and Kaminski (2001, Chapter 6). For second-order differential equations, see Olde Daalhuis and Olver (1995a), Olde Daalhuis (1995, 1996), and Murphy and Wood (1997).

For higher-order differential equations, see Olde Daalhuis (1998a,b). The first of these two references also provides an introduction to the powerful Borel transform theory. In this connection see also Byatt-Smith (2000).

For nonlinear differential equations see Olde Daalhuis (2005a,b).

2.11(vi) Direct Numerical Transformations

The transformations in §3.9 for summing slowly convergent series can also be very effective when applied to divergent asymptotic series.

A simple example is provided by Euler's transformation (§3.9(ii)) applied to the asymptotic expansion for the exponential integral (§6.12(i)):

2.11.24 $\quad e^x E_1(x) \sim \sum_{s=0}^{\infty} (-1)^s \dfrac{s!}{x^{s+1}}, \quad x \to +\infty.$

Taking $x = 5$ and rounding to 5D, we obtain

2.11.25
$$e^5 E_1(5) = 0.20000 - 0.04000 + 0.01600 - 0.00960 \\ + 0.00768 - 0.00768 + 0.00922 - 0.01290 \\ + 0.02064 - 0.03716 + 0.07432 - \cdots.$$

The numerically smallest terms are the 5th and 6th. Truncation after 5 terms yields 0.17408, compared with the correct value

2.11.26 $\quad e^5 E_1(5) = 0.17042\ldots.$

We now compute the forward differences Δ^j, $j = 0, 1, 2, \ldots$, of the moduli of the rounded values of the first 6 neglected terms:

2.11.27
$$\Delta^0 = 0.00768, \quad \Delta^1 = 0.00154, \\ \Delta^2 = 0.00214, \quad \Delta^3 = 0.00192, \\ \Delta^4 = 0.00280, \quad \Delta^5 = 0.00434.$$

Multiplying these differences by $(-1)^j 2^{-j-1}$ and summing, we obtain

2.11.28
$$0.00384 - 0.00038 + 0.00027 - 0.00012 \\ + 0.00009 - 0.00007 = 0.00363.$$

Subtraction of this result from the sum of the first 5 terms in (2.11.25) yields 0.17045, which is much closer to the true value.

The process just used is equivalent to re-expanding the remainder term of the original asymptotic series (2.11.24) in powers of $1/(x+5)$ and truncating the new series optimally. Further improvements in accuracy can be realized by making a second application of the Euler transformation; see Olver (1997b, pp. 540–543).

Similar improvements are achievable by Aitken's Δ^2-process, Wynn's ϵ-algorithm, and other acceleration transformations. For a comprehensive survey see Weniger (1989).

The following example, based on Weniger (1996), illustrates their power.

For large $|z|$, with $|\operatorname{ph} z| \leq \frac{3}{2}\pi - \delta \;(<\frac{3}{2}\pi)$, the Whittaker function of the second kind has the asymptotic expansion (§13.19)

2.11.29 $\quad W_{\kappa,\mu}(z) \sim \sum_{n=0}^{\infty} a_n,$

in which

2.11.30 $\quad a_n = \dfrac{e^{-z/2}}{z^{n-\kappa} n!} \left(\mu^2 - (\kappa - \tfrac{1}{2})^2\right) \left(\mu^2 - (\kappa - \tfrac{3}{2})^2\right) \\ \cdots \left(\mu^2 - (\kappa - n + \tfrac{1}{2})^2\right).$

With $z = 1.0$, $\kappa = 2.3$, $\mu = 0.5$, the values of a_n to 8D are supplied in the second column of Table 2.11.1.

Table 2.11.1: Whittaker functions with Levin's transformation.

n	a_n	s_n	d_n
0	0.60653 066	0.60653 066	0.60653 066
1	−1.81352 667	−1.20699 601	−0.91106 488
2	0.35363 770	−0.85335 831	−0.82413 405
3	0.02475 464	−0.82860 367	−0.83323 429
4	−0.00736 451	−0.83596 818	−0.83303 750
5	0.00676 062	−0.82920 756	−0.83298 901
6	−0.01125 643	−0.84046 399	−0.83299 429
7	0.02796 418	−0.81249 981	−0.83299 530
8	−0.09364 504	−0.90614 485	−0.83299 504
9	0.39736 710	−0.50877 775	−0.83299 501
10	−2.05001 686	−2.55879 461	−0.83299 503

The next column lists the partial sums $s_n = a_0 + a_1 + \cdots + a_n$. Optimum truncation occurs just prior to the numerically smallest term, that is, at s_4. Comparison with the true value

2.11.31 $\quad W_{2.3, 0.5}(1.0) = -0.83299\,50268\,27526\cdots$

shows that this direct estimate is correct to almost 3D.

The fourth column of Table 2.11.1 gives the results of applying the following variant of *Levin's transformation*:

2.11.32 $\quad d_n = \dfrac{\sum_{j=0}^{n} (-1)^j \binom{n}{j} (j+1)^{n-1} \dfrac{s_j}{a_{j+1}}}{\sum_{j=0}^{n} (-1)^j \binom{n}{j} (j+1)^{n-1} \dfrac{1}{a_{j+1}}}.$

By $n = 10$ we already have 8 correct decimals. Furthermore, on proceeding to higher values of n with higher precision, much more accuracy is achievable. For example, using double precision d_{20} is found to agree with (2.11.31) to 13D.

However, direct numerical transformations need to be used with care. Their extrapolation is based on assumed forms of remainder terms that may not always be appropriate for asymptotic expansions. For example, extrapolated values may converge to an accurate value on one side of a Stokes line (§2.11(iv)), and converge to a quite inaccurate value on the other.

References

General References

The main references used in writing this chapter are Olver (1997b) and Wong (1989).

For additional bibliographic reading see Bender (1974), Bleistein and Handelsman (1975), Copson (1965), de Bruijn (1961), Dingle (1973), Erdélyi (1956), Jones (1972, 1997), Lauwerier (1974), Odlyzko (1995), Paris and Kaminski (2001), Slavyanov and Lay (2000), Temme (1995c), and Wasow (1965).

Sources

The following list gives the references or other indications of proofs that were used in constructing the various sections of this chapter. These sources supplement the references that are quoted in the text.

§2.1 Olver (1997b, Chapter 1).

§2.2 Olver (1997b, pp. 11–16), Fabijonas and Olver (1999).

§2.3 Olver (1997b, Chapter 3). For (2.3.9) see Wong (1989, §2.2). For (2.3.12) use termwise integration in an analogous manner to that used to prove Watson's lemma (Olver (1997b, pp. 71–72). (2.3.18) follows from (1.10.15) and (1.10.17) with $f(t) = p(t)$, $g(t) = q(t)/\left(p'(t)(p(t)-p(a))^{(\lambda/\mu)-1}\right)$, using Cauchy's integral formula for the residue, and integrating by parts. See also Cicuta and Montaldi (1975).

§2.4 Olver (1997b, Chapter 4 and pp. 315–320), Wong (1989, p. 31).

§2.5 Wong (1989, Chapter 3), Doetsch (1955, §6.5).

§2.6 Wong (1989, Chapter 6).

§2.7 Olver (1997b, Chapters 5–7), Olver (1994a), Olde Daalhuis and Olver (1994), Olde Daalhuis (1998a).

§2.8 Olver (1997b, Chapters 10–12).

§2.10 Olver (1997b, Chapter 8).

§2.11 Olver (1997b, pp. 76–78 and 540–543), Olver (1991a), Weniger (1996). The computations in the example in §2.11(vi) were carried out at NIST.

Chapter 3

Numerical Methods

N. M. Temme[1]

Areas **72**

3.1 Arithmetics and Error Measures 72
3.2 Linear Algebra 73
3.3 Interpolation 75
3.4 Differentiation 77
3.5 Quadrature 78
3.6 Linear Difference Equations 85
3.7 Ordinary Differential Equations 88
3.8 Nonlinear Equations 90
3.9 Acceleration of Convergence 93
3.10 Continued Fractions 94
3.11 Approximation Techniques 96
3.12 Mathematical Constants 100

References **100**

---

[1] Centrum voor Wiskunde en Informatica, Department MAS, Amsterdam, The Netherlands.
Acknowledgments: The author thanks W. Gautschi and C. Brezinski for their valuable contributions, and F. W. J. Olver for assisting with the writing of §§3.6, 3.7, 3.8, and 3.11.
Copyright © 2009 National Institute of Standards and Technology. All rights reserved.

Areas

3.1 Arithmetics and Error Measures

3.1(i) Floating-Point Arithmetic

Computer arithmetic is described for the *binary* based system with base 2; another frequently used system is the *hexadecimal* system with base 16.

A nonzero *normalized binary floating-point machine number* x is represented as

3.1.1 $\qquad x = (-1)^s \cdot (b_0.b_1 b_2 \ldots b_{p-1}) \cdot 2^E, \qquad b_0 = 1,$

where s is equal to 1 or 0, each b_j, $j \geq 1$, is either 0 or 1, b_1 is the *most significant bit*, p ($\in \mathbb{N}$) is the number of significant bits b_j, b_{p-1} is the *least significant bit*, E is an integer called the *exponent*, $b_0.b_1 b_2 \ldots b_{p-1}$ is the *significand*, and $f = .b_1 b_2 \ldots b_{p-1}$ is the *fractional part*.

The set of *machine numbers* $\mathbb{R}_{\text{fl}}$ is the union of 0 and the set

3.1.2 $\qquad (-1)^s 2^E \sum_{j=0}^{p-1} b_j 2^{-j},$

with $b_0 = 1$ and all allowable choices of E, p, s, and b_j.

Let $E_{\min} \leq E \leq E_{\max}$ with $E_{\min} < 0$ and $E_{\max} > 0$. For given values of $E_{\min}$, $E_{\max}$, and p, the *format width in bits* N of a computer word is the total number of bits: the sign (one bit), the significant bits $b_1, b_2, \ldots, b_{p-1}$ ($p-1$ bits), and the bits allocated to the exponent (the remaining $N - p$ bits). The integers p, $E_{\min}$, and $E_{\max}$ are characteristics of the machine. The *machine epsilon* ϵ_M, that is, the distance between 1 and the next larger machine number with $E = 0$ is given by $\epsilon_M = 2^{-p+1}$. The *machine precision* is $\frac{1}{2}\epsilon_M = 2^{-p}$. The lower and upper bounds for the absolute values of the nonzero machine numbers are given by

3.1.3 $\qquad N_{\min} \equiv 2^{E_{\min}} \leq |x| \leq 2^{E_{\max}+1}\left(1 - 2^{-p}\right) \equiv N_{\max}.$

Underflow (overflow) after computing $x \neq 0$ occurs when $|x|$ is smaller (larger) than $N_{\min}$ ($N_{\max}$).

IEEE Standard

The current standard is the ANSI/IEEE Standard 754; see IEEE (1985, §§1–4). In the case of normalized binary representation the memory positions for *single precision* ($N = 32$, $p = 24$, $E_{\min} = -126$, $E_{\max} = 127$) and *double precision* ($N = 64$, $p = 53$, $E_{\min} = -1022$, $E_{\max} = 1023$) are as in Figure 3.1.1. The respective machine precisions are $\frac{1}{2}\epsilon_M = 0.596 \times 10^{-7}$ and $\frac{1}{2}\epsilon_M = 0.111 \times 10^{-15}$.

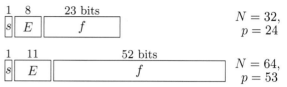

Figure 3.1.1: Floating-point arithmetic. Memory positions in single and double precision, in the case of binary representation.

Rounding

Let x be any positive number with

3.1.4 $\qquad x = (1.b_1 b_2 \ldots b_{p-1} b_p b_{p+1} \ldots) \cdot 2^E,$

$N_{\min} \leq x \leq N_{\max}$, and

3.1.5 $\qquad \begin{aligned} x_- &= (1.b_1 b_2 \ldots b_{p-1}) \cdot 2^E, \\ x_+ &= ((1.b_1 b_2 \ldots b_{p-1}) + \epsilon_M) \cdot 2^E. \end{aligned}$

Then *rounding by chopping* or *rounding down* of x gives x_-, with maximum relative error ϵ_M. *Symmetric rounding* or *rounding to nearest* of x gives x_- or x_+, whichever is nearer to x, with maximum relative error equal to the machine precision $\frac{1}{2}\epsilon_M = 2^{-p}$.

Negative numbers x are rounded in the same way as $-x$.

For further information see Goldberg (1991) and Overton (2001).

3.1(ii) Interval Arithmetic

Interval arithmetic is intended for bounding the total effect of rounding errors of calculations with machine numbers. With this arithmetic the computed result can be proved to lie in a certain interval, which leads to *validated computing* with guaranteed and rigorous inclusion regions for the results.

Let G be the set of closed intervals $\{[a, b]\}$. The elementary arithmetical operations on intervals are defined as follows:

3.1.6 $\qquad I * J = \{x * y \mid x \in I, y \in J\}, \qquad I, J \in G,$

where $* \in \{+, -, \cdot, /\}$, with appropriate roundings of the end points of $I * J$ when machine numbers are being used. Division is possible only if the divisor interval does not contain zero.

A basic text on interval arithmetic and analysis is Alefeld and Herzberger (1983), and for applications and further information see Moore (1979) and Petković and Petković (1998). The last reference includes analogs for arithmetic in the complex plane $\mathbb{C}$.

3.1(iii) Rational Arithmetics

Computer algebra systems use *exact rational arithmetic* with rational numbers p/q, where p and q are multi-length integers. During the calculations common divisors are removed from the rational numbers, and the final results can be converted to decimal representations of arbitrary length. For further information see Matula and Kornerup (1980).

3.1(iv) Level-Index Arithmetic

To eliminate overflow or underflow in finite-precision arithmetic numbers are represented by using *generalized logarithms* $\ln_\ell(x)$ given by

3.1.7 $\quad \ln_0(x) = x, \quad \ln_\ell(x) = \ln(\ln_{\ell-1}(x)), \quad \ell = 1, 2, \ldots,$

with $x \geq 0$ and ℓ the unique nonnegative integer such that $a \equiv \ln_\ell(x) \in [0, 1)$. In *level-index arithmetic* x is represented by $\ell + a$ (or $-(\ell + a)$ for negative numbers). Also in this arithmetic *generalized precision* can be defined, which includes absolute error and relative precision (§3.1(v)) as special cases.

For further information see Clenshaw and Olver (1984) and Clenshaw et al. (1989). For applications see Lozier (1993).

For further references on level-index arithmetic (and also other arithmetics) see Anuta et al. (1996). See also Hayes (2009).

3.1(v) Error Measures

If x^* is an approximation to a real or complex number x, then the *absolute error* is

3.1.8 $\quad\quad\quad\quad \epsilon_a = |x^* - x|.$

If $x \neq 0$, the *relative error* is

3.1.9 $\quad\quad\quad\quad \epsilon_r = \left|\dfrac{x^* - x}{x}\right| = \dfrac{\epsilon_a}{|x|}.$

The *relative precision* is

3.1.10 $\quad\quad\quad\quad \epsilon_{rp} = |\ln(x^*/x)|,$

where $xx^* > 0$ for real variables, and $xx^* \neq 0$ for complex variables (with the principal value of the logarithm).

The *mollified error* is

3.1.11 $\quad\quad\quad\quad \epsilon_m = \dfrac{|x^* - x|}{\max(|x|, 1)}.$

For error measures for complex arithmetic see Olver (1983).

3.2 Linear Algebra

3.2(i) Gaussian Elimination

To solve the system

3.2.1 $\quad\quad\quad\quad\quad \mathbf{A}\mathbf{x} = \mathbf{b},$

with Gaussian elimination, where $\mathbf{A}$ is a nonsingular $n \times n$ matrix and $\mathbf{b}$ is an $n \times 1$ vector, we start with the *augmented matrix*

3.2.2 $\quad \begin{bmatrix} a_{11} & \cdots & a_{1n} & b_1 \\ \vdots & \ddots & \vdots & \vdots \\ a_{n1} & \cdots & a_{nn} & b_n \end{bmatrix}.$

By repeatedly subtracting multiples of each row from the subsequent rows we obtain a matrix of the form

3.2.3 $\quad \begin{bmatrix} u_{11} & u_{12} & \cdots & u_{1n} & y_1 \\ 0 & u_{22} & \cdots & u_{2n} & y_2 \\ \vdots & \ddots & \ddots & \vdots & \vdots \\ 0 & \cdots & 0 & u_{nn} & y_n \end{bmatrix}.$

During this reduction process we store the *multipliers* ℓ_{jk} that are used in each column to eliminate other elements in that column. This yields a *lower triangular matrix* of the form

3.2.4 $\quad \mathbf{L} = \begin{bmatrix} 1 & 0 & \cdots & 0 \\ \ell_{21} & 1 & \cdots & 0 \\ \vdots & \ddots & \ddots & \vdots \\ \ell_{n1} & \cdots & \ell_{n,n-1} & 1 \end{bmatrix}.$

If we denote by $\mathbf{U}$ the upper triangular matrix comprising the elements u_{jk} in (3.2.3), then we have the factorization, or *triangular decomposition*,

3.2.5 $\quad\quad\quad\quad\quad \mathbf{A} = \mathbf{L}\mathbf{U}.$

With $\mathbf{y} = [y_1, y_2, \ldots, y_n]^{\mathrm{T}}$ the process of solution can then be regarded as first solving the equation $\mathbf{L}\mathbf{y} = \mathbf{b}$ for $\mathbf{y}$ (*forward elimination*), followed by the solution of $\mathbf{U}\mathbf{x} = \mathbf{y}$ for $\mathbf{x}$ (*back substitution*).

For more details see Golub and Van Loan (1996, pp. 87–100).

Example

3.2.6 $\quad \begin{bmatrix} 1 & 2 & 3 \\ 2 & 3 & 1 \\ 3 & 1 & 2 \end{bmatrix} = \begin{bmatrix} 1 & 0 & 0 \\ 2 & 1 & 0 \\ 3 & 5 & 1 \end{bmatrix} \begin{bmatrix} 1 & 2 & 3 \\ 0 & -1 & -5 \\ 0 & 0 & 18 \end{bmatrix}.$

In solving $\mathbf{A}\mathbf{x} = [1, 1, 1]^{\mathrm{T}}$, we obtain by forward elimination $\mathbf{y} = [1, -1, 3]^{\mathrm{T}}$, and by back substitution $\mathbf{x} = [\frac{1}{6}, \frac{1}{6}, \frac{1}{6}]^{\mathrm{T}}$.

In practice, if any of the multipliers ℓ_{jk} are unduly large in magnitude compared with unity, then Gaussian elimination is unstable. To avoid instability the rows are interchanged at each elimination step in such a way that the absolute value of the element that is used as a divisor, the *pivot element*, is not less than that of the other available elements in its column. Then $|\ell_{jk}| \leq 1$ in all cases. This modification is called *Gaussian elimination with partial pivoting*.

For more information on pivoting see Golub and Van Loan (1996, pp. 109–123).

Iterative Refinement

When the factorization (3.2.5) is available, the accuracy of the computed solution $\mathbf{x}$ can be improved with little extra computation. Because of rounding errors, the *residual vector* $\mathbf{r} = \mathbf{b} - \mathbf{A}\mathbf{x}$ is nonzero as a rule. We solve the system $\mathbf{A}\delta\mathbf{x} = \mathbf{r}$ for $\delta\mathbf{x}$, taking advantage of the existing triangular decomposition of $\mathbf{A}$ to obtain an improved solution $\mathbf{x} + \delta\mathbf{x}$.

3.2(ii) Gaussian Elimination for a Tridiagonal Matrix

Tridiagonal matrices are ones in which the only nonzero elements occur on the main diagonal and two adjacent diagonals. Thus

$$3.2.7 \quad \mathbf{A} = \begin{bmatrix} b_1 & c_1 & & & 0 \\ a_2 & b_2 & c_2 & & \\ & \ddots & \ddots & \ddots & \\ & & a_{n-1} & b_{n-1} & c_{n-1} \\ 0 & & & a_n & b_n \end{bmatrix}.$$

Assume that $\mathbf{A}$ can be factored as in (3.2.5), but without partial pivoting. Then

$$3.2.8 \quad \mathbf{L} = \begin{bmatrix} 1 & 0 & & & 0 \\ \ell_2 & 1 & 0 & & \\ & \ddots & \ddots & \ddots & \\ & & \ell_{n-1} & 1 & 0 \\ 0 & & & \ell_n & 1 \end{bmatrix},$$

$$3.2.9 \quad \mathbf{U} = \begin{bmatrix} d_1 & u_1 & & & 0 \\ 0 & d_2 & u_2 & & \\ & \ddots & \ddots & \ddots & \\ & & 0 & d_{n-1} & u_{n-1} \\ 0 & & & 0 & d_n \end{bmatrix},$$

where $u_j = c_j$, $j = 1, 2, \ldots, n-1$, $d_1 = b_1$, and

3.2.10 $\quad \ell_j = a_j/d_{j-1}, \quad d_j = b_j - \ell_j c_{j-1}, \quad j = 2, \ldots, n.$

Forward elimination for solving $\mathbf{A}\mathbf{x} = \mathbf{f}$ then becomes $y_1 = f_1$,

3.2.11 $\quad y_j = f_j - \ell_j y_{j-1}, \quad j = 2, \ldots, n,$

and back substitution is $x_n = y_n/d_n$, followed by

3.2.12 $\quad x_j = (y_j - u_j x_{j+1})/d_j, \quad j = n-1, \ldots, 1.$

For more information on solving tridiagonal systems see Golub and Van Loan (1996, pp. 152–160).

3.2(iii) Condition of Linear Systems

The *p-norm of a vector* $\mathbf{x} = [x_1, \ldots, x_n]^{\mathrm{T}}$ is given by

$$3.2.13 \quad \|\mathbf{x}\|_p = \left(\sum_{j=1}^{n} |x_j|^p\right)^{1/p}, \quad p = 1, 2, \ldots,$$
$$\|\mathbf{x}\|_\infty = \max_{1 \leq j \leq n} |x_j|.$$

The *Euclidean norm* is the case $p = 2$.

The *p-norm of a matrix* $\mathbf{A} = [a_{jk}]$ is

$$3.2.14 \quad \|\mathbf{A}\|_p = \max_{\mathbf{x} \neq \mathbf{0}} \frac{\|\mathbf{A}\mathbf{x}\|_p}{\|\mathbf{x}\|_p}.$$

The cases $p = 1, 2$, and ∞ are the most important:

$$\|\mathbf{A}\|_1 = \max_{1 \leq k \leq n} \sum_{j=1}^{n} |a_{jk}|,$$

$$3.2.15 \quad \|\mathbf{A}\|_\infty = \max_{1 \leq j \leq n} \sum_{k=1}^{n} |a_{jk}|,$$

$$\|\mathbf{A}\|_2 = \sqrt{\rho(\mathbf{A}\mathbf{A}^{\mathrm{T}})},$$

where $\rho(\mathbf{A}\mathbf{A}^{\mathrm{T}})$ is the largest of the absolute values of the eigenvalues of the matrix $\mathbf{A}\mathbf{A}^{\mathrm{T}}$; see §3.2(iv). (We are assuming that the matrix $\mathbf{A}$ is real; if not $\mathbf{A}^{\mathrm{T}}$ is replaced by $\mathbf{A}^{\mathrm{H}}$, the transpose of the complex conjugate of $\mathbf{A}$.)

The sensitivity of the solution vector $\mathbf{x}$ in (3.2.1) to small perturbations in the matrix $\mathbf{A}$ and the vector $\mathbf{b}$ is measured by the *condition number*

3.2.16 $\quad \kappa(\mathbf{A}) = \|\mathbf{A}\|_p \|\mathbf{A}^{-1}\|_p,$

where $\|\cdot\|_p$ is one of the matrix norms. For any norm (3.2.14) we have $\kappa(\mathbf{A}) \geq 1$. The larger the value $\kappa(\mathbf{A})$, the more ill-conditioned the system.

Let $\mathbf{x}^*$ denote a computed solution of the system (3.2.1), with $\mathbf{r} = \mathbf{b} - \mathbf{A}\mathbf{x}^*$ again denoting the residual. Then we have the *a posteriori* error bound

$$3.2.17 \quad \frac{\|\mathbf{x}^* - \mathbf{x}\|_p}{\|\mathbf{x}\|_p} \leq \kappa(\mathbf{A}) \frac{\|\mathbf{r}\|_p}{\|\mathbf{b}\|_p}.$$

For further information see Brezinski (1999) and Trefethen and Bau (1997, Chapter 3).

3.2(iv) Eigenvalues and Eigenvectors

If $\mathbf{A}$ is an $n \times n$ matrix, then a real or complex number λ is called an *eigenvalue* of $\mathbf{A}$, and a nonzero vector $\mathbf{x}$ a corresponding (*right*) *eigenvector*, if

3.2.18 $\quad \mathbf{A}\mathbf{x} = \lambda \mathbf{x}.$

A nonzero vector $\mathbf{y}$ is called a *left eigenvector* of $\mathbf{A}$ corresponding to the eigenvalue λ if $\mathbf{y}^{\mathrm{T}}\mathbf{A} = \lambda \mathbf{y}^{\mathrm{T}}$ or, equivalently, $\mathbf{A}^{\mathrm{T}}\mathbf{y} = \lambda\mathbf{y}$. A *normalized* eigenvector has Euclidean norm 1; compare (3.2.13) with $p = 2$.

The polynomial

3.2.19 $\quad p_n(\lambda) = \det[\lambda \mathbf{I} - \mathbf{A}]$

is called the *characteristic polynomial* of $\mathbf{A}$ and its zeros are the eigenvalues of $\mathbf{A}$. The *multiplicity* of an eigenvalue is its multiplicity as a zero of the characteristic polynomial (§3.8(i)). To an eigenvalue of multiplicity m, there correspond m linearly independent eigenvectors provided that $\mathbf{A}$ is *nondefective*, that is, $\mathbf{A}$ has a complete set of n linearly independent eigenvectors.

3.2(v) Condition of Eigenvalues

If $\mathbf{A}$ is nondefective and λ is a simple zero of $p_n(\lambda)$, then the sensitivity of λ to small perturbations in the matrix $\mathbf{A}$ is measured by the *condition number*

3.2.20 $$\kappa(\lambda) = \frac{1}{|\mathbf{y}^\mathrm{T}\mathbf{x}|},$$

where $\mathbf{x}$ and $\mathbf{y}$ are the normalized right and left eigenvectors of $\mathbf{A}$ corresponding to the eigenvalue λ. Because $|\mathbf{y}^\mathrm{T}\mathbf{x}| = |\cos\theta|$, where θ is the angle between $\mathbf{y}^\mathrm{T}$ and $\mathbf{x}$ we always have $\kappa(\lambda) \geq 1$. When $\mathbf{A}$ is a symmetric matrix, the left and right eigenvectors coincide, yielding $\kappa(\lambda) = 1$, and the calculation of its eigenvalues is a well-conditioned problem.

3.2(vi) Lanczos Tridiagonalization of a Symmetric Matrix

Define the *Lanczos vectors* $\mathbf{v}_j$ by $\mathbf{v}_0 = \mathbf{0}$, a normalized vector $\mathbf{v}_1$ (perhaps chosen randomly), and for $j = 1, 2, \ldots, n-1$,

3.2.21 $$\beta_{j+1}\mathbf{v}_{j+1} = \mathbf{A}\mathbf{v}_j - \alpha_j\mathbf{v}_j - \beta_j\mathbf{v}_{j-1},$$
$$\alpha_j = \mathbf{v}_j^\mathrm{T}\mathbf{A}\mathbf{v}_j, \quad \beta_{j+1} = \mathbf{v}_{j+1}^\mathrm{T}\mathbf{A}\mathbf{v}_j.$$

Then all $\mathbf{v}_j$, $1 \leq j \leq n$, are normalized and $\mathbf{v}_j^\mathrm{T}\mathbf{v}_k = 0$ for $j, k = 1, 2, \ldots, n$, $j \neq k$. The tridiagonal matrix

3.2.22 $$\mathbf{B} = \begin{bmatrix} \alpha_1 & \beta_2 & & & 0 \\ \beta_2 & \alpha_2 & \beta_3 & & \\ & \ddots & \ddots & \ddots & \\ & & \beta_{n-1} & \alpha_{n-1} & \beta_n \\ 0 & & & \beta_n & \alpha_n \end{bmatrix}$$

has the same eigenvalues as $\mathbf{A}$. Its characteristic polynomial can be obtained from the recursion

3.2.23 $$p_{k+1}(\lambda) = (\lambda - \alpha_{k+1})p_k(\lambda) - \beta_{k+1}^2 p_{k-1}(\lambda),$$
$$k = 0, 1, \ldots, n-1,$$

with $p_{-1}(\lambda) = 0$, $p_0(\lambda) = 1$.

For numerical information see Stewart (2001, pp. 347–368).

3.2(vii) Computation of Eigenvalues

Many methods are available for computing eigenvalues; see Golub and Van Loan (1996, Chapters 7, 8), Trefethen and Bau (1997, Chapter 5), and Wilkinson (1988, Chapters 8, 9).

3.3 Interpolation

3.3(i) Lagrange Interpolation

The *nodes* or *abscissas* z_k are real or complex; function values are $f_k = f(z_k)$. Given $n+1$ distinct points z_k and $n+1$ corresponding function values f_k, the *Lagrange interpolation polynomial* is the unique polynomial $P_n(z)$ of degree not exceeding n such that $P_n(z_k) = f_k$, $k = 0, 1, \ldots, n$. It is given by

3.3.1 $$P_n(z) = \sum_{k=0}^{n} \ell_k(z) f_k = \sum_{k=0}^{n} \frac{\omega_{n+1}(z)}{(z - z_k)\omega'_{n+1}(z_k)} f_k,$$

where

3.3.2 $$\ell_k(z) = \prod_{j=0}^{n}{}' \frac{z - z_j}{z_k - z_j}, \quad \ell_k(z_j) = \delta_{k,j}.$$

Here the prime signifies that the factor for $j = k$ is to be omitted, $\delta_{k,j}$ is the Kronecker symbol, and ω_{n+1} is the *nodal polynomial*

3.3.3 $$\omega_{n+1}(z) = \prod_{k=0}^{n}(z - z_k).$$

With an error term the *Lagrange interpolation formula* for f is given by

3.3.4 $$f(z) = \sum_{k=0}^{n} \ell_k(z) f_k + R_n(z).$$

If f, x ($= z$), and the nodes x_k are real, and $f^{(n+1)}$ is continuous on the smallest closed interval I containing $x, x_0, x_1, \ldots, x_n$, then the error can be expressed

3.3.5 $$R_n(x) = \frac{f^{(n+1)}(\xi)}{(n+1)!} \omega_{n+1}(x),$$

for some $\xi \in I$. If f is analytic in a simply-connected domain D (§1.13(i)), then for $z \in D$,

3.3.6 $$R_n(z) = \frac{\omega_{n+1}(z)}{2\pi i} \int_C \frac{f(\zeta)}{(\zeta - z)\omega_{n+1}(\zeta)} d\zeta,$$

where C is a simple closed contour in D described in the positive rotational sense and enclosing the points $z, z_1, z_2, \ldots, z_n$.

3.3(ii) Lagrange Interpolation with Equally-Spaced Nodes

The $(n+1)$-point formula (3.3.4) can be written in the form

3.3.7 $$f_t = f(x_0 + th) = \sum_{k=n_0}^{n_1} A_k^n f_k + R_{n,t}, \quad n_0 < t < n_1,$$

where the nodes $x_k = x_0 + kh$ ($h > 0$) and function f are real,

3.3.8 $$n_0 = -\tfrac{1}{2}(n - \sigma), \quad n_1 = \tfrac{1}{2}(n + \sigma),$$

3.3.9 $$\sigma = \tfrac{1}{2}(1 - (-1)^n),$$

and A_k^n are the *Lagrangian interpolation coefficients* defined by

3.3.10 $$A_k^n = \frac{(-1)^{n_1+k}}{(k - n_0)!(n_1 - k)!(t - k)} \prod_{m=n_0}^{n_1}(t - m).$$

The remainder is given by

3.3.11
$$R_{n,t} = R_n(x_0 + th) = \frac{h^{n+1}}{(n+1)!} f^{(n+1)}(\xi) \prod_{k=n_0}^{n_1} (t-k),$$

where ξ is as in §3.3(i).

Let c_n be defined by

3.3.12
$$c_n = \frac{1}{(n+1)!} \max \prod_{k=n_0}^{n_1} |t-k|,$$

where the maximum is taken over t-intervals given in the formulas below. Then for these t-intervals,

3.3.13 $\quad |R_{n,t}| \leq c_n h^{n+1} \left| f^{(n+1)}(\xi) \right|.$

Linear Interpolation

3.3.14 $\quad f_t = (1-t)f_0 + t f_1 + R_{1,t}, \qquad 0 < t < 1,$

3.3.15 $\quad c_1 = \frac{1}{8}, \qquad 0 < t < 1.$

Three-Point Formula

3.3.16 $\quad f_t = \sum_{k=-1}^{1} A_k^2 f_k + R_{2,t}, \qquad |t| < 1,$

3.3.17 $\quad A_{-1}^2 = \frac{1}{2}t(t-1), \quad A_0^2 = 1 - t^2, \quad A_1^2 = \frac{1}{2}t(t+1),$

3.3.18 $\quad c_2 = 1/(9\sqrt{3}) = 0.0641\ldots, \qquad |t| < 1.$

For four-point to eight-point formulas see http://dlmf.nist.gov/3.3.ii.

3.3(iii) Divided Differences

The *divided differences* of f relative to a sequence of distinct points $z_0, z_1, z_2, \ldots$ are defined by

3.3.34
$$\begin{aligned}[z_0]f &= f_0, \\ [z_0, z_1]f &= ([z_1]f - [z_0]f)/(z_1 - z_0), \\ [z_0, z_1, z_2]f &= ([z_1, z_2]f - [z_0, z_1]f)/(z_2 - z_0),\end{aligned}$$

and so on. Explicitly, the *divided difference of order n* is given by

3.3.35
$$[z_0, z_1, \ldots, z_n]f = \sum_{k=0}^{n} \left(f(z_k) \bigg/ \prod_{\substack{0 \leq j \leq n \\ j \neq k}} (z_k - z_j) \right).$$

If f and the z_k ($= x_k$) are real, and f is n times continuously differentiable on a closed interval containing the x_k, then

3.3.36 $\quad [x_0, x_1, \ldots, x_n]f = \dfrac{f^{(n)}(\xi)}{n!}$

and again ξ is as in §3.3(i). If f is analytic in a simply-connected domain D, then for $z \in D$,

3.3.37 $\quad [z_0, z_1, \ldots, z_n]f = \dfrac{1}{2\pi i} \int_C \dfrac{f(\zeta)}{\omega_{n+1}(\zeta)} d\zeta,$

where $\omega_{n+1}(\zeta)$ is given by (3.3.3), and C is a simple closed contour in D described in the positive rotational sense and enclosing $z_0, z_1, \ldots, z_n$.

3.3(iv) Newton's Interpolation Formula

This represents the Lagrange interpolation polynomial in terms of divided differences:

3.3.38
$$\begin{aligned}f(z) &= [z_0]f + (z - z_0)[z_0, z_1]f \\ &\quad + (z-z_0)(z-z_1)[z_0, z_1, z_2]f + \cdots \\ &\quad + (z-z_0)(z-z_1)\cdots(z-z_{n-1})[z_0, z_1, \ldots, z_n]f \\ &\quad + R_n(z).\end{aligned}$$

The interpolation error $R_n(z)$ is as in §3.3(i). Newton's formula has the advantage of allowing easy updating: incorporation of a new point z_{n+1} requires only addition of the term with $[z_0, z_1, \ldots, z_{n+1}]f$ to (3.3.38), plus the computation of this divided difference. Another advantage is its robustness with respect to confluence of the set of points $z_0, z_1, \ldots, z_n$. For example, for $k+1$ coincident points the limiting form is given by $[z_0, z_0, \ldots, z_0]f = f^{(k)}(z_0)/k!$.

3.3(v) Inverse Interpolation

In this method we interchange the roles of the points z_k and the function values f_k. It can be used for solving a nonlinear scalar equation $f(z) = 0$ approximately. Another approach is to combine the methods of §3.8 with direct interpolation and §3.4.

Example

To compute the first negative zero $a_1 = -2.33810\,7410\ldots$ of the Airy function $f(x) = \mathrm{Ai}(x)$ (§9.2). The inverse interpolation polynomial is given by

3.3.39 $\quad \begin{aligned}x(f) &= [f_0]x + (f - f_0)[f_0, f_1]x \\ &\quad + (f - f_0)(f - f_1)[f_0, f_1, f_2]x;\end{aligned}$

compare (3.3.38). With $x_0 = -2.2$, $x_1 = -2.3$, $x_2 = -2.4$, we obtain

3.3.40
$$\begin{aligned}x &= -2.2 \\ &\quad + 1.44011\,1973(f - 0.09614\,53780) + 0.08865\,85832 \\ &\quad \times (f - 0.09614\,53780)(f - 0.02670\,63331),\end{aligned}$$

and with $f = 0$ we find that $x = -2.33823\,2462$, with 4 correct digits. By using this approximation to x as a new point, $x_3 = x$, and evaluating $[f_0, f_1, f_2, f_3]x = 1.12388\,6190$, we find that $x = -2.33810\,7409$, with 9 correct digits.

For comparison, we use Newton's interpolation formula (3.3.38)

3.3.41 $\quad \begin{aligned}f(x) &= 0.09614\,53780 + 0.69439\,04495(x + 2.1) \\ &\quad - 0.03007\,14275(x + 2.2)(x + 2.3),\end{aligned}$

with the derivative

3.3.42 $\quad f'(x) = 0.55906\,90257 - 0.06014\,28550 x,$

and compute an approximation to a_1 by using Newton's rule (§3.8(ii)) with starting value $x = -2.5$. This gives the new point $x_3 = -2.33934\ 0514$. Then by using x_3 in Newton's interpolation formula, evaluating $[x_0, x_1, x_2, x_3]f = -0.26608\ 28233$ and recomputing $f'(x)$, another application of Newton's rule with starting value x_3 gives the approximation $x = 2.33810\ 7373$, with 8 correct digits.

3.3(vi) Other Interpolation Methods

For Hermite interpolation, trigonometric interpolation, spline interpolation, rational interpolation (by using continued fractions), interpolation based on Chebyshev points, and bivariate interpolation, see Bulirsch and Rutishauser (1968), Davis (1975, pp. 27–31), and Mason and Handscomb (2003, Chapter 6). These references also describe convergence properties of the interpolation formulas.

For interpolation of a bounded function f on $\mathbb{R}$ the *cardinal function* of f is defined by

3.3.43 $$C(f,h)(x) = \sum_{k=-\infty}^{\infty} f(kh) S(k,h)(x),$$

where

3.3.44 $$S(k,h)(x) = \frac{\sin(\pi(x-kh)/h)}{\pi(x-kh)/h},$$

is called the *Sinc function*. For theory and applications see Stenger (1993, Chapter 3).

3.4 Differentiation

3.4(i) Equally-Spaced Nodes

The Lagrange $(n+1)$-point formula is

3.4.1 $$hf'_t = hf'(x_0 + th) = \sum_{k=n_0}^{n_1} B_k^n f_k + hR'_{n,t}, \quad n_0 < t < n_1,$$

and follows from the differentiated form of (3.3.4). The B_k^n are the *differentiated Lagrangian interpolation coefficients*:

3.4.2 $$B_k^n = dA_k^n/dt,$$

where A_k^n is as in (3.3.10).

If $f^{(n+2)}(x)$ is continuous on the interval I defined in §3.3(i), then the remainder in (3.4.1) is given by

3.4.3 $$hR'_{n,t} = \frac{h^{n+1}}{(n+1)!} \left(f^{(n+1)}(\xi_0) \frac{d}{dt} \prod_{k=n_0}^{n_1}(t-k) + f^{(n+2)}(\xi_1) \prod_{k=n_0}^{n_1}(t-k) \right),$$

where ξ_0 and $\xi_1 \in I$.

For the values of n_0 and n_1 used in the formulas below

3.4.4 $$h\left|R'_{n,t}\right| \leq h^{n+1}\left(c_n \left|f^{(n+2)}(\xi_1)\right| + \frac{1}{n+1}\left|f^{(n+1)}(\xi_0)\right|\right),$$
$$n_0 < t < n_1,$$

where c_n is defined by (3.3.12), with numerical values as in §3.3(ii).

Two-Point Formula

3.4.5 $$hf'_t = -f_0 + f_1 + hR'_{1,t}, \qquad 0 < t < 1.$$

Three-Point Formula

3.4.6 $$hf'_t = -\tfrac{1}{2}(1-2t)f_{-1} - 2tf_0 + \tfrac{1}{2}(1+2t)f_1 + hR'_{2,t},$$
$$|t| < 1.$$

For four-point to eight-point formulas see http://dlmf.nist.gov/3.4.i.

For corresponding formulas for second, third, and fourth derivatives, with $t = 0$, see Collatz (1960, Table III, pp. 538–539). For formulas for derivatives with equally-spaced real nodes and based on Sinc approximations (§3.3(vi)), see Stenger (1993, §3.5).

3.4(ii) Analytic Functions

If f can be extended analytically into the complex plane, then from Cauchy's integral formula (§1.9(iii))

3.4.17 $$\frac{1}{k!}f^{(k)}(x_0) = \frac{1}{2\pi i} \int_C \frac{f(\zeta)}{(\zeta - x_0)^{k+1}}\,d\zeta,$$

where C is a simple closed contour described in the positive rotational sense such that C and its interior lie in the domain of analyticity of f, and x_0 is interior to C. Taking C to be a circle of radius r centered at x_0, we obtain

3.4.18 $$\frac{1}{k!}f^{(k)}(x_0) = \frac{1}{2\pi r^k}\int_0^{2\pi} f(x_0 + re^{i\theta})e^{-ik\theta}\,d\theta.$$

The integral on the right-hand side can be approximated by the composite trapezoidal rule (3.5.2).

Example

$f(z) = e^z$, $x_0 = 0$. The integral (3.4.18) becomes

3.4.19 $$\frac{1}{k!} = \frac{1}{2\pi r^k}\int_0^{2\pi} e^{r\cos\theta}\cos(r\sin\theta - k\theta)\,d\theta.$$

With the choice $r = k$ (which is crucial when k is large because of numerical cancellation) the integrand equals e^k at the dominant points $\theta = 0, 2\pi$, and in combination with the factor k^{-k} in front of the integral sign this gives a rough approximation to $1/k!$. The choice $r = k$ is motivated by saddle-point analysis; see §2.4(iv) or examples in §3.5(ix). As explained in §§3.5(i) and 3.5(ix) the composite trapezoidal rule can be very efficient for computing integrals with analytic periodic integrands.

3.4(iii) Partial Derivatives

First-Order

For partial derivatives we use the notation $u_{t,s} = u(x_0 + th, y_0 + sh)$.

3.4.20 $\quad \dfrac{\partial u_{0,0}}{\partial x} = \dfrac{1}{2h}(u_{1,0} - u_{-1,0}) + O(h^2),$

3.4.21
$$\dfrac{\partial u_{0,0}}{\partial x} = \dfrac{1}{4h}(u_{1,1} - u_{-1,1} + u_{1,-1} - u_{-1,-1}) + O(h^2).$$

Second-Order

3.4.22 $\quad \dfrac{\partial^2 u_{0,0}}{\partial x^2} = \dfrac{1}{h^2}(u_{1,0} - 2u_{0,0} + u_{-1,0}) + O(h^2),$

3.4.23 $\quad \dfrac{\partial^2 u_{0,0}}{\partial x^2} = \dfrac{1}{12h^2}(-u_{2,0} + 16u_{1,0} - 30u_{0,0}$
$\qquad\qquad\qquad + 16u_{-1,0} - u_{-2,0}) + O(h^4),$

3.4.24
$$\dfrac{\partial^2 u_{0,0}}{\partial x^2} = \dfrac{1}{3h^2}(u_{1,1} - 2u_{0,1} + u_{-1,1} + u_{1,0} - 2u_{0,0} + u_{-1,0}$$
$$+ u_{1,-1} - 2u_{0,-1} + u_{-1,-1}) + O(h^2).$$

3.4.25 $\quad \dfrac{\partial^2 u_{0,0}}{\partial x\,\partial y} = \dfrac{1}{4h^2}(u_{1,1} - u_{1,-1} - u_{-1,1} + u_{-1,-1}) + O(h^2),$

3.4.26 $\quad \dfrac{\partial^2 u_{0,0}}{\partial x\,\partial y} = -\dfrac{1}{2h^2}(u_{1,0} + u_{-1,0} + u_{0,1} + u_{0,-1} - 2u_{0,0}$
$\qquad\qquad\qquad - u_{1,1} - u_{-1,-1}) + O(h^2).$

Laplacian

3.4.27 $\quad \nabla^2 u = \dfrac{\partial^2 u}{\partial x^2} + \dfrac{\partial^2 u}{\partial y^2}.$

3.4.28 $\quad \nabla^2 u_{0,0} = \dfrac{1}{h^2}(u_{1,0} + u_{0,1} + u_{-1,0} + u_{0,-1} - 4u_{0,0})$
$\qquad\qquad + O(h^2),$

3.4.29
$$\nabla^2 u_{0,0}$$
$$= \dfrac{1}{12h^2}(-60u_{0,0} + 16(u_{1,0} + u_{0,1} + u_{-1,0} + u_{0,-1})$$
$$- (u_{2,0} + u_{0,2} + u_{-2,0} + u_{0,-2})) + O(h^4).$$

For fourth-order formulas and the biharmonic operator see http://dlmf.nist.gov/3.4.iii.

The results in this subsection for the partial derivatives follow from Panow (1955, Table 10). Those for the Laplacian and the biharmonic operator follow from the formulas for the partial derivatives.

For additional formulas involving values of $\nabla^2 u$ and $\nabla^4 u$ on square, triangular, and cubic grids, see Collatz (1960, Table VI, pp. 542–546).

3.5 Quadrature

3.5(i) Trapezoidal Rules

The *elementary trapezoidal rule* is given by

3.5.1 $\quad \displaystyle\int_a^b f(x)\,dx = \tfrac{1}{2}h(f(a) + f(b)) - \tfrac{1}{12}h^3 f''(\xi),$

where $h = b - a$, $f \in C^2[a,b]$, and $a < \xi < b$.

The *composite trapezoidal rule* is

3.5.2
$$\int_a^b f(x)\,dx = h(\tfrac{1}{2}f_0 + f_1 + \cdots + f_{n-1} + \tfrac{1}{2}f_n) + E_n(f),$$

where $h = (b-a)/n$, $x_k = a + kh$, $f_k = f(x_k)$, $k = 0, 1, \ldots, n$, and

3.5.3 $\quad E_n(f) = -\dfrac{b-a}{12}h^2 f''(\xi), \qquad a < \xi < b.$

If in addition f is periodic, $f \in C^k(\mathbb{R})$, and the integral is taken over a period, then

3.5.4 $\quad E_n(f) = O(h^k), \qquad h \to 0.$

In particular, when $k = \infty$ the error term is an exponentially-small function of $1/h$, and in these circumstances the composite trapezoidal rule is exceptionally efficient. For an example see §3.5(ix).

Similar results hold for the trapezoidal rule in the form

3.5.5 $\quad \displaystyle\int_{-\infty}^{\infty} f(t)\,dt = h \sum_{k=-\infty}^{\infty} f(kh) + E_h(f),$

with a function f that is analytic in a strip containing $\mathbb{R}$. For further information and examples, see Goodwin (1949a). In Stenger (1993, Chapter 3) the rule (3.5.5) is considered in the framework of Sinc approximations (§3.3(vi)). See also Poisson's summation formula (§1.8(iv)).

If k in (3.5.4) is not arbitrarily large, and if odd-order derivatives of f are known at the end points a and b, then the composite trapezoidal rule can be improved by means of the Euler–Maclaurin formula (§2.10(i)). See Davis and Rabinowitz (1984, pp. 134–142) and Temme (1996a, p. 25).

3.5(ii) Simpson's Rule

Let $h = \tfrac{1}{2}(b-a)$ and $f \in C^4[a,b]$. Then the *elementary Simpson's rule* is

3.5.6 $\quad \displaystyle\int_a^b f(x)\,dx = \tfrac{1}{3}h(f(a) + 4f(\tfrac{1}{2}(a+b)) + f(b))$
$\qquad\qquad - \tfrac{1}{90}h^5 f^{(4)}(\xi),$

where $a < \xi < b$.

3.5 Quadrature

Now let $h = (b-a)/n$, $x_k = a + kh$, and $f_k = f(x_k)$, $k = 0, 1, \ldots, n$. Then the *composite Simpson's rule* is

3.5.7 $$\int_a^b f(x)\,dx = \tfrac{1}{3}h(f_0 + 4f_1 + 2f_2 + 4f_3 + 2f_4 + \cdots + 4f_{n-1} + f_n) + E_n(f),$$

where n is even and

3.5.8 $$E_n(f) = -\frac{b-a}{180}h^4 f^{(4)}(\xi), \quad a < \xi < b.$$

Simpson's rule can be regarded as a combination of two trapezoidal rules, one with step size h and one with step size $h/2$ to refine the error term.

3.5(iii) Romberg Integration

Further refinements are achieved by *Romberg integration*. If $f \in C^{2m+2}[a,b]$, then the remainder $E_n(f)$ in (3.5.2) can be expanded in the form

3.5.9 $$E_n(f) = c_1 h^2 + c_2 h^4 + \cdots + c_m h^{2m} + O(h^{2m+2}),$$

where $h = (b-a)/n$. As in Simpson's rule, by combining the rule for h with that for $h/2$, the first error term $c_1 h^2$ in (3.5.9) can be eliminated. With the Romberg scheme successive terms $c_1 h^2, c_2 h^4, \ldots,$ in (3.5.9) are eliminated, according to the formula

3.5.10
$$G_k(\tfrac{1}{2}h) = G_{k-1}(\tfrac{1}{2}h) + \frac{G_{k-1}(\tfrac{1}{2}h) - G_{k-1}(h)}{4^k - 1}, \quad k \geq 1,$$

beginning with

3.5.11 $G_0(h) = h(\tfrac{1}{2}f_0 + f_1 + \cdots + f_{n-1} + \tfrac{1}{2}f_n),$

although we may also start with the elementary rule with $G_0(h) = \tfrac{1}{2}h(f(a) + f(b))$ and $h = b - a$. To generate $G_k(h)$ the quantities $G_0(h), G_0(h/2), \ldots, G_0(h/2^k)$ are needed. These can be found by means of the recursion

3.5.12 $G_0(\tfrac{1}{2}h) = \tfrac{1}{2}G_0(h) + \tfrac{1}{2}h \sum_{k=0}^{n-1} f\left(x_0 + (k + \tfrac{1}{2})h\right),$

which depends on function values computed previously.

If $f \in C^{2k+2}(a,b)$, then for $j, k = 0, 1, \ldots,$

3.5.13 $$\int_a^b f(x)\,dx - G_k\left(\frac{b-a}{2^j}\right) = -\frac{(b-a)^{2k+3}}{2^{k(k+1)}} \frac{4^{-j(k+1)}}{(2k+2)!} |B_{2k+2}| f^{(2k+2)}(\xi),$$

for some $\xi \in (a,b)$. For the Bernoulli numbers B_m see §24.2(i).

When $f \in C^\infty$, the Romberg method affords a means of obtaining high accuracy in many cases with a relatively simple adaptive algorithm. However, as illustrated by the next example, other methods may be more efficient.

Example

With $J_0(t)$ denoting the Bessel function (§10.2(ii)) the integral

3.5.14 $$\int_0^\infty e^{-pt} J_0(t)\,dt = \frac{1}{\sqrt{p^2+1}}$$

is computed with $p = 1$ on the interval $[0, 30]$. Using (3.5.10) with $h = 30/4 = 7.5$ we obtain $G_7(h)$ with 14 correct digits. About $2^9 = 512$ function evaluations are needed. (With the 20-point Gauss–Laguerre formula (§3.5(v)) the same precision can be achieved with 15 function evaluations.) With $j = 2$ and $k = 7$, the coefficient of the derivative $f^{(16)}(\xi)$ in (3.5.13) is found to be $(0.14\ldots) \times 10^{-13}$.

See Davis and Rabinowitz (1984, pp. 440–441) for modifications of the Romberg method when the function f is singular.

3.5(iv) Interpolatory Quadrature Rules

An *interpolatory quadrature rule*

3.5.15 $$\int_a^b f(x)w(x)\,dx = \sum_{k=1}^n w_k f(x_k) + E_n(f),$$

with *weight function* $w(x)$, is one for which $E_n(f) = 0$ whenever f is a polynomial of degree $\leq n - 1$. The *nodes* $x_1, x_2, \ldots, x_n$ are prescribed, and the *weights* w_k and *error term* $E_n(f)$ are found by integrating the product of the Lagrange interpolation polynomial of degree $n - 1$ and $w(x)$.

If the extreme members of the set of nodes $x_1, x_2, \ldots, x_n$ are the endpoints a and b, then the quadrature rule is said to be *closed*. Or if the set $x_1, x_2, \ldots, x_n$ lies in the open interval (a,b), then the quadrature rule is said to be *open*.

Rules of closed type include the *Newton–Cotes formulas* such as the trapezoidal rules and Simpson's rule. Examples of open rules are the Gauss formulas (§3.5(v)), the *midpoint rule*, and *Fejér's quadrature rule*. For the latter $a = -1$, $b = 1$, and the nodes x_k are the extrema of the Chebyshev polynomial $T_n(x)$ (§3.11(ii) and §18.3). If we add -1 and 1 to this set of x_k, then the resulting closed formula is the frequently-used *Clenshaw–Curtis formula*, whose weights are positive and given by

3.5.16 $$w_k = \frac{g_k}{n}\left(1 - \sum_{j=1}^{\lfloor n/2 \rfloor} \frac{b_j}{4j^2 - 1}\cos(2jk\pi/n)\right),$$

where $x_k = \cos(k\pi/n), k = 0, 1, \ldots, n$, and

3.5.17 $g_k = \begin{cases} 1, & k = 0, n, \\ 2, & \text{otherwise,} \end{cases} \quad b_j = \begin{cases} 1, & j = \tfrac{1}{2}n, \\ 2, & \text{otherwise.} \end{cases}$

For further information, see Mason and Handscomb (2003, Chapter 8), Davis and Rabinowitz (1984, pp. 74–92), and Clenshaw and Curtis (1960).

For a detailed comparison of the Clenshaw–Curtis formula with Gauss quadrature (§3.5(v)), see Trefethen (2008).

3.5(v) Gauss Quadrature

Let $\{p_n\}$ denote the set of monic polynomials p_n of degree n (coefficient of x^n equal to 1) that are orthogonal with respect to a positive weight function w on a finite or infinite interval (a,b); compare §18.2(i). In *Gauss quadrature* (also known as *Gauss–Christoffel quadrature*) we use (3.5.15) with nodes x_k the zeros of p_n, and weights w_k given by

3.5.18 $$w_k = \int_a^b \frac{p_n(x)}{(x-x_k)p_n'(x_k)} w(x)\,dx.$$

The w_k are also known as *Christoffel coefficients* or *Christoffel numbers* and they are all positive. The remainder is given by

3.5.19 $$E_n(f) = \gamma_n f^{(2n)}(\xi)/(2n)!,$$

where

3.5.20 $$\gamma_n = \int_a^b p_n^2(x) w(x)\,dx,$$

and ξ is some point in (a,b). As a consequence, the rule is exact for polynomials of degree $\leq 2n-1$.

In practical applications the weight function $w(x)$ is chosen to simulate the asymptotic behavior of the integrand as the endpoints are approached. For C^∞ functions Gauss quadrature can be very efficient. In adaptive algorithms the evaluation of the nodes and weights may cause difficulties, unless exact values are known.

For the derivation of Gauss quadrature formulas see Gautschi (2004, pp. 22–32), Gil et al. (2007a, §5.3), and Davis and Rabinowitz (1984, §§2.7 and 3.6). Stroud and Secrest (1966) includes computational methods and extensive tables. For further extensions, applications, and computation of orthogonal polynomials and Gauss-type formulas, see Gautschi (1994, 1996, 2004). For effective testing of Gaussian quadrature rules see Gautschi (1983).

For the classical orthogonal polynomials related to the following Gauss rules, see §18.3. The given quantities γ_n follow from (18.2.5), (18.2.7), Table 18.3.1, and the relation $\gamma_n = h_n/k_n^2$.

Gauss–Legendre Formula

3.5.21 $$[a,b]=[-1,1], \quad w(x)=1, \quad \gamma_n = \frac{2^{2n+1}}{2n+1}\frac{(n!)^4}{((2n)!)^2}.$$

The nodes x_k and weights w_k for $n=5,10$ are shown in Tables 3.5.1 and 3.5.2. The $p_n(x)$ are the *monic* Legendre polynomials, that is, the polynomials $P_n(x)$ (§18.3) scaled so that the coefficient of the highest power of x in their explicit forms is unity.

Table 3.5.1: Nodes and weights for the 5-point Gauss–Legendre formula.

$\pm x_k$	w_k
0.00000 00000 00000	0.56888 88888 88889
0.53846 93101 05683	0.47862 86704 99366
0.90617 98459 38664	0.23692 68850 56189

Table 3.5.2: Nodes and weights for the 10-point Gauss–Legendre formula.

$\pm x_k$	w_k
0.14887 43389 81631 211	0.29552 42247 14752 870
0.43339 53941 29247 191	0.26926 67193 09996 355
0.67940 95682 99024 406	0.21908 63625 15982 044
0.86506 33666 88984 511	0.14945 13491 50580 593
0.97390 65285 17171 720	0.06667 13443 08688 138

For corresponding results for $n=20,40,80$; see http://dlmf.nist.gov/3.5.v.

Gauss–Chebyshev Formula

3.5.22 $$[a,b]=[-1,1], \quad w(x)=(1-x^2)^{-1/2}, \quad \gamma_n = \frac{\pi}{2^{2n-1}}.$$

The nodes x_k and weights w_k are known explicitly:

3.5.23 $$x_k = \cos\left(\frac{2k-1}{2n}\pi\right), \quad w_k = \frac{\pi}{n}, \quad k=1,2,\ldots,n.$$

Nodes and weights are also known explicitly for the other three weight functions in the set $w(x) = (1-x)^{\pm 1/2}(1+x)^{\pm 1/2}$; see http://dlmf.nist.gov/3.5.v.

Gauss–Jacobi Formula

3.5.26 $$[a,b]=[-1,1], \quad w(x)=(1-x)^\alpha(1+x)^\beta, \quad \gamma_n = \frac{\Gamma(n+\alpha+1)\,\Gamma(n+\beta+1)\,\Gamma(n+\alpha+\beta+1)}{(2n+\alpha+\beta+1)(\Gamma(2n+\alpha+\beta+1))^2} 2^{2n+\alpha+\beta+1} n!,$$
$$\alpha > -1, \beta > -1.$$

The $p_n(x)$ are the monic Jacobi polynomials $P_n^{(\alpha,\beta)}(x)$ (§18.3).

Gauss–Laguerre Formula

3.5.27 $$[a,b)=[0,\infty), \quad w(x)=x^\alpha e^{-x}, \quad \gamma_n = n!\,\Gamma(n+\alpha+1), \quad \alpha > -1.$$

3.5 QUADRATURE

If $\alpha \neq 0$ this is called the *generalized Gauss–Laguerre formula*.

The nodes x_k and weights w_k for $\alpha = 0$ and $n = 5, 10$ are shown in Tables 3.5.6 and 3.5.7. The $p_n(x)$ are the monic Laguerre polynomials $L_n(x)$ (§18.3).

Table 3.5.6: Nodes and weights for the 5-point Gauss–Laguerre formula.

x_k	w_k
0.26356 03197 18141	0.52175 56105 82809
0.14134 03059 10652×10^1	0.39866 68110 83176
0.35964 25771 04072×10^1	0.75942 44968 17076×10^{-1}
0.70858 10005 85884×10^1	0.36117 58679 92205×10^{-2}
0.12640 80084 42758×10^2	0.23369 97238 57762×10^{-4}

Table 3.5.7: Nodes and weights for the 10-point Gauss–Laguerre formula.

x_k	w_k
0.13779 34705 40492 431	0.30844 11157 65020 141
0.72945 45495 03170 498	0.40111 99291 55273 552
0.18083 42901 74031 605×10^1	0.21806 82876 11809 422
0.34014 33697 85489 951×10^1	0.62087 45609 86777 475×10^{-1}
0.55524 96140 06380 363×10^1	0.95015 16975 18110 055×10^{-2}
0.83301 52746 76449 670×10^1	0.75300 83885 87538 775×10^{-3}
0.11843 78583 79000 656×10^2	0.28259 23349 59956 557×10^{-4}
0.16279 25783 13781 021×10^2	0.42493 13984 96268 637×10^{-6}
0.21996 58581 19807 620×10^2	0.18395 64823 97963 078×10^{-8}
0.29920 69701 22738 916×10^2	0.99118 27219 60900 856×10^{-12}

For the corresponding results for $n = 15, 20$ see http://dlmf.nist.gov/3.5.v.

Gauss–Hermite Formula

3.5.28 $$(a,b) = (-\infty, \infty), \quad w(x) = e^{-x^2}, \quad \gamma_n = \sqrt{\pi}\frac{n!}{2^n}.$$

The nodes x_k and weights w_k for $n = 5, 10$ are shown in Tables 3.5.10 and 3.5.11. The $p_n(x)$ are the monic Hermite polynomials $H_n(x)$ (§18.3).

Table 3.5.10: Nodes and weights for the 5-point Gauss–Hermite formula.

$\pm x_k$	w_k
0.00000 00000 00000	0.94530 87204 82942
0.95857 24646 13819	0.39361 93231 52241
0.20201 82870 45609×10^1	0.19953 24205 90459×10^{-1}

Table 3.5.11: Nodes and weights for the 10-point Gauss–Hermite formula.

$\pm x_k$	w_k
0.34290 13272 23704 609	0.61086 26337 35325 799
0.10366 10829 78951 365×10^1	0.24013 86110 82314 686
0.17566 83649 29988 177×10^1	0.33874 39445 54810 631×10^{-1}
0.25327 31674 23278 980×10^1	0.13436 45746 78123 269×10^{-2}
0.34361 59118 83773 760×10^1	0.76404 32855 23262 063×10^{-5}

For the corresponding results for $n = 15, 20$ see http://dlmf.nist.gov/3.5.v.

Gauss Formula for a Logarithmic Weight Function

3.5.29 $$[a,b] = [0,1], \quad w(x) = \ln(1/x).$$

The nodes x_k and weights w_k for $n = 5, 10$ are shown in Tables 3.5.14 and 3.5.15.

Table 3.5.14: Nodes and weights for the 5-point Gauss formula for the logarithmic weight function.

x_k	w_k
$0.29134\ 47215\ 19721 \times 10^{-1}$	$0.29789\ 34717\ 82894$
$0.17397\ 72133\ 20898$	$0.34977\ 62265\ 13224$
$0.41170\ 25202\ 84902$	$0.23448\ 82900\ 44052$
$0.67731\ 41745\ 82820$	$0.98930\ 45951\ 66331 \times 10^{-1}$
$0.89477\ 13610\ 31008$	$0.18911\ 55214\ 31958 \times 10^{-1}$

Table 3.5.15: Nodes and weights for the 10-point Gauss formula for the logarithmic weight function.

x_k	w_k
$0.90426\ 30962\ 19965\ 064 \times 10^{-2}$	$0.12095\ 51319\ 54570\ 515$
$0.53971\ 26622\ 25006\ 295 \times 10^{-1}$	$0.18636\ 35425\ 64071\ 870$
$0.13531\ 18246\ 39250\ 775$	$0.19566\ 08732\ 77759\ 983$
$0.24705\ 24162\ 87159\ 824$	$0.17357\ 71421\ 82906\ 921$
$0.38021\ 25396\ 09332\ 334$	$0.13569\ 56729\ 95484\ 202$
$0.52379\ 23179\ 71843\ 201$	$0.93646\ 75853\ 81105\ 260 \times 10^{-1}$
$0.66577\ 52055\ 16424\ 597$	$0.55787\ 72735\ 14158\ 741 \times 10^{-1}$
$0.79419\ 04160\ 11966\ 217$	$0.27159\ 81089\ 92333\ 311 \times 10^{-1}$
$0.89816\ 10912\ 19003\ 538$	$0.95151\ 82602\ 84851\ 500 \times 10^{-2}$
$0.96884\ 79887\ 18633\ 539$	$0.16381\ 57633\ 59826\ 325 \times 10^{-2}$

For the corresponding results for $n = 15, 20$ see http://dlmf.nist.gov/3.5.v.

3.5(vi) Eigenvalue/Eigenvector Characterization of Gauss Quadrature Formulas

All the monic orthogonal polynomials $\{p_n\}$ used with Gauss quadrature satisfy a three-term recurrence relation (§18.2(iv)):

3.5.30
$$p_{k+1}(x) = (x - \alpha_k)p_k(x) - \beta_k p_{k-1}(x), \quad k = 0, 1, \ldots,$$
with $\beta_k > 0$, $p_{-1}(x) = 0$, and $p_0(x) = 1$.

The Gauss nodes x_k (the zeros of p_n) are the eigenvalues of the (symmetric tridiagonal) *Jacobi matrix* of order $n \times n$:

3.5.31
$$\mathbf{J}_n = \begin{bmatrix} \alpha_0 & \sqrt{\beta_1} & & & 0 \\ \sqrt{\beta_1} & \alpha_1 & \sqrt{\beta_2} & & \\ & \ddots & \ddots & \ddots & \\ & & \sqrt{\beta_{n-2}} & \alpha_{n-2} & \sqrt{\beta_{n-1}} \\ 0 & & & \sqrt{\beta_{n-1}} & \alpha_{n-1} \end{bmatrix}.$$

Let $\mathbf{v}_k$ denote the normalized eigenvector of $\mathbf{J}_n$ corresponding to the eigenvalue x_k. Then the weights are given by

3.5.32 $\qquad w_k = \beta_0 v_{k,1}^2, \qquad k = 1, 2, \ldots, n,$

where $\beta_0 = \int_a^b w(x)\,dx$ and $v_{k,1}$ is the first element of $\mathbf{v}_k$. Also, the error constant (3.5.20) is given by

3.5.33 $\qquad \gamma_n = \beta_0 \beta_1 \cdots \beta_n.$

Tables 3.5.1, 3.5.2, 3.5.6, 3.5.7, 3.5.10, and 3.5.11 can be verified by application of the results given in the present subsection. In these cases the coefficients α_k and β_k are obtainable explicitly from results given in §18.9(i).

3.5(vii) Oscillatory Integrals

Integrals of the form

3.5.34 $\qquad \int_a^b f(x)\cos(\omega x)\,dx, \quad \int_a^b f(x)\sin(\omega x)\,dx,$

can be computed by *Filon's rule*. See Davis and Rabinowitz (1984, pp. 146–168).

Oscillatory integral transforms are treated in Wong (1982) by a method based on Gaussian quadrature. A comparison of several methods, including an extension of the Clenshaw–Curtis formula (§3.5(iv)), is given in Evans and Webster (1999).

For computing infinite oscillatory integrals, *Longman's method* may be used. The integral is written as an alternating series of positive and negative subintegrals that are computed individually; see Longman (1956). Convergence acceleration schemes, for example Levin's transformation (§3.9(v)), can be used when evaluating the series. Further methods are given in Clendenin (1966) and Lyness (1985).

For a comprehensive survey of quadrature of highly oscillatory integrals, including multidimensional integrals, see Iserles et al. (2006).

3.5(viii) Complex Gauss Quadrature

For the *Bromwich integral*

3.5.35
$$I(f) = \frac{1}{2\pi i} \int_{c-i\infty}^{c+i\infty} e^{\zeta} \zeta^{-s} f(\zeta) \, d\zeta, \quad s > 0, \ c > c_0 > 0,$$

a *complex Gauss quadrature formula* is available. Here $f(\zeta)$ is assumed analytic in the half-plane $\Re\zeta > c_0$ and bounded as $\zeta \to \infty$ in $|\text{ph}\,\zeta| \leq \frac{1}{2}\pi$. The quadrature rule for (3.5.35) is

3.5.36
$$I(f) = \sum_{k=1}^{n} w_k f(\zeta_k) + E_n(f),$$

where $E_n(f) = 0$ if $f(\zeta)$ is a polynomial of degree $\leq 2n - 1$ in $1/\zeta$. *Complex orthogonal polynomials* $p_n(1/\zeta)$ of degree $n = 0, 1, 2, \ldots$, in $1/\zeta$ that satisfy the orthogonality condition

3.5.37
$$\int_{c-i\infty}^{c+i\infty} e^{\zeta} \zeta^{-s} p_k(1/\zeta) p_\ell(1/\zeta) \, d\zeta = 0, \quad k \neq \ell,$$

are related to Bessel polynomials (§§10.49(ii) and 18.34). The complex Gauss nodes ζ_k have positive real part for all $s > 0$.

The nodes and weights of the 5-point complex Gauss quadrature formula (3.5.36) for $s = 1$ are shown in Table 3.5.18. Extensive tables of quadrature nodes and weights can be found in Krylov and Skoblya (1985).

Table 3.5.18: Nodes and weights for the 5-point complex Gauss quadrature formula with $s = 1$.

ζ_k	w_k
3.65569 4325+6.54373 6899i	3.83966 1630−0.27357 03863i
3.65569 4325−6.54373 6899i	3.83966 1630+0.27357 03863i
5.70095 3299+3.21026 5600i	−25.07945 221 +2.18725 2294i
5.70095 3299−3.21026 5600i	−25.07945 221 −2.18725 2294i
6.28670 4752+0.00000 0000i	43.47958 116 +0.00000 0000i

Example. Laplace Transform Inversion

From §1.14(iii)

3.5.38
$$G(p) = \int_0^\infty e^{-pt} g(t) \, dt,$$

3.5.39
$$g(t) = \frac{1}{2\pi i} \int_{\sigma-i\infty}^{\sigma+i\infty} e^{tp} G(p) \, dp,$$

with appropriate conditions. The pair

3.5.40
$$g(t) = J_0(t), \quad G(p) = \frac{1}{\sqrt{p^2 + 1}},$$

where $J_0(t)$ is the Bessel function (§10.2(ii)), satisfy these conditions, provided that $\sigma > 0$. The integral (3.5.39) has the form (3.5.35) if we set $\zeta = tp$, $c = t\sigma$, and $f(\zeta) = t^{-1} \zeta^s G(\zeta/t)$. We choose $s = 1$ so that $f(\zeta) = O(1)$ at infinity. Equation (3.5.36), without the error term, becomes

3.5.41
$$g(t) = \sum_{k=1}^{n} \frac{w_k \zeta_k}{\sqrt{\zeta_k^2 + t^2}},$$

approximately.

Using Table 3.5.18 we compute $g(t)$ for $n = 5$. The results are given in the middle column of Table 3.5.19, accompanied by the actual 10D values in the last column. Agreement is very good for small values of t, but not for larger values. For these cases the integration path may need to be deformed; see §3.5(ix).

Table 3.5.19: Laplace transform inversion.

t	$g(t)$	$J_0(t)$
0.0	1.00000 00000	1.00000 00000
0.5	0.93846 98072	0.93846 98072
1.0	0.76519 76866	0.76519 76865
2.0	0.22389 07791	0.22389 10326
5.0	−0.17759 67713	−0.17902 54097
10.0	−0.24593 57645	−0.07540 53543

3.5(ix) Other Contour Integrals

A frequent problem with contour integrals is heavy cancellation, which occurs especially when the value of the integral is exponentially small compared with the maximum absolute value of the integrand. To avoid cancellation we try to deform the path to pass through a saddle point in such a way that the maximum contribution of the integrand is derived from the neighborhood of the saddle point. For example, steepest descent paths can be used; see §2.4(iv).

Example

In (3.5.35) take $s = 1$ and $f(\zeta) = e^{-2\lambda\sqrt{\zeta}}$, with $\lambda > 0$. When λ is large the integral becomes exponentially small, and application of the quadrature rule of §3.5(viii) is useless. In fact from (7.14.4) and the inversion formula for the Laplace transform (§1.14(iii)) we have

$$3.5.42 \qquad \operatorname{erfc} \lambda = \frac{1}{2\pi i} \int_{c-i\infty}^{c+i\infty} e^{\zeta - 2\lambda\sqrt{\zeta}} \frac{d\zeta}{\zeta}, \qquad c > 0,$$

where $\operatorname{erfc} z$ is the complementary error function, and from (7.12.1) it follows that

$$3.5.43 \qquad \operatorname{erfc} \lambda \sim \frac{e^{-\lambda^2}}{\sqrt{\pi}\lambda}, \qquad \lambda \to \infty.$$

With the transformation $\zeta = \lambda^2 t$, (3.5.42) becomes

$$3.5.44 \qquad \operatorname{erfc} \lambda = \frac{1}{2\pi i} \int_{c-i\infty}^{c+i\infty} e^{\lambda^2 (t - 2\sqrt{t})} \frac{dt}{t}, \qquad c > 0,$$

with saddle point at $t = 1$, and when $c = 1$ the vertical path intersects the real axis at the saddle point. The steepest descent path is given by $\Im(t - 2\sqrt{t}) = 0$, or in polar coordinates $t = re^{i\theta}$ we have $r = \sec^2(\frac{1}{2}\theta)$. Thus

$$3.5.45 \qquad \operatorname{erfc} \lambda = \frac{e^{-\lambda^2}}{2\pi} \int_{-\pi}^{\pi} e^{-\lambda^2 \tan^2(\frac{1}{2}\theta)} d\theta.$$

The integrand can be extended as a periodic C^∞ function on $\mathbb{R}$ with period 2π and as noted in §3.5(i), the trapezoidal rule is exceptionally efficient in this case.

Table 3.5.20 gives the results of applying the composite trapezoidal rule (3.5.2) with step size h; n indicates the number of function values in the rule that are larger than 10^{-15} (we exploit the fact that the integrand is even). All digits shown in the approximation in the final row are correct.

Table 3.5.20: Composite trapezoidal rule for the integral (3.5.45) with $\lambda = 10$.

h	$\operatorname{erfc} \lambda$	n
0.25	$0.20949\ 49432\ 96679 \times 10^{-44}$	5
0.20	$0.20886\ 11645\ 34559 \times 10^{-44}$	6
0.15	$0.20884\ 87588\ 72946 \times 10^{-44}$	8
0.10	$0.20884\ 87583\ 76254 \times 10^{-44}$	11

A second example is provided in Gil et al. (2001), where the method of contour integration is used to evaluate Scorer functions of complex argument (§9.12). See also Gil et al. (2003b).

If f is meromorphic, with poles near the saddle point, then the foregoing method can be modified. A special case is the rule for Hilbert transforms (§1.14(v)):

$$3.5.46 \qquad \mathcal{H}(f; x) = \frac{1}{\pi} \fint_{-\infty}^{\infty} \frac{f(t)}{t - x} dt, \qquad x \in \mathbb{R},$$

where the integral is the Cauchy principal value. See Kress and Martensen (1970).

Other contour integrals occur in standard integral transforms or their inverses, for example, Hankel transforms (§10.22(v)), Kontorovich–Lebedev transforms (§10.43(v)), and Mellin transforms (§1.14(iv)).

3.5(x) Cubature Formulas

Table 3.5.21 supplies cubature rules, including *weights* w_j, for the disk D, given by $x^2 + y^2 \leq h^2$:

$$3.5.47 \qquad \frac{1}{\pi h^2} \iint_D f(x, y) \, dx \, dy = \sum_{j=1}^n w_j f(x_j, y_j) + R,$$

and the square S, given by $|x| \leq h$, $|y| \leq h$:

$$3.5.48 \qquad \frac{1}{4h^2} \iint_S f(x, y) \, dx \, dy = \sum_{j=1}^n w_j f(x_j, y_j) + R.$$

For these results and further information on cubature formulas see Cools (2003).

For integrals in higher dimensions, *Monte Carlo methods* are another—often the only—alternative. The standard Monte Carlo method samples points uniformly from the integration region to estimate the integral and its error. In more advanced methods points are sampled from a probability distribution, so that they are concentrated in regions that make the largest contribution to the integral. With N function values, the Monte Carlo method aims at an error of order $1/\sqrt{N}$, independently of the dimension of the domain of integration. See Davis and Rabinowitz (1984, pp. 384–417) and Schürer (2004).

Table 3.5.21: Cubature formulas for disk and square.

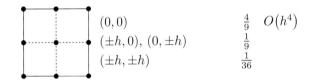

Diagram	(x_j, y_j)	w_j	R
	$(0,0)$	$\tfrac{1}{2}$	$O(h^4)$
	$(\pm h, 0)$	$\tfrac{1}{8}$	
	$(0, \pm h)$	$\tfrac{1}{8}$	
	$(\pm\tfrac{1}{2}h, \pm\tfrac{1}{2}h)$	$\tfrac{1}{4}$	$O(h^4)$
	$(0,0)$	$\tfrac{1}{6}$	$O(h^6)$
	$(\pm h, 0), (0, \pm h)$	$\tfrac{1}{24}$	
	$(\pm\tfrac{1}{2}h, \pm\tfrac{1}{2}h)$	$\tfrac{1}{6}$	
	$(0,0)$	$\tfrac{1}{4}$	$O(h^6)$
	$(\pm\tfrac{1}{3}\sqrt{6}h, 0)$	$\tfrac{1}{8}$	
	$(\pm\tfrac{1}{6}\sqrt{6}h, \pm\tfrac{1}{2}\sqrt{2}h)$	$\tfrac{1}{8}$	
	$(0,0)$	$\tfrac{4}{9}$	$O(h^4)$
	$(\pm h, 0), (0, \pm h)$	$\tfrac{1}{9}$	
	$(\pm h, \pm h)$	$\tfrac{1}{36}$	
	$(\pm\tfrac{1}{3}\sqrt{3}h, \pm\tfrac{1}{3}\sqrt{3}h)$	$\tfrac{1}{4}$	$O(h^4)$
	$(0,0)$	$\tfrac{16}{81}$	$O(h^6)$
	$(\pm\sqrt{\tfrac{3}{5}}h, 0), (0, \pm\sqrt{\tfrac{3}{5}}h)$	$\tfrac{10}{81}$	
	$(\pm\sqrt{\tfrac{3}{5}}h, 0), (\pm\sqrt{\tfrac{3}{5}}h, 0)$	$\tfrac{25}{324}$	

3.6 Linear Difference Equations

3.6(i) Introduction

Many special functions satisfy second-order recurrence relations, or difference equations, of the form

3.6.1 $$a_n w_{n+1} - b_n w_n + c_n w_{n-1} = d_n,$$

or equivalently,

3.6.2 $$a_n \Delta^2 w_{n-1} + (2a_n - b_n)\Delta w_{n-1} + (a_n - b_n + c_n)w_{n-1} = d_n,$$

where $\Delta w_{n-1} = w_n - w_{n-1}$, $\Delta^2 w_{n-1} = \Delta w_n - \Delta w_{n-1}$, and $n \in \mathbb{Z}$. If $d_n = 0$, $\forall n$, then the difference equation is *homogeneous*; otherwise it is *inhomogeneous*.

Difference equations are simple and attractive for computation. In practice, however, problems of severe instability often arise and in §§3.6(ii)–3.6(vii) we show how these difficulties may be overcome.

3.6(ii) Homogeneous Equations

Given numerical values of w_0 and w_1, the solution w_n of the equation

3.6.3 $$a_n w_{n+1} - b_n w_n + c_n w_{n-1} = 0,$$

with $a_n \neq 0$, $\forall n$, can be computed recursively for $n = 2, 3, \ldots$. Unless exact arithmetic is being used, however, each step of the calculation introduces rounding errors. These errors have the effect of perturbing the solution by unwanted small multiples of w_n and of an independent solution g_n, say. This is of little consequence if the wanted solution is growing in magnitude at least as fast as any other solution of (3.6.3), and the recursion process is *stable*.

But suppose that w_n is a nontrivial solution such that

3.6.4 $$w_n/g_n \to 0, \qquad n \to \infty.$$

Then w_n is said to be a *recessive* (equivalently, *minimal* or *distinguished*) solution as $n \to \infty$, and it is unique except for a constant factor. In this situation the unwanted multiples of g_n grow more rapidly than the wanted solution, and the computations are *unstable*. Stability can be restored, however, by *backward recursion*, provided that $c_n \neq 0$, $\forall n$: starting from w_N and w_{N+1}, with N large, equation (3.6.3) is applied to generate in succession $w_{N-1}, w_{N-2}, \ldots, w_0$. The unwanted multiples of g_n now decay in comparison with w_n, hence are of little consequence.

The values of w_N and w_{N+1} needed to begin the backward recursion may be available, for example, from asymptotic expansions (§2.9). However, there are alternative procedures that do not require w_N and w_{N+1} to be known in advance. These are described in §§ 3.6(iii) and 3.6(v).

3.6(iii) Miller's Algorithm

Because the recessive solution of a homogeneous equation is the fastest growing solution in the backward direction, it occurred to J.C.P. Miller (Bickley *et al.* (1952, pp. xvi–xvii)) that arbitrary "trial values" can be assigned to w_N and w_{N+1}, for example, 1 and 0. A "trial solution" is then computed by backward recursion, in the course of which the original components of the unwanted solution g_n die away. It therefore remains to apply a normalizing factor Λ. The process is then repeated with a higher value of N, and the normalized solutions compared. If agreement is not within a prescribed tolerance the cycle is continued.

The normalizing factor Λ can be the true value of w_0 divided by its trial value, or Λ can be chosen to satisfy a known property of the wanted solution of the form

$$3.6.5 \qquad \sum_{n=0}^{\infty} \lambda_n w_n = 1,$$

where the λ's are constants. The latter method is usually superior when the true value of w_0 is zero or pathologically small.

For further information on Miller's algorithm, including examples, convergence proofs, and error analyses, see Wimp (1984, Chapter 4), Gautschi (1967, 1997a), and Olver (1964a). See also Gautschi (1967) and Gil et al. (2007a, Chapter 4) for the computation of recessive solutions via continued fractions.

3.6(iv) Inhomogeneous Equations

Similar principles apply to equation (3.6.1) when $a_n c_n \neq 0$, $\forall n$, and $d_n \neq 0$ for some, or all, values of n. If, as $n \to \infty$, the wanted solution w_n grows (decays) in magnitude at least as fast as any solution of the corresponding homogeneous equation, then forward (backward) recursion is stable.

A new problem arises, however, if, as $n \to \infty$, the asymptotic behavior of w_n is intermediate to those of two independent solutions f_n and g_n of the corresponding inhomogeneous equation (the complementary functions). More precisely, assume that $f_0 \neq 0$, $g_n \neq 0$ for all sufficiently large n, and as $n \to \infty$

$$3.6.6 \qquad f_n/g_n \to 0, \quad w_n/g_n \to 0.$$

Then computation of w_n by forward recursion is unstable. If it also happens that $f_n/w_n \to 0$ as $n \to \infty$, then computation of w_n by backward recursion is unstable as well. However, w_n can be computed successfully in these circumstances by *boundary-value methods*, as follows.

Let us assume the normalizing condition is of the form $w_0 = \lambda$, where λ is a constant, and then solve the following tridiagonal system of algebraic equations for the unknowns $w_1^{(N)}, w_2^{(N)}, \ldots, w_{N-1}^{(N)}$; see §3.2(ii). Here N is an arbitrary positive integer.

$$3.6.7 \qquad \begin{bmatrix} -b_1 & a_1 & & & 0 \\ c_2 & -b_2 & a_2 & & \\ & \ddots & \ddots & \ddots & \\ & & c_{N-2} & -b_{N-2} & a_{N-2} \\ 0 & & & c_{N-1} & -b_{N-1} \end{bmatrix} \begin{bmatrix} w_1^{(N)} \\ w_2^{(N)} \\ \vdots \\ w_{N-2}^{(N)} \\ w_{N-1}^{(N)} \end{bmatrix} = \begin{bmatrix} d_1 - c_1 \lambda \\ d_2 \\ \vdots \\ d_{N-2} \\ d_{N-1} \end{bmatrix}.$$

Then as $N \to \infty$ with n fixed, $w_n^{(N)} \to w_n$.

3.6(v) Olver's Algorithm

To apply the method just described a succession of values can be prescribed for the arbitrary integer N and the results compared. However, a more powerful procedure combines the solution of the algebraic equations with the determination of the optimum value of N. It is applicable equally to the computation of the recessive solution of the homogeneous equation (3.6.3) or the computation of any solution w_n of the inhomogeneous equation (3.6.1) for which the conditions of §3.6(iv) are satisfied.

Suppose again that $f_0 \neq 0$, w_0 is given, and we wish to calculate $w_1, w_2, \ldots, w_M$ to a prescribed relative accuracy ϵ for a given value of M. We first compute, by forward recurrence, the solution p_n of the homogeneous equation (3.6.3) with initial values $p_0 = 0$, $p_1 = 1$. At the same time we construct a sequence e_n, $n = 0, 1, \ldots$, defined by

$$3.6.8 \qquad a_n e_n = c_n e_{n-1} - d_n p_n,$$

beginning with $e_0 = w_0$. (This part of the process is equivalent to forward elimination.) The computation is continued until a value N ($\geq M$) is reached for which

$$3.6.9 \qquad \left| \frac{e_N}{p_N p_{N+1}} \right| \leq \epsilon \min_{1 \leq n \leq M} \left| \frac{e_n}{p_n p_{n+1}} \right|.$$

Then w_n is generated by backward recursion from

$$3.6.10 \qquad p_{n+1} w_n = p_n w_{n+1} + e_n,$$

starting with $w_N = 0$. (This part of the process is back substitution.)

An example is included in the next subsection. For further information, including a more general form of normalizing condition, other examples, convergence proofs, and error analyses, see Olver (1967a), Olver and Sookne (1972), and Wimp (1984, Chapter 6).

3.6(vi) Examples

Example 1. Bessel Functions

The difference equation

3.6.11 $\qquad w_{n+1} - 2nw_n + w_{n-1} = 0, \quad n = 1, 2, \ldots,$

is satisfied by $J_n(1)$ and $Y_n(1)$, where $J_n(x)$ and $Y_n(x)$ are the Bessel functions of the first kind. For large n,

3.6.12 $\qquad J_n(1) \sim \dfrac{1}{(2\pi n)^{1/2}} \left(\dfrac{e}{2n}\right)^n,$

3.6.13 $\qquad Y_n(1) \sim \left(\dfrac{2}{\pi n}\right)^{1/2} \left(\dfrac{2n}{e}\right)^n,$

(§10.19(i)). Thus $Y_n(1)$ is dominant and can be computed by forward recursion, whereas $J_n(1)$ is recessive and has to be computed by backward recursion. The backward recursion can be carried out using independently computed values of $J_N(1)$ and $J_{N+1}(1)$ or by use of Miller's algorithm (§3.6(iii)) or Olver's algorithm (§3.6(v)).

Example 2. Weber Function

The Weber function $\mathbf{E}_n(1)$ satisfies

3.6.14 $\quad w_{n+1} - 2nw_n + w_{n-1} = -(2/\pi)(1-(-1)^n),$

for $n = 1, 2, \ldots,$ and as $n \to \infty$

3.6.15 $\qquad \mathbf{E}_{2n}(1) \sim \dfrac{2}{(4n^2-1)\pi},$

3.6.16 $\qquad \mathbf{E}_{2n+1}(1) \sim \dfrac{2}{(2n+1)\pi};$

see §11.11(ii). Thus the asymptotic behavior of the particular solution $\mathbf{E}_n(1)$ is intermediate to those of the complementary functions $J_n(1)$ and $Y_n(1)$; moreover, the conditions for Olver's algorithm are satisfied. We apply the algorithm to compute $\mathbf{E}_n(1)$ to 8S for the range $n = 1, 2, \ldots, 10$, beginning with the value $\mathbf{E}_0(1) = -0.56865\ 663$ obtained from the Maclaurin series expansion (§11.10(iii)).

In the notation of §3.6(v) we have $M = 10$ and $\epsilon = \tfrac{1}{2} \times 10^{-8}$. The least value of N that satisfies (3.6.9) is found to be 16. The results of the computations are displayed in Table 3.6.1. The values of w_n for $n = 1, 2, \ldots, 10$ are the wanted values of $\mathbf{E}_n(1)$. (It should be observed that for $n > 10$, however, the w_n are progressively poorer approximations to $\mathbf{E}_n(1)$: the underlined digits are in error.)

Table 3.6.1: Weber function $w_n = \mathbf{E}_n(1)$ computed by Olver's algorithm.

n	p_n	e_n	$e_n/(p_n p_{n+1})$	w_n
0	$0.00000\ 000$	$-0.56865\ 663$		$-0.56865\ 663$
1	$0.10000\ 000 \times 10^1$	$0.70458\ 291$	$0.35229\ 146$	$0.43816\ 243$
2	$0.20000\ 000 \times 10^1$	$0.70458\ 291$	$0.50327\ 351 \times 10^{-1}$	$0.17174\ 195$
3	$0.70000\ 000 \times 10^1$	$0.96172\ 597 \times 10^1$	$0.34347\ 356 \times 10^{-1}$	$0.24880\ 538$
4	$0.40000\ 000 \times 10^2$	$0.96172\ 597 \times 10^1$	$0.76815\ 174 \times 10^{-3}$	$0.47850\ 795 \times 10^{-1}$
5	$0.31300\ 000 \times 10^3$	$0.40814\ 124 \times 10^3$	$0.42199\ 534 \times 10^{-3}$	$0.13400\ 098$
6	$0.30900\ 000 \times 10^4$	$0.40814\ 124 \times 10^3$	$0.35924\ 754 \times 10^{-5}$	$0.18919\ 443 \times 10^{-1}$
7	$0.36767\ 000 \times 10^5$	$0.47221\ 340 \times 10^5$	$0.25102\ 029 \times 10^{-5}$	$0.93032\ 343 \times 10^{-1}$
8	$0.51164\ 800 \times 10^6$	$0.47221\ 340 \times 10^5$	$0.11324\ 804 \times 10^{-7}$	$0.10293\ 811 \times 10^{-1}$
9	$0.81496\ 010 \times 10^7$	$0.10423\ 616 \times 10^8$	$0.87496\ 485 \times 10^{-8}$	$0.71668\ 638 \times 10^{-1}$
10	$0.14618\ 117 \times 10^9$	$0.10423\ 616 \times 10^8$	$0.24457\ 824 \times 10^{-10}$	$0.65021\ 292 \times 10^{-2}$
11	$0.29154\ 738 \times 10^{10}$	$0.37225\ 201 \times 10^{10}$	$0.19952\ 026 \times 10^{-10}$	$0.58373\ 946 \times 10^{-1}$
12	$0.63994\ 242 \times 10^{11}$	$0.37225\ 201 \times 10^{10}$	$0.37946\ 279 \times 10^{-13}$	$0.44851\ 3\underline{87} \times 10^{-2}$
13	$0.15329\ 463 \times 10^{13}$	$0.19555\ 304 \times 10^{13}$	$0.32057\ 909 \times 10^{-13}$	$0.49269\ \underline{383} \times 10^{-1}$
14	$0.39792\ 611 \times 10^{14}$	$0.19555\ 304 \times 10^{13}$	$0.44167\ 174 \times 10^{-16}$	$0.327\underline{92\ 861} \times 10^{-2}$
15	$0.11126\ 602 \times 10^{16}$	$0.14186\ 384 \times 10^{16}$	$0.38242\ 250 \times 10^{-16}$	$0.425\underline{50\ 628} \times 10^{-1}$
16	$0.33340\ 012 \times 10^{17}$	$0.14186\ 384 \times 10^{16}$	$0.39924\ 861 \times 10^{-19}$	$\underline{0.00000\ 000}$

3.6(vii) Linear Difference Equations of Other Orders

Similar considerations apply to the first-order equation

3.6.17 $\qquad a_n w_{n+1} - b_n w_n = d_n.$

Thus in the inhomogeneous case it may sometimes be necessary to recur backwards to achieve stability. For analyses and examples see Gautschi (1997a).

For a difference equation of order $k\ (\geq 3)$,

3.6.18 $\quad a_{n,k} w_{n+k} + a_{n,k-1} w_{n+k-1} + \cdots + a_{n,0} w_n = d_n,$

or for systems of k first-order inhomogeneous equations, boundary-value methods are the rule rather than the exception. Typically $k - \ell$ conditions are prescribed at the beginning of the range, and ℓ conditions at the end.

Here $\ell \in [0, k]$, and its actual value depends on the asymptotic behavior of the wanted solution in relation to those of the other solutions. Within this framework forward and backward recursion may be regarded as the special cases $\ell = 0$ and $\ell = k$, respectively.

For further information see Wimp (1984, Chapters 7–8), Cash and Zahar (1994), and Lozier (1980).

3.7 Ordinary Differential Equations

3.7(i) Introduction

Consideration will be limited to *ordinary linear second-order differential equations*

3.7.1 $$\frac{d^2 w}{dz^2} + f(z) \frac{dw}{dz} + g(z) w = h(z),$$

where f, g, and h are analytic functions in a domain $D \subset \mathbb{C}$. If $h = 0$ the differential equation is *homogeneous*, otherwise it is *inhomogeneous*. For applications to special functions f, g, and h are often simple rational functions.

For general information on solutions of equation (3.7.1) see §1.13. For classification of singularities of (3.7.1) and expansions of solutions in the neighborhoods of singularities, see §2.7. For an introduction to numerical methods for ordinary differential equations, see Ascher and Petzold (1998), Hairer et al. (1993), and Iserles (1996).

3.7(ii) Taylor-Series Method: Initial-Value Problems

Assume that we wish to integrate (3.7.1) along a finite path $\mathscr{P}$ from $z = a$ to $z = b$ in a domain D. The path is partitioned at $P + 1$ points labeled successively $z_0, z_1, \ldots, z_P$, with $z_0 = a$, $z_P = b$.

By repeated differentiation of (3.7.1) all derivatives of $w(z)$ can be expressed in terms of $w(z)$ and $w'(z)$ as follows. Write

3.7.2 $$w^{(s)}(z) = f_s(z) w(z) + g_s(z) w'(z) + h_s(z), \quad s = 0, 1, 2, \ldots,$$

with

3.7.3 $$\begin{array}{lll} f_0(z) = 1, & g_0(z) = 0, & h_0(z) = 0, \\ f_1(z) = 0, & g_1(z) = 1, & h_1(z) = 0. \end{array}$$

Then for $s = 2, 3, \ldots,$

3.7.4 $$\begin{aligned} f_s(z) &= f'_{s-1}(z) - g(z) g_{s-1}(z), \\ g_s(z) &= f_{s-1}(z) - f(z) g_{s-1}(z) + g'_{s-1}(z), \\ h_s(z) &= h(z) g_{s-1}(z) + h'_{s-1}(z). \end{aligned}$$

Write $\tau_j = z_{j+1} - z_j$, $j = 0, 1, \ldots, P$, expand $w(z)$ and $w'(z)$ in Taylor series (§1.10(i)) centered at $z = z_j$, and apply (3.7.2). Then

3.7.5 $$\begin{bmatrix} w(z_{j+1}) \\ w'(z_{j+1}) \end{bmatrix} = \mathbf{A}(\tau_j, z_j) \begin{bmatrix} w(z_j) \\ w'(z_j) \end{bmatrix} + \mathbf{b}(\tau_j, z_j),$$

where $\mathbf{A}(\tau, z)$ is the matrix

3.7.6 $$\mathbf{A}(\tau, z) = \begin{bmatrix} A_{11}(\tau, z) & A_{12}(\tau, z) \\ A_{21}(\tau, z) & A_{22}(\tau, z) \end{bmatrix},$$

and $\mathbf{b}(\tau, z)$ is the vector

3.7.7 $$\mathbf{b}(\tau, z) = \begin{bmatrix} b_1(\tau, z) \\ b_2(\tau, z) \end{bmatrix},$$

with

3.7.8 $$\begin{aligned} A_{11}(\tau, z) &= \sum_{s=0}^{\infty} \frac{\tau^s}{s!} f_s(z), \\ A_{12}(\tau, z) &= \sum_{s=0}^{\infty} \frac{\tau^s}{s!} g_s(z), \\ A_{21}(\tau, z) &= \sum_{s=0}^{\infty} \frac{\tau^s}{s!} f_{s+1}(z), \\ A_{22}(\tau, z) &= \sum_{s=0}^{\infty} \frac{\tau^s}{s!} g_{s+1}(z), \end{aligned}$$

3.7.9 $$b_1(\tau, z) = \sum_{s=0}^{\infty} \frac{\tau^s}{s!} h_s(z), \quad b_2(\tau, z) = \sum_{s=0}^{\infty} \frac{\tau^s}{s!} h_{s+1}(z).$$

If the solution $w(z)$ that we are seeking grows in magnitude at least as fast as all other solutions of (3.7.1) as we pass along $\mathscr{P}$ from a to b, then $w(z)$ and $w'(z)$ may be computed in a stable manner for $z = z_0, z_1, \ldots, z_P$ by successive application of (3.7.5) for $j = 0, 1, \ldots, P - 1$, beginning with initial values $w(a)$ and $w'(a)$.

Similarly, if $w(z)$ is decaying at least as fast as all other solutions along $\mathscr{P}$, then we may reverse the labeling of the z_j along $\mathscr{P}$ and begin with initial values $w(b)$ and $w'(b)$.

3.7(iii) Taylor-Series Method: Boundary-Value Problems

Now suppose the path $\mathscr{P}$ is such that the rate of growth of $w(z)$ along $\mathscr{P}$ is intermediate to that of two other solutions. (This can happen only for inhomogeneous equations.) Then to compute $w(z)$ in a stable manner we solve the set of equations (3.7.5) simultaneously for $j = 0, 1, \ldots, P$, as follows. Let $\mathbf{A}$ be the $(2P) \times (2P+2)$ band matrix

3.7.10
$$\mathbf{A} = \begin{bmatrix} -\mathbf{A}(\tau_0, z_0) & \mathbf{I} & \mathbf{0} & \cdots & \mathbf{0} & \mathbf{0} \\ \mathbf{0} & -\mathbf{A}(\tau_1, z_1) & \mathbf{I} & \cdots & \mathbf{0} & \mathbf{0} \\ \vdots & \vdots & \ddots & \ddots & \vdots & \vdots \\ \mathbf{0} & \mathbf{0} & \cdots & -\mathbf{A}(\tau_{P-2}, z_{P-2}) & \mathbf{I} & \mathbf{0} \\ \mathbf{0} & \mathbf{0} & \cdots & \mathbf{0} & -\mathbf{A}(\tau_{P-1}, z_{P-1}) & \mathbf{I} \end{bmatrix}$$

($\mathbf{I}$ and $\mathbf{0}$ being the identity and zero matrices of order 2×2.) Also let $\mathbf{w}$ denote the $(2P+2) \times 1$ vector

3.7.11
$$\mathbf{w} = [w(z_0), w'(z_0), w(z_1), w'(z_1), \ldots, w(z_P), w'(z_P)]^{\mathrm{T}},$$

and $\mathbf{b}$ the $(2P) \times 1$ vector

3.7.12
$$\mathbf{b} = [b_1(\tau_0, z_0), b_2(\tau_0, z_0), b_1(\tau_1, z_1), b_2(\tau_1, z_1), \ldots, b_1(\tau_{P-1}, z_{P-1}), b_2(\tau_{P-1}, z_{P-1})]^{\mathrm{T}}.$$

Then

3.7.13
$$\mathbf{A}\mathbf{w} = \mathbf{b}.$$

This is a set of $2P$ equations for the $2P + 2$ unknowns, $w(z_j)$ and $w'(z_j)$, $j = 0, 1, \ldots, P$. The remaining two equations are supplied by boundary conditions of the form

3.7.14
$$\alpha_0 w(z_0) + \beta_0 w'(z_0) = \gamma_0,$$
$$\alpha_1 w(z_P) + \beta_1 w'(z_P) = \gamma_1,$$

where the α's, β's, and γ's are constants.

If, for example, $\beta_0 = \beta_1 = 0$, then on moving the contributions of $w(z_0)$ and $w(z_P)$ to the right-hand side of (3.7.13) the resulting system of equations is not tridiagonal, but can readily be made tridiagonal by annihilating the elements of $\mathbf{A}$ that lie below the main diagonal and its two adjacent diagonals. The equations can then be solved by the method of §3.2(ii), if the differential equation is homogeneous, or by Olver's algorithm (§3.6(v)). The latter is especially useful if the endpoint b of $\mathscr{P}$ is at ∞, or if the differential equation is inhomogeneous.

It will be observed that the present formulation of the Taylor-series method permits considerable parallelism in the computation, both for initial-value and boundary-value problems.

For further information and examples, see Olde Daalhuis and Olver (1998, §7) and Lozier and Olver (1993). General methods for boundary-value problems for ordinary differential equations are given in Ascher et al. (1995).

3.7(iv) Sturm–Liouville Eigenvalue Problems

Let (a, b) be a finite or infinite interval and $q(x)$ be a real-valued continuous (or piecewise continuous) function on the closure of (a, b). The *Sturm–Liouville eigenvalue problem* is the construction of a nontrivial solution of the system

3.7.15
$$\frac{d^2 w_k}{dx^2} + (\lambda_k - q(x)) w_k = 0,$$

3.7.16
$$w_k(a) = w_k(b) = 0,$$

with limits taken in (3.7.16) when a or b, or both, are infinite. The values λ_k are the *eigenvalues* and the corresponding solutions w_k of the differential equation are the *eigenfunctions*. The eigenvalues λ_k are simple, that is, there is only one corresponding eigenfunction (apart from a normalization factor), and when ordered increasingly the eigenvalues satisfy

3.7.17
$$\lambda_1 < \lambda_2 < \lambda_3 < \cdots, \quad \lim_{k \to \infty} \lambda_k = \infty.$$

If $q(x)$ is C^∞ on the closure of (a, b), then the discretized form (3.7.13) of the differential equation can be used. This converts the problem into a tridiagonal matrix problem in which the elements of the matrix are polynomials in λ; compare §3.2(vi). The larger the absolute values of the eigenvalues λ_k that are being sought, the smaller the integration steps $|\tau_j|$ need to be.

For further information, including other methods and examples, see Pryce (1993, §2.5.1).

3.7(v) Runge–Kutta Method

The Runge–Kutta method applies to linear or nonlinear differential equations. The method consists of a set of rules each of which is equivalent to a truncated Taylor-series expansion, but the rules avoid the need for analytic differentiations of the differential equation.

First-Order Equations

For $w' = f(z, w)$ the standard fourth-order rule reads

3.7.18
$$w_{n+1} = w_n + \tfrac{1}{6}(k_1 + 2k_2 + 2k_3 + k_4) + O(h^5),$$

where $h = z_{n+1} - z_n$ and

3.7.19
$$k_1 = hf(z_n, w_n),$$
$$k_2 = hf(z_n + \tfrac{1}{2}h, w_n + \tfrac{1}{2}k_1),$$
$$k_3 = hf(z_n + \tfrac{1}{2}h, w_n + \tfrac{1}{2}k_2),$$
$$k_4 = hf(z_n + h, w_n + k_3).$$

The order estimate $O(h^5)$ holds if the solution $w(z)$ has five continuous derivatives.

Second-Order Equations

For $w'' = f(z, w, w')$ the standard fourth-order rule reads

3.7.20
$$w_{n+1} = w_n + \tfrac{1}{6}h(6w'_n + k_1 + k_2 + k_3) + O(h^5),$$
$$w'_{n+1} = w'_n + \tfrac{1}{6}(k_1 + 2k_2 + 2k_3 + k_4) + O(h^5),$$

where

3.7.21
$$k_1 = hf(z_n, w_n, w'_n),$$
$$k_2 = hf(z_n + \tfrac{1}{2}h, w_n + \tfrac{1}{2}hw'_n + \tfrac{1}{8}hk_1, w'_n + \tfrac{1}{2}k_1),$$
$$k_3 = hf(z_n + \tfrac{1}{2}h, w_n + \tfrac{1}{2}hw'_n + \tfrac{1}{8}hk_2, w'_n + \tfrac{1}{2}k_2),$$
$$k_4 = hf(z_n + h, w_n + hw'_n + \tfrac{1}{2}hk_3, w'_n + k_3).$$

The order estimates $O(h^5)$ hold if the solution $w(z)$ has five continuous derivatives.

An extensive literature exists on the numerical solution of ordinary differential equations by Runge–Kutta, multistep, or other methods. See, for example, Butcher (1987), Dekker and Verwer (1984, Chapter 3), Hairer et al. (1993, Chapter 2), and Hairer and Wanner (1996, Chapter 4).

3.8 Nonlinear Equations

3.8(i) Introduction

The equation to be solved is

3.8.1
$$f(z) = 0,$$

where z is a real or complex variable and the function f is nonlinear. Solutions are called *roots* of the equation, or *zeros* of f. If $f(z_0) = 0$ and $f'(z_0) \neq 0$, then z_0 is a *simple zero* of f. If $f(z_0) = f'(z_0) = \cdots = f^{(m-1)}(z_0) = 0$ and $f^{(m)}(z_0) \neq 0$, then z_0 is a zero of f of *multiplicity* m; compare §1.10(i).

Sometimes the equation takes the form

3.8.2
$$z = \phi(z),$$

and the solutions are called *fixed points* of ϕ.

Equations (3.8.1) and (3.8.2) are usually solved by iterative methods. Let $z_1, z_2, \ldots$ be a sequence of approximations to a root, or fixed point, ζ. If

3.8.3
$$|z_{n+1} - \zeta| < A |z_n - \zeta|^p$$

for all n sufficiently large, where A and p are independent of n, then the sequence is said to have *convergence of the pth order*. (More precisely, p is the largest of the possible set of indices for (3.8.3).) If $p = 1$ and $A < 1$, then the convergence is said to be *linear* or *geometric*. If $p = 2$, then the convergence is *quadratic*; if $p = 3$, then the convergence is *cubic*, and so on.

An iterative method converges *locally* to a solution ζ if there exists a neighborhood N of ζ such that $z_n \to \zeta$ whenever the initial approximation z_0 lies within N.

3.8(ii) Newton's Rule

This is an iterative method for real twice-continuously differentiable, or complex analytic, functions:

3.8.4
$$z_{n+1} = z_n - \frac{f(z_n)}{f'(z_n)}, \qquad n = 0, 1, \ldots.$$

If ζ is a simple zero, then the iteration converges locally and quadratically. For multiple zeros the convergence is linear, but if the multiplicity m is known then quadratic convergence can be restored by multiplying the ratio $f(z_n)/f'(z_n)$ in (3.8.4) by m.

For real functions $f(x)$ the sequence of approximations to a real zero ξ will always converge (and converge quadratically) if either:

(a) $f(x_0)f''(x_0) > 0$ and $f'(x), f''(x)$ do not change sign between x_0 and ξ (monotonic convergence).

(b) $f(x_0)f''(x_0) < 0$, $f'(x), f''(x)$ do not change sign in the interval (x_0, x_1), and $\xi \in [x_0, x_1]$ (monotonic convergence after the first iteration).

Example

$f(x) = x - \tan x$. The first positive zero of $f(x)$ lies in the interval $(\pi, \tfrac{3}{2}\pi)$; see Figure 4.15.3. From this graph we estimate an initial value $x_0 = 4.65$. Newton's rule is given by

3.8.5 $x_{n+1} = \phi(x_n), \quad \phi(x) = x + x \cot^2 x - \cot x.$

Results appear in Table 3.8.1. The choice of x_0 here is critical. When $x_0 \leq 4.2875$ or $x_0 \geq 4.7125$, Newton's rule does not converge to the required zero. The convergence is faster when we use instead the function $f(x) = x \cos x - \sin x$; in addition, the successful interval for the starting value x_0 is larger.

Table 3.8.1: Newton's rule for $x - \tan x = 0$.

n	x_n
0	4.65000 00000 000
1	4.60567 66065 900
2	4.55140 53475 751
3	4.50903 76975 617
4	4.49455 61600 185
5	4.49341 56569 391
6	4.49340 94580 903
7	4.49340 94579 091
8	4.49340 94579 091

3.8(iii) Other Methods

Bisection Method

If $f(a)f(b) < 0$ with $a < b$, then the interval $[a,b]$ contains one or more zeros of f. Bisection of this interval is used to decide where at least one zero is located. All zeros of f in the original interval $[a,b]$ can be computed to any predetermined accuracy. Convergence is slow however; see Kaufman and Lenker (1986) and Nievergelt (1995).

Regula Falsi

Let x_0 and x_1 be such that $f_0 = f(x_0)$ and $f_1 = f(x_1)$ have opposite signs. Inverse linear interpolation (§3.3(v)) is used to obtain the first approximation:

$$3.8.6 \qquad x_2 = x_1 - \frac{x_1 - x_0}{f_1 - f_0} f_1 = \frac{f_1 x_0 - f_0 x_1}{f_1 - f_0}.$$

We continue with x_2 and either x_0 or x_1, depending which of f_0 and f_1 is of opposite sign to $f(x_2)$, and so on. The convergence is linear, and again more than one zero may occur in the original interval $[x_0, x_1]$.

Secant Method

Whether or not f_0 and f_1 have opposite signs, x_2 is computed as in (3.8.6). If the wanted zero ξ is simple, then the method converges locally with order of convergence $p = \frac{1}{2}(1 + \sqrt{5}) = 1.618\ldots$. Because the method requires only one function evaluation per iteration, its numerical efficiency is ultimately higher than that of Newton's method. There is no guaranteed convergence: the first approximation x_2 may be outside $[x_0, x_1]$.

Steffensen's Method

This iterative method for solving $z = \phi(z)$ is given by

$$3.8.7 \qquad z_{n+1} = z_n - \frac{(\phi(z_n) - z_n)^2}{\phi(\phi(z_n)) - 2\phi(z_n) + z_n}, \quad n = 0, 1, 2, \ldots.$$

It converges locally and quadratically for both $\mathbb{R}$ and $\mathbb{C}$.

For other efficient derivative-free methods, see Le (1985).

Eigenvalue Methods

For the computation of zeros of orthogonal polynomials as eigenvalues of finite tridiagonal matrices (§3.5(vi)), see Gil et al. (2007a, pp. 205–207). For the computation of zeros of Bessel functions, Coulomb functions, and conical functions as eigenvalues of finite parts of infinite tridiagonal matrices, see Grad and Zakrajšek (1973), Ikebe (1975), Ikebe et al. (1991), Ball (2000), and Gil et al. (2007a, pp. 205–213).

3.8(iv) Zeros of Polynomials

The polynomial

$$3.8.8 \qquad p(z) = a_n z^n + a_{n-1} z^{n-1} + \cdots + a_0, \quad a_n \neq 0,$$

has n zeros in $\mathbb{C}$, counting each zero according to its multiplicity. Explicit formulas for the zeros are available if $n \leq 4$; see §§1.11(iii) and 4.43. No explicit general formulas exist when $n \geq 5$.

After a zero ζ has been computed, the factor $z - \zeta$ is factored out of $p(z)$ as a by-product of Horner's scheme (§1.11(i)) for the computation of $p(\zeta)$. In this way polynomials of successively lower degree can be used to find the remaining zeros. (This process is called *deflation*.) However, to guard against the accumulation of rounding errors, a final iteration for each zero should also be performed on the original polynomial $p(z)$.

Example

$p(z) = z^4 - 1$. The zeros are ± 1 and $\pm i$. Newton's method is given by

$$3.8.9 \qquad z_{n+1} = \phi(z_n), \quad \phi(z) = \frac{3z^4 + 1}{4z^3}.$$

The results for $z_0 = 1.5$ are given in Table 3.8.2.

Table 3.8.2: Newton's rule for $z^4 - 1 = 0$.

n	z_n
0	1.50000 00000 000
1	1.19907 40740 741
2	1.04431 68969 414
3	1.00274 20038 676
4	1.00001 12265 490
5	1.00000 00001 891
6	1.00000 00000 000

As in the case of Table 3.8.1 the quadratic nature of convergence is clearly evident: as the zero is approached, the number of correct decimal places doubles at each iteration.

Newton's rule can also be used for complex zeros of $p(z)$. However, when the coefficients are all real, complex arithmetic can be avoided by the following iterative process.

Bairstow's Method

Let $z^2 - sz - t$ be an approximation to the real quadratic factor of $p(z)$ that corresponds to a pair of conjugate complex zeros or to a pair of real zeros. We construct sequences q_j and r_j, $j = n+1, n, \ldots, 0$, from $q_{n+1} = r_{n+1} = 0$, $q_n = r_n = a_n$, and for $j \leq n - 1$,

$$3.8.10 \qquad q_j = a_j + s q_{j+1} + t q_{j+2}, \quad r_j = q_j + s r_{j+1} + t r_{j+2}.$$

Then the next approximation to the quadratic factor is $z^2 - (s + \Delta s)z - (t + \Delta t)$, where

$$3.8.11 \qquad \Delta s = \frac{r_3 q_0 - r_2 q_1}{r_2^2 - \ell r_3}, \quad \Delta t = \frac{\ell q_1 - r_2 q_0}{r_2^2 - \ell r_3}, \quad \ell = s r_2 + t r_3.$$

The method converges locally and quadratically, except when the wanted quadratic factor is a multiple factor of $q(z)$. On the last iteration $q_n z^{n-2} + q_{n-1} z^{n-3} + \cdots + q_2$ is the quotient on dividing $p(z)$ by $z^2 - sz - t$.

Example

$p(z) = z^4 - 2z^2 + 1$. With the starting values $s_0 = \frac{7}{4}$, $t_0 = -\frac{1}{2}$, an approximation to the quadratic factor $z^2 - 2z + 1 = (z-1)^2$ is computed ($s = 2$, $t = -1$). Table 3.8.3 gives the successive values of s and t. The quadratic nature of the convergence is evident.

Table 3.8.3: Bairstow's method for factoring $z^4 - 2z^2 + 1$.

n	s_n	t_n
0	1.75000 00000 000	-0.50000 00000 000
1	2.13527 29454 109	-1.21235 75284 943
2	2.01786 10488 956	-1.02528 61401 539
3	2.00036 06329 466	-1.00047 63067 522
4	2.00000 01474 803	-1.00000 01858 298
5	2.00000 00000 000	-1.00000 00000 000

This example illustrates the fact that the method succeeds even if the two zeros of the wanted quadratic factor are real and the same.

For further information on the computation of zeros of polynomials see McNamee (2007).

3.8(v) Zeros of Analytic Functions

Newton's rule is the most frequently used iterative process for accurate computation of real or complex zeros of analytic functions $f(z)$. Another iterative method is *Halley's rule*:

3.8.12 $\quad z_{n+1} = z_n - \dfrac{f(z_n)}{f'(z_n) - (f''(z_n) f(z_n)/(2 f'(z_n)))}$.

This is useful when $f(z)$ satisfies a second-order linear differential equation because of the ease of computing $f''(z_n)$. The rule converges locally and is cubically convergent.

Initial approximations to the zeros can often be found from asymptotic or other approximations to $f(z)$, or by application of the phase principle or Rouché's theorem; see §1.10(iv). These results are also useful in ensuring that no zeros are overlooked when the complex plane is being searched.

For an example involving the Airy functions, see Fabijonas and Olver (1999).

For fixed-point methods for computing zeros of special functions, see Segura (2002), Gil and Segura (2003), and Gil et al. (2007a, Chapter 7).

3.8(vi) Conditioning of Zeros

Suppose $f(z)$ also depends on a parameter α, denoted by $f(z, \alpha)$. Then the sensitivity of a simple zero z to changes in α is given by

3.8.13 $\quad \dfrac{dz}{d\alpha} = -\dfrac{\partial f}{\partial \alpha} \bigg/ \dfrac{\partial f}{\partial z}$.

Thus if f is the polynomial (3.8.8) and α is the coefficient a_j, say, then

3.8.14 $\quad \dfrac{dz}{da_j} = -\dfrac{z^j}{f'(z)}$.

For moderate or large values of n it is not uncommon for the magnitude of the right-hand side of (3.8.14) to be very large compared with unity, signifying that the computation of zeros of polynomials is often an ill-posed problem.

Example. Wilkinson's Polynomial

The zeros of

3.8.15 $\quad p(x) = (x-1)(x-2) \cdots (x-20)$

are well separated but extremely ill-conditioned. Consider $x = 20$ and $j = 19$. We have $p'(20) = 19!$ and $a_{19} = 1 + 2 + \cdots + 20 = 210$. The perturbation factor (3.8.14) is given by

3.8.16 $\quad \dfrac{dx}{da_{19}} = -\dfrac{20^{19}}{19!} = (-4.30\ldots) \times 10^7$.

Corresponding numerical factors in this example for other zeros and other values of j are obtained in Gautschi (1984, §4).

3.8(vii) Systems of Nonlinear Equations

For fixed-point iterations and Newton's method for solving systems of nonlinear equations, see Gautschi (1997b, Chapter 4, §9) and Ortega and Rheinboldt (1970).

3.8(viii) Fixed-Point Iterations: Fractals

The convergence of iterative methods

3.8.17 $\quad z_{n+1} = \phi(z_n), \qquad n = 0, 1, \ldots,$

for solving fixed-point problems (3.8.2) cannot always be predicted, especially in the complex plane.

Consider, for example, (3.8.9). Starting this iteration in the neighborhood of one of the four zeros $\pm 1, \pm i$, sequences $\{z_n\}$ are generated that converge to these zeros. For an arbitrary starting point $z_0 \in \mathbb{C}$, convergence cannot be predicted, and the boundary of the set of points z_0 that generate a sequence converging to a particular zero has a very complicated structure. It is called a *Julia set*. In general the Julia set of an analytic function $f(z)$ is a *fractal*, that is, a set that is self-similar. See Julia (1918) and Devaney (1986).

3.9 Acceleration of Convergence

3.9(i) Sequence Transformations

All sequences (series) in this section are sequences (series) of real or complex numbers.

A transformation of a convergent sequence $\{s_n\}$ with limit σ into a sequence $\{t_n\}$ is called *limit-preserving* if $\{t_n\}$ converges to the same limit σ.

The transformation is *accelerating* if it is limit-preserving and if

$$3.9.1 \qquad \lim_{n \to \infty} \frac{t_n - \sigma}{s_n - \sigma} = 0.$$

Similarly for convergent series if we regard the sum as the limit of the sequence of partial sums.

It should be borne in mind that a sequence (series) transformation can be effective for one type of sequence (series) but may not accelerate convergence for another type. It may even fail altogether by not being limit-preserving.

3.9(ii) Euler's Transformation of Series

If $S = \sum_{k=0}^{\infty}(-1)^k a_k$ is a convergent series, then

$$3.9.2 \qquad S = \sum_{k=0}^{\infty}(-1)^k 2^{-k-1} \Delta^k a_0,$$

provided that the right-hand side converges. Here Δ is the *forward difference operator*:

$$3.9.3 \qquad \Delta^k a_0 = \Delta^{k-1} a_1 - \Delta^{k-1} a_0, \quad k = 1, 2, \dots.$$

Thus

$$3.9.4 \qquad \Delta^k a_0 = \sum_{m=0}^{k}(-1)^m \binom{k}{m} a_{k-m}.$$

Euler's transformation is usually applied to alternating series. Examples are provided by the following analytic transformations of slowly-convergent series into rapidly convergent ones:

$$3.9.5 \quad \ln 2 = 1 - \frac{1}{2} + \frac{1}{3} - \frac{1}{4} + \cdots = \frac{1}{1 \cdot 2^1} + \frac{1}{2 \cdot 2^2} + \frac{1}{3 \cdot 2^3} + \cdots,$$

$$3.9.6 \quad \begin{aligned}\frac{\pi}{4} &= 1 - \frac{1}{3} + \frac{1}{5} - \frac{1}{7} + \cdots \\ &= \frac{1}{2}\left(1 + \frac{1!}{1 \cdot 3} + \frac{2!}{3 \cdot 5} + \frac{3!}{3 \cdot 5 \cdot 7} + \cdots\right).\end{aligned}$$

3.9(iii) Aitken's Δ^2-Process

$$3.9.7 \quad t_n = s_n - \frac{(\Delta s_n)^2}{\Delta^2 s_n} = s_n - \frac{(s_{n+1} - s_n)^2}{s_{n+2} - 2s_{n+1} + s_n}.$$

This transformation is accelerating if $\{s_n\}$ is a *linearly convergent sequence*, i.e., a sequence for which

$$3.9.8 \qquad \lim_{n \to \infty} \frac{s_{n+1} - \sigma}{s_n - \sigma} = \rho, \qquad |\rho| < 1.$$

When applied repeatedly, Aitken's process is known as the *iterated Δ^2-process*. See Brezinski and Redivo Zaglia (1991, pp. 39–42).

3.9(iv) Shanks' Transformation

Shanks' transformation is a generalization of Aitken's Δ^2-process. Let k be a fixed positive integer. Then the transformation of the sequence $\{s_n\}$ into a sequence $\{t_{n,2k}\}$ is given by

$$3.9.9 \qquad t_{n,2k} = \frac{H_{k+1}(s_n)}{H_k(\Delta^2 s_n)}, \quad n = 0, 1, 2, \dots,$$

where H_m is the *Hankel determinant*

$$3.9.10 \quad H_m(u_n) = \begin{vmatrix} u_n & u_{n+1} & \cdots & u_{n+m-1} \\ u_{n+1} & u_{n+2} & \cdots & u_{n+m} \\ \vdots & \vdots & \ddots & \vdots \\ u_{n+m-1} & u_{n+m} & \cdots & u_{n+2m-2} \end{vmatrix}.$$

The ratio of the Hankel determinants in (3.9.9) can be computed recursively by *Wynn's epsilon algorithm*:

3.9.11
$$\varepsilon_{-1}^{(n)} = 0, \quad \varepsilon_0^{(n)} = s_n, \qquad n = 0, 1, 2, \dots,$$
$$\varepsilon_{m+1}^{(n)} = \varepsilon_{m-1}^{(n+1)} + \frac{1}{\varepsilon_m^{(n+1)} - \varepsilon_m^{(n)}}, \quad n, m = 0, 1, 2, \dots.$$

Then $t_{n,2k} = \varepsilon_{2k}^{(n)}$. Aitken's Δ^2-process is the case $k = 1$.

If s_n is the nth partial sum of a power series f, then $t_{n,2k} = \varepsilon_{2k}^{(n)}$ is the Padé approximant $[(n+k)/k]_f$ (§3.11(iv)).

For further information on the epsilon algorithm see Brezinski and Redivo Zaglia (1991, pp. 78–95).

Example

In Table 3.9.1 values of the transforms $t_{n,2k}$ are supplied for

$$3.9.12 \qquad s_n = \sum_{j=1}^{n} \frac{(-1)^{j+1}}{j^2},$$

with $s_\infty = \frac{1}{12}\pi^2 = 0.82246\,70334\,24\dots$.

Table 3.9.1: Shanks' transformation for $s_n = \sum_{j=1}^{n}(-1)^{j+1}j^{-2}$.

n	$t_{n,2}$	$t_{n,4}$	$t_{n,6}$	$t_{n,8}$	$t_{n,10}$
0	0.80000 00000 00	0.82182 62806 24	0.82244 84501 47	0.82246 64909 60	0.82246 70175 41
1	0.82692 30769 23	0.82259 02017 65	0.82247 05346 57	0.82246 71342 06	0.82246 70363 45
2	0.82111 11111 11	0.82243 44785 14	0.82246 61821 45	0.82246 70102 48	0.82246 70327 79
3	0.82300 13550 14	0.82247 78118 35	0.82246 72851 83	0.82246 70397 56	0.82246 70335 90
4	0.82221 76684 88	0.82246 28314 41	0.82246 69467 93	0.82246 70314 36	0.82246 70333 75
5	0.82259 80392 16	0.82246 88857 22	0.82246 70670 21	0.82246 70341 24	0.82246 70334 40
6	0.82239 19390 77	0.82246 61352 37	0.82246 70190 76	0.82246 70331 54	0.82246 70334 18
7	0.82251 30483 23	0.82246 75033 13	0.82246 70400 56	0.82246 70335 37	0.82246 70334 26
8	0.82243 73137 33	0.82246 67719 32	0.82246 70301 49	0.82246 70333 73	0.82246 70334 23
9	0.82248 70624 89	0.82246 71865 91	0.82246 70351 34	0.82246 70334 48	0.82246 70334 24
10	0.82245 30535 15	0.82246 69397 57	0.82246 70324 88	0.82246 70334 12	0.82246 70334 24

3.9(v) Levin's and Weniger's Transformations

We give a special form of *Levin's transformation* in which the sequence $s = \{s_n\}$ of partial sums $s_n = \sum_{j=0}^{n} a_j$ is transformed into:

3.9.13
$$\mathcal{L}_k^{(n)}(s) = \frac{\sum_{j=0}^{k}(-1)^j \binom{k}{j} c_{j,k,n}\, s_{n+j}/a_{n+j+1}}{\sum_{j=0}^{k}(-1)^j \binom{k}{j} c_{j,k,n}/a_{n+j+1}},$$

where k is a fixed nonnegative integer, and

3.9.14
$$c_{j,k,n} = \frac{(n+j+1)^{k-1}}{(n+k+1)^{k-1}}.$$

Sequences that are accelerated by Levin's transformation include *logarithmically convergent* sequences, i.e., sequences s_n converging to σ such that

3.9.15
$$\lim_{n \to \infty} \frac{s_{n+1} - \sigma}{s_n - \sigma} = 1.$$

For further information see Brezinski and Redivo Zaglia (1991, pp. 39–42).

In *Weniger's transformations* the numbers $c_{j,k,n}$ in (3.9.13) are chosen as follows:

3.9.16
$$c_{j,k,n} = \frac{(\beta + n + j)_{k-1}}{(\beta + n + k)_{k-1}},$$

or

3.9.17
$$c_{j,k,n} = \frac{(-\gamma - n - j)_{k-1}}{(-\gamma - n - k)_{k-1}},$$

where $(a)_0 = 1$ and $(a)_j = a(a+1)\cdots(a+j-1)$ are Pochhammer symbols (§5.2(iii)), and the constants β and γ are chosen arbitrarily subject to certain conditions. See Weniger (1989).

3.9(vi) Applications and Further Transformations

For examples and other transformations for convergent sequences and series, see Wimp (1981, pp. 156–199), Brezinski and Redivo Zaglia (1991, pp. 55–72), and Sidi (2003, Chapters 6, 12–13, 15–16, 19–24, and pp. 483–492).

For applications to asymptotic expansions, see §2.11(vi), Olver (1997b, pp. 540–543), and Weniger (1989, 2003).

3.10 Continued Fractions

3.10(i) Introduction

See §1.12 for relevant properties of continued fractions, including the following definitions:

3.10.1
$$C = b_0 + \frac{a_1}{b_1 +} \frac{a_2}{b_2 +} \cdots, \qquad a_n \neq 0,$$

3.10.2
$$C_n = b_0 + \frac{a_1}{b_1 +} \frac{a_2}{b_2 +} \cdots \frac{a_n}{b_n} = \frac{A_n}{B_n}.$$

C_n is the nth *approximant* or *convergent* to C.

3.10(ii) Relations to Power Series

Every convergent, asymptotic, or formal series

3.10.3
$$u_0 + u_1 + u_2 + \cdots$$

can be converted into a continued fraction C of type (3.10.1), and with the property that the nth convergent $C_n = A_n/B_n$ to C is equal to the nth partial sum of the series in (3.10.3), that is,

3.10.4
$$\frac{A_n}{B_n} = u_0 + u_1 + \cdots + u_n, \qquad n = 0, 1, \ldots.$$

For instance, if none of the u_n vanish, then we can define

$$b_0 = u_0, \quad b_1 = 1, \quad a_1 = u_1,$$

3.10.5
$$b_n = 1 + \frac{u_n}{u_{n-1}}, \quad a_n = -\frac{u_n}{u_{n-1}}, \qquad n \geq 2.$$

However, other continued fractions with the same limit may converge in a much larger domain of the complex plane than the fraction given by (3.10.4) and (3.10.5). For example, by converting the Maclaurin expansion of arctan z (4.24.3), we obtain a continued fraction with the same region of convergence ($|z| \leq 1$, $z \neq \pm i$), whereas the continued fraction (4.25.4) converges for all $z \in \mathbb{C}$ except on the branch cuts from i to $i\infty$ and $-i$ to $-i\infty$.

Stieltjes Fractions

A continued fraction of the form

3.10.6 $$C = \frac{a_0}{1-} \frac{a_1 z}{1-} \frac{a_2 z}{1-} \cdots$$

is called a *Stieltjes fraction (S-fraction)*. We say that it *corresponds* to the formal power series

3.10.7 $$f(z) = c_0 + c_1 z + c_2 z^2 + \cdots$$

if the expansion of its nth convergent C_n in ascending powers of z agrees with (3.10.7) up to and including the term in z^{n-1}, $n = 1, 2, 3, \ldots$.

Quotient-Difference Algorithm

For several special functions the S-fractions are known explicitly, but in any case the coefficients a_n can always be calculated from the power-series coefficients by means of the *quotient-difference algorithm*; see Table 3.10.1.

Table 3.10.1: Quotient-difference scheme.

	q_1^0			
e_0^1		e_1^0		
	q_1^1		q_2^0	
e_0^2		e_1^1		e_2^0
	q_1^2		q_2^1		q_3^0
e_0^3		e_1^2		e_2^1		e_3^0
	q_1^3		q_2^2		q_3^1
e_0^4		e_1^3		e_2^2		e_3^1

The first two columns in this table are defined by

3.10.8 $$e_0^n = 0, \quad n = 1, 2, \ldots,$$
$$q_1^n = c_{n+1}/c_n, \quad n = 0, 1, \ldots,$$

where the c_n ($\neq 0$) appear in (3.10.7). We continue by means of the *rhombus rule*

3.10.9 $$e_j^k = e_{j-1}^{k+1} + q_j^{k+1} - q_j^k, \quad j \geq 1, k \geq 0,$$
$$q_{j+1}^k = q_j^{k+1} e_j^{k+1}/e_j^k, \quad j \geq 1, k \geq 0.$$

Then the coefficients a_n of the S-fraction (3.10.6) are given by

3.10.10 $a_0 = c_0, \quad a_1 = q_1^0, \quad a_2 = e_1^0, \quad a_3 = q_2^0, \quad a_4 = e_2^0, \quad \ldots$

The quotient-difference algorithm is frequently unstable and may require high-precision arithmetic or exact arithmetic. A more stable version of the algorithm is discussed in Stokes (1980). For applications to Bessel functions and Whittaker functions (Chapters 10 and 13), see Gargantini and Henrici (1967).

Jacobi Fractions

A continued fraction of the form

3.10.11 $$C = \frac{\beta_0}{1 - \alpha_0 z -} \frac{\beta_1 z^2}{1 - \alpha_1 z -} \frac{\beta_2 z^2}{1 - \alpha_2 z -} \cdots$$

is called a *Jacobi fraction (J-fraction)*. We say that it is *associated* with the formal power series $f(z)$ in (3.10.7) if the expansion of its nth convergent C_n in ascending powers of z, agrees with (3.10.7) up to and including the term in z^{2n-1}, $n = 1, 2, 3, \ldots$. For the same function $f(z)$, the convergent C_n of the Jacobi fraction (3.10.11) equals the convergent C_{2n} of the Stieltjes fraction (3.10.6).

Examples of S- and J-Fractions

For elementary functions, see §§4.9 and 4.35.

For special functions see §5.10 (gamma function), §7.9 (error function), §8.9 (incomplete gamma functions), §8.17(v) (incomplete beta function), §8.19(vii) (generalized exponential integral), §§10.10 and 10.33 (quotients of Bessel functions), §13.6 (quotients of confluent hypergeometric functions), §13.19 (quotients of Whittaker functions), and §15.7 (quotients of hypergeometric functions).

For further information and examples see Lorentzen and Waadeland (1992, pp. 292–330, 560–599) and Cuyt et al. (2008).

3.10(iii) Numerical Evaluation of Continued Fractions

Forward Recurrence Algorithm

The A_n and B_n of (3.10.2) can be computed by means of three-term recurrence relations (1.12.5). However, this may be unstable; also overflow and underflow may occur when evaluating A_n and B_n (making it necessary to re-scale from time to time).

Backward Recurrence Algorithm

To compute the C_n of (3.10.2) we perform the iterated divisions

3.10.12 $$u_n = b_n, \quad u_k = b_k + \frac{a_{k+1}}{u_{k+1}}, \quad k = n-1, n-2, \ldots, 0.$$

Then $u_0 = C_n$. To achieve a prescribed accuracy, either *a priori* knowledge is needed of the value of n, or n is determined by trial and error. In general this algorithm is more stable than the forward algorithm; see Jones and Thron (1974).

Forward Series Recurrence Algorithm

The continued fraction

3.10.13 $$C = \frac{a_0}{1-}\frac{a_1}{1-}\frac{a_2}{1-}\cdots$$

can be written in the form

3.10.14 $$C = \sum_{k=0}^{\infty} t_k,$$

where

3.10.15 $$t_0 = a_0, \quad t_k = \rho_k t_{k-1}, \quad \rho_0 = 0,$$
$$\rho_k = \frac{a_k(1+\rho_{k-1})}{1-a_k(1+\rho_{k-1})}, \quad k=1,2,3,\ldots.$$

The nth partial sum $t_0 + t_1 + \cdots + t_{n-1}$ equals the nth convergent of (3.10.13), $n=1,2,3,\ldots$. In contrast to the preceding algorithms in this subsection no scaling problems arise and no *a priori* information is needed.

In Gautschi (1979b) the forward series algorithm is used for the evaluation of a continued fraction of an incomplete gamma function (see §8.9).

Steed's Algorithm

This forward algorithm achieves efficiency and stability in the computation of the convergents $C_n = A_n/B_n$, and is related to the forward series recurrence algorithm. Again, no scaling problems arise and no *a priori* information is needed.

Let

3.10.16 $$C_0 = b_0, \quad D_1 = 1/b_1, \quad \nabla C_1 = a_1 D_1, \quad C_1 = C_0 + \nabla C_1.$$

(∇ is the *backward difference operator*.) Then for $n \geq 2$,

3.10.17 $$D_n = \frac{1}{D_{n-1}a_n + b_n},$$
$$\nabla C_n = (b_n D_n - 1)\nabla C_{n-1},$$
$$C_n = C_{n-1} + \nabla C_n.$$

The recurrences are continued until $(\nabla C_n)/C_n$ is within a prescribed relative precision.

For further information on the preceding algorithms, including convergence in the complex plane and methods for accelerating convergence, see Blanch (1964) and Lorentzen and Waadeland (1992, Chapter 3). For the evaluation of special functions by using continued fractions see Cuyt et al. (2008), Gautschi (1967, §1), Gil et al. (2007a, Chapter 6), and Wimp (1984, Chapter 4, §5). See also §§6.18(i), 7.22(i), 8.25(iv), 10.74(v), 14.32, 28.34(ii), 29.20(i), 30.16(i), 33.23(v).

3.11 Approximation Techniques

3.11(i) Minimax Polynomial Approximations

Let $f(x)$ be continuous on a closed interval $[a,b]$. Then there exists a unique nth degree polynomial $p_n(x)$, called the *minimax* (or *best uniform*) polynomial approximation to $f(x)$ on $[a,b]$, that minimizes $\max_{a \leq x \leq b} |\epsilon_n(x)|$, where $\epsilon_n(x) = f(x) - p_n(x)$.

A sufficient condition for $p_n(x)$ to be the minimax polynomial is that $|\epsilon_n(x)|$ attains its maximum at $n+2$ distinct points in $[a,b]$ and $\epsilon_n(x)$ changes sign at these consecutive maxima.

If we have a sufficiently close approximation

3.11.1 $$p_n(x) = a_n x^n + a_{n-1} x^{n-1} + \cdots + a_0$$

to $f(x)$, then the coefficients a_k can be computed iteratively. Assume that $f'(x)$ is continuous on $[a,b]$ and let $x_0 = a$, $x_{n+1} = b$, and $x_1, x_2, \ldots, x_n$ be the zeros of $\epsilon'_n(x)$ in (a,b) arranged so that

3.11.2 $$x_0 < x_1 < x_2 < \cdots < x_n < x_{n+1}.$$

Also, let

3.11.3 $$m_j = (-1)^j \epsilon_n(x_j), \quad j = 0,1,\ldots,n+1.$$

(Thus the m_j are approximations to m, where $\pm m$ is the maximum value of $|\epsilon_n(x)|$ on $[a,b]$.)

Then (in general) a better approximation to $p_n(x)$ is given by

3.11.4 $$\sum_{k=0}^{n}(a_k + \delta a_k)x^k,$$

where

3.11.5 $$\sum_{k=0}^{n} x_j^k \delta a_k = (-1)^j(m_j - m), \quad j=0,1,\ldots,n+1.$$

This is a set of $n+2$ equations for the $n+2$ unknowns $\delta a_0, \delta a_1, \ldots, \delta a_n$ and m.

The iterative process converges locally and quadratically (§3.8(i)).

A method for obtaining a sufficiently accurate first approximation is described in the next subsection.

For the theory of minimax approximations see Meinardus (1967). For examples of minimax polynomial approximations to elementary and special functions see Hart et al. (1968). See also Cody (1970) and Ralston (1965).

3.11(ii) Chebyshev-Series Expansions

The Chebyshev polynomials T_n are given by

3.11.6 $$T_n(x) = \cos(n \arccos x), \quad -1 \leq x \leq 1.$$

They satisfy the recurrence relation

3.11.7 $$T_{n+1}(x) - 2x\,T_n(x) + T_{n-1}(x) = 0, \quad n=1,2,\ldots,$$

with initial values $T_0(x) = 1$, $T_1(x) = x$. They enjoy an orthogonal property with respect to integrals:

3.11.8 $$\int_{-1}^{1} \frac{T_j(x)\,T_k(x)}{\sqrt{1-x^2}}\,dx = \begin{cases} \pi, & j=k=0, \\ \frac{1}{2}\pi, & j=k\neq 0, \\ 0, & j\neq k, \end{cases}$$

as well as an orthogonal property with respect to sums, as follows. When $n > 0$ and $0 \leq j \leq n$, $0 \leq k \leq n$,

3.11.9 $$\sum_{\ell=0}^{n}{}'' T_j(x_\ell) T_k(x_\ell) = \begin{cases} n, & j = k = 0 \text{ or } n, \\ \frac{1}{2}n, & j = k \neq 0 \text{ or } n, \\ 0, & j \neq k, \end{cases}$$

where $x_\ell = \cos(\pi\ell/n)$ and the double prime means that the first and last terms are to be halved.

For these and further properties of Chebyshev polynomials, see Chapter 18, Gil et al. (2007a, Chapter 3), and Mason and Handscomb (2003).

Chebyshev Expansions

If f is continuously differentiable on $[-1, 1]$, then with

3.11.10 $$c_n = \frac{2}{\pi} \int_0^\pi f(\cos\theta) \cos(n\theta) \, d\theta, \quad n = 0, 1, 2, \ldots,$$

the expansion

3.11.11 $$f(x) = \sum_{n=0}^{\infty}{}' c_n T_n(x), \quad -1 \leq x \leq 1,$$

converges uniformly. Here the single prime on the summation symbol means that the first term is to be halved. In fact, (3.11.11) is the Fourier-series expansion of $f(\cos\theta)$; compare (3.11.6) and §1.8(i).

Furthermore, if $f \in C^\infty[-1, 1]$, then the convergence of (3.11.11) is usually very rapid; compare (1.8.7) with k arbitrary.

For general intervals $[a, b]$ we rescale:

3.11.12 $$f(x) = \sum_{n=0}^{\infty}{}' d_n T_n\left(\frac{2x - a - b}{b - a}\right).$$

Because the series (3.11.12) converges rapidly we obtain a very good first approximation to the minimax polynomial $p_n(x)$ for $[a, b]$ if we truncate (3.11.12) at its $(n + 1)$th term. This is because in the notation of §3.11(i)

3.11.13 $$\epsilon_n(x) = d_{n+1} T_{n+1}\left(\frac{2x - a - b}{b - a}\right),$$

approximately, and the right-hand side enjoys exactly those properties concerning its maxima and minima that are required for the minimax approximation; compare Figure 18.4.3.

More precisely, it is known that for the interval $[a, b]$, the ratio of the maximum value of the remainder

3.11.14 $$\left| \sum_{k=n+1}^{\infty} d_k T_k\left(\frac{2x - a - b}{b - a}\right) \right|$$

to the maximum error of the minimax polynomial $p_n(x)$ is bounded by $1 + L_n$, where L_n is the nth *Lebesgue constant* for Fourier series; see §1.8(i). Since $L_0 = 1$, L_n is a monotonically increasing function of n, and (for example) $L_{1000} = 4.07\ldots$, this means that in practice the gain in replacing a truncated Chebyshev-series expansion by the corresponding minimax polynomial approximation is hardly worthwhile. Moreover, the set of minimax approximations $p_0(x), p_1(x), p_2(x), \ldots, p_n(x)$ requires the calculation and storage of $\frac{1}{2}(n+1)(n+2)$ coefficients, whereas the corresponding set of Chebyshev-series approximations requires only $n + 1$ coefficients.

Calculation of Chebyshev Coefficients

The c_n in (3.11.11) can be calculated from (3.11.10), but in general it is more efficient to make use of the orthogonal property (3.11.9). Also, in cases where $f(x)$ satisfies a linear ordinary differential equation with polynomial coefficients, the expansion (3.11.11) can be substituted in the differential equation to yield a recurrence relation satisfied by the c_n.

For details and examples of these methods, see Clenshaw (1957, 1962) and Miller (1966). See also Mason and Handscomb (2003, Chapter 10) and Fox and Parker (1968, Chapter 5).

Summation of Chebyshev Series: Clenshaw's Algorithm

For the expansion (3.11.11), numerical values of the Chebyshev polynomials $T_n(x)$ can be generated by application of the recurrence relation (3.11.7). A more efficient procedure is as follows. Let $c_n T_n(x)$ be the last term retained in the truncated series. Beginning with $u_{n+1} = 0$, $u_n = c_n$, we apply

3.11.15 $$u_k = 2xu_{k+1} - u_{k+2} + c_k, \quad k = n - 1, n - 2, \ldots, 0.$$

Then the sum of the truncated expansion equals $\frac{1}{2}(u_0 - u_2)$. For error analysis and modifications of Clenshaw's algorithm, see Oliver (1977).

Complex Variables

If x is replaced by a complex variable z and $f(z)$ is analytic, then the expansion (3.11.11) converges within an ellipse. However, in general (3.11.11) affords no advantage in $\mathbb{C}$ for numerical purposes compared with the Maclaurin expansion of $f(z)$.

For further details on Chebyshev-series expansions in the complex plane, see Mason and Handscomb (2003, §5.10).

3.11(iii) Minimax Rational Approximations

Let f be continuous on a closed interval $[a, b]$ and w be a continuous nonvanishing function on $[a, b]$: w is called a *weight function*. Then the *minimax* (or *best uniform*) rational approximation

3.11.16 $$R_{k,\ell}(x) = \frac{p_0 + p_1 x + \cdots + p_k x^k}{1 + q_1 x + \cdots + q_\ell x^\ell}$$

of *type* $[k, \ell]$ to f on $[a, b]$ minimizes the maximum value of $|\epsilon_{k,\ell}(x)|$ on $[a, b]$, where

3.11.17 $$\epsilon_{k,\ell}(x) = \frac{R_{k,\ell}(x) - f(x)}{w(x)}.$$

The theory of polynomial minimax approximation given in §3.11(i) can be extended to the case when $p_n(x)$ is replaced by a rational function $R_{k,\ell}(x)$. There exists a unique solution of this minimax problem and there are at least $k+\ell+2$ values x_j, $a \leq x_0 < x_1 < \cdots < x_{k+\ell+1} \leq b$, such that $m_j = m$, where

3.11.18 $\quad m_j = (-1)^j \epsilon_{k,\ell}(x_j), \quad j = 0, 1, \ldots, k+\ell+1,$

and $\pm m$ is the maximum of $|\epsilon_{k,\ell}(x)|$ on $[a,b]$.

A collection of minimax rational approximations to elementary and special functions can be found in Hart et al. (1968).

A widely implemented and used algorithm for calculating the coefficients p_j and q_j in (3.11.16) is *Remez's second algorithm*. See Remez (1957), Werner et al. (1967), and Johnson and Blair (1973).

Example

With $w(x) = 1$ and 14-digit computation, we obtain the following rational approximation of type $[3,3]$ to the Bessel function $J_0(x)$ (§10.2(ii)) on the interval $0 \leq x \leq j_{0,1}$, where $j_{0,1}$ is the first positive zero of $J_0(x)$:

3.11.19 $\quad R_{3,3}(x) = \dfrac{p_0 + p_1 x + p_2 x^2 + p_3 x^3}{1 + q_1 x + q_2 x^2 + q_3 x^3},$

with coefficients given in Table 3.11.1.

Table 3.11.1: Coefficients p_j, q_j for the minimax rational approximation $R_{3,3}(x)$.

j	p_j	q_j
0	0.99999 99891 7854	
1	$-0.34038\ 93820\ 9347$	$-0.34039\ 05233\ 8838$
2	$-0.18915\ 48376\ 3222$	$0.06086\ 50162\ 9812$
3	$0.06658\ 31942\ 0166$	$-0.01864\ 47680\ 9090$

The error curve is shown in Figure 3.11.1.

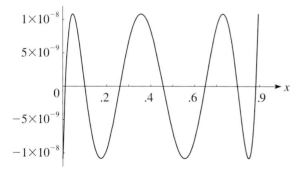

Figure 3.11.1: Error $R_{3,3}(x) - J_0(x)$ of the minimax rational approximation $R_{3,3}(x)$ to the Bessel function $J_0(x)$ for $0 \leq x \leq j_{0,1}$ ($= 0.89357\ldots$).

3.11(iv) Padé Approximations

Let

3.11.20 $\quad f(z) = c_0 + c_1 z + c_2 z^2 + \cdots$

be a formal power series. The rational function

3.11.21 $\quad \dfrac{N_{p,q}(z)}{D_{p,q}(z)} = \dfrac{a_0 + a_1 z + \cdots + a_p z^p}{b_0 + b_1 z + \cdots + b_q z^q}$

is called a *Padé approximant at zero* of f if

3.11.22 $\quad N_{p,q}(z) - f(z) D_{p,q}(z) = O\!\left(z^{p+q+1}\right), \quad z \to 0.$

It is denoted by $[p/q]_f(z)$. Thus if $b_0 \neq 0$, then the Maclaurin expansion of (3.11.21) agrees with (3.11.20) up to, and including, the term in z^{p+q}.

The requirement (3.11.22) implies

3.11.23
$$\begin{aligned}
a_0 &= c_0 b_0, \\
a_1 &= c_1 b_0 + c_0 b_1, \\
&\vdots \\
a_p &= c_p b_0 + c_{p-1} b_1 + \cdots + c_{p-q} b_q, \\
0 &= c_{p+1} b_0 + c_p b_1 + \cdots + c_{p-q+1} b_q, \\
&\vdots \\
0 &= c_{p+q} b_0 + c_{p+q-1} b_1 + \cdots + c_p b_q,
\end{aligned}$$

where $c_j = 0$ if $j < 0$. With $b_0 = 1$, the last q equations give $b_1, \ldots, b_q$ as the solution of a system of linear equations. The first $p+1$ equations then yield $a_0, \ldots, a_p$.

The array of Padé approximants

3.11.24
$$\begin{matrix}
[0/0]_f & [0/1]_f & [0/2]_f & \cdots \\
[1/0]_f & [1/1]_f & [1/2]_f & \cdots \\
[2/0]_f & [2/1]_f & [2/2]_f & \cdots \\
\vdots & \vdots & \vdots & \ddots
\end{matrix}$$

is called a *Padé table*. Approximants with the same denominator degree are located in the same column of the table.

For convergence results for Padé approximants, and the connection with continued fractions and Gaussian quadrature, see Baker and Graves-Morris (1996, §4.7).

The Padé approximants can be computed by *Wynn's cross rule*. Any five approximants arranged in the Padé table as

$$\begin{matrix} & N & \\ W & C & E \\ & S & \end{matrix}$$

satisfy

3.11.25 $\quad (N-C)^{-1} + (S-C)^{-1} = (W-C)^{-1} + (E-C)^{-1}.$

Starting with the first column $[n/0]_f$, $n = 0,1,2,\ldots$, and initializing the preceding column by $[n/-1]_f = \infty$, $n = 1,2,\ldots$, we can compute the lower triangular part of the table via (3.11.25). Similarly, the upper triangular part follows from the first row $[0/n]_f$, $n = 0,1,2,\ldots$, by initializing $[-1/n]_f = 0$, $n = 1,2,\ldots$.

For the recursive computation of $[n+k/k]_f$ by Wynn's epsilon algorithm, see (3.9.11) and the subsequent text.

Laplace Transform Inversion

Numerical inversion of the Laplace transform (§1.14(iii))

3.11.26 $\quad F(s) = \mathscr{L}(f;s) = \int_0^\infty e^{-st} f(t)\, dt$

requires $f = \mathscr{L}^{-1} F$ to be obtained from numerical values of F. A general procedure is to approximate F by a rational function R (vanishing at infinity) and then approximate f by $r = \mathscr{L}^{-1} R$. When F has an explicit power-series expansion a possible choice of R is a Padé approximation to F. See Luke (1969b, §16.4) for several examples involving special functions.

For further information on Padé approximations, see Baker and Graves-Morris (1996, §4.7), Brezinski (1980, pp. 9–39 and 126–177), and Lorentzen and Waadeland (1992, pp. 367–395).

3.11(v) Least Squares Approximations

Suppose a function $f(x)$ is approximated by the polynomial

3.11.27 $\quad p_n(x) = a_n x^n + a_{n-1} x^{n-1} + \cdots + a_0$

that minimizes

3.11.28 $\quad S = \sum_{j=1}^J (f(x_j) - p_n(x_j))^2.$

Here x_j, $j = 1, 2, \ldots, J$, is a given set of distinct real points and $J \geq n+1$. From the equations $\partial S/\partial a_k = 0$, $k = 0, 1, \ldots, n$, we derive the *normal equations*

3.11.29 $\quad \begin{bmatrix} X_0 & X_1 & \cdots & X_n \\ X_1 & X_2 & \cdots & X_{n+1} \\ \vdots & \vdots & \ddots & \vdots \\ X_n & X_{n+1} & \cdots & X_{2n} \end{bmatrix} \begin{bmatrix} a_0 \\ a_1 \\ \vdots \\ a_n \end{bmatrix} = \begin{bmatrix} F_0 \\ F_1 \\ \vdots \\ F_n \end{bmatrix},$

where

3.11.30 $\quad X_k = \sum_{j=1}^J x_j^k, \quad F_k = \sum_{j=1}^J f(x_j) x_j^k.$

(3.11.29) is a system of $n+1$ linear equations for the coefficients $a_0, a_1, \ldots, a_n$. The matrix is symmetric and positive definite, but the system is ill-conditioned when n is large because the lower rows of the matrix are approximately proportional to one another. If $J = n+1$, then $p_n(x)$ is the Lagrange interpolation polynomial for the set $x_1, x_2, \ldots, x_J$ (§3.3(i)).

More generally, let $f(x)$ be approximated by a linear combination

3.11.31 $\quad \Phi_n(x) = a_n \phi_n(x) + a_{n-1} \phi_{n-1}(x) + \cdots + a_0 \phi_0(x)$

of given functions $\phi_k(x)$, $k = 0, 1, \ldots, n$, that minimizes

3.11.32 $\quad \sum_{j=1}^J w(x_j) (f(x_j) - \Phi_n(x_j))^2,$

$w(x)$ being a given positive *weight function*, and again $J \geq n+1$. Then (3.11.29) is replaced by

3.11.33 $\quad \begin{bmatrix} X_{00} & X_{01} & \cdots & X_{0n} \\ X_{10} & X_{11} & \cdots & X_{1n} \\ \vdots & \vdots & \ddots & \vdots \\ X_{n0} & X_{n1} & \cdots & X_{nn} \end{bmatrix} \begin{bmatrix} a_0 \\ a_1 \\ \vdots \\ a_n \end{bmatrix} = \begin{bmatrix} F_0 \\ F_1 \\ \vdots \\ F_n \end{bmatrix},$

with

3.11.34 $\quad X_{k\ell} = \sum_{j=1}^J w(x_j) \phi_k(x_j) \phi_\ell(x_j),$

and

3.11.35 $\quad F_k = \sum_{j=1}^J w(x_j) f(x_j) \phi_k(x_j).$

Since $X_{k\ell} = X_{\ell k}$, the matrix is again symmetric.

If the functions $\phi_k(x)$ are linearly independent on the set $x_1, x_2, \ldots, x_J$, that is, the only solution of the system of equations

3.11.36 $\quad \sum_{k=0}^n c_k \phi_k(x_j) = 0, \quad j = 1, 2, \ldots, J,$

is $c_0 = c_1 = \cdots = c_n = 0$, then the approximation $\Phi_n(x)$ is determined uniquely.

Now suppose that $X_{k\ell} = 0$ when $k \neq \ell$, that is, the functions $\phi_k(x)$ *are orthogonal with respect to weighted summation on the discrete set* $x_1, x_2, \ldots, x_J$. Then the system (3.11.33) is diagonal and hence well-conditioned.

A set of functions $\phi_0(x), \phi_1(x), \ldots, \phi_n(x)$ that is linearly independent on the set $x_1, x_2, \ldots, x_J$ (compare (3.11.36)) can always be orthogonalized in the sense given in the preceding paragraph by the *Gram–Schmidt* procedure; see Gautschi (1997b).

Example. The Discrete Fourier Transform

We take n complex exponentials $\phi_k(x) = e^{ikx}$, $k = 0, 1, \ldots, n-1$, and approximate $f(x)$ by the linear combination (3.11.31). The functions $\phi_k(x)$ are orthogonal on the set $x_0, x_1, \ldots, x_{n-1}$, $x_j = 2\pi j/n$, with respect to the weight function $w(x) = 1$, in the sense that

3.11.37 $\quad \sum_{j=0}^{n-1} \phi_k(x_j) \overline{\phi_\ell(x_j)} = n \delta_{k,\ell}, \quad k, \ell = 0, 1, \ldots, n-1,$

$\delta_{k,\ell}$ being Kronecker's symbol and the bar denoting complex conjugate. In consequence we can solve the system

3.11.38 $\quad f_j = \sum_{k=0}^{n-1} a_k \phi_k(x_j), \quad j = 0, 1, \ldots, n-1,$

and obtain

$$a_k = \frac{1}{n} \sum_{j=0}^{n-1} f_j \overline{\phi_k(x_j)}, \quad k = 0, 1, \ldots, n-1. \quad \text{3.11.39}$$

With this choice of a_k and $f_j = f(x_j)$, the corresponding sum (3.11.32) vanishes.

The pair of vectors $\{\mathbf{f}, \mathbf{a}\}$

$$\begin{aligned}\mathbf{f} &= [f_0, f_1, \ldots, f_{n-1}]^{\mathrm{T}}, \\ \mathbf{a} &= [a_0, a_1, \ldots, a_{n-1}]^{\mathrm{T}},\end{aligned} \quad \text{3.11.40}$$

is called a *discrete Fourier transform pair*.

The Fast Fourier Transform

The direct computation of the discrete Fourier transform (3.11.38), that is, of

$$f_j = \sum_{k=0}^{n-1} a_k \omega_n^{jk}, \quad \omega_n = e^{2\pi i/n}, \quad j = 0, 1, \ldots, n-1, \quad \text{3.11.41}$$

requires approximately n^2 multiplications. The method of the *fast Fourier transform* (FFT) exploits the structure of the matrix $\mathbf{\Omega}$ with elements ω_n^{jk}, $j, k = 0, 1, \ldots, n-1$. If $n = 2^m$, then $\mathbf{\Omega}$ can be factored into m matrices, the rows of which contain only a few nonzero entries and the nonzero entries are equal apart from signs. In consequence of this structure the number of operations can be reduced to $nm = n \log_2 n$ operations.

The property

$$\omega_n^{2(k-(n/2))} = \omega_{n/2}^k \quad \text{3.11.42}$$

is of fundamental importance in the FFT algorithm. If n is not a power of 2, then modifications are possible. For the original reference see Cooley and Tukey (1965). For further details and algorithms, see Van Loan (1992).

For further information on least squares approximations, including examples, see Gautschi (1997b, Chapter 2) and Björck (1996, Chapters 1 and 2).

3.11(vi) Splines

Splines are defined piecewise and usually by low-degree polynomials. Given $n+1$ distinct points x_k in the real interval $[a, b]$, with $(a =)x_0 < x_1 < \cdots < x_{n-1} < x_n (= b)$, on each subinterval $[x_k, x_{k+1}]$, $k = 0, 1, \ldots, n-1$, a low-degree polynomial is defined with coefficients determined by, for example, values f_k and f'_k of a function f and its derivative at the nodes x_k and x_{k+1}. The set of all the polynomials defines a function, the *spline*, on $[a, b]$. By taking more derivatives into account, the smoothness of the spline will increase.

For splines based on Bernoulli and Euler polynomials, see §24.17(ii).

For many applications a spline function is a more adaptable approximating tool than the Lagrange interpolation polynomial involving a comparable number of parameters; see §3.3(i), where a single polynomial is used for interpolating $f(x)$ on the complete interval $[a, b]$. Multivariate functions can also be approximated in terms of multivariate polynomial splines. See de Boor (2001), Chui (1988), and Schumaker (1981) for further information.

In computer graphics a special type of spline is used which produces a *Bézier curve*. A cubic Bézier curve is defined by four points. Two are endpoints: (x_0, y_0) and (x_3, y_3); the other points (x_1, y_1) and (x_2, y_2) are control points. The slope of the curve at (x_0, y_0) is tangent to the line between (x_0, y_0) and (x_1, y_1); similarly the slope at (x_3, y_3) is tangent to the line between x_2, y_2 and x_3, y_3. The curve is described by $x(t)$ and $y(t)$, which are cubic polynomials with $t \in [0, 1]$. A complete spline results by composing several Bézier curves. A special applications area of Bézier curves is mathematical typography and the design of type fonts. See Knuth (1986, pp. 116-136).

3.12 Mathematical Constants

The fundamental constant

$$\pi = 3.14159\,26535\,89793\,23846\ldots \quad \text{3.12.1}$$

can be defined analytically in numerous ways, for example,

$$\pi = 4 \int_0^1 \frac{dt}{1+t^2}. \quad \text{3.12.2}$$

Other constants that appear in this Handbook include the base e of natural logarithms

$$e = 2.71828\,18284\,59045\,23536\ldots, \quad \text{3.12.3}$$

see §4.2(ii), and Euler's constant γ

$$\gamma = 0.57721\,56649\,01532\,86060\ldots, \quad \text{3.12.4}$$

see §5.2(ii).

For access to online high-precision numerical values of mathematical constants see Sloane (2003). For historical and other information see Finch (2003).

References

General References

Lozier and Olver (1994) gives an overview of the numerical evaluation of special functions. For more detailed information see Gautschi (1997b), Gil et al. (2007a), Henrici (1974, 1977, 1986), Hildebrand (1974), Luke (1969a,b).

Sources

The following list gives the references or other indications of proofs that were used in constructing the various sections of this chapter. These sources supplement the references that are quoted in the text.

§**3.2** Young and Gregory (1988, pp. 741–743), Wilkinson (1988, Chapter 2, §§8 10, and pp. 394–395, 423).

§**3.3** Davis (1975, Chapters 2–4), National Bureau of Standards (1944, pp. xv–xvii), Hildebrand (1974, Chapter 2), Ostrowski (1973, pp. 18–26).

§**3.4** Hildebrand (1974, pp. 85–89). The coefficients B_k^n are obtained by differentiation of the A_k^n; compare (3.4.2).

§**3.5** Davis and Rabinowitz (1984, pp. 54–58, 118–120, 137, 434–436), Bauer et al. (1963), Golub and Welsch (1969), Salzer (1955). For (3.5.18)–(3.5.19) see Waldvogel (2006). For Table 3.5.21 see Stroud (1971, pp. 243–249, 278–279).

In §3.5(v) all numerical values of the nodes x_k and corresponding weights w_k that appear in the tables in the text and on the Web site can be computed, for example, by means of the quadruple-precision analogs of the softwares **recur** and **gauss** given in Gautschi (1994), or in the case of the tables for the logarithmic weight function with **recur** replaced by **cheb**, also provided in Gautschi (1994). The three softwares can be used for other values of n, and other values of the parameters α and β that appear in some of the weight functions.

§**3.6** Olver (1967a).

§**3.8** Gautschi (1997b, pp. 217–225, 230–234), Ostrowski (1973, Chapters 3–11), Traub (1964, pp. 268–269), National Physical Laboratory (1961, pp. 57–59), Hildebrand (1974, p. 582).

§**3.9** Knopp (1964, pp. 253–255).

§**3.10** Blanch (1964), Rutishauser (1957), Wall (1948, pp. 17–19), Barnett et al. (1974), Barnett (1981a).

§**3.11** Powell (1967), Meinardus (1967, §3), Wynn (1966).

§**3.12** For more digits in (3.12.1), (3.12.3), and (3.12.4) see OEIS Sequences A000796, A001113, and A001620. See also Sloane (2003).

Chapter 4
Elementary Functions
R. Roy[1] and F. W. J. Olver[2]

Notation   104
    4.1   Special Notation 104

Logarithm, Exponential, Powers   104
    4.2   Definitions 104
    4.3   Graphics 106
    4.4   Special Values and Limits 107
    4.5   Inequalities 108
    4.6   Power Series 108
    4.7   Derivatives and Differential Equations . . 108
    4.8   Identities 109
    4.9   Continued Fractions 109
    4.10   Integrals 110
    4.11   Sums 110
    4.12   Generalized Logarithms and Exponentials 111
    4.13   Lambert W-Function 111

Trigonometric Functions   111
    4.14   Definitions and Periodicity 112
    4.15   Graphics 112
    4.16   Elementary Properties 115
    4.17   Special Values and Limits 116
    4.18   Inequalities 116
    4.19   Maclaurin Series and Laurent Series . . . 116
    4.20   Derivatives and Differential Equations . . 117
    4.21   Identities 117
    4.22   Infinite Products and Partial Fractions . . 118
    4.23   Inverse Trigonometric Functions 118
    4.24   Inverse Trigonometric Functions: Further Properties 121
    4.25   Continued Fractions 121
    4.26   Integrals 122
    4.27   Sums 123

Hyperbolic Functions   123
    4.28   Definitions and Periodicity 123
    4.29   Graphics 123
    4.30   Elementary Properties 124
    4.31   Special Values and Limits 125
    4.32   Inequalities 125
    4.33   Maclaurin Series and Laurent Series . . . 125
    4.34   Derivatives and Differential Equations . . 125
    4.35   Identities 125
    4.36   Infinite Products and Partial Fractions . . 126
    4.37   Inverse Hyperbolic Functions 127
    4.38   Inverse Hyperbolic Functions: Further Properties 129
    4.39   Continued Fractions 129
    4.40   Integrals 129
    4.41   Sums 130

Applications   130
    4.42   Solution of Triangles 130
    4.43   Cubic Equations 131
    4.44   Other Applications 131

Computation   131
    4.45   Methods of Computation 131
    4.46   Tables 132
    4.47   Approximations 132
    4.48   Software 133

References   133

---

[1]Department of Mathematics and Computer Science, Beloit College, Beloit, Wisconsin.
[2]Institute for Physical Science and Technology and Department of Mathematics, University of Maryland, College Park, Maryland.
  Acknowledgments: The authors are grateful to Steven G. Krantz and Peter R. Turner for advice on early drafts of this chapter.
  Copyright © 2009 National Institute of Standards and Technology. All rights reserved.

Notation

4.1 Special Notation

(For other notation see pp. xiv and 873.)

k, m, n integers.
a, c real or complex constants.
x, y real variables.
$z = x + iy$ complex variable.
e base of natural logarithms.

It is assumed the user is familiar with the definitions and properties of elementary functions of real arguments x. The main purpose of the present chapter is to extend these definitions and properties to complex arguments z.

The main functions treated in this chapter are the logarithm $\ln z$, $\operatorname{Ln} z$; the exponential $\exp z$, e^z; the circular trigonometric (or just trigonometric) functions $\sin z$, $\cos z$, $\tan z$, $\csc z$, $\sec z$, $\cot z$; the inverse trigonometric functions $\arcsin z$, $\operatorname{Arcsin} z$, etc.; the hyperbolic trigonometric (or just hyperbolic) functions $\sinh z$, $\cosh z$, $\tanh z$, $\operatorname{csch} z$, $\operatorname{sech} z$, $\coth z$; the inverse hyperbolic functions $\operatorname{arcsinh} z$, $\operatorname{Arcsinh} z$, etc.

Sometimes in the literature the meanings of $\ln$ and Ln are interchanged; similarly for $\arcsin z$ and $\operatorname{Arcsin} z$, etc. Sometimes "arc" is replaced by the index "-1", e.g. $\sin^{-1} z$ for $\arcsin z$ and $\operatorname{Sin}^{-1} z$ for $\operatorname{Arcsin} z$.

Logarithm, Exponential, Powers

4.2 Definitions

4.2(i) The Logarithm

The *general logarithm function* $\operatorname{Ln} z$ is defined by

4.2.1 $$\operatorname{Ln} z = \int_1^z \frac{dt}{t}, \qquad z \neq 0,$$

where the integration path does not intersect the origin. This is a multivalued function of z with branch point at $z = 0$.

The *principal value*, or *principal branch*, is defined by

4.2.2 $$\ln z = \int_1^z \frac{dt}{t},$$

where the path does not intersect $(-\infty, 0]$; see Figure 4.2.1. $\ln z$ is a single-valued analytic function on $\mathbb{C} \setminus (-\infty, 0]$ and real-valued when z ranges over the positive real numbers.

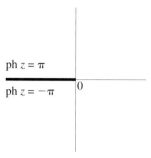

Figure 4.2.1: z-plane: Branch cut for $\ln z$ and z^α.

The real and imaginary parts of $\ln z$ are given by

4.2.3 $$\ln z = \ln |z| + i \operatorname{ph} z, \qquad -\pi < \operatorname{ph} z < \pi.$$

For $\operatorname{ph} z$ see §1.9(i).

The only zero of $\ln z$ is at $z = 1$.

Most texts extend the definition of the principal value to include the *branch cut*

4.2.4 $$z = x, \qquad -\infty < x < 0,$$

by replacing (4.2.3) with

4.2.5 $$\ln z = \ln |z| + i \operatorname{ph} z, \qquad -\pi < \operatorname{ph} z \leq \pi.$$

With this definition the general logarithm is given by

4.2.6 $$\operatorname{Ln} z = \ln z + 2k\pi i,$$

where k is the excess of the number of times the path in (4.2.1) crosses the negative real axis in the positive sense over the number of times in the negative sense.

In this Handbook we allow a further extension by regarding the cut as representing two sets of points, one set corresponding to the "upper side" and denoted by $z = x + i0$, the other set corresponding to the "lower side" and denoted by $z = x - i0$. Again see Figure 4.2.1. Then

4.2.7 $$\ln(x \pm i0) = \ln |x| \pm i\pi, \qquad -\infty < x < 0,$$

with either upper signs or lower signs taken throughout. Consequently $\ln z$ is two-valued on the cut, and discontinuous across the cut. We regard this as the *closed definition of the principal value*.

In contrast to (4.2.5) the closed definition is symmetric. As a consequence, it has the advantage of extending regions of validity of properties of principal values. For example, with the definition (4.2.5) the identity (4.8.7) is valid only when $|\operatorname{ph} z| < \pi$, but with the closed definition the identity (4.8.7) is valid when $|\operatorname{ph} z| \leq \pi$. For another example see (4.2.37).

In this Handbook it is usually clear from the context which definition of principal value is being used. *However, in the absence of any indication to the contrary it is assumed that the definition is the closed one.* For other examples in this chapter see §§4.23, 4.24, 4.37, and 4.38.

4.2(ii) Logarithms to a General Base a

With $a, b \neq 0$ or 1,

4.2.8 $$\log_a z = \ln z / \ln a,$$

4.2.9 $$\log_a z = \frac{\log_b z}{\log_b a},$$

4.2.10 $$\log_a b = \frac{1}{\log_b a}.$$

Natural logarithms have as base the unique positive number

4.2.11 $$e = 2.71828\ 18284\ 59045\ 23536\ldots$$

such that

4.2.12 $$\ln e = 1.$$

Equivalently,

4.2.13 $$\int_1^e \frac{dt}{t} = 1.$$

Thus

4.2.14 $$\log_e z = \ln z,$$

4.2.15 $\log_{10} z = (\ln z)/(\ln 10) = (\log_{10} e)\ln z,$

4.2.16 $$\ln z = (\ln 10)\log_{10} z,$$

4.2.17 $\log_{10} e = 0.43429\ 44819\ 03251\ 82765\ldots,$

4.2.18 $\ln 10 = 2.30258\ 50929\ 94045\ 68401\ldots.$

$\log_e x = \ln x$ is also called the *Napierian* or *hyperbolic* logarithm. $\log_{10} x$ is the *common* or *Briggs* logarithm.

4.2(iii) The Exponential Function

4.2.19 $$\exp z = 1 + \frac{z}{1!} + \frac{z^2}{2!} + \frac{z^3}{3!} + \cdots.$$

The function $\exp$ is an entire function of z, with no real or complex zeros. It has period $2\pi i$:

4.2.20 $$\exp(z + 2\pi i) = \exp z.$$

Also,

4.2.21 $$\exp(-z) = 1/\exp(z).$$

4.2.22 $$|\exp z| = \exp(\Re z).$$

The general value of the phase is given by

4.2.23 $$\mathrm{ph}(\exp z) = \Im z + 2k\pi, \qquad k \in \mathbb{Z}.$$

If $z = x + iy$, then

4.2.24 $$\exp z = e^x \cos y + i e^x \sin y.$$

If $\zeta \neq 0$ then

4.2.25 $$\exp z = \zeta \iff z = \mathrm{Ln}\,\zeta.$$

4.2(iv) Powers

Powers with General Bases

The general a^{th} power of z is defined by

4.2.26 $$z^a = \exp(a\,\mathrm{Ln}\,z), \qquad z \neq 0.$$

In particular, $z^0 = 1$, and if $a = n = 1, 2, 3, \ldots$, then

4.2.27 $$z^a = \underbrace{z \cdot z \cdots z}_{n\text{ times}} = 1/z^{-a}.$$

In all other cases, z^a is a multivalued function with branch point at $z = 0$. The *principal value* is

4.2.28 $$z^a = \exp(a \ln z).$$

This is an analytic function of z on $\mathbb{C} \setminus (-\infty, 0]$, and is two-valued and discontinuous on the cut shown in Figure 4.2.1, unless $a \in \mathbb{Z}$.

4.2.29 $$|z^a| = |z|^{\Re a} \exp(-(\Im a)\,\mathrm{ph}\,z),$$

4.2.30 $$\mathrm{ph}(z^a) = (\Re a)\,\mathrm{ph}\,z + (\Im a)\ln|z|,$$

where $\mathrm{ph}\,z \in [-\pi, \pi]$ for the principal value of z^a, and is unrestricted in the general case. When a is real

4.2.31 $$|z^a| = |z|^a, \quad \mathrm{ph}(z^a) = a\,\mathrm{ph}\,z.$$

Unless indicated otherwise, it is assumed throughout this Handbook that a power assumes its principal value. With this convention,

4.2.32 $$e^z = \exp z,$$

but the general value of e^z is

4.2.33 $$e^z = (\exp z)\exp(2k z\pi i), \qquad k \in \mathbb{Z}.$$

For $z = 1$

4.2.34 $$e = 1 + \frac{1}{1!} + \frac{1}{2!} + \frac{1}{3!} + \cdots.$$

If z^a has its general value, with $a \neq 0$, and if $w \neq 0$, then

4.2.35 $$z^a = w \iff z = \exp\left(\frac{1}{a}\mathrm{Ln}\,w\right).$$

This result is also valid when z^a has its principal value, provided that the branch of $\mathrm{Ln}\,w$ satisfies

4.2.36 $$-\pi \leq \Im\left(\frac{1}{a}\mathrm{Ln}\,w\right) \leq \pi.$$

Another example of a principal value is provided by

4.2.37 $$\sqrt{z^2} = \begin{cases} z, & \Re z \geq 0, \\ -z, & \Re z \leq 0. \end{cases}$$

Again, without the closed definition the $\geq$ and $\leq$ signs would have to be replaced by $>$ and $<$, respectively.

4.3 Graphics

4.3(i) Real Arguments

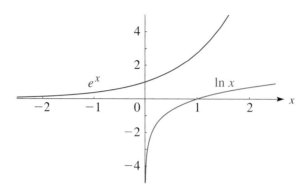

Figure 4.3.1: $\ln x$ and e^x.

4.3(ii) Complex Arguments: Conformal Maps

Figure 4.3.2 illustrates the conformal mapping of the strip $-\pi < \Im z < \pi$ onto the whole w-plane cut along the negative real axis, where $w = e^z$ and $z = \ln w$ (principal value). Corresponding points share the same letters, with bars signifying complex conjugates. Lines parallel to the real axis in the z-plane map onto rays in the w-plane, and lines parallel to the imaginary axis in the z-plane map onto circles centered at the origin in the w-plane. In the labeling of corresponding points r is a real parameter that can lie anywhere in the interval $(0, \infty)$.

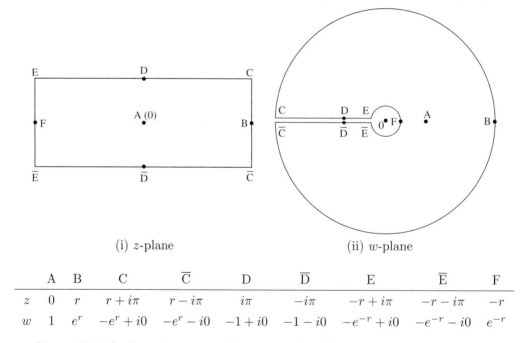

	A	B	C	$\overline{C}$	D	$\overline{D}$	E	$\overline{E}$	F
z	0	r	$r+i\pi$	$r-i\pi$	$i\pi$	$-i\pi$	$-r+i\pi$	$-r-i\pi$	$-r$
w	1	e^r	$-e^r+i0$	$-e^r-i0$	$-1+i0$	$-1-i0$	$-e^{-r}+i0$	$-e^{-r}-i0$	e^{-r}

Figure 4.3.2: Conformal mapping of exponential and logarithm. $w = e^z$, $z = \ln w$.

4.4 Special Values and Limits

4.3(iii) Complex Arguments: Surfaces

In the graphics shown in this subsection height corresponds to the absolute value of the function and color to the phase. See also p. xiv.

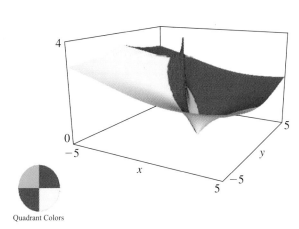

Figure 4.3.3: $\ln(x+iy)$ (principal value). There is a branch cut along the negative real axis.

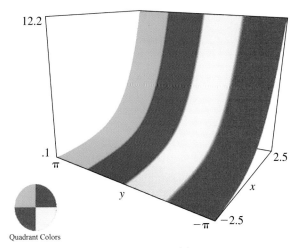

Figure 4.3.4: e^{x+iy}.

4.4 Special Values and Limits

4.4(i) Logarithms

4.4.1 $$\ln 1 = 0,$$

4.4.2 $$\ln(-1 \pm i0) = \pm \pi i,$$

4.4.3 $$\ln(\pm i) = \pm \tfrac{1}{2}\pi i.$$

4.4(ii) Powers

4.4.4 $$e^0 = 1,$$

4.4.5 $$e^{\pm \pi i} = -1,$$

4.4.6 $$e^{\pm \pi i/2} = \pm i,$$

4.4.7 $$e^{2\pi k i} = 1, \qquad k \in \mathbb{Z},$$

4.4.8 $$e^{\pm \pi i/3} = \tfrac{1}{2} \pm i\tfrac{\sqrt{3}}{2},$$

4.4.9 $$e^{\pm 2\pi i/3} = -\tfrac{1}{2} \pm i\tfrac{\sqrt{3}}{2},$$

4.4.10 $$e^{\pm \pi i/4} = \tfrac{1}{\sqrt{2}} \pm i\tfrac{1}{\sqrt{2}},$$

4.4.11 $$e^{\pm 3\pi i/4} = -\tfrac{1}{\sqrt{2}} \pm i\tfrac{1}{\sqrt{2}},$$

4.4.12 $$i^{\pm i} = e^{\mp \pi/2}.$$

4.4(iii) Limits

4.4.13 $$\lim_{x \to \infty} x^{-a} \ln x = 0, \qquad \Re a > 0,$$

4.4.14 $$\lim_{x \to 0} x^a \ln x = 0, \qquad \Re a > 0,$$

4.4.15 $$\lim_{x \to \infty} x^a e^{-x} = 0,$$

4.4.16 $$\lim_{z \to \infty} z^a e^{-z} = 0, \quad |\operatorname{ph} z| \leq \tfrac{1}{2}\pi - \delta \; (< \tfrac{1}{2}\pi),$$

where $a \; (\in \mathbb{C})$ and $\delta \; (\in (0, \tfrac{1}{2}\pi])$ are constants.

4.4.17 $$\lim_{n \to \infty} \left(1 + \frac{z}{n}\right)^n = e^z, \quad z = \text{constant}.$$

4.4.18 $$\lim_{n \to \infty} \left(1 + \frac{1}{n}\right)^n = e.$$

4.4.19 $$\lim_{n \to \infty} \left(\left(\sum_{k=1}^{n} \frac{1}{k}\right) - \ln n\right) = \gamma = 0.57721\,56649\,01532\,86060\ldots,$$

where γ is Euler's constant; see (5.2.3).

4.5 Inequalities

4.5(i) Logarithms

4.5.1 $$\frac{x}{1+x} < \ln(1+x) < x, \quad x > -1, x \neq 0,$$

4.5.2 $$x < -\ln(1-x) < \frac{x}{1-x}, \quad x < 1, x \neq 0,$$

4.5.3 $$|\ln(1-x)| < \tfrac{3}{2}x, \quad 0 < x \leq 0.5828\ldots,$$

4.5.4 $$\ln x \leq x - 1, \quad x > 0,$$

4.5.5 $$\ln x \leq a(x^{1/a} - 1), \quad a, x > 0,$$

4.5.6 $$|\ln(1+z)| \leq -\ln(1-|z|), \quad |z| < 1.$$

For more inequalities involving the logarithm function see Mitrinović (1964, pp. 75–77), Mitrinović (1970, pp. 272–276), and Bullen (1998, pp. 159–160).

4.5(ii) Exponentials

In (4.5.7)–(4.5.12) it is assumed that $x \neq 0$. (When $x = 0$ the inequalities become equalities.)

4.5.7 $$e^{-x/(1-x)} < 1 - x < e^{-x}, \quad x < 1,$$

4.5.8 $$1 + x < e^x, \quad -\infty < x < \infty,$$

4.5.9 $$e^x < \frac{1}{1-x}, \quad x < 1,$$

4.5.10 $$\frac{x}{1+x} < 1 - e^{-x} < x, \quad x > -1,$$

4.5.11 $$x < e^x - 1 < \frac{x}{1-x}, \quad x < 1,$$

4.5.12 $$e^{x/(1+x)} < 1 + x, \quad x > -1,$$

4.5.13 $$e^{xy/(x+y)} < \left(1 + \frac{x}{y}\right)^y < e^x, \quad x > 0, y > 0,$$

4.5.14 $$e^{-x} < 1 - \tfrac{1}{2}x, \quad 0 < x \leq 1.5936\ldots,$$

4.5.15 $$\tfrac{1}{4}|z| < |e^z - 1| < \tfrac{7}{4}|z|, \quad 0 < |z| < 1,$$

4.5.16 $$|e^z - 1| \leq e^{|z|} - 1 \leq |z|e^{|z|}, \quad z \in \mathbb{C}.$$

For more inequalities involving the exponential function see Mitrinović (1964, pp. 73–77), Mitrinović (1970, pp. 266–271), and Bullen (1998, pp. 81–83).

4.6 Power Series

4.6(i) Logarithms

4.6.1 $$\ln(1+z) = z - \tfrac{1}{2}z^2 + \tfrac{1}{3}z^3 - \cdots, \quad |z| \leq 1, z \neq -1,$$

4.6.2 $$\ln z = \left(\frac{z-1}{z}\right) + \frac{1}{2}\left(\frac{z-1}{z}\right)^2 + \frac{1}{3}\left(\frac{z-1}{z}\right)^3 + \cdots, \quad \Re z \geq \tfrac{1}{2},$$

4.6.3 $$\ln z = (z-1) - \tfrac{1}{2}(z-1)^2 + \tfrac{1}{3}(z-1)^3 - \cdots, \quad |z-1| \leq 1, z \neq 0,$$

4.6.4 $$\ln z = 2\left(\left(\frac{z-1}{z+1}\right) + \frac{1}{3}\left(\frac{z-1}{z+1}\right)^3 + \frac{1}{5}\left(\frac{z-1}{z+1}\right)^5 + \cdots\right), \quad \Re z \geq 0, z \neq 0,$$

4.6.5 $$\ln\left(\frac{z+1}{z-1}\right) = 2\left(\frac{1}{z} + \frac{1}{3z^3} + \frac{1}{5z^5} + \cdots\right), \quad |z| \geq 1, z \neq \pm 1,$$

4.6.6 $$\ln(z+a) = \ln a + 2\left(\left(\frac{z}{2a+z}\right) + \frac{1}{3}\left(\frac{z}{2a+z}\right)^3 + \frac{1}{5}\left(\frac{z}{2a+z}\right)^5 + \cdots\right), \quad a > 0, \Re z \geq -a, z \neq -a.$$

4.6(ii) Powers

Binomial Expansion

4.6.7 $$(1+z)^a = 1 + \frac{a}{1!}z + \frac{a(a-1)}{2!}z^2 + \frac{a(a-1)(a-2)}{3!}z^3 + \cdots,$$

valid when a is any real or complex constant and $|z| < 1$. If $a = 0, 1, 2, \ldots$, then the series terminates and z is unrestricted.

4.7 Derivatives and Differential Equations

4.7(i) Logarithms

4.7.1 $$\frac{d}{dz}\ln z = \frac{1}{z},$$

4.7.2 $$\frac{d}{dz}\operatorname{Ln} z = \frac{1}{z},$$

4.7.3 $$\frac{d^n}{dz^n}\ln z = (-1)^{n-1}(n-1)!z^{-n},$$

4.7.4 $$\frac{d^n}{dz^n}\operatorname{Ln} z = (-1)^{n-1}(n-1)!z^{-n}.$$

4.8 Identities

For a nonvanishing analytic function $f(z)$, the general solution of the differential equation

4.7.5 $$\frac{dw}{dz} = \frac{f'(z)}{f(z)}$$

is

4.7.6 $$w(z) = \operatorname{Ln}(f(z)) + \text{constant}.$$

4.7(ii) Exponentials and Powers

4.7.7 $$\frac{d}{dz} e^z = e^z,$$

4.7.8 $$\frac{d}{dz} e^{az} = a e^{az},$$

4.7.9 $$\frac{d}{dz} a^z = a^z \ln a, \qquad a \ne 0.$$

When a^z is a general power, $\ln a$ is replaced by the branch of $\operatorname{Ln} a$ used in constructing a^z.

4.7.10 $$\frac{d}{dz} z^a = a z^{a-1},$$

4.7.11 $$\frac{d^n}{dz^n} z^a = a(a-1)(a-2)\cdots(a-n+1) z^{a-n}.$$

The general solution of the differential equation

4.7.12 $$\frac{dw}{dz} = f(z) w$$

is

4.7.13 $$w = \exp\left(\int f(z)\, dz \right) + \text{constant}.$$

The general solution of the differential equation

4.7.14 $$\frac{d^2 w}{dz^2} = a w, \qquad a \ne 0,$$

is

4.7.15 $$w = A e^{\sqrt{a} z} + B e^{-\sqrt{a} z},$$

where A and B are arbitrary constants.

For other differential equations see Kamke (1977, pp. 396–413).

4.8 Identities

4.8(i) Logarithms

In (4.8.1)–(4.8.4) $z_1 z_2 \ne 0$.

4.8.1 $$\operatorname{Ln}(z_1 z_2) = \operatorname{Ln} z_1 + \operatorname{Ln} z_2.$$

This is interpreted that every value of $\operatorname{Ln}(z_1 z_2)$ is one of the values of $\operatorname{Ln} z_1 + \operatorname{Ln} z_2$, and vice versa.

4.8.2 $\ln(z_1 z_2) = \ln z_1 + \ln z_2, \quad -\pi \le \operatorname{ph} z_1 + \operatorname{ph} z_2 \le \pi,$

4.8.3 $$\operatorname{Ln} \frac{z_1}{z_2} = \operatorname{Ln} z_1 - \operatorname{Ln} z_2,$$

4.8.4 $\ln \frac{z_1}{z_2} = \ln z_1 - \ln z_2, \quad -\pi \le \operatorname{ph} z_1 - \operatorname{ph} z_2 \le \pi.$

In (4.8.5)–(4.8.7) and (4.8.10) $z \ne 0$.

4.8.5 $$\operatorname{Ln}(z^n) = n \operatorname{Ln} z, \qquad n \in \mathbb{Z},$$

4.8.6 $$\ln(z^n) = n \ln z, \quad n \in \mathbb{Z}, \; -\pi \le n \operatorname{ph} z \le \pi,$$

4.8.7 $$\ln \frac{1}{z} = -\ln z, \qquad |\operatorname{ph} z| \le \pi.$$

4.8.8 $$\operatorname{Ln}(\exp z) = z + 2 k \pi i, \qquad k \in \mathbb{Z},$$

4.8.9 $$\ln(\exp z) = z, \qquad -\pi \le \Im z \le \pi,$$

4.8.10 $$\exp(\ln z) = \exp(\operatorname{Ln} z) = z.$$

If $a \ne 0$ and a^z has its general value, then

4.8.11 $$\operatorname{Ln}(a^z) = z \operatorname{Ln} a + 2 k \pi i, \qquad k \in \mathbb{Z}.$$

If $a \ne 0$ and a^z has its principal value, then

4.8.12 $$\ln(a^z) = z \ln a + 2 k \pi i,$$

where the integer k is chosen so that $\Re(-i z \ln a) + 2 k \pi \in [-\pi, \pi]$.

4.8.13 $$\ln(a^x) = x \ln a, \qquad a > 0.$$

4.8(ii) Powers

4.8.14 $$a^{z_1} a^{z_2} = a^{z_1 + z_2},$$

4.8.15 $$a^z b^z = (ab)^z, \qquad -\pi \le \operatorname{ph} a + \operatorname{ph} b \le \pi,$$

4.8.16 $$e^{z_1} e^{z_2} = e^{z_1 + z_2},$$

4.8.17 $$(e^{z_1})^{z_2} = e^{z_1 z_2}, \qquad -\pi \le \Im z_1 \le \pi.$$

The restriction on z_1 can be removed when z_2 is an integer.

4.9 Continued Fractions

4.9(i) Logarithms

4.9.1 $$\ln(1+z) = \frac{z}{1+} \frac{z}{2+} \frac{z}{3+} \frac{4z}{4+} \frac{4z}{5+} \frac{9z}{6+} \frac{9z}{7+} \cdots,$$
$$|\operatorname{ph}(1+z)| < \pi.$$

4.9.2 $$\ln\left(\frac{1+z}{1-z}\right) = \frac{2z}{1-} \frac{z^2}{3-} \frac{4z^2}{5-} \frac{9z^2}{7-} \frac{16z^2}{9-} \cdots,$$

valid when $z \in \mathbb{C} \setminus (-\infty, -1] \cup [1, \infty)$; see Figure 4.23.1(i).

For other continued fractions involving logarithms see Lorentzen and Waadeland (1992, pp. 566–568). See also Cuyt et al. (2008, pp. 196–200).

4.9(ii) Exponentials

For $z \in \mathbb{C}$,

4.9.3
$$e^z = \cfrac{1}{1-} \cfrac{z}{1+} \cfrac{z}{2-} \cfrac{z}{3+} \cfrac{z}{2-} \cfrac{z}{5+} \cfrac{z}{2-} \cdots$$
$$= 1 + \cfrac{z}{1-} \cfrac{z}{2+} \cfrac{z}{3-} \cfrac{z}{2+} \cfrac{z}{5-} \cfrac{z}{2+} \cfrac{z}{7-} \cdots$$
$$= 1 + \cfrac{z}{1-(z/2)+} \cfrac{z^2/(4\cdot 3)}{1+} \cfrac{z^2/(4\cdot 15)}{1+} \cfrac{z^2/(4\cdot 35)}{1+} \cdots \cfrac{z^2/(4(4n^2-1))}{1+} \cdots$$

4.9.4
$$e^z - e_{n-1}(z) = \cfrac{z^n}{n!-} \cfrac{n!z}{(n+1)+} \cfrac{z}{(n+2)-} \cfrac{(n+1)z}{(n+3)+} \cfrac{2z}{(n+4)-} \cfrac{(n+2)z}{(n+5)+} \cfrac{3z}{(n+6)-} \cdots,$$

where

4.9.5
$$e_n(z) = \sum_{k=0}^{n} \frac{z^k}{k!}.$$

For other continued fractions involving the exponential function see Lorentzen and Waadeland (1992, pp. 563–564). See also Cuyt et al. (2008, pp. 193–195).

4.9(iii) Powers

See Cuyt et al. (2008, pp. 217–220).

4.10 Integrals

4.10(i) Logarithms

4.10.1 $\displaystyle\int \frac{dz}{z} = \ln z,$

4.10.2 $\displaystyle\int \ln z \, dz = z \ln z - z,$

4.10.3 $\displaystyle\int z^n \ln z \, dz = \frac{z^{n+1}}{n+1} \ln z - \frac{z^{n+1}}{(n+1)^2}, \quad n \neq -1,$

4.10.4 $\displaystyle\int \frac{dz}{z \ln z} = \ln(\ln z),$

4.10.5 $\displaystyle\int_0^1 \frac{\ln t}{1-t} dt = -\frac{\pi^2}{6},$

4.10.6 $\displaystyle\int_0^1 \frac{\ln t}{1+t} dt = -\frac{\pi^2}{12},$

4.10.7 $\displaystyle\fint_0^x \frac{dt}{\ln t} = \mathrm{li}(x), \qquad x > 1.$

The left-hand side of (4.10.7) is a Cauchy principal value (§1.4(v)). For $\mathrm{li}(x)$ see §6.2(i).

4.10(ii) Exponentials

For $a, b \neq 0$,

4.10.8 $\displaystyle\int e^{az} \, dz = \frac{e^{az}}{a},$

4.10.9 $\displaystyle\int \frac{dz}{e^{az}+b} = \frac{1}{ab}(az - \ln(e^{az}+b)),$

4.10.10 $\displaystyle\int \frac{e^{az}-1}{e^{az}+1} dz = \frac{2}{a} \ln\left(e^{az/2} + e^{-az/2}\right),$

4.10.11 $\displaystyle\int_{-\infty}^{\infty} e^{-cx^2} dx = \sqrt{\frac{\pi}{c}}, \qquad \Re c > 0,$

4.10.12 $\displaystyle\int_0^{\ln 2} \frac{xe^x}{e^x-1} dx = \frac{\pi^2}{12},$

4.10.13 $\displaystyle\int_0^{\infty} \frac{dx}{e^x+1} = \ln 2.$

4.10(iii) Compendia

Extensive compendia of indefinite and definite integrals of logarithms and exponentials include Apelblat (1983, pp. 16–47), Bierens de Haan (1939), Gröbner and Hofreiter (1949, pp. 107–116), Gröbner and Hofreiter (1950, pp. 52–90), Gradshteyn and Ryzhik (2000, Chapters 2–4), and Prudnikov et al. (1986a, §§1.3, 1.6, 2.3, 2.6).

4.11 Sums

For infinite series involving logarithms and/or exponentials, see Gradshteyn and Ryzhik (2000, Chapter 1), Hansen (1975, §44), and Prudnikov et al. (1986a, Chapter 5).

4.12 Generalized Logarithms and Exponentials

A *generalized exponential function* $\phi(x)$ satisfies the equations

4.12.1 $\qquad \phi(x+1) = e^{\phi(x)}, \qquad -1 < x < \infty,$

4.12.2 $\qquad \phi(0) = 0,$

and is strictly increasing when $0 \leq x \leq 1$. Its inverse $\psi(x)$ is called a *generalized logarithm*. It, too, is strictly increasing when $0 \leq x \leq 1$, and

4.12.3 $\qquad \psi(e^x) = 1 + \psi(x), \qquad -\infty < x < \infty,$

4.12.4 $\qquad \psi(0) = 0.$

These functions are not unique. The simplest choice is given by

4.12.5 $\qquad \phi(x) = \psi(x) = x, \qquad 0 \leq x \leq 1.$

Then

4.12.6 $\qquad \phi(x) = \ln(x+1), \qquad -1 < x < 0,$

and

4.12.7 $\qquad \phi(x) = \exp\exp\cdots\exp(x - \lfloor x \rfloor), \qquad x > 1,$

where the exponentiations are carried out $\lfloor x \rfloor$ times. Correspondingly,

4.12.8 $\qquad \psi(x) = e^x - 1, \qquad -\infty < x < 0,$

and

4.12.9 $\qquad \psi(x) = \ell + \ln^{(\ell)} x, \qquad x > 1,$

where $\ln^{(\ell)} x$ denotes the ℓ-th repeated logarithm of x, and ℓ is the positive integer determined by the condition

4.12.10 $\qquad 0 \leq \ln^{(\ell)} x < 1.$

Both $\phi(x)$ and $\psi(x)$ are continuously differentiable.

For further information, see Clenshaw *et al.* (1986). For C^∞ generalized logarithms, see Walker (1991). For analytic generalized logarithms, see Kneser (1950).

4.13 Lambert W-Function

The Lambert W-function $W(x)$ is the solution of the equation

4.13.1 $\qquad We^W = x.$

On the x-interval $[0, \infty)$ there is one real solution, and it is nonnegative and increasing. On the x-interval $(-1/e, 0)$ there are two real solutions, one increasing and the other decreasing. We call the solution for which $W(x) \geq W(-1/e)$ the *principal branch* and denote it by $\mathrm{Wp}(x)$. The other solution is denoted by $\mathrm{Wm}(x)$. See Figure 4.13.1.

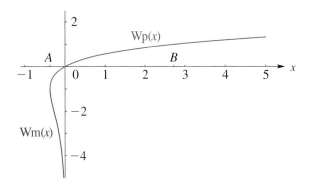

Figure 4.13.1: Branches $\mathrm{Wp}(x)$ and $\mathrm{Wm}(x)$ of the Lambert W-function. A and B denote the points $-1/e$ and e, respectively, on the x-axis.

Properties include:

4.13.2 $\qquad \begin{aligned} &\mathrm{Wp}(-1/e) = \mathrm{Wm}(-1/e) = -1, \\ &\mathrm{Wp}(0) = 0, \quad \mathrm{Wp}(e) = 1. \end{aligned}$

4.13.3 $\qquad U + \ln U = x, \quad U = U(x) = W(e^x).$

4.13.4 $\qquad \dfrac{dW}{dx} = \dfrac{e^{-W}}{1+W}, \qquad x \neq -\dfrac{1}{e}.$

4.13.5 $\qquad \mathrm{Wp}(x) = \sum_{n=1}^{\infty} (-1)^{n-1} \dfrac{n^{n-2}}{(n-1)!} x^n, \qquad |x| < \dfrac{1}{e}.$

4.13.6
$$W\left(-e^{-1-(t^2/2)}\right) = \sum_{n=0}^{\infty} (-1)^{n-1} c_n t^n, \quad |t| < 2\sqrt{\pi},$$

where $t \geq 0$ for Wp, $t \leq 0$ for Wm,

4.13.7 $\qquad c_0 = 1, c_1 = 1, c_2 = \tfrac{1}{3}, c_3 = \tfrac{1}{36}, c_4 = -\tfrac{1}{270},$

4.13.8 $\qquad c_n = \dfrac{1}{n+1}\left(c_{n-1} - \sum_{k=2}^{n-1} k c_k c_{n+1-k}\right), \quad n \geq 2,$

and

4.13.9 $\qquad 1 \cdot 3 \cdot 5 \cdots (2n+1) c_{2n+1} = g_n,$

where g_n is defined in §5.11(i).

As $x \to +\infty$

4.13.10
$$\mathrm{Wp}(x) = \xi - \ln \xi + \dfrac{\ln \xi}{\xi} + \dfrac{(\ln \xi)^2}{2\xi^2} - \dfrac{\ln \xi}{\xi^2} + O\left(\dfrac{(\ln \xi)^3}{\xi^3}\right),$$

where $\xi = \ln x$. As $x \to 0-$

4.13.11
$$\mathrm{Wm}(x) = -\eta - \ln \eta - \dfrac{\ln \eta}{\eta} - \dfrac{(\ln \eta)^2}{2\eta^2} - \dfrac{\ln \eta}{\eta^2} + O\left(\dfrac{(\ln \eta)^3}{\eta^3}\right),$$

where $\eta = \ln(-1/x)$.

For the foregoing results and further information see Borwein and Corless (1999), Corless *et al.* (1996), de Bruijn (1961, pp. 25–28), Olver (1997b, pp. 12–13), and Siewert and Burniston (1973).

For integral representations of all branches of the Lambert W-function see Kheyfits (2004).

Trigonometric Functions

4.14 Definitions and Periodicity

$$4.14.1 \qquad \sin z = \frac{e^{iz} - e^{-iz}}{2i},$$

$$4.14.2 \qquad \cos z = \frac{e^{iz} + e^{-iz}}{2},$$

$$4.14.3 \qquad \cos z \pm i \sin z = e^{\pm iz},$$

$$4.14.4 \qquad \tan z = \frac{\sin z}{\cos z},$$

$$4.14.5 \qquad \csc z = \frac{1}{\sin z},$$

$$4.14.6 \qquad \sec z = \frac{1}{\cos z},$$

$$4.14.7 \qquad \cot z = \frac{\cos z}{\sin z} = \frac{1}{\tan z}.$$

The functions $\sin z$ and $\cos z$ are entire. In $\mathbb{C}$ the zeros of $\sin z$ are $z = k\pi$, $k \in \mathbb{Z}$; the zeros of $\cos z$ are $z = \left(k + \tfrac{1}{2}\right)\pi$, $k \in \mathbb{Z}$. The functions $\tan z$, $\csc z$, $\sec z$, and $\cot z$ are meromorphic, and the locations of their zeros and poles follow from (4.14.4) to (4.14.7).

For $k \in \mathbb{Z}$

$$4.14.8 \qquad \sin(z + 2k\pi) = \sin z,$$

$$4.14.9 \qquad \cos(z + 2k\pi) = \cos z,$$

$$4.14.10 \qquad \tan(z + k\pi) = \tan z.$$

4.15 Graphics

4.15(i) Real Arguments

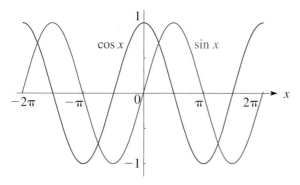

Figure 4.15.1: $\sin x$ and $\cos x$.

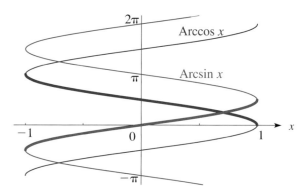

Figure 4.15.2: $\operatorname{Arcsin} x$ and $\operatorname{Arccos} x$. Principal values are shown with thickened lines.

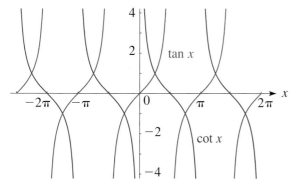

Figure 4.15.3: $\tan x$ and $\cot x$.

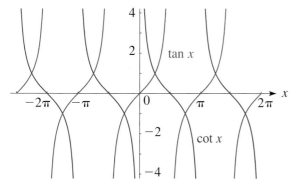

Figure 4.15.4: $\arctan x$ and $\operatorname{arccot} x$. Only principal values are shown. $\operatorname{arccot} x$ is discontinuous at $x = 0$.

4.15 Graphics

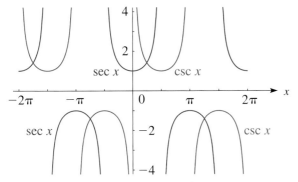

Figure 4.15.5: $\csc x$ and $\sec x$.

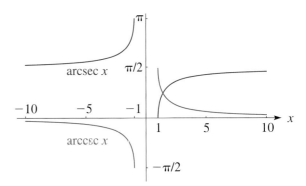

Figure 4.15.6: $\operatorname{arccsc} x$ and $\operatorname{arcsec} x$. Only principal values are shown. (Both functions are complex when $-1 < x < 1$.)

4.15(ii) Complex Arguments: Conformal Maps

Figure 4.15.7 illustrates the conformal mapping of the strip $-\frac{1}{2}\pi < \Re z < \frac{1}{2}\pi$ onto the whole w-plane cut along the real axis from $-\infty$ to -1 and 1 to ∞, where $w = \sin z$ and $z = \arcsin w$ (principal value). Corresponding points share the same letters, with bars signifying complex conjugates. Lines parallel to the real axis in the z-plane map onto ellipses in the w-plane with foci at $w = \pm 1$, and lines parallel to the imaginary axis in the z-plane map onto rectangular hyperbolas confocal with the ellipses. In the labeling of corresponding points r is a real parameter that can lie anywhere in the interval $(0, \infty)$.

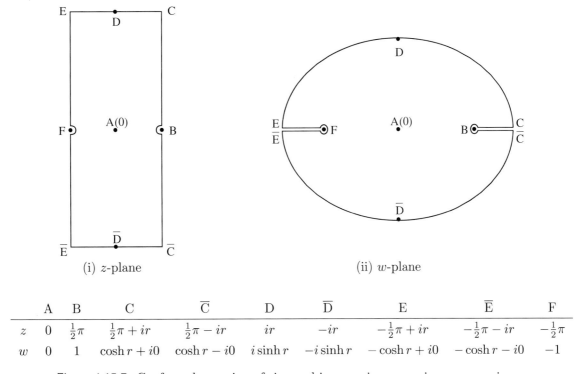

	A	B	C	$\overline{C}$	D	$\overline{D}$	E	$\overline{E}$	F
z	0	$\frac{1}{2}\pi$	$\frac{1}{2}\pi + ir$	$\frac{1}{2}\pi - ir$	ir	$-ir$	$-\frac{1}{2}\pi + ir$	$-\frac{1}{2}\pi - ir$	$-\frac{1}{2}\pi$
w	0	1	$\cosh r + i0$	$\cosh r - i0$	$i\sinh r$	$-i\sinh r$	$-\cosh r + i0$	$-\cosh r - i0$	-1

Figure 4.15.7: Conformal mapping of sine and inverse sine. $w = \sin z$, $z = \arcsin w$.

4.15(iii) Complex Arguments: Surfaces

In the graphics shown in this subsection height corresponds to the absolute value of the function and color to the phase. See also p. xiv.

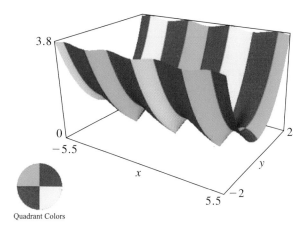

Figure 4.15.8: $\sin(x+iy)$.

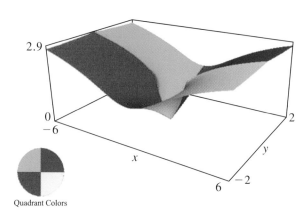

Figure 4.15.9: $\arcsin(x+iy)$ (principal value). There are branch cuts along the real axis from $-\infty$ to -1 and 1 to ∞.

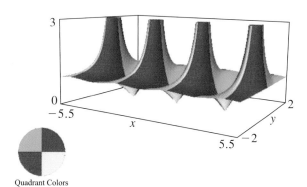

Figure 4.15.10: $\tan(x+iy)$.

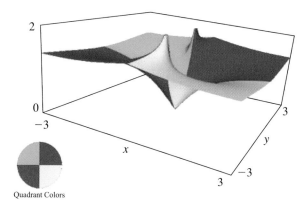

Figure 4.15.11: $\arctan(x+iy)$ (principal value). There are branch cuts along the imaginary axis from $-i\infty$ to $-i$ and i to $i\infty$.

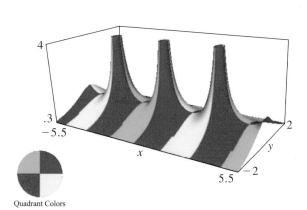

Figure 4.15.12: $\csc(x+iy)$.

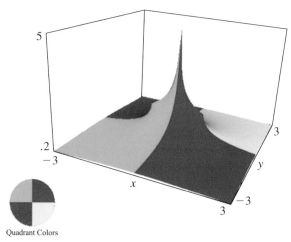

Figure 4.15.13: $\operatorname{arccsc}(x+iy)$ (principal value). There is a branch cut along the real axis from -1 to 1.

The corresponding surfaces for $\cos(x+iy)$, $\cot(x+iy)$, and $\sec(x+iy)$ are similar. In consequence of the identities

4.15.1 $\qquad \cos(x+iy) = \sin\left(x + \tfrac{1}{2}\pi + iy\right),$

4.15.2 $\qquad \cot(x+iy) = -\tan\left(x + \tfrac{1}{2}\pi + iy\right),$

4.15.3 $\qquad \sec(x+iy) = \csc\left(x + \tfrac{1}{2}\pi + iy\right),$

they can be obtained by translating the surfaces shown in Figures 4.15.8, 4.15.10, 4.15.12 by $-\tfrac{1}{2}\pi$ parallel to the x-axis, and adjusting the phase coloring in the case of Figure 4.15.10.

The corresponding surfaces for $\arccos(x+iy)$, $\text{arccot}(x+iy)$, $\text{arcsec}(x+iy)$ can be visualized from Figures 4.15.9, 4.15.11, 4.15.13 with the aid of equations (4.23.16)–(4.23.18).

4.16 Elementary Properties

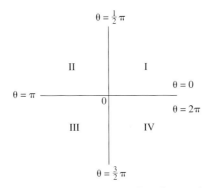

Figure 4.16.1: Quadrants for the angle θ.

Table 4.16.1: Signs of the trigonometric functions in the four quadrants.

Quadrant	$\sin\theta, \csc\theta$	$\cos\theta, \sec\theta$	$\tan\theta, \cot\theta$
I	+	+	+
II	+	−	−
III	−	−	+
IV	−	+	−

Table 4.16.2: Trigonometric functions: quarter periods and change of sign.

x	$-\theta$	$\tfrac{1}{2}\pi \pm \theta$	$\pi \pm \theta$	$\tfrac{3}{2}\pi \pm \theta$	$2\pi \pm \theta$
$\sin x$	$-\sin\theta$	$\cos\theta$	$\mp\sin\theta$	$-\cos\theta$	$\pm\sin\theta$
$\cos x$	$\cos\theta$	$\mp\sin\theta$	$-\cos\theta$	$\pm\sin\theta$	$\cos\theta$
$\tan x$	$-\tan\theta$	$\mp\cot\theta$	$\pm\tan\theta$	$\mp\cot\theta$	$\pm\tan\theta$
$\csc x$	$-\csc\theta$	$\sec\theta$	$\mp\csc\theta$	$-\sec\theta$	$\pm\csc\theta$
$\sec x$	$\sec\theta$	$\mp\csc\theta$	$-\sec\theta$	$\pm\csc\theta$	$\sec\theta$
$\cot x$	$-\cot\theta$	$\mp\tan\theta$	$\pm\cot\theta$	$\mp\tan\theta$	$\pm\cot\theta$

Table 4.16.3: Trigonometric functions: interrelations. All square roots have their principal values when the functions are real, nonnegative, and finite.

	$\sin\theta = a$	$\cos\theta = a$	$\tan\theta = a$	$\csc\theta = a$	$\sec\theta = a$	$\cot\theta = a$
$\sin\theta$	a	$(1-a^2)^{1/2}$	$a(1+a^2)^{-1/2}$	a^{-1}	$a^{-1}(a^2-1)^{1/2}$	$(1+a^2)^{-1/2}$
$\cos\theta$	$(1-a^2)^{1/2}$	a	$(1+a^2)^{-1/2}$	$a^{-1}(a^2-1)^{1/2}$	a^{-1}	$a(1+a^2)^{-1/2}$
$\tan\theta$	$a(1-a^2)^{-1/2}$	$a^{-1}(1-a^2)^{1/2}$	a	$(a^2-1)^{-1/2}$	$(a^2-1)^{1/2}$	a^{-1}
$\csc\theta$	a^{-1}	$(1-a^2)^{-1/2}$	$a^{-1}(1+a^2)^{1/2}$	a	$a(a^2-1)^{-1/2}$	$(1+a^2)^{1/2}$
$\sec\theta$	$(1-a^2)^{-1/2}$	a^{-1}	$(1+a^2)^{1/2}$	$a(a^2-1)^{-1/2}$	a	$a^{-1}(1+a^2)^{1/2}$
$\cot\theta$	$a^{-1}(1-a^2)^{1/2}$	$a(1-a^2)^{-1/2}$	a^{-1}	$(a^2-1)^{1/2}$	$(a^2-1)^{-1/2}$	a

4.17 Special Values and Limits

Table 4.17.1: Trigonometric functions: values at multiples of $\frac{1}{12}\pi$.

θ	$\sin\theta$	$\cos\theta$	$\tan\theta$	$\csc\theta$	$\sec\theta$	$\cot\theta$
0	0	1	0	∞	1	∞
$\pi/12$	$\frac{1}{4}\sqrt{2}(\sqrt{3}-1)$	$\frac{1}{4}\sqrt{2}(\sqrt{3}+1)$	$2-\sqrt{3}$	$\sqrt{2}(\sqrt{3}+1)$	$\sqrt{2}(\sqrt{3}-1)$	$2+\sqrt{3}$
$\pi/6$	$\frac{1}{2}$	$\frac{1}{2}\sqrt{3}$	$\frac{1}{3}\sqrt{3}$	2	$\frac{2}{3}\sqrt{3}$	$\sqrt{3}$
$\pi/4$	$\frac{1}{2}\sqrt{2}$	$\frac{1}{2}\sqrt{2}$	1	$\sqrt{2}$	$\sqrt{2}$	1
$\pi/3$	$\frac{1}{2}\sqrt{3}$	$\frac{1}{2}$	$\sqrt{3}$	$\frac{2}{3}\sqrt{3}$	2	$\frac{1}{3}\sqrt{3}$
$5\pi/12$	$\frac{1}{4}\sqrt{2}(\sqrt{3}+1)$	$\frac{1}{4}\sqrt{2}(\sqrt{3}-1)$	$2+\sqrt{3}$	$\sqrt{2}(\sqrt{3}-1)$	$\sqrt{2}(\sqrt{3}+1)$	$2-\sqrt{3}$
$\pi/2$	1	0	∞	1	∞	0
$7\pi/12$	$\frac{1}{4}\sqrt{2}(\sqrt{3}+1)$	$-\frac{1}{4}\sqrt{2}(\sqrt{3}-1)$	$-(2+\sqrt{3})$	$\sqrt{2}(\sqrt{3}-1)$	$-\sqrt{2}(\sqrt{3}+1)$	$-(2-\sqrt{3})$
$2\pi/3$	$\frac{1}{2}\sqrt{3}$	$-\frac{1}{2}$	$-\sqrt{3}$	$\frac{2}{3}\sqrt{3}$	-2	$-\frac{1}{3}\sqrt{3}$
$3\pi/4$	$\frac{1}{2}\sqrt{2}$	$-\frac{1}{2}\sqrt{2}$	-1	$\sqrt{2}$	$-\sqrt{2}$	-1
$5\pi/6$	$\frac{1}{2}$	$-\frac{1}{2}\sqrt{3}$	$-\frac{1}{3}\sqrt{3}$	2	$-\frac{2}{3}\sqrt{3}$	$-\sqrt{3}$
$11\pi/12$	$\frac{1}{4}\sqrt{2}(\sqrt{3}-1)$	$-\frac{1}{4}\sqrt{2}(\sqrt{3}+1)$	$-(2-\sqrt{3})$	$\sqrt{2}(\sqrt{3}+1)$	$-\sqrt{2}(\sqrt{3}-1)$	$-(2+\sqrt{3})$
π	0	-1	0	∞	-1	∞

4.17.1 $$\lim_{z\to 0}\frac{\sin z}{z}=1,$$

4.17.2 $$\lim_{z\to 0}\frac{\tan z}{z}=1.$$

4.17.3 $$\lim_{z\to 0}\frac{1-\cos z}{z^2}=\frac{1}{2}.$$

4.18 Inequalities

Jordan's Inequality

4.18.1 $$\frac{2x}{\pi}\leq\sin x\leq x, \qquad 0\leq x\leq\tfrac{1}{2}\pi.$$

4.18.2 $$x\leq\tan x, \qquad 0\leq x<\tfrac{1}{2}\pi,$$

4.18.3 $$\cos x\leq\frac{\sin x}{x}\leq 1, \qquad 0\leq x\leq\pi,$$

4.18.4 $$\pi<\frac{\sin(\pi x)}{x(1-x)}\leq 4, \qquad 0<x<1.$$

With $z=x+iy$,

4.18.5 $$|\sinh y|\leq|\sin z|\leq\cosh y,$$

4.18.6 $$|\sinh y|\leq|\cos z|\leq\cosh y,$$

4.18.7 $$|\csc z|\leq\operatorname{csch}|y|,$$

4.18.8 $$|\cos z|\leq\cosh|z|,$$

4.18.9 $$|\sin z|\leq\sinh|z|,$$

4.18.10 $$|\cos z|<2, \quad |\sin z|\leq\tfrac{6}{5}|z|, \qquad |z|<1.$$

For more inequalities see Mitrinović (1964, pp. 101–111), Mitrinović (1970, pp. 235–265), and Bullen (1998, pp. 250–254).

4.19 Maclaurin Series and Laurent Series

4.19.1 $$\sin z = z - \frac{z^3}{3!} + \frac{z^5}{5!} - \frac{z^7}{7!} + \cdots,$$

4.19.2 $$\cos z = 1 - \frac{z^2}{2!} + \frac{z^4}{4!} - \frac{z^6}{6!} + \cdots.$$

In (4.19.3)–(4.19.9), B_n are the Bernoulli numbers and E_n are the Euler numbers (§§24.2(i)–24.2(ii)).

4.19.3 $$\tan z = z + \frac{z^3}{3} + \frac{2}{15}z^5 + \frac{17}{315}z^7 + \cdots$$
$$+ \frac{(-1)^{n-1}2^{2n}(2^{2n}-1)B_{2n}}{(2n)!}z^{2n-1}+\cdots,$$
$$|z|<\tfrac{1}{2}\pi,$$

4.19.4 $$\csc z = \frac{1}{z} + \frac{z}{6} + \frac{7}{360}z^3 + \frac{31}{15120}z^5 + \cdots$$
$$+ \frac{(-1)^{n-1}2(2^{2n-1}-1)B_{2n}}{(2n)!}z^{2n-1}+\cdots,$$
$$0<|z|<\pi,$$

4.19.5
$$\sec z = 1 + \frac{z^2}{2} + \frac{5}{24}z^4 + \frac{61}{720}z^6 + \cdots$$
$$+ \frac{(-1)^n E_{2n}}{(2n)!} z^{2n} + \cdots, \qquad |z| < \tfrac{1}{2}\pi,$$

4.19.6
$$\cot z = \frac{1}{z} - \frac{z}{3} - \frac{z^3}{45} - \frac{2}{945}z^5 - \cdots$$
$$- \frac{(-1)^{n-1} 2^{2n} B_{2n}}{(2n)!} z^{2n-1} - \cdots, \qquad 0 < |z| < \pi,$$

4.19.7
$$\ln\left(\frac{\sin z}{z}\right) = \sum_{n=1}^{\infty} \frac{(-1)^n 2^{2n-1} B_{2n}}{n(2n)!} z^{2n}, \quad |z| < \pi,$$

4.19.8
$$\ln(\cos z) = \sum_{n=1}^{\infty} \frac{(-1)^n 2^{2n-1}(2^{2n}-1) B_{2n}}{n(2n)!} z^{2n}, \quad |z| < \tfrac{1}{2}\pi,$$

4.19.9
$$\ln\left(\frac{\tan z}{z}\right) = \sum_{n=1}^{\infty} \frac{(-1)^{n-1} 2^{2n}(2^{2n-1}-1) B_{2n}}{n(2n)!} z^{2n},$$
$$|z| < \tfrac{1}{2}\pi.$$

4.20 Derivatives and Differential Equations

4.20.1 $\quad \dfrac{d}{dz}\sin z = \cos z,$

4.20.2 $\quad \dfrac{d}{dz}\cos z = -\sin z,$

4.20.3 $\quad \dfrac{d}{dz}\tan z = \sec^2 z,$

4.20.4 $\quad \dfrac{d}{dz}\csc z = -\csc z \cot z,$

4.20.5 $\quad \dfrac{d}{dz}\sec z = \sec z \tan z,$

4.20.6 $\quad \dfrac{d}{dz}\cot z = -\csc^2 z,$

4.20.7 $\quad \dfrac{d^n}{dz^n}\sin z = \sin(z + \tfrac{1}{2}n\pi),$

4.20.8 $\quad \dfrac{d^n}{dz^n}\cos z = \cos(z + \tfrac{1}{2}n\pi).$

With $a \neq 0$, the general solutions of the differential equations

4.20.9 $\quad \dfrac{d^2 w}{dz^2} + a^2 w = 0,$

4.20.10 $\quad \left(\dfrac{dw}{dz}\right)^2 + a^2 w^2 = 1,$

4.20.11 $\quad \dfrac{dw}{dz} - a^2 w^2 = 1,$

are respectively

4.20.12 $\quad w = A\cos(az) + B\sin(az),$

4.20.13 $\quad w = (1/a)\sin(az + c),$

4.20.14 $\quad w = (1/a)\tan(az + c),$

where A, B, c are arbitrary constants.

For other differential equations see Kamke (1977, pp. 355–358 and 396–400).

4.21 Identities

4.21(i) Addition Formulas

4.21.1 $\quad \sin u \pm \cos u = \sqrt{2}\sin\bigl(u \pm \tfrac{1}{4}\pi\bigr) = \sqrt{2}\cos\bigl(u \mp \tfrac{1}{4}\pi\bigr).$

4.21.2 $\quad \sin(u \pm v) = \sin u \cos v \pm \cos u \sin v,$

4.21.3 $\quad \cos(u \pm v) = \cos u \cos v \mp \sin u \sin v,$

4.21.4 $\quad \tan(u \pm v) = \dfrac{\tan u \pm \tan v}{1 \mp \tan u \tan v},$

4.21.5 $\quad \cot(u \pm v) = \dfrac{\pm \cot u \cot v - 1}{\cot u \pm \cot v}.$

4.21.6 $\quad \sin u + \sin v = 2 \sin\left(\dfrac{u+v}{2}\right)\cos\left(\dfrac{u-v}{2}\right),$

4.21.7 $\quad \sin u - \sin v = 2 \cos\left(\dfrac{u+v}{2}\right)\sin\left(\dfrac{u-v}{2}\right),$

4.21.8 $\quad \cos u + \cos v = 2 \cos\left(\dfrac{u+v}{2}\right)\cos\left(\dfrac{u-v}{2}\right),$

4.21.9 $\quad \cos u - \cos v = -2 \sin\left(\dfrac{u+v}{2}\right)\sin\left(\dfrac{u-v}{2}\right).$

4.21.10 $\quad \tan u \pm \tan v = \dfrac{\sin(u \pm v)}{\cos u \cos v},$

4.21.11 $\quad \cot u \pm \cot v = \dfrac{\sin(v \pm u)}{\sin u \sin v}.$

4.21(ii) Squares and Products

4.21.12 $\quad \sin^2 z + \cos^2 z = 1,$

4.21.13 $\quad \sec^2 z = 1 + \tan^2 z,$

4.21.14 $\quad \csc^2 z = 1 + \cot^2 z.$

4.21.15 $\quad 2\sin u \sin v = \cos(u-v) - \cos(u+v),$

4.21.16 $\quad 2\cos u \cos v = \cos(u-v) + \cos(u+v),$

4.21.17 $\quad 2\sin u \cos v = \sin(u-v) + \sin(u+v).$

4.21.18 $\quad \sin^2 u - \sin^2 v = \sin(u+v)\sin(u-v),$

4.21.19 $\quad \cos^2 u - \cos^2 v = -\sin(u+v)\sin(u-v),$

4.21.20 $\quad \cos^2 u - \sin^2 v = \cos(u+v)\cos(u-v).$

4.21(iii) Multiples of the Argument

4.21.21
$$\sin \frac{z}{2} = \pm \left(\frac{1 - \cos z}{2} \right)^{1/2},$$

4.21.22
$$\cos \frac{z}{2} = \pm \left(\frac{1 + \cos z}{2} \right)^{1/2},$$

4.21.23
$$\tan \frac{z}{2} = \pm \left(\frac{1 - \cos z}{1 + \cos z} \right)^{1/2} = \frac{1 - \cos z}{\sin z} = \frac{\sin z}{1 + \cos z}.$$

In (4.21.21)–(4.21.23) Table 4.16.1 and analytic continuation will assist in resolving sign ambiguities.

4.21.24
$$\sin(-z) = -\sin z,$$

4.21.25
$$\cos(-z) = \cos z,$$

4.21.26
$$\tan(-z) = -\tan z.$$

4.21.27
$$\sin(2z) = 2 \sin z \cos z = \frac{2 \tan z}{1 + \tan^2 z},$$

4.21.28
$$\cos(2z) = 2\cos^2 z - 1 = 1 - 2\sin^2 z$$
$$= \cos^2 z - \sin^2 z = \frac{1 - \tan^2 z}{1 + \tan^2 z},$$

4.21.29
$$\tan(2z) = \frac{2 \tan z}{1 - \tan^2 z} = \frac{2 \cot z}{\cot^2 z - 1} = \frac{2}{\cot z - \tan z}.$$

4.21.30
$$\sin(3z) = 3 \sin z - 4 \sin^3 z,$$

4.21.31
$$\cos(3z) = -3 \cos z + 4 \cos^3 z,$$

4.21.32
$$\sin(4z) = 8 \cos^3 z \sin z - 4 \cos z \sin z,$$

4.21.33
$$\cos(4z) = 8 \cos^4 z - 8 \cos^2 z + 1.$$

De Moivre's Theorem

When $n \in \mathbb{Z}$

4.21.34
$$\cos(nz) + i \sin(nz) = (\cos z + i \sin z)^n.$$

This result is also valid when n is fractional or complex, provided that $-\pi \leq \Re z \leq \pi$.

4.21.35
$$\sin(nz) = 2^{n-1} \prod_{k=0}^{n-1} \sin \left(z + \frac{k\pi}{n} \right), \quad n = 1, 2, 3, \ldots.$$

If $t = \tan(\tfrac{1}{2} z)$, then

4.21.36
$$\sin z = \frac{2t}{1 + t^2}, \quad \cos z = \frac{1 - t^2}{1 + t^2}, \quad dz = \frac{2}{1 + t^2} dt.$$

4.21(iv) Real and Imaginary Parts; Moduli

With $z = x + iy$

4.21.37
$$\sin z = \sin x \cosh y + i \cos x \sinh y,$$

4.21.38
$$\cos z = \cos x \cosh y - i \sin x \sinh y,$$

4.21.39
$$\tan z = \frac{\sin(2x) + i \sinh(2y)}{\cos(2x) + \cosh(2y)},$$

4.21.40
$$\cot z = \frac{\sin(2x) - i \sinh(2y)}{\cosh(2y) - \cos(2x)}.$$

4.21.41
$$|\sin z| = (\sin^2 x + \sinh^2 y)^{1/2} = \left(\tfrac{1}{2} (\cosh(2y) - \cos(2x)) \right)^{1/2},$$

4.21.42
$$|\cos z| = (\cos^2 x + \sinh^2 y)^{1/2} = \left(\tfrac{1}{2} (\cosh(2y) + \cos(2x)) \right)^{1/2},$$

4.21.43
$$|\tan z| = \left(\frac{\cosh(2y) - \cos(2x)}{\cosh(2y) + \cos(2x)} \right)^{1/2}.$$

4.22 Infinite Products and Partial Fractions

4.22.1
$$\sin z = z \prod_{n=1}^{\infty} \left(1 - \frac{z^2}{n^2 \pi^2} \right),$$

4.22.2
$$\cos z = \prod_{n=1}^{\infty} \left(1 - \frac{4z^2}{(2n-1)^2 \pi^2} \right).$$

When $z \neq n\pi$, $n \in \mathbb{Z}$,

4.22.3
$$\cot z = \frac{1}{z} + 2z \sum_{n=1}^{\infty} \frac{1}{z^2 - n^2 \pi^2},$$

4.22.4
$$\csc^2 z = \sum_{n=-\infty}^{\infty} \frac{1}{(z - n\pi)^2},$$

4.22.5
$$\csc z = \frac{1}{z} + 2z \sum_{n=1}^{\infty} \frac{(-1)^n}{z^2 - n^2 \pi^2}.$$

4.23 Inverse Trigonometric Functions

4.23(i) General Definitions

The general values of the inverse trigonometric functions are defined by

4.23.1
$$\operatorname{Arcsin} z = \int_0^z \frac{dt}{(1 - t^2)^{1/2}},$$

4.23.2
$$\operatorname{Arccos} z = \int_z^1 \frac{dt}{(1 - t^2)^{1/2}},$$

4.23.3
$$\operatorname{Arctan} z = \int_0^z \frac{dt}{1 + t^2}, \qquad z \neq \pm i,$$

4.23.4
$$\operatorname{Arccsc} z = \operatorname{Arcsin}(1/z),$$

4.23.5
$$\operatorname{Arcsec} z = \operatorname{Arccos}(1/z),$$

4.23.6
$$\operatorname{Arccot} z = \operatorname{Arctan}(1/z).$$

In (4.23.1) and (4.23.2) the integration paths may not pass through either of the points $t = \pm 1$. The function $(1-t^2)^{1/2}$ assumes its principal value when $t \in (-1,1)$; elsewhere on the integration paths the branch is determined by continuity. In (4.23.3) the integration path may not intersect $\pm i$. Each of the six functions is a multivalued function of z. Arctan z and Arccot z have branch points at $z = \pm i$; the other four functions have branch points at $z = \pm 1$.

4.23(ii) Principal Values

The *principal values* (or *principal branches*) of the inverse sine, cosine, and tangent are obtained by introducing cuts in the z-plane as indicated in Figures 4.23.1(i) and 4.23.1(ii), and requiring the integration paths in (4.23.1)–(4.23.3) not to cross these cuts. Compare the principal value of the logarithm (§4.2(i)). The principal branches are denoted by arcsin z, arccos z, arctan z, respectively. Each is two-valued on the corresponding cuts, and each is real on the part of the real axis that remains after deleting the intersections with the corresponding cuts.

The principal values of the inverse cosecant, secant, and cotangent are given by

4.23.7 $\qquad \operatorname{arccsc} z = \arcsin(1/z),$

4.23.8 $\qquad \operatorname{arcsec} z = \arccos(1/z).$

4.23.9 $\qquad \operatorname{arccot} z = \arctan(1/z), \qquad z \neq \pm i.$

These functions are analytic in the cut plane depicted in Figures 4.23.1(iii) and 4.23.1(iv).

Except where indicated otherwise, it is assumed throughout this Handbook that the inverse trigonometric functions assume their principal values.

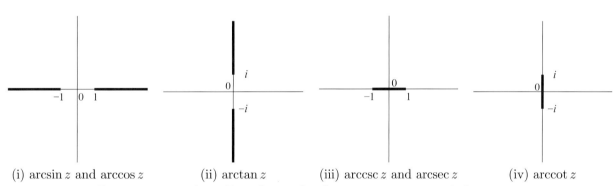

(i) arcsin z and arccos z (ii) arctan z (iii) arccsc z and arcsec z (iv) arccot z

Figure 4.23.1: z-plane. Branch cuts for the inverse trigonometric functions.

Graphs of the principal values for real arguments are given in §4.15. This section also includes conformal mappings, and surface plots for complex arguments.

4.23(iii) Reflection Formulas

4.23.10 $\quad \arcsin(-z) = -\arcsin z,$

4.23.11 $\quad \arccos(-z) = \pi - \arccos z.$

4.23.12 $\quad \arctan(-z) = -\arctan z, \qquad z \neq \pm i.$

4.23.13 $\quad \operatorname{arccsc}(-z) = -\operatorname{arccsc} z,$

4.23.14 $\quad \operatorname{arcsec}(-z) = \pi - \operatorname{arcsec} z.$

4.23.15 $\quad \operatorname{arccot}(-z) = -\operatorname{arccot} z, \qquad z \neq \pm i.$

4.23.16 $\qquad \arccos z = \tfrac{1}{2}\pi - \arcsin z,$

4.23.17 $\qquad \operatorname{arcsec} z = \tfrac{1}{2}\pi - \operatorname{arccsc} z.$

4.23.18 $\qquad \operatorname{arccot} z = \pm \tfrac{1}{2}\pi - \arctan z, \qquad \Re z \gtrless 0.$

4.23(iv) Logarithmic Forms

Throughout this subsection *all* quantities assume their principal values.

Inverse Sine

4.23.19
$$\arcsin z = -i \ln\!\left((1-z^2)^{1/2} + iz\right),$$
$$z \in \mathbb{C} \setminus (-\infty, -1) \cup (1, \infty);$$

compare Figure 4.23.1(i). On the cuts

4.23.20
$$\arcsin x = \tfrac{1}{2}\pi \pm i \ln\!\left((x^2-1)^{1/2} + x\right), \qquad x \in [1, \infty),$$

4.23.21
$$\arcsin x = -\tfrac{1}{2}\pi \pm i \ln\!\left((x^2-1)^{1/2} - x\right),$$
$$x \in (-\infty, -1],$$

upper signs being taken on upper sides, and lower signs on lower sides.

Inverse Cosine

4.23.22
$$\arccos z = \tfrac{1}{2}\pi + i\ln\left((1-z^2)^{1/2} + iz\right),$$
$$z \in \mathbb{C} \setminus (-\infty, -1) \cup (1, \infty);$$

compare Figure 4.23.1(i). An equivalent definition is

4.23.23
$$\arccos z = -2i\ln\left(\left(\frac{1+z}{2}\right)^{1/2} + i\left(\frac{1-z}{2}\right)^{1/2}\right),$$
$$z \in \mathbb{C} \setminus (-\infty, -1) \cup (1, \infty);$$

see Kahan (1987).

On the cuts

4.23.24 $\quad \arccos x = \mp i\ln\left((x^2-1)^{1/2} + x\right), \quad x \in [1, \infty),$

4.23.25
$$\arccos x = \pi \mp i\ln\left((x^2-1)^{1/2} - x\right),$$
$$x \in (-\infty, -1],$$

the upper/lower signs corresponding to the upper/lower sides.

Inverse Tangent

4.23.26
$$\arctan z = \frac{i}{2}\ln\left(\frac{i+z}{i-z}\right), \quad z/i \in \mathbb{C} \setminus (-\infty, -1] \cup [1, \infty);$$

compare Figure 4.23.1(ii). On the cuts

4.23.27
$$\arctan(iy) = \pm\frac{1}{2}\pi + \frac{i}{2}\ln\left(\frac{y+1}{y-1}\right),$$
$$y \in (-\infty, -1) \cup (1, \infty),$$

the upper/lower sign corresponding to the right/left side.

Other Inverse Functions

For the corresponding results for $\operatorname{arccsc} z$, $\operatorname{arcsec} z$, and $\operatorname{arccot} z$, use (4.23.7)–(4.23.9). Care needs to be taken on the cuts, for example, if $0 < x < \infty$ then $1/(x+i0) = (1/x) - i0$.

4.23(v) Fundamental Property

With $k \in \mathbb{Z}$, the general solutions of the equations

4.23.28 $\quad z = \sin w,$

4.23.29 $\quad z = \cos w,$

4.23.30 $\quad z = \tan w,$

are respectively

4.23.31 $\quad w = \operatorname{Arcsin} z = (-1)^k \arcsin z + k\pi,$

4.23.32 $\quad w = \operatorname{Arccos} z = \pm \arccos z + 2k\pi,$

4.23.33 $\quad w = \operatorname{Arctan} z = \arctan z + k\pi, \qquad z \neq \pm i.$

4.23(vi) Real and Imaginary Parts

4.23.34 $\quad \arcsin z = \arcsin \beta + i\ln\left(\alpha + (\alpha^2-1)^{1/2}\right),$

4.23.35 $\quad \arccos z = \arccos \beta - i\ln\left(\alpha + (\alpha^2-1)^{1/2}\right),$

4.23.36
$$\arctan z = \tfrac{1}{2}\arctan\left(\frac{2x}{1-x^2-y^2}\right)$$
$$+ \tfrac{1}{4}i\ln\left(\frac{x^2+(y+1)^2}{x^2+(y-1)^2}\right),$$

where $z = x + iy$ and $x \in [-1, 1]$ in (4.23.34) and (4.23.35), and $|z| < 1$ in (4.23.36). Also,

4.23.37 $\quad \alpha = \tfrac{1}{2}\left((x+1)^2 + y^2\right)^{1/2} + \tfrac{1}{2}\left((x-1)^2 + y^2\right)^{1/2},$

4.23.38 $\quad \beta = \tfrac{1}{2}\left((x+1)^2 + y^2\right)^{1/2} - \tfrac{1}{2}\left((x-1)^2 + y^2\right)^{1/2}.$

4.23(vii) Special Values and Interrelations

Table 4.23.1: Inverse trigonometric functions: principal values at 0, ± 1, $\pm \infty$.

x	$\arcsin x$	$\arccos x$	$\arctan x$	$\operatorname{arccsc} x$	$\operatorname{arcsec} x$	$\operatorname{arccot} x$
$-\infty$	—	—	$-\tfrac{1}{2}\pi$	0	$\tfrac{1}{2}\pi$	0
-1	$-\tfrac{1}{2}\pi$	π	$-\tfrac{1}{4}\pi$	$-\tfrac{1}{2}\pi$	π	$-\tfrac{1}{4}\pi$
0	0	$\tfrac{1}{2}\pi$	0	—	—	$\mp\tfrac{1}{2}\pi$
1	$\tfrac{1}{2}\pi$	0	$\tfrac{1}{4}\pi$	$\tfrac{1}{2}\pi$	0	$\tfrac{1}{4}\pi$
∞	—	—	$\tfrac{1}{2}\pi$	0	$\tfrac{1}{2}\pi$	0

For interrelations see Table 4.16.3. For example, from the heading and last entry in the penultimate column we have $\operatorname{arcsec} a = \operatorname{arccot}((a^2-1)^{-1/2})$.

4.23(viii) Gudermannian Function

The *Gudermannian* $\operatorname{gd}(x)$ is defined by

4.23.39 $$\operatorname{gd}(x) = \int_0^x \operatorname{sech} t\, dt, \quad -\infty < x < \infty.$$

Equivalently,

4.23.40
$$\begin{aligned}\operatorname{gd}(x) &= 2\arctan(e^x) - \tfrac{1}{2}\pi \\ &= \operatorname{arcsin}(\tanh x) = \operatorname{arccsc}(\coth x) \\ &= \operatorname{arccos}(\operatorname{sech} x) = \operatorname{arcsec}(\cosh x) \\ &= \arctan(\sinh x) = \operatorname{arccot}(\operatorname{csch} x).\end{aligned}$$

The inverse Gudermannian function is given by

4.23.41 $$\operatorname{gd}^{-1}(x) = \int_0^x \sec t\, dt, \quad -\tfrac{1}{2}\pi < x < \tfrac{1}{2}\pi.$$

Equivalently, and again when $-\tfrac{1}{2}\pi < x < \tfrac{1}{2}\pi$,

4.23.42
$$\begin{aligned}\operatorname{gd}^{-1}(x) &= \ln\tan\!\left(\tfrac{1}{2}x + \tfrac{1}{4}\pi\right) = \ln(\sec x + \tan x)\\ &= \operatorname{arcsinh}(\tan x) = \operatorname{arccsch}(\cot x) \\ &= \operatorname{arccosh}(\sec x) = \operatorname{arcsech}(\cos x) \\ &= \operatorname{arctanh}(\sin x) = \operatorname{arccoth}(\csc x).\end{aligned}$$

4.24 Inverse Trigonometric Functions: Further Properties

4.24(i) Power Series

4.24.1
$$\arcsin z = z + \frac{1}{2}\frac{z^3}{3} + \frac{1\cdot 3}{2\cdot 4}\frac{z^5}{5} + \frac{1\cdot 3\cdot 5}{2\cdot 4\cdot 6}\frac{z^7}{7} + \cdots, \quad |z| \le 1.$$

4.24.2
$$\arccos z = (2(1-z))^{1/2} \\ \times \left(1 + \sum_{n=1}^{\infty} \frac{1\cdot 3\cdot 5\cdots(2n-1)}{2^{2n}(2n+1)n!}(1-z)^n\right), \\ |1-z| \le 2.$$

4.24.3
$$\arctan z = z - \frac{z^3}{3} + \frac{z^5}{5} - \frac{z^7}{7} + \cdots, \quad |z|\le 1,\ z\ne \pm i.$$

4.24.4
$$\arctan z = \pm\frac{\pi}{2} - \frac{1}{z} + \frac{1}{3z^3} - \frac{1}{5z^5} + \cdots, \quad \Re z \gtreqless 0,\ |z| \ge 1.$$

4.24.5
$$\arctan z = \frac{z}{z^2 + 1} \\ \times \left(1 + \frac{2}{3}\frac{z^2}{1+z^2} + \frac{2\cdot 4}{3\cdot 5}\left(\frac{z^2}{1+z^2}\right)^2 + \cdots\right), \\ \Re(z^2) > -\tfrac{1}{2}.$$

which requires $z\,(=x+iy)$ to lie between the two rectangular hyperbolas given by

4.24.6 $$x^2 - y^2 = -\tfrac{1}{2}.$$

4.24(ii) Derivatives

4.24.7 $$\frac{d}{dz}\arcsin z = (1-z^2)^{-1/2},$$

4.24.8 $$\frac{d}{dz}\arccos z = -(1-z^2)^{-1/2},$$

4.24.9 $$\frac{d}{dz}\arctan z = \frac{1}{1+z^2}.$$

4.24.10 $$\frac{d}{dz}\operatorname{arccsc} z = \mp\frac{1}{z(z^2-1)^{1/2}}, \qquad \Re z \gtreqless 0.$$

4.24.11 $$\frac{d}{dz}\operatorname{arcsec} z = \pm\frac{1}{z(z^2-1)^{1/2}}, \qquad \Re z \gtreqless 0.$$

4.24.12 $$\frac{d}{dz}\operatorname{arccot} z = -\frac{1}{1+z^2}.$$

4.24(iii) Addition Formulas

4.24.13
$$\operatorname{Arcsin} u \pm \operatorname{Arcsin} v \\ = \operatorname{Arcsin}\!\left(u(1-v^2)^{1/2} \pm v(1-u^2)^{1/2}\right),$$

4.24.14
$$\operatorname{Arccos} u \pm \operatorname{Arccos} v \\ = \operatorname{Arccos}\!\left(uv \mp ((1-u^2)(1-v^2))^{1/2}\right),$$

4.24.15 $$\operatorname{Arctan} u \pm \operatorname{Arctan} v = \operatorname{Arctan}\!\left(\frac{u\pm v}{1\mp uv}\right),$$

4.24.16
$$\operatorname{Arcsin} u \pm \operatorname{Arccos} v \\ = \operatorname{Arcsin}\!\left(uv \pm ((1-u^2)(1-v^2))^{1/2}\right) \\ = \operatorname{Arccos}\!\left(v(1-u^2)^{1/2} \mp u(1-v^2)^{1/2}\right),$$

4.24.17
$$\operatorname{Arctan} u \pm \operatorname{Arccot} v = \operatorname{Arctan}\!\left(\frac{uv\pm 1}{v\mp u}\right) \\ = \operatorname{Arccot}\!\left(\frac{v\mp u}{uv\pm 1}\right).$$

The above equations are interpreted in the sense that every value of the left-hand side is a value of the right-hand side and vice versa. All square roots have either possible value.

4.25 Continued Fractions

4.25.1
$$\tan z = \cfrac{z}{1-}\cfrac{z^2}{3-}\cfrac{z^2}{5-}\cfrac{z^2}{7-}\cdots,\quad z\ne \pm\tfrac{1}{2}\pi,\ \pm\tfrac{3}{2}\pi,\ \ldots.$$

4.25.2 $$\tan(az) = \cfrac{a\tan z}{1+} \cfrac{(1-a^2)\tan^2 z}{3+} \cfrac{(4-a^2)\tan^2 z}{5+} \cfrac{(9-a^2)\tan^2 z}{7+} \cdots, \quad |\Re z| < \tfrac{1}{2}\pi, \, az \neq \pm\tfrac{1}{2}\pi, \pm\tfrac{3}{2}\pi, \ldots.$$

4.25.3 $$\frac{\arcsin z}{\sqrt{1-z^2}} = \cfrac{z}{1-} \cfrac{1\cdot 2z^2}{3-} \cfrac{1\cdot 2z^2}{5-} \cfrac{3\cdot 4z^2}{7-} \cfrac{3\cdot 4z^2}{9-} \cdots,$$

valid when z lies in the open cut plane shown in Figure 4.23.1(i).

4.25.4 $$\arctan z = \cfrac{z}{1+} \cfrac{z^2}{3+} \cfrac{4z^2}{5+} \cfrac{9z^2}{7+} \cfrac{16z^2}{9+} \cdots,$$

valid when z lies in the open cut plane shown in Figure 4.23.1(ii).

4.25.5
$$e^{2a\arctan(1/z)} = 1 + \cfrac{2a}{z-a+} \cfrac{a^2+1}{3z+} \cfrac{a^2+4}{5z+} \cfrac{a^2+9}{7z+} \cdots,$$

valid when z lies in the open cut plane shown in Figure 4.23.1(iv).

See Lorentzen and Waadeland (1992, pp. 560–571) for other continued fractions involving inverse trigonometric functions. See also Cuyt et al. (2008, pp. 201–203, 205–210).

4.26 Integrals

4.26(i) Introduction

Throughout this section the variables are assumed to be real. The results in §§4.26(ii) and 4.26(iv) can be extended to the complex plane by using continuous branches and avoiding singularities.

4.26(ii) Indefinite Integrals

4.26.1 $\int \sin x \, dx = -\cos x,$

4.26.2 $\int \cos x \, dx = \sin x.$

4.26.3 $\int \tan x \, dx = -\ln(\cos x), \qquad -\tfrac{1}{2}\pi < x < \tfrac{1}{2}\pi.$

4.26.4 $\int \csc x \, dx = \ln(\tan \tfrac{1}{2}x), \qquad 0 < x < \pi.$

4.26.5 $\int \sec x \, dx = \operatorname{gd}^{-1}(x), \qquad -\tfrac{1}{2}\pi < x < \tfrac{1}{2}\pi.$

For the right-hand side see (4.23.41) and (4.23.42).

4.26.6 $\int \cot x \, dx = \ln(\sin x), \qquad 0 < x < \pi.$

4.26.7 $\int e^{ax}\sin(bx)\,dx = \dfrac{e^{ax}}{a^2+b^2}(a\sin(bx) - b\cos(bx)),$

4.26.8 $\int e^{ax}\cos(bx)\,dx = \dfrac{e^{ax}}{a^2+b^2}(a\cos(bx) + b\sin(bx)).$

4.26(iii) Definite Integrals

Throughout this subsection m and n are integers.

Orthogonality Properties

4.26.9 $\displaystyle\int_0^\pi \sin(mt)\sin(nt)\,dt = 0, \qquad m \neq n,$

4.26.10 $\displaystyle\int_0^\pi \cos(mt)\cos(nt)\,dt = 0, \qquad m \neq n,$

4.26.11 $\displaystyle\int_0^\pi \sin^2(nt)\,dt = \int_0^\pi \cos^2(nt)\,dt = \tfrac{1}{2}\pi, \quad n \neq 0.$

4.26.12 $\displaystyle\int_0^\infty \frac{\sin(mt)}{t}\,dt = \begin{cases} \tfrac{1}{2}\pi, & m > 0, \\ 0, & m = 0, \\ -\tfrac{1}{2}\pi, & m < 0. \end{cases}$

4.26.13 $\displaystyle\int_0^\infty \sin(t^2)\,dt = \int_0^\infty \cos(t^2)\,dt = \dfrac{1}{2}\sqrt{\dfrac{\pi}{2}}.$

4.26(iv) Inverse Trigonometric Functions

4.26.14 $\displaystyle\int \arcsin x \, dx = x\arcsin x + (1-x^2)^{1/2}, \quad -1 < x < 1,$

4.26.15 $\displaystyle\int \arccos x \, dx = x\arccos x - (1-x^2)^{1/2}, \quad -1 < x < 1.$

4.26.16 $\displaystyle\int \arctan x \, dx = x\arctan x - \tfrac{1}{2}\ln(1+x^2), \quad -\infty < x < \infty,$

4.26.17 $\displaystyle\int \operatorname{arccsc} x \, dx = x\operatorname{arccsc} x + \ln\!\left(x + (x^2-1)^{1/2}\right), \quad 1 < x < \infty,$

4.26.18 $\displaystyle\int \operatorname{arcsec} x \, dx = x\operatorname{arcsec} x - \ln\!\left(x + (x^2-1)^{1/2}\right), \quad 1 < x < \infty,$

4.26.19 $\displaystyle\int \operatorname{arccot} x \, dx = x\operatorname{arccot} x + \tfrac{1}{2}\ln(1+x^2), \quad 0 < x < \infty.$

4.26.20 $\displaystyle\int x \arcsin x \, dx = \left(\dfrac{x^2}{2} - \dfrac{1}{4}\right)\arcsin x + \dfrac{x}{4}(1-x^2)^{1/2}, \quad -1 < x < 1,$

4.26.21 $\displaystyle\int x \arccos x \, dx = \left(\dfrac{x^2}{2} - \dfrac{1}{4}\right)\arccos x - \dfrac{x}{4}(1-x^2)^{1/2}, \quad -1 < x < 1.$

4.26(v) Compendia

Extensive compendia of indefinite and definite integrals of trigonometric and inverse trigonometric functions include Apelblat (1983, pp. 48–109), Bierens de Haan (1939), Gradshteyn and Ryzhik (2000, Chapters 2–4), Gröbner and Hofreiter (1949, pp. 116–139), Gröbner and Hofreiter (1950, pp. 94–160), and Prudnikov et al. (1986a, §§1.5, 1.7, 2.5, 2.7).

4.27 Sums

For sums of trigonometric and inverse trigonometric functions see Gradshteyn and Ryzhik (2000, Chapter 1), Hansen (1975, §§14–42), Oberhettinger (1973), and Prudnikov et al. (1986a, Chapter 5).

Hyperbolic Functions

4.28 Definitions and Periodicity

4.28.1
$$\sinh z = \frac{e^z - e^{-z}}{2},$$

4.28.2
$$\cosh z = \frac{e^z + e^{-z}}{2},$$

4.28.3
$$\cosh z \pm \sinh z = e^{\pm z},$$

4.28.4
$$\tanh z = \frac{\sinh z}{\cosh z},$$

4.28.5
$$\operatorname{csch} z = \frac{1}{\sinh z},$$

4.28.6
$$\operatorname{sech} z = \frac{1}{\cosh z},$$

4.28.7
$$\coth z = \frac{1}{\tanh z}.$$

Relations to Trigonometric Functions

4.28.8
$$\sin(iz) = i \sinh z,$$

4.28.9
$$\cos(iz) = \cosh z,$$

4.28.10
$$\tan(iz) = i \tanh z,$$

4.28.11
$$\csc(iz) = -i \operatorname{csch} z,$$

4.28.12
$$\sec(iz) = \operatorname{sech} z,$$

4.28.13
$$\cot(iz) = -i \coth z.$$

As a consequence, many properties of the hyperbolic functions follow immediately from the corresponding properties of the trigonometric functions.

Periodicity and Zeros

The functions $\sinh z$ and $\cosh z$ have period $2\pi i$, and $\tanh z$ has period πi. The zeros of $\sinh z$ and $\cosh z$ are $z = ik\pi$ and $z = i\left(k + \frac{1}{2}\right)\pi$, respectively, $k \in \mathbb{Z}$.

4.29 Graphics

4.29(i) Real Arguments

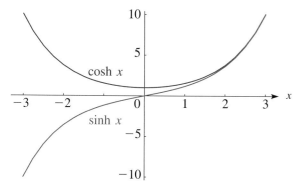

Figure 4.29.1: $\sinh x$ and $\cosh x$.

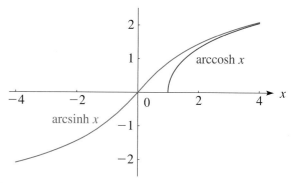

Figure 4.29.2: Principal values of $\operatorname{arcsinh} x$ and $\operatorname{arccosh} x$. ($\operatorname{arccosh} x$ is complex when $x < 1$.)

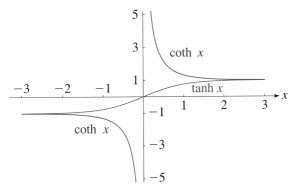

Figure 4.29.3: $\tanh x$ and $\coth x$.

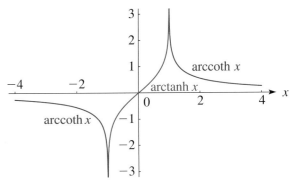

Figure 4.29.4: Principal values of $\operatorname{arctanh} x$ and $\operatorname{arccoth} x$. ($\operatorname{arctanh} x$ is complex when $x < -1$ or $x > 1$, and $\operatorname{arccoth} x$ is complex when $-1 < x < 1$.)

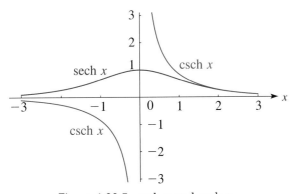

Figure 4.29.5: $\operatorname{csch} x$ and $\operatorname{sech} x$.

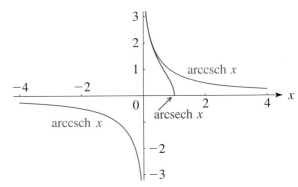

Figure 4.29.6: Principal values of $\operatorname{arccsch} x$ and $\operatorname{arcsech} x$. ($\operatorname{arcsech} x$ is complex when $x < 0$ and $x > 1$.)

4.29(ii) Complex Arguments

The conformal mapping $w = \sinh z$ is obtainable from Figure 4.15.7 by rotating both the w-plane and the z-plane through an angle $\tfrac{1}{2}\pi$, compare (4.28.8).

The surfaces for the complex hyperbolic and inverse hyperbolic functions are similar to the surfaces depicted in §4.15(iii) for the trigonometric and inverse trigonometric functions. They can be visualized with the aid of equations (4.28.8)–(4.28.13).

4.30 Elementary Properties

Table 4.30.1: Hyperbolic functions: interrelations. All square roots have their principal values when the functions are real, nonnegative, and finite.

	$\sinh\theta = a$	$\cosh\theta = a$	$\tanh\theta = a$	$\operatorname{csch}\theta = a$	$\operatorname{sech}\theta = a$	$\coth\theta = a$
$\sinh\theta$	a	$(a^2-1)^{1/2}$	$a(1-a^2)^{-1/2}$	a^{-1}	$a^{-1}(1-a^2)^{1/2}$	$(a^2-1)^{-1/2}$
$\cosh\theta$	$(1+a^2)^{1/2}$	a	$(1-a^2)^{-1/2}$	$a^{-1}(1+a^2)^{1/2}$	a^{-1}	$a(a^2-1)^{-1/2}$
$\tanh\theta$	$a(1+a^2)^{-1/2}$	$a^{-1}(a^2-1)^{1/2}$	a	$(1+a^2)^{-1/2}$	$(1-a^2)^{1/2}$	a^{-1}
$\operatorname{csch}\theta$	a^{-1}	$(a^2-1)^{-1/2}$	$a^{-1}(1-a^2)^{1/2}$	a	$a(1-a^2)^{-1/2}$	$(a^2-1)^{1/2}$
$\operatorname{sech}\theta$	$(1+a^2)^{-1/2}$	a^{-1}	$(1-a^2)^{1/2}$	$a(1+a^2)^{-1/2}$	a	$a^{-1}(a^2-1)^{1/2}$
$\coth\theta$	$a^{-1}(a^2+1)^{1/2}$	$a(a^2-1)^{-1/2}$	a^{-1}	$(1+a^2)^{1/2}$	$(1-a^2)^{-1/2}$	a

4.31 Special Values and Limits

Table 4.31.1: Hyperbolic functions: values at multiples of $\frac{1}{2}\pi i$.

z	0	$\frac{1}{2}\pi i$	πi	$\frac{3}{2}\pi i$	∞
$\sinh z$	0	i	0	$-i$	∞
$\cosh z$	1	0	-1	0	∞
$\tanh z$	0	∞i	0	$-\infty i$	1
$\csch z$	∞	$-i$	∞	i	0
$\sech z$	1	∞	-1	∞	0
$\coth z$	∞	0	∞	0	1

4.31.1 $$\lim_{z \to 0} \frac{\sinh z}{z} = 1,$$

4.31.2 $$\lim_{z \to 0} \frac{\tanh z}{z} = 1,$$

4.31.3 $$\lim_{z \to 0} \frac{\cosh z - 1}{z^2} = \frac{1}{2}.$$

4.32 Inequalities

For x real,

4.32.1 $$\cosh x \leq \left(\frac{\sinh x}{x}\right)^3,$$

4.32.2 $$\sin x \cos x < \tanh x < x, \qquad x > 0,$$

4.32.3 $$|\cosh x - \cosh y| \geq |x - y|\sqrt{\sinh x \sinh y}, \quad x > 0, y > 0,$$

4.32.4 $$\arctan x \leq \tfrac{1}{2}\pi \tanh x, \qquad x \geq 0.$$

For these and other inequalities involving hyperbolic functions see Mitrinović (1964, pp. 61, 76, 159) and Mitrinović (1970, p. 270).

4.33 Maclaurin Series and Laurent Series

4.33.1 $$\sinh z = z + \frac{z^3}{3!} + \frac{z^5}{5!} + \cdots,$$

4.33.2 $$\cosh z = 1 + \frac{z^2}{2!} + \frac{z^4}{4!} + \cdots.$$

4.33.3 $$\tanh z = z - \frac{z^3}{3} + \frac{2}{15}z^5 - \frac{17}{315}z^7 + \cdots + \frac{2^{2n}(2^{2n}-1)B_{2n}}{(2n)!}z^{2n-1} + \cdots, \qquad |z| < \tfrac{1}{2}\pi.$$

For B_{2n} see §24.2(i). For expansions that correspond to (4.19.4)–(4.19.9), change z to iz and use (4.28.8)–(4.28.13).

4.34 Derivatives and Differential Equations

4.34.1 $$\frac{d}{dz}\sinh z = \cosh z,$$

4.34.2 $$\frac{d}{dz}\cosh z = \sinh z,$$

4.34.3 $$\frac{d}{dz}\tanh z = \sech^2 z,$$

4.34.4 $$\frac{d}{dz}\csch z = -\csch z \coth z,$$

4.34.5 $$\frac{d}{dz}\sech z = -\sech z \tanh z,$$

4.34.6 $$\frac{d}{dz}\coth z = -\csch^2 z.$$

With $a \neq 0$, the general solutions of the differential equations

4.34.7 $$\frac{d^2 w}{dz^2} - a^2 w = 0,$$

4.34.8 $$\left(\frac{dw}{dz}\right)^2 - a^2 w^2 = 1,$$

4.34.9 $$\left(\frac{dw}{dz}\right)^2 - a^2 w^2 = -1,$$

4.34.10 $$\frac{dw}{dz} + a^2 w^2 = 1,$$

are respectively

4.34.11 $$w = A\cosh(az) + B\sinh(az),$$

4.34.12 $$w = (1/a)\sinh(az + c),$$

4.34.13 $$w = (1/a)\cosh(az + c),$$

4.34.14 $$w = (1/a)\coth(az + c),$$

where A, B, c are arbitrary constants.

For other differential equations see Kamke (1977, pp. 289–400).

4.35 Identities

4.35(i) Addition Formulas

4.35.1 $$\sinh(u \pm v) = \sinh u \cosh v \pm \cosh u \sinh v,$$

4.35.2 $$\cosh(u \pm v) = \cosh u \cosh v \pm \sinh u \sinh v,$$

4.35.3 $$\tanh(u \pm v) = \frac{\tanh u \pm \tanh v}{1 \pm \tanh u \tanh v},$$

4.35.4 $$\coth(u \pm v) = \frac{\pm \coth u \coth v + 1}{\coth u \pm \coth v}.$$

4.35.5 $\quad \sinh u + \sinh v = 2\sinh\left(\dfrac{u+v}{2}\right)\cosh\left(\dfrac{u-v}{2}\right),$

4.35.6 $\quad \sinh u - \sinh v = 2\cosh\left(\dfrac{u+v}{2}\right)\sinh\left(\dfrac{u-v}{2}\right),$

4.35.7 $\quad \cosh u + \cosh v = 2\cosh\left(\dfrac{u+v}{2}\right)\cosh\left(\dfrac{u-v}{2}\right),$

4.35.8 $\quad \cosh u - \cosh v = 2\sinh\left(\dfrac{u+v}{2}\right)\sinh\left(\dfrac{u-v}{2}\right),$

4.35.9 $\quad \tanh u \pm \tanh v = \dfrac{\sinh(u \pm v)}{\cosh u \cosh v},$

4.35.10 $\quad \coth u \pm \coth v = \dfrac{\sinh(v \pm u)}{\sinh u \sinh v}.$

4.35(ii) Squares and Products

4.35.11 $\quad \cosh^2 z - \sinh^2 z = 1,$

4.35.12 $\quad \operatorname{sech}^2 z = 1 - \tanh^2 z,$

4.35.13 $\quad \operatorname{csch}^2 z = \coth^2 z - 1.$

4.35.14 $\quad 2\sinh u \sinh v = \cosh(u+v) - \cosh(u-v),$

4.35.15 $\quad 2\cosh u \cosh v = \cosh(u+v) + \cosh(u-v),$

4.35.16 $\quad 2\sinh u \cosh v = \sinh(u+v) + \sinh(u-v).$

4.35.17 $\quad \sinh^2 u - \sinh^2 v = \sinh(u+v)\sinh(u-v),$

4.35.18 $\quad \cosh^2 u - \cosh^2 v = \sinh(u+v)\sinh(u-v),$

4.35.19 $\quad \sinh^2 u + \cosh^2 v = \cosh(u+v)\cosh(u-v).$

4.35(iii) Multiples of the Argument

4.35.20 $\quad \sinh\dfrac{z}{2} = \left(\dfrac{\cosh z - 1}{2}\right)^{1/2},$

4.35.21 $\quad \cosh\dfrac{z}{2} = \left(\dfrac{\cosh z + 1}{2}\right)^{1/2},$

4.35.22
$$\tanh\dfrac{z}{2} = \left(\dfrac{\cosh z - 1}{\cosh z + 1}\right)^{1/2} = \dfrac{\cosh z - 1}{\sinh z} = \dfrac{\sinh z}{\cosh z + 1}.$$

The square roots assume their principal value on the positive real axis, and are determined by continuity elsewhere.

4.35.23 $\quad \sinh(-z) = -\sinh z,$

4.35.24 $\quad \cosh(-z) = \cosh z,$

4.35.25 $\quad \tanh(-z) = -\tanh z.$

4.35.26 $\quad \sinh(2z) = 2\sinh z \cosh z = \dfrac{2\tanh z}{1 - \tanh^2 z},$

4.35.27 $\quad \cosh(2z) = 2\cosh^2 z - 1 = 2\sinh^2 z + 1$
$\qquad\qquad = \cosh^2 z + \sinh^2 z,$

4.35.28 $\quad \tanh(2z) = \dfrac{2\tanh z}{1 + \tanh^2 z},$

4.35.29 $\quad \sinh(3z) = 3\sinh z + 4\sinh^3 z,$

4.35.30 $\quad \cosh(3z) = -3\cosh z + 4\cosh^3 z,$

4.35.31 $\quad \sinh(4z) = 4\sinh^3 z \cosh z + 4\cosh^3 z \sinh z,$

4.35.32 $\quad \cosh(4z) = \cosh^4 z + 6\sinh^2 z \cosh^2 z + \sinh^4 z.$

4.35.33
$$\cosh(nz) \pm \sinh(nz) = (\cosh z \pm \sinh z)^n, \quad n \in \mathbb{Z}.$$

4.35(iv) Real and Imaginary Parts; Moduli

With $z = x + iy$

4.35.34 $\quad \sinh z = \sinh x \cos y + i\cosh x \sin y,$

4.35.35 $\quad \cosh z = \cosh x \cos y + i\sinh x \sin y,$

4.35.36 $\quad \tanh z = \dfrac{\sinh(2x) + i\sin(2y)}{\cosh(2x) + \cos(2y)},$

4.35.37 $\quad \coth z = \dfrac{\sinh(2x) - i\sin(2y)}{\cosh(2x) - \cos(2y)}.$

4.35.38 $\quad |\sinh z| = (\sinh^2 x + \sin^2 y)^{1/2}$
$\qquad\qquad = \left(\tfrac{1}{2}(\cosh(2x) - \cos(2y))\right)^{1/2},$

4.35.39 $\quad |\cosh z| = (\sinh^2 x + \cos^2 y)^{1/2}$
$\qquad\qquad = \left(\tfrac{1}{2}(\cosh(2x) + \cos(2y))\right)^{1/2},$

4.35.40 $\quad |\tanh z| = \left(\dfrac{\cosh(2x) - \cos(2y)}{\cosh(2x) + \cos(2y)}\right)^{1/2}.$

4.36 Infinite Products and Partial Fractions

4.36.1 $\quad \sinh z = z\displaystyle\prod_{n=1}^{\infty}\left(1 + \dfrac{z^2}{n^2\pi^2}\right),$

4.36.2 $\quad \cosh z = \displaystyle\prod_{n=1}^{\infty}\left(1 + \dfrac{4z^2}{(2n-1)^2\pi^2}\right).$

When $z \neq n\pi i,\ n \in \mathbb{Z}$,

4.36.3 $\quad \coth z = \dfrac{1}{z} + 2z\displaystyle\sum_{n=1}^{\infty}\dfrac{1}{z^2 + n^2\pi^2},$

4.36.4 $\quad \operatorname{csch}^2 z = \displaystyle\sum_{n=-\infty}^{\infty}\dfrac{1}{(z - n\pi i)^2},$

4.36.5 $\quad \operatorname{csch} z = \dfrac{1}{z} + 2z\displaystyle\sum_{n=1}^{\infty}\dfrac{(-1)^n}{z^2 + n^2\pi^2}.$

4.37 Inverse Hyperbolic Functions

4.37(i) General Definitions

The general values of the inverse hyperbolic functions are defined by

4.37.1 $\qquad \operatorname{Arcsinh} z = \int_0^z \frac{dt}{(1+t^2)^{1/2}},$

4.37.2 $\qquad \operatorname{Arccosh} z = \int_1^z \frac{dt}{(t^2-1)^{1/2}},$

4.37.3 $\qquad \operatorname{Arctanh} z = \int_0^z \frac{dt}{1-t^2}, \qquad z \neq \pm 1,$

4.37.4 $\qquad \operatorname{Arccsch} z = \operatorname{Arcsinh}(1/z),$

4.37.5 $\qquad \operatorname{Arcsech} z = \operatorname{Arccosh}(1/z),$

4.37.6 $\qquad \operatorname{Arccoth} z = \operatorname{Arctanh}(1/z).$

In (4.37.1) the integration path may not pass through either of the points $t = \pm i$, and the function $(1+t^2)^{1/2}$ assumes its principal value when t is real. In (4.37.2) the integration path may not pass through either of the points ± 1, and the function $(t^2 - 1)^{1/2}$ assumes its principal value when $t \in (1, \infty)$. Elsewhere on the integration paths in (4.37.1) and (4.37.2) the branches are determined by continuity. In (4.37.3) the integration path may not intersect ± 1. Each of the six functions is a multivalued function of z. $\operatorname{Arcsinh} z$ and $\operatorname{Arccsch} z$ have branch points at $z = \pm i$; the other four functions have branch points at $z = \pm 1$.

4.37(ii) Principal Values

The *principal values* (or *principal branches*) of the inverse sinh, cosh, and tanh are obtained by introducing cuts in the z-plane as indicated in Figure 4.37.1(i)-(iii), and requiring the integration paths in (4.37.1)–(4.37.3) not to cross these cuts. Compare the principal value of the logarithm (§4.2(i)). The principal branches are denoted by arcsinh, arccosh, arctanh respectively. Each is two-valued on the corresponding cut(s), and each is real on the part of the real axis that remains after deleting the intersections with the corresponding cuts.

The principal values of the inverse hyperbolic cosecant, hyperbolic secant, and hyperbolic tangent are given by

4.37.7 $\qquad \operatorname{arccsch} z = \operatorname{arcsinh}(1/z),$

4.37.8 $\qquad \operatorname{arcsech} z = \operatorname{arccosh}(1/z).$

4.37.9 $\qquad \operatorname{arccoth} z = \operatorname{arctanh}(1/z), \qquad z \neq \pm 1.$

These functions are analytic in the cut plane depicted in Figure 4.37.1(iv), (v), (vi), respectively.

Except where indicated otherwise, it is assumed throughout this Handbook that the inverse hyperbolic functions assume their principal values.

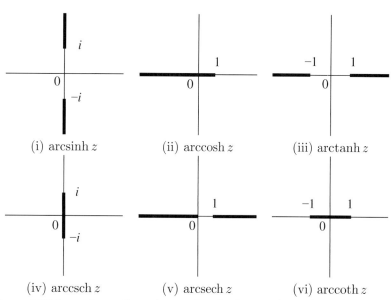

Figure 4.37.1: z-plane. Branch cuts for the inverse hyperbolic functions.

Graphs of the principal values for real arguments are given in §4.29. This section also indicates conformal mappings, and surface plots for complex arguments.

4.37(iii) Reflection Formulas

4.37.10 $\quad \operatorname{arcsinh}(-z) = -\operatorname{arcsinh} z.$

4.37.11 $\quad \operatorname{arccosh}(-z) = \pm \pi i + \operatorname{arccosh} z, \qquad \Im z \gtrless 0.$

4.37.12 $\quad \operatorname{arctanh}(-z) = -\operatorname{arctanh} z, \qquad z \neq \pm 1.$

4.37.13 $\quad \operatorname{arccsch}(-z) = -\operatorname{arccsch} z.$

4.37.14 $\quad \operatorname{arcsech}(-z) = \mp \pi i + \operatorname{arcsech} z, \qquad \Im z \gtrless 0.$

4.37.15 $\quad \operatorname{arccoth}(-z) = -\operatorname{arccoth} z, \qquad z \neq \pm 1.$

4.37(iv) Logarithmic Forms

Throughout this subsection *all* quantities assume their principal values.

Inverse Hyperbolic Sine

4.37.16
$$\operatorname{arcsinh} z = \ln\left((z^2+1)^{1/2} + z\right),$$
$$z/i \in \mathbb{C} \setminus (-\infty, -1) \cup (1, \infty);$$

compare Figure 4.37.1(i). On the cuts

4.37.17
$$\operatorname{arcsinh}(iy) = \tfrac{1}{2}\pi i \pm \ln\left((y^2-1)^{1/2} + y\right), \quad y \in [1, \infty),$$

4.37.18
$$\operatorname{arcsinh}(iy) = -\tfrac{1}{2}\pi i \pm \ln\left((y^2-1)^{1/2} - y\right),$$
$$y \in (-\infty, -1],$$

the upper/lower signs corresponding to the right/left sides.

Inverse Hyperbolic Cosine

4.37.19
$$\operatorname{arccosh} z = \ln\left(\pm(z^2-1)^{1/2} + z\right), \quad z \in \mathbb{C} \setminus (-\infty, 1),$$

the upper or lower sign being taken according as $\Re z \gtrless 0$; compare Figure 4.37.1(ii). Also,

4.37.20
$$\operatorname{arccosh}(iy) = \pm \tfrac{1}{2}\pi i + \ln\left((y^2+1)^{1/2} \pm y\right), \quad y \gtrless 0.$$

It should be noted that the imaginary axis is not a cut; the function defined by (4.37.19) and (4.37.20) is analytic everywhere except on $(-\infty, 1]$. Compare Figure 4.37.1(ii).

An equivalent definition is

4.37.21
$$\operatorname{arccosh} z = 2\ln\left(\left(\frac{z+1}{2}\right)^{1/2} + \left(\frac{z-1}{2}\right)^{1/2}\right),$$
$$z \in \mathbb{C} \setminus (-\infty, 1);$$

see Kahan (1987).

On the part of the cuts from -1 to 1

4.37.22
$$\operatorname{arccosh} x = \pm \ln\left(i(1-x^2)^{1/2} + x\right), \quad x \in (-1, 1],$$

the upper/lower sign corresponding to the upper/lower side.

On the part of the cut from $-\infty$ to -1

4.37.23
$$\operatorname{arccosh} x = \pm \pi i + \ln\left((x^2-1)^{1/2} - x\right), \quad x \in (-\infty, -1],$$

the upper/lower sign corresponding to the upper/lower side.

Inverse Hyperbolic Tangent

4.37.24
$$\operatorname{arctanh} z = \tfrac{1}{2}\ln\left(\frac{1+z}{1-z}\right), \quad z \in \mathbb{C} \setminus (-\infty, -1] \cup [1, \infty);$$

compare Figure 4.37.1(iii). On the cuts

4.37.25 $\quad \operatorname{arctanh} x = \pm \tfrac{1}{2}\pi i + \tfrac{1}{2}\ln\left(\frac{x+1}{x-1}\right),$
$$x \in (-\infty, -1) \cup (1, \infty),$$

the upper/lower sign corresponding to the upper/lower sides.

Other Inverse Functions

For the corresponding results for $\operatorname{arccsch} z$, $\operatorname{arcsech} z$, and $\operatorname{arccoth} z$, use (4.37.7)–(4.37.9); compare §4.23(iv).

4.37(v) Fundamental Property

With $k \in \mathbb{Z}$, the general solutions of the equations

4.37.26 $\qquad z = \sinh w,$

4.37.27 $\qquad z = \cosh w,$

4.37.28 $\qquad z = \tanh w,$

are respectively given by

4.37.29 $\quad w = \operatorname{Arcsinh} z = (-1)^k \operatorname{arcsinh} z + k\pi i,$

4.37.30 $\quad w = \operatorname{Arccosh} z = \pm \operatorname{arccosh} z + 2k\pi i,$

4.37.31 $\quad w = \operatorname{Arctanh} z = \operatorname{arctanh} z + k\pi i, \qquad z \neq \pm 1.$

4.37(vi) Interrelations

Table 4.30.1 can also be used to find interrelations between inverse hyperbolic functions. For example, $\operatorname{arcsech} a = \operatorname{arccoth}\left((1-a^2)^{-1/2}\right)$.

4.38 Inverse Hyperbolic Functions: Further Properties

4.38(i) Power Series

4.38.1 $\quad \operatorname{arcsinh} z = z - \dfrac{1}{2}\dfrac{z^3}{3} + \dfrac{1\cdot 3}{2\cdot 4}\dfrac{z^5}{5} - \dfrac{1\cdot 3\cdot 5}{2\cdot 4\cdot 6}\dfrac{z^7}{7} + \cdots,\quad |z|<1.$

4.38.2
$$\operatorname{arcsinh} z = \ln(2z) + \dfrac{1}{2}\dfrac{1}{2z^2} - \dfrac{1\cdot 3}{2\cdot 4}\dfrac{1}{4z^4} + \dfrac{1\cdot 3\cdot 5}{2\cdot 4\cdot 6}\dfrac{1}{6z^6} - \cdots,$$
$\Re z > 0,\ |z|>1.$

4.38.3
$$\operatorname{arccosh} z = \ln(2z) - \dfrac{1}{2}\dfrac{1}{2z^2} - \dfrac{1\cdot 3}{2\cdot 4}\dfrac{1}{4z^4} - \dfrac{1\cdot 3\cdot 5}{2\cdot 4\cdot 6}\dfrac{1}{6z^6} - \cdots,\quad |z|>1.$$

4.38.4
$$\operatorname{arccosh} z = (2(z-1))^{1/2} \times \left(1 + \sum_{n=1}^{\infty} (-1)^n \dfrac{1\cdot 3\cdot 5\cdots(2n-1)}{2^{2n} n!(2n+1)}(z-1)^n\right),$$
$\Re z > 0,\ |z-1|\le 2.$

4.38.5
$$\operatorname{arctanh} z = z + \dfrac{z^3}{3} + \dfrac{z^5}{5} + \dfrac{z^7}{7} + \cdots,\quad |z|\le 1,\ z\ne \pm 1.$$

4.38.6
$$\operatorname{arctanh} z = \pm i\dfrac{\pi}{2} + \dfrac{1}{z} + \dfrac{1}{3z^3} + \dfrac{1}{5z^5} + \cdots,\quad \Im z \gtreqless 0,\ |z|\ge 1.$$

4.38.7
$$\operatorname{arctanh} z = \dfrac{z}{1-z^2}\times \left(1 + \dfrac{2}{3}\dfrac{z^2}{z^2-1} + \dfrac{2\cdot 4}{3\cdot 5}\left(\dfrac{z^2}{z^2-1}\right)^2 + \cdots\right),$$
$\Re(z^2) < \tfrac{1}{2},$

which requires $z\,(=x+iy)$ to lie between the two rectangular hyperbolas given by

4.38.8 $\qquad x^2 - y^2 = \tfrac{1}{2}.$

4.38(ii) Derivatives

In the following equations square roots have their principal values.

4.38.9 $\quad \dfrac{d}{dz}\operatorname{arcsinh} z = (1+z^2)^{-1/2}.$

4.38.10 $\quad \dfrac{d}{dz}\operatorname{arccosh} z = \pm(z^2-1)^{-1/2},\qquad \Re z \gtreqless 0.$

4.38.11 $\quad \dfrac{d}{dz}\operatorname{arctanh} z = \dfrac{1}{1-z^2}.$

4.38.12 $\quad \dfrac{d}{dz}\operatorname{arccsch} z = \mp \dfrac{1}{z(1+z^2)^{1/2}},\qquad \Re z \gtreqless 0.$

4.38.13 $\quad \dfrac{d}{dz}\operatorname{arcsech} z = -\dfrac{1}{z(1-z^2)^{1/2}}.$

4.38.14 $\quad \dfrac{d}{dz}\operatorname{arccoth} z = \dfrac{1}{1-z^2}.$

4.38(iii) Addition Formulas

4.38.15
$$\operatorname{Arcsinh} u \pm \operatorname{Arcsinh} v = \operatorname{Arcsinh}\!\left(u(1+v^2)^{1/2} \pm v(1+u^2)^{1/2}\right),$$

4.38.16
$$\operatorname{Arccosh} u \pm \operatorname{Arccosh} v = \operatorname{Arccosh}\!\left(uv \pm ((u^2-1)(v^2-1))^{1/2}\right),$$

4.38.17 $\quad \operatorname{Arctanh} u \pm \operatorname{Arctanh} v = \operatorname{Arctanh}\!\left(\dfrac{u\pm v}{1\pm uv}\right),$

4.38.18
$$\operatorname{Arcsinh} u \pm \operatorname{Arccosh} v = \operatorname{Arcsinh}\!\left(uv \pm ((1+u^2)(v^2-1))^{1/2}\right) = \operatorname{Arccosh}\!\left(v(1+u^2)^{1/2} \pm u(v^2-1)^{1/2}\right),$$

4.38.19
$$\operatorname{Arctanh} u \pm \operatorname{Arccoth} v = \operatorname{Arctanh}\!\left(\dfrac{uv\pm 1}{v\pm u}\right) = \operatorname{Arccoth}\!\left(\dfrac{v\pm u}{uv\pm 1}\right).$$

The above equations are interpreted in the sense that every value of the left-hand side is a value of the right-hand side and vice-versa. All square roots have either possible value.

4.39 Continued Fractions

4.39.1
$$\tanh z = \dfrac{z}{1+}\,\dfrac{z^2}{3+}\,\dfrac{z^2}{5+}\,\dfrac{z^2}{7+}\cdots,\quad z\ne \pm\tfrac{1}{2}\pi i, \pm\tfrac{3}{2}\pi i,\ldots.$$

4.39.2
$$\dfrac{\operatorname{arcsinh} z}{\sqrt{1+z^2}} = \dfrac{z}{1+}\,\dfrac{1\cdot 2 z^2}{3+}\,\dfrac{1\cdot 2 z^2}{5+}\,\dfrac{3\cdot 4 z^2}{7+}\,\dfrac{3\cdot 4 z^2}{9+}\cdots,$$

where z is in the open cut plane of Figure 4.37.1(i).

4.39.3 $\quad \operatorname{arctanh} z = \dfrac{z}{1-}\,\dfrac{z^2}{3-}\,\dfrac{4z^2}{5-}\,\dfrac{9z^2}{7-}\cdots,$

where z is in the open cut plane of Figure 4.37.1(iii).

For these and other continued fractions involving inverse hyperbolic functions see Lorentzen and Waadeland (1992, pp. 569–571). See also Cuyt et al. (2008, pp. 211–217).

4.40 Integrals

4.40(i) Introduction

Throughout this section the variables are assumed to be real. The results in §§4.40(ii) and 4.40(iv) can be extended to the complex plane by using continuous branches and avoiding singularities.

4.40(ii) Indefinite Integrals

4.40.1 $$\int \sinh x \, dx = \cosh x,$$

4.40.2 $$\int \cosh x \, dx = \sinh x,$$

4.40.3 $$\int \tanh x \, dx = \ln(\cosh x).$$

4.40.4 $$\int \operatorname{csch} x \, dx = \ln\bigl(\tanh\bigl(\tfrac{1}{2}x\bigr)\bigr), \qquad 0 < x < \infty.$$

4.40.5 $$\int \operatorname{sech} x \, dx = \operatorname{gd}(x).$$

For the right-hand side see (4.23.39) and (4.23.40).

4.40.6 $$\int \coth x \, dx = \ln(\sinh x), \qquad 0 < x < \infty.$$

4.40(iii) Definite Integrals

4.40.7
$$\int_0^\infty e^{-x} \frac{\sin(ax)}{\sinh x} \, dx = \tfrac{1}{2}\pi \coth\bigl(\tfrac{1}{2}\pi a\bigr) - \frac{1}{a}, \quad a \neq 0,$$

4.40.8 $$\int_0^\infty \frac{\sinh(ax)}{\sinh(\pi x)} \, dx = \tfrac{1}{2}\tan\bigl(\tfrac{1}{2}a\bigr), \quad -\pi < a < \pi,$$

4.40.9 $$\int_{-\infty}^\infty \frac{e^{ax}}{\bigl(\cosh\bigl(\tfrac{1}{2}x\bigr)\bigr)^2} \, dx = \frac{4\pi a}{\sin(\pi a)}, \quad -1 < a < 1,$$

4.40.10
$$\int_0^\infty \frac{\tanh(ax) - \tanh(bx)}{x} \, dx = \ln\left(\frac{a}{b}\right), \quad a > 0, b > 0.$$

4.40(iv) Inverse Hyperbolic Functions

4.40.11 $$\int \operatorname{arcsinh} x \, dx = x \operatorname{arcsinh} x - (1+x^2)^{1/2}.$$

4.40.12
$$\int \operatorname{arccosh} x \, dx = x \operatorname{arccosh} x - (x^2-1)^{1/2}, \quad 1 < x < \infty,$$

4.40.13 $$\int \operatorname{arctanh} x \, dx = x \operatorname{arctanh} x + \tfrac{1}{2}\ln(1-x^2),$$
$$-1 < x < 1,$$

4.40.14
$$\int \operatorname{arccsch} x \, dx = x \operatorname{arccsch} x + \operatorname{arcsinh} x, \quad 0 < x < \infty,$$

4.40.15
$$\int \operatorname{arcsech} x \, dx = x \operatorname{arcsech} x + \arcsin x, \quad 0 < x < 1,$$

4.40.16 $$\int \operatorname{arccoth} x \, dx = x \operatorname{arccoth} x + \tfrac{1}{2}\ln(x^2-1),$$
$$1 < x < \infty.$$

4.40(v) Compendia

Extensive compendia of indefinite and definite integrals of hyperbolic functions include Apelblat (1983, pp. 96–109), Bierens de Haan (1939), Gröbner and Hofreiter (1949, pp. 139–160), Gröbner and Hofreiter (1950, pp. 160–167), Gradshteyn and Ryzhik (2000, Chapters 2–4), and Prudnikov et al. (1986a, §§1.4, 1.8, 2.4, 2.8).

4.41 Sums

For sums of hyperbolic functions see Gradshteyn and Ryzhik (2000, Chapter 1), Hansen (1975, §43), Prudnikov et al. (1986a, §5.3), and Zucker (1979).

Applications

4.42 Solution of Triangles

4.42(i) Planar Right Triangles

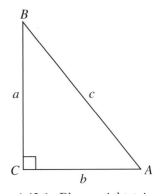

Figure 4.42.1: Planar right triangle.

4.42.1 $$\sin A = \frac{a}{c} = \frac{1}{\csc A},$$

4.42.2 $$\cos A = \frac{b}{c} = \frac{1}{\sec A},$$

4.42.3 $$\tan A = \frac{a}{b} = \frac{1}{\cot A}.$$

4.42(ii) Planar Triangles

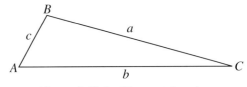

Figure 4.42.2: Planar triangle.

4.42.4 $$\frac{a}{\sin A} = \frac{b}{\sin B} = \frac{c}{\sin C},$$

4.42.5
$$c^2 = a^2 + b^2 - 2ab\cos C,$$

4.42.6
$$a = b\cos C + c\cos B$$

4.42.7 area $= \frac{1}{2}bc\sin A = (s(s-a)(s-b)(s-c))^{1/2}$, where $s = \frac{1}{2}(a+b+c)$ (the semiperimeter).

4.42(iii) Spherical Triangles

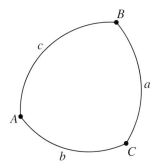

Figure 4.42.3: Spherical triangle.

4.42.8 $\quad \cos a = \cos b \cos c + \sin b \sin c \cos A,$

4.42.9
$$\frac{\sin A}{\sin a} = \frac{\sin B}{\sin b} = \frac{\sin C}{\sin c},$$

4.42.10 $\quad \sin a \cos B = \cos b \sin c - \sin b \cos c \cos A,$

4.42.11 $\quad \cos a \cos C = \sin a \cot b - \sin C \cot B,$

4.42.12 $\quad \cos A = -\cos B \cos C + \sin B \sin C \cos a.$

For these and other formulas see Smart (1962, Chapter 1).

4.43 Cubic Equations

Let

4.43.1
$$A = \left(-\tfrac{4}{3}p\right)^{1/2}, \quad B = \left(\tfrac{4}{3}p\right)^{1/2},$$
$$C = \left(-\frac{27q^2}{4p^3}\right)^{1/2}, \quad D = -\left(\frac{27q^2}{4p^3}\right)^{1/2},$$

where $p(\neq 0)$ and q are real constants. The roots of

4.43.2
$$z^3 + pz + q = 0$$

are:

(a) $A\sin a$, $A\sin\left(a + \frac{2}{3}\pi\right)$, and $A\sin\left(a + \frac{4}{3}\pi\right)$, with $\sin(3a) = C$, when $p < 0$ and $C \leq 1$.

(b) $A\cosh a$, $A\cosh\left(a + \frac{2}{3}\pi i\right)$, and $A\cosh\left(a + \frac{4}{3}\pi i\right)$, with $\cosh(3a) = C$, when $p < 0$ and $C > 1$.

(c) $B\sinh a$, $B\sinh\left(a + \frac{2}{3}\pi i\right)$, and $B\sinh\left(a + \frac{4}{3}\pi i\right)$, with $\sinh(3a) = D$, when $p > 0$.

Note that in Case (a) all the roots are real, whereas in Cases (b) and (c) there is one real root and a conjugate pair of complex roots. See also §1.11(iii).

4.44 Other Applications

For applications of generalized exponentials and generalized logarithms to computer arithmetic see §3.1(iv).

For an application of the Lambert W-function to generalized Gaussian noise see Chapeau-Blondeau and Monir (2002).

Computation

4.45 Methods of Computation

4.45(i) Real Variables

Logarithms

The function $\ln x$ can always be computed from its ascending power series after preliminary scaling. Suppose first $1/10 \leq x \leq 10$. Then we take square roots repeatedly until $|y|$ is sufficiently small, where

4.45.1
$$y = x^{2^{-m}} - 1.$$

After computing $\ln(1+y)$ from (4.6.1)

4.45.2
$$\ln x = 2^m \ln(1+y).$$

For other values of x set $x = 10^m \xi$, where $1/10 \leq \xi \leq 10$ and $m \in \mathbb{Z}$. Then

4.45.3
$$\ln x = \ln \xi + m \ln 10.$$

Exponentials

Let x have any real value. First, rescale via

4.45.4
$$m = \left\lfloor \frac{x}{\ln 10} + \frac{1}{2} \right\rfloor, \quad y = x - m\ln 10.$$

Then

4.45.5
$$e^x = 10^m e^y,$$

and since $|y| \leq \frac{1}{2}\ln 10 = 1.15\ldots$, e^y can be computed straightforwardly from (4.2.19).

Trigonometric Functions

Let x have any real value. We first compute $\xi = x/\pi$, followed by

4.45.6
$$m = \left\lfloor \xi + \tfrac{1}{2} \right\rfloor, \quad \theta = \pi(\xi - m).$$

Then

4.45.7 $\quad \sin x = (-1)^m \sin\theta, \quad \cos x = (-1)^m \cos\theta,$

and since $|\theta| \leq \frac{1}{2}\pi = 1.57\ldots$, $\sin\theta$ and $\cos\theta$ can be computed straightforwardly from (4.19.1) and (4.19.2).

The other trigonometric functions can be found from the definitions (4.14.4)–(4.14.7).

Inverse Trigonometric Functions

The function $\arctan x$ can always be computed from its ascending power series after preliminary transformations to reduce the size of x. From (4.24.15) with $u = v = ((1+x^2)^{1/2} - 1)/x$, we have

4.45.8
$$2\arctan\frac{(1+x^2)^{1/2} - 1}{x} = \arctan x, \quad 0 < x < \infty.$$

Beginning with $x_0 = x$, generate the sequence

4.45.9
$$x_n = \frac{(1+x_{n-1}^2)^{1/2} - 1}{x_{n-1}}, \quad n = 1, 2, 3, \ldots,$$

until x_n is sufficiently small. We then compute $\arctan x_n$ from (4.24.3), followed by

4.45.10
$$\arctan x = 2^n \arctan x_n.$$

Another method, when x is large, is to sum

4.45.11
$$\arctan x = \frac{\pi}{2} - \frac{1}{x} + \frac{1}{3x^3} - \frac{1}{5x^5} + \cdots;$$

compare (4.24.4).

As an example, take $x = 9.47376$. Then

4.45.12
$$x_1 = 0.90000\ldots, \quad x_2 = 0.38373\ldots,$$
$$x_3 = 0.18528\ldots, \quad x_4 = 0.09185\ldots.$$

From (4.24.3) $\arctan x_4 = 0.09160\ldots$. From (4.45.10)

4.45.13 $\quad \arctan x = 16 \arctan x_4 = 1.46563\ldots$.

As a check, from (4.45.11)

4.45.14
$$\arctan x = 1.57079\ldots - 0.10555\ldots + 0.00039\ldots - \cdots$$
$$= 1.46563\ldots.$$

For the remaining inverse trigonometric functions, we may use the identities provided by the fourth row of Table 4.16.3. For example, $\arcsin x = \arctan(x(1-x^2)^{-1/2})$.

Hyperbolic and Inverse Hyperbolic Functions

The hyperbolic functions can be computed directly from the definitions (4.28.1)–(4.28.7). The inverses arcsinh, arccosh, and arctanh can be computed from the logarithmic forms given in §4.37(iv), with real arguments. For arccsch, arcsech, and arccoth we have (4.37.7)–(4.37.9).

Other Methods

See Luther (1995), Ziv (1991), Cody and Waite (1980), Rosenberg and McNamee (1976), Carlson (1972a). For interval-arithmetic algorithms, see Markov (1981). For Shift-and-Add and CORDIC algorithms, see Muller (1997), Merrheim (1994), Schelin (1983). For multi-precision methods, see Smith (1989), Brent (1976).

4.45(ii) Complex Variables

For $\ln z$ and e^z

4.45.15 $\quad \ln z = \ln|z| + i\operatorname{ph} z, \quad -\pi \leq \operatorname{ph} z \leq \pi,$

4.45.16 $\quad e^z = e^{\Re z}(\cos(\Im z) + i\sin(\Im z)).$

See §1.9(i) for the precise relationship of $\operatorname{ph} z$ to the arctangent function.

The trigonometric functions may be computed from the definitions (4.14.1)–(4.14.7), and their inverses from the logarithmic forms in §4.23(iv), followed by (4.23.7)–(4.23.9). Similarly for the hyperbolic and inverse hyperbolic functions; compare (4.28.1)–(4.28.7), §4.37(iv), and (4.37.7)–(4.37.9).

For other methods see Miel (1981).

4.45(iii) Lambert W-Function

For $x \in [-1/e, \infty)$ the principal branch $\operatorname{Wp}(x)$ can be computed by solving the defining equation $We^W = x$ numerically, for example, by Newton's rule (§3.8(ii)). Initial approximations are obtainable, for example, from the power series (4.13.6) (with $t \geq 0$) when x is close to $-1/e$, from the asymptotic expansion (4.13.10) when x is large, and by numerical integration of the differential equation (4.13.4) (§3.7) for other values of x.

Similarly for $\operatorname{Wm}(x)$ in the interval $[-1/e, 0)$.

See also Barry *et al.* (1995) and Chapeau-Blondeau and Monir (2002).

4.46 Tables

Extensive numerical tables of all the elementary functions for real values of their arguments appear in Abramowitz and Stegun (1964, Chapter 4). This handbook also includes lists of references for earlier tables, as do Fletcher *et al.* (1962) and Lebedev and Fedorova (1960).

For 40D values of the first 500 roots of $\tan x = x$, see Robinson (1972). (These roots are zeros of the Bessel function $J_{3/2}(x)$; see §10.21.)

For 10S values of the first five complex roots of $\sin z = az$, $\cos z = az$, and $\cosh z = az$, for selected positive values of a, see Fettis (1976).

See also Luther (1995).

4.47 Approximations

4.47(i) Chebyshev-Series Expansions

Clenshaw (1962) and Luke (1975, Chapter 3) give 20D coefficients for ln, exp, sin, cos, tan, cot, arcsin, arctan, arcsinh. Schonfelder (1980) gives 40D coefficients for sin, cos, tan.

4.47(ii) Rational Functions

Hart et al. (1968) give ln, exp, sin, cos, tan, cot, arcsin, arccos, arctan, sinh, cosh, tanh, arcsinh, arccosh. Precision is variable.

4.47(iii) Padé Approximations

Luke (1975, Chapter 3) supplies real and complex approximations for ln, exp, sin, cos, tan, arctan, arcsinh. Precision is variable.

4.47(iv) Additional References

See Luke (1975, pp. 288–289) and Luke (1969b, pp.74–76).

4.48 Software

See http://dlmf.nist.gov/4.48.

References

General References

The main references used in writing this chapter are Levinson and Redheffer (1970), Hobson (1928), Wall (1948), and Whittaker and Watson (1927). For additional bibliographic reading see Copson (1935) and Silverman (1967).

Sources

The following list gives the references or other indications of proofs that were used in constructing the various sections of this chapter. These sources supplement the references that are quoted in the text.

§4.2 Levinson and Redheffer (1970, pp. 62–67), Hobson (1928, pp. 289–301).

§4.3 These graphics were produced at NIST.

§4.4 Levinson and Redheffer (1970, pp. 62–63, 69), Hardy (1952, pp. 403–420).

§4.5 (4.5.1) and (4.5.5) can be verified by the methods of Hardy et al. (1967, pp. 106–107). (4.5.2) and (4.5.4) follow from (4.5.1). (4.5.3) follows from the fact that $x = 0$ and $x = 0.5828\ldots$ are successive zeros of $\frac{3}{2}x + \ln(1-x)$. (4.5.6) is obtained from the Maclaurin expansion of $\ln(1+z)$. (4.5.7) to (4.5.12) are obtained by exponentiating the inequalities (4.5.1) and (4.5.2). For (4.5.13), see Hardy et al. (1967, p. 102). (4.5.14) follows from the fact that $1 - \frac{1}{2}x - e^{-x}$ has 0 and $1.5936\ldots$ as consecutive zeros. (4.5.15) and (4.5.16) can be derived from the Maclaurin expansion of e^z.

§4.6 For (4.6.1) see Hardy (1952, pp. 471–473). (4.6.2)–(4.6.6) are variations of this. For (4.6.7) see Hardy (1952, pp. 476–477).

§4.7 Levinson and Redheffer (1970, pp. 53–54, 62–69).

§4.8 Levinson and Redheffer (1970, pp. 62–66), Hobson (1928, pp. 297–299).

§4.10 (4.10.1)–(4.10.4) and (4.10.8)–(4.10.10) can be verified by differentiation. For (4.10.5) and (4.10.6), expand by the geometric series and integrate term by term to get a series which can be summed by Andrews et al. (1999, p. 12). For (4.10.11) apply (5.4.6) and (5.9.1). To evaluate (4.10.12) and (4.10.13), expand by the geometric series and integrate term by term. The dilogarithm series which appears from (4.10.12) can be summed by Andrews et al. (1999, p. 105).

§4.13 To verify the radius of convergence of the series (4.13.6) map the plane of W onto the plane of t via $t = (-2v)^{1/2}$, where $v = W + \ln W + 1 - i\pi$. Then W is analytic at $t = 0$, and its nearest singularities to the origin are located at $t = 2\sqrt{\pi}e^{\pm\pi i/4}$. Figure 4.13.1 was produced at NIST.

§4.14 Levinson and Redheffer (1970, pp. 55–57).

§4.15 These graphics were produced at NIST.

§4.16 Hobson (1928, pp. 19–24).

§4.17 Hobson (1928, pp. 29–32, 53–75).

§4.18 For (4.18.1) see Copson (1935, p. 136). (4.18.3) follows by the same method and (4.18.2) is a consequence. (4.18.5) to (4.18.9) are straightforward and (4.18.10) is obtained from the Maclaurin expansions of $\cos z$ and $\sin z$. For the second inequality in (4.18.4), it is sufficient to show $f(x) \equiv 4x(1-x) - \sin(\pi x) \geq 0$ for $0 \leq x \leq 1/2$. The function $f(x)$ is zero at $x = 0$ and $x = 1/2$ and it has no zeros in $(0, 1/2)$, because $f'(x) = 4(1 - 2x) - \pi\cos(\pi x)$ can have only one zero in $(0, 1/2)$ where $y = 4(1 - 2x)$ intersects $y = \pi\cos(\pi x)$. The first inequality is proved similarly.

§4.19 Hobson (1928, pp. 288–293, 360–367).

§4.20 Levinson and Redheffer (1970, pp. 53–60).

§4.21 Hobson (1928, Chapter 4 and pp. 19, 21, 45, 52–53, 60, 63–69, 237–239, 331). For (4.21.35) see Walker (1996, §1.9).

§4.22 Hobson (1928, Chapter 17), Levinson and Redheffer (1970, pp. 387–389).

§4.23 Levinson and Redheffer (1970, pp. 68–70), Hobson (1928, pp. 32–33, 332–333), Fletcher *et al.* (1962, §§12.1, 12.2). (4.23.10)–(4.23.18) and also Table 4.23.1 follow from the definitions in §§4.23(i), 4.23(ii). To verify (4.23.19), denote the right-hand side by $\phi(z)$, and the domain $\mathbb{C} \setminus (-\infty, -1] \cup [1, \infty)$ by D. If $z = x \in (-1, 1)$, then $\phi'(x) = (1 - x^2)^{-1/2}$ and $\phi(0) = 0$. Hence (4.23.19) applies; compare (4.23.1) with Arcsin replaced by arcsin. We may now extend (4.23.19) to the rest of D simply by showing that $\phi(z)$ is analytic on D; compare §1.10(ii). Since the principal value of $(1 - z^2)^{1/2}$ is analytic on D, the only possible singularities of $\phi(z)$ occur on the branch cut of the logarithm, that is, when $(1 - z^2)^{1/2} = -iz - t$ with $t \in [0, \infty)$. By squaring the last equation we see that $(1 - z^2)^{1/2} + iz$ is real only when z lies on the imaginary axis, and it is then positive. The proofs of (4.23.22), (4.23.23), (4.23.26) are similar, or in the case of (4.23.22) we may simply refer to (4.23.16). (4.23.40) and (4.23.42) may be verified by differentiation plus comparison of values as $x \to 0$.

§4.24 Hobson (1928, pp. 54–55, 279–280, 321), Levinson and Redheffer (1970, pp. 68–70). For (4.24.10) and (4.24.11) note that the principal value of $(z^2 - 1)^{1/2}$ is discontinuous on the imaginary axis, hence we switch to the other branch when crossing this axis. This accounts for the two signs.

§4.25 Jones and Thron (1980, pp. 202–203), Wall (1948, pp. 343–349).

§4.26 (4.26.1)–(4.26.8) and (4.26.14)–(4.26.21) may be verified by differentiation. For (4.26.12) and (4.26.13) see Copson (1935, pp. 137 and 227).

§4.28 Hobson (1928, pp. 322–326), Levinson and Redheffer (1970, pp. 56–57).

§4.29 These graphics were produced at NIST.

§4.30 Hobson (1928, pp. 323–326).

§4.31 Hobson (1928, p. 326), Levinson and Redheffer (1970, p. 61).

§4.35 Hobson (1928, pp. 323–325, 331).

§4.36 For (4.36.1)–(4.36.5) replace z by iz in (4.22.1)–(4.22.5) and apply (4.28.8)–(4.28.13).

§4.37 Levinson and Redheffer (1970, pp. 68–69). (4.37.11) follows from (4.37.19). The equations in §4.37(iv) may be verified in a similar manner to those of §4.23(iv). The only new feature is that in (4.37.19) the principal value of $(z^2 - 1)^{1/2}$ is discontinuous on the imaginary axis, hence to continue $(z^2 - 1)^{1/2}$ analytically we switch to the other branch. This accounts for the $\pm$ sign in (4.37.19).

§4.38 For (4.38.1) expand $(1 + z^2)^{-1/2}$ by the binomial theorem and integrate term by term. For (4.38.2), write $(1 + z^2)^{-1/2} = z^{-1}(1 + (1/z^2))^{-1/2}$, $\Re z > 0$, and then expand and integrate. To find the constant of integration note that for large z, arcsinh z behaves like $\ln(2z)$ and the constant is $\ln 2$. (4.38.3) is proved similarly. (4.38.4)–(4.38.7) follow from the corresponding series for inverse trigonometric functions in §4.24. For (4.38.9)–(4.38.14) take the derivatives of the logarithmic forms of inverse hyperbolic functions in §4.37(iv). For (4.38.15)–(4.38.19) use similar analysis to that for §4.24(iii).

§4.40 (4.40.1)–(4.40.6) and (4.40.11)–(4.40.16) may be verified by differentiation. For (4.40.7)–(4.40.10) see Copson (1935, p. 155).

§4.42 Hobson (1928, p. 18 and Chapter 10).

§4.43 Hobson (1928, p. 335).

Chapter 5
Gamma Function

R. A. Askey[1] and R. Roy[2]

Notation   **136**
- 5.1 Special Notation 136

Properties   **136**
- 5.2 Definitions 136
- 5.3 Graphics 136
- 5.4 Special Values and Extrema 137
- 5.5 Functional Relations 138
- 5.6 Inequalities 138
- 5.7 Series Expansions 139
- 5.8 Infinite Products 139
- 5.9 Integral Representations 139
- 5.10 Continued Fractions 140
- 5.11 Asymptotic Expansions 140
- 5.12 Beta Function 142
- 5.13 Integrals 143
- 5.14 Multidimensional Integrals 143
- 5.15 Polygamma Functions 144
- 5.16 Sums 144
- 5.17 Barnes' G-Function (Double Gamma Function) 144
- 5.18 q-Gamma and Beta Functions 145

Applications   **145**
- 5.19 Mathematical Applications 145
- 5.20 Physical Applications 145

Computation   **146**
- 5.21 Methods of Computation 146
- 5.22 Tables 146
- 5.23 Approximations 146
- 5.24 Software 147

References   **147**

---

[1]Department of Mathematics, University of Wisconsin, Madison, Wisconsin.
[2]Department of Mathematics and Computer Science, Beloit College, Beloit, Wisconsin.
Acknowledgments: This chapter is based in part on Abramowitz and Stegun (1964, Chapter 6) by P. J. Davis.
Copyright © 2009 National Institute of Standards and Technology. All rights reserved.

Notation

5.1 Special Notation

(For other notation see pp. xiv and 873.)

j, m, n	nonnegative integers.
k	nonnegative integer, except in §5.20.
x, y	real variables.
$z = x + iy$	complex variable.
a, b, q, s, w	real or complex variables with $|q| < 1$.
δ	arbitrary small positive constant.
γ	Euler's constant (§5.2(ii)).
primes	derivatives with respect to the variable.

The main functions treated in this chapter are the gamma function $\Gamma(z)$, the psi function (or digamma function) $\psi(z)$, the beta function $\mathrm{B}(a,b)$, and the q-gamma function $\Gamma_q(z)$.

The notation $\Gamma(z)$ is due to Legendre. Alternative notations for this function are: $\Pi(z-1)$ (Gauss) and $(z-1)!$. Alternative notations for the psi function are: $\Psi(z-1)$ (Gauss) Jahnke and Emde (1945); $\Psi(z)$ Davis (1933); $\mathsf{F}(z-1)$ Pairman (1919).

Properties

5.2 Definitions

5.2(i) Gamma and Psi Functions

Euler's Integral

5.2.1 $$\Gamma(z) = \int_0^\infty e^{-t} t^{z-1}\, dt, \qquad \Re z > 0.$$

When $\Re z \leq 0$, $\Gamma(z)$ is defined by analytic continuation. It is a meromorphic function with no zeros, and with simple poles of residue $(-1)^n/n!$ at $z = -n$. $1/\Gamma(z)$ is entire, with simple zeros at $z = -n$.

5.2.2 $$\psi(z) = \Gamma'(z)/\Gamma(z), \quad z \neq 0, -1, -2, \ldots.$$

$\psi(z)$ is meromorphic with simple poles of residue -1 at $z = -n$.

5.2(ii) Euler's Constant

5.2.3 $$\gamma = \lim_{n\to\infty}\left(1 + \frac{1}{2} + \frac{1}{3} + \cdots + \frac{1}{n} - \ln n\right)$$
$$= 0.57721\,56649\,01532\,86060\ldots.$$

5.2(iii) Pochhammer's Symbol

5.2.4 $(a)_0 = 1, \quad (a)_n = a(a+1)(a+2)\cdots(a+n-1),$

5.2.5 $(a)_n = \Gamma(a+n)/\Gamma(a), \qquad a \neq 0, -1, -2, \ldots.$

5.3 Graphics

5.3(i) Real Argument

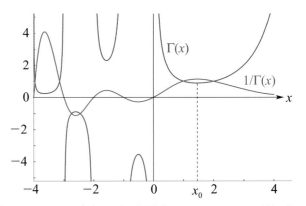

Figure 5.3.1: $\Gamma(x)$ and $1/\Gamma(x)$. $x_0 = 1.46\ldots$, $\Gamma(x_0) = 0.88\ldots$; see §5.4(iii).

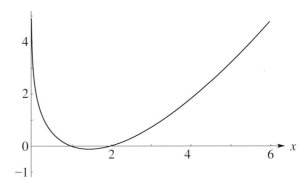

Figure 5.3.2: $\ln \Gamma(x)$. This function is convex on $(0, \infty)$; compare §5.5(iv).

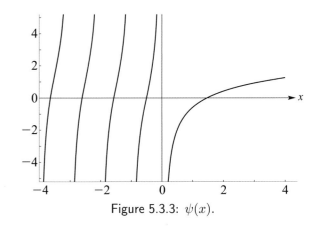

Figure 5.3.3: $\psi(x)$.

5.3(ii) Complex Argument

In the graphics shown in this subsection, both the height and color correspond to the absolute value of the function. See also p. xiv.

5.4 Special Values and Extrema

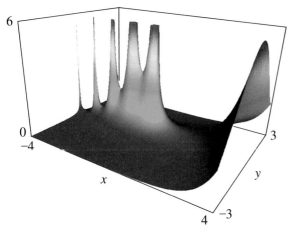

Figure 5.3.4: $|\Gamma(x+iy)|$.

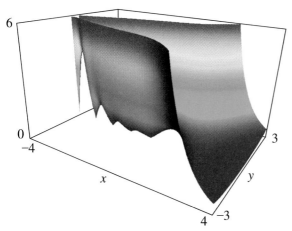

Figure 5.3.5: $1/|\Gamma(x+iy)|$.

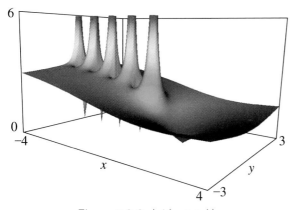

Figure 5.3.6: $|\psi(x+iy)|$.

5.4 Special Values and Extrema

5.4(i) Gamma Function

5.4.1 $\quad \Gamma(1) = 1, \quad n! = \Gamma(n+1).$

5.4.2 $\quad n!! = \begin{cases} 2^{\frac{1}{2}n}\,\Gamma\!\left(\tfrac{1}{2}n+1\right), & n \text{ even,} \\ \pi^{-\frac{1}{2}}2^{\frac{1}{2}n+\frac{1}{2}}\,\Gamma\!\left(\tfrac{1}{2}n+1\right), & n \text{ odd.} \end{cases}$

(The second line of Formula (5.4.2) also applies when $n = -1$.)

5.4.3 $\quad |\Gamma(iy)| = \left(\dfrac{\pi}{y\sinh(\pi y)}\right)^{1/2},$

5.4.4 $\quad \Gamma\!\left(\tfrac{1}{2}+iy\right)\Gamma\!\left(\tfrac{1}{2}-iy\right) = \left|\Gamma\!\left(\tfrac{1}{2}+iy\right)\right|^2 = \dfrac{\pi}{\cosh(\pi y)},$

5.4.5 $\quad \Gamma\!\left(\tfrac{1}{4}+iy\right)\Gamma\!\left(\tfrac{3}{4}-iy\right) = \dfrac{\pi\sqrt{2}}{\cosh(\pi y)+i\sinh(\pi y)}.$

5.4.6 $\quad \Gamma\!\left(\tfrac{1}{2}\right) = \pi^{1/2}$
$\qquad\qquad = 1.77245\,38509\,05516\,02729\ldots,$

5.4.7 $\quad \Gamma\!\left(\tfrac{1}{3}\right) = 2.67893\,85347\,07747\,63365\ldots,$

5.4.8 $\quad \Gamma\!\left(\tfrac{2}{3}\right) = 1.35411\,79394\,26400\,41694\ldots,$

5.4.9 $\quad \Gamma\!\left(\tfrac{1}{4}\right) = 3.62560\,99082\,21908\,31193\ldots,$

5.4.10 $\quad \Gamma\!\left(\tfrac{3}{4}\right) = 1.22541\,67024\,65177\,64512\ldots.$

5.4.11 $\quad \Gamma'(1) = -\gamma.$

5.4(ii) Psi Function

5.4.12 $\quad \psi(1) = -\gamma, \quad \psi'(1) = \tfrac{1}{6}\pi^2,$

5.4.13 $\quad \psi\!\left(\tfrac{1}{2}\right) = -\gamma - 2\ln 2, \quad \psi'\!\left(\tfrac{1}{2}\right) = \tfrac{1}{2}\pi^2.$

For higher derivatives of $\psi(z)$ at $z = 1$ and $z = \tfrac{1}{2}$, see §5.15.

5.4.14 $\quad \psi(n+1) = \sum_{k=1}^{n} \dfrac{1}{k} - \gamma,$

5.4.15 $\quad \psi\!\left(n+\tfrac{1}{2}\right) = -\gamma - 2\ln 2 + 2\left(1 + \tfrac{1}{3} + \cdots + \tfrac{1}{2n-1}\right),$
$\qquad\qquad\qquad\qquad\qquad\qquad n = 1, 2, \ldots.$

5.4.16 $\quad \Im\psi(iy) = \dfrac{1}{2y} + \dfrac{\pi}{2}\coth(\pi y),$

5.4.17 $\quad \Im\psi\!\left(\tfrac{1}{2}+iy\right) = \dfrac{\pi}{2}\tanh(\pi y),$

5.4.18 $\quad \Im\psi(1+iy) = -\dfrac{1}{2y} + \dfrac{\pi}{2}\coth(\pi y).$

If p, q are integers with $0 < p < q$, then

5.4.19
$$\psi\!\left(\dfrac{p}{q}\right) = -\gamma - \ln q - \dfrac{\pi}{2}\cot\!\left(\dfrac{\pi p}{q}\right) + \dfrac{1}{2}\sum_{k=1}^{q-1}\cos\!\left(\dfrac{2\pi k p}{q}\right)\ln\!\left(2 - 2\cos\!\left(\dfrac{2\pi k}{q}\right)\right).$$

5.4(iii) Extrema

Table 5.4.1: $\Gamma'(x_n) = \psi(x_n) = 0$.

n	x_n	$\Gamma(x_n)$
0	1.46163 21449	0.88560 31944
1	−0.50408 30083	−3.54464 36112
2	−1.57349 84732	2.30240 72583
3	−2.61072 08875	−0.88813 63584
4	−3.63529 33665	0.24512 75398
5	−4.65323 77626	−0.05277 96396
6	−5.66716 24513	0.00932 45945
7	−6.67841 82649	−0.00139 73966
8	−7.68778 83250	0.00018 18784
9	−8.69576 41633	−0.00002 09253
10	−9.70267 25406	0.00000 21574

Compare Figure 5.3.1.

As $n \to \infty$,

5.4.20 $\quad x_n = -n + \dfrac{1}{\pi}\arctan\left(\dfrac{\pi}{\ln n}\right) + O\left(\dfrac{1}{n(\ln n)^2}\right).$

For error bounds for this estimate see Walker (2007, Theorem 5).

5.5 Functional Relations

5.5(i) Recurrence

5.5.1 $\qquad \Gamma(z+1) = z\,\Gamma(z),$

5.5.2 $\qquad \psi(z+1) = \psi(z) + \dfrac{1}{z}.$

5.5(ii) Reflection

5.5.3 $\quad \Gamma(z)\,\Gamma(1-z) = \pi/\sin(\pi z), \quad z \neq 0, \pm 1, \ldots,$

5.5.4 $\quad \psi(z) - \psi(1-z) = -\pi/\tan(\pi z), \quad z \neq 0, \pm 1, \ldots.$

5.5(iii) Multiplication

Duplication Formula

For $2z \neq 0, -1, -2, \ldots,$

5.5.5 $\qquad \Gamma(2z) = \pi^{-1/2} 2^{2z-1}\,\Gamma(z)\,\Gamma\!\left(z + \tfrac{1}{2}\right).$

Gauss's Multiplication Formula

For $nz \neq 0, -1, -2, \ldots,$

5.5.6 $\quad \Gamma(nz) = (2\pi)^{(1-n)/2} n^{nz-(1/2)} \displaystyle\prod_{k=0}^{n-1} \Gamma\!\left(z + \dfrac{k}{n}\right).$

5.5.7 $\qquad \displaystyle\prod_{k=1}^{n-1} \Gamma\!\left(\dfrac{k}{n}\right) = (2\pi)^{(n-1)/2} n^{-1/2}.$

5.5.8 $\qquad \psi(2z) = \tfrac{1}{2}\left(\psi(z) + \psi\!\left(z + \tfrac{1}{2}\right)\right) + \ln 2,$

5.5.9 $\qquad \psi(nz) = \dfrac{1}{n}\displaystyle\sum_{k=0}^{n-1} \psi\!\left(z + \dfrac{k}{n}\right) + \ln n.$

5.5(iv) Bohr–Mollerup Theorem

If a positive function $f(x)$ on $(0, \infty)$ satisfies $f(x+1) = xf(x)$, $f(1) = 1$, and $\ln f(x)$ is convex (see §1.4(viii)), then $f(x) = \Gamma(x)$.

5.6 Inequalities

5.6(i) Real Variables

Throughout this subsection $x > 0$.

5.6.1 $\quad 1 < (2\pi)^{-1/2} x^{(1/2)-x} e^x\,\Gamma(x) < e^{1/(12x)},$

5.6.2 $\qquad \dfrac{1}{\Gamma(x)} + \dfrac{1}{\Gamma(1/x)} \leq 2,$

5.6.3 $\qquad \dfrac{1}{(\Gamma(x))^2} + \dfrac{1}{(\Gamma(1/x))^2} \leq 2,$

Gautschi's Inequality

5.6.4 $\quad x^{1-s} < \dfrac{\Gamma(x+1)}{\Gamma(x+s)} < (x+1)^{1-s}, \quad 0 < s < 1.$

5.6.5 $\quad \begin{aligned}&\exp\!\left((1-s)\,\psi\!\left(x + s^{1/2}\right)\right) \\ &\leq \dfrac{\Gamma(x+1)}{\Gamma(x+s)} \leq \exp\!\left((1-s)\,\psi\!\left(x + \tfrac{1}{2}(s+1)\right)\right),\end{aligned}$
$\hfill 0 < s < 1.$

5.6(ii) Complex Variables

5.6.6 $\qquad |\Gamma(x+iy)| \leq |\Gamma(x)|,$

5.6.7 $\quad |\Gamma(x+iy)| \geq (\operatorname{sech}(\pi y))^{1/2}\,\Gamma(x), \qquad x \geq \tfrac{1}{2}.$

For $b - a \geq 1$, $a \geq 0$, and $z = x + iy$ with $x > 0$,

5.6.8 $\qquad \left|\dfrac{\Gamma(z+a)}{\Gamma(z+b)}\right| \leq \dfrac{1}{|z|^{b-a}}.$

For $x \geq 0$,

5.6.9 $\quad |\Gamma(z)| \leq (2\pi)^{1/2} |z|^{x-(1/2)} e^{-\pi|y|/2} \exp\!\left(\tfrac{1}{6}|z|^{-1}\right).$

5.7 Series Expansions

5.7(i) Maclaurin and Taylor Series

Throughout this subsection $\zeta(k)$ is as in Chapter 25.

5.7.1
$$\frac{1}{\Gamma(z)} = \sum_{k=1}^{\infty} c_k z^k,$$

where $c_1 = 1$, $c_2 = \gamma$, and

5.7.2
$$(k-1)c_k = \gamma c_{k-1} - \zeta(2)c_{k-2} + \zeta(3)c_{k-3} - \cdots \\ + (-1)^k \zeta(k-1)c_1, \qquad k \geq 3.$$

For 15D numerical values of c_k see Abramowitz and Stegun (1964, p. 256), and for 31D values see Wrench (1968).

5.7.3
$$\ln \Gamma(1+z) = -\ln(1+z) \\ + z(1-\gamma) + \sum_{k=2}^{\infty} (-1)^k (\zeta(k) - 1) \frac{z^k}{k},$$
$$|z| < 2.$$

5.7.4
$$\psi(1+z) = -\gamma + \sum_{k=2}^{\infty} (-1)^k \zeta(k) z^{k-1}, \qquad |z| < 1,$$

5.7.5
$$\psi(1+z) = \frac{1}{2z} - \frac{\pi}{2} \cot(\pi z) + \frac{1}{z^2 - 1} + 1 \\ - \gamma - \sum_{k=1}^{\infty} (\zeta(2k+1) - 1) z^{2k},$$
$$|z| < 2, z \neq 0, \pm 1.$$

For 20D numerical values of the coefficients of the Maclaurin series for $\Gamma(z+3)$ see Luke (1969b, p. 299).

5.7(ii) Other Series

When $z \neq 0, -1, -2, \ldots,$

5.7.6
$$\psi(z) = -\gamma - \frac{1}{z} + \sum_{k=1}^{\infty} \frac{z}{k(k+z)} \\ = -\gamma + \sum_{k=0}^{\infty} \left(\frac{1}{k+1} - \frac{1}{k+z} \right),$$

and

5.7.7
$$\psi\left(\frac{z+1}{2}\right) - \psi\left(\frac{z}{2}\right) = 2 \sum_{k=0}^{\infty} \frac{(-1)^k}{k+z}.$$

Also,

5.7.8
$$\Im \psi(1+iy) = \sum_{k=1}^{\infty} \frac{y}{k^2 + y^2}.$$

5.8 Infinite Products

5.8.1
$$\Gamma(z) = \lim_{k \to \infty} \frac{k! k^z}{z(z+1) \cdots (z+k)}, \quad z \neq 0, -1, -2, \ldots,$$

5.8.2
$$\frac{1}{\Gamma(z)} = z e^{\gamma z} \prod_{k=1}^{\infty} \left(1 + \frac{z}{k}\right) e^{-z/k},$$

5.8.3
$$\left| \frac{\Gamma(x)}{\Gamma(x+iy)} \right|^2 = \prod_{k=0}^{\infty} \left(1 + \frac{y^2}{(x+k)^2}\right), \quad x \neq 0, -1, \ldots.$$

If

5.0.4
$$\sum_{k=1}^{m} a_k = \sum_{k=1}^{m} b_k,$$

then

5.8.5
$$\prod_{k=0}^{\infty} \frac{(a_1+k)(a_2+k) \cdots (a_m+k)}{(b_1+k)(b_2+k) \cdots (b_m+k)} \\ = \frac{\Gamma(b_1) \Gamma(b_2) \cdots \Gamma(b_m)}{\Gamma(a_1) \Gamma(a_2) \cdots \Gamma(a_m)},$$

provided that none of the b_k is zero or a negative integer.

5.9 Integral Representations

5.9(i) Gamma Function

5.9.1
$$\frac{1}{\mu} \Gamma\left(\frac{\nu}{\mu}\right) \frac{1}{z^{\nu/\mu}} = \int_0^{\infty} \exp(-zt^\mu) t^{\nu-1} \, dt,$$

$\Re \nu > 0$, $\mu > 0$, and $\Re z > 0$. (The fractional powers have their principal values.)

Hankel's Loop Integral

5.9.2
$$\frac{1}{\Gamma(z)} = \frac{1}{2\pi i} \int_{-\infty}^{(0+)} e^t t^{-z} \, dt,$$

where the contour begins at $-\infty$, circles the origin once in the positive direction, and returns to $-\infty$. t^{-z} has its principal value where t crosses the positive real axis, and is continuous. See Figure 5.9.1.

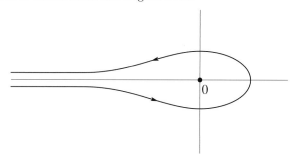

Figure 5.9.1: t-plane. Contour for Hankel's loop integral.

5.9.3
$$c^{-z} \Gamma(z) = \int_{-\infty}^{\infty} |t|^{2z-1} e^{-ct^2} \, dt, \quad c > 0, \Re z > 0,$$

where the path is the real axis.

5.9.4
$$\Gamma(z) = \int_1^{\infty} t^{z-1} e^{-t} \, dt + \sum_{k=0}^{\infty} \frac{(-1)^k}{(z+k)k!},$$
$$z \neq 0, -1, -2, \ldots.$$

5.9.5 $$\Gamma(z) = \int_0^\infty t^{z-1}\left(e^{-t} - \sum_{k=0}^n \frac{(-1)^k t^k}{k!}\right) dt,$$
$$-n-1 < \Re z < -n.$$

5.9.6 $\Gamma(z)\cos\left(\tfrac{1}{2}\pi z\right) = \int_0^\infty t^{z-1}\cos t\, dt,\quad 0 < \Re z < 1,$

5.9.7 $\Gamma(z)\sin\left(\tfrac{1}{2}\pi z\right) = \int_0^\infty t^{z-1}\sin t\, dt,\quad -1 < \Re z < 1.$

5.9.8
$$\Gamma\left(1+\frac{1}{n}\right)\cos\left(\frac{\pi}{2n}\right) = \int_0^\infty \cos(t^n)\, dt,\quad n=2,3,4,\ldots,$$

5.9.9
$$\Gamma\left(1+\frac{1}{n}\right)\sin\left(\frac{\pi}{2n}\right) = \int_0^\infty \sin(t^n)\, dt,\quad n=2,3,4,\ldots.$$

Binet's Formula

5.9.10
$$\ln\Gamma(z) = \left(z-\tfrac{1}{2}\right)\ln z - z + \tfrac{1}{2}\ln(2\pi) + 2\int_0^\infty \frac{\arctan(t/z)}{e^{2\pi t}-1}\, dt,$$

where $|\operatorname{ph} z| < \pi/2$ and the inverse tangent has its principal value.

5.9.11
$$\ln\Gamma(z+1) = -\gamma z - \frac{1}{2\pi i}\int_{-c-\infty i}^{-c+\infty i} \frac{\pi z^{-s}}{s\sin(\pi s)}\zeta(-s)\, ds,$$

where $|\operatorname{ph} z| \leq \pi - \delta\,(<\pi)$, $1 < c < 2$, and $\zeta(s)$ is as in Chapter 25.

For additional representations see Whittaker and Watson (1927, §§12.31–12.32).

5.9(ii) Psi Function, Euler's Constant, and Derivatives

For $\Re z > 0$,

5.9.12 $\psi(z) = \int_0^\infty \left(\dfrac{e^{-t}}{t} - \dfrac{e^{-zt}}{1-e^{-t}}\right) dt,$

5.9.13 $\psi(z) = \ln z + \int_0^\infty \left(\dfrac{1}{t} - \dfrac{1}{1-e^{-t}}\right) e^{-tz}\, dt,$

5.9.14 $\psi(z) = \int_0^\infty \left(e^{-t} - \dfrac{1}{(1+t)^z}\right) \dfrac{dt}{t},$

5.9.15 $\psi(z) = \ln z - \dfrac{1}{2z} - 2\int_0^\infty \dfrac{t\, dt}{(t^2+z^2)(e^{2\pi t}-1)}.$

5.9.16 $\psi(z) + \gamma = \int_0^\infty \dfrac{e^{-t}-e^{-zt}}{1-e^{-t}}\, dt = \int_0^1 \dfrac{1-t^{z-1}}{1-t}\, dt.$

5.9.17 $\psi(z+1) = -\gamma + \dfrac{1}{2\pi i}\int_{-c-\infty i}^{-c+\infty i} \dfrac{\pi z^{-s-1}}{\sin(\pi s)}\zeta(-s)\, ds,$

where $|\operatorname{ph} z| \leq \pi - \delta(<\pi)$ and $1 < c < 2$.

5.9.18
$$\gamma = -\int_0^\infty e^{-t}\ln t\, dt = \int_0^\infty \left(\frac{1}{1+t} - e^{-t}\right)\frac{dt}{t}$$
$$= \int_0^1 (1-e^{-t})\frac{dt}{t} - \int_1^\infty e^{-t}\frac{dt}{t}$$
$$= \int_0^\infty \left(\frac{e^{-t}}{1-e^{-t}} - \frac{e^{-t}}{t}\right) dt.$$

5.9.19 $\Gamma^{(n)}(z) = \int_0^\infty (\ln t)^n e^{-t} t^{z-1}\, dt,\quad n\geq 0,\ \Re z > 0.$

5.10 Continued Fractions

For $\Re z > 0$,

5.10.1
$$\ln\Gamma(z) + z - \left(z-\tfrac{1}{2}\right)\ln z - \tfrac{1}{2}\ln(2\pi)$$
$$= \frac{a_0}{z+}\frac{a_1}{z+}\frac{a_2}{z+}\frac{a_3}{z+}\frac{a_4}{z+}\frac{a_5}{z+}\cdots,$$

where

5.10.2
$a_0 = \tfrac{1}{12},\quad a_1 = \tfrac{1}{30},\quad a_2 = \tfrac{53}{210},\quad a_3 = \tfrac{195}{371},$
$a_4 = \tfrac{22999}{22737},\quad a_5 = \tfrac{299\,44523}{197\,33142},\quad a_6 = \tfrac{10\,95352\,41009}{4\,82642\,75462}.$

For exact values of a_7 to a_{11} and 40S values of a_0 to a_{40}, see Char (1980). Also see Cuyt et al. (2008, pp. 223–228), Jones and Thron (1980, pp. 348–350), and Lorentzen and Waadeland (1992, pp. 221–224) for further information.

5.11 Asymptotic Expansions

5.11(i) Poincaré-Type Expansions

As $z \to \infty$ in the sector $|\operatorname{ph} z| \leq \pi - \delta\,(<\pi)$,

5.11.1
$$\ln\Gamma(z) \sim \left(z-\tfrac{1}{2}\right)\ln z - z + \tfrac{1}{2}\ln(2\pi) + \sum_{k=1}^\infty \frac{B_{2k}}{2k(2k-1)z^{2k-1}}$$

and

5.11.2 $\psi(z) \sim \ln z - \dfrac{1}{2z} - \sum_{k=1}^\infty \dfrac{B_{2k}}{2k z^{2k}}.$

For the Bernoulli numbers B_{2k}, see §24.2(i).

With the same conditions,

5.11.3 $\Gamma(z) \sim e^{-z} z^z \left(\dfrac{2\pi}{z}\right)^{1/2}\left(\sum_{k=0}^\infty \dfrac{g_k}{z^k}\right),$

where

5.11.4
$g_0 = 1,\quad g_1 = \tfrac{1}{12},\quad g_2 = \tfrac{1}{288},\quad g_3 = -\tfrac{139}{51840},$
$g_4 = -\tfrac{571}{24\,88320},\quad g_5 = \tfrac{1\,63879}{2090\,18880},\quad g_6 = \tfrac{52\,46819}{7\,52467\,96800}.$

Also,

5.11.5 $g_k = \sqrt{2}\left(\tfrac{1}{2}\right)_k a_{2k},$

where $a_0 = \frac{1}{2}\sqrt{2}$ and

5.11.6
$$a_0 a_k + \frac{1}{2} a_1 a_{k-1} + \frac{1}{3} a_2 a_{k-2} + \cdots + \frac{1}{k+1} a_k a_0 = \frac{1}{k} a_{k-1}, \qquad k \geq 1.$$

Wrench (1968) gives exact values of g_k up to g_{20}. Spira (1971) corrects errors in Wrench's results and also supplies exact and 45D values of g_k for $k = 21, 22, \ldots, 30$. For an asymptotic expansion of g_k as $k \to \infty$ see Boyd (1994).

Terminology

The expansion (5.11.1) is called *Stirling's series* (Whittaker and Watson (1927, §12.33)), whereas the expansion (5.11.3), or sometimes just its leading term, is known as *Stirling's formula* (Abramowitz and Stegun (1964, §6.1), Olver (1997b, p. 88)).

Next, and again with the same conditions,

5.11.7 $\quad \Gamma(az + b) \sim \sqrt{2\pi} e^{-az} (az)^{az+b-(1/2)},$

where $a\ (>0)$ and $b\ (\in \mathbb{C})$ are both fixed, and

5.11.8
$$\ln \Gamma(z+h) \sim \left(z + h - \tfrac{1}{2}\right) \ln z - z + \tfrac{1}{2} \ln(2\pi) + \sum_{k=2}^{\infty} \frac{(-1)^k B_k(h)}{k(k-1) z^{k-1}},$$

where $h\ (\in [0,1])$ is fixed, and $B_k(h)$ is the Bernoulli polynomial defined in §24.2(i).

Lastly, as $y \to \pm\infty$,

5.11.9 $\quad |\Gamma(x + iy)| \sim \sqrt{2\pi} |y|^{x-(1/2)} e^{-\pi |y|/2},$

uniformly for bounded real values of x.

5.11(ii) Error Bounds and Exponential Improvement

If the sums in the expansions (5.11.1) and (5.11.2) are terminated at $k = n-1$ ($k \geq 0$) and z is real and positive, then the remainder terms are bounded in magnitude by the first neglected terms and have the same sign. If z is complex, then the remainder terms are bounded in magnitude by $\sec^{2n}\left(\tfrac{1}{2} \operatorname{ph} z\right)$ for (5.11.1), and $\sec^{2n+1}\left(\tfrac{1}{2} \operatorname{ph} z\right)$ for (5.11.2), times the first neglected terms.

For the remainder term in (5.11.3) write

5.11.10 $\quad \Gamma(z) = e^{-z} z^z \left(\frac{2\pi}{z}\right)^{1/2} \left(\sum_{k=0}^{K-1} \frac{g_k}{z^k} + R_K(z)\right),$
$$K = 1, 2, 3, \ldots.$$

Then

5.11.11
$$|R_K(z)| \leq \frac{(1 + \zeta(K)) \Gamma(K)}{2(2\pi)^{K+1} |z|^K} \left(1 + \min(\sec(\operatorname{ph} z), 2K^{\frac{1}{2}})\right),$$
$$|\operatorname{ph} z| \leq \tfrac{1}{2}\pi,$$

where $\zeta(K)$ is as in Chapter 25. For this result and a similar bound for the sector $\tfrac{1}{2}\pi \leq \operatorname{ph} z \leq \pi$ see Boyd (1994).

For further information see Olver (1997b, pp. 293–295), and for other error bounds see Whittaker and Watson (1927, §12.33), Spira (1971), and Schäfke and Finsterer (1990).

For re-expansions of the remainder terms in (5.11.1) and (5.11.3) in series of incomplete gamma functions with exponential improvement (§2.11(iii)) in the asymptotic expansions, see Berry (1991), Boyd (1994), and Paris and Kaminski (2001, §6.4).

5.11(iii) Ratios

In this subsection a, b, and c are real or complex constants.

If $z \to \infty$ in the sector $|\operatorname{ph} z| \leq \pi - \delta\ (<\pi)$, then

5.11.12 $\quad \dfrac{\Gamma(z+a)}{\Gamma(z+b)} \sim z^{a-b},$

5.11.13 $\quad \dfrac{\Gamma(z+a)}{\Gamma(z+b)} \sim z^{a-b} \sum_{k=0}^{\infty} \frac{G_k(a,b)}{z^k}.$

Also, with the added condition $\Re(b-a) > 0$,

5.11.14
$$\frac{\Gamma(z+a)}{\Gamma(z+b)} \sim \left(z + \frac{a+b-1}{2}\right)^{a-b} \sum_{k=0}^{\infty} \frac{H_k(a,b)}{\left(z + \tfrac{1}{2}(a+b-1)\right)^{2k}}.$$

Here

5.11.15
$$G_0(a,b) = 1, \quad G_1(a,b) = \tfrac{1}{2}(a-b)(a+b-1),$$
$$G_2(a,b) = \frac{1}{12} \binom{a-b}{2} (3(a+b-1)^2 - (a-b+1)),$$

5.11.16
$$H_0(a,b) = 1, \quad H_1(a,b) = -\frac{1}{12} \binom{a-b}{2}(a-b+1),$$
$$H_2(a,b) = \frac{1}{240} \binom{a-b}{4} (2(a-b+1) + 5(a-b+1)^2).$$

In terms of generalized Bernoulli polynomials $B_n^{(\ell)}(x)$ (§24.16(i)), we have for $k = 0, 1, \ldots,$

5.11.17 $\quad G_k(a,b) = \binom{a-b}{k} B_k^{(a-b+1)}(a),$

5.11.18 $\quad H_k(a,b) = \binom{a-b}{2k} B_{2k}^{(a-b+1)}\left(\frac{a-b+1}{2}\right).$

Lastly, and again if $z \to \infty$ in the sector $|\operatorname{ph} z| \leq$

$\pi - \delta$ ($< \pi$), then

5.11.19
$$\frac{\Gamma(z+a)\,\Gamma(z+b)}{\Gamma(z+c)}$$
$$\sim \sum_{k=0}^{\infty}(-1)^k \frac{(c-a)_k(c-b)_k}{k!}\,\Gamma(a+b-c+z-k).$$

For the error term in (5.11.19) in the case $z = x$ (> 0) and $c = 1$, see Olver (1995).

5.12 Beta Function

In this section all fractional powers have their principal values, except where noted otherwise. In (5.12.1)–(5.12.4) it is assumed $\Re a > 0$ and $\Re b > 0$.

Euler's Beta Integral

5.12.1 $\quad \mathrm{B}(a,b) = \int_0^1 t^{a-1}(1-t)^{b-1}\,dt = \dfrac{\Gamma(a)\,\Gamma(b)}{\Gamma(a+b)}.$

5.12.2 $\quad \int_0^{\pi/2} \sin^{2a-1}\theta \cos^{2b-1}\theta\,d\theta = \tfrac{1}{2}\mathrm{B}(a,b).$

5.12.3 $\quad \int_0^\infty \dfrac{t^{a-1}\,dt}{(1+t)^{a+b}} = \mathrm{B}(a,b).$

5.12.4
$$\int_0^1 \frac{t^{a-1}(1-t)^{b-1}}{(t+z)^{a+b}}\,dt = \mathrm{B}(a,b)(1+z)^{-a}z^{-b},\ |\mathrm{ph}\,z|<\pi.$$

5.12.5
$$\int_0^{\pi/2} (\cos t)^{a-1}\cos(bt)\,dt$$
$$= \frac{\pi}{2^a}\frac{1}{a\,\mathrm{B}\!\left(\tfrac{1}{2}(a+b+1),\tfrac{1}{2}(a-b+1)\right)},\quad \Re a>0.$$

5.12.6
$$\int_0^\pi (\sin t)^{a-1} e^{ibt}\,dt$$
$$= \frac{\pi}{2^{a-1}}\frac{e^{i\pi b/2}}{a\,\mathrm{B}\!\left(\tfrac{1}{2}(a+b+1),\tfrac{1}{2}(a-b+1)\right)},\quad \Re a>0.$$

5.12.7 $\quad \int_0^\infty \dfrac{\cosh(2bt)}{(\cosh t)^{2a}}\,dt = 4^{a-1}\mathrm{B}(a+b,a-b),\ \Re a>|\Re b|.$

5.12.8
$$\frac{1}{2\pi}\int_{-\infty}^\infty \frac{dt}{(w+it)^a(z-it)^b} = \frac{(w+z)^{1-a-b}}{(a+b-1)\mathrm{B}(a,b)},$$
$$\Re(a+b)>1,\ \Re w>0,\ \Re z>0.$$

In (5.12.8) the fractional powers have their principal values when $w > 0$ and $z > 0$, and are continued via continuity.

5.12.9 $\quad \dfrac{1}{2\pi i}\int_{c-\infty i}^{c+\infty i} t^{-a}(1-t)^{-1-b}\,dt = \dfrac{1}{b\,\mathrm{B}(a,b)},$
$$0<c<1,\ \Re(a+b)>0.$$

5.12.10 $\quad \dfrac{1}{2\pi i}\int_0^{(1+)} t^{a-1}(t-1)^{b-1}\,dt = \dfrac{\sin(\pi b)}{\pi}\mathrm{B}(a,b),\ \Re a > 0,$

with the contour as shown in Figure 5.12.1.

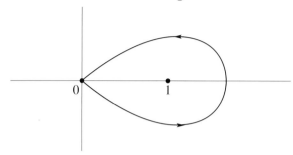

Figure 5.12.1: t-plane. Contour for first loop integral for the beta function.

In (5.12.11) and (5.12.12) the fractional powers are continuous on the integration paths and take their principal values at the beginning.

5.12.11 $\quad \dfrac{1}{e^{2\pi i a}-1}\int_\infty^{(0+)} t^{a-1}(1+t)^{-a-b}\,dt = \mathrm{B}(a,b),$

when $\Re b > 0$, a is not an integer and the contour cuts the real axis between -1 and the origin. See Figure 5.12.2.

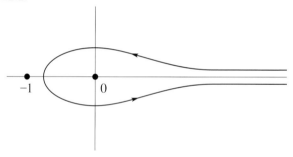

Figure 5.12.2: t-plane. Contour for second loop integral for the beta function.

Pochhammer's Integral

When $a, b \in \mathbb{C}$

5.12.12 $\quad \int_P^{(1+,0+,1-,0-)} t^{a-1}(1-t)^{b-1}\,dt$
$$= -4e^{\pi i(a+b)}\sin(\pi a)\sin(\pi b)\,\mathrm{B}(a,b),$$

where the contour starts from an arbitrary point P in the interval $(0,1)$, circles 1 and then 0 in the positive sense, circles 1 and then 0 in the negative sense, and returns to P. It can always be deformed into the contour shown in Figure 5.12.3.

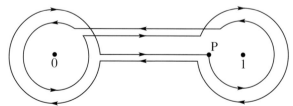

Figure 5.12.3: t-plane. Contour for Pochhammer's integral.

5.13 Integrals

In (5.13.1) the integration path is a straight line parallel to the imaginary axis.

5.13.1 $$\frac{1}{2\pi i}\int_{c-i\infty}^{c+i\infty}\Gamma(s+a)\,\Gamma(b-s)z^{-s}\,ds = \frac{\Gamma(a+b)z^a}{(1+z)^{a+b}}, \quad \Re(a+b) > 0,\ -\Re a < c < \Re b,\ |\operatorname{ph} z| < \pi.$$

5.13.2 $$\frac{1}{2\pi}\int_{-\infty}^{\infty}|\Gamma(a+it)|^2 e^{(2b-\pi)t}\,dt = \frac{\Gamma(2a)}{(2\sin b)^{2a}}, \quad a > 0,\ 0 < b < \pi.$$

Barnes' Beta Integral

5.13.3 $$\frac{1}{2\pi}\int_{-\infty}^{\infty}\Gamma(a+it)\,\Gamma(b+it)\,\Gamma(c-it)\,\Gamma(d-it)\,dt = \frac{\Gamma(a+c)\,\Gamma(a+d)\,\Gamma(b+c)\,\Gamma(b+d)}{\Gamma(a+b+c+d)}, \quad \Re a, \Re b, \Re c, \Re d > 0.$$

Ramanujan's Beta Integral

5.13.4 $$\int_{-\infty}^{\infty}\frac{dt}{\Gamma(a+t)\,\Gamma(b+t)\,\Gamma(c-t)\,\Gamma(d-t)} = \frac{\Gamma(a+b+c+d-3)}{\Gamma(a+c-1)\,\Gamma(a+d-1)\,\Gamma(b+c-1)\,\Gamma(b+d-1)}, \quad \Re(a+b+c+d) > 3.$$

de Branges–Wilson Beta Integral

5.13.5 $$\frac{1}{4\pi}\int_{-\infty}^{\infty}\frac{\prod_{k=1}^{4}\Gamma(a_k+it)\,\Gamma(a_k-it)}{\Gamma(2it)\,\Gamma(-2it)}\,dt = \frac{\prod_{1\le j<k\le 4}\Gamma(a_j+a_k)}{\Gamma(a_1+a_2+a_3+a_4)}, \quad \Re(a_k) > 0,\ k=1,2,3,4.$$

For compendia of integrals of gamma functions see Apelblat (1983, pp. 124–127 and 129–130), Erdélyi et al. (1954a,b), Gradshteyn and Ryzhik (2000, pp. 644–652), Oberhettinger (1974, pp. 191–204), Oberhettinger and Badii (1973, pp. 307–316), Prudnikov et al. (1986b, pp. 57–64), Prudnikov et al. (1992a, pp. 127–130), and Prudnikov et al. (1992b, pp. 113–123).

5.14 Multidimensional Integrals

Let V_n be the simplex: $t_1 + t_2 + \cdots + t_n \le 1$, $t_k \ge 0$. Then for $\Re z_k > 0$, $k = 1, 2, \ldots, n+1$,

5.14.1 $$\int_{V_n} t_1^{z_1-1}t_2^{z_2-1}\cdots t_n^{z_n-1}\,dt_1\,dt_2\cdots dt_n = \frac{\Gamma(z_1)\,\Gamma(z_2)\cdots\Gamma(z_n)}{\Gamma(1+z_1+z_2+\cdots+z_n)},$$

5.14.2 $$\int_{V_n}\left(1-\sum_{k=1}^{n}t_k\right)^{z_{n+1}-1}\prod_{k=1}^{n}t_k^{z_k-1}\,dt_k = \frac{\Gamma(z_1)\,\Gamma(z_2)\cdots\Gamma(z_{n+1})}{\Gamma(z_1+z_2+\cdots+z_{n+1})}.$$

Selberg-type Integrals

Let

5.14.3 $$\Delta(t_1,t_2,\ldots,t_n) = \prod_{1\le j<k\le n}(t_j - t_k).$$

Then

5.14.4 $$\int_{[0,1]^n} t_1 t_2\cdots t_m |\Delta(t_1,\ldots,t_n)|^{2c}\prod_{k=1}^{n}t_k^{a-1}(1-t_k)^{b-1}\,dt_k = \frac{1}{(\Gamma(1+c))^n}\prod_{k=1}^{m}\frac{a+(n-k)c}{a+b+(2n-k-1)c}$$
$$\times \prod_{k=1}^{n}\frac{\Gamma(a+(n-k)c)\,\Gamma(b+(n-k)c)\,\Gamma(1+kc)}{\Gamma(a+b+(2n-k-1)c)},$$

provided that $\Re a, \Re b > 0$, $\Re c > -\min(1/n, \Re a/(n-1), \Re b/(n-1))$.

Secondly,

5.14.5 $$\int_{[0,\infty)^n} t_1 t_2 \cdots t_m |\Delta(t_1,\ldots,t_n)|^{2c}\prod_{k=1}^{n}t_k^{a-1}e^{-t_k}\,dt_k = \prod_{k=1}^{m}(a+(n-k)c)\frac{\prod_{k=1}^{n}\Gamma(a+(n-k)c)\,\Gamma(1+kc)}{(\Gamma(1+c))^n},$$

when $\Re a > 0$, $\Re c > -\min(1/n, \Re a/(n-1))$.

Thirdly,

5.14.6 $$\frac{1}{(2\pi)^{n/2}}\int_{(-\infty,\infty)^n}|\Delta(t_1,\ldots,t_n)|^{2c}\prod_{k=1}^{n}\exp\!\left(-\tfrac{1}{2}t_k^2\right)dt_k = \frac{\prod_{k=1}^{n}\Gamma(1+kc)}{(\Gamma(1+c))^n}, \quad \Re c > -1/n.$$

Dyson's Integral

5.14.7
$$\frac{1}{(2\pi)^n} \int_{[-\pi,\pi]^n} \prod_{1\leq j<k\leq n} |e^{i\theta_j} - e^{i\theta_k}|^{2b} \, d\theta_1 \cdots d\theta_n = \frac{\Gamma(1+bn)}{(\Gamma(1+b))^n}, \qquad \Re b > -1/n.$$

5.15 Polygamma Functions

The functions $\psi^{(n)}(z)$, $n = 1, 2, \ldots$, are called the *polygamma functions*. In particular, $\psi'(z)$ is the *trigamma function*; ψ'', $\psi^{(3)}$, $\psi^{(4)}$ are the *tetra-, penta-,* and *hexagamma functions* respectively. Most properties of these functions follow straightforwardly by differentiation of properties of the psi function. This includes asymptotic expansions: compare §§2.1(ii)–2.1(iii).

In (5.15.2)–(5.15.7) $n, m = 1, 2, 3, \ldots$, and for $\zeta(n+1)$ see §25.6(i).

5.15.1
$$\psi'(z) = \sum_{k=0}^{\infty} \frac{1}{(k+z)^2}, \quad z \neq 0, -1, -2, \ldots,$$

5.15.2
$$\psi^{(n)}(1) = (-1)^{n+1} n! \, \zeta(n+1),$$

5.15.3
$$\psi^{(n)}\left(\tfrac{1}{2}\right) = (-1)^{n+1} n! (2^{n+1} - 1) \zeta(n+1),$$

5.15.4
$$\psi'\left(n - \tfrac{1}{2}\right) = \tfrac{1}{2}\pi^2 - 4 \sum_{k=1}^{n-1} \frac{1}{(2k-1)^2},$$

5.15.5
$$\psi^{(n)}(z+1) = \psi^{(n)}(z) + (-1)^n n! z^{-n-1},$$

5.15.6
$$\psi^{(n)}(1-z) + (-1)^{n-1} \psi^{(n)}(z) = (-1)^n \pi \frac{d^n}{dz^n} \cot(\pi z),$$

5.15.7
$$\psi^{(n)}(mz) = \frac{1}{m^{n+1}} \sum_{k=0}^{m-1} \psi^{(n)}\left(z + \frac{k}{m}\right).$$

As $z \to \infty$ in $|\operatorname{ph} z| \leq \pi - \delta \, (< \pi)$

5.15.8
$$\psi'(z) \sim \frac{1}{z} + \frac{1}{2z^2} + \sum_{k=1}^{\infty} \frac{B_{2k}}{z^{2k+1}}.$$

For B_{2k} see §24.2(i).

For continued fractions for $\psi'(z)$ and $\psi''(z)$ see Cuyt et al. (2008, pp. 231–238).

5.16 Sums

5.16.1
$$\sum_{k=1}^{\infty} (-1)^k \psi'(k) = -\frac{\pi^2}{8},$$

5.16.2
$$\sum_{k=1}^{\infty} \frac{1}{k} \psi'(k+1) = \zeta(3) = -\tfrac{1}{2} \psi''(1).$$

For further sums involving the psi function see Hansen (1975, pp. 360–367). For sums of gamma functions see Andrews et al. (1999, Chapters 2 and 3) and §§15.2(i), 16.2.

For related sums involving finite field analogs of the gamma and beta functions (Gauss and Jacobi sums) see Andrews et al. (1999, Chapter 1) and Terras (1999, pp. 90, 149).

5.17 Barnes' G-Function (Double Gamma Function)

5.17.1
$$G(z+1) = \Gamma(z)\, G(z), \quad G(1) = 1,$$

5.17.2
$$G(n) = (n-2)!(n-3)! \cdots 1!, \quad n = 2, 3, \ldots.$$

5.17.3
$$G(z+1) = (2\pi)^{z/2} \exp\left(-\tfrac{1}{2}z(z+1) - \tfrac{1}{2}\gamma z^2\right) \times \prod_{k=1}^{\infty} \left(\left(1 + \frac{z}{k}\right)^k \exp\left(-z + \frac{z^2}{2k}\right)\right).$$

5.17.4
$$\operatorname{Ln} G(z+1) = \tfrac{1}{2} z \ln(2\pi) - \tfrac{1}{2} z(z+1) + z \operatorname{Ln} \Gamma(z+1) - \int_0^z \operatorname{Ln} \Gamma(t+1) \, dt.$$

In this equation (and in (5.17.5) below), the Ln's have their principal values on the positive real axis and are continued via continuity, as in §4.2(i).

When $z \to \infty$ in $|\operatorname{ph} z| \leq \pi - \delta \, (< \pi)$,

5.17.5
$$\operatorname{Ln} G(z+1) \sim \tfrac{1}{4} z^2 + z \Gamma(z+1) - \left(\tfrac{1}{2}z(z+1) + \tfrac{1}{12}\right) \operatorname{Ln} z - \ln A + \sum_{k=1}^{\infty} \frac{B_{2k+2}}{2k(2k+1)(2k+2) z^{2k}};$$

see Ferreira and López (2001). This reference also provides bounds for the error term. Here B_{2k+2} is the Bernoulli number (§24.2(i)), and A is *Glaisher's constant*, given by

5.17.6
$$A = e^C = 1.28242\,71291\,00622\,63687\ldots,$$

where

5.17.7
$$C = \lim_{n\to\infty} \left(\sum_{k=1}^{n} k \ln k - \left(\tfrac{1}{2}n^2 + \tfrac{1}{2}n + \tfrac{1}{12}\right) \ln n + \tfrac{1}{4}n^2 \right)$$
$$= \frac{\gamma + \ln(2\pi)}{12} - \frac{\zeta'(2)}{2\pi^2} = \frac{1}{12} - \zeta'(-1),$$

and ζ' is the derivative of the zeta function (Chapter 25).

For Glaisher's constant see also Greene and Knuth (1982, p. 100) and §2.10(i).

5.18 q-Gamma and Beta Functions

5.18(i) q-Factorials

5.18.1 $$(a;q)_n = \prod_{k=0}^{n-1}(1-aq^k), \quad n=0,1,2,\ldots,$$

5.18.2 $$n!_q = 1(1+q)\cdots(1+q+\cdots+q^{n-1}) = (q;q)_n (1-q)^{-n}.$$

When $|q| < 1$,

5.18.3 $$(a;q)_\infty = \prod_{k=0}^{\infty}(1-aq^k).$$

See also §17.2(i).

5.18(ii) q-Gamma Function

When $0 < q < 1$,

5.18.4 $$\Gamma_q(z) = (q;q)_\infty (1-q)^{1-z} / (q^z;q)_\infty,$$

5.18.5 $$\Gamma_q(1) = \Gamma_q(2) = 1,$$

5.18.6 $$n!_q = \Gamma_q(n+1),$$

5.18.7 $$\Gamma_q(z+1) = \frac{1-q^z}{1-q}\Gamma_q(z).$$

Also, $\ln\Gamma_q(x)$ is convex for $x > 0$, and the analog of the Bohr-Mollerup theorem (§5.5(iv)) holds.

If $0 < q < r < 1$, then

5.18.8 $$\Gamma_q(x) < \Gamma_r(x),$$

when $0 < x < 1$ or when $x > 2$, and

5.18.9 $$\Gamma_q(x) > \Gamma_r(x),$$

when $1 < x < 2$.

5.18.10 $$\lim_{q \to 1-} \Gamma_q(z) = \Gamma(z).$$

For generalized asymptotic expansions of $\ln\Gamma_q(z)$ as $|z| \to \infty$ see Olde Daalhuis (1994) and Moak (1984).

5.18(iii) q-Beta Function

5.18.11 $$\mathrm{B}_q(a,b) = \frac{\Gamma_q(a)\Gamma_q(b)}{\Gamma_q(a+b)}.$$

5.18.12 $$\mathrm{B}_q(a,b) = \int_0^1 \frac{t^{a-1}(tq;q)_\infty}{(tq^b;q)_\infty} d_q t,$$
$$0 < q < 1, \Re a > 0, \Re b > 0.$$

For q-integrals see §17.2(v).

Applications

5.19 Mathematical Applications

5.19(i) Summation of Rational Functions

As shown in Temme (1996a, §3.4), the results given in §5.7(ii) can be used to sum infinite series of rational functions.

Example

5.19.1 $$S = \sum_{k=0}^{\infty} a_k, \quad a_k = \frac{k}{(3k+2)(2k+1)(k+1)}.$$

By decomposition into partial fractions (§1.2(iii))

5.19.2 $$a_k = \frac{2}{k+\frac{2}{3}} - \frac{1}{k+\frac{1}{2}} - \frac{1}{k+1}$$
$$= \left(\frac{1}{k+1} - \frac{1}{k+\frac{1}{2}}\right) - 2\left(\frac{1}{k+1} - \frac{1}{k+\frac{2}{3}}\right).$$

Hence from (5.7.6), (5.4.13), and (5.4.19)

5.19.3 $$S = \psi\left(\tfrac{1}{2}\right) - 2\psi\left(\tfrac{2}{3}\right) - \gamma = 3\ln 3 - 2\ln 2 - \tfrac{1}{3}\pi\sqrt{3}.$$

5.19(ii) Mellin–Barnes Integrals

Many special functions $f(z)$ can be represented as a *Mellin–Barnes integral*, that is, an integral of a product of gamma functions, reciprocals of gamma functions, and a power of z, the integration contour being doubly-infinite and eventually parallel to the imaginary axis at both ends. The left-hand side of (5.13.1) is a typical example. By translating the contour parallel to itself and summing the residues of the integrand, asymptotic expansions of $f(z)$ for large $|z|$, or small $|z|$, can be obtained complete with an integral representation of the error term. For further information and examples see §2.5 and Paris and Kaminski (2001, Chapters 5, 6, and 8).

5.19(iii) n-Dimensional Sphere

The volume V and surface area S of the n-dimensional sphere of radius r are given by

5.19.4 $$V = \frac{\pi^{\frac{1}{2}n} r^n}{\Gamma(\frac{1}{2}n+1)}, \quad S = \frac{2\pi^{\frac{1}{2}n} r^{n-1}}{\Gamma(\frac{1}{2}n)} = \frac{n}{r}V.$$

5.20 Physical Applications

Rutherford Scattering

In nonrelativistic quantum mechanics, collisions between two charged particles are described with the aid of the Coulomb phase shift $\mathrm{ph}\,\Gamma(\ell+1+i\eta)$; see (33.2.10) and Clark (1979).

Solvable Models of Statistical Mechanics

Suppose the potential energy of a gas of n point charges with positions $x_1, x_2, \ldots, x_n$ and free to move on the infinite line $-\infty < x < \infty$, is given by

5.20.1 $$W = \frac{1}{2}\sum_{\ell=1}^{n} x_\ell^2 - \sum_{1 \le \ell < j \le n} \ln|x_\ell - x_j|.$$

The probability density of the positions when the gas is in thermodynamic equilibrium is:

5.20.2 $$P(x_1, \ldots, x_n) = C \exp(-W/(kT)),$$

where k is the Boltzmann constant, T the temperature and C a constant. Then the partition function (with $\beta = 1/(kT)$) is given by

5.20.3 $$\begin{aligned}\psi_n(\beta) &= \int_{\mathbb{R}^n} e^{-\beta W}\, dx \\ &= (2\pi)^{n/2} \beta^{-(n/2)-(\beta n(n-1)/4)} \\ &\quad \times (\Gamma(1 + \tfrac{1}{2}\beta))^{-n} \prod_{j=1}^{n} \Gamma(1 + \tfrac{1}{2}j\beta).\end{aligned}$$

See (5.14.6).

For n charges free to move on a circular wire of radius 1,

5.20.4 $$W = -\sum_{1 \le \ell < j \le n} \ln|e^{i\theta_\ell} - e^{i\theta_j}|,$$

and the partition function is given by

5.20.5 $$\begin{aligned}\psi_n(\beta) &= \frac{1}{(2\pi)^n} \int_{[-\pi,\pi]^n} e^{-\beta W}\, d\theta_1 \cdots d\theta_n \\ &= \Gamma(1 + \tfrac{1}{2}n\beta)(\Gamma(1 + \tfrac{1}{2}\beta))^{-n}.\end{aligned}$$

See (5.14.7).

For further information see Mehta (2004).

Elementary Particles

Veneziano (1968) identifies relationships between particle scattering amplitudes described by the beta function, an important early development in string theory. Carlitz (1972) describes the partition function of dense hadronic matter in terms of a gamma function.

Computation

5.21 Methods of Computation

An effective way of computing $\Gamma(z)$ in the right half-plane is backward recurrence, beginning with a value generated from the asymptotic expansion (5.11.3). Or we can use forward recurrence, with an initial value obtained e.g. from (5.7.3). For the left half-plane we can continue the backward recurrence or make use of the reflection formula (5.5.3).

Similarly for $\ln\Gamma(z)$, $\psi(z)$, and the polygamma functions.

Another approach is to apply numerical quadrature (§3.5) to the integral (5.9.2), using paths of steepest descent for the contour. See Schmelzer and Trefethen (2007).

For a comprehensive survey see van der Laan and Temme (1984, Chapter III). See also Borwein and Zucker (1992).

5.22 Tables

5.22(i) Introduction

For early tables for both real and complex variables see Fletcher et al. (1962), Lebedev and Fedorova (1960), and Luke (1975, p. 21).

5.22(ii) Real Variables

Abramowitz and Stegun (1964, Chapter 6) tabulates $\Gamma(x)$, $\ln\Gamma(x)$, $\psi(x)$, and $\psi'(x)$ for $x = 1(.005)2$ to 10D; $\psi''(x)$ and $\psi^{(3)}(x)$ for $x = 1(.01)2$ to 10D; $\Gamma(n)$, $1/\Gamma(n)$, $\Gamma(n + \tfrac{1}{2})$, $\psi(n)$, $\log_{10}\Gamma(n)$, $\log_{10}\Gamma(n + \tfrac{1}{3})$, $\log_{10}\Gamma(n + \tfrac{1}{2})$, and $\log_{10}\Gamma(n + \tfrac{2}{3})$ for $n = 1(1)101$ to 8–11S; $\Gamma(n+1)$ for $n = 100(100)1000$ to 20S. Zhang and Jin (1996, pp. 67–69 and 72) tabulates $\Gamma(x)$, $1/\Gamma(x)$, $\Gamma(-x)$, $\ln\Gamma(x)$, $\psi(x)$, $\psi(-x)$, $\psi'(x)$, and $\psi'(-x)$ for $x = 0(.1)5$ to 8D or 8S; $\Gamma(n+1)$ for $n = 0(1)100(10)250(50)500(100)3000$ to 51S.

5.22(iii) Complex Variables

Abramov (1960) tabulates $\ln\Gamma(x+iy)$ for $x = 1$ (.01) 2, $y = 0$ (.01) 4 to 6D. Abramowitz and Stegun (1964, Chapter 6) tabulates $\ln\Gamma(x+iy)$ for $x = 1$ (.1) 2, $y = 0$ (.1) 10 to 12D. This reference also includes $\psi(x+iy)$ for the same arguments to 5D. Zhang and Jin (1996, pp. 70, 71, and 73) tabulates the real and imaginary parts of $\Gamma(x+iy)$, $\ln\Gamma(x+iy)$, and $\psi(x+iy)$ for $x = 0.5, 1, 5, 10$, $y = 0(.5)10$ to 8S.

5.23 Approximations

5.23(i) Rational Approximations

Cody and Hillstrom (1967) gives minimax rational approximations for $\ln\Gamma(x)$ for the ranges $0.5 \le x \le 1.5$, $1.5 \le x \le 4$, $4 \le x \le 12$; precision is variable. Hart et al. (1968) gives minimax polynomial and rational approximations to $\Gamma(x)$ and $\ln\Gamma(x)$ in the intervals $0 \le x \le 1$, $8 \le x \le 1000$, $12 \le x \le 1000$; precision is variable. Cody et al. (1973) gives minimax rational approximations for $\psi(x)$ for the ranges $0.5 \le x \le 3$ and $3 \le x < \infty$; precision is variable.

For additional approximations see Hart et al. (1968, Appendix B), Luke (1975, pp. 22–23), and Weniger (2003).

5.23(ii) Expansions in Chebyshev Series

Luke (1969b) gives the coefficients to 20D for the Chebyshev-series expansions of $\Gamma(1+x)$, $1/\Gamma(1+x)$, $\Gamma(x+3)$, $\ln\Gamma(x+3)$, $\psi(x+3)$, and the first six derivatives of $\psi(x+3)$ for $0 \le x \le 1$. These coefficients are reproduced in Luke (1975). Clenshaw (1962) also gives 20D Chebyshev-series coefficients for $\Gamma(1+x)$ and its reciprocal for $0 \le x \le 1$. See Luke (1975, pp. 22–23) for additional expansions.

5.23(iii) Approximations in the Complex Plane

See Schmelzer and Trefethen (2007) for a survey of rational approximations to various scaled versions of $\Gamma(z)$.

For rational approximations to $\psi(z) + \gamma$ see Luke (1975, pp. 13–16).

5.24 Software

See http://dlmf.nist.gov/5.24.

References

General References

The main references used in writing this chapter are Andrews *et al.* (1999), Carlson (1977b), Erdélyi *et al.* (1953a), Nielsen (1906a), Olver (1997b), Paris and Kaminski (2001), Temme (1996a), and Whittaker and Watson (1927).

Sources

The following list gives the references or other indications of proofs that were used in constructing the various sections of this chapter. These sources supplement the references that are quoted in the text.

§**5.2** Olver (1997b, Chapter 2, §§1 and 2), Temme (1996a, Chapters 1 and 3).

§**5.3** These graphics were computed at NIST.

§**5.4** Olver (1997b, Chapter 2, §§1 and 2), Andrews *et al.* (1999, §1.2). For (5.4.2) use (5.4.6) and (5.5.1). For (5.4.20) use (5.11.2) to solve $\psi(1-x) = \pi\cot(\pi x)$ with $x = -n + u$ and n large.

§**5.5** Olver (1997b, Chapter 2, §§1 and 2), Temme (1996a, Chapter 3), Andrews *et al.* (1999, §1.9). (5.5.9) follows from (5.5.6).

§**5.6** Gautschi (1959b, 1974), Alzer (1997a), Laforgia (1984), Kershaw (1983), Lorch (2002), Carlson (1977b, §3.10), Paris and Kaminski (2001, §2.1.3). For (5.6.1) see §5.11(ii).

§**5.7** Wrench (1968) (errors on p. 621 are corrected here), Erdélyi *et al.* (1953a, §1.17), Olver (1997b, Chapter 2, §2), Temme (1996a, §3.4). (5.7.7) follows from (5.7.6) and (5.5.2).

§**5.8** Olver (1997b, Chapter 2, §1), Whittaker and Watson (1927, §12.13).

§**5.9** Olver (1997b, Chapter 2, §§1 and 2), Temme (1996a, Chapter 3), Whittaker and Watson (1927, §§12.3–12.32 and §13.6). (5.9.3) follows from (5.2.1) by a change of variables. (5.9.8) and (5.9.9) follow from (5.9.6) and (5.9.7). (5.9.15) and (5.9.17) are the differentiated forms of (5.9.10) and (5.9.11). (5.9.19) is the differentiated form of (5.2.1).

§**5.10** Wall (1948, Chapter 19).

§**5.11** Olver (1997b, Chapter 3, §8, Chapter 4, §5, and Chapter 8, §4), Temme (1996a, §3.6.2), Paris and Kaminski (2001, §2.2.5). (5.11.7) and (5.11.9) are derived from (5.11.3).

§**5.12** Carlson (1977b, §4.2 and p. 70), Nielsen (1906a, §64), Temme (1996a, §3.8: an error in Ex.3.13 is corrected here), Olver (1997b, p. 38). (5.12.11) follows from (5.12.3).

§**5.13** Paris and Kaminski (2001, §3.3.4), Titchmarsh (1986a, pp. 188 and 194), Andrews *et al.* (1999, §3.6).

§**5.14** Andrews *et al.* (1999, §§1.8, 8.1–8.3, and 8.7), Mehta (2004, pp. 224–227).

§**5.16** Jordan (1939, pp. 344–345).

§**5.17** Whittaker and Watson (1927, p. 264), Olver (1997b, Chapter 8, §3.3), and the differentiated form of (25.4.1).

§**5.18** Andrews *et al.* (1999, §§10.1–10.3).

§**5.19** Stein and Shakarchi (2003, pp. 208–209) and Robnik (1980). The formula for V can also be verified by setting $t_k = (x_k/r)^2$ and $z_k = \frac{1}{2}, k = 1, 2, \ldots, n$, in (5.14.1). The formula for S can be verified in a similar way from (5.14.2), or derived by differentiating the formula for V.

§**5.20** Andrews *et al.* (1999, §8.2), Mehta (2004, Chapters 4 and 11).

Chapter 6
Exponential, Logarithmic, Sine, and Cosine Integrals

N. M. Temme[1]

Notation   **150**
6.1 Special Notation 150

Properties   **150**
6.2 Definitions and Interrelations 150
6.3 Graphics 151
6.4 Analytic Continuation 151
6.5 Further Interrelations 151
6.6 Power Series 151
6.7 Integral Representations 152
6.8 Inequalities 152
6.9 Continued Fraction 153
6.10 Other Series Expansions 153
6.11 Relations to Other Functions 153
6.12 Asymptotic Expansions 153

6.13 Zeros 154
6.14 Integrals 154
6.15 Sums 154

Applications   **154**
6.16 Mathematical Applications 154
6.17 Physical Applications 155

Computation   **155**
6.18 Methods of Computation 155
6.19 Tables 156
6.20 Approximations 156
6.21 Software 157

References   **157**

---

[1]Centrum voor Wiskunde en Informatica, Department MAS, Amsterdam, The Netherlands.
Acknowledgments: This chapter is based in part on Abramowitz and Stegun (1964, Chapter 5) by Walter Gautschi and William F. Cahill. Walter Gautschi provided the author with a list of references and comments collected since the original publication.
Copyright © 2009 National Institute of Standards and Technology. All rights reserved.

Notation

6.1 Special Notation

(For other notation see pp. xiv and 873.)

- x real variable.
- z complex variable.
- n nonnegative integer.
- δ arbitrary small positive constant.
- γ Euler's constant (§5.2(ii)).

Unless otherwise noted, primes indicate derivatives with respect to the argument.

The main functions treated in this chapter are the exponential integrals $\mathrm{Ei}(x)$, $E_1(z)$, and $\mathrm{Ein}(z)$; the logarithmic integral $\mathrm{li}(x)$; the sine integrals $\mathrm{Si}(z)$ and $\mathrm{si}(z)$; the cosine integrals $\mathrm{Ci}(z)$ and $\mathrm{Cin}(z)$.

Properties

6.2 Definitions and Interrelations

6.2(i) Exponential and Logarithmic Integrals

The *principal value* of the exponential integral $E_1(z)$ is defined by

6.2.1 $$E_1(z) = \int_z^\infty \frac{e^{-t}}{t}\,dt, \qquad z \neq 0,$$

where the path does not cross the negative real axis or pass through the origin. As in the case of the logarithm (§4.2(i)) there is a cut along the interval $(-\infty, 0]$ and the principal value is two-valued on $(-\infty, 0)$.

Unless indicated otherwise, it is assumed throughout this Handbook that $E_1(z)$ assumes its principal value. This is also true of the functions $\mathrm{Ci}(z)$ and $\mathrm{Chi}(z)$ defined in §6.2(ii).

6.2.2 $$E_1(z) = e^{-z}\int_0^\infty \frac{e^{-t}}{t+z}\,dt, \qquad |\mathrm{ph}\,z| < \pi.$$

6.2.3 $$\mathrm{Ein}(z) = \int_0^z \frac{1-e^{-t}}{t}\,dt.$$

$\mathrm{Ein}(z)$ is sometimes called the *complementary exponential integral*. It is entire.

6.2.4 $$E_1(z) = \mathrm{Ein}(z) - \ln z - \gamma.$$

In the next three equations $x > 0$.

6.2.5 $$\mathrm{Ei}(x) = -\fint_{-x}^\infty \frac{e^{-t}}{t}\,dt = \fint_{-\infty}^x \frac{e^t}{t}\,dt,$$

6.2.6 $$\mathrm{Ei}(-x) = -\int_x^\infty \frac{e^{-t}}{t}\,dt = -E_1(x),$$

6.2.7 $$\mathrm{Ei}(\pm x) = -\mathrm{Ein}(\mp x) + \ln x + \gamma.$$

$(\mathrm{Ei}(x)$ is undefined when $x = 0$, or when x is not real.)

The *logarithmic integral* is defined by

6.2.8 $$\mathrm{li}(x) = \fint_0^x \frac{dt}{\ln t} = \mathrm{Ei}(\ln x), \qquad x > 1.$$

The generalized exponential integral $E_p(z)$, $p \in \mathbb{C}$, is treated in Chapter 8.

6.2(ii) Sine and Cosine Integrals

6.2.9 $$\mathrm{Si}(z) = \int_0^z \frac{\sin t}{t}\,dt.$$

$\mathrm{Si}(z)$ is an odd entire function.

6.2.10 $$\mathrm{si}(z) = -\int_z^\infty \frac{\sin t}{t}\,dt = \mathrm{Si}(z) - \tfrac{1}{2}\pi.$$

6.2.11 $$\mathrm{Ci}(z) = -\int_z^\infty \frac{\cos t}{t}\,dt,$$

where the path does not cross the negative real axis or pass through the origin. This is the *principal value*; compare (6.2.1).

6.2.12 $$\mathrm{Cin}(z) = \int_0^z \frac{1-\cos t}{t}\,dt.$$

$\mathrm{Cin}(z)$ is an even entire function.

6.2.13 $$\mathrm{Ci}(z) = -\mathrm{Cin}(z) + \ln z + \gamma.$$

Values at Infinity

6.2.14 $$\lim_{x\to\infty}\mathrm{Si}(x) = \tfrac{1}{2}\pi, \qquad \lim_{x\to\infty}\mathrm{Ci}(x) = 0.$$

Hyperbolic Analogs of the Sine and Cosine Integrals

6.2.15 $$\mathrm{Shi}(z) = \int_0^z \frac{\sinh t}{t}\,dt,$$

6.2.16 $$\mathrm{Chi}(z) = \gamma + \ln z + \int_0^z \frac{\cosh t - 1}{t}\,dt.$$

6.2(iii) Auxiliary Functions

6.2.17 $$\mathrm{f}(z) = \mathrm{Ci}(z)\sin z - \mathrm{si}(z)\cos z,$$

6.2.18 $$\mathrm{g}(z) = -\mathrm{Ci}(z)\cos z - \mathrm{si}(z)\sin z.$$

6.2.19 $$\mathrm{Si}(z) = \tfrac{1}{2}\pi - \mathrm{f}(z)\cos z - \mathrm{g}(z)\sin z,$$

6.2.20 $$\mathrm{Ci}(z) = \mathrm{f}(z)\sin z - \mathrm{g}(z)\cos z.$$

6.2.21 $$\frac{d\mathrm{f}(z)}{dz} = -\mathrm{g}(z), \qquad \frac{d\mathrm{g}(z)}{dz} = \mathrm{f}(z) - \frac{1}{z}.$$

6.3 Graphics

6.3(i) Real Variable

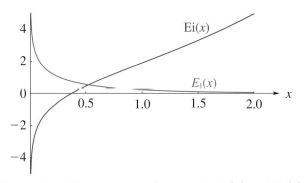

Figure 6.3.1: The exponential integrals $E_1(x)$ and $\mathrm{Ei}(x)$, $0 < x \leq 2$.

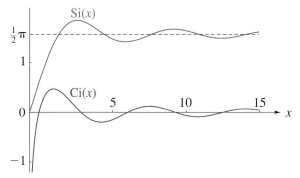

Figure 6.3.2: The sine and cosine integrals $\mathrm{Si}(x)$, $\mathrm{Ci}(x)$, $0 \leq x \leq 15$.

For a graph of $\mathrm{li}(x)$ see Figure 6.16.2.

6.3(ii) Complex Variable

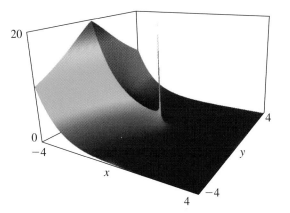

Figure 6.3.3: $|E_1(x+iy)|$, $-4 \leq x \leq 4$, $-4 \leq y \leq 4$. Principal value. There is a cut along the negative real axis. Also, $|E_1(z)| \to \infty$ logarithmically as $z \to 0$.

6.4 Analytic Continuation

Analytic continuation of the principal value of $E_1(z)$ yields a multi-valued function with branch points at $z = 0$ and $z = \infty$. The general value of $E_1(z)$ is given by

6.4.1 $$E_1(z) = \mathrm{Ein}(z) - \mathrm{Ln}\, z - \gamma;$$

compare (6.2.4) and (4.2.6). Thus

6.4.2 $$E_1(ze^{2m\pi i}) = E_1(z) - 2m\pi i, \qquad m \in \mathbb{Z},$$

and

6.4.3
$$E_1(ze^{\pm \pi i}) = \mathrm{Ein}(-z) - \ln z - \gamma \mp \pi i, \quad |\mathrm{ph}\, z| \leq \pi.$$

The general values of the other functions are defined in a similar manner, and

6.4.4 $$\mathrm{Ci}(ze^{\pm \pi i}) = \pm \pi i + \mathrm{Ci}(z),$$

6.4.5 $$\mathrm{Chi}(ze^{\pm \pi i}) = \pm \pi i + \mathrm{Chi}(z),$$

6.4.6 $$\mathrm{f}(ze^{\pm \pi i}) = \pi e^{\mp iz} - \mathrm{f}(z),$$

6.4.7 $$\mathrm{g}(ze^{\pm \pi i}) = \mp \pi i e^{\mp iz} + \mathrm{g}(z).$$

Unless indicated otherwise, in the rest of this chapter and elsewhere in this Handbook the functions $E_1(z)$, $\mathrm{Ci}(z)$, $\mathrm{Chi}(z)$, $\mathrm{f}(z)$, and $\mathrm{g}(z)$ assume their principal values, that is, the branches that are real on the positive real axis and two-valued on the negative real axis.

6.5 Further Interrelations

When $x > 0$,

6.5.1 $$E_1(-x \pm i0) = -\mathrm{Ei}(x) \mp i\pi,$$

6.5.2 $$\mathrm{Ei}(x) = -\tfrac{1}{2}(E_1(-x+i0) + E_1(-x-i0)),$$

6.5.3 $$\tfrac{1}{2}(\mathrm{Ei}(x) + E_1(x)) = \mathrm{Shi}(x) = -i\,\mathrm{Si}(ix),$$

6.5.4 $$\tfrac{1}{2}(\mathrm{Ei}(x) - E_1(x)) = \mathrm{Chi}(x) = \mathrm{Ci}(ix) - \tfrac{1}{2}\pi i.$$

When $|\mathrm{ph}\, z| < \tfrac{1}{2}\pi$,

6.5.5 $$\mathrm{Si}(z) = \tfrac{1}{2}i(E_1(-iz) - E_1(iz)) + \tfrac{1}{2}\pi,$$

6.5.6 $$\mathrm{Ci}(z) = -\tfrac{1}{2}(E_1(iz) + E_1(-iz)),$$

6.5.7 $$\mathrm{g}(z) \pm i\,\mathrm{f}(z) = E_1(\mp iz)e^{\mp iz}.$$

6.6 Power Series

6.6.1 $$\mathrm{Ei}(x) = \gamma + \ln x + \sum_{n=1}^{\infty} \frac{x^n}{n!\, n}, \qquad x > 0.$$

6.6.2 $$E_1(z) = -\gamma - \ln z - \sum_{n=1}^{\infty} \frac{(-1)^n z^n}{n!\, n}.$$

6.6.3 $$E_1(z) = -\ln z + e^{-z} \sum_{n=0}^{\infty} \frac{z^n}{n!} \psi(n+1),$$

where ψ denotes the logarithmic derivative of the gamma function (§5.2(i)).

6.6.4 $$\operatorname{Ein}(z) = \sum_{n=1}^{\infty} \frac{(-1)^{n-1} z^n}{n!\, n},$$

6.6.5 $$\operatorname{Si}(z) = \sum_{n=0}^{\infty} \frac{(-1)^n z^{2n+1}}{(2n+1)!(2n+1)},$$

6.6.6 $$\operatorname{Ci}(z) = \gamma + \ln z + \sum_{n=1}^{\infty} \frac{(-1)^n z^{2n}}{(2n)!(2n)}.$$

The series in this section converge for all finite values of x and $|z|$.

6.7 Integral Representations

6.7(i) Exponential Integrals

6.7.1 $$\int_0^\infty \frac{e^{-at}}{t+b}\, dt = \int_0^\infty \frac{e^{iat}}{t+ib}\, dt = e^{ab} E_1(ab),\ a>0,\ b>0,$$

6.7.2 $$e^x \int_0^\alpha \frac{e^{-xt}}{1-t}\, dt = \operatorname{Ei}(x) - \operatorname{Ei}((1-\alpha)x),$$
$$0 \le \alpha < 1,\ x > 0.$$

6.7.3 $$\int_x^\infty \frac{e^{it}}{a^2+t^2}\, dt = \frac{i}{2a}\left(e^a E_1(a-ix) - e^{-a} E_1(-a-ix)\right),$$
$$a>0,\ x>0,$$

6.7.4 $$\int_x^\infty \frac{t e^{it}}{a^2+t^2}\, dt = \tfrac{1}{2}\left(e^a E_1(a-ix) + e^{-a} E_1(-a-ix)\right),$$
$$a>0,\ x>0.$$

6.7.5 $$\int_x^\infty \frac{e^{-t}}{a^2+t^2}\, dt = -\frac{1}{2ai}\left(e^{ia} E_1(x+ia) - e^{-ia} E_1(x-ia)\right),$$
$$a>0,\ x\in\mathbb{R},$$

6.7.6 $$\int_x^\infty \frac{t e^{-t}}{a^2+t^2}\, dt = \tfrac{1}{2}\left(e^{ia} E_1(x+ia) + e^{-ia} E_1(x-ia)\right),$$
$$a>0,\ x\in\mathbb{R}.$$

6.7.7 $$\int_0^1 \frac{e^{-at}\sin(bt)}{t}\, dt = \Im\operatorname{Ein}(a+ib),\ a,b\in\mathbb{R},$$

6.7.8 $$\int_0^1 \frac{e^{-at}(1-\cos(bt))}{t}\, dt = \Re\operatorname{Ein}(a+ib) - \operatorname{Ein}(a),$$
$$a,b\in\mathbb{R}.$$

Many integrals with exponentials and rational functions, for example, integrals of the type $\int e^z R(z)\, dz$, where $R(z)$ is an arbitrary rational function, can be represented in finite form in terms of the function $E_1(z)$ and elementary functions; see Lebedev (1965, p. 42).

6.7(ii) Sine and Cosine Integrals

When $z \in \mathbb{C}$

6.7.9 $$\operatorname{si}(z) = -\int_0^{\pi/2} e^{-z\cos t} \cos(z\sin t)\, dt,$$

6.7.10 $$\operatorname{Ein}(z) - \operatorname{Cin}(z) = \int_0^{\pi/2} e^{-z\cos t}\sin(z\sin t)\, dt,$$

6.7.11 $$\int_0^1 \frac{(1-e^{-at})\cos(bt)}{t}\, dt = \Re\operatorname{Ein}(a+ib) - \operatorname{Cin}(b),$$
$$a,b\in\mathbb{R}.$$

6.7(iii) Auxiliary Functions

6.7.12 $$\operatorname{g}(z) + i\operatorname{f}(z) = e^{-iz}\int_z^\infty \frac{e^{it}}{t}\, dt,\ |\operatorname{ph} z|\le \pi.$$

The path of integration does not cross the negative real axis or pass through the origin.

6.7.13 $$\operatorname{f}(z) = \int_0^\infty \frac{\sin t}{t+z}\, dt = \int_0^\infty \frac{e^{-zt}}{t^2+1}\, dt,$$

6.7.14 $$\operatorname{g}(z) = \int_0^\infty \frac{\cos t}{t+z}\, dt = \int_0^\infty \frac{t e^{-zt}}{t^2+1}\, dt.$$

The first integrals on the right-hand sides apply when $|\operatorname{ph} z|<\pi$; the second ones when $\Re z\ge 0$ and (in the case of (6.7.14)) $z\ne 0$.

When $|\operatorname{ph} z|<\pi$

6.7.15 $$\operatorname{f}(z) = 2\int_0^\infty K_0\!\left(2\sqrt{zt}\right)\cos t\, dt,$$

6.7.16 $$\operatorname{g}(z) = 2\int_0^\infty K_0\!\left(2\sqrt{zt}\right)\sin t\, dt.$$

For K_0 see §10.25(ii).

6.7(iv) Compendia

For collections of integral representations see Bierens de Haan (1939, pp. 56–59, 72–73, 82–84, 121, 133–136, 155, 179–181, 223, 225–227, 230, 259–260, 374, 377, 397–398, 408, 416, 424, 431, 438–439, 442–444, 488, 496–500, 567–571, 585, 602, 638, 675–677), Corrington (1961), Erdélyi et al. (1954a, vol. 1, pp. 267–270), Geller and Ng (1969), Nielsen (1906b), Oberhettinger (1974, pp. 244–246), Oberhettinger and Badii (1973, pp. 364–371), and Watrasiewicz (1967).

6.8 Inequalities

In this section $x>0$.

6.8.1 $$\tfrac{1}{2}\ln\!\left(1+\frac{2}{x}\right) < e^x E_1(x) < \ln\!\left(1+\frac{1}{x}\right),$$

6.8.2 $$\frac{x}{x+1} < x e^x E_1(x) < \frac{x+1}{x+2},$$

6.8.3 $$\frac{x(x+3)}{x^2+4x+2} < x e^x E_1(x) < \frac{x^2+5x+2}{x^2+6x+6}.$$

6.9 Continued Fraction

$$6.9.1 \quad E_1(z) = \frac{e^{-z}}{z+}\frac{1}{1+}\frac{1}{z+}\frac{2}{1+}\frac{2}{z+}\frac{3}{1+}\frac{3}{z+}\cdots,$$
$$|\operatorname{ph} z| < \pi.$$

See also Cuyt *et al.* (2008, pp. 287–290).

6.10 Other Series Expansions

6.10(i) Inverse Factorial Series

$$6.10.1 \quad E_1(z) = e^{-z}\left(\frac{c_0}{z} + \frac{c_1}{z(z+1)} + \frac{2!c_2}{z(z+1)(z+2)} + \frac{3!c_3}{z(z+1)(z+2)(z+3)} + \cdots\right),$$
$$\Re z > 0,$$

where

6.10.2 $\quad c_0 = 1, \quad c_1 = -1, \quad c_2 = \tfrac{1}{2}, \quad c_3 = -\tfrac{1}{3}, \quad c_4 = \tfrac{1}{6},$

and

$$6.10.3 \quad c_k = -\sum_{j=0}^{k-1}\frac{c_j}{k-j}, \qquad k \geq 1.$$

For a more general result (incomplete gamma function), and also for a result for the logarithmic integral, see Nielsen (1906a, p. 283: Formula (3) is incorrect).

6.10(ii) Expansions in Series of Spherical Bessel Functions

For the notation see §10.47(ii).

$$6.10.4 \quad \operatorname{Si}(z) = z\sum_{n=0}^{\infty}\left(\mathsf{j}_n(\tfrac{1}{2}z)\right)^2,$$

$$6.10.5 \quad \operatorname{Cin}(z) = \sum_{n=1}^{\infty} a_n \left(\mathsf{j}_n(\tfrac{1}{2}z)\right)^2,$$

$$6.10.6 \quad \operatorname{Ei}(x) = \gamma + \ln|x| + \sum_{n=0}^{\infty}(-1)^n(x-a_n)\left(\mathsf{i}_n^{(1)}(\tfrac{1}{2}x)\right)^2,$$
$$x \neq 0,$$

where

6.10.7 $\quad a_n = (2n+1)\left(1 - (-1)^n + \psi(n+1) - \psi(1)\right),$

and ψ denotes the logarithmic derivative of the gamma function (§5.2(i)).

6.10.8
$$\operatorname{Ein}(z) = ze^{-z/2}\left(\mathsf{i}_0^{(1)}(\tfrac{1}{2}z) + \sum_{n=1}^{\infty}\frac{2n+1}{n(n+1)}\mathsf{i}_n^{(1)}(\tfrac{1}{2}z)\right).$$

For (6.10.4)–(6.10.8) and further results see Harris (2000) and Luke (1969b, pp. 56–57). An expansion for $E_1(z)$ can be obtained by combining (6.2.4) and (6.10.8).

6.11 Relations to Other Functions

For the notation see §§8.2(i) and 13.2(i).

Incomplete Gamma Function

6.11.1 $\quad E_1(z) = \Gamma(0, z).$

Confluent Hypergeometric Function

6.11.2 $\quad E_1(z) = e^{-z}U(1, 1, z),$

6.11.3 $\quad \mathrm{g}(z) + i\,\mathrm{f}(z) = U(1, 1, -iz).$

6.12 Asymptotic Expansions

6.12(i) Exponential and Logarithmic Integrals

$$6.12.1 \quad E_1(z) \sim \frac{e^{-z}}{z}\left(1 - \frac{1!}{z} + \frac{2!}{z^2} - \frac{3!}{z^3} + \cdots\right),$$
$$z \to \infty, \ |\operatorname{ph} z| \leq \tfrac{3}{2}\pi - \delta(<\tfrac{3}{2}\pi).$$

When $|\operatorname{ph} z| \leq \tfrac{1}{2}\pi$ the remainder is bounded in magnitude by the first neglected term, and has the same sign when $\operatorname{ph} z = 0$. When $\tfrac{1}{2}\pi \leq |\operatorname{ph} z| < \pi$ the remainder term is bounded in magnitude by $\csc(|\operatorname{ph} z|)$ times the first neglected term. For these and other error bounds see Olver (1997b, pp. 109–112) with $\alpha = 0$.

For re-expansions of the remainder term leading to larger sectors of validity, exponential improvement, and a smooth interpretation of the Stokes phenomenon, see §§2.11(ii)–2.11(iv), with $p = 1$.

$$6.12.2 \quad \operatorname{Ei}(x) \sim \frac{e^x}{x}\left(1 + \frac{1!}{x} + \frac{2!}{x^2} + \frac{3!}{x^3} + \cdots\right), \quad x \to +\infty.$$

If the expansion is terminated at the nth term, then the remainder term is bounded by $1 + \chi(n+1)$ times the next term. For the function χ see §9.7(i).

The asymptotic expansion of $\operatorname{li}(x)$ as $x \to \infty$ is obtainable from (6.2.8) and (6.12.2).

6.12(ii) Sine and Cosine Integrals

The asymptotic expansions of $\operatorname{Si}(z)$ and $\operatorname{Ci}(z)$ are given by (6.2.19), (6.2.20), together with

$$6.12.3 \quad \mathrm{f}(z) \sim \frac{1}{z}\left(1 - \frac{2!}{z^2} + \frac{4!}{z^4} - \frac{6!}{z^6} + \cdots\right),$$

$$6.12.4 \quad \mathrm{g}(z) \sim \frac{1}{z^2}\left(1 - \frac{3!}{z^2} + \frac{5!}{z^4} - \frac{7!}{z^6} + \cdots\right),$$

as $z \to \infty$ in $|\operatorname{ph} z| \leq \pi - \delta(<\pi)$.

The remainder terms are given by

$$6.12.5 \quad \mathrm{f}(z) = \frac{1}{z}\sum_{m=0}^{n-1}(-1)^m\frac{(2m)!}{z^{2m}} + R_n^{(\mathrm{f})}(z),$$

$$6.12.6 \quad \mathrm{g}(z) = \frac{1}{z^2}\sum_{m=0}^{n-1}(-1)^m\frac{(2m+1)!}{z^{2m}} + R_n^{(\mathrm{g})}(z),$$

where, for $n = 0, 1, 2, \ldots$,

6.12.7 $\quad R_n^{(\mathrm{f})}(z) = (-1)^n \int_0^\infty \frac{e^{-zt} t^{2n}}{t^2 + 1} \, dt,$

6.12.8 $\quad R_n^{(\mathrm{g})}(z) = (-1)^n \int_0^\infty \frac{e^{-zt} t^{2n+1}}{t^2 + 1} \, dt.$

When $|\operatorname{ph} z| \leq \frac{1}{4}\pi$, these remainders are bounded in magnitude by the first neglected terms in (6.12.3) and (6.12.4), respectively, and have the same signs as these terms when $\operatorname{ph} z = 0$. When $\frac{1}{4}\pi \leq |\operatorname{ph} z| < \frac{1}{2}\pi$ the remainders are bounded in magnitude by $\csc(2|\operatorname{ph} z|)$ times the first neglected terms.

For other phase ranges use (6.4.6) and (6.4.7). For exponentially-improved asymptotic expansions, use (6.5.5), (6.5.6), and §6.12(i).

6.13 Zeros

The function $\operatorname{Ei}(x)$ has one real zero x_0, given by

6.13.1 $\quad x_0 = 0.37250\,74107\,81366\,63446\,19918\,66580\ldots.$

$\operatorname{Ci}(x)$ and $\operatorname{si}(x)$ each have an infinite number of positive real zeros, which are denoted by c_k, s_k, respectively, arranged in ascending order of absolute value for $k = 0, 1, 2, \ldots$. Values of c_1 and c_2 to 30D are given by MacLeod (1996).

As $k \to \infty$,

6.13.2
$c_k, s_k \sim \alpha + \frac{1}{\alpha} - \frac{16}{3}\frac{1}{\alpha^3} + \frac{1673}{15}\frac{1}{\alpha^5} - \frac{5\,07746}{105}\frac{1}{\alpha^7} + \cdots,$

where $\alpha = k\pi$ for c_k, and $\alpha = (k + \frac{1}{2})\pi$ for s_k. For these results, together with the next three terms in (6.13.2), see MacLeod (2002a). See also Riekstyņš (1991, pp. 176–177).

6.14 Integrals

6.14(i) Laplace Transforms

6.14.1 $\quad \int_0^\infty e^{-at} E_1(t) \, dt = \frac{1}{a} \ln(1 + a), \quad \Re a > -1,$

6.14.2 $\quad \int_0^\infty e^{-at} \operatorname{Ci}(t) \, dt = -\frac{1}{2a} \ln(1 + a^2), \quad \Re a > 0,$

6.14.3 $\quad \int_0^\infty e^{-at} \operatorname{si}(t) \, dt = -\frac{1}{a} \arctan a, \quad \Re a > 0.$

6.14(ii) Other Integrals

6.14.4 $\quad \int_0^\infty E_1^2(t) \, dt = 2 \ln 2,$

6.14.5 $\quad \int_0^\infty \cos t \operatorname{Ci}(t) \, dt = \int_0^\infty \sin t \operatorname{si}(t) \, dt = -\frac{1}{4}\pi,$

6.14.6 $\quad \int_0^\infty \operatorname{Ci}^2(t) \, dt = \int_0^\infty \operatorname{si}^2(t) \, dt = \frac{1}{2}\pi,$

6.14.7 $\quad \int_0^\infty \operatorname{Ci}(t) \operatorname{si}(t) \, dt = \ln 2.$

6.14(iii) Compendia

For collections of integrals, see Apelblat (1983, pp. 110–123), Bierens de Haan (1939, pp. 373–374, 409, 479, 571–572, 637, 664–673, 680–682, 685–697), Erdélyi et al. (1954a, vol. 1, pp. 40–42, 96–98, 177–178, 325), Geller and Ng (1969), Gradshteyn and Ryzhik (2000, §§5.2–5.3 and 6.2–6.27), Marichev (1983, pp. 182–184), Nielsen (1906b), Oberhettinger (1974, pp. 139–141), Oberhettinger (1990, pp. 53–55 and 158–160), Oberhettinger and Badii (1973, pp. 172–179), Prudnikov et al. (1986b, vol. 2, pp. 24–29 and 64–92), Prudnikov et al. (1992a, §§3.4–3.6), Prudnikov et al. (1992b, §§3.4–3.6), and Watrasiewicz (1967).

6.15 Sums

6.15.1 $\quad \sum_{n=1}^\infty \operatorname{Ci}(\pi n) = \frac{1}{2}(\ln 2 - \gamma),$

6.15.2 $\quad \sum_{n=1}^\infty \frac{\operatorname{si}(\pi n)}{n} = \frac{1}{2}\pi(\ln \pi - 1),$

6.15.3 $\quad \sum_{n=1}^\infty (-1)^n \operatorname{Ci}(2\pi n) = 1 - \ln 2 - \frac{1}{2}\gamma,$

6.15.4 $\quad \sum_{n=1}^\infty (-1)^n \frac{\operatorname{si}(2\pi n)}{n} = \pi(\frac{3}{2} \ln 2 - 1).$

For further sums see Fempl (1960), Hansen (1975, pp. 423–424), Harris (2000), Prudnikov et al. (1986b, vol. 2, pp. 649–650), and Slavić (1974).

Applications

6.16 Mathematical Applications

6.16(i) The Gibbs Phenomenon

Consider the Fourier series

6.16.1
$\sin x + \frac{1}{3}\sin(3x) + \frac{1}{5}\sin(5x) + \cdots$
$= \begin{cases} \frac{1}{4}\pi, & 0 < x < \pi, \\ 0, & x = 0, \\ -\frac{1}{4}\pi, & -\pi < x < 0. \end{cases}$

The nth partial sum is given by

6.16.2 $\quad S_n(x) = \sum_{k=0}^{n-1} \frac{\sin((2k+1)x)}{2k+1} = \frac{1}{2}\int_0^x \frac{\sin(2nt)}{\sin t} \, dt$
$= \frac{1}{2}\operatorname{Si}(2nx) + R_n(x),$

6.17 Physical Applications

where

6.16.3 $\quad R_n(x) = \dfrac{1}{2} \displaystyle\int_0^x \left(\dfrac{1}{\sin t} - \dfrac{1}{t} \right) \sin(2nt)\, dt.$

By integration by parts

6.16.4 $\quad R_n(x) = O(n^{-1}), \qquad n \to \infty,$

uniformly for $x \in [-\pi, \pi]$. Hence, if x is fixed and $n \to \infty$, then $S_n(x) \to \tfrac{1}{4}\pi$, 0, or $-\tfrac{1}{4}\pi$ according as $0 < x < \pi$, $x = 0$, or $-\pi < x < 0$; compare (6.2.14).

These limits are not approached uniformly, however. The first maximum of $\tfrac{1}{2}\operatorname{Si}(x)$ for positive x occurs at $x = \pi$ and equals $(1.1789\ldots) \times \tfrac{1}{4}\pi$; compare Figure 6.3.2. Hence if $x = \pi/(2n)$ and $n \to \infty$, then the limiting value of $S_n(x)$ overshoots $\tfrac{1}{4}\pi$ by approximately 18%. Similarly if $x = \pi/n$, then the limiting value of $S_n(x)$ undershoots $\tfrac{1}{4}\pi$ by approximately 10%, and so on. Compare Figure 6.16.1.

This nonuniformity of convergence is an illustration of the *Gibbs phenomenon*. It occurs with Fourier-series expansions of all piecewise continuous functions. See Carslaw (1930) for additional graphs and information.

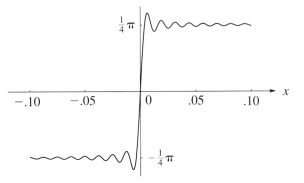

Figure 6.16.1: Graph of $S_n(x)$, $n = 250$, $-0.1 \le x \le 0.1$, illustrating the Gibbs phenomenon.

6.16(ii) Number-Theoretic Significance of li(x)

If we assume Riemann's hypothesis that all nonreal zeros of $\zeta(s)$ have real part of $\tfrac{1}{2}$ (§25.10(i)), then

6.16.5 $\quad \operatorname{li}(x) - \pi(x) = O(\sqrt{x} \ln x), \qquad x \to \infty,$

where $\pi(x)$ is the number of primes less than or equal to x. Compare §27.12 and Figure 6.16.2. See also Bays and Hudson (2000).

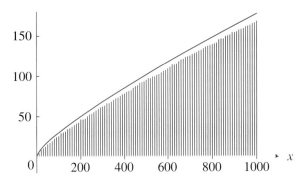

Figure 6.16.2: The logarithmic integral $\operatorname{li}(x)$, together with vertical bars indicating the value of $\pi(x)$ for $x = 10, 20, \ldots, 1000$.

6.17 Physical Applications

Geller and Ng (1969) cites work with applications from diffusion theory, transport problems, the study of the radiative equilibrium of stellar atmospheres, and the evaluation of exchange integrals occurring in quantum mechanics. For applications in astrophysics, see also van de Hulst (1980). Lebedev (1965) gives an application to electromagnetic theory (radiation of a linear half-wave oscillator), in which sine and cosine integrals are used.

Computation

6.18 Methods of Computation

6.18(i) Main Functions

For small or moderate values of x and $|z|$, the expansion in power series (§6.6) or in series of spherical Bessel functions (§6.10(ii)) can be used. For large x or $|z|$ these series suffer from slow convergence or cancellation (or both). However, this problem is less severe for the series of spherical Bessel functions because of their more rapid rate of convergence, and also (except in the case of (6.10.6)) absence of cancellation when $z = x\ (>0)$.

For large x and $|z|$, expansions in inverse factorial series (§6.10(i)) or asymptotic expansions (§6.12) are available. The attainable accuracy of the asymptotic expansions can be increased considerably by exponential improvement. Also, other ranges of $\operatorname{ph} z$ can be covered by use of the continuation formulas of §6.4.

Quadrature of the integral representations is another effective method. For example, the Gauss-Laguerre formula (§3.5(v)) can be applied to (6.2.2); see Todd (1954) and Tseng and Lee (1998). For an application of the Gauss-Legendre formula (§3.5(v)) see Tooper and Mark (1968).

Lastly, the continued fraction (6.9.1) can be used if $|z|$ is bounded away from the origin. Convergence becomes slow when z is near the negative real axis, however.

6.18(ii) Auxiliary Functions

Power series, asymptotic expansions, and quadrature can also be used to compute the functions $\mathrm{f}(z)$ and $\mathrm{g}(z)$. In addition, Acton (1974) developed a recurrence procedure, as follows. For $n = 0, 1, 2, \ldots$, define

6.18.1
$$A_n = \int_0^\infty \frac{te^{-zt}}{1+t^2}\left(\frac{t^2}{1+t^2}\right)^n dt,$$
$$B_n = \int_0^\infty \frac{e^{-zt}}{1+t^2}\left(\frac{t^2}{1+t^2}\right)^n dt,$$
$$C_n = \int_0^\infty e^{-zt}\left(\frac{t^2}{1+t^2}\right)^n dt.$$

Then $\mathrm{f}(z) = B_0$, $\mathrm{g}(z) = A_0$, and

6.18.2
$$A_{n-1} = A_n + \frac{z}{2n}C_n, \quad B_{n-1} = \frac{2nB_n + zA_{n-1}}{2n-1},$$
$$C_{n-1} = C_n + B_{n-1}, \quad n = 1, 2, 3, \ldots.$$

A_0, B_0, and C_0 can be computed by Miller's algorithm (§3.6(iii)), starting with initial values $(A_N, B_N, C_N) = (1, 0, 0)$, say, where N is an arbitrary large integer, and normalizing via $C_0 = 1/z$.

6.18(iii) Zeros

Zeros of $\mathrm{Ci}(x)$ and $\mathrm{si}(x)$ can be computed to high precision by Newton's rule (§3.8(ii)), using values supplied by the asymptotic expansion (6.13.2) as initial approximations.

6.18(iv) Other References

For a comprehensive survey of computational methods for the functions treated in this chapter, see van der Laan and Temme (1984, Ch. IV).

6.19 Tables

6.19(i) Introduction

Lebedev and Fedorova (1960) and Fletcher et al. (1962) give comprehensive indexes of mathematical tables. This section lists relevant tables that appeared later.

6.19(ii) Real Variables

- Abramowitz and Stegun (1964, Chapter 5) includes $x^{-1}\mathrm{Si}(x)$, $-x^{-2}\mathrm{Cin}(x)$, $x^{-1}\mathrm{Ein}(x)$, $-x^{-1}\mathrm{Ein}(-x)$, $x = 0(.01)0.5$; $\mathrm{Si}(x)$, $\mathrm{Ci}(x)$, $\mathrm{Ei}(x)$, $E_1(x)$, $x = 0.5(.01)2$; $\mathrm{Si}(x)$, $\mathrm{Ci}(x)$, $xe^{-x}\mathrm{Ei}(x)$, $xe^x E_1(x)$, $x = 2(.1)10$; $x\mathrm{f}(x)$, $x^2\mathrm{g}(x)$, $xe^{-x}\mathrm{Ei}(x)$, $xe^x E_1(x)$, $x^{-1} = 0(.005)0.1$; $\mathrm{Si}(\pi x)$, $\mathrm{Cin}(\pi x)$, $x = 0(.1)10$. Accuracy varies but is within the range 8S–11S.

- Zhang and Jin (1996, pp. 652, 689) includes $\mathrm{Si}(x)$, $\mathrm{Ci}(x)$, $x = 0(.5)20(2)30$, 8D; $\mathrm{Ei}(x)$, $E_1(x)$, $x = [0, 100]$, 8S.

6.19(iii) Complex Variables, $z = x + iy$

- Abramowitz and Stegun (1964, Chapter 5) includes the real and imaginary parts of $ze^z E_1(z)$, $x = -19(1)20$, $y = 0(1)20$, 6D; $e^z E_1(z)$, $x = -4(.5) - 2$, $y = 0(.2)1$, 6D; $E_1(z) + \ln z$, $x = -2(.5)2.5$, $y = 0(.2)1$, 6D.

- Zhang and Jin (1996, pp. 690–692) includes the real and imaginary parts of $E_1(z)$, $\pm x = 0.5, 1, 3, 5, 10, 15, 20, 50, 100$, $y = 0(.5)1(1)5(5)30, 50, 100$, 8S.

6.20 Approximations

6.20(i) Approximations in Terms of Elementary Functions

- Hastings (1955) gives several minimax polynomial and rational approximations for $E_1(x) + \ln x$, $xe^x E_1(x)$, and the auxiliary functions $\mathrm{f}(x)$ and $\mathrm{g}(x)$. These are included in Abramowitz and Stegun (1964, Ch. 5).

- Cody and Thacher (1968) provides minimax rational approximations for $E_1(x)$, with accuracies up to 20S.

- Cody and Thacher (1969) provides minimax rational approximations for $\mathrm{Ei}(x)$, with accuracies up to 20S.

- MacLeod (1996) provides rational approximations for the sine and cosine integrals and for the auxiliary functions f and g, with accuracies up to 20S.

6.20(ii) Expansions in Chebyshev Series

- Clenshaw (1962) gives Chebyshev coefficients for $-E_1(x) - \ln|x|$ for $-4 \leq x \leq 4$ and $e^x E_1(x)$ for $x \geq 4$ (20D).

- Luke and Wimp (1963) covers $\mathrm{Ei}(x)$ for $x \leq -4$ (20D), and $\mathrm{Si}(x)$ and $\mathrm{Ci}(x)$ for $x \geq 4$ (20D).

- Luke (1969b, pp. 41–42) gives Chebyshev expansions of $\mathrm{Ein}(ax)$, $\mathrm{Si}(ax)$, and $\mathrm{Cin}(ax)$ for $-1 \leq x \leq 1$, $a \in \mathbb{C}$. The coefficients are given in terms of series of Bessel functions.

- Luke (1969b, pp. 321–322) covers $\text{Ein}(x)$ and $-\text{Ein}(-x)$ for $0 \leq x \leq 8$ (the Chebyshev coefficients are given to 20D); $E_1(x)$ for $x \geq 5$ (20D), and $\text{Ei}(x)$ for $x \geq 8$ (15D). Coefficients for the sine and cosine integrals are given on pp. 325–327.

- Luke (1969b, p. 25) gives a Chebyshev expansion near infinity for the confluent hypergeometric U-function (§13.2(i)) from which Chebyshev expansions near infinity for $E_1(z)$, $f(z)$, and $g(z)$ follow by using (6.11.2) and (6.11.3). Luke also includes a recursion scheme for computing the coefficients in the expansions of the U functions. If $|\operatorname{ph} z| < \pi$ the scheme can be used in backward direction.

6.20(iii) Padé-Type and Rational Expansions

- Luke (1969b, pp. 402, 410, and 415–421) gives main diagonal Padé approximations for $\text{Ein}(z)$, $\text{Si}(z)$, $\text{Cin}(z)$ (valid near the origin), and $E_1(z)$ (valid for large $|z|$); approximate errors are given for a selection of z-values.

- Luke (1969b, pp. 411–414) gives rational approximations for $\text{Ein}(z)$.

6.21 Software

See http://dlmf.nist.gov/6.21.

References

General References

For general bibliographic reading see Andrews *et al.* (1999), Jeffreys and Jeffreys (1956), Lebedev (1965), Olver (1997b), and Temme (1996a).

Sources

The following list gives the references or other indications of proofs that were used in constructing the various sections of this chapter. These sources supplement the references that are quoted in the text.

§6.2 Olver (1997b, pp. 40–42).

§6.3 These graphics were produced at NIST.

§6.4 For (6.4.1) see Olver (1997b, p. 40). (6.4.3) follows from (6.6.2) and (6.6.4). (6.4.4) and (6.4.5) follow from (6.2.13) and (6.2.16). (6.4.6) and (6.4.7) follow from (6.2.17), (6.2.18), and (6.4.4).

§6.5 For (6.5.1) and (6.5.2) see Olver (1997b, p. 41). (6.5.3) and (6.5.4) follow from (6.6.1), (6.6.2), (6.6.5), and (6.6.6). For (6.5.5) and (6.5.6) see Olver (1997b, p. 42). (6.5.7) follows from (6.2.10), (6.2.17), (6.2.18), (6.5.5), and (6.5.6).

§6.6 Olver (1997b, pp. 40–43). (6.6.3) follows from (6.11.2) and (13.2.9).

§6.7 (6.7.1) and (6.7.2) follow from the definitions (§6.2(i)). (6.7.3)–(6.7.6) follow from differentiation with respect to x. (6.7.7) and (6.7.8) follow from replacing the trigonometric functions by exponentials. For (6.7.9)–(6.7.11) see Nielsen (1906b, p. 13: there are sign errors in Eq. (27)). (6.7.12)–(6.7.14) follow from (6.5.7), (6.2.1), and (6.2.2); for the second equations in (6.7.13) and (6.7.14) see Temme (1996a, pp. 187–188). For (6.7.15) and (6.7.16) use (10.32.10).

§6.8 See Gautschi (1959b) for (6.8.1), and Luke (1969b, p. 201) for (6.8.2) and (6.8.3).

§6.9 Nielsen (1906b, pp. 42–44), or Lorentzen and Waadeland (1992, p. 577).

§6.10 Nielsen (1906a, p. 283). (6.10.3) follows from $1/(1 - \ln(1-t)) = \sum_{k=0}^{\infty} c_k t^k$.

§6.11 Temme (1996a, pp. 180 and 187). For (6.11.3) use (6.5.7).

§6.12 For (6.12.2) see Olver (1997b, p. 227). (6.12.3) and (6.12.4) follow from (6.7.13) and (6.7.14) by applying Watson's lemma (§2.4(i)). (6.12.5)–(6.12.8) follow from (6.7.13), (6.7.14), and the identity $(t^2 + 1)^{-1} = \sum_{m=0}^{n-1}(-1)^m t^{2m} + (-1)^n t^{2n}(t^2+1)^{-1}$. The error bounds are obtained by setting $t = \sqrt{\tau}$ in (6.12.7) and (6.12.8), rotating the integration path in the τ-plane through an angle $-2\operatorname{ph} z$, and then replacing $|\tau + 1|$ by its minimum value on the path.

§6.13 See Cody and Thacher (1969) for x_0 in (6.13.1).

§6.14 Nielsen (1906b, pp. 48–50, 53, and 54: there is a $\frac{1}{2}$ missing in the formula that corresponds to (6.14.2) and a sign error in the formula that corresponds to (6.14.7)).

§6.15 Slavić (1974).

§6.16 Temme (1996a, pp. 181–182: the numerical value $1.089490\ldots$ on p. 182 should be replaced by $1.1789\ldots$). Gibbs reported this phenomenon in a letter to *Nature*, **59** (1899, p. 606). Figures 6.16.1 and 6.16.2 were produced at NIST.

Chapter 7
Error Functions, Dawson's and Fresnel Integrals

N. M. Temme[1]

Notation   **160**
- 7.1 Special Notation 160

Properties   **160**
- 7.2 Definitions 160
- 7.3 Graphics 160
- 7.4 Symmetry 161
- 7.5 Interrelations 162
- 7.6 Series Expansions 162
- 7.7 Integral Representations 162
- 7.8 Inequalities 163
- 7.9 Continued Fractions 163
- 7.10 Derivatives 163
- 7.11 Relations to Other Functions 164
- 7.12 Asymptotic Expansions 164
- 7.13 Zeros 165
- 7.14 Integrals 166
- 7.15 Sums 166
- 7.16 Generalized Error Functions 166
- 7.17 Inverse Error Functions 166
- 7.18 Repeated Integrals of the Complementary Error Function 167
- 7.19 Voigt Functions 167

Applications   **168**
- 7.20 Mathematical Applications 168
- 7.21 Physical Applications 169

Computation   **169**
- 7.22 Methods of Computation 169
- 7.23 Tables 169
- 7.24 Approximations 170
- 7.25 Software 170

References   **170**

---

[1]Centrum voor Wiskunde en Informatica, Department MAS, Amsterdam, The Netherlands.
Acknowledgments: This chapter is based in part on Abramowitz and Stegun (1964, Chapter 7) by Walter Gautschi. Walter Gautschi provided the author with a list of references and comments collected since the original publication.
Copyright © 2009 National Institute of Standards and Technology. All rights reserved.

Notation

7.1 Special Notation

(For other notation see pp. xiv and 873.)

- x real variable.
- z complex variable.
- n nonnegative integer.
- δ arbitrary small positive constant.
- γ Euler's constant (§5.2(ii)).

Unless otherwise noted, primes indicate derivatives with respect to the argument.

The main functions treated in this chapter are the error function $\operatorname{erf} z$; the complementary error functions $\operatorname{erfc} z$ and $w(z)$; Dawson's integral $F(z)$; the Fresnel integrals $\mathcal{F}(z)$, $C(z)$, and $S(z)$; the Goodwin–Staton integral $G(z)$; the repeated integrals of the complementary error function $\mathrm{i}^n\operatorname{erfc}(z)$; the Voigt functions $\mathsf{U}(x,t)$ and $\mathsf{V}(x,t)$.

Alternative notations are $P(z) = \tfrac{1}{2}\operatorname{erfc}(-z/\sqrt{2})$, $Q(z) = \Phi(z) = \tfrac{1}{2}\operatorname{erfc}(z/\sqrt{2})$, $\operatorname{Erf} z = \tfrac{1}{2}\sqrt{\pi}\operatorname{erf} z$, $\operatorname{Erfi} z = e^{z^2} F(z)$, $C_1(z) = C(\sqrt{2/\pi}\,z)$, $S_1(z) = S(\sqrt{2/\pi}\,z)$, $C_2(z) = C(\sqrt{2z/\pi})$, $S_2(z) = S(\sqrt{2z/\pi})$.

The notations $P(z)$, $Q(z)$, and $\Phi(z)$ are used in mathematical statistics, where these functions are called the *normal* or *Gaussian probability functions*.

Properties

7.2 Definitions

7.2(i) Error Functions

7.2.1
$$\operatorname{erf} z = \frac{2}{\sqrt{\pi}} \int_0^z e^{-t^2}\, dt,$$

7.2.2
$$\operatorname{erfc} z = \frac{2}{\sqrt{\pi}} \int_z^\infty e^{-t^2}\, dt = 1 - \operatorname{erf} z,$$

7.2.3
$$w(z) = e^{-z^2}\left(1 + \frac{2i}{\sqrt{\pi}}\int_0^z e^{t^2}\,dt\right) = e^{-z^2}\operatorname{erfc}(-iz).$$

$\operatorname{erf} z$, $\operatorname{erfc} z$, and $w(z)$ are entire functions of z, as is $F(z)$ in the next subsection.

Values at Infinity

7.2.4
$$\lim_{z\to\infty}\operatorname{erf} z = 1, \quad \lim_{z\to\infty}\operatorname{erfc} z = 0,$$
$$|\operatorname{ph} z| \le \tfrac{1}{4}\pi - \delta(<\tfrac{1}{4}\pi).$$

7.2(ii) Dawson's Integral

7.2.5
$$F(z) = e^{-z^2} \int_0^z e^{t^2}\, dt.$$

7.2(iii) Fresnel Integrals

7.2.6
$$\mathcal{F}(z) = \int_z^\infty e^{\frac{1}{2}\pi i t^2}\, dt,$$

7.2.7
$$C(z) = \int_0^z \cos\!\left(\tfrac{1}{2}\pi t^2\right) dt,$$

7.2.8
$$S(z) = \int_0^z \sin\!\left(\tfrac{1}{2}\pi t^2\right) dt,$$

$\mathcal{F}(z)$, $C(z)$, and $S(z)$ are entire functions of z, as are $\mathrm{f}(z)$ and $\mathrm{g}(z)$ in the next subsection.

Values at Infinity

7.2.9
$$\lim_{x\to\infty} C(x) = \tfrac{1}{2}, \quad \lim_{x\to\infty} S(x) = \tfrac{1}{2}.$$

7.2(iv) Auxiliary Functions

7.2.10
$$\mathrm{f}(z) = \left(\tfrac{1}{2} - S(z)\right)\cos\!\left(\tfrac{1}{2}\pi z^2\right) - \left(\tfrac{1}{2} - C(z)\right)\sin\!\left(\tfrac{1}{2}\pi z^2\right),$$

7.2.11
$$\mathrm{g}(z) = \left(\tfrac{1}{2} - C(z)\right)\cos\!\left(\tfrac{1}{2}\pi z^2\right) + \left(\tfrac{1}{2} - S(z)\right)\sin\!\left(\tfrac{1}{2}\pi z^2\right).$$

7.2(v) Goodwin–Staton Integral

7.2.12
$$G(z) = \int_0^\infty \frac{e^{-t^2}}{t+z}\, dt, \qquad |\operatorname{ph} z| < \pi.$$

7.3 Graphics

7.3(i) Real Variable

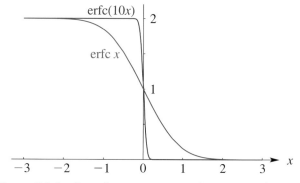

Figure 7.3.1: Complementary error functions $\operatorname{erfc} x$ and $\operatorname{erfc}(10x)$, $-3 \le x \le 3$.

7.4 SYMMETRY

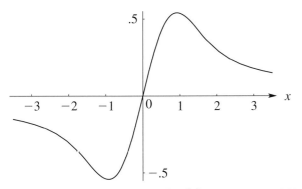

Figure 7.3.2: Dawson's integral $F(x)$, $-3.5 \leq x \leq 3.5$.

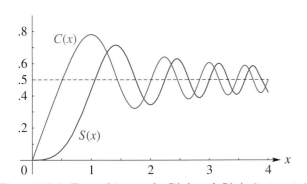

Figure 7.3.3: Fresnel integrals $C(x)$ and $S(x)$, $0 \leq x \leq 4$.

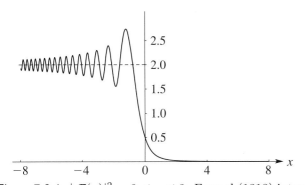

Figure 7.3.4: $|\mathcal{F}(x)|^2$, $-8 \leq x \leq 8$. Fresnel (1818) introduced the integral $\mathcal{F}(x)$ in his study of the interference pattern at the edge of a shadow. He observed that the intensity distribution is given by $|\mathcal{F}(x)|^2$.

7.3(ii) Complex Variable

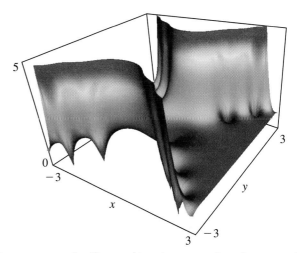

Figure 7.3.5: $|\operatorname{erf}(x+iy)|$, $-3 \leq x \leq 3$, $-3 \leq y \leq 3$. Compare §7.13(i).

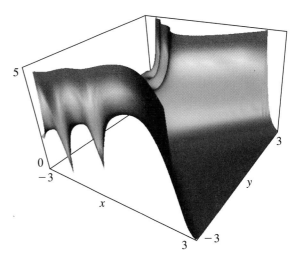

Figure 7.3.6: $|\operatorname{erfc}(x+iy)|$, $-3 \leq x \leq 3$, $-3 \leq y \leq 3$. Compare §§7.12(i) and 7.13(ii).

7.4 Symmetry

7.4.1 $$\operatorname{erf}(-z) = -\operatorname{erf}(z),$$

7.4.2 $$\operatorname{erfc}(-z) = 2 - \operatorname{erfc}(z),$$

7.4.3 $$w(-z) = 2e^{-z^2} - w(z).$$

7.4.4 $$F(-z) = -F(z).$$

7.4.5 $$C(-z) = -C(z), \quad S(-z) = -S(z),$$

7.4.6 $$C(iz) = iC(z), \quad S(iz) = -iS(z).$$

7.4.7
$$\operatorname{f}(iz) = (1/\sqrt{2})e^{\frac{1}{4}\pi i - \frac{1}{2}\pi i z^2} - i\operatorname{f}(z),$$
$$\operatorname{g}(iz) = (1/\sqrt{2})e^{-\frac{1}{4}\pi i - \frac{1}{2}\pi i z^2} + i\operatorname{g}(z).$$

7.4.8
$$\mathrm{f}(-z) = \sqrt{2}\cos\left(\tfrac{1}{4}\pi + \tfrac{1}{2}\pi z^2\right) - \mathrm{f}(z),$$
$$\mathrm{g}(-z) = \sqrt{2}\sin\left(\tfrac{1}{4}\pi + \tfrac{1}{2}\pi z^2\right) - \mathrm{g}(z).$$

7.5 Interrelations

7.5.1 $F(z) = \tfrac{1}{2}i\sqrt{\pi}\left(e^{-z^2} - w(z)\right) = -\tfrac{1}{2}i\sqrt{\pi}e^{-z^2}\operatorname{erf}(iz).$

7.5.2 $C(z) + iS(z) = \tfrac{1}{2}(1+i) - \mathcal{F}(z).$

7.5.3 $C(z) = \tfrac{1}{2} + \mathrm{f}(z)\sin\left(\tfrac{1}{2}\pi z^2\right) - \mathrm{g}(z)\cos\left(\tfrac{1}{2}\pi z^2\right),$

7.5.4 $S(z) = \tfrac{1}{2} - \mathrm{f}(z)\cos\left(\tfrac{1}{2}\pi z^2\right) - \mathrm{g}(z)\sin\left(\tfrac{1}{2}\pi z^2\right).$

7.5.5 $e^{-\tfrac{1}{2}\pi i z^2}\mathcal{F}(z) = \mathrm{g}(z) + i\,\mathrm{f}(z).$

7.5.6 $e^{\pm\tfrac{1}{2}\pi i z^2}(\mathrm{g}(z) \pm i\,\mathrm{f}(z)) = \tfrac{1}{2}(1 \pm i) - (C(z) \pm iS(z)).$

In (7.5.8)–(7.5.10)

7.5.7 $\zeta = \tfrac{1}{2}\sqrt{\pi}(1 \mp i)z,$

and either all upper signs or all lower signs are taken throughout.

7.5.8 $C(z) \pm iS(z) = \tfrac{1}{2}(1 \pm i)\operatorname{erf}\zeta.$

7.5.9 $C(z) \pm iS(z) = \tfrac{1}{2}(1 \pm i)\left(1 - e^{\pm\tfrac{1}{2}\pi i z^2}w(i\zeta)\right).$

7.5.10 $\mathrm{g}(z) \pm i\,\mathrm{f}(z) = \tfrac{1}{2}(1 \pm i)e^{\zeta^2}\operatorname{erfc}\zeta.$

7.5.11 $|\mathcal{F}(x)|^2 = \mathrm{f}^2(x) + \mathrm{g}^2(x), \qquad x \geq 0,$

7.5.12
$$|\mathcal{F}(x)|^2 = 2 + \mathrm{f}^2(-x) + \mathrm{g}^2(-x)$$
$$- 2\sqrt{2}\cos\left(\tfrac{1}{4}\pi + \tfrac{1}{2}\pi x^2\right)\mathrm{f}(-x)$$
$$- 2\sqrt{2}\cos\left(\tfrac{1}{4}\pi - \tfrac{1}{2}\pi x^2\right)\mathrm{g}(-x),$$
$$x \leq 0.$$

See Figure 7.3.4.

7.5.13 $G(x) = \sqrt{\pi}F(x) - \tfrac{1}{2}e^{-x^2}\operatorname{Ei}(x^2), \qquad x > 0.$

For $\operatorname{Ei}(x)$ see §6.2(i).

7.6 Series Expansions

7.6(i) Power Series

7.6.1 $\operatorname{erf} z = \dfrac{2}{\sqrt{\pi}}\sum_{n=0}^{\infty}\dfrac{(-1)^n z^{2n+1}}{n!(2n+1)},$

7.6.2 $\operatorname{erf} z = \dfrac{2}{\sqrt{\pi}}e^{-z^2}\sum_{n=0}^{\infty}\dfrac{2^n z^{2n+1}}{1\cdot 3\cdots(2n+1)},$

7.6.3 $w(z) = \sum_{n=0}^{\infty}\dfrac{(iz)^n}{\Gamma(\tfrac{1}{2}n+1)}.$

7.6.4 $C(z) = \sum_{n=0}^{\infty}\dfrac{(-1)^n(\tfrac{1}{2}\pi)^{2n}}{(2n)!(4n+1)}z^{4n+1},$

7.6.5
$$C(z) = \cos\left(\tfrac{1}{2}\pi z^2\right)\sum_{n=0}^{\infty}\dfrac{(-1)^n\pi^{2n}}{1\cdot 3\cdots(4n+1)}z^{4n+1}$$
$$+ \sin\left(\tfrac{1}{2}\pi z^2\right)\sum_{n=0}^{\infty}\dfrac{(-1)^n\pi^{2n+1}}{1\cdot 3\cdots(4n+3)}z^{4n+3}.$$

7.6.6 $S(z) = \sum_{n=0}^{\infty}\dfrac{(-1)^n(\tfrac{1}{2}\pi)^{2n+1}}{(2n+1)!(4n+3)}z^{4n+3},$

7.6.7
$$S(z) = -\cos\left(\tfrac{1}{2}\pi z^2\right)\sum_{n=0}^{\infty}\dfrac{(-1)^n\pi^{2n+1}}{1\cdot 3\cdots(4n+3)}z^{4n+3}$$
$$+ \sin\left(\tfrac{1}{2}\pi z^2\right)\sum_{n=0}^{\infty}\dfrac{(-1)^n\pi^{2n}}{1\cdot 3\cdots(4n+1)}z^{4n+1}.$$

The series in this subsection and in §7.6(ii) converge for all finite values of $|z|$.

7.6(ii) Expansions in Series of Spherical Bessel Functions

For the notation see §§10.47(ii) and 18.3.

7.6.8 $\operatorname{erf} z = \dfrac{2z}{\sqrt{\pi}}\sum_{n=0}^{\infty}(-1)^n\left(\mathrm{i}^{(1)}_{2n}(z^2) - \mathrm{i}^{(1)}_{2n+1}(z^2)\right),$

7.6.9 $\operatorname{erf}(az) = \dfrac{2z}{\sqrt{\pi}}e^{(\tfrac{1}{2}-a^2)z^2}\sum_{n=0}^{\infty}T_{2n+1}(a)\,\mathrm{i}^{(1)}_n\left(\tfrac{1}{2}z^2\right),$
$$-1 \leq a \leq 1.$$

7.6.10 $C(z) = z\sum_{n=0}^{\infty}\mathrm{j}_{2n}\left(\tfrac{1}{2}\pi z^2\right),$

7.6.11 $S(z) = z\sum_{n=0}^{\infty}\mathrm{j}_{2n+1}\left(\tfrac{1}{2}\pi z^2\right).$

For further results see Luke (1969b, pp. 57–58).

7.7 Integral Representations

7.7(i) Error Functions and Dawson's Integral

Integrals of the type $\int e^{-z^2}R(z)\,dz$, where $R(z)$ is an arbitrary rational function, can be written in closed form in terms of the error functions and elementary functions.

7.7.1 $\operatorname{erfc} z = \dfrac{2}{\pi}e^{-z^2}\int_0^{\infty}\dfrac{e^{-z^2 t^2}}{t^2+1}\,dt, \quad |\operatorname{ph} z| \leq \tfrac{1}{4}\pi,$

7.7.2 $w(z) = \dfrac{1}{\pi i}\int_{-\infty}^{\infty}\dfrac{e^{-t^2}\,dt}{t-z} = \dfrac{2z}{\pi i}\int_0^{\infty}\dfrac{e^{-t^2}\,dt}{t^2-z^2}, \quad \Im z > 0.$

7.7.3 $\int_0^{\infty}e^{-at^2+2izt}\,dt = \tfrac{1}{2}\sqrt{\dfrac{\pi}{a}}e^{-z^2/a} + \dfrac{i}{\sqrt{a}}F\left(\dfrac{z}{\sqrt{a}}\right),$
$$\Re a > 0.$$

7.7.4 $\int_0^{\infty}\dfrac{e^{-at}}{\sqrt{t+z^2}}\,dt = \sqrt{\dfrac{\pi}{a}}e^{az^2}\operatorname{erfc}(\sqrt{az}), \quad \Re a > 0,\ \Re z > 0.$

7.8 Inequalities

7.7.5 $\quad \int_0^1 \frac{e^{-at^2}}{t^2+1} dt = \frac{\pi}{4} e^a \left(1 - (\operatorname{erf} \sqrt{a})^2\right), \quad \Re a > 0.$

7.7.6
$\int_x^\infty e^{-(at^2+2bt+c)} dt$
$= \frac{1}{2}\sqrt{\frac{\pi}{a}} e^{(b^2-ac)/a} \operatorname{erfc}\left(\sqrt{a}x + \frac{b}{\sqrt{a}}\right), \quad \Re a > 0.$

7.7.7
$\int_x^\infty e^{-a^2t^2 - (b^2/t^2)} dt = \frac{\sqrt{\pi}}{4a} \left(e^{2ab} \operatorname{erfc}(ax + (b/x)) + e^{-2ab} \operatorname{erfc}(ax - (b/x))\right),$
$x > 0, |\operatorname{ph} a| < \tfrac{1}{4}\pi.$

7.7.8 $\quad \int_0^\infty e^{-a^2t^2 - (b^2/t^2)} dt = \frac{\sqrt{\pi}}{2a} e^{-2ab},$
$|\operatorname{ph} a| < \tfrac{1}{4}\pi, |\operatorname{ph} b| < \tfrac{1}{4}\pi.$

7.7.9 $\quad \int_0^x \operatorname{erf} t \, dt = x \operatorname{erf} x + \frac{1}{\sqrt{\pi}} \left(e^{-x^2} - 1\right).$

7.7(ii) Auxiliary Functions

7.7.10 $\quad \operatorname{f}(z) = \frac{1}{\pi\sqrt{2}} \int_0^\infty \frac{e^{-\pi z^2 t/2}}{\sqrt{t}(t^2+1)} dt, \quad |\operatorname{ph} z| \leq \tfrac{1}{4}\pi,$

7.7.11 $\quad \operatorname{g}(z) = \frac{1}{\pi\sqrt{2}} \int_0^\infty \frac{\sqrt{t} e^{-\pi z^2 t/2}}{t^2+1} dt, \quad |\operatorname{ph} z| \leq \tfrac{1}{4}\pi,$

7.7.12 $\quad \operatorname{g}(z) + i\operatorname{f}(z) = e^{-\pi i z^2/2} \int_z^\infty e^{\pi i t^2/2} dt.$

Mellin–Barnes Integrals

7.7.13
$\operatorname{f}(z) = \frac{(2\pi)^{-3/2}}{2\pi i} \int_{c-i\infty}^{c+i\infty} \zeta^{-s} \Gamma(s) \Gamma\left(s + \tfrac{1}{2}\right)$
$\times \Gamma\left(s + \tfrac{3}{4}\right) \Gamma\left(\tfrac{1}{4} - s\right) ds,$

7.7.14
$\operatorname{g}(z) = \frac{(2\pi)^{-3/2}}{2\pi i} \int_{c-i\infty}^{c+i\infty} \zeta^{-s} \Gamma(s) \Gamma\left(s + \tfrac{1}{2}\right)$
$\times \Gamma\left(s + \tfrac{1}{4}\right) \Gamma\left(\tfrac{3}{4} - s\right) ds.$

In (7.7.13) and (7.7.14) the integration paths are straight lines, $\zeta = \tfrac{1}{16}\pi^2 z^4$, and c is a constant such that $0 < c < \tfrac{1}{4}$ in (7.7.13), and $0 < c < \tfrac{3}{4}$ in (7.7.14).

7.7.15 $\quad \int_0^\infty e^{-at} \cos(t^2) dt = \sqrt{\frac{\pi}{2}} \operatorname{f}\left(\frac{a}{\sqrt{2\pi}}\right), \quad \Re a > 0,$

7.7.16 $\quad \int_0^\infty e^{-at} \sin(t^2) dt = \sqrt{\frac{\pi}{2}} \operatorname{g}\left(\frac{a}{\sqrt{2\pi}}\right), \quad \Re a > 0.$

7.7(iii) Compendia

For other integral representations see Erdélyi et al. (1954a, vol. 1, pp. 265–267, 270), Ng and Geller (1969), Oberhettinger (1974, pp. 246–248), and Oberhettinger and Badii (1973, pp. 371–377).

7.8 Inequalities

Let $\operatorname{M}(x)$ denote *Mills' ratio*:

7.8.1 $\quad \operatorname{M}(x) = \frac{\int_x^\infty e^{-t^2} dt}{e^{-x^2}} = e^{x^2} \int_x^\infty e^{-t^2} dt.$

(Other notations are often used.) Then

7.8.2 $\quad \frac{1}{x + \sqrt{x^2+2}} < \operatorname{M}(x) \leq \frac{1}{x + \sqrt{x^2+(4/\pi)}}, \quad x \geq 0,$

7.8.3 $\quad \frac{\sqrt{\pi}}{2\sqrt{\pi}x + 2} \leq \operatorname{M}(x) < \frac{1}{x+1}, \quad x \geq 0,$

7.8.4 $\quad \operatorname{M}(x) < \frac{2}{3x + \sqrt{x^2+4}}, \quad x > -\tfrac{1}{2}\sqrt{2},$

7.8.5
$\frac{x^2}{2x^2+1} \leq \frac{x^2(2x^2+5)}{4x^4+12x^2+3} \leq x \operatorname{M}(x)$
$< \frac{2x^4+9x^2+4}{4x^4+20x^2+15} < \frac{x^2+1}{2x^2+3},$
$x \geq 0.$

Next,

7.8.6 $\quad \int_0^x e^{at^2} dt < \frac{1}{3ax}\left(2e^{ax^2} + ax^2 - 2\right), \quad a, x > 0.$

7.8.7 $\quad \int_0^x e^{t^2} dt < \frac{e^{x^2}-1}{x}, \quad x > 0.$

7.9 Continued Fractions

7.9.1
$\sqrt{\pi} e^{z^2} \operatorname{erfc} z = \frac{z}{z^2+} \frac{\tfrac{1}{2}}{1+} \frac{1}{z^2+} \frac{\tfrac{3}{2}}{1+} \frac{2}{z^2+} \cdots,$
$\Re z > 0,$

7.9.2
$\sqrt{\pi} e^{z^2} \operatorname{erfc} z = \frac{2z}{2z^2+1-} \frac{1 \cdot 2}{2z^2+5-} \frac{3 \cdot 4}{2z^2+9-} \cdots,$
$\Re z > 0,$

7.9.3
$w(z) = \frac{i}{\sqrt{\pi}} \frac{1}{z-} \frac{\tfrac{1}{2}}{z-} \frac{1}{z-} \frac{\tfrac{3}{2}}{z-} \frac{2}{z-} \cdots, \quad \Im z > 0.$

See also Cuyt et al. (2008, pp. 255–260, 263–267, 270–273).

7.10 Derivatives

7.10.1
$\frac{d^{n+1} \operatorname{erf} z}{dz^{n+1}} = (-1)^n \frac{2}{\sqrt{\pi}} H_n(z) e^{-z^2}, \quad n = 0, 1, 2, \ldots.$

For the Hermite polynomial $H_n(z)$ see §18.3.

7.10.2 $\quad w'(z) = -2z\, w(z) + (2i/\sqrt{\pi}),$

7.10.3
$$w^{(n+2)}(z) + 2z\,w^{(n+1)}(z) + 2(n+1)\,w^{(n)}(z) = 0,$$
$$n = 0, 1, 2, \ldots.$$

7.10.4 $\quad \dfrac{d\mathrm{f}(z)}{dz} = -\pi z\,\mathrm{g}(z), \quad \dfrac{d\mathrm{g}(z)}{dz} = \pi z\,\mathrm{f}(z) - 1.$

7.11 Relations to Other Functions

Incomplete Gamma Functions and Generalized Exponential Integral

For the notation see §§8.2(i) and 8.19(i).

7.11.1 $\quad \operatorname{erf} z = \dfrac{1}{\sqrt{\pi}}\,\gamma\!\left(\tfrac{1}{2}, z^2\right),$

7.11.2 $\quad \operatorname{erfc} z = \dfrac{1}{\sqrt{\pi}}\,\Gamma\!\left(\tfrac{1}{2}, z^2\right),$

7.11.3 $\quad \operatorname{erfc} z = \dfrac{z}{\sqrt{\pi}}\,E_{\frac{1}{2}}\!\left(z^2\right).$

Confluent Hypergeometric Functions

For the notation see §13.2(i).

7.11.4 $\quad \operatorname{erf} z = \dfrac{2z}{\sqrt{\pi}}\,M\!\left(\tfrac{1}{2}, \tfrac{3}{2}, -z^2\right) = \dfrac{2z}{\sqrt{\pi}}\,e^{-z^2} M\!\left(1, \tfrac{3}{2}, z^2\right),$

7.11.5
$$\operatorname{erfc} z = \dfrac{1}{\sqrt{\pi}}\,e^{-z^2}\,U\!\left(\tfrac{1}{2}, \tfrac{1}{2}, z^2\right) = \dfrac{z}{\sqrt{\pi}}\,e^{-z^2}\,U\!\left(1, \tfrac{3}{2}, z^2\right).$$

7.11.6
$$C(z) + i\,S(z) = z\,M\!\left(\tfrac{1}{2}, \tfrac{3}{2}, \tfrac{1}{2}\pi i z^2\right)$$
$$= z e^{\pi i z^2/2}\,M\!\left(1, \tfrac{3}{2}, -\tfrac{1}{2}\pi i z^2\right).$$

Generalized Hypergeometric Functions

For the notation see §§16.2(i) and 16.2(ii).

7.11.7 $\quad C(z) = z\,_1F_2\!\left(\tfrac{1}{4};\tfrac{5}{4},\tfrac{1}{2};-\tfrac{1}{16}\pi^2 z^4\right),$

7.11.8 $\quad S(z) = \tfrac{1}{6}\pi z^3\,_1F_2\!\left(\tfrac{3}{4};\tfrac{7}{4},\tfrac{3}{2};-\tfrac{1}{16}\pi^2 z^4\right).$

7.12 Asymptotic Expansions

7.12(i) Complementary Error Function

As $z \to \infty$

7.12.1
$$\operatorname{erfc} z \sim \dfrac{e^{-z^2}}{\sqrt{\pi}\,z} \sum_{m=0}^{\infty} (-1)^m \dfrac{1\cdot 3\cdot 5 \cdots (2m-1)}{(2z^2)^m},$$
$$\operatorname{erfc}(-z) \sim 2 - \dfrac{e^{-z^2}}{\sqrt{\pi}\,z} \sum_{m=0}^{\infty} (-1)^m \dfrac{1\cdot 3\cdot 5 \cdots (2m-1)}{(2z^2)^m},$$

both expansions being valid when $|\operatorname{ph} z| \le \tfrac{3}{4}\pi - \delta$ $(<\tfrac{3}{4}\pi)$.

When $|\operatorname{ph} z| \le \tfrac{1}{4}\pi$ the remainder terms are bounded in magnitude by the first neglected terms, and have the same sign as these terms when $\operatorname{ph} z = 0$. When $\tfrac{1}{4}\pi \le |\operatorname{ph} z| < \tfrac{1}{2}\pi$ the remainder terms are bounded in magnitude by $\csc(2|\operatorname{ph} z|)$ times the first neglected terms. For these and other error bounds see Olver (1997b, pp. 109–112), with $\alpha = \tfrac{1}{2}$ and z replaced by z^2; compare (7.11.2).

For re-expansions of the remainder terms leading to larger sectors of validity, exponential improvement, and a smooth interpretation of the Stokes phenomenon, see §§2.11(ii)–2.11(iv) and use (7.11.3). (Note that some of these re-expansions themselves involve the complementary error function.)

7.12(ii) Fresnel Integrals

The asymptotic expansions of $C(z)$ and $S(z)$ are given by (7.5.3), (7.5.4), and

7.12.2 $\quad \mathrm{f}(z) \sim \dfrac{1}{\pi z} \sum_{m=0}^{\infty} (-1)^m \dfrac{1\cdot 3\cdot 5 \cdots (4m-1)}{(\pi z^2)^{2m}},$

7.12.3 $\quad \mathrm{g}(z) \sim \dfrac{1}{\pi^2 z^3} \sum_{m=0}^{\infty} (-1)^m \dfrac{1\cdot 3\cdot 5 \cdots (4m+1)}{(\pi z^2)^{2m}},$

as $z \to \infty$ in $|\operatorname{ph} z| \le \tfrac{1}{2}\pi - \delta (<\tfrac{1}{2}\pi)$. The remainder terms are given by

7.12.4 $\quad \mathrm{f}(z) = \dfrac{1}{\pi z} \sum_{m=0}^{n-1} (-1)^m \dfrac{1\cdot 3\cdots (4m-1)}{(\pi z^2)^{2m}} + R_n^{(\mathrm{f})}(z),$

7.12.5
$$\mathrm{g}(z) = \dfrac{1}{\pi^2 z^3} \sum_{m=0}^{n-1} (-1)^m \dfrac{1\cdot 3 \cdots (4m+1)}{(\pi z^2)^{2m}} + R_n^{(\mathrm{g})}(z),$$

where, for $n = 0, 1, 2, \ldots$ and $|\operatorname{ph} z| < \tfrac{1}{4}\pi$,

7.12.6 $\quad R_n^{(\mathrm{f})}(z) = \dfrac{(-1)^n}{\pi\sqrt{2}} \int_0^{\infty} \dfrac{e^{-\pi z^2 t/2}\, t^{2n-(1/2)}}{t^2 + 1}\, dt,$

7.12.7 $\quad R_n^{(\mathrm{g})}(z) = \dfrac{(-1)^n}{\pi\sqrt{2}} \int_0^{\infty} \dfrac{e^{-\pi z^2 t/2}\, t^{2n+(1/2)}}{t^2 + 1}\, dt.$

When $|\operatorname{ph} z| \le \tfrac{1}{8}\pi$, $R_n^{(\mathrm{f})}(z)$ and $R_n^{(\mathrm{g})}(z)$ are bounded in magnitude by the first neglected terms in (7.12.2) and (7.12.3), respectively, and have the same signs as these terms when $\operatorname{ph} z = 0$. They are bounded by $|\csc(4\operatorname{ph} z)|$ times the first neglected terms when $\tfrac{1}{8}\pi \le |\operatorname{ph} z| < \tfrac{1}{4}\pi$.

For other phase ranges use (7.4.7) and (7.4.8). For exponentially-improved expansions use (7.5.7), (7.5.10), and §7.12(i).

7.12(iii) Goodwin–Staton Integral

See Olver (1997b, p. 115) for an expansion of $G(z)$ with bounds for the remainder for real and complex values of z.

7.13 Zeros

7.13(i) Zeros of erf z

erf z has a simple zero at $z = 0$, and in the first quadrant of $\mathbb{C}$ there is an infinite set of zeros $z_n = x_n + iy_n$, $n = 1, 2, 3, \ldots$, arranged in order of increasing absolute value. The other zeros of erf z are $-z_n$, $\overline{z}_n$, $-\overline{z}_n$.

Table 7.13.1 gives 10D values of the first five x_n and y_n. For graphical illustration see Figure 7.3.5.

Table 7.13.1: Zeros $x_n + iy_n$ of erf z.

n	x_n	y_n
1	1.45061 61632	1.88094 30002
2	2.24465 92738	2.61657 51407
3	2.83974 10469	3.17562 80996
4	3.33546 07354	3.64617 43764
5	3.76900 55670	4.06069 72339

As $n \to \infty$

7.13.1
$$x_n \sim \lambda - \tfrac{1}{4}\mu\lambda^{-1} + \tfrac{1}{16}(1 - \mu + \tfrac{1}{2}\mu^2)\lambda^{-3} - \cdots,$$
$$y_n \sim \lambda + \tfrac{1}{4}\mu\lambda^{-1} + \tfrac{1}{16}(1 - \mu + \tfrac{1}{2}\mu^2)\lambda^{-3} + \cdots,$$
where

7.13.2 $\quad \lambda = \sqrt{(n - \tfrac{1}{8})\pi}, \quad \mu = \ln\left(\lambda\sqrt{2\pi}\right).$

7.13(ii) Zeros of erfc z

In the second quadrant of $\mathbb{C}$, erfc z has an infinite set of zeros $z_n = x_n + iy_n$, $n = 1, 2, 3, \ldots$, arranged in order of increasing absolute value. The other zeros of erfc z are $\overline{z}_n$. The zeros of $w(z)$ are iz_n and $i\overline{z}_n$.

Table 7.13.2 gives 10D values of the first five x_n and y_n. For graphical illustration see Figure 7.3.6.

Table 7.13.2: Zeros $x_n + iy_n$ of erfc z.

n	x_n	y_n
1	$-1.35481\ 01281$	1.99146 68428
2	$-2.17704\ 49061$	2.69114 90243
3	$-2.78438\ 76132$	3.23533 08684
4	$-3.28741\ 07894$	3.69730 97025
5	$-3.72594\ 87194$	4.10610 72847

As $n \to \infty$

7.13.3
$$x_n \sim -\lambda + \tfrac{1}{4}\mu\lambda^{-1} - \tfrac{1}{16}(1 - \mu + \tfrac{1}{2}\mu^2)\lambda^{-3} + \cdots,$$
$$y_n \sim \lambda + \tfrac{1}{4}\mu\lambda^{-1} + \tfrac{1}{16}(1 - \mu + \tfrac{1}{2}\mu^2)\lambda^{-3} + \cdots,$$
where

7.13.4 $\quad \lambda = \sqrt{(n - \tfrac{1}{8})\pi}, \quad \mu = \ln\left(2\lambda\sqrt{2\pi}\right).$

7.13(iii) Zeros of the Fresnel Integrals

At $z = 0$, $C(z)$ has a simple zero and $S(z)$ has a triple zero. In the first quadrant of $\mathbb{C}$ $C(z)$ has an infinite set of zeros $z_n = x_n + iy_n$, $n = 1, 2, 3, \ldots$, arranged in order of increasing absolute value. Similarly for $S(z)$. Let z_n be a zero of one of the Fresnel integrals. Then $-z_n$, $\overline{z}_n$, $-\overline{z}_n$, iz_n, $-iz_n$, $i\overline{z}_n$, $-i\overline{z}_n$ are also zeros of the same integral.

Tables 7.13.3 and 7.13.4 give 10D values of the first five x_n and y_n of $C(z)$ and $S(z)$, respectively.

Table 7.13.3: Complex zeros $x_n + iy_n$ of $C(z)$.

n	x_n	y_n
1	1.74366 74862	0.30573 50636
2	2.65145 95973	0.25290 39555
3	3.32035 93363	0.22395 34581
4	3.87573 44884	0.20474 74706
5	4.36106 35170	0.19066 97324

As $n \to \infty$ the x_n and y_n corresponding to the zeros of $C(z)$ satisfy

7.13.5 $\quad x_n \sim \lambda + \dfrac{\alpha(\alpha\pi - 4)}{8\pi\lambda^3} + \cdots, \quad y_n \sim \dfrac{\alpha}{2\lambda} + \cdots,$

with

7.13.6 $\quad \lambda = \sqrt{4n - 1}, \quad \alpha = (2/\pi)\ln(\pi\lambda).$

Table 7.13.4: Complex zeros $x_n + iy_n$ of $S(z)$.

n	x_n	y_n
1	2.00925 70118	0.28854 78973
2	2.83347 72325	0.24428 52408
3	3.46753 30835	0.21849 26805
4	4.00257 82433	0.20085 10251
5	4.47418 92952	0.18768 85891

As $n \to \infty$ the x_n and y_n corresponding to the zeros of $S(z)$ satisfy (7.13.5) with

7.13.7 $\quad \lambda = 2\sqrt{n}, \quad \alpha = (2/\pi)\ln(\pi\lambda).$

7.13(iv) Zeros of $\mathcal{F}(z)$

In consequence of (7.5.5) and (7.5.10), zeros of $\mathcal{F}(z)$ are related to zeros of erfc z. Thus if z_n is a zero of erfc z (§7.13(ii)), then $(1 + i)z_n/\sqrt{\pi}$ is a zero of $\mathcal{F}(z)$.

For an asymptotic expansion of the zeros of $\int_0^z \exp(\tfrac{1}{2}\pi i t^2)\, dt\ (= \mathcal{F}(0) - \mathcal{F}(z) = C(z) + iS(z))$ see Tuẑilin (1971).

7.14 Integrals

7.14(i) Error Functions

Fourier Transform

7.14.1
$$\int_0^\infty e^{2iat} \operatorname{erfc}(bt)\,dt = \frac{1}{a\sqrt{\pi}} F\left(\frac{a}{b}\right) + \frac{i}{2a}\left(1 - e^{-(a/b)^2}\right),$$
$$a \in \mathbb{C},\ |\operatorname{ph} b| < \tfrac{1}{4}\pi.$$

When $a = 0$ the limit is taken.

Laplace Transforms

7.14.2
$$\int_0^\infty e^{-at} \operatorname{erf}(bt)\,dt = \frac{1}{a} e^{a^2/(4b^2)} \operatorname{erfc}\left(\frac{a}{2b}\right),$$
$$\Re a > 0,\ |\operatorname{ph} b| < \tfrac{1}{4}\pi,$$

7.14.3
$$\int_0^\infty e^{-at} \operatorname{erf} \sqrt{bt}\,dt = \frac{1}{a}\sqrt{\frac{b}{a+b}},\quad \Re a > 0,\ \Re b > 0,$$

7.14.4
$$\int_0^\infty e^{(a-b)t} \operatorname{erfc}\left(\sqrt{at} + \sqrt{\frac{c}{t}}\right) dt$$
$$= \frac{e^{-2(\sqrt{ac}+\sqrt{bc})}}{\sqrt{b}(\sqrt{a}+\sqrt{b})},\quad |\operatorname{ph} a| < \tfrac{1}{2}\pi,\ \Re b > 0,\ \Re c \geq 0.$$

7.14(ii) Fresnel Integrals

Laplace Transforms

7.14.5 $\quad\displaystyle\int_0^\infty e^{-at} C(t)\,dt = \frac{1}{a}\operatorname{f}\left(\frac{a}{\pi}\right),\quad \Re a > 0,$

7.14.6 $\quad\displaystyle\int_0^\infty e^{-at} S(t)\,dt = \frac{1}{a}\operatorname{g}\left(\frac{a}{\pi}\right),\quad \Re a > 0,$

7.14.7
$$\int_0^\infty e^{-at} C\left(\sqrt{\frac{2t}{\pi}}\right) dt = \frac{(\sqrt{a^2+1}+a)^{\frac{1}{2}}}{2a\sqrt{a^2+1}},\quad \Re a > 0,$$

7.14.8
$$\int_0^\infty e^{-at} S\left(\sqrt{\frac{2t}{\pi}}\right) dt = \frac{(\sqrt{a^2+1}-a)^{\frac{1}{2}}}{2a\sqrt{a^2+1}},\quad \Re a > 0.$$

7.14(iii) Compendia

For collections of integrals see Apelblat (1983, pp. 131–146), Erdélyi et al. (1954a, vol. 1, pp. 40, 96, 176–177), Geller and Ng (1971), Gradshteyn and Ryzhik (2000, §§5.4 and 6.28–6.32), Marichev (1983, pp. 184–189), Ng and Geller (1969), Oberhettinger (1974, pp. 138–139, 142–143), Oberhettinger (1990, pp. 48–52, 155–158), Oberhettinger and Badii (1973, pp. 171–172, 179–181), Prudnikov et al. (1986b, vol. 2, pp. 30–36, 93–143), Prudnikov et al. (1992a, §§3.7–3.8), and Prudnikov et al. (1992b, §§3.7–3.8). In a series of ten papers Hadži (1968, 1969, 1970, 1972, 1973, 1975a,b, 1976a,b, 1978) gives many integrals containing error functions and Fresnel integrals, also in combination with the hypergeometric function, confluent hypergeometric functions, and generalized hypergeometric functions.

7.15 Sums

For sums involving the error function see Hansen (1975, p. 423) and Prudnikov et al. (1986b, vol. 2, pp. 650–651).

7.16 Generalized Error Functions

Generalizations of the error function and Dawson's integral are $\int_0^x e^{-t^p}\,dt$ and $\int_0^x e^{t^p}\,dt$. These functions can be expressed in terms of the incomplete gamma function $\gamma(a,z)$ (§8.2(i)) by change of integration variable.

7.17 Inverse Error Functions

7.17(i) Notation

The inverses of the functions $x = \operatorname{erf} y$, $x = \operatorname{erfc} y$, $y \in \mathbb{R}$, are denoted by

7.17.1 $\qquad y = \operatorname{inverf} x,\quad y = \operatorname{inverfc} x,$

respectively.

7.17(ii) Power Series

With $t = \tfrac{1}{2}\sqrt{\pi}x$,

7.17.2 $\quad \operatorname{inverf} x = t + \tfrac{1}{3}t^3 + \tfrac{7}{30}t^5 + \tfrac{127}{630}t^7 + \cdots,\quad |x|<1.$

For 25S values of the first 200 coefficients see Strecok (1968).

7.17(iii) Asymptotic Expansion of inverfc x for Small x

As $x \to 0$

7.17.3 $\quad \operatorname{inverfc} x \sim u^{-1/2} + a_2 u^{3/2} + a_3 u^{5/2} + a_4 u^{7/2} + \cdots,$

where

7.17.4
$$a_2 = \tfrac{1}{8}v,\quad a_3 = -\tfrac{1}{32}(v^2 + 6v - 6),$$
$$a_4 = \tfrac{1}{384}(4v^3 + 27v^2 + 108v - 300),$$

7.17.5 $\qquad u = -2/\ln(\pi x^2 \ln(1/x)),$

and

7.17.6 $\qquad v = \ln(\ln(1/x)) - 2 + \ln\pi.$

7.18 Repeated Integrals of the Complementary Error Function

7.18(i) Definition

7.18.1 $\quad i^{-1}\operatorname{erfc}(z) = \dfrac{2}{\sqrt{\pi}} e^{-z^2}, \quad i^0\operatorname{erfc}(z) = \operatorname{erfc} z,$

and for $n = 0, 1, 2, \ldots,$

7.18.2
$$i^n\operatorname{erfc}(z) = \int_z^\infty i^{n-1}\operatorname{erfc}(t)\,dt = \frac{2}{\sqrt{\pi}} \int_z^\infty \frac{(t-z)^n}{n!} e^{-t^2}\,dt.$$

7.18(ii) Graphics

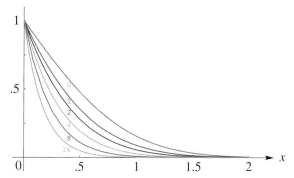

Figure 7.18.1: Repeated integrals of the scaled complementary error function $2^n \Gamma\left(\tfrac{1}{2}n+1\right) i^n\operatorname{erfc}(x)$, $n = 0, 1, 2, 4, 8, 16$.

7.18(iii) Properties

7.18.3 $\quad \dfrac{d}{dz} i^n\operatorname{erfc}(z) = -i^{n-1}\operatorname{erfc}(z), \quad n = 0, 1, 2, \ldots,$

7.18.4 $\quad \dfrac{d^n}{dz^n}\left(e^{z^2}\operatorname{erfc} z\right) = (-1)^n 2^n n! e^{z^2} i^n\operatorname{erfc}(z),$
$\hspace{8em} n = 0, 1, 2, \ldots.$

7.18.5
$$\frac{d^2 W}{dz^2} + 2z \frac{dW}{dz} - 2nW = 0,$$
$$W(z) = A\, i^n\operatorname{erfc}(z) + B\, i^n\operatorname{erfc}(-z),$$

where $n = 1, 2, 3, \ldots,$ and A, B are arbitrary constants.

7.18.6 $\quad i^n\operatorname{erfc}(z) = \sum_{k=0}^\infty \dfrac{(-1)^k z^k}{2^{n-k} k!\, \Gamma\left(1 + \tfrac{1}{2}(n-k)\right)}.$

7.18.7 $\quad i^n\operatorname{erfc}(z) = -\dfrac{z}{n} i^{n-1}\operatorname{erfc}(z) + \dfrac{1}{2n} i^{n-2}\operatorname{erfc}(z),$
$\hspace{10em} n = 1, 2, 3, \ldots.$

7.18(iv) Relations to Other Functions

For the notation see §§18.3, 13.2(i), and 12.2.

Hermite Polynomials

7.18.8 $\quad (-1)^n i^n\operatorname{erfc}(z) + i^n\operatorname{erfc}(-z) = \dfrac{i^{-n}}{2^{n-1} n!} H_n(iz).$

Confluent Hypergeometric Functions

7.18.9
$$i^n\operatorname{erfc}(z) = e^{-z^2}\left(\frac{1}{2^n\,\Gamma\left(\tfrac{1}{2}n+1\right)} M\left(\tfrac{1}{2}n + \tfrac{1}{2}, \tfrac{1}{2}, z^2\right)\right.$$
$$\left. -\frac{z}{2^{n-1}\,\Gamma\left(\tfrac{1}{2}n+\tfrac{1}{2}\right)} M\left(\tfrac{1}{2}n+1, \tfrac{3}{2}, z^2\right)\right),$$

7.18.10 $\quad i^n\operatorname{erfc}(z) = \dfrac{e^{-z^2}}{2^n\sqrt{\pi}} U\left(\tfrac{1}{2}n+\tfrac{1}{2}, \tfrac{1}{2}, z^2\right).$

Parabolic Cylinder Functions

7.18.11 $\quad i^n\operatorname{erfc}(z) = \dfrac{e^{-z^2/2}}{\sqrt{2^{n-1}\pi}} U\left(n+\tfrac{1}{2}, z\sqrt{2}\right).$

Probability Functions

7.18.12 $\quad i^n\operatorname{erfc}(z) = \dfrac{1}{\sqrt{2^{n-1}\pi}} Hh_n\left(\sqrt{2}z\right).$

See Jeffreys and Jeffreys (1956, §§23.081–23.09).

7.18(v) Continued Fraction

7.18.13
$$\frac{i^n\operatorname{erfc}(z)}{i^{n-1}\operatorname{erfc}(z)} = \frac{1/2}{z+} \frac{(n+1)/2}{z+} \frac{(n+2)/2}{z+}\cdots, \quad \Re z > 0.$$

See also Cuyt et al. (2008, p. 269).

7.18(vi) Asymptotic Expansion

7.18.14 $\quad i^n\operatorname{erfc}(z) \sim \dfrac{2}{\sqrt{\pi}} \dfrac{e^{-z^2}}{(2z)^{n+1}} \sum_{m=0}^\infty \dfrac{(-1)^m (2m+n)!}{n!\, m!\, (2z)^{2m}},$
$\hspace{6em} z \to \infty,\ |\operatorname{ph} z| \leq \tfrac{3}{4}\pi - \delta (< \tfrac{3}{4}\pi).$

7.19 Voigt Functions

7.19(i) Definitions

For $x \in \mathbb{R}$ and $t > 0$,

7.19.1 $\quad \mathsf{U}(x, t) = \dfrac{1}{\sqrt{4\pi t}} \displaystyle\int_{-\infty}^\infty \dfrac{e^{-(x-y)^2/(4t)}}{1+y^2}\,dy,$

7.19.2 $\quad \mathsf{V}(x, t) = \dfrac{1}{\sqrt{4\pi t}} \displaystyle\int_{-\infty}^\infty \dfrac{y e^{-(x-y)^2/(4t)}}{1+y^2}\,dy.$

7.19.3
$$\mathsf{U}(x,t) + i\mathsf{V}(x,t) = \sqrt{\dfrac{\pi}{4t}} e^{z^2} \operatorname{erfc} z, \quad z = (1-ix)/(2\sqrt{t}).$$

7.19.4
$$H(a,u) = \dfrac{a}{\pi} \int_{-\infty}^\infty \dfrac{e^{-t^2}\,dt}{(u-t)^2 + a^2} = \dfrac{1}{a\sqrt{\pi}} \mathsf{U}\left(\dfrac{u}{a}, \dfrac{1}{4a^2}\right).$$

$H(a, u)$ is sometimes called the *line broadening function*; see, for example, Finn and Mugglestone (1965).

7.19(ii) Graphics

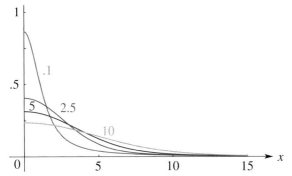

Figure 7.19.1: Voigt function $\mathsf{U}(x,t)$, $t = 0.1, 2.5, 5, 10$.

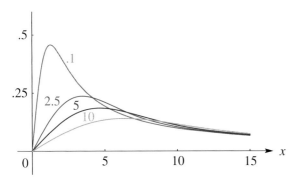

Figure 7.19.2: Voigt function $\mathsf{V}(x,t)$, $t = 0.1, 2.5, 5, 10$.

7.19(iii) Properties

7.19.5 $\quad \lim_{t \to 0} \mathsf{U}(x,t) = \dfrac{1}{1+x^2}, \quad \lim_{t \to 0} \mathsf{V}(x,t) = \dfrac{x}{1+x^2}.$

7.19.6 $\quad \mathsf{U}(-x,t) = \mathsf{U}(x,t), \quad \mathsf{V}(-x,t) = -\mathsf{V}(x,t).$

7.19.7 $\quad 0 < \mathsf{U}(x,t) \leq 1, \quad -1 \leq \mathsf{V}(x,t) \leq 1.$

7.19.8 $\quad \mathsf{V}(x,t) = x\,\mathsf{U}(x,t) + 2t\dfrac{\partial \mathsf{U}(x,t)}{\partial x},$

7.19.9 $\quad \mathsf{U}(x,t) = 1 - x\,\mathsf{V}(x,t) - 2t\dfrac{\partial \mathsf{V}(x,t)}{\partial x}.$

7.19(iv) Other Integral Representations

7.19.10 $\quad \mathsf{U}\!\left(\dfrac{u}{a}, \dfrac{1}{4a^2}\right) = a \displaystyle\int_0^\infty e^{-at-\frac{1}{4}t^2} \cos(ut)\,dt,$

7.19.11 $\quad \mathsf{V}\!\left(\dfrac{u}{a}, \dfrac{1}{4a^2}\right) = a \displaystyle\int_0^\infty e^{-at-\frac{1}{4}t^2} \sin(ut)\,dt.$

Applications

7.20 Mathematical Applications

7.20(i) Asymptotics

For applications of the complementary error function in uniform asymptotic approximations of integrals—saddle point coalescing with a pole or saddle point coalescing with an endpoint—see Wong (1989, Chapter 7), Olver (1997b, Chapter 9), and van der Waerden (1951).

The complementary error function also plays a ubiquitous role in constructing exponentially-improved asymptotic expansions and providing a smooth interpretation of the Stokes phenomenon; see §§2.11(iii) and 2.11(iv).

7.20(ii) Cornu's Spiral

Let the set $\{x(t), y(t), t\}$ be defined by $x(t) = C(t)$, $y(t) = S(t)$, $t \geq 0$. Then the set $\{x(t), y(t)\}$ is called *Cornu's spiral*: it is the projection of the corkscrew on the $\{x,y\}$-plane. See Figure 7.20.1. The spiral has several special properties (see Temme (1996a, p. 184)). Let $P(t) = P(x(t), y(t))$ be any point on the projected spiral. Then the arc length between the origin and $P(t)$ equals t, and is directly proportional to the curvature at $P(t)$, which equals πt. Furthermore, because $dy/dx = \tan(\tfrac{1}{2}\pi t^2)$, the angle between the x-axis and the tangent to the spiral at $P(t)$ is given by $\tfrac{1}{2}\pi t^2$.

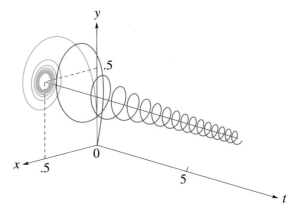

Figure 7.20.1: Cornu's spiral, formed from Fresnel integrals, is defined parametrically by $x = C(t)$, $y = S(t)$, $t \in [0, \infty)$.

7.20(iii) Statistics

The normal distribution function with mean m and standard deviation σ is given by

7.20.1
$$\frac{1}{\sigma\sqrt{2\pi}} \int_{-\infty}^{x} e^{-(t-m)^2/(2\sigma^2)}\, dt$$
$$= \frac{1}{2}\operatorname{erfc}\left(\frac{m-x}{\sigma\sqrt{2}}\right) = Q\left(\frac{m-x}{\sigma}\right) = P\left(\frac{x-m}{\sigma}\right).$$

For applications in statistics and probability theory, also for the role of the normal distribution functions (the error functions and probability integrals) in the asymptotics of arbitrary probability density functions, see Johnson et al. (1994, Chapter 13) and Patel and Read (1982, Chapters 2 and 3).

7.21 Physical Applications

The error functions, Fresnel integrals, and related functions occur in a variety of physical applications. Fresnel integrals and Cornu's spiral occurred originally in the analysis of the diffraction of light; see Born and Wolf (1999, §8.7). More recently, Cornu's spiral appears in the design of highways and railroad tracks, robot trajectory planning, and computer-aided design; see Meek and Walton (1992).

Carslaw and Jaeger (1959) gives many applications and points out the importance of the repeated integrals of the complementary error function $i^n\operatorname{erfc}(z)$. Fried and Conte (1961) mentions the role of $w(z)$ in the theory of linearized waves or oscillations in a hot plasma; $w(z)$ is called the *plasma dispersion function* or *Faddeeva function*; see Faddeeva and Terent'ev (1954). Ng and Geller (1969) cites work with applications from atomic physics and astrophysics.

Voigt functions can be regarded as the convolution of a Gaussian and a Lorentzian, and appear when the analysis of light (or particulate) absorption (or emission) involves thermal motion effects. These applications include astrophysics, plasma diagnostics, neutron diffraction, laser spectroscopy, and surface scattering. See Mitchell and Zemansky (1961, §IV.2), Armstrong (1967), and Ahn et al. (2001). Dawson's integral appears in de-convolving even more complex motional effects; see Pratt (2007).

Computation

7.22 Methods of Computation

7.22(i) Main Functions

The methods available for computing the main functions in this chapter are analogous to those described in §§6.18(i)–6.18(iv) for the exponential integral and sine and cosine integrals, and similar comments apply. Additional references are Matta and Reichel (1971) for the application of the trapezoidal rule, for example, to the first of (7.7.2), and Gautschi (1970) and Cuyt et al. (2008) for continued fractions.

7.22(ii) Goodwin–Staton Integral

See Goodwin and Staton (1948).

7.22(iii) Repeated Integrals of the Complementary Error Function

The recursion scheme given by (7.18.1) and (7.18.7) can be used for computing $i^n\operatorname{erfc}(x)$. See Gautschi (1977a), where forward and backward recursions are used; see also Gautschi (1961).

7.22(iv) Voigt Functions

The computation of these functions can be based on algorithms for the complementary error function with complex argument; compare (7.19.3).

7.22(v) Other References

For a comprehensive survey of computational methods for the functions treated in this chapter, see van der Laan and Temme (1984, Ch. V).

7.23 Tables

7.23(i) Introduction

Lebedev and Fedorova (1960) and Fletcher et al. (1962) give comprehensive indexes of mathematical tables. This section lists relevant tables that appeared later.

7.23(ii) Real Variables

- Abramowitz and Stegun (1964, Chapter 7) includes $\operatorname{erf} x$, $(2/\sqrt{\pi})e^{-x^2}$, $x \in [0,2]$, 10D; $(2/\sqrt{\pi})e^{-x^2}$, $x \in [2,10]$, 8S; $xe^{x^2}\operatorname{erfc} x$, $x^{-2} \in [0, 0.25]$, 7D; $2^n\,\Gamma(\tfrac{1}{2}n+1)\,i^n\operatorname{erfc}(x)$, $n = 1(1)6, 10, 11$, $x \in [0, 5]$, 6S; $F(x)$, $x \in [0,2]$, 10D; $xF(x)$, $x^{-2} \in [0, 0.25]$, 9D; $C(x)$, $S(x)$, $x \in [0, 5]$, 7D; f(x), g(x), $x \in [0,1]$, $x^{-1} \in [0,1]$, 15D.

- Abramowitz and Stegun (1964, Table 27.6) includes the Goodwin–Staton integral $G(x)$, $x = 1(.1)3(.5)8$, 4D; also $G(x) + \ln x$, $x = 0(.05)1$, 4D.

- Finn and Mugglestone (1965) includes the Voigt function $H(a, u)$, $u \in [0, 22]$, $a \in [0, 1]$, 6S.

- Zhang and Jin (1996, pp. 637, 639) includes $(2/\sqrt{\pi})e^{-x^2}$, $\operatorname{erf} x$, $x = 0(.02)1(.04)3$, 8D; $C(x)$, $S(x)$, $x = 0(.2)10(2)100(100)500$, 8D.

7.23(iii) Complex Variables, $z = x + iy$

- Abramowitz and Stegun (1964, Chapter 7) includes $w(z)$, $x = 0(.1)3.9$, $y = 0(.1)3$, 6D.

- Zhang and Jin (1996, pp. 638, 640–641) includes the real and imaginary parts of $\operatorname{erf} z$, $x \in [0,5]$, $y = 0.5(.5)3$, 7D and 8D, respectively; the real and imaginary parts of $\int_x^\infty e^{\pm it^2} dt$, $(1/\sqrt{\pi})e^{\mp i(x^2 + (\pi/4))} \int_x^\infty e^{\pm it^2} dt$, $x = 0(.5)20(1)25$, 8D, together with the corresponding modulus and phase to 8D and 6D (degrees), respectively.

7.23(iv) Zeros

- Fettis et al. (1973) gives the first 100 zeros of $\operatorname{erf} z$ and $w(z)$ (the table on page 406 of this reference is for $w(z)$, not for $\operatorname{erfc} z$), 11S.

- Zhang and Jin (1996, p. 642) includes the first 10 zeros of $\operatorname{erf} z$, 9D; the first 25 distinct zeros of $C(z)$ and $S(z)$, 8S.

7.24 Approximations

7.24(i) Approximations in Terms of Elementary Functions

- Hastings (1955) gives several minimax polynomial and rational approximations for $\operatorname{erf} x$, $\operatorname{erfc} x$ and the auxiliary functions $f(x)$ and $g(x)$.

- Cody (1969) provides minimax rational approximations for $\operatorname{erf} x$ and $\operatorname{erfc} x$. The maximum relative precision is about 20S.

- Cody (1968) gives minimax rational approximations for the Fresnel integrals (maximum relative precision 19S); for a Fortran algorithm and comments see Snyder (1993).

- Cody et al. (1970) gives minimax rational approximations to Dawson's integral $F(x)$ (maximum relative precision 20S–22S).

7.24(ii) Expansions in Chebyshev Series

- Luke (1969b, pp. 323–324) covers $\frac{1}{2}\sqrt{\pi}\operatorname{erf} x$ and $e^{x^2} F(x)$ for $-3 \leq x \leq 3$ (the Chebyshev coefficients are given to 20D); $\sqrt{\pi} x e^{x^2} \operatorname{erfc} x$ and $2x F(x)$ for $x \geq 3$ (the Chebyshev coefficients are given to 20D and 15D, respectively). Coefficients for the Fresnel integrals are given on pp. 328–330 (20D).

- Bulirsch (1967) provides Chebyshev coefficients for the auxiliary functions $f(x)$ and $g(x)$ for $x \geq 3$ (15D).

- Schonfelder (1978) gives coefficients of Chebyshev expansions for $x^{-1}\operatorname{erf} x$ on $0 \leq x \leq 2$, for $xe^{x^2}\operatorname{erfc} x$ on $[2,\infty)$, and for $e^{x^2}\operatorname{erfc} x$ on $[0,\infty)$ (30D).

- Shepherd and Laframboise (1981) gives coefficients of Chebyshev series for $(1 + 2x)e^{x^2}\operatorname{erfc} x$ on $(0,\infty)$ (22D).

7.24(iii) Padé-Type Expansions

- Luke (1969b, vol. 2, pp. 422–435) gives main diagonal Padé approximations for $F(z)$, $\operatorname{erf} z$, $\operatorname{erfc} z$, $C(z)$, and $S(z)$; approximate errors are given for a selection of z-values.

7.25 Software

See http://dlmf.nist.gov/7.25.

References

General References

For general bibliographic reading see Carslaw and Jaeger (1959), Lebedev (1965), Olver (1997b), and Temme (1996a).

Sources

The following list gives the references or other indications of proofs that were used in constructing the various sections of this chapter. These sources supplement the references that are quoted in the text.

§7.2 Olver (1997b, pp. 43–44) and Temme (1996a, pp. 180, 182–183, 275–276).

§7.3 These graphics were produced at NIST.

§7.4 (7.4.7) follows from (7.2.10), (7.2.11), and (7.4.6). (7.4.8) follows from (7.4.7).

§7.5 (7.5.1) follows from (7.2.1)–(7.2.3). (7.5.2) follows from (7.2.6)–(7.2.9). (7.5.3) and (7.5.4) follow from (7.2.10) and (7.2.11). (7.5.5) and (7.5.6) follow from (7.2.10), (7.2.11), and (7.5.2). (7.5.8) follows from (7.2.1), (7.2.7), and (7.2.8). (7.5.9) follows from (7.2.2), (7.2.3), (7.5.7), and (7.5.8). (7.5.10) follows from (7.2.2), (7.5.6), and (7.5.8). (7.5.11) follows from (7.5.5). (7.5.12) follows from (7.4.8) and (7.5.11). For (7.5.13) see Olver (1997b, p. 44).

§7.6 For (7.6.1)–(7.6.3) see van der Laan and Temme (1984, pp. 185–186). (7.6.4) and (7.6.6) follow from (7.2.7) and (7.2.8). (7.6.5) and (7.6.7) follow from (7.5.8) and (7.6.2). For (7.6.8) differentiate: use (7.2.1) for the left-hand side, and (10.47.7) and the second of (10.29.1) for the right-hand side. The same method can be used for (7.6.10) and (7.6.11). For (7.6.9) write the coefficients in the Chebyshev-series expansion of $\exp(a^2 z^2)\operatorname{erf}(az)$ as integrals (§3.11(ii)), then apply (5.12.5), (7.6.2), and (13.6.9).

§7.7 (7.7.1), (7.7.2), and (7.7.4) are given in van der Laan and Temme (1984, pp. 185–186). (7.7.3) follows from integrating from 0 to iz/a and from iz/a to ∞. (7.7.5) follows by differentiating with respect to a (after multiplying the equation by e^{-a}). (7.7.6), (7.7.7), and (7.7.9) follow by differentiating with respect to x. For (7.7.8) let $x \to 0+$ in (7.7.7) and use (7.2.2), (7.2.4). For (7.7.10) and (7.7.11) see van der Laan and Temme (1984, Chapter V). (7.7.12) follows from (7.5.5) and (7.2.6). (7.7.13) and (7.7.14) follow by taking Mellin transforms (§1.14(iv)), and applying (7.7.10), (7.7.11), (5.12.1).

§7.8 See Mills (1926), Mitrinović (1970, p. 177), and Gautschi (1959b) for (7.8.2); Kesavan and Vasudevamurthy (1985) for the lower bound in (7.8.3); Laforgia and Sismondi (1988) for the upper bound in (7.8.3); Gupta (1970) for (7.8.4); Luke (1969b, p. 201) for (7.8.5); Martić (1978) for (7.8.6); Crstici and Tudor (1975) for (7.8.7). See also Wu (1982).

§7.9 Nielsen (1906a, p. 217) and Lorentzen and Waadeland (1992, pp. 576–577). (7.9.2) is the even part of (7.9.1) (compare §1.12(iv)).

§7.10 These results may be verified by differentiation of the definitions given in §7.2.

§7.11 These results may be verified by comparing the power-series expansions of both sides of each equation. For (7.11.5) use (7.7.1) and (13.4.4).

§7.12 (7.12.2) and (7.12.3) follow from (7.7.10) and (7.7.11) by applying Watson's lemma in its extended form (§2.4(i)). (7.12.4)–(7.12.7) follow from (7.7.10), (7.7.11), and the identity $(t^2+1)^{-1} = \sum_{m=0}^{n-1}(-1)^m t^{2m} + (-1)^n t^{2n}(t^2+1)^{-1}$. The error bounds are obtained by setting $t = \sqrt{\tau}$ in (7.12.6) and (7.12.7), rotating the integration path in the τ-plane through an angle $-4\operatorname{ph} z$, and then replacing $|\tau+1|$ by its minimum value on the path.

§7.13 Fettis et al. (1973), Fettis and Caslin (1973), and Kreyszig (1957).

§7.14 For (7.14.1) and (7.14.2) integrate by parts and apply (7.7.3), (7.7.6). (7.14.3) follows from (7.14.4) with $c = 0$. For (7.14.4) integrate by parts and apply (10.32.10), (10.39.2). For (7.14.5) and (7.14.6) integrate by parts, and use (7.7.15) and (7.7.16). For (7.14.7) and (7.14.8) consider the integrals $\int_0^\infty e^{-at}\left(C\left(\sqrt{2t/\pi}\right) \pm i S\left(\sqrt{2t/\pi}\right)\right) dt$ and integrate by parts. The results are $1\big/\left(a\sqrt{2(a \mp i)}\right)$, from which (7.14.7) and (7.14.8) follow.

§7.17 For (7.17.2) see Carlitz (1963). (7.17.3) follows from Blair et al. (1976) after modifications.

§7.18 Hartree (1936) and Lorentzen and Waadeland (1992, p. 577). The graphs were produced at NIST.

§7.19 (7.19.3) follows from (7.2.3) and (7.7.2). (7.19.5) follows from the definitions (7.19.1), (7.19.2), together with (1.17.6) or §2.3(iii). For the first of (7.19.7) use (7.19.1) for the lower bound, and (7.19.10) for the upper bound. For the second of (7.19.7) use (7.19.11). For (7.19.8) and (7.19.9) again use the definitions (7.19.1) and (7.19.2). For (7.19.10) and (7.19.11) see Armstrong (1967). The graphs were produced at NIST.

§7.20 The diagram was produced by the author.

Chapter 8
Incomplete Gamma and Related Functions

R. B. Paris[1]

Notation	**174**
8.1 Special Notation	174

Incomplete Gamma Functions	**174**
8.2 Definitions and Basic Properties	174
8.3 Graphics	175
8.4 Special Values	176
8.5 Confluent Hypergeometric Representations	177
8.6 Integral Representations	177
8.7 Series Expansions	178
8.8 Recurrence Relations and Derivatives	178
8.9 Continued Fractions	179
8.10 Inequalities	179
8.11 Asymptotic Approximations and Expansions	179
8.12 Uniform Asymptotic Expansions for Large Parameter	181
8.13 Zeros	182
8.14 Integrals	182
8.15 Sums	183
8.16 Generalizations	183

Related Functions	**183**
8.17 Incomplete Beta Functions	183
8.18 Asymptotic Expansions of $I_x(a,b)$	184
8.19 Generalized Exponential Integral	185
8.20 Asymptotic Expansions of $E_p(z)$	187
8.21 Generalized Sine and Cosine Integrals	188

Applications	**189**
8.22 Mathematical Applications	189
8.23 Statistical Applications	189
8.24 Physical Applications	189

Computation	**190**
8.25 Methods of Computation	190
8.26 Tables	190
8.27 Approximations	191
8.28 Software	191

References	**191**

---

[1] Division of Mathematical Sciences, University of Abertay Dundee, Dundee, United Kingdom.
Acknowledgments: This chapter is based in part on Abramowitz and Stegun (1964, Chapters 5 and 6), by W. Gautschi and F. Cahill, and P. J. Davis, respectively.
Copyright © 2009 National Institute of Standards and Technology. All rights reserved.

Notation

8.1 Special Notation

(For other notation see pp. xiv and 873.)

x	real variable.
z	complex variable.
a, p	real or complex parameters.
k, n	nonnegative integers.
δ	arbitrary small positive constant.
$\Gamma(z)$	gamma function (§5.2(i)).
$\psi(z)$	$\Gamma'(z)/\Gamma(z)$.

Unless otherwise indicated, primes denote derivatives with respect to the argument.

The functions treated in this chapter are the incomplete gamma functions $\gamma(a,z)$, $\Gamma(a,z)$, $\gamma^*(a,z)$, $P(a,z)$, and $Q(a,z)$; the incomplete beta functions $\mathrm{B}_x(a,b)$ and $I_x(a,b)$; the generalized exponential integral $E_p(z)$; the generalized sine and cosine integrals $\mathrm{si}(a,z)$, $\mathrm{ci}(a,z)$, $\mathrm{Si}(a,z)$, and $\mathrm{Ci}(a,z)$.

Alternative notations include: *Prym's functions* $P_z(a) = \gamma(a,z)$, $Q_z(a) = \Gamma(a,z)$, Nielsen (1906a, pp. 25–26), Batchelder (1967, p. 63); $(a,z)! = \gamma(a+1,z)$, $[a,z]! = \Gamma(a+1,z)$, Dingle (1973); $B(a,b,x) = \mathrm{B}_x(a,b)$, $I(a,b,x) = I_x(a,b)$, Magnus *et al.* (1966); $\mathrm{Si}(a,x) \to \mathrm{Si}(1-a,x)$, $\mathrm{Ci}(a,x) \to \mathrm{Ci}(1-a,x)$, Luke (1975).

Incomplete Gamma Functions

8.2 Definitions and Basic Properties

8.2(i) Definitions

The *general values* of the *incomplete gamma functions* $\gamma(a,z)$ and $\Gamma(a,z)$ are defined by

8.2.1 $$\gamma(a,z) = \int_0^z t^{a-1} e^{-t}\, dt, \qquad \Re a > 0,$$

8.2.2 $$\Gamma(a,z) = \int_z^\infty t^{a-1} e^{-t}\, dt,$$

without restrictions on the integration paths. However, when the integration paths do not cross the negative real axis, and in the case of (8.2.2) exclude the origin, $\gamma(a,z)$ and $\Gamma(a,z)$ take their *principal values*; compare §4.2(i). *Except where indicated otherwise* in this Handbook these principal values are assumed. For example,

8.2.3 $$\gamma(a,z) + \Gamma(a,z) = \Gamma(a), \quad a \neq 0, -1, -2, \dots.$$

Normalized functions are:

8.2.4 $$P(a,z) = \frac{\gamma(a,z)}{\Gamma(a)}, \quad Q(a,z) = \frac{\Gamma(a,z)}{\Gamma(a)},$$

8.2.5 $$P(a,z) + Q(a,z) = 1.$$

In addition,

8.2.6 $$\gamma^*(a,z) = z^{-a} P(a,z) = \frac{z^{-a}}{\Gamma(a)}\gamma(a,z).$$

8.2.7 $$\gamma^*(a,z) = \frac{1}{\Gamma(a)} \int_0^1 t^{a-1} e^{-zt}\, dt, \qquad \Re a > 0.$$

8.2(ii) Analytic Continuation

In this subsection the functions γ and Γ have their general values.

The function $\gamma^*(a,z)$ is entire in z and a. When $z \neq 0$, $\Gamma(a,z)$ is an entire function of a, and $\gamma(a,z)$ is meromorphic with simple poles at $a = -n$, $n = 0, 1, 2, \dots$, with residue $(-1)^n/n!$.

For $m \in \mathbb{Z}$,

8.2.8 $$\gamma\bigl(a, z e^{2\pi m i}\bigr) = e^{2\pi m i a} \gamma(a,z), \quad a \neq 0, -1, -2, \dots,$$

8.2.9 $$\Gamma\bigl(a, z e^{2\pi m i}\bigr) = e^{2\pi m i a} \Gamma(a,z) + (1 - e^{2\pi m i a}) \Gamma(a).$$

(8.2.9) also holds when a is zero or a negative integer, provided that the right-hand side is replaced by its limiting value. For example, in the case $m = -1$ we have

8.2.10 $$e^{-\pi i a} \Gamma\bigl(a, z e^{\pi i}\bigr) - e^{\pi i a} \Gamma\bigl(a, z e^{-\pi i}\bigr) = -\frac{2\pi i}{\Gamma(1-a)},$$

without restriction on a.

Lastly,

8.2.11 $$\Gamma\bigl(a, z e^{\pm \pi i}\bigr) = \Gamma(a)(1 - z^a e^{\pm \pi i a} \gamma^*(a,-z)).$$

8.2(iii) Differential Equations

If $w = \gamma(a,z)$ or $\Gamma(a,z)$, then

8.2.12 $$\frac{d^2 w}{dz^2} + \left(1 + \frac{1-a}{z}\right) \frac{dw}{dz} = 0.$$

If $w = e^z z^{1-a} \Gamma(a,z)$, then

8.2.13 $$\frac{d^2 w}{dz^2} - \left(1 + \frac{1-a}{z}\right) \frac{dw}{dz} + \frac{1-a}{z^2} w = 0.$$

Also,

8.2.14 $$z \frac{d^2 \gamma^*}{dz^2} + (a + 1 + z) \frac{d\gamma^*}{dz} + a\, \gamma^* = 0.$$

8.3 Graphics

8.3(i) Real Variables

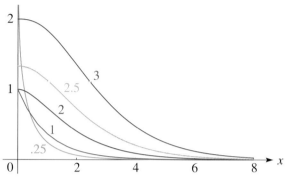

Figure 8.3.1: $\Gamma(a,x)$, $a = 0.25, 1, 2, 2.5, 3$.

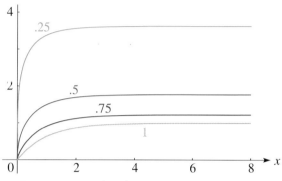

Figure 8.3.2: $\gamma(a,x)$, $a = 0.25, 0.5, 0.75, 1$.

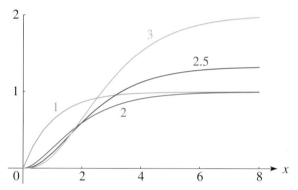

Figure 8.3.3: $\gamma(a,x)$, $a = 1, 2, 2.5, 3$.

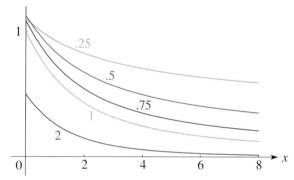

Figure 8.3.4: $\gamma^*(a,x)$ $(= x^{-a} P(a,x))$, $a = 0.25, 0.5, 0.75, 1, 2$.

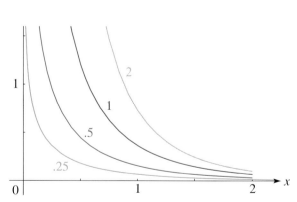

Figure 8.3.5: $x^{-a} - \gamma^*(a,x)$ $(= x^{-a} Q(a,x))$, $a = 0.25, 0.5, 1, 2$.

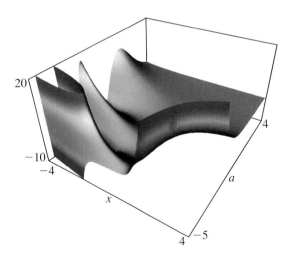

Figure 8.3.6: $\gamma^*(a,x)$ $(= x^{-a} P(a,x))$, $-4 \leq x \leq 4$, $-5 \leq a \leq 4$.

Some monotonicity properties of $\gamma^*(a,x)$ and $\Gamma(a,x)$ in the four quadrants of the (a,x)-plane in Figure 8.3.6 are given in Erdélyi et al. (1953b, §9.6).

8.3(ii) Complex Argument

In the graphics shown in this subsection, height corresponds to the absolute value of the function and color to the phase. See p. xiv.

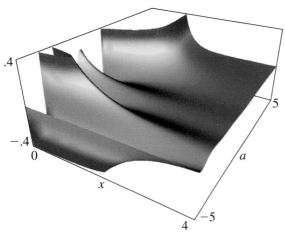

Figure 8.3.7: $x^{-a} - \gamma^*(a,x)$ $(= x^{-a} Q(a,x))$, $0 \le x \le 4$, $-5 \le a \le 5$.

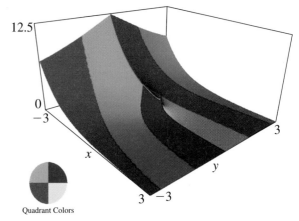

Figure 8.3.8: $\Gamma(0.25, x+iy)$, $-3 \le x \le 3$, $-3 \le y \le 3$. Principal value. There is a cut along the negative real axis. When $x = y = 0$, $\Gamma(0.25, 0) = \Gamma(0.25) = 3.625\ldots$.

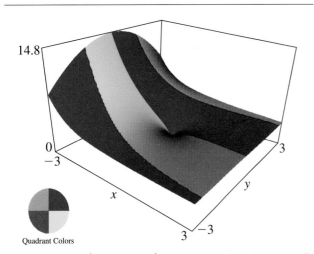

Figure 8.3.9: $\gamma(0.25, x+iy)$, $-3 \le x \le 3$, $-3 \le y \le 3$. Principal value. There is a cut along the negative real axis.

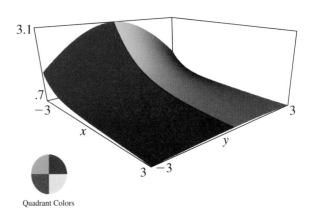

Figure 8.3.10: $\gamma^*(0.25, x+iy)$, $-3 \le x \le 3$, $-3 \le y \le 3$.

For additional graphics see http://dlmf.nist.gov/8.3.ii.

8.4 Special Values

For $\operatorname{erf}(z)$, $\operatorname{erfc}(z)$, and $F(z)$, see §§7.2(i), 7.2(ii). For $E_n(z)$ see §8.19(i).

8.4.1 $$\gamma\left(\tfrac{1}{2}, z^2\right) = 2 \int_0^z e^{-t^2}\, dt = \sqrt{\pi}\, \operatorname{erf}(z),$$

8.4.2 $$\gamma^*(a, 0) = \frac{1}{\Gamma(a+1)},$$

8.4.3 $$\gamma^*\left(\tfrac{1}{2}, -z^2\right) = \frac{2 e^{z^2}}{z\sqrt{\pi}}\, F(z).$$

8.4.4 $$\Gamma(0, z) = \int_z^\infty t^{-1} e^{-t}\, dt = E_1(z),$$

8.4.5 $$\Gamma(1, z) = e^{-z},$$

8.4.6 $$\Gamma\left(\tfrac{1}{2}, z^2\right) = 2\int_z^\infty e^{-t^2}\,dt = \sqrt{\pi}\,\operatorname{erfc}(z).$$

For $n = 0, 1, 2, \ldots$,

8.4.7 $$\gamma(n+1, z) = n!(1 - e^{-z} e_n(z)),$$

8.4.8 $$\Gamma(n+1, z) = n! e^{-z} e_n(z),$$

8.4.9 $$P(n+1, z) = 1 - e^{-z} e_n(z),$$

8.4.10 $$Q(n+1, z) = e^{-z} e_n(z),$$

where

8.4.11 $$e_n(z) = \sum_{k=0}^n \frac{z^k}{k!}.$$

Also

8.4.12 $$\gamma^*(-n, z) = z^n,$$

8.4.13 $$\Gamma(1-n, z) = z^{1-n} E_n(z),$$

8.4.14 $$Q\left(n+\tfrac{1}{2}, z^2\right) = \operatorname{erfc}(z) + \frac{e^{-z^2}}{\sqrt{\pi}} \sum_{k=1}^n \frac{z^{2k-1}}{\left(\tfrac{1}{2}\right)_k},$$

8.4.15
$$\Gamma(-n, z) = \frac{(-1)^n}{n!}\left(E_1(z) - e^{-z}\sum_{k=0}^{n-1} \frac{(-1)^k k!}{z^{k+1}}\right)$$
$$= \frac{(-1)^n}{n!}(\psi(n+1) - \ln z) - z^{-n}\sum_{\substack{k=0\\k\neq n}}^{\infty} \frac{(-z)^k}{k!(k-n)}.$$

8.5 Confluent Hypergeometric Representations

For the confluent hypergeometric functions M, $\mathbf{M}$, U, and the Whittaker functions $M_{\kappa,\mu}$ and $W_{\kappa,\mu}$, see §§13.2(i) and 13.14(i).

8.5.1
$$\gamma(a, z) = a^{-1} z^a e^{-z} M(1, 1+a, z)$$
$$= a^{-1} z^a M(a, 1+a, -z), \quad a \neq 0, -1, -2, \ldots.$$

8.5.2 $$\gamma^*(a, z) = e^{-z}\mathbf{M}(1, 1+a, z) = \mathbf{M}(a, 1+a, -z).$$

8.5.3 $$\Gamma(a, z) = e^{-z} U(1-a, 1-a, z) = z^a e^{-z} U(1, 1+a, z).$$

8.5.4 $$\gamma(a, z) = a^{-1} z^{\frac{1}{2}a - \frac{1}{2}} e^{-\frac{1}{2}z} M_{\frac{1}{2}a - \frac{1}{2}, \frac{1}{2}a}(z).$$

8.5.5 $$\Gamma(a, z) = e^{-\frac{1}{2}z} z^{\frac{1}{2}a - \frac{1}{2}} W_{\frac{1}{2}a - \frac{1}{2}, \frac{1}{2}a}(z).$$

8.6 Integral Representations

8.6(i) Integrals Along the Real Line

For the Bessel function $J_\nu(z)$ and modified Bessel function $K_\nu(z)$, see §§10.2(ii) and 10.25(ii).

8.6.1 $$\gamma(a, z) = \frac{z^a}{\sin(\pi a)}\int_0^\pi e^{z\cos t}\cos(at + z\sin t)\,dt, \quad a \notin \mathbb{Z},$$

8.6.2 $$\gamma(a, z) = z^{\frac{1}{2}a}\int_0^\infty e^{-t} t^{\frac{1}{2}a - 1} J_a\left(2\sqrt{zt}\right)\,dt, \quad \Re a > 0.$$

8.6.3 $$\gamma(a, z) = z^a \int_0^\infty \exp\left(-at - z e^{-t}\right)\,dt, \quad \Re a > 0.$$

8.6.4 $$\Gamma(a, z) = \frac{z^a e^{-z}}{\Gamma(1-a)}\int_0^\infty \frac{t^{-a} e^{-t}}{z+t}\,dt, \quad |\operatorname{ph} z| < \pi,\ \Re a < 1,$$

8.6.5 $$\Gamma(a, z) = z^a e^{-z}\int_0^\infty \frac{e^{-zt}}{(1+t)^{1-a}}\,dt, \quad \Re z > 0,$$

8.6.6 $$\Gamma(a, z) = \frac{2 z^{\frac{1}{2}a} e^{-z}}{\Gamma(1-a)}\int_0^\infty e^{-t} t^{-\frac{1}{2}a} K_a\left(2\sqrt{zt}\right)\,dt, \quad \Re a < 1,$$

8.6.7 $$\Gamma(a, z) = z^a \int_0^\infty \exp\left(at - z e^t\right)\,dt, \quad \Re z > 0.$$

8.6(ii) Contour Integrals

8.6.8
$$\gamma(a, z) = \frac{-iz^a}{2\sin(\pi a)}\int_{-1}^{(0+)} t^{a-1} e^{zt}\,dt, \quad z \neq 0,\ a \notin \mathbb{Z};$$

t^{a-1} takes its principal value where the path intersects the positive real axis, and is continuous elsewhere on the path.

8.6.9 $$\Gamma\left(-a, z e^{\pm\pi i}\right) = \frac{e^z e^{\mp\pi i a}}{\Gamma(1+a)}\int_0^\infty \frac{t^a e^{-zt}}{t - 1}\,dt, \quad \Re z > 0,\ \Re a > -1,$$

where the integration path passes above or below the pole at $t = 1$, according as upper or lower signs are taken.

Mellin–Barnes Integrals

In (8.6.10)–(8.6.12), c is a real constant and the path of integration is indented (if necessary) so that it separates the poles of the gamma function from the other pole in the integrand, in the case of (8.6.10) and (8.6.11), and from the poles at $s = 0, 1, 2, \ldots$ in the case of (8.6.12).

8.6.10 $$\gamma(a, z) = \frac{1}{2\pi i}\int_{c-i\infty}^{c+i\infty} \frac{\Gamma(s)}{a - s} z^{a-s}\,ds, \quad |\operatorname{ph} z| < \tfrac{1}{2}\pi,\ a \neq 0, -1, -2, \ldots,$$

8.6.11
$$\Gamma(a,z) = \frac{1}{2\pi i} \int_{c-i\infty}^{c+i\infty} \Gamma(s+a) \frac{z^{-s}}{s} \, ds, \qquad |\mathrm{ph}\, z| < \tfrac{1}{2}\pi,$$

8.6.12
$$\Gamma(a,z) = -\frac{z^{a-1}e^{-z}}{\Gamma(1-a)}$$
$$\times \frac{1}{2\pi i} \int_{c-i\infty}^{c+i\infty} \Gamma(s+1-a) \frac{\pi z^{-s}}{\sin(\pi s)} \, ds,$$
$$|\mathrm{ph}\, z| < \tfrac{3}{2}\pi, \; a \neq 1, 2, 3, \ldots.$$

8.6(iii) Compendia

For collections of integral representations of $\gamma(a,z)$ and $\Gamma(a,z)$ see Erdélyi et al. (1953b, §9.3), Oberhettinger (1972, pp. 68–69), Oberhettinger and Badii (1973, pp. 309–312), Prudnikov et al. (1992b, §3.10), and Temme (1996a, pp. 282–283).

8.7 Series Expansions

For the functions $e_n(z)$, $\mathrm{i}_n^{(1)}(z)$, and $L_n^{(\alpha)}(x)$ see (8.4.11), §§10.47(ii), and 18.3, respectively.

8.7.1
$$\gamma^*(a,z) = e^{-z} \sum_{k=0}^{\infty} \frac{z^k}{\Gamma(a+k+1)} = \frac{1}{\Gamma(a)} \sum_{k=0}^{\infty} \frac{(-z)^k}{k!(a+k)}.$$

8.7.2
$$\gamma(a, x+y) - \gamma(a, x)$$
$$= \Gamma(a, x) - \Gamma(a, x+y)$$
$$= e^{-x} x^{a-1} \sum_{n=0}^{\infty} \frac{(1-a)_n}{(-x)^n} (1 - e^{-y} e_n(y)),$$
$$|y| < |x|.$$

8.7.3
$$\Gamma(a,z) = \Gamma(a) - \sum_{k=0}^{\infty} \frac{(-1)^k z^{a+k}}{k!(a+k)}$$
$$= \Gamma(a) \left(1 - z^a e^{-z} \sum_{k=0}^{\infty} \frac{z^k}{\Gamma(a+k+1)} \right),$$
$$a \neq 0, -1, -2, \ldots.$$

8.7.4
$$\gamma(a,x) = \Gamma(a) x^{\frac{1}{2}a} e^{-x} \sum_{n=0}^{\infty} e_n(-1) x^{\frac{1}{2}n} I_{n+a}(2x^{1/2}),$$
$$a \neq 0, -1, -2, \ldots.$$

8.7.5
$$\gamma^*(a,z) = e^{-\frac{1}{2}z} \sum_{n=0}^{\infty} \frac{(1-a)_n}{\Gamma(n+a+1)} (2n+1) \mathrm{i}_n^{(1)}(\tfrac{1}{2}z).$$

8.7.6
$$\Gamma(a,x) = x^a e^{-x} \sum_{n=0}^{\infty} \frac{L_n^{(a)}(x)}{n+1}, \qquad x > 0.$$

For an expansion for $\gamma(a, ix)$ in series of Bessel functions $J_n(x)$ that converges rapidly when $a > 0$ and x (≥ 0) is small or moderate in magnitude see Barakat (1961).

8.8 Recurrence Relations and Derivatives

8.8.1 $\qquad \gamma(a+1, z) = a\,\gamma(a,z) - z^a e^{-z},$

8.8.2 $\qquad \Gamma(a+1, z) = a\,\Gamma(a,z) + z^a e^{-z}.$

If $w(a,z) = \gamma(a,z)$ or $\Gamma(a,z)$, then

8.8.3 $w(a+2, z) - (a+1+z)w(a+1, z) + az w(a,z) = 0.$

8.8.4 $\qquad z\,\gamma^*(a+1, z) = \gamma^*(a,z) - \dfrac{e^{-z}}{\Gamma(a+1)}.$

8.8.5 $\qquad P(a+1, z) = P(a,z) - \dfrac{z^a e^{-z}}{\Gamma(a+1)},$

8.8.6 $\qquad Q(a+1, z) = Q(a,z) + \dfrac{z^a e^{-z}}{\Gamma(a+1)}.$

For $n = 0, 1, 2, \ldots,$

8.8.7
$$\gamma(a+n, z) = (a)_n\, \gamma(a,z) - z^a e^{-z} \sum_{k=0}^{n-1} \frac{\Gamma(a+n)}{\Gamma(a+k+1)} z^k,$$

8.8.8
$$\gamma(a,z)$$
$$= \frac{\Gamma(a)}{\Gamma(a-n)} \gamma(a-n, z) - z^{a-1} e^{-z} \sum_{k=0}^{n-1} \frac{\Gamma(a)}{\Gamma(a-k)} z^{-k},$$

8.8.9
$$\Gamma(a+n, z) = (a)_n\, \Gamma(a,z) + z^a e^{-z} \sum_{k=0}^{n-1} \frac{\Gamma(a+n)}{\Gamma(a+k+1)} z^k,$$

8.8.10
$$\Gamma(a,z)$$
$$= \frac{\Gamma(a)}{\Gamma(a-n)} \Gamma(a-n, z) + z^{a-1} e^{-z} \sum_{k=0}^{n-1} \frac{\Gamma(a)}{\Gamma(a-k)} z^{-k},$$

8.8.11 $P(a+n, z) = P(a,z) - z^a e^{-z} \sum_{k=0}^{n-1} \dfrac{z^k}{\Gamma(a+k+1)},$

8.8.12 $Q(a+n, z) = Q(a,z) + z^a e^{-z} \sum_{k=0}^{n-1} \dfrac{z^k}{\Gamma(a+k+1)}.$

8.8.13 $\qquad \dfrac{d}{dz} \gamma(a,z) = -\dfrac{d}{dz} \Gamma(a,z) = z^{a-1} e^{-z},$

8.8.14 $\qquad \dfrac{\partial}{\partial a} \gamma^*(a,z) \bigg|_{a=0} = -E_1(z) - \ln z.$

For $E_1(z)$ see §8.19(i).

For $n = 0, 1, 2, \ldots,$

8.8.15 $\dfrac{d^n}{dz^n} (z^{-a} \gamma(a,z)) = (-1)^n z^{-a-n} \gamma(a+n, z),$

8.8.16 $\dfrac{d^n}{dz^n} (z^{-a} \Gamma(a,z)) = (-1)^n z^{-a-n} \Gamma(a+n, z),$

8.8.17 $\dfrac{d^n}{dz^n} (e^z \gamma(a,z)) = (-1)^n (1-a)_n e^z \gamma(a-n, z),$

8.8.18 $\dfrac{d^n}{dz^n} (z^a e^z \gamma^*(a,z)) = z^{a-n} e^z \gamma^*(a-n, z),$

8.8.19 $\dfrac{d^n}{dz^n} (e^z \Gamma(a,z)) = (-1)^n (1-a)_n e^z \Gamma(a-n, z).$

8.9 Continued Fractions

$$8.9.1 \quad \Gamma(a+1)e^z \gamma^*(a,z) = \cfrac{1}{1-} \cfrac{z}{a+1+} \cfrac{z}{a+2-} \cfrac{(a+1)z}{a+3+} \cfrac{2z}{a+4-} \cfrac{(a+2)z}{a+5+} \cfrac{3z}{a+6-} \cdots, \quad a \ne -1, -2, \dots,$$

$$8.9.2 \quad z^{-a} e^z \Gamma(a,z) = \cfrac{z^{-1}}{1+} \cfrac{(1-a)z^{-1}}{1+} \cfrac{z^{-1}}{1+} \cfrac{(2-a)z^{-1}}{1+} \cfrac{2z^{-1}}{1+} \cfrac{(3-a)z^{-1}}{1+} \cfrac{3z^{-1}}{1+} \cdots, \quad |\mathrm{ph}\, z| < \pi.$$

For these expansions and further information see Jones and Thron (1985). See also Cuyt *et al.* (2008, pp. 240–251).

8.10 Inequalities

$$8.10.1 \quad x^{1-a} e^x \Gamma(a,x) \le 1, \quad x > 0, \; 0 < a \le 1,$$

$$8.10.2 \quad \gamma(a,x) \ge \frac{x^{a-1}}{a}(1 - e^{-x}), \quad x > 0, \; 0 < a \le 1.$$

The inequalities in (8.10.1) and (8.10.2) are reversed when $a \ge 1$. If ϑ is defined by

$$8.10.3 \quad x^{1-a} e^x \Gamma(a,x) = 1 + \frac{a-1}{x} \vartheta,$$

then $\vartheta \to 1$ as $x \to \infty$, and

$$8.10.4 \quad 0 < \vartheta \le 1, \quad x > 0, \; a \le 2.$$

For further inequalities of these types see Qi and Mei (1999).

Padé Approximants

For $n = 1, 2, \ldots$,

$$8.10.5 \quad A_n < x^{1-a} e^x \Gamma(a,x) < B_n, \quad x > 0, \; a < 1,$$

where

$$8.10.6 \quad \begin{aligned} A_1 &= \frac{x}{x+1-a}, & B_1 &= \frac{x+1}{x+2-a}, \\ A_2 &= \frac{x(x+3-a)}{x^2 + 2(2-a)x + (1-a)(2-a)}, \\ B_2 &= \frac{x^2 + (5-a)x + 2}{x^2 + 2(3-a)x + (2-a)(3-a)}. \end{aligned}$$

For hypergeometric polynomial representations of A_n and B_n, see Luke (1969b, §14.6).

Next, define

$$8.10.7 \quad I = \int_0^x t^{a-1} e^t \, dt = \Gamma(a) x^a \gamma^*(a, -x), \quad \Re a > 0.$$

Then

$$8.10.8 \quad \frac{(a+1)(a+2) - x}{(a+1)(a+2+x)} < a x^{-a} e^{-x} I < \frac{a+1}{a+1+x}, \quad x > 0, \; a \ge 0.$$

Also, define

$$8.10.9 \quad c_a = (\Gamma(1+a))^{1/(a-1)}, \quad d_a = (\Gamma(1+a))^{-1/a}.$$

Then

$$8.10.10 \quad \begin{aligned} \frac{x}{2a}\left(\left(1+\frac{2}{x}\right)^a - 1\right) &< x^{1-a} e^x \Gamma(a,x) \\ &\le \frac{x}{ac_a}\left(\left(1+\frac{c_a}{x}\right)^a - 1\right), \end{aligned} \quad x \ge 0, \; 0 < a < 1,$$

and

$$8.10.11 \quad (1 - e^{-\alpha_a x})^a \le P(a,x) \le (1 - e^{-\beta_a x})^a, \quad x \ge 0, \; a > 0,$$

where

$$8.10.12 \quad \alpha_a = \begin{cases} 1, & 0 < a < 1, \\ d_a, & a > 1, \end{cases} \quad \beta_a = \begin{cases} d_a, & 0 < a < 1, \\ 1, & a > 1. \end{cases}$$

Equalities in (8.10.11) apply only when $a = 1$.

Lastly,

$$8.10.13 \quad \frac{\Gamma(n,n)}{\Gamma(n)} < \frac{1}{2} < \frac{\Gamma(n, n-1)}{\Gamma(n)}, \quad n = 1, 2, 3, \ldots.$$

8.11 Asymptotic Approximations and Expansions

8.11(i) Large z, Fixed a

Define

$$8.11.1 \quad u_k = (-1)^k (1-a)_k = (a-1)(a-2) \cdots (a-k),$$

$$8.11.2 \quad \Gamma(a,z) = z^{a-1} e^{-z} \left(\sum_{k=0}^{n-1} \frac{u_k}{z^k} + R_n(a,z) \right), \quad n = 1, 2, \ldots.$$

Then as $z \to \infty$ with a and n fixed

$$8.11.3 \quad R_n(a,z) = O(z^{-n}), \quad |\mathrm{ph}\, z| \le \tfrac{3}{2}\pi - \delta,$$

where δ denotes an arbitrary small positive constant.

If a is real and $z\,(=x)$ is positive, then $R_n(a,x)$ is bounded in absolute value by the first neglected term u_n/x^n and has the same sign provided that $n \ge a - 1$. For bounds on $R_n(a,z)$ when a is real and z is complex see Olver (1997b, pp. 109–112). For an exponentially-improved asymptotic expansion (§2.11(iii)) see Olver (1991a).

8.11(ii) Large a, Fixed z

$$8.11.4 \quad \gamma(a,z) = z^a e^{-z} \sum_{k=0}^{\infty} \frac{z^k}{(a)_{k+1}}, \quad a \neq 0,-1,-2,\ldots.$$

This expansion is absolutely convergent for all finite z, and it can also be regarded as a generalized asymptotic expansion (§2.1(v)) of $\gamma(a,z)$ as $a \to \infty$ in $|\operatorname{ph} a| \leq \pi - \delta$.

Also,

$$8.11.5 \quad P(a,z) \sim \frac{z^a e^{-z}}{\Gamma(1+a)} \sim (2\pi a)^{-\frac{1}{2}} e^{a-z} (z/a)^a,$$
$$a \to \infty, \ |\operatorname{ph} a| \leq \pi - \delta.$$

8.11(iii) Large a, Fixed x/a

If $x = \lambda a$, with λ fixed, then as $a \to +\infty$

$$8.11.6 \quad \gamma(a,x) \sim -x^a e^{-x} \sum_{k=0}^{\infty} \frac{(-a)^k b_k(\lambda)}{(x-a)^{2k+1}}, \quad 0 < \lambda < 1,$$

$$8.11.7 \quad \Gamma(a,x) \sim x^a e^{-x} \sum_{k=0}^{\infty} \frac{(-a)^k b_k(\lambda)}{(x-a)^{2k+1}}, \quad \lambda > 1,$$

where

$$8.11.8 \quad b_0(\lambda) = 1, \quad b_1(\lambda) = \lambda, \quad b_2(\lambda) = \lambda(2\lambda+1),$$

and for $k = 1, 2, \ldots$,

$$8.11.9 \quad b_k(\lambda) = \lambda(1-\lambda) b'_{k-1}(\lambda) + (2k-1)\lambda b_{k-1}(\lambda).$$

The expansion (8.11.7) also applies when $a \to -\infty$ with $\lambda < 0$, and in this case Gautschi (1959a) supplies numerical bounds for the remainders in the truncated expansion (8.11.7). For extensions to complex variables see Temme (1994a, §4), and also Mahler (1930), Tricomi (1950b), and Paris (2002b).

8.11(iv) Large a, Bounded $(x-a)/(2a)^{\frac{1}{2}}$

If $x = a + (2a)^{\frac{1}{2}} y$ and $a \to +\infty$, then

$$8.11.10$$
$$P(a+1,x) = \tfrac{1}{2}\operatorname{erfc}(-y) - \tfrac{1}{3}\sqrt{\tfrac{2}{\pi a}}(1+y^2)e^{-y^2} + O(a^{-1}),$$

$$8.11.11 \quad \gamma^*(1-a,-x) = x^{a-1}\left(-\cos(\pi a) + \frac{\sin(\pi a)}{\pi}\left(2\sqrt{\pi}\,F(y) + \frac{2}{3}\sqrt{\frac{2\pi}{a}}(1-y^2)\right)e^{y^2} + O(a^{-1})\right),$$

in both cases uniformly with respect to bounded real values of y. For Dawson's integral $F(y)$ see §7.2(ii). See Tricomi (1950b) for these approximations, together with higher terms and extensions to complex variables. For related expansions involving Hermite polynomials see Pagurova (1965).

8.11(v) Other Approximations

As $z \to \infty$,

$$8.11.12$$
$$\Gamma(z,z) \sim z^{z-1} e^{-z}\left(\sqrt{\frac{\pi}{2}}z^{\frac{1}{2}} - \frac{1}{3} + \frac{\sqrt{2\pi}}{24 z^{\frac{1}{2}}} - \frac{4}{135 z}\right.$$
$$\left. + \frac{\sqrt{2\pi}}{576 z^{\frac{3}{2}}} + \frac{8}{2835 z^2} + \ldots\right),$$
$$|\operatorname{ph} z| \leq \pi - \delta.$$

For the function $e_n(z)$ defined by (8.4.11),

$$8.11.13 \quad \lim_{n \to \infty} \frac{e_n(nx)}{e^{nx}} = \begin{cases} 0, & x > 1, \\ \tfrac{1}{2}, & x = 1, \\ 1, & 0 \leq x < 1. \end{cases}$$

With $x = 1$, an asymptotic expansion of $e_n(nx)/e^{nx}$ follows from (8.11.14) and (8.11.16).

If $S_n(x)$ is defined by

$$8.11.14 \quad e^{nx} = e_n(nx) + \frac{(nx)^n}{n!} S_n(x),$$

then

$$8.11.15 \quad S_n(x) = \frac{\gamma(n+1,nx)}{(nx)^n e^{-nx}}.$$

As $n \to \infty$

$$8.11.16$$
$$S_n(1) - \tfrac{1}{2}\frac{n! e^n}{n^n} \sim -\tfrac{2}{3} + \tfrac{4}{135}n^{-1} - \tfrac{8}{2835}n^{-2} - \tfrac{16}{8505}n^{-3} + \ldots,$$

$$8.11.17$$
$$S_n(-1) \sim -\tfrac{1}{2} + \tfrac{1}{8}n^{-1} + \tfrac{1}{32}n^{-2} - \tfrac{1}{128}n^{-3} - \tfrac{13}{512}n^{-4} + \ldots.$$

Also,

$$8.11.18 \quad S_n(x) \sim \sum_{k=0}^{\infty} d_k(x) n^{-k}, \quad n \to \infty,$$

uniformly for $x \in (-\infty, 1-\delta]$, with

$$8.11.19 \quad d_k(x) = \frac{(-1)^k b_k(x)}{(1-x)^{2k+1}}, \quad k = 0,1,2,\ldots,$$

and $b_k(x)$ as in §8.11(iii).

For (8.11.18) and extensions to complex values of x see Buckholtz (1963). For a uniformly valid expansion for $n \to \infty$ and $x \in [\delta, 1]$, see Wong (1973b).

8.12 Uniform Asymptotic Expansions for Large Parameter

Define

8.12.1 $\quad \lambda = z/a, \quad \eta = (2(\lambda - 1 - \ln \lambda))^{1/2},$

where the branch of the square root is continuous and satisfies $\eta(\lambda) \sim \lambda - 1$ as $\lambda \to 1$. Then

8.12.2 $\quad \tfrac{1}{2}\eta^2 = \lambda - 1 - \ln \lambda, \quad \dfrac{d\eta}{d\lambda} = \dfrac{\lambda - 1}{\lambda \eta}.$

Also, denote

8.12.3 $\quad P(a, z) = \tfrac{1}{2}\operatorname{erfc}\left(-\eta\sqrt{a/2}\right) - S(a, \eta),$

8.12.4 $\quad Q(a, z) = \tfrac{1}{2}\operatorname{erfc}\left(\eta\sqrt{a/2}\right) + S(a, \eta),$

8.12.5
$$\Gamma(a+1) \dfrac{e^{\pm \pi i a}}{2\pi i} \Gamma(-a, z e^{\pm \pi i})$$
$$= \mp \tfrac{1}{2}\operatorname{erfc}\left(\pm i\eta\sqrt{a/2}\right) + iT(a, \eta),$$

and

8.12.6
$$z^{-a}\gamma^*(-a, -z)$$
$$= \cos(\pi a) - 2\sin(\pi a)\left(\dfrac{e^{\frac{1}{2}a\eta^2}}{\sqrt{\pi}}F\left(\eta\sqrt{a/2}\right) + T(a, \eta)\right),$$

where $F(x)$ is Dawson's integral; see §7.2(ii). Then as $a \to \infty$ in the sector $|\operatorname{ph} a| \leq \pi - \delta (<\pi)$,

8.12.7 $\quad S(a, \eta) \sim \dfrac{e^{-\frac{1}{2}a\eta^2}}{\sqrt{2\pi a}} \displaystyle\sum_{k=0}^{\infty} c_k(\eta) a^{-k},$

8.12.8 $\quad T(a, \eta) \sim \dfrac{e^{\frac{1}{2}a\eta^2}}{\sqrt{2\pi a}} \displaystyle\sum_{k=0}^{\infty} c_k(\eta)(-a)^{-k},$

in each case uniformly with respect to λ in the sector $|\operatorname{ph} \lambda| \leq 2\pi - \delta (<2\pi)$.

With $\mu = \lambda - 1$, the coefficients $c_k(\eta)$ are given by

8.12.9 $\quad c_0(\eta) = \dfrac{1}{\mu} - \dfrac{1}{\eta}, \quad c_1(\eta) = \dfrac{1}{\eta^3} - \dfrac{1}{\mu^3} - \dfrac{1}{\mu^2} - \dfrac{1}{12\mu},$

8.12.10 $\quad c_k(\eta) = \dfrac{1}{\eta}\dfrac{d}{d\eta}c_{k-1}(\eta) + (-1)^k \dfrac{g_k}{\mu}, \quad k = 1, 2, \ldots,$

where g_k, $k = 0, 1, 2, \ldots$, are the coefficients that appear in the asymptotic expansion (5.11.3) of $\Gamma(z)$. The right-hand sides of equations (8.12.9), (8.12.10) have removable singularities at $\eta = 0$, and the Maclaurin series expansion of $c_k(\eta)$ is given by

8.12.11 $\quad c_k(\eta) = \displaystyle\sum_{n=0}^{\infty} d_{k,n}\eta^n, \quad |\eta| < 2\sqrt{\pi},$

where $d_{0,0} = -\tfrac{1}{3}$,

8.12.12
$$d_{0,n} = (n+2)\alpha_{n+2}, \quad n \geq 1,$$
$$d_{k,n} = (-1)^k g_k d_{0,n} + (n+2)d_{k-1,n+2}, \quad n \geq 0, k \geq 1,$$

and $\alpha_3, \alpha_4, \ldots$ are defined by

8.12.13 $\quad \lambda - 1 = \eta + \tfrac{1}{3}\eta^2 + \displaystyle\sum_{n=3}^{\infty} \alpha_n \eta^n, \quad |\eta| < 2\sqrt{\pi}.$

In particular,

8.12.14
$\alpha_3 = \tfrac{1}{36}, \quad \alpha_4 = -\tfrac{1}{270}, \quad \alpha_5 = \tfrac{1}{4320},$
$\alpha_6 = \tfrac{1}{17010}, \quad \alpha_7 = -\tfrac{139}{54\,43200}, \quad \alpha_8 = \tfrac{1}{2\,04120}.$

For numerical values of $d_{k,n}$ to 30D for $k = 0(1)9$ and $n = 0(1)N_k$, where $N_k = 28 - 4\lfloor k/2 \rfloor$, see DiDonato and Morris (1986).

Special cases are given by

8.12.15 $\quad Q(a, a) \sim \dfrac{1}{2} + \dfrac{1}{\sqrt{2\pi a}} \displaystyle\sum_{k=0}^{\infty} c_k(0) a^{-k}, \quad |\operatorname{ph} a| \leq \pi - \delta,$

8.12.16
$$\dfrac{e^{\pm \pi i a}}{2i \sin(\pi a)} Q(-a, a e^{\pm \pi i})$$
$$\sim \pm\dfrac{1}{2} - \dfrac{i}{\sqrt{2\pi a}} \sum_{k=0}^{\infty} c_k(0)(-a)^{-k}, \quad |\operatorname{ph} a| \leq \pi - \delta,$$

where

8.12.17
$c_0(0) = -\tfrac{1}{3}, \quad c_1(0) = -\tfrac{1}{540},$
$c_2(0) = \tfrac{25}{6048}, \quad c_3(0) = \tfrac{101}{1\,55520},$
$c_4(0) = -\tfrac{31\,84811}{36951\,55200}, \quad c_5(0) = -\tfrac{27\,45493}{81517\,36320}.$

For error bounds for (8.12.7) see Paris (2002a). For the asymptotic behavior of $c_k(\eta)$ as $k \to \infty$ see Dunster et al. (1998) and Olde Daalhuis (1998c). The last reference also includes an exponentially-improved version (§2.11(iii)) of the expansions (8.12.4) and (8.12.7) for $Q(a, z)$.

A different type of uniform expansion with coefficients that do not possess a removable singularity at $z = a$ is given by

8.12.18
$$\left.\begin{array}{l}Q(a, z) \\ P(a, z)\end{array}\right\} \sim \dfrac{z^{a-\frac{1}{2}}e^{-z}}{\Gamma(a)}\left(d(\pm\chi)\sum_{k=0}^{\infty}\dfrac{A_k(\chi)}{z^{k/2}} \pm \sum_{k=1}^{\infty}\dfrac{B_k(\chi)}{z^{k/2}}\right),$$

for $z \to \infty$ in $|\operatorname{ph} z| < \tfrac{1}{2}\pi$, with $\Re(z - a) \leq 0$ for $P(a, z)$ and $\Re(z - a) \geq 0$ for $Q(a, z)$. Here

8.12.19 $\quad \chi = (z - a)/\sqrt{z}, \quad d(\pm\chi) = \sqrt{\tfrac{1}{2}\pi}e^{\chi^2/2}\operatorname{erfc}(\pm\chi/\sqrt{2}),$

and

8.12.20 $\quad A_0(\chi) = 1, \quad A_1(\chi) = \tfrac{1}{2}\chi + \tfrac{1}{6}\chi^3, \quad B_1(\chi) = \tfrac{1}{3} + \tfrac{1}{6}\chi^2.$

Higher coefficients $A_k(\chi)$, $B_k(\chi)$, up to $k = 8$, are given in Paris (2002b).

Lastly, a uniform approximation for $\Gamma(a, ax)$ for large a, with error bounds, can be found in Dunster (1996a).

For other uniform asymptotic approximations of the incomplete gamma functions in terms of the function erfc see Paris (2002b) and Dunster (1996a).

Inverse Function

For asymptotic expansions, as $a \to \infty$, of the *inverse function* $x = x(a, q)$ that satisfies the equation

8.12.21
$$Q(a, x) = q$$

see Temme (1992a). These expansions involve the inverse error function $\operatorname{inverfc}(x)$ (§7.17), and are uniform with respect to $q \in [0, 1]$. As a special case,

8.12.22
$$x(a, \tfrac{1}{2}) \sim a - \tfrac{1}{3} + \tfrac{8}{405}a^{-1} + \tfrac{184}{25515}a^{-2} + \tfrac{2248}{34\,44525}a^{-3} + \cdots, \qquad a \to \infty.$$

8.13 Zeros

8.13(i) x-Zeros of $\gamma^*(a, x)$

The function $\gamma^*(a, x)$ has no real zeros for $a \geq 0$. For $a < 0$ and $n = 1, 2, 3, \ldots$, there exist:

(a) one negative zero $x_-(a)$ and no positive zeros when $1 - 2n < a < 2 - 2n$;

(b) one negative zero $x_-(a)$ and one positive zero $x_+(a)$ when $-2n < a < 1 - 2n$.

The negative zero $x_-(a)$ decreases monotonically in the interval $-1 < a < 0$, and satisfies

8.13.1
$$1 + a^{-1} < x_-(a) < \ln|a|, \quad -1 < a < 0.$$

When $-5 \leq a \leq 4$ the behavior of the x-zeros as functions of a can be seen by taking the slice $\gamma^*(a, x) = 0$ of the surface depicted in Figure 8.3.6. Note that from (8.4.12) $\gamma^*(-n, 0) = 0$, $n = 1, 2, 3, \ldots$.

For asymptotic approximations for $x_+(a)$ and $x_-(a)$ as $a \to -\infty$ see Tricomi (1950b), with corrections by Kölbig (1972b).

8.13(ii) λ-Zeros of $\gamma(a, \lambda a)$ and $\Gamma(a, \lambda a)$

For information on the distribution and computation of zeros of $\gamma(a, \lambda a)$ and $\Gamma(a, \lambda a)$ in the complex λ-plane for large values of the positive real parameter a see Temme (1995a).

8.13(iii) a-Zeros of $\gamma^*(a, x)$

For fixed x and $n = 1, 2, 3, \ldots$, $\gamma^*(a, x)$ has:

(a) two zeros in each of the intervals $-2n < a < 2-2n$ when $x < 0$;

(b) two zeros in each of the intervals $-2n < a < 1-2n$ when $0 < x \leq x_n^*$;

(c) zeros at $a = -n$ when $x = 0$.

As x increases the positive zeros coalesce to form a double zero at (a_n^*, x_n^*). The values of the first six double zeros are given to 5D in Table 8.13.1. For values up to $n = 10$ see Kölbig (1972b). Approximations to a_n^*, x_n^* for large n can be found in Kölbig (1970). When $x > x_n^*$ a pair of conjugate trajectories emanate from the point $a = a_n^*$ in the complex a-plane. See Kölbig (1970, 1972b) for further information.

Table 8.13.1: Double zeros (a_n^*, x_n^*) of $\gamma^*(a, x)$.

n	a_n^*	x_n^*
1	-1.64425	0.30809
2	-3.63887	0.77997
3	-5.63573	1.28634
4	-7.63372	1.80754
5	-9.63230	2.33692
6	-11.63126	2.87150

8.14 Integrals

8.14.1
$$\int_0^\infty e^{-ax} \frac{\gamma(b, x)}{\Gamma(b)} \, dx = \frac{(1+a)^{-b}}{a}, \quad \Re a > 0, \Re b > -1,$$

8.14.2
$$\int_0^\infty e^{-ax} \Gamma(b, x) \, dx = \Gamma(b) \frac{1 - (1+a)^{-b}}{a}, \quad \Re a > -1, \Re b > -1.$$

In (8.14.1) and (8.14.2) limiting values are used when $b = 0$.

8.14.3
$$\int_0^\infty x^{a-1} \gamma(b, x) \, dx = -\frac{\Gamma(a+b)}{a}, \quad \Re a < 0, \Re(a+b) > 0,$$

8.14.4
$$\int_0^\infty x^{a-1} \Gamma(b, x) \, dx = \frac{\Gamma(a+b)}{a}, \quad \Re a > 0, \Re(a+b) > 0,$$

8.14.5
$$\int_0^\infty x^{a-1} e^{-sx} \gamma(b, x) \, dx = \frac{\Gamma(a+b)}{b(1+s)^{a+b}} F(1, a+b; 1+b; 1/(1+s)),$$
$$\Re s > 0, \Re(a+b) > 0,$$

8.14.6
$$\int_0^\infty x^{a-1} e^{-sx} \Gamma(b, x) \, dx = \frac{\Gamma(a+b)}{a(1+s)^{a+b}} F(1, a+b; 1+a; s/(1+s)),$$
$$\Re s > -1, \Re(a+b) > 0, \Re a > 0.$$

For the hypergeometric function $F(a, b; c; z)$ see §15.2(i).

For additional integrals see Apelblat (1983, §8.2), Erdélyi et al. (1953b, §9.3), Erdélyi et al. (1954a,b), Gradshteyn and Ryzhik (2000, §6.45), Marichev (1983, pp.189–190), Oberhettinger (1972, pp. 68–69), Prudnikov et al. (1986b, §§1.2, 2.10), and Prudnikov et al. (1992a, §3.10).

8.15 Sums

8.15.1 $\quad \gamma(a, \lambda x) = \lambda^a \sum_{k=0}^{\infty} \gamma(a+k, x) \frac{(1-\lambda)^k}{k!}.$

For sums of infinite series whose terms include incomplete gamma functions, see Prudnikov et al. (1986b, §5.2).

8.16 Generalizations

For a generalization of the incomplete gamma function, including asymptotic approximations, see Chaudhry and Zubair (1994, 2001) and Chaudhry et al. (1996). Other generalizations are considered in Guthmann (1991) and Paris (2003).

Related Functions

8.17 Incomplete Beta Functions

8.17(i) Definitions and Basic Properties

Throughout §§8.17 and 8.18 we assume that $a > 0$, $b > 0$, and $0 \le x \le 1$. However, in the case of §8.17 it is straightforward to continue most results analytically to other real values of a, b, and x, and also to complex values.

8.17.1 $\quad B_x(a,b) = \int_0^x t^{a-1}(1-t)^{b-1}\, dt,$

8.17.2 $\quad I_x(a,b) = B_x(a,b)/B(a,b),$

where, as in §5.12, $B(a,b)$ denotes the Beta function:

8.17.3 $\quad B(a,b) = \frac{\Gamma(a)\,\Gamma(b)}{\Gamma(a+b)}.$

8.17.4 $\quad I_x(a,b) = 1 - I_{1-x}(b,a),$

8.17.5 $\quad I_x(m, n-m+1) = \sum_{j=m}^{n} \binom{n}{j} x^j (1-x)^{n-j},$

8.17.6 $\quad I_x(a,a) = \tfrac{1}{2} I_{4x(1-x)}\left(a, \tfrac{1}{2}\right), \quad 0 \le x \le \tfrac{1}{2}.$

For a historical profile of $B_x(a,b)$ see Dutka (1981).

8.17(ii) Hypergeometric Representations

8.17.7 $\quad B_x(a,b) = \frac{x^a}{a} F(a, 1-b; a+1; x),$

8.17.8 $\quad B_x(a,b) = \frac{x^a(1-x)^b}{a} F(a+b, 1; a+1; x),$

8.17.9 $\quad B_x(a,b) = \frac{x^a(1-x)^{b-1}}{a} F\left(\genfrac{}{}{0pt}{}{1, 1-b}{a+1}; \frac{x}{x-1}\right).$

For the hypergeometric function $F(a,b;c;z)$ see §15.2(i).

8.17(iii) Integral Representation

With $a > 0$, $b > 0$, and $0 < x < 1$,

8.17.10 $\quad I_x(a,b) = \frac{x^a(1-x)^b}{2\pi i} \int_{c-i\infty}^{c+i\infty} s^{-a}(1-s)^{-b} \frac{ds}{s-x},$

where $x < c < 1$ and the branches of s^{-a} and $(1-s)^{-b}$ are continuous on the path and assume their principal values when $s = c$.

Further integral representations can be obtained by combining the results given in §8.17(ii) with §15.6.

8.17(iv) Recurrence Relations

With

8.17.11 $\quad x' = 1 - x, \quad c = a + b - 1,$

8.17.12 $\quad I_x(a,b) = x\, I_x(a-1, b) + x'\, I_x(a, b-1),$

8.17.13 $\quad (a+b)\, I_x(a,b) = a\, I_x(a+1, b) + b\, I_x(a, b+1),$

8.17.14 $\quad (a + bx)\, I_x(a,b) = xb\, I_x(a-1, b+1) + a\, I_x(a+1, b),$

8.17.15 $\quad (b + ax')\, I_x(a,b) = ax'\, I_x(a+1, b-1) + b\, I_x(a, b+1),$

8.17.16 $\quad a\, I_x(a+1, b) = (a + cx)\, I_x(a,b) - cx\, I_x(a-1, b),$

8.17.17 $\quad b\, I_x(a, b+1) = (b + cx')\, I_x(a,b) - cx'\, I_x(a, b-1),$

8.17.18 $\quad I_x(a,b) = I_x(a+1, b-1) + \frac{x^a (x')^{b-1}}{a\, B(a,b)},$

8.17.19 $\quad I_x(a,b) = I_x(a-1, b+1) - \frac{x^{a-1}(x')^b}{b\, B(a,b)},$

8.17.20 $\quad I_x(a,b) = I_x(a+1, b) + \frac{x^a(x')^b}{a\, B(a,b)},$

8.17.21 $\quad I_x(a,b) = I_x(a, b+1) - \frac{x^a(x')^b}{b\, B(a,b)}.$

8.17(v) Continued Fraction

8.17.22 $I_x(a,b) = \dfrac{x^a(1-x)^b}{a\,\mathrm{B}(a,b)}\left(\dfrac{1}{1+}\,\dfrac{d_1}{1+}\,\dfrac{d_2}{1+}\,\dfrac{d_3}{1+}\cdots\right),$

where

8.17.23
$$d_{2m} = \frac{m(b-m)x}{(a+2m-1)(a+2m)},$$
$$d_{2m+1} = -\frac{(a+m)(a+b+m)x}{(a+2m)(a+2m+1)}.$$

The $4m$ and $4m+1$ convergents are less than $I_x(a,b)$, and the $4m+2$ and $4m+3$ convergents are greater than $I_x(a,b)$.

See also Cuyt et al. (2008, pp. 385–389).

The expansion (8.17.22) converges rapidly for $x < (a+1)/(a+b+2)$. For $x > (a+1)/(a+b+2)$ or $1-x < (b+1)/(a+b+2)$, more rapid convergence is obtained by computing $I_{1-x}(b,a)$ and using (8.17.4).

8.17(vi) Sums

For sums of infinite series whose terms involve the incomplete Beta function see Hansen (1975, §62).

8.18 Asymptotic Expansions of $I_x(a,b)$

8.18(i) Large Parameters, Fixed x

If b and x are fixed, with $b > 0$ and $0 < x < 1$, then as $a \to \infty$

8.18.1
$$I_x(a,b) = \Gamma(a+b)x^a(1-x)^{b-1}$$
$$\times\left(\sum_{k=0}^{n-1}\frac{1}{\Gamma(a+k+1)\,\Gamma(b-k)}\left(\frac{x}{1-x}\right)^k + O\!\left(\frac{1}{\Gamma(a+n+1)}\right)\right),$$

for each $n = 0, 1, 2, \ldots$. If $b = 1, 2, 3, \ldots$ and $n \geq b$, then the O-term can be omitted and the result is exact.

If $b \to \infty$ and a and x are fixed, with $a > 0$ and $0 < x < 1$, then (8.18.1), with a and b interchanged and x replaced by $1 - x$, can be combined with (8.17.4).

8.18(ii) Large Parameters: Uniform Asymptotic Expansions

Large a, Fixed b

Let

8.18.2 $\xi = -\ln x.$

Then as $a \to \infty$, with $b\,(> 0)$ fixed,

8.18.3 $I_x(a,b) \sim \dfrac{\Gamma(a+b)}{\Gamma(a)}\displaystyle\sum_{k=0}^{\infty} d_k F_k,$

uniformly for $x \in (0,1)$. The functions F_k are defined by

8.18.4 $aF_{k+1} = (k + b - a\xi)F_k + k\xi F_{k-1},$

with

8.18.5 $F_0 = a^{-b}\,Q(b,a\xi),\quad F_1 = \dfrac{b-a\xi}{a}F_0 + \dfrac{\xi^b e^{-a\xi}}{a\,\Gamma(b)},$

and $Q(a,z)$ as in §8.2(i). The coefficients d_k are defined by the generating function

8.18.6 $\left(\dfrac{1-e^{-t}}{t}\right)^{b-1} = \displaystyle\sum_{k=0}^{\infty} d_k(t-\xi)^k.$

In particular,

8.18.7 $d_0 = \left(\dfrac{1-x}{\xi}\right)^{b-1},\quad d_1 = \dfrac{x\xi + x - 1}{(1-x)\xi}(b-1)d_0.$

Compare also §24.16(i).

Symmetric Case

Let

8.18.8 $x_0 = a/(a+b).$

Then as $a + b \to \infty$,

8.18.9
$$I_x(a,b) \sim \tfrac{1}{2}\operatorname{erfc}\!\left(-\eta\sqrt{b/2}\right) + \frac{1}{\sqrt{2\pi(a+b)}}$$
$$\times\left(\frac{x}{x_0}\right)^a\left(\frac{1-x}{1-x_0}\right)^b\sum_{k=0}^{\infty}\frac{(-1)^k c_k(\eta)}{(a+b)^k},$$

uniformly for $x \in (0,1)$ and $a/(a+b),\,b/(a+b) \in [\delta, 1-\delta]$, where δ again denotes an arbitrary small positive constant. For erfc see §7.2(i). Also,

8.18.10 $-\tfrac{1}{2}\eta^2 = x_0 \ln\!\left(\dfrac{x}{x_0}\right) + (1-x_0)\ln\!\left(\dfrac{1-x}{1-x_0}\right),$

with $\eta/(x - x_0) > 0$, and

8.18.11 $c_0(\eta) = \dfrac{1}{\eta} - \dfrac{\sqrt{x_0(1-x_0)}}{x - x_0},$

with limiting value

8.18.12 $c_0(0) = \dfrac{1-2x_0}{3\sqrt{x_0(1-x_0)}}.$

For this result, and for higher coefficients $c_k(\eta)$ see Temme (1996a, §11.3.3.2). All of the $c_k(\eta)$ are analytic at $\eta = 0$.

8.19 Generalized Exponential Integral

General Case

Let $\widetilde{\Gamma}(z)$ denote the scaled gamma function

8.18.13 $\quad \widetilde{\Gamma}(z) = (2\pi)^{-1/2} e^z z^{(1/2)-z} \Gamma(z),$

$\mu = b/a$, and x_0 again be as in (8.18.8). Then as $a \to \infty$

8.18.14
$$I_x(a,b) \sim Q(b, a\zeta) \\ - \frac{(2\pi b)^{-1/2}}{\widetilde{\Gamma}(b)} \left(\frac{x}{x_0}\right)^a \left(\frac{1-x}{1-x_0}\right)^b \sum_{k=0}^{\infty} \frac{h_k(\zeta, \mu)}{a^k},$$

uniformly for $b \in (0, \infty)$ and $x \in (0, 1)$. Here

8.18.15
$$\mu \ln \zeta - \zeta = \ln x + \mu \ln(1-x) + (1+\mu) \ln(1+\mu) - \mu,$$

with $(\zeta - \mu)/(x_0 - x) > 0$, and

8.18.16 $\quad h_0(\zeta, \mu) = \mu \left(\frac{1}{\zeta - \mu} - \frac{(1+\mu)^{-3/2}}{x_0 - x}\right),$

with limiting value

8.18.17 $\quad h_0(\mu, \mu) = \frac{1}{3} \left(\frac{1-\mu}{\sqrt{1+\mu}} - 1\right).$

For this result and higher coefficients $h_k(\zeta, \mu)$ see Temme (1996a, §11.3.3.3). All of the $h_k(\zeta, \mu)$ are analytic at $\zeta = \mu$ (corresponding to $x = x_0$).

Inverse Function

For asymptotic expansions for large values of a and/or b of the x-solution of the equation

8.18.18 $\quad I_x(a,b) = p, \qquad 0 \le p \le 1,$

see Temme (1992b).

8.19 Generalized Exponential Integral

8.19(i) Definition and Integral Representations

For $p, z \in \mathbb{C}$

8.19.1 $\quad E_p(z) = z^{p-1} \Gamma(1-p, z).$

Most properties of $E_p(z)$ follow straightforwardly from those of $\Gamma(a, z)$. For an extensive treatment of $E_1(z)$ see Chapter 6.

8.19.2 $\quad E_p(z) = z^{p-1} \int_z^{\infty} \frac{e^{-t}}{t^p} dt.$

When the path of integration excludes the origin and does not cross the negative real axis (8.19.2) defines the *principal value* of $E_p(z)$, and *unless indicated otherwise* in this Handbook principal values are assumed.

Other Integral Representations

8.19.3 $\quad E_p(z) = \int_1^{\infty} \frac{e^{-zt}}{t^p} dt, \qquad |\operatorname{ph} z| < \tfrac{1}{2}\pi,$

8.19.4 $\quad E_p(z) = \frac{z^{p-1} e^{-z}}{\Gamma(p)} \int_0^{\infty} \frac{t^{p-1} e^{-zt}}{1+t} dt,$
$$|\operatorname{ph} z| < \tfrac{1}{2}\pi, \Re p > 0.$$

Integral representations of Mellin–Barnes type for $E_p(z)$ follow immediately from (8.6.11), (8.6.12), and (8.19.1).

8.19(ii) Graphics

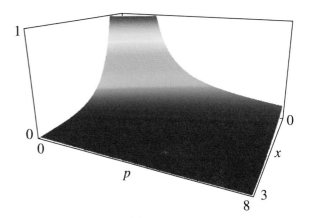

Figure 8.19.1: $E_p(x)$, $0 \le x \le 3$, $0 \le p \le 8$.

In Figures 8.19.2 and 8.19.3, height corresponds to the absolute value of the function and color to the phase. See p. xiv.

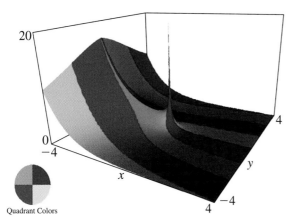

Figure 8.19.2: $E_{\frac{1}{2}}(x+iy)$, $-4 \leq x \leq 4$, $-4 \leq y \leq 4$. Principal value. There is a branch cut along the negative real axis.

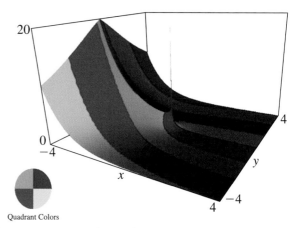

Figure 8.19.3: $E_1(x+iy)$, $-4 \leq x \leq 4$, $-4 \leq y \leq 4$. Principal value. There is a branch cut along the negative real axis.

For additional graphics see http://dlmf.nist.gov/8.19.ii.

8.19(iii) Special Values

8.19.5
$$E_0(z) = z^{-1} e^{-z}, \qquad z \neq 0,$$

8.19.6
$$E_p(0) = \frac{1}{p-1}, \qquad \Re p > 1,$$

8.19.7
$$E_n(z) = \frac{(-z)^{n-1}}{(n-1)!} E_1(z) + \frac{e^{-z}}{(n-1)!} \sum_{k=0}^{n-2} (n-k-2)!(-z)^k,$$
$$n = 2, 3, \ldots.$$

8.19(iv) Series Expansions

For $n = 1, 2, 3, \ldots$,

8.19.8
$$E_n(z) = \frac{(-z)^{n-1}}{(n-1)!} (\psi(n) - \ln z) - \sum_{\substack{k=0 \\ k \neq n-1}}^{\infty} \frac{(-z)^k}{k!(1-n+k)},$$

and

8.19.9
$$E_n(z) = \frac{(-1)^n z^{n-1}}{(n-1)!} \ln z + \frac{e^{-z}}{(n-1)!} \sum_{k=1}^{n-1} (-z)^{k-1} \Gamma(n-k)$$
$$+ \frac{e^{-z}(-z)^{n-1}}{(n-1)!} \sum_{k=0}^{\infty} \frac{z^k}{k!} \psi(k+1),$$

with $|\operatorname{ph} z| \leq \pi$ in both equations. For $\psi(x)$ see §5.2(i). When $p \in \mathbb{C}$

8.19.10 $\quad E_p(z) = z^{p-1} \Gamma(1-p) - \sum_{k=0}^{\infty} \frac{(-z)^k}{k!(1-p+k)},$

8.19.11
$$E_p(z) = \Gamma(1-p) \left(z^{p-1} - e^{-z} \sum_{k=0}^{\infty} \frac{z^k}{\Gamma(2-p+k)} \right),$$

again with $|\operatorname{ph} z| \leq \pi$ in both equations. The right-hand sides are replaced by their limiting forms when $p = 1, 2, 3, \ldots$.

8.19(v) Recurrence Relation and Derivatives

8.19.12 $\quad p\, E_{p+1}(z) + z\, E_p(z) = e^{-z}.$

8.19.13 $\quad \dfrac{d}{dz} E_p(z) = -E_{p-1}(z),$

8.19.14 $\quad \dfrac{d}{dz}(e^z E_p(z)) = e^z E_p(z) \left(1 + \dfrac{p-1}{z}\right) - \dfrac{1}{z}.$

p-Derivatives

For $j = 1, 2, 3, \ldots$,

8.19.15
$$\frac{\partial^j E_p(z)}{\partial p^j} = (-1)^j \int_1^{\infty} (\ln t)^j t^{-p} e^{-zt}\, dt, \quad \Re z > 0.$$

For properties and numerical tables see Milgram (1985), and also (when $p = 1$) MacLeod (2002b).

8.19(vi) Relation to Confluent Hypergeometric Function

8.19.16 $\quad E_p(z) = z^{p-1} e^{-z} U(p, p, z).$

For $U(a, b, z)$ see §13.2(i).

8.19(vii) Continued Fraction

8.19.17
$$E_p(z) = e^{-z}\left(\cfrac{1}{z+}\cfrac{p}{1+}\cfrac{1}{z+}\cfrac{p+1}{1+}\cfrac{2}{z+}\cdots\right), \quad |\operatorname{ph} z| < \pi.$$

See also Cuyt et al. (2008, pp. 277–285).

8.19(viii) Analytic Continuation

The general function $E_p(z)$ is attained by extending the path in (8.19.2) across the negative real axis. Unless p is a nonpositive integer, $E_p(z)$ has a branch point at $z = 0$. For $z \neq 0$ each branch of $E_p(z)$ is an entire function of p.

8.19.18
$$E_p(ze^{2m\pi i}) = \frac{2\pi i e^{m p \pi i}}{\Gamma(p)} \frac{\sin(mp\pi)}{\sin(p\pi)} z^{p-1} + E_p(z),$$
$$m \in \mathbb{Z}, \, z \neq 0.$$

8.19(ix) Inequalities

For $n = 1, 2, 3, \ldots$ and $x > 0$,

8.19.19
$$\frac{n-1}{n} E_n(x) < E_{n+1}(x) < E_n(x),$$

8.19.20
$$(E_n(x))^2 < E_{n-1}(x) E_{n+1}(x),$$

8.19.21
$$\frac{1}{x+n} < e^x E_n(x) \leq \frac{1}{x+n-1},$$

8.19.22
$$\frac{d}{dx}\frac{E_n(x)}{E_{n-1}(x)} > 0.$$

8.19(x) Integrals

8.19.23
$$\int_z^\infty E_{p-1}(t)\, dt = E_p(z), \quad |\operatorname{ph} z| < \pi,$$

8.19.24
$$\int_0^\infty e^{-at} E_n(t)\, dt$$
$$= \frac{(-1)^{n-1}}{a^n}\left(\ln(1+a) + \sum_{k=1}^{n-1}\frac{(-1)^k a^k}{k}\right),$$
$$n = 1, 2, \ldots, \Re a > -1,$$

8.19.25
$$\int_0^\infty e^{-at} t^{b-1} E_p(t)\, dt = \frac{\Gamma(b)(1+a)^{-b}}{p+b-1}$$
$$\times F(1, b; p+b; a/(1+a)),$$
$$\Re a > -1, \Re(p+b) > 1.$$

8.19.26
$$\int_0^\infty E_p(t) E_q(t)\, dt = \frac{L(p) + L(q)}{p+q-1},$$
$$p > 0, q > 0, p+q > 1,$$

where

8.19.27
$$L(p) = \int_0^\infty e^{-t} E_p(t)\, dt = \frac{1}{2p} F(1,1;1+p;\tfrac{1}{2}), \quad p > 0.$$

For the hypergeometric function $F(a, b; c; z)$ see §15.2(i). When $p = 1, 2, 3, \ldots$, $L(p)$ can also be evaluated via (8.19.24).

For collections of integrals involving $E_p(z)$, especially for integer p, see Apelblat (1983, §§7.1–7.2) and LeCaine (1945).

8.19(xi) Further Generalizations

For higher-order generalized exponential integrals see Meijer and Baken (1987) and Milgram (1985).

8.20 Asymptotic Expansions of $E_p(z)$

8.20(i) Large z

8.20.1
$$E_p(z) = \frac{e^{-z}}{z}\left(\sum_{k=0}^{n-1}(-1)^k\frac{(p)_k}{z^k} + (-1)^n\frac{(p)_n e^z}{z^{n-1}} E_{n+p}(z)\right),$$
$$n = 1, 2, 3, \ldots.$$

As $z \to \infty$

8.20.2
$$E_p(z) \sim \frac{e^{-z}}{z}\sum_{k=0}^\infty (-1)^k\frac{(p)_k}{z^k}, \quad |\operatorname{ph} z| \leq \tfrac{3}{2}\pi - \delta,$$

and

8.20.3
$$E_p(z) \sim \pm\frac{2\pi i}{\Gamma(p)} e^{\mp p\pi i} z^{p-1} + \frac{e^{-z}}{z}\sum_{k=0}^\infty \frac{(-1)^k (p)_k}{z^k},$$
$$\tfrac{1}{2}\pi + \delta \leq \pm\operatorname{ph} z \leq \tfrac{7}{2}\pi - \delta,$$

δ again denoting an arbitrary small positive constant. Where the sectors of validity of (8.20.2) and (8.20.3) overlap the contribution of the first term on the right-hand side of (8.20.3) is exponentially small compared to the other contribution; compare §2.11(ii).

For an exponentially-improved asymptotic expansion of $E_p(z)$ see §2.11(iii).

8.20(ii) Large p

For $x \geq 0$ and $p > 1$ let $x = \lambda p$ and define $A_0(\lambda) = 1$,

8.20.4
$$A_{k+1}(\lambda) = (1 - 2k\lambda)A_k(\lambda) + \lambda(\lambda+1)\frac{dA_k(\lambda)}{d\lambda},$$
$$k = 0, 1, 2, \ldots,$$

so that $A_k(\lambda)$ is a polynomial in λ of degree $k-1$ when $k \geq 1$. In particular,

8.20.5
$$A_1(\lambda) = 1, \quad A_2(\lambda) = 1 - 2\lambda, \quad A_3(\lambda) = 1 - 8\lambda + 6\lambda^2.$$

Then as $p \to \infty$

8.20.6
$$E_p(\lambda p) \sim \frac{e^{-\lambda p}}{(\lambda+1)p}\sum_{k=0}^\infty \frac{A_k(\lambda)}{(\lambda+1)^{2k}}\frac{1}{p^k},$$

uniformly for $\lambda \in [0, \infty)$.

For further information, including extensions to complex values of x and p, see Temme (1994a, §4) and Dunster (1996b, 1997).

8.21 Generalized Sine and Cosine Integrals

8.21(i) Definitions: General Values

With γ and Γ denoting here the general values of the incomplete gamma functions (§8.2(i)), we define

8.21.1 $\quad \operatorname{ci}(a,z) \pm i\operatorname{si}(a,z) = e^{\pm\frac{1}{2}\pi i a} \Gamma\left(a, ze^{\mp\frac{1}{2}\pi i}\right),$

8.21.2 $\quad \operatorname{Ci}(a,z) \pm i\operatorname{Si}(a,z) = e^{\pm\frac{1}{2}\pi i a} \gamma\left(a, ze^{\mp\frac{1}{2}\pi i}\right).$

From §§8.2(i) and 8.2(ii) it follows that each of the four functions $\operatorname{si}(a,z)$, $\operatorname{ci}(a,z)$, $\operatorname{Si}(a,z)$, and $\operatorname{Ci}(a,z)$ is a multivalued function of z with branch point at $z=0$. Furthermore, $\operatorname{si}(a,z)$ and $\operatorname{ci}(a,z)$ are entire functions of a, and $\operatorname{Si}(a,z)$ and $\operatorname{Ci}(a,z)$ are meromorphic functions of a with simple poles at $a = -1, -3, -5, \ldots$ and $a = 0, -2, -4, \ldots$, respectively.

8.21(ii) Definitions: Principal Values

When $\operatorname{ph} z = 0$ (and when $a \neq -1, -3, -5, \ldots$, in the case of $\operatorname{Si}(a,z)$, or $a \neq 0, -2, -4, \ldots$, in the case of $\operatorname{Ci}(a,z)$) the *principal values* of $\operatorname{si}(a,z)$, $\operatorname{ci}(a,z)$, $\operatorname{Si}(a,z)$, and $\operatorname{Ci}(a,z)$ are defined by (8.21.1) and (8.21.2) with the incomplete gamma functions assuming their principal values (§8.2(i)). Elsewhere in the sector $|\operatorname{ph} z| \leq \pi$ the principal values are defined by analytic continuation from $\operatorname{ph} z = 0$; compare §4.2(i).

From here on it is assumed that *unless indicated otherwise* the functions $\operatorname{si}(a,z)$, $\operatorname{ci}(a,z)$, $\operatorname{Si}(a,z)$, and $\operatorname{Ci}(a,z)$ have their principal values.

Properties of the four functions that are stated below in §§8.21(iii) and 8.21(iv) follow directly from the definitions given above, together with properties of the incomplete gamma functions given earlier in this chapter. In the case of §8.21(iv) the equation

8.21.3 $\quad \int_0^\infty t^{a-1} e^{\pm it} \, dt = e^{\pm\frac{1}{2}\pi i a} \Gamma(a), \quad 0 < \Re a < 1,$

(obtained from (5.2.1) by rotation of the integration path) is also needed.

8.21(iii) Integral Representations

8.21.4 $\quad \operatorname{si}(a,z) = \int_z^\infty t^{a-1} \sin t \, dt, \qquad \Re a < 1,$

8.21.5 $\quad \operatorname{ci}(a,z) = \int_z^\infty t^{a-1} \cos t \, dt, \qquad \Re a < 1,$

8.21.6 $\quad \operatorname{Si}(a,z) = \int_0^z t^{a-1} \sin t \, dt, \qquad \Re a > -1,$

8.21.7 $\quad \operatorname{Ci}(a,z) = \int_0^z t^{a-1} \cos t \, dt, \qquad \Re a > 0.$

In these representations the integration paths do not cross the negative real axis, and in the case of (8.21.4) and (8.21.5) the paths also exclude the origin.

8.21(iv) Interrelations

8.21.8
$\operatorname{Si}(a,z) = \Gamma(a) \sin\left(\tfrac{1}{2}\pi a\right) - \operatorname{si}(a,z), \quad a \neq -1, -3, -5, \ldots,$

8.21.9
$\operatorname{Ci}(a,z) = \Gamma(a) \cos\left(\tfrac{1}{2}\pi a\right) - \operatorname{ci}(a,z), \quad a \neq 0, -2, -4, \ldots.$

8.21(v) Special Values

8.21.10 $\quad \operatorname{si}(0,z) = -\operatorname{si}(z), \quad \operatorname{ci}(0,z) = -\operatorname{Ci}(z),$

8.21.11 $\quad \operatorname{Si}(0,z) = \operatorname{Si}(z).$

For the functions on the right-hand sides of (8.21.10) and (8.21.11) see §6.2(ii).

8.21.12 $\quad \operatorname{Si}(a,\infty) = \Gamma(a) \sin\left(\tfrac{1}{2}\pi a\right), \quad a \neq -1, -3, -5, \ldots,$

8.21.13 $\quad \operatorname{Ci}(a,\infty) = \Gamma(a) \cos\left(\tfrac{1}{2}\pi a\right), \quad a \neq 0, -2, -4, \ldots.$

8.21(vi) Series Expansions

Power-Series Expansions

8.21.14 $\quad \operatorname{Si}(a,z) = z^a \sum_{k=0}^\infty \frac{(-1)^k z^{2k+1}}{(2k+a+1)(2k+1)!},$
$$a \neq -1, -3, -5, \ldots,$$

8.21.15
$$\operatorname{Ci}(a,z) = z^a \sum_{k=0}^\infty \frac{(-1)^k z^{2k}}{(2k+a)(2k)!}, \quad a \neq 0, -2, -4, \ldots.$$

Spherical-Bessel-Function Expansions

8.21.16 $\quad \operatorname{Si}(a,z) = z^a \sum_{k=0}^\infty \frac{\left(2k+\tfrac{3}{2}\right)\left(1-\tfrac{1}{2}a\right)_k}{\left(\tfrac{1}{2}+\tfrac{1}{2}a\right)_{k+1}} \mathsf{j}_{2k+1}(z),$
$$a \neq -1, -3, -5, \ldots,$$

8.21.17 $\quad \operatorname{Ci}(a,z) = z^a \sum_{k=0}^\infty \frac{\left(2k+\tfrac{1}{2}\right)\left(\tfrac{1}{2}-\tfrac{1}{2}a\right)_k}{\left(\tfrac{1}{2}a\right)_{k+1}} \mathsf{j}_{2k}(z),$
$$a \neq 0, -2, -4, \ldots.$$

For $\mathsf{j}_n(z)$ see §10.47(ii). For (8.21.16), (8.21.17), and further expansions in series of Bessel functions see Luke (1969b, pp. 56–57).

8.21(vii) Auxiliary Functions

8.21.18 $\quad f(a,z) = \operatorname{si}(a,z)\cos z - \operatorname{ci}(a,z)\sin z,$

8.21.19 $\quad g(a,z) = \operatorname{si}(a,z)\sin z + \operatorname{ci}(a,z)\cos z.$

8.21.20 $\quad \operatorname{si}(a,z) = f(a,z)\cos z + g(a,z)\sin z,$

8.21.21 $\quad \operatorname{ci}(a,z) = -f(a,z)\sin z + g(a,z)\cos z.$

When $|\operatorname{ph} z| < \pi$ and $\Re a < 1$,

8.21.22 $\quad f(a,z) = \int_0^\infty \frac{\sin t}{(t+z)^{1-a}}\, dt,$

8.21.23 $\quad g(a,z) = \int_0^\infty \frac{\cos t}{(t+z)^{1-a}}\, dt.$

When $|\operatorname{ph} z| < \frac{1}{2}\pi$,

8.21.24
$$f(a,z) = \frac{z^a}{2}\int_0^\infty \left((1+it)^{a-1} + (1-it)^{a-1}\right) e^{-zt}\, dt,$$

8.21.25
$$g(a,z) = \frac{z^a}{2i}\int_0^\infty \left((1-it)^{a-1} - (1+it)^{a-1}\right) e^{-zt}\, dt.$$

8.21(viii) Asymptotic Expansions

When $z \to \infty$ with $|\operatorname{ph} z| \le \pi - \delta\ (<\pi)$,

8.21.26 $\quad f(a,z) \sim z^{a-1}\sum_{k=0}^\infty \frac{(-1)^k (1-a)_{2k}}{z^{2k}},$

8.21.27 $\quad g(a,z) \sim z^{a-1}\sum_{k=0}^\infty \frac{(-1)^k (1-a)_{2k+1}}{z^{2k+1}}.$

For the corresponding expansions for $\operatorname{si}(a,z)$ and $\operatorname{ci}(a,z)$ apply (8.21.20) and (8.21.21).

Applications

8.22 Mathematical Applications

8.22(i) Terminant Function

The so-called *terminant function* $F_p(z)$, defined by

8.22.1 $\quad F_p(z) = \frac{\Gamma(p)}{2\pi}z^{1-p}E_p(z) = \frac{\Gamma(p)}{2\pi}\Gamma(1-p,z),$

plays a fundamental role in re-expansions of remainder terms in asymptotic expansions, including exponentially-improved expansions and a smooth interpretation of the Stokes phenomenon. See §§2.11(ii)–2.11(v) and the references supplied in these subsections.

8.22(ii) Riemann Zeta Function and Incomplete Riemann Zeta Function

The function $\Gamma(a,z)$, with $|\operatorname{ph} a| \le \frac{1}{2}\pi$ and $\operatorname{ph} z = \frac{1}{2}\pi$, has an intimate connection with the Riemann zeta function $\zeta(s)$ (§25.2(i)) on the critical line $\Re s = \frac{1}{2}$. See Paris and Cang (1997).

If $\zeta_x(s)$ denotes the *incomplete Riemann zeta function* defined by

8.22.2 $\quad \zeta_x(s) = \frac{1}{\Gamma(s)}\int_0^x \frac{t^{s-1}}{e^t - 1}\, dt, \qquad \Re s > 1,$

so that $\lim_{x\to\infty} \zeta_x(s) = \zeta(s)$, then

8.22.3 $\quad \zeta_x(s) = \sum_{k=1}^\infty k^{-s} P(s, kx), \qquad \Re s > 1.$

For further information on $\zeta_x(s)$, including zeros and uniform asymptotic approximations, see Kölbig (1970, 1972a) and Dunster (2006).

8.23 Statistical Applications

The functions $P(a,x)$ and $Q(a,x)$ are used extensively in statistics as the probability integrals of the gamma distribution; see Johnson et al. (1994, pp. 337–414). Particular forms are the chi-square distribution functions; see Johnson et al. (1994, pp. 415–493). The function $B_x(a,b)$ and its normalization $I_x(a,b)$ play a similar role in statistics in connection with the beta distribution; see Johnson et al. (1995, pp. 210–275). In queueing theory the Erlang loss function is used, which can be expressed in terms of the reciprocal of $Q(a,x)$; see Jagerman (1974) and Cooper (1981, pp. 80, 316–319).

8.24 Physical Applications

8.24(i) Incomplete Gamma Functions

The function $\gamma(a,x)$ appears in: discussions of power-law relaxation times in complex physical systems (Sornette (1998)); logarithmic oscillations in relaxation times for proteins (Metzler et al. (1999)); Gaussian orbitals and exponential (Slater) orbitals in quantum chemistry (Shavitt (1963), Shavitt and Karplus (1965)); population biology and ecological systems (Camacho et al. (2002)).

8.24(ii) Incomplete Beta Functions

The function $I_x(a,b)$ appears in: Monte Carlo sampling in statistical mechanics (Kofke (2004)); analysis of packings of soft or granular objects (Prellberg and Owczarek (1995)); growth formulas in cosmology (Hamilton (2001)).

8.24(iii) Generalized Exponential Integral

The function $E_p(x)$, with $p > 0$, appears in theories of transport and radiative equilibrium (Hopf (1934), Kourganoff (1952), Altaç (1996)).

With more general values of p, $E_p(x)$ supplies fundamental auxiliary functions that are used in the computation of molecular electronic integrals in quantum chemistry (Harris (2002), Shavitt (1963)), and also wave acoustics of overlapping sound beams (Ding (2000)).

Computation

8.25 Methods of Computation

8.25(i) Series Expansions

Although the series expansions in §§8.7, 8.19(iv), and 8.21(vi) converge for all finite values of z, they are cumbersome to use when $|z|$ is large owing to slowness of convergence and cancellation. For large $|z|$ the corresponding asymptotic expansions (generally divergent) are used instead. See also Luke (1975, pp. 101–102) and Temme (1994a).

8.25(ii) Quadrature

See Allasia and Besenghi (1987a) for the numerical computation of $\Gamma(a, z)$ from (8.6.4) by means of the trapezoidal rule.

8.25(iii) Asymptotic Expansions

DiDonato and Morris (1986) describes an algorithm for computing $P(a, x)$ and $Q(a, x)$ for $a \geq 0$, $x \geq 0$, and $a + x \neq 0$ from the uniform expansions in §8.12. The algorithm supplies 14S accuracy. A numerical inversion procedure is also given for calculating the value of x (with 10S accuracy), when a and $P(a, x)$ are specified, based on Newton's rule (§3.8(ii)). See also Temme (1987, 1994a).

8.25(iv) Continued Fractions

The computation of $\gamma(a, z)$ and $\Gamma(a, z)$ by means of continued fractions is described in Jones and Thron (1985) and Gautschi (1979a, §§4.3, 5). See also Jacobsen et al. (1986) and Temme (1996a, p. 280).

8.25(v) Recurrence Relations

Expansions involving incomplete gamma functions often require the generation of sequences $P(a+n, x)$, $Q(a+n, x)$, or $\gamma^*(a+n, x)$ for fixed a and $n = 0, 1, 2, \ldots$. An efficient procedure, based partly on the recurrence relations (8.8.5) and (8.8.6), is described in Gautschi (1979a, 1999).

Stable recursive schemes for the computation of $E_p(x)$ are described in Miller (1960) for $x > 0$ and integer p. For $x > 0$ and real p see Amos (1980) and Chiccoli et al. (1987, 1988). See also Chiccoli et al. (1990) and Stegun and Zucker (1974).

8.26 Tables

8.26(i) Introduction

For tables published before 1961 see Fletcher et al. (1962) and Lebedev and Fedorova (1960).

8.26(ii) Incomplete Gamma Functions

- Khamis (1965) tabulates $P(a, x)$ for $a = 0.05(.05)10(.1)20(.25)70$, $0.0001 \leq x \leq 250$ to 10D.

- Pagurova (1963) tabulates $P(a, x)$ and $Q(a, x)$ (with different notation) for $a = 0(.05)3$, $x = 0(.05)1$ to 7D.

- Pearson (1965) tabulates the function $I(u, p)$ ($= P(p+1, u)$) for $p = -1(.05)0(.1)5(.2)50$, $u = 0(.1)u_p$ to 7D, where $I(u, u_p)$ rounds off to 1 to 7D; also $I(u, p)$ for $p = -0.75(.01)-1$, $u = 0(.1)6$ to 5D.

- Zhang and Jin (1996, Table 3.8) tabulates $\gamma(a, x)$ for $a = 0.5, 1, 3, 5, 10, 25, 50, 100$, $x = 0(.1)1(1)3, 5(5)30, 50, 100$ to 8D or 8S.

8.26(iii) Incomplete Beta Functions

- Pearson (1968) tabulates $I_x(a, b)$ for $x = 0.01(.01)1$, $a, b = 0.5(.5)11(1)50$, with $b \leq a$, to 7D.

- Zhang and Jin (1996, Table 3.9) tabulates $I_x(a, b)$ for $x = 0(.05)1$, $a = 0.5, 1, 3, 5, 10$, $b = 1, 10$ to 8D.

8.26(iv) Generalized Exponential Integral

- Abramowitz and Stegun (1964, pp. 245–248) tabulates $E_n(x)$ for $n = 2, 3, 4, 10, 20$, $x = 0(.01)2$ to 7D; also $(x+n)e^x E_n(x)$ for $n = 2, 3, 4, 10, 20$, $x^{-1} = 0(.01)0.1(.05)0.5$ to 6S.

- Chiccoli *et al.* (1988) presents a short table of $E_p(x)$ for $p = -\frac{9}{2}(1) - \frac{1}{2}$, $0 \le x \le 200$ to 14S.

- Pagurova (1961) tabulates $E_n(x)$ for $n = 0(1)20$, $x = 0(.01)2(.1)10$ to 4-9S; $e^x E_n(x)$ for $n = 2(1)10$, $x = 10(.1)20$ to 7D; $e^x E_p(x)$ for $p = 0(.1)1$, $x = 0.01(.01)7(.05)12(.1)20$ to 7S or 7D.

- Stankiewicz (1968) tabulates $E_n(x)$ for $n = 1(1)10$, $x = 0.01(.01)5$ to 7D.

- Zhang and Jin (1996, Table 19.1) tabulates $E_n(x)$ for $n = 1, 2, 3, 5, 10, 15, 20$, $x = 0(.1)1, 1.5, 2, 3, 5, 10, 20, 30, 50, 100$ to 7D or 8S.

8.27 Approximations

8.27(i) Incomplete Gamma Functions

- DiDonato (1978) gives a simple approximation for the function $F(p,x) = x^{-p}e^{x^2/2}\int_x^\infty e^{-t^2/2}t^p\,dt$ (which is related to the incomplete gamma function by a change of variables) for real p and large positive x. This takes the form $F(p,x) = 4x/h(p,x)$, approximately, where $h(p,x) = 3(x^2 - p) + \sqrt{(x^2-p)^2 + 8(x^2+p)}$ and is shown to produce an absolute error $O(x^{-7})$ as $x \to \infty$.

- Luke (1975, §4.3) gives Padé approximation methods, combined with a detailed analysis of the error terms, valid for real and complex variables except on the negative real z-axis. See also Temme (1994a, §3).

- Luke (1969b, pp. 25, 40–41) gives Chebyshev-series expansions for $\Gamma(a, \omega z)$ (by specifying parameters) with $1 \le \omega < \infty$, and $\gamma(a, \omega z)$ with $0 \le \omega \le 1$; see also Temme (1994a, §3).

- Luke (1969b, p. 186) gives hypergeometric polynomial representations that converge uniformly on compact subsets of the z-plane that exclude $z = 0$ and are valid for $|\text{ph } z| < \pi$.

8.27(ii) Generalized Exponential Integral

- Luke (1975, p. 103) gives Chebyshev-series expansions for $E_1(x)$ and related functions for $x \ge 5$.

- Luke (1975, p. 106) gives rational and Padé approximations, with remainders, for $E_1(z)$ and $z^{-1}\int_0^z t^{-1}(1-e^{-t})\,dt$ for complex z with $|\text{ph } z| \le \pi$.

- Verbeeck (1970) gives polynomial and rational approximations for $E_p(x) = (e^{-x}/x)P(z)$, approximately, where $P(z)$ denotes a quotient of polynomials of equal degree in $z = x^{-1}$.

8.28 Software

See http://dlmf.nist.gov/8.28.

References

General References

The main references used in writing this chapter are Erdélyi *et al.* (1953b), Luke (1969b), and Temme (1996a). For additional bibliographic reading see Gautschi (1998), Olver (1997b), and Wong (1989).

Sources

The following list gives the references or other indications of proofs that were used in constructing the various sections of this chapter. These sources supplement the references that are quoted in the text.

§8.2 Olver (1997b, p. 45), Temme (1996b). (8.2.12)–(8.2.14) follow from the definitions in §8.2(i).

§8.3 The graphics were produced at NIST.

§8.4 Erdélyi *et al.* (1953b, Chapter 9). (8.4.14) follows from (8.4.6) and (8.8.6). (8.4.15) follows from the first series in (8.7.3) by combining $\Gamma(a)$ with the term $k = n$ of the series, then taking the limit as $a \to -n$.

§8.5 Slater (1960, §5.6) and §13.2(vii). For (8.5.4) see (13.18.4).

§8.6 For (8.6.1) use (8.6.8), replacing t by e^{it} with $-\pi \le t \le \pi$. For (8.6.2) substitute for $J_a(2\sqrt{zt})$ by (10.2.2), integrate term by term and refer to (8.7.1) and (8.2.6). (8.6.6) may be proved in a similar manner with the aid also of (10.25.2), (10.27.4), (8.2.3), and analytic continuation when $a = -2, -3, -4, \ldots$. For (8.6.3) and (8.6.7) apply (8.2.1) and (8.2.2), taking new integration variables $ze^{\mp t}$. For (8.6.4) and (8.6.5) see Temme (1996a, §§11.2.1–11.2.2). For (8.6.8) assume temporarily $\Re a > 0$, collapse the integration path onto the interval $[-1, 0]$ and use (8.2.1). For (8.6.9) see Temme (1996b). For (8.6.10)–(8.6.12) see Paris and Kaminski (2001, §3.4.3).

§8.7 Erdélyi *et al.* (1953b, Chapter 9). (8.7.3) follows from (8.7.1) and (8.2.3). For (8.7.5) use (8.5.2) and (13.11.1).

§8.8 Erdélyi *et al.* (1953b, Chapter 9). These results also follow straightforwardly from §8.2(i).

§8.10 Olver (1997b, pp. 66–67), Luke (1969b, pp. 195, 201). For (8.10.10)–(8.10.13), see Gautschi (1959b), Alzer (1997b), and Vietoris (1983).

§8.11 Olver (1997b, pp. 66, 109–112), Temme (1996a, p. 280). For (8.11.4) see (8.5.1) or (8.7.1). (8.11.5) follows from the leading terms of (8.11.4) and (5.11.3). For (8.11.6) and (8.11.7) see Gautschi (1959a) and Temme (1994a). (8.11.12) can be obtained from (8.12.15), (8.2.4), and (5.11.3). (8.11.15) follows from (8.4.7) with $z = nx$. For (8.11.16) see Ramanujan (1962, pp. 323–324). For (8.11.17) see Copson (1933).

§8.12 Temme (1979b, 1992a, 1996b), Paris (2002b), and Ferreira et al. (2005).

§8.13 Erdélyi et al. (1953b, §9.6), Tricomi (1950b), Lew (1994), and Kölbig (1972b).

§8.14 (8.14.1)–(8.14.2) are obtained by term-by-term integration using (8.7.1) and (8.7.3). (8.14.3)–(8.14.6) are obtained by specializing (13.10.10), (13.10.11), (13.10.3), (13.10.4) by means of (8.5.1)–(8.5.3).

§8.15 Tricomi (1950b).

§8.17 Temme (1996a, §§11.3–11.3.2). For (8.17.5) combine (8.17.8) and (15.8.1). For (8.17.6) combine (8.17.8), (15.8.18), and (15.8.1). For the last paragraph of §8.17(v) see Zhang and Jin (1996, p. 65).

§8.18 Temme (1996a, §§11.3.3.1–11.3.3.3). For (8.18.1) use (8.17.9) and apply §15.12(ii).

§8.19 For (8.19.1)–(8.19.4) see Temme (1996a, p. 180). For (8.19.5)–(8.19.7) use (8.19.1), (8.19.3), (8.4.15). (8.19.8) follows from (8.4.13) and (8.4.15). (8.19.9) follows from (6.6.3), (8.19.1), and (8.4.15). (8.19.10) and (8.19.11) follow from (8.19.1) and (8.7.3). For (8.19.12)–(8.19.16) combine (8.19.1) with (8.8.2), (8.8.16), (8.8.19), and (8.5.3). For (8.19.17) combine (8.9.2) and (8.19.1). For (8.19.18) see Olver (1994b). For (8.19.19)–(8.19.22) see Hopf (1934, pp. 26–27). For (8.19.23) use (8.19.13). For (8.19.24)–(8.19.27) see Kourganoff (1952, Appendix 1). The graphics were produced at NIST.

§8.20 Olver (1991a), Gautschi (1959a).

§8.21 For §8.21(iii) follow the prescription given in the final paragraph of §8.21(ii). Thus for (8.21.4) and (8.21.5) replace z by iz with $\operatorname{ph} z = 0$ in (8.2.2), deform the path of integration to run along the positive imaginary axis, and replace t by it. Then extend to the sector $|\operatorname{ph} z| \leq \pi$ by analytic continuation. Similarly for (8.21.6) and (8.21.7). For §8.21(iv) temporarily restrict $0 < \Re a < 1$. Then (8.21.8) and (8.21.9) follow immediately from (8.21.3)–(8.21.7). Subsequently, ease the restrictions on a by analytic continuation with respect to a; compare §8.21(i). For (8.21.12) and (8.21.13) use (8.21.8) and (8.21.9), and also (8.21.4) and (8.21.5). (8.21.14) and (8.21.15) are obtained by expansion of the trigonometric functions in (8.21.6), (8.21.7), and termwise integration. See also Luke (1975, p. 115). (8.21.22) and (8.21.23) follow from (8.21.4), (8.21.5), (8.21.18), and (8.21.19). For (8.21.24) and (8.21.25) assume $\operatorname{ph} z = 0$, and in the integrals for $\operatorname{ci}(a,z) \pm i\operatorname{si}(a,z)$ obtained from (8.21.4) and (8.21.5) set $t = (1+\tau)z$, rotate the integration paths in the τ-plane through $\pm \frac{1}{2}\pi$, and apply (8.21.18) and (8.21.19). The restriction $\operatorname{ph} z = 0$ is eased to $|\operatorname{ph} z| < \frac{1}{2}\pi$ by analytic continuation. For (8.21.26) and (8.21.27) apply Watson's lemma to (8.21.24) and (8.21.25), and then extend the sector of validity from $|\operatorname{ph} z| \leq \frac{1}{2}\pi - \delta$ to $|\operatorname{ph} z| \leq \pi - \delta$; see §2.4(i).

Chapter 9
Airy and Related Functions

F. W. J. Olver[1]

Notation   194
    9.1   Special Notation 194

Airy Functions   194
    9.2   Differential Equation 194
    9.3   Graphics 195
    9.4   Maclaurin Series 196
    9.5   Integral Representations 196
    9.6   Relations to Other Functions 196
    9.7   Asymptotic Expansions 198
    9.8   Modulus and Phase 199
    9.9   Zeros 200
    9.10   Integrals 202
    9.11   Products 203

Related Functions   204
    9.12   Scorer Functions 204
    9.13   Generalized Airy Functions 206
    9.14   Incomplete Airy Functions 208

Applications   208
    9.15   Mathematical Applications 208
    9.16   Physical Applications 209

Computation   209
    9.17   Methods of Computation 209
    9.18   Tables 210
    9.19   Approximations 211
    9.20   Software 212

References   212

---

[1] Institute for Physical Science and Technology and Department of Mathematics, University of Maryland, College Park, Maryland.
    Acknowledgments: This chapter is based in part on Abramowitz and Stegun (1964, Chapter 10) by H. A. Antosiewicz. The author is pleased to acknowledge the assistance of Bruce R. Fabijonas for computing the numerical tables in §9.9, and of Leonard Maximon for writing §9.16.
    Copyright © 2009 National Institute of Standards and Technology. All rights reserved.

Notation

9.1 Special Notation

(For other notation see pp. xiv and 873.)

k	nonnegative integer, except in §9.9(iii).
x	real variable.
$z (= x + iy)$	complex variable.
δ	arbitrary small positive constant.
primes	derivatives with respect to argument.

The main functions treated in this chapter are the Airy functions $\mathrm{Ai}(z)$ and $\mathrm{Bi}(z)$, and the Scorer functions $\mathrm{Gi}(z)$ and $\mathrm{Hi}(z)$ (also known as inhomogeneous Airy functions).

Other notations that have been used are as follows: $\mathrm{Ai}(-x)$ and $\mathrm{Bi}(-x)$ for $\mathrm{Ai}(x)$ and $\mathrm{Bi}(x)$ (Jeffreys (1928), later changed to $\mathrm{Ai}(x)$ and $\mathrm{Bi}(x)$); $U(x) = \sqrt{\pi}\,\mathrm{Bi}(x)$, $V(x) = \sqrt{\pi}\,\mathrm{Ai}(x)$ (Fock (1945)); $A(x) = 3^{-1/3}\pi\,\mathrm{Ai}(-3^{-1/3}x)$ (Szegö (1967, §1.81)); $e_0(x) = \pi\,\mathrm{Hi}(-x)$, $\widetilde{e}_0(x) = -\pi\,\mathrm{Gi}(-x)$ (Tumarkin (1959)).

Airy Functions

9.2 Differential Equation

9.2(i) Airy's Equation

9.2.1
$$\frac{d^2 w}{dz^2} = zw.$$

All solutions are entire functions of z.
Standard solutions are:

9.2.2 $w = \mathrm{Ai}(z),\ \mathrm{Bi}(z),\ \mathrm{Ai}\!\left(ze^{\mp 2\pi i/3}\right).$

9.2(ii) Initial Values

9.2.3 $\mathrm{Ai}(0) = \dfrac{1}{3^{2/3}\,\Gamma\!\left(\frac{2}{3}\right)} = 0.35502\,80538\ldots,$

9.2.4 $\mathrm{Ai}'(0) = -\dfrac{1}{3^{1/3}\,\Gamma\!\left(\frac{1}{3}\right)} = -0.25881\,94037\ldots,$

9.2.5 $\mathrm{Bi}(0) = \dfrac{1}{3^{1/6}\,\Gamma\!\left(\frac{2}{3}\right)} = 0.61492\,66274\ldots,$

9.2.6 $\mathrm{Bi}'(0) = \dfrac{3^{1/6}}{\Gamma\!\left(\frac{1}{3}\right)} = 0.44828\,83573\ldots.$

9.2(iii) Numerically Satisfactory Pairs of Solutions

Table 9.2.1 lists numerically satisfactory pairs of solutions of (9.2.1) for the stated regions; compare §2.7(iv).

Table 9.2.1: Numerically satisfactory solutions of Airy's equation.

Pair	Region
$\mathrm{Ai}(x), \mathrm{Bi}(x)$	$-\infty < x < \infty$
$\mathrm{Ai}(z), \mathrm{Bi}(z)$	$\begin{cases} \lvert\mathrm{ph}\,z\rvert \le \tfrac{1}{3}\pi \\ -\infty < z \le 0 \end{cases}$
$\mathrm{Ai}(z), \mathrm{Ai}\!\left(ze^{-2\pi i/3}\right)$	$-\tfrac{1}{3}\pi \le \mathrm{ph}\,z \le \pi$
$\mathrm{Ai}(z), \mathrm{Ai}\!\left(ze^{2\pi i/3}\right)$	$-\pi \le \mathrm{ph}\,z \le \tfrac{1}{3}\pi$
$\mathrm{Ai}\!\left(ze^{\mp 2\pi i/3}\right)$	$\lvert\mathrm{ph}(-z)\rvert \le \tfrac{2}{3}\pi$

9.2(iv) Wronskians

9.2.7 $\mathscr{W}\{\mathrm{Ai}(z), \mathrm{Bi}(z)\} = \dfrac{1}{\pi},$

9.2.8 $\mathscr{W}\!\left\{\mathrm{Ai}(z), \mathrm{Ai}\!\left(ze^{\mp 2\pi i/3}\right)\right\} = \dfrac{e^{\pm \pi i/6}}{2\pi},$

9.2.9 $\mathscr{W}\!\left\{\mathrm{Ai}\!\left(ze^{-2\pi i/3}\right), \mathrm{Ai}\!\left(ze^{2\pi i/3}\right)\right\} = \dfrac{1}{2\pi i}.$

9.2(v) Connection Formulas

9.2.10 $\mathrm{Bi}(z) = e^{-\pi i/6}\,\mathrm{Ai}\!\left(ze^{-2\pi i/3}\right) + e^{\pi i/6}\,\mathrm{Ai}\!\left(ze^{2\pi i/3}\right).$

9.2.11 $\mathrm{Ai}\!\left(ze^{\mp 2\pi i/3}\right) = \tfrac{1}{2} e^{\mp \pi i/3}\left(\mathrm{Ai}(z) \pm i\,\mathrm{Bi}(z)\right).$

9.2.12
$$\mathrm{Ai}(z) + e^{-2\pi i/3}\,\mathrm{Ai}\!\left(ze^{-2\pi i/3}\right) + e^{2\pi i/3}\,\mathrm{Ai}\!\left(ze^{2\pi i/3}\right) = 0,$$

9.2.13
$$\mathrm{Bi}(z) + e^{-2\pi i/3}\,\mathrm{Bi}\!\left(ze^{-2\pi i/3}\right) + e^{2\pi i/3}\,\mathrm{Bi}\!\left(ze^{2\pi i/3}\right) = 0.$$

9.2.14 $\mathrm{Ai}(-z) = e^{\pi i/3}\,\mathrm{Ai}\!\left(ze^{\pi i/3}\right) + e^{-\pi i/3}\,\mathrm{Ai}\!\left(ze^{-\pi i/3}\right),$

9.2.15 $\mathrm{Bi}(-z) = e^{-\pi i/6}\,\mathrm{Ai}\!\left(ze^{\pi i/3}\right) + e^{\pi i/6}\,\mathrm{Ai}\!\left(ze^{-\pi i/3}\right).$

9.2(vi) Riccati Form of Differential Equation

9.2.16
$$\frac{dW}{dz} + W^2 = z,$$

$W = (1/w)\,dw/dz$, where w is any nontrivial solution of (9.2.1). See also Smith (1990).

9.3 Graphics

9.3(i) Real Variable

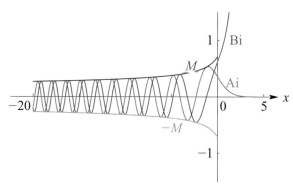

Figure 9.3.1: $\operatorname{Ai}(x)$, $\operatorname{Bi}(x)$, $M(x)$. For $M(x)$ see §9.8(i).

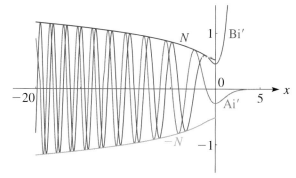

Figure 9.3.2: $\operatorname{Ai}'(x)$, $\operatorname{Bi}'(x)$, $N(x)$. For $N(x)$ see §9.8(i).

9.3(ii) Complex Variable

In the graphics shown in this subsection, height corresponds to the absolute value of the function and color to the phase. See also p. xiv.

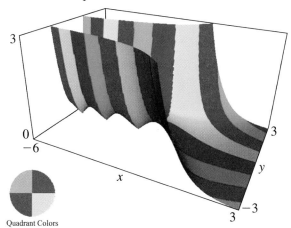

Figure 9.3.3: $\operatorname{Ai}(x+iy)$.

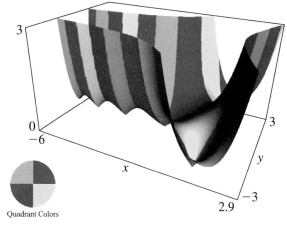

Figure 9.3.4: $\operatorname{Bi}(x+iy)$.

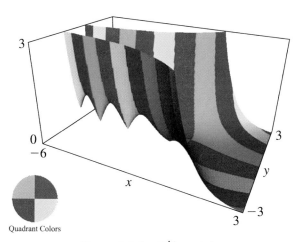

Figure 9.3.5: $\operatorname{Ai}'(x+iy)$.

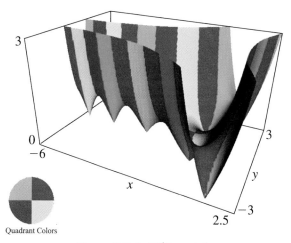

Figure 9.3.6: $\operatorname{Bi}'(x+iy)$.

9.4 Maclaurin Series

For $z \in \mathbb{C}$

9.4.1 $\quad \operatorname{Ai}(z) = \operatorname{Ai}(0)\left(1 + \frac{1}{3!}z^3 + \frac{1\cdot 4}{6!}z^6 + \frac{1\cdot 4\cdot 7}{9!}z^9 + \cdots\right) + \operatorname{Ai}'(0)\left(z + \frac{2}{4!}z^4 + \frac{2\cdot 5}{7!}z^7 + \frac{2\cdot 5\cdot 8}{10!}z^{10} + \cdots\right),$

9.4.2 $\quad \operatorname{Ai}'(z) = \operatorname{Ai}'(0)\left(1 + \frac{2}{3!}z^3 + \frac{2\cdot 5}{6!}z^6 + \frac{2\cdot 5\cdot 8}{9!}z^9 + \cdots\right) + \operatorname{Ai}(0)\left(\frac{1}{2!}z^2 + \frac{1\cdot 4}{5!}z^5 + \frac{1\cdot 4\cdot 7}{8!}z^8 + \cdots\right),$

9.4.3 $\quad \operatorname{Bi}(z) = \operatorname{Bi}(0)\left(1 + \frac{1}{3!}z^3 + \frac{1\cdot 4}{6!}z^6 + \frac{1\cdot 4\cdot 7}{9!}z^9 + \cdots\right) + \operatorname{Bi}'(0)\left(z + \frac{2}{4!}z^4 + \frac{2\cdot 5}{7!}z^7 + \frac{2\cdot 5\cdot 8}{10!}z^{10} + \cdots\right),$

9.4.4 $\quad \operatorname{Bi}'(z) = \operatorname{Bi}'(0)\left(1 + \frac{2}{3!}z^3 + \frac{2\cdot 5}{6!}z^6 + \frac{2\cdot 5\cdot 8}{9!}z^9 + \cdots\right) + \operatorname{Bi}(0)\left(\frac{1}{2!}z^2 + \frac{1\cdot 4}{5!}z^5 + \frac{1\cdot 4\cdot 7}{8!}z^8 + \cdots\right).$

9.5 Integral Representations

9.5(i) Real Variable

9.5.1 $\quad \operatorname{Ai}(x) = \frac{1}{\pi}\int_0^\infty \cos\left(\tfrac{1}{3}t^3 + xt\right) dt.$

9.5.2
$$\operatorname{Ai}(-x) = \frac{x^{1/2}}{\pi}\int_{-1}^\infty \cos\left(x^{3/2}\left(\tfrac{1}{3}t^3 + t^2 - \tfrac{2}{3}\right)\right) dt, \; x > 0.$$

9.5.3
$\operatorname{Bi}(x)$
$$= \frac{1}{\pi}\int_0^\infty \exp\left(-\tfrac{1}{3}t^3 + xt\right) dt + \frac{1}{\pi}\int_0^\infty \sin\left(\tfrac{1}{3}t^3 + xt\right) dt.$$

See also (9.10.19), (9.11.3), (36.9.2), and Vallée and Soares (2004, §2.1.3).

9.5(ii) Complex Variable

9.5.4 $\quad \operatorname{Ai}(z) = \frac{1}{2\pi i}\int_{\infty e^{-\pi i/3}}^{\infty e^{\pi i/3}} \exp\left(\tfrac{1}{3}t^3 - zt\right) dt,$

9.5.5
$$\operatorname{Bi}(z) = \frac{1}{2\pi}\int_{-\infty}^{\infty e^{\pi i/3}} \exp\left(\tfrac{1}{3}t^3 - zt\right) dt$$
$$+ \frac{1}{2\pi}\int_{-\infty}^{\infty e^{-\pi i/3}} \exp\left(\tfrac{1}{3}t^3 - zt\right) dt.$$

9.5.6 $\quad \operatorname{Ai}(z) = \frac{\sqrt{3}}{2\pi}\int_0^\infty \exp\left(-\frac{t^3}{3} - \frac{z^3}{3t^3}\right) dt.$

9.5.7
$$\operatorname{Ai}(z) = \frac{e^{-\zeta}}{\pi}\int_0^\infty \exp\left(-z^{1/2}t^2\right)\cos\left(\tfrac{1}{3}t^3\right) dt, \; |\operatorname{ph} z| < \pi.$$

9.5.8
$$\operatorname{Ai}(z) = \frac{e^{-\zeta}\zeta^{-1/6}}{\sqrt{\pi}(48)^{1/6}\Gamma\!\left(\tfrac{5}{6}\right)}\int_0^\infty e^{-t}t^{-1/6}\left(2 + \frac{t}{\zeta}\right)^{-1/6} dt,$$
$$|\operatorname{ph} z| < \tfrac{2}{3}\pi.$$

In (9.5.7) and (9.5.8) $\zeta = \tfrac{2}{3}z^{3/2}$.
See also (9.10.18) and (9.11.4).

9.6 Relations to Other Functions

9.6(i) Airy Functions as Bessel Functions, Hankel Functions, and Modified Bessel Functions

For the notation see §§10.2(ii) and 10.25(ii). With

9.6.1 $\quad \zeta = \tfrac{2}{3}z^{3/2},$

9.6.2
$$\operatorname{Ai}(z) = \pi^{-1}\sqrt{z/3}\, K_{\pm 1/3}(\zeta)$$
$$= \tfrac{1}{3}\sqrt{z}\left(I_{-1/3}(\zeta) - I_{1/3}(\zeta)\right)$$
$$= \tfrac{1}{2}\sqrt{z/3}\, e^{2\pi i/3} H^{(1)}_{1/3}\!\left(\zeta e^{\pi i/2}\right)$$
$$= \tfrac{1}{2}\sqrt{z/3}\, e^{\pi i/3} H^{(1)}_{-1/3}\!\left(\zeta e^{\pi i/2}\right)$$
$$= \tfrac{1}{2}\sqrt{z/3}\, e^{-2\pi i/3} H^{(2)}_{1/3}\!\left(\zeta e^{-\pi i/2}\right)$$
$$= \tfrac{1}{2}\sqrt{z/3}\, e^{-\pi i/3} H^{(2)}_{-1/3}\!\left(\zeta e^{-\pi i/2}\right),$$

9.6.3
$$\operatorname{Ai}'(z) = -\pi^{-1}(z/\sqrt{3})\, K_{\pm 2/3}(\zeta)$$
$$= (z/3)\left(I_{2/3}(\zeta) - I_{-2/3}(\zeta)\right)$$
$$= \tfrac{1}{2}(z/\sqrt{3})e^{-\pi i/6} H^{(1)}_{2/3}\!\left(\zeta e^{\pi i/2}\right)$$
$$= \tfrac{1}{2}(z/\sqrt{3})e^{-5\pi i/6} H^{(1)}_{-2/3}\!\left(\zeta e^{\pi i/2}\right)$$
$$= \tfrac{1}{2}(z/\sqrt{3})e^{\pi i/6} H^{(2)}_{2/3}\!\left(\zeta e^{-\pi i/2}\right)$$
$$= \tfrac{1}{2}(z/\sqrt{3})e^{5\pi i/6} H^{(2)}_{-2/3}\!\left(\zeta e^{-\pi i/2}\right),$$

9.6.4
$$\operatorname{Bi}(z) = \sqrt{z/3}\left(I_{1/3}(\zeta) + I_{-1/3}(\zeta)\right)$$
$$= \tfrac{1}{2}\sqrt{z/3}\left(e^{\pi i/6} H^{(1)}_{1/3}\!\left(\zeta e^{-\pi i/2}\right)\right.$$
$$\left. + e^{-\pi i/6} H^{(2)}_{1/3}\!\left(\zeta e^{\pi i/2}\right)\right)$$
$$= \tfrac{1}{2}\sqrt{z/3}\left(e^{-\pi i/6} H^{(1)}_{-1/3}\!\left(\zeta e^{-\pi i/2}\right)\right.$$
$$\left. + e^{\pi i/6} H^{(2)}_{-1/3}\!\left(\zeta e^{\pi i/2}\right)\right),$$

9.6 Relations to Other Functions

$$\text{Bi}'(z) = (z/\sqrt{3})\left(I_{2/3}(\zeta) + I_{-2/3}(\zeta)\right)$$

9.6.5
$$= \tfrac{1}{2}(z/\sqrt{3})\left(e^{\pi i/3} H^{(1)}_{2/3}(\zeta e^{-\pi i/2}) + e^{-\pi i/3} H^{(2)}_{2/3}(\zeta e^{\pi i/2})\right)$$
$$= \tfrac{1}{2}(z/\sqrt{3})\left(e^{-\pi i/3} H^{(1)}_{-2/3}(\zeta e^{-\pi i/2}) + e^{\pi i/3} H^{(2)}_{-2/3}(\zeta e^{\pi i/2})\right),$$

9.6.6
$$\text{Ai}(-z) = (\sqrt{z}/3)\left(J_{1/3}(\zeta) + J_{-1/3}(\zeta)\right)$$
$$= \tfrac{1}{2}\sqrt{z/3}\left(e^{\pi i/6} H^{(1)}_{1/3}(\zeta) + e^{-\pi i/6} H^{(2)}_{1/3}(\zeta)\right)$$
$$= \tfrac{1}{2}\sqrt{z/3}\left(e^{-\pi i/6} H^{(1)}_{-1/3}(\zeta) + e^{\pi i/6} H^{(2)}_{-1/3}(\zeta)\right),$$

9.6.7
$$\text{Ai}'(-z) = (z/3)\left(J_{2/3}(\zeta) - J_{-2/3}(\zeta)\right)$$
$$= \tfrac{1}{2}(z/\sqrt{3})\left(e^{-\pi i/6} H^{(1)}_{2/3}(\zeta) + e^{\pi i/6} H^{(2)}_{2/3}(\zeta)\right)$$
$$= \tfrac{1}{2}(z/\sqrt{3})\left(e^{-5\pi i/6} H^{(1)}_{-2/3}(\zeta) + e^{5\pi i/6} H^{(2)}_{-2/3}(\zeta)\right),$$

9.6.8
$$\text{Bi}(-z) = \sqrt{z/3}\left(J_{-1/3}(\zeta) - J_{1/3}(\zeta)\right)$$
$$= \tfrac{1}{2}\sqrt{z/3}\left(e^{2\pi i/3} H^{(1)}_{1/3}(\zeta) + e^{-2\pi i/3} H^{(2)}_{1/3}(\zeta)\right)$$
$$= \tfrac{1}{2}\sqrt{z/3}\left(e^{\pi i/3} H^{(1)}_{-1/3}(\zeta) + e^{-\pi i/3} H^{(2)}_{-1/3}(\zeta)\right),$$

9.6.9
$$\text{Bi}'(-z) = (z/\sqrt{3})\left(J_{-2/3}(\zeta) + J_{2/3}(\zeta)\right)$$
$$= \tfrac{1}{2}(z/\sqrt{3})\left(e^{\pi i/3} H^{(1)}_{2/3}(\zeta) + e^{-\pi i/3} H^{(2)}_{2/3}(\zeta)\right)$$
$$= \tfrac{1}{2}(z/\sqrt{3})\left(e^{-\pi i/3} H^{(1)}_{-2/3}(\zeta) + e^{\pi i/3} H^{(2)}_{-2/3}(\zeta)\right).$$

9.6(ii) Bessel Functions, Hankel Functions, and Modified Bessel Functions as Airy Functions

Again, for the notation see §§10.2(ii) and 10.25(ii). With

9.6.10
$$z = (\tfrac{3}{2}\zeta)^{2/3},$$

9.6.11 $\quad J_{\pm 1/3}(\zeta) = \tfrac{1}{2}\sqrt{3/z}\left(\sqrt{3}\,\text{Ai}(-z) \mp \text{Bi}(-z)\right),$

9.6.12 $\quad J_{\pm 2/3}(\zeta) = \tfrac{1}{2}(\sqrt{3}/z)\left(\pm\sqrt{3}\,\text{Ai}'(-z) + \text{Bi}'(-z)\right),$

9.6.13 $\quad I_{\pm 1/3}(\zeta) = \tfrac{1}{2}\sqrt{3/z}\left(\mp\sqrt{3}\,\text{Ai}(z) + \text{Bi}(z)\right),$

9.6.14 $\quad I_{\pm 2/3}(\zeta) = \tfrac{1}{2}(\sqrt{3}/z)\left(\pm\sqrt{3}\,\text{Ai}'(z) + \text{Bi}'(z)\right),$

9.6.15 $\quad K_{\pm 1/3}(\zeta) = \pi\sqrt{3/z}\,\text{Ai}(z),$

9.6.16 $\quad K_{\pm 2/3}(\zeta) = -\pi(\sqrt{3}/z)\,\text{Ai}'(z),$

9.6.17
$$H^{(1)}_{1/3}(\zeta) = e^{-\pi i/3} H^{(1)}_{-1/3}(\zeta)$$
$$= e^{-\pi i/6}\sqrt{3/z}\left(\text{Ai}(-z) - i\,\text{Bi}(-z)\right),$$

9.6.18
$$H^{(1)}_{2/3}(\zeta) = e^{-2\pi i/3} H^{(1)}_{-2/3}(\zeta)$$
$$= e^{\pi i/6}(\sqrt{3}/z)\left(\text{Ai}'(-z) - i\,\text{Bi}'(-z)\right),$$

9.6.19
$$H^{(2)}_{1/3}(\zeta) = e^{\pi i/3} H^{(2)}_{-1/3}(\zeta)$$
$$= e^{\pi i/6}\sqrt{3/z}\left(\text{Ai}(-z) + i\,\text{Bi}(-z)\right),$$

9.6.20
$$H^{(2)}_{2/3}(\zeta) = e^{2\pi i/3} H^{(2)}_{-2/3}(\zeta)$$
$$= e^{-\pi i/6}(\sqrt{3}/z)\left(\text{Ai}'(-z) + i\,\text{Bi}'(-z)\right).$$

9.6(iii) Airy Functions as Confluent Hypergeometric Functions

For the notation see §§13.1, 13.2, and 13.14(i). With ζ as in (9.6.1),

9.6.21
$$\text{Ai}(z) = \tfrac{1}{2}\pi^{-1/2} z^{-1/4} W_{0,1/3}(2\zeta)$$
$$= 3^{-1/6}\pi^{-1/2}\zeta^{2/3} e^{-\zeta} U\left(\tfrac{5}{6}, \tfrac{5}{3}, 2\zeta\right),$$

9.6.22
$$\text{Ai}'(z) = -\tfrac{1}{2}\pi^{-1/2} z^{1/4} W_{0,2/3}(2\zeta)$$
$$= -3^{1/6}\pi^{-1/2}\zeta^{4/3} e^{-\zeta} U\left(\tfrac{7}{6}, \tfrac{7}{3}, 2\zeta\right),$$

9.6.23
$$\text{Bi}(z) = \frac{1}{2^{1/3}\,\Gamma\!\left(\tfrac{2}{3}\right)} z^{-1/4} M_{0,-1/3}(2\zeta)$$
$$+ \frac{3}{2^{5/3}\,\Gamma\!\left(\tfrac{1}{3}\right)} z^{-1/4} M_{0,1/3}(2\zeta),$$

9.6.24
$$\text{Bi}'(z) = \frac{2^{1/3}}{\Gamma\!\left(\tfrac{1}{3}\right)} z^{1/4} M_{0,-2/3}(2\zeta)$$
$$+ \frac{3}{2^{10/3}\,\Gamma\!\left(\tfrac{2}{3}\right)} z^{1/4} M_{0,2/3}(2\zeta),$$

9.6.25
$$\text{Bi}(z) = \frac{1}{3^{1/6}\,\Gamma\!\left(\tfrac{2}{3}\right)} e^{-\zeta}\,{}_1F_1\!\left(\tfrac{1}{6};\tfrac{1}{3};2\zeta\right)$$
$$+ \frac{3^{5/6}}{2^{2/3}\,\Gamma\!\left(\tfrac{1}{3}\right)}\zeta^{2/3} e^{-\zeta}\,{}_1F_1\!\left(\tfrac{5}{6};\tfrac{5}{3};2\zeta\right),$$

9.6.26
$$\text{Bi}'(z) = \frac{3^{1/6}}{\Gamma\!\left(\tfrac{1}{3}\right)} e^{-\zeta}\,{}_1F_1\!\left(-\tfrac{1}{6};-\tfrac{1}{3};2\zeta\right)$$
$$+ \frac{3^{7/6}}{2^{7/3}\,\Gamma\!\left(\tfrac{2}{3}\right)}\zeta^{4/3} e^{-\zeta}\,{}_1F_1\!\left(\tfrac{7}{6};\tfrac{7}{3};\zeta\right).$$

9.7 Asymptotic Expansions

9.7(i) Notation

Here δ denotes an arbitrary small positive constant and

9.7.1
$$\zeta = \tfrac{2}{3} z^{3/2}.$$

Also $u_0 = v_0 = 1$ and for $k = 1, 2, \ldots,$

9.7.2
$$u_k = \frac{(2k+1)(2k+3)(2k+5)\cdots(6k-1)}{(216)^k (k)!},$$

$$v_k = \frac{6k+1}{1-6k} u_k.$$

Lastly,

9.7.3
$$\chi(n) = \pi^{1/2}\, \Gamma\!\left(\tfrac{1}{2}n+1\right) / \Gamma\!\left(\tfrac{1}{2}n+\tfrac{1}{2}\right).$$

Numerical values of this function are given in Table 9.7.1 for $n = 1(1)20$ to 2D. For large n,

9.7.4
$$\chi(n) \sim \left(\tfrac{1}{2}\pi n\right)^{1/2}.$$

Table 9.7.1: $\chi(n)$.

n	$\chi(n)$	n	$\chi(n)$	n	$\chi(n)$	n	$\chi(n)$
1	1.57	6	3.20	11	4.25	16	5.09
2	2.00	7	3.44	12	4.43	17	5.24
3	2.36	8	3.66	13	4.61	18	5.39
4	2.67	9	3.87	14	4.77	19	5.54
5	2.95	10	4.06	15	4.94	20	5.68

9.7(ii) Poincaré-Type Expansions

As $z \to \infty$ the following asymptotic expansions are valid uniformly in the stated sectors.

9.7.5 $\quad \mathrm{Ai}(z) \sim \dfrac{e^{-\zeta}}{2\sqrt{\pi}\, z^{1/4}} \displaystyle\sum_{k=0}^{\infty} (-1)^k \dfrac{u_k}{\zeta^k}, \quad |\mathrm{ph}\, z| \leq \pi - \delta,$

9.7.6 $\quad \mathrm{Ai}'(z) \sim -\dfrac{z^{1/4} e^{-\zeta}}{2\sqrt{\pi}} \displaystyle\sum_{k=0}^{\infty} (-1)^k \dfrac{v_k}{\zeta^k}, \quad |\mathrm{ph}\, z| \leq \pi - \delta,$

9.7.7 $\quad \mathrm{Bi}(z) \sim \dfrac{e^{\zeta}}{\sqrt{\pi}\, z^{1/4}} \displaystyle\sum_{k=0}^{\infty} \dfrac{u_k}{\zeta^k}, \quad |\mathrm{ph}\, z| \leq \tfrac{1}{3}\pi - \delta,$

9.7.8 $\quad \mathrm{Bi}'(z) \sim \dfrac{z^{1/4} e^{\zeta}}{\sqrt{\pi}} \displaystyle\sum_{k=0}^{\infty} \dfrac{v_k}{\zeta^k}, \quad |\mathrm{ph}\, z| \leq \tfrac{1}{3}\pi - \delta.$

9.7.9 $\quad \mathrm{Ai}(-z) \sim \dfrac{1}{\sqrt{\pi}\, z^{1/4}} \left(\cos\!\left(\zeta - \tfrac{1}{4}\pi\right) \displaystyle\sum_{k=0}^{\infty} (-1)^k \dfrac{u_{2k}}{\zeta^{2k}} + \sin\!\left(\zeta - \tfrac{1}{4}\pi\right) \displaystyle\sum_{k=0}^{\infty} (-1)^k \dfrac{u_{2k+1}}{\zeta^{2k+1}} \right), \quad |\mathrm{ph}\, z| \leq \tfrac{2}{3}\pi - \delta,$

9.7.10 $\quad \mathrm{Ai}'(-z) \sim \dfrac{z^{1/4}}{\sqrt{\pi}} \left(\sin\!\left(\zeta - \tfrac{1}{4}\pi\right) \displaystyle\sum_{k=0}^{\infty} (-1)^k \dfrac{v_{2k}}{\zeta^{2k}} - \cos\!\left(\zeta - \tfrac{1}{4}\pi\right) \displaystyle\sum_{k=0}^{\infty} (-1)^k \dfrac{v_{2k+1}}{\zeta^{2k+1}} \right), \quad |\mathrm{ph}\, z| \leq \tfrac{2}{3}\pi - \delta,$

9.7.11 $\quad \mathrm{Bi}(-z) \sim \dfrac{1}{\sqrt{\pi}\, z^{1/4}} \left(-\sin\!\left(\zeta - \tfrac{1}{4}\pi\right) \displaystyle\sum_{k=0}^{\infty} (-1)^k \dfrac{u_{2k}}{\zeta^{2k}} + \cos\!\left(\zeta - \tfrac{1}{4}\pi\right) \displaystyle\sum_{k=0}^{\infty} (-1)^k \dfrac{u_{2k+1}}{\zeta^{2k+1}} \right), \quad |\mathrm{ph}\, z| \leq \tfrac{2}{3}\pi - \delta,$

9.7.12 $\quad \mathrm{Bi}'(-z) \sim \dfrac{z^{1/4}}{\sqrt{\pi}} \left(\cos\!\left(\zeta - \tfrac{1}{4}\pi\right) \displaystyle\sum_{k=0}^{\infty} (-1)^k \dfrac{v_{2k}}{\zeta^{2k}} + \sin\!\left(\zeta - \tfrac{1}{4}\pi\right) \displaystyle\sum_{k=0}^{\infty} (-1)^k \dfrac{v_{2k+1}}{\zeta^{2k+1}} \right), \quad |\mathrm{ph}\, z| \leq \tfrac{2}{3}\pi - \delta.$

9.7.13 $\quad \mathrm{Bi}\!\left(z e^{\pm \pi i/3}\right) \sim \sqrt{\dfrac{2}{\pi}} \dfrac{e^{\pm \pi i/6}}{z^{1/4}} \left(\cos\!\left(\zeta - \tfrac{1}{4}\pi \mp \tfrac{1}{2} i \ln 2\right) \displaystyle\sum_{k=0}^{\infty} (-1)^k \dfrac{u_{2k}}{\zeta^{2k}} + \sin\!\left(\zeta - \tfrac{1}{4}\pi \mp \tfrac{1}{2} i \ln 2\right) \displaystyle\sum_{k=0}^{\infty} (-1)^k \dfrac{u_{2k+1}}{\zeta^{2k+1}} \right),$

$$|\mathrm{ph}\, z| \leq \tfrac{2}{3}\pi - \delta,$$

9.7.14
$$\mathrm{Bi}'\!\left(z e^{\pm \pi i/3}\right) \sim \sqrt{\dfrac{2}{\pi}}\, e^{\mp \pi i/6} z^{1/4} \left(-\sin\!\left(\zeta - \tfrac{1}{4}\pi \mp \tfrac{1}{2} i \ln 2\right) \sum_{k=0}^{\infty} (-1)^k \dfrac{v_{2k}}{\zeta^{2k}} + \cos\!\left(\zeta - \tfrac{1}{4}\pi \mp \tfrac{1}{2} i \ln 2\right) \sum_{k=0}^{\infty} (-1)^k \dfrac{v_{2k+1}}{\zeta^{2k+1}} \right),$$

$$|\mathrm{ph}\, z| \leq \tfrac{2}{3}\pi - \delta.$$

9.7(iii) Error Bounds for Real Variables

In (9.7.5) and (9.7.6) the nth error term, that is, the error on truncating the expansion at n terms, is bounded in magnitude by the first neglected term and has the same sign, provided that the following term is of opposite sign, that is, if $n \geq 0$ for (9.7.5) and $n \geq 1$ for (9.7.6).

In (9.7.7) and (9.7.8) the nth error term is bounded in magnitude by the first neglected term multiplied by $2\chi(n)\exp(\sigma\pi/(72\zeta))$ where $\sigma = 5$ for (9.7.7) and $\sigma = 7$ for (9.7.8), provided that $n \geq 1$ in both cases.

In (9.7.9)–(9.7.12) the nth error term in each infinite series is bounded in magnitude by the first neglected term and has the same sign, provided that the following term in the series is of opposite sign.

As special cases, when $0 < x < \infty$

9.7.15
$$\operatorname{Ai}(x) \leq \frac{e^{-\xi}}{2\sqrt{\pi}x^{1/4}}, \quad |\operatorname{Ai}'(x)| \leq \frac{x^{1/4}e^{-\xi}}{2\sqrt{\pi}}\left(1 + \frac{7}{72\xi}\right),$$

9.7.16
$$\operatorname{Bi}(x) \leq \frac{e^{\xi}}{\sqrt{\pi}x^{1/4}}\left(1 + \frac{5\pi}{72\xi}\exp\left(\frac{5\pi}{72\xi}\right)\right),$$
$$\operatorname{Bi}'(x) \leq \frac{x^{1/4}e^{\xi}}{\sqrt{\pi}}\left(1 + \frac{7\pi}{72\xi}\exp\left(\frac{7\pi}{72\xi}\right)\right),$$

where $\xi = \frac{2}{3}x^{3/2}$.

9.7(iv) Error Bounds for Complex Variables

When $n \geq 1$ the nth error term in (9.7.5) and (9.7.6) is bounded in magnitude by the first neglected term multiplied by

9.7.17
$$2\exp\left(\frac{\sigma}{36|\zeta|}\right), \quad 2\chi(n)\exp\left(\frac{\sigma\pi}{72|\zeta|}\right)$$
$$\text{or} \quad \frac{4\chi(n)}{|\cos(\operatorname{ph}\zeta)|^n}\exp\left(\frac{\sigma\pi}{36|\Re\zeta|}\right),$$

according as $|\operatorname{ph}z| \leq \frac{1}{3}\pi$, $\frac{1}{3}\pi \leq |\operatorname{ph}z| \leq \frac{2}{3}\pi$, or $\frac{2}{3}\pi \leq |\operatorname{ph}z| \leq \pi$. Here $\sigma = 5$ for (9.7.5) and $\sigma = 7$ for (9.7.6).

Corresponding bounds for the errors in (9.7.7) to (9.7.14) may be obtained by use of these results and those of §9.2(v) and their differentiated forms.

For other error bounds see Boyd (1993).

9.7(v) Exponentially-Improved Expansions

In (9.7.5) and (9.7.6) let

9.7.18 $\quad \operatorname{Ai}(z) = \dfrac{e^{-\zeta}}{2\sqrt{\pi}z^{1/4}}\left(\displaystyle\sum_{k=0}^{n-1}(-1)^k\dfrac{u_k}{\zeta^k} + R_n(z)\right),$

9.7.19 $\quad \operatorname{Ai}'(z) = -\dfrac{z^{1/4}e^{-\zeta}}{2\sqrt{\pi}}\left(\displaystyle\sum_{k=0}^{n-1}(-1)^k\dfrac{v_k}{\zeta^k} + S_n(z)\right),$

with $n = \lfloor 2|\zeta|\rfloor$. Then

9.7.20
$$R_n(z) = (-1)^n\sum_{k=0}^{m-1}(-1)^k u_k\frac{G_{n-k}(2\zeta)}{\zeta^k} + R_{m,n}(z),$$

9.7.21
$$S_n(z) = (-1)^{n-1}\sum_{k=0}^{m-1}(-1)^k v_k\frac{G_{n-k}(2\zeta)}{\zeta^k} + S_{m,n}(z),$$

where

9.7.22 $\quad G_p(z) = \dfrac{e^z}{2\pi}\Gamma(p)\Gamma(1-p,z).$

(For the notation see §8.2(i).) And as $z \to \infty$ with m fixed

9.7.23
$$R_{m,n}(z), S_{m,n}(z) = O\!\left(e^{-2|\zeta|}\zeta^{-m}\right), \quad |\operatorname{ph}z| \leq \tfrac{2}{3}\pi.$$

For re-expansions of the remainder terms in (9.7.7)–(9.7.14) combine the results of this section with those of §9.2(v) and their differentiated forms, as in §9.7(iv).

For higher re-expansions of the remainder terms see Olde Daalhuis (1995, 1996), and Olde Daalhuis and Olver (1995a).

9.8 Modulus and Phase

9.8(i) Definitions

Throughout this section x is real and nonpositive.

9.8.1 $\quad \operatorname{Ai}(x) = M(x)\sin\theta(x),$

9.8.2 $\quad \operatorname{Bi}(x) = M(x)\cos\theta(x),$

9.8.3 $\quad M(x) = \sqrt{\operatorname{Ai}^2(x) + \operatorname{Bi}^2(x)},$

9.8.4 $\quad \theta(x) = \arctan(\operatorname{Ai}(x)/\operatorname{Bi}(x)).$

9.8.5 $\quad \operatorname{Ai}'(x) = N(x)\sin\phi(x),$

9.8.6 $\quad \operatorname{Bi}'(x) = N(x)\cos\phi(x),$

9.8.7 $\quad N(x) = \sqrt{\operatorname{Ai}'^2(x) + \operatorname{Bi}'^2(x)},$

9.8.8 $\quad \phi(x) = \arctan(\operatorname{Ai}'(x)/\operatorname{Bi}'(x)).$

Graphs of $M(x)$ and $N(x)$ are included in §9.3(i). The branches of $\theta(x)$ and $\phi(x)$ are continuous and fixed by $\theta(0) = -\phi(0) = \frac{1}{6}\pi$. (These definitions of $\theta(x)$ and $\phi(x)$ differ from Abramowitz and Stegun (1964, Chapter 10), and agree more closely with those used in Miller (1946) and Olver (1997b, Chapter 11).)

In terms of Bessel functions, and with $\xi = \frac{2}{3}|x|^{3/2}$,

9.8.9 $\quad |x|^{1/2}M^2(x) = \tfrac{1}{2}\xi\left(J_{1/3}^2(\xi) + Y_{1/3}^2(\xi)\right),$

9.8.10 $\quad |x|^{-1/2}N^2(x) = \tfrac{1}{2}\xi\left(J_{2/3}^2(\xi) + Y_{2/3}^2(\xi)\right),$

9.8.11 $\quad \theta(x) = \tfrac{2}{3}\pi + \arctan(Y_{1/3}(\xi)/J_{1/3}(\xi)),$

9.8.12 $\quad \phi(x) = \tfrac{1}{3}\pi + \arctan(Y_{2/3}(\xi)/J_{2/3}(\xi)).$

9.8(ii) Identities

Primes denote differentiations with respect to x, which is continued to be assumed real and nonpositive.

9.8.13 $\quad M(x) N(x) \sin(\theta(x) - \phi(x)) = \pi^{-1},$

9.8.14 $\quad M^2(x) \theta'(x) = -\pi^{-1}, \quad N^2(x) \phi'(x) = \pi^{-1} x,$
$\quad N(x) N'(x) = x M(x) M'(x),$

9.8.15 $\quad \begin{aligned} N^2(x) &= M'^2(x) + M^2(x) \theta'^2(x) \\ &= M'^2(x) + \pi^{-2} M^{-2}(x), \end{aligned}$

9.8.16 $\quad \begin{aligned} x^2 M^2(x) &= N'^2(x) + N^2(x) \phi'^2(x) \\ &= N'^2(x) + \pi^{-2} x^2 N^{-2}(x), \end{aligned}$

9.8.17 $\quad \begin{aligned} \tan(\theta(x) - \phi(x)) &= 1/(\pi M(x) M'(x)) \\ &= -M(x) \theta'(x) / M'(x), \end{aligned}$

9.8.18 $\quad \begin{aligned} M''(x) &= x M(x) + \pi^{-2} M^{-3}(x), \\ (M^2)'''(x) - 4x (M^2)'(x) - 2 M^2(x) &= 0, \end{aligned}$

9.8.19 $\quad \theta'^2(x) + \frac{1}{2}(\theta'''(x)/\theta'(x)) - \frac{3}{4}(\theta''(x)/\theta'(x))^2 = -x.$

9.8(iii) Monotonicity

As x increases from $-\infty$ to 0 each of the functions $M(x)$, $M'(x)$, $|x|^{-1/4} N(x)$, $M(x) N(x)$, $\theta'(x)$, $\phi'(x)$ is increasing, and each of the functions $|x|^{1/4} M(x)$, $\theta(x)$, $\phi(x)$ is decreasing.

9.8(iv) Asymptotic Expansions

As $x \to -\infty$

9.8.20
$$M^2(x) \sim \frac{1}{\pi(-x)^{1/2}} \sum_{k=0}^{\infty} \frac{1 \cdot 3 \cdot 5 \cdots (6k-1)}{k!(96)^k} \frac{1}{x^{3k}},$$

9.8.21
$$N^2(x) \sim \frac{(-x)^{1/2}}{\pi} \sum_{k=0}^{\infty} \frac{1 \cdot 3 \cdot 5 \cdots (6k-1)}{k!(96)^k} \frac{1+6k}{1-6k} \frac{1}{x^{3k}},$$

9.8.22
$$\theta(x) \sim \frac{\pi}{4} + \frac{2}{3}(-x)^{3/2} \left(1 + \frac{5}{32}\frac{1}{x^3} + \frac{1105}{6144}\frac{1}{x^6} \right.$$
$$\left. + \frac{82825}{65536}\frac{1}{x^9} + \frac{12820\,31525}{587\,20256}\frac{1}{x^{12}} + \cdots \right),$$

9.8.23
$$\phi(x) \sim -\frac{\pi}{4} + \frac{2}{3}(-x)^{3/2} \left(1 - \frac{7}{32}\frac{1}{x^3} - \frac{1463}{6144}\frac{1}{x^6} \right.$$
$$\left. - \frac{4\,95271}{3\,27680}\frac{1}{x^9} - \frac{2065\,30429}{83\,88608}\frac{1}{x^{12}} - \cdots \right).$$

In (9.8.20) and (9.8.21) the remainder after n terms does not exceed the $(n+1)$th term in absolute value and is of the same sign, provided that $n \geq 0$ for (9.8.20) and $n \geq 1$ for (9.8.21).

For higher terms in (9.8.22) and (9.8.23) see Fabijonas et al. (2004). Also, approximate values (25S) of the coefficients of the powers $x^{-15}, x^{-18}, \ldots, x^{-56}$ are available in Sherry (1959).

9.9 Zeros

9.9(i) Distribution and Notation

On the real line, $\operatorname{Ai}(x)$, $\operatorname{Ai}'(x)$, $\operatorname{Bi}(x)$, $\operatorname{Bi}'(x)$ each have an infinite number of zeros, all of which are negative. They are denoted by a_k, a'_k, b_k, b'_k, respectively, arranged in ascending order of absolute value for $k = 1, 2, \ldots$.

$\operatorname{Ai}(z)$ and $\operatorname{Ai}'(z)$ have no other zeros. However, $\operatorname{Bi}(z)$ and $\operatorname{Bi}'(z)$ each have an infinite number of complex zeros. They lie in the sectors $\frac{1}{3}\pi < \operatorname{ph} z < \frac{1}{2}\pi$ and $-\frac{1}{2}\pi < \operatorname{ph} z < -\frac{1}{3}\pi$, and are denoted by β_k, β'_k, respectively, in the former sector, and by $\bar{\beta}_k$, $\bar{\beta}'_k$, in the conjugate sector, again arranged in ascending order of absolute value (modulus) for $k = 1, 2, \ldots$. See §9.3(ii) for visualizations.

For the distribution in $\mathbb{C}$ of the zeros of $\operatorname{Ai}'(z) - \sigma \operatorname{Ai}(z)$, where σ is an arbitrary complex constant, see Muraveĭ (1976).

9.9(ii) Relation to Modulus and Phase

9.9.1 $\quad \theta(a_k) = \phi(a'_{k+1}) = k\pi,$

9.9.2 $\quad \theta(b_k) = \phi(b'_k) = (k - \tfrac{1}{2})\pi.$

9.9.3 $\quad \operatorname{Ai}'(a_k) = \dfrac{(-1)^{k-1}}{\pi M(a_k)}, \quad \operatorname{Bi}'(b_k) = \dfrac{(-1)^{k-1}}{\pi M(b_k)},$

9.9.4 $\quad \operatorname{Ai}(a'_k) = \dfrac{(-1)^{k-1}}{\pi N(a'_k)}, \quad \operatorname{Bi}(b'_k) = \dfrac{(-1)^k}{\pi N(b'_k)}.$

9.9(iii) Derivatives With Respect to k

If k is regarded as a continuous variable, then

9.9.5
$$\operatorname{Ai}'(a_k) = (-1)^{k-1} \left(-\frac{da_k}{dk}\right)^{-1/2},$$
$$\operatorname{Ai}(a'_k) = (-1)^{k-1} \left(a'_k \frac{da'_k}{dk}\right)^{-1/2}.$$

See Olver (1954, Appendix).

9.9(iv) Asymptotic Expansions

For large k

9.9.6 $$a_k = -T\left(\tfrac{3}{8}\pi(4k-1)\right),$$

9.9.7 $$\operatorname{Ai}'(a_k) = (-1)^{k-1} V\left(\tfrac{3}{8}\pi(4k-1)\right),$$

9.9.8 $$a'_k = -U\left(\tfrac{3}{8}\pi(4k-3)\right),$$

9.9.9 $$\operatorname{Ai}(a'_k) = (-1)^{k-1} W\left(\tfrac{3}{8}\pi(4k-3)\right).$$

9.9.10 $$b_k = -T\left(\tfrac{3}{8}\pi(4k-3)\right),$$

9.9.11 $$\operatorname{Bi}'(b_k) = (-1)^{k-1} V\left(\tfrac{3}{8}\pi(4k-3)\right),$$

9.9.12 $$b'_k = -U\left(\tfrac{3}{8}\pi(4k-1)\right),$$

9.9.13 $$\operatorname{Bi}(b'_k) = (-1)^k W\left(\tfrac{3}{8}\pi(4k-1)\right).$$

9.9.14 $$\beta_k = e^{\pi i/3} T\left(\tfrac{3}{8}\pi(4k-1) + \tfrac{3}{4} i \ln 2\right),$$

9.9.15 $$\operatorname{Bi}'(\beta_k) = (-1)^k \sqrt{2} e^{-\pi i/6} V\left(\tfrac{3}{8}\pi(4k-1) + \tfrac{3}{4} i \ln 2\right),$$

9.9.16 $$\beta'_k = e^{\pi i/3} U\left(\tfrac{3}{8}\pi(4k-3) + \tfrac{3}{4} i \ln 2\right),$$

9.9.17 $$\operatorname{Bi}(\beta'_k) = (-1)^{k-1} \sqrt{2} e^{\pi i/6} W\left(\tfrac{3}{8}\pi(4k-3) + \tfrac{3}{4} i \ln 2\right).$$

Here

9.9.18 $$T(t) \sim t^{2/3} \left(1 + \frac{5}{48} t^{-2} - \frac{5}{36} t^{-4} + \frac{77125}{82944} t^{-6} - \frac{1080\,56875}{69\,67296} t^{-8} + \frac{16\,23755\,96875}{3344\,30208} t^{-10} - \cdots\right),$$

9.9.19 $$U(t) \sim t^{2/3} \left(1 - \frac{7}{48} t^{-2} + \frac{35}{288} t^{-4} - \frac{1\,81223}{2\,07360} t^{-6} + \frac{186\,83371}{12\,44160} t^{-8} - \frac{9\,11458\,84361}{1911\,02976} t^{-10} + \cdots\right),$$

9.9.20 $$V(t) \sim \pi^{-1/2} t^{1/6} \left(1 + \frac{5}{48} t^{-2} - \frac{1525}{4608} t^{-4} + \frac{23\,97875}{6\,63552} t^{-6} - \frac{7\,48989\,40625}{8918\,13888} t^{-8} + \frac{14419\,83037\,34375}{4\,28070\,66624} t^{-10} - \cdots\right),$$

9.9.21
$$W(t) \sim \pi^{-1/2} t^{-1/6} \left(1 - \frac{7}{96} t^{-2} + \frac{1673}{6144} t^{-4} - \frac{843\,94709}{265\,42080} t^{-6} + \frac{78\,02771\,35421}{1\,01921\,58720} t^{-8} - \frac{20444\,90510\,51945}{6\,52298\,15808} t^{-10} + \cdots\right).$$

For higher terms see Fabijonas and Olver (1999).

For error bounds for the asymptotic expansions of a_k, b_k, a'_k, and b'_k see Pittaluga and Sacripante (1991), and a conjecture given in Fabijonas and Olver (1999).

9.9(v) Tables

Tables 9.9.1 and 9.9.2 give 10D values of the first five real zeros of Ai, Ai$'$, Bi, Bi$'$, together with the associated values of the derivative or the function. Tables 9.9.3 and 9.9.4 give the corresponding results for the first five complex zeros of Bi and Bi$'$ in the upper half plane.

For versions of Tables 9.9.1–9.9.4 that cover $k = 1(1)10$ see http://dlmf.nist.gov/9.9.v.

Table 9.9.1: Zeros of Ai and Ai$'$.

k	a_k	$\operatorname{Ai}'(a_k)$	a'_k	$\operatorname{Ai}(a'_k)$
1	$-2.33810\,74105$	$0.70121\,08227$	$-1.01879\,29716$	$0.53565\,66560$
2	$-4.08794\,94441$	$-0.80311\,13697$	$-3.24819\,75822$	$-0.41901\,54780$
3	$-5.52055\,98281$	$0.86520\,40259$	$-4.82009\,92112$	$0.38040\,64686$
4	$-6.78670\,80901$	$-0.91085\,07370$	$-6.16330\,73556$	$-0.35790\,79437$
5	$-7.94413\,35871$	$0.94733\,57094$	$-7.37217\,72550$	$0.34230\,12444$

Table 9.9.2: Real zeros of Bi and Bi$'$.

k	b_k	$\operatorname{Bi}'(b_k)$	b'_k	$\operatorname{Bi}(b'_k)$
1	$-1.17371\,32227$	$0.60195\,78880$	$-2.29443\,96826$	$-0.45494\,43836$
2	$-3.27109\,33028$	$-0.76031\,01415$	$-4.07315\,50891$	$0.39652\,28361$
3	$-4.83073\,78417$	$0.83699\,10126$	$-5.51239\,57297$	$-0.36796\,91615$
4	$-6.16985\,21283$	$-0.88947\,99014$	$-6.78129\,44460$	$0.34949\,91168$
5	$-7.37676\,20794$	$0.92998\,36386$	$-7.94017\,86892$	$-0.33602\,62401$

Table 9.9.3: Complex zeros of Bi.

	$e^{-\pi i/3}\beta_k$		$\mathrm{Bi}'(\beta_k)$	
k	modulus	phase	modulus	phase
1	2.35387 33809	0.09533 49591	0.99310 68457	2.64060 02521
2	4.09328 73094	0.04178 55604	1.13612 83345	$-0.51328\ 28720$
3	5.52350 35011	0.02668 05442	1.22374 37881	2.62462 83591
4	6.78865 95301	0.01958 69751	1.28822 92493	$-0.51871\ 63829$
5	7.94555 90160	0.01547 08228	1.33979 47726	2.62185 44560

Table 9.9.4: Complex zeros of Bi$'$.

	$e^{-\pi i/3}\beta'_k$		$\mathrm{Bi}(\beta'_k)$	
k	modulus	phase	modulus	phase
1	1.12139 32942	0.33072 66208	0.75004 14897	0.46597 78930
2	3.25690 82266	0.05938 99367	0.59221 66315	$-2.63235\ 40329$
3	4.82400 26102	0.03278 56423	0.53787 06321	0.51549 32992
4	6.16568 66408	0.02266 24588	0.50611 02160	$-2.62362\ 85920$
5	7.37383 79870	0.01731 96481	0.48406 00643	0.51928 28169

9.10 Integrals

9.10(i) Indefinite Integrals

9.10.1 $\quad \int_z^\infty \mathrm{Ai}(t)\,dt = \pi\left(\mathrm{Ai}(z)\,\mathrm{Gi}'(z) - \mathrm{Ai}'(z)\,\mathrm{Gi}(z)\right),$

9.10.2 $\quad \int_{-\infty}^z \mathrm{Ai}(t)\,dt = \pi\left(\mathrm{Ai}(z)\,\mathrm{Hi}'(z) - \mathrm{Ai}'(z)\,\mathrm{Hi}(z)\right),$

9.10.3
$$\int_{-\infty}^z \mathrm{Bi}(t)\,dt = \int_0^z \mathrm{Bi}(t)\,dt$$
$$= \pi\left(\mathrm{Bi}'(z)\,\mathrm{Gi}(z) - \mathrm{Bi}(z)\,\mathrm{Gi}'(z)\right)$$
$$= \pi\left(\mathrm{Bi}(z)\,\mathrm{Hi}'(z) - \mathrm{Bi}'(z)\,\mathrm{Hi}(z)\right).$$

For the functions Gi and Hi see §9.12.

9.10(ii) Asymptotic Approximations

9.10.4
$$\int_x^\infty \mathrm{Ai}(t)\,dt \sim \tfrac{1}{2}\pi^{-1/2}x^{-3/4}\exp\!\left(-\tfrac{2}{3}x^{3/2}\right), \qquad x \to \infty,$$

9.10.5
$$\int_0^x \mathrm{Bi}(t)\,dt \sim \pi^{-1/2}x^{-3/4}\exp\!\left(\tfrac{2}{3}x^{3/2}\right), \qquad x \to \infty.$$

9.10.6
$$\int_{-\infty}^x \mathrm{Ai}(t)\,dt = \pi^{-1/2}(-x)^{-3/4}\cos\!\left(\tfrac{2}{3}(-x)^{3/2} + \tfrac{1}{4}\pi\right)$$
$$+ O\!\left(|x|^{-9/4}\right), \qquad x \to -\infty,$$

9.10.7
$$\int_{-\infty}^x \mathrm{Bi}(t)\,dt = \pi^{-1/2}(-x)^{-3/4}\sin\!\left(\tfrac{2}{3}(-x)^{3/2} + \tfrac{1}{4}\pi\right)$$
$$+ O\!\left(|x|^{-9/4}\right), \qquad x \to -\infty.$$

For higher terms in (9.10.4)–(9.10.7) see Vallée and Soares (2004, §3.1.3). For error bounds see Boyd (1993). See also Muldoon (1970).

9.10(iii) Other Indefinite Integrals

Let $w(z)$ be any solution of Airy's equation (9.2.1). Then

9.10.8 $\quad \int z w(z)\,dz = w'(z),$

9.10.9 $\quad \int z^2 w(z)\,dz = z w'(z) - w(z),$

9.10.10
$$\int z^{n+3} w(z)\,dz = z^{n+2} w'(z) - (n+2)z^{n+1} w(z)$$
$$+ (n+1)(n+2)\int z^n w(z)\,dz,$$
$$n = 0, 1, 2, \ldots.$$

See also §9.11(iv).

9.10(iv) Definite Integrals

9.10.11 $\quad \int_0^\infty \mathrm{Ai}(t)\,dt = \tfrac{1}{3}, \qquad \int_{-\infty}^0 \mathrm{Ai}(t)\,dt = \tfrac{2}{3},$

9.11 Products

9.10.12
$$\int_{-\infty}^{0} \operatorname{Bi}(t)\,dt = 0.$$

9.10(v) Laplace Transforms

9.10.13
$$\int_{-\infty}^{\infty} e^{pt}\operatorname{Ai}(t)\,dt = e^{p^3/3}, \qquad \Re p > 0.$$

9.10.14
$$\int_{0}^{\infty} e^{-pt}\operatorname{Ai}(t)\,dt = e^{-p^3/3}\left(\frac{1}{3} - \frac{p\,{}_1F_1\!\left(\frac{1}{3};\frac{4}{3};\frac{1}{3}p^3\right)}{3^{4/3}\,\Gamma\!\left(\frac{4}{3}\right)}\right.$$
$$\left. + \frac{p^2\,{}_1F_1\!\left(\frac{2}{3};\frac{5}{3};\frac{1}{3}p^3\right)}{3^{5/3}\,\Gamma\!\left(\frac{5}{3}\right)}\right),$$
$$p \in \mathbb{C}.$$

9.10.15
$$\int_{0}^{\infty} e^{-pt}\operatorname{Ai}(-t)\,dt$$
$$= \frac{1}{3}e^{p^3/3}\left(\frac{\Gamma\!\left(\frac{1}{3},\frac{1}{3}p^3\right)}{\Gamma\!\left(\frac{1}{3}\right)} + \frac{\Gamma\!\left(\frac{2}{3},\frac{1}{3}p^3\right)}{\Gamma\!\left(\frac{2}{3}\right)}\right), \quad \Re p > 0,$$

9.10.16
$$\int_{0}^{\infty} e^{-pt}\operatorname{Bi}(-t)\,dt$$
$$= \frac{1}{\sqrt{3}}e^{p^3/3}\left(\frac{\Gamma\!\left(\frac{2}{3},\frac{1}{3}p^3\right)}{\Gamma\!\left(\frac{2}{3}\right)} - \frac{\Gamma\!\left(\frac{1}{3},\frac{1}{3}p^3\right)}{\Gamma\!\left(\frac{1}{3}\right)}\right),$$
$$\Re p > 0.$$

For the confluent hypergeometric function ${}_1F_1$ and the incomplete gamma function Γ see §§13.1, 13.2, and 8.2(i).

For Laplace transforms of products of Airy functions see Shawagfeh (1992).

9.10(vi) Mellin Transform

9.10.17
$$\int_{0}^{\infty} t^{\alpha-1}\operatorname{Ai}(t)\,dt = \frac{\Gamma(\alpha)}{3^{(\alpha+2)/3}\,\Gamma\!\left(\frac{1}{3}\alpha + \frac{2}{3}\right)}, \quad \Re\alpha > 0.$$

9.10(vii) Stieltjes Transforms

9.10.18
$$\operatorname{Ai}(z) = \frac{z^{5/4}e^{-(2/3)z^{3/2}}}{2^{7/2}\pi}\int_{0}^{\infty}\frac{t^{-1/2}e^{-(2/3)t^{3/2}}\operatorname{Ai}(t)}{z^{3/2}+t^{3/2}}\,dt,$$
$$|\operatorname{ph} z| < \tfrac{2}{3}\pi.$$

9.10.19
$$\operatorname{Bi}(x) = \frac{x^{5/4}e^{(2/3)x^{3/2}}}{2^{5/2}\pi}\fint_{0}^{\infty}\frac{t^{-1/2}e^{-(2/3)t^{3/2}}\operatorname{Ai}(t)}{x^{3/2}-t^{3/2}}\,dt,$$
$$x > 0,$$

where the last integral is a Cauchy principal value (§1.4(v)).

9.10(viii) Repeated Integrals

9.10.20
$$\int_{0}^{x}\int_{0}^{v}\operatorname{Ai}(t)\,dt\,dv = x\int_{0}^{x}\operatorname{Ai}(t)\,dt - \operatorname{Ai}'(x) + \operatorname{Ai}'(0),$$

9.10.21
$$\int_{0}^{x}\int_{0}^{v}\operatorname{Bi}(t)\,dt\,dv = x\int_{0}^{x}\operatorname{Bi}(t)\,dt - \operatorname{Bi}'(x) + \operatorname{Bi}'(0),$$

9.10.22
$$\int_{0}^{\infty}\int_{t}^{\infty}\cdots\int_{t}^{\infty}\operatorname{Ai}(-t)(dt)^n = \frac{2\cos\!\left(\frac{1}{3}(n-1)\pi\right)}{3^{(n+2)/3}\,\Gamma\!\left(\frac{1}{3}n + \frac{2}{3}\right)},$$
$$n = 1, 2, \ldots.$$

9.10(ix) Compendia

For further integrals, including the Airy transform, see §9.11(iv), Widder (1979), Prudnikov et al. (1990, §1.8.1), Prudnikov et al. (1992a, pp. 405–413), Prudnikov et al. (1992b, §4.3.25), Vallée and Soares (2004, Chapters 3, 4).

9.11 Products

9.11(i) Differential Equation

9.11.1
$$\frac{d^3w}{dz^3} - 4z\frac{dw}{dz} - 2w = 0, \qquad w = w_1 w_2,$$

where w_1 and w_2 are any solutions of (9.2.1). For example, $w = \operatorname{Ai}^2(z)$, $\operatorname{Ai}(z)\operatorname{Bi}(z)$, $\operatorname{Ai}(z)\operatorname{Ai}(ze^{\mp 2\pi i/3})$, $M^2(z)$. Numerically satisfactory triads of solutions can be constructed where needed on $\mathbb{R}$ or $\mathbb{C}$ by inspection of the asymptotic expansions supplied in §9.7.

9.11(ii) Wronskian

9.11.2
$$\mathscr{W}\{\operatorname{Ai}^2(z), \operatorname{Ai}(z)\operatorname{Bi}(z), \operatorname{Bi}^2(z)\} = 2\pi^{-3}.$$

9.11(iii) Integral Representations

9.11.3
$$\operatorname{Ai}^2(x) = \frac{1}{4\pi\sqrt{3}}\int_{0}^{\infty} J_0\!\left(\tfrac{1}{12}t^3 + xt\right)t\,dt, \quad x \geq 0,$$

where J_0 is the Bessel function (§10.2(ii)).

9.11.4
$$\operatorname{Ai}^2(z) + \operatorname{Bi}^2(z) = \frac{1}{\pi^{3/2}}\int_{0}^{\infty} \exp\!\left(zt - \tfrac{1}{12}t^3\right)t^{-1/2}\,dt.$$

For an integral representation of the Dirac delta involving a product of two Ai functions see §1.17(ii).

For further integral representations see Reid (1995, 1997a,b).

9.11(iv) Indefinite Integrals

Let w_1, w_2 be any solutions of (9.2.1), not necessarily distinct. Then

9.11.5
$$\int w_1 w_2 \, dz = -w_1' w_2' + z w_1 w_2,$$

9.11.6
$$\int w_1 w_2' \, dz = \tfrac{1}{2}(w_1 w_2 + z \mathscr{W}\{w_1, w_2\}),$$

9.11.7
$$\int w_1' w_2' \, dz = \tfrac{1}{3}(w_1 w_2' + w_1' w_2 + z w_1' w_2' - z^2 w_1 w_2),$$

9.11.8
$$\int z w_1 w_2 \, dz = \tfrac{1}{6}(w_1 w_2' + w_1' w_2) - \tfrac{1}{3}(z w_1' w_2' - z^2 w_1 w_2),$$

9.11.9
$$\int z w_1 w_2' \, dz = \tfrac{1}{2} w_1' w_2' + \tfrac{1}{4} z^2 \mathscr{W}\{w_1, w_2\},$$

9.11.10
$$\int z w_1' w_2' \, dz = \tfrac{3}{10}(-w_1 w_2 + z w_1 w_2' + z w_1' w_2) + \tfrac{1}{5}(z^2 w_1' w_2' - z^3 w_1 w_2).$$

For $\int z^n w_1 w_2 \, dz$, $\int z^n w_1 w_2' \, dz$, $\int z^n w_1' w_2' \, dz$, where n is any positive integer, see Albright (1977). For related integrals see Gordon (1969, Appendix B).

For any continuously-differentiable function f

9.11.11
$$\int \frac{1}{w_1^2} f'\left(\frac{w_2}{w_1}\right) dz = \frac{1}{\mathscr{W}\{w_1, w_2\}} f\left(\frac{w_2}{w_1}\right).$$

Examples

9.11.12
$$\int \frac{dz}{\operatorname{Ai}^2(z)} = \pi \frac{\operatorname{Bi}(z)}{\operatorname{Ai}(z)},$$

9.11.13
$$\int \frac{dz}{\operatorname{Ai}(z) \operatorname{Bi}(z)} = \pi \ln\left(\frac{\operatorname{Bi}(z)}{\operatorname{Ai}(z)}\right),$$

9.11.14
$$\int \frac{\operatorname{Ai}(z) \operatorname{Bi}(z)}{(\operatorname{Ai}^2(z) + \operatorname{Bi}^2(z))^2} dz = \frac{\pi}{2} \frac{\operatorname{Bi}^2(z)}{\operatorname{Ai}^2(z) + \operatorname{Bi}^2(z)}.$$

9.11(v) Definite Integrals

9.11.15
$$\int_0^\infty t^{\alpha-1} \operatorname{Ai}^2(t) \, dt = \frac{2\,\Gamma(\alpha)}{\pi^{1/2} 12^{(2\alpha+5)/6} \Gamma\!\left(\tfrac{1}{3}\alpha + \tfrac{5}{6}\right)},$$
$$\Re \alpha > 0.$$

9.11.16
$$\int_{-\infty}^\infty \operatorname{Ai}^3(t) \, dt = \frac{\Gamma^2\!\left(\tfrac{1}{3}\right)}{4\pi^2},$$

9.11.17
$$\int_{-\infty}^\infty \operatorname{Ai}^2(t) \operatorname{Bi}(t) \, dt = \frac{\Gamma^2\!\left(\tfrac{1}{3}\right)}{4\sqrt{3}\pi^2}.$$

9.11.18
$$\int_0^\infty \operatorname{Ai}^4(t) \, dt = \frac{\ln 3}{24 \pi^2}.$$

9.11.19
$$\int_0^\infty \frac{dt}{\operatorname{Ai}^2(t) + \operatorname{Bi}^2(t)} = \int_0^\infty \frac{t \, dt}{\operatorname{Ai}'^2(t) + \operatorname{Bi}'^2(t)} = \frac{\pi^2}{6}.$$

For further definite integrals see Prudnikov et al. (1990, §1.8.2), Laurenzi (1993), Reid (1995, 1997a,b), and Vallée and Soares (2004, Chapters 3, 4).

Related Functions

9.12 Scorer Functions

9.12(i) Differential Equation

9.12.1
$$\frac{d^2 w}{dz^2} - zw = \frac{1}{\pi}.$$

Solutions of this equation are the *Scorer functions* and can be found by the method of variation of parameters (§1.13(iii)). The general solution is given by

9.12.2
$$w(z) = A w_1(z) + B w_2(z) + p(z),$$

where A and B are arbitrary constants, $w_1(z)$ and $w_2(z)$ are any two linearly independent solutions of Airy's equation (9.2.1), and $p(z)$ is any particular solution of (9.12.1). Standard particular solutions are

9.12.3
$$-\operatorname{Gi}(z), \quad \operatorname{Hi}(z), \quad e^{\mp 2\pi i/3} \operatorname{Hi}\!\left(z e^{\mp 2\pi i/3}\right),$$

where

9.12.4
$$\operatorname{Gi}(z) = \operatorname{Bi}(z) \int_z^\infty \operatorname{Ai}(t) \, dt + \operatorname{Ai}(z) \int_0^z \operatorname{Bi}(t) \, dt,$$

9.12.5
$$\operatorname{Hi}(z) = \operatorname{Bi}(z) \int_{-\infty}^z \operatorname{Ai}(t) \, dt - \operatorname{Ai}(z) \int_{-\infty}^z \operatorname{Bi}(t) \, dt.$$

$\operatorname{Gi}(z)$ and $\operatorname{Hi}(z)$ are entire functions of z.

9.12(ii) Graphs

See Figures 9.12.1 and 9.12.2.

9.12(iii) Initial Values

9.12.6
$$\operatorname{Gi}(0) = \tfrac{1}{2} \operatorname{Hi}(0) = \tfrac{1}{3} \operatorname{Bi}(0)$$
$$= 1\big/\!\left(3^{7/6} \Gamma\!\left(\tfrac{2}{3}\right)\right) = 0.20497\,55424\ldots,$$

9.12.7
$$\operatorname{Gi}'(0) = \tfrac{1}{2} \operatorname{Hi}'(0) = \tfrac{1}{3} \operatorname{Bi}'(0) = 1\big/\!\left(3^{5/6} \Gamma\!\left(\tfrac{1}{3}\right)\right)$$
$$= 0.14942\,94524\ldots.$$

9.12(iv) Numerically Satisfactory Solutions

$-\operatorname{Gi}(x)$ is a numerically satisfactory companion to the complementary functions $\operatorname{Ai}(x)$ and $\operatorname{Bi}(x)$ on the interval $0 \le x < \infty$. $\operatorname{Hi}(x)$ is a numerically satisfactory companion to $\operatorname{Ai}(x)$ and $\operatorname{Bi}(x)$ on the interval $-\infty < x \le 0$.

In $\mathbb{C}$, numerically satisfactory sets of solutions are given by

9.12.8
$$-\operatorname{Gi}(z), \operatorname{Ai}(z), \operatorname{Bi}(z), \qquad |\operatorname{ph} z| \le \tfrac{1}{3}\pi,$$

9.12 Scorer Functions

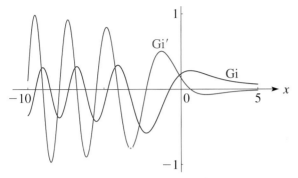

Figure 9.12.1: Gi(x), Gi$'(x)$.

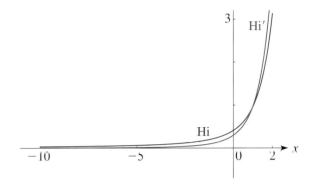

Figure 9.12.2: Hi(x), Hi$'(x)$.

9.12.9
$$\text{Hi}(z), \text{Ai}\!\left(ze^{-2\pi i/3}\right), \text{Ai}\!\left(ze^{2\pi i/3}\right), \quad |\text{ph}(-z)| \leq \tfrac{2}{3}\pi,$$
and

9.12.10
$$e^{\mp 2\pi i/3}\,\text{Hi}\!\left(ze^{\mp 2\pi i/3}\right), \text{Ai}(z), \text{Ai}\!\left(ze^{\pm 2\pi i/3}\right),$$
$$-\pi \leq \pm \text{ph}\,z \leq \tfrac{1}{3}\pi.$$

9.12(v) Connection Formulas

9.12.11
$$\text{Gi}(z) + \text{Hi}(z) = \text{Bi}(z),$$

9.12.12
$$\text{Gi}(z) = \tfrac{1}{2}e^{\pi i/3}\,\text{Hi}\!\left(ze^{-2\pi i/3}\right) + \tfrac{1}{2}e^{-\pi i/3}\,\text{Hi}\!\left(ze^{2\pi i/3}\right),$$

9.12.13 $\text{Gi}(z) = e^{\mp \pi i/3}\,\text{Hi}\!\left(ze^{\pm 2\pi i/3}\right) \pm i\,\text{Ai}(z),$

9.12.14
$$\text{Hi}(z) = e^{\pm 2\pi i/3}\,\text{Hi}\!\left(ze^{\pm 2\pi i/3}\right) + 2e^{\mp \pi i/6}\,\text{Ai}\!\left(ze^{\mp 2\pi i/3}\right).$$

9.12(vi) Maclaurin Series

9.12.15
$$\text{Gi}(z) = \frac{3^{-2/3}}{\pi}\sum_{k=0}^{\infty}\cos\!\left(\frac{2k-1}{3}\pi\right)\Gamma\!\left(\frac{k+1}{3}\right)\frac{(3^{1/3}z)^k}{k!},$$

9.12.16
$$\text{Gi}'(z) = \frac{3^{-1/3}}{\pi}\sum_{k=0}^{\infty}\cos\!\left(\frac{2k+1}{3}\pi\right)\Gamma\!\left(\frac{k+2}{3}\right)\frac{(3^{1/3}z)^k}{k!}.$$

9.12.17 $\text{Hi}(z) = \dfrac{3^{-2/3}}{\pi}\displaystyle\sum_{k=0}^{\infty}\Gamma\!\left(\dfrac{k+1}{3}\right)\dfrac{(3^{1/3}z)^k}{k!},$

9.12.18 $\text{Hi}'(z) = \dfrac{3^{-1/3}}{\pi}\displaystyle\sum_{k=0}^{\infty}\Gamma\!\left(\dfrac{k+2}{3}\right)\dfrac{(3^{1/3}z)^k}{k!}.$

9.12(vii) Integral Representations

9.12.19 $\text{Gi}(x) = \dfrac{1}{\pi}\displaystyle\int_0^{\infty}\sin\!\left(\tfrac{1}{3}t^3 + xt\right)dt, \quad x\in\mathbb{R}.$

9.12.20 $\text{Hi}(z) = \dfrac{1}{\pi}\displaystyle\int_0^{\infty}\exp\!\left(-\tfrac{1}{3}t^3 + zt\right)dt,$

9.12.21
$$\text{Gi}(z) = -\frac{1}{\pi}\int_0^{\infty}\exp\!\left(-\tfrac{1}{3}t^3 - \tfrac{1}{2}zt\right)\cos\!\left(\tfrac{1}{2}\sqrt{3}zt + \tfrac{2}{3}\pi\right)dt.$$

If $\zeta = \tfrac{2}{3}z^{3/2}$ or $\tfrac{2}{3}x^{3/2}$, and $K_{1/3}$ is the modified Bessel function (§10.25(ii)), then

9.12.22 $\text{Hi}(-z) = \dfrac{4z^2}{3^{3/2}\pi^2}\displaystyle\int_0^{\infty}\dfrac{K_{1/3}(t)}{\zeta^2 + t^2}\,dt, \quad |\text{ph}\,z| < \tfrac{1}{3}\pi,$

9.12.23 $\text{Gi}(x) = \dfrac{4x^2}{3^{3/2}\pi^2}\displaystyle\fint_0^{\infty}\dfrac{K_{1/3}(t)}{\zeta^2 - t^2}\,dt, \quad x > 0,$

where the last integral is a Cauchy principal value (§1.4(v)).

Mellin–Barnes Type Integral

9.12.24
$$\text{Hi}(z) = \frac{3^{-2/3}}{2\pi^2 i}\int_{-i\infty}^{i\infty}\Gamma\!\left(\tfrac{1}{3} + \tfrac{1}{3}t\right)\Gamma(-t)(3^{1/3}e^{\pi i}z)^t\,dt,$$

where the integration contour separates the poles of $\Gamma\!\left(\tfrac{1}{3} + \tfrac{1}{3}t\right)$ from those of $\Gamma(-t)$.

9.12(viii) Asymptotic Expansions

Functions and Derivatives

As $z \to \infty$, and with δ denoting an arbitrary small positive constant,

9.12.25 $\text{Gi}(z) \sim \dfrac{1}{\pi z}\displaystyle\sum_{k=0}^{\infty}\dfrac{(3k)!}{k!(3z^3)^k}, \quad |\text{ph}\,z| \leq \tfrac{1}{3}\pi - \delta,$

9.12.26 $\text{Gi}'(z) \sim -\dfrac{1}{\pi z^2}\displaystyle\sum_{k=0}^{\infty}\dfrac{(3k+1)!}{k!(3z^3)^k}, \quad |\text{ph}\,z| \leq \tfrac{1}{3}\pi - \delta.$

9.12.27
$$\text{Hi}(z) \sim -\frac{1}{\pi z}\sum_{k=0}^{\infty}\frac{(3k)!}{k!(3z^3)^k}, \quad |\text{ph}(-z)| \leq \tfrac{2}{3}\pi - \delta,$$

9.12.28
$$\text{Hi}'(z) \sim \frac{1}{\pi z^2} \sum_{k=0}^{\infty} \frac{(3k+1)!}{k!(3z^3)^k}, \quad |\text{ph}(-z)| \leq \tfrac{2}{3}\pi - \delta.$$

For other phase ranges combine these results with the connection formulas (9.12.11)–(9.12.14) and the asymptotic expansions given in §9.7. For example, with the notation of §9.7(i).

9.12.29
$$\text{Hi}(z) \sim -\frac{1}{\pi z}\sum_{k=0}^{\infty}\frac{(3k)!}{k!(3z^3)^k} + \frac{e^\zeta}{\sqrt{\pi}z^{1/4}}\sum_{k=0}^{\infty}\frac{u_k}{\zeta^k},$$
$$|\text{ph}\,z| \leq \pi - \delta.$$

Integrals

9.12.30
$$\int_0^z \text{Gi}(t)\,dt \sim \frac{1}{\pi}\ln z + \frac{2\gamma + \ln 3}{3\pi} - \frac{1}{\pi}\sum_{k=1}^{\infty}\frac{(3k-1)!}{k!(3z^3)^k},$$
$$|\text{ph}\,z| \leq \tfrac{1}{3}\pi - \delta.$$

9.12.31
$$\int_0^z \text{Hi}(-t)\,dt \sim \frac{1}{\pi}\ln z + \frac{2\gamma + \ln 3}{3\pi}$$
$$+ \frac{1}{\pi}\sum_{k=1}^{\infty}(-1)^{k-1}\frac{(3k-1)!}{k!(3z^3)^k},$$
$$|\text{ph}\,z| \leq \tfrac{2}{3}\pi - \delta,$$

where γ is Euler's constant (§5.2(ii)).

9.12(ix) Zeros

All zeros, real or complex, of $\text{Gi}(z)$ and $\text{Hi}(z)$ are simple.

Neither $\text{Hi}(z)$ nor $\text{Hi}'(z)$ has real zeros.

$\text{Gi}(z)$ has no nonnegative real zeros and $\text{Gi}'(z)$ has exactly one nonnegative real zero, given by $z = 0.60907\,54170\,7\ldots$. Both $\text{Gi}(z)$ and $\text{Gi}'(z)$ have an infinity of negative real zeros, and they are interlaced.

For the above properties and further results, including the distribution of complex zeros, asymptotic approximations for the numerically large real or complex zeros, and numerical tables see Gil et al. (2003c).

For graphical illustration of the real zeros see Figures 9.12.1 and 9.12.2.

9.13 Generalized Airy Functions

9.13(i) Generalizations from the Differential Equation

Equations of the form

9.13.1
$$\frac{d^2w}{dz^2} = z^n w, \qquad n = 1, 2, 3, \ldots,$$

are used in approximating solutions to differential equations with multiple turning points; see §2.8(v). The general solution of (9.13.1) is given by

9.13.2
$$w = z^{1/2}\mathscr{Z}_p(\zeta),$$

where

9.13.3 $\quad p = \dfrac{1}{n+2}, \quad \zeta = \dfrac{2}{n+2}z^{(n+2)/2} = 2pz^{1/(2p)},$

and $\mathscr{Z}_p$ is any linear combination of the modified Bessel functions I_p and $e^{p\pi i}K_p$ (§10.25(ii)).

Swanson and Headley (1967) define independent solutions $A_n(z)$ and $B_n(z)$ of (9.13.1) by

9.13.4
$$A_n(z) = (2p/\pi)\sin(p\pi)z^{1/2}K_p(\zeta),$$
$$B_n(z) = (pz)^{1/2}(I_{-p}(\zeta) + I_p(\zeta)),$$

when z is real and positive, and by analytic continuation elsewhere. (All solutions of (9.13.1) are entire functions of z.) When $n = 1$, $A_n(z)$ and $B_n(z)$ become $\text{Ai}(z)$ and $\text{Bi}(z)$, respectively.

Properties of $A_n(z)$ and $B_n(z)$ follow from the corresponding properties of the modified Bessel functions. They include:

9.13.5
$$A_n(0) = p^{1/2}B_n(0) = \frac{p^{1-p}}{\Gamma(1-p)},$$
$$-A_n'(0) = p^{1/2}B_n'(0) = \frac{p^p}{\Gamma(p)}.$$

9.13.6 $\quad A_n(-z) = \begin{cases} pz^{1/2}(J_{-p}(\zeta) + J_p(\zeta)), & n \text{ odd}, \\ p^{1/2}B_n(z), & n \text{ even}, \end{cases}$

9.13.7 $\quad B_n(-z) = \begin{cases} (pz)^{1/2}(J_{-p}(\zeta) - J_p(\zeta)), & n \text{ odd}, \\ p^{-1/2}A_n(z), & n \text{ even}. \end{cases}$

9.13.8 $\quad \mathscr{W}\{A_n(z), B_n(z)\} = \dfrac{2}{\pi}p^{1/2}\sin(p\pi).$

As $z \to \infty$

9.13.9 $\quad A_n(z) = \sqrt{p/\pi}\sin(p\pi)z^{-n/4}e^{-\zeta}(1 + O(\zeta^{-1})), \qquad |\text{ph}\,z| \leq 3p\pi - \delta,$

9.13.10 $\quad A_n(-z) = \begin{cases} 2\sqrt{p/\pi}\cos(\tfrac{1}{2}p\pi)z^{-n/4}\left(\cos(\zeta - \tfrac{1}{4}\pi) + e^{|\Im\zeta|}O(\zeta^{-1})\right), & |\text{ph}\,z| \leq 2p\pi - \delta, n \text{ odd}, \\ \sqrt{p/\pi}z^{-n/4}e^\zeta(1 + O(\zeta^{-1})), & |\text{ph}\,z| \leq p\pi - \delta, n \text{ even}, \end{cases}$

9.13.11 $\quad B_n(z) = \pi^{-1/2}z^{-n/4}e^\zeta(1 + O(\zeta^{-1})), \qquad |\text{ph}\,z| \leq p\pi - \delta,$

9.13.12 $\quad B_n(-z) = \begin{cases} -(2/\sqrt{\pi})\sin(\tfrac{1}{2}p\pi)z^{-n/4}\left(\sin(\zeta - \tfrac{1}{4}\pi) + e^{|\Im\zeta|}O(\zeta^{-1})\right), & |\text{ph}\,z| \leq 2p\pi - \delta, n \text{ odd}, \\ (1/\sqrt{\pi})\sin(p\pi)z^{-n/4}e^{-\zeta}(1 + O(\zeta^{-1})), & |\text{ph}\,z| \leq 3p\pi - \delta, n \text{ even}. \end{cases}$

The distribution in $\mathbb{C}$ and asymptotic properties of the zeros of $A_n(z)$, $A'_n(z)$, $B_n(z)$, and $B'_n(z)$ are investigated in Swanson and Headley (1967) and Headley and Barwell (1975).

In Olver (1977a, 1978) a different normalization is used. In place of (9.13.1) we have

9.13.13 $$\frac{d^2w}{dt^2} = \tfrac{1}{4}m^2 t^{m-2} w,$$

where $m = 3, 4, 5, \ldots$. For real variables the solutions of (9.13.13) are denoted by $U_m(t)$, $U_m(-t)$ when m is even, and by $V_m(t)$, $\overline{V}_m(t)$ when m is odd. (The overbar has nothing to do with complex conjugates.) Their relations to the functions $A_n(z)$ and $B_n(z)$ are given by

9.13.14 $\quad m = n+2 = 1/p, \quad t = (\tfrac{1}{2}m)^{-2/m} z = \zeta^{2/m},$

9.13.15
$$\sqrt{2\pi} \left(\tfrac{1}{2}m\right)^{(m-1)/m} \csc(\pi/m) A_n(z)$$
$$= \begin{cases} U_m(t), & m \text{ even}, \\ V_m(t), & m \text{ odd}, \end{cases}$$

9.13.16
$$\sqrt{\pi} \left(\tfrac{1}{2}m\right)^{(m-2)/(2m)} \csc(\pi/m) B_n(z)$$
$$= \begin{cases} U_m(-t), & m \text{ even}, \\ \overline{V}_m(t), & m \text{ odd}. \end{cases}$$

Properties and graphs of $U_m(t)$, $V_m(t)$, $\overline{V}_m(t)$ are included in Olver (1977a) together with properties and graphs of real solutions of the equation

9.13.17 $\quad \dfrac{d^2w}{dt^2} = -\tfrac{1}{4}m^2 t^{m-2} w, \quad m$ even,

which are denoted by $W_m(t)$, $W_m(-t)$.

In $\mathbb{C}$, the solutions of (9.13.13) used in Olver (1978) are

9.13.18 $\quad w = U_m(t e^{-2j\pi i/m}), \quad j = 0, \pm 1, \pm 2, \ldots.$

The function on the right-hand side is recessive in the sector $-(2j-1)\pi/m \leq \operatorname{ph} z \leq (2j+1)\pi/m$, and is therefore an essential member of any numerically satisfactory pair of solutions in this region.

Another normalization of (9.13.17) is used in Smirnov (1960), given by

9.13.19 $\quad \dfrac{d^2w}{dx^2} + x^\alpha w = 0,$

where $\alpha > -2$ and $x > 0$. Solutions are $w = U_1(x, \alpha)$, $U_2(x, \alpha)$, where

9.13.20
$$U_1(x, \alpha) = \frac{1}{(\alpha+2)^{1/(\alpha+2)}}$$
$$\times \Gamma\!\left(\frac{\alpha+1}{\alpha+2}\right) x^{1/2} J_{-1/(\alpha+2)}\!\left(\frac{2}{\alpha+2} x^{(\alpha+2)/2}\right),$$

9.13.21
$$U_2(x, \alpha) = (\alpha+2)^{1/(\alpha+2)}$$
$$\times \Gamma\!\left(\frac{\alpha+3}{\alpha+2}\right) x^{1/2} J_{1/(\alpha+2)}\!\left(\frac{2}{\alpha+2} x^{(\alpha+2)/2}\right),$$

and J denotes the Bessel function (§10.2(ii)).

When α is a positive integer the relation of these functions to $W_m(t)$, $W_m(-t)$ is as follows:

9.13.22 $\quad \alpha = m-2, \quad x = (m/2)^{2/m} t,$

9.13.23
$$U_1(x, \alpha) = \frac{\pi^{1/2}}{2^{(m+2)/(2m)} \Gamma(1/m)} \left(W_m(t) + W_m(-t)\right),$$

9.13.24
$$U_2(x, \alpha) = \frac{\pi^{1/2} m^{2/m}}{2^{(m+2)/(2m)} \Gamma(-1/m)} \left(W_m(t) - W_m(-t)\right).$$

For properties of the zeros of the functions defined in this subsection see Laforgia and Muldoon (1988) and references given therein.

9.13(ii) Generalizations from Integral Representations

Reid (1972) and Drazin and Reid (1981, Appendix) introduce the following contour integrals in constructing approximate solutions to the Orr–Sommerfeld equation for fluid flow:

9.13.25 $\quad A_k(z, p) = \dfrac{1}{2\pi i} \displaystyle\int_{\mathscr{L}_k} t^{-p} \exp\!\left(zt - \tfrac{1}{3}t^3\right) dt,$
$$k = 1, 2, 3, \, p \in \mathbb{C},$$

9.13.26 $\quad B_0(z, p) = \dfrac{1}{2\pi i} \displaystyle\int_{\mathscr{L}_0} t^{-p} \exp\!\left(zt - \tfrac{1}{3}t^3\right) dt,$
$$p = 0, \pm 1, \pm 2, \ldots,$$

9.13.27 $\quad B_k(z, p) = \displaystyle\int_{\mathscr{I}_k} t^{-p} \exp\!\left(zt - \tfrac{1}{3}t^3\right) dt,$
$$k = 1, 2, 3, \, p = 0, \pm 1, \pm 2, \ldots,$$

with $z \in \mathbb{C}$ in all cases. The integration paths $\mathscr{L}_0$, $\mathscr{L}_1$, $\mathscr{L}_2$, $\mathscr{L}_3$ are depicted in Figure 9.13.1. $\mathscr{I}_1$, $\mathscr{I}_2$, $\mathscr{I}_3$ are depicted in Figure 9.13.2. When p is not an integer the branch of t^{-p} in (9.13.25) is usually chosen to be $\exp(-p(\ln|t| + i \operatorname{ph} t))$ with $0 \leq \operatorname{ph} t < 2\pi$.

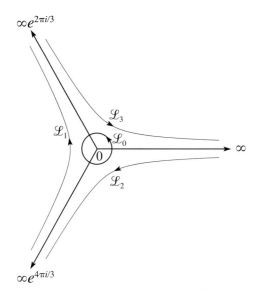

Figure 9.13.1: t-plane. Paths $\mathscr{L}_0, \mathscr{L}_1, \mathscr{L}_2, \mathscr{L}_3$.

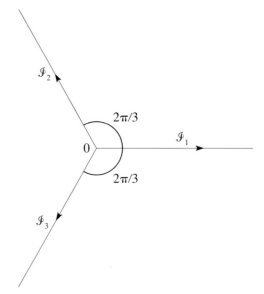

Figure 9.13.2: t-plane. Paths $\mathscr{I}_1, \mathscr{I}_2, \mathscr{I}_3$.

When $p = 0$

9.13.28 $$A_1(z,0) = \operatorname{Ai}(z),$$

9.13.29
$$A_2(z,0) = e^{2\pi i/3}\operatorname{Ai}\!\left(ze^{2\pi i/3}\right),$$
$$A_3(z,0) = e^{-2\pi i/3}\operatorname{Ai}\!\left(ze^{-2\pi i/3}\right),$$

and

9.13.30 $$B_0(z,0) = 0, \quad B_1(z,0) = \pi\operatorname{Hi}(z).$$

Each of the functions $A_k(z,p)$ and $B_k(z,p)$ satisfies the differential equation

9.13.31 $$\frac{d^3w}{dz^3} - z\frac{dw}{dz} + (p-1)w = 0,$$

and the difference equation

9.13.32 $$f(p-3) - zf(p-1) + (p-1)f(p) = 0.$$

The $A_k(z,p)$ are related by

9.13.33
$$A_2(z,p) = e^{-2(p-1)\pi i/3}A_1\!\left(ze^{2\pi i/3},p\right),$$
$$A_3(z,p) = e^{2(p-1)\pi i/3}A_1\!\left(ze^{-2\pi i/3},p\right).$$

Connection formulas for the solutions of (9.13.31) include

9.13.34 $A_1(z,p) + A_2(z,p) + A_3(z,p) + B_0(z,p) = 0,$

9.13.35 $B_2(z,p) - B_3(z,p) = 2\pi i\, A_1(z,p),$

9.13.36 $B_3(z,p) - B_1(z,p) = 2\pi i\, A_2(z,p),$

9.13.37 $B_1(z,p) - B_2(z,p) = 2\pi i\, A_3(z,p).$

Further properties of these functions, and also of similar contour integrals containing an additional factor $(\ln t)^q$, $q = 1, 2, \ldots$, in the integrand, are derived in Reid (1972), Drazin and Reid (1981, Appendix), and Baldwin (1985). These properties include Wronskians, asymptotic expansions, and information on zeros.

For further generalizations via integral representations see Chin and Hedstrom (1978), Janson et al. (1993, §10), and Kamimoto (1998).

9.14 Incomplete Airy Functions

Incomplete Airy functions are defined by the contour integral (9.5.4) when one of the integration limits is replaced by a variable real or complex parameter. For information, including asymptotic approximations, computation, and applications, see Levey and Felsen (1969), Constantinides and Marhefka (1993), and Michaeli (1996).

Applications

9.15 Mathematical Applications

Airy functions play an indispensable role in the construction of uniform asymptotic expansions for contour integrals with coalescing saddle points, and for solutions of linear second-order ordinary differential equations with a simple turning point. For descriptions of, and references to, the underlying theory see §§2.4(v) and 2.8(iii).

9.16 Physical Applications

Airy functions are applied in many branches of both classical and quantum physics. The function Ai(x) first appears as an integral in two articles by G.B. Airy on the intensity of light in the neighborhood of a caustic (Airy (1838, 1849)). Details of the Airy theory are given in van de Hulst (1957) in the chapter on the optics of a raindrop. See also Berry (1966, 1969).

The frequent appearances of the Airy functions in both classical and quantum physics is associated with wave equations with turning points, for which asymptotic (WKBJ) solutions are exponential on one side and oscillatory on the other. The Airy functions constitute uniform approximations whose region of validity includes the turning point and its neighborhood. Within classical physics, they appear prominently in physical optics, electromagnetism, radiative transfer, fluid mechanics, and nonlinear wave propagation. Examples dealing with the propagation of light and with radiation of electromagnetic waves are given in Landau and Lifshitz (1962). Extensive use is made of Airy functions in investigations in the theory of electromagnetic diffraction and radiowave propagation (Fock (1965)). A quite different application is made in the study of the diffraction of sound pulses by a circular cylinder (Friedlander (1958)).

In fluid dynamics, Airy functions enter several topics. In the study of the stability of a two-dimensional viscous fluid, the flow is governed by the Orr–Sommerfeld equation (a fourth-order differential equation). Again, the quest for asymptotic approximations that are uniformly valid solutions to this equation in the neighborhoods of critical points leads (after choosing solvable equations with similar asymptotic properties) to Airy functions. Other applications appear in the study of instability of Couette flow of an inviscid fluid. These examples of transitions to turbulence are presented in detail in Drazin and Reid (1981) with the problem of hydrodynamic stability. The investigation of the transition between subsonic and supersonic of a two-dimensional gas flow leads to the Euler–Tricomi equation (Landau and Lifshitz (1987)). An application of Airy functions to the solution of this equation is given in Gramtcheff (1981).

Airy functions play a prominent role in problems defined by nonlinear wave equations. These first appeared in connection with the equation governing the evolution of long shallow water waves of permanent form, generally called solitons, and are predicted by the Korteweg–de Vries (KdV) equation (a third-order nonlinear partial differential equation). The KdV equation and solitons have applications in many branches of physics, including plasma physics lattice dynamics, and quantum mechanics. (Ablowitz and Segur (1981), Ablowitz and Clarkson (1991), and Whitham (1974).)

Reference to many of these applications as well as to the theory of elasticity and to the heat equation are given in Vallée and Soares (2004): a book devoted specifically to the Airy and Scorer functions and their applications in physics.

An example from quantum mechanics is given in Landau and Lifshitz (1965), in which the exact solution of the Schrödinger equation for the motion of a particle in a homogeneous external field is expressed in terms of Ai(x). Solutions of the Schrödinger equation involving the Airy functions are given for other potentials in Vallée and Soares (2004). This reference provides several examples of applications to problems in quantum mechanics in which Airy functions give uniform asymptotic approximations, valid in the neighborhood of a turning point. A study of the semiclassical description of quantum-mechanical scattering is given in Ford and Wheeler (1959a,b). In the case of the rainbow, the scattering amplitude is expressed in terms of Ai(x), the analysis being similar to that given originally by Airy (1838) for the corresponding problem in optics.

An application of the Scorer functions is to the problem of the uniform loading of infinite plates (Rothman (1954a,b)).

Computation

9.17 Methods of Computation

9.17(i) Maclaurin Expansions

Although the Maclaurin-series expansions of §§9.4 and 9.12(vi) converge for all finite values of z, they are cumbersome to use when $|z|$ is large owing to slowness of convergence and cancellation. For large $|z|$ the asymptotic expansions of §§9.7 and 9.12(viii) should be used instead. Since these expansions diverge, the accuracy they yield is limited by the magnitude of $|z|$. However, in the case of Ai(z) and Bi(z) this accuracy can be increased considerably by use of the exponentially-improved forms of expansion supplied in §9.7(v).

9.17(ii) Differential Equations

A comprehensive and powerful approach is to integrate the defining differential equation (9.2.1) by direct numerical methods. As described in §3.7(ii), to ensure stability the integration path must be chosen in such a way that as we proceed along it the wanted solution grows at least as fast as all other solutions of the differential equation. In the case of Ai(z), for example, this means that in the sectors $\frac{1}{3}\pi < |\operatorname{ph} z| < \pi$ we may integrate along outward rays from the origin with initial values obtained from §9.2(ii). But when $|\operatorname{ph} z| < \frac{1}{3}\pi$ the integration has to be towards the origin, with starting values of $\operatorname{Ai}(z)$ and $\operatorname{Ai}'(z)$ computed from their asymptotic expansions. On the remaining rays, given by $\operatorname{ph} z = \pm\frac{1}{3}\pi$ and π, integration can proceed in either direction.

For further information see Lozier and Olver (1993) and Fabijonas et al. (2004). The former reference includes a parallelized version of the method.

In the case of the Scorer functions, integration of the differential equation (9.12.1) is more difficult than (9.2.1), because in some regions stable directions of integration do not exist. An example is provided by $\operatorname{Gi}(x)$ on the positive real axis. In these cases boundary-value methods need to be used instead; see §3.7(iii).

9.17(iii) Integral Representations

Among the integral representations of the Airy functions the Stieltjes transform (9.10.18) furnishes a way of computing $\operatorname{Ai}(z)$ in the complex plane, once values of this function can be generated on the positive real axis. For details, including the application of a generalized form of Gaussian quadrature, see Gordon (1969, Appendix A) and Schulten et al. (1979).

Gil et al. (2002a) describes two methods for the computation of $\operatorname{Ai}(z)$ and $\operatorname{Ai}'(z)$ for $z \in \mathbb{C}$. In the first method the integration path for the contour integral (9.5.4) is deformed to coincide with paths of steepest descent (§2.4(iv)). The trapezoidal rule (§3.5(i)) is then applied. The second method is to apply generalized Gauss–Laguerre quadrature (§3.5(v)) to the integral (9.5.8). For the second method see also Gautschi (2002a). The methods for $\operatorname{Ai}'(z)$ are similar.

For quadrature methods for Scorer functions see Gil et al. (2001), Lee (1980), and Gordon (1970, Appendix A); but see also Gautschi (1983).

9.17(iv) Via Bessel Functions

In consequence of §9.6(i), algorithms for generating Bessel functions, Hankel functions, and modified Bessel functions (§10.74) can also be applied to $\operatorname{Ai}(z)$, $\operatorname{Bi}(z)$, and their derivatives.

9.17(v) Zeros

Zeros of the Airy functions, and their derivatives, can be computed to high precision via Newton's rule (§3.8(ii)) or Halley's rule (§3.8(v)), using values supplied by the asymptotic expansions of §9.9(iv) as initial approximations. This method was used in the computation of the tables in §9.9(v). See also Fabijonas et al. (2004).

For the computation of the zeros of the Scorer functions and their derivatives see Gil et al. (2003c).

9.18 Tables

9.18(i) Introduction

Additional listings of early tables of the functions treated in this chapter are given in Fletcher et al. (1962) and Lebedev and Fedorova (1960).

9.18(ii) Real Variables

- Miller (1946) tabulates $\operatorname{Ai}(x)$, $\operatorname{Ai}'(x)$ for $x = -20(.01)2$; $\log_{10}\operatorname{Ai}(x)$, $\operatorname{Ai}'(x)/\operatorname{Ai}(x)$ for $x = 0(.1)25(1)75$; $\operatorname{Bi}(x)$, $\operatorname{Bi}'(x)$ for $x = -10(.1)2.5$; $\log_{10}\operatorname{Bi}(x)$, $\operatorname{Bi}'(x)/\operatorname{Bi}(x)$ for $x = 0(.1)10$; $M(x)$, $N(x)$, $\theta(x)$, $\phi(x)$ (respectively $F(x)$, $G(x)$, $\chi(x)$, $\psi(x)$) for $x = -80(1)-30(.1)0$. Precision is generally 8D; slightly less for some of the auxiliary functions. Extracts from these tables are included in Abramowitz and Stegun (1964, Chapter 10), together with some auxiliary functions for large arguments.

- Fox (1960, Table 3) tabulates $2\pi^{1/2}x^{1/4} \times \exp(\frac{2}{3}x^{3/2})\operatorname{Ai}(x)$, $2\pi^{1/2}x^{-1/4}\exp(\frac{2}{3}x^{3/2})\operatorname{Ai}'(x)$, $\pi^{1/2}x^{1/4}\exp(-\frac{2}{3}x^{3/2})\operatorname{Bi}(x)$, and $\pi^{1/2}x^{-1/4} \times \exp(-\frac{2}{3}x^{3/2})\operatorname{Bi}'(x)$ for $\frac{3}{2}x^{-3/2} = 0(.001)0.05$, together with similar auxiliary functions for negative values of x. Precision is 10D.

- Zhang and Jin (1996, p. 337) tabulates $\operatorname{Ai}(x)$, $\operatorname{Ai}'(x)$, $\operatorname{Bi}(x)$, $\operatorname{Bi}'(x)$ for $x = 0(1)20$ to 8S and for $x = -20(1)0$ to 9D.

- Yakovleva (1969) tabulates Fock's functions $U(x) \equiv \sqrt{\pi}\operatorname{Bi}(x)$, $U'(x) \equiv \sqrt{\pi}\operatorname{Bi}'(x)$, $V(x) \equiv \sqrt{\pi}\operatorname{Ai}(x)$, $V'(x) \equiv \sqrt{\pi}\operatorname{Ai}'(x)$ for $x = -9(.001)9$. Precision is 7S.

9.18(iii) Complex Variables

- Woodward and Woodward (1946) tabulates the real and imaginary parts of $\operatorname{Ai}(z)$, $\operatorname{Ai}'(z)$, $\operatorname{Bi}(z)$, $\operatorname{Bi}'(z)$ for $\Re z = -2.4(.2)2.4$, $\Im z = -2.4(.2)0$. Precision is 4D.

- Harvard (1945) tabulates the real and imaginary parts of $h_1(z)$, $h_1'(z)$, $h_2(z)$, $h_2'(z)$ for $-x_0 \leq \Re z \leq x_0$, $0 \leq \Im z \leq y_0$, $|x_0 + iy_0| < 6.1$, with interval 0.1 in $\Re z$ and $\Im z$. Precision is 8D. Here $h_1(z) = -2^{4/3}3^{1/6}i\operatorname{Ai}(e^{-\pi i/3}z)$, $h_2(z) = 2^{4/3}3^{1/6}i\operatorname{Ai}(e^{\pi i/3}z)$.

9.18(iv) Zeros

- Miller (1946) tabulates a_k, $\text{Ai}'(a_k)$, a_k', $\text{Ai}(a_k')$, $k = 1(1)50$; b_k, $\text{Bi}'(b_k)$, b_k', $\text{Bi}(b_k')$, $k = 1(1)20$. Precision is 8D. Entries for $k = 1(1)20$ are reproduced in Abramowitz and Stegun (1964, Chapter 10).

- Sherry (1959) tabulates a_k, $\text{Ai}'(a_k)$, a_k', $\text{Ai}(a_k')$, $k = 1(1)50$; 20S.

- Zhang and Jin (1996, p. 339) tabulates a_k, $\text{Ai}'(a_k)$, a_k', $\text{Ai}(a_k')$, b_k, $\text{Bi}'(b_k)$, b_k', $\text{Bi}(b_k')$, $k = 1(1)20$; 8D.

- Corless et al. (1992) gives the real and imaginary parts of β_k for $k = 1(1)13$; 14S.

- See also §9.9(v).

9.18(v) Integrals

- Rothman (1954a) tabulates $\int_0^x \text{Ai}(t)\,dt$ and $\int_0^x \text{Bi}(t)\,dt$ for $x = -10(.1)\infty$ and $-10(.1)2$, respectively; 7D. The entries in the columns headed $\int_0^x \text{Ai}(-x)\,dx$ and $\int_0^x \text{Bi}(-x)\,dx$ all have the wrong sign. The tables are reproduced in Abramowitz and Stegun (1964, Chapter 10), and the sign errors are corrected in later reprintings.

- NBS (1958) tabulates $\int_0^x \text{Ai}(-t)\,dt$ and $\int_0^x \int_0^v \text{Ai}(-t)\,dt\,dv$ (see (9.10.20)) for $x = -2(.01)5$ to 8D and 7D, respectively.

- Zhang and Jin (1996, p. 338) tabulates $\int_0^x \text{Ai}(t)\,dt$ and $\int_0^x \text{Bi}(t)\,dt$ for $x = -10(.2)10$ to 8D or 8S.

9.18(vi) Scorer Functions

- Scorer (1950) tabulates $\text{Gi}(x)$ and $\text{Hi}(-x)$ for $x = 0(.1)10$; 7D.

- Rothman (1954b) tabulates $\int_0^x \text{Gi}(t)\,dt$, $\text{Gi}'(x)$, $\int_0^x \text{Hi}(-t)\,dt$, $-\text{Hi}'(-x)$ for $x = 0(.1)10$; 7D.

- NBS (1958) tabulates $A_0(x) \equiv \pi\text{Hi}(-x)$ and $-A_0'(x) \equiv \pi\text{Hi}'(-x)$ for $x = 0(.01)1(.02)5(.05)11$ and $1/x = 0.01(.01)0.1$; $\int_0^x A_0(t)\,dt$ for $x = 0.5, 1(1)11$. Precision is 8D.

- Nosova and Tumarkin (1965) tabulates $e_0(x) \equiv \pi\text{Hi}(-x)$, $e_0'(x) \equiv -\pi\text{Hi}'(-x)$, $\tilde{e}_0(-x) \equiv -\pi\text{Gi}(-x)$, $\tilde{e}_0'(-x) \equiv \pi\text{Gi}'(x)$ for $x = -1(.01)10$; 7D. Also included are the real and imaginary parts of $e_0(z)$ and $ie_0'(z)$, where $z = iy$ and $y = 0(.01)9$; 6-7D.

- Gil et al. (2003c) tabulates the only positive zero of $\text{Gi}'(z)$, the first 10 negative real zeros of $\text{Gi}(z)$ and $\text{Gi}'(z)$, and the first 10 complex zeros of $\text{Gi}(z)$, $\text{Gi}'(z)$, $\text{Hi}(z)$, and $\text{Hi}'(z)$. Precision is 11 or 12S.

9.18(vii) Generalized Airy Functions

- Smirnov (1960) tabulates $U_1(x,\alpha)$, $U_2(x,\alpha)$, defined by (9.13.20), (9.13.21), and also $\partial U_1(x,\alpha)/\partial x$, $\partial U_2(x,\alpha)/\partial x$, for $\alpha = 1$, $x = -6(.01)10$ to 5D or 5S, and also for $\alpha = \pm\frac{1}{4}, \pm\frac{1}{3}, \pm\frac{1}{2}, \pm\frac{2}{3}, \pm\frac{3}{4}, \frac{5}{4}, \frac{4}{3}, \frac{3}{2}, \frac{5}{3}, \frac{7}{4}, 2$, $x = 0(.01)6$; 4D.

9.19 Approximations

9.19(i) Approximations in Terms of Elementary Functions

- Martín et al. (1992) provides two simple formulas for approximating $\text{Ai}(x)$ to graphical accuracy, one for $-\infty < x \leq 0$, the other for $0 \leq x < \infty$.

- Moshier (1989, §6.14) provides minimax rational approximations for calculating $\text{Ai}(x)$, $\text{Ai}'(x)$, $\text{Bi}(x)$, $\text{Bi}'(x)$. They are in terms of the variable ζ, where $\zeta = \frac{2}{3}x^{3/2}$ when x is positive, $\zeta = \frac{2}{3}(-x)^{3/2}$ when x is negative, and $\zeta = 0$ when $x = 0$. The approximations apply when $2 \leq \zeta < \infty$, that is, when $3^{2/3} \leq x < \infty$ or $-\infty < x \leq -3^{2/3}$. The precision in the coefficients is 21S.

9.19(ii) Expansions in Chebyshev Series

These expansions are for real arguments x and are supplied in sets of four for each function, corresponding to intervals $-\infty < x \leq a$, $a \leq x \leq 0$, $0 \leq x \leq b$, $b \leq x < \infty$. The constants a and b are chosen numerically, with a view to equalizing the effort required for summing the series.

- Prince (1975) covers $\text{Ai}(x)$, $\text{Ai}'(x)$, $\text{Bi}(x)$, $\text{Bi}'(x)$. The Chebyshev coefficients are given to 10-11D. Fortran programs are included. See also Razaz and Schonfelder (1981).

- Németh (1992, Chapter 8) covers $\text{Ai}(x)$, $\text{Ai}'(x)$, $\text{Bi}(x)$, $\text{Bi}'(x)$, and integrals $\int_0^x \text{Ai}(t)\,dt$, $\int_0^x \text{Bi}(t)\,dt$, $\int_0^x \int_0^v \text{Ai}(t)\,dt\,dv$, $\int_0^x \int_0^v \text{Bi}(t)\,dt\,dv$ (see also (9.10.20) and (9.10.21)). The Chebyshev coefficients are given to 15D. Chebyshev coefficients are also given for expansions of the second and higher (real) zeros of $\text{Ai}(x)$, $\text{Ai}'(x)$, $\text{Bi}(x)$, $\text{Bi}'(x)$, again to 15D.

- Razaz and Schonfelder (1980) covers $\text{Ai}(x)$, $\text{Ai}'(x)$, $\text{Bi}(x)$, $\text{Bi}'(x)$. The Chebyshev coefficients are given to 30D.

9.19(iii) Approximations in the Complex Plane

- Corless et al. (1992) describe a method of approximation based on subdividing $\mathbb{C}$ into a triangular mesh, with values of $\mathrm{Ai}(z)$, $\mathrm{Ai}'(z)$ stored at the nodes. $\mathrm{Ai}(z)$ and $\mathrm{Ai}'(z)$ are then computed from Taylor-series expansions centered at one of the nearest nodes. The Taylor coefficients are generated by recursion, starting from the stored values of $\mathrm{Ai}(z)$, $\mathrm{Ai}'(z)$ at the node. Similarly for $\mathrm{Bi}(z)$, $\mathrm{Bi}'(z)$.

9.19(iv) Scorer Functions

- MacLeod (1994) supplies Chebyshev-series expansions to cover $\mathrm{Gi}(x)$ for $0 \le x < \infty$ and $\mathrm{Hi}(x)$ for $-\infty < x \le 0$. The Chebyshev coefficients are given to 20D.

9.20 Software

See http://dlmf.nist.gov/9.20.

References

General References

The main references used in writing this chapter are Miller (1946) and Olver (1997b). For additional bibliographic reading see Bleistein and Handelsman (1975), Jeffreys and Jeffreys (1956), Lebedev (1965), Temme (1996a), Wasow (1965, 1985), and Wong (1989).

Sources

The following list gives the references or other indications of proofs that were used in constructing the various sections of this chapter. These sources supplement the references that are quoted in the text.

§9.2 Miller (1946), Olver (1997b, Chapter 11).

§9.3 These graphics were produced by NIST.

§9.4 Miller (1946, p. B17), Olver (1997b, p. 54).

§9.5 Miller (1946, p. B17), Olver (1997b, p. 103). For (9.5.4) see Olver (1997b, p. 53). For (9.5.5) combine (9.2.10) and (9.5.4). For (9.5.6) see Reid (1995). For (9.5.7) see Copson (1963). (9.5.8) follows from the first of (9.6.2) and (10.32.9).

§9.6 Miller (1946, p. B17), Olver (1997b, pp. 392–393). For (9.6.21)–(9.6.24) combine (9.6.1)–(9.6.5) and (10.39.8)–(10.39.10). For (9.6.25), (9.6.26) combine (9.6.23), (9.6.24) with (13.14.2) and refer to §13.1.

§9.7 For (9.7.4) and Table 9.7.1 see Olver (1997b, p. 225). For (9.7.5)–(9.7.14) and §9.7(iii) see Olver (1997b, pp. 392–393 and 413–414). For §9.7(iv) see (9.6.1)–(9.6.5) and Olver (1997b, pp. 266–267). For (9.7.18)–(9.7.23) see Olver (1991b, 1993a).

§9.8 For (9.8.9)–(9.8.12) combine (9.6.17) and (9.6.18) with (10.4.3). For (9.8.13) use (9.2.7). For (9.8.14)–(9.8.19) combine (9.8.9)–(9.8.12) with §10.18(ii); see also Olver (1997b, p. 404), Miller (1946, p. B10). For §9.8(iii) see Olver (1997b, p. 404). For §9.8(iv) combine (9.8.9)–(9.8.12) with §10.18(iii); see also Miller (1946, p. B48).

§9.9 Olver (1997b, pp. 404, 414–415), Miller (1946, p. B48), Olver (1954, Appendix). For the computation of Tables 9.9.1–9.9.4 see §9.17(v).

§9.10 For (9.10.1) combine (9.12.4) and its differentiated form with the first of (9.10.11). For (9.10.2) combine (9.12.11) and its differentiated form with (9.10.1), then apply (9.2.7) and (9.10.11). (9.10.3) is proved in a similar manner. For (9.10.4)–(9.10.7) integrate the leading terms of the asymptotic expansions given in §9.7(ii) and use (9.10.11), (9.10.12). To verify (9.10.8)–(9.10.10)—and also (9.10.20), (9.10.21)—differentiate, and refer to (9.2.1). For (9.10.11) and (9.10.12) see Olver (1997b, p. 431). For (9.10.13) see Widder (1979). For (9.10.14)–(9.10.16) see Gibbs (1973, problem 72–21). For (9.10.17) see Olver (1997b, p. 338). For (9.10.18) and (9.10.19) see Schulten et al. (1979). To verify (9.10.20) and (9.10.21) differentiate, and refer to (9.2.1). For (9.10.22) see Olver (1997b, pp. 342–344).

§9.11 For (9.11.1) see §1.13(v). For (9.11.2) use (9.2.1) and (9.2.7). For (9.11.3) and (9.11.4) see Lebedev (1965, p. 142) and Muldoon (1977). For (9.11.5)–(9.11.14) see Albright (1977) and Albright and Gavathas (1986). For (9.11.15) see Reid (1995). For (9.11.16) and (9.11.17) see Reid (1997a). For (9.11.18) see Laurenzi (1993).

For (9.11.19) extend the definitions of §9.8(i) to positive values of x, obtain the indefinite integrals of $1/M^2(x)$ and $x/N^2(x)$ via the first two of (9.8.14), then combine the values of $\theta(0)$ and $\phi(0)$ given in §9.8(i) with $\theta(+\infty) = \phi(+\infty) = 0$ obtained from (9.8.4), (9.8.8), and §9.7(ii). (Communicated by M.E. Muldoon.)

§9.12 For (9.12.1)–(9.12.7) see Olver (1997b, pp. 430–431). For §9.12(iv) refer to the asymptotic expansions given in §§9.7(ii) and 9.12(viii). (9.12.11)–(9.12.14) can be verified with the aid of §§9.2(ii) and 9.12(iii). For (9.12.17) expand the integral in (9.12.20) in powers of zt and integrate term-by-term by means of (5.2.1). For (9.12.15) substitute into (9.12.11) by means of (9.12.17), (9.4.3), (9.2.5), (9.2.6), and use (5.5.3). For (9.12.16) and (9.12.18) use differentiation. (9.12.20) can be verified by showing that the right-hand side satisfies the differential equation (9.12.1) and the initial conditions given in §9.12(iii). For (9.12.19) combine (9.5.3) and (9.12.11). For (9.12.21), see Lee (1980). For (9.12.22), (9.12.23) see Gordon (1970, Appendix A). For (9.12.24) see Exton (1983). For (9.12.25), (9.12.27) see Olver (1997b, pp. 431–432). For (9.12.26), (9.12.28) refer to §2.1(ii).

Except for the constant term, (9.12.31) can be verified by termwise integration of (9.12.27). To evaluate the constant term replace z by $-x$ (≤ 0) in (9.12.20) and integrate (§1.5(v)) to obtain $\pi \int_0^x \mathrm{Hi}(-t)\,dt = \int_0^\infty (1 - e^{-xt}) e^{-\frac{1}{3}t^3} t^{-1}\,dt$. Next, integrate the right-hand side of this equation by parts—integrating the factor t^{-1} and differentiating the rest. As $x \to \infty$ the asymptotic expansions of $\int_0^\infty x e^{-xt} e^{-\frac{1}{3}t^3} (\ln t)\,dt$ and $\int_0^\infty e^{-xt} t^2 e^{-\frac{1}{3}t^3} (\ln t)\,dt$ follow from (2.3.9). Also, $\int_0^\infty t^2 e^{-\frac{1}{3}t^3} (\ln t)\,dt$ can be found by replacing $\frac{1}{3}t^3$ by t and referring to the first of (5.9.18). For (9.12.30) integrate (9.12.25) and obtain the constant term by combining (9.12.12) and (9.12.31). (Equations (9.12.30) and (9.12.31) first appeared in Rothman (1954b). As noted in this reference these results were derived by the author of the present DLMF chapter, but the proof was not included.) The graphs were produced by NIST.

Chapter 10
Bessel Functions

F. W. J. Olver[1] and L. C. Maximon[2]

Notation — **217**
- 10.1 Special Notation 217

Bessel and Hankel Functions — **217**
- 10.2 Definitions 217
- 10.3 Graphics 218
- 10.4 Connection Formulas 222
- 10.5 Wronskians and Cross-Products 222
- 10.6 Recurrence Relations and Derivatives . . 222
- 10.7 Limiting Forms 223
- 10.8 Power Series 223
- 10.9 Integral Representations 223
- 10.10 Continued Fractions 226
- 10.11 Analytic Continuation 226
- 10.12 Generating Function and Associated Series 226
- 10.13 Other Differential Equations 226
- 10.14 Inequalities; Monotonicity 227
- 10.15 Derivatives with Respect to Order 227
- 10.16 Relations to Other Functions 228
- 10.17 Asymptotic Expansions for Large Argument 228
- 10.18 Modulus and Phase Functions 230
- 10.19 Asymptotic Expansions for Large Order . 231
- 10.20 Uniform Asymptotic Expansions for Large Order 232
- 10.21 Zeros 235
- 10.22 Integrals 240
- 10.23 Sums 246
- 10.24 Functions of Imaginary Order 248

Modified Bessel Functions — **248**
- 10.25 Definitions 248
- 10.26 Graphics 249
- 10.27 Connection Formulas 251
- 10.28 Wronskians and Cross-Products 251
- 10.29 Recurrence Relations and Derivatives . . 251
- 10.30 Limiting Forms 252
- 10.31 Power Series 252
- 10.32 Integral Representations 252
- 10.33 Continued Fractions 253
- 10.34 Analytic Continuation 253
- 10.35 Generating Function and Associated Series 254
- 10.36 Other Differential Equations 254
- 10.37 Inequalities; Monotonicity 254
- 10.38 Derivatives with Respect to Order 254
- 10.39 Relations to Other Functions 254
- 10.40 Asymptotic Expansions for Large Argument 255
- 10.41 Asymptotic Expansions for Large Order . 256
- 10.42 Zeros 258
- 10.43 Integrals 258
- 10.44 Sums 260
- 10.45 Functions of Imaginary Order 261
- 10.46 Generalized and Incomplete Bessel Functions; Mittag-Leffler Function 261

Spherical Bessel Functions — **262**
- 10.47 Definitions and Basic Properties 262
- 10.48 Graphs 262
- 10.49 Explicit Formulas 264
- 10.50 Wronskians and Cross-Products 265
- 10.51 Recurrence Relations and Derivatives . . 265
- 10.52 Limiting Forms 265
- 10.53 Power Series 265
- 10.54 Integral Representations 266
- 10.55 Continued Fractions 266
- 10.56 Generating Functions 266
- 10.57 Uniform Asymptotic Expansions for Large Order 266
- 10.58 Zeros 266
- 10.59 Integrals 267
- 10.60 Sums 267

Kelvin Functions — **267**
- 10.61 Definitions and Basic Properties 267
- 10.62 Graphs 268
- 10.63 Recurrence Relations and Derivatives . . 269
- 10.64 Integral Representations 269
- 10.65 Power Series 269
- 10.66 Expansions in Series of Bessel Functions . 270
- 10.67 Asymptotic Expansions for Large Argument 271
- 10.68 Modulus and Phase Functions 272
- 10.69 Uniform Asymptotic Expansions for Large Order 273
- 10.70 Zeros 273
- 10.71 Integrals 274

Applications — **274**
- 10.72 Mathematical Applications 274
- 10.73 Physical Applications 275

[1]Institute for Physical Science and Technology and Department of Mathematics, University of Maryland, College Park, Maryland.
[2]Center for Nuclear Studies, Department of Physics, The George Washington University, Washington, D.C.

Computation **276**
   10.74 Methods of Computation 276
   10.75 Tables 278
   10.76 Approximations 281

   10.77 Software 281

References **281**

Acknowledgments: This chapter is based in part on Abramowitz and Stegun (1964, Chapters 9, 10, and 11) by F. W. J. Olver, H. A. Antosiewicz, and Y. L. Luke, respectively. The authors are pleased to acknowledge assistance of Martin E. Muldoon with §§10.21 and 10.42, Adri Olde Daalhuis with the verification of Eqs. (10.15.6)–(10.15.9), (10.38.6), (10.38.7), (10.60.7)–(10.60.9), and (10.61.9)–(10.61.12), Peter Paule and Frédéric Chyzak for the verification of Eqs. (10.15.6)–(10.15.9), (10.38.6), (10.38.7), (10.56.1)–(10.56.5), (10.60.4), (10.60.6), (10.60.10), and (10.60.11) by application of computer algebra, Nico Temme with the verification of Eqs. (10.15.6)–(10.15.9), (10.38.6), and (10.38.7), and Roderick Wong with the verification of §§10.22(v) and 10.43(v).

   Copyright © 2009 National Institute of Standards and Technology. All rights reserved.

Notation

10.1 Special Notation

(For other notation see pp. xiv and 873.)

m, n	integers. In §§10.47–10.71 n is nonnegative.
k	nonnegative integer (except in §10.73).
x, y	real variables.
z	complex variable.
ν	real or complex parameter (the order).
δ	arbitrary small positive constant.
ϑ	$z(d/dz)$.
$\psi(x)$	$\Gamma'(x)/\Gamma(x)$: logarithmic derivative of the gamma function (§5.2(i)).
primes	derivatives with respect to argument, except where indicated otherwise.

The main functions treated in this chapter are the Bessel functions $J_\nu(z)$, $Y_\nu(z)$; Hankel functions $H_\nu^{(1)}(z)$, $H_\nu^{(2)}(z)$; modified Bessel functions $I_\nu(z)$, $K_\nu(z)$; spherical Bessel functions $\mathsf{j}_n(z)$, $\mathsf{y}_n(z)$, $\mathsf{h}_n^{(1)}(z)$, $\mathsf{h}_n^{(2)}(z)$; modified spherical Bessel functions $\mathsf{i}_n^{(1)}(z)$, $\mathsf{i}_n^{(2)}(z)$, $\mathsf{k}_n(z)$; Kelvin functions $\mathrm{ber}_\nu(x)$, $\mathrm{bei}_\nu(x)$, $\mathrm{ker}_\nu(x)$, $\mathrm{kei}_\nu(x)$. For the spherical Bessel functions and modified spherical Bessel functions the order n is a nonnegative integer. For the other functions when the order ν is replaced by n, it can be any integer. For the Kelvin functions the order ν is always assumed to be real.

A common alternative notation for $Y_\nu(z)$ is $N_\nu(z)$. Other notations that have been used are as follows.

Abramowitz and Stegun (1964): $j_n(z)$, $y_n(z)$, $h_n^{(1)}(z)$, $h_n^{(2)}(z)$, for $\mathsf{j}_n(z)$, $\mathsf{y}_n(z)$, $\mathsf{h}_n^{(1)}(z)$, $\mathsf{h}_n^{(2)}(z)$, respectively, when $n \geq 0$.

Jeffreys and Jeffreys (1956): $\mathrm{Hs}_\nu(z)$ for $H_\nu^{(1)}(z)$, $\mathrm{Hi}_\nu(z)$ for $H_\nu^{(2)}(z)$, $\mathrm{Kh}_\nu(z)$ for $(2/\pi) K_\nu(z)$.

Whittaker and Watson (1927): $K_\nu(z)$ for $\cos(\nu\pi) K_\nu(z)$.

For older notations see British Association for the Advancement of Science (1937, pp. xix–xx) and Watson (1944, Chapters 1–3).

Bessel and Hankel Functions

10.2 Definitions

10.2(i) Bessel's Equation

$$10.2.1 \qquad z^2 \frac{d^2 w}{dz^2} + z \frac{dw}{dz} + (z^2 - \nu^2) w = 0.$$

This differential equation has a regular singularity at $z = 0$ with indices $\pm \nu$, and an irregular singularity at $z = \infty$ of rank 1; compare §§2.7(i) and 2.7(ii).

10.2(ii) Standard Solutions

Bessel Function of the First Kind

$$10.2.2 \qquad J_\nu(z) = (\tfrac{1}{2}z)^\nu \sum_{k=0}^\infty (-1)^k \frac{(\tfrac{1}{4}z^2)^k}{k!\,\Gamma(\nu+k+1)}.$$

This solution of (10.2.1) is an analytic function of $z \in \mathbb{C}$, except for a branch point at $z = 0$ when ν is not an integer. The *principal branch* of $J_\nu(z)$ corresponds to the principal value of $(\tfrac{1}{2}z)^\nu$ (§4.2(iv)) and is analytic in the z-plane cut along the interval $(-\infty, 0]$.

When $\nu = n$ ($\in \mathbb{Z}$), $J_\nu(z)$ is entire in z.

For fixed z ($\neq 0$) each branch of $J_\nu(z)$ is entire in ν.

Bessel Function of the Second Kind (Weber's Function)

$$10.2.3 \qquad Y_\nu(z) = \frac{J_\nu(z) \cos(\nu\pi) - J_{-\nu}(z)}{\sin(\nu\pi)}.$$

When ν is an integer the right-hand side is replaced by its limiting value:

$$10.2.4 \qquad Y_n(z) = \frac{1}{\pi} \left. \frac{\partial J_\nu(z)}{\partial \nu} \right|_{\nu=n} + \frac{(-1)^n}{\pi} \left. \frac{\partial J_\nu(z)}{\partial \nu} \right|_{\nu=-n},$$
$$n = 0, \pm 1, \pm 2, \ldots.$$

Whether or not ν is an integer $Y_\nu(z)$ has a branch point at $z = 0$. The *principal branch* corresponds to the principal branches of $J_{\pm\nu}(z)$ in (10.2.3) and (10.2.4), with a cut in the z-plane along the interval $(-\infty, 0]$.

Except in the case of $J_{\pm n}(z)$, the principal branches of $J_\nu(z)$ and $Y_\nu(z)$ are two-valued and discontinuous on the cut $\mathrm{ph}\, z = \pm\pi$; compare §4.2(i).

Both $J_\nu(z)$ and $Y_\nu(z)$ are real when ν is real and $\mathrm{ph}\, z = 0$.

For fixed z ($\neq 0$) each branch of $Y_\nu(z)$ is entire in ν.

Bessel Functions of the Third Kind (Hankel Functions)

These solutions of (10.2.1) are denoted by $H_\nu^{(1)}(z)$ and $H_\nu^{(2)}(z)$, and their defining properties are given by

$$10.2.5 \qquad H_\nu^{(1)}(z) \sim \sqrt{2/(\pi z)}\, e^{i(z - \frac{1}{2}\nu\pi - \frac{1}{4}\pi)}$$

as $z \to \infty$ in $-\pi + \delta \leq \mathrm{ph}\, z \leq 2\pi - \delta$, and

$$10.2.6 \qquad H_\nu^{(2)}(z) \sim \sqrt{2/(\pi z)}\, e^{-i(z - \frac{1}{2}\nu\pi - \frac{1}{4}\pi)}$$

as $z \to \infty$ in $-2\pi + \delta \leq \mathrm{ph}\, z \leq \pi - \delta$, where δ is an arbitrary small positive constant. Each solution has a branch point at $z = 0$ for all $\nu \in \mathbb{C}$. The *principal branches* correspond to principal values of the square roots in (10.2.5) and (10.2.6), again with a cut in the z-plane along the interval $(-\infty, 0]$.

The principal branches of $H_\nu^{(1)}(z)$ and $H_\nu^{(2)}(z)$ are two-valued and discontinuous on the cut $\mathrm{ph}\, z = \pm\pi$.

For fixed z ($\neq 0$) each branch of $H_\nu^{(1)}(z)$ and $H_\nu^{(2)}(z)$ is entire in ν.

Branch Conventions

Except where indicated otherwise, it is assumed throughout this Handbook that the symbols $J_\nu(z)$, $Y_\nu(z)$, $H_\nu^{(1)}(z)$, and $H_\nu^{(2)}(z)$ denote the principal values of these functions.

Cylinder Functions

The notation $\mathscr{C}_\nu(z)$ denotes $J_\nu(z)$, $Y_\nu(z)$, $H_\nu^{(1)}(z)$, $H_\nu^{(2)}(z)$, or any nontrivial linear combination of these functions, the coefficients in which are independent of z and ν.

10.2(iii) Numerically Satisfactory Pairs of Solutions

Table 10.2.1 lists numerically satisfactory pairs of solutions (§2.7(iv)) of (10.2.1) for the stated intervals or regions in the case $\Re\nu \geq 0$. When $\Re\nu < 0$, ν is replaced by $-\nu$ throughout.

Table 10.2.1: Numerically satisfactory pairs of solutions of Bessel's equation.

Pair	Interval or Region
$J_\nu(x), Y_\nu(x)$	$0 < x < \infty$
$J_\nu(z), Y_\nu(z)$	neighborhood of 0 in $|\operatorname{ph} z| \leq \pi$
$J_\nu(z), H_\nu^{(1)}(z)$	$0 \leq \operatorname{ph} z \leq \pi$
$J_\nu(z), H_\nu^{(2)}(z)$	$-\pi \leq \operatorname{ph} z \leq 0$
$H_\nu^{(1)}(z), H_\nu^{(2)}(z)$	neighborhood of ∞ in $|\operatorname{ph} z| \leq \pi$

10.3 Graphics

10.3(i) Real Order and Variable

See Figures 10.3.1–10.3.8. For the modulus and phase functions $M_\nu(x)$, $\theta_\nu(x)$, $N_\nu(x)$, and $\phi_\nu(x)$ see §10.18.

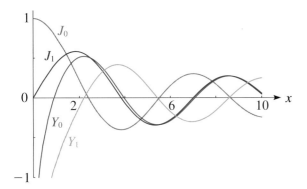

Figure 10.3.1: $J_0(x), Y_0(x), J_1(x), Y_1(x), 0 \leq x \leq 10$.

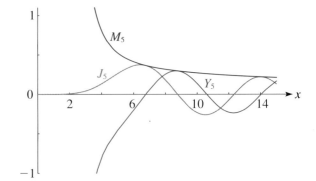

Figure 10.3.2: $J_5(x), Y_5(x), M_5(x), 0 \leq x \leq 15$.

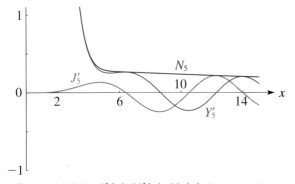

Figure 10.3.3: $J_5'(x), Y_5'(x), N_5(x), 0 \leq x \leq 15$.

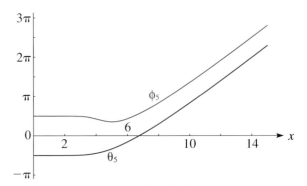

Figure 10.3.4: $\theta_5(x), \phi_5(x), 0 \leq x \leq 15$.

10.3 Graphics

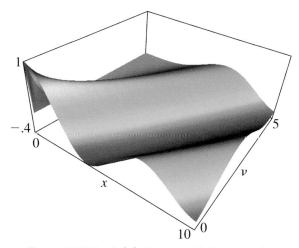

Figure 10.3.5: $J_\nu(x), 0 \le x \le 10, 0 \le \nu \le 5$.

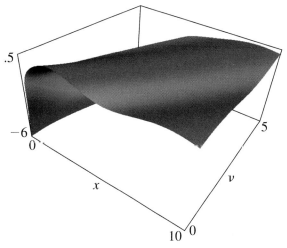

Figure 10.3.6: $Y_\nu(x), 0 < x \le 10, 0 \le \nu \le 5$.

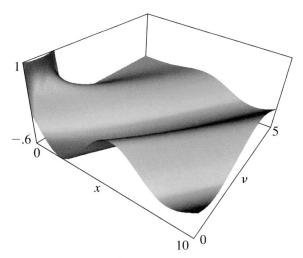

Figure 10.3.7: $J'_\nu(x), 0 \le x \le 10, 0 \le \nu \le 5$.

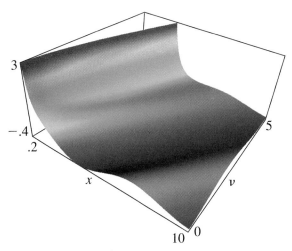

Figure 10.3.8: $Y'_\nu(x), 0.2 \le x \le 10, 0 \le \nu \le 5$.

10.3(ii) Real Order, Complex Variable

See Figures 10.3.9–10.3.16. In these graphics, height corresponds to the absolute value of the function and color to the phase. See also p. xiv.

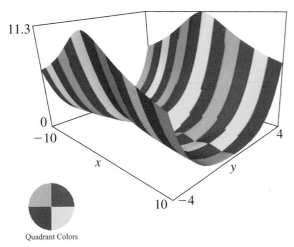

Figure 10.3.9: $J_0(x+iy)$, $-10 \leq x \leq 10, -4 \leq y \leq 4$.

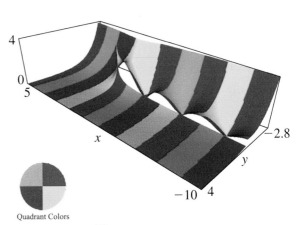

Figure 10.3.10: $H_0^{(1)}(x+iy)$, $-10 \leq x \leq 5, -2.8 \leq y \leq 4$. Principal value. There is a cut along the negative real axis.

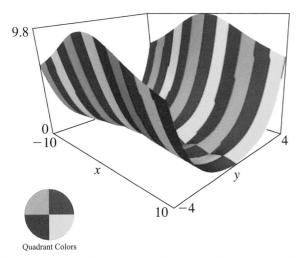

Figure 10.3.11: $J_1(x+iy)$, $-10 \leq x \leq 10, -4 \leq y \leq 4$.

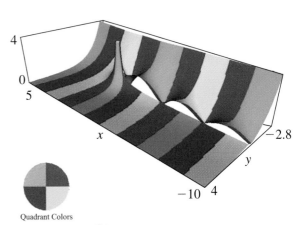

Figure 10.3.12: $H_1^{(1)}(x+iy)$, $-10 \leq x \leq 5, -2.8 \leq y \leq 4$. Principal value. There is a cut along the negative real axis.

10.3 GRAPHICS

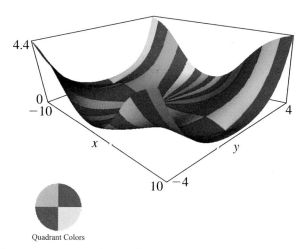

Figure 10.3.13: $J_5(x+iy)$, $-10 \leq x \leq 10$, $-4 \leq y \leq 4$.

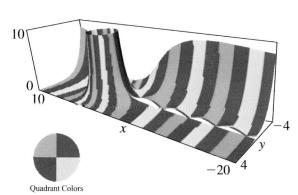

Figure 10.3.14: $H_5^{(1)}(x+iy)$, $-20 \leq x \leq 10$, $-4 \leq y \leq 4$. Principal value. There is a cut along the negative real axis.

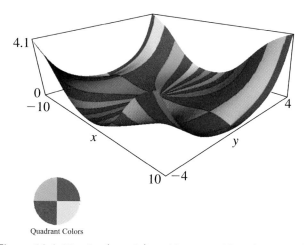

Figure 10.3.15: $J_{5.5}(x+iy)$, $-10 \leq x \leq 10$, $-4 \leq y \leq 4$. Principal value. There is a cut along the negative real axis.

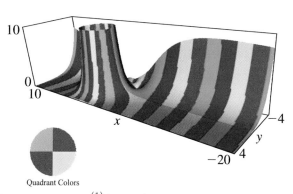

Figure 10.3.16: $H_{5.5}^{(1)}(x+iy)$, $-20 \leq x \leq 10$, $-4 \leq y \leq 4$. Principal value. There is a cut along the negative real axis.

10.3(iii) Imaginary Order, Real Variable

See Figures 10.3.17–10.3.19. For the notation see §10.24.

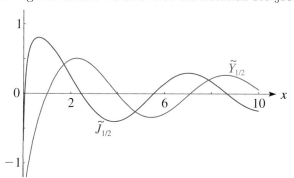

Figure 10.3.17: $\widetilde{J}_{1/2}(x), \widetilde{Y}_{1/2}(x), 0.01 \leq x \leq 10$.

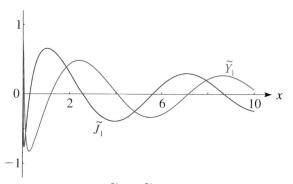

Figure 10.3.18: $\widetilde{J}_1(x), \widetilde{Y}_1(x), 0.01 \leq x \leq 10$.

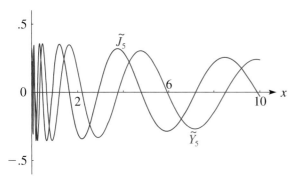

Figure 10.3.19: $\widetilde{J}_5(x), \widetilde{Y}_5(x), 0.01 \le x \le 10$.

10.4 Connection Formulas

Other solutions of (10.2.1) include $J_{-\nu}(z)$, $Y_{-\nu}(z)$, $H^{(1)}_{-\nu}(z)$, and $H^{(2)}_{-\nu}(z)$.

10.4.1
$$J_{-n}(z) = (-1)^n J_n(z),$$
$$Y_{-n}(z) = (-1)^n Y_n(z),$$

10.4.2
$$H^{(1)}_{-n}(z) = (-1)^n H^{(1)}_n(z),$$
$$H^{(2)}_{-n}(z) = (-1)^n H^{(2)}_n(z).$$

10.4.3
$$H^{(1)}_\nu(z) = J_\nu(z) + i Y_\nu(z),$$
$$H^{(2)}_\nu(z) = J_\nu(z) - i Y_\nu(z),$$

10.4.4
$$J_\nu(z) = \frac{1}{2}\left(H^{(1)}_\nu(z) + H^{(2)}_\nu(z)\right),$$
$$Y_\nu(z) = \frac{1}{2i}\left(H^{(1)}_\nu(z) - H^{(2)}_\nu(z)\right).$$

10.4.5 $\quad J_\nu(z) = \csc(\nu\pi)\left(Y_{-\nu}(z) - Y_\nu(z)\cos(\nu\pi)\right).$

10.4.6
$$H^{(1)}_{-\nu}(z) = e^{\nu\pi i} H^{(1)}_\nu(z),$$
$$H^{(2)}_{-\nu}(z) = e^{-\nu\pi i} H^{(2)}_\nu(z).$$

10.4.7
$$H^{(1)}_\nu(z) = i\csc(\nu\pi)\left(e^{-\nu\pi i} J_\nu(z) - J_{-\nu}(z)\right)$$
$$= \csc(\nu\pi)\left(Y_{-\nu}(z) - e^{-\nu\pi i} Y_\nu(z)\right),$$

10.4.8
$$H^{(2)}_\nu(z) = i\csc(\nu\pi)\left(J_{-\nu}(z) - e^{\nu\pi i} J_\nu(z)\right)$$
$$= \csc(\nu\pi)\left(Y_{-\nu}(z) - e^{\nu\pi i} Y_\nu(z)\right).$$

In (10.4.5), (10.4.7), and (10.4.8) limiting values are taken when $\nu = n$; compare (10.2.3) and (10.2.4).

See also §10.11.

10.5 Wronskians and Cross-Products

10.5.1
$$\mathscr{W}\{J_\nu(z), J_{-\nu}(z)\} = J_{\nu+1}(z) J_{-\nu}(z) + J_\nu(z) J_{-\nu-1}(z)$$
$$= -2\sin(\nu\pi)/(\pi z),$$

10.5.2
$$\mathscr{W}\{J_\nu(z), Y_\nu(z)\} = J_{\nu+1}(z) Y_\nu(z) - J_\nu(z) Y_{\nu+1}(z)$$
$$= 2/(\pi z),$$

10.5.3
$$\mathscr{W}\{J_\nu(z), H^{(1)}_\nu(z)\} = J_{\nu+1}(z) H^{(1)}_\nu(z) - J_\nu(z) H^{(1)}_{\nu+1}(z)$$
$$= 2i/(\pi z),$$

10.5.4
$$\mathscr{W}\{J_\nu(z), H^{(2)}_\nu(z)\} = J_{\nu+1}(z) H^{(2)}_\nu(z) - J_\nu(z) H^{(2)}_{\nu+1}(z)$$
$$= -2i/(\pi z),$$

10.5.5
$$\mathscr{W}\left\{H^{(1)}_\nu(z), H^{(2)}_\nu(z)\right\} = H^{(1)}_{\nu+1}(z) H^{(2)}_\nu(z) - H^{(1)}_\nu(z) H^{(2)}_{\nu+1}(z)$$
$$= -4i/(\pi z).$$

10.6 Recurrence Relations and Derivatives

10.6(i) Recurrence Relations

With $\mathscr{C}_\nu(z)$ defined as in §10.2(ii),

10.6.1
$$\mathscr{C}_{\nu-1}(z) + \mathscr{C}_{\nu+1}(z) = (2\nu/z)\mathscr{C}_\nu(z),$$
$$\mathscr{C}_{\nu-1}(z) - \mathscr{C}_{\nu+1}(z) = 2\mathscr{C}'_\nu(z).$$

10.6.2
$$\mathscr{C}'_\nu(z) = \mathscr{C}_{\nu-1}(z) - (\nu/z)\mathscr{C}_\nu(z),$$
$$\mathscr{C}'_\nu(z) = -\mathscr{C}_{\nu+1}(z) + (\nu/z)\mathscr{C}_\nu(z).$$

10.6.3
$$J'_0(z) = -J_1(z), \qquad Y'_0(z) = -Y_1(z),$$
$$H^{(1)'}_0(z) = -H^{(1)}_1(z), \quad H^{(2)'}_0(z) = -H^{(2)}_1(z).$$

If $f_\nu(z) = z^p \mathscr{C}_\nu(\lambda z^q)$, where p, q, and $\lambda\ (\ne 0)$ are real or complex constants, then

10.6.4
$$f_{\nu-1}(z) + f_{\nu+1}(z) = (2\nu/\lambda)z^{-q}f_\nu(z),$$
$$(p+\nu q)f_{\nu-1}(z) + (p-\nu q)f_{\nu+1}(z)$$
$$= (2\nu/\lambda)z^{1-q}f'_\nu(z).$$

10.6.5
$$zf'_\nu(z) = \lambda q z^q f_{\nu-1}(z) + (p-\nu q)f_\nu(z),$$
$$zf'_\nu(z) = -\lambda q z^q f_{\nu+1}(z) + (p+\nu q)f_\nu(z).$$

10.6(ii) Derivatives

For $k = 0, 1, 2, \ldots$,

10.6.6
$$\left(\frac{1}{z}\frac{d}{dz}\right)^k (z^\nu \mathscr{C}_\nu(z)) = z^{\nu-k}\mathscr{C}_{\nu-k}(z),$$
$$\left(\frac{1}{z}\frac{d}{dz}\right)^k (z^{-\nu}\mathscr{C}_\nu(z)) = (-1)^k z^{-\nu-k}\mathscr{C}_{\nu+k}(z).$$

10.6.7 $\quad \mathscr{C}^{(k)}_\nu(z) = \dfrac{1}{2^k}\displaystyle\sum_{n=0}^{k}(-1)^n\binom{k}{n}\mathscr{C}_{\nu-k+2n}(z).$

10.6(iii) Cross-Products

Let

10.6.8
$$p_\nu = J_\nu(a) Y_\nu(b) - J_\nu(b) Y_\nu(a),$$
$$q_\nu = J_\nu(a) Y'_\nu(b) - J'_\nu(b) Y_\nu(a),$$
$$r_\nu = J'_\nu(a) Y_\nu(b) - J_\nu(b) Y'_\nu(a),$$
$$s_\nu = J'_\nu(a) Y'_\nu(b) - J'_\nu(b) Y'_\nu(a),$$

where a and b are independent of ν. Then

10.6.9
$$p_{\nu+1} - p_{\nu-1} = -\frac{2\nu}{a} q_\nu - \frac{2\nu}{b} r_\nu,$$
$$q_{\nu+1} + r_\nu = \frac{\nu}{a} p_\nu - \frac{\nu+1}{b} p_{\nu+1},$$
$$r_{\nu+1} + q_\nu = \frac{\nu}{b} p_\nu - \frac{\nu+1}{a} p_{\nu+1},$$
$$s_\nu = \tfrac{1}{2} p_{\nu+1} + \tfrac{1}{2} p_{\nu-1} - \frac{\nu^2}{ab} p_\nu,$$

and

10.6.10
$$p_\nu s_\nu - q_\nu r_\nu = 4/(\pi^2 ab).$$

10.7 Limiting Forms

10.7(i) $z \to 0$

When ν is fixed and $z \to 0$,

10.7.1 $\quad J_0(z) \to 1, \quad Y_0(z) \sim (2/\pi) \ln z,$

10.7.2 $\quad H_0^{(1)}(z) \sim -H_0^{(2)}(z) \sim (2i/\pi) \ln z,$

10.7.3 $\quad J_\nu(z) \sim (\tfrac{1}{2}z)^\nu / \Gamma(\nu+1), \quad \nu \neq -1, -2, -3, \ldots,$

10.7.4
$$Y_\nu(z) \sim -(1/\pi) \Gamma(\nu) (\tfrac{1}{2}z)^{-\nu},$$
$$\Re \nu > 0 \text{ or } \nu = -\tfrac{1}{2}, -\tfrac{3}{2}, -\tfrac{5}{2}, \ldots,$$

10.7.5
$$Y_{-\nu}(z) \sim -(1/\pi) \cos(\nu\pi) \Gamma(\nu) (\tfrac{1}{2}z)^{-\nu},$$
$$\Re \nu > 0, \; \nu \neq \tfrac{1}{2}, \tfrac{3}{2}, \tfrac{5}{2}, \ldots,$$

10.7.6
$$Y_{i\nu}(z) = \frac{i \operatorname{csch}(\nu\pi)}{\Gamma(1-i\nu)} (\tfrac{1}{2}z)^{-i\nu} - \frac{i \coth(\nu\pi)}{\Gamma(1+i\nu)} (\tfrac{1}{2}z)^{i\nu}$$
$$+ e^{|\nu \operatorname{ph} z|} o(1), \qquad \nu \in \mathbb{R} \text{ and } \nu \neq 0.$$

See also §10.24 when $z = x \; (> 0)$.

10.7.7
$$H_\nu^{(1)}(z) \sim -H_\nu^{(2)}(z) \sim -(i/\pi) \Gamma(\nu) (\tfrac{1}{2}z)^{-\nu}, \quad \Re\nu > 0.$$

For $H_{-\nu}^{(1)}(z)$ and $H_{-\nu}^{(2)}(z)$ when $\Re\nu > 0$ combine (10.4.6) and (10.7.7). For $H_{i\nu}^{(1)}(z)$ and $H_{i\nu}^{(2)}(z)$ when $\nu \in \mathbb{R}$ and $\nu \neq 0$ combine (10.4.3), (10.7.3), and (10.7.6).

10.7(ii) $z \to \infty$

When ν is fixed and $z \to \infty$,

10.7.8
$$J_\nu(z) = \sqrt{2/(\pi z)} \left(\cos\left(z - \tfrac{1}{2}\nu\pi - \tfrac{1}{4}\pi\right) + e^{|\Im z|} o(1) \right),$$
$$Y_\nu(z) = \sqrt{2/(\pi z)} \left(\sin\left(z - \tfrac{1}{2}\nu\pi - \tfrac{1}{4}\pi\right) + e^{|\Im z|} o(1) \right),$$
$$|\operatorname{ph} z| \leq \pi - \delta(< \pi).$$

For the corresponding results for $H_\nu^{(1)}(z)$ and $H_\nu^{(2)}(z)$ see (10.2.5) and (10.2.6).

10.8 Power Series

For $J_\nu(z)$ see (10.2.2) and (10.4.1). When ν is not an integer the corresponding expansions for $Y_\nu(z)$, $H_\nu^{(1)}(z)$, and $H_\nu^{(2)}(z)$ are obtained by combining (10.2.2) with (10.2.3), (10.4.7), and (10.4.8).

When $n = 0, 1, 2, \ldots,$

10.8.1
$$Y_n(z) = -\frac{(\tfrac{1}{2}z)^{-n}}{\pi} \sum_{k=0}^{n-1} \frac{(n-k-1)!}{k!} \left(\tfrac{1}{4}z^2\right)^k$$
$$+ \frac{2}{\pi} \ln\left(\tfrac{1}{2}z\right) J_n(z) - \frac{(\tfrac{1}{2}z)^n}{\pi} \sum_{k=0}^{\infty} (\psi(k+1)$$
$$+ \psi(n+k+1)) \frac{(-\tfrac{1}{4}z^2)^k}{k!(n+k)!},$$

where $\psi(x) = \Gamma'(x)/\Gamma(x)$ (§5.2(i)). In particular,

10.8.2
$$Y_0(z) = \frac{2}{\pi} \left(\ln\left(\tfrac{1}{2}z\right) + \gamma \right) J_0(z) + \frac{2}{\pi} \left(\frac{\tfrac{1}{4}z^2}{(1!)^2} - \left(1 + \tfrac{1}{2}\right) \frac{(\tfrac{1}{4}z^2)^2}{(2!)^2} \right.$$
$$\left. + \left(1 + \tfrac{1}{2} + \tfrac{1}{3}\right) \frac{(\tfrac{1}{4}z^2)^3}{(3!)^2} - \cdots \right),$$

where γ is Euler's constant (§5.2(ii)).

For negative values of n use (10.4.1).

The corresponding results for $H_n^{(1)}(z)$ and $H_n^{(2)}(z)$ are obtained via (10.4.3) with $\nu = n$.

10.8.3
$$J_\nu(z) J_\mu(z) = (\tfrac{1}{2}z)^{\nu+\mu} \sum_{k=0}^{\infty} \frac{(\nu+\mu+k+1)_k (-\tfrac{1}{4}z^2)^k}{k! \, \Gamma(\nu+k+1) \, \Gamma(\mu+k+1)}.$$

10.9 Integral Representations

10.9(i) Integrals along the Real Line

Bessel's Integral

10.9.1
$$J_0(z) = \frac{1}{\pi} \int_0^\pi \cos(z \sin\theta) \, d\theta = \frac{1}{\pi} \int_0^\pi \cos(z \cos\theta) \, d\theta,$$

10.9.2
$$J_n(z) = \frac{1}{\pi} \int_0^\pi \cos(z \sin\theta - n\theta) \, d\theta$$
$$= \frac{i^{-n}}{\pi} \int_0^\pi e^{iz \cos\theta} \cos(n\theta) \, d\theta, \qquad n \in \mathbb{Z}.$$

Neumann's Integral

10.9.3
$$Y_0(z) = \frac{4}{\pi^2} \int_0^{\frac{1}{2}\pi} \cos(z\cos\theta)\left(\gamma + \ln(2z\sin^2\theta)\right)d\theta,$$
where γ is Euler's constant (§5.2(ii)).

Poisson's and Related Integrals

10.9.4
$$J_\nu(z) = \frac{(\frac{1}{2}z)^\nu}{\pi^{\frac{1}{2}}\Gamma(\nu+\frac{1}{2})} \int_0^\pi \cos(z\cos\theta)(\sin\theta)^{2\nu}d\theta$$
$$= \frac{2(\frac{1}{2}z)^\nu}{\pi^{\frac{1}{2}}\Gamma(\nu+\frac{1}{2})} \int_0^1 (1-t^2)^{\nu-\frac{1}{2}}\cos(zt)dt,$$
$$\Re\nu > -\tfrac{1}{2}.$$

10.9.5
$$Y_\nu(z) = \frac{2(\frac{1}{2}z)^\nu}{\pi^{\frac{1}{2}}\Gamma(\nu+\frac{1}{2})}\left(\int_0^1 (1-t^2)^{\nu-\frac{1}{2}}\sin(zt)dt - \int_0^\infty e^{-zt}(1+t^2)^{\nu-\frac{1}{2}}dt\right),$$
$$\Re\nu > -\tfrac{1}{2}, |\operatorname{ph}z| < \tfrac{1}{2}\pi.$$

Schläfli's and Related Integrals

10.9.6
$$J_\nu(z) = \frac{1}{\pi}\int_0^\pi \cos(z\sin\theta - \nu\theta)d\theta$$
$$-\frac{\sin(\nu\pi)}{\pi}\int_0^\infty e^{-z\sinh t - \nu t}dt, \quad |\operatorname{ph}z| < \tfrac{1}{2}\pi,$$

10.9.7
$$Y_\nu(z) = \frac{1}{\pi}\int_0^\pi \sin(z\sin\theta - \nu\theta)d\theta$$
$$-\frac{1}{\pi}\int_0^\infty \left(e^{\nu t} + e^{-\nu t}\cos(\nu\pi)\right)e^{-z\sinh t}dt,$$
$$|\operatorname{ph}z| < \tfrac{1}{2}\pi.$$

Mehler–Sonine and Related Integrals

10.9.8
$$J_\nu(x) = \frac{2}{\pi}\int_0^\infty \sin(x\cosh t - \tfrac{1}{2}\nu\pi)\cosh(\nu t)dt,$$
$$Y_\nu(x) = -\frac{2}{\pi}\int_0^\infty \cos(x\cosh t - \tfrac{1}{2}\nu\pi)\cosh(\nu t)dt,$$
$$|\Re\nu| < 1, x > 0.$$

In particular,

10.9.9
$$J_0(x) = \frac{2}{\pi}\int_0^\infty \sin(x\cosh t)dt, \quad x > 0,$$
$$Y_0(x) = -\frac{2}{\pi}\int_0^\infty \cos(x\cosh t)dt, \quad x > 0.$$

10.9.10
$$H_\nu^{(1)}(z) = \frac{e^{-\frac{1}{2}\nu\pi i}}{\pi i}\int_{-\infty}^\infty e^{iz\cosh t - \nu t}dt, \quad 0 < \operatorname{ph}z < \pi,$$

10.9.11
$$H_\nu^{(2)}(z) = -\frac{e^{\frac{1}{2}\nu\pi i}}{\pi i}\int_{-\infty}^\infty e^{-iz\cosh t - \nu t}dt, \quad -\pi < \operatorname{ph}z < 0.$$

10.9.12
$$J_\nu(x) = \frac{2(\frac{1}{2}x)^{-\nu}}{\pi^{\frac{1}{2}}\Gamma(\frac{1}{2}-\nu)}\int_1^\infty \frac{\sin(xt)dt}{(t^2-1)^{\nu+\frac{1}{2}}},$$
$$Y_\nu(x) = -\frac{2(\frac{1}{2}x)^{-\nu}}{\pi^{\frac{1}{2}}\Gamma(\frac{1}{2}-\nu)}\int_1^\infty \frac{\cos(xt)dt}{(t^2-1)^{\nu+\frac{1}{2}}},$$
$$|\Re\nu| < \tfrac{1}{2}, x > 0.$$

10.9.13
$$\left(\frac{z+\zeta}{z-\zeta}\right)^{\frac{1}{2}\nu}J_\nu\left((z^2-\zeta^2)^{\frac{1}{2}}\right)$$
$$= \frac{1}{\pi}\int_0^\pi e^{\zeta\cos\theta}\cos(z\sin\theta - \nu\theta)d\theta$$
$$-\frac{\sin(\nu\pi)}{\pi}\int_0^\infty e^{-\zeta\cosh t - z\sinh t - \nu t}dt,$$
$$\Re(z+\zeta) > 0,$$

10.9.14
$$\left(\frac{z+\zeta}{z-\zeta}\right)^{\frac{1}{2}\nu}Y_\nu\left((z^2-\zeta^2)^{\frac{1}{2}}\right)$$
$$= \frac{1}{\pi}\int_0^\pi e^{\zeta\cos\theta}\sin(z\sin\theta - \nu\theta)d\theta$$
$$-\frac{1}{\pi}\int_0^\infty \left(e^{\nu t + \zeta\cosh t} + e^{-\nu t - \zeta\cosh t}\cos(\nu\pi)\right)$$
$$\times e^{-z\sinh t}dt, \quad \Re(z\pm\zeta) > 0.$$

10.9.15
$$\left(\frac{z+\zeta}{z-\zeta}\right)^{\frac{1}{2}\nu}H_\nu^{(1)}\left((z^2-\zeta^2)^{\frac{1}{2}}\right)$$
$$= \frac{1}{\pi i}e^{-\frac{1}{2}\nu\pi i}\int_{-\infty}^\infty e^{iz\cosh t + i\zeta\sinh t - \nu t}dt,$$
$$\Im(z\pm\zeta) > 0,$$

10.9.16
$$\left(\frac{z+\zeta}{z-\zeta}\right)^{\frac{1}{2}\nu}H_\nu^{(2)}\left((z^2-\zeta^2)^{\frac{1}{2}}\right)$$
$$= -\frac{1}{\pi i}e^{\frac{1}{2}\nu\pi i}\int_{-\infty}^\infty e^{-iz\cosh t - i\zeta\sinh t - \nu t}dt,$$
$$\Im(z\pm\zeta) < 0.$$

10.9(ii) Contour Integrals

Schläfli–Sommerfeld Integrals

When $|\operatorname{ph}z| < \tfrac{1}{2}\pi$,

10.9.17
$$J_\nu(z) = \frac{1}{2\pi i}\int_{\infty-\pi i}^{\infty+\pi i} e^{z\sinh t - \nu t}dt,$$

and

10.9.18
$$H_\nu^{(1)}(z) = \frac{1}{\pi i}\int_{-\infty}^{\infty+\pi i} e^{z\sinh t - \nu t}dt,$$
$$H_\nu^{(2)}(z) = -\frac{1}{\pi i}\int_{-\infty}^{\infty-\pi i} e^{z\sinh t - \nu t}dt.$$

10.9 Integral Representations

Schläfli's Integral

10.9.19 $\quad J_\nu(z) = \dfrac{(\frac{1}{2}z)^\nu}{2\pi i} \displaystyle\int_{-\infty}^{(0+)} \exp\left(t - \dfrac{z^2}{4t}\right) \dfrac{dt}{t^{\nu+1}},$

where the integration path is a simple loop contour, and $t^{\nu+1}$ is continuous on the path and takes its principal value at the intersection with the positive real axis.

Hankel's Integrals

In (10.9.20) and (10.9.21) the integration paths are simple loop contours not enclosing $t = -1$. Also, $(t^2-1)^{\nu-\frac{1}{2}}$ is continuous on the path, and takes its principal value at the intersection with the interval $(1, \infty)$.

10.9.20
$$J_\nu(z) = \frac{\Gamma(\frac{1}{2}-\nu)(\frac{1}{2}z)^\nu}{\pi^{\frac{3}{2}} i} \int_0^{(1+)} \cos(zt)(t^2-1)^{\nu-\frac{1}{2}} dt,$$
$$\nu \neq \tfrac{1}{2}, \tfrac{3}{2}, \ldots.$$

10.9.21
$$H_\nu^{(1)}(z) = \frac{\Gamma(\frac{1}{2}-\nu)(\frac{1}{2}z)^\nu}{\pi^{\frac{3}{2}} i} \int_{1+i\infty}^{(1+)} e^{izt}(t^2-1)^{\nu-\frac{1}{2}} dt,$$
$$H_\nu^{(2)}(z) = \frac{\Gamma(\frac{1}{2}-\nu)(\frac{1}{2}z)^\nu}{\pi^{\frac{3}{2}} i} \int_{1-i\infty}^{(1+)} e^{-izt}(t^2-1)^{\nu-\frac{1}{2}} dt,$$
$$\nu \neq \tfrac{1}{2}, \tfrac{3}{2}, \ldots, |\operatorname{ph} z| < \tfrac{1}{2}\pi.$$

Mellin–Barnes Type Integrals

10.9.22
$$J_\nu(x) = \frac{1}{2\pi i} \int_{-i\infty}^{i\infty} \frac{\Gamma(-t)(\frac{1}{2}x)^{\nu+2t}}{\Gamma(\nu+t+1)} dt, \quad \Re\nu > 0, x > 0,$$

where the integration path passes to the left of $t = 0, 1, 2, \ldots$.

10.9.23
$$J_\nu(z) = \frac{1}{2\pi i} \int_{-\infty-ic}^{-\infty+ic} \frac{\Gamma(t)}{\Gamma(\nu-t+1)} (\tfrac{1}{2}z)^{\nu-2t} dt,$$

where c is a positive constant and the integration path encloses the points $t = 0, -1, -2, \ldots$.

In (10.9.24) and (10.9.25) c is any constant exceeding $\max(\Re\nu, 0)$.

10.9.24
$$H_\nu^{(1)}(z) = -\frac{e^{-\frac{1}{2}\nu\pi i}}{2\pi^2} \int_{c-i\infty}^{c+i\infty} \Gamma(t)\Gamma(t-\nu)(-\tfrac{1}{2}iz)^{\nu-2t} dt,$$
$$0 < \operatorname{ph} z < \pi,$$

10.9.25
$$H_\nu^{(2)}(z) = \frac{e^{\frac{1}{2}\nu\pi i}}{2\pi^2} \int_{c-i\infty}^{c+i\infty} \Gamma(t)\Gamma(t-\nu)(\tfrac{1}{2}iz)^{\nu-2t} dt,$$
$$-\pi < \operatorname{ph} z < 0.$$

For (10.9.22)–(10.9.25) and further integrals of this type see Paris and Kaminski (2001, pp. 114–116).

10.9(iii) Products

10.9.26 $\quad J_\mu(z) J_\nu(z) = \dfrac{2}{\pi} \displaystyle\int_0^{\pi/2} J_{\mu+\nu}(2z\cos\theta) \cos(\mu-\nu)\theta \, d\theta, \qquad \Re(\mu+\nu) > -1.$

10.9.27 $\quad J_\nu(z) J_\nu(\zeta) = \dfrac{2}{\pi} \displaystyle\int_0^{\pi/2} J_{2\nu}\left(2(z\zeta)^{\frac{1}{2}} \sin\theta\right) \cos\left((z-\zeta)\cos\theta\right) d\theta, \qquad \Re\nu > -\tfrac{1}{2},$

where the square root has its principal value.

10.9.28 $\quad J_\nu(z) J_\nu(\zeta) = \dfrac{1}{2\pi i} \displaystyle\int_{c-i\infty}^{c+i\infty} \exp\left(\dfrac{1}{2}t - \dfrac{z^2+\zeta^2}{2t}\right) I_\nu\left(\dfrac{z\zeta}{t}\right) \dfrac{dt}{t}, \qquad \Re\nu > -1,$

where c is a positive constant. For the function I_ν see §10.25(ii).

Mellin–Barnes Type

10.9.29 $\quad J_\mu(x) J_\nu(x) = \dfrac{1}{2\pi i} \displaystyle\int_{-i\infty}^{i\infty} \dfrac{\Gamma(-t)\,\Gamma(2t+\mu+\nu+1)(\frac{1}{2}x)^{\mu+\nu+2t}}{\Gamma(t+\mu+1)\Gamma(t+\nu+1)\Gamma(t+\mu+\nu+1)} dt, \qquad x > 0,$

where the path of integration separates the poles of $\Gamma(-t)$ from those of $\Gamma(2t+\mu+\nu+1)$. See Paris and Kaminski (2001, p. 116) for related results.

Nicholson's Integral

10.9.30 $\quad J_\nu^2(z) + Y_\nu^2(z) = \dfrac{8}{\pi^2} \displaystyle\int_0^\infty \cosh(2\nu t) K_0(2z \sinh t)\, dt, \qquad |\operatorname{ph} z| < \tfrac{1}{2}\pi.$

For the function K_0 see §10.25(ii).

10.9(iv) Compendia

For collections of integral representations of Bessel and Hankel functions see Erdélyi et al. (1953b, §§7.3 and 7.12), Erdélyi et al. (1954a, pp. 43–48, 51–60, 99–105, 108–115, 123–124, 272–276, and 356–357), Gröbner and Hofreiter (1950, pp. 189–192), Marichev (1983, pp. 191–192 and 196–210), Magnus et al. (1966, §3.6), and Watson (1944, Chapter 6).

10.10 Continued Fractions

Assume $J_{\nu-1}(z) \neq 0$. Then

10.10.1
$$\frac{J_\nu(z)}{J_{\nu-1}(z)} = \frac{1}{2\nu z^{-1} -} \; \frac{1}{2(\nu+1)z^{-1} -} \; \frac{1}{2(\nu+2)z^{-1} -} \cdots, \quad z \neq 0,$$

10.10.2
$$\frac{J_\nu(z)}{J_{\nu-1}(z)} = \frac{\frac{1}{2}z/\nu}{1 -} \; \frac{\frac{1}{4}z^2/(\nu(\nu+1))}{1 -} \; \frac{\frac{1}{4}z^2/((\nu+1)(\nu+2))}{1 -} \cdots, \quad \nu \neq 0, -1, -2, \ldots.$$

See also Cuyt et al. (2008, pp. 349–356).

10.11 Analytic Continuation

When $m \in \mathbb{Z}$,

10.11.1 $\qquad J_\nu(ze^{m\pi i}) = e^{m\nu\pi i} J_\nu(z),$

10.11.2
$$Y_\nu(ze^{m\pi i}) = e^{-m\nu\pi i} Y_\nu(z) + 2i \sin(m\nu\pi) \cot(\nu\pi) J_\nu(z).$$

10.11.3
$$\sin(\nu\pi) H_\nu^{(1)}(ze^{m\pi i}) = -\sin((m-1)\nu\pi) H_\nu^{(1)}(z) - e^{-\nu\pi i} \sin(m\nu\pi) H_\nu^{(2)}(z),$$

10.11.4
$$\sin(\nu\pi) H_\nu^{(2)}(ze^{m\pi i}) = e^{\nu\pi i} \sin(m\nu\pi) H_\nu^{(1)}(z) + \sin((m+1)\nu\pi) H_\nu^{(2)}(z).$$

10.11.5
$$H_\nu^{(1)}(ze^{\pi i}) = -e^{-\nu\pi i} H_\nu^{(2)}(z),$$
$$H_\nu^{(2)}(ze^{-\pi i}) = -e^{\nu\pi i} H_\nu^{(1)}(z).$$

If $\nu = n \; (\in \mathbb{Z})$, then limiting values are taken in (10.11.2)–(10.11.4):

10.11.6 $Y_n(ze^{m\pi i}) = (-1)^{mn}(Y_n(z) + 2im J_n(z)),$

10.11.7
$$H_n^{(1)}(ze^{m\pi i}) = (-1)^{mn-1}((m-1) H_n^{(1)}(z) + m H_n^{(2)}(z)),$$

10.11.8
$$H_n^{(2)}(ze^{m\pi i}) = (-1)^{mn}(m H_n^{(1)}(z) + (m+1) H_n^{(2)}(z)).$$

For real ν,

10.11.9
$$J_\nu(\bar{z}) = \overline{J_\nu(z)}, \qquad Y_\nu(\bar{z}) = \overline{Y_\nu(z)},$$
$$H_\nu^{(1)}(\bar{z}) = \overline{H_\nu^{(2)}(z)}, \quad H_\nu^{(2)}(\bar{z}) = \overline{H_\nu^{(1)}(z)}.$$

For complex ν replace ν by $\bar{\nu}$ on the right-hand sides.

10.12 Generating Function and Associated Series

For $z \in \mathbb{C}$ and $t \in \mathbb{C} \setminus \{0\}$,

10.12.1 $\qquad e^{\frac{1}{2}z(t-t^{-1})} = \sum_{m=-\infty}^{\infty} t^m J_m(z).$

For $z, \theta \in \mathbb{C}$,

10.12.2
$$\cos(z\sin\theta) = J_0(z) + 2\sum_{k=1}^{\infty} J_{2k}(z)\cos(2k\theta),$$
$$\sin(z\sin\theta) = 2\sum_{k=0}^{\infty} J_{2k+1}(z)\sin((2k+1)\theta),$$

10.12.3
$$\cos(z\cos\theta) = J_0(z) + 2\sum_{k=1}^{\infty} (-1)^k J_{2k}(z)\cos(2k\theta),$$
$$\sin(z\cos\theta) = 2\sum_{k=0}^{\infty} (-1)^k J_{2k+1}(z)\cos((2k+1)\theta).$$

10.12.4 $1 = J_0(z) + 2J_2(z) + 2J_4(z) + 2J_6(z) + \cdots,$

10.12.5 $\cos z = J_0(z) - 2J_2(z) + 2J_4(z) - 2J_6(z) + \cdots,$
$\sin z = 2J_1(z) - 2J_3(z) + 2J_5(z) - \cdots,$

10.12.6 $\frac{1}{2}z\cos z = J_1(z) - 9J_3(z) + 25J_5(z) - 49J_7(z) + \cdots,$
$\frac{1}{2}z\sin z = 4J_2(z) - 16J_4(z) + 36J_6(z) - \cdots.$

10.13 Other Differential Equations

In the following equations $\nu, \lambda, p, q,$ and r are real or complex constants with $\lambda \neq 0$, $p \neq 0$, and $q \neq 0$.

10.13.1 $\quad w'' + \left(\lambda^2 - \frac{\nu^2 - \frac{1}{4}}{z^2}\right)w = 0, \quad w = z^{\frac{1}{2}}\mathscr{C}_\nu(\lambda z),$

10.13.2 $\quad w'' + \left(\frac{\lambda^2}{4z} - \frac{\nu^2 - 1}{4z^2}\right)w = 0, \quad w = z^{\frac{1}{2}}\mathscr{C}_\nu\left(\lambda z^{\frac{1}{2}}\right),$

10.13.3 $\quad w'' + \lambda^2 z^{p-2} w = 0, \quad w = z^{\frac{1}{2}}\mathscr{C}_{1/p}\left(2\lambda z^{\frac{1}{2}p}/p\right),$

10.13.4 $\quad w'' - \frac{2\nu - 1}{z}w' + \lambda^2 w = 0, \quad w = z^\nu \mathscr{C}_\nu(\lambda z),$

10.13.5 $\quad z^2 w'' + (1-2r)zw' + (\lambda^2 q^2 z^{2q} + r^2 - \nu^2 q^2)w = 0, \qquad w = z^r \mathscr{C}_\nu(\lambda z^q),$

10.13.6 $\qquad w'' + (\lambda^2 e^{2z} - \nu^2)w = 0, \quad w = \mathscr{C}_\nu(\lambda e^z),$

10.14 Inequalities; Monotonicity

10.13.7
$$z^2(z^2 - \nu^2)w'' + z(z^2 - 3\nu^2)w'$$
$$+ ((z^2 - \nu^2)^2 - (z^2 + \nu^2))w = 0,$$
$$w = \mathscr{C}'_\nu(z),$$

10.13.8
$$w^{(2n)} = (-1)^n \lambda^{2n} z^{-n} w,$$
$$w = z^{\frac{1}{2}n} \mathscr{C}_n\left(2\lambda e^{k\pi i/n} z^{\frac{1}{2}}\right), \quad k = 0, 1, \ldots, 2n - 1.$$

In (10.13.9)–(10.13.11) $\mathscr{C}_\nu(z), \mathscr{D}_\mu(z)$ are any cylinder functions of orders ν, μ, respectively, and $\vartheta = z(d/dz)$.

10.13.9
$$z^2 w''' + 3zw'' + (4z^2 + 1 - 4\nu^2)w' + 4zw = 0,$$
$$w = \mathscr{C}_\nu(z)\mathscr{D}_\nu(z),$$

10.13.10
$$z^3 w''' + z(4z^2 + 1 - 4\nu^2)w' + (4\nu^2 - 1)w = 0,$$
$$w = z\mathscr{C}_\nu(z)\mathscr{D}_\nu(z),$$

10.13.11
$$\left(\vartheta^4 - 2(\nu^2 + \mu^2)\vartheta^2 + (\nu^2 - \mu^2)^2\right)w$$
$$+ 4z^2(\vartheta + 1)(\vartheta + 2)w = 0, \quad w = \mathscr{C}_\nu(z)\mathscr{D}_\mu(z).$$

For further differential equations see Kamke (1977, pp. 440–451). See also Watson (1944, pp. 95–100).

10.14 Inequalities; Monotonicity

10.14.1
$$|J_\nu(x)| \leq 1, \qquad \nu \geq 0, x \in \mathbb{R},$$
$$|J_\nu(x)| \leq 2^{-\frac{1}{2}}, \qquad \nu \geq 1, x \in \mathbb{R}.$$

10.14.2
$$0 < J_\nu(\nu) < \frac{2^{\frac{1}{3}}}{3^{\frac{2}{3}} \Gamma\left(\frac{2}{3}\right) \nu^{\frac{1}{3}}}, \qquad \nu > 0.$$

For monotonicity properties of $J_\nu(\nu)$ and $J'_\nu(\nu)$ see Lorch (1992).

10.14.3
$$|J_n(z)| \leq e^{|\Im z|}, \qquad n \in \mathbb{Z}.$$

10.14.4
$$|J_\nu(z)| \leq \frac{|\frac{1}{2}z|^\nu e^{|\Im z|}}{\Gamma(\nu + 1)}, \qquad \nu \geq -\frac{1}{2}.$$

10.14.5
$$|J_\nu(\nu x)| \leq \frac{x^\nu \exp\left(\nu(1 - x^2)^{\frac{1}{2}}\right)}{\left(1 + (1 - x^2)^{\frac{1}{2}}\right)^\nu}, \quad \nu \geq 0, 0 < x \leq 1;$$

see Siegel (1953).

10.14.6
$$|J'_\nu(\nu x)| \leq \frac{(1+x^2)^{\frac{1}{4}}}{x(2\pi\nu)^{\frac{1}{2}}} \frac{x^\nu \exp\left(\nu(1-x^2)^{\frac{1}{2}}\right)}{\left(1 + (1 - x^2)^{\frac{1}{2}}\right)^\nu},$$
$$\nu > 0, 0 < x \leq 1;$$

see Watson (1944, p. 255). For a related bound for $Y_\nu(\nu x)$ see Siegel and Sleator (1954).

10.14.7
$$1 \leq \frac{J_\nu(\nu x)}{x^\nu J_\nu(\nu)} \leq e^{\nu(1-x)}, \quad \nu \geq 0, 0 < x \leq 1;$$

see Paris (1984). For similar bounds for $\mathscr{C}_\nu(x)$ (§10.2(ii)) see Laforgia (1986).

Kapteyn's Inequality

10.14.8
$$|J_n(nz)| \leq \frac{\left|z^n \exp\left(n(1-z^2)^{\frac{1}{2}}\right)\right|}{\left|1 + (1-z^2)^{\frac{1}{2}}\right|^n}, \quad n = 0, 1, 2, \ldots,$$

where $(1 - z^2)^{\frac{1}{2}}$ has its principal value.

10.14.9
$$|J_n(nz)| \leq 1, \quad n = 0, 1, 2, \ldots, z \in \mathbf{K},$$

where $\mathbf{K}$ is defined in §10.20(ii).

For inequalities for the function $\Gamma(\nu+1)(2/x)^\nu J_\nu(x)$ with $\nu > -\frac{1}{2}$ see Neuman (2004).

For further monotonicity properties see Landau (1999, 2000).

10.15 Derivatives with Respect to Order

Noninteger Values of ν

10.15.1
$$\frac{\partial J_\nu(z)}{\partial \nu} = J_\nu(z) \ln\left(\tfrac{1}{2}z\right)$$
$$- \left(\tfrac{1}{2}z\right)^\nu \sum_{k=0}^\infty (-1)^k \frac{\psi(\nu+k+1)}{\Gamma(\nu+k+1)} \frac{\left(\tfrac{1}{4}z^2\right)^k}{k!},$$

10.15.2
$$\frac{\partial Y_\nu(z)}{\partial \nu} = \cot(\nu\pi)\left(\frac{\partial J_\nu(z)}{\partial \nu} - \pi Y_\nu(z)\right)$$
$$- \csc(\nu\pi) \frac{\partial J_{-\nu}(z)}{\partial \nu} - \pi J_\nu(z).$$

Integer Values of ν

10.15.3
$$\left.\frac{\partial J_\nu(z)}{\partial \nu}\right|_{\nu=n} = \frac{\pi}{2} Y_n(z) + \frac{n!}{2(\tfrac{1}{2}z)^n} \sum_{k=0}^{n-1} \frac{(\tfrac{1}{2}z)^k J_k(z)}{k!(n-k)},$$

10.15.4
$$\left.\frac{\partial Y_\nu(z)}{\partial \nu}\right|_{\nu=n} = -\frac{\pi}{2} J_n(z) + \frac{n!}{2(\tfrac{1}{2}z)^n} \sum_{k=0}^{n-1} \frac{(\tfrac{1}{2}z)^k Y_k(z)}{k!(n-k)},$$

10.15.5
$$\left.\frac{\partial J_\nu(z)}{\partial \nu}\right|_{\nu=0} = \frac{\pi}{2} Y_0(z), \quad \left.\frac{\partial Y_\nu(z)}{\partial \nu}\right|_{\nu=0} = -\frac{\pi}{2} J_0(z).$$

Half-Integer Values of ν

For the notations Ci and Si see §6.2(ii). When $x > 0$,

10.15.6
$$\left.\frac{\partial J_\nu(x)}{\partial \nu}\right|_{\nu=\frac{1}{2}} = \sqrt{\frac{2}{\pi x}}\left(\mathrm{Ci}(2x)\sin x - \mathrm{Si}(2x)\cos x\right),$$

10.15.7
$$\left.\frac{\partial J_\nu(x)}{\partial \nu}\right|_{\nu=-\frac{1}{2}} = \sqrt{\frac{2}{\pi x}}\left(\mathrm{Ci}(2x)\cos x + \mathrm{Si}(2x)\sin x\right),$$

10.15.8
$$\left.\frac{\partial Y_\nu(x)}{\partial \nu}\right|_{\nu=\frac{1}{2}} = \sqrt{\frac{2}{\pi x}}\big(\mathrm{Ci}(2x)\cos x \\ + (\mathrm{Si}(2x) - \pi)\sin x\big),$$

10.15.9
$$\left.\frac{\partial Y_\nu(x)}{\partial \nu}\right|_{\nu=-\frac{1}{2}} = -\sqrt{\frac{2}{\pi x}}\big(\mathrm{Ci}(2x)\sin x \\ - (\mathrm{Si}(2x) - \pi)\cos x\big).$$

For further results see Brychkov and Geddes (2005) and Landau (1999, 2000).

10.16 Relations to Other Functions

Elementary Functions

10.16.1
$$J_{\frac{1}{2}}(z) = Y_{-\frac{1}{2}}(z) = \left(\frac{2}{\pi z}\right)^{\frac{1}{2}}\sin z,$$
$$J_{-\frac{1}{2}}(z) = -Y_{\frac{1}{2}}(z) = \left(\frac{2}{\pi z}\right)^{\frac{1}{2}}\cos z,$$

10.16.2
$$H^{(1)}_{\frac{1}{2}}(z) = -i H^{(1)}_{-\frac{1}{2}}(z) = -i\left(\frac{2}{\pi z}\right)^{\frac{1}{2}}e^{iz},$$
$$H^{(2)}_{\frac{1}{2}}(z) = i H^{(2)}_{-\frac{1}{2}}(z) = i\left(\frac{2}{\pi z}\right)^{\frac{1}{2}}e^{-iz}.$$

For these and general results when ν is half an odd integer see §§10.47(ii) and 10.49(i).

Airy Functions

See §§9.6(i) and 9.6(ii).

Parabolic Cylinder Functions

With the notation of §12.14(i),

10.16.3
$$J_{\frac{1}{4}}(z) = -2^{-\frac{1}{4}}\pi^{-\frac{1}{2}}z^{-\frac{1}{4}}\left(W\big(0, 2z^{\frac{1}{2}}\big) - W\big(0, -2z^{\frac{1}{2}}\big)\right),$$
$$J_{-\frac{1}{4}}(z) = 2^{-\frac{1}{4}}\pi^{-\frac{1}{2}}z^{-\frac{1}{4}}\left(W\big(0, 2z^{\frac{1}{2}}\big) + W\big(0, -2z^{\frac{1}{2}}\big)\right).$$

10.16.4
$$J_{\frac{3}{4}}(z) = -2^{-\frac{1}{4}}\pi^{-\frac{1}{2}}z^{-\frac{3}{4}}\left(W'\big(0, 2z^{\frac{1}{2}}\big) - W'\big(0, -2z^{\frac{1}{2}}\big)\right),$$
$$J_{-\frac{3}{4}}(z) = -2^{-\frac{1}{4}}\pi^{-\frac{1}{2}}z^{-\frac{3}{4}}\left(W'\big(0, 2z^{\frac{1}{2}}\big) + W'\big(0, -2z^{\frac{1}{2}}\big)\right).$$

Principal values on each side of these equations correspond.

Confluent Hypergeometric Functions

10.16.5
$$J_\nu(z) = \frac{(\frac{1}{2}z)^\nu e^{\mp iz}}{\Gamma(\nu+1)}M\big(\nu + \tfrac{1}{2}, 2\nu+1, \pm 2iz\big),$$

10.16.6
$$\left.\begin{array}{l}H^{(1)}_\nu(z)\\ H^{(2)}_\nu(z)\end{array}\right\} = \mp 2\pi^{-\frac{1}{2}}ie^{\mp\nu\pi i}(2z)^\nu \\ \times e^{\pm iz}U\big(\nu + \tfrac{1}{2}, 2\nu+1, \mp 2iz\big).$$

For the functions M and U see §13.2(i).

10.16.7
$$J_\nu(z) = \frac{e^{\mp(2\nu+1)\pi i/4}}{2^{2\nu}\,\Gamma(\nu+1)}(2z)^{-\frac{1}{2}}M_{0,\nu}(\pm 2iz),$$
$$2\nu \neq -1, -2-3, \ldots,$$

10.16.8
$$\left.\begin{array}{l}H^{(1)}_\nu(z)\\ H^{(2)}_\nu(z)\end{array}\right\} = e^{\mp(2\nu+1)\pi i/4}\left(\frac{2}{\pi z}\right)^{\frac{1}{2}}W_{0,\nu}(\mp 2iz).$$

For the functions $M_{0,\nu}$ and $W_{0,\nu}$ see §13.14(i).

In all cases principal branches correspond at least when $|\mathrm{ph}\,z| \le \tfrac{1}{2}\pi$.

Generalized Hypergeometric Functions

10.16.9
$$J_\nu(z) = \frac{(\tfrac{1}{2}z)^\nu}{\Gamma(\nu+1)}\,_0F_1\big(-;\nu+1;-\tfrac{1}{4}z^2\big).$$

For $_0F_1$ see (16.2.1).

With $\mathbf{F}$ as in §15.2(i), and with z and ν fixed,

10.16.10
$$J_\nu(z) = (\tfrac{1}{2}z)^\nu \lim \mathbf{F}\big(\lambda, \mu; \nu+1; -z^2/(4\lambda\mu)\big),$$

as λ and $\mu \to \infty$ in $\mathbb{C}$. For this result see Watson (1944, §5.7).

10.17 Asymptotic Expansions for Large Argument

10.17(i) Hankel's Expansions

Define $a_0(\nu) = 1$,

10.17.1
$$a_k(\nu) = \frac{(4\nu^2 - 1^2)(4\nu^2 - 3^2)\cdots(4\nu^2 - (2k-1)^2)}{k!\,8^k},$$
$$k \ge 1,$$

10.17.2
$$\omega = z - \tfrac{1}{2}\nu\pi - \tfrac{1}{4}\pi,$$

and let δ denote an arbitrary small positive constant. Then as $z \to \infty$, with ν fixed,

10.17.3
$$J_\nu(z) \sim \left(\frac{2}{\pi z}\right)^{\frac{1}{2}}\bigg(\cos\omega\sum_{k=0}^\infty (-1)^k \frac{a_{2k}(\nu)}{z^{2k}} \\ - \sin\omega\sum_{k=0}^\infty (-1)^k \frac{a_{2k+1}(\nu)}{z^{2k+1}}\bigg),$$
$$|\mathrm{ph}\,z| \le \pi - \delta,$$

10.17.4
$$Y_\nu(z) \sim \left(\frac{2}{\pi z}\right)^{\frac{1}{2}} \left(\sin\omega \sum_{k=0}^{\infty} (-1)^k \frac{a_{2k}(\nu)}{z^{2k}} + \cos\omega \sum_{k=0}^{\infty} (-1)^k \frac{a_{2k+1}(\nu)}{z^{2k+1}}\right), \quad |\operatorname{ph} z| \leq \pi - \delta,$$

10.17.5
$$H_\nu^{(1)}(z) \sim \left(\frac{2}{\pi z}\right)^{\frac{1}{2}} e^{i\omega} \sum_{k=0}^{\infty} i^k \frac{a_k(\nu)}{z^k}, \quad -\pi + \delta \leq \operatorname{ph} z \leq 2\pi - \delta,$$

10.17.6
$$H_\nu^{(2)}(z) \sim \left(\frac{2}{\pi z}\right)^{\frac{1}{2}} e^{-i\omega} \sum_{k=0}^{\infty} (-i)^k \frac{a_k(\nu)}{z^k}, \quad -2\pi + \delta \leq \operatorname{ph} z \leq \pi - \delta,$$

where the branch of $z^{\frac{1}{2}}$ is determined by

10.17.7
$$z^{\frac{1}{2}} = \exp\left(\tfrac{1}{2}\ln|z| + \tfrac{1}{2}i\operatorname{ph} z\right).$$

Corresponding expansions for other ranges of $\operatorname{ph} z$ can be obtained by combining (10.17.3), (10.17.5), (10.17.6) with the continuation formulas (10.11.1), (10.11.3), (10.11.4) (or (10.11.7), (10.11.8)), and also the connection formula given by the second of (10.4.4).

10.17(ii) Asymptotic Expansions of Derivatives

We continue to use the notation of §10.17(i). Also, $b_0(\nu) = 1$, $b_1(\nu) = (4\nu^2 + 3)/8$, and for $k \geq 2$,

10.17.8
$$b_k(\nu) = \frac{\left((4\nu^2 - 1^2)(4\nu^2 - 3^2)\cdots(4\nu^2 - (2k-3)^2)\right)(4\nu^2 + 4k^2 - 1)}{k!\,8^k}.$$

Then as $z \to \infty$ with ν fixed,

10.17.9
$$J'_\nu(z) \sim -\left(\frac{2}{\pi z}\right)^{\frac{1}{2}} \left(\sin\omega \sum_{k=0}^{\infty} (-1)^k \frac{b_{2k}(\nu)}{z^{2k}} + \cos\omega \sum_{k=0}^{\infty} (-1)^k \frac{b_{2k+1}(\nu)}{z^{2k+1}}\right), \quad |\operatorname{ph} z| \leq \pi - \delta,$$

10.17.10
$$Y'_\nu(z) \sim \left(\frac{2}{\pi z}\right)^{\frac{1}{2}} \left(\cos\omega \sum_{k=0}^{\infty} (-1)^k \frac{b_{2k}(\nu)}{z^{2k}} - \sin\omega \sum_{k=0}^{\infty} (-1)^k \frac{b_{2k+1}(\nu)}{z^{2k+1}}\right), \quad |\operatorname{ph} z| \leq \pi - \delta,$$

10.17.11
$$H_\nu^{(1)\prime}(z) \sim i\left(\frac{2}{\pi z}\right)^{\frac{1}{2}} e^{i\omega} \sum_{k=0}^{\infty} i^k \frac{b_k(\nu)}{z^k}, \quad -\pi + \delta \leq \operatorname{ph} z \leq 2\pi - \delta,$$

10.17.12
$$H_\nu^{(2)\prime}(z) \sim -i\left(\frac{2}{\pi z}\right)^{\frac{1}{2}} e^{-i\omega} \sum_{k=0}^{\infty} (-i)^k \frac{b_k(\nu)}{z^k}, \quad -2\pi + \delta \leq \operatorname{ph} z \leq \pi - \delta.$$

10.17(iii) Error Bounds for Real Argument and Order

In the expansions (10.17.3) and (10.17.4) assume that $\nu \geq 0$ and $z > 0$. Then the remainder associated with the sum $\sum_{k=0}^{\ell-1}(-1)^k a_{2k}(\nu)z^{-2k}$ does not exceed the first neglected term in absolute value and has the same sign provided that $\ell \geq \max(\tfrac{1}{2}\nu - \tfrac{1}{4}, 1)$. Similarly for $\sum_{k=0}^{\ell-1}(-1)^k a_{2k+1}(\nu)z^{-2k-1}$, provided that $\ell \geq \max(\tfrac{1}{2}\nu - \tfrac{3}{4}, 1)$.

In the expansions (10.17.5) and (10.17.6) assume that $\nu > -\tfrac{1}{2}$ and $z > 0$. If these expansions are terminated when $k = \ell - 1$, then the remainder term is bounded in absolute value by the first neglected term, provided that $\ell \geq \max(\nu - \tfrac{1}{2}, 1)$.

10.17(iv) Error Bounds for Complex Argument and Order

For (10.17.5) and (10.17.6) write

10.17.13
$$\left.\begin{array}{r}H_\nu^{(1)}(z)\\ H_\nu^{(2)}(z)\end{array}\right\} = \left(\frac{2}{\pi z}\right)^{\frac{1}{2}} e^{\pm i\omega}\left(\sum_{k=0}^{\ell-1}(\pm i)^k \frac{a_k(\nu)}{z^k} + R_\ell^\pm(\nu, z)\right), \quad \ell = 1, 2, \ldots.$$

Then

10.17.14
$$|R_\ell^\pm(\nu, z)| \leq 2|a_\ell(\nu)|\mathcal{V}_{z,\pm i\infty}(t^{-\ell}) \times \exp\left(|\nu^2 - \tfrac{1}{4}|\mathcal{V}_{z,\pm i\infty}(t^{-\ell})\right),$$

where $\mathcal{V}$ denotes the variational operator (2.3.6), and the paths of variation are subject to the condition that $|\Im t|$ changes monotonically. Bounds for $\mathcal{V}_{z,i\infty}(t^{-\ell})$ are given by

$$
\text{10.17.15} \quad \mathcal{V}_{z,i\infty}(t^{-\ell}) \leq \begin{cases} |z|^{-\ell}, & 0 \leq \mathrm{ph}\, z \leq \pi, \\ \chi(\ell)|z|^{-\ell}, & -\tfrac{1}{2}\pi \leq \mathrm{ph}\, z \leq 0 \text{ or } \pi \leq \mathrm{ph}\, z \leq \tfrac{3}{2}\pi, \\ 2\chi(\ell)|\Im z|^{-\ell}, & -\pi < \mathrm{ph}\, z \leq -\tfrac{1}{2}\pi \text{ or } \tfrac{3}{2}\pi \leq \mathrm{ph}\, z < 2\pi, \end{cases}
$$

where $\chi(\ell) = \pi^{\frac{1}{2}}\,\Gamma(\tfrac{1}{2}\ell+1)/\Gamma(\tfrac{1}{2}\ell+\tfrac{1}{2})$; see §9.7(i). The bounds (10.17.15) also apply to $\mathcal{V}_{z,-i\infty}(t^{-\ell})$ in the conjugate sectors.

Corresponding error bounds for (10.17.3) and (10.17.4) are obtainable by combining (10.17.13) and (10.17.14) with (10.4.4).

10.17(v) Exponentially-Improved Expansions

As in §9.7(v) denote

$$\text{10.17.16} \quad G_p(z) = \frac{e^z}{2\pi}\,\Gamma(p)\,\Gamma(1-p,z),$$

where $\Gamma(1-p,z)$ is the incomplete gamma function (§8.2(i)). Then in (10.17.13) as $z \to \infty$ with $|\ell - 2|z||$ bounded and $m\,(\geq 0)$ fixed,

$$
\text{10.17.17} \quad \begin{aligned} R_\ell^\pm(\nu,z) = (-1)^\ell 2\cos(\nu\pi) \\ \times\left(\sum_{k=0}^{m-1}(\pm i)^k \frac{a_k(\nu)}{z^k}\, G_{\ell-k}(\mp 2iz) \right. \\ \left. + R_{m,\ell}^\pm(\nu,z)\right), \end{aligned}
$$

where

$$\text{10.17.18} \quad R_{m,\ell}^\pm(\nu,z) = O\!\left(e^{-2|z|} z^{-m}\right), \quad |\mathrm{ph}(z e^{\mp \frac{1}{2}\pi i})| \leq \pi.$$

For higher re-expansions of the remainder terms see Olde Daalhuis and Olver (1995a) and Olde Daalhuis (1995, 1996).

10.18 Modulus and Phase Functions

10.18(i) Definitions

For $\nu \geq 0$ and $x > 0$

$$\text{10.18.1} \quad M_\nu(x) e^{i\theta_\nu(x)} = H_\nu^{(1)}(x),$$

$$\text{10.18.2} \quad N_\nu(x) e^{i\phi_\nu(x)} = H_\nu^{(1)\prime}(x),$$

where $M_\nu(x)\,(>0)$, $N_\nu(x)\,(>0)$, $\theta_\nu(x)$, and $\phi_\nu(x)$ are continuous real functions of ν and x, with the branches of $\theta_\nu(x)$ and $\phi_\nu(x)$ fixed by

$$\text{10.18.3} \quad \theta_\nu(x) \to -\tfrac{1}{2}\pi, \quad \phi_\nu(x) \to \tfrac{1}{2}\pi, \quad x \to 0+.$$

10.18(ii) Basic Properties

$$\text{10.18.4} \quad \begin{aligned} J_\nu(x) &= M_\nu(x)\cos\theta_\nu(x), \\ Y_\nu(x) &= M_\nu(x)\sin\theta_\nu(x), \end{aligned}$$

$$\text{10.18.5} \quad \begin{aligned} J_\nu'(x) &= N_\nu(x)\cos\phi_\nu(x), \\ Y_\nu'(x) &= N_\nu(x)\sin\phi_\nu(x), \end{aligned}$$

$$\text{10.18.6} \quad \begin{aligned} M_\nu(x) &= \left(J_\nu^2(x) + Y_\nu^2(x)\right)^{\frac{1}{2}}, \\ N_\nu(x) &= \left(J_\nu^{\prime 2}(x) + Y_\nu^{\prime 2}(x)\right)^{\frac{1}{2}}, \end{aligned}$$

$$\text{10.18.7} \quad \begin{aligned} \theta_\nu(x) &= \mathrm{Arctan}(Y_\nu(x)/J_\nu(x)), \\ \phi_\nu(x) &= \mathrm{Arctan}(Y_\nu'(x)/J_\nu'(x)). \end{aligned}$$

$$\text{10.18.8} \quad M_\nu^2(x)\,\theta_\nu'(x) = \frac{2}{\pi x}, \quad N_\nu^2(x)\,\phi_\nu'(x) = \frac{2(x^2-\nu^2)}{\pi x^3},$$

10.18.9
$$N_\nu^2(x) = M_\nu^{\prime 2}(x) + M_\nu^2(x)\,\theta_\nu^{\prime 2}(x) = M_\nu^{\prime 2}(x) + \frac{4}{(\pi x\, M_\nu(x))^2},$$

10.18.10
$$(x^2-\nu^2)\,M_\nu(x)\,M_\nu'(x) + x^2\,N_\nu(x)\,N_\nu'(x) + x\,N_\nu^2(x) = 0.$$

10.18.11
$$\tan(\phi_\nu(x) - \theta_\nu(x)) = \frac{M_\nu(x)\,\theta_\nu'(x)}{M_\nu'(x)} = \frac{2}{\pi x\, M_\nu(x)\, M_\nu'(x)},$$

$$\text{10.18.12} \quad M_\nu(x)\,N_\nu(x)\sin(\phi_\nu(x) - \theta_\nu(x)) = \frac{2}{\pi x}.$$

10.18.13
$$x^2\,M_\nu''(x) + x\,M_\nu'(x) + (x^2-\nu^2)\,M_\nu(x) = \frac{4}{\pi^2 M_\nu^3(x)},$$

10.18.14
$$w'' + \left(1 + \frac{\tfrac{1}{4}-\nu^2}{x^2}\right)w = \frac{4}{\pi^2 w^3}, \quad w = x^{\frac{1}{2}} M_\nu(x),$$

10.18.15
$$x^3 w''' + x(4x^2 + 1 - 4\nu^2)w' + (4\nu^2-1)w = 0, \quad w = x\, M_\nu^2(x).$$

$$\text{10.18.16} \quad \theta_\nu^{\prime 2}(x) + \frac{1}{2}\frac{\theta_\nu'''(x)}{\theta_\nu'(x)} - \frac{3}{4}\left(\frac{\theta_\nu''(x)}{\theta_\nu'(x)}\right)^2 = 1 - \frac{\nu^2 - \tfrac{1}{4}}{x^2}.$$

10.18(iii) Asymptotic Expansions for Large Argument

As $x \to \infty$, with ν fixed and $\mu = 4\nu^2$,

10.18.17 $\quad M_\nu^2(x) \sim \dfrac{2}{\pi x}\left(1 + \dfrac{1}{2}\dfrac{\mu-1}{(2x)^2} + \dfrac{1\cdot 3}{2\cdot 4}\dfrac{(\mu-1)(\mu-9)}{(2x)^4} + \dfrac{1\cdot 3\cdot 5}{2\cdot 4\cdot 6}\dfrac{(\mu-1)(\mu-9)(\mu-25)}{(2x)^6} + \cdots\right),$

10.18.18
$$\theta_\nu(x) \sim x - \left(\dfrac{1}{2}\nu + \dfrac{1}{4}\right)\pi + \dfrac{\mu-1}{2(4x)} + \dfrac{(\mu-1)(\mu-25)}{6(4x)^3} + \dfrac{(\mu-1)(\mu^2 - 114\mu + 1073)}{5(4x)^5}$$
$$+ \dfrac{(\mu-1)(5\mu^3 - 1535\mu^2 + 54703\mu - 3\,75733)}{14(4x)^7} + \cdots.$$

Also,

10.18.19 $\quad N_\nu^2(x) \sim \dfrac{2}{\pi x}\left(1 - \dfrac{1}{2}\dfrac{\mu-3}{(2x)^2} - \dfrac{1}{2\cdot 4}\dfrac{(\mu-1)(\mu-45)}{(2x)^4} - \cdots\right),$

the general term in this expansion being

10.18.20 $\quad -\dfrac{(2k-3)!!}{(2k)!!}\dfrac{(\mu-1)(\mu-9)\cdots(\mu-(2k-3)^2)(\mu-(2k+1)(2k-1)^2)}{(2x)^{2k}}, \quad k \geq 2,$

and

10.18.21 $\quad \phi_\nu(x) \sim x - \left(\dfrac{1}{2}\nu - \dfrac{1}{4}\right)\pi + \dfrac{\mu+3}{2(4x)} + \dfrac{\mu^2 + 46\mu - 63}{6(4x)^3} + \dfrac{\mu^3 + 185\mu^2 - 2053\mu + 1899}{5(4x)^5} + \cdots.$

The remainder after k terms in (10.18.17) does not exceed the $(k+1)$th term in absolute value and is of the same sign, provided that $k > \nu - \tfrac{1}{2}$.

10.19 Asymptotic Expansions for Large Order

10.19(i) Asymptotic Forms

If $\nu \to \infty$ through positive real values, with $z\,(\neq 0)$ fixed, then

10.19.1 $\quad J_\nu(z) \sim \dfrac{1}{\sqrt{2\pi\nu}}\left(\dfrac{ez}{2\nu}\right)^\nu,$

10.19.2
$$Y_\nu(z) \sim -i\,H_\nu^{(1)}(z) \sim i\,H_\nu^{(2)}(z) \sim -\sqrt{\dfrac{2}{\pi\nu}}\left(\dfrac{ez}{2\nu}\right)^{-\nu}.$$

10.19(ii) Debye's Expansions

If $\nu \to \infty$ through positive real values with $\alpha\,(>0)$ fixed, then

10.19.3
$$J_\nu(\nu\,\text{sech}\,\alpha) \sim \dfrac{e^{\nu(\tanh\alpha-\alpha)}}{(2\pi\nu\tanh\alpha)^{\frac{1}{2}}}\sum_{k=0}^\infty \dfrac{U_k(\coth\alpha)}{\nu^k},$$
$$Y_\nu(\nu\,\text{sech}\,\alpha) \sim -\dfrac{e^{\nu(\alpha-\tanh\alpha)}}{(\tfrac{1}{2}\pi\nu\tanh\alpha)^{\frac{1}{2}}}\sum_{k=0}^\infty (-1)^k\dfrac{U_k(\coth\alpha)}{\nu^k},$$

10.19.4
$$J'_\nu(\nu\,\text{sech}\,\alpha) \sim \left(\dfrac{\sinh(2\alpha)}{4\pi\nu}\right)^{\frac{1}{2}} e^{\nu(\tanh\alpha-\alpha)}\sum_{k=0}^\infty \dfrac{V_k(\coth\alpha)}{\nu^k},$$
$$Y'_\nu(\nu\,\text{sech}\,\alpha) \sim \left(\dfrac{\sinh(2\alpha)}{\pi\nu}\right)^{\frac{1}{2}} e^{\nu(\alpha-\tanh\alpha)}\sum_{k=0}^\infty (-1)^k\dfrac{V_k(\coth\alpha)}{\nu^k}.$$

If $\nu \to \infty$ through positive real values with $\beta\,(\in (0, \tfrac{1}{2}\pi))$ fixed, and

10.19.5 $\quad \xi = \nu(\tan\beta - \beta) - \tfrac{1}{4}\pi,$

then

10.19.6
$$J_\nu(\nu\sec\beta) \sim \left(\dfrac{2}{\pi\nu\tan\beta}\right)^{\frac{1}{2}}\left(\cos\xi\sum_{k=0}^\infty \dfrac{U_{2k}(i\cot\beta)}{\nu^{2k}}\right.$$
$$\left. -i\sin\xi\sum_{k=0}^\infty \dfrac{U_{2k+1}(i\cot\beta)}{\nu^{2k+1}}\right),$$
$$Y_\nu(\nu\sec\beta) \sim \left(\dfrac{2}{\pi\nu\tan\beta}\right)^{\frac{1}{2}}\left(\sin\xi\sum_{k=0}^\infty \dfrac{U_{2k}(i\cot\beta)}{\nu^{2k}}\right.$$
$$\left. +i\cos\xi\sum_{k=0}^\infty \dfrac{U_{2k+1}(i\cot\beta)}{\nu^{2k+1}}\right),$$

10.19.7
$$J'_\nu(\nu\sec\beta) \sim \left(\dfrac{\sin(2\beta)}{\pi\nu}\right)^{\frac{1}{2}}\left(-\sin\xi\sum_{k=0}^\infty \dfrac{V_{2k}(i\cot\beta)}{\nu^{2k}}\right.$$
$$\left. -i\cos\xi\sum_{k=0}^\infty \dfrac{V_{2k+1}(i\cot\beta)}{\nu^{2k+1}}\right),$$
$$Y'_\nu(\nu\sec\beta) \sim \left(\dfrac{\sin(2\beta)}{\pi\nu}\right)^{\frac{1}{2}}\left(\cos\xi\sum_{k=0}^\infty \dfrac{V_{2k}(i\cot\beta)}{\nu^{2k}}\right.$$
$$\left. -i\sin\xi\sum_{k=0}^\infty \dfrac{V_{2k+1}(i\cot\beta)}{\nu^{2k+1}}\right).$$

In these expansions $U_k(p)$ and $V_k(p)$ are the polynomials in p of degree $3k$ defined in §10.41(ii).

For error bounds for the first of (10.19.3) see Olver (1997b, p. 382).

10.19(iii) Transition Region

As $\nu \to \infty$, with $a(\in \mathbb{C})$ fixed,

10.19.8
$$J_\nu\left(\nu + a\nu^{\frac{1}{3}}\right) \sim \frac{2^{\frac{1}{3}}}{\nu^{\frac{1}{3}}} \operatorname{Ai}\left(-2^{\frac{1}{3}}a\right) \sum_{k=0}^\infty \frac{P_k(a)}{\nu^{2k/3}} + \frac{2^{\frac{2}{3}}}{\nu} \operatorname{Ai}'\left(-2^{\frac{1}{3}}a\right) \sum_{k=0}^\infty \frac{Q_k(a)}{\nu^{2k/3}}, \qquad |\operatorname{ph}\nu| \leq \tfrac{1}{2}\pi - \delta,$$

$$Y_\nu\left(\nu + a\nu^{\frac{1}{3}}\right) \sim -\frac{2^{\frac{1}{3}}}{\nu^{\frac{1}{3}}} \operatorname{Bi}\left(-2^{\frac{1}{3}}a\right) \sum_{k=0}^\infty \frac{P_k(a)}{\nu^{2k/3}} - \frac{2^{\frac{2}{3}}}{\nu} \operatorname{Bi}'\left(-2^{\frac{1}{3}}a\right) \sum_{k=0}^\infty \frac{Q_k(a)}{\nu^{2k/3}}, \qquad |\operatorname{ph}\nu| \leq \tfrac{1}{2}\pi - \delta.$$

Also,

10.19.9
$$\left.\begin{array}{l} H_\nu^{(1)}\left(\nu + a\nu^{\frac{1}{3}}\right) \\ H_\nu^{(2)}\left(\nu + a\nu^{\frac{1}{3}}\right) \end{array}\right\} \sim \frac{2^{\frac{4}{3}}}{\nu^{\frac{1}{3}}} e^{\mp \pi i/3} \operatorname{Ai}\left(e^{\mp \pi i/3}2^{\frac{1}{3}}a\right) \sum_{k=0}^\infty \frac{P_k(a)}{\nu^{2k/3}} + \frac{2^{\frac{5}{3}}}{\nu} e^{\pm \pi i/3} \operatorname{Ai}'\left(e^{\mp \pi i/3}2^{\frac{1}{3}}a\right) \sum_{k=0}^\infty \frac{Q_k(a)}{\nu^{2k/3}},$$

with sectors of validity $-\tfrac{1}{2}\pi + \delta \leq \pm \operatorname{ph}\nu \leq \tfrac{3}{2}\pi - \delta$. Here Ai and Bi are the Airy functions (§9.2), and

10.19.10
$$P_0(a) = 1, \quad P_1(a) = -\tfrac{1}{5}a, \quad P_2(a) = -\tfrac{9}{100}a^5 + \tfrac{3}{35}a^2,$$
$$P_3(a) = \tfrac{957}{7000}a^6 - \tfrac{173}{3150}a^3 - \tfrac{1}{225},$$
$$P_4(a) = \tfrac{27}{20000}a^{10} - \tfrac{23573}{147000}a^7 + \tfrac{5903}{138600}a^4 + \tfrac{947}{346500}a,$$
$$Q_0(a) = \tfrac{3}{10}a^2, \quad Q_1(a) = -\tfrac{17}{70}a^3 + \tfrac{1}{70},$$

10.19.11
$$Q_2(a) = -\tfrac{9}{1000}a^7 + \tfrac{611}{3150}a^4 - \tfrac{37}{3150}a,$$
$$Q_3(a) = -\tfrac{549}{28000}a^8 - \tfrac{110767}{693000}a^5 + \tfrac{79}{12375}a^2.$$

For corresponding expansions for derivatives see http://dlmf.nist.gov/10.19.iii.

For proofs and also for the corresponding expansions for second derivatives see Olver (1952).

For higher coefficients in (10.19.8) in the case $a = 0$ (that is, in the expansions of $J_\nu(\nu)$ and $Y_\nu(\nu)$), see Watson (1944, §8.21) and Temme (1997).

10.20 Uniform Asymptotic Expansions for Large Order

10.20(i) Real Variables

Define $\zeta = \zeta(z)$ to be the solution of the differential equation

10.20.1
$$\left(\frac{d\zeta}{dz}\right)^2 = \frac{1-z^2}{\zeta z^2}$$

that is infinitely differentiable on the interval $0 < z < \infty$, including $z = 1$. Then

10.20.2
$$\tfrac{2}{3}\zeta^{\frac{3}{2}} = \int_z^1 \frac{\sqrt{1-t^2}}{t} dt = \ln\left(\frac{1+\sqrt{1-z^2}}{z}\right) - \sqrt{1-z^2},$$
$$0 < z \leq 1,$$

10.20.3
$$\tfrac{2}{3}(-\zeta)^{\frac{3}{2}} = \int_1^z \frac{\sqrt{t^2-1}}{t} dt = \sqrt{z^2-1} - \operatorname{arcsec} z,$$
$$1 \leq z < \infty,$$

all functions taking their principal values, with $\zeta = \infty, 0, -\infty$, corresponding to $z = 0, 1, \infty$, respectively.

As $\nu \to \infty$ through positive real values

10.20.4
$$J_\nu(\nu z) \sim \left(\frac{4\zeta}{1-z^2}\right)^{\frac{1}{4}} \left(\frac{\operatorname{Ai}\left(\nu^{\frac{2}{3}}\zeta\right)}{\nu^{\frac{1}{3}}} \sum_{k=0}^\infty \frac{A_k(\zeta)}{\nu^{2k}} + \frac{\operatorname{Ai}'\left(\nu^{\frac{2}{3}}\zeta\right)}{\nu^{\frac{5}{3}}} \sum_{k=0}^\infty \frac{B_k(\zeta)}{\nu^{2k}} \right),$$

10.20.5
$$Y_\nu(\nu z) \sim -\left(\frac{4\zeta}{1-z^2}\right)^{\frac{1}{4}} \left(\frac{\operatorname{Bi}\left(\nu^{\frac{2}{3}}\zeta\right)}{\nu^{\frac{1}{3}}} \sum_{k=0}^\infty \frac{A_k(\zeta)}{\nu^{2k}} + \frac{\operatorname{Bi}'\left(\nu^{\frac{2}{3}}\zeta\right)}{\nu^{\frac{5}{3}}} \sum_{k=0}^\infty \frac{B_k(\zeta)}{\nu^{2k}} \right),$$

10.20.6
$$\left.\begin{array}{l} H_\nu^{(1)}(\nu z) \\ H_\nu^{(2)}(\nu z) \end{array}\right\} \sim 2e^{\mp \pi i/3} \left(\frac{4\zeta}{1-z^2}\right)^{\frac{1}{4}} \left(\frac{\operatorname{Ai}\left(e^{\pm 2\pi i/3}\nu^{\frac{2}{3}}\zeta\right)}{\nu^{\frac{1}{3}}} \sum_{k=0}^\infty \frac{A_k(\zeta)}{\nu^{2k}} + \frac{e^{\pm \pi i/3}\operatorname{Ai}'\left(e^{\pm 2\pi i/3}\nu^{\frac{2}{3}}\zeta\right)}{\nu^{\frac{5}{3}}} \sum_{k=0}^\infty \frac{B_k(\zeta)}{\nu^{2k}} \right),$$

10.20 Uniform Asymptotic Expansions for Large Order

10.20.7
$$J'_\nu(\nu z) \sim -\frac{2}{z}\left(\frac{1-z^2}{4\zeta}\right)^{\frac{1}{4}}\left(\frac{\operatorname{Ai}\left(\nu^{\frac{2}{3}}\zeta\right)}{\nu^{\frac{4}{3}}}\sum_{k=0}^\infty \frac{C_k(\zeta)}{\nu^{2k}} + \frac{\operatorname{Ai}'\left(\nu^{\frac{2}{3}}\zeta\right)}{\nu^{\frac{2}{3}}}\sum_{k=0}^\infty \frac{D_k(\zeta)}{\nu^{2k}}\right),$$

10.20.8
$$Y'_\nu(\nu z) \sim \frac{2}{z}\left(\frac{1-z^2}{4\zeta}\right)^{\frac{1}{4}}\left(\frac{\operatorname{Bi}\left(\nu^{\frac{2}{3}}\zeta\right)}{\nu^{\frac{4}{3}}}\sum_{k=0}^\infty \frac{C_k(\zeta)}{\nu^{2k}} + \frac{\operatorname{Bi}'\left(\nu^{\frac{2}{3}}\zeta\right)}{\nu^{\frac{2}{3}}}\sum_{k=0}^\infty \frac{D_k(\zeta)}{\nu^{2k}}\right),$$

10.20.9
$$\left.\begin{array}{r}H_\nu^{(1)'}(\nu z)\\ H_\nu^{(2)'}(\nu z)\end{array}\right\} \sim \frac{4e^{\mp 2\pi i/3}}{z}\left(\frac{1-z^2}{4\zeta}\right)^{\frac{1}{4}}\left(\frac{e^{\mp 2\pi i/3}\operatorname{Ai}\left(e^{\pm 2\pi i/3}\nu^{\frac{2}{3}}\zeta\right)}{\nu^{\frac{4}{3}}}\sum_{k=0}^\infty \frac{C_k(\zeta)}{\nu^{2k}} + \frac{\operatorname{Ai}'\left(e^{\pm 2\pi i/3}\nu^{\frac{2}{3}}\zeta\right)}{\nu^{\frac{2}{3}}}\sum_{k=0}^\infty \frac{D_k(\zeta)}{\nu^{2k}}\right),$$

uniformly for $z \in (0,\infty)$ in all cases, where Ai and Bi are the Airy functions (§9.2).

In the following formulas for the coefficients $A_k(\zeta)$, $B_k(\zeta)$, $C_k(\zeta)$, and $D_k(\zeta)$, u_k, v_k are the constants defined in §9.7(i), and $U_k(p)$, $V_k(p)$ are the polynomials in p of degree $3k$ defined in §10.41(ii).

Interval $0 < z < 1$

10.20.10 $A_k(\zeta) = \sum_{j=0}^{2k}(\tfrac{3}{2})^j v_j \zeta^{-3j/2} U_{2k-j}\left((1-z^2)^{-\frac{1}{2}}\right),$

10.20.11
$B_k(\zeta) = -\zeta^{-\frac{1}{2}}\sum_{j=0}^{2k+1}(\tfrac{3}{2})^j u_j \zeta^{-3j/2} U_{2k-j+1}\left((1-z^2)^{-\frac{1}{2}}\right),$

10.20.12
$C_k(\zeta) = -\zeta^{\frac{1}{2}}\sum_{j=0}^{2k+1}(\tfrac{3}{2})^j v_j \zeta^{-3j/2} V_{2k-j+1}\left((1-z^2)^{-\frac{1}{2}}\right),$

10.20.13 $D_k(\zeta) = \sum_{j=0}^{2k}(\tfrac{3}{2})^j u_j \zeta^{-3j/2} V_{2k-j}\left((1-z^2)^{-\frac{1}{2}}\right).$

Interval $1 < z < \infty$

In formulas (10.20.10)–(10.20.13) replace $\zeta^{\frac{1}{2}}$, $\zeta^{-\frac{1}{2}}$, $\zeta^{-3j/2}$, and $(1-z^2)^{-\frac{1}{2}}$ by $-i(-\zeta)^{\frac{1}{2}}$, $i(-\zeta)^{-\frac{1}{2}}$, $i^{3j}(-\zeta)^{-3j/2}$, and $i(z^2-1)^{-\frac{1}{2}}$, respectively.

Note: Another way of arranging the above formulas for the coefficients $A_k(\zeta), B_k(\zeta), C_k(\zeta)$, and $D_k(\zeta)$ would be by analogy with (12.10.42) and (12.10.46). In this way there is less usage of many-valued functions.

Values at $\zeta = 0$

10.20.14
$\begin{aligned}&A_0(0) = 1,\quad A_1(0) = -\tfrac{1}{225},\\ &A_2(0) = \tfrac{1\,51439}{2182\,95000},\quad A_3(0) = -\tfrac{8872\,78009}{250\,49351\,25000},\\ &B_0(0) = \tfrac{1}{70}2^{\frac{1}{3}},\quad B_1(0) = -\tfrac{1213}{10\,23750}2^{\frac{1}{3}},\\ &B_2(0) = \tfrac{1\,65425\,37833}{3774\,32055\,00000}2^{\frac{1}{3}},\\ &B_3(0) = -\tfrac{430\,99056\,39368\,59253}{5\,68167\,34399\,42500\,00000}2^{\frac{1}{3}}.\end{aligned}$

Each of the coefficients $A_k(\zeta), B_k(\zeta), C_k(\zeta),$ and $D_k(\zeta)$, $k = 0, 1, 2, \ldots$, is real and infinitely differentiable on the interval $-\infty < \zeta < \infty$. For (10.20.14) and further information on the coefficients see Temme (1997).

For numerical tables of $\zeta = \zeta(z)$, $(4\zeta/(1-z^2))^{\frac{1}{4}}$ and $A_k(\zeta), B_k(\zeta), C_k(\zeta)$, and $D_k(\zeta)$ see Olver (1962, pp. 28–42).

10.20(ii) Complex Variables

The function $\zeta = \zeta(z)$ given by (10.20.2) and (10.20.3) can be continued analytically to the z-plane cut along the negative real axis. Corresponding points of the mapping are shown in Figures 10.20.1 and 10.20.2.

The equations of the curved boundaries D_1E_1 and D_2E_2 in the ζ-plane are given parametrically by

10.20.15 $\zeta = (\tfrac{3}{2})^{\frac{2}{3}}(\tau \mp i\pi)^{\frac{2}{3}},\qquad 0 \le \tau < \infty,$

respectively.

The curves BP_1E_1 and BP_2E_2 in the z-plane are the inverse maps of the line segments

10.20.16 $\zeta = e^{\mp i\pi/3}\tau,\qquad 0 \le \tau \le (\tfrac{3}{2}\pi)^{\frac{2}{3}},$

respectively. They are given parametrically by

10.20.17
$z = \pm(\tau\coth\tau - \tau^2)^{\frac{1}{2}} \pm i(\tau^2 - \tau\tanh\tau)^{\frac{1}{2}},\ 0 \le \tau \le \tau_0,$
where $\tau_0 = 1.19968\ldots$ is the positive root of the equation $\tau = \coth\tau$. The points P_1, P_2 where these curves intersect the imaginary axis are $\pm ic$, where

10.20.18 $c = (\tau_0^2 - 1)^{\frac{1}{2}} = 0.66274\ldots.$

The eye-shaped closed domain in the uncut z-plane that is bounded by BP_1E_1 and BP_2E_2 is denoted by **K**; see Figure 10.20.3.

As $\nu \to \infty$ through positive real values the expansions (10.20.4)–(10.20.9) apply uniformly for $|\operatorname{ph} z| \le \pi - \delta$, the coefficients $A_k(\zeta), B_k(\zeta), C_k(\zeta),$ and $D_k(\zeta)$, being the analytic continuations of the functions defined in §10.20(i) when ζ is real.

For proofs of the above results and for error bounds and extensions of the regions of validity see Olver (1997b, pp. 419–425). For extensions to complex ν see Olver (1954). For resurgence properties of the coefficients (§2.7(ii)) see Howls and Olde Daalhuis (1999). For further results see Dunster (2001a), Wang and Wong (2002), and Paris (2004).

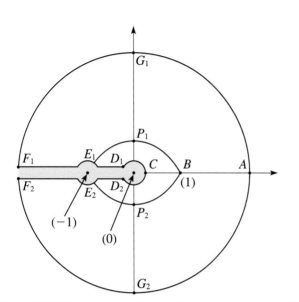

Figure 10.20.1: z-plane. P_1 and P_2 are the points $\pm ic$. $c = 0.66274\ldots$.

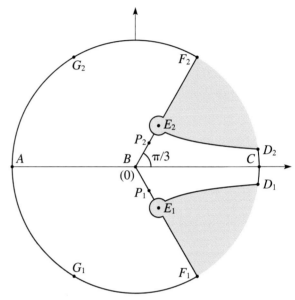

Figure 10.20.2: ζ-plane. E_1 and E_2 are the points $e^{\mp \pi i/3}(3\pi/2)^{2/3}$.

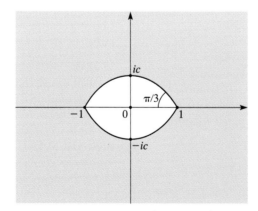

Figure 10.20.3: z-plane. Domain **K** (unshaded). $c = 0.66274\ldots$.

10.20(iii) Double Asymptotic Properties

For asymptotic properties of the expansions (10.20.4)–(10.20.6) with respect to large values of z see §10.41(v).

10.21 Zeros

10.21(i) Distribution

The zeros of any cylinder function or its derivative are simple, with the possible exceptions of $z = 0$ in the case of the functions, and $z = 0, \pm\nu$ in the case of the derivatives.

If ν is real, then $J_\nu(z)$, $J'_\nu(z)$, $Y_\nu(z)$, and $Y'_\nu(z)$, each have an infinite number of positive real zeros. All of these zeros are simple, provided that $\nu \geq -1$ in the case of $J'_\nu(z)$, and $\nu \geq -\frac{1}{2}$ in the case of $Y'_\nu(z)$. When all of their zeros are simple, the mth positive zeros of these functions are denoted by $j_{\nu,m}$, $j'_{\nu,m}$, $y_{\nu,m}$, and $y'_{\nu,m}$ respectively, except that $z = 0$ is counted as the first zero of $J'_0(z)$. Since $J'_0(z) = -J_1(z)$ we have

10.21.1 $\quad j'_{0,1} = 0, \quad j'_{0,m} = j_{1,m-1}, \quad m = 2, 3, \ldots.$

When $\nu \geq 0$, the zeros interlace according to the inequalities

10.21.2 $\quad j_{\nu,1} < j_{\nu+1,1} < j_{\nu,2} < j_{\nu+1,2} < j_{\nu,3} < \cdots,$
$\quad y_{\nu,1} < y_{\nu+1,1} < y_{\nu,2} < y_{\nu+1,2} < y_{\nu,3} < \cdots,$

10.21.3 $\quad \nu \leq j'_{\nu,1} < y_{\nu,1} < y'_{\nu,1} < j_{\nu,1} < j'_{\nu,2} < y_{\nu,2} < \cdots.$

The positive zeros of any two real distinct cylinder functions of the same order are interlaced, as are the positive zeros of any real cylinder function $\mathscr{C}_\nu(z)$ and the contiguous function $\mathscr{C}_{\nu+1}(z)$. See also Elbert and Laforgia (1994).

When $\nu \geq -1$ the zeros of $J_\nu(z)$ are all real. If $\nu < -1$ and ν is not an integer, then the number of complex zeros of $J_\nu(z)$ is $2\lfloor -\nu \rfloor$. If $\lfloor -\nu \rfloor$ is odd, then two of these zeros lie on the imaginary axis.

If $\nu \geq 0$, then the zeros of $J'_\nu(z)$ are all real.

For information on the real double zeros of $J'_\nu(z)$ and $Y'_\nu(z)$ when $\nu < -1$ and $\nu < -\frac{1}{2}$, respectively, see Döring (1971) and Kerimov and Skorokhodov (1986). The latter reference also has information on double zeros of the second and third derivatives of $J_\nu(z)$ and $Y_\nu(z)$.

No two of the functions $J_0(z), J_1(z), J_2(z), \ldots$, have any common zeros other than $z = 0$; see Watson (1944, §15.28).

10.21(ii) Analytic Properties

If ρ_ν is a zero of the cylinder function

10.21.4 $\quad \mathscr{C}_\nu(z) = J_\nu(z)\cos(\pi t) + Y_\nu(z)\sin(\pi t),$

where t is a parameter, then

10.21.5 $\quad \mathscr{C}'_\nu(\rho_\nu) = \mathscr{C}_{\nu-1}(\rho_\nu) = -\mathscr{C}_{\nu+1}(\rho_\nu).$

If σ_ν is a zero of $\mathscr{C}'_\nu(z)$, then

10.21.6 $\quad \mathscr{C}_\nu(\sigma_\nu) = \dfrac{\sigma_\nu}{\nu}\mathscr{C}_{\nu-1}(\sigma_\nu) = \dfrac{\sigma_\nu}{\nu}\mathscr{C}_{\nu+1}(\sigma_\nu).$

The parameter t may be regarded as a continuous variable and ρ_ν, σ_ν as functions $\rho_\nu(t)$, $\sigma_\nu(t)$ of t. If $\nu \geq 0$ and these functions are fixed by

10.21.7 $\quad \rho_\nu(0) = 0, \quad \sigma_\nu(0) = j'_{\nu,1},$

then

10.21.8
$$j_{\nu,m} = \rho_\nu(m), \quad y_{\nu,m} = \rho_\nu(m - \tfrac{1}{2}), \quad m = 1, 2, \ldots,$$

10.21.9
$$j'_{\nu,m} = \sigma_\nu(m - 1), \quad y'_{\nu,m} = \sigma_\nu(m - \tfrac{1}{2}),$$
$$m = 1, 2, \ldots.$$

10.21.10
$$\mathscr{C}'_\nu(\rho_\nu) = \left(\frac{\rho_\nu}{2}\frac{d\rho_\nu}{dt}\right)^{-\frac{1}{2}}, \quad \mathscr{C}_\nu(\sigma_\nu) = \left(\frac{\sigma_\nu^2 - \nu^2}{2\sigma_\nu}\frac{d\sigma_\nu}{dt}\right)^{-\frac{1}{2}},$$

10.21.11
$$2\rho_\nu^2 \frac{d\rho_\nu}{dt}\frac{d^3\rho_\nu}{dt^3} - 3\rho_\nu^2\left(\frac{d^2\rho_\nu}{dt^2}\right)^2 - 4\pi^2\rho_\nu^2\left(\frac{d\rho_\nu}{dt}\right)^2$$
$$+ (4\rho_\nu^2 + 1 - 4\nu^2)\left(\frac{d\rho_\nu}{dt}\right)^4 = 0.$$

The functions $\rho_\nu(t)$ and $\sigma_\nu(t)$ are related to the inverses of the phase functions $\theta_\nu(x)$ and $\phi_\nu(x)$ defined in §10.18(i): if $\nu \geq 0$, then

10.21.12 $\quad \theta_\nu(j_{\nu,m}) = (m - \tfrac{1}{2})\pi, \quad \theta_\nu(y_{\nu,m}) = (m-1)\pi,$
$\quad m = 1, 2, \ldots,$

10.21.13 $\quad \phi_\nu(j'_{\nu,m}) = (m - \tfrac{1}{2})\pi, \quad \phi_\nu(y'_{\nu,m}) = m\pi,$
$\quad m = 1, 2, \ldots.$

For sign properties of the forward differences that are defined by

10.21.14 $\quad \Delta\rho_\nu(t) = \rho_\nu(t+1) - \rho_\nu(t),$
$\quad \Delta^2\rho_\nu(t) = \Delta\rho_\nu(t+1) - \Delta\rho_\nu(t), \ldots,$

when $t = 1, 2, 3, \ldots$, and similarly for $\sigma_\nu(t)$, see Lorch and Szego (1963, 1964), Lorch et al. (1970, 1972), and Muldoon (1977).

10.21(iii) Infinite Products

10.21.15 $\quad J_\nu(z) = \dfrac{(\tfrac{1}{2}z)^\nu}{\Gamma(\nu+1)}\prod_{k=1}^{\infty}\left(1 - \dfrac{z^2}{j_{\nu,k}^2}\right), \quad \nu \geq 0,$

10.21.16 $\quad J'_\nu(z) = \dfrac{(\tfrac{1}{2}z)^{\nu-1}}{2\,\Gamma(\nu)}\prod_{k=1}^{\infty}\left(1 - \dfrac{z^2}{j'^{\,2}_{\nu,k}}\right), \quad \nu > 0.$

10.21(iv) Monotonicity Properties

Any positive zero c of the cylinder function $\mathscr{C}_\nu(x)$ and any positive zero c' of $\mathscr{C}_\nu'(x)$ such that $c' > |\nu|$ are definable as continuous and increasing functions of ν:

10.21.17 $\quad \dfrac{dc}{d\nu} = 2c \int_0^\infty K_0(2c \sinh t) e^{-2\nu t}\, dt,$

10.21.18 $\quad \dfrac{dc'}{d\nu} = \dfrac{2c'}{c'^2 - \nu^2} \int_0^\infty (c'^2 \cosh(2t) - \nu^2) \\ \times K_0(2c' \sinh t) e^{-2\nu t}\, dt,$

where K_0 is defined in §10.25(ii).

In particular, $j_{\nu,m}$, $y_{\nu,m}$, $j'_{\nu,m}$, and $y'_{\nu,m}$ are increasing functions of ν when $\nu \geq 0$. It is also true that the positive zeros j''_ν and j'''_ν of $J''_\nu(x)$ and $J'''_\nu(x)$, respectively, are increasing functions of ν when $\nu > 0$, provided that in the latter case $j'''_\nu > \sqrt{3}$ when $0 < \nu < 1$.

$j_{\nu,m}/\nu$ and $j'_{\nu,m}/\nu$ are decreasing functions of ν when $\nu > 0$ for $m = 1, 2, 3, \ldots$.

For further monotonicity properties see Elbert (2001), Lorch (1990, 1993, 1995), Lorch and Szego (1990, 1995), and Muldoon (1981). For inequalities for zeros arising from monotonicity properties see Laforgia and Muldoon (1983).

10.21(v) Inequalities

For bounds for the smallest real or purely imaginary zeros of $J_\nu(x)$ when ν is real see Ismail and Muldoon (1995).

10.21(vi) McMahon's Asymptotic Expansions for Large Zeros

If ν (≥ 0) is fixed, $\mu = 4\nu^2$, and $m \to \infty$, then

10.21.19
$$j_{\nu,m}, y_{\nu,m} \sim a - \dfrac{\mu - 1}{8a} - \dfrac{4(\mu - 1)(7\mu - 31)}{3(8a)^3} - \dfrac{32(\mu - 1)(83\mu^2 - 982\mu + 3779)}{15(8a)^5} \\ - \dfrac{64(\mu - 1)(6949\mu^3 - 1\,53855\mu^2 + 15\,85743\mu - 62\,77237)}{105(8a)^7} - \cdots,$$

where $a = (m + \tfrac{1}{2}\nu - \tfrac{1}{4})\pi$ for $j_{\nu,m}$, $a = (m + \tfrac{1}{2}\nu - \tfrac{3}{4})\pi$ for $y_{\nu,m}$. With $a = (t + \tfrac{1}{2}\nu - \tfrac{1}{4})\pi$, the right-hand side is the asymptotic expansion of $\rho_\nu(t)$ for large t.

10.21.20
$$j'_{\nu,m}, y'_{\nu,m} \sim b - \dfrac{\mu + 3}{8b} - \dfrac{4(7\mu^2 + 82\mu - 9)}{3(8b)^3} - \dfrac{32(83\mu^3 + 2075\mu^2 - 3039\mu + 3537)}{15(8b)^5} \\ - \dfrac{64(6949\mu^4 + 2\,96492\mu^3 - 12\,48002\mu^2 + 74\,14380\mu - 58\,53627)}{105(8b)^7} - \cdots,$$

where $b = (m + \tfrac{1}{2}\nu - \tfrac{3}{4})\pi$ for $j'_{\nu,m}$, $b = (m + \tfrac{1}{2}\nu - \tfrac{1}{4})\pi$ for $y'_{\nu,m}$, and $b = (t + \tfrac{1}{2}\nu + \tfrac{1}{4})\pi$ for $\sigma_\nu(t)$.

For the next three terms in (10.21.19) and the next two terms in (10.21.20) see Bickley et al. (1952, p. xxxvii) or Olver (1960, pp. xvii–xviii).

For error bounds see Wong and Lang (1990), Wong (1995), and Elbert and Laforgia (2000). See also Laforgia (1979).

For the mth positive zero $j''_{\nu,m}$ of $J''_\nu(x)$ Wong and Lang (1990) gives the corresponding expansion

10.21.21 $\quad j''_{\nu,m} \sim c - \dfrac{\mu + 7}{8c} - \dfrac{28\mu^2 + 424\mu + 1724}{3(8c)^3} - \cdots,$

where $c = (m + \tfrac{1}{2}\nu - \tfrac{1}{4})\pi$ if $0 < \nu < 1$, and $c = (m + \tfrac{1}{2}\nu - \tfrac{5}{4})\pi$ if $\nu > 1$. An error bound is included for the case $\nu \geq \tfrac{3}{2}$.

10.21(vii) Asymptotic Expansions for Large Order

Let $\mathscr{C}_\nu(x)$, $\rho_\nu(t)$, and $\sigma_\nu(t)$ be defined as in §10.21(ii) and $M(x)$, $\theta(x)$, $N(x)$, and $\phi(x)$ denote the modulus and phase functions for the Airy functions and their derivatives as in §9.8.

As $\nu \to \infty$ with t (>0) fixed,

10.21.22 $\quad \rho_\nu(t) \sim \nu \sum_{k=0}^\infty \dfrac{\alpha_k}{\nu^{2k/3}},$

10.21.23 $\quad \mathscr{C}_\nu'(\rho_\nu(t)) \sim \dfrac{(2/\nu)^{\tfrac{2}{3}}}{\pi M\!\left(-2^{\tfrac{1}{3}}\alpha\right)} \sum_{k=0}^\infty \dfrac{\beta_k}{\nu^{2k/3}},$

where α is given by

10.21.24 $\quad \theta\!\left(-2^{\tfrac{1}{3}}\alpha\right) = \pi t,$

and

10.21.25 $\quad \alpha_0 = 1, \quad \alpha_1 = \alpha, \quad \alpha_2 = \tfrac{3}{10}\alpha^2, \\ \alpha_3 = -\tfrac{1}{350}\alpha^3 + \tfrac{1}{70}, \quad \alpha_4 = -\tfrac{479}{63000}\alpha^4 - \tfrac{1}{3150}\alpha, \\ \alpha_5 = \tfrac{20231}{80\,85000}\alpha^5 - \tfrac{551}{1\,61700}\alpha^2,$

10.21.26
$\beta_0 = 1, \quad \beta_1 = -\tfrac{4}{5}\alpha, \quad \beta_2 = \tfrac{18}{35}\alpha^2, \\ \beta_3 = -\tfrac{88}{315}\alpha^3 - \tfrac{11}{1575}, \quad \beta_4 = \tfrac{79586}{6\,06375}\alpha^4 + \tfrac{9824}{6\,06375}\alpha.$

10.21 ZEROS

As $\nu \to \infty$ with $t\ (> -\tfrac{1}{6})$ fixed,

10.21.27 $\qquad \sigma_\nu(t) \sim \nu \sum_{k=0}^{\infty} \dfrac{\alpha'_k}{\nu^{2k/3}},$

10.21.28 $\quad \mathscr{C}_\nu(\sigma_\nu(t)) \sim \dfrac{(2/\nu)^{\tfrac{1}{3}}}{\pi\, N\!\left(-2^{\tfrac{1}{3}}\alpha'\right)} \sum_{k=0}^{\infty} \dfrac{\beta'_k}{\nu^{2k/3}},$

where α' is given by

10.21.29 $\qquad \phi\!\left(-2^{\tfrac{1}{3}}\alpha'\right) = \pi t,$

and

10.21.30
$\alpha'_0 = 1, \quad \alpha'_1 = \alpha', \quad \alpha'_2 = \tfrac{3}{10}\alpha'^2 - \tfrac{1}{10}\alpha'^{-1},$
$\alpha'_3 = -\tfrac{1}{350}\alpha'^3 - \tfrac{1}{25} - \tfrac{1}{200}\alpha'^{-3},$
$\alpha'_4 = -\tfrac{479}{63000}\alpha'^4 + \tfrac{509}{31500}\alpha' + \tfrac{1}{1500}\alpha'^{-2} - \tfrac{1}{2000}\alpha'^{-5},$

10.21.31
$\beta'_0 = 1, \quad \beta'_1 = -\tfrac{1}{5}\alpha', \quad \beta'_2 = \tfrac{9}{350}\alpha'^2 + \tfrac{1}{100}\alpha'^{-1},$
$\beta'_3 = \tfrac{89}{15750}\alpha'^3 - \tfrac{47}{4500} + \tfrac{1}{3000}\alpha'^{-3}.$

In particular, with the notation as below,

10.21.32 $\qquad j_{\nu,m} \sim \nu \sum_{k=0}^{\infty} \dfrac{\alpha_k}{\nu^{2k/3}},$

10.21.33 $\qquad y_{\nu,m} \sim \nu \sum_{k=0}^{\infty} \dfrac{\alpha_k}{\nu^{2k/3}},$

10.21.34 $\quad J'_\nu(j_{\nu,m}) \sim (-1)^m \dfrac{(2/\nu)^{\tfrac{2}{3}}}{\pi\, M(a_m)} \sum_{k=0}^{\infty} \dfrac{\beta_k}{\nu^{2k/3}},$

10.21.35 $\quad Y'_\nu(y_{\nu,m}) \sim (-1)^{m-1} \dfrac{(2/\nu)^{\tfrac{2}{3}}}{\pi\, M(b_m)} \sum_{k=0}^{\infty} \dfrac{\beta_k}{\nu^{2k/3}},$

and

10.21.36 $\qquad j'_{\nu,m} \sim \nu \sum_{k=0}^{\infty} \dfrac{\alpha'_k}{\nu^{2k/3}},$

10.21.37 $\qquad y'_{\nu,m} \sim \nu \sum_{k=0}^{\infty} \dfrac{\alpha'_k}{\nu^{2k/3}},$

10.21.38 $\quad J_\nu(j'_{\nu,m}) \sim (-1)^{m-1} \dfrac{(2/\nu)^{\tfrac{1}{3}}}{\pi\, N(a'_m)} \sum_{k=0}^{\infty} \dfrac{\beta'_k}{\nu^{2k/3}},$

10.21.39 $\quad Y_\nu(y'_{\nu,m}) \sim (-1)^{m-1} \dfrac{(2/\nu)^{\tfrac{1}{3}}}{\pi\, N(b'_m)} \sum_{k=0}^{\infty} \dfrac{\beta'_k}{\nu^{2k/3}}.$

Here a_m, b_m, a'_m, b'_m are the mth negative zeros of $\mathrm{Ai}(x)$, $\mathrm{Bi}(x)$, $\mathrm{Ai}'(x)$, $\mathrm{Bi}'(x)$, respectively (§9.9), α_k, β_k, α'_k, β'_k are given by (10.21.25), (10.21.26), (10.21.30), and (10.21.31), with $\alpha = -2^{-\tfrac{1}{3}} a_m$ in the case of $j_{\nu,m}$ and $J'_\nu(j_{\nu,m})$, $\alpha = -2^{-\tfrac{1}{3}} b_m$ in the case of $y_{\nu,m}$ and $Y'_\nu(y_{\nu,m})$, $\alpha' = -2^{-\tfrac{1}{3}} a'_m$ in the case of $j'_{\nu,m}$ and $J_\nu(j'_{\nu,m})$, $\alpha' = -2^{-\tfrac{1}{3}} b'_m$ in the case of $y'_{\nu,m}$ and $Y_\nu(y'_{\nu,m})$.

For error bounds for (10.21.32) see Qu and Wong (1999); for (10.21.36) and (10.21.37) see Elbert and Laforgia (1997). See also Spigler (1980).

For the first zeros rounded numerical values of the coefficients are given by

10.21.40
$j_{\nu,1} \sim \nu + 1.85575\,71\nu^{\tfrac{1}{3}} + 1.03315\,0\nu^{-\tfrac{1}{3}} - 0.00397\nu^{-1} - 0.0908\nu^{-\tfrac{5}{3}} + 0.043\nu^{-\tfrac{7}{3}} + \cdots,$
$y_{\nu,1} \sim \nu + 0.93157\,68\nu^{\tfrac{1}{3}} + 0.26035\,1\nu^{-\tfrac{1}{3}} + 0.01198\nu^{-1} - 0.0060\nu^{-\tfrac{5}{3}} - 0.001\nu^{-\tfrac{7}{3}} + \cdots,$
$J'_\nu(j_{\nu,1}) \sim -1.11310\,28\nu^{-\tfrac{2}{3}} \div (1 + 1.48460\,6\nu^{-\tfrac{2}{3}} + 0.43294\nu^{-\tfrac{4}{3}} - 0.1943\nu^{-2} + 0.019\nu^{-\tfrac{8}{3}} + \cdots),$
$Y'_\nu(y_{\nu,1}) \sim 0.95554\,86\nu^{-\tfrac{2}{3}} \div (1 + 0.74526\,1\nu^{-\tfrac{2}{3}} + 0.10910\nu^{-\tfrac{4}{3}} - 0.0185\nu^{-2} - 0.003\nu^{-\tfrac{8}{3}} + \cdots),$
$j'_{\nu,1} \sim \nu + 0.80861\,65\nu^{\tfrac{1}{3}} + 0.07249\,0\nu^{-\tfrac{1}{3}} - 0.05097\nu^{-1} + 0.0094\nu^{-\tfrac{5}{3}} + \cdots,$
$y'_{\nu,1} \sim \nu + 1.82109\,80\nu^{\tfrac{1}{3}} + 0.94000\,7\nu^{-\tfrac{1}{3}} - 0.05808\nu^{-1} - 0.0540\nu^{-\tfrac{5}{3}} + \cdots.$
$J_\nu(j'_{\nu,1}) \sim 0.67488\,51\nu^{-\tfrac{1}{3}}(1 - 0.16172\,3\nu^{-\tfrac{2}{3}} + 0.02918\nu^{-\tfrac{4}{3}} - 0.0068\nu^{-2} + \cdots),$
$Y_\nu(y'_{\nu,1}) \sim 0.57319\,40\nu^{-\tfrac{1}{3}}(1 - 0.36422\,0\nu^{-\tfrac{2}{3}} + 0.09077\nu^{-\tfrac{4}{3}} + 0.0237\nu^{-2} + \cdots).$

For numerical coefficients for $m = 2, 3, 4, 5$ see Olver (1951, Tables 3–6).

The expansions (10.21.32)–(10.21.39) become progressively weaker as m increases. The approximations that follow in §10.21(viii) do not suffer from this drawback.

10.21(viii) Uniform Asymptotic Approximations for Large Order

As $\nu \to \infty$ the following four approximations hold uniformly for $m = 1, 2, \ldots$:

10.21.41 $\quad j_{\nu,m} = \nu z(\zeta) + \dfrac{z(\zeta)(h(\zeta))^2 B_0(\zeta)}{2\nu} + O\!\left(\dfrac{1}{\nu^3}\right),$

$\zeta = \nu^{-\tfrac{2}{3}} a_m,$

10.21.42
$$J'_\nu(j_{\nu,m}) = -\frac{2}{\nu^{\frac{2}{3}}} \frac{\operatorname{Ai}'(a_m)}{z(\zeta)h(\zeta)}\left(1 + O\left(\frac{1}{\nu^2}\right)\right), \quad \zeta = \nu^{-\frac{2}{3}} a_m,$$

10.21.43
$$j'_{\nu,m} = \nu z(\zeta) + \frac{z(\zeta)(h(\zeta))^2 C_0(\zeta)}{2\zeta\nu} + O\left(\frac{1}{\nu}\right), \quad \zeta = \nu^{-\frac{2}{3}} a'_m,$$

10.21.44
$$J_\nu(j'_{\nu,m}) = \frac{h(\zeta)\operatorname{Ai}(a'_m)}{\nu^{\frac{1}{3}}}\left(1 + O\left(\frac{1}{\nu^{\frac{4}{3}}}\right)\right), \quad \zeta = \nu^{-\frac{2}{3}} a'_m.$$

Here a_m and a'_m denote respectively the zeros of the Airy function $\operatorname{Ai}(z)$ and its derivative $\operatorname{Ai}'(z)$; see §9.9. Next, $z(\zeta)$ is the inverse of the function $\zeta = \zeta(z)$ defined by (10.20.3). $B_0(\zeta)$ and $C_0(\zeta)$ are defined by (10.20.11) and (10.20.12) with $k = 0$. Lastly,

10.21.45
$$h(\zeta) = \left(4\zeta/(1-z^2)\right)^{\frac{1}{4}}.$$

(Note: If the term $z(\zeta)(h(\zeta))^2 C_0(\zeta)/(2\zeta\nu)$ in (10.21.43) is omitted, then the uniform character of the error term $O(1/\nu)$ is destroyed.)

Corresponding uniform approximations for $y_{\nu,m}$, $Y'_\nu(y_{\nu,m})$, $y'_{\nu,m}$, and $Y_\nu(y'_{\nu,m})$, are obtained from (10.21.41)–(10.21.44) by changing the symbols j, J, Ai, Ai', a_m, and a'_m to y, Y, $-\operatorname{Bi}$, $-\operatorname{Bi}'$, b_m, and b'_m, respectively.

For derivations and further information, including extensions to uniform asymptotic expansions, see Olver (1954, 1960). The latter reference includes numerical tables of the first few coefficients in the uniform asymptotic expansions.

10.21(ix) Complex Zeros

This subsection describes the distribution in $\mathbb{C}$ of the zeros of the principal branches of the Bessel functions of the second and third kinds, and their derivatives, in the case when the order is a positive integer n. For further information, including uniform asymptotic expansions, extensions to other branches of the functions and their derivatives, and extensions to half-integer values of the order, see Olver (1954). (There is an inaccuracy in Figures 11 and 14 in this reference. Each curve that represents an infinite string of nonreal zeros should be located on the opposite side of its straight line asymptote. This inaccuracy was repeated in Abramowitz and Stegun (1964, Figures 9.5 and 9.6). See Kerimov and Skorokhodov (1985a,b) and Figures 10.21.3–10.21.6.)

See also Cruz and Sesma (1982); Cruz et al. (1991), Kerimov and Skorokhodov (1984c, 1987, 1988), Kokologiannaki et al. (1992), and references supplied in §10.75(iii).

Zeros of $Y_n(nz)$ and $Y'_n(nz)$

In Figures 10.21.1, 10.21.3, and 10.21.5 the two continuous curves that join the points ± 1 are the boundaries of $\mathbf{K}$, that is, the eye-shaped domain depicted in Figure 10.20.3. These curves therefore intersect the imaginary axis at the points $z = \pm ic$, where $c = 0.66274\ldots$.

The first set of zeros of the principal value of $Y_n(nz)$ are the points $z = y_{n,m}/n$, $m = 1, 2, \ldots$, on the positive real axis (§10.21(i)). Secondly, there is a conjugate pair of infinite strings of zeros with asymptotes $\Im z = \pm ia/n$, where

10.21.46
$$a = \tfrac{1}{2}\ln 3 = 0.54931\ldots.$$

Lastly, there are two conjugate sets, with n zeros in each set, that are asymptotically close to the boundary of $\mathbf{K}$ as $n \to \infty$. Figures 10.21.1, 10.21.3, and 10.21.5 plot the actual zeros for $n = 1, 5$, and 10, respectively.

The zeros of $Y'_n(nz)$ have a similar pattern to those of $Y_n(nz)$.

Zeros of $H_n^{(1)}(nz)$, $H_n^{(2)}(nz)$, $H_n^{(1)'}(nz)$, $H_n^{(2)'}(nz)$

In Figures 10.21.2, 10.21.4, and 10.21.6 the continuous curve that joins the points ± 1 is the lower boundary of $\mathbf{K}$.

The first set of zeros of the principal value of $H_n^{(1)}(nz)$ is an infinite string with asymptote $\Im z = -id/n$, where

10.21.47
$$d = \tfrac{1}{2}\ln 2 = 0.34657\ldots.$$

The only other set comprises n zeros that are asymptotically close to the lower boundary of $\mathbf{K}$ as $n \to \infty$. Figures 10.21.2, 10.21.4, and 10.21.6 plot the actual zeros for $n = 1, 5$, and 10, respectively.

The zeros of $H_n^{(1)'}(nz)$ have a similar pattern to those of $H_n^{(1)}(nz)$. The zeros of $H_n^{(2)}(nz)$ and $H_n^{(2)'}(nz)$ are the complex conjugates of the zeros of $H_n^{(1)}(nz)$ and $H_n^{(1)'}(nz)$, respectively.

Zeros of $J_0(z) - iJ_1(z)$ and $J_n(z) - iJ_{n+1}(z)$

For information see Synolakis (1988), MacDonald (1989, 1997), and Ikebe et al. (1993).

10.21(x) Cross-Products

Throughout this subsection we assume $\nu \geq 0$, $x > 0$, $\lambda > 1$, and we denote $4\nu^2$ by μ.

The zeros of the functions

10.21.48
$$J_\nu(x) Y_\nu(\lambda x) - Y_\nu(x) J_\nu(\lambda x)$$

and

10.21.49
$$J'_\nu(x) Y'_\nu(\lambda x) - Y'_\nu(x) J'_\nu(\lambda x)$$

are simple and the asymptotic expansion of the mth positive zero as $m \to \infty$ is given by

10.21.50
$$\alpha + \frac{p}{\alpha} + \frac{q - p^2}{\alpha^3} + \frac{r - 4pq + 2p^3}{\alpha^5} + \cdots,$$

10.21 ZEROS

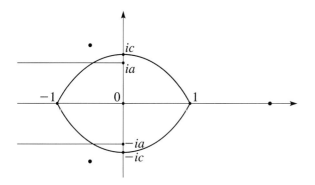

Figure 10.21.1: Zeros ••• of $Y_n(nz)$ in $|\operatorname{ph} z| \leq \pi$. Case $n = 1$, $-1.6 \leq \Re z \leq 2.6$.

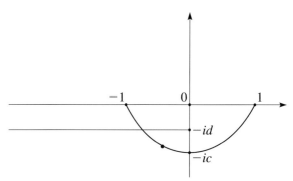

Figure 10.21.2: Zeros ••• of $H_n^{(1)}(nz)$ in $|\operatorname{ph} z| \leq \pi$. Case $n = 1$, $-2.8 \leq \Re z \leq 1.4$.

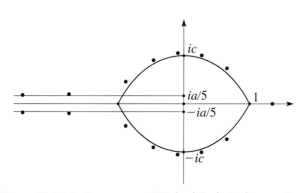

Figure 10.21.3: Zeros ••• of $Y_n(nz)$ in $|\operatorname{ph} z| \leq \pi$. Case $n = 5$, $-2.6 \leq \Re z \leq 1.6$.

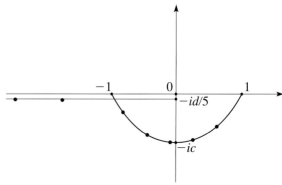

Figure 10.21.4: Zeros ••• of $H_n^{(1)}(nz)$ in $|\operatorname{ph} z| \leq \pi$. Case $n = 5$, $-2.6 \leq \Re z \leq 1.6$.

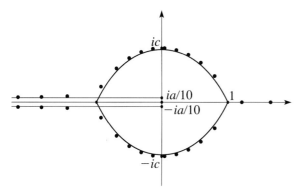

Figure 10.21.5: Zeros ••• of $Y_n(nz)$ in $|\operatorname{ph} z| \leq \pi$. Case $n = 10$, $-2.3 \leq \Re z \leq 1.9$.

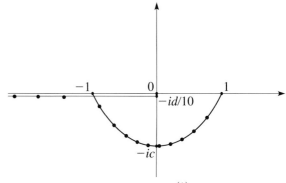

Figure 10.21.6: Zeros ••• of $H_n^{(1)}(nz)$ in $|\operatorname{ph} z| \leq \pi$. Case $n = 10$, $-2.3 \leq \Re z \leq 1.9$.

where, in the case of (10.21.48),

10.21.51

$$\alpha = \frac{m\pi}{\lambda - 1}, \quad p = \frac{\mu - 1}{8\lambda}, \quad q = \frac{(\mu - 1)(\mu - 25)(\lambda^3 - 1)}{6(4\lambda)^3(\lambda - 1)},$$

$$r = \frac{(\mu - 1)(\mu^2 - 114\mu + 1073)(\lambda^5 - 1)}{5(4\lambda)^5(\lambda - 1)},$$

10.21.54
$$\alpha = \frac{(m - \tfrac{1}{2})\pi}{\lambda - 1}, \quad p = \frac{(\mu + 3)\lambda - (\mu - 1)}{8\lambda(\lambda - 1)}, \quad q = \frac{(\mu^2 + 46\mu - 63)\lambda^3 - (\mu - 1)(\mu - 25)}{6(4\lambda)^3(\lambda - 1)},$$
$$r = \frac{(\mu^3 + 185\mu^2 - 2053\mu + 1899)\lambda^5 - (\mu - 1)(\mu^2 - 114\mu + 1073)}{5(4\lambda)^5(\lambda - 1)}.$$

Higher coefficients in the asymptotic expansions in this subsection can be obtained by expressing the cross-products in terms of the modulus and phase functions (§10.18), and then reverting the asymptotic expansion for the difference of the phase functions.

For further information see Cochran (1963, 1964, 1966a,b), Kalähne (1907), Martinek et al. (1966), Muldoon (1979), and Salchev and Popov (1976).

10.21(xi) Riccati–Bessel Functions

The Riccati–Bessel functions are $(\tfrac{1}{2}\pi x)^{\tfrac{1}{2}} J_\nu(x)$ and $(\tfrac{1}{2}\pi x)^{\tfrac{1}{2}} Y_\nu(x)$. Except possibly for $x = 0$ their zeros are the same as those of $J_\nu(x)$ and $Y_\nu(x)$, respectively. For information on the zeros of the derivatives of Riccati–Bessel functions, and also on zeros of their cross-products, see Boyer (1969). This information includes asymptotic approximations analogous to those given in §§10.21(vi), 10.21(vii), and 10.21(x).

and, in the case of (10.21.49),

10.21.52
$$\alpha = \frac{(m - 1)\pi}{\lambda - 1}, \quad p = \frac{\mu + 3}{8\lambda},$$
$$q = \frac{(\mu^2 + 46\mu - 63)(\lambda^3 - 1)}{6(4\lambda)^3(\lambda - 1)},$$
$$r = \frac{(\mu^3 + 185\mu^2 - 2053\mu + 1899)(\lambda^5 - 1)}{5(4\lambda)^5(\lambda - 1)}.$$

The asymptotic expansion of the large positive zeros (not necessarily the mth) of the function

10.21.53 $$J'_\nu(x) Y_\nu(\lambda x) - Y'_\nu(x) J_\nu(\lambda x)$$

is given by (10.21.50), where

10.21(xii) Zeros of $\alpha J_\nu(x) + x J'_\nu(x)$

For properties of the positive zeros of the function $\alpha J_\nu(x) + x J'_\nu(x)$, with α and ν real, see Landau (1999).

10.21(xiii) Rayleigh Function

The *Rayleigh function* $\sigma_n(\nu)$ is defined by

10.21.55 $$\sigma_n(\nu) = \sum_{m=1}^\infty (j_{\nu,m})^{-2n}, \quad n = 1, 2, 3, \ldots.$$

For properties, computation, and generalizations see Kapitsa (1951a), Kerimov (1999), and Gupta and Muldoon (2000). See also Watson (1944, §§15.5, 15.51).

10.21(xiv) ν-Zeros

For information on zeros of Bessel and Hankel functions as functions of the order, see Cochran (1965), Cochran and Hoffspiegel (1970), Hethcote (1970), and Conde and Kalla (1979).

10.22 Integrals

10.22(i) Indefinite Integrals

In this subsection $\mathscr{C}_\nu(z)$ and $\mathscr{D}_\mu(z)$ denote cylinder functions (§10.2(ii)) of orders ν and μ, respectively, not necessarily distinct.

10.22.1 $$\int z^{\nu+1} \mathscr{C}_\nu(z)\, dz = z^{\nu+1} \mathscr{C}_{\nu+1}(z), \quad \int z^{-\nu+1} \mathscr{C}_\nu(z)\, dz = -z^{-\nu+1} \mathscr{C}_{\nu-1}(z).$$

10.22.2 $$\int z^\nu \mathscr{C}_\nu(z)\, dz = \pi^{\tfrac{1}{2}} 2^{\nu-1} \Gamma\!\left(\nu + \tfrac{1}{2}\right) z \left(\mathscr{C}_\nu(z) \mathbf{H}_{\nu-1}(z) - \mathscr{C}_{\nu-1}(z) \mathbf{H}_\nu(z)\right), \qquad \nu \neq -\tfrac{1}{2}.$$

10.22 Integrals

For the Struve function $\mathbf{H}_\nu(z)$ see §11.2(i).

10.22.3
$$\int e^{iz} z^\nu \mathscr{C}_\nu(z)\,dz = \frac{e^{iz} z^{\nu+1}}{2\nu+1}(\mathscr{C}_\nu(z) - i\mathscr{C}_{\nu+1}(z)), \qquad \nu \neq -\tfrac{1}{2},$$

$$\int e^{iz} z^{-\nu} \mathscr{C}_\nu(z)\,dz = \frac{e^{iz} z^{-\nu+1}}{1-2\nu}(\mathscr{C}_\nu(z) + i\mathscr{C}_{\nu-1}(z)), \qquad \nu \neq \tfrac{1}{2}.$$

Products

10.22.4
$$\int z\,\mathscr{C}_\mu(az)\mathscr{D}_\mu(bz)\,dz = \frac{z\,(a\mathscr{C}_{\mu+1}(az)\mathscr{D}_\mu(bz) - b\mathscr{C}_\mu(az)\mathscr{D}_{\mu+1}(bz))}{a^2 - b^2}, \qquad a^2 \neq b^2,$$

10.22.5
$$\int z\,\mathscr{C}_\mu(az)\mathscr{D}_\mu(az)\,dz = \tfrac{1}{4}z^2\left(2\mathscr{C}_\mu(az)\mathscr{D}_\mu(az) - \mathscr{C}_{\mu-1}(az)\mathscr{D}_{\mu+1}(az) - \mathscr{C}_{\mu+1}(az)\mathscr{D}_{\mu-1}(az)\right),$$

10.22.6
$$\int \mathscr{C}_\mu(az)\mathscr{D}_\nu(az)\,\frac{dz}{z} = -\frac{az(\mathscr{C}_{\mu+1}(az)\mathscr{D}_\nu(az) - \mathscr{C}_\mu(az)\mathscr{D}_{\nu+1}(az))}{\mu^2 - \nu^2} + \frac{\mathscr{C}_\mu(az)\mathscr{D}_\nu(az)}{\mu+\nu}, \qquad \mu^2 \neq \nu^2,$$

10.22.7
$$\int z^{\mu+\nu+1}\mathscr{C}_\mu(az)\mathscr{D}_\nu(az)\,dz = \frac{z^{\mu+\nu+2}}{2(\mu+\nu+1)}\left(\mathscr{C}_\mu(az)\mathscr{D}_\nu(az) + \mathscr{C}_{\mu+1}(az)\mathscr{D}_{\nu+1}(az)\right), \qquad \mu+\nu \neq -1,$$

$$\int z^{-\mu-\nu+1}\mathscr{C}_\mu(az)\mathscr{D}_\nu(az)\,dz = \frac{z^{-\mu-\nu+2}}{2(1-\mu-\nu)}\left(\mathscr{C}_\mu(az)\mathscr{D}_\nu(az) + \mathscr{C}_{\mu-1}(az)\mathscr{D}_{\nu-1}(az)\right), \qquad \mu+\nu \neq 1.$$

10.22(ii) Integrals over Finite Intervals

Throughout this subsection $x > 0$.

10.22.8
$$\int_0^x J_\nu(t)\,dt = 2\sum_{k=0}^\infty J_{\nu+2k+1}(x), \qquad \Re\nu > -1.$$

10.22.9
$$\int_0^x J_{2n}(t)\,dt = \int_0^x J_0(t)\,dt - 2\sum_{k=0}^{n-1} J_{2k+1}(x), \qquad \int_0^x J_{2n+1}(t)\,dt = 1 - J_0(x) - 2\sum_{k=1}^n J_{2k}(x), \quad n = 0,1,\ldots.$$

10.22.10
$$\int_0^x t^\mu J_\nu(t)\,dt = x^\mu \frac{\Gamma(\tfrac{1}{2}\nu + \tfrac{1}{2}\mu + \tfrac{1}{2})}{\Gamma(\tfrac{1}{2}\nu - \tfrac{1}{2}\mu + \tfrac{1}{2})}\sum_{k=0}^\infty \frac{(\nu+2k+1)\Gamma(\tfrac{1}{2}\nu - \tfrac{1}{2}\mu + \tfrac{1}{2} + k)}{\Gamma(\tfrac{1}{2}\nu + \tfrac{1}{2}\mu + \tfrac{3}{2} + k)} J_{\nu+2k+1}(x), \quad \Re(\mu+\nu+1) > 0.$$

10.22.11
$$\int_0^x \frac{1-J_0(t)}{t}\,dt = \frac{1}{2}\sum_{k=1}^\infty \frac{\psi(k+1) - \psi(1)}{k!}(\tfrac{1}{2}x)^k J_k(x),$$

10.22.12
$$x\int_0^x \frac{1-J_0(t)}{t}\,dt = 2\sum_{k=0}^\infty (2k+3)(\psi(k+2) - \psi(1))J_{2k+3}(x)$$
$$= x - 2J_1(x) + 2\sum_{k=0}^\infty (2k+5)\left(\psi(k+3) - \psi(1) - 1\right) J_{2k+5}(x),$$

where $\psi(x) = \Gamma'(x)/\Gamma(x)$ (§5.2(i)). See also (10.22.39).

Trigonometric Arguments

10.22.13
$$\int_0^{\frac{1}{2}\pi} J_{2\nu}(2z\cos\theta)\cos(2\mu\theta)\,d\theta = \tfrac{1}{2}\pi J_{\nu+\mu}(z) J_{\nu-\mu}(z), \qquad \Re\nu > -\tfrac{1}{2},$$

10.22.14
$$\int_0^\pi J_{2\nu}(2z\sin\theta)\cos(2\mu\theta)\,d\theta = \pi\cos(\mu\pi) J_{\nu+\mu}(z) J_{\nu-\mu}(z), \qquad \Re\nu > -\tfrac{1}{2},$$

10.22.15
$$\int_0^\pi J_{2\nu}(2z\sin\theta)\sin(2\mu\theta)\,d\theta = \pi\sin(\mu\pi) J_{\nu+\mu}(z) J_{\nu-\mu}(z), \qquad \Re\nu > -1.$$

10.22.16
$$\int_0^{\frac{1}{2}\pi} J_0(2z\sin\theta)\cos(2n\theta)\,d\theta = \tfrac{1}{2}\pi J_n^2(z), \qquad n = 0,1,2,\ldots.$$

10.22.17
$$\int_0^{\frac{1}{2}\pi} Y_{2\nu}(2z\cos\theta)\cos(2\mu\theta)\,d\theta = \tfrac{1}{2}\pi\cot(2\nu\pi)\,J_{\nu+\mu}(z)\,J_{\nu-\mu}(z) - \tfrac{1}{2}\pi\csc(2\nu\pi)\,J_{\mu-\nu}(z)\,J_{-\mu-\nu}(z), \quad -\tfrac{1}{2}<\Re\nu<\tfrac{1}{2},$$

10.22.18
$$\int_0^{\frac{1}{2}\pi} Y_0(2z\sin\theta)\cos(2n\theta)\,d\theta = \tfrac{1}{2}\pi\,J_n(z)\,Y_n(z), \qquad n=0,1,2,\ldots.$$

10.22.19
$$\int_0^{\frac{1}{2}\pi} J_\mu(z\sin\theta)(\sin\theta)^{\mu+1}(\cos\theta)^{2\nu+1}\,d\theta = 2^\nu\,\Gamma(\nu+1)z^{-\nu-1}\,J_{\mu+\nu+1}(z), \quad \Re\mu>-1,\ \Re\nu>-1,$$

10.22.20
$$\int_0^{\frac{1}{2}\pi} J_\mu(z\sin\theta)(\sin\theta)^\mu(\cos\theta)^{2\mu}\,d\theta = \pi^{\frac{1}{2}}2^{\mu-1}z^{-\mu}\,\Gamma\!\left(\mu+\tfrac{1}{2}\right)J_\mu^2\!\left(\tfrac{1}{2}z\right), \qquad \Re\mu>-\tfrac{1}{2},$$

10.22.21
$$\int_0^{\frac{1}{2}\pi} Y_\mu(z\sin\theta)(\sin\theta)^\mu(\cos\theta)^{2\mu}\,d\theta = \pi^{\frac{1}{2}}2^{\mu-1}z^{-\mu}\,\Gamma\!\left(\mu+\tfrac{1}{2}\right)J_\mu\!\left(\tfrac{1}{2}z\right)Y_\mu\!\left(\tfrac{1}{2}z\right), \qquad \Re\mu>-\tfrac{1}{2}.$$

10.22.22
$$\int_0^{\frac{1}{2}\pi} J_\mu(z\sin^2\theta)\,J_\nu(z\cos^2\theta)(\sin\theta)^{2\mu+1}(\cos\theta)^{2\nu+1}\,d\theta = \frac{\Gamma\!\left(\mu+\tfrac{1}{2}\right)\Gamma\!\left(\nu+\tfrac{1}{2}\right)J_{\mu+\nu+\frac{1}{2}}(z)}{(8\pi z)^{\frac{1}{2}}\Gamma(\mu+\nu+1)}, \quad \Re\mu>-\tfrac{1}{2},\Re\nu>-\tfrac{1}{2}.$$

10.22.23
$$\int_0^{\frac{1}{2}\pi} J_\mu(z\sin^2\theta)\,J_\nu(z\cos^2\theta)(\sin\theta)^{2\alpha-1}\sec\theta\,d\theta = \frac{(\mu+\nu+\alpha)\,\Gamma(\mu+\alpha)2^{\alpha-1}}{\nu\,\Gamma(\mu+1)z^\alpha}\,J_{\mu+\nu+\alpha}(z),$$
$$\Re(\mu+\alpha)>0,\ \Re\nu>0.$$

10.22.24
$$\int_0^{\frac{1}{2}\pi} J_\mu(z\sin^2\theta)\,J_\nu(z\cos^2\theta)\cot\theta\,d\theta = \tfrac{1}{2}\mu^{-1}\,J_{\mu+\nu}(z), \qquad \Re\mu>0,\Re\nu>-1.$$

10.22.25
$$\int_0^{\frac{1}{2}\pi} J_\mu(z\sin\theta)\,I_\nu(z\cos\theta)(\tan\theta)^{\mu+1}\,d\theta = \frac{\Gamma\!\left(\tfrac{1}{2}\nu-\tfrac{1}{2}\mu\right)(\tfrac{1}{2}z)^\mu}{2\Gamma\!\left(\tfrac{1}{2}\nu+\tfrac{1}{2}\mu+1\right)}\,J_\nu(z), \qquad \Re\nu>\Re\mu>-1.$$

For I_ν see §10.25(ii).

10.22.26
$$\int_0^{\frac{1}{2}\pi} J_\mu(z\sin\theta)\,J_\nu(\zeta\cos\theta)(\sin\theta)^{\mu+1}(\cos\theta)^{\nu+1}\,d\theta = \frac{z^\mu\zeta^\nu\,J_{\mu+\nu+1}\!\left(\sqrt{\zeta^2+z^2}\right)}{(\zeta^2+z^2)^{\frac{1}{2}(\mu+\nu+1)}}, \quad \Re\mu>-1,\Re\nu>-1.$$

Products

10.22.27
$$\int_0^x t\,J_{\nu-1}^2(t)\,dt = 2\sum_{k=0}^\infty (\nu+2k)\,J_{\nu+2k}^2(x), \qquad \Re\nu>0,$$

10.22.28
$$\int_0^x t\,\left(J_{\nu-1}^2(t)-J_{\nu+1}^2(t)\right)dt = 2\nu\,J_\nu^2(x), \qquad \Re\nu>0,$$

10.22.29
$$\int_0^x t\,J_0^2(t)\,dt = \tfrac{1}{2}x^2\left(J_0^2(x)+J_1^2(x)\right).$$

10.22.30
$$\int_0^x J_n(t)\,J_{n+1}(t)\,dt = \tfrac{1}{2}\left(1-J_0^2(x)\right) - \sum_{k=1}^n J_k^2(x) = \sum_{k=n+1}^\infty J_k^2(x), \qquad n=0,1,2,\ldots.$$

Convolutions

10.22.31
$$\int_0^x J_\mu(t)\,J_\nu(x-t)\,dt = 2\sum_{k=0}^\infty (-1)^k\,J_{\mu+\nu+2k+1}(x), \qquad \Re\mu>-1,\Re\nu>-1.$$

10.22.32
$$\int_0^x J_\nu(t)\,J_{1-\nu}(x-t)\,dt = J_0(x)-\cos x, \qquad -1<\Re\nu<2.$$

10.22.33
$$\int_0^x J_\nu(t)\,J_{-\nu}(x-t)\,dt = \sin x, \qquad |\Re\nu|<1.$$

10.22.34
$$\int_0^x t^{-1}\,J_\mu(t)\,J_\nu(x-t)\,dt = \frac{J_{\mu+\nu}(x)}{\mu}, \qquad \Re\mu>0,\Re\nu>-1.$$

10.22 Integrals

10.22.35
$$\int_0^x \frac{J_\mu(t)\, J_\nu(x-t)}{t(x-t)}\,dt = \frac{(\mu+\nu)\, J_{\mu+\nu}(x)}{\mu\nu x}, \qquad \Re\mu > 0, \Re\nu > 0.$$

Fractional Integral

10.22.36
$$\frac{1}{\Gamma(\alpha)}\int_0^x (x-t)^{\alpha-1} J_\nu(t)\,dt = 2^\alpha \sum_{k=0}^\infty \frac{(\alpha)_k}{k!}\, J_{\nu+\alpha+2k}(x), \qquad \Re\alpha > 0, \Re\nu \geq 0.$$

When $\alpha = m = 1,2,3,\ldots$ the left-hand side of (10.22.36) is the mth repeated integral of $J_\nu(x)$ (§§1.4(v) and 1.15(vi)).

Orthogonality

If $\nu > -1$, then

10.22.37
$$\int_0^1 t\, J_\nu(j_{\nu,\ell} t)\, J_\nu(j_{\nu,m} t)\,dt = \tfrac{1}{2}\delta_{\ell,m}\, (J'_\nu(j_{\nu,\ell}))^2,$$

where $j_{\nu,\ell}$ and $j_{\nu,m}$ are zeros of $J_\nu(x)$ (§10.21(i)), and $\delta_{\ell,m}$ is Kronecker's symbol.

Also, if a, b, ν are real constants with $b \neq 0$ and $\nu > -1$, then

10.22.38
$$\int_0^1 t\, J_\nu(\alpha_\ell t)\, J_\nu(\alpha_m t)\,dt = \delta_{\ell,m}\left(\frac{a^2}{b^2} + \alpha_\ell^2 - \nu^2\right) \frac{(J_\nu(\alpha_\ell))^2}{2\alpha_\ell^2},$$

where α_ℓ and α_m are positive zeros of $a J_\nu(x) + bx J'_\nu(x)$. (Compare (10.22.55)).

10.22(iii) Integrals over the Interval (x, ∞)

When $x > 0$

10.22.39
$$\int_x^\infty \frac{J_0(t)}{t}\,dt + \gamma + \ln(\tfrac{1}{2}x) = \int_0^x \frac{1 - J_0(t)}{t}\,dt = \sum_{k=1}^\infty (-1)^{k-1} \frac{(\tfrac{1}{2}x)^{2k}}{2k(k!)^2},$$

10.22.40
$$\int_x^\infty \frac{Y_0(t)}{t}\,dt = -\frac{1}{\pi}\left(\ln(\tfrac{1}{2}x) + \gamma\right)^2 + \frac{\pi}{6} + \frac{2}{\pi}\sum_{k=1}^\infty (-1)^k \left(\psi(k+1) + \frac{1}{2k} - \ln(\tfrac{1}{2}x)\right) \frac{(\tfrac{1}{2}x)^{2k}}{2k(k!)^2},$$

where γ is Euler's constant (§5.2(ii)).

10.22(iv) Integrals over the Interval $(0, \infty)$

10.22.41
$$\int_0^\infty J_\nu(t)\,dt = 1, \qquad \Re\nu > -1,$$

10.22.42
$$\int_0^\infty Y_\nu(t)\,dt = -\tan(\tfrac{1}{2}\nu\pi), \qquad |\Re\nu| < 1.$$

10.22.43
$$\int_0^\infty t^\mu J_\nu(t)\,dt = 2^\mu \frac{\Gamma(\tfrac{1}{2}\nu + \tfrac{1}{2}\mu + \tfrac{1}{2})}{\Gamma(\tfrac{1}{2}\nu - \tfrac{1}{2}\mu + \tfrac{1}{2})}, \qquad \Re(\mu+\nu) > -1, \Re\mu < \tfrac{1}{2},$$

10.22.44
$$\int_0^\infty t^\mu Y_\nu(t)\,dt = \frac{2^\mu}{\pi}\, \Gamma(\tfrac{1}{2}\mu + \tfrac{1}{2}\nu + \tfrac{1}{2})\, \Gamma(\tfrac{1}{2}\mu - \tfrac{1}{2}\nu + \tfrac{1}{2}) \sin(\tfrac{1}{2}\mu - \tfrac{1}{2}\nu)\pi, \qquad \Re(\mu \pm \nu) > -1, \Re\mu < \tfrac{1}{2}.$$

10.22.45
$$\int_0^\infty \frac{1 - J_0(t)}{t^\mu}\,dt = -\frac{\pi \sec(\tfrac{1}{2}\mu\pi)}{2^\mu\, \Gamma^2(\tfrac{1}{2}\mu + \tfrac{1}{2})}, \qquad 1 < \Re\mu < 3.$$

10.22.46
$$\int_0^\infty \frac{t^{\nu+1} J_\nu(at)}{(t^2 + b^2)^{\mu+1}}\,dt = \frac{a^\mu b^{\nu-\mu}}{2^\mu\, \Gamma(\mu+1)}\, K_{\nu-\mu}(ab), \quad a > 0,\, \Re b > 0,\, -1 < \Re\nu < 2\Re\mu + \tfrac{3}{2}.$$

10.22.47
$$\int_0^\infty \frac{t^\nu Y_\nu(at)}{t^2 + b^2}\,dt = -b^{\nu-1} K_\nu(ab), \qquad a > 0, \Re b > 0, -\tfrac{1}{2} < \Re\nu < \tfrac{5}{2}.$$

For K_ν see §10.25(ii).

10.22.48
$$\int_0^\infty J_\mu(x\cosh\phi)(\cosh\phi)^{1-\mu}(\sinh\phi)^{2\nu+1}\,d\phi = 2^\nu\, \Gamma(\nu+1)\, x^{-\nu-1}\, J_{\mu-\nu-1}(x), \quad x > 0, \Re\nu > -1, \Re\mu > 2\Re\nu + \tfrac{1}{2}.$$

10.22.49 $\quad\int_0^\infty t^{\mu-1} e^{-at} J_\nu(bt)\, dt = \frac{(\tfrac12 b)^\nu}{a^{\mu+\nu}}\, \Gamma(\mu+\nu)\, \mathbf{F}\!\left(\frac{\mu+\nu}{2}, \frac{\mu+\nu+1}{2}; \nu+1; -\frac{b^2}{a^2}\right), \quad \Re(\mu+\nu)>0, \Re(a\pm ib)>0,$

10.22.50
$$\int_0^\infty t^{\mu-1} e^{-at} Y_\nu(bt)\, dt = \cot(\nu\pi) \frac{(\tfrac12 b)^\nu \Gamma(\mu+\nu)}{(a^2+b^2)^{\frac12(\mu+\nu)}} \, \mathbf{F}\!\left(\frac{\mu+\nu}{2}, \frac{1-\mu+\nu}{2}; \nu+1; \frac{b^2}{a^2+b^2}\right)$$
$$- \csc(\nu\pi) \frac{(\tfrac12 b)^{-\nu}\Gamma(\mu-\nu)}{(a^2+b^2)^{\frac12(\mu-\nu)}}\, \mathbf{F}\!\left(\frac{\mu-\nu}{2}, \frac{1-\mu-\nu}{2}; 1-\nu; \frac{b^2}{a^2+b^2}\right),$$
$$\Re\mu > |\Re\nu|,\ \Re(a\pm ib)>0.$$

For the hypergeometric function **F** see §15.2(i).

10.22.51 $\quad\int_0^\infty J_\nu(bt)\exp(-p^2 t^2) t^{\nu+1}\, dt = \dfrac{b^\nu}{(2p^2)^{\nu+1}}\exp\!\left(-\dfrac{b^2}{4p^2}\right), \qquad \Re\nu>-1,\ \Re(p^2)>0,$

10.22.52 $\quad\int_0^\infty J_\nu(bt)\exp(-p^2 t^2)\, dt = \dfrac{\sqrt\pi}{2p}\exp\!\left(-\dfrac{b^2}{8p^2}\right) I_{\nu/2}\!\left(\dfrac{b^2}{8p^2}\right), \qquad \Re\nu>-1,\ \Re(p^2)>0,$

10.22.53
$$\int_0^\infty Y_{2\nu}(bt)\exp(-p^2 t^2)\, dt = -\frac{\sqrt\pi}{2p}\exp\!\left(-\frac{b^2}{8p^2}\right)\!\left(I_\nu\!\left(\frac{b^2}{8p^2}\right)\tan(\nu\pi) + \frac{1}{\pi} K_\nu\!\left(\frac{b^2}{8p^2}\right)\sec(\nu\pi)\right), \quad |\Re\nu|<\tfrac12,\ \Re(p^2)>0.$$

For I and K see §10.25(ii).

10.22.54 $\quad\int_0^\infty J_\nu(bt)\exp(-p^2 t^2) t^{\mu-1}\, dt = \dfrac{(\tfrac12 b/p)^\nu \Gamma(\tfrac12\nu+\tfrac12\mu)}{2p^\mu}\exp\!\left(-\dfrac{b^2}{4p^2}\right)\mathbf{M}\!\left(\tfrac12\nu-\tfrac12\mu+1, \nu+1, \dfrac{b^2}{4p^2}\right),$
$$\Re(\mu+\nu)>0,\ \Re(p^2)>0.$$

For the confluent hypergeometric function **M** see §13.2(i).

Orthogonality

10.22.55 $\quad\int_0^\infty t^{-1} J_{\nu+2\ell+1}(t) J_{\nu+2m+1}(t)\, dt = \dfrac{\delta_{\ell,m}}{2(2\ell+\nu+1)}, \qquad \nu+\ell+m>-1.$

Weber–Schafheitlin Discontinuous Integrals, including Special Cases

10.22.56 $\quad\int_0^\infty \dfrac{J_\mu(at) J_\nu(bt)}{t^\lambda}\, dt = \dfrac{a^\mu \Gamma\!\left(\tfrac12\nu+\tfrac12\mu-\tfrac12\lambda+\tfrac12\right)}{2^\lambda b^{\mu-\lambda+1}\Gamma\!\left(\tfrac12\nu-\tfrac12\mu+\tfrac12\lambda+\tfrac12\right)}\, \mathbf{F}\!\left(\tfrac12(\mu+\nu-\lambda+1), \tfrac12(\mu-\nu-\lambda+1); \mu+1; \dfrac{a^2}{b^2}\right),$
$$0<a<b,\ \Re(\mu+\nu+1)>\Re\lambda>-1.$$

If $0<b<a$, then interchange a and b, and also μ and ν. If $b=a$, then

10.22.57 $\quad\int_0^\infty \dfrac{J_\mu(at) J_\nu(at)}{t^\lambda}\, dt = \dfrac{(\tfrac12 a)^{\lambda-1}\Gamma\!\left(\tfrac12\mu+\tfrac12\nu-\tfrac12\lambda+\tfrac12\right)\Gamma(\lambda)}{2\,\Gamma\!\left(\tfrac12\lambda+\tfrac12\nu-\tfrac12\mu+\tfrac12\right)\Gamma\!\left(\tfrac12\lambda+\tfrac12\mu-\tfrac12\nu+\tfrac12\right)\Gamma\!\left(\tfrac12\lambda+\tfrac12\mu+\tfrac12\nu+\tfrac12\right)},$
$$\Re(\mu+\nu+1)>\Re\lambda>0.$$

10.22.58 $\quad\int_0^\infty \dfrac{J_\nu(at) J_\nu(bt)}{t^\lambda}\, dt = \dfrac{(ab)^\nu \Gamma\!\left(\nu-\tfrac12\lambda+\tfrac12\right)}{2^\lambda (a^2+b^2)^{\nu-\frac12\lambda+\frac12}\Gamma\!\left(\tfrac12\lambda+\tfrac12\right)}\, \mathbf{F}\!\left(\dfrac{2\nu+1-\lambda}{4}, \dfrac{2\nu+3-\lambda}{4}; \nu+1; \dfrac{4a^2 b^2}{(a^2+b^2)^2}\right),$
$$a\neq b,\ \Re(2\nu+1)>\Re\lambda>-1.$$

When $\Re\mu>-1$

10.22.59 $\quad\int_0^\infty e^{ibt} J_\mu(at)\, dt = \begin{cases} \dfrac{\exp(i\mu\arcsin(b/a))}{(a^2-b^2)^{\frac12}}, & 0\leq b<a, \\[6pt] \dfrac{ia^\mu \exp(\tfrac12\mu\pi i)}{(b^2-a^2)^{\frac12}\!\left(b+(b^2-a^2)^{\frac12}\right)^\mu}, & 0<a<b. \end{cases}$

10.22.60 $\quad\int_0^\infty e^{ibt} Y_0(at)\, dt = \begin{cases} (2i/\pi)(a^2-b^2)^{-\frac12}\arcsin(b/a), & 0\leq b<a, \\[6pt] (b^2-a^2)^{-\frac12}\!\left(-1+\dfrac{2i}{\pi}\ln\!\left(\dfrac{a}{b+(b^2-a^2)^{\frac12}}\right)\right), & 0<a<b. \end{cases}$

10.22 INTEGRALS

When $\Re\mu > 0$,

10.22.61 $$\int_0^\infty t^{-1} e^{ibt} J_\mu(at)\,dt = \begin{cases} (1/\mu)\exp(i\mu \arcsin(b/a)), & 0 \le b \le a, \\ \dfrac{a^\mu \exp\left(\frac{1}{2}\mu\pi i\right)}{\mu\left(b + (b^2-a^2)^{\frac{1}{2}}\right)^\mu}, & 0 < a \le b. \end{cases}$$

When $\Re\nu > \Re\mu > -1$,

10.22.62 $$\int_0^\infty t^{\mu-\nu+1} J_\mu(at) J_\nu(bt)\,dt = \begin{cases} 0, & 0 < b < a, \\ \dfrac{2^{\mu-\nu+1} a^\mu (b^2-a^2)^{\nu-\mu-1}}{b^\nu \Gamma(\nu-\mu)}, & 0 < a \le b. \end{cases}$$

When $\Re\mu > 0$,

10.22.63 $$\int_0^\infty J_\mu(at) J_{\mu-1}(bt)\,dt = \begin{cases} b^{\mu-1} a^{-\mu}, & 0 < b < a, \\ (2b)^{-1}, & b = a(>0), \\ 0, & 0 < a < b. \end{cases}$$

When $n = 0, 1, 2, \ldots$ and $\Re\mu > -n-1$,

10.22.64 $$\int_0^\infty J_{\mu+2n+1}(at) J_\mu(bt)\,dt = \begin{cases} \dfrac{b^\mu \Gamma(\mu+n+1)}{a^{\mu+1} n!}\,\mathbf{F}\!\left(-n, \mu+n+1; \mu+1; \dfrac{b^2}{a^2}\right), & 0 < b < a, \\ (-1)^n/(2a), & b = a(>0), \\ 0, & 0 < a < b. \end{cases}$$

10.22.65 $$\int_0^\infty J_0(at)\left(J_0(bt) - J_0(ct)\right) \dfrac{dt}{t} = \begin{cases} 0, & 0 \le b < a,\, 0 < c \le a, \\ \ln(c/a), & 0 \le b < a \le c. \end{cases}$$

Other Double Products

In (10.22.66)–(10.22.70) a, b, c are positive constants.

10.22.66 $$\int_0^\infty e^{-at} J_\nu(bt) J_\nu(ct)\,dt = \dfrac{1}{\pi(bc)^{\frac{1}{2}}}\, Q_{\nu-\frac{1}{2}}\!\left(\dfrac{a^2+b^2+c^2}{2bc}\right), \qquad \Re\nu > -\tfrac{1}{2}.$$

10.22.67 $$\int_0^\infty t \exp(-p^2 t^2)\, J_\nu(at) J_\nu(bt)\,dt = \dfrac{1}{2p^2} \exp\!\left(-\dfrac{a^2+b^2}{4p^2}\right) I_\nu\!\left(\dfrac{ab}{2p^2}\right), \qquad \Re\nu > -1, \Re(p^2) > 0.$$

10.22.68 $$\int_0^\infty t \exp(-p^2 t^2)\, J_0(at) Y_0(at)\,dt = -\dfrac{1}{2\pi p^2} \exp\!\left(-\dfrac{a^2}{2p^2}\right) K_0\!\left(\dfrac{a^2}{2p^2}\right), \qquad \Re(p^2) > 0.$$

For the associated Legendre function Q see §14.3(ii) with $\mu = 0$. For I and K see §10.25(ii).

10.22.69 $$\int_0^\infty J_\nu(at) J_\nu(bt) \dfrac{t\,dt}{t^2 - z^2} = \begin{cases} \tfrac{1}{2}\pi i\, J_\nu(bz) H_\nu^{(1)}(az), & a > b \\ \tfrac{1}{2}\pi i\, J_\nu(az) H_\nu^{(1)}(bz), & b > a \end{cases}, \qquad \Re\nu > -1, \Im z > 0.$$

10.22.70 $$\int_0^\infty Y_\nu(at) J_{\nu+1}(bt) \dfrac{t\,dt}{t^2 - z^2} = \tfrac{1}{2}\pi\, J_{\nu+1}(bz) H_\nu^{(1)}(az), \qquad a \ge b > 0, \Re\nu > -\tfrac{3}{2}, \Im z > 0.$$

Equation (10.22.70) also remains valid if the order $\nu + 1$ of the J functions on both sides is replaced by $\nu + 2n - 3$, $n = 1, 2, \ldots$, and the constraint $\Re\nu > -\tfrac{3}{2}$ is replaced by $\Re\nu > -n + \tfrac{1}{2}$.

See also §1.17(ii) for an integral representation of the Dirac delta in terms of a product of Bessel functions.

Triple Products

In (10.22.71) and (10.22.72) a, b, c are positive constants.

10.22.71 $$\int_0^\infty J_\mu(at) J_\nu(bt) J_\nu(ct) t^{1-\mu}\,dt = \dfrac{(bc)^{\mu-1}(\sin\phi)^{\mu-\frac{1}{2}}}{(2\pi)^{\frac{1}{2}} a^\mu}\, \mathsf{P}_{\nu-\frac{1}{2}}^{\frac{1}{2}-\mu}(\cos\phi),$$
$$\Re\mu > -\tfrac{1}{2},\, \Re\nu > -1,\, |b-c| < a < b+c,\, \cos\phi = (b^2+c^2-a^2)/(2bc).$$

10.22.72 $$\int_0^\infty J_\mu(at) J_\nu(bt) J_\nu(ct) t^{1-\mu}\,dt = \dfrac{(bc)^{\mu-1} \cos(\nu\pi)(\sinh\chi)^{\mu-\frac{1}{2}}}{(\tfrac{1}{2}\pi^3)^{\frac{1}{2}} a^\mu}\, Q_{\nu-\frac{1}{2}}^{\frac{1}{2}-\mu}(\cosh\chi),$$
$$\Re\mu > -\tfrac{1}{2},\, \Re\nu > -1,\, a > b+c,\, \cosh\chi = (a^2-b^2-c^2)/(2bc).$$

For the Ferrers function P and the associated Legendre function Q, see §§14.3(i) and 14.3(ii), respectively.

In (10.22.74) and (10.22.75), a, b, c are positive constants and

10.22.73 $$A = s(s-a)(s-b)(s-c), \quad s = \tfrac{1}{2}(a+b+c).$$

(Thus if a, b, c are the sides of a triangle, then $A^{\frac{1}{2}}$ is the area of the triangle.)

If $\Re\nu > -\tfrac{1}{2}$, then

10.22.74 $$\int_0^\infty J_\nu(at)\, J_\nu(bt)\, J_\nu(ct) t^{1-\nu}\, dt = \begin{cases} \dfrac{2^{\nu-1} A^{\nu-\frac{1}{2}}}{\pi^{\frac{1}{2}} (abc)^\nu\, \Gamma(\nu+\tfrac{1}{2})}, & A > 0, \\ 0, & A \le 0. \end{cases}$$

If $|\nu| < \tfrac{1}{2}$, then

10.22.75 $$\int_0^\infty Y_\nu(at)\, J_\nu(bt)\, J_\nu(ct) t^{1+\nu}\, dt = \begin{cases} -\dfrac{(abc)^\nu (-A)^{-\nu-\frac{1}{2}}}{\pi^{\frac{1}{2}} 2^{\nu+1}\, \Gamma(\tfrac{1}{2}-\nu)}, & 0 < a < |b-c|, \\ 0, & |b-c| < a < b+c, \\ \dfrac{(abc)^\nu (-A)^{-\nu-\frac{1}{2}}}{\pi^{\frac{1}{2}} 2^{\nu+1}\, \Gamma(\tfrac{1}{2}-\nu)}, & a > b+c. \end{cases}$$

Additional infinite integrals over the product of three Bessel functions (including modified Bessel functions) are given in Gervois and Navelet (1984, 1985a,b, 1986a,b).

10.22(v) Hankel Transform

The *Hankel transform* (or *Bessel transform*) of a function $f(x)$ is defined as

10.22.76 $$g(y) = \int_0^\infty f(x)\, J_\nu(xy)(xy)^{\frac{1}{2}}\, dx.$$

Hankel's inversion theorem is given by

10.22.77 $$f(y) = \int_0^\infty g(x)\, J_\nu(xy)(xy)^{\frac{1}{2}}\, dx.$$

Sufficient conditions for the validity of (10.22.77) are that $\int_0^\infty |f(x)|\, dx < \infty$ when $\nu \ge -\tfrac{1}{2}$, or that $\int_0^\infty |f(x)|\, dx < \infty$ and $\int_0^1 x^{\nu+\frac{1}{2}} |f(x)|\, dx < \infty$ when $-1 < \nu < -\tfrac{1}{2}$; see Titchmarsh (1986a, Theorem 135, Chapter 8) and Akhiezer (1988, p. 62).

For asymptotic expansions of Hankel transforms see Wong (1976, 1977) and Frenzen and Wong (1985).

For collections of Hankel transforms see Erdélyi et al. (1954b, Chapter 8) and Oberhettinger (1972).

10.22(vi) Compendia

For collections of integrals of the functions $J_\nu(z)$, $Y_\nu(z)$, $H_\nu^{(1)}(z)$, and $H_\nu^{(2)}(z)$, including integrals with respect to the order, see Andrews et al. (1999, pp. 216–225), Apelblat (1983, §12), Erdélyi et al. (1953b, §§7.7.1–7.7.7 and 7.14–7.14.2), Erdélyi et al. (1954a,b), Gradshteyn and Ryzhik (2000, §§5.5 and 6.5–6.7), Gröbner and Hofreiter (1950, pp. 196–204), Luke (1962), Magnus et al. (1966, §3.8), Marichev (1983, pp. 191–216), Oberhettinger (1974, §§1.10 and 2.7), Oberhettinger (1990, §§1.13–1.16 and 2.13–2.16), Oberhettinger and Badii (1973, §§1.14 and 2.12), Okui (1974, 1975), Prudnikov et al. (1986b, §§1.8–1.10, 2.12–2.14, 3.2.4–3.2.7, 3.3.2, and 3.4.1), Prudnikov et al. (1992a, §§3.12–3.14), Prudnikov et al. (1992b, §§3.12–3.14), Watson (1944, Chapters 5, 12, 13, and 14), and Wheelon (1968).

10.23 Sums

10.23(i) Multiplication Theorem

10.23.1
$$\mathscr{C}_\nu(\lambda z) = \lambda^{\pm\nu} \sum_{k=0}^\infty \frac{(\mp 1)^k (\lambda^2-1)^k (\tfrac{1}{2} z)^k}{k!} \mathscr{C}_{\nu\pm k}(z),$$
$$|\lambda^2 - 1| < 1.$$

If $\mathscr{C} = J$ and the upper signs are taken, then the restriction on λ is unnecessary.

10.23(ii) Addition Theorems

Neumann's Addition Theorem

10.23.2 $$\mathscr{C}_\nu(u \pm v) = \sum_{k=-\infty}^\infty \mathscr{C}_{\nu \mp k}(u)\, J_k(v), \quad |v| < |u|.$$

The restriction $|v| < |u|$ is unnecessary when $\mathscr{C} = J$ and ν is an integer. Special cases are:

10.23.3 $$J_0^2(z) + 2\sum_{k=1}^\infty J_k^2(z) = 1,$$

10.23.4 $$\sum_{k=0}^{2n} (-1)^k J_k(z) J_{2n-k}(z) + 2\sum_{k=1}^\infty J_k(z) J_{2n+k}(z) = 0, \quad n \ge 1,$$

10.23.5 $$\sum_{k=0}^n J_k(z) J_{n-k}(z) + 2\sum_{k=1}^\infty (-1)^k J_k(z) J_{n+k}(z) = J_n(2z).$$

10.23 Sums

Graf's and Gegenbauer's Addition Theorems

Define

10.23.6 $w = \sqrt{u^2 + v^2 - 2uv\cos\alpha}$,
$u - v\cos\alpha = w\cos\chi$, $v\sin\alpha = w\sin\chi$,

the branches being continuous and chosen so that $w \to u$ and $\chi \to 0$ as $v \to 0$. If u, v are real and positive and $0 < \alpha < \pi$, then w and χ are real and nonnegative, and the geometrical relationship is shown in Figure 10.23.1.

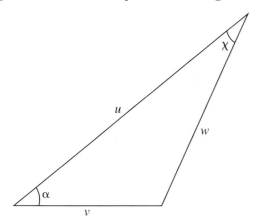

Figure 10.23.1: Graf's and Gegenbauer's addition theorems.

10.23.7 $\mathscr{C}_\nu(w) {\cos \atop \sin} (\nu\chi) = \sum_{k=-\infty}^{\infty} \mathscr{C}_{\nu+k}(u) J_k(v) {\cos \atop \sin} (k\alpha)$,

$$|ve^{\pm i\alpha}| < |u|.$$

10.23.8
$$\frac{\mathscr{C}_\nu(w)}{w^\nu} = 2^\nu \Gamma(\nu)$$
$$\times \sum_{k=0}^{\infty} (\nu+k) \frac{\mathscr{C}_{\nu+k}(u)}{u^\nu} \frac{J_{\nu+k}(v)}{v^\nu} C_k^{(\nu)}(\cos\alpha),$$
$$\nu \neq 0, -1, \ldots, |ve^{\pm i\alpha}| < |u|,$$

where $C_k^{(\nu)}(\cos\alpha)$ is Gegenbauer's polynomial (§18.3). The restriction $|ve^{\pm i\alpha}| < |u|$ is unnecessary in (10.23.7) when $\mathscr{C} = J$ and ν is an integer, and in (10.23.8) when $\mathscr{C} = J$.

The degenerate form of (10.23.8) when $u = \infty$ is given by

10.23.9
$$e^{iv\cos\alpha} = \frac{\Gamma(\nu)}{(\tfrac{1}{2}v)^\nu} \sum_{k=0}^{\infty} (\nu+k) i^k J_{\nu+k}(v) C_k^{(\nu)}(\cos\alpha),$$
$$\nu \neq 0, -1, \ldots.$$

Partial Fractions

For expansions of products of Bessel functions of the first kind in partial fractions see Rogers (2005).

10.23(iii) Series Expansions of Arbitrary Functions

Neumann's Expansion

10.23.10 $f(z) = a_0 J_0(z) + 2\sum_{k=1}^{\infty} a_k J_k(z)$, $|z| < c$,

where c is the distance of the nearest singularity of the analytic function $f(z)$ from $z = 0$,

10.23.11 $a_k = \frac{1}{2\pi i} \int_{|z|=c'} f(t) O_k(t) \, dt$, $0 < c' < c$,

and $O_k(t)$ is *Neumann's polynomial*, defined by the generating function:

10.23.12
$$\frac{1}{t-z} = J_0(z) O_0(t) + 2\sum_{k=1}^{\infty} J_k(z) O_k(t),\quad |z| < |t|.$$

$O_n(t)$ is a polynomial of degree $n+1$ in $1/t$: $O_0(t) = 1/t$ and

10.23.13
$$O_n(t) = \frac{1}{4} \sum_{k=0}^{\lfloor n/2 \rfloor} \frac{(n-k-1)! n}{k!} \left(\frac{2}{t}\right)^{n-2k+1},\quad n=1,2,\ldots.$$

For the more general form of expansion

10.23.14 $z^\nu f(z) = a_0 J_\nu(z) + 2\sum_{k=1}^{\infty} a_k J_{\nu+k}(z)$

see Watson (1944, §16.13), and for further generalizations see Watson (1944, Chapter 16) and (Erdélyi et al., 1953b, §7.10.1).

Examples

10.23.15 $(\tfrac{1}{2}z)^\nu = \sum_{k=0}^{\infty} \frac{(\nu+2k)\Gamma(\nu+k)}{k!} J_{\nu+2k}(z)$,
$$\nu \neq 0, -1, -2, \ldots,$$

10.23.16
$$Y_0(z) = \frac{2}{\pi}\left(\ln(\tfrac{1}{2}z) + \gamma\right) J_0(z) - \frac{4}{\pi} \sum_{k=1}^{\infty} (-1)^k \frac{J_{2k}(z)}{k},$$

10.23.17
$$Y_n(z) = -\frac{n!(\tfrac{1}{2}z)^{-n}}{\pi} \sum_{k=0}^{n-1} \frac{(\tfrac{1}{2}z)^k J_k(z)}{k!(n-k)}$$
$$+ \frac{2}{\pi}\left(\ln(\tfrac{1}{2}z) - \psi(n+1)\right) J_n(z)$$
$$- \frac{2}{\pi} \sum_{k=1}^{\infty} (-1)^k \frac{(n+2k) J_{n+2k}(z)}{k(n+k)},$$

where γ is Euler's constant and $\psi(n+1) = \Gamma'(n+1)/\Gamma(n+1)$ (§5.2).

Other examples are provided by (10.12.1)–(10.12.6), (10.23.2), and (10.23.7).

Fourier–Bessel Expansion

Assume $f(t)$ satisfies

10.23.18
$$\int_0^1 t^{\frac{1}{2}} |f(t)|\, dt < \infty,$$

and define

10.23.19
$$a_m = \frac{2}{(J_{\nu+1}(j_{\nu,m}))^2} \int_0^1 t f(t) J_\nu(j_{\nu,m} t)\, dt, \quad \nu \geq -\tfrac{1}{2},$$

where $j_{\nu,m}$ is as in §10.21(i). If $0 < x < 1$, then

10.23.20 $\quad \tfrac{1}{2} f(x-) + \tfrac{1}{2} f(x+) = \sum_{m=1}^{\infty} a_m J_\nu(j_{\nu,m} x),$

provided that $f(t)$ is of bounded variation (§1.4(v)) on an interval $[a,b]$ with $0 < a < x < b < 1$. This result is proved in Watson (1944, Chapter 18) and further information is provided in this reference, including the behavior of the series near $x = 0$ and $x = 1$.

As an example,

10.23.21 $\quad x^\nu = \sum_{m=1}^{\infty} \frac{2 J_\nu(j_{\nu,m} x)}{j_{\nu,m} J_{\nu+1}(j_{\nu,m})}, \quad \nu > 0,\, 0 \leq x < 1.$

(Note that when $x = 1$ the left-hand side is 1 and the right-hand side is 0.)

Other Series Expansions

For other types of expansions of arbitrary functions in series of Bessel functions, see Watson (1944, Chapters 17–19) and Erdélyi et al. (1953b, §§ 7.10.2–7.10.4). See also Schäfke (1960, 1961b).

10.23(iv) Compendia

For collections of sums of series involving Bessel or Hankel functions see Erdélyi et al. (1953b, §7.15), Gradshteyn and Ryzhik (2000, §§8.51–8.53), Hansen (1975), Luke (1969b, §9.4), Prudnikov et al. (1986b, pp. 651–691 and 697–700), and Wheelon (1968, pp. 48–51).

10.24 Functions of Imaginary Order

With $z = x$ and ν replaced by $i\nu$, Bessel's equation (10.2.1) becomes

10.24.1 $\quad x^2 \dfrac{d^2 w}{dx^2} + x \dfrac{dw}{dx} + (x^2 + \nu^2) w = 0.$

For $\nu \in \mathbb{R}$ and $x \in (0, \infty)$ define

10.24.2
$$\widetilde{J}_\nu(x) = \operatorname{sech}\left(\tfrac{1}{2}\pi\nu\right) \Re(J_{i\nu}(x)),$$
$$\widetilde{Y}_\nu(x) = \operatorname{sech}\left(\tfrac{1}{2}\pi\nu\right) \Re(Y_{i\nu}(x)),$$

and

10.24.3 $\quad \Gamma(1 + i\nu) = \left(\dfrac{\pi\nu}{\sinh(\pi\nu)}\right)^{\frac{1}{2}} e^{i\gamma_\nu},$

where γ_ν is real and continuous with $\gamma_0 = 0$; compare (5.4.3). Then

10.24.4 $\quad \widetilde{J}_{-\nu}(x) = \widetilde{J}_\nu(x), \quad \widetilde{Y}_{-\nu}(x) = \widetilde{Y}_\nu(x),$

and $\widetilde{J}_\nu(x)$, $\widetilde{Y}_\nu(x)$ are linearly independent solutions of (10.24.1):

10.24.5 $\quad \mathscr{W}\{\widetilde{J}_\nu(x), \widetilde{Y}_\nu(x)\} = 2/(\pi x).$

As $x \to +\infty$, with ν fixed,

10.24.6
$$\widetilde{J}_\nu(x) = \sqrt{2/(\pi x)} \cos\left(x - \tfrac{1}{4}\pi\right) + O\left(x^{-\frac{3}{2}}\right),$$
$$\widetilde{Y}_\nu(x) = \sqrt{2/(\pi x)} \sin\left(x - \tfrac{1}{4}\pi\right) + O\left(x^{-\frac{3}{2}}\right).$$

As $x \to 0+$, with ν fixed,

10.24.7
$$\widetilde{J}_\nu(x) = \left(\dfrac{2 \tanh(\tfrac{1}{2}\pi\nu)}{\pi\nu}\right)^{\frac{1}{2}} \cos\left(\nu \ln(\tfrac{1}{2} x) - \gamma_\nu\right) + O(x^2),$$

10.24.8
$$\widetilde{Y}_\nu(x) = \left(\dfrac{2 \coth(\tfrac{1}{2}\pi\nu)}{\pi\nu}\right)^{\frac{1}{2}} \sin\left(\nu \ln(\tfrac{1}{2} x) - \gamma_\nu\right) + O(x^2), \quad \nu > 0,$$

and

10.24.9 $\quad \widetilde{Y}_0(x) = Y_0(x) = \dfrac{2}{\pi}\left(\ln(\tfrac{1}{2} x) + \gamma\right) + O(x^2 \ln x),$

where γ denotes Euler's constant §5.2(ii).

In consequence of (10.24.6), when x is large $\widetilde{J}_\nu(x)$ and $\widetilde{Y}_\nu(x)$ comprise a numerically satisfactory pair of solutions of (10.24.1); compare §2.7(iv). Also, in consequence of (10.24.7)–(10.24.9), when x is small either $\widetilde{J}_\nu(x)$ and $\tanh(\tfrac{1}{2}\pi\nu)\widetilde{Y}_\nu(x)$ or $\widetilde{J}_\nu(x)$ and $\widetilde{Y}_\nu(x)$ comprise a numerically satisfactory pair depending whether $\nu \neq 0$ or $\nu = 0$.

For graphs of $\widetilde{J}_\nu(x)$ and $\widetilde{Y}_\nu(x)$ see §10.3(iii).

For mathematical properties and applications of $\widetilde{J}_\nu(x)$ and $\widetilde{Y}_\nu(x)$, including zeros and uniform asymptotic expansions for large ν, see Dunster (1990a). In this reference $\widetilde{J}_\nu(x)$ and $\widetilde{Y}_\nu(x)$ are denoted respectively by $F_{i\nu}(x)$ and $G_{i\nu}(x)$.

Modified Bessel Functions

10.25 Definitions

10.25(i) Modified Bessel's Equation

10.25.1 $\quad z^2 \dfrac{d^2 w}{dz^2} + z \dfrac{dw}{dz} - (z^2 + \nu^2) w = 0.$

This equation is obtained from Bessel's equation (10.2.1) on replacing z by $\pm i z$, and it has the same kinds of singularities. Its solutions are called *modified Bessel functions* or *Bessel functions of imaginary argument*.

10.25(ii) Standard Solutions

10.25.2 $$I_\nu(z) = (\tfrac{1}{2}z)^\nu \sum_{k=0}^{\infty} \frac{(\tfrac{1}{4}z^2)^k}{k!\,\Gamma(\nu+k+1)}.$$

This solution has properties analogous to those of $J_\nu(z)$, defined in §10.2(ii). In particular, the *principal branch* of $I_\nu(z)$ is defined in a similar way: it corresponds to the principal value of $(\tfrac{1}{2}z)^\nu$, is analytic in $\mathbb{C}\setminus(-\infty,0]$, and two-valued and discontinuous on the cut $\mathrm{ph}\,z = \pm\pi$.

The defining property of the second standard solution $K_\nu(z)$ of (10.25.1) is

10.25.3 $$K_\nu(z) \sim \sqrt{\pi/(2z)}\,e^{-z},$$

as $z \to \infty$ in $|\mathrm{ph}\,z| \leq \tfrac{3}{2}\pi - \delta$ ($< \tfrac{3}{2}\pi$). It has a branch point at $z=0$ for all $\nu \in \mathbb{C}$. The *principal branch* corresponds to the principal value of the square root in (10.25.3), is analytic in $\mathbb{C}\setminus(-\infty,0]$, and two-valued and discontinuous on the cut $\mathrm{ph}\,z = \pm\pi$.

Both $I_\nu(z)$ and $K_\nu(z)$ are real when ν is real and $\mathrm{ph}\,z = 0$.

For fixed z ($\neq 0$) each branch of $I_\nu(z)$ and $K_\nu(z)$ is entire in ν.

Branch Conventions

Except where indicated otherwise it is assumed throughout this Handbook that the symbols $I_\nu(z)$ and $K_\nu(z)$ denote the principal values of these functions.

Symbol $\mathscr{Z}_\nu(z)$

Corresponding to the symbol $\mathscr{C}_\nu$ introduced in §10.2(ii), we sometimes use $\mathscr{Z}_\nu(z)$ to denote $I_\nu(z)$, $e^{\nu\pi i}K_\nu(z)$, or any nontrivial linear combination of these functions, the coefficients in which are independent of z and ν.

10.25(iii) Numerically Satisfactory Pairs of Solutions

Table 10.25.1 lists numerically satisfactory pairs of solutions (§2.7(iv)) of (10.25.1). It is assumed that $\Re\nu \geq 0$. When $\Re\nu < 0$, $I_\nu(z)$ is replaced by $I_{-\nu}(z)$.

Table 10.25.1: Numerically satisfactory pairs of solutions of the modified Bessel's equation.

Pair	Region
$I_\nu(z), K_\nu(z)$	$\|\mathrm{ph}\,z\| \leq \tfrac{1}{2}\pi$
$I_\nu(z), K_\nu(ze^{\mp\pi i})$	$\tfrac{1}{2}\pi \leq \pm\mathrm{ph}\,z \leq \tfrac{3}{2}\pi$

10.26 Graphics

10.26(i) Real Order and Variable

See Figures 10.26.1–10.26.6.

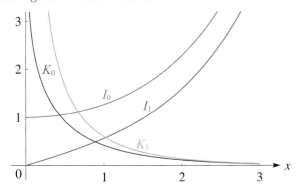

Figure 10.26.1: $I_0(x), I_1(x), K_0(x), K_1(x)$, $0 \leq x \leq 3$.

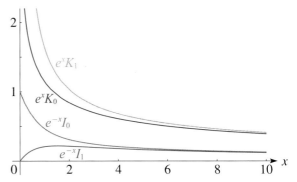

Figure 10.26.2: $e^{-x}I_0(x), e^{-x}I_1(x), e^x K_0(x), e^x K_1(x)$, $0 \leq x \leq 10$.

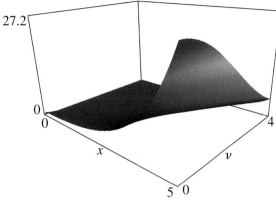

Figure 10.26.3: $I_\nu(x)$, $0 \le x \le 5, 0 \le \nu \le 4$.

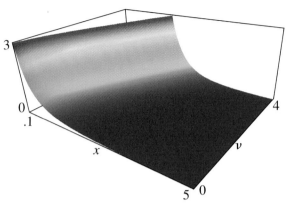

Figure 10.26.4: $K_\nu(x)$, $0.1 \le x \le 5, 0 \le \nu \le 4$.

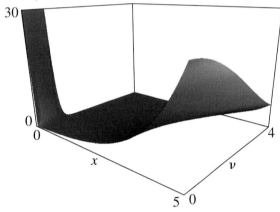

Figure 10.26.5: $I'_\nu(x)$, $0 \le x \le 5, 0 \le \nu \le 4$.

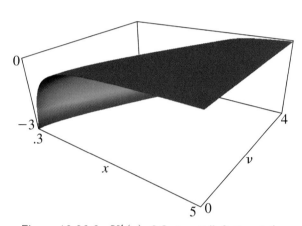

Figure 10.26.6: $K'_\nu(x)$, $0.3 \le x \le 5, 0 \le \nu \le 4$.

10.26(ii) Real Order, Complex Variable

Apply (10.27.6) and (10.27.8) to §10.3(ii).

10.26(iii) Imaginary Order, Real Variable

See Figures 10.26.7–10.26.10. For the notation, see §10.45.

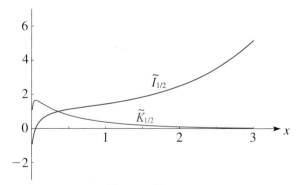

Figure 10.26.7: $\widetilde{I}_{1/2}(x), \widetilde{K}_{1/2}(x), 0.01 \le x \le 3$.

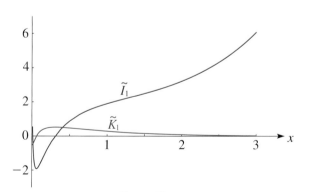

Figure 10.26.8: $\widetilde{I}_1(x), \widetilde{K}_1(x), 0.01 \le x \le 3$.

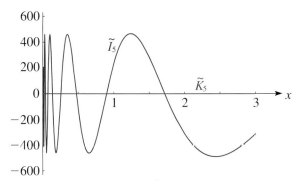

Figure 10.26.9: $\widetilde{I}_5(x), \widetilde{K}_5(x), 0.01 \leq x \leq 3$.

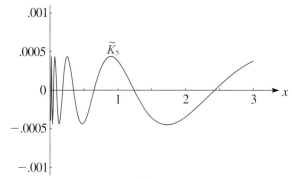

Figure 10.26.10: $\widetilde{K}_5(x), 0.01 \leq x \leq 3$.

10.27 Connection Formulas

Other solutions of (10.25.1) are $I_{-\nu}(z)$ and $K_{-\nu}(z)$.

10.27.1 $$I_{-n}(z) = I_n(z),$$

10.27.2 $$I_{-\nu}(z) = I_\nu(z) + (2/\pi)\sin(\nu\pi)\,K_\nu(z),$$

10.27.3 $$K_{-\nu}(z) = K_\nu(z).$$

10.27.4 $$K_\nu(z) = \tfrac{1}{2}\pi \frac{I_{-\nu}(z) - I_\nu(z)}{\sin(\nu\pi)}.$$

When ν is an integer limiting values are taken:

10.27.5 $$K_n(z) = \frac{(-1)^{n-1}}{2}\left(\left.\frac{\partial I_\nu(z)}{\partial \nu}\right|_{\nu=n} + \left.\frac{\partial I_\nu(z)}{\partial \nu}\right|_{\nu=-n}\right),$$
$$n = 0, \pm 1, \pm 2, \ldots.$$

In terms of the solutions of (10.2.1),

10.27.6 $$I_\nu(z) = e^{\mp \nu\pi i/2} J_\nu\!\left(ze^{\pm\pi i/2}\right), \quad -\pi \leq \pm\operatorname{ph} z \leq \tfrac{1}{2}\pi,$$

10.27.7 $$I_\nu(z) = \tfrac{1}{2}e^{\mp \nu\pi i/2}\left(H_\nu^{(1)}\!\left(ze^{\pm\pi i/2}\right) + H_\nu^{(2)}\!\left(ze^{\pm\pi i/2}\right)\right),$$
$$-\pi \leq \pm\operatorname{ph} z \leq \tfrac{1}{2}\pi.$$

10.27.8
$$K_\nu(z) = \begin{cases} \tfrac{1}{2}\pi i e^{\nu\pi i/2} H_\nu^{(1)}\!\left(ze^{\pi i/2}\right), & -\pi \leq \operatorname{ph} z \leq \tfrac{1}{2}\pi, \\ -\tfrac{1}{2}\pi i e^{-\nu\pi i/2} H_\nu^{(2)}\!\left(ze^{-\pi i/2}\right), & -\tfrac{1}{2}\pi \leq \operatorname{ph} z \leq \pi. \end{cases}$$

10.27.9 $$\pi i\, J_\nu(z) = e^{-\nu\pi i/2} K_\nu\!\left(ze^{-\pi i/2}\right) - e^{\nu\pi i/2} K_\nu\!\left(ze^{\pi i/2}\right), \quad |\operatorname{ph} z| \leq \tfrac{1}{2}\pi.$$

10.27.10 $$-\pi Y_\nu(z) = e^{-\nu\pi i/2} K_\nu\!\left(ze^{-\pi i/2}\right) + e^{\nu\pi i/2} K_\nu\!\left(ze^{\pi i/2}\right),$$
$$|\operatorname{ph} z| \leq \tfrac{1}{2}\pi.$$

10.27.11 $$Y_\nu(z) = e^{\pm(\nu+1)\pi i/2} I_\nu\!\left(ze^{\mp\pi i/2}\right) - (2/\pi)e^{\mp\nu\pi i/2} K_\nu\!\left(ze^{\mp\pi i/2}\right),$$
$$-\tfrac{1}{2}\pi \leq \pm\operatorname{ph} z \leq \pi.$$

See also §10.34.

Many properties of modified Bessel functions follow immediately from those of ordinary Bessel functions by application of (10.27.6)–(10.27.8).

10.28 Wronskians and Cross-Products

10.28.1 $$\mathscr{W}\{I_\nu(z), I_{-\nu}(z)\} = I_\nu(z)\,I_{-\nu-1}(z) - I_{\nu+1}(z)\,I_{-\nu}(z)$$
$$= -2\sin(\nu\pi)/(\pi z),$$

10.28.2 $$\mathscr{W}\{K_\nu(z), I_\nu(z)\} = I_\nu(z)\,K_{\nu+1}(z) + I_{\nu+1}(z)\,K_\nu(z)$$
$$= 1/z.$$

10.29 Recurrence Relations and Derivatives

10.29(i) Recurrence Relations

With $\mathscr{Z}_\nu(z)$ defined as in §10.25(ii),

10.29.1 $$\mathscr{Z}_{\nu-1}(z) - \mathscr{Z}_{\nu+1}(z) = (2\nu/z)\,\mathscr{Z}_\nu(z),$$
$$\mathscr{Z}_{\nu-1}(z) + \mathscr{Z}_{\nu+1}(z) = 2\mathscr{Z}_\nu'(z).$$

10.29.2 $$\mathscr{Z}_\nu'(z) = \mathscr{Z}_{\nu-1}(z) - (\nu/z)\,\mathscr{Z}_\nu(z),$$
$$\mathscr{Z}_\nu'(z) = \mathscr{Z}_{\nu+1}(z) + (\nu/z)\,\mathscr{Z}_\nu(z).$$

10.29.3 $$I_0'(z) = I_1(z), \quad K_0'(z) = -K_1(z).$$

10.29(ii) Derivatives

For $k = 0, 1, 2, \ldots,$

10.29.4
$$\left(\frac{1}{z}\frac{d}{dz}\right)^k (z^\nu \mathscr{Z}_\nu(z)) = z^{\nu-k} \mathscr{Z}_{\nu-k}(z),$$
$$\left(\frac{1}{z}\frac{d}{dz}\right)^k (z^{-\nu} \mathscr{Z}_\nu(z)) = z^{-\nu-k} \mathscr{Z}_{\nu+k}(z).$$

10.29.5
$$\mathscr{Z}_\nu^{(k)}(z) = \frac{1}{2^k}\left(\mathscr{Z}_{\nu-k}(z) + \binom{k}{1}\mathscr{Z}_{\nu-k+2}(z) \right.$$
$$\left. + \binom{k}{2}\mathscr{Z}_{\nu-k+4}(z) + \cdots + \mathscr{Z}_{\nu+k}(z)\right).$$

10.30 Limiting Forms

10.30(i) $z \to 0$

When ν is fixed and $z \to 0$,

10.30.1 $\quad I_\nu(z) \sim (\tfrac{1}{2}z)^\nu / \Gamma(\nu+1), \quad \nu \neq -1, -2, -3, \ldots,$

10.30.2 $\quad K_\nu(z) \sim \tfrac{1}{2}\Gamma(\nu)(\tfrac{1}{2}z)^{-\nu}, \quad\quad \Re\nu > 0,$

10.30.3 $\quad K_0(z) \sim -\ln z.$

For $K_\nu(x)$, when ν is purely imaginary and $x \to 0+$, see (10.45.2) and (10.45.7).

10.30(ii) $z \to \infty$

When ν is fixed and $z \to \infty$,

10.30.4 $\quad I_\nu(z) \sim e^z/\sqrt{2\pi z}, \quad\quad |\operatorname{ph} z| \leq \tfrac{1}{2}\pi - \delta,$

10.30.5
$$I_\nu(z) \sim e^{\pm(\nu+\frac{1}{2})\pi i} e^{-z}/\sqrt{2\pi z},$$
$$\tfrac{1}{2}\pi + \delta \leq \pm \operatorname{ph} z \leq \tfrac{3}{2}\pi - \delta.$$

For $K_\nu(z)$ see (10.25.3).

10.31 Power Series

For $I_\nu(z)$ see (10.25.2) and (10.27.1). When ν is not an integer the corresponding expansion for $K_\nu(z)$ is obtained from (10.25.2) and (10.27.4).
When $n = 0, 1, 2, \ldots,$

10.31.1
$$K_n(z) = \tfrac{1}{2}(\tfrac{1}{2}z)^{-n} \sum_{k=0}^{n-1} \frac{(n-k-1)!}{k!}(-\tfrac{1}{4}z^2)^k$$
$$+ (-1)^{n+1}\ln(\tfrac{1}{2}z) I_n(z)$$
$$+ (-1)^n \tfrac{1}{2}(\tfrac{1}{2}z)^n \sum_{k=0}^{\infty} (\psi(k+1)$$
$$+ \psi(n+k+1)) \frac{(\tfrac{1}{4}z^2)^k}{k!(n+k)!},$$

where $\psi(x) = \Gamma'(x)/\Gamma(x)$ (§5.2(i)). In particular,

10.31.2
$$K_0(z) = -\left(\ln(\tfrac{1}{2}z) + \gamma\right) I_0(z) + \frac{\tfrac{1}{4}z^2}{(1!)^2}$$
$$+ (1 + \tfrac{1}{2})\frac{(\tfrac{1}{4}z^2)^2}{(2!)^2} + (1 + \tfrac{1}{2} + \tfrac{1}{3})\frac{(\tfrac{1}{4}z^2)^3}{(3!)^2} + \cdots.$$

For negative values of n use (10.27.3).

10.31.3
$$I_\nu(z) I_\mu(z) = (\tfrac{1}{2}z)^{\nu+\mu} \sum_{k=0}^{\infty} \frac{(\nu+\mu+k+1)_k (\tfrac{1}{4}z^2)^k}{k!\,\Gamma(\nu+k+1)\,\Gamma(\mu+k+1)}.$$

10.32 Integral Representations

10.32(i) Integrals along the Real Line

10.32.1
$$I_0(z) = \frac{1}{\pi}\int_0^\pi e^{\pm z\cos\theta}\,d\theta = \frac{1}{\pi}\int_0^\pi \cosh(z\cos\theta)\,d\theta.$$

10.32.2
$$I_\nu(z) = \frac{(\tfrac{1}{2}z)^\nu}{\pi^{\frac{1}{2}}\Gamma(\nu+\tfrac{1}{2})}\int_0^\pi e^{\pm z\cos\theta}(\sin\theta)^{2\nu}\,d\theta$$
$$= \frac{(\tfrac{1}{2}z)^\nu}{\pi^{\frac{1}{2}}\Gamma(\nu+\tfrac{1}{2})}\int_{-1}^1 (1-t^2)^{\nu-\frac{1}{2}} e^{\pm zt}\,dt,$$
$$\Re\nu > -\tfrac{1}{2}.$$

10.32.3 $\quad I_n(z) = \dfrac{1}{\pi}\int_0^\pi e^{z\cos\theta} \cos(n\theta)\,d\theta.$

10.32.4
$$I_\nu(z) = \frac{1}{\pi}\int_0^\pi e^{z\cos\theta}\cos(\nu\theta)\,d\theta$$
$$- \frac{\sin(\nu\pi)}{\pi}\int_0^\infty e^{-z\cosh t - \nu t}\,dt, \quad |\operatorname{ph} z| < \tfrac{1}{2}\pi.$$

10.32.5
$$K_0(z) = -\frac{1}{\pi}\int_0^\pi e^{\pm z\cos\theta}\left(\gamma + \ln\bigl(2z(\sin\theta)^2\bigr)\right) d\theta.$$

10.32.6
$$K_0(x) = \int_0^\infty \cos(x\sinh t)\,dt = \int_0^\infty \frac{\cos(xt)}{\sqrt{t^2+1}}\,dt, \quad x > 0.$$

10.32.7
$$K_\nu(x) = \sec(\tfrac{1}{2}\nu\pi)\int_0^\infty \cos(x\sinh t)\cosh(\nu t)\,dt$$
$$= \csc(\tfrac{1}{2}\nu\pi)\int_0^\infty \sin(x\sinh t)\sinh(\nu t)\,dt,$$
$$|\Re\nu| < 1, x > 0.$$

10.32.8
$$K_\nu(z) = \frac{\pi^{\frac{1}{2}}(\tfrac{1}{2}z)^\nu}{\Gamma(\nu+\tfrac{1}{2})}\int_0^\infty e^{-z\cosh t}(\sinh t)^{2\nu}\,dt$$
$$= \frac{\pi^{\frac{1}{2}}(\tfrac{1}{2}z)^\nu}{\Gamma(\nu+\tfrac{1}{2})}\int_1^\infty e^{-zt}(t^2-1)^{\nu-\frac{1}{2}}\,dt,$$
$$\Re\nu > -\tfrac{1}{2}, |\operatorname{ph} z| < \tfrac{1}{2}\pi.$$

10.32.9
$$K_\nu(z) = \int_0^\infty e^{-z\cosh t}\cosh(\nu t)\,dt, \quad |\operatorname{ph} z| < \tfrac{1}{2}\pi.$$

10.33 Continued Fractions

10.32.10
$$K_\nu(z) = \tfrac{1}{2}(\tfrac{1}{2}z)^\nu \int_0^\infty \exp\left(-t - \frac{z^2}{4t}\right) \frac{dt}{t^{\nu+1}}, \quad |\mathrm{ph}\, z| < \tfrac{1}{4}\pi.$$

Basset's Integral

10.32.11
$$K_\nu(xz) = \frac{\Gamma(\nu + \tfrac{1}{2})(2z)^\nu}{\pi^{\frac{1}{2}} x^\nu} \int_0^\infty \frac{\cos(xt)\, dt}{(t^2 + z^2)^{\nu + \frac{1}{2}}},$$
$$\Re\nu > -\tfrac{1}{2},\, x > 0,\, |\mathrm{ph}\, z| < \tfrac{1}{2}\pi.$$

10.32(ii) Contour Integrals

10.32.12
$$I_\nu(z) = \frac{1}{2\pi i} \int_{\infty - i\pi}^{\infty + i\pi} e^{z \cosh t - \nu t}\, dt, \quad |\mathrm{ph}\, z| < \tfrac{1}{2}\pi.$$

Mellin–Barnes Type

10.32.13
$$K_\nu(z) = \frac{(\tfrac{1}{2}z)^\nu}{4\pi i} \int_{c - i\infty}^{c + i\infty} \Gamma(t)\, \Gamma(t - \nu)(\tfrac{1}{2}z)^{-2t}\, dt,$$
$$c > \max(\Re\nu, 0), |\mathrm{ph}\, z| < \pi.$$

10.32.14
$$K_\nu(z) = \frac{1}{2\pi^2 i} \left(\frac{\pi}{2z}\right)^{\frac{1}{2}} e^{-z} \cos(\nu\pi)$$
$$\times \int_{-i\infty}^{i\infty} \Gamma(t)\, \Gamma(\tfrac{1}{2} - t - \nu)\, \Gamma(\tfrac{1}{2} - t + \nu)(2z)^t\, dt,$$
$$\nu - \tfrac{1}{2} \notin \mathbb{Z}, |\mathrm{ph}\, z| < \tfrac{3}{2}\pi.$$

In (10.32.14) the integration contour separates the poles of $\Gamma(t)$ from the poles of $\Gamma(\tfrac{1}{2} - t - \nu)\, \Gamma(\tfrac{1}{2} - t + \nu)$.

10.32(iii) Products

10.32.15
$$I_\mu(z)\, I_\nu(z) = \frac{2}{\pi} \int_0^{\frac{1}{2}\pi} I_{\mu+\nu}(2z \cos\theta) \cos((\mu - \nu)\theta)\, d\theta,$$
$$\Re(\mu + \nu) > -1.$$

10.32.16
$$I_\mu(x)\, K_\nu(x) = \int_0^\infty J_{\mu \pm \nu}(2x \sinh t) e^{(-\mu \pm \nu)t}\, dt,$$
$$\Re(\mu \mp \nu) > -\tfrac{1}{2},\, \Re(\mu \pm \nu) > -1,\, x > 0.$$

10.32.17
$$K_\mu(z)\, K_\nu(z) = 2 \int_0^\infty K_{\mu \pm \nu}(2z \cosh t) \cosh((\mu \mp \nu)t)\, dt,$$
$$|\mathrm{ph}\, z| < \tfrac{1}{2}\pi.$$

10.32.18
$$K_\nu(z)\, K_\nu(\zeta)$$
$$= \tfrac{1}{2} \int_0^\infty \exp\left(-\frac{t}{2} - \frac{z^2 + \zeta^2}{2t}\right) K_\nu\left(\frac{z\zeta}{t}\right) \frac{dt}{t},$$
$$|\mathrm{ph}\, z| < \pi, |\mathrm{ph}\, \zeta| < \pi, |\mathrm{ph}(z+\zeta)| < \tfrac{1}{4}\pi.$$

Mellin–Barnes Type

10.32.19
$$K_\mu(z)\, K_\nu(z) = \frac{1}{8\pi i} \int_{c-i\infty}^{c+i\infty} \frac{\Gamma(t + \tfrac{1}{2}\mu + \tfrac{1}{2}\nu)\, \Gamma(t + \tfrac{1}{2}\mu - \tfrac{1}{2}\nu)\, \Gamma(t - \tfrac{1}{2}\mu + \tfrac{1}{2}\nu)\, \Gamma(t - \tfrac{1}{2}\mu - \tfrac{1}{2}\nu)}{\Gamma(2t)} (\tfrac{1}{2}z)^{-2t}\, dt,$$
$$c > \tfrac{1}{2}(|\Re\mu| + |\Re\nu|), |\mathrm{ph}\, z| < \tfrac{1}{2}\pi.$$

For similar integrals for $J_\nu(z)\, K_\nu(z)$ and $I_\nu(z)\, K_\nu(z)$ see Paris and Kaminski (2001, p. 116).

10.32(iv) Compendia

For collections of integral representations of modified Bessel functions, or products of modified Bessel functions, see Erdélyi et al. (1953b, §§7.3, 7.12, and 7.14.2), Erdélyi et al. (1954a, pp. 48–60, 105–115, 276–285, and 357–359), Gröbner and Hofreiter (1950, pp. 193–194), Magnus et al. (1966, §3.7), Marichev (1983, pp. 191–216), and Watson (1944, Chapters 6, 12, and 13).

10.33 Continued Fractions

Assume $I_{\nu-1}(z) \neq 0$. Then

10.33.1
$$\frac{I_\nu(z)}{I_{\nu-1}(z)} = \frac{1}{2\nu z^{-1}+} \frac{1}{2(\nu+1)z^{-1}+} \frac{1}{2(\nu+2)z^{-1}+} \cdots,$$
$$z \neq 0,$$

10.33.2
$$\frac{I_\nu(z)}{I_{\nu-1}(z)} = \frac{\tfrac{1}{2}z/\nu}{1+} \frac{\tfrac{1}{4}z^2/(\nu(\nu+1))}{1+} \frac{\tfrac{1}{4}z^2/((\nu+1)(\nu+2))}{1+} \cdots,$$
$$\nu \neq 0, -1, -2, \ldots.$$

See also Cuyt et al. (2008, pp. 361–367).

10.34 Analytic Continuation

When $m \in \mathbb{Z}$,

10.34.1
$$I_\nu(z e^{m\pi i}) = e^{m\nu\pi i} I_\nu(z),$$

10.34.2
$$K_\nu(z e^{m\pi i}) = e^{-m\nu\pi i} K_\nu(z) - \pi i \sin(m\nu\pi) \csc(\nu\pi) I_\nu(z).$$

10.34.3
$$I_\nu(z e^{m\pi i}) = (i/\pi) \left(\pm e^{m\nu\pi i} K_\nu(z e^{\pm\pi i}) \right.$$
$$\left. \mp e^{(m \mp 1)\nu\pi i} K_\nu(z) \right),$$

10.34.4
$$K_\nu(z e^{m\pi i}) = \csc(\nu\pi) \left(\pm \sin(m\nu\pi) K_\nu(z e^{\pm\pi i}) \right.$$
$$\left. \mp \sin((m \mp 1)\nu\pi) K_\nu(z) \right).$$

If $\nu = n(\in \mathbb{Z})$, then limiting values are taken in (10.34.2) and (10.34.4):

10.34.5
$$K_n(ze^{m\pi i}) = (-1)^{mn} K_n(z) + (-1)^{n(m-1)-1} m\pi i\, I_n(z),$$

10.34.6
$$K_n(ze^{m\pi i}) = \pm(-1)^{n(m-1)} m\, K_n(ze^{\pm\pi i})$$
$$\mp (-1)^{nm}(m \mp 1) K_n(z).$$

For real ν,

10.34.7 $\quad I_\nu(\overline{z}) = \overline{I_\nu(z)}, \quad K_\nu(\overline{z}) = \overline{K_\nu(z)}.$

For complex ν replace ν by $\overline{\nu}$ on the right-hand sides.

10.35 Generating Function and Associated Series

For $z \in \mathbb{C}$ and $t \in \mathbb{C}\setminus\{0\}$,

10.35.1 $\quad e^{\frac{1}{2}z(t+t^{-1})} = \sum_{m=-\infty}^{\infty} t^m I_m(z).$

For $z, \theta \in \mathbb{C}$,

10.35.2 $\quad e^{z\cos\theta} = I_0(z) + 2\sum_{k=1}^{\infty} I_k(z)\cos(k\theta),$

10.35.3
$$e^{z\sin\theta} = I_0(z) + 2\sum_{k=0}^{\infty}(-1)^k I_{2k+1}(z)\sin((2k+1)\theta)$$
$$+ 2\sum_{k=1}^{\infty}(-1)^k I_{2k}(z)\cos(2k\theta).$$

10.35.4 $\quad 1 = I_0(z) - 2I_2(z) + 2I_4(z) - 2I_6(z) + \cdots,$

10.35.5 $\quad e^{\pm z} = I_0(z) \pm 2I_1(z) + 2I_2(z) \pm 2I_3(z) + \cdots,$

10.35.6
$$\cosh z = I_0(z) + 2I_2(z) + 2I_4(z) + 2I_6(z) + \ldots,$$
$$\sinh z = 2I_1(z) + 2I_3(z) + 2I_5(z) + \ldots.$$

10.36 Other Differential Equations

The quantity λ^2 in (10.13.1)–(10.13.6) and (10.13.8) can be replaced by $-\lambda^2$ if at the same time the symbol $\mathscr{C}$ in the given solutions is replaced by $\mathscr{Z}$. Also,

10.36.1
$$z^2(z^2+\nu^2)w'' + z(z^2+3\nu^2)w'$$
$$- ((z^2+\nu^2)^2 + z^2 - \nu^2) w = 0, \quad w = \mathscr{Z}'_\nu(z),$$

10.36.2
$$z^2 w'' + z(1 \pm 2z)w' + (\pm z - \nu^2)w = 0,$$
$$w = e^{\mp z}\mathscr{Z}_\nu(z).$$

Differential equations for products can be obtained from (10.13.9)–(10.13.11) by replacing z by iz.

10.37 Inequalities; Monotonicity

If $\nu\ (\geq 0)$ is fixed, then throughout the interval $0 < x < \infty$, $I_\nu(x)$ is positive and increasing, and $K_\nu(x)$ is positive and decreasing.

If $x\ (> 0)$ is fixed, then throughout the interval $0 < \nu < \infty$, $I_\nu(x)$ is decreasing, and $K_\nu(x)$ is increasing.

For sharper inequalities when the variables are real see Paris (1984) and Laforgia (1991).

If $0 \leq \nu < \mu$ and $|\operatorname{ph} z| < \pi$, then

10.37.1 $\quad |K_\nu(z)| < |K_\mu(z)|.$

See also Pal'tsev (1999) and Petropoulou (2000).

10.38 Derivatives with Respect to Order

10.38.1
$$\frac{\partial I_\nu(z)}{\partial \nu} = I_\nu(z)\ln\!\left(\tfrac{1}{2}z\right) - \left(\tfrac{1}{2}z\right)^\nu \sum_{k=0}^{\infty} \frac{\psi(\nu+k+1)}{\Gamma(\nu+k+1)} \frac{(\tfrac{1}{4}z^2)^k}{k!},$$

10.38.2
$$\frac{\partial K_\nu(z)}{\partial \nu} = \tfrac{1}{2}\pi\csc(\nu\pi)\left(\frac{\partial I_{-\nu}(z)}{\partial \nu} - \frac{\partial I_\nu(z)}{\partial \nu}\right)$$
$$- \pi\cot(\nu\pi) K_\nu(z), \qquad \nu \notin \mathbb{Z}.$$

Integer Values of ν

10.38.3
$$(-1)^n \frac{\partial I_\nu(z)}{\partial \nu}\bigg|_{\nu=n} = -K_n(z)$$
$$+ \frac{n!}{2(\tfrac{1}{2}z)^n}\sum_{k=0}^{n-1}(-1)^k \frac{(\tfrac{1}{2}z)^k I_k(z)}{k!(n-k)},$$

10.38.4 $\quad \dfrac{\partial K_\nu(z)}{\partial \nu}\bigg|_{\nu=n} = \dfrac{n!}{2(\tfrac{1}{2}z)^n}\sum_{k=0}^{n-1}\dfrac{(\tfrac{1}{2}z)^k K_k(z)}{k!(n-k)}.$

10.38.5 $\quad \dfrac{\partial I_\nu(z)}{\partial \nu}\bigg|_{\nu=0} = -K_0(z), \quad \dfrac{\partial K_\nu(z)}{\partial \nu}\bigg|_{\nu=0} = 0.$

Half-Integer Values of ν

For the notations E_1 and Ei see §6.2(i). When $x > 0$,

10.38.6
$$\frac{\partial I_\nu(x)}{\partial \nu}\bigg|_{\nu=\pm\frac{1}{2}} = -\frac{1}{\sqrt{2\pi x}}\left(E_1(2x)e^x \pm \operatorname{Ei}(2x)e^{-x}\right),$$

10.38.7 $\quad \dfrac{\partial K_\nu(x)}{\partial \nu}\bigg|_{\nu=\pm\frac{1}{2}} = \pm\sqrt{\dfrac{\pi}{2x}}\, E_1(2x) e^x.$

For further results see Brychkov and Geddes (2005).

10.39 Relations to Other Functions

Elementary Functions

10.39.1
$$I_{\frac{1}{2}}(z) = \left(\frac{2}{\pi z}\right)^{\frac{1}{2}}\sinh z, \quad I_{-\frac{1}{2}}(z) = \left(\frac{2}{\pi z}\right)^{\frac{1}{2}}\cosh z,$$

10.39.2 $\quad K_{\frac{1}{2}}(z) = K_{-\frac{1}{2}}(z) = \left(\dfrac{\pi}{2z}\right)^{\frac{1}{2}} e^{-z}.$

For these and general results when ν is half an odd integer see §§10.47(ii) and 10.49(ii).

Airy Functions

See §§9.6(i) and 9.6(ii).

Parabolic Cylinder Functions

With the notation of §12.2(i),

10.39.3 $\quad K_{\frac{1}{4}}(z) = \pi^{\frac{1}{2}} z^{-\frac{1}{4}} U\left(0, 2z^{\frac{1}{2}}\right),$

10.39.4
$$K_{\frac{3}{4}}(z) = \tfrac{1}{2}\pi^{\frac{1}{2}} z^{-\frac{3}{4}} \left(\tfrac{1}{2} U\left(1, 2z^{\frac{1}{2}}\right) + U\left(-1, 2z^{\frac{1}{2}}\right)\right).$$

Principal values on each side of these equations correspond. For these and further results see Miller (1955, pp. 42–43 and 77–79).

Confluent Hypergeometric Functions

10.39.5 $\quad I_\nu(z) = \dfrac{(\tfrac{1}{2}z)^\nu e^{\pm z}}{\Gamma(\nu+1)} M\left(\nu+\tfrac{1}{2}, 2\nu+1, \mp 2z\right),$

10.39.6 $\quad K_\nu(z) = \pi^{\frac{1}{2}} (2z)^\nu e^{-z} U\left(\nu+\tfrac{1}{2}, 2\nu+1, 2z\right),$

10.39.7
$$I_\nu(z) = \frac{(2z)^{-\frac{1}{2}} M_{0,\nu}(2z)}{2^{2\nu}\, \Gamma(\nu+1)}, \quad 2\nu \neq -1, -2, -3, \ldots,$$

10.39.8 $\quad K_\nu(z) = \left(\dfrac{\pi}{2z}\right)^{\frac{1}{2}} W_{0,\nu}(2z).$

For the functions M, U, $M_{0,\nu}$, and $W_{0,\nu}$ see §§13.2(i) and 13.14(i).

Generalized Hypergeometric Functions and Hypergeometric Function

10.39.9 $\quad I_\nu(z) = \dfrac{(\tfrac{1}{2}z)^\nu}{\Gamma(\nu+1)}\, {}_0F_1\left(-;\nu+1;\tfrac{1}{4}z^2\right),$

10.39.10 $\quad I_\nu(z) = (\tfrac{1}{2}z)^\nu \lim \mathbf{F}\left(\lambda, \mu; \nu+1; z^2/(4\lambda\mu)\right),$

as λ and $\mu \to \infty$ in $\mathbb{C}$, with z and ν fixed. For the functions ${}_0F_1$ and $\mathbf{F}$ see (16.2.1) and §15.2(i).

10.40 Asymptotic Expansions for Large Argument

10.40(i) Hankel's Expansions

With the notation of §§10.17(i) and 10.17(ii), as $z \to \infty$ with ν fixed,

10.40.1
$$I_\nu(z) \sim \frac{e^z}{(2\pi z)^{\frac{1}{2}}} \sum_{k=0}^{\infty} (-1)^k \frac{a_k(\nu)}{z^k}, \quad |\mathrm{ph}\, z| \leq \tfrac{1}{2}\pi - \delta,$$

10.40.2
$$K_\nu(z) \sim \left(\frac{\pi}{2z}\right)^{\frac{1}{2}} e^{-z} \sum_{k=0}^{\infty} \frac{a_k(\nu)}{z^k}, \quad |\mathrm{ph}\, z| \leq \tfrac{3}{2}\pi - \delta,$$

10.40.3
$$I'_\nu(z) \sim \frac{e^z}{(2\pi z)^{\frac{1}{2}}} \sum_{k=0}^{\infty} (-1)^k \frac{b_k(\nu)}{z^k}, \quad |\mathrm{ph}\, z| \leq \tfrac{1}{2}\pi - \delta,$$

10.40.4
$$K'_\nu(z) \sim -\left(\frac{\pi}{2z}\right)^{\frac{1}{2}} e^{-z} \sum_{k=0}^{\infty} \frac{b_k(\nu)}{z^k}, \quad |\mathrm{ph}\, z| \leq \tfrac{3}{2}\pi - \delta.$$

Corresponding expansions for $I_\nu(z)$, $K_\nu(z)$, $I'_\nu(z)$, and $K'_\nu(z)$ for other ranges of $\mathrm{ph}\, z$ are obtainable by combining (10.34.3), (10.34.4), (10.34.6), and their differentiated forms, with (10.40.2) and (10.40.4). In particular, use of (10.34.3) with $m = 0$ yields the following more general (and more accurate) version of (10.40.1):

10.40.5
$$I_\nu(z) \sim \frac{e^z}{(2\pi z)^{\frac{1}{2}}} \sum_{k=0}^{\infty} (-1)^k \frac{a_k(\nu)}{z^k}$$
$$\pm i e^{\pm \nu \pi i} \frac{e^{-z}}{(2\pi z)^{\frac{1}{2}}} \sum_{k=0}^{\infty} \frac{a_k(\nu)}{z^k},$$
$$-\tfrac{1}{2}\pi + \delta \leq \pm \mathrm{ph}\, z \leq \tfrac{3}{2}\pi - \delta.$$

Products

With $\mu = 4\nu^2$ and fixed,

10.40.6
$$I_\nu(z) K_\nu(z) \sim \frac{1}{2z}\left(1 - \frac{1}{2}\frac{\mu-1}{(2z)^2} + \frac{1\cdot 3}{2\cdot 4}\frac{(\mu-1)(\mu-9)}{(2z)^4} - \cdots\right),$$

10.40.7
$$I'_\nu(z) K'_\nu(z) \sim -\frac{1}{2z}\left(1 + \frac{1}{2}\frac{\mu-3}{(2z)^2} - \frac{1}{2\cdot 4}\frac{(\mu-1)(\mu-45)}{(2z)^4} + \cdots\right),$$

as $z \to \infty$ in $|\mathrm{ph}\, z| \leq \tfrac{1}{2}\pi - \delta$. The general terms in (10.40.6) and (10.40.7) can be written down by analogy with (10.18.17), (10.18.19), and (10.18.20).

ν-Derivative

For fixed ν,

10.40.8
$$\frac{\partial K_\nu(z)}{\partial \nu} \sim \left(\frac{\pi}{2z}\right)^{\frac{1}{2}} \frac{\nu e^{-z}}{z} \sum_{k=0}^{\infty} \frac{\alpha_k(\nu)}{(8z)^k},$$

as $z \to \infty$ in $|\mathrm{ph}\, z| \leq \tfrac{3}{2}\pi - \delta$. Here $\alpha_0(\nu) = 1$ and

10.40.9
$$\alpha_k(\nu) = \frac{(4\nu^2 - 1^2)(4\nu^2 - 3^2)\cdots(4\nu^2 - (2k+1)^2)}{(k+1)!}$$
$$\times \left(\frac{1}{4\nu^2 - 1^2} + \frac{1}{4\nu^2 - 3^2} + \cdots + \frac{1}{4\nu^2 - (2k+1)^2}\right).$$

10.40(ii) Error Bounds for Real Argument and Order

In the expansion (10.40.2) assume that $z > 0$ and the sum is truncated when $k = \ell - 1$. Then the remainder term does not exceed the first neglected term in absolute value and has the same sign provided that $\ell \geq \max(|\nu| - \frac{1}{2}, 1)$.

For the error term in (10.40.1) see §10.40(iii).

10.40(iii) Error Bounds for Complex Argument and Order

For (10.40.2) write

10.40.10
$$K_\nu(z) = \left(\frac{\pi}{2z}\right)^{\frac{1}{2}} e^{-z} \left(\sum_{k=0}^{\ell-1} \frac{a_k(\nu)}{z^k} + R_\ell(\nu, z)\right), \quad \ell = 1, 2, \ldots.$$

Then

10.40.11
$$|R_\ell(\nu, z)| \leq 2|a_\ell(\nu)| \mathcal{V}_{z,\infty}(t^{-\ell}) \exp\left(|\nu^2 - \tfrac{1}{4}| \mathcal{V}_{z,\infty}(t^{-1})\right),$$

where $\mathcal{V}$ denotes the variational operator (§2.3(i)), and the paths of variation are subject to the condition that $|\Re t|$ changes monotonically. Bounds for $\mathcal{V}_{z,\infty}(t^{-\ell})$ are given by

10.40.12
$$\mathcal{V}_{z,\infty}(t^{-\ell}) \leq \begin{cases} |z|^{-\ell}, & |\operatorname{ph} z| \leq \tfrac{1}{2}\pi, \\ \chi(\ell)|z|^{-\ell}, & \tfrac{1}{2}\pi \leq |\operatorname{ph} z| \leq \pi, \\ 2\chi(\ell)|\Re z|^{-\ell}, & \pi \leq |\operatorname{ph} z| \leq \tfrac{3}{2}\pi, \end{cases}$$

where $\chi(\ell) = \pi^{\frac{1}{2}} \Gamma\left(\tfrac{1}{2}\ell + 1\right) / \Gamma\left(\tfrac{1}{2}\ell + \tfrac{1}{2}\right)$; see §9.7(i).

A similar result for (10.40.1) is obtained by combining (10.34.3), with $m = 0$, and (10.40.10)–(10.40.12); see Olver (1997b, p. 269).

10.40(iv) Exponentially-Improved Expansions

In (10.40.10)

10.40.13
$$R_\ell(\nu, z) = (-1)^\ell 2 \cos(\nu\pi) \times \left(\sum_{k=0}^{m-1} \frac{a_k(\nu)}{z^k} G_{\ell-k}(2z) + R_{m,\ell}(\nu, z)\right),$$

where $G_p(z)$ is given by (10.17.16). If $z \to \infty$ with $|\ell - 2|z||$ bounded and $m (\geq 0)$ fixed, then

10.40.14
$$R_{m,\ell}(\nu, z) = O\left(e^{-2|z|} z^{-m}\right), \quad |\operatorname{ph} z| \leq \pi.$$

For higher re-expansions of the remainder term see Olde Daalhuis and Olver (1995a), Olde Daalhuis (1995, 1996), and Paris (2001a,b).

10.41 Asymptotic Expansions for Large Order

10.41(i) Asymptotic Forms

If $\nu \to \infty$ through positive real values with $z (\neq 0)$ fixed, then

10.41.1
$$I_\nu(z) \sim \frac{1}{\sqrt{2\pi\nu}} \left(\frac{ez}{2\nu}\right)^\nu,$$

10.41.2
$$K_\nu(z) \sim \sqrt{\frac{\pi}{2\nu}} \left(\frac{ez}{2\nu}\right)^{-\nu}.$$

10.41(ii) Uniform Expansions for Real Variable

As $\nu \to \infty$ through positive real values,

10.41.3
$$I_\nu(\nu z) \sim \frac{e^{\nu\eta}}{(2\pi\nu)^{\frac{1}{2}} (1+z^2)^{\frac{1}{4}}} \sum_{k=0}^{\infty} \frac{U_k(p)}{\nu^k},$$

10.41.4
$$K_\nu(\nu z) \sim \left(\frac{\pi}{2\nu}\right)^{\frac{1}{2}} \frac{e^{-\nu\eta}}{(1+z^2)^{\frac{1}{4}}} \sum_{k=0}^{\infty} (-1)^k \frac{U_k(p)}{\nu^k},$$

10.41.5
$$I'_\nu(\nu z) \sim \frac{(1+z^2)^{\frac{1}{4}} e^{\nu\eta}}{(2\pi\nu)^{\frac{1}{2}} z} \sum_{k=0}^{\infty} \frac{V_k(p)}{\nu^k},$$

10.41.6
$$K'_\nu(\nu z) \sim -\left(\frac{\pi}{2\nu}\right)^{\frac{1}{2}} \frac{(1+z^2)^{\frac{1}{4}} e^{-\nu\eta}}{z} \sum_{k=0}^{\infty} (-1)^k \frac{V_k(p)}{\nu^k},$$

uniformly for $0 < z < \infty$. Here

10.41.7
$$\eta = (1+z^2)^{\frac{1}{2}} + \ln \frac{z}{1 + (1+z^2)^{\frac{1}{2}}},$$

10.41.8
$$p = (1+z^2)^{-\frac{1}{2}},$$

where the branches assume their principal values. Also, $U_k(p)$ and $V_k(p)$ are polynomials in p of degree $3k$, given by $U_0(p) = V_0(p) = 1$, and

10.41.9
$$U_{k+1}(p) = \tfrac{1}{2} p^2 (1-p^2) U'_k(p) + \frac{1}{8} \int_0^p (1-5t^2) U_k(t)\, dt,$$
$$V_{k+1}(p) = U_{k+1}(p) - \tfrac{1}{2} p(1-p^2) U_k(p) - p^2(1-p^2) U'_k(p), \quad k = 0, 1, 2, \ldots.$$

For $k = 1, 2, 3$,

10.41.10
$$U_1(p) = \tfrac{1}{24}(3p - 5p^3), \quad U_2(p) = \tfrac{1}{1152}(81p^2 - 462p^4 + 385p^6),$$
$$U_3(p) = \tfrac{1}{4\,14720} (30375p^3 - 3\,69603p^5 + 7\,65765p^7 - 4\,25425p^9),$$

10.41.11
$$V_1(p) = \tfrac{1}{24}(-9p + 7p^3), \quad V_2(p) = \tfrac{1}{1152}(-135p^2 + 594p^4 - 455p^6),$$
$$V_3(p) = \tfrac{1}{4\,14720} (-42525p^3 + 4\,51737p^5 - 8\,83575p^7 + 4\,75475p^9).$$

10.41 Asymptotic Expansions for Large Order

For $U_4(p)$, $U_5(p)$, $U_6(p)$, see Bickley et al. (1952, p. xxxv).

For numerical tables of $\eta = \eta(z)$ and the coefficients $U_k(p)$, $V_k(p)$, see Olver (1962, pp. 43–51).

10.41(iii) Uniform Expansions for Complex Variable

The expansions (10.41.3)–(10.41.6) also hold uniformly in the sector $|\operatorname{ph} z| \leq \frac{1}{2}\pi - \delta \ (< \frac{1}{2}\pi)$, with the branches of the fractional powers in (10.41.3)–(10.41.8) extended by continuity from the positive real z-axis.

Figures 10.41.1 and 10.41.2 show corresponding points of the mapping of the z-plane and the η-plane. The curve E_1BE_2 in the z-plane is the upper boundary of the domain **K** depicted in Figure 10.20.3 and rotated through an angle $-\frac{1}{2}\pi$. Thus B is the point $z = c$, where c is given by (10.20.18).

For derivations of the results in this subsection, and also error bounds, see Olver (1997b, pp. 374–378). For extensions of the regions of validity in the z-plane and extensions to complex values of ν see Olver (1997b, pp. 378–382).

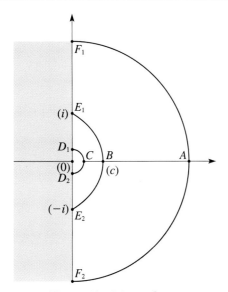

Figure 10.41.1: z-plane.

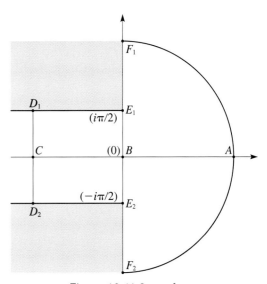

Figure 10.41.2: η-plane.

For expansions in inverse factorial series see Dunster et al. (1993).

10.41(iv) Double Asymptotic Properties

The series (10.41.3)–(10.41.6) can also be regarded as generalized asymptotic expansions for large $|z|$. Thus as $z \to \infty$ with $\ell \ (\geq 1)$ and $\nu \ (> 0)$ both fixed,

10.41.12
$$I_\nu(\nu z) = \frac{e^{\nu\eta}}{(2\pi\nu)^{\frac{1}{2}}(1+z^2)^{\frac{1}{4}}} \left(\sum_{k=0}^{\ell-1} \frac{U_k(p)}{\nu^k} + O\left(\frac{1}{z^\ell}\right) \right),$$
$$|\operatorname{ph} z| \leq \tfrac{1}{2}\pi - \delta,$$

10.41.13
$$K_\nu(\nu z) = \left(\frac{\pi}{2\nu}\right)^{\frac{1}{2}} \frac{e^{-\nu\eta}}{(1+z^2)^{\frac{1}{4}}}$$
$$\times \left(\sum_{k=0}^{\ell-1} (-1)^k \frac{U_k(p)}{\nu^k} + O\left(\frac{1}{z^\ell}\right) \right),$$
$$|\operatorname{ph} z| \leq \tfrac{3}{2}\pi - \delta.$$

Similarly for (10.41.5) and (10.41.6).

In the case of (10.41.13) with positive real values of z the result is a consequence of the error bounds given in Olver (1997b, pp. 377–378). Then by expanding the quantities η, $(1+z^2)^{-\frac{1}{4}}$, and $U_k(p)$, $k = 0, 1, \ldots, \ell - 1$, and rearranging, we arrive at an expansion of the right-hand side of (10.41.13) in powers of z^{-1}. Moreover, because of the uniqueness property of asymptotic expansions (§2.1(iii)) this expansion must agree with (10.40.2), with z replaced by νz, up to and including the term in $z^{-(\ell-1)}$. It also enjoys the same sector of validity.

To establish (10.41.12) we substitute into (10.34.3), with $m = 0$ and z replaced by νz, by means of (10.41.13) observing that when $|z|$ is large the effect of replacing z by $ze^{\pm\pi i}$ is to replace η, $(1+z^2)^{\frac{1}{4}}$, and p by $-\eta$, $\pm i(1+z^2)^{\frac{1}{4}}$, and $-p$, respectively.

10.41(v) Double Asymptotic Properties (Continued)

Similar analysis can be developed for the uniform asymptotic expansions in terms of Airy functions given in §10.20. We first prove that for the expansions (10.20.6) for the Hankel functions $H_\nu^{(1)}(\nu z)$ and $H_\nu^{(2)}(\nu z)$ the z-asymptotic property applies when $z \to \pm i\infty$, respectively. This is a consequence of the error bounds associated with these expansions. We then extend the validity of this property from $z \to \pm i\infty$ to $z \to \infty$ in the sector $-\pi + \delta \leq \mathrm{ph}\, z \leq 2\pi - \delta$ in the case of $H_\nu^{(1)}(\nu z)$, and to $z \to \infty$ in the sector $-2\pi + \delta \leq \mathrm{ph}\, z \leq \pi - \delta$ in the case of $H_\nu^{(2)}(\nu z)$. This is done by re-expansion with the aid of (10.20.10), (10.20.11), and §10.41(ii), followed by comparison with (10.17.5) and (10.17.6), with z replaced by νz. Lastly, we substitute into (10.4.4), again with z replaced by νz. The final results are:

10.41.14
$$J_\nu(\nu z) = \left(\frac{4\zeta}{1-z^2}\right)^{\frac{1}{4}} \left(\frac{\mathrm{Ai}\left(\nu^{\frac{2}{3}}\zeta\right)}{\nu^{\frac{1}{3}}} \left(\sum_{k=0}^{\ell} \frac{A_k(\zeta)}{\nu^{2k}} + O\left(\frac{1}{\zeta^{3\ell+3}}\right)\right) \right.$$
$$\left. + \frac{\mathrm{Ai}'\left(\nu^{\frac{2}{3}}\zeta\right)}{\nu^{\frac{5}{3}}} \left(\sum_{k=0}^{\ell-1} \frac{B_k(\zeta)}{\nu^{2k}} + O\left(\frac{1}{\zeta^{3\ell+1}}\right)\right)\right),$$

10.41.15
$$Y_\nu(\nu z) = -\left(\frac{4\zeta}{1-z^2}\right)^{\frac{1}{4}} \left(\frac{\mathrm{Bi}\left(\nu^{\frac{2}{3}}\zeta\right)}{\nu^{\frac{1}{3}}} \left(\sum_{k=0}^{\ell} \frac{A_k(\zeta)}{\nu^{2k}} + O\left(\frac{1}{\zeta^{3\ell+3}}\right)\right) \right.$$
$$\left. + \frac{\mathrm{Bi}'\left(\nu^{\frac{2}{3}}\zeta\right)}{\nu^{\frac{5}{3}}} \left(\sum_{k=0}^{\ell-1} \frac{B_k(\zeta)}{\nu^{2k}} + O\left(\frac{1}{\zeta^{3\ell+1}}\right)\right)\right),$$

as $z \to \infty$ in $|\mathrm{ph}\, z| \leq \pi - \delta$, or equivalently as $\zeta \to \infty$ in $|\mathrm{ph}(-\zeta)| \leq \frac{2}{3}\pi - \delta$, for fixed ℓ (≥ 0) and fixed ν (> 0).

It needs to be noted that the results (10.41.14) and (10.41.15) do *not* apply when $z \to 0+$ or equivalently $\zeta \to +\infty$. This is because $A_k(\zeta)$ and $\zeta^{-\frac{1}{2}} B_k(\zeta)$, $k = 0, 1, \ldots$, do not form an asymptotic scale (§2.1(v)) as $\zeta \to +\infty$; see Olver (1997b, pp. 422–425).

10.42 Zeros

Properties of the zeros of $I_\nu(z)$ and $K_\nu(z)$ may be deduced from those of $J_\nu(z)$ and $H_\nu^{(1)}(z)$, respectively, by application of the transformations (10.27.6) and (10.27.8).

For example, if ν is real, then the zeros of $I_\nu(z)$ are all complex unless $-2\ell < \nu < -(2\ell-1)$ for some positive integer ℓ, in which event $I_\nu(z)$ has two real zeros.

The distribution of the zeros of $K_n(nz)$ in the sector $-\frac{3}{2}\pi \leq \mathrm{ph}\, z \leq \frac{1}{2}\pi$ in the cases $n = 1, 5, 10$ is obtained on rotating Figures 10.21.2, 10.21.4, 10.21.6, respectively, through an angle $-\frac{1}{2}\pi$ so that in each case the cut lies along the positive imaginary axis. The zeros in the sector $-\frac{1}{2}\pi \leq \mathrm{ph}\, z \leq \frac{3}{2}\pi$ are their conjugates.

$K_n(z)$ has no zeros in the sector $|\mathrm{ph}\, z| \leq \frac{1}{2}\pi$; this result remains true when n is replaced by any real number ν. For the number of zeros of $K_\nu(z)$ in the sector $|\mathrm{ph}\, z| \leq \pi$, when ν is real, see Watson (1944, pp. 511–513).

See also Kerimov and Skorokhodov (1984b,a).

10.43 Integrals

10.43(i) Indefinite Integrals

Let $\mathscr{Z}_\nu(z)$ be defined as in §10.25(ii). Then

10.43.1
$$\int z^{\nu+1} \mathscr{Z}_\nu(z)\, dz = z^{\nu+1} \mathscr{Z}_{\nu+1}(z),$$
$$\int z^{-\nu+1} \mathscr{Z}_\nu(z)\, dz = z^{-\nu+1} \mathscr{Z}_{\nu-1}(z).$$

10.43.2
$$\int z^\nu \mathscr{Z}_\nu(z)\, dz = \pi^{\frac{1}{2}} 2^{\nu-1} \Gamma\left(\nu + \tfrac{1}{2}\right) z$$
$$\times \left(\mathscr{Z}_\nu(z) \mathbf{L}_{\nu-1}(z) - \mathscr{Z}_{\nu-1}(z) \mathbf{L}_\nu(z)\right),$$
$$\nu \neq -\tfrac{1}{2}.$$

For the modified Struve function $\mathbf{L}_\nu(z)$ see §11.2(i).

10.43.3
$$\int e^{\pm z} z^\nu \mathscr{Z}_\nu(z)\, dz = \frac{e^{\pm z} z^{\nu+1}}{2\nu+1} \left(\mathscr{Z}_\nu(z) \mp \mathscr{Z}_{\nu+1}(z)\right),$$
$$\nu \neq -\tfrac{1}{2},$$
$$\int e^{\pm z} z^{-\nu} \mathscr{Z}_\nu(z)\, dz = \frac{e^{\pm z} z^{-\nu+1}}{1-2\nu} \left(\mathscr{Z}_\nu(z) \mp \mathscr{Z}_{\nu-1}(z)\right),$$
$$\nu \neq \tfrac{1}{2}.$$

10.43(ii) Integrals over the Intervals $(0, x)$ and (x, ∞)

10.43.4
$$\int_0^x \frac{I_0(t) - 1}{t}\, dt$$
$$= \frac{1}{2} \sum_{k=1}^\infty (-1)^{k-1} \frac{\psi(k+1) - \psi(1)}{k!} \left(\tfrac{1}{2}x\right)^k I_k(x)$$
$$= \frac{2}{x} \sum_{k=0}^\infty (-1)^k (2k+3)(\psi(k+2) - \psi(1)) I_{2k+3}(x).$$

10.43.5
$$\int_x^\infty \frac{K_0(t)}{t}\, dt = \frac{1}{2}\left(\ln\left(\tfrac{1}{2}x\right) + \gamma\right)^2 + \frac{\pi^2}{24} - \sum_{k=1}^\infty \left(\psi(k+1)\right.$$
$$\left. + \frac{1}{2k} - \ln\left(\tfrac{1}{2}x\right)\right) \frac{\left(\tfrac{1}{2}x\right)^{2k}}{2k(k!)^2},$$

where $\psi = \Gamma'/\Gamma$ and γ is Euler's constant (§5.2).

10.43.6
$$\int_0^x e^{-t} I_n(t)\, dt = xe^{-x}(I_0(x) + I_1(x)) + n(e^{-x} I_0(x) - 1)$$
$$+ 2e^{-x} \sum_{k=1}^{n-1} (n-k)\, I_k(x),$$
$$n = 0, 1, 2, \ldots.$$

10.43.7
$$\int_0^x e^{\pm t} t^\nu I_\nu(t)\, dt = \frac{e^{\pm x} x^{\nu+1}}{2\nu+1} (I_\nu(x) \mp I_{\nu+1}(x)),$$
$$\Re\nu > -\tfrac{1}{2},$$

10.43.8
$$\int_0^x e^{\pm t} t^{-\nu} I_\nu(t)\, dt = -\frac{e^{\pm x} x^{-\nu+1}}{2\nu-1} (I_\nu(x) \mp I_{\nu-1}(x))$$
$$\mp \frac{2^{-\nu+1}}{(2\nu-1)\Gamma(\nu)}, \qquad \nu \neq \tfrac{1}{2}.$$

10.43.9
$$\int_0^x e^{\pm t} t^\nu K_\nu(t)\, dt = \frac{e^{\pm x} x^{\nu+1}}{2\nu+1} (K_\nu(x) \pm K_{\nu+1}(x))$$
$$\mp \frac{2^\nu \Gamma(\nu+1)}{2\nu+1}, \qquad \Re\nu > -\tfrac{1}{2},$$

10.43.10
$$\int_x^\infty e^t t^{-\nu} K_\nu(t)\, dt = \frac{e^x x^{-\nu+1}}{2\nu-1} (K_\nu(x) + K_{\nu-1}(x)),$$
$$\Re\nu > \tfrac{1}{2}.$$

10.43(iii) Fractional Integrals

The *Bickley function* $\operatorname{Ki}_\alpha(x)$ is defined by

10.43.11 $\operatorname{Ki}_\alpha(x) = \dfrac{1}{\Gamma(\alpha)} \displaystyle\int_x^\infty (t-x)^{\alpha-1} K_0(t)\, dt,$

when $\Re\alpha > 0$ and $x > 0$, and by analytic continuation elsewhere. Equivalently,

10.43.12
$$\operatorname{Ki}_\alpha(x) = \int_0^\infty \frac{e^{-x\cosh t}}{(\cosh t)^\alpha}\, dt, \qquad x > 0.$$

Properties

10.43.13 $\operatorname{Ki}_\alpha(x) = \displaystyle\int_x^\infty \operatorname{Ki}_{\alpha-1}(t)\, dt,$

10.43.14 $\operatorname{Ki}_0(x) = K_0(x),$

10.43.15 $\operatorname{Ki}_{-n}(x) = (-1)^n \dfrac{d^n}{dx^n} K_0(x), \quad n = 1, 2, 3, \ldots.$

10.43.16 $\operatorname{Ki}_\alpha(0) = \dfrac{\sqrt{\pi}\,\Gamma\left(\tfrac{1}{2}\alpha\right)}{2\,\Gamma\left(\tfrac{1}{2}\alpha + \tfrac{1}{2}\right)}, \quad \alpha \neq 0, -2, -4, \ldots.$

10.43.17
$$\alpha \operatorname{Ki}_{\alpha+1}(x) + x \operatorname{Ki}_\alpha(x)$$
$$+ (1-\alpha)\operatorname{Ki}_{\alpha-1}(x) - x \operatorname{Ki}_{\alpha-2}(x) = 0.$$

For further properties of the Bickley function, including asymptotic expansions and generalizations, see Amos (1983, 1989) and Luke (1962, Chapter 8).

10.43(iv) Integrals over the Interval $(0, \infty)$

10.43.18 $\displaystyle\int_0^\infty K_\nu(t)\, dt = \tfrac{1}{2}\pi \sec\left(\tfrac{1}{2}\pi\nu\right), \qquad |\Re\nu| < 1.$

10.43.19
$$\int_0^\infty t^{\mu-1} K_\nu(t)\, dt = 2^{\mu-2}\, \Gamma\left(\tfrac{1}{2}\mu - \tfrac{1}{2}\nu\right) \Gamma\left(\tfrac{1}{2}\mu + \tfrac{1}{2}\nu\right),$$
$$|\Re\nu| < \Re\mu.$$

10.43.20 $\displaystyle\int_0^\infty \cos(at) K_0(t)\, dt = \dfrac{\pi}{2(1+a^2)^{\frac{1}{2}}}, \quad |\Im a| < 1,$

10.43.21 $\displaystyle\int_0^\infty \sin(at) K_0(t)\, dt = \dfrac{\operatorname{arcsinh} a}{(1+a^2)^{\frac{1}{2}}}, \quad |\Im a| < 1.$

When $\Re\mu > |\Re\nu|$,

10.43.22
$$\int_0^\infty t^{\mu-1} e^{-at} K_\nu(t)\, dt = \begin{cases} \left(\tfrac{1}{2}\pi\right)^{\frac{1}{2}} \Gamma(\mu-\nu)\Gamma(\mu+\nu)(1-a^2)^{-\frac{1}{2}\mu+\frac{1}{4}} \mathsf{P}_{\nu-\frac{1}{2}}^{-\mu+\frac{1}{2}}(a), & -1 < a < 1, \\ \left(\tfrac{1}{2}\pi\right)^{\frac{1}{2}} \Gamma(\mu-\nu)\Gamma(\mu+\nu)(a^2-1)^{-\frac{1}{2}\mu+\frac{1}{4}} P_{\nu-\frac{1}{2}}^{-\mu+\frac{1}{2}}(a), & \Re a \geq 0, a \neq 1. \end{cases}$$

For the second equation there is a cut in the a-plane along the interval $[0,1]$, and all quantities assume their principal values (§4.2(i)). For the Ferrers function P and the associated Legendre function P, see §§14.3(i) and 14.21(i).

10.43.23 $\displaystyle\int_0^\infty t^{\nu+1} I_\nu(bt) \exp(-p^2 t^2)\, dt = \dfrac{b^\nu}{(2p^2)^{\nu+1}} \exp\left(\dfrac{b^2}{4p^2}\right), \qquad \Re\nu > -1,\ \Re(p^2) > 0,$

10.43.24 $\displaystyle\int_0^\infty I_\nu(bt) \exp(-p^2 t^2)\, dt = \dfrac{\sqrt{\pi}}{2p} \exp\left(\dfrac{b^2}{8p^2}\right) I_{\frac{1}{2}\nu}\left(\dfrac{b^2}{8p^2}\right), \qquad \Re\nu > -1,\ \Re(p^2) > 0,$

10.43.25 $\displaystyle\int_0^\infty K_\nu(bt) \exp(-p^2 t^2)\, dt = \dfrac{\sqrt{\pi}}{4p} \sec\left(\tfrac{1}{2}\pi\nu\right) \exp\left(\dfrac{b^2}{8p^2}\right) K_{\frac{1}{2}\nu}\left(\dfrac{b^2}{8p^2}\right), \qquad |\Re\nu| < 1,\ \Re(p^2) > 0.$

$$\text{10.43.26} \quad \int_0^\infty \frac{K_\mu(at)\, J_\nu(bt)}{t^\lambda}\, dt = \frac{b^\nu\, \Gamma\left(\tfrac{1}{2}\nu - \tfrac{1}{2}\lambda + \tfrac{1}{2}\mu + \tfrac{1}{2}\right) \Gamma\left(\tfrac{1}{2}\nu - \tfrac{1}{2}\lambda - \tfrac{1}{2}\mu + \tfrac{1}{2}\right)}{2^{\lambda+1} a^{\nu-\lambda+1}}$$
$$\times \mathbf{F}\left(\frac{\nu-\lambda+\mu+1}{2}, \frac{\nu-\lambda-\mu+1}{2}; \nu+1; -\frac{b^2}{a^2}\right),$$
$$\Re(\nu+1-\lambda) > |\Re\mu|,\ \Re a > |\Im b|.$$

For the hypergeometric function $\mathbf{F}$ see §15.2(i).

$$\text{10.43.27} \quad \int_0^\infty t^{\mu+\nu+1} K_\mu(at)\, J_\nu(bt)\, dt = \frac{(2a)^\mu (2b)^\nu \Gamma(\mu+\nu+1)}{(a^2+b^2)^{\mu+\nu+1}}, \qquad \Re(\nu+1) > |\Re\mu|,\ \Re a > |\Im b|.$$

$$\text{10.43.28} \quad \int_0^\infty t \exp(-p^2 t^2)\, I_\nu(at)\, I_\nu(bt)\, dt = \frac{1}{2p^2} \exp\left(\frac{a^2+b^2}{4p^2}\right) I_\nu\left(\frac{ab}{2p^2}\right), \qquad \Re\nu > -1,\ \Re(p^2) > 0,$$

$$\text{10.43.29} \quad \int_0^\infty t \exp(-p^2 t^2)\, I_0(at)\, K_0(at)\, dt = \frac{1}{4p^2} \exp\left(\frac{a^2}{2p^2}\right) K_0\left(\frac{a^2}{2p^2}\right), \qquad \Re(p^2) > 0.$$

For infinite integrals of triple products of modified and unmodified Bessel functions, see Gervois and Navelet (1984, 1985a,b, 1986a,b).

10.43(v) Kontorovich–Lebedev Transform

The *Kontorovich–Lebedev transform* of a function $g(x)$ is defined as

$$\text{10.43.30} \quad f(y) = \frac{2y}{\pi^2} \sinh(\pi y) \int_0^\infty \frac{g(x)}{x} K_{iy}(x)\, dx.$$

Then

$$\text{10.43.31} \quad g(x) = \int_0^\infty f(y)\, K_{iy}(x)\, dy,$$

provided that either of the following sets of conditions is satisfied:

(a) On the interval $0 < x < \infty$, $x^{-1} g(x)$ is continuously differentiable and each of $xg(x)$ and $x\, d(x^{-1}g(x))/dx$ is absolutely integrable.

(b) $g(x)$ is piecewise continuous and of bounded variation on every compact interval in $(0,\infty)$, and each of the following integrals

$$\text{10.43.32} \quad \int_0^{\tfrac{1}{2}} \frac{g(x)}{x} \ln\left(\frac{1}{x}\right) dx, \quad \int_{\tfrac{1}{2}}^\infty \frac{|g(x)|}{x^{\tfrac{1}{2}}}\, dx,$$

converges.

For asymptotic expansions of the direct transform (10.43.30) see Wong (1981), and for asymptotic expansions of the inverse transform (10.43.31) see Naylor (1990, 1996).

For collections of the Kontorovich–Lebedev transform, see Erdélyi et al. (1954b, Chapter 12), Prudnikov et al. (1986b, pp. 404–412), and Oberhettinger (1972, Chapter 5).

10.43(vi) Compendia

For collections of integrals of the functions $I_\nu(z)$ and $K_\nu(z)$, including integrals with respect to the order, see Apelblat (1983, §12), Erdélyi et al. (1953b, §§7.7.1–7.7.7 and 7.14–7.14.2), Erdélyi et al. (1954a,b), Gradshteyn and Ryzhik (2000, §§5.5, 6.5–6.7), Gröbner and Hofreiter (1950, pp. 197–203), Luke (1962), Magnus et al. (1966, §3.8), Marichev (1983, pp. 191–216), Oberhettinger (1972), Oberhettinger (1974, §§1.11 and 2.7), Oberhettinger (1990, §§1.17–1.20 and 2.17–2.20), Oberhettinger and Badii (1973, §§1.15 and 2.13), Okui (1974, 1975), Prudnikov et al. (1986b, §§1.11–1.12, 2.15–2.16, 3.2.8–3.2.10, and 3.4.1), Prudnikov et al. (1992a, §§3.15, 3.16), Prudnikov et al. (1992b, §§3.15, 3.16), Watson (1944, Chapter 13), and Wheelon (1968).

10.44 Sums

10.44(i) Multiplication Theorem

$$\text{10.44.1} \quad \mathscr{L}_\nu(\lambda z) = \lambda^{\pm\nu} \sum_{k=0}^\infty \frac{(\lambda^2-1)^k (\tfrac{1}{2}z)^k}{k!}\, \mathscr{L}_{\nu\pm k}(z),$$
$$|\lambda^2 - 1| < 1.$$

If $\mathscr{L} = I$ and the upper signs are taken, then the restriction on λ is unnecessary.

Examples

$$\text{10.44.2} \quad I_\nu(z) = \sum_{k=0}^\infty \frac{z^k}{k!} J_{\nu+k}(z), \quad J_\nu(z) = \sum_{k=0}^\infty (-1)^k \frac{z^k}{k!} I_{\nu+k}(z).$$

10.44(ii) Addition Theorems

Neumann's Addition Theorem

$$\text{10.44.3} \quad \mathscr{L}_\nu(u \pm v) = \sum_{k=-\infty}^\infty (\pm 1)^k \mathscr{L}_{\nu+k}(u)\, I_k(v), \quad |v| < |u|.$$

The restriction $|v| < |u|$ is unnecessary when $\mathscr{L} = I$ and ν is an integer.

Graf's and Gegenbauer's Addition Theorems

For results analogous to (10.23.7) and (10.23.8) see Watson (1944, §§11.3 and 11.41).

10.44(iii) Neumann-Type Expansions

10.44.4
$$\left(\tfrac{1}{2}z\right)^\nu = \sum_{k=0}^\infty (-1)^k \frac{(\nu+2k)\,\Gamma(\nu+k)}{k!} I_{\nu+2k}(z),$$
$$\nu \neq 0, -1, -2, \ldots.$$

10.44.5
$$K_0(z) = -\left(\ln\left(\tfrac{1}{2}z\right) + \gamma\right) I_0(z) + 2\sum_{k=1}^\infty \frac{I_{2k}(z)}{k},$$

10.44.6
$$K_n(z) = \frac{n!\left(\tfrac{1}{2}z\right)^{-n}}{2} \sum_{k=0}^{n-1} (-1)^k \frac{\left(\tfrac{1}{2}z\right)^k I_k(z)}{k!(n-k)}$$
$$+ (-1)^{n-1}\left(\ln\left(\tfrac{1}{2}z\right) - \psi(n+1)\right) I_n(z)$$
$$+ (-1)^n \sum_{k=1}^\infty \frac{(n+2k)\,I_{n+2k}(z)}{k(n+k)},$$

where γ is Euler's constant and $\psi = \Gamma'/\Gamma$ (§5.2).

10.44(iv) Compendia

For collections of sums and series involving modified Bessel functions see Erdélyi et al. (1953b, §7.15), Hansen (1975), and Prudnikov et al. (1986b, pp. 691–700).

10.45 Functions of Imaginary Order

With $z = x$, and ν replaced by $i\nu$, the modified Bessel's equation (10.25.1) becomes

10.45.1
$$x^2 \frac{d^2w}{dx^2} + x \frac{dw}{dx} + (\nu^2 - x^2)w = 0.$$

For $\nu \in \mathbb{R}$ and $x \in (0, \infty)$ define

10.45.2
$$\widetilde{I}_\nu(x) = \Re(I_{i\nu}(x)), \qquad \widetilde{K}_\nu(x) = K_{i\nu}(x).$$

Then

10.45.3
$$\widetilde{I}_{-\nu}(x) = \widetilde{I}_\nu(x), \qquad \widetilde{K}_{-\nu}(x) = \widetilde{K}_\nu(x),$$

and $\widetilde{I}_\nu(x)$, $\widetilde{K}_\nu(x)$ are real and linearly independent solutions of (10.45.1):

10.45.4
$$\mathscr{W}\{\widetilde{K}_\nu(x), \widetilde{I}_\nu(x)\} = 1/x.$$

As $x \to +\infty$

10.45.5
$$\widetilde{I}_\nu(x) = (2\pi x)^{-\frac{1}{2}} e^x \left(1 + O(x^{-1})\right),$$
$$\widetilde{K}_\nu(x) = (\pi/(2x))^{\frac{1}{2}} e^{-x} \left(1 + O(x^{-1})\right).$$

As $x \to 0+$

10.45.6
$$\widetilde{I}_\nu(x) = \left(\frac{\sinh(\pi\nu)}{\pi\nu}\right)^{\frac{1}{2}} \cos\left(\nu \ln\left(\tfrac{1}{2}x\right) - \gamma_\nu\right) + O(x^2),$$

where γ_ν is as in §10.24. The corresponding result for $\widetilde{K}_\nu(x)$ is given by

10.45.7
$$\widetilde{K}_\nu(x) = -\left(\frac{\pi}{\nu \sinh(\pi\nu)}\right)^{\frac{1}{2}} \sin\left(\nu \ln\left(\tfrac{1}{2}x\right) - \gamma_\nu\right) + O(x^2),$$

when $\nu > 0$, and

10.45.8 $\quad \widetilde{K}_0(x) = K_0(x) = -\ln\left(\tfrac{1}{2}x\right) - \gamma + O(x^2 \ln x)$,

where γ again denotes Euler's constant (§5.2(ii)).

In consequence of (10.45.5)–(10.45.7), $\widetilde{I}_\nu(x)$ and $\widetilde{K}_\nu(x)$ comprise a numerically satisfactory pair of solutions of (10.45.1) when x is large, and either $\widetilde{I}_\nu(x)$ and $(1/\pi)\sinh(\pi\nu)\widetilde{K}_\nu(x)$, or $\widetilde{I}_\nu(x)$ and $\widetilde{K}_\nu(x)$, comprise a numerically satisfactory pair when x is small, depending whether $\nu \neq 0$ or $\nu = 0$.

For graphs of $\widetilde{I}_\nu(x)$ and $\widetilde{K}_\nu(x)$ see §10.26(iii).

For properties of $\widetilde{I}_\nu(x)$ and $\widetilde{K}_\nu(x)$, including uniform asymptotic expansions for large ν and zeros, see Dunster (1990a). In this reference $\widetilde{I}_\nu(x)$ is denoted by $(1/\pi)\sinh(\pi\nu)L_{i\nu}(x)$. See also Gil et al. (2003a) and Balogh (1967).

10.46 Generalized and Incomplete Bessel Functions; Mittag-Leffler Function

The function $\phi(\rho, \beta; z)$ is defined by

10.46.1
$$\phi(\rho, \beta; z) = \sum_{k=0}^\infty \frac{z^k}{k!\,\Gamma(\rho k + \beta)}, \qquad \rho > -1.$$

From (10.25.2)

10.46.2 $\quad I_\nu(z) = \left(\tfrac{1}{2}z\right)^\nu \phi\left(1, \nu+1; \tfrac{1}{4}z^2\right).$

For asymptotic expansions of $\phi(\rho, \beta; z)$ as $z \to \infty$ in various sectors of the complex z-plane for fixed real values of ρ and fixed real or complex values of β, see Wright (1935) when $\rho > 0$, and Wright (1940b) when $-1 < \rho < 0$. For exponentially-improved asymptotic expansions in the same circumstances, together with smooth interpretations of the corresponding Stokes phenomenon (§§2.11(iii)–2.11(v)) see Wong and Zhao (1999a) when $\rho > 0$, and Wong and Zhao (1999b) when $-1 < \rho < 0$.

The Laplace transform of $\phi(\rho, \beta; z)$ can be expressed in terms of the *Mittag-Leffler function*:

10.46.3
$$E_{a,b}(z) = \sum_{k=0}^\infty \frac{z^k}{\Gamma(ak+b)}, \qquad a > 0.$$

See Paris (2002c). This reference includes exponentially-improved asymptotic expansions for $E_{a,b}(z)$ when $|z| \to \infty$, together with a smooth interpretation of Stokes phenomena. See also Wong and Zhao (2002a), and for further information on the Mittag-Leffler function see Erdélyi et al. (1955, §18.1) and Paris and Kaminski (2001, §5.1.4).

For incomplete modified Bessel functions and Hankel functions, including applications, see Cicchetti and Faraone (2004).

Spherical Bessel Functions

10.47 Definitions and Basic Properties

10.47(i) Differential Equations

10.47.1 $\quad z^2 \dfrac{d^2w}{dz^2} + 2z\dfrac{dw}{dz} + \left(z^2 - n(n+1)\right)w = 0,$

10.47.2 $\quad z^2 \dfrac{d^2w}{dz^2} + 2z\dfrac{dw}{dz} - \left(z^2 + n(n+1)\right)w = 0.$

Here, and throughout the remainder of §§10.47–10.60, *n is a nonnegative integer*. (This is in contrast to other treatments of spherical Bessel functions, including Abramowitz and Stegun (1964, Chapter 10), in which n can be any integer. However, there is a gain in symmetry, without any loss of generality in applications, on restricting $n \geq 0$.)

Equations (10.47.1) and (10.47.2) each have a regular singularity at $z = 0$ with indices $n, -n-1$, and an irregular singularity at $z = \infty$ of rank 1; compare §§2.7(i)–2.7(ii).

10.47(ii) Standard Solutions

Equation (10.47.1)

10.47.3
$$\mathsf{j}_n(z) = \sqrt{\tfrac{1}{2}\pi/z}\, J_{n+\frac{1}{2}}(z) = (-1)^n \sqrt{\tfrac{1}{2}\pi/z}\, Y_{-n-\frac{1}{2}}(z),$$

10.47.4
$$\mathsf{y}_n(z) = \sqrt{\tfrac{1}{2}\pi/z}\, Y_{n+\frac{1}{2}}(z) = (-1)^{n+1} \sqrt{\tfrac{1}{2}\pi/z}\, J_{-n-\frac{1}{2}}(z),$$

10.47.5
$$\mathsf{h}_n^{(1)}(z) = \sqrt{\tfrac{1}{2}\pi/z}\, H_{n+\frac{1}{2}}^{(1)}(z) = (-1)^{n+1} i\sqrt{\tfrac{1}{2}\pi/z}\, H_{-n-\frac{1}{2}}^{(1)}(z),$$

10.47.6
$$\mathsf{h}_n^{(2)}(z) = \sqrt{\tfrac{1}{2}\pi/z}\, H_{n+\frac{1}{2}}^{(2)}(z) = (-1)^n i\sqrt{\tfrac{1}{2}\pi/z}\, H_{-n-\frac{1}{2}}^{(2)}(z).$$

$\mathsf{j}_n(z)$ and $\mathsf{y}_n(z)$ are the *spherical Bessel functions of the first and second kinds*, respectively; $\mathsf{h}_n^{(1)}(z)$ and $\mathsf{h}_n^{(2)}(z)$ are the *spherical Bessel functions of the third kind*.

Equation (10.47.2)

10.47.7 $\quad \mathsf{i}_n^{(1)}(z) = \sqrt{\tfrac{1}{2}\pi/z}\, I_{n+\frac{1}{2}}(z)$

10.47.8 $\quad \mathsf{i}_n^{(2)}(z) = \sqrt{\tfrac{1}{2}\pi/z}\, I_{-n-\frac{1}{2}}(z)$

10.47.9 $\quad \mathsf{k}_n(z) = \sqrt{\tfrac{1}{2}\pi/z}\, K_{n+\frac{1}{2}}(z) = \sqrt{\tfrac{1}{2}\pi/z}\, K_{-n-\frac{1}{2}}(z).$

$\mathsf{i}_n^{(1)}(z)$, $\mathsf{i}_n^{(2)}(z)$, and $\mathsf{k}_n(z)$ are the *modified spherical Bessel functions*.

Many properties of $\mathsf{j}_n(z)$, $\mathsf{y}_n(z)$, $\mathsf{h}_n^{(1)}(z)$, $\mathsf{h}_n^{(2)}(z)$, $\mathsf{i}_n^{(1)}(z)$, $\mathsf{i}_n^{(2)}(z)$, and $\mathsf{k}_n(z)$ follow straightforwardly from the above definitions and results given in preceding sections of this chapter. For example, $z^{-n}\mathsf{j}_n(z)$, $z^{n+1}\mathsf{y}_n(z)$, $z^{n+1}\mathsf{h}_n^{(1)}(z)$, $z^{n+1}\mathsf{h}_n^{(2)}(z)$, $z^{-n}\mathsf{i}_n^{(1)}(z)$, $z^{n+1}\mathsf{i}_n^{(2)}(z)$, and $z^{n+1}\mathsf{k}_n(z)$ are all entire functions of z.

10.47(iii) Numerically Satisfactory Pairs of Solutions

For (10.47.1) numerically satisfactory pairs of solutions are given by Table 10.2.1 with the symbols J, Y, H, and ν replaced by j, y, h, and n, respectively.

For (10.47.2) numerically satisfactory pairs of solutions are $\mathsf{i}_n^{(1)}(z)$ and $\mathsf{k}_n(z)$ in the right half of the z-plane, and $\mathsf{i}_n^{(1)}(z)$ and $\mathsf{k}_n(-z)$ in the left half of the z-plane.

10.47(iv) Interrelations

10.47.10
$$\mathsf{h}_n^{(1)}(z) = \mathsf{j}_n(z) + i\,\mathsf{y}_n(z), \quad \mathsf{h}_n^{(2)}(z) = \mathsf{j}_n(z) - i\,\mathsf{y}_n(z).$$

10.47.11 $\quad \mathsf{k}_n(z) = (-1)^{n+1}\tfrac{1}{2}\pi \left(\mathsf{i}_n^{(1)}(z) - \mathsf{i}_n^{(2)}(z)\right).$

10.47.12 $\quad \mathsf{i}_n^{(1)}(z) = i^{-n}\,\mathsf{j}_n(iz), \quad \mathsf{i}_n^{(2)}(z) = i^{-n-1}\,\mathsf{y}_n(iz).$

10.47.13 $\quad \mathsf{k}_n(z) = -\tfrac{1}{2}\pi i^n\, \mathsf{h}_n^{(1)}(iz) = -\tfrac{1}{2}\pi i^{-n}\, \mathsf{h}_n^{(2)}(-iz).$

10.47(v) Reflection Formulas

10.47.14
$$\mathsf{j}_n(-z) = (-1)^n\,\mathsf{j}_n(z), \quad \mathsf{y}_n(-z) = (-1)^{n+1}\,\mathsf{y}_n(z),$$

10.47.15
$$\mathsf{h}_n^{(1)}(-z) = (-1)^n\,\mathsf{h}_n^{(2)}(z), \quad \mathsf{h}_n^{(2)}(-z) = (-1)^n\,\mathsf{h}_n^{(1)}(z).$$

10.47.16
$$\mathsf{i}_n^{(1)}(-z) = (-1)^n\,\mathsf{i}_n^{(1)}(z), \quad \mathsf{i}_n^{(2)}(-z) = (-1)^{n+1}\,\mathsf{i}_n^{(2)}(z),$$

10.47.17 $\quad \mathsf{k}_n(-z) = -\tfrac{1}{2}\pi\left(\mathsf{i}_n^{(1)}(z) + \mathsf{i}_n^{(2)}(z)\right).$

10.48 Graphs

For unmodified spherical Bessel functions see Figures 10.48.1–10.48.4. For modified spherical Bessel functions see Figures 10.48.5–10.48.7.

10.48 Graphs

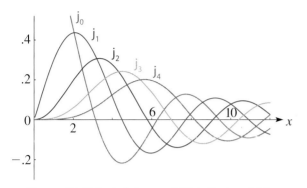

Figure 10.48.1: $\mathsf{j}_n(x), n = 0(1)4, 0 \leq x \leq 12$.

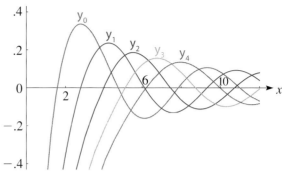

Figure 10.48.2: $\mathsf{y}_n(x), n = 0(1)4, 0 < x \leq 12$.

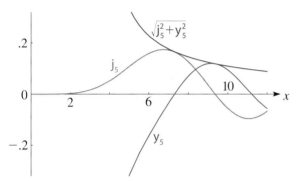

Figure 10.48.3: $\mathsf{j}_5(x), \mathsf{y}_5(x), \sqrt{\mathsf{j}_5^2(x) + \mathsf{y}_5^2(x)}, 0 \leq x \leq 12$.

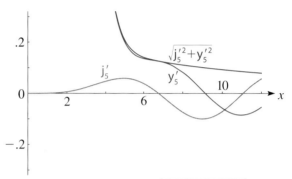

Figure 10.48.4: $\mathsf{j}_5'(x), \mathsf{y}_5'(x), \sqrt{\mathsf{j}_5'^2(x) + \mathsf{y}_5'^2(x)}, 0 \leq x \leq 12$.

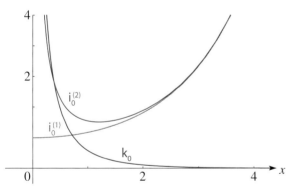

Figure 10.48.5: $\mathsf{i}_0^{(1)}(x), \mathsf{i}_0^{(2)}(x), \mathsf{k}_0(x), 0 \leq x \leq 4$.

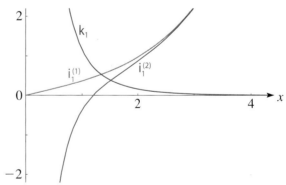

Figure 10.48.6: $\mathsf{i}_1^{(1)}(x), \mathsf{i}_1^{(2)}(x), \mathsf{k}_1(x), 0 \leq x \leq 4$.

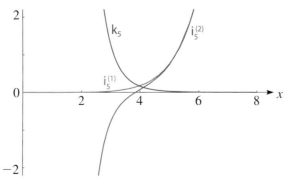

Figure 10.48.7: $\mathsf{i}_5^{(1)}(x), \mathsf{i}_5^{(2)}(x), \mathsf{k}_5(x), 0 \leq x \leq 8$.

10.49 Explicit Formulas

10.49(i) Unmodified Functions

Define $a_k(\nu)$ as in (10.17.1). Then

10.49.1
$$a_k(n + \tfrac{1}{2}) = \begin{cases} \dfrac{(n+k)!}{2^k k!(n-k)!}, & k = 0, 1, \ldots, n, \\ 0, & k = n+1, n+2, \ldots. \end{cases}$$

10.49.2
$$\mathsf{j}_n(z) = \sin(z - \tfrac{1}{2}n\pi) \sum_{k=0}^{\lfloor n/2 \rfloor} (-1)^k \frac{a_{2k}(n+\tfrac{1}{2})}{z^{2k+1}}$$
$$+ \cos(z - \tfrac{1}{2}n\pi) \sum_{k=0}^{\lfloor (n-1)/2 \rfloor} (-1)^k \frac{a_{2k+1}(n+\tfrac{1}{2})}{z^{2k+2}}.$$

10.49.3
$$\mathsf{j}_0(z) = \frac{\sin z}{z}, \quad \mathsf{j}_1(z) = \frac{\sin z}{z^2} - \frac{\cos z}{z},$$
$$\mathsf{j}_2(z) = \left(-\frac{1}{z} + \frac{3}{z^3}\right) \sin z - \frac{3}{z^2} \cos z.$$

10.49.4
$$\mathsf{y}_n(z) = -\cos(z - \tfrac{1}{2}n\pi) \sum_{k=0}^{\lfloor n/2 \rfloor} (-1)^k \frac{a_{2k}(n+\tfrac{1}{2})}{z^{2k+1}}$$
$$+ \sin(z - \tfrac{1}{2}n\pi) \sum_{k=0}^{\lfloor (n-1)/2 \rfloor} (-1)^k \frac{a_{2k+1}(n+\tfrac{1}{2})}{z^{2k+2}}.$$

10.49.5
$$\mathsf{y}_0(z) = -\frac{\cos z}{z}, \quad \mathsf{y}_1(z) = -\frac{\cos z}{z^2} - \frac{\sin z}{z},$$
$$\mathsf{y}_2(z) = \left(\frac{1}{z} - \frac{3}{z^3}\right) \cos z - \frac{3}{z^2} \sin z.$$

10.49.6
$$\mathsf{h}_n^{(1)}(z) = e^{iz} \sum_{k=0}^n i^{k-n-1} \frac{a_k(n+\tfrac{1}{2})}{z^{k+1}},$$

10.49.7
$$\mathsf{h}_n^{(2)}(z) = e^{-iz} \sum_{k=0}^n (-i)^{k-n-1} \frac{a_k(n+\tfrac{1}{2})}{z^{k+1}}.$$

10.49(ii) Modified Functions

Again, with $a_k(n+\tfrac{1}{2})$ as in (10.49.1),

10.49.8
$$\mathsf{i}_n^{(1)}(z) = \tfrac{1}{2} e^z \sum_{k=0}^n (-1)^k \frac{a_k(n+\tfrac{1}{2})}{z^{k+1}}$$
$$+ (-1)^{n+1} \tfrac{1}{2} e^{-z} \sum_{k=0}^n \frac{a_k(n+\tfrac{1}{2})}{z^{k+1}}.$$

10.49.9
$$\mathsf{i}_0^{(1)}(z) = \frac{\sinh z}{z}, \quad \mathsf{i}_1^{(1)}(z) = -\frac{\sinh z}{z^2} + \frac{\cosh z}{z},$$
$$\mathsf{i}_2^{(1)}(z) = \left(\frac{1}{z} + \frac{3}{z^3}\right) \sinh z - \frac{3}{z^2} \cosh z.$$

10.49.10
$$\mathsf{i}_n^{(2)}(z) = \tfrac{1}{2} e^z \sum_{k=0}^n (-1)^k \frac{a_k(n+\tfrac{1}{2})}{z^{k+1}}$$
$$+ (-1)^n \tfrac{1}{2} e^{-z} \sum_{k=0}^n \frac{a_k(n+\tfrac{1}{2})}{z^{k+1}}.$$

10.49.11
$$\mathsf{i}_0^{(2)}(z) = \frac{\cosh z}{z}, \quad \mathsf{i}_1^{(2)}(z) = -\frac{\cosh z}{z^2} + \frac{\sinh z}{z},$$
$$\mathsf{i}_2^{(2)}(z) = \left(\frac{1}{z} + \frac{3}{z^3}\right) \cosh z - \frac{3}{z^2} \sinh z.$$

10.49.12
$$\mathsf{k}_n(z) = \tfrac{1}{2} \pi e^{-z} \sum_{k=0}^n \frac{a_k(n+\tfrac{1}{2})}{z^{k+1}}.$$

10.49.13
$$\mathsf{k}_0(z) = \tfrac{1}{2}\pi \frac{e^{-z}}{z}, \quad \mathsf{k}_1(z) = \tfrac{1}{2}\pi e^{-z}\left(\frac{1}{z} + \frac{1}{z^2}\right),$$
$$\mathsf{k}_2(z) = \tfrac{1}{2}\pi e^{-z}\left(\frac{1}{z} + \frac{3}{z^2} + \frac{3}{z^3}\right).$$

$\sum_{k=0}^n a_k(n+\tfrac{1}{2}) z^{n-k}$ is sometimes called the *Bessel polynomial of degree n*. For a survey of properties of these polynomials and their generalizations see Grosswald (1978). See also §18.34, de Bruin *et al.* (1981a,b), and Dunster (2001c).

10.49(iii) Rayleigh's Formulas

10.49.14
$$\mathsf{j}_n(z) = z^n \left(-\frac{1}{z}\frac{d}{dz}\right)^n \frac{\sin z}{z},$$
$$\mathsf{y}_n(z) = -z^n \left(-\frac{1}{z}\frac{d}{dz}\right)^n \frac{\cos z}{z}.$$

10.49.15
$$\mathsf{i}_n^{(1)}(z) = z^n \left(\frac{1}{z}\frac{d}{dz}\right)^n \frac{\sinh z}{z},$$
$$\mathsf{i}_n^{(2)}(z) = z^n \left(\frac{1}{z}\frac{d}{dz}\right)^n \frac{\cosh z}{z}.$$

10.49.16
$$\mathsf{k}_n(z) = (-1)^n \tfrac{1}{2}\pi z^n \left(\frac{1}{z}\frac{d}{dz}\right)^n \frac{e^{-z}}{z}.$$

10.49(iv) Sums or Differences of Squares

Denote

10.49.17
$$s_k(n+\tfrac{1}{2}) = \frac{(2k)!(n+k)!}{2^{2k}(k!)^2(n-k)!}, \quad k = 0, 1, \ldots, n.$$

Then

10.49.18
$$\mathsf{j}_n^2(z) + \mathsf{y}_n^2(z) = \sum_{k=0}^n \frac{s_k(n+\tfrac{1}{2})}{z^{2k+2}}.$$

10.49.19
$$\mathsf{j}_0^2(z) + \mathsf{y}_0^2(z) = z^{-2}, \quad \mathsf{j}_1^2(z) + \mathsf{y}_1^2(z) = z^{-2} + z^{-4},$$
$$\mathsf{j}_2^2(z) + \mathsf{y}_2^2(z) = z^{-2} + 3z^{-4} + 9z^{-6}.$$

10.49.20
$$\left(\mathsf{i}_n^{(1)}(z)\right)^2 - \left(\mathsf{i}_n^{(2)}(z)\right)^2 = (-1)^{n+1} \sum_{k=0}^n (-1)^k \frac{s_k(n+\tfrac{1}{2})}{z^{2k+2}}.$$

10.49.21
$$\left(\mathsf{i}_0^{(1)}(z)\right)^2 - \left(\mathsf{i}_0^{(2)}(z)\right)^2 = -z^{-2},$$
$$\left(\mathsf{i}_1^{(1)}(z)\right)^2 - \left(\mathsf{i}_1^{(2)}(z)\right)^2 = z^{-2} - z^{-4},$$
$$\left(\mathsf{i}_2^{(1)}(z)\right)^2 - \left(\mathsf{i}_2^{(2)}(z)\right)^2 = -z^{-2} + 3z^{-4} - 9z^{-6}.$$

10.50 Wronskians and Cross Products

10.50.1
$$\mathscr{W}\{\mathsf{j}_n(z), \mathsf{y}_n(z)\} = z^{-2},$$
$$\mathscr{W}\left\{\mathsf{h}_n^{(1)}(z), \mathsf{h}_n^{(2)}(z)\right\} = -2iz^{-2}.$$

10.50.2
$$\mathscr{W}\left\{\mathsf{i}_n^{(1)}(z), \mathsf{i}_n^{(2)}(z)\right\} = (-1)^{n+1} z^{-2},$$
$$\mathscr{W}\left\{\mathsf{i}_n^{(1)}(z), \mathsf{k}_n(z)\right\} = \mathscr{W}\left\{\mathsf{i}_n^{(2)}(z), \mathsf{k}_n(z)\right\}$$
$$= -\tfrac{1}{2}\pi z^{-2}.$$

10.50.3
$$\mathsf{j}_{n+1}(z)\,\mathsf{y}_n(z) - \mathsf{j}_n(z)\,\mathsf{y}_{n+1}(z) = z^{-2},$$
$$\mathsf{j}_{n+2}(z)\,\mathsf{y}_n(z) - \mathsf{j}_n(z)\,\mathsf{y}_{n+2}(z) = (2n+3)z^{-3}.$$

10.50.4
$$\mathsf{j}_0(z)\,\mathsf{j}_n(z) + \mathsf{y}_0(z)\,\mathsf{y}_n(z)$$
$$= \cos\!\left(\tfrac{1}{2}n\pi\right) \sum_{k=0}^{\lfloor n/2 \rfloor} (-1)^k \frac{a_{2k}(n+\tfrac{1}{2})}{z^{2k+2}}$$
$$+ \sin\!\left(\tfrac{1}{2}n\pi\right) \sum_{k=0}^{\lfloor (n-1)/2 \rfloor} (-1)^k \frac{a_{2k+1}(n+\tfrac{1}{2})}{z^{2k+3}},$$

where $a_k(n+\tfrac{1}{2})$ is given by (10.49.1).

Results corresponding to (10.50.3) and (10.50.4) for $\mathsf{i}_n^{(1)}(z)$ and $\mathsf{i}_n^{(2)}(z)$ are obtainable via (10.47.12).

10.51 Recurrence Relations and Derivatives

10.51(i) Unmodified Functions

Let $f_n(z)$ denote any of $\mathsf{j}_n(z)$, $\mathsf{y}_n(z)$, $\mathsf{h}_n^{(1)}(z)$, or $\mathsf{h}_n^{(2)}(z)$. Then

10.51.1
$$f_{n-1}(z) + f_{n+1}(z) = ((2n+1)/z) f_n(z),$$
$$nf_{n-1}(z) - (n+1) f_{n+1}(z) = (2n+1) f_n'(z),$$
$$n = 1, 2, \ldots,$$

10.51.2
$$f_n'(z) = f_{n-1}(z) - ((n+1)/z) f_n(z), \ n = 1, 2, \ldots,$$
$$f_n'(z) = -f_{n+1}(z) + (n/z) f_n(z), \quad n = 0, 1, \ldots.$$

10.51.3
$$\left(\frac{1}{z}\frac{d}{dz}\right)^m (z^{n+1} f_n(z)) = z^{n-m+1} f_{n-m}(z),$$
$$m = 0, 1, \ldots, n,$$
$$\left(\frac{1}{z}\frac{d}{dz}\right)^m (z^{-n} f_n(z)) = (-1)^m z^{-n-m} f_{n+m}(z),$$
$$m = 0, 1, \ldots.$$

10.51(ii) Modified Functions

Let $g_n(z)$ denote $\mathsf{i}_n^{(1)}(z)$, $\mathsf{i}_n^{(2)}(z)$, or $(-1)^n\,\mathsf{k}_n(z)$. Then
$$g_{n-1}(z) - g_{n+1}(z) = ((2n+1)/z) g_n(z)$$

10.51.4
$$ng_{n-1}(z) + (n+1) g_{n+1}(z) = (2n+1) g_n'(z),$$
$$n = 1, 2, \ldots,$$

10.51.5
$$g_n'(z) = g_{n-1}(z) - ((n+1)/z) g_n(z), \ n = 1, 2, \ldots,$$
$$g_n'(z) = g_{n+1}(z) + (n/z) g_n(z), \quad n = 0, 1, \ldots.$$

10.51.6
$$\left(\frac{1}{z}\frac{d}{dz}\right)^m (z^{n+1} g_n(z)) = z^{n-m+1} g_{n-m}(z),$$
$$m = 0, 1, \ldots, n,$$
$$\left(\frac{1}{z}\frac{d}{dz}\right)^m (z^{-n} g_n(z)) = z^{-n-m} g_{n+m}(z),$$
$$m = 0, 1, \ldots.$$

10.52 Limiting Forms

10.52(i) $z \to 0$

10.52.1
$$\mathsf{j}_n(z), \mathsf{i}_n^{(1)}(z) \sim z^n/(2n+1)!!,$$

10.52.2
$$-\mathsf{y}_n(z), i\,\mathsf{h}_n^{(1)}(z), -i\,\mathsf{h}_n^{(2)}(z), (-1)^n\,\mathsf{i}_n^{(2)}(z), (2/\pi)\,\mathsf{k}_n(z)$$
$$\sim (2n-1)!!/z^{n+1}.$$

10.52(ii) $z \to \infty$

10.52.3
$$\mathsf{j}_n(z) = z^{-1} \sin(z - \tfrac{1}{2}n\pi) + e^{|\Im z|} O(z^{-2}),$$
$$\mathsf{y}_n(z) = -z^{-1} \cos(z - \tfrac{1}{2}n\pi) + e^{|\Im z|} O(z^{-2}),$$

10.52.4
$$\mathsf{h}_n^{(1)}(z) \sim i^{-n-1} z^{-1} e^{iz}, \quad \mathsf{h}_n^{(2)}(z) \sim i^{n+1} z^{-1} e^{-iz},$$

10.52.5
$$\mathsf{i}_n^{(1)}(z) \sim \mathsf{i}_n^{(2)}(z) \sim \tfrac{1}{2} z^{-1} e^z, \ |\operatorname{ph} z| \leq \tfrac{1}{2}\pi - \delta (<\tfrac{1}{2}\pi),$$

10.52.6
$$\mathsf{k}_n(z) \sim \tfrac{1}{2}\pi z^{-1} e^{-z}.$$

10.53 Power Series

10.53.1
$$\mathsf{j}_n(z) = z^n \sum_{k=0}^{\infty} \frac{(-\tfrac{1}{2} z^2)^k}{k!(2n+2k+1)!!},$$

10.53.2
$$\mathsf{y}_n(z) = -\frac{1}{z^{n+1}} \sum_{k=0}^{n} \frac{(2n-2k-1)!!(\tfrac{1}{2} z^2)^k}{k!}$$
$$+ \frac{(-1)^{n+1}}{z^{n+1}} \sum_{k=n+1}^{\infty} \frac{(-\tfrac{1}{2} z^2)^k}{k!(2k-2n-1)!!}.$$

10.53.3 $\quad \mathsf{i}_n^{(1)}(z) = z^n \sum_{k=0}^{\infty} \frac{(\frac{1}{2}z^2)^k}{k!(2n+2k+1)!!},$

10.53.4
$$\mathsf{i}_n^{(2)}(z) = \frac{(-1)^n}{z^{n+1}} \sum_{k=0}^{n} \frac{(2n-2k-1)!!(-\frac{1}{2}z^2)^k}{k!}$$
$$+ \frac{1}{z^{n+1}} \sum_{k=n+1}^{\infty} \frac{(\frac{1}{2}z^2)^k}{k!(2k-2n-1)!!}.$$

For $\mathsf{h}_n^{(1)}(z)$ and $\mathsf{h}_n^{(2)}(z)$ combine (10.47.10), (10.53.1), and (10.53.2). For $\mathsf{k}_n(z)$ combine (10.47.11), (10.53.3), and (10.53.4).

10.54 Integral Representations

10.54.1 $\quad \mathsf{j}_n(z) = \frac{z^n}{2^{n+1} n!} \int_0^{\pi} \cos(z \cos\theta)(\sin\theta)^{2n+1} \, d\theta.$

10.54.2 $\quad \mathsf{j}_n(z) = \frac{(-i)^n}{2} \int_0^{\pi} e^{iz\cos\theta} P_n(\cos\theta) \sin\theta \, d\theta.$

10.54.3 $\quad \mathsf{k}_n(z) = \frac{\pi}{2} \int_1^{\infty} e^{-zt} P_n(t) \, dt, \qquad |\mathrm{ph}\, z| < \tfrac{1}{2}\pi.$

10.54.4 $\quad \mathsf{j}_n(z) = \frac{(-i)^{n+1}}{2\pi} \int_{i\infty}^{(-1+,1+)} e^{izt} Q_n(t) \, dt,$
$$|\mathrm{ph}\, z| < \tfrac{1}{2}\pi.$$

10.54.5
$$\mathsf{h}_n^{(1)}(z) = \frac{(-i)^{n+1}}{\pi} \int_{i\infty}^{(1+)} e^{izt} Q_n(t) \, dt,$$
$$\mathsf{h}_n^{(2)}(z) = \frac{(-i)^{n+1}}{\pi} \int_{i\infty}^{(-1+)} e^{izt} Q_n(t) \, dt,$$
$$|\mathrm{ph}\, z| < \tfrac{1}{2}\pi.$$

For the Legendre polynomial P_n and the associated Legendre function Q_n see §§18.3 and 14.21(i), with $\mu = 0$ and $\nu = n$.

Additional integral representations can be obtained by combining the definitions (10.47.3)–(10.47.9) with the results given in §10.9 and §10.32.

10.55 Continued Fractions

For continued fractions for $\mathsf{j}_{n+1}(z)/\mathsf{j}_n(z)$ and $\mathsf{i}_{n+1}^{(1)}(z)/\mathsf{i}_n^{(1)}(z)$ see Cuyt et al. (2008, pp. 350, 353, 362, 363, 367–369).

10.56 Generating Functions

When $2|t| < |z|$,

10.56.1 $\quad \frac{\cos\sqrt{z^2 - 2zt}}{z} = \frac{\cos z}{z} + \sum_{n=1}^{\infty} \frac{t^n}{n!} \mathsf{j}_{n-1}(z),$

10.56.2 $\quad \frac{\sin\sqrt{z^2 - 2zt}}{z} = \frac{\sin z}{z} + \sum_{n=1}^{\infty} \frac{t^n}{n!} \mathsf{y}_{n-1}(z).$

10.56.3 $\quad \frac{\cosh\sqrt{z^2 + 2izt}}{z} = \frac{\cosh z}{z} + \sum_{n=1}^{\infty} \frac{(it)^n}{n!} \mathsf{i}_{n-1}^{(1)}(z),$

10.56.4 $\quad \frac{\sinh\sqrt{z^2 + 2izt}}{z} = \frac{\sinh z}{z} + \sum_{n=1}^{\infty} \frac{(it)^n}{n!} \mathsf{i}_{n-1}^{(2)}(z),$

10.56.5
$$\frac{\exp(-\sqrt{z^2 + 2izt})}{z} = \frac{e^{-z}}{z} + \frac{2}{\pi} \sum_{n=1}^{\infty} \frac{(-it)^n}{n!} \mathsf{k}_{n-1}(z).$$

10.57 Uniform Asymptotic Expansions for Large Order

Asymptotic expansions for $\mathsf{j}_n((n+\tfrac{1}{2})z)$, $\mathsf{y}_n((n+\tfrac{1}{2})z)$, $\mathsf{h}_n^{(1)}((n+\tfrac{1}{2})z)$, $\mathsf{h}_n^{(2)}((n+\tfrac{1}{2})z)$, $\mathsf{i}_n^{(1)}((n+\tfrac{1}{2})z)$, and $\mathsf{k}_n((n+\tfrac{1}{2})z)$ as $n \to \infty$ that are uniform with respect to z can be obtained from the results given in §§10.20 and 10.41 by use of the definitions (10.47.3)–(10.47.7) and (10.47.9). Subsequently, for $\mathsf{i}_n^{(2)}((n+\tfrac{1}{2})z)$ the connection formula (10.47.11) is available.

For the corresponding expansion for $\mathsf{j}_n'((n+\tfrac{1}{2})z)$ use

10.57.1
$$\mathsf{j}_n'((n+\tfrac{1}{2})z) = \frac{\pi^{\frac{1}{2}}}{((2n+1)z)^{\frac{1}{2}}} J_{n+\frac{1}{2}}'((n+\tfrac{1}{2})z)$$
$$- \frac{\pi^{\frac{1}{2}}}{((2n+1)z)^{\frac{3}{2}}} J_{n+\frac{1}{2}}((n+\tfrac{1}{2})z).$$

Similarly for the expansions of the derivatives of the other six functions.

10.58 Zeros

For $n \geq 0$ the mth positive zeros of $\mathsf{j}_n(x)$, $\mathsf{j}_n'(x)$, $\mathsf{y}_n(x)$, and $\mathsf{y}_n'(x)$ are denoted by $a_{n,m}$, $a_{n,m}'$, $b_{n,m}$, and $b_{n,m}'$, respectively, except that for $n = 0$ we count $x = 0$ as the first zero of $\mathsf{j}_0'(x)$.

With the notation of §10.21(i),

10.58.1 $\quad a_{n,m} = j_{n+\frac{1}{2},m}, \qquad b_{n,m} = y_{n+\frac{1}{2},m},$

10.58.2
$$\mathsf{j}_n'(a_{n,m}) = \sqrt{\frac{\pi}{2\, j_{n+\frac{1}{2},m}}}\, J_{n+\frac{1}{2}}'\left(j_{n+\frac{1}{2},m}\right),$$
$$\mathsf{y}_n'(b_{n,m}) = \sqrt{\frac{\pi}{2\, y_{n+\frac{1}{2},m}}}\, Y_{n+\frac{1}{2}}'\left(y_{n+\frac{1}{2},m}\right).$$

Hence properties of $a_{n,m}$ and $b_{n,m}$ are derivable straightforwardly from results given in §§10.21(i)–10.21(iii), 10.21(vi)–10.21(viii), and 10.21(x). However, there are no simple relations that connect the zeros of the derivatives. For some properties of $a_{n,m}'$ and $b_{n,m}'$, including asymptotic expansions, see Olver (1960, pp. xix–xxi).

See also Davies (1973), de Bruin et al. (1981a,b), and Gottlieb (1985).

10.59 Integrals

10.59.1 $\int_{-\infty}^{\infty} e^{ibt} \mathsf{j}_n(t)\, dt = \begin{cases} \pi i^n P_n(b), & -1 < b < 1, \\ \frac{1}{2}\pi(\pm i)^n, & b = \pm 1, \\ 0, & \pm b > 1, \end{cases}$

where P_n is the Legendre polynomial (§18.3).

For an integral representation of the Dirac delta in terms of a product of spherical Bessel functions of the first kind see §1.17(ii), and for a generalization see Maximon (1991).

Additional integrals can be obtained by combining the definitions (10.47.3)–(10.47.9) with the results given in §10.22 and §10.43. For integrals of products see also Mehrem et al. (1991).

10.60 Sums

10.60(i) Addition Theorems

Define u, v, w, and α as in §10.23(ii). Then with P_n again denoting the Legendre polynomial of degree n,

10.60.1 $\dfrac{\cos w}{w} = -\sum_{n=0}^{\infty}(2n+1)\mathsf{j}_n(v)\mathsf{y}_n(u) P_n(\cos\alpha),$

$$|ve^{\pm i\alpha}| < |u|.$$

10.60.2 $\dfrac{\sin w}{w} = \sum_{n=0}^{\infty}(2n+1)\mathsf{j}_n(v)\mathsf{j}_n(u) P_n(\cos\alpha).$

10.60.3 $\dfrac{e^{-w}}{w} = \dfrac{2}{\pi}\sum_{n=0}^{\infty}(2n+1)\mathsf{i}_n^{(1)}(v)\mathsf{k}_n(u) P_n(\cos\alpha),$

$$|ve^{\pm i\alpha}| < |u|.$$

10.60(ii) Duplication Formulas

10.60.4
$\mathsf{j}_n(2z) = -n!z^{n+1}\sum_{k=0}^{n}\dfrac{2n-2k+1}{k!(2n-k+1)!}\mathsf{j}_{n-k}(z)\mathsf{y}_{n-k}(z),$

10.60.5
$\mathsf{y}_n(2z)$
$= n!z^{n+1}\sum_{k=0}^{n}\dfrac{n-k+\frac{1}{2}}{k!(2n-k+1)!}(\mathsf{j}_{n-k}^2(z) - \mathsf{y}_{n-k}^2(z)),$

10.60.6
$\mathsf{k}_n(2z) = \dfrac{1}{\pi}n!z^{n+1}\sum_{k=0}^{n}(-1)^k\dfrac{2n-2k+1}{k!(2n-k+1)!}\mathsf{k}_{n-k}^2(z).$

10.60(iii) Other Series

10.60.7 $e^{iz\cos\alpha} = \sum_{n=0}^{\infty}(2n+1)i^n \mathsf{j}_n(z) P_n(\cos\alpha),$

10.60.8 $e^{z\cos\alpha} = \sum_{n=0}^{\infty}(2n+1)\mathsf{i}_n^{(1)}(z) P_n(\cos\alpha),$

10.60.9 $e^{-z\cos\alpha} = \sum_{n=0}^{\infty}(-1)^n(2n+1)\mathsf{i}_n^{(1)}(z) P_n(\cos\alpha).$

10.60.10
$J_0(z\sin\alpha) = \sum_{n=0}^{\infty}(4n+1)\dfrac{(2n)!}{2^{2n}(n!)^2}\mathsf{j}_{2n}(z) P_{2n}(\cos\alpha).$

10.60.11 $\sum_{n=0}^{\infty}\mathsf{j}_n^2(z) = \dfrac{\mathrm{Si}(2z)}{2z}.$

For Si see §6.2(ii).

10.60.12 $\sum_{n=0}^{\infty}(2n+1)\mathsf{j}_n^2(z) = 1,$

10.60.13 $\sum_{n=0}^{\infty}(-1)^n(2n+1)\mathsf{j}_n^2(z) = \dfrac{\sin(2z)}{2z},$

10.60.14 $\sum_{n=0}^{\infty}(2n+1)(\mathsf{j}_n'(z))^2 = \tfrac{1}{3}.$

For further sums of series of spherical Bessel functions, or modified spherical Bessel functions, see §6.10(ii), Luke (1969b, pp. 55–58), Vavreck and Thompson (1984), Harris (2000), and Rottbrand (2000).

10.60(iv) Compendia

For collections of sums of series relevant to spherical Bessel functions or Bessel functions of half odd integer order see Erdélyi et al. (1953b, pp. 43–45 and 98–105), Gradshteyn and Ryzhik (2000, §§8.51, 8.53), Hansen (1975), Magnus et al. (1966, pp. 106–108 and 123–138), and Prudnikov et al. (1986b, pp. 635–637 and 651–700). See also Watson (1944, Chapters 11 and 16).

Kelvin Functions

10.61 Definitions and Basic Properties

10.61(i) Definitions

Throughout §§10.61–§10.71 it is assumed that $x \geq 0$, $\nu \in \mathbb{R}$, and n is a nonnegative integer.

10.61.1
$\mathrm{ber}_\nu x + i\,\mathrm{bei}_\nu x = J_\nu\!\left(xe^{3\pi i/4}\right) = e^{\nu\pi i} J_\nu\!\left(xe^{-\pi i/4}\right)$
$= e^{\nu\pi i/2} I_\nu\!\left(xe^{\pi i/4}\right)$
$= e^{3\nu\pi i/2} I_\nu\!\left(xe^{-3\pi i/4}\right),$

$$\ker_\nu x + i\kei_\nu x = e^{-\nu\pi i/2} K_\nu\!\left(xe^{\pi i/4}\right)$$

10.61.2
$$= \tfrac{1}{2}\pi i\, H^{(1)}_\nu\!\left(xe^{3\pi i/4}\right)$$
$$= -\tfrac{1}{2}\pi i e^{-\nu\pi i}\, H^{(2)}_\nu\!\left(xe^{-\pi i/4}\right).$$

When $\nu = 0$ suffices on ber, bei, ker, and kei are usually suppressed.

Most properties of $\ber_\nu x$, $\bei_\nu x$, $\ker_\nu x$, and $\kei_\nu x$ follow straightforwardly from the above definitions and results given in preceding sections of this chapter.

10.61(ii) Differential Equations

10.61.3
$$x^2\frac{d^2w}{dx^2} + x\frac{dw}{dx} - (ix^2 + \nu^2)w = 0,$$
$$w = \ber_\nu x + i\bei_\nu x, \quad \ber_{-\nu} x + i\bei_{-\nu} x$$
$$\ker_\nu x + i\kei_\nu x, \quad \ker_{-\nu} x + i\kei_{-\nu} x.$$

10.61.4
$$x^4\frac{d^4w}{dx^4} + 2x^3\frac{d^3w}{dx^3} - (1+2\nu^2)\left(x^2\frac{d^2w}{dx^2} - x\frac{dw}{dx}\right)$$
$$+ (\nu^4 - 4\nu^2 + x^4)w = 0,$$
$$w = \ber_{\pm\nu} x, \bei_{\pm\nu} x, \ker_{\pm\nu} x, \kei_{\pm\nu} x.$$

10.61(iii) Reflection Formulas for Arguments

In general, Kelvin functions have a branch point at $x = 0$ and functions with arguments $xe^{\pm\pi i}$ are complex. The branch point is absent, however, in the case of $\ber_\nu$ and $\bei_\nu$ when ν is an integer. In particular,

10.61.5
$$\ber_n(-x) = (-1)^n \ber_n x, \quad \bei_n(-x) = (-1)^n \bei_n x.$$

10.61(iv) Reflection Formulas for Orders

10.61.6
$$\ber_{-\nu} x = \cos(\nu\pi)\ber_\nu x + \sin(\nu\pi)\bei_\nu x + (2/\pi)\sin(\nu\pi)\ker_\nu x,$$
$$\bei_{-\nu} x = -\sin(\nu\pi)\ber_\nu x + \cos(\nu\pi)\bei_\nu x + (2/\pi)\sin(\nu\pi)\kei_\nu x.$$

10.61.7
$$\ker_{-\nu} x = \cos(\nu\pi)\ker_\nu x - \sin(\nu\pi)\kei_\nu x,$$
$$\kei_{-\nu} x = \sin(\nu\pi)\ker_\nu x + \cos(\nu\pi)\kei_\nu x.$$

10.61.8
$$\ber_{-n} x = (-1)^n \ber_n x, \quad \bei_{-n} x = (-1)^n \bei_n x,$$
$$\ker_{-n} x = (-1)^n \ker_n x, \quad \kei_{-n} x = (-1)^n \kei_n x.$$

10.61(v) Orders $\pm\tfrac{1}{2}$

10.61.9
$$\ber_{\tfrac{1}{2}}\!\left(x\sqrt{2}\right) = \frac{2^{-\tfrac{3}{4}}}{\sqrt{\pi x}}\left(e^x\cos\!\left(x + \frac{\pi}{8}\right) - e^{-x}\cos\!\left(x - \frac{\pi}{8}\right)\right),$$
$$\bei_{\tfrac{1}{2}}\!\left(x\sqrt{2}\right) = \frac{2^{-\tfrac{3}{4}}}{\sqrt{\pi x}}\left(e^x\sin\!\left(x + \frac{\pi}{8}\right) + e^{-x}\sin\!\left(x - \frac{\pi}{8}\right)\right).$$

10.61.10
$$\ber_{-\tfrac{1}{2}}\!\left(x\sqrt{2}\right) = \frac{2^{-\tfrac{3}{4}}}{\sqrt{\pi x}}\left(e^x\sin\!\left(x + \frac{\pi}{8}\right) - e^{-x}\sin\!\left(x - \frac{\pi}{8}\right)\right),$$
$$\bei_{-\tfrac{1}{2}}\!\left(x\sqrt{2}\right) = -\frac{2^{-\tfrac{3}{4}}}{\sqrt{\pi x}}\left(e^x\cos\!\left(x + \frac{\pi}{8}\right) + e^{-x}\cos\!\left(x - \frac{\pi}{8}\right)\right).$$

10.61.11
$$\ker_{\tfrac{1}{2}}\!\left(x\sqrt{2}\right) = \kei_{-\tfrac{1}{2}}\!\left(x\sqrt{2}\right)$$
$$= -2^{-\tfrac{3}{4}}\sqrt{\frac{\pi}{x}}e^{-x}\sin\!\left(x - \frac{\pi}{8}\right),$$

10.61.12
$$\kei_{\tfrac{1}{2}}\!\left(x\sqrt{2}\right) = -\ker_{-\tfrac{1}{2}}\!\left(x\sqrt{2}\right)$$
$$= -2^{-\tfrac{3}{4}}\sqrt{\frac{\pi}{x}}e^{-x}\cos\!\left(x - \frac{\pi}{8}\right).$$

10.62 Graphs

See Figures 10.62.1–10.62.4. For the modulus functions $M(x)$ and $N(x)$ see §10.68(i) with $\nu = 0$.

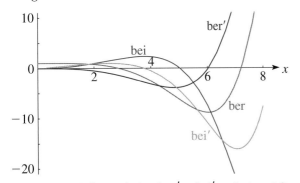

Figure 10.62.1: $\ber x, \bei x, \ber' x, \bei' x$, $0 \leq x \leq 8$.

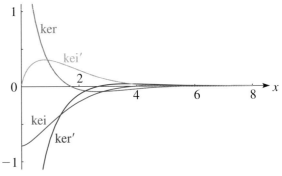

Figure 10.62.2: $\ker x, \kei x, \ker' x, \kei' x$, $0 \leq x \leq 8$.

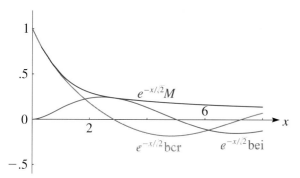

Figure 10.62.3: $e^{-x/\sqrt{2}}\operatorname{ber} x$, $e^{-x/\sqrt{2}}\operatorname{bei} x$, $e^{-x/\sqrt{2}}M(x)$, $0 \le x \le 8$.

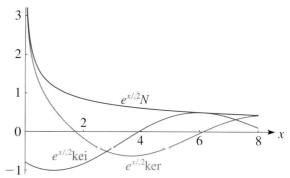

Figure 10.62.4: $e^{x/\sqrt{2}}\ker x$, $e^{x/\sqrt{2}}\ker x$, $e^{x/\sqrt{2}}N(x)$, $0 \le x \le 8$.

10.63 Recurrence Relations and Derivatives

10.63(i) $\operatorname{ber}_\nu x$, $\operatorname{bei}_\nu x$, $\ker_\nu x$, $\ker_\nu x$

Let $f_\nu(x)$, $g_\nu(x)$ denote any one of the ordered pairs:

10.63.1
$$\operatorname{ber}_\nu x, \operatorname{bei}_\nu x; \quad \operatorname{bei}_\nu x, -\operatorname{ber}_\nu x;$$
$$\ker_\nu x, \ker_\nu x; \quad \ker_\nu x, -\ker_\nu x.$$

Then

10.63.2
$$f_{\nu-1}(x) + f_{\nu+1}(x) = -(\nu\sqrt{2}/x)\left(f_\nu(x) - g_\nu(x)\right),$$
$$f_{\nu+1}(x) + g_{\nu+1}(x) - f_{\nu-1}(x) - g_{\nu-1}(x) = 2\sqrt{2}f'_\nu(x),$$
$$f'_\nu(x) = -(1/\sqrt{2})\left(f_{\nu-1}(x) + g_{\nu-1}(x)\right) - (\nu/x)f_\nu(x),$$
$$f'_\nu(x) = (1/\sqrt{2})\left(f_{\nu+1}(x) + g_{\nu+1}(x)\right) + (\nu/x)f_\nu(x).$$

10.63.3
$$\sqrt{2}\operatorname{ber}' x = \operatorname{ber}_1 x + \operatorname{bei}_1 x,$$
$$\sqrt{2}\operatorname{bei}' x = -\operatorname{ber}_1 x + \operatorname{bei}_1 x.$$

10.63.4
$$\sqrt{2}\ker' x = \ker_1 x + \ker_1 x,$$
$$\sqrt{2}\ker' x = -\ker_1 x + \ker_1 x.$$

10.63(ii) Cross-Products

Let

10.63.5
$$p_\nu = \operatorname{ber}_\nu^2 x + \operatorname{bei}_\nu^2 x, \quad q_\nu = \operatorname{ber}_\nu x \operatorname{bei}'_\nu x - \operatorname{ber}'_\nu x \operatorname{bei}_\nu x,$$
$$r_\nu = \operatorname{ber}_\nu x \operatorname{ber}'_\nu x + \operatorname{bei}_\nu x \operatorname{bei}'_\nu x,$$
$$s_\nu = \left(\operatorname{ber}'_\nu x\right)^2 + \left(\operatorname{bei}'_\nu x\right)^2.$$

Then

10.63.6
$$p_{\nu+1} = p_{\nu-1} - (4\nu/x)r_\nu,$$
$$q_{\nu+1} = -(\nu/x)p_\nu + r_\nu = -q_{\nu-1} + 2r_\nu,$$
$$r_{\nu+1} = -((\nu+1)/x)p_{\nu+1} + q_\nu,$$
$$s_\nu = \tfrac{1}{2}p_{\nu+1} + \tfrac{1}{2}p_{\nu-1} - (\nu^2/x^2)p_\nu,$$

and

10.63.7
$$p_\nu s_\nu = r_\nu^2 + q_\nu^2.$$

Equations (10.63.6) and (10.63.7) also hold when the symbols ber and bei in (10.63.5) are replaced throughout by ker and kei, respectively.

10.64 Integral Representations

Schläfli-Type Integrals

10.64.1
$$\operatorname{ber}_n\left(x\sqrt{2}\right) = \frac{(-1)^n}{\pi}\int_0^\pi \cos(x\sin t - nt)\cosh(x\sin t)\,dt,$$

10.64.2
$$\operatorname{bei}_n\left(x\sqrt{2}\right) = \frac{(-1)^n}{\pi}\int_0^\pi \sin(x\sin t - nt)\sinh(x\sin t)\,dt.$$

See Apelblat (1991) for these results, and also for similar representations for $\operatorname{ber}_\nu(x\sqrt{2})$, $\operatorname{bei}_\nu(x\sqrt{2})$, and their ν-derivatives.

10.65 Power Series

10.65(i) $\operatorname{ber}_\nu x$ and $\operatorname{bei}_\nu x$

10.65.1
$$\operatorname{ber}_\nu x = (\tfrac{1}{2}x)^\nu \sum_{k=0}^\infty \frac{\cos\left(\tfrac{3}{4}\nu\pi + \tfrac{1}{2}k\pi\right)}{k!\,\Gamma(\nu+k+1)}(\tfrac{1}{4}x^2)^k,$$
$$\operatorname{bei}_\nu x = (\tfrac{1}{2}x)^\nu \sum_{k=0}^\infty \frac{\sin\left(\tfrac{3}{4}\nu\pi + \tfrac{1}{2}k\pi\right)}{k!\,\Gamma(\nu+k+1)}(\tfrac{1}{4}x^2)^k.$$

10.65.2
$$\operatorname{ber} x = 1 - \frac{(\tfrac{1}{4}x^2)^2}{(2!)^2} + \frac{(\tfrac{1}{4}x^2)^4}{(4!)^2} - \cdots,$$
$$\operatorname{bei} x = \tfrac{1}{4}x^2 - \frac{(\tfrac{1}{4}x^2)^3}{(3!)^2} + \frac{(\tfrac{1}{4}x^2)^5}{(5!)^2} - \cdots.$$

10.65(ii) $\ker_\nu x$ and $\kei_\nu x$

When ν is not an integer combine (10.65.1) with (10.61.6). Also, with $\psi(x) = \Gamma'(x)/\Gamma(x)$,

10.65.3
$$\ker_n x = \tfrac{1}{2}(\tfrac{1}{2}x)^{-n} \sum_{k=0}^{n-1} \frac{(n-k-1)!}{k!} \cos(\tfrac{3}{4}n\pi + \tfrac{1}{2}k\pi)(\tfrac{1}{4}x^2)^k - \ln(\tfrac{1}{2}x) \ber_n x$$
$$+ \tfrac{1}{4}\pi \bei_n x + \tfrac{1}{2}(\tfrac{1}{2}x)^n \sum_{k=0}^{\infty} \frac{\psi(k+1) + \psi(n+k+1)}{k!(n+k)!} \cos(\tfrac{3}{4}n\pi + \tfrac{1}{2}k\pi)(\tfrac{1}{4}x^2)^k,$$

10.65.4
$$\kei_n x = -\tfrac{1}{2}(\tfrac{1}{2}x)^{-n} \sum_{k=0}^{n-1} \frac{(n-k-1)!}{k!} \sin(\tfrac{3}{4}n\pi + \tfrac{1}{2}k\pi)(\tfrac{1}{4}x^2)^k - \ln(\tfrac{1}{2}x) \bei_n x$$
$$- \tfrac{1}{4}\pi \ber_n x + \tfrac{1}{2}(\tfrac{1}{2}x)^n \sum_{k=0}^{\infty} \frac{\psi(k+1) + \psi(n+k+1)}{k!(n+k)!} \sin(\tfrac{3}{4}n\pi + \tfrac{1}{2}k\pi)(\tfrac{1}{4}x^2)^k.$$

10.65.5
$$\ker x = -\ln(\tfrac{1}{2}x) \ber x + \tfrac{1}{4}\pi \bei x + \sum_{k=0}^{\infty} (-1)^k \frac{\psi(2k+1)}{((2k)!)^2} (\tfrac{1}{4}x^2)^{2k},$$
$$\kei x = -\ln(\tfrac{1}{2}x) \bei x - \tfrac{1}{4}\pi \ber x + \sum_{k=0}^{\infty} (-1)^k \frac{\psi(2k+2)}{((2k+1)!)^2} (\tfrac{1}{4}x^2)^{2k+1}.$$

10.65(iii) Cross-Products and Sums of Squares

10.65.6
$$\ber_\nu^2 x + \bei_\nu^2 x = (\tfrac{1}{2}x)^{2\nu} \sum_{k=0}^{\infty} \frac{1}{\Gamma(\nu+k+1)\Gamma(\nu+2k+1)} \frac{(\tfrac{1}{4}x^2)^{2k}}{k!},$$

10.65.7
$$\ber_\nu x \bei_\nu' x - \ber_\nu' x \bei_\nu x = (\tfrac{1}{2}x)^{2\nu+1} \sum_{k=0}^{\infty} \frac{1}{\Gamma(\nu+k+1)\Gamma(\nu+2k+2)} \frac{(\tfrac{1}{4}x^2)^{2k}}{k!},$$

10.65.8
$$\ber_\nu x \ber_\nu' x + \bei_\nu x \bei_\nu' x = \tfrac{1}{2}(\tfrac{1}{2}x)^{2\nu-1} \sum_{k=0}^{\infty} \frac{1}{\Gamma(\nu+k+1)\Gamma(\nu+2k)} \frac{(\tfrac{1}{4}x^2)^{2k}}{k!},$$

10.65.9
$$(\ber_\nu' x)^2 + (\bei_\nu' x)^2 = (\tfrac{1}{2}x)^{2\nu-2} \sum_{k=0}^{\infty} \frac{2k^2 + 2\nu k + \tfrac{1}{4}\nu^2}{\Gamma(\nu+k+1)\Gamma(\nu+2k+1)} \frac{(\tfrac{1}{4}x^2)^{2k}}{k!}.$$

10.65(iv) Compendia

For further power series summable in terms of Kelvin functions and their derivatives see Hansen (1975).

10.66 Expansions in Series of Bessel Functions

10.66.1
$$\ber_\nu x + i \bei_\nu x = \sum_{k=0}^{\infty} \frac{e^{(3\nu+k)\pi i/4} x^k J_{\nu+k}(x)}{2^{k/2} k!} = \sum_{k=0}^{\infty} \frac{e^{(3\nu+3k)\pi i/4} x^k I_{\nu+k}(x)}{2^{k/2} k!}.$$

10.66.2
$$\ber_n(x\sqrt{2}) = \sum_{k=-\infty}^{\infty} (-1)^{n+k} J_{n+2k}(x) I_{2k}(x), \quad \bei_n(x\sqrt{2}) = \sum_{k=-\infty}^{\infty} (-1)^{n+k} J_{n+2k+1}(x) I_{2k+1}(x).$$

10.67 Asymptotic Expansions for Large Argument

10.67(i) $\text{ber}_\nu x, \text{bei}_\nu x, \text{ker}_\nu x, \text{kei}_\nu x$, and Derivatives

Define $a_k(\nu)$ and $b_k(\nu)$ as in §§10.17(i) and 10.17(ii). Then as $x \to \infty$ with ν fixed,

10.67.1
$$\text{ker}_\nu x \sim e^{-x/\sqrt{2}} \left(\frac{\pi}{2x}\right)^{\frac{1}{2}} \sum_{k=0}^{\infty} \frac{a_k(\nu)}{x^k} \cos\left(\frac{x}{\sqrt{2}} + \left(\frac{\nu}{2} + \frac{k}{4} + \frac{1}{8}\right)\pi\right),$$

10.67.2
$$\text{kei}_\nu x \sim -e^{-x/\sqrt{2}} \left(\frac{\pi}{2x}\right)^{\frac{1}{2}} \sum_{k=0}^{\infty} \frac{a_k(\nu)}{x^k} \sin\left(\frac{x}{\sqrt{2}} + \left(\frac{\nu}{2} + \frac{k}{4} + \frac{1}{8}\right)\pi\right).$$

10.67.3
$$\text{ber}_\nu x \sim \frac{e^{x/\sqrt{2}}}{(2\pi x)^{\frac{1}{2}}} \sum_{k=0}^{\infty} \frac{a_k(\nu)}{x^k} \cos\left(\frac{x}{\sqrt{2}} + \left(\frac{\nu}{2} + \frac{3k}{4} - \frac{1}{8}\right)\pi\right) - \frac{1}{\pi}(\sin(2\nu\pi)\text{ker}_\nu x + \cos(2\nu\pi)\text{kei}_\nu x),$$

10.67.4
$$\text{bei}_\nu x \sim \frac{e^{x/\sqrt{2}}}{(2\pi x)^{\frac{1}{2}}} \sum_{k=0}^{\infty} \frac{a_k(\nu)}{x^k} \sin\left(\frac{x}{\sqrt{2}} + \left(\frac{\nu}{2} + \frac{3k}{4} - \frac{1}{8}\right)\pi\right) + \frac{1}{\pi}(\cos(2\nu\pi)\text{ker}_\nu x - \sin(2\nu\pi)\text{kei}_\nu x).$$

10.67.5
$$\text{ker}'_\nu x \sim -e^{-x/\sqrt{2}} \left(\frac{\pi}{2x}\right)^{\frac{1}{2}} \sum_{k=0}^{\infty} \frac{b_k(\nu)}{x^k} \cos\left(\frac{x}{\sqrt{2}} + \left(\frac{\nu}{2} + \frac{k}{4} - \frac{1}{8}\right)\pi\right),$$

10.67.6
$$\text{kei}'_\nu x \sim e^{-x/\sqrt{2}} \left(\frac{\pi}{2x}\right)^{\frac{1}{2}} \sum_{k=0}^{\infty} \frac{b_k(\nu)}{x^k} \sin\left(\frac{x}{\sqrt{2}} + \left(\frac{\nu}{2} + \frac{k}{4} - \frac{1}{8}\right)\pi\right).$$

10.67.7
$$\text{ber}'_\nu x \sim \frac{e^{x/\sqrt{2}}}{(2\pi x)^{\frac{1}{2}}} \sum_{k=0}^{\infty} \frac{b_k(\nu)}{x^k} \cos\left(\frac{x}{\sqrt{2}} + \left(\frac{\nu}{2} + \frac{3k}{4} + \frac{1}{8}\right)\pi\right) - \frac{1}{\pi}(\sin(2\nu\pi)\text{ker}'_\nu x + \cos(2\nu\pi)\text{kei}'_\nu x),$$

10.67.8
$$\text{bei}'_\nu x \sim \frac{e^{x/\sqrt{2}}}{(2\pi x)^{\frac{1}{2}}} \sum_{k=0}^{\infty} \frac{b_k(\nu)}{x^k} \sin\left(\frac{x}{\sqrt{2}} + \left(\frac{\nu}{2} + \frac{3k}{4} + \frac{1}{8}\right)\pi\right) + \frac{1}{\pi}(\cos(2\nu\pi)\text{ker}'_\nu x - \sin(2\nu\pi)\text{kei}'_\nu x).$$

The contributions of the terms in $\text{ker}_\nu x$, $\text{kei}_\nu x$, $\text{ker}'_\nu x$, and $\text{kei}'_\nu x$ on the right-hand sides of (10.67.3), (10.67.4), (10.67.7), and (10.67.8) are exponentially small compared with the other terms, and hence can be neglected in the sense of Poincaré asymptotic expansions (§2.1(iii)). However, their inclusion improves numerical accuracy.

10.67(ii) Cross-Products and Sums of Squares in the Case $\nu = 0$

As $x \to \infty$

10.67.9
$$\text{ber}^2 x + \text{bei}^2 x \sim \frac{e^{x\sqrt{2}}}{2\pi x}\left(1 + \frac{1}{4\sqrt{2}}\frac{1}{x} + \frac{1}{64}\frac{1}{x^2} - \frac{33}{256\sqrt{2}}\frac{1}{x^3} - \frac{1797}{8192}\frac{1}{x^4} + \cdots\right),$$

10.67.10
$$\text{ber}\, x\, \text{bei}'\, x - \text{ber}'\, x\, \text{bei}\, x \sim \frac{e^{x\sqrt{2}}}{2\pi x}\left(\frac{1}{\sqrt{2}} + \frac{1}{8}\frac{1}{x} + \frac{9}{64\sqrt{2}}\frac{1}{x^2} + \frac{39}{512}\frac{1}{x^3} + \frac{75}{8192\sqrt{2}}\frac{1}{x^4} + \cdots\right),$$

10.67.11
$$\text{ber}\, x\, \text{ber}'\, x + \text{bei}\, x\, \text{bei}'\, x \sim \frac{e^{x\sqrt{2}}}{2\pi x}\left(\frac{1}{\sqrt{2}} - \frac{3}{8}\frac{1}{x} - \frac{15}{64\sqrt{2}}\frac{1}{x^2} - \frac{45}{512}\frac{1}{x^3} + \frac{315}{8192\sqrt{2}}\frac{1}{x^4} + \cdots\right),$$

10.67.12
$$(\text{ber}'\, x)^2 + (\text{bei}'\, x)^2 \sim \frac{e^{x\sqrt{2}}}{2\pi x}\left(1 - \frac{3}{4\sqrt{2}}\frac{1}{x} + \frac{9}{64}\frac{1}{x^2} + \frac{75}{256\sqrt{2}}\frac{1}{x^3} + \frac{2475}{8192}\frac{1}{x^4} + \cdots\right).$$

10.67.13
$$\text{ker}^2 x + \text{kei}^2 x \sim \frac{\pi}{2x}e^{-x\sqrt{2}}\left(1 - \frac{1}{4\sqrt{2}}\frac{1}{x} + \frac{1}{64}\frac{1}{x^2} + \frac{33}{256\sqrt{2}}\frac{1}{x^3} - \frac{1797}{8192}\frac{1}{x^4} + \cdots\right),$$

10.67.14
$$\text{ker}\, x\, \text{kei}'\, x - \text{ker}'\, x\, \text{kei}\, x \sim -\frac{\pi}{2x}e^{-x\sqrt{2}}\left(\frac{1}{\sqrt{2}} - \frac{1}{8}\frac{1}{x} + \frac{9}{64\sqrt{2}}\frac{1}{x^2} - \frac{39}{512}\frac{1}{x^3} + \frac{75}{8192\sqrt{2}}\frac{1}{x^4} + \cdots\right),$$

10.67.15
$$\text{ker}\, x\, \text{ker}'\, x + \text{kei}\, x\, \text{kei}'\, x \sim -\frac{\pi}{2x}e^{-x\sqrt{2}}\left(\frac{1}{\sqrt{2}} + \frac{3}{8}\frac{1}{x} - \frac{15}{64\sqrt{2}}\frac{1}{x^2} + \frac{45}{512}\frac{1}{x^3} + \frac{315}{8192\sqrt{2}}\frac{1}{x^4} + \cdots\right),$$

10.67.16
$$(\text{ker}'\, x)^2 + (\text{kei}'\, x)^2 \sim \frac{\pi}{2x}e^{-x\sqrt{2}}\left(1 + \frac{3}{4\sqrt{2}}\frac{1}{x} + \frac{9}{64}\frac{1}{x^2} - \frac{75}{256\sqrt{2}}\frac{1}{x^3} + \frac{2475}{8192}\frac{1}{x^4} + \cdots\right).$$

10.68 Modulus and Phase Functions

10.68(i) Definitions

10.68.1
$$M_\nu(x)e^{i\theta_\nu(x)} = \text{ber}_\nu x + i\,\text{bei}_\nu x,$$

10.68.2
$$N_\nu(x)e^{i\phi_\nu(x)} = \text{ker}_\nu x + i\,\text{kei}_\nu x,$$

where $M_\nu(x)\,(>0)$, $N_\nu(x)\,(>0)$, $\theta_\nu(x)$, and $\phi_\nu(x)$ are continuous real functions of x and ν, with the branches of $\theta_\nu(x)$ and $\phi_\nu(x)$ chosen to satisfy (10.68.18) and (10.68.21) as $x \to \infty$. (See also §10.68(iv).)

10.68(ii) Basic Properties

10.68.3
$$\text{ber}_\nu x = M_\nu(x)\cos\theta_\nu(x), \quad \text{bei}_\nu x = M_\nu(x)\sin\theta_\nu(x),$$

10.68.4
$$\text{ker}_\nu x = N_\nu(x)\cos\phi_\nu(x), \quad \text{kei}_\nu x = N_\nu(x)\sin\phi_\nu(x).$$

10.68.5
$$M_\nu(x) = (\text{ber}_\nu^2 x + \text{bei}_\nu^2 x)^{1/2}, \quad N_\nu(x) = (\text{ker}_\nu^2 x + \text{kei}_\nu^2 x)^{1/2},$$

10.68.6
$$\theta_\nu(x) = \text{Arctan}(\text{bei}_\nu x/\text{ber}_\nu x), \quad \phi_\nu(x) = \text{Arctan}(\text{kei}_\nu x/\text{ker}_\nu x).$$

10.68.7
$$M_{-n}(x) = M_n(x), \quad \theta_{-n}(x) = \theta_n(x) - n\pi.$$

With arguments (x) suppressed,

10.68.8
$$\text{ber}'_\nu x = \tfrac{1}{2}M_{\nu+1}\cos(\theta_{\nu+1} - \tfrac{1}{4}\pi) - \tfrac{1}{2}M_{\nu-1}\cos(\theta_{\nu-1} - \tfrac{1}{4}\pi)$$
$$= (\nu/x)M_\nu\cos\theta_\nu + M_{\nu+1}\cos(\theta_{\nu+1} - \tfrac{1}{4}\pi) = -(\nu/x)M_\nu\cos\theta_\nu - M_{\nu-1}\cos(\theta_{\nu-1} - \tfrac{1}{4}\pi),$$

10.68.9
$$\text{bei}'_\nu x = \tfrac{1}{2}M_{\nu+1}\sin(\theta_{\nu+1} - \tfrac{1}{4}\pi) - \tfrac{1}{2}M_{\nu-1}\sin(\theta_{\nu-1} - \tfrac{1}{4}\pi)$$
$$= (\nu/x)M_\nu\sin\theta_\nu + M_{\nu+1}\sin(\theta_{\nu+1} - \tfrac{1}{4}\pi) = -(\nu/x)M_\nu\sin\theta_\nu - M_{\nu-1}\sin(\theta_{\nu-1} - \tfrac{1}{4}\pi).$$

10.68.10
$$\text{ber}' x = M_1\cos(\theta_1 - \tfrac{1}{4}\pi), \quad \text{bei}' x = M_1\sin(\theta_1 - \tfrac{1}{4}\pi).$$

10.68.11
$$M'_\nu = (\nu/x)M_\nu + M_{\nu+1}\cos(\theta_{\nu+1} - \theta_\nu - \tfrac{1}{4}\pi) = -(\nu/x)M_\nu - M_{\nu-1}\cos(\theta_{\nu-1} - \theta_\nu - \tfrac{1}{4}\pi),$$

10.68.12
$$\theta'_\nu = (M_{\nu+1}/M_\nu)\sin(\theta_{\nu+1} - \theta_\nu - \tfrac{1}{4}\pi) = -(M_{\nu-1}/M_\nu)\sin(\theta_{\nu-1} - \theta_\nu - \tfrac{1}{4}\pi).$$

10.68.13
$$M'_0 = M_1\cos(\theta_1 - \theta_0 - \tfrac{1}{4}\pi), \quad \theta'_0 = (M_1/M_0)\sin(\theta_1 - \theta_0 - \tfrac{1}{4}\pi).$$

10.68.14
$$d(xM_\nu^2\theta'_\nu)/dx = xM_\nu^2, \quad x^2M''_\nu + xM'_\nu - \nu^2 M_\nu = x^2 M_\nu {\theta'_\nu}^2.$$

Equations (10.68.8)–(10.68.14) also hold with the symbols ber, bei, M, and θ replaced throughout by ker, kei, N, and ϕ, respectively. In place of (10.68.7),

10.68.15
$$N_{-\nu}(x) = N_\nu(x), \quad \phi_{-\nu}(x) = \phi_\nu(x) + \nu\pi.$$

10.68(iii) Asymptotic Expansions for Large Argument

When ν is fixed, $\mu = 4\nu^2$, and $x \to \infty$

10.68.16
$$M_\nu(x) = \frac{e^{x/\sqrt{2}}}{(2\pi x)^{\frac{1}{2}}}\left(1 - \frac{\mu-1}{8\sqrt{2}}\frac{1}{x} + \frac{(\mu-1)^2}{256}\frac{1}{x^2} - \frac{(\mu-1)(\mu^2+14\mu-399)}{6144\sqrt{2}}\frac{1}{x^3} + O\!\left(\frac{1}{x^4}\right)\right),$$

10.68.17
$$\ln M_\nu(x) = \frac{x}{\sqrt{2}} - \frac{1}{2}\ln(2\pi x) - \frac{\mu-1}{8\sqrt{2}}\frac{1}{x} - \frac{(\mu-1)(\mu-25)}{384\sqrt{2}}\frac{1}{x^3} - \frac{(\mu-1)(\mu-13)}{128}\frac{1}{x^4} + O\!\left(\frac{1}{x^5}\right),$$

10.68.18
$$\theta_\nu(x) = \frac{x}{\sqrt{2}} + \left(\frac{1}{2}\nu - \frac{1}{8}\right)\pi + \frac{\mu-1}{8\sqrt{2}}\frac{1}{x} + \frac{\mu-1}{16}\frac{1}{x^2} - \frac{(\mu-1)(\mu-25)}{384\sqrt{2}}\frac{1}{x^3} + O\!\left(\frac{1}{x^5}\right).$$

10.68.19
$$N_\nu(x) = e^{-x/\sqrt{2}}\left(\frac{\pi}{2x}\right)^{\frac{1}{2}}\left(1 + \frac{\mu-1}{8\sqrt{2}}\frac{1}{x} + \frac{(\mu-1)^2}{256}\frac{1}{x^2} + \frac{(\mu-1)(\mu^2+14\mu-399)}{6144\sqrt{2}}\frac{1}{x^3} + O\!\left(\frac{1}{x^4}\right)\right),$$

10.68.20
$$\ln N_\nu(x) = -\frac{x}{\sqrt{2}} + \frac{1}{2}\ln\!\left(\frac{\pi}{2x}\right) + \frac{\mu-1}{8\sqrt{2}}\frac{1}{x} + \frac{(\mu-1)(\mu-25)}{384\sqrt{2}}\frac{1}{x^3} - \frac{(\mu-1)(\mu-13)}{128}\frac{1}{x^4} + O\!\left(\frac{1}{x^5}\right),$$

10.68.21
$$\phi_\nu(x) = -\frac{x}{\sqrt{2}} - \left(\frac{1}{2}\nu + \frac{1}{8}\right)\pi - \frac{\mu-1}{8\sqrt{2}}\frac{1}{x} + \frac{\mu-1}{16}\frac{1}{x^2} + \frac{(\mu-1)(\mu-25)}{384\sqrt{2}}\frac{1}{x^3} + O\!\left(\frac{1}{x^5}\right).$$

10.68(iv) Further Properties

Additional properties of the modulus and phase functions are given in Young and Kirk (1964, pp. xi–xv). However, care needs to be exercised with the branches of the phases. Thus this reference gives $\phi_1(0) = \frac{5}{4}\pi$ (Eq. (6.10)), and $\lim_{x\to\infty}(\phi_1(x) + (x/\sqrt{2})) = -\frac{5}{8}\pi$ (Eqs. (10.20) and (Eqs. (10.26b)). However, numerical tabulations show that if the second of these equations applies and $\phi_1(x)$ is continuous, then $\phi_1(0) = -\frac{3}{4}\pi$; compare Abramowitz and Stegun (1964, p. 433).

10.69 Uniform Asymptotic Expansions for Large Order

Let $U_k(p)$ and $V_k(p)$ be the polynomials defined in §10.41(ii), and

10.69.1
$$\xi = (1 + ix^2)^{1/2}.$$

Then as $\nu \to +\infty$,

10.69.2
$$\operatorname{ber}_\nu(\nu x) + i\operatorname{bei}_\nu(\nu x) \sim \frac{e^{\nu\xi}}{(2\pi\nu\xi)^{1/2}} \left(\frac{xe^{3\pi i/4}}{1+\xi}\right)^\nu \sum_{k=0}^{\infty} \frac{U_k(\xi^{-1})}{\nu^k},$$

10.69.3
$$\operatorname{ker}_\nu(\nu x) + i\operatorname{kei}_\nu(\nu x) \sim e^{-\nu\xi} \left(\frac{\pi}{2\nu\xi}\right)^{1/2} \left(\frac{xe^{3\pi i/4}}{1+\xi}\right)^{-\nu} \sum_{k=0}^{\infty} (-1)^k \frac{U_k(\xi^{-1})}{\nu^k},$$

10.69.4
$$\operatorname{ber}'_\nu(\nu x) + i\operatorname{bei}'_\nu(\nu x) \sim \frac{e^{\nu\xi}}{x} \left(\frac{\xi}{2\pi\nu}\right)^{1/2} \left(\frac{xe^{3\pi i/4}}{1+\xi}\right)^\nu \sum_{k=0}^{\infty} \frac{V_k(\xi^{-1})}{\nu^k},$$

10.69.5
$$\operatorname{ker}'_\nu(\nu x) + i\operatorname{kei}'_\nu(\nu x) \sim -\frac{e^{-\nu\xi}}{x} \left(\frac{\pi\xi}{2\nu}\right)^{1/2} \left(\frac{xe^{3\pi i/4}}{1+\xi}\right)^{-\nu} \sum_{k=0}^{\infty} (-1)^k \frac{V_k(\xi^{-1})}{\nu^k},$$

uniformly for $x \in (0, \infty)$. All fractional powers take their principal values.

All four expansions also enjoy the same kind of double asymptotic property described in §10.41(iv).

Accuracy in (10.69.2) and (10.69.4) can be increased by including exponentially-small contributions as in (10.67.3), (10.67.4), (10.67.7), and (10.67.8) with x replaced by νx.

10.70 Zeros

Asymptotic approximations for large zeros are as follows. Let $\mu = 4\nu^2$ and $f(t)$ denote the formal series

10.70.1
$$\frac{\mu-1}{16t} + \frac{\mu-1}{32t^2} + \frac{(\mu-1)(5\mu+19)}{1536t^3} + \frac{3(\mu-1)^2}{512t^4} + \cdots.$$

If m is a large positive integer, then

10.70.2
$$\begin{aligned}
\text{zeros of } \operatorname{ber}_\nu x &\sim \sqrt{2}(t - f(t)), & t &= (m - \tfrac{1}{2}\nu - \tfrac{3}{8})\pi, \\
\text{zeros of } \operatorname{bei}_\nu x &\sim \sqrt{2}(t - f(t)), & t &= (m - \tfrac{1}{2}\nu + \tfrac{1}{8})\pi, \\
\text{zeros of } \operatorname{ker}_\nu x &\sim \sqrt{2}(t + f(-t)), & t &= (m - \tfrac{1}{2}\nu - \tfrac{5}{8})\pi, \\
\text{zeros of } \operatorname{kei}_\nu x &\sim \sqrt{2}(t + f(-t)), & t &= (m - \tfrac{1}{2}\nu - \tfrac{1}{8})\pi.
\end{aligned}$$

In the case $\nu = 0$, numerical tabulations (Abramowitz and Stegun (1964, Table 9.12)) indicate that each of (10.70.2) corresponds to the mth zero of the function on the left-hand side. For the next six terms in the series (10.70.1) see MacLeod (2002a).

10.71 Integrals

10.71(i) Indefinite Integrals

In the following equations f_ν, g_ν is any one of the four ordered pairs given in (10.63.1), and $\widehat{f}_\nu, \widehat{g}_\nu$ is either the same ordered pair or any other ordered pair in (10.63.1).

10.71.1
$$\int x^{1+\nu} f_\nu \, dx = -\frac{x^{1+\nu}}{\sqrt{2}}(f_{\nu+1} - g_{\nu+1}) = -x^{1+\nu}\left(\frac{\nu}{x}g_\nu - g'_\nu\right),$$

10.71.2
$$\int x^{1-\nu} f_\nu \, dx = \frac{x^{1-\nu}}{\sqrt{2}}(f_{\nu-1} - g_{\nu-1}) = x^{1-\nu}\left(\frac{\nu}{x}g_\nu + g'_\nu\right).$$

10.71.3
$$\int x(f_\nu \widehat{g}_\nu - g_\nu \widehat{f}_\nu) \, dx = \frac{x}{2\sqrt{2}}\left(\widehat{f}_\nu(f_{\nu+1} + g_{\nu+1}) - \widehat{g}_\nu(f_{\nu+1} - g_{\nu+1}) - f_\nu(\widehat{f}_{\nu+1} + \widehat{g}_{\nu+1}) + g_\nu(\widehat{f}_{\nu+1} - \widehat{g}_{\nu+1})\right)$$
$$= \tfrac{1}{2}x(f'_\nu \widehat{f}_\nu - f_\nu \widehat{f}'_\nu + g'_\nu \widehat{g}_\nu - g_\nu \widehat{g}'_\nu),$$

10.71.4
$$\int x(f_\nu \widehat{g}_\nu + g_\nu \widehat{f}_\nu) \, dx = \tfrac{1}{4}x^2(2f_\nu \widehat{g}_\nu - f_{\nu-1}\widehat{g}_{\nu+1} - f_{\nu+1}\widehat{g}_{\nu-1} + 2g_\nu \widehat{f}_\nu - g_{\nu-1}\widehat{f}_{\nu+1} - g_{\nu+1}\widehat{f}_{\nu-1}).$$

10.71.5
$$\int x(f_\nu^2 + g_\nu^2) \, dx = x(f_\nu g'_\nu - f'_\nu g_\nu) = -\frac{x}{\sqrt{2}}(f_\nu f_{\nu+1} + g_\nu g_{\nu+1} - f_\nu g_{\nu+1} + f_{\nu+1}g_\nu),$$

10.71.6
$$\int x f_\nu g_\nu \, dx = \tfrac{1}{4}x^2 \left(2f_\nu g_\nu - f_{\nu-1}g_{\nu+1} - f_{\nu+1}g_{\nu-1}\right),$$

10.71.7
$$\int x(f_\nu^2 - g_\nu^2) \, dx = \tfrac{1}{2}x^2 \left(f_\nu^2 - f_{\nu-1}f_{\nu+1} - g_\nu^2 + g_{\nu-1}g_{\nu+1}\right).$$

Examples

10.71.8
$$\int x M_\nu^2(x) \, dx = x(\mathrm{ber}_\nu x \, \mathrm{bei}'_\nu x - \mathrm{ber}'_\nu x \, \mathrm{bei}_\nu x), \qquad \int x N_\nu^2(x) \, dx = x(\mathrm{ker}_\nu x \, \mathrm{kei}'_\nu x - \mathrm{ker}'_\nu x \, \mathrm{kei}_\nu x),$$

where $M_\nu(x)$ and $N_\nu(x)$ are the modulus functions introduced in §10.68(i).

10.71(ii) Definite Integrals

See Kerr (1978) and Glasser (1979).

10.71(iii) Compendia

For infinite double integrals involving Kelvin functions see Prudnikov et al. (1986b, pp. 630–631).

For direct and inverse Laplace transforms of Kelvin functions see Prudnikov et al. (1992a, §3.19) and Prudnikov et al. (1992b, §3.19).

Applications

10.72 Mathematical Applications

10.72(i) Differential Equations with Turning Points

Bessel functions and modified Bessel functions are often used as approximants in the construction of uniform asymptotic approximations and expansions for solutions of linear second-order differential equations containing a parameter. The canonical form of differential equation for these problems is given by

10.72.1
$$\frac{d^2w}{dz^2} = \left(u^2 f(z) + g(z)\right)w,$$

where z is a real or complex variable and u is a large real or complex parameter.

Simple Turning Points

In regions in which (10.72.1) has a simple turning point z_0, that is, $f(z)$ and $g(z)$ are analytic (or with weaker conditions if $z = x$ is a real variable) and z_0 is a simple zero of $f(z)$, asymptotic expansions of the solutions w for large u can be constructed in terms of Airy functions or equivalently Bessel functions or modified Bessel functions of order $\tfrac{1}{3}$ (§9.6(i)). These expansions are uniform with respect to z, including the turning point z_0 and its neighborhood, and the region of validity often includes cut neighborhoods (§1.10(vi)) of other singularities of the differential equation, especially irregular singularities.

For further information and references see §§2.8(i) and 2.8(iii).

Multiple or Fractional Turning Points

If $f(z)$ has a double zero z_0, or more generally z_0 is a zero of order m, $m = 2, 3, 4, \ldots$, then uniform asymptotic approximations (but *not* expansions) can be constructed in terms of Bessel functions, or modified Bessel functions, of order $1/(m+2)$. The number m can also be replaced by any real constant λ (> -2) in the sense that $(z - z_0)^{-\lambda} f(z)$ is analytic and nonvanishing at z_0; moreover, $g(z)$ is permitted to have a single or double pole at z_0. The order of the approximating Bessel functions, or modified Bessel functions, is $1/(\lambda + 2)$, except in the case when $g(z)$ has a double pole at z_0. See §2.8(v) for references.

10.72(ii) Differential Equations with Poles

In regions in which the function $f(z)$ has a simple pole at $z = z_0$ and $(z - z_0)^2 g(z)$ is analytic at $z = z_0$ (the case $\lambda = -1$ in §10.72(i)), asymptotic expansions of the solutions w of (10.72.1) for large u can be constructed in terms of Bessel functions and modified Bessel functions of order $\pm\sqrt{1+4\rho}$, where ρ is the limiting value of $(z - z_0)^2 g(z)$ as $z \to z_0$. These asymptotic expansions are uniform with respect to z, including cut neighborhoods of z_0, and again the region of uniformity often includes cut neighborhoods of other singularities of the differential equation.

For further information and references see §§2.8(i) and 2.8(iv).

10.72(iii) Differential Equations with a Double Pole and a Movable Turning Point

In (10.72.1) assume $f(z) = f(z, \alpha)$ and $g(z) = g(z, \alpha)$ depend continuously on a real parameter α, $f(z, \alpha)$ has a simple zero $z = z_0(\alpha)$ and a double pole $z = 0$, except for a critical value $\alpha = a$, where $z_0(a) = 0$. Assume that whether or not $\alpha = a$, $z^2 g(z, \alpha)$ is analytic at $z = 0$. Then for large u asymptotic approximations of the solutions w can be constructed in terms of Bessel functions, or modified Bessel functions, of variable order (in fact the order depends on u and α). These approximations are uniform with respect to both z and α, including $z = z_0(a)$, the cut neighborhood of $z = 0$, and $\alpha = a$. See §2.8(vi) for references.

10.73 Physical Applications

10.73(i) Bessel and Modified Bessel Functions

Bessel functions first appear in the investigation of a physical problem in Daniel Bernoulli's analysis of the small oscillations of a uniform heavy flexible chain. For this problem and its further generalizations, see Korenev (2002, Chapter 4, §37) and Gray *et al.* (1922, Chapter I, §1, Chapter XVI, §4).

Bessel functions of the first kind, $J_n(x)$, arise naturally in applications having cylindrical symmetry in which the physics is described either by Laplace's equation $\nabla^2 V = 0$, or by the Helmholtz equation $(\nabla^2 + k^2)\psi = 0$.

Laplace's equation governs problems in heat conduction, in the distribution of potential in an electrostatic field, and in hydrodynamics in the irrotational motion of an incompressible fluid. See Jackson (1999, Chapter 3, §§3.7, 3.8, 3.11, 3.13), Lamb (1932, Chapter V, §§100–102; Chapter VIII, §§186, 191–193; Chapter X, §§303, 304), Happel and Brenner (1973, Chapter 3, §3.3; Chapter 7, §7.3), Korenev (2002, Chapter 4, §43), and Gray *et al.* (1922, Chapter XI). In cylindrical coordinates r, ϕ, z, (§1.5(ii)) we have

$$10.73.1 \quad \nabla^2 V = \frac{1}{r}\frac{\partial}{\partial r}\left(r\frac{\partial V}{\partial r}\right) + \frac{1}{r^2}\frac{\partial^2 V}{\partial \phi^2} + \frac{\partial^2 V}{\partial z^2} = 0,$$

and on separation of variables we obtain solutions of the form $e^{\pm in\phi} e^{\pm \kappa z} J_n(\kappa r)$, from which a solution satisfying prescribed boundary conditions may be constructed.

The Helmholtz equation, $(\nabla^2 + k^2)\psi = 0$, follows from the wave equation

$$10.73.2 \quad \nabla^2 \psi = \frac{1}{c^2}\frac{\partial^2 \psi}{\partial t^2},$$

on assuming a time dependence of the form $e^{\pm ikt}$. This equation governs problems in acoustic and electromagnetic wave propagation. See Jackson (1999, Chapter 9, §9.6), Jones (1986, Chapters 7, 8), and Lord Rayleigh (1945, Vol. I, Chapter IX, §§200–211, 218, 219, 221a; Vol. II, Chapter XIII, §272a; Chapter XV, §302; Chapter XVIII; Chapter XIX, §350; Chapter XX, §357; Chapter XXI, §369). It is fundamental in the study of electromagnetic wave transmission. Consequently, Bessel functions $J_n(x)$, and modified Bessel functions $I_n(x)$, are central to the analysis of microwave and optical transmission in waveguides, including coaxial and fiber. See Krivoshlykov (1994, Chapter 2, §2.2.10; Chapter 5, §5.2.2), Kapany and Burke (1972, Chapters 4–6; Chapter 7, §A.1), and Slater (1942, Chapter 4, §§20, 25).

Bessel functions enter in the study of the scattering of light and other electromagnetic radiation, not only from cylindrical surfaces but also in the statistical analysis involved in scattering from rough surfaces. See Smith (1997, Chapter 3, §3.7; Chapter 6, §6.4), Beckmann and Spizzichino (1963, Chapter 4, §§4.2, 4.3; Chapter 5, §§5.2, 5.3; Chapter 6, §6.1; Chapter 7, §7.1.), Kerker (1969, Chapter 5, §5.6.4; Chapter 7, §7.5.6), and Bayvel and Jones (1981, Chapter 1, §§1.6.5, 1.6.6).

More recently, Bessel functions appear in the inverse problem in wave propagation, with applications in medicine, astronomy, and acoustic imaging. See Colton and Kress (1998, Chapter 2, §§2.4, 2.5; Chapter 3, §3.4).

In the theory of plates and shells, the oscillations of a circular plate are determined by the differential equation

10.73.3
$$\nabla^4 W + \lambda^2 \frac{\partial^2 W}{\partial t^2} = 0.$$

See Korenev (2002). On separation of variables into cylindrical coordinates, the Bessel functions $J_n(x)$, and modified Bessel functions $I_n(x)$ and $K_n(x)$, all appear.

10.73(ii) Spherical Bessel Functions

The functions $\mathsf{j}_n(x)$, $\mathsf{y}_n(x)$, $\mathsf{h}_n^{(1)}(x)$, and $\mathsf{h}_n^{(2)}(x)$ arise in the solution (again by separation of variables) of the Helmholtz equation in spherical coordinates ρ, θ, ϕ (§1.5(ii)):

10.73.4
$$(\nabla^2 + k^2)f = \frac{1}{\rho^2}\frac{\partial}{\partial \rho}\left(\rho^2 \frac{\partial f}{\partial \rho}\right) + \frac{1}{\rho^2 \sin \theta}\frac{\partial}{\partial \theta}\left(\sin \theta \frac{\partial f}{\partial \theta}\right) + \frac{1}{\rho^2 \sin^2 \theta}\frac{\partial^2 f}{\partial \phi^2} + k^2 f.$$

With the spherical harmonic $Y_{\ell,m}(\theta, \phi)$ defined as in §14.30(i), the solutions are of the form $f = g_\ell(k\rho) Y_{\ell,m}(\theta, \phi)$ with $g_\ell = \mathsf{j}_\ell, \mathsf{y}_\ell, \mathsf{h}_\ell^{(1)}$, or $\mathsf{h}_\ell^{(2)}$, depending on the boundary conditions. Accordingly, the spherical Bessel functions appear in all problems in three dimensions with spherical symmetry involving the scattering of electromagnetic radiation. See Jackson (1999, Chapter 9, §9.6), Bayvel and Jones (1981, Chapter 1, §1.5.1), and Konopinski (1981, Chapter 9, §9.1). In quantum mechanics the spherical Bessel functions arise in the solution of the Schrödinger wave equation for a particle in a central potential. See Messiah (1961, Chapter IX, §§7–10).

10.73(iii) Kelvin Functions

The analysis of the current distribution in circular conductors leads to the Kelvin functions $\operatorname{ber} x$, $\operatorname{bei} x$, $\operatorname{ker} x$, and $\operatorname{kei} x$. See Relton (1965, Chapter X, §§10.2, 10.3), Bowman (1958, Chapter III, §§51–53), McLachlan (1961, Chapters VIII and IX), and Russell (1909). The McLachlan reference also includes other applications of Kelvin functions.

10.73(iv) Bickley Functions

See Bickley (1935) and Altaç (1996).

10.73(v) Rayleigh Function

For applications of the Rayleigh function $\sigma_n(\nu)$ (§10.21(xiii)) to problems of heat conduction and diffusion in liquids see Kapitsa (1951b).

Computation

10.74 Methods of Computation

10.74(i) Series Expansions

The power-series expansions given in §§10.2 and 10.8, together with the connection formulas of §10.4, can be used to compute the Bessel and Hankel functions when the argument x or z is sufficiently small in absolute value. In the case of the modified Bessel function $K_\nu(z)$ see especially Temme (1975).

In other circumstances the power series are prone to slow convergence and heavy numerical cancellation.

If x or $|z|$ is large compared with $|\nu|^2$, then the asymptotic expansions of §§10.17(i)–10.17(iv) are available. Furthermore, the attainable accuracy can be increased substantially by use of the exponentially-improved expansions given in §10.17(v), even more so by application of the hyperasymptotic expansions to be found in the references in that subsection.

For large positive real values of ν the uniform asymptotic expansions of §§10.20(i) and 10.20(ii) can be used. Moreover, because of their double asymptotic properties (§10.41(v)) these expansions can also be used for large x or $|z|$, whether or not ν is large. It should be noted, however, that there is a difficulty in evaluating the coefficients $A_k(\zeta)$, $B_k(\zeta)$, $C_k(\zeta)$, and $D_k(\zeta)$, from the explicit expressions (10.20.10)–(10.20.13) when z is close to 1 owing to severe cancellation. Temme (1997) shows how to overcome this difficulty by use of the Maclaurin expansions for these coefficients or by use of auxiliary functions.

Similar observations apply to the computation of modified Bessel functions, spherical Bessel functions, and Kelvin functions. In the case of the spherical Bessel functions the explicit formulas given in §§10.49(i) and 10.49(ii) are terminating cases of the asymptotic expansions given in §§10.17(i) and 10.40(i) for the Bessel functions and modified Bessel functions. And since there are no error terms they could, in theory, be used for all values of z; however, there may be severe cancellation when $|z|$ is not large compared with n^2.

10.74(ii) Differential Equations

A comprehensive and powerful approach is to integrate the differential equations (10.2.1) and (10.25.1) by direct numerical methods. As described in §3.7(ii), to insure stability the integration path must be chosen in such a way that as we proceed along it the wanted solution grows in magnitude at least as fast as all other solutions of the differential equation.

In the interval $0 < x < \nu$, $J_\nu(x)$ needs to be integrated in the forward direction and $Y_\nu(x)$ in the backward direction, with initial values for the former obtained from the power-series expansion (10.2.2) and for the latter from asymptotic expansions (§§10.17(i) and 10.20(i)). In the interval $\nu < x < \infty$ either direction of integration can be used for both functions.

Similarly, to maintain stability in the interval $0 < x < \infty$ the integration direction has to be forwards in the case of $I_\nu(x)$ and backwards in the case of $K_\nu(x)$, with initial values obtained in an analogous manner to those for $J_\nu(x)$ and $Y_\nu(x)$.

For $z \in \mathbb{C}$ the function $H_\nu^{(1)}(z)$, for example, can always be computed in a stable manner in the sector $0 \leq \operatorname{ph} z \leq \pi$ by integrating along rays towards the origin.

Similar considerations apply to the spherical Bessel functions and Kelvin functions.

For further information, including parallel methods for solving the differential equations, see Lozier and Olver (1993).

10.74(iii) Integral Representations

For evaluation of the Hankel functions $H_\nu^{(1)}(z)$ and $H_\nu^{(2)}(z)$ for complex values of ν and z based on the integral representations (10.9.18) see Remenets (1973).

For applications of generalized Gauss–Laguerre quadrature (§3.5(v)) to the evaluation of the modified Bessel functions $K_\nu(z)$ for $0 < \nu < 1$ and $0 < x < \infty$ see Gautschi (2002a). The integral representation used is based on (10.32.8).

For evaluation of $K_\nu(z)$ from (10.32.14) with $\nu = n$ and z complex, see Mechel (1966).

10.74(iv) Recurrence Relations

If values of the Bessel functions $J_\nu(z)$, $Y_\nu(z)$, or the other functions treated in this chapter, are needed for integer-spaced ranges of values of the order ν, then a simple and powerful procedure is provided by recurrence relations typified by the first of (10.6.1).

Suppose, for example, $\nu = n \in 0, 1, 2, \ldots$, and $x \in (0, \infty)$. Then $J_n(x)$ and $Y_n(x)$ can be generated by either forward or backward recurrence on n when $n < x$, but if $n > x$ then to maintain stability $J_n(x)$ has to be generated by backward recurrence on n, and $Y_n(x)$ has to be generated by forward recurrence on n. In the case of $J_n(x)$, the need for initial values can be avoided by application of Olver's algorithm (§3.6(v)) in conjunction with Equation (10.12.4) used as a normalizing condition, or in the case of noninteger orders, (10.23.15).

For further information see Gautschi (1967), Olver and Sookne (1972), Temme (1975), Campbell (1980), and Kerimov and Skorokhodov (1984a).

10.74(v) Continued Fractions

For applications of the continued-fraction expansions (10.10.1), (10.10.2), (10.33.1), and (10.33.2) to the computation of Bessel functions and modified Bessel functions see Gargantini and Henrici (1967), Amos (1974), Gautschi and Slavik (1978), Tretter and Walster (1980), Thompson and Barnett (1986), and Cuyt et al. (2008).

10.74(vi) Zeros and Associated Values

Newton's rule (§3.8(i)) or Halley's rule (§3.8(v)) can be used to compute to arbitrarily high accuracy the real or complex zeros of all the functions treated in this chapter. Necessary values of the first derivatives of the functions are obtained by the use of (10.6.2), for example. Newton's rule is quadratically convergent and Halley's rule is cubically convergent. See also Segura (1998, 2001).

Methods for obtaining initial approximations to the zeros include asymptotic expansions (§§10.21(vi)-10.21(ix)), graphical intersection of $2D$ graphs in $\mathbb{R}$ (e.g., §10.3(i)) with the x-axis, or graphical intersection of $3D$ complex-variable surfaces (e.g., §10.3(ii)) with the plane $z = 0$.

To ensure that no zeros are overlooked, standard tools are the phase principle and Rouché's theorem; see §1.10(iv).

Real Zeros

See Olver (1960, pp. xvi–xxix), Grad and Zakrajšek (1973), Temme (1979a), Ikebe et al. (1991), Zafiropoulos et al. (1996), Vrahatis et al. (1997a), Ball (2000), and Gil and Segura (2003).

Complex Zeros

See Leung and Ghaderpanah (1979), Kerimov and Skorokhodov (1984b,c, 1985a,b), Skorokhodov (1985), Modenov and Filonov (1986), and Vrahatis et al. (1997b).

Multiple Zeros

See Kerimov and Skorokhodov (1985c, 1986, 1987, 1988).

10.74(vii) Integrals

Hankel Transform

See Cornille (1972), Johansen and Sørensen (1979), Gabutti (1979), Gabutti and Minetti (1981), Candel (1981), Wong (1982), Lund (1985), Piessens and Branders (1985), Hansen (1985), Bezvoda et al. (1986), Puoskari (1988), Christensen (1990), Campos (1995), Lucas and Stone (1995), Barakat and Parshall (1996), Sidi (1997), Secada (1999).

Fourier–Bessel Expansion

For the computation of the integral (10.23.19) see Piessens and Branders (1983, 1985), Lewanowicz (1991), and Zhileĭkin and Kukarkin (1995).

Spherical Bessel Transform

The *spherical Bessel transform* is the Hankel transform (10.22.76) in the case when ν is half an odd positive integer.

See Lehman et al. (1981), Puoskari (1988), and Sharafeddin et al. (1992).

Kontorovich–Lebedev Transform

See Ehrenmark (1995).

Products

For infinite integrals involving products of two Bessel functions of the first kind, see Linz and Kropp (1973), Gabutti (1980), Ikonomou et al. (1995), and Lucas (1995).

10.74(viii) Functions of Imaginary Order

For the computation of the functions $\widetilde{I}_\nu(x)$ and $\widetilde{K}_\nu(x)$ defined by (10.45.2) see Temme (1994b) and Gil et al. (2002b, 2003a, 2004a).

10.75 Tables

10.75(i) Introduction

Comprehensive listings and descriptions of tables of the functions treated in this chapter are provided in Bateman and Archibald (1944), Lebedev and Fedorova (1960), Fletcher et al. (1962), and Luke (1975, §9.13.2). Only a few of the more comprehensive of these early tables are included in the listings in the following subsections. Also, for additional listings of tables pertaining to complex arguments see Babushkina et al. (1997).

10.75(ii) Bessel Functions and their Derivatives

- British Association for the Advancement of Science (1937) tabulates $J_0(x)$, $J_1(x)$, $x = 0(.001)16(.01)25$, 10D; $Y_0(x)$, $Y_1(x)$, $x = 0.01(.01)25$, 8–9S or 8D. Also included are auxiliary functions to facilitate interpolation of the tables of $Y_0(x)$, $Y_1(x)$ for small values of x, as well as auxiliary functions to compute all four functions for large values of x.

- Bickley et al. (1952) tabulates $J_n(x)$, $Y_n(x)$ or $x^n Y_n(x)$, $n = 2(1)20$, $x = 0(.01$ or $.1)10(.1)25$, 8D (for $J_n(x)$), 8S (for $Y_n(x)$ or $x^n Y_n(x)$); $J_n(x)$, $Y_n(x)$, $n = 0(1)20$, $x = 0$ or $0.1(.1)25$, 10D (for $J_n(x)$), 10S (for $Y_n(x)$).

- Olver (1962) provides tables for the uniform asymptotic expansions given in §10.20(i), including ζ and $(4\zeta/(1-x^2))^{\frac{1}{4}}$ as functions of x ($= z$) and the coefficients $A_k(\zeta)$, $B_k(\zeta)$, $C_k(\zeta)$, $D_k(\zeta)$ as functions of ζ. These enable $J_\nu(\nu x)$, $Y_\nu(\nu x)$, $J'_\nu(\nu x)$, $Y'_\nu(\nu x)$ to be computed to 10S when $\nu \geq 15$, except in the neighborhoods of zeros.

- The main tables in Abramowitz and Stegun (1964, Chapter 9) give $J_0(x)$ to 15D, $J_1(x)$, $J_2(x)$, $Y_0(x)$, $Y_1(x)$ to 10D, $Y_2(x)$ to 8D, $x = 0(.1)17.5$; $Y_n(x) - (2/\pi) J_n(x) \ln x$, $n = 0, 1$, $x = 0(.1)2$, 8D; $J_n(x)$, $Y_n(x)$, $n = 3(1)9$, $x = 0(.2)20$, 5D or 5S; $J_n(x)$, $Y_n(x)$, $n = 0(1)20(10)50, 100$, $x = 1, 2, 5, 10, 50, 100$, 10S; modulus and phase functions $\sqrt{x} M_n(x)$, $\theta_n(x) - x$, $n = 0, 1, 2$, $1/x = 0(.01)0.1$, 8D.

- Achenbach (1986) tabulates $J_0(x)$, $J_1(x)$, $Y_0(x)$, $Y_1(x)$, $x = 0(.1)8$, 20D or 18–20S.

- Zhang and Jin (1996, pp. 185–195) tabulates $J_n(x)$, $J'_n(x)$, $Y_n(x)$, $Y'_n(x)$, $n = 0(1)10(10)50, 100$, $x = 1, 5, 10, 25, 50, 100$, 9S; $J_{n+\alpha}(x)$, $J'_{n+\alpha}(x)$, $Y_{n+\alpha}(x)$, $Y'_{n+\alpha}(x)$, $n = 0(1)5, 10, 30, 50, 100$, $\alpha = \frac{1}{4}, \frac{1}{3}, \frac{1}{2}, \frac{2}{3}, \frac{3}{4}$, $x = 1, 5, 10, 50$, 8S; real and imaginary parts of $J_{n+\alpha}(z)$, $J'_{n+\alpha}(z)$, $Y_{n+\alpha}(z)$, $Y'_{n+\alpha}(z)$, $n = 0(1)15, 20(10)50, 100$, $\alpha = 0, \frac{1}{2}$, $z = 4+2i$, $20+10i$, 8S.

10.75(iii) Zeros and Associated Values of the Bessel Functions, Hankel Functions, and their Derivatives

Real Zeros

- British Association for the Advancement of Science (1937) tabulates $j_{0,m}$, $J_1(j_{0,m})$, $j_{1,m}$, $J_0(j_{1,m})$, $m = 1(1)150$, 10D; $y_{0,m}$, $Y_1(y_{0,m})$, $y_{1,m}$, $Y_0(y_{1,m})$, $m = 1(1)50$, 8D.

- Olver (1960) tabulates $j_{n,m}$, $J'_n(j_{n,m})$, $j'_{n,m}$, $J_n(j'_{n,m})$, $y_{n,m}$, $Y'_n(y_{n,m})$, $y'_{n,m}$, $Y_n(y'_{n,m})$, $n = 0(\frac{1}{2})20\frac{1}{2}$, $m = 1(1)50$, 8D. Also included are tables of the coefficients in the uniform asymptotic expansions of these zeros and associated values as $n \to \infty$; see §10.21(viii), and more fully Olver (1954).

- Morgenthaler and Reismann (1963) tabulates $j'_{n,m}$ for $n = 21(1)51$ and $j'_{n,m} < 100$, 7-10S.

- Abramowitz and Stegun (1964, Chapter 9) tabulates $j_{n,m}$, $J'_n(j_{n,m})$, $j'_{n,m}$, $J_n(j'_{n,m})$, $n = 0(1)8$, $m = 1(1)20$, 5D (10D for $n = 0$), $y_{n,m}$, $Y'_n(y_{n,m})$, $y'_{n,m}$, $Y_n(y'_{n,m})$, $n = 0(1)8$, $m = 1(1)20$, 5D

(8D for $n = 0$), $J_0(j_{0,m} x)$, $m = 1(1)5$, $x = 0(.02)1$, 5D. Also included are the first 5 zeros of the functions $x J_1(x) - \lambda J_0(x)$, $J_1(x) - \lambda x J_0(x)$, $J_0(x) Y_0(\lambda x) - Y_0(x) J_0(\lambda x)$, $J_1(x) Y_1(\lambda x) - Y_1(x) J_1(\lambda x)$, $J_1(x) Y_0(\lambda x) - Y_1(x) J_0(\lambda x)$ for various values of λ and λ^{-1} in the interval $[0, 1]$, 4–8D.

- Abramowitz and Stegun (1964, Chapter 10) tabulates $j_{\nu,m}$, $J'_\nu(j_{\nu,m})$, $j'_{\nu,m}$, $J_\nu(j'_{\nu,m})$, $y_{\nu,m}$, $Y'_\nu(y_{\nu,m})$, $y'_{\nu,m}$, $Y_\nu(y'_{\nu,m})$, $\nu = \frac{1}{2}(1)19\frac{1}{2}$, $m = 1(1)m_\nu$, where m_ν ranges from 8 at $\nu = \frac{1}{2}$ down to 1 at $\nu = 19\frac{1}{2}$, 6–7D.

- Makinouchi (1966) tabulates all values of $j_{\nu,m}$ and $y_{\nu,m}$ in the interval $(0, 100)$, with at least 29S. These are for $\nu = 0(1)5, 10, 20$; $\nu = \frac{3}{2}, \frac{5}{2}$; $\nu = m/n$ with $m = 1(1)n - 1$ and $n = 3(1)8$, except for $\nu = \frac{1}{2}$.

- Döring (1971) tabulates the first 100 values of ν (> 1) for which $J'_{-\nu}(x)$ has the double zero $x = \nu$, 10D.

- Heller (1976) tabulates $j_{0,m}$, $J_1(j_{0,m})$, $j_{1,m}$, $J_0(j_{1,m})$, $j'_{1,m}$, $J_1(j'_{1,m})$ for $m = 1(1)100$, 25D.

- Wills et al. (1982) tabulates $j_{0,m}$, $j_{1,m}$, $y_{0,m}$, $y_{1,m}$ for $m = 1(1)30$, 35D.

- Kerimov and Skorokhodov (1985c) tabulates 201 double zeros of $J''_{-\nu}(x)$, 10 double zeros of $J'''_{-\nu}(x)$, 101 double zeros of $Y'_{-\nu}(x)$, 201 double zeros of $Y''_{-\nu}(x)$, and 10 double zeros of $Y'''_{-\nu}(x)$, all to 8 or 9D.

- Zhang and Jin (1996, pp. 196–198) tabulates $j_{n,m}$, $j'_{n,m}$, $y_{n,m}$, $y'_{n,m}$, $n = 0(1)3$, $m = 1(1)10$, 8D; the first five zeros of $J_n(x) Y_n(\lambda x) - J_n(\lambda x) Y_n(x)$, $J'_n(x) Y'_n(\lambda x) - J'_n(\lambda x) Y'_n(x)$, $n = 0, 1, 2$, $\lambda = 1.1(.1)1.6, 1.8, 2(.5)5$, 7D.

Complex Zeros

- Abramowitz and Stegun (1964, p. 373) tabulates the three smallest zeros of $Y_0(z)$, $Y_1(z)$, $Y'_1(z)$ in the sector $0 < \operatorname{ph} z \leq \pi$, together with the corresponding values of $Y_1(z)$, $Y_0(z)$, $Y_1(z)$, respectively, to 9D. (There is an error in the value of $Y_0(z)$ at the 3rd zero of $Y_1(z)$: the last four digits should be 2533; see Amos (1985).)

- Döring (1966) tabulates all zeros of $Y_0(z)$, $Y_1(z)$, $H_0^{(1)}(z)$, $H_1^{(1)}(z)$, that lie in the sector $|z| < 158$, $|\operatorname{ph} z| \leq \pi$, to 10D. Some of the smaller zeros of $Y_n(z)$ and $H_n^{(1)}(z)$ for $n = 2, 3, 4, 5, 15$ are also included.

- Kerimov and Skorokhodov (1985a) tabulates 5 (nonreal) complex conjugate pairs of zeros of the principal branches of $Y_n(z)$ and $Y'_n(z)$ for $n = 0(1)5$, 8D.

- Kerimov and Skorokhodov (1985b) tabulates 50 zeros of the principal branches of $H_0^{(1)}(z)$ and $H_1^{(1)}(z)$, 8D.

- Kerimov and Skorokhodov (1987) tabulates 100 complex double zeros ν of $Y'_\nu(ze^{-\pi i})$ and $H_\nu^{(1)'}(ze^{-\pi i})$, 8D.

- MacDonald (1989) tabulates the first 30 zeros, in ascending order of absolute value in the fourth quadrant, of the function $J_0(z) - i J_1(z)$, 6D. (Other zeros of this function can be obtained by reflection in the imaginary axis).

- Zhang and Jin (1996, p. 199) tabulates the real and imaginary parts of the first 15 conjugate pairs of complex zeros of $Y_0(z)$, $Y_1(z)$, $Y'_1(z)$ and the corresponding values of $Y_1(z)$, $Y_0(z)$, $Y_1(z)$, respectively, 10D.

10.75(iv) Integrals of Bessel Functions

- Abramowitz and Stegun (1964, Chapter 11) tabulates $\int_0^x J_0(t)\,dt$, $\int_0^x Y_0(t)\,dt$, $x = 0(.1)10$, 10D; $\int_0^x t^{-1}(1 - J_0(t))\,dt$, $\int_x^\infty t^{-1} Y_0(t)\,dt$, $x = 0(.1)5$, 8D.

- Zhang and Jin (1996, p. 270) tabulates $\int_0^x J_0(t)\,dt$, $\int_0^x t^{-1}(1 - J_0(t))\,dt$, $\int_0^x Y_0(t)\,dt$, $\int_x^\infty t^{-1} Y_0(t)\,dt$, $x = 0(.1)1(.5)20$, 8D.

10.75(v) Modified Bessel Functions and their Derivatives

- British Association for the Advancement of Science (1937) tabulates $I_0(x)$, $I_1(x)$, $x = 0(.001)5$, 7–8D; $K_0(x)$, $K_1(x)$, $x = 0.01(.01)5$, 7–10D; $e^{-x} I_0(x)$, $e^{-x} I_1(x)$, $e^x K_0(x)$, $e^x K_1(x)$, $x = 5(.01)10(.1)20$, 8D. Also included are auxiliary functions to facilitate interpolation of the tables of $K_0(x)$, $K_1(x)$ for small values of x.

- Bickley et al. (1952) tabulates $x^{-n} I_n(x)$ or $e^{-x} I_n(x)$, $x^n K_n(x)$ or $e^x K_n(x)$, $n = 2(1)20$, $x = 0(.01$ or $.1)\, 10(.1)\,20$, 8S; $I_n(x)$, $K_n(x)$, $n = 0(1)20$, $x = 0$ or $0.1(.1)20$, 10S.

- Olver (1962) provides tables for the uniform asymptotic expansions given in §10.41(ii), including η and the coefficients $U_k(p)$, $V_k(p)$ as functions of $p = (1 + x^2)^{-\frac{1}{2}}$. These enable $I_\nu(\nu x)$, $K_\nu(\nu x)$, $I'_\nu(\nu x)$, $K'_\nu(\nu x)$ to be computed to 10S when $\nu \geq 16$.

- The main tables in Abramowitz and Stegun (1964, Chapter 9) give $e^{-x}I_n(x)$, $e^x K_n(x)$, $n = 0, 1, 2$, $x = 0(.1)10(.2)20$, 8D–10D or 10S; $\sqrt{x}e^{-x}I_n(x)$, $(\sqrt{x}/\pi) e^x K_n(x)$, $n = 0, 1, 2$, $1/x = 0(.002)0.05$; $K_0(x) + I_0(x)\ln x$, $x(K_1(x) - I_1(x)\ln x)$, $x = 0(.1)2$, 8D; $e^{-x}I_n(x)$, $e^x K_n(x)$, $n = 3(1)9$, $x = 0(.2)10(.5)20$, 5S; $I_n(x)$, $K_n(x)$, $n = 0(1)20(10)50, 100$, $x = 1, 2, 5, 10, 50, 100$, 9–10S.

- Achenbach (1986) tabulates $I_0(x)$, $I_1(x)$, $K_0(x)$, $K_1(x)$, $x = 0(.1)8$, 19D or 19–21S.

- Zhang and Jin (1996, pp. 240–250) tabulates $I_n(x)$, $I'_n(x)$, $K_n(x)$, $K'_n(x)$, $n = 0(1)10(10)50, 100$, $x = 1, 5, 10, 25, 50, 100$, 9S; $I_{n+\alpha}(x)$, $I'_{n+\alpha}(x)$, $K_{n+\alpha}(x)$, $K'_{n+\alpha}(x)$, $n = 0(1)5$, 10, 30, 50, 100, $\alpha = \frac{1}{4}, \frac{1}{3}, \frac{1}{2}, \frac{2}{3}, \frac{3}{4}$, $x = 1, 5, 10, 50$, 8S; real and imaginary parts of $I_{n+\alpha}(z)$, $I'_{n+\alpha}(z)$, $K_{n+\alpha}(z)$, $K'_{n+\alpha}(z)$, $n = 0(1)15$, $20(10)50, 100$, $\alpha = 0, \frac{1}{2}$, $z = 4 + 2i, 20 + 10i$, 8S.

10.75(vi) Zeros of Modified Bessel Functions and their Derivatives

- Parnes (1972) tabulates all zeros of the principal value of $K_n(z)$, for $n = 2(1)10$, 9D.

- Leung and Ghaderpanah (1979), tabulates all zeros of the principal value of $K_n(z)$, for $n = 2(1)10$, 29S.

- Kerimov and Skorokhodov (1984b) tabulates all zeros of the principal values of $K_n(z)$ and $K'_n(z)$, for $n = 2(1)20$, 9S.

- Kerimov and Skorokhodov (1984c) tabulates all zeros of $I_{-n-\frac{1}{2}}(z)$ and $I'_{-n-\frac{1}{2}}(z)$ in the sector $0 \leq \mathrm{ph}\, z \leq \frac{1}{2}\pi$ for $n = 1(1)20$, 9S.

- Kerimov and Skorokhodov (1985b) tabulates all zeros of $K_n(z)$ and $K'_n(z)$ in the sector $-\frac{1}{2}\pi < \mathrm{ph}\, z \leq \frac{3}{2}\pi$ for $n = 0(1)5$, 8D.

10.75(vii) Integrals of Modified Bessel Functions

- Abramowitz and Stegun (1964, Chapter 11) tabulates $e^{-x}\int_0^x I_0(t)\,dt$, $e^x \int_x^\infty K_0(t)\,dt$, $x = 0(.1)10$, 7D; $e^{-x}\int_0^x t^{-1}(I_0(t) - 1)\,dt$, $xe^x \int_x^\infty t^{-1} K_0(t)\,dt$, $x = 0(.1)5$, 6D.

- Bickley and Nayler (1935) tabulates $\mathrm{Ki}_n(x)$ (§10.43(iii)) for $n = 1(1)16$, $x = 0(.05)0.2(.1)\,2, 3$, 9D.

- Zhang and Jin (1996, p. 271) tabulates $e^{-x}\int_0^x I_0(t)\,dt$, $e^{-x}\int_0^x t^{-1}(I_0(t) - 1)\,dt$, $e^x \int_x^\infty K_0(t)\,dt$, $xe^x \int_x^\infty t^{-1} K_0(t)\,dt$, $x = 0(.1)1(.5)20$, 8D.

10.75(viii) Modified Bessel Functions of Imaginary or Complex Order

For the notation see §10.45.

- Žurina and Karmazina (1967) tabulates $\widetilde{K}_\nu(x)$ for $\nu = 0.01(.01)10$, $x = 0.1(.1)10.2$, 7S.

- Rappoport (1979) tabulates the real and imaginary parts of $K_{\frac{1}{2} + i\tau}(x)$ for $\tau = 0.01(.01)10$, $x = 0.1(.2)9.5$, 7S.

10.75(ix) Spherical Bessel Functions, Modified Spherical Bessel Functions, and their Derivatives

- The main tables in Abramowitz and Stegun (1964, Chapter 10) give $\mathsf{j}_n(x)$, $\mathsf{y}_n(x)$ $n = 0(1)8$, $x = 0(.1)10$, 5–8S; $\mathsf{j}_n(x)$, $\mathsf{y}_n(x)$ $n = 0(1)20(10)50$, 100, $x = 1, 2, 5, 10, 50, 100$, 10S; $\mathsf{i}_n^{(1)}(x)$, $\mathsf{k}_n(x)$, $n = 0, 1, 2$, $x = 0(.1)5$, 4–9D; $\mathsf{i}_n^{(1)}(x)$, $\mathsf{k}_n(x)$, $n = 0(1)20(10)50, 100$, $x = 1, 2, 5, 10, 50, 100$, 10S. (For the notation see §10.1 and §10.47(ii).)

- Zhang and Jin (1996, pp. 296–305) tabulates $\mathsf{j}_n(x)$, $\mathsf{j}'_n(x)$, $\mathsf{y}_n(x)$, $\mathsf{y}'_n(x)$, $\mathsf{i}_n^{(1)}(x)$, $\mathsf{i}_n^{(1)'}(x)$, $\mathsf{k}_n(x)$, $\mathsf{k}'_n(x)$, $n = 0(1)10(10)30, 50, 100$, $x = 1, 5, 10, 25, 50, 100$, 8S; $x\mathsf{j}_n(x)$, $(x\mathsf{j}_n(x))'$, $x\mathsf{y}_n(x)$, $(x\mathsf{y}_n(x))'$ (Riccati–Bessel functions and their derivatives), $n = 0(1)10(10)30, 50, 100$, $x = 1, 5, 10, 25, 50, 100$, 8S; real and imaginary parts of $\mathsf{j}_n(z)$, $\mathsf{j}'_n(z)$, $\mathsf{y}_n(z)$, $\mathsf{y}'_n(z)$, $\mathsf{i}_n^{(1)}(z)$, $\mathsf{i}_n^{(1)'}(z)$, $\mathsf{k}_n(z)$, $\mathsf{k}'_n(z)$, $n = 0(1)15$, $20(10)50, 100$, $z = 4 + 2i, 20 + 10i$, 8S. (For the notation replace j, y, i, k by $\mathsf{j}, \mathsf{y}, \mathsf{i}^{(1)}, \mathsf{k}$, respectively.)

10.75(x) Zeros and Associated Values of Derivatives of Spherical Bessel Functions

For the notation see §10.58.

- Olver (1960) tabulates $a'_{n,m}$, $\mathsf{j}_n(a'_{n,m})$, $b'_{n,m}$, $\mathsf{y}_n(b'_{n,m})$, $n = 1(1)20$, $m = 1(1)50$, 8D. Also included are tables of the coefficients in the uniform asymptotic expansions of these zeros and associated values as $n \to \infty$.

10.75(xi) Kelvin Functions and their Derivatives

- Young and Kirk (1964) tabulates $\text{ber}_n x$, $\text{bei}_n x$, $\text{ker}_n x$, $\text{kei}_n x$, $n = 0, 1$, $x = 0(.1)10$, 15D; $\text{ber}_n x$, $\text{bei}_n x$, $\text{ker}_n x$, $\text{kei}_n x$, modulus and phase functions $M_n(x)$, $\theta_n(x)$, $N_n(x)$, $\phi_n(x)$, $n = 0, 1, 2$, $x = 0(.01)2.5$, 8S, and $n = 0(1)10$, $x = 0(.1)10$, 7S. Also included are auxiliary functions to facilitate interpolation of the tables for $n = 0(1)10$ for small values of x. (Concerning the phase functions see §10.68(iv).)

- Abramowitz and Stegun (1964, Chapter 9) tabulates $\text{ber}_n x$, $\text{bei}_n x$, $\text{ker}_n x$, $\text{kei}_n x$, $n = 0, 1$, $x = 0(.1)5$, 9–10D; $x^n(\text{ker}_n x + (\text{ber}_n x)(\ln x))$, $x^n(\text{kei}_n x + (\text{bei}_n x)(\ln x))$, $n = 0, 1$, $x = 0(.1)1$, 9D; modulus and phase functions $M_n(x)$, $\theta_n(x)$, $N_n(x)$, $\phi_n(x)$, $n = 0, 1$, $x = 0(.2)7$, 6D; $\sqrt{x}e^{-x/\sqrt{2}}M_n(x)$, $\theta_n(x) - (x/\sqrt{2})$, $\sqrt{x}e^{x/\sqrt{2}}N_n(x)$, $\phi_n(x) + (x/\sqrt{2})$, $n = 0, 1$, $1/x = 0(.01)0.15$, 5D.

- Zhang and Jin (1996, p. 322) tabulates $\text{ber}\, x$, $\text{ber}'x$, $\text{bei}\, x$, $\text{bei}'x$, $\text{ker}\, x$, $\text{ker}'x$, $\text{kei}\, x$, $\text{kei}'x$, $x = 0(1)20$, 7S.

10.75(xii) Zeros of Kelvin Functions and their Derivatives

- Zhang and Jin (1996, p. 323) tabulates the first 20 real zeros of $\text{ber}\, x$, $\text{ber}'x$, $\text{bei}\, x$, $\text{bei}'x$, $\text{ker}\, x$, $\text{ker}'x$, $\text{kei}\, x$, $\text{kei}'x$, 8D.

10.76 Approximations

10.76(i) Introduction

Because of the comprehensive nature of more recent software packages (§10.77), the following subsections include only references that give representative examples of the kind of approximations that can be used to generate the functions that appear in the present chapter. For references to other approximations, see for example, Luke (1975, §9.13.3).

10.76(ii) Bessel Functions, Hankel Functions, and Modified Bessel Functions

Real Variable and Order : Functions

Luke (1971a,b, 1972), Luke (1975, Tables 9.1, 9.2, 9.5, 9.6, 9.11–9.15, 9.17–9.21), Weniger and Čížek (1990), Németh (1992, Chapters 4–6).

Real Variable and Order : Zeros

Piessens (1984, 1990), Piessens and Ahmed (1986), Németh (1992, Chapter 7).

Real Variable and Order : Integrals

Luke (1975, Tables 9.3, 9.4, 9.7–9.9, 9.16, 9.22), Németh (1992, Chapter 10).

Complex Variable; Real Order

Luke (1975, Tables 9.23–9.28), Coleman and Monaghan (1983), Coleman (1987), Zhang (1996), Zhang and Belward (1997).

Real Variable; Imaginary Order

Poquérusse and Alexiou (1999).

10.76(iii) Other Functions

Bickley Functions

Blair et al. (1978).

Spherical Bessel Functions

Delic (1979).

Kelvin Functions

Luke (1975, Table 9.10), Németh (1992, Chapter 9).

10.77 Software

See http://dlmf.nist.gov/10.77.

References

General References

The main references used in writing this chapter are Watson (1944) and Olver (1997b).

Sources

The following list gives the references or other indications of proofs that were used in constructing the various sections of this chapter. These sources supplement the references that are quoted in the text.

§10.2 Olver (1997b, pp. 57, 237–238, 242–243) and Watson (1944, pp. 38–45, 57–64, 196–198). The conclusions in §10.2(iii) follow from §2.7(iv) and the limiting forms of the solutions as $z \to 0$ and as $z \to \infty$; see §10.7.

§10.3 These graphics were produced at NIST.

§10.4 Olver (1997b, pp. 56, 238–239, 242–243) and Watson (1944, pp. 74–75).

§10.5 For the Wronskians use (1.13.5) and the limiting forms in §10.7. Then for the cross-products apply (10.6.2).

§10.6 For (10.6.1) and (10.6.2) see Olver (1997b, pp. 58–59, 240–242) or Watson (1944, pp. 45, 66, 73–74). (10.6.3) are special cases, and (10.6.4), (10.6.5) follow by straightforward substitution. For (10.6.6) see Watson (1944, pp. 46). For (10.6.7) use induction combined with the second of (10.6.1). For (10.6.8)–(10.6.10) see Goodwin (1949b).

§10.7 For (10.7.1) and (10.7.3) use (10.2.2) and (10.8.2). For (10.7.2) use (10.4.3) and (10.7.1). For (10.7.4) and (10.7.5) use (10.2.3) and (10.7.3) when ν is not an integer; (10.4.1), (10.8.1) otherwise. For (10.7.6) use (10.2.3) and (10.7.3). For (10.7.7) use (10.4.3), (10.7.3), and (10.7.4). For (10.7.8) see (10.17.3) and (10.17.4).

§10.8 Olver (1997b, p. 243) and Watson (1944, p. 147).

§10.9 Watson (1944, pp. 19–21, 47–48, 68–71, 150, 160–170, 174–180, 436, 438, 441–444). For (10.9.3) see Olver (1997b, p. 244) (with "Exercises 2.2 and 9.5" corrected to "Exercises 2.3 and 9.5"). For (10.9.5), (10.9.10), (10.9.11), (10.9.13), (10.9.14) see Erdélyi et al. (1953b, pp. 18, 21, 82). (The condition $\Re(z \pm \zeta) > 0$ in (10.9.14) is weaker than the corresponding condition in Erdélyi et al. (1953b, p. 82, Eq. (18)).) (10.9.15), (10.9.16) follow from (10.9.10), (10.9.11) by change of variables $z = \zeta \cosh\phi$, $t \to t - \ln\tanh(\frac{1}{2}\phi)$, $\phi > 0$. For (10.9.27) see Erdélyi et al. (1953b, p. 47). See also Olver (1997b, pp. 340–341).

§10.10 Watson (1944, §§5.6, 9.65).

§10.11 For (10.11.1)–(10.11.5) use (10.2.2), (10.2.3), (10.4.3). For (10.11.6)–(10.11.8) take limits. For (10.11.9) use the Schwarz Reflection Principle (§1.10(ii)).

§10.12 For (10.12.1) see Olver (1997b, pp. 55–56). For (10.12.2)–(10.12.6) set $t = e^{i\theta}$ and $ie^{i\theta}$, and apply other straighforward substitutions, including differentiations with respect to θ in the case of (10.12.6). See also Watson (1944, pp. 22–23).

§10.13 These results are obtainable from (10.2.1) by straightforward substitutions. See also §1.13(v).

§10.14 Watson (1944, pp. 49, 258–259, 268–270, 406) and Olver (1997b, pp. 59, 426).

§10.15 For (10.15.1) see Watson (1944, pp. 61–62) or Olver (1997b, p. 243). For (10.15.2) use (10.2.3). For (10.15.3)–(10.15.5) see Olver (1997b, p. 244). (10.15.6)–(10.15.9) appear without proof in Magnus et al. (1966, §3.3.3). To derive (10.15.6) the left-hand side satisfies the differential equation $x^2(d^2W/dx^2) + x(dW/dx) + (x^2 - \frac{1}{4})W = \sqrt{2/(\pi x)}\sin x$, obtained by differentiating (10.2.1) with respect to ν, setting $\nu = \frac{1}{2}$, and referring to (10.16.1) for w. This inhomogeneous equation for W can be solved by variation of parameters (§1.13(ii)), using the fact that independent solutions of the corresponding homogeneous equation are $J_{\frac{1}{2}}(x)$ and $Y_{\frac{1}{2}}(x)$ with Wronskian $2/(\pi x)$, and subsequently referring to (6.2.9) and (6.2.11). Similarly for (10.15.7). (10.15.8) and (10.15.9) follow from (10.15.2), (10.15.6), (10.15.7), and (10.16.1).

§10.16 For (10.16.3), (10.16.4) see Miller (1955, p. 43). For (10.16.5) and (10.16.6) see Olver (1997b, pp. 255, 259) and apply (10.27.8). For (10.16.7) and (10.16.8) apply (13.14.4) and (13.14.5). For (10.16.9) combine (10.2.2) and (16.2.1).

§10.17 Olver (1997b, pp. 237–242, 266–269), Watson (1944, pp. 205–206). (10.17.8)–(10.17.12) follow by differentiation of the corresponding expansions in §10.17(i); compare §2.1(iii). For (10.17.16)–(10.17.18) see Olver (1991b, Theorem 1) or Olver (1993a, Theorem 1.1), and (10.16.6).

§10.18 For (10.18.3) see §10.7(i). (10.18.4)–(10.18.16) are verifiable by straightforward substitutions. For (10.18.17), and also the concluding paragraph of §10.18(iii), see Watson (1944, pp. 448–449). For (10.18.19) substitute into $N_\nu^2(x) = {H_\nu^{(1)}}'(x){H_\nu^{(2)}}'(x)$ by means of (10.17.11), (10.17.12). The general term in (10.18.20) can be verified via (10.18.10). For (10.18.18) the first two terms can be found from (10.18.7), (10.17.3), (10.17.4), except for an arbitrary integer multiple of π. Higher terms can be calculated via (10.18.8), (10.18.17). By continuity, the multiple of π is independent of ν, hence it may be determined, e.g. by setting $\nu = \frac{1}{2}$ and referring to (10.16.1). Similar methods can be used for (10.18.21), together with the interlacing properties of the zeros of $J_{1/2}(z)$, $Y_{1/2}(z)$, and their derivatives (§10.21(i)). See also Bickley et al. (1952, p. xxxiv).

§10.19 (10.19.1), (10.19.2) follow from (10.2.2), (10.2.3), (10.4.3), (10.8.1), (5.5.3), (5.11.3). For (10.19.3) and (10.19.6) see Watson (1944, pp. 241–245) and Bickley et al. (1952, p. xxxv). The expansions for the derivatives are established in a similar manner, with the coefficients calculated by term-by-term differentiation; compare §2.1(iii).

§10.20 Olver (1997b, pp. 419–425), Olver (1954).

§10.21 For §10.21(i) see Watson (1944, pp. 477–487), Olver (1997b, pp. 244–249), Döring (1971), and Kerimov and Skorokhodov (1985a). For §10.21(ii) see Watson (1944, pp. 508 and 510) and Olver (1950). (In the latter reference t in (10.21.4) is replaced by $-t$.) (10.21.5) and (10.21.6) follow from (10.6.2). (10.21.12) and (10.21.13) follow from (10.18.3), (10.21.2), (10.21.3), and the fact that $\theta_\nu(x)$ is increasing when $x > 0$, whereas $\phi_\nu(x)$ is decreasing when $0 < x < \nu$ and increasing when $x > \nu$; compare (10.18.8). For (10.21.15), (10.21.16) see Watson (1944, pp. 497–498). For §10.21(iv) see Watson (1944, pp. 508–510), Lorch (1990, 1995), Wong and Lang (1991), McCann (1977), Lewis and Muldoon (1977), and Mercer (1992). For (10.21.19) see Watson (1944, pp. 503–507) or Olver (1997b, pp. 247–248). Similar methods can be used for (10.21.20). For (10.21.22)–(10.21.40) see Olver (1951, 1952). The zeros depicted in Figures 10.21.1–10.21.6 were computed at NIST using methods referred to in §10.74(vi). For (10.21.48)–(10.21.54) see McMahon (9495), Gray et al. (1922, p. 261), and Cochran (1964).

§10.22 For (10.22.1)–(10.22.3) differentiate and use (10.6.2), (11.4.27), (11.4.28). For (10.22.4)–(10.22.7) see Watson (1944, pp. 132–136). For (10.22.8)–(10.22.12) see Luke (1962, pp. 51–53). To verify (10.22.13) construct the expansion of the left-hand side in powers of z by use of (10.2.2), followed by term-by-term integration with the aid of (5.12.5) and (5.12.1). Then compare the result with the corresponding expansion of the right-hand side obtained from (10.8.3). Next, the result $\int_0^{2\pi} J_{2\nu}(2z \sin\theta) e^{\pm 2i\mu\theta} d\theta = \pi e^{\pm i\mu\pi} J_{\nu+\mu}(z) J_{\nu-\mu}(z)$, $\Re \nu > -\frac{1}{2}$, is proved in a similar manner with the aid of (5.12.6) in place of (5.12.5)—from which (10.22.14) and (10.22.15) both follow. (10.22.17) follows by combining (10.22.13) and (10.2.3); (10.22.16) is a special case of (10.22.14). For (10.22.18) replace θ by $\frac{1}{2}\pi - \theta$ and set $\mu = n$ in (10.22.17); then apply (10.2.3) and let $\nu \to 0$. For (10.22.19), (10.22.22), (10.22.25), (10.22.26) see Watson (1944, Chapter 12). (In the case of (10.22.25), page 374 of this reference lacks a factor $\frac{1}{2}$ on the right-hand side.) The verification of (10.22.20) is similar to that of (10.22.13), the role of (5.12.5) now being played by (5.12.2). For (10.22.21) combine (10.2.3) and (10.22.20). For (10.22.23) and (10.22.24) see Luke (1962, p. 302 (36) and p. 303 (39), respectively). For (10.22.27) see Watson (1944, p. 151). For (10.22.28), (10.22.29) differentiate and use (10.6.2). For (10.22.30) with $n \geq 1$ it follows by differentiation and use of (10.6.2) that the left-hand side equals $\int_0^x t^{-1} J_n^2(t) dt - \frac{1}{2} J_n^2(x)$; application of Watson (1944, p. 152) yields the second result, then for the first result refer to (10.23.3). Some modifications of the proof of (10.22.30) are needed when $n = 0$. For (10.22.31)–(10.22.35) see Watson (1944, p. 380). For (10.22.36) replace t by $z - t$, substitute for t^α via (10.23.15) (with z replaced by t, and ν replaced by α), and then apply (10.22.34). For (10.22.37) use (10.22.4) and (10.22.5); a similar proof applies to (10.22.38) after replacing $\mathscr{C}_{\mu+1}(az)$ and $\mathscr{D}_{\mu+1}(bz)$ by $\mp \mathscr{C}'_\mu(az)$ and $\mp \mathscr{D}'_\mu(bz)$, respectively, by means of (10.6.2). For the first result in (10.22.39) use (10.22.43) with $\nu = 0$ and μ replaced by $\mu - 1$, split the integration range at $t = x$ and take limits as $\mu \to 0$; for the second result substitute into the first result by (10.2.2) and integrate term by term. (10.22.40), is proved in a similar manner, starting from (10.22.44) and substituting by means of (10.8.2) and (10.2.2) with $\nu = 0$ for the term-by-term integration. For (10.22.41)–(10.22.45) see Luke (1962, pp. 56–57). For (10.22.46) see Erdélyi et al. (1953b, p. 96). (10.22.47) is the special case of Eq. (6) of Watson (1944, §13.53) obtained by setting $\mu = b = 0$, $\rho = \nu + 1$, and subsequently replacing k by b. For (10.22.48) see Sneddon (1966, Eq. (2.1.32)). For (10.22.49)–(10.22.59) see Watson (1944, pp. 385, 394, 403–405, 407; there is an error in Eq. (1), p. 407). For (10.22.60) differentiate (10.22.59) with respect to μ and use (10.2.4) with $n = 0$. For (10.22.61) see Watson (1944, p. 405). (10.22.62) follows from (10.22.56) with $\lambda = \nu - \mu - 1$ and (15.4.6). For (10.22.63), (10.22.64) see Watson (1944, p. 404). For (10.22.65) apply (10.22.56) with $\mu = \nu = 0$, then let $\lambda \to 1$. For (10.22.66), (10.22.67) see Watson (1944, pp. 389, 395). For (10.22.68) set $a = b$ in (10.22.67), differentiate with respect to ν and apply (10.2.4) and (10.27.5) with $n = 0$. For (10.22.69), (10.22.70), see Watson (1944, p. 429, Eqs. (3),(4), with $\mu = \nu + 1$ in (3)). For (10.22.71), (10.22.72) see Watson (1944, pp. 411, 412). For (10.22.74), (10.22.75) see Watson (1944, pp. 411) and Askey et al. (1986).

§10.23 Watson (1944, §§5.22, 11.3, 11.4, 16.11 and pp. 64, 67, 71, 138). (10.23.2) is obtained from (10.23.7) by taking $\chi = 0$ and $\alpha = 0, \pi$. For (10.23.21) see Temme (1996a, p. 247).

§10.24 (10.24.6)–(10.24.9) follow from (10.24.2)–(10.24.4) combined with (10.2.2), (10.2.3), (10.8.2), (10.17.3), and (10.17.4). (10.24.5) can be verified from (1.13.5) and either (10.24.6) or

(10.24.7)–(10.24.9) and their differentiated forms.

§10.25 Olver (1997b, pp. 60, 236–237, 250). The conclusions in §10.25(iii) follow from §2.7(iv) and the limiting forms of the solutions as $z \to 0$ and $z \to \infty$; see (10.25.3) and §10.30. See also (10.27.3).

§10.26 These graphics were produced at NIST.

§10.27 For (10.27.1)–(10.27.6) and (10.27.8) see Olver (1997b, pp. 60–61 and 250–252), Watson (1944, pp. 77–79), and (10.11.5). For (10.27.7), (10.27.9),–(10.27.11) combine these results with (10.4.4), and also use (10.34.2) with $m = 1$.

§10.28 For the Wronskians use (1.13.5) and the limiting forms in §10.30. For the cross-products apply (10.29.2).

§10.29 Watson (1944, p. 79). For (10.29.5) use induction combined with the second of (10.29.1).

§10.30 For (10.30.1) use (10.25.2). For (10.30.2) and (10.30.3) use (10.27.4) when ν is not an integer; (10.27.3), (10.31.1) otherwise. For (10.30.4), (10.30.5) use (10.40.1) and (10.34.1) with $m = \pm 1$.

§10.31 Olver (1997b, p. 253) or Watson (1944, p. 80). For (10.31.3) combine (10.8.3) and (10.27.6).

§10.32 Watson (1944, pp. 79, 80, 172, 181–183, 191, 193, 439–441), Erdélyi et al. (1953b, p. 82, 97–98), Paris and Kaminski (2001, p. 114). Also use (10.27.8). For (10.32.16) see Dixon and Ferrar (1930). (An error in the conditions has been corrected.) For (10.32.19) see Titchmarsh (1986a, Eq. (7.10.2)).

§10.33 Combine (10.10.1), (10.10.2) with (10.27.6).

§10.34 Watson (1944, p. 80) and Olver (1997b, pp. 253, 381). For (10.34.3) take $m = \pm 1$ in (10.34.2), and combine with (10.34.1).

§10.35 For (10.35.1) replace z and t in (10.12.1) by iz and $-it$, respectively, and apply (10.27.6). (10.35.2)–(10.35.6) are obtained by setting $t = e^{i\theta}$, $t = -ie^{i\theta}$, together with other straightforward substitutions.

§10.37 Olver (1997b, pp. 251–252). For (10.37.1) see Everitt and Jones (1977).

§10.38 (10.38.1) is obtained by differentiation of (10.25.2); compare (10.15.1). For (10.38.2) use (10.27.4). (10.38.3)–(10.38.5) are proved in a similar way to (10.15.3)–(10.15.5). (10.38.6) and (10.38.7) are stated without proof and in a slightly different notation in Magnus et al. (1966, §3.3.3).

Both cases of (10.38.6) can be derived by a method analogous to that used for (10.15.6) and (10.15.7). (10.38.7) follows from (10.38.2) and (10.38.6).

§10.39 For (10.39.5)–(10.39.10) combine (10.16.5)–(10.16.10) with (10.27.6) and (10.27.8).

§10.40 Watson (1944, pp. 202–203, 206–207), Olver (1997b, pp. 250–251, 266–269, 325). Also use (10.27.8). (10.40.3) and (10.40.4) are obtained by differentiation of (10.40.1) and (10.40.2); compare §2.1(iii). (10.40.6) and (10.40.7) are obtained by multiplication of (10.40.1)–(10.40.4): that the coefficients are the same as in (10.18.17) and (10.18.19) is a consequence of the fact that $I_\nu(x) K_\nu(x)$ and $I'_\nu(x) K'_\nu(x)$ satisfy the same differential equations as $M_\nu^2(x) = |H_\nu^{(1)}(x)|^2 = H_\nu^{(1)}(x) H_\nu^{(2)}(x)$ and $N_\nu^2(x) = |H_\nu^{(1)'}(x)|^2 = H_\nu^{(1)'}(x) H_\nu^{(2)'}(x)$, respectively, except for replacement of x by ix. For the statement concerning the accuracy of (10.40.5) use the error bounds given by (10.40.10)–(10.40.12). For (10.40.14) see Olver (1991b) together with (10.39.6).

§10.41 Olver (1997b, pp. 374–378). For (10.41.1), (10.41.2) combine (10.19.1), (10.19.2) with (10.27.6), (10.27.8).

§10.42 Watson (1944, pp. 511–513) and Olver (1997b, p. 254).

§10.43 For (10.43.1)–(10.43.3) differentiate, apply (10.29.2), and also (11.4.29) and (11.4.30) in the case of (10.43.2). For (10.43.4) replace x by ix in (10.22.11), (10.22.12) and use (10.27.6). For (10.43.5) combine (10.22.39) and (10.22.40) by means of (10.4.3) to obtain an expansion for $\int_x^\infty (H_0^{(1)}(t)/t) \, dt$; then replace x by ix and use (10.27.8). For (10.43.6)–(10.43.10) differentiate, apply §10.29(i) and also verify the limiting behavior as $x \to 0$ or $x \to \infty$. For (10.43.12) substitute into (10.43.11) by means of (10.32.9) with $\nu = 0$, invert the order of integration and apply (5.2.1). (10.43.13)–(10.43.16) follow from (10.43.12), and in the case of (10.43.16), (5.12.1). For (10.43.17) see Bickley and Nayler (1935). For §10.43(iv) see Watson (1944, pp. 388, 394–395, 410). For some results it is necessary to use the connection formulas (10.27.6); for example, to obtain (10.43.23) set $a = ib$ in Watson (1944, p. 394, Eq. (4)). Equations (10.43.22) follow from Eq. (7) of Watson (1944, §13.21). For (10.43.25) see Erdélyi et al. (1953b, p. 51). For (10.43.29) combine (10.22.68),

(10.27.6), (10.27.10). In §10.43(v), for Conditions (a) see Sneddon (1972, pp. 359–361). For Conditions (b) see Lebedev et al. (1965, pp. 194–196).

§10.44 For (10.44.1) combine (10.23.1) with (10.27.6) or with (10.27.8). Equations (10.44.2) are special cases of (10.23.1) and (10.44.1) with $\lambda = i$. For (10.44.3) combine (10.23.2) and (10.27.1) with (10.27.6) or with (10.27.8). For (10.44.4)–(10.44.6) combine (10.23.15)–(10.23.17) with (10.27.6), (10.27.8), and (10.4.3).

§10.45 Equations (10.45.5)–(10.45.8) follow from (10.25.2), (10.27.4), (10.31.2), (10.40.1), and (10.40.2). The Wronskian (10.45.4) can be verified from (1.13.5) and either (10.45.5) or (10.45.6)–(10.45.8) and their differentiated forms.

§10.47 For (10.47.3)–(10.47.9) use (10.2.3), (10.4.6), (10.27.3). For §10.47(iii) use §10.52. For (10.47.10)–(10.47.13) use (10.4.3), (10.27.4), (10.27.6), (10.27.8), and the definitions (10.47.3)–(10.47.9). For (10.47.14)–(10.47.16) use (10.11.1), (10.11.2), (10.34.1), with $m = 1$ in each case, and the definitions (10.47.3)–(10.47.9). For (10.47.17) use (10.47.11) and (10.47.16).

§10.48 These graphs were produced at NIST.

§10.49 For (10.49.1)–(10.49.7) observe that when $\nu = n + \frac{1}{2}$ the asymptotic expansions (10.17.3)–(10.17.6) terminate, and as a consequence of the error bounds of §10.17(iv) they represent the left-hand sides exactly. For (10.49.8)–(10.49.13) use the same method as for (10.49.1)–(10.49.7), or combine the results of §10.49(i) with (10.47.12) and (10.47.13). For the first of (10.49.14) combine the second of (10.51.3), with $n = 0$ and $m = n$, and the first of (10.49.3). Similarly for the second of (10.49.14) and also (10.49.15), (10.49.16). For (10.49.18) observe that from (10.18.6), (10.47.3), and (10.47.4), $j_n^2(z) + y_n^2(z) = (\pi/(2z))M_{n+\frac{1}{2}}^2(z)$. Then apply (10.18.17). To derive (10.49.20) combine (10.47.12) and (10.49.18).

§10.50 That the Wronskians are constant multiples of z^{-2} follows from (1.13.5). The constants can be found from the limiting forms (and their derivatives) given in §§10.52(i) or 10.52(ii). For (10.50.3) combine (10.50.1) with (10.51.1) and (10.51.2). For (10.50.4) use (10.49.2)–(10.49.5).

§10.51 For (10.51.1) and (10.51.2) combine (10.6.1) and (10.6.2) with the definitions (10.47.3)–(10.47.5). For (10.51.3) apply induction with the aid of (10.51.2). For (10.51.4) and (10.51.5) combine (10.29.1) and (10.29.2) with the definitions (10.47.7) and (10.47.9). For (10.51.6) apply induction with the aid of (10.51.5).

§10.52 For (10.52.1), (10.52.2) use §10.53. For (10.52.3)–(10.52.6) use (10.49.2), (10.49.4), (10.49.6)–(10.49.8), (10.49.10), and (10.49.12).

§10.53 Combine (10.2.2) and (10.25.2) with (10.47.3), (10.47.4), and (10.47.7).

§10.54 Watson (1944, pp. 50 and 174–175). For (10.54.1) use (10.9.4).

§10.56 To verify (10.56.1) and (10.56.2) show that each side of both equations satisfies the differential equation $(2t-z)(d^2w/dt^2) + (dw/dt) = zw$ via the first of (10.51.1) and (10.49.3), (10.49.5). Then check the initial conditions at $t = 0$. (10.56.3) and (10.56.4) follow from (10.56.1) and (10.56.2) via (10.47.12); then (10.56.5) follows from (10.47.11).

§10.57 For (10.57.1) use the differentiated form of the first of (10.47.3).

§10.59 For (10.59.1) suppose first $b \neq 0$. The left-hand side is $2i \int_0^\infty \sin(bt) j_n(t)\, dt$ or $2\int_0^\infty \cos(bt) j_n(t)\, dt$ according as n is odd or even, see (10.47.14). Next, apply (10.22.64) with $a = 1$, $\mu = \frac{1}{2}$ or $-\frac{1}{2}$, and subsequently replace $2n+1$ or $2n$ by n. For $J_{\pm(1/2)}(bt)$ and $J_{n+(1/2)}(t)$ we have (10.16.1) and (10.47.3); also the function $_2F_1$ is interpreted as a Legendre polynomial for both odd and even n via (14.3.11), (14.3.13), and (14.3.14). When $b = 0$, use (10.22.43), (10.47.3), and also $P_n(0) = (-1)^{\frac{1}{2}n} \left(\frac{1}{2}\right)_{\frac{1}{2}n} \big/ \left(\frac{1}{2}n\right)!$ or 0, according as the nonnegative integer n is even or odd; see (14.5.1) and §5.5.

§10.60 For (10.60.1)–(10.60.3) use (10.23.8) with $\nu = \frac{1}{2}$ and $\mathscr{C} = Y, J, H^{(1)}$; subsequently apply (10.47.12) and (10.47.13) in the case of (10.60.3). For (10.60.4) set $\mathscr{C}_\nu = Y_\nu$, $u = v = z$, $\nu = -n-\frac{1}{2}$, and $\alpha = \pi$ in (10.23.8). Then refer to (10.47.3), (10.47.4), and also apply the following results obtained from Table 18.6.1: $C_k^{(-n-\frac{1}{2})}(-1)$ equals $(2n+1)!/(k!(2n+1-k)!)$ when $k = 0, 1, \ldots, 2n+1$, and equals 0 when $k = 2n+2, 2n+3, \ldots$. For (10.60.5) use the same procedure, but with $\mathscr{C}_\nu = J_\nu$. (10.60.6) follows by combining (10.60.4) and (10.60.5) with §10.47(iv). For (10.60.7)–(10.60.9) see Watson (1944, pp. 368–369). For (10.60.10) use Watson (1944, p. 370, Eq. (9)) with $\nu = \frac{1}{2}$, $\phi = \alpha$, $\phi' = \frac{1}{2}\pi$; also

Eq. (18.7.9). For (10.60.11) see Watson (1944, p. 152). For (10.60.12) and (10.60.13) substitute $u = v = z$, with $\alpha = 0$ and π, into (10.60.2). For (10.60.14) see Vavreck and Thompson (1984).

§10.61 For (10.61.3) set $z = xe^{3\pi i/4}$ in (10.2.1). (10.61.4) follows by taking real and imaginary parts, and straightforward substitutions. For (10.61.5)–(10.61.8) see Whitehead (1911). (10.61.11) and (10.61.12) follow from the terminating forms of (10.67.1) and (10.67.2). Then (10.61.9) and (10.61.10) follow from these results and the terminating forms of (10.67.3) and (10.67.4). (Compare the derivation of the results given in §10.49(i) from (10.17.3)–(10.17.6).) The version of (10.61.9)–(10.61.10) given in Apelblat (1991) contains two sign errors.

§10.62 These graphs were produced at NIST.

§10.63 For (10.63.1)–(10.63.4) set $z = xe^{3\pi i/4}$ in (10.6.1) and (10.6.2). For (10.63.5)–(10.63.7) set $a = xe^{3\pi i/4}$. Then from (10.61.1) and (10.63.5) $J_\nu(a) J_\nu(\bar{a}) = p_\nu$, $J'_\nu(a) J'_\nu(\bar{a}) = s_\nu$, $J_\nu(a) J'_\nu(\bar{a}) = e^{3\pi i/4}(r_\nu - iq_\nu)$, $J_\nu(\bar{a}) J'_\nu(a) = e^{-3\pi i/4}(r_\nu + iq_\nu)$. Combine these results with (10.6.2) and eliminate the derivatives. See also Petiau (1955, pp. 266–267) (but this reference contains errors). For the functions $\ker_\nu x$ and $\kei_\nu x$ use the second of (10.61.2).

§10.65 Whitehead (1911). For (10.65.1), (10.65.2) combine (10.2.2), (10.61.1). For (10.65.3)–(10.65.5) combine (10.31.1), (10.61.1), and (10.61.2); see also Young and Kirk (1964, p. x).

§10.66 For (10.66.1) apply (10.23.1) with $\mathscr{C} = J$ and $\lambda = e^{3\pi i/4}$; also (10.44.1) with $\mathscr{L} = I$ and $\lambda = e^{\pi i/4}$. For (10.66.2) apply (10.23.2) with $\mathscr{C} = J$, $\nu = n$, $u = -x$, $v = ix$, and equate real and imaginary parts.

§10.67 For (10.67.1)–(10.67.8) combine (10.61.1), (10.61.2), and their differentiated forms with (10.40.1)–(10.40.4). To obtain the exponentially-small terms in (10.67.3), (10.67.4), (10.67.7), and (10.67.8), use the identity $\pi i I_\nu(xe^{\pi i/4}) = K_\nu(xe^{-3\pi i/4}) - e^{\nu\pi i} K_\nu(xe^{\pi i/4})$, obtained from (10.27.6) and (10.27.9). The final sentence in §10.67(i) is justified by error bounds obtained as in §10.40(iii). For (10.67.9)–(10.67.16), first replace the cos and sin functions in (10.67.1)–(10.67.4) by exponential functions by constructing the corresponding expansions for $\ber_\nu x \pm i \bei_\nu x$ and $\ker_\nu x \pm i \kei_\nu x$ and discarding the exponentially-small terms. Then set $\nu = 0$ and apply straightforward manipulations.

§10.68 (10.68.3)–(10.68.15) are derived from the definitions §10.68(i), the differential equation (10.61.3), the reflection formulas in §10.61(iv), and recurrence relations in §10.63(i) by straightforward manipulations. For (10.68.16)–(10.68.21) combine (10.68.5) and (10.68.6) with (10.67.1)–(10.67.4), ignoring the exponentially-small terms in (10.67.3) and (10.67.4). See also Whitehead (1911) and Young and Kirk (1964, pp. xiv–xv).

§10.69 Combine the results given in §§10.41(ii) and 10.41(iii) with the definitions (10.61.1) and (10.61.2).

§10.70 Revert (10.68.18) and (10.68.21) (§2.2).

§10.71 Differentiate and use (10.63.2) and (10.68.5). See also Young and Kirk (1964, pp. xvi–xvii).

Chapter 11
Struve and Related Functions
R. B. Paris[1]

Notation — 288
- 11.1 Special Notation 288

Struve and Modified Struve Functions — 288
- 11.2 Definitions 288
- 11.3 Graphics 289
- 11.4 Basic Properties 291
- 11.5 Integral Representations 292
- 11.6 Asymptotic Expansions 293
- 11.7 Integrals and Sums 293
- 11.8 Analogs to Kelvin Functions 294

Related Functions — 294
- 11.9 Lommel Functions 294

- 11.10 Anger–Weber Functions 295
- 11.11 Asymptotic Expansions of Anger–Weber Functions 297

Applications — 298
- 11.12 Physical Applications 298

Computation — 298
- 11.13 Methods of Computation 299
- 11.14 Tables 299
- 11.15 Approximations 300
- 11.16 Software 300

References — 300

---

[1]Division of Mathematical Sciences, University of Abertay Dundee, Dundee, United Kingdom.
Acknowledgments: This chapter is based in part on Abramowitz and Stegun (1964, Chapter 12) by M. Abramowitz. The author is indebted to Adri Olde Daalhuis for correcting a long-standing error in Eq. (11.10.23) in previous literature.
Copyright © 2009 National Institute of Standards and Technology. All rights reserved.

Notation

11.1 Special Notation

(For other notation see pp. xiv and 873.)

- x real variable.
- z complex variable.
- ν real or complex order.
- n integer order.
- k nonnegative integer.
- δ arbitrary small positive constant.

Unless indicated otherwise, primes denote derivatives with respect to the argument. For the functions $J_\nu(z)$, $Y_\nu(z)$, $H_\nu^{(1)}(z)$, $H_\nu^{(2)}(z)$, $I_\nu(z)$, and $K_\nu(z)$ see §§10.2(ii), 10.25(ii).

The functions treated in this chapter are the Struve functions $\mathbf{H}_\nu(z)$ and $\mathbf{K}_\nu(z)$, the modified Struve functions $\mathbf{L}_\nu(z)$ and $\mathbf{M}_\nu(z)$, the Lommel functions $s_{\mu,\nu}(z)$ and $S_{\mu,\nu}(z)$, the Anger function $\mathbf{J}_\nu(z)$, the Weber function $\mathbf{E}_\nu(z)$, and the associated Anger–Weber function $\mathbf{A}_\nu(z)$.

Struve and Modified Struve Functions

11.2 Definitions

11.2(i) Power-Series Expansions

11.2.1 $\mathbf{H}_\nu(z) = (\tfrac{1}{2}z)^{\nu+1} \sum_{n=0}^{\infty} \dfrac{(-1)^n (\tfrac{1}{2}z)^{2n}}{\Gamma(n+\tfrac{3}{2})\,\Gamma(n+\nu+\tfrac{3}{2})}$,

11.2.2
$$\mathbf{L}_\nu(z) = -i e^{-\tfrac{1}{2}\pi i \nu} \mathbf{H}_\nu(iz)$$
$$= (\tfrac{1}{2}z)^{\nu+1} \sum_{n=0}^{\infty} \dfrac{(\tfrac{1}{2}z)^{2n}}{\Gamma(n+\tfrac{3}{2})\,\Gamma(n+\nu+\tfrac{3}{2})}.$$

Principal values correspond to principal values of $(\tfrac{1}{2}z)^{\nu+1}$; compare §4.2(i).

The expansions (11.2.1) and (11.2.2) are absolutely convergent for all finite values of z. The functions $z^{-\nu-1}\mathbf{H}_\nu(z)$ and $z^{-\nu-1}\mathbf{L}_\nu(z)$ are entire functions of z and ν.

11.2.3 $\mathbf{H}_0(z) = \dfrac{2}{\pi}\left(z - \dfrac{z^3}{1^2 \cdot 3^2} + \dfrac{z^5}{1^2 \cdot 3^2 \cdot 5^2} - \cdots\right)$,

11.2.4 $\mathbf{L}_0(z) = \dfrac{2}{\pi}\left(z + \dfrac{z^3}{1^2 \cdot 3^2} + \dfrac{z^5}{1^2 \cdot 3^2 \cdot 5^2} + \cdots\right)$.

11.2.5 $\mathbf{K}_\nu(z) = \mathbf{H}_\nu(z) - Y_\nu(z)$,

11.2.6 $\mathbf{M}_\nu(z) = \mathbf{L}_\nu(z) - I_\nu(z)$.

Principal values of $\mathbf{K}_\nu(z)$ and $\mathbf{M}_\nu(z)$ correspond to principal values of the functions on the right-hand sides of (11.2.5) and (11.2.6).

Unless indicated otherwise, $\mathbf{H}_\nu(z)$, $\mathbf{K}_\nu(z)$, $\mathbf{L}_\nu(z)$, and $\mathbf{M}_\nu(z)$ assume their principal values throughout this Handbook.

11.2(ii) Differential Equations

Struve's Equation

11.2.7 $\dfrac{d^2w}{dz^2} + \dfrac{1}{z}\dfrac{dw}{dz} + \left(1 - \dfrac{\nu^2}{z^2}\right)w = \dfrac{(\tfrac{1}{2}z)^{\nu-1}}{\sqrt{\pi}\,\Gamma(\nu+\tfrac{1}{2})}$.

Particular solutions:

11.2.8 $w = \mathbf{H}_\nu(z), \mathbf{K}_\nu(z)$.

Modified Struve's Equation

11.2.9 $\dfrac{d^2w}{dz^2} + \dfrac{1}{z}\dfrac{dw}{dz} - \left(1 + \dfrac{\nu^2}{z^2}\right)w = \dfrac{(\tfrac{1}{2}z)^{\nu-1}}{\sqrt{\pi}\,\Gamma(\nu+\tfrac{1}{2})}$.

Particular solutions:

11.2.10 $w = \mathbf{L}_\nu(z), \mathbf{M}_\nu(z)$.

11.2(iii) Numerically Satisfactory Solutions

In this subsection A and B are arbitrary constants.

When $z = x$, $0 < x < \infty$, and $\Re\nu \geq 0$, numerically satisfactory general solutions of (11.2.7) are given by

11.2.11 $w = \mathbf{H}_\nu(x) + A J_\nu(x) + B Y_\nu(x)$,

11.2.12 $w = \mathbf{K}_\nu(x) + A J_\nu(x) + B Y_\nu(x)$.

(11.2.11) applies when x is bounded, and (11.2.12) applies when x is bounded away from the origin.

When $z \in \mathbb{C}$ and $\Re\nu \geq 0$, numerically satisfactory general solutions of (11.2.7) are given by

11.2.13 $w = \mathbf{H}_\nu(z) + A J_\nu(z) + B H_\nu^{(1)}(z)$,

11.2.14 $w = \mathbf{H}_\nu(z) + A J_\nu(z) + B H_\nu^{(2)}(z)$,

11.2.15 $w = \mathbf{K}_\nu(z) + A H_\nu^{(1)}(z) + B H_\nu^{(2)}(z)$.

(11.2.13) applies when $0 \leq \mathrm{ph}\,z \leq \pi$ and $|z|$ is bounded. (11.2.14) applies when $-\pi \leq \mathrm{ph}\,z \leq 0$ and $|z|$ is bounded. (11.2.15) applies when $|\mathrm{ph}\,z| \leq \pi$ and z is bounded away from the origin.

When $\Re\nu \geq 0$, numerically satisfactory general solutions of (11.2.9) are given by

11.2.16 $w = \mathbf{L}_\nu(z) + A K_\nu(z) + B I_\nu(z)$,

11.2.17 $w = \mathbf{M}_\nu(z) + A K_\nu(z) + B I_\nu(z)$.

(11.2.16) applies when $|\mathrm{ph}\,z| \leq \tfrac{1}{2}\pi$ with $|z|$ bounded. (11.2.17) applies when $|\mathrm{ph}\,z| \leq \tfrac{1}{2}\pi$ with z bounded away from the origin.

11.3 Graphics

11.3(i) Struve Functions

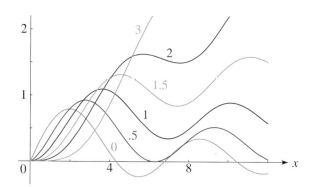

Figure 11.3.1: $\mathbf{H}_\nu(x)$ for $0 \leq x \leq 12$ and $\nu = 0, \frac{1}{2}, 1, \frac{3}{2}, 2, 3$.

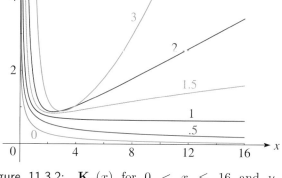

Figure 11.3.2: $\mathbf{K}_\nu(x)$ for $0 < x \leq 16$ and $\nu = 0, \frac{1}{2}, 1, \frac{3}{2}, 2, 3$.

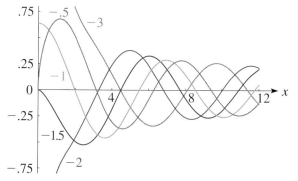

Figure 11.3.3: $\mathbf{H}_\nu(x)$ for $0 \leq x \leq 12$ and $\nu = -3, -2, -\frac{3}{2}, -1, -\frac{1}{2}$.

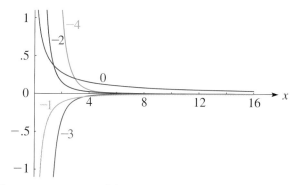

Figure 11.3.4: $\mathbf{K}_\nu(x)$ for $0 < x \leq 16$ and $\nu = -4, -3, -2, -1, 0$. If $\nu = -\frac{1}{2}, -\frac{3}{2}, \ldots$, then $\mathbf{K}_\nu(x)$ is identically zero.

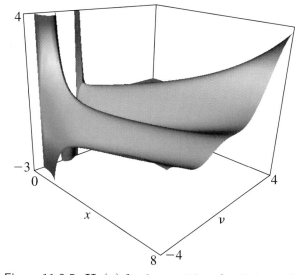

Figure 11.3.5: $\mathbf{H}_\nu(x)$ for $0 \leq x \leq 8$ and $-4 \leq \nu \leq 4$.

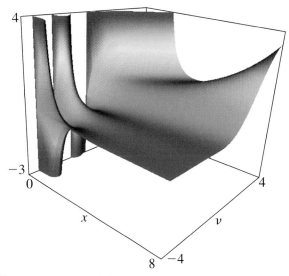

Figure 11.3.6: $\mathbf{K}_\nu(x)$ for $0 \leq x \leq 8$ and $-4 \leq \nu \leq 4$.

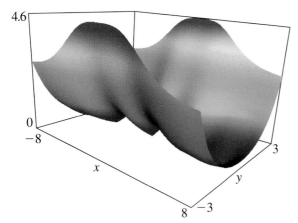

Figure 11.3.7: $|\mathbf{H}_0(x+iy)|$ for $-8 \leq x \leq 8$ and $-3 \leq y \leq 3$.

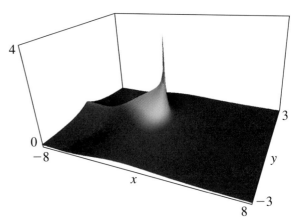

Figure 11.3.8: $|\mathbf{K}_0(x+iy)|$ (principal value) for $-8 \leq x \leq 8$ and $-3 \leq y \leq 3$. There is a cut along the negative real axis.

For further graphics see http://dlmf.nist.gov/11.3.i.

11.3(ii) Modified Struve Functions

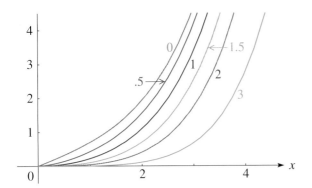

Figure 11.3.13: $\mathbf{L}_\nu(x)$ for $0 \leq x < 4.38$ and $\nu = 0, \frac{1}{2}, 1, \frac{3}{2}, 2, 3$.

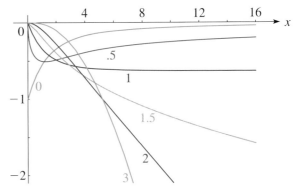

Figure 11.3.14: $\mathbf{M}_\nu(x)$ for $0 \leq x \leq 16$ and $\nu = 0, \frac{1}{2}, 1, \frac{3}{2}, 2, 3$.

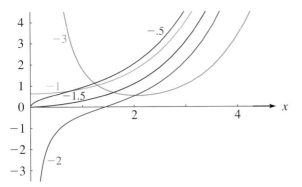

Figure 11.3.15: $\mathbf{L}_\nu(x)$ for $0 \leq x < 4.25$ and $\nu = -3, -2, -\frac{3}{2}, -1, -\frac{1}{2}$.

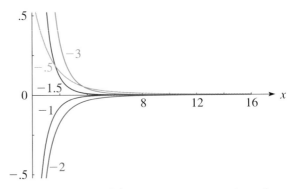

Figure 11.3.16: $\mathbf{M}_\nu(x)$ for $0 < x \leq 16$ and $\nu = -3, -2, -\frac{3}{2}, -1, -\frac{1}{2}$.

11.4 Basic Properties

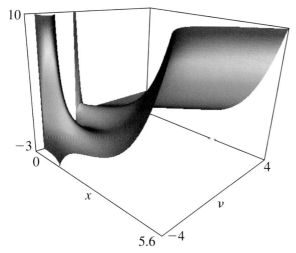

Figure 11.3.17: $\mathbf{L}_\nu(x)$ for $0 \leq x \leq 5.6$ and $-4 \leq \nu \leq 4$.

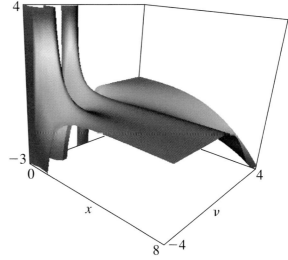

Figure 11.3.18: $\mathbf{M}_\nu(x)$ for $0 \leq x \leq 8$ and $-4 \leq \nu \leq 4$.

For further graphics see http://dlmf.nist.gov/11.3.ii.

11.4 Basic Properties

11.4(i) Half-Integer Orders

For $n = 0, 1, 2, \ldots$,

11.4.1 $\quad \mathbf{K}_{n+\frac{1}{2}}(z) = \left(\dfrac{2}{\pi z}\right)^{\frac{1}{2}} \sum_{m=0}^{n} \dfrac{(2m)! \, 2^{-2m}}{m!(n-m)!} (\tfrac{1}{2}z)^{n-2m},$

11.4.2
$$\mathbf{L}_{n+\frac{1}{2}}(z) = I_{-n-\frac{1}{2}}(z)$$
$$- \left(\dfrac{2}{\pi z}\right)^{\frac{1}{2}} \sum_{m=0}^{n} \dfrac{(-1)^m (2m)! \, 2^{-2m}}{m!(n-m)!} (\tfrac{1}{2}z)^{n-2m},$$

11.4.3 $\quad \mathbf{H}_{-n-\frac{1}{2}}(z) = (-1)^n J_{n+\frac{1}{2}}(z),$

11.4.4 $\quad \mathbf{L}_{-n-\frac{1}{2}}(z) = I_{n+\frac{1}{2}}(z).$

11.4.5 $\quad \mathbf{H}_{\frac{1}{2}}(z) = \left(\dfrac{2}{\pi z}\right)^{\frac{1}{2}} (1 - \cos z),$

11.4.6 $\quad \mathbf{H}_{-\frac{1}{2}}(z) = \left(\dfrac{2}{\pi z}\right)^{\frac{1}{2}} \sin z,$

11.4.7 $\quad \mathbf{L}_{\frac{1}{2}}(z) = \left(\dfrac{2}{\pi z}\right)^{\frac{1}{2}} (\cosh z - 1),$

11.4.8 $\quad \mathbf{L}_{-\frac{1}{2}}(z) = \left(\dfrac{2}{\pi z}\right)^{\frac{1}{2}} \sinh z,$

11.4.9
$$\mathbf{H}_{\frac{3}{2}}(z) = \left(\dfrac{z}{2\pi}\right)^{\frac{1}{2}} \left(1 + \dfrac{2}{z^2}\right) - \left(\dfrac{2}{\pi z}\right)^{\frac{1}{2}} \left(\sin z + \dfrac{\cos z}{z}\right),$$

11.4.10 $\quad \mathbf{H}_{-\frac{3}{2}}(z) = \left(\dfrac{2}{\pi z}\right)^{\frac{1}{2}} \left(\cos z - \dfrac{\sin z}{z}\right),$

11.4.11
$$\mathbf{L}_{\frac{3}{2}}(z) = -\left(\dfrac{z}{2\pi}\right)^{\frac{1}{2}} \left(1 - \dfrac{2}{z^2}\right) + \left(\dfrac{2}{\pi z}\right)^{\frac{1}{2}} \left(\sinh z - \dfrac{\cosh z}{z}\right),$$

11.4.12 $\quad \mathbf{L}_{-\frac{3}{2}}(z) = \left(\dfrac{2}{\pi z}\right)^{\frac{1}{2}} \left(\cosh z - \dfrac{\sinh z}{z}\right).$

11.4(ii) Inequalities

11.4.13 $\quad \mathbf{H}_\nu(x) \geq 0, \qquad x > 0, \nu \geq \tfrac{1}{2}.$

11.4.14
$$\mathbf{H}_\nu(z) = \dfrac{2(\tfrac{1}{2}z)^{\nu+1}}{\sqrt{\pi}\,\Gamma(\nu+\tfrac{3}{2})} (1 + \vartheta), \quad \nu \neq -\tfrac{3}{2}, -\tfrac{5}{2}, -\tfrac{7}{2}, \ldots,$$

where

11.4.15 $\quad |\vartheta| < \dfrac{2}{3} \exp\!\left(\dfrac{\tfrac{1}{4}|z|^2}{|\nu_0 + \tfrac{3}{2}|} - 1\right),$

and $|\nu_0 + \tfrac{3}{2}|$ is the smallest of the numbers $|\nu + \tfrac{3}{2}|$, $|\nu + \tfrac{5}{2}|$, $|\nu + \tfrac{9}{2}|, \ldots$.

11.4(iii) Analytic Continuation

11.4.16 $\quad \mathbf{H}_\nu(ze^{m\pi i}) = e^{m\pi i(\nu+1)} \mathbf{H}_\nu(z), \qquad m \in \mathbb{Z},$

11.4.17 $\quad \mathbf{L}_\nu(ze^{m\pi i}) = e^{m\pi i(\nu+1)} \mathbf{L}_\nu(z), \qquad m \in \mathbb{Z}.$

11.4(iv) Expansions in Series of Bessel Functions

11.4.18
$$\mathbf{H}_\nu(z) = \frac{4}{\pi^{1/2}\,\Gamma\!\left(\nu+\tfrac{1}{2}\right)}$$
$$\times \sum_{k=0}^{\infty} \frac{(2k+\nu+1)\,\Gamma(k+\nu+1)}{k!(2k+1)(2k+2\nu+1)} J_{2k+\nu+1}(z),$$
$$\nu \ne -1,-2,-3,\dots,$$

11.4.19 $\quad \mathbf{H}_\nu(z) = \left(\dfrac{z}{2\pi}\right)^{1/2} \displaystyle\sum_{k=0}^{\infty} \dfrac{(\tfrac{1}{2}z)^k}{k!(k+\tfrac{1}{2})} J_{k+\nu+\tfrac{1}{2}}(z),$

11.4.20 $\quad \mathbf{H}_\nu(z) = \dfrac{(\tfrac{1}{2}z)^{\nu+\tfrac{1}{2}}}{\Gamma(\nu+\tfrac{1}{2})} \displaystyle\sum_{k=0}^{\infty} \dfrac{(\tfrac{1}{2}z)^k}{k!(k+\nu+\tfrac{1}{2})} J_{k+\tfrac{1}{2}}(z),$

11.4.21
$$\mathbf{H}_0(z) = \frac{4}{\pi}\sum_{k=0}^{\infty}\frac{J_{2k+1}(z)}{2k+1} = 2\sum_{k=0}^{\infty}(-1)^k J^2_{k+\tfrac{1}{2}}(\tfrac{1}{2}z),$$

$$\mathbf{H}_1(z) = \frac{2}{\pi}(1-J_0(z)) + \frac{4}{\pi}\sum_{k=1}^{\infty}\frac{J_{2k}(z)}{4k^2-1}$$

11.4.22
$$= 4\sum_{k=0}^{\infty} J_{2k+\tfrac{1}{2}}(\tfrac{1}{2}z)\,J_{2k+\tfrac{3}{2}}(\tfrac{1}{2}z).$$

For these and further results see Luke (1969b, §9.4.5), and §10.23(iii).

11.4(v) Recurrence Relations and Derivatives

11.4.23
$$\mathbf{H}_{\nu-1}(z) + \mathbf{H}_{\nu+1}(z) = \frac{2\nu}{z}\mathbf{H}_\nu(z) + \frac{(\tfrac{1}{2}z)^\nu}{\sqrt{\pi}\,\Gamma(\nu+\tfrac{3}{2})},$$

11.4.24
$$\mathbf{H}_{\nu-1}(z) - \mathbf{H}_{\nu+1}(z) = 2\mathbf{H}'_\nu(z) - \frac{(\tfrac{1}{2}z)^\nu}{\sqrt{\pi}\,\Gamma(\nu+\tfrac{3}{2})},$$

11.4.25
$$\mathbf{L}_{\nu-1}(z) - \mathbf{L}_{\nu+1}(z) = \frac{2\nu}{z}\mathbf{L}_\nu(z) + \frac{(\tfrac{1}{2}z)^\nu}{\sqrt{\pi}\,\Gamma(\nu+\tfrac{3}{2})},$$

11.4.26
$$\mathbf{L}_{\nu-1}(z) + \mathbf{L}_{\nu+1}(z) = 2\mathbf{L}'_\nu(z) - \frac{(\tfrac{1}{2}z)^\nu}{\sqrt{\pi}\,\Gamma(\nu+\tfrac{3}{2})}.$$

11.4.27 $\quad \dfrac{d}{dz}\left(z^\nu \mathbf{H}_\nu(z)\right) = z^\nu \mathbf{H}_{\nu-1}(z),$

11.4.28 $\quad \dfrac{d}{dz}\left(z^{-\nu} \mathbf{H}_\nu(z)\right) = \dfrac{2^{-\nu}}{\sqrt{\pi}\,\Gamma(\nu+\tfrac{3}{2})} - z^{-\nu}\mathbf{H}_{\nu+1}(z),$

11.4.29 $\quad \dfrac{d}{dz}\left(z^\nu \mathbf{L}_\nu(z)\right) = z^\nu \mathbf{L}_{\nu-1}(z),$

11.4.30 $\quad \dfrac{d}{dz}\left(z^{-\nu} \mathbf{L}_\nu(z)\right) = \dfrac{2^{-\nu}}{\sqrt{\pi}\,\Gamma(\nu+\tfrac{3}{2})} + z^{-\nu}\mathbf{L}_{\nu+1}(z).$

11.4.31
$$\mathcal{H}_{\nu-m}(z) = z^{m-\nu}\left(\frac{1}{z}\frac{d}{dz}\right)^m (z^\nu \mathcal{H}_\nu(z)),\; m=1,2,3,\dots,$$

where $\mathcal{H}_\nu(z)$ denotes either $\mathbf{H}_\nu(z)$ or $\mathbf{L}_\nu(z)$.

11.4.32 $\quad \mathbf{H}'_0(z) = \dfrac{2}{\pi} - \mathbf{H}_1(z),\quad \dfrac{d}{dz}(z\,\mathbf{H}_1(z)) = z\,\mathbf{H}_0(z),$

11.4.33 $\quad \mathbf{L}'_0(z) = \dfrac{2}{\pi} + \mathbf{L}_1(z),\quad \dfrac{d}{dz}(z\,\mathbf{L}_1(z)) = z\,\mathbf{L}_0(z).$

11.4(vi) Derivatives with Respect to Order

For derivatives with respect to the order ν, see Apelblat (1989) and Brychkov and Geddes (2005).

11.4(vii) Zeros

For properties of zeros of $\mathbf{H}_\nu(x)$ see Steinig (1970).

For asymptotic expansions of zeros of $\mathbf{H}_0(x)$ see MacLeod (2002a).

11.5 Integral Representations

11.5(i) Integrals Along the Real Line

11.5.1
$$\mathbf{H}_\nu(z) = \frac{2(\tfrac{1}{2}z)^\nu}{\sqrt{\pi}\,\Gamma(\nu+\tfrac{1}{2})} \int_0^1 (1-t^2)^{\nu-\tfrac{1}{2}} \sin(zt)\,dt$$
$$= \frac{2(\tfrac{1}{2}z)^\nu}{\sqrt{\pi}\,\Gamma(\nu+\tfrac{1}{2})} \int_0^{\pi/2} \sin(z\cos\theta)(\sin\theta)^{2\nu}\,d\theta,$$
$$\Re\nu > -\tfrac{1}{2},$$

11.5.2 $\quad \mathbf{K}_\nu(z) = \dfrac{2(\tfrac{1}{2}z)^\nu}{\sqrt{\pi}\,\Gamma(\nu+\tfrac{1}{2})}\displaystyle\int_0^\infty e^{-zt}(1+t^2)^{\nu-\tfrac{1}{2}}\,dt,\; \Re z > 0,$

11.5.3 $\quad \mathbf{K}_0(z) = \dfrac{2}{\pi}\displaystyle\int_0^\infty e^{-z\sinh t}\,dt,\quad \Re z > 0,$

11.5.4 $\quad \mathbf{M}_\nu(z) = -\dfrac{2(\tfrac{1}{2}z)^\nu}{\sqrt{\pi}\,\Gamma(\nu+\tfrac{1}{2})}\displaystyle\int_0^1 e^{-zt}(1-t^2)^{\nu-\tfrac{1}{2}}\,dt,$
$$\Re\nu > -\tfrac{1}{2},$$

11.5.5 $\quad \mathbf{M}_0(z) = -\dfrac{2}{\pi}\displaystyle\int_0^{\pi/2} e^{-z\cos\theta}\,d\theta,$

11.5.6
$$\mathbf{L}_\nu(z) = \frac{2(\tfrac{1}{2}z)^\nu}{\sqrt{\pi}\,\Gamma(\nu+\tfrac{1}{2})}\int_0^{\pi/2} \sinh(z\cos\theta)(\sin\theta)^{2\nu}\,d\theta,$$
$$\Re\nu > -\tfrac{1}{2},$$

11.5.7
$$I_{-\nu}(x) - \mathbf{L}_\nu(x)$$
$$= \frac{2(\tfrac{1}{2}x)^\nu}{\sqrt{\pi}\,\Gamma(\nu+\tfrac{1}{2})}\int_0^\infty (1+t^2)^{\nu-\tfrac{1}{2}} \sin(xt)\,dt,$$
$$x > 0,\; \Re\nu < \tfrac{1}{2}.$$

11.5(ii) Contour Integrals

For loop-integral versions of (11.5.1), (11.5.2), (11.5.4), and (11.5.7) see Babister (1967, §§3.3 and 3.14).

Mellin–Barnes Integrals

11.5.8
$$(\tfrac{1}{2}x)^{-\nu-1}\mathbf{H}_\nu(x) = -\frac{1}{2\pi i}\int_{-i\infty}^{i\infty}\frac{\pi\csc(\pi s)}{\Gamma(\tfrac{3}{2}+s)\Gamma(\tfrac{3}{2}+\nu+s)}(\tfrac{1}{4}x^2)^s\,ds,$$
$$x>0,\ \Re\nu>-1,$$

11.5.9
$$(\tfrac{1}{2}z)^{-\nu-1}\mathbf{L}_\nu(z) = \frac{1}{2\pi i}\int_\infty^{(0+)}\frac{\pi\csc(\pi s)}{\Gamma(\tfrac{3}{2}+s)\Gamma(\tfrac{3}{2}+\nu+s)}(-\tfrac{1}{4}z^2)^s\,ds.$$

In (11.5.8) and (11.5.9) the path of integration separates the poles of the integrand at $s=0,1,2,\ldots$ from those at $s=-1,-2,-3,\ldots$.

11.5(iii) Compendia

For further integral representations see Babister (1967, §§3.3, 3.14), Erdélyi et al. (1954a, §§5.17, 15.3), Magnus et al. (1966, p. 114), Oberhettinger (1972), Oberhettinger (1974, §2.7), Oberhettinger and Badii (1973, §2.14), and Watson (1944, pp. 330, 374, and 426).

11.6 Asymptotic Expansions

11.6(i) Large $|z|$, Fixed ν

11.6.1
$$\mathbf{K}_\nu(z) \sim \frac{1}{\pi}\sum_{k=0}^\infty \frac{\Gamma(k+\tfrac{1}{2})(\tfrac{1}{2}z)^{\nu-2k-1}}{\Gamma(\nu+\tfrac{1}{2}-k)},\quad |\operatorname{ph} z|\le \pi-\delta,$$

where δ is an arbitrary small positive constant. If the series on the right-hand side of (11.6.1) is truncated after $m(\ge 0)$ terms, then the remainder term $R_m(z)$ is $O(z^{\nu-2m-1})$. If ν is real, z is positive, and $m+\tfrac{1}{2}-\nu\ge 0$, then $R_m(z)$ is of the same sign and numerically less than the first neglected term.

11.6.2
$$\mathbf{M}_\nu(z) \sim \frac{1}{\pi}\sum_{k=0}^\infty (-1)^{k+1}\frac{\Gamma(k+\tfrac{1}{2})(\tfrac{1}{2}z)^{\nu-2k-1}}{\Gamma(\nu+\tfrac{1}{2}-k)},$$
$$|\operatorname{ph} z|\le \tfrac{1}{2}\pi-\delta.$$

For re-expansions of the remainder terms in (11.6.1) and (11.6.2), see Dingle (1973, p. 445).

For the corresponding expansions for $\mathbf{H}_\nu(z)$ and $\mathbf{L}_\nu(z)$ combine (11.6.1), (11.6.2) with (11.2.5), (11.2.6), (10.17.4), and (10.40.1).

11.6.3
$$\int_0^z \mathbf{K}_0(t)\,dt - \frac{2}{\pi}(\ln(2z)+\gamma)$$
$$\sim \frac{2}{\pi}\sum_{k=1}^\infty (-1)^{k+1}\frac{(2k)!(2k-1)!}{(k!)^2(2z)^{2k}},\quad |\operatorname{ph} z|\le \pi-\delta,$$

11.6.4
$$\int_0^z \mathbf{M}_0(t)\,dt + \frac{2}{\pi}(\ln(2z)+\gamma)$$
$$\sim \frac{2}{\pi}\sum_{k=1}^\infty \frac{(2k)!(2k-1)!}{(k!)^2(2z)^{2k}},\quad |\operatorname{ph} z|\le \tfrac{1}{2}\pi-\delta,$$

where γ is Euler's constant (§5.2(ii)).

11.6(ii) Large $|\nu|$, Fixed z

11.6.5
$$\mathbf{H}_\nu(z),\mathbf{L}_\nu(z) \sim \frac{z}{\pi\nu\sqrt{2}}\left(\frac{ez}{2\nu}\right)^\nu,\quad |\operatorname{ph}\nu|\le \pi-\delta.$$

More fully, the series (11.2.1) and (11.2.2) can be regarded as generalized asymptotic expansions (§2.1(v)).

11.6(iii) Large $|\nu|$, Fixed z/ν

For fixed $\lambda(>1)$

11.6.6
$$\mathbf{K}_\nu(\lambda\nu) \sim \frac{(\tfrac{1}{2}\lambda\nu)^{\nu-1}}{\sqrt{\pi}\,\Gamma(\nu+\tfrac{1}{2})}\sum_{k=0}^\infty \frac{k!c_k(\lambda)}{\nu^k},\quad |\operatorname{ph}\nu|\le \tfrac{1}{2}\pi-\delta,$$

and for fixed $\lambda\ (>0)$

11.6.7
$$\mathbf{M}_\nu(\lambda\nu) \sim -\frac{(\tfrac{1}{2}\lambda\nu)^{\nu-1}}{\sqrt{\pi}\,\Gamma(\nu+\tfrac{1}{2})}\sum_{k=0}^\infty \frac{k!c_k(i\lambda)}{\nu^k},$$
$$|\operatorname{ph}\nu|\le \tfrac{1}{2}\pi-\delta.$$

Here

11.6.8
$$c_0(\lambda)=1,\quad c_1(\lambda)=2\lambda^{-2},$$
$$c_2(\lambda)=6\lambda^{-4}-\tfrac{1}{2}\lambda^{-2},\quad c_3(\lambda)=20\lambda^{-6}-4\lambda^{-4},$$
$$c_4(\lambda)=70\lambda^{-8}-\tfrac{45}{2}\lambda^{-6}+\tfrac{3}{8}\lambda^{-4},$$

and for higher coefficients $c_k(\lambda)$ see Dingle (1973, p. 203).

For the corresponding result for $\mathbf{H}_\nu(\lambda\nu)$ use (11.2.5) and (10.19.6). See also Watson (1944, p. 336).

For fixed $\lambda\ (>0)$

11.6.9
$$\mathbf{L}_\nu(\lambda\nu) \sim I_\nu(\lambda\nu),\quad |\operatorname{ph}\nu|\le \tfrac{1}{2}\pi-\delta,$$

and for an estimate of the relative error in this approximation see Watson (1944, p. 336).

11.7 Integrals and Sums

11.7(i) Indefinite Integrals

11.7.1
$$\int z^\nu \mathbf{H}_{\nu-1}(z)\,dz = z^\nu \mathbf{H}_\nu(z),$$

11.7.2
$$\int z^{-\nu}\mathbf{H}_{\nu+1}(z)\,dz = -z^{-\nu}\mathbf{H}_\nu(z) + \frac{2^{-\nu}z}{\sqrt{\pi}\,\Gamma(\nu+\tfrac{3}{2})},$$

11.7.3
$$\int z^\nu \mathbf{L}_{\nu-1}(z)\,dz = z^\nu \mathbf{L}_\nu(z),$$

11.7.4 $\int z^{-\nu} \mathbf{L}_{\nu+1}(z)\, dz = z^{-\nu} \mathbf{L}_\nu(z) - \dfrac{2^{-\nu} z}{\sqrt{\pi}\, \Gamma(\nu + \frac{3}{2})}.$

If

11.7.5 $f_\nu(z) = \displaystyle\int_0^z t^\nu\, \mathbf{H}_\nu(t)\, dt,$

then

11.7.6 $f_{\nu+1}(z) = (2\nu + 1) f_\nu(z) - z^{\nu+1} \mathbf{H}_\nu(z) + \dfrac{(\frac{1}{2} z^2)^{\nu+1}}{(\nu+1)\sqrt{\pi}\, \Gamma(\nu + \frac{3}{2})}, \quad \Re\nu > -1.$

11.7(ii) Definite Integrals

11.7.7
$$\int_0^{\pi/2} \mathbf{H}_\nu(z \sin\theta) \dfrac{(\sin\theta)^{\nu+1}}{(\cos\theta)^{2\nu}}\, d\theta = \dfrac{2^{-\nu}}{\sqrt{\pi}} \Gamma(\tfrac{1}{2} - \nu) z^{\nu-1}(1 - \cos z), \quad -\tfrac{3}{2} < \Re\nu < \tfrac{1}{2},$$

11.7.8 $\displaystyle\int_0^\infty \mathbf{H}_0(t)\, \dfrac{dt}{t} = \tfrac{1}{2}\pi, \quad \int_0^\infty \mathbf{H}_1(t)\, \dfrac{dt}{t^2} = \tfrac{1}{4}\pi,$

11.7.9 $\displaystyle\int_0^\infty \mathbf{H}_\nu(t)\, dt = -\cot(\tfrac{1}{2}\pi\nu), \quad -2 < \Re\nu < 0,$

11.7.10 $\displaystyle\int_0^\infty t^{-\nu-1} \mathbf{H}_\nu(t)\, dt = \dfrac{\pi}{2^{\nu+1}\, \Gamma(\nu + 1)}, \quad \Re\nu > -\tfrac{3}{2},$

11.7.11 $\displaystyle\int_0^\infty t^{\mu-\nu-1} \mathbf{H}_\nu(t)\, dt = \dfrac{\Gamma(\frac{1}{2}\mu)\, 2^{\mu-\nu-1} \tan(\frac{1}{2}\pi\mu)}{\Gamma(\nu - \frac{1}{2}\mu + 1)}, \quad |\Re\mu| < 1,\ \Re\nu > \Re\mu - \tfrac{3}{2},$

11.7.12
$$\int_0^\infty t^{-\mu-\nu} \mathbf{H}_\mu(t)\, \mathbf{H}_\nu(t)\, dt = \dfrac{\sqrt{\pi}\, \Gamma(\mu + \nu)}{2^{\mu+\nu}\, \Gamma(\mu + \nu + \frac{1}{2})\, \Gamma(\mu + \frac{1}{2})\, \Gamma(\nu + \frac{1}{2})}, \quad \Re(\mu + \nu) > 0.$$

For other integrals involving products of Struve functions see Zanovello (1978, 1995). For integrals involving products of $\mathbf{M}_\nu(t)$ functions, see Paris and Sy (1983, Appendix).

11.7(iii) Laplace Transforms

The following Laplace transforms of $\mathbf{H}_\nu(t)$ require $\Re a > 0$ for convergence, while those of $\mathbf{L}_\nu(t)$ require $\Re a > 1$.

11.7.13 $\displaystyle\int_0^\infty e^{-at} \mathbf{H}_0(t)\, dt = \dfrac{2}{\pi\sqrt{1+a^2}} \ln\left(\dfrac{1 + \sqrt{1+a^2}}{a}\right),$

11.7.14 $\displaystyle\int_0^\infty e^{-at} \mathbf{H}_1(t)\, dt = \dfrac{2}{\pi a} - \dfrac{2a}{\pi\sqrt{1+a^2}} \ln\left(\dfrac{1 + \sqrt{1+a^2}}{a}\right),$

11.7.15 $\displaystyle\int_0^\infty e^{-at} \mathbf{L}_0(t)\, dt = \dfrac{2}{\pi\sqrt{a^2-1}} \arcsin\left(\dfrac{1}{a}\right),$

11.7.16 $\displaystyle\int_0^\infty e^{-at} \mathbf{L}_1(t)\, dt = \dfrac{2a}{\pi\sqrt{a^2-1}} \arctan\left(\dfrac{1}{\sqrt{a^2-1}}\right) - \dfrac{2}{\pi a}.$

11.7(iv) Integrals with Respect to Order

For integrals of $\mathbf{H}_\nu(x)$ and $\mathbf{L}_\nu(x)$ with respect to the order ν, see Apelblat (1989).

11.7(v) Compendia

For further integrals see Apelblat (1983, §12.16), Babister (1967, Chapter 3), Erdélyi et al. (1954a, §§4.19, 6.8, 8.15, 9.4, 10.3, 11.3, and 15.3), Luke (1962, Chapters 9, 11), Gradshteyn and Ryzhik (2000, §6.8), Marichev (1983, pp. 192–193 and 215–216), Oberhettinger (1972), Oberhettinger (1974, §1.12), Oberhettinger (1990, §§1.21 and 2.21), Oberhettinger and Badii (1973, §1.16), Prudnikov et al. (1990, §§1.4 and 2.7), Prudnikov et al. (1992a, §3.17), and Prudnikov et al. (1992b, §3.17).

For sums of Struve functions see Hansen (1975, p. 456) and Prudnikov et al. (1990, §6.4.1).

11.8 Analogs to Kelvin Functions

For properties of Struve functions of argument $xe^{\pm 3\pi i/4}$ see McLachlan and Meyers (1936).

Related Functions

11.9 Lommel Functions

11.9(i) Definitions

The inhomogeneous Bessel differential equation

11.9.1 $\dfrac{d^2 w}{dz^2} + \dfrac{1}{z} \dfrac{dw}{dz} + \left(1 - \dfrac{\nu^2}{z^2}\right) w = z^{\mu-1}$

can be regarded as a generalization of (11.2.7). Provided that $\mu \pm \nu \neq -1, -3, -5, \ldots$, (11.9.1) has the general solution

11.9.2 $w = s_{\mu,\nu}(z) + A J_\nu(z) + B Y_\nu(z),$

where A, B are arbitrary constants, $s_{\mu,\nu}(z)$ is the *Lommel function* defined by

11.9.3 $s_{\mu,\nu}(z) = z^{\mu+1} \displaystyle\sum_{k=0}^\infty (-1)^k \dfrac{z^{2k}}{a_{k+1}(\mu, \nu)},$

and

11.9.4 $a_k(\mu, \nu) = \displaystyle\prod_{m=1}^k \left((\mu + 2m - 1)^2 - \nu^2\right), \quad k = 0, 1, 2, \ldots.$

Another solution of (11.9.1) that is defined for all values of μ and ν is $S_{\mu,\nu}(z)$, where

11.9.5 $\quad S_{\mu,\nu}(z) = s_{\mu,\nu}(z) + 2^{\mu-1}\Gamma\left(\tfrac{1}{2}\mu + \tfrac{1}{2}\nu + \tfrac{1}{2}\right)\Gamma\left(\tfrac{1}{2}\mu - \tfrac{1}{2}\nu + \tfrac{1}{2}\right)\left(\sin\left(\tfrac{1}{2}(\mu-\nu)\pi\right)J_\nu(z) - \cos\left(\tfrac{1}{2}(\mu-\nu)\pi\right)Y_\nu(z)\right)$,

the right-hand side being replaced by its limiting form when $\mu \pm \nu$ is an odd negative integer.

Reflection Formulas

11.9.6 $\qquad\qquad\qquad s_{\mu,-\nu}(z) = s_{\mu,\nu}(z), \quad S_{\mu,-\nu}(z) = S_{\mu,\nu}(z).$

For the foregoing results and further information see Watson (1944, §§10.7–10.73) and Babister (1967, §3.16).

11.9(ii) Expansions in Series of Bessel Functions

When $\mu \pm \nu \neq -1, -2, -3, \ldots,$

11.9.7 $\qquad s_{\mu,\nu}(z) = 2^{\mu+1}\sum_{k=0}^{\infty}\frac{(2k+\mu+1)\,\Gamma(k+\mu+1)}{k!(2k+\mu-\nu+1)(2k+\mu+\nu+1)}J_{2k+\mu+1}(z),$

11.9.8 $\qquad s_{\mu,\nu}(z) = 2^{(\mu+\nu-1)/2}\Gamma\!\left(\tfrac{1}{2}\mu + \tfrac{1}{2}\nu + \tfrac{1}{2}\right)z^{(\mu+1-\nu)/2}\sum_{k=0}^{\infty}\frac{(\tfrac{1}{2}z)^k}{k!(2k+\mu-\nu+1)}J_{k+\frac{1}{2}(\mu+\nu+1)}(z).$

For these and further results see Luke (1969b, §9.4.5).

11.9(iii) Asymptotic Expansion

For fixed μ and ν,

11.9.9 $\quad S_{\mu,\nu}(z) \sim z^{\mu-1}\sum_{k=0}^{\infty}(-1)^k a_k(-\mu,\nu)z^{-2k},$
$\qquad\qquad z \to \infty,\ |\operatorname{ph} z| \leq \pi - \delta(<\pi).$

For $a_k(\mu,\nu)$ see (11.9.4). If either of $\mu \pm \nu$ equals an odd positive integer, then the right-hand side of (11.9.9) terminates and represents $S_{\mu,\nu}(z)$ exactly.

For uniform asymptotic expansions, for large ν and fixed $\mu = -1, 0, 1, 2, \ldots,$ of solutions of the inhomogeneous modified Bessel differential equation that corresponds to (11.9.1) see Olver (1997b, pp. 388–390).

11.9(iv) References

For further information on Lommel functions see Watson (1944, §§10.7–10.75) and Babister (1967, Chapter 3). For descriptive properties of $s_{\mu,\nu}(x)$ see Steinig (1972).

For collections of integral representations and integrals see Apelblat (1983, §12.17), Babister (1967, p. 85), Erdélyi et al. (1954a, §§4.19 and 5.17), Gradshteyn and Ryzhik (2000, §6.86), Marichev (1983, p. 193), Oberhettinger (1972, pp. 127–128, 168–169, and 188–189), Oberhettinger (1974, §§1.12 and 2.7), Oberhettinger (1990, pp. 105–106 and 191–192), Oberhettinger and Badii (1973, §2.14), Prudnikov et al. (1990, §§1.6 and 2.9), Prudnikov et al. (1992a, §3.34), and Prudnikov et al. (1992b, §3.32).

11.10 Anger–Weber Functions

11.10(i) Definitions

The Anger function $\mathbf{J}_\nu(z)$ and Weber function $\mathbf{E}_\nu(z)$ are defined by

11.10.1 $\qquad \mathbf{J}_\nu(z) = \frac{1}{\pi}\int_0^\pi \cos(\nu\theta - z\sin\theta)\,d\theta,$

11.10.2 $\qquad \mathbf{E}_\nu(z) = \frac{1}{\pi}\int_0^\pi \sin(\nu\theta - z\sin\theta)\,d\theta.$

Each is an entire function of z and ν. Also,

11.10.3
$$\frac{1}{\pi}\int_0^{2\pi}\cos(\nu\theta - z\sin\theta)\,d\theta = (1 + \cos(2\pi\nu))\,\mathbf{J}_\nu(z) + \sin(2\pi\nu)\,\mathbf{E}_\nu(z).$$

The associated Anger–Weber function $\mathbf{A}_\nu(z)$ is defined by

11.10.4 $\qquad \mathbf{A}_\nu(z) = \frac{1}{\pi}\int_0^\infty e^{-\nu t - z\sinh t}\,dt, \qquad \Re z > 0.$

(11.10.4) also applies when $\Re z = 0$ and $\Re\nu > 0$.

11.10(ii) Differential Equations

The Anger and Weber functions satisfy the inhomogeneous Bessel differential equation

11.10.5 $\qquad \frac{d^2w}{dz^2} + \frac{1}{z}\frac{dw}{dz} + \left(1 - \frac{\nu^2}{z^2}\right)w = f(\nu, z),$

where

11.10.6 $\qquad f(\nu, z) = \frac{(z-\nu)}{\pi z^2}\sin(\pi\nu), \quad w = \mathbf{J}_\nu(z),$

or

11.10.7 $\qquad f(\nu, z) = -\frac{1}{\pi z^2}(z + \nu + (z-\nu)\cos(\pi\nu)), \quad w = \mathbf{E}_\nu(z).$

11.10(iii) Maclaurin Series

11.10.8 $\mathbf{J}_\nu(z) = \cos\left(\tfrac{1}{2}\pi\nu\right) S_1(\nu,z) + \sin\left(\tfrac{1}{2}\pi\nu\right) S_2(\nu,z),$

11.10.9 $\mathbf{E}_\nu(z) = \sin\left(\tfrac{1}{2}\pi\nu\right) S_1(\nu,z) - \cos\left(\tfrac{1}{2}\pi\nu\right) S_2(\nu,z),$

where

11.10.10 $S_1(\nu,z) = \sum_{k=0}^{\infty} \dfrac{(-1)^k (\tfrac{1}{2}z)^{2k}}{\Gamma\left(k+\tfrac{1}{2}\nu+1\right)\Gamma\left(k-\tfrac{1}{2}\nu+1\right)},$

11.10.11 $S_2(\nu,z) = \sum_{k=0}^{\infty} \dfrac{(-1)^k (\tfrac{1}{2}z)^{2k+1}}{\Gamma\left(k+\tfrac{1}{2}\nu+\tfrac{3}{2}\right)\Gamma\left(k-\tfrac{1}{2}\nu+\tfrac{3}{2}\right)}.$

These expansions converge absolutely for all finite values of z.

11.10(iv) Graphics

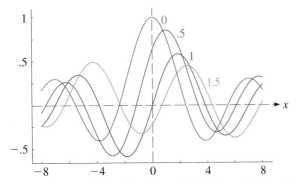

Figure 11.10.1: Anger function $\mathbf{J}_\nu(x)$ for $-8 \leq x \leq 8$ and $\nu = 0, \tfrac{1}{2}, 1, \tfrac{3}{2}$.

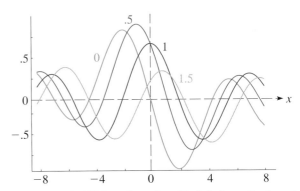

Figure 11.10.2: Weber function $\mathbf{E}_\nu(x)$ for $-8 \leq x \leq 8$ and $\nu = 0, \tfrac{1}{2}, 1, \tfrac{3}{2}$.

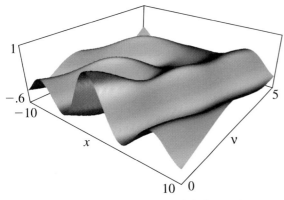

Figure 11.10.3: Anger function $\mathbf{J}_\nu(x)$ for $-10 \leq x \leq 10$ and $0 \leq \nu \leq 5$.

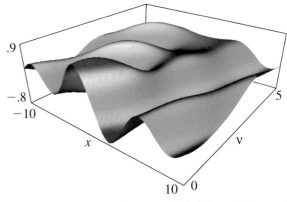

Figure 11.10.4: Weber function $\mathbf{E}_\nu(x)$ for $-10 \leq x \leq 10$ and $0 \leq \nu \leq 5$.

11.10(v) Interrelations

11.10.12 $\mathbf{J}_\nu(-z) = \mathbf{J}_{-\nu}(z), \quad \mathbf{E}_\nu(-z) = -\mathbf{E}_{-\nu}(z).$

11.10.13 $\sin(\pi\nu)\,\mathbf{J}_\nu(z) = \cos(\pi\nu)\,\mathbf{E}_\nu(z) - \mathbf{E}_{-\nu}(z),$

11.10.14 $\sin(\pi\nu)\,\mathbf{E}_\nu(z) = \mathbf{J}_{-\nu}(z) - \cos(\pi\nu)\,\mathbf{J}_\nu(z).$

11.10.15 $\mathbf{J}_\nu(z) = J_\nu(z) + \sin(\pi\nu)\,\mathbf{A}_\nu(z),$

11.10.16 $\mathbf{E}_\nu(z) = -Y_\nu(z) - \cos(\pi\nu)\,\mathbf{A}_\nu(z) - \mathbf{A}_{-\nu}(z).$

11.10(vi) Relations to Other Functions

11.10.17 $\mathbf{J}_\nu(z) = \dfrac{\sin(\pi\nu)}{\pi}\left(s_{0,\nu}(z) - \nu\, s_{-1,\nu}(z)\right),$

11.10.18 $\begin{aligned}\mathbf{E}_\nu(z) = &-\dfrac{1}{\pi}(1+\cos(\pi\nu))\,s_{0,\nu}(z) \\ &-\dfrac{\nu}{\pi}(1-\cos(\pi\nu))\,s_{-1,\nu}(z).\end{aligned}$

11.10.19
$$\mathbf{J}_{-\frac{1}{2}}(z) = \mathbf{E}_{\frac{1}{2}}(z)$$
$$= (\tfrac{1}{2}\pi z)^{-\frac{1}{2}}(A_+(\chi)\cos z - A_-(\chi)\sin z),$$

11.10.20
$$\mathbf{J}_{\frac{1}{2}}(z) = -\mathbf{E}_{-\frac{1}{2}}(z)$$
$$= (\tfrac{1}{2}\pi z)^{-\frac{1}{2}}(A_+(\chi)\sin z + A_-(\chi)\cos z),$$

where

11.10.21 $\quad A_\pm(\chi) = C(\chi) \pm S(\chi), \quad \chi = (2z/\pi)^{\frac{1}{2}}$

For the Fresnel integrals C and S see §7.2(iii).

For $n = 1, 2, 3, \ldots,$

11.10.22
$$\mathbf{E}_n(z) = -\mathbf{H}_n(z) + \frac{1}{\pi}\sum_{k=0}^{m_1}\frac{\Gamma(k+\tfrac{1}{2})}{\Gamma(n+\tfrac{1}{2}-k)}(\tfrac{1}{2}z)^{n-2k-1},$$

and

11.10.23
$$\mathbf{E}_{-n}(z) = -\mathbf{H}_{-n}(z)$$
$$+ \frac{(-1)^{n+1}}{\pi}\sum_{k=0}^{m_2}\frac{\Gamma(n-k-\tfrac{1}{2})}{\Gamma(k+\tfrac{3}{2})}(\tfrac{1}{2}z)^{-n+2k+1},$$

where

11.10.24 $\quad m_1 = \lfloor \tfrac{1}{2}n - \tfrac{1}{2} \rfloor, \quad m_2 = \lceil \tfrac{1}{2}n - \tfrac{3}{2} \rceil.$

11.10(vii) Special Values

11.10.25 $\quad \mathbf{J}_\nu(0) = \dfrac{\sin(\pi\nu)}{\pi\nu}, \quad \mathbf{E}_\nu(0) = \dfrac{1-\cos(\pi\nu)}{\pi\nu}.$

11.10.26 $\quad \mathbf{E}_0(z) = -\mathbf{H}_0(z), \quad \mathbf{E}_1(z) = \dfrac{2}{\pi} - \mathbf{H}_1(z).$

11.10.27 $\quad \left.\dfrac{\partial}{\partial\nu}\mathbf{J}_\nu(z)\right|_{\nu=0} = \tfrac{1}{2}\pi\,\mathbf{H}_0(z),$

11.10.28 $\quad \left.\dfrac{\partial}{\partial\nu}\mathbf{E}_\nu(z)\right|_{\nu=0} = \tfrac{1}{2}\pi\,J_0(z).$

11.10.29 $\quad \mathbf{J}_n(z) = J_n(z), \qquad n \in \mathbb{Z}.$

11.10(viii) Expansions in Series of Products of Bessel Functions

11.10.30
$$\mathbf{J}_\nu(z) =$$
$$2\sin\!\left(\tfrac{1}{2}\nu\pi\right)\sum_{k=0}^{\infty}(-1)^k J_{k-\frac{1}{2}\nu+\frac{1}{2}}\!\left(\tfrac{1}{2}z\right)J_{k+\frac{1}{2}\nu+\frac{1}{2}}\!\left(\tfrac{1}{2}z\right)$$
$$+ 2\cos\!\left(\tfrac{1}{2}\nu\pi\right){\sum_{k=0}^{\infty}}'(-1)^k J_{k-\frac{1}{2}\nu}\!\left(\tfrac{1}{2}z\right)J_{k+\frac{1}{2}\nu}\!\left(\tfrac{1}{2}z\right),$$

11.10.31
$$\mathbf{E}_\nu(z) =$$
$$-2\cos\!\left(\tfrac{1}{2}\nu\pi\right)\sum_{k=0}^{\infty}(-1)^k J_{k-\frac{1}{2}\nu+\frac{1}{2}}\!\left(\tfrac{1}{2}z\right)J_{k+\frac{1}{2}\nu+\frac{1}{2}}\!\left(\tfrac{1}{2}z\right)$$
$$+ 2\sin\!\left(\tfrac{1}{2}\nu\pi\right){\sum_{k=0}^{\infty}}'(-1)^k J_{k-\frac{1}{2}\nu}\!\left(\tfrac{1}{2}z\right)J_{k+\frac{1}{2}\nu}\!\left(\tfrac{1}{2}z\right),$$

where the prime on the second summation symbols means that the first term is to be halved.

11.10(ix) Recurrence Relations and Derivatives

11.10.32 $\quad \mathbf{J}_{\nu-1}(z) + \mathbf{J}_{\nu+1}(z) = \dfrac{2\nu}{z}\mathbf{J}_\nu(z) - \dfrac{2}{\pi z}\sin(\pi\nu),$

11.10.33
$$\mathbf{E}_{\nu-1}(z) + \mathbf{E}_{\nu+1}(z) = \dfrac{2\nu}{z}\mathbf{E}_\nu(z) - \dfrac{2}{\pi z}(1-\cos(\pi\nu)).$$

11.10.34 $\quad 2\mathbf{J}'_\nu(z) = \mathbf{J}_{\nu-1}(z) - \mathbf{J}_{\nu+1}(z),$

11.10.35 $\quad 2\mathbf{E}'_\nu(z) = \mathbf{E}_{\nu-1}(z) - \mathbf{E}_{\nu+1}(z),$

11.10.36 $\quad z\mathbf{J}'_\nu(z) \pm \nu\mathbf{J}_\nu(z) = \pm z\mathbf{J}_{\nu\mp 1}(z) \pm \dfrac{\sin(\pi\nu)}{\pi},$

11.10.37
$$z\mathbf{E}'_\nu(z) \pm \nu\mathbf{E}_\nu(z) = \pm z\mathbf{E}_{\nu\mp 1}(z) \pm \dfrac{(1-\cos(\pi\nu))}{\pi}.$$

11.10(x) Integrals and Sums

For collections of integral representations and integrals see Erdélyi et al. (1954a, §§4.19 and 5.17), Marichev (1983, pp. 194–195 and 214–215), Oberhettinger (1972, p. 128), Oberhettinger (1974, §§1.12 and 2.7), Oberhettinger (1990, pp. 105 and 189–190), Prudnikov et al. (1990, §§1.5 and 2.8), Prudnikov et al. (1992a, §3.18), Prudnikov et al. (1992b, §3.18), and Zanovello (1977).

For sums see Hansen (1975, pp. 456–457) and Prudnikov et al. (1990, §§6.4.2–6.4.3).

11.11 Asymptotic Expansions of Anger–Weber Functions

11.11(i) Large $|z|$, Fixed ν

Let $F_0(\nu) = G_0(\nu) = 1$, and for $k = 1, 2, 3, \ldots,$

11.11.1
$$F_k(\nu) = (\nu^2 - 1^2)(\nu^2 - 3^2)\cdots(\nu^2 - (2k-1)^2),$$
$$G_k(\nu) = (\nu^2 - 2^2)(\nu^2 - 4^2)\cdots(\nu^2 - (2k)^2).$$

Then as $z \to \infty$ in $|\operatorname{ph} z| \le \pi - \delta \,(< \pi)$

11.11.2
$$\mathbf{J}_\nu(z) \sim J_\nu(z)$$
$$+ \frac{\sin(\pi\nu)}{\pi z}\!\left(\sum_{k=0}^\infty \frac{F_k(\nu)}{z^{2k}} - \frac{\nu}{z}\sum_{k=0}^\infty \frac{G_k(\nu)}{z^{2k}}\right),$$

11.11.3
$$\mathbf{E}_\nu(z) \sim -Y_\nu(z) - \frac{1+\cos(\pi\nu)}{\pi z}\sum_{k=0}^\infty \frac{F_k(\nu)}{z^{2k}}$$
$$- \frac{\nu(1-\cos(\pi\nu))}{\pi z^2}\sum_{k=0}^\infty \frac{G_k(\nu)}{z^{2k}},$$

11.11.4 $\quad \mathbf{A}_\nu(z) \sim \dfrac{1}{\pi z}\sum_{k=0}^\infty \dfrac{F_k(\nu)}{z^{2k}} - \dfrac{\nu}{\pi z^2}\sum_{k=0}^\infty \dfrac{G_k(\nu)}{z^{2k}}.$

11.11(ii) Large $|\nu|$, Fixed z

If z is fixed, and $\nu \to \infty$ in $|\operatorname{ph}\nu| \leq \pi$ in such a way that ν is bounded away from the set of all integers, then

11.11.5 $\quad \mathbf{J}_\nu(z) = \frac{\sin(\pi\nu)}{\pi\nu}\left(1 - \frac{\nu z}{\nu^2-1} + O\!\left(\frac{1}{\nu^2}\right)\right),$

11.11.6 $\quad \mathbf{E}_\nu(z) = \frac{2}{\pi\nu}\left(\sin^2\!\left(\tfrac{1}{2}\pi\nu\right) + \frac{\nu z}{\nu^2-1}\cos^2\!\left(\tfrac{1}{2}\pi\nu\right) + O\!\left(\frac{1}{\nu^2}\right)\right).$

If $\nu = n(\in \mathbb{Z})$, then (11.10.29) applies for $\mathbf{J}_n(z)$, and

11.11.7 $\quad \begin{aligned}\mathbf{E}_{2n}(z) &\sim \frac{2z}{(4n^2-1)\pi},\\ \mathbf{E}_{2n+1}(z) &\sim \frac{2}{(2n+1)\pi},\quad n \to \pm\infty.\end{aligned}$

11.11(iii) Large ν, Fixed z/ν

For fixed $\lambda \,(>0)$,

11.11.8 $\quad \mathbf{A}_\nu(\lambda\nu) \sim \frac{1}{\pi}\sum_{k=0}^{\infty}\frac{(2k)!\,a_k(\lambda)}{\nu^{2k+1}},$
$\nu \to \infty,\ |\operatorname{ph}\nu| \leq \pi - \delta\,(<\pi),$

where

11.11.9 $\quad \begin{aligned} a_0 &= \frac{1}{1+\lambda},\quad a_1 = -\frac{\lambda}{2(1+\lambda)^4},\\ a_2 &= \frac{9\lambda^2-\lambda}{24(1+\lambda)^7},\quad a_3 = -\frac{225\lambda^3-54\lambda^2+\lambda}{720(1+\lambda)^{10}}.\end{aligned}$

For fixed $\lambda(>1)$,

11.11.10 $\quad \mathbf{A}_{-\nu}(\lambda\nu) \sim -\frac{1}{\pi}\sum_{k=0}^{\infty}\frac{(2k)!\,a_k(-\lambda)}{\nu^{2k+1}},\ \nu \to +\infty.$

For fixed λ, $0 < \lambda < 1$,

11.11.11 $\quad \mathbf{A}_{-\nu}(\lambda\nu) \sim \sqrt{\frac{2}{\pi\nu}}\,e^{-\nu\mu}\sum_{k=0}^{\infty}\frac{(\tfrac{1}{2})_k b_k(\lambda)}{\nu^k},\ \nu \to +\infty,$

where

11.11.12 $\quad \mu = \sqrt{1-\lambda^2} - \ln\!\left(\frac{1+\sqrt{1-\lambda^2}}{\lambda}\right),$

and

11.11.13 $\quad \begin{aligned} b_0(\lambda) &= \frac{1}{(1-\lambda^2)^{1/4}},\quad b_1(\lambda) = \frac{2+3\lambda^2}{12(1-\lambda^2)^{7/4}},\\ b_2(\lambda) &= \frac{4+300\lambda^2+81\lambda^4}{864(1-\lambda^2)^{13/4}}.\end{aligned}$

In particular, as $\nu \to +\infty$,

11.11.14 $\quad \mathbf{A}_{-\nu}(\lambda\nu) \sim \frac{1}{\pi\nu(\lambda-1)},\qquad \lambda > 1,$

11.11.15 $\quad \mathbf{A}_{-\nu}(\lambda\nu) \sim \left(\frac{2}{\pi\nu}\right)^{1/2}\!\left(\frac{1+\sqrt{1-\lambda^2}}{\lambda}\right)^\nu \frac{e^{-\nu\sqrt{1-\lambda^2}}}{(1-\lambda^2)^{1/4}},$
$0 < \lambda < 1.$

Also, as $\nu \to +\infty$,

11.11.16 $\quad \mathbf{A}_{-\nu}(\nu) \sim \frac{2^{4/3}}{3^{7/6}\,\Gamma(\tfrac{2}{3})\nu^{1/3}},$

and

11.11.17 $\quad \mathbf{A}_{-\nu}\!\left(\nu + a\nu^{1/3}\right) = 2^{1/3}\nu^{-1/3}\operatorname{Hi}\!\left(-2^{1/3}a\right) + O(\nu^{-1}),$

uniformly for bounded real values of a. For the Scorer function Hi see §9.12(i).

All of (11.11.10)–(11.11.17) can be regarded as special cases of two asymptotic expansions given in Olver (1997b, pp. 352–357) for $\mathbf{A}_{-\nu}(\lambda\nu)$ as $\nu \to +\infty$, one being uniform for $\delta \leq \lambda \leq 1$, where δ again denotes an arbitrary small positive constant, and the other being uniform for $1 \leq \lambda < \infty$. (Note that Olver's definition of $\mathbf{A}_\nu(z)$ omits the factor $1/\pi$ in (11.10.4).) See also Watson (1944, §10.15).

Lastly, corresponding asymptotic approximations and expansions for $\mathbf{J}_\nu(\lambda\nu)$ and $\mathbf{E}_\nu(\lambda\nu)$ follow from (11.10.15) and (11.10.16) and the corresponding asymptotic expansions for the Bessel functions $J_\nu(z)$ and $Y_\nu(z)$; see §10.19(ii). In particular,

11.11.18 $\quad \mathbf{J}_\nu(\nu) \sim \frac{2^{1/3}}{3^{2/3}\,\Gamma(\tfrac{2}{3})\nu^{1/3}},\qquad \nu \to +\infty,$

11.11.19 $\quad \mathbf{E}_\nu(\nu) \sim \frac{2^{1/3}}{3^{7/6}\,\Gamma(\tfrac{2}{3})\nu^{1/3}},\qquad \nu \to +\infty.$

Applications

11.12 Physical Applications

Applications of Struve functions occur in water-wave and surface-wave problems (Hirata (1975) and Ahmadi and Widnall (1985)), unsteady aerodynamics (Shaw (1985) and Wehausen and Laitone (1960)), distribution of fluid pressure over a vibrating disk (McLachlan (1934)), resistive MHD instability theory (Paris and Sy (1983)), and optical diffraction (Levine and Schwinger (1948)). More recently Struve functions have appeared in many particle quantum dynamical studies of spin decoherence (Shao and Hänggi (1998)) and nanotubes (Pedersen (2003)).

Computation

11.13 Methods of Computation

11.13(i) Introduction

Subsequent subsections treat the computation of Struve functions. The treatment of Lommel and Anger–Weber functions is similar. For a review of methods for the computation of $\mathbf{H}_\nu(z)$ see Zanovello (1975).

11.13(ii) Series Expansions

Although the power-series expansions (11.2.1) and (11.2.2), and the Bessel-function expansions of §11.4(iv) converge for all finite values of z, they are cumbersome to use when $|z|$ is large owing to slowness of convergence and cancellation. For large $|z|$ and/or $|\nu|$ the asymptotic expansions given in §11.6 should be used instead.

11.13(iii) Quadrature

For numerical purposes the most convenient of the representations given in §11.5, at least for real variables, include the integrals (11.5.2)–(11.5.5) for $\mathbf{K}_\nu(z)$ and $\mathbf{M}_\nu(z)$. Subsequently $\mathbf{H}_\nu(z)$ and $\mathbf{L}_\nu(z)$ are obtainable via (11.2.5) and (11.2.6). Other integrals that appear in §11.5(i) have highly oscillatory integrands unless z is small.

For complex variables the methods described in §§3.5(viii) and 3.5(ix) are available.

11.13(iv) Differential Equations

A comprehensive approach is to integrate the defining inhomogeneous differential equations (11.2.7) and (11.2.9) numerically, using methods described in §3.7. To insure stability the integration path must be chosen so that as we proceed along it the wanted solution grows in magnitude at least as rapidly as the complementary solutions.

Suppose $\nu \geq 0$ and x is real and positive. Then from the limiting forms for small argument (§§11.2(i), 10.7(i), 10.30(i)), limiting forms for large argument (§§11.6(i), 10.7(ii), 10.30(ii)), and the connection formulas (11.2.5) and (11.2.6), it is seen that $\mathbf{H}_\nu(x)$ and $\mathbf{L}_\nu(x)$ can be computed in a stable manner by integrating forwards, that is, from the origin toward infinity. The solution $\mathbf{K}_\nu(x)$ needs to be integrated backwards for small x, and either forwards or backwards for large x depending whether or not ν exceeds $\frac{1}{2}$. For $\mathbf{M}_\nu(x)$ both forward and backward integration are unstable, and boundary-value methods are required (§3.7(iii)).

11.13(v) Difference Equations

Sequences of values of $\mathbf{H}_\nu(z)$ and $\mathbf{L}_\nu(z)$, with z fixed, can be computed by application of the inhomogeneous difference equations (11.4.23) and (11.4.25). There are similar problems to those described in §11.13(iv) concerning stability. In consequence forward recurrence, backward recurrence, or boundary-value methods may be necessary. See §3.6 for implementation of these methods, and with the Weber function $\mathbf{E}_n(x)$ as an example.

11.14 Tables

11.14(i) Introduction

For tables before 1961 see Fletcher et al. (1962) and Lebedev and Fedorova (1960). Tables listed in these Indices are omitted from the subsections that follow.

11.14(ii) Struve Functions

- Abramowitz and Stegun (1964, Chapter 12) tabulates $\mathbf{H}_n(x)$, $\mathbf{H}_n(x) - Y_n(x)$, and $I_n(x) - \mathbf{L}_n(x)$ for $n = 0, 1$ and $x = 0(.1)5$, $x^{-1} = 0(.01)0.2$ to 6D or 7D.

- Agrest et al. (1982) tabulates $\mathbf{H}_n(x)$ and $e^{-x}\mathbf{L}_n(x)$ for $n = 0, 1$ and $x = 0(.001)5(.005)15(.01)100$ to 11D.

- Barrett (1964) tabulates $\mathbf{L}_n(x)$ for $n = 0, 1$ and $x = 0.2(.005)4(.05)10(.1)19.2$ to 5 or 6S, $x = 6(.25)59.5(.5)100$ to 2S.

- Zanovello (1975) tabulates $\mathbf{H}_n(x)$ for $n = -4(1)15$ and $x = 0.5(.5)26$ to 8D or 9S.

- Zhang and Jin (1996) tabulates $\mathbf{H}_n(x)$ and $\mathbf{L}_n(x)$ for $n = -4(1)3$ and $x = 0(1)20$ to 8D or 7S.

11.14(iii) Integrals

- Abramowitz and Stegun (1964, Chapter 12) tabulates $\int_0^x (I_0(t) - \mathbf{L}_0(t))\,dt$ and $(2/\pi)\int_x^\infty t^{-1}\mathbf{H}_0(t)\,dt$ for $x = 0(.1)5$ to 5D or 7D; $\int_0^x (\mathbf{H}_0(t) - Y_0(t))\,dt - (2/\pi)\ln x$, $\int_0^x (I_0(t) - \mathbf{L}_0(t))\,dt - (2/\pi)\ln x$, and $\int_x^\infty t^{-1}(\mathbf{H}_0(t) - Y_0(t))\,dt$ for $x^{-1} = 0(.01)0.2$ to 6D.

- Agrest et al. (1982) tabulates $\int_0^x \mathbf{H}_0(t)\,dt$ and $e^{-x}\int_0^x \mathbf{L}_0(t)\,dt$ for $x = 0(.001)5(.005)15(.01)100$ to 11D.

11.14(iv) Anger–Weber Functions

- Bernard and Ishimaru (1962) tabulates $\mathbf{J}_\nu(x)$ and $\mathbf{E}_\nu(x)$ for $\nu = -10(.1)10$ and $x = 0(.1)10$ to 5D.

- Jahnke and Emde (1945) tabulates $\mathbf{E}_n(x)$ for $n = 1, 2$ and $x = 0(.01)14.99$ to 4D.

11.14(v) Incomplete Functions

- Agrest and Maksimov (1971, Chapter 11) defines *incomplete* Struve, Anger, and Weber functions and includes tables of an incomplete Struve function $\mathbf{H}_n(x,\alpha)$ for $n = 0, 1$, $x = 0(.2)10$, and $\alpha = 0(.2)1.4, \frac{1}{2}\pi$, together with surface plots.

11.15 Approximations

11.15(i) Expansions in Chebyshev Series

- Luke (1975, pp. 416–421) gives Chebyshev-series expansions for $\mathbf{H}_n(x)$, $\mathbf{L}_n(x)$, $0 \le |x| \le 8$, and $\mathbf{H}_n(x) - Y_n(x)$, $x \ge 8$, for $n = 0, 1$; $\int_0^x t^{-m} \mathbf{H}_0(t)\,dt$, $\int_0^x t^{-m} \mathbf{L}_0(t)\,dt$, $0 \le |x| \le 8$, $m = 0, 1$ and $\int_0^x (\mathbf{H}_0(t) - Y_0(t))\,dt$, $\int_x^\infty t^{-1}(\mathbf{H}_0(t) - Y_0(t))\,dt$, $x \ge 8$; the coefficients are to 20D.

- MacLeod (1993) gives Chebyshev-series expansions for $\mathbf{L}_0(x)$, $\mathbf{L}_1(x)$, $0 \le x \le 16$, and $I_0(x) - \mathbf{L}_0(x)$, $I_1(x) - \mathbf{L}_1(x)$, $x \ge 16$; the coefficients are to 20D.

11.15(ii) Rational and Polynomial Approximations

- Newman (1984) gives polynomial approximations for $\mathbf{H}_n(x)$ for $n = 0, 1$, $0 \le x \le 3$, and rational-fraction approximations for $\mathbf{H}_n(x) - Y_n(x)$ for $n = 0, 1$, $x \ge 3$. The maximum errors do not exceed 1.2×10^{-8} for the former and 2.5×10^{-8} for the latter.

11.16 Software

See http://dlmf.nist.gov/11.16.

References

General References

The main references used in writing this chapter are Babister (1967, Chapter 3) and Watson (1944, Chapter 10). For additional bibliographic reading see Erdélyi et al. (1953b, §7.5), Luke (1969b), Luke (1975, Chapter 10), Magnus et al. (1966, §3.10), and Olver (1997b).

Sources

The following list gives the references or other indications of proofs that were used in constructing the various sections of this chapter. These sources supplement the references that are quoted in the text.

§11.2 Watson (1944, pp. 328–329), Olver (1997b, pp. 274–277). The notation $\mathbf{M}_\nu(z)$ is new and this function has been introduced to play a similar role to $\mathbf{L}_\nu(z)$ that $\mathbf{K}_\nu(z)$ does to $\mathbf{H}_\nu(z)$. For §11.2(iii) see §2.7(iv) and Olver (1997b, pp. 274–277). The last reference restricts (11.2.15) to the sector $|\mathrm{ph}\,z| \le \frac{1}{2}\pi$, and instead covers the sector $\frac{1}{2}\pi \le \mathrm{ph}\,z \le \frac{3}{2}\pi$ with another set of solutions. (Similarly for the conjugate sector $-\frac{3}{2}\pi \le \mathrm{ph}\,z \le -\frac{1}{2}\pi$.)

§11.3 The graphics were produced at NIST.

§11.4 Watson (1944, §§10.4, 10.45). (11.4.1), (11.4.2) both follow from Erdélyi et al. (1953b, p. 39, Eq. (64)): for (11.4.1) set $\xi = z$ and use (11.2.5); for (11.4.2) replace $Y_{n+\frac{1}{2}}(\xi)$ by $(-1)^{n+1} J_{-n-\frac{1}{2}}(\xi)$ in consequence of (10.2.3), then set $\xi = iz$ and use (11.2.2), (10.27.6). For (11.4.3), (11.4.4) see Babister (1967, pp. 64, 75). For (11.4.5)–(11.4.12) combine (11.4.3), (11.4.4) with (10.47.3), (10.47.7), §10.49(i), §10.49(ii), (11.4.23), (11.4.25). (11.4.16), (11.4.17) follow from (11.2.1), (11.2.2). For (11.4.31) see Babister (1967, pp. 60, 74).

§11.5 For (11.5.1)–(11.5.3), (11.5.6), and (11.5.7) see Watson (1944, pp. 328, 331, 332), together with (11.2.5) in the case of (11.5.2) and (11.5.3), and (11.2.2) in the case of (11.5.6). For (11.5.4) and (11.5.5) see Babister (1967, Eq. (3.102)) and collapse the integration path on to the interval $[0, 1]$. For (11.5.8) see Babister (1967, §3.7) with modified convergence conditions. For (11.5.9) deform the integration path in (11.5.8) into a loop and use (11.2.2).

§11.6 For (11.6.1) apply Watson's lemma (§2.4(i)) to (11.5.2), or combine Watson (1944, p. 333, Eq. (2)) with (11.2.5). See also the subsequent text in this reference, and Olver (1997b, p. 277, Ex. 15.5). For (11.6.2) convert (11.5.4) to a loop integral $\int_0^{(1+)}$ to remove the restriction $\Re\nu > -\frac{1}{2}$, extend the loop to pass through the point $t = \infty$ on the positive real axis, then apply Laplace's method (§2.4(iii)) to each of the two integrals with paths from $t = 0$ to $t = \infty$, one passing below $t = 1$ and the other passing above $t = 1$. For (11.6.3) write the integrals over the intervals $[0,\infty)$ and $[z,\infty)$; use (11.6.1) with the first term extracted, and a limiting procedure on the integral over $[0,\infty)$. For (11.6.4) replace z by iz in (11.6.3) and apply (11.2.5), (11.2.6), (10.27.11). For (11.6.5) apply (5.11.7) to (11.2.1), (11.2.2). For (11.6.6) and (11.6.9) see Watson (1944, §10.43): a similar method can be used for (11.6.7), starting from (11.5.4).

§11.7 For (11.7.1)–(11.7.6) use §11.4(v). For (11.7.7)–(11.7.12) see Babister (1967, pp. 68, 71–72), Watson (1944, pp. 392, 397). For (11.7.13)–(11.7.16) see Babister (1967, §§3.13, 3.15).

§11.9 Watson (1944, §10.75).

§11.10 Watson (1944, pp. 308–312). The notation $\mathbf{A}_\nu(z)$, without the factor $1/\pi$, was introduced in Olver (1997b, p. 84). For (11.10.12) use (11.10.1), (11.10.2). For (11.10.16) combine (11.10.14), (11.10.15), and (10.2.3). For (11.10.19), (11.10.20), use (11.10.8)–(11.10.11) with $\nu = \pm\frac{1}{2}$ and identify the resulting sums with those associated with the right-hand sides via (7.6.5), (7.6.7). For (11.10.22), (11.10.23) see Watson (1944, pp. 336–337) or Erdélyi *et al.* (1953b, p. 40). The upper summation limit in (11.10.23) is given incorrectly in Watson (1944, p. 337), and this error is reproduced in Erdélyi *et al.* (1953b), as well as in later printings of Abramowitz and Stegun (1964, Chapter 12)—earlier printings contained a different error. (11.10.23) can be derived by combining (11.2.1) with (11.10.12), (11.10.22). For (11.10.25) use (11.10.1) and (11.10.2). For (11.10.26) use (11.10.22). (11.10.27) and (11.10.28) can be obtained by differentiation of (11.10.1) and (11.10.2), followed by straightforward manipulation of the integrals and comparison with (11.5.1) and (11.10.1). For (11.10.29) use (11.10.1) and (10.9.2). For (11.10.30)–(11.10.31) see Luke (1969b, p. 55).

The graphics were produced at NIST.

§11.11 Watson (1944, §§10.14–10.15). (11.11.2), (11.11.3) follow from (11.10.15), (11.10.16). (11.11.5), (11.11.6) follow from (11.10.8)–(11.10.11). Eqs. (11.11.7) follow from (11.6.5). For (11.11.11), see Dingle (1973, p. 388). For (11.11.8)–(11.11.19), see Olver (1997b, pp. 103 and 352).

Chapter 12
Parabolic Cylinder Functions

N. M. Temme[1]

Notation   **304**
    12.1   Special Notation 304

Properties   **304**
    12.2   Differential Equations 304
    12.3   Graphics 305
    12.4   Power-Series Expansions 307
    12.5   Integral Representations 307
    12.6   Continued Fraction 308
    12.7   Relations to Other Functions 308
    12.8   Recurrence Relations and Derivatives . . 309
    12.9   Asymptotic Expansions for Large Variable 309
    12.10   Uniform Asymptotic Expansions for Large Parameter 309
    12.11   Zeros 312
    12.12   Integrals 313
    12.13   Sums 313
    12.14   The Function $W(a,x)$ 314
    12.15   Generalized Parabolic Cylinder Functions 317

Applications   **317**
    12.16   Mathematical Applications 317
    12.17   Physical Applications 317

Computation   **317**
    12.18   Methods of Computation 318
    12.19   Tables 318
    12.20   Approximations 318
    12.21   Software 318

References   **318**

---

[1]Centrum voor Wiskunde en Informatica, Department MAS, Amsterdam, The Netherlands.
  Acknowledgments: This chapter is based in part on Abramowitz and Stegun (1964, Chapter 19) by J. C. P. Miller.
  Copyright © 2009 National Institute of Standards and Technology. All rights reserved.

Notation

12.1 Special Notation

(For other notation see pp. xiv and 873.)

x, y real variables.
z complex variable.
n, s nonnegative integers.
a, ν real or complex parameters.
δ arbitrary small positive constant.

Unless otherwise noted, primes indicate derivatives with respect to the variable, and fractional powers take their principal values.

The main functions treated in this chapter are the parabolic cylinder functions (PCFs), also known as Weber parabolic cylinder functions: $U(a, z)$, $V(a, z)$, $\overline{U}(a, z)$, and $W(a, z)$. These notations are due to Miller (1952, 1955). An older notation, due to Whittaker (1902), for $U(a, z)$ is $D_\nu(z)$. The notations are related by $U(a, z) = D_{-a-\frac{1}{2}}(z)$. Whittaker's notation $D_\nu(z)$ is useful when ν is a nonnegative integer (Hermite polynomial case).

Properties

12.2 Differential Equations

12.2(i) Introduction

PCFs are solutions of the differential equation

$$12.2.1 \qquad \frac{d^2w}{dz^2} + \left(az^2 + bz + c\right)w = 0,$$

with three distinct standard forms

$$12.2.2 \qquad \frac{d^2w}{dz^2} - \left(\tfrac{1}{4}z^2 + a\right)w = 0,$$

$$12.2.3 \qquad \frac{d^2w}{dz^2} + \left(\tfrac{1}{4}z^2 - a\right)w = 0,$$

$$12.2.4 \qquad \frac{d^2w}{dz^2} + \left(\nu + \tfrac{1}{2} - \tfrac{1}{4}z^2\right)w = 0.$$

Each of these equations is transformable into the others. Standard solutions are $U(a, \pm z)$, $V(a, \pm z)$, $\overline{U}(a, \pm x)$ (*not* complex conjugate), $U(-a, \pm iz)$ for (12.2.2); $W(a, \pm x)$ for (12.2.3); $D_\nu(\pm z)$ for (12.2.4), where

$$12.2.5 \qquad D_\nu(z) = U\left(-\tfrac{1}{2} - \nu, z\right).$$

All solutions are entire functions of z and entire functions of a or ν.

For real values of z $(= x)$, numerically satisfactory pairs of solutions (§2.7(iv)) of (12.2.2) are $U(a, x)$ and $V(a, x)$ when x is positive, or $U(a, -x)$ and $V(a, -x)$ when x is negative. For (12.2.3) $W(a, x)$ and $W(a, -x)$ comprise a numerically satisfactory pair, for all $x \in \mathbb{R}$. The solutions $W(a, \pm x)$ are treated in §12.14.

In $\mathbb{C}$, for $j = 0, 1, 2, 3$, $U\big((-1)^{j-1}a, (-i)^{j-1}z\big)$ and $U\big((-1)^j a, (-i)^j z\big)$ comprise a numerically satisfactory pair of solutions in the half-plane $\tfrac{1}{4}(2j-3)\pi \leq \mathrm{ph}\, z \leq \tfrac{1}{4}(2j+1)\pi$.

12.2(ii) Values at $z = 0$

$$12.2.6 \qquad U(a, 0) = \frac{\sqrt{\pi}}{2^{\frac{1}{2}a + \frac{1}{4}} \Gamma\left(\frac{3}{4} + \frac{1}{2}a\right)},$$

$$12.2.7 \qquad U'(a, 0) = -\frac{\sqrt{\pi}}{2^{\frac{1}{2}a - \frac{1}{4}} \Gamma\left(\frac{1}{4} + \frac{1}{2}a\right)},$$

$$12.2.8 \qquad V(a, 0) = \frac{\pi 2^{\frac{1}{2}a + \frac{1}{4}}}{\left(\Gamma\left(\frac{3}{4} - \frac{1}{2}a\right)\right)^2 \Gamma\left(\frac{1}{4} + \frac{1}{2}a\right)},$$

$$12.2.9 \qquad V'(a, 0) = \frac{\pi 2^{\frac{1}{2}a + \frac{3}{4}}}{\left(\Gamma\left(\frac{1}{4} - \frac{1}{2}a\right)\right)^2 \Gamma\left(\frac{3}{4} + \frac{1}{2}a\right)}.$$

12.2(iii) Wronskians

$$12.2.10 \qquad \mathscr{W}\{U(a, z), V(a, z)\} = \sqrt{2/\pi},$$

$$12.2.11 \qquad \mathscr{W}\{U(a, z), U(a, -z)\} = \frac{\sqrt{2\pi}}{\Gamma\left(\frac{1}{2} + a\right)},$$

$$12.2.12 \qquad \mathscr{W}\{U(a, z), U(-a, \pm iz)\} = \mp i e^{\pm i\pi\left(\frac{1}{2}a + \frac{1}{4}\right)}.$$

12.2(iv) Reflection Formulas

For $n = 0, 1, \ldots$,

$$12.2.13 \quad U\left(-n - \tfrac{1}{2}, -z\right) = (-1)^n U\left(-n - \tfrac{1}{2}, z\right),$$

$$12.2.14 \quad V\left(n + \tfrac{1}{2}, -z\right) = (-1)^n V\left(n + \tfrac{1}{2}, z\right).$$

12.2(v) Connection Formulas

$$12.2.15 \qquad U(a, -z) = -\sin(\pi a)\, U(a, z) + \frac{\pi}{\Gamma\left(\frac{1}{2} + a\right)} V(a, z),$$

$$12.2.16 \qquad V(a, -z) = \frac{\cos(\pi a)}{\Gamma\left(\frac{1}{2} - a\right)} U(a, z) + \sin(\pi a)\, V(a, z).$$

$$12.2.17 \qquad \sqrt{2\pi}\, U(-a, \pm iz) = \Gamma\left(\tfrac{1}{2} + a\right)\left(e^{\mp i\pi\left(\frac{1}{2}a - \frac{1}{4}\right)} U(a, z) + e^{\pm i\pi\left(\frac{1}{2}a - \frac{1}{4}\right)} U(a, -z)\right).$$

$$12.2.18 \qquad \sqrt{2\pi}\, U(a, z) = \Gamma\left(\tfrac{1}{2} - a\right)\left(e^{\mp i\pi\left(\frac{1}{2}a + \frac{1}{4}\right)} U(-a, \pm iz) + e^{\pm i\pi\left(\frac{1}{2}a + \frac{1}{4}\right)} U(-a, \mp iz)\right),$$

12.3 Graphics

12.2.19
$$U(a, z) = \pm i e^{\pm i\pi a} U(a, -z) \\ + \frac{\sqrt{2\pi}}{\Gamma(\frac{1}{2} + a)} e^{\pm i\pi(\frac{1}{2}a - \frac{1}{4})} U(-a, \pm iz).$$

12.2.20
$$V(a, z) \\ = \frac{\mp i}{\Gamma(\frac{1}{2} - a)} U(a, z) + \sqrt{\frac{2}{\pi}} e^{\mp i\pi(\frac{1}{2}a - \frac{1}{4})} U(-a, \pm iz).$$

12.2(vi) Solution $\overline{U}(a, x)$; Modulus and Phase Functions

When $z\ (=x)$ is real the solution $\overline{U}(a, x)$ is defined by

12.2.21
$$\overline{U}(a, x) = \Gamma(\tfrac{1}{2} - a)\, V(a, x),$$

unless $a = \frac{1}{2}, \frac{3}{2}, \ldots$, in which case $\overline{U}(a, x)$ is undefined. Its importance is that when a is negative and $|a|$ is large, $U(a, x)$ and $\overline{U}(a, x)$ asymptotically have the same envelope (modulus) and are $\frac{1}{2}\pi$ out of phase in the oscillatory interval $-2\sqrt{-a} < x < 2\sqrt{-a}$. Properties of $\overline{U}(a, x)$ follow immediately from those of $V(a, x)$ via (12.2.21).

In the oscillatory interval we define

12.2.22
$$U(a, x) + i\overline{U}(a, x) = F(a, x) e^{i\theta(a, x)},$$

12.2.23
$$U'(a, x) + i\overline{U}'(a, x) = -G(a, x) e^{i\psi(a, x)},$$

where $F(a, x)\ (>0)$, $\theta(a, x)$, $G(a, x)\ (>0)$, and $\psi(a, x)$ are real. F or G is the *modulus* and θ or ψ is the corresponding *phase*.

For properties of the modulus and phase functions, including differential equations, see Miller (1955, pp. 72–73). For graphs of the modulus functions see §12.3(i).

12.3 Graphics

12.3(i) Real Variables

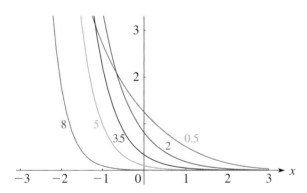

Figure 12.3.1: $U(a, x)$, $a = 0.5, 2, 3.5, 5, 8$.

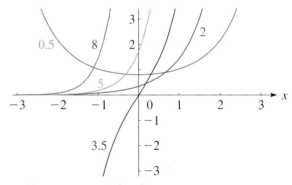

Figure 12.3.2: $V(a, x)$, $a = 0.5, 2, 3.5, 5, 8$.

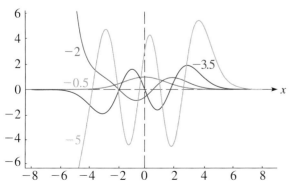

Figure 12.3.3: $U(a, x)$, $a = -0.5, -2, -3.5, -5$.

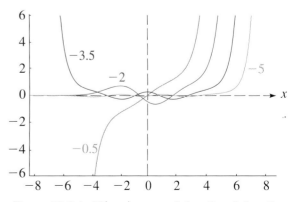

Figure 12.3.4: $V(a, x)$, $a = -0.5, -2, -3.5, -5$.

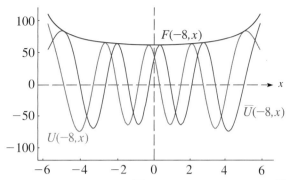

Figure 12.3.5: $U(-8,x)$, $\overline{U}(-8,x)$, $F(-8,x)$, $-4\sqrt{2} \leq x \leq 4\sqrt{2}$.

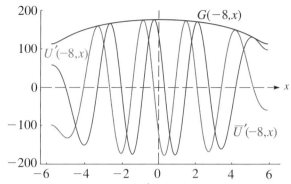

Figure 12.3.6: $U'(-8,x)$, $\overline{U}'(-8,x)$, $G(-8,x)$, $-4\sqrt{2} \leq x \leq 4\sqrt{2}$.

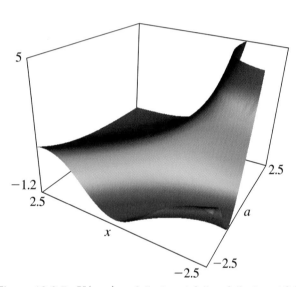

Figure 12.3.7: $U(a,x)$, $-2.5 \leq a \leq 2.5$, $-2.5 \leq x \leq 2.5$.

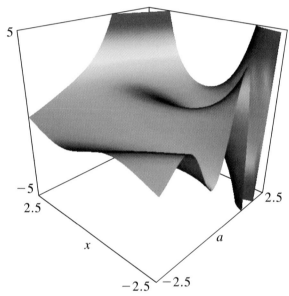

Figure 12.3.8: $V(a,x)$, $-2.5 \leq a \leq 2.5$, $-2.5 \leq x \leq 2.5$.

12.3(ii) Complex Variables

In the graphics shown in this subsection, height corresponds to the absolute value of the function and color to the phase. See also p. xiv.

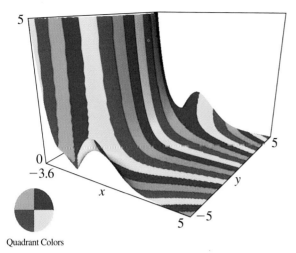

Figure 12.3.9: $U(3.5, x+iy)$, $-3.6 \leq x \leq 5$, $-5 \leq y \leq 5$.

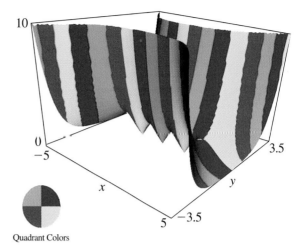

Figure 12.3.10: $U(-3.5, x+iy)$, $-5 \leq x \leq 5$, $-3.5 \leq y \leq 3.5$.

12.4 Power-Series Expansions

12.4.1 $\quad U(a,z) = U(a,0)u_1(a,z) + U'(a,0)u_2(a,z)$,

12.4.2 $\quad V(a,z) = V(a,0)u_1(a,z) + V'(a,0)u_2(a,z)$,

where the initial values are given by (12.2.6)–(12.2.9), and $u_1(a,z)$ and $u_2(a,z)$ are the even and odd solutions of (12.2.2) given by

12.4.3
$$u_1(a,z) = e^{-\frac{1}{4}z^2}\left(1 + (a+\tfrac{1}{2})\frac{z^2}{2!} + (a+\tfrac{1}{2})(a+\tfrac{5}{2})\frac{z^4}{4!} + \cdots\right),$$

12.4.4
$$u_2(a,z) = e^{-\frac{1}{4}z^2}\left(z + (a+\tfrac{3}{2})\frac{z^3}{3!} + (a+\tfrac{3}{2})(a+\tfrac{7}{2})\frac{z^5}{5!} + \cdots\right).$$

Equivalently,

12.4.5
$$u_1(a,z) = e^{\frac{1}{4}z^2}\left(1 + (a-\tfrac{1}{2})\frac{z^2}{2!} + (a-\tfrac{1}{2})(a-\tfrac{5}{2})\frac{z^4}{4!} + \cdots\right),$$

12.4.6
$$u_2(a,z) = e^{\frac{1}{4}z^2}\left(z + (a-\tfrac{3}{2})\frac{z^3}{3!} + (a-\tfrac{3}{2})(a-\tfrac{7}{2})\frac{z^5}{5!} + \cdots\right).$$

These series converge for all values of z.

12.5 Integral Representations

12.5(i) Integrals Along the Real Line

12.5.1
$$U(a,z) = \frac{e^{-\frac{1}{4}z^2}}{\Gamma(\tfrac{1}{2}+a)}\int_0^\infty t^{a-\frac{1}{2}}e^{-\frac{1}{2}t^2-zt}\,dt, \quad \Re a > -\tfrac{1}{2},$$

12.5.2
$$U(a,z) = \frac{ze^{-\frac{1}{4}z^2}}{\Gamma(\tfrac{1}{4}+\tfrac{1}{2}a)}\int_0^\infty t^{\frac{1}{2}a-\frac{3}{4}}e^{-t}\left(z^2+2t\right)^{-\frac{1}{2}a-\frac{3}{4}}dt,$$
$$|\mathrm{ph}\,z| < \tfrac{1}{2}\pi, \Re a > -\tfrac{1}{2},$$

12.5.3
$$U(a,z) = \frac{e^{-\frac{1}{4}z^2}}{\Gamma(\tfrac{3}{4}+\tfrac{1}{2}a)}\int_0^\infty t^{\frac{1}{2}a-\frac{1}{4}}e^{-t}\left(z^2+2t\right)^{-\frac{1}{2}a-\frac{1}{4}}dt,$$
$$|\mathrm{ph}\,z| < \tfrac{1}{2}\pi, \Re a > -\tfrac{3}{2},$$

12.5.4
$$U(a,z) = \sqrt{\frac{2}{\pi}}e^{\frac{1}{4}z^2}$$
$$\times \int_0^\infty t^{-a-\frac{1}{2}}e^{-\frac{1}{2}t^2}\cos\!\left(zt + \left(\tfrac{1}{2}a+\tfrac{1}{4}\right)\pi\right)dt,$$
$$\Re a < \tfrac{1}{2}.$$

12.5(ii) Contour Integrals

The following integrals correspond to those of §12.5(i).

12.5.5
$$U(a,z) = \frac{\Gamma(\tfrac{1}{2}-a)}{2\pi i}e^{-\frac{1}{4}z^2}\int_{-\infty}^{(0+)}e^{zt-\frac{1}{2}t^2}t^{a-\frac{1}{2}}\,dt,$$
$$a \neq \tfrac{1}{2},\tfrac{3}{2},\tfrac{5}{2},\ldots, -\pi < \mathrm{ph}\,t < \pi.$$

Restrictions on a are not needed in the following two representations:

12.5.6
$$U(a,z) = \frac{e^{\frac{1}{4}z^2}}{i\sqrt{2\pi}}\int_{c-i\infty}^{c+i\infty}e^{-zt+\frac{1}{2}t^2}t^{-a-\frac{1}{2}}\,dt,$$
$$-\tfrac{1}{2}\pi < \mathrm{ph}\,t < \tfrac{1}{2}\pi, c > 0,$$

12.5.7
$$V(a,z) = \frac{e^{-\frac{1}{4}z^2}}{2\pi}\left(\int_{-ic-\infty}^{-ic+\infty} + \int_{ic-\infty}^{ic+\infty}\right)e^{zt-\frac{1}{2}t^2}t^{a-\frac{1}{2}}\,dt,$$
$$-\pi < \mathrm{ph}\,t < \pi, c > 0.$$

For proofs and further results see Miller (1955, §4) and Whittaker (1902).

12.5(iii) Mellin–Barnes Integrals

12.5.8
$$U(a, z) = \frac{e^{-\frac{1}{4}z^2} z^{-a-\frac{1}{2}}}{2\pi i \, \Gamma\left(\frac{1}{2} + a\right)}$$
$$\times \int_{-i\infty}^{i\infty} \Gamma(t)\,\Gamma\left(\tfrac{1}{2} + a - 2t\right) 2^t z^{2t}\, dt,$$
$$a \neq -\tfrac{1}{2}, -\tfrac{3}{2}, -\tfrac{5}{2}, \ldots, \; |\mathrm{ph}\, z| < \tfrac{3}{4}\pi,$$
where the contour separates the poles of $\Gamma(t)$ from those of $\Gamma\left(\tfrac{1}{2} + a - 2t\right)$.

12.5.9
$$V(a, z) = \sqrt{\frac{2}{\pi}}\,\frac{e^{\frac{1}{4}z^2} z^{a-\frac{1}{2}}}{2\pi i \, \Gamma\left(\frac{1}{2} - a\right)}$$
$$\times \int_{-i\infty}^{i\infty} \Gamma(t)\,\Gamma\left(\tfrac{1}{2} - a - 2t\right) 2^t z^{2t} \cos(\pi t)\, dt,$$
$$a \neq \tfrac{1}{2}, \tfrac{3}{2}, \tfrac{5}{2}, \ldots, \; |\mathrm{ph}\, z| < \tfrac{1}{4}\pi,$$
where the contour separates the poles of $\Gamma(t)$ from those of $\Gamma\left(\tfrac{1}{2} - a - 2t\right)$.

12.5(iv) Compendia

For further collections of integral representations see Apelblat (1983, pp. 427-436), Erdélyi et al. (1953b, v. 2, pp. 119–120), Erdélyi et al. (1954a, pp. 289–291 and 362), Gradshteyn and Ryzhik (2000, §§9.24–9.25), Magnus et al. (1966, pp. 328–330), Oberhettinger (1974, pp. 251–252), and Oberhettinger and Badii (1973, pp. 378–384).

12.6 Continued Fraction

For a continued-fraction expansion of the ratio $U(a, x)/U(a-1, x)$ see Cuyt et al. (2008, pp. 340–341).

12.7 Relations to Other Functions

12.7(i) Hermite Polynomials

For the notation see §18.3.

12.7.1 $\quad U\left(-\tfrac{1}{2}, z\right) = D_0(z) = e^{-\frac{1}{4}z^2},$

12.7.2
$$U\left(-n - \tfrac{1}{2}, z\right) = D_n(z) = e^{-\frac{1}{4}z^2} \mathit{He}_n(z)$$
$$= 2^{-n/2} e^{-\frac{1}{4}z^2} H_n\left(z/\sqrt{2}\right),$$
$$n = 0, 1, 2, \ldots,$$

12.7.3
$$V\left(n + \tfrac{1}{2}, z\right) = \sqrt{2/\pi}\, e^{\frac{1}{4}z^2} (-i)^n \mathit{He}_n(iz)$$
$$= \sqrt{2/\pi}\, e^{\frac{1}{4}z^2} (-i)^n 2^{-\frac{1}{2}n} H_n\left(iz/\sqrt{2}\right),$$
$$n = 0, 1, 2, \ldots.$$

12.7(ii) Error Functions, Dawson's Integral, and Probability Function

For the notation see §§7.2 and 7.18.

12.7.4 $\quad V\left(-\tfrac{1}{2}, z\right) = (2/\sqrt{\pi})\, e^{\frac{1}{4}z^2} F\left(z/\sqrt{2}\right),$

12.7.5 $\quad U\left(\tfrac{1}{2}, z\right) = D_{-1}(z) = \sqrt{\tfrac{1}{2}\pi}\, e^{\frac{1}{4}z^2} \mathrm{erfc}\left(z/\sqrt{2}\right),$

12.7.6
$$U\left(n + \tfrac{1}{2}, z\right) = D_{-n-1}(z)$$
$$= \sqrt{\frac{\pi}{2}}\,\frac{(-1)^n}{n!}\, e^{-\frac{1}{4}z^2} \frac{d^n\left(e^{\frac{1}{2}z^2} \mathrm{erfc}\left(z/\sqrt{2}\right)\right)}{dz^n},$$
$$n = 0, 1, 2, \ldots,$$

12.7.7
$$U\left(n + \tfrac{1}{2}, z\right) = e^{\frac{1}{4}z^2} \mathit{Hh}_n(z)$$
$$= \sqrt{\pi}\, 2^{\frac{1}{2}(n-1)} e^{\frac{1}{4}z^2} \mathrm{i}^n\mathrm{erfc}\left(z/\sqrt{2}\right),$$
$$n = -1, 0, 1, \ldots.$$

12.7(iii) Modified Bessel Functions

For the notation see §10.25(ii).

12.7.8
$$U(-2, z) = \frac{z^{5/2}}{4\sqrt{2\pi}}\left(2 K_{\frac{1}{4}}\!\left(\tfrac{1}{4}z^2\right) + 3 K_{\frac{3}{4}}\!\left(\tfrac{1}{4}z^2\right) - K_{\frac{5}{4}}\!\left(\tfrac{1}{4}z^2\right)\right),$$

12.7.9 $\quad U(-1, z) = \dfrac{z^{3/2}}{2\sqrt{2\pi}}\left(K_{\frac{1}{4}}\!\left(\tfrac{1}{4}z^2\right) + K_{\frac{3}{4}}\!\left(\tfrac{1}{4}z^2\right)\right),$

12.7.10 $\quad U(0, z) = \sqrt{\dfrac{z}{2\pi}}\, K_{\frac{1}{4}}\!\left(\tfrac{1}{4}z^2\right),$

12.7.11 $\quad U(1, z) = \dfrac{z^{3/2}}{\sqrt{2\pi}}\left(K_{\frac{3}{4}}\!\left(\tfrac{1}{4}z^2\right) - K_{\frac{1}{4}}\!\left(\tfrac{1}{4}z^2\right)\right).$

For these, the corresponding results for $U(a, z)$ with $a = 2, \pm 3, -\tfrac{1}{2}, -\tfrac{3}{2}, -\tfrac{5}{2}$, and the corresponding results for $V(a, z)$ with $a = 0, \pm 1, \pm 2, \pm 3, \tfrac{1}{2}, \tfrac{3}{2}, \tfrac{5}{2}$, see Miller (1955, pp. 42–43 and 77–79).

12.7(iv) Confluent Hypergeometric Functions

For the notation see §§13.2(i) and 13.14(i).

The even and odd solutions of (12.2.2) (see (12.4.3)–(12.4.6)) are given by

12.7.12
$$u_1(a, z) = e^{-\frac{1}{4}z^2} M\left(\tfrac{1}{2}a + \tfrac{1}{4}, \tfrac{1}{2}, \tfrac{1}{2}z^2\right)$$
$$= e^{\frac{1}{4}z^2} M\left(-\tfrac{1}{2}a + \tfrac{1}{4}, \tfrac{1}{2}, -\tfrac{1}{2}z^2\right),$$

12.7.13
$$u_2(a, z) = z e^{-\frac{1}{4}z^2} M\left(\tfrac{1}{2}a + \tfrac{3}{4}, \tfrac{3}{2}, \tfrac{1}{2}z^2\right)$$
$$= z e^{\frac{1}{4}z^2} M\left(-\tfrac{1}{2}a + \tfrac{3}{4}, \tfrac{3}{2}, -\tfrac{1}{2}z^2\right).$$

Also,

12.7.14
$$U(a, z) = 2^{-\frac{1}{4} - \frac{1}{2}a} e^{-\frac{1}{4}z^2} U\left(\tfrac{1}{2}a + \tfrac{1}{4}, \tfrac{1}{2}, \tfrac{1}{2}z^2\right)$$
$$= 2^{-\frac{3}{4} - \frac{1}{2}a} z e^{-\frac{1}{4}z^2} U\left(\tfrac{1}{2}a + \tfrac{3}{4}, \tfrac{3}{2}, \tfrac{1}{2}z^2\right)$$
$$= 2^{-\frac{1}{2}a} z^{-\frac{1}{2}} W_{-\frac{1}{2}a, \pm\frac{1}{4}}\!\left(\tfrac{1}{2}z^2\right).$$

(It should be observed that the functions on the right-hand sides of (12.7.14) are multivalued; hence, for example, z cannot be replaced simply by $-z$.)

12.8 Recurrence Relations and Derivatives

12.8(i) Recurrence Relations

12.8.1 $\quad zU(a,z) - U(a-1,z) + (a+\tfrac{1}{2})U(a+1,z) = 0,$

12.8.2 $\quad U'(a,z) + \tfrac{1}{2}zU(a,z) + (a+\tfrac{1}{2})U(a+1,z) = 0,$

12.8.3 $\quad U'(a,z) - \tfrac{1}{2}zU(a,z) + U(a-1,z) = 0,$

12.8.4 $\quad 2U'(a,z) + U(a-1,z) + (a+\tfrac{1}{2})U(a+1,z) = 0.$

(12.8.1)–(12.8.4) are also satisfied by $\overline{U}(a,z)$.

12.8.5 $\quad zV(a,z) - V(a+1,z) + (a-\tfrac{1}{2})V(a-1,z) = 0,$

12.8.6 $\quad V'(a,z) - \tfrac{1}{2}zV(a,z) - (a-\tfrac{1}{2})V(a-1,z) = 0,$

12.8.7 $\quad V'(a,z) + \tfrac{1}{2}zV(a,z) - V(a+1,z) = 0,$

12.8.8 $\quad 2V'(a,z) - V(a+1,z) - (a-\tfrac{1}{2})V(a-1,z) = 0.$

12.8(ii) Derivatives

For $m = 0, 1, 2, \ldots,$

12.8.9
$$\frac{d^m}{dz^m}\left(e^{\frac{1}{4}z^2} U(a,z)\right) = (-1)^m \left(\tfrac{1}{2}+a\right)_m e^{\frac{1}{4}z^2} U(a+m,z),$$

12.8.10
$$\frac{d^m}{dz^m}\left(e^{-\frac{1}{4}z^2} U(a,z)\right) = (-1)^m e^{-\frac{1}{4}z^2} U(a-m,z),$$

12.8.11
$$\frac{d^m}{dz^m}\left(e^{\frac{1}{4}z^2} V(a,z)\right) = e^{\frac{1}{4}z^2} V(a+m,z),$$

12.8.12
$$\frac{d^m}{dz^m}\left(e^{-\frac{1}{4}z^2} V(a,z)\right) = (-1)^m \left(\tfrac{1}{2}-a\right)_m e^{-\frac{1}{4}z^2} V(a-m,z).$$

12.9 Asymptotic Expansions for Large Variable

12.9(i) Poincaré-Type Expansions

Throughout this subsection δ is an arbitrary small positive constant.

As $z \to \infty$

12.9.1
$$U(a,z) \sim e^{-\frac{1}{4}z^2} z^{-a-\frac{1}{2}} \sum_{s=0}^{\infty} (-1)^s \frac{\left(\tfrac{1}{2}+a\right)_{2s}}{s!(2z^2)^s},$$
$$|\operatorname{ph} z| \leq \tfrac{3}{4}\pi - \delta (<\tfrac{3}{4}\pi),$$

12.9.2
$$V(a,z) \sim \sqrt{\tfrac{2}{\pi}} e^{\frac{1}{4}z^2} z^{a-\frac{1}{2}} \sum_{s=0}^{\infty} \frac{\left(\tfrac{1}{2}-a\right)_{2s}}{s!(2z^2)^s},$$
$$|\operatorname{ph} z| \leq \tfrac{1}{4}\pi - \delta (<\tfrac{1}{4}\pi).$$

12.9.3
$$U(a,z) \sim e^{-\frac{1}{4}z^2} z^{-a-\frac{1}{2}} \sum_{s=0}^{\infty} (-1)^s \frac{\left(\tfrac{1}{2}+a\right)_{2s}}{s!(2z^2)^s}$$
$$\pm i \frac{\sqrt{2\pi}}{\Gamma\left(\tfrac{1}{2}+a\right)} e^{\mp i\pi a} e^{\frac{1}{4}z^2} z^{a-\frac{1}{2}} \sum_{s=0}^{\infty} \frac{\left(\tfrac{1}{2}-a\right)_{2s}}{s!(2z^2)^s},$$
$$\tfrac{1}{4}\pi + \delta \leq \pm \operatorname{ph} z \leq \tfrac{5}{4}\pi - \delta,$$

12.9.4
$$V(a,z) \sim \sqrt{\tfrac{2}{\pi}} e^{\frac{1}{4}z^2} z^{a-\frac{1}{2}} \sum_{s=0}^{\infty} \frac{\left(\tfrac{1}{2}-a\right)_{2s}}{s!(2z^2)^s}$$
$$\pm \frac{i}{\Gamma\left(\tfrac{1}{2}-a\right)} e^{-\frac{1}{4}z^2} z^{-a-\frac{1}{2}} \sum_{s=0}^{\infty} (-1)^s \frac{\left(\tfrac{1}{2}+a\right)_{2s}}{s!(2z^2)^s},$$
$$-\tfrac{1}{4}\pi + \delta \leq \pm \operatorname{ph} z \leq \tfrac{3}{4}\pi - \delta.$$

12.9(ii) Bounds and Re-Expansions for the Remainder Terms

Bounds and re-expansions for the error term in (12.9.1) can be obtained by use of (12.7.14) and §§13.7(ii), 13.7(iii). Corresponding results for (12.9.2) can be obtained via (12.2.20).

12.10 Uniform Asymptotic Expansions for Large Parameter

12.10(i) Introduction

In this section we give asymptotic expansions of PCFs for large values of the parameter a that are uniform with respect to the variable z, when both a and z ($=x$) are real. These expansions follow from Olver (1959), where detailed information is also given for complex variables.

With the transformations

12.10.1 $\quad a = \pm\tfrac{1}{2}\mu^2, \quad x = \mu t\sqrt{2},$

(12.2.2) becomes

12.10.2 $\quad \dfrac{d^2 w}{dt^2} = \mu^4(t^2 \pm 1)w.$

With the upper sign in (12.10.2), expansions can be constructed for large μ in terms of elementary functions that are uniform for $t \in (-\infty, \infty)$ (§2.8(ii)). With the lower sign there are turning points at $t = \pm 1$, which need to be excluded from the regions of validity. These cases are treated in §§12.10(ii)–12.10(vi).

The turning points can be included if expansions in terms of Airy functions are used instead of elementary functions (§2.8(iii)). These cases are treated in §§12.10(vii)–12.10(viii).

Throughout this section the symbol δ again denotes an arbitrary small positive constant.

12.10(ii) Negative a, $2\sqrt{-a} < x < \infty$

As $a \to -\infty$

12.10.3 $\quad U\left(-\tfrac{1}{2}\mu^2, \mu t\sqrt{2}\right) \sim \dfrac{g(\mu)e^{-\mu^2\xi}}{(t^2-1)^{\frac{1}{4}}} \displaystyle\sum_{s=0}^{\infty} \dfrac{\mathcal{A}_s(t)}{\mu^{2s}}$,

12.10.4
$$U'\left(-\tfrac{1}{2}\mu^2, \mu t\sqrt{2}\right) \sim -\dfrac{\mu}{\sqrt{2}}g(\mu)(t^2-1)^{\frac{1}{4}}e^{-\mu^2\xi}\sum_{s=0}^{\infty}\dfrac{\mathcal{B}_s(t)}{\mu^{2s}},$$

12.10.5
$$V\left(-\tfrac{1}{2}\mu^2, \mu t\sqrt{2}\right) \sim \dfrac{2g(\mu)}{\Gamma(\tfrac{1}{2}+\tfrac{1}{2}\mu^2)}\dfrac{e^{\mu^2\xi}}{(t^2-1)^{\frac{1}{4}}} \\ \times \sum_{s=0}^{\infty}(-1)^s\dfrac{\mathcal{A}_s(t)}{\mu^{2s}},$$

12.10.6
$$V'\left(-\tfrac{1}{2}\mu^2, \mu t\sqrt{2}\right) \sim \dfrac{\sqrt{2}\mu g(\mu)}{\Gamma(\tfrac{1}{2}+\tfrac{1}{2}\mu^2)}(t^2-1)^{\frac{1}{4}} \\ \times e^{\mu^2\xi}\sum_{s=0}^{\infty}(-1)^s\dfrac{\mathcal{B}_s(t)}{\mu^{2s}},$$

uniformly for $t \in [1+\delta, \infty)$, where

12.10.7 $\quad \xi = \tfrac{1}{2}t\sqrt{t^2-1} - \tfrac{1}{2}\ln\left(t+\sqrt{t^2-1}\right)$.

The coefficients are given by

12.10.8 $\quad \mathcal{A}_s(t) = \dfrac{u_s(t)}{(t^2-1)^{\frac{3}{2}s}}, \quad \mathcal{B}_s(t) = \dfrac{v_s(t)}{(t^2-1)^{\frac{3}{2}s}}$,

where $u_s(t)$ and $v_s(t)$ are polynomials in t of degree $3s$, (s odd), $3s-2$ (s even, $s \geq 2$). For $s = 0, 1, 2$,

12.10.9
$$u_0(t) = 1, \quad u_1(t) = \dfrac{t(t^2-6)}{24}, \\ u_2(t) = \dfrac{-9t^4+249t^2+145}{1152},$$

12.10.10
$$v_0(t) = 1, \quad v_1(t) = \dfrac{t(t^2+6)}{24}, \\ v_2(t) = \dfrac{15t^4-327t^2-143}{1152}.$$

Higher polynomials $u_s(t)$ can be calculated from the recurrence relation

12.10.11 $\quad (t^2-1)u'_s(t) - 3stu_s(t) = r_{s-1}(t)$,

where

12.10.12
$$8r_s(t) = (3t^2+2)u_s(t) - 12(s+1)tr_{s-1}(t) \\ + 4(t^2-1)r'_{s-1}(t),$$

and the $v_s(t)$ then follow from

12.10.13 $\quad v_s(t) = u_s(t) + \tfrac{1}{2}tu_{s-1}(t) - r_{s-2}(t)$.

Lastly, the function $g(\mu)$ in (12.10.3) and (12.10.4) has the asymptotic expansion:

12.10.14 $\quad g(\mu) \sim h(\mu)\left(1 + \dfrac{1}{2}\displaystyle\sum_{s=1}^{\infty}\dfrac{\gamma_s}{(\tfrac{1}{2}\mu^2)^s}\right)$,

where

12.10.15 $\quad h(\mu) = 2^{-\frac{1}{4}\mu^2 - \frac{1}{4}}e^{-\frac{1}{4}\mu^2}\mu^{\frac{1}{2}\mu^2-\frac{1}{2}}$,

and the coefficients γ_s are defined by

12.10.16 $\quad \Gamma\left(\tfrac{1}{2}+z\right) \sim \sqrt{2\pi}e^{-z}z^z\displaystyle\sum_{s=0}^{\infty}\dfrac{\gamma_s}{z^s}$;

compare (5.11.8). For $s \leq 4$

12.10.17 $\quad \gamma_0 = 1, \quad \gamma_1 = -\tfrac{1}{24}, \quad \gamma_2 = \tfrac{1}{1152}, \\ \gamma_3 = \tfrac{1003}{4\,14720}, \quad \gamma_4 = -\tfrac{4027}{398\,13120}.$

12.10(iii) Negative a, $-\infty < x < -2\sqrt{-a}$

When $\mu \to \infty$, asymptotic expansions for the functions $U\left(-\tfrac{1}{2}\mu^2, -\mu t\sqrt{2}\right)$ and $V\left(-\tfrac{1}{2}\mu^2, -\mu t\sqrt{2}\right)$ that are uniform for $t \in [1+\delta, \infty)$ are obtainable by substitution into (12.2.15) and (12.2.16) by means of (12.10.3) and (12.10.5). Similarly for $U'\left(-\tfrac{1}{2}\mu^2, -\mu t\sqrt{2}\right)$ and $V'\left(-\tfrac{1}{2}\mu^2, -\mu t\sqrt{2}\right)$.

12.10(iv) Negative a, $-2\sqrt{-a} < x < 2\sqrt{-a}$

As $a \to -\infty$

12.10.18 $\quad U\left(-\tfrac{1}{2}\mu^2, \mu t\sqrt{2}\right) \sim \dfrac{2g(\mu)}{(1-t^2)^{\frac{1}{4}}}\left(\cos\kappa\displaystyle\sum_{s=0}^{\infty}(-1)^s\dfrac{\widetilde{\mathcal{A}}_{2s}(t)}{\mu^{4s}} - \sin\kappa\sum_{s=0}^{\infty}(-1)^s\dfrac{\widetilde{\mathcal{A}}_{2s+1}(t)}{\mu^{4s+2}}\right)$,

12.10.19 $\quad U'\left(-\tfrac{1}{2}\mu^2, \mu t\sqrt{2}\right) \sim \mu\sqrt{2}g(\mu)(1-t^2)^{\frac{1}{4}}\left(\sin\kappa\displaystyle\sum_{s=0}^{\infty}(-1)^s\dfrac{\widetilde{\mathcal{B}}_{2s}(t)}{\mu^{4s}} + \cos\kappa\sum_{s=0}^{\infty}(-1)^s\dfrac{\widetilde{\mathcal{B}}_{2s+1}(t)}{\mu^{4s+2}}\right)$,

12.10.20 $\quad V\left(-\tfrac{1}{2}\mu^2, \mu t\sqrt{2}\right) \sim \dfrac{2g(\mu)}{\Gamma(\tfrac{1}{2}+\tfrac{1}{2}\mu^2)(1-t^2)^{\frac{1}{4}}}\left(\cos\chi\displaystyle\sum_{s=0}^{\infty}(-1)^s\dfrac{\widetilde{\mathcal{A}}_{2s}(t)}{\mu^{4s}} - \sin\chi\sum_{s=0}^{\infty}(-1)^s\dfrac{\widetilde{\mathcal{A}}_{2s+1}(t)}{\mu^{4s+2}}\right)$,

12.10.21 $\quad V'\left(-\tfrac{1}{2}\mu^2, \mu t\sqrt{2}\right) \sim \dfrac{\mu\sqrt{2}g(\mu)(1-t^2)^{\frac{1}{4}}}{\Gamma(\tfrac{1}{2}+\tfrac{1}{2}\mu^2)}\left(\sin\chi\displaystyle\sum_{s=0}^{\infty}(-1)^s\dfrac{\widetilde{\mathcal{B}}_{2s}(t)}{\mu^{4s}} + \cos\chi\sum_{s=0}^{\infty}(-1)^s\dfrac{\widetilde{\mathcal{B}}_{2s+1}(t)}{\mu^{4s+2}}\right)$, uniformly for $t \in [-1+\delta, 1-\delta]$. The quantities κ and χ are defined by

12.10.22 $\quad \kappa = \mu^2 \eta - \tfrac{1}{4}\pi, \quad \chi = \mu^2 \eta + \tfrac{1}{4}\pi,$

where

12.10.23 $\quad \eta = \tfrac{1}{2}\arccos t - \tfrac{1}{2} t\sqrt{1-t^2},$

and the coefficients $\widetilde{\mathcal{A}}_s(t)$ and $\widetilde{\mathcal{B}}_s(t)$ are given by

12.10.24 $\quad \widetilde{\mathcal{A}}_s(t) = \dfrac{u_s(t)}{(1-t^2)^{\frac{3}{2}s}}, \quad \widetilde{\mathcal{B}}_s(t) = \dfrac{v_s(t)}{(1-t^2)^{\frac{3}{2}s}};$

compare (12.10.8).

12.10(v) Positive a, $-\infty < x < \infty$

As $a \to \infty$

12.10.25
$$U\left(\tfrac{1}{2}\mu^2, \mu t \sqrt{2}\right) \sim \dfrac{\overline{g}(\mu) e^{-\mu^2 \overline{\xi}}}{(t^2+1)^{\frac{1}{4}}} \sum_{s=0}^{\infty} \dfrac{\overline{u}_s(t)}{(t^2+1)^{\frac{3}{2}s}} \dfrac{1}{\mu^{2s}},$$

uniformly for $t \in \mathbb{R}$. Here bars do not denote complex conjugates; instead

12.10.26 $\quad \overline{\xi} = \tfrac{1}{2} t\sqrt{t^2+1} + \tfrac{1}{2}\ln\!\left(t + \sqrt{t^2+1}\right),$

12.10.27 $\quad \overline{u}_s(t) = i^s u_s(-it),$

and the function $\overline{g}(\mu)$ has the asymptotic expansion

12.10.28 $\quad \overline{g}(\mu) \sim \dfrac{1}{\mu \sqrt{2} h(\mu)} \left(1 + \tfrac{1}{2} \sum_{s=1}^{\infty} (-1)^s \dfrac{\gamma_s}{(\tfrac{1}{2}\mu^2)^s}\right),$

where $h(\mu)$ and γ_s are as in §12.10(ii).

With the same conditions

12.10.29
$$U'\left(\tfrac{1}{2}\mu^2, \mu t \sqrt{2}\right) \\ \sim -\dfrac{\mu}{\sqrt{2}} \overline{g}(\mu)(t^2+1)^{\frac{1}{4}} e^{-\mu^2 \overline{\xi}} \sum_{s=0}^{\infty} \dfrac{\overline{v}_s(t)}{(t^2+1)^{\frac{3}{2}s}} \dfrac{1}{\mu^{2s}},$$

where

12.10.30 $\quad \overline{v}_s(t) = i^s v_s(-it).$

12.10(vi) Modifications of Expansions in Elementary Functions

In Temme (2000) modifications are given of Olver's expansions. An example is the following modification of (12.10.3)

12.10.31 $\quad U\left(-\tfrac{1}{2}\mu^2, \mu t \sqrt{2}\right) \sim \dfrac{h(\mu) e^{-\mu^2 \xi}}{(t^2-1)^{\frac{1}{4}}} \sum_{s=0}^{\infty} \dfrac{\mathsf{A}_s(\tau)}{\mu^{2s}},$

where ξ and $h(\mu)$ are as in (12.10.7) and (12.10.15),

12.10.32 $\quad \tau = \dfrac{1}{2}\left(\dfrac{t}{\sqrt{t^2-1}} - 1\right),$

and the coefficients $\mathsf{A}_s(\tau)$ are the product of τ^s and a polynomial in τ of degree $2s$. They satisfy the recursion

12.10.33
$$\mathsf{A}_{s+1}(\tau) = -4\tau^2(\tau+1)^2 \dfrac{d}{d\tau} \mathsf{A}_s(\tau) \\ -\dfrac{1}{4}\int_0^\tau (20u^2 + 20u + 3)\mathsf{A}_s(u)\, du,$$
$s = 0, 1, 2, \ldots,$

starting with $\mathsf{A}_o(\tau) = 1$. Explicitly,

$\mathsf{A}_1(\tau) = -\tfrac{1}{12}\tau(20\tau^2 + 30\tau + 9),$

12.10.34 $\quad \mathsf{A}_2(\tau) = \tfrac{1}{288}\tau^2(6160\tau^4 + 18480\tau^3 + 19404\tau^2 + 8028\tau + 945).$

The modified expansion (12.10.31) shares the property of (12.10.3) that it applies when $\mu \to \infty$ uniformly with respect to $t \in [1+\delta, \infty)$. In addition, it enjoys a double asymptotic property: it holds if either or both μ and t tend to infinity. Observe that if $t \to \infty$, then $\mathsf{A}_s(\tau) = O(t^{-2s})$, whereas $\mathcal{A}_s(t) = O(1)$ or $O(t^{-2})$ according as s is even or odd. The proof of the double asymptotic property then follows with the aid of error bounds; compare §10.41(iv).

For additional information see Temme (2000). See also Olver (1997b, pp. 206–208) and Jones (2006).

12.10(vii) Negative a, $-2\sqrt{-a} < x < \infty$. Expansions in Terms of Airy Functions

The following expansions hold for large positive real values of μ, uniformly for $t \in [-1+\delta, \infty)$. (For complex values of μ and t see Olver (1959).)

12.10.35
$$U\left(-\tfrac{1}{2}\mu^2, \mu t \sqrt{2}\right) \sim 2\pi^{\frac{1}{2}} \mu^{\frac{1}{3}} g(\mu) \phi(\zeta) \left(\operatorname{Ai}\!\left(\mu^{\frac{4}{3}}\zeta\right) \sum_{s=0}^{\infty} \dfrac{A_s(\zeta)}{\mu^{4s}} + \dfrac{\operatorname{Ai}'\!\left(\mu^{\frac{4}{3}}\zeta\right)}{\mu^{\frac{8}{3}}} \sum_{s=0}^{\infty} \dfrac{B_s(\zeta)}{\mu^{4s}} \right),$$

12.10.36
$$U'\left(-\tfrac{1}{2}\mu^2, \mu t \sqrt{2}\right) \sim \dfrac{(2\pi)^{\frac{1}{2}} \mu^{\frac{2}{3}} g(\mu)}{\phi(\zeta)} \left(\dfrac{\operatorname{Ai}\!\left(\mu^{\frac{4}{3}}\zeta\right)}{\mu^{\frac{4}{3}}} \sum_{s=0}^{\infty} \dfrac{C_s(\zeta)}{\mu^{4s}} + \operatorname{Ai}'\!\left(\mu^{\frac{4}{3}}\zeta\right) \sum_{s=0}^{\infty} \dfrac{D_s(\zeta)}{\mu^{4s}} \right),$$

12.10.37
$$V\left(-\tfrac{1}{2}\mu^2, \mu t\sqrt{2}\right) \sim \frac{2\pi^{\frac{1}{2}}\mu^{\frac{1}{3}}g(\mu)\phi(\zeta)}{\Gamma\left(\tfrac{1}{2}+\tfrac{1}{2}\mu^2\right)}\left(\operatorname{Bi}\left(\mu^{\frac{4}{3}}\zeta\right)\sum_{s=0}^{\infty}\frac{A_s(\zeta)}{\mu^{4s}} + \frac{\operatorname{Bi}'\left(\mu^{\frac{4}{3}}\zeta\right)}{\mu^{\frac{8}{3}}}\sum_{s=0}^{\infty}\frac{B_s(\zeta)}{\mu^{4s}}\right),$$

12.10.38
$$V'\left(-\tfrac{1}{2}\mu^2, \mu t\sqrt{2}\right) \sim \frac{(2\pi)^{\frac{1}{2}}\mu^{\frac{2}{3}}g(\mu)}{\phi(\zeta)\,\Gamma\left(\tfrac{1}{2}+\tfrac{1}{2}\mu^2\right)}\left(\frac{\operatorname{Bi}\left(\mu^{\frac{4}{3}}\zeta\right)}{\mu^{\frac{4}{3}}}\sum_{s=0}^{\infty}\frac{C_s(\zeta)}{\mu^{4s}} + \operatorname{Bi}'\left(\mu^{\frac{4}{3}}\zeta\right)\sum_{s=0}^{\infty}\frac{D_s(\zeta)}{\mu^{4s}}\right).$$

The variable ζ is defined by

12.10.39
$$\tfrac{2}{3}\zeta^{\frac{3}{2}} = \xi, \quad 1 \le t, (\zeta \ge 0);$$
$$\tfrac{2}{3}(-\zeta)^{\frac{3}{2}} = \eta, \quad -1 < t \le 1, (\zeta \le 0),$$

where ξ, η are given by (12.10.7), (12.10.23), respectively, and

12.10.40
$$\phi(\zeta) = \left(\frac{\zeta}{t^2-1}\right)^{\frac{1}{4}}.$$

The function $\zeta = \zeta(t)$ is real for $t > -1$ and analytic at $t = 1$. Inversely, with $w = 2^{-\frac{1}{3}}\zeta$,

12.10.41
$$t = 1 + w - \tfrac{1}{10}w^2 + \tfrac{11}{350}w^3 - \tfrac{823}{63000}w^4 + \tfrac{1\,50653}{242\,55000}w^5 + \cdots, \quad |\zeta| < \left(\tfrac{3}{4}\pi\right)^{\frac{2}{3}}.$$

For $g(\mu)$ see (12.10.14). The coefficients $A_s(\zeta)$ and $B_s(\zeta)$ are given by

12.10.42
$$A_s(\zeta) = \zeta^{-3s}\sum_{m=0}^{2s}\beta_m(\phi(\zeta))^{6(2s-m)}u_{2s-m}(t),$$
$$\zeta^2 B_s(\zeta) = -\zeta^{-3s}\sum_{m=0}^{2s+1}\alpha_m(\phi(\zeta))^{6(2s-m+1)}u_{2s-m+1}(t),$$

where $\phi(\zeta)$ is as in (12.10.40), $u_k(t)$ is as in §12.10(ii), $\alpha_0 = 1$, and

12.10.43
$$\alpha_m = \frac{(2m+1)(2m+3)\cdots(6m-1)}{m!(144)^m},$$
$$\beta_m = -\frac{6m+1}{6m-1}\alpha_m.$$

The coefficients $C_s(\zeta)$ and $D_s(\zeta)$ in (12.10.36) and (12.10.38) are given by

12.10.44
$$C_s(\zeta) = \chi(\zeta)A_s(\zeta) + A'_s(\zeta) + \zeta B_s(\zeta),$$
$$D_s(\zeta) = A_s(\zeta) + \chi(\zeta)B_{s-1}(\zeta) + B'_{s-1}(\zeta),$$

where

12.10.45
$$\chi(\zeta) = \frac{\phi'(\zeta)}{\phi(\zeta)} = \frac{1-2t(\phi(\zeta))^6}{4\zeta}.$$

Explicitly,

12.10.46
$$\zeta C_s(\zeta) = -\zeta^{-3s}\sum_{m=0}^{2s+1}\beta_m(\phi(\zeta))^{6(2s-m+1)}v_{2s-m+1}(t),$$
$$D_s(\zeta) = \zeta^{-3s}\sum_{m=0}^{2s}\alpha_m(\phi(\zeta))^{6(2s-m)}v_{2s-m}(t),$$

where $v_k(t)$ is as in §12.10(ii).

Modified Expansions

The expansions (12.10.35)–(12.10.38) can be modified, again see Temme (2000), and the new expansions hold if either or both μ and t tend to infinity. This is provable by the methods used in §10.41(v).

12.10(viii) Negative a, $-\infty < x < 2\sqrt{-a}$. Expansions in Terms of Airy Functions

When $\mu \to \infty$, asymptotic expansions for $U\left(-\tfrac{1}{2}\mu^2, -\mu t\sqrt{2}\right)$ and $V\left(-\tfrac{1}{2}\mu^2, -\mu t\sqrt{2}\right)$ that are uniform for $t \in [-1+\delta, \infty)$ are obtained by substitution into (12.2.15) and (12.2.16) by means of (12.10.35) and (12.10.37). Similarly for $U'\left(-\tfrac{1}{2}\mu^2, -\mu t\sqrt{2}\right)$ and $V'\left(-\tfrac{1}{2}\mu^2, -\mu t\sqrt{2}\right)$.

12.11 Zeros

12.11(i) Distribution of Real Zeros

If $a \ge -\tfrac{1}{2}$, then $U(a,x)$ has no real zeros. If $-\tfrac{3}{2} < a < -\tfrac{1}{2}$, then $U(a,x)$ has no positive real zeros. If $-2n-\tfrac{3}{2} < a < -2n+\tfrac{1}{2}$, $n = 1, 2, \ldots$, then $U(a,x)$ has n positive real zeros. Lastly, when $a = -n-\tfrac{1}{2}$, $n = 1, 2, \ldots$ (Hermite polynomial case) $U(a,x)$ has n zeros and they lie in the interval $[-2\sqrt{-a}, 2\sqrt{-a}]$. For further information on these cases see Dean (1966).

If $a > -\tfrac{1}{2}$, then $V(a,x)$ has no positive real zeros, and if $a = \tfrac{3}{2} - 2n$, $n \in \mathbb{Z}$, then $V(a,x)$ has a zero at $x = 0$.

12.11(ii) Asymptotic Expansions of Large Zeros

When $a > -\tfrac{1}{2}$, $U(a,z)$ has a string of complex zeros that approaches the ray $\operatorname{ph} z = \tfrac{3}{4}\pi$ as $z \to \infty$, and a conjugate string. When $a > -\tfrac{1}{2}$ the zeros are asymptotically given by $z_{a,s}$ and $\bar{z}_{a,s}$, where s is a large positive integer and

12.11.1
$$z_{a,s} = e^{\tfrac{3}{4}\pi i}\sqrt{2\tau_s}\left(1 - \frac{ia\lambda_s}{2\tau_s} + \frac{2a^2\lambda_s^2 - 8a^2\lambda_s + 4a^2 + 3}{16\tau_s^2} + O\left(\lambda_s^3\tau_s^{-3}\right)\right),$$

with

12.11.2
$$\tau_s = \left(2s+\tfrac{1}{2}-a\right)\pi + i\ln\left(\pi^{-\tfrac{1}{2}}2^{-a-\tfrac{1}{2}}\Gamma\left(\tfrac{1}{2}+a\right)\right),$$

and

12.11.3 $\qquad \lambda_s = \ln \tau_s - \frac{1}{2}\pi i.$

When $a = \frac{1}{2}$ these zeros are the same as the zeros of the complementary error function $\operatorname{erfc}(z/\sqrt{2})$; compare (12.7.5). Numerical calculations in this case show that $z_{\frac{1}{2},s}$ corresponds to the sth zero on the string; compare §7.13(ii).

12.11(iii) Asymptotic Expansions for Large Parameter

For large negative values of a the real zeros of $U(a, x)$, $U'(a, x)$, $V(a, x)$, and $V'(a, x)$ can be approximated by reversion of the Airy-type asymptotic expansions of §§12.10(vii) and 12.10(viii). For example, let the sth real zeros of $U(a, x)$ and $U'(a, x)$, counted in descending order away from the point $z = 2\sqrt{-a}$, be denoted by $u_{a,s}$ and $u'_{a,s}$, respectively. Then

12.11.4 $\quad u_{a,s} \sim 2^{\frac{1}{2}}\mu \left(p_0(\alpha) + \frac{p_1(\alpha)}{\mu^4} + \frac{p_2(\alpha)}{\mu^8} + \cdots \right),$

as $\mu \, (= \sqrt{-2a}) \to \infty$, s fixed. Here $\alpha = \mu^{-\frac{4}{3}} a_s$, a_s denoting the sth negative zero of the function Ai (see §9.9(i)). The first two coefficients are given by

12.11.5 $\qquad p_0(\zeta) = t(\zeta),$

where $t(\zeta)$ is the function inverse to $\zeta(t)$, defined by (12.10.39) (see also (12.10.41)), and

12.11.6 $\quad p_1(\zeta) = \frac{t^3 - 6t}{24(t^2 - 1)^2} + \frac{5}{48((t^2 - 1)\zeta^3)^{\frac{1}{2}}}.$

Similarly, for the zeros of $U'(a, x)$ we have

12.11.7 $\quad u'_{a,s} \sim 2^{\frac{1}{2}}\mu \left(q_0(\beta) + \frac{q_1(\beta)}{\mu^4} + \frac{q_2(\beta)}{\mu^8} + \cdots \right),$

where $\beta = \mu^{-\frac{4}{3}} a'_s$, a'_s denoting the sth negative zero of the function Ai$'$ and

12.11.8 $\qquad q_0(\zeta) = t(\zeta).$

For the first zero of $U(a, x)$ we also have

12.11.9
$u_{a,1} \sim 2^{\frac{1}{2}}\mu \Big(1 - 1.85575\,708\mu^{-4/3} - 0.34438\,34\mu^{-8/3}$
$\qquad - 0.16871\,5\mu^{-4} - 0.11414\mu^{-16/3} - 0.0808\mu^{-20/3}$
$\qquad - \cdots \Big),$

where the numerical coefficients have been rounded off.

For further information, including associated functions, see Olver (1959).

12.12 Integrals

12.12.1
$$\int_0^\infty e^{-\frac{1}{4}t^2} t^{\mu-1} U(a,t) \, dt = \frac{\sqrt{\pi}\, 2^{-\frac{1}{2}(\mu+a+\frac{1}{2})} \Gamma(\mu)}{\Gamma\!\left(\frac{1}{2}(\mu+a+\frac{3}{2})\right)},$$
$$\Re\mu > 0,$$

12.12.2
$$\int_0^\infty e^{-\frac{3}{4}t^2} t^{-a-\frac{3}{2}} U(a,t) \, dt$$
$$= 2^{\frac{1}{4}+\frac{1}{2}a} \Gamma\!\left(-a-\tfrac{1}{2}\right) \cos\!\left((\tfrac{1}{4}a+\tfrac{1}{8})\pi\right), \quad \Re a < -\tfrac{1}{2},$$

12.12.3
$$\int_0^\infty e^{-\frac{1}{4}t^2} t^{-a-\frac{1}{2}} (x^2 + t^2)^{-1} U(a,t) \, dt$$
$$= \sqrt{\pi/2}\, \Gamma\!\left(\tfrac{1}{2} - a\right) x^{-a-\frac{3}{2}} e^{\frac{1}{4}x^2} U(-a, x),$$
$$\Re a < \tfrac{1}{2}, x > 0.$$

Nicholson-type Integral

12.12.4
$(U(a,z))^2 + (\overline{U}(a,z))^2$
$= \frac{2^{\frac{3}{2}}}{\pi} \Gamma\!\left(\tfrac{1}{2} - a\right) \int_0^\infty \frac{e^{2at + \frac{1}{2} z^2 \tanh t}}{\sqrt{\sinh(2t)}}\, dt, \quad \Re a < \tfrac{1}{2}.$

When $z \, (= x)$ is real the left-hand side equals $(F(a, x))^2$; compare (12.2.22).

For further integrals see §§13.10, 13.23, and use (12.7.14).

For compendia of integrals see Erdélyi et al. (1953b, v. 2, pp. 121–122), Erdélyi et al. (1954a,b, v. 1, pp. 60–61, 115, 210–211, and 336; v. 2, pp. 76–80, 115, 151, 171, and 395–398), Gradshteyn and Ryzhik (2000, §7.7), Magnus et al. (1966, pp. 330–331), Marichev (1983, pp. 190–191), Oberhettinger (1974, pp. 144–145), Oberhettinger (1990, pp. 106–108 and 192), Oberhettinger and Badii (1973, pp. 181–185), Prudnikov et al. (1986b, pp. 36–37, 155–168, 243–246, 289–290, 327–328, 419–420, and 619), Prudnikov et al. (1992a, §3.11), and Prudnikov et al. (1992b, §3.11).

See also Barr (1968) and Lowdon (1970).

12.13 Sums

12.13(i) Addition Theorems

12.13.1
$$U(a, x+y) = e^{\frac{1}{2}xy + \frac{1}{4}y^2} \sum_{m=0}^\infty \frac{(-y)^m}{m!} U(a-m, x),$$

12.13.2
$U(a, x+y)$
$= e^{-\frac{1}{2}xy - \frac{1}{4}y^2} \sum_{m=0}^\infty \binom{-a-\frac{1}{2}}{m} y^m U(a+m, x),$

12.13.3
$$V(a, x+y) = e^{\frac{1}{2}xy + \frac{1}{4}y^2} \sum_{m=0}^\infty \binom{a-\frac{1}{2}}{m} y^m V(a-m, x),$$

12.13.4 $\quad V(a, x+y) = e^{-\frac{1}{2}xy - \frac{1}{4}y^2} \sum_{m=0}^\infty \frac{y^m}{m!} V(a+m, x).$

12.13.5
$$U(a, x\cos t + y\sin t)$$
$$= e^{\frac{1}{4}(x\sin t - y\cos t)^2}$$
$$\times \sum_{m=0}^{\infty} \binom{-a-\frac{1}{2}}{m} (\tan t)^m U(m+a,x) U(-m-\tfrac{1}{2},y),$$
$$\Re a \leq -\tfrac{1}{2}, 0 \leq t \leq \tfrac{1}{4}\pi.$$

12.13.6
$$n!\,U(n+\tfrac{1}{2},z) = i^n e^{-\frac{1}{2}z^2}\operatorname{erfc}(z/\sqrt{2})\,U(-n-\tfrac{1}{2},iz)$$
$$+ \sum_{m=1}^{\lfloor\frac{1}{2}n+\frac{1}{2}\rfloor} U(2m-n-\tfrac{1}{2},z),$$
$$n = 0,1,2,\ldots.$$

For erfc see §7.2(i).

12.13(ii) Other Series

For other series see Dhar (1940), Hansen (1975, pp. 421–422), Hillion (1997), Miller (1974), Prudnikov et al. (1986b, p. 651), Shanker (1940b,a,c), and Varma (1941).

12.14 The Function $W(a,x)$

12.14(i) Introduction

In this section solutions of equation (12.2.3) are considered. This equation is important when a and z $(=x)$ are real, and we shall assume this to be the case. In other cases the general theory of (12.2.2) is available. $W(a,x)$ and $W(a,-x)$ form a numerically satisfactory pair of solutions when $-\infty < x < \infty$.

12.14(ii) Values at $z=0$ and Wronskian

12.14.1 $\qquad W(a,0) = 2^{-\frac{3}{4}} \left|\dfrac{\Gamma(\frac{1}{4}+\frac{1}{2}ia)}{\Gamma(\frac{3}{4}+\frac{1}{2}ia)}\right|^{\frac{1}{2}},$

12.14.2 $\qquad W'(a,0) = -2^{-\frac{1}{4}} \left|\dfrac{\Gamma(\frac{3}{4}+\frac{1}{2}ia)}{\Gamma(\frac{1}{4}+\frac{1}{2}ia)}\right|^{\frac{1}{2}}.$

12.14.3 $\qquad \mathscr{W}\{W(a,x), W(a,-x)\} = 1.$

12.14(iii) Graphs

For the modulus functions $\widetilde{F}(a,x)$ and $\widetilde{G}(a,x)$ see §12.14(x).

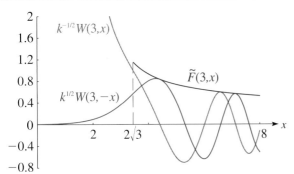

Figure 12.14.1: $k^{-1/2}\,W(3,x)$, $k^{1/2}\,W(3,-x)$, $\widetilde{F}(3,x)$, $0 \leq x \leq 8$.

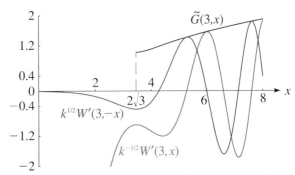

Figure 12.14.2: $k^{-1/2}\,W'(3,x)$, $k^{1/2}\,W'(3,-x)$, $\widetilde{G}(3,x)$, $0 \leq x \leq 8$.

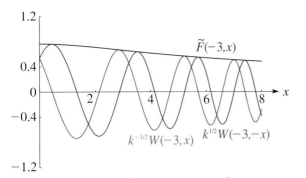

Figure 12.14.3: $k^{-1/2}\,W(-3,x)$, $k^{1/2}\,W(-3,-x)$, $\widetilde{F}(-3,x)$, $0 \leq x \leq 8$.

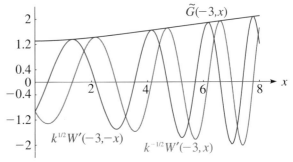

Figure 12.14.4: $k^{-1/2}\,W'(-3,x)$, $k^{1/2}\,W'(-3,-x)$, $\widetilde{G}(-3,x)$, $0 \leq x \leq 8$.

12.14(iv) Connection Formula

$$W(a,x) = \sqrt{k/2}\, e^{\frac{1}{4}\pi a} \left(e^{i\rho} U\left(ia, xe^{-\pi i/4}\right) \right.$$
12.14.4
$$\left. + e^{-i\rho} U\left(-ia, xe^{\pi i/4}\right) \right),$$

where

12.14.5 $\quad k = \sqrt{1+e^{2\pi a}} - e^{\pi a}, \quad 1/k = \sqrt{1+e^{2\pi a}} + e^{\pi a},$

12.14.6 $\quad\quad\quad\quad \rho = \tfrac{1}{8}\pi + \tfrac{1}{2}\phi_2,$

12.14.7 $\quad\quad\quad\quad \phi_2 = \mathrm{ph}\,\Gamma\!\left(\tfrac{1}{2} + ia\right),$

the branch of ph being zero when $a=0$ and defined by continuity elsewhere.

12.14(v) Power-Series Expansions

12.14.8 $\quad W(a,x) = W(a,0) w_1(a,x) + W'(a,0) w_2(a,x).$

Here $w_1(a,x)$ and $w_2(a,x)$ are the even and odd solutions of (12.2.3):

12.14.9 $\quad\quad w_1(a,x) = \sum_{n=0}^{\infty} \alpha_n(a) \frac{x^{2n}}{(2n)!},$

12.14.10 $\quad\quad w_2(a,x) = \sum_{n=0}^{\infty} \beta_n(a) \frac{x^{2n+1}}{(2n+1)!},$

where $\alpha_n(a)$ and $\beta_n(a)$ satisfy the recursion relations

12.14.11 $\quad \begin{aligned} \alpha_{n+2} &= a\alpha_{n+1} - \tfrac{1}{2}(n+1)(2n+1)\alpha_n, \\ \beta_{n+2} &= a\beta_{n+1} - \tfrac{1}{2}(n+1)(2n+3)\beta_n, \end{aligned}$

with

12.14.12 $\quad \alpha_0(a) = 1, \quad \alpha_1(a) = a, \quad \beta_0(a) = 1, \quad \beta_1(a) = a.$

Other expansions, involving $\cos\!\left(\tfrac{1}{4}x^2\right)$ and $\sin\!\left(\tfrac{1}{4}x^2\right)$, can be obtained from (12.4.3) to (12.4.6) by replacing a by $-ia$ and z by $xe^{\pi i/4}$; see Miller (1955, p. 80), and also (12.14.15) and (12.14.16).

12.14(vi) Integral Representations

These follow from the contour integrals of §12.5(ii), which are valid for general complex values of the argument z and parameter a. See Miller (1955, p. 26).

12.14(vii) Relations to Other Functions

Bessel Functions

For the notation see §10.2(ii). When $x > 0$

12.14.13 $\quad W(0, \pm x) = 2^{-\frac{5}{4}} \sqrt{\pi x} \left(J_{-\frac{1}{4}}\!\left(\tfrac{1}{4}x^2\right) \mp J_{\frac{1}{4}}\!\left(\tfrac{1}{4}x^2\right) \right),$

12.14.14
$$\frac{d}{dx} W(0, \pm x) = -2^{-\frac{9}{4}} x \sqrt{\pi x} \left(J_{\frac{3}{4}}\!\left(\tfrac{1}{4}x^2\right) \pm J_{-\frac{3}{4}}\!\left(\tfrac{1}{4}x^2\right) \right).$$

Confluent Hypergeometric Functions

For the notation see §13.2(i).

The even and odd solutions of (12.2.3) (see §12.14(v)) are given by

12.14.15 $\quad \begin{aligned} w_1(a,x) &= e^{-\frac{1}{4}ix^2} M\!\left(\tfrac{1}{4} - \tfrac{1}{2}ia, \tfrac{1}{2}, \tfrac{1}{2}ix^2\right) \\ &= e^{\frac{1}{4}ix^2} M\!\left(\tfrac{1}{4} + \tfrac{1}{2}ia, \tfrac{1}{2}, -\tfrac{1}{2}ix^2\right), \end{aligned}$

12.14.16 $\quad \begin{aligned} w_2(a,x) &= x e^{-\frac{1}{4}ix^2} M\!\left(\tfrac{3}{4} - \tfrac{1}{2}ia, \tfrac{3}{2}, \tfrac{1}{2}ix^2\right) \\ &= x e^{\frac{1}{4}ix^2} M\!\left(\tfrac{3}{4} + \tfrac{1}{2}ia, \tfrac{3}{2}, -\tfrac{1}{2}ix^2\right). \end{aligned}$

12.14(viii) Asymptotic Expansions for Large Variable

Write

12.14.17 $\quad W(a,x) = \sqrt{\frac{2k}{x}} \left(s_1(a,x) \cos\omega - s_2(a,x) \sin\omega \right),$

12.14.18
$$W(a,-x) = \sqrt{\frac{2}{kx}} \left(s_1(a,x) \sin\omega + s_2(a,x) \cos\omega \right),$$

where

12.14.19 $\quad\quad \omega = \tfrac{1}{4}x^2 - a\ln x + \tfrac{1}{4}\pi + \tfrac{1}{2}\phi_2,$

with ϕ_2 given by (12.14.7). Then as $x \to \infty$

12.14.20
$$s_1(a,x) \sim 1 + \frac{d_2}{1!\,2x^2} - \frac{c_4}{2!\,2^2 x^4} - \frac{d_6}{3!\,2^3 x^6} + \frac{c_8}{4!\,2^4 x^8} + \cdots,$$

12.14.21
$$s_2(a,x) \sim -\frac{c_2}{1!\,2x^2} - \frac{d_4}{2!\,2^2 x^4} + \frac{c_6}{3!\,2^3 x^6} + \frac{d_8}{4!\,2^4 x^8} - \cdots.$$

The coefficients c_{2r} and d_{2r} are obtainable by equating real and imaginary parts in

12.14.22 $\quad\quad c_{2r} + i d_{2r} = \frac{\Gamma\!\left(2r + \tfrac{1}{2} + ia\right)}{\Gamma\!\left(\tfrac{1}{2} + ia\right)}.$

Equivalently,

12.14.23 $\quad s_1(a,x) + i s_2(a,x) \sim \sum_{r=0}^{\infty} (-i)^r \frac{\left(\tfrac{1}{2} + ia\right)_{2r}}{2^r r!\, x^{2r}}.$

12.14(ix) Uniform Asymptotic Expansions for Large Parameter

The differential equation

12.14.24 $\quad\quad \dfrac{d^2 w}{dt^2} = \mu^4 (1 - t^2) w$

follows from (12.2.3), and has solutions $W\!\left(\tfrac{1}{2}\mu^2, \pm\mu t\sqrt{2}\right)$. For real μ and t oscillations occur outside the t-interval $[-1, 1]$. Airy-type uniform asymptotic expansions can be used to include either one of the turning points ± 1. In the following expansions, obtained from Olver (1959), μ is large and positive, and δ is again an arbitrary small positive constant.

Positive a, $2\sqrt{a} < x < \infty$

12.14.25
$$W\left(\tfrac{1}{2}\mu^2, \mu t\sqrt{2}\right) \sim \frac{2^{-\frac{1}{2}} e^{-\frac{1}{4}\pi\mu^2} l(\mu)}{(t^2-1)^{\frac{1}{4}}} \left(\cos\sigma \sum_{s=0}^{\infty} (-1)^s \frac{\mathcal{A}_{2s}(t)}{\mu^{4s}} - \sin\sigma \sum_{s=0}^{\infty} (-1)^s \frac{\mathcal{A}_{2s+1}(t)}{\mu^{4s+2}}\right),$$

12.14.26
$$W\left(\tfrac{1}{2}\mu^2, -\mu t\sqrt{2}\right) \sim \frac{2^{\frac{1}{2}} e^{\frac{1}{4}\pi\mu^2} l(\mu)}{(t^2-1)^{\frac{1}{4}}} \left(\sin\sigma \sum_{s=0}^{\infty} (-1)^s \frac{\mathcal{A}_{2s}(t)}{\mu^{4s}} + \cos\sigma \sum_{s=0}^{\infty} (-1)^s \frac{\mathcal{A}_{2s+1}(t)}{\mu^{4s+2}}\right),$$

uniformly for $t \in [1+\delta, \infty)$. Here $\mathcal{A}_s(t)$ is as in §12.10(ii), σ is defined by

12.14.27 $\qquad \sigma = \mu^2 \xi + \tfrac{1}{4}\pi,$

with ξ given by (12.10.7), and

12.14.28 $\quad l(\mu) = \sqrt{2} e^{\frac{1}{8}\pi\mu^2} e^{i(\frac{1}{2}\phi_2 - \frac{1}{8}\pi)} g(\mu e^{-\frac{1}{4}\pi i}),$

with $g(\mu)$ as in §12.10(ii). The function $l(\mu)$ has the asymptotic expansion

12.14.29 $\qquad l(\mu) \sim \frac{2^{\frac{1}{4}}}{\mu^{\frac{1}{2}}} \sum_{s=0}^{\infty} \frac{l_s}{\mu^{4s}},$

with

12.14.30 $\quad l_0 = 1, \quad l_1 = -\tfrac{1}{1152}, \quad l_2 = -\tfrac{16123}{398\,13120}.$

Positive a, $-2\sqrt{a} < x < 2\sqrt{a}$

12.14.31
$$W\left(\tfrac{1}{2}\mu^2, \mu t\sqrt{2}\right) \sim \frac{l(\mu) e^{\mu^2 \eta}}{2^{\frac{1}{2}} e^{\frac{1}{4}\pi\mu^2} (1-t^2)^{\frac{1}{4}}} \sum_{s=0}^{\infty} (-1)^s \frac{\widetilde{\mathcal{A}}_s(t)}{\mu^{2s}},$$

uniformly for $t \in [-1+\delta, 1-\delta]$, with η given by (12.10.23) and $\widetilde{\mathcal{A}}_s(t)$ given by (12.10.24).

The expansions for the derivatives corresponding to (12.14.25), (12.14.26), and (12.14.31) may be obtained by formal term-by-term differentiation with respect to t; compare the analogous results in §§12.10(ii)–12.10(v).

Airy-type Uniform Expansions

12.14.32 $\quad W\left(\tfrac{1}{2}\mu^2, \mu t\sqrt{2}\right) \sim \frac{\pi^{\frac{1}{2}} \mu^{\frac{1}{3}} l(\mu)}{2^{\frac{1}{2}} e^{\frac{1}{4}\pi\mu^2}} \phi(\zeta) \left(\operatorname{Bi}\left(-\mu^{\frac{4}{3}}\zeta\right) \sum_{s=0}^{\infty} (-1)^s \frac{A_s(\zeta)}{\mu^{4s}} + \frac{\operatorname{Bi}'\left(-\mu^{\frac{4}{3}}\zeta\right)}{\mu^{\frac{8}{3}}} \sum_{s=0}^{\infty} (-1)^s \frac{B_s(\zeta)}{\mu^{4s}}\right),$

12.14.33 $\quad W\left(\tfrac{1}{2}\mu^2, -\mu t\sqrt{2}\right) \sim \frac{\pi^{\frac{1}{2}} \mu^{\frac{1}{3}} l(\mu)}{2^{-\frac{1}{2}} e^{-\frac{1}{4}\pi\mu^2}} \phi(\zeta) \left(\operatorname{Ai}\left(-\mu^{\frac{4}{3}}\zeta\right) \sum_{s=0}^{\infty} (-1)^s \frac{A_s(\zeta)}{\mu^{4s}} + \frac{\operatorname{Ai}'\left(-\mu^{\frac{4}{3}}\zeta\right)}{\mu^{\frac{8}{3}}} \sum_{s=0}^{\infty} (-1)^s \frac{B_s(\zeta)}{\mu^{4s}}\right),$

uniformly for $t \in [-1+\delta, \infty)$, with ζ, $\phi(\zeta)$, $A_s(\zeta)$, and $B_s(\zeta)$ as in §12.10(vii). For the corresponding expansions for the derivatives see Olver (1959).

Negative a, $-\infty < x < \infty$

In this case there are no real turning points, and the solutions of (12.2.3), with z replaced by x, oscillate on the entire real x-axis.

12.14.34 $\quad W\left(-\tfrac{1}{2}\mu^2, \mu t\sqrt{2}\right) \sim \frac{l(\mu)}{(t^2+1)^{\frac{1}{4}}} \left(\cos\overline{\sigma} \sum_{s=0}^{\infty} \frac{(-1)^s \overline{u}_{2s}(t)}{(t^2+1)^{3s} \mu^{4s}} - \sin\overline{\sigma} \sum_{s=0}^{\infty} \frac{(-1)^s \overline{u}_{2s+1}(t)}{(t^2+1)^{3s+\frac{3}{2}} \mu^{4s+2}}\right),$

12.14.35 $\quad W'\left(-\tfrac{1}{2}\mu^2, \mu t\sqrt{2}\right) \sim -\frac{\mu}{\sqrt{2}} l(\mu)(t^2+1)^{\frac{1}{4}} \left(\sin\overline{\sigma} \sum_{s=0}^{\infty} \frac{(-1)^s \overline{v}_{2s}(t)}{(t^2+1)^{3s} \mu^{4s}} + \cos\overline{\sigma} \sum_{s=0}^{\infty} \frac{(-1)^s \overline{v}_{2s+1}(t)}{(t^2+1)^{3s+\frac{3}{2}} \mu^{4s+2}}\right),$

uniformly for $t \in \mathbb{R}$, where

12.14.36 $\qquad \overline{\sigma} = \mu^2 \overline{\xi} + \tfrac{1}{4}\pi,$

and $\overline{\xi}$ and the coefficients $\overline{u}_s(t)$ and $\overline{v}_s(t)$ as in §12.10(v).

12.14(x) Modulus and Phase Functions

As noted in §12.14(ix), when a is negative the solutions of (12.2.3), with z replaced by x, are oscillatory on the whole real line; also, when a is positive there is a central interval $-2\sqrt{a} < x < 2\sqrt{a}$ in which the solutions are exponential in character. In the oscillatory intervals we write

12.14.37
$$k^{-1/2} W(a,x) + i k^{1/2} W(a,-x) = \widetilde{F}(a,x) e^{i\widetilde{\theta}(a,x)},$$

12.14.38
$$k^{-1/2} W'(a,x) + ik^{1/2} W'(a,-x) = -\widetilde{G}(a,x)e^{i\widetilde{\psi}(a,x)},$$

where k is defined in (12.14.5), and $\widetilde{F}(a,x)$ (>0), $\widetilde{\theta}(a,x)$, $\widetilde{G}(a,x)$ (>0), and $\widetilde{\psi}(a,x)$ are real. $\widetilde{F}$ or $\widetilde{G}$ is the *modulus* and $\widetilde{\theta}$ or $\widetilde{\psi}$ is the corresponding *phase*. Compare §12.2(vi).

For properties of the modulus and phase functions, including differential equations and asymptotic expansions for large x, see Miller (1955, pp. 87–88). For graphs of the modulus functions see §12.14(iii).

12.14(xi) Zeros of $W(a,x)$, $W'(a,x)$

For asymptotic expansions of the zeros of $W(a,x)$ and $W'(a,x)$, see Olver (1959).

12.15 Generalized Parabolic Cylinder Functions

The equation

12.15.1
$$\frac{d^2w}{dz^2} + \left(\nu + \lambda^{-1} - \lambda^{-2}z^\lambda\right)w = 0$$

can be viewed as a generalization of (12.2.4). This equation arises in the study of non-self-adjoint elliptic boundary-value problems involving an indefinite weight function. See Faierman (1992) for power series and asymptotic expansions of a solution of (12.15.1).

Applications

12.16 Mathematical Applications

PCFs are used as basic approximating functions in the theory of contour integrals with a coalescing saddle point and an algebraic singularity, and in the theory of differential equations with two coalescing turning points; see §§2.4(vi) and 2.8(vi). For examples see §§13.20(iii), 13.20(iv), 14.15(v), and 14.26.

Sleeman (1968b) considers certain orthogonality properties of the PCFs and corresponding eigenvalues. In Brazel et al. (1992) exponential asymptotics are considered in connection with an eigenvalue problem involving PCFs.

PCFs are also used in integral transforms with respect to the parameter, and inversion formulas exist for kernels containing PCFs. See Erdélyi (1941a), Cherry (1948), and Lowdon (1970). Integral transforms and sampling expansions are considered in Jerri (1982).

12.17 Physical Applications

The main applications of PCFs in mathematical physics arise when solving the Helmholtz equation

12.17.1
$$\nabla^2 w + k^2 w = 0,$$

where k is a constant, and ∇^2 is the Laplacian

12.17.2
$$\nabla^2 = \frac{\partial^2}{\partial x^2} + \frac{\partial^2}{\partial y^2} + \frac{\partial^2}{\partial z^2}$$

in Cartesian coordinates x, y, z of three-dimensional space (§1.5(ii)). By using instead coordinates of the parabolic cylinder ξ, η, ζ, defined by

12.17.3
$$x = \xi\eta, \quad y = \tfrac{1}{2}\xi^2 - \tfrac{1}{2}\eta^2, \quad z = \zeta,$$

(12.17.1) becomes

12.17.4
$$\frac{1}{\xi^2 + \eta^2}\left(\frac{\partial^2 w}{\partial \xi^2} + \frac{\partial^2 w}{\partial \eta^2}\right) + \frac{\partial^2 w}{\partial \zeta^2} + k^2 w = 0.$$

Setting $w = U(\xi)V(\eta)W(\zeta)$ and separating variables, we obtain

12.17.5
$$\frac{d^2 U}{d\xi^2} + \left(\sigma\xi^2 + \lambda\right)U = 0,$$
$$\frac{d^2 V}{d\eta^2} + \left(\sigma\eta^2 - \lambda\right)V = 0,$$
$$\frac{d^2 W}{d\zeta^2} + \left(k^2 - \sigma\right)W = 0,$$

with arbitrary constants σ, λ. The first two equations can be transformed into (12.2.2) or (12.2.3).

In a similar manner coordinates of the paraboloid of revolution transform the Helmholtz equation into equations related to the differential equations considered in this chapter. See Buchholz (1969, §4) and Morse and Feshbach (1953a, pp. 515 and 553).

Buchholz (1969) collects many results on boundary-value problems involving PCFs. Miller (1974) treats separation of variables by group theoretic methods. Dean (1966) describes the role of PCFs in quantum mechanical systems closely related to the one-dimensional harmonic oscillator.

Problems on high-frequency scattering in homogeneous media by parabolic cylinders lead to asymptotic methods for integrals involving PCFs. For this topic and other boundary-value problems see Boyd (1973), Hillion (1997), Magnus (1941), Morse and Feshbach (1953a,b), Müller (1988), Ott (1985), Rice (1954), and Shanmugam (1978).

Lastly, parabolic cylinder functions arise in the description of ultra cold atoms in harmonic trapping potentials; see Busch et al. (1998) and Edwards et al. (1999).

Computation

12.18 Methods of Computation

Because PCFs are special cases of confluent hypergeometric functions, the methods of computation described in §13.29 are applicable to PCFs. These include the use of power-series expansions, recursion, integral representations, differential equations, asymptotic expansions, and expansions in series of Bessel functions. See, especially, Temme (2000) and Gil et al. (2004b, 2006b,c).

12.19 Tables

- Abramowitz and Stegun (1964, Chapter 19) includes $U(a,x)$ and $V(a,x)$ for $\pm a = 0(.1)1(.5)5$, $x = 0(.1)5$, 5S; $W(a, \pm x)$ for $\pm a = 0(.1)1(1)5$, $x = 0(.1)5$, 4-5D or 4-5S.

- Miller (1955) includes $W(a,x)$, $W(a,-x)$, and reduced derivatives for $a = -10(1)10$, $x = 0(.1)10$, 8D or 8S. Modulus and phase functions, and also other auxiliary functions are tabulated.

- Fox (1960) includes modulus and phase functions for $W(a,x)$ and $W(a,-x)$, and several auxiliary functions for $x^{-1} = 0(.005)0.1$, $a = -10(1)10$, 8S.

- Kireyeva and Karpov (1961) includes $D_p(x(1+i))$ for $\pm x = 0(.1)5$, $p = 0(.1)2$, and $\pm x = 5(.01)10$, $p = 0(.5)2$, 7D.

- Karpov and Čistova (1964) includes $D_p(x)$ for $p = -2(.1)0$, $\pm x = 0(.01)5$; $p = -2(.05)0$, $\pm x = 5(.01)10$, 6D.

- Karpov and Čistova (1968) includes $e^{-\frac{1}{4}x^2} D_p(-x)$ and $e^{-\frac{1}{4}x^2} D_p(ix)$ for $x = 0(.01)5$ and $x^{-1} = 0(.001 \text{ or } .0001)5$, $p = -1(.1)1$, 7D or 8S.

- Murzewski and Sowa (1972) includes $D_{-n}(x)$ $\left(= U\left(n - \frac{1}{2}, x\right)\right)$ for $n = 1(1)20$, $x = 0(.05)3$, 7S.

- Zhang and Jin (1996, pp. 455–473) includes $U\left(\pm n - \frac{1}{2}, x\right)$, $V\left(\pm n - \frac{1}{2}, x\right)$, $U\left(\pm \nu - \frac{1}{2}, x\right)$, $V\left(\pm \nu - \frac{1}{2}, x\right)$, and derivatives, $\nu = n + \frac{1}{2}$, $n = 0(1)10(10)30$, $x = 0.5, 1, 5, 10, 30, 50$, 8S; $W(a, \pm x)$, $W(-a, \pm x)$, and derivatives, $a = h(1)5 + h$, $x = 0.5, 1$ and $a = h(1)5 + h$, $x = 5$, $h = 0, 0.5$, 8S. Also, first zeros of $U(a,x)$, $V(a,x)$, and of derivatives, $a = -6(.5)-1$, 6D; first three zeros of $W(a,-x)$ and of derivative, $a = 0(.5)4$, 6D; first three zeros of $W(-a, \pm x)$ and of derivative, $a = 0.5(.5)5.5$, 6D; real and imaginary parts of $U(a,z)$, $a = -1.5(1)1.5$, $z = x + iy$, $x = 0.5, 1, 5, 10$, $y = 0(.5)10$, 8S.

For other tables prior to 1961 see Fletcher et al. (1962) and Lebedev and Fedorova (1960).

12.20 Approximations

Luke (1969b, pp. 25 and 35) gives Chebyshev-series expansions for the confluent hypergeometric functions $U(a,b,x)$ and $M(a,b,x)$ (§13.2(i)) whose regions of validity include intervals with endpoints $x = \infty$ and $x = 0$, respectively. As special cases of these results a Chebyshev-series expansion for $U(a,x)$ valid when $\lambda \leq x < \infty$ follows from (12.7.14), and Chebyshev-series expansions for $U(a,x)$ and $V(a,x)$ valid when $0 \leq x \leq \lambda$ follow from (12.4.1), (12.4.2), (12.7.12), and (12.7.13). Here λ denotes an arbitrary positive constant.

12.21 Software

See http://dlmf.nist.gov/12.21.

References

General References

The main references used in writing this chapter are Erdélyi et al. (1953b, v. 2), Miller (1955), and Olver (1959). For additional bibliographic reading see Buchholz (1969), Lebedev (1965), Magnus et al. (1966), Olver (1997b), and Temme (1996a).

Sources

The following list gives the references or other indications of proofs that were used in constructing the various sections of this chapter. These sources supplement the references that are quoted in the text.

§12.2 See Miller (1955, pp. 9–10, 17, 63–64, and 72), Miller (1952), Miller (1950), and Olver (1997b, Chapter 5, §3.3).

§12.3 These graphics were produced at NIST.

§12.4 See Miller (1955, pp. 61–63).

§12.5 See Whittaker (1902) and Miller (1955, pp. 19 and 25–26). For (12.5.4) combine (12.2.18) and (12.5.1). In Miller (1955, p. 26) the conditions on a given in Eqs. (12.5.8) and (12.5.9) are missing. These conditions are needed to ensure that in each integrand no poles of the two gamma functions coincide.

§12.7 See Miller (1955, pp. 40–43, 73–74, 76, and 77–79). For (12.7.7) combine (7.18.11) and (7.18.12).

§12.8 See Miller (1955, p. 65). (12.8.9), (12.8.10), (12.8.11), and (12.8.12) can be obtained from (12.5.1), (12.5.6), (12.5.7), and (12.5.9), respectively.

§12.9 (12.9.1) is obtained from (12.7.14) and (13.7.3). (12.9.2)–(12.9.4) follow from (12.2.18) and (12.2.20). See also Whittaker (1902) and Whittaker and Watson (1927, pp. 348–349).

§12.10 See Olver (1959). Equations (12.10.42)–(12.10.46) are rearrangements of Olver's results and have the advantage of avoiding the many-valued functions in the explicit expressions for $A_s(\zeta)$, $B_s(\zeta)$, $C_s(\zeta)$, and $D_s(\zeta)$.

§12.11 See Whittaker and Watson (1927, p. 354), Dean (1966), Riekstyņš (1991, p. 195), and Olver (1959). (12.11.9) is obtained by truncating (12.11.4) at its second term, and applying (12.10.41) with terms up to and including w^5.

§12.12 See Erdélyi et al. (1953b, Chapter 8). For (12.12.4) see Durand (1975).

§12.13 (12.13.1)–(12.13.4) follow from the results in §12.8(ii) and Taylor's theorem (§1.10(i)). For (12.13.5) see Shanker (1939) or Erdélyi et al. (1953b, Chapter 8). For (12.13.6) see Lepe (1985).

§12.14 See Miller (1955, pp. 17–18, 26, 43, 80–82, 87, 89), Miller (1952), and Olver (1959). The graphs were produced at NIST.

§12.17 See Jeffreys and Jeffreys (1956, §§18.04 and 23.08) and Morse and Feshbach (1953a,b, pp. 553, 1403–1405).

Chapter 13
Confluent Hypergeometric Functions

A. B. Olde Daalhuis[1]

Notation — **322**
13.1 Special Notation 322

Kummer Functions — **322**
13.2 Definitions and Basic Properties 322
13.3 Recurrence Relations and Derivatives . . 325
13.4 Integral Representations 326
13.5 Continued Fractions 327
13.6 Relations to Other Functions 327
13.7 Asymptotic Expansions for Large Argument 328
13.8 Asymptotic Approximations for Large Parameters 330
13.9 Zeros 331
13.10 Integrals 332
13.11 Series 333
13.12 Products 333
13.13 Addition and Multiplication Theorems . . 333

Whittaker Functions — **334**
13.14 Definitions and Basic Properties 334
13.15 Recurrence Relations and Derivatives . . 336
13.16 Integral Representations 337
13.17 Continued Fractions 338
13.18 Relations to Other Functions 338
13.19 Asymptotic Expansions for Large Argument 339
13.20 Uniform Asymptotic Approximations for Large μ 339
13.21 Uniform Asymptotic Approximations for Large κ 341
13.22 Zeros 342
13.23 Integrals 343
13.24 Series 344
13.25 Products 345
13.26 Addition and Multiplication Theorems . . 345

Applications — **345**
13.27 Mathematical Applications 345
13.28 Physical Applications 346

Computation — **346**
13.29 Methods of Computation 346
13.30 Tables 347
13.31 Approximations 347
13.32 Software 347

References — **347**

---

[1]School of Mathematics, Edinburgh University, Edinburgh, United Kingdom.
Acknowledgments: This chapter is based in part on Abramowitz and Stegun (1964, Chapter 13) by L.J. Slater. The author is indebted to J. Wimp for several references.
Copyright © 2009 National Institute of Standards and Technology. All rights reserved.

Notation

13.1 Special Notation

(For other notation see pp. xiv and 873.)

m	integer.
n, s	nonnegative integers.
x, y	real variables.
z	complex variable.
δ	arbitrary small positive constant.
γ	Euler's constant (§5.2(ii)).
$\Gamma(x)$	Gamma function (§5.2(i)).
$\psi(x)$	$\Gamma'(x)/\Gamma(x)$.

The main functions treated in this chapter are the Kummer functions $M(a,b,z)$ and $U(a,b,z)$, Olver's function $\mathbf{M}(a,b,z)$, and the Whittaker functions $M_{\kappa,\mu}(z)$ and $W_{\kappa,\mu}(z)$.

Other notations are: ${}_1F_1(a;b;z)$ (§16.2(i)) and $\Phi(a;b;z)$ (Humbert (1920)) for $M(a,b,z)$; $\Psi(a;b;z)$ (Erdélyi et al. (1953a, §6.5)) for $U(a,b,z)$; $V(b-a,b,z)$ (Olver (1997b, p. 256)) for $e^z U(a,b,-z)$; $\Gamma(1+2\mu)\mathscr{M}_{\kappa,\mu}$ (Buchholz (1969, p. 12)) for $M_{\kappa,\mu}(z)$.

For an historical account of notations see Slater (1960, Chapter 1).

Kummer Functions

13.2 Definitions and Basic Properties

13.2(i) Differential Equation

Kummer's Equation

13.2.1 $$z\frac{d^2w}{dz^2} + (b-z)\frac{dw}{dz} - aw = 0.$$

This equation has a regular singularity at the origin with indices 0 and $1-b$, and an irregular singularity at infinity of rank one. It can be regarded as the limiting form of the hypergeometric differential equation (§15.10(i)) that is obtained on replacing z by z/b, letting $b \to \infty$, and subsequently replacing the symbol c by b. In effect, the regular singularities of the hypergeometric differential equation at b and ∞ coalesce into an irregular singularity at ∞.

Standard Solutions

The first two standard solutions are:

13.2.2
$$M(a,b,z) = \sum_{s=0}^{\infty} \frac{(a)_s}{(b)_s s!} z^s = 1 + \frac{a}{b}z + \frac{a(a+1)}{b(b+1)2!}z^2 + \cdots,$$

and

13.2.3 $$\mathbf{M}(a,b,z) = \sum_{s=0}^{\infty} \frac{(a)_s}{\Gamma(b+s)s!} z^s,$$

except that $M(a,b,z)$ does not exist when b is a nonpositive integer. In other cases

13.2.4 $$M(a,b,z) = \Gamma(b)\mathbf{M}(a,b,z).$$

The series (13.2.2) and (13.2.3) converge for all $z \in \mathbb{C}$. $M(a,b,z)$ is entire in z and a, and is a meromorphic function of b. $\mathbf{M}(a,b,z)$ is entire in z, a, and b.

Although $M(a,b,z)$ does not exist when $b = -n$, $n = 0,1,2,\ldots$, many formulas containing $M(a,b,z)$ continue to apply in their limiting form. In particular,

13.2.5
$$\lim_{b \to -n} \frac{M(a,b,z)}{\Gamma(b)} = \mathbf{M}(a,-n,z)$$
$$= \frac{(a)_{n+1}}{(n+1)!} z^{n+1} M(a+n+1, n+2, z).$$

When $a = -n$, $n = 0,1,2,\ldots$, $\mathbf{M}(a,b,z)$ is a polynomial in z of degree not exceeding n; this is also true of $M(a,b,z)$ provided that b is not a nonpositive integer.

Another standard solution of (13.2.1) is $U(a,b,z)$, which is determined uniquely by the property

13.2.6 $$U(a,b,z) \sim z^{-a}, \quad z \to \infty, \; |\operatorname{ph} z| \leq \tfrac{3}{2}\pi - \delta,$$

where δ is an arbitrary small positive constant. In general, $U(a,b,z)$ has a branch point at $z=0$. The *principal branch* corresponds to the *principal value* of z^{-a} in (13.2.6), and has a cut in the z-plane along the interval $(-\infty, 0]$; compare §4.2(i).

When $a = -n$, $n = 0,1,2,\ldots$, $U(a,b,z)$ is a polynomial in z of degree n:

13.2.7 $$U(-n,b,z) = (-1)^n \sum_{s=0}^{n} \binom{n}{s} (b+s)_{n-s} (-z)^s.$$

Similarly, when $a - b + 1 = -n$, $n = 0,1,2,\ldots$,

13.2.8 $$U(a, a+n+1, z) = z^{-a} \sum_{s=0}^{n} \binom{n}{s} (a)_s z^{-s}.$$

When $b = n+1$, $n = 0, 1, 2, \ldots$,

13.2.9
$$U(a, n+1, z) = \frac{(-1)^{n+1}}{n!\, \Gamma(a-n)} \sum_{k=0}^{\infty} \frac{(a)_k}{(n+1)_k k!} z^k \left(\ln z + \psi(a+k) - \psi(1+k) - \psi(n+k+1) \right) + \frac{1}{\Gamma(a)} \sum_{k=1}^{n} \frac{(k-1)!(1-a+k)_{n-k}}{(n-k)!} z^{-k},$$

if $a \neq 0, -1, -2, \ldots$, or

13.2.10
$$U(a, n+1, z) = (-1)^a \sum_{k=0}^{-a} \binom{-a}{k} (n+k+1)_{-a-k} (-z)^k,$$

if $a = 0, -1, -2, \ldots$.

When $b = -n$, $n = 0, 1, 2, \ldots$, the following equation can be combined with (13.2.9) and (13.2.10):

13.2.11 $\quad U(a, -n, z) = z^{n+1} U(a+n+1, n+2, z).$

13.2(ii) Analytic Continuation

When $m \in \mathbb{Z}$,

13.2.12
$$U\left(a, b, z e^{2\pi i m}\right) = \frac{2\pi i e^{-\pi i b m} \sin(\pi b m)}{\Gamma(1+a-b)\sin(\pi b)} \mathbf{M}(a, b, z) + e^{-2\pi i b m} U(a, b, z).$$

Except when $z = 0$ each branch of $U(a, b, z)$ is entire in a and b. Unless specified otherwise, however, $U(a, b, z)$ is assumed to have its principal value.

13.2(iii) Limiting Forms as $z \to 0$

13.2.13 $\quad M(a, b, z) = 1 + O(z).$

Next, in cases when $a = -n$ or $-n + b - 1$, where n is a nonnegative integer,

13.2.14 $\quad U(-n, b, z) = (-1)^n (b)_n + O(z),$

13.2.15
$$U(-n + b - 1, b, z) = (-1)^n (2-b)_n z^{1-b} + O(z^{2-b}).$$

In all other cases

13.2.16
$$U(a, b, z) = \frac{\Gamma(b-1)}{\Gamma(a)} z^{1-b} + O(z^{2-\Re b}), \quad \Re b \geq 2, \; b \neq 2,$$

13.2.17
$$U(a, 2, z) = \frac{1}{\Gamma(a)} z^{-1} + O(\ln z),$$

13.2.18
$$U(a, b, z) = \frac{\Gamma(b-1)}{\Gamma(a)} z^{1-b} + \frac{\Gamma(1-b)}{\Gamma(a-b+1)} + O(z^{2-\Re b}), \\ 1 \leq \Re b < 2, \; b \neq 1,$$

13.2.19
$$U(a, 1, z) = -\frac{1}{\Gamma(a)} \left(\ln z + \psi(a) + 2\gamma \right) + O(z \ln z),$$

13.2.20
$$U(a, b, z) = \frac{\Gamma(1-b)}{\Gamma(a-b+1)} + O(z^{1-\Re b}), \quad 0 < \Re b < 1,$$

13.2.21
$$U(a, 0, z) = \frac{1}{\Gamma(a+1)} + O(z \ln z),$$

13.2.22
$$U(a, b, z) = \frac{\Gamma(1-b)}{\Gamma(a-b+1)} + O(z), \quad \Re b \leq 0, \; b \neq 0.$$

13.2(iv) Limiting Forms as $z \to \infty$

Except when $a = 0, -1, \ldots$ (polynomial cases),

13.2.23 $\quad \mathbf{M}(a, b, z) \sim e^z z^{a-b} / \Gamma(a), \quad |\mathrm{ph}\, z| \leq \tfrac{1}{2}\pi - \delta,$

where δ is an arbitrary small positive constant.

For $U(a, b, z)$ see (13.2.6).

13.2(v) Numerically Satisfactory Solutions

Fundamental pairs of solutions of (13.2.1) that are numerically satisfactory (§2.7(iv)) in the neighborhood of infinity are

13.2.24
$$U(a, b, z), \quad e^z U(b-a, b, e^{-\pi i} z), \\ -\tfrac{1}{2}\pi \leq \mathrm{ph}\, z \leq \tfrac{3}{2}\pi,$$

13.2.25
$$U(a, b, z), \quad e^z U(b-a, b, e^{\pi i} z), \\ -\tfrac{3}{2}\pi \leq \mathrm{ph}\, z \leq \tfrac{1}{2}\pi.$$

A fundamental pair of solutions that is numerically satisfactory near the origin is

13.2.26
$$M(a, b, z), \quad z^{1-b} M(a-b+1, 2-b, z), \quad b \notin \mathbb{Z}.$$

When $b = n+1 = 1, 2, 3, \ldots$, a fundamental pair that is numerically satisfactory near the origin is $M(a, n+1, z)$ and

13.2.27
$$\sum_{k=1}^{n} \frac{n!(k-1)!}{(n-k)!(1-a)_k} z^{-k} - \sum_{k=0}^{\infty} \frac{(a)_k}{(n+1)_k k!} z^k \left(\ln z + \psi(a+k) - \psi(1+k) - \psi(n+k+1) \right),$$

if $a - n \neq 0, -1, -2, \ldots$, or $M(a, n+1, z)$ and

13.2.28
$$\sum_{k=1}^{n} \frac{n!(k-1)!}{(n-k)!(1-a)_k} z^{-k}$$
$$- \sum_{k=0}^{-a} \frac{(a)_k}{(n+1)_k k!} z^k \left(\ln z + \psi(1-a-k) - \psi(1+k) - \psi(n+k+1) \right) + (-1)^{1-a}(-a)! \sum_{k=1-a}^{\infty} \frac{(k-1+a)!}{(n+1)_k k!} z^k,$$

if $a = 0, -1, -2, \ldots$, or $M(a, n+1, z)$ and

13.2.29
$$\sum_{k=a}^{n} \frac{(k-1)!}{(n-k)!(k-a)!} z^{-k},$$

if $a = 1, 2, \ldots, n$.

When $b = -n = 0, -1, -2, \ldots$, a fundamental pair that is numerically satisfactory near the origin is $z^{n+1} \times M(a+n+1, n+2, z)$ and

13.2.30
$$\sum_{k=1}^{n+1} \frac{(n+1)!(k-1)!}{(n-k+1)!(-a-n)_k} z^{n-k+1} - \sum_{k=0}^{\infty} \frac{(a+n+1)_k}{(n+2)_k k!} z^{n+k+1} \left(\ln z + \psi(a+n+k+1) - \psi(1+k) - \psi(n+k+2) \right),$$

if $a \neq 0, -1, -2, \ldots$, or $z^{n+1} M(a+n+1, n+2, z)$ and

13.2.31
$$\sum_{k=1}^{n+1} \frac{(n+1)!(k-1)!}{(n-k+1)!(-a-n)_k} z^{n-k+1}$$
$$- \sum_{k=0}^{-a-n-1} \frac{(a+n+1)_k}{(n+2)_k k!} z^{n+k+1} \left(\ln z + \psi(-a-n-k) - \psi(1+k) - \psi(n+k+2) \right)$$
$$+ (-1)^{n-a}(-a-n-1)! \sum_{k=-a-n}^{\infty} \frac{(k+a+n)!}{(n+2)_k k!} z^{n+k+1},$$

if $a = -n-1, -n-2, -n-3, \ldots$, or $z^{n+1} M(a+n+1, n+2, z)$ and

13.2.32
$$\sum_{k=a+n+1}^{n+1} \frac{(k-1)!}{(n-k+1)!(k-a-n-1)!} z^{n-k+1},$$

if $a = 0, -1, \ldots, -n$.

13.2(vi) Wronskians

13.2.33 $\quad \mathscr{W}\left\{\mathbf{M}(a,b,z), z^{1-b}\mathbf{M}(a-b+1, 2-b, z)\right\} = \sin(\pi b) z^{-b} e^z / \pi,$

13.2.34 $\quad \mathscr{W}\left\{\mathbf{M}(a,b,z), U(a,b,z)\right\} = -z^{-b} e^z / \Gamma(a),$

13.2.35 $\quad \mathscr{W}\left\{\mathbf{M}(a,b,z), e^z U(b-a, b, e^{\pm \pi i} z)\right\} = e^{\mp b \pi i} z^{-b} e^z / \Gamma(b-a),$

13.2.36 $\quad \mathscr{W}\left\{z^{1-b}\mathbf{M}(a-b+1, 2-b, z), U(a,b,z)\right\} = -z^{-b} e^z / \Gamma(a-b+1),$

13.2.37 $\quad \mathscr{W}\left\{z^{1-b}\mathbf{M}(a-b+1, 2-b, z), e^z U(b-a, b, e^{\pm \pi i} z)\right\} = -z^{-b} e^z / \Gamma(1-a),$

13.2.38 $\quad \mathscr{W}\left\{U(a,b,z), e^z U(b-a, b, e^{\pm \pi i} z)\right\} = e^{\pm(a-b)\pi i} z^{-b} e^z.$

13.2(vii) Connection Formulas

Kummer's Transformations

13.2.39
$$M(a,b,z) = e^z M(b-a, b, -z),$$

13.2.40
$$U(a,b,z) = z^{1-b} U(a-b+1, 2-b, z).$$

13.2.41
$$\frac{1}{\Gamma(b)} M(a,b,z) = \frac{e^{\mp a\pi i}}{\Gamma(b-a)} U(a,b,z) + \frac{e^{\pm(b-a)\pi i}}{\Gamma(a)} e^z U(b-a, b, e^{\pm\pi i} z).$$

Also, when b is not an integer

13.2.42
$$U(a,b,z) = \frac{\Gamma(1-b)}{\Gamma(a-b+1)} M(a,b,z) + \frac{\Gamma(b-1)}{\Gamma(a)} z^{1-b} M(a-b+1, 2-b, z).$$

13.3 Recurrence Relations and Derivatives

13.3(i) Recurrence Relations

13.3.1
$$(b-a) M(a-1, b, z) + (2a-b+z) M(a,b,z) - a M(a+1, b, z) = 0,$$

13.3.2
$$b(b-1) M(a, b-1, z) + b(1-b-z) M(a,b,z) + z(b-a) M(a, b+1, z) = 0,$$

13.3.3
$$(a-b+1) M(a,b,z) - a M(a+1, b, z) + (b-1) M(a, b-1, z) = 0,$$

13.3.4
$$b M(a,b,z) - b M(a-1, b, z) - z M(a, b+1, z) = 0,$$

13.3.5
$$b(a+z) M(a,b,z) + z(a-b) M(a, b+1, z) - ab M(a+1, b, z) = 0,$$

13.3.6
$$(a-1+z) M(a,b,z) + (b-a) M(a-1, b, z) + (1-b) M(a, b-1, z) = 0.$$

13.3.7
$$U(a-1, b, z) + (b-2a-z) U(a,b,z) + a(a-b+1) U(a+1, b, z) = 0,$$

13.3.8
$$(b-a-1) U(a, b-1, z) + (1-b-z) U(a,b,z) + z U(a, b+1, z) = 0,$$

13.3.9
$$U(a,b,z) - a U(a+1, b, z) - U(a, b-1, z) = 0,$$

13.3.10
$$(b-a) U(a,b,z) + U(a-1, b, z) - z U(a, b+1, z) = 0,$$

13.3.11
$$(a+z) U(a,b,z) - z U(a, b+1, z) + a(b-a-1) U(a+1, b, z) = 0,$$

13.3.12
$$(a-1+z) U(a,b,z) - U(a-1, b, z) + (a-b+1) U(a, b-1, z) = 0.$$

Kummer's differential equation (13.2.1) is equivalent to

13.3.13
$$(a+1) z M(a+2, b+2, z) + (b+1)(b-z) M(a+1, b+1, z) - b(b+1) M(a,b,z) = 0,$$

and

13.3.14
$$(a+1) z U(a+2, b+2, z) + (z-b) U(a+1, b+1, z) - U(a,b,z) = 0.$$

13.3(ii) Differentiation Formulas

13.3.15
$$\frac{d}{dz} M(a,b,z) = \frac{a}{b} M(a+1, b+1, z),$$

13.3.16
$$\frac{d^n}{dz^n} M(a,b,z) = \frac{(a)_n}{(b)_n} M(a+n, b+n, z),$$

13.3.17
$$\left(z \frac{d}{dz} z \right)^n \left(z^{a-1} M(a,b,z) \right) = (a)_n z^{a+n-1} M(a+n, b, z),$$

13.3.18
$$\frac{d^n}{dz^n} \left(z^{b-1} M(a,b,z) \right) = (b-n)_n z^{b-n-1} M(a, b-n, z),$$

13.3.19
$$\left(z \frac{d}{dz} z \right)^n \left(z^{b-a-1} e^{-z} M(a,b,z) \right) = (b-a)_n z^{b-a+n-1} e^{-z} M(a-n, b, z),$$

13.3.20
$$\frac{d^n}{dz^n} \left(e^{-z} M(a,b,z) \right) = (-1)^n \frac{(b-a)_n}{(b)_n} e^{-z} M(a, b+n, z),$$

13.3.21
$$\frac{d^n}{dz^n} \left(z^{b-1} e^{-z} M(a,b,z) \right) = (b-n)_n z^{b-n-1} e^{-z} M(a-n, b-n, z).$$

13.3.22 $\quad \dfrac{d}{dz} U(a,b,z) = -a\, U(a+1,b+1,z),$

13.3.23 $\quad \dfrac{d^n}{dz^n} U(a,b,z) = (-1)^n (a)_n\, U(a+n,b+n,z),$

13.3.24
$$\left(z\dfrac{d}{dz}z\right)^n \left(z^{a-1} U(a,b,z)\right)$$
$$= (a)_n (a-b+1)_n\, z^{a+n-1} U(a+n,b,z),$$

13.3.25
$$\dfrac{d^n}{dz^n}\left(z^{b-1} U(a,b,z)\right)$$
$$= (-1)^n (a-b+1)_n\, z^{b-n-1} U(a,b-n,z),$$

13.3.26
$$\left(z\dfrac{d}{dz}z\right)^n \left(z^{b-a-1} e^{-z} U(a,b,z)\right)$$
$$= (-1)^n z^{b-a+n-1} e^{-z} U(a-n,b,z),$$

13.3.27 $\quad \dfrac{d^n}{dz^n}\left(e^{-z} U(a,b,z)\right) = (-1)^n e^{-z} U(a,b+n,z),$

13.3.28
$$\dfrac{d^n}{dz^n}\left(z^{b-1} e^{-z} U(a,b,z)\right)$$
$$= (-1)^n z^{b-n-1} e^{-z} U(a-n,b-n,z).$$

Other versions of several of the identities in this subsection can be constructed with the aid of the operator identity

13.3.29 $\quad \left(z\dfrac{d}{dz}\right)^n = z^n \dfrac{d^n}{dz^n} z^n, \quad n = 1, 2, 3, \ldots.$

13.4 Integral Representations

13.4(i) Integrals Along the Real Line

13.4.1
$$\mathbf{M}(a,b,z) = \dfrac{1}{\Gamma(a)\,\Gamma(b-a)} \int_0^1 e^{zt} t^{a-1} (1-t)^{b-a-1}\, dt,$$
$$\Re b > \Re a > 0,$$

13.4.2
$$\mathbf{M}(a,b,z) = \dfrac{1}{\Gamma(b-c)} \int_0^1 \mathbf{M}(a,c,zt)\, t^{c-1} (1-t)^{b-c-1}\, dt,$$
$$\Re b > \Re c > 0,$$

13.4.3
$$\mathbf{M}(a,b,-z) = \dfrac{z^{\frac{1}{2} - \frac{1}{2}b}}{\Gamma(a)} \int_0^\infty e^{-t} t^{a - \frac{1}{2}b - \frac{1}{2}} J_{b-1}\!\left(2\sqrt{zt}\right) dt,$$
$$\Re a > 0.$$

For the function J_{b-1} see §10.2(ii).

13.4.4
$$U(a,b,z) = \dfrac{1}{\Gamma(a)} \int_0^\infty e^{-zt} t^{a-1} (1+t)^{b-a-1}\, dt,$$
$$\Re a > 0,\ |\mathrm{ph}\, z| < \tfrac{1}{2}\pi,$$

13.4.5
$$U(a,b,z)$$
$$= \dfrac{z^{1-a}}{\Gamma(a)\,\Gamma(1+a-b)} \int_0^\infty \dfrac{U(b-a,b,t)\, e^{-t} t^{a-1}}{t+z}\, dt,$$
$$|\mathrm{ph}\, z| < \pi,\ \Re a > \max(\Re b - 1, 0),$$

13.4.6
$$U(a,b,z)$$
$$= \dfrac{(-1)^n z^{1-b-n}}{\Gamma(1+a-b)} \int_0^\infty \dfrac{\mathbf{M}(b-a,b,t)\, e^{-t} t^{b+n-1}}{t+z}\, dt,$$
$$|\mathrm{ph}\, z| < \pi,\ n = 0,1,2,\ldots,\ -\Re b < n < 1 + \Re(a-b),$$

13.4.7
$$U(a,b,z) = \dfrac{2 z^{\frac{1}{2} - \frac{1}{2}b}}{\Gamma(a)\,\Gamma(a-b+1)}$$
$$\times \int_0^\infty e^{-t} t^{a - \frac{1}{2}b - \frac{1}{2}} K_{b-1}\!\left(2\sqrt{zt}\right) dt,$$
$$\Re a > \max(\Re b - 1, 0),$$

13.4.8
$$U(a,b,z) = z^{c-a}$$
$$\times \int_0^\infty e^{-zt} t^{c-1}\, {}_2\mathbf{F}_1(a, a-b+1; c; -t)\, dt,$$
$$|\mathrm{ph}\, z| < \tfrac{1}{2}\pi,$$

where c is arbitrary, $\Re c > 0$. For the functions K_{b-1} and ${}_2\mathbf{F}_1$ see §10.25(ii) and §§15.1, 15.2(i).

13.4(ii) Contour Integrals

13.4.9
$$\mathbf{M}(a,b,z) = \dfrac{\Gamma(1+a-b)}{2\pi i\, \Gamma(a)} \int_0^{(1+)} e^{zt} t^{a-1} (t-1)^{b-a-1}\, dt,$$
$$b - a \neq 1, 2, 3, \ldots,\ \Re a > 0.$$

13.4.10
$$\mathbf{M}(a,b,z)$$
$$= e^{-a\pi i}\, \dfrac{\Gamma(1-a)}{2\pi i\, \Gamma(b-a)} \int_1^{(0+)} e^{zt} t^{a-1} (1-t)^{b-a-1}\, dt,$$
$$a \neq 1, 2, 3, \ldots,\ \Re(b-a) > 0.$$

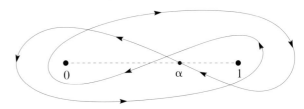

Figure 13.4.1: Contour of integration in (13.4.11). (Compare Figure 5.12.3.)

13.4.11
$$\mathbf{M}(a,b,z)$$
$$= e^{-b\pi i}\, \Gamma(1-a)\, \Gamma(1+a-b)$$
$$\times \dfrac{1}{4\pi^2} \int_\alpha^{(0+,1+,0-,1-)} e^{zt} t^{a-1} (1-t)^{b-a-1}\, dt,$$
$$a, b - a \neq 1, 2, 3, \ldots.$$

The contour of integration starts and terminates at a point α on the real axis between 0 and 1. It encircles $t = 0$ and $t = 1$ once in the positive sense, and then once in the negative sense. See Figure 13.4.1. The fractional powers are continuous and assume their principal values at $t = \alpha$. Similar conventions also apply to the remaining integrals in this subsection.

13.4.12
$$\mathbf{M}(a, c, z) = \frac{\Gamma(b)}{2\pi i} z^{1-b} \int_{-\infty}^{(0+,1+)} e^{zt} t^{-b} {}_2\mathbf{F}_1(a, b; c; 1/t) \, dt,$$
$$b \neq 0, -1, -2, \ldots, |\operatorname{ph} z| < \tfrac{1}{2}\pi.$$

At the point where the contour crosses the interval $(1, \infty)$, t^{-b} and the ${}_2\mathbf{F}_1$ function assume their principal values; compare §§15.1 and 15.2(i). A special case is

13.4.13
$$\mathbf{M}(a, b, z) = \frac{z^{1-b}}{2\pi i} \int_{-\infty}^{(0+,1+)} e^{zt} t^{-b} \left(1 - \frac{1}{t}\right)^{-a} dt,$$
$$|\operatorname{ph} z| < \tfrac{1}{2}\pi.$$

13.4.14
$$U(a, b, z) = e^{-a\pi i} \frac{\Gamma(1-a)}{2\pi i} \int_{\infty}^{(0+)} e^{-zt} t^{a-1} (1+t)^{b-a-1} \, dt,$$
$$a \neq 1, 2, 3, \ldots, |\operatorname{ph} z| < \tfrac{1}{2}\pi.$$

The contour cuts the real axis between -1 and 0. At this point the fractional powers are determined by $\operatorname{ph} t = \pi$ and $\operatorname{ph}(1+t) = 0$.

13.4.15
$$\frac{U(a, b, z)}{\Gamma(c)\Gamma(c-b+1)} = \frac{z^{1-c}}{2\pi i} \int_{-\infty}^{(0+)} e^{zt} t^{-c} {}_2\mathbf{F}_1\left(a, c; a+c-b+1; 1-\frac{1}{t}\right) dt, \qquad |\operatorname{ph} z| < \tfrac{1}{2}\pi.$$

Again, t^{-c} and the ${}_2\mathbf{F}_1$ function assume their principal values where the contour intersects the positive real axis.

13.4(iii) Mellin–Barnes Integrals

If $a \neq 0, -1, -2, \ldots$, then

13.4.16
$$\mathbf{M}(a, b, -z) = \frac{1}{2\pi i \, \Gamma(a)} \int_{-i\infty}^{i\infty} \frac{\Gamma(a+t)\Gamma(-t)}{\Gamma(b+t)} z^t \, dt,$$
$$|\operatorname{ph} z| < \tfrac{1}{2}\pi,$$

where the contour of integration separates the poles of $\Gamma(a+t)$ from those of $\Gamma(-t)$.

If a and $a - b + 1 \neq 0, -1, -2, \ldots$, then

13.4.17
$$U(a, b, z) = \frac{z^{-a}}{2\pi i} \int_{-i\infty}^{i\infty} \frac{\Gamma(a+t)\Gamma(1+a-b+t)\Gamma(-t)}{\Gamma(a)\Gamma(1+a-b)} z^{-t} \, dt,$$
$$|\operatorname{ph} z| < \tfrac{3}{2}\pi,$$

where the contour of integration separates the poles of $\Gamma(a+t)\Gamma(1+a-b+t)$ from those of $\Gamma(-t)$.

13.4.18
$$U(a, b, z) = \frac{z^{1-b} e^z}{2\pi i} \int_{-i\infty}^{i\infty} \frac{\Gamma(b-1+t)\Gamma(t)}{\Gamma(a+t)} z^{-t} \, dt,$$
$$|\operatorname{ph} z| < \tfrac{1}{2}\pi,$$

where the contour of integration passes all the poles of $\Gamma(b-1+t)\Gamma(t)$ on the right-hand side.

13.5 Continued Fractions

If $a, b \in \mathbb{C}$ such that $a \neq -1, -2, -3, \ldots$, and $a - b \neq 0, 1, 2, \ldots$, then

13.5.1
$$\frac{M(a, b, z)}{M(a+1, b+1, z)} = 1 + \frac{u_1 z}{1+} \frac{u_2 z}{1+} \cdots,$$

where

13.5.2
$$u_{2n+1} = \frac{a - b - n}{(b+2n)(b+2n+1)},$$
$$u_{2n} = \frac{a+n}{(b+2n-1)(b+2n)}.$$

This continued fraction converges to the meromorphic function of z on the left-hand side everywhere in $\mathbb{C}$. For more details on how a continued fraction converges to a meromorphic function see Jones and Thron (1980).

If $a, b \in \mathbb{C}$ such that $a \neq 0, -1, -2, \ldots$, and $b - a \neq 2, 3, 4, \ldots$, then

13.5.3
$$\frac{U(a, b, z)}{U(a, b-1, z)} = 1 + \frac{v_1/z}{1+} \frac{v_2/z}{1+} \cdots,$$

where

13.5.4
$$v_{2n+1} = a + n, \qquad v_{2n} = a - b + n + 1.$$

This continued fraction converges to the meromorphic function of z on the left-hand side throughout the sector $|\operatorname{ph} z| < \pi$.

See also Cuyt et al. (2008, pp. 322–330).

13.6 Relations to Other Functions

13.6(i) Elementary Functions

13.6.1
$$M(a, a, z) = e^z,$$

13.6.2
$$M(1, 2, 2z) = \frac{e^z}{z} \sinh z,$$

13.6.3
$$M(0, b, z) = U(0, b, z) = 1,$$

13.6.4
$$U(a, a+1, z) = z^{-a}.$$

13.6(ii) Incomplete Gamma Functions

For the notation see §§6.2(i), 7.2(i), 8.2(i), and 8.19(i). When $a - b$ is an integer or a is a positive integer the Kummer functions can be expressed as incomplete gamma functions (or generalized exponential integrals). For example,

13.6.5
$$M(a, a+1, -z) = e^{-z} M(1, a+1, z) = az^{-a} \gamma(a, z),$$

13.6.6
$$U(a, a, z) = z^{1-a} U(1, 2-a, z)$$
$$= z^{1-a} e^z E_a(z) = e^z \Gamma(1-a, z).$$

Special cases are the error functions

13.6.7 $\quad M\left(\tfrac{1}{2}, \tfrac{3}{2}, -z^2\right) = \dfrac{\sqrt{\pi}}{2z} \operatorname{erf}(z),$

13.6.8 $\quad U\left(\tfrac{1}{2}, \tfrac{1}{2}, z^2\right) = \sqrt{\pi} e^{z^2} \operatorname{erfc}(z).$

13.6(iii) Modified Bessel Functions

When $b = 2a$ the Kummer functions can be expressed as modified Bessel functions. For the notation see §§10.25(ii) and 9.2(i).

13.6.9
$$M\left(\nu + \tfrac{1}{2}, 2\nu+1, 2z\right) = \Gamma(1+\nu)e^z (z/2)^{-\nu} I_\nu(z),$$

13.6.10
$$U\left(\nu + \tfrac{1}{2}, 2\nu+1, 2z\right) = \frac{1}{\sqrt{\pi}} e^z (2z)^{-\nu} K_\nu(z),$$

13.6.11 $\quad U\left(\tfrac{5}{6}, \tfrac{5}{3}, \tfrac{4}{3} z^{3/2}\right) = \sqrt{\pi} \dfrac{3^{5/6} \exp\left(\tfrac{2}{3} z^{3/2}\right)}{2^{2/3} z} \operatorname{Ai}(z).$

13.6(iv) Parabolic Cylinder Functions

For the notation see §12.2.

13.6.12 $\quad U\left(\tfrac{1}{2}a + \tfrac{1}{4}, \tfrac{1}{2}, \tfrac{1}{2}z^2\right) = 2^{\tfrac{1}{2}a + \tfrac{1}{4}} e^{\tfrac{1}{4}z^2} U(a, z),$

13.6.13 $\quad U\left(\tfrac{1}{2}a + \tfrac{3}{4}, \tfrac{3}{2}, \tfrac{1}{2}z^2\right) = 2^{\tfrac{1}{2}a + \tfrac{3}{4}} \dfrac{e^{\tfrac{1}{4}z^2}}{z} U(a, z).$

13.6.14
$$M\left(\tfrac{1}{2}a + \tfrac{1}{4}, \tfrac{1}{2}, \tfrac{1}{2}z^2\right) = \frac{2^{\tfrac{1}{2}a - \tfrac{3}{4}} \Gamma\left(\tfrac{1}{2}a + \tfrac{3}{4}\right) e^{\tfrac{1}{4}z^2}}{\sqrt{\pi}}$$
$$\times (U(a, z) + U(a, -z)),$$

13.6.15
$$M\left(\tfrac{1}{2}a + \tfrac{3}{4}, \tfrac{3}{2}, \tfrac{1}{2}z^2\right) = \frac{2^{\tfrac{1}{2}a - \tfrac{5}{4}} \Gamma\left(\tfrac{1}{2}a + \tfrac{1}{4}\right) e^{\tfrac{1}{4}z^2}}{z\sqrt{\pi}}$$
$$\times (U(a, -z) - U(a, z)).$$

13.6(v) Orthogonal Polynomials

Special cases of §13.6(iv) are as follows. For the notation see §§18.3, 18.19.

Hermite Polynomials

13.6.16 $\quad M\left(-n, \tfrac{1}{2}, z^2\right) = (-1)^n \dfrac{n!}{(2n)!} H_{2n}(z),$

13.6.17 $\quad M\left(-n, \tfrac{3}{2}, z^2\right) = (-1)^n \dfrac{n!}{(2n+1)! 2z} H_{2n+1}(z),$

13.6.18 $\quad U\left(\tfrac{1}{2} - \tfrac{1}{2}n, \tfrac{3}{2}, z^2\right) = 2^{-n} z^{-1} H_n(z).$

Laguerre Polynomials

13.6.19
$$U(-n, \alpha+1, z) = (-1)^n (\alpha+1)_n M(-n, \alpha+1, z)$$
$$= (-1)^n n! L_n^{(\alpha)}(z).$$

Charlier Polynomials

13.6.20
$$U(-n, z-n+1, a) = (-z)_n M(-n, z-n+1, a)$$
$$= a^n C_n(z, a).$$

13.6(vi) Generalized Hypergeometric Functions

13.6.21 $\quad U(a, b, z) = z^{-a} {}_2F_0\left(a, a-b+1; -; -z^{-1}\right).$

For the definition of ${}_2F_0\left(a, a-b+1; -; -z^{-1}\right)$ when neither a nor $a - b + 1$ is a nonpositive integer see §16.5.

13.7 Asymptotic Expansions for Large Argument

13.7(i) Poincaré-Type Expansions

As $x \to \infty$

13.7.1 $\quad \mathbf{M}(a, b, x) \sim \dfrac{e^x x^{a-b}}{\Gamma(a)} \sum_{s=0}^{\infty} \dfrac{(1-a)_s (b-a)_s}{s!} x^{-s},$

provided that $a \neq 0, -1, \ldots$.

As $z \to \infty$

13.7.2
$$\mathbf{M}(a, b, z) \sim \frac{e^z z^{a-b}}{\Gamma(a)} \sum_{s=0}^{\infty} \frac{(1-a)_s (b-a)_s}{s!} z^{-s} + \frac{e^{\pm \pi i a} z^{-a}}{\Gamma(b-a)} \sum_{s=0}^{\infty} \frac{(a)_s (a-b+1)_s}{s!} (-z)^{-s}, \quad -\tfrac{1}{2}\pi + \delta \leq \pm \operatorname{ph} z \leq \tfrac{3}{2}\pi - \delta,$$

unless $a = 0, -1, \ldots$ and $b - a = 0, -1, \ldots$. Here δ denotes an arbitrary small positive constant. Also,

13.7.3
$$U(a, b, z) \sim z^{-a} \sum_{s=0}^{\infty} \frac{(a)_s (a-b+1)_s}{s!} (-z)^{-s}, \qquad |\operatorname{ph} z| \leq \tfrac{3}{2}\pi - \delta.$$

13.7(ii) Error Bounds

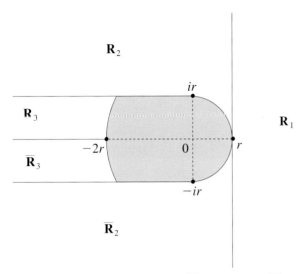

Figure 13.7.1: Regions $\mathbf{R}_1$, $\mathbf{R}_2$, $\overline{\mathbf{R}}_2$, $\mathbf{R}_3$, and $\overline{\mathbf{R}}_3$ are the closures of the indicated unshaded regions bounded by the straight lines and circular arcs centered at the origin, with $r = |b - 2a|$.

13.7.4
$$U(a,b,z) = z^{-a} \sum_{s=0}^{n-1} \frac{(a)_s (a-b+1)_s}{s!} (-z)^{-s} + \varepsilon_n(z),$$

where

13.7.5
$$|\varepsilon_n(z)|, \; \beta^{-1}|\varepsilon_n'(z)| \leq 2\alpha C_n \left| \frac{(a)_n (a-b+1)_n}{n! z^{a+n}} \right| \exp\left(\frac{2\alpha\rho C_1}{|z|}\right),$$

and with the notation of Figure 13.7.1

13.7.6 $\quad C_n = 1, \quad \chi(n), \quad (\chi(n) + \sigma\nu^2 n)\nu^n,$

according as

13.7.7 $\quad z \in \mathbf{R}_1, \quad z \in \mathbf{R}_2 \cup \overline{\mathbf{R}}_2, \quad z \in \mathbf{R}_3 \cup \overline{\mathbf{R}}_3,$

respectively, with

13.7.8
$$\sigma = |(b-2a)/z|, \quad \nu = \left(\tfrac{1}{2} + \tfrac{1}{2}\sqrt{1-4\sigma^2}\right)^{-1/2},$$
$$\chi(n) = \sqrt{\pi}\,\Gamma(\tfrac{1}{2}n+1)/\Gamma(\tfrac{1}{2}n+\tfrac{1}{2}).$$

Also, when $z \in \mathbf{R}_1 \cup \mathbf{R}_2 \cup \overline{\mathbf{R}}_2$

13.7.9
$$\alpha = \frac{1}{1-\sigma}, \quad \beta = \frac{1-\sigma^2 + \sigma|z|^{-1}}{2(1-\sigma)},$$
$$\rho = \tfrac{1}{2}|2a^2 - 2ab + b| + \frac{\sigma(1+\tfrac{1}{4}\sigma)}{(1-\sigma)^2},$$

and when $z \in \mathbf{R}_3 \cup \overline{\mathbf{R}}_3$ σ is replaced by $\nu\sigma$ and $|z|^{-1}$ is replaced by $\nu|z|^{-1}$ everywhere in (13.7.9).

For numerical values of $\chi(n)$ see Table 9.7.1.

Corresponding error bounds for (13.7.2) can be constructed by combining (13.2.41) with (13.7.4)–(13.7.9).

13.7(iii) Exponentially-Improved Expansion

Let

13.7.10
$$U(a,b,z) = z^{-a} \sum_{s=0}^{n-1} \frac{(a)_s (a-b+1)_s}{s!} (-z)^{-s} + R_n(a,b,z),$$

and

13.7.11
$$R_n(a,b,z) = \frac{(-1)^n 2\pi z^{a-b}}{\Gamma(a)\,\Gamma(a-b+1)} \left(\sum_{s=0}^{m-1} \frac{(1-a)_s (b-a)_s}{s!} (-z)^{-s} G_{n+2a-b-s}(z) + (1-a)_m (b-a)_m R_{m,n}(a,b,z) \right),$$

where m is an arbitrary nonnegative integer, and

13.7.12
$$G_p(z) = \frac{e^z}{2\pi} \Gamma(p)\,\Gamma(1-p,z).$$

(For the notation see §8.2(i).) Then as $z \to \infty$ with $||z| - n|$ bounded and a,b,m fixed

13.7.13
$$R_{m,n}(a,b,z) = \begin{cases} O(e^{-|z|} z^{-m}), & |\operatorname{ph} z| \leq \pi, \\ O(e^{z} z^{-m}), & \pi \leq |\operatorname{ph} z| \leq \tfrac{5}{2}\pi - \delta. \end{cases}$$

For proofs see Olver (1991b, 1993a). For extensions to hyperasymptotic expansions see Olde Daalhuis and Olver (1995a).

13.8 Asymptotic Approximations for Large Parameters

13.8(i) Large $|b|$, Fixed a and z

If $b \to \infty$ in $\mathbb{C}$ in such a way that $|b+n| \geq \delta > 0$ for all $n = 0, 1, 2, \ldots$, then

13.8.1 $\quad M(a,b,z) = \sum_{s=0}^{n-1} \frac{(a)_s}{(b)_s s!} z^s + O\big(|b|^{-n}\big).$

For fixed a and z in $\mathbb{C}$

13.8.2 $\quad M(a,b,z) \sim \frac{\Gamma(b)}{\Gamma(b-a)} \sum_{s=0}^{\infty} (a)_s q_s(z,a) b^{-s-a},$

as $b \to \infty$ in $|\operatorname{ph} b| \leq \pi - \delta$, where $q_0(z,a) = 1$ and

13.8.3 $\quad (e^t - 1)^{a-1} \exp\big(t + z(1 - e^{-t})\big) = \sum_{s=0}^{\infty} q_s(z,a) t^{s+a-1}.$

When the foregoing results are combined with Kummer's transformation (13.2.39), an approximation is obtained for the case when $|b|$ is large, and $|b-a|$ and $|z|$ are bounded.

13.8(ii) Large b and z, Fixed a and b/z

Let $\lambda = z/b > 0$ and $\zeta = \sqrt{2(\lambda - 1 - \ln \lambda)}$ with $\operatorname{sign}(\zeta) = \operatorname{sign}(\lambda - 1)$. Then

13.8.4 $\quad M(a,b,z) \sim b^{\frac{1}{2}a} e^{\frac{1}{4}\zeta^2 b} \left(\lambda \left(\frac{\lambda-1}{\zeta}\right)^{a-1} U\big(a - \tfrac{1}{2}, -\zeta\sqrt{b}\big) + \left(\lambda \left(\frac{\lambda-1}{\zeta}\right)^{a-1} - \left(\frac{\zeta}{\lambda-1}\right)^a \right) \frac{U\big(a - \tfrac{3}{2}, -\zeta\sqrt{b}\big)}{\zeta\sqrt{b}} \right)$

and

13.8.5 $\quad U(a,b,z) \sim b^{-\frac{1}{2}a} e^{\frac{1}{4}\zeta^2 b} \left(\lambda \left(\frac{\lambda-1}{\zeta}\right)^{a-1} U\big(a - \tfrac{1}{2}, \zeta\sqrt{b}\big) - \left(\lambda \left(\frac{\lambda-1}{\zeta}\right)^{a-1} - \left(\frac{\zeta}{\lambda-1}\right)^a \right) \frac{U\big(a - \tfrac{3}{2}, \zeta\sqrt{b}\big)}{\zeta\sqrt{b}} \right)$

as $b \to \infty$, uniformly in compact λ-intervals of $(0, \infty)$ and compact real a-intervals. For the parabolic cylinder function U see §12.2, and for an extension to an asymptotic expansion see Temme (1978).

Special cases are

13.8.6 $\quad M(a,b,b) = \sqrt{\pi} \left(\frac{b}{2}\right)^{\frac{1}{2}a} \left(\frac{1}{\Gamma\big(\tfrac{1}{2}(a+1)\big)} + \frac{(a+1)\sqrt{8/b}}{3\Gamma\big(\tfrac{1}{2}a\big)} + O\left(\frac{1}{b}\right) \right),$

and

13.8.7 $\quad U(a,b,b) = \sqrt{\pi} (2b)^{-\frac{1}{2}a} \left(\frac{1}{\Gamma\big(\tfrac{1}{2}(a+1)\big)} - \frac{(a+1)\sqrt{8/b}}{3\Gamma\big(\tfrac{1}{2}a\big)} + O\left(\frac{1}{b}\right) \right).$

To obtain approximations for $M(a,b,z)$ and $U(a,b,z)$ that hold as $b \to \infty$, with $a > \tfrac{1}{2} - b$ and $z > 0$ combine (13.14.4), (13.14.5) with §13.20(i).

Also, more complicated—but more powerful—uniform asymptotic approximations can be obtained by combining (13.14.4), (13.14.5) with §§13.20(iii) and 13.20(iv).

13.8(iii) Large a

For the notation see §§10.2(ii), 10.25(ii), and 2.8(iv).

When $a \to +\infty$ with b (≤ 1) fixed,

13.8.8 $\quad U(a,b,x) = \frac{2e^{\frac{1}{2}x}}{\Gamma(a)} \left(\sqrt{\frac{2}{\beta}} \tanh\left(\frac{w}{2}\right) \left(\frac{1 - e^{-w}}{\beta}\right)^{-b} \beta^{1-b} K_{1-b}(2\beta a) + a^{-1} \left(\frac{a^{-1} + \beta}{1 + \beta}\right)^{1-b} e^{-2\beta a} O(1) \right),$

where $w = \operatorname{arccosh}\big(1 + (2a)^{-1} x\big)$, and $\beta = (w + \sinh w)/2$. (13.8.8) holds uniformly with respect to $x \in [0, \infty)$. For the case $b > 1$ the transformation (13.2.40) can be used.

For an extension to an asymptotic expansion complete with error bounds see Temme (1990b), and for related results see §13.21(i).

When $a \to -\infty$ with b (≥ 1) fixed,

13.8.9 $\quad M(a,b,x) = \Gamma(b) e^{\frac{1}{2}x} \big((\tfrac{1}{2}b - a)x\big)^{\frac{1}{2} - \frac{1}{2}b} \left(J_{b-1}\big(\sqrt{2x(b-2a)}\big) + \operatorname{env} J_{b-1}\big(\sqrt{2x(b-2a)}\big) O\big(|a|^{-\frac{1}{2}}\big) \right),$

and

$$U(a,b,x) = \Gamma\left(\tfrac{1}{2}b - a + \tfrac{1}{2}\right) e^{\frac{1}{2}x} x^{\frac{1}{2} - \frac{1}{2}b} \left(\cos(a\pi) J_{b-1}\left(\sqrt{2x(b-2a)}\right) - \sin(a\pi) Y_{b-1}\left(\sqrt{2x(b-2a)}\right) \right.$$
$$\left. + \mathrm{env} Y_{b-1}\left(\sqrt{2x(b-2a)}\right) O\left(|a|^{-\frac{1}{2}}\right)\right),$$

13.8.10

uniformly with respect to bounded positive values of x in each case.

For asymptotic approximations to $M(a,b,x)$ and $U(a,b,x)$ as $a \to -\infty$ that hold uniformly with respect to $x \in (0, \infty)$ and bounded positive values of $(b-1)/|a|$, combine (13.14.4), (13.14.5) with §§13.21(ii), 13.21(iii).

13.9 Zeros

13.9(i) Zeros of $M(a,b,z)$

If a and $b - a \neq 0, -1, -2, \ldots$, then $M(a,b,z)$ has infinitely many z-zeros in $\mathbb{C}$. When $a, b \in \mathbb{R}$ the number of real zeros is finite. Let $p(a,b)$ be the number of positive zeros. Then

13.9.1 $\quad p(a,b) = \lceil -a \rceil, \qquad a < 0, b \geq 0,$

13.9.2 $\quad p(a,b) = 0, \qquad a \geq 0, b \geq 0,$

13.9.3 $\quad p(a,b) = 1, \qquad a \geq 0, -1 < b < 0,$

13.9.4 $\quad p(a,b) = \lfloor -\tfrac{1}{2}b \rfloor - \lfloor -\tfrac{1}{2}(b+1) \rfloor, \quad a \geq 0, b \leq -1.$

13.9.5
$$p(a,b) = \lceil -a \rceil - \lceil -b \rceil, \quad \lceil -a \rceil \geq \lceil -b \rceil, a < 0, b < 0,$$

13.9.6
$$p(a,b) = \lfloor \tfrac{1}{2}(\lceil -b \rceil - \lceil -a \rceil + 1) \rfloor - \lfloor \tfrac{1}{2}(\lceil -b \rceil - \lceil -a \rceil) \rfloor,$$
$$\lceil -b \rceil > \lceil -a \rceil > 0.$$

The number of negative real zeros $n(a,b)$ is given by

13.9.7 $\quad n(a,b) = p(b-a, b).$

When $a < 0$ and $b > 0$ let ϕ_r, $r = 1, 2, 3, \ldots$, be the positive zeros of $M(a,b,x)$ arranged in increasing order of magnitude, and let $j_{b-1,r}$ be the rth positive zero of the Bessel function $J_{b-1}(x)$ (§10.21(i)). Then

13.9.8
$$\phi_r = \frac{j_{b-1,r}^2}{2b - 4a}\left(1 + \frac{2b(b-2) + j_{b-1,r}^2}{3(2b-4a)^2}\right) + O\left(\frac{1}{a^5}\right),$$

as $a \to -\infty$ with r fixed.

Inequalities for ϕ_r are given in Gatteschi (1990), and identities involving infinite series of all of the complex zeros of $M(a,b,x)$ are given in Ahmed and Muldoon (1980).

For fixed $a, b \in \mathbb{C}$ the large z-zeros of $M(a,b,z)$ satisfy

13.9.9
$$z = \pm(2n + a)\pi i + \ln\left(-\frac{\Gamma(a)}{\Gamma(b-a)}(\pm 2n\pi i)^{b-2a}\right) + O(n^{-1} \ln n),$$

where n is a large positive integer, and the logarithm takes its principal value (§4.2(i)).

Let P_α denote the closure of the domain that is bounded by the parabola $y^2 = 4\alpha(x + \alpha)$ and contains the origin. Then $M(a,b,z)$ has no zeros in the regions $P_{b/a}$, if $0 < b \leq a$; P_1, if $1 \leq a \leq b$; P_α, where $\alpha = (2a - b + ab)/(a(a+1))$, if $0 < a < 1$ and $a \leq b < 2a/(1-a)$. The same results apply for the nth partial sums of the Maclaurin series (13.2.2) of $M(a,b,z)$.

More information on the location of real zeros can be found in Zarzo et al. (1995).

For fixed b and z in $\mathbb{C}$ the large a-zeros of $M(a,b,z)$ are given by

13.9.10
$$a = -\frac{\pi^2}{4z}\left(n^2 + (b - \tfrac{3}{2})n\right)$$
$$- \frac{1}{16z}\left((b - \tfrac{3}{2})^2\pi^2 + \tfrac{4}{3}z^2 - 8b(z-1) - 4b^2 - 3\right)$$
$$+ O(n^{-1}),$$

where n is a large positive integer.

For fixed a and z in $\mathbb{C}$ the function $M(a,b,z)$ has only a finite number of b-zeros.

13.9(ii) Zeros of $U(a,b,z)$

For fixed a and b in $\mathbb{C}$, $U(a,b,z)$ has a finite number of z-zeros in the sector $|\mathrm{ph}\, z| \leq \tfrac{3}{2}\pi - \delta(< \tfrac{3}{2}\pi)$. Let $T(a,b)$ be the total number of zeros in the sector $|\mathrm{ph}\, z| < \pi$, $P(a,b)$ be the corresponding number of positive zeros, and a, b, and $a - b + 1$ be nonintegers. For the case $b \leq 1$

13.9.11 $\quad T(a,b) = \lfloor -a \rfloor + 1, \quad a < 0, \Gamma(a)\Gamma(a-b+1) > 0,$

13.9.12 $\quad T(a,b) = \lfloor -a \rfloor, \quad a < 0, \Gamma(a)\Gamma(a-b+1) < 0,$

13.9.13 $\quad T(a,b) = 0, \qquad a > 0,$

and

13.9.14 $\quad P(a,b) = \lceil b - a - 1 \rceil, \quad a + 1 < b,$

13.9.15 $\quad P(a,b) = 0, \qquad a + 1 \geq b.$

For the case $b \geq 1$ we can use $T(a,b) = T(a-b+1, 2-b)$ and $P(a,b) = P(a-b+1, 2-b)$.

In Wimp (1965) it is shown that if $a, b \in \mathbb{R}$ and $2a - b > -1$, then $U(a,b,z)$ has no zeros in the sector $|\mathrm{ph}\, z| \leq \tfrac{1}{2}\pi$.

Inequalities for the zeros of $U(a,b,x)$ are given in Gatteschi (1990).

For fixed b and z in $\mathbb{C}$ the large a-zeros of $U(a,b,z)$ are given by

13.9.16
$$a \sim -n - \frac{2}{\pi}\sqrt{zn} - \frac{2z}{\pi^2} + \tfrac{1}{2}b + \tfrac{1}{4} + \frac{z^2\left(\tfrac{1}{3} - 4\pi^{-2}\right) + z - (b-1)^2 + \tfrac{1}{4}}{4\pi\sqrt{zn}} + O\left(\frac{1}{n}\right),$$

where n is a large positive integer.

For fixed a and z in $\mathbb{C}$, $U(a,b,z)$ has two infinite strings of b-zeros that are asymptotic to the imaginary axis as $|b| \to \infty$.

13.10 Integrals

13.10(i) Indefinite Integrals

When $a \neq 1$,

13.10.1 $\quad \int \mathbf{M}(a,b,z)\,dz = \frac{1}{a-1}\mathbf{M}(a-1,b-1,z),$

13.10.2 $\quad \int U(a,b,z)\,dz = -\frac{1}{a-1}U(a-1,b-1,z).$

Other formulas of this kind can be constructed by inversion of the differentiation formulas given in §13.3(ii).

13.10(ii) Laplace Transforms

For the notation see §§15.1, 15.2(i), and 10.25(ii).

13.10.3
$$\int_0^\infty e^{-zt}t^{b-1}\mathbf{M}(a,c,kt)\,dt = \Gamma(b)z^{-b}\,{}_2\mathbf{F}_1(a,b;c;k/z),$$
$$\Re b > 0, \Re z > \max(\Re k, 0),$$

13.10.4
$$\int_0^\infty e^{-zt}t^{b-1}\mathbf{M}(a,b,t)\,dt = z^{-b}\left(1 - \frac{1}{z}\right)^{-a},$$
$$\Re b > 0, \Re z > 1,$$

13.10.5
$$\int_0^\infty e^{-t}t^{b-1}\mathbf{M}(a,c,t)\,dt = \frac{\Gamma(b)\Gamma(c-a-b)}{\Gamma(c-a)\Gamma(c-b)},$$
$$\Re(c-a) > \Re b > 0,$$

13.10.6
$$\int_0^\infty e^{-zt-t^2}t^{2b-2}\mathbf{M}(a,b,t^2)\,dt$$
$$= \tfrac{1}{2}\pi^{-\tfrac{1}{2}}\Gamma\left(b-\tfrac{1}{2}\right)U\left(b-\tfrac{1}{2}, a+\tfrac{1}{2}, \tfrac{1}{4}z^2\right),$$
$$\Re b > \tfrac{1}{2}, \Re z > 0,$$

13.10.7
$$\int_0^\infty e^{-zt}t^{b-1}U(a,c,t)\,dt$$
$$= \Gamma(b)\Gamma(b-c+1)$$
$$\times z^{-b}\,{}_2\mathbf{F}_1\left(a,b;a+b-c+1;1-\frac{1}{z}\right),$$
$$\Re b > \max(\Re c - 1, 0), \Re z > 0.$$

Loop Integrals

13.10.8
$$\frac{1}{2\pi i}\int_{-\infty}^{(0+)} e^{tz}t^{-a}\mathbf{M}(a,b,y/t)\,dt$$
$$= \frac{1}{\Gamma(a)}z^{\tfrac{1}{2}(2a-b-1)}y^{\tfrac{1}{2}(1-b)}I_{b-1}(2\sqrt{zy}),$$
$$\Re z > 0.$$

13.10.9
$$\frac{1}{2\pi i}\int_{-\infty}^{(0+)} e^{tz}t^{-a}U(a,b,y/t)\,dt$$
$$= \frac{2z^{\tfrac{1}{2}(2a-b-1)}y^{\tfrac{1}{2}(1-b)}}{\Gamma(a)\Gamma(a-b+1)}K_{b-1}(2\sqrt{zy}), \quad \Re z > 0.$$

For additional Laplace transforms see Erdélyi et al. (1954a, §§4.22, 5.20), Oberhettinger and Badii (1973, §1.17), and Prudnikov et al. (1992a, §§3.34, 3.35). Inverse Laplace transforms are given in Oberhettinger and Badii (1973, §2.16) and Prudnikov et al. (1992b, §§3.33, 3.34).

13.10(iii) Mellin Transforms

13.10.10
$$\int_0^\infty t^{\lambda-1}\mathbf{M}(a,b,-t)\,dt = \frac{\Gamma(\lambda)\Gamma(a-\lambda)}{\Gamma(a)\Gamma(b-\lambda)}, \quad 0 < \Re\lambda < \Re a,$$

13.10.11
$$\int_0^\infty t^{\lambda-1}U(a,b,t)\,dt = \frac{\Gamma(\lambda)\Gamma(a-\lambda)\Gamma(\lambda-b+1)}{\Gamma(a)\Gamma(a-b+1)},$$
$$\max(\Re b - 1, 0) < \Re\lambda < \Re a.$$

For additional Mellin transforms see Erdélyi et al. (1954a, §§6.9, 7.5), Marichev (1983, pp. 283–287), and Oberhettinger (1974, §§1.13, 2.8).

13.10(iv) Fourier Transforms

13.10.12
$$\int_0^\infty \cos(2xt)\mathbf{M}(a,b,-t^2)\,dt$$
$$= \frac{\sqrt{\pi}}{2\Gamma(a)}x^{2a-1}e^{-x^2}U\left(b-\tfrac{1}{2}, a+\tfrac{1}{2}, x^2\right),$$
$$\Re a > 0.$$

For additional Fourier transforms see Erdélyi et al. (1954a, §§1.14, 2.14, 3.3) and Oberhettinger (1990, §§1.22, 2.22).

13.10(v) Hankel Transforms

For the notation see §10.2(ii).

13.10.13
$$\int_0^\infty e^{-t}t^{b-1-\tfrac{1}{2}\nu}\mathbf{M}(a,b,t)J_\nu\left(2\sqrt{xt}\right)dt$$
$$= x^{-a+\tfrac{1}{2}\nu}e^{-x}\mathbf{M}(\nu-b+1, \nu-a+1, x),$$
$$x > 0, 2\Re a < \Re\nu + \tfrac{5}{2}, \Re b > 0,$$

13.10.14
$$\int_0^\infty e^{-t}t^{\tfrac{1}{2}\nu}\mathbf{M}(a,b,t)J_\nu\left(2\sqrt{xt}\right)dt$$
$$= \frac{x^{\tfrac{1}{2}\nu}e^{-x}}{\Gamma(b-a)}U(a, a-b+\nu+2, x),$$
$$x > 0, -1 < \Re\nu < 2\Re(b-a) - \tfrac{1}{2},$$

13.10.15
$$\int_0^\infty t^{\frac{1}{2}\nu} U(a,b,t) J_\nu\left(2\sqrt{xt}\right) dt$$
$$= \frac{\Gamma(\nu-b+2)}{\Gamma(a)} x^{\frac{1}{2}\nu} U(\nu-b+2, \nu-a+2, x),$$
$$x > 0, \max(\Re b - 2, -1) < \Re\nu < 2\Re a + \tfrac{1}{2}.$$

13.10.16
$$\int_0^\infty e^{-t} t^{\frac{1}{2}\nu} U(a,b,t) J_\nu\left(2\sqrt{xt}\right) dt$$
$$= \Gamma(\nu-b+2) x^{\frac{1}{2}\nu} e^{-x} \mathbf{M}(a, a-b+\nu+2, x),$$
$$x > 0, \max(\Re b - 2, -1) < \Re\nu.$$

For additional Hankel transforms and also other Bessel transforms see Erdélyi et al. (1954b, §8.18) and Oberhettinger (1972, §§1.16 and 3.4.42–46, 4.4.45–47, 5.94–97).

13.10(vi) Other Integrals

For integral transforms in terms of Whittaker functions see §13.23(iv). Additional integrals can be found in Apelblat (1983, pp. 388–392), Erdélyi et al. (1954b), Gradshteyn and Ryzhik (2000, §7.6), Magnus et al. (1966, §6.1.2), Prudnikov et al. (1990, §§1.13, 1.14, 2.19, 4.2.2), Prudnikov et al. (1992a, §§3.35, 3.36), and Prudnikov et al. (1992b, §§3.33, 3.34). See also (13.4.2), (13.4.5), (13.4.6).

13.11 Series

For $z \in \mathbb{C}$,

13.11.1
$$M(a,b,z) = \Gamma\left(a-\tfrac{1}{2}\right) e^{\frac{1}{2}z} \left(\tfrac{1}{4}z\right)^{\frac{1}{2}-a}$$
$$\times \sum_{s=0}^\infty \frac{(2a-1)_s (2a-b)_s}{(b)_s s!}$$
$$\times \left(a - \tfrac{1}{2} + s\right) I_{a-\frac{1}{2}+s}\left(\tfrac{1}{2}z\right),$$
$$a + \tfrac{1}{2}, b \neq 0, -1, -2, \ldots.$$

(13.6.9) is a special case.

For additional expansions combine (13.14.4), (13.14.5), and §13.24. For other series expansions see Hansen (1975, §§66 and 87), Prudnikov et al. (1990, §6.6), and Tricomi (1954, §1.8). See also §13.13.

13.12 Products

13.12.1
$$M(a,b,z) M(-a,-b,-z)$$
$$+ \frac{a(a-b)z^2}{b^2(1-b^2)} M(1+a, 2+b, z) M(1-a, 2-b, -z) = 1.$$

For generalizations of this quadratic relation see Majima et al. (2000).

For integral representations, integrals, and series containing products of $M(a,b,z)$ and $U(a,b,z)$ see Erdélyi et al. (1953a, §6.15.3).

13.13 Addition and Multiplication Theorems

13.13(i) Addition Theorems for $M(a,b,z)$

The function $M(a,b,x+y)$ has the following expansions:

13.13.1
$$\sum_{n=0}^\infty \frac{(a)_n y^n}{(b)_n n!} M(a+n, b+n, x),$$

13.13.2
$$\left(\frac{x+y}{x}\right)^{1-b} \sum_{n=0}^\infty \frac{(1-b)_n (-y/x)^n}{n!} M(a, b-n, x),$$
$$|y| < |x|,$$

13.13.3
$$\left(\frac{x}{x+y}\right)^a \sum_{n=0}^\infty \frac{(a)_n y^n}{n!(x+y)^n} M(a+n, b, x), \quad \Re(y/x) > -\tfrac{1}{2},$$

13.13.4
$$e^y \sum_{n=0}^\infty \frac{(b-a)_n (-y)^n}{(b)_n n!} M(a, b+n, x),$$

13.13.5
$$e^y \left(\frac{x}{x+y}\right)^{b-a} \sum_{n=0}^\infty \frac{(b-a)_n y^n}{n!(x+y)^n}$$
$$\times M(a-n, b, x), \quad \Re((y+x)/x) > \tfrac{1}{2},$$

13.13.6
$$e^y \left(\frac{x+y}{x}\right)^{1-b} \sum_{n=0}^\infty \frac{(1-b)_n (-y)^n}{n! x^n}$$
$$\times M(a-n, b-n, x), \quad |y| < |x|.$$

13.13(ii) Addition Theorems for $U(a,b,z)$

The function $U(a,b,x+y)$ has the following expansions:

13.13.7
$$\sum_{n=0}^\infty \frac{(a)_n (-y)^n}{n!} U(a+n, b+n, x), \quad |y| < |x|,$$

13.13.8
$$\left(\frac{x+y}{x}\right)^{1-b} \sum_{n=0}^\infty \frac{(1+a-b)_n (-y/x)^n}{n!} U(a, b-n, x),$$
$$|y| < |x|,$$

13.13.9
$$\left(\frac{x}{x+y}\right)^a \sum_{n=0}^\infty \frac{(a)_n (1+a-b)_n y^n}{n!(x+y)^n} U(a+n, b, x),$$
$$\Re(y/x) > -\tfrac{1}{2},$$

13.13.10
$$e^y \sum_{n=0}^\infty \frac{(-y)^n}{n!} U(a, b+n, x), \quad |y| < |x|,$$

13.13.11
$$e^y \left(\frac{x}{x+y}\right)^{b-a} \sum_{n=0}^\infty \frac{(-y)^n}{n!(x+y)^n} U(a-n, b, x),$$
$$\Re(y/x) > -\tfrac{1}{2},$$

13.13.12
$$e^y \left(\frac{x+y}{x}\right)^{1-b} \sum_{n=0}^\infty \frac{(-y)^n}{n! x^n} U(a-n, b-n, x), \quad |y| < |x|.$$

13.13(iii) Multiplication Theorems for $M(a,b,z)$ and $U(a,b,z)$

To obtain similar expansions for $M(a,b,xy)$ and $U(a,b,xy)$, replace y in the previous two subsections by $(y-1)x$.

Whittaker Functions

13.14 Definitions and Basic Properties

13.14(i) Differential Equation

Whittaker's Equation

13.14.1 $$\frac{d^2W}{dz^2} + \left(-\frac{1}{4} + \frac{\kappa}{z} + \frac{\frac{1}{4}-\mu^2}{z^2}\right)W = 0.$$

This equation is obtained from Kummer's equation (13.2.1) via the substitutions $W = e^{-\frac{1}{2}z}z^{\frac{1}{2}+\mu}w$, $\kappa = \frac{1}{2}b - a$, and $\mu = \frac{1}{2}b - \frac{1}{2}$. It has a regular singularity at the origin with indices $\frac{1}{2} \pm \mu$, and an irregular singularity at infinity of rank one.

Standard Solutions

Standard solutions are:

13.14.2 $$M_{\kappa,\mu}(z) = e^{-\frac{1}{2}z}z^{\frac{1}{2}+\mu}M\left(\tfrac{1}{2}+\mu-\kappa, 1+2\mu, z\right),$$

13.14.3 $$W_{\kappa,\mu}(z) = e^{-\frac{1}{2}z}z^{\frac{1}{2}+\mu}U\left(\tfrac{1}{2}+\mu-\kappa, 1+2\mu, z\right),$$

except that $M_{\kappa,\mu}(z)$ does not exist when $2\mu = -1, -2, -3, \ldots$.

Conversely,

13.14.4 $$M(a,b,z) = e^{\frac{1}{2}z}z^{-\frac{1}{2}b}M_{\frac{1}{2}b-a,\frac{1}{2}b-\frac{1}{2}}(z),$$

13.14.5 $$U(a,b,z) = e^{\frac{1}{2}z}z^{-\frac{1}{2}b}W_{\frac{1}{2}b-a,\frac{1}{2}b-\frac{1}{2}}(z).$$

The series

13.14.6
$$M_{\kappa,\mu}(z) = e^{-\frac{1}{2}z}z^{\frac{1}{2}+\mu}\sum_{s=0}^{\infty}\frac{\left(\frac{1}{2}+\mu-\kappa\right)_s}{(1+2\mu)_s s!}z^s$$
$$= z^{\frac{1}{2}+\mu}\sum_{n=0}^{\infty} {}_2F_1\left(\begin{array}{c}-n,\frac{1}{2}+\mu-\kappa\\1+2\mu\end{array};2\right)\frac{\left(-\frac{1}{2}z\right)^n}{n!},$$
$$2\mu \neq -1,-2,-3,\ldots,$$

converge for all $z \in \mathbb{C}$.

In general $M_{\kappa,\mu}(z)$ and $W_{\kappa,\mu}(z)$ are many-valued functions of z with branch points at $z = 0$ and $z = \infty$. The *principal branches* correspond to the principal branches of the functions $z^{\frac{1}{2}+\mu}$ and $U\left(\frac{1}{2}+\mu-\kappa, 1+2\mu, z\right)$ on the right-hand sides of the equations (13.14.2) and (13.14.3); compare §4.2(i).

Although $M_{\kappa,\mu}(z)$ does not exist when $2\mu = -1, -2, -3, \ldots$, many formulas containing $M_{\kappa,\mu}(z)$ continue to apply in their limiting form. For example, if $n = 0, 1, 2, \ldots$, then

13.14.7
$$\lim_{2\mu \to -n-1}\frac{M_{\kappa,\mu}(z)}{\Gamma(2\mu+1)} = \frac{\left(-\frac{1}{2}n-\kappa\right)_{n+1}}{(n+1)!}M_{\kappa,\frac{1}{2}(n+1)}(z)$$
$$= e^{-\frac{1}{2}z}z^{-\frac{1}{2}n}\sum_{s=n+1}^{\infty}\frac{\left(-\frac{1}{2}n-\kappa\right)_s}{\Gamma(s-n)s!}z^s.$$

If $2\mu = \pm n$, where $n = 0, 1, 2, \ldots$, then

13.14.8
$$W_{\kappa,\pm\frac{1}{2}n}(z) = \frac{(-1)^n e^{-\frac{1}{2}z}z^{\frac{1}{2}n+\frac{1}{2}}}{n!\,\Gamma\left(\frac{1}{2}-\frac{1}{2}n-\kappa\right)}\left(\sum_{k=1}^{n}\frac{n!(k-1)!}{(n-k)!\left(\kappa+\frac{1}{2}-\frac{1}{2}n\right)_k}z^{-k}\right.$$
$$\left.-\sum_{k=0}^{\infty}\frac{\left(\frac{1}{2}n+\frac{1}{2}-\kappa\right)_k}{(n+1)_k k!}z^k\left(\ln z + \psi\left(\tfrac{1}{2}n+\tfrac{1}{2}-\kappa+k\right) - \psi(1+k) - \psi(n+1+k)\right)\right),$$
$$\kappa - \tfrac{1}{2}n - \tfrac{1}{2} \neq 0, 1, 2, \ldots,$$

or

13.14.9
$$W_{\kappa,\pm\frac{1}{2}n}(z) = (-1)^{\kappa-\frac{1}{2}n-\frac{1}{2}}e^{-\frac{1}{2}z}z^{\frac{1}{2}n+\frac{1}{2}}\sum_{k=0}^{\kappa-\frac{1}{2}n-\frac{1}{2}}\binom{\kappa-\frac{1}{2}n-\frac{1}{2}}{k}(n+1+k)_{\kappa-k-\frac{1}{2}n-\frac{1}{2}}(-z)^k, \quad \kappa - \tfrac{1}{2}n - \tfrac{1}{2} = 0, 1, 2, \ldots.$$

13.14(ii) Analytic Continuation

13.14.10 $$M_{\kappa,\mu}\!\left(ze^{\pm\pi i}\right) = \pm ie^{\pm\mu\pi i}M_{-\kappa,\mu}(z).$$

In (13.14.11)–(13.14.13) m is any integer.

13.14.11 $$M_{\kappa,\mu}\!\left(ze^{2m\pi i}\right) = (-1)^m e^{2m\mu\pi i}M_{\kappa,\mu}(z).$$

$$W_{\kappa,\mu}\bigl(ze^{2m\pi i}\bigr) = \frac{(-1)^{m+1} 2\pi i \sin(2\pi\mu m)}{\Gamma\bigl(\tfrac{1}{2}-\mu-\kappa\bigr)\Gamma(1+2\mu)\sin(2\pi\mu)} M_{\kappa,\mu}(z) + (-1)^m e^{-2m\mu\pi i} W_{\kappa,\mu}(z). \quad \text{13.14.12}$$

$$\begin{aligned}\text{13.14.13}\quad (-1)^m W_{\kappa,\mu}\bigl(ze^{2m\pi i}\bigr) &= -\frac{e^{2\kappa\pi i}\sin(2m\mu\pi) + \sin((2m-2)\mu\pi)}{\sin(2\mu\pi)} W_{\kappa,\mu}(z) \\ &\quad - \frac{\sin(2m\mu\pi) 2\pi i e^{\kappa\pi i}}{\sin(2\mu\pi)\Gamma\bigl(\tfrac{1}{2}+\mu-\kappa\bigr)\Gamma\bigl(\tfrac{1}{2}-\mu-\kappa\bigr)} W_{-\kappa,\mu}\bigl(ze^{\pi i}\bigr).\end{aligned}$$

Except when $z = 0$, each branch of the functions $M_{\kappa,\mu}(z)/\Gamma(2\mu+1)$ and $W_{\kappa,\mu}(z)$ is entire in κ and μ. Also, unless specified otherwise $M_{\kappa,\mu}(z)$ and $W_{\kappa,\mu}(z)$ are assumed to have their principal values.

13.14(iii) Limiting Forms as $z \to 0$

13.14.14
$$M_{\kappa,\mu}(z) = z^{\mu+\tfrac{1}{2}}(1+O(z)), \quad 2\mu \neq -1,-2,-3,\dots.$$

In cases when $\tfrac{1}{2}-\kappa\pm\mu = -n$, where n is a nonnegative integer,

13.14.15
$$W_{\tfrac{1}{2}\pm\mu+n,\mu}(z) = (-1)^n (1\pm 2\mu)_n z^{\tfrac{1}{2}\pm\mu} + O\bigl(z^{\tfrac{3}{2}\pm\mu}\bigr).$$

In all other cases

13.14.16
$$W_{\kappa,\mu}(z) = \frac{\Gamma(2\mu)}{\Gamma\bigl(\tfrac{1}{2}+\mu-\kappa\bigr)} z^{\tfrac{1}{2}-\mu} + O\bigl(z^{\tfrac{3}{2}-\Re\mu}\bigr),$$
$$\Re\mu \geq \tfrac{1}{2},\ \mu \neq \tfrac{1}{2},$$

13.14.17
$$W_{\kappa,\tfrac{1}{2}}(z) = \frac{1}{\Gamma(1-\kappa)} + O(z\ln z),$$

13.14.18
$$\begin{aligned}W_{\kappa,\mu}(z) &= \frac{\Gamma(2\mu)}{\Gamma\bigl(\tfrac{1}{2}+\mu-\kappa\bigr)} z^{\tfrac{1}{2}-\mu} + \frac{\Gamma(-2\mu)}{\Gamma\bigl(\tfrac{1}{2}-\mu-\kappa\bigr)} z^{\tfrac{1}{2}+\mu} \\ &\quad + O\bigl(z^{\tfrac{3}{2}-\Re\mu}\bigr), \quad 0 \leq \Re\mu < \tfrac{1}{2},\ \mu \neq 0,\end{aligned}$$

13.14.19
$$\begin{aligned}W_{\kappa,0}(z) &= -\frac{\sqrt{z}}{\Gamma\bigl(\tfrac{1}{2}-\kappa\bigr)}\Bigl(\ln z + \psi\bigl(\tfrac{1}{2}-\kappa\bigr) + 2\gamma\Bigr) \\ &\quad + O\bigl(z^{3/2}\ln z\bigr).\end{aligned}$$

For $W_{\kappa,\mu}(z)$ with $\Re\mu < 0$ use (13.14.31).

13.14(iv) Limiting Forms as $z \to \infty$

Except when $\mu-\kappa = -\tfrac{1}{2},-\tfrac{3}{2},\dots$ (polynomial cases),

13.14.20
$$M_{\kappa,\mu}(z) \sim \Gamma(1+2\mu) e^{\tfrac{1}{2}z} z^{-\kappa}\big/\Gamma\bigl(\tfrac{1}{2}+\mu-\kappa\bigr),$$
$$|\operatorname{ph} z| \leq \tfrac{1}{2}\pi - \delta,$$

where δ is an arbitrary small positive constant. Also,

13.14.21
$$W_{\kappa,\mu}(z) \sim e^{-\tfrac{1}{2}z} z^{\kappa}, \quad |\operatorname{ph} z| \leq \tfrac{3}{2}\pi - \delta.$$

13.14(v) Numerically Satisfactory Solutions

Fundamental pairs of solutions of (13.14.1) that are numerically satisfactory (§2.7(iv)) in the neighborhood of infinity are

13.14.22 $\quad W_{\kappa,\mu}(z),\quad W_{-\kappa,\mu}\bigl(e^{-\pi i}z\bigr),\quad -\tfrac{1}{2}\pi \leq \operatorname{ph} z \leq \tfrac{3}{2}\pi,$

13.14.23 $\quad W_{\kappa,\mu}(z),\quad W_{-\kappa,\mu}\bigl(e^{\pi i}z\bigr),\quad -\tfrac{3}{2}\pi \leq \operatorname{ph} z \leq \tfrac{1}{2}\pi.$

A fundamental pair of solutions that is numerically satisfactory in the sector $|\operatorname{ph} z| \leq \pi$ near the origin is

13.14.24 $\quad M_{\kappa,\mu}(z),\quad M_{\kappa,-\mu}(z),\quad 2\mu \notin \mathbb{Z}.$

When 2μ is an integer we may use the results of §13.2(v) with the substitutions $b = 2\mu+1$, $a = \mu-\kappa+\tfrac{1}{2}$, and $W = e^{-\tfrac{1}{2}z}z^{\tfrac{1}{2}+\mu}w$, where W is the solution of (13.14.1) corresponding to the solution w of (13.2.1).

13.14(vi) Wronskians

13.14.25 $\quad \mathscr{W}\{M_{\kappa,\mu}(z), M_{\kappa,-\mu}(z)\} = -2\mu,$

13.14.26 $\quad \mathscr{W}\{M_{\kappa,\mu}(z), W_{\kappa,\mu}(z)\} = -\dfrac{\Gamma(1+2\mu)}{\Gamma\bigl(\tfrac{1}{2}+\mu-\kappa\bigr)},$

13.14.27
$$\mathscr{W}\bigl\{M_{\kappa,\mu}(z), W_{-\kappa,\mu}\bigl(e^{\pm\pi i}z\bigr)\bigr\} = \frac{\Gamma(1+2\mu)}{\Gamma\bigl(\tfrac{1}{2}+\mu+\kappa\bigr)} e^{\mp(\tfrac{1}{2}+\mu)\pi i},$$

13.14.28 $\quad \mathscr{W}\{M_{\kappa,-\mu}(z), W_{\kappa,\mu}(z)\} = -\dfrac{\Gamma(1-2\mu)}{\Gamma\bigl(\tfrac{1}{2}-\mu-\kappa\bigr)},$

13.14.29
$$\begin{aligned}&\mathscr{W}\bigl\{M_{\kappa,-\mu}(z), W_{-\kappa,\mu}\bigl(e^{\pm\pi i}z\bigr)\bigr\} \\ &= \frac{\Gamma(1-2\mu)}{\Gamma\bigl(\tfrac{1}{2}-\mu+\kappa\bigr)} e^{\mp(\tfrac{1}{2}-\mu)\pi i},\end{aligned}$$

13.14.30 $\quad \mathscr{W}\bigl\{W_{\kappa,\mu}(z), W_{-\kappa,\mu}\bigl(e^{\pm\pi i}z\bigr)\bigr\} = e^{\mp\kappa\pi i}.$

13.14(vii) Connection Formulas

13.14.31 $\quad W_{\kappa,\mu}(z) = W_{\kappa,-\mu}(z).$

13.14.32
$$\begin{aligned}\frac{1}{\Gamma(1+2\mu)} M_{\kappa,\mu}(z) &= \frac{e^{\pm(\kappa-\mu-\tfrac{1}{2})\pi i}}{\Gamma\bigl(\tfrac{1}{2}+\mu+\kappa\bigr)} W_{\kappa,\mu}(z) \\ &\quad + \frac{e^{\pm\kappa\pi i}}{\Gamma\bigl(\tfrac{1}{2}+\mu-\kappa\bigr)} W_{-\kappa,\mu}\bigl(e^{\pm\pi i}z\bigr).\end{aligned}$$

When 2μ is not an integer

13.14.33
$$\begin{aligned}W_{\kappa,\mu}(z) &= \frac{\Gamma(-2\mu)}{\Gamma\bigl(\tfrac{1}{2}-\mu-\kappa\bigr)} M_{\kappa,\mu}(z) \\ &\quad + \frac{\Gamma(2\mu)}{\Gamma\bigl(\tfrac{1}{2}+\mu-\kappa\bigr)} M_{\kappa,-\mu}(z).\end{aligned}$$

13.15 Recurrence Relations and Derivatives

13.15(i) Recurrence Relations

13.15.1 $\quad (\kappa - \mu - \tfrac{1}{2}) M_{\kappa-1,\mu}(z) + (z - 2\kappa) M_{\kappa,\mu}(z) + (\kappa + \mu + \tfrac{1}{2}) M_{\kappa+1,\mu}(z) = 0,$

13.15.2 $\quad 2\mu(1+2\mu)\sqrt{z}\, M_{\kappa-\frac{1}{2},\mu-\frac{1}{2}}(z) - (z+2\mu)(1+2\mu) M_{\kappa,\mu}(z) + (\kappa+\mu+\tfrac{1}{2})\sqrt{z}\, M_{\kappa+\frac{1}{2},\mu+\frac{1}{2}}(z) = 0,$

13.15.3 $\quad (\kappa-\mu-\tfrac{1}{2}) M_{\kappa-\frac{1}{2},\mu+\frac{1}{2}}(z) + (1+2\mu)\sqrt{z}\, M_{\kappa,\mu}(z) - (\kappa+\mu+\tfrac{1}{2}) M_{\kappa+\frac{1}{2},\mu+\frac{1}{2}}(z) = 0,$

13.15.4 $\quad 2\mu M_{\kappa-\frac{1}{2},\mu-\frac{1}{2}}(z) - 2\mu M_{\kappa+\frac{1}{2},\mu-\frac{1}{2}}(z) - \sqrt{z}\, M_{\kappa,\mu}(z) = 0,$

13.15.5 $\quad 2\mu(1+2\mu) M_{\kappa,\mu}(z) - 2\mu(1+2\mu)\sqrt{z}\, M_{\kappa-\frac{1}{2},\mu-\frac{1}{2}}(z) - (\kappa-\mu-\tfrac{1}{2})\sqrt{z}\, M_{\kappa-\frac{1}{2},\mu+\frac{1}{2}}(z) = 0,$

13.15.6 $\quad 2\mu(1+2\mu)\sqrt{z}\, M_{\kappa+\frac{1}{2},\mu-\frac{1}{2}}(z) + (z-2\mu)(1+2\mu) M_{\kappa,\mu}(z) + (\kappa-\mu-\tfrac{1}{2})\sqrt{z}\, M_{\kappa-\frac{1}{2},\mu+\frac{1}{2}}(z) = 0,$

13.15.7 $\quad 2\mu(1+2\mu)\sqrt{z}\, M_{\kappa+\frac{1}{2},\mu-\frac{1}{2}}(z) - 2\mu(1+2\mu) M_{\kappa,\mu}(z) + (\kappa+\mu+\tfrac{1}{2})\sqrt{z}\, M_{\kappa+\frac{1}{2},\mu+\frac{1}{2}}(z) = 0.$

13.15.8 $\quad W_{\kappa+\frac{1}{2},\mu+\frac{1}{2}}(z) - \sqrt{z}\, W_{\kappa,\mu}(z) + (\kappa-\mu-\tfrac{1}{2}) W_{\kappa-\frac{1}{2},\mu+\frac{1}{2}}(z) = 0,$

13.15.9 $\quad W_{\kappa+\frac{1}{2},\mu-\frac{1}{2}}(z) - \sqrt{z}\, W_{\kappa,\mu}(z) + (\kappa+\mu-\tfrac{1}{2}) W_{\kappa-\frac{1}{2},\mu-\frac{1}{2}}(z) = 0,$

13.15.10 $\quad 2\mu W_{\kappa,\mu}(z) - \sqrt{z}\, W_{\kappa+\frac{1}{2},\mu+\frac{1}{2}}(z) + \sqrt{z}\, W_{\kappa+\frac{1}{2},\mu-\frac{1}{2}}(z) = 0,$

13.15.11 $\quad W_{\kappa+1,\mu}(z) + (2\kappa - z) W_{\kappa,\mu}(z) + (\kappa-\mu-\tfrac{1}{2})(\kappa+\mu-\tfrac{1}{2}) W_{\kappa-1,\mu}(z) = 0,$

13.15.12 $\quad (\kappa-\mu-\tfrac{1}{2})\sqrt{z}\, W_{\kappa-\frac{1}{2},\mu+\frac{1}{2}}(z) + 2\mu W_{\kappa,\mu}(z) - (\kappa+\mu-\tfrac{1}{2})\sqrt{z}\, W_{\kappa-\frac{1}{2},\mu-\frac{1}{2}}(z) = 0,$

13.15.13 $\quad (\kappa+\mu-\tfrac{1}{2})\sqrt{z}\, W_{\kappa-\frac{1}{2},\mu-\frac{1}{2}}(z) - (z+2\mu) W_{\kappa,\mu}(z) + \sqrt{z}\, W_{\kappa+\frac{1}{2},\mu+\frac{1}{2}}(z) = 0,$

13.15.14 $\quad (\kappa-\mu-\tfrac{1}{2})\sqrt{z}\, W_{\kappa-\frac{1}{2},\mu+\frac{1}{2}}(z) - (z-2\mu) W_{\kappa,\mu}(z) + \sqrt{z}\, W_{\kappa+\frac{1}{2},\mu-\frac{1}{2}}(z) = 0.$

13.15(ii) Differentiation Formulas

13.15.15 $\quad \dfrac{d^n}{dz^n}\left(e^{\frac{1}{2}z} z^{\mu-\frac{1}{2}} M_{\kappa,\mu}(z)\right) = (-1)^n (-2\mu)_n e^{\frac{1}{2}z} z^{\mu-\frac{1}{2}(n+1)} M_{\kappa-\frac{1}{2}n,\mu-\frac{1}{2}n}(z),$

13.15.16 $\quad \dfrac{d^n}{dz^n}\left(e^{\frac{1}{2}z} z^{-\mu-\frac{1}{2}} M_{\kappa,\mu}(z)\right) = \dfrac{\left(\tfrac{1}{2}+\mu-\kappa\right)_n}{(1+2\mu)_n} e^{\frac{1}{2}z} z^{-\mu-\frac{1}{2}(n+1)} M_{\kappa-\frac{1}{2}n,\mu+\frac{1}{2}n}(z),$

13.15.17 $\quad \left(z\dfrac{d}{dz}z\right)^n \left(e^{\frac{1}{2}z} z^{-\kappa-1} M_{\kappa,\mu}(z)\right) = \left(\tfrac{1}{2}+\mu-\kappa\right)_n e^{\frac{1}{2}z} z^{n-\kappa-1} M_{\kappa-n,\mu}(z),$

13.15.18 $\quad \dfrac{d^n}{dz^n}\left(e^{-\frac{1}{2}z} z^{\mu-\frac{1}{2}} M_{\kappa,\mu}(z)\right) = (-1)^n (-2\mu)_n e^{-\frac{1}{2}z} z^{\mu-\frac{1}{2}(n+1)} M_{\kappa+\frac{1}{2}n,\mu-\frac{1}{2}n}(z),$

13.15.19 $\quad \dfrac{d^n}{dz^n}\left(e^{-\frac{1}{2}z} z^{-\mu-\frac{1}{2}} M_{\kappa,\mu}(z)\right) = (-1)^n \dfrac{\left(\tfrac{1}{2}+\mu+\kappa\right)_n}{(1+2\mu)_n} e^{-\frac{1}{2}z} z^{-\mu-\frac{1}{2}(n+1)} M_{\kappa+\frac{1}{2}n,\mu+\frac{1}{2}n}(z),$

13.15.20 $\quad \left(z\dfrac{d}{dz}z\right)^n \left(e^{-\frac{1}{2}z} z^{\kappa-1} M_{\kappa,\mu}(z)\right) = \left(\tfrac{1}{2}+\mu+\kappa\right)_n e^{-\frac{1}{2}z} z^{\kappa+n-1} M_{\kappa+n,\mu}(z).$

13.15.21 $\quad \dfrac{d^n}{dz^n}\left(e^{\frac{1}{2}z} z^{-\mu-\frac{1}{2}} W_{\kappa,\mu}(z)\right) = (-1)^n \left(\tfrac{1}{2}+\mu-\kappa\right)_n e^{\frac{1}{2}z} z^{-\mu-\frac{1}{2}(n+1)} W_{\kappa-\frac{1}{2}n,\mu+\frac{1}{2}n}(z),$

13.15.22 $\quad \dfrac{d^n}{dz^n}\left(e^{\frac{1}{2}z} z^{\mu-\frac{1}{2}} W_{\kappa,\mu}(z)\right) = (-1)^n \left(\tfrac{1}{2}-\mu-\kappa\right)_n e^{\frac{1}{2}z} z^{\mu-\frac{1}{2}(n+1)} W_{\kappa-\frac{1}{2}n,\mu-\frac{1}{2}n}(z),$

13.15.23 $\quad \left(z\dfrac{d}{dz}z\right)^n \left(e^{\frac{1}{2}z} z^{-\kappa-1} W_{\kappa,\mu}(z)\right) = \left(\tfrac{1}{2}+\mu-\kappa\right)_n \left(\tfrac{1}{2}-\mu-\kappa\right)_n e^{\frac{1}{2}z} z^{n-\kappa-1} W_{\kappa-n,\mu}(z),$

13.15.24 $\quad \dfrac{d^n}{dz^n}\left(e^{-\frac{1}{2}z} z^{-\mu-\frac{1}{2}} W_{\kappa,\mu}(z)\right) = (-1)^n e^{-\frac{1}{2}z} z^{-\mu-\frac{1}{2}(n+1)} W_{\kappa+\frac{1}{2}n,\mu+\frac{1}{2}n}(z),$

13.15.25 $\quad \dfrac{d^n}{dz^n}\left(e^{-\frac{1}{2}z} z^{\mu-\frac{1}{2}} W_{\kappa,\mu}(z)\right) = (-1)^n e^{-\frac{1}{2}z} z^{\mu-\frac{1}{2}(n+1)} W_{\kappa+\frac{1}{2}n,\mu-\frac{1}{2}n}(z),$

13.15.26 $\quad \left(z\dfrac{d}{dz}z\right)^n \left(e^{-\frac{1}{2}z} z^{\kappa-1} W_{\kappa,\mu}(z)\right) = (-1)^n e^{-\frac{1}{2}z} z^{\kappa+n-1} W_{\kappa+n,\mu}(z).$

Other versions of several of the identities in this subsection can be constructed by use of (13.3.29).

13.16 Integral Representations

13.16(i) Integrals Along the Real Line

In this subsection see §§10.2(ii), 10.25(ii) for the functions $J_{2\mu}$, $I_{2\mu}$, and $K_{2\mu}$, and §§15.1, 15.2(i) for ${}_2\mathbf{F}_1$.

13.16.1 $\quad M_{\kappa,\mu}(z) = \dfrac{\Gamma(1+2\mu)z^{\mu+\frac{1}{2}}2^{-2\mu}}{\Gamma\left(\frac{1}{2}+\mu-\kappa\right)\Gamma\left(\frac{1}{2}+\mu+\kappa\right)} \displaystyle\int_{-1}^{1} e^{\frac{1}{2}zt}(1+t)^{\mu-\frac{1}{2}-\kappa}(1-t)^{\mu-\frac{1}{2}+\kappa}\,dt, \qquad \Re\mu+\tfrac{1}{2}>|\Re\kappa|,$

13.16.2 $\quad M_{\kappa,\mu}(z) = \dfrac{\Gamma(1+2\mu)z^{\lambda}}{\Gamma(1+2\mu-2\lambda)\Gamma(2\lambda)} \displaystyle\int_{0}^{1} M_{\kappa-\lambda,\mu-\lambda}(zt)e^{\frac{1}{2}z(t-1)}t^{\mu-\lambda-\frac{1}{2}}(1-t)^{2\lambda-1}\,dt, \qquad \Re\mu+\tfrac{1}{2}>\Re\lambda>0,$

13.16.3 $\quad \dfrac{1}{\Gamma(1+2\mu)}M_{\kappa,\mu}(z) = \dfrac{\sqrt{z}e^{\frac{1}{2}z}}{\Gamma\left(\frac{1}{2}+\mu+\kappa\right)} \displaystyle\int_{0}^{\infty} e^{-t}t^{\kappa-\frac{1}{2}}J_{2\mu}\left(2\sqrt{zt}\right)\,dt, \qquad \Re(\kappa+\mu)+\tfrac{1}{2}>0,$

13.16.4 $\quad \dfrac{1}{\Gamma(1+2\mu)}M_{\kappa,\mu}(z) = \dfrac{\sqrt{z}e^{-\frac{1}{2}z}}{\Gamma\left(\frac{1}{2}+\mu-\kappa\right)} \displaystyle\int_{0}^{\infty} e^{-t}t^{-\kappa-\frac{1}{2}}I_{2\mu}\left(2\sqrt{zt}\right)\,dt, \qquad \Re(\kappa-\mu)-\tfrac{1}{2}>0.$

13.16.5 $\quad W_{\kappa,\mu}(z) = \dfrac{z^{\mu+\frac{1}{2}}2^{-2\mu}}{\Gamma\left(\frac{1}{2}+\mu-\kappa\right)} \displaystyle\int_{1}^{\infty} e^{-\frac{1}{2}zt}(t-1)^{\mu-\frac{1}{2}-\kappa}(t+1)^{\mu-\frac{1}{2}+\kappa}\,dt, \quad \Re\mu+\tfrac{1}{2}>\Re\kappa,\ |\operatorname{ph} z|<\tfrac{1}{2}\pi,$

13.16.6
$$W_{\kappa,\mu}(z) = \dfrac{e^{-\frac{1}{2}z}z^{\kappa+1}}{\Gamma\left(\frac{1}{2}+\mu-\kappa\right)\Gamma\left(\frac{1}{2}-\mu-\kappa\right)} \int_{0}^{\infty} \dfrac{W_{-\kappa,\mu}(t)e^{-\frac{1}{2}t}t^{-\kappa-1}}{t+z}\,dt, \quad |\operatorname{ph} z|<\pi,\ \Re(\tfrac{1}{2}+\mu-\kappa)>\max(2\Re\mu,0),$$

13.16.7 $\quad W_{\kappa,\mu}(z) = \dfrac{(-1)^n e^{-\frac{1}{2}z}z^{\frac{1}{2}-\mu-n}}{\Gamma(1+2\mu)\Gamma\left(\frac{1}{2}-\mu-\kappa\right)} \displaystyle\int_{0}^{\infty} \dfrac{M_{-\kappa,\mu}(t)e^{-\frac{1}{2}t}t^{n+\mu-\frac{1}{2}}}{t+z}\,dt,$
$$|\operatorname{ph} z|<\pi,\ n=0,1,2,\ldots,\ -\Re(1+2\mu)<n<|\Re\mu|+\Re\kappa<\tfrac{1}{2},$$

13.16.8 $\quad W_{\kappa,\mu}(z) = \dfrac{2\sqrt{z}e^{-\frac{1}{2}z}}{\Gamma\left(\frac{1}{2}+\mu-\kappa\right)\Gamma\left(\frac{1}{2}-\mu-\kappa\right)} \displaystyle\int_{0}^{\infty} e^{-t}t^{-\kappa-\frac{1}{2}}K_{2\mu}\left(2\sqrt{zt}\right)\,dt, \qquad \Re(\mu-\kappa)+\tfrac{1}{2}>0,$

13.16.9 $\quad W_{\kappa,\mu}(z) = e^{-\frac{1}{2}z}z^{\kappa+c} \displaystyle\int_{0}^{\infty} e^{-zt}t^{c-1}\,{}_2\mathbf{F}_1\!\left(\begin{array}{c}\frac{1}{2}+\mu-\kappa,\frac{1}{2}-\mu-\kappa\\c\end{array};-t\right)dt, \qquad |\operatorname{ph} z|<\tfrac{1}{2}\pi,$

where c is arbitrary, $\Re c>0$.

13.16(ii) Contour Integrals

For contour integral representations combine (13.14.2) and (13.14.3) with §13.4(ii). See Buchholz (1969, §2.3), Erdélyi et al. (1953a, §6.11.3), and Slater (1960, Chapter 3). See also §13.16(iii).

13.16(iii) Mellin–Barnes Integrals

If $\frac{1}{2}+\mu-\kappa\neq 0,-1,-2,\ldots,$ then

13.16.10 $\quad \dfrac{1}{\Gamma(1+2\mu)}M_{\kappa,\mu}\!\left(e^{\pm\pi i}z\right) = \dfrac{e^{\frac{1}{2}z\pm(\frac{1}{2}+\mu)\pi i}}{2\pi i\,\Gamma\left(\frac{1}{2}+\mu-\kappa\right)} \displaystyle\int_{-i\infty}^{i\infty} \dfrac{\Gamma(t-\kappa)\Gamma\left(\frac{1}{2}+\mu-t\right)}{\Gamma\left(\frac{1}{2}+\mu+t\right)}z^t\,dt, \qquad |\operatorname{ph} z|<\tfrac{1}{2}\pi,$

where the contour of integration separates the poles of $\Gamma(t-\kappa)$ from those of $\Gamma\!\left(\frac{1}{2}+\mu-t\right)$.

If $\frac{1}{2}\pm\mu-\kappa\neq 0,-1,-2,\ldots,$ then

13.16.11 $\quad W_{\kappa,\mu}(z) = \dfrac{e^{-\frac{1}{2}z}}{2\pi i} \displaystyle\int_{-i\infty}^{i\infty} \dfrac{\Gamma\left(\frac{1}{2}+\mu+t\right)\Gamma\left(\frac{1}{2}-\mu+t\right)\Gamma(-\kappa-t)}{\Gamma\left(\frac{1}{2}+\mu-\kappa\right)\Gamma\left(\frac{1}{2}-\mu-\kappa\right)}z^{-t}\,dt, \qquad |\operatorname{ph} z|<\tfrac{3}{2}\pi,$

where the contour of integration separates the poles of $\Gamma\!\left(\frac{1}{2}+\mu+t\right)\Gamma\!\left(\frac{1}{2}-\mu+t\right)$ from those of $\Gamma(-\kappa-t)$.

13.16.12 $\quad W_{\kappa,\mu}(z) = \dfrac{e^{\frac{1}{2}z}}{2\pi i} \displaystyle\int_{-i\infty}^{i\infty} \dfrac{\Gamma\left(\frac{1}{2}+\mu+t\right)\Gamma\left(\frac{1}{2}-\mu+t\right)}{\Gamma(1-\kappa+t)}z^{-t}\,dt, \qquad |\operatorname{ph} z|<\tfrac{1}{2}\pi,$

where the contour of integration passes all the poles of $\Gamma\!\left(\frac{1}{2}+\mu+t\right)\Gamma\!\left(\frac{1}{2}-\mu+t\right)$ on the right-hand side.

13.17 Continued Fractions

If $\kappa, \mu \in \mathbb{C}$ such that $\mu \pm (\kappa - \tfrac{1}{2}) \neq -1, -2, -3, \ldots$, then

13.17.1
$$\frac{\sqrt{z}\, M_{\kappa,\mu}(z)}{M_{\kappa-\frac{1}{2},\mu+\frac{1}{2}}(z)} = 1 + \frac{u_1 z}{1+} \frac{u_2 z}{1+} \cdots,$$

where

13.17.2
$$u_{2n+1} = -\frac{\tfrac{1}{2} + \mu + \kappa + n}{(2\mu + 2n + 1)(2\mu + 2n + 2)},$$
$$u_{2n} = \frac{\tfrac{1}{2} + \mu - \kappa + n}{(2\mu + 2n)(2\mu + 2n + 1)}.$$

This continued fraction converges to the meromorphic function of z on the left-hand side for all $z \in \mathbb{C}$. For more details on how a continued fraction converges to a meromorphic function see Jones and Thron (1980).

If $\kappa, \mu \in \mathbb{C}$ such that $\mu + \tfrac{1}{2} \pm (\kappa+1) \neq -1, -2, -3, \ldots$, then

13.17.3
$$\frac{W_{\kappa,\mu}(z)}{\sqrt{z}\, W_{\kappa-\frac{1}{2},\mu-\frac{1}{2}}(z)} = 1 + \frac{v_1/z}{1+} \frac{v_2/z}{1+} \cdots,$$

where

13.17.4 $v_{2n+1} = \tfrac{1}{2} + \mu - \kappa + n$, $v_{2n} = \tfrac{1}{2} - \mu - \kappa + n$.

This continued fraction converges to the meromorphic function of z on the left-hand side throughout the sector $|\operatorname{ph} z| < \pi$.

See also Cuyt et al. (2008, pp. 336–337).

13.18 Relations to Other Functions

13.18(i) Elementary Functions

13.18.1 $\qquad M_{0,\frac{1}{2}}(2z) = 2\sinh z,$

13.18.2
$$M_{\kappa,\kappa-\frac{1}{2}}(z) = W_{\kappa,\kappa-\frac{1}{2}}(z) = W_{\kappa,-\kappa+\frac{1}{2}}(z) = e^{-\frac{1}{2}z} z^\kappa,$$

13.18.3 $\qquad M_{\kappa,-\kappa-\frac{1}{2}}(z) = e^{\frac{1}{2}z} z^{-\kappa}.$

13.18(ii) Incomplete Gamma Functions

For the notation see §§6.2(i), 7.2(i), and 8.2(i). When $\tfrac{1}{2} - \kappa \pm \mu$ is an integer the Whittaker functions can be expressed as incomplete gamma functions (or generalized exponential integrals). For example,

13.18.4 $\qquad M_{\mu-\frac{1}{2},\mu}(z) = 2\mu e^{\frac{1}{2}z} z^{\frac{1}{2}-\mu}\, \gamma(2\mu, z),$

13.18.5 $\qquad W_{\mu-\frac{1}{2},\mu}(z) = e^{\frac{1}{2}z} z^{\frac{1}{2}-\mu}\, \Gamma(2\mu, z).$

Special cases are the error functions

13.18.6 $\qquad M_{-\frac{1}{4},\frac{1}{4}}(z^2) = \tfrac{1}{2} e^{\frac{1}{2}z^2} \sqrt{\pi z}\, \operatorname{erf}(z),$

13.18.7 $\qquad W_{-\frac{1}{4},-\frac{1}{4}}(z^2) = e^{\frac{1}{2}z^2} \sqrt{\pi z}\, \operatorname{erfc}(z).$

13.18(iii) Modified Bessel Functions

When $\kappa = 0$ the Whittaker functions can be expressed as modified Bessel functions. For the notation see §§10.25(ii) and 9.2(i).

13.18.8 $\qquad M_{0,\nu}(2z) = 2^{2\nu+\frac{1}{2}}\, \Gamma(1+\nu)\sqrt{z}\, I_\nu(z),$

13.18.9 $\qquad W_{0,\nu}(2z) = \sqrt{2z/\pi}\, K_\nu(z),$

13.18.10 $\qquad W_{0,\frac{1}{3}}\!\left(\tfrac{4}{3} z^{\frac{3}{2}}\right) = 2\sqrt{\pi}\, z^{\frac{1}{4}}\, \operatorname{Ai}(z).$

13.18(iv) Parabolic Cylinder Functions

For the notation see §12.2.

13.18.11 $\qquad W_{-\frac{1}{2}a,\pm\frac{1}{4}}\!\left(\tfrac{1}{2} z^2\right) = 2^{\frac{1}{2}a} \sqrt{z}\, U(a,z),$

13.18.12
$$M_{-\frac{1}{2}a,-\frac{1}{4}}\!\left(\tfrac{1}{2} z^2\right) = 2^{\frac{1}{2}a-1}\, \Gamma\!\left(\tfrac{1}{2}a + \tfrac{3}{4}\right) \sqrt{z/\pi}$$
$$\times (U(a,z) + U(a,-z)),$$

13.18.13
$$M_{-\frac{1}{2}a,\frac{1}{4}}\!\left(\tfrac{1}{2} z^2\right) = 2^{\frac{1}{2}a-2}\, \Gamma\!\left(\tfrac{1}{2}a + \tfrac{1}{4}\right) \sqrt{z/\pi}$$
$$\times (U(a,-z) - U(a,z)).$$

13.18(v) Orthogonal Polynomials

Special cases of §13.18(iv) are as follows. For the notation see §18.3.

Hermite Polynomials

13.18.14 $\qquad M_{\frac{1}{4}+n,-\frac{1}{4}}(z^2) = (-1)^n \dfrac{n!}{(2n)!}\, e^{-\frac{1}{2}z^2} \sqrt{z}\, H_{2n}(z),$

13.18.15
$$M_{\frac{3}{4}+n,\frac{1}{4}}(z^2) = (-1)^n \frac{n!}{(2n+1)!} \frac{e^{-\frac{1}{2}z^2}\sqrt{z}}{2}\, H_{2n+1}(z),$$

13.18.16 $\qquad W_{\frac{1}{4}+\frac{1}{2}n,\frac{1}{4}}(z^2) = 2^{-n} e^{-\frac{1}{2}z^2} \sqrt{z}\, H_n(z).$

Laguerre Polynomials

13.18.17
$$W_{\frac{1}{2}\alpha+\frac{1}{2}+n,\frac{1}{2}\alpha}(z) = (-1)^n (\alpha+1)_n\, M_{\frac{1}{2}\alpha+\frac{1}{2}+n,\frac{1}{2}\alpha}(z)$$
$$= (-1)^n n!\, e^{-\frac{1}{2}z} z^{\frac{1}{2}\alpha+\frac{1}{2}} L_n^{(\alpha)}(z).$$

13.19 Asymptotic Expansions for Large Argument

As $x \to \infty$

13.19.1
$$M_{\kappa,\mu}(x) \sim \frac{\Gamma(1+2\mu)}{\Gamma(\frac{1}{2}+\mu-\kappa)} e^{\frac{1}{2}x} x^{-\kappa} \sum_{s=0}^{\infty} \frac{\left(\frac{1}{2}-\mu+\kappa\right)_s \left(\frac{1}{2}+\mu+\kappa\right)_s}{s!} x^{-s}, \qquad \mu-\kappa \neq -\tfrac{1}{2},-\tfrac{3}{2},\dots.$$

As $z \to \infty$

13.19.2
$$M_{\kappa,\mu}(z) \sim \frac{\Gamma(1+2\mu)}{\Gamma(\frac{1}{2}+\mu-\kappa)} e^{\frac{1}{2}z} z^{-\kappa} \sum_{s=0}^{\infty} \frac{\left(\frac{1}{2}-\mu+\kappa\right)_s \left(\frac{1}{2}+\mu+\kappa\right)_s}{s!} z^{-s}$$
$$+ \frac{\Gamma(1+2\mu)}{\Gamma(\frac{1}{2}+\mu+\kappa)} e^{-\frac{1}{2}z \pm (\frac{1}{2}+\mu-\kappa)\pi i} z^{\kappa} \sum_{s=0}^{\infty} \frac{\left(\frac{1}{2}+\mu-\kappa\right)_s \left(\frac{1}{2}-\mu-\kappa\right)_s}{s!} (-z)^{-s},$$
$$-\tfrac{1}{2}\pi + \delta \leq \pm\operatorname{ph} z \leq \tfrac{3}{2}\pi - \delta,$$

provided that both $\mu \mp \kappa \neq -\tfrac{1}{2}, -\tfrac{3}{2}, \dots$. Again, δ denotes an arbitrary small positive constant. Also,

13.19.3
$$W_{\kappa,\mu}(z) \sim e^{-\frac{1}{2}z} z^{\kappa} \sum_{s=0}^{\infty} \frac{\left(\frac{1}{2}+\mu-\kappa\right)_s \left(\frac{1}{2}-\mu-\kappa\right)_s}{s!} (-z)^{-s}, \qquad |\operatorname{ph} z| \leq \tfrac{3}{2}\pi - \delta.$$

Error bounds and exponentially-improved expansions are derivable by combining §§13.7(ii) and 13.7(iii) with (13.14.2) and (13.14.3). See also Olver (1965).

For an asymptotic expansion of $W_{\kappa,\mu}(z)$ as $z \to \infty$ that is valid in the sector $|\operatorname{ph} z| \leq \pi - \delta$ and where the real parameters κ, μ are subject to the growth conditions $\kappa = o(z)$, $\mu = o(\sqrt{z})$, see Wong (1973a).

13.20 Uniform Asymptotic Approximations for Large μ

13.20(i) Large μ, Fixed κ

When $\mu \to \infty$ in the sector $|\operatorname{ph} \mu| \leq \tfrac{1}{2}\pi - \delta (< \tfrac{1}{2}\pi)$, with $\kappa (\in \mathbb{C})$ fixed

13.20.1 $M_{\kappa,\mu}(z) = z^{\mu+\frac{1}{2}} \left(1 + O(\mu^{-1})\right),$

uniformly for bounded values of $|z|$; also

13.20.2
$$W_{\kappa,\mu}(x) = \pi^{-\frac{1}{2}} \Gamma(\kappa+\mu) \left(\tfrac{1}{4}x\right)^{\frac{1}{2}-\mu} \left(1 + O(\mu^{-1})\right),$$
uniformly for bounded positive values of x. For an extension of (13.20.1) to an asymptotic expansion, together with error bounds, see Olver (1997b, Chapter 10, Ex. 3.4).

13.20(ii) Large μ, $0 \leq \kappa \leq (1-\delta)\mu$

Let

13.20.3 $X = \sqrt{4\mu^2 - 4\kappa x + x^2}.$

Then as $\mu \to \infty$

13.20.4
$$M_{\kappa,\mu}(x) = \sqrt{\frac{2\mu x}{X}} \left(\frac{4\mu^2 x}{2\mu^2 - \kappa x + \mu X}\right)^{\mu}$$
$$\times \left(\frac{2(\mu-\kappa)}{X+x-2\kappa}\right)^{\kappa} e^{\frac{1}{2}X-\mu} \left(1 + O\left(\frac{1}{\mu}\right)\right),$$

13.20.5
$$W_{\kappa,\mu}(x) = \sqrt{\frac{x}{X}} \left(\frac{2\mu^2 - \kappa x + \mu X}{(\mu-\kappa)x}\right)^{\mu} \left(\frac{X+x-2\kappa}{2}\right)^{\kappa}$$
$$\times e^{-\frac{1}{2}X-\kappa} \left(1 + O\left(\frac{1}{\mu}\right)\right),$$

uniformly with respect to $x \in (0,\infty)$ and $\kappa \in [0, (1-\delta)\mu]$, where δ again denotes an arbitrary small positive constant.

13.20(iii) Large μ, $-(1-\delta)\mu \leq \kappa \leq \mu$

Let

13.20.6 $\alpha = \sqrt{2|\kappa-\mu|/\mu},$

13.20.7 $X = \sqrt{|x^2 - 4\kappa x + 4\mu^2|},$

13.20.8 $\Phi(\kappa,\mu,x) = \left(\frac{\mu^2 \zeta^2 - 2\kappa\mu + 2\mu^2}{x^2 - 4\kappa x + 4\mu^2}\right)^{\frac{1}{4}} \sqrt{x},$

with the variable ζ defined implicitly as follows:

(a) In the case $-\mu < \kappa < \mu$

13.20.9
$$\zeta\sqrt{\zeta^2 + \alpha^2} + \alpha^2 \operatorname{arcsinh}\left(\frac{\zeta}{\alpha}\right)$$
$$= \frac{X}{\mu} - \frac{2\kappa}{\mu} \ln\left(\frac{X+x-2\kappa}{2\sqrt{\mu^2-\kappa^2}}\right) - 2\ln\left(\frac{\mu X + 2\mu^2 - \kappa x}{x\sqrt{\mu^2-\kappa^2}}\right).$$

(b) In the case $\mu = \kappa$

13.20.10 $\zeta = \pm\sqrt{\dfrac{x}{\mu} - 2 - 2\ln\left(\dfrac{x}{2\mu}\right)},$

the upper or lower sign being taken according as $x \gtrless 2\mu$.

(In both cases (a) and (b) the x-interval $(0,\infty)$ is mapped one-to-one onto the ζ-interval $(-\infty,\infty)$, with

$x = 0$ and ∞ corresponding to $\zeta = -\infty$ and ∞, respectively.) Then as $\mu \to \infty$

13.20.11
$$W_{\kappa,\mu}(x) = \left(\tfrac{1}{2}\mu\right)^{-\frac{1}{4}} \left(\frac{\kappa+\mu}{e}\right)^{\frac{1}{2}(\kappa+\mu)} \Phi(\kappa,\mu,x)$$
$$\times U\left(\mu-\kappa, \zeta\sqrt{2\mu}\right)\left(1 + O(\mu^{-1}\ln\mu)\right),$$

13.20.12
$$M_{\kappa,\mu}(x) = (8\mu)^{\frac{1}{4}} \left(\frac{2\mu}{e}\right)^{2\mu} \left(\frac{e}{\kappa+\mu}\right)^{\frac{1}{2}(\kappa+\mu)} \Phi(\kappa,\mu,x)$$
$$\times U\left(\mu-\kappa, -\zeta\sqrt{2\mu}\right)\left(1 + O(\mu^{-1}\ln\mu)\right),$$

13.20.13
$$\zeta\sqrt{\zeta^2-\alpha^2} - \alpha^2 \operatorname{arccosh}\left(\frac{\zeta}{\alpha}\right) = \frac{X}{\mu} - \frac{2\kappa}{\mu}\ln\left(\frac{X+x-2\kappa}{2\sqrt{\kappa^2-\mu^2}}\right) - 2\ln\left(\frac{\kappa x - \mu X - 2\mu^2}{x\sqrt{\kappa^2-\mu^2}}\right), \quad x \geq 2\kappa + 2\sqrt{\kappa^2-\mu^2},$$

13.20.14
$$\zeta\sqrt{\alpha^2-\zeta^2} + \alpha^2 \arcsin\left(\frac{\zeta}{\alpha}\right) = \frac{X}{\mu} + \frac{2\kappa}{\mu}\arctan\left(\frac{x-2\kappa}{X}\right) - 2\arctan\left(\frac{\kappa x - 2\mu^2}{\mu X}\right),$$
$$2\kappa - 2\sqrt{\kappa^2-\mu^2} \leq x \leq 2\kappa + 2\sqrt{\kappa^2-\mu^2},$$

13.20.15
$$-\zeta\sqrt{\zeta^2-\alpha^2} - \alpha^2 \operatorname{arccosh}\left(-\frac{\zeta}{\alpha}\right) = -\frac{X}{\mu} + \frac{2\kappa}{\mu}\ln\left(\frac{2\kappa - X - x}{2\sqrt{\kappa^2-\mu^2}}\right) + 2\ln\left(\frac{\mu X + 2\mu^2 - \kappa x}{x\sqrt{\kappa^2-\mu^2}}\right),$$
$$0 < x \leq 2\kappa - 2\sqrt{\kappa^2-\mu^2},$$

when $\mu < \kappa$, and by (13.20.10) when $\mu = \kappa$. (As in §13.20(iii) $x = 0$ and ∞ correspond to $\zeta = -\infty$ and ∞, respectively). Then as $\mu \to \infty$

13.20.16 $W_{\kappa,\mu}(x) = \left(\tfrac{1}{2}\mu\right)^{-\frac{1}{4}} \left(\dfrac{\kappa+\mu}{e}\right)^{\frac{1}{2}(\kappa+\mu)} \Phi(\kappa,\mu,x) \left(U\left(\mu-\kappa, \zeta\sqrt{2\mu}\right) + \operatorname{env}U\left(\mu-\kappa, \zeta\sqrt{2\mu}\right) O\left(\mu^{-\frac{2}{3}}\right)\right),$

13.20.17 $M_{\kappa,\mu}(x) = (8\mu)^{\frac{1}{4}} \left(\dfrac{2\mu}{e}\right)^{2\mu} \left(\dfrac{e}{\kappa+\mu}\right)^{\frac{1}{2}(\kappa+\mu)} \Phi(\kappa,\mu,x) \left(U\left(\mu-\kappa, -\zeta\sqrt{2\mu}\right) + \operatorname{env}\overline{U}\left(\mu-\kappa, \zeta\sqrt{2\mu}\right) O\left(\mu^{-\frac{2}{3}}\right)\right),$

uniformly with respect to $\zeta \in [0,\infty)$ and $\kappa \in [\mu, \mu/\delta]$.

Also,

13.20.18 $W_{\kappa,\mu}(x) = \left(\tfrac{1}{2}\mu\right)^{-\frac{1}{4}} \left(\dfrac{\kappa+\mu}{e}\right)^{\frac{1}{2}(\kappa+\mu)} \Phi(\kappa,\mu,x) \left(U\left(\mu-\kappa, \zeta\sqrt{2\mu}\right) + \operatorname{env}\overline{U}\left(\mu-\kappa, -\zeta\sqrt{2\mu}\right) O\left(\mu^{-\frac{2}{3}}\right)\right),$

13.20.19
$$M_{\kappa,\mu}(x) = (8\mu)^{\frac{1}{4}} \left(\frac{2\mu}{e}\right)^{2\mu} \left(\frac{e}{\kappa+\mu}\right)^{\frac{1}{2}(\kappa+\mu)} \Phi(\kappa,\mu,x) \left(U\left(\mu-\kappa, -\zeta\sqrt{2\mu}\right) + \operatorname{env}U\left(\mu-\kappa, -\zeta\sqrt{2\mu}\right) O\left(\mu^{-\frac{2}{3}}\right)\right),$$

uniformly with respect to $\zeta \in (-\infty, 0]$ and $\kappa \in [\mu, \mu/\delta]$.

For the parabolic cylinder functions U and $\overline{U}$ see §12.2, and for the env functions associated with U and $\overline{U}$ see §14.15(v).

These results are proved in Olver (1980b). Equations (13.20.17) and (13.20.18) are simpler than (6.10) and (6.11) in this reference. Olver (1980b) also supplies error bounds and corresponding approximations when

uniformly with respect to $x \in (0,\infty)$ and $\kappa \in [-(1-\delta)\mu, \mu]$. For the parabolic cylinder function U see §12.2.

These results are proved in Olver (1980b). This reference also supplies error bounds and corresponding approximations when x, κ, and μ are replaced by ix, $i\kappa$, and $i\mu$, respectively.

13.20(iv) Large μ, $\mu \leq \kappa \leq \mu/\delta$

Again define α, X, and $\Phi(\kappa,\mu,x)$ by (13.20.6)–(13.20.8), but with ζ now defined by

x, κ, and μ are replaced by ix, $i\kappa$, and $i\mu$, respectively.

It should be noted that (13.20.11), (13.20.16), and (13.20.18) differ only in the common error terms. Hence without the error terms the approximation holds for $-(1-\delta)\mu \leq \kappa \leq \mu/\delta$. Similarly for (13.20.12), (13.20.17), and (13.20.19).

13.20(v) Large μ, Other Expansions

For uniform approximations valid when μ is large, $x/i \in (0, \infty)$, and $\kappa/i \in [0, \mu/\delta]$, see Olver (1997b, pp. 401–403). These approximations are in terms of Airy functions.

For uniform approximations of $M_{\kappa,i\mu}(z)$ and $W_{\kappa,i\mu}(z)$, κ and μ real, one or both large, see Dunster (2003a).

13.21 Uniform Asymptotic Approximations for Large κ

13.21(i) Large κ, Fixed μ

For the notation see §§10.2(ii), 10.25(ii), and 2.8(iv).

When $\kappa \to \infty$ through positive real values with μ (≥ 0) fixed

$$13.21.1 \quad M_{\kappa,\mu}(x) = \sqrt{x}\,\Gamma(2\mu+1)\kappa^{-\mu}\left(J_{2\mu}(2\sqrt{x\kappa}) + \mathrm{env}J_{2\mu}(2\sqrt{x\kappa})\,O\!\left(\kappa^{-\frac{1}{2}}\right)\right),$$

$$13.21.2 \quad W_{\kappa,\mu}(x) = \sqrt{x}\,\Gamma\!\left(\kappa+\tfrac{1}{2}\right)\left(\sin(\kappa\pi - \mu\pi)\,J_{2\mu}(2\sqrt{x\kappa}) - \cos(\kappa\pi - \mu\pi)\,Y_{2\mu}(2\sqrt{x\kappa}) + \mathrm{env}Y_{2\mu}(2\sqrt{x\kappa})\,O\!\left(\kappa^{-\frac{1}{2}}\right)\right),$$

$$13.21.3 \quad W_{-\kappa,\mu}(xe^{-\pi i}) = \frac{\pi\sqrt{x}}{\Gamma(\kappa+\tfrac{1}{2})}e^{\mu\pi i}\left(H^{(1)}_{2\mu}(2\sqrt{x\kappa}) + \mathrm{env}Y_{2\mu}(2\sqrt{x\kappa})\,O\!\left(\kappa^{-\frac{1}{2}}\right)\right),$$

$$13.21.4 \quad W_{-\kappa,\mu}(xe^{\pi i}) = \frac{\pi\sqrt{x}}{\Gamma(\kappa+\tfrac{1}{2})}e^{-\mu\pi i}\left(H^{(2)}_{2\mu}(2\sqrt{x\kappa}) + \mathrm{env}Y_{2\mu}(2\sqrt{x\kappa})\,O\!\left(\kappa^{-\frac{1}{2}}\right)\right),$$

uniformly with respect to $x \in (0, A]$ in each case, where A is an arbitrary positive constant.

Other types of approximations when $\kappa \to \infty$ through positive real values with μ (≥ 0) fixed are as follows. Define

$$13.21.5 \quad 2\sqrt{\zeta} = \sqrt{x+x^2} + \ln(\sqrt{x} + \sqrt{1+x}).$$

Then

$$13.21.6 \quad M_{-\kappa,\mu}(4\kappa x) = \frac{2\,\Gamma(2\mu+1)}{\kappa^{\mu-\frac{1}{2}}}\left(\frac{x\zeta}{1+x}\right)^{\frac{1}{4}} I_{2\mu}\!\left(4\kappa\zeta^{\frac{1}{2}}\right)\!\left(1+O(\kappa^{-1})\right),$$

$$13.21.7 \quad W_{-\kappa,\mu}(4\kappa x) = \frac{\sqrt{8/\pi}\,e^{\kappa}}{\kappa^{\kappa-\frac{1}{2}}}\left(\frac{x\zeta}{1+x}\right)^{\frac{1}{4}} K_{2\mu}\!\left(4\kappa\zeta^{\frac{1}{2}}\right)\!\left(1+O(\kappa^{-1})\right),$$

uniformly with respect to $x \in (0, \infty)$.

For (13.21.6), (13.21.7), and extensions to asymptotic expansions and error bounds, see Olver (1997b, Chapter 12, Exs. 12.4.5, 12.4.6). For extensions to complex values of x see López (1999).

13.21(ii) Large κ, $0 \leq \mu \leq (1-\delta)\kappa$

Let

$$13.21.8 \quad c(\kappa,\mu) = e^{\mu\pi i}\sqrt{\tfrac{1}{2}\pi}\left(\frac{\kappa-\mu}{\kappa+\mu}\right)^{\frac{1}{2}\mu}\left(\frac{e^2}{\kappa^2-\mu^2}\right)^{\frac{1}{2}\kappa},$$

$$13.21.9 \quad X = \sqrt{|x^2 - 4\kappa x + 4\mu^2|},$$

$$13.21.10 \quad \Psi(\kappa,\mu,x) = \left(\frac{4\mu^2 - \kappa\zeta}{x^2 - 4\kappa x + 4\mu^2}\right)^{\frac{1}{4}}\sqrt{x},$$

with the variable ζ defined implicitly by

$$13.21.11 \quad \sqrt{4\mu^2 - \kappa\zeta} - \mu\ln\!\left(\frac{2\mu + \sqrt{4\mu^2 - \kappa\zeta}}{2\mu - \sqrt{4\mu^2 - \kappa\zeta}}\right) = \tfrac{1}{2}X + \mu\ln\!\left(\frac{x\sqrt{\kappa^2-\mu^2}}{2\mu^2 - \kappa x + \mu X}\right) + \kappa\ln\!\left(\frac{2\sqrt{\kappa^2-\mu^2}}{2\kappa - x - X}\right),$$
$$0 < x \leq 2\kappa - 2\sqrt{\kappa^2-\mu^2},$$

and

$$13.21.12 \quad \sqrt{\kappa\zeta - 4\mu^2} - 2\mu\arctan\!\left(\frac{\sqrt{\kappa\zeta - 4\mu^2}}{2\mu}\right) = \tfrac{1}{2}(X - \pi\mu) - \mu\arctan\!\left(\frac{x\kappa - 2\mu^2}{\mu X}\right) + \kappa\arcsin\!\left(\frac{X}{2\sqrt{\kappa^2-\mu^2}}\right),$$
$$2\kappa - 2\sqrt{\kappa^2-\mu^2} \leq x < 2\kappa + 2\sqrt{\kappa^2-\mu^2}.$$

Then as $\kappa \to \infty$

$$13.21.13 \quad M_{\kappa,\mu}(x) = \Gamma(2\mu+1)\left(\frac{e^2}{\kappa^2-\mu^2}\right)^{\frac{1}{2}\mu}\left(\frac{\kappa-\mu}{\kappa+\mu}\right)^{\frac{1}{2}\kappa}\Psi(\kappa,\mu,x)\left(J_{2\mu}\!\left(\sqrt{\zeta\kappa}\right) + \mathrm{env}J_{2\mu}\!\left(\sqrt{\zeta\kappa}\right)O(\kappa^{-1})\right),$$

$$13.21.14 \quad W_{\kappa,\mu}(x) = \frac{e^{-\mu\pi i}}{\pi}\Gamma\!\left(\kappa+\mu+\tfrac{1}{2}\right)\Gamma\!\left(\kappa-\mu+\tfrac{1}{2}\right)c(\kappa,\mu)\Psi(\kappa,\mu,x)$$
$$\times\left(\sin(\kappa\pi-\mu\pi)J_{2\mu}\!\left(\sqrt{\zeta\kappa}\right) - \cos(\kappa\pi-\mu\pi)Y_{2\mu}\!\left(\sqrt{\zeta\kappa}\right) + \mathrm{env}Y_{2\mu}\!\left(\sqrt{\zeta\kappa}\right)O(\kappa^{-1})\right),$$

13.21.15
$$W_{-\kappa,\mu}(xe^{-\pi i}) = c(\kappa,\mu)\Psi(\kappa,\mu,x)\left(H^{(1)}_{2\mu}\left(\sqrt{\zeta\kappa}\right) + \text{env}Y_{2\mu}\left(\sqrt{\zeta\kappa}\right)O(\kappa^{-1})\right),$$

13.21.16
$$W_{-\kappa,\mu}(xe^{\pi i}) = e^{-2\mu\pi i}c(\kappa,\mu)\Psi(\kappa,\mu,x)\left(H^{(2)}_{2\mu}\left(\sqrt{\zeta\kappa}\right) + \text{env}Y_{2\mu}\left(\sqrt{\zeta\kappa}\right)O(\kappa^{-1})\right),$$

uniformly with respect to $\mu \in [0,(1-\delta)\kappa]$ and $x \in \left(0,(1-\delta)(2\kappa+2\sqrt{\kappa^2-\mu^2})\right]$, where δ again denotes an arbitrary small positive constant. For the functions $J_{2\mu}$, $Y_{2\mu}$, $H^{(1)}_{2\mu}$, and $H^{(2)}_{2\mu}$ see §10.2(ii), and for the env functions associated with $J_{2\mu}$ and $Y_{2\mu}$ see §2.8(iv).

These approximations are proved in Dunster (1989). This reference also includes error bounds and extensions to asymptotic expansions and complex values of x.

13.21(iii) Large κ, $0 \leq \mu \leq (1-\delta)\kappa$ (Continued)

Let

13.21.17 $\quad \widehat{c}(\kappa,\mu) = \sqrt{2\pi}\kappa^{\frac{1}{6}}\left(\dfrac{\kappa-\mu}{\kappa+\mu}\right)^{\frac{1}{2}\mu}\left(\dfrac{e^2}{\kappa^2-\mu^2}\right)^{\frac{1}{2}\kappa},$

13.21.18 $\quad X = \sqrt{|x^2-4\kappa x+4\mu^2|},$

13.21.19 $\quad \widehat{\Psi}(\kappa,\mu,x) = \left(\dfrac{\widehat{\zeta}}{x^2-4\kappa x+4\mu^2}\right)^{\frac{1}{4}}\sqrt{2x},$

and define the variable $\widehat{\zeta}$ implicitly by

13.21.20
$$\widehat{\zeta} = -\left(\frac{3}{2\kappa}\left(-\frac{1}{2}X + 2\mu\arctan\left(\frac{x\kappa - x\sqrt{\kappa^2-\mu^2} - 2\mu^2}{\mu X}\right) + \kappa\arccos\left(\frac{x-2\kappa}{2\sqrt{\kappa^2-\mu^2}}\right)\right)\right)^{2/3},$$
$$2\kappa - 2\sqrt{\kappa^2-\mu^2} < x \leq 2\kappa + 2\sqrt{\kappa^2-\mu^2},$$

and

13.21.21
$$\widehat{\zeta} = \left(\frac{3}{2\kappa}\left(\frac{1}{2}X + \mu\ln\left(\frac{x\sqrt{\kappa^2-\mu^2}}{\kappa x - 2\mu^2 - \mu X}\right) + \kappa\ln\left(\frac{2\sqrt{\kappa^2-\mu^2}}{x-2\kappa+X}\right)\right)\right)^{2/3}, \quad x \geq 2\kappa+2\sqrt{\kappa^2-\mu^2}.$$

Then as $\kappa \to \infty$

13.21.22
$$M_{\kappa,\mu}(x) = \frac{1}{2\pi}\,\Gamma(2\mu+1)\,\Gamma\!\left(\kappa-\mu+\tfrac{1}{2}\right)\widehat{c}(\kappa,\mu)\widehat{\Psi}(\kappa,\mu,x)$$
$$\times\left(\sin(\kappa\pi-\mu\pi)\,\text{Ai}\!\left(\kappa^{\frac{2}{3}}\widehat{\zeta}\right) + \cos(\kappa\pi-\mu\pi)\,\text{Bi}\!\left(\kappa^{\frac{2}{3}}\widehat{\zeta}\right) + \text{envBi}\!\left(\kappa^{\frac{2}{3}}\widehat{\zeta}\right)O(\kappa^{-1})\right),$$

13.21.23 $\quad W_{\kappa,\mu}(x) = \sqrt{2\pi}\kappa^{\frac{1}{6}}\left(\dfrac{\kappa+\mu}{\kappa-\mu}\right)^{\frac{1}{2}\mu}\left(\dfrac{\kappa^2-\mu^2}{e^2}\right)^{\frac{1}{2}\kappa}\widehat{\Psi}(\kappa,\mu,x)\left(\text{Ai}\!\left(\kappa^{\frac{2}{3}}\widehat{\zeta}\right) + \text{envAi}\!\left(\kappa^{\frac{2}{3}}\widehat{\zeta}\right)O(\kappa^{-1})\right),$

13.21.24 $\quad W_{-\kappa,\mu}(xe^{-\pi i}) = e^{(\kappa-\frac{1}{6})\pi i}\widehat{c}(\kappa,\mu)\widehat{\Psi}(\kappa,\mu,x)\left(\text{Ai}\!\left(\kappa^{\frac{2}{3}}\widehat{\zeta}e^{-\frac{2}{3}\pi i}\right) + \text{envBi}\!\left(\kappa^{\frac{2}{3}}\widehat{\zeta}\right)O(\kappa^{-1})\right),$

13.21.25 $\quad W_{-\kappa,\mu}(xe^{\pi i}) = e^{-(\kappa-\frac{1}{6})\pi i}\widehat{c}(\kappa,\mu)\widehat{\Psi}(\kappa,\mu,x)\left(\text{Ai}\!\left(\kappa^{\frac{2}{3}}\widehat{\zeta}e^{\frac{2}{3}\pi i}\right) + \text{envBi}\!\left(\kappa^{\frac{2}{3}}\widehat{\zeta}\right)O(\kappa^{-1})\right),$

uniformly with respect to $\mu \in [0,(1-\delta)\kappa]$ and $x \in \left[(1+\delta)(2\kappa-2\sqrt{\kappa^2-\mu^2}),\infty\right)$. For the functions Ai and Bi see §9.2(i), and for the env functions associated with Ai and Bi see §2.8(iii).

These approximations are proved in Dunster (1989). This reference also includes error bounds and extensions to asymptotic expansions and complex values of x.

13.21(iv) Large κ, Other Expansions

For a uniform asymptotic expansion in terms of Airy functions for $W_{\kappa,\mu}(4\kappa x)$ when κ is large and positive, μ is real with $|\mu|$ bounded, and $x \in [\delta,\infty)$ see Olver (1997b, Chapter 11, Ex. 7.3). This expansion is simpler in form than the expansions of Dunster (1989) that correspond to the approximations given in §13.21(iii), but the conditions on μ are more restrictive.

For asymptotic expansions having double asymptotic properties see Skovgaard (1966).

See also §13.20(v).

13.22 Zeros

From (13.14.2) and (13.14.3) $M_{\kappa,\mu}(z)$ has the same zeros as $M\!\left(\tfrac{1}{2}+\mu-\kappa, 1+2\mu, z\right)$ and $W_{\kappa,\mu}(z)$ has the same zeros as $U\!\left(\tfrac{1}{2}+\mu-\kappa, 1+2\mu, z\right)$, hence the results given in §13.9 can be adopted.

Asymptotic approximations to the zeros when the parameters κ and/or μ are large can be found by reversion of the uniform approximations provided in §§13.20 and 13.21. For example, if $\mu(\geq 0)$ is fixed and $\kappa(>0)$ is large, then the rth positive zero ϕ_r of $M_{\kappa,\mu}(z)$ is given by

13.22.1 $$\phi_r = \frac{j_{2\mu,r}^2}{4\kappa} + j_{2\mu,r}\, O\!\left(\kappa^{-\frac{3}{2}}\right),$$

where $j_{2\mu,r}$ is the rth positive zero of the Bessel function $J_{2\mu}(x)$ (§10.21(i)). (13.22.1) is a weaker version of (13.9.8).

13.23 Integrals

13.23(i) Laplace and Mellin Transforms

For the notation see §§15.1, 15.2(i), and 10.25(ii).

13.23.1 $$\int_0^\infty e^{-zt} t^{\nu-1} M_{\kappa,\mu}(t)\, dt = \frac{\Gamma\!\left(\mu+\nu+\tfrac{1}{2}\right)}{\left(z+\tfrac{1}{2}\right)^{\mu+\nu+\frac{1}{2}}}\, {}_2F_1\!\left(\begin{array}{c}\tfrac{1}{2}+\mu-\kappa,\tfrac{1}{2}+\mu+\nu\\1+2\mu\end{array}; \frac{1}{z+\tfrac{1}{2}}\right), \quad \Re(\mu+\nu+\tfrac{1}{2})>0,\ \Re z > \tfrac{1}{2}.$$

13.23.2 $$\int_0^\infty e^{-zt} t^{\mu-\frac{1}{2}} M_{\kappa,\mu}(t)\, dt = \Gamma(2\mu+1)\left(z+\tfrac{1}{2}\right)^{-\kappa-\mu-\frac{1}{2}}\left(z-\tfrac{1}{2}\right)^{\kappa-\mu-\frac{1}{2}}, \quad \Re\mu > -\tfrac{1}{2},\ \Re z > \tfrac{1}{2},$$

13.23.3 $$\frac{1}{\Gamma(1+2\mu)}\int_0^\infty e^{-\frac{1}{2}t} t^{\nu-1} M_{\kappa,\mu}(t)\, dt = \frac{\Gamma\!\left(\mu+\nu+\tfrac{1}{2}\right)\Gamma(\kappa-\nu)}{\Gamma\!\left(\tfrac{1}{2}+\mu+\kappa\right)\Gamma\!\left(\tfrac{1}{2}+\mu-\nu\right)}, \quad -\tfrac{1}{2}-\Re\mu < \Re\nu < \Re\kappa.$$

13.23.4 $$\int_0^\infty e^{-zt} t^{\nu-1} W_{\kappa,\mu}(t)\, dt = \Gamma\!\left(\tfrac{1}{2}+\mu+\nu\right)\Gamma\!\left(\tfrac{1}{2}-\mu+\nu\right)\, {}_2F_1\!\left(\begin{array}{c}\tfrac{1}{2}-\mu+\nu,\tfrac{1}{2}+\mu+\nu\\\nu-\kappa+1\end{array}; \tfrac{1}{2}-z\right),$$
$$\Re(\nu+\tfrac{1}{2}) > |\Re\mu|,\ \Re z > -\tfrac{1}{2},$$

13.23.5 $$\int_0^\infty e^{\frac{1}{2}t} t^{\nu-1} W_{\kappa,\mu}(t)\, dt = \frac{\Gamma\!\left(\tfrac{1}{2}+\mu+\nu\right)\Gamma\!\left(\tfrac{1}{2}-\mu+\nu\right)\Gamma(-\kappa-\nu)}{\Gamma\!\left(\tfrac{1}{2}+\mu-\kappa\right)\Gamma\!\left(\tfrac{1}{2}-\mu-\kappa\right)}, \quad |\Re\mu|-\tfrac{1}{2} < \Re\nu < -\Re\kappa.$$

13.23.6 $$\frac{1}{\Gamma(1+2\mu)2\pi i}\int_{-\infty}^{(0+)} e^{zt+\frac{1}{2}t^{-1}} t^\kappa M_{\kappa,\mu}(t^{-1})\, dt = \frac{z^{-\kappa-\frac{1}{2}}}{\Gamma\!\left(\tfrac{1}{2}+\mu-\kappa\right)}\, I_{2\mu}(2\sqrt{z}), \quad \Re z > 0.$$

13.23.7 $$\frac{1}{2\pi i}\int_{-\infty}^{(0+)} e^{zt+\frac{1}{2}t^{-1}} t^\kappa W_{\kappa,\mu}(t^{-1})\, dt = \frac{2z^{-\kappa-\frac{1}{2}}}{\Gamma\!\left(\tfrac{1}{2}+\mu-\kappa\right)\Gamma\!\left(\tfrac{1}{2}-\mu-\kappa\right)}\, K_{2\mu}(2\sqrt{z}), \quad \Re z > 0.$$

For additional Laplace and Mellin transforms see Erdélyi et al. (1954a, §§4.22, 5.20, 6.9, 7.5), Marichev (1983, pp. 283–287), Oberhettinger and Badii (1973, §1.17), Oberhettinger (1974, §§1.13, 2.8), and Prudnikov et al. (1992a, §§3.34, 3.35). Inverse Laplace transforms are given in Oberhettinger and Badii (1973, §2.16) and Prudnikov et al. (1992b, §§3.33, 3.34).

13.23(ii) Fourier Transforms

13.23.8 $$\frac{1}{\Gamma(1+2\mu)}\int_0^\infty \cos(2xt) e^{-\frac{1}{2}t^2} t^{-2\mu-1} M_{\kappa,\mu}(t^2)\, dt = \frac{\sqrt{\pi}\, e^{-\frac{1}{2}x^2} x^{\mu+\kappa-1}}{2\,\Gamma\!\left(\tfrac{1}{2}+\mu+\kappa\right)}\, W_{\frac{1}{2}\kappa-\frac{3}{2}\mu,\frac{1}{2}\kappa+\frac{1}{2}\mu}(x^2),\quad \Re(\kappa+\mu) > -\tfrac{1}{2}.$$

For additional Fourier transforms see Erdélyi et al. (1954a, §§1.14, 2.14, 3.3) and Oberhettinger (1990, §§1.22, 2.22).

13.23(iii) Hankel Transforms

For the notation see §10.2(ii).

13.23.9
$$\int_0^\infty e^{-\frac{1}{2}t} t^{\mu-\frac{1}{2}(\nu+1)} M_{\kappa,\mu}(t)\, J_\nu(2\sqrt{xt})\, dt = \frac{\Gamma(1+2\mu)}{\Gamma\!\left(\tfrac{1}{2}-\mu+\kappa+\nu\right)}\, e^{-\frac{1}{2}x} x^{\frac{1}{2}(\kappa-\mu-\frac{3}{2})}\, M_{\frac{1}{2}(\kappa+3\mu-\nu+\frac{1}{2}),\frac{1}{2}(\kappa-\mu+\nu-\frac{1}{2})}(x),$$
$$x > 0,\ -\tfrac{1}{2} < \Re\mu < \Re(\kappa+\tfrac{1}{2}\nu)+\tfrac{3}{4},$$

13.23.10 $$\frac{1}{\Gamma(1+2\mu)} \int_0^\infty e^{-\frac{1}{2}t} t^{\frac{1}{2}(\nu-1)-\mu} M_{\kappa,\mu}(t) J_\nu\left(2\sqrt{xt}\right) dt = \frac{e^{-\frac{1}{2}x} x^{\frac{1}{2}(\kappa+\mu-\frac{3}{2})}}{\Gamma\left(\frac{1}{2}+\mu+\kappa\right)} W_{\frac{1}{2}(\kappa-3\mu+\nu+\frac{1}{2}),\frac{1}{2}(\kappa+\mu-\nu-\frac{1}{2})}(x),$$
$$x > 0, \; -1 < \Re\nu < 2\Re(\mu+\kappa) + \tfrac{1}{2}.$$

13.23.11 $$\int_0^\infty e^{\frac{1}{2}t} t^{\frac{1}{2}(\nu-1)-\mu} W_{\kappa,\mu}(t) J_\nu\left(2\sqrt{xt}\right) dt = \frac{\Gamma(\nu-2\mu+1)}{\Gamma\left(\frac{1}{2}+\mu-\kappa\right)} e^{\frac{1}{2}x} x^{\frac{1}{2}(\mu-\kappa-\frac{3}{2})} W_{\frac{1}{2}(\kappa+3\mu-\nu-\frac{1}{2}),\frac{1}{2}(\kappa-\mu+\nu+\frac{1}{2})}(x),$$
$$x > 0, \; \max(2\Re\mu-1,-1) < \Re\nu < 2\Re(\mu-\kappa) + \tfrac{3}{2},$$

13.23.12 $$\int_0^\infty e^{-\frac{1}{2}t} t^{\frac{1}{2}(\nu-1)-\mu} W_{\kappa,\mu}(t) J_\nu\left(2\sqrt{xt}\right) dt = \frac{\Gamma(\nu-2\mu+1)}{\Gamma\left(\frac{3}{2}-\mu-\kappa+\nu\right)} e^{-\frac{1}{2}x} x^{\frac{1}{2}(\mu+\kappa-\frac{3}{2})} M_{\frac{1}{2}(\kappa-3\mu+\nu+\frac{1}{2}),\frac{1}{2}(\nu-\mu-\kappa+\frac{1}{2})}(x),$$
$$x > 0, \; \max(2\Re\mu-1,-1) < \Re\nu.$$

For additional Hankel transforms and also other Bessel transforms see Erdélyi et al. (1954b, §8.18) and Oberhettinger (1972, §1.16 and 3.4.42–46, 4.4.45–47, 5.94–97).

13.23(iv) Integral Transforms in terms of Whittaker Functions

Let $f(x)$ be absolutely integrable on the interval $[r, R]$ for all positive $r < R$, $f(x) = O(x^{\rho_0})$ as $x \to 0+$, and $f(x) = O(e^{-\rho_1 x})$ as $x \to +\infty$, where $\rho_1 > \frac{1}{2}$. Then for μ in the half-plane $\Re\mu \geq \mu_1 > \max\left(-\rho_0, \Re\kappa - \frac{1}{2}\right)$

13.23.13 $$g(\mu) = \frac{1}{\Gamma(1+2\mu)} \int_0^\infty f(x) x^{-\frac{3}{2}} M_{\kappa,\mu}(x) \, dx,$$

13.23.14 $$f(x) = \frac{1}{\pi i \sqrt{x}} \int_{\mu_1-i\infty}^{\mu_1+i\infty} \mu g(\mu) \Gamma\left(\tfrac{1}{2}+\mu-\kappa\right) W_{\kappa,\mu}(x) \, d\mu.$$

For additional integral transforms see Magnus et al. (1966, p. 189), Prudnikov et al. (1992b, §§4.3.39–4.3.42), and Wimp (1964).

13.23(v) Other Integrals

Additional integrals involving confluent hypergeometric functions can be found in Apelblat (1983, pp. 388–392), Erdélyi et al. (1954b), Gradshteyn and Ryzhik (2000, §7.6), and Prudnikov et al. (1990, §§1.13, 1.14, 2.19, 4.2.2). See also (13.16.2), (13.16.6), (13.16.7).

13.24 Series

13.24(i) Expansions in Series of Whittaker Functions

For expansions of arbitrary functions in series of $M_{\kappa,\mu}(z)$ functions see Schäfke (1961b).

13.24(ii) Expansions in Series of Bessel Functions

For $z \in \mathbb{C}$, and again with the notation of §§10.2(ii) and 10.25(ii),

13.24.1 $$M_{\kappa,\mu}(z) = \Gamma(\kappa+\mu) 2^{2\kappa+2\mu} z^{\frac{1}{2}-\kappa} \sum_{s=0}^\infty (-1)^s \frac{(2\kappa+2\mu)_s (2\kappa)_s}{(1+2\mu)_s s!} (\kappa+\mu+s) I_{\kappa+\mu+s}\left(\tfrac{1}{2}z\right), \quad 2\mu, \kappa+\mu \neq -1,-2,-3,\ldots,$$

and

13.24.2 $$\frac{1}{\Gamma(1+2\mu)} M_{\kappa,\mu}(z) = 2^{2\mu} z^{\mu+\frac{1}{2}} \sum_{s=0}^\infty p_s^{(\mu)}(z) \left(2\sqrt{\kappa z}\right)^{-2\mu-s} J_{2\mu+s}\left(2\sqrt{\kappa z}\right),$$

where $p_0^{(\mu)}(z) = 1$, $p_1^{(\mu)}(z) = \frac{1}{6} z^2$, and higher polynomials $p_s^{(\mu)}(z)$ are defined by

13.24.3 $$\exp\left(-\tfrac{1}{2} z \left(\coth t - \tfrac{1}{t}\right)\right) \left(\frac{t}{\sinh t}\right)^{1-2\mu} = \sum_{s=0}^\infty p_s^{(\mu)}(z) \left(-\frac{t}{z}\right)^s.$$

(13.18.8) is a special case of (13.24.1).

Additional expansions in terms of Bessel functions are given in Luke (1959). See also López (1999).

For other series expansions see Prudnikov et al. (1990, §6.6). See also §13.26.

13.25 Products

13.25.1
$$M_{\kappa,\mu}(z) M_{\kappa,-\mu-1}(z) + \frac{(\frac{1}{2}+\mu+\kappa)(\frac{1}{2}+\mu-\kappa)}{4\mu(1+\mu)(1+2\mu)^2} M_{\kappa,\mu+1}(z) M_{\kappa,-\mu}(z) = 1.$$

For integral representations, integrals, and series containing products of $M_{\kappa,\mu}(z)$ and $W_{\kappa,\mu}(z)$ see Erdélyi et al. (1053a, §6.15.3).

13.26 Addition and Multiplication Theorems

13.26(i) Addition Theorems for $M_{\kappa,\mu}(z)$

The function $M_{\kappa,\mu}(x+y)$ has the following expansions:

13.26.1
$$e^{-\frac{1}{2}y}\left(\frac{x}{x+y}\right)^{\mu-\frac{1}{2}} \sum_{n=0}^{\infty} \frac{(-2\mu)_n}{n!}\left(\frac{-y}{\sqrt{x}}\right)^n \times M_{\kappa-\frac{1}{2}n,\mu-\frac{1}{2}n}(x), \qquad |y|<|x|,$$

13.26.2
$$e^{-\frac{1}{2}y}\left(\frac{x+y}{x}\right)^{\mu+\frac{1}{2}} \sum_{n=0}^{\infty} \frac{(\frac{1}{2}+\mu-\kappa)_n}{(1+2\mu)_n n!}\left(\frac{y}{\sqrt{x}}\right)^n \times M_{\kappa-\frac{1}{2}n,\mu+\frac{1}{2}n}(x),$$

13.26.3
$$e^{-\frac{1}{2}y}\left(\frac{x+y}{x}\right)^{\kappa} \sum_{n=0}^{\infty} \frac{(\frac{1}{2}+\mu-\kappa)_n y^n}{n!(x+y)^n} M_{\kappa-n,\mu}(x), \qquad \Re(y/x) > -\tfrac{1}{2},$$

13.26.4
$$e^{\frac{1}{2}y}\left(\frac{x}{x+y}\right)^{\mu-\frac{1}{2}} \sum_{n=0}^{\infty} \frac{(-2\mu)_n}{n!}\left(\frac{-y}{\sqrt{x}}\right)^n \times M_{\kappa+\frac{1}{2}n,\mu-\frac{1}{2}n}(x), \qquad |y|<|x|,$$

13.26.5
$$e^{\frac{1}{2}y}\left(\frac{x+y}{x}\right)^{\mu+\frac{1}{2}} \sum_{n=0}^{\infty} \frac{(\frac{1}{2}+\mu+\kappa)_n}{(1+2\mu)_n n!}\left(\frac{-y}{\sqrt{x}}\right)^n \times M_{\kappa+\frac{1}{2}n,\mu+\frac{1}{2}n}(x),$$

13.26.6
$$e^{\frac{1}{2}y}\left(\frac{x}{x+y}\right)^{\kappa} \sum_{n=0}^{\infty} \frac{(\frac{1}{2}+\mu+\kappa)_n y^n}{n!(x+y)^n} M_{\kappa+n,\mu}(x), \qquad \Re((y+x)/x) > \tfrac{1}{2}.$$

13.26(ii) Addition Theorems for $W_{\kappa,\mu}(z)$

The function $W_{\kappa,\mu}(x+y)$ has the following expansions:

13.26.7
$$e^{-\frac{1}{2}y}\left(\frac{x}{x+y}\right)^{\mu-\frac{1}{2}} \sum_{n=0}^{\infty} \frac{(\frac{1}{2}-\mu-\kappa)_n}{n!}\left(\frac{-y}{\sqrt{x}}\right)^n \times W_{\kappa-\frac{1}{2}n,\mu-\frac{1}{2}n}(x), \qquad |y|<|x|,$$

13.26.8
$$e^{-\frac{1}{2}y}\left(\frac{x+y}{x}\right)^{\mu+\frac{1}{2}} \sum_{n=0}^{\infty} \frac{(\frac{1}{2}+\mu-\kappa)_n}{n!}\left(\frac{-y}{\sqrt{x}}\right)^n \times W_{\kappa-\frac{1}{2}n,\mu+\frac{1}{2}n}(x), \qquad |y|<|x|,$$

13.26.9
$$e^{-\frac{1}{2}y}\left(\frac{x+y}{x}\right)^{\kappa} \sum_{n=0}^{\infty} \frac{(\frac{1}{2}+\mu-\kappa)_n(\frac{1}{2}-\mu-\kappa)_n}{n!} \times \left(\frac{y}{x+y}\right)^n W_{\kappa-n,\mu}(x), \qquad \Re(y/x) > -\tfrac{1}{2},$$

13.26.10
$$e^{\frac{1}{2}y}\left(\frac{x}{x+y}\right)^{\mu-\frac{1}{2}} \sum_{n=0}^{\infty} \frac{1}{n!}\left(\frac{-y}{\sqrt{x}}\right)^n \times W_{\kappa+\frac{1}{2}n,\mu-\frac{1}{2}n}(x), \qquad |y|<|x|,$$

13.26.11
$$e^{\frac{1}{2}y}\left(\frac{x+y}{x}\right)^{\mu+\frac{1}{2}} \sum_{n=0}^{\infty} \frac{1}{n!}\left(\frac{-y}{\sqrt{x}}\right)^n \times W_{\kappa+\frac{1}{2}n,\mu+\frac{1}{2}n}(x), \qquad |y|<|x|,$$

13.26.12
$$e^{\frac{1}{2}y}\left(\frac{x}{x+y}\right)^{\kappa} \sum_{n=0}^{\infty} \frac{1}{n!}\left(\frac{-y}{x+y}\right)^n W_{\kappa+n,\mu}(x), \qquad \Re(y/x) > -\tfrac{1}{2}.$$

13.26(iii) Multiplication Theorems for $M_{\kappa,\mu}(z)$ and $W_{\kappa,\mu}(z)$

To obtain similar expansions for $M_{\kappa,\mu}(xy)$ and $W_{\kappa,\mu}(xy)$, replace y in the previous two subsections by $(y-1)x$.

Applications

13.27 Mathematical Applications

Confluent hypergeometric functions are connected with representations of the group of third-order triangular matrices. The elements of this group are of the form

13.27.1
$$g = \begin{pmatrix} 1 & \alpha & \beta \\ 0 & \gamma & \delta \\ 0 & 0 & 1 \end{pmatrix},$$

where α, β, γ, δ are real numbers, and $\gamma > 0$. Vilenkin (1968, Chapter 8) constructs irreducible representations of this group, in which the diagonal matrices correspond to operators of multiplication by an exponential function. The other group elements correspond to integral operators whose kernels can be expressed in terms of Whittaker functions. This identification can be used to obtain various properties of the Whittaker functions, including recurrence relations and derivatives.

For applications of Whittaker functions to the uniform asymptotic theory of differential equations with a coalescing turning point and simple pole see §§2.8(vi) and 18.15(i).

13.28 Physical Applications

13.28(i) Exact Solutions of the Wave Equation

The reduced wave equation $\nabla^2 w = k^2 w$ in paraboloidal coordinates, $x = 2\sqrt{\xi\eta}\cos\phi$, $y = 2\sqrt{\xi\eta}\sin\phi$, $z = \xi - \eta$, can be solved via separation of variables $w = f_1(\xi)f_2(\eta)e^{ip\phi}$, where

13.28.1
$$f_1(\xi) = \xi^{-\frac{1}{2}} V^{(1)}_{\kappa,\frac{1}{2}p}(2ik\xi), \quad f_2(\eta) = \eta^{-\frac{1}{2}} V^{(2)}_{\kappa,\frac{1}{2}p}(-2ik\eta),$$

and $V^{(j)}_{\kappa,\mu}(z)$, $j = 1, 2$, denotes any pair of solutions of Whittaker's equation (13.14.1). See Hochstadt (1971, Chapter 7).

For potentials in quantum mechanics that are solvable in terms of confluent hypergeometric functions see Negro et al. (2000).

13.28(ii) Coulomb Functions

See Chapter 33.

13.28(iii) Other Applications

For dynamics of many-body systems see Meden and Schönhammer (1992); for tomography see D'Ariano et al. (1994); for generalized coherent states see Barut and Girardello (1971); for relativistic cosmology see Crisóstomo et al. (2004).

Computation

13.29 Methods of Computation

13.29(i) Series Expansions

Although the Maclaurin series expansion (13.2.2) converges for all finite values of z, it is cumbersome to use when $|z|$ is large owing to slowness of convergence and cancellation. For large $|z|$ the asymptotic expansions of §13.7 should be used instead. Accuracy is limited by the magnitude of $|z|$. However, this accuracy can be increased considerably by use of the exponentially-improved forms of expansion supplied by the combination of (13.7.10) and (13.7.11), or by use of the hyperasymptotic expansions given in Olde Daalhuis and Olver (1995a). For large values of the parameters a and b the approximations in §13.8 are available.

Similarly for the Whittaker functions.

13.29(ii) Differential Equations

A comprehensive and powerful approach is to integrate the differential equations (13.2.1) and (13.14.1) by direct numerical methods. As described in §3.7(ii), to insure stability the integration path must be chosen in such a way that as we proceed along it the wanted solution grows in magnitude at least as fast as all other solutions of the differential equation.

For $M(a, b, z)$ and $M_{\kappa,\mu}(z)$ this means that in the sector $|\operatorname{ph} z| \leq \pi$ we may integrate along outward rays from the origin with initial values obtained from (13.2.2) and (13.14.2).

For $U(a, b, z)$ and $W_{\kappa,\mu}(z)$ we may integrate along outward rays from the origin in the sectors $\frac{1}{2}\pi < |\operatorname{ph} z| < \frac{3}{2}\pi$, with initial values obtained from connection formulas in §13.2(vii), §13.14(vii). In the sector $|\operatorname{ph} z| < \frac{1}{2}\pi$ the integration has to be towards the origin, with starting values computed from asymptotic expansions (§§13.7 and 13.19). On the rays $\operatorname{ph} z = \pm\frac{1}{2}\pi$, integration can proceed in either direction.

13.29(iii) Integral Representations

The integral representations (13.4.1) and (13.4.4) can be used to compute the Kummer functions, and (13.16.1) and (13.16.5) for the Whittaker functions. In Allasia and Besenghi (1991) and Allasia and Besenghi (1987b) the high accuracy of the trapezoidal rule for the computation of Kummer functions is described. Gauss quadrature methods are discussed in Gautschi (2002b).

13.29(iv) Recurrence Relations

The recurrence relations in §§13.3(i) and 13.15(i) can be used to compute the confluent hypergeometric functions in an efficient way. In the following two examples Olver's algorithm (§3.6(v)) can be used.

Example 1

We assume $2\mu \neq -1, -2, -3, \ldots$. Then we have

13.29.1
$$\frac{z^2(n+\mu-\frac{1}{2})\left((n+\mu+\frac{1}{2})^2 - \kappa^2\right)}{(n+\mu)(n+\mu+\frac{1}{2})(n+\mu+1)} y(n+1)$$
$$+ 16\left((n+\mu)^2 - \frac{1}{2}\kappa z - \frac{1}{4}\right) y(n)$$
$$- 16\left((n+\mu)^2 - \frac{1}{4}\right) y(n-1) = 0,$$

with recessive solution

13.29.2 $\qquad y(n) = z^{-n-\mu-\frac{1}{2}} M_{\kappa,n+\mu}(z),$

normalizing relation

13.29.3 $\quad e^{-\frac{1}{2}z} = \sum_{s=0}^{\infty} \frac{(2\mu)_s (\frac{1}{2}+\mu-\kappa)_s}{(2\mu)_{2s} s!} (-z)^s y(s),$

and estimate

13.29.4 $\qquad y(n) = 1 + O(n^{-1}), \qquad n \to \infty.$

13.30 Tables

Example 2

We assume $a, a+1-b \neq 0, -1, -2, \ldots$. Then we have

$$13.29.5 \quad \begin{aligned}(n+a)w(n) - (2(n+a+1)+z-b)\,w(n+1) \\ + (n+a-b+2)w(n+2) = 0,\end{aligned}$$

with recessive solution

$$13.29.6 \qquad w(n) = (a)_n\, U(n+a,b,z),$$

normalizing relation

$$13.29.7 \qquad z^{-a} = \sum_{s=0}^{\infty} \frac{(a-b+1)_s}{s!} w(s),$$

and estimate

$$13.29.8 \quad w(n) \sim \frac{\sqrt{\pi}\, e^{\frac{1}{2}z} z^{\frac{1}{4}(4a-2b+1)}}{\Gamma(a)\,\Gamma(a+1-b)} n^{\frac{1}{4}(4a-2b-3)} e^{-2\sqrt{nz}},$$

as $n \to \infty$. See Temme (1983), and also Wimp (1984, Chapter 5).

13.30 Tables

- Žurina and Osipova (1964) tabulates $M(a,b,x)$ and $U(a,b,x)$ for $b = 2$, $a = -0.98(.02)1.10$, $x = 0(.01)4$, 7D or 7S.

- Slater (1960) tabulates $M(a,b,x)$ for $a = -1(.1)1$, $b = 0.1(.1)1$, and $x = 0.1(.1)10$, 7–9S; $M(a,b,1)$ for $a = -11(.2)2$ and $b = -4(.2)1$, 7D; the smallest positive x-zero of $M(a,b,x)$ for $a = -4(.1)-0.1$ and $b = 0.1(.1)2.5$, 7D.

- Abramowitz and Stegun (1964, Chapter 13) tabulates $M(a,b,x)$ for $a = -1(.1)1$, $b = 0.1(.1)1$, and $x = 0.1(.1)1(1)10$, 8S. Also the smallest positive x-zero of $M(a,b,x)$ for $a = -1(.1)-0.1$ and $b = 0.1(.1)1$, 7D.

- Zhang and Jin (1996, pp. 411–423) tabulates $M(a,b,x)$ and $U(a,b,x)$ for $a = -5(.5)5$, $b = 0.5(.5)5$, and $x = 0.1, 1, 5, 10, 20, 30$, 8S (for $M(a,b,x)$) and 7S (for $U(a,b,x)$).

For other tables prior to 1961 see Fletcher et al. (1962) and Lebedev and Fedorova (1960).

13.31 Approximations

13.31(i) Chebyshev-Series Expansions

Luke (1969b, pp. 35 and 25) provides Chebyshev-series expansions of $M(a,b,x)$ and $U(a,b,x)$ that include the intervals $0 \le x \le \alpha$ and $\alpha \le x < \infty$, respectively, where α is an arbitrary positive constant.

13.31(ii) Padé Approximations

For a discussion of the convergence of the Padé approximants that are related to the continued fraction (13.5.1) see Wimp (1985).

13.31(iii) Rational Approximations

In Luke (1977a) the following rational approximation is given, together with its rate of convergence. For the notation see §16.2(i).

Let $a, a+1-b \neq 0, -1, -2, \ldots$, $|\mathrm{ph}\, z| < \pi$,

$$13.31.1 \quad \begin{aligned} A_n(z) = \sum_{s=0}^{n} & \frac{(-n)_s (n+1)_s (a)_s (b)_s}{(a+1)_s (b+1)_s (n!)^2} \\ & \times {}_3F_3\!\left(\begin{matrix}-n+s, n+1+s, 1 \\ 1+s, a+1+s, b+1+s\end{matrix}; -z\right),\end{aligned}$$

and

$$13.31.2 \qquad B_n(z) = {}_2F_2\!\left(\begin{matrix}-n, n+1 \\ a+1, b+1\end{matrix}; -z\right).$$

Then

$$13.31.3 \qquad z^a\, U(a, 1+a-b, z) = \lim_{n\to\infty} \frac{A_n(z)}{B_n(z)}.$$

13.32 Software

See http://dlmf.nist.gov/13.32.

References

General References

The main references used in writing this chapter are Buchholz (1969), Erdélyi et al. (1953a), Olver (1997b), Slater (1960), and Temme (1996a). For additional bibliographic reading see Andrews et al. (1999), Hochstadt (1971), Luke (1969a,b), Wang and Guo (1989), and Whittaker and Watson (1927).

Sources

The following list gives the references or other indications of proofs that were used in constructing the various sections this chapter. These sources supplement the references that are quoted in the text.

§13.2 Olver (1997b, Chapter 7, §§3, 9, 10), Slater (1960, §§1.5, 1.5.1, 2.1.2), and Temme (1996a, §§7.1, 7.2). (13.2.7) and (13.2.8) are terminating forms of the asymptotic expansion (13.7.3) (that the $\sim$ sign can be replaced by $=$ in these circumstances follows from (13.7.4) and (13.7.5).)

To verify (13.2.12) replace the U functions by **M** functions by means of (13.2.42) and (13.2.4), then recall that each **M** function is an entire function of z. For (13.2.13)–(13.2.22) see Temme (1996a, Ex. 7.6: an error in the equation that corresponds to (13.2.19) has been corrected). For (13.2.23) see (13.7.2). (13.2.27)–(13.2.32) are obtained by considering limiting forms of the connection formulas in §13.2(vii); see Olver (1997b, Chapter 7, Ex. 10.6) and Slater (1960, §§1.5–1.5.1).

§13.3 Slater (1960, §§2.1, 2.2). Note that in this reference (2.2.7) and (2.1.32) contain errors. The correct versions are (13.3.13) and (13.3.28), respectively. To see that (13.3.13) and (13.3.14) are equivalent to (13.2.1) use (13.3.16) and (13.3.23). For the operator identity (13.3.29) see Fleury and Turbiner (1994).

§13.4 Buchholz (1969, §1.4), Erdélyi et al. (1953a, §6.11), and Slater (1960, Chapter 3). For (13.4.5) use (13.2.41); compare the proof of Lemma 3.1 in Olde Daalhuis and Olver (1994). For (13.4.16)–(13.4.18) combine the results of Buchholz (1969, §5.4) with (13.14.2), (13.14.3).

§13.5 Jones and Thron (1980, Theorems 6.3 and 6.5).

§13.6 See Temme (1996a, §§7.2–7.3 and p. 254) and Buchholz (1969, §3.3). In the former reference each of the equations on p. 180 that correspond to (13.6.7) and (13.6.8) contains an error. For (13.6.16)–(13.6.18) combine §13.6(iv) with §12.7(i) and (18.5.18). (The last equation is needed to illustrate that $x^n H_n(x)$ is an even function of x.) For (13.6.19) see (18.11.2). For (13.6.20) combine (18.20.8) with (16.2.3), replace the ${}_1F_1$ notation by M (§13.1), and then use (13.2.42) (in which the final term vanishes). Alternatively, for (13.6.20) combine (16.2.3) with Andrews et al. (1999, p. 347). When neither a nor $a - b + 1$ is a nonpositive integer (13.6.21) can be verified by comparison of (13.4.17) and (16.5.1). If a is a nonpositive integer, then both sides of (13.6.21) reduce to a polynomial in z (compare §§13.2(i) and 16.2(iv)), and (13.6.21) follows by comparing coefficients. Similarly if $a - b + 1$ is a nonpositive integer, then both sides of (13.6.21) reduce to z^{1-b} times a polynomial in z with identical coefficients.

§13.7 Temme (1996a, §7.2) or Olver (1997b, pp. 256–258), and Olver (1965).

§13.8 Slater (1960, §4.3), Temme (1978), and Temme (1990b). For (13.8.2) and (13.8.3) use Watson's lemma for loop integrals (Olver (1997b, §4.5)) and (13.4.10).

§13.9 Buchholz (1969, Chapter 17), Erdélyi et al. (1953a, §6.16), Slater (1960, Chapter 6), and Tricomi (1950a). The proof of (13.9.8) is given in Tricomi (1947). (13.9.9) follows from (13.7.2). For the paragraph following (13.9.9) see Andrews et al. (1999, §4.16). For (13.9.10) and (13.9.16) use (13.8.9), (13.8.10), and the asymptotics of zeros of Bessel functions (§10.21(vi)). For the final paragraph of §13.9(ii) apply Kummer's transformation (13.2.39) to the final term in the connection relation (13.2.42) and then use the asymptotic relation (13.8.1).

§13.10 Erdélyi et al. (1953a, §§6.10, 6.15.2) and Slater (1960, Chapter 3). Also Buchholz (1969, §11.1), including the references given there. (13.10.14) and (13.10.16) are from Erdélyi et al. (1954b, §8.18). For (13.10.14) substitute the integral (13.4.1) for $\mathbf{M}(a, b, t)$, interchange the order of integration, then apply (13.4.3) and (13.6.1) followed by (13.4.4). For (13.10.16) interchange the roles of (13.4.1) and (13.4.4).

§13.11 Slater (1960, §2.7.3: Eq. (2.7.14) has errors).

§13.12 (13.12.1) follows from the fact that its left-hand side is bounded at infinity: use (13.2.39), (13.2.41), and (13.7.3).

§13.13 Erdélyi et al. (1953a, §6.14) and Slater (1960, §§2.3–2.3.3). In the first reference Equation (2) needs the constraint $|\lambda - 1| < 1$ and Equation (6) should have no constraint. In the second reference Eq. (2.3.6) contains an error: $(x + y)^n$ should be replaced by $(x + y)^{-n}$.

§13.14 Olver (1997b, Chapter 7, §§9–11, and Ex. 11.2), Buchholz (1969, §2.3a), Slater (1960, §§1.7.1, 2.4.2), Temme (1996a, §7.2). For (13.14.8) and (13.14.9) take limiting values in (13.14.33), using (13.14.2) and (13.2.2). For (13.14.14)–(13.14.19) combine §13.2(iii) and (13.14.4), (13.14.5). For (13.14.20), (13.14.21) use (13.19.2), (13.19.3). For (13.14.32) and (13.14.33) combine (13.2.41) and (13.2.42) with (13.14.4) and (13.14.5). (13.14.31) follows from (13.14.33).

§13.15 Slater (1960, §§2.4, 2.4.1, 2.5). Note that (2.5.4) and (2.5.10) contain errors: the correct versions are (13.15.2) and (13.15.10), respectively.

§13.16 Buchholz (1969, §5.4), Erdélyi et al. (1953a, §6.11), and Slater (1960, Chapter 3). For §13.16(i) combine §13.4(i) with (13.14.4) and (13.14.5).

§**13.17** Jones and Thron (1980, Theorems 6.3 and 6.5).

§**13.18** Combine §13.6 with (13.14.4) and (13.14.5).

§**13.19** Temme (1996a, §7.2).

§**13.20** Olver (1997b, Chapter 7, §11.2). For (13.20.2) use (13.14.3) and apply the method of steepest descents (§2.4(iv)) to the integral representation (13.4.14).

§**13.21** Slater (1960, §4.4.3). The asymptotic approximations (13.21.2)–(13.21.4) follow from §13.21(ii) and (5.11.7).

§**13.22** Olver (1997b, Chapter 12, (7.05)).

§**13.23** Buchholz (1969, §§10, 11.1), Erdélyi et al. (1953a, §§6.10, 6.15.2), Slater (1960, Chapter 3), and Snow (1952, Chapter XI). (13.23.3) and (13.23.5) can also be derived as limiting forms of (13.23.1) and (13.23.4), respectively; compare (15.4.20). For (13.23.9)–(13.23.12) combine §13.10(v) with (13.14.4) and (13.14.5).

§**13.24** Slater (1960, §2.7.3) and Buchholz (1969, §7.4). In the first reference (2.7.16) contains an error.

§**13.25** For (13.25.1) combine (13.12.1) with (13.14.4).

§**13.26** Slater (1960, §§2.6–2.6.3). Note that (2.6.3) and (2.6.6) contain errors; the correct versions are (13.26.3) and (13.26.6), respectively.

Chapter 14

Legendre and Related Functions

T. M. Dunster[1]

Notation — 352
- 14.1 Special Notation — 352

Real Arguments — 352
- 14.2 Differential Equations — 352
- 14.3 Definitions and Hypergeometric Representations — 353
- 14.4 Graphics — 355
- 14.5 Special Values — 359
- 14.6 Integer Order — 360
- 14.7 Integer Degree and Order — 360
- 14.8 Behavior at Singularities — 361
- 14.9 Connection Formulas — 362
- 14.10 Recurrence Relations and Derivatives — 362
- 14.11 Derivatives with Respect to Degree or Order — 363
- 14.12 Integral Representations — 363
- 14.13 Trigonometric Expansions — 364
- 14.14 Continued Fractions — 364
- 14.15 Uniform Asymptotic Approximations — 365
- 14.16 Zeros — 368
- 14.17 Integrals — 368
- 14.18 Sums — 370
- 14.19 Toroidal (or Ring) Functions — 371
- 14.20 Conical (or Mehler) Functions — 372

Complex Arguments — 375
- 14.21 Definitions and Basic Properties — 375
- 14.22 Graphics — 375
- 14.23 Values on the Cut — 376
- 14.24 Analytic Continuation — 376
- 14.25 Integral Representations — 377
- 14.26 Uniform Asymptotic Expansions — 377
- 14.27 Zeros — 377
- 14.28 Sums — 377
- 14.29 Generalizations — 377

Applications — 378
- 14.30 Spherical and Spheroidal Harmonics — 378
- 14.31 Other Applications — 379

Computation — 379
- 14.32 Methods of Computation — 379
- 14.33 Tables — 380
- 14.34 Software — 380

References — 380

[1]Department of Mathematics and Statistics, San Diego State University, San Diego, California.
Acknowledgments: This chapter is based in part on Abramowitz and Stegun (1964, Chapter 8) by Irene A. Stegun. The author is indebted to Richard Paris for correcting a long-standing error in Eq. (14.18.3) in previous literature.
Copyright © 2009 National Institute of Standards and Technology. All rights reserved.

Notation

14.1 Special Notation

(For other notation see pp. xiv and 873.)

x, y, τ	real variables.
$z = x + iy$	complex variable.
m, n	nonnegative integers used for order and degree, respectively.
μ, ν	general order and degree, respectively.
$-\tfrac{1}{2} + i\tau$	complex degree, $\tau \in \mathbb{R}$.
γ	Euler's constant (§5.2(ii)).
δ	arbitrary small positive constant.
$\psi(x)$	logarithmic derivative of gamma function (§5.2(i)).
$\psi'(x)$	$d\psi(x)/dx$.
$\mathbf{F}(a, b; c; z)$	Olver's scaled hypergeometric function: $F(a,b;c;z)/\Gamma(c)$.

Multivalued functions take their principal values (§4.2(i)) unless indicated otherwise.

The main functions treated in this chapter are the Legendre functions $\mathsf{P}_\nu(x)$, $\mathsf{Q}_\nu(x)$, $P_\nu(z)$, $Q_\nu(z)$; Ferrers functions $\mathsf{P}_\nu^\mu(x)$, $\mathsf{Q}_\nu^\mu(x)$ (also known as the Legendre functions on the cut); associated Legendre functions $P_\nu^\mu(z)$, $Q_\nu^\mu(z)$, $\boldsymbol{Q}_\nu^\mu(z)$; conical functions $\mathsf{P}_{-\frac{1}{2}+i\tau}^\mu(x)$, $\mathsf{Q}_{-\frac{1}{2}+i\tau}^\mu(x)$, $\widehat{\mathsf{Q}}_{-\frac{1}{2}+i\tau}^\mu(x)$, $P_{-\frac{1}{2}+i\tau}^\mu(x)$, $Q_{-\frac{1}{2}+i\tau}^\mu(x)$ (also known as Mehler functions).

Among other notations commonly used in the literature Erdélyi et al. (1953a) and Olver (1997b) denote $\mathsf{P}_\nu^\mu(x)$ and $\mathsf{Q}_\nu^\mu(x)$ by $P_\nu^\mu(x)$ and $Q_\nu^\mu(x)$, respectively. Magnus et al. (1966) denotes $\mathsf{P}_\nu^\mu(x)$, $\mathsf{Q}_\nu^\mu(x)$, $P_\nu^\mu(z)$, and $Q_\nu^\mu(z)$ by $P_\nu^\mu(x)$, $Q_\nu^\mu(x)$, $\mathfrak{P}_\nu^\mu(z)$, and $\mathfrak{Q}_\nu^\mu(z)$, respectively. Hobson (1931) denotes both $\mathsf{P}_\nu^\mu(x)$ and $P_\nu^\mu(x)$ by $P_\nu^\mu(x)$; similarly for $\mathsf{Q}_\nu^\mu(x)$ and $Q_\nu^\mu(x)$.

Real Arguments

14.2 Differential Equations

14.2(i) Legendre's Equation

14.2.1 $$(1 - x^2)\frac{d^2w}{dx^2} - 2x\frac{dw}{dx} + \nu(\nu + 1)w = 0.$$

Standard solutions: $\mathsf{P}_\nu(\pm x)$, $\mathsf{Q}_\nu(\pm x)$, $\mathsf{Q}_{-\nu-1}(\pm x)$, $P_\nu(\pm x)$, $Q_\nu(\pm x)$, $Q_{-\nu-1}(\pm x)$. $\mathsf{P}_\nu(x)$ and $\mathsf{Q}_\nu(x)$ are real when $\nu \in \mathbb{R}$ and $x \in (-1, 1)$, and $P_\nu(x)$ and $Q_\nu(x)$ are real when $\nu \in \mathbb{R}$ and $x \in (1, \infty)$.

14.2(ii) Associated Legendre Equation

14.2.2 $$(1 - x^2)\frac{d^2w}{dx^2} - 2x\frac{dw}{dx} + \left(\nu(\nu+1) - \frac{\mu^2}{1-x^2}\right)w = 0.$$

Standard solutions: $\mathsf{P}_\nu^\mu(\pm x)$, $\mathsf{P}_\nu^{-\mu}(\pm x)$, $\mathsf{Q}_\nu^\mu(\pm x)$, $\mathsf{Q}_{-\nu-1}^\mu(\pm x)$, $P_\nu^\mu(\pm x)$, $P_\nu^{-\mu}(\pm x)$, $\boldsymbol{Q}_\nu^\mu(\pm x)$, $\boldsymbol{Q}_{-\nu-1}^\mu(\pm x)$.

(14.2.2) reduces to (14.2.1) when $\mu = 0$. Ferrers functions and the associated Legendre functions are related to the Legendre functions by the equations $\mathsf{P}_\nu^0(x) = \mathsf{P}_\nu(x)$, $\mathsf{Q}_\nu^0(x) = \mathsf{Q}_\nu(x)$, $P_\nu^0(x) = P_\nu(x)$, $Q_\nu^0(x) = Q_\nu(x)$, $\boldsymbol{Q}_\nu^0(x) = \boldsymbol{Q}_\nu(x) = Q_\nu(x)/\Gamma(\nu+1)$.

$\mathsf{P}_\nu^\mu(x)$, $\mathsf{P}_{-\frac{1}{2}+i\tau}^\mu(x)$, and $\mathsf{Q}_\nu^\mu(x)$ are real when ν, μ, and $\tau \in \mathbb{R}$, and $x \in (-1, 1)$; $P_\nu^\mu(x)$, $Q_\nu^\mu(x)$, and $\boldsymbol{Q}_\nu^\mu(x)$ are real when ν and $\mu \in \mathbb{R}$, and $x \in (1, \infty)$.

Unless stated otherwise in §§14.2–14.20 it is assumed that the arguments of the functions $\mathsf{P}_\nu^\mu(x)$ and $\mathsf{Q}_\nu^\mu(x)$ lie in the interval $(-1, 1)$, and the arguments of the functions $P_\nu^\mu(x)$, $Q_\nu^\mu(x)$, and $\boldsymbol{Q}_\nu^\mu(x)$ lie in the interval $(1, \infty)$. For extensions to complex arguments see §§14.21–14.28.

14.2(iii) Numerically Satisfactory Solutions

Equation (14.2.2) has regular singularities at $x = 1, -1$, and ∞, with exponent pairs $\{-\tfrac{1}{2}\mu, \tfrac{1}{2}\mu\}$, $\{-\tfrac{1}{2}\mu, \tfrac{1}{2}\mu\}$, and $\{-\nu-1, \nu\}$, respectively; compare §2.7(i).

When $\mu - \nu \neq 0, -1, -2, \ldots$, and $\mu + \nu \neq -1, -2, -3, \ldots$, $\mathsf{P}_\nu^{-\mu}(x)$ and $\mathsf{P}_\nu^{-\mu}(-x)$ are linearly independent, and when $\Re\mu \geq 0$ they are recessive at $x = 1$ and $x = -1$, respectively. Hence they comprise a numerically satisfactory pair of solutions (§2.7(iv)) of (14.2.2) in the interval $-1 < x < 1$. When $\mu - \nu = 0, -1, -2, \ldots$, or $\mu + \nu = -1, -2, -3, \ldots$, $\mathsf{P}_\nu^{-\mu}(x)$ and $\mathsf{P}_\nu^{-\mu}(-x)$ are linearly dependent, and in these cases either may be paired with almost any linearly independent solution to form a numerically satisfactory pair.

When $\Re\mu \geq 0$ and $\Re\nu \geq -\tfrac{1}{2}$, $P_\nu^{-\mu}(x)$ and $\boldsymbol{Q}_\nu^\mu(x)$ are linearly independent, and recessive at $x = 1$ and $x = \infty$, respectively. Hence they comprise a numerically satisfactory pair of solutions of (14.2.2) in the interval $1 < x < \infty$. With the same conditions, $P_\nu^{-\mu}(-x)$ and $\boldsymbol{Q}_\nu^\mu(-x)$ comprise a numerically satisfactory pair of solutions in the interval $-\infty < x < -1$.

14.2(iv) Wronskians and Cross-Products

14.2.3 $$\mathscr{W}\left\{\mathsf{P}_\nu^{-\mu}(x), \mathsf{P}_\nu^{-\mu}(-x)\right\} = \frac{2}{\Gamma(\mu-\nu)\Gamma(\nu+\mu+1)(1-x^2)},$$

14.2.4 $$\mathscr{W}\left\{\mathsf{P}_\nu^\mu(x), \mathsf{Q}_\nu^\mu(x)\right\} = \frac{\Gamma(\nu+\mu+1)}{\Gamma(\nu-\mu+1)(1-x^2)},$$

14.2.5 $$\mathsf{P}_{\nu+1}^\mu(x)\mathsf{Q}_\nu^\mu(x) - \mathsf{P}_\nu^\mu(x)\mathsf{Q}_{\nu+1}^\mu(x) = \frac{\Gamma(\nu+\mu+1)}{\Gamma(\nu-\mu+2)},$$

14.2.6 $\quad \mathscr{W}\left\{\mathsf{P}_\nu^{-\mu}(x), \mathsf{Q}_\nu^\mu(x)\right\} = \dfrac{\cos(\mu\pi)}{1-x^2}$,

14.2.7 $\quad \mathscr{W}\left\{P_\nu^{-\mu}(x), P_\nu^\mu(x)\right\} = -\dfrac{2\sin(\mu\pi)}{\pi(x^2-1)}$,

14.2.8 $\quad \mathscr{W}\left\{P_\nu^{-\mu}(x), \boldsymbol{Q}_\nu^\mu(x)\right\} = -\dfrac{1}{\Gamma(\nu+\mu+1)(x^2-1)}$,

14.2.9 $\quad \mathscr{W}\left\{\boldsymbol{Q}_\nu^\mu(x), \boldsymbol{Q}_{-\nu-1}^\mu(x)\right\} = \dfrac{\cos(\nu\pi)}{x^2-1}$,

14.2.10
$$\mathscr{W}\left\{P_\nu^\mu(x), Q_\nu^\mu(x)\right\} = -e^{\mu\pi i}\dfrac{\Gamma(\nu+\mu+1)}{\Gamma(\nu-\mu+1)(x^2-1)},$$

14.2.11
$$P_{\nu+1}^\mu(x)\,Q_\nu^\mu(x) - P_\nu^\mu(x)\,Q_{\nu+1}^\mu(x) = e^{\mu\pi i}\dfrac{\Gamma(\nu+\mu+1)}{\Gamma(\nu-\mu+2)}.$$

14.3 Definitions and Hypergeometric Representations

14.3(i) Interval $-1 < x < 1$

The following are real-valued solutions of (14.2.2) when $\mu, \nu \in \mathbb{R}$ and $x \in (-1,1)$.

Ferrers Function of the First Kind

14.3.1
$$\mathsf{P}_\nu^\mu(x) = \left(\frac{1+x}{1-x}\right)^{\mu/2} \mathbf{F}\!\left(\nu+1, -\nu; 1-\mu; \tfrac{1}{2}-\tfrac{1}{2}x\right).$$

Ferrers Function of the Second Kind

14.3.2
$$\mathsf{Q}_\nu^\mu(x) = \frac{\pi}{2\sin(\mu\pi)}\left(\cos(\mu\pi)\left(\frac{1+x}{1-x}\right)^{\mu/2} \mathbf{F}\!\left(\nu+1,-\nu;1-\mu;\tfrac{1}{2}-\tfrac{1}{2}x\right)\right.$$
$$\left.- \frac{\Gamma(\nu+\mu+1)}{\Gamma(\nu-\mu+1)}\left(\frac{1-x}{1+x}\right)^{\mu/2}\mathbf{F}\!\left(\nu+1,-\nu;1+\mu;\tfrac{1}{2}-\tfrac{1}{2}x\right)\right).$$

Here and elsewhere in this chapter

14.3.3
$$\mathbf{F}(a,b;c;x) = \frac{1}{\Gamma(c)}F(a,b;c;x)$$

is Olver's hypergeometric function (§15.1).

$\mathsf{P}_\nu^\mu(x)$ exists for all values of μ and ν. $\mathsf{Q}_\nu^\mu(x)$ is undefined when $\mu+\nu = -1,-2,-3,\ldots$.

When $\mu = m = 0,1,2,\ldots$, (14.3.1) reduces to

14.3.4
$$\mathsf{P}_\nu^m(x) = (-1)^m \frac{\Gamma(\nu+m+1)}{2^m\,\Gamma(\nu-m+1)}\left(1-x^2\right)^{m/2}\mathbf{F}\!\left(\nu+m+1, m-\nu; m+1; \tfrac{1}{2}-\tfrac{1}{2}x\right);$$

equivalently,

14.3.5
$$\mathsf{P}_\nu^m(x) = (-1)^m \frac{\Gamma(\nu+m+1)}{\Gamma(\nu-m+1)}\left(\frac{1-x}{1+x}\right)^{m/2}\mathbf{F}\!\left(\nu+1,-\nu;m+1;\tfrac{1}{2}-\tfrac{1}{2}x\right).$$

When $\mu = m\ (\in \mathbb{Z})$ (14.3.2) is replaced by its limiting value; see Hobson (1931, §132) for details. See also (14.3.12)–(14.3.14) for this case.

14.3(ii) Interval $1 < x < \infty$

Associated Legendre Function of the First Kind

14.3.6
$$P_\nu^\mu(x) = \left(\frac{x+1}{x-1}\right)^{\mu/2} \mathbf{F}\!\left(\nu+1,-\nu;1-\mu;\tfrac{1}{2}-\tfrac{1}{2}x\right).$$

Associated Legendre Function of the Second Kind

14.3.7
$$Q_\nu^\mu(x) = e^{\mu\pi i}\frac{\pi^{1/2}\,\Gamma(\nu+\mu+1)\,(x^2-1)^{\mu/2}}{2^{\nu+1}x^{\nu+\mu+1}}\mathbf{F}\left(\tfrac{1}{2}\nu+\tfrac{1}{2}\mu+1,\tfrac{1}{2}\nu+\tfrac{1}{2}\mu+\tfrac{1}{2};\nu+\tfrac{3}{2};\frac{1}{x^2}\right),\quad \mu+\nu\neq -1,-2,-3,\ldots.$$

When $\mu = m = 1,2,3,\ldots$, (14.3.6) reduces to

14.3.8
$$P_\nu^m(x) = \frac{\Gamma(\nu+m+1)}{2^m\,\Gamma(\nu-m+1)}(x^2-1)^{m/2}\,\mathbf{F}\left(\nu+m+1,m-\nu;m+1;\tfrac{1}{2}-\tfrac{1}{2}x\right).$$

As standard solutions of (14.2.2) we take the pair $P_\nu^{-\mu}(x)$ and $\boldsymbol{Q}_\nu^\mu(x)$, where

14.3.9
$$P_\nu^{-\mu}(x) = \left(\frac{x-1}{x+1}\right)^{\mu/2}\mathbf{F}\left(\nu+1,-\nu;\mu+1;\tfrac{1}{2}-\tfrac{1}{2}x\right),$$

and

14.3.10
$$\boldsymbol{Q}_\nu^\mu(x) = e^{-\mu\pi i}\frac{Q_\nu^\mu(x)}{\Gamma(\nu+\mu+1)}.$$

Like $P_\nu^\mu(x)$, but unlike $Q_\nu^\mu(x)$, $\boldsymbol{Q}_\nu^\mu(x)$ is real-valued when $\nu,\mu\in\mathbb{R}$ and $x\in(1,\infty)$, and is defined for all values of ν and μ. The notation $\boldsymbol{Q}_\nu^\mu(x)$ is due to Olver (1997b, pp. 170 and 178).

14.3(iii) Alternative Hypergeometric Representations

14.3.11
$$\mathsf{P}_\nu^\mu(x) = \cos\left(\tfrac{1}{2}(\nu+\mu)\pi\right)w_1(\nu,\mu,x) + \sin\left(\tfrac{1}{2}(\nu+\mu)\pi\right)w_2(\nu,\mu,x),$$

14.3.12
$$\mathsf{Q}_\nu^\mu(x) = -\tfrac{1}{2}\pi\sin\left(\tfrac{1}{2}(\nu+\mu)\pi\right)w_1(\nu,\mu,x) + \tfrac{1}{2}\pi\cos\left(\tfrac{1}{2}(\nu+\mu)\pi\right)w_2(\nu,\mu,x),$$

where

14.3.13
$$w_1(\nu,\mu,x) = \frac{2^\mu\,\Gamma\!\left(\tfrac{1}{2}\nu+\tfrac{1}{2}\mu+\tfrac{1}{2}\right)}{\Gamma\!\left(\tfrac{1}{2}\nu-\tfrac{1}{2}\mu+1\right)}(1-x^2)^{-\mu/2}\,\mathbf{F}\!\left(-\tfrac{1}{2}\nu-\tfrac{1}{2}\mu,\tfrac{1}{2}\nu-\tfrac{1}{2}\mu+\tfrac{1}{2};\tfrac{1}{2};x^2\right),$$

14.3.14
$$w_2(\nu,\mu,x) = \frac{2^\mu\,\Gamma\!\left(\tfrac{1}{2}\nu+\tfrac{1}{2}\mu+1\right)}{\Gamma\!\left(\tfrac{1}{2}\nu-\tfrac{1}{2}\mu+\tfrac{1}{2}\right)}x(1-x^2)^{-\mu/2}\,\mathbf{F}\!\left(\tfrac{1}{2}-\tfrac{1}{2}\nu-\tfrac{1}{2}\mu,\tfrac{1}{2}\nu-\tfrac{1}{2}\mu+1;\tfrac{3}{2};x^2\right).$$

14.3.15
$$P_\nu^{-\mu}(x) = 2^{-\mu}(x^2-1)^{\mu/2}\,\mathbf{F}\!\left(\mu-\nu,\nu+\mu+1;\mu+1;\tfrac{1}{2}-\tfrac{1}{2}x\right),$$

14.3.16
$$\cos(\nu\pi)\,P_\nu^{-\mu}(x) = \frac{2^\nu\pi^{1/2}x^{\nu-\mu}(x^2-1)^{\mu/2}}{\Gamma(\nu+\mu+1)}\mathbf{F}\!\left(\tfrac{1}{2}\mu-\tfrac{1}{2}\nu,\tfrac{1}{2}\mu-\tfrac{1}{2}\nu+\tfrac{1}{2};\tfrac{1}{2}-\nu;\frac{1}{x^2}\right)$$
$$-\frac{\pi^{1/2}(x^2-1)^{\mu/2}}{2^{\nu+1}\,\Gamma(\mu-\nu)x^{\nu+\mu+1}}\mathbf{F}\!\left(\tfrac{1}{2}\nu+\tfrac{1}{2}\mu+1,\tfrac{1}{2}\nu+\tfrac{1}{2}\mu+\tfrac{1}{2};\nu+\tfrac{3}{2};\frac{1}{x^2}\right),$$

14.3.17
$$P_\nu^{-\mu}(x) = \frac{\pi(x^2-1)^{\mu/2}}{2^\mu}\left(\frac{\mathbf{F}\!\left(\tfrac{1}{2}\mu-\tfrac{1}{2}\nu,\tfrac{1}{2}\nu+\tfrac{1}{2}\mu+\tfrac{1}{2};\tfrac{1}{2};x^2\right)}{\Gamma\!\left(\tfrac{1}{2}\mu-\tfrac{1}{2}\nu+\tfrac{1}{2}\right)\Gamma\!\left(\tfrac{1}{2}\nu+\tfrac{1}{2}\mu+1\right)} - \frac{x\,\mathbf{F}\!\left(\tfrac{1}{2}\mu-\tfrac{1}{2}\nu+\tfrac{1}{2},\tfrac{1}{2}\nu+\tfrac{1}{2}\mu+1;\tfrac{3}{2};x^2\right)}{\Gamma\!\left(\tfrac{1}{2}\mu-\tfrac{1}{2}\nu\right)\Gamma\!\left(\tfrac{1}{2}\nu+\tfrac{1}{2}\mu+\tfrac{1}{2}\right)}\right),$$

14.3.18
$$P_\nu^{-\mu}(x) = 2^{-\mu}x^{\nu-\mu}(x^2-1)^{\mu/2}\,\mathbf{F}\!\left(\tfrac{1}{2}\mu-\tfrac{1}{2}\nu,\tfrac{1}{2}\mu-\tfrac{1}{2}\nu+\tfrac{1}{2};\mu+1;1-\frac{1}{x^2}\right),$$

14.3.19
$$\boldsymbol{Q}_\nu^\mu(x) = \frac{2^\nu\,\Gamma(\nu+1)(x+1)^{\mu/2}}{(x-1)^{(\mu/2)+\nu+1}}\mathbf{F}\!\left(\nu+1,\nu+\mu+1;2\nu+2;\frac{2}{1-x}\right),$$

14.3.20
$$\frac{2\sin(\mu\pi)}{\pi}\boldsymbol{Q}_\nu^\mu(x) = \frac{(x+1)^{\mu/2}}{\Gamma(\nu+\mu+1)(x-1)^{\mu/2}}\mathbf{F}\!\left(\nu+1,-\nu;1-\mu;\tfrac{1}{2}-\tfrac{1}{2}x\right)$$
$$-\frac{(x-1)^{\mu/2}}{\Gamma(\nu-\mu+1)(x+1)^{\mu/2}}\mathbf{F}\!\left(\nu+1,-\nu;\mu+1;\tfrac{1}{2}-\tfrac{1}{2}x\right).$$

For further hypergeometric representations of $P_\nu^\mu(x)$ and $Q_\nu^\mu(x)$ see Erdélyi et al. (1953a, pp. 123–139), Andrews et al. (1999, §3.1), Magnus et al. (1966, pp. 153–163), and §15.8(iv).

14.3(iv) Relations to Other Functions

In terms of the *Gegenbauer function* $C_\alpha^{(\beta)}(x)$ and the *Jacobi function* $\phi_\lambda^{(\alpha,\beta)}(t)$ (§§15.9(iii), 15.9(ii)):

14.3.21
$$\mathsf{P}_\nu^\mu(x) = \frac{2^\mu \, \Gamma(1-2\mu) \, \Gamma(\nu+\mu+1)}{\Gamma(\nu-\mu+1) \, \Gamma(1-\mu) \, (1-x^2)^{\mu/2}} C_{\nu+\mu}^{(\frac{1}{2}-\mu)}(x).$$

14.3.22
$$P_\nu^\mu(x) = \frac{2^\mu \, \Gamma(1-2\mu) \, \Gamma(\nu+\mu+1)}{\Gamma(\nu-\mu+1) \, \Gamma(1-\mu) \, (x^2-1)^{\mu/2}} C_{\nu+\mu}^{(\frac{1}{2}-\mu)}(x).$$

14.3.23
$$P_\nu^\mu(x) = \frac{1}{\Gamma(1-\mu)} \left(\frac{x+1}{x-1}\right)^{\mu/2} \phi_{-i(2\nu+1)}^{(-\mu,\mu)}\left(\operatorname{arcsinh}\left((\tfrac{1}{2}x-\tfrac{1}{2})^{1/2}\right)\right).$$

Compare also (18.11.1).

14.4 Graphics

14.4(i) Ferrers Functions: 2D Graphs

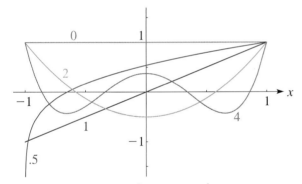

Figure 14.4.1: $\mathsf{P}_\nu^0(x), \nu = 0, \tfrac{1}{2}, 1, 2, 4$.

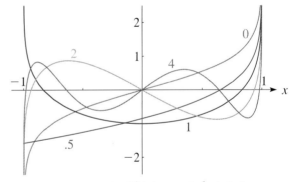

Figure 14.4.2: $\mathsf{Q}_\nu^0(x), \nu = 0, \tfrac{1}{2}, 1, 2, 4$.

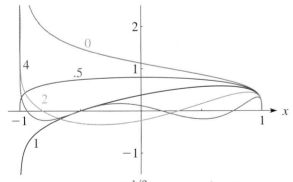

Figure 14.4.3: $\mathsf{P}_\nu^{-1/2}(x), \nu = 0, \tfrac{1}{2}, 1, 2, 4$.

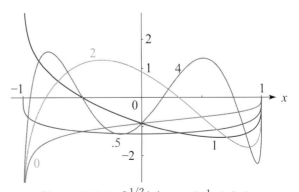

Figure 14.4.4: $\mathsf{Q}_\nu^{1/2}(x), \nu = 0, \tfrac{1}{2}, 1, 2, 4$.

For additional graphs see http://dlmf.nist.gov/14.4.i.

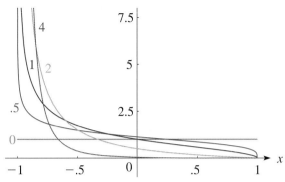

Figure 14.4.7: $\mathsf{P}_0^{-\mu}(x), \mu = 0, \frac{1}{2}, 1, 2, 4.$

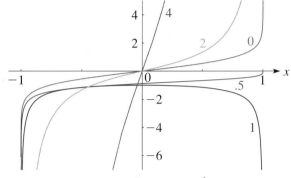

Figure 14.4.8: $\mathsf{Q}_0^{\mu}(x), \mu = 0, \frac{1}{2}, 1, 2, 4.$

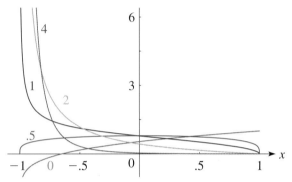

Figure 14.4.9: $\mathsf{P}_{1/2}^{-\mu}(x), \mu = 0, \frac{1}{2}, 1, 2, 4.$

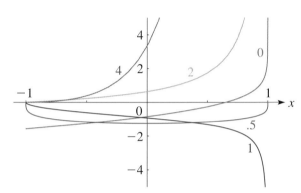

Figure 14.4.10: $\mathsf{Q}_{1/2}^{\mu}(x), \mu = 0, \frac{1}{2}, 1, 2, 4.$

For additional graphs see http://dlmf.nist.gov/14.4.i.

14.4(ii) Ferrers Functions: 3D Surfaces

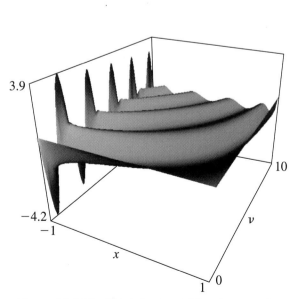

Figure 14.4.13: $\mathsf{P}_\nu^0(x), 0 \le \nu \le 10, -1 < x < 1.$

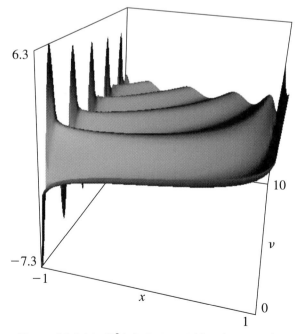

Figure 14.4.14: $\mathsf{Q}_\nu^0(x), 0 \le \nu \le 10, -1 < x < 1.$

14.4 Graphics

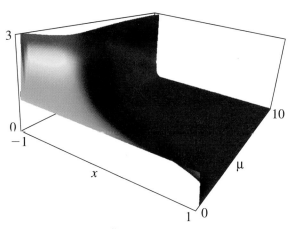

Figure 14.4.15: $\mathsf{P}_0^{-\mu}(x), 0 \leq \mu \leq 10, -1 < x < 1$.

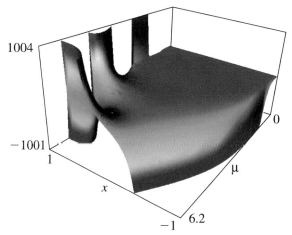

Figure 14.4.16: $\mathsf{Q}_0^{\mu}(x), 0 \leq \mu \leq 6.2, -1 < x < 1$.

14.4(iii) Associated Legendre Functions: 2D Graphs

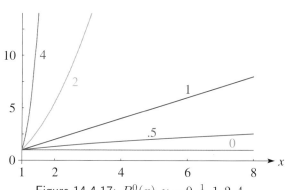

Figure 14.4.17: $P_\nu^0(x), \nu = 0, \tfrac{1}{2}, 1, 2, 4$.

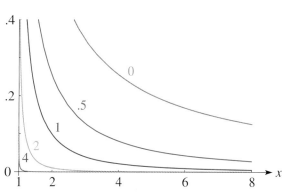

Figure 14.4.18: $\boldsymbol{Q}_\nu^0(x), \nu = 0, \tfrac{1}{2}, 1, 2, 4$.

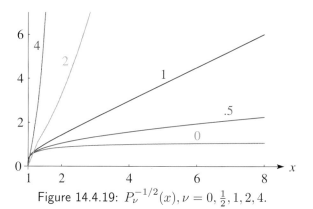

Figure 14.4.19: $P_\nu^{-1/2}(x), \nu = 0, \tfrac{1}{2}, 1, 2, 4$.

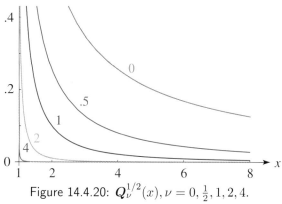

Figure 14.4.20: $\boldsymbol{Q}_\nu^{1/2}(x), \nu = 0, \tfrac{1}{2}, 1, 2, 4$.

For additional graphs see http://dlmf.nist.gov/14.4.iii.

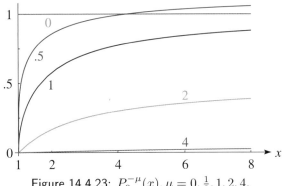

Figure 14.4.23: $P_0^{-\mu}(x), \mu = 0, \tfrac{1}{2}, 1, 2, 4$.

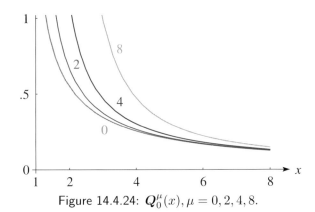

Figure 14.4.24: $\boldsymbol{Q}_0^{\mu}(x), \mu = 0, 2, 4, 8$.

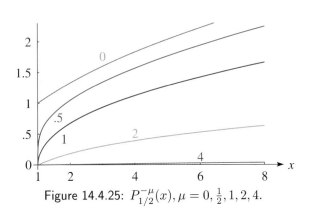

Figure 14.4.25: $P_{1/2}^{-\mu}(x), \mu = 0, \tfrac{1}{2}, 1, 2, 4$.

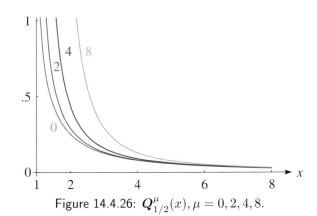

Figure 14.4.26: $\boldsymbol{Q}_{1/2}^{\mu}(x), \mu = 0, 2, 4, 8$.

For additional graphs see http://dlmf.nist.gov/14.4.iii.

14.4(iv) Associated Legendre Functions: 3D Surfaces

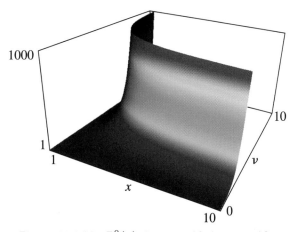

Figure 14.4.29: $P_\nu^0(x), 0 \leq \nu \leq 10, 1 < x < 10$.

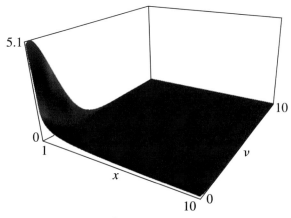

Figure 14.4.30: $\boldsymbol{Q}_\nu^0(x), 0 \leq \nu \leq 10, 1 < x < 10$.

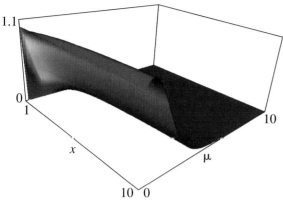

Figure 14.4.31: $\mathsf{P}_0^{-\mu}(x), 0 \leq \mu \leq 10, 1 < x < 10$.

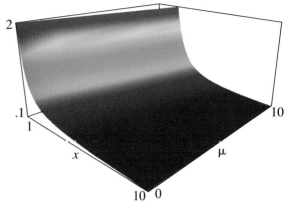

Figure 14.4.32: $\boldsymbol{Q}_0^\mu(x), 0 \leq \mu \leq 10, 1 < x < 10$.

14.5 Special Values

14.5(i) $x = 0$

14.5.1 $\quad \mathsf{P}_\nu^\mu(0) = \dfrac{2^\mu \pi^{1/2}}{\Gamma\left(\frac{1}{2}\nu - \frac{1}{2}\mu + 1\right) \Gamma\left(\frac{1}{2} - \frac{1}{2}\nu - \frac{1}{2}\mu\right)}$,

14.5.2 $\quad \left.\dfrac{d\mathsf{P}_\nu^\mu(x)}{dx}\right|_{x=0} = -\dfrac{2^{\mu+1}\pi^{1/2}}{\Gamma\left(\frac{1}{2}\nu - \frac{1}{2}\mu + \frac{1}{2}\right) \Gamma\left(-\frac{1}{2}\nu - \frac{1}{2}\mu\right)}$,

14.5.3
$$\mathsf{Q}_\nu^\mu(0) = -\dfrac{2^{\mu-1} \pi^{1/2} \sin\left(\frac{1}{2}(\nu+\mu)\pi\right) \Gamma\left(\frac{1}{2}\nu + \frac{1}{2}\mu + \frac{1}{2}\right)}{\Gamma\left(\frac{1}{2}\nu - \frac{1}{2}\mu + 1\right)},$$
$$\nu + \mu \neq -1, -3, -5, \dots,$$

14.5.4
$$\left.\dfrac{d\mathsf{Q}_\nu^\mu(x)}{dx}\right|_{x=0} = \dfrac{2^\mu \pi^{1/2} \cos\left(\frac{1}{2}(\nu+\mu)\pi\right) \Gamma\left(\frac{1}{2}\nu + \frac{1}{2}\mu + 1\right)}{\Gamma\left(\frac{1}{2}\nu - \frac{1}{2}\mu + \frac{1}{2}\right)},$$
$$\nu + \mu \neq -2, -4, -6, \dots.$$

14.5(ii) $\mu = 0$, $\nu = 0, 1$

14.5.5 $\quad \mathsf{P}_0(x) = P_0(x) = 1$,

14.5.6 $\quad \mathsf{P}_1(x) = P_1(x) = x$.

14.5.7 $\quad \mathsf{Q}_0(x) = \dfrac{1}{2} \ln\left(\dfrac{1+x}{1-x}\right)$,

14.5.8 $\quad \mathsf{Q}_1(x) = \dfrac{x}{2} \ln\left(\dfrac{1+x}{1-x}\right) - 1$.

14.5.9 $\quad \boldsymbol{Q}_0(x) = \dfrac{1}{2} \ln\left(\dfrac{x+1}{x-1}\right)$,

14.5.10 $\quad \boldsymbol{Q}_1(x) = \dfrac{x}{2} \ln\left(\dfrac{x+1}{x-1}\right) - 1$.

14.5(iii) $\mu = \pm\frac{1}{2}$

In this subsection and the next two, $0 < \theta < \pi$ and $\xi > 0$.

14.5.11 $\quad \mathsf{P}_\nu^{1/2}(\cos\theta) = \left(\dfrac{2}{\pi \sin\theta}\right)^{1/2} \cos\left(\left(\nu + \tfrac{1}{2}\right)\theta\right)$,

14.5.12 $\quad \mathsf{P}_\nu^{-1/2}(\cos\theta) = \left(\dfrac{2}{\pi \sin\theta}\right)^{1/2} \dfrac{\sin\left(\left(\nu + \tfrac{1}{2}\right)\theta\right)}{\nu + \tfrac{1}{2}}$,

14.5.13 $\quad \mathsf{Q}_\nu^{1/2}(\cos\theta) = -\left(\dfrac{\pi}{2 \sin\theta}\right)^{1/2} \sin\left(\left(\nu + \tfrac{1}{2}\right)\theta\right)$,

14.5.14 $\quad \mathsf{Q}_\nu^{-1/2}(\cos\theta) = -\left(\dfrac{\pi}{2 \sin\theta}\right)^{1/2} \dfrac{\cos\left(\left(\nu + \tfrac{1}{2}\right)\theta\right)}{\nu + \tfrac{1}{2}}$.

14.5.15
$$P_\nu^{1/2}(\cosh\xi) = \left(\dfrac{2}{\pi \sinh\xi}\right)^{1/2} \cosh\left(\left(\nu + \tfrac{1}{2}\right)\xi\right),$$

14.5.16
$$P_\nu^{-1/2}(\cosh\xi) = \left(\dfrac{2}{\pi \sinh\xi}\right)^{1/2} \dfrac{\sinh\left(\left(\nu + \tfrac{1}{2}\right)\xi\right)}{\nu + \tfrac{1}{2}},$$

14.5.17
$$\boldsymbol{Q}_\nu^{\pm 1/2}(\cosh\xi) = \left(\dfrac{\pi}{2 \sinh\xi}\right)^{1/2} \dfrac{\exp\left(-\left(\nu + \tfrac{1}{2}\right)\xi\right)}{\Gamma\left(\nu + \tfrac{3}{2}\right)}.$$

14.5(iv) $\mu = -\nu$

14.5.18 $\quad \mathsf{P}_\nu^{-\nu}(\cos\theta) = \dfrac{(\sin\theta)^\nu}{2^\nu \, \Gamma(\nu+1)}$,

14.5.19 $\quad P_\nu^{-\nu}(\cosh\xi) = \dfrac{(\sinh\xi)^\nu}{2^\nu \, \Gamma(\nu+1)}.$

14.5(v) $\mu = 0, \nu = \pm\frac{1}{2}$

In this subsection $K(k)$ and $E(k)$ denote the complete elliptic integrals of the first and second kinds; see §19.2(ii).

14.5.20 $\quad \mathsf{P}_{\frac{1}{2}}(\cos\theta) = \frac{2}{\pi}\left(2E\left(\sin\left(\tfrac{1}{2}\theta\right)\right) - K\left(\sin\left(\tfrac{1}{2}\theta\right)\right)\right),$

14.5.21 $\quad \mathsf{P}_{-\frac{1}{2}}(\cos\theta) = \frac{2}{\pi} K\left(\sin\left(\tfrac{1}{2}\theta\right)\right),$

14.5.22 $\quad \mathsf{Q}_{\frac{1}{2}}(\cos\theta) = K\left(\cos\left(\tfrac{1}{2}\theta\right)\right) - 2E\left(\cos\left(\tfrac{1}{2}\theta\right)\right),$

14.5.23 $\quad \mathsf{Q}_{-\frac{1}{2}}(\cos\theta) = K\left(\cos\left(\tfrac{1}{2}\theta\right)\right).$

14.5.24 $\quad P_{\frac{1}{2}}(\cosh\xi) = \frac{2}{\pi} e^{\xi/2} E\left((1 - e^{-2\xi})^{1/2}\right),$

14.5.25 $\quad P_{-\frac{1}{2}}(\cosh\xi) = \frac{2}{\pi\cosh\left(\tfrac{1}{2}\xi\right)} K\left(\tanh\left(\tfrac{1}{2}\xi\right)\right),$

14.5.26
$$\boldsymbol{Q}_{\frac{1}{2}}(\cosh\xi) = 2\pi^{-1/2}\cosh\xi\,\operatorname{sech}\left(\tfrac{1}{2}\xi\right) K\left(\operatorname{sech}\left(\tfrac{1}{2}\xi\right)\right) \\ - 4\pi^{-1/2}\cosh\left(\tfrac{1}{2}\xi\right) E\left(\operatorname{sech}\left(\tfrac{1}{2}\xi\right)\right),$$

14.5.27 $\quad \boldsymbol{Q}_{-\frac{1}{2}}(\cosh\xi) = 2\pi^{-1/2} e^{-\xi/2} K\left(e^{-\xi}\right).$

14.6 Integer Order

14.6(i) Nonnegative Integer Orders

For $m = 0, 1, 2, \ldots,$

14.6.1 $\quad \mathsf{P}_\nu^m(x) = (-1)^m \left(1 - x^2\right)^{m/2} \frac{d^m \mathsf{P}_\nu(x)}{dx^m},$

14.6.2 $\quad \mathsf{Q}_\nu^m(x) = (-1)^m \left(1 - x^2\right)^{m/2} \frac{d^m \mathsf{Q}_\nu(x)}{dx^m}.$

14.6.3 $\quad P_\nu^m(x) = \left(x^2 - 1\right)^{m/2} \frac{d^m P_\nu(x)}{dx^m},$

14.6.4 $\quad Q_\nu^m(x) = \left(x^2 - 1\right)^{m/2} \frac{d^m Q_\nu(x)}{dx^m},$

14.6.5 $\quad (\nu+1)_m \boldsymbol{Q}_\nu^m(x) = (-1)^m \left(x^2 - 1\right)^{m/2} \frac{d^m \boldsymbol{Q}_\nu(x)}{dx^m}.$

14.6(ii) Negative Integer Orders

For $m = 1, 2, 3, \ldots,$

14.6.6 $\quad \mathsf{P}_\nu^{-m}(x) = \left(1 - x^2\right)^{-m/2} \int_x^1 \cdots \int_x^1 \mathsf{P}_\nu(x) (dx)^m.$

14.6.7 $\quad P_\nu^{-m}(x) = \left(x^2 - 1\right)^{-m/2} \int_1^x \cdots \int_1^x P_\nu(x) (dx)^m,$

14.6.8
$$Q_\nu^{-m}(x) = (-1)^m \left(x^2 - 1\right)^{-m/2} \\ \times \int_x^\infty \cdots \int_x^\infty Q_\nu(x) (dx)^m.$$

For connections between positive and negative integer orders see (14.9.3), (14.9.4), and (14.9.13).

14.7 Integer Degree and Order

14.7(i) $\mu = 0$

For $n = 0, 1, 2, \ldots,$

14.7.1 $\quad \mathsf{P}_n^0(x) = \mathsf{P}_n(x) = P_n^0(x) = P_n(x), \qquad x \in \mathbb{R},$

where $P_n(x)$ is the *Legendre polynomial of degree n*. For additional properties of $P_n(x)$ see Chapter 18.

14.7.2
$$\mathsf{Q}_n^0(x) = \mathsf{Q}_n(x) = \tfrac{1}{2} P_n(x) \ln\left(\frac{1+x}{1-x}\right) - W_{n-1}(x),$$

where $W_{-1}(x) = 0$, and for $n \geq 1$,

14.7.3
$$W_{n-1}(x) = \sum_{s=0}^{n-1} \frac{(n+s)!(\psi(n+1) - \psi(s+1))}{2^s (n-s)!(s!)^2} (x-1)^s;$$

equivalently,

14.7.4 $\quad W_{n-1}(x) = \sum_{k=1}^n \frac{1}{k} P_{k-1}(x) P_{n-k}(x).$

14.7.5 $\quad W_0(x) = 1, \quad W_1(x) = \tfrac{3}{2}x, \quad W_2(x) = \tfrac{5}{2}x^2 - \tfrac{2}{3}.$

Next,

14.7.6 $\quad Q_n^0(x) = Q_n(x) = n!\, \boldsymbol{Q}_n^0(x) = n!\, \boldsymbol{Q}_n(x),$

where

14.7.7
$$Q_n(x) = \tfrac{1}{2} P_n(x) \ln\left(\frac{x+1}{x-1}\right) - W_{n-1}(x), \, n = 0, 1, 2, \ldots.$$

14.7(ii) Rodrigues-Type Formulas

For $m = 0, 1, 2, \ldots,$ and $n = 0, 1, 2, \ldots,$

14.7.8 $\quad \mathsf{P}_n^m(x) = (-1)^m \left(1 - x^2\right)^{m/2} \frac{d^m}{dx^m} \mathsf{P}_n(x),$

14.7.9 $\quad \mathsf{Q}_n^m(x) = (-1)^m \left(1 - x^2\right)^{m/2} \frac{d^m}{dx^m} \mathsf{Q}_n(x),$

14.7.10
$$\mathsf{P}_n^m(x) = (-1)^{m+n} \frac{\left(1 - x^2\right)^{m/2}}{2^n n!} \frac{d^{m+n}}{dx^{m+n}} \left(1 - x^2\right)^n.$$

14.7.11 $\quad P_n^m(x) = \left(x^2 - 1\right)^{m/2} \frac{d^m}{dx^m} P_n(x),$

14.7.12 $\quad Q_n^m(x) = \left(x^2 - 1\right)^{m/2} \frac{d^m}{dx^m} Q_n(x),$

14.7.13 $\quad P_n(x) = \frac{1}{2^n n!} \frac{d^n}{dx^n} \left(x^2 - 1\right)^n,$

14.7.14 $\quad P_n^m(x) = \frac{\left(x^2 - 1\right)^{m/2}}{2^n n!} \frac{d^{m+n}}{dx^{m+n}} \left(x^2 - 1\right)^n,$

14.7.15 $\quad P_m^m(x) = \frac{(2m)!}{2^m m!} \left(x^2 - 1\right)^{m/2}.$

When m is even and $m \leq n$, $\mathsf{P}_n^m(x)$ and $P_n^m(x)$ are polynomials of degree n. Also,

14.7.16 $\quad \mathsf{P}_n^m(x) = P_n^m(x) = 0, \qquad m > n.$

14.7(iii) Reflection Formulas

14.7.17 $\quad \mathsf{P}_n^m(-x) = (-1)^{n-m}\, \mathsf{P}_n^m(x),$

14.7.18 $\quad \mathsf{Q}_n^{\pm m}(-x) = (-1)^{n-m-1}\, \mathsf{Q}_n^{\pm m}(x).$

14.7(iv) Generating Functions

When $-1 < x < 1$ and $|h| < 1$,

14.7.19 $\quad \displaystyle\sum_{n=0}^{\infty} \mathsf{P}_n(x) h^n = \left(1 - 2xh + h^2\right)^{-1/2},$

14.7.20
$$\sum_{n=0}^{\infty} \mathsf{Q}_n(x) h^n = \frac{1}{(1 - 2xh + h^2)^{1/2}} \\ \times \ln\left(\frac{x - h + (1 - 2xh + h^2)^{1/2}}{(1 - x^2)^{1/2}}\right).$$

When $-1 < x < 1$ and $|h| > 1$,

14.7.21 $\quad \displaystyle\sum_{n=0}^{\infty} \mathsf{P}_n(x) h^{-n-1} = \left(1 - 2xh + h^2\right)^{-1/2}.$

When $x > 1$, (14.7.19) applies with $|h| < x - (x^2 - 1)^{1/2}$. Also, with the same conditions

14.7.22
$$\sum_{n=0}^{\infty} Q_n(x) h^n = \frac{1}{(1 - 2xh + h^2)^{1/2}} \\ \times \ln\left(\frac{x - h + (1 - 2xh + h^2)^{1/2}}{(x^2 - 1)^{1/2}}\right).$$

Lastly, when $x > 1$, (14.7.21) applies with $|h| > x + (x^2 - 1)^{1/2}$.

For other generating functions see Magnus *et al.* (1966, pp. 232–233) and Rainville (1960, pp. 163–165, 168, 170–171, 184).

14.8 Behavior at Singularities

14.8(i) $x \to 1-$ or $x \to -1+$

As $x \to 1-$,

14.8.1 $\quad \mathsf{P}_\nu^\mu(x) \sim \dfrac{1}{\Gamma(1-\mu)} \left(\dfrac{2}{1-x}\right)^{\mu/2}, \quad \mu \ne 1, 2, 3, \ldots,$

14.8.2 $\quad \mathsf{P}_\nu^m(x) \sim (-1)^m \dfrac{(\nu - m + 1)_{2m}}{m!} \left(\dfrac{1-x}{2}\right)^{m/2},$
$m = 1, 2, 3, \ldots, \nu \ne m-1, m-2, \ldots, -m,$

14.8.3 $\quad \mathsf{Q}_\nu(x) = \dfrac{1}{2} \ln\left(\dfrac{2}{1-x}\right) - \gamma - \psi(\nu + 1) + O(1-x),$
$\nu \ne -1, -2, -3, \ldots,$

where γ is Euler's constant (§5.2(ii)). In the next three relations $\Re\mu > 0$.

14.8.4 $\quad \mathsf{Q}_\nu^\mu(x) \sim \dfrac{1}{2} \cos(\mu\pi)\, \Gamma(\mu) \left(\dfrac{2}{1-x}\right)^{\mu/2}, \quad \mu \ne \tfrac{1}{2}, \tfrac{3}{2}, \tfrac{5}{2}, \ldots,$

14.8.5
$$\mathsf{Q}_\nu^\mu(x) \\ \sim (-1)^{\mu + (1/2)} \frac{\pi\, \Gamma(\nu + \mu + 1)}{2\, \Gamma(\mu + 1)\, \Gamma(\nu - \mu + 1)} \left(\frac{1-x}{2}\right)^{\mu/2},$$
$\mu = \tfrac{1}{2}, \tfrac{3}{2}, \tfrac{5}{2}, \ldots, \nu \pm \mu \ne -1, -2, -3, \ldots,$

14.8.6 $\quad \mathsf{Q}_\nu^{-\mu}(x) \sim \dfrac{\Gamma(\mu)\, \Gamma(\nu - \mu + 1)}{2\, \Gamma(\nu + \mu + 1)} \left(\dfrac{2}{1-x}\right)^{\mu/2},$
$\nu \pm \mu \ne -1, -2, -3, \ldots.$

The behavior of $P_\nu^\mu(x)$ and $Q_\nu^\mu(x)$ as $x \to -1+$ follows from the above results and the connection formulas (14.9.8) and (14.9.10).

14.8(ii) $x \to 1+$

14.8.7 $\quad P_\nu^\mu(x) \sim \dfrac{1}{\Gamma(1-\mu)} \left(\dfrac{2}{x-1}\right)^{\mu/2}, \quad \mu \ne 1, 2, 3, \ldots,$

14.8.8 $\quad P_\nu^m(x) \sim \dfrac{\Gamma(\nu + m + 1)}{m!\, \Gamma(\nu - m + 1)} \left(\dfrac{x-1}{2}\right)^{m/2},$
$m = 1, 2, 3, \ldots, \nu \pm m \ne -1, -2, -3, \ldots,$

14.8.9 $\quad \boldsymbol{Q}_\nu(x) = -\dfrac{\ln(x-1)}{2\,\Gamma(\nu+1)} + \dfrac{\tfrac{1}{2}\ln 2 - \gamma - \psi(\nu+1)}{\Gamma(\nu+1)} + O(x-1), \quad \nu \ne -1, -2, -3, \ldots,$

14.8.10 $\quad \boldsymbol{Q}_{-n}(x) \to (-1)^{n+1}(n-1)!, \quad n = 1, 2, 3, \ldots,$

14.8.11 $\quad \boldsymbol{Q}_\nu^\mu(x) \sim \dfrac{\Gamma(\mu)}{2\,\Gamma(\nu + \mu + 1)} \left(\dfrac{2}{x-1}\right)^{\mu/2},$
$\Re\mu > 0, \nu + \mu \ne -1, -2, -3, \ldots.$

14.8(iii) $x \to \infty$

14.8.12 $\quad P_\nu^\mu(x) \sim \dfrac{\Gamma(\nu + \tfrac{1}{2})}{\pi^{1/2}\, \Gamma(\nu - \mu + 1)} (2x)^\nu,$
$\Re\nu > -\tfrac{1}{2}, \mu - \nu \ne 1, 2, 3, \ldots,$

14.8.13 $\quad P_\nu^\mu(x) \sim \dfrac{\Gamma(-\nu - \tfrac{1}{2})}{\pi^{1/2}\, \Gamma(-\mu - \nu)(2x)^{\nu+1}},$
$\Re\nu < -\tfrac{1}{2}, \nu + \mu \ne 0, 1, 2, \ldots,$

14.8.14 $\quad P_{-1/2}^\mu(x) \sim \dfrac{1}{\Gamma(\tfrac{1}{2} - \mu)} \left(\dfrac{2}{\pi x}\right)^{1/2} \ln x,$
$\mu \ne \tfrac{1}{2}, \tfrac{3}{2}, \tfrac{5}{2}, \ldots,$

14.8.15
$$Q_\nu^\mu(x) \sim \frac{\pi^{1/2}}{\Gamma(\nu+\frac{3}{2})(2x)^{\nu+1}}, \quad \nu \neq -\frac{3}{2}, -\frac{5}{2}, -\frac{7}{2}, \ldots,$$

14.8.16
$$Q_{-n-(1/2)}^\mu(x) \sim \frac{\pi^{1/2}\,\Gamma(\mu+n+\frac{1}{2})}{n!\,\Gamma(\mu-n+\frac{1}{2})(2x)^{n+(1/2)}},$$
$$n = 1, 2, 3, \ldots, \mu - n + \tfrac{1}{2} \neq 0, -1, -2, \ldots.$$

14.9 Connection Formulas

14.9(i) Connections Between $\mathsf{P}_\nu^{\pm\mu}(x)$, $\mathsf{P}_{-\nu-1}^{\pm\mu}(x)$, $\mathsf{Q}_\nu^{\pm\mu}(x)$, $\mathsf{Q}_{-\nu-1}^\mu(x)$

14.9.1
$$\frac{\pi \sin(\mu\pi)}{2\,\Gamma(\nu-\mu+1)}\,\mathsf{P}_\nu^{-\mu}(x) = -\frac{1}{\Gamma(\nu+\mu+1)}\,\mathsf{Q}_\nu^\mu(x) + \frac{\cos(\mu\pi)}{\Gamma(\nu-\mu+1)}\,\mathsf{Q}_\nu^{-\mu}(x).$$

14.9.2
$$\frac{2\sin(\mu\pi)}{\pi\,\Gamma(\nu-\mu+1)}\,\mathsf{Q}_\nu^{-\mu}(x) = \frac{1}{\Gamma(\nu+\mu+1)}\,\mathsf{P}_\nu^\mu(x) - \frac{\cos(\mu\pi)}{\Gamma(\nu-\mu+1)}\,\mathsf{P}_\nu^{-\mu}(x),$$

14.9.3
$$\mathsf{P}_\nu^{-m}(x) = (-1)^m \frac{\Gamma(\nu-m+1)}{\Gamma(\nu+m+1)}\,\mathsf{P}_\nu^m(x),$$

14.9.4
$$\mathsf{Q}_\nu^{-m}(x) = (-1)^m \frac{\Gamma(\nu-m+1)}{\Gamma(\nu+m+1)}\,\mathsf{Q}_\nu^m(x),$$
$$\nu \neq m-1, m-2, \ldots.$$

14.9.5
$$\mathsf{P}_{-\nu-1}^\mu(x) = \mathsf{P}_\nu^\mu(x), \quad \mathsf{P}_{-\nu-1}^{-\mu}(x) = \mathsf{P}_\nu^{-\mu}(x).$$

14.9.6
$$\pi\cos(\nu\pi)\cos(\mu\pi)\,\mathsf{P}_\nu^\mu(x) = \sin((\nu+\mu)\pi)\,\mathsf{Q}_\nu^\mu(x) - \sin((\nu-\mu)\pi)\,\mathsf{Q}_{-\nu-1}^\mu(x).$$

14.9(ii) Connections Between $\mathsf{P}_\nu^{\pm\mu}(\pm x)$, $\mathsf{Q}_\nu^{-\mu}(\pm x)$, $\mathsf{Q}_\nu^\mu(x)$

14.9.7
$$\frac{\sin((\nu-\mu)\pi)}{\Gamma(\nu+\mu+1)}\,\mathsf{P}_\nu^\mu(x) = \frac{\sin(\nu\pi)}{\Gamma(\nu-\mu+1)}\,\mathsf{P}_\nu^{-\mu}(x) - \frac{\sin(\mu\pi)}{\Gamma(\nu-\mu+1)}\,\mathsf{P}_\nu^{-\mu}(-x),$$

14.9.8
$$\tfrac{1}{2}\pi\sin((\nu-\mu)\pi)\,\mathsf{P}_\nu^{-\mu}(x) = -\cos((\nu-\mu)\pi)\,\mathsf{Q}_\nu^{-\mu}(x) - \mathsf{Q}_\nu^{-\mu}(-x),$$

14.9.9
$$\frac{2}{\Gamma(\nu+\mu+1)\,\Gamma(\mu-\nu)}\,\mathsf{Q}_\nu^\mu(x) = -\cos(\nu\pi)\,\mathsf{P}_\nu^{-\mu}(x) + \cos(\mu\pi)\,\mathsf{P}_\nu^{-\mu}(-x),$$

14.9.10
$$(2/\pi)\sin((\nu-\mu)\pi)\,\mathsf{Q}_\nu^{-\mu}(x) = \cos((\nu-\mu)\pi)\,\mathsf{P}_\nu^{-\mu}(x) - \mathsf{P}_\nu^{-\mu}(-x).$$

14.9(iii) Connections Between $P_\nu^{\pm\mu}(x)$, $P_{-\nu-1}^{\pm\mu}(x)$, $Q_\nu^{\pm\mu}(x)$, $Q_{-\nu-1}^\mu(x)$

14.9.11
$$P_{-\nu-1}^{-\mu}(x) = P_\nu^{-\mu}(x), \quad P_{-\nu-1}^\mu(x) = P_\nu^\mu(x),$$

14.9.12
$$\cos(\nu\pi)\,P_\nu^{-\mu}(x) = -\frac{Q_\nu^\mu(x)}{\Gamma(\mu-\nu)} + \frac{Q_{-\nu-1}^\mu(x)}{\Gamma(\nu+\mu+1)}.$$

14.9.13
$$P_\nu^{-m}(x) = \frac{\Gamma(\nu-m+1)}{\Gamma(\nu+m+1)}\,P_\nu^m(x), \quad \nu \neq m-1, m-2, \ldots.$$

14.9.14
$$Q_\nu^{-\mu}(x) = Q_\nu^\mu(x),$$

14.9.15
$$\frac{2\sin(\mu\pi)}{\pi}\,Q_\nu^\mu(x) = \frac{P_\nu^\mu(x)}{\Gamma(\nu+\mu+1)} - \frac{P_\nu^{-\mu}(x)}{\Gamma(\nu-\mu+1)}.$$

14.9(iv) Whipple's Formula

14.9.16
$$Q_\nu^\mu(x) = \left(\tfrac{1}{2}\pi\right)^{1/2}(x^2-1)^{-1/4}\,P_{-\mu-(1/2)}^{-\nu-(1/2)}\!\left(x(x^2-1)^{-1/2}\right).$$

Equivalently,

14.9.17
$$P_\nu^\mu(x) = (2/\pi)^{1/2}(x^2-1)^{-1/4}\,Q_{-\mu-(1/2)}^{\nu+(1/2)}\!\left(x(x^2-1)^{-1/2}\right).$$

14.10 Recurrence Relations and Derivatives

14.10.1
$$\mathsf{P}_\nu^{\mu+2}(x) + 2(\mu+1)x(1-x^2)^{-1/2}\,\mathsf{P}_\nu^{\mu+1}(x) + (\nu-\mu)(\nu+\mu+1)\,\mathsf{P}_\nu^\mu(x) = 0,$$

14.10.2
$$(1-x^2)^{1/2}\,\mathsf{P}_\nu^{\mu+1}(x) - (\nu-\mu+1)\,\mathsf{P}_{\nu+1}^\mu(x) + (\nu+\mu+1)x\,\mathsf{P}_\nu^\mu(x) = 0,$$

14.10.3
$$(\nu-\mu+2)\,\mathsf{P}_{\nu+2}^\mu(x) - (2\nu+3)x\,\mathsf{P}_{\nu+1}^\mu(x) + (\nu+\mu+1)\,\mathsf{P}_\nu^\mu(x) = 0,$$

14.10.4
$$(1-x^2)\frac{d\mathsf{P}_\nu^\mu(x)}{dx} = (\mu-\nu-1)\,\mathsf{P}_{\nu+1}^\mu(x) + (\nu+1)x\,\mathsf{P}_\nu^\mu(x),$$

14.10.5
$$(1-x^2)\frac{d\mathsf{P}_\nu^\mu(x)}{dx} = (\nu+\mu)\,\mathsf{P}_{\nu-1}^\mu(x) - \nu x\,\mathsf{P}_\nu^\mu(x).$$

$\mathsf{Q}_\nu^\mu(x)$ also satisfies (14.10.1)–(14.10.5).

14.10.6
$$P_\nu^{\mu+2}(x) + 2(\mu+1)x(x^2-1)^{-1/2}\,P_\nu^{\mu+1}(x) - (\nu-\mu)(\nu+\mu+1)\,P_\nu^\mu(x) = 0,$$

14.10.7
$$(x^2-1)^{1/2}\,P_\nu^{\mu+1}(x) - (\nu-\mu+1)\,P_{\nu+1}^\mu(x) + (\nu+\mu+1)x\,P_\nu^\mu(x) = 0.$$

$Q_\nu^\mu(x)$ also satisfies (14.10.6) and (14.10.7). In addition, $P_\nu^\mu(x)$ and $Q_\nu^\mu(x)$ satisfy (14.10.3)–(14.10.5).

14.11 Derivatives with Respect to Degree or Order

14.11.1
$$\frac{\partial}{\partial \nu} \mathsf{P}^\mu_\nu(x) = \pi \cot(\nu\pi) \mathsf{P}^\mu_\nu(x) - \frac{1}{\pi}\mathsf{A}^\mu_\nu(x),$$

14.11.2
$$\frac{\partial}{\partial \nu} \mathsf{Q}^\mu_\nu(x) = -\tfrac{1}{2}\pi^2 \mathsf{P}^\mu_\nu(x) + \frac{\pi \sin(\mu\pi)}{\sin(\nu\pi)\sin((\nu+\mu)\pi)} \mathsf{Q}^\mu_\nu(x) - \tfrac{1}{2}\cot((\nu+\mu)\pi)\mathsf{A}^\mu_\nu(x) + \tfrac{1}{2}\csc((\nu+\mu)\pi)\mathsf{A}^\mu_\nu(-x),$$

where

14.11.3
$$\mathsf{A}^\mu_\nu(x) = \sin(\nu\pi) \left(\frac{1+x}{1-x}\right)^{\mu/2} \sum_{k=0}^\infty \frac{\left(\tfrac{1}{2}-\tfrac{1}{2}x\right)^k \Gamma(k-\nu)\,\Gamma(k+\nu+1)}{k!\,\Gamma(k-\mu+1)} \left(\psi(k+\nu+1) - \psi(k-\nu)\right).$$

14.11.4
$$\left.\frac{\partial}{\partial \mu}\mathsf{P}^\mu_\nu(x)\right|_{\mu=0} = (\psi(-\nu) - \pi\cot(\nu\pi))\,\mathsf{P}_\nu(x) + \mathsf{Q}_\nu(x),$$

14.11.5
$$\left.\frac{\partial}{\partial \mu}\mathsf{Q}^\mu_\nu(x)\right|_{\mu=0} = -\tfrac{1}{4}\pi^2 \mathsf{P}_\nu(x) + (\psi(-\nu) - \pi\cot(\nu\pi))\,\mathsf{Q}_\nu(x).$$

(14.11.1) holds if $\mathsf{P}^\mu_\nu(x)$ is replaced by $P^\mu_\nu(x)$, provided that the factor $((1+x)/(1-x))^{\mu/2}$ in (14.11.3) is replaced by $((x+1)/(x-1))^{\mu/2}$. (14.11.4) holds if $\mathsf{P}^\mu_\nu(x)$, $\mathsf{P}_\nu(x)$, and $\mathsf{Q}_\nu(x)$ are replaced by $P^\mu_\nu(x)$, $P_\nu(x)$, and $Q_\nu(x)$, respectively.

For further results see Magnus et al. (1966, pp. 177–178).

14.12 Integral Representations

14.12(i) $-1 < x < 1$

Mehler–Dirichlet Formula

14.12.1
$$\mathsf{P}^\mu_\nu(\cos\theta) = \frac{2^{1/2}(\sin\theta)^\mu}{\pi^{1/2}\,\Gamma\!\left(\tfrac{1}{2}-\mu\right)} \int_0^\theta \frac{\cos\!\left((\nu+\tfrac{1}{2})t\right)}{(\cos t - \cos\theta)^{\mu+(1/2)}}\,dt, \qquad 0 < \theta < \pi,\ \Re\mu < \tfrac{1}{2}.$$

14.12.2
$$\mathsf{P}^{-\mu}_\nu(x) = \frac{(1-x^2)^{-\mu/2}}{\Gamma(\mu)} \int_x^1 \mathsf{P}_\nu(t)(t-x)^{\mu-1}\,dt, \qquad \Re\mu > 0;$$

compare (14.6.6).

14.12.3
$$\mathsf{Q}^\mu_\nu(\cos\theta) = \frac{\pi^{1/2}\,\Gamma(\nu+\mu+1)(\sin\theta)^\mu}{2^{\mu+1}\,\Gamma\!\left(\mu+\tfrac{1}{2}\right)\Gamma(\nu-\mu+1)} \left(\int_0^\infty \frac{(\sinh t)^{2\mu}}{(\cos\theta + i\sin\theta\cosh t)^{\nu+\mu+1}}\,dt + \int_0^\infty \frac{(\sinh t)^{2\mu}}{(\cos\theta - i\sin\theta\cosh t)^{\nu+\mu+1}}\,dt\right),$$
$$0 < \theta < \pi,\ \Re\mu > -\tfrac{1}{2},\ \Re(\nu\pm\mu) > -1.$$

14.12(ii) $1 < x < \infty$

14.12.4
$$P^{-\mu}_\nu(x) = \frac{2^{1/2}\,\Gamma\!\left(\mu+\tfrac{1}{2}\right)(x^2-1)^{\mu/2}}{\pi^{1/2}\,\Gamma(\nu+\mu+1)\,\Gamma(\mu-\nu)} \int_0^\infty \frac{\cosh\!\left((\nu+\tfrac{1}{2})t\right)}{(x+\cosh t)^{\mu+(1/2)}}\,dt, \qquad \nu+\mu \neq -1,-2,-3,\ldots,\ \Re(\mu-\nu) > 0.$$

14.12.5
$$P^{-\mu}_\nu(x) = \frac{(x^2-1)^{-\mu/2}}{\Gamma(\mu)} \int_1^x P_\nu(t)(x-t)^{\mu-1}\,dt, \qquad \Re\mu > 0.$$

14.12.6
$$\boldsymbol{Q}^\mu_\nu(x) = \frac{\pi^{1/2}\,(x^2-1)^{\mu/2}}{2^\mu\,\Gamma\!\left(\mu+\tfrac{1}{2}\right)\Gamma(\nu-\mu+1)} \int_0^\infty \frac{(\sinh t)^{2\mu}}{(x+(x^2-1)^{1/2}\cosh t)^{\nu+\mu+1}}\,dt, \qquad \Re(\nu+1) > \Re\mu > -\tfrac{1}{2}.$$

14.12.7
$$P^m_\nu(x) = \frac{(\nu+1)_m}{\pi} \int_0^\pi \left(x + (x^2-1)^{1/2}\cos\phi\right)^\nu \cos(m\phi)\,d\phi,$$

14.12.8
$$P^m_n(x) = \frac{2^m m!\,(n+m)!\,(x^2-1)^{m/2}}{(2m)!\,(n-m)!\,\pi} \int_0^\pi \left(x + (x^2-1)^{1/2}\cos\phi\right)^{n-m} (\sin\phi)^{2m}\,d\phi, \qquad n \geq m.$$

14.12.9
$$\boldsymbol{Q}_n^m(x) = \frac{1}{n!} \int_0^u \left(x - (x^2-1)^{1/2} \cosh t\right)^n \cosh(mt)\,dt,$$

where

14.12.10
$$u = \frac{1}{2}\ln\left(\frac{x+1}{x-1}\right).$$

14.12.11
$$\boldsymbol{Q}_n^m(x) = \frac{(x^2-1)^{m/2}}{2^{n+1} n!} \int_{-1}^1 \frac{(1-t^2)^n}{(x-t)^{n+m+1}}\,dt,$$

14.12.12
$$\boldsymbol{Q}_n^m(x) = \frac{1}{(n-m)!} P_n^m(x) \int_x^\infty \frac{dt}{(t^2-1)(P_n^m(t))^2}, \qquad n \geq m.$$

Neumann's Integral

14.12.13
$$\boldsymbol{Q}_n(x) = \frac{1}{2(n!)} \int_{-1}^1 \frac{P_n(t)}{x-t}\,dt.$$

Heine's Integral

14.12.14
$$\boldsymbol{Q}_n(x) = \frac{1}{n!} \int_0^\infty \frac{dt}{\left(x + (x^2-1)^{1/2}\cosh t\right)^{n+1}}.$$

For further integral representations see Erdélyi et al. (1953a, pp. 158–159) and Magnus et al. (1966, pp. 184–190), and for contour integrals and other representations see §14.25.

14.13 Trigonometric Expansions

When $0 < \theta < \pi$,

14.13.1
$$\mathsf{P}_\nu^\mu(\cos\theta) = \frac{2^{\mu+1}(\sin\theta)^\mu}{\pi^{1/2}} \sum_{k=0}^\infty \frac{\Gamma(\nu+\mu+k+1)}{\Gamma(\nu+k+\frac{3}{2})} \frac{(\mu+\frac{1}{2})_k}{k!} \sin((\nu+\mu+2k+1)\theta),$$

14.13.2
$$\mathsf{Q}_\nu^\mu(\cos\theta) = \pi^{1/2} 2^\mu (\sin\theta)^\mu \sum_{k=0}^\infty \frac{\Gamma(\nu+\mu+k+1)}{\Gamma(\nu+k+\frac{3}{2})} \frac{(\mu+\frac{1}{2})_k}{k!} \cos((\nu+\mu+2k+1)\theta),$$

14.13.3
$$\mathsf{P}_n(\cos\theta) = \frac{2^{2n+2}(n!)^2}{\pi(2n+1)!} \sum_{k=0}^\infty \frac{1\cdot 3\cdots(2k-1)}{k!} \frac{(n+1)(n+2)\cdots(n+k)}{(2n+3)(2n+5)\cdots(2n+2k+1)} \sin((n+2k+1)\theta),$$

14.13.4
$$\mathsf{Q}_n(\cos\theta) = \frac{2^{2n+1}(n!)^2}{(2n+1)!} \sum_{k=0}^\infty \frac{1\cdot 3\cdots(2k-1)}{k!} \frac{(n+1)(n+2)\cdots(n+k)}{(2n+3)(2n+5)\cdots(2n+2k+1)} \cos((n+2k+1)\theta).$$

For these and other trigonometric expansions see Erdélyi et al. (1953a, pp. 146–147).

14.14 Continued Fractions

14.14.1
$$\tfrac{1}{2}(x^2-1)^{1/2} \frac{P_\nu^\mu(x)}{P_\nu^{\mu-1}(x)} = \frac{x_0}{y_0+} \frac{x_1}{y_1+} \frac{x_2}{y_2+} \cdots,$$

where

14.14.2
$$x_k = \tfrac{1}{4}(\nu-\mu-k+1)(\nu+\mu+k)(x^2-1), \qquad y_k = (\mu+k)x,$$

provided that x_{k+1} and y_k do not vanish simultaneously for any $k = 0, 1, 2, \ldots$.

14.14.3
$$(\nu-\mu)\frac{Q_\nu^\mu(x)}{Q_{\nu-1}^\mu(x)} = \frac{x_0}{y_0-} \frac{x_1}{y_1-} \frac{x_2}{y_2-} \cdots, \qquad \nu \neq \mu,$$

where now

14.14.4
$$x_k = (\nu+\mu+k)(\nu-\mu+k), \qquad y_k = (2\nu+2k+1)x,$$

again provided x_{k+1} and y_k do not vanish simultaneously for any $k = 0, 1, 2, \ldots$.

14.15 Uniform Asymptotic Approximations

14.15(i) Large μ, Fixed ν

For the interval $-1 < x < 1$ with fixed ν, real μ, and arbitrary fixed values of the nonnegative integer J,

14.15.1 $$\mathsf{P}_\nu^{-\mu}(\pm x) = \left(\frac{1 \mp x}{1 \pm x}\right)^{\mu/2} \left(\sum_{j=0}^{J-1} \frac{(\nu+1)_j(-\nu)_j}{j!\,\Gamma(j+1+\mu)} \left(\frac{1 \mp x}{2}\right)^j + O\left(\frac{1}{\Gamma(J+1+\mu)}\right)\right)$$

as $\mu \to \infty$, uniformly with respect to x. In other words, the convergent hypergeometric series expansions of $\mathsf{P}_\nu^{-\mu}(\pm x)$ are also generalized (and uniform) asymptotic expansions as $\mu \to \infty$, with scale $1/\Gamma(j+1+\mu)$, $j = 0, 1, 2, \ldots$; compare §2.1(v).

Provided that $\mu - \nu \notin \mathbb{Z}$ the corresponding expansions for $\mathsf{P}_\nu^\mu(x)$ and $\mathsf{Q}_\nu^{\mp\mu}(x)$ can be obtained from the connection formulas (14.9.7), (14.9.9), and (14.9.10).

For the interval $1 < x < \infty$ the following asymptotic approximations hold when $\mu \to \infty$, with ν ($\geq -\frac{1}{2}$) fixed, uniformly with respect to x:

14.15.2 $$P_\nu^{-\mu}(x) = \frac{1}{\Gamma(\mu+1)} \left(\frac{2\mu u}{\pi}\right)^{1/2} K_{\nu+\frac{1}{2}}(\mu u) \left(1 + O\left(\frac{1}{\mu}\right)\right),$$

14.15.3 $$\boldsymbol{Q}_\nu^\mu(x) = \frac{1}{\mu^{\nu+(1/2)}} \left(\frac{\pi u}{2}\right)^{1/2} I_{\nu+\frac{1}{2}}(\mu u) \left(1 + O\left(\frac{1}{\mu}\right)\right),$$

where u is given by (14.12.10). Here I and K are the modified Bessel functions (§10.25(ii)).

For asymptotic expansions and explicit error bounds, see Dunster (2003b) and Gil et al. (2000).

14.15(ii) Large μ, $0 \leq \nu + \frac{1}{2} \leq (1-\delta)\mu$

In this and subsequent subsections δ denotes an arbitrary constant such that $0 < \delta < 1$.

As $\mu \to \infty$,

14.15.4 $$\mathsf{P}_\nu^{-\mu}(x) = \frac{1}{\Gamma(\mu+1)} (1-\alpha^2)^{-\mu/2} \left(\frac{1-\alpha}{1+\alpha}\right)^{(\nu/2)+(1/4)} \left(\frac{p}{x}\right)^{1/2} e^{-\mu\rho} \left(1 + O\left(\frac{1}{\mu}\right)\right),$$

uniformly with respect to $x \in (-1, 1)$ and $\nu + \frac{1}{2} \in [0, (1-\delta)\mu]$, where

14.15.5 $$\alpha = \frac{\nu + \frac{1}{2}}{\mu} \quad (< 1),$$

14.15.6 $$p = \frac{x}{(\alpha^2 x^2 + 1 - \alpha^2)^{1/2}},$$

and

14.15.7 $$\rho = \frac{1}{2} \ln\left(\frac{1+p}{1-p}\right) + \frac{1}{2}\alpha \ln\left(\frac{1-\alpha p}{1+\alpha p}\right).$$

With the same conditions, the corresponding approximation for $\mathsf{P}_\nu^{-\mu}(-x)$ is obtained by replacing $e^{-\mu\rho}$ by $e^{\mu\rho}$ on the right-hand side of (14.15.4). Approximations for $\mathsf{P}_\nu^\mu(x)$ and $\mathsf{Q}_\nu^{\mp\mu}(x)$ can then be achieved via (14.9.7), (14.9.9), and (14.9.10).

Next,

14.15.8 $$P_\nu^{-\mu}(x) = \left(\frac{2\mu}{\pi}\right)^{1/2} \frac{1}{\Gamma(\mu+1)} \left(\frac{1-\alpha}{1+\alpha}\right)^{(\nu/2)+(1/4)} (1-\alpha^2)^{-\mu/2} \left(\frac{\alpha^2 + \eta^2}{\alpha^2(x^2-1)+1}\right)^{1/4} K_{\nu+\frac{1}{2}}(\mu\eta) \left(1 + O\left(\frac{1}{\mu}\right)\right),$$

14.15.9 $$\boldsymbol{Q}_\nu^\mu(x) = \left(\frac{\pi}{2}\right)^{1/2} \left(\frac{e}{\mu}\right)^{\nu+(1/2)} \left(\frac{1-\alpha}{1+\alpha}\right)^{\mu/2} (1-\alpha^2)^{-(\nu/2)-(1/4)} \left(\frac{\alpha^2 + \eta^2}{\alpha^2(x^2-1)+1}\right)^{1/4} I_{\nu+\frac{1}{2}}(\mu\eta) \left(1 + O\left(\frac{1}{\mu}\right)\right),$$

uniformly with respect to $x \in (1, \infty)$ and $\nu + \frac{1}{2} \in [0, (1-\delta)\mu]$. Here α is again given by (14.15.5), and η is defined implicitly by

14.15.10 $$\alpha \ln\left((\alpha^2+\eta^2)^{1/2} + \alpha\right) - \alpha \ln \eta - (\alpha^2+\eta^2)^{1/2} = \frac{1}{2}\ln\left(\frac{(1+\alpha^2)x^2 + 1 - \alpha^2 - 2x(\alpha^2 x^2 - \alpha^2 + 1)^{1/2}}{(x^2-1)(1-\alpha^2)}\right) + \frac{1}{2}\alpha \ln\left(\frac{\alpha^2(2x^2-1) + 1 + 2\alpha x(\alpha^2 x^2 - \alpha^2 + 1)^{1/2}}{1-\alpha^2}\right).$$

The interval $1 < x < \infty$ is mapped one-to-one to the interval $0 < \eta < \infty$, with the points $x = 1$ and $x = \infty$ corresponding to $\eta = \infty$ and $\eta = 0$, respectively. For asymptotic expansions and explicit error bounds, see Dunster (2003b).

14.15(iii) Large ν, Fixed μ

For $\nu \to \infty$ and fixed μ (≥ 0),

14.15.11 $$\mathsf{P}_\nu^{-\mu}(\cos\theta) = \frac{1}{\nu^\mu}\left(\frac{\theta}{\sin\theta}\right)^{1/2}\left(J_\mu\big((\nu+\tfrac{1}{2})\theta\big) + O\left(\frac{1}{\nu}\right)\mathrm{env}J_\mu\big((\nu+\tfrac{1}{2})\theta\big)\right),$$

14.15.12 $$\mathsf{Q}_\nu^{-\mu}(\cos\theta) = -\frac{\pi}{2\nu^\mu}\left(\frac{\theta}{\sin\theta}\right)^{1/2}\left(Y_\mu\big((\nu+\tfrac{1}{2})\theta\big) + O\left(\frac{1}{\nu}\right)\mathrm{env}Y_\mu\big((\nu+\tfrac{1}{2})\theta\big)\right),$$

uniformly for $\theta \in (0, \pi-\delta]$. For the Bessel functions J and Y see §10.2(ii), and for the env functions associated with J and Y see §2.8(iv).

Next,

14.15.13 $$P_\nu^{-\mu}(\cosh\xi) = \frac{1}{\nu^\mu}\left(\frac{\xi}{\sinh\xi}\right)^{1/2} I_\mu\big((\nu+\tfrac{1}{2})\xi\big)\left(1 + O\left(\frac{1}{\nu}\right)\right),$$

14.15.14 $$\boldsymbol{Q}_\nu^\mu(\cosh\xi) = \frac{\nu^\mu}{\Gamma(\nu+\mu+1)}\left(\frac{\xi}{\sinh\xi}\right)^{1/2} K_\mu\big((\nu+\tfrac{1}{2})\xi\big)\left(1 + O\left(\frac{1}{\nu}\right)\right),$$

uniformly for $\xi \in (0, \infty)$.

For asymptotic expansions and explicit error bounds, see Olver (1997b, Chapter 12, §§12, 13) and Jones (2001). For convergent series expansions see Dunster (2004).

See also Olver (1997b, pp. 311–313) and §18.15(iii) for a generalized asymptotic expansion in terms of elementary functions for Legendre polynomials $P_n(\cos\theta)$ as $n \to \infty$ with θ fixed.

14.15(iv) Large ν, $0 \leq \mu \leq (1-\delta)(\nu+\tfrac{1}{2})$

As $\nu \to \infty$,

14.15.15 $$\mathsf{P}_\nu^{-\mu}(x) = \beta\left(\frac{y-\alpha^2}{1-\alpha^2-x^2}\right)^{1/4}\left(J_\mu\big((\nu+\tfrac{1}{2})y^{1/2}\big) + O\left(\frac{1}{\nu}\right)\mathrm{env}J_\mu\big((\nu+\tfrac{1}{2})y^{1/2}\big)\right),$$

14.15.16 $$\mathsf{Q}_\nu^{-\mu}(x) = -\frac{\pi\beta}{2}\left(\frac{y-\alpha^2}{1-\alpha^2-x^2}\right)^{1/4}\left(Y_\mu\big((\nu+\tfrac{1}{2})y^{1/2}\big) + O\left(\frac{1}{\nu}\right)\mathrm{env}Y_\mu\big((\nu+\tfrac{1}{2})y^{1/2}\big)\right),$$

uniformly with respect to $x \in [0,1)$ and $\mu \in [0, (1-\delta)(\nu+\tfrac{1}{2})]$. For α, β, and y see below.

Next,

14.15.17 $$P_\nu^{-\mu}(x) = \beta\left(\frac{\alpha^2-y}{x^2-1+\alpha^2}\right)^{1/4} I_\mu\big((\nu+\tfrac{1}{2})|y|^{1/2}\big)\left(1 + O\left(\frac{1}{\nu}\right)\right),$$

14.15.18 $$\boldsymbol{Q}_\nu^\mu(x) = \frac{1}{\beta\,\Gamma(\nu+\mu+1)}\left(\frac{\alpha^2-y}{x^2-1+\alpha^2}\right)^{1/4} K_\mu\big((\nu+\tfrac{1}{2})|y|^{1/2}\big)\left(1 + O\left(\frac{1}{\nu}\right)\right),$$

uniformly with respect to $x \in (1, \infty)$ and $\mu \in [0, (1-\delta)(\nu+\tfrac{1}{2})]$. In (14.15.15)–(14.15.18)

14.15.19 $$\alpha = \frac{\mu}{\nu+\tfrac{1}{2}}\ (<1),$$

14.15.20 $$\beta = e^\mu \left(\frac{\nu-\mu+\tfrac{1}{2}}{\nu+\mu+\tfrac{1}{2}}\right)^{(\nu/2)+(1/4)} \left((\nu+\tfrac{1}{2})^2-\mu^2\right)^{-\mu/2},$$

and the variable y is defined implicitly by

14.15.21 $$(y-\alpha^2)^{1/2} - \alpha\arctan\left(\frac{(y-\alpha^2)^{1/2}}{\alpha}\right) = \arccos\left(\frac{x}{(1-\alpha^2)^{1/2}}\right) - \frac{\alpha}{2}\arccos\left(\frac{(1+\alpha^2)x^2-1+\alpha^2}{(1-\alpha^2)(1-x^2)}\right),$$

$$x \leq (1-\alpha^2)^{1/2},\ y \geq \alpha^2,$$

14.15 Uniform Asymptotic Approximations

and

$$(\alpha^2 - y)^{1/2} + \tfrac{1}{2}\alpha \ln|y| - \alpha \ln\left((\alpha^2 - y)^{1/2} + \alpha\right)$$

14.15.22
$$= \ln\left(\frac{x + (x^2 - 1 + \alpha^2)^{1/2}}{(1-\alpha^2)^{1/2}}\right) + \frac{\alpha}{2}\ln\left(\frac{(1-\alpha^2)|1-x^2|}{(1+\alpha^2)x^2 - 1 + \alpha^2 + 2\alpha x(x^2-1+\alpha^2)^{1/2}}\right),$$
$$x \geq (1-\alpha^2)^{1/2},\ y \leq \alpha^2,$$

where the inverse trigonometric functions take their principal values (§4.23(ii)). The points $x = (1-\alpha^2)^{1/2}$, $x = 1$, and $x = \infty$ are mapped to $y = \alpha^2$, $y = 0$, and $y = -\infty$, respectively. The interval $0 \leq x < \infty$ is mapped one to one to the interval $-\infty < y \leq y_0$, where $y = y_0$ is the (positive) solution of (14.15.21) when $x = 0$.

For asymptotic expansions and explicit error bounds, see Boyd and Dunster (1986).

14.15(v) Large ν, $(\nu + \tfrac{1}{2})\delta \leq \mu \leq (\nu + \tfrac{1}{2})/\delta$

Here we introduce the envelopes of the parabolic cylinder functions $U(-c, x)$, $\overline{U}(-c, x)$, which are defined in §12.2. For $f(x) = U(-c, x)$ or $\overline{U}(-c, x)$, with c and x nonnegative,

14.15.23
$$\operatorname{env} f(x) = \begin{cases} \left((U(-c,x))^2 + (\overline{U}(-c,x))^2\right)^{1/2}, & 0 \leq x \leq X_c, \\ \sqrt{2} f(x), & X_c \leq x < \infty, \end{cases}$$

where $x = X_c$ denotes the largest positive root of the equation $U(-c, x) = \overline{U}(-c, x)$.

As $\nu \to \infty$,

14.15.24
$$\mathsf{P}_\nu^{-\mu}(x) = \frac{1}{(\nu + \tfrac{1}{2})^{1/4} 2^{(\nu+\mu)/2} \Gamma(\tfrac{1}{2}\nu + \tfrac{1}{2}\mu + \tfrac{3}{4})}\left(\frac{\zeta^2 - a^2}{x^2 - a^2}\right)^{1/4}$$
$$\times \left(U\left(\mu - \nu - \tfrac{1}{2}, (2\nu+1)^{1/2}\zeta\right) + O(\nu^{-2/3})\operatorname{env}U\left(\mu - \nu - \tfrac{1}{2}, (2\nu+1)^{1/2}\zeta\right)\right),$$

14.15.25
$$\mathsf{Q}_\nu^{-\mu}(x) = \frac{\pi}{(\nu + \tfrac{1}{2})^{1/4} 2^{(\nu+\mu+2)/2} \Gamma(\tfrac{1}{2}\nu + \tfrac{1}{2}\mu + \tfrac{3}{4})}\left(\frac{\zeta^2 - a^2}{x^2 - a^2}\right)^{1/4}$$
$$\times \left(\overline{U}\left(\mu - \nu - \tfrac{1}{2}, (2\nu+1)^{1/2}\zeta\right) + O(\nu^{-2/3})\operatorname{env}\overline{U}\left(\mu - \nu - \tfrac{1}{2}, (2\nu+1)^{1/2}\zeta\right)\right),$$

uniformly with respect to $x \in [0, 1)$ and $\mu \in [\delta(\nu + \tfrac{1}{2}), \nu + \tfrac{1}{2}]$. Here

14.15.26
$$a = \frac{\left((\nu + \mu + \tfrac{1}{2})|\nu - \mu + \tfrac{1}{2}|\right)^{1/2}}{\nu + \tfrac{1}{2}}, \qquad \alpha = \left(\frac{2|\nu - \mu + \tfrac{1}{2}|}{\nu + \tfrac{1}{2}}\right)^{1/2},$$

and the variable ζ is defined implicitly by

14.15.27
$$\tfrac{1}{2}\zeta(\zeta^2 - \alpha^2)^{1/2} - \tfrac{1}{2}\alpha^2 \operatorname{arccosh}\left(\frac{\zeta}{\alpha}\right) = (1-a^2)^{1/2}\operatorname{arctanh}\left(\frac{1}{x}\left(\frac{x^2-a^2}{1-a^2}\right)^{1/2}\right) - \operatorname{arccosh}\left(\frac{x}{a}\right), \quad a \leq x < 1,\ \alpha \leq \zeta < \infty,$$

and

14.15.28
$$\tfrac{1}{2}\alpha^2 \arcsin\left(\frac{\zeta}{\alpha}\right) + \tfrac{1}{2}\zeta(\alpha^2 - \zeta^2)^{1/2} = \arcsin\left(\frac{x}{a}\right) - (1-a^2)^{1/2}\arctan\left(x\left(\frac{1-a^2}{a^2-x^2}\right)^{1/2}\right), \quad -a \leq x \leq a,\ -\alpha \leq \zeta \leq \alpha,$$

when $a > 0$, and

14.15.29
$$\zeta^2 = -\ln(1-x^2), \qquad -1 < x < 1,$$

when $a = 0$. The inverse hyperbolic and trigonometric functions take their principal values (§§4.23(ii), 4.37(ii)).

When $a > 0$ the interval $-a \leq x < 1$ is mapped one-to-one to the interval $-\alpha \leq \zeta < \infty$, with the points $x = -a$, $x = a$, and $x = 1$ corresponding to $\zeta = -\alpha$, $\zeta = \alpha$, and $\zeta = \infty$, respectively. When $a = 0$ the interval $-1 < x < 1$ is mapped one-to-one to the interval $-\infty < \zeta < \infty$, with the points $x = -1$, 0, and 1 corresponding to $\zeta = -\infty$, 0, and ∞, respectively.

Next, as $\nu \to \infty$,

14.15.30 $\quad \mathsf{P}_\nu^{-\mu}(x) = \dfrac{1}{(\nu+\frac{1}{2})^{1/4} 2^{(\nu+\mu)/2} \Gamma(\frac{1}{2}\nu + \frac{1}{2}\mu + \frac{3}{4})} \left(\dfrac{\zeta^2 + \alpha^2}{x^2 + a^2}\right)^{1/4} U\!\left(\mu - \nu - \tfrac{1}{2}, (2\nu+1)^{1/2} \zeta\right) \left(1 + O(\nu^{-1} \ln \nu)\right),$

uniformly with respect to $x \in (-1, 1)$ and $\mu \in [\nu + \frac{1}{2}, (1/\delta)(\nu + \frac{1}{2})]$. Here ζ is defined implicitly by

14.15.31 $\quad \dfrac{1}{2}\zeta\left(\zeta^2 + \alpha^2\right)^{1/2} + \dfrac{1}{2}\alpha^2 \operatorname{arcsinh}\!\left(\dfrac{\zeta}{\alpha}\right) = (1+a^2)^{1/2} \operatorname{arctanh}\!\left(x \left(\dfrac{1+a^2}{x^2 + a^2}\right)^{1/2}\right) - \operatorname{arcsinh}\!\left(\dfrac{x}{a}\right),$
$-1 < x < 1, \ -\infty < \zeta < \infty,$

when $a > 0$, which maps the interval $-1 < x < 1$ one-to-one to the interval $-\infty < \zeta < \infty$: the points $x = -1$ and $x = 1$ correspond to $\zeta = -\infty$ and $\zeta = \infty$, respectively. When $a = 0$ (14.15.29) again applies. (The inverse hyperbolic functions again take their principal values.)

Since (14.15.30) holds for negative x, corresponding approximations for $\mathsf{Q}_\nu^{\mp\mu}(x)$, uniformly valid in the interval $-1 < x < 1$, can be obtained from (14.9.9) and (14.9.10).

For error bounds and other extensions see Olver (1975b).

14.16 Zeros

14.16(i) Notation

Throughout this section we assume that μ and ν are real, and when they are not integers we write

14.16.1 $\quad \mu = m + \delta_\mu, \quad \nu = n + \delta_\nu,$

where $m, n \in \mathbb{Z}$ and $\delta_\mu, \delta_\nu \in (0, 1)$. For all cases concerning $\mathsf{P}_\nu^\mu(x)$ and $P_\nu^\mu(x)$ we assume that $\nu \geq -\frac{1}{2}$ without loss of generality (see (14.9.5) and (14.9.11)).

14.16(ii) Interval $-1 < x < 1$

The number of zeros of $\mathsf{P}_\nu^\mu(x)$ in the interval $(-1, 1)$ is $\max(\lceil \nu - |\mu| \rceil, 0)$ if any of the following sets of conditions hold:

(a) $\mu \leq 0$.

(b) $\mu > 0$, $n \geq m$, and $\delta_\nu > \delta_\mu$.

(c) $\mu > 0$, $n < m$, and $m - n$ is odd.

(d) $\nu = 0, 1, 2, 3, \ldots$.

The number of zeros of $\mathsf{P}_\nu^\mu(x)$ in the interval $(-1, 1)$ is $\max(\lceil \nu - |\mu| \rceil, 0) + 1$ if either of the following sets of conditions holds:

(a) $\mu > 0$, $n > m$, and $\delta_\nu \leq \delta_\mu$.

(b) $\mu > 0$, $n < m$, and $m - n$ is even.

The zeros of $\mathsf{Q}_\nu^\mu(x)$ in the interval $(-1, 1)$ interlace those of $\mathsf{P}_\nu^\mu(x)$. $\mathsf{Q}_\nu^\mu(x)$ has $\max(\lceil \nu - |\mu| \rceil, 0) + k$ zeros in the interval $(-1, 1)$, where k can take one of the values $-1, 0, 1, 2$, subject to $\max(\lceil \nu - |\mu| \rceil, 0) + k$ being even or odd according as $\cos(\nu\pi)$ and $\cos(\mu\pi)$ have opposite signs or the same sign. In the special case $\mu = 0$ and $\nu = n = 0, 1, 2, 3, \ldots$, $\mathsf{Q}_n(x)$ has $n + 1$ zeros in the interval $-1 < x < 1$.

For uniform asymptotic approximations for the zeros of $\mathsf{P}_n^{-m}(x)$ in the interval $-1 < x < 1$ when $n \to \infty$ with $m \ (\geq 0)$ fixed, see Olver (1997b, p. 469).

14.16(iii) Interval $1 < x < \infty$

$P_\nu^\mu(x)$ has exactly one zero in the interval $(1, \infty)$ if either of the following sets of conditions holds:

(a) $\mu > 0$, $\mu > \nu$, $\mu \notin \mathbb{Z}$, and $\sin((\mu - \nu)\pi)$ and $\sin(\mu\pi)$ have opposite signs.

(b) $\mu \leq \nu$, $\mu \notin \mathbb{Z}$, and $\lfloor \mu \rfloor$ is odd.

For all other values of μ and ν (with $\nu \geq -\frac{1}{2}$) $P_\nu^\mu(x)$ has no zeros in the interval $(1, \infty)$.

$Q_\nu^\mu(x)$ has no zeros in the interval $(1, \infty)$ when $\nu > -1$, and at most one zero in the interval $(1, \infty)$ when $\nu < -1$.

14.17 Integrals

14.17(i) Indefinite Integrals

14.17.1
$$\int (1 - x^2)^{-\mu/2} \mathsf{P}_\nu^\mu(x) \, dx = -(1 - x^2)^{-(\mu-1)/2} \mathsf{P}_\nu^{\mu-1}(x).$$

14.17.2
$$\int (1 - x^2)^{\mu/2} \mathsf{P}_\nu^\mu(x) \, dx = \dfrac{(1 - x^2)^{(\mu+1)/2}}{(\nu - \mu)(\nu + \mu + 1)} \mathsf{P}_\nu^{\mu+1}(x),$$
$\mu \neq \nu$ or $-\nu - 1$.

14.17 Integrals

14.17.3

$$\int x \, \mathsf{P}_\nu^\mu(x) \, \mathsf{Q}_\nu^\mu(x) \, dx = \frac{1}{2\nu(\nu+1)} \Big((\mu^2 - (\nu+1)(\nu+x^2)) \, \mathsf{P}_\nu^\mu(x) \, \mathsf{Q}_\nu^\mu(x)$$
$$+ (\nu+1)(\nu-\mu+1)x(\mathsf{P}_\nu^\mu(x) \, \mathsf{Q}_{\nu+1}^\mu(x) + \mathsf{P}_{\nu+1}^\mu(x) \, \mathsf{Q}_\nu^\mu(x)) - (\nu-\mu+1)^2 \, \mathsf{P}_{\nu+1}^\mu(x) \, \mathsf{Q}_{\nu+1}^\mu(x) \Big),$$
$$\nu \neq 0, -1.$$

14.17.4

$$\int \frac{x}{(1-x^2)^{3/2}} \, \mathsf{P}_\nu^\mu(x) \, \mathsf{Q}_\nu^\mu(x) \, dx$$
$$= \frac{1}{(1-4\mu^2)(1-x^2)^{1/2}} \Big((1 - 2\mu^2 + 2\nu(\nu+1)) \, \mathsf{P}_\nu^\mu(x) \, \mathsf{Q}_\nu^\mu(x) + (2\nu+1)(\mu-\nu-1)x(\mathsf{P}_\nu^\mu(x) \, \mathsf{Q}_{\nu+1}^\mu(x) + \mathsf{P}_{\nu+1}^\mu(x) \, \mathsf{Q}_\nu^\mu(x))$$
$$+ 2(\mu-\nu-1)^2 \, \mathsf{P}_{\nu+1}^\mu(x) \, \mathsf{Q}_{\nu+1}^\mu(x) \Big),$$
$$\mu \neq \pm \tfrac{1}{2}.$$

In (14.17.1)–(14.17.4), P may be replaced by Q, and in (14.17.3) and (14.17.4), Q may be replaced by P.

For further results, see Maximon (1955) and Prudnikov et al. (1990, pp. 37–39). See also (14.12.2), (14.12.5), and (14.12.12).

14.17(ii) Barnes' Integral

14.17.5
$$\int_0^1 x^\sigma (1-x^2)^{\mu/2} \, \mathsf{P}_\nu^{-\mu}(x) \, dx = \frac{\Gamma\!\left(\tfrac{1}{2}\sigma + \tfrac{1}{2}\right) \Gamma\!\left(\tfrac{1}{2}\sigma + 1\right)}{2^{\mu+1} \Gamma\!\left(\tfrac{1}{2}\sigma - \tfrac{1}{2}\nu + \tfrac{1}{2}\mu + 1\right) \Gamma\!\left(\tfrac{1}{2}\sigma + \tfrac{1}{2}\nu + \tfrac{1}{2}\mu + \tfrac{3}{2}\right)}, \quad \Re\sigma > -1, \, \Re\mu > -1.$$

14.17(iii) Orthogonality Properties

For $l, m, n = 0, 1, 2, \ldots$,

14.17.6
$$\int_{-1}^1 \mathsf{P}_l^m(x) \, \mathsf{P}_n^m(x) \, dx = \delta_{l,n} \frac{(n+m)!}{(n-m)! \left(n+\tfrac{1}{2}\right)},$$

14.17.7
$$\int_{-1}^1 \mathsf{P}_l^m(x) \, \mathsf{P}_n^{-m}(x) \, dx = (-1)^m \delta_{l,n} \frac{1}{l + \tfrac{1}{2}},$$

14.17.8
$$\int_{-1}^1 \frac{\mathsf{P}_n^l(x) \, \mathsf{P}_n^m(x)}{1-x^2} \, dx = \delta_{l,m} \frac{(n+m)!}{(n-m)! \, m}, \qquad m > 0,$$

14.17.9
$$\int_{-1}^1 \frac{\mathsf{P}_n^l(x) \, \mathsf{P}_n^{-m}(x)}{1-x^2} \, dx = (-1)^l \delta_{l,m} \frac{1}{l}, \qquad l > 0.$$

14.17(iv) Definite Integrals of Products

With $\psi(x) = \Gamma'(x)/\Gamma(x)$ (§5.2(i)),

14.17.10
$$\int_{-1}^1 \mathsf{P}_\nu(x) \, \mathsf{P}_\lambda(x) \, dx = \frac{2(2\sin(\nu\pi)\sin(\lambda\pi)(\psi(\nu+1) - \psi(\lambda+1)) + \pi\sin((\lambda-\nu)\pi))}{\pi^2 (\lambda-\nu)(\lambda+\nu+1)}, \quad \lambda \neq \nu \text{ or } -\nu-1.$$

14.17.11
$$\int_{-1}^1 (\mathsf{P}_\nu(x))^2 \, dx = \frac{\pi^2 - 2\sin^2(\nu\pi)\psi'(\nu+1)}{\pi^2 \left(\nu+\tfrac{1}{2}\right)}, \qquad \nu \neq -\tfrac{1}{2}.$$

14.17.12
$$\int_{-1}^1 \mathsf{Q}_\nu(x) \, \mathsf{Q}_\lambda(x) \, dx = \frac{((\psi(\nu+1) - \psi(\lambda+1))(1 + \cos(\nu\pi)\cos(\lambda\pi)) + \tfrac{1}{2}\pi \sin((\lambda-\nu)\pi))}{(\lambda-\nu)(\lambda+\nu+1)},$$
$$\lambda \neq \nu \text{ or } -\nu-1, \, \lambda \text{ and } \nu \neq -1, -2, -3, \ldots.$$

14.17.13
$$\int_{-1}^1 (\mathsf{Q}_\nu(x))^2 \, dx = \frac{\pi^2 - 2(1 + \cos^2(\nu\pi))\psi'(\nu+1)}{2(2\nu+1)}, \qquad \nu \neq -\tfrac{1}{2} \text{ or } -1, -2, -3, \ldots.$$

14.17.14
$$\int_{-1}^1 \mathsf{P}_\nu(x) \, \mathsf{Q}_\lambda(x) \, dx = \frac{2\sin(\nu\pi)\cos(\lambda\pi)(\psi(\nu+1) - \psi(\lambda+1)) + \pi\cos((\lambda-\nu)\pi) - \pi}{\pi(\lambda-\nu)(\lambda+\nu+1)}, \quad \Re\lambda > 0, \, \Re\nu > 0, \, \lambda \neq \nu.$$

14.17.15
$$\int_{-1}^{1} \mathsf{P}_\nu(x)\,\mathsf{Q}_\nu(x)\,dx = -\frac{\sin(2\nu\pi)\,\psi'(\nu+1)}{\pi(2\nu+1)}, \quad \Re\nu > 0.$$

14.17.16
$$\int_{-1}^{1} \mathsf{P}_l^m(x)\,\mathsf{Q}_n^m(x)\,dx = \frac{\left(1-(-1)^{l+n}\right)(l+m)!}{(l-n)(l+n+1)(l-m)!},$$
$$l,m,n = 0,1,2,\dots,\ l \neq n.$$

14.17.17
$$\int_0^\pi \mathsf{Q}_l(\cos\theta)\,\mathsf{P}_m(\cos\theta)\,\mathsf{P}_n(\cos\theta)\sin\theta\,d\theta$$
$$= 0, \quad l,m,n = 1,2,3,\dots,\ |m-n| < l < m+n.$$

(When $l+m+n$ is even the condition $|m-n| < l < m+n$ is not needed.) Next,

14.17.18
$$\int_1^\infty P_\nu(x)\,Q_\lambda(x)\,dx = \frac{1}{(\lambda-\nu)(\nu+\lambda+1)},$$
$$\Re\lambda > \Re\nu > 0.$$

14.17.19
$$\int_1^\infty Q_\nu(x)\,Q_\lambda(x)\,dx$$
$$= \frac{\psi(\lambda+1) - \psi(\nu+1)}{(\lambda-\nu)(\lambda+\nu+1)},$$
$$\Re(\lambda+\nu) > -1,\ \lambda \neq \nu,\ \lambda \text{ and } \nu \neq -1,-2,-3,\dots.$$

14.17.20
$$\int_1^\infty (Q_\nu(x))^2\,dx = \frac{\psi'(\nu+1)}{2\nu+1}, \quad \Re\nu > -\tfrac{1}{2}.$$

For further results, see Prudnikov et al. (1990, pp. 194–240); also (34.3.21).

14.17(v) Laplace Transforms

For Laplace transforms and inverse Laplace transforms involving associated Legendre functions, see Erdélyi et al. (1954a, pp. 179–181, 270–272), Oberhettinger and Badii (1973, pp. 113–118, 317–324), Prudnikov et al. (1992a, §§3.22, 3.32, and 3.33), and Prudnikov et al. (1992b, §§3.20, 3.30, and 3.31).

14.17(vi) Mellin Transforms

For Mellin transforms involving associated Legendre functions see Oberhettinger (1974, pp. 69–82) and Marichev (1983, pp. 247–283), and for inverse transforms see Oberhettinger (1974, pp. 205–215).

14.18 Sums

14.18(i) Expansion Theorem

For expansions of arbitrary functions in series of Legendre polynomials see §18.18(i), and for expansions of arbitrary functions in series of associated Legendre functions see Schäfke (1961b).

14.18(ii) Addition Theorems

In (14.18.1) and (14.18.2), θ_1, θ_2, and $\theta_1 + \theta_2$ all lie in $[0,\pi)$, and ϕ is real.

14.18.1
$$\mathsf{P}_\nu(\cos\theta_1\cos\theta_2 + \sin\theta_1\sin\theta_2\cos\phi) = \mathsf{P}_\nu(\cos\theta_1)\,\mathsf{P}_\nu(\cos\theta_2) + 2\sum_{m=1}^\infty (-1)^m\,\mathsf{P}_\nu^{-m}(\cos\theta_1)\,\mathsf{P}_\nu^m(\cos\theta_2)\cos(m\phi),$$

14.18.2
$$\mathsf{P}_n(\cos\theta_1\cos\theta_2 + \sin\theta_1\sin\theta_2\cos\phi) = \sum_{m=-n}^n (-1)^m\,\mathsf{P}_n^{-m}(\cos\theta_1)\,\mathsf{P}_n^m(\cos\theta_2)\cos(m\phi).$$

In (14.18.3), θ_1 lies in $(0,\tfrac{1}{2}\pi)$, θ_2 and $\theta_1 + \theta_2$ both lie in $(0,\pi)$, $\theta_1 < \theta_2$, ϕ is real, and $\nu \neq -1,-2,-3,\dots$.

14.18.3
$$\mathsf{Q}_\nu(\cos\theta_1\cos\theta_2 + \sin\theta_1\sin\theta_2\cos\phi) = \mathsf{P}_\nu(\cos\theta_1)\,\mathsf{Q}_\nu(\cos\theta_2) + 2\sum_{m=1}^\infty (-1)^m\,\mathsf{P}_\nu^{-m}(\cos\theta_1)\,\mathsf{Q}_\nu^m(\cos\theta_2)\cos(m\phi).$$

In (14.18.4) and (14.18.5), ξ_1 and ξ_2 are positive, and ϕ is real; also in (14.18.5) $\xi_1 < \xi_2$ and $\nu \neq -1,-2,-3,\dots$.

14.18.4
$$P_\nu(\cosh\xi_1\cosh\xi_2 - \sinh\xi_1\sinh\xi_2\cos\phi) = P_\nu(\cosh\xi_1)\,P_\nu(\cosh\xi_2) + 2\sum_{m=1}^\infty (-1)^m\,P_\nu^{-m}(\cosh\xi_1)\,P_\nu^m(\cosh\xi_2)\cos(m\phi),$$

14.18.5
$$Q_\nu(\cosh\xi_1\cosh\xi_2 - \sinh\xi_1\sinh\xi_2\cos\phi) = P_\nu(\cosh\xi_1)\,Q_\nu(\cosh\xi_2) + 2\sum_{m=1}^\infty (-1)^m\,P_\nu^{-m}(\cosh\xi_1)\,Q_\nu^m(\cosh\xi_2)\cos(m\phi).$$

14.18(iii) Other Sums

14.18.6
$$(x-y)\sum_{k=0}^{n}(2k+1)P_k(x)P_k(y) = (n+1)\left(P_{n+1}(x)P_n(y) - P_n(x)P_{n+1}(y)\right),$$

14.18.7
$$(x-y)\sum_{k=0}^{n}(2k+1)P_k(x)Q_k(y) = (n+1)\left(P_{n+1}(x)Q_n(y) - P_n(x)Q_{n+1}(y)\right) - 1.$$

Zonal Harmonic Series

14.18.8
$$\mathsf{P}_\nu(-x) = \frac{\sin(\nu\pi)}{\pi}\sum_{n=0}^{\infty}\frac{2n+1}{(\nu-n)(\nu+n+1)}\mathsf{P}_n(x), \qquad \nu \notin \mathbb{Z}.$$

Dougall's Expansion

14.18.9
$$\mathsf{P}_\nu^{-\mu}(x) = \frac{\sin(\nu\pi)}{\pi}\sum_{n=0}^{\infty}(-1)^n\frac{2n+1}{(\nu-n)(\nu+n+1)}\mathsf{P}_n^{-\mu}(x), \qquad -1 < x \le 1,\, \mu \ge 0,\, \nu \notin \mathbb{Z}.$$

For a series representation of the Dirac delta in terms of products of Legendre polynomials see (1.17.22).

14.18(iv) Compendia

For collections of sums involving associated Legendre functions, see Hansen (1975, pp. 367–377, 457–460, and 475), Erdélyi et al. (1953a, §3.10), Gradshteyn and Ryzhik (2000, §8.92), Magnus et al. (1966, pp. 178–184), and Prudnikov et al. (1990, §§5.2, 6.5). See also §18.18 and (34.3.19).

14.19 Toroidal (or Ring) Functions

14.19(i) Introduction

When $\nu = n - \frac{1}{2}$, $n = 0, 1, 2, \ldots$, $\mu \in \mathbb{R}$, and $x \in (1, \infty)$ solutions of (14.2.2) are known as *toroidal* or *ring functions*. This form of the differential equation arises when Laplace's equation is transformed into *toroidal coordinates* (η, θ, ϕ), which are related to Cartesian coordinates (x, y, z) by

14.19.1
$$x = \frac{c\sinh\eta\cos\phi}{\cosh\eta - \cos\theta}, \quad y = \frac{c\sinh\eta\sin\phi}{\cosh\eta - \cos\theta}, \quad z = \frac{c\sin\theta}{\cosh\eta - \cos\theta},$$

where the constant c is a scaling factor. Most required properties of toroidal functions come directly from the results for $P_\nu^\mu(x)$ and $\boldsymbol{Q}_\nu^\mu(x)$. In particular, for $\mu = 0$ and $\nu = \pm\frac{1}{2}$ see §14.5(v).

14.19(ii) Hypergeometric Representations

With **F** as in §14.3 and $\xi > 0$,

14.19.2
$$P_{\nu-\frac{1}{2}}^\mu(\cosh\xi) = \frac{\Gamma(1-2\mu)2^{2\mu}}{\Gamma(1-\mu)\left(1 - e^{-2\xi}\right)^\mu e^{(\nu+(1/2))\xi}}\,\mathbf{F}\!\left(\tfrac{1}{2} - \mu, \tfrac{1}{2} + \nu - \mu; 1 - 2\mu; e^{-2\xi}\right), \qquad \mu \ne \tfrac{1}{2}.$$

14.19.3
$$\boldsymbol{Q}_{\nu-\frac{1}{2}}^\mu(\cosh\xi) = \frac{\pi^{1/2}\left(1 - e^{-2\xi}\right)^\mu}{e^{(\nu+(1/2))\xi}}\,\mathbf{F}\!\left(\mu + \tfrac{1}{2}, \nu + \mu + \tfrac{1}{2}; \nu + 1; e^{-2\xi}\right).$$

14.19(iii) Integral Representations

With $\xi > 0$,

14.19.4
$$P_{n-\frac{1}{2}}^m(\cosh\xi) = \frac{\Gamma(n + m + \tfrac{1}{2})(\sinh\xi)^m}{2^m \pi^{1/2}\,\Gamma(n - m + \tfrac{1}{2})\,\Gamma(m + \tfrac{1}{2})}\int_0^\pi \frac{(\sin\phi)^{2m}}{(\cosh\xi + \cos\phi\sinh\xi)^{n+m+(1/2)}}\,d\phi,$$

14.19.5
$$\boldsymbol{Q}_{n-\frac{1}{2}}^m(\cosh\xi) = \frac{\Gamma(n + \tfrac{1}{2})}{\Gamma(n + m + \tfrac{1}{2})\,\Gamma(n - m + \tfrac{1}{2})}\int_0^\infty \frac{\cosh(mt)}{(\cosh\xi + \cosh t\sinh\xi)^{n+(1/2)}}\,dt, \qquad m < n + \tfrac{1}{2}.$$

14.19(iv) Sums

With $\xi > 0$,

14.19.6 $$\boldsymbol{Q}^\mu_{-\frac{1}{2}}(\cosh\xi) + 2\sum_{n=1}^{\infty}\frac{\Gamma(\mu+n+\frac{1}{2})}{\Gamma(\mu+\frac{1}{2})}\boldsymbol{Q}^\mu_{n-\frac{1}{2}}(\cosh\xi)\cos(n\phi) = \frac{(\frac{1}{2}\pi)^{1/2}(\sinh\xi)^\mu}{(\cosh\xi - \cos\phi)^{\mu+(1/2)}}, \qquad \Re\mu > -\tfrac{1}{2}.$$

14.19(v) Whipple's Formula for Toroidal Functions

With $\xi > 0$,

14.19.7 $$P^m_{n-\frac{1}{2}}(\cosh\xi) = \frac{\Gamma(n+m+\frac{1}{2})}{\Gamma(n-m+\frac{1}{2})}\left(\frac{2}{\pi\sinh\xi}\right)^{1/2}\boldsymbol{Q}^n_{m-\frac{1}{2}}(\coth\xi),$$

14.19.8 $$\boldsymbol{Q}^m_{n-\frac{1}{2}}(\cosh\xi) = \frac{\Gamma(m-n+\frac{1}{2})}{\Gamma(m+n+\frac{1}{2})}\left(\frac{\pi}{2\sinh\xi}\right)^{1/2}P^n_{m-\frac{1}{2}}(\coth\xi).$$

14.20 Conical (or Mehler) Functions

14.20(i) Definitions and Wronskians

Throughout §14.20 we assume that $\nu = -\frac{1}{2}+i\tau$, with $\mu \geq 0$ and $\tau \geq 0$. (14.2.2) takes the form

14.20.1 $$(1-x^2)\frac{d^2w}{dx^2} - 2x\frac{dw}{dx} - \left(\tau^2 + \frac{1}{4} + \frac{\mu^2}{1-x^2}\right)w = 0.$$

Solutions are known as *conical* or *Mehler functions*. For $-1 < x < 1$ and $\tau > 0$, a numerically satisfactory pair of real conical functions is $\mathsf{P}^{-\mu}_{-\frac{1}{2}+i\tau}(x)$ and $\mathsf{P}^{-\mu}_{-\frac{1}{2}+i\tau}(-x)$.

Another real-valued solution $\widehat{\mathsf{Q}}^{-\mu}_{-\frac{1}{2}+i\tau}(x)$ of (14.20.1) was introduced in Dunster (1991). This is defined by

14.20.2 $$\widehat{\mathsf{Q}}^{-\mu}_{-\frac{1}{2}+i\tau}(x) = \Re\left(e^{\mu\pi i}\mathsf{Q}^{-\mu}_{-\frac{1}{2}+i\tau}(x)\right) - \tfrac{1}{2}\pi\sin(\mu\pi)\mathsf{P}^{-\mu}_{-\frac{1}{2}+i\tau}(x).$$

Equivalently,

14.20.3 $$\widehat{\mathsf{Q}}^{-\mu}_{-\frac{1}{2}+i\tau}(x) = \frac{\pi e^{-\tau\pi}\sin(\mu\pi)\sinh(\tau\pi)}{2(\cosh^2(\tau\pi) - \sin^2(\mu\pi))}\mathsf{P}^{-\mu}_{-\frac{1}{2}+i\tau}(x) + \frac{\pi(e^{-\tau\pi}\cos^2(\mu\pi) + \sinh(\tau\pi))}{2(\cosh^2(\tau\pi) - \sin^2(\mu\pi))}\mathsf{P}^{-\mu}_{-\frac{1}{2}+i\tau}(-x).$$

$\widehat{\mathsf{Q}}^{-\mu}_{-\frac{1}{2}+i\tau}(x)$ exists except when $\mu = \tfrac{1}{2}, \tfrac{3}{2}, \ldots$ and $\tau = 0$; compare §14.3(i). It is an important companion solution to $\mathsf{P}^{-\mu}_{-\frac{1}{2}+i\tau}(x)$ when τ is large; compare §§14.20(vii), 14.20(viii), and 10.25(iii).

14.20.4 $$\mathscr{W}\left\{\mathsf{P}^{-\mu}_{-\frac{1}{2}+i\tau}(x), \mathsf{P}^{-\mu}_{-\frac{1}{2}+i\tau}(-x)\right\} = \frac{2}{|\Gamma(\mu+\frac{1}{2}+i\tau)|^2(1-x^2)}.$$

14.20.5 $$\mathscr{W}\left\{\mathsf{P}^{-\mu}_{-\frac{1}{2}+i\tau}(x), \widehat{\mathsf{Q}}^{-\mu}_{-\frac{1}{2}+i\tau}(x)\right\} = \frac{\pi(e^{-\tau\pi}\cos^2(\mu\pi) + \sinh(\tau\pi))}{|\Gamma(\mu+\frac{1}{2}+i\tau)|^2(\cosh^2(\tau\pi) - \sin^2(\mu\pi))(1-x^2)},$$

provided that $\widehat{\mathsf{Q}}^{-\mu}_{-\frac{1}{2}+i\tau}(x)$ exists.

Lastly, for the range $1 < x < \infty$, $P^{-\mu}_{-\frac{1}{2}+i\tau}(x)$ is a real-valued solution of (14.20.1); in terms of $Q^\mu_{-\frac{1}{2}\pm i\tau}(x)$ (which are complex-valued in general):

14.20.6 $$P^{-\mu}_{-\frac{1}{2}+i\tau}(x) = \frac{ie^{-\mu\pi i}}{\sinh(\tau\pi)\left|\Gamma(\mu+\frac{1}{2}+i\tau)\right|^2}\left(Q^\mu_{-\frac{1}{2}+i\tau}(x) - Q^\mu_{-\frac{1}{2}-i\tau}(x)\right), \qquad \tau \neq 0.$$

14.20(ii) Graphics

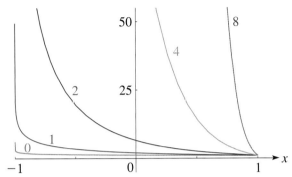

Figure 14.20.1: $\mathsf{P}^0_{-\frac{1}{2}+i\tau}(x), \tau = 0, 1, 2, 4, 8$.

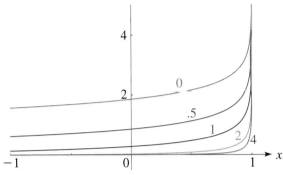

Figure 14.20.2: $\widehat{\mathsf{Q}}^0_{-\frac{1}{2}+i\tau}(x), \tau = 0, \frac{1}{2}, 1, 2, 4$.

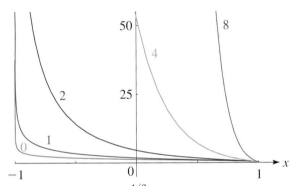

Figure 14.20.3: $\mathsf{P}^{-1/2}_{-\frac{1}{2}+i\tau}(x), \tau = 0, 1, 2, 4, 8$.

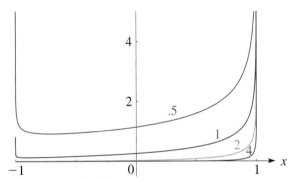

Figure 14.20.4: $\widehat{\mathsf{Q}}^{-1/2}_{-\frac{1}{2}+i\tau}(x), \tau = \frac{1}{2}, 1, 2, 4$. (This function does not exist when $\tau = 0$.)

For additional graphs see http://dlmf.nist.gov/14.20.ii.

14.20(iii) Behavior as $x \to 1$

The behavior of $\mathsf{P}^{-\mu}_{-\frac{1}{2}+i\tau}(\pm x)$ as $x \to 1-$ is given in §14.8(i). For $\mu > 0$ and $x \to 1-$,

14.20.7 $\quad \widehat{\mathsf{Q}}^\mu_{-\frac{1}{2}+i\tau}(x) \sim \frac{1}{2}\,\Gamma(\mu) \left(\frac{2}{1-x}\right)^{\mu/2},$

14.20.8
$$\widehat{\mathsf{Q}}^{-\mu}_{-\frac{1}{2}+i\tau}(x) \sim \frac{\pi\,\Gamma(\mu)(e^{-\tau\pi}\cos^2(\mu\pi) + \sinh(\tau\pi))}{2(\cosh^2(\tau\pi) - \sin^2(\mu\pi))\left|\Gamma\!\left(\mu+\frac{1}{2}+i\tau\right)\right|^2}$$
$$\times \left(\frac{2}{1-x}\right)^{\mu/2}.$$

14.20(iv) Integral Representation

When $0 < \theta < \pi$,

14.20.9 $\quad \mathsf{P}_{-\frac{1}{2}+i\tau}(\cos\theta) = \frac{2}{\pi} \int_0^\theta \frac{\cosh(\tau\phi)}{\sqrt{2(\cos\phi - \cos\theta)}}\,d\phi.$

14.20(v) Trigonometric Expansion

14.20.10
$$\mathsf{P}_{-\frac{1}{2}+i\tau}(\cos\theta) = 1 + \frac{4\tau^2 + 1^2}{2^2}\sin^2\!\left(\tfrac{1}{2}\theta\right)$$
$$+ \frac{(4\tau^2+1^2)(4\tau^2+3^2)}{2^2 \cdot 4^2}\sin^4\!\left(\tfrac{1}{2}\theta\right)$$
$$+ \cdots, \qquad 0 \le \theta \le \pi.$$

From (14.20.9) or (14.20.10) it is evident that $\mathsf{P}_{-\frac{1}{2}+i\tau}(\cos\theta)$ is positive for real θ.

14.20(vi) Generalized Mehler–Fock Transformation

14.20.11
$$f(\tau) = \frac{\tau}{\pi}\sinh(\tau\pi)\,\Gamma\!\left(\tfrac{1}{2}-\mu+i\tau\right)$$
$$\times \Gamma\!\left(\tfrac{1}{2}-\mu-i\tau\right) \int_1^\infty P^\mu_{-\frac{1}{2}+i\tau}(x)g(x)\,dx,$$

where

14.20.12 $\quad g(x) = \int_0^\infty P^\mu_{-\frac{1}{2}+i\tau}(x)f(\tau)\,d\tau.$

Special cases:

14.20.13 $\quad P_{-\frac{1}{2}+i\tau}(x) = \dfrac{\cosh(\tau\pi)}{\pi} \displaystyle\int_1^\infty \dfrac{P_{-\frac{1}{2}+i\tau}(t)}{x+t}\,dt,$

14.20.14
$$\pi \int_0^\infty \dfrac{\tau\tanh(\tau\pi)}{\cosh(\tau\pi)} P_{-\frac{1}{2}+i\tau}(x)\, P_{-\frac{1}{2}+i\tau}(y)\,d\tau = \dfrac{1}{y+x}.$$

14.20(vii) Asymptotic Approximations: Large τ, Fixed μ

For $\tau \to \infty$ and fixed μ,

14.20.15 $\quad \mathsf{P}^{-\mu}_{-\frac{1}{2}+i\tau}(\cos\theta) = \dfrac{1}{\tau^\mu}\left(\dfrac{\theta}{\sin\theta}\right)^{1/2} I_\mu(\tau\theta)$
$$\times(1 + O(1/\tau)),$$

14.20.16 $\quad \widehat{\mathsf{Q}}^{-\mu}_{-\frac{1}{2}+i\tau}(\cos\theta) = \dfrac{1}{\tau^\mu}\left(\dfrac{\theta}{\sin\theta}\right)^{1/2} K_\mu(\tau\theta)$
$$\times(1 + O(1/\tau)),$$

uniformly for $\theta \in (0, \pi - \delta]$, where I and K are the modified Bessel functions (§10.25(ii)) and δ is an arbitrary constant such that $0 < \delta < \pi$. For asymptotic expansions and explicit error bounds, see Olver (1997b, pp. 473–474). See also Žurina and Karmazina (1966).

14.20(viii) Asymptotic Approximations: Large τ, $0 \le \mu \le A\tau$

In this subsection and §14.20(ix), A and δ denote arbitrary constants such that $A > 0$ and $0 < \delta < 2$.

As $\tau \to \infty$,

14.20.17
$$\mathsf{P}^{-\mu}_{-\frac{1}{2}+i\tau}(x) = \sigma(\mu, \tau)\left(\dfrac{\alpha^2 + \eta}{1 + \alpha^2 - x^2}\right)^{1/4} I_\mu(\tau\eta^{1/2})$$
$$\times(1 + O(1/\tau)),$$

14.20.18
$$\widehat{\mathsf{Q}}^{-\mu}_{-\frac{1}{2}+i\tau}(x) = \sigma(\mu, \tau)\left(\dfrac{\alpha^2 + \eta}{1 + \alpha^2 - x^2}\right)^{1/4} K_\mu(\tau\eta^{1/2})$$
$$\times(1 + O(1/\tau)),$$

uniformly for $x \in [-1 + \delta, 1)$ and $\mu \in [0, A\tau]$. Here

14.20.19 $\quad\quad\quad \alpha = \mu/\tau,$

14.20.20 $\quad\quad \sigma(\mu,\tau) = \dfrac{\exp(\mu - \tau\arctan\alpha)}{(\mu^2 + \tau^2)^{\mu/2}}.$

The variable η is defined implicitly by

14.20.21
$$\left(\alpha^2 + \eta\right)^{1/2} + \tfrac{1}{2}\alpha\ln\eta - \alpha\ln\!\left(\left(\alpha^2+\eta\right)^{1/2} + \alpha\right)$$
$$= \arccos\!\left(\dfrac{x}{(1+\alpha^2)^{1/2}}\right) + \dfrac{\alpha}{2}\ln\!\left(\dfrac{1+\alpha^2 + (\alpha^2-1)x^2 - 2\alpha x(1+\alpha^2-x^2)^{1/2}}{(1+\alpha^2)(1-x^2)}\right),$$

where the inverse trigonometric functions take their principal values. The interval $-1 < x < 1$ is mapped one-to-one to the interval $0 < \eta < \infty$, with the points $x = -1$ and $x = 1$ corresponding to $\eta = \infty$ and $\eta = 0$, respectively.

For extensions to complex arguments (including the range $1 < x < \infty$), asymptotic expansions, and explicit error bounds, see Dunster (1991).

14.20(ix) Asymptotic Approximations: Large μ, $0 \le \tau \le A\mu$

As $\mu \to \infty$,

14.20.22 $\quad \mathsf{P}^{-\mu}_{-\frac{1}{2}+i\tau}(x) = \dfrac{\beta\exp(\mu\beta\arctan\beta)}{\Gamma(\mu+1)(1+\beta^2)^{\mu/2}}\dfrac{e^{-\mu\rho}}{(1+\beta^2 - x^2\beta^2)^{1/4}}\left(1 + O\!\left(\dfrac{1}{\mu}\right)\right),$

uniformly for $x \in (-1, 1)$ and $\tau \in [0, A\mu]$. Here

14.20.23 $\quad\quad\quad\quad \beta = \tau/\mu,$

and the variable ρ is defined by

14.20.24 $\quad \rho = \dfrac{1}{2}\ln\!\left(\dfrac{(1-\beta^2)x^2 + 1 + \beta^2 + 2x(1 + \beta^2 - \beta^2 x^2)^{1/2}}{1 - x^2}\right) + \beta\arctan\!\left(\dfrac{\beta x}{\sqrt{1 + \beta^2 - \beta^2 x^2}}\right) - \dfrac{1}{2}\ln(1+\beta^2),$

with the inverse tangent taking its principal value. The interval $-1 < x < 1$ is mapped one-to-one to the interval $-\infty < \rho < \infty$, with the points $x = -1$, $x = 0$, and $x = 1$ corresponding to $\rho = -\infty$, $\rho = 0$, and $\rho = \infty$, respectively.

With the same conditions, the corresponding approximation for $\mathsf{P}^{-\mu}_{-\frac{1}{2}+i\tau}(-x)$ is obtainable by replacing $e^{-\mu\rho}$ by $e^{\mu\rho}$ on the right-hand side of (14.20.22). Approximations for $\mathsf{P}^{\mu}_{-\frac{1}{2}+i\tau}(x)$ and $\widehat{\mathsf{Q}}^{-\mu}_{-\frac{1}{2}+i\tau}(x)$ can then be achieved via (14.9.7) and (14.20.3).

For extensions to complex arguments (including the range $1 < x < \infty$), asymptotic expansions, and explicit error bounds, see Dunster (1991).

14.20(x) Zeros and Integrals

For zeros of $\mathsf{P}_{-\frac{1}{2}+i\tau}(x)$ see Hobson (1931, §237).

For integrals with respect to τ involving $\mathsf{P}_{-\frac{1}{2}+i\tau}(x)$, see Prudnikov et al. (1990, pp. 218–228).

Complex Arguments

14.21 Definitions and Basic Properties

14.21(i) Associated Legendre Equation

14.21.1
$$\left(1 - z^2\right)\frac{d^2w}{dz^2} - 2z\frac{dw}{dz} + \left(\nu(\nu+1) - \frac{\mu^2}{1-z^2}\right)w = 0.$$

Standard solutions: the associated Legendre functions $P^{\mu}_{\nu}(z)$, $P^{-\mu}_{\nu}(z)$, $Q^{\mu}_{\nu}(z)$, and $\boldsymbol{Q}^{\mu}_{-\nu-1}(z)$. $P^{\pm\mu}_{\nu}(z)$ and $Q^{\mu}_{\nu}(z)$ exist for all values of ν, μ, and z, except possibly $z = \pm 1$ and ∞, which are branch points (or poles) of the functions, in general. When z is complex $P^{\pm\mu}_{\nu}(z)$, $Q^{\mu}_{\nu}(z)$, and $\boldsymbol{Q}^{\mu}_{\nu}(z)$ are defined by (14.3.6)–(14.3.10) with x replaced by z: the principal branches are obtained by taking the principal values of all the multivalued functions appearing in these representations when $z \in (1, \infty)$, and by continuity elsewhere in the z-plane with a cut along the interval $(-\infty, 1]$; compare §4.2(i). The principal branches of $P^{\pm\mu}_{\nu}(z)$ and $\boldsymbol{Q}^{\mu}_{\nu}(z)$ are real when $\nu, \mu \in \mathbb{R}$ and $z \in (1, \infty)$.

14.21(ii) Numerically Satisfactory Solutions

When $\Re\nu \geq -\frac{1}{2}$ and $\Re\mu \geq 0$, a numerically satisfactory pair of solutions of (14.21.1) in the half-plane $|\operatorname{ph} z| < \frac{1}{2}\pi$ is given by $P^{-\mu}_{\nu}(z)$ and $\boldsymbol{Q}^{\mu}_{\nu}(z)$.

14.21(iii) Properties

Many of the properties stated in preceding sections extend immediately from the x-interval $(1, \infty)$ to the cut z-plane $\mathbb{C}\setminus(-\infty, 1]$. This includes, for example, the Wronskian relations (14.2.7)–(14.2.11); hypergeometric representations (14.3.6)–(14.3.10) and (14.3.15)–(14.3.20); results for integer orders (14.6.3)–(14.6.5), (14.6.7), (14.6.8), (14.7.6), (14.7.7), and (14.7.11)–(14.7.16); behavior at singularities (14.8.7)–(14.8.16); connection formulas (14.9.11)–(14.9.16); recurrence relations (14.10.3)–(14.10.7). The generating function expansions (14.7.19) (with P replaced by P) and (14.7.22) apply when $|h| < \min\left|z \pm \left(z^2 - 1\right)^{1/2}\right|$; (14.7.21) (with P replaced by P) applies when $|h| > \max\left|z \pm \left(z^2 - 1\right)^{1/2}\right|$.

14.22 Graphics

In the graphics shown in this section, height corresponds to the absolute value of the function and color to the phase. See also p. xiv.

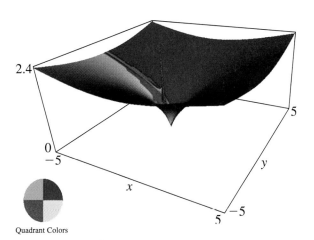

Figure 14.22.1: $P^0_{1/2}(x + iy)$, $-5 \leq x \leq 5$, $-5 \leq y \leq 5$. There is a cut along the real axis from $-\infty$ to -1.

Figure 14.22.2: $P^{-1/2}_{1/2}(x + iy)$, $-5 \leq x \leq 5$, $-5 \leq y \leq 5$. There is a cut along the real axis from $-\infty$ to 1.

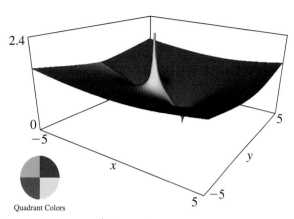

Figure 14.22.3: $P_{1/2}^{-1}(x+iy), -5 \le x \le 5, -5 \le y \le 5$. There is a cut along the real axis from $-\infty$ to 1.

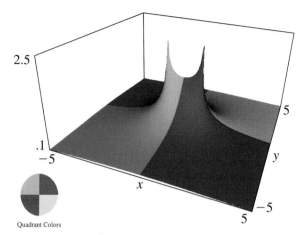

Figure 14.22.4: $\boldsymbol{Q}_0^0(x+iy), -5 \le x \le 5, -5 \le y \le 5$. There is a cut along the real axis from -1 to 1.

14.23 Values on the Cut

When $-1 < x < 1$,

14.23.1
$$P_\nu^\mu(x \pm i0) = e^{\mp\mu\pi i/2}\, \mathsf{P}_\nu^\mu(x),$$

14.23.2
$$\boldsymbol{Q}_\nu^\mu(x \pm i0) = \frac{e^{\pm\mu\pi i/2}}{\Gamma(\nu+\mu+1)}\left(\mathsf{Q}_\nu^\mu(x) \mp \tfrac{1}{2}\pi i\, \mathsf{P}_\nu^\mu(x)\right).$$

In terms of the hypergeometric function $\mathbf{F}$ (§14.3(i))

14.23.3
$$\boldsymbol{Q}_\nu^\mu(x \pm i0) = \frac{e^{\mp\nu\pi i/2}\pi^{3/2}(1-x^2)^{\mu/2}}{2^{\nu+1}} \left(\frac{x\,\mathbf{F}\!\left(\tfrac{1}{2}\mu - \tfrac{1}{2}\nu + \tfrac{1}{2}, \tfrac{1}{2}\nu + \tfrac{1}{2}\mu + 1; \tfrac{3}{2}; x^2\right)}{\Gamma\!\left(\tfrac{1}{2}\nu - \tfrac{1}{2}\mu + \tfrac{1}{2}\right)\Gamma\!\left(\tfrac{1}{2}\nu + \tfrac{1}{2}\mu + \tfrac{1}{2}\right)} \mp i \frac{\mathbf{F}\!\left(\tfrac{1}{2}\mu - \tfrac{1}{2}\nu, \tfrac{1}{2}\nu + \tfrac{1}{2}\mu + \tfrac{1}{2}; \tfrac{1}{2}; x^2\right)}{\Gamma\!\left(\tfrac{1}{2}\nu - \tfrac{1}{2}\mu + 1\right)\Gamma\!\left(\tfrac{1}{2}\nu + \tfrac{1}{2}\mu + 1\right)} \right).$$

Conversely,

14.23.4 $\mathsf{P}_\nu^\mu(x) = e^{\pm\mu\pi i/2} P_\nu^\mu(x \pm i0),$

14.23.5
$$\mathsf{Q}_\nu^\mu(x) = \tfrac{1}{2}\Gamma(\nu+\mu+1)\left(e^{-\mu\pi i/2}\boldsymbol{Q}_\nu^\mu(x+i0) + e^{\mu\pi i/2}\boldsymbol{Q}_\nu^\mu(x-i0)\right),$$

or equivalently,

14.23.6
$$\mathsf{Q}_\nu^\mu(x) = e^{\mp\mu\pi i/2}\Gamma(\nu+\mu+1)\,\boldsymbol{Q}_\nu^\mu(x \pm i0) \pm \tfrac{1}{2}\pi i e^{\pm\mu\pi i/2} P_\nu^\mu(x \pm i0).$$

If cuts are introduced along the intervals $(-\infty, -1]$ and $[1, \infty)$, then (14.23.4) and (14.23.6) could be used to extend the definitions of $\mathsf{P}_\nu^\mu(x)$ and $\mathsf{Q}_\nu^\mu(x)$ to complex x.

The conical function defined by (14.20.2) can be represented similarly by

14.23.7
$$\widehat{\mathsf{Q}}_{-\frac{1}{2}+i\tau}^{-\mu}(x) = \tfrac{1}{2}e^{3\mu\pi i/2} Q_{-\frac{1}{2}+i\tau}^{-\mu}(x-i0) + \tfrac{1}{2}e^{-3\mu\pi i/2} Q_{-\frac{1}{2}-i\tau}^{-\mu}(x+i0).$$

14.24 Analytic Continuation

Let s be an arbitrary integer, and $P_\nu^{-\mu}(ze^{s\pi i})$ and $\boldsymbol{Q}_\nu^\mu(ze^{s\pi i})$ denote the branches obtained from the principal branches by making $\tfrac{1}{2}s$ circuits, in the positive sense, of the ellipse having ± 1 as foci and passing through z. Then

14.24.1
$$P_\nu^{-\mu}(ze^{s\pi i}) = e^{s\nu\pi i} P_\nu^{-\mu}(z) + \frac{2i\sin\!\left(\left(\nu+\tfrac{1}{2}\right)s\pi\right)e^{-s\pi i/2}}{\cos(\nu\pi)\,\Gamma(\mu-\nu)}\boldsymbol{Q}_\nu^\mu(z),$$

14.24.2 $\boldsymbol{Q}_\nu^\mu(ze^{s\pi i}) = (-1)^s e^{-s\nu\pi i}\boldsymbol{Q}_\nu^\mu(z),$

the limiting value being taken in (14.24.1) when 2ν is an odd integer.

Next, let $P_{\nu,s}^{-\mu}(z)$ and $\boldsymbol{Q}_{\nu,s}^\mu(z)$ denote the branches obtained from the principal branches by encircling the branch point 1 (but not the branch point -1) s times in the positive sense. Then

14.24.3 $P_{\nu,s}^{-\mu}(z) = e^{s\mu\pi i} P_\nu^{-\mu}(z),$

$$\boldsymbol{Q}_{\nu,s}^{\mu}(z) = e^{-s\mu\pi i}\,\boldsymbol{Q}_{\nu}^{\mu}(z)$$
14.24.4
$$-\frac{\pi i \sin(s\mu\pi)}{\sin(\mu\pi)\,\Gamma(\nu-\mu+1)}\,P_{\nu}^{-\mu}(z),$$

the limiting value being taken in (14.24.4) when $\mu \in \mathbb{Z}$.

For fixed z, other than ± 1 or ∞, each branch of $P_{\nu}^{-\mu}(z)$ and $\boldsymbol{Q}_{\nu}^{\mu}(z)$ is an entire function of each parameter ν and μ.

The behavior of $P_{\nu}^{-\mu}(z)$ and $\boldsymbol{Q}_{\nu}^{\mu}(z)$ as $z \to -1$ from the left on the upper or lower side of the cut from $-\infty$ to 1 can be deduced from (14.8.7)–(14.8.11), combined with (14.24.1) and (14.24.2) with $s = \pm 1$.

14.25 Integral Representations

The principal values of $P_{\nu}^{-\mu}(z)$ and $\boldsymbol{Q}_{\nu}^{\mu}(z)$ (§14.21(i)) are given by

14.25.1
$$P_{\nu}^{-\mu}(z) = \frac{(z^2-1)^{\mu/2}}{2^{\nu}\,\Gamma(\mu-\nu)\,\Gamma(\nu+1)} \int_0^{\infty} \frac{(\sinh t)^{2\nu+1}}{(z+\cosh t)^{\nu+\mu+1}}\,dt,$$
$$\Re\mu > \Re\nu > -1,$$

14.25.2
$$\boldsymbol{Q}_{\nu}^{\mu}(z) = \frac{\pi^{1/2}\,(z^2-1)^{\mu/2}}{2^{\mu}\,\Gamma\!\left(\mu+\tfrac{1}{2}\right)\Gamma(\nu-\mu+1)}$$
$$\times \int_0^{\infty} \frac{(\sinh t)^{2\mu}}{(z+(z^2-1)^{1/2}\cosh t)^{\nu+\mu+1}}\,dt,$$
$$\Re(\nu+1) > \Re\mu > -\tfrac{1}{2},$$

where the multivalued functions have their principal values when $1 < z < \infty$ and are continuous in $\mathbb{C} \setminus (-\infty, 1]$.

For corresponding contour integrals, with less restrictions on μ and ν, see Olver (1997b, pp. 174–179), and for further integral representations see Magnus et al. (1966, §4.6.1).

14.26 Uniform Asymptotic Expansions

The uniform asymptotic approximations given in §14.15 for $P_{\nu}^{-\mu}(x)$ and $\boldsymbol{Q}_{\nu}^{\mu}(x)$ for $1 < x < \infty$ are extended to domains in the complex plane in the following references: §§14.15(i) and 14.15(ii), Dunster (2003b); §14.15(iii), Olver (1997b, Chapter 12); §14.15(iv), Boyd and Dunster (1986). For an extension of §14.15(iv) to complex argument and imaginary parameters, see Dunster (1990b).

See also Frenzen (1990), Gil et al. (2000), Shivakumar and Wong (1988), Ursell (1984), and Wong (1989) for uniform asymptotic approximations obtained from integral representations.

14.27 Zeros

$P_{\nu}^{\mu}(x \pm i0)$ (either side of the cut) has exactly one zero in the interval $(-\infty, -1)$ if either of the following sets of conditions holds:

(a) $\mu < 0$, $\mu \notin \mathbb{Z}$, $\nu \in \mathbb{Z}$, and $\sin((\mu-\nu)\pi)$ and $\sin(\mu\pi)$ have opposite signs.

(b) $\mu, \nu \in \mathbb{Z}$, $\mu + \nu < 0$, and ν is odd.

For all other values of the parameters $P_{\nu}^{\mu}(x \pm i0)$ has no zeros in the interval $(-\infty, -1)$.

For complex zeros of $P_{\nu}^{\mu}(z)$ see Hobson (1931, §§233, 234, and 238).

14.28 Sums

14.28(i) Addition Theorem

When $\Re z_1 > 0$, $\Re z_2 > 0$, $|\mathrm{ph}(z_1-1)| < \pi$, and $|\mathrm{ph}(z_2-1)| < \pi$,

14.28.1
$$P_{\nu}\!\left(z_1 z_2 - (z_1^2-1)^{1/2}(z_2^2-1)^{1/2}\cos\phi\right)$$
$$= P_{\nu}(z_1)\,P_{\nu}(z_2) + 2\sum_{m=1}^{\infty} (-1)^m \frac{\Gamma(\nu-m+1)}{\Gamma(\nu+m+1)}$$
$$\times P_{\nu}^{m}(z_1)\,P_{\nu}^{m}(z_2)\cos(m\phi),$$

where the branches of the square roots have their principal values when $z_1, z_2 \in (1, \infty)$ and are continuous when $z_1, z_2 \in \mathbb{C} \setminus (0, 1]$. For this and similar results see Erdélyi et al. (1953a, §3.11).

14.28(ii) Heine's Formula

14.28.2
$$\sum_{n=0}^{\infty} (2n+1)\,Q_n(z_1)\,P_n(z_2) = \frac{1}{z_1-z_2}, \quad z_1 \in \mathcal{E}_1,\ z_2 \in \mathcal{E}_2,$$

where $\mathcal{E}_1$ and $\mathcal{E}_2$ are ellipses with foci at ± 1, $\mathcal{E}_2$ being properly interior to $\mathcal{E}_1$. The series converges uniformly for z_1 outside or on $\mathcal{E}_1$, and z_2 within or on $\mathcal{E}_2$.

14.28(iii) Other Sums

See §14.18(iv).

14.29 Generalizations

Solutions of the equation

14.29.1
$$(1-z^2)\frac{d^2 w}{dz^2} - 2z\frac{dw}{dz}$$
$$+ \left(\nu(\nu+1) - \frac{\mu_1^2}{2(1-z)} - \frac{\mu_2^2}{2(1+z)}\right)w$$
$$= 0$$

are called *Generalized Associated Legendre Functions*. As in the case of (14.21.1), the solutions are hypergeometric functions, and (14.29.1) reduces to (14.21.1) when $\mu_1 = \mu_2 = \mu$. For properties see Virchenko and Fedotova (2001) and Braaksma and Meulenbeld (1967).

For inhomogeneous versions of the associated Legendre equation, and properties of their solutions, see Babister (1967, pp. 252–264).

Applications

14.30 Spherical and Spheroidal Harmonics

14.30(i) Definitions

With l and m integers such that $0 \leq m \leq l$, and θ and ϕ angles such that $0 \leq \theta \leq \pi$, $0 \leq \phi \leq 2\pi$,

14.30.1
$$Y_{l,m}(\theta,\phi) = \left(\frac{(l-m)!(2l+1)}{4\pi(l+m)!}\right)^{1/2} e^{im\phi}\, \mathsf{P}_l^m(\cos\theta),$$

14.30.2
$$Y_l^m(\theta,\phi) = \cos(m\phi)\,\mathsf{P}_l^m(\cos\theta) \text{ or } \sin(m\phi)\,\mathsf{P}_l^m(\cos\theta).$$

$Y_{l,m}(\theta,\phi)$ are known as *spherical harmonics*. $Y_l^m(\theta,\phi)$ are known as *surface harmonics of the first kind*: *tesseral* for $m < l$ and *sectorial* for $m = l$. Sometimes $Y_{l,m}(\theta,\phi)$ is denoted by $i^{-l}\mathfrak{D}_{lm}(\theta,\phi)$; also the definition of $Y_{l,m}(\theta,\phi)$ can differ from (14.30.1), for example, by inclusion of a factor $(-1)^m$.

$P_n^m(x)$ and $Q_n^m(x)$ $(x > 1)$ are often referred to as the *prolate spheroidal harmonics of the first and second kinds*, respectively. $P_n^m(ix)$ and $Q_n^m(ix)$ $(x > 0)$ are known as *oblate spheroidal harmonics of the first and second kinds*, respectively. Segura and Gil (1999) introduced the scaled oblate spheroidal harmonics $R_n^m(x) = e^{-i\pi n/2} P_n^m(ix)$ and $T_n^m(x) = ie^{i\pi n/2} Q_n^m(ix)$ which are real when $x > 0$ and $n = 0, 1, 2, \ldots$.

14.30(ii) Basic Properties

Most mathematical properties of $Y_{l,m}(\theta,\phi)$ can be derived directly from (14.30.1) and the properties of the Ferrers function of the first kind given earlier in this chapter.

Explicit Representation

14.30.3
$$Y_{l,m}(\theta,\phi) = \frac{(-1)^{l+m}}{2^l l!}\left(\frac{(l-m)!(2l+1)}{4\pi(l+m)!}\right)^{1/2} e^{im\phi}(\sin\theta)^m \left(\frac{d}{d(\cos\theta)}\right)^{l+m}(\sin\theta)^{2l}.$$

Special Values

14.30.4
$$Y_{l,m}(0,\phi) = \begin{cases} \left(\dfrac{2l+1}{4\pi}\right)^{1/2}, & m = 0, \\ 0, & m = 1, 2, 3, \ldots, \end{cases}$$

14.30.5
$$Y_{l,m}\!\left(\tfrac{1}{2}\pi,\phi\right) = \begin{cases} \dfrac{(-1)^{(l+m)/2}e^{im\phi}}{2^l\left(\tfrac{1}{2}l-\tfrac{1}{2}m\right)!\left(\tfrac{1}{2}l+\tfrac{1}{2}m\right)!}\left(\dfrac{(l-m)!(l+m)!(2l+1)}{4\pi}\right)^{1/2}, & \tfrac{1}{2}l+\tfrac{1}{2}m \in \mathbb{Z}, \\ 0, & \tfrac{1}{2}l+\tfrac{1}{2}m \notin \mathbb{Z}. \end{cases}$$

Symmetry

14.30.6
$$Y_{l,-m}(\theta,\phi) = (-1)^m\, Y_{l,m}^*(\theta,\phi).$$

Parity Operation

14.30.7
$$Y_{l,m}(\pi-\theta,\phi+\pi) = (-1)^l\, Y_{l,m}(\theta,\phi).$$

Orthogonality

14.30.8
$$\int_0^{2\pi}\!\!\int_0^\pi Y_{l_1,m_1}^*(\theta,\phi)\,Y_{l_2,m_2}(\theta,\phi)\sin\theta\, d\theta\, d\phi = \delta_{l_1,l_2}\delta_{m_1,m_2};$$

here and elsewhere in this section the asterisk (*) denotes complex conjugate.

See also (34.3.22), and for further related integrals see Askey *et al.* (1986).

14.30(iii) Sums

Distributional Completeness

For a series representation of the product of two Dirac deltas in terms of products of spherical harmonics see §1.17(iii).

Addition Theorem

14.30.9
$$P_l(\cos\theta_1 \cos\theta_2 + \sin\theta_1 \sin\theta_2 \cos(\phi_1 - \phi_2)) = \frac{4\pi}{2l+1} \sum_{m=-l}^{l} Y_{l,m}^*(\theta_1, \phi_1) Y_{l,m}(\theta_2, \phi_2).$$

See also (18.18.9) and (34.3.19).

14.30(iv) Applications

In general, *spherical harmonics* are defined as the class of homogeneous harmonic polynomials. See Andrews et al. (1999, Chapter 9). The special class of spherical harmonics $Y_{l,m}(\theta,\phi)$, defined by (14.30.1), appear in many physical applications. As an example, Laplace's equation $\nabla^2 W = 0$ in spherical coordinates (§1.5(ii)):

14.30.10
$$\frac{1}{\rho^2}\frac{\partial}{\partial \rho}\left(\rho^2 \frac{\partial W}{\partial \rho}\right) + \frac{1}{\rho^2 \sin\theta}\frac{\partial}{\partial \theta}\left(\sin\theta \frac{\partial W}{\partial \theta}\right) + \frac{1}{\rho^2 \sin^2\theta}\frac{\partial^2 W}{\partial \phi^2} = 0,$$

has solutions $W(\rho,\theta,\phi) = \rho^l Y_{l,m}(\theta,\phi)$, which are everywhere one-valued and continuous.

In the quantization of angular momentum the spherical harmonics $Y_{l,m}(\theta,\phi)$ are normalized solutions of the eigenvalue equation

14.30.11
$$L^2 Y_{l,m} = \hbar^2 l(l+1) Y_{l,m},$$

where $\hbar$ is the reduced Planck's constant, and L^2 is the *angular momentum operator* in spherical coordinates:

14.30.12
$$L^2 = -\hbar^2 \left(\frac{1}{\sin\theta}\frac{\partial}{\partial \theta}\left(\sin\theta \frac{\partial}{\partial \theta}\right) + \frac{1}{\sin^2\theta}\frac{\partial^2}{\partial \phi^2}\right);$$

see Edmonds (1974, §2.5).

For applications in geophysics see Stacey (1977, §§4.2, 6.3, and 8.1).

14.31 Other Applications

14.31(i) Toroidal Functions

Applications of toroidal functions include expansion of vacuum magnetic fields in stellarators and tokamaks (van Milligen and López Fraguas (1994)), analytic solutions of Poisson's equation in channel-like geometries (Hoyles et al. (1998)), and Dirichlet problems with toroidal symmetry (Gil et al. (2000)).

14.31(ii) Conical Functions

The conical functions $P^m_{-\frac{1}{2}+i\tau}(x)$ appear in boundary-value problems for the Laplace equation in toroidal coordinates (§14.19(i)) for regions bounded by cones, by two intersecting spheres, or by one or two confocal hyperboloids of revolution (Kölbig (1981)). These functions are also used in the Mehler–Fock integral transform (§14.20(vi)) for problems in potential and heat theory, and in elementary particle physics (Sneddon (1972, Chapter 7) and Braaksma and Meulenbeld (1967)). The conical functions and Mehler–Fock transform generalize to Jacobi functions and the Jacobi transform; see Koornwinder (1984a) and references therein.

14.31(iii) Miscellaneous

Many additional physical applications of Legendre polynomials and associated Legendre functions include solution of the Helmholtz equation, as well as the Laplace equation, in spherical coordinates (Temme (1996a)), quantum mechanics (Edmonds (1974)), and high-frequency scattering by a sphere (Nussenzveig (1965)). See also §18.39.

Legendre functions $P_\nu(x)$ of complex degree ν appear in the application of complex angular momentum techniques to atomic and molecular scattering (Connor and Mackay (1979)).

Computation

14.32 Methods of Computation

Essentially the same comments that are made in §15.19 concerning the computation of hypergeometric functions apply to the functions described in the present chapter. In particular, for small or moderate values of the parameters μ and ν the power-series expansions of the various hypergeometric function representations given in §§14.3(i)–14.3(iii), 14.19(ii), and 14.20(i) can be selected in such a way that convergence is stable, and reasonably rapid, especially when the argument of the functions is real. In other cases recurrence relations (§14.10) provide a powerful method when applied in a stable direction (§3.6); see Olver and Smith (1983) and Gautschi (1967).

Other methods include:

- Application of the uniform asymptotic expansions for large values of the parameters given in §§14.15 and 14.20(vii)–14.20(ix).

- Numerical integration (§3.7) of the defining differential equations (14.2.2), (14.20.1), and (14.21.1).

- Quadrature (§3.5) of the integral representations given in §§14.12, 14.19(iii), 14.20(iv), and 14.25; see Segura and Gil (1999) and Gil et al. (2000).

- Evaluation (§3.10) of the continued fractions given in §14.14. See Gil and Segura (2000).

14.33 Tables

- Abramowitz and Stegun (1964, Chapter 8) tabulates $\mathsf{P}_n(x)$ for $n = 0(1)3, 9, 10$, $x = 0(.01)1$, 5–8D; $\mathsf{P}'_n(x)$ for $n = 1(1)4, 9, 10$, $x = 0(.01)1$, 5–7D; $\mathsf{Q}_n(x)$ and $\mathsf{Q}'_n(x)$ for $n = 0(1)3, 9, 10$, $x = 0(.01)1$, 6–8D; $P_n(x)$ and $P'_n(x)$ for $n = 0(1)5, 9, 10$, $x = 1(.2)10$, 6S; $Q_n(x)$ and $Q'_n(x)$ for $n = 0(1)3, 9, 10$, $x = 1(.2)10$, 6S. (Here primes denote derivatives with respect to x.)

- Zhang and Jin (1996, Chapter 4) tabulates $\mathsf{P}_n(x)$ for $n = 2(1)5, 10$, $x = 0(.1)1$, 7D; $\mathsf{P}_n(\cos\theta)$ for $n = 1(1)4, 10$, $\theta = 0(5°)90°$, 8D; $\mathsf{Q}_n(x)$ for $n = 0(1)2, 10$, $x = 0(.1)0.9$, 8S; $\mathsf{Q}_n(\cos\theta)$ for $n = 0(1)3, 10$, $\theta = 0(5°)90°$, 8D; $\mathsf{P}_n^m(x)$ for $m = 1(1)4$, $n - m = 0(1)2$, $n = 10$, $x = 0, 0.5$, 8S; $\mathsf{Q}_n^m(x)$ for $m = 1(1)4$, $n = 0(1)2, 10$, 8S; $\mathsf{P}_\nu^m(\cos\theta)$ for $m = 0(1)3$, $\nu = 0(.25)5$, $\theta = 0(15°)90°$, 5D; $P_n(x)$ for $n = 2(1)5, 10$, $x = 1(1)10$, 7S; $Q_n(x)$ for $n = 0(1)2, 10$, $x = 2(1)10$, 8S. Corresponding values of the derivative of each function are also included, as are 6D values of the first 5 ν-zeros of $\mathsf{P}_\nu^m(\cos\theta)$ and of its derivative for $m = 0(1)4$, $\theta = 10°, 30°, 150°$.

- Belousov (1962) tabulates $\mathsf{P}_n^m(\cos\theta)$ (normalized) for $m = 0(1)36$, $n - m = 0(1)56$, $\theta = 0(2.5°)90°$, 6D.

- Žurina and Karmazina (1964, 1965) tabulate the conical functions $\mathsf{P}_{-\frac{1}{2}+i\tau}(x)$ for $\tau = 0(.01)50$, $x = -0.9(.1)0.9$, 7S; $\hat{P}_{-\frac{1}{2}+i\tau}(x)$ for $\tau = 0(.01)50$, $x = 1.1(.1)2(.2)5(.5)10(10)60$, 7D. Auxiliary tables are included to facilitate computation for larger values of τ when $-1 < x < 1$.

- Žurina and Karmazina (1963) tabulates the conical functions $\mathsf{P}^1_{-\frac{1}{2}+i\tau}(x)$ for $\tau = 0(.01)25$, $x = -0.9(.1)0.9$, 7S; $\hat{P}^1_{-\frac{1}{2}+i\tau}(x)$ for $\tau = 0(.01)25$, $x = 1.1(.1)2(.2)5(.5)10(10)60$, 7S. Auxiliary tables are included to assist computation for larger values of τ when $-1 < x < 1$.

For tables prior to 1961 see Fletcher et al. (1962) and Lebedev and Fedorova (1960).

14.34 Software

See http://dlmf.nist.gov/14.34.

References

General References

The main reference used in writing this chapter is Olver (1997b). For additional bibliographic reading see Erdélyi et al. (1953a, Chapter III), Hobson (1931), Jeffreys and Jeffreys (1956), MacRobert (1967), Magnus et al. (1966), Robin (1957, 1958, 1959), Snow (1952), Szegö (1967), Temme (1996a), and Wong (1989).

Sources

The following list gives the references or other indications of proofs that were used in constructing the various sections of this chapter. These sources supplement the references that are quoted in the text.

§14.2 For §§14.2(i), 14.2(ii) see Olver (1997b, pp. 169). For §14.2(iii) see Olver (1997b, p. 172) for the pair $P_\nu^{-\mu}(x)$ and $\boldsymbol{Q}_\nu^\mu(x)$. The result for $\mathsf{P}_\nu^{-\mu}(x)$ and $\mathsf{P}_\nu^{-\mu}(-x)$ follows from the fact that when $\Re\mu \geq 0$, $\mathsf{P}_\nu^{-\mu}(x)$ is recessive as $x \to 1-$ and $\mathsf{P}_\nu^{-\mu}(-x)$ is recessive as $x \to -1+$; see §14.8(i). For §14.2(iv) see Olver (1997b, p. 172) for (14.2.4), (14.2.5); the other results may be derived in a similar manner, or by application of the connection formulas in §14.9.

§14.3 For (14.3.1)–(14.3.4) see Olver (1997b, pp. 159, 186). The version of (14.3.4) given in Hobson (1931, p. 386) has an error. For (14.3.5) use (14.3.1) and (14.9.3). For (14.3.8) see Olver (1997b, p. 159, Eq. (9.05)). For (14.3.11) and (14.3.12) see Olver (1997b, p. 187). For (14.3.16)–(14.3.20) see Erdélyi et al. (1953a, pp. 123–139). For (14.3.21) combine (14.3.22) and (14.23.4). For (14.3.22) see Erdélyi et al. (1953a, p. 175). For (14.3.23) see (14.3.6) and Olver (1997b, p. 167).

§14.4 These graphics were produced at NIST.

§14.5 (14.5.1)–(14.5.4) may be derived from (14.3.11)–(14.3.14) and also (14.10.4) with P replaced by Q. For (14.5.5)–(14.5.10) use §14.7(i). For (14.5.11)–(14.5.19) see Erdélyi et al. (1953a, p. 150). (14.5.20)–(14.5.27) are given in Magnus et al. (1966, p. 173): to verify these compare the hypergeometric representations of the Legendre functions and elliptic integrals (§§14.3 and 19.5).

§14.6 See Erdélyi et al. (1953a, pp. 148–149). For (14.6.5) combine (14.3.10) and (14.6.4).

REFERENCES

§14.7 Olver (1997b, pp. 174, 181–182, 188). For (14.7.17), (14.7.18) use (14.9.8), (14.9.10). For (14.7.19)–(14.7.22) see Erdélyi et al. (1953a, p. 154) and Olver (1997b, pp. 51, 85).

§14.8 Olver (1997b, pp. 171, 173, 186), Erdélyi et al. (1953a, p. 163). (14.8.5) may be derived from (14.8.1), (14.9.2). (14.8.10) may be derived from (14.8.7), (14.9.12). (14.8.16) may be derived from (14.8.15), (14.9.12).

§14.9 Olver (1997b, pp. 171, 174, 186, 188). For (14.9.3), (14.9.4) use (14.9.1), (14.9.2). For (14.9.17) use (14.9.14), (14.9.16).

§14.10 Erdélyi et al. (1953a, pp. 160–161).

§14.11 (14.11.1) may be derived from (14.3.1). (14.11.2) may be derived from (14.9.10) and (14.11.1). (14.11.4) may be derived from (14.3.1) and the hypergeometric expansion for $Q_\nu(x)$ (Hobson (1931, §132)). (14.11.5) may be derived from (14.9.8), (14.9.10), and (14.11.4).

§14.12 Erdélyi et al. (1953a, pp. 155–159), Olver (1997b, pp. 181–183, 185).

§14.13 Erdélyi et al. (1953a, pp. 146, 151).

§14.14 (14.14.1) follows from (14.10.6). (14.14.3) follows from (14.10.3), with $P_\nu^\mu(x)$ replaced by $Q_\nu^\mu(x)$. For further details see Gil et al. (2000).

§14.15 Dunster (2003b), Olver (1997b, pp. 463–469), Boyd and Dunster (1986). (14.15.1) may be derived from (14.3.1) and §15.12(ii). For (14.15.24)–(14.15.31) see Olver (1975b).

§14.16 Hobson (1931, pp. 386–389, 399–401).

§14.17 Erdélyi et al. (1953a, pp. 170–172), Olver (1997b, pp. 188–189). (14.17.1)–(14.17.4) may be verified by differentiation and using the recurrence relations (§14.10). (14.17.7), (14.17.9) may be derived from (14.9.3), (14.17.6), (14.17.8). The version of (14.17.16) given in Erdélyi et al. (1953a, p. 171, Eq. (18)) is incorrect. For (14.17.17) see Din (1981).

§14.18 Erdélyi et al. (1953a, pp. 162, 167–169), Olver (1997b, p. 183). Errors in Erdélyi et al. (1953a, pp. 168–169) have been corrected. (14.18.2) may be derived from (14.7.16), (14.9.3), (14.18.1). (14.18.8) may be derived from (14.7.17), (14.18.9).

§14.19 Erdélyi et al. (1953a, pp. 156–157, 166, 173). For (14.19.7), (14.19.8) combine (14.9.16), (14.9.17) with (14.9.11)–(14.9.13).

§14.20 (14.20.3) follows from (14.9.10), (14.20.2). (14.20.4), (14.20.5) follow from (14.2.3), (14.20.3). (14.20.6) follows from (14.3.10), (14.9.12). (14.20.7), (14.20.8) follow from §14.8(i) and (14.20.3). (14.20.9) follows from (14.12.1). For (14.20.10) see Erdélyi et al. (1953a, p. 174). For (14.20.11)–(14.20.14) see Braaksma and Meulenbeld (1967). For (14.20.15) see Olver (1997b, p. 473). (14.20.16) may be derived from (14.20.18). For §§14.20(viii), 14.20(ix) see Dunster (1991, Eqs. (5.11), (5.14) have been corrected). The graphs were produced at NIST.

§14.21 Olver (1997b, pp. 169–185), Erdélyi et al. (1953a, Chapter 3).

§14.22 These graphics were produced at NIST.

§14.23 For (14.23.1), (14.23.5) see Olver (1997b, p. 185). For (14.23.7) see Dunster (1991). (14.23.2) may be derived from (14.23.4), (14.24.2). (14.23.3) may be derived from (14.3.11), (14.3.12), (14.9.14). (14.23.4) may be derived from (14.23.1). (14.23.6) may be derived from (14.23.1), (14.23.2).

§14.24 Olver (1997b, p, 179).

§14.25 For (14.25.1) see Olver (1997b, p. 179). For (14.25.2) see Erdélyi et al. (1953a, p. 155).

§14.27 Hobson (1931, pp. 391–399).

§14.28 Erdélyi et al. (1953a, p. 168) or Olver (1997b, p. 473).

§14.30 Edmonds (1974, pp. 20–24, 63). (Note that Edmonds' $P_l^m(x)$ differs by a factor $(-1)^m$ from $\mathsf{P}_l^m(x)$.) (14.30.4) may be derived from (14.8.1), (14.8.2), (14.30.1). (14.30.5) may be derived from (14.5.1), (14.30.1). (14.30.7) may be derived from (14.7.17), (14.30.1). (14.30.3) also follows from (14.30.1), (14.7.10).

Chapter 15

Hypergeometric Function

A. B. Olde Daalhuis[1]

Notation **384**
    15.1   Special Notation 384

Properties **384**
    15.2   Definitions and Analytical Properties . . . 384
    15.3   Graphics 385
    15.4   Special Cases 386
    15.5   Derivatives and Contiguous Functions . . 387
    15.6   Integral Representations 388
    15.7   Continued Fractions 389
    15.8   Transformations of Variable 390
    15.9   Relations to Other Functions 393
    15.10 Hypergeometric Differential Equation . . 394
    15.11 Riemann's Differential Equation 396

    15.12 Asymptotic Approximations 396
    15.13 Zeros 398
    15.14 Integrals 398
    15.15 Sums 399
    15.16 Products 399

Applications **399**
    15.17 Mathematical Applications 399
    15.18 Physical Applications 400

Computation **400**
    15.19 Methods of Computation 400
    15.20 Software 401

References **401**

---

[1]School of Mathematics, Edinburgh University, Edinburgh, United Kingdom.
    Acknowledgments: This chapter is based in part on Chapter 15 of Abramowitz and Stegun (1964) by Fritz Oberhettinger. The author thanks Richard Askey and Simon Ruijsenaars for many helpful recommendations.
    Copyright © 2009 National Institute of Standards and Technology. All rights reserved.

Notation

15.1 Special Notation

(For other notation see pp. xiv and 873.)

x	real variable.
$z = x + iy$	complex variable.
a, b, c	real or complex parameters.
k, ℓ, m, n	integers.
s	nonnegative integer.
δ	arbitrary small positive constant.
$\Gamma(z)$	gamma function (§5.2(i)).
$\psi(z)$	$\Gamma'(z)/\Gamma(z)$.

Unless indicated otherwise primes denote derivatives with respect to the variable.

We use the following notations for the hypergeometric function:

15.1.1 $\quad {}_2F_1(a,b;c;z) = F(a,b;c;z) = F\left(\begin{matrix} a,b \\ c \end{matrix}; z\right),$

and also

15.1.2
$$\frac{F(a,b;c;z)}{\Gamma(c)} = \mathbf{F}(a,b;c;z) = \mathbf{F}\left(\begin{matrix} a,b \\ c \end{matrix}; z\right) = {}_2\mathbf{F}_1(a,b;c;z),$$

(Olver (1997b, Chapter 5)).

Properties

15.2 Definitions and Analytical Properties

15.2(i) Gauss Series

The hypergeometric function $F(a,b;c;z)$ is defined by the *Gauss series*

15.2.1
$$\begin{aligned} F(a,b;c;z) &= \sum_{s=0}^{\infty} \frac{(a)_s (b)_s}{(c)_s s!} z^s \\ &= 1 + \frac{ab}{c} z + \frac{a(a+1)b(b+1)}{c(c+1)2!} z^2 + \cdots \\ &= \frac{\Gamma(c)}{\Gamma(a)\Gamma(b)} \sum_{s=0}^{\infty} \frac{\Gamma(a+s)\Gamma(b+s)}{\Gamma(c+s)s!} z^s, \end{aligned}$$

on the disk $|z| < 1$, and by analytic continuation elsewhere. In general, $F(a,b;c;z)$ does not exist when $c = 0, -1, -2, \ldots$. The branch obtained by introducing a cut from 1 to $+\infty$ on the real z-axis, that is, the branch in the sector $|\operatorname{ph}(1-z)| \leq \pi$, is the *principal branch* (or *principal value*) of $F(a,b;c;z)$.

For all values of c

15.2.2 $\quad \mathbf{F}(a,b;c;z) = \sum_{s=0}^{\infty} \frac{(a)_s(b)_s}{\Gamma(c+s)s!} z^s, \qquad |z| < 1,$

again with analytic continuation for other values of z, and with the principal branch defined in a similar way.

Except where indicated otherwise principal branches of $F(a,b;c;z)$ and $\mathbf{F}(a,b;c;z)$ are assumed throughout this Handbook.

The difference between the principal branches on the two sides of the branch cut (§4.2(i)) is given by

15.2.3
$$\mathbf{F}\left(\begin{matrix} a,b \\ c \end{matrix}; x+i0\right) - \mathbf{F}\left(\begin{matrix} a,b \\ c \end{matrix}; x-i0\right)$$
$$= \frac{2\pi i}{\Gamma(a)\Gamma(b)}(x-1)^{c-a-b}\mathbf{F}\left(\begin{matrix} c-a, c-b \\ c-a-b+1 \end{matrix}; 1-x\right),$$
$$x > 1.$$

On the circle of convergence, $|z| = 1$, the Gauss series:

(a) Converges absolutely when $\Re(c-a-b) > 0$.

(b) Converges conditionally when $-1 < \Re(c-a-b) \leq 0$ and $z=1$ is excluded.

(c) Diverges when $\Re(c-a-b) \leq -1$.

For the case $z=1$ see also §15.4(ii).

15.2(ii) Analytic Properties

The principal branch of $\mathbf{F}(a,b;c;z)$ is an entire function of a, b, and c. The same is true of other branches, provided that $z=0,1,$ and ∞ are excluded. As a multi-valued function of z, $\mathbf{F}(a,b;c;z)$ is analytic everywhere except for possible branch points at $z=0, 1,$ and ∞. The same properties hold for $F(a,b;c;z)$, except that as a function of c, $F(a,b;c;z)$ in general has poles at $c = 0, -1, -2, \ldots$.

Because of the analytic properties with respect to a, b, and c, it is usually legitimate to take limits in formulas involving functions that are undefined for certain values of the parameters.

For example, when $a = -m$, $m = 0, 1, 2, \ldots$, and $c \neq 0, -1, -2, \ldots$, $F(a, b; c; z)$ is a polynomial:

15.2.4
$$F(-m, b; c; z) = \sum_{n=0}^{m} \frac{(-m)_n (b)_n}{(c)_n n!} z^n = \sum_{n=0}^{m} (-1)^n \binom{m}{n} \frac{(b)_n}{(c)_n} z^n.$$

This formula is also valid when $c = -m - \ell$, $\ell = 0, 1, 2, \ldots$, provided that we use the interpretation

15.2.5 $\quad F\left(\begin{matrix} -m, b \\ -m - \ell \end{matrix}; z\right) = \lim_{c \to -m-\ell} \left(\lim_{a \to -m} F\left(\begin{matrix} a, b \\ c \end{matrix}; z\right) \right),$

and not

15.2.6 $\quad F\left(\begin{matrix} -m, b \\ -m - \ell \end{matrix}; z\right) = \lim_{a \to -m} F\left(\begin{matrix} a, b \\ a - \ell \end{matrix}; z\right),$

which is sometimes used in the literature. (Both interpretations give solutions of the hypergeometric differential equation (15.10.1), as does $\mathbf{F}(a, b; c; z)$, which is analytic at $c = 0, -1, -2, \ldots$.) For illustration see Figures 15.3.6 and 15.3.7.

In the case $c = -m$ the right-hand side of (15.2.4) becomes the first $m + 1$ terms of the Maclaurin series for $(1 - z)^{-b}$.

15.3 Graphics

15.3(i) Graphs

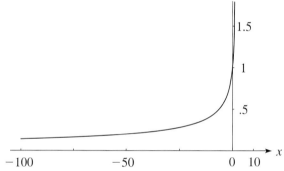

Figure 15.3.1: $F\left(\frac{4}{3}, \frac{9}{16}; \frac{14}{5}; x\right)$, $-100 \leq x \leq 1$.

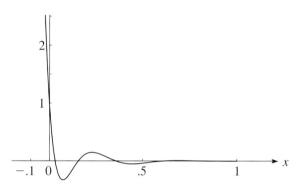

Figure 15.3.2: $F(5, -10; 1; x)$, $-0.023 \leq x \leq 1$.

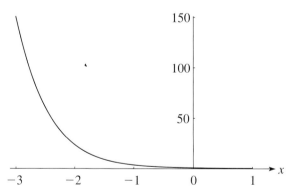

Figure 15.3.3: $F(1, -10; 10; x)$, $-3 \leq x \leq 1$.

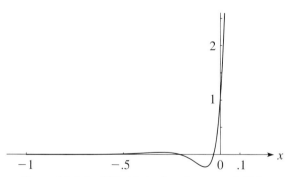

Figure 15.3.4: $F(5, 10; 1; x)$, $-1 \leq x \leq 0.022$.

15.3(ii) Surfaces

In Figures 15.3.5 and 15.3.6, height corresponds to the absolute value of the function and color to the phase. See also p. xiv.

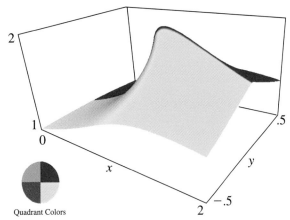

Figure 15.3.5: $F\left(\frac{4}{3}, \frac{9}{16}; \frac{14}{5}; x+iy\right), 0 \leq x \leq 2, -0.5 \leq y \leq 0.5$. (There is a cut along the real axis from 1 to ∞.)

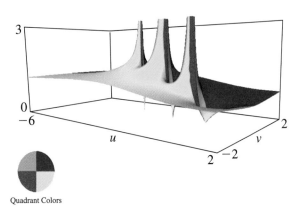

Figure 15.3.6: $F\left(-3, \frac{3}{5}; u+iv; \frac{1}{2}\right), -6 \leq u \leq 2, -2 \leq v \leq 2$. (With $c = u+iv$ the only poles occur at $c = 0, -1, -2$; compare §15.2(ii).)

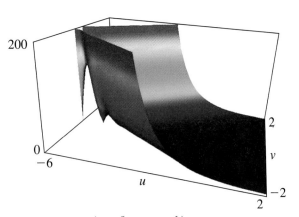

Figure 15.3.7: $|\mathbf{F}\left(-3, \frac{3}{5}; u+iv; \frac{1}{2}\right)|, -6 \leq u \leq 2, -2 \leq v \leq 2$.

15.4 Special Cases

15.4(i) Elementary Functions

The following results hold for principal branches when $|z| < 1$, and by analytic continuation elsewhere. Exceptions are (15.4.8) and (15.4.10), that hold for $|z| < \pi/4$, and (15.4.12), (15.4.14), and (15.4.16), that hold for $|z| < \pi/2$.

15.4.1 $\quad F(1,1;2;z) = -z^{-1}\ln(1-z),$

15.4.2 $\quad F\left(\frac{1}{2}, 1; \frac{3}{2}; z^2\right) = \frac{1}{2z}\ln\left(\frac{1+z}{1-z}\right),$

15.4.3 $\quad F\left(\frac{1}{2}, 1; \frac{3}{2}; -z^2\right) = z^{-1}\arctan z,$

15.4.4 $\quad F\left(\frac{1}{2}, \frac{1}{2}; \frac{3}{2}; z^2\right) = z^{-1}\arcsin z,$

15.4.5 $\quad F\left(\frac{1}{2}, \frac{1}{2}; \frac{3}{2}; -z^2\right) = z^{-1}\ln\left(z+\sqrt{1+z^2}\right).$

15.4.6 $\quad F(a,b;b;z) = (1-z)^{-a};$

compare §15.2(ii).

15.4.7 $\quad F\left(a, \frac{1}{2}+a; \frac{1}{2}; z^2\right) = \frac{1}{2}\left((1+z)^{-2a} + (1-z)^{-2a}\right),$

15.4.8 $\quad F\left(a, \frac{1}{2}+a; \frac{1}{2}; -\tan^2 z\right) = (\cos z)^{2a}\cos(2az).$

15.4.9
$$F\left(a, \frac{1}{2}+a; \frac{3}{2}; z^2\right) = \frac{1}{(2-4a)z}\left((1+z)^{1-2a} - (1-z)^{1-2a}\right),$$

15.4.10
$$F\left(a, \frac{1}{2}+a; \frac{3}{2}; -\tan^2 z\right) = (\cos z)^{2a}\frac{\sin((1-2a)z)}{(1-2a)\sin z}.$$

15.4.11
$$F\left(-a, a; \frac{1}{2}; -z^2\right) = \frac{1}{2}\left(\left(\sqrt{1+z^2}+z\right)^{2a} + \left(\sqrt{1+z^2}-z\right)^{2a}\right),$$

15.4.12 $\quad F\left(-a, a; \frac{1}{2}; \sin^2 z\right) = \cos(2az).$

15.4.13
$$F\left(a, 1-a; \frac{1}{2}; -z^2\right) = \frac{1}{2\sqrt{1+z^2}}\left(\left(\sqrt{1+z^2}+z\right)^{2a-1} + \left(\sqrt{1+z^2}-z\right)^{2a-1}\right),$$

15.4.14 $\quad F\left(a, 1-a; \frac{1}{2}; \sin^2 z\right) = \dfrac{\cos((2a-1)z)}{\cos z}.$

15.4.15
$$F\left(a, 1-a; \tfrac{3}{2}; -z^2\right) = \frac{1}{(2-4a)z}\left(\left(\sqrt{1+z^2}+z\right)^{1-2a} - \left(\sqrt{1+z^2}-z\right)^{1-2a}\right),$$

15.4.16 $F\left(a, 1-a; \tfrac{3}{2}; \sin^2 z\right) = \dfrac{\sin((2a-1)z)}{(2a-1)\sin z}.$

15.4.17 $F\left(a, \tfrac{1}{2}+a; 1+2a; z\right) = \left(\tfrac{1}{2}+\tfrac{1}{2}\sqrt{1-z}\right)^{-2a},$

15.4.18 $F\left(a, \tfrac{1}{2}+a; 2a; z\right) = \dfrac{1}{\sqrt{1-z}}\left(\tfrac{1}{2}+\tfrac{1}{2}\sqrt{1-z}\right)^{1-2a}.$

15.4.19 $F(a+1, b; a; z) = (1-(1-(b/a))z)(1-z)^{-1-b}.$

For an extensive list of elementary representations see Prudnikov et al. (1990, pp. 468–488).

15.4(ii) Argument Unity

If $\Re(c-a-b) > 0$, then

15.4.20 $F(a, b; c; 1) = \dfrac{\Gamma(c)\,\Gamma(c-a-b)}{\Gamma(c-a)\,\Gamma(c-b)}.$

If $c = a+b$, then

15.4.21 $\lim\limits_{z \to 1-} \dfrac{F(a, b; a+b; z)}{-\ln(1-z)} = \dfrac{\Gamma(a+b)}{\Gamma(a)\,\Gamma(b)}.$

If $\Re(c-a-b) = 0$ and $c \neq a+b$, then

15.4.22
$$\lim_{z \to 1-} (1-z)^{a+b-c}\left(F(a,b;c;z) - \dfrac{\Gamma(c)\,\Gamma(c-a-b)}{\Gamma(c-a)\,\Gamma(c-b)}\right) = \dfrac{\Gamma(c)\,\Gamma(a+b-c)}{\Gamma(a)\,\Gamma(b)}.$$

If $\Re(c-a-b) < 0$, then

15.4.23 $\lim\limits_{z \to 1-} \dfrac{F(a,b;c;z)}{(1-z)^{c-a-b}} = \dfrac{\Gamma(c)\,\Gamma(a+b-c)}{\Gamma(a)\,\Gamma(b)}.$

Chu–Vandermonde Identity

15.4.24 $F(-n, b; c; 1) = \dfrac{(c-b)_n}{(c)_n}, \quad n = 0, 1, 2, \ldots.$

Dougall's Bilateral Sum

This is a generalization of (15.4.20). If a, b are not integers and $\Re(c+d-a-b) > 1$, then

15.4.25
$$\sum_{n=-\infty}^{\infty} \frac{\Gamma(a+n)\,\Gamma(b+n)}{\Gamma(c+n)\,\Gamma(d+n)} = \frac{\pi^2}{\sin(\pi a)\sin(\pi b)} \frac{\Gamma(c+d-a-b-1)}{\Gamma(c-a)\,\Gamma(d-a)\,\Gamma(c-b)\,\Gamma(d-b)}.$$

15.4(iii) Other Arguments

15.4.26 $F(a, b; a-b+1; -1) = \dfrac{\Gamma(a-b+1)\,\Gamma(\tfrac{1}{2}a+1)}{\Gamma(a+1)\,\Gamma(\tfrac{1}{2}a-b+1)}.$

15.4.27 $F(1, a; a+1; -1) = \tfrac{1}{2}a\left(\psi\left(\tfrac{1}{2}a+\tfrac{1}{2}\right) - \psi\left(\tfrac{1}{2}a\right)\right).$

15.4.28
$$F\left(a, b; \tfrac{1}{2}a+\tfrac{1}{2}b+\tfrac{1}{2}; \tfrac{1}{2}\right) = \sqrt{\pi}\,\frac{\Gamma\left(\tfrac{1}{2}a+\tfrac{1}{2}b+\tfrac{1}{2}\right)}{\Gamma\left(\tfrac{1}{2}a+\tfrac{1}{2}\right)\,\Gamma\left(\tfrac{1}{2}b+\tfrac{1}{2}\right)}.$$

15.4.29
$$F\left(a, b; \tfrac{1}{2}a+\tfrac{1}{2}b+1; \tfrac{1}{2}\right) = \frac{2\sqrt{\pi}}{a-b}\Gamma\left(\tfrac{1}{2}a+\tfrac{1}{2}b+1\right) \times \left(\frac{1}{\Gamma\left(\tfrac{1}{2}a\right)\Gamma\left(\tfrac{1}{2}b+\tfrac{1}{2}\right)} - \frac{1}{\Gamma\left(\tfrac{1}{2}a+\tfrac{1}{2}\right)\Gamma\left(\tfrac{1}{2}b\right)}\right).$$

15.4.30 $F\left(a, 1-a; b; \tfrac{1}{2}\right) = \dfrac{2^{1-b}\sqrt{\pi}\,\Gamma(b)}{\Gamma\left(\tfrac{1}{2}a+\tfrac{1}{2}b\right)\,\Gamma\left(\tfrac{1}{2}b-\tfrac{1}{2}a+\tfrac{1}{2}\right)}.$

15.4.31
$$F\left(a, \tfrac{1}{2}+a; \tfrac{3}{2}-2a; -\tfrac{1}{3}\right) = \left(\tfrac{8}{9}\right)^{-2a} \frac{\Gamma\left(\tfrac{4}{3}\right)\,\Gamma\left(\tfrac{3}{2}-2a\right)}{\Gamma\left(\tfrac{3}{2}\right)\,\Gamma\left(\tfrac{4}{3}-2a\right)}.$$

15.4.32
$$F\left(a, \tfrac{1}{2}+a; \tfrac{5}{6}+\tfrac{2}{3}a; \tfrac{1}{9}\right) = \sqrt{\pi}\left(\tfrac{3}{4}\right)^a \frac{\Gamma\left(\tfrac{5}{6}+\tfrac{2}{3}a\right)}{\Gamma\left(\tfrac{1}{2}+\tfrac{1}{3}a\right)\,\Gamma\left(\tfrac{5}{6}+\tfrac{1}{3}a\right)}.$$

15.4.33
$$F\left(3a, \tfrac{1}{3}+a; \tfrac{2}{3}+2a; e^{i\pi/3}\right) = \sqrt{\pi}\,e^{i\pi a/2}\left(\tfrac{16}{27}\right)^{(3a+1)/6} \frac{\Gamma\left(\tfrac{5}{6}+a\right)}{\Gamma\left(\tfrac{2}{3}+a\right)\,\Gamma\left(\tfrac{2}{3}\right)}.$$

15.5 Derivatives and Contiguous Functions

15.5(i) Differentiation Formulas

15.5.1 $\dfrac{d}{dz} F(a, b; c; z) = \dfrac{ab}{c} F(a+1, b+1; c+1; z),$

15.5.2 $\dfrac{d^n}{dz^n} F(a, b; c; z) = \dfrac{(a)_n (b)_n}{(c)_n} F(a+n, b+n; c+n; z).$

15.5.3
$$\left(z\frac{d}{dz}z\right)^n \left(z^{a-1} F(a, b; c; z)\right) = (a)_n z^{a+n-1} F(a+n, b; c; z).$$

15.5.4
$$\frac{d^n}{dz^n}\left(z^{c-1} F(a, b; c; z)\right) = (c-n)_n z^{c-n-1} F(a, b; c-n; z).$$

15.5.5
$$\left(z\frac{d}{dz}z\right)^n \left(z^{c-a-1}(1-z)^{a+b-c} F(a, b; c; z)\right) = (c-a)_n z^{c-a+n-1}(1-z)^{a-n+b-c} F(a-n, b; c; z).$$

15.5.6
$$\frac{d^n}{dz^n}\left((1-z)^{a+b-c}F(a,b;c;z)\right)$$
$$=\frac{(c-a)_n(c-b)_n}{(c)_n}(1-z)^{a+b-c-n}F(a,b;c+n;z).$$

15.5.7
$$\left((1-z)\frac{d}{dz}(1-z)\right)^n\left((1-z)^{a-1}F(a,b;c;z)\right)$$
$$=(-1)^n\frac{(a)_n(c-b)_n}{(c)_n}(1-z)^{a+n-1}$$
$$\times F(a+n,b;c+n;z).$$

15.5.8
$$\left((1-z)\frac{d}{dz}(1-z)\right)^n\left(z^{c-1}(1-z)^{b-c}F(a,b;c;z)\right)$$
$$=(c-n)_n z^{c-n-1}(1-z)^{b-c+n}F(a-n,b;c-n;z).$$

15.5.9
$$\frac{d^n}{dz^n}\left(z^{c-1}(1-z)^{a+b-c}F(a,b;c;z)\right)$$
$$=(c-n)_n z^{c-n-1}(1-z)^{a+b-c-n}$$
$$\times F(a-n,b-n;c-n;z).$$

Other versions of several of the identities in this subsection can be constructed with the aid of the operator identity

15.5.10
$$\left(z\frac{d}{dz}z\right)^n = z^n\frac{d^n}{dz^n}z^n,\quad n=1,2,3,\ldots.$$

See Erdélyi et al. (1953a, pp. 102–103).

15.5(ii) Contiguous Functions

The six functions $F(a\pm 1,b;c;z)$, $F(a,b\pm 1;c;z)$, $F(a,b;c\pm 1;z)$ are said to be *contiguous* to $F(a,b;c;z)$.

15.5.11
$$(c-a)F(a-1,b;c;z)$$
$$+(2a-c+(b-a)z)F(a,b;c;z)$$
$$+a(z-1)F(a+1,b;c;z)=0,$$

15.5.12
$$(b-a)F(a,b;c;z)+aF(a+1,b;c;z)$$
$$-bF(a,b+1;c;z)=0,$$

15.5.13
$$(c-a-b)F(a,b;c;z)$$
$$+a(1-z)F(a+1,b;c;z)$$
$$-(c-b)F(a,b-1;c;z)=0,$$

15.5.14
$$c(a+(b-c)z)F(a,b;c;z)$$
$$-ac(1-z)F(a+1,b;c;z)$$
$$+(c-a)(c-b)zF(a,b;c+1;z)=0,$$

15.5.15
$$(c-a-1)F(a,b;c;z)+aF(a+1,b;c;z)$$
$$-(c-1)F(a,b;c-1;z)=0,$$

15.5.16
$$c(1-z)F(a,b;c;z)-cF(a-1,b;c;z)$$
$$+(c-b)zF(a,b;c+1;z)=0,$$

15.5.17
$$(a-1+(b+1-c)z)F(a,b;c;z)$$
$$+(c-a)F(a-1,b;c;z)$$
$$-(c-1)(1-z)F(a,b;c-1;z)=0,$$

15.5.18
$$c(c-1)(z-1)F(a,b;c-1;z)$$
$$+c(c-1-(2c-a-b-1)z)F(a,b;c;z)$$
$$+(c-a)(c-b)zF(a,b;c+1;z)=0.$$

By repeated applications of (15.5.11)–(15.5.18) any function $F(a+k,b+\ell;c+m;z)$, in which k,ℓ,m are integers, can be expressed as a linear combination of $F(a,b;c;z)$ and any one of its contiguous functions, with coefficients that are rational functions of a,b,c, and z.

An equivalent equation to the hypergeometric differential equation (15.10.1) is

15.5.19
$$z(1-z)(a+1)(b+1)F(a+2,b+2;c+2;z)$$
$$+(c-(a+b+1)z)(c+1)F(a+1,b+1;c+1;z)$$
$$-c(c+1)F(a,b;c;z)=0.$$

Further contiguous relations include:

15.5.20
$$z(1-z)(dF(a,b;c;z)/dz)$$
$$=(c-a)F(a-1,b;c;z)+(a-c+bz)F(a,b;c;z)$$
$$=(c-b)F(a,b-1;c;z)+(b-c+az)F(a,b;c;z),$$

15.5.21
$$c(1-z)(dF(a,b;c;z)/dz)$$
$$=(c-a)(c-b)F(a,b;c+1;z)+c(a+b-c)F(a,b;c;z).$$

15.6 Integral Representations

The function $\mathbf{F}(a,b;c;z)$ (not $F(a,b;c;z)$) has the following integral representations:

15.6.1
$$\frac{1}{\Gamma(b)\Gamma(c-b)}\int_0^1\frac{t^{b-1}(1-t)^{c-b-1}}{(1-zt)^a}dt,\quad \Re c>\Re b>0.$$

15.6.2
$$\frac{\Gamma(1+b-c)}{2\pi i\,\Gamma(b)}\int_0^{(1+)}\frac{t^{b-1}(t-1)^{c-b-1}}{(1-zt)^a}dt,$$
$$c-b\neq 1,2,3,\ldots,\ \Re b>0.$$

15.6.3
$$e^{-b\pi i}\frac{\Gamma(1-b)}{2\pi i\,\Gamma(c-b)}\int_\infty^{(0+)}\frac{t^{b-1}(t+1)^{a-c}}{(t-zt+1)^a}dt,$$
$$b\neq 1,2,3,\ldots,\ \Re(c-b)>0.$$

15.6.4
$$e^{-b\pi i}\frac{\Gamma(1-b)}{2\pi i\,\Gamma(c-b)}\int_1^{(0+)}\frac{t^{b-1}(1-t)^{c-b-1}}{(1-zt)^a}dt,$$
$$b\neq 1,2,3,\ldots,\ \Re(c-b)>0.$$

15.6.5
$$e^{-c\pi i}\Gamma(1-b)\Gamma(1+b-c)$$
$$\times\frac{1}{4\pi^2}\int_A^{(0+,1+,0-,1-)}\frac{t^{b-1}(1-t)^{c-b-1}}{(1-zt)^a}dt,$$
$$b,c-b\neq 1,2,3,\ldots.$$

15.6.6
$$\frac{1}{2\pi i\,\Gamma(a)\Gamma(b)}\int_{-i\infty}^{i\infty}\frac{\Gamma(a+t)\Gamma(b+t)\Gamma(-t)}{\Gamma(c+t)}(-z)^t dt,$$
$$a,b\neq 0,-1,-2,\ldots.$$

15.6.7
$$\frac{1}{2\pi i\,\Gamma(a)\,\Gamma(b)\,\Gamma(c-a)\,\Gamma(c-b)}\int_{-i\infty}^{i\infty}\Gamma(a+t)\,\Gamma(b+t)\,\Gamma(c-a-b-t)\,\Gamma(-t)(1-z)^t\,dt,$$
$$a, b, c-a, c-b \neq 0, -1, -2, \ldots.$$

15.6.8
$$\frac{1}{\Gamma(c-d)}\int_0^1 \mathbf{F}(a,b;d;zt) t^{d-1}(1-t)^{c-d-1}\,dt, \qquad \Re c > \Re d > 0.$$

15.6.9
$$\int_0^1 \frac{t^{d-1}(1-t)^{c-d-1}}{(1-zt)^{a+b-\lambda}} \mathbf{F}\!\left(\begin{matrix}\lambda-a,\lambda-b\\d\end{matrix};zt\right)\mathbf{F}\!\left(\begin{matrix}a+b-\lambda,\lambda-d\\c-d\end{matrix};\frac{(1-t)z}{1-zt}\right)dt, \qquad \Re c > \Re d > 0.$$

These representations are valid when $|\mathrm{ph}(1-z)|<\pi$, except (15.6.6) which holds for $|\mathrm{ph}(-z)|<\pi$. In all cases the integrands are continuous functions of t on the integration paths, except possibly at the endpoints. In addition:

In (15.6.1) all functions in the integrand assume their principal values.

In (15.6.2) the point $1/z$ lies outside the integration contour, t^{b-1} and $(t-1)^{c-b-1}$ assume their principal values where the contour cuts the interval $(1,\infty)$, and $(1-zt)^a = 1$ at $t=0$.

In (15.6.3) the point $1/(z-1)$ lies outside the integration contour, the contour cuts the real axis between $t=-1$ and 0, at which point $\mathrm{ph}\,t = \pi$ and $\mathrm{ph}(1+t) = 0$.

In (15.6.4) the point $1/z$ lies outside the integration contour, and at the point where the contour cuts the negative real axis $\mathrm{ph}\,t = \pi$ and $\mathrm{ph}(1-t) = 0$.

In (15.6.5) the integration contour starts and terminates at a point A on the real axis between 0 and 1. It encircles $t=0$ and $t=1$ once in the positive direction, and then once in the negative direction. See Figure 15.6.1. At the starting point $\mathrm{ph}\,t$ and $\mathrm{ph}(1-t)$ are zero. Compare Figure 5.12.3.

In (15.6.6) the integration contour separates the poles of $\Gamma(a+t)$ and $\Gamma(b+t)$ from those of $\Gamma(-t)$, and $(-z)^t$ has its principal value.

In (15.6.7) the integration contour separates the poles of $\Gamma(a+t)$ and $\Gamma(b+t)$ from those of $\Gamma(c-a-b-t)$ and $\Gamma(-t)$, and $(1-z)^t$ has its principal value.

In each of (15.6.8) and (15.6.9) all functions in the integrand assume their principal values.

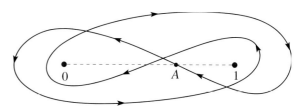

Figure 15.6.1: t-plane. Contour of integration in (15.6.5).

15.7 Continued Fractions

If $|\mathrm{ph}(1-z)|<\pi$, then

15.7.1
$$\frac{\mathbf{F}(a,b;c;z)}{\mathbf{F}(a,b+1;c+1;z)} = t_0 - \frac{u_1 z}{t_1 -} \frac{u_2 z}{t_2 -} \frac{u_3 z}{t_3 -} \cdots,$$

where

15.7.2
$$t_n = c+n, \quad u_{2n+1} = (a+n)(c-b+n),$$
$$u_{2n} = (b+n)(c-a+n).$$

If $|z|<1$, then

15.7.3
$$\frac{\mathbf{F}(a,b;c;z)}{\mathbf{F}(a,b+1;c+1;z)} = v_0 - \frac{w_1}{v_1 -} \frac{w_2}{v_2 -} \frac{w_3}{v_3 -} \cdots,$$

where

15.7.4
$$v_n = c+n+(b-a+n+1)z,$$
$$w_n = (b+n)(c-a+n)z.$$

If $\Re z < \tfrac{1}{2}$, then

15.7.5
$$\frac{\mathbf{F}(a,b;c;z)}{\mathbf{F}(a+1,b+1;c+1;z)} = x_0 + \frac{y_1}{x_1 +} \frac{y_2}{x_2 +} \frac{y_3}{x_3 +} \cdots,$$

where

15.7.6
$$x_n = c+n-(a+b+2n+1)z,$$
$$y_n = (a+n)(b+n)z(1-z).$$

See also Cuyt et al. (2008, pp. 295–309).

15.8 Transformations of Variable

15.8(i) Linear Transformations

All functions in this subsection and §15.8(ii) assume their principal values.

15.8.1
$$\mathbf{F}\left(\begin{matrix}a,b\\c\end{matrix};z\right) = (1-z)^{-a}\mathbf{F}\left(\begin{matrix}a,c-b\\c\end{matrix};\frac{z}{z-1}\right) = (1-z)^{-b}\mathbf{F}\left(\begin{matrix}c-a,b\\c\end{matrix};\frac{z}{z-1}\right) = (1-z)^{c-a-b}\mathbf{F}\left(\begin{matrix}c-a,c-b\\c\end{matrix};z\right),$$
$$|\mathrm{ph}(1-z)| < \pi.$$

15.8.2
$$\frac{\sin(\pi(b-a))}{\pi}\mathbf{F}\left(\begin{matrix}a,b\\c\end{matrix};z\right) = \frac{(-z)^{-a}}{\Gamma(b)\Gamma(c-a)}\mathbf{F}\left(\begin{matrix}a,a-c+1\\a-b+1\end{matrix};\frac{1}{z}\right) - \frac{(-z)^{-b}}{\Gamma(a)\Gamma(c-b)}\mathbf{F}\left(\begin{matrix}b,b-c+1\\b-a+1\end{matrix};\frac{1}{z}\right), \quad |\mathrm{ph}(-z)| < \pi.$$

15.8.3
$$\frac{\sin(\pi(b-a))}{\pi}\mathbf{F}\left(\begin{matrix}a,b\\c\end{matrix};z\right) = \frac{(1-z)^{-a}}{\Gamma(b)\Gamma(c-a)}\mathbf{F}\left(\begin{matrix}a,c-b\\a-b+1\end{matrix};\frac{1}{1-z}\right) - \frac{(1-z)^{-b}}{\Gamma(a)\Gamma(c-b)}\mathbf{F}\left(\begin{matrix}b,c-a\\b-a+1\end{matrix};\frac{1}{1-z}\right), \quad |\mathrm{ph}(-z)| < \pi.$$

15.8.4
$$\frac{\sin(\pi(c-a-b))}{\pi}\mathbf{F}\left(\begin{matrix}a,b\\c\end{matrix};z\right) = \frac{1}{\Gamma(c-a)\Gamma(c-b)}\mathbf{F}\left(\begin{matrix}a,b\\a+b-c+1\end{matrix};1-z\right) - \frac{(1-z)^{c-a-b}}{\Gamma(a)\Gamma(b)}\mathbf{F}\left(\begin{matrix}c-a,c-b\\c-a-b+1\end{matrix};1-z\right),$$
$$|\mathrm{ph}\,z| < \pi, |\mathrm{ph}(1-z)| < \pi.$$

15.8.5
$$\frac{\sin(\pi(c-a-b))}{\pi}\mathbf{F}\left(\begin{matrix}a,b\\c\end{matrix};z\right) = \frac{z^{-a}}{\Gamma(c-a)\Gamma(c-b)}\mathbf{F}\left(\begin{matrix}a,a-c+1\\a+b-c+1\end{matrix};1-\frac{1}{z}\right)$$
$$-\frac{(1-z)^{c-a-b}z^{a-c}}{\Gamma(a)\Gamma(b)}\mathbf{F}\left(\begin{matrix}c-a,1-a\\c-a-b+1\end{matrix};1-\frac{1}{z}\right), \quad |\mathrm{ph}\,z| < \pi, |\mathrm{ph}(1-z)| < \pi.$$

15.8(ii) Linear Transformations: Limiting Cases

With $m = 0, 1, 2, \ldots$, polynomial cases of (15.8.2)–(15.8.5) are given by

15.8.6
$$F\left(\begin{matrix}-m,b\\c\end{matrix};z\right) = \frac{(b)_m}{(c)_m}(-z)^m F\left(\begin{matrix}-m,1-c-m\\1-b-m\end{matrix};\frac{1}{z}\right) = \frac{(b)_m}{(c)_m}(1-z)^m F\left(\begin{matrix}-m,c-b\\1-b-m\end{matrix};\frac{1}{1-z}\right),$$

15.8.7
$$F\left(\begin{matrix}-m,b\\c\end{matrix};z\right) = \frac{(c-b)_m}{(c)_m} F\left(\begin{matrix}-m,b\\b-c-m+1\end{matrix};1-z\right) = \frac{(c-b)_m}{(c)_m}z^m F\left(\begin{matrix}-m,1-c-m\\b-c-m+1\end{matrix};1-\frac{1}{z}\right),$$

with the understanding that if $b = -\ell$, $\ell = 0, 1, 2, \ldots$, then $m \leq \ell$.

When $b - a$ is an integer limits are taken in (15.8.2) and (15.8.3) as follows.
If $b - a$ is a nonnegative integer, then

15.8.8
$$\mathbf{F}\left(\begin{matrix}a,a+m\\c\end{matrix};z\right) = \frac{(-z)^{-a}}{\Gamma(a+m)}\sum_{k=0}^{m-1}\frac{(a)_k(m-k-1)!}{k!\,\Gamma(c-a-k)}z^{-k} + \frac{(-z)^{-a}}{\Gamma(a)}\sum_{k=0}^{\infty}\frac{(a+m)_k}{k!(k+m)!\,\Gamma(c-a-k-m)}(-1)^k z^{-k-m}$$
$$\times (\ln(-z) + \psi(k+1) + \psi(k+m+1) - \psi(a+k+m) - \psi(c-a-k-m)),$$
$$|z| > 1, |\mathrm{ph}(-z)| < \pi,$$

15.8.9
$$\mathbf{F}\left(\begin{matrix}a,a+m\\c\end{matrix};z\right) = \frac{(1-z)^{-a}}{\Gamma(a+m)\Gamma(c-a)}\sum_{k=0}^{m-1}\frac{(a)_k(c-a-m)_k(m-k-1)!}{k!}(z-1)^{-k}$$
$$+ \frac{(-1)^m(1-z)^{-a-m}}{\Gamma(a)\Gamma(c-a-m)}\sum_{k=0}^{\infty}\frac{(a+m)_k(c-a)_k}{k!(k+m)!}(1-z)^{-k}$$
$$\times (\ln(1-z) + \psi(k+1) + \psi(k+m+1) - \psi(a+k+m) - \psi(c-a+k)),$$
$$|z-1| > 1, |\mathrm{ph}(1-z)| < \pi.$$

In (15.8.8) when $c - a - k - m$ is a nonpositive integer $\psi(c-a-k-m)/\Gamma(c-a-k-m)$ is interpreted as $(-1)^{m+k+a-c+1}(m+k+a-c)!$. Also, if a is a nonpositive integer, then (15.8.6) applies.

15.8 Transformations of Variable

Alternatively, if $b - a$ is a negative integer, then we interchange a and b in $\mathbf{F}(a,b;c;z)$.

In a similar way, when $c - a - b$ is an integer limits are taken in (15.8.4) and (15.8.5) as follows.

If $c - a - b$ is a nonnegative integer, then

15.8.10
$$\mathbf{F}\!\left(\begin{matrix}a,b\\a+b+m\end{matrix};z\right) = \frac{1}{\Gamma(a+m)\,\Gamma(b+m)}\sum_{k=0}^{m-1}\frac{(a)_k(b)_k(m-k-1)!}{k!}(z-1)^k - \frac{(z-1)^m}{\Gamma(a)\,\Gamma(b)}\sum_{k=0}^{\infty}\frac{(a+m)_k(b+m)_k}{k!(k+m)!}(1-z)^k$$
$$\times\,(\ln(1-z) - \psi(k+1) - \psi(k+m+1) + \psi(a+k+m) + \psi(b+k+m)),$$
$$|z-1|<1,\,|\mathrm{ph}(1-z)|<\pi,$$

15.8.11
$$\mathbf{F}\!\left(\begin{matrix}a,b\\a+b+m\end{matrix};z\right) = \frac{z^{-a}}{\Gamma(a+m)}\sum_{k=0}^{m-1}\frac{(a)_k(m-k-1)!}{k!\,\Gamma(b+m-k)}\left(1-\frac{1}{z}\right)^k - \frac{z^{-a}}{\Gamma(a)}\sum_{k=0}^{\infty}\frac{(a+m)_k}{k!(k+m)!\,\Gamma(b-k)}(-1)^k\left(1-\frac{1}{z}\right)^{k+m}$$
$$\times\,\left(\ln\!\left(\frac{1-z}{z}\right) - \psi(k+1) - \psi(k+m+1) + \psi(a+k+m) + \psi(b-k)\right),$$
$$\Re z > \tfrac{1}{2},\,|\mathrm{ph}\,z|<\pi,\,|\mathrm{ph}(1-z)|<\pi.$$

In (15.8.11) when $b - k$ is a nonpositive integer, $\psi(b-k)/\Gamma(b-k)$ is interpreted as $(-1)^{k-b+1}(k-b)!$. Also, if a or b or both are nonpositive integers, then (15.8.7) applies.

Lastly, if $c - a - b$ is a negative integer, then we first apply the transformation

15.8.12
$$\mathbf{F}(a,b;a+b-m;z) = (1-z)^{-m}\,\mathbf{F}\!\left(\tilde{a},\tilde{b};\tilde{a}+\tilde{b}+m;z\right),$$
$$\tilde{a} = a-m,\ \tilde{b} = b-m.$$

15.8(iii) Quadratic Transformations

A quadratic transformation relates two hypergeometric functions, with the variable in one a quadratic function of the variable in the other, possibly combined with a fractional linear transformation.

A necessary and sufficient condition that there exists a quadratic transformation is that at least one of the equations shown in Table 15.8.1 is satisfied.

Table 15.8.1: Quadratic transformations of the hypergeometric function.

Group 1	Group 2	Group 3	Group 4
	$c = a-b+1$	$a = b+\tfrac{1}{2}$	
$c = 2a$	$c = b-a+1$	$b = a+\tfrac{1}{2}$	$c = \tfrac{1}{2}$
$c = 2b$	$c = \tfrac{1}{2}(a+b+1)$	$c = a+b+\tfrac{1}{2}$	$c = \tfrac{3}{2}$
	$a+b = 1$	$c = a+b-\tfrac{1}{2}$	

The hypergeometric functions that correspond to Groups 1 and 2 have z as variable. The hypergeometric functions that correspond to Groups 3 and 4 have a nonlinear function of z as variable. The transformation formulas between two hypergeometric functions in Group 2, or two hypergeometric functions in Group 3, are the linear transformations (15.8.1).

In the equations that follow in this subsection all functions take their principal values.

Group 1 $\longrightarrow$ Group 3

15.8.13
$$F\!\left(\begin{matrix}a,b\\2b\end{matrix};z\right) = \left(1 - \tfrac{1}{2}z\right)^{-a} F\!\left(\begin{matrix}\tfrac{1}{2}a,\tfrac{1}{2}a+\tfrac{1}{2}\\b+\tfrac{1}{2}\end{matrix};\left(\frac{z}{2-z}\right)^{2}\right),\qquad |\mathrm{ph}(1-z)|<\pi,$$

15.8.14
$$F\!\left(\begin{matrix}a,b\\2b\end{matrix};z\right) = (1-z)^{-a/2} F\!\left(\begin{matrix}\tfrac{1}{2}a,b-\tfrac{1}{2}a\\b+\tfrac{1}{2}\end{matrix};\frac{z^2}{4z-4}\right),\qquad |\mathrm{ph}(1-z)|<\pi.$$

Group 2 $\longrightarrow$ Group 3

15.8.15
$$F\!\left(\begin{matrix}a,b\\a-b+1\end{matrix};z\right) = (1+z)^{-a} F\!\left(\begin{matrix}\tfrac{1}{2}a,\tfrac{1}{2}a+\tfrac{1}{2}\\a-b+1\end{matrix};\frac{4z}{(1+z)^2}\right),\qquad |z|<1,$$

15.8.16
$$F\!\left(\begin{matrix}a,b\\a-b+1\end{matrix};z\right) = (1-z)^{-a} F\!\left(\begin{matrix}\tfrac{1}{2}a,\tfrac{1}{2}a-b+\tfrac{1}{2}\\a-b+1\end{matrix};\frac{-4z}{(1-z)^2}\right),\qquad |z|<1.$$

15.8.17 $$F\left({a,b \atop \tfrac{1}{2}(a+b+1)};z\right) = (1-2z)^{-a} F\left({\tfrac{1}{2}a, \tfrac{1}{2}a+\tfrac{1}{2} \atop \tfrac{1}{2}(a+b+1)};\frac{4z(z-1)}{(1-2z)^2}\right), \qquad \Re z < \tfrac{1}{2},$$

15.8.18 $$F\left({a,b \atop \tfrac{1}{2}(a+b+1)};z\right) = F\left({\tfrac{1}{2}a, \tfrac{1}{2}b \atop \tfrac{1}{2}(a+b+1)}; 4z(1-z)\right), \qquad \Re z < \tfrac{1}{2}.$$

15.8.19 $$F\left({a, 1-a \atop c};z\right) = (1-2z)^{1-a-c}(1-z)^{c-1} F\left({\tfrac{1}{2}(a+c), \tfrac{1}{2}(a+c-1) \atop c};\frac{4z(z-1)}{(1-2z)^2}\right), \qquad \Re z < \tfrac{1}{2},$$

15.8.20 $$F\left({a, 1-a \atop c};z\right) = (1-z)^{c-1} F\left({\tfrac{1}{2}(c-a), \tfrac{1}{2}(a+c-1) \atop c}; 4z(1-z)\right), \qquad \Re z < \tfrac{1}{2}.$$

Group 2 $\longrightarrow$ Group 1

15.8.21 $$F\left({a,b \atop a-b+1};z\right) = (1+\sqrt{z})^{-2a} F\left({a, a-b+\tfrac{1}{2} \atop 2a-2b+1};\frac{4\sqrt{z}}{(1+\sqrt{z})^2}\right), \qquad |\operatorname{ph} z| < \pi, |z| < 1.$$

15.8.22 $$F\left({a,b \atop \tfrac{1}{2}(a+b+1)};z\right) = \left(\frac{\sqrt{1-z^{-1}}-1}{\sqrt{1-z^{-1}}+1}\right)^a F\left({a, \tfrac{1}{2}(a+b) \atop a+b};\frac{4\sqrt{1-z^{-1}}}{(\sqrt{1-z^{-1}}+1)^2}\right), \qquad |\operatorname{ph}(-z)| < \pi, \Re z < \tfrac{1}{2}.$$

15.8.23 $$F\left({a, 1-a \atop c};z\right) = \left(\sqrt{1-z^{-1}}-1\right)^{1-a} \left(\sqrt{1-z^{-1}}+1\right)^{a-2c+1} (1-z^{-1})^{c-1} F\left({c-a, c-\tfrac{1}{2} \atop 2c-1};\frac{4\sqrt{1-z^{-1}}}{(\sqrt{1-z^{-1}}+1)^2}\right),$$
$$|\operatorname{ph}(-z)| < \pi, \Re z < \tfrac{1}{2}.$$

Group 2 $\longrightarrow$ Group 4

15.8.24 $$F\left({a,b \atop a-b+1};z\right) = (1-z)^{-a} \frac{\Gamma(a-b+1)\Gamma(\tfrac{1}{2})}{\Gamma(\tfrac{1}{2}a+\tfrac{1}{2})\Gamma(\tfrac{1}{2}a-b+1)} F\left({\tfrac{1}{2}a, \tfrac{1}{2}a-b+\tfrac{1}{2} \atop \tfrac{1}{2}};\left(\frac{z+1}{z-1}\right)^2\right)$$
$$+ (1+z)(1-z)^{-a-1} \frac{\Gamma(a-b+1)\Gamma(-\tfrac{1}{2})}{\Gamma(\tfrac{1}{2}a)\Gamma(\tfrac{1}{2}a-b+\tfrac{1}{2})} F\left({\tfrac{1}{2}a+\tfrac{1}{2}, \tfrac{1}{2}a-b+1 \atop \tfrac{3}{2}};\left(\frac{z+1}{z-1}\right)^2\right),$$
$$|\operatorname{ph}(-z)| < \pi.$$

15.8.25 $$F\left({a,b \atop \tfrac{1}{2}(a+b+1)};z\right) = \frac{\Gamma(\tfrac{1}{2}(a+b+1))\Gamma(\tfrac{1}{2})}{\Gamma(\tfrac{1}{2}a+\tfrac{1}{2})\Gamma(\tfrac{1}{2}b+\tfrac{1}{2})} F\left({\tfrac{1}{2}a, \tfrac{1}{2}b \atop \tfrac{1}{2}};(1-2z)^2\right)$$
$$+ (1-2z)\frac{\Gamma(\tfrac{1}{2}(a+b+1))\Gamma(-\tfrac{1}{2})}{\Gamma(\tfrac{1}{2}a)\Gamma(\tfrac{1}{2}b)} F\left({\tfrac{1}{2}a+\tfrac{1}{2}, \tfrac{1}{2}b+\tfrac{1}{2} \atop \tfrac{3}{2}};(1-2z)^2\right),$$
$$|\operatorname{ph} z| < \pi, |\operatorname{ph}(1-z)| < \pi.$$

15.8.26 $$F\left({a, 1-a \atop c};z\right) = (1-z)^{c-1}\frac{\Gamma(c)\Gamma(\tfrac{1}{2})}{\Gamma(\tfrac{1}{2}(c-a+1))\Gamma(\tfrac{1}{2}c+\tfrac{1}{2}a)} F\left({\tfrac{1}{2}c-\tfrac{1}{2}a, \tfrac{1}{2}c+\tfrac{1}{2}a-\tfrac{1}{2} \atop \tfrac{1}{2}};(1-2z)^2\right)$$
$$+ (1-2z)(1-z)^{c-1}\frac{\Gamma(c)\Gamma(-\tfrac{1}{2})}{\Gamma(\tfrac{1}{2}c-\tfrac{1}{2}a)\Gamma(\tfrac{1}{2}(c+a-1))} F\left({\tfrac{1}{2}c-\tfrac{1}{2}a+\tfrac{1}{2}, \tfrac{1}{2}c+\tfrac{1}{2}a \atop \tfrac{3}{2}};(1-2z)^2\right),$$
$$|\operatorname{ph} z| < \pi, |\operatorname{ph}(1-z)| < \pi.$$

Group 4 $\longrightarrow$ Group 2

15.8.27 $$\frac{2\,\Gamma(\tfrac{1}{2})\,\Gamma(a+b+\tfrac{1}{2})}{\Gamma(a+\tfrac{1}{2})\,\Gamma(b+\tfrac{1}{2})} F(a,b;\tfrac{1}{2};z) = F(2a, 2b; a+b+\tfrac{1}{2}; \tfrac{1}{2}-\tfrac{1}{2}\sqrt{z}) + F(2a, 2b; a+b+\tfrac{1}{2}; \tfrac{1}{2}+\tfrac{1}{2}\sqrt{z}),$$
$$|\operatorname{ph} z| < \pi, |\operatorname{ph}(1-z)| < \pi.$$

15.8.28
$$\frac{2\sqrt{z}\,\Gamma\left(-\tfrac{1}{2}\right)\Gamma\left(a+b-\tfrac{1}{2}\right)}{\Gamma\left(a-\tfrac{1}{2}\right)\Gamma\left(b-\tfrac{1}{2}\right)} F\left(a,b;\tfrac{3}{2};z\right) = F\left(2a-1,2b-1;a+b-\tfrac{1}{2};\tfrac{1}{2}-\tfrac{1}{2}\sqrt{z}\right)$$
$$- F\left(2a-1,2b-1;a+b-\tfrac{1}{2};\tfrac{1}{2}+\tfrac{1}{2}\sqrt{z}\right), \qquad |\operatorname{ph} z| < \pi, |\operatorname{ph}(1-z)| < \pi.$$

15.8(iv) Quadratic Transformations (Continued)

When the intersection of two groups in Table 15.8.1 is not empty there exist special quadratic transformations, with only one free parameter, between two hypergeometric functions in the same group.

Examples
$b = \tfrac{1}{3}a + \tfrac{1}{3}$, $c = 2b = a - b + 1$ in Groups 1 and 2.
(15.8.21) becomes

15.8.29
$$F\left(\begin{matrix} a, \tfrac{1}{3}a + \tfrac{1}{3} \\ \tfrac{2}{3}a + \tfrac{2}{3} \end{matrix}; z\right) = (1+\sqrt{z})^{-2a}\, F\left(\begin{matrix} a, \tfrac{2}{3}a + \tfrac{1}{6} \\ \tfrac{4}{3}a + \tfrac{1}{3} \end{matrix}; \frac{4\sqrt{z}}{(1+\sqrt{z})^2}\right).$$

This is a quadratic transformation between two cases in Group 1.

We can also use (15.8.13), followed by the inverse of (15.8.15), and obtain

15.8.30 $\left(1 - \tfrac{1}{2}z\right)^{-a} F\left(\begin{matrix} \tfrac{1}{2}a, \tfrac{1}{2}a + \tfrac{1}{2} \\ \tfrac{1}{3}a + \tfrac{5}{6} \end{matrix}; \left(\tfrac{z}{2-z}\right)^2\right) = F\left(\begin{matrix} a, \tfrac{1}{3}a + \tfrac{1}{3} \\ \tfrac{2}{3}a + \tfrac{2}{3} \end{matrix}; z\right) = (1+z)^{-a} F\left(\begin{matrix} \tfrac{1}{2}a, \tfrac{1}{2}a + \tfrac{1}{2} \\ \tfrac{2}{3}a + \tfrac{2}{3} \end{matrix}; \frac{4z}{(1+z)^2}\right),$

which is a quadratic transformation between two cases in Group 3.

For further examples see Andrews et al. (1999, pp. 130–132 and 176–177).

15.8(v) Cubic Transformations

Examples

15.8.31
$$F\left(\begin{matrix} 3a, 3a + \tfrac{1}{2} \\ 4a + \tfrac{2}{3} \end{matrix}; z\right) = \left(1 - \tfrac{9}{8}z\right)^{-2a} F\left(\begin{matrix} a, a + \tfrac{1}{2} \\ 2a + \tfrac{5}{6} \end{matrix}; \frac{27z^2(z-1)}{(9z-8)^2}\right), \qquad \Re z < \tfrac{8}{9}.$$

With $\zeta = e^{2\pi i/3}(1-z)/\left(z - e^{4\pi i/3}\right)$

15.8.32
$$\frac{(1-z^3)^a}{(-z)^{3a}}\left(\frac{1}{\Gamma\left(a+\tfrac{2}{3}\right)\Gamma\left(\tfrac{2}{3}\right)} F\left(\begin{matrix} a, a + \tfrac{1}{3} \\ \tfrac{2}{3} \end{matrix}; z^{-3}\right) + \frac{e^{\tfrac{1}{3}\pi i}}{z\,\Gamma(a)\Gamma\left(\tfrac{4}{3}\right)} F\left(\begin{matrix} a + \tfrac{1}{3}, a + \tfrac{2}{3} \\ \tfrac{4}{3} \end{matrix}; z^{-3}\right)\right)$$
$$= \frac{3^{\tfrac{3}{2}a + \tfrac{1}{2}} e^{\tfrac{1}{2}a\pi i}\,\Gamma\left(a + \tfrac{1}{3}\right)(1-\zeta)^a}{2\pi\,\Gamma\left(2a + \tfrac{2}{3}\right)(-\zeta)^{2a}} F\left(\begin{matrix} a + \tfrac{1}{3}, 3a \\ 2a + \tfrac{2}{3} \end{matrix}; \zeta^{-1}\right), \qquad |z| > 1, |\operatorname{ph}(-z)| < \tfrac{1}{3}\pi.$$

Ramanujan's Cubic Transformation

15.8.33
$$F\left(\begin{matrix} \tfrac{1}{3}, \tfrac{2}{3} \\ 1 \end{matrix}; 1 - \left(\frac{1-z}{1+2z}\right)^3\right) = (1+2z)\, F\left(\begin{matrix} \tfrac{1}{3}, \tfrac{2}{3} \\ 1 \end{matrix}; z^3\right),$$

provided that z lies in the intersection of the open disks $\left|z - \tfrac{1}{4} \pm \tfrac{1}{4}\sqrt{3}i\right| < \tfrac{1}{2}\sqrt{3}$, or equivalently, $|\operatorname{ph}((1-z)/(1+2z))| < \pi/3$. This is used in a cubic analog of the arithmetic-geometric mean. See Borwein and Borwein (1991), and also Berndt et al. (1995).

For further examples and higher-order transformations see Goursat (1881), Watson (1910), and Vidūnas (2005); see also Erdélyi et al. (1953a, pp. 67 and 113–114).

15.9 Relations to Other Functions

15.9(i) Orthogonal Polynomials

For the notation see §§18.3 and 18.19.

Jacobi

15.9.1
$$P_n^{(\alpha,\beta)}(x) = \frac{(\alpha+1)_n}{n!}\, F\left(\begin{matrix} -n, n + \alpha + \beta + 1 \\ \alpha + 1 \end{matrix}; \frac{1-x}{2}\right).$$

Gegenbauer (or Ultraspherical)

15.9.2 $C_n^{(\lambda)}(x) = \dfrac{(2\lambda)_n}{n!}\, F\left(\begin{matrix} -n, n + 2\lambda \\ \lambda + \tfrac{1}{2} \end{matrix}; \dfrac{1-x}{2}\right).$

15.9.3 $C_n^{(\lambda)}(x) = (2x)^n \dfrac{(\lambda)_n}{n!}\, F\left(\begin{matrix} -\tfrac{1}{2}n, \tfrac{1}{2}(1-n) \\ 1 - \lambda - n \end{matrix}; \dfrac{1}{x^2}\right).$

15.9.4 $\quad C_n^{(\lambda)}(\cos\theta) = e^{ni\theta} \dfrac{(\lambda)_n}{n!} F\left(\begin{matrix}-n,\lambda\\1-\lambda-n\end{matrix};e^{-2i\theta}\right).$

Chebyshev

15.9.5 $\quad T_n(x) = F\left(\begin{matrix}-n,n\\\tfrac{1}{2}\end{matrix};\dfrac{1-x}{2}\right).$

15.9.6 $\quad U_n(x) = (n+1) F\left(\begin{matrix}-n,n+2\\\tfrac{3}{2}\end{matrix};\dfrac{1-x}{2}\right).$

Legendre

15.9.7 $\quad P_n(x) = F\left(\begin{matrix}-n,n+1\\1\end{matrix};\dfrac{1-x}{2}\right).$

Krawtchouk

15.9.8 $\quad K_n(x;p,N) = F\left(\begin{matrix}-n,-x\\-N\end{matrix};\dfrac{1}{p}\right),\quad n=0,1,2,\ldots,N;$

compare also §15.2(ii).

Meixner

15.9.9 $\quad M_n(x;\beta,c) = F\left(\begin{matrix}-n,-x\\\beta\end{matrix};1-\dfrac{1}{c}\right).$

Meixner–Pollaczek

15.9.10 $\quad P_n^{(\lambda)}(x;\phi) = \dfrac{(2\lambda)_n}{n!} e^{ni\phi} F\left(\begin{matrix}-n,\lambda+ix\\2\lambda\end{matrix};1-e^{-2i\phi}\right).$

15.9(ii) Jacobi Function

This is a generalization of Jacobi polynomials (§18.3) and has the representation

15.9.11 $\quad \phi_\lambda^{(\alpha,\beta)}(t) = F\left(\begin{matrix}\tfrac{1}{2}(\alpha+\beta+1-i\lambda),\tfrac{1}{2}(\alpha+\beta+1+i\lambda)\\\alpha+1\end{matrix};-\sinh^2 t\right).$

The *Jacobi transform* is defined as

15.9.12 $\quad \widetilde{f}(\lambda) = \int_0^\infty f(t)\phi_\lambda^{(\alpha,\beta)}(t)(2\sinh t)^{2\alpha+1}(2\cosh t)^{2\beta+1}\,dt,$

with inverse

15.9.13 $\quad f(t) = \dfrac{1}{2\pi i}\int_{-i\infty}^{i\infty} \widetilde{f}(i\lambda)\,\Phi_{i\lambda}^{(\alpha,\beta)}(t)\dfrac{\Gamma\left(\tfrac{1}{2}(\alpha+\beta+1+\lambda)\right)\Gamma\left(\tfrac{1}{2}(\alpha-\beta+1+\lambda)\right)}{\Gamma(\alpha+1)\Gamma(\lambda)2^{\alpha+\beta+1-\lambda}}\,d\lambda,$

where the contour of integration is located to the right of the poles of the gamma functions in the integrand, and

15.9.14 $\quad \Phi_\lambda^{(\alpha,\beta)}(t) = (2\cosh t)^{i\lambda-\alpha-\beta-1} F\left(\begin{matrix}\tfrac{1}{2}(\alpha+\beta+1-i\lambda),\tfrac{1}{2}(\alpha-\beta+1-i\lambda)\\1-i\lambda\end{matrix};\operatorname{sech}^2 t\right).$

For this result, together with restrictions on the functions $f(t)$ and $\widetilde{f}(\lambda)$, see Koornwinder (1984a).

15.9(iii) Gegenbauer Function

This is a generalization of Gegenbauer (or ultraspherical) polynomials (§18.3). It is defined by:

15.9.15 $\quad C_\alpha^{(\lambda)}(z) = \dfrac{\Gamma(\alpha+2\lambda)}{\Gamma(2\lambda)\Gamma(\alpha+1)} F\left(\begin{matrix}-\alpha,\alpha+2\lambda\\\lambda+\tfrac{1}{2}\end{matrix};\dfrac{1-z}{2}\right).$

15.9(iv) Associated Legendre Functions; Ferrers Functions

Any hypergeometric function for which a quadratic transformation exists can be expressed in terms of associated Legendre functions or Ferrers functions. For examples see §§14.3(i)–14.3(iii) and 14.21(iii).

For further examples see http://dlmf.nist.gov/15.9.iv.

15.10 Hypergeometric Differential Equation

15.10(i) Fundamental Solutions

15.10.1 $\quad z(1-z)\dfrac{d^2 w}{dz^2} + (c-(a+b+1)z)\dfrac{dw}{dz} - abw = 0.$

This is the *hypergeometric differential equation*. It has regular singularities at $z=0,1,\infty$, with corresponding exponent pairs $\{0,1-c\}$, $\{0,c-a-b\}$, $\{a,b\}$, respectively. When none of the exponent pairs differ by an integer, that is, when none of c, $c-a-b$, $a-b$ is an integer, we have the following pairs $f_1(z)$, $f_2(z)$ of fundamental solutions. They are also numerically satisfactory (§2.7(iv)) in the neighborhood of the corresponding singularity.

15.10 Hypergeometric Differential Equation

Singularity $z = 0$

15.10.2
$$f_1(z) = F\left(\begin{matrix} a, b \\ c \end{matrix}; z\right),$$
$$f_2(z) = z^{1-c} F\left(\begin{matrix} a-c+1, b-c+1 \\ 2-c \end{matrix}; z\right),$$

15.10.3 $\mathscr{W}\{f_1(z), f_2(z)\} = (1-c)z^{-c}(1-z)^{c-a-b-1}.$

Singularity $z = 1$

15.10.4
$$f_1(z) = F\left(\begin{matrix} a, b \\ a+b+1-c \end{matrix}; 1-z\right),$$
$$f_2(z) = (1-z)^{c-a-b} F\left(\begin{matrix} c-a, c-b \\ c-a-b+1 \end{matrix}; 1-z\right),$$

15.10.5 $\mathscr{W}\{f_1(z), f_2(z)\} = (a+b-c)z^{-c}(1-z)^{c-a-b-1}.$

Singularity $z = \infty$

15.10.6
$$f_1(z) = z^{-a} F\left(\begin{matrix} a, a-c+1 \\ a-b+1 \end{matrix}; \frac{1}{z}\right),$$
$$f_2(z) = z^{-b} F\left(\begin{matrix} b, b-c+1 \\ b-a+1 \end{matrix}; \frac{1}{z}\right),$$

15.10.7 $\mathscr{W}\{f_1(z), f_2(z)\} = (a-b)z^{-c}(z-1)^{c-a-b-1}.$

(a) If c equals $n = 1, 2, 3, \ldots,$ and $a = 1, 2, \ldots, n-1$, then fundamental solutions in the neighborhood of $z = 0$ are given by (15.10.2) with the interpretation (15.2.5) for $f_2(z)$.

(b) If c equals $n = 1, 2, 3, \ldots,$ and $a \neq 1, 2, \ldots, n-1$, then fundamental solutions in the neighborhood of $z = 0$ are given by $F(a, b; n; z)$ and

15.10.8
$$F\left(\begin{matrix} a, b \\ n \end{matrix}; z\right) \ln z - \sum_{k=1}^{n-1} \frac{(n-1)!(k-1)!}{(n-k-1)!(1-a)_k(1-b)_k}(-z)^{-k}$$
$$+ \sum_{k=0}^{\infty} \frac{(a)_k(b)_k}{(n)_k k!} z^k \left(\psi(a+k) + \psi(b+k) - \psi(1+k) - \psi(n+k)\right), \quad a, b \neq n-1, n-2, \ldots, 0, -1, -2, \ldots,$$

or

15.10.9
$$F\left(\begin{matrix} -m, b \\ n \end{matrix}; z\right) \ln z - \sum_{k=1}^{n-1} \frac{(n-1)!(k-1)!}{(n-k-1)!(m+1)_k(1-b)_k}(-z)^{-k}$$
$$+ \sum_{k=0}^{m} \frac{(-m)_k(b)_k}{(n)_k k!} z^k \left(\psi(1+m-k) + \psi(b+k) - \psi(1+k) - \psi(n+k)\right)$$
$$+ (-1)^m m! \sum_{k=m+1}^{\infty} \frac{(k-1-m)!(b)_k}{(n)_k k!} z^k, \quad a = -m, m = 0, 1, 2, \ldots; b \neq n-1, n-2, \ldots, 0, -1, -2, \ldots,$$

or

15.10.10
$$F\left(\begin{matrix} -m, -\ell \\ n \end{matrix}; z\right) \ln z - \sum_{k=1}^{n-1} \frac{(n-1)!(k-1)!}{(n-k-1)!(m+1)_k(\ell+1)_k}(-z)^{-k}$$
$$+ \sum_{k=0}^{\ell} \frac{(-m)_k(-\ell)_k}{(n)_k k!} z^k \left(\psi(1+m-k) + \psi(1+\ell-k) - \psi(1+k) - \psi(n+k)\right)$$
$$+ (-1)^\ell \ell! \sum_{k=\ell+1}^{m} \frac{(k-1-\ell)!(-m)_k}{(n)_k k!} z^k, \quad a = -m, m = 0, 1, 2, \ldots; b = -\ell, \ell = 0, 1, 2, \ldots, m.$$

Moreover, in (15.10.9) and (15.10.10) the symbols a and b are interchangeable.

(c) If c equals $2-n = 0, -1, -2, \ldots,$ then fundamental solutions in the neighborhood of $z = 0$ are given by z^{n-1} times those in (a) and (b) with a and b replaced by $a+n-1$ and $b+n-1$, respectively.

(d) If $a + b + 1 - c$ equals $n = 1, 2, 3, \ldots,$ or $2 - n = 0, -1, -2, \ldots,$ then fundamental solutions in the neighborhood of $z = 1$ are given by those in (a), (b), and (c) with z replaced by $1 - z$.

(e) Finally, if $a - b + 1$ equals $n = 1, 2, 3, \ldots,$ or $2-n = 0, -1, -2, \ldots,$ then fundamental solutions in the neighborhood of $z = \infty$ are given by z^{-a} times those in (a), (b), and (c) with b and z replaced by $a - c + 1$ and $1/z$, respectively.

15.10(ii) Kummer's 24 Solutions and Connection Formulas

The three pairs of fundamental solutions given by (15.10.2), (15.10.4), and (15.10.6) can be transformed into 18 other solutions by means of (15.8.1), leading to a total of 24 solutions known as *Kummer's solutions*. See http://dlmf.nist.gov/15.10.ii for Kummer's solutions and their connection formulas.

15.11 Riemann's Differential Equation

15.11(i) Equations with Three Singularities

The importance of (15.10.1) is that any homogeneous linear differential equation of the second order with at most three distinct singularities, all regular, in the extended plane can be transformed into (15.10.1). The most general form is given by

$$15.11.1 \quad \frac{d^2w}{dz^2} + \left(\frac{1-a_1-a_2}{z-\alpha} + \frac{1-b_1-b_2}{z-\beta} + \frac{1-c_1-c_2}{z-\gamma}\right)\frac{dw}{dz} + \left(\frac{(\alpha-\beta)(\alpha-\gamma)a_1a_2}{z-\alpha} + \frac{(\beta-\alpha)(\beta-\gamma)b_1b_2}{z-\beta} + \frac{(\gamma-\alpha)(\gamma-\beta)c_1c_2}{z-\gamma}\right)\frac{w}{(z-\alpha)(z-\beta)(z-\gamma)} = 0,$$

with

$$15.11.2 \quad a_1 + a_2 + b_1 + b_2 + c_1 + c_2 = 1.$$

Here $\{a_1, a_2\}$, $\{b_1, b_2\}$, $\{c_1, c_2\}$ are the exponent pairs at the points α, β, γ, respectively. Cases in which there are fewer than three singularities are included automatically by allowing the choice $\{0, 1\}$ for exponent pairs. Also, if any of α, β, γ, is at infinity, then we take the corresponding limit in (15.11.1).

The complete set of solutions of (15.11.1) is denoted by *Riemann's P-symbol*:

$$15.11.3 \quad w = P\begin{Bmatrix} \alpha & \beta & \gamma & \\ a_1 & b_1 & c_1 & z \\ a_2 & b_2 & c_2 & \end{Bmatrix}.$$

In particular,

$$15.11.4 \quad w = P\begin{Bmatrix} 0 & 1 & \infty & \\ 0 & 0 & a & z \\ 1-c & c-a-b & b & \end{Bmatrix}$$

denotes the set of solutions of (15.10.1).

15.11(ii) Transformation Formulas

A conformal mapping of the extended complex plane onto itself has the form

$$15.11.5 \quad t = (\kappa z + \lambda)/(\mu z + \nu),$$

where $\kappa, \lambda, \mu, \nu$ are real or complex constants such that $\kappa\nu - \lambda\mu = 1$. These constants can be chosen to map any two sets of three distinct points $\{\alpha, \beta, \gamma\}$ and $\{\tilde{\alpha}, \tilde{\beta}, \tilde{\gamma}\}$ onto each other. Symbolically:

$$15.11.6 \quad P\begin{Bmatrix} \alpha & \beta & \gamma & \\ a_1 & b_1 & c_1 & z \\ a_2 & b_2 & c_2 & \end{Bmatrix} = P\begin{Bmatrix} \tilde{\alpha} & \tilde{\beta} & \tilde{\gamma} & \\ a_1 & b_1 & c_1 & t \\ a_2 & b_2 & c_2 & \end{Bmatrix}.$$

The reduction of a general homogeneous linear differential equation of the second order with at most three regular singularities to the hypergeometric differential equation is given by

$$15.11.7 \quad P\begin{Bmatrix} \alpha & \beta & \gamma & \\ a_1 & b_1 & c_1 & z \\ a_2 & b_2 & c_2 & \end{Bmatrix} = \left(\frac{z-\alpha}{z-\gamma}\right)^{a_1}\left(\frac{z-\beta}{z-\gamma}\right)^{b_1} P\begin{Bmatrix} 0 & 1 & \infty & \\ 0 & 0 & a_1+b_1+c_1 & \frac{(z-\alpha)(\beta-\gamma)}{(z-\gamma)(\beta-\alpha)} \\ a_2-a_1 & b_2-b_1 & a_1+b_1+c_2 & \end{Bmatrix}.$$

We also have

$$15.11.8 \quad z^\lambda(1-z)^\mu P\begin{Bmatrix} 0 & 1 & \infty & \\ a_1 & b_1 & c_1 & z \\ a_2 & b_2 & c_2 & \end{Bmatrix} = P\begin{Bmatrix} 0 & 1 & \infty & \\ a_1+\lambda & b_1+\mu & c_1-\lambda-\mu & z \\ a_2+\lambda & b_2+\mu & c_2-\lambda-\mu & \end{Bmatrix},$$

for arbitrary λ and μ.

15.12 Asymptotic Approximations

15.12(i) Large Variable

For the asymptotic behavior of $\mathbf{F}(a,b;c;z)$ as $z \to \infty$ with a, b, c fixed, combine (15.2.2) with (15.8.2) or (15.8.8).

15.12(ii) Large c

Let δ denote an arbitrary small positive constant. Also let a, b, z be real or complex and fixed, and at least one of the following conditions be satisfied:

(a) a and/or $b \in \{0, -1, -2, \ldots\}$.

(b) $\Re z < \frac{1}{2}$ and $|c+n| \geq \delta$ for all $n \in \{0, 1, 2, \ldots\}$.

15.12 ASYMPTOTIC APPROXIMATIONS

(c) $\Re z = \frac{1}{2}$ and $|\operatorname{ph} c| \le \pi - \delta$.

(d) $\Re z > \frac{1}{2}$ and $\alpha_- - \frac{1}{2}\pi + \delta \le \operatorname{ph} c \le \alpha_+ + \frac{1}{2}\pi - \delta$, where

15.12.1
$$\alpha_\pm = \arctan\left(\frac{\operatorname{ph} z - \operatorname{ph}(1-z) \mp \pi}{\ln|1 - z^{-1}|}\right),$$

with z restricted so that $\pm\alpha_\pm \in [0, \frac{1}{2}\pi)$.

Then for fixed $m \in \{0, 1, 2, \dots\}$,

15.12.2
$$F(a, b; c; z) = \sum_{s=0}^{m-1} \frac{(a)_s (b)_s}{(c)_s s!} z^s + O(c^{-m}), \quad |c| \to \infty.$$

Similar results for other sectors are given in Wagner (1988). For the more general case in which $a^2 = o(c)$ and $b^2 = o(c)$ see Wagner (1990).

15.12(iii) Other Large Parameters

Again, throughout this subsection δ denotes an arbitrary small positive constant, and a, b, c, z are real or complex and fixed.

As $\lambda \to \infty$,

15.12.3
$$F\left(\begin{matrix} a, b \\ c + \lambda \end{matrix}; z\right) \sim \frac{\Gamma(c + \lambda)}{\Gamma(c - b + \lambda)} \sum_{s=0}^\infty q_s(z)(b)_s \lambda^{-s-b},$$

where $q_0(z) = 1$ and $q_s(z)$, $s = 1, 2, \dots$, are defined by the generating function

15.12.4
$$\left(\frac{e^t - 1}{t}\right)^{b-1} e^{t(1-c)} (1 - z + ze^{-t})^{-a} = \sum_{s=0}^\infty q_s(z) t^s.$$

If $|\operatorname{ph}(1-z)| < \pi$, then (15.12.3) applies when $|\operatorname{ph} \lambda| \le \frac{1}{2}\pi - \delta$. If $\Re z \le \frac{1}{2}$, then (15.12.3) applies when $|\operatorname{ph} \lambda| \le \pi - \delta$.

If $|\operatorname{ph}(z-1)| < \pi$, then as $\lambda \to \infty$ with $|\operatorname{ph} \lambda| \le \pi - \delta$,

15.12.5
$$\mathbf{F}\left(\begin{matrix} a + \lambda, b - \lambda \\ c \end{matrix}; \frac{1}{2} - \frac{1}{2}z\right)$$
$$= 2^{(a+b-1)/2} \frac{(z+1)^{(c-a-b-1)/2}}{(z-1)^{c/2}} \sqrt{\zeta \sinh \zeta} \left(\lambda + \tfrac{1}{2}a - \tfrac{1}{2}b\right)^{1-c} \left(I_{c-1}\big((\lambda + \tfrac{1}{2}a - \tfrac{1}{2}b)\zeta\big)(1 + O(\lambda^{-2}))\right.$$
$$\left. + \frac{I_{c-2}\big((\lambda + \tfrac{1}{2}a - \tfrac{1}{2}b)\zeta\big)}{2\lambda + a - b} \left(\big(c - \tfrac{1}{2}\big)\big(c - \tfrac{3}{2}\big)\left(\frac{1}{\zeta} - \coth \zeta\right) + \tfrac{1}{2}(2c - a - b - 1)(a + b - 1)\tanh\big(\tfrac{1}{2}\zeta\big) + O(\lambda^{-2})\right)\right),$$

where

15.12.6
$$\zeta = \operatorname{arccosh} z.$$

For $I_\nu(z)$ see §10.25(ii). For this result and an extension to an asymptotic expansion with error bounds see Jones (2001).

See also Dunster (1999) where the asymptotics of Jacobi polynomials is described; compare (15.9.1).

If $|\operatorname{ph} z| < \pi$, then as $\lambda \to \infty$ with $|\operatorname{ph} \lambda| \le \pi - \delta$,

15.12.7
$$F\left(\begin{matrix} a, b - \lambda \\ c + \lambda \end{matrix}; -z\right)$$
$$= 2^{b-c+(1/2)} \left(\frac{z+1}{2\sqrt{z}}\right)^\lambda \left(\lambda^{a/2} U\big(a - \tfrac{1}{2}, -\alpha\sqrt{\lambda}\big)\left((1+z)^{c-a-b} z^{1-c}\left(\frac{\alpha}{z-1}\right)^{1-a} + O(\lambda^{-1})\right)\right.$$
$$\left. + \frac{\lambda^{(a-1)/2}}{\alpha} U\big(a - \tfrac{3}{2}, -\alpha\sqrt{\lambda}\big)\left((1+z)^{c-a-b} z^{1-c}\left(\frac{\alpha}{z-1}\right)^{1-a} - 2^{c-b-(1/2)}\left(\frac{\alpha}{z-1}\right)^a + O(\lambda^{-1})\right)\right),$$

where

15.12.8
$$\alpha = \left(-2\ln\left(1 - \left(\frac{z-1}{z+1}\right)^2\right)\right)^{1/2},$$

with the branch chosen to be continuous and $\Re \alpha > 0$ when $\Re((z-1)/(z+1)) > 0$. For $U(a, z)$ see §12.2, and for an extension to an asymptotic expansion see Olde Daalhuis (2003a).

If $|\operatorname{ph} z| < \pi$, then as $\lambda \to \infty$ with $|\operatorname{ph} \lambda| \le \frac{1}{2}\pi - \delta$,

15.12.9
$$(z+1)^{3\lambda/2}(2\lambda)^{c-1} \mathbf{F}\!\left(\begin{matrix} a+\lambda, b+2\lambda \\ c \end{matrix}; -z\right)$$
$$= \lambda^{-1/3}\left(e^{\pi i(a-c+\lambda+(1/3))}\operatorname{Ai}\!\left(e^{-2\pi i/3}\lambda^{2/3}\beta^2\right) + e^{\pi i(c-a-\lambda-(1/3))}\operatorname{Ai}\!\left(e^{2\pi i/3}\lambda^{2/3}\beta^2\right)\right)\!\left(a_0(\zeta) + O(\lambda^{-1})\right)$$
$$+ \lambda^{-2/3}\left(e^{\pi i(a-c+\lambda+(2/3))}\operatorname{Ai}'\!\left(e^{-2\pi i/3}\lambda^{2/3}\beta^2\right) + e^{\pi i(c-a-\lambda-(2/3))}\operatorname{Ai}'\!\left(e^{2\pi i/3}\lambda^{2/3}\beta^2\right)\right)\!\left(a_1(\zeta) + O(\lambda^{-1})\right),$$

where

15.12.10
$$\zeta = \operatorname{arccosh}\!\left(\tfrac{1}{4}z - 1\right),$$

15.12.11
$$\beta = \left(-\frac{3}{2}\zeta + \frac{9}{4}\ln\!\left(\frac{2+e^{\zeta}}{2+e^{-\zeta}}\right)\right)^{1/3},$$

with the branch chosen to be continuous and $\beta > 0$ when $\zeta > 0$. Also,

15.12.12
$$a_0(\zeta) = \tfrac{1}{2}G_0(\beta) + \tfrac{1}{2}G_0(-\beta), \quad a_1(\zeta) = \left(\tfrac{1}{2}G_0(\beta) - \tfrac{1}{2}G_0(-\beta)\right)/\beta,$$

where

15.12.13
$$G_0(\pm\beta) = \left(2+e^{\pm\zeta}\right)^{c-b-(1/2)}\left(1+e^{\pm\zeta}\right)^{a-c+(1/2)}\left(z-1-e^{\pm\zeta}\right)^{-a+(1/2)}\sqrt{\frac{\beta}{e^{\zeta}-e^{-\zeta}}}.$$

For $\operatorname{Ai}(z)$ see §9.2, and for further information and an extension to an asymptotic expansion see Olde Daalhuis (2003b). (Two errors in this reference are corrected in (15.12.9).)

By combination of the foregoing results of this subsection with the linear transformations of §15.8(i) and the connection formulas of §15.10(ii), similar asymptotic approximations for $F(a+e_1\lambda, b+e_2\lambda; c+e_3\lambda; z)$ can be obtained with $e_j = \pm 1$ or 0, $j=1,2,3$. For more details see Olde Daalhuis (2010). For other extensions, see Wagner (1986) and Temme (2003).

15.13 Zeros

Let $N(a,b,c)$ denote the number of zeros of $F(a,b;c;z)$ in the sector $|\operatorname{ph}(1-z)| < \pi$. If a,b,c are real, $a,b,c,c-a,c-b \ne 0,-1,-2,\ldots$, and, without loss of generality, $b \ge a$, $c \ge a+b$ (compare (15.8.1)), then

15.13.1
$$N(a,b,c) = \begin{cases} 0, & a > 0, \\ \lfloor -a \rfloor + \tfrac{1}{2}(1+S), & a < 0, c-a > 0, \\ \lfloor -a \rfloor + \tfrac{1}{2}(1+S) + \lfloor a-c+1 \rfloor S, & a < 0, c-a < 0, \end{cases}$$

where $S = \operatorname{sign}(\Gamma(a)\,\Gamma(b)\,\Gamma(c-a)\,\Gamma(c-b))$.

If a, b, c, $c-a$, or $c-b \in \{0,-1,-2,\ldots\}$, then $F(a,b;c;z)$ is not defined, or reduces to a polynomial, or reduces to $(1-z)^{c-a-b}$ times a polynomial.

For further information on the location of real zeros see Zarzo et al. (1995). A small table of zeros is given in Conde and Kalla (1981).

15.14 Integrals

The Mellin transform of the hypergeometric function of negative argument is given by

15.14.1
$$\int_0^{\infty} x^{s-1}\mathbf{F}\!\left(\begin{matrix} a,b \\ c \end{matrix}; -x\right) dx = \frac{\Gamma(s)\,\Gamma(a-s)\,\Gamma(b-s)}{\Gamma(a)\,\Gamma(b)\,\Gamma(c-s)},$$
$$\min(\Re a, \Re b) > \Re s > 0.$$

Integrals of the form $\int x^{\alpha}(x+t)^{\beta} F(a,b;c;x)\,dx$ and more complicated forms are given in Apelblat (1983, pp. 370–387), Prudnikov et al. (1990, §§1.15 and 2.21), and Gradshteyn and Ryzhik (2000, §7.5).

Fourier transforms of hypergeometric functions are given in Erdélyi et al. (1954a, §§1.14 and 2.14). Laplace transforms of hypergeometric functions are given in Erdélyi et al. (1954a, §4.21), Oberhettinger and Badii (1973, §1.19), and Prudnikov et al. (1992a, §3.37). Inverse Laplace transforms of hypergeometric functions are given in Erdélyi et al. (1954a, §5.19), Oberhettinger and Badii (1973, §2.18), and Prudnikov et al. (1992b, §3.35). Mellin transforms of hypergeometric functions are given in Erdélyi et al. (1954a, §6.9), Oberhettinger (1974, §1.15), and Marichev (1983, pp. 288–299). Inverse Mellin transforms are given in Erdélyi et al. (1954a, §7.5). Hankel transforms of hypergeometric functions are given in Oberhettinger (1972, §1.17) and Erdélyi et al. (1954b, §8.17).

15.15 Sums

15.15.1
$$\mathbf{F}\left(\begin{matrix}a,b\\c\end{matrix};\frac{1}{z}\right) = \left(1-\frac{z_0}{z}\right)^{-a} \sum_{s=0}^{\infty} \frac{(a)_s}{s!} \times \mathbf{F}\left(\begin{matrix}-s,b\\c\end{matrix};\frac{1}{z_0}\right)\left(1-\frac{z}{z_0}\right)^{-s}.$$

Here z_0 ($\neq 0$) is an arbitrary complex constant and the expansion converges when $|z - z_0| > \max(|z_0|, |z_0 - 1|)$. For further information see Bühring (1987a) and Kalla (1992).

For other integral transforms see Erdélyi et al. (1954b), Prudnikov et al. (1992b, §4.3.43), and also §15.9(ii).

For compendia of finite sums and infinite series involving hypergeometric functions see Prudnikov et al. (1990, §§5.3 and 6.7) and Hansen (1975).

15.16 Products

15.16.1
$$F\left(\begin{matrix}a,b\\c-\frac{1}{2}\end{matrix};z\right) F\left(\begin{matrix}c-a,c-b\\c+\frac{1}{2}\end{matrix};z\right) = \sum_{s=0}^{\infty} \frac{(c)_s}{(c+\frac{1}{2})_s} A_s z^s, \qquad |z| < 1,$$

where $A_0 = 1$ and A_s, $s = 1, 2, \ldots$, are defined by the generating function

15.16.2
$$(1-z)^{a+b-c} F(2a, 2b; 2c-1; z) = \sum_{s=0}^{\infty} A_s z^s, \quad |z| < 1.$$

Also,

15.16.3
$$F\left(\begin{matrix}a,b\\c\end{matrix};z\right) F\left(\begin{matrix}a,b\\c\end{matrix};\zeta\right) = \sum_{s=0}^{\infty} \frac{(a)_s(b)_s(c-a)_s(c-b)_s}{(c)_s(c)_{2s}s!} (z\zeta)^s F\left(\begin{matrix}a+s,b+s\\c+2s\end{matrix};z+\zeta-z\zeta\right),$$
$$|z| < 1, |\zeta| < 1, |z+\zeta-z\zeta| < 1.$$

15.16.4
$$F\left(\begin{matrix}a,b\\c\end{matrix};z\right) F\left(\begin{matrix}-a,-b\\-c\end{matrix};z\right) + \frac{ab(a-c)(b-c)}{c^2(1-c^2)}z^2 F\left(\begin{matrix}1+a,1+b\\2+c\end{matrix};z\right) F\left(\begin{matrix}1-a,1-b\\2-c\end{matrix};z\right) = 1.$$

Generalized Legendre's Relation

15.16.5
$$F\left(\begin{matrix}\frac{1}{2}+\lambda,-\frac{1}{2}-\nu\\1+\lambda+\mu\end{matrix};z\right) F\left(\begin{matrix}\frac{1}{2}-\lambda,\frac{1}{2}+\nu\\1+\nu+\mu\end{matrix};1-z\right) + F\left(\begin{matrix}\frac{1}{2}+\lambda,\frac{1}{2}-\nu\\1+\lambda+\mu\end{matrix};z\right) F\left(\begin{matrix}-\frac{1}{2}-\lambda,\frac{1}{2}+\nu\\1+\nu+\mu\end{matrix};1-z\right)$$
$$- F\left(\begin{matrix}\frac{1}{2}+\lambda,\frac{1}{2}-\nu\\1+\lambda+\mu\end{matrix};z\right) F\left(\begin{matrix}\frac{1}{2}-\lambda,\frac{1}{2}+\nu\\1+\nu+\mu\end{matrix};1-z\right) = \frac{\Gamma(1+\lambda+\mu)\Gamma(1+\nu+\mu)}{\Gamma(\lambda+\mu+\nu+\frac{3}{2})\Gamma(\frac{1}{2}+\nu)},$$
$$|\operatorname{ph} z| < \pi, |\operatorname{ph}(1-z)| < \pi.$$

For further results of this kind, and also series of products of hypergeometric functions, see Erdélyi et al. (1953a, §2.5.2).

Applications

15.17 Mathematical Applications

15.17(i) Differential Equations

This topic is treated in §§15.10 and 15.11.

The logarithmic derivatives of some hypergeometric functions for which quadratic transformations exist (§15.8(iii)) are solutions of Painlevé equations. See §32.10(vi).

15.17(ii) Conformal Mappings

The quotient of two solutions of (15.10.1) maps the closed upper half-plane $\Im z \geq 0$ conformally onto a curvilinear triangle. See Klein (1894) and Hochstadt (1971). Hypergeometric functions, especially complete elliptic integrals, also play an important role in quasi-conformal mapping. See Anderson et al. (1997).

15.17(iii) Group Representations

For harmonic analysis it is more natural to represent hypergeometric functions as a Jacobi function (§15.9(ii)). For special values of α and β there are many group-theoretic interpretations. First, as spherical functions on noncompact Riemannian symmetric spaces of rank one, but also as associated spherical functions, intertwining functions, matrix elements of $SL(2,\mathbb{R})$, and spherical functions on certain nonsymmetric Gelfand pairs. Harmonic analysis can be developed for the Jacobi transform either as a generalization of the Fourier-cosine transform (§1.14(ii)) or as a specialization of a group Fourier transform. For further information see Koornwinder (1984a).

15.17(iv) Combinatorics

In combinatorics, hypergeometric identities classify single sums of products of binomial coefficients. See Egorychev (1984, §2.3).

Quadratic transformations give insight into the relation of elliptic integrals to the arithmetic-geometric mean (§19.22(ii)). See Andrews et al. (1999, §3.2).

15.17(v) Monodromy Groups

The three singular points in Riemann's differential equation (15.11.1) lead to an interesting Riemann sheet structure. By considering, as a group, all analytic transformations of a basis of solutions under analytic continuation around all paths on the Riemann sheet, we obtain the monodromy group. These monodromy groups are finite iff the solutions of Riemann's differential equation are all algebraic. For a survey of this topic see Gray (2000).

15.18 Physical Applications

The hypergeometric function has allowed the development of "solvable" models for one-dimensional quantum scattering through and over barriers (Eckart (1930), Bhattacharjie and Sudarshan (1962)), and generalized to include position-dependent effective masses (Dekar et al. (1999)).

More varied applications include photon scattering from atoms (Gavrila (1967)), energy distributions of particles in plasmas (Mace and Hellberg (1995)), conformal field theory of critical phenomena (Burkhardt and Xue (1991)), quantum chromo-dynamics (Atkinson and Johnson (1988)), and general parametrization of the effective potentials of interaction between atoms in diatomic molecules (Herrick and O'Connor (1998)).

Computation

15.19 Methods of Computation

15.19(i) Maclaurin Expansions

The Gauss series (15.2.1) converges for $|z| < 1$. For $z \in \mathbb{R}$ it is always possible to apply one of the linear transformations in §15.8(i) in such a way that the hypergeometric function is expressed in terms of hypergeometric functions with an argument in the interval $[0, \frac{1}{2}]$.

For $z \in \mathbb{C}$ it is possible to use the linear transformations in such a way that the new arguments lie within the unit circle, except when $z = e^{\pm \pi i/3}$. This is because the linear transformations map the pair $\{e^{\pi i/3}, e^{-\pi i/3}\}$ onto itself. However, by appropriate choice of the constant z_0 in (15.15.1) we can obtain an infinite series that converges on a disk containing $z = e^{\pm \pi i/3}$. Moreover, it is also possible to accelerate convergence by appropriate choice of z_0.

Large values of $|a|$ or $|b|$, for example, delay convergence of the Gauss series, and may also lead to severe cancellation.

For further information see Bühring (1987a), Forrey (1997), and Kalla (1992).

15.19(ii) Differential Equation

A comprehensive and powerful approach is to integrate the hypergeometric differential equation (15.10.1) by direct numerical methods. As noted in §3.7(ii), the integration path should be chosen so that the wanted solution grows in magnitude at least as fast as all other solutions. However, since the growth near the singularities of the differential equation is algebraic rather than exponential, the resulting instabilities in the numerical integration might be tolerable in some cases.

15.19(iii) Integral Representations

The representation (15.6.1) can be used to compute the hypergeometric function in the sector $|\mathrm{ph}(1-z)| < \pi$. Gauss quadrature approximations are discussed in Gautschi (2002b).

15.19(iv) Recurrence Relations

The relations in §15.5(ii) can be used to compute $F(a, b; c; z)$, provided that care is taken to apply these relations in a stable manner; see §3.6(ii). Initial values for moderate values of $|a|$ and $|b|$ can be obtained by the methods of §15.19(i), and for large values of $|a|$, $|b|$, or $|c|$ via the asymptotic expansions of §§15.12(ii) and 15.12(iii).

For example, in the half-plane $\Re z \leq \frac{1}{2}$ we can use (15.12.2) or (15.12.3) to compute $F(a, b; c + N + 1; z)$ and $F(a, b; c + N; z)$, where N is a large positive integer, and then apply (15.5.18) in the backward direction. When $\Re z > \frac{1}{2}$ it is better to begin with one of the linear transformations (15.8.4), (15.8.7), or (15.8.8). For further information see Gil et al. (2006a, 2007b).

15.20 Software

See http://dlmf.nist.gov/15.20.

References

General References

The main references used in writing this chapter are Andrews et al. (1999) and Temme (1996a). For additional bibliographic reading see Erdélyi et al. (1953a), Hochstadt (1971), Luke (1969a), Olver (1997b), Slater (1966), Wang and Guo (1989), and Whittaker and Watson (1927).

Sources

The following list gives the references or other indications of proofs that were used in constructing the various sections of this chapter. These sources supplement the references that are quoted in the text.

§15.2 Andrews et al. (1999, §2.1), Olver (1997b, Chapter 5, Theorem 9.1), Temme (1996a, §5.1). (15.2.3) is a consequence of (15.8.4).

§15.3 These graphics were produced at NIST.

§15.4 Andrews et al. (1999, §2.2). For (15.4.1)–(15.4.6) see Temme (1996a, §5.1). For (15.4.7) use (15.8.27), (15.4.6), and (5.5.5). For (15.4.9) use (15.8.28), (15.4.6), and (5.5.5). For (15.4.11) use (15.8.1) and (15.4.7). For (15.4.13) use (15.8.1) and (15.4.11). For (15.4.15) use (15.8.1) and (15.4.9). For (15.4.17) and (15.4.18) use (15.8.15) and (15.4.6). For (15.4.19) use (15.5.11) and (15.4.6). For (15.4.21) use (15.8.10). For (15.4.22) and (15.4.23) use (15.8.4), (5.5.3), and (5.5.1). For (15.4.25) see Dougall (1907) or Andrews et al. (1999, §2.8). For (15.4.26) use (15.8.24) and (5.5.5). For (15.4.27) use (15.2.1) and (5.7.7). For (15.4.28) use (15.8.1), (15.4.26), and (5.5.5). For (15.4.29) use (15.8.1), (15.5.15), (15.4.26), (5.5.5), and (5.5.1). For (15.4.30) use (15.8.1) and (15.4.28). For (15.4.31) use (15.8.1), Erdélyi et al. (1953a, Eq. (2.11.41)), (15.4.20), and (5.5.5). For (15.4.32) use (15.8.30), (15.4.28), (5.5.5), and (5.5.6). (Note that Erdélyi et al. (1953a, Eq. (2.8.54)) contains an error.) For (15.4.33) let $z \to -\infty$ in (15.8.32) and use (5.5.5).

§15.5 Andrews et al. (1999, §2.5). For (15.5.2)–(15.5.9) use induction. For (15.5.10) see Fleury and Turbiner (1994).

§15.6 Andrews et al. (1999, §§2.2, 2.4, 2.9), Erdélyi et al. (1953a, §2.1.3), and Whittaker and Watson (1927, pp. 290–291). (15.6.3) follows from (15.6.4).

§15.7 Andrews et al. (1999, pp. 94, 97–98, and Ex. 26 on p. 119), Lorentzen and Waadeland (1992, §6.1), and Berndt (1989, pp. 134–137). These references contain several restrictions on the parameters a, b, and c. This is because they use the function $F(a,b;c;z)$. No restrictions are needed for $\mathbf{F}(a,b;c;z)$.

§15.8 For (15.8.1)–(15.8.4) see Olver (1997b, Chapter 5, §10). For (15.8.5) combine (15.8.1) and (15.8.2). (15.8.6) and (15.8.7) are obtained as limits of (15.8.2)–(15.8.5) as $a \to -m$, together with (5.5.3). For (15.8.8) and (15.8.10) see Erdélyi et al. (1953a, §§2.1.4 and 2.3.1). (15.8.9) and (15.8.11) follow from (15.8.10) and (15.8.8), respectively, via (15.8.1). For §15.8(iii) see Andrews et al. (1999, §3.1). For (15.8.31) and (15.8.32) see Goursat (1881, Eq. (110)) and Watson (1910). The version of (15.8.31) given in Erdélyi et al. (1953a, p. 114 (40)) contains a typographical error. For (15.8.33) see Chan (1998).

§15.9 For (15.9.1)–(15.9.10) see §§18.5(ii) and 18.20(ii). For (15.9.15) see Erdélyi et al. (1953a, §§3.15.1 and 3.15.2).

§15.10 Andrews et al. (1999, §2.3), Olver (1997b, pp. 163–168), and Luke (1969a, Chapter III). (15.10.9) is a corrected version of (2.3.20) in the first reference.

§15.11 Andrews et al. (1999, §2.3) or Olver (1997b, pp. 156–158).

§15.12 For (15.12.2) see Wagner (1988). The region of validity given in Luke (1969a, p. 235) is incorrect. For (15.12.3) see Luke (1969a, §7.2) and Olver (1997b, p. 162). The sector of validity given in the first reference is incorrect. See the third footnote in the second reference.

§15.13 Runckel (1971).

§15.14 Andrews et al. (1999, §2.4).

§15.16 Burchnall and Chaundy (1940, 1948) and Elliott (1903). For (15.16.4) use (15.8.2) and (15.8.4), combined with (15.8.1) to show that the left-hand side of (15.16.4) is an entire function of z. Then apply Liouville's theorem (1.9(iii)).

Chapter 16

Generalized Hypergeometric Functions and Meijer G-Function

R. A. Askey[1] and A. B. Olde Daalhuis[2]

Notation   **404**
- 16.1   Special Notation 404

Generalized Hypergeometric Functions   **404**
- 16.2   Definition and Analytic Properties 404
- 16.3   Derivatives and Contiguous Functions . . 405
- 16.4   Argument Unity 405
- 16.5   Integral Representations and Integrals . . 408
- 16.6   Transformations of Variable 408
- 16.7   Relations to Other Functions 409
- 16.8   Differential Equations 409
- 16.9   Zeros 410
- 16.10   Expansions in Series of ${}_pF_q$ Functions . . 410
- 16.11   Asymptotic Expansions 411
- 16.12   Products 412

Two-Variable Hypergeometric Functions   **412**
- 16.13   Appell Functions 412
- 16.14   Partial Differential Equations 413
- 16.15   Integral Representations and Integrals . . 414
- 16.16   Transformations of Variables 414

Meijer G-Function   **415**
- 16.17   Definition 415
- 16.18   Special Cases 416
- 16.19   Identities 416
- 16.20   Integrals and Series 416
- 16.21   Differential Equation 417
- 16.22   Asymptotic Expansions 417

Applications   **417**
- 16.23   Mathematical Applications 417
- 16.24   Physical Applications 417

Computation   **418**
- 16.25   Methods of Computation 418
- 16.26   Approximations 418
- 16.27   Software 418

References   **418**

---

[1]Department of Mathematics, University of Wisconsin, Madison, Wisconsin.
[2]School of Mathematics, Edinburgh University, Edinburgh, United Kingdom.
Acknowledgments: The authors are pleased to acknowledge the assistance of B. L. J. Braaksma with §§16.5 and 16.11.
Copyright © 2009 National Institute of Standards and Technology. All rights reserved.

Notation

16.1 Special Notation

(For other notation see pp. xiv and 873.)

p, q	nonnegative integers.
k, n	nonnegative integers, unless stated otherwise.
z	complex variable.
$a_1, a_2, \ldots, a_p$ $b_1, b_2, \ldots, b_q$	real or complex parameters.
δ	arbitrary small positive constant.
$\mathbf{a}$	vector $(a_1, a_2, \ldots, a_p)$.
$\mathbf{b}$	vector $(b_1, b_2, \ldots, b_q)$.
$(\mathbf{a})_k$	$(a_1)_k (a_2)_k \cdots (a_p)_k$.
$(\mathbf{b})_k$	$(b_1)_k (b_2)_k \cdots (b_q)_k$.
D	d/dz.
ϑ	$z\, d/dz$.

The main functions treated in this chapter are the generalized hypergeometric function ${}_pF_q\!\left({a_1,\ldots,a_p \atop b_1,\ldots,b_q};z\right)$, the Appell (two-variable hypergeometric) functions $F_1(\alpha;\beta,\beta';\gamma;x,y)$, $F_2(\alpha;\beta,\beta';\gamma,\gamma';x,y)$, $F_3(\alpha,\alpha';\beta,\beta';\gamma;x,y)$, $F_4(\alpha;\beta;\gamma,\gamma';x,y)$, and the Meijer G-function $G_{p,q}^{m,n}\!\left(z;{a_1,\ldots,a_p \atop b_1,\ldots,b_q}\right)$. Alternative notations are ${}_pF_q\!\left({\mathbf{a} \atop \mathbf{b}};z\right)$, ${}_pF_q(a_1,\ldots,a_p;b_1,\ldots,b_q;z)$, and ${}_pF_q(\mathbf{a};\mathbf{b};z)$ for the generalized hypergeometric function, $F_1(\alpha,\beta,\beta';\gamma;x,y)$, $F_2(\alpha,\beta,\beta';\gamma,\gamma';x,y)$, $F_3(\alpha,\alpha',\beta,\beta';\gamma;x,y)$, $F_4(\alpha,\beta;\gamma,\gamma';x,y)$, for the Appell functions, and $G_{p,q}^{m,n}(z;\mathbf{a};\mathbf{b})$ for the Meijer G-function.

Generalized Hypergeometric Functions

16.2 Definition and Analytic Properties

16.2(i) Generalized Hypergeometric Series

Throughout this chapter it is assumed that none of the bottom parameters $b_1, b_2, \ldots, b_q$ is a nonpositive integer, *unless stated otherwise*. Then formally

16.2.1 $\quad {}_pF_q\!\left({a_1,\ldots,a_p \atop b_1,\ldots,b_q};z\right) = \sum_{k=0}^{\infty} \frac{(a_1)_k \cdots (a_p)_k}{(b_1)_k \cdots (b_q)_k} \frac{z^k}{k!}.$

Equivalently, the function is denoted by ${}_pF_q\!\left({\mathbf{a} \atop \mathbf{b}};z\right)$ or ${}_pF_q(\mathbf{a};\mathbf{b};z)$, and sometimes, for brevity, by ${}_pF_q(z)$.

16.2(ii) Case $p \leq q$

When $p \leq q$ the series (16.2.1) converges for all finite values of z and defines an entire function.

16.2(iii) Case $p = q + 1$

Suppose first one or more of the top parameters a_j is a nonpositive integer. Then the series (16.2.1) terminates and the generalized hypergeometric function is a polynomial in z.

If none of the a_j is a nonpositive integer, then the radius of convergence of the series (16.2.1) is 1, and outside the open disk $|z| < 1$ the generalized hypergeometric function is defined by analytic continuation with respect to z. The branch obtained by introducing a cut from 1 to $+\infty$ on the real axis, that is, the branch in the sector $|\mathrm{ph}(1-z)| \leq \pi$, is the *principal branch* (or *principal value*) of ${}_{q+1}F_q(\mathbf{a};\mathbf{b};z)$; compare §4.2(i). Elsewhere the generalized hypergeometric function is a multivalued function that is analytic except for possible branch points at $z = 0, 1$, and ∞. *Unless indicated otherwise* it is assumed that in this Handbook generalized hypergeometric functions assume their principal values.

On the circle $|z| = 1$ the series (16.2.1) is absolutely convergent if $\Re\gamma_q > 0$, convergent except at $z = 1$ if $-1 < \Re\gamma_q \leq 0$, and divergent if $\Re\gamma_q \leq -1$, where

16.2.2 $\quad \gamma_q = (b_1 + \cdots + b_q) - (a_1 + \cdots + a_{q+1}).$

16.2(iv) Case $p > q + 1$

Polynomials

In general the series (16.2.1) diverges for all nonzero values of z. However, when one or more of the top parameters a_j is a nonpositive integer the series terminates and the generalized hypergeometric function is a polynomial in z. Note that if $-m$ is the value of the numerically largest a_j that is a nonpositive integer, then the identity

16.2.3

$${}_{p+1}F_q\!\left({-m, \mathbf{a} \atop \mathbf{b}};z\right)$$
$$= \frac{(\mathbf{a})_m (-z)^m}{(\mathbf{b})_m} \, {}_{q+1}F_p\!\left({-m, 1-m-\mathbf{b} \atop 1-m-\mathbf{a}};\frac{(-1)^{p+q}}{z}\right)$$

can be used to interchange p and q.

Note also that any partial sum of the generalized hypergeometric series can be represented as a generalized hypergeometric function via

16.2.4

$$\sum_{k=0}^{m} \frac{(\mathbf{a})_k}{(\mathbf{b})_k} \frac{z^k}{k!}$$
$$= \frac{(\mathbf{a})_m z^m}{(\mathbf{b})_m m!} \, {}_{q+2}F_p\!\left({-m, 1, 1-m-\mathbf{b} \atop 1-m-\mathbf{a}};\frac{(-1)^{p+q+1}}{z}\right).$$

Non-Polynomials

See §16.5 for the definition of ${}_pF_q(\mathbf{a};\mathbf{b};z)$ as a contour integral when $p > q+1$ and none of the a_k is a nonpositive integer. (However, *except where indicated otherwise* in this Handbook we assume that when $p > q+1$ at least one of the a_k is a nonpositive integer.)

16.2(v) Behavior with Respect to Parameters

Let

16.2.5
$$_p\mathbf{F}_q(\mathbf{a};\mathbf{b};z) = {_pF_q}\!\left(\begin{matrix}a_1,\ldots,a_p\\b_1,\ldots,b_q\end{matrix};z\right)\!\bigg/(\Gamma(b_1)\cdots\Gamma(b_q)) = \sum_{k=0}^{\infty}\frac{(a_1)_k\cdots(a_p)_k}{\Gamma(b_1+k)\cdots\Gamma(b_q+k)}\frac{z^k}{k!};$$

compare (15.2.2) in the case $p = 2$, $q = 1$. When $p \leq q+1$ and z is fixed and not a branch point, any branch of $_p\mathbf{F}_q(\mathbf{a};\mathbf{b};z)$ is an entire function of each of the parameters $a_1,\ldots,a_p,b_1,\ldots,b_q$.

16.3 Derivatives and Contiguous Functions

16.3(i) Differentiation Formulas

16.3.1
$$\frac{d^n}{dz^n}{_pF_q}\!\left(\begin{matrix}a_1,\ldots,a_p\\b_1,\ldots,b_q\end{matrix};z\right) = \frac{(\mathbf{a})_n}{(\mathbf{b})_n}{_pF_q}\!\left(\begin{matrix}a_1+n,\ldots,a_p+n\\b_1+n,\ldots,b_q+n\end{matrix};z\right),$$

16.3.2
$$\frac{d^n}{dz^n}\!\left(z^\gamma\,{_pF_q}\!\left(\begin{matrix}a_1,\ldots,a_p\\b_1,\ldots,b_q\end{matrix};z\right)\right) = (\gamma-n+1)_n z^{\gamma-n}\,{_{p+1}F_{q+1}}\!\left(\begin{matrix}\gamma+1,a_1,\ldots,a_p\\\gamma+1-n,b_1,\ldots,b_q\end{matrix};z\right),$$

16.3.3
$$\left(z\frac{d}{dz}z\right)^{\!n}\!\left(z^{\gamma-1}\,{_{p+1}F_q}\!\left(\begin{matrix}\gamma,a_1,\ldots,a_p\\b_1,\ldots,b_q\end{matrix};z\right)\right) = (\gamma)_n z^{\gamma+n-1}\,{_{p+1}F_q}\!\left(\begin{matrix}\gamma+n,a_1,\ldots,a_p\\b_1,\ldots,b_q\end{matrix};z\right),$$

16.3.4
$$\frac{d^n}{dz^n}\!\left(z^{\gamma-1}\,{_pF_{q+1}}\!\left(\begin{matrix}a_1,\ldots,a_p\\\gamma,b_1,\ldots,b_q\end{matrix};z\right)\right) = (\gamma-n)_n z^{\gamma-n-1}\,{_pF_{q+1}}\!\left(\begin{matrix}a_1,\ldots,a_p\\\gamma-n,b_1,\ldots,b_q\end{matrix};z\right).$$

Other versions of these identities can be constructed with the aid of the operator identity

16.3.5
$$\left(z\frac{d}{dz}z\right)^{\!n} = z^n\frac{d^n}{dz^n}z^n, \qquad\qquad n = 1, 2, \ldots.$$

16.3(ii) Contiguous Functions

Two generalized hypergeometric functions $_pF_q(\mathbf{a};\mathbf{b};z)$ are *(generalized) contiguous* if they have the same pair of values of p and q, and corresponding parameters differ by integers. If $p \leq q+1$, then any $q+2$ distinct contiguous functions are linearly related. Examples are provided by the following recurrence relations:

16.3.6
$$z\,{_0F_1}(-;b+1;z) + b(b-1)\,{_0F_1}(-;b;z) - b(b-1)\,{_0F_1}(-;b-1;z) = 0,$$

16.3.7
$$_3F_2\!\left(\begin{matrix}a_1+2,a_2,a_3\\b_1,b_2\end{matrix};z\right)a_1(a_1+1)(1-z) + {_3F_2}\!\left(\begin{matrix}a_1+1,a_2,a_3\\b_1,b_2\end{matrix};z\right)a_1\left(b_1+b_2-3a_1-2+z(2a_1-a_2-a_3+1)\right)$$
$$+ {_3F_2}\!\left(\begin{matrix}a_1,a_2,a_3\\b_1,b_2\end{matrix};z\right)\left((2a_1-b_1)(2a_1-b_2)+a_1-a_1^2-z(a_1-a_2)(a_1-a_3)\right)$$
$$- {_3F_2}\!\left(\begin{matrix}a_1-1,a_2,a_3\\b_1,b_2\end{matrix};z\right)(a_1-b_1)(a_1-b_2) = 0.$$

For further examples see §§13.3(i), 15.5(ii), and the following references: Rainville (1960, §48), Wimp (1968), and Luke (1975, §5.13).

16.4 Argument Unity

16.4(i) Classification

The function $_{q+1}F_q(\mathbf{a};\mathbf{b};z)$ is *well-poised* if

16.4.1
$$a_1 + b_1 = \cdots = a_q + b_q = a_{q+1} + 1.$$

It is *very well-poised* if it is well-poised and $a_1 = b_1 + 1$.

The special case $_{q+1}F_q(\mathbf{a};\mathbf{b};1)$ is *k-balanced* if a_{q+1} is a nonpositive integer and

16.4.2
$$a_1 + \cdots + a_{q+1} + k = b_1 + \cdots + b_q.$$

When $k = 1$ the function is said to be *balanced* or *Saalschützian*.

16.4(ii) Examples

The function $_{q+1}F_q$ with argument unity and general values of the parameters is discussed in Bühring (1992). Special cases are as follows:

Pfaff–Saalschütz Balanced Sum

16.4.3
$$_3F_2\left(\begin{matrix} -n, a, b \\ c, d \end{matrix}; 1\right) = \frac{(c-a)_n (c-b)_n}{(c)_n (c-a-b)_n},$$

when $c + d = a + b + 1 - n$, $n = 0, 1, \ldots$. See Erdélyi et al. (1953a, §4.4(4)) for a non-terminating balanced identity.

Dixon's Well-Poised Sum

16.4.4
$$_3F_2\left(\begin{matrix} a, b, c \\ a-b+1, a-c+1 \end{matrix}; 1\right) = \frac{\Gamma\left(\tfrac{1}{2}a+1\right)\Gamma(a-b+1)\Gamma(a-c+1)\Gamma\left(\tfrac{1}{2}a-b-c+1\right)}{\Gamma(a+1)\Gamma\left(\tfrac{1}{2}a-b+1\right)\Gamma\left(\tfrac{1}{2}a-c+1\right)\Gamma(a-b-c+1)},$$

when $\Re(a - 2b - 2c) > -2$, or when the series terminates with $a = -n$:

16.4.5
$$_3F_2\left(\begin{matrix} -n, b, c \\ 1-b-n, 1-c-n \end{matrix}; 1\right) = \begin{cases} 0, & n = 2k+1, \\ \dfrac{(2k)!\, \Gamma(b+k)\, \Gamma(c+k)\, \Gamma(b+c+2k)}{k!\, \Gamma(b+2k)\, \Gamma(c+2k)\, \Gamma(b+c+k)}, & n = 2k, \end{cases}$$

where $k = 0, 1, \ldots$.

Watson's Sum

16.4.6
$$_3F_2\left(\begin{matrix} a, b, c \\ \tfrac{1}{2}(a+b+1), 2c \end{matrix}; 1\right) = \frac{\Gamma\left(\tfrac{1}{2}\right)\Gamma\left(c+\tfrac{1}{2}\right)\Gamma\left(\tfrac{1}{2}(a+b+1)\right)\Gamma\left(c+\tfrac{1}{2}(1-a-b)\right)}{\Gamma\left(\tfrac{1}{2}(a+1)\right)\Gamma\left(\tfrac{1}{2}(b+1)\right)\Gamma\left(c+\tfrac{1}{2}(1-a)\right)\Gamma\left(c+\tfrac{1}{2}(1-b)\right)},$$

when $\Re(2c - a - b) > -1$, or when the series terminates with $a = -n$.

Whipple's Sum

16.4.7
$$_3F_2\left(\begin{matrix} a, 1-a, c \\ d, 2c-d+1 \end{matrix}; 1\right) = \frac{\pi\, \Gamma(d)\, \Gamma(2c-d+1)\, 2^{1-2c}}{\Gamma\left(c+\tfrac{1}{2}(a-d+1)\right)\Gamma\left(c+1-\tfrac{1}{2}(a+d)\right)\Gamma\left(\tfrac{1}{2}(a+d)\right)\Gamma\left(\tfrac{1}{2}(d-a+1)\right)},$$

when $\Re c > 0$ or when a is an integer.

Džrbasjan's Sum

This is (16.4.7) in the case $c = -n$:

16.4.8
$$_3F_2\left(\begin{matrix} -n, a, 1-a \\ d, 1-d-2n \end{matrix}; 1\right) = \frac{\left(\tfrac{1}{2}(a+d)\right)_n \left(\tfrac{1}{2}(d-a+1)\right)_n}{\left(\tfrac{1}{2}d\right)_n \left(\tfrac{1}{2}(d+1)\right)_n}, \qquad n = 0, 1, \ldots.$$

Rogers–Dougall Very Well-Poised Sum

16.4.9
$$_5F_4\left(\begin{matrix} a, \tfrac{1}{2}a+1, b, c, d \\ \tfrac{1}{2}a, a-b+1, a-c+1, a-d+1 \end{matrix}; 1\right) = \frac{\Gamma(a-b+1)\Gamma(a-c+1)\Gamma(a-d+1)\Gamma(a-b-c-d+1)}{\Gamma(a+1)\Gamma(a-b-c+1)\Gamma(a-b-d+1)\Gamma(a-c-d+1)},$$

when $\Re(b + c + d - a) < 1$, or when the series terminates with $d = -n$.

Dougall's Very Well-Poised Sum

16.4.10
$$_7F_6\left(\begin{matrix} a, \tfrac{1}{2}a+1, b, c, d, f, -n \\ \tfrac{1}{2}a, a-b+1, a-c+1, a-d+1, a-f+1, a+n+1 \end{matrix}; 1\right)$$
$$= \frac{(a+1)_n (a-b-c+1)_n (a-b-d+1)_n (a-c-d+1)_n}{(a-b+1)_n (a-c+1)_n (a-d+1)_n (a-b-c-d+1)_n}, \qquad n = 0, 1, \ldots,$$

when $2a + 1 = b + c + d + f - n$. The last condition is equivalent to the sum of the top parameters plus 2 equals the sum of the bottom parameters, that is, the series is 2-balanced.

16.4(iii) Identities

16.4.11
$$_3F_2\left(\begin{matrix}a,b,c\\d,e\end{matrix};1\right)=\frac{\Gamma(e)\,\Gamma(d+e-a-b-c)}{\Gamma(e-a)\,\Gamma(d+e-b-c)}\,_3F_2\left(\begin{matrix}a,d-b,d-c\\d,d+e-b-c\end{matrix};1\right),$$

when $\Re(d+e-a-b-c)>0$ and $\Re(e-a)>0$. The function $_3F_2(a,b,c;d,e;1)$ is analytic in the parameters a,b,c,d,e when its series expansion converges and the bottom parameters are not negative integers or zero. (16.4.11) provides a partial analytic continuation to the region when the only restrictions on the parameters are $\Re(e-a)>0$, and d,e, and $d+e-b-c\neq 0,-1,\ldots$. A detailed treatment of analytic continuation in (16.4.11) and asymptotic approximations as the variables a,b,c,d,e approach infinity is given by Aomoto (1987).

There are two types of three-term identities for $_3F_2$'s. The first are recurrence relations that extend those for $_2F_1$'s; see §15.5(ii). Examples are (16.3.7) with $z=1$. Also,

16.4.12
$$(a-d)(b-d)(c-d)\left(_3F_2\left(\begin{matrix}a,b,c\\d+1,e\end{matrix};1\right)-{}_3F_2\left(\begin{matrix}a,b,c\\d,e\end{matrix};1\right)\right)+abc\,_3F_2\left(\begin{matrix}a,b,c\\d,e\end{matrix};1\right)$$
$$=d(d-1)(a+b+c-d-e+1)\left(_3F_2\left(\begin{matrix}a,b,c\\d,e\end{matrix};1\right)-{}_3F_2\left(\begin{matrix}a,b,c\\d-1,e\end{matrix};1\right)\right),$$

and

16.4.13
$$_3F_2\left(\begin{matrix}a,b,c\\d,e\end{matrix};1\right)=\frac{c(e-a)}{de}\,_3F_2\left(\begin{matrix}a,b+1,c+1\\d+1,e+1\end{matrix};1\right)+\frac{d-c}{d}\,_3F_2\left(\begin{matrix}a,b+1,c\\d+1,e\end{matrix};1\right).$$

Methods of deriving such identities are given by Bailey (1964), Rainville (1960), Raynal (1979), and Wilson (1978). Lists are given by Raynal (1979) and Wilson (1978). See Raynal (1979) for a statement in terms of $3j$ symbols (Chapter 34). Also see Wilf and Zeilberger (1992a,b) for information on the Wilf–Zeilberger algorithm which can be used to find such relations.

The other three-term relations are extensions of Kummer's relations for $_2F_1$'s given in §15.10(ii). See Bailey (1964, pp. 19–22).

Balanced $_4F_3(1)$ series have transformation formulas and three-term relations. The basic transformation is given by

16.4.14
$$_4F_3\left(\begin{matrix}-n,a,b,c\\d,e,f\end{matrix};1\right)=\frac{(e-a)_n(f-a)_n}{(e)_n(f)_n}\,_4F_3\left(\begin{matrix}-n,a,d-b,d-c\\d,a-e-n+1,a-f-n+1\end{matrix};1\right),$$

when $a+b+c-n+1=d+e+f$. These series contain $6j$ symbols as special cases when the parameters are integers; compare §34.4.

The characterizing properties (18.22.2), (18.22.10), (18.22.19), (18.22.20), and (18.26.14) of the Hahn and Wilson class polynomials are examples of the contiguous relations mentioned in the previous three paragraphs.

Contiguous balanced series have parameters shifted by an integer but still balanced. One example of such a three-term relation is the recurrence relation (18.26.16) for Racah polynomials. See Raynal (1979), Wilson (1978), and Bailey (1964).

A different type of transformation is that of Whipple:

16.4.15
$$_7F_6\left(\begin{matrix}a,\tfrac{1}{2}a+1,b,c,d,e,f\\\tfrac{1}{2}a,a-b+1,a-c+1,a-d+1,a-e+1,a-f+1\end{matrix};1\right)$$
$$=\frac{\Gamma(a-d+1)\,\Gamma(a-e+1)\,\Gamma(a-f+1)\,\Gamma(a-d-e-f+1)}{\Gamma(a+1)\,\Gamma(a-d-e+1)\,\Gamma(a-d-f+1)\,\Gamma(a-e-f+1)}\,_4F_3\left(\begin{matrix}a-b-c+1,d,e,f\\a-b+1,a-c+1,d+e+f-a\end{matrix};1\right),$$

when the series on the right terminates and the series on the left converges. When the series on the right does not terminate, a second term appears. See Bailey (1964, §4.4(4)).

Transformations for both balanced $_4F_3(1)$ and very well-poised $_7F_6(1)$ are included in Bailey (1964, pp. 56–63). A similar theory is available for very well-poised $_9F_8(1)$'s which are 2-balanced. See Bailey (1964, §§4.3(7) and 7.6(1)) for the transformation formulas and Wilson (1978) for contiguous relations.

Relations between three solutions of three-term recurrence relations are given by Masson (1991). See also Lewanowicz (1985) (with corrections in Lewanowicz (1987)) for further examples of recurrence relations.

16.4(iv) Continued Fractions

For continued fractions for ratios of $_3F_2$ functions with argument unity, see Cuyt et al. (2008, pp. 315–317).

16.4(v) Bilateral Series

Denote, formally, the bilateral hypergeometric function

16.4.16
$$_pH_q\left(\begin{array}{c}a_1,\ldots,a_p\\b_1,\ldots,b_q\end{array};z\right) = \sum_{k=-\infty}^{\infty} \frac{(a_1)_k\cdots(a_p)_k}{(b_1)_k\cdots(b_q)_k} z^k.$$

Then

16.4.17
$$_2H_2\left(\begin{array}{c}a,b\\c,d\end{array};1\right) = \frac{\Gamma(c)\,\Gamma(d)\,\Gamma(1-a)\,\Gamma(1-b)\,\Gamma(c+d-a-b-1)}{\Gamma(c-a)\,\Gamma(d-a)\,\Gamma(c-b)\,\Gamma(d-b)}, \qquad \Re(c+d-a-b) > 1.$$

This is *Dougall's bilateral sum*; see Andrews et al. (1999, §2.8).

16.5 Integral Representations and Integrals

When $z \neq 0$ and $a_k \neq 0, -1, -2, \ldots$, $k = 1, 2, \ldots, p$,

16.5.1
$$\left(\prod_{k=1}^{p}\Gamma(a_k)\bigg/\prod_{k=1}^{q}\Gamma(b_k)\right) {}_pF_q\left(\begin{array}{c}a_1,\ldots,a_p\\b_1,\ldots,b_q\end{array};z\right) = \frac{1}{2\pi i}\int_L \left(\prod_{k=1}^{p}\Gamma(a_k+s)\bigg/\prod_{k=1}^{q}\Gamma(b_k+s)\right)\Gamma(-s)(-z)^s\,ds,$$

where the contour of integration separates the poles of $\Gamma(a_k+s)$, $k = 1, \ldots, p$, from those of $\Gamma(-s)$.

Suppose first that L is a contour that starts at infinity on a line parallel to the positive real axis, encircles the nonnegative integers in the negative sense, and ends at infinity on another line parallel to the positive real axis. Then the integral converges when $p < q+1$ provided that $z \neq 0$, or when $p = q+1$ provided that $0 < |z| < 1$, and provides an integral representation of the left-hand side with these conditions.

Secondly, suppose that L is a contour from $-i\infty$ to $i\infty$. Then the integral converges when $q < p+1$ and $|\mathrm{ph}(-z)| < (p+1-q)\pi/2$. In the case $p = q$ the left-hand side of (16.5.1) is an entire function, and the right-hand side supplies an integral representation valid when $|\mathrm{ph}(-z)| < \pi/2$. In the case $p = q+1$ the right-hand side of (16.5.1) supplies the analytic continuation of the left-hand side from the open unit disk to the sector $|\mathrm{ph}(1-z)| < \pi$; compare §16.2(iii). Lastly, when $p > q+1$ the right-hand side of (16.5.1) can be regarded as the definition of the (customarily undefined) left-hand side. In this event, the formal power-series expansion of the left-hand side (obtained from (16.2.1)) is the asymptotic expansion of the right-hand side as $z \to 0$ in the sector $|\mathrm{ph}(-z)| \leq (p+1-q-\delta)\pi/2$, where δ is an arbitrary small positive constant.

Next, when $p \leq q$,

16.5.2
$$_{p+1}F_{q+1}\left(\begin{array}{c}a_0,\ldots,a_p\\b_0,\ldots,b_q\end{array};z\right) = \frac{\Gamma(b_0)}{\Gamma(a_0)\,\Gamma(b_0-a_0)}\int_0^1 t^{a_0-1}(1-t)^{b_0-a_0-1}\,{}_pF_q\left(\begin{array}{c}a_1,\ldots,a_p\\b_1,\ldots,b_q\end{array};zt\right)dt, \quad \Re b_0 > \Re a_0 > 0,$$

16.5.3
$$_{p+1}F_q\left(\begin{array}{c}a_0,\ldots,a_p\\b_1,\ldots,b_q\end{array};z\right) = \frac{1}{\Gamma(a_0)}\int_0^{\infty} e^{-t}t^{a_0-1}\,{}_pF_q\left(\begin{array}{c}a_1,\ldots,a_p\\b_1,\ldots,b_q\end{array};zt\right)dt, \qquad \Re z < 1, \Re a_0 > 0,$$

16.5.4
$$_pF_{q+1}\left(\begin{array}{c}a_1,\ldots,a_p\\b_0,\ldots,b_q\end{array};z\right) = \frac{\Gamma(b_0)}{2\pi i}\int_{c-i\infty}^{c+i\infty} e^t t^{-b_0}\,{}_pF_q\left(\begin{array}{c}a_1,\ldots,a_p\\b_1,\ldots,b_q\end{array};\frac{z}{t}\right)dt, \qquad c > 0, \Re b_0 > 0.$$

In (16.5.2)–(16.5.4) all many-valued functions in the integrands assume their principal values, and all integration paths are straight lines.

(16.5.2) also holds when $p = q+1$, provided that $|\mathrm{ph}(1-z)| < \pi$. In (16.5.3) the restriction $\Re z < 1$ can be removed when $p < q$. (16.5.4) also holds when $p = q+1$, provided that $\max(0, \Re z) < c$. Lastly, the restrictions on the parameters can be eased by replacing the integration paths with loop contours; see Luke (1969a, §3.6).

Laplace transforms and inverse Laplace transforms of generalized hypergeometric functions are given in Prudnikov et al. (1992a, §3.38) and Prudnikov et al. (1992b, §3.36). For further integral representations and integrals see Apelblat (1983, §16), Erdélyi et al. (1953a, §4.6), Erdélyi et al. (1954a, §§6.9 and 7.5), Luke (1969a, §3.6), and Prudnikov et al. (1990, §§2.22, 4.2.4, and 4.3.1).

16.6 Transformations of Variable

Quadratic

16.6.1
$$_3F_2\left(\begin{array}{c}a,b,c\\a-b+1,a-c+1\end{array};z\right) = (1-z)^{-a}\,{}_3F_2\left(\begin{array}{c}a-b-c+1,\tfrac{1}{2}a,\tfrac{1}{2}(a+1)\\a-b+1,a-c+1\end{array};\frac{-4z}{(1-z)^2}\right).$$

Cubic

16.6.2
$$_3F_2\left(\begin{matrix}a, 2b-a-1, 2-2b+a\\b, a-b+\frac{3}{2}\end{matrix}; \frac{z}{4}\right) = (1-z)^{-a} {}_3F_2\left(\begin{matrix}\frac{1}{3}a, \frac{1}{3}a+\frac{1}{3}, \frac{1}{3}a+\frac{2}{3}\\b, a-b+\frac{3}{2}\end{matrix}; \frac{-27z}{4(1-z)^3}\right).$$

For Kummer-type transformations of ${}_2F_2$ functions see Miller (2003) and Paris (2005a), and for further transformations see Erdélyi et al. (1953a, §4.5).

16.7 Relations to Other Functions

For orthogonal polynomials see Chapter 18. For $3j$, $6j$, $9j$ symbols see Chapter 34. Further representations of special functions in terms of ${}_pF_q$ functions are given in Luke (1969a, §§6.2–6.3), and an extensive list of ${}_{q+1}F_q$ functions with rational numbers as parameters is given in Krupnikov and Kölbig (1997).

16.8 Differential Equations

16.8(i) Classification of Singularities

An *ordinary point* of the differential equation

16.8.1
$$\frac{d^n w}{dz^n} + f_{n-1}(z)\frac{d^{n-1}w}{dz^{n-1}} + f_{n-2}(z)\frac{d^{n-2}w}{dz^{n-2}} + \cdots + f_1(z)\frac{dw}{dz} + f_0(z)w = 0$$

is a value z_0 of z at which all the coefficients $f_j(z)$, $j = 0, 1, \ldots, n-1$, are analytic. If z_0 is not an ordinary point but $(z-z_0)^{n-j}f_j(z)$, $j = 0, 1, \ldots, n-1$, are analytic at $z = z_0$, then z_0 is a *regular singularity*. All other singularities are *irregular*. Compare §2.7(i) in the case $n = 2$. Similar definitions apply in the case $z_0 = \infty$: we transform ∞ into the origin by replacing z in (16.8.1) by $1/z$; again compare §2.7(i).

For further information see Hille (1976, pp. 360–370).

16.8(ii) The Generalized Hypergeometric Differential Equation

With the notation

16.8.2
$$D = \frac{d}{dz}, \quad \vartheta = z\frac{d}{dz},$$

the function $w = {}_pF_q(\mathbf{a}; \mathbf{b}; z)$ satisfies the differential equation

16.8.3
$$(\vartheta(\vartheta+b_1-1)\cdots(\vartheta+b_q-1) - z(\vartheta+a_1)\cdots(\vartheta+a_p))w = 0.$$

Equivalently,

16.8.4
$$z^q D^{q+1}w + \sum_{j=1}^{q} z^{j-1}(\alpha_j z + \beta_j)D^j w + \alpha_0 w = 0, \qquad p \le q,$$

or

16.8.5
$$z^q(1-z)D^{q+1}w + \sum_{j=1}^{q} z^{j-1}(\alpha_j z + \beta_j)D^j w + \alpha_0 w = 0, \qquad p = q+1,$$

where α_j and β_j are constants. Equation (16.8.4) has a regular singularity at $z = 0$, and an irregular singularity at $z = \infty$, whereas (16.8.5) has regular singularities at $z = 0, 1$, and ∞. In each case there are no other singularities. Equation (16.8.3) is of order $\max(p, q+1)$. In Letessier et al. (1994) examples are discussed in which the generalized hypergeometric function satisfies a differential equation that is of order 1 or even 2 less than might be expected.

When no b_j is an integer, and no two b_j differ by an integer, a fundamental set of solutions of (16.8.3) is given by

16.8.6
$$w_0(z) = {}_pF_q\left(\begin{matrix}a_1, \ldots, a_p\\b_1, \ldots, b_q\end{matrix}; z\right), \quad w_j(z) = z^{1-b_j} {}_pF_q\left(\begin{matrix}1+a_1-b_j, \ldots, 1+a_p-b_j\\2-b_j, 1+b_1-b_j, \ldots * \ldots, 1+b_q-b_j\end{matrix}; z\right), \quad j = 1, \ldots, q,$$

where $*$ indicates that the entry $1 + b_j - b_j$ is omitted. For other values of the b_j, series solutions in powers of z (possibly involving also $\ln z$) can be constructed via a limiting process; compare §2.7(i) in the case of second-order differential equations. For details see Smith (1939a,b), and Nørlund (1955).

When $p = q+1$, and no two a_j differ by an integer, another fundamental set of solutions of (16.8.3) is given by

16.8.7
$$\widetilde{w}_j(z) = (-z)^{-a_j} {}_{q+1}F_q\left(\begin{matrix}a_j, 1-b_1+a_j, \ldots, 1-b_q+a_j\\1-a_1+a_j, \ldots * \ldots, 1-a_{q+1}+a_j\end{matrix}; \frac{1}{z}\right), \qquad j = 1, \ldots, q+1,$$

where $*$ indicates that the entry $1 - a_j + a_j$ is omitted. We have the connection formula

16.8.8
$$_{q+1}F_q\left(\begin{matrix}a_1,\ldots,a_{q+1}\\b_1,\ldots,b_q\end{matrix};z\right) = \sum_{j=1}^{q+1}\left(\prod_{\substack{k=1\\k\neq j}}^{q+1}\frac{\Gamma(a_k-a_j)}{\Gamma(a_k)}\bigg/\prod_{k=1}^{q}\frac{\Gamma(b_k-a_j)}{\Gamma(b_k)}\right)\widetilde{w}_j(z), \qquad |\operatorname{ph}(-z)|\leq\pi.$$

More generally if z_0 ($\in\mathbb{C}$) is an arbitrary constant, $|z-z_0| > \max(|z_0|, |z_0-1|)$, and $|\operatorname{ph}(z_0-z)| < \pi$, then

16.8.9
$$\left(\prod_{k=1}^{q+1}\Gamma(a_k)\bigg/\prod_{k=1}^{q}\Gamma(b_k)\right)\,_{q+1}F_q\left(\begin{matrix}a_1,\ldots,a_{q+1}\\b_1,\ldots,b_q\end{matrix};z\right)$$
$$= \sum_{j=1}^{q+1}(z_0-z)^{-a_j}\sum_{n=0}^{\infty}\frac{\Gamma(a_j+n)}{n!}\left(\prod_{\substack{k=1\\k\neq j}}^{q+1}\Gamma(a_k-a_j-n)\bigg/\prod_{k=1}^{q}\Gamma(b_k-a_j-n)\right)$$
$$\times\,_{q+1}F_q\left(\begin{matrix}a_1-a_j-n,\ldots,a_{q+1}-a_j-n\\b_1-a_j-n,\ldots,b_q-a_j-n\end{matrix};z_0\right)(z-z_0)^{-n}.$$

(Note that the generalized hypergeometric functions on the right-hand side are polynomials in z_0.)

When $p = q+1$ and some of the a_j differ by an integer a limiting process can again be applied. For details see Nørlund (1955). In this reference it is also explained that in general when $q > 1$ no simple representations in terms of generalized hypergeometric functions are available for the fundamental solutions near $z = 1$. Analytical continuation formulas for $_{q+1}F_q(\mathbf{a};\mathbf{b};z)$ near $z = 1$ are given in Bühring (1987b) for the case $q = 2$, and in Bühring (1992) for the general case.

16.8(iii) Confluence of Singularities

If $p \leq q$, then

16.8.10
$$\lim_{|\alpha|\to\infty}\,_{p+1}F_q\left(\begin{matrix}a_1,\ldots,a_p,\alpha\\b_1,\ldots,b_q\end{matrix};\frac{z}{\alpha}\right) = \,_pF_q\left(\begin{matrix}a_1,\ldots,a_p\\b_1,\ldots,b_q\end{matrix};z\right).$$

Thus in the case $p = q$ the regular singularities of the function on the left-hand side at α and ∞ coalesce into an irregular singularity at ∞.

Next, if $p \leq q+1$ and $|\operatorname{ph}\beta| \leq \pi - \delta\ (<\pi)$, then

16.8.11
$$\lim_{|\beta|\to\infty}\,_pF_{q+1}\left(\begin{matrix}a_1,\ldots,a_p\\b_1,\ldots,b_q,\beta\end{matrix};\beta z\right) = \,_pF_q\left(\begin{matrix}a_1,\ldots,a_p\\b_1,\ldots,b_q\end{matrix};z\right),$$

provided that in the case $p = q+1$ we have $|z| < 1$ when $|\operatorname{ph}\beta| \leq \frac{1}{2}\pi$, and $|z| < |\sin(\operatorname{ph}\beta)|$ when $\frac{1}{2}\pi \leq |\operatorname{ph}\beta| \leq \pi - \delta$ ($<\pi$).

16.9 Zeros

Assume that $p = q$ and none of the a_j is a nonpositive integer. Then $_pF_p(\mathbf{a};\mathbf{b};z)$ has at most finitely many zeros if and only if the a_j can be re-indexed for $j = 1,\ldots,p$ in such a way that $a_j - b_j$ is a nonnegative integer.

Next, assume that $p = q$ and that the a_j and the quotients $(\mathbf{a})_j/(\mathbf{b})_j$ are all real. Then $_pF_p(\mathbf{a};\mathbf{b};z)$ has at most finitely many real zeros.

These results are proved in Ki and Kim (2000). For further information on zeros see Hille (1929).

16.10 Expansions in Series of $_pF_q$ Functions

The following expansion, with appropriate conditions and together with similar results, is given in Fields and Wimp (1961):

16.10.1
$$_{p+r}F_{q+s}\left(\begin{matrix}a_1,\ldots,a_p,c_1,\ldots,c_r\\b_1,\ldots,b_q,d_1,\ldots,d_s\end{matrix};z\zeta\right)$$
$$= \sum_{k=0}^{\infty}\frac{(\mathbf{a})_k(\alpha)_k(\beta)_k(-z)^k}{(\mathbf{b})_k(\gamma+k)_k k!}\,_{p+2}F_{q+1}\left(\begin{matrix}\alpha+k,\beta+k,a_1+k,\ldots,a_p+k\\\gamma+2k+1,b_1+k,\ldots,b_q+k\end{matrix};z\right)\,_{r+2}F_{s+2}\left(\begin{matrix}-k,\gamma+k,c_1,\ldots,c_r\\\alpha,\beta,d_1,\ldots,d_s\end{matrix};\zeta\right).$$

Here α, β, and γ are free real or complex parameters.

The next expansion is given in Nørlund (1955, equation (1.21)):

16.10.2 $\quad {}_{p+1}F_p\!\left(\begin{array}{c}a_1,\ldots,a_{p+1}\\ b_1,\ldots,b_p\end{array};z\zeta\right)=(1-z)^{-a_1}\sum_{k=0}^{\infty}\frac{(a_1)_k}{k!}\,{}_{p+1}F_p\!\left(\begin{array}{c}-k,a_2,\ldots,a_{p+1}\\ b_1,\ldots,b_p\end{array};\zeta\right)\left(\frac{z}{z-1}\right)^k.$

When $|\zeta-1|<1$ the series on the right-hand side converges in the half-plane $\Re z<\tfrac{1}{2}$.

Expansions of the form $\sum_{n=1}^{\infty}(\pm 1)^n\,{}_pF_{p+1}(\mathbf{a};\mathbf{b};-n^2 z^2)$ are discussed in Miller (1997), and further series of generalized hypergeometric functions are given in Luke (1969b, Chapter 9), Luke (1975, §§5.10.2 and 5.11), and Prudnikov et al. (1990, §§5.3, 6.8–6.9).

16.11 Asymptotic Expansions

16.11(i) Formal Series

For subsequent use we define two formal infinite series, $E_{p,q}(z)$ and $H_{p,q}(z)$, as follows:

16.11.1 $\quad E_{p,q}(z)=(2\pi)^{(p-q)/2}\kappa^{-\nu-(1/2)}e^{\kappa z^{1/\kappa}}\sum_{k=0}^{\infty}c_k\left(\kappa z^{1/\kappa}\right)^{\nu-k},\qquad p<q+1,$

16.11.2 $\quad H_{p,q}(z)=\sum_{m=1}^{p}\sum_{k=0}^{\infty}\frac{(-1)^k}{k!}\,\Gamma(a_m+k)\left(\prod_{\substack{\ell=1\\ \ell\neq m}}^{p}\Gamma(a_\ell-a_m-k)\bigg/\prod_{\ell=1}^{q}\Gamma(b_\ell-a_m-k)\right)z^{-a_m-k}.$

In (16.11.1)

16.11.3 $\quad \kappa=q-p+1,\quad \nu=a_1+\cdots+a_p-b_1-\cdots-b_q+\tfrac{1}{2}(q-p),$

and

16.11.4 $\quad c_0=1,\quad c_k=-\dfrac{1}{k\kappa^\kappa}\sum_{m=0}^{k-1}c_m e_{k,m},\qquad k\geq 1,$

where

16.11.5 $\quad e_{k,m}=\sum_{j=1}^{q+1}(1-\nu-\kappa b_j+m)_{\kappa+k-m}\left(\prod_{\ell=1}^{p}(a_\ell-b_j)\bigg/\prod_{\substack{\ell=1\\ \ell\neq j}}^{q+1}(b_\ell-b_j)\right),$

and $b_{q+1}=1$.

It may be observed that $H_{p,q}(z)$ represents the sum of the residues of the poles of the integrand in (16.5.1) at $s=-a_j,-a_j-1,\ldots,\;j=1,\ldots,p$, provided that these poles are all simple, that is, no two of the a_j differ by an integer. (If this condition is violated, then the definition of $H_{p,q}(z)$ has to be modified so that the residues are those associated with the multiple poles. In consequence, logarithmic terms may appear. See (15.8.8) for an example.)

16.11(ii) Expansions for Large Variable

In this subsection we assume that none of $a_1, a_2, \ldots, a_p$ is a nonpositive integer.

Case $p=q+1$

The formal series (16.11.2) for $H_{q+1,q}(z)$ converges if $|z|>1$, and

16.11.6 $\quad \left(\prod_{\ell=1}^{q+1}\Gamma(a_\ell)\bigg/\prod_{\ell=1}^{q}\Gamma(b_\ell)\right){}_{q+1}F_q\!\left(\begin{array}{c}a_1,\ldots,a_{q+1}\\ b_1,\ldots,b_q\end{array};z\right)=H_{q+1,q}(-z),\qquad |\mathrm{ph}(-z)|\leq\pi;$

compare (16.8.8).

Case $p=q$

As $z\to\infty$ in $|\mathrm{ph}\,z|\leq\pi$,

16.11.7 $\quad \left(\prod_{\ell=1}^{q}\Gamma(a_\ell)\bigg/\prod_{\ell=1}^{q}\Gamma(b_\ell)\right){}_qF_q\!\left(\begin{array}{c}a_1,\ldots,a_q\\ b_1,\ldots,b_q\end{array};z\right)\sim H_{q,q}(ze^{\mp\pi i})+E_{q,q}(z),$

where upper or lower signs are chosen according as z lies in the upper or lower half-plane. (Either sign may be used when $\mathrm{ph}\,z=0$ since the first term on the right-hand side becomes exponentially small compared with the second term.)

For the special case $a_1=1$, $p=q=2$ explicit representations for the right-hand side of (16.11.7) in terms of generalized hypergeometric functions are given in Kim (1972).

Case $p = q - 1$

As $z \to \infty$ in $|\operatorname{ph} z| \leq \pi$,

16.11.8
$$\left(\prod_{\ell=1}^{q-1} \Gamma(a_\ell) \Big/ \prod_{\ell=1}^{q} \Gamma(b_\ell)\right) {}_{q-1}F_q\left(\begin{matrix} a_1, \ldots, a_{q-1} \\ b_1, \ldots, b_q \end{matrix}; -z\right) \sim H_{q-1,q}(z) + E_{q-1,q}(ze^{-\pi i}) + E_{q-1,q}(ze^{\pi i}).$$

Case $p \leq q - 2$

As $z \to \infty$ in $|\operatorname{ph} z| \leq \pi$,

16.11.9
$$\left(\prod_{\ell=1}^{p} \Gamma(a_\ell) \Big/ \prod_{\ell=1}^{q} \Gamma(b_\ell)\right) {}_{p}F_q\left(\begin{matrix} a_1, \ldots, a_p \\ b_1, \ldots, b_q \end{matrix}; -z\right) \sim E_{p,q}(ze^{-\pi i}) + E_{p,q}(ze^{\pi i}).$$

16.11(iii) Expansions for Large Parameters

If z is fixed and $|\operatorname{ph}(1-z)| < \pi$, then for each nonnegative integer m

16.11.10
$$_{p+1}F_p\left(\begin{matrix} a_1+r, \ldots, a_{k-1}+r, a_k, \ldots, a_{p+1} \\ b_1+r, \ldots, b_k+r, b_{k+1}, \ldots, b_p \end{matrix}; z\right) = \sum_{n=0}^{m-1} \frac{(a_1+r)_n \cdots (a_{k-1}+r)_n (a_k)_n \cdots (a_{p+1})_n}{(b_1+r)_n \cdots (b_k+r)_n (b_{k+1})_n \cdots (b_p)_n} \frac{z^n}{n!} + O\left(\frac{1}{r^m}\right),$$

as $r \to +\infty$. Here k can have any integer value from 1 to p. Also if $p < q$, then

16.11.11
$$_{p}F_q\left(\begin{matrix} a_1+r, \ldots, a_p+r \\ b_1+r, \ldots, b_q+r \end{matrix}; z\right) = \sum_{n=0}^{m-1} \frac{(a_1+r)_n \cdots (a_p+r)_n}{(b_1+r)_n \cdots (b_q+r)_n} \frac{z^n}{n!} + O\left(\frac{1}{r^{(q-p)m}}\right),$$

again as $r \to +\infty$. For these and other results see Knottnerus (1960). See also Luke (1969a, §7.3).

Asymptotic expansions for the polynomials ${}_{p+2}F_q(-r, r+a_0, \mathbf{a}; \mathbf{b}; z)$ as $r \to \infty$ through integer values are given in Fields and Luke (1963a,b) and Fields (1965).

16.12 Products

16.12.1
$${}_0F_1(-;a;z)\,{}_0F_1(-;b;z) = {}_2F_3\left(\begin{matrix} \frac{1}{2}(a+b), \frac{1}{2}(a+b-1) \\ a, b, a+b-1 \end{matrix}; 4z\right).$$

16.12.2
$$\left({}_2F_1\left(\begin{matrix} a, b \\ a+b+\frac{1}{2} \end{matrix}; z\right)\right)^2 = {}_3F_2\left(\begin{matrix} 2a, 2b, a+b \\ a+b+\frac{1}{2}, 2a+2b \end{matrix}; z\right).$$

More generally,

16.12.3
$$\left({}_2F_1\left(\begin{matrix} a, b \\ c \end{matrix}; z\right)\right)^2 = \sum_{k=0}^{\infty} \frac{(2a)_k (2b)_k (c-\frac{1}{2})_k}{(c)_k (2c-1)_k k!} \, {}_4F_3\left(\begin{matrix} -\frac{1}{2}k, \frac{1}{2}(1-k), a+b-c+\frac{1}{2}, \frac{1}{2} \\ a+\frac{1}{2}, b+\frac{1}{2}, \frac{3}{2}-k-c \end{matrix}; 1\right) z^k, \qquad |z| < 1.$$

For further identities see Goursat (1883) and Erdélyi et al. (1953a, §4.3).

Two-Variable Hypergeometric Functions

16.13 Appell Functions

The following four functions of two real or complex variables x and y cannot be expressed as a product of two ${}_2F_1$ functions, in general, but they satisfy partial differential equations that resemble the hypergeometric differential equation (15.10.1):

16.13.1 $$F_1(\alpha;\beta,\beta';\gamma;x,y) = \sum_{m,n=0}^{\infty} \frac{(\alpha)_{m+n}(\beta)_m(\beta')_n}{(\gamma)_{m+n} m! n!} x^m y^n, \qquad \max(|x|,|y|) < 1,$$

16.13.2 $$F_2(\alpha;\beta,\beta';\gamma,\gamma';x,y) = \sum_{m,n=0}^{\infty} \frac{(\alpha)_{m+n}(\beta)_m(\beta')_n}{(\gamma)_m(\gamma')_n m! n!} x^m y^n, \qquad |x|+|y| < 1,$$

16.13.3 $$F_3(\alpha,\alpha';\beta,\beta';\gamma;x,y) = \sum_{m,n=0}^{\infty} \frac{(\alpha)_m(\alpha')_n(\beta)_m(\beta')_n}{(\gamma)_{m+n} m! n!} x^m y^n, \qquad \max(|x|,|y|) < 1,$$

16.13.4 $$F_4(\alpha;\beta;\gamma,\gamma';x,y) = \sum_{m,n=0}^{\infty} \frac{(\alpha)_{m+n}(\beta)_{m+n}}{(\gamma)_m(\gamma')_n m! n!} x^m y^n, \qquad \sqrt{|x|}+\sqrt{|y|} < 1.$$

Here and elsewhere it is assumed that neither of the bottom parameters γ and γ' is a nonpositive integer.

16.14 Partial Differential Equations

16.14(i) Appell Functions

16.14.1
$$x(1-x)\frac{\partial^2 F_1}{\partial x^2} + y(1-x)\frac{\partial^2 F_1}{\partial x \partial y} + (\gamma - (\alpha+\beta+1)x)\frac{\partial F_1}{\partial x} - \beta y\frac{\partial F_1}{\partial y} - \alpha\beta F_1 = 0,$$
$$y(1-y)\frac{\partial^2 F_1}{\partial y^2} + x(1-y)\frac{\partial^2 F_1}{\partial x \partial y} + (\gamma - (\alpha+\beta'+1)y)\frac{\partial F_1}{\partial y} - \beta' x\frac{\partial F_1}{\partial x} - \alpha\beta' F_1 = 0,$$

16.14.2
$$x(1-x)\frac{\partial^2 F_2}{\partial x^2} - xy\frac{\partial^2 F_2}{\partial x \partial y} + (\gamma - (\alpha+\beta+1)x)\frac{\partial F_2}{\partial x} - \beta y\frac{\partial F_2}{\partial y} - \alpha\beta F_2 = 0,$$
$$y(1-y)\frac{\partial^2 F_2}{\partial y^2} - xy\frac{\partial^2 F_2}{\partial x \partial y} + (\gamma' - (\alpha+\beta'+1)y)\frac{\partial F_2}{\partial y} - \beta' x\frac{\partial F_2}{\partial x} - \alpha\beta' F_2 = 0,$$

16.14.3
$$x(1-x)\frac{\partial^2 F_3}{\partial x^2} + y\frac{\partial^2 F_3}{\partial x \partial y} + (\gamma - (\alpha+\beta+1)x)\frac{\partial F_3}{\partial x} - \alpha\beta F_3 = 0,$$
$$y(1-y)\frac{\partial^2 F_3}{\partial y^2} + x\frac{\partial^2 F_3}{\partial x \partial y} + (\gamma - (\alpha'+\beta'+1)y)\frac{\partial F_3}{\partial y} - \alpha'\beta' F_3 = 0,$$

16.14.4
$$x(1-x)\frac{\partial^2 F_4}{\partial x^2} - 2xy\frac{\partial^2 F_4}{\partial x \partial y} - y^2\frac{\partial^2 F_4}{\partial y^2} + (\gamma - (\alpha+\beta+1)x)\frac{\partial F_4}{\partial x} - (\alpha+\beta+1)y\frac{\partial F_4}{\partial y} - \alpha\beta F_4 = 0,$$
$$y(1-y)\frac{\partial^2 F_4}{\partial y^2} - 2xy\frac{\partial^2 F_4}{\partial x \partial y} - x^2\frac{\partial^2 F_4}{\partial x^2} + (\gamma' - (\alpha+\beta+1)y)\frac{\partial F_4}{\partial y} - (\alpha+\beta+1)x\frac{\partial F_4}{\partial x} - \alpha\beta F_4 = 0.$$

16.14(ii) Other Functions

In addition to the four Appell functions there are 24 other sums of double series that cannot be expressed as a product of two $_2F_1$ functions, and which satisfy pairs of linear partial differential equations of the second order. Two examples are provided by

16.14.5 $$G_2(\alpha,\alpha';\beta,\beta';x,y) = \sum_{m,n=0}^{\infty} \frac{\Gamma(\alpha+m)\,\Gamma(\alpha'+n)\,\Gamma(\beta+n-m)\,\Gamma(\beta'+m-n)}{\Gamma(\alpha)\,\Gamma(\alpha')\,\Gamma(\beta)\,\Gamma(\beta')} \frac{x^m y^n}{m!n!}, \qquad |x|<1,\,|y|<1,$$

16.14.6 $$G_3(\alpha,\alpha';x,y) = \sum_{m,n=0}^{\infty} \frac{\Gamma(\alpha+2n-m)\,\Gamma(\alpha'+2m-n)}{\Gamma(\alpha)\,\Gamma(\alpha')} \frac{x^m y^n}{m!n!}, \qquad |x|+|y|<\tfrac{1}{4}.$$

(The region of convergence $|x|+|y|<\tfrac{1}{4}$ is not quite maximal.) See Erdélyi et al. (1953a, §§5.7.1–5.7.2) for further information.

16.15 Integral Representations and Integrals

16.15.1
$$F_1(\alpha;\beta,\beta';\gamma;x,y) = \frac{\Gamma(\gamma)}{\Gamma(\alpha)\,\Gamma(\gamma-\alpha)} \int_0^1 \frac{u^{\alpha-1}(1-u)^{\gamma-\alpha-1}}{(1-ux)^\beta(1-uy)^{\beta'}}\,du, \quad \Re\alpha > 0, \Re(\gamma-\alpha) > 0,$$

16.15.2
$$F_2(\alpha;\beta,\beta';\gamma,\gamma';x,y) = \frac{\Gamma(\gamma)\,\Gamma(\gamma')}{\Gamma(\beta)\,\Gamma(\beta')\,\Gamma(\gamma-\beta)\,\Gamma(\gamma'-\beta')} \int_0^1\!\!\int_0^1 \frac{u^{\beta-1}v^{\beta'-1}(1-u)^{\gamma-\beta-1}(1-v)^{\gamma'-\beta'-1}}{(1-ux-vy)^\alpha}\,du\,dv,$$
$$\Re\gamma > \Re\beta > 0, \Re\gamma' > \Re\beta' > 0,$$

16.15.3
$$F_3(\alpha,\alpha';\beta,\beta';\gamma;x,y) = \frac{\Gamma(\gamma)}{\Gamma(\beta)\,\Gamma(\beta')\,\Gamma(\gamma-\beta-\beta')} \iint_\Delta \frac{u^{\beta-1}v^{\beta'-1}(1-u-v)^{\gamma-\beta-\beta'-1}}{(1-ux)^\alpha(1-vy)^{\alpha'}}\,du\,dv,$$
$$\Re(\gamma-\beta-\beta') > 0, \Re\beta > 0, \Re\beta' > 0,$$

where Δ is the triangle defined by $u \geq 0$, $v \geq 0$, $u+v \leq 1$.

16.15.4
$$F_4(\alpha;\beta;\gamma,\gamma';x(1-y),y(1-x))$$
$$= \frac{\Gamma(\gamma)\,\Gamma(\gamma')}{\Gamma(\alpha)\,\Gamma(\beta)\,\Gamma(\gamma-\alpha)\,\Gamma(\gamma'-\beta)} \int_0^1\!\!\int_0^1 \frac{u^{\alpha-1}v^{\beta-1}(1-u)^{\gamma-\alpha-1}(1-v)^{\gamma'-\beta-1}}{(1-ux)^{\gamma+\gamma'-\alpha-1}(1-vy)^{\gamma+\gamma'-\beta-1}(1-ux-vy)^{\alpha+\beta-\gamma-\gamma'+1}}\,du\,dv,$$
$$\Re\gamma > \Re\alpha > 0, \Re\gamma' > \Re\beta > 0.$$

For these and other formulas, including double Mellin–Barnes integrals, see Erdélyi *et al.* (1953a, §5.8). These representations can be used to derive analytic continuations of the Appell functions, including convergent series expansions for large x, large y, or both. For inverse Laplace transforms of Appell functions see Prudnikov *et al.* (1992b, §3.40).

16.16 Transformations of Variables

16.16(i) Reduction Formulas

16.16.1
$$F_1(\alpha;\beta,\beta';\beta+\beta';x,y) = (1-y)^{-\alpha}\,{}_2F_1\!\left(\begin{matrix}\alpha,\beta\\\beta+\beta'\end{matrix};\frac{x-y}{1-y}\right),$$

16.16.2
$$F_2(\alpha;\beta,\beta';\gamma,\beta';x,y) = (1-y)^{-\alpha}\,{}_2F_1\!\left(\begin{matrix}\alpha,\beta\\\gamma\end{matrix};\frac{x}{1-y}\right),$$

16.16.3
$$F_2(\alpha;\beta,\beta';\gamma,\alpha;x,y) = (1-y)^{-\beta'}\,F_1\!\left(\beta;\alpha-\beta',\beta';\gamma;x,\frac{x}{1-y}\right),$$

16.16.4
$$F_3(\alpha,\gamma-\alpha;\beta,\beta';\gamma;x,y) = (1-y)^{-\beta'}\,F_1\!\left(\alpha;\beta,\beta';\gamma;x,\frac{y}{y-1}\right),$$

16.16.5
$$F_3(\alpha,\gamma-\alpha;\beta,\gamma-\beta;\gamma;x,y) = (1-y)^{\alpha+\beta-\gamma}\,{}_2F_1\!\left(\begin{matrix}\alpha,\beta\\\gamma\end{matrix};x+y-xy\right),$$

16.16.6
$$F_4(\alpha;\beta;\gamma,\alpha+\beta-\gamma+1;x(1-y),y(1-x)) = {}_2F_1\!\left(\begin{matrix}\alpha,\beta\\\gamma\end{matrix};x\right){}_2F_1\!\left(\begin{matrix}\alpha,\beta\\\alpha+\beta-\gamma+1\end{matrix};y\right).$$

See Erdélyi *et al.* (1953a, §5.10) for these and further reduction formulas. An extension of (16.16.6) is given by

16.16.7
$$F_4(\alpha;\beta;\gamma,\gamma';x(1-y),y(1-x))$$
$$= \sum_{k=0}^\infty \frac{(\alpha)_k(\beta)_k(\alpha+\beta-\gamma-\gamma'+1)_k}{(\gamma)_k(\gamma')_k k!} x^k y^k\,{}_2F_1\!\left(\begin{matrix}\alpha+k,\beta+k\\\gamma+k\end{matrix};x\right){}_2F_1\!\left(\begin{matrix}\alpha+k,\beta+k\\\gamma'+k\end{matrix};y\right);$$

see Burchnall and Chaundy (1940, 1941).

16.16(ii) Other Transformations

16.16.8
$$F_1(\alpha;\beta,\beta';\gamma;x,y) = (1-x)^{-\beta}(1-y)^{-\beta'}\,F_1\!\left(\gamma-\alpha;\beta,\beta';\gamma;\frac{x}{x-1},\frac{y}{y-1}\right)$$
$$= (1-x)^{-\alpha}\,F_1\!\left(\alpha;\gamma-\beta-\beta',\beta';\gamma;\frac{x}{x-1},\frac{y-x}{1-x}\right),$$

MEIJER G-FUNCTION

16.16.9
$$F_2(\alpha;\beta,\beta';\gamma,\gamma';x,y) = (1-x)^{-\alpha} F_2\left(\alpha;\gamma-\beta,\beta';\gamma,\gamma';\frac{x}{x-1},\frac{y}{1-x}\right),$$

16.16.10
$$F_4(\alpha;\beta;\gamma,\gamma';x,y) = \frac{\Gamma(\gamma')\Gamma(\beta-\alpha)}{\Gamma(\gamma'-\alpha)\Gamma(\beta)}(-y)^{-\alpha} F_4\left(\alpha;\alpha-\gamma'+1;\gamma,\alpha-\beta+1;\frac{x}{y},\frac{1}{y}\right)$$
$$+ \frac{\Gamma(\gamma')\Gamma(\alpha-\beta)}{\Gamma(\gamma'-\beta)\Gamma(\alpha)}(-y)^{-\beta} F_4\left(\beta;\beta-\gamma'+1;\gamma,\beta-\alpha+1;\frac{x}{y},\frac{1}{y}\right).$$

For quadratic transformations of Appell functions see Carlson (1976).

Meijer G-Function

16.17 Definition

Again assume $a_1, a_2, \ldots, a_p$ and $b_1, b_2, \ldots, b_q$ are real or complex parameters. Assume also that m and n are integers such that $0 \leq m \leq q$ and $0 \leq n \leq p$, and none of $a_k - b_j$ is a positive integer when $1 \leq k \leq n$ and $1 \leq j \leq m$. Then the *Meijer G-function* is defined via the Mellin–Barnes integral representation:

16.17.1
$$G_{p,q}^{m,n}(z;\mathbf{a};\mathbf{b}) = G_{p,q}^{m,n}\left(z;\begin{matrix}a_1,\ldots,a_p\\b_1,\ldots,b_q\end{matrix}\right)$$
$$= \frac{1}{2\pi i}\int_L \left(\prod_{\ell=1}^{m}\Gamma(b_\ell-s)\prod_{\ell=1}^{n}\Gamma(1-a_\ell+s)\Big/\left(\prod_{\ell=m}^{q-1}\Gamma(1-b_{\ell+1}+s)\prod_{\ell=n}^{p-1}\Gamma(a_{\ell+1}-s)\right)\right)z^s\,ds,$$

where the integration path L separates the poles of the factors $\Gamma(b_\ell - s)$ from those of the factors $\Gamma(1 - a_\ell + s)$. There are three possible choices for L, illustrated in Figure 16.17.1 in the case $m = 1$, $n = 2$:

(i) L goes from $-i\infty$ to $i\infty$. The integral converges if $p + q < 2(m+n)$ and $|\text{ph } z| < (m+n-\tfrac{1}{2}(p+q))\pi$.

(ii) L is a loop that starts at infinity on a line parallel to the positive real axis, encircles the poles of the $\Gamma(b_\ell - s)$ once in the negative sense and returns to infinity on another line parallel to the positive real axis. The integral converges for all z ($\neq 0$) if $p < q$, and for $0 < |z| < 1$ if $p = q \geq 1$.

(iii) L is a loop that starts at infinity on a line parallel to the negative real axis, encircles the poles of the $\Gamma(1 - a_\ell + s)$ once in the positive sense and returns to infinity on another line parallel to the negative real axis. The integral converges for all z if $p > q$, and for $|z| > 1$ if $p = q \geq 1$.

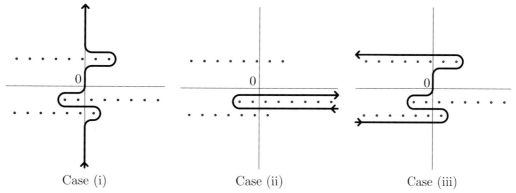

Case (i)   Case (ii)   Case (iii)

Figure 16.17.1: s-plane. Path L for the integral representation (16.17.1) of the Meijer G-function.

When more than one of Cases (i), (ii), and (iii) is applicable the same value is obtained for the Meijer G-function.

Assume $p \leq q$, no two of the bottom parameters b_j, $j = 1, \ldots, m$, differ by an integer, and $a_j - b_k$ is not a positive integer when $j = 1, 2, \ldots, n$ and $k = 1, 2, \ldots, m$. Then

16.17.2
$$G_{p,q}^{m,n}\left(z;\begin{matrix}a_1,\ldots,a_p\\b_1,\ldots,b_q\end{matrix}\right) = \sum_{k=1}^{m} A_{p,q,k}^{m,n}(z)\,_pF_{q-1}\left(\begin{matrix}1+b_k-a_1,\ldots,1+b_k-a_p\\1+b_k-b_1,\ldots*\ldots,1+b_k-b_q\end{matrix};(-1)^{p-m-n}z\right),$$

where $*$ indicates that the entry $1 + b_k - b_k$ is omitted. Also,

16.17.3 $\quad A_{p,q,k}^{m,n}(z) = \prod_{\substack{\ell=1 \\ \ell \neq k}}^{m} \Gamma(b_\ell - b_k) \prod_{\ell=1}^{n} \Gamma(1 + b_k - a_\ell) z^{b_k} \bigg/ \left(\prod_{\ell=m}^{q-1} \Gamma(1 + b_k - b_{\ell+1}) \prod_{\ell=n}^{p-1} \Gamma(a_{\ell+1} - b_k) \right).$

16.18 Special Cases

The $_1F_1$ and $_2F_1$ functions introduced in Chapters 13 and 15, as well as the more general $_pF_q$ functions introduced in the present chapter, are all special cases of the Meijer G-function. This is a consequence of the following relations:

16.18.1
$$_pF_q\left(\begin{matrix} a_1, \ldots, a_p \\ b_1, \ldots, b_q \end{matrix}; z \right) = \left(\prod_{k=1}^{q} \Gamma(b_k) \bigg/ \prod_{k=1}^{p} \Gamma(a_k) \right) G_{p,q+1}^{1,p}\left(-z; \begin{matrix} 1 - a_1, \ldots, 1 - a_p \\ 0, 1 - b_1, \ldots, 1 - b_q \end{matrix} \right)$$
$$= \left(\prod_{k=1}^{q} \Gamma(b_k) \bigg/ \prod_{k=1}^{p} \Gamma(a_k) \right) G_{q+1,p}^{p,1}\left(-\frac{1}{z}; \begin{matrix} 1, b_1, \ldots, b_q \\ a_1, \ldots, a_p \end{matrix} \right).$$

As a corollary, special cases of the $_1F_1$ and $_2F_1$ functions, including Airy functions, Bessel functions, parabolic cylinder functions, Ferrers functions, associated Legendre functions, and many orthogonal polynomials, are all special cases of the Meijer G-function. Representations of special functions in terms of the Meijer G-function are given in Erdélyi et al. (1953a, §5.6), Luke (1969a, §§6.4–6.5), and Mathai (1993, §3.10).

16.19 Identities

16.19.1 $\quad G_{p,q}^{m,n}\left(\frac{1}{z}; \begin{matrix} a_1, \ldots, a_p \\ b_1, \ldots, b_q \end{matrix} \right) = G_{q,p}^{n,m}\left(z; \begin{matrix} 1 - b_1, \ldots, 1 - b_q \\ 1 - a_1, \ldots, 1 - a_p \end{matrix} \right),$

16.19.2 $\quad z^\mu G_{p,q}^{m,n}\left(z; \begin{matrix} a_1, \ldots, a_p \\ b_1, \ldots, b_q \end{matrix} \right) = G_{p,q}^{m,n}\left(z; \begin{matrix} a_1 + \mu, \ldots, a_p + \mu \\ b_1 + \mu, \ldots, b_q + \mu \end{matrix} \right),$

16.19.3 $\quad G_{p+1,q+1}^{m,n+1}\left(z; \begin{matrix} a_0, \ldots, a_p \\ b_1, \ldots, b_q, a_0 \end{matrix} \right) = G_{p,q}^{m,n}\left(z; \begin{matrix} a_1, \ldots, a_p \\ b_1, \ldots, b_q \end{matrix} \right),$

16.19.4 $\quad G_{p,q}^{m,n}\left(z; \begin{matrix} a_1, \ldots, a_p \\ b_1, \ldots, b_q \end{matrix} \right) = \frac{2^{p+1+b_1+\cdots+b_q-m-n-a_1-\cdots-a_p}}{\pi^{m+n-\frac{1}{2}(p+q)}} G_{2p,2q}^{2m,2n}\left(2^{2p-2q} z^2; \begin{matrix} \frac{1}{2}a_1, \frac{1}{2}a_1 + \frac{1}{2}, \ldots, \frac{1}{2}a_p, \frac{1}{2}a_p + \frac{1}{2} \\ \frac{1}{2}b_1, \frac{1}{2}b_1 + \frac{1}{2}, \ldots, \frac{1}{2}b_q, \frac{1}{2}b_q + \frac{1}{2} \end{matrix} \right),$

16.19.5 $\quad \vartheta G_{p,q}^{m,n}\left(z; \begin{matrix} a_1, \ldots, a_p \\ b_1, \ldots, b_q \end{matrix} \right) = G_{p,q}^{m,n}\left(z; \begin{matrix} a_1 - 1, a_2, \ldots, a_p \\ b_1, \ldots, b_q \end{matrix} \right) + (a_1 - 1) G_{p,q}^{m,n}\left(z; \begin{matrix} a_1, \ldots, a_p \\ b_1, \ldots, b_q \end{matrix} \right),$

16.19.6 $\quad \int_0^1 t^{-a_0} (1-t)^{a_0 - b_{q+1} - 1} G_{p,q}^{m,n}\left(zt; \begin{matrix} a_1, \ldots, a_p \\ b_1, \ldots, b_q \end{matrix} \right) dt = \Gamma(a_0 - b_{q+1}) G_{p+1,q+1}^{m,n+1}\left(z; \begin{matrix} a_0, \ldots, a_p \\ b_1, \ldots, b_{q+1} \end{matrix} \right),$

where again $\vartheta = z\, d/dz$. For conditions for (16.19.6) see Luke (1969a, Chapter 5). This reference and Mathai (1993, §§2.2 and 2.4) also supply additional identities.

16.20 Integrals and Series

Integrals of the Meijer G-function are given in Apelblat (1983, §19), Erdélyi et al. (1953a, §5.5.2), Erdélyi et al. (1954a, §§6.9 and 7.5), Luke (1969a, §3.6), Luke (1975, §5.6), Mathai (1993, §3.10), and Prudnikov et al. (1990, §2.24). Extensive lists of Laplace transforms and inverse Laplace transforms of the Meijer G-function are given in Prudnikov et al. (1992a, §3.40) and Prudnikov et al. (1992b, §3.38).

Series of the Meijer G-function are given in Erdélyi et al. (1953a, §5.5.1), Luke (1975, §5.8), and Prudnikov et al. (1990, §6.11).

16.21 Differential Equation

$w = G^{m,n}_{p,q}(z; \mathbf{a}; \mathbf{b})$ satisfies the differential equation

16.21.1
$$\left((-1)^{p-m-n}z(\vartheta - a_1 + 1)\cdots(\vartheta - a_p + 1) - (\vartheta - b_1)\cdots(\vartheta - b_q)\right)w = 0,$$

where again $\vartheta = z\,d/dz$. This equation is of order $\max(p,q)$. In consequence of (16.19.1) we may assume, without loss of generality, that $p \le q$. With the classification of §16.8(i), when $p < q$ the only singularities of (16.21.1) are a regular singularity at $z = 0$ and an irregular singularity at $z = \infty$. When $p = q$ the only singularities of (16.21.1) are regular singularities at $z = 0$, $(-1)^{p-m-n}$, and ∞.

A fundamental set of solutions of (16.21.1) is given by

16.21.2
$$G^{1,p}_{p,q}\left(ze^{(p-m-n-1)\pi i}; \begin{matrix} a_1, \ldots, a_p \\ b_j, b_1, \ldots, b_{j-1}, b_{j+1}, \ldots, b_q \end{matrix}\right), \qquad j = 1, \ldots, q.$$

For other fundamental sets see Erdélyi et al. (1953a, §5.4) and Marichev (1984).

16.22 Asymptotic Expansions

Asymptotic expansions of $G^{m,n}_{p,q}(z; \mathbf{a}; \mathbf{b})$ for large z are given in Luke (1969a, §§5.7 and 5.10) and Luke (1975, §5.9). For asymptotic expansions of Meijer G-functions with large parameters see Fields (1973, 1983).

Applications

16.23 Mathematical Applications

16.23(i) Differential Equations

A variety of problems in classical mechanics and mathematical physics lead to Picard–Fuchs equations. These equations are frequently solvable in terms of generalized hypergeometric functions, and the monodromy of generalized hypergeometric functions plays an important role in describing properties of the solutions. See, for example, Berglund et al. (1994).

16.23(ii) Random Graphs

A substantial transition occurs in a random graph of n vertices when the number of edges becomes approximately $\frac{1}{2}n$. In Janson et al. (1993) limiting distributions are discussed for the sparse connected components of these graphs, and the asymptotics of three $_2F_2$ functions are applied to compute the expected value of the excess.

16.23(iii) Conformal Mapping

The Bieberbach conjecture states that if $\sum_{n=0}^{\infty} a_n z^n$ is a conformal map of the unit disk to any complex domain, then $|a_n| \le n|a_1|$. In the proof of this conjecture de Branges (1985) uses the inequality

16.23.1
$$_3F_2\left(\begin{matrix} -n, n+\alpha+2, \frac{1}{2}(\alpha+1) \\ \alpha+1, \frac{1}{2}(\alpha+3) \end{matrix}; x\right) > 0,$$

when $0 \le x < 1$, $\alpha > -2$, and $n = 0, 1, 2, \ldots$. The proof of this inequality is given in Askey and Gasper (1976). See also Kazarinoff (1988).

16.23(iv) Combinatorics and Number Theory

Many combinatorial identities, especially ones involving binomial and related coefficients, are special cases of hypergeometric identities. In Petkovšek et al. (1996) tools are given for automated proofs of these identities.

16.24 Physical Applications

16.24(i) Random Walks

Generalized hypergeometric functions and Appell functions appear in the evaluation of the so-called Watson integrals which characterize the simplest possible lattice walks. They are also potentially useful for the solution of more complicated restricted lattice walk problems, and the 3D Ising model; see Barber and Ninham (1970, pp. 147–148).

16.24(ii) Loop Integrals in Feynman Diagrams

Appell functions are used for the evaluation of one-loop integrals in Feynman diagrams. See Cabral-Rosetti and Sanchis-Lozano (2000).

For an extension to two-loop integrals see Moch et al. (2002).

16.24(iii) $3j$, $6j$, and $9j$ Symbols

The $3j$ symbols, or Clebsch–Gordan coefficients, play an important role in the decomposition of reducible representations of the rotation group into irreducible representations. They can be expressed as ${}_3F_2$ functions with unit argument. The coefficients of transformations between different coupling schemes of three angular momenta are related to the Wigner $6j$ symbols. These are balanced ${}_4F_3$ functions with unit argument. Lastly, special cases of the $9j$ symbols are ${}_5F_4$ functions with unit argument. For further information see Chapter 34 and Varshalovich et al. (1988, §§8.2.5, 8.8, and 9.2.3).

Computation

16.25 Methods of Computation

Methods for computing the functions of the present chapter include power series, asymptotic expansions, integral representations, differential equations, and recurrence relations. They are similar to those described for confluent hypergeometric functions, and hypergeometric functions in §§13.29 and 15.19. There is, however, an added feature in the numerical solution of differential equations and difference equations (recurrence relations). This occurs when the wanted solution is intermediate in asymptotic growth compared with other solutions. In these cases integration, or recurrence, in either a forward or a backward direction is unstable. Instead a boundary-value problem needs to be formulated and solved. See §§3.6(vii), 3.7(iii), Olde Daalhuis and Olver (1998), Lozier (1980), and Wimp (1984, Chapters 7, 8).

16.26 Approximations

For discussions of the approximation of generalized hypergeometric functions and the Meijer G-function in terms of polynomials, rational functions, and Chebyshev polynomials see Luke (1975, §§5.12 - 5.13) and Luke (1977b, Chapters 1 and 9).

16.27 Software

See http://dlmf.nist.gov/16.27.

References

General References

The main references used in writing this chapter are Erdélyi et al. (1953a), Luke (1969a, 1975). For additional bibliographic reading see Andrews et al. (1999) and Slater (1966).

Sources

The following list gives the references or other indications of proofs that were used in constructing the various sections of this chapter. These sources supplement the references that are quoted in the text.

§16.2 Luke (1975, Chapter 5), Slater (1966, Chapter 2). The statement that follows (16.2.5) follows from the uniform convergence of (16.2.5) when $p \leq q$, and also when $p = q + 1$ provided that $|z| < 1$. For other values of z, apply the straightforward generalization (to higher-order differential equations) of Theorem 3.2 in Olver (1997b, Chapter 5).

§16.3 Luke (1975, §5.2.2), Rainville (1960, §48). For (16.3.5) see Fleury and Turbiner (1994).

§16.4 Andrews et al. (1999, Chapters 2 and 3), Slater (1966, Chapters 2 and 6). For (16.4.13) see Donovan et al. (1999).

§16.5 Luke (1969a, §3.6). To justify the last sentence in the third paragraph, translate the integration contour L in (16.5.1) to the right, then apply the residue theorem (§1.10(iv)) making use of the estimate (5.11.9) for the gamma function; compare §2.4(ii).

§16.6 For (16.6.1) see Whipple (1927). For (16.6.2) see Bailey (1929).

§16.8 Luke (1969a, §§3.5 and 5.1). For (16.8.9) see Bühring (1988).

§16.11 Paris and Kaminski (2001, §2.3), Wright (1940a, p. 391), Meijer (1946, p. 1172), Luke (1969a, Chapter 7).

§16.12 Erdélyi et al. (1953a, §4.3). For (16.12.1) see Bailey (1928). For (16.12.2) see Clausen (1828). For (16.12.3) see Chaundy (1969, Chapter 12, Problem 12).

§16.13 Erdélyi et al. (1953a, §5.7).

§16.14 Erdélyi et al. (1953a, §5.9).

§16.16 Erdélyi et al. (1953a, §5.11).

§16.17 Luke (1969a, Chapter 5).

§16.18 Luke (1969a, Chapter 5).

§16.19 These results are straightforward consequences of the definition (16.17.1).

§16.21 Luke (1969a, Chapter 5).

Chapter 17
q-Hypergeometric and Related Functions

G. E. Andrews[1]

Notation — 420
17.1 Special Notation — 420

Properties — 420
17.2 Calculus — 420
17.3 q-Elementary and q-Special Functions — 422
17.4 Basic Hypergeometric Functions — 423
17.5 $_0\phi_0, {}_1\phi_0, {}_1\phi_1$ Functions — 423
17.6 $_2\phi_1$ Function — 424
17.7 Special Cases of Higher $_r\phi_s$ Functions — 426
17.8 Special Cases of $_r\psi_r$ Functions — 427
17.9 Transformations of Higher $_r\phi_r$ Functions — 428
17.10 Transformations of $_r\psi_r$ Functions — 429
17.11 Transformations of q-Appell Functions — 430
17.12 Bailey Pairs — 430
17.13 Integrals — 431
17.14 Constant Term Identities — 431
17.15 Generalizations — 432

Applications — 432
17.16 Mathematical Applications — 432
17.17 Physical Applications — 432

Computation — 432
17.18 Methods of Computation — 432
17.19 Software — 432

References — 432

---

[1] Department of Mathematics, Pennsylvania State University, University Park, Pennsylvania.
Copyright © 2009 National Institute of Standards and Technology. All rights reserved.

Notation

17.1 Special Notation

(For other notation see pp. xiv and 873.)

k, j, m, n, r, s nonnegative integers.
z complex variable.
x real variable.
$q \;(\in \mathbb{C})$ base: unless stated otherwise $|q| < 1$.
$(a;q)_n$ q-shifted factorial:
     $(1-a)(1-aq)\cdots(1-aq^{n-1})$.

The main functions treated in this chapter are the basic hypergeometric (or q-hypergeometric) function ${}_r\phi_s(a_1, a_2, \ldots, a_r; b_1, b_2, \ldots, b_s; q, z)$, the bilateral basic hypergeometric (or bilateral q-hypergeometric) function ${}_r\psi_s(a_1, a_2, \ldots, a_r; b_1, b_2, \ldots, b_s; q, z)$, and the q-analogs of the Appell functions $\Phi^{(1)}(a; b, b'; c; x, y)$, $\Phi^{(2)}(a; b, b'; c, c'; x, y)$, $\Phi^{(3)}(a, a'; b, b'; c; x, y)$, and $\Phi^{(4)}(a; b; c, c'; x, y)$.

Another function notation used is the "idem" function:

$$f(\chi_1; \chi_2, \ldots, \chi_n) + \operatorname{idem}(\chi_1; \chi_2, \ldots, \chi_n)$$
$$= \sum_{j=1}^{n} f(\chi_j; \chi_1, \chi_2, \ldots, \chi_{j-1}, \chi_{j+1}, \ldots, \chi_n).$$

These notations agree with Gasper and Rahman (2004) (except for the q-Appell functions which are not considered in this reference). A slightly different notation is that in Bailey (1935) and Slater (1966); see §17.4(i). Fine (1988) uses $F(a, b; t : q)$ for a particular specialization of a ${}_2\phi_1$ function.

Properties

17.2 Calculus

17.2(i) q-Calculus

For $n = 0, 1, 2, \ldots$,

17.2.1 $(a;q)_n = (1-a)(1-aq)\cdots(1-aq^{n-1})$,

17.2.2 $(a;q)_{-n} = \dfrac{1}{(aq^{-n};q)_n} = \dfrac{(-q/a)^n q^{\binom{n}{2}}}{(q/a;q)_n}$.

For $\nu \in \mathbb{C}$

17.2.3 $(a;q)_\nu = \displaystyle\prod_{j=0}^{\infty}\left(\dfrac{1-aq^j}{1-aq^{\nu+j}}\right)$,

when this product converges.

17.2.4 $(a;q)_\infty = \displaystyle\prod_{j=0}^{\infty}(1-aq^j)$,

17.2.5 $(a_1, a_2, \ldots, a_r; q)_n = \displaystyle\prod_{j=1}^{r}(a_j;q)_n$,

17.2.6 $(a_1, a_2, \ldots, a_r; q)_\infty = \displaystyle\prod_{j=1}^{r}(a_j;q)_\infty$.

17.2.7 $(a;q^{-1})_n = (a^{-1};q)_n (-a)^n q^{-\binom{n}{2}}$,

17.2.8 $\dfrac{(a;q^{-1})_n}{(b;q^{-1})_n} = \dfrac{(a^{-1};q)_n}{(b^{-1};q)_n}\left(\dfrac{a}{b}\right)^n$,

17.2.9 $(a;q)_n = (q^{1-n}/a;q)_n (-a)^n q^{\binom{n}{2}}$,

17.2.10 $\dfrac{(a;q)_n}{(b;q)_n} = \dfrac{(q^{1-n}/a;q)_n}{(q^{1-n}/b;q)_n}\left(\dfrac{a}{b}\right)^n$,

17.2.11 $(aq^{-n};q)_n = (q/a;q)_n \left(-\dfrac{a}{q}\right)^n q^{-\binom{n}{2}}$,

17.2.12 $\dfrac{(aq^{-n};q)_n}{(bq^{-n};q)_n} = \dfrac{(q/a;q)_n}{(q/b;q)_n}\left(\dfrac{a}{b}\right)^n$.

17.2.13 $(a;q)_{n-k} = \dfrac{(a;q)_n}{(q^{1-n}/a;q)_k}\left(-\dfrac{q}{a}\right)^k q^{\binom{k}{2}-nk}$,

17.2.14 $\dfrac{(a;q)_{n-k}}{(b;q)_{n-k}} = \dfrac{(a;q)_n}{(b;q)_n}\dfrac{(q^{1-n}/b;q)_k}{(q^{1-n}/a;q)_k}\left(\dfrac{b}{a}\right)^k$,

17.2.15 $(aq^{-n};q)_k = \dfrac{(a;q)_k (q/a;q)_n}{(q^{1-k}/a;q)_n}q^{-nk}$,

17.2.16 $(aq^{-n};q)_{n-k} = \dfrac{(q/a;q)_n}{(q/a;q)_k}\left(-\dfrac{a}{q}\right)^{n-k} q^{\binom{k}{2}-\binom{n}{2}}$,

17.2.17 $(aq^n;q)_k = \dfrac{(a;q)_k (aq^k;q)_n}{(a;q)_n}$,

17.2.18 $(aq^k;q)_{n-k} = \dfrac{(a;q)_n}{(a;q)_k}$.

17.2.19 $(a;q)_{2n} = (a, aq; q^2)_n$,

more generally,

17.2.20 $(a;q)_{kn} = (a, aq, \ldots, aq^{k-1}; q^k)_n$.

17.2.21 $(a^2; q^2)_n = (a;q)_n (-a;q)_n$,

17.2.22 $\dfrac{\left(qa^{\frac{1}{2}}, -aq^{\frac{1}{2}}; q\right)_n}{\left(a^{\frac{1}{2}}, -a^{\frac{1}{2}}; q\right)_n} = \dfrac{(aq^2; q^2)_n}{(a; q^2)_n} = \dfrac{1-aq^{2n}}{1-a}$,

more generally,

17.2.23 $\dfrac{\left(aq^{\frac{1}{k}}, q\omega_k a^{\frac{1}{k}}, \ldots, q\omega_k^{k-1} a^{\frac{1}{k}}; q\right)_n}{\left(a^{\frac{1}{k}}, \omega_k a^{\frac{1}{k}}, \ldots, \omega_k^{k-1} a^{\frac{1}{k}}; q\right)_n}$
$= \dfrac{(aq^k; q^k)_n}{(a; q^k)_n} = \dfrac{1-aq^{kn}}{1-a}$,

17.2 CALCULUS

where $\omega_k = e^{2\pi i/k}$.

17.2.24
$$\lim_{\tau \to 0} (a/\tau;q)_n \tau^n = \lim_{\sigma \to \infty} (a\sigma;q)_n \sigma^{-n} = (-a)^n q^{\binom{n}{2}},$$

17.2.25 $\displaystyle\lim_{\tau \to 0} \frac{(a/\tau;q)_n}{(b/\tau;q)_n} = \lim_{\sigma \to \infty} \frac{(a\sigma;q)_n}{(b\sigma;q)_n} = \left(\frac{a}{b}\right)^n,$

17.2.26 $\displaystyle\lim_{\tau \to 0} \frac{(a/\tau;q)_n (b/\tau;q)_n}{(c/\tau^2;q)_n} = (-1)^n \left(\frac{ab}{c}\right)^n q^{\binom{n}{2}}.$

17.2(ii) Binomial Coefficients

17.2.27
$$\begin{bmatrix} n \\ m \end{bmatrix}_q = \frac{(q;q)_n}{(q;q)_m (q;q)_{n-m}}$$
$$= \frac{(q^{-n};q)_m (-1)^m q^{nm-\binom{m}{2}}}{(q;q)_m},$$

17.2.28 $\displaystyle\lim_{q \to 1} \begin{bmatrix} n \\ m \end{bmatrix}_q = \binom{n}{m} = \frac{n!}{m!(n-m)!},$

17.2.29 $\displaystyle \begin{bmatrix} m+n \\ m \end{bmatrix}_q = \frac{(q^{n+1};q)_m}{(q;q)_m},$

17.2.30 $\displaystyle \begin{bmatrix} -n \\ m \end{bmatrix}_q = \begin{bmatrix} m+n-1 \\ m \end{bmatrix}_q (-1)^m q^{-mn-\binom{m}{2}},$

17.2.31 $\displaystyle \begin{bmatrix} n \\ m \end{bmatrix}_q = \begin{bmatrix} n-1 \\ m-1 \end{bmatrix}_q + q^m \begin{bmatrix} n-1 \\ m \end{bmatrix}_q,$

17.2.32 $\displaystyle \begin{bmatrix} n \\ m \end{bmatrix}_q = \begin{bmatrix} n-1 \\ m \end{bmatrix}_q + q^{n-m} \begin{bmatrix} n-1 \\ m-1 \end{bmatrix}_q,$

17.2.33
$$\lim_{n \to \infty} \begin{bmatrix} n \\ m \end{bmatrix}_q = \frac{1}{(q;q)_m} = \frac{1}{(1-q)(1-q^2)\cdots(1-q^m)},$$

17.2.34 $\displaystyle\lim_{n \to \infty} \begin{bmatrix} rn+u \\ sn+t \end{bmatrix}_q = \frac{1}{(q;q)_\infty} = \prod_{j=1}^{\infty} \frac{1}{(1-q^j)},$

provided that $r > s$.

17.2(iii) Binomial Theorem

17.2.35
$$\sum_{j=0}^{n} \begin{bmatrix} n \\ j \end{bmatrix}_q (-z)^j q^{\binom{j}{2}} = (z;q)_n$$
$$= (1-z)(1-zq)\cdots(1-zq^{n-1}).$$

In the limit as $q \to 1$, (17.2.35) reduces to the standard binomial theorem

17.2.36 $\displaystyle \sum_{j=0}^{n} \binom{n}{j}(-z)^j = (1-z)^n.$

Also,

17.2.37 $\displaystyle \sum_{n=0}^{\infty} \frac{(a;q)_n}{(q;q)_n} z^n = \frac{(az;q)_\infty}{(z;q)_\infty},$

provided that $|z| < 1$. When $a = q^{m+1}$, where m is a nonnegative integer, (17.2.37) reduces to the q-binomial series

17.2.38 $\displaystyle \sum_{n=0}^{\infty} \begin{bmatrix} n+m \\ n \end{bmatrix}_q z^n = \frac{1}{(z;q)_{m+1}}.$

17.2.39 $\displaystyle \sum_{j=0}^{n} \begin{bmatrix} n \\ j \end{bmatrix}_{q^2} q^j = (-q;q)_n,$

17.2.40 $\displaystyle \sum_{j=0}^{2n} (-1)^j \begin{bmatrix} 2n \\ j \end{bmatrix}_q = (q;q^2)_n.$

When $n \to \infty$ in (17.2.35), and when $m \to \infty$ in (17.2.38), the results become convergent infinite series and infinite products (see (17.5.1) and (17.5.4)).

17.2(iv) Derivatives

The q-derivatives of $f(z)$ are defined by

17.2.41 $\mathcal{D}_q f(z) = \begin{cases} \dfrac{f(z) - f(zq)}{(1-q)z}, & z \neq 0, \\ f'(0), & z = 0, \end{cases}$

and

17.2.42 $f^{[n]}(z) = \mathcal{D}_q^n f(z) = \begin{cases} z^{-n}(1-q)^{-n} \sum_{j=0}^{n} q^{-nj+\binom{j+1}{2}} (-1)^j \begin{bmatrix} n \\ j \end{bmatrix}_q f(zq^j), & z \neq 0, \\ \dfrac{f^{(n)}(0)(q;q)_n}{n!(1-q)^n}, & z = 0. \end{cases}$

When $q \to 1$ the q-derivatives converge to the corresponding ordinary derivatives.

Product Rule

17.2.43 $\mathcal{D}_q(f(z)g(z)) = g(z)f^{[1]}(z) + f(zq)g^{[1]}(z).$

Leibniz Rule

17.2.44 $\mathcal{D}_q^n(f(z)g(z)) = \displaystyle\sum_{j=0}^{n} \begin{bmatrix} n \\ j \end{bmatrix}_q f^{[n-j]}(zq^j)g^{[j]}(z).$

q-differential equations are considered in §17.6(iv).

17.2(v) Integrals

If $f(x)$ is continuous at $x=0$, then

17.2.45
$$\int_0^1 f(x)\,d_q x = (1-q)\sum_{j=0}^\infty f(q^j)q^j,$$

and more generally,

17.2.46
$$\int_0^a f(x)\,d_q x = a(1-q)\sum_{j=0}^\infty f(aq^j)q^j.$$

If $f(x)$ is continuous on $[0,a]$, then

17.2.47
$$\lim_{q\to 1-}\int_0^a f(x)\,d_q x = \int_0^a f(x)\,dx.$$

Infinite Range

17.2.48
$$\int_0^\infty f(x)\,d_q x = \lim_{n\to\infty}\int_0^{q^{-n}} f(x)\,d_q x = (1-q)\sum_{j=-\infty}^\infty f(q^j)q^j,$$

provided that $\sum_{j=-\infty}^\infty f(q^j)q^j$ converges.

17.2(vi) Rogers–Ramanujan Identities

17.2.49
$$1 + \sum_{n=1}^\infty \frac{q^{n^2}}{(1-q)(1-q^2)\cdots(1-q^n)} = \prod_{n=0}^\infty \frac{1}{(1-q^{5n+1})(1-q^{5n+4})},$$

17.2.50
$$1 + \sum_{n=1}^\infty \frac{q^{n^2+n}}{(1-q)(1-q^2)\cdots(1-q^n)} = \prod_{n=0}^\infty \frac{1}{(1-q^{5n+2})(1-q^{5n+3})}.$$

These identities are the first in a large collection of similar results. See §17.14.

17.3 q-Elementary and q-Special Functions

17.3(i) Elementary Functions

q-Exponential Functions

17.3.1
$$e_q(x) = \sum_{n=0}^\infty \frac{(1-q)^n x^n}{(q;q)_n} = \frac{1}{((1-q)x;q)_\infty},$$

17.3.2
$$E_q(x) = \sum_{n=0}^\infty \frac{(1-q)^n q^{\binom{n}{2}} x^n}{(q;q)_n} = (-(1-q)x;q)_\infty.$$

q-Sine Functions

17.3.3
$$\sin_q(x) = \frac{1}{2i}(e_q(ix) - e_q(-ix)) = \sum_{n=0}^\infty \frac{(1-q)^{2n+1}(-1)^n x^{2n+1}}{(q;q)_{2n+1}},$$

17.3.4
$$\mathrm{Sin}_q(x) = \frac{1}{2i}(E_q(ix) - E_q(-ix)) = \sum_{n=0}^\infty \frac{(1-q)^{2n+1} q^{n(2n+1)}(-1)^n x^{2n+1}}{(q;q)_{2n+1}}.$$

q-Cosine Functions

17.3.5
$$\cos_q(x) = \frac{1}{2}(e_q(ix) + e_q(-ix)) = \sum_{n=0}^\infty \frac{(1-q)^{2n}(-1)^n x^{2n}}{(q;q)_{2n}},$$

17.3.6
$$\mathrm{Cos}_q(x) = \frac{1}{2}(E_q(ix) + E_q(-ix)) = \sum_{n=0}^\infty \frac{(1-q)^{2n} q^{n(2n-1)}(-1)^n x^{2n}}{(q;q)_{2n}}.$$

See also Suslov (2003).

17.3(ii) Gamma and Beta Functions

See §5.18.

17.3(iii) Bernoulli Polynomials; Euler and Stirling Numbers

q-Bernoulli Polynomials

17.3.7
$$\beta_n(x,q) = (1-q)^{1-n}\sum_{r=0}^n (-1)^r \binom{n}{r}\frac{r+1}{(1-q^{r+1})}q^{rx}.$$

q-Euler Numbers

17.3.8
$$A_{m,s}(q) = q^{\binom{s-m}{2}+\binom{s}{2}}\sum_{j=0}^s (-1)^j q^{\binom{j}{2}}\begin{bmatrix}m+1\\j\end{bmatrix}_q \frac{(1-q^{s-j})^m}{(1-q)^m}.$$

q-Stirling Numbers

17.3.9
$$a_{m,s}(q) = \frac{q^{-\binom{s}{2}}(1-q)^s}{(q;q)_s}\sum_{j=0}^s (-1)^j q^{\binom{j}{2}}\begin{bmatrix}s\\j\end{bmatrix}_q \frac{(1-q^{s-j})^m}{(1-q)^m}.$$

These were introduced in Carlitz (1954b, 1958). The $\beta_n(x,q)$ are, in fact, rational functions of q, and not necessarily polynomials. The $A_{m,s}(q)$ are always polynomials in q, and the $a_{m,s}(q)$ are polynomials in q for $0 \le s \le m$.

17.3(iv) Theta Functions

See §§17.8 and 20.5.

17.3(v) Orthogonal Polynomials

See §§18.27–18.29.

17.4 Basic Hypergeometric Functions

17.4(i) $_r\phi_s$ Functions

17.4.1
$$_{r+1}\phi_s\left(\begin{matrix}a_0,a_1,a_2,\ldots,a_r\\b_1,b_2,\ldots,b_s\end{matrix};q,z\right)$$
$$= {}_{r+1}\phi_s(a_0,a_1,\ldots,a_r;b_1,b_2,\ldots,b_s;q,z)$$
$$= \sum_{n=0}^{\infty}\frac{(a_0;q)_n(a_1;q)_n\cdots(a_r;q)_n}{(q;q)_n(b_1;q)_n\cdots(b_s;q)_n}\left((-1)^n q^{\binom{n}{2}}\right)^{s-r}z^n.$$

Here and elsewhere it is assumed that the b_j do not take any of the values q^{-n}. The infinite series converges for all z when $s > r$, and for $|z| < 1$ when $s = r$.

17.4.2
$$\lim_{q\to 1^-}{}_{r+1}\phi_r\left(\begin{matrix}q^{a_0},q^{a_1},\ldots,q^{a_r}\\q^{b_1},\ldots,q^{b_r}\end{matrix};q,z\right)$$
$$= {}_{r+1}F_r\left(\begin{matrix}a_0,a_1,\ldots,a_r\\b_1,\ldots,b_r\end{matrix};z\right).$$

For the function on the right-hand side see §16.2(i).

This notation is from Gasper and Rahman (2004). It is slightly at variance with the notation in Bailey (1935) and Slater (1966). In these references the factor $\left((-1)^n q^{\binom{n}{2}}\right)^{s-r}$ is not included in the sum. In practice this discrepancy does not usually cause serious problems because the case most often considered is $r = s$.

17.4(ii) $_r\psi_s$ Functions

17.4.3
$$_r\psi_s\left(\begin{matrix}a_1,a_2,\ldots,a_r\\b_1,b_2,\ldots,b_s\end{matrix};q,z\right)$$
$$= {}_r\psi_s(a_1,a_2,\ldots,a_r;b_1,b_2,\ldots,b_s;q,z)$$
$$= \sum_{n=-\infty}^{\infty}\frac{(a_1,a_2,\ldots,a_r;q)_n(-1)^{(s-r)n}q^{(s-r)\binom{n}{2}}z^n}{(b_1,b_2,\ldots,b_s;q)_n}$$
$$= \sum_{n=0}^{\infty}\frac{(a_1,a_2,\ldots,a_r;q)_n(-1)^{(s-r)n}q^{(s-r)\binom{n}{2}}z^n}{(b_1,b_2,\ldots,b_s;q)_n}$$
$$+\sum_{n=1}^{\infty}\frac{(q/b_1,q/b_2,\ldots,q/b_s;q)_n}{(q/a_1,q/a_2,\ldots,q/a_r;q)_n}\left(\frac{b_1b_2\cdots b_s}{a_1a_2\cdots a_r z}\right)^n.$$

Here and elsewhere the b_j must not take any of the values q^{-n}, and the a_j must not take any of the values q^{n+1}. The infinite series converge when $s \geq r$ provided that $|(b_1\cdots b_s)/(a_1\cdots a_r z)| < 1$ and also, in the case $s = r$, $|z| < 1$.

17.4.4
$$\lim_{q\to 1^-}{}_r\psi_r\left(\begin{matrix}q^{a_1},q^{a_2},\ldots,q^{a_r}\\q^{b_1},q^{b_2},\ldots,q^{b_r}\end{matrix};q,z\right)$$
$$= {}_rH_r\left(\begin{matrix}a_1,a_2,\ldots,a_r\\b_1,b_2,\ldots,b_r\end{matrix};z\right).$$

For the function $_rH_r$ see §16.4(v).

17.4(iii) Appell Functions

The following definitions apply when $|x| < 1$ and $|y| < 1$:

17.4.5
$$\Phi^{(1)}(a;b,b';c;x,y)$$
$$= \sum_{m,n\geq 0}\frac{(a;q)_{m+n}(b;q)_m(b';q)_n x^m y^n}{(q;q)_m(q;q)_n(c;q)_{m+n}},$$

17.4.6
$$\Phi^{(2)}(a;b,b';c,c';x,y)$$
$$= \sum_{m,n\geq 0}\frac{(a;q)_{m+n}(b;q)_m(b';q)_n x^m y^n}{(q;q)_m(q;q)_n(c;q)_m(c';q)_n},$$

17.4.7
$$\Phi^{(3)}(a,a';b,b';c;x,y)$$
$$= \sum_{m,n\geq 0}\frac{(a,b;q)_m(a',b';q)_n x^m y^n}{(q;q)_m(q;q)_n(c;q)_{m+n}},$$

17.4.8 $\Phi^{(4)}(a;b;c,c';x,y) = \displaystyle\sum_{m,n\geq 0}\frac{(a,b;q)_{m+n}x^m y^n}{(q,c;q)_m(q,c';q)_n}.$

17.4(iv) Classification

The series (17.4.1) is said to be *balanced* or *Saalschützian* when it terminates, $r = s$, $z = q$, and

17.4.9 $qa_0a_1\cdots a_s = b_1b_2\cdots b_s.$

The series (17.4.1) is said to be *k-balanced* when $r = s$ and

17.4.10 $q^k a_0 a_1\cdots a_s = b_1 b_2\cdots b_s.$

The series (17.4.1) is said to be *well-poised* when $r = s$ and

17.4.11 $a_0 q = a_1 b_1 = a_2 b_2 = \cdots = a_s b_s.$

The series (17.4.1) is said to be *very-well-poised* when $r = s$, (17.4.11) is satisfied, and

17.4.12 $b_1 = -b_2 = \sqrt{a_0}.$

The series (17.4.1) is said to be *nearly-poised* when $r = s$ and

17.4.13 $a_0 q = a_1 b_1 = a_2 b_2 = \cdots = a_{s-1}b_{s-1}.$

17.5 $_0\phi_0, {}_1\phi_0, {}_1\phi_1$ Functions

Euler's Second Sum

17.5.1
$$_0\phi_0(-;-;q,z) = \sum_{n=0}^{\infty}\frac{(-1)^n q^{\binom{n}{2}}z^n}{(q;q)_n} = (z;q)_\infty, \quad |z| < 1;$$

compare (17.3.2).

q-Binomial Series

17.5.2 $_1\phi_0(a;-;q,z) = \dfrac{(az;q)_\infty}{(z;q)_\infty}, \quad |z| < 1;$

compare (17.2.37).

q-Binomial Theorem

17.5.3 $\quad {}_1\phi_0\big(q^{-n};-;q,z\big) = \big(zq^{-n};q\big)_n.$

This is (17.2.35) reformulated.

Euler's First Sum

17.5.4 $\quad {}_1\phi_0(0;-;q,z) = \sum_{n=0}^{\infty} \dfrac{z^n}{(q;q)_n} = \dfrac{1}{(z;q)_\infty}, \quad |z|<1;$

compare (17.3.1).

Cauchy's Sum

17.5.5 $\quad {}_1\phi_1\!\left(\begin{matrix} a \\ c \end{matrix}; q, c/a\right) = \dfrac{(c/a;q)_\infty}{(c;q)_\infty}, \qquad |c|<|a|.$

17.6 ${}_2\phi_1$ Function

17.6(i) Special Values

q-Gauss Sum

17.6.1 $\quad {}_2\phi_1\!\left(\begin{matrix} a,b \\ c \end{matrix}; q, c/(ab)\right) = \dfrac{(c/a, c/b; q)_\infty}{(c, c/(ab); q)_\infty}.$

First q-Chu–Vandermonde Sum

17.6.2 $\quad {}_2\phi_1\!\left(\begin{matrix} a, q^{-n} \\ c \end{matrix}; q, cq^n/a\right) = \dfrac{(c/a;q)_n}{(c;q)_n}.$

Second q-Chu–Vandermonde Sum

This reverses the order of summation in (17.6.2):

17.6.3 $\quad {}_2\phi_1\!\left(\begin{matrix} a, q^{-n} \\ c \end{matrix}; q, q\right) = \dfrac{a^n(c/a;q)_n}{(c;q)_n}.$

Andrews–Askey Sum

17.6.4
$${}_2\phi_1\!\left(\begin{matrix} b^2, b^2/c \\ c \end{matrix}; q^2, cq/b^2\right)$$
$$= \dfrac{1}{2} \dfrac{(b^2, q; q^2)_\infty}{(c, cq/b^2; q^2)_\infty} \left(\dfrac{(c/b;q)_\infty}{(b;q)_\infty} + \dfrac{(-c/b;q)_\infty}{(-b;q)_\infty} \right),$$
$$|cq| < |b^2|.$$

Bailey–Daum q-Kummer Sum

17.6.5 $\quad {}_2\phi_1\!\left(\begin{matrix} a,b \\ aq/b \end{matrix}; q, -q/b\right) = \dfrac{(-q;q)_\infty \big(aq, aq^2/b^2; q^2\big)_\infty}{(-q/b, aq/b; q)_\infty},$
$$|b| > |q|.$$

17.6(ii) ${}_2\phi_1$ Transformations

Heine's First Transformation

17.6.6 $\quad {}_2\phi_1\!\left(\begin{matrix} a,b \\ c \end{matrix}; q, z\right) = \dfrac{(b, az; q)_\infty}{(c, z; q)_\infty} \, {}_2\phi_1\!\left(\begin{matrix} c/b, z \\ az \end{matrix}; q, b\right),$
$$|z|<1, |b|<1.$$

Heine's Second Tranformation

17.6.7 $\quad {}_2\phi_1\!\left(\begin{matrix} a,b \\ c \end{matrix}; q, z\right) = \dfrac{(c/b, bz; q)_\infty}{(c, z; q)_\infty} \, {}_2\phi_1\!\left(\begin{matrix} abz/c, b \\ bz \end{matrix}; q, c/b\right),$
$$|z|<1, |c|<|b|.$$

Heine's Third Transformation

17.6.8
$${}_2\phi_1\!\left(\begin{matrix} a,b \\ c \end{matrix}; q, z\right)$$
$$= \dfrac{(abz/c; q)_\infty}{(z;q)_\infty} \, {}_2\phi_1\!\left(\begin{matrix} c/a, c/b \\ c \end{matrix}; q, abz/c\right),$$
$$|z|<1, |abz|<|c|.$$

Fine's First Transformation

17.6.9 $\quad {}_2\phi_1\!\left(\begin{matrix} q, aq \\ bq \end{matrix}; q, z\right) = -\dfrac{(1-b)(aq/b)}{(1-(aq/b))} \sum_{n=0}^{\infty} \dfrac{(aq, azq/b; q)_n q^n}{(azq^2/b; q)_n} + \dfrac{(aq, azq/b;q)_\infty}{(aq/b; q)_\infty} \, {}_2\phi_1\!\left(\begin{matrix} q, 0 \\ bq \end{matrix}; q, z\right), \qquad |z|<1.$

Fine's Second Transformation

17.6.10 $\quad (1-z) \, {}_2\phi_1\!\left(\begin{matrix} q, aq \\ bq \end{matrix}; q, z\right) = \sum_{n=0}^{\infty} \dfrac{(b/a;q)_n (-az)^n q^{(n^2+n)/2}}{(bq, zq; q)_n}, \qquad |z|<1.$

Fine's Third Transformation

17.6.11 $\quad \dfrac{1-z}{1-b} \, {}_2\phi_1\!\left(\begin{matrix} q, aq \\ bq \end{matrix}; q, z\right) = \sum_{n=0}^{\infty} \dfrac{(aq;q)_n (azq/b;q)_{2n} b^n}{(zq, aq/b; q)_n} - aq \sum_{n=0}^{\infty} \dfrac{(aq;q)_n (azq/b;q)_{2n+1} (bq)^n}{(zq;q)_n (aq/b;q)_{n+1}}, \quad |z|<1, |b|<1.$

Rogers–Fine Identity

17.6.12 $\quad (1-z) \, {}_2\phi_1\!\left(\begin{matrix} q, aq \\ bq \end{matrix}; q, z\right) = \sum_{n=0}^{\infty} \dfrac{(aq, azq/b; q)_n}{(bq, zq; q)_n}(1-azq^{2n+1})(bz)^n q^{n^2}, \qquad |z|<1.$

Nonterminating Form of the q-Vandermonde Sum

17.6.13
$$_2\phi_1(a,b;c;q,q) + \frac{(q/c,a,b;q)_\infty}{(c/q,aq/c,bq/c;q)_\infty}\,_2\phi_1\big(aq/c,bq/c;q^2/c;q,q\big) = \frac{(q/c,abq/c;q)_\infty}{(aq/c,bq/c;q)_\infty},$$

17.6.14
$$\sum_{n=0}^\infty \frac{(a;q)_n\,(b;q^2)_n\,z^n}{(q;q)_n\,(azb;q^2)_n} = \frac{(az,bz;q^2)_\infty}{(z,azb;q^2)_\infty}\,_2\phi_1\!\left(\genfrac{}{}{0pt}{}{a,b}{bz};q^2,zq\right).$$

Three-Term $_2\phi_1$ Transformations

17.6.15
$$_2\phi_1\!\left(\genfrac{}{}{0pt}{}{a,b}{c};q,z\right) = \frac{(abz/c,q/c;q)_\infty}{(az/c,q/a;q)_\infty}\,_2\phi_1\!\left(\genfrac{}{}{0pt}{}{c/a,cq/(abz)}{cq/(az)};q,bq/c\right)$$
$$-\frac{(b,q/c,c/a,az/q,q^2/(az);q)_\infty}{(c/q,bq/c,q/a,az/c,cq/(az);q)_\infty}\,_2\phi_1\!\left(\genfrac{}{}{0pt}{}{aq/c,bq/c}{q^2/c};q,z\right), \qquad |z|<1, |bq|<|c|.$$

17.6.16
$$_2\phi_1\!\left(\genfrac{}{}{0pt}{}{a,b}{c};q,z\right) = \frac{(b,c/a,az,q/(az);q)_\infty}{(c,b/a,z,q/z;q)_\infty}\,_2\phi_1\!\left(\genfrac{}{}{0pt}{}{a,aq/c}{aq/b};q,cq/(abz)\right)$$
$$+\frac{(a,c/b,bz,q/(bz);q)_\infty}{(c,a/b,z,q/z;q)_\infty}\,_2\phi_1\!\left(\genfrac{}{}{0pt}{}{b,bq/c}{bq/a};q,cq/(abz)\right), \qquad |z|<1,\, |abz|<|cq|.$$

17.6(iii) Contiguous Relations

Heine's Contiguous Relations

17.6.17
$$_2\phi_1\!\left(\genfrac{}{}{0pt}{}{a,b}{c/q};q,z\right) - {}_2\phi_1\!\left(\genfrac{}{}{0pt}{}{a,b}{c};q,z\right) = cz\frac{(1-a)(1-b)}{(q-c)(1-c)}\,_2\phi_1\!\left(\genfrac{}{}{0pt}{}{aq,bq}{cq};q,z\right),$$

17.6.18
$$_2\phi_1\!\left(\genfrac{}{}{0pt}{}{aq,b}{c};q,z\right) - {}_2\phi_1\!\left(\genfrac{}{}{0pt}{}{a,b}{c};q,z\right) = az\frac{1-b}{1-c}\,_2\phi_1\!\left(\genfrac{}{}{0pt}{}{aq,bq}{cq};q,z\right),$$

17.6.19
$$_2\phi_1\!\left(\genfrac{}{}{0pt}{}{aq,b}{cq};q,z\right) - {}_2\phi_1\!\left(\genfrac{}{}{0pt}{}{a,b}{c};q,z\right) = az\frac{(1-b)(1-(c/a))}{(1-c)(1-cq)}\,_2\phi_1\!\left(\genfrac{}{}{0pt}{}{aq,bq}{cq^2};q,z\right),$$

17.6.20
$$_2\phi_1\!\left(\genfrac{}{}{0pt}{}{aq,b/q}{c};q,z\right) - {}_2\phi_1\!\left(\genfrac{}{}{0pt}{}{a,b}{c};q,z\right) = az\frac{(1-b/(aq))}{1-c}\,_2\phi_1\!\left(\genfrac{}{}{0pt}{}{aq,b}{cq};q,z\right),$$

17.6.21
$$b(1-a)\,_2\phi_1\!\left(\genfrac{}{}{0pt}{}{aq,b}{c};q,z\right) - a(1-b)\,_2\phi_1\!\left(\genfrac{}{}{0pt}{}{a,bq}{c};q,z\right) = (b-a)\,_2\phi_1\!\left(\genfrac{}{}{0pt}{}{a,b}{c};q,z\right),$$

17.6.22
$$a\left(1-\frac{b}{c}\right)\,_2\phi_1\!\left(\genfrac{}{}{0pt}{}{a,b/q}{c};q,z\right) - b\left(1-\frac{a}{c}\right)\,_2\phi_1\!\left(\genfrac{}{}{0pt}{}{a/q,b}{c};q,z\right) = (a-b)\left(1-\frac{abz}{cq}\right)\,_2\phi_1\!\left(\genfrac{}{}{0pt}{}{a,b}{c};q,z\right),$$

17.6.23
$$q\left(1-\frac{a}{c}\right)\,_2\phi_1\!\left(\genfrac{}{}{0pt}{}{a/q,b}{c};q,z\right) + (1-a)\left(1-\frac{abz}{c}\right)\,_2\phi_1\!\left(\genfrac{}{}{0pt}{}{aq,b}{c};q,z\right) = \left(1+q-a-\frac{aq}{c}+\frac{a^2z}{c}-\frac{abz}{c}\right)\,_2\phi_1\!\left(\genfrac{}{}{0pt}{}{a,b}{c};q,z\right),$$

17.6.24
$$(1-c)(q-c)(abz-c)\,_2\phi_1\!\left(\genfrac{}{}{0pt}{}{a,b}{c/q};q,z\right) + z(c-a)(c-b)\,_2\phi_1\!\left(\genfrac{}{}{0pt}{}{a,b}{cq};q,z\right)$$
$$= (c-1)(c(q-c) + z(ca+cb-ab-abq))\,_2\phi_1\!\left(\genfrac{}{}{0pt}{}{a,b}{c};q,z\right).$$

17.6(iv) Differential Equations

Iterations of $\mathcal{D}$

17.6.25
$$\mathcal{D}_q^n\,_2\phi_1\!\left(\genfrac{}{}{0pt}{}{a,b}{c};q,zd\right) = \frac{(a,b;q)_n\,d^n}{(c;q)_n(1-q)^n}\,_2\phi_1\!\left(\genfrac{}{}{0pt}{}{aq^n,bq^n}{cq^n};q,dz\right),$$

17.6.26
$$\mathcal{D}_q^n\left(\frac{(z;q)_\infty}{(abz/c;q)_\infty}\,_2\phi_1\!\left(\genfrac{}{}{0pt}{}{a,b}{c};q,z\right)\right) = \frac{(c/a,c/b;q)_n}{(c;q)_n(1-q)^n}\left(\frac{ab}{c}\right)^n\frac{(zq^n;q)_\infty}{(abz/c;q)_\infty}\,_2\phi_1\!\left(\genfrac{}{}{0pt}{}{a,b}{cq^n};q,zq^n\right).$$

q-Differential Equation

17.6.27
$$z(c - abqz)\mathcal{D}_q^2 \, {}_2\phi_1\!\left(\begin{matrix}a,b\\c\end{matrix};q,z\right) + \left(\frac{1-c}{1-q} + \frac{(1-a)(1-b)-(1-abq)}{1-q}z\right)\mathcal{D}_q \, {}_2\phi_1\!\left(\begin{matrix}a,b\\c\end{matrix};q,z\right)$$
$$- \frac{(1-a)(1-b)}{(1-q)^2} \, {}_2\phi_1\!\left(\begin{matrix}a,b\\c\end{matrix};q,z\right) = 0.$$

(17.6.27) reduces to the hypergeometric equation (15.10.1) with the substitutions $a \to q^a$, $b \to q^b$, $c \to q^c$, followed by $\lim_{q \to 1-}$.

17.6(v) Integral Representations

17.6.28
$${}_2\phi_1\!\left(\begin{matrix}q^\alpha, q^\beta\\q^\gamma\end{matrix};q,z\right) = \frac{\Gamma_q(\gamma)}{\Gamma_q(\beta)\,\Gamma_q(\gamma-\beta)} \int_0^1 \frac{t^{\beta-1}\,(tq;q)_{\gamma-\beta-1}}{(xt;q)_\alpha}\,d_q t.$$

17.6.29
$${}_2\phi_1\!\left(\begin{matrix}a,b\\c\end{matrix};q,z\right) = \left(\frac{-1}{2\pi i}\right)\frac{(a,b;q)_\infty}{(q,c;q)_\infty} \int_{-i\infty}^{i\infty} \frac{(q^{1+\zeta}, cq^\zeta;q)_\infty}{(aq^\zeta, bq^\zeta;q)_\infty}\frac{\pi(-z)^\zeta}{\sin(\pi\zeta)}\,d\zeta,$$

where $|z| < 1$, $|\mathrm{ph}(-z)| < \pi$, and the contour of integration separates the poles of $(q^{1+\zeta}, cq^\zeta;q)_\infty / \sin(\pi\zeta)$ from those of $1/(aq^\zeta, bq^\zeta;q)_\infty$, and the infimum of the distances of the poles from the contour is positive.

17.6(vi) Continued Fractions

For continued-fraction representations of the ${}_2\phi_1$ function, see Cuyt et al. (2008, pp. 395–399).

17.7 Special Cases of Higher ${}_r\phi_s$ Functions

17.7(i) ${}_2\phi_2$ Functions

q-Analog of Bailey's ${}_2F_1(-1)$ Sum

17.7.1
$${}_2\phi_2\!\left(\begin{matrix}a, q/a\\-q, b\end{matrix};q,-b\right) = \frac{(ab, bq/a; q^2)_\infty}{(b;q)_\infty}, \quad |b| < 1.$$

q-Analog of Gauss's ${}_2F_1(-1)$ Sum

17.7.2
$${}_2\phi_2\!\left(\begin{matrix}a^2, b^2\\abq^{\frac{1}{2}}, -abq^{\frac{1}{2}}\end{matrix};q,-q\right) = \frac{(a^2q, b^2q; q^2)_\infty}{(q, a^2b^2q; q^2)_\infty}.$$

Sum Related to (17.6.4)

17.7.3
$${}_2\phi_2\!\left(\begin{matrix}c^2/b^2, b^2\\c, cq\end{matrix};q^2,q\right)$$
$$= \frac{1}{2}\frac{(b^2, q; q^2)_\infty}{(c, cq; q^2)_\infty}\left(\frac{(c/b;q)_\infty}{(b;q)_\infty} + \frac{(-c/b;q)_\infty}{(-b;q)_\infty}\right).$$

17.7(ii) ${}_3\phi_2$ Functions

q-Pfaff–Saalschütz Sum

17.7.4
$${}_3\phi_2\!\left(\begin{matrix}a, b, q^{-n}\\c, abq^{1-n}/c\end{matrix};q,q\right) = \frac{(c/a, c/b;q)_n}{(c, c/(ab);q)_n}.$$

Nonterminating Form of the q-Saalschütz Sum

17.7.5
$${}_3\phi_2\!\left(\begin{matrix}a,b,c\\e,f\end{matrix};q,q\right) + \frac{(q/e, a, b, c, qf/e;q)_\infty}{(e/q, aq/e, bq/e, cq/e, f;q)_\infty}$$
$$\times {}_3\phi_2\!\left(\begin{matrix}aq/e, bq/e, cq/e\\q^2/e, qf/e\end{matrix};q,q\right)$$
$$= \frac{(q/e, f/a, f/b, f/c;q)_\infty}{(aq/e, bq/e, cq/e, f;q)_\infty},$$

where $ef = abcq$.

F. H. Jackson's Terminating q-Analog of Dixon's Sum

17.7.6
$${}_3\phi_2\!\left(\begin{matrix}q^{-2n}, b, c\\q^{1-2n}/b, q^{1-2n}/c\end{matrix};q,\frac{q^{2-n}}{bc}\right) = \frac{(b,c;q)_n\,(q,bc;q)_{2n}}{(q,bc;q)_n\,(b,c;q)_{2n}}.$$

Continued Fractions

For continued-fraction representations of a ratio of ${}_3\phi_2$ functions, see Cuyt et al. (2008, pp. 399–400).

17.7(iii) Other ${}_r\phi_s$ Functions

q-Analog of Dixon's ${}_3F_2(1)$ Sum

17.7.7
$${}_4\phi_3\!\left(\begin{matrix}a, -qa^{\frac{1}{2}}, b, c\\-a^{\frac{1}{2}}, aq/b, aq/c\end{matrix};q,\frac{qa^{\frac{1}{2}}}{bc}\right)$$
$$= \frac{\left(aq, qa^{\frac{1}{2}}/b, qa^{\frac{1}{2}}/c, aq/(bc);q\right)_\infty}{\left(aq/b, aq/c, qa^{\frac{1}{2}}, qa^{\frac{1}{2}}/(bc);q\right)_\infty}.$$

Gasper–Rahman q-Analog of Watson's ${}_3F_2$ Sum

17.7.8
$${}_8\phi_7\!\left(\begin{matrix}\lambda, q\lambda^{\frac{1}{2}}, -q\lambda^{\frac{1}{2}}, a, b, c, -c, \lambda q/c^2\\ \lambda^{\frac{1}{2}}, -\lambda^{\frac{1}{2}}, \lambda q/a, \lambda q/b, \lambda q/c, -\lambda q/c, c^2\end{matrix};q,-\frac{\lambda q}{ab}\right)$$
$$= \frac{(\lambda q, c^2/\lambda;q)_\infty\,(aq, bq, c^2 q/a, c^2 q/b; q^2)_\infty}{(\lambda q/a, \lambda q/b;q)_\infty\,(q, abq, c^2 q, c^2 q/(ab); q^2)_\infty},$$

where $\lambda = -c(ab/q)^{\frac{1}{2}}$.

Andrews' Terminating q-Analog of (17.7.8)

17.7.9
$$_4\phi_3\left(\begin{matrix} q^{-n}, aq^n, c, -c \\ (aq)^{\frac{1}{2}}, -(aq)^{\frac{1}{2}}, c^2 \end{matrix}; q, q\right)$$
$$= \begin{cases} 0, & n \text{ odd,} \\ \dfrac{c^n \left(q, aq/c^2; q^2\right)_{n/2}}{(aq, c^2q; q^2)_{n/2}}, & n \text{ even.} \end{cases}$$

Gasper–Rahman q-Analog of Whipple's $_3F_2$ Sum

17.7.10
$$_8\phi_7\left(\begin{matrix} -c, q(-c)^{\frac{1}{2}}, -q(-c)^{\frac{1}{2}}, a, q/a, c, -d, -q/d \\ (-c)^{\frac{1}{2}}, -(-c)^{\frac{1}{2}}, -cq/a, -ac, -q, cq/d, cd \end{matrix}; q, c\right)$$
$$= \frac{(-c, -cq; q)_\infty \left(acd, acq/d, cdq/a, cq^2/(ad); q^2\right)_\infty}{(cd, cq/d, -ac, -cq/a; q)_\infty}.$$

Andrews' Terminating q-Analog

17.7.11
$$_4\phi_3\left(\begin{matrix} q^{-n}, q^{n+1}, c, -c \\ e, c^2q/e, -q \end{matrix}; q, q\right)$$
$$= \frac{\left(eq^{-n}, eq^{n+1}, c^2q^{1-n}/e, c^2q^{n+2}/e; q^2\right)_\infty}{(e, c^2q/e; q)_\infty}.$$

First q-Analog of Bailey's $_4F_3(1)$ Sum

17.7.12
$$_4\phi_3\left(\begin{matrix} a, aq, b^2q^{2n}, q^{-2n} \\ b, bq, a^2q^2 \end{matrix}; q^2, q^2\right) = \frac{a^n (-q, b/a; q)_n}{(-aq, b; q)_n}.$$

Second q-Analog of Bailey's $_4F_3(1)$ Sum

17.7.13
$$_4\phi_3\left(\begin{matrix} a, aq, b^2q^{2n-2}, q^{-2n} \\ b, bq, a^2 \end{matrix}; q^2, q^2\right)$$
$$= \frac{a^n (-q, b/a; q)_n (1 - bq^{n-1})}{(-a, b; q)_n (1 - bq^{2n-1})}.$$

F. H. Jackson's q-Analog of Dougall's $_7F_6(1)$ Sum

17.7.14
$$_8\phi_7\left(\begin{matrix} a, qa^{\frac{1}{2}}, -qa^{\frac{1}{2}}, b, c, d, e, q^{-n} \\ a^{\frac{1}{2}}, -a^{\frac{1}{2}}, aq/b, aq/c, aq/d, aq/e, aq^{n+1} \end{matrix}; q, q\right)$$
$$= \frac{(aq, aq/(bc), aq/(bd), aq/(cd); q)_n}{(aq/b, aq/c, aq/d, aq/(bcd); q)_n},$$

where $a^2q = bcdeq^{-n}$.

Limiting Cases of (17.7.14)

17.7.15
$$_6\phi_5\left(\begin{matrix} a, qa^{\frac{1}{2}}, -qa^{\frac{1}{2}}, b, c, d \\ a^{\frac{1}{2}}, -a^{\frac{1}{2}}, aq/b, aq/c, aq/d \end{matrix}; q, \frac{aq}{bcd}\right)$$
$$= \frac{(aq, aq/(bc), aq/(bd), aq/(cd); q)_\infty}{(aq/b, aq/c, aq/d, aq/(bcd); q)_\infty},$$

and when $d = q^{-n}$,

17.7.16
$$_6\phi_5\left(\begin{matrix} a, qa^{\frac{1}{2}}, -qa^{\frac{1}{2}}, b, c, q^{-n} \\ a^{\frac{1}{2}}, -a^{\frac{1}{2}}, aq/b, aq/c, aq^{n+1} \end{matrix}; q, \frac{aq^{n+1}}{bc}\right)$$
$$= \frac{(aq, aq/(bc); q)_n}{(aq/b, aq/c; q)_n}.$$

See http://dlmf.nist.gov/17.7.iii for additional results.

17.8 Special Cases of $_r\psi_r$ Functions

Jacobi's Triple Product

17.8.1
$$\sum_{n=-\infty}^{\infty} (-z)^n q^{n(n-1)/2} = (q, z, q/z; q)_\infty;$$

compare (20.5.9).

Ramanujan's $_1\psi_1$ Summation

17.8.2
$$_1\psi_1\left(\begin{matrix} a \\ b \end{matrix}; q, z\right) = \frac{(q, b/a, az, q/(az); q)_\infty}{(b, q/a, z, b/(az); q)_\infty}.$$

Quintuple Product Identity

17.8.3
$$\sum_{n=-\infty}^{\infty} (-1)^n q^{n(3n-1)/2} z^{3n} (1 + zq^n)$$
$$= (q, -z, -q/z; q)_\infty \left(qz^2, q/z^2; q^2\right)_\infty.$$

Bailey's Bilateral Summations

17.8.4
$$_2\psi_2(b, c; aq/b, aq/c; q, -aq/(bc)) = \frac{(aq/(bc); q)_\infty \left(aq^2/b^2, aq^2/c^2, q^2, aq, q/a; q^2\right)_\infty}{(aq/b, aq/c, q/b, q/c, -aq/(bc); q)_\infty},$$

17.8.5
$$_3\psi_3\left(\begin{matrix} b, c, d \\ q/b, q/c, q/d \end{matrix}; q, \frac{q}{bcd}\right) = \frac{(q, q/(bc), q/(bd), q/(cd); q)_\infty}{(q/b, q/c, q/d, q/(bcd); q)_\infty},$$

17.8.6
$$_4\psi_4\left(\begin{matrix} -qa^{\frac{1}{2}}, b, c, d \\ -a^{\frac{1}{2}}, aq/b, aq/c, aq/d \end{matrix}; q, \frac{qa^{\frac{3}{2}}}{bcd}\right) = \frac{\left(aq, aq/(bc), aq/(bd), aq/(cd), qa^{\frac{1}{2}}/b, qa^{\frac{1}{2}}/c, qa^{\frac{1}{2}}/d, q, q/a; q\right)_\infty}{\left(aq/b, aq/c, aq/d, q/b, q/c, q/d, qa^{\frac{1}{2}}, qa^{-\frac{1}{2}}, qa^{\frac{3}{2}}/(bcd); q\right)_\infty},$$

17.8.7
$$_6\psi_6\left(\begin{matrix} qa^{\frac{1}{2}}, -qa^{\frac{1}{2}}, b, c, d, e \\ a^{\frac{1}{2}}, -a^{\frac{1}{2}}, aq/b, aq/c, aq/d, aq/e \end{matrix}; q, \frac{qa^2}{bcde}\right) = \frac{(aq, aq/(bc), aq/(bd), aq/(be), aq/(cd), aq/(ce), aq/(de), q, q/a; q)_\infty}{(aq/b, aq/c, aq/d, aq/e, q/b, q/c, q/d, q/e, qa^2/(bcde); q)_\infty}.$$

17.9 Transformations of Higher $_r\phi_r$ Functions

17.9(i) $_2\phi_1 \to {}_2\phi_2$, $_3\phi_1$, or $_3\phi_2$

F. H. Jackson's Transformations

17.9.1
$$_2\phi_1\left(\begin{matrix} a, b \\ c \end{matrix}; q, z\right) = \frac{(za; q)_\infty}{(z; q)_\infty} {}_2\phi_2\left(\begin{matrix} a, c/b \\ c, az \end{matrix}; q, bz\right),$$

17.9.2
$$_2\phi_1\left(\begin{matrix} q^{-n}, b \\ c \end{matrix}; q, z\right) = \frac{(c/b; q)_n}{(c; q)_n} b^n {}_3\phi_1\left(\begin{matrix} q^{-n}, b, q/c \\ bq^{1-n}/c \end{matrix}; q, z/c\right),$$

17.9.3
$$_2\phi_1\left(\begin{matrix} a, b \\ c \end{matrix}; q, z\right) = \frac{(abz/c; q)_\infty}{(bz/c; q)_\infty} {}_3\phi_2\left(\begin{matrix} a, c/b, 0 \\ c, cq/bz \end{matrix}; q, q\right),$$

17.9.4
$$_2\phi_1\left(\begin{matrix} q^{-n}, b \\ c \end{matrix}; q, z\right) = \frac{(c/b; q)_n}{(c; q)_n} \left(\frac{bz}{q}\right)^n {}_3\phi_2\left(\begin{matrix} q^{-n}, q/z, q^{1-n}/c \\ bq^{1-n}/c, 0 \end{matrix}; q, q\right),$$

17.9.5
$$_2\phi_1\left(\begin{matrix} q^{-n}, b \\ c \end{matrix}; q, z\right) = \frac{(c/b; q)_n}{(c; q)_n} {}_3\phi_2\left(\begin{matrix} q^{-n}, b, bzq^{-n}/c \\ bq^{1-n}/c, 0 \end{matrix}; q, q\right).$$

17.9(ii) $_3\phi_2 \to {}_3\phi_2$

Transformations of $_3\phi_2$-Series

17.9.6
$$_3\phi_2\left(\begin{matrix} a, b, c \\ d, e \end{matrix}; q, de/(abc)\right) = \frac{(e/a, de/(bc); q)_\infty}{(e, de/(abc); q)_\infty} {}_3\phi_2\left(\begin{matrix} a, d/b, d/c \\ d, de/(bc) \end{matrix}; q, e/a\right),$$

17.9.7
$$_3\phi_2\left(\begin{matrix} a, b, c \\ d, e \end{matrix}; q, de/(abc)\right) = \frac{(b, de/(ab), de/(bc); q)_\infty}{(d, e, de/(abc); q)_\infty} {}_3\phi_2\left(\begin{matrix} d/b, e/b, de/(abc) \\ de/(ab), de/(bc) \end{matrix}; q, b\right),$$

17.9.8
$$_3\phi_2\left(\begin{matrix} q^{-n}, b, c \\ d, e \end{matrix}; q, q\right) = \frac{(de/(bc); q)_n}{(e; q)_n} \left(\frac{bc}{d}\right)^n {}_3\phi_2\left(\begin{matrix} q^{-n}, d/b, d/c \\ d, de/(bc) \end{matrix}; q, q\right),$$

17.9.9
$$_3\phi_2\left(\begin{matrix} q^{-n}, b, c \\ d, e \end{matrix}; q, q\right) = \frac{(e/c; q)_n}{(e; q)_n} c^n {}_3\phi_2\left(\begin{matrix} q^{-n}, c, d/b \\ d, cq^{1-n}/e \end{matrix}; q, \frac{bq}{e}\right),$$

17.9.10
$$_3\phi_2\left(\begin{matrix} q^{-n}, b, c \\ d, e \end{matrix}; q, \frac{deq^n}{bc}\right) = \frac{(e/c; q)_n}{(e; q)_n} {}_3\phi_2\left(\begin{matrix} q^{-n}, c, d/b \\ d, cq^{1-n}/e \end{matrix}; q, q\right).$$

q-Sheppard Identity

17.9.11
$$_3\phi_2\left(\begin{matrix} q^{-n}, b, c \\ d, e \end{matrix}; q, q\right) = \frac{(e/c, d/c; q)_n}{(e, d; q)_n} c^n {}_3\phi_2\left(\begin{matrix} q^{-n}, c, cbq^{1-n}/(de) \\ cq^{1-n}/e, cq^{1-n}/d \end{matrix}; q, q\right),$$

For further results see http://dlmf.nist.gov/17.9.ii.

17.9(iii) Further $_r\phi_s$ Functions

Sears' Balanced $_4\phi_3$ Transformations

With $def = abcq^{1-n}$

17.9.14
$$_4\phi_3\left(\begin{matrix} q^{-n}, a, b, c \\ d, e, f \end{matrix}; q, q\right) = \frac{(e/a, f/a; q)_n}{(e, f; q)_n} a^n {}_4\phi_3\left(\begin{matrix} q^{-n}, a, d/b, d/c \\ d, aq^{1-n}/e, aq^{1-n}/f \end{matrix}; q, q\right)$$
$$= \frac{(a, ef/(ab), ef/(ac); q)_n}{(e, f, ef/(abc); q)_n} {}_4\phi_3\left(\begin{matrix} q^{-n}, e/a, f/a, ef/(abc) \\ ef/(ab), ef/(ac), q^{1-n}/a \end{matrix}; q, q\right).$$

Watson's q-Analog of Whipple's Theorem

With n a nonnegative integer

17.9.15 $\quad \dfrac{(aq, aq/(de); q)_n}{(aq/d, aq/e; q)_n} {}_4\phi_3 \left(\begin{matrix} aq/(bc), d, e, q^{-n} \\ aq/b, aq/c, deq^{-n}/a \end{matrix} ; q, q \right) = {}_8\phi_7 \left(\begin{matrix} a, qa^{\frac{1}{2}}, -qa^{\frac{1}{2}}, b, c, d, e, q^{-n} \\ a^{\frac{1}{2}}, -a^{\frac{1}{2}}, aq/b, aq/c, aq/d, aq/e, aq^{n+1} \end{matrix} ; q, \dfrac{a^2 q^{2+n}}{bcde} \right).$

Bailey's Transformation of Very-Well-Poised ${}_8\phi_7$

17.9.16

$${}_8\phi_7 \left(\begin{matrix} a, qa^{\frac{1}{2}}, -qa^{\frac{1}{2}}, b, c, d, e, f \\ a^{\frac{1}{2}}, -a^{\frac{1}{2}}, aq/b, aq/c, aq/d, aq/e, aq/f \end{matrix} ; q, \dfrac{a^2 q^2}{bcdef} \right)$$

$$= \dfrac{(aq, aq/(de), aq/(df), aq/(ef); q)_\infty}{(aq/d, aq/e, aq/f, aq/(def); q)_\infty} {}_4\phi_3 \left(\begin{matrix} aq/(bc), d, e, f \\ aq/b, aq/c, def/a \end{matrix} ; q, q \right)$$

$$+ \dfrac{(aq, aq/(bc), d, e, f, a^2 q^2/(bdef), a^2 q^2/(cdef); q)_\infty}{(aq/b, aq/c, aq/d, aq/e, aq/f, a^2 q^2/(bcdef), def/(aq); q)_\infty} {}_4\phi_3 \left(\begin{matrix} aq/(de), aq/(df), aq/(ef), a^2 q^2/(bcdef) \\ a^2 q^2/(bdef), a^2 q^2/(cdef), aq^2/(def) \end{matrix} ; q, q \right).$$

For additional results see http://dlmf.nist.gov/17.9.iii and Gasper and Rahman (2004, Appendix III and Chapter 2).

17.9(iv) Bibasic Series

Mixed-Base Heine-Type Transformations

17.9.19
$$\sum_{n=0}^{\infty} \dfrac{(a; q^2)_n (b; q)_n}{(q^2; q^2)_n (c; q)_n} z^n = \dfrac{(b; q)_\infty (az; q^2)_\infty}{(c; q)_\infty (z; q^2)_\infty} \sum_{n=0}^{\infty} \dfrac{(c/b; q)_{2n} (z; q^2)_n b^{2n}}{(q; q)_{2n} (az; q^2)_n}$$
$$+ \dfrac{(b; q)_\infty (azq; q^2)_\infty}{(c; q)_\infty (zq; q^2)_\infty} \sum_{n=0}^{\infty} \dfrac{(c/b; q)_{2n+1} (zq; q^2)_n b^{2n+1}}{(q; q)_{2n+1} (azq; q^2)_n}.$$

17.9.20 $\quad \displaystyle\sum_{n=0}^{\infty} \dfrac{(a; q^k)_n (b; q)_{kn}}{(q^k; q^k)_n (c; q)_{kn}} z^n = \dfrac{(b; q)_\infty (az; q^k)_\infty}{(c; q)_\infty (z; q^k)_\infty} \sum_{n=0}^{\infty} \dfrac{(c/b; q)_n (z; q^k)_n b^n}{(q; q)_n (az; q^k)_n}, \qquad k = 1, 2, 3, \ldots.$

17.10 Transformations of ${}_r\psi_r$ Functions

Bailey's ${}_2\psi_2$ Transformations

17.10.1 $\quad {}_2\psi_2 \left(\begin{matrix} a, b \\ c, d \end{matrix} ; q, z \right) = \dfrac{(az, d/a, c/b, dq/(abz); q)_\infty}{(z, d, q/b, cd/(abz); q)_\infty} {}_2\psi_2 \left(\begin{matrix} a, abz/d \\ az, c \end{matrix} ; q, \dfrac{d}{a} \right),$

17.10.2 $\quad {}_2\psi_2 \left(\begin{matrix} a, b \\ c, d \end{matrix} ; q, z \right) = \dfrac{(az, bz, cq/(abz), dq/(abz); q)_\infty}{(q/a, q/b, c, d; q)_\infty} {}_2\psi_2 \left(\begin{matrix} abz/c, abz/d \\ az, bz \end{matrix} ; q, \dfrac{cd}{abz} \right).$

Other Transformations

17.10.3
$${}_8\psi_8 \left(\begin{matrix} qa^{\frac{1}{2}}, -qa^{\frac{1}{2}}, c, d, e, f, aq^{-n}, q^{-n} \\ a^{\frac{1}{2}}, -a^{\frac{1}{2}}, aq/c, aq/d, aq/e, aq/f, q^{n+1}, aq^{n+1} \end{matrix} ; q, \dfrac{a^2 q^{2n+2}}{cdef} \right)$$
$$= \dfrac{(aq, q/a, aq/(cd), aq/(ef); q)_n}{(q/c, q/d, aq/e, aq/f; q)_n} {}_4\psi_4 \left(\begin{matrix} e, f, aq^{n+1}/(cd), q^{-n} \\ aq/c, aq/d, q^{n+1}, ef/(aq^n) \end{matrix} ; q, q \right),$$

17.10.4
$${}_2\psi_2 \left(\begin{matrix} e, f \\ aq/c, aq/d \end{matrix} ; q, \dfrac{aq}{ef} \right) = \dfrac{(q/c, q/d, aq/e, aq/f; q)_\infty}{(aq, q/a, aq/(cd), aq/(ef); q)_\infty} \sum_{n=-\infty}^{\infty} \dfrac{(1 - aq^{2n})(c, d, e, f; q)_n}{(1-a)(aq/c, aq/d, aq/e, aq/f; q)_n} \left(\dfrac{qa^3}{cdef} \right)^n q^{n^2}.$$

17.10.5
$$\dfrac{(aq/b, aq/c, aq/d, aq/e, q/(ab), q/(ac), q/(ad), q/(ae); q)_\infty}{(fa, ga, f/a, g/a, qa^2, q/a^2; q)_\infty} {}_8\psi_8 \left(\begin{matrix} qa, -qa, ba, ca, da, ea, fa, ga \\ a, -a, aq/b, aq/c, aq/d, aq/e, aq/f, aq/g \end{matrix} ; q, \dfrac{q^2}{bcdefg} \right)$$
$$= \dfrac{(q, q/(bf), q/(cf), q/(df), q/(ef), qf/b, qf/c, qf/d, qf/e; q)_\infty}{(fa, q/(fa), aq/f, f/a, g/f, fg, qf^2; q)_\infty}$$
$$\times {}_8\phi_7 \left(\begin{matrix} f^2, qf, -qf, fb, fc, fd, fe, fg \\ f, -f, fq/b, fq/c, fq/d, fq/e, fq/g \end{matrix} ; q, \dfrac{q^2}{bcdefg} \right) + \text{idem}(f; g).$$

17.10.6
$$\frac{(aq/b,aq/c,aq/d,aq/e,aq/f,q/(ab),q/(ac),q/(ad),q/(ae),q/(af);q)_\infty}{(ag,ah,ak,g/a,h/a,k/a,qa^2,q/a^2;q)_\infty}$$
$$\times {}_{10}\psi_{10}\left(\begin{array}{c}qa,-qa,ba,ca,da,ea,fa,ga,ha,ka\\a,-a,aq/b,aq/c,aq/d,aq/e,aq/f,aq/g,aq/h,aq/k\end{array};q,\frac{q^2}{bcdefghk}\right)$$
$$=\frac{(q,q/(bg),q/(cg),q/(dg),q/(eg),q/(fg),qg/b,qg/c,qg/d,qg/e,qg/f;q)_\infty}{(gh,gk,h/g,ag,q/(ag),g/a,aq/g,qg^2;q)_\infty}$$
$$\times {}_{10}\phi_9\left(\begin{array}{c}g^2,qg,-qg,gb,gc,gd,ge,gf,gh,gk\\g,-g,qg/b,qg/c,qg/d,qg/e,qg/f,qg/h,qg/k\end{array};q,\frac{q^2}{bcdefghk}\right)+\text{idem}(g;h,k).$$

17.11 Transformations of q-Appell Functions

17.11.1 $$\Phi^{(1)}(a;b,b';c;x,y)=\frac{(a,bx,b'y;q)_\infty}{(c,x,y;q)_\infty}{}_3\phi_2\left(\begin{array}{c}c/a,x,y\\bx,b'y\end{array};q,a\right),$$

17.11.2 $$\Phi^{(2)}(a;b,b';c,c';x,y)=\frac{(b,ax;q)_\infty}{(c,x;q)_\infty}\sum_{n,r\geq 0}\frac{(a,b';q)_n\,(c/b,x;q)_r\,b^r y^n}{(q,c';q)_n\,(q)_r\,(ax;q)_{n+r}},$$

17.11.3 $$\Phi^{(3)}(a,a';b,b';c;x,y)=\frac{(a,bx;q)_\infty}{(c,x;q)_\infty}\sum_{n,r\geq 0}\frac{(a',b';q)_n\,(x;q)_r\,(c/a;q)_{n+r}\,a^r y^n}{(q,c/a;q)_n\,(q,bx;q)_r}.$$

Of (17.11.1)–(17.11.3) only (17.11.1) has a natural generalization: the following sum reduces to (17.11.1) when $n=2$.

17.11.4
$$\sum_{m_1,\ldots,m_n\geq 0}\frac{(a;q)_{m_1+m_2+\cdots+m_n}(b_1;q)_{m_1}(b_2;q)_{m_2}\cdots(b_n;q)_{m_n}x_1^{m_1}x_2^{m_2}\cdots x_n^{m_n}}{(q;q)_{m_1}(q;q)_{m_2}\cdots(q;q)_{m_n}(c;q)_{m_1+m_2+\cdots+m_n}}$$
$$=\frac{(a,b_1x_1,b_2x_2,\ldots,b_nx_n;q)_\infty}{(c,x_1,x_2,\ldots,x_n;q)_\infty}{}_{n+1}\phi_n\left(\begin{array}{c}c/a,x_1,x_2,\ldots,x_n\\b_1x_1,b_2x_2,\ldots,b_nx_n\end{array};q,a\right).$$

17.12 Bailey Pairs

Bailey Transform

17.12.1 $$\sum_{n=0}^\infty \alpha_n\gamma_n=\sum_{n=0}^\infty \beta_n\delta_n,$$

where

17.12.2 $$\beta_n=\sum_{j=0}^n \alpha_j u_{n-j}v_{n+j},\quad \gamma_n=\sum_{j=n}^\infty \delta_j u_{j-n}v_{j+n}.$$

Bailey Pairs

A sequence of pairs of rational functions of several variables (α_n,β_n), $n=0,1,2,\ldots$, is called a *Bailey pair* provided that for each $n\geq 0$

17.12.3 $$\beta_n=\sum_{j=0}^n \frac{\alpha_j}{(q;q)_{n-j}(aq;q)_{n+j}}.$$

Weak Bailey Lemma

If (α_n,β_n) is a Bailey pair, then

17.12.4 $$\sum_{n=0}^\infty q^{n^2}a^n\beta_n=\frac{1}{(aq;q)_\infty}\sum_{n=0}^\infty q^{n^2}a^n\alpha_n.$$

Strong Bailey Lemma

If (α_n,β_n) is a Bailey pair, then so is (α'_n,β'_n), where

17.12.5
$$\left(\frac{aq}{\rho_1},\frac{aq}{\rho_2};q\right)_n \alpha'_n=(\rho_1,\rho_2;q)_n\left(\frac{aq}{\rho_1\rho_2}\right)^n \alpha_n$$
$$\left(\frac{aq}{\rho_1},\frac{aq}{\rho_2};q\right)_n \beta'_n$$
$$=\sum_{j=0}^n (\rho_1,\rho_2;q)_j\left(\frac{aq}{\rho_1\rho_2};q\right)_{n-j}\left(\frac{aq}{\rho_1\rho_2}\right)^j \frac{\beta_j}{(q;q)_{n-j}}$$

When (17.12.5) is iterated the resulting infinite sequence of Bailey pairs is called a *Bailey Chain*.

The Bailey pair that implies the Rogers–Ramanujan identities §17.2(vi) is:

17.12.6
$$\alpha_n=\frac{(a;q)_n\,(1-aq^{2n})(-1)^n q^{n(3n-1)/2}a^n}{(q;q)_n\,(1-a)},$$
$$\beta_n=\frac{1}{(q;q)_n}.$$

The Bailey pair and Bailey chain concepts have been extended considerably. See Andrews (2000, 2001), Andrews and Berkovich (1998), Andrews et al. (1999),

Milne and Lilly (1992), Spiridonov (2002), and Warnaar (1998).

17.13 Integrals

In this section, for the function Γ_q see §5.18(ii).

17.13.1
$$\int_{-c}^{d} \frac{(-qx/c;q)_\infty (qx/d;q)_\infty}{(-ax/c;q)_\infty (bx/d;q)_\infty} d_q x$$
$$= \frac{(1-q)(q;q)_\infty (ab;q)_\infty cd(-c/d;q)_\infty (-d/c;q)_\infty}{(a;q)_\infty (b;q)_\infty (c+d)(-bc/d;q)_\infty (-ad/c;q)_\infty},$$
or, when $0 < q < 1$,

17.13.2
$$\int_{-c}^{d} \frac{(-qx/c;q)_\infty (qx/d;q)_\infty}{(-xq^\alpha/c;q)_\infty (xq^\beta/d;q)_\infty} d_q x$$
$$= \frac{\Gamma_q(\alpha)\Gamma_q(\beta)}{\Gamma_q(\alpha+\beta)} \frac{cd}{c+d} \frac{(-c/d;q)_\infty (-d/c;q)_\infty}{(-q^\beta c/d;q)_\infty (-q^\alpha d/c;q)_\infty}.$$

Ramanujan's Integrals

17.13.3
$$\int_0^\infty t^{\alpha-1} \frac{(-tq^{\alpha+\beta};q)_\infty}{(-t;q)_\infty} d_q t = \frac{\Gamma(\alpha)\Gamma(1-\alpha)\Gamma_q(\beta)}{\Gamma_q(1-\alpha)\Gamma_q(\alpha+\beta)},$$

17.13.4
$$\int_0^\infty t^{\alpha-1} \frac{(-ctq^{\alpha+\beta};q)_\infty}{(-ct;q)_\infty} d_q t$$
$$= \frac{\Gamma_q(\alpha)\Gamma_q(\beta)(-cq^\alpha;q)_\infty (-q^{1-\alpha}/c;q)_\infty}{\Gamma_q(\alpha+\beta)(-c;q)_\infty (-q/c;q)_\infty}.$$

Askey (1980) conjectured extensions of the foregoing integrals that are closely related to Macdonald (1982). These conjectures are proved independently in Habsieger (1988) and Kadell (1988).

17.14 Constant Term Identities

Zeilberger–Bressoud Theorem (Andrews' q-Dyson Conjecture)

17.14.1
$$\frac{(q;q)_{a_1+a_2+\cdots+a_n}}{(q;q)_{a_1}(q;q)_{a_2}\cdots(q;q)_{a_n}} = \text{coeff. of } x_1^0 x_2^0 \cdots x_n^0 \text{ in } \prod_{1 \leq j < k \leq n} \left(\frac{x_j}{x_k};q\right)_{a_j} \left(\frac{qx_k}{x_j};q\right)_{a_k}.$$

Rogers–Ramanujan Constant Term Identities

In the following, $G(q)$ and $H(q)$ denote the left-hand sides of (17.2.49) and (17.2.50), respectively.

17.14.2
$$\sum_{n=0}^\infty \frac{q^{n(n+1)}}{(q^2;q^2)_n (-q;q^2)_{n+1}} = \text{coeff. of } z^0 \text{ in } \frac{(-zq;q^2)_\infty (-z^{-1}q;q^2)_\infty (q^2;q^2)_\infty}{(z^{-1}q^2;q^2)_\infty (-q;q^2)_\infty (z^{-1}q;q^2)_\infty}$$
$$= \frac{1}{(-q;q^2)_\infty} \text{coeff. of } z^0 \text{ in } \frac{(-zq;q^2)_\infty (-z^{-1}q;q^2)_\infty (q^2;q^2)_\infty}{(z^{-1}q;q)_\infty} = \frac{H(q)}{(-q;q^2)_\infty},$$

17.14.3
$$\sum_{n=0}^\infty \frac{q^{n(n+1)}}{(q^2;q^2)_n (-q;q^2)_{n+1}} = \text{coeff. of } z^0 \text{ in } \frac{(-zq;q^2)_\infty (-z^{-1}q;q^2)_\infty (q^2;q^2)_\infty}{(z^{-1};q^2)_\infty (-q;q^2)_\infty (z^{-1}q;q^2)_\infty}$$
$$= \frac{1}{(-q;q^2)_\infty} \text{coeff. of } z^0 \text{ in } \frac{(-zq;q^2)_\infty (-z^{-1}q;q^2)_\infty (q^2;q^2)_\infty}{(z^{-1};q)_\infty} = \frac{G(q)}{(-q;q^2)_\infty},$$

17.14.4
$$\sum_{n=0}^\infty \frac{q^{n^2}}{(q^2;q^2)_n (q;q^2)_n} = \text{coeff. of } z^0 \text{ in } \frac{(-zq;q^2)_\infty (-z^{-1}q;q^2)_\infty (q^2;q^2)_\infty}{(-z^{-1};q^2)_\infty (q;q^2)_\infty (z^{-1};q^2)_\infty}$$
$$= \frac{1}{(q;q^2)_\infty} \text{coeff. of } z^0 \text{ in } \frac{(-zq;q^2)_\infty (-z^{-1}q;q^2)_\infty (q^2;q^2)_\infty}{(z^{-2};q^4)_\infty} = \frac{G(q^4)}{(q;q^2)_\infty},$$

17.14.5
$$\sum_{n=0}^\infty \frac{q^{n^2+2n}}{(q^2;q^2)_n (q;q^2)_{n+1}} = \text{coeff. of } z^0 \text{ in } \frac{(-zq;q^2)_\infty (-z^{-1}q;q^2)_\infty (q^2;q^2)_\infty}{(-q^2 z^{-1};q^2)_\infty (q;q^2)_\infty (z^{-1}q^2;q^2)_\infty}$$
$$= \frac{1}{(q;q^2)_\infty} \text{coeff. of } z^0 \text{ in } \frac{(-zq;q^2)_\infty (-z^{-1}q;q^2)_\infty (q^2;q^2)_\infty}{(q^4 z^{-2};q^4)_\infty} = \frac{H(q^4)}{(q;q^2)_\infty}.$$

Macdonald (1982) includes extensive conjectures on generalizations of (17.14.1) to root systems. These conjectures were proved in Cherednik (1995), Habsieger (1986), and Kadell (1994); see also Macdonald (1998). For additional results of the type (17.14.2)–(17.14.5) see Andrews (1986, Chapter 4).

17.15 Generalizations

For higher-dimensional basic hypergometric functions, see Milne (1985b,c,d,a, 1988, 1994, 1997) and Gustafson (1987).

Applications

17.16 Mathematical Applications

Many special cases of q-series arise in the theory of partitions, a topic treated in §§27.14(i) and 26.9. In Lie algebras Lepowsky and Milne (1978) and Lepowsky and Wilson (1982) laid foundations for extensive interaction with q-series. These and other applications are described in the surveys Andrews (1974, 1986). More recent applications are given in Gasper and Rahman (2004, Chapter 8) and Fine (1988, Chapters 1 and 2).

17.17 Physical Applications

In exactly solved models in statistical mechanics (Baxter (1981, 1982)) the methods and identities of §17.12 play a substantial role. See Berkovich and McCoy (1998) and Bethuel (1998) for recent surveys.

Quantum groups also apply q-series extensively. Quantum groups are really not groups at all but certain Hopf algebras. They were given this name because they play a role in quantum physics analogous to the role of Lie groups and special functions in classical mechanics. See Kassel (1995).

A substantial literature on q-deformed quantum-mechanical Schrödinger equations has developed recently. It involves q-generalizations of exponentials and Laguerre polynomials, and has been applied to the problems of the harmonic oscillator and Coulomb potentials. See Micu and Papp (2005), where many earlier references are cited.

Computation

17.18 Methods of Computation

The two main methods for computing basic hypergeometric functions are: (1) numerical summation of the defining series given in §§17.4(i) and 17.4(ii); (2) modular transformations. Method (1) is applicable within the circles of convergence of the defining series, although it is often cumbersome owing to slowness of convergence and/or severe cancellation. Method (2) is very powerful when applicable (Andrews (1976, Chapter 5)); however, it is applicable only rarely. Lehner (1941) uses Method (2) in connection with the Rogers–Ramanujan identities.

Method (1) can sometimes be improved by application of convergence acceleration procedures; see §3.9. Shanks (1955) applies such methods in several q-series problems; see Andrews et al. (1986).

17.19 Software

See http://dlmf.nist.gov/17.19.

References

General References

The main reference used in writing this chapter is Gasper and Rahman (2004). For additional bibliographic reading see Andrews (1974, 1976, 1986), Andrews et al. (1999), Bailey (1935), Fine (1988), Kac and Cheung (2002), and Slater (1966).

Sources

The following list gives the references or other indications of proofs that were used in constructing the various sections of this chapter. These sources supplement the references that are quoted in the text.

§**17.2** Andrews (1976, pp. 17, 36, 37, 49 and §7.3). (17.2.43) is derived from (17.2.41); (17.2.42) and (17.2.44) follow by induction.

§**17.5** Andrews (1976, pp. 17, 19, 36), Gasper and Rahman (2004, pp. 25–26).

§**17.6** Gasper and Rahman (2004, pp. 13–15, 18, 23, 26–28, 115, 356, 363–364). For (17.6.9)–(17.6.12) see Fine (1988, pp. 12–15), and for (17.6.14) see Andrews (1966a).

§**17.7** Gasper and Rahman (2004, pp. 17, 19, 28–29, 42–44, 58, 61–62, 81–83, 355–356), Andrews (1996). For (17.7.3) combine Gasper and Rahman (2004, p. 359, Equation (III.4)) with (17.6.4).

§**17.8** Gasper and Rahman (2004, pp. 15–16, 52, 140–141, 146–147, 149–150, 153).

§**17.9** Gasper and Rahman (2004, pp. 43, 50, 63–64, 70–73, 359, 361), Andrews (1966b). For (17.9.11) apply (17.9.10) to (17.9.9) and interchange the roles of d and e.

§**17.10** Gasper and Rahman (2004, pp. 147–150, 364–366).

§**17.11** Andrews (1972).

§**17.12** Andrews (1984).

§**17.13** Gasper and Rahman (2004, p. 52), Berndt (1991, p. 29), Askey (1980).

§**17.14** Zeilberger and Bressoud (1985), Andrews (1986, p. 34).

Chapter 18
Orthogonal Polynomials

T. H. Koornwinder[1], R. Wong[2], R. Koekoek[3] and R. F. Swarttouw[4]

Notation   **436**
18.1   Notation 436

General Orthogonal Polynomials   **437**
18.2   General Orthogonal Polynomials 437

Classical Orthogonal Polynomials   **438**
18.3   Definitions 438
18.4   Graphics 440
18.5   Explicit Representations 442
18.6   Symmetry, Special Values, and Limits to Monomials 443
18.7   Interrelations and Limit Relations 444
18.8   Differential Equations 445
18.9   Recurrence Relations and Derivatives . . 446
18.10   Integral Representations 447
18.11   Relations to Other Functions 448
18.12   Generating Functions 449
18.13   Continued Fractions 450
18.14   Inequalities 450
18.15   Asymptotic Approximations 451
18.16   Zeros 454
18.17   Integrals 455
18.18   Sums 459

Askey Scheme   **462**
18.19   Hahn Class: Definitions 462
18.20   Hahn Class: Explicit Representations . . 462
18.21   Hahn Class: Interrelations 463
18.22   Hahn Class: Recurrence Relations and Differences 464
18.23   Hahn Class: Generating Functions . . . 466
18.24   Hahn Class: Asymptotic Approximations . 466
18.25   Wilson Class: Definitions 467
18.26   Wilson Class: Continued 468

Other Orthogonal Polynomials   **470**
18.27 q-Hahn Class 470
18.28   Askey–Wilson Class 472
18.29   Asymptotic Approximations for q-Hahn and Askey–Wilson Classes 474
18.30   Associated OP's 474
18.31   Bernstein–Szegö Polynomials 474
18.32   OP's with Respect to Freud Weights . . . 475
18.33   Polynomials Orthogonal on the Unit Circle 475
18.34   Bessel Polynomials 476
18.35   Pollaczek Polynomials 476
18.36   Miscellaneous Polynomials 477
18.37   Classical OP's in Two or More Variables . 477

Applications   **478**
18.38   Mathematical Applications 478
18.39   Physical Applications 479

Computation   **479**
18.40   Methods of Computation 479
18.41   Tables 480
18.42   Software 480

References   **480**

---

[1]University of Amsterdam, Korteweg–de Vries Institute, Amsterdam, The Netherlands.
[2]City University of Hong Kong, Liu Bie Ju Centre for Mathematical Sciences, Kowloon, Hong Kong.
[3]Delft University of Technology, Delft Institute of Applied Mathematics, Delft, The Netherlands.
[4]Vrije Universiteit Amsterdam, Department of Mathematics, Amsterdam, The Netherlands.
Copyright © 2009 National Institute of Standards and Technology. All rights reserved.

Notation

18.1 Notation

18.1(i) Special Notation

(For other notation see pp. xiv and 873.)

x, y	real variables.
$z(= x + iy)$	complex variable.
q	real variable such that $0 < q < 1$, unless stated otherwise.
ℓ, m	nonnegative integers.
n	nonnegative integer, except in §18.30.
N	positive integer.
$\delta(x-a)$	Dirac delta (§1.17).
δ	arbitrary small positive constant.
$p_n(x)$	polynomial in x of degree n.
$p_{-1}(x)$	0.
$w(x)$	weight function (≥ 0) on an open interval (a,b).
w_x	weights (> 0) at points $x \in X$ of a finite or countably infinite subset of $\mathbb{R}$.
OP's	orthogonal polynomials.

x-Differences

Forward differences:
$$\Delta_x\left(f(x)\right) = f(x+1) - f(x),$$
$$\Delta_x^{n+1}\left(f(x)\right) = \Delta_x\left(\Delta_x^n(f(x))\right).$$

Backward differences:
$$\nabla_x\left(f(x)\right) = f(x) - f(x-1),$$
$$\nabla_x^{n+1}\left(f(x)\right) = \nabla_x\left(\nabla_x^n(f(x))\right).$$

Central differences in imaginary direction:
$$\delta_x\left(f(x)\right) = \left(f(x+\tfrac{1}{2}i) - f(x-\tfrac{1}{2}i)\right)/i,$$
$$\delta_x^{n+1}\left(f(x)\right) = \delta_x\left(\delta_x^n(f(x))\right).$$

q-Pochhammer Symbol

$$(z;q)_0 = 1, \quad (z;q)_n = (1-z)(1-zq)\cdots(1-zq^{n-1}),$$
$$(z_1,\ldots,z_k;q)_n = (z_1;q)_n \cdots (z_k;q)_n.$$

Infinite q-Product

$$(z;q)_\infty = \prod_{j=0}^\infty (1-zq^j),$$
$$(z_1,\ldots,z_k;q)_\infty = (z_1;q)_\infty \cdots (z_k;q)_\infty.$$

18.1(ii) Main Functions

The main functions treated in this chapter are:

Classical OP's

Jacobi: $P_n^{(\alpha,\beta)}(x)$.

Ultraspherical (or Gegenbauer): $C_n^{(\lambda)}(x)$.

Chebyshev of first, second, third, and fourth kinds: $T_n(x), U_n(x), V_n(x), W_n(x)$.

Shifted Chebyshev of first and second kinds: $T_n^*(x), U_n^*(x)$.

Legendre: $P_n(x)$.

Shifted Legendre: $P_n^*(x)$.

Laguerre: $L_n^{(\alpha)}(x)$ and $L_n(x) = L_n^{(0)}(x)$. ($L_n^{(\alpha)}(x)$ with $\alpha \neq 0$ is also called Generalized Laguerre.)

Hermite: $H_n(x), \mathit{He}_n(x)$.

Hahn Class OP's

Hahn: $Q_n(x;\alpha,\beta,N)$.

Krawtchouk: $K_n(x;p,N)$.

Meixner: $M_n(x;\beta,c)$.

Charlier: $C_n(x,a)$.

Continuous Hahn: $p_n(x;a,b,\overline{a},\overline{b})$.

Meixner–Pollaczek: $P_n^{(\lambda)}(x;\phi)$.

Wilson Class OP's

Wilson: $W_n(x;a,b,c,d)$.

Racah: $R_n(x;\alpha,\beta,\gamma,\delta)$.

Continuous Dual Hahn: $S_n(x;a,b,c)$.

Dual Hahn: $R_n(x;\gamma,\delta,N)$.

q-Hahn Class OP's

q-Hahn: $Q_n(x;\alpha,\beta,N;q)$.

Big q-Jacobi: $P_n(x;a,b,c;q)$.

Little q-Jacobi: $p_n(x;a,b;q)$.

q-Laguerre: $L_n^{(\alpha)}(x;q)$.

Stieltjes–Wigert: $S_n(x;q)$.

Discrete q-Hermite I: $h_n(x;q)$.

Discrete q-Hermite II: $\tilde{h}_n(x;q)$.

Askey–Wilson Class OP's

Askey–Wilson: $p_n(x;a,b,c,d\,|\,q)$.

Al-Salam–Chihara: $Q_n(x;a,b\,|\,q)$.

Continuous q-Ultraspherical: $C_n(x;\beta\,|\,q)$.

Continuous q-Hermite: $H_n(x\,|\,q)$.

Continuous q^{-1}-Hermite: $h_n(x\,|\,q)$.

q-Racah: $R_n(x;\alpha,\beta,\gamma,\delta\,|\,q)$.

Other OP's

Bessel: $y_n(x;a)$.

Pollaczek: $P_n^{(\lambda)}(x;a,b)$.

Classical OP's in Two Variables

Disk: $R_{m,n}^{(\alpha)}(z)$.

Triangle: $P_{m,n}^{\alpha,\beta,\gamma}(x,y)$.

18.1(iii) Other Notations

In Szegö (1975, §4.7) the ultraspherical polynomials $C_n^{(\lambda)}(x)$ are denoted by $P_n^{(\lambda)}(x)$. The ultraspherical polynomials will not be considered for $\lambda = 0$. They are defined in the literature by $C_0^{(0)}(x) = 1$ and

18.1.1 $$C_n^{(0)}(x) = \frac{2}{n} T_n(x) = \frac{2(n-1)!}{\left(\frac{1}{2}\right)_n} P_n^{(-\frac{1}{2},-\frac{1}{2})}(x), \quad n = 1,2,3,\ldots.$$

Nor do we consider the shifted Jacobi polynomials:

18.1.2 $$G_n(p,q,x) = \frac{n!}{(n+p)_n} P_n^{(p-q,q-1)}(2x-1),$$

or the dilated Chebyshev polynomials of the first and second kinds:

18.1.3 $$C_n(x) = 2T_n\left(\tfrac{1}{2}x\right), \quad S_n(x) = U_n\left(\tfrac{1}{2}x\right).$$

In Koekoek and Swarttouw (1998) δ_x denotes the operator $i\delta_x$.

General Orthogonal Polynomials

18.2 General Orthogonal Polynomials

18.2(i) Definition

Orthogonality on Intervals

Let (a,b) be a finite or infinite open interval in $\mathbb{R}$. A system (or set) of polynomials $\{p_n(x)\}$, $n = 0,1,2,\ldots$, is said to be *orthogonal on (a,b) with respect to the weight function* $w(x)$ (≥ 0) if

18.2.1 $$\int_a^b p_n(x)p_m(x)w(x)\,dx = 0, \qquad n \neq m.$$

Here $w(x)$ is continuous or piecewise continuous or integrable, and such that $0 < \int_a^b x^{2n}w(x)\,dx < \infty$ for all n.

It is assumed throughout this chapter that for each polynomial $p_n(x)$ that is orthogonal on an open interval (a,b) the variable x is confined to the closure of (a,b) *unless indicated otherwise*. (However, under appropriate conditions almost all equations given in the chapter can be continued analytically to various complex values of the variables.)

Orthogonality on Finite Point Sets

Let X be a finite set of distinct points on $\mathbb{R}$, or a countable infinite set of distinct points on $\mathbb{R}$, and w_x, $x \in X$, be a set of positive constants. Then a system of polynomials $\{p_n(x)\}$, $n = 0,1,2,\ldots$, is said to be *orthogonal* on X with respect to the *weights* w_x if

18.2.2 $$\sum_{x \in X} p_n(x)p_m(x)w_x = 0, \qquad n \neq m,$$

when X is infinite, or

18.2.3 $$\sum_{x \in X} p_n(x)p_m(x)w_x = 0, \quad n,m = 0,1,\ldots,N; n \neq m,$$

when X is a finite set of $N+1$ distinct points. In the former case we also require

18.2.4 $$\sum_{x \in X} x^{2n}w_x < \infty, \qquad n = 0,1,\ldots,$$

whereas in the latter case the system $\{p_n(x)\}$ is finite: $n = 0,1,\ldots,N$.

More generally than (18.2.1)–(18.2.3), $w(x)\,dx$ may be replaced in (18.2.1) by a positive measure $d\alpha(x)$, where $\alpha(x)$ is a bounded nondecreasing function on the closure of (a,b) with an infinite number of points of increase, and such that $0 < \int_a^b x^{2n}\,d\alpha(x) < \infty$ for all n. See McDonald and Weiss (1999, Chapters 3, 4) and Szegö (1975, §1.4).

18.2(ii) x-Difference Operators

If the orthogonality discrete set X is $\{0,1,\ldots,N\}$ or $\{0,1,2,\ldots\}$, then the role of the differentiation operator d/dx in the case of classical OP's (§18.3) is played by Δ_x, the forward-difference operator, or by ∇_x, the backward-difference operator; compare §18.1(i). This happens, for example, with the Hahn class OP's (§18.20(i)).

If the orthogonality interval is $(-\infty,\infty)$ or $(0,\infty)$, then the role of d/dx can be played by δ_x, the central-difference operator in the imaginary direction (§18.1(i)). This happens, for example, with the continuous Hahn polynomials and Meixner–Pollaczek polynomials (§18.20(i)).

18.2(iii) Normalization

The orthogonality relations (18.2.1)–(18.2.3) each determine the polynomials $p_n(x)$ uniquely up to constant factors, which may be fixed by suitable normalization.

If we define

18.2.5 $$h_n = \int_a^b (p_n(x))^2 w(x)\,dx \text{ or } \sum_{x \in X} (p_n(x))^2 w_x,$$

18.2.6
$$\tilde{h}_n = \int_a^b x\,(p_n(x))^2\,w(x)\,dx \text{ or } \sum_{x\in X} x\,(p_n(x))^2\,w_x,$$

and

18.2.7 $p_n(x) = k_n x^n + \tilde{k}_n x^{n-1} + \tilde{\tilde{k}}_n x^{n-2} + \cdots,$

then two special normalizations are: (i) *orthonormal OP's*: $h_n = 1$, $k_n > 0$; (ii) *monic OP's*: $k_n = 1$.

18.2(iv) Recurrence Relations

As in §18.1(i) we assume that $p_{-1}(x) \equiv 0$.

First Form

18.2.8
$$p_{n+1}(x) = (A_n x + B_n) p_n(x) - C_n p_{n-1}(x), \quad n \geq 0.$$

Here A_n, B_n ($n \geq 0$), and C_n ($n \geq 1$) are real constants, and $A_{n-1} A_n C_n > 0$ for $n \geq 1$. Then

18.2.9
$$A_n = \frac{k_{n+1}}{k_n}, \quad B_n = \left(\frac{\tilde{k}_{n+1}}{k_{n+1}} - \frac{\tilde{k}_n}{k_n}\right) A_n = -\frac{\tilde{h}_n}{h_n} A_n,$$
$$C_n = \frac{A_n \tilde{\tilde{k}}_n + B_n \tilde{k}_n - \tilde{\tilde{k}}_{n+1}}{k_{n-1}} = \frac{A_n}{A_{n-1}} \frac{h_n}{h_{n-1}}.$$

Second Form

18.2.10
$$x p_n(x) = a_n p_{n+1}(x) + b_n p_n(x) + c_n p_{n-1}(x), \quad n \geq 0.$$

Here a_n, b_n ($n \geq 0$), c_n ($n \geq 1$) are real constants, and $a_{n-1} c_n > 0$ ($n \geq 1$). Then

18.2.11
$$a_n = \frac{k_n}{k_{n+1}}, \quad b_n = \frac{\tilde{k}_n}{k_n} - \frac{\tilde{k}_{n+1}}{k_{n+1}} = \frac{\tilde{h}_n}{h_n},$$
$$c_n = \frac{\tilde{\tilde{k}}_n - a_n \tilde{\tilde{k}}_{n+1} - b_n \tilde{k}_n}{k_{n-1}} = a_{n-1} \frac{h_n}{h_{n-1}}.$$

If the OP's are orthonormal, then $c_n = a_{n-1}$ ($n \geq 1$). If the OP's are monic, then $a_n = 1$ ($n \geq 0$).

Conversely, if a system of polynomials $\{p_n(x)\}$ satisfies (18.2.10) with $a_{n-1} c_n > 0$ ($n \geq 1$), then $\{p_n(x)\}$ is orthogonal with respect to some positive measure on $\mathbb{R}$ (Favard's theorem). The measure is not necessarily of the form $w(x)\,dx$ nor is it necessarily unique.

18.2(v) Christoffel–Darboux Formula

18.2.12
$$\sum_{\ell=0}^n \frac{p_\ell(x) p_\ell(y)}{h_\ell} = \frac{k_n}{h_n k_{n+1}} \frac{p_{n+1}(x) p_n(y) - p_n(x) p_{n+1}(y)}{x - y},$$
$$x \neq y.$$

Confluent Form

18.2.13
$$\sum_{\ell=0}^n \frac{(p_\ell(x))^2}{h_\ell} = \frac{k_n}{h_n k_{n+1}} \left(p'_{n+1}(x) p_n(x) - p'_n(x) p_{n+1}(x)\right).$$

18.2(vi) Zeros

All n zeros of an OP $p_n(x)$ are simple, and they are located in the interval of orthogonality (a, b). The zeros of $p_n(x)$ and $p_{n+1}(x)$ separate each other, and if $m < n$ then between any two zeros of $p_m(x)$ there is at least one zero of $p_n(x)$.

For illustrations of these properties see Figures 18.4.1–18.4.7.

Classical Orthogonal Polynomials

18.3 Definitions

Table 18.3.1 provides the definitions of Jacobi, Laguerre, and Hermite polynomials via orthogonality and normalization (§§18.2(i) and 18.2(iii)). This table also includes the following special cases of Jacobi polynomials: ultraspherical, Chebyshev, and Legendre.

18.3   Definitions

Table 18.3.1: Orthogonality properties for classical OP's: intervals, weight functions, normalizations, leading coefficients, and parameter constraints. In the second row $\mathcal{A}$ denotes $2^{\alpha+\beta+1}\Gamma(n+\alpha+1)\Gamma(n+\beta+1)/((2n+\alpha+\beta+1)\Gamma(n+\alpha+\beta+1)n!)$. For further implications of the parameter constraints see the *Note* in §18.5(iii).

Name	$p_n(x)$	(a,b)	$w(x)$	h_n	k_n	$\tilde{k}_n/k_n$	Constraints
Jacobi	$P_n^{(\alpha,\beta)}(x)$	$(-1,1)$	$(1-x)^\alpha(1+x)^\beta$	$\mathcal{A}$	$\dfrac{(n+\alpha+\beta+1)_n}{2^n n!}$	$\dfrac{n(\alpha-\beta)}{2n+\alpha+\beta}$	$\alpha,\beta > -1$
Ultraspherical (Gegenbauer)	$C_n^{(\lambda)}(x)$	$(-1,1)$	$(1-x^2)^{\lambda-\frac{1}{2}}$	$\dfrac{2^{1-2\lambda}\pi\,\Gamma(n+2\lambda)}{(n+\lambda)(\Gamma(\lambda))^2 n!}$	$\dfrac{2^n(\lambda)_n}{n!}$	0	$\lambda > -\tfrac{1}{2}, \lambda \neq 0$
Chebyshev of first kind	$T_n(x)$	$(-1,1)$	$(1-x^2)^{-\frac{1}{2}}$	$\begin{cases}\tfrac{1}{2}\pi, & n>0\\ \pi, & n=0\end{cases}$	$\begin{cases}2^{n-1}, & n>0\\ 1, & n=0\end{cases}$	0	
Chebyshev of second kind	$U_n(x)$	$(-1,1)$	$(1-x^2)^{\frac{1}{2}}$	$\tfrac{1}{2}\pi$	2^n	0	
Chebyshev of third kind	$V_n(x)$	$(-1,1)$	$(1-x)^{\frac{1}{2}}(1+x)^{-\frac{1}{2}}$	π	2^n	$\tfrac{1}{2}$	
Chebyshev of fourth kind	$W_n(x)$	$(-1,1)$	$(1-x)^{-\frac{1}{2}}(1+x)^{\frac{1}{2}}$	π	2^n	$-\tfrac{1}{2}$	
Shifted Chebyshev of first kind	$T_n^*(x)$	$(0,1)$	$(x-x^2)^{-\frac{1}{2}}$	$\begin{cases}\tfrac{1}{2}\pi, & n>0\\ \pi, & n=0\end{cases}$	$\begin{cases}2^{2n-1}, & n>0\\ 1, & n=0\end{cases}$	$-\tfrac{1}{2}n$	
Shifted Chebyshev of second kind	$U_n^*(x)$	$(0,1)$	$(x-x^2)^{\frac{1}{2}}$	$\tfrac{1}{8}\pi$	2^{2n}	$-\tfrac{1}{2}n$	
Legendre	$P_n(x)$	$(-1,1)$	1	$2/(2n+1)$	$2^n(\tfrac{1}{2})_n/n!$	0	
Shifted Legendre	$P_n^*(x)$	$(0,1)$	1	$1/(2n+1)$	$2^{2n}(\tfrac{1}{2})_n/n!$	$-\tfrac{1}{2}n$	
Laguerre	$L_n^{(\alpha)}(x)$	$(0,\infty)$	$e^{-x}x^\alpha$	$\Gamma(n+\alpha+1)/n!$	$(-1)^n/n!$	$-n(n+\alpha)$	$\alpha > -1$
Hermite	$H_n(x)$	$(-\infty,\infty)$	e^{-x^2}	$\pi^{\frac{1}{2}}2^n n!$	2^n	0	
Hermite	$He_n(x)$	$(-\infty,\infty)$	$e^{-\frac{1}{2}x^2}$	$(2\pi)^{\frac{1}{2}}n!$	1	0	

For exact values of the coefficients of the Jacobi polynomials $P_n^{(\alpha,\beta)}(x)$, the ultraspherical polynomials $C_n^{(\lambda)}(x)$, the Chebyshev polynomials $T_n(x)$ and $U_n(x)$, the Legendre polynomials $P_n(x)$, the Laguerre polynomials $L_n(x)$, and the Hermite polynomials $H_n(x)$, see Abramowitz and Stegun (1964, pp. 793–801). The Jacobi polynomials are in powers of $x-1$ for $n = 0,1,\ldots,6$. The ultraspherical polynomials are in powers of x for $n = 0,1,\ldots,6$. The other polynomials are in powers of x for $n = 0,1,\ldots,12$. See also §18.5(iv).

Chebyshev

In this chapter, formulas for the Chebyshev polynomials of the second, third, and fourth kinds will not be given as extensively as those of the first kind. However, most of these formulas can be obtained by specialization of formulas for Jacobi polynomials, via (18.7.4)–(18.7.6).

In addition to the orthogonal property given by Table 18.3.1, the Chebyshev polynomials $T_n(x)$, $n = 0,1,\ldots,N$, are orthogonal on the discrete point set comprising the zeros $x_{N+1,n}$, $n = 1, 2, \ldots, N+1$, of $T_{N+1}(x)$:

18.3.1
$$\sum_{n=1}^{N+1} T_j(x_{N+1,n}) T_k(x_{N+1,n}) = 0,$$
$$0 \le j \le N, \ 0 \le k \le N, \ j \ne k,$$

where

18.3.2 $\quad x_{N+1,n} = \cos\bigl((n-\tfrac{1}{2})\pi/(N+1)\bigr).$

When $j = k \ne 0$ the sum in (18.3.1) is $\tfrac{1}{2}(N+1)$. When $j = k = 0$ the sum in (18.3.1) is $N+1$.

For proofs of these results and for similar properties of the Chebyshev polynomials of the second, third, and fourth kinds see Mason and Handscomb (2003, §4.6).

For another version of the discrete orthogonality property of the polynomials $T_n(x)$ see (3.11.9).

Legendre

Legendre polynomials are special cases of Legendre functions, Ferrers functions, and associated Legendre functions (§14.7(i)). In consequence, additional properties are included in Chapter 14.

18.4 Graphics

18.4(i) Graphs

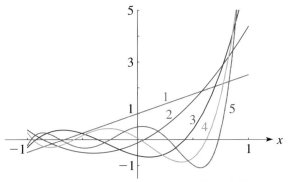

Figure 18.4.1: Jacobi polynomials $P_n^{(1.5,-0.5)}(x)$, $n = 1, 2, 3, 4, 5$.

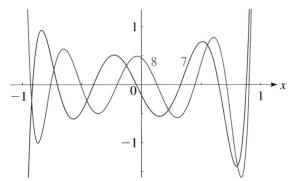

Figure 18.4.2: Jacobi polynomials $P_n^{(1.25,0.75)}(x)$, $n = 7, 8$. This illustrates inequalities for extrema of a Jacobi polynomial; see (18.14.16). See also Askey (1990).

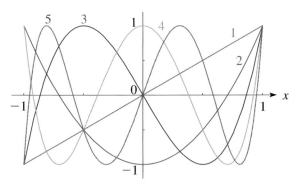

Figure 18.4.3: Chebyshev polynomials $T_n(x)$, $n = 1, 2, 3, 4, 5$.

18.4 GRAPHICS

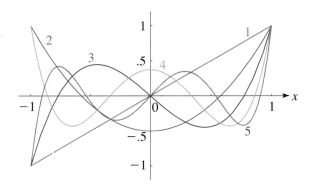

Figure 18.4.4: Legendre polynomials $P_n(x)$, $n = 1, 2, 3, 4, 5$.

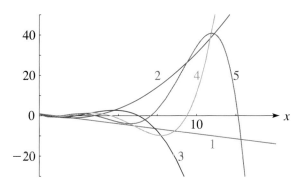

Figure 18.4.5: Laguerre polynomials $L_n(x)$, $n = 1, 2, 3, 4, 5$.

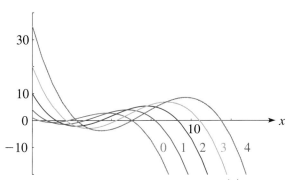

Figure 18.4.6: Laguerre polynomials $L_3^{(\alpha)}(x)$, $\alpha = 0, 1, 2, 3, 4$.

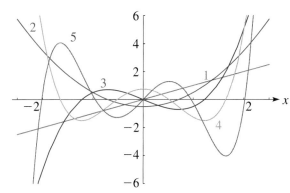

Figure 18.4.7: Monic Hermite polynomials $h_n(x) = 2^{-n} H_n(x)$, $n = 1, 2, 3, 4, 5$.

18.4(ii) Surfaces

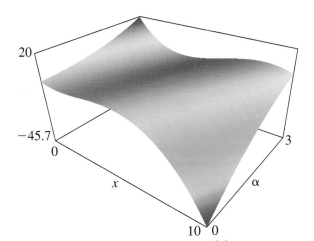

Figure 18.4.8: Laguerre polynomials $L_3^{(\alpha)}(x)$, $0 \leq \alpha \leq 3$, $0 \leq x \leq 10$.

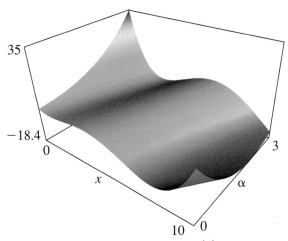

Figure 18.4.9: Laguerre polynomials $L_4^{(\alpha)}(x)$, $0 \leq \alpha \leq 3$, $0 \leq x \leq 10$.

18.5 Explicit Representations

18.5(i) Trigonometric Functions

Chebyshev

With $x = \cos\theta$,

18.5.1 $\quad T_n(x) = \cos(n\theta),$

18.5.2 $\quad U_n(x) = (\sin(n+1)\theta)/\sin\theta,$

18.5.3 $\quad V_n(x) = (\sin(n+\tfrac{1}{2})\theta)/\sin(\tfrac{1}{2}\theta),$

18.5.4 $\quad W_n(x) = (\cos(n+\tfrac{1}{2})\theta)/\cos(\tfrac{1}{2}\theta).$

18.5(ii) Rodrigues Formulas

18.5.5 $\quad p_n(x) = \dfrac{1}{\kappa_n w(x)} \dfrac{d^n}{dx^n} \left(w(x)(F(x))^n\right).$

In this equation $w(x)$ is as in Table 18.3.1, and $F(x)$, κ_n are as in Table 18.5.1.

Table 18.5.1: Classical OP's: Rodrigues formulas (18.5.5).

$p_n(x)$	$F(x)$	κ_n
$P_n^{(\alpha,\beta)}(x)$	$1-x^2$	$(-2)^n n!$
$C_n^{(\lambda)}(x)$	$1-x^2$	$\dfrac{(-2)^n\left(\lambda+\tfrac{1}{2}\right)_n n!}{(2\lambda)_n}$
$T_n(x)$	$1-x^2$	$(-2)^n\left(\tfrac{1}{2}\right)_n$
$U_n(x)$	$1-x^2$	$\dfrac{(-2)^n\left(\tfrac{3}{2}\right)_n}{n+1}$
$V_n(x)$	$1-x^2$	$\dfrac{(-2)^n\left(\tfrac{3}{2}\right)_n}{2n+1}$
$W_n(x)$	$1-x^2$	$(-2)^n\left(\tfrac{1}{2}\right)_n$
$P_n(x)$	$1-x^2$	$(-2)^n n!$
$L_n^{(\alpha)}(x)$	x	$n!$
$H_n(x)$	1	$(-1)^n$
$He_n(x)$	1	$(-1)^n$

Related formula:

18.5.6 $\quad L_n^{(\alpha)}\left(\dfrac{1}{x}\right) = \dfrac{(-1)^n}{n!} x^{n+\alpha+1} e^{1/x} \dfrac{d^n}{dx^n}\left(x^{-\alpha-1} e^{-1/x}\right).$

18.5(iii) Finite Power Series, the Hypergeometric Function, and Generalized Hypergeometric Functions

For the definitions of $_2F_1$, $_1F_1$, and $_2F_0$ see §16.2.

Jacobi

18.5.7
$$P_n^{(\alpha,\beta)}(x) = \sum_{\ell=0}^{n} \frac{(n+\alpha+\beta+1)_\ell (\alpha+\ell+1)_{n-\ell}}{\ell!\,(n-\ell)!} \left(\frac{x-1}{2}\right)^\ell$$
$$= \frac{(\alpha+1)_n}{n!}\,{}_2F_1\!\left(\begin{matrix}-n, n+\alpha+\beta+1\\ \alpha+1\end{matrix};\frac{1-x}{2}\right),$$

18.5.8
$$P_n^{(\alpha,\beta)}(x) = 2^{-n}\sum_{\ell=0}^{n}\binom{n+\alpha}{\ell}\binom{n+\beta}{n-\ell}(x-1)^{n-\ell}(x+1)^\ell$$
$$= \frac{(\alpha+1)_n}{n!}\left(\frac{x+1}{2}\right)^n {}_2F_1\!\left(\begin{matrix}-n,-n-\beta\\ \alpha+1\end{matrix};\frac{x-1}{x+1}\right),$$

and two similar formulas by symmetry; compare the second row in Table 18.6.1.

Ultraspherical

18.5.9 $\quad C_n^{(\lambda)}(x) = \dfrac{(2\lambda)_n}{n!}\,{}_2F_1\!\left(\begin{matrix}-n, n+2\lambda\\ \lambda+\tfrac{1}{2}\end{matrix};\dfrac{1-x}{2}\right),$

18.5.10
$$C_n^{(\lambda)}(x) = \sum_{\ell=0}^{\lfloor n/2\rfloor}\frac{(-1)^\ell(\lambda)_{n-\ell}}{\ell!\,(n-2\ell)!}(2x)^{n-2\ell}$$
$$= (2x)^n\frac{(\lambda)_n}{n!}\,{}_2F_1\!\left(\begin{matrix}-\tfrac{1}{2}n,-\tfrac{1}{2}n+\tfrac{1}{2}\\ 1-\lambda-n\end{matrix};\frac{1}{x^2}\right),$$

18.5.11
$$C_n^{(\lambda)}(\cos\theta) = \sum_{\ell=0}^{n}\frac{(\lambda)_\ell(\lambda)_{n-\ell}}{\ell!\,(n-\ell)!}\cos((n-2\ell)\theta)$$
$$= e^{in\theta}\frac{(\lambda)_n}{n!}\,{}_2F_1\!\left(\begin{matrix}-n,\lambda\\ 1-\lambda-n\end{matrix};e^{-2i\theta}\right).$$

Laguerre

18.5.12
$$L_n^{(\alpha)}(x) = \sum_{\ell=0}^{n} \frac{(\alpha+\ell+1)_{n-\ell}}{(n-\ell)!\,\ell!}(-x)^\ell$$
$$= \frac{(\alpha+1)_n}{n!}\,{}_1F_1\!\left(\begin{array}{c}-n\\ \alpha+1\end{array};x\right).$$

Hermite

18.5.13
$$H_n(x) = n!\sum_{\ell=0}^{\lfloor n/2\rfloor}\frac{(-1)^\ell (2x)^{n-2\ell}}{\ell!\,(n-2\ell)!}$$
$$= (2x)^n\,{}_2F_0\!\left(\begin{array}{c}-\tfrac{1}{2}n,-\tfrac{1}{2}n+\tfrac{1}{2}\\ -\end{array};-\frac{1}{x^2}\right).$$

For corresponding formulas for Chebyshev, Legendre, and the Hermite He_n polynomials apply (18.7.3)–(18.7.6), (18.7.9), and (18.7.11).

Note. The first of each of equations (18.5.7) and (18.5.8) can be regarded as definitions of $P_n^{(\alpha,\beta)}(x)$ when the conditions $\alpha > -1$ and $\beta > -1$ are not satisfied. However, in these circumstances the orthogonality property (18.2.1) disappears. For this reason, and also in the interest of simplicity, in the case of the Jacobi polynomials $P_n^{(\alpha,\beta)}(x)$ we assume throughout this chapter that $\alpha > -1$ and $\beta > -1$, *unless stated otherwise*. Similarly in the cases of the ultraspherical polynomials $C_n^{(\lambda)}(x)$ and the Laguerre polynomials $L_n^{(\alpha)}(x)$ we assume that $\lambda > -\tfrac{1}{2}, \lambda \neq 0$, and $\alpha > -1$, *unless stated otherwise*.

18.5(iv) Numerical Coefficients

Chebyshev

18.5.14
$T_0(x) = 1, \quad T_1(x) = x, \quad T_2(x) = 2x^2 - 1,$
$T_3(x) = 4x^3 - 3x, \quad T_4(x) = 8x^4 - 8x^2 + 1,$
$T_5(x) = 16x^5 - 20x^3 + 5x,$
$T_6(x) = 32x^6 - 48x^4 + 18x^2 - 1.$

18.5.15
$U_0(x) = 1, \quad U_1(x) = 2x, \quad U_2(x) = 4x^2 - 1,$
$U_3(x) = 8x^3 - 4x, \quad U_4(x) = 16x^4 - 12x^2 + 1,$
$U_5(x) = 32x^5 - 32x^3 + 6x,$
$U_6(x) = 64x^6 - 80x^4 + 24x^2 - 1.$

Legendre

18.5.16
$P_0(x) = 1, \quad P_1(x) = x, \quad P_2(x) = \tfrac{3}{2}x^2 - \tfrac{1}{2},$
$P_3(x) = \tfrac{5}{2}x^3 - \tfrac{3}{2}x, \quad P_4(x) = \tfrac{35}{8}x^4 - \tfrac{15}{4}x^2 + \tfrac{3}{8},$
$P_5(x) = \tfrac{63}{8}x^5 - \tfrac{35}{4}x^3 + \tfrac{15}{8}x,$
$P_6(x) = \tfrac{231}{16}x^6 - \tfrac{315}{16}x^4 + \tfrac{105}{16}x^2 - \tfrac{5}{16}.$

Laguerre

18.5.17
$L_0(x) = 1, \quad L_1(x) = -x + 1, \quad L_2(x) = \tfrac{1}{2}x^2 - 2x + 1,$
$L_3(x) = -\tfrac{1}{6}x^3 + \tfrac{3}{2}x^2 - 3x + 1,$
$L_4(x) = \tfrac{1}{24}x^4 - \tfrac{2}{3}x^3 + 3x^2 - 4x + 1,$
$L_5(x) = -\tfrac{1}{120}x^5 + \tfrac{5}{24}x^4 - \tfrac{5}{3}x^3 + 5x^2 - 5x + 1,$
$L_6(x) = \tfrac{1}{720}x^6 - \tfrac{1}{20}x^5 + \tfrac{5}{8}x^4 - \tfrac{10}{3}x^3 + \tfrac{15}{2}x^2 - 6x + 1.$

Hermite

18.5.18
$H_0(x) = 1, \quad H_1(x) = 2x, \quad H_2(x) = 4x^2 - 2,$
$H_3(x) = 8x^3 - 12x, \quad H_4(x) = 16x^4 - 48x^2 + 12,$
$H_5(x) = 32x^5 - 160x^3 + 120x,$
$H_6(x) = 64x^6 - 480x^4 + 720x^2 - 120.$

18.5.19
$He_0(x) = 1, \quad He_1(x) = x, \quad He_2(x) = x^2 - 1,$
$He_3(x) = x^3 - 3x, \quad He_4(x) = x^4 - 6x^2 + 3,$
$He_5(x) = x^5 - 10x^3 + 15x,$
$He_6(x) = x^6 - 15x^4 + 45x^2 - 15.$

For the corresponding polynomials of degrees 7 through 12 see Abramowitz and Stegun (1964, Tables 22.3, 22.5, 22.9, 22.10, 22.12).

18.6 Symmetry, Special Values, and Limits to Monomials

18.6(i) Symmetry and Special Values

For Jacobi, ultraspherical, Chebyshev, Legendre, and Hermite polynomials, see Table 18.6.1.

Laguerre

18.6.1
$$L_n^{(\alpha)}(0) = \frac{(\alpha+1)_n}{n!}.$$

Table 18.6.1: Classical OP's: symmetry and special values.

$p_n(x)$	$p_n(-x)$	$p_n(1)$	$p_{2n}(0)$	$p'_{2n+1}(0)$
$P_n^{(\alpha,\beta)}(x)$	$(-1)^n P_n^{(\beta,\alpha)}(x)$	$(\alpha+1)_n/n!$		
$P_n^{(\alpha,\alpha)}(x)$	$(-1)^n P_n^{(\alpha,\alpha)}(x)$	$(\alpha+1)_n/n!$	$(-\tfrac{1}{4})^n (n+\alpha+1)_n/n!$	$(-\tfrac{1}{4})^n (n+\alpha+1)_{n+1}/n!$
$C_n^{(\lambda)}(x)$	$(-1)^n C_n^{(\lambda)}(x)$	$(2\lambda)_n/n!$	$(-1)^n (\lambda)_n/n!$	$2(-1)^n (\lambda)_{n+1}/n!$
$T_n(x)$	$(-1)^n T_n(x)$	1	$(-1)^n$	$(-1)^n(2n+1)$
$U_n(x)$	$(-1)^n U_n(x)$	$n+1$	$(-1)^n$	$(-1)^n(2n+2)$
$V_n(x)$	$(-1)^n W_n(x)$	$2n+1$	$(-1)^n$	$(-1)^n(2n+2)$
$W_n(x)$	$(-1)^n V_n(x)$	1	$(-1)^n$	$(-1)^n(2n+2)$
$P_n(x)$	$(-1)^n P_n(x)$	1	$(-1)^n \left(\tfrac{1}{2}\right)_n/n!$	$2(-1)^n \left(\tfrac{1}{2}\right)_{n+1}/n!$
$H_n(x)$	$(-1)^n H_n(x)$		$(-1)^n(n+1)_n$	$2(-1)^n(n+1)_{n+1}$
$He_n(x)$	$(-1)^n He_n(x)$		$(-\tfrac{1}{2})^n(n+1)_n$	$(-\tfrac{1}{2})^n(n+1)_{n+1}$

18.6(ii) Limits to Monomials

18.6.2 $$\lim_{\alpha\to\infty} \frac{P_n^{(\alpha,\beta)}(x)}{P_n^{(\alpha,\beta)}(1)} = \left(\frac{1+x}{2}\right)^n,$$

18.6.3 $$\lim_{\beta\to\infty} \frac{P_n^{(\alpha,\beta)}(x)}{P_n^{(\alpha,\beta)}(-1)} = \left(\frac{1-x}{2}\right)^n,$$

18.6.4 $$\lim_{\lambda\to\infty} \frac{C_n^{(\lambda)}(x)}{C_n^{(\lambda)}(1)} = x^n,$$

18.6.5 $$\lim_{\alpha\to\infty} \frac{L_n^{(\alpha)}(\alpha x)}{L_n^{(\alpha)}(0)} = (1-x)^n.$$

18.7 Interrelations and Limit Relations

18.7(i) Linear Transformations

Ultraspherical and Jacobi

18.7.1 $\quad C_n^{(\lambda)}(x) = \dfrac{(2\lambda)_n}{(\lambda+\tfrac{1}{2})_n} P_n^{(\lambda-\tfrac{1}{2},\lambda-\tfrac{1}{2})}(x),$

18.7.2 $\quad P_n^{(\alpha,\alpha)}(x) = \dfrac{(\alpha+1)_n}{(2\alpha+1)_n} C_n^{(\alpha+\tfrac{1}{2})}(x).$

Chebyshev, Ultraspherical, and Jacobi

18.7.3 $\quad T_n(x) = P_n^{(-\tfrac{1}{2},-\tfrac{1}{2})}(x) \big/ P_n^{(-\tfrac{1}{2},-\tfrac{1}{2})}(1),$

18.7.4
$$U_n(x) = C_n^{(1)}(x) = (n+1)\, P_n^{(\tfrac{1}{2},\tfrac{1}{2})}(x) \big/ P_n^{(\tfrac{1}{2},\tfrac{1}{2})}(1),$$

18.7.5 $\quad V_n(x) = (2n+1)\, P_n^{(\tfrac{1}{2},-\tfrac{1}{2})}(x) \big/ P_n^{(\tfrac{1}{2},-\tfrac{1}{2})}(1),$

18.7.6 $\quad W_n(x) = P_n^{(-\tfrac{1}{2},\tfrac{1}{2})}(x) \big/ P_n^{(-\tfrac{1}{2},\tfrac{1}{2})}(1).$

18.7.7 $\quad T_n^*(x) = T_n(2x-1),$

18.7.8 $\quad U_n^*(x) = U_n(2x-1).$

See also (18.9.9)–(18.9.12).

Legendre, Ultraspherical, and Jacobi

18.7.9 $\quad P_n(x) = C_n^{(\tfrac{1}{2})}(x) = P_n^{(0,0)}(x).$

18.7.10 $\quad P_n^*(x) = P_n(2x-1).$

Hermite

18.7.11 $\quad He_n(x) = 2^{-\tfrac{1}{2}n} H_n\!\left(2^{-\tfrac{1}{2}} x\right),$

18.7.12 $\quad H_n(x) = 2^{\tfrac{1}{2}n} He_n\!\left(2^{\tfrac{1}{2}} x\right).$

18.7(ii) Quadratic Transformations

$$18.7.13 \quad \frac{P_{2n}^{(\alpha,\alpha)}(x)}{P_{2n}^{(\alpha,\alpha)}(1)} = \frac{P_n^{(\alpha,-\frac{1}{2})}(2x^2-1)}{P_n^{(\alpha,-\frac{1}{2})}(1)},$$

$$18.7.14 \quad \frac{P_{2n+1}^{(\alpha,\alpha)}(x)}{P_{2n+1}^{(\alpha,\alpha)}(1)} = \frac{x\,P_n^{(\alpha,\frac{1}{2})}(2x^2-1)}{P_n^{(\alpha,\frac{1}{2})}(1)}.$$

$$18.7.15 \quad C_{2n}^{(\lambda)}(x) = \frac{(\lambda)_n}{(\frac{1}{2})_n}\,P_n^{(\lambda-\frac{1}{2},-\frac{1}{2})}(2x^2-1),$$

$$18.7.16 \quad C_{2n+1}^{(\lambda)}(x) = \frac{(\lambda)_{n+1}}{(\frac{1}{2})_{n+1}}\,x\,P_n^{(\lambda-\frac{1}{2},\frac{1}{2})}(2x^2-1).$$

$$18.7.17 \quad U_{2n}(x) = V_n(2x^2-1),$$

$$18.7.18 \quad T_{2n+1}(x) = x\,W_n(2x^2-1).$$

$$18.7.19 \quad H_{2n}(x) = (-1)^n 2^{2n} n!\, L_n^{(-\frac{1}{2})}(x^2),$$

$$18.7.20 \quad H_{2n+1}(x) = (-1)^n 2^{2n+1} n!\, x\, L_n^{(\frac{1}{2})}(x^2).$$

18.7(iii) Limit Relations

Jacobi → Laguerre

$$18.7.21 \quad \lim_{\beta\to\infty} P_n^{(\alpha,\beta)}(1-(2x/\beta)) = L_n^{(\alpha)}(x).$$

$$18.7.22 \quad \lim_{\alpha\to\infty} P_n^{(\alpha,\beta)}((2x/\alpha)-1) = (-1)^n L_n^{(\beta)}(x).$$

Jacobi → Hermite

$$18.7.23 \quad \lim_{\alpha\to\infty} \alpha^{-\frac{1}{2}n}\, P_n^{(\alpha,\alpha)}\!\left(\alpha^{-\frac{1}{2}}x\right) = \frac{H_n(x)}{2^n n!}.$$

Ultraspherical → Hermite

$$18.7.24 \quad \lim_{\lambda\to\infty} \lambda^{-\frac{1}{2}n}\, C_n^{(\lambda)}\!\left(\lambda^{-\frac{1}{2}}x\right) = \frac{H_n(x)}{n!}.$$

$$18.7.25 \quad \lim_{\lambda\to 0} \frac{1}{\lambda} C_n^{(\lambda)}(x) = \frac{2}{n} T_n(x), \qquad n \ge 1.$$

Laguerre → Hermite

18.7.26
$$\lim_{\alpha\to\infty} \left(\frac{2}{\alpha}\right)^{\frac{1}{2}n} L_n^{(\alpha)}\!\left((2\alpha)^{\frac{1}{2}}x + \alpha\right) = \frac{(-1)^n}{n!} H_n(x).$$

See Figure 18.21.1 for the Askey schematic representation of most of these limits.

18.8 Differential Equations

See Table 18.8.1 and also Table 22.6 of Abramowitz and Stegun (1964).

Table 18.8.1: Classical OP's: differential equations $A(x)f''(x) + B(x)f'(x) + C(x)f(x) + \lambda_n f(x) = 0$.

$f(x)$	$A(x)$	$B(x)$	$C(x)$	λ_n
$P_n^{(\alpha,\beta)}(x)$	$1-x^2$	$\beta-\alpha-(\alpha+\beta+2)x$	0	$n(n+\alpha+\beta+1)$
$(\sin\tfrac{1}{2}x)^{\alpha+\frac{1}{2}}(\cos\tfrac{1}{2}x)^{\beta+\frac{1}{2}}$ $\times P_n^{(\alpha,\beta)}(\cos x)$	1	0	$\dfrac{\frac{1}{4}-\alpha^2}{4\sin^2\frac{1}{2}x} + \dfrac{\frac{1}{4}-\beta^2}{4\cos^2\frac{1}{2}x}$	$\left(n+\tfrac{1}{2}(\alpha+\beta+1)\right)^2$
$(\sin x)^{\alpha+\frac{1}{2}} P_n^{(\alpha,\alpha)}(\cos x)$	1	0	$(\tfrac{1}{4}-\alpha^2)/\sin^2 x$	$(n+\alpha+\tfrac{1}{2})^2$
$C_n^{(\lambda)}(x)$	$1-x^2$	$-(2\lambda+1)x$	0	$n(n+2\lambda)$
$T_n(x)$	$1-x^2$	$-x$	0	n^2
$U_n(x)$	$1-x^2$	$-3x$	0	$n(n+2)$
$P_n(x)$	$1-x^2$	$-2x$	0	$n(n+1)$
$L_n^{(\alpha)}(x)$	x	$\alpha+1-x$	0	n
$e^{-\frac{1}{2}x^2} x^{\alpha+\frac{1}{2}} L_n^{(\alpha)}(x^2)$	1	0	$-x^2 + (\tfrac{1}{4}-\alpha^2)x^{-2}$	$4n+2\alpha+2$
$H_n(x)$	1	$-2x$	0	$2n$
$e^{-\frac{1}{2}x^2} H_n(x)$	1	0	$-x^2$	$2n+1$
$He_n(x)$	1	$-x$	0	n

18.9 Recurrence Relations and Derivatives

18.9(i) Recurrence Relations

18.9.1 $p_{n+1}(x) = (A_n x + B_n) p_n(x) - C_n p_{n-1}(x).$

For $p_n(x) = P_n^{(\alpha,\beta)}(x),$

18.9.2
$$A_n = \frac{(2n+\alpha+\beta+1)(2n+\alpha+\beta+2)}{2(n+1)(n+\alpha+\beta+1)},$$
$$B_n = \frac{(\alpha^2-\beta^2)(2n+\alpha+\beta+1)}{2(n+1)(n+\alpha+\beta+1)(2n+\alpha+\beta)},$$
$$C_n = \frac{(n+\alpha)(n+\beta)(2n+\alpha+\beta+2)}{(n+1)(n+\alpha+\beta+1)(2n+\alpha+\beta)}.$$

For the other classical OP's see Table 18.9.1; compare also §18.2(iv).

Table 18.9.1: Classical OP's: recurrence relations (18.9.1).

$p_n(x)$	A_n	B_n	C_n
$C_n^{(\lambda)}(x)$	$\frac{2n+\lambda}{n+1}$	0	$\frac{n+2\lambda-1}{n+1}$
$T_n(x)$	$2 - \delta_{n,0}$	0	1
$U_n(x)$	2	0	1
$T_n^*(x)$	$4 - 2\delta_{n,0}$	$-2 + \delta_{n,0}$	1
$U_n^*(x)$	4	-2	1
$P_n(x)$	$\frac{2n+1}{n+1}$	0	$\frac{n}{n+1}$
$P_n^*(x)$	$\frac{4n+2}{n+1}$	$-\frac{2n+1}{n+1}$	$\frac{n}{n+1}$
$L_n^{(\alpha)}(x)$	$-\frac{1}{n+1}$	$\frac{2n+\alpha+1}{n+1}$	$\frac{n+\alpha}{n+1}$
$H_n(x)$	2	0	$2n$
$He_n(x)$	1	0	n

18.9(ii) Contiguous Relations in the Parameters and the Degree

Jacobi

18.9.3 $P_n^{(\alpha,\beta-1)}(x) - P_n^{(\alpha-1,\beta)}(x) = P_{n-1}^{(\alpha,\beta)}(x),$

18.9.4
$(1-x)P_n^{(\alpha+1,\beta)}(x) + (1+x)P_n^{(\alpha,\beta+1)}(x) = 2P_n^{(\alpha,\beta)}(x).$

18.9.5
$(2n+\alpha+\beta+1)P_n^{(\alpha,\beta)}(x)$
$= (n+\alpha+\beta+1)P_n^{(\alpha,\beta+1)}(x) + (n+\alpha)P_{n-1}^{(\alpha,\beta+1)}(x),$

18.9.6
$(n+\tfrac{1}{2}\alpha+\tfrac{1}{2}\beta+1)(1+x)P_n^{(\alpha,\beta+1)}(x)$
$= (n+1)P_{n+1}^{(\alpha,\beta)}(x) + (n+\beta+1)P_n^{(\alpha,\beta)}(x),$

and a similar pair to (18.9.5) and (18.9.6) by symmetry; compare the second row in Table 18.6.1.

Ultraspherical

18.9.7 $(n+\lambda)C_n^{(\lambda)}(x) = \lambda\left(C_n^{(\lambda+1)}(x) - C_{n-2}^{(\lambda+1)}(x)\right),$

18.9.8
$4\lambda(n+\lambda+1)(1-x^2)C_n^{(\lambda+1)}(x)$
$= -(n+1)(n+2)C_{n+2}^{(\lambda)}(x)$
$+ (n+2\lambda)(n+2\lambda+1)C_n^{(\lambda)}(x).$

Chebyshev

18.9.9 $T_n(x) = \tfrac{1}{2}\left(U_n(x) - U_{n-2}(x)\right),$

18.9.10 $(1-x^2)U_n(x) = -\tfrac{1}{2}\left(T_{n+2}(x) - T_n(x)\right).$

18.9.11 $W_n(x) + W_{n-1}(x) = 2T_n(x),$

18.9.12 $T_{n+1}(x) + T_n(x) = (1+x)W_n(x).$

Laguerre

18.9.13 $L_n^{(\alpha)}(x) = L_n^{(\alpha+1)}(x) - L_{n-1}^{(\alpha+1)}(x),$

18.9.14
$x L_n^{(\alpha+1)}(x) = -(n+1)L_{n+1}^{(\alpha)}(x)$
$+ (n+\alpha+1)L_n^{(\alpha)}(x).$

18.9(iii) Derivatives

Jacobi

18.9.15 $\frac{d}{dx}P_n^{(\alpha,\beta)}(x) = \tfrac{1}{2}(n+\alpha+\beta+1)P_{n-1}^{(\alpha+1,\beta+1)}(x),$

18.9.16
$\frac{d}{dx}\left((1-x)^\alpha(1+x)^\beta P_n^{(\alpha,\beta)}(x)\right)$
$= -2(n+1)(1-x)^{\alpha-1}(1+x)^{\beta-1}P_{n+1}^{(\alpha-1,\beta-1)}(x).$

18.9.17
$(2n+\alpha+\beta)(1-x^2)\frac{d}{dx}P_n^{(\alpha,\beta)}(x)$
$= n(\alpha-\beta-(2n+\alpha+\beta)x)P_n^{(\alpha,\beta)}(x)$
$+ 2(n+\alpha)(n+\beta)P_{n-1}^{(\alpha,\beta)}(x),$

18.9.18
$(2n+\alpha+\beta+2)(1-x^2)\frac{d}{dx}P_n^{(\alpha,\beta)}(x)$
$= (n+\alpha+\beta+1)(\alpha-\beta+(2n+\alpha+\beta+2)x)P_n^{(\alpha,\beta)}(x)$
$- 2(n+1)(n+\alpha+\beta+1)P_{n+1}^{(\alpha,\beta)}(x).$

Ultraspherical

18.9.19 $\frac{d}{dx}C_n^{(\lambda)}(x) = 2\lambda C_{n-1}^{(\lambda+1)}(x),$

18.9.20
$\frac{d}{dx}\left((1-x^2)^{\lambda-\tfrac{1}{2}}C_n^{(\lambda)}(x)\right)$
$= -\frac{(n+1)(n+2\lambda-1)}{2(\lambda-1)}(1-x^2)^{\lambda-\tfrac{3}{2}}C_{n+1}^{(\lambda-1)}(x).$

18.10   Integral Representations

Chebyshev

18.9.21
$$\frac{d}{dx} T_n(x) = n\, U_{n-1}(x),$$

18.9.22
$$\frac{d}{dx}\left((1-x^2)^{\frac{1}{2}} U_n(x)\right) = -(n+1)(1-x^2)^{-\frac{1}{2}} T_{n+1}(x).$$

Laguerre

18.9.23
$$\frac{d}{dx} L_n^{(\alpha)}(x) = -L_{n-1}^{(\alpha+1)}(x),$$

18.9.24
$$\frac{d}{dx}\left(e^{-x} x^\alpha L_n^{(\alpha)}(x)\right) = (n+1)e^{-x} x^{\alpha-1} L_{n+1}^{(\alpha-1)}(x).$$

Hermite

18.9.25
$$\frac{d}{dx} H_n(x) = 2n\, H_{n-1}(x),$$

18.9.26
$$\frac{d}{dx}\left(e^{-x^2} H_n(x)\right) = -e^{-x^2} H_{n+1}(x).$$

18.9.27
$$\frac{d}{dx} He_n(x) = n\, He_{n-1}(x),$$

18.9.28
$$\frac{d}{dx}\left(e^{-\frac{1}{2}x^2} He_n(x)\right) = -e^{-\frac{1}{2}x^2} He_{n+1}(x).$$

18.10 Integral Representations

18.10(i) Dirichlet–Mehler-Type Integral Representations

Ultraspherical

18.10.1
$$\frac{P_n^{(\alpha,\alpha)}(\cos\theta)}{P_n^{(\alpha,\alpha)}(1)} = \frac{C_n^{(\alpha+\frac{1}{2})}(\cos\theta)}{C_n^{(\alpha+\frac{1}{2})}(1)} = \frac{2^{\alpha+\frac{1}{2}}\Gamma(\alpha+1)}{\pi^{\frac{1}{2}}\Gamma(\alpha+\frac{1}{2})}(\sin\theta)^{-2\alpha}\int_0^\theta \frac{\cos\bigl((n+\alpha+\frac{1}{2})\phi\bigr)}{(\cos\phi - \cos\theta)^{-\alpha+\frac{1}{2}}}\,d\phi, \quad 0 < \theta < \pi,\ \alpha > -\tfrac{1}{2}.$$

Legendre

18.10.2
$$P_n(\cos\theta) = \frac{2^{\frac{1}{2}}}{\pi}\int_0^\theta \frac{\cos\bigl((n+\frac{1}{2})\phi\bigr)}{(\cos\phi - \cos\theta)^{\frac{1}{2}}}\,d\phi, \qquad 0 < \theta < \pi.$$

Generalizations of (18.10.1) are given in Gasper (1975, (6),(8)) and Koornwinder (1975b, (5.7),(5.8)).

18.10(ii) Laplace-Type Integral Representations

Jacobi

18.10.3
$$\frac{P_n^{(\alpha,\beta)}(\cos\theta)}{P_n^{(\alpha,\beta)}(1)} = \frac{2\,\Gamma(\alpha+1)}{\pi^{\frac{1}{2}}\Gamma(\alpha-\beta)\Gamma(\beta+\frac{1}{2})}$$
$$\times \int_0^1\int_0^\pi \bigl((\cos\tfrac{1}{2}\theta)^2 - r^2(\sin\tfrac{1}{2}\theta)^2 + ir\sin\theta\cos\phi\bigr)^n (1-r^2)^{\alpha-\beta-1} r^{2\beta+1}(\sin\phi)^{2\beta}\,d\phi\,dr,$$
$$\alpha > \beta > -\tfrac{1}{2}.$$

Ultraspherical

18.10.4
$$\frac{P_n^{(\alpha,\alpha)}(\cos\theta)}{P_n^{(\alpha,\alpha)}(1)} = \frac{C_n^{(\alpha+\frac{1}{2})}(\cos\theta)}{C_n^{(\alpha+\frac{1}{2})}(1)} = \frac{\Gamma(\alpha+1)}{\pi^{\frac{1}{2}}\Gamma(\alpha+\frac{1}{2})}\int_0^\pi (\cos\theta + i\sin\theta\cos\phi)^n (\sin\phi)^{2\alpha}\,d\phi, \qquad \alpha > -\tfrac{1}{2}.$$

Legendre

18.10.5
$$P_n(\cos\theta) = \frac{1}{\pi}\int_0^\pi (\cos\theta + i\sin\theta\cos\phi)^n\,d\phi.$$

Laguerre

18.10.6
$$L_n^{(\alpha)}(x^2) = \frac{2(-1)^n}{\pi^{\frac{1}{2}}\Gamma(\alpha+\frac{1}{2})n!}\int_0^\infty\int_0^\pi (x^2 - r^2 + 2ixr\cos\phi)^n e^{-r^2} r^{2\alpha+1}(\sin\phi)^{2\alpha}\,d\phi\,dr, \qquad \alpha > -\tfrac{1}{2}.$$

Hermite

18.10.7
$$H_n(x) = \frac{2^n}{\pi^{\frac{1}{2}}}\int_{-\infty}^\infty (x+it)^n e^{-t^2}\,dt.$$

18.10(iii) Contour Integral Representations

Table 18.10.1 gives contour integral representations of the form

18.10.8
$$p_n(x) = \frac{g_0(x)}{2\pi i} \int_C (g_1(z,x))^n g_2(z,x)(z-c)^{-1}\, dz$$

for the Jacobi, Laguerre, and Hermite polynomials. Here C is a simple closed contour encircling $z = c$ once in the positive sense.

Table 18.10.1: Classical OP's: contour integral representations (18.10.8).

$p_n(x)$	$g_0(x)$	$g_1(z,x)$	$g_2(z,x)$	c	Conditions
$P_n^{(\alpha,\beta)}(x)$	$(1-x)^{-\alpha}(1+x)^{-\beta}$	$\dfrac{z^2-1}{2(z-x)}$	$(1-z)^\alpha(1+z)^\beta$	x	± 1 outside C.
$C_n^{(\lambda)}(x)$	1	z^{-1}	$(1-2xz+z^2)^{-\lambda}$	0	
$T_n(x)$	1	z^{-1}	$\dfrac{1-xz}{1-2xz+z^2}$	0	$e^{\pm i\theta}$ outside C (where $x = \cos\theta$).
$U_n(x)$	1	z^{-1}	$(1-2xz+z^2)^{-1}$	0	
$P_n(x)$	1	z^{-1}	$(1-2xz+z^2)^{-\frac{1}{2}}$	0	
$P_n(x)$	1	$\dfrac{z^2-1}{2(z-x)}$	1	x	
$L_n^{(\alpha)}(x)$	$e^x x^{-\alpha}$	$z(z-x)^{-1}$	$z^\alpha e^{-z}$	x	0 outside C.
$H_n(x)/n!$	1	z^{-1}	e^{2xz-z^2}	0	
$\mathit{He}_n(x)/n!$	1	z^{-1}	$e^{xz-\frac{1}{2}z^2}$	0	

18.10(iv) Other Integral Representations

Laguerre

18.10.9 $L_n^{(\alpha)}(x) = \dfrac{e^x x^{-\frac{1}{2}\alpha}}{n!} \int_0^\infty e^{-t} t^{n+\frac{1}{2}\alpha} J_\alpha(2\sqrt{xt})\, dt,$
$\alpha > -1.$

For the Bessel function $J_\nu(z)$ see §10.2(ii).

Hermite

18.10.10
$$H_n(x) = \frac{(-2i)^n e^{x^2}}{\pi^{\frac{1}{2}}} \int_{-\infty}^\infty e^{-t^2} t^n e^{2ixt}\, dt$$
$$= \frac{2^{n+1}}{\pi^{\frac{1}{2}}} e^{x^2} \int_0^\infty e^{-t^2} t^n \cos(2xt - \tfrac{1}{2}n\pi)\, dt.$$

See also §18.17.

18.11 Relations to Other Functions

18.11(i) Explicit Formulas

See §§18.5(i) and 18.5(iii) for relations to trigonometric functions, the hypergeometric function, and generalized hypergeometric functions.

Ultraspherical

18.11.1
$$\mathsf{P}_n^m(x) = (\tfrac{1}{2})_m (-2)^m (1-x^2)^{\frac{1}{2}m} C_{n-m}^{(m+\frac{1}{2})}(x)$$
$$= (n+1)_m (-2)^{-m} (1-x^2)^{\frac{1}{2}m} P_{n-m}^{(m,m)}(x),$$
$$0 \le m \le n.$$

For the Ferrers function $\mathsf{P}_n^m(x)$, see §14.3(i).

Compare also (14.3.21) and (14.3.22).

Laguerre

18.11.2
$$L_n^{(\alpha)}(x) = \frac{(\alpha+1)_n}{n!} M(-n, \alpha+1, x)$$
$$= \frac{(-1)^n}{n!} U(-n, \alpha+1, x)$$
$$= \frac{(\alpha+1)_n}{n!} z^{-\frac{1}{2}(\alpha+1)} e^{\frac{1}{2}z} M_{n+\frac{1}{2}(\alpha+1),\frac{1}{2}\alpha}(z)$$
$$= \frac{(-1)^n}{n!} z^{-\frac{1}{2}(\alpha+1)} e^{\frac{1}{2}z} W_{n+\frac{1}{2}(\alpha+1),\frac{1}{2}\alpha}(z).$$

For the confluent hypergeometric functions $M(a,b,z)$ and $U(a,b,z)$, see §13.2(i), and for the Whittaker functions $M_{\kappa,\mu}(z)$ and $W_{\kappa,\mu}(z)$ see §13.14(i).

Hermite

$$18.11.3 \quad \begin{aligned} H_n(x) &= 2^n\, U\!\left(-\tfrac{1}{2}n, \tfrac{1}{2}, x^2\right) \\ &= 2^n x\, U\!\left(-\tfrac{1}{2}n + \tfrac{1}{2}, \tfrac{3}{2}, x^2\right) \\ &= 2^{\frac{1}{2}n} e^{\frac{1}{2}x^2} U\!\left(-n - \tfrac{1}{2}, 2^{\frac{1}{2}} x\right). \end{aligned}$$

$$18.11.4 \quad \begin{aligned} He_n(x) &= 2^{\frac{1}{2}n}\, U\!\left(-\tfrac{1}{2}n, \tfrac{1}{2}, \tfrac{1}{2}x^2\right) \\ &= 2^{\frac{1}{2}(n-1)} x\, U\!\left(-\tfrac{1}{2}n + \tfrac{1}{2}, \tfrac{3}{2}, \tfrac{1}{2}x^2\right) \\ &= e^{\frac{1}{4}x^2}\, U\!\left(-n - \tfrac{1}{2}, x\right). \end{aligned}$$

For the parabolic cylinder function $U(a,z)$, see §12.2.

18.11(ii) Formulas of Mehler–Heine Type

Jacobi

18.11.5
$$\lim_{n\to\infty} \frac{1}{n^\alpha} P_n^{(\alpha,\beta)}\!\left(1 - \frac{z^2}{2n^2}\right) = \lim_{n\to\infty} \frac{1}{n^\alpha} P_n^{(\alpha,\beta)}\!\left(\cos \frac{z}{n}\right) = \frac{2^\alpha}{z^\alpha} J_\alpha(z).$$

Laguerre

18.11.6 $\quad \displaystyle \lim_{n\to\infty} \frac{1}{n^\alpha} L_n^{(\alpha)}\!\left(\frac{z}{n}\right) = \frac{1}{z^{\frac{1}{2}\alpha}} J_\alpha\!\left(2z^{\frac{1}{2}}\right).$

Hermite

18.11.7 $\quad \displaystyle \lim_{n\to\infty} \frac{(-1)^n n^{\frac{1}{2}}}{2^{2n} n!} H_{2n}\!\left(\frac{z}{2n^{\frac{1}{2}}}\right) = \frac{1}{\pi^{\frac{1}{2}}} \cos z,$

18.11.8 $\quad \displaystyle \lim_{n\to\infty} \frac{(-1)^n}{2^{2n} n!} H_{2n+1}\!\left(\frac{z}{2n^{\frac{1}{2}}}\right) = \frac{2}{\pi^{\frac{1}{2}}} \sin z.$

For the Bessel function $J_\nu(z)$, see §10.2(ii). The limits (18.11.5)–(18.11.8) hold uniformly for z in any bounded subset of $\mathbb{C}$.

18.12 Generating Functions

With the notation of §§10.2(ii), 10.25(ii), and 15.2,

Jacobi

18.12.1
$$\frac{2^{\alpha+\beta}}{R(1+R-z)^\alpha (1+R+z)^\beta} = \sum_{n=0}^\infty P_n^{(\alpha,\beta)}(x) z^n, \quad R = \sqrt{1 - 2xz + z^2},\ |z| < 1.$$

18.12.2
$$\begin{aligned} &\left(\tfrac{1}{2}(1-x)z\right)^{-\frac{1}{2}\alpha} J_\alpha\!\left(\sqrt{2(1-x)z}\right) \\ &\quad \times \left(\tfrac{1}{2}(1+x)z\right)^{-\frac{1}{2}\beta} I_\beta\!\left(\sqrt{2(1+x)z}\right) \\ &= \sum_{n=0}^\infty \frac{P_n^{(\alpha,\beta)}(x)}{\Gamma(n+\alpha+1)\,\Gamma(n+\beta+1)} z^n. \end{aligned}$$

18.12.3
$$(1+z)^{-\alpha-\beta-1} \times {}_2F_1\!\left(\begin{array}{c}\tfrac{1}{2}(\alpha+\beta+1), \tfrac{1}{2}(\alpha+\beta+2) \\ \beta+1\end{array}; \frac{2(x+1)z}{(1+z)^2}\right)$$
$$= \sum_{n=0}^\infty \frac{(\alpha+\beta+1)_n}{(\beta+1)_n} P_n^{(\alpha,\beta)}(x) z^n, \quad |z| < 1,$$

and a similar formula by symmetry; compare the second row in Table 18.6.1. For the hypergeometric function ${}_2F_1$ see §§15.1, 15.2(i).

Ultraspherical

18.12.4
$$(1 - 2xz + z^2)^{-\lambda} = \sum_{n=0}^\infty C_n^{(\lambda)}(x) z^n$$
$$= \sum_{n=0}^\infty \frac{(2\lambda)_n}{(\lambda+\tfrac{1}{2})_n} P_n^{(\lambda-\frac{1}{2},\lambda-\frac{1}{2})}(x) z^n, \quad |z| < 1.$$

18.12.5
$$\frac{1 - xz}{(1-2xz+z^2)^{\lambda+1}} = \sum_{n=0}^\infty \frac{n+2\lambda}{2\lambda} C_n^{(\lambda)}(x) z^n, \quad |z| < 1.$$

18.12.6
$$\Gamma\!\left(\lambda+\tfrac{1}{2}\right) e^{z\cos\theta} \left(\tfrac{1}{2} z \sin\theta\right)^{\frac{1}{2}-\lambda} J_{\lambda-\frac{1}{2}}(z\sin\theta) = \sum_{n=0}^\infty \frac{C_n^{(\lambda)}(\cos\theta)}{(2\lambda)_n} z^n, \quad 0 \le \theta \le \pi.$$

Chebyshev

18.12.7 $\quad \displaystyle \frac{1 - z^2}{1 - 2xz + z^2} = 1 + 2\sum_{n=1}^\infty T_n(x) z^n, \quad |z| < 1.$

18.12.8 $\quad \displaystyle \frac{1 - xz}{1 - 2xz + z^2} = \sum_{n=0}^\infty T_n(x) z^n, \quad |z| < 1.$

18.12.9 $\quad \displaystyle -\ln(1 - 2xz + z^2) = 2\sum_{n=1}^\infty \frac{T_n(x)}{n} z^n, \quad |z| < 1.$

18.12.10 $\quad \displaystyle \frac{1}{1 - 2xz + z^2} = \sum_{n=0}^\infty U_n(x) z^n, \quad |z| < 1.$

Legendre

18.12.11 $\quad \displaystyle \frac{1}{\sqrt{1 - 2xz + z^2}} = \sum_{n=0}^\infty P_n(x) z^n, \quad |z| < 1.$

18.12.12 $\quad \displaystyle e^{xz} J_0\!\left(z\sqrt{1-x^2}\right) = \sum_{n=0}^\infty \frac{P_n(x)}{n!} z^n.$

Laguerre

18.12.13
$$(1-z)^{-\alpha-1} \exp\!\left(\frac{xz}{z-1}\right) = \sum_{n=0}^\infty L_n^{(\alpha)}(x) z^n, \quad |z| < 1.$$

18.12.14
$$\Gamma(\alpha+1)(xz)^{-\frac{1}{2}\alpha} e^z J_\alpha\!\left(2\sqrt{xz}\right) = \sum_{n=0}^\infty \frac{L_n^{(\alpha)}(x)}{(\alpha+1)_n} z^n.$$

Hermite

18.12.15
$$e^{2xz-z^2} = \sum_{n=0}^{\infty} \frac{H_n(x)}{n!} z^n,$$

18.12.16
$$e^{xz-\frac{1}{2}z^2} = \sum_{n=0}^{\infty} \frac{He_n(x)}{n!} z^n.$$

18.13 Continued Fractions

We use the terminology of §1.12(ii).

Chebyshev

$T_n(x)$ is the denominator of the nth approximant to:

18.13.1
$$\frac{-1}{x+}\frac{-1}{2x+}\frac{-1}{2x+}\cdots,$$

and $U_n(x)$ is the denominator of the nth approximant to:

18.13.2
$$\frac{-1}{2x+}\frac{-1}{2x+}\frac{-1}{2x+}\cdots.$$

Legendre

$P_n(x)$ is the denominator of the nth approximant to:

18.13.3
$$\frac{a_1}{x+}\frac{-\frac{1}{2}}{\frac{3}{2}x+}\frac{-\frac{2}{3}}{\frac{5}{3}x+}\frac{-\frac{3}{4}}{\frac{7}{4}x+}\cdots,$$

where a_1 is an arbitrary nonzero constant.

Laguerre

$L_n(x)$ is the denominator of the nth approximant to:

18.13.4
$$\frac{a_1}{1-x+}\frac{-\frac{1}{2}}{\frac{1}{2}(3-x)+}\frac{-\frac{2}{3}}{\frac{1}{3}(5-x)+}\frac{-\frac{3}{4}}{\frac{1}{4}(7-x)+}\cdots,$$

where a_1 is again an arbitrary nonzero constant.

Hermite

$H_n(x)$ is the denominator of the nth approximant to:

18.13.5
$$\frac{1}{2x+}\frac{-2}{2x+}\frac{-4}{2x+}\frac{-6}{2x+}\cdots.$$

See also Cuyt et al. (2008, pp. 91–99).

18.14 Inequalities

18.14(i) Upper Bounds

Jacobi

18.14.1
$$|P_n^{(\alpha,\beta)}(x)| \leq P_n^{(\alpha,\beta)}(1) = \frac{(\alpha+1)_n}{n!},$$
$$-1 \leq x \leq 1, \alpha \geq \beta > -1, \alpha \geq -\tfrac{1}{2},$$

18.14.2
$$|P_n^{(\alpha,\beta)}(x)| \leq |P_n^{(\alpha,\beta)}(-1)| = \frac{(\beta+1)_n}{n!},$$
$$-1 \leq x \leq 1, \beta \geq \alpha > -1, \beta \geq -\tfrac{1}{2}.$$

18.14.3
$$\left(\tfrac{1}{2}(1-x)\right)^{\frac{1}{2}\alpha+\frac{1}{4}} \left(\tfrac{1}{2}(1+x)\right)^{\frac{1}{2}\beta+\frac{1}{4}} |P_n^{(\alpha,\beta)}(x)|$$
$$\leq \frac{\Gamma(\max(\alpha,\beta)+n+1)}{\pi^{\frac{1}{2}} n! \left(n+\frac{1}{2}(\alpha+\beta+1)\right)^{\max(\alpha,\beta)+\frac{1}{2}}},$$
$$-1 \leq x \leq 1, -\tfrac{1}{2} \leq \alpha \leq \tfrac{1}{2}, -\tfrac{1}{2} \leq \beta \leq \tfrac{1}{2}.$$

Ultraspherical

18.14.4
$$|C_n^{(\lambda)}(x)| \leq C_n^{(\lambda)}(1) = \frac{(2\lambda)_n}{n!}, \quad -1 \leq x \leq 1, \lambda > 0.$$

18.14.5
$$|C_{2m}^{(\lambda)}(x)| \leq |C_{2m}^{(\lambda)}(0)| = \left|\frac{(\lambda)_m}{m!}\right|,$$
$$-1 \leq x \leq 1, -\tfrac{1}{2} < \lambda < 0,$$

18.14.6
$$|C_{2m+1}^{(\lambda)}(x)| < \frac{-2(\lambda)_{m+1}}{((2m+1)(2\lambda+2m+1))^{\frac{1}{2}} m!},$$
$$-1 \leq x \leq 1, -\tfrac{1}{2} < \lambda < 0.$$

18.14.7
$$(n+\lambda)^{1-\lambda}(1-x^2)^{\frac{1}{2}\lambda}|C_n^{(\lambda)}(x)| < \frac{2^{1-\lambda}}{\Gamma(\lambda)},$$
$$-1 \leq x \leq 1, 0 < \lambda < 1.$$

Laguerre

18.14.8
$$e^{-\frac{1}{2}x}\left|L_n^{(\alpha)}(x)\right| \leq L_n^{(\alpha)}(0) = \frac{(\alpha+1)_n}{n!},$$
$$0 \leq x < \infty, \alpha \geq 0.$$

Hermite

18.14.9
$$\frac{1}{(2^n n!)^{\frac{1}{2}}} e^{-\frac{1}{2}x^2} |H_n(x)| \leq 1, \quad -\infty < x < \infty.$$

For further inequalities see Abramowitz and Stegun (1964, §22.14).

18.14(ii) Turan-Type Inequalities

Legendre

18.14.10
$$(P_n(x))^2 \geq P_{n-1}(x) P_{n+1}(x), \quad -1 \leq x \leq 1.$$

Jacobi

Let $R_n(x) = P_n^{(\alpha,\beta)}(x)/P_n^{(\alpha,\beta)}(1)$. Then

18.14.11
$$(R_n(x))^2 \geq R_{n-1}(x)R_{n+1}(x), -1 \leq x \leq 1, \beta \geq \alpha > -1.$$

Laguerre

18.14.12
$$(L_n^{(\alpha)}(x))^2 \geq L_{n-1}^{(\alpha)}(x) L_{n+1}^{(\alpha)}(x), \quad 0 \leq x < \infty, \alpha \geq 0.$$

Hermite

18.14.13
$$(H_n(x))^2 \geq H_{n-1}(x) H_{n+1}(x), \quad -\infty < x < \infty.$$

18.14(iii) Local Maxima and Minima

Jacobi

Let the maxima $x_{n,m}$, $m = 0, 1, \ldots, n$, of $|P_n^{(\alpha,\beta)}(x)|$ in $[-1, 1]$ be arranged so that

18.14.14
$$-1 = x_{n,0} < x_{n,1} < \cdots < x_{n,n-1} < x_{n,n} = 1.$$

When $(\alpha + \tfrac{1}{2})(\beta + \tfrac{1}{2}) > 0$ choose m so that

18.14.15
$$x_{n,m} \leq (\beta-\alpha)/(\alpha+\beta+1) \leq x_{n,m+1}.$$

Then

18.14.16
$$|P_n^{(\alpha,\beta)}(x_{n,0})| > |P_n^{(\alpha,\beta)}(x_{n,1})| > \cdots > |P_n^{(\alpha,\beta)}(x_{n,m})|,$$
$$|P_n^{(\alpha,\beta)}(x_{n,n})| > |P_n^{(\alpha,\beta)}(x_{n,n-1})| > \cdots > |P_n^{(\alpha,\beta)}(x_{n,m+1})|, \qquad \alpha > -\tfrac{1}{2}, \beta > -\tfrac{1}{2}.$$

18.14.17
$$|P_n^{(\alpha,\beta)}(x_{n,0})| < |P_n^{(\alpha,\beta)}(x_{n,1})| < \cdots < |P_n^{(\alpha,\beta)}(x_{n,m})|,$$
$$|P_n^{(\alpha,\beta)}(x_{n,n})| < |P_n^{(\alpha,\beta)}(x_{n,n-1})| < \cdots < |P_n^{(\alpha,\beta)}(x_{n,m+1})|, \qquad -1 < \alpha < -\tfrac{1}{2}, -1 < \beta < -\tfrac{1}{2}.$$

Also,

18.14.18
$$|P_n^{(\alpha,\beta)}(x_{n,0})| < |P_n^{(\alpha,\beta)}(x_{n,1})| < \cdots < |P_n^{(\alpha,\beta)}(x_{n,n})|,$$
$$\alpha \geq -\tfrac{1}{2},\ -1 < \beta \leq -\tfrac{1}{2},$$

18.14.19
$$|P_n^{(\alpha,\beta)}(x_{n,0})| > |P_n^{(\alpha,\beta)}(x_{n,1})| > \cdots > |P_n^{(\alpha,\beta)}(x_{n,n})|,$$
$$\beta \geq -\tfrac{1}{2},\ -1 < \alpha \leq -\tfrac{1}{2},$$

except that when $\alpha = \beta = -\tfrac{1}{2}$ (Chebyshev case) $|P_n^{(\alpha,\beta)}(x_{n,m})|$ is constant.

Szegö–Szász Inequality

18.14.20
$$\left|\frac{P_n^{(\alpha,\beta)}(x_{n,n-m})}{P_n^{(\alpha,\beta)}(1)}\right| > \left|\frac{P_{n+1}^{(\alpha,\beta)}(x_{n+1,n-m+1})}{P_{n+1}^{(\alpha,\beta)}(1)}\right|,$$
$$\alpha = \beta > -\tfrac{1}{2},\ m = 1,2,\ldots,n.$$

For extensions of (18.14.20) see Askey (1990) and Wong and Zhang (1994a,b).

Laguerre

Let the maxima $x_{n,m}$, $m = 0, 1, \ldots, n-1$, of $|L_n^{(\alpha)}(x)|$ in $[0, \infty)$ be arranged so that

18.14.21 $\quad 0 = x_{n,0} < x_{n,1} < \cdots < x_{n,n-1} < x_{n,n} = \infty$.

When $\alpha > -\tfrac{1}{2}$ choose m so that

18.14.22 $\qquad x_{n,m} \leq \alpha + \tfrac{1}{2} \leq x_{n,m+1}$.

Then

18.14.23
$$|L_n^{(\alpha)}(x_{n,0})| > |L_n^{(\alpha)}(x_{n,1})| > \cdots > |L_n^{(\alpha)}(x_{n,m})|,$$
$$|L_n^{(\alpha)}(x_{n,n-1})| > |L_n^{(\alpha)}(x_{n,n-2})| > \cdots > |L_n^{(\alpha)}(x_{n,m+1})|.$$

Also, when $\alpha \leq -\tfrac{1}{2}$

18.14.24
$$|L_n^{(\alpha)}(x_{n,0})| < |L_n^{(\alpha)}(x_{n,1})| < \cdots < |L_n^{(\alpha)}(x_{n,n-1})|.$$

Hermite

The successive maxima of $|H_n(x)|$ form a decreasing sequence for $x \leq 0$, and an increasing sequence for $x \geq 0$.

18.15 Asymptotic Approximations

18.15(i) Jacobi

With the exception of the penultimate paragraph, we assume throughout this subsection that α, β, and M ($= 0, 1, 2, \ldots$) are all fixed.

18.15.1
$$(\sin \tfrac{1}{2}\theta)^{\alpha+\frac{1}{2}} (\cos \tfrac{1}{2}\theta)^{\beta+\frac{1}{2}} P_n^{(\alpha,\beta)}(\cos\theta)$$
$$= \pi^{-1} 2^{2n+\alpha+\beta+1} \mathrm{B}(n+\alpha+1, n+\beta+1)$$
$$\times \left(\sum_{m=0}^{M-1} \frac{f_m(\theta)}{2^m(2n+\alpha+\beta+2)_m} + O(n^{-M})\right),$$

as $n \to \infty$, uniformly with respect to $\theta \in [\delta, \pi-\delta]$. Here, and elsewhere in §18.15, δ is an arbitrary small positive constant. Also, $\mathrm{B}(a,b)$ is the beta function (§5.12) and

18.15.2 $\quad f_m(\theta) = \sum_{\ell=0}^{m} \frac{C_{m,\ell}(\alpha,\beta)}{\ell!(m-\ell)!} \frac{\cos\theta_{n,m,\ell}}{(\sin\tfrac{1}{2}\theta)^\ell (\cos\tfrac{1}{2}\theta)^{m-\ell}},$

where

18.15.3
$$C_{m,\ell}(\alpha,\beta) = (\tfrac{1}{2}+\alpha)_\ell (\tfrac{1}{2}-\alpha)_\ell (\tfrac{1}{2}+\beta)_{m-\ell} (\tfrac{1}{2}-\beta)_{m-\ell},$$

and

18.15.4 $\quad \theta_{n,m,\ell} = \tfrac{1}{2}(2n+\alpha+\beta+m+1)\theta - \tfrac{1}{2}(\alpha+\ell+\tfrac{1}{2})\pi.$

When $\alpha, \beta \in (-\tfrac{1}{2}, \tfrac{1}{2})$, the error term in (18.15.1) is less than twice the first neglected term in absolute value. See Hahn (1980), where corresponding results are given when x is replaced by a complex variable z that is bounded away from the orthogonality interval $[-1,1]$.

Next, let

18.15.5 $\qquad \rho = n + \tfrac{1}{2}(\alpha+\beta+1).$

Then as $n \to \infty$,

18.15.6
$$(\sin \tfrac{1}{2}\theta)^{\alpha+\frac{1}{2}} (\cos \tfrac{1}{2}\theta)^{\beta+\frac{1}{2}} P_n^{(\alpha,\beta)}(\cos\theta)$$
$$= \frac{\Gamma(n+\alpha+1)}{2^{\frac{1}{2}} \rho^\alpha n!} \left(\theta^{\frac{1}{2}} J_\alpha(\rho\theta) \sum_{m=0}^{M} \frac{A_m(\theta)}{\rho^{2m}}\right.$$
$$\left.+ \theta^{\frac{3}{2}} J_{\alpha+1}(\rho\theta) \sum_{m=0}^{M-1} \frac{B_m(\theta)}{\rho^{2m+1}} + \varepsilon_M(\rho,\theta)\right),$$

where $J_\nu(z)$ is the Bessel function (§10.2(ii)), and

18.15.7
$$\varepsilon_M(\rho,\theta) = \begin{cases} \theta\, O(\rho^{-2M-(3/2)}), & c\rho^{-1} \leq \theta \leq \pi-\delta, \\ \theta^{\alpha+(5/2)} O(\rho^{-2M+\alpha}), & 0 \leq \theta \leq c\rho^{-1}, \end{cases}$$

with c denoting an arbitrary positive constant. Also,

18.15.8
$$A_0(\theta) = 1, \quad \theta B_0(\theta) = \tfrac{1}{4} g(\theta),$$
$$A_1(\theta) = \tfrac{1}{8} g'(\theta) - \frac{1+2\alpha}{8} \frac{g(\theta)}{\theta} - \frac{1}{32}(g(\theta))^2,$$

where

18.15.9

$$g(\theta) = \left(\tfrac{1}{4} - \alpha^2\right)\left(\cot\left(\tfrac{1}{2}\theta\right) - \left(\tfrac{1}{2}\theta\right)^{-1}\right) - \left(\tfrac{1}{4} - \beta^2\right)\tan\left(\tfrac{1}{2}\theta\right).$$

For higher coefficients see Baratella and Gatteschi (1988), and for another estimate of the error term see Wong and Zhao (2003).

For large β, fixed α, and $0 \le n/\beta \le c$, Dunster (1999) gives asymptotic expansions of $P_n^{(\alpha,\beta)}(z)$ that are uniform in unbounded complex z-domains containing $z = \pm 1$. These expansions are in terms of Whittaker functions (§13.14). This reference also supplies asymptotic expansions of $P_n^{(\alpha,\beta)}(z)$ for large n, fixed α, and $0 \le \beta/n \le c$. The latter expansions are in terms of Bessel functions, and are uniform in complex z-domains not containing neighborhoods of 1. For a complementary result, see Wong and Zhao (2004). By using the symmetry property given in the second row of Table 18.6.1, the roles of α and β can be interchanged.

For an asymptotic expansion of $P_n^{(\alpha,\beta)}(z)$ as $n \to \infty$ that holds uniformly for complex z bounded away from $[-1, 1]$, see Elliott (1971). The first term of this expansion also appears in Szegö (1975, Theorem 8.21.7).

18.15(ii) Ultraspherical

For fixed $\lambda \in (0, 1)$ and fixed $M = 0, 1, 2, \ldots,$

18.15.10

$$C_n^{(\lambda)}(\cos\theta) = \frac{2^{2\lambda}\,\Gamma\!\left(\lambda + \tfrac{1}{2}\right)}{\pi^{\frac{1}{2}}\,\Gamma(\lambda + 1)} \frac{(2\lambda)_n}{(\lambda + 1)_n}$$
$$\times \left(\sum_{m=0}^{M-1} \frac{(\lambda)_m (1-\lambda)_m}{m!\,(n+\lambda+1)_m} \frac{\cos\theta_{n,m}}{(2\sin\theta)^{m+\lambda}}\right.$$
$$\left. + O\!\left(\frac{1}{n^M}\right)\right),$$

as $n \to \infty$ uniformly with respect to $\theta \in [\delta, \pi - \delta]$, where

18.15.11 $\theta_{n,m} = (n + m + \lambda)\theta - \tfrac{1}{2}(m + \lambda)\pi.$

For a bound on the error term in (18.15.10) see Szegö (1975, Theorem 8.21.11).

Asymptotic expansions for $C_n^{(\lambda)}(\cos\theta)$ can be obtained from the results given in §18.15(i) by setting $\alpha = \beta = \lambda - \tfrac{1}{2}$ and referring to (18.7.1). See also Szegö (1933) and Szegö (1975, Eq. (8.21.14)).

18.15(iii) Legendre

For fixed $M = 0, 1, 2, \ldots,$

18.15.12

$$P_n(\cos\theta) = \left(\frac{2}{\sin\theta}\right)^{\frac{1}{2}} \sum_{m=0}^{M-1} \binom{-\tfrac{1}{2}}{m}\binom{m - \tfrac{1}{2}}{n} \frac{\cos\alpha_{n,m}}{(2\sin\theta)^m}$$
$$+ O\!\left(\frac{1}{n^{M+\frac{1}{2}}}\right),$$

as $n \to \infty$, uniformly with respect to $\theta \in [\delta, \pi - \delta]$, where

18.15.13 $\alpha_{n,m} = (n - m + \tfrac{1}{2})\theta + (n - \tfrac{1}{2}m - \tfrac{1}{4})\pi.$

Also, when $\tfrac{1}{6}\pi < \theta < \tfrac{5}{6}\pi$, the right-hand side of (18.15.12) with $M = \infty$ converges; paradoxically, however, the sum is $2P_n(\cos\theta)$ and not $P_n(\cos\theta)$ as stated erroneously in Szegö (1975, §8.4(3)).

For these results and further information see Olver (1997b, pp. 311–313). For another form of the asymptotic expansion, complete with error bound, see Szegö (1975, Theorem 8.21.5).

For asymptotic expansions of $P_n(\cos\theta)$ and $P_n(\cosh\xi)$ that are uniformly valid when $0 \le \theta \le \pi - \delta$ and $0 \le \xi < \infty$ see §14.15(iii) with $\mu = 0$ and $\nu = n$. These expansions are in terms of Bessel functions and modified Bessel functions, respectively.

18.15(iv) Laguerre

In Terms of Elementary Functions

For fixed $M = 0, 1, 2, \ldots,$ and fixed α,

18.15.14 $L_n^{(\alpha)}(x) = \dfrac{n^{\frac{1}{2}\alpha - \frac{1}{4}} e^{\frac{1}{2}x}}{\pi^{\frac{1}{2}} x^{\frac{1}{2}\alpha + \frac{1}{4}}} \left(\cos\theta_n^{(\alpha)}(x) \left(\sum_{m=0}^{M-1} \dfrac{a_m(x)}{n^{\frac{1}{2}m}} + O\!\left(\dfrac{1}{n^{\frac{1}{2}M}}\right)\right) + \sin\theta_n^{(\alpha)}(x)\left(\sum_{m=1}^{M-1} \dfrac{b_m(x)}{n^{\frac{1}{2}m}} + O\!\left(\dfrac{1}{n^{\frac{1}{2}M}}\right)\right)\right),$

as $n \to \infty$, uniformly on compact x-intervals in $(0, \infty)$, where

18.15.15 $\theta_n^{(\alpha)}(x) = 2(nx)^{\frac{1}{2}} - \left(\tfrac{1}{2}\alpha + \tfrac{1}{4}\right)\pi.$

The leading coefficients are given by

18.15.16 $a_0(x) = 1, \quad a_1(x) = 0, \quad b_1(x) = \dfrac{1}{48 x^{\frac{1}{2}}}\left(4x^2 - 12\alpha^2 - 24\alpha x - 24x + 3\right).$

In Terms of Bessel Functions

Define

18.15.17 $\nu = 4n + 2\alpha + 2,$

18.15 Asymptotic Approximations

18.15.18
$$\xi = \tfrac{1}{2}\left(\sqrt{x-x^2} + \arcsin(\sqrt{x})\right), \qquad 0 \le x \le 1.$$

Then for fixed $M = 0, 1, 2, \ldots$, and fixed α,

18.15.19
$$L_n^{(\alpha)}(\nu x) = \frac{e^{\frac{1}{2}\nu x}}{2^\alpha x^{\frac{1}{2}\alpha+\frac{1}{4}}(1-x)^{\frac{1}{4}}}\left(\xi^{\frac{1}{2}} J_\alpha(\nu\xi)\sum_{m=0}^{M-1}\frac{A_m(\xi)}{\nu^{2m}} + \xi^{-\frac{1}{2}} J_{\alpha+1}(\nu\xi)\sum_{m=0}^{M-1}\frac{B_m(\xi)}{\nu^{2m+1}} + \xi^{\frac{1}{2}}\operatorname{env}J_\alpha(\nu\xi)\,O\!\left(\frac{1}{\nu^{2M-1}}\right)\right),$$

as $n \to \infty$ uniformly for $0 \le x \le 1-\delta$. Here $J_\nu(z)$ denotes the Bessel function (§10.2(ii)), $\operatorname{env}J_\nu(z)$ denotes its envelope (§2.8(iv)), and δ is again an arbitrary small positive constant. The leading coefficients are given by $A_0(\xi) = 1$ and

18.15.20
$$B_0(\xi) = -\frac{1}{2}\left(\frac{1-4\alpha^2}{8} + \xi\left(\frac{1-x}{x}\right)^{\frac{1}{2}}\left(\frac{4\alpha^2-1}{8} + \frac{1}{4}\frac{x}{1-x} + \frac{5}{24}\left(\frac{x}{1-x}\right)^2\right)\right).$$

In Terms of Airy Functions

Again define ν as in (18.15.17); also,

18.15.21
$$\zeta = -\left(\tfrac{3}{4}\left(\arccos(\sqrt{x}) - \sqrt{x-x^2}\right)\right)^{\frac{2}{3}}, \qquad 0 \le x \le 1,$$
$$\zeta = \left(\tfrac{3}{4}\left(\sqrt{x^2-x} - \operatorname{arccosh}(\sqrt{x})\right)\right)^{\frac{2}{3}}, \qquad x \ge 1.$$

Then for fixed $M = 0, 1, 2, \ldots$, and fixed α,

18.15.22
$$L_n^{(\alpha)}(\nu x) \sim (-1)^n \frac{e^{\frac{1}{2}\nu x}}{2^{\alpha-\frac{1}{2}}x^{\frac{1}{2}\alpha+\frac{1}{4}}}$$
$$\times \left(\frac{\zeta}{x-1}\right)^{\frac{1}{4}}\left(\frac{\operatorname{Ai}\left(\nu^{\frac{2}{3}}\zeta\right)}{\nu^{\frac{1}{3}}}\sum_{m=0}^{M-1}\frac{E_m(\zeta)}{\nu^{2m}} + \frac{\operatorname{Ai}'\left(\nu^{\frac{2}{3}}\zeta\right)}{\nu^{\frac{5}{3}}}\sum_{m=0}^{M-1}\frac{F_m(\zeta)}{\nu^{2m}} + \operatorname{envAi}\left(\nu^{\frac{2}{3}}\zeta\right)O\!\left(\frac{1}{\nu^{2M-\frac{2}{3}}}\right)\right),$$

as $n \to \infty$ uniformly for $\delta \le x < \infty$. Here Ai denotes the Airy function (§9.2), Ai' denotes its derivative, and envAi denotes its envelope (§2.8(iii)). The leading coefficients are given by $E_0(\zeta) = 1$ and

18.15.23
$$F_0(\zeta) = -\frac{5}{48\zeta^2} + \left(\frac{x-1}{x\zeta}\right)^{\frac{1}{2}}\left(\frac{1}{2}\alpha^2 - \frac{1}{8} - \frac{1}{4}\frac{x}{x-1} + \frac{5}{24}\left(\frac{x}{x-1}\right)^2\right), \qquad 0 \le x < \infty.$$

18.15(v) Hermite

Define

18.15.24
$$\mu = 2n+1,$$

18.15.25
$$\lambda_n = \begin{cases} \Gamma(n+1)/\Gamma(\tfrac{1}{2}n+1), & n \text{ even}, \\ \Gamma(n+2)\big/\!\left(\mu^{\frac{1}{2}}\Gamma(\tfrac{1}{2}n+\tfrac{3}{2})\right), & n \text{ odd}, \end{cases}$$

and

18.15.26
$$\omega_{n,m}(x) = \mu^{\frac{1}{2}}x - \tfrac{1}{2}(m+n)\pi.$$

Then for fixed $M = 0, 1, 2, \ldots$,

18.15.27
$$H_n(x) = \lambda_n e^{\frac{1}{2}x^2}\left(\sum_{m=0}^{M-1}\frac{u_m(x)\cos\omega_{n,m}(x)}{\mu^{\frac{1}{2}m}} + O\!\left(\frac{1}{\mu^{\frac{1}{2}M}}\right)\right),$$

as $n \to \infty$, uniformly on compact x-intervals on $\mathbb{R}$. The coefficients $u_m(x)$ are polynomials in x, and $u_0(x) = 1$, $u_1(x) = \tfrac{1}{6}x^3$.

For more powerful asymptotic expansions as $n \to \infty$ in terms of elementary functions that apply uniformly when $1 + \delta \le t < \infty$, $-1 + \delta \le t \le 1 - \delta$, or $-\infty < t \le -1 - \delta$, where $t = x/\sqrt{2n+1}$ and δ is again an arbitrary small positive constant, see §§12.10(i)–12.10(iv) and 12.10(vi). And for asymptotic expansions as $n \to \infty$ in terms of Airy functions that apply uniformly when $-1 + \delta \le t < \infty$ or $-\infty < t \le 1 - \delta$, see §§12.10(vii) and 12.10(viii). With $\mu = \sqrt{2n+1}$ the expansions in Chapter 12 are for the parabolic cylinder function $U\!\left(-\tfrac{1}{2}\mu^2, \mu t\sqrt{2}\right)$, which is related to the Hermite polynomials via

18.15.28 $H_n(x) = 2^{\frac{1}{4}(\mu^2-1)}e^{\frac{1}{2}\mu^2 t^2}U\!\left(-\tfrac{1}{2}\mu^2, \mu t\sqrt{2}\right);$

compare (18.11.3).

For an error bound for the first term in the Airy-function expansions see Olver (1997b, p. 403).

18.15(vi) Other Approximations

The asymptotic behavior of the classical OP's as $x \to \pm\infty$ with the degree and parameters fixed is evident from their explicit polynomial forms; see, for example, (18.2.7) and the last two columns of Table 18.3.1.

For asymptotic approximations of Jacobi, ultraspherical, and Laguerre polynomials in terms of Hermite polynomials, see López and Temme (1999a). These approximations apply when the parameters are large, namely α and β (subject to restrictions) in the case of Jacobi polynomials, λ in the case of ultraspherical polynomials, and $|\alpha| + |x|$ in the case of Laguerre polynomials. See also Dunster (1999).

18.16 Zeros

18.16(i) Distribution

See §18.2(vi).

18.16(ii) Jacobi

Let $\theta_{n,m}$, $m = 1, 2, \ldots, n$, denote the zeros of $P_n^{(\alpha,\beta)}(\cos\theta)$ with

18.16.1 $\quad 0 < \theta_{n,1} < \theta_{n,2} < \cdots < \theta_{n,n} < \pi.$

Then $\theta_{n,m}$ is strictly increasing in α and strictly decreasing in β; furthermore, if $\alpha = \beta$, then $\theta_{n,m}$ is strictly increasing in α.

Inequalities

18.16.2 $\quad \dfrac{(m - \tfrac{1}{2})\pi}{n + \tfrac{1}{2}} \le \theta_{n,m} \le \dfrac{m\pi}{n + \tfrac{1}{2}}, \quad \alpha, \beta \in [-\tfrac{1}{2}, \tfrac{1}{2}],$

18.16.3
$$\dfrac{(m - \tfrac{1}{2})\pi}{n} \le \theta_{n,m} \le \dfrac{m\pi}{n+1},$$
$\alpha = \beta,\ \alpha \in [-\tfrac{1}{2}, \tfrac{1}{2}],\ m = 1, 2, \ldots, \lfloor \tfrac{1}{2} n \rfloor.$

Also, with ρ defined as in (18.15.5)

18.16.4
$$\dfrac{\left(m + \tfrac{1}{2}(\alpha + \beta - 1)\right)\pi}{\rho} < \theta_{n,m} < \dfrac{m\pi}{\rho}, \quad \alpha, \beta \in [-\tfrac{1}{2}, \tfrac{1}{2}],$$
except when $\alpha^2 = \beta^2 = \tfrac{1}{4}$.

18.16.5
$$\theta_{n,m} > \dfrac{\left(m + \tfrac{1}{2}\alpha - \tfrac{1}{4}\right)\pi}{n + \alpha + \tfrac{1}{2}},$$
$\alpha = \beta,\ \alpha \in (-\tfrac{1}{2}, \tfrac{1}{2}),\ m = 1, 2, \ldots, \lfloor \tfrac{1}{2} n \rfloor.$

Let $j_{\alpha,m}$ be the mth positive zero of the Bessel function $J_\alpha(x)$ (§10.21(i)). Then

18.16.6
$$\theta_{n,m} \le \dfrac{j_{\alpha,m}}{\left(\rho^2 + \tfrac{1}{12}(1 - \alpha^2 - 3\beta^2)\right)^{\tfrac{1}{2}}}, \quad \alpha, \beta \in [-\tfrac{1}{2}, \tfrac{1}{2}],$$

18.16.7
$$\theta_{n,m} \ge \dfrac{j_{\alpha,m}}{\left(\rho^2 + \tfrac{1}{4} - \tfrac{1}{2}(\alpha^2 + \beta^2) - \pi^{-2}(1 - 4\alpha^2)\right)^{\tfrac{1}{2}}},$$
$\alpha, \beta \in [-\tfrac{1}{2}, \tfrac{1}{2}],\ m = 1, 2, \ldots, \lfloor \tfrac{1}{2} n \rfloor.$

Asymptotic Behavior

Let $\phi_m = j_{\alpha,m}/\rho$. Then as $n \to \infty$, with $\alpha\ (> -\tfrac{1}{2})$ and $\beta\ (\ge -1 - \alpha)$ fixed,

18.16.8
$$\theta_{n,m} = \phi_m + \left(\left(\alpha^2 - \tfrac{1}{4}\right)\dfrac{1 - \phi_m \cot\phi_m}{2\phi_m} - \tfrac{1}{4}(\alpha^2 - \beta^2)\tan(\tfrac{1}{2}\phi_m)\right)\dfrac{1}{\rho^2} + \phi_m^2\, O\!\left(\dfrac{1}{\rho^3}\right),$$

uniformly for $m = 1, 2, \ldots, \lfloor cn \rfloor$, where c is an arbitrary constant such that $0 < c < 1$.

18.16(iii) Ultraspherical and Legendre

For ultraspherical and Legendre polynomials, set $\alpha = \beta$ and $\alpha = \beta = 0$, respectively, in the results given in §18.16(ii).

18.16(iv) Laguerre

The zeros of $L_n^{(\alpha)}(x)$ are denoted by $x_{n,m}$, $m = 1, 2, \ldots, n$, with

18.16.9 $\quad 0 < x_{n,1} < x_{n,2} < \cdots < x_{n,n}.$

Also, ν is again defined by (18.15.17).

Inequalities

For $n = 1, 2, \ldots, m$, and with $j_{\alpha,m}$ as in §18.16(ii),

18.16.10 $\quad x_{n,m} > j_{\alpha,m}^2 / \nu,$

18.16.11
$$x_{n,m} < (4m + 2\alpha + 2)\left(2m + \alpha + 1 + \left((2m + \alpha + 1)^2 + \tfrac{1}{4} - \alpha^2\right)^{\tfrac{1}{2}}\right)\big/\nu.$$

The constant $j_{\alpha,m}^2$ in (18.16.10) is the best possible since the ratio of the two sides of this inequality tends to 1 as $n \to \infty$.

For the smallest and largest zeros we have

18.16.12 $\quad x_{n,1} > 2n + \alpha - 2 - (1 + 4(n-1)(n + \alpha - 1))^{\tfrac{1}{2}},$

18.16.13 $\quad x_{n,n} < 2n + \alpha - 2 + (1 + 4(n-1)(n + \alpha - 1))^{\tfrac{1}{2}}.$

Asymptotic Behavior

As $n \to \infty$, with α and m fixed,

18.16.14
$$x_{n, n-m+1} = \nu + 2^{\tfrac{2}{3}} a_m \nu^{\tfrac{1}{3}} + \tfrac{1}{5} 2^{\tfrac{4}{3}} a_m^2 \nu^{-\tfrac{1}{3}} + O(n^{-1}),$$

where a_m is the mth negative zero of $\operatorname{Ai}(x)$ (§9.9(i)). For three additional terms in this expansion see Gatteschi (2002). Also,

18.16.15 $\quad x_{n,m} < \nu + 2^{\tfrac{2}{3}} a_m \nu^{\tfrac{1}{3}} + 2^{-\tfrac{2}{3}} a_m^2 \nu^{-\tfrac{1}{3}},$

when $\alpha \notin (-\tfrac{1}{2}, \tfrac{1}{2})$.

18.16(v) Hermite

All zeros of $H_n(x)$ lie in the open interval $(-\sqrt{2n+1}, \sqrt{2n+1})$. In view of the reflection formula, given in Table 18.6.1, we may consider just the positive zeros $x_{n,m}$, $m = 1, 2, \ldots, \lfloor \tfrac{1}{2}n \rfloor$. Arrange them in decreasing order:

18.16.16 $\quad (2n+1)^{\frac{1}{2}} > x_{n,1} > x_{n,2} > \cdots > x_{n,\lfloor n/2 \rfloor} > 0.$

Then

18.16.17 $\quad x_{n,m} = (2n+1)^{\frac{1}{2}} + 2^{-\frac{1}{3}}(2n+1)^{-\frac{1}{6}} a_m + \epsilon_{n,m},$

where a_m is the mth negative zero of $\mathrm{Ai}(x)$ (§9.9(i)), $\epsilon_{n,m} < 0$, and as $n \to \infty$ with m fixed

18.16.18 $\quad \epsilon_{n,m} = O\!\left(n^{-\frac{5}{6}}\right).$

For an asymptotic expansion of $x_{n,m}$ as $n \to \infty$ that applies uniformly for $m = 1, 2, \ldots, \lfloor \tfrac{1}{2}n \rfloor$, see Olver (1959, §14(i)). In the notation of this reference $x_{n,m} = u_{a,m}$, $\mu = \sqrt{2n+1}$, and $\alpha = \mu^{-\frac{4}{3}} a_m$. For an error bound for the first approximation yielded by this expansion see Olver (1997b, p. 408).

Lastly, in view of (18.7.19) and (18.7.20), results for the zeros of $L_n^{(\pm\frac{1}{2})}(x)$ lead immediately to results for the zeros of $H_n(x)$.

18.16(vi) Additional References

For further information on the zeros of the classical orthogonal polynomials, see Szegö (1975, Chapter VI), Erdélyi et al. (1953b, §§10.16 and 10.17), Gatteschi (1987, 2002), López and Temme (1999a), and Temme (1990a).

18.17 Integrals

18.17(i) Indefinite Integrals

Jacobi

18.17.1
$$2n \int_0^x (1-y)^\alpha (1+y)^\beta P_n^{(\alpha,\beta)}(y)\, dy \\ = P_{n-1}^{(\alpha+1,\beta+1)}(0) - (1-x)^{\alpha+1}(1+x)^{\beta+1} P_{n-1}^{(\alpha+1,\beta+1)}(x).$$

Laguerre

18.17.2
$$\int_0^x L_m(y) L_n(x-y)\, dy = \int_0^x L_{m+n}(y)\, dy \\ = L_{m+n}(x) - L_{m+n+1}(x).$$

Hermite

18.17.3 $\quad \displaystyle\int_0^x H_n(y)\, dy = \frac{1}{2(n+1)}(H_{n+1}(x) - H_{n+1}(0)),$

18.17.4 $\quad \displaystyle\int_0^x e^{-y^2} H_n(y)\, dy = H_{n-1}(0) - e^{-x^2} H_{n-1}(x).$

18.17(ii) Integral Representations for Products

Ultraspherical

18.17.5 $\quad \displaystyle\frac{C_n^{(\lambda)}(\cos\theta_1)}{C_n^{(\lambda)}(1)} \frac{C_n^{(\lambda)}(\cos\theta_2)}{C_n^{(\lambda)}(1)} = \frac{\Gamma\!\left(\lambda+\tfrac{1}{2}\right)}{\pi^{\frac{1}{2}} \Gamma(\lambda)} \int_0^\pi \frac{C_n^{(\lambda)}(\cos\theta_1 \cos\theta_2 + \sin\theta_1 \sin\theta_2 \cos\phi)}{C_n^{(\lambda)}(1)} (\sin\phi)^{2\lambda-1}\, d\phi, \quad \lambda > 0.$

Legendre

18.17.6 $\quad \displaystyle P_n(\cos\theta_1) P_n(\cos\theta_2) = \frac{1}{\pi} \int_0^\pi P_n(\cos\theta_1 \cos\theta_2 + \sin\theta_1 \sin\theta_2 \cos\phi)\, d\phi.$

For formulas for Jacobi and Laguerre polynomials analogous to (18.17.5) and (18.17.6), see Koornwinder (1974, 1977).

18.17(iii) Nicholson-Type Integrals

Legendre

18.17.7 $\quad (P_n(x))^2 + 4\pi^{-2}(\mathsf{Q}_n(x))^2 = 4\pi^{-2} \displaystyle\int_1^\infty Q_n\!\left(x^2 + (1-x^2)t\right)(t^2 - 1)^{-\frac{1}{2}}\, dt, \quad -1 < x < 1.$

For the Ferrers function $\mathsf{Q}_n(x)$ and Legendre function $Q_n(x)$ see §§14.3(i) and 14.3(ii), with $\mu = 0$ and $\nu = n$.

Hermite

18.17.8 $\quad (H_n(x))^2 + 2^n (n!)^2 e^{x^2} \left(V\!\left(-n-\tfrac{1}{2}, 2^{\frac{1}{2}} x\right)\right)^2 = \dfrac{2^{n+\frac{3}{2}} n!\, e^{x^2}}{\pi} \displaystyle\int_0^\infty \dfrac{e^{-(2n+1)t + x^2 \tanh t}}{(\sinh 2t)^{\frac{1}{2}}}\, dt.$

For the parabolic cylinder function $V(a, z)$ see §12.2. For similar formulas for ultraspherical polynomials see Durand (1975), and for Jacobi and Laguerre polynomials see Durand (1978).

18.17(iv) Fractional Integrals

Jacobi

18.17.9
$$\frac{(1-x)^{\alpha+\mu} P_n^{(\alpha+\mu,\beta-\mu)}(x)}{\Gamma(\alpha+\mu+n+1)} = \int_x^1 \frac{(1-y)^\alpha P_n^{(\alpha,\beta)}(y)}{\Gamma(\alpha+n+1)} \frac{(y-x)^{\mu-1}}{\Gamma(\mu)} dy, \qquad \mu>0,\ -1<x<1,$$

18.17.10
$$\frac{x^{\beta+\mu}(x+1)^n}{\Gamma(\beta+\mu+n+1)} P_n^{(\alpha,\beta+\mu)}\left(\frac{x-1}{x+1}\right) = \int_0^x \frac{y^\beta(y+1)^n}{\Gamma(\beta+n+1)} P_n^{(\alpha,\beta)}\left(\frac{y-1}{y+1}\right) \frac{(x-y)^{\mu-1}}{\Gamma(\mu)} dy,\ \mu>0,\ x>0,$$

18.17.11
$$\frac{\Gamma(n+\alpha+\beta-\mu+1)}{x^{n+\alpha+\beta-\mu+1}} P_n^{(\alpha,\beta-\mu)}(1-2x^{-1}) = \int_x^\infty \frac{\Gamma(n+\alpha+\beta+1)}{y^{n+\alpha+\beta+1}} P_n^{(\alpha,\beta)}(1-2y^{-1}) \frac{(y-x)^{\mu-1}}{\Gamma(\mu)} dy,$$
$$\alpha+\beta+1>\mu>0,\ x>1,$$

and three formulas similar to (18.17.9)–(18.17.11) by symmetry; compare the second row in Table 18.6.1.

Ultraspherical

18.17.12
$$\frac{\Gamma(\lambda-\mu) C_n^{(\lambda-\mu)}\left(x^{-\frac{1}{2}}\right)}{x^{\lambda-\mu+\frac{1}{2}n}} = \int_x^\infty \frac{\Gamma(\lambda) C_n^{(\lambda)}\left(y^{-\frac{1}{2}}\right)}{y^{\lambda+\frac{1}{2}n}} \frac{(y-x)^{\mu-1}}{\Gamma(\mu)} dy, \qquad \lambda>\mu>0,\ x>0,$$

18.17.13
$$\frac{x^{\frac{1}{2}n}(x-1)^{\lambda+\mu-\frac{1}{2}}}{\Gamma(\lambda+\mu+\frac{1}{2})} \frac{C_n^{(\lambda+\mu)}\left(x^{-\frac{1}{2}}\right)}{C_n^{(\lambda+\mu)}(1)} = \int_1^x \frac{y^{\frac{1}{2}n}(y-1)^{\lambda-\frac{1}{2}}}{\Gamma(\lambda+\frac{1}{2})} \frac{C_n^{(\lambda)}\left(y^{-\frac{1}{2}}\right)}{C_n^{(\lambda)}(1)} \frac{(x-y)^{\mu-1}}{\Gamma(\mu)} dy, \qquad \mu>0,\ x>1.$$

Laguerre

18.17.14
$$\frac{x^{\alpha+\mu} L_n^{(\alpha+\mu)}(x)}{\Gamma(\alpha+\mu+n+1)} = \int_0^x \frac{y^\alpha L_n^{(\alpha)}(y)}{\Gamma(\alpha+n+1)} \frac{(x-y)^{\mu-1}}{\Gamma(\mu)} dy, \qquad \mu>0,\ x>0.$$

18.17.15
$$e^{-x} L_n^{(\alpha)}(x) = \int_x^\infty e^{-y} L_n^{(\alpha+\mu)}(y) \frac{(y-x)^{\mu-1}}{\Gamma(\mu)} dy, \qquad \mu>0.$$

18.17(v) Fourier Transforms

Throughout this subsection we assume $y>0$; sometimes however, this restriction can be eased by analytic continuation.

Jacobi

18.17.16
$$\int_{-1}^1 (1-x)^\alpha (1+x)^\beta P_n^{(\alpha,\beta)}(x) e^{ixy} dx$$
$$= \frac{(iy)^n e^{iy}}{n!} 2^{n+\alpha+\beta+1} \mathrm{B}(n+\alpha+1, n+\beta+1)\, {}_1F_1(n+\alpha+1; 2n+\alpha+\beta+2; -2iy).$$

For the beta function $\mathrm{B}(a,b)$ see §5.12, and for the confluent hypergeometric function ${}_1F_1$ see (16.2.1) and Chapter 13.

Ultraspherical

18.17.17
$$\int_0^1 (1-x^2)^{\lambda-\frac{1}{2}} C_{2n}^{(\lambda)}(x) \cos(xy) dx = \frac{(-1)^n \pi\, \Gamma(2n+2\lambda) J_{\lambda+2n}(y)}{(2n)!\, \Gamma(\lambda)(2y)^\lambda},$$

18.17.18
$$\int_0^1 (1-x^2)^{\lambda-\frac{1}{2}} C_{2n+1}^{(\lambda)}(x) \sin(xy) dx = \frac{(-1)^n \pi\, \Gamma(2n+2\lambda+1) J_{2n+\lambda+1}(y)}{(2n+1)!\, \Gamma(\lambda)(2y)^\lambda}.$$

For the Bessel function J_ν see §10.2(ii).

Legendre

18.17.19
$$\int_{-1}^1 P_n(x) e^{ixy} dx = i^n \sqrt{\frac{2\pi}{y}} J_{n+\frac{1}{2}}(y),$$

18.17.20
$$\int_0^1 P_n(1-2x^2) \cos(xy) dx = (-1)^n \tfrac{1}{2}\pi\, J_{n+\frac{1}{2}}(\tfrac{1}{2}y) J_{-n-\frac{1}{2}}(\tfrac{1}{2}y),$$

18.17.21
$$\int_0^1 P_n(1-2x^2) \sin(xy) dx = \tfrac{1}{2}\pi \left(J_{n+\frac{1}{2}}(\tfrac{1}{2}y)\right)^2.$$

18.17   INTEGRALS

Hermite

18.17.22
$$\frac{1}{2\sqrt{\pi}} \int_{-\infty}^{\infty} e^{-\frac{1}{4}x^2} He_n(x) e^{\frac{1}{2}ixy} \, dx = i^n e^{-\frac{1}{4}y^2} He_n(y),$$

18.17.23
$$\int_0^{\infty} e^{-\frac{1}{2}x^2} He_{2n}(x) \cos(xy) \, dx = (-1)^n \sqrt{\tfrac{1}{2}\pi} y^{2n} e^{-\frac{1}{2}y^2},$$

18.17.24
$$\int_0^{\infty} e^{-x^2} He_{2n}(2x) \cos(xy) \, dx = (-1)^n \tfrac{1}{2}\sqrt{\pi} e^{-\frac{1}{4}y^2} He_{2n}(y).$$

18.17.25
$$\int_0^{\infty} e^{-\frac{1}{2}x^2} He_n(x) He_{n+2m}(x) \cos(xy) \, dx = (-1)^m \sqrt{\tfrac{1}{2}\pi} n! \, y^{2m} e^{-\frac{1}{2}y^2} L_n^{(2m)}(y^2),$$

18.17.26
$$\int_0^{\infty} e^{-\frac{1}{2}x^2} He_n(x) He_{n+2m+1}(x) \sin(xy) \, dx = (-1)^m \sqrt{\tfrac{1}{2}\pi} n! \, y^{2m+1} e^{-\frac{1}{2}y^2} L_n^{(2m+1)}(y^2).$$

18.17.27
$$\int_0^{\infty} e^{-\frac{1}{2}x^2} He_{2n+1}(x) \sin(xy) \, dx = (-1)^n \sqrt{\tfrac{1}{2}\pi} y^{2n+1} e^{-\frac{1}{2}y^2},$$

18.17.28
$$\int_0^{\infty} e^{-x^2} He_{2n+1}(2x) \sin(xy) \, dx = (-1)^n \tfrac{1}{2}\sqrt{\pi} e^{-\frac{1}{4}y^2} He_{2n+1}(y).$$

Laguerre

18.17.29
$$\int_0^{\infty} x^{2m} e^{-\frac{1}{2}x^2} L_n^{(2m)}(x^2) \cos(xy) \, dx = (-1)^m \sqrt{\tfrac{1}{2}\pi} \frac{1}{n!} e^{-\frac{1}{2}y^2} He_n(y) He_{n+2m}(y).$$

18.17.30
$$\int_0^{\infty} x^{2n} e^{-\frac{1}{2}x^2} L_n^{(n-\frac{1}{2})}\!\left(\tfrac{1}{2}x^2\right) \cos(xy) \, dx = \sqrt{\tfrac{1}{2}\pi}\, y^{2n} e^{-\frac{1}{2}y^2} L_n^{(n-\frac{1}{2})}\!\left(\tfrac{1}{2}y^2\right).$$

18.17.31
$$\int_0^{\infty} e^{-ax} x^{\nu-2n} L_{2n-1}^{(\nu-2n)}(ax) \cos(xy) \, dx = i\frac{(-1)^n \Gamma(\nu)}{2(2n-1)!} y^{2n-1} \left((a+iy)^{-\nu} - (a-iy)^{-\nu}\right), \quad \nu > 2n-1, \, a > 0,$$

18.17.32
$$\int_0^{\infty} e^{-ax} x^{\nu-1-2n} L_{2n}^{(\nu-1-2n)}(ax) \cos(xy) \, dx = \frac{(-1)^n \Gamma(\nu)}{2(2n)!} y^{2n} \left((a+iy)^{-\nu} + (a-iy)^{-\nu}\right), \quad \nu > 2n, \, a > 0.$$

18.17(vi) Laplace Transforms

Jacobi

18.17.33
$$\int_{-1}^{1} e^{-(x+1)z} P_n^{(\alpha,\beta)}(x)(1-x)^{\alpha}(1+x)^{\beta} \, dx$$
$$= \frac{(-1)^n 2^{\alpha+\beta+n+1} \Gamma(\alpha+n+1)\Gamma(\beta+n+1)}{\Gamma(\alpha+\beta+2n+2) n!} z^n {}_1F_1\!\left(\begin{matrix} \beta+n+1 \\ \alpha+\beta+2n+2 \end{matrix}; -2z\right), \qquad z \in \mathbb{C}.$$

For the confluent hypergeometric function ${}_1F_1$ see (16.2.1) and Chapter 13.

Laguerre

18.17.34
$$\int_0^{\infty} e^{-xz} L_n^{(\alpha)}(x) e^{-x} x^{\alpha} \, dx = \frac{\Gamma(\alpha+n+1) z^n}{n!(z+1)^{\alpha+n+1}}, \qquad \Re z > -1.$$

Hermite

18.17.35
$$\int_{-\infty}^{\infty} e^{-xz} H_n(x) e^{-x^2} \, dx = \pi^{\frac{1}{2}} (-z)^n e^{\frac{1}{4}z^2}, \qquad z \in \mathbb{C}.$$

18.17(vii) Mellin Transforms

Jacobi

18.17.36
$$\int_{-1}^{1} (1-x)^{z-1}(1+x)^{\beta} P_n^{(\alpha,\beta)}(x) \, dx = \frac{2^{\beta+z} \Gamma(z) \Gamma(1+\beta+n)(1+\alpha-z)_n}{n! \, \Gamma(1+\beta+z+n)}, \qquad \Re z > 0.$$

Ultraspherical

18.17.37
$$\int_0^1 (1-x^2)^{\lambda-\frac{1}{2}} C_n^{(\lambda)}(x) x^{z-1}\, dx = \frac{\pi\, 2^{1-2\lambda-z}\, \Gamma(n+2\lambda)\, \Gamma(z)}{n!\, \Gamma(\lambda)\, \Gamma\!\left(\frac{1}{2}+\frac{1}{2}n+\lambda+\frac{1}{2}z\right) \Gamma\!\left(\frac{1}{2}+\frac{1}{2}z-\frac{1}{2}n\right)}, \qquad \Re z > 0.$$

Legendre

18.17.38
$$\int_0^1 P_{2n}(x) x^{z-1}\, dx = \frac{(-1)^n \left(\frac{1}{2}-\frac{1}{2}z\right)_n}{2\left(\frac{1}{2}z\right)_{n+1}}, \qquad \Re z > 0,$$

18.17.39
$$\int_0^1 P_{2n+1}(x) x^{z-1}\, dx = \frac{(-1)^n \left(1-\frac{1}{2}z\right)_n}{2\left(\frac{1}{2}+\frac{1}{2}z\right)_{n+1}}, \qquad \Re z > -1.$$

Laguerre

18.17.40
$$\int_0^\infty e^{-ax} L_n^{(\alpha)}(bx) x^{z-1}\, dx = \frac{\Gamma(z+n)}{n!}(a-b)^n a^{-n-z}\, {}_2F_1\!\left(\begin{matrix}-n, 1+\alpha-z\\ 1-n-z\end{matrix}; \frac{a}{a-b}\right), \quad \Re a > 0, \Re z > 0.$$

For the hypergeometric function ${}_2F_1$ see §§15.1 and 15.2(i).

Hermite

18.17.41
$$\int_0^\infty e^{-ax} He_n(x) x^{z-1}\, dx = \Gamma(z+n) a^{-n-z}\, {}_2F_2\!\left(\begin{matrix}-\frac{1}{2}n, -\frac{1}{2}n+\frac{1}{2}\\ -\frac{1}{2}z-\frac{1}{2}n, -\frac{1}{2}z-\frac{1}{2}n+\frac{1}{2}\end{matrix}; -\frac{1}{2}a^2\right),$$

$\Re a > 0$. Also, $\Re z > 0$, n even; $\Re z > -1$, n odd.

For the generalized hypergeometric function ${}_2F_2$ see (16.2.1).

18.17(viii) Other Integrals

Chebyshev

18.17.42
$$\fint_{-1}^1 T_n(y) \frac{(1-y^2)^{-\frac{1}{2}}}{y-x}\, dy = \pi U_{n-1}(x),$$

18.17.43
$$\fint_{-1}^1 U_{n-1}(y) \frac{(1-y^2)^{\frac{1}{2}}}{y-x}\, dy = -\pi T_n(x).$$

These integrals are Cauchy principal values (§1.4(v)).

Legendre

18.17.44
$$\int_{-1}^1 \frac{P_n(x)-P_n(t)}{|x-t|}\, dt = 2\left(1+\frac{1}{2}+\cdots+\frac{1}{n}\right) P_n(x), \qquad -1 \le x \le 1.$$

The case $x=1$ is a limit case of an integral for Jacobi polynomials; see Askey and Razban (1972).

18.17.45
$$(n+\tfrac{1}{2})(1+x)^{\frac{1}{2}} \int_{-1}^x (x-t)^{-\frac{1}{2}} P_n(t)\, dt = T_n(x) + T_{n+1}(x),$$

18.17.46
$$(n+\tfrac{1}{2})(1-x)^{\frac{1}{2}} \int_x^1 (t-x)^{-\frac{1}{2}} P_n(t)\, dt = T_n(x) - T_{n+1}(x).$$

Laguerre

18.17.47
$$\int_0^x t^\alpha \frac{L_m^{(\alpha)}(t)}{L_m^{(\alpha)}(0)} (x-t)^\beta \frac{L_n^{(\beta)}(x-t)}{L_n^{(\beta)}(0)}\, dt = \frac{\Gamma(\alpha+1)\Gamma(\beta+1)}{\Gamma(\alpha+\beta+2)} x^{\alpha+\beta+1} \frac{L_{m+n}^{(\alpha+\beta+1)}(x)}{L_{m+n}^{(\alpha+\beta+1)}(0)}.$$

Hermite

18.17.48
$$\int_{-\infty}^\infty H_m(y) e^{-y^2} H_n(x-y) e^{-(x-y)^2}\, dy = \pi^{\frac{1}{2}} 2^{-\frac{1}{2}(m+n+1)} H_{m+n}\!\left(2^{-\frac{1}{2}} x\right) e^{-\frac{1}{2}x^2}.$$

18.17.49
$$\int_{-\infty}^\infty H_\ell(x) H_m(x) H_n(x) e^{-x^2}\, dx = \frac{2^{\frac{1}{2}(\ell+m+n)} \ell!\, m!\, n!\, \sqrt{\pi}}{\left(\frac{1}{2}\ell+\frac{1}{2}m-\frac{1}{2}n\right)!\left(\frac{1}{2}m+\frac{1}{2}n-\frac{1}{2}\ell\right)!\left(\frac{1}{2}n+\frac{1}{2}\ell-\frac{1}{2}m\right)!},$$

provided that $\ell+m+n$ is even and the sum of any two of ℓ, m, n is not less than the third; otherwise the integral is zero.

18.17(ix) Compendia

For further integrals, see Apelblat (1983, pp. 189–204), Erdélyi et al. (1954a, pp. 38–39, 94–95, 170–176, 259–261, 324), Erdélyi et al. (1954b, pp. 42–44, 271–294), Gradshteyn and Ryzhik (2000, pp. 788–806), Gröbner and Hofreiter (1950, pp. 23–30), Marichev (1983, pp. 216–247), Oberhettinger (1972, pp. 64–67), Oberhettinger (1974, pp. 83–92), Oberhettinger (1990, pp. 44–47 and 152–154), Oberhettinger and Badii (1973, pp. 103–112), Prudnikov et al. (1986b, pp. 420–617), Prudnikov et al. (1992a, pp. 419–476), and Prudnikov et al. (1992b, pp. 280–308).

18.18 Sums

18.18(i) Series Expansions of Arbitrary Functions

Jacobi

Let $f(z)$ be analytic within an ellipse E with foci $z = \pm 1$, and

18.18.1
$$a_n = \frac{n!(2n+\alpha+\beta+1)\Gamma(n+\alpha+\beta+1)}{2^{\alpha+\beta+1}\Gamma(n+\alpha+1)\Gamma(n+\beta+1)} \times \int_{-1}^{1} f(x) P_n^{(\alpha,\beta)}(x)(1-x)^\alpha (1+x)^\beta \, dx.$$

Then

18.18.2
$$f(z) = \sum_{n=0}^{\infty} a_n P_n^{(\alpha,\beta)}(z),$$

when z lies in the interior of E. Moreover, the series (18.18.2) converges uniformly on any compact domain within E.

Alternatively, assume $f(x)$ is real and continuous and $f'(x)$ is piecewise continuous on $(-1,1)$. Assume also the integrals $\int_{-1}^{1}(f(x))^2 (1-x)^\alpha (1+x)^\beta \, dx$ and $\int_{-1}^{1}(f'(x))^2 (1-x)^{\alpha+1}(1+x)^{\beta+1}\, dx$ converge. Then (18.18.2), with z replaced by x, applies when $-1 < x < 1$; moreover, the convergence is uniform on any compact interval within $(-1, 1)$.

Chebyshev

See §3.11(ii), or set $\alpha = \beta = \pm\frac{1}{2}$ in the above results for Jacobi and refer to (18.7.3)–(18.7.6).

Legendre

This is the case $\alpha = \beta = 0$ of Jacobi. Equation (18.18.1) becomes

18.18.3
$$a_n = \left(n + \tfrac{1}{2}\right) \int_{-1}^{1} f(x) P_n(x) \, dx.$$

Laguerre

Assume $f(x)$ is real and continuous and $f'(x)$ is piecewise continuous on $(0, \infty)$. Assume also $\int_0^\infty (f(x))^2 e^{-x} x^\alpha \, dx$ converges. Then

18.18.4
$$f(x) = \sum_{n=0}^{\infty} b_n L_n^{(\alpha)}(x), \qquad 0 < x < \infty,$$

where

18.18.5
$$b_n = \frac{n!}{\Gamma(n+\alpha+1)} \int_0^\infty f(x) L_n^{(\alpha)}(x) e^{-x} x^\alpha \, dx.$$

The convergence of the series (18.18.4) is uniform on any compact interval in $(0, \infty)$.

Hermite

Assume $f(x)$ is real and continuous and $f'(x)$ is piecewise continuous on $(-\infty, \infty)$. Assume also $\int_{-\infty}^{\infty}(f(x))^2 e^{-x^2}\, dx$ converges. Then

18.18.6
$$f(x) = \sum_{n=0}^{\infty} d_n H_n(x), \qquad -\infty < x < \infty,$$

where

18.18.7
$$d_n = \frac{1}{\sqrt{\pi} 2^n n!} \int_{-\infty}^{\infty} f(x) H_n(x) e^{-x^2}\, dx.$$

The convergence of the series (18.18.6) is uniform on any compact interval in $(-\infty, \infty)$.

18.18(ii) Addition Theorems

Ultraspherical

18.18.8
$$C_n^{(\lambda)}(\cos\theta_1 \cos\theta_2 + \sin\theta_1 \sin\theta_2 \cos\phi)$$
$$= \sum_{\ell=0}^{n} 2^{2\ell}(n-\ell)! \frac{2\lambda+2\ell-1}{2\lambda-1} \frac{((\lambda)_\ell)^2}{(2\lambda)_{n+\ell}} (\sin\theta_1)^\ell C_{n-\ell}^{(\lambda+\ell)}(\cos\theta_1) (\sin\theta_2)^\ell C_{n-\ell}^{(\lambda+\ell)}(\cos\theta_2) C_\ell^{(\lambda-\frac{1}{2})}(\cos\phi),$$
$$\lambda > 0, \lambda \neq \tfrac{1}{2}.$$

For the case $\lambda = \frac{1}{2}$ use (18.18.9); compare (18.7.9).

Legendre

18.18.9
$$P_n(\cos\theta_1 \cos\theta_2 + \sin\theta_1 \sin\theta_2 \cos\phi)$$
$$= P_n(\cos\theta_1) P_n(\cos\theta_2) + 2 \sum_{\ell=1}^{n} \frac{(n-\ell)!\,(n+\ell)!}{2^{2\ell}(n!)^2} (\sin\theta_1)^\ell P_{n-\ell}^{(\ell,\ell)}(\cos\theta_1) (\sin\theta_2)^\ell P_{n-\ell}^{(\ell,\ell)}(\cos\theta_2) \cos(\ell\phi).$$

For (18.18.8), (18.18.9), and the corresponding formula for Jacobi polynomials see Koornwinder (1975a). See also (14.30.9).

Laguerre

18.18.10 $$L_n^{(\alpha_1+\cdots+\alpha_r+r-1)}(x_1+\cdots+x_r) = \sum_{m_1+\cdots+m_r=n} L_{m_1}^{(\alpha_1)}(x_1)\cdots L_{m_r}^{(\alpha_r)}(x_r).$$

Hermite

18.18.11 $$\frac{(a_1^2+\cdots+a_r^2)^{\frac{1}{2}n}}{n!} H_n\left(\frac{a_1x_1+\cdots+a_rx_r}{(a_1^2+\cdots+a_r^2)^{\frac{1}{2}}}\right) = \sum_{m_1+\cdots+m_r=n} \frac{a_1^{m_1}\cdots a_r^{m_r}}{m_1!\cdots m_r!} H_{m_1}(x_1)\cdots H_{m_r}(x_r).$$

18.18(iii) Multiplication Theorems

Laguerre

18.18.12 $$\frac{L_n^{(\alpha)}(\lambda x)}{L_n^{(\alpha)}(0)} = \sum_{\ell=0}^{n} \binom{n}{\ell} \lambda^\ell (1-\lambda)^{n-\ell} \frac{L_\ell^{(\alpha)}(x)}{L_\ell^{(\alpha)}(0)}.$$

Hermite

18.18.13 $$H_n(\lambda x) = \lambda^n \sum_{\ell=0}^{\lfloor n/2 \rfloor} \frac{(-n)_{2\ell}}{\ell!} (1-\lambda^{-2})^\ell H_{n-2\ell}(x).$$

18.18(iv) Connection Formulas

Jacobi

18.18.14 $$P_n^{(\gamma,\beta)}(x) = \frac{(\beta+1)_n}{(\alpha+\beta+2)_n} \sum_{\ell=0}^{n} \frac{\alpha+\beta+2\ell+1}{\alpha+\beta+1} \frac{(\alpha+\beta+1)_\ell (n+\beta+\gamma+1)_\ell}{(\beta+1)_\ell (n+\alpha+\beta+2)_\ell} \frac{(\gamma-\alpha)_{n-\ell}}{(n-\ell)!} P_\ell^{(\alpha,\beta)}(x),$$

18.18.15 $$\left(\frac{1+x}{2}\right)^n = \frac{(\beta+1)_n}{(\alpha+\beta+2)_n} \sum_{\ell=0}^{n} \frac{\alpha+\beta+2\ell+1}{\alpha+\beta+1} \frac{(\alpha+\beta+1)_\ell (n-\ell+1)_\ell}{(\beta+1)_\ell (n+\alpha+\beta+2)_\ell} P_\ell^{(\alpha,\beta)}(x),$$

and a similar pair of equations by symmetry; compare the second row in Table 18.6.1.

Ultraspherical

18.18.16
$$C_n^{(\mu)}(x) = \sum_{\ell=0}^{\lfloor n/2 \rfloor} \frac{\lambda+n-2\ell}{\lambda} \frac{(\mu)_{n-\ell}}{(\lambda+1)_{n-\ell}} \frac{(\mu-\lambda)_\ell}{\ell!} C_{n-2\ell}^{(\lambda)}(x),$$

18.18.17
$$(2x)^n = n! \sum_{\ell=0}^{\lfloor n/2 \rfloor} \frac{\lambda+n-2\ell}{\lambda} \frac{1}{(\lambda+1)_{n-\ell} \ell!} C_{n-2\ell}^{(\lambda)}(x).$$

Laguerre

18.18.18 $$L_n^{(\beta)}(x) = \sum_{\ell=0}^{n} \frac{(\beta-\alpha)_{n-\ell}}{(n-\ell)!} L_\ell^{(\alpha)}(x),$$

18.18.19 $$x^n = (\alpha+1)_n \sum_{\ell=0}^{n} \frac{(-n)_\ell}{(\alpha+1)_\ell} L_\ell^{(\alpha)}(x).$$

Hermite

18.18.20 $$(2x)^n = \sum_{\ell=0}^{\lfloor n/2 \rfloor} \frac{(-n)_{2\ell}}{\ell!} H_{n-2\ell}(x).$$

18.18(v) Linearization Formulas

Chebyshev

18.18.21 $$T_m(x) T_n(x) = \tfrac{1}{2}(T_{m+n}(x) + T_{m-n}(x)).$$

Ultraspherical

18.18.22
$$C_m^{(\lambda)}(x) C_n^{(\lambda)}(x) = \sum_{\ell=0}^{\min(m,n)} \frac{(m+n+\lambda-2\ell)(m+n-2\ell)!}{(m+n+\lambda-\ell)\ell!(m-\ell)!(n-\ell)!} \times \frac{(\lambda)_\ell (\lambda)_{m-\ell} (\lambda)_{n-\ell} (2\lambda)_{m+n-\ell}}{(\lambda)_{m+n-\ell} (2\lambda)_{m+n-2\ell}} C_{m+n-2\ell}^{(\lambda)}(x).$$

Hermite

18.18.23
$$H_m(x)\,H_n(x) = \sum_{\ell=0}^{\min(m,n)} \binom{m}{\ell}\binom{n}{\ell} 2^\ell \ell!\, H_{m+n-2\ell}(x).$$

The coefficients in the expansions (18.18.22) and (18.18.23) are positive, provided that in the former case $\lambda > 0$.

18.18(vi) Bateman-Type Sums

Jacobi

With

18.18.24 $\quad b_{n,\ell} = \binom{n}{\ell}\dfrac{(n+\alpha+\beta+1)_\ell(-\beta-n)_{n-\ell}}{2^\ell(\alpha+1)_n},$

18.18.25
$$\frac{P_n^{(\alpha,\beta)}(x)}{P_n^{(\alpha,\beta)}(1)}\frac{P_n^{(\alpha,\beta)}(y)}{P_n^{(\alpha,\beta)}(1)} = \sum_{\ell=0}^{n} b_{n,\ell}(x+y)^\ell \\ \times \frac{P_\ell^{(\alpha,\beta)}\!\left((1+xy)/(x+y)\right)}{P_\ell^{(\alpha,\beta)}(1)},$$

18.18.26 $\quad \dfrac{P_n^{(\alpha,\beta)}(x)}{P_n^{(\alpha,\beta)}(1)} = \sum_{\ell=0}^{n} b_{n,\ell}(x+1)^\ell.$

18.18(vii) Poisson Kernels

Laguerre

18.18.27
$$\sum_{n=0}^{\infty} \frac{n!\,L_n^{(\alpha)}(x)\,L_n^{(\alpha)}(y)}{(\alpha+1)_n} z^n \\ = \frac{\Gamma(\alpha+1)(xyz)^{-\frac{1}{2}\alpha}}{1-z} \\ \times \exp\!\left(\frac{-(x+y)z}{1-z}\right) I_\alpha\!\left(\frac{2(xyz)^{\frac{1}{2}}}{1-z}\right),$$
$$|z| < 1.$$

For the modified Bessel function $I_\nu(z)$ see §10.25(ii).

Hermite

18.18.28
$$\sum_{n=0}^{\infty} \frac{H_n(x)\,H_n(y)}{2^n n!} z^n \\ = (1-z^2)^{-\frac{1}{2}} \exp\!\left(\frac{2xyz - (x^2+y^2)z^2}{1-z^2}\right),$$
$$|z| < 1.$$

These Poisson kernels are positive, provided that x, y are real, $0 \le z < 1$, and in the case of (18.18.27) $x, y \ge 0$.

18.18(viii) Other Sums

In this subsection the variables x and y are not confined to the closures of the intervals of orthogonality; compare §18.2(i).

Ultraspherical

18.18.29 $\quad \displaystyle\sum_{\ell=0}^{n} C_\ell^{(\lambda)}(x)\,C_{n-\ell}^{(\mu)}(x) = C_n^{(\lambda+\mu)}(x).$

18.18.30 $\quad \displaystyle\sum_{\ell=0}^{n} \frac{\ell+2\lambda}{2\lambda} C_\ell^{(\lambda)}(x) x^{n-\ell} = C_n^{(\lambda+1)}(x).$

Chebyshev

18.18.31 $\quad \displaystyle\sum_{\ell=0}^{n} T_\ell(x) x^{n-\ell} = U_n(x).$

18.18.32 $\quad 2\displaystyle\sum_{\ell=0}^{n} T_{2\ell}(x) = 1 + U_{2n}(x),$

18.18.33 $\quad 2\displaystyle\sum_{\ell=0}^{n} T_{2\ell+1}(x) = U_{2n+1}(x).$

18.18.34 $\quad 2(1-x^2)\displaystyle\sum_{\ell=0}^{n} U_{2\ell}(x) = 1 - T_{2n+2}(x),$

18.18.35 $\quad 2(1-x^2)\displaystyle\sum_{\ell=0}^{n} U_{2\ell+1}(x) = x - T_{2n+3}(x).$

Legendre and Chebyshev

18.18.36 $\quad \displaystyle\sum_{\ell=0}^{n} P_\ell(x)\,P_{n-\ell}(x) = U_n(x).$

Laguerre

18.18.37 $\quad \displaystyle\sum_{\ell=0}^{n} L_\ell^{(\alpha)}(x) = L_n^{(\alpha+1)}(x),$

18.18.38 $\quad \displaystyle\sum_{\ell=0}^{n} L_\ell^{(\alpha)}(x)\,L_{n-\ell}^{(\beta)}(y) = L_n^{(\alpha+\beta+1)}(x+y).$

Hermite and Laguerre

18.18.39
$$\sum_{\ell=0}^{n} \binom{n}{\ell} H_\ell\!\left(2^{\frac{1}{2}}x\right) H_{n-\ell}\!\left(2^{\frac{1}{2}}y\right) = 2^{\frac{1}{2}n}\,H_n(x+y),$$

18.18.40
$$\sum_{\ell=0}^{n} \binom{n}{\ell} H_{2\ell}(x)\,H_{2n-2\ell}(y) = (-1)^n 2^{2n} n!\,L_n\!\left(x^2+y^2\right).$$

18.18(ix) Compendia

For further sums see Hansen (1975, pp. 292-330), Gradshteyn and Ryzhik (2000, pp. 978–993), and Prudnikov et al. (1986b, pp. 637-644 and 700-718).

Askey Scheme

18.19 Hahn Class: Definitions

Hahn, Krawtchouk, Meixner, and Charlier

Tables 18.19.1 and 18.19.2 provide definitions via orthogonality and normalization (§§18.2(i), 18.2(iii)) for the Hahn polynomials $Q_n(x;\alpha,\beta,N)$, Krawtchouk polynomials $K_n(x;p,N)$, Meixner polynomials $M_n(x;\beta,c)$, and Charlier polynomials $C_n(x,a)$.

Table 18.19.1: Orthogonality properties for Hahn, Krawtchouk, Meixner, and Charlier OP's: discrete sets, weight functions, normalizations, and parameter constraints.

$p_n(x)$	X	w_x	h_n
$Q_n(x;\alpha,\beta,N)$, $n=0,1,\ldots,N$	$\{0,1,\ldots,N\}$	$\dfrac{(\alpha+1)_x(\beta+1)_{N-x}}{x!(N-x)!}$, $\alpha,\beta > -1$ or $\alpha,\beta < -N$	$\dfrac{(-1)^n(n+\alpha+\beta+1)_{N+1}(\beta+1)_n n!}{(2n+\alpha+\beta+1)(\alpha+1)_n(-N)_n N!}$ If $\alpha,\beta < -N$, then $(-1)^N w_x > 0$ and $(-1)^N h_n > 0$.
$K_n(x;p,N)$, $n=0,1,\ldots,N$	$\{0,1,\ldots,N\}$	$\binom{N}{x}p^x(1-p)^{N-x}$, $0 < p < 1$	$\left(\dfrac{1-p}{p}\right)^n \Big/ \binom{N}{n}$
$M_n(x;\beta,c)$	$\{0,1,2,\ldots\}$	$(\beta)_x c^x/x!$, $\beta > 0$, $0 < c < 1$	$\dfrac{c^{-n} n!}{(\beta)_n (1-c)^\beta}$
$C_n(x,a)$	$\{0,1,2,\ldots\}$	$a^x/x!$, $a > 0$	$a^{-n} e^a n!$

Table 18.19.2: Hahn, Krawtchouk, Meixner, and Charlier OP's: leading coefficients.

$p_n(x)$	k_n
$Q_n(x;\alpha,\beta,N)$	$\dfrac{(n+\alpha+\beta+1)_n}{(\alpha+1)_n(-N)_n}$
$K_n(x;p,N)$	$p^{-n}/(-N)_n$
$M_n(x;\beta,c)$	$(1-c^{-1})^n/(\beta)_n$
$C_n(x,a)$	$(-a)^{-n}$

Continuous Hahn

These polynomials are orthogonal on $(-\infty,\infty)$, and with $\Re a > 0$, $\Re b > 0$ are defined as follows.

18.19.1 $$p_n(x) = p_n(x;a,b,\overline{a},\overline{b}),$$

18.19.2 $$w(z;a,b,\overline{a},\overline{b}) = \Gamma(a+iz)\,\Gamma(b+iz)\,\Gamma(\overline{a}-iz)\,\Gamma(\overline{b}-iz),$$

18.19.3 $$w(x) = w(x;a,b,\overline{a},\overline{b}) = |\Gamma(a+ix)\,\Gamma(b+ix)|^2,$$

18.19.4 $$h_n = \frac{2\pi\,\Gamma(n+a+\overline{a})\,\Gamma(n+b+\overline{b})\,|\Gamma(n+a+\overline{b})|^2}{(2n+2\Re(a+b)-1)\,\Gamma(n+2\Re(a+b)-1)\,n!},$$

18.19.5 $$k_n = \frac{(n+2\Re(a+b)-1)_n}{n!}.$$

Meixner–Pollaczek

These polynomials are orthogonal on $(-\infty,\infty)$, and are defined as follows.

18.19.6 $$p_n(x) = P_n^{(\lambda)}(x;\phi),$$

18.19.7 $$w^{(\lambda)}(z;\phi) = \Gamma(\lambda+iz)\,\Gamma(\lambda-iz)e^{(2\phi-\pi)z},$$

18.19.8 $$w(x) = w^{(\lambda)}(x;\phi) = |\Gamma(\lambda+ix)|^2 \, e^{(2\phi-\pi)x},$$
$$\lambda > 0, \ 0 < \phi < \pi,$$

18.19.9 $$h_n = \frac{2\pi\,\Gamma(n+2\lambda)}{(2\sin\phi)^{2\lambda} n!}, \quad k_n = \frac{(2\sin\phi)^n}{n!}.$$

18.20 Hahn Class: Explicit Representations

18.20(i) Rodrigues Formulas

For comments on the use of the forward-difference operator Δ_x, the backward-difference operator ∇_x, and the central-difference operator δ_x, see §18.2(ii).

Hahn, Krawtchouk, Meixner, and Charlier

18.20.1
$$p_n(x) = \frac{1}{\kappa_n w_x} \nabla_x^n \left(w_x \prod_{\ell=0}^{n-1} F(x+\ell) \right), \quad x \in X.$$

In (18.20.1) X and w_x are as in Table 18.19.1. For the Hahn polynomials $p_n(x) = Q_n(x;\alpha,\beta,N)$ and

18.20.2
$$F(x) = (x+\alpha+1)(x-N), \quad \kappa_n = (-N)_n(\alpha+1)_n.$$

For the Krawtchouk, Meixner, and Charlier polynomials, $F(x)$ and κ_n are as in Table 18.20.1.

Table 18.20.1: Krawtchouk, Meixner, and Charlier OP's: Rodrigues formulas (18.20.1).

$p_n(x)$	$F(x)$	κ_n
$K_n(x;p,N)$	$x-N$	$(-N)_n$
$M_n(x;\beta,c)$	$x+\beta$	$(\beta)_n$
$C_n(x,a)$	1	1

Continuous Hahn

18.20.3
$$w(x;a,b,\bar{a},\bar{b})\, p_n(x;a,b,\bar{a},\bar{b})$$
$$= \frac{1}{n!} \delta_x^n \left(w(x; a+\tfrac{1}{2}n, b+\tfrac{1}{2}n, \bar{a}+\tfrac{1}{2}n, \bar{b}+\tfrac{1}{2}n) \right).$$

Meixner–Pollaczek

18.20.4
$$w^{(\lambda)}(x;\phi)\, P_n^{(\lambda)}(x;\phi) = \frac{1}{n!} \delta_x^n \left(w^{(\lambda+\frac{1}{2}n)}(x;\phi) \right).$$

18.20(ii) Hypergeometric Function and Generalized Hypergeometric Functions

For the definition of hypergeometric and generalized hypergeometric functions see §16.2.

18.20.5
$$Q_n(x;\alpha,\beta,N) = {}_3F_2\!\left(\begin{matrix}-n, n+\alpha+\beta+1, -x\\ \alpha+1, -N\end{matrix}; 1\right),$$
$$n = 0, 1, \ldots, N.$$

18.20.6
$$K_n(x;p,N) = {}_2F_1\!\left(\begin{matrix}-n,-x\\ -N\end{matrix}; p^{-1}\right),$$
$$n = 0, 1, \ldots, N.$$

18.20.7
$$M_n(x;\beta,c) = {}_2F_1\!\left(\begin{matrix}-n,-x\\ \beta\end{matrix}; 1-c^{-1}\right).$$

18.20.8
$$C_n(x,a) = {}_2F_0\!\left(\begin{matrix}-n,-x\\ -\end{matrix}; -a^{-1}\right).$$

18.20.9
$$p_n(x;a,b,\bar{a},\bar{b})$$
$$= \frac{i^n(a+\bar{a})_n(a+\bar{b})_n}{n!}$$
$$\times {}_3F_2\!\left(\begin{matrix}-n, n+2\Re(a+b)-1, a+ix\\ a+\bar{a}, a+\bar{b}\end{matrix}; 1\right).$$

(For symmetry properties of $p_n(x;a,b,\bar{a},\bar{b})$ with respect to a, b, $\bar{a}$, $\bar{b}$ see Andrews et al. (1999, Corollary 3.3.4).)

18.20.10
$$P_n^{(\lambda)}(x;\phi) = \frac{(2\lambda)_n}{n!} e^{in\phi} {}_2F_1\!\left(\begin{matrix}-n, \lambda+ix\\ 2\lambda\end{matrix}; 1-e^{-2i\phi}\right).$$

18.21 Hahn Class: Interrelations

18.21(i) Dualities

Duality of Hahn and Dual Hahn

18.21.1
$$Q_n(x;\alpha,\beta,N) = R_x(n(n+\alpha+\beta+1);\alpha,\beta,N),$$
$$n,x = 0,1,\ldots,N.$$

For the dual Hahn polynomial $R_n(x;\gamma,\delta,N)$ see §18.25.

Self-Dualities

18.21.2
$$K_n(x;p,N) = K_x(n;p,N), \quad n,x = 0,1,\ldots,N.$$
$$M_n(x;\beta,c) = M_x(n;\beta,c), \quad n,x = 0,1,2,\ldots.$$
$$C_n(x,a) = C_x(n,a), \quad n,x = 0,1,2,\ldots.$$

18.21(ii) Limit Relations and Special Cases

Hahn → Krawtchouk

18.21.3 $\quad \lim_{t \to \infty} Q_n(x;pt,(1-p)t,N) = K_n(x;p,N).$

Hahn → Meixner

18.21.4
$$\lim_{N \to \infty} Q_n(x;b-1, N(c^{-1}-1), N) = M_n(x;b,c).$$

Hahn → Jacobi

18.21.5 $\quad \lim_{N \to \infty} Q_n(Nx;\alpha,\beta,N) = \dfrac{P_n^{(\alpha,\beta)}(1-2x)}{P_n^{(\alpha,\beta)}(1)}.$

Krawtchouk → Charlier

18.21.6 $\quad \lim_{N \to \infty} K_n(x; N^{-1}a, N) = C_n(x,a).$

Meixner → Charlier

18.21.7 $\quad \lim_{\beta \to \infty} M_n(x;\beta, a(a+\beta)^{-1}) = C_n(x,a).$

Meixner → Laguerre

18.21.8 $\quad \lim_{c \to 1} M_n((1-c)^{-1}x; \alpha+1, c) = \dfrac{L_n^{(\alpha)}(x)}{L_n^{(\alpha)}(0)}.$

Charlier → Hermite

18.21.9 $\quad \lim_{a \to \infty} (2a)^{\frac{1}{2}n} C_n\!\left((2a)^{\frac{1}{2}}x+a, a\right) = (-1)^n H_n(x).$

Continuous Hahn → Meixner–Pollaczek

18.21.10
$$\lim_{t \to \infty} t^{-n} p_n(x-t; \lambda+it, -t\tan\phi, \lambda-it, -t\tan\phi)$$
$$= \frac{(-1)^n}{(\cos\phi)^n} P_n^{(\lambda)}(x;\phi).$$

18.21.11
$$p_n(x;a, a+\tfrac{1}{2}, a, a+\tfrac{1}{2}) = 2^{-2n}(4a+n)_n P_n^{(2a)}(2x; \tfrac{1}{2}\pi).$$

Meixner–Pollaczek → Laguerre

18.21.12 $\lim_{\phi \to 0} P_n^{(\frac{1}{2}\alpha+\frac{1}{2})}(-(2\phi)^{-1}x;\phi) = L_n^{(\alpha)}(x).$

A graphical representation of limits in §§18.7(iii), 18.21(ii), and 18.26(ii) is provided by the *Askey scheme* depicted in Figure 18.21.1.

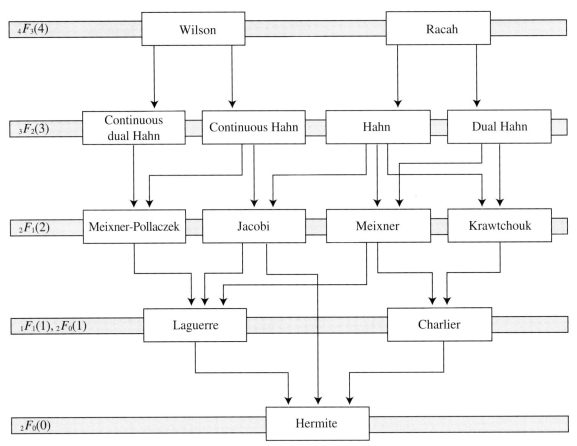

Figure 18.21.1: Askey scheme. The number of free real parameters is zero for Hermite polynomials. It increases by one for each row ascended in the scheme, culminating with four free real parameters for the Wilson and Racah polynomials. (This is with the convention that the real and imaginary parts of the parameters are counted separately in the case of the continuous Hahn polynomials.)

18.22 Hahn Class: Recurrence Relations and Differences

18.22(i) Recurrence Relations in n

Hahn

With

18.22.1 $\qquad p_n(x) = Q_n(x;\alpha,\beta,N),$

18.22.2
$-xp_n(x) = A_n p_{n+1}(x) - (A_n + C_n) p_n(x) + C_n p_{n-1}(x),$

where

18.22.3 $A_n = \dfrac{(n+\alpha+\beta+1)(n+\alpha+1)(N-n)}{(2n+\alpha+\beta+1)(2n+\alpha+\beta+2)},$

$C_n = \dfrac{n(n+\alpha+\beta+N+1)(n+\beta)}{(2n+\alpha+\beta)(2n+\alpha+\beta+1)}.$

Krawtchouk, Meixner, and Charlier

These polynomials satisfy (18.22.2) with $p_n(x)$, A_n, and C_n as in Table 18.22.1.

Table 18.22.1: Recurrence relations (18.22.2) for Krawtchouk, Meixner, and Charlier polynomials.

$p_n(x)$	A_n	C_n
$K_n(x;p,N)$	$p(N-n)$	$n(1-p)$
$M_n(x;\beta,c)$	$\dfrac{c(n+\beta)}{1-c}$	$\dfrac{n}{1-c}$
$C_n(x,a)$	a	n

18.22 Hahn Class: Recurrence Relations and Differences

Continuous Hahn

With

18.22.4 $\quad q_n(x) = p_n(x; a, b, \overline{a}, \overline{b})/p_n(ia; a, b, \overline{a}, \overline{b})$,

18.22.5 $\quad \begin{aligned}(a+ix)q_n(x) \\ = \tilde{A}_n q_{n+1}(x) - (\tilde{A}_n + \tilde{C}_n)q_n(x) + \tilde{C}_n q_{n-1}(x),\end{aligned}$

where

18.22.6
$$\tilde{A}_n = -\frac{(n+2\Re(a+b)-1)(n+a+\overline{a})(n+a+\overline{b})}{(2n+2\Re(a+b)-1)(2n+2\Re(a+b))},$$
$$\tilde{C}_n = \frac{n(n+b+\overline{a}-1)(n+b+\overline{b}-1)}{(2n+2\Re(a+b)-2)(2n+2\Re(a+b)-1)}.$$

Meixner–Pollaczek

With

18.22.7 $\quad p_n(x) = P_n^{(\lambda)}(x; \phi)$,

18.22.8 $\quad \begin{aligned}(n+1)p_{n+1}(x) = 2\left(x\sin\phi + (n+\lambda)\cos\phi\right)p_n(x) \\ - (n+2\lambda-1)p_{n-1}(x).\end{aligned}$

18.22(ii) Difference Equations in x

Hahn

With

18.22.9 $\quad p_n(x) = Q_n(x; \alpha, \beta, N)$,

18.22.10 $\quad \begin{aligned}A(x)p_n(x+1) - (A(x)+C(x))p_n(x) \\ + C(x)p_n(x-1) \\ - n(n+\alpha+\beta+1)p_n(x) = 0,\end{aligned}$

where

18.22.11 $\quad \begin{aligned}A(x) &= (x+\alpha+1)(x-N), \\ C(x) &= x(x-\beta-N-1).\end{aligned}$

Krawtchouk, Meixner, and Charlier

18.22.12 $\quad \begin{aligned}A(x)p_n(x+1) - (A(x)+C(x))p_n(x) \\ + C(x)p_n(x-1) + \lambda_n p_n(x) = 0.\end{aligned}$

For $A(x)$, $C(x)$, and λ_n in (18.22.12) see Table 18.22.2.

Table 18.22.2: Difference equations (18.22.12) for Krawtchouk, Meixner, and Charlier polynomials.

$p_n(x)$	$A(x)$	$C(x)$	λ_n
$K_n(x; p, N)$	$p(x-N)$	$(p-1)x$	$-n$
$M_n(x; \beta, c)$	$c(x+\beta)$	x	$n(1-c)$
$C_n(x, a)$	a	x	n

Continuous Hahn

With

18.22.13 $\quad p_n(x) = p_n(x; a, b, \overline{a}, \overline{b})$,

18.22.14 $\quad \begin{aligned}A(x)p_n(x+i) - (A(x)+C(x))p_n(x) \\ + C(x)p_n(x-i) \\ + n(n+2\Re(a+b)-1)p_n(x) = 0,\end{aligned}$

where

18.22.15 $\quad A(x) = (x+i\overline{a})(x+i\overline{b}), \quad C(x) = (x-ia)(x-ib).$

Meixner–Pollaczek

With

18.22.16 $\quad p_n(x) = P_n^{(\lambda)}(x; \phi)$,

18.22.17 $\quad \begin{aligned}A(x)p_n(x+i) - (A(x)+C(x))p_n(x) \\ + C(x)p_n(x-i) + 2n\sin\phi\, p_n(x) = 0,\end{aligned}$

where

18.22.18 $\quad A(x) = e^{i\phi}(x+i\lambda), \quad C(x) = e^{-i\phi}(x-i\lambda).$

18.22(iii) x-Differences

Hahn

18.22.19
$$\Delta_x Q_n(x; \alpha, \beta, N)$$
$$= -\frac{n(n+\alpha+\beta+1)}{(\alpha+1)N} Q_{n-1}(x; \alpha+1, \beta+1, N-1),$$

18.22.20
$$\nabla_x \left(\frac{(\alpha+1)_x (\beta+1)_{N-x}}{x!\,(N-x)!} Q_n(x; \alpha, \beta, N)\right)$$
$$= \frac{N+1}{\beta} \frac{(\alpha)_x (\beta)_{N+1-x}}{x!\,(N+1-x)!}$$
$$\times Q_{n+1}(x; \alpha-1, \beta-1, N+1).$$

Krawtchouk

18.22.21 $\quad \Delta_x K_n(x; p, N) = -\dfrac{n}{pN} K_{n-1}(x; p, N-1),$

18.22.22
$$\nabla_x \left(\binom{N}{x} p^x (1-p)^{N-x} K_n(x; p, N)\right)$$
$$= \binom{N+1}{x} p^x (1-p)^{N-x} K_{n+1}(x; p, N+1).$$

Meixner

18.22.23 $\quad \Delta_x M_n(x; \beta, c) = -\dfrac{n(1-c)}{\beta c} M_{n-1}(x; \beta+1, c),$

18.22.24
$$\nabla_x \left(\frac{(\beta)_x c^x}{x!} M_n(x; \beta, c)\right)$$
$$= \frac{(\beta-1)_x c^x}{x!} M_{n+1}(x; \beta-1, c).$$

Charlier

18.22.25 $\quad \Delta_x C_n(x, a) = -\dfrac{n}{a} C_{n-1}(x, a),$

18.22.26 $\quad \nabla_x \left(\dfrac{a^x}{x!} C_n(x, a)\right) = \dfrac{a^x}{x!} C_{n+1}(x, a).$

Continuous Hahn

18.22.27 $\qquad \delta_x\left(p_n(x;a,b,\overline{a},\overline{b})\right) = (n + 2\Re(a+b) - 1)\, p_{n-1}(x; a+\tfrac{1}{2}, b+\tfrac{1}{2}, \overline{a}+\tfrac{1}{2}, \overline{b}+\tfrac{1}{2})$,

18.22.28 $\quad \delta_x\left(w(x; a+\tfrac{1}{2}, b+\tfrac{1}{2}, \overline{a}+\tfrac{1}{2}, \overline{b}+\tfrac{1}{2}) p_n(x; a+\tfrac{1}{2}, b+\tfrac{1}{2}, \overline{a}+\tfrac{1}{2}, \overline{b}+\tfrac{1}{2})\right) = -(n+1) w(x; a,b,\overline{a},\overline{b}) p_{n+1}(x; a,b,\overline{a},\overline{b})$.

Meixner–Pollaczek

18.22.29 $\quad \delta_x\left(P_n^{(\lambda)}(x;\phi)\right) = 2\sin\phi\, P_{n-1}^{(\lambda+\tfrac{1}{2})}(x;\phi)$,

18.22.30
$$\delta_x\left(w^{(\lambda+\tfrac{1}{2})}(x;\phi)\, P_n^{(\lambda+\tfrac{1}{2})}(x;\phi)\right) = -(n+1) w^{(\lambda)}(x;\phi)\, P_{n+1}^{(\lambda)}(x;\phi).$$

18.23 Hahn Class: Generating Functions

For the definition of generalized hypergeometric functions see §16.2.

Hahn

18.23.1
$${}_1F_1\!\left(\begin{matrix}-x\\ \alpha+1\end{matrix}; -z\right) {}_1F_1\!\left(\begin{matrix}x-N\\ \beta+1\end{matrix}; z\right) = \sum_{n=0}^{N} \frac{(-N)_n}{(\beta+1)_n n!} Q_n(x;\alpha,\beta,N) z^n, \quad x = 0,1,\ldots,N.$$

18.23.2
$${}_2F_0\!\left(\begin{matrix}-x, -x+\beta+N+1\\ -\end{matrix}; -z\right) \times {}_2F_0\!\left(\begin{matrix}x-N, x+\alpha+1\\ -\end{matrix}; z\right) = \sum_{n=0}^{N} \frac{(-N)_n(\alpha+1)_n}{n!} Q_n(x;\alpha,\beta,N) z^n, \quad x = 0,1,\ldots,N.$$

Krawtchouk

18.23.3
$$\left(1 - \frac{1-p}{p} z\right)^x (1+z)^{N-x} = \sum_{n=0}^{N} \binom{N}{n} K_n(x; p, N) z^n, \quad x = 0,1,\ldots,N.$$

Meixner

18.23.4
$$\left(1 - \frac{z}{c}\right)^x (1-z)^{-x-\beta} = \sum_{n=0}^{\infty} \frac{(\beta)_n}{n!} M_n(x;\beta,c) z^n, \quad x = 0,1,2,\ldots,\ |z|<1.$$

Charlier

18.23.5 $\quad e^z \left(1 - \frac{z}{a}\right)^x = \sum_{n=0}^{\infty} \frac{C_n(x,a)}{n!} z^n, \quad x = 0,1,2,\ldots.$

Continuous Hahn

18.23.6
$${}_1F_1\!\left(\begin{matrix}a+ix\\ 2\Re a\end{matrix}; -iz\right) {}_1F_1\!\left(\begin{matrix}\overline{b}-ix\\ 2\Re b\end{matrix}; iz\right) = \sum_{n=0}^{\infty} \frac{p_n(x; a,b,\overline{a},\overline{b})}{(2\Re a)_n (2\Re b)_n} z^n.$$

Meixner–Pollaczek

18.23.7
$$(1 - e^{i\phi} z)^{-\lambda+ix} (1 - e^{-i\phi} z)^{-\lambda-ix} = \sum_{n=0}^{\infty} P_n^{(\lambda)}(x;\phi) z^n, \qquad |z| < 1.$$

18.24 Hahn Class: Asymptotic Approximations

Krawtchouk

With $x = \lambda N$ and $\nu = n/N$, Li and Wong (2000) gives an asymptotic expansion for $K_n(x;p,N)$ as $n \to \infty$, that holds uniformly for λ and ν in compact subintervals of $(0,1)$. This expansion is in terms of the parabolic cylinder function and its derivative.

With $\mu = N/n$ and x fixed, Qiu and Wong (2004) gives an asymptotic expansion for $K_n(x;p,N)$ as $n \to \infty$, that holds uniformly for $\mu \in [1,\infty)$. This expansion is in terms of confluent hypergeometric functions. Asymptotic approximations are also provided for the zeros of $K_n(x;p,N)$ in various cases depending on the values of p and μ.

Meixner

For two asymptotic expansions of $M_n(nx;\beta,c)$ as $n \to \infty$, with β and c fixed, see Jin and Wong (1998). The first expansion holds uniformly for $\delta \leq x \leq 1 + \delta$, and the second for $1 - \delta \leq x \leq 1 + \delta^{-1}$, δ being an arbitrary small positive constant. Both expansions are in terms of parabolic cylinder functions.

For asymptotic approximations for the zeros of $M_n(nx;\beta,c)$ in terms of zeros of $\mathrm{Ai}(x)$ (§9.9(i)), see Jin and Wong (1999).

Charlier

Dunster (2001b) provides various asymptotic expansions for $C_n(x,a)$ as $n \to \infty$, in terms of elementary functions or in terms of Bessel functions. Taken together, these expansions are uniformly valid for $-\infty < x < \infty$ and for a in unbounded intervals—each of which contains $[0, (1-\delta)n]$, where δ again denotes an arbitrary

18.25 Wilson Class: Definitions

18.25(i) Preliminaries

For the Wilson class OP's $p_n(x)$ with $x = \lambda(y)$: if the y-orthogonality set is $\{0, 1, \ldots, N\}$, then the role of the differentiation operator d/dx in the Jacobi, Laguerre, and Hermite cases is played by the operator Δ_y followed by division by $\Delta_y(\lambda(y))$, or by the operator ∇_y followed by division by $\nabla_y(\lambda(y))$. Alternatively if the y-orthogonality interval is $(0, \infty)$, then the role of d/dx is played by the operator δ_y followed by division by $\delta_y(\lambda(y))$.

Table 18.25.1 lists the transformations of variable, orthogonality ranges, and parameter constraints that are needed in §18.2(i) for the Wilson polynomials $W_n(x; a, b, c, d)$, continuous dual Hahn polynomials $S_n(x; a, b, c)$, Racah polynomials $R_n(x; \alpha, \beta, \gamma, \delta)$, and dual Hahn polynomials $R_n(x; \gamma, \delta, N)$.

Meixner–Pollaczek

For an asymptotic expansion of $P_n^{(\lambda)}(nx; \phi)$ as $n \to \infty$, with ϕ fixed, see Li and Wong (2001). This expansion is uniformly valid in any compact x-interval on the real line and is in terms of parabolic cylinder functions. Corresponding approximations are included for the zeros of $P_n^{(\lambda)}(nx; \phi)$.

Approximations in Terms of Laguerre Polynomials

For asymptotic approximations to $P_n^{(\lambda)}(x; \phi)$ as $|x + i\lambda| \to \infty$, with n fixed, see Temme and López (2001). These approximations are in terms of Laguerre polynomials and hold uniformly for $\mathrm{ph}(x + i\lambda) \in [0, \pi]$. Compare also (18.21.12). Similar approximations are included for Jacobi, Krawtchouk, and Meixner polynomials.

small positive constant. See also Bo and Wong (1994) and Goh (1998).

Table 18.25.1: Wilson class OP's: transformations of variable, orthogonality ranges, and parameter constraints.

$p_n(x)$	$x = \lambda(y)$	Orthogonality range for y	Constraints
$W_n(x; a, b, c, d)$	y^2	$(0, \infty)$	$\Re(a, b, c, d) > 0$; nonreal parameters in conjugate pairs
$S_n(x; a, b, c)$	y^2	$(0, \infty)$	$\Re(a, b, c) > 0$; nonreal parameters in conjugate pairs
$R_n(x; \alpha, \beta, \gamma, \delta)$	$y(y + \gamma + \delta + 1)$	$\{0, 1, \ldots, N\}$	$\alpha + 1$ or $\beta + \delta + 1$ or $\gamma + 1 = -N$; for further constraints see (18.25.1)
$R_n(x; \gamma, \delta, N)$	$y(y + \gamma + \delta + 1)$	$\{0, 1, \ldots, N\}$	$\gamma, \delta > -1$ or $< -N$

Further Constraints for Racah Polynomials

If $\alpha + 1 = -N$, then the weights will be positive iff one of the following eight sets of inequalities holds:

18.25.1
$$-\delta - 1 < \beta < \gamma + 1 < -N + 1.$$
$$N - 1 < -\delta - 1 < \beta < \gamma + 1.$$
$$\gamma, \delta > -1, \quad \beta > N + \gamma.$$
$$\gamma, \delta > -1, \quad \beta < -N - \delta.$$
$$N - 1 < N + \gamma < \beta < -N - \delta.$$
$$N + \gamma < \beta < -N - \delta < -N - 1.$$
$$\gamma, \delta < -N, \quad \beta > -1 - \delta.$$
$$\gamma, \delta < -N, \quad \beta < \gamma + 1.$$

The first four sets imply $\gamma + \delta > -2$, and the last four imply $\gamma + \delta < -2N$.

18.25(ii) Weights and Normalizations: Continuous Cases

18.25.2
$$\int_0^\infty p_n(x) p_m(x) w(x)\, dx = h_n \delta_{n,m}.$$

Wilson

18.25.3
$$p_n(x) = W_n(x; a_1, a_2, a_3, a_4),$$

18.25.4
$$w(y^2) = \frac{1}{2y} \left| \frac{\prod_j \Gamma(a_j + iy)}{\Gamma(2iy)} \right|^2,$$

18.25.5
$$h_n = \frac{n!\, 2\pi \prod_{j<\ell} \Gamma(n + a_j + a_\ell)}{(2n - 1 + \sum_j a_j)\, \Gamma\!\left(n - 1 + \sum_j a_j\right)}.$$

Continuous Dual Hahn

18.25.6 $\quad p_n(x) = S_n(x; a_1, a_2, a_3),$

18.25.7 $\quad w(y^2) = \dfrac{1}{2y} \left| \dfrac{\prod_j \Gamma(a_j + iy)}{\Gamma(2iy)} \right|^2,$

18.25.8 $\quad h_n = n! \, 2\pi \prod_{j<\ell} \Gamma(n + a_j + a_\ell).$

18.25(iii) Weights and Normalizations: Discrete Cases

18.25.9
$$\sum_{y=0}^{N} p_n(y(y+\gamma+\delta+1)) p_m(y(y+\gamma+\delta+1))$$
$$\times \dfrac{\gamma + \delta + 1 + 2y}{\gamma + \delta + 1 + y} \omega_y = h_n \delta_{n,m}.$$

Racah

18.25.10 $\quad p_n(x) = R_n(x; \alpha, \beta, \gamma, \delta), \quad \alpha + 1 = -N,$

18.25.11
$$\omega_y = \dfrac{(\alpha+1)_y (\beta+\delta+1)_y (\gamma+1)_y (\gamma+\delta+2)_y}{(-\alpha+\gamma+\delta+1)_y (-\beta+\gamma+1)_y (\delta+1)_y y!},$$

18.25.12
$$h_n = \dfrac{(-\beta)_N (\gamma+\delta+2)_N}{(-\beta+\gamma+1)_N (\delta+1)_N} \dfrac{(n+\alpha+\beta+1)_n n!}{(\alpha+\beta+2)_{2n}}$$
$$\times \dfrac{(\alpha+\beta-\gamma+1)_n (\alpha-\delta+1)_n (\beta+1)_n}{(\alpha+1)_n (\beta+\delta+1)_n (\gamma+1)_n}.$$

Dual Hahn

18.25.13 $\quad p_n(x) = R_n(x; \gamma, \delta, N),$

18.25.14 $\quad \omega_y = \dfrac{(-1)^y (-N)_y (\gamma+1)_y (\gamma+\delta+1)_2}{(N+\gamma+\delta+2)_y (\delta+1)_y y!},$

18.25.15 $\quad h_n = \dfrac{n! \, (N-n)! \, (\gamma+\delta+2)_N}{N! \, (\gamma+1)_n (\delta+1)_{N-n}}.$

18.25(iv) Leading Coefficients

Table 18.25.2 provides the leading coefficients k_n (§18.2(iii)) for the Wilson, continuous dual Hahn, Racah, and dual Hahn polynomials.

Table 18.25.2: Wilson class OP's: leading coefficients.

$p_n(x)$	k_n
$W_n(x; a, b, c, d)$	$(-1)^n (n + a + b + c + d - 1)_n$
$S_n(x; a, b, c)$	$(-1)^n$
$R_n(x; \alpha, \beta, \gamma, \delta)$	$\dfrac{(n+\alpha+\beta+1)_n}{(\alpha+1)_n (\beta+\delta+1)_n (\gamma+1)_n}$
$R_n(x; \gamma, \delta, N)$	$\dfrac{1}{(\gamma+1)_n (-N)_n}$

18.26 Wilson Class: Continued

18.26(i) Representations as Generalized Hypergeometric Functions

For the definition of generalized hypergeometric functions see §16.2.

18.26.1 $\quad W_n(y^2; a, b, c, d) = (a+b)_n (a+c)_n (a+d)_n \; {}_4F_3\!\left(\begin{array}{c} -n, n+a+b+c+d-1, a+iy, a-iy \\ a+b, a+c, a+d \end{array}; 1\right).$

18.26.2 $\quad \dfrac{S_n(y^2; a, b, c)}{(a+b)_n (a+c)_n} = {}_3F_2\!\left(\begin{array}{c} -n, a+iy, a-iy \\ a+b, a+c \end{array}; 1\right).$

18.26.3 $\quad R_n(y(y+\gamma+\delta+1); \alpha, \beta, \gamma, \delta) = {}_4F_3\!\left(\begin{array}{c} -n, n+\alpha+\beta+1, -y, y+\gamma+\delta+1 \\ \alpha+1, \beta+\delta+1, \gamma+1 \end{array}; 1\right),$
$\alpha + 1$ or $\beta + \delta + 1$ or $\gamma + 1 = -N;\ n = 0, 1, \ldots, N.$

18.26.4 $\quad R_n(y(y+\gamma+\delta+1); \gamma, \delta, N) = {}_3F_2\!\left(\begin{array}{c} -n, -y, y+\gamma+\delta+1 \\ \gamma+1, -N \end{array}; 1\right), \qquad n = 0, 1, \ldots, N.$

18.26(ii) Limit Relations

Wilson → Continuous Dual Hahn

18.26.5 $\quad \lim\limits_{d \to \infty} \dfrac{W_n(x; a, b, c, d)}{(a+d)_n} = S_n(x; a, b, c).$

18.26 Wilson Class: Continued

Wilson → Continuous Hahn

18.26.6
$$\lim_{t\to\infty} \frac{W_n\big((x+t)^2; a-it, b-it, \overline{a}+it, \overline{b}+it\big)}{(-2t)^n n!} = p_n(x; a, b, \overline{a}, \overline{b}).$$

Wilson → Jacobi

18.26.7
$$\lim_{t\to\infty} \frac{W_n\big(\tfrac{1}{2}(1-x)t^2; \tfrac{1}{2}\alpha+\tfrac{1}{2}, \tfrac{1}{2}\alpha+\tfrac{1}{2}, \tfrac{1}{2}\beta+\tfrac{1}{2}+it, \tfrac{1}{2}\beta+\tfrac{1}{2}-it\big)}{t^{2n} n!} = P_n^{(\alpha,\beta)}(x).$$

Continuous Dual Hahn → Meixner–Pollaczek

18.26.8
$$\lim_{t\to\infty} S_n\big((x-t)^2; \lambda+it, \lambda-it, t\cot\phi\big)/t^n = n!(\csc\phi)^n P_n^{(\lambda)}(x; \phi).$$

Racah → Dual Hahn

18.26.9
$$\lim_{\beta\to\infty} R_n(x; -N-1, \beta, \gamma, \delta) = R_n(x; \gamma, \delta, N).$$

Racah → Hahn

18.26.10
$$\lim_{\delta\to\infty} R_n(x(x+\gamma+\delta+1); \alpha, \beta, -N-1, \delta) = Q_n(x; \alpha, \beta, N).$$

Dual Hahn → Krawtchouk

18.26.11
$$\lim_{t\to\infty} R_n(x(x+t+1); pt, (1-p)t, N) = K_n(x; p, N).$$

Dual Hahn → Meixner

With

18.26.12
$$r(x; \beta, c, N) = x(x+\beta+c^{-1}(1-c)N),$$

18.26.13
$$\lim_{N\to\infty} R_n\big(r(x; \beta, c, N); \beta-1, c^{-1}(1-c)N, N\big) = M_n(x; \beta, c).$$

See also Figure 18.21.1.

18.26(iii) Difference Relations

For comments on the use of the forward-difference operator Δ_x, the backward-difference operator ∇_x, and the central-difference operator δ_x, see §18.2(ii).

For each family only the y-difference that lowers n is given. See Koekoek and Swarttouw (1998, Chapter 1) for further formulas.

18.26.14
$$\delta_y\big(W_n(y^2; a, b, c, d)\big)/\delta_y(y^2) = -n(n+a+b+c+d-1)\, W_{n-1}\big(y^2; a+\tfrac{1}{2}, b+\tfrac{1}{2}, c+\tfrac{1}{2}, d+\tfrac{1}{2}\big).$$

18.26.15
$$\delta_y\big(S_n(y^2; a, b, c)\big)/\delta_y(y^2) = -n\, S_{n-1}\big(y^2; a+\tfrac{1}{2}, b+\tfrac{1}{2}, c+\tfrac{1}{2}\big).$$

18.26.16
$$\frac{\Delta_y\big(R_n(y(y+\gamma+\delta+1); \alpha, \beta, \gamma, \delta)\big)}{\Delta_y(y(y+\gamma+\delta+1))} = \frac{n(n+\alpha+\beta+1)}{(\alpha+1)(\beta+\delta+1)(\gamma+1)} R_{n-1}(y(y+\gamma+\delta+2); \alpha+1, \beta+1, \gamma+1, \delta).$$

18.26.17
$$\frac{\Delta_y\big(R_n(y(y+\gamma+\delta+1); \gamma, \delta, N)\big)}{\Delta_y(y(y+\gamma+\delta+1))} = -\frac{n}{(\gamma+1)N} R_{n-1}(y(y+\gamma+\delta+2); \gamma+1, \delta, N-1).$$

18.26(iv) Generating Functions

For the hypergeometric function ${}_2F_1$ see §§15.1 and 15.2(i).

Wilson

18.26.18
$${}_2F_1\!\left(\begin{array}{c} a+iy, d+iy \\ a+d \end{array}; z\right) {}_2F_1\!\left(\begin{array}{c} b-iy, c-iy \\ b+c \end{array}; z\right) = \sum_{n=0}^{\infty} \frac{W_n(y^2; a, b, c, d)}{(a+d)_n (b+c)_n n!} z^n, \qquad |z| < 1.$$

Continuous Dual Hahn

18.26.19
$$(1-z)^{-c+iy} {}_2F_1\left(\begin{matrix} a+iy, b+iy \\ a+b \end{matrix}; z\right) = \sum_{n=0}^{\infty} \frac{S_n(y^2; a, b, c)}{(a+b)_n n!} z^n, \qquad |z| < 1.$$

Racah

18.26.20
$${}_2F_1\left(\begin{matrix} -y, -y+\beta-\gamma \\ \beta+\delta+1 \end{matrix}; z\right) {}_2F_1\left(\begin{matrix} y-N, y+\gamma+1 \\ -\delta-N \end{matrix}; z\right) = \sum_{n=0}^{N} \frac{(-N)_n(\gamma+1)_n}{(-\delta-N)_n n!} R_n(y(y+\gamma+\delta+1); -N-1, \beta, \gamma, \delta) z^n.$$

Dual Hahn

18.26.21
$$(1-z)^y {}_2F_1\left(\begin{matrix} y-N, y+\gamma+1 \\ -\delta-N \end{matrix}; z\right) = \sum_{n=0}^{N} \frac{(\gamma+1)_n(-N)_n}{(-\delta-N)_n n!} R_n(y(y+\gamma+\delta+1); \gamma, \delta, N) z^n.$$

18.26(v) Asymptotic Approximations

For asymptotic expansions of Wilson polynomials of large degree see Wilson (1991), and for asymptotic approximations to their largest zeros see Chen and Ismail (1998).

Other Orthogonal Polynomials

18.27 q-Hahn Class

18.27(i) Introduction

The q-hypergeometric OP's comprise the q-Hahn class OP's and the Askey–Wilson class OP's (§18.28). For the notation of q-hypergeometric functions see §§17.2 and 17.4(i).

The q-Hahn class OP's comprise systems of OP's $\{p_n(x)\}$, $n = 0, 1, \ldots, N$, or $n = 0, 1, 2, \ldots$, that are eigenfunctions of a second-order q-difference operator. Thus

18.27.1
$$A(x) p_n(qx) + B(x) p_n(x) + C(x) p_n(q^{-1}x) = \lambda_n p_n(x),$$

where $A(x)$, $B(x)$, and $C(x)$ are independent of n, and where the λ_n are the eigenvalues. In the q-Hahn class OP's the role of the operator d/dx in the Jacobi, Laguerre, and Hermite cases is played by the q-derivative $\mathcal{D}_q$, as defined in (17.2.41). A (nonexhaustive) classification of such systems of OP's was made by Hahn (1949). There are 18 families of OP's of q-Hahn class. These families depend on further parameters, in addition to q. The generic (top level) cases are the q-Hahn polynomials and the big q-Jacobi polynomials, each of which depends on three further parameters.

All these systems of OP's have orthogonality properties of the form

18.27.2
$$\sum_{x \in X} p_n(x) p_m(x) |x| v_x = h_n \delta_{n,m},$$

where X is given by $X = \{aq^y\}_{y \in I_+}$ or $X = \{aq^y\}_{y \in I_+} \cup \{-bq^y\}_{y \in I_-}$. Here a, b are fixed positive real numbers, and I_+ and I_- are sequences of successive integers, finite or unbounded in one direction, or unbounded in both directions. If I_+ and I_- are both nonempty, then they are both unbounded to the right. Some of the systems of OP's that occur in the classification do not have a unique orthogonality property. Thus in addition to a relation of the form (18.27.2), such systems may also satisfy orthogonality relations with respect to a continuous weight function on some interval.

Here only a few families are mentioned. They are defined by their q-hypergeometric representations, followed by their orthogonality properties. For other formulas, including q-difference equations, recurrence relations, duality formulas, special cases, and limit relations, see Koekoek and Swarttouw (1998, Chapter 3). See also Gasper and Rahman (2004, pp. 195–199, 228–230) and Ismail (2005, Chapters 13, 18, 21).

18.27(ii) q-Hahn Polynomials

18.27.3
$$Q_n(x) = Q_n(x; \alpha, \beta, N; q) = {}_3\phi_2\left(\begin{matrix} q^{-n}, \alpha\beta q^{n+1}, x \\ \alpha q, q^{-N} \end{matrix}; q, q\right), \qquad n = 0, 1, \ldots, N.$$

18.27.4
$$\sum_{y=0}^{N} Q_n(q^{-y}) Q_m(q^{-y}) \frac{(\alpha q, q^{-N}; q)_y (\alpha\beta q)^{-y}}{(q, \beta^{-1} q^{-N}; q)_y} = h_n \delta_{n,m}, \qquad n, m = 0, 1, \ldots, N.$$

For h_n see Koekoek and Swarttouw (1998, Eq. (3.6.2)).

18.27(iii) Big q-Jacobi Polynomials

18.27.5 $\quad P_n(x;a,b,c;q) = {}_3\phi_2\left(\begin{matrix} q^{-n}, abq^{n+1}, x \\ aq, cq \end{matrix}; q, q\right),$

and

18.27.6
$$\begin{aligned} &P_n^{(\alpha,\beta)}(x;c,d;q) \\ &= \frac{c^n q^{-(\alpha+1)n}\left(q^{\alpha+1}, -q^{\alpha+1}c^{-1}d; q\right)_n}{(q, -q; q)_n} \\ &\quad \times P_n\!\left(q^{\alpha+1}c^{-1}dx; q^\alpha, q^\beta, -q^\alpha c^{-1}d; q\right). \end{aligned}$$

The orthogonality relations are given by (18.27.2), with

18.27.7 $\quad p_n(x) = P_n(x;a,b,c;q),$

18.27.8 $\quad X = \{aq^{\ell+1}\}_{\ell=0,1,2,\ldots} \cup \{cq^{\ell+1}\}_{\ell=0,1,2,\ldots},$

18.27.9 $\quad v_x = \frac{(a^{-1}x, c^{-1}x; q)_\infty}{(x, bc^{-1}x; q)_\infty},$
$$0 < a < q^{-1},\ 0 < b < q^{-1},\ c < 0,$$

and

18.27.10 $\quad p_n(x) = P_n^{(\alpha,\beta)}(x;c,d;q)$

18.27.11 $\quad X = \{cq^\ell\}_{\ell=0,1,2,\ldots} \cup \{-dq^\ell\}_{\ell=0,1,2,\ldots},$

18.27.12 $\quad v_x = \frac{(qx/c, -qx/d; q)_\infty}{(q^{\alpha+1}x/c, -q^{\beta+1}x/d; q)_\infty},\quad \alpha,\beta > -1,\ c,d > 0.$

For h_n see Koekoek and Swarttouw (1998, Eq. (3.5.2)).

18.27(iv) Little q-Jacobi Polynomials

18.27.13
$$p_n(x) = p_n(x;a,b;q) = {}_2\phi_1\left(\begin{matrix} q^{-n}, abq^{n+1} \\ aq \end{matrix}; q, qx\right).$$

18.27.14
$$\sum_{y=0}^\infty p_n(q^y) p_m(q^y) \frac{(bq;q)_y (aq)^y}{(q;q)_y} = h_n \delta_{n,m},\quad 0 < a < q^{-1},\ b < q^{-1}.$$

For h_n see Koekoek and Swarttouw (1998, Eq. (3.12.2)).

18.27(v) q-Laguerre Polynomials

18.27.15
$$L_n^{(\alpha)}(x;q) = \frac{(q^{\alpha+1};q)_n}{(q;q)_n} {}_1\phi_1\!\left(\begin{matrix} q^{-n} \\ q^{\alpha+1} \end{matrix}; q, -xq^{n+\alpha+1}\right).$$

The measure is not uniquely determined:

18.27.16
$$\int_0^\infty L_n^{(\alpha)}(x;q) L_m^{(\alpha)}(x;q) \frac{x^\alpha}{(-x;q)_\infty} dx = \frac{(q^{\alpha+1};q)_n}{(q;q)_n q^n} h_0^{(1)} \delta_{n,m},\quad \alpha > -1,$$

where $h_0^{(1)}$ is given in Koekoek and Swarttouw (1998, Eq. (3.21.2), and

18.27.17
$$\sum_{y=-\infty}^\infty L_n^{(\alpha)}(cq^y;q) L_m^{(\alpha)}(cq^y;q) \frac{q^{y(\alpha+1)}}{(-cq^y;q)_\infty} = \frac{(q^{\alpha+1};q)_n}{(q;q)_n q^n} h_0^{(2)} \delta_{n,m},\quad \alpha > -1,\ c > 0,$$

where $h_0^{(2)}$ is given in Koekoek and Swarttouw (1998, Eq. (3.21.3)).

18.27(vi) Stieltjes–Wigert Polynomials

18.27.18
$$\begin{aligned} S_n(x;q) &= \sum_{\ell=0}^n \frac{q^{\ell^2}(-x)^\ell}{(q;q)_\ell (q;q)_{n-\ell}} \\ &= \frac{1}{(q;q)_n} {}_1\phi_1\!\left(\begin{matrix} q^{-n} \\ 0 \end{matrix}; q, -q^{n+1}x\right). \end{aligned}$$

(Sometimes in the literature x is replaced by $q^{\frac{1}{2}}x$.)

The measure is not uniquely determined:

18.27.19
$$\int_0^\infty \frac{S_n(x;q) S_m(x;q)}{(-x, -qx^{-1};q)_\infty} dx = \frac{\ln(q^{-1})}{q^n} \frac{(q;q)_\infty}{(q;q)_n} \delta_{n,m},$$

and

18.27.20
$$\int_0^\infty S_n\!\left(q^{\frac{1}{2}}x;q\right) S_m\!\left(q^{\frac{1}{2}}x;q\right) \exp\!\left(-\frac{(\ln x)^2}{2\ln(q^{-1})}\right) dx = \frac{\sqrt{2\pi q^{-1} \ln(q^{-1})}}{q^n (q;q)_n} \delta_{n,m}.$$

18.27(vii) Discrete q-Hermite I and II Polynomials

Discrete q-Hermite I

18.27.21
$$\begin{aligned} h_n(x;q) &= (q;q)_n \sum_{\ell=0}^{\lfloor n/2 \rfloor} \frac{(-1)^\ell q^{\ell(\ell-1)} x^{n-2\ell}}{(q^2;q^2)_\ell (q;q)_{n-2\ell}} \\ &= x^n {}_2\phi_0\!\left(\begin{matrix} q^{-n}, q^{-n+1} \\ - \end{matrix}; q^2, x^{-2}q^{2n-1}\right). \end{aligned}$$

18.27.22
$$\begin{aligned} &\sum_{\ell=0}^\infty \bigl(h_n(q^\ell;q) h_m(q^\ell;q) + h_n(-q^\ell;q) h_m(-q^\ell;q)\bigr) \\ &\quad \times (q^{\ell+1}, -q^{\ell+1};q)_\infty q^\ell \\ &= (q;q)_n (q,-1,-q;q)_\infty q^{n(n-1)/2} \delta_{n,m}. \end{aligned}$$

Discrete q-Hermite II

18.27.23
$$\tilde{h}_n(x;q) = (q;q)_n \sum_{\ell=0}^{\lfloor n/2 \rfloor} \frac{(-1)^\ell q^{-2n\ell} q^{\ell(2\ell+1)} x^{n-2\ell}}{(q^2;q^2)_\ell (q;q)_{n-2\ell}}$$
$$= x^n\, {}_2\phi_1\!\left(\begin{matrix}q^{-n}, q^{-n+1}\\ 0\end{matrix};q^2, -x^{-2}q^2\right).$$

18.27.24
$$\sum_{\ell=-\infty}^{\infty} \Big(\tilde{h}_n(cq^\ell;q)\,\tilde{h}_m(cq^\ell;q)$$
$$+ \tilde{h}_n(-cq^\ell;q)\,\tilde{h}_m(-cq^\ell;q)\Big) \frac{q^\ell}{(-c^2 q^{2\ell};q^2)_\infty}$$
$$= 2\frac{(q^2,-c^2q,-c^{-2}q;q^2)_\infty}{(q,-c^2,-c^{-2}q^2;q^2)_\infty} \frac{(q;q)_n}{q^{n^2}} \delta_{n,m},$$
$$c > 0.$$

(For discrete q-Hermite II polynomials the measure is not uniquely determined.)

18.28 Askey–Wilson Class

18.28(i) Introduction

The Askey–Wilson class OP's comprise the four-parameter families of Askey–Wilson polynomials and of q-Racah polynomials, and cases of these families obtained by specialization of parameters. The Askey–Wilson polynomials form a system of OP's $\{p_n(x)\}$, $n = 0, 1, 2, \ldots$, that are orthogonal with respect to a weight function on a bounded interval, possibly supplemented with discrete weights on a finite set. The q-Racah polynomials form a system of OP's $\{p_n(x)\}$, $n = 0, 1, 2, \ldots, N$, that are orthogonal with respect to a weight function on a sequence $\{q^{-y} + cq^{y+1}\}$, $y = 0, 1, \ldots, N$, with c a constant. Both the Askey–Wilson polynomials and the q-Racah polynomials can best be described as functions of z (resp. y) such that $P_n(z) = p_n(\tfrac{1}{2}(z+z^{-1}))$ in the Askey–Wilson case, and $P_n(y) = p_n(q^{-y} + cq^{y+1})$ in the q-Racah case, and both are eigenfunctions of a second-order q-difference operator similar to (18.27.1).

In the remainder of this section the Askey–Wilson class OP's are defined by their q-hypergeometric representations, followed by their orthogonal properties. For further properties see Koekoek and Swarttouw (1998, Chapter 3). See also Gasper and Rahman (2004, pp. 180–199) and Ismail (2005, Chapter 15). For the notation of q-hypergeometric functions see §§17.2 and 17.4(i).

18.28(ii) Askey–Wilson Polynomials

18.28.1
$$p_n(\cos\theta)$$
$$= p_n(\cos\theta; a,b,c,d \,|\, q)$$
$$= a^{-n} \sum_{\ell=0}^{n} q^\ell \left(abq^\ell, acq^\ell, adq^\ell; q\right)_{n-\ell}$$
$$\times \frac{(q^{-n}, abcdq^{n-1};q)_\ell}{(q;q)_\ell} \prod_{j=0}^{\ell-1}(1 - 2aq^j\cos\theta + a^2q^{2j})$$
$$= a^{-n}(ab,ac,ad;q)_n$$
$$\times {}_4\phi_3\!\left(\begin{matrix}q^{-n}, abcdq^{n-1}, ae^{i\theta}, ae^{-i\theta}\\ ab, ac, ad\end{matrix}; q, q\right).$$

Assume a, b, c, d are all real, or two of them are real and two form a conjugate pair, or none of them are real but they form two conjugate pairs. Furthermore, $|ab|$, $|ac|$, $|ad|$, $|bc|$, $|bd|$, $|cd| < 1$. Then

18.28.2
$$\int_{-1}^{1} p_n(x) p_m(x) w(x)\, dx = h_n \delta_{n,m}, \quad |a|,|b|,|c|,|d| \leq 1,$$
where

18.28.3 $\quad 2\pi \sin\theta\, w(\cos\theta) = \left|\frac{(e^{2i\theta};q)_\infty}{(ae^{i\theta}, be^{i\theta}, ce^{i\theta}, de^{i\theta}; q)_\infty}\right|^2,$

18.28.4 $\quad h_0 = \dfrac{(abcd;q)_\infty}{(q, ab, ac, ad, bc, bd, cd; q)_\infty},$

18.28.5
$$h_n = h_0 \frac{(1 - abcdq^{n-1})\,(q, ab, ac, ad, bc, bd, cd; q)_n}{(1 - abcdq^{2n-1})(abcd;q)_n},$$
$$n = 1, 2, \ldots.$$

More generally, without the constraints in (18.28.2),

18.28.6
$$\int_{-1}^{1} p_n(x) p_m(x) w(x)\, dx + \sum_{\ell} p_n(x_\ell) p_m(x_\ell) \omega_\ell = h_n \delta_{n,m},$$

with $w(x)$ and h_n as above. Also, x_ℓ are the points $\tfrac{1}{2}(\alpha q^\ell + \alpha^{-1} q^{-\ell})$ with α any of the a, b, c, d whose absolute value exceeds 1, and the sum is over the $\ell = 0, 1, 2, \ldots$ with $|\alpha q^\ell| > 1$. See Koekoek and Swarttouw (1998, Eq. (3.1.3)) for the value of ω_ℓ when $\alpha = a$.

18.28(iii) Al-Salam–Chihara Polynomials

18.28.7
$$Q_n(\cos\theta; a, b \mid q)$$
$$= p_n(\cos\theta; a, b, 0, 0 \mid q)$$
$$= a^{-n} \sum_{\ell=0}^{n} q^\ell \frac{(abq^\ell; q)_{n-\ell} (q^{-n}; q)_\ell}{(q; q)_\ell}$$
$$\times \prod_{j=0}^{\ell-1}(1 - 2aq^j \cos\theta + a^2 q^{2j})$$
$$= \frac{(ab; q)_n}{a^n} {}_3\phi_2\!\left(\begin{matrix} q^{-n}, ae^{i\theta}, ae^{-i\theta} \\ ab, 0 \end{matrix}; q, q\right)$$
$$= (be^{-i\theta}; q)_n e^{in\theta} {}_2\phi_1\!\left(\begin{matrix} q^{-n}, ae^{i\theta} \\ b^{-1}q^{1-n}e^{i\theta} \end{matrix}; q, b^{-1}qe^{-i\theta}\right).$$

18.28.8
$$\frac{1}{2\pi}\int_0^\pi Q_n(\cos\theta; a, b \mid q) Q_m(\cos\theta; a, b \mid q)$$
$$\times \left|\frac{(e^{2i\theta}; q)_\infty}{(ae^{i\theta}, be^{i\theta}; q)_\infty}\right|^2 d\theta = \frac{\delta_{n,m}}{(q^{n+1}, abq^n; q)_\infty},$$
$$a, b \in \mathbb{R} \text{ or } a = \bar{b}; |ab| < 1; |a|, |b| \le 1.$$

More generally, without the constraints $|a|, |b| \le 1$ discrete terms need to be added to the right-hand side of (18.28.8); see Koekoek and Swarttouw (1998, Eq. (3.8.3)).

18.28(iv) q^{-1}-Al-Salam–Chihara Polynomials

18.28.9
$$Q_n\!\left(\tfrac{1}{2}(aq^{-y} + a^{-1}q^y); a, b \mid q^{-1}\right)$$
$$= (-1)^n b^n q^{-\frac{1}{2}n(n-1)}$$
$$\times ((ab)^{-1}; q)_n {}_3\phi_1\!\left(\begin{matrix} q^{-n}, q^{-y}, a^{-2}q^y \\ (ab)^{-1} \end{matrix}; q, q^n ab^{-1}\right).$$

18.28.10
$$\sum_{y=0}^{\infty} \frac{(1 - q^{2y}a^{-2})(a^{-2}, (ab)^{-1}; q)_y}{(1 - a^{-2})(q, bqa^{-1}; q)_y} (ba^{-1})^y q^{y^2}$$
$$\times Q_n\!\left(\tfrac{1}{2}(aq^{-y} + a^{-1}q^y); a, b \mid q^{-1}\right)$$
$$\times Q_m\!\left(\tfrac{1}{2}(aq^{-y} + a^{-1}q^y); a, b \mid q^{-1}\right)$$
$$= \frac{(qa^{-2}; q)_\infty}{(ba^{-1}q; q)_\infty} (q, (ab)^{-1}; q)_n (ab)^n q^{-n^2} \delta_{n,m}.$$

Eq. (18.28.10) is valid when either

18.28.11 $\quad 0 < q < 1, a, b \in \mathbb{R}, ab > 1, a^{-1}b < q^{-1},$
or
18.28.12 $\quad 0 < q < 1, a/i, b/i \in \mathbb{R}, (\Im a)(\Im b) > 0, a^{-1}b < q^{-1}.$

If, in addition to (18.28.11) or (18.28.12), we have $a^{-1}b \le q$, then the measure in (18.28.10) is uniquely determined. Also, if $q < a^{-1}b < q^{-1}$, then (18.28.10) holds with a, b interchanged. For further nondegenerate cases see Chihara and Ismail (1993) and Christiansen and Ismail (2006).

18.28(v) Continuous q-Ultraspherical Polynomials

18.28.13
$$C_n(\cos\theta; \beta \mid q)$$
$$= \sum_{\ell=0}^{n} \frac{(\beta; q)_\ell (\beta; q)_{n-\ell}}{(q; q)_\ell (q; q)_{n-\ell}} e^{i(n-2\ell)\theta}$$
$$= \frac{(\beta; q)_n}{(q; q)_n} e^{in\theta} {}_2\phi_1\!\left(\begin{matrix} q^{-n}, \beta \\ \beta^{-1}q^{1-n} \end{matrix}; q, \beta^{-1}qe^{-2i\theta}\right).$$

18.28.14
$$C_n(\cos\theta; \beta \mid q)$$
$$= \frac{(\beta^2; q)_n}{(q; q)_n \beta^{\frac{1}{2}n}} {}_4\phi_3\!\left(\begin{matrix} q^{-n}, \beta^2 q^n, \beta^{\frac{1}{2}}e^{i\theta}, \beta^{\frac{1}{2}}e^{-i\theta} \\ \beta q^{\frac{1}{2}}, -\beta, -\beta q^{\frac{1}{2}} \end{matrix}; q, q\right).$$

18.28.15
$$\frac{1}{2\pi}\int_0^\pi C_n(\cos\theta; \beta \mid q) C_m(\cos\theta; \beta \mid q) \left|\frac{(e^{2i\theta}; q)_\infty}{(\beta e^{2i\theta}; q)_\infty}\right|^2 d\theta$$
$$= \frac{(\beta, \beta q; q)_\infty}{(\beta^2, q; q)_\infty} \frac{(1-\beta)(\beta^2; q)_n}{(1 - \beta q^n)(q; q)_n} \delta_{n,m}, \quad -1 < \beta < 1.$$

These polynomials are also called *Rogers polynomials*.

18.28(vi) Continuous q-Hermite Polynomials

18.28.16
$$H_n(\cos\theta \mid q) = \sum_{\ell=0}^{n} \frac{(q; q)_n}{(q; q)_\ell (q; q)_{n-\ell}} e^{i(n-2\ell)\theta}$$
$$= e^{in\theta} {}_2\phi_0\!\left(\begin{matrix} q^{-n}, 0 \\ - \end{matrix}; q, q^n e^{-2i\theta}\right).$$

18.28.17
$$\frac{1}{2\pi}\int_0^\pi H_n(\cos\theta \mid q) H_m(\cos\theta \mid q) \left|(e^{2i\theta}; q)_\infty\right|^2 d\theta$$
$$= \frac{\delta_{n,m}}{(q^{n+1}; q)_\infty}.$$

18.28(vii) Continuous q^{-1}-Hermite Polynomials

18.28.18
$$h_n(\sinh t \mid q) = \sum_{\ell=0}^{n} q^{\frac{1}{2}\ell(\ell+1)} \frac{(q^{-n}; q)_\ell}{(q; q)_\ell} e^{(n-2\ell)t}$$
$$= e^{nt} {}_1\phi_1\!\left(\begin{matrix} q^{-n} \\ 0 \end{matrix}; q, -qe^{-2t}\right)$$
$$= i^{-n} H_n(i \sinh t \mid q^{-1}).$$

For continuous q^{-1}-Hermite polynomials the orthogonality measure is not unique. See Askey (1989) and Ismail and Masson (1994) for examples.

18.28(viii) q-Racah Polynomials

With $x = q^{-y} + \gamma\delta q^{y+1}$,

18.28.19
$$R_n(x) = R_n(x;\alpha,\beta,\gamma,\delta \mid q)$$
$$= \sum_{\ell=0}^n \frac{q^\ell\left(q^{-n},\alpha\beta q^{n+1};q\right)_\ell}{(\alpha q,\beta\delta q,\gamma q,q;q)_\ell} \prod_{j=0}^{\ell-1}(1-q^j x + \gamma\delta q^{2j+1})$$
$$= {}_4\phi_3\left(\begin{matrix}q^{-n},\alpha\beta q^{n+1},q^{-y},\gamma\delta q^{y+1}\\ \alpha q,\beta\delta q,\gamma q\end{matrix};q,q\right),$$
αq, $\beta\delta q$, or $\gamma q = q^{-N}$; $n = 0,1,\ldots,N$.

18.28.20
$$\sum_{y=0}^N R_n(q^{-y}+\gamma\delta q^{y+1})R_m(q^{-y}+\gamma\delta q^{y+1})\omega_y$$
$$= h_n\delta_{n,m}, \qquad n,m = 0,1,\ldots,N.$$

For ω_y and h_n see Koekoek and Swarttouw (1998, Eq. (3.2.2)).

18.29 Asymptotic Approximations for q-Hahn and Askey–Wilson Classes

Ismail (1986) gives asymptotic expansions as $n \to \infty$, with x and other parameters fixed, for continuous q-ultraspherical, big and little q-Jacobi, and Askey–Wilson polynomials. These asymptotic expansions are in fact convergent expansions. For Askey–Wilson $p_n(\cos\theta;a,b,c,d \mid q)$ the leading term is given by

18.29.1
$$(bc,bd,cd;q)_n \left(Q_n(e^{i\theta};a,b,c,d \mid q) + Q_n(e^{-i\theta};a,b,c,d \mid q)\right),$$

where with $z = e^{\pm i\theta}$,

18.29.2
$$Q_n(z;a,b,c,d \mid q) \sim \frac{z^n\left(az^{-1},bz^{-1},cz^{-1},dz^{-1};q\right)_\infty}{(z^{-2},bc,bd,cd;q)_\infty},$$
$n \to \infty$; z,a,b,c,d,q fixed.

For a uniform asymptotic expansion of the Stieltjes–Wigert polynomials, see Wang and Wong (2006).

For asymptotic approximations to the largest zeros of the q-Laguerre and continuous q^{-1}-Hermite polynomials see Chen and Ismail (1998).

18.30 Associated OP's

In the recurrence relation (18.2.8) assume that the coefficients A_n, B_n, and C_{n+1} are defined when n is a continuous nonnegative real variable, and let c be an arbitrary positive constant. Assume also

18.30.1 $\qquad A_n A_{n+1} C_{n+1} > 0, \qquad n \geq 0.$

Then the *associated orthogonal polynomials* $p_n(x;c)$ are defined by

18.30.2 $\qquad p_{-1}(x;c) = 0, \quad p_0(x;c) = 1,$

and

18.30.3 $\qquad p_{n+1}(x;c) = (A_{n+c}x + B_{n+c})p_n(x;c)$
$$- C_{n+c}p_{n-1}(x;c), \qquad n = 0,1,\ldots.$$

Assume also that Eq. (18.30.3) continues to hold, except that when $n = 0$, B_c is replaced by an arbitrary real constant. Then the polynomials $p_n(x,c)$ generated in this manner are called *corecursive associated OP's*.

Associated Jacobi Polynomials

These are defined by

18.30.4 $\qquad P_n^{(\alpha,\beta)}(x;c) = p_n(x;c), \qquad n = 0,1,\ldots,$

where $p_n(x;c)$ is given by (18.30.2) and (18.30.3), with A_n, B_n, and C_n as in (18.9.2). Explicitly,

18.30.5
$$\frac{(-1)^n(\alpha+\beta+c+1)_n n!\, P_n^{(\alpha,\beta)}(x;c)}{(\alpha+\beta+2c+1)_n(\beta+c+1)_n}$$
$$= \sum_{\ell=0}^n \frac{(-n)_\ell(n+\alpha+\beta+2c+1)_\ell}{(c+1)_\ell(\beta+c+1)_\ell}\left(\tfrac{1}{2}x+\tfrac{1}{2}\right)^\ell$$
$$\times {}_4F_3\left(\begin{matrix}\ell-n,n+\ell+\alpha+\beta+2c+1,\beta+c,c\\ \beta+\ell+c+1,\ell+c+1,\alpha+\beta+2c\end{matrix};1\right),$$

where the generalized hypergeometric function ${}_4F_3$ is defined by (16.2.1).

For corresponding corecursive associated Jacobi polynomials see Letessier (1995).

Associated Legendre Polynomials

These are defined by

18.30.6 $\qquad P_n(x;c) = P_n^{(0,0)}(x;c), \qquad n = 0,1,\ldots.$

Explicitly,

18.30.7 $\qquad P_n(x;c) = \sum_{\ell=0}^n \frac{c}{\ell+c}\, P_\ell(x)\, P_{n-\ell}(x).$

(These polynomials are not to be confused with associated Legendre functions §14.3(ii).)

For further results on associated Legendre polynomials see Chihara (1978, Chapter VI, §12); on associated Jacobi polynomials, see Wimp (1987) and Ismail and Masson (1991). For associated Pollaczek polynomials (compare §18.35) see Erdélyi et al. (1953b, §10.21). For associated Askey–Wilson polynomials see Rahman (2001).

18.31 Bernstein–Szegö Polynomials

Let $\rho(x)$ be a polynomial of degree ℓ and positive when $-1 \leq x \leq 1$. The *Bernstein–Szegö polynomials* $\{p_n(x)\}$, $n = 0,1,\ldots$, are orthogonal on $(-1,1)$ with respect to three types of weight function: $(1-x^2)^{-\frac{1}{2}}(\rho(x))^{-1}$, $(1-x^2)^{\frac{1}{2}}(\rho(x))^{-1}$, $(1-x)^{\frac{1}{2}}(1+x)^{-\frac{1}{2}}(\rho(x))^{-1}$. In consequence, $p_n(\cos\theta)$ can be given explicitly in terms of $\rho(\cos\theta)$ and sines and cosines, provided that $\ell < 2n$ in the first case, $\ell < 2n+2$ in the second case, and $\ell < 2n+1$ in the third case. See Szegö (1975, §2.6).

18.32 OP's with Respect to Freud Weights

A *Freud weight* is a weight function of the form

18.32.1 $$w(x) = \exp(-Q(x)), \quad -\infty < x < \infty,$$

where $Q(x)$ is real, even, nonnegative, and continuously differentiable. Of special interest are the cases $Q(x) = x^{2m}$, $m = 1, 2, \ldots$. No explicit expressions for the corresponding OP's are available. However, for asymptotic approximations in terms of elementary functions for the OP's, and also for their largest zeros, see Levin and Lubinsky (2001) and Nevai (1986). For a uniform asymptotic expansion in terms of Airy functions (§9.2) for the OP's in the case $Q(x) = x^4$ see Bo and Wong (1999).

18.33 Polynomials Orthogonal on the Unit Circle

18.33(i) Definition

A system of polynomials $\{\phi_n(z)\}$, $n = 0, 1, \ldots$, where $\phi_n(z)$ is of proper degree n, is *orthonormal on the unit circle* with respect to the weight function $w(z)$ (≥ 0) if

18.33.1 $$\frac{1}{2\pi i} \int_{|z|=1} \phi_n(z) \overline{\phi_m(z)} w(z) \frac{dz}{z} = \delta_{n,m},$$

where the bar signifies complex conjugate. See Simon (2005a,b) for general theory.

18.33(ii) Recurrence Relations

Denote

18.33.2 $$\phi_n(z) = \kappa_n z^n + \sum_{\ell=1}^{n} \kappa_{n,n-\ell} z^{n-\ell},$$

where $\kappa_n(> 0)$, and $\kappa_{n,n-\ell}(\in \mathbb{C})$ are constants. Also denote

18.33.3 $$\phi_n^*(z) = \kappa_n z^n + \sum_{\ell=1}^{n} \overline{\kappa}_{n,n-\ell} z^{n-\ell},$$

where the bar again signifies compex conjugate. Then

18.33.4 $\kappa_n z \phi_n(z) = \kappa_{n+1} \phi_{n+1}(z) - \phi_{n+1}(0) \phi_{n+1}^*(z),$

18.33.5 $\kappa_n \phi_{n+1}(z) = \kappa_{n+1} z \phi_n(z) + \phi_{n+1}(0) \phi_n^*(z),$

18.33.6 $$\begin{aligned}&\kappa_n \phi_n(0) \phi_{n+1}(z) + \kappa_{n-1} \phi_{n+1}(0) z \phi_{n-1}(z) \\ &= (\kappa_n \phi_{n+1}(0) + \kappa_{n+1} \phi_n(0) z) \phi_n(z).\end{aligned}$$

18.33(iii) Connection with OP's on the Line

Assume that $w(e^{i\phi}) = w(e^{-i\phi})$. Set

18.33.7 $$\begin{aligned}w_1(x) &= (1-x^2)^{-\frac{1}{2}} w\left(x + i(1-x^2)^{\frac{1}{2}}\right), \\ w_2(x) &= (1-x^2)^{\frac{1}{2}} w\left(x + i(1-x^2)^{\frac{1}{2}}\right).\end{aligned}$$

Let $\{p_n(x)\}$ and $\{q_n(x)\}$, $n = 0, 1, \ldots$, be OP's with weight functions $w_1(x)$ and $w_2(x)$, respectively, on $(-1, 1)$. Then

18.33.8
$$\begin{aligned}&p_n\left(\tfrac{1}{2}(z + z^{-1})\right) \\ &= (\text{const.}) \times \left(z^{-n} \phi_{2n}(z) + z^n \phi_{2n}(z^{-1})\right) \\ &= (\text{const.}) \times \left(z^{-n+1} \phi_{2n-1}(z) + z^{n-1} \phi_{2n-1}(z^{-1})\right),\end{aligned}$$

18.33.9
$$\begin{aligned}&q_n\left(\tfrac{1}{2}(z + z^{-1})\right) \\ &= (\text{const.}) \times \frac{z^{-n-1} \phi_{2n+2}(z) - z^{n+1} \phi_{2n+2}(z^{-1})}{z - z^{-1}} \\ &= (\text{const.}) \times \frac{z^{-n} \phi_{2n+1}(z) - z^n \phi_{2n+1}(z^{-1})}{z - z^{-1}}.\end{aligned}$$

Conversely,

18.33.10
$$\begin{aligned}&z^{-n} \phi_{2n}(z) \\ &= A_n p_n\left(\tfrac{1}{2}(z + z^{-1})\right) + B_n (z - z^{-1}) q_{n-1}\left(\tfrac{1}{2}(z + z^{-1})\right),\end{aligned}$$

18.33.11
$$\begin{aligned}&z^{-n+1} \phi_{2n-1}(z) \\ &= C_n p_n\left(\tfrac{1}{2}(z + z^{-1})\right) + D_n (z - z^{-1}) q_{n-1}\left(\tfrac{1}{2}(z + z^{-1})\right),\end{aligned}$$

where A_n, B_n, C_n, and D_n are independent of z.

18.33(iv) Special Cases

Trivial

18.33.12 $$\phi_n(z) = z^n, \quad w(z) = 1.$$

Szegö–Askey

18.33.13
$$\phi_n(z) = \sum_{\ell=0}^{n} \frac{(\lambda+1)_\ell (\lambda)_{n-\ell}}{\ell!\,(n-\ell)!} z^\ell = \frac{(\lambda)_n}{n!} {}_2F_1\left(\begin{matrix}-n, \lambda+1 \\ -\lambda-n+1\end{matrix}; z\right),$$

with

$$w(z) = \left(1 - \tfrac{1}{2}(z + z^{-1})\right)^\lambda,$$

18.33.14 $$\begin{aligned}w_1(x) &= (1-x)^{\lambda-\frac{1}{2}}(1+x)^{-\frac{1}{2}}, \\ w_2(x) &= (1-x)^{\lambda+\frac{1}{2}}(1+x)^{\frac{1}{2}}, \qquad \lambda > -\tfrac{1}{2}.\end{aligned}$$

For the hypergeometric function ${}_2F_1$ see §§15.1 and 15.2(i).

Askey

18.33.15
$$\begin{aligned}\phi_n(z) &= \sum_{\ell=0}^{n} \frac{(aq^2; q^2)_\ell (a; q^2)_{n-\ell}}{(q^2; q^2)_\ell (q^2; q^2)_{n-\ell}} (q^{-1} z)^\ell \\ &= \frac{(a; q^2)_n}{(q^2; q^2)_n} {}_2\phi_1\left(\begin{matrix}aq^2, q^{-2n} \\ a^{-1} q^{2-2n}\end{matrix}; q^2, \frac{qz}{a}\right),\end{aligned}$$

with

18.33.16 $$w(z) = \left|(qz; q^2)_\infty \big/ (aqz; q^2)_\infty\right|^2, \quad a^2 q^2 < 1.$$

For the notation, including the basic hypergeometric function ${}_2\phi_1$, see §§17.2 and 17.4(i).

When $a = 0$ the Askey case is also known as the *Rogers–Szegö case*.

18.33(v) Biorthogonal Polynomials on the Unit Circle

See Baxter (1961) for general theory. See Askey (1982) and Pastro (1985) for special cases extending (18.33.13)–(18.33.14) and (18.33.15)–(18.33.16), respectively. See Gasper (1981) and Hendriksen and van Rossum (1986) for relations with Laurent polynomials orthogonal on the unit circle. See Al-Salam and Ismail (1994) for special biorthogonal rational functions on the unit circle.

18.34 Bessel Polynomials

18.34(i) Definitions and Recurrence Relation

For the confluent hypergeometric function $_1F_1$ and the generalized hypergeometric function $_2F_0$ see §16.2(ii) and §16.2(iv).

18.34.1
$$y_n(x;a) = {}_2F_0\left(\begin{matrix}-n, n+a-1 \\ -\end{matrix}; -\frac{x}{2}\right)$$
$$= (n+a-1)_n \left(\frac{x}{2}\right)^n {}_1F_1\left(\begin{matrix}-n \\ -2n-a+2\end{matrix}; \frac{2}{x}\right).$$

Other notations in use are given by

18.34.2 $y_n(x) = y_n(x;2), \quad \theta_n(x) = x^n y_n(x^{-1}),$

and

18.34.3 $y_n(x;a,b) = y_n(2x/b;a), \quad \theta_n(x;a,b) = x^n y_n(x^{-1};a,b).$

Often only the polynomials (18.34.2) are called *Bessel polynomials*, while the polynomials (18.34.1) and (18.34.3) are called *generalized Bessel polynomials*. See also §10.49(ii).

18.34.4
$$y_{n+1}(x;a) = (A_n x + B_n) y_n(x;a) - C_n y_{n-1}(x;a),$$
where

18.34.5
$$A_n = \frac{(2n+a)(2n+a-1)}{2(n+a-1)},$$
$$B_n = \frac{(a-2)(2n+a-1)}{(n+a-1)(2n+a-2)},$$
$$C_n = \frac{-n(2n+a)}{(n+a-1)(2n+a-2)}.$$

18.34(ii) Orthogonality

Because the coefficients C_n in (18.34.4) are not all positive, the polynomials $y_n(x;a)$ cannot be orthogonal on the line with respect to a positive weight function. There is orthogonality on the unit circle, however:

18.34.6
$$\frac{1}{2\pi i}\int_{|z|=1} z^{a-2} y_n(z;a) y_m(z;a) e^{-2/z}\,dz$$
$$= \frac{(-1)^{n+a-1} n!\, 2^{a-1}}{(n+a-2)!(2n+a-1)}\delta_{n,m}, \qquad a = 1, 2, \ldots,$$

the integration path being taken in the positive rotational sense.

Orthogonality can also be expressed in terms of *moment functionals*; see Durán (1993), Evans et al. (1993), and Maroni (1995).

18.34(iii) Other Properties

18.34.7
$$x^2 y_n''(x;a) + (ax+2)y_n'(x;a) - n(n+a-1)y_n(x;a) = 0,$$

where primes denote derivatives with respect to x.

18.34.8 $\displaystyle\lim_{\alpha\to\infty} \frac{P_n^{(\alpha,a-\alpha-2)}(1+\alpha x)}{P_n^{(\alpha,a-\alpha-2)}(1)} = y_n(x;a).$

For uniform asymptotic expansions of $y_n(x;a)$ as $n\to\infty$ in terms of Airy functions (§9.2) see Wong and Zhang (1997) and Dunster (2001c). For uniform asymptotic expansions in terms of Hermite polynomials see López and Temme (1999b).

For further information on Bessel polynomials see §10.49(ii).

18.35 Pollaczek Polynomials

18.35(i) Definition and Hypergeometric Representation

18.35.1 $P_{-1}^{(\lambda)}(x;a,b) = 0, \quad P_0^{(\lambda)}(x;a,b) = 1,$

and

18.35.2
$$(n+1)P_{n+1}^{(\lambda)}(x;a,b) = 2((n+\lambda+a)x + b)P_n^{(\lambda)}(x;a,b)$$
$$- (n+2\lambda-1)P_{n-1}^{(\lambda)}(x;a,b),$$
$$n = 0, 1, \ldots.$$

Next, let

18.35.3 $\tau_{a,b}(\theta) = \dfrac{a\cos\theta+b}{\sin\theta}, \qquad 0 < \theta < \pi.$

Then

18.35.4
$$P_n^{(\lambda)}(\cos\theta;a,b)$$
$$= \frac{(\lambda - i\tau_{a,b}(\theta))_n}{n!}e^{in\theta}$$
$$\times {}_2F_1\left(\begin{matrix}-n, \lambda + i\tau_{a,b}(\theta)\\-n-\lambda+1+i\tau_{a,b}(\theta)\end{matrix}; e^{-2i\theta}\right)$$
$$= \sum_{\ell=0}^n \frac{(\lambda+i\tau_{a,b}(\theta))_\ell}{\ell!}\frac{(\lambda-i\tau_{a,b}(\theta))_{n-\ell}}{(n-\ell)!}e^{i(n-2\ell)\theta}.$$

For the hypergeometric function $_2F_1$ see §§15.1, 15.2(i).

18.35(ii) Orthogonality

18.35.5
$$\int_{-1}^{1} P_n^{(\lambda)}(x;a,b)\, P_m^{(\lambda)}(x;a,b) w^{(\lambda)}(x;a,b)\, dx = 0,$$
$$n \neq m,$$

where

18.35.6
$$w^{(\lambda)}(\cos\theta;a,b) = \pi^{-1} 2^{2\lambda-1} e^{(2\theta-\pi)\tau_{a,b}(\theta)}$$
$$\times (\sin\theta)^{2\lambda-1} |\Gamma(\lambda + i\tau_{a,b}(\theta))|^2,$$
$$a \geq b \geq -a,\ \lambda > -\tfrac{1}{2},\ 0 < \theta < \pi.$$

18.35(iii) Other Properties

18.35.7
$$(1 - ze^{i\theta})^{-\lambda + i\tau_{a,b}(\theta)} (1 - ze^{-i\theta})^{-\lambda - i\tau_{a,b}(\theta)}$$
$$= \sum_{n=0}^{\infty} P_n^{(\lambda)}(\cos\theta;a,b) z^n,\quad |z|<1,\ 0<\theta<\pi.$$

18.35.8
$$P_n^{(\lambda)}(x;0,0) = C_n^{(\lambda)}(x),$$

18.35.9
$$P_n^{(\lambda)}(\cos\phi;0, x\sin\phi) = P_n^{(\lambda)}(x;\phi).$$

For the polynomials $C_n^{(\lambda)}(x)$ and $P_n^{(\lambda)}(x;\phi)$ see §§18.3 and 18.19, respectively.

See Bo and Wong (1996) for an asymptotic expansion of $P_n^{(\frac{1}{2})}\!\left(\cos(n^{-\frac{1}{2}}\theta);a,b\right)$ as $n \to \infty$, with a and b fixed. This expansion is in terms of the Airy function $\mathrm{Ai}(x)$ and its derivative (§9.2), and is uniform in any compact θ-interval in $(0,\infty)$. Also included is an asymptotic approximation for the zeros of $P_n^{(\frac{1}{2})}\!\left(\cos(n^{-\frac{1}{2}}\theta);a,b\right)$.

18.36 Miscellaneous Polynomials

18.36(i) Jacobi-Type Polynomials

These are OP's on the interval $(-1,1)$ with respect to an orthogonality measure obtained by adding constant multiples of "Dirac delta weights" at -1 and 1 to the weight function for the Jacobi polynomials. For further information see Koornwinder (1984a) and Kwon et al. (2006).

Similar OP's can also be constructed for the Laguerre polynomials; see Koornwinder (1984b, (4.8)).

18.36(ii) Sobolev OP's

Sobolev OP's are orthogonal with respect to an inner product involving derivatives. For an introductory survey to this subject, see Marcellán et al. (1993). Other relevant references include Iserles et al. (1991) and Koekoek et al. (1998).

18.36(iii) Multiple OP's

These are polynomials in one variable that are orthogonal with respect to a number of different measures. They are related to Hermite-Padé approximation and can be used for proofs of irrationality or transcendence of interesting numbers. For further information see Ismail (2005, Chapter 23).

18.36(iv) Orthogonal Matrix Polynomials

These are matrix-valued polynomials that are orthogonal with respect to a square matrix of measures on the real line. Classes of such polynomials have been found that generalize the classical OP's in the sense that they satisfy second-order matrix differential equations with coefficients independent of the degree. For further information see Durán and Grünbaum (2005).

18.37 Classical OP's in Two or More Variables

18.37(i) Disk Polynomials

Definition in Terms of Jacobi Polynomials

18.37.1
$$R_{m,n}^{(\alpha)}(re^{i\theta}) = e^{i(m-n)\theta} r^{|m-n|} \frac{P_{\min(m,n)}^{(\alpha,|m-n|)}(2r^2-1)}{P_{\min(m,n)}^{(\alpha,|m-n|)}(1)},$$
$$r \geq 0,\ \theta \in \mathbb{R},\ \alpha > -1.$$

Orthogonality

18.37.2
$$\iint_{x^2+y^2<1} R_{m,n}^{(\alpha)}(x+iy)\, R_{j,\ell}^{(\alpha)}(x-iy)(1-x^2-y^2)^\alpha\, dx\, dy$$
$$= 0, \qquad m \neq j \text{ and/or } n \neq \ell.$$

Equivalent Definition

The following three conditions, taken together, determine $R_{m,n}^{(\alpha)}(z)$ uniquely:

18.37.3
$$R_{m,n}^{(\alpha)}(z) = \sum_{j=0}^{\min(m,n)} c_j z^{m-j} \overline{z}^{n-j},$$

where c_j are real or complex constants, with $c_0 \neq 0$;

18.37.4
$$\iint_{x^2+y^2<1} R_{m,n}^{(\alpha)}(x+iy)(x-iy)^{m-j}(x+iy)^{n-j}$$
$$\times (1-x^2-y^2)^\alpha\, dx\, dy = 0,$$
$$j = 1, 2, \ldots, \min(m,n);$$

18.37.5
$$R_{m,n}^{(\alpha)}(1) = 1.$$

Explicit Representation

18.37.6
$$R_{m,n}^{(\alpha)}(z) = \sum_{j=0}^{\min(m,n)} \frac{(-1)^j (\alpha+1)_{m+n-j}(-m)_j(-n)_j}{(\alpha+1)_m (\alpha+1)_n j!} \times z^{m-j} \bar{z}^{n-j}.$$

18.37(ii) OP's on the Triangle

Definition in Terms of Jacobi Polynomials

18.37.7
$$P_{m,n}^{\alpha,\beta,\gamma}(x,y) = P_{m-n}^{(\alpha,\beta+\gamma+2n+1)}(2x-1) \times x^n P_n^{(\beta,\gamma)}(2x^{-1}y - 1),$$
$$m \geq n \geq 0, \alpha, \beta, \gamma > -1.$$

Orthogonality

18.37.8
$$\iint_{0<y<x<1} P_{m,n}^{\alpha,\beta,\gamma}(x,y) P_{j,\ell}^{\alpha,\beta,\gamma}(x,y) \times (1-x)^\alpha (x-y)^\beta y^\gamma \, dx \, dy = 0,$$
$$m \neq j \text{ and/or } n \neq \ell.$$

See Dunkl and Xu (2001, §2.3.3) for analogs of (18.37.1) and (18.37.7) on a d-dimensional simplex.

18.37(iii) OP's Associated with Root Systems

Orthogonal polynomials associated with root systems are certain systems of trigonometric polynomials in several variables, symmetric under a certain finite group (Weyl group), and orthogonal on a torus. In one variable they are essentially ultraspherical, Jacobi, continuous q-ultraspherical, or Askey–Wilson polynomials. In several variables they occur, for $q=1$, as *Jack polynomials* and also as *Jacobi polynomials associated with root systems*; see Macdonald (1995, Chapter VI, §10), Stanley (1989), Kuznetsov and Sahi (2006, Part 1), Heckman (1991). For general q they occur as *Macdonald polynomials for root system A_n*, as *Macdonald polynomials for general root systems*, and as *Macdonald-Koornwinder polynomials*; see Macdonald (1995, Chapter VI), Macdonald (2000, 2003), Koornwinder (1992).

Applications

18.38 Mathematical Applications

18.38(i) Classical OP's: Numerical Analysis

Approximation Theory

The scaled Chebyshev polynomial $2^{1-n} T_n(x)$, $n \geq 1$, enjoys the "minimax" property on the interval $[-1,1]$, that is, $|2^{1-n} T_n(x)|$ has the least maximum value among all monic polynomials of degree n. In consequence, expansions of functions that are infinitely differentiable on $[-1,1]$ in series of Chebyshev polynomials usually converge extremely rapidly. For these results and applications in approximation theory see §3.11(ii) and Mason and Handscomb (2003, Chapter 3), Cheney (1982, p. 108), and Rivlin (1969, p. 31).

Quadrature

Classical OP's play a fundamental role in Gaussian quadrature. If the nodes in a quadrature formula with a positive weight function are chosen to be the zeros of the nth degree OP with the same weight function, and the interval of orthogonality is the same as the integration range, then the weights in the quadrature formula can be chosen in such a way that the formula is exact for all polynomials of degree not exceeding $2n-1$. See §3.5(v).

Differential Equations

Linear ordinary differential equations can be solved directly in series of Chebyshev polynomials (or other OP's) by a method originated by Clenshaw (1957). This process has been generalized to spectral methods for solving partial differential equations. For further information see Mason and Handscomb (2003, Chapters 10 and 11), Gottlieb and Orszag (1977, pp. 7–19), and Guo (1998, pp. 120–151).

18.38(ii) Classical OP's: Other Applications

Integrable Systems

The Toda equation provides an important model of a completely integrable system. It has elegant structures, including N-soliton solutions, Lax pairs, and Bäcklund transformations. While the Toda equation is an important model of nonlinear systems, the special functions of mathematical physics are usually regarded as solutions to linear equations. However, by using Hirota's technique of bilinear formalism of soliton theory, Nakamura (1996) shows that a wide class of exact solutions of the Toda equation can be expressed in terms of various special functions, and in particular classical OP's. For instance,

18.38.1 $V_n(x) = 2n \, H_{n+1}(x) \, H_{n-1}(x) / (H_n(x))^2$,

with $H_n(x)$ as in §18.3, satisfies the Toda equation

18.38.2
$$(d^2/dx^2) \ln V_n(x) = V_{n+1}(x) + V_{n-1}(x) - 2V_n(x), \quad n = 1, 2, \ldots.$$

Complex Function Theory

The Askey–Gasper inequality

18.38.3
$$\sum_{m=0}^n P_m^{(\alpha,0)}(x) \geq 0, \quad -1 \leq x \leq 1, \alpha > -1, n = 0, 1, \ldots,$$

was used in de Branges' proof of the long-standing Bieberbach conjecture concerning univalent functions on the unit disk in the complex plane. See de Branges (1985).

Zonal Spherical Harmonics

Ultraspherical polynomials are zonal spherical harmonics. As such they have many applications. See, for example, Andrews et al. (1999, Chapter 9). See also §14.30.

Random Matrix Theory

Hermite polynomials (and their Freud-weight analogs (§18.32)) play an important role in random matrix theory. See Fyodorov (2005) and Deift (1998, Chapter 5).

Riemann–Hilbert Problems

See Deift (1998, Chapter 7) and Ismail (2005, Chapter 22).

Radon Transform

See Deans (1983, Chapters 4, 7).

18.38(iii) Other OP's

Group Representations

For group-theoretic interpretations of OP's see Vilenkin and Klimyk (1991, 1992, 1993).

Coding Theory

For applications of Krawtchouk polynomials $K_n(x; p, N)$ and q-Racah polynomials $R_n(x; \alpha, \beta, \gamma, \delta \,|\, q)$ to coding theory see Bannai (1990, pp. 38–43), Leonard (1982), and Chihara (1987).

18.39 Physical Applications

18.39(i) Quantum Mechanics

Classical OP's appear when the time-dependent Schrödinger equation is solved by separation of variables. Consider, for example, the one-dimensional form of this equation for a particle of mass m with potential energy $V(x)$:

18.39.1 $\quad \left(\dfrac{-\hbar^2}{2m} \dfrac{\partial^2}{\partial x^2} + V(x) \right) \psi(x, t) = i\hbar \dfrac{\partial}{\partial t} \psi(x, t),$

where $\hbar$ is the reduced Planck's constant. On substituting $\psi(x, t) = \eta(x) \zeta(t)$, we obtain two ordinary differential equations, each of which involve the same constant E. The equation for $\eta(x)$ is

18.39.2 $\quad \dfrac{d^2 \eta}{dx^2} + \dfrac{2m}{\hbar^2} (E - V(x)) \eta = 0.$

For a harmonic oscillator, the potential energy is given by

18.39.3 $\quad V(x) = \tfrac{1}{2} m \omega^2 x^2,$

where ω is the angular frequency. For (18.39.2) to have a nontrivial bounded solution in the interval $-\infty < x < \infty$, the constant E (the total energy of the particle) must satisfy

18.39.4 $\quad E = E_n = \left(n + \tfrac{1}{2} \right) \hbar \omega, \quad n = 0, 1, 2, \ldots.$

The corresponding eigenfunctions are

18.39.5 $\quad \eta_n(x) = \pi^{-\frac{1}{4}} 2^{-\frac{1}{2} n} (n!\, b)^{-\frac{1}{2}} H_n(x/b) e^{-x^2/2b^2},$

where $b = (\hbar/m\omega)^{1/2}$, and H_n is the Hermite polynomial. For further details, see Seaborn (1991, p. 224) or Nikiforov and Uvarov (1988, pp. 71-72).

A second example is provided by the three-dimensional time-independent Schrödinger equation

18.39.6 $\quad \nabla^2 \psi + \dfrac{2m}{\hbar^2} (E - V(\mathbf{x})) \psi = 0,$

when this is solved by separation of variables in spherical coordinates (§1.5(ii)). The eigenfunctions of one of the separated ordinary differential equations are Legendre polynomials. See Seaborn (1991, pp. 69-75).

For a third example, one in which the eigenfunctions are Laguerre polynomials, see Seaborn (1991, pp. 87-93) and Nikiforov and Uvarov (1988, pp. 76-80 and 320-323).

18.39(ii) Other Applications

For applications of Legendre polynomials in fluid dynamics to study the flow around the outside of a puff of hot gas rising through the air, see Paterson (1983).

For applications and an extension of the Szegö–Szász inequality (18.14.20) for Legendre polynomials ($\alpha = \beta = 0$) to obtain global bounds on the variation of the phase of an elastic scattering amplitude, see Cornille and Martin (1972, 1974).

For physical applications of q-Laguerre polynomials see §17.17.

For interpretations of zeros of classical OP's as equilibrium positions of charges in electrostatic problems (assuming logarithmic interaction), see Ismail (2000a,b).

Computation

18.40 Methods of Computation

Orthogonal polynomials can be computed from their explicit polynomial form by Horner's scheme (§1.11(i)). Usually, however, other methods are more efficient, especially the numerical solution of difference equations (§3.6) and the application of uniform asymptotic expansions (when available) for OP's of large degree.

However, for applications in which the OP's appear only as terms in series expansions (compare §18.18(i)) the need to compute them can be avoided altogether by use instead of Clenshaw's algorithm (§3.11(ii)) and its straightforward generalization to OP's other than Chebyshev. For further information see Clenshaw (1955), Gautschi (2004, §§2.1, 8.1), and Mason and Handscomb (2003, §2.4).

18.41 Tables

18.41(i) Polynomials

For $P_n(x)$ ($=\mathsf{P}_n(x)$) see §14.33.

Abramowitz and Stegun (1964, Tables 22.4, 22.6, 22.11, and 22.13) tabulates $T_n(x)$, $U_n(x)$, $L_n(x)$, and $H_n(x)$ for $n = 0(1)12$. The ranges of x are $0.2(.2)1$ for $T_n(x)$ and $U_n(x)$, and $0.5, 1, 3, 5, 10$ for $L_n(x)$ and $H_n(x)$. The precision is 10D, except for $H_n(x)$ which is 6-11S.

18.41(ii) Zeros

For $P_n(x)$, $L_n(x)$, and $H_n(x)$ see §3.5(v). See also Abramowitz and Stegun (1964, Tables 25.4, 25.9, and 25.10).

18.41(iii) Other Tables

For tables prior to 1961 see Fletcher et al. (1962) and Lebedev and Fedorova (1960).

18.42 Software

See http://dlmf.nist.gov/18.42.

References

General References

The main references for writing this chapter are Andrews et al. (1999), Askey and Wilson (1985), Chihara (1978), Koekoek and Swarttouw (1998), and Szegö (1975).

Sources

The following list gives the references or other indications of proofs that were used in constructing the various sections of this chapter. These sources supplement the references that are quoted in the text.

§18.2 Andrews et al. (1999, Chapter 5), Szegö (1975, §§2.2(i), 3.2), Chihara (1978, Chapter 1, Theorem 3.3 and p. 21).

§18.3 In Table 18.3.1, for Row 2 see Szegö (1975, §2.4, Item 1, (4.3.3), and (4.21.6)): the entry in the last column follows from (18.5.7); for Row 3 see Szegö (1975, (4.7.1), (4.7.14), and (4.7.9)): the entry in the last column follows from the symmetry in the fourth row of Table 18.6.1; for Rows 4–7 see Andrews et al. (1999, §5.1 and Remark 2.5.3); for Row 10 specialize Row 2 to $\alpha = \beta = 0$; for Row 12 see Szegö (1975, §2.4, Item 2, (5.1.1), and (5.1.8)): the entry in the last column follows from (18.5.12); for Row 13 see Szegö (1975, §2.4, Item 3, (5.5.1), and (5.5.6)): the entry in the last column follows from the symmetry in the tenth row of Table 18.6.1.

§18.4 These graphics were produced at NIST.

§18.5 To verify (18.5.1)–(18.5.4) substitute them in the fourth through seventh rows, respectively, of Table 18.3.1. Alternatively, combine Szegö (1975, (4.1.7), (4.1.8)) with (18.7.5), (18.7.6). In Table 18.5.1, for Rows 2, 3, 9, 10 see Szegö (1975, (4.3.1), (4.7.12), (5.1.5), (5.5.3)); Rows 4–7 follow from Row 2 combined with (18.5.1)–(18.5.4) and Table 18.6.1, second row; Row 8 is the case $\alpha = \beta = 0$ of Row 2. For (18.5.6) see Truesdell (1948, §18, (5)). For (18.5.7) see Szegö (1975, (4.21.2)). For (18.5.8) see Szegö (1975, (4.3.2)). For (18.5.9) see Szegö (1975, (4.7.6)). For (18.5.10) see Szegö (1975, (4.7.31)). For (18.5.11) see Andrews et al. (1999, (6.4.11)). For (18.5.12) see Szegö (1975, (5.3.3)). For (18.5.13) see Szegö (1975, (5.5.4)). For (18.5.14)–(18.5.19) apply the recurrence relations (§18.9(i)), with initial values obtained from the values of k_n and $\tilde{k}_n/k_n$ given in Table 18.3.1 with $n = 0, 1$.

§18.6 For (18.6.1) see Szegö (1975, (5.1.7)). For Table 18.6.1, Rows 2, 4, 10, see Szegö (1975, (4.1.3), (4.1.1), (4.7.4), (2.3.3), (4.7.3), (5.5.5)); the entries in the fourth and fifth columns of Rows 3 and 4 of Table 18.6.1 follow from (18.5.9) combined (for Row 3) with (18.7.2); the other entries of Row 3 of Table 18.6.1 are the case $\alpha = \beta$ of Row 2; Rows 5–8 follow from (18.5.1)–(18.5.4); Row 9 is the case $\alpha = 0$ of Row 3. (18.6.2) follows from (18.6.3) by the second row of Table 18.6.1. (18.6.3) follows from (18.5.7) together with the second row of Table 18.6.1. (18.6.4) follows from (18.5.10) together with the fourth row of Table 18.6.1. (18.6.5) follows from (18.5.12) together with (18.6.1).

§18.7 For (18.7.1)–(18.7.6) and (18.7.13), (18.7.14) see Szegö (1975, (4.1.5), (4.1.7), (4.1.8), (4.7.1)). (18.7.7)–(18.7.12) follow from the definitions given by Table 18.3.1. (18.7.15) and (18.7.16) follow from (18.7.13) and (18.7.14), combined with (18.7.1). (18.7.17) follows from (18.7.4), (18.7.5), and (18.7.13). (18.7.18) follows from (18.7.3), (18.7.6), and (18.7.14). For (18.7.19) and (18.7.20) see Szegö (1975, (5.6.1)). For (18.7.21) see Szegö (1975, (5.3.4)). (18.7.22) follows from (18.7.21) and the symmetry in Row 2 of Table 18.6.1. (18.7.23) follows from (18.7.24) and (18.7.1). For (18.7.24) see Szegö (1975, (5.6.3)). For (18.7.25) see Szegö (1975, (4.7.8)). For (18.7.26) see Calogero (1978).

§18.8 For Table 18.8.1, Rows 2, 3, 5, 9–10, 11–12, see Szegö (1975, (4.2.4), (4.24.2), (4.7.5), (5.1.2), (5.5.2)), respectively; Row 4 is the special case $\alpha = \beta$ of Row 2; Rows 6, 7, 8 are the special cases $\alpha = \beta = -\frac{1}{2}$, $\alpha = \beta = \frac{1}{2}$, $\alpha = \beta = 0$, respectively, of Row 2.

§18.9 For (18.9.1), (18.9.2) see Szegö (1975, (4.5.1)). For Table 18.9.1, Rows 2, 9, 10, see Szegö (1975, (4.7.17), (5.1.10), (5.5.8)), respectively; Rows 3 and 4 are rewritings of elementary trigonometric identities in view of (18.5.1), (18.5.2); Row 7 is the special case $\alpha = \beta = 0$ of (18.9.2). For (18.9.3)–(18.9.5) see Rainville (1960, §138, (17), (16), (14)). For (18.9.6) see Szegö (1975, (4.5.4)). For (18.9.7) see Szegö (1975, (4.7.29)). For (18.9.8) substitute (18.7.15) or (18.7.16); the resulting formula is a special case of Rainville (1960, §138, (11)). (18.9.9)–(18.9.12) are rewritings of elementary trigonometric identities in view of (18.5.1)–(18.5.4). For (18.9.13), (18.9.14) see Szegö (1975, (5.1.13), (5.1.14)). For (18.9.15) see Szegö (1975, (4.21.7)). (18.9.16) is an immediate corollary of (18.5.5) and Table 18.5.1, Row 2. For (18.9.17) and (18.9.18) see Koornwinder (2006, §4). For (18.9.19) see Szegö (1975, (4.7.14)). (18.9.20) is an immediate corollary of (18.5.5) and Table 18.5.1, Row 3. (18.9.21) and (18.9.22) are rewritings of elementary trigonometric differentiation formulas. For (18.9.23) see Szegö (1975, (5.1.14)). (18.9.24) is an immediate corollary of (18.5.5) and Table 18.5.1, Row 9. For (18.9.25) see Szegö (1975, (5.5.10)). (18.9.26) is an immediate corollary of (18.5.5) and Table 18.5.1, Row 10.

§18.10 For (18.10.1) combine (14.12.1) and (14.3.21). For (18.10.2) see Szegö (1975, (4.8.6)). For (18.10.3) see Askey (1975, (4.20)). For (18.10.4) see Andrews *et al.* (1999, Theorem 6.7.4). For (18.10.5) see Szegö (1975, (4.8.10)). (18.10.6) can be obtained as a limit case of (18.10.3) in view of (18.7.21). (18.10.7) can be obtained as a limit case of (18.10.4) in view of (18.7.23). Table 18.10.1 follows from the corresponding Rodrigues formulas (§18.5(ii)) or generating functions (§18.12); see Szegö (1975, (4.4.6), (4.82.1), (4.8.16), (4.8.1), (5.4.8)). For (18.10.9), (18.10.10) see Andrews *et al.* (1999, (6.2.15), (6.1.4)).

§18.11 For (18.11.1) see Andrews *et al.* (1999, (9.6.7)). (18.11.2) is a rewriting of (18.5.12). For (18.11.3) see Temme (1990a, (3.1)). For (18.11.5) see Szegö (1975, Theorem 8.1.1). For (18.11.6) see Szegö (1975, Theorem 8.22.4). For (18.11.7) and (18.11.8) see Szegö (1975, Theorem 8.22.8).

§18.12 For (18.12.1) see Andrews *et al.* (1999, (6.4.3)). For (18.12.2) see Bateman (1905, pp. 113–114) and Koornwinder (1974, p. 128). For (18.12.3) see Andrews *et al.* (1999, (6.4.7)). For (18.12.4) see Szegö (1975, (4.7.23)). (18.12.5) follows by combining (18.12.4) and its z-differentiated form. For (18.12.6) see Rainville (1960, §144, (7)). For (18.12.7) see Andrews *et al.* (1999, (5.1.16)). (18.12.8) is an immediate consequence of (18.12.7). For (18.12.9) see Szegö (1975, (4.7.25)). (18.12.10) is the special case $\lambda = 1$ of (18.12.4) in view of (18.7.4). (18.12.11) is the special case $\lambda = \frac{1}{2}$ of (18.12.4). (18.12.12) is the special case $\lambda = \frac{1}{2}$ of (18.12.6). For (18.12.13) see Andrews *et al.* (1999, (6.2.4)). For (18.12.14) see Szegö (1975, (5.1.16)). For (18.12.15) see Andrews *et al.* (1999, (6.1.7)).

§18.13 Lorentzen and Waadeland (1992, pp. 446–448).

§18.14 For (18.14.1) and (18.14.2) see Szegö (1975, Theorem 7.32.1). For (18.14.3) see Chow *et al.* (1994). For (18.14.4), (18.14.5), and (18.14.6) see Szegö (1975, Theorem 7.33.1). For (18.14.7) see Lorch (1984). For (18.14.8) see Koornwinder (1977, Remark 4.1). For (18.14.9) see Szász (1951). For (18.14.10) see Szegö (1948). For (18.14.11) see Gasper (1972). For (18.14.12), (18.14.13) see Skovgaard (1954). For (18.14.14)–(18.14.19) see Szegö (1975, discussion following Theorem 7.32.1). For (18.14.20) see Szász (1950). For (18.14.21)–(18.14.24) see Szegö (1975, Theorem 7.6.1). For the last statement about the successive maxima of $|H_n(x)|$ see Szegö (1975, Theorem 7.6.3).

§18.15 Frenzen and Wong (1985, 1988), Szegö (1975, Theorems 8.21.11, 8.22.2, 8.22.6).

§18.16 Szegö (1975, Theorems 6.21.1, 6.21.2, 6.21.3, 6.3.2). For (18.16.6), (18.16.7) see Gatteschi (1987). For (18.16.8) see Frenzen and Wong (1985). For (18.16.10) and (18.16.11) see Szegö (1975, Theorem 6.31.3). For (18.16.12) and (18.16.13) see Ismail and Li (1992). For (18.16.14) see Tricomi (1949). For (18.16.15) see Szegö (1975, Theorem 6.32). See also Gatteschi (2002). For (18.16.16)–(18.16.18) see Szegö (1975, Theorem 6.32) and Sun (1996).

§18.17 For (18.17.1) see Erdélyi et al. (1953b, §10.8(38)). For the first equation in (18.17.2) apply the convolution property of the Laplace transform (§1.14(iii)) to (18.17.34) with $\alpha = 0$. For the second equation combine (18.9.23), (18.9.13), and (18.6.1). For (18.17.3) and (18.17.4) use (18.9.25) and (18.9.26). For (18.17.5) see Ismail (2005, (9.6.2)). (18.17.6) is the case $\alpha = 0$ of (18.17.5). For (18.17.7), (18.17.8) see Durand (1975). For (18.17.9) and (18.17.10) see Andrews et al. (1999, Theorem 6.7.2). For (18.17.11)–(18.17.15) see Askey and Fitch (1969). For (18.17.16) use (18.5.5), integrate by parts n times, expand $e^{-iy(1-x)}$ in a Maclaurin series, and integrate term by term. For (18.17.17) use (18.5.5), integrate repeatedly by parts, expand $\cos(xy)$ in a Maclaurin series, and integrate term by term; the proofs of (18.17.18) and (18.17.19) are similar. For (18.17.20) expand $\cos(xy)$ in a Maclaurin series, make the change of integration variable $1 - 2x^2 = t$, apply (18.5.5), integrate by parts n times, and use (10.8.3); the proof of (18.17.21) is similar. For (18.17.22) see Strichartz (1994, §7.6). For (18.17.23) use (18.12.16) and the fact that the Fourier transform of $e^{-\frac{1}{2}x^2}$ is $e^{-\frac{1}{2}y^2}$; the proofs of (18.17.24), (18.17.27), and (18.17.28) are similar, except that in the case of (18.17.24), (18.12.15) replaces (18.12.16). For (18.17.25) use (18.18.23); similarly for (18.17.26). For (18.17.29) take the inverse Fourier transform and apply (18.17.25). For (18.17.30) consider the Fourier transform of this function instead of the cosine transform, and replace $L_n^{(n-\frac{1}{2})}(\frac{1}{2}x^2)$ by its explicit form (18.5.12); then integrate term by term and rearrange the the consequential finite double sum into a single sum. For (18.17.31) use (18.5.5) and then integrate by parts; similarly for (18.17.32). For (18.17.33) expand the exponential in the integral as a power series in z and interchange integration and summation. The resulting integral can be evaluated by considering the term $\ell = 0$ in (18.18.14). (18.17.33) may also be verified by applying Kummer's transformation (13.2.39) to (18.17.16). (18.17.34) follows by substituting (18.5.12) into the integrand and performing termwise integration. (18.17.35) follows by use of (18.5.5) and integration by parts. For (18.17.36) use (18.5.7) and apply (16.4.3). For (18.17.37) use (18.5.5) and integrate by parts n times. For (18.17.38) use the first equality in (18.5.10), with $\lambda = \frac{1}{2}$ and n replaced by $2n$, integrate term by term, then apply (16.4.3); the proof of (18.17.39) is similar. For (18.17.40) use (18.5.12), integrate term by term, then apply (15.8.6). For (18.17.41) use (18.5.13) and integrate term by term. (18.17.42) and (18.17.43) follow from the case $\alpha = \beta = \pm\frac{1}{2}$ of Szegö (1975, Theorem 4.61.2), where the hypergeometric function on the right-hand side is rewritten as in Erdélyi et al. (1953b, 19.8(19)). For (18.17.44) see Tuck (1964). (18.17.46) is obtained from (18.17.9) for $\alpha = \beta = 0$, $\mu = \frac{1}{2}$, together with (18.7.5), (18.7.3), and the second row of Table 18.6.1. (18.17.45) is obtained from (18.17.46) by symmetry; compare Rows 5 and 9 of Table 18.6.1. (18.17.47) and (18.17.48) follow by use of (18.17.34) and (18.17.35). For (18.17.49) see Andrews et al. (1999, p. 328).

§18.18 For (18.18.2) see Szegö (1975, Theorem 9.1.2 and the Remarks on p. 248). For (18.18.3)–(18.18.7) see Lebedev (1965, pp. 68–71 and 88–89) and Nikiforov and Uvarov (1988, pp. 21 and 59). For (18.18.8) see Carlson (1971). (18.18.9) is the case $\alpha = 0$ of (18.18.8). (18.18.10) follows from (18.12.13). (18.18.11) follows from (18.12.15). (18.18.12) follows by computing $\int_0^\infty L_n^{(\alpha)}(\lambda x) L_\ell^{(\alpha)}(x) e^{-x} x^\alpha \, dx$ with use of the Rodrigues formula (Table 18.5.1), integration by parts, and (18.5.12). (18.18.13) follows from (18.18.12) for $\alpha = \pm\frac{1}{2}$ by (18.7.19), (18.7.20). For (18.18.14) see Andrews et al. (1999, Theorem 7.1.3). For (18.18.15) see Askey (1974). For (18.18.16) see Andrews et al. (1999, Theorem 7.1.4′). (18.18.17) follows from (18.18.16), (18.6.4), and the fourth row of Table 18.6.1. (18.18.18) follows from (18.12.13). (18.18.19) follows from (18.18.12) by dividing both sides by λ^n and letting $\lambda \to \infty$. (18.18.20) is the case $\beta = \pm\frac{1}{2}$ of (18.18.19) in view of (18.7.19), (18.7.20). For (18.18.21)–(18.18.23) see Andrews et al. (1999, (5.1.6), Theorems 6.8.2 and 6.8.1 and Remarks 6.8.2 and 6.8.1). For (18.18.25), (18.18.26) see Koornwinder (1974). For (18.18.27), (18.18.28) see Andrews et al. (1999, (6.2.25), (6.1.13)). For the positivity of the Poisson kernels see Askey (1975, p. 16). (18.18.29) follows from (18.12.4).

(18.18.30) follows from (18.12.4) and (18.12.5). (18.18.31) is the limiting case of (18.18.30) as $\lambda \to 0$. Each of the formulas (18.18.32)–(18.18.35) is equivalent to a difference formula together with a trivial $n = 0$ case, and each difference formula can be rewritten via (18.5.1), (18.5.2) as a well-known trigonometric identity. (18.18.36) is the special case $\lambda = \mu = \frac{1}{2}$ of (18.18.29). For (18.18.37) see Szegö (1975, (5.1.13)). (18.18.38) follows from (18.12.13), and is the special case $r = 2$ of (18.18.10). For (18.18.39) see Szegö (1975, (5.5.11)). (18.18.40) is the special case $\alpha = -\frac{1}{2}$ of (18.18.38) in view of (18.7.19).

§18.19 For Table 18.19.1 see Ismail (2005, (6.2.4), (6.2.35), (6.1.4), (6.1.21)). For Table 18.19.2, Rows 2, 4, see Ismail (2005, (6.2.7), (6.1.7)); Row 3 follows from (18.20.6); Row 5 follows from (18.20.8). For (18.19.1)–(18.19.4) see Askey (1985, (4), (5)). (18.19.5) follows from (18.20.9). For (18.19.6)–(18.19.9) see Ismail (2005, (5.9.8), (5.9.9)). The formula for k_n in (18.19.9) follows from (18.20.10).

§18.20 For (18.20.2) see Karlin and McGregor (1961, (1.8)). For Table 18.20.1, Rows 2, 3, see Ismail (2005, (6.2.42), (6.1.17)); for Row 4 see Chihara (1978, Chapter V, (3.2)). (18.20.3) follows by iteration of (18.22.28). (18.20.4) follows by iteration of (18.22.30). For (18.20.5)–(18.20.8) and (18.20.10) see Ismail (2005, (6.2.3), (6.2.34), (6.1.3), (6.1.20), (5.9.5)). For (18.20.9) see Askey (1985).

§18.21 For (18.21.3), (18.21.5), (18.21.7), (18.21.8) see Ismail (2005, §6.2, unnumbered formula after (6.2.34), also (6.2.17), (6.1.19), (6.1.18)). For (18.21.1) see Karlin and McGregor (1961, (1.19)). The three identities in (18.21.2) follow from (18.20.6), (18.20.7), (18.20.8). (18.21.4) follows from (18.20.5) and (18.20.7). (18.21.6) follows from (18.20.6) and (18.20.8). (18.21.9) follows from (18.22.2), Row 4 in Table 18.22.1, (18.9.1), and Row 10 in Table 18.9.1. (18.21.10) follows from (18.20.9) and (18.20.10). For (18.21.11) see Koornwinder (1989, (2.6)). (18.21.12) follows from (18.20.10) and (18.5.12). For Figure 18.21.1 see Askey and Wilson (1985, p. 46), together with correction in Askey (1985).

§18.22 For (18.22.1)–(18.22.3) see Ismail (2005, (6.2.8) and (6.2.9)). For Table 18.22.1 see Ismail (2005, (6.2.36), (6.1.5), and (6.1.25)). (18.22.4)–(18.22.6) is a limiting case of Andrews *et al.* (1999, (3.8.2)), in view of (18.26.6). For (18.22.7)–(18.22.8) see Ismail (2005, (5.9.1)). For (18.22.9)–(18.22.11) see Ismail (2005, (6.2.16)). For Table 18.22.2, Rows 2 and 3, see Ismail (2005, (6.2.38), (6.1.15)); Row 4 follows from Table 18.22.1, Row 4, and the third identity in (18.21.2). (18.22.13)–(18.22.15) is a limiting case of Koekoek and Swarttouw (1998, (1.1.6)) in view of (18.26.6). Koekoek and Swarttouw (1998, (1.1.6)) is a limiting case of Askey and Wilson (1985, (5.7)) in view of Koekoek and Swarttouw (1998, (5.1.1)). (18.22.16)–(18.22.17) follow from (18.22.14) in view of (18.21.10). For (18.22.19)–(18.22.25) see Ismail (2005, (6.2.5), (6.2.13), (6.2.39), (6.2.40), (6.1.13), §6.1, unnumbered formula following (6.1.15), also (6.1.23)). (18.22.26) follows from (18.22.24) and (18.21.7). (18.22.27) follows from (18.20.9). (18.22.28) follows from (18.22.14) and (18.22.27). (18.22.29) follows from (18.20.10). (18.22.30) follows from (18.22.28) in view of (18.21.10).

§18.23 For (18.23.3)–(18.23.5), (18.23.7) see Ismail (2005, (6.2.43), (6.1.8), (6.1.22), (5.9.3)). (18.23.1) and (18.23.2) follow by expanding the factors on the left as power series in z, and substituting (18.20.5) on the right. (18.23.6) is a limiting case of (18.26.18) via (18.26.6).

§18.25 For Table 18.25.1, Rows 2, 3, 4, see Wilson (1980); for Row 5 see Ismail (2005, (6.2.20)). (18.25.1) follows from (18.25.11). For (18.25.3)–(18.25.5) see Wilson (1980) and Andrews *et al.* (1999, (3.8.3)). For (18.25.6)–(18.25.8) see Wilson (1980). For (18.25.10)–(18.25.12) see Wilson (1980). For (18.25.13)–(18.25.15) see Ismail (2005, (6.2.20)). Table 18.25.2 follows from §18.26(i).

§18.26 For (18.26.1) see Andrews *et al.* (1999, Definition 3.8.1). For (18.26.2) and (18.26.3) see Wilson (1980). For (18.26.4) see Ismail (2005, (6.2.19)). For (18.26.5) and (18.26.7) see Wilson (1980). (18.26.6) follows from (18.26.1) and (18.20.9). (18.26.8) follows from (18.26.2) and (18.20.10). (18.26.9) follows from (18.26.3) and (18.26.4). (18.26.10) follows from (18.26.3) and (18.20.5). For (18.26.11) see Karlin and McGregor (1961, (1.21)). (18.26.12), (18.26.13) follow from (18.26.4) and (18.20.7). (18.26.14)–(18.26.17) follow from §18.26(i). For (18.26.18) see Ismail *et al.* (1990, (6.1)). (18.26.19) follows by expanding both factors on the left as power series in z, and substituting (18.26.2) on the right. (18.26.20) follows from (18.26.18), (18.26.1), (18.26.3). For (18.26.21) see Ismail (2005, (6.2.31)).

§18.27 For (18.27.3), (18.27.4) see Gasper and Rahman (2004, (7.2.21), (7.2.22)) and Ismail (2005,

(18.5.1), (18.5.2)). For (18.27.8)–(18.27.11) see Gasper and Rahman (2004, (7.3.10), (7.3.12)) and Ismail (2005, (18.4.7), (18.4.14)). For (18.27.14) see Gasper and Rahman (2004, (7.3.1), (7.3.3)). For (18.27.16), (18.27.17) see Ismail (2005, (21.8.2), (21.8.4)) and Moak (1981, Theorem 2). For (18.27.18)–(18.27.20) see Ismail (2005, (21.8.3), (21.8.46)). For (18.27.21)–(18.27.24) see Al-Salam and Carlitz (1965).

§**18.28** For (18.28.1)–(18.28.6) see Askey and Wilson (1985), Gasper and Rahman (2004, (7.5.2), (7.5.15), (7.5.21)), Ismail (2005, (15.2.4), (15.2.5)). For (18.28.7), (18.28.8) see Ismail (2005, (15.1.5), (15.1.6), (15.1.11)). For (18.28.9)–(18.28.12) see Askey and Ismail (1984, Chapter 3). For (18.28.13)–(18.28.15) see Gasper and Rahman (2004, (7.4.2), (7.4.14)–(7.4.16)) and Ismail (2005, (13.2.3)–(13.2.5), (13.2.11)). For (18.28.16)–(18.28.18) see Ismail (2005, (13.1.7), (13.1.11), (21.2.1), (21.2.5)). For (18.28.19), (18.28.20) see Gasper and Rahman (2004, (7.2.11)) and Ismail (2005, (15.6.1), (15.6.7)).

§**18.30** For (18.30.5) see Wimp (1987, Theorem 1). (18.30.7) is mentioned in Chihara (1978, Chapter VI, (12.6)), and proved in Barrucand and Dickinson (1968).

§**18.33** For (18.33.1), (18.33.4), (18.33.8), and (18.33.9) see Szegö (1975, (11.1.8), (11.4.6), (11.4.7), (11.5.2)). (18.33.6) follows from (18.33.4), (18.33.5). (18.33.10), (18.33.11) follow from (18.33.8), (18.33.9). For (18.33.13)–(18.33.16) see Askey (1982) and Pastro (1985).

§**18.34** Ismail (2005, Chapter 4).

§**18.35** Ismail (2005, Chapter 5).

§**18.37** Dunkl and Xu (2001, §2.4.3), Koornwinder (1975c).

Chapter 19

Elliptic Integrals

B. C. Carlson[1]

Notation   **486**
- 19.1 Special Notation 486

Legendre's Integrals   **486**
- 19.2 Definitions 486
- 19.3 Graphics 488
- 19.4 Derivatives and Differential Equations . . 490
- 19.5 Maclaurin and Related Expansions 490
- 19.6 Special Cases 491
- 19.7 Connection Formulas 491
- 19.8 Quadratic Transformations 492
- 19.9 Inequalities 494
- 19.10 Relations to Other Functions 494
- 19.11 Addition Theorems 495
- 19.12 Asymptotic Approximations 495
- 19.13 Integrals of Elliptic Integrals 496
- 19.14 Reduction of General Elliptic Integrals . . 496

Symmetric Integrals   **497**
- 19.15 Advantages of Symmetry 497
- 19.16 Definitions 497
- 19.17 Graphics 499
- 19.18 Derivatives and Differential Equations . . 500
- 19.19 Taylor and Related Series 501
- 19.20 Special Cases 502

- 19.21 Connection Formulas 503
- 19.22 Quadratic Transformations 504
- 19.23 Integral Representations 506
- 19.24 Inequalities 506
- 19.25 Relations to Other Functions 507
- 19.26 Addition Theorems 509
- 19.27 Asymptotic Approximations and Expansions 510
- 19.28 Integrals of Elliptic Integrals 511
- 19.29 Reduction of General Elliptic Integrals . . 512

Applications   **514**
- 19.30 Lengths of Plane Curves 514
- 19.31 Probability Distributions 515
- 19.32 Conformal Map onto a Rectangle 515
- 19.33 Triaxial Ellipsoids 515
- 19.34 Mutual Inductance of Coaxial Circles . . . 516
- 19.35 Other Applications 516

Computation   **517**
- 19.36 Methods of Computation 517
- 19.37 Tables 518
- 19.38 Approximations 519
- 19.39 Software 519

References   **519**

---

[1]Mathematics Department and Ames Laboratory (U.S. Department of Energy), Iowa State University, Ames, Iowa.
    Acknowledgments: The parts of this chapter that deal with Legendre's integrals are based in part on Abramowitz and Stegun (1964, Chapter 17) by L. M. Milne-Thomson. I am greatly indebted to R. C. Winther for indispensable technical support and to F. W. J. Olver for long-sustained encouragement of a new approach to elliptic integrals. I thank E. Neuman for improvements to §§19.9 and 19.24(i).
    Copyright © 2009 National Institute of Standards and Technology. All rights reserved.

Notation

19.1 Special Notation

(For other notation see pp. xiv and 873.)

l, m, n nonnegative integers.
ϕ real or complex argument (or amplitude).
k real or complex modulus.
k' complementary real or complex modulus, $k^2 + k'^2 = 1$.
α^2 real or complex parameter.
$\mathrm{B}(a,b)$ beta function (§5.12).

All square roots have their principal values. All derivatives are denoted by differentials, not by primes.

The first set of main functions treated in this chapter are Legendre's complete integrals

19.1.1 $$K(k), \quad E(k), \quad \Pi(\alpha^2, k),$$

of the first, second, and third kinds, respectively, and Legendre's incomplete integrals

19.1.2 $$F(\phi, k), \quad E(\phi, k), \quad \Pi(\phi, \alpha^2, k),$$

of the first, second, and third kinds, respectively. This notation follows Byrd and Friedman (1971, 110). We use also the function $D(\phi, k)$, introduced by Jahnke et al. (1966, p. 43). The functions (19.1.1) and (19.1.2) are used in Erdélyi et al. (1953b, Chapter 13), except that $\Pi(\alpha^2, k)$ and $\Pi(\phi, \alpha^2, k)$ are denoted by $\Pi_1(\nu, k)$ and $\Pi(\phi, \nu, k)$, respectively, where $\nu = -\alpha^2$.

In Abramowitz and Stegun (1964, Chapter 17) the functions (19.1.1) and (19.1.2) are denoted, in order, by $K(\alpha), E(\alpha), \Pi(n\backslash\alpha), F(\phi\backslash\alpha), E(\phi\backslash\alpha)$, and $\Pi(n;\phi\backslash\alpha)$, where $\alpha = \arcsin k$ and n is the α^2 (not related to k) in (19.1.1) and (19.1.2). Also, frequently in this reference α is replaced by m and $\backslash\alpha$ by $|m$, where $m = k^2$. However, it should be noted that in Chapter 8 of Abramowitz and Stegun (1964) the notation used for elliptic integrals differs from Chapter 17 and is consistent with that used in the present chapter and the rest of the NIST Handbook and DLMF.

The second set of main functions treated in this chapter is

19.1.3 $$\begin{array}{c} R_C(x,y), \quad R_F(x,y,z), \quad R_G(x,y,z), \\ R_J(x,y,z,p), \quad R_D(x,y,z), \\ R_{-a}(b_1, b_2, \ldots, b_n; z_1, z_2, \ldots, z_n). \end{array}$$

$R_F(x,y,z), R_G(x,y,z)$, and $R_J(x,y,z,p)$ are the symmetric (in x, y, and z) integrals of the first, second, and third kinds; they are complete if exactly one of x, y, and z is identically 0.

$R_{-a}(b_1, b_2, \ldots, b_n; z_1, z_2, \ldots, z_n)$ is a multivariate hypergeometric function that includes all the functions in (19.1.3).

A third set of functions, introduced by Bulirsch (1965a,b, 1969a), is

19.1.4 $$\begin{array}{c} \mathrm{el1}(x, k_c), \quad \mathrm{el2}(x, k_c, a, b), \\ \mathrm{el3}(x, k_c, p), \quad \mathrm{cel}(k_c, p, a, b). \end{array}$$

The first three functions are incomplete integrals of the first, second, and third kinds, and the cel function includes complete integrals of all three kinds.

Legendre's Integrals

19.2 Definitions

19.2(i) General Elliptic Integrals

Let $s^2(t)$ be a cubic or quartic polynomial in t with simple zeros, and let $r(s,t)$ be a rational function of s and t containing at least one odd power of s. Then

19.2.1 $$\int r(s,t)\, dt$$

is called an *elliptic integral*. Because s^2 is a polynomial, we have

19.2.2 $$r(s,t) = \frac{(p_1 + p_2 s)(p_3 - p_4 s)s}{(p_3 + p_4 s)(p_3 - p_4 s)s} = \frac{\rho}{s} + \sigma,$$

where p_j is a polynomial in t while ρ and σ are rational functions of t. Thus the *elliptic part* of (19.2.1) is

19.2.3 $$\int \frac{\rho(t)}{s(t)}\, dt.$$

19.2(ii) Legendre's Integrals

Assume $1 - \sin^2\phi \in \mathbb{C}\setminus(-\infty, 0]$ and $1 - k^2\sin^2\phi \in \mathbb{C}\setminus(-\infty, 0]$, except that one of them may be 0, and $1 - \alpha^2\sin^2\phi \in \mathbb{C}\setminus\{0\}$. Then

19.2.4
$$F(\phi, k) = \int_0^\phi \frac{d\theta}{\sqrt{1 - k^2\sin^2\theta}}$$
$$= \int_0^{\sin\phi} \frac{dt}{\sqrt{1-t^2}\sqrt{1-k^2 t^2}},$$

19.2.5
$$E(\phi, k) = \int_0^\phi \sqrt{1 - k^2\sin^2\theta}\, d\theta$$
$$= \int_0^{\sin\phi} \frac{\sqrt{1-k^2 t^2}}{\sqrt{1-t^2}}\, dt.$$

19.2.6
$$D(\phi, k) = \int_0^\phi \frac{\sin^2\theta\, d\theta}{\sqrt{1 - k^2\sin^2\theta}}$$
$$= \int_0^{\sin\phi} \frac{t^2\, dt}{\sqrt{1-t^2}\sqrt{1-k^2 t^2}}$$
$$= (F(\phi, k) - E(\phi, k))/k^2.$$

19.2.7
$$\Pi(\phi, \alpha^2, k) = \int_0^\phi \frac{d\theta}{\sqrt{1 - k^2 \sin^2 \theta}(1 - \alpha^2 \sin^2 \theta)}$$
$$= \int_0^{\sin \phi} \frac{dt}{\sqrt{1 - t^2}\sqrt{1 - k^2 t^2}(1 - \alpha^2 t^2)}.$$

The paths of integration are the line segments connecting the limits of integration. The integral for $E(\phi, k)$ is well defined if $k^2 = \sin^2 \phi = 1$, and the Cauchy principal value (§1.4(v)) of $\Pi(\phi, \alpha^2, k)$ is taken if $1 - \alpha^2 \sin^2 \phi$ vanishes at an interior point of the integration path. Also, if k^2 and α^2 are real, then $\Pi(\phi, \alpha^2, k)$ is called a *circular* or *hyperbolic case* according as $\alpha^2(\alpha^2 - k^2)(\alpha^2 - 1)$ is negative or positive. The circular and hyperbolic cases alternate in the four intervals of the real line separated by the points $\alpha^2 = 0, k^2, 1$.

The cases with $\phi = \pi/2$ are the *complete integrals*:

19.2.8
$$K(k) = F(\pi/2, k), \quad E(k) = E(\pi/2, k),$$
$$D(k) = D(\pi/2, k) = (K(k) - E(k))/k^2,$$
$$\Pi(\alpha^2, k) = \Pi(\pi/2, \alpha^2, k),$$

19.2.9 $\quad K'(k) = K(k'), \quad E'(k) = E(k'), \quad k' = \sqrt{1 - k^2}.$

If m is an integer, then

19.2.10
$$F(m\pi \pm \phi, k) = 2m\,K(k) \pm F(\phi, k),$$
$$E(m\pi \pm \phi, k) = 2m\,E(k) \pm E(\phi, k),$$
$$D(m\pi \pm \phi, k) = 2m\,D(k) \pm D(\phi, k).$$

19.2(iii) Bulirsch's Integrals

Bulirsch's integrals are linear combinations of Legendre's integrals that are chosen to facilitate computational application of Bartky's transformation (Bartky (1938)). Two are defined by

19.2.11
$$\mathrm{cel}(k_c, p, a, b) = \int_0^{\pi/2} \frac{a \cos^2 \theta + b \sin^2 \theta}{\cos^2 \theta + p \sin^2 \theta} \frac{d\theta}{\sqrt{\cos^2 \theta + k_c^2 \sin^2 \theta}},$$

19.2.12
$$\mathrm{el2}(x, k_c, a, b) = \int_0^{\arctan x} \frac{a + b \tan^2 \theta}{\sqrt{(1 + \tan^2 \theta)(1 + k_c^2 \tan^2 \theta)}} d\theta.$$

Here a, b, p are real parameters, and k_c and x are real or complex variables, with $p \neq 0$, $k_c \neq 0$. If $-\infty < p < 0$, then the integral in (19.2.11) is a Cauchy principal value.

With

19.2.13 $\quad k_c = k', \quad p = 1 - \alpha^2, \quad x = \tan \phi,$

special cases include

19.2.14
$$K(k) = \mathrm{cel}(k_c, 1, 1, 1),$$
$$E(k) = \mathrm{cel}(k_c, 1, 1, k_c^2), \quad D(k) = \mathrm{cel}(k_c, 1, 0, 1),$$
$$(E(k) - k'^2 K(k))/k^2 = \mathrm{cel}(k_c, 1, 1, 0),$$
$$\Pi(\alpha^2, k) = \mathrm{cel}(k_c, p, 1, 1),$$

and

19.2.15
$$F(\phi, k) = \mathrm{el1}(x, k_c) = \mathrm{el2}(x, k_c, 1, 1),$$
$$E(\phi, k) = \mathrm{el2}(x, k_c, 1, k_c^2),$$
$$D(\phi, k) = \mathrm{el2}(x, k_c, 0, 1).$$

The integrals are *complete* if $x = \infty$. If $1 < k \leq 1/\sin \phi$, then k_c is pure imaginary.

Lastly, corresponding to Legendre's incomplete integral of the third kind we have

19.2.16
$$\mathrm{el3}(x, k_c, p)$$
$$= \int_0^{\arctan x} \frac{d\theta}{(\cos^2 \theta + p \sin^2 \theta)\sqrt{\cos^2 \theta + k_c^2 \sin^2 \theta}}$$
$$= \Pi(\arctan x, 1 - p, k), \qquad x^2 \neq -1/p.$$

19.2(iv) $R_C(x, y)$

Let $x \in \mathbb{C} \setminus (-\infty, 0)$ and $y \in \mathbb{C} \setminus \{0\}$. We define

19.2.17 $\quad R_C(x, y) = \frac{1}{2} \int_0^\infty \frac{dt}{\sqrt{t + x}(t + y)},$

where the Cauchy principal value is taken if $y < 0$. Formulas involving $\Pi(\phi, \alpha^2, k)$ that are customarily different for circular cases, ordinary hyperbolic cases, and (hyperbolic) Cauchy principal values, are united in a single formula by using $R_C(x, y)$.

In (19.2.18)–(19.2.22) the inverse trigonometric and hyperbolic functions assume their principal values (§§4.23(ii) and 4.37(ii)). When x and y are positive, $R_C(x, y)$ is an inverse circular function if $x < y$ and an inverse hyperbolic function (or logarithm) if $x > y$:

19.2.18
$$R_C(x, y) = \frac{1}{\sqrt{y - x}} \arctan \sqrt{\frac{y - x}{x}}$$
$$= \frac{1}{\sqrt{y - x}} \arccos \sqrt{x/y}, \quad 0 \leq x < y,$$

19.2.19
$$R_C(x, y) = \frac{1}{\sqrt{x - y}} \mathrm{arctanh} \sqrt{\frac{x - y}{x}}$$
$$= \frac{1}{\sqrt{x - y}} \ln \frac{\sqrt{x} + \sqrt{x - y}}{\sqrt{y}}, \quad 0 < y < x.$$

The Cauchy principal value is hyperbolic:

19.2.20
$$R_C(x, y) = \sqrt{\frac{x}{x - y}} R_C(x - y, -y)$$
$$= \frac{1}{\sqrt{x - y}} \mathrm{arctanh} \sqrt{\frac{x}{x - y}}$$
$$= \frac{1}{\sqrt{x - y}} \ln \frac{\sqrt{x} + \sqrt{x - y}}{\sqrt{-y}}, \quad y < 0 \leq x.$$

If the line segment with endpoints x and y lies in $\mathbb{C}\setminus(-\infty, 0]$, then

19.2.21 $\quad R_C(x, y) = \int_0^1 (v^2 x + (1-v^2)y)^{-1/2}\, dv,$

19.2.22
$$R_C(x, y) = \frac{2}{\pi} \int_0^{\pi/2} R_C\bigl(y, x\cos^2\theta + y\sin^2\theta\bigr)\, d\theta.$$

19.3 Graphics

19.3(i) Real Variables

See Figures 19.3.1–19.3.6 for complete and incomplete Legendre's elliptic integrals.

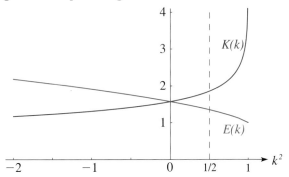

Figure 19.3.1: $K(k)$ and $E(k)$ as functions of k^2 for $-2 \le k^2 \le 1$. Graphs of $K'(k)$ and $E'(k)$ are the mirror images in the vertical line $k^2 = \frac{1}{2}$.

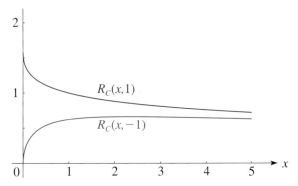

Figure 19.3.2: $R_C(x, 1)$ and the Cauchy principal value of $R_C(x, -1)$ for $0 \le x \le 5$. Both functions are asymptotic to $\ln(4x)/\sqrt{4x}$ as $x \to \infty$; see (19.2.19) and (19.2.20). Note that $R_C(x, \pm y) = y^{-1/2} R_C(x/y, \pm 1)$, $y > 0$.

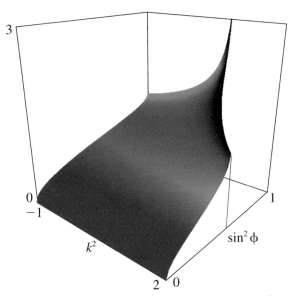

Figure 19.3.3: $F(\phi, k)$ as a function of k^2 and $\sin^2 \phi$ for $-1 \le k^2 \le 2$, $0 \le \sin^2 \phi \le 1$. If $\sin^2 \phi = 1$ ($\ge k^2$), then the function reduces to $K(k)$, becoming infinite when $k^2 = 1$. If $\sin^2 \phi = 1/k^2$ (< 1), then it has the value $K(1/k)/k$: put $c = k^2$ in (19.25.5) and use (19.25.1).

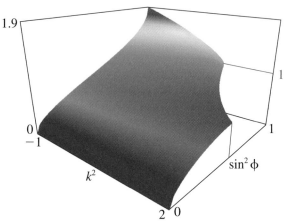

Figure 19.3.4: $E(\phi, k)$ as a function of k^2 and $\sin^2 \phi$ for $-1 \le k^2 \le 2$, $0 \le \sin^2 \phi \le 1$. If $\sin^2 \phi = 1$ ($\ge k^2$), then the function reduces to $E(k)$, with value 1 at $k^2 = 1$. If $\sin^2 \phi = 1/k^2$ (< 1), then it has the value $k\,E(1/k) + (k'^2/k)\,K(1/k)$, with limit 1 as $k^2 \to 1+$: put $c = k^2$ in (19.25.7) and use (19.25.1).

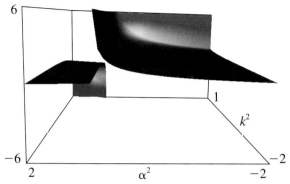

Figure 19.3.5: $\Pi(\alpha^2, k)$ as a function of k^2 and α^2 for $-2 \leq k^2 < 1$, $-2 \leq \alpha^2 \leq 2$. Cauchy principal values are shown when $\alpha^2 > 1$. The function is unbounded as $\alpha^2 \to 1-$, and also (with the same sign as $1-\alpha^2$) as $k^2 \to 1-$. As $\alpha^2 \to 1+$ it has the limit $K(k)-(E(k)/k'^2)$. If $\alpha^2 = 0$, then it reduces to $K(k)$. If $k^2 = 0$, then it has the value $\tfrac{1}{2}\pi/\sqrt{1-\alpha^2}$ when $\alpha^2 < 1$, and 0 when $\alpha^2 > 1$. See §19.6(i).

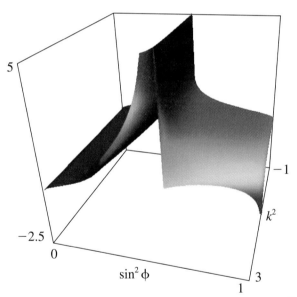

Figure 19.3.6: $\Pi(\phi, 2, k)$ as a function of k^2 and $\sin^2\phi$ for $-1 \leq k^2 \leq 3$, $0 \leq \sin^2\phi < 1$. Cauchy principal values are shown when $\sin^2\phi > \tfrac{1}{2}$. The function tends to $+\infty$ as $\sin^2\phi \to \tfrac{1}{2}$, except in the last case below. If $\sin^2\phi = 1$ $(> k^2)$, then the function reduces to $\Pi(2, k)$ with Cauchy principal value $K(k) - \Pi(\tfrac{1}{2}k^2, k)$, which tends to $-\infty$ as $k^2 \to 1-$. See (19.6.5) and (19.6.6). If $\sin^2\phi = 1/k^2$ (< 1), then by (19.7.4) it reduces to $\Pi(2/k^2, 1/k)/k$, $k^2 \neq 2$, with Cauchy principal value $(K(1/k) - \Pi(\tfrac{1}{2}, 1/k))/k$, $1 < k^2 < 2$, by (19.6.5). Its value tends to $-\infty$ as $k^2 \to 1+$ by (19.6.6), and to the negative of the second lemniscate constant (see (19.20.22)) as $k^2 (= \csc^2\phi) \to 2-$.

19.3(ii) Complex Variables

In Figures 19.3.7 and 19.3.8 for complete Legendre's elliptic integrals with complex arguments, height corresponds to the absolute value of the function and color to the phase. See also p. xiv.

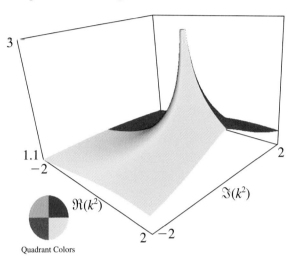

Figure 19.3.7: $K(k)$ as a function of complex k^2 for $-2 \leq \Re(k^2) \leq 2$, $-2 \leq \Im(k^2) \leq 2$. There is a branch cut where $1 < k^2 < \infty$.

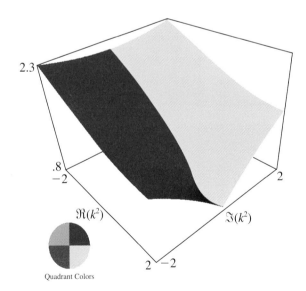

Figure 19.3.8: $E(k)$ as a function of complex k^2 for $-2 \leq \Re(k^2) \leq 2$, $-2 \leq \Im(k^2) \leq 2$. There is a branch cut where $1 < k^2 < \infty$.

For further graphics, see http://dlmf.nist.gov/19.3.ii.

19.4 Derivatives and Differential Equations

19.4(i) Derivatives

19.4.1
$$\frac{dK(k)}{dk} = \frac{E(k) - k'^2 K(k)}{kk'^2},$$
$$\frac{d(E(k) - k'^2 K(k))}{dk} = k K(k),$$

19.4.2
$$\frac{dE(k)}{dk} = \frac{E(k) - K(k)}{k}, \quad \frac{d(E(k) - K(k))}{dk} = -\frac{k E(k)}{k'^2},$$

19.4.3
$$\frac{d^2 E(k)}{dk^2} = -\frac{1}{k}\frac{dK(k)}{dk} = \frac{k'^2 K(k) - E(k)}{k^2 k'^2},$$

19.4.4
$$\frac{\partial \Pi(\alpha^2, k)}{\partial k} = \frac{k}{k'^2(k^2 - \alpha^2)}(E(k) - k'^2 \Pi(\alpha^2, k)).$$

19.4.5
$$\frac{\partial F(\phi, k)}{\partial k} = \frac{E(\phi, k) - k'^2 F(\phi, k)}{kk'^2} - \frac{k \sin\phi \cos\phi}{k'^2 \sqrt{1 - k^2 \sin^2\phi}},$$

19.4.6
$$\frac{\partial E(\phi, k)}{\partial k} = \frac{E(\phi, k) - F(\phi, k)}{k},$$

19.4.7
$$\frac{\partial \Pi(\phi, \alpha^2, k)}{\partial k} = \frac{k}{k'^2(k^2 - \alpha^2)} \left(E(\phi, k) - k'^2 \Pi(\phi, \alpha^2, k) - \frac{k^2 \sin\phi \cos\phi}{\sqrt{1 - k^2 \sin^2\phi}} \right).$$

19.4(ii) Differential Equations

Let $D_k = \partial/\partial k$. Then

19.4.8
$$(kk'^2 D_k^2 + (1 - 3k^2)D_k - k) F(\phi, k) = \frac{-k \sin\phi \cos\phi}{(1 - k^2 \sin^2\phi)^{3/2}},$$

19.4.9
$$(kk'^2 D_k^2 + k'^2 D_k + k) E(\phi, k) = \frac{k \sin\phi \cos\phi}{\sqrt{1 - k^2 \sin^2\phi}}.$$

If $\phi = \pi/2$, then these two equations become hypergeometric differential equations (15.10.1) for $K(k)$ and $E(k)$. An analogous differential equation of third order for $\Pi(\phi, \alpha^2, k)$ is given in Byrd and Friedman (1971, 118.03).

19.5 Maclaurin and Related Expansions

If $|k| < 1$ and $|\alpha| < 1$, then

19.5.1
$$K(k) = \frac{\pi}{2} \sum_{m=0}^{\infty} \frac{\left(\frac{1}{2}\right)_m \left(\frac{1}{2}\right)_m}{m!\, m!} k^{2m} = \frac{\pi}{2} {}_2F_1\left(\tfrac{1}{2}, \tfrac{1}{2}; 1; k^2\right),$$

where ${}_2F_1$ is the Gauss hypergeometric function (§§15.1 and 15.2(i)).

19.5.2
$$E(k) = \frac{\pi}{2} \sum_{m=0}^{\infty} \frac{\left(-\frac{1}{2}\right)_m \left(\frac{1}{2}\right)_m}{m!\, m!} k^{2m} = \frac{\pi}{2} {}_2F_1\left(-\tfrac{1}{2}, \tfrac{1}{2}; 1; k^2\right),$$

19.5.3
$$D(k) = \frac{\pi}{4} \sum_{m=0}^{\infty} \frac{\left(\frac{3}{2}\right)_m \left(\frac{1}{2}\right)_m}{(m+1)!\, m!} k^{2m} = \frac{\pi}{4} {}_2F_1\left(\tfrac{3}{2}, \tfrac{1}{2}; 2; k^2\right),$$

19.5.4
$$\Pi(\alpha^2, k) = \frac{\pi}{2} \sum_{n=0}^{\infty} \frac{\left(\frac{1}{2}\right)_n}{n!} \sum_{m=0}^{n} \frac{\left(\frac{1}{2}\right)_m}{m!} k^{2m} \alpha^{2n-2m}$$
$$= \frac{\pi}{2} F_1\left(\tfrac{1}{2}; \tfrac{1}{2}, 1; 1; k^2, \alpha^2\right),$$

where $F_1(\alpha; \beta, \beta'; \gamma; x, y)$ is an Appell function (§16.13).

For *Jacobi's nome q*:

19.5.5
$$q = \exp(-\pi K'(k)/K(k)) = r + 8r^2 + 84r^3 + 992r^4 + \cdots,$$
$$r = \tfrac{1}{16}k^2, \quad 0 \leq k \leq 1.$$

Also,

19.5.6
$$q = \lambda + 2\lambda^5 + 15\lambda^9 + 150\lambda^{13} + 1707\lambda^{17} + \cdots, \quad 0 \leq k \leq 1,$$

where

19.5.7
$$\lambda = (1 - \sqrt{k'})/(2(1 + \sqrt{k'})).$$

Coefficients of terms up to λ^{49} are given in Lee (1990), along with tables of fractional errors in $K(k)$ and $E(k)$, $0.1 \leq k^2 \leq 0.9999$, obtained by using 12 different truncations of (19.5.6) in (19.5.8) and (19.5.9).

19.5.8
$$K(k) = \frac{\pi}{2}\left(1 + 2\sum_{n=1}^{\infty} q^{n^2}\right)^2, \quad |q| < 1,$$

19.5.9
$$E(k) = K(k) + \frac{2\pi^2}{K(k)} \frac{\sum_{n=1}^{\infty}(-1)^n n^2 q^{n^2}}{1 + 2\sum_{n=1}^{\infty}(-1)^n q^{n^2}}, \quad |q| < 1.$$

An infinite series for $\ln K(k)$ is equivalent to the infinite product

19.5.10
$$K(k) = \frac{\pi}{2} \prod_{m=1}^{\infty}(1 + k_m),$$

where $k_0 = k$ and

19.5.11
$$k_{m+1} = \frac{1 - \sqrt{1 - k_m^2}}{1 + \sqrt{1 - k_m^2}}, \quad m = 0, 1, \ldots.$$

Series expansions of $F(\phi, k)$ and $E(\phi, k)$ are surveyed and improved in Van de Vel (1969), and the case of $F(\phi, k)$ is summarized in Gautschi (1975, §1.3.2). For series expansions of $\Pi(\phi, \alpha^2, k)$ when $|\alpha^2| < 1$ see Erdélyi et al. (1953b, §13.6(9)). See also Karp et al. (2007).

19.6 Special Cases

19.6(i) Complete Elliptic Integrals

19.6.1
$$K(0) = E(0) = K'(1) = E'(1) = \tfrac{1}{2}\pi,$$
$$K(1) = K'(0) = \infty, \quad E(1) = E'(0) = 1.$$

19.6.2
$$\Pi(k^2, k) = E(k)/k'^2, \qquad k^2 < 1,$$
$$\Pi(-k, k) = \tfrac{1}{4}\pi(1+k)^{-1} + \tfrac{1}{2}K(k), \quad 0 \le k^2 < 1.$$

19.6.3 $\quad \Pi(\alpha^2, 0) = \pi/(2\sqrt{1-\alpha^2}), \quad \Pi(0, k) = K(k),$
$$-\infty < \alpha^2 < 1.$$

19.6.4
$$\Pi(\alpha^2, k) \to +\infty, \qquad \alpha^2 \to 1-,$$
$$\Pi(\alpha^2, k) \to \infty \operatorname{sign}(1-\alpha^2), \qquad k^2 \to 1-.$$

If $1 < \alpha^2 < \infty$, then the Cauchy principal value satisfies

19.6.5 $\quad \Pi(\alpha^2, k) = K(k) - \Pi(k^2/\alpha^2, k),$

and

$$\Pi(\alpha^2, 0) = 0,$$

19.6.6 $\quad \Pi(\alpha^2, k) \to K(k) - \left(E(k)/k'^2\right), \qquad \alpha^2 \to 1+,$
$$\Pi(\alpha^2, k) \to -\infty, \qquad k^2 \to 1-.$$

Exact values of $K(k)$ and $E(k)$ for various special values of k are given in Byrd and Friedman (1971, 111.10 and 111.11) and Cooper et al. (2006).

19.6(ii) $F(\phi, k)$

19.6.7
$$F(0, k) = 0, \quad F(\phi, 0) = \phi, \quad F\left(\tfrac{1}{2}\pi, 1\right) = \infty,$$
$$F\left(\tfrac{1}{2}\pi, k\right) = K(k), \quad \lim_{\phi \to 0} F(\phi, k)/\phi = 1.$$

19.6.8 $\quad F(\phi, 1) = (\sin \phi)\, R_C\left(1, \cos^2 \phi\right) = \operatorname{gd}^{-1}(\phi).$

For the inverse Gudermannian function $\operatorname{gd}^{-1}(\phi)$ see §4.23(viii). Compare also (19.10.2).

19.6(iii) $E(\phi, k)$

19.6.9
$$E(0, k) = 0, \quad E(\phi, 0) = \phi, \quad E\left(\tfrac{1}{2}\pi, 1\right) = 1,$$
$$E(\phi, 1) = \sin \phi, \quad E\left(\tfrac{1}{2}\pi, k\right) = E(k).$$

19.6.10 $\quad \lim_{\phi \to 0} E(\phi, k)/\phi = 1.$

19.6(iv) $\Pi(\phi, \alpha^2, k)$

Circular and hyperbolic cases, including Cauchy principal values, are unified by using $R_C(x, y)$. Let $c = \csc^2 \phi \ne \alpha^2$ and $\Delta = \sqrt{1 - k^2 \sin^2 \phi}$. Then

19.6.11
$$\Pi(0, \alpha^2, k) = 0, \quad \Pi(\phi, 0, 0) = \phi, \quad \Pi(\phi, 1, 0) = \tan \phi.$$

19.6.12
$$\Pi(\phi, \alpha^2, 0) = R_C(c-1, c-\alpha^2),$$
$$\Pi(\phi, \alpha^2, 1) = \frac{1}{1-\alpha^2}\left(R_C(c, c-1) - \alpha^2 R_C(c, c-\alpha^2)\right),$$
$$\Pi(\phi, 1, 1) = \tfrac{1}{2}(R_C(c, c-1) + \sqrt{c}(c-1)^{-1}).$$

19.6.13
$$\Pi(\phi, 0, k) = F(\phi, k),$$
$$\Pi(\phi, k^2, k) = \frac{1}{k'^2}\left(E(\phi, k) - \frac{k^2}{\Delta}\sin\phi\cos\phi\right),$$
$$\Pi(\phi, 1, k) = F(\phi, k) - \frac{1}{k'^2}(E(\phi, k) - \Delta \tan \phi).$$

19.6.14
$$\Pi\left(\tfrac{1}{2}\pi, \alpha^2, k\right) = \Pi(\alpha^2, k), \quad \lim_{\phi \to 0} \Pi(\phi, \alpha^2, k)/\phi = 1.$$

For the Cauchy principal value of $\Pi(\phi, \alpha^2, k)$ when $\alpha^2 > c$, see §19.7(iii).

19.6(v) $R_C(x, y)$

19.6.15
$$R_C(x, x) = x^{-1/2}, \quad R_C(\lambda x, \lambda y) = \lambda^{-1/2} R_C(x, y),$$
$$R_C(x, y) \to +\infty, \qquad y \to 0+ \text{ or } y \to 0-, x > 0,$$
$$R_C(0, y) = \tfrac{1}{2}\pi y^{-1/2}, \qquad |\operatorname{ph} y| < \pi,$$
$$R_C(0, y) = 0, \qquad y < 0.$$

19.7 Connection Formulas

19.7(i) Complete Integrals of the First and Second Kinds

Legendre's Relation

19.7.1 $\quad E(k)\, K'(k) + E'(k)\, K(k) - K(k)\, K'(k) = \tfrac{1}{2}\pi.$

Also,

19.7.2
$$K(ik/k') = k'\, K(k), \quad K(k'/ik) = k\, K(k'),$$
$$E(ik/k') = (1/k')\, E(k), \quad E(k'/ik) = (1/k)\, E(k').$$

19.7.3
$$K(1/k) = k(K(k) \mp i\, K(k')),$$
$$K(1/k') = k'(K(k') \pm i\, K(k)),$$
$$E(1/k) = (1/k)\left(E(k) \pm i\, E(k') - k'^2\, K(k) \mp i k^2\, K(k')\right),$$
$$E(1/k') = (1/k')\left(E(k') \mp i\, E(k) - k^2\, K(k')\right.$$
$$\left. \pm i k'^2\, K(k)\right),$$

where upper signs apply if $\Im k^2 > 0$ and lower signs if $\Im k^2 < 0$. This dichotomy of signs (missing in several references) is due to Fettis (1970).

19.7(ii) Change of Modulus and Amplitude

Reciprocal-Modulus Transformation

19.7.4
$$F(\phi, k_1) = k F(\beta, k),$$
$$E(\phi, k_1) = (E(\beta, k) - k'^2 F(\beta, k))/k,$$
$$\Pi(\phi, \alpha^2, k_1) = k \Pi(\beta, k^2 \alpha^2, k),$$
$$k_1 = 1/k, \sin \beta = k_1 \sin \phi \leq 1.$$

Imaginary-Modulus Transformation

19.7.5
$$F(\phi, ik) = \kappa' F(\theta, \kappa),$$
$$E(\phi, ik) = (1/\kappa') \Big(E(\theta, \kappa) - \kappa^2 (\sin \theta \cos \theta)$$
$$\times (1 - \kappa^2 \sin^2 \theta)^{-1/2} \Big),$$
$$\Pi(\phi, \alpha^2, ik) = (\kappa'/\alpha_1^2) \Big(\kappa^2 F(\theta, \kappa) + \kappa'^2 \alpha^2 \Pi(\theta, \alpha_1^2, \kappa) \Big),$$

where

19.7.6
$$\kappa = \frac{k}{\sqrt{1+k^2}}, \quad \kappa' = \frac{1}{\sqrt{1+k^2}},$$
$$\sin \theta = \frac{\sqrt{1+k^2} \sin \phi}{\sqrt{1+k^2 \sin^2 \phi}}, \quad \alpha_1^2 = \frac{\alpha^2 + k^2}{1+k^2}.$$

Imaginary-Argument Transformation

With $\sinh \phi = \tan \psi$,

19.7.7
$$F(i\phi, k) = i F(\psi, k'),$$
$$E(i\phi, k) = i \Big(F(\psi, k') - E(\psi, k')$$
$$+ (\tan \psi)\sqrt{1 - k'^2 \sin^2 \psi} \Big),$$
$$\Pi(i\phi, \alpha^2, k) = i \big(F(\psi, k')$$
$$- \alpha^2 \Pi(\psi, 1 - \alpha^2, k') \big) / (1 - \alpha^2).$$

For two further transformations of this type see Erdélyi et al. (1953b, p. 316).

19.7(iii) Change of Parameter of $\Pi(\phi, \alpha^2, k)$

There are three relations connecting $\Pi(\phi, \alpha^2, k)$ and $\Pi(\phi, \omega^2, k)$, where ω^2 is a rational function of α^2. If k^2 and α^2 are real, then both integrals are circular cases or both are hyperbolic cases (see §19.2(ii)).

The first of the three relations maps each circular region onto itself and each hyperbolic region onto the other; in particular, it gives the Cauchy principal value of $\Pi(\phi, \alpha^2, k)$ when $\alpha^2 > \csc^2 \phi$ (see (19.6.5) for the complete case). Let $c = \csc^2 \phi \neq \alpha^2$. Then

19.7.8
$$\Pi(\phi, \alpha^2, k) + \Pi(\phi, \omega^2, k)$$
$$= F(\phi, k) + \sqrt{c}\, R_C\big((c-1)(c-k^2), (c-\alpha^2)(c-\omega^2)\big),$$
$$\alpha^2 \omega^2 = k^2.$$

Since $k^2 \leq c$ we have $\alpha^2 \omega^2 \leq c$; hence $\alpha^2 > c$ implies $\omega^2 < 1 \leq c$.

The second relation maps each hyperbolic region onto itself and each circular region onto the other:

19.7.9
$$(k^2 - \alpha^2) \Pi(\phi, \alpha^2, k) + (k^2 - \omega^2) \Pi(\phi, \omega^2, k)$$
$$= k^2 F(\phi, k)$$
$$- \alpha^2 \omega^2 \sqrt{c-1}\, R_C\big(c(c-k^2), (c-\alpha^2)(c-\omega^2)\big),$$
$$(1 - \alpha^2)(1 - \omega^2) = 1 - k^2.$$

The third relation (missing from the literature of Legendre's integrals) maps each circular region onto the other and each hyperbolic region onto the other:

19.7.10
$$(1 - \alpha^2) \Pi(\phi, \alpha^2, k) + (1 - \omega^2) \Pi(\phi, \omega^2, k)$$
$$= F(\phi, k) + (1 - \alpha^2 - \omega^2)\sqrt{c - k^2}$$
$$\times R_C\big(c(c-1), (c-\alpha^2)(c-\omega^2)\big),$$
$$(k^2 - \alpha^2)(k^2 - \omega^2) = k^2(k^2 - 1).$$

19.8 Quadratic Transformations

19.8(i) Gauss's Arithmetic-Geometric Mean (AGM)

When a_0 and g_0 are positive numbers, define

19.8.1
$$a_{n+1} = \frac{a_n + g_n}{2}, \quad g_{n+1} = \sqrt{a_n g_n}, \quad n = 0, 1, 2, \ldots.$$

As $n \to \infty$, a_n and g_n converge to a common limit $M(a_0, g_0)$ called the AGM (*Arithmetic-Geometric Mean*) of a_0 and g_0. By symmetry in a_0 and g_0 we may assume $a_0 \geq g_0$ and define

19.8.2
$$c_n = \sqrt{a_n^2 - g_n^2}.$$

Then

19.8.3
$$c_{n+1} = \frac{a_n - g_n}{2} = \frac{c_n^2}{4 a_{n+1}},$$

showing that the convergence of c_n to 0 and of a_n and g_n to $M(a_0, g_0)$ is quadratic in each case.

The AGM has the integral representations

19.8.4
$$\frac{1}{M(a_0, g_0)} = \frac{2}{\pi} \int_0^{\pi/2} \frac{d\theta}{\sqrt{a_0^2 \cos^2 \theta + g_0^2 \sin^2 \theta}}$$
$$= \frac{1}{\pi} \int_0^\infty \frac{dt}{\sqrt{t(t + a_0^2)(t + g_0^2)}}.$$

The first of these shows that

19.8.5
$$K(k) = \frac{\pi}{2 M(1, k')}, \quad -\infty < k^2 < 1.$$

19.8 Quadratic Transformations

The AGM appears in

$$E(k) = \frac{\pi}{2M(1,k')}\left(a_0^2 - \sum_{n=0}^{\infty} 2^{n-1} c_n^2\right)$$

19.8.6
$$= K(k)\left(a_1^2 - \sum_{n=2}^{\infty} 2^{n-1} c_n^2\right),$$
$$-\infty < k^2 < 1,\ a_0 = 1,\ g_0 = k',$$

and in

19.8.7
$$\Pi(\alpha^2, k) = \frac{\pi}{4M(1,k')}\left(2 + \frac{\alpha^2}{1-\alpha^2}\sum_{n=0}^{\infty} Q_n\right),$$
$$-\infty < k^2 < 1,\ -\infty < \alpha^2 < 1,$$

where $a_0 = 1,\ g_0 = k',\ p_0^2 = 1 - \alpha^2,\ Q_0 = 1$, and

19.8.8
$$p_{n+1} = \frac{p_n^2 + a_n g_n}{2 p_n},\quad \varepsilon_n = \frac{p_n^2 - a_n g_n}{p_n^2 + a_n g_n},$$
$$Q_{n+1} = \tfrac{1}{2} Q_n \varepsilon_n,\qquad n = 0, 1, \ldots.$$

Again, p_n and ε_n converge quadratically to $M(a_0, g_0)$ and 0, respectively, and Q_n converges to 0 faster than quadratically. If $\alpha^2 > 1$, then the Cauchy principal value is

19.8.9
$$\Pi(\alpha^2, k) = \frac{\pi}{4M(1,k')}\frac{k^2}{k^2 - \alpha^2}\sum_{n=0}^{\infty} Q_n,$$
$$-\infty < k^2 < 1,\ 1 < \alpha^2 < \infty,$$

where (19.8.8) still applies, but with

19.8.10 $$p_0^2 = 1 - (k^2/\alpha^2).$$

19.8(ii) Landen Transformations

Descending Landen Transformation

Let
$$k_1 = \frac{1-k'}{1+k'},$$

19.8.11 $$\phi_1 = \phi + \arctan(k' \tan\phi)$$
$$= \arcsin\left((1+k')\frac{\sin\phi\cos\phi}{\sqrt{1-k^2\sin^2\phi}}\right).$$

(Note that $0 < k < 1$ and $0 < \phi < \pi/2$ imply $k_1 < k$ and $\phi < \phi_1 < 2\phi$, and also that $\phi = \pi/2$ implies $\phi_1 = \pi$.) Then

19.8.12
$$K(k) = (1+k_1)K(k_1),$$
$$E(k) = (1+k')E(k_1) - k' K(k).$$

19.8.13
$$F(\phi, k) = \tfrac{1}{2}(1+k_1) F(\phi_1, k_1),$$
$$E(\phi, k) = \tfrac{1}{2}(1+k') E(\phi_1, k_1) - k' F(\phi, k) + \tfrac{1}{2}(1-k')\sin\phi_1.$$

19.8.14
$$2(k^2 - \alpha^2)\Pi(\phi, \alpha^2, k) = \frac{\omega^2 - \alpha^2}{1+k'}\Pi(\phi_1, \alpha_1^2, k_1)$$
$$+ k^2 F(\phi, k)$$
$$- (1+k')\alpha_1^2 R_C(c_1, c_1 - \alpha_1^2),$$

where

19.8.15 $$\omega^2 = \frac{k^2 - \alpha^2}{1 - \alpha^2},\quad \alpha_1^2 = \frac{\alpha^2 \omega^2}{(1+k')^2},\quad c_1 = \csc^2\phi_1.$$

Ascending Landen Transformation

Let

19.8.16
$$k_2 = 2\sqrt{k}/(1+k),$$
$$2\phi_2 = \phi + \arcsin(k\sin\phi).$$

(Note that $0 < k < 1$ and $0 < \phi \le \pi/2$ imply $k < k_2 < 1$ and $\phi_2 < \phi$.) Then

19.8.17
$$F(\phi, k) = \frac{2}{1+k} F(\phi_2, k_2),$$
$$E(\phi, k) = (1+k) E(\phi_2, k_2) + (1-k) F(\phi_2, k_2) - k\sin\phi.$$

19.8(iii) Gauss Transformation

We consider only the descending Gauss transformation because its (ascending) inverse moves $F(\phi, k)$ closer to the singularity at $k = \sin\phi = 1$. Let

19.8.18
$$k_1 = (1-k')/(1+k'),$$
$$\sin\psi_1 = \frac{(1+k')\sin\phi}{1+\Delta},\quad \Delta = \sqrt{1 - k^2\sin^2\phi}.$$

(Note that $0 < k < 1$ and $0 < \phi < \pi/2$ imply $k_1 < k$ and $\psi_1 < \phi$, and also that $\phi = \pi/2$ implies $\psi_1 = \pi/2$, thus preserving completeness.) Then

19.8.19
$$F(\phi, k) = (1+k_1) F(\psi_1, k_1),$$
$$E(\phi, k) = (1+k') E(\psi_1, k_1) - k' F(\phi, k) + (1-\Delta)\cot\phi,$$

19.8.20
$$\rho\,\Pi(\phi, \alpha^2, k) = \frac{4}{1+k'}\Pi(\psi_1, \alpha_1^2, k_1)$$
$$+ (\rho - 1) F(\phi, k) - R_C(c-1, c-\alpha^2),$$

where

19.8.21
$$\rho = \sqrt{1 - (k^2/\alpha^2)},$$
$$\alpha_1^2 = \alpha^2(1+\rho)^2/(1+k')^2,\quad c = \csc^2\phi.$$

If $0 < \alpha^2 < k^2$, then ρ is pure imaginary.

19.9 Inequalities

19.9(i) Complete Integrals

Throughout this subsection $0 < k < 1$, except in (19.9.4).

19.9.1
$$\ln 4 \leq K(k) + \ln k' \leq \pi/2, \quad 1 \leq E(k) \leq \pi/2.$$
$$1 \leq (2/\pi)\sqrt{1-\alpha^2}\,\Pi(\alpha^2, k) \leq 1/k', \qquad \alpha^2 < 1.$$

19.9.2
$$1 + \frac{k'^2}{8} < \frac{K(k)}{\ln(4/k')} < 1 + \frac{k'^2}{4},$$

19.9.3
$$9 + \frac{k^2 k'^2}{8} < \frac{(8+k^2)K(k)}{\ln(4/k')} < 9.096.$$

The left-hand inequalities in (19.9.2) and (19.9.3) are equivalent, but the right-hand inequality of (19.9.3) is sharper than that of (19.9.2) when $0 < k^2 \leq 0.922$.

19.9.4
$$\left(\frac{1 + k'^{3/2}}{2}\right)^{2/3} \leq \frac{2}{\pi} E(k) \leq \left(\frac{1+k'^2}{2}\right)^{1/2}$$

for $0 \leq k \leq 1$. The lower bound in (19.9.4) is sharper than $2/\pi$ when $0 \leq k^2 \leq 0.9960$.

19.9.5
$$\ln \frac{(1+\sqrt{k'})^2}{k} < \frac{\pi K'(k)}{2K(k)} < \ln \frac{2(1+k')}{k}.$$

For a sharper, but more complicated, version of (19.9.5) see Anderson et al. (1990).

Other inequalities are:

19.9.6
$$(1 - \tfrac{3}{4}k^2)^{-1/2} < \frac{4}{\pi k^2}(K(k) - E(k)) < (k')^{-3/4},$$

19.9.7
$$(1 - \tfrac{1}{4}k^2)^{-1/2} < \frac{4}{\pi k^2}(E(k) - k'^2 K(k))$$
$$< \min((k')^{-1/4}, 4/\pi),$$

19.9.8
$$k' < \frac{E(k)}{K(k)} < \left(\frac{1+k'}{2}\right)^2.$$

Further inequalities for $K(k)$ and $E(k)$ can be found in Alzer and Qiu (2004), Anderson et al. (1992a,b, 1997), and Qiu and Vamanamurthy (1996).

The perimeter $L(a, b)$ of an ellipse with semiaxes a, b is given by

19.9.9
$$L(a, b) = 4a\,E(k), \quad k^2 = 1 - (b^2/a^2), a > b.$$

Almkvist and Berndt (1988) list thirteen approximations to $L(a, b)$ that have been proposed by various authors. The earliest is due to Kepler and the most accurate to Ramanujan. Ramanujan's approximation and its leading error term yield the following approximation to $L(a, b)/(\pi(a + b))$:

19.9.10
$$1 + \frac{3\lambda^2}{10 + \sqrt{4 - 3\lambda^2}} + \frac{3\lambda^{10}}{2^{17}}, \quad \lambda = \frac{a-b}{a+b}.$$

Even for the extremely eccentric ellipse with $a = 99$ and $b = 1$, this is correct within 0.023%. Barnard et al. (2000) shows that nine of the thirteen approximations, including Ramanujan's, are from below and four are from above. See also Barnard et al. (2001).

19.9(ii) Incomplete Integrals

Throughout this subsection we assume that $0 < k < 1$, $0 \leq \phi \leq \pi/2$, and $\Delta = \sqrt{1 - k^2 \sin^2 \phi} > 0$.

Simple inequalities for incomplete integrals follow directly from the defining integrals (§19.2(ii)) together with (19.6.12):

19.9.11
$$\phi \leq F(\phi, k) \leq \min(\phi/\Delta, \mathrm{gd}^{-1}(\phi)),$$

where $\mathrm{gd}^{-1}(\phi)$ is given by (4.23.41) and (4.23.42). Also,

19.9.12
$$\max(\sin \phi, \phi\Delta) \leq E(\phi, k) \leq \phi,$$

19.9.13
$$\Pi(\phi, \alpha^2, 0) \leq \Pi(\phi, \alpha^2, k)$$
$$\leq \min(\Pi(\phi, \alpha^2, 0)/\Delta, \Pi(\phi, \alpha^2, 1)).$$

Sharper inequalities for $F(\phi, k)$ are:

19.9.14
$$\frac{3}{1 + \Delta + \cos \phi} < \frac{F(\phi, k)}{\sin \phi} < \frac{1}{(\Delta \cos \phi)^{1/3}},$$

19.9.15
$$1 < F(\phi, k) \bigg/ \left((\sin \phi)\ln\left(\frac{4}{\Delta + \cos \phi}\right)\right)$$
$$< \frac{4}{2 + (1+k^2)\sin^2 \phi}.$$

19.9.16
$$F(\phi, k) = \frac{2}{\pi} K(k') \ln\left(\frac{4}{\Delta + \cos \phi}\right) - \theta \Delta^2,$$
$$(\sin \phi)/8 < \theta < (\ln 2)/(k^2 \sin \phi).$$

(19.9.15) is useful when k^2 and $\sin^2 \phi$ are both close to 1, since the bounds are then nearly equal; otherwise (19.9.14) is preferable.

Inequalities for both $F(\phi, k)$ and $E(\phi, k)$ involving inverse circular or inverse hyperbolic functions are given in Carlson (1961b, §4). For example,

19.9.17
$$L \leq F(\phi, k) \leq \sqrt{UL} \leq \tfrac{1}{2}(U+L) \leq U,$$

where

19.9.18
$$L = (1/\sigma) \operatorname{arctanh}(\sigma \sin \phi), \quad \sigma = \sqrt{(1+k^2)/2},$$
$$U = \tfrac{1}{2}\operatorname{arctanh}(\sin \phi) + \tfrac{1}{2}k^{-1}\operatorname{arctanh}(k \sin \phi).$$

Other inequalities for $F(\phi, k)$ can be obtained from inequalities for $R_F(x, y, z)$ given in Carlson (1966, (2.15)) and Carlson (1970) via (19.25.5).

19.10 Relations to Other Functions

19.10(i) Theta and Elliptic Functions

For relations of Legendre's integrals to theta functions, Jacobian functions, and Weierstrass functions, see §§20.9(i), 22.15(ii), and 23.6(iv), respectively. See also Erdélyi et al. (1953b, Chapter 13).

19.10(ii) Elementary Functions

If $y > 0$ is assumed (without loss of generality), then

19.10.1
$$\ln(x/y) = (x-y)\, R_C\big(\tfrac{1}{4}(x+y)^2, xy\big),$$
$$\arctan(x/y) = x\, R_C(y^2, y^2+x^2),$$
$$\operatorname{arctanh}(x/y) = x\, R_C(y^2, y^2-x^2),$$
$$\arcsin(x/y) = x\, R_C(y^2-x^2, y^2),$$
$$\operatorname{arcsinh}(x/y) = x\, R_C(y^2+x^2, y^2),$$
$$\arccos(x/y) = (y^2-x^2)^{1/2}\, R_C(x^2, y^2),$$
$$\operatorname{arccosh}(x/y) = (x^2-y^2)^{1/2}\, R_C(x^2, y^2).$$

In each case when $y = 1$, the quantity multiplying R_C supplies the asymptotic behavior of the left-hand side as the left-hand side tends to 0.

For relations to the Gudermannian function $\operatorname{gd}(x)$ and its inverse $\operatorname{gd}^{-1}(x)$ (§4.23(viii)), see (19.6.8) and

19.10.2
$$(\sinh\phi)\, R_C(1, \cosh^2\phi) = \operatorname{gd}(\phi).$$

19.11 Addition Theorems

19.11(i) General Formulas

19.11.1
$$F(\theta, k) + F(\phi, k) = F(\psi, k),$$

19.11.2
$$E(\theta, k) + E(\phi, k) = E(\psi, k) + k^2 \sin\theta \sin\phi \sin\psi.$$

Here

19.11.3
$$\sin\psi = \frac{(\sin\theta\cos\phi)\Delta(\phi) + (\sin\phi\cos\theta)\Delta(\theta)}{1 - k^2 \sin^2\theta \sin^2\phi},$$
$$\Delta(\theta) = \sqrt{1 - k^2 \sin^2\theta}.$$

Also,

19.11.4
$$\cos\psi = \frac{\cos\theta\cos\phi - (\sin\theta\sin\phi)\Delta(\theta)\Delta(\phi)}{1 - k^2 \sin^2\theta \sin^2\phi},$$
$$\tan(\tfrac{1}{2}\psi) = \frac{(\sin\theta)\Delta(\phi) + (\sin\phi)\Delta(\theta)}{\cos\theta + \cos\phi}.$$

Lastly,

19.11.5
$$\Pi(\theta, \alpha^2, k) + \Pi(\phi, \alpha^2, k) = \Pi(\psi, \alpha^2, k) - \alpha^2 R_C(\gamma - \delta, \gamma),$$

where

19.11.6
$$\gamma = ((\csc^2\theta) - \alpha^2)((\csc^2\phi) - \alpha^2)((\csc^2\psi) - \alpha^2),$$
$$\delta = \alpha^2(1-\alpha^2)(\alpha^2 - k^2).$$

19.11(ii) Case $\psi = \pi/2$

19.11.7
$$F(\phi, k) = K(k) - F(\theta, k),$$

19.11.8 $E(\phi, k) = E(k) - E(\theta, k) + k^2 \sin\theta \sin\phi,$

where

19.11.9
$$\tan\theta = 1/(k' \tan\phi).$$

19.11.10
$$\Pi(\phi, \alpha^2, k) = \Pi(\alpha^2, k) - \Pi(\theta, \alpha^2, k) - \alpha^2 R_C(\gamma - \delta, \gamma),$$

where

19.11.11
$$\gamma = (1-\alpha^2)((\csc^2\theta) - \alpha^2)((\csc^2\phi) - \alpha^2),$$
$$\delta = \alpha^2(1-\alpha^2)(\alpha^2 - k^2).$$

19.11(iii) Duplication Formulas

If $\phi = \theta$ in §19.11(i) and $\Delta(\theta)$ is again defined by (19.11.3), then

19.11.12
$$F(\psi, k) = 2F(\theta, k),$$

19.11.13 $E(\psi, k) = 2E(\theta, k) - k^2 \sin^2\theta \sin\psi,$

19.11.14 $\sin\psi = (\sin 2\theta)\Delta(\theta)/(1 - k^2 \sin^4\theta),$

19.11.15
$$\cos\psi = (\cos(2\theta) + k^2 \sin^4\theta)/(1 - k^2 \sin^4\theta),$$
$$\tan(\tfrac{1}{2}\psi) = (\tan\theta)\Delta(\theta),$$
$$\sin\theta = (\sin\psi)/\sqrt{(1+\cos\psi)(1+\Delta(\psi))},$$
$$\cos\theta = \sqrt{\frac{(\cos\psi) + \Delta(\psi)}{1 + \Delta(\psi)}},$$
$$\tan\theta = \tan(\tfrac{1}{2}\psi)\sqrt{\frac{1 + \cos\psi}{(\cos\psi) + \Delta(\psi)}},$$

19.11.16 $\Pi(\psi, \alpha^2, k) = 2\,\Pi(\theta, \alpha^2, k) + \alpha^2 R_C(\gamma - \delta, \gamma),$

19.11.17
$$\gamma = ((\csc^2\theta) - \alpha^2)^2((\csc^2\psi) - \alpha^2),$$
$$\delta = \alpha^2(1-\alpha^2)(\alpha^2 - k^2).$$

19.12 Asymptotic Approximations

With $\psi(x)$ denoting the digamma function (§5.2(i)) in this subsection, the asymptotic behavior of $K(k)$ and $E(k)$ near the singularity at $k = 1$ is given by the following convergent series:

19.12.1
$$K(k) = \sum_{m=0}^{\infty} \frac{(\tfrac{1}{2})_m (\tfrac{1}{2})_m}{m!\, m!} k'^{2m} \left(\ln\left(\frac{1}{k'}\right) + d(m)\right),$$
$$0 < |k'| < 1,$$

19.12.2
$$E(k) = 1 + \frac{1}{2}\sum_{m=0}^{\infty} \frac{\left(\frac{1}{2}\right)_m \left(\frac{3}{2}\right)_m}{(2)_m m!} k'^{2m+2}$$
$$\times \left(\ln\left(\frac{1}{k'}\right) + d(m) - \frac{1}{(2m+1)(2m+2)}\right),$$
$$|k'| < 1,$$

where

19.12.3
$$d(m) = \psi(1+m) - \psi\left(\tfrac{1}{2}+m\right),$$
$$d(m+1) = d(m) - \frac{2}{(2m+1)(2m+2)}, \quad m = 0, 1, \ldots.$$

For the asymptotic behavior of $F(\phi,k)$ and $E(\phi,k)$ as $\phi \to \tfrac{1}{2}\pi-$ and $k \to 1-$ see Kaplan (1948, §2), Van de Vel (1969), and Karp and Sitnik (2007).

As $k^2 \to 1-$

19.12.4
$$(1-\alpha^2)\,\Pi(\alpha^2, k)$$
$$= \left(\ln\frac{4}{k'}\right)\left(1 + O(k'^2)\right) - \alpha^2 R_C(1, 1-\alpha^2),$$
$$-\infty < \alpha^2 < 1,$$

19.12.5
$$(1-\alpha^2)\,\Pi(\alpha^2, k)$$
$$= \left(\ln\left(\frac{4}{k'}\right) - R_C(1, 1-\alpha^{-2})\right)\left(1 + O(k'^2)\right),$$
$$1 < \alpha^2 < \infty.$$

Asymptotic approximations for $\Pi(\phi,\alpha^2,k)$, with different variables, are given in Karp et al. (2007). They are useful primarily when $(1-k)/(1-\sin\phi)$ is either small or large compared with 1.

If $x \geq 0$ and $y > 0$, then

19.12.6
$$R_C(x,y) = \frac{\pi}{2\sqrt{y}} - \frac{\sqrt{x}}{y}\left(1 + O\left(\sqrt{\frac{x}{y}}\right)\right), \quad x/y \to 0,$$

19.12.7
$$R_C(x,y) = \frac{1}{2\sqrt{x}}\left(\left(1 + \frac{y}{2x}\right)\ln\left(\frac{4x}{y}\right) - \frac{y}{2x}\right)$$
$$\times (1 + O(y^2/x^2)), \quad y/x \to 0.$$

19.13 Integrals of Elliptic Integrals

19.13(i) Integration with Respect to the Modulus

For definite and indefinite integrals of complete elliptic integrals see Byrd and Friedman (1971, pp. 610–612, 615), Prudnikov et al. (1990, §§1.11, 2.16), Glasser (1976), Bushell (1987), and Cvijović and Klinowski (1999).

For definite and indefinite integrals of incomplete elliptic integrals see Byrd and Friedman (1971, pp. 613, 616), Prudnikov et al. (1990, §§1.10.2, 2.15.2), and Cvijović and Klinowski (1994).

19.13(ii) Integration with Respect to the Amplitude

Various integrals are listed by Byrd and Friedman (1971, p. 630) and Prudnikov et al. (1990, §§1.10.1, 2.15.1). Cvijović and Klinowski (1994) contains fractional integrals (with free parameters) for $F(\phi,k)$ and $E(\phi,k)$, together with special cases.

19.13(iii) Laplace Transforms

For direct and inverse Laplace transforms for the complete elliptic integrals $K(k)$, $E(k)$, and $D(k)$ see Prudnikov et al. (1992a, §3.31) and Prudnikov et al. (1992b, §§3.29 and 4.3.33), respectively.

19.14 Reduction of General Elliptic Integrals

19.14(i) Examples

In (19.14.1)–(19.14.3) both the integrand and $\cos\phi$ are assumed to be nonnegative. Cases in which $\cos\phi < 0$ can be included by application of (19.2.10).

19.14.1
$$\int_1^x \frac{dt}{\sqrt{t^3-1}} = 3^{-1/4}F(\phi,k),$$
$$\cos\phi = \frac{\sqrt{3}+1-x}{\sqrt{3}-1+x}, \quad k^2 = \frac{2-\sqrt{3}}{4}.$$

19.14.2
$$\int_x^1 \frac{dt}{\sqrt{1-t^3}} = 3^{-1/4}F(\phi,k),$$
$$\cos\phi = \frac{\sqrt{3}-1+x}{\sqrt{3}+1-x}, \quad k^2 = \frac{2+\sqrt{3}}{4}.$$

19.14.3
$$\int_0^x \frac{dt}{\sqrt{1+t^4}} = \frac{\operatorname{sign}(x)}{2} F(\phi,k),$$
$$\cos\phi = \frac{1-x^2}{1+x^2}, \quad k^2 = \frac{1}{2}.$$

19.14.4
$$\int_y^x \frac{dt}{\sqrt{(a_1+b_1 t^2)(a_2+b_2 t^2)}} = \frac{1}{\sqrt{\gamma-\alpha}} F(\phi,k),$$
$$k^2 = (\gamma-\beta)/(\gamma-\alpha).$$

In (19.14.4) $0 \leq y < x$, each quadratic polynomial is positive on the interval (y,x), and α, β, γ is a permutation of $0, a_1 b_2, a_2 b_1$ (not all 0 by assumption) such that $\alpha \leq \beta \leq \gamma$. More generally in (19.14.4),

19.14.5
$$\sin^2\phi = \frac{\gamma-\alpha}{U^2+\gamma},$$

where

19.14.6
$$(x^2-y^2)U = x\sqrt{(a_1+b_1 y^2)(a_2+b_2 y^2)}$$
$$+ y\sqrt{(a_1+b_1 x^2)(a_2+b_2 x^2)}.$$

There are four important special cases of (19.14.4)–(19.14.6), as follows. If $y = 0$, then

19.14.7 $$\sin^2 \phi = \frac{(\gamma - \alpha)x^2}{a_1 a_2 + \gamma x^2}.$$

If $x = \infty$, then

19.14.8 $$\sin^2 \phi = \frac{\gamma - \alpha}{b_1 b_2 y^2 + \gamma}.$$

If $a_1 + b_1 y^2 = 0$, then

19.14.9 $$\sin^2 \phi = \frac{(\gamma - \alpha)(x^2 - y^2)}{\gamma(x^2 - y^2) - a_1(a_2 + b_2 x^2)}.$$

If $a_1 + b_1 x^2 = 0$, then

19.14.10 $$\sin^2 \phi = \frac{(\gamma - \alpha)(y^2 - x^2)}{\gamma(y^2 - x^2) - a_1(a_2 + b_2 y^2)}.$$

(These four cases include 12 integrals in Abramowitz and Stegun (1964, p. 596).)

19.14(ii) General Case

Legendre (1825–1832) showed that every elliptic integral can be expressed in terms of the three integrals in (19.1.2) supplemented by algebraic, logarithmic, and trigonometric functions. The classical method of reducing (19.2.3) to Legendre's integrals is described in many places, especially Erdélyi et al. (1953b, §13.5), Abramowitz and Stegun (1964, Chapter 17), and Labahn and Mutrie (1997, §3). The last reference gives a clear summary of the various steps involving linear fractional transformations, partial-fraction decomposition, and recurrence relations. It then improves the classical method by first applying Hermite reduction to (19.2.3) to arrive at integrands without multiple poles and uses implicit full partial-fraction decomposition and implicit root finding to minimize computing with algebraic extensions. The choice among 21 transformations for final reduction to Legendre's normal form depends on inequalities involving the limits of integration and the zeros of the cubic or quartic polynomial. A similar remark applies to the transformations given in Erdélyi et al. (1953b, §13.5) and to the choice among explicit reductions in the extensive table of Byrd and Friedman (1971), in which one limit of integration is assumed to be a branch point of the integrand at which the integral converges. If no such branch point is accessible from the interval of integration (for example, if the integrand is $(t(3-t)(4-t))^{-3/2}$ and the interval is $[1,2]$), then no method using this assumption succeeds.

Symmetric Integrals

19.15 Advantages of Symmetry

Elliptic integrals are special cases of a particular multivariate hypergeometric function called *Lauricella's* F_D (Carlson (1961b)). The function $R_{-a}(b_1, b_2, \ldots, b_n; z_1, z_2, \ldots, z_n)$ (Carlson (1963)) reveals the full permutation symmetry that is partially hidden in F_D, and leads to symmetric standard integrals that simplify many aspects of theory, applications, and numerical computation.

Symmetry in x, y, z of $R_F(x, y, z)$, $R_G(x, y, z)$, and $R_J(x, y, z, p)$ replaces the five transformations (19.7.2), (19.7.4)–(19.7.7) of Legendre's integrals; compare (19.25.17). Symmetry unifies the Landen transformations of §19.8(ii) with the Gauss transformations of §19.8(iii), as indicated following (19.22.22) and (19.36.9). (19.21.12) unifies the three transformations in §19.7(iii) that change the parameter of Legendre's third integral.

Symmetry allows the expansion (19.19.7) in a series of elementary symmetric functions that gives high precision with relatively few terms and provides the most efficient method of computing the incomplete integral of the third kind (§19.36(i)).

Symmetry makes possible the reduction theorems of §19.29(i), permitting remarkable compression of tables of integrals while generalizing the interval of integration. (Compare (19.14.4)–(19.14.10) with (19.29.19), and see the last paragraph of §19.29(i) and the text following (19.29.15).) These reduction theorems, unknown in the Legendre theory, allow symbolic integration without imposing conditions on the parameters and the limits of integration (see §19.29(ii)).

For the many properties of ellipses and triaxial ellipsoids that can be represented by elliptic integrals, any symmetry in the semiaxes remains obvious when symmetric integrals are used (see (19.30.5) and §19.33). For example, the computation of depolarization factors for solid ellipsoids is simplified considerably; compare (19.33.7) with Cronemeyer (1991).

19.16 Definitions

19.16(i) Symmetric Integrals

19.16.1 $$R_F(x, y, z) = \frac{1}{2} \int_0^\infty \frac{dt}{s(t)},$$

19.16.2 $$R_J(x, y, z, p) = \frac{3}{2} \int_0^\infty \frac{dt}{s(t)(t+p)},$$

19.16.3
$$R_G(x,y,z) = \frac{1}{4\pi} \int_0^{2\pi} \int_0^\pi \left(x\sin^2\theta\cos^2\phi + y\sin^2\theta\sin^2\phi + z\cos^2\theta\right)^{\frac{1}{2}} \sin\theta\, d\theta\, d\phi,$$

where $p\ (\neq 0)$ is a real or complex constant, and

19.16.4
$$s(t) = \sqrt{t+x}\sqrt{t+y}\sqrt{t+z}.$$

In (19.16.1) and (19.16.2), $x, y, z \in \mathbb{C}\setminus(-\infty, 0]$ except that one or more of x, y, z may be 0 when the corresponding integral converges. In (19.16.2) the Cauchy principal value is taken when p is real and negative. In (19.16.3) $\Re x, \Re y, \Re z \geq 0$. It should be noted that the integrals (19.16.1)–(19.16.3) have been normalized so that $R_F(1,1,1) = R_J(1,1,1,1) = R_G(1,1,1) = 1$.

A fourth integral that is symmetric in only two variables is defined by

19.16.5 $\quad R_D(x,y,z) = R_J(x,y,z,z) = \dfrac{3}{2}\displaystyle\int_0^\infty \dfrac{dt}{s(t)(t+z)},$

with the same conditions on x, y, z as for (19.16.1), but now $z \neq 0$.

Just as the elementary function $R_C(x, y)$ (§19.2(iv)) is the degenerate case

19.16.6 $\quad R_C(x,y) = R_F(x,y,y),$

and R_D is a degenerate case of R_J, so is R_J a degenerate case of the *hyperelliptic integral*,

19.16.7 $\quad \dfrac{3}{2}\displaystyle\int_0^\infty \dfrac{dt}{\prod_{j=1}^5 \sqrt{t+x_j}}.$

19.16(ii) $R_{-a}(\mathbf{b}; \mathbf{z})$

All elliptic integrals of the form (19.2.3) and many multiple integrals, including (19.16.3), are special cases of a multivariate hypergeometric function

19.16.8 $\quad R_{-a}(\mathbf{b}; \mathbf{z}) = R_{-a}(b_1, \ldots, b_n; z_1, \ldots, z_n),$

which is homogeneous and of *degree* $-a$ in the z's, and symmetric when the same permutation is applied to both sets of subscripts $1, \ldots, n$. Thus $R_{-a}(\mathbf{b}; \mathbf{z})$ is symmetric in the variables z_j and z_ℓ if the parameters b_j and b_ℓ are equal. The R-function is often used to make a unified statement of a property of several elliptic integrals. Before 1969 $R_{-a}(\mathbf{b}; \mathbf{z})$ was denoted by $R(a; \mathbf{b}; \mathbf{z})$.

19.16.9
$$R_{-a}(\mathbf{b}; \mathbf{z}) = \frac{1}{\mathrm{B}(a,a')} \int_0^\infty t^{a'-1} \prod_{j=1}^n (t+z_j)^{-b_j} dt$$
$$= \frac{1}{\mathrm{B}(a,a')} \int_0^\infty t^{a-1} \prod_{j=1}^n (1+tz_j)^{-b_j} dt,$$
$$a, a' > 0,\ z_j \in \mathbb{C}\setminus(-\infty, 0],$$

where $\mathrm{B}(x,y)$ is the beta function (§5.12) and

19.16.10 $\quad a' = -a + \displaystyle\sum_{j=1}^n b_j.$

19.16.11
$$R_{-a}(\mathbf{b}; \lambda\mathbf{z}) = \lambda^{-a} R_{-a}(\mathbf{b}; \mathbf{z}),$$
$$R_{-a}(\mathbf{b}; x\mathbf{1}) = x^{-a}, \qquad \mathbf{1} = (1,\ldots,1).$$

When $n = 4$ a useful version of (19.16.9) is given by

19.16.12
$$R_{-a}(b_1, \ldots, b_4; c-1, c-k^2, c, c-\alpha^2)$$
$$= \frac{2(\sin^2\phi)^{1-a'}}{\mathrm{B}(a,a')} \int_0^\phi (\sin\theta)^{2a-1} (\sin^2\phi - \sin^2\theta)^{a'-1}$$
$$\times (\cos\theta)^{1-2b_1}(1-k^2\sin^2\theta)^{-b_2}(1-\alpha^2\sin^2\theta)^{-b_4} d\theta,$$

where

19.16.13 $\quad c = \csc^2\phi;\quad a, a' > 0;\quad b_3 = a + a' - b_1 - b_2 - b_4.$

For further information, especially representation of the R-function as a Dirichlet average, see Carlson (1977b).

19.16(iii) Elliptic Cases of $R_{-a}(\mathbf{b}; \mathbf{z})$

$R_{-a}(\mathbf{b}; \mathbf{z})$ is an elliptic integral if the z's are distinct and exactly four of the parameters $a, a', b_1, \ldots, b_n$ are half-odd-integers, the rest are integers, and none of a, a', $a + a'$ is zero or a negative integer. The only cases that are integrals of the first kind are the four in which each of a and a' is either $\frac{1}{2}$ or 1 and each b_j is $\frac{1}{2}$. The only cases that are integrals of the third kind are those in which at least one b_j is a positive integer. All other elliptic cases are integrals of the second kind.

19.16.14 $\quad R_F(x,y,z) = R_{-\frac{1}{2}}\left(\tfrac{1}{2},\tfrac{1}{2},\tfrac{1}{2}; x,y,z\right),$

19.16.15 $\quad R_D(x,y,z) = R_{-\frac{3}{2}}\left(\tfrac{1}{2},\tfrac{1}{2},\tfrac{3}{2}; x,y,z\right),$

19.16.16 $\quad R_J(x,y,z,p) = R_{-\frac{3}{2}}\left(\tfrac{1}{2},\tfrac{1}{2},\tfrac{1}{2},1; x,y,z,p\right),$

19.16.17 $\quad R_G(x,y,z) = R_{\frac{1}{2}}\left(\tfrac{1}{2},\tfrac{1}{2},\tfrac{1}{2}; x,y,z\right),$

19.16.18 $\quad R_C(x,y) = R_{-\frac{1}{2}}\left(\tfrac{1}{2},1; x,y\right).$

When one variable is 0 without destroying convergence, any one of (19.16.14)–(19.16.17) is said to be *complete* and can be written as an R-function with one less variable:

19.16.19
$$R_{-a}(b_1, \ldots, b_n; 0, z_2, \ldots, z_n)$$
$$= \frac{\mathrm{B}(a, a' - b_1)}{\mathrm{B}(a, a')} R_{-a}(b_2, \ldots, b_n; z_2, \ldots, z_n),$$
$$a + a' > 0,\ a' > b_1.$$

19.17 Graphics

Thus

19.16.20 $\quad R_F(0,y,z) = \tfrac{1}{2}\pi R_{-\frac{1}{2}}\left(\tfrac{1}{2},\tfrac{1}{2};y,z\right),$

19.16.21 $\quad R_D(0,y,z) = \tfrac{3}{4}\pi R_{-\frac{3}{2}}\left(\tfrac{1}{2},\tfrac{3}{2};y,z\right),$

19.16.22 $\quad R_J(0,y,z,p) = \tfrac{3}{4}\pi R_{-\frac{3}{2}}\left(\tfrac{1}{2},\tfrac{1}{2},1;y,z,p\right),$

19.16.23
$$R_G(0,y,z) = \tfrac{1}{4}\pi R_{\frac{1}{2}}\left(\tfrac{1}{2},\tfrac{1}{2};y,z\right)$$
$$= \tfrac{1}{4}\pi z\, R_{-\frac{1}{2}}\left(-\tfrac{1}{2},\tfrac{3}{2};y,z\right).$$

The last R-function has $a = a' = \tfrac{1}{2}$.

Each of the four complete integrals (19.16.20)–(19.16.23) can be integrated to recover the incomplete integral:

19.16.24
$$R_{-a}(\mathbf{b};\mathbf{z}) = \frac{z_1^{a'-b_1}}{\mathrm{B}(b_1, a'-b_1)}\int_0^\infty t^{b_1-1}(t+z_1)^{-a'}$$
$$\times R_{-a}(\mathbf{b};0,t+z_2,\ldots,t+z_n)\,dt,$$
$$a' > b_1,\ a+a' > b_1 > 0.$$

19.17 Graphics

See Figures 19.17.1–19.17.8 for symmetric elliptic integrals with real arguments.

Because the R-function is homogeneous, there is no loss of generality in giving one variable the value 1 or -1 (as in Figure 19.3.2). For R_F, R_G, and R_J, which are symmetric in x,y,z, we may further assume that z is the largest of x,y,z if the variables are real, then choose $z=1$, and consider only $0 \le x \le 1$ and $0 \le y \le 1$. The cases $x=0$ or $y=0$ correspond to the complete integrals. The case $y=1$ corresponds to elementary functions.

To view $R_F(0,y,1)$ and $2R_G(0,y,1)$ for complex y, put $y = 1-k^2$, use (19.25.1), and see Figures 19.3.7–19.3.8.

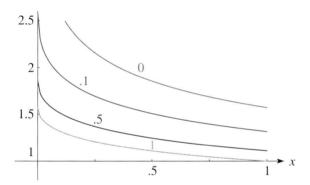

Figure 19.17.1: $R_F(x,y,1)$ for $0 \le x \le 1$, $y = 0, 0.1, 0.5, 1$. $y = 1$ corresponds to $R_C(x,1)$.

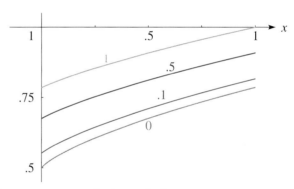

Figure 19.17.2: $R_G(x,y,1)$ for $0 \le x \le 1$, $y = 0, 0.1, 0.5, 1$. $y = 1$ corresponds to $\tfrac{1}{2}(R_C(x,1)+\sqrt{x})$.

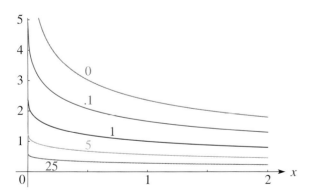

Figure 19.17.3: $R_D(x,y,1)$ for $0 \le x \le 2$, $y = 0, 0.1, 1, 5, 25$. $y = 1$ corresponds to $\tfrac{3}{2}(R_C(x,1) - \sqrt{x})/(1-x)$, $x \ne 1$.

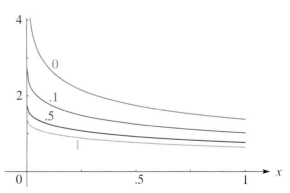

Figure 19.17.4: $R_J(x,y,1,2)$ for $0 \le x \le 1$, $y = 0, 0.1, 0.5, 1$. $y = 1$ corresponds to $3(R_C(x,1) - R_C(x,2))$.

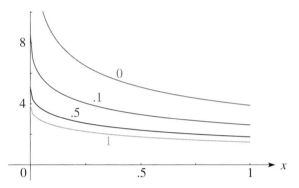

Figure 19.17.5: $R_J(x,y,1,0.5)$ for $0 \leq x \leq 1$, $y = 0, 0.1, 0.5, 1$. $y = 1$ corresponds to $6(R_C(x,0.5) - R_C(x,1))$.

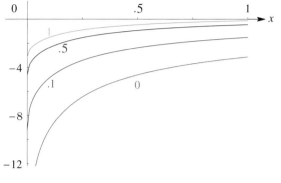

Figure 19.17.6: Cauchy principal value of $R_J(x,y,1,-0.5)$ for $0 \leq x \leq 1$, $y = 0, 0.1, 0.5, 1$. $y = 1$ corresponds to $2(R_C(x,-0.5) - R_C(x,1))$.

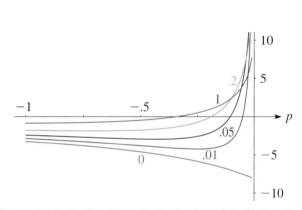

Figure 19.17.7: Cauchy principal value of $R_J(0.5,y,1,p)$ for $y = 0, 0.01, 0.05, 0.2, 1$, $-1 \leq p < 0$. $y = 1$ corresponds to $3(R_C(0.5,p) - (\pi/\sqrt{8}))/(1-p)$. As $p \to 0$ the curve for $y = 0$ has the finite limit $-8.10386\ldots$; see (19.20.10).

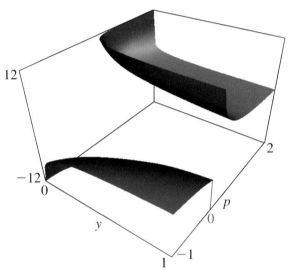

Figure 19.17.8: $R_J(0,y,1,p)$, $0 \leq y \leq 1$, $-1 \leq p \leq 2$. Cauchy principal values are shown when $p < 0$. The function is asymptotic to $\frac{3}{2}\pi/\sqrt{yp}$ as $p \to 0+$, and to $(\frac{3}{2}/p)\ln(16/y)$ as $y \to 0+$. As $p \to 0-$ it has the limit $(-6/y)\,R_G(0,y,1)$. When $p = 1$, it reduces to $R_D(0,y,1)$. If $y = 1$, then it has the value $\frac{3}{2}\pi/(p+\sqrt{p})$ when $p > 0$, and $\frac{3}{2}\pi/(p-1)$ when $p < 0$. See (19.20.10), (19.20.11), and (19.20.8) for the cases $p \to 0\pm$, $y \to 0+$, and $y = 1$, respectively.

19.18 Derivatives and Differential Equations

19.18(i) Derivatives

19.18.1 $$\frac{\partial R_F(x,y,z)}{\partial z} = -\tfrac{1}{6} R_D(x,y,z),$$

19.18.2 $$\frac{d}{dx} R_G(x+a, x+b, x+c) = \tfrac{1}{2} R_F(x+a, x+b, x+c).$$

Let $\partial_j = \partial/\partial z_j$, and $\mathbf{e}_j$ be an n-tuple with 1 in the jth place and 0's elsewhere. Also define

19.18.3 $$w_j = b_j \bigg/ \sum_{j=1}^{n} b_j, \quad a' = -a + \sum_{j=1}^{n} b_j.$$

The next two equations apply to (19.16.14)–(19.16.18) and (19.16.20)–(19.16.23).

19.18.4 $$\partial_j R_{-a}(\mathbf{b}; \mathbf{z}) = -a w_j R_{-a-1}(\mathbf{b} + \mathbf{e}_j; \mathbf{z}),$$

19.18.5 $(z_j\partial_j + b_j)R_{-a}(\mathbf{b};\mathbf{z}) = w_j a' R_{-a}(\mathbf{b}+\mathbf{e}_j;\mathbf{z})$.

19.18(ii) Differential Equations

19.18.6 $\left(\dfrac{\partial}{\partial x} + \dfrac{\partial}{\partial y} + \dfrac{\partial}{\partial z}\right) R_F(x,y,z) = \dfrac{-1}{2\sqrt{xyz}}$,

19.18.7 $\left(\dfrac{\partial}{\partial x} + \dfrac{\partial}{\partial y} + \dfrac{\partial}{\partial z}\right) R_G(x,y,z) = \tfrac{1}{2} R_F(x,y,z)$.

19.18.8 $\sum_{j=1}^{n} \partial_j R_{-a}(\mathbf{b};\mathbf{z}) = -a\, R_{-a-1}(\mathbf{b};\mathbf{z})$.

19.18.9
$\left(x\dfrac{\partial}{\partial x} + y\dfrac{\partial}{\partial y} + z\dfrac{\partial}{\partial z}\right) R_F(x,y,z) = -\tfrac{1}{2} R_F(x,y,z)$,

19.18.10
$\left((x-y)\dfrac{\partial^2}{\partial x\,\partial y} + \tfrac{1}{2}\left(\dfrac{\partial}{\partial y} - \dfrac{\partial}{\partial x}\right)\right) R_F(x,y,z) = 0$,

and two similar equations obtained by permuting x,y,z in (19.18.10).

More concisely, if $v = R_{-a}(\mathbf{b};\mathbf{z})$, then each of (19.16.14)–(19.16.18) and (19.16.20)–(19.16.23) satisfies *Euler's homogeneity relation*:

19.18.11 $\sum_{j=1}^{n} z_j\partial_j v = -av$,

and also a system of $n(n-1)/2$ *Euler–Poisson differential equations* (of which only $n-1$ are independent):

19.18.12 $(z_j\partial_j + b_j)\partial_l v = (z_l\partial_l + b_l)\partial_j v$,

or equivalently,

19.18.13 $((z_j - z_l)\partial_j\partial_l + b_j\partial_l - b_l\partial_j)v = 0$.

Here $j,l = 1,2,\ldots,n$ and $j \neq l$. For group-theoretical aspects of this system see Carlson (1963, §VI). If $n=2$, then elimination of $\partial_2 v$ between (19.18.11) and (19.18.12), followed by the substitution $(b_1,b_2,z_1,z_2) = (b, c-b, 1-z, 1)$, produces the Gauss hypergeometric equation (15.10.1).

The next four differential equations apply to the complete case of R_F and R_G in the form $R_{-a}(\tfrac{1}{2},\tfrac{1}{2};z_1,z_2)$ (see (19.16.20) and (19.16.23)).

The function $w = R_{-a}(\tfrac{1}{2},\tfrac{1}{2};x+y,x-y)$ satisfies an *Euler–Poisson–Darboux equation*:

19.18.14 $\dfrac{\partial^2 w}{\partial x^2} = \dfrac{\partial^2 w}{\partial y^2} + \dfrac{1}{y}\dfrac{\partial w}{\partial y}$.

Also $W = R_{-a}(\tfrac{1}{2},\tfrac{1}{2}; t+r, t-r)$, with $r = \sqrt{x^2+y^2}$, satisfies a *wave equation*:

19.18.15 $\dfrac{\partial^2 W}{\partial t^2} = \dfrac{\partial^2 W}{\partial x^2} + \dfrac{\partial^2 W}{\partial y^2}$.

Similarly, the function $u = R_{-a}(\tfrac{1}{2},\tfrac{1}{2};x+iy,x-iy)$ satisfies an equation of *axially symmetric potential theory*:

19.18.16 $\dfrac{\partial^2 u}{\partial x^2} + \dfrac{\partial^2 u}{\partial y^2} + \dfrac{1}{y}\dfrac{\partial u}{\partial y} = 0$,

and $U = R_{-a}(\tfrac{1}{2},\tfrac{1}{2};z+i\rho,z-i\rho)$, with $\rho = \sqrt{x^2+y^2}$, satisfies *Laplace's equation*:

19.18.17 $\dfrac{\partial^2 U}{\partial x^2} + \dfrac{\partial^2 U}{\partial y^2} + \dfrac{\partial^2 U}{\partial z^2} = 0$.

19.19 Taylor and Related Series

For $N = 0,1,2,\ldots$ define the homogeneous hypergeometric polynomial

19.19.1 $T_N(\mathbf{b},\mathbf{z}) = \sum \dfrac{(b_1)_{m_1}\cdots(b_n)_{m_n}}{m_1!\cdots m_n!} z_1^{m_1}\cdots z_n^{m_n}$,

where the summation extends over all nonnegative integers $m_1,\ldots,m_n$ whose sum is N. The following two multivariate hypergeometric series apply to each of the integrals (19.16.14)–(19.16.18) and (19.16.20)–(19.16.23):

19.19.2 $R_{-a}(\mathbf{b};\mathbf{z}) = \sum_{N=0}^{\infty} \dfrac{(a)_N}{(c)_N} T_N(\mathbf{b}, \mathbf{1}-\mathbf{z})$,
$c = \sum_{j=1}^n b_j,\ |1-z_j| < 1$,

19.19.3 $R_{-a}(\mathbf{b};\mathbf{z}) = z_n^{-a} \sum_{N=0}^{\infty} \dfrac{(a)_N}{(c)_N} T_N(b_1,\ldots,b_{n-1}; 1-(z_1/z_n),\ldots,1-(z_{n-1}/z_n))$, $c = \sum_{j=1}^n b_j,\ |1-(z_j/z_n)| < 1$.

If $n=2$, then (19.19.3) is a Gauss hypergeometric series (see (19.25.43) and (15.2.1)).

Define the *elementary symmetric function* $E_s(\mathbf{z})$ by

19.19.4 $\prod_{j=1}^{n}(1+tz_j) = \sum_{s=0}^{n} t^s E_s(\mathbf{z})$,

and define the n-tuple $\tfrac{1}{2} = (\tfrac{1}{2},\ldots,\tfrac{1}{2})$. Then

19.19.5
$T_N(\tfrac{1}{2},\mathbf{z}) = \sum (-1)^{M+N} (\tfrac{1}{2})_M \dfrac{E_1^{m_1}(\mathbf{z})\cdots E_n^{m_n}(\mathbf{z})}{m_1!\cdots m_n!}$,

where $M = \sum_{j=1}^n m_j$ and the summation extends over all nonnegative integers $m_1,\ldots,m_n$ such that $\sum_{j=1}^n j m_j = N$.

This form of T_N can be applied to (19.16.14)–(19.16.18) and (19.16.20)–(19.16.23) if we use

19.19.6 $R_J(x,y,z,p) = R_{-\frac{3}{2}}\left(\frac{1}{2},\frac{1}{2},\frac{1}{2},\frac{1}{2},\frac{1}{2};x,y,z,p,p\right)$

as well as (19.16.5) and (19.16.6). The number of terms in T_N can be greatly reduced by using variables $\mathbf{Z} = \mathbf{1} - (\mathbf{z}/A)$ with A chosen to make $E_1(\mathbf{Z}) = 0$. Then T_N has at most one term if $N \leq 5$ in the series for R_F. For R_J and R_D, T_N has at most one term if $N \leq 3$, and two terms if $N = 4$ or 5.

19.19.7 $R_{-a}\left(\frac{1}{2};\mathbf{z}\right) = A^{-a} \sum_{N=0}^{\infty} \frac{(a)_N}{\left(\frac{1}{2}n\right)_N} T_N\left(\frac{1}{2},\mathbf{Z}\right),$

where

19.19.8
$$A = \frac{1}{n}\sum_{j=1}^n z_j, \quad Z_j = 1 - (z_j/A),$$
$$E_1(\mathbf{Z}) = 0, \qquad |Z_j| < 1.$$

A special case is given in (19.36.1).

19.20 Special Cases

19.20(i) $R_F(x,y,z)$

In this subsection, and also §§19.20(ii)–19.20(v), the variables of all R-functions satisfy the constraints specified in §19.16(i) unless other conditions are stated.

19.20.1
$$R_F(x,x,x) = x^{-1/2},$$
$$R_F(\lambda x, \lambda y, \lambda z) = \lambda^{-1/2} R_F(x,y,z),$$
$$R_F(x,y,y) = R_C(x,y),$$
$$R_F(0,y,y) = \tfrac{1}{2}\pi y^{-1/2},$$
$$R_F(0,0,z) = \infty.$$

The *first lemniscate constant* is given by

19.20.2 $\int_0^1 \dfrac{dt}{\sqrt{1-t^4}} = R_F(0,1,2) = \dfrac{\left(\Gamma\left(\frac{1}{4}\right)\right)^2}{4(2\pi)^{1/2}}$
$= 1.31102\,87771\,46059\,90523\dots.$

Todd (1975) refers to a proof by T. Schneider that this is a transcendental number. The *general lemniscatic case* is

19.20.3 $R_F(x,a,y) = R_{-\frac{1}{4}}\left(\frac{3}{4},\frac{1}{2};a^2,xy\right), \quad a = \tfrac{1}{2}(x+y).$

19.20(ii) $R_G(x,y,z)$

19.20.4
$$R_G(x,x,x) = x^{1/2},$$
$$R_G(\lambda x, \lambda y, \lambda z) = \lambda^{1/2} R_G(x,y,z),$$
$$R_G(0,y,y) = \tfrac{1}{4}\pi y^{1/2},$$
$$R_G(0,0,z) = \tfrac{1}{2}z^{1/2},$$

19.20.5 $2R_G(x,y,y) = y\,R_C(x,y) + \sqrt{x}.$

19.20(iii) $R_J(x,y,z,p)$

19.20.6
$$R_J(x,x,x,x) = x^{-3/2},$$
$$R_J(\lambda x, \lambda y, \lambda z, \lambda p) = \lambda^{-3/2} R_J(x,y,z,p),$$
$$R_J(x,y,z,z) = R_D(x,y,z),$$
$$R_J(0,0,z,p) = \infty,$$
$$R_J(x,x,x,p) = R_D(p,p,x) = \frac{3}{x-p}\left(R_C(x,p) - \frac{1}{\sqrt{x}}\right),$$
$$x \neq p,\ xp \neq 0.$$

19.20.7
$$R_J(x,y,z,p) \to +\infty, \quad p \to 0+ \text{ or } 0-;\ x,y,z > 0.$$

19.20.8
$$R_J(0,y,y,p) = \frac{3\pi}{2(y\sqrt{p} + p\sqrt{y})}, \qquad p > 0,$$
$$R_J(0,y,y,-q) = \frac{-3\pi}{2\sqrt{y}(y+q)}, \qquad q > 0,$$
$$R_J(x,y,y,p) = \frac{3}{p-y}(R_C(x,y) - R_C(x,p)), \quad p \neq y,$$
$$R_J(x,y,y,y) = R_D(x,y,y).$$

19.20.9 $R_J(0,y,z,\pm\sqrt{yz}) = \pm\dfrac{3}{2\sqrt{yz}} R_F(0,y,z).$

19.20.10
$$\lim_{p\to 0+} \sqrt{p}\,R_J(0,y,z,p) = \frac{3\pi}{2\sqrt{yz}},$$
$$\lim_{p\to 0-} R_J(0,y,z,p) = -R_D(0,y,z) - R_D(0,z,y)$$
$$= \frac{-6}{yz} R_G(0,y,z).$$

19.20.11 $R_J(0,y,z,p) \sim \dfrac{3}{2p\sqrt{z}}\ln\left(\dfrac{16z}{y}\right), \quad y \to 0+;\ p\,(\neq 0)\text{ real}.$

19.20.12 $\lim_{p\to\pm\infty} p\,R_J(x,y,z,p) = 3R_F(x,y,z).$

19.20.13
$2(p-x)R_J(x,y,z,p) = 3R_F(x,y,z) - 3\sqrt{x}\,R_C(yz,p^2),$
$p = x \pm \sqrt{(y-x)(z-x)},$

where x,y,z may be permuted.

When the variables are real and distinct, the various cases of $R_J(x,y,z,p)$ are called *circular (hyperbolic) cases* if $(p-x)(p-y)(p-z)$ is positive (negative), because they typically occur in conjunction with inverse circular (hyperbolic) functions. Cases encountered in dynamical problems are usually circular; hyperbolic cases include Cauchy principal values. If x,y,z are permuted so that

$0 \leq x < y < z$, then the Cauchy principal value of R_J is given by

19.20.14
$$(q + z) R_J(x, y, z, -q)$$
$$= (p - z) R_J(x, y, z, p) - 3R_F(x, y, z)$$
$$+ 3\left(\frac{xyz}{xy + pq}\right)^{1/2} R_C(xy + pq, pq),$$

valid when

19.20.15 $\quad q > 0, \quad p = \dfrac{z(x + y + q) - xy}{z + q},$

or

19.20.16 $\quad p = wy + (1 - w)z, \quad w = \dfrac{z - x}{z + q}, \quad 0 < w < 1.$

Since $x < y < p < z$, p is in a hyperbolic region. In the complete case ($x = 0$) (19.20.14) reduces to

19.20.17
$$(q + z) R_J(0, y, z, -q)$$
$$= (p - z) R_J(0, y, z, p) - 3R_F(0, y, z),$$
$$p = z(y + q)/(z + q), \ w = z/(z + q).$$

19.20(iv) $R_D(x, y, z)$

19.20.18
$$R_D(x, x, x) = x^{-3/2},$$
$$R_D(\lambda x, \lambda y, \lambda z) = \lambda^{-3/2} R_D(x, y, z),$$
$$R_D(0, y, y) = \tfrac{3}{4}\pi\, y^{-3/2},$$
$$R_D(0, 0, z) = \infty.$$

19.20.19 $\quad R_D(x, y, z) \sim 3(xyz)^{-1/2}, \quad z/\sqrt{xy} \to 0.$

19.20.20 $\quad R_D(x, y, y) = \dfrac{3}{2(y - x)}\left(R_C(x, y) - \dfrac{\sqrt{x}}{y}\right),$
$$x \neq y,\ y \neq 0,$$

19.20.21
$$R_D(x, x, z) = \frac{3}{z - x}\left(R_C(z, x) - \frac{1}{\sqrt{z}}\right), \quad x \neq z,\ xz \neq 0.$$

The *second lemniscate constant* is given by

19.20.22
$$\int_0^1 \frac{t^2\, dt}{\sqrt{1 - t^4}} = \tfrac{1}{3} R_D(0, 2, 1) = \frac{(\Gamma(\tfrac{3}{4}))^2}{(2\pi)^{1/2}}$$
$$= 0.59907\,01173\,67796\,10371\ldots.$$

Todd (1975) refers to a proof by T. Schneider that this is a transcendental number. Compare (19.20.2). The *general lemniscatic case* is

19.20.23
$$R_D(x, y, a) = R_{-\frac{3}{4}}\left(\tfrac{5}{4}, \tfrac{1}{2}; a^2, xy\right), \quad a = \tfrac{1}{2}x + \tfrac{1}{2}y.$$

19.20(v) $R_{-a}(\mathbf{b}; \mathbf{z})$

Define $c = \sum_{j=1}^n b_j$. Then

19.20.24 $\quad R_0(\mathbf{b}; \mathbf{z}) = 1, \quad R_N(\mathbf{b}; \mathbf{z}) = \dfrac{N!}{(c)_N} T_N(\mathbf{b}, \mathbf{z}),$
$$N = 0, 1, 2, \ldots,$$

where T_N is defined by (19.19.1). Also,

19.20.25 $\quad R_{-c}(\mathbf{b}; \mathbf{z}) = \displaystyle\prod_{j=1}^n z_j^{-b_j},$

19.20.26
$$R_{-a}(\mathbf{b}; \mathbf{z}) = \prod_{j=1}^n z_j^{-b_j}\, R_{-a'}\left(\mathbf{b}; \mathbf{z^{-1}}\right),$$
$$a + a' = c,\ \mathbf{z^{-1}} = (z_1^{-1}, \ldots, z_n^{-1}).$$

See also (19.16.11) and (19.16.19).

19.21 Connection Formulas

19.21(i) Complete Integrals

Legendre's relation (19.7.1) can be written

19.21.1
$$R_F(0, z + 1, z)\, R_D(0, z + 1, 1)$$
$$+ R_D(0, z + 1, z)\, R_F(0, z + 1, 1) = 3\pi/(2z),$$
$$z \in \mathbb{C}\setminus(-\infty, 0].$$

The case $z = 1$ shows that the product of the two lemniscate constants, (19.20.2) and (19.20.22), is $\pi/4$.

19.21.2 $\quad 3R_F(0, y, z) = z\, R_D(0, y, z) + y\, R_D(0, z, y).$

19.21.3
$$6R_G(0, y, z) = yz(R_D(0, y, z) + R_D(0, z, y))$$
$$= 3z\, R_F(0, y, z) + z(y - z)\, R_D(0, y, z).$$

The complete cases of R_F and R_G have connection formulas resulting from those for the Gauss hypergeometric function (Erdélyi et al. (1953a, §2.9)). Upper signs apply if $0 < \mathrm{ph}\,z < \pi$, and lower signs if $-\pi < \mathrm{ph}\,z < 0$:

19.21.4 $\quad R_F(0, z - 1, z) = R_F(0, 1 - z, 1) \mp i R_F(0, z, 1),$

19.21.5
$$2R_G(0, z - 1, z) = 2R_G(0, 1 - z, 1) \pm i2R_G(0, z, 1)$$
$$+ (z - 1) R_F(0, 1 - z, 1)$$
$$\mp i z R_F(0, z, 1).$$

Let y, z, and p be positive and distinct, and permute y and z to ensure that y does not lie between z and p. The complete case of R_J can be expressed in terms of R_F and R_D:

19.21.6
$$(\sqrt{rp}/z)\, R_J(0, y, z, p) = (r - 1) R_F(0, y, z)\, R_D(p, rz, z)$$
$$+ R_D(0, y, z)\, R_F(p, rz, z),$$
$$r = (y - p)/(y - z) > 0.$$

If $0 < p < z$ and $y = z + 1$, then as $p \to 0$ (19.21.6) reduces to Legendre's relation (19.21.1).

19.21(ii) Incomplete Integrals

$R_D(x,y,z)$ is symmetric only in x and y, but either (nonzero) x or (nonzero) y can be moved to the third position by using

19.21.7
$$(x-y)R_D(y,z,x) + (z-y)R_D(x,y,z) = 3R_F(x,y,z) - 3\sqrt{y/(xz)},$$

or the corresponding equation with x and y interchanged.

19.21.8
$$R_D(y,z,x) + R_D(z,x,y) + R_D(x,y,z) = 3(xyz)^{-1/2},$$

19.21.9
$$xR_D(y,z,x) + yR_D(z,x,y) + zR_D(x,y,z) = 3R_F(x,y,z).$$

19.21.10
$$2R_G(x,y,z) = zR_F(x,y,z) - \tfrac{1}{3}(x-z)(y-z)R_D(x,y,z) + \sqrt{xy/z}, \quad z \neq 0.$$

Because R_G is completely symmetric, x, y, z can be permuted on the right-hand side of (19.21.10) so that $(x-z)(y-z) \leq 0$ if the variables are real, thereby avoiding cancellations when R_G is calculated from R_F and R_D (see §19.36(i)).

19.21.11
$$6R_G(x,y,z) = 3(x+y+z)R_F(x,y,z) - \sum x^2 R_D(y,z,x) = \sum x(y+z)R_D(y,z,x),$$

where both summations extend over the three cyclic permutations of x, y, z.

Connection formulas for $R_{-a}(\mathbf{b};\mathbf{z})$ are given in Carlson (1977b, pp. 99, 101, and 123–124).

19.21(iii) Change of Parameter of R_J

Let x, y, z be real and nonnegative, with at most one of them 0. Change-of-parameter relations can be used to shift the parameter p of R_J from either circular region to the other, or from either hyperbolic region to the other (§19.20(iii)). The latter case allows evaluation of Cauchy principal values (see (19.20.14)).

19.21.12
$$(p-x)R_J(x,y,z,p) + (q-x)R_J(x,y,z,q) = 3R_F(x,y,z) - 3R_C(\xi,\eta),$$

where

19.21.13
$$(p-x)(q-x) = (y-x)(z-x), \quad \xi = yz/x, \quad \eta = pq/x,$$

and x, y, z may be permuted. Also,

19.21.14
$$\eta - \xi = p + q - y - z = \frac{(p-y)(p-z)}{p-x} = \frac{(q-y)(q-z)}{q-x}$$
$$= \frac{(p-y)(q-y)}{x-y} = \frac{(p-z)(q-z)}{x-z}.$$

For each value of p, permutation of x, y, z produces three values of q, one of which lies in the same region as p and two lie in the other region of the same type. In (19.21.12), if x is the largest (smallest) of x, y, and z, then p and q lie in the same region if it is circular (hyperbolic); otherwise p and q lie in different regions, both circular or both hyperbolic. If $x = 0$, then $\xi = \eta = \infty$ and $R_C(\xi,\eta) = 0$; hence

19.21.15
$$pR_J(0,y,z,p) + qR_J(0,y,z,q) = 3R_F(0,y,z), \quad pq = yz.$$

19.22 Quadratic Transformations

19.22(i) Complete Integrals

Let $\Re x > 0$, $\Re y > 0$, $a = (x+y)/2$, and $p \neq 0$. Then

19.22.1
$$R_F(0,x^2,y^2) = R_F(0,xy,a^2),$$

19.22.2
$$2R_G(0,x^2,y^2) = 4R_G(0,xy,a^2) - xy R_F(0,xy,a^2),$$

19.22.3
$$2y^2 R_D(0,x^2,y^2) = \tfrac{1}{4}(y^2-x^2)R_D(0,xy,a^2) + 3R_F(0,xy,a^2).$$

19.22.4
$$(p_\pm^2 - p_\mp^2)R_J(0,x^2,y^2,p^2) = 2(p_\pm^2 - a^2)R_J(0,xy,a^2,p_\pm^2) - 3R_F(0,xy,a^2) + 3\pi/(2p),$$

where

19.22.5
$$2p_\pm = \sqrt{(p+x)(p+y)} \pm \sqrt{(p-x)(p-y)},$$

and hence

19.22.6
$$p_+ p_- = pa, \quad p_+^2 + p_-^2 = p^2 + xy,$$
$$p_+^2 - p_-^2 = \sqrt{(p^2-x^2)(p^2-y^2)},$$
$$4(p_\pm^2 - a^2) = (\sqrt{p^2-x^2} \pm \sqrt{p^2-y^2})^2.$$

Bartky's Transformation

19.22.7
$$2p^2 R_J(0,x^2,y^2,p^2) = v_+ v_- R_J(0,xy,a^2,v_+^2) + 3R_F(0,xy,a^2),$$
$$v_\pm = (p^2 \pm xy)/(2p).$$

If $p = y$, then (19.22.7) reduces to (19.22.3), but if $p = x$ or $p = y$, then both sides of (19.22.4) are 0 by (19.20.9). If $x < p < y$ or $y < p < x$, then p_+ and p_- are complex conjugates.

19.22(ii) Gauss's Arithmetic-Geometric Mean (AGM)

The AGM, $M(a_0, g_0)$, of two positive numbers a_0 and g_0 is defined in §19.8(i). Again, we assume that $a_0 \geq g_0$ (except in (19.22.10)), and define $c_n = \sqrt{a_n^2 - g_n^2}$. Then

$$19.22.8 \qquad \frac{2}{\pi} R_F(0, a_0^2, g_0^2) = \frac{1}{M(a_0, g_0)},$$

$$19.22.9 \qquad \begin{aligned} \frac{4}{\pi} R_G(0, a_0^2, g_0^2) &= \frac{1}{M(a_0, g_0)} \left(a_0^2 - \sum_{n=0}^{\infty} 2^{n-1} c_n^2 \right) \\ &= \frac{1}{M(a_0, g_0)} \left(a_1^2 - \sum_{n=2}^{\infty} 2^{n-1} c_n^2 \right), \end{aligned}$$

and

$$19.22.10 \qquad R_D(0, g_0^2, a_0^2) = \frac{3\pi}{4M(a_0, g_0) a_0^2} \sum_{n=0}^{\infty} Q_n,$$

where

$$19.22.11 \qquad Q_0 = 1, \quad Q_{n+1} = \tfrac{1}{2} Q_n \frac{a_n - g_n}{a_n + g_n}.$$

Q_n has the same sign as $a_0 - g_0$ for $n \geq 1$.

$$19.22.12 \qquad R_J(0, g_0^2, a_0^2, p_0^2) = \frac{3\pi}{4M(a_0, g_0) p_0^2} \sum_{n=0}^{\infty} Q_n,$$

where $p_0 > 0$ and

$$19.22.13 \qquad \begin{aligned} p_{n+1} &= \frac{p_n^2 + a_n g_n}{2 p_n}, \quad \varepsilon_n = \frac{p_n^2 - a_n g_n}{p_n^2 + a_n g_n}, \\ Q_0 &= 1, \quad Q_{n+1} = \tfrac{1}{2} Q_n \varepsilon_n. \end{aligned}$$

(If $p_0 = a_0$, then $p_n = a_n$ and (19.22.13) reduces to (19.22.11).) As $n \to \infty$, p_n and ε_n converge quadratically to $M(a_0, g_0)$ and 0, respectively, and Q_n converges to 0 faster than quadratically. If the last variable of R_J is negative, then the Cauchy principal value is

$$19.22.14 \qquad \begin{aligned} R_J(0, g_0^2, a_0^2, -q_0^2) &= \frac{-3\pi}{4M(a_0, g_0)(q_0^2 + a_0^2)} \\ &\quad \times \left(2 + \frac{a_0^2 - g_0^2}{q_0^2 + g_0^2} \sum_{n=0}^{\infty} Q_n \right), \end{aligned}$$

and (19.22.13) still applies, provided that

$$19.22.15 \qquad p_0^2 = a_0^2 (q_0^2 + g_0^2) / (q_0^2 + a_0^2).$$

19.22(iii) Incomplete Integrals

Let x, y, and z have positive real parts, assume $p \neq 0$, and retain (19.22.5) and (19.22.6). Define

$$19.22.16 \qquad \begin{aligned} a &= (x+y)/2, \\ 2z_{\pm} &= \sqrt{(z+x)(z+y)} \pm \sqrt{(z-x)(z-y)}, \end{aligned}$$

so that

$$19.22.17 \qquad \begin{aligned} z_+ z_- &= za, \quad z_+^2 + z_-^2 = z^2 + xy, \\ z_+^2 - z_-^2 &= \sqrt{(z^2 - x^2)(z^2 - y^2)}, \\ 4(z_{\pm}^2 - a^2) &= (\sqrt{z^2 - x^2} \pm \sqrt{z^2 - y^2})^2. \end{aligned}$$

Then

$$19.22.18 \qquad R_F(x^2, y^2, z^2) = R_F(a^2, z_-^2, z_+^2),$$

$$19.22.19 \qquad \begin{aligned} (z_{\pm}^2 - z_{\mp}^2) R_D(x^2, y^2, z^2) &= 2(z_{\pm}^2 - a^2) R_D(a^2, z_{\mp}^2, z_{\pm}^2) \\ &\quad - 3 R_F(x^2, y^2, z^2) + (3/z), \end{aligned}$$

$$19.22.20 \qquad \begin{aligned} (p_{\pm}^2 - p_{\mp}^2) R_J(x^2, y^2, z^2, p^2) &= 2(p_{\pm}^2 - a^2) R_J(a^2, z_+^2, z_-^2, p_{\pm}^2) \\ &\quad - 3 R_F(x^2, y^2, z^2) + 3 R_C(z^2, p^2), \end{aligned}$$

$$19.22.21 \qquad \begin{aligned} 2 R_G(x^2, y^2, z^2) &= 4 R_G(a^2, z_+^2, z_-^2) \\ &\quad - xy\, R_F(x^2, y^2, z^2) - z, \end{aligned}$$

$$19.22.22 \qquad R_C(x^2, y^2) = R_C(a^2, ay).$$

If x, y, z are real and positive, then (19.22.18)–(19.22.21) are ascending Landen transformations when $x, y < z$ (implying $a < z_- < z_+$), and descending Gauss transformations when $z < x, y$ (implying $z_+ < z_- < a$). Ascent and descent correspond respectively to increase and decrease of k in Legendre's notation. Descending Gauss transformations include, as special cases, transformations of complete integrals into complete integrals; ascending Landen transformations do not.

If $p = x$ or $p = y$, then (19.22.20) reduces to $0 = 0$ by (19.20.13), and if $z = x$ or $z = y$ then (19.22.19) reduces to $0 = 0$ by (19.20.20) and (19.22.22). If $x < z < y$ or $y < z < x$, then z_+ and z_- are complex conjugates. However, if x and y are complex conjugates and z and p are real, then the right-hand sides of all transformations in §§19.22(i) and 19.22(iii)—except (19.22.3) and (19.22.22)—are free of complex numbers and $p_{\pm}^2 - p_{\mp}^2 = \pm |p^2 - x^2| \neq 0$.

The transformations inverse to the ones just described are the descending Landen transformations and the ascending Gauss transformations. The equations inverse to (19.22.5) and (19.22.16) are given by

$$19.22.23 \qquad \begin{aligned} x + y &= 2a, \quad x - y = (2/a)\sqrt{(a^2 - z_+^2)(a^2 - z_-^2)}, \\ z &= z_+ z_- / a, \end{aligned}$$

and the corresponding equations with z, z_+, and z_- replaced by p, p_+, and p_-, respectively. These relations need to be used with caution because y is negative when $0 < a < z_+ z_- (z_+^2 + z_-^2)^{-1/2}$.

19.23 Integral Representations

In (19.23.1)–(19.23.3) we assume $\Re y > 0$ and $\Re z > 0$.

19.23.1 $\quad R_F(0,y,z) = \int_0^{\pi/2} (y\cos^2\theta + z\sin^2\theta)^{-1/2}\,d\theta,$

19.23.2 $\quad R_G(0,y,z) = \frac{1}{2}\int_0^{\pi/2} (y\cos^2\theta + z\sin^2\theta)^{1/2}\,d\theta,$

19.23.3
$$R_D(0,y,z) = 3\int_0^{\pi/2} (y\cos^2\theta + z\sin^2\theta)^{-3/2}\sin^2\theta\,d\theta.$$

19.23.4
$$R_F(0,y,z) = \frac{2}{\pi}\int_0^{\pi/2} R_C(y, z\cos^2\theta)\,d\theta$$
$$= \frac{2}{\pi}\int_0^{\infty} R_C(y\cosh^2 t, z)\,dt.$$

19.23.5
$$R_F(x,y,z) = \frac{2}{\pi}\int_0^{\pi/2} R_C(x, y\cos^2\theta + z\sin^2\theta)\,d\theta,$$
$$\Re y > 0, \Re z > 0,$$

19.23.6
$4\pi R_F(x,y,z)$
$$= \int_0^{2\pi}\int_0^{\pi} \frac{\sin\theta\,d\theta\,d\phi}{(x\sin^2\theta\cos^2\phi + y\sin^2\theta\sin^2\phi + z\cos^2\theta)^{1/2}},$$

where x, y, and z have positive real parts—except that at most one of them may be 0.

In (19.23.7)–(19.23.10) one or more of the variables may be 0 if the integral converges. In (19.23.8) $n = 2$, and in (19.23.9) $n = 3$. Also, in (19.23.8) and (19.23.10) B denotes the beta function (§5.12).

19.23.7 $\quad R_G(x,y,z) = \frac{1}{4}\int_0^{\infty} \frac{1}{\sqrt{t+x}\sqrt{t+y}\sqrt{t+z}}\left(\frac{x}{t+x} + \frac{y}{t+y} + \frac{z}{t+z}\right)t\,dt, \quad x,y,z \in \mathbb{C}\backslash(-\infty, 0].$

19.23.8 $\quad R_{-a}(\mathbf{b};\mathbf{z}) = \frac{2}{B(b_1, b_2)}\int_0^{\pi/2}(z_1\cos^2\theta + z_2\sin^2\theta)^{-a}(\cos\theta)^{2b_1-1}(\sin\theta)^{2b_2-1}\,d\theta, \quad b_1, b_2 > 0; \Re z_1, \Re z_2 > 0.$

With l_1, l_2, l_3 denoting any permutation of $\sin\theta\cos\phi$, $\sin\theta\sin\phi$, $\cos\theta$,

19.23.9 $\quad R_{-a}(\mathbf{b};\mathbf{z}) = \frac{4\,\Gamma(b_1+b_2+b_3)}{\Gamma(b_1)\Gamma(b_2)\Gamma(b_3)}\int_0^{\pi/2}\int_0^{\pi/2}\left(\sum_{j=1}^{3} z_j l_j^2\right)^{-a}\prod_{j=1}^{3} l_j^{2b_j-1}\sin\theta\,d\theta\,d\phi, \quad b_j > 0, \Re z_j > 0.$

19.23.10
$$R_{-a}(\mathbf{b};\mathbf{z}) = \frac{1}{B(a,a')}\int_0^1 u^{a-1}(1-u)^{a'-1}\prod_{j=1}^{n}(1-u+uz_j)^{-b_j}\,du, \quad a, a' > 0; a+a' = \sum_{j=1}^{n} b_j; z_j \in \mathbb{C}\backslash(-\infty, 0].$$

For generalizations of (19.16.3) and (19.23.8) see Carlson (1964, (6.2), (6.12), and (6.1)).

19.24 Inequalities

19.24(i) Complete Integrals

The condition $y \leq z$ for (19.24.1) and (19.24.2) serves only to identify y as the smaller of the two nonzero variables of a symmetric function; it does not restrict validity.

19.24.1
$$\ln 4 \leq \sqrt{z}\,R_F(0,y,z) + \ln\sqrt{y/z} \leq \tfrac{1}{2}\pi, \quad 0 < y \leq z,$$

19.24.2 $\quad \tfrac{1}{2} \leq z^{-1/2} R_G(0,y,z) \leq \tfrac{1}{4}\pi, \quad 0 \leq y \leq z,$

19.24.3
$$\left(\frac{y^{3/2}+z^{3/2}}{2}\right)^{2/3} \leq \frac{4}{\pi} R_G(0, y^2, z^2) \leq \left(\frac{y^2+z^2}{2}\right)^{1/2},$$
$$y > 0, z > 0.$$

If y, z, and p are positive, then

19.24.4
$$\frac{2}{\sqrt{p}}(2yz+yp+zp)^{-1/2} \leq \frac{4}{3\pi} R_J(0,y,z,p) \leq (yzp^2)^{-3/8}.$$

Inequalities for $R_D(0,y,z)$ are included as the case $p = z$.

A series of successively sharper inequalities is obtained from the AGM process (§19.8(i)) with $a_0 \geq g_0 > 0$:

19.24.5 $\quad \dfrac{1}{a_n} \leq \dfrac{2}{\pi} R_F(0, a_0^2, g_0^2) \leq \dfrac{1}{g_n}, \quad n = 0, 1, 2, \ldots,$

where

19.24.6 $\quad a_{n+1} = (a_n + g_n)/2, \quad g_{n+1} = \sqrt{a_n g_n}.$

Other inequalities can be obtained by applying Carlson (1966, Theorems 2 and 3) to (19.16.20)–(19.16.23). Approximations and one-sided inequalities for $R_G(0,y,z)$ follow from those given in §19.9(i) for the length $L(a,b)$ of an ellipse with semiaxes a and b, since

19.24.7 $\quad L(a,b) = 8 R_G(0, a^2, b^2).$

For $x > 0$, $y > 0$, and $x \neq y$, the complete cases of R_F and R_G satisfy

19.24.8
$$R_F(x,y,0)\, R_G(x,y,0) > \tfrac{1}{8}\pi^2,$$
$$R_F(x,y,0) + 2R_G(x,y,0) > \pi.$$

Also, with the notation of (19.24.6),

19.24.9
$$\tfrac{1}{2} g_1^2 \leq \frac{R_G(a_0^2, g_0^2, 0)}{R_F(a_0^2, g_0^2, 0)} \leq \tfrac{1}{2} a_1^2,$$

with equality iff $a_0 = g_0$.

19.24(ii) Incomplete Integrals

Inequalities for $R_{-a}(\mathbf{b}; \mathbf{z})$ in Carlson (1966, Theorems 2 and 3) can be applied to (19.16.14)–(19.16.17). All variables are positive, and equality occurs iff all variables are equal.

Examples

19.24.10
$$\frac{3}{\sqrt{x} + \sqrt{y} + \sqrt{z}} \leq R_F(x,y,z) \leq \frac{1}{(xyz)^{1/6}},$$

19.24.11
$$\left(\frac{5}{\sqrt{x} + \sqrt{y} + \sqrt{z} + 2\sqrt{p}}\right)^3 \leq R_J(x,y,z,p)$$
$$\leq (xyzp^2)^{-3/10},$$

19.24.12
$$\tfrac{1}{3}(\sqrt{x} + \sqrt{y} + \sqrt{z})$$
$$\leq R_G(x,y,z) \leq \min\left(\sqrt{\frac{x+y+z}{3}}, \frac{x^2+y^2+z^2}{3\sqrt{xyz}}\right).$$

Inequalities for $R_C(x,y)$ and $R_D(x,y,z)$ are included as special cases (see (19.16.6) and (19.16.5)).

Other inequalities for $R_F(x,y,z)$ are given in Carlson (1970).

If $a\ (\neq 0)$ is real, all components of $\mathbf{b}$ and $\mathbf{z}$ are positive, and the components of $\mathbf{z}$ are not all equal, then

19.24.13
$$R_a(\mathbf{b};\mathbf{z})\, R_{-a}(\mathbf{b};\mathbf{z}) > 1, \quad R_a(\mathbf{b};\mathbf{z}) + R_{-a}(\mathbf{b};\mathbf{z}) > 2;$$

see Neuman (2003, (2.13)). Special cases with $a = \pm\tfrac{1}{2}$ are (19.24.8) (because of (19.16.20), (19.16.23)), and

19.24.14
$$R_F(x,y,z)\, R_G(x,y,z) > 1,$$
$$R_F(x,y,z) + R_G(x,y,z) > 2.$$

The same reference also gives upper and lower bounds for symmetric integrals in terms of their elementary degenerate cases. These bounds include a sharper but more complicated lower bound than that supplied in the next result:

19.24.15
$$R_C\bigl(x, \tfrac{1}{2}(y+z)\bigr) \leq R_F(x,y,z) \leq R_C(x, \sqrt{yz}), \quad x \geq 0,$$

with equality iff $y = z$.

19.25 Relations to Other Functions

19.25(i) Legendre's Integrals as Symmetric Integrals

Let $k'^2 = 1 - k^2$ and $c = \csc^2\phi$. Then

19.25.1
$$K(k) = R_F(0, k'^2, 1), \quad E(k) = 2R_G(0, k'^2, 1),$$
$$E(k) = \tfrac{1}{3}k'^2\bigl(R_D(0, k'^2, 1) + R_D(0, 1, k'^2)\bigr),$$
$$K(k) - E(k) = k^2 D(k) = \tfrac{1}{3}k^2 R_D(0, k'^2, 1),$$
$$E(k) - k'^2 K(k) = \tfrac{1}{3}k^2 k'^2 R_D(0, 1, k'^2).$$

19.25.2 $\Pi(\alpha^2, k) - K(k) = \tfrac{1}{3}\alpha^2 R_J(0, k'^2, 1, 1-\alpha^2).$

19.25.3 $\Pi(\alpha^2, k) = \tfrac{1}{2}\pi R_{-\frac{1}{2}}\bigl(\tfrac{1}{2}, -\tfrac{1}{2}, 1; k'^2, 1, 1-\alpha^2\bigr),$

with Cauchy principal value

19.25.4
$$\Pi(\alpha^2, k) = -\tfrac{1}{3}(k^2/\alpha^2) R_J\bigl(0, 1-k^2, 1, 1-(k^2/\alpha^2)\bigr),$$
$$-\infty < k^2 < 1 < \alpha^2.$$

19.25.5 $F(\phi, k) = R_F(c-1, c-k^2, c),$

19.25.6 $\dfrac{\partial F(\phi, k)}{\partial k} = \tfrac{1}{3} k\, R_D(c-1, c, c-k^2).$

19.25.7
$$E(\phi, k) = 2R_G(c-1, c-k^2, c)$$
$$\quad - (c-1)\, R_F(c-1, c-k^2, c)$$
$$\quad - \sqrt{(c-1)(c-k^2)/c},$$

19.25.8 $E(\phi, k) = R_{-\frac{1}{2}}\bigl(\tfrac{1}{2}, -\tfrac{1}{2}, \tfrac{3}{2}; c-1, c-k^2, c\bigr),$

19.25.9
$$E(\phi, k) = R_F(c-1, c-k^2, c) - \tfrac{1}{3}k^2 R_D(c-1, c-k^2, c),$$

19.25.10
$$E(\phi, k) = k'^2\, R_F(c-1, c-k^2, c)$$
$$\quad + \tfrac{1}{3}k^2 k'^2 R_D(c-1, c, c-k^2)$$
$$\quad + k^2 \sqrt{(c-1)/(c(c-k^2))}, \quad c > k^2,$$

19.25.11
$$E(\phi, k) = -\tfrac{1}{3}k'^2 R_D(c-k^2, c, c-1)$$
$$\quad + \sqrt{(c-k^2)/(c(c-1))}, \quad \phi \neq \tfrac{1}{2}\pi.$$

Equations (19.25.9)–(19.25.11) correspond to three (nonzero) choices for the last variable of R_D; see (19.21.7). All terms on the right-hand sides are nonnegative when $k^2 \leq 0$, $0 \leq k^2 \leq 1$, or $1 \leq k^2 \leq c$, respectively.

19.25.12 $\dfrac{\partial E(\phi, k)}{\partial k} = -\tfrac{1}{3} k\, R_D(c-1, c-k^2, c).$

19.25.13 $D(\phi, k) = \tfrac{1}{3} R_D(c-1, c-k^2, c).$

19.25.14
$$\Pi(\phi, \alpha^2, k) - F(\phi, k) = \tfrac{1}{3}\alpha^2 R_J(c-1, c-k^2, c, c-\alpha^2),$$

19.25.15
$$\Pi(\phi, \alpha^2, k) = R_{-\frac{1}{2}}\bigl(\tfrac{1}{2}, \tfrac{1}{2}, -\tfrac{1}{2}, 1; c-1, c-k^2, c, c-\alpha^2\bigr).$$

If $\alpha^2 > c$, then the Cauchy principal value is

19.25.16
$$\Pi(\phi, \alpha^2, k) = -\tfrac{1}{3}\omega^2 R_J(c-1, c-k^2, c, c-\omega^2) \\ + \sqrt{\frac{(c-1)(c-k^2)}{(\alpha^2-1)(1-\omega^2)}} \\ \times R_C(c(\alpha^2-1)(1-\omega^2), (\alpha^2-c)(c-\omega^2)), \\ \omega^2 = k^2/\alpha^2.$$

The transformations in §19.7(ii) result from the symmetry and homogeneity of functions on the right-hand sides of (19.25.5), (19.25.7), and (19.25.14). For example, if we write (19.25.5) as

19.25.17 $\quad F(\phi, k) = R_F(x, y, z),$

with

19.25.18 $\quad (x, y, z) = (c-1, c-k^2, c),$

then the five nontrivial permutations of x, y, z that leave R_F invariant change $k^2 \,(= (z-y)/(z-x))$ into $1/k^2$, k'^2, $1/k'^2$, $-k^2/k'^2$, $-k'^2/k^2$, and $\sin\phi \,(= \sqrt{(z-x)/z})$ into $k\sin\phi$, $-i\tan\phi$, $-ik'\tan\phi$, $(k'\sin\phi)/\sqrt{1-k^2\sin^2\phi}$, $-ik\sin\phi/\sqrt{1-k^2\sin^2\phi}$. Thus the five permutations induce five transformations of Legendre's integrals (and also of the Jacobian elliptic functions).

The three changes of parameter of $\Pi(\phi, \alpha^2, k)$ in §19.7(iii) are unified in (19.21.12) by way of (19.25.14).

19.25(ii) Bulirsch's Integrals as Symmetric Integrals

Let $r = 1/x^2$. Then

19.25.19
$$\operatorname{cel}(k_c, p, a, b) = a\, R_F(0, k_c^2, 1) \\ + \tfrac{1}{3}(b - pa) R_J(0, k_c^2, 1, p),$$

19.25.20
$$\operatorname{el1}(x, k_c) = R_F(r, r + k_c^2, r + 1),$$

19.25.21
$$\operatorname{el2}(x, k_c, a, b) = a\operatorname{el1}(x, k_c) \\ + \tfrac{1}{3}(b-a) R_D(r, r+k_c^2, r+1),$$

19.25.22
$$\operatorname{el3}(x, k_c, p) = \operatorname{el1}(x, k_c) \\ + \tfrac{1}{3}(1-p) R_J(r, r+k_c^2, r+1, r+p).$$

19.25(iii) Symmetric Integrals as Legendre's Integrals

Assume $0 \leq x \leq y \leq z$, $x < z$, and $p > 0$. Let

19.25.23
$$\phi = \arccos\sqrt{x/z} = \arcsin\sqrt{(z-x)/z}, \\ k = \sqrt{\frac{z-y}{z-x}}, \quad \alpha^2 = \frac{z-p}{z-x},$$

with $\alpha \neq 0$. Then

19.25.24 $\quad (z-x)^{1/2} R_F(x, y, z) = F(\phi, k),$

19.25.25
$$(z-x)^{3/2} R_D(x, y, z) = (3/k^2)(F(\phi, k) - E(\phi, k)),$$

19.25.26
$$(z-x)^{3/2} R_J(x, y, z, p) = (3/\alpha^2)(\Pi(\phi, \alpha^2, k) - F(\phi, k)),$$

19.25.27
$$2(z-x)^{-1/2} R_G(x, y, z) = E(\phi, k) + (\cot\phi)^2 F(\phi, k) \\ + (\cot\phi)\sqrt{1 - k^2 \sin^2\phi}.$$

19.25(iv) Theta Functions

For relations of symmetric integrals to theta functions, see §20.9(i).

19.25(v) Jacobian Elliptic Functions

For the notation see §§22.2, 22.15, and 22.16(i).

With $0 \leq k^2 \leq 1$ and p, q, r any permutation of the letters c, d, n, define

19.25.28 $\quad \Delta(\mathrm{p},\mathrm{q}) = \operatorname{ps}^2(u,k) - \operatorname{qs}^2(u,k) = -\Delta(\mathrm{q},\mathrm{p}),$

which implies

19.25.29 $\quad \Delta(\mathrm{n},\mathrm{d}) = k^2, \quad \Delta(\mathrm{d},\mathrm{c}) = k'^2, \quad \Delta(\mathrm{n},\mathrm{c}) = 1.$

If $\operatorname{cs}^2(u,k) \geq 0$, then

19.25.30 $\quad \operatorname{am}(u,k) = R_C(\operatorname{cs}^2(u,k), \operatorname{ns}^2(u,k)),$

19.25.31 $\quad u = R_F(\operatorname{ps}^2(u,k), \operatorname{qs}^2(u,k), \operatorname{rs}^2(u,k));$

compare (19.25.35) and (20.9.3).

19.25.32
$$\operatorname{arcps}(x,k) = R_F(x^2, x^2 + \Delta(\mathrm{q},\mathrm{p}), x^2 + \Delta(\mathrm{r},\mathrm{p})),$$

19.25.33
$$\operatorname{arcsp}(x,k) = x R_F(1, 1 + \Delta(\mathrm{q},\mathrm{p})x^2, 1 + \Delta(\mathrm{r},\mathrm{p})x^2),$$

19.25.34
$$\operatorname{arcpq}(x,k) = \sqrt{w} R_F(x^2, 1, 1 + \Delta(\mathrm{r},\mathrm{q})w), \\ w = (1-x^2)/\Delta(\mathrm{q},\mathrm{p}),$$

where we assume $0 \leq x^2 \leq 1$ if $x = \operatorname{sn}$, cn, or cd; $x^2 \geq 1$ if $x = \operatorname{ns}$, nc, or dc; x real if $x = \operatorname{cs}$ or sc; $k' \leq x \leq 1$ if $x = \operatorname{dn}$; $1 \leq x \leq 1/k'$ if $x = \operatorname{nd}$; $x^2 \geq k'^2$ if $x = \operatorname{ds}$; $0 \leq x^2 \leq 1/k'^2$ if $x = \operatorname{sd}$.

For the use of R-functions with $\Delta(\mathrm{p},\mathrm{q})$ in unifying other properties of Jacobian elliptic functions, see Carlson (2004, 2006a,b, 2008).

Inversions of 12 elliptic integrals of the first kind, producing the 12 Jacobian elliptic functions, are combined and simplified by using the properties of $R_F(x, y, z)$. See (19.29.19), Carlson (2005), and (22.15.11), and compare with Abramowitz and Stegun (1964, Eqs. (17.4.41)–(17.4.52)). For analogous integrals of the second kind, which are not invertible in terms of single-valued functions, see (19.29.20) and (19.29.21) and compare with Gradshteyn and Ryzhik (2000, §3.153,1–10 and §3.156,1–9).

19.25(vi) Weierstrass Elliptic Functions

For the notation see §23.2.

19.25.35 $\quad z = R_F(\wp(z) - e_1, \wp(z) - e_2, \wp(z) - e_3),$

provided that

19.25.36 $\quad \wp(z) - e_j \in \mathbb{C} \setminus (-\infty, 0], \qquad j = 1, 2, 3,$

and the left-hand side does not vanish for more than one value of j. Also,

19.25.37
$$\zeta(z) + z\wp(z) = 2R_G(\wp(z) - e_1, \wp(z) - e_2, \wp(z) - e_3).$$

In (19.25.38) and (19.25.39) j, k, ℓ is any permutation of the numbers $1, 2, 3$.

19.25.38 $\quad \omega_j = R_F(0, e_j - e_k, e_j - e_\ell),$

19.25.39 $\quad \eta_j + \omega_j e_j = 2R_G(0, e_j - e_k, e_j - e_\ell).$

Lastly,

19.25.40 $\quad z = \sigma(z) R_F\left(\sigma_1^2(z), \sigma_2^2(z), \sigma_3^2(z)\right),$

where

19.25.41 $\quad \sigma_j(z) = \exp(-\eta_j z) \sigma(z + \omega_j) / \sigma(\omega_j), \quad j = 1, 2, 3.$

19.25(vii) Hypergeometric Function

19.25.42 $\quad {}_2F_1(a, b; c; z) = R_{-a}(b, c - b; 1 - z, 1),$

19.25.43
$R_{-a}(b_1, b_2; z_1, z_2) = z_2^{-a} {}_2F_1(a, b_1; b_1 + b_2; 1 - (z_1/z_2)).$

For these results and extensions to the Appell function F_1 (§16.13) and Lauricella's function F_D see Carlson (1963). (F_1 and F_D are equivalent to the R-function of 3 and n variables, respectively, but lack full symmetry.)

19.26 Addition Theorems

19.26(i) General Formulas

In this subsection, and also §§19.26(ii) and 19.26(iii), we assume that λ, x, y, z are positive, except that at most one of x, y, z can be 0.

19.26.1
$$R_F(x + \lambda, y + \lambda, z + \lambda) + R_F(x + \mu, y + \mu, z + \mu) = R_F(x, y, z),$$

where $\mu > 0$ and

19.26.2
$$x + \mu = \lambda^{-2} \left(\sqrt{(x + \lambda)yz} + \sqrt{x(y + \lambda)(z + \lambda)} \right)^2,$$

with corresponding equations for $y + \mu$ and $z + \mu$ obtained by permuting x, y, z. Also,

19.26.3 $\quad \sqrt{z} = \dfrac{\xi\zeta' + \eta'\zeta - \xi\eta'}{\sqrt{\xi\eta\zeta'} + \sqrt{\xi'\eta'\zeta}},$

where

19.26.4
$$\begin{aligned}(\xi, \eta, \zeta) &= (x + \lambda, y + \lambda, z + \lambda), \\ (\xi', \eta', \zeta') &= (x + \mu, y + \mu, z + \mu),\end{aligned}$$

with $\sqrt{x}$ and $\sqrt{y}$ obtained by permuting $x, y,$ and z. (Note that $\xi\zeta' + \eta'\zeta - \xi\eta' = \xi'\zeta + \eta\zeta' - \xi'\eta$.) Equivalent forms of (19.26.2) are given by

19.26.5
$$\mu = \lambda^{-2} \left(\sqrt{xyz} + \sqrt{(x + \lambda)(y + \lambda)(z + \lambda)} \right)^2 - \lambda - x - y - z,$$

and

19.26.6 $\quad (\lambda\mu - xy - xz - yz)^2 = 4xyz(\lambda + \mu + x + y + z).$

Also,

19.26.7
$$\begin{aligned}&R_D(x + \lambda, y + \lambda, z + \lambda) + R_D(x + \mu, y + \mu, z + \mu) \\ &= R_D(x, y, z) - \dfrac{3}{\sqrt{z(z + \lambda)(z + \mu)}},\end{aligned}$$

19.26.8
$$\begin{aligned}&2R_G(x + \lambda, y + \lambda, z + \lambda) + 2R_G(x + \mu, y + \mu, z + \mu) \\ &= 2R_G(x, y, z) + \lambda R_F(x + \lambda, y + \lambda, z + \lambda) \\ &\quad + \mu R_F(x + \mu, y + \mu, z + \mu) + \sqrt{\lambda + \mu + x + y + z}.\end{aligned}$$

19.26.9
$$\begin{aligned}&R_J(x + \lambda, y + \lambda, z + \lambda, p + \lambda) \\ &\quad + R_J(x + \mu, y + \mu, z + \mu, p + \mu) \\ &= R_J(x, y, z, p) - 3R_C(\gamma - \delta, \gamma),\end{aligned}$$

where

19.26.10 $\quad \gamma = p(p + \lambda)(p + \mu), \quad \delta = (p - x)(p - y)(p - z).$

Lastly,

19.26.11 $\quad R_C(x + \lambda, y + \lambda) + R_C(x + \mu, y + \mu) = R_C(x, y),$

where $\lambda > 0, y > 0, x \geq 0,$ and

19.26.12
$$\begin{aligned}x + \mu &= \lambda^{-2}(\sqrt{x + \lambda}\, y + \sqrt{x}(y + \lambda))^2, \\ y + \mu &= (y(y + \lambda)/\lambda^2)(\sqrt{x} + \sqrt{x + \lambda})^2.\end{aligned}$$

Equivalent forms of (19.26.11) are given by

19.26.13
$$\begin{aligned}&R_C(\alpha^2, \alpha^2 - \theta) + R_C(\beta^2, \beta^2 - \theta) \\ &= R_C(\sigma^2, \sigma^2 - \theta), \quad \sigma = (\alpha\beta + \theta)/(\alpha + \beta),\end{aligned}$$

where $0 < \gamma^2 - \theta < \gamma^2$ for $\gamma = \alpha, \beta, \sigma$, except that $\sigma^2 - \theta$ can be 0, and

19.26.14
$$\begin{aligned}&(p - y) R_C(x, p) + (q - y) R_C(x, q) \\ &= (\eta - \xi) R_C(\xi, \eta), \quad x \geq 0, y \geq 0; p, q \in \mathbb{R} \setminus \{0\},\end{aligned}$$

where

19.26.15
$$\begin{aligned}(p - x)(q - x) &= (y - x)^2, \quad \xi = y^2/x, \\ \eta &= pq/x, \quad \eta - \xi = p + q - 2y.\end{aligned}$$

19.26(ii) Case $x = 0$

If $x = 0$, then $\lambda\mu = yz$. For example,

19.26.16
$$R_F(\lambda, y+\lambda, z+\lambda) = R_F(0, y, z) - R_F(\mu, y+\mu, z+\mu), \quad \lambda\mu = yz.$$

An equivalent version for R_C is

19.26.17
$$\sqrt{\alpha}\, R_C(\beta, \alpha+\beta) + \sqrt{\beta}\, R_C(\alpha, \alpha+\beta) = \pi/2, \quad \alpha, \beta \in \mathbb{C}\setminus(-\infty, 0), \; \alpha+\beta > 0.$$

19.26(iii) Duplication Formulas

19.26.18
$$R_F(x, y, z) = 2R_F(x+\lambda, y+\lambda, z+\lambda) = R_F\left(\frac{x+\lambda}{4}, \frac{y+\lambda}{4}, \frac{z+\lambda}{4}\right),$$

where

19.26.19 $\quad \lambda = \sqrt{x}\sqrt{y} + \sqrt{y}\sqrt{z} + \sqrt{z}\sqrt{x}.$

19.26.20
$$R_D(x, y, z) = 2R_D(x+\lambda, y+\lambda, z+\lambda) + \frac{3}{\sqrt{z}(z+\lambda)}.$$

19.26.21
$$2R_G(x, y, z) = 4R_G(x+\lambda, y+\lambda, z+\lambda) - \lambda R_F(x, y, z) - \sqrt{x} - \sqrt{y} - \sqrt{z}.$$

19.26.22
$$R_J(x, y, z, p) = 2R_J(x+\lambda, y+\lambda, z+\lambda, p+\lambda) + 3R_C(\alpha^2, \beta^2),$$

where

19.26.23
$$\alpha = p(\sqrt{x}+\sqrt{y}+\sqrt{z}) + \sqrt{x}\sqrt{y}\sqrt{z}, \quad \beta = \sqrt{p}(p+\lambda),$$
$$\beta \pm \alpha = (\sqrt{p} \pm \sqrt{x})(\sqrt{p} \pm \sqrt{y})(\sqrt{p} \pm \sqrt{z}),$$
$$\beta^2 - \alpha^2 = (p-x)(p-y)(p-z),$$

either upper or lower signs being taken throughout.

The equations inverse to $z+\lambda = (\sqrt{z}+\sqrt{x})(\sqrt{z}+\sqrt{y})$ and the two other equations obtained by permuting x, y, z (see (19.26.19)) are

19.26.24
$$z = (\xi\zeta + \eta\zeta - \xi\eta)^2/(4\xi\eta\zeta),$$
$$(\xi, \eta, \zeta) = (x+\lambda, y+\lambda, z+\lambda),$$

and two similar equations obtained by exchanging z with x (and ζ with ξ), or z with y (and ζ with η).

Next,

19.26.25
$$R_C(x, y) = 2R_C(x+\lambda, y+\lambda), \quad \lambda = y + 2\sqrt{x}\sqrt{y}.$$

Equivalent forms are given by (19.22.22). Also,

19.26.26
$$R_C(x^2, y^2) = R_C(a^2, ay),$$
$$a = (x+y)/2, \; \Re x \geq 0, \Re y > 0,$$

and

19.26.27
$$R_C(x^2, x^2 - \theta) = 2R_C(s^2, s^2 - \theta),$$
$$s = x + \sqrt{x^2 - \theta}, \; \theta \neq x^2 \text{ or } s^2.$$

19.27 Asymptotic Approximations and Expansions

19.27(i) Notation

Throughout this section

19.27.1
$$a = \tfrac{1}{2}(x+y), \quad b = \tfrac{1}{2}(y+z), \quad c = \tfrac{1}{3}(x+y+z),$$
$$f = (xyz)^{1/3}, \quad g = (xy)^{1/2}, \quad h = (yz)^{1/2}.$$

19.27(ii) $R_F(x, y, z)$

Assume x, y, and z are real and nonnegative and at most one of them is 0. Then

19.27.2
$$R_F(x, y, z) = \frac{1}{2\sqrt{z}}\left(\ln\frac{8z}{a+g}\right)\left(1 + O\left(\frac{a}{z}\right)\right), \quad a/z \to 0.$$

19.27.3
$$R_F(x, y, z) = R_F(0, y, z) - \frac{1}{\sqrt{h}}\left(\sqrt{\frac{x}{h}} + O\left(\frac{x}{h}\right)\right), \quad x/h \to 0.$$

19.27(iii) $R_G(x, y, z)$

Assume x, y, and z are real and nonnegative and at most one of them is 0. Then

19.27.4 $\quad R_G(x, y, z) = \frac{\sqrt{z}}{2}\left(1 + O\left(\frac{a}{z}\ln\frac{z}{a}\right)\right), \quad a/z \to 0.$

19.27.5
$$R_G(x, y, z) = R_G(0, y, z) + \sqrt{x}\, O\left(\sqrt{x/h}\right), \quad x/h \to 0.$$

19.27.6
$$R_G(0, y, z) = \frac{\sqrt{z}}{2} + \frac{y}{8\sqrt{z}}\left(\ln\left(\frac{16z}{y}\right) - 1\right)\left(1 + O\left(\frac{y}{z}\right)\right),$$
$$y/z \to 0.$$

19.27(iv) $R_D(x, y, z)$

Assume x and y are real and nonnegative, at most one of them is 0, and $z > 0$. Then

19.27.7
$$R_D(x, y, z) = \frac{3}{2z^{3/2}}\left(\ln\left(\frac{8z}{a+g}\right) - 2\right)\left(1 + O\left(\frac{a}{z}\right)\right),$$
$$a/z \to 0.$$

19.27.8
$$R_D(x, y, z) = \frac{3}{\sqrt{xyz}} - \frac{6}{xy}R_G(x, y, 0)\left(1 + O\left(\frac{z}{g}\right)\right),$$
$$z/g \to 0.$$

19.27.9
$$R_D(x, y, z) = \frac{3}{\sqrt{xz}(\sqrt{y}+\sqrt{z})}\left(1 + O\left(\frac{b}{x}\ln\frac{x}{b}\right)\right),$$
$$b/x \to 0.$$

19.27.10
$$R_D(x,y,z) = R_D(0,y,z) - \frac{3\sqrt{x}}{hz}\left(1 + O\left(\sqrt{\frac{x}{h}}\right)\right),$$
$$x/h \to 0.$$

19.27(v) $R_J(x,y,z,p)$

Assume x, y, and z are real and nonnegative, at most one of them is 0, and $p > 0$. Then

19.27.11
$$R_J(x,y,z,p) = \frac{3}{p} R_F(x,y,z) - \frac{3\pi}{2p^{3/2}}\left(1 + O\left(\sqrt{\frac{c}{p}}\right)\right),$$
$$c/p \to 0.$$

19.27.12
$$R_J(x,y,z,p) = \frac{3}{2\sqrt{xyz}}\left(\ln\left(\frac{4f}{p}\right) - 2\right)\left(1 + O\left(\frac{p}{f}\right)\right),$$
$$p/f \to 0.$$

19.27.13
$$R_J(x,y,z,p) = \frac{3}{2\sqrt{zp}}\left(\ln\left(\frac{8z}{a+g}\right) - 2R_C\left(1,\frac{p}{z}\right)\right.$$
$$\left. + O\left(\left(\frac{a}{z} + \frac{a}{p}\right)\ln\frac{p}{a}\right)\right),$$
$$\max(x,y)/\min(z,p) \to 0.$$

19.27.14
$$R_J(x,y,z,p) = \frac{3}{\sqrt{yz}} R_C(x,p) - \frac{6}{yz} R_G(0,y,z)$$
$$+ O\left(\frac{\sqrt{x+2p}}{yz}\right),$$
$$\max(x,p)/\min(y,z) \to 0.$$

19.27.15
$$R_J(x,y,z,p) = R_J(0,y,z,p)$$
$$- \frac{3\sqrt{x}}{hp}\left(1 + O\left(\left(\frac{b}{h}+\frac{h}{p}\right)\sqrt{\frac{x}{h}}\right)\right),$$
$$x/\min(y,z,p) \to 0.$$

19.27.16
$$R_J(x,y,z,p) = (3/\sqrt{x}) R_C\big((h+p)^2, 2(b+h)p\big)$$
$$+ O\left(\frac{1}{x^{3/2}} \ln\frac{x}{b+h}\right),$$
$$\max(y,z,p)/x \to 0.$$

19.27(vi) Asymptotic Expansions

The approximations in §§19.27(i)–19.27(v) are furnished with upper and lower bounds by Carlson and Gustafson (1994), sometimes with two or three approximations of differing accuracies. Although they are obtained (with some exceptions) by approximating uniformly the integrand of each elliptic integral, some occur also as the leading terms of known asymptotic series with error bounds (Wong (1983, §4), Carlson and Gustafson (1985), López (2000, 2001)). These series converge but not fast enough, given the complicated nature of their terms, to be very useful in practice.

A similar (but more general) situation prevails for $R_{-a}(\mathbf{b};\mathbf{z})$ when some of the variables $z_1,\ldots,z_n$ are smaller in magnitude than the rest; see Carlson (1985, (4.16)–(4.19) and (2.26)–(2.29)).

19.28 Integrals of Elliptic Integrals

In (19.28.1)–(19.28.3) we assume $\Re\sigma > 0$. Also, B again denotes the beta function (§5.12).

19.28.1 $\int_0^1 t^{\sigma-1} R_F(0,t,1)\, dt = \tfrac{1}{2}\left(\mathrm{B}(\sigma,\tfrac{1}{2})\right)^2,$

19.28.2 $\int_0^1 t^{\sigma-1} R_G(0,t,1)\, dt = \dfrac{\sigma}{4\sigma+2}\left(\mathrm{B}(\sigma,\tfrac{1}{2})\right)^2,$

19.28.3
$$\int_0^1 t^{\sigma-1}(1-t) R_D(0,t,1)\, dt = \frac{3}{4\sigma+2}\left(\mathrm{B}(\sigma,\tfrac{1}{2})\right)^2.$$

19.28.4
$$\int_0^1 t^{\sigma-1}(1-t)^{c-1} R_{-a}(b_1,b_2;t,1)\, dt$$
$$= \frac{\Gamma(c)\,\Gamma(\sigma)\,\Gamma(\sigma+b_2-a)}{\Gamma(\sigma+c-a)\,\Gamma(\sigma+b_2)},$$
$$c = b_1 + b_2 > 0,\ \Re\sigma > \max(0, a-b_2).$$

In (19.28.5)–(19.28.9) we assume x, y, z, and p are real and positive.

19.28.5 $\int_z^\infty R_D(x,y,t)\, dt = 6 R_F(x,y,z),$

19.28.6
$$\int_0^1 R_D\big(x, y, v^2 z + (1-v^2)p\big)\, dv = R_J(x,y,z,p).$$

19.28.7 $\int_0^\infty R_J(x,y,z,r^2)\, dr = \tfrac{3}{2}\pi R_F(xy, xz, yz),$

19.28.8
$$\int_0^\infty R_J(tx, y, z, tp)\, dt = \frac{6}{\sqrt{p}} R_C(p,x)\, R_F(0,y,z).$$

19.28.9
$$\int_0^{\pi/2} R_F\big(\sin^2\theta\cos^2(x+y), \sin^2\theta\cos^2(x-y), 1\big)\, d\theta$$
$$= R_F(0, \cos^2 x, 1)\, R_F(0, \cos^2 y, 1),$$

19.28.10
$$\int_0^\infty R_F\big((ac+bd)^2, (ad+bc)^2, 4abcd\cosh^2 z\big)\, dz$$
$$= \tfrac{1}{2} R_F(0, a^2, b^2)\, R_F(0, c^2, d^2), \quad a,b,c,d > 0.$$

See also (19.16.24). To replace a single component of $\mathbf{z}$ in $R_{-a}(\mathbf{b};\mathbf{z})$ by several different variables (as in (19.28.6)), see Carlson (1963, (7.9)).

19.29 Reduction of General Elliptic Integrals

19.29(i) Reduction Theorems

These theorems reduce integrals over a real interval (y, x) of certain integrands containing the square root of a quartic or cubic polynomial to symmetric integrals over $(0, \infty)$ containing the square root of a cubic polynomial (compare §19.16(i)). Let

19.29.1 $\quad X_\alpha = \sqrt{a_\alpha + b_\alpha x}, \quad Y_\alpha = \sqrt{a_\alpha + b_\alpha y},$
$$x > y, \; 1 \leq \alpha \leq 5,$$

19.29.2 $\quad d_{\alpha\beta} = a_\alpha b_\beta - a_\beta b_\alpha, \quad d_{\alpha\beta} \neq 0 \text{ if } \alpha \neq \beta,$

and assume that the line segment with endpoints $a_\alpha + b_\alpha x$ and $a_\alpha + b_\alpha y$ lies in $\mathbb{C}\backslash(-\infty, 0)$ for $1 \leq \alpha \leq 4$. If

19.29.3 $\quad s(t) = \prod_{\alpha=1}^{4} \sqrt{a_\alpha + b_\alpha t}$

and $\alpha, \beta, \gamma, \delta$ is any permutation of the numbers $1, 2, 3, 4$, then

19.29.4 $\quad \int_y^x \frac{dt}{s(t)} = 2R_F\left(U_{12}^2, U_{13}^2, U_{23}^2\right),$

where

19.29.5
$U_{\alpha\beta} = (X_\alpha X_\beta Y_\gamma Y_\delta + Y_\alpha Y_\beta X_\gamma X_\delta)/(x-y),$
$U_{\alpha\beta} = U_{\beta\alpha} = U_{\gamma\delta} = U_{\delta\gamma}, \quad U_{\alpha\beta}^2 - U_{\alpha\gamma}^2 = d_{\alpha\delta} d_{\beta\gamma}.$

There are only three distinct U's with subscripts ≤ 4, and at most one of them can be 0 because the d's are nonzero. Then

19.29.6
$U_{\alpha\beta} = \sqrt{b_\alpha}\sqrt{b_\beta} Y_\gamma Y_\delta + Y_\alpha Y_\beta \sqrt{b_\gamma}\sqrt{b_\delta}, \quad x = \infty,$
$U_{\alpha\beta} = X_\alpha X_\beta \sqrt{-b_\gamma}\sqrt{-b_\delta} + \sqrt{-b_\alpha}\sqrt{-b_\beta} X_\gamma X_\delta,$
$$y = -\infty.$$

19.29.7 $\quad \int_y^x \frac{a_\alpha + b_\alpha t}{a_\delta + b_\delta t} \frac{dt}{s(t)} = \frac{2}{3} d_{\alpha\beta} d_{\alpha\gamma} R_D\left(U_{\alpha\beta}^2, U_{\alpha\gamma}^2, U_{\alpha\delta}^2\right)$
$$+ \frac{2 X_\alpha Y_\alpha}{X_\delta Y_\delta U_{\alpha\delta}}, \quad U_{\alpha\delta} \neq 0.$$

$\int_y^x \frac{a_\alpha + b_\alpha t}{a_5 + b_5 t} \frac{dt}{s(t)}$

19.29.8
$= \frac{2}{3} \frac{d_{\alpha\beta} d_{\alpha\gamma} d_{\alpha\delta}}{d_{\alpha 5}} R_J\left(U_{12}^2, U_{13}^2, U_{23}^2, U_{\alpha 5}^2\right)$
$+ 2 R_C\left(S_{\alpha 5}^2, Q_{\alpha 5}^2\right), \quad S_{\alpha 5}^2 \in \mathbb{C}\backslash(-\infty, 0),$

where

19.29.9
$U_{\alpha 5}^2 = U_{\alpha\beta}^2 - \frac{d_{\alpha\gamma} d_{\alpha\delta} d_{\beta 5}}{d_{\alpha 5}} = U_{\beta\gamma}^2 - \frac{d_{\alpha\beta} d_{\alpha\gamma} d_{\delta 5}}{d_{\alpha 5}} \neq 0,$
$S_{\alpha 5} = \frac{1}{x-y}\left(\frac{X_\beta X_\gamma X_\delta}{X_\alpha} Y_5^2 + \frac{Y_\beta Y_\gamma Y_\delta}{Y_\alpha} X_5^2\right),$
$Q_{\alpha 5} = \frac{X_5 Y_5}{X_\alpha Y_\alpha} U_{\alpha 5} \neq 0, \quad S_{\alpha 5}^2 - Q_{\alpha 5}^2 = \frac{d_{\beta 5} d_{\gamma 5} d_{\delta 5}}{d_{\alpha 5}}.$

The Cauchy principal value is taken when $U_{\alpha 5}^2$ or $Q_{\alpha 5}^2$ is real and negative. Cubic cases of these formulas are obtained by setting one of the factors in (19.29.3) equal to 1.

The advantages of symmetric integrals for tables of integrals and symbolic integration are illustrated by (19.29.4) and its cubic case, which replace the $8 + 8 + 12 = 28$ formulas in Gradshteyn and Ryzhik (2000, 3.147, 3.131, 3.152) after taking x^2 as the variable of integration in 3.152. Moreover, the requirement that one limit of integration be a branch point of the integrand is eliminated without doubling the number of standard integrals in the result. (19.29.7) subsumes all 72 formulas in Gradshteyn and Ryzhik (2000, 3.168), and its cubic cases similarly replace the $18 + 36 + 18 = 72$ formulas in Gradshteyn and Ryzhik (2000, 3.133, 3.142, and 3.141(1-18)). For example, 3.142(2) is included as

19.29.10
$\int_u^b \sqrt{\frac{a-t}{(b-t)(t-c)^3}} \, dt = -\frac{2}{3}(a-b)(b-u)^{3/2} R_D$
$$+ \frac{2}{b-c}\sqrt{\frac{(a-u)(b-u)}{u-c}},$$
$$a > b > u > c,$$

where the arguments of the R_D function are, in order, $(a-b)(u-c), (b-c)(a-u), (a-b)(b-c)$.

19.29(ii) Reduction to Basic Integrals

(19.2.3) can be written

19.29.11
$$I(\mathbf{m}) = \int_y^x \prod_{\alpha=1}^{h}(a_\alpha + b_\alpha t)^{-1/2} \prod_{j=1}^{n}(a_j + b_j t)^{m_j} \, dt,$$

where $x > y$, $h = 3$ or 4, $n \geq h$, and m_j is an integer. Define

19.29.12 $\quad \mathbf{m} = (m_1, \ldots, m_n) = \sum_{j=1}^{n} m_j \mathbf{e}_j,$

where $\mathbf{e}_j$ is an n-tuple with 1 in the jth position and 0's elsewhere. Define also $\mathbf{0} = (0, \ldots, 0)$ and retain the notation and conditions associated with (19.29.1) and (19.29.2). The integrals in (19.29.4), (19.29.7), and (19.29.8) are $I(\mathbf{0})$, $I(\mathbf{e}_\alpha - \mathbf{e}_\delta)$, and $I(\mathbf{e}_\alpha - \mathbf{e}_5)$, respectively.

The only cases of $I(\mathbf{m})$ that are integrals of the *first kind* are the two ($h = 3$ or 4) with $\mathbf{m} = \mathbf{0}$. The only cases that are integrals of the *third kind* are those in which at least one m_j with $j > h$ is a negative integer and those in which $h = 4$ and $\sum_{j=1}^{n} m_j$ is a positive integer. All other cases are integrals of the *second kind*.

$I(\mathbf{m})$ can be reduced to a linear combination of *basic integrals* and algebraic functions. In the cubic case ($h = 3$) the basic integrals are

19.29.13 $\qquad I(\mathbf{0}); \quad I(-\mathbf{e}_j), \qquad 1 \leq j \leq n.$

19.29 Reduction of General Elliptic Integrals

In the quartic case ($h = 4$) the basic integrals are

19.29.14
$$I(\mathbf{0}); \quad I(-\mathbf{e}_j), \quad 1 \le j \le n;$$
$$I(\mathbf{e}_\alpha), \quad 1 \le \alpha \le 4.$$

Basic integrals of type $I(-\mathbf{e}_j)$, $1 \le j \le h$, are not linearly independent, nor are those of type $I(\mathbf{e}_j)$, $1 \le j \le 4$.

The reduction of $I(\mathbf{m})$ is carried out by a relation derived from partial fractions and by use of two recurrence relations. These are given in Carlson (1999, (2.19), (3.5), (3.11)) and simplified in Carlson (2002, (1.10), (1.7), (1.8)) by means of modified definitions. Partial fractions provide a reduction to integrals in which $\mathbf{m}$ has at most one nonzero component, and these are then reduced to basic integrals by the recurrence relations. A special case of Carlson (1999, (2.19)) is given by

19.29.15
$$b_j I(\mathbf{e}_l - \mathbf{e}_j) = d_{lj} I(-\mathbf{e}_j) + b_l I(\mathbf{0}), \quad j, l = 1, 2, \ldots, n,$$

which shows how to express the basic integral $I(-\mathbf{e}_j)$ in terms of symmetric integrals by using (19.29.4) and either (19.29.7) or (19.29.8). The first choice gives a formula that includes the $18+9+18 = 45$ formulas in Gradshteyn and Ryzhik (2000, 3.133, 3.156, 3.158), and the second choice includes the $8+8+8+12 = 36$ formulas in Gradshteyn and Ryzhik (2000, 3.151, 3.149, 3.137, 3.157) (after setting $x^2 = t$ in some cases).

If $h = 3$, then the recurrence relation (Carlson (1999, (3.5))) has the special case

19.29.16
$$b_\beta b_\gamma I(\mathbf{e}_\alpha) = d_{\alpha\beta} d_{\alpha\gamma} I(-\mathbf{e}_\alpha) + 2b_\alpha \left(\frac{s(x)}{a_\alpha + b_\alpha x} - \frac{s(y)}{a_\alpha + b_\alpha y} \right),$$

where α, β, γ is any permutation of the numbers $1, 2, 3$, and

19.29.17
$$s(t) = \prod_{\alpha=1}^{3} \sqrt{a_\alpha + b_\alpha t}.$$

(This shows why $I(\mathbf{e}_\alpha)$ is not needed as a basic integral in the cubic case.) In the quartic case this recurrence relation has an extra term in $I(2\mathbf{e}_\alpha)$, and hence $I(\mathbf{e}_\alpha)$, $1 \le \alpha \le 4$, is a basic integral. It can be expressed in terms of symmetric integrals by setting $a_5 = 1$ and $b_5 = 0$ in (19.29.8).

The other recurrence relation is

19.29.18
$$b_j^q I(q\mathbf{e}_l) = \sum_{r=0}^{q} \binom{q}{r} b_l^r d_{lj}^{q-r} I(r\mathbf{e}_j), \quad j, l = 1, 2, \ldots, n;$$

see Carlson (1999, (3.11)). An example that uses (19.29.15)–(19.29.18) is given in §19.34.

For an implementation by James FitzSimons of the method for reducing $I(\mathbf{m})$ to basic integrals and extensive tables of such reductions, see Carlson (1999) and Carlson and FitzSimons (2000).

Another method of reduction is given in Gray (2002). It depends primarily on multivariate recurrence relations that replace one integral by two or more.

19.29(iii) Examples

The first formula replaces (19.14.4)–(19.14.10). Define $Q_j(t) = a_j + b_j t^2$, $j = 1, 2$, and assume both Q's are positive for $0 \le y < t < x$. Then

19.29.19
$$\int_y^x \frac{dt}{\sqrt{Q_1(t)Q_2(t)}} = R_F(U^2 + a_1 b_2, U^2 + a_2 b_1, U^2),$$

19.29.20
$$\int_y^x \frac{t^2 \, dt}{\sqrt{Q_1(t)Q_2(t)}}$$
$$= \tfrac{1}{3} a_1 a_2 R_D(U^2 + a_1 b_2, U^2 + a_2 b_1, U^2) + (xy/U),$$

and

19.29.21
$$\int_y^x \frac{dt}{t^2 \sqrt{Q_1(t)Q_2(t)}}$$
$$= \tfrac{1}{3} b_1 b_2 R_D(U^2 + a_1 b_2, U^2 + a_2 b_1, U^2) + (xyU)^{-1},$$

where

19.29.22
$$(x^2 - y^2)U = x\sqrt{Q_1(y)Q_2(y)} + y\sqrt{Q_1(x)Q_2(x)}.$$

If both square roots in (19.29.22) are 0, then the indeterminacy in the two preceding equations can be removed by using (19.27.8) to evaluate the integral as $R_G(a_1 b_2, a_2 b_1, 0)$ multiplied either by $-2/(b_1 b_2)$ or by $-2/(a_1 a_2)$ in the cases of (19.29.20) or (19.29.21), respectively. If $x = \infty$, then U is found by taking the limit. For example,

19.29.23
$$\int_y^\infty \frac{dt}{\sqrt{(t^2 + a^2)(t^2 - b^2)}} = R_F(y^2 + a^2, y^2 - b^2, y^2).$$

Next, for $j = 1, 2$, define $Q_j(t) = f_j + g_j t + h_j t^2$, and assume both Q's are positive for $y < t < x$. If each has real zeros, then (19.29.4) may be simpler than

19.29.24
$$\int_y^x \frac{dt}{\sqrt{Q_1(t)Q_2(t)}}$$
$$= 4R_F(U, U + D_{12} + V, U + D_{12} - V),$$

where

19.29.25
$$(x - y)^2 U = S_1 S_2,$$
$$S_j = \left(\sqrt{Q_j(x)} + \sqrt{Q_j(y)}\right)^2 - h_j(x - y)^2,$$
$$D_{jl} = 2f_j h_l + 2h_j f_l - g_j g_l, \quad V = \sqrt{D_{12}^2 - D_{11} D_{22}}.$$

(The variables of R_F are real and nonnegative unless both Q's have real zeros and those of Q_1 interlace those of Q_2.) If $Q_1(t) = (a_1 + b_1 t)(a_2 + b_2 t)$, where both linear factors are positive for $y < t < x$, and

$Q_2(t) = f_2 + g_2 t + h_2 t^2$, then (19.29.25) is modified so that

19.29.26
$$S_1 = (X_1 Y_2 + Y_1 X_2)^2,$$
$$X_j = \sqrt{a_j + b_j x}, \quad Y_j = \sqrt{a_j + b_j y},$$
$$D_{12} = 2a_1 a_2 h_2 + 2 b_1 b_2 f_2 - (a_1 b_2 + a_2 b_1) g_2,$$
$$D_{11} = -(a_1 b_2 - a_2 b_1)^2 = -d_{12}^2,$$

with other quantities remaining as in (19.29.25). In the cubic case, in which $a_2 = 1$, $b_2 = 0$, (19.29.26) reduces further to

19.29.27
$$S_1 = (X_1 + Y_1)^2, \quad D_{12} = 2a_1 h_2 - b_1 g_2, \quad D_{11} = -b_1^2.$$

For example, because $t^3 - a^3 = (t-a)(t^2 + at + a^2)$, we find that when $0 \le a \le y < x$

19.29.28
$$\int_y^x \frac{dt}{\sqrt{t^3 - a^3}}$$
$$= 4 R_F\left(U, U - 3a + 2\sqrt{3}a, U - 3a - 2\sqrt{3}a\right),$$

where

19.29.29
$$(x-y)^2 U = \left(\sqrt{x-a} + \sqrt{y-a}\right)^2 \left((\xi + \eta)^2 - (x-y)^2\right),$$
$$\xi = \sqrt{x^2 + ax + a^2}, \quad \eta = \sqrt{y^2 + ay + a^2}.$$

Lastly, define $Q(t^2) = f + gt^2 + ht^4$ and assume $Q(t^2)$ is positive and monotonic for $y < t < x$. Then

19.29.30
$$\int_y^x \frac{dt}{\sqrt{Q(t^2)}}$$
$$= 2 R_F\left(U, U - g + 2\sqrt{fh}, U - g - 2\sqrt{fh}\right),$$

where

19.29.31
$$(x-y)^2 U = \left(\sqrt{Q(x^2)} + \sqrt{Q(y^2)}\right)^2 - h(x^2 - y^2)^2.$$

For example, if $0 \le y \le x$ and $a^4 \ge 0$, then

19.29.32
$$\int_y^x \frac{dt}{\sqrt{t^4 + a^4}} = 2 R_F\left(U, U + 2a^2, U - 2a^2\right),$$

where

19.29.33
$$(x-y)^2 U = \left(\sqrt{x^4 + a^4} + \sqrt{y^4 + a^4}\right)^2 - (x^2 - y^2)^2.$$

Applications

19.30 Lengths of Plane Curves

19.30(i) Ellipse

The arclength s of the ellipse

19.30.1
$$x = a \sin \phi, \quad y = b \cos \phi, \quad 0 \le \phi \le 2\pi,$$

with $a > b$, is given by

19.30.2
$$s = a \int_0^\phi \sqrt{1 - k^2 \sin^2 \theta}\, d\theta.$$

When $0 \le \phi \le \tfrac{1}{2}\pi$,

19.30.3
$$s/a = E(\phi, k)$$
$$= R_F(c-1, c-k^2, c) - \tfrac{1}{3}k^2 R_D(c-1, c-k^2, c),$$

where

19.30.4
$$k^2 = 1 - (b^2/a^2), \quad c = \csc^2 \phi.$$

Cancellation on the second right-hand side of (19.30.3) can be avoided by use of (19.25.10).

The length of the ellipse is

19.30.5
$$L(a,b) = 4a\, E(k) = 8a\, R_G(0, b^2/a^2, 1)$$
$$= 8 R_G(0, a^2, b^2) = 8ab\, R_G(0, a^{-2}, b^{-2}),$$

showing the symmetry in a and b. Approximations and inequalities for $L(a,b)$ are given in §19.9(i).

Let a^2 and b^2 be replaced respectively by $a^2 + \lambda$ and $b^2 + \lambda$, where $\lambda \in (-b^2, \infty)$, to produce a family of confocal ellipses. As λ increases, the eccentricity k decreases and the rate of change of arclength for a fixed value of ϕ is given by

19.30.6
$$\frac{\partial s}{\partial (1/k)} = \sqrt{a^2 - b^2}\, F(\phi, k)$$
$$= \sqrt{a^2 - b^2}\, R_F(c-1, c-k^2, c),$$
$$k^2 = (a^2 - b^2)/(a^2 + \lambda), \quad c = \csc^2 \phi.$$

19.30(ii) Hyperbola

The arclength s of the hyperbola

19.30.7
$$x = a\sqrt{t+1}, \quad y = b\sqrt{t}, \quad 0 \le t < \infty,$$

is given by

19.30.8
$$s = \frac{1}{2} \int_0^{y^2/b^2} \sqrt{\frac{(a^2 + b^2)t + b^2}{t(t+1)}}\, dt.$$

From (19.29.7), with $a_\delta = 1$ and $b_\delta = 0$,

19.30.9
$$s = \tfrac{1}{2} I(\mathbf{e}_1) = -\tfrac{1}{3} a^2 b^2 R_D(r, r + b^2 + a^2, r + b^2)$$
$$+ y\sqrt{\frac{r + b^2 + a^2}{r + b^2}}, \quad r = b^4/y^2.$$

For s in terms of $E(\phi, k)$, $F(\phi, k)$, and an algebraic term, see Byrd and Friedman (1971, p. 3). See Carlson (1977b, Ex. 9.4-1 and 9.4-4)) for arclengths of hyperbolas and ellipses in terms of R_{-a} that differ only in the sign of b^2.

19.30(iii) Bernoulli's Lemniscate

For $0 \leq \theta \leq \frac{1}{4}\pi$, the arclength s of Bernoulli's lemniscate

19.30.10 $$r^2 = 2a^2 \cos(2\theta), \qquad 0 \leq \theta \leq 2\pi,$$

is given by

19.30.11
$$s = 2a^2 \int_0^r \frac{dt}{\sqrt{4a^4 - t^4}} = \sqrt{2a^2}\, R_F(q-1, q, q+1),$$
$$q = 2a^2/r^2 = \sec(2\theta),$$

or equivalently,

19.30.12
$$s = a\, F\!\left(\phi, 1/\sqrt{2}\right),$$
$$\phi = \arcsin\sqrt{2/(q+1)} = \arccos(\tan\theta).$$

The perimeter length P of the lemniscate is given by

19.30.13
$$P = 4\sqrt{2a^2}\, R_F(0,1,2) = \sqrt{2a^2} \times 5.24411\,51\ldots$$
$$= 4a\, K\!\left(1/\sqrt{2}\right) = a \times 7.41629\,87\ldots.$$

For other plane curves with arclength representable by an elliptic integral see Greenhill (1892, p. 190) and Bowman (1953, pp. 32–33).

19.31 Probability Distributions

$R_G(x, y, z)$ and $R_F(x, y, z)$ occur as the expectation values, relative to a normal probability distribution in $\mathbb{R}^2$ or $\mathbb{R}^3$, of the square root or reciprocal square root of a quadratic form. More generally, let $\mathbf{A}\ (= [a_{r,s}])$ and $\mathbf{B}\ (= [b_{r,s}])$ be real positive-definite matrices with n rows and n columns, and let $\lambda_1, \ldots, \lambda_n$ be the eigenvalues of $\mathbf{A}\mathbf{B}^{-1}$. If $\mathbf{x}$ is a column vector with elements $x_1, x_2, \ldots, x_n$ and transpose $\mathbf{x}^{\mathrm{T}}$, then

19.31.1
$$\mathbf{x}^{\mathrm{T}} \mathbf{A} \mathbf{x} = \sum_{r=1}^n \sum_{s=1}^n a_{r,s} x_r x_s,$$

and

19.31.2
$$\int_{\mathbb{R}^n} (\mathbf{x}^{\mathrm{T}}\mathbf{A}\mathbf{x})^\mu \exp\!\left(-\mathbf{x}^{\mathrm{T}}\mathbf{B}\mathbf{x}\right) dx_1 \cdots dx_n$$
$$= \frac{\pi^{n/2}\, \Gamma\!\left(\mu + \tfrac{1}{2}n\right)}{\sqrt{\det \mathbf{B}}\, \Gamma\!\left(\tfrac{1}{2}n\right)}\, R_\mu\!\left(\tfrac{1}{2}, \ldots, \tfrac{1}{2}; \lambda_1, \ldots, \lambda_n\right),$$
$$\mu > -\tfrac{1}{2}n.$$

§19.16(iii) shows that for $n = 3$ the incomplete cases of R_F and R_G occur when $\mu = -1/2$ and $\mu = 1/2$, respectively, while their complete cases occur when $n = 2$.

For (19.31.2) and generalizations see Carlson (1972b).

19.32 Conformal Map onto a Rectangle

The function

19.32.1 $$z(p) = R_F(p - x_1, p - x_2, p - x_3),$$

with x_1, x_2, x_3 real constants, has differential

19.32.2
$$dz = -\frac{1}{2}\left(\prod_{j=1}^3 (p - x_j)^{-1/2}\right) dp,$$
$$\Im p > 0;\ 0 < \mathrm{ph}(p - x_j) < \pi,\ j = 1, 2, 3.$$

If

19.32.3 $$x_1 > x_2 > x_3,$$

then $z(p)$ is a Schwartz–Christoffel mapping of the open upper-half p-plane onto the interior of the rectangle in the z-plane with vertices

19.32.4
$$z(\infty) = 0,$$
$$z(x_1) = R_F(0, x_1 - x_2, x_1 - x_3) \quad (> 0),$$
$$z(x_2) = z(x_1) + z(x_3),$$
$$z(x_3) = R_F(x_3 - x_1, x_3 - x_2, 0)$$
$$= -i\, R_F(0, x_1 - x_3, x_2 - x_3).$$

As p proceeds along the entire real axis with the upper half-plane on the right, z describes the rectangle in the clockwise direction; hence $z(x_3)$ is negative imaginary.

For further connections between elliptic integrals and conformal maps, see Bowman (1953, pp. 44–85).

19.33 Triaxial Ellipsoids

19.33(i) Surface Area

The surface area of an ellipsoid with semiaxes a, b, c, and volume $V = 4\pi abc/3$ is given by

19.33.1 $$S = 3V\, R_G\!\left(a^{-2}, b^{-2}, c^{-2}\right),$$

or equivalently,

19.33.2
$$\frac{S}{2\pi} = c^2 + \frac{ab}{\sin\phi}\left(E(\phi, k)\sin^2\phi + F(\phi, k)\cos^2\phi\right),$$
$$a \geq b \geq c,$$

where

19.33.3 $$\cos\phi = \frac{c}{a}, \quad k^2 = \frac{a^2(b^2 - c^2)}{b^2(a^2 - c^2)}.$$

Application of (19.16.23) transforms the last quantity in (19.30.5) into a two-dimensional analog of (19.33.1).

For additional geometrical properties of ellipsoids (and ellipses), see Carlson (1964, p. 417).

19.33(ii) Potential of a Charged Conducting Ellipsoid

If a conducting ellipsoid with semiaxes a, b, c bears an electric charge Q, then the equipotential surfaces in the exterior region are confocal ellipsoids:

19.33.4 $\quad \dfrac{x^2}{a^2+\lambda} + \dfrac{y^2}{b^2+\lambda} + \dfrac{z^2}{c^2+\lambda} = 1, \qquad \lambda \geq 0.$

The potential is

19.33.5 $\quad V(\lambda) = Q\, R_F\!\left(a^2+\lambda, b^2+\lambda, c^2+\lambda\right),$

and the electric capacity $C = Q/V(0)$ is given by

19.33.6 $\quad 1/C = R_F\!\left(a^2, b^2, c^2\right).$

A conducting elliptic disk is included as the case $c = 0$.

19.33(iii) Depolarization Factors

Let a homogeneous magnetic ellipsoid with semiaxes a, b, c, volume $V = 4\pi abc/3$, and susceptibility χ be placed in a previously uniform magnetic field H parallel to the principal axis with semiaxis c. The external field and the induced magnetization together produce a uniform field inside the ellipsoid with strength $H/(1+L_c\chi)$, where L_c is the demagnetizing factor, given in cgs units by

19.33.7
$$L_c = 2\pi abc \int_0^\infty \frac{d\lambda}{\sqrt{(a^2+\lambda)(b^2+\lambda)(c^2+\lambda)^3}}$$
$$= V\, R_D\!\left(a^2, b^2, c^2\right).$$

The same result holds for a homogeneous dielectric ellipsoid in an electric field. By (19.21.8),

19.33.8 $\qquad L_a + L_b + L_c = 4\pi,$

where L_a and L_b are obtained from L_c by permutation of a, b, and c. Expressions in terms of Legendre's integrals, numerical tables, and further references are given by Cronemeyer (1991).

19.33(iv) Self-Energy of an Ellipsoidal Distribution

Ellipsoidal distributions of charge or mass are used to model certain atomic nuclei and some elliptical galaxies. Let the density of charge or mass be

19.33.9
$$\rho(x,y,z) = f\!\left(\sqrt{(x^2/\alpha^2)+(y^2/\beta^2)+(z^2/\gamma^2)}\right),$$

where α, β, γ are dimensionless positive constants. The contours of constant density are a family of similar, rather than confocal, ellipsoids. In suitable units the self-energy of the distribution is given by

19.33.10
$$U = \frac{1}{2}\int_{\mathbb{R}^6} \frac{\rho(x,y,z)\rho(x',y',z')\,dx\,dy\,dz\,dx'\,dy'\,dz'}{\sqrt{(x-x')^2+(y-y')^2+(z-z')^2}}.$$

Subject to mild conditions on f this becomes

19.33.11 $\quad U = \tfrac{1}{2}(\alpha\beta\gamma)^2\, R_F\!\left(\alpha^2,\beta^2,\gamma^2\right) \int_0^\infty (g(r))^2\,dr,$

where

19.33.12 $\qquad g(r) = 4\pi \int_r^\infty f(t)\,t\,dt.$

19.34 Mutual Inductance of Coaxial Circles

The mutual inductance M of two coaxial circles of radius a and b with centers at a distance h apart is given in cgs units by

19.34.1
$$\frac{c^2 M}{2\pi} = ab\int_0^{2\pi}(h^2+a^2+b^2-2ab\cos\theta)^{-1/2}\cos\theta\,d\theta$$
$$= 2ab\int_{-1}^{1}\frac{t\,dt}{\sqrt{(1+t)(1-t)(a_3-2abt)}} = 2ab\,I(\mathbf{e}_5),$$

where c is the speed of light, and in (19.29.11),

19.34.2 $\qquad a_3 = h^2+a^2+b^2, \quad a_5 = 0, \quad b_5 = 1.$

The method of §19.29(ii) uses (19.29.18), (19.29.16), and (19.29.15) to produce

19.34.3
$$2ab\,I(\mathbf{e}_5) = a_3 I(\mathbf{0}) - I(\mathbf{e}_3) = a_3 I(\mathbf{0}) - r_+^2 r_-^2 I(-\mathbf{e}_3)$$
$$= 2ab(I(\mathbf{0}) - r_-^2 I(\mathbf{e}_1-\mathbf{e}_3)),$$

where $a_1 + b_1 t = 1 + t$ and

19.34.4 $\qquad r_\pm^2 = a_3 \pm 2ab = h^2+(a\pm b)^2$

is the square of the maximum (upper signs) or minimum (lower signs) distance between the circles. Application of (19.29.4) and (19.29.7) with $\alpha = 1$, $a_\beta + b_\beta t = 1-t$, $\delta = 3$, and $a_\gamma + b_\gamma t = 1$ yields

19.34.5 $\quad \dfrac{3c^2}{8\pi ab} M = 3R_F\!\left(0, r_+^2, r_-^2\right) - 2r_-^2\, R_D\!\left(0, r_+^2, r_-^2\right),$

or, by (19.21.3),

19.34.6
$$\frac{c^2}{2\pi}M = (r_+^2+r_-^2)\,R_F\!\left(0,r_+^2,r_-^2\right) - 4R_G\!\left(0,r_+^2,r_-^2\right).$$

A simpler form of the result is

19.34.7 $\quad M = (2/c^2)(\pi a^2)(\pi b^2)\, R_{-\frac{3}{2}}\!\left(\tfrac{3}{2},\tfrac{3}{2};r_+^2,r_-^2\right).$

References for other inductance problems solvable in terms of elliptic integrals are given in Grover (1946, pp. 8 and 283).

19.35 Other Applications

19.35(i) Mathematical

Generalizations of elliptic integrals appear in analysis of modular theorems of Ramanujan (Anderson *et al.* (2000)); analysis of Selberg integrals (Van Diejen and Spiridonov (2001)); use of Legendre's relation (19.7.1) to compute π to high precision (Borwein and Borwein (1987, p. 26)).

19.35(ii) Physical

Elliptic integrals appear in lattice models of critical phenomena (Guttmann and Prellberg (1993)); theories of layered materials (Parkinson (1969)); fluid dynamics (Kida (1981)); string theory (Arutyunov and Staudacher (2004)); astrophysics (Dexter and Agol (2009)).

Computation

19.36 Methods of Computation

19.36(i) Duplication Method

Numerical differences between the variables of a symmetric integral can be reduced in magnitude by successive factors of 4 by repeated applications of the duplication theorem, as shown by (19.26.18). When the differences are moderately small, the iteration is stopped, the elementary symmetric functions of certain differences are calculated, and a polynomial consisting of a fixed number of terms of the sum in (19.19.7) is evaluated. For R_F the polynomial of degree 7, for example, is

19.36.1
$$1 - \tfrac{1}{10}E_2 + \tfrac{1}{14}E_3 + \tfrac{1}{24}E_2^2 - \tfrac{3}{44}E_2E_3 \\ - \tfrac{5}{208}E_2^3 + \tfrac{3}{104}E_3^2 + \tfrac{1}{16}E_2^2E_3,$$

where the elementary symmetric functions E_s are defined by (19.19.4). If (19.36.1) is used instead of its first five terms, then the factor $(3r)^{-1/6}$ in Carlson (1995, (2.2)) is changed to $(3r)^{-1/8}$.

For a polynomial for both R_D and R_J see http://dlmf.nist.gov/19.36.i.

Example

Three applications of (19.26.18) yield

19.36.3 $\qquad R_F(1,2,4) = R_F(z_1, z_2, z_3),$

where, in the notation of (19.19.7) with $a = -\tfrac{1}{2}$ and $n = 3$,

19.36.4
$z_1 = 2.10985\ 99098\ 8, \quad z_2 = 2.12548\ 49098\ 8,$
$z_3 = 2.15673\ 49098\ 8, \quad A = 2.13069\ 32432\ 1,$
$Z_1 = 0.00977\ 77253\ 5, \quad Z_2 = 0.00244\ 44313\ 4,$
$Z_3 = -Z_1 - Z_2 = -0.01222\ 21566\ 9,$
$E_2 = -1.25480\ 14 \times 10^{-4}, \quad E_3 = -2.9212 \times 10^{-7}.$

The first five terms of (19.36.1) suffice for

19.36.5 $\qquad R_F(1,2,4) = 0.68508\ 58166\ldots.$

All cases of R_F, R_C, R_J, and R_D are computed by essentially the same procedure (after transforming Cauchy principal values by means of (19.20.14) and (19.2.20)). Complex values of the variables are allowed, with some restrictions in the case of R_J that are sufficient but not always necessary. The computation is slowest for complete cases. For details see Carlson (1995, 2002) and Carlson and FitzSimons (2000). In the Appendix of the last reference it is shown how to compute R_J without computing R_C more than once. Because of cancellations in (19.26.21) it is advisable to compute R_G from R_F and R_D by (19.21.10) or else to use §19.36(ii).

Legendre's integrals can be computed from symmetric integrals by using the relations in §19.25(i). Note the remark following (19.25.11). If (19.25.9) is used when $0 \leq k^2 \leq 1$, cancellations may lead to loss of significant figures when k^2 is close to 1 and $\phi > \pi/4$, as shown by Reinsch and Raab (2000). The cancellations can be eliminated, however, by using (19.25.10).

Accurate values of $F(\phi, k) - E(\phi, k)$ for k^2 near 0 can be obtained from R_D by (19.2.6) and (19.25.13).

19.36(ii) Quadratic Transformations

Complete cases of Legendre's integrals and symmetric integrals can be computed with quadratic convergence by the AGM method (including Bartky transformations), using the equations in §19.8(i) and §19.22(ii), respectively.

The incomplete integrals $R_F(x,y,z)$ and $R_G(x,y,z)$ can be computed by successive transformations in which two of the three variables converge quadratically to a common value and the integrals reduce to R_C, accompanied by two quadratically convergent series in the case of R_G; compare Carlson (1965, §§5,6). (In Legendre's notation the modulus k approaches 0 or 1.) Let

19.36.6
$2a_{n+1} = a_n + \sqrt{a_n^2 - c_n^2},$
$2c_{n+1} = a_n - \sqrt{a_n^2 - c_n^2} = c_n^2/(2a_{n+1}),$
$2t_{n+1} = t_n + \sqrt{t_n^2 + \theta c_n^2},$

where $n = 0, 1, 2, \ldots$, and

19.36.7 $\quad 0 < c_0 < a_0, \quad t_0 \geq 0, \quad t_0^2 + \theta a_0^2 \geq 0, \quad \theta = \pm 1.$

Then (19.22.18) implies that

19.36.8 $\qquad R_F\!\left(t_n^2, t_n^2 + \theta c_n^2, t_n^2 + \theta a_n^2\right)$

is independent of n. As $n \to \infty$, c_n, a_n, and t_n converge quadratically to limits 0, M, and T, respectively; hence

19.36.9
$R_F\!\left(t_0^2, t_0^2 + \theta c_0^2, t_0^2 + \theta a_0^2\right) = R_F(T^2, T^2, T^2 + \theta M^2)$
$= R_C(T^2 + \theta M^2, T^2).$

If $t_0 = a_0$ and $\theta = -1$, so that $t_n = a_n$, then this procedure reduces to the AGM method for the complete integral.

The step from n to $n+1$ is an ascending Landen transformation if $\theta = 1$ (leading ultimately to a hyperbolic case of R_C) or a descending Gauss transformation if $\theta = -1$ (leading to a circular case of R_C). If x, y, and z are permuted so that $0 \leq x < y < z$, then the computation of $R_F(x,y,z)$ is fastest if we make $c_0^2 \leq a_0^2/2$ by choosing $\theta = 1$ when $y < (x+z)/2$ or $\theta = -1$ when $y \geq (x+z)/2$.

Example

We compute $R_F(1,2,4)$ by setting $\theta = 1$, $t_0 = c_0 = 1$, and $a_0 = \sqrt{3}$. Then

19.36.10
$$c_3^2 = 6.65 \times 10^{-12}, \quad a_3^2 = 2.46209\,30206\,0 = M^2,$$
$$t_3^2 = 1.46971\,53173\,1 = T^2.$$

Hence

19.36.11
$$R_F(1,2,4) = R_C(T^2 + M^2, T^2) = 0.68508\,58166,$$

in agreement with (19.36.5). Here R_C is computed either by the duplication algorithm in Carlson (1995) or via (19.2.19).

For an error estimate and the corresponding procedure for $R_G(x,y,z)$, see http://dlmf.nist.gov/19.36.ii.

$F(\phi, k)$ can be evaluated by using (19.25.5). $E(\phi, k)$ can be evaluated by using (19.25.7), and R_D by using (19.21.10), but cancellations may become significant. Thompson (1997, pp. 499, 504) uses descending Landen transformations for both $F(\phi, k)$ and $E(\phi, k)$. A summary for $F(\phi, k)$ is given in Gautschi (1975, §3). For computation of $K(k)$ and $E(k)$ with complex k see Fettis and Caslin (1969) and Morita (1978).

(19.22.20) reduces to $0 = 0$ if $p = x$ or $p = y$, and (19.22.19) reduces to $0 = 0$ if $z = x$ or $z = y$. Near these points there will be loss of significant figures in the computation of R_J or R_D.

Descending Gauss transformations of $\Pi(\phi, \alpha^2, k)$ (see (19.8.20)) are used in Fettis (1965) to compute a large table (see §19.37(iii)). This method loses significant figures in ρ if α^2 and k^2 are nearly equal unless they are given exact values—as they can be for tables. If $\alpha^2 = k^2$, then the method fails, but the function can be expressed by (19.6.13) in terms of $E(\phi, k)$, for which Neuman (1969) uses ascending Landen transformations.

Computation of Legendre's integrals of all three kinds by quadratic transformation is described by Cazenave (1969, pp. 128–159, 208–230).

Quadratic transformations can be applied to compute Bulirsch's integrals (§19.2(iii)). The function $\mathrm{cel}(k_c, p, a, b)$ is computed by successive Bartky transformations (Bulirsch and Stoer (1968), Bulirsch (1969b)). The function $\mathrm{el2}(x, k_c, a, b)$ is computed by descending Landen transformations if x is real, or by descending Gauss transformations if x is complex (Bulirsch (1965a)). Remedies for cancellation when x is real and near 0 are supplied in Midy (1975). See also Bulirsch (1969a) and Reinsch and Raab (2000).

Bulirsch (1969a,b) extend Bartky's transformation to $\mathrm{el3}(x, k_c, p)$ by expressing it in terms of the first incomplete integral, a complete integral of the third kind, and a more complicated integral to which Bartky's method can be applied. The cases $k_c^2/2 \leq p < \infty$ and $-\infty < p < k_c^2/2$ require different treatment for numerical purposes, and again precautions are needed to avoid cancellations.

19.36(iii) Via Theta Functions

Lee (1990) compares the use of theta functions for computation of $K(k)$, $E(k)$, and $K(k) - E(k)$, $0 \leq k^2 \leq 1$, with four other methods. Also, see Todd (1975) for a special case of $K(k)$. For computation of Legendre's integral of the third kind, see Abramowitz and Stegun (1964, §§17.7 and 17.8, Examples 15, 17, 19, and 20). For integrals of the second and third kinds see Lawden (1989, §§3.4–3.7).

19.36(iv) Other Methods

Numerical quadrature is slower than most methods for the standard integrals but can be useful for elliptic integrals that have complicated representations in terms of standard integrals. See §3.5.

For series expansions of Legendre's integrals see §19.5. Faster convergence of power series for $K(k)$ and $E(k)$ can be achieved by using (19.5.1) and (19.5.2) in the right-hand sides of (19.8.12). A three-part computational procedure for $\Pi(\phi, \alpha^2, k)$ is described by Franke (1965) for $\alpha^2 < 1$.

When the values of complete integrals are known, addition theorems with $\psi = \pi/2$ (§19.11(ii)) ease the computation of functions such as $F(\phi, k)$ when $\frac{1}{2}\pi - \phi$ is small and positive. Similarly, §19.26(ii) eases the computation of functions such as $R_F(x, y, z)$ when $x\,(>0)$ is small compared with $\min(y, z)$. These special theorems are also useful for checking computer codes.

19.37 Tables

19.37(i) Introduction

Only tables published since 1960 are included. For earlier tables see Fletcher (1948), Lebedev and Fedorova (1960), and Fletcher et al. (1962).

19.37(ii) Legendre's Complete Integrals

Functions $K(k)$ and $E(k)$

Tabulated for $k^2 = 0(.01)1$ to 6D by Byrd and Friedman (1971), to 15D for $K(k)$ and 9D for $E(k)$ by Abramowitz and Stegun (1964, Chapter 17), and to 10D by Fettis and Caslin (1964).

Tabulated for $k = 0(.01)1$ to 10D by Fettis and Caslin (1964), and for $k = 0(.02)1$ to 7D by Zhang and Jin (1996, p. 673).

Tabulated for $\arcsin k = 0(1°)90°$ to 6D by Byrd and Friedman (1971) and to 15D by Abramowitz and Stegun (1964, Chapter 17).

Functions $K(k)$, $K'(k)$, and $iK'(k)/K(k)$

Tabulated with $k = Re^{i\theta}$ for $R = 0(.01)1$ and $\theta = 0(1°)90°$ to 11D by Fettis and Caslin (1969).

Function $\exp(-\pi K'(k)/K(k))(= q(k))$

Tabulated for $k^2 = 0(.01)1$ to 6D by Byrd and Friedman (1971) and to 15D by Abramowitz and Stegun (1964, Chapter 17).

Tabulated for $\arcsin k = 0(1°)90°$ to 6D by Byrd and Friedman (1971) and to 15D by Abramowitz and Stegun (1964, Chapter 17).

Tabulated for $k^2 = 0(.001)1$ to 8D by Beliakov et al. (1962).

19.37(iii) Legendre's Incomplete Integrals

Functions $F(\phi, k)$ and $E(\phi, k)$

Tabulated for $\phi = 0(5°)90°$, $k^2 = 0(.01)1$ to 10D by Fettis and Caslin (1964).

Tabulated for $\phi = 0(1°)90°$, $k^2 = 0(.01)1$ to 7S by Beliakov et al. (1962). ($F(\phi, k)$ is presented as $\Pi(\phi, 0, k)$.)

Tabulated for $\phi = 0(5°)90°$, $k = 0(.01)1$ to 10D by Fettis and Caslin (1964).

Tabulated for $\phi = 0(5°)90°$, $\arcsin k = 0(1°)90°$ to 6D by Byrd and Friedman (1971), for $\phi = 0(5°)90°$, $\arcsin k = 0(2°)90°$ and $5°(10°)85°$ to 8D by Abramowitz and Stegun (1964, Chapter 17), and for $\phi = 0(10°)90°$, $\arcsin k = 0(5°)90°$ to 9D by Zhang and Jin (1996, pp. 674–675).

Function $\Pi(\phi, \alpha^2, k)$

Tabulated (with different notation) for $\phi = 0(15°)90°$, $\alpha^2 = 0(.1)1$, $\arcsin k = 0(15°)90°$ to 5D by Abramowitz and Stegun (1964, Chapter 17), and for $\phi = 0(15°)90°$, $\alpha^2 = 0(.1)1$, $\arcsin k = 0(15°)90°$ to 7D by Zhang and Jin (1996, pp. 676–677).

Tabulated for $\phi = 5°(5°)80°(2.5°)90°$, $\alpha^2 = -1(.1) - 0.1, 0.1(.1)1$, $k^2 = 0(.05)0.9(.02)1$ to 10D by Fettis and Caslin (1964) (and warns of inaccuracies in Selfridge and Maxfield (1958) and Paxton and Rollin (1959)).

Tabulated for $\phi = 0(1°)90°$, $\alpha^2 = 0(.05)0.85, 0.88(.02)0.94(.01)0.98(.005)1$, $k^2 = 0(.01)1$ to 7S by Beliakov et al. (1962).

19.37(iv) Symmetric Integrals

Functions $R_F(x^2, 1, y^2)$ and $R_G(x^2, 1, y^2)$

Tabulated for $x = 0(.1)1$, $y = 1(.2)6$ to 3D by Nellis and Carlson (1966).

Function $R_F(a^2, b^2, c^2)$ with $abc = 1$

Tabulated for $\sigma = 0(.05)0.5(.1)1(.2)2(.5)5$, $\cos(3\gamma) = -1(.2)1$ to 5D by Carlson (1961a). Here $\sigma^2 = \frac{2}{3}((\ln a)^2 + (\ln b)^2 + (\ln c)^2)$, $\cos(3\gamma) = (4/\sigma^3)(\ln a)(\ln b)(\ln c)$, and a, b, c are semiaxes of an ellipsoid with the same volume as the unit sphere.

Check Values

For check values of symmetric integrals with real or complex variables to 14S see Carlson (1995).

19.38 Approximations

Minimax polynomial approximations (§3.11(i)) for $K(k)$ and $E(k)$ in terms of $m = k^2$ with $0 \le m < 1$ can be found in Abramowitz and Stegun (1964, §17.3) with maximum absolute errors ranging from 4×10^{-5} to 2×10^{-8}. Approximations of the same type for $K(k)$ and $E(k)$ for $0 < k \le 1$ are given in Cody (1965a) with maximum absolute errors ranging from 4×10^{-5} to 4×10^{-18}. Cody (1965b) gives Chebyshev-series expansions (§3.11(ii)) with maximum precision 25D.

Approximations for Legendre's complete or incomplete integrals of all three kinds, derived by Padé approximation of the square root in the integrand, are given in Luke (1968, 1970). They are valid over parts of the complex k and ϕ planes. The accuracy is controlled by the number of terms retained in the approximation; for real variables the number of significant figures appears to be roughly twice the number of terms retained, perhaps even for ϕ near $\pi/2$ with the improvements made in the 1970 reference.

19.39 Software

See http://dlmf.nist.gov/19.39.

References

General References

The main references used for writing this chapter are Erdélyi et al. (1953b), Byerly (1888), Cazenave (1969), and Byrd and Friedman (1971) for Legendre's integrals, and Carlson (1977b) for symmetric integrals. For additional bibliographic reading see Cayley (1895), Greenhill (1892), Legendre (1825–1832), Tricomi (1951), and Whittaker and Watson (1927).

Sources

The following list gives the references or other indications of proofs that were used in constructing the various sections of this chapter. These sources supplement the references that are quoted in the text.

§19.2 Bulirsch (1965a, 1969a,b), Bulirsch and Stoer (1968). To prove (19.2.20) evaluate the two parts of the Cauchy principal value (intervals $(0, -y-\delta)$ and $(-y+\delta, \infty)$) using Carlson (1977b, (8.2-2)), and reduce the first part to R_C by Carlson (1977b, (9.8-4)) with $B = C$. Apply (19.12.7) to both parts as $\delta \to 0$ and combine the two logarithms. For (19.2.21) see (19.16.18) and put $\cos\theta = v$ in (19.23.8). For (19.2.22) put $z = x$ in (19.23.5) and interchange x and y.

§19.3 The graphics were produced at NIST.

§19.4 Cazenave (1969, p. 175). (19.4.1)–(19.4.7) follow by differentiation of the definitions in §19.2(ii). (19.4.8) agrees also with Edwards (1954, vol. 1, p. 402) and with expansion to first order in k. The term on the right side in Byrd and Friedman (1971, 118.01) has the wrong sign.

§19.5 For (19.5.1)–(19.5.4) put $\sin\phi = 1$ and $t = \sqrt{x}$ in (19.2.4)–(19.2.7). Then compare with Erdélyi et al. (1953a, 2.1.3(10) and 2.1.1(2)) in the first three cases, and with Erdélyi et al. (1953a, 5.8.2(5) and 5.7.1(6)) in the fourth case. For (19.5.5) and (19.5.6) see Kneser (1927, (12) and p. 218); Byrd and Friedman (1971, 901.00) is incorrect. (19.5.8) and (19.5.9) follow from Borwein and Borwein (1987, (2.1.13) and (2.3.17), respectively). For (19.5.10) iterate (19.8.12).

§19.6 For the first line of (19.6.2) put $\alpha = k$ in the first line of (19.25.2) and use the last line of (19.25.1). For the second line of (19.6.2), and also for (19.6.5), use (19.7.8) and (19.6.15). For the first line of (19.6.6) use (19.6.5) and (19.6.2). For more detail as $k^2 \to 1-$ see §19.12. For (19.6.7), (19.6.8) use (19.2.4), (19.16.6), and (19.25.5). For (19.6.9), (19.6.10) use (19.2.5). For (19.6.11)–(19.6.14) Byrd and Friedman (1971, 111.01 and 111.04, p. 10) also needs $\alpha\sin\phi < 1$. Start with (19.25.14). For the second equation of (19.6.12) use (19.20.8). For (19.6.13) use (19.16.5) with (19.25.10) and (19.25.11).

§19.7 Three proofs of (19.7.1) are given in Duren (1991). To prove it from (19.21.1) put $z + 1 = 1/k^2$, use homogeneity, and apply the penultimate equation in (19.25.1) twice. For (19.7.4)–(19.7.7) see the penultimate paragraph in §19.25(i). (19.7.8)–(19.7.10) follow from the change of parameter for the symmetric integral of the third kind; see §19.21(iii) and (19.25.14).

§19.8 Cox (1984, 1985), Borwein and Borwein (1987, Chapter 1), Cazenave (1969, pp. 114–127). To prove the second equality in (19.8.4), put $\tan\theta = \sqrt{t}/g_0$. (19.8.7) is derived from (19.22.12) and (19.25.14), and (19.8.9) is derived from (19.6.5) and (19.8.7); see also Carlson (2002). For (19.8.16) and (19.8.17) replace (ϕ, k) by (ϕ_2, k_2), and then (ϕ_1, k_1) by (ϕ, k) in (19.8.11) and (19.8.13). See also Hancock (1958, pp. 74–77) for proof of (19.8.13) and (19.8.17).

§19.9 For (19.9.1) see Erdélyi et al. (1953b, §13.8(9),(11)), (19.9.13), (19.6.12), and (19.6.15). For (19.9.2) and (19.9.3) see Qiu and Vamanamurthy (1996). For (19.9.4) see Barnard et al. (2000, (6)); the first inequality was given earlier by Qiu and Shen (1997, Theorem 2). For (19.9.5) see Lehto and Virtanen (1973, p. 62). For (19.9.6) and (19.9.7) see (19.25.1) and (19.16.21) and then apply Carlson (1966, (2.15)), in which $H < H'$ for $0 < k \leq 1$ in both cases. In (19.9.7) the upper bound $4/\pi$, which is the smaller of the two when $k^2 \geq 0.855\ldots$, is given by Anderson and Vamanamurthy (1985). For (19.9.8) see (19.25.1), Neuman (2003, (4.2)), and (19.24.9). For (19.9.9) see (19.30.5). For (19.9.14) see (19.24.10) and (19.25.5). For (19.9.15) and (19.9.16) see Carlson and Gustafson (1985, (1.2), (1.22)).

§19.10 For (19.10.1) see (19.2.17). For (19.10.2) use (19.6.8).

§19.11 Byerly (1888, pp. 243–245, 256–258), Edwards (1954, v. 2, pp. 511–513), Cazenave (1969, pp. 83–85). (19.11.5) can be derived from (19.26.9), (19.25.26), and (19.11.1).

§19.12 For (19.12.1) and (19.12.2) see Cayley (1895, p. 54) and Cazenave (1969, pp. 165–169). For (19.12.4) and (19.12.5) use (19.25.2), (19.27.13), and (19.6.5). For (19.12.6) and (19.12.7) see Carlson and Gustafson (1994, (22),(24)).

§19.14 For (19.14.1)–(19.14.3) see Cazenave (1969, pp. 286,276). For (19.14.4) use (19.29.19) and (19.25.24).

§19.16 See Carlson (1977b, (6.8-6), Ex. 6.8-8, and (5.9-1)). To prove (19.16.12) put $t = \csc^2\theta - \csc^2\phi$ in the first integral in (19.16.9). For (19.16.19) and (19.16.23) see Carlson (1977b, (5.9-19) and (8.3-4)). To derive (19.16.24) exchange subscripts 1 and n in Carlson (1963, (7.4)), put $t = s/z_1$, and use (19.16.19).

§19.17 The graphics were produced at NIST.

§19.18 (19.18.1) is derived from (19.16.1), (19.16.5), and (19.18.4). (19.18.2) follows from (19.18.8). For (19.18.4) and (19.18.5) put $t = -a$ and $c = a + a'$ in Carlson (1977b, (5.9-9),(5.9-10)). (19.18.6) comes from (19.18.8) and (19.20.25). For (19.18.8) and (19.18.11) see Carlson (1977b, (5.9-2)). For (19.18.12)–(19.18.17) see Carlson (1977b, §5.4).

§19.19 To prove (19.19.2) expand the product in (19.23.10) in powers of u. (19.19.3) is derived from (19.16.11) and (19.19.2). For (19.19.5) see Carlson (1979, (A.12)). For (19.19.6) compare (19.16.2) and (19.16.9).

§19.20 In (19.20.2) put $t = 1/\sqrt{s+1}$; alternatively use (19.29.19). For the second equality replace t^4 by t and apply (5.12.1). For (19.20.3) use Carlson (1977b, Ex. 6.9-5 and p. 309) and (19.25.42). For (19.20.4) use (19.20.5) and (19.16.3). For (19.20.5) put $z = y$ in (19.21.10). For (19.20.6) substitute in (19.16.2) and (19.16.5). In (19.20.7) see (19.27.12) for $p \to 0+$; for $p \to 0-$ use (19.20.17) and (19.6.15). In (19.20.8) the third equation is proved by partial fractions, and also implies the first two equations by (19.6.15). For (19.20.9) put $x = 0$ in (19.20.13). For (19.20.10) interchange x and z in (19.27.14) and use (19.6.15). For (19.20.11) use (19.27.13), (19.20.17), and (19.27.2). For (19.20.12) see (19.27.11) and (19.21.12). For (19.20.13) let $q = p$ in (19.21.12). For (19.20.14) exchange x and z in (19.21.12) and use (19.2.20). For the third equation in (19.20.18) put $t = y \tan^2 \theta$ in (19.16.5); for the fourth equation see (19.27.7). For (19.20.19) see (19.27.8). For (19.20.20) and (19.20.21) use (19.16.15), (19.16.9), and Carlson (1977b, Table 8.5-1). In (19.20.22) put $t = 1/\sqrt{s+1}$; alternatively use (19.29.20). For the second equality replace t^4 by t and apply (5.12.1). For (19.20.23) use Carlson (1977b, Ex. 6.9-5 and p. 309) and (19.25.42). For (19.20.24)–(19.20.26) see Carlson (1977b, (6.2-1),(6.8-15)).

§19.21 To prove (19.21.1) see the text following (19.21.6), use (19.20.10), and analytic continuation. For (19.21.2) put $x = 0$ in (19.21.9). For (19.21.3) put $x = 0$ in (19.21.11) and (19.21.10). (19.21.6) is equivalent to Zill and Carlson (1970, (7.15)). For (19.21.8) and (19.21.9) see Carlson (1977b, (5.9-5),(5.9-6)) and (19.20.25). To obtain (19.21.7) eliminate $R_D(z,x,y)$ between (19.21.8) and (19.21.9), which follow from Carlson (1977b, (5.9-5), (6.6-5), and (5.9-6)). For (19.21.10) see Carlson (1977b, Table 9.3-1). To prove (19.21.11) write $xt/(t+x) = x - (x^2/(t+x))$ in (19.23.7) and similarly for y and z. Then use (19.21.9). For (19.21.12)–(19.21.15) see Zill and Carlson (1970, (4.6)).

§19.22 In (19.22.18), (19.22.21), and (19.22.20), put $z = 0$ to obtain (19.22.1), (19.22.2), and (19.22.4), respectively. (19.22.3) is derivable from (19.22.2) and (19.21.3), or more directly by putting $p = y$ in (19.22.7). For (19.22.7) see Carlson (1976, (4.14),(4.13)), where $(\pi/4)R_L(y,z,p) = R_F(0,y,z) - (p/3) R_J(0,y,z,p)$. For (19.22.8)–(19.22.15) iterate the results given in §19.22(i); see also (19.16.20), (19.16.23), and Carlson (2002, Section 2). For (19.22.18) see Carlson (1964, (5.13)). For (19.22.19) put $p = z$ in (19.22.20). For (19.22.20) see Zill and Carlson (1970, (5.7)) and Carlson (1990, (8.5)). For (19.22.21) see Carlson (1964, (5.16)). For (19.22.22) put $z = y$ in (19.22.18). In the ascending Landen case let $k^2 = (z_+^2 - z_-^2)/(z_+^2 - a^2)$ and $k_1^2 = (z^2 - y^2)/(z^2 - x^2)$ to get the second equation in (19.8.11). In the descending Gauss case let $k_1^2 = (a^2 - z_-^2)/(a^2 - z_+^2)$ and $k^2 = (z^2 - y^2)/(z^2 - x^2)$ to get the first equation in (19.8.11).

§19.23 For (19.23.8) and (19.23.9) see Carlson (1977b, Exercises 5.9-19, 5.9-20, and p. 306). By §19.16(iii), (19.23.8) implies (19.23.1)–(19.23.3), and (19.23.9) implies (19.23.6). Use (19.23.8) to integrate over θ in (19.23.6) and then permute variables to prove (19.23.5). To prove (19.23.4) put $z = 0$ in (19.23.5), relabel variables, and substitute $\cos \theta = \operatorname{sech} t$. For (19.23.7) and (19.23.10) see Carlson (1977b, (9.1-9) and (6.8-2), respectively).

§19.24 For (19.24.1)–(19.24.3) use (19.9.1) and (19.9.4). For (19.24.4) see (19.16.22) and Carlson (1966, (2.15)). For (19.24.3) see (19.30.5). (19.24.8) is a special case of (19.24.13). For (19.24.9) see Neuman (2003, (4.2)).

§19.25 (19.25.1), (19.25.2), and (19.25.3) are derived from the incomplete cases. For (19.25.4) put $c = 1$ in (19.25.16). (19.25.5) and (19.25.7) come from Carlson (1977b, (9.3-2) and (9.3-3)). For (19.25.6) and (19.25.12) apply (19.18.4) to (19.25.5) and (19.25.8), respectively. (19.25.8) and (19.25.15) are special cases of (19.16.12). To get (19.25.9), (19.25.10), and (19.25.11), let $(c-1, c-k^2, c) = (x, y, z)$ and eliminate R_G between (19.25.7) and each of the three forms of (19.25.10) obtained by permuting x, y and z. For (19.25.13) combine (19.2.6) and (19.25.9). For (19.25.14) see Zill and Carlson (1970, (2.5)). For (19.25.16) substitute (19.25.14) in (19.7.8) and use (19.2.20). For (19.25.19)–(19.25.22) rewrite Bulirsch's integrals (§19.2(iii)) in terms of Legendre's integrals, then use §19.25(i) to convert them to R-functions. For (19.25.24)–(19.25.27) define $c = \csc^2 \phi$, write $(x, y, z, p)/(z - x) = (c - 1, c - k^2, c, c - \alpha^2)$, then use (19.25.5), (19.25.9), (19.25.14), and (19.25.7) to prove (19.25.24), (19.25.25), (19.25.26), and (19.25.27), respectively. To prove (19.25.29) use $(\operatorname{cs}, \operatorname{ds}, \operatorname{ns}) = (\operatorname{cn}, \operatorname{dn}, 1)/\operatorname{sn}$ (suppressing variables (u,k)). For (19.25.30) see Carlson (2006a, Comments following proof of Proposition 4.1). For (19.25.31) see Carlson (2004, (1.8)). In (19.25.32), (19.25.33), and (19.25.34), substitute $x = \operatorname{ps}(u, k)$, $\operatorname{sp}(u, k)$, and $\operatorname{pq}(u, k)$, respectively, to recover (19.25.31). To prove (19.25.35) use (23.6.36), with $z = \wp(w)$ as prescribed in the text that follows (23.6.36), substitute $u = t + \wp(w)$ and compare with (19.16.1). Then put $z = \omega_j$ to obtain (19.25.38). For (19.25.37) and (19.25.39) see Carlson (1964, (3.10) and (3.2)). For (19.25.40) combine Erdélyi et al. (1953b, §§13.12(22), 13.13(22)) and (19.25.35).

§19.26 Addition theorems (and therefore duplication theorems) for the symmetric integrals are proved by Zill and Carlson (1970, §8). For other proofs of (19.26.1) see Carlson (1977b, §9.7) and Carlson (1978, Theorem 3). To prove (19.26.13) use (19.2.9) to show that $2\sqrt{\theta} R_C(\sigma^2, \sigma^2 - \theta) = \ln\left((\sigma + \sqrt{\theta})/(\sigma - \sqrt{\theta})\right)$, then apply this to all three terms. To prove (19.26.14) put $z = y$ in (19.21.12) and use (19.20.8). For (19.26.17) put $\theta = -\alpha\beta$ in (19.26.13) and use homogeneity. For (19.26.18)–(19.26.27) put $\mu = \lambda$ in the formulas of §19.26(i). For proofs of (19.26.18) not invoking the addition theorem, see Carlson (1977b, §9.6) and Carlson (1998, §2). Equations (19.26.25) and (19.26.20) are degenerate cases of (19.26.18) and (19.26.22), respectively.

§19.27 Carlson and Gustafson (1994). For (19.27.2) see Carlson and Gustafson (1985).

§19.28 To prove (19.28.1)–(19.28.3) from (19.28.4) use §19.16(iii). To prove (19.28.4) expand the R-function in powers of $1 - t$ by (19.19.3), integrate term by term, and use Erdélyi et al. (1953a, 2.8(46)). (19.28.5) is equivalent to (19.18.1). In (19.28.6) let $v = \sqrt{u}$ and use Carlson (1963, (7.9)). In (19.28.7) substitute (19.16.2), change the order of integration, and use (19.29.4). Use Carlson (1963, (7.11)) and (19.16.20) to prove (19.28.8) and (19.28.10). In the first case Carlson (1977b, (5.9-21)) is needed; in the second case put $(z_1, z_2, \zeta_1, \zeta_2) = (a^2, b^2, c^2, d^2)$, use Carlson (1977b, (9.8-4)), and substitute $t = (ab/cd) \exp(2z)$. To prove (19.28.9) from (19.28.10), put $a = \exp(ix) = 1/b$, $c = \exp(iy) = 1/d$, $\cosh z = 1/\sin \theta$, and on the right-hand side use (19.22.1).

§19.29 For (19.29.4) see Carlson (1998, (3.6)). For (19.29.7), a special case of (19.29.8), see also Carlson (1987, (4.14)). For (19.29.8) see Carlson (1999, (4.10)) and Carlson (1988, (5.6)). For (19.29.10) see Byrd and Friedman (1971, p. 76, Eq. (234.13), and p. 74) for notation. Then use Carlson (2006b, (3.2)) with $(p, q, r) = (n, d, c)$ for reduction to R_D. For (19.29.19)–(19.29.33) take t^2 as a new variable where appropriate. Then factor quadratic polynomials, use (19.29.4), and apply (19.22.18) to remove any complex quantities. For (19.29.20) use (19.29.7) with $a_\alpha + b_\alpha t = t$ and $a_\delta + b_\delta t = 1$. For (19.29.21) use (19.29.7) with $a_\alpha + b_\alpha t = 1$ and $a_\delta + b_\delta t = t$. With regard to (19.29.28) see Carlson (1977a, p. 238).

§19.30 Carlson (1977b, §9.4 and Ex. 8.3-7, with solution on p. 312). For (19.30.5) see (19.25.1). For (19.30.6) use (19.4.6).

§19.32 Carlson (1977b, pp. 234–235). For (19.32.2) use (19.18.6).

§19.33 Carlson (1977b, pp. 271, 313, (9.4-10), and Ex. 9.4-3) and Carlson (1961a). For other proofs of (19.33.1) and (19.33.2) see Watson (1935b), Bowman (1953, pp. 31–32), and Carlson (1964, p. 417). For the first equality in (19.33.7) see Becker and Sauter (1964, p. 106).

§19.34 For (19.34.1) see Becker and Sauter (1964, p. 194). For (19.34.7) see Carlson (1977b, Ex. 9.3-2 and p. 313); alternatively, substitute Carlson (1977b, (9.2-3) and (9.2-2)) in (19.34.6) and use Carlson (1977b, Table 9.3-2).

§19.36 For the quadratic transformations see Carlson (1965, (3.1), (3.2), Sections 5, 6). To obtain (19.36.6) and (19.36.8) from (19.22.18), let $(x^2, y^2, z^2) = (t_n^2, t_n^2 + \theta c_n^2, t_n^2 + \theta a_n^2)$ and $(a^2, z_-^2, z_+^2) = (t_{n+1}^2, t_{n+1}^2 + \theta c_{n+1}^2, t_{n+1}^2 + \theta a_{n+1}^2)$. Then use the expression for $z_\pm^2 - a^2$ from (19.22.17) and the definition of a from (19.22.16).

Chapter 20
Theta Functions
W. P. Reinhardt[1] and P. L. Walker[2]

Notation		**524**
20.1	Special Notation	524
Properties		**524**
20.2	Definitions and Periodic Properties	524
20.3	Graphics	525
20.4	Values at $z = 0$	529
20.5	Infinite Products and Related Results	529
20.6	Power Series	530
20.7	Identities	530
20.8	Watson's Expansions	531
20.9	Relations to Other Functions	532
20.10	Integrals	532
20.11	Generalizations and Analogs	532
Applications		**533**
20.12	Mathematical Applications	533
20.13	Physical Applications	533
Computation		**534**
20.14	Methods of Computation	534
20.15	Tables	534
20.16	Software	534
References		**534**

---

[1] University of Washington, Seattle, Washington.
[2] American University of Sharjah, Sharjah, United Arab Emirates.
Acknowledgments: This chapter is based in part on Abramowitz and Stegun (1964, Chapter 16), by L. M. Milne-Thomson.
Copyright © 2009 National Institute of Standards and Technology. All rights reserved.

Notation

20.1 Special Notation

(For other notation see pp. xiv and 873.)

m, n	integers.
$z \ (\in \mathbb{C})$	the argument.
$\tau \ (\in \mathbb{C})$	the lattice parameter, $\Im\tau > 0$.
$q \ (\in \mathbb{C})$	the nome, $q = e^{i\pi\tau}$, $0 < |q| < 1$. Since τ is not a single-valued function of q, it is assumed that τ is known, even when q is specified. Most applications concern the rectangular case $\Re\tau = 0$, $\Im\tau > 0$, so that $0 < q < 1$ and τ and q are uniquely related.
q^α	$e^{i\alpha\pi\tau}$ for $\alpha \in \mathbb{R}$ (resolving issues of choice of branch).
S_1/S_2	set of all elements of S_1, modulo elements of S_2. Thus two elements of S_1/S_2 are equivalent if they are both in S_1 and their difference is in S_2. (For an example see §20.12(ii).)

The main functions treated in this chapter are the theta functions $\theta_j(z|\tau) = \theta_j(z,q)$ where $j = 1,2,3,4$ and $q = e^{i\pi\tau}$. When τ is fixed the notation is often abbreviated in the literature as $\theta_j(z)$, or even as simply θ_j, it being then understood that the argument is the primary variable. Sometimes the theta functions are called the Jacobian or classical theta functions to distinguish them from generalizations; compare Chapter 21.

Primes on the θ symbols indicate derivatives with respect to the argument of the θ function.

Other Notations

Jacobi's original notation: $\Theta(z|\tau)$, $\Theta_1(z|\tau)$, $H(z|\tau)$, $H_1(z|\tau)$, respectively, for $\theta_4(u|\tau)$, $\theta_3(u|\tau)$, $\theta_1(u|\tau)$, $\theta_2(u|\tau)$, where $u = z/\theta_3^2(0|\tau)$. Here the symbol H denotes capital eta. See, for example, Whittaker and Watson (1927, p. 479) and Copson (1935, pp. 405, 411).

Neville's notation: $\theta_s(z|\tau)$, $\theta_c(z|\tau)$, $\theta_d(z|\tau)$, $\theta_n(z|\tau)$, respectively, for $\theta_3^2(0|\tau)\theta_1(u|\tau)/\theta_1'(0|\tau)$, $\theta_2(u|\tau)/\theta_2(0|\tau)$, $\theta_3(u|\tau)/\theta_3(0|\tau)$, $\theta_4(u|\tau)/\theta_4(0|\tau)$, where again $u = z/\theta_3^2(0|\tau)$. This notation simplifies the relationship of the theta functions to Jacobian elliptic functions (§22.2); see Neville (1951).

McKean and Moll's notation: $\vartheta_j(z|\tau) = \theta_j(\pi z|\tau)$, $j = 1,2,3,4$. See McKean and Moll (1999, p. 125).

Additional notations that have been used in the literature are summarized in Whittaker and Watson (1927, p. 487).

Properties

20.2 Definitions and Periodic Properties

20.2(i) Fourier Series

20.2.1
$$\theta_1(z|\tau) = \theta_1(z,q)$$
$$= 2\sum_{n=0}^{\infty}(-1)^n q^{(n+\frac{1}{2})^2}\sin((2n+1)z),$$

20.2.2 $\theta_2(z|\tau) = \theta_2(z,q) = 2\sum_{n=0}^{\infty} q^{(n+\frac{1}{2})^2}\cos((2n+1)z),$

20.2.3 $\theta_3(z|\tau) = \theta_3(z,q) = 1 + 2\sum_{n=1}^{\infty} q^{n^2}\cos(2nz),$

20.2.4 $\theta_4(z|\tau) = \theta_4(z,q) = 1 + 2\sum_{n=1}^{\infty}(-1)^n q^{n^2}\cos(2nz).$

Corresponding expansions for $\theta_j'(z|\tau)$, $j = 1,2,3,4$, can be found by differentiating (20.2.1)–(20.2.4) with respect to z.

20.2(ii) Periodicity and Quasi-Periodicity

For fixed τ, each $\theta_j(z|\tau)$ is an entire function of z with period 2π; $\theta_1(z|\tau)$ is odd in z and the others are even. For fixed z, each of $\theta_1(z|\tau)/\sin z$, $\theta_2(z|\tau)/\cos z$, $\theta_3(z|\tau)$, and $\theta_4(z|\tau)$ is an analytic function of τ for $\Im\tau > 0$, with a natural boundary $\Im\tau = 0$, and correspondingly, an analytic function of q for $|q| < 1$ with a natural boundary $|q| = 1$.

The four points $(0, \pi, \pi+\tau\pi, \tau\pi)$ are the vertices of the *fundamental parallelogram* in the z-plane; see Figure 20.2.1. The points

20.2.5 $\qquad z_{m,n} = (m+n\tau)\pi, \qquad m,n \in \mathbb{Z},$

are the *lattice points*. The theta functions are quasi-periodic on the lattice:

20.2.6
$$\theta_1(z+(m+n\tau)\pi|\tau) = (-1)^{m+n}q^{-n^2}e^{-2inz}\theta_1(z|\tau),$$

20.2.7
$$\theta_2(z+(m+n\tau)\pi|\tau) = (-1)^m q^{-n^2}e^{-2inz}\theta_2(z|\tau),$$

20.2.8
$$\theta_3(z+(m+n\tau)\pi|\tau) = q^{-n^2}e^{-2inz}\theta_3(z|\tau),$$

20.2.9
$$\theta_4(z+(m+n\tau)\pi|\tau) = (-1)^n q^{-n^2}e^{-2inz}\theta_4(z|\tau).$$

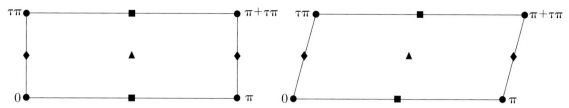

Figure 20.2.1: z-plane. Fundamental parallelogram. Left-hand diagram is the rectangular case (τ purely imaginary); right-hand diagram is the general case. ● zeros of $\theta_1(z|\tau)$, ■ zeros of $\theta_2(z|\tau)$, ▲ zeros of $\theta_3(z|\tau)$, ◆ zeros of $\theta_4(z|\tau)$.

20.2(iii) Translation of the Argument by Half-Periods

With

20.2.10 $\qquad M \equiv M(z|\tau) = e^{iz+(i\pi\tau/4)},$

20.2.11 $\qquad \begin{aligned}\theta_1(z|\tau) &= -\theta_2\bigl(z+\tfrac{1}{2}\pi\big|\tau\bigr) = -iM\,\theta_4\bigl(z+\tfrac{1}{2}\pi\tau\big|\tau\bigr)\\ &= -iM\,\theta_3\bigl(z+\tfrac{1}{2}\pi+\tfrac{1}{2}\pi\tau\big|\tau\bigr),\end{aligned}$

20.2.12 $\qquad \begin{aligned}\theta_2(z|\tau) &= \theta_1\bigl(z+\tfrac{1}{2}\pi\big|\tau\bigr) = M\,\theta_3\bigl(z+\tfrac{1}{2}\pi\tau\big|\tau\bigr)\\ &= M\,\theta_4\bigl(z+\tfrac{1}{2}\pi+\tfrac{1}{2}\pi\tau\big|\tau\bigr),\end{aligned}$

20.2.13 $\qquad \begin{aligned}\theta_3(z|\tau) &= \theta_4\bigl(z+\tfrac{1}{2}\pi\big|\tau\bigr) = M\,\theta_2\bigl(z+\tfrac{1}{2}\pi\tau\big|\tau\bigr)\\ &= M\,\theta_1\bigl(z+\tfrac{1}{2}\pi+\tfrac{1}{2}\pi\tau\big|\tau\bigr),\end{aligned}$

20.2.14 $\qquad \begin{aligned}\theta_4(z|\tau) &= \theta_3\bigl(z+\tfrac{1}{2}\pi\big|\tau\bigr) = -iM\,\theta_1\bigl(z+\tfrac{1}{2}\pi\tau\big|\tau\bigr)\\ &= iM\,\theta_2\bigl(z+\tfrac{1}{2}\pi+\tfrac{1}{2}\pi\tau\big|\tau\bigr).\end{aligned}$

20.2(iv) z-Zeros

For $m, n \in \mathbb{Z}$, the z-zeros of $\theta_j(z|\tau)$, $j = 1, 2, 3, 4$, are $(m+n\tau)\pi$, $(m+\tfrac{1}{2}+n\tau)\pi$, $(m+\tfrac{1}{2}+(n+\tfrac{1}{2})\tau)\pi$, $(m+(n+\tfrac{1}{2})\tau)\pi$ respectively.

20.3 Graphics

20.3(i) θ-Functions: Real Variable and Real Nome

See Figures 20.3.1–20.3.13.

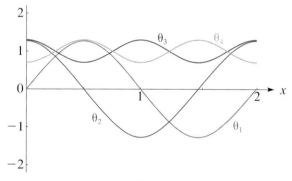

Figure 20.3.1: $\theta_j(\pi x, 0.15)$, $0 \leq x \leq 2$, $j = 1, 2, 3, 4$.

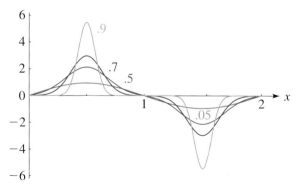

Figure 20.3.2: $\theta_1(\pi x, q)$, $0 \leq x \leq 2$, $q = 0.05$, 0.5, 0.7, 0.9. For $q \leq q^{\text{Dedekind}}$, $\theta_1(\pi x, q)$ is convex in x for $0 < x < 1$. Here $q^{\text{Dedekind}} = e^{-\pi y_0} = 0.19$ approximately, where $y = y_0$ corresponds to the maximum value of Dedekind's eta function $\eta(iy)$ as depicted in Figure 23.16.1.

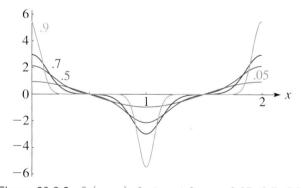

Figure 20.3.3: $\theta_2(\pi x, q)$, $0 \leq x \leq 2$, $q = 0.05$, 0.5, 0.7, 0.9.

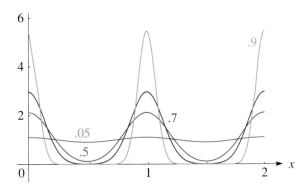

Figure 20.3.4: $\theta_3(\pi x, q)$, $0 \le x \le 2$, $q = 0.05, 0.5, 0.7, 0.9$.

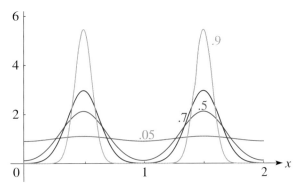

Figure 20.3.5: $\theta_4(\pi x, q)$, $0 \le x \le 2$, $q = 0.05, 0.5, 0.7, 0.9$.

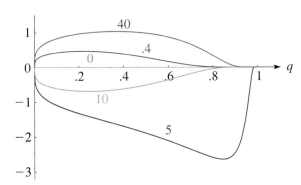

Figure 20.3.6: $\theta_1(x, q)$, $0 \le q \le 1$, $x = 0, 0.4, 5, 10, 40$.

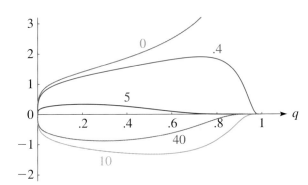

Figure 20.3.7: $\theta_2(x, q)$, $0 \le q \le 1$, $x = 0, 0.4, 5, 10, 40$.

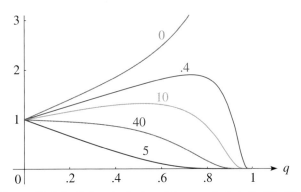

Figure 20.3.8: $\theta_3(x, q)$, $0 \le q \le 1$, $x = 0, 0.4, 5, 10, 40$.

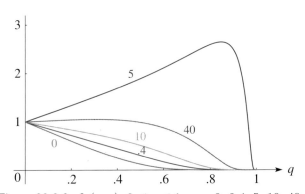

Figure 20.3.9: $\theta_4(x, q)$, $0 \le q \le 1$, $x = 0, 0.4, 5, 10, 40$.

20.3 GRAPHICS

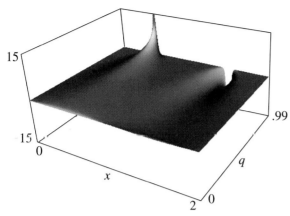

Figure 20.3.10: $\theta_1(\pi x, q)$, $0 \leq x \leq 2$, $0 \leq q \leq 0.99$.

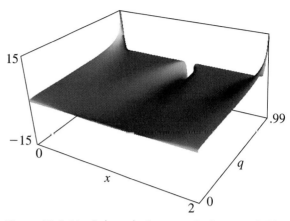

Figure 20.3.11: $\theta_2(\pi x, q)$, $0 \leq x \leq 2$, $0 \leq q \leq 0.99$.

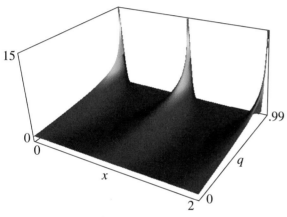

Figure 20.3.12: $\theta_3(\pi x, q)$, $0 \leq x \leq 2$, $0 \leq q \leq 0.99$.

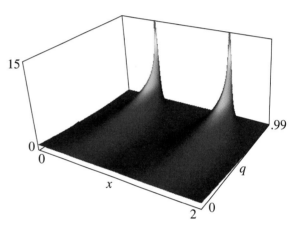

Figure 20.3.13: $\theta_4(\pi x, q)$, $0 \leq x \leq 2$, $0 \leq q \leq 0.99$.

20.3(ii) θ-Functions: Complex Variable and Real Nome

See Figures 20.3.14–20.3.17. In these graphics, height corresponds to the absolute value of the function and color to the phase. See also p. xiv.

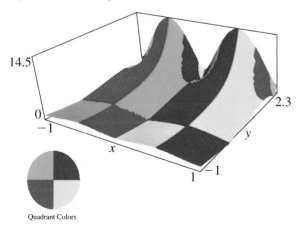

Figure 20.3.14: $\theta_1(\pi x + iy, 0.12)$, $-1 \leq x \leq 1$, $-1 \leq y \leq 2.3$.

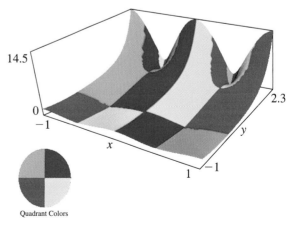

Figure 20.3.15: $\theta_2(\pi x + iy, 0.12)$, $-1 \leq x \leq 1$, $-1 \leq y \leq 2.3$.

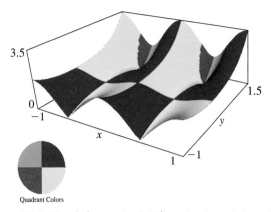

Figure 20.3.16: $\theta_3(\pi x + iy, 0.12)$, $-1 \leq x \leq 1$, $-1 \leq y \leq 1.5$.

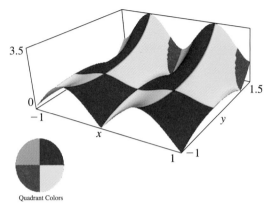

Figure 20.3.17: $\theta_4(\pi x + iy, 0.12)$, $-1 \leq x \leq 1$, $-1 \leq y \leq 1.5$.

20.3(iii) θ-Functions: Real Variable and Complex Lattice Parameter

See Figures 20.3.18–20.3.21. In these graphics this subsection, height corresponds to the absolute value of the function and color to the phase. See also p. xiv.

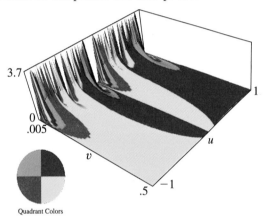

Figure 20.3.18: $\theta_1(0.1|u+iv)$, $-1 \leq u \leq 1$, $0.005 \leq v \leq 0.5$. The value 0.1 of z is chosen arbitrarily since θ_1 vanishes identically when $z=0$.

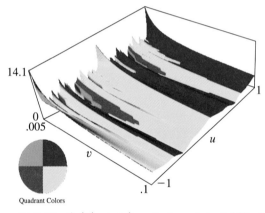

Figure 20.3.19: $\theta_2(0|u+iv)$, $-1 \leq u \leq 1$, $0.005 \leq v \leq 0.1$.

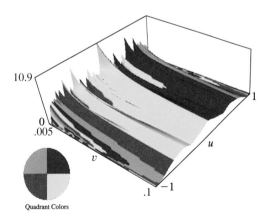

Figure 20.3.20: $\theta_3(0|u+iv)$, $-1 \leq u \leq 1$, $0.005 \leq v \leq 0.1$.

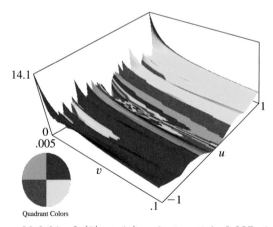

Figure 20.3.21: $\theta_4(0|u+iv)$, $-1 \leq u \leq 1$, $0.005 \leq v \leq 0.1$.

20.4 Values at $z = 0$

20.4(i) Functions and First Derivatives

20.4.1 $\quad \theta_1(0,q) = \theta_2'(0,q) = \theta_3'(0,q) = \theta_4'(0,q) = 0,$

20.4.2 $\quad \theta_1'(0,q) = 2q^{1/4} \prod_{n=1}^{\infty} \left(1 - q^{2n}\right)^3,$

20.4.3 $\quad \theta_2(0,q) = 2q^{1/4} \prod_{n=1}^{\infty} \left(1 - q^{2n}\right) \left(1 + q^{2n}\right)^2,$

20.4.4 $\quad \theta_3(0,q) = \prod_{n=1}^{\infty} \left(1 - q^{2n}\right) \left(1 + q^{2n-1}\right)^2,$

20.4.5 $\quad \theta_4(0,q) = \prod_{n=1}^{\infty} \left(1 - q^{2n}\right) \left(1 - q^{2n-1}\right)^2.$

Jacobi's Identity

20.4.6 $\quad \theta_1'(0,q) = \theta_2(0,q)\,\theta_3(0,q)\,\theta_4(0,q).$

20.4(ii) Higher Derivatives

20.4.7 $\quad \theta_1''(0,q) = \theta_2'''(0,q) = \theta_3'''(0,q) = \theta_4'''(0,q) = 0.$

20.4.8 $\quad \dfrac{\theta_1'''(0,q)}{\theta_1'(0,q)} = -1 + 24 \sum_{n=1}^{\infty} \dfrac{q^{2n}}{(1 - q^{2n})^2}.$

20.4.9 $\quad \dfrac{\theta_2''(0,q)}{\theta_2(0,q)} = -1 - 8 \sum_{n=1}^{\infty} \dfrac{q^{2n}}{(1 + q^{2n})^2},$

20.4.10 $\quad \dfrac{\theta_3''(0,q)}{\theta_3(0,q)} = -8 \sum_{n=1}^{\infty} \dfrac{q^{2n-1}}{(1 + q^{2n-1})^2},$

20.4.11 $\quad \dfrac{\theta_4''(0,q)}{\theta_4(0,q)} = 8 \sum_{n=1}^{\infty} \dfrac{q^{2n-1}}{(1 - q^{2n-1})^2}.$

20.4.12 $\quad \dfrac{\theta_1'''(0,q)}{\theta_1'(0,q)} = \dfrac{\theta_2''(0,q)}{\theta_2(0,q)} + \dfrac{\theta_3''(0,q)}{\theta_3(0,q)} + \dfrac{\theta_4''(0,q)}{\theta_4(0,q)}.$

20.5 Infinite Products and Related Results

20.5(i) Single Products

20.5.1
$$\theta_1(z,q) = 2q^{1/4} \sin z \prod_{n=1}^{\infty} \left(1 - q^{2n}\right)\left(1 - 2q^{2n}\cos(2z) + q^{4n}\right),$$

20.5.2
$$\theta_2(z,q) = 2q^{1/4} \cos z \prod_{n=1}^{\infty} \left(1 - q^{2n}\right)\left(1 + 2q^{2n}\cos(2z) + q^{4n}\right),$$

20.5.3
$$\theta_3(z,q) = \prod_{n=1}^{\infty} \left(1 - q^{2n}\right)\left(1 + 2q^{2n-1}\cos(2z) + q^{4n-2}\right),$$

20.5.4
$$\theta_4(z,q) = \prod_{n=1}^{\infty} \left(1 - q^{2n}\right)\left(1 - 2q^{2n-1}\cos(2z) + q^{4n-2}\right).$$

20.5.5
$$\theta_1(z|\tau) = \theta_1'(0|\tau)\sin z \prod_{n=1}^{\infty} \dfrac{\sin(n\pi\tau + z)\sin(n\pi\tau - z)}{\sin^2(n\pi\tau)},$$

20.5.6
$$\theta_2(z|\tau) = \theta_2(0|\tau)\cos z \prod_{n=1}^{\infty} \dfrac{\cos(n\pi\tau + z)\cos(n\pi\tau - z)}{\cos^2(n\pi\tau)},$$

20.5.7
$$\theta_3(z|\tau) = \theta_3(0|\tau) \prod_{n=1}^{\infty} \dfrac{\cos\!\left((n-\tfrac{1}{2})\pi\tau + z\right)\cos\!\left((n-\tfrac{1}{2})\pi\tau - z\right)}{\cos^2\!\left((n-\tfrac{1}{2})\pi\tau\right)},$$

20.5.8
$$\theta_4(z|\tau) = \theta_4(0|\tau) \prod_{n=1}^{\infty} \dfrac{\sin\!\left((n-\tfrac{1}{2})\pi\tau + z\right)\sin\!\left((n-\tfrac{1}{2})\pi\tau - z\right)}{\sin^2\!\left((n-\tfrac{1}{2})\pi\tau\right)}.$$

Jacobi's Triple Product

20.5.9
$$\theta_3(\pi z|\tau) = \sum_{n=-\infty}^{\infty} p^{2n} q^{n^2}$$
$$= \prod_{n=1}^{\infty} \left(1 - q^{2n}\right)\left(1 + q^{2n-1}p^2\right)\left(1 + q^{2n-1}p^{-2}\right),$$

where $p = e^{i\pi z}$, $q = e^{i\pi\tau}$.

20.5(ii) Logarithmic Derivatives

When $|\Im z| < \pi\Im\tau$,

20.5.10
$$\dfrac{\theta_1'(z,q)}{\theta_1(z,q)} - \cot z = 4\sin(2z)\sum_{n=1}^{\infty}\dfrac{q^{2n}}{1 - 2q^{2n}\cos(2z) + q^{4n}}$$
$$= 4\sum_{n=1}^{\infty}\dfrac{q^{2n}}{1 - q^{2n}}\sin(2nz),$$

20.5.11
$$\dfrac{\theta_2'(z,q)}{\theta_2(z,q)} + \tan z = -4\sin(2z)\sum_{n=1}^{\infty}\dfrac{q^{2n}}{1 + 2q^{2n}\cos(2z) + q^{4n}}$$
$$= 4\sum_{n=1}^{\infty}(-1)^n\dfrac{q^{2n}}{1 - q^{2n}}\sin(2nz).$$

The left-hand sides of (20.5.10) and (20.5.11) are replaced by their limiting values when $\cot z$ or $\tan z$ are undefined.

When $|\Im z| < \tfrac{1}{2}\pi\Im\tau$,

20.5.12
$$\dfrac{\theta_3'(z,q)}{\theta_3(z,q)} = -4\sin(2z)\sum_{n=1}^{\infty}\dfrac{q^{2n-1}}{1 + 2q^{2n-1}\cos(2z) + q^{4n-2}}$$
$$= 4\sum_{n=1}^{\infty}(-1)^n\dfrac{q^n}{1 - q^{2n}}\sin(2nz),$$

20.5.13
$$\frac{\theta_4'(z,q)}{\theta_4(z,q)} = 4\sin(2z) \sum_{n=1}^{\infty} \frac{q^{2n-1}}{1 - 2q^{2n-1}\cos(2z) + q^{4n-2}}$$
$$= 4 \sum_{n=1}^{\infty} \frac{q^n}{1-q^{2n}} \sin(2nz).$$

With the given conditions the infinite series in (20.5.10)–(20.5.13) converge absolutely and uniformly in compact sets in the z-plane.

20.5(iii) Double Products

20.5.14
$$\theta_1(z|\tau) = z\,\theta_1'(0|\tau) \lim_{N\to\infty} \prod_{n=-N}^{N} \lim_{M\to\infty} \prod_{\substack{m=-M \\ |m|+|n|\neq 0}}^{M} \left(1 + \frac{z}{(m+n\tau)\pi}\right),$$

20.5.15
$$\theta_2(z|\tau) = \theta_2(0|\tau) \lim_{N\to\infty} \prod_{n=-N}^{N} \lim_{M\to\infty} \prod_{m=1-M}^{M} \left(1 + \frac{z}{(m-\tfrac{1}{2}+n\tau)\pi}\right),$$

20.5.16
$$\theta_3(z|\tau) = \theta_3(0|\tau) \lim_{N\to\infty} \prod_{n=1-N}^{N} \lim_{M\to\infty} \prod_{m=1-M}^{M} \left(1 + \frac{z}{(m-\tfrac{1}{2}+(n-\tfrac{1}{2})\tau)\pi}\right),$$

20.5.17
$$\theta_4(z|\tau) = \theta_4(0|\tau) \lim_{N\to\infty} \prod_{n=1-N}^{N} \lim_{M\to\infty} \prod_{m=-M}^{M} \left(1 + \frac{z}{(m+(n-\tfrac{1}{2})\tau)\pi}\right).$$

These double products are not absolutely convergent; hence the order of the limits is important. The order shown is in accordance with the Eisenstein convention (Walker (1996, §0.3)).

20.6 Power Series

Assume

20.6.1 $\qquad |\pi z| < \min |z_{m,n}|,$

where $z_{m,n}$ is given by (20.2.5) and the minimum is for $m, n \in \mathbb{Z}$, except $m = n = 0$. Then

20.6.2 $\quad \theta_1(\pi z|\tau) = \pi z\, \theta_1'(0|\tau) \exp\left(-\sum_{j=1}^{\infty} \frac{1}{2j} \delta_{2j}(\tau) z^{2j}\right),$

20.6.3 $\quad \theta_2(\pi z|\tau) = \theta_2(0|\tau) \exp\left(-\sum_{j=1}^{\infty} \frac{1}{2j} \alpha_{2j}(\tau) z^{2j}\right),$

20.6.4 $\quad \theta_3(\pi z|\tau) = \theta_3(0|\tau) \exp\left(-\sum_{j=1}^{\infty} \frac{1}{2j} \beta_{2j}(\tau) z^{2j}\right),$

20.6.5 $\quad \theta_4(\pi z|\tau) = \theta_4(0|\tau) \exp\left(-\sum_{j=1}^{\infty} \frac{1}{2j} \gamma_{2j}(\tau) z^{2j}\right).$

Here the coefficients are given by

20.6.6 $\quad \delta_{2j}(\tau) = \sum_{n=-\infty}^{\infty} \sum_{\substack{m=-\infty \\ |m|+|n|\neq 0}}^{\infty} (m+n\tau)^{-2j},$

20.6.7 $\quad \alpha_{2j}(\tau) = \sum_{n=-\infty}^{\infty} \sum_{m=-\infty}^{\infty} (m - \tfrac{1}{2} + n\tau)^{-2j},$

20.6.8 $\quad \beta_{2j}(\tau) = \sum_{n=-\infty}^{\infty} \sum_{m=-\infty}^{\infty} (m - \tfrac{1}{2} + (n-\tfrac{1}{2})\tau)^{-2j},$

20.6.9 $\quad \gamma_{2j}(\tau) = \sum_{n=-\infty}^{\infty} \sum_{m=-\infty}^{\infty} (m + (n-\tfrac{1}{2})\tau)^{-2j},$

and satisfy

20.6.10
$$\alpha_{2j}(\tau) = 2^{2j}\delta_{2j}(2\tau) - \delta_{2j}(\tau),$$
$$\beta_{2j}(\tau) = 2^{2j}\gamma_{2j}(2\tau) - \gamma_{2j}(\tau).$$

In the double series the order of summation is important only when $j = 1$. For further information on δ_{2j} see §23.9: since the double sums in (20.6.6) and (23.9.1) are the same, we have $\delta_{2n} = c_n/(2n-1)$ when $n \geq 2$.

20.7 Identities

20.7(i) Sums of Squares

20.7.1
$$\theta_3^2(0,q)\,\theta_3^2(z,q) = \theta_4^2(0,q)\,\theta_4^2(z,q) + \theta_2^2(0,q)\,\theta_2^2(z,q),$$

20.7.2
$$\theta_3^2(0,q)\,\theta_4^2(z,q) = \theta_2^2(0,q)\,\theta_1^2(z,q) + \theta_4^2(0,q)\,\theta_3^2(z,q),$$

20.7.3
$$\theta_2^2(0,q)\,\theta_4^2(z,q) = \theta_3^2(0,q)\,\theta_1^2(z,q) + \theta_4^2(0,q)\,\theta_2^2(z,q),$$

20.7.4
$$\theta_2^2(0,q)\,\theta_3^2(z,q) = \theta_4^2(0,q)\,\theta_1^2(z,q) + \theta_3^2(0,q)\,\theta_2^2(z,q).$$

Also

20.7.5 $\qquad \theta_3^4(0,q) = \theta_2^4(0,q) + \theta_4^4(0,q).$

20.7(ii) Addition Formulas

20.7.6
$$\theta_4^2(0,q)\,\theta_1(w+z,q)\,\theta_1(w-z,q) \\ = \theta_3^2(w,q)\,\theta_2^2(z,q) - \theta_2^2(w,q)\,\theta_3^2(z,q),$$

20.7.7
$$\theta_4^2(0,q)\,\theta_2(w+z,q)\,\theta_2(w-z,q) \\ = \theta_4^2(w,q)\,\theta_2^2(z,q) - \theta_1^2(w,q)\,\theta_3^2(z,q),$$

20.7.8
$$\theta_4^2(0,q)\,\theta_3(w+z,q)\,\theta_3(w-z,q) \\ = \theta_4^2(w,q)\,\theta_3^2(z,q) - \theta_1^2(w,q)\,\theta_2^2(z,q),$$

20.7.9
$$\theta_4^2(0,q)\,\theta_4(w+z,q)\,\theta_4(w-z,q) \\ = \theta_3^2(w,q)\,\theta_3^2(z,q) - \theta_2^2(w,q)\,\theta_2^2(z,q).$$

For these and similar formulas see Lawden (1989, §1.4) and Whittaker and Watson (1927, pp. 487–488).

20.7(iii) Duplication Formula

20.7.10 $\quad \theta_1(2z,q) = 2\,\dfrac{\theta_1(z,q)\,\theta_2(z,q)\,\theta_3(z,q)\,\theta_4(z,q)}{\theta_2(0,q)\,\theta_3(0,q)\,\theta_4(0,q)}.$

20.7(iv) Transformations of Nome

20.7.11 $\quad \dfrac{\theta_1(z,q)\,\theta_2(z,q)}{\theta_1(2z,q^2)} = \dfrac{\theta_3(z,q)\,\theta_4(z,q)}{\theta_4(2z,q^2)} = \theta_4(0,q^2),$

20.7.12 $\quad \dfrac{\theta_1(z,q^2)\,\theta_4(z,q^2)}{\theta_1(z,q)} = \dfrac{\theta_2(z,q^2)\,\theta_3(z,q^2)}{\theta_2(z,q)} = \tfrac{1}{2}\theta_2(0,q).$

20.7(v) Watson's Identities

20.7.13
$$\theta_1(z,q)\,\theta_1(w,q) = \theta_3(z+w,q^2)\,\theta_2(z-w,q^2) \\ - \theta_2(z+w,q^2)\,\theta_3(z-w,q^2),$$

20.7.14
$$\theta_3(z,q)\,\theta_3(w,q) = \theta_3(z+w,q^2)\,\theta_3(z-w,q^2) \\ + \theta_2(z+w,q^2)\,\theta_2(z-w,q^2).$$

20.7(vi) Landen Transformations

With

20.7.15 $\quad A \equiv A(\tau) = 1/\theta_4(0|2\tau),$

20.7.16 $\quad \theta_1(2z|2\tau) = A\,\theta_1(z|\tau)\,\theta_2(z|\tau),$

20.7.17 $\quad \theta_2(2z|2\tau) = A\,\theta_1\!\left(\tfrac{1}{4}\pi - z|\tau\right)\theta_1\!\left(\tfrac{1}{4}\pi + z|\tau\right),$

20.7.18 $\quad \theta_3(2z|2\tau) = A\,\theta_3\!\left(\tfrac{1}{4}\pi - z|\tau\right)\theta_3\!\left(\tfrac{1}{4}\pi + z|\tau\right),$

20.7.19 $\quad \theta_4(2z|2\tau) = A\,\theta_3(z|\tau)\,\theta_4(z|\tau).$

Next, with

20.7.20 $\quad B \equiv B(\tau) = 1/\!\left(\theta_3(0|\tau)\,\theta_4(0|\tau)\,\theta_3\!\left(\tfrac{1}{4}\pi|\tau\right)\right),$

20.7.21 $\quad \theta_1(4z|4\tau) = B\,\theta_1(z|\tau)\,\theta_1\!\left(\tfrac{1}{4}\pi - z|\tau\right)\theta_1\!\left(\tfrac{1}{4}\pi + z|\tau\right)\theta_2(z|\tau),$

20.7.22 $\quad \theta_2(4z|4\tau) = B\,\theta_2\!\left(\tfrac{1}{8}\pi - z|\tau\right)\theta_2\!\left(\tfrac{1}{8}\pi + z|\tau\right)\theta_2\!\left(\tfrac{3}{8}\pi - z|\tau\right)\theta_2\!\left(\tfrac{3}{8}\pi + z|\tau\right),$

20.7.23 $\quad \theta_3(4z|4\tau) = B\,\theta_3\!\left(\tfrac{1}{8}\pi - z|\tau\right)\theta_3\!\left(\tfrac{1}{8}\pi + z|\tau\right)\theta_3\!\left(\tfrac{3}{8}\pi - z|\tau\right)\theta_3\!\left(\tfrac{3}{8}\pi + z|\tau\right),$

20.7.24 $\quad \theta_4(4z|4\tau) = B\,\theta_4(z|\tau)\,\theta_4\!\left(\tfrac{1}{4}\pi - z|\tau\right)\theta_4\!\left(\tfrac{1}{4}\pi + z|\tau\right)\theta_3(z|\tau).$

20.7(vii) Derivatives of Ratios of Theta Functions

20.7.25 $\quad \dfrac{d}{dz}\!\left(\dfrac{\theta_2(z|\tau)}{\theta_4(z|\tau)}\right) = -\dfrac{\theta_3^2(0|\tau)\,\theta_1(z|\tau)\,\theta_3(z|\tau)}{\theta_4^2(z|\tau)}.$

See Lawden (1989, pp. 19–20). This reference also gives ten additional identities involving permutations of the four theta functions.

20.7(viii) Transformations of Lattice Parameter

20.7.26 $\quad \theta_1(z|\tau+1) = e^{i\pi/4}\,\theta_1(z|\tau),$

20.7.27 $\quad \theta_2(z|\tau+1) = e^{i\pi/4}\,\theta_2(z|\tau),$

20.7.28 $\quad \theta_3(z|\tau+1) = \theta_4(z|\tau),$

20.7.29 $\quad \theta_4(z|\tau+1) = \theta_3(z|\tau).$

In the following equations $\tau' = -1/\tau$, and all square roots assume their principal values.

20.7.30 $\quad (-i\tau)^{1/2}\,\theta_1(z|\tau) = -i\exp\!\left(i\tau' z^2/\pi\right)\theta_1(z\tau'|\tau'),$

20.7.31 $\quad (-i\tau)^{1/2}\,\theta_2(z|\tau) = \exp\!\left(i\tau' z^2/\pi\right)\theta_4(z\tau'|\tau'),$

20.7.32 $\quad (-i\tau)^{1/2}\,\theta_3(z|\tau) = \exp\!\left(i\tau' z^2/\pi\right)\theta_3(z\tau'|\tau'),$

20.7.33 $\quad (-i\tau)^{1/2}\,\theta_4(z|\tau) = \exp\!\left(i\tau' z^2/\pi\right)\theta_2(z\tau'|\tau').$

These are examples of modular transformations; see §23.15.

20.8 Watson's Expansions

20.8.1
$$\frac{\theta_2(0,q)\,\theta_3(z,q)\,\theta_4(z,q)}{\theta_2(z,q)} = 2\sum_{n=-\infty}^{\infty}\frac{(-1)^n q^{n^2} e^{i2nz}}{q^{-n}e^{-iz}+q^n e^{iz}}.$$

See Watson (1935a). This reference and Bellman (1961, pp. 46–47) include other expansions of this type.

20.9 Relations to Other Functions

20.9(i) Elliptic Integrals

With k defined by

20.9.1
$$k = \theta_2^2(0|\tau)/\theta_3^2(0|\tau)$$

and the notation of §19.2(ii), the complete Legendre integrals of the first kind may be expressed as theta functions:

20.9.2 $\quad K(k) = \tfrac{1}{2}\pi\,\theta_3^2(0|\tau), \quad K'(k) = -i\tau\,K(k),$

together with (22.2.1).

In the case of the symetric integrals, with the notation of §19.16(i) we have

20.9.3 $\quad R_F\left(\dfrac{\theta_2^2(z,q)}{\theta_2^2(0,q)}, \dfrac{\theta_3^2(z,q)}{\theta_3^2(0,q)}, \dfrac{\theta_4^2(z,q)}{\theta_4^2(0,q)}\right) = \dfrac{\theta_1'(0,q)}{\theta_1(z,q)}z,$

20.9.4 $\quad R_F\left(0, \theta_3^4(0,q), \theta_4^4(0,q)\right) = \tfrac{1}{2}\pi,$

20.9.5 $\quad \exp\left(-\dfrac{\pi\,R_F(0, k^2, 1)}{R_F(0, k'^2, 1)}\right) = q.$

20.9(ii) Elliptic Functions and Modular Functions

See §§22.2 and 23.6(i) for the relations of Jacobian and Weierstrass elliptic functions to theta functions.

The relations (20.9.1) and (20.9.2) between k and τ (or q) are solutions of *Jacobi's inversion problem*; see Baker (1995) and Whittaker and Watson (1927, pp. 480–485).

As a function of τ, k^2 is the *elliptic modular function*; see Walker (1996, Chapter 7) and (23.15.2), (23.15.6).

20.9(iii) Riemann Zeta Function

See Koblitz (1993, Ch. 2, §4) and Titchmarsh (1986b, pp. 21–22). See also §§20.10(i) and 25.2.

20.10 Integrals

20.10(i) Mellin Transforms with respect to the Lattice Parameter

Let s be a constant such that $\Re s > 2$. Then

20.10.1
$$\int_0^\infty x^{s-1}\theta_2(0|ix^2)\,dx = 2^s(1-2^{-s})\pi^{-s/2}\Gamma\!\left(\tfrac{1}{2}s\right)\zeta(s),$$

20.10.2 $\quad \displaystyle\int_0^\infty x^{s-1}(\theta_3(0|ix^2)-1)\,dx = \pi^{-s/2}\Gamma\!\left(\tfrac{1}{2}s\right)\zeta(s),$

20.10.3
$$\int_0^\infty x^{s-1}(1-\theta_4(0|ix^2))\,dx$$
$$= (1-2^{1-s})\pi^{-s/2}\Gamma\!\left(\tfrac{1}{2}s\right)\zeta(s).$$

Here $\zeta(s)$ again denotes the Riemann zeta function (§25.2).

For further results see Oberhettinger (1974, pp. 157–159).

20.10(ii) Laplace Transforms with respect to the Lattice Parameter

Let s, ℓ, and β be constants such that $\Re s > 0$, $\ell > 0$, and $\sinh|\beta| \leq \ell$. Then

20.10.4
$$\int_0^\infty e^{-st}\theta_1\!\left(\dfrac{\beta\pi}{2\ell}\bigg|\dfrac{i\pi t}{\ell^2}\right)dt$$
$$= \int_0^\infty e^{-st}\theta_2\!\left(\dfrac{(1+\beta)\pi}{2\ell}\bigg|\dfrac{i\pi t}{\ell^2}\right)dt$$
$$= -\dfrac{\ell}{\sqrt{s}}\sinh(\beta\sqrt{s})\operatorname{sech}(\ell\sqrt{s}),$$

20.10.5
$$\int_0^\infty e^{-st}\theta_3\!\left(\dfrac{(1+\beta)\pi}{2\ell}\bigg|\dfrac{i\pi t}{\ell^2}\right)dt$$
$$= \int_0^\infty e^{-st}\theta_4\!\left(\dfrac{\beta\pi}{2\ell}\bigg|\dfrac{i\pi t}{\ell^2}\right)dt$$
$$= \dfrac{\ell}{\sqrt{s}}\cosh(\beta\sqrt{s})\operatorname{csch}(\ell\sqrt{s}).$$

For corresponding results for argument derivatives of the theta functions see Erdélyi et al. (1954a, pp. 224–225) or Oberhettinger and Badii (1973, p. 193).

20.10(iii) Compendia

For further integrals of theta functions see Erdélyi et al. (1954a, pp. 61–62 and 339), Prudnikov et al. (1990, pp. 356–358), Prudnikov et al. (1992a, §3.41), and Gradshteyn and Ryzhik (2000, pp. 627–628).

20.11 Generalizations and Analogs

20.11(i) Gauss Sum

For relatively prime integers m, n with $n > 0$ and mn even, the *Gauss sum* $G(m, n)$ is defined by

20.11.1
$$G(m,n) = \sum_{k=0}^{n-1} e^{-\pi i k^2 m/n};$$

see Lerch (1903). It is a discrete analog of theta functions. If both m, n are positive, then $G(m, n)$ allows inversion of its arguments as a modular transformation (compare (23.15.3) and (23.15.4)):

20.11.2
$$\dfrac{1}{\sqrt{n}}G(m,n) = \dfrac{1}{\sqrt{n}}\sum_{k=0}^{n-1} e^{-\pi i k^2 m/n}$$
$$= \dfrac{e^{-\pi i/4}}{\sqrt{m}}\sum_{j=0}^{m-1} e^{\pi i j^2 n/m} = \dfrac{e^{-\pi i/4}}{\sqrt{m}}G(-n,m).$$

This is the discrete analog of the Poisson identity (§1.8(iv)).

20.11(ii) Ramanujan's Theta Function and q-Series

Ramanujan's theta function $f(a,b)$ is defined by

20.11.3 $$f(a,b) = \sum_{n=-\infty}^{\infty} a^{n(n+1)/2} b^{n(n-1)/2},$$

where $a, b \in \mathbb{C}$ and $|ab| < 1$. With the substitutions $a = qe^{2iz}$, $b = qe^{-2iz}$, with $q = e^{i\pi\tau}$, we have

20.11.4 $$f(a,b) = \theta_3(z|\tau).$$

In the case $z = 0$ identities for theta functions become identities in the complex variable q, with $|q| < 1$, that involve rational functions, power series, and continued fractions; see Adiga et al. (1985), McKean and Moll (1999, pp. 156–158), and Andrews et al. (1988, §10.7).

20.11(iii) Ramanujan's Change of Base

As in §20.11(ii), the modulus k of elliptic integrals (§19.2(ii)), Jacobian elliptic functions (§22.2), and Weierstrass elliptic functions (§23.6(ii)) can be expanded in q-series via (20.9.1). However, in this case q is no longer regarded as an independent complex variable within the unit circle, because k is related to the variable $\tau = \tau(k)$ of the theta functions via (20.9.2). This is Jacobi's inversion problem of §20.9(ii).

The first of equations (20.9.2) can also be written

20.11.5 $$_2F_1\left(\tfrac{1}{2}, \tfrac{1}{2}; 1; k^2\right) = \theta_3^2(0|\tau);$$

see §19.5. Similar identities can be constructed for $_2F_1\left(\tfrac{1}{3}, \tfrac{2}{3}; 1; k^2\right)$, $_2F_1\left(\tfrac{1}{4}, \tfrac{3}{4}; 1; k^2\right)$, and $_2F_1\left(\tfrac{1}{6}, \tfrac{5}{6}; 1; k^2\right)$. These results are called *Ramanujan's changes of base*. Each provides an extension of Jacobi's inversion problem. See Berndt et al. (1995) and Shen (1998). For applications to rapidly convergent expansions for π see Chudnovsky and Chudnovsky (1988), and for applications in the construction of *elliptic-hypergeometric series* see Rosengren (2004).

20.11(iv) Theta Functions with Characteristics

Multidimensional *theta functions with characteristics* are defined in §21.2(ii) and their properties are described in §§21.3(ii), 21.5(ii), and 21.6. For specialization to the one-dimensional theta functions treated in the present chapter, see Rauch and Lebowitz (1973) and §21.7(iii).

Applications

20.12 Mathematical Applications

20.12(i) Number Theory

For applications of $\theta_3(0,q)$ to problems involving sums of squares of integers see §27.13(iv), and for extensions see Estermann (1959), Serre (1973, pp. 106–109), Koblitz (1993, pp. 176–177), and McKean and Moll (1999, pp. 142–143).

For applications of Jacobi's triple product (20.5.9) to Ramanujan's $\tau(n)$ function and Euler's pentagonal numbers see Hardy and Wright (1979, pp. 132–160) and McKean and Moll (1999, pp. 143–145). For an application of a generalization in affine root systems see Macdonald (1972).

20.12(ii) Uniformization and Embedding of Complex Tori

For the terminology and notation see McKean and Moll (1999, pp. 48–53).

The space of complex tori $\mathbb{C}/(\mathbb{Z} + \tau\mathbb{Z})$ (that is, the set of complex numbers z in which two of these numbers z_1 and z_2 are regarded as equivalent if there exist integers m, n such that $z_1 - z_2 = m + \tau n$) is mapped into the projective space P^3 via the identification $z \to (\theta_1(2z|\tau), \theta_2(2z|\tau), \theta_3(2z|\tau), \theta_4(2z|\tau))$. Thus theta functions "uniformize" the complex torus. This ability to uniformize multiply-connected spaces (manifolds), or multi-sheeted functions of a complex variable (Riemann (1899), Rauch and Lebowitz (1973), Siegel (1988)) has led to applications in string theory (Green et al. (1988a,b), Krichever and Novikov (1989)), and also in statistical mechanics (Baxter (1982)).

20.13 Physical Applications

The functions $\theta_j(z|\tau)$, $j = 1, 2, 3, 4$, provide periodic solutions of the partial differential equation

20.13.1 $$\partial \theta(z|\tau)/\partial \tau = \kappa\, \partial^2 \theta(z|\tau)/\partial z^2,$$

with $\kappa = -i\pi/4$.

For $\tau = it$, with α, t, z real, (20.13.1) takes the form of a real-time t diffusion equation

20.13.2 $$\partial \theta/\partial t = \alpha\, \partial^2 \theta/\partial z^2,$$

with diffusion constant $\alpha = \pi/4$. Let $z, \alpha, t \in \mathbb{R}$. Then the nonperiodic Gaussian

20.13.3 $$g(z,t) = \sqrt{\frac{\pi}{4\alpha t}} \exp\left(-\frac{z^2}{4\alpha t}\right)$$

is also a solution of (20.13.2), and it approaches a Dirac delta (§1.17) at $t = 0$. These two apparently different solutions differ only in their normalization and boundary conditions. From (20.2.3), (20.2.4), (20.7.32), and (20.7.33),

20.13.4 $$\sqrt{\frac{\pi}{4\alpha t}} \sum_{n=-\infty}^{\infty} e^{-(n\pi+z)^2/(4\alpha t)} = \theta_3(z|i4\alpha t/\pi),$$

and

20.13.5 $$\sqrt{\frac{\pi}{4\alpha t}} \sum_{n=-\infty}^{\infty} (-1)^n e^{-(n\pi+z)^2/(4\alpha t)} = \theta_4(z|i4\alpha t/\pi).$$

Thus the classical theta functions are "periodized", or "anti-periodized", Gaussians; see Bellman (1961, pp. 18, 19). Theta-function solutions to the heat diffusion equation with simple boundary conditions are discussed in Lawden (1989, pp. 1–3), and with more general boundary conditions in Körner (1989, pp. 274–281).

In the singular limit $\Im\tau \to 0+$, the functions $\theta_j(z|\tau)$, $j = 1, 2, 3, 4$, become integral kernels of Feynman path integrals (distribution-valued Green's functions); see Schulman (1981, pp. 194–195). This allows analytic time propagation of quantum wave-packets in a box, or on a ring, as closed-form solutions of the time-dependent Schrödinger equation.

Computation

20.14 Methods of Computation

The Fourier series of §20.2(i) usually converge rapidly because of the factors $q^{(n+\frac{1}{2})^2}$ or q^{n^2}, and provide a convenient way of calculating values of $\theta_j(z|\tau)$. Similarly, their z-differentiated forms provide a convenient way of calculating the corresponding derivatives. For instance, the first three terms of (20.2.1) give the value of $\theta_1(2-i|i)$ $(= \theta_1(2-i,e^{-\pi}))$ to 12 decimal places.

For values of $|q|$ near 1 the transformations of §20.7(viii) can be used to replace τ with a value that has a larger imaginary part and hence a smaller value of $|q|$. For instance, to find $\theta_3(z, 0.9)$ we use (20.7.32) with $q = 0.9 = e^{i\pi\tau}$, $\tau = -i\ln(0.9)/\pi$. Then $\tau' = -1/\tau = -i\pi/\ln(0.9)$ and $q' = e^{i\pi\tau'} = \exp(\pi^2/\ln(0.9)) = (2.07\ldots) \times 10^{-41}$. Hence the first term of the series (20.2.3) for $\theta_3(z\tau'|\tau')$ suffices for most purposes. In theory, starting from any value of τ, a finite number of applications of the transformations $\tau \to \tau + 1$ and $\tau \to -1/\tau$ will result in a value of τ with $\Im\tau \geq \sqrt{3}/2$; see §23.18. In practice a value with, say, $\Im\tau \geq 1/2$, $|q| \leq 0.2$, is found quickly and is satisfactory for numerical evaluation.

20.15 Tables

Theta functions are tabulated in Jahnke and Emde (1945, p. 45). This reference gives $\theta_j(x,q)$, $j = 1, 2, 3, 4$, and their logarithmic x-derivatives to 4D for $x/\pi = 0(.1)1$, $\alpha = 0(9°)90°$, where α is the modular angle given by

20.15.1 $$\sin\alpha = \theta_2^2(0,q)/\theta_3^2(0,q) = k.$$

Spenceley and Spenceley (1947) tabulates $\theta_1(x,q)/\theta_2(0,q)$, $\theta_2(x,q)/\theta_2(0,q)$, $\theta_3(x,q)/\theta_4(0,q)$, $\theta_4(x,q)/\theta_4(0,q)$ to 12D for $u = 0(1°)90°$, $\alpha = 0(1°)89°$, where $u = 2x/(\pi\theta_3^2(0,q))$ and α is defined by (20.15.1), together with the corresponding values of $\theta_2(0,q)$ and $\theta_4(0,q)$.

Lawden (1989, pp. 270–279) tabulates $\theta_j(x,q)$, $j = 1, 2, 3, 4$, to 5D for $x = 0(1°)90°$, $q = 0.1(.1)0.9$, and also q to 5D for $k^2 = 0(.01)1$.

Tables of Neville's theta functions $\theta_s(x,q)$, $\theta_c(x,q)$, $\theta_d(x,q)$, $\theta_n(x,q)$ (see §20.1) and their logarithmic x-derivatives are given in Abramowitz and Stegun (1964, pp. 582–585) to 9D for $\varepsilon, \alpha = 0(5°)90°$, where (in radian measure) $\varepsilon = x/\theta_3^2(0,q) = \pi x/(2K(k))$, and α is defined by (20.15.1).

For other tables prior to 1961 see Fletcher et al. (1962, pp. 508–514) and Lebedev and Fedorova (1960, pp. 227–230).

20.16 Software

See http://dlmf.nist.gov/20.16.

References

General References

The main references used in writing this chapter are Whittaker and Watson (1927), Lawden (1989), and Walker (1996). For further bibliographic reading see McKean and Moll (1999).

Sources

The following list gives the references or other indications of proofs that were used in constructing the various sections of this chapter. These sources supplement the references that are quoted in the text.

§20.2 Whittaker and Watson (1927, pp. 463–465) and Lawden (1989, Chapter 1).

§20.3 These graphics were produced at NIST.

§20.4 Lawden (1989, pp. 12–23), Walker (1996, pp. 90–92), and Whittaker and Watson (1927, pp. 470–473). (20.4.1)–(20.4.5) are special cases of (20.5.1)–(20.5.4).

§20.5 Lawden (1989, pp. 12–23), Walker (1996, pp. 86–98), Whittaker and Watson (1927, pp. 469–473), and Bellman (1961, p. 44). Equations (20.5.14)–(20.5.17) follow from (20.5.5)–(20.5.8) by use of the infinite products for the sine and cosine (§4.22).

§20.6 Walker (1996, §3.2) and §23.9. (20.6.2)–(20.6.5) may be derived by termwise expansion in (20.5.14)–(20.5.17). (20.6.10) may be derived from (20.6.6) and (20.6.8) by subtraction of terms with even j, in a similar manner to $\sum_{n=1}^{\infty}(-1)^{n-1}n^{-j} = (1-2^{1-j})\sum_{n=1}^{\infty} n^{-j}$.

§20.7 Lawden (1989, pp. 5–23), Whittaker and Watson (1927, pp. 466–477), McKean and Moll (1999, pp. 129–130), Watson (1935a), Bellman (1961, p. 61), and Serre (1973, p. 109). The first equalities in (20.7.11) and (20.7.12) follow by translation of z by $\frac{1}{2}\pi$ as in (20.2.11)–(20.2.14). The second equalities follow from (20.5.5)–(20.5.8) and the identity $\prod_{n=1}^{\infty}(1+q^n)(1-q^{2n-1}) = 1$ (Walker (1996, p. 90)).

§20.9 Walker (1996, p. 156), Whittaker and Watson (1927, pp. 480–485), Serre (1973, p. 109), and McKean and Moll (1999, §§3.3, 3.9). For (20.9.3) combination of (20.4.6) and (23.6.5)–(23.6.7) yields $\wp(z) - e_j = \left(\frac{v\theta_1'(0,q)\,\theta_{j+1}(v,q)}{z\theta_1(v,q)\,\theta_{j+1}(0,q)}\right)^2$, $j = 1, 2, 3$, where $v = \pi z/(2\omega_1)$. Then by application of (19.25.35) and use of the properties that R_F is homogenous and of degree $-\frac{1}{2}$ in its three variables (§§19.16(ii), 19.16(iii)), we derive $z = \frac{z\,\theta_1(v,q)}{v\,\theta_1'(0,q)} R_F\left(\frac{\theta_2^2(v,q)}{\theta_2^2(0,q)}, \frac{\theta_3^2(v,q)}{\theta_3^2(0,q)}, \frac{\theta_4^2(v,q)}{\theta_4^2(0,q)}\right)$. This equation becomes (20.9.3) when the z's are cancelled and v is renamed z. For (20.9.4), from (19.25.1) and Erdélyi et al. (1953b, 13.20(11)) we have $K(k) = R_F\left(0, \frac{\theta_4^4(0,q)}{\theta_3^4(0,q)}, 1\right) = \theta_3^2(0,q)\,R_F\left(0, \theta_3^4(0,q), \theta_4^4(0,q)\right)$, where the second equality uses the homogeneity and symmetry of R_F. Comparison with (20.9.2) proves (20.9.4). For (20.9.5), by (19.25.1) the left side is $\exp(-\pi K(k')/K(k))$, which equals q by Erdélyi et al. (1953b, 13.19(4)).

§20.10 Bellman (1961, pp. 20–24). For (20.10.1) and (20.10.3) use §20.7(viii) with appropriate changes of integration variable. For (20.10.2) use (20.2.3) with $z = 0$, $\tau = it$, Bellman (1961, pp. 28–32), Koblitz (1993, pp. 70–75), and/or Titchmarsh (1986b, §2.6).

§20.11 Bellman (1961, pp. 38–39), Walker (1996, pp. 181–182), and McKean and Moll (1999, pp. 140–147 and 151–152).

§20.13 Whittaker and Watson (1927, p. 470).

Chapter 21

Multidimensional Theta Functions

B. Deconinck[1]

Notation — **538**
- 21.1 Special Notation 538

Properties — **538**
- 21.2 Definitions 538
- 21.3 Symmetry and Quasi-Periodicity 539
- 21.4 Graphics 539
- 21.5 Modular Transformations 541
- 21.6 Products 542

Applications — **543**
- 21.7 Riemann Surfaces 543
- 21.8 Abelian Functions 545
- 21.9 Integrable Equations 545

Computation — **546**
- 21.10 Methods of Computation 546
- 21.11 Software 546

References — **546**

---

[1] Department of Applied Mathematics, University of Washington, Seattle, Washington.
Copyright © 2009 National Institute of Standards and Technology. All rights reserved.

Notation

21.1 Special Notation

(For other notation see pp. xiv and 873.)

g, h	positive integers.
$\mathbb{Z}^g$	$\mathbb{Z} \times \mathbb{Z} \times \cdots \times \mathbb{Z}$ (g times).
$\mathbb{R}^g$	$\mathbb{R} \times \mathbb{R} \times \cdots \times \mathbb{R}$ (g times).
$\mathbb{Z}^{g \times h}$	set of all $g \times h$ matrices with integer elements.
$\mathbf{\Omega}$	$g \times g$ complex, symmetric matrix with $\Im \mathbf{\Omega}$ strictly positive definite, i.e., a Riemann matrix.
$\boldsymbol{\alpha}, \boldsymbol{\beta}$	g-dimensional vectors, with all elements in $[0, 1)$, unless stated otherwise.
a_j	jth element of vector $\mathbf{a}$.
A_{jk}	(j, k)th element of matrix $\mathbf{A}$.
$\mathbf{a} \cdot \mathbf{b}$	scalar product of the vectors $\mathbf{a}$ and $\mathbf{b}$.
$\mathbf{a} \cdot \mathbf{\Omega} \cdot \mathbf{b}$	$[\mathbf{\Omega a}] \cdot \mathbf{b} = [\mathbf{\Omega b}] \cdot \mathbf{a}$.
$\mathbf{0}_g$	$g \times g$ zero matrix.
$\mathbf{I}_g$	$g \times g$ identity matrix.
$\mathbf{J}_{2g}$	$\begin{bmatrix} \mathbf{0}_g & \mathbf{I}_g \\ -\mathbf{I}_g & \mathbf{0}_g \end{bmatrix}$.
S^g	set of g-dimensional vectors with elements in S.
$\|S\|$	number of elements of the set S.
$S_1 S_2$	set of all elements of the form "element of $S_1 \times$ element of S_2".
S_1/S_2	set of all elements of S_1, modulo elements of S_2. Thus two elements of S_1/S_2 are equivalent if they are both in S_1 and their difference is in S_2. (For an example see §20.12(ii).)
$a \circ b$	intersection index of a and b, two cycles lying on a closed surface. $a \circ b = 0$ if a and b do not intersect. Otherwise $a \circ b$ gets an additive contribution from every intersection point. This contribution is 1 if the basis of the tangent vectors of the a and b cycles (§21.7(i)) at the point of intersection is positively oriented; otherwise it is -1.
$\oint_a \omega$	line integral of the differential ω over the cycle a.

Lowercase boldface letters or numbers are g-dimensional real or complex vectors, either row or column depending on the context. Uppercase boldface letters are $g \times g$ real or complex matrices.

The main functions treated in this chapter are the Riemann theta functions $\theta(\mathbf{z}|\mathbf{\Omega})$, and the Riemann theta functions with characteristics $\theta\begin{bmatrix}\boldsymbol{\alpha}\\\boldsymbol{\beta}\end{bmatrix}(\mathbf{z}|\mathbf{\Omega})$.

The function $\Theta(\boldsymbol{\phi}|\mathbf{B}) = \theta(\boldsymbol{\phi}/(2\pi i)|\mathbf{B}/(2\pi i))$ is also commonly used; see, for example, Belokolos et al. (1994, §2.5), Dubrovin (1981), and Fay (1973, Chapter 1).

Properties

21.2 Definitions

21.2(i) Riemann Theta Functions

21.2.1 $$\theta(\mathbf{z}|\mathbf{\Omega}) = \sum_{\mathbf{n} \in \mathbb{Z}^g} e^{2\pi i \left(\frac{1}{2}\mathbf{n} \cdot \mathbf{\Omega} \cdot \mathbf{n} + \mathbf{n} \cdot \mathbf{z}\right)}.$$

This g-tuple Fourier series converges absolutely and uniformly on compact sets of the $\mathbf{z}$ and $\mathbf{\Omega}$ spaces; hence $\theta(\mathbf{z}|\mathbf{\Omega})$ is an analytic function of (each element of) $\mathbf{z}$ and (each element of) $\mathbf{\Omega}$. $\theta(\mathbf{z}|\mathbf{\Omega})$ is also referred to as a theta function with g components, a g-dimensional theta function or as a genus g theta function.

For numerical purposes we use the *scaled Riemann theta function* $\hat{\theta}(\mathbf{z}|\mathbf{\Omega})$, defined by (Deconinck et al. (2004)),

21.2.2 $$\hat{\theta}(\mathbf{z}|\mathbf{\Omega}) = e^{-\pi[\Im \mathbf{z}] \cdot [\Im \mathbf{\Omega}]^{-1} \cdot [\Im \mathbf{z}]} \theta(\mathbf{z}|\mathbf{\Omega}).$$

$\hat{\theta}(\mathbf{z}|\mathbf{\Omega})$ is a bounded nonanalytic function of $\mathbf{z}$. Many applications involve quotients of Riemann theta functions: the exponential factor then disappears.

Example

21.2.3
$$\theta\left(z_1, z_2 \middle| \begin{bmatrix} i & -\frac{1}{2} \\ -\frac{1}{2} & i \end{bmatrix}\right)$$
$$= \sum_{n_1=-\infty}^{\infty} \sum_{n_2=-\infty}^{\infty} e^{-\pi(n_1^2+n_2^2)} e^{-i\pi n_1 n_2} e^{2\pi i(n_1 z_1 + n_2 z_2)}.$$

With $z_1 = x_1 + iy_1$, $z_2 = x_2 + iy_2$,

$$\hat{\theta}\left(x_1 + iy_1, x_2 + iy_2 \middle| \begin{bmatrix} i & -\frac{1}{2} \\ -\frac{1}{2} & i \end{bmatrix}\right)$$

21.2.4
$$= \sum_{n_1=-\infty}^{\infty} \sum_{n_2=-\infty}^{\infty} e^{-\pi(n_1+y_1)^2 - \pi(n_2+y_2)^2}$$
$$\times e^{\pi i(2n_1 x_1 + 2n_2 x_2 - n_1 n_2)}.$$

21.2(ii) Riemann Theta Functions with Characteristics

Let $\boldsymbol{\alpha}, \boldsymbol{\beta} \in \mathbb{R}^g$. Define

21.2.5
$$\theta\begin{bmatrix}\boldsymbol{\alpha}\\\boldsymbol{\beta}\end{bmatrix}(\mathbf{z}|\boldsymbol{\Omega}) = \sum_{\mathbf{n}\in\mathbb{Z}^g} e^{2\pi i\left(\frac{1}{2}[\mathbf{n}+\boldsymbol{\alpha}]\cdot\boldsymbol{\Omega}\cdot[\mathbf{n}+\boldsymbol{\alpha}]+[\mathbf{n}+\boldsymbol{\alpha}]\cdot[\mathbf{z}+\boldsymbol{\beta}]\right)}.$$

This function is referred to as a *Riemann theta function with characteristics* $\begin{bmatrix}\boldsymbol{\alpha}\\\boldsymbol{\beta}\end{bmatrix}$. It is a translation of the Riemann theta function (21.2.1), multiplied by an exponential factor:

21.2.6
$$\theta\begin{bmatrix}\boldsymbol{\alpha}\\\boldsymbol{\beta}\end{bmatrix}(\mathbf{z}|\boldsymbol{\Omega}) = e^{2\pi i\left(\frac{1}{2}\boldsymbol{\alpha}\cdot\boldsymbol{\Omega}\cdot\boldsymbol{\alpha}+\boldsymbol{\alpha}\cdot[\mathbf{z}+\boldsymbol{\beta}]\right)} \theta(\mathbf{z}+\boldsymbol{\Omega}\boldsymbol{\alpha}+\boldsymbol{\beta}|\boldsymbol{\Omega}),$$

and

21.2.7
$$\theta\begin{bmatrix}\mathbf{0}\\\mathbf{0}\end{bmatrix}(\mathbf{z}|\boldsymbol{\Omega}) = \theta(\mathbf{z}|\boldsymbol{\Omega}).$$

Characteristics whose elements are either 0 or $\frac{1}{2}$ are called *half-period characteristics*. For given $\boldsymbol{\Omega}$, there are 2^{2g} g-dimensional Riemann theta functions with half-period characteristics.

21.2(iii) Relation to Classical Theta Functions

For $g=1$, and with the notation of §20.2(i),

21.2.8
$$\theta(z|\Omega) = \theta_3(\pi z|\Omega),$$

21.2.9
$$\theta_1(\pi z|\Omega) = -\theta\begin{bmatrix}\frac{1}{2}\\\frac{1}{2}\end{bmatrix}(z|\Omega),$$

21.2.10
$$\theta_2(\pi z|\Omega) = \theta\begin{bmatrix}\frac{1}{2}\\0\end{bmatrix}(z|\Omega),$$

21.2.11
$$\theta_3(\pi z|\Omega) = \theta\begin{bmatrix}0\\0\end{bmatrix}(z|\Omega),$$

21.2.12
$$\theta_4(\pi z|\Omega) = \theta\begin{bmatrix}0\\\frac{1}{2}\end{bmatrix}(z|\Omega).$$

21.3 Symmetry and Quasi-Periodicity

21.3(i) Riemann Theta Functions

21.3.1
$$\theta(-\mathbf{z}|\boldsymbol{\Omega}) = \theta(\mathbf{z}|\boldsymbol{\Omega}),$$

21.3.2
$$\theta(\mathbf{z}+\mathbf{m}_1|\boldsymbol{\Omega}) = \theta(\mathbf{z}|\boldsymbol{\Omega}),$$

when $\mathbf{m}_1 \in \mathbb{Z}^g$. Thus $\theta(\mathbf{z}|\boldsymbol{\Omega})$ is periodic, with period 1, in each element of $\mathbf{z}$. More generally,

21.3.3
$$\theta(\mathbf{z}+\mathbf{m}_1+\boldsymbol{\Omega}\mathbf{m}_2|\boldsymbol{\Omega}) = e^{-2\pi i\left(\frac{1}{2}\mathbf{m}_2\cdot\boldsymbol{\Omega}\cdot\mathbf{m}_2+\mathbf{m}_2\cdot\mathbf{z}\right)}\theta(\mathbf{z}|\boldsymbol{\Omega}),$$

with $\mathbf{m}_1, \mathbf{m}_2 \in \mathbb{Z}^g$. This is the *quasi-periodicity* property of the Riemann theta function. It determines the Riemann theta function up to a constant factor. The set of points $\mathbf{m}_1 + \boldsymbol{\Omega}\mathbf{m}_2$ form a g-dimensional lattice, the *period lattice* of the Riemann theta function.

21.3(ii) Riemann Theta Functions with Characteristics

Again, with $\mathbf{m}_1, \mathbf{m}_2 \in \mathbb{Z}^g$

21.3.4
$$\theta\begin{bmatrix}\boldsymbol{\alpha}+\mathbf{m}_1\\\boldsymbol{\beta}+\mathbf{m}_2\end{bmatrix}(\mathbf{z}|\boldsymbol{\Omega}) = e^{2\pi i\boldsymbol{\alpha}\cdot\mathbf{m}_1}\theta\begin{bmatrix}\boldsymbol{\alpha}\\\boldsymbol{\beta}\end{bmatrix}(\mathbf{z}|\boldsymbol{\Omega}).$$

Because of this property, the elements of $\boldsymbol{\alpha}$ and $\boldsymbol{\beta}$ are usually restricted to $[0,1)$, without loss of generality.

21.3.5
$$\theta\begin{bmatrix}\boldsymbol{\alpha}\\\boldsymbol{\beta}\end{bmatrix}(\mathbf{z}+\mathbf{m}_1+\boldsymbol{\Omega}\mathbf{m}_2|\boldsymbol{\Omega})$$
$$= e^{2\pi i\left(\boldsymbol{\alpha}\cdot\mathbf{m}_1-\boldsymbol{\beta}\cdot\mathbf{m}_2-\frac{1}{2}\mathbf{m}_2\cdot\boldsymbol{\Omega}\cdot\mathbf{m}_2-\mathbf{m}_2\cdot\mathbf{z}\right)}\theta\begin{bmatrix}\boldsymbol{\alpha}\\\boldsymbol{\beta}\end{bmatrix}(\mathbf{z}|\boldsymbol{\Omega}).$$

For Riemann theta functions with half-period characteristics,

21.3.6
$$\theta\begin{bmatrix}\boldsymbol{\alpha}\\\boldsymbol{\beta}\end{bmatrix}(-\mathbf{z}|\boldsymbol{\Omega}) = (-1)^{4\boldsymbol{\alpha}\cdot\boldsymbol{\beta}}\theta\begin{bmatrix}\boldsymbol{\alpha}\\\boldsymbol{\beta}\end{bmatrix}(\mathbf{z}|\boldsymbol{\Omega}).$$

See also §20.2(iii) for the case $g=1$ and classical theta functions.

21.4 Graphics

Figure 21.4.1 provides surfaces of the scaled Riemann theta function $\hat{\theta}(\mathbf{z}|\boldsymbol{\Omega})$, with

21.4.1
$$\boldsymbol{\Omega} = \begin{bmatrix} 1.69098\,3006 + 0.95105\,6516\,i & 1.5 + 0.36327\,1264\,i \\ 1.5 + 0.36327\,1264\,i & 1.30901\,6994 + 0.95105\,6516\,i \end{bmatrix}.$$

This Riemann matrix originates from the Riemann surface represented by the algebraic curve $\mu^3 - \lambda^7 + 2\lambda^3\mu = 0$; compare §21.7(i).

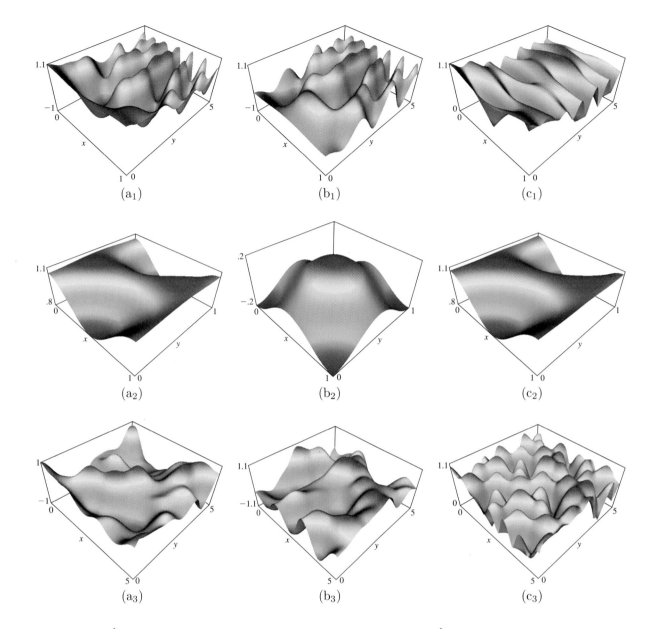

Figure 21.4.1: $\hat{\theta}(\mathbf{z}|\mathbf{\Omega})$ parametrized by (21.4.1). The surface plots are of $\hat{\theta}(x+iy, 0|\mathbf{\Omega})$, $0 \leq x \leq 1$, $0 \leq y \leq 5$ (suffix 1); $\hat{\theta}(x, y|\mathbf{\Omega})$, $0 \leq x \leq 1$, $0 \leq y \leq 1$ (suffix 2); $\hat{\theta}(ix, iy|\mathbf{\Omega})$, $0 \leq x \leq 5$, $0 \leq y \leq 5$ (suffix 3). Shown are the real part (a), the imaginary part (b), and the modulus (c).

For the scaled Riemann theta functions depicted in Figures 21.4.2–21.4.5

21.4.2
$$\mathbf{\Omega}_1 = \begin{bmatrix} i & -\frac{1}{2} \\ -\frac{1}{2} & i \end{bmatrix},$$

and

21.4.3
$$\mathbf{\Omega}_2 = \begin{bmatrix} -\frac{1}{2}+i & \frac{1}{2}-\frac{1}{2}i & -\frac{1}{2}-\frac{1}{2}i \\ \frac{1}{2}-\frac{1}{2}i & i & 0 \\ -\frac{1}{2}-\frac{1}{2}i & 0 & i \end{bmatrix}.$$

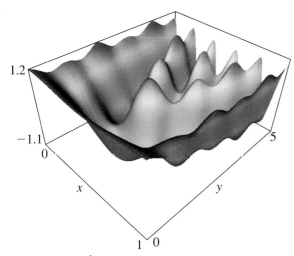

Figure 21.4.2: $\Re\hat{\theta}(x+iy,0|\mathbf{\Omega}_1)$, $0 \leq x \leq 1$, $0 \leq y \leq 5$. (The imaginary part looks very similar.)

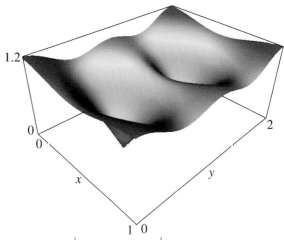

Figure 21.4.3: $\left|\hat{\theta}(x+iy,0|\mathbf{\Omega}_1)\right|$, $0 \leq x \leq 1$, $0 \leq y \leq 2$.

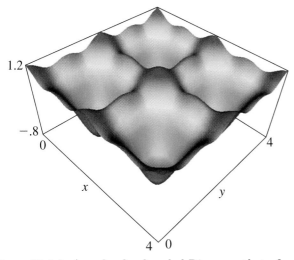

Figure 21.4.4: A real-valued scaled Riemann theta function: $\hat{\theta}(ix,iy|\mathbf{\Omega}_1)$, $0 \leq x \leq 4$, $0 \leq y \leq 4$. In this case, the quasi-periods are commensurable, resulting in a doubly-periodic configuration.

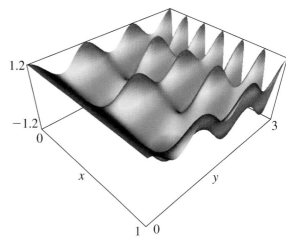

Figure 21.4.5: The real part of a genus 3 scaled Riemann theta function: $\Re\hat{\theta}(x+iy,0,0|\mathbf{\Omega}_2)$, $0 \leq x \leq 1$, $0 \leq y \leq 3$. This Riemann matrix originates from the genus 3 Riemann surface represented by the algebraic curve $\mu^3 + 2\mu - \lambda^4 = 0$; compare §21.7(i).

21.5 Modular Transformations

21.5(i) Riemann Theta Functions

Let **A**, **B**, **C**, and **D** be $g \times g$ matrices with integer elements such that

21.5.1
$$\mathbf{\Gamma} = \begin{bmatrix} \mathbf{A} & \mathbf{B} \\ \mathbf{C} & \mathbf{D} \end{bmatrix}$$

is a *symplectic matrix*, that is,

21.5.2
$$\mathbf{\Gamma} \mathbf{J}_{2g} \mathbf{\Gamma}^{\mathrm{T}} = \mathbf{J}_{2g}.$$

Then

21.5.3
$$\det \mathbf{\Gamma} = 1,$$

and

21.5.4
$$\theta\left([\mathbf{C}\mathbf{\Omega}+\mathbf{D}]^{-1}]^{\mathrm{T}} \mathbf{z} \,\Big|\, [\mathbf{A}\mathbf{\Omega}+\mathbf{B}][\mathbf{C}\mathbf{\Omega}+\mathbf{D}]^{-1}\right)$$
$$= \xi(\mathbf{\Gamma})\sqrt{\det[\mathbf{C}\mathbf{\Omega}+\mathbf{D}]} e^{\pi i \mathbf{z} \cdot [[\mathbf{C}\mathbf{\Omega}+\mathbf{D}]^{-1}\mathbf{C}] \cdot \mathbf{z}} \theta(\mathbf{z}|\mathbf{\Omega}).$$

Here $\xi(\mathbf{\Gamma})$ is an eighth root of unity, that is, $(\xi(\mathbf{\Gamma}))^8 = 1$. For general $\mathbf{\Gamma}$, it is difficult to decide which root needs to be used. The choice depends on $\mathbf{\Gamma}$, but is independent of $\mathbf{z}$ and $\boldsymbol{\Omega}$. Equation (21.5.4) is the *modular transformation* property for Riemann theta functions.

The modular transformations form a group under the composition of such transformations, the *modular group*, which is generated by simpler transformations, for which $\xi(\boldsymbol{\Gamma})$ is determinate:

21.5.5 $\quad \boldsymbol{\Gamma} = \begin{bmatrix} \mathbf{A} & \mathbf{0}_g \\ \mathbf{0}_g & [\mathbf{A}^{-1}]^{\mathrm{T}} \end{bmatrix} \Rightarrow \theta(\mathbf{Az}|\mathbf{A\Omega A}^{\mathrm{T}}) = \theta(\mathbf{z}|\boldsymbol{\Omega})$.

($\mathbf{A}$ invertible with integer elements.)

21.5.6 $\quad \boldsymbol{\Gamma} = \begin{bmatrix} \mathbf{I}_g & \mathbf{B} \\ \mathbf{0}_g & \mathbf{I}_g \end{bmatrix} \Rightarrow \theta(\mathbf{z}|\boldsymbol{\Omega}+\mathbf{B}) = \theta(\mathbf{z}|\boldsymbol{\Omega})$.

($\mathbf{B}$ symmetric with integer elements and even diagonal elements.)

21.5.7

$$\boldsymbol{\Gamma} = \begin{bmatrix} \mathbf{I}_g & \mathbf{B} \\ \mathbf{0}_g & \mathbf{I}_g \end{bmatrix} \Rightarrow \theta(\mathbf{z}|\boldsymbol{\Omega}+\mathbf{B}) = \theta\!\left(\mathbf{z}+\tfrac{1}{2}\operatorname{diag}\mathbf{B}\big|\boldsymbol{\Omega}\right).$$

($\mathbf{B}$ symmetric with integer elements.) See Heil (1995, p. 24).

21.5.8

$$\boldsymbol{\Gamma} = \begin{bmatrix} \mathbf{0}_g & -\mathbf{I}_g \\ \mathbf{I}_g & \mathbf{0}_g \end{bmatrix}$$

$$\Rightarrow \quad \theta(\boldsymbol{\Omega}^{-1}\mathbf{z}|-\boldsymbol{\Omega}^{-1}) = \sqrt{\det[-i\boldsymbol{\Omega}]}\, e^{\pi i \mathbf{z}\cdot \boldsymbol{\Omega}^{-1}\cdot \mathbf{z}}\, \theta(\mathbf{z}|\boldsymbol{\Omega}),$$

where the square root assumes its principal value.

21.5(ii) Riemann Theta Functions with Characteristics

21.5.9
$$\theta\!\begin{bmatrix} \mathbf{D}\boldsymbol{\alpha} - \mathbf{C}\boldsymbol{\beta} + \tfrac{1}{2}\operatorname{diag}[\mathbf{CD}^{\mathrm{T}}] \\ -\mathbf{B}\boldsymbol{\alpha} + \mathbf{A}\boldsymbol{\beta} + \tfrac{1}{2}\operatorname{diag}[\mathbf{AB}^{\mathrm{T}}] \end{bmatrix}\!\left([[\mathbf{C}\boldsymbol{\Omega}+\mathbf{D}]^{-1}]^{\mathrm{T}}\mathbf{z}\,\Big|\,[\mathbf{A}\boldsymbol{\Omega}+\mathbf{B}][\mathbf{C}\boldsymbol{\Omega}+\mathbf{D}]^{-1}\right)$$
$$= \kappa(\boldsymbol{\alpha},\boldsymbol{\beta},\boldsymbol{\Gamma})\sqrt{\det[\mathbf{C}\boldsymbol{\Omega}+\mathbf{D}]}\, e^{\pi i \mathbf{z}\cdot[[\mathbf{C}\boldsymbol{\Omega}+\mathbf{D}]^{-1}\mathbf{C}]\cdot \mathbf{z}}\, \theta\!\begin{bmatrix}\boldsymbol{\alpha}\\ \boldsymbol{\beta}\end{bmatrix}\!(\mathbf{z}|\boldsymbol{\Omega}),$$

where $\kappa(\boldsymbol{\alpha},\boldsymbol{\beta},\boldsymbol{\Gamma})$ is a complex number that depends on $\boldsymbol{\alpha}$, $\boldsymbol{\beta}$, and $\boldsymbol{\Gamma}$. However, $\kappa(\boldsymbol{\alpha},\boldsymbol{\beta},\boldsymbol{\Gamma})$ is independent of $\mathbf{z}$ and $\boldsymbol{\Omega}$. For explicit results in the case $g=1$, see §20.7(viii).

21.6 Products

21.6(i) Riemann Identity

Let $\mathbf{T} = [T_{jk}]$ be an arbitrary $h \times h$ orthogonal matrix (that is, $\mathbf{TT}^{\mathrm{T}} = \mathbf{I}$) with rational elements. Also, let $\mathbf{Z}$ be an arbitrary $g \times h$ matrix. Define

21.6.1 $\quad \mathcal{K} = \mathbb{Z}^{g\times h}\mathbf{T}/(\mathbb{Z}^{g\times h}\mathbf{T} \cap \mathbb{Z}^{g\times h})$,

that is, $\mathcal{K}$ is the set of all $g \times h$ matrices that are obtained by premultiplying $\mathbf{T}$ by any $g \times h$ matrix with integer elements; two such matrices in $\mathcal{K}$ are considered *equivalent* if their difference is a matrix with integer elements. Also, let

21.6.2 $\quad \mathcal{D} = |\mathbf{T}^{\mathrm{T}}\mathbb{Z}^h/(\mathbf{T}^{\mathrm{T}}\mathbb{Z}^h \cap \mathbb{Z}^h)|,$

that is, $\mathcal{D}$ is the number of elements in the set containing all h-dimensional vectors obtained by multiplying $\mathbf{T}^{\mathrm{T}}$ on the right by a vector with integer elements. Two such vectors are considered *equivalent* if their difference is a vector with integer elements. Then

21.6.3
$$\prod_{j=1}^{h} \theta\!\left(\sum_{k=1}^{h} T_{jk}\mathbf{z}_k \,\Big|\, \boldsymbol{\Omega}\right)$$
$$= \frac{1}{\mathcal{D}^g} \sum_{\mathbf{A}\in\mathcal{K}}\sum_{\mathbf{B}\in\mathcal{K}} e^{2\pi i \operatorname{tr}\left[\frac{1}{2}\mathbf{A}^{\mathrm{T}}\boldsymbol{\Omega}\mathbf{A} + \mathbf{A}^{\mathrm{T}}[\mathbf{Z}+\mathbf{B}]\right]}$$
$$\times \prod_{j=1}^{h} \theta(\mathbf{z}_j + \boldsymbol{\Omega}\mathbf{a}_j + \mathbf{b}_j | \boldsymbol{\Omega}),$$

where $\mathbf{z}_j$, $\mathbf{a}_j$, $\mathbf{b}_j$ denote respectively the jth columns of $\mathbf{Z}$, $\mathbf{A}$, $\mathbf{B}$. This is the *Riemann identity*. On using theta functions with characteristics, it becomes

21.6.4
$$\prod_{j=1}^{h} \theta\!\begin{bmatrix}\sum_{k=1}^{h} T_{jk}\mathbf{c}_k \\ \sum_{k=1}^{h} T_{jk}\mathbf{d}_k\end{bmatrix}\!\left(\sum_{k=1}^{h} T_{jk}\mathbf{z}_k\,\Big|\,\boldsymbol{\Omega}\right)$$
$$= \frac{1}{\mathcal{D}^g} \sum_{\mathbf{A}\in\mathcal{K}}\sum_{\mathbf{B}\in\mathcal{K}} e^{-2\pi i \sum_{j=1}^{h} \mathbf{b}_j\cdot \mathbf{c}_j} \prod_{j=1}^{h} \theta\!\begin{bmatrix}\mathbf{a}_j + \mathbf{c}_j \\ \mathbf{b}_j + \mathbf{d}_j\end{bmatrix}\!(\mathbf{z}_j|\boldsymbol{\Omega}),$$

where $\mathbf{c}_j$ and $\mathbf{d}_j$ are arbitrary h-dimensional vectors. Many identities involving products of theta functions can be established using these formulas.

Example

Let $h = 4$ and

21.6.5 $\quad \mathbf{T} = \begin{bmatrix} 1 & 1 & 1 & 1 \\ 1 & 1 & -1 & -1 \\ 1 & -1 & 1 & -1 \\ 1 & -1 & -1 & 1 \end{bmatrix}.$

Then

$$\theta\left(\frac{\mathbf{x}+\mathbf{y}+\mathbf{u}+\mathbf{v}}{2}\bigg|\mathbf{\Omega}\right)\theta\left(\frac{\mathbf{x}+\mathbf{y}-\mathbf{u}-\mathbf{v}}{2}\bigg|\mathbf{\Omega}\right)\theta\left(\frac{\mathbf{x}-\mathbf{y}+\mathbf{u}-\mathbf{v}}{2}\bigg|\mathbf{\Omega}\right)\theta\left(\frac{\mathbf{x}-\mathbf{y}-\mathbf{u}+\mathbf{v}}{2}\bigg|\mathbf{\Omega}\right)$$

21.6.6
$$=\frac{1}{2^g}\sum_{\boldsymbol{\alpha}\in\frac{1}{2}\mathbb{Z}^g/\mathbb{Z}^g}\sum_{\boldsymbol{\beta}\in\frac{1}{2}\mathbb{Z}^g/\mathbb{Z}^g}e^{2\pi i(2\boldsymbol{\alpha}\cdot\boldsymbol{\Omega}\cdot\boldsymbol{\alpha}+\boldsymbol{\alpha}\cdot[\mathbf{x}+\mathbf{y}+\mathbf{u}+\mathbf{v}])}$$
$$\times\theta(\mathbf{x}+\mathbf{\Omega}\boldsymbol{\alpha}+\boldsymbol{\beta}|\mathbf{\Omega})\,\theta(\mathbf{y}+\mathbf{\Omega}\boldsymbol{\alpha}+\boldsymbol{\beta}|\mathbf{\Omega})\,\theta(\mathbf{u}+\mathbf{\Omega}\boldsymbol{\alpha}+\boldsymbol{\beta}|\mathbf{\Omega})\,\theta(\mathbf{v}+\mathbf{\Omega}\boldsymbol{\alpha}+\boldsymbol{\beta}|\mathbf{\Omega}),$$

and

21.6.7
$$\theta\!\begin{bmatrix}\frac{1}{2}[\mathbf{c}_1+\mathbf{c}_2+\mathbf{c}_3+\mathbf{c}_4]\\\frac{1}{2}[\mathbf{d}_1+\mathbf{d}_2+\mathbf{d}_3+\mathbf{d}_4]\end{bmatrix}\!\left(\frac{\mathbf{x}+\mathbf{y}+\mathbf{u}+\mathbf{v}}{2}\bigg|\mathbf{\Omega}\right)\theta\!\begin{bmatrix}\frac{1}{2}[\mathbf{c}_1+\mathbf{c}_2-\mathbf{c}_3-\mathbf{c}_4]\\\frac{1}{2}[\mathbf{d}_1+\mathbf{d}_2-\mathbf{d}_3-\mathbf{d}_4]\end{bmatrix}\!\left(\frac{\mathbf{x}+\mathbf{y}-\mathbf{u}-\mathbf{v}}{2}\bigg|\mathbf{\Omega}\right)$$
$$\times\theta\!\begin{bmatrix}\frac{1}{2}[\mathbf{c}_1-\mathbf{c}_2+\mathbf{c}_3-\mathbf{c}_4]\\\frac{1}{2}[\mathbf{d}_1-\mathbf{d}_2+\mathbf{d}_3-\mathbf{d}_4]\end{bmatrix}\!\left(\frac{\mathbf{x}-\mathbf{y}+\mathbf{u}-\mathbf{v}}{2}\bigg|\mathbf{\Omega}\right)\theta\!\begin{bmatrix}\frac{1}{2}[\mathbf{c}_1-\mathbf{c}_2-\mathbf{c}_3+\mathbf{c}_4]\\\frac{1}{2}[\mathbf{d}_1-\mathbf{d}_2-\mathbf{d}_3+\mathbf{d}_4]\end{bmatrix}\!\left(\frac{\mathbf{x}-\mathbf{y}-\mathbf{u}+\mathbf{v}}{2}\bigg|\mathbf{\Omega}\right)$$
$$=\frac{1}{2^g}\sum_{\boldsymbol{\alpha}\in\frac{1}{2}\mathbb{Z}^g/\mathbb{Z}^g}\sum_{\boldsymbol{\beta}\in\frac{1}{2}\mathbb{Z}^g/\mathbb{Z}^g}e^{-2\pi i\boldsymbol{\beta}\cdot[\mathbf{c}_1+\mathbf{c}_2+\mathbf{c}_3+\mathbf{c}_4]}\theta\!\begin{bmatrix}\mathbf{c}_1+\boldsymbol{\alpha}\\\mathbf{d}_1+\boldsymbol{\beta}\end{bmatrix}\!(\mathbf{x}|\mathbf{\Omega})\,\theta\!\begin{bmatrix}\mathbf{c}_2+\boldsymbol{\alpha}\\\mathbf{d}_2+\boldsymbol{\beta}\end{bmatrix}\!(\mathbf{y}|\mathbf{\Omega})\,\theta\!\begin{bmatrix}\mathbf{c}_3+\boldsymbol{\alpha}\\\mathbf{d}_3+\boldsymbol{\beta}\end{bmatrix}\!(\mathbf{u}|\mathbf{\Omega})\,\theta\!\begin{bmatrix}\mathbf{c}_4+\boldsymbol{\alpha}\\\mathbf{d}_4+\boldsymbol{\beta}\end{bmatrix}\!(\mathbf{v}|\mathbf{\Omega}).$$

21.6(ii) Addition Formulas

Let $\boldsymbol{\alpha},\boldsymbol{\beta},\boldsymbol{\gamma},\boldsymbol{\delta}\in\mathbb{R}^g$. Then

21.6.8
$$\theta\!\begin{bmatrix}\boldsymbol{\alpha}\\\boldsymbol{\gamma}\end{bmatrix}\!(\mathbf{z}_1|\mathbf{\Omega})\,\theta\!\begin{bmatrix}\boldsymbol{\beta}\\\boldsymbol{\delta}\end{bmatrix}\!(\mathbf{z}_2|\mathbf{\Omega})$$
$$=\sum_{\boldsymbol{\nu}\in\mathbb{Z}^g/(2\mathbb{Z}^g)}\theta\!\begin{bmatrix}\frac{1}{2}[\boldsymbol{\alpha}+\boldsymbol{\beta}+\boldsymbol{\nu}]\\\boldsymbol{\gamma}+\boldsymbol{\delta}\end{bmatrix}\!(\mathbf{z}_1+\mathbf{z}_2|2\mathbf{\Omega})$$
$$\times\theta\!\begin{bmatrix}\frac{1}{2}[\boldsymbol{\alpha}-\boldsymbol{\beta}+\boldsymbol{\nu}]\\\boldsymbol{\gamma}-\boldsymbol{\delta}\end{bmatrix}\!(\mathbf{z}_1-\mathbf{z}_2|2\mathbf{\Omega}).$$

Thus $\boldsymbol{\nu}$ is a g-dimensional vector whose entries are either 0 or 1. For this result and a generalization see Koizumi (1976) and Belokolos et al. (1994, pp. 38–41). For addition formulas for classical theta functions see §20.7(ii).

Applications

21.7 Riemann Surfaces

21.7(i) Connection of Riemann Theta Functions to Riemann Surfaces

In almost all applications, a Riemann theta function is associated with a compact Riemann surface. Although there are other ways to represent Riemann surfaces (see e.g. Belokolos et al. (1994, §2.1)), they are obtainable from *plane algebraic curves* (Springer (1957), or Riemann (1851)). Consider the set of points in $\mathbb{C}^2$ that satisfy the equation

21.7.1
$$P(\lambda,\mu)=0,$$

where $P(\lambda,\mu)$ is a polynomial in λ and μ that does not factor over $\mathbb{C}^2$. Equation (21.7.1) determines a plane algebraic curve in $\mathbb{C}^2$, which is made compact by adding its points at infinity. To accomplish this we write (21.7.1) in terms of homogeneous coordinates:

21.7.2
$$\tilde{P}(\tilde{\lambda},\tilde{\mu},\tilde{\eta})=0,$$

by setting $\lambda=\tilde{\lambda}/\tilde{\eta}$, $\mu=\tilde{\mu}/\tilde{\eta}$, and then clearing fractions. This compact curve may have singular points, that is, points at which the gradient of $\tilde{P}$ vanishes. Removing the singularities of this curve gives rise to a two-dimensional connected manifold with a complex-analytic structure, that is, a *Riemann surface*. All compact Riemann surfaces can be obtained this way.

Since a Riemann surface Γ is a two-dimensional manifold that is orientable (owing to its analytic structure), its only topological invariant is its *genus g* (the number of *handles* in the surface). On this surface, we choose $2g$ *cycles* (that is, closed oriented curves, each with at most a finite number of singular points) $a_j, b_j, j=1,2,\ldots,g$, such that their *intersection indices* satisfy

21.7.3
$$a_j\circ a_k=0,\quad b_j\circ b_k=0,\quad a_j\circ b_k=\delta_{j,k}.$$

For example, Figure 21.7.1 depicts a genus 2 surface.

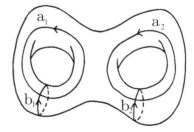

Figure 21.7.1: A basis of cycles for a genus 2 surface.

On a Riemann surface of genus g, there are g linearly independent *holomorphic differentials* ω_j, $j=$

$1, 2, \ldots, g$. If a local coordinate z is chosen on the Riemann surface, then the local coordinate representation of these holomorphic differentials is given by

21.7.4 $\qquad \omega_j = f_j(z)\, dz, \qquad j = 1, 2, \ldots, g,$

where $f_j(z)$, $j = 1, 2, \ldots, g$ are analytic functions. Thus the differentials ω_j, $j = 1, 2, \ldots, g$ have no singularities on Γ. Note that for the purposes of integrating these holomorphic differentials, all cycles on the surface are a linear combination of the cycles a_j, b_j, $j = 1, 2, \ldots, g$. The ω_j are normalized so that

21.7.5 $\qquad \oint_{a_k} \omega_j = \delta_{j,k}, \qquad j, k = 1, 2, \ldots, g.$

Then the matrix defined by

21.7.6 $\qquad \Omega_{jk} = \oint_{b_k} \omega_j, \qquad j, k = 1, 2, \ldots, g,$

is a Riemann matrix and it is used to define the corresponding Riemann theta function. *In this way, we associate a Riemann theta function with every compact Riemann surface Γ.*

Riemann theta functions originating from Riemann surfaces are special in the sense that a general g-dimensional Riemann theta function depends on $g(g+1)/2$ complex parameters. In contrast, a g-dimensional Riemann theta function arising from a compact Riemann surface of genus g (> 1) depends on at most $3g - 3$ complex parameters (one complex parameter for the case $g = 1$). These special Riemann theta functions satisfy many special identities, two of which appear in the following subsections. For more information, see Dubrovin (1981), Brieskorn and Knörrer (1986, §9.3), Belokolos et al. (1994, Chapter 2), and Mumford (1984, §2.2–2.3).

21.7(ii) Fay's Trisecant Identity

Let $\boldsymbol{\alpha}$, $\boldsymbol{\beta}$ be such that

21.7.7
$$\left(\frac{\partial}{\partial z_1} \theta\!\begin{bmatrix}\boldsymbol{\alpha}\\ \boldsymbol{\beta}\end{bmatrix}\!(\mathbf{z}|\boldsymbol{\Omega})\Big|_{\mathbf{z}=\mathbf{0}}, \ldots, \frac{\partial}{\partial z_g} \theta\!\begin{bmatrix}\boldsymbol{\alpha}\\ \boldsymbol{\beta}\end{bmatrix}\!(\mathbf{z}|\boldsymbol{\Omega})\Big|_{\mathbf{z}=\mathbf{0}} \right) \neq \mathbf{0}.$$

Define the holomorphic differential

21.7.8 $\qquad \zeta = \sum_{j=1}^{g} \omega_j \frac{\partial}{\partial z_j} \theta\!\begin{bmatrix}\boldsymbol{\alpha}\\ \boldsymbol{\beta}\end{bmatrix}\!(\mathbf{z}|\boldsymbol{\Omega})\Big|_{\mathbf{z}=\mathbf{0}}.$

Then the *prime form* on the corresponding compact Riemann surface Γ is defined by

21.7.9
$$E(P_1, P_2) = \theta\!\begin{bmatrix}\boldsymbol{\alpha}\\ \boldsymbol{\beta}\end{bmatrix}\!\left(\int_{P_1}^{P_2} \boldsymbol{\omega}\,\Big|\,\boldsymbol{\Omega}\right) \Big/ \left(\sqrt{\zeta(P_1)} \sqrt{\zeta(P_2)} \right),$$

where P_1 and P_2 are points on Γ, $\boldsymbol{\omega} = (\omega_1, \omega_2, \ldots, \omega_g)$, and the path of integration on Γ from P_1 to P_2 is identical for all components. Here $\sqrt{\zeta(P)}$ is such that $\sqrt{\zeta(P)}^2 = \zeta(P)$, $P \in \Gamma$. Either branch of the square roots may be chosen, as long as the branch is consistent across Γ. For all $\mathbf{z} \in \mathbb{C}^g$, and all P_1, P_2, P_3, P_4 on Γ, Fay's identity is given by

21.7.10
$$\theta\!\left(\mathbf{z} + \int_{P_1}^{P_3}\boldsymbol{\omega}\,\Big|\,\boldsymbol{\Omega}\right) \theta\!\left(\mathbf{z} + \int_{P_2}^{P_4}\boldsymbol{\omega}\,\Big|\,\boldsymbol{\Omega}\right) E(P_3, P_2) E(P_1, P_4) + \theta\!\left(\mathbf{z} + \int_{P_2}^{P_3}\boldsymbol{\omega}\,\Big|\,\boldsymbol{\Omega}\right) \theta\!\left(\mathbf{z} + \int_{P_1}^{P_4}\boldsymbol{\omega}\,\Big|\,\boldsymbol{\Omega}\right) E(P_3, P_1) E(P_4, P_2)$$
$$= \theta(\mathbf{z}|\boldsymbol{\Omega})\, \theta\!\left(\mathbf{z} + \int_{P_1}^{P_3}\boldsymbol{\omega} + \int_{P_2}^{P_4}\boldsymbol{\omega}\,\Big|\,\boldsymbol{\Omega}\right) E(P_1, P_2) E(P_3, P_4),$$

where again all integration paths are identical for all components. Generalizations of this identity are given in Fay (1973, Chapter 2). Fay derives (21.7.10) as a special case of a more general class of addition theorems for Riemann theta functions on Riemann surfaces.

21.7(iii) Frobenius' Identity

Let Γ be a *hyperelliptic Riemann surface*. These are Riemann surfaces that may be obtained from algebraic curves of the form

21.7.11 $\qquad \mu^2 = Q(\lambda),$

where $Q(\lambda)$ is a polynomial in λ of odd degree $2g + 1$ (≥ 5). The genus of this surface is g. The zeros λ_j, $j = 1, 2, \ldots, 2g + 1$ of $Q(\lambda)$ specify the finite branch points P_j, that is, points at which $\mu_j = 0$, on the Riemann surface. Denote the set of all branch points by $B = \{P_1, P_2, \ldots, P_{2g+1}, P_\infty\}$. Consider a fixed subset U of B, such that the number of elements $|U|$ in the set U is $g+1$, and $P_\infty \notin U$. Next, define an isomorphism $\boldsymbol{\eta}$ which maps every subset T of B with an even number of elements to a $2g$-dimensional vector $\boldsymbol{\eta}(T)$ with elements either 0 or $\tfrac{1}{2}$. Define the operation

21.7.12 $\qquad T_1 \ominus T_2 = (T_1 \cup T_2) \backslash (T_1 \cap T_2).$

Also, $T^c = B \backslash T$, $\boldsymbol{\eta}^1(T) = (\eta_1(T), \eta_2(T), \ldots, \eta_g(T))$, and $\boldsymbol{\eta}^2(T) = (\eta_{g+1}(T), \eta_{g+2}(T), \ldots, \eta_{2g}(T))$. Then the isomorphism is determined completely by:

21.7.13 $\quad\quad\quad \boldsymbol{\eta}(T) = \boldsymbol{\eta}(T^c),$

21.7.14 $\quad\quad \boldsymbol{\eta}(T_1 \ominus T_2) = \boldsymbol{\eta}(T_1) + \boldsymbol{\eta}(T_2),$

21.7.15 $\quad 4\boldsymbol{\eta}^1(T) \cdot \boldsymbol{\eta}^2(T) = \tfrac{1}{2}(|T \ominus U| - g - 1) \pmod 2$

21.7.16
$4(\boldsymbol{\eta}^1(T_1) \cdot \boldsymbol{\eta}^2(T_2) - \boldsymbol{\eta}^2(T_1) \cdot \boldsymbol{\eta}^1(T_2)) \equiv |T_1 \cap T_2| \pmod{2}.$

Furthermore, let $\boldsymbol{\eta}(P_\infty) = \mathbf{0}$ and $\boldsymbol{\eta}(P_j) = \boldsymbol{\eta}(\{P_j, P_\infty\})$. Then for all $\mathbf{z}_j \in \mathbb{C}^g$, $j = 1, 2, 3, 4$, such that $\mathbf{z}_1 + \mathbf{z}_2 + \mathbf{z}_3 + \mathbf{z}_4 = 0$, and for all $\boldsymbol{\alpha}_j, \boldsymbol{\beta}_j \in \mathbb{R}^g$, such that $\boldsymbol{\alpha}_1 + \boldsymbol{\alpha}_2 + \boldsymbol{\alpha}_3 + \boldsymbol{\alpha}_4 = 0$ and $\boldsymbol{\beta}_1 + \boldsymbol{\beta}_2 + \boldsymbol{\beta}_3 + \boldsymbol{\beta}_4 = 0$, we have *Frobenius' identity*:

21.7.17
$$\sum_{P_j \in U} \prod_{k=1}^4 \theta\begin{bmatrix} \boldsymbol{\alpha}_k + \boldsymbol{\eta}^1(P_j) \\ \boldsymbol{\beta}_k + \boldsymbol{\eta}^2(P_j) \end{bmatrix}(\mathbf{z}_k | \boldsymbol{\Omega})$$
$$= \sum_{P_j \in U^c} \prod_{k=1}^4 \theta\begin{bmatrix} \boldsymbol{\alpha}_k + \boldsymbol{\eta}^1(P_j) \\ \boldsymbol{\beta}_k + \boldsymbol{\eta}^2(P_j) \end{bmatrix}(\mathbf{z}_k | \boldsymbol{\Omega}).$$

21.8 Abelian Functions

An Abelian function is a $2g$-fold periodic, meromorphic function of g complex variables. In consequence, Abelian functions are generalizations of elliptic functions (§23.2(iii)) to more than one complex variable. For every Abelian function, there is a positive integer n, such that the Abelian function can be expressed as a ratio of linear combinations of products with n factors of Riemann theta functions with characteristics that share a common period lattice. For further information see Igusa (1972, pp. 132–135) and Markushevich (1992).

21.9 Integrable Equations

Riemann theta functions arise in the study of *integrable differential equations* that have applications in many areas, including fluid mechanics (Ablowitz and Segur (1981, Chapter 4)), magnetic monopoles (Ercolani and Sinha (1989)), and string theory (Deligne et al. (1999, Part 3)). Typical examples of such equations are the Korteweg–de Vries equation

21.9.1 $\quad\quad\quad 4u_t = 6uu_x + u_{xxx},$

and the nonlinear Schrödinger equations

21.9.2 $\quad\quad\quad iu_t = -\tfrac{1}{2}u_{xx} \pm |u|^2 u.$

Here, and in what follows, $x, y,$ and t suffixes indicate partial derivatives.

Particularly important for the use of Riemann theta functions is the Kadomtsev–Petviashvili (KP) equation, which describes the propagation of two-dimensional, long-wave length surface waves in shallow water (Ablowitz and Segur (1981, Chapter 4)):

21.9.3 $\quad\quad (-4u_t + 6uu_x + u_{xxx})_x + 3u_{yy} = 0.$

Here x and y are spatial variables, t is time, and $u(x, y, t)$ is the elevation of the surface wave. All quantities are made dimensionless by a suitable scaling transformation. The KP equation has a class of quasi-periodic solutions described by Riemann theta functions, given by

21.9.4 $\quad u(x, y, t) = c + 2\dfrac{\partial^2}{\partial x^2} \ln(\theta(\mathbf{k}x + \mathbf{l}y + \boldsymbol{\omega}t + \boldsymbol{\phi}|\boldsymbol{\Omega})),$

where c is a complex constant and $\mathbf{k}$, $\mathbf{l}$, $\boldsymbol{\omega}$, and $\boldsymbol{\phi}$ are g-dimensional complex vectors; see Krichever (1976). These parameters, including $\boldsymbol{\Omega}$, are not free: they are determined by a compact, connected Riemann surface (Krichever (1976)), or alternatively by an appropriate initial condition $u(x, y, 0)$ (Deconinck and Segur (1998)). These solutions have been compared successfully with physical experiments for $g = 1, 2$ (Wiegel (1960), Hammack et al. (1989), and Hammack et al. (1995)). See Figures 21.9.1 and 21.9.2.

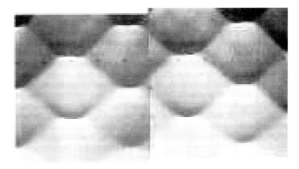

Figure 21.9.1: Two-dimensional periodic waves in a shallow water wave tank, taken from Hammack et al. (1995, p. 97) by permission of Cambridge University Press. The original caption reads "Mosaic of two overhead photographs, showing surface patterns of waves in shallow water."

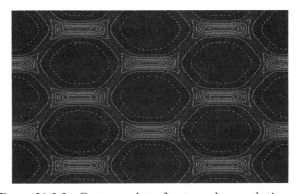

Figure 21.9.2: Contour plot of a two-phase solution of Equation (21.9.3). Such a solution is given in terms of a Riemann theta function with two phases; see Krichever (1976), Dubrovin (1981), and Hammack et al. (1995).

Furthermore, the solutions of the KP equation solve the *Schottky problem*: this is the question concerning conditions that a Riemann matrix needs to satisfy in order to be associated with a Riemann surface (Schottky (1903)). Following the work of Krichever (1976), Novikov conjectured that the Riemann theta function in (21.9.4) gives rise to a solution of the KP equation (21.9.3) if, and only if, the theta function originates from a Riemann surface; see Dubrovin (1981, §IV.4). The first part of this conjecture was established in Krichever (1976); the second part was proved in Shiota (1986).

Computation

21.10 Methods of Computation

21.10(i) General Riemann Theta Functions

Although the defining Fourier series (21.2.1) is uniformly convergent on compact sets, its evaluation is cumbersome when one or more of the eigenvalues of $\Im(\Omega)$ is near zero. Furthermore, for fixed Ω different terms of the Fourier series dominate for different values of $\mathbf{z}$.

To overcome these obstacles, we compute instead the scaled function $\hat{\theta}(\mathbf{z}|\Omega)$ (§21.2(i)) from the expansion

21.10.1
$$\hat{\theta}(\mathbf{z}|\Omega) = \sum_{\mathbf{n} \in S(\epsilon)} e^{\pi i [\mathbf{n} - [\mathbf{Y}^{-1}\mathbf{y}]] \cdot \mathbf{X} \cdot [\mathbf{n} - [\mathbf{Y}^{-1}\mathbf{y}]]}$$
$$\times e^{2\pi i [\mathbf{n} - [\mathbf{Y}^{-1}\mathbf{y}]] \cdot \mathbf{x}} e^{-\pi [\mathbf{n} + [\mathbf{Y}^{-1}\mathbf{y}]] \cdot \mathbf{Y} \cdot [\mathbf{n} + [\mathbf{Y}^{-1}\mathbf{y}]]},$$

where ϵ is the tolerated maximum absolute error for $\hat{\theta}(\mathbf{z}|\Omega)$. Here $\mathbf{X} = \Re(\Omega)$, $\mathbf{Y} = \Im(\Omega)$, $\mathbf{x} = \Re(\mathbf{z})$, $\mathbf{y} = \Im(\mathbf{z})$, and

21.10.2
$$S(\epsilon) = \left\{ \mathbf{m} \in \mathbb{Z}^g \,\middle|\, \pi \left[\mathbf{m} + [\mathbf{Y}^{-1}\mathbf{y}]\right] \cdot \mathbf{Y} \cdot \left[\mathbf{m} + [\mathbf{Y}^{-1}\mathbf{y}]\right] \leq R(\epsilon) \right\}.$$

Thus $S(\epsilon)$ is the set of all integer vectors that are contained in an ellipsoid centered at the fractional part of $\mathbf{Y}^{-1}\mathbf{y}$, and whose size is determined by the allowed absolute error. The value of $R(\epsilon)$ is determined as follows. Let r be the length of the shortest vector of the lattice $\Lambda = \{\sqrt{\pi}\,\mathbf{Tm} | \mathbf{m} \in \mathbb{Z}^g\}$, and $\mathbf{T}^T\mathbf{T} = \mathbf{Y}$ be the Cholesky decomposition of $\mathbf{Y}$ (Atkinson (1989, p. 254)). Then $R(\epsilon)$ is the greater of $\sqrt{g/2} + r$ and the smallest positive root of the equation

21.10.3 $\quad \Gamma\!\left(\tfrac{1}{2}g, R^2\right)/(2gr^g) = \epsilon.$

For the incomplete gamma function $\Gamma(a,z)$, see §8.2(i).

The construction (21.10.2) amounts to determining all integer vectors in a g-dimensional ellipsoid. For this purpose it is convenient to have the ellipsoid as spherical as possible (Siegel (1973, pp. 144–159), Heil (1995)).

Usually, (21.10.1) can also be used for the efficient evaluation of $\hat{\theta}(\mathbf{z}|\Omega)$ for fixed Ω and varying $\mathbf{z}$, by addition of a few vectors to the set $S(\epsilon)$.

21.10(ii) Riemann Theta Functions Associated with a Riemann Surface

In addition to evaluating the Fourier series, the main problem here is to compute a Riemann matrix originating from a Riemann surface. Various approaches are considered in the following references:

- Belokolos *et al.* (1994, Chapter 5) and references therein. Here the Riemann surface is represented by the action of a Schottky group on a region of the complex plane. The same representation is used in Gianni *et al.* (1998).

- Tretkoff and Tretkoff (1984). Here a Hurwitz system is chosen to represent the Riemann surface.

- Deconinck and van Hoeij (2001). Here a plane algebraic curve representation of the Riemann surface is used.

21.11 Software

See http://dlmf.nist.gov/21.11.

References

General References

The main references used in writing this chapter are Mumford (1983, 1984), Igusa (1972), and Belokolos *et al.* (1994). For additional bibliographic reading see Dubrovin (1981), Siegel (1971, 1973), and Fay (1973).

Sources

The following list gives the references or other indications of proofs that were used in constructing the various sections of this chapter. These sources supplement the references that are quoted in the text.

§21.3 Mumford (1983, pp. 120–122).

§21.4 These graphics were computed by the author, using the algorithms described in Deconinck *et al.* (2004).

§21.5 Arnol'd (1997, p. 222), Mumford (1983, pp. 189–210), Igusa (1972, pp. 78–85).

§**21.6** Mumford (1983, pp. 211–216), Dubrovin (1981, pp. 22–23).

§**21.7** Mumford (1984, pp. 106–120 and 207–260).

§**21.9** Dubrovin (1981), Belokolos *et al.* (1994).

§**21.10** Deconinck *et al.* (2004).

Chapter 22
Jacobian Elliptic Functions

W. P. Reinhardt[1] and P. L. Walker[2]

Notation — **550**
22.1 Special Notation 550

Properties — **550**
22.2 Definitions 550
22.3 Graphics 550
22.4 Periods, Poles, and Zeros 553
22.5 Special Values 554
22.6 Elementary Identities 556
22.7 Landen Transformations 556
22.8 Addition Theorems 557
22.9 Cyclic Identities 558
22.10 Maclaurin Series 558
22.11 Fourier and Hyperbolic Series 559
22.12 Expansions in Other Trigonometric Series and Doubly-Infinite Partial Fractions: Eisenstein Series 559

22.13 Derivatives and Differential Equations . . 560
22.14 Integrals 560
22.15 Inverse Functions 561
22.16 Related Functions 561
22.17 Moduli Outside the Interval [0,1] 563

Applications — **563**
22.18 Mathematical Applications 563
22.19 Physical Applications 564

Computation — **566**
22.20 Methods of Computation 566
22.21 Tables 567
22.22 Software 567

References — **567**

---

[1] University of Washington, Seattle, Washington.
[2] American University of Sharjah, Sharjah, United Arab Emirates.
Acknowledgments: This chapter is based in part on Abramowitz and Stegun (1964, Chapters 16,18) by L. M. Milne-Thomson and T. H. Southard respectively.
Copyright © 2009 National Institute of Standards and Technology. All rights reserved.

Notation

22.1 Special Notation

(For other notation see pp. xiv and 873.)

x, y	real variables.
z	complex variable.
k	modulus. Except in §§22.3(iv), 22.17, and 22.19, $0 \leq k \leq 1$.
k'	complementary modulus, $k^2 + k'^2 = 1$. If $k \in [0,1]$, then $k' \in [0,1]$.
K, K'	$K(k), K'(k) = K(k')$ (complete elliptic integrals of the first kind (§19.2(ii))).
q	nome. $0 \leq q < 1$ except in §22.17; see also §20.1.
τ	iK'/K.

All derivatives are denoted by differentials, not primes.

The functions treated in this chapter are the three principal Jacobian elliptic functions $\operatorname{sn}(z,k)$, $\operatorname{cn}(z,k)$, $\operatorname{dn}(z,k)$; the nine subsidiary Jacobian elliptic functions $\operatorname{cd}(z,k)$, $\operatorname{sd}(z,k)$, $\operatorname{nd}(z,k)$, $\operatorname{dc}(z,k)$, $\operatorname{nc}(z,k)$, $\operatorname{sc}(z,k)$, $\operatorname{ns}(z,k)$, $\operatorname{ds}(z,k)$, $\operatorname{cs}(z,k)$; the amplitude function $\operatorname{am}(x,k)$; Jacobi's epsilon and zeta functions $\mathcal{E}(x,k)$ and $\mathrm{Z}(x|k)$.

The notation $\operatorname{sn}(z,k)$, $\operatorname{cn}(z,k)$, $\operatorname{dn}(z,k)$ is due to Gudermann (1838), following Jacobi (1827); that for the subsidiary functions is due to Glaisher (1882). Other notations for $\operatorname{sn}(z,k)$ are $\operatorname{sn}(z|m)$ and $\operatorname{sn}(z,m)$ with $m = k^2$; see Abramowitz and Stegun (1964) and Walker (1996). Similarly for the other functions.

Properties

22.2 Definitions

The *nome* q is given in terms of the *modulus* k by

22.2.1 $$q = \exp(-\pi K'(k)/K(k)),$$

where $K(k)$, $K'(k)$ are defined in §19.2(ii). Inversely,

22.2.2 $$k = \frac{\theta_2^2(0,q)}{\theta_3^2(0,q)}, \quad k' = \frac{\theta_4^2(0,q)}{\theta_3^2(0,q)}, \quad K(k) = \frac{\pi}{2}\theta_3^2(0,q),$$

where $k' = \sqrt{1-k^2}$ and the theta functions are defined in §20.2(i).

With

22.2.3 $$\zeta = \frac{\pi z}{2K(k)},$$

22.2.4 $$\operatorname{sn}(z,k) = \frac{\theta_3(0,q)}{\theta_2(0,q)}\frac{\theta_1(\zeta,q)}{\theta_4(\zeta,q)} = \frac{1}{\operatorname{ns}(z,k)},$$

22.2.5 $$\operatorname{cn}(z,k) = \frac{\theta_4(0,q)}{\theta_2(0,q)}\frac{\theta_2(\zeta,q)}{\theta_4(\zeta,q)} = \frac{1}{\operatorname{nc}(z,k)},$$

22.2.6 $$\operatorname{dn}(z,k) = \frac{\theta_4(0,q)}{\theta_3(0,q)}\frac{\theta_3(\zeta,q)}{\theta_4(\zeta,q)} = \frac{1}{\operatorname{nd}(z,k)},$$

22.2.7 $$\operatorname{sd}(z,k) = \frac{\theta_3^2(0,q)}{\theta_2(0,q)\theta_4(0,q)}\frac{\theta_1(\zeta,q)}{\theta_3(\zeta,q)} = \frac{1}{\operatorname{ds}(z,k)},$$

22.2.8 $$\operatorname{cd}(z,k) = \frac{\theta_3(0,q)}{\theta_2(0,q)}\frac{\theta_2(\zeta,q)}{\theta_3(\zeta,q)} = \frac{1}{\operatorname{dc}(z,k)},$$

22.2.9 $$\operatorname{sc}(z,k) = \frac{\theta_3(0,q)}{\theta_4(0,q)}\frac{\theta_1(\zeta,q)}{\theta_2(\zeta,q)} = \frac{1}{\operatorname{cs}(z,k)}.$$

As a function of z, with fixed k, each of the 12 Jacobian elliptic functions is doubly periodic, having two periods whose ratio is not real. Each is meromorphic in z for fixed k, with simple poles and simple zeros, and each is meromorphic in k for fixed z. For $k \in [0,1]$, all functions are real for $z \in \mathbb{R}$.

Glaisher's Notation

The Jacobian functions are related in the following way. Let p, q, r be any three of the letters s, c, d, n. Then

22.2.10 $$\operatorname{pq}(z,k) = \frac{\operatorname{pr}(z,k)}{\operatorname{qr}(z,k)} = \frac{1}{\operatorname{qp}(z,k)},$$

with the convention that functions with the same two letters are replaced by unity; e.g. $\operatorname{ss}(z,k) = 1$.

The six functions containing the letter s in their two-letter name are odd in z; the other six are even in z.

In terms of Neville's theta functions (§20.1)

22.2.11 $$\operatorname{pq}(z,k) = \theta_p(z|\tau)/\theta_q(z|\tau),$$

where

22.2.12 $$\tau = iK'(k)/K(k),$$

and p, q are any pair of the letters s, c, d, n.

22.3 Graphics

22.3(i) Real Variables: Line Graphs

See Figures 22.3.1–22.3.4 for line graphs of the functions $\operatorname{sn}(x,k)$, $\operatorname{cn}(x,k)$, $\operatorname{dn}(x,k)$, and $\operatorname{nd}(x,k)$ for representative values of real x and real k illustrating the near trigonometric ($k=0$), and near hyperbolic ($k=1$) limits. For corresponding graphs for the other 8 Jacobian elliptic functions see http://dlmf.nist.gov/22.3.i.

22.3 GRAPHICS

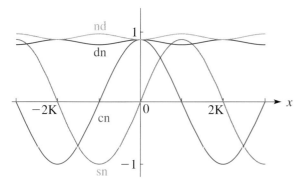

Figure 22.3.1: $k = 0.4$, $-3K \leq x \leq 3K$, $K = 1.6399\ldots$.

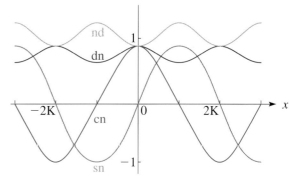

Figure 22.3.2: $k = 0.7$, $-3K \leq x \leq 3K$, $K = 1.8456\ldots$. For $\operatorname{cn}(x, k)$ the curve for $k = 1/\sqrt{2} = 0.70710\ldots$ is a boundary between the curves that have an inflection point in the interval $0 \leq x \leq 2K(k)$, and its translates, and those that do not; see Walker (1996, p. 146).

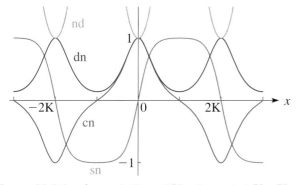

Figure 22.3.3: $k = 0.99$, $-3K \leq x \leq 3K$, $K = 3.3566\ldots$.

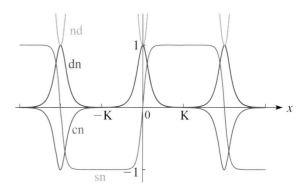

Figure 22.3.4: $k = 0.999999$, $-3K \leq x \leq 3K$, $K = 7.9474\ldots$.

22.3(ii) Real Variables: Surfaces

See Figure 22.3.13 for $\operatorname{sn}(x, k)$ as a function of real arguments x and k. The period diverges logarithmically as $k \to 1-$; see §19.12. For the corresponding surfaces for $\operatorname{cn}(x, k)$ and $\operatorname{dn}(x, k)$ see http://dlmf.nist.gov/22.3.ii.

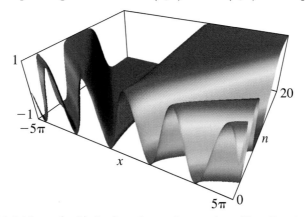

Figure 22.3.13: $\operatorname{sn}(x, k)$ for $k = 1 - e^{-n}$, $n = 0$ to 20, $-5\pi \leq x \leq 5\pi$.

22.3(iii) Complex z; Real k

22.3(iv) Complex k

In Figure 22.3.16 height corresponds to the absolute value of the function and color to the phase. See p. xiv.

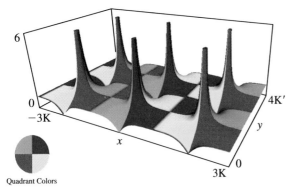

Figure 22.3.16: $\operatorname{sn}(x+iy, k)$ for $k = 0.99$, $-3K \leq x \leq 3K$, $0 \leq y \leq 4K'$. $K = 3.3566\ldots$, $K' = 1.5786\ldots$.

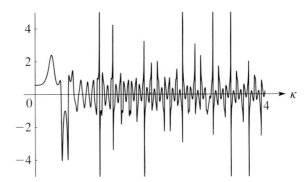

Figure 22.3.22: $\Re\operatorname{sn}(x, k)$, $x = 120$, as a function of $k^2 = i\kappa$, $0 \leq \kappa \leq 4$.

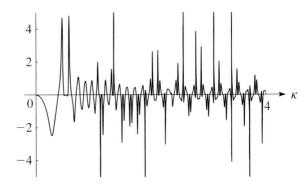

Figure 22.3.23: $\Im\operatorname{sn}(x, k)$, $x = 120$, as a function of $k^2 = i\kappa$, $0 \leq \kappa \leq 4$.

For the corresponding surfaces for the copolar functions $\operatorname{cn}(z, k)$ and $\operatorname{dn}(z, k)$ and the coperiodic functions $\operatorname{cd}(z, k)$, $\operatorname{dc}(z, k)$, and $\operatorname{ns}(z, k)$ with $z = x + iy$ see http://dlmf.nist.gov/22.3.iii.

In Figures 22.3.24 and 22.3.25, height corresponds to the absolute value of the function and color to the phase. See p. xiv.

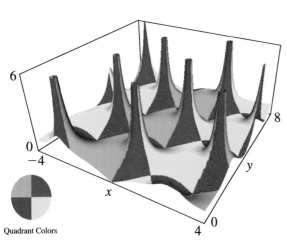

Figure 22.3.24: $\operatorname{sn}(x+iy, k)$ for $-4 \leq x \leq 4$, $0 \leq y \leq 8$, $k = 1 + \frac{1}{2}i$. $K = 1.5149\ldots + i0.5235\ldots$, $K' = 1.4620\ldots - i0.3552\ldots$.

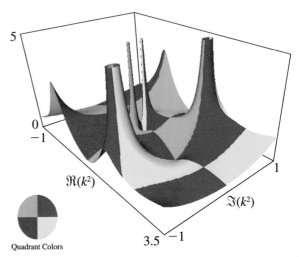

Figure 22.3.25: $\operatorname{sn}(5, k)$ as a function of complex k^2, $-1 \leq \Re(k^2) \leq 3.5$, $-1 \leq \Im(k^2) \leq 1$. Compare §22.17(ii).

22.4 PERIODS, POLES, AND ZEROS

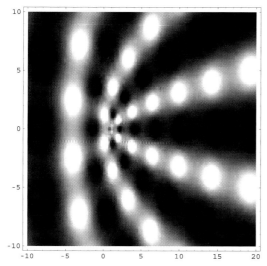

Figure 22.3.26: Density plot of $|\operatorname{sn}(5,k)|$ as a function of complex k^2, $-10 \le \Re(k^2) \le 20$, $-10 \le \Im(k^2) \le 10$. Grayscale, running from 0 (black) to 10 (white), with $|(\operatorname{sn}(5,k))| > 10$ truncated to 10. White spots correspond to poles.

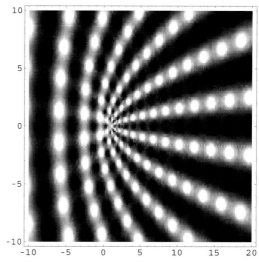

Figure 22.3.27: Density plot of $|\operatorname{sn}(10,k)|$ as a function of complex k^2, $-10 \le \Re(k^2) \le 20$, $-10 \le \Im(k^2) \le 10$. Grayscale, running from 0 (black) to 10 (white), with $|\operatorname{sn}(10,k)| > 10$ truncated to 10. White spots correspond to poles.

For corresponding density plots with arguments 20 and 30 see http://dlmf.nist.gov/22.3.iv.

22.4 Periods, Poles, and Zeros

22.4(i) Distribution

For each Jacobian function, Table 22.4.1 gives its periods in the z-plane in the left column, and the position of one of its poles in the second row. The other poles are at *congruent points*, which is the set of points obtained by making translations by $2mK + 2niK'$, where $m, n \in \mathbb{Z}$. For example, the poles of $\operatorname{sn}(z,k)$, abbreviated as sn in the following tables, are at $z = 2mK + (2n+1)iK'$.

Table 22.4.1: Periods and poles of Jacobian elliptic functions.

Periods	z-Poles			
	iK'	$K+iK'$	K	0
$4K, 2iK'$	sn	cd	dc	ns
$4K, 2K+2iK'$	cn	sd	nc	ds
$2K, 4iK'$	dn	nd	sc	cs

Three functions in the same column of Table 22.4.1 are *copolar*, and four functions in the same row are *cope-*riodic.

Table 22.4.2 displays the periods and zeros of the functions in the z-plane in a similar manner to Table 22.4.1. Again, one member of each congruent set of zeros appears in the second row; all others are generated by translations of the form $2mK + 2niK'$, where $m, n \in \mathbb{Z}$.

Table 22.4.2: Periods and zeros of Jacobian elliptic functions.

Periods	z-Zeros			
	0	K	$K+iK'$	iK'
$4K, 2iK'$	sn	cd	dc	ns
$4K, 2K+2iK'$	sd	cn	ds	nc
$2K, 4iK'$	sc	cs	dn	nd

Figure 22.4.1 illustrates the locations in the z-plane of the poles and zeros of the three principal Jacobian functions in the rectangle with vertices 0, $2K$, $2K + 2iK'$, $2iK'$. The other poles and zeros are at the congruent points.

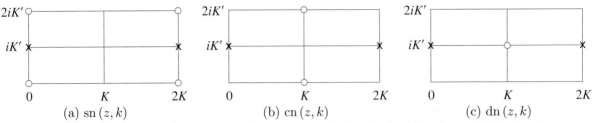

Figure 22.4.1: z-plane. Poles × × × and zeros ∘ ∘ ∘ of the principal Jacobian elliptic functions.

22.4(ii) Graphical Interpretation via Glaisher's Notation

Figure 22.4.2 depicts the *fundamental unit cell* in the z-plane, with vertices s $= 0$, c $= K$, d $= K + iK'$, n $= iK'$. The set of points $z = mK + niK'$, $m, n \in \mathbb{Z}$, comprise the *lattice* for the 12 Jacobian functions; all other lattice unit cells are generated by translation of the fundamental unit cell by $mK + niK'$, where again $m, n \in \mathbb{Z}$.

Figure 22.4.2: z-plane. Fundamental unit cell.

Using the p,q notation of (22.2.10), Figure 22.4.2 serves as a mnemonic for the poles, zeros, periods, and half-periods of the 12 Jacobian elliptic functions as follows. Let p,q be any two distinct letters from the set s,c,d,n which appear in counterclockwise orientation at the corners of all lattice unit cells. Then: (a) In any lattice unit cell pq(z, k) has a simple zero at $z =$ p and a simple pole at $z =$ q. (b) The difference between p and the nearest q is a half-period of pq(z, k). This half-period will be plus or minus a member of the triple $K, iK', K + iK'$; the other two members of this triple are quarter periods of pq(z, k).

22.4(iii) Translation by Half or Quarter Periods

See Table 22.4.3.

For example, $\operatorname{sn}(z + K, k) = \operatorname{cd}(z, k)$. (The modulus k is suppressed throughout the table.)

For the other nine functions see http://dlmf.nist.gov/22.4.iii.

Table 22.4.3: Half- or quarter-period shifts of variable for the Jacobian elliptic functions.

	u					
	$z + K$	$z + K + iK'$	$z + iK'$	$z + 2K$	$z + 2K + 2iK'$	$z + 2iK'$
sn u	cd z	k^{-1} dc z	k^{-1} ns z	$-$ sn z	$-$ sn z	sn z
cn u	$-k'$ sd z	$-ik'k^{-1}$ nc z	$-ik^{-1}$ ds z	$-$ cn z	cn z	$-$ cn z
dn u	k' nd z	ik' sc z	$-i$ cs z	dn z	$-$ dn z	$-$ dn z

22.5 Special Values

22.5(i) Special Values of z

Table 22.5.1 gives the value of each of the functions $\operatorname{sn}(z, k)$, $\operatorname{cn}(z, k)$, $\operatorname{dn}(z, k)$, together with its z-derivative (or at a pole, the residue), for values of z that are integer multiples of K, iK'. For example, at $z = K + iK'$, $\operatorname{sn}(z, k) = 1/k$, $d\operatorname{sn}(z, k)/dz = 0$. (The modulus k is suppressed throughout the table.)

For the other nine functions see http://dlmf.nist.gov/22.5.i.

22.5 Special Values

Table 22.5.1: Jacobian elliptic function values, together with derivatives or residues, for special values of the variable.

	0	K	$K+iK'$	iK'	$2K$	$2K+2iK'$	$2iK'$
$\operatorname{sn} z$	$0,1$	$1,0$	$1/k,0$	$\infty,1/k$	$0,-1$	$0,-1$	$0,1$
$\operatorname{cn} z$	$1,0$	$0,-k'$	$-ik'/k,0$	$\infty,-i/k$	$-1,0$	$1,0$	$-1,0$
$\operatorname{dn} z$	$1,0$	$k',0$	$0,ik'$	$\infty,-i$	$1,0$	$-1,0$	$-1,0$

Table 22.5.2 gives $\operatorname{sn}(z,k)$, $\operatorname{cn}(z,k)$, $\operatorname{dn}(z,k)$ for other special values of z. For example, $\operatorname{sn}\left(\tfrac{1}{2}K,k\right)=(1+k')^{-1/2}$. For the other nine functions ratios can be taken; compare (22.2.10).

Table 22.5.2: Other special values of Jacobian elliptic functions.

	$\tfrac{1}{2}K$	$\tfrac{1}{2}(K+iK')$	$\tfrac{1}{2}iK'$
$\operatorname{sn} z$	$(1+k')^{-1/2}$	$\left((1+k)^{1/2}+i(1-k)^{1/2}\right)/(2k)^{1/2}$	$ik^{-1/2}$
$\operatorname{cn} z$	$(k'/(1+k'))^{1/2}$	$(1-i)k'^{1/2}/(2k)^{1/2}$	$(1+k)^{1/2}k^{-1/2}$
$\operatorname{dn} z$	$k'^{1/2}$	$k'^{1/2}((1+k')^{1/2}-i(1-k')^{1/2})/2^{1/2}$	$(1+k)^{1/2}$

	$\tfrac{3}{2}K$	$\tfrac{3}{2}(K+iK')$	$\tfrac{3}{2}iK'$
$\operatorname{sn} z$	$(1+k')^{-1/2}$	$(1+i)((1+k)^{1/2}-i(1-k)^{1/2})/(2k^{1/2})$	$-ik^{-1/2}$
$\operatorname{cn} z$	$-(k'/(1+k'))^{1/2}$	$(1-i)k'^{1/2}/(2k)^{1/2}$	$-(1+k)^{1/2}k^{-1/2}$
$\operatorname{dn} z$	$k'^{1/2}$	$(-1+i)k'^{1/2}((1+k')^{1/2}+i(1-k')^{1/2})/2$	$-(1+k)^{1/2}$

22.5(ii) Limiting Values of k

If $k \to 0+$, then $K \to \pi/2$ and $K' \to \infty$; if $k \to 1-$, then $K \to \infty$ and $K' \to \pi/2$. In these cases the elliptic functions degenerate into elementary trigonometric and hyperbolic functions, respectively. See Tables 22.5.3 and 22.5.4.

Table 22.5.3: Limiting forms of Jacobian elliptic functions as $k \to 0$.

$\operatorname{sn}(z,k) \to \sin z$	$\operatorname{cd}(z,k) \to \cos z$	$\operatorname{dc}(z,k) \to \sec z$	$\operatorname{ns}(z,k) \to \csc z$
$\operatorname{cn}(z,k) \to \cos z$	$\operatorname{sd}(z,k) \to \sin z$	$\operatorname{nc}(z,k) \to \sec z$	$\operatorname{ds}(z,k) \to \csc z$
$\operatorname{dn}(z,k) \to 1$	$\operatorname{nd}(z,k) \to 1$	$\operatorname{sc}(z,k) \to \tan z$	$\operatorname{cs}(z,k) \to \cot z$

Table 22.5.4: Limiting forms of Jacobian elliptic functions as $k \to 1$.

$\operatorname{sn}(z,k) \to \tanh z$	$\operatorname{cd}(z,k) \to 1$	$\operatorname{dc}(z,k) \to 1$	$\operatorname{ns}(z,k) \to \coth z$
$\operatorname{cn}(z,k) \to \operatorname{sech} z$	$\operatorname{sd}(z,k) \to \sinh z$	$\operatorname{nc}(z,k) \to \cosh z$	$\operatorname{ds}(z,k) \to \operatorname{csch} z$
$\operatorname{dn}(z,k) \to \operatorname{sech} z$	$\operatorname{nd}(z,k) \to \cosh z$	$\operatorname{sc}(z,k) \to \sinh z$	$\operatorname{cs}(z,k) \to \operatorname{csch} z$

Expansions for K, K' as $k \to 0$ or 1 are given in §§19.5, 19.12.

For values of K, K' when $k^2 = \tfrac{1}{2}$ (lemniscatic case) see §23.5(iii), and for $k^2 = e^{i\pi/3}$ (equianharmonic case) see §23.5(v).

22.6 Elementary Identities

22.6(i) Sums of Squares

22.6.1
$$\operatorname{sn}^2(z,k) + \operatorname{cn}^2(z,k) = k^2 \operatorname{sn}^2(z,k) + \operatorname{dn}^2(z,k) = 1,$$

22.6.2 $\quad 1 + \operatorname{cs}^2(z,k) = k^2 + \operatorname{ds}^2(z,k) = \operatorname{ns}^2(z,k),$

22.6.3 $\quad k'^2 \operatorname{sc}^2(z,k) + 1 = \operatorname{dc}^2(z,k) = k'^2 \operatorname{nc}^2(z,k) + k^2,$

22.6.4
$$-k^2 k'^2 \operatorname{sd}^2(z,k) = k^2 (\operatorname{cd}^2(z,k) - 1) = k'^2 (1 - \operatorname{nd}^2(z,k)).$$

22.6(ii) Double Argument

22.6.5 $\quad \operatorname{sn}(2z,k) = \dfrac{2 \operatorname{sn}(z,k) \operatorname{cn}(z,k) \operatorname{dn}(z,k)}{1 - k^2 \operatorname{sn}^4(z,k)},$

22.6.6
$$\operatorname{cn}(2z,k) = \frac{\operatorname{cn}^2(z,k) - \operatorname{sn}^2(z,k) \operatorname{dn}^2(z,k)}{1 - k^2 \operatorname{sn}^4(z,k)}$$
$$= \frac{\operatorname{cn}^4(z,k) - k'^2 \operatorname{sn}^4(z,k)}{1 - k^2 \operatorname{sn}^4(z,k)},$$

22.6.7
$$\operatorname{dn}(2z,k) = \frac{\operatorname{dn}^2(z,k) - k^2 \operatorname{sn}^2(z,k) \operatorname{dn}^2(z,k)}{1 - k^2 \operatorname{sn}^4(z,k)}$$
$$= \frac{\operatorname{dn}^4(z,k) + k^2 k'^2 \operatorname{sn}^4(z,k)}{1 - k^2 \operatorname{sn}^4(z,k)}.$$

For corresponding results for the other nine functions see http://dlmf.nist.gov/22.6.ii. See also Carlson (2004).

22.6.17 $\quad \dfrac{1 - \operatorname{cn}(2z,k)}{1 + \operatorname{cn}(2z,k)} = \dfrac{\operatorname{sn}^2(z,k) \operatorname{dn}^2(z,k)}{\operatorname{cn}^2(z,k)},$

22.6.18 $\quad \dfrac{1 - \operatorname{dn}(2z,k)}{1 + \operatorname{dn}(2z,k)} = \dfrac{k^2 \operatorname{sn}^2(z,k) \operatorname{cn}^2(z,k)}{\operatorname{dn}^2(z,k)}.$

22.6(iii) Half Argument

22.6.19
$$\operatorname{sn}^2\left(\tfrac{1}{2}z,k\right) = \frac{1 - \operatorname{cn}(z,k)}{1 + \operatorname{dn}(z,k)} = \frac{1 - \operatorname{dn}(z,k)}{k^2(1 + \operatorname{cn}(z,k))}$$
$$= \frac{\operatorname{dn}(z,k) - k^2 \operatorname{cn}(z,k) - k'^2}{k^2(\operatorname{dn}(z,k) - \operatorname{cn}(z,k))},$$

22.6.20
$$\operatorname{cn}^2\left(\tfrac{1}{2}z,k\right) = \frac{-k'^2 + \operatorname{dn}(z,k) + k^2 \operatorname{cn}(z,k)}{k^2(1 + \operatorname{cn}(z,k))}$$
$$= \frac{k'^2(1 - \operatorname{dn}(z,k))}{k^2(\operatorname{dn}(z,k) - \operatorname{cn}(z,k))}$$
$$= \frac{k'^2(1 + \operatorname{cn}(z,k))}{k'^2 + \operatorname{dn}(z,k) - k^2 \operatorname{cn}(z,k)},$$

22.6.21
$$\operatorname{dn}^2\left(\tfrac{1}{2}z,k\right) = \frac{k^2 \operatorname{cn}(z,k) + \operatorname{dn}(z,k) + k'^2}{1 + \operatorname{dn}(z,k)}$$
$$= \frac{k'^2(1 - \operatorname{cn}(z,k))}{\operatorname{dn}(z,k) - \operatorname{cn}(z,k)}$$
$$= \frac{k'^2(1 + \operatorname{dn}(z,k))}{k'^2 + \operatorname{dn}(z,k) - k^2 \operatorname{cn}(z,k)}.$$

If {p,q,r} is any permutation of {c,d,n}, then

22.6.22
$$\operatorname{pq}^2\left(\tfrac{1}{2}z,k\right) = \frac{\operatorname{ps}(z,k) + \operatorname{rs}(z,k)}{\operatorname{qs}(z,k) + \operatorname{rs}(z,k)}$$
$$= \frac{\operatorname{pq}(z,k) + \operatorname{rq}(z,k)}{1 + \operatorname{rq}(z,k)} = \frac{\operatorname{pr}(z,k) + 1}{\operatorname{qr}(z,k) + 1}.$$

For (22.6.22) and similar results, see Carlson (2004).

22.6(iv) Rotation of Argument (Jacobi's Imaginary Transformation)

Table 22.6.1: Jacobi's imaginary transformation of Jacobian elliptic functions.

$\operatorname{sn}(iz,k) = i \operatorname{sc}(z,k')$	$\operatorname{dc}(iz,k) = \operatorname{dn}(z,k')$
$\operatorname{cn}(iz,k) = \operatorname{nc}(z,k')$	$\operatorname{nc}(iz,k) = \operatorname{cn}(z,k')$
$\operatorname{dn}(iz,k) = \operatorname{dc}(z,k')$	$\operatorname{sc}(iz,k) = i \operatorname{sn}(z,k')$
$\operatorname{cd}(iz,k) = \operatorname{nd}(z,k')$	$\operatorname{ns}(iz,k) = -i \operatorname{cs}(z,k')$
$\operatorname{sd}(iz,k) = i \operatorname{sd}(z,k')$	$\operatorname{ds}(iz,k) = -i \operatorname{ds}(z,k')$
$\operatorname{nd}(iz,k) = \operatorname{cd}(z,k')$	$\operatorname{cs}(iz,k) = -i \operatorname{ns}(z,k')$

22.6(v) Change of Modulus

See §22.17.

22.7 Landen Transformations

22.7(i) Descending Landen Transformation

With

22.7.1 $\quad k_1 = \dfrac{1 - k'}{1 + k'},$

22.7.2 $\quad \operatorname{sn}(z,k) = \dfrac{(1 + k_1) \operatorname{sn}(z/(1+k_1), k_1)}{1 + k_1 \operatorname{sn}^2(z/(1+k_1), k_1)},$

22.7.3
$$\operatorname{cn}(z,k) = \frac{\operatorname{cn}(z/(1+k_1), k_1) \operatorname{dn}(z/(1+k_1), k_1)}{1 + k_1 \operatorname{sn}^2(z/(1+k_1), k_1)},$$

22.7.4 $\quad \operatorname{dn}(z,k) = \dfrac{\operatorname{dn}^2(z/(1+k_1), k_1) - (1 - k_1)}{1 + k_1 - \operatorname{dn}^2(z/(1+k_1), k_1)}.$

22.7(ii) Ascending Landen Transformation

With

$$22.7.5 \qquad k_2 = \frac{2\sqrt{k}}{1+k}, \quad k_2' = \frac{1-k}{1+k},$$

22.7.6
$$\operatorname{sn}(z,k) = \frac{(1+k_2')\operatorname{sn}(z/(1+k_2'),k_2)\operatorname{cn}(z/(1+k_2'),k_2)}{\operatorname{dn}(z/(1+k_2'),k_2)},$$

$$22.7.7 \quad \operatorname{cn}(z,k) = \frac{(1+k_2')(\operatorname{dn}^2(z/(1+k_2'),k_2) - k_2')}{k_2^2 \operatorname{dn}(z/(1+k_2'),k_2)},$$

$$22.7.8 \quad \operatorname{dn}(z,k) = \frac{(1-k_2')(\operatorname{dn}^2(z/(1+k_2'),k_2) + k_2')}{k_2^2 \operatorname{dn}(z/(1+k_2'),k_2)}.$$

22.7(iii) Generalized Landen Transformations

See Khare and Sukhatme (2004).

22.8 Addition Theorems

22.8(i) Sum of Two Arguments

For $u, v \in \mathbb{C}$, and with the common modulus k suppressed:

$$22.8.1 \quad \operatorname{sn}(u+v) = \frac{\operatorname{sn} u \operatorname{cn} v \operatorname{dn} v + \operatorname{sn} v \operatorname{cn} u \operatorname{dn} u}{1 - k^2 \operatorname{sn}^2 u \operatorname{sn}^2 v},$$

$$22.8.2 \quad \operatorname{cn}(u+v) = \frac{\operatorname{cn} u \operatorname{cn} v - \operatorname{sn} u \operatorname{dn} u \operatorname{sn} v \operatorname{dn} v}{1 - k^2 \operatorname{sn}^2 u \operatorname{sn}^2 v},$$

$$22.8.3 \quad \operatorname{dn}(u+v) = \frac{\operatorname{dn} u \operatorname{dn} v - k^2 \operatorname{sn} u \operatorname{cn} u \operatorname{sn} v \operatorname{cn} v}{1 - k^2 \operatorname{sn}^2 u \operatorname{sn}^2 v}.$$

See also Carlson (2004).

For the other nine functions see `http://dlmf.nist.gov/22.8.i`.

22.8(ii) Alternative Forms for Sum of Two Arguments

For $u, v \in \mathbb{C}$, and with the common modulus k suppressed:

$$22.8.13 \quad \operatorname{sn}(u+v) = \frac{\operatorname{sn}^2 u - \operatorname{sn}^2 v}{\operatorname{sn} u \operatorname{cn} v \operatorname{dn} v - \operatorname{sn} v \operatorname{cn} u \operatorname{dn} u},$$

$$22.8.14 \quad \operatorname{sn}(u+v) = \frac{\operatorname{sn} u \operatorname{cn} u \operatorname{dn} v + \operatorname{sn} v \operatorname{cn} v \operatorname{dn} u}{\operatorname{cn} u \operatorname{cn} v + \operatorname{sn} u \operatorname{dn} u \operatorname{sn} v \operatorname{dn} v},$$

$$22.8.15 \quad \operatorname{cn}(u+v) = \frac{\operatorname{sn} u \operatorname{cn} u \operatorname{dn} v - \operatorname{sn} v \operatorname{cn} v \operatorname{dn} u}{\operatorname{sn} u \operatorname{cn} v \operatorname{dn} v - \operatorname{sn} v \operatorname{cn} u \operatorname{dn} u},$$

$$22.8.16 \quad \operatorname{cn}(u+v) = \frac{1 - \operatorname{sn}^2 u - \operatorname{sn}^2 v + k^2 \operatorname{sn}^2 u \operatorname{sn}^2 v}{\operatorname{cn} u \operatorname{cn} v + \operatorname{sn} u \operatorname{dn} u \operatorname{sn} v \operatorname{dn} v},$$

$$22.8.17 \quad \operatorname{dn}(u+v) = \frac{\operatorname{sn} u \operatorname{cn} v \operatorname{dn} u - \operatorname{sn} v \operatorname{cn} u \operatorname{dn} v}{\operatorname{sn} u \operatorname{cn} v \operatorname{dn} v - \operatorname{sn} v \operatorname{cn} u \operatorname{dn} u},$$

$$22.8.18 \quad \operatorname{dn}(u+v) = \frac{\operatorname{cn} u \operatorname{dn} u \operatorname{cn} v \operatorname{dn} v + k'^2 \operatorname{sn} u \operatorname{sn} v}{\operatorname{cn} u \operatorname{cn} v + \operatorname{sn} u \operatorname{dn} u \operatorname{sn} v \operatorname{dn} v}.$$

See also Carlson (2004).

22.8(iii) Special Relations Between Arguments

In the following equations the common modulus k is again suppressed.

Let

$$22.8.19 \qquad z_1 + z_2 + z_3 + z_4 = 0.$$

Then

$$22.8.20 \quad \begin{vmatrix} \operatorname{sn} z_1 & \operatorname{cn} z_1 & \operatorname{dn} z_1 & 1 \\ \operatorname{sn} z_2 & \operatorname{cn} z_2 & \operatorname{dn} z_2 & 1 \\ \operatorname{sn} z_3 & \operatorname{cn} z_3 & \operatorname{dn} z_3 & 1 \\ \operatorname{sn} z_4 & \operatorname{cn} z_4 & \operatorname{dn} z_4 & 1 \end{vmatrix} = 0,$$

and

$$22.8.21 \quad \begin{aligned} & k'^2 - k'^2 k^2 \operatorname{sn} z_1 \operatorname{sn} z_2 \operatorname{sn} z_3 \operatorname{sn} z_4 \\ & + k^2 \operatorname{cn} z_1 \operatorname{cn} z_2 \operatorname{cn} z_3 \operatorname{cn} z_4 \\ & - \operatorname{dn} z_1 \operatorname{dn} z_2 \operatorname{dn} z_3 \operatorname{dn} z_4 = 0. \end{aligned}$$

A geometric interpretation of (22.8.20) analogous to that of (23.10.5) is given in Whittaker and Watson (1927, p. 530).

Next, let

$$22.8.22 \qquad z_1 + z_2 + z_3 + z_4 = 2K(k).$$

Then

$$22.8.23 \quad \begin{vmatrix} \operatorname{sn} z_1 \operatorname{cn} z_1 & \operatorname{cn} z_1 \operatorname{dn} z_1 & \operatorname{cn} z_1 & \operatorname{dn} z_1 \\ \operatorname{sn} z_2 \operatorname{cn} z_2 & \operatorname{cn} z_2 \operatorname{dn} z_2 & \operatorname{cn} z_2 & \operatorname{dn} z_2 \\ \operatorname{sn} z_3 \operatorname{cn} z_3 & \operatorname{cn} z_3 \operatorname{dn} z_3 & \operatorname{cn} z_3 & \operatorname{dn} z_3 \\ \operatorname{sn} z_4 \operatorname{cn} z_4 & \operatorname{cn} z_4 \operatorname{dn} z_4 & \operatorname{cn} z_4 & \operatorname{dn} z_4 \end{vmatrix} = 0.$$

For these and related identities see Copson (1935, pp. 415–416).

If sums/differences of the z_j's are rational multiples of $K(k)$, then further relations follow. For instance, if

$$22.8.24 \qquad z_1 - z_2 = z_2 - z_3 = \tfrac{2}{3} K(k),$$

then

$$22.8.25 \quad \frac{(\operatorname{dn} z_2 + \operatorname{dn} z_3)(\operatorname{dn} z_3 + \operatorname{dn} z_1)(\operatorname{dn} z_1 + \operatorname{dn} z_2)}{\operatorname{dn} z_1 + \operatorname{dn} z_2 + \operatorname{dn} z_3}$$

is independent of z_1, z_2, z_3. Similarly, if

$$22.8.26 \quad z_1 - z_2 = z_2 - z_3 = z_3 - z_4 = \tfrac{1}{2} K(k),$$

then

$$22.8.27 \qquad \operatorname{dn} z_1 \operatorname{dn} z_3 = \operatorname{dn} z_2 \operatorname{dn} z_4 = k'.$$

Greenhill (1959, pp. 121–130) reviews these results in terms of the geometric *poristic polygon* constructions of Poncelet. Generalizations are given in §22.9.

22.9 Cyclic Identities

22.9(i) Notation

The following notation is a generalization of that of Khare and Sukhatme (2002).

Throughout this subsection m and p are positive integers with $1 \leq m \leq p$.

22.9.1 $\quad s_{m,p}^{(2)} = \operatorname{sn}\left(z + 2p^{-1}(m-1)K(k), k\right),$

22.9.2 $\quad c_{m,p}^{(2)} = \operatorname{cn}\left(z + 2p^{-1}(m-1)K(k), k\right),$

22.9.3 $\quad d_{m,p}^{(2)} = \operatorname{dn}\left(z + 2p^{-1}(m-1)K(k), k\right),$

22.9.4 $\quad s_{m,p}^{(4)} = \operatorname{sn}\left(z + 4p^{-1}(m-1)K(k), k\right),$

22.9.5 $\quad c_{m,p}^{(4)} = \operatorname{cn}\left(z + 4p^{-1}(m-1)K(k), k\right),$

22.9.6 $\quad d_{m,p}^{(4)} = \operatorname{dn}\left(z + 4p^{-1}(m-1)K(k), k\right).$

In the remainder of this section the *rank* of an identity is the maximum number of elliptic function factors in each term in the identity. The value of p determines the number of *points* in the identity. The argument z is suppressed in the above notation, as all cyclic identities are independent of z.

22.9(ii) Typical Identities of Rank 2

In this subsection $1 \leq m \leq p$ and $1 \leq n \leq p$.

Three Points

With

22.9.7 $\quad \kappa = \operatorname{dn}(2K(k)/3, k),$

22.9.8 $\quad s_{1,2}^{(4)} s_{2,2}^{(4)} + s_{2,2}^{(4)} s_{3,2}^{(4)} + s_{3,2}^{(4)} s_{1,2}^{(4)} = \dfrac{\kappa^2 - 1}{k^2},$

22.9.9 $\quad c_{1,2}^{(4)} c_{2,2}^{(4)} + c_{2,2}^{(4)} c_{3,2}^{(4)} + c_{3,2}^{(4)} c_{1,2}^{(4)} = -\dfrac{\kappa(\kappa+2)}{(1+\kappa)^2},$

22.9.10 $\quad \begin{aligned} & d_{1,2}^{(2)} d_{2,2}^{(2)} + d_{2,2}^{(2)} d_{3,2}^{(2)} + d_{3,2}^{(2)} d_{1,2}^{(2)} \\ & = d_{1,2}^{(4)} d_{2,2}^{(4)} + d_{2,2}^{(4)} d_{3,2}^{(4)} + d_{3,2}^{(4)} d_{1,2}^{(4)} = \kappa(\kappa+2). \end{aligned}$

These identities are *cyclic* in the sense that each of the indices m, n in the first product of, for example, the form $s_{m,2}^{(4)} s_{n,2}^{(4)}$ are *simultaneously permuted* in the cyclic order: $m \to m+1 \to m+2 \to \cdots p \to 1 \to 2 \to \cdots m-1$; $n \to n+1 \to n+2 \to \cdots p \to 1 \to 2 \to \cdots n-1$. Many of the identities that follow also have this property.

22.9(iii) Typical Identities of Rank 3

Two Points

22.9.11 $\quad \left(d_{1,2}^{(2)}\right)^2 d_{2,2}^{(2)} \pm \left(d_{2,2}^{(2)}\right)^2 d_{1,2}^{(2)} = k'\left(d_{1,2}^{(2)} \pm d_{2,2}^{(2)}\right),$

22.9.12 $\quad c_{1,2}^{(2)} s_{1,2}^{(2)} d_{2,2}^{(2)} + c_{2,2}^{(2)} s_{2,2}^{(2)} d_{1,2}^{(2)} = 0.$

Three Points

With κ defined as in (22.9.7),

22.9.13 $\quad s_{1,3}^{(4)} s_{2,3}^{(4)} s_{3,3}^{(4)} = -\dfrac{1}{1-\kappa^2}\left(s_{1,3}^{(4)} + s_{2,3}^{(4)} + s_{3,3}^{(4)}\right),$

22.9.14 $\quad c_{1,3}^{(4)} c_{2,3}^{(4)} c_{3,3}^{(4)} = \dfrac{\kappa^2}{1-\kappa^2}\left(c_{1,3}^{(4)} + c_{2,3}^{(4)} + c_{3,3}^{(4)}\right),$

22.9.15 $\quad \begin{aligned} & d_{1,3}^{(2)} d_{2,3}^{(2)} d_{3,3}^{(2)} \\ & = \dfrac{\kappa^2 + k^2 - 1}{1-\kappa^2}\left(d_{1,3}^{(2)} + d_{2,3}^{(2)} + d_{3,3}^{(2)}\right), \end{aligned}$

22.9.16 $\quad \begin{aligned} & s_{1,3}^{(4)} c_{2,3}^{(4)} c_{3,3}^{(4)} + s_{2,3}^{(4)} c_{3,3}^{(4)} c_{1,3}^{(4)} + s_{3,3}^{(4)} c_{1,3}^{(4)} c_{2,3}^{(4)} \\ & = \dfrac{\kappa(\kappa+2)}{1-\kappa^2}\left(s_{1,3}^{(4)} + s_{2,3}^{(4)} + s_{3,3}^{(4)}\right). \end{aligned}$

Four Points

22.9.17 $\quad \begin{aligned} & d_{1,4}^{(2)} d_{2,4}^{(2)} d_{3,4}^{(2)} \pm d_{2,4}^{(2)} d_{3,4}^{(2)} d_{4,4}^{(2)} + d_{3,4}^{(2)} d_{4,4}^{(2)} d_{1,4}^{(2)} \pm d_{4,4}^{(2)} d_{1,4}^{(2)} d_{2,4}^{(2)} \\ & = k'\left(\pm d_{1,4}^{(2)} + d_{2,4}^{(2)} \pm d_{3,4}^{(2)} + d_{4,4}^{(2)}\right), \end{aligned}$

22.9.18 $\quad \begin{aligned} & \left(d_{1,4}^{(2)}\right)^2 d_{3,4}^{(2)} \pm \left(d_{2,4}^{(2)}\right)^2 d_{4,4}^{(2)} + \left(d_{3,4}^{(2)}\right)^2 d_{1,4}^{(2)} \\ & \pm \left(d_{4,4}^{(2)}\right)^2 d_{2,4}^{(2)} = k'\left(d_{1,4}^{(2)} \pm d_{2,4}^{(2)} + d_{3,4}^{(2)} \pm d_{4,4}^{(2)}\right), \end{aligned}$

22.9.19 $\quad \begin{aligned} & c_{1,4}^{(2)} s_{1,4}^{(2)} d_{3,4}^{(2)} + c_{3,4}^{(2)} s_{3,4}^{(2)} d_{1,4}^{(2)} \\ & = c_{2,4}^{(2)} s_{2,4}^{(2)} d_{4,4}^{(2)} + c_{4,4}^{(2)} s_{4,4}^{(2)} d_{2,4}^{(2)} = 0. \end{aligned}$

For identities of rank 4 and higher see http://dlmf.nist.gov/22.9.iv.

22.10 Maclaurin Series

22.10(i) Maclaurin Series in z

Initial terms are given by

22.10.1
$$\operatorname{sn}(z,k) = z - \left(1+k^2\right)\dfrac{z^3}{3!} + \left(1+14k^2+k^4\right)\dfrac{z^5}{5!} - \left(1+135k^2+135k^4+k^6\right)\dfrac{z^7}{7!} + O(z^9),$$

22.10.2
$$\operatorname{cn}(z,k) = 1 - \dfrac{z^2}{2!} + \left(1+4k^2\right)\dfrac{z^4}{4!} - \left(1+44k^2+16k^4\right)\dfrac{z^6}{6!} + O(z^8),$$

22.10.3
$$\operatorname{dn}(z,k) = 1 - k^2\dfrac{z^2}{2!} + k^2\left(4+k^2\right)\dfrac{z^4}{4!} - k^2\left(16+44k^2+k^4\right)\dfrac{z^6}{6!} + O(z^8).$$

Further terms may be derived by substituting in the differential equations (22.13.13), (22.13.14), (22.13.15). The full expansions converge when $|z| < \min(K(k), K'(k))$.

22.10(ii) Maclaurin Series in k and k'

Initial terms are given by

22.10.4
$$\operatorname{sn}(z,k) = \sin z - \frac{k^2}{4}(z - \sin z \cos z)\cos z + O(k^4),$$

22.10.5
$$\operatorname{cn}(z,k) = \cos z + \frac{k^2}{4}(z - \sin z \cos z)\sin z + O(k^4),$$

22.10.6
$$\operatorname{dn}(z,k) = 1 - \frac{k^2}{2}\sin^2 z + O(k^4),$$

22.10.7
$$\operatorname{sn}(z,k) = \tanh z - \frac{k'^2}{4}(z - \sinh z \cosh z)\operatorname{sech}^2 z + O(k'^4),$$

22.10.8
$$\operatorname{cn}(z,k) = \operatorname{sech} z + \frac{k'^2}{4}(z - \sinh z \cosh z)\tanh z \operatorname{sech} z + O(k'^4),$$

22.10.9
$$\operatorname{dn}(z,k) = \operatorname{sech} z + \frac{k'^2}{4}(z + \sinh z \cosh z)\tanh z \operatorname{sech} z + O(k'^4).$$

Further terms may be derived from the differential equations (22.13.13), (22.13.14), (22.13.15), or from the integral representations of the inverse functions in §22.15(ii). The radius of convergence is the distance to the origin from the nearest pole in the complex k-plane in the case of (22.10.4)–(22.10.6), or complex k'-plane in the case of (22.10.7)–(22.10.9); see §22.17.

22.11 Fourier and Hyperbolic Series

Throughout this section q and ζ are defined as in §22.2. If $q\exp(2|\Im\zeta|) < 1$, then

22.11.1
$$\operatorname{sn}(z,k) = \frac{2\pi}{Kk}\sum_{n=0}^{\infty}\frac{q^{n+\frac{1}{2}}\sin((2n+1)\zeta)}{1-q^{2n+1}},$$

22.11.2
$$\operatorname{cn}(z,k) = \frac{2\pi}{Kk}\sum_{n=0}^{\infty}\frac{q^{n+\frac{1}{2}}\cos((2n+1)\zeta)}{1+q^{2n+1}},$$

22.11.3
$$\operatorname{dn}(z,k) = \frac{\pi}{2K} + \frac{2\pi}{K}\sum_{n=1}^{\infty}\frac{q^n\cos(2n\zeta)}{1+q^{2n}}.$$

For the other nine functions see http://dlmf.nist.gov/22.11.

Next, with $E = E(k)$ denoting the complete elliptic integral of the second kind (§19.2(ii)) and $q\exp(2|\Im\zeta|) < 1$,

22.11.13
$$\operatorname{sn}^2(z,k) = \frac{1}{k^2}\left(1 - \frac{E}{K}\right) - \frac{2\pi^2}{k^2K^2}\sum_{n=1}^{\infty}\frac{nq^n}{1-q^{2n}}\cos(2n\zeta).$$

Similar expansions for $\operatorname{cn}^2(z,k)$ and $\operatorname{dn}^2(z,k)$ follow immediately from (22.6.1).

For further Fourier series see Oberhettinger (1973, pp. 23–27).

A related hyperbolic series is

22.11.14
$$k^2\operatorname{sn}^2(z,k) = \frac{E'}{K'} - \left(\frac{\pi}{2K'}\right)^2\sum_{n=-\infty}^{\infty}\left(\operatorname{sech}^2\left(\frac{\pi}{2K'}(z-2nK)\right)\right),$$

where $E' = E'(k)$ is defined by §19.2.9. Again, similar expansions for $\operatorname{cn}^2(z,k)$ and $\operatorname{dn}^2(z,k)$ may be derived via (22.6.1). See Dunne and Rao (2000).

22.12 Expansions in Other Trigonometric Series and Doubly-Infinite Partial Fractions: Eisenstein Series

With $t \in \mathbb{C}$ and

22.12.1
$$\tau = iK'(k)/K(k),$$

22.12.2
$$2Kk\operatorname{sn}(2Kt,k) = \sum_{n=-\infty}^{\infty}\frac{\pi}{\sin(\pi(t-(n+\frac{1}{2})\tau))}$$
$$= \sum_{n=-\infty}^{\infty}\left(\sum_{m=-\infty}^{\infty}\frac{(-1)^m}{t-m-(n+\frac{1}{2})\tau}\right),$$

22.12.3
$$2iKk\operatorname{cn}(2Kt,k) = \sum_{n=-\infty}^{\infty}\frac{(-1)^n\pi}{\sin(\pi(t-(n+\frac{1}{2})\tau))}$$
$$= \sum_{n=-\infty}^{\infty}\left(\sum_{m=-\infty}^{\infty}\frac{(-1)^{m+n}}{t-m-(n+\frac{1}{2})\tau}\right),$$

22.12.4
$$2iK\operatorname{dn}(2Kt,k)$$
$$= \lim_{N\to\infty}\sum_{n=-N}^{N}(-1)^n\frac{\pi}{\tan(\pi(t-(n+\frac{1}{2})\tau))}$$
$$= \lim_{N\to\infty}\sum_{n=-N}^{N}(-1)^n\left(\lim_{M\to\infty}\sum_{m=-M}^{M}\frac{1}{t-m-(n+\frac{1}{2})\tau}\right).$$

The double sums in (22.12.2)–(22.12.4) are convergent but not absolutely convergent, hence the order of the summations is important. Compare §20.5(iii).

For corresponding expansions for the subsidiary functions see http://dlmf.nist.gov/22.12.

22.13 Derivatives and Differential Equations

22.13(i) Derivatives

Table 22.13.1: Derivatives of Jacobian elliptic functions with respect to variable.

$\frac{d}{dz}(\operatorname{sn} z) = \operatorname{cn} z \operatorname{dn} z$	$\frac{d}{dz}(\operatorname{dc} z) = k'^2 \operatorname{sc} z \operatorname{nc} z$
$\frac{d}{dz}(\operatorname{cn} z) = -\operatorname{sn} z \operatorname{dn} z$	$\frac{d}{dz}(\operatorname{nc} z) = \operatorname{sc} z \operatorname{dc} z$
$\frac{d}{dz}(\operatorname{dn} z) = -k^2 \operatorname{sn} z \operatorname{cn} z$	$\frac{d}{dz}(\operatorname{sc} z) = \operatorname{dc} z \operatorname{nc} z$
$\frac{d}{dz}(\operatorname{cd} z) = -k'^2 \operatorname{sd} z \operatorname{nd} z$	$\frac{d}{dz}(\operatorname{ns} z) = -\operatorname{ds} z \operatorname{cs} z$
$\frac{d}{dz}(\operatorname{sd} z) = \operatorname{cd} z \operatorname{nd} z$	$\frac{d}{dz}(\operatorname{ds} z) = -\operatorname{cs} z \operatorname{ns} z$
$\frac{d}{dz}(\operatorname{nd} z) = k^2 \operatorname{sd} z \operatorname{cd} z$	$\frac{d}{dz}(\operatorname{cs} z) = -\operatorname{ns} z \operatorname{ds} z$

Note that each derivative in Table 22.13.1 is a constant multiple of the product of the corresponding copolar functions. (The modulus k is suppressed throughout the table.)

For alternative, and symmetric, formulations of these results see Carlson (2004, 2006a).

22.13(ii) First-Order Differential Equations

22.13.1
$$\left(\frac{d}{dz} \operatorname{sn}(z,k)\right)^2 = \left(1 - \operatorname{sn}^2(z,k)\right)\left(1 - k^2 \operatorname{sn}^2(z,k)\right),$$

22.13.2
$$\left(\frac{d}{dz} \operatorname{cn}(z,k)\right)^2 = \left(1 - \operatorname{cn}^2(z,k)\right)\left(k'^2 + k^2 \operatorname{cn}^2(z,k)\right),$$

22.13.3
$$\left(\frac{d}{dz} \operatorname{dn}(z,k)\right)^2 = \left(1 - \operatorname{dn}^2(z,k)\right)\left(\operatorname{dn}^2(z,k) - k'^2\right).$$

For corresponding equations for the subsidiary functions see http://dlmf.nist.gov/22.13.ii.

For alternative, and symmetric, formulations of these results see Carlson (2006a).

22.13(iii) Second-Order Differential Equations

22.13.13
$$\frac{d^2}{dz^2} \operatorname{sn}(z,k) = -(1+k^2)\operatorname{sn}(z,k) + 2k^2 \operatorname{sn}^3(z,k),$$

22.13.14
$$\frac{d^2}{dz^2} \operatorname{cn}(z,k) = -(k'^2 - k^2)\operatorname{cn}(z,k) - 2k^2 \operatorname{cn}^3(z,k),$$

22.13.15
$$\frac{d^2}{dz^2} \operatorname{dn}(z,k) = (1+k'^2)\operatorname{dn}(z,k) - 2\operatorname{dn}^3(z,k).$$

For corresponding equations for the subsidiary functions see http://dlmf.nist.gov/22.13.iii.

For alternative, and symmetric, formulations of these results see Carlson (2006a).

22.14 Integrals

22.14(i) Indefinite Integrals of Jacobian Elliptic Functions

With $x \in \mathbb{R}$,

22.14.1
$$\int \operatorname{sn}(x,k)\,dx = k^{-1}\ln(\operatorname{dn}(x,k) - k\operatorname{cn}(x,k)),$$

22.14.2
$$\int \operatorname{cn}(x,k)\,dx = k^{-1}\operatorname{Arccos}(\operatorname{dn}(x,k)),$$

22.14.3
$$\int \operatorname{dn}(x,k)\,dx = \operatorname{Arcsin}(\operatorname{sn}(x,k)) = \operatorname{am}(x,k).$$

The branches of the inverse trigonometric functions are chosen so that they are continuous. See §22.16(i) for $\operatorname{am}(z,k)$.

For alternative, and symmetric, formulations of these results see Carlson (2006a).

For the corresponding results for the subsidiary functions see http://dlmf.nist.gov/22.14.i.

22.14(ii) Indefinite Integrals of Powers of Jacobian Elliptic Functions

See §22.16(ii). The indefinite integral of the 3rd power of a Jacobian function can be expressed as an elementary function of Jacobian functions and a product of Jacobian functions. The indefinite integral of a 4th power can be expressed as a complete elliptic integral, a polynomial in Jacobian functions, and the integration variable. See Lawden (1989, pp. 87–88). See also Gradshteyn and Ryzhik (2000, pp. 618–619) and Carlson (2006a).

For indefinite integrals of squares and products of even powers of Jacobian functions in terms of symmetric elliptic integrals, see Carlson (2006b).

22.14(iii) Other Indefinite Integrals

In (22.14.13)–(22.14.15), $0 < x < 2K$.

22.14.13
$$\int \frac{dx}{\operatorname{sn}(x,k)} = \ln\left(\frac{\operatorname{sn}(x,k)}{\operatorname{cn}(x,k) + \operatorname{dn}(x,k)}\right),$$

22.14.14
$$\int \frac{\operatorname{cn}(x,k)\,dx}{\operatorname{sn}(x,k)} = \frac{1}{2}\ln\left(\frac{1 - \operatorname{dn}(x,k)}{1 + \operatorname{dn}(x,k)}\right),$$

22.14.15
$$\int \frac{\operatorname{cn}(x,k)\,dx}{\operatorname{sn}^2(x,k)} = -\frac{\operatorname{dn}(x,k)}{\operatorname{sn}(x,k)}.$$

For additional results see Gradshteyn and Ryzhik (2000, pp. 619–622) and Lawden (1989, Chapter 3).

22.14(iv) Definite Integrals

22.14.16 $\quad \int_0^{K(k)} \ln(\operatorname{sn}(t,k))\,dt = -\tfrac{1}{4}K'(k) - \tfrac{1}{2}K(k)\ln k,$

22.14.17 $\quad \int_0^{K(k)} \ln(\operatorname{cn}(t,k))\,dt = -\tfrac{1}{4}K'(k) + \tfrac{1}{2}K(k)\ln(k'/k),$

22.14.18 $\quad \int_0^{K(k)} \ln(\operatorname{dn}(t,k))\,dt = \tfrac{1}{2}K(k)\ln k'.$

Corresponding results for the subsidiary functions follow by subtraction; compare (22.2.10).

22.15 Inverse Functions

22.15(i) Definitions

The inverse Jacobian elliptic functions can be defined in an analogous manner to the inverse trigonometric functions (§4.23). With real variables, the solutions of the equations

22.15.1 $\qquad \operatorname{sn}(\xi, k) = x, \qquad -1 \le x \le 1,$

22.15.2 $\qquad \operatorname{cn}(\eta, k) = x, \qquad -1 \le x \le 1,$

22.15.3 $\qquad \operatorname{dn}(\zeta, k) = x, \qquad k' \le x \le 1,$

are denoted respectively by

22.15.4
$$\xi = \operatorname{arcsn}(x,k), \quad \eta = \operatorname{arccn}(x,k), \quad \zeta = \operatorname{arcdn}(x,k).$$

Each of these inverse functions is multivalued. The *principal values* satisfy

22.15.5 $\qquad -K \le \operatorname{arcsn}(x,k) \le K,$

22.15.6 $\qquad 0 \le \operatorname{arccn}(x,k) \le 2K,$

22.15.7 $\qquad 0 \le \operatorname{arcdn}(x,k) \le K,$

and *unless stated otherwise* it is assumed that the inverse functions assume their principal values. The general solutions of (22.15.1), (22.15.2), (22.15.3) are, respectively,

22.15.8 $\qquad \xi = (-1)^m \operatorname{arcsn}(x,k) + 2mK,$

22.15.9 $\qquad \eta = \pm \operatorname{arccn}(x,k) + 4mK,$

22.15.10 $\qquad \zeta = \pm \operatorname{arcdn}(x,k) + 2mK,$

where $m \in \mathbb{Z}$.

Equations (22.15.1) and (22.15.4), for $\operatorname{arcsn}(x,k)$, are equivalent to (22.15.12) and also to

22.15.11 $\quad x = \int_0^{\operatorname{sn}(x,k)} \frac{dt}{\sqrt{(1-t^2)(1-k^2t^2)}},$
$$-1 \le x \le 1, 0 \le k \le 1.$$

Similarly with (22.15.13)–(22.15.14) and also the other nine Jacobian elliptic functions.

22.15(ii) Representations as Elliptic Integrals

22.15.12 $\quad \operatorname{arcsn}(x,k) = \int_0^x \frac{dt}{\sqrt{(1-t^2)(1-k^2t^2)}}, \quad -1 \le x \le 1,$

22.15.13 $\quad \operatorname{arccn}(x,k) = \int_x^1 \frac{dt}{\sqrt{(1-t^2)(k'^2 + k^2 t^2)}}, \quad -1 \le x \le 1,$

22.15.14 $\quad \operatorname{arcdn}(x,k) = \int_x^1 \frac{dt}{\sqrt{(1-t^2)(t^2 - k'^2)}}, \quad k' \le x \le 1.$

For the corresponding results for the subsidiary functions see http://dlmf.nist.gov/22.15.ii.

The integrals (22.15.12)–(22.15.14) can be regarded as *normal forms* for representing the inverse functions. Other integrals, for example,
$$\int_x^b \frac{dt}{\sqrt{(a^2 + t^2)(b^2 - t^2)}}$$
can be transformed into normal form by elementary change of variables. Comprehensive treatments are given by Carlson (2005), Lawden (1989, pp. 52–55), Bowman (1953, Chapter IX), and Erdélyi et al. (1953b, pp. 296–301). See also Abramowitz and Stegun (1964, p. 596).

For representations of the inverse functions as symmetric elliptic integrals see §19.25(v). For power-series expansions see Carlson (2008).

22.16 Related Functions

22.16(i) Jacobi's Amplitude (am) Function

Definition

22.16.1 $\qquad \operatorname{am}(x,k) = \operatorname{Arcsin}(\operatorname{sn}(x,k)), \qquad x \in \mathbb{R},$

where the inverse sine has its principal value when $-K \le x \le K$ and is defined by continuity elsewhere. See Figure 22.16.1. $\operatorname{am}(x,k)$ is an infinitely differentiable function of x.

Quasi-Periodicity

22.16.2 $\qquad \operatorname{am}(x + 2K, k) = \operatorname{am}(x,k) + \pi.$

Integral Representation

22.16.3 $\qquad \operatorname{am}(x,k) = \int_0^x \operatorname{dn}(t,k)\,dt.$

Special Values

22.16.4 $\qquad \operatorname{am}(x,0) = x,$

22.16.5 $\qquad \operatorname{am}(x,1) = \operatorname{gd}(x).$

For the Gudermannian function $\operatorname{gd}(x)$ see §4.23(viii).

Approximation for Small x

22.16.6 $\quad \operatorname{am}(x,k) = x - k^2 \frac{x^3}{3!} + k^2(4+k^2)\frac{x^5}{5!} + O(x^7).$

Approximations for Small k, k'

22.16.7 $\quad \operatorname{am}(x, k) = x - \frac{1}{4}k^2(x - \sin x \cos x) + O(k^4)$,

22.16.8 $\quad \operatorname{am}(x, k) = \operatorname{gd} x - \frac{1}{4}k'^2(x - \sinh x \cosh x) \operatorname{sech} x + O(k'^4)$.

Fourier Series

With q as in (22.2.1) and $\zeta = \pi x/(2K)$,

22.16.9 $\quad \operatorname{am}(x, k) = \frac{\pi}{2K}x + 2\sum_{n=1}^{\infty} \frac{q^n \sin(2n\zeta)}{n(1+q^{2n})}$.

Relation to Elliptic Integrals

If $-K \leq x \leq K$, then the following four equations are equivalent:

22.16.10 $\quad x = F(\phi, k)$,

22.16.11 $\quad \operatorname{am}(x, k) = \phi$,

22.16.12 $\quad \operatorname{sn}(x, k) = \sin \phi = \sin(\operatorname{am}(x, k))$,

22.16.13 $\quad \operatorname{cn}(x, k) = \cos \phi = \cos(\operatorname{am}(x, k))$.

For $F(\phi, k)$ see §19.2(ii).

22.16(ii) Jacobi's Epsilon Function

Definition

For $x \in \mathbb{R}$

22.16.14 $\quad \mathcal{E}(x, k) = \int_0^x \sqrt{\frac{1-k^2 t^2}{1-t^2}}\, dt$;

compare (19.2.5). See Figure 22.16.2.

Other Integral Representations

22.16.15 $\quad \mathcal{E}(x, k) = x - k^2 \int_0^x \operatorname{sn}^2(t, k)\, dt$,

22.16.16 $\quad \mathcal{E}(x, k) = k'^2 x + k^2 \int_0^x \operatorname{cn}^2(t, k)\, dt$,

22.16.17 $\quad \mathcal{E}(x, k) = \int_0^x \operatorname{dn}^2(t, k)\, dt$.

For corresponding formulas for the subsidiary functions see http://dlmf.nist.gov/22.16.ii.

Quasi-Addition and Quasi-Periodic Formulas

22.16.27
$\mathcal{E}(x_1 + x_2, k) = \mathcal{E}(x_1, k) + \mathcal{E}(x_2, k) - k^2 \operatorname{sn}(x_1, k) \operatorname{sn}(x_2, k) \operatorname{sn}(x_1 + x_2, k)$,

22.16.28
$\mathcal{E}(x + K, k) = \mathcal{E}(x, k) + E(k) - k^2 \operatorname{sn}(x, k) \operatorname{cd}(x, k)$,

22.16.29 $\quad \mathcal{E}(x + 2K, k) = \mathcal{E}(x, k) + 2E(k)$.

For $E(k)$ see §19.2(ii).

Relation to Theta Functions

22.16.30 $\quad \mathcal{E}(x, k) = \frac{1}{\theta_3^2(0, q)\, \theta_4(\xi, q)} \frac{d}{d\xi}\theta_4(\xi, q) + \frac{E(k)}{K(k)}x$,

where $\xi = x/\theta_3^2(0, q)$. For θ_j see §20.2(i). For $E(k)$ see §19.2(ii).

Relation to the Elliptic Integral $E(\phi, k)$

22.16.31 $\quad E(\operatorname{am}(x, k), k) = \mathcal{E}(x, k), \quad -K \leq x \leq K$.

For $E(\phi, k)$ see §19.2(ii). See also (22.16.14).

22.16(iii) Jacobi's Zeta Function

Definition

With $E(k)$ and $K(k)$ as in §19.2(ii) and $x \in \mathbb{R}$,

22.16.32 $\quad \operatorname{Z}(x|k) = \mathcal{E}(x, k) - (E(k)/K(k))x$.

See Figure 22.16.3. (Sometimes in the literature $\operatorname{Z}(x|k)$ is denoted by $\operatorname{Z}(\operatorname{am}(x, k), k^2)$.)

Properties

$\operatorname{Z}(x|k)$ satisfies the same quasi-addition formula as the function $\mathcal{E}(x, k)$, given by (22.16.27). Also,

22.16.33 $\quad \operatorname{Z}(x + K|k) = \operatorname{Z}(x|k) - k^2 \operatorname{sn}(x, k) \operatorname{cd}(x, k)$,

22.16.34 $\quad \operatorname{Z}(x + 2K|k) = \operatorname{Z}(x|k)$.

22.16(iv) Graphs

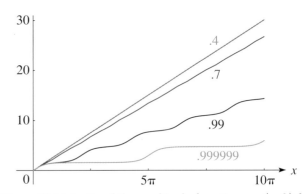

Figure 22.16.1: Jacobi's amplitude function $\operatorname{am}(x, k)$ for $0 \leq x \leq 10\pi$ and $k = 0.4, 0.7, 0.99, 0.999999$. Values of k greater than 1 are illustrated in Figure 22.19.1.

22.17 Moduli Outside the Interval [0,1]

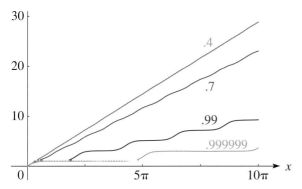

Figure 22.16.2: Jacobi's epsilon function $\mathcal{E}(x,k)$ for $0 \leq x \leq 10\pi$ and $k = 0.4, 0.7, 0.99, 0.999999$. (These graphs are similar to those in Figure 22.16.1; compare (22.16.3), (22.16.17), and the graphs of $\mathrm{dn}\,(x,k)$ in §22.3(i).)

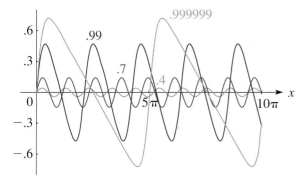

Figure 22.16.3: Jacobi's zeta function $\mathrm{Z}(x|k)$ for $0 \leq x \leq 10\pi$ and $k = 0.4, 0.7, 0.99, 0.999999$.

22.17 Moduli Outside the Interval [0,1]

22.17(i) Real or Purely Imaginary Moduli

Jacobian elliptic functions with real moduli in the intervals $(-\infty, 0)$ and $(1, \infty)$, or with purely imaginary moduli are related to functions with moduli in the interval $[0, 1]$ by the following formulas.

First

22.17.1 $$\mathrm{pq}\,(z,k) = \mathrm{pq}\,(z,-k),$$

for all twelve functions.
Secondly,

22.17.2 $$\mathrm{sn}\,(z, 1/k) = k\,\mathrm{sn}\,(z/k, k),$$

22.17.3 $$\mathrm{cn}\,(z, 1/k) = \mathrm{dn}\,(z/k, k),$$

22.17.4 $$\mathrm{dn}\,(z, 1/k) = \mathrm{cn}\,(z/k, k).$$

Thirdly, with

22.17.5 $$k_1 = \frac{k}{\sqrt{1+k^2}}, \quad k_1 k_1' = \frac{k}{1+k^2},$$

22.17.6 $$\mathrm{sn}\,(z, ik) = k_1'\,\mathrm{sd}\,(z/k_1', k_1),$$

22.17.7 $$\mathrm{cn}\,(z, ik) = \mathrm{cd}\,(z/k_1', k_1),$$

22.17.8 $$\mathrm{dn}\,(z, ik) = \mathrm{nd}\,(z/k_1', k_1).$$

In terms of the coefficients of the power series of §22.10(i), the above equations are polynomial identities in k. In (22.17.5) either value of the square root can be chosen.

22.17(ii) Complex Moduli

When z is fixed each of the twelve Jacobian elliptic functions is a meromorphic function of k^2. For illustrations see Figures 22.3.25–22.3.27. In consequence, the formulas in this chapter remain valid when k is complex. In particular, the Landen transformations in §§22.7(i) and 22.7(ii) are valid for all complex values of k, irrespective of which values of $\sqrt{k}$ and $k' = \sqrt{1-k^2}$ are chosen—as long as they are used consistently. For proofs of these results and further information see Walker (2003).

Applications

22.18 Mathematical Applications

22.18(i) Lengths and Parametrization of Plane Curves

Ellipse

22.18.1 $$(x^2/a^2) + (y^2/b^2) = 1,$$

with $a \geq b > 0$, is parametrized by

22.18.2 $$x = a\,\mathrm{sn}\,(u,k), \quad y = b\,\mathrm{cn}\,(u,k),$$

where $k = \sqrt{1-(b^2/a^2)}$ is the eccentricity, and $0 \leq u \leq 4K(k)$. The arc length $l(u)$ in the first quadrant, measured from $u = 0$, is

22.18.3 $$l(u) = a\,\mathcal{E}(u,k),$$

where $\mathcal{E}(u,k)$ is Jacobi's epsilon function (§22.16(ii)).

Lemniscate

In polar coordinates, $x = r\cos\phi$, $y = r\sin\phi$, the lemniscate is given by $r^2 = \cos(2\phi)$, $0 \leq \phi \leq 2\pi$. The arc length $l(r)$, measured from $\phi = 0$, is

22.18.4 $$l(r) = (1/\sqrt{2})\,\mathrm{arccn}\left(r, 1/\sqrt{2}\right).$$

Inversely:

22.18.5 $$r = \mathrm{cn}\left(\sqrt{2}l, 1/\sqrt{2}\right),$$

and

22.18.6 $$\begin{aligned} x &= \mathrm{cn}\left(\sqrt{2}l, 1/\sqrt{2}\right)\mathrm{dn}\left(\sqrt{2}l, 1/\sqrt{2}\right), \\ y &= \mathrm{cn}\left(\sqrt{2}l, 1/\sqrt{2}\right)\mathrm{sn}\left(\sqrt{2}l, 1/\sqrt{2}\right)\Big/\sqrt{2}. \end{aligned}$$

For these and other examples see Lawden (1989, Chapter 4), Whittaker and Watson (1927, §22.8), and Siegel (1988, pp. 1–7).

22.18(ii) Conformal Mapping

With $k \in [0,1]$ the mapping $z \to w = \operatorname{sn}(z,k)$ gives a conformal map of the closed rectangle $[-K,K] \times [0,K']$ onto the half-plane $\Im w \geq 0$, with $0, \pm K, \pm K + iK', iK'$ mapping to $0, \pm 1, \pm k^{-2}, \infty$ respectively. The half-open rectangle $(-K,K) \times [-K',K']$ maps onto $\mathbb{C}$ cut along the intervals $(-\infty, -1]$ and $[1, \infty)$. See Akhiezer (1990, Chapter 8) and McKean and Moll (1999, Chapter 2) for discussions of the inverse mapping. Bowman (1953, Chapters V–VI) gives an overview of the use of Jacobian elliptic functions in conformal maps for engineering applications.

22.18(iii) Uniformization and Other Parametrizations

By use of the functions sn and cn, parametrizations of algebraic equations, such as

22.18.7
$$ax^2y^2 + b(x^2y + xy^2) + c(x^2 + y^2) + 2dxy + e(x+y) + f = 0,$$

in which a,b,c,d,e,f are real constants, can be achieved in terms of single-valued functions. This circumvents the cumbersome branch structure of the multivalued functions $x(y)$ or $y(x)$, and constitutes the process of *uniformization*; see Siegel (1988, Chapter II). See Baxter (1982, p. 471) for an example from statistical mechanics. Discussion of parametrization of the angles of spherical trigonometry in terms of Jacobian elliptic functions is given in Greenhill (1959, p. 131) and Lawden (1989, §4.4).

22.18(iv) Elliptic Curves and the Jacobi–Abel Addition Theorem

Algebraic curves of the form $y^2 = P(x)$, where P is a nonsingular polynomial of degree 3 or 4 (see McKean and Moll (1999, §1.10)), are *elliptic curves*, which are also considered in §23.20(ii). The special case $y^2 = (1-x^2)(1-k^2x^2)$ is in *Jacobian normal form*. For any two points (x_1, y_1) and (x_2, y_2) on this curve, their *sum* (x_3, y_3), always a third point on the curve, is defined by the Jacobi–Abel addition law

22.18.8
$$x_3 = \frac{x_1 y_2 + x_2 y_1}{1 - k^2 x_1^2 x_2^2},$$
$$y_3 = \frac{y_1 y_2 + x_2(-(1+k^2)x_1 + 2k^2 x_1^3)}{1 - k^2 x_1^2 x_2^2} + x_3 \frac{2k^2 x_1 y_1 x_2^2}{1 - k^2 x_1^2 x_2^2},$$

a construction due to Abel; see Whittaker and Watson (1927, pp. 442, 496–497). This provides an abelian group structure, and leads to important results in number theory, discussed in an elementary manner by Silverman and Tate (1992), and more fully by Koblitz (1993, Chapter 1, especially §1.7) and McKean and Moll (1999, Chapter 3). The existence of this group structure is connected to the Jacobian elliptic functions via the differential equation (22.13.1). With the identification $x = \operatorname{sn}(z,k)$, $y = d(\operatorname{sn}(z,k))/dz$, the addition law (22.18.8) is transformed into the addition theorem (22.8.1); see Akhiezer (1990, pp. 42, 45, 73–74) and McKean and Moll (1999, §§2.14, 2.16). The theory of elliptic functions brings together complex analysis, algebraic curves, number theory, and geometry: Lang (1987), Siegel (1988), and Serre (1973).

22.19 Physical Applications

22.19(i) Classical Dynamics: The Pendulum

With appropriate scalings, Newton's equation of motion for a pendulum with a mass in a gravitational field constrained to move in a vertical plane at a fixed distance from a fulcrum is

22.19.1
$$\frac{d^2\theta(t)}{dt^2} = -\sin\theta(t),$$

θ being the angular displacement from the point of stable equilibrium, $\theta = 0$. The bounded ($-\pi \leq \theta \leq \pi$) oscillatory solution of (22.19.1) is traditionally written

22.19.2
$$\sin\left(\tfrac{1}{2}\theta(t)\right) = \sin\left(\tfrac{1}{2}\alpha\right) \operatorname{sn}\left(t, \sin\left(\tfrac{1}{2}\alpha\right)\right),$$

for an initial angular displacement α, with $d\theta/dt = 0$ at time 0; see Lawden (1989, pp. 114–117). The period is $4K\left(\sin\left(\tfrac{1}{2}\alpha\right)\right)$. The angle $\alpha = \pi$ is a *separatrix*, separating oscillatory and unbounded motion. With the same initial conditions, if the sign of gravity is reversed then the new period is $4K'\left(\sin\left(\tfrac{1}{2}\alpha\right)\right)$; see Whittaker (1964, §44).

Alternatively, Sala (1989) writes:

22.19.3
$$\theta(t) = 2\operatorname{am}\left(t, \sqrt{2/E}\right),$$

for the initial conditions $\theta(0) = 0$, the point of stable equilibrium for $E = 0$, and $d\theta(t)/dt = \sqrt{2E}$. Here $E = \tfrac{1}{2}(d\theta(t)/dt)^2 + 1 - \cos\theta(t)$ is the *energy*, which is a first integral of the motion. This formulation gives the bounded and unbounded solutions from the same formula (22.19.3), for $k \geq 1$ and $k \leq 1$, respectively. Also, $\theta(t)$ is not restricted to the principal range $-\pi \leq \theta \leq \pi$. Figure 22.19.1 shows the nature of the solutions $\theta(t)$ of (22.19.3) by graphing $\operatorname{am}(x,k)$ for both $0 \leq k \leq 1$, as in Figure 22.16.1, and $k \geq 1$, where it is periodic.

22.19 Physical Applications

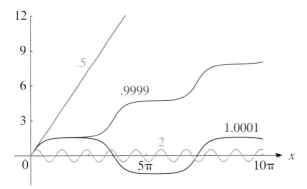

Figure 22.19.1: Jacobi's amplitude function am (x, k) for $0 \leq x \leq 10\pi$ and $k = 0.5, 0.9999, 1.0001, 2$. When $k < 1$, am (x, k) increases monotonically indicating that the motion of the pendulum is unbounded in θ, corresponding to free rotation about the fulcrum; compare Figure 22.16.1. As $k \to 1-$, plateaus are seen as the motion approaches the separatrix where $\theta = n\pi$, $n = \pm 1, \pm 2, \ldots$, at which points the motion is time independent for $k = 1$. This corresponds to the pendulum being "upside down" at a point of unstable equilibrium. For $k > 1$, the motion is periodic in x, corresponding to bounded oscillatory motion.

22.19(ii) Classical Dynamics: The Quartic Oscillator

Classical motion in one dimension is described by Newton's equation

22.19.4 $$\frac{d^2 x(t)}{dt^2} = -\frac{dV(x)}{dx},$$

where $V(x)$ is the potential energy, and $x(t)$ is the coordinate as a function of time t. The potential

22.19.5 $$V(x) = \pm \tfrac{1}{2} x^2 \pm \tfrac{1}{4} \beta x^4$$

plays a prototypal role in classical mechanics (Lawden (1989, §5.2)), quantum mechanics (Schulman (1981, Chapter 29)), and quantum field theory (Pokorski (1987, p. 203), Parisi (1988, §14.6)). Its dynamics for purely imaginary time is connected to the theory of instantons (Itzykson and Zuber (1980, p. 572), Schäfer and Shuryak (1998)), to WKB theory, and to large-order perturbation theory (Bender and Wu (1973), Simon (1982)).

For β real and positive, three of the four possible combinations of signs give rise to bounded oscillatory motions. We consider the case of a particle of mass 1, initially held at rest at displacement a from the origin and then released at time $t = 0$. The subsequent position as a function of time, $x(t)$, for the three cases is given with results expressed in terms of a and the dimensionless parameter $\eta = \tfrac{1}{2} \beta a^2$.

Case I: $V(x) = \tfrac{1}{2} x^2 + \tfrac{1}{4} \beta x^4$

This is an example of *Duffing's equation*; see Ablowitz and Clarkson (1991, pp. 150–152) and Lawden (1989, pp. 117–119). The subsequent time evolution is always oscillatory with period $4K(k)/\sqrt{1+\eta}$:

22.19.6 $$x(t) = a \operatorname{cn}\left(\sqrt{1+\eta}\, t, 1/\sqrt{2+\eta^{-1}}\right).$$

Case II: $V(x) = \tfrac{1}{2} x^2 - \tfrac{1}{4} \beta x^4$

There is bounded oscillatory motion near $x = 0$, with period $4K(k)/\sqrt{1-\eta}$, for initial displacements with $|a| \leq \sqrt{1/\beta}$:

22.19.7 $$x(t) = a \operatorname{sn}\left(\sqrt{1-\eta}\, t, 1/\sqrt{\eta^{-1}-1}\right).$$

As $a \to \sqrt{1/\beta}$ from below the period diverges since $a = \pm\sqrt{1/\beta}$ are points of unstable equilibrium.

Case III: $V(x) = -\tfrac{1}{2} x^2 + \tfrac{1}{4} \beta x^4$

Two types of oscillatory motion are possible. For an initial displacement with $\sqrt{1/\beta} \leq |a| \leq \sqrt{2/\beta}$, bounded oscillations take place near one of the two points of stable equilibrium $x = \pm\sqrt{1/\beta}$. Such oscillations, of period $4K(k)/\sqrt{\eta}$, are given by:

22.19.8 $$x(t) = a \operatorname{dn}\left(\sqrt{\eta}\, t, \sqrt{2-\eta^{-1}}\right).$$

As $a \to \sqrt{2/\beta}$ from below the period diverges since $x = 0$ is a point of unstable equilibrium. For initial displacement with $|a| \geq \sqrt{2/\beta}$ the motion extends over the full range $-a \leq x \leq a$:

22.19.9 $$x(t) = a \operatorname{cn}\left(\sqrt{2\eta-1}\, t, 1/\sqrt{2-\eta^{-1}}\right),$$

with period $4K(k)/\sqrt{2\eta-1}$. As $|a| \to \sqrt{2/\beta}$ from above the period again diverges. Both the dn and cn solutions approach $a \operatorname{sech} t$ as $a \to \sqrt{2/\beta}$ from the appropriate directions.

22.19(iii) Nonlinear ODEs and PDEs

Many nonlinear ordinary and partial differential equations have solutions that may be expressed in terms of Jacobian elliptic functions. These include the time dependent, and time independent, nonlinear Schrödinger equations (NLSE) (Drazin and Johnson (1993, Chapter 2), Ablowitz and Clarkson (1991, pp. 42, 99)), the Korteweg–de Vries (KdV) equation (Kruskal (1974), Li and Olver (2000)), the sine-Gordon equation, and others; see Drazin and Johnson (1993, Chapter 2) for an overview. Such solutions include standing or stationary waves, periodic cnoidal waves, and single and multi-solitons occurring in diverse physical situations such as water waves, optical pulses, quantum fluids, and electrical impulses (Hasegawa (1989), Carr et al. (2000), Kivshar and Luther-Davies (1998), and Boyd (1998, Appendix D2.2)).

22.19(iv) Tops

The classical rotation of rigid bodies in free space or about a fixed point may be described in terms of elliptic, or *hyperelliptic*, functions if the motion is *integrable* (Audin (1999, Chapter 1)). Hyperelliptic functions $u(z)$ are solutions of the equation $z = \int_0^u (f(x))^{-1/2} dx$, where $f(x)$ is a polynomial of degree higher than 4. Elementary discussions of this topic appear in Lawden (1989, §5.7), Greenhill (1959, pp. 101–103), and Whittaker (1964, Chapter VI). A more abstract overview is Audin (1999, Chapters III and IV), and a complete discussion of analytical solutions in the elliptic and hyperelliptic cases appears in Golubev (1960, Chapters V and VII), the original hyperelliptic investigation being due to Kowalevski (1889).

22.19(v) Other Applications

Numerous other physical or engineering applications involving Jacobian elliptic functions, and their inverses, to problems of classical dynamics, electrostatics, and hydrodynamics appear in Bowman (1953, Chapters VII and VIII) and Lawden (1989, Chapter 5). Whittaker (1964, Chapter IV) enumerates the complete class of one-body classical mechanical problems that are solvable this way.

Computation

22.20 Methods of Computation

22.20(i) Via Theta Functions

A powerful way of computing the twelve Jacobian elliptic functions for real or complex values of both the argument z and the modulus k is to use the definitions in terms of theta functions given in §22.2, obtaining the theta functions via methods described in §20.14.

22.20(ii) Arithmetic-Geometric Mean

Given real or complex numbers a_0, b_0, with b_0/a_0 not real and negative, define

22.20.1
$$a_n = \tfrac{1}{2}(a_{n-1} + b_{n-1}), \quad b_n = (a_{n-1}b_{n-1})^{1/2},$$
$$c_n = \tfrac{1}{2}(a_{n-1} - b_{n-1}),$$

for $n \geq 1$, where the square root is chosen so that $\operatorname{ph} b_n = \tfrac{1}{2}(\operatorname{ph} a_{n-1} + \operatorname{ph} b_{n-1})$, where $\operatorname{ph} a_{n-1}$ and $\operatorname{ph} b_{n-1}$ are chosen so that their difference is numerically less than π. Then as $n \to \infty$ sequences $\{a_n\}$, $\{b_n\}$ converge to a common limit $M = M(a_0, b_0)$, the *arithmetic-geometric mean* of a_0, b_0. And since

22.20.2 $\max(|a_n - M|, |b_n - M|, |c_n|) \leq (\text{const.}) \times 2^{-2^n}$,

convergence is very rapid.

For x real and $k \in (0, 1)$, use (22.20.1) with $a_0 = 1$, $b_0 = k' \in (0, 1)$, $c_0 = k$, and continue until c_N is zero to the required accuracy. Next, compute $\phi_N, \phi_{N-1}, \ldots, \phi_0$, where

22.20.3 $\phi_N = 2^N a_N x$,

22.20.4 $\phi_{n-1} = \tfrac{1}{2}\left(\phi_n + \arcsin\left(\dfrac{c_n}{a_n}\sin\phi_n\right)\right),$

and the inverse sine has its principal value (§4.23(ii)). Then

22.20.5 $\operatorname{sn}(x, k) = \sin\phi_0, \quad \operatorname{cn}(x, k) = \cos\phi_0,$
$$\operatorname{dn}(x, k) = \dfrac{\cos\phi_0}{\cos(\phi_1 - \phi_0)},$$

and the subsidiary functions can be found using (22.2.10).

See also Wachspress (2000).

Example

To compute sn, cn, dn to 10D when $x = 0.8$, $k = 0.65$.

Four iterations of (22.20.1) lead to $c_4 = 6.5 \times 10^{-12}$. From (22.20.3) and (22.20.4) we obtain $\phi_1 = 1.40213\,91827$ and $\phi_0 = 0.76850\,92170$. Then from (22.20.5), $\operatorname{sn}(0.8, 0.65) = 0.69506\,42165$, $\operatorname{cn}(0.8, 0.65) = 0.71894\,76580$, $\operatorname{dn}(0.8, 0.65) = 0.89212\,34349$.

22.20(iii) Landen Transformations

By application of the transformations given in §§22.7(i) and 22.7(ii), k or k' can always be made sufficiently small to enable the approximations given in §22.10(ii) to be applied. The rate of convergence is similar to that for the arithmetic-geometric mean.

Example

To compute $\operatorname{dn}(x, k)$ to 6D for $x = 0.2$, $k^2 = 0.19$, $k' = 0.9$.

From (22.7.1), $k_1 = \tfrac{1}{19}$ and $x/(1 + k_1) = 0.19$. From the first two terms in (22.10.6) we find $\operatorname{dn}\left(0.19, \tfrac{1}{19}\right) = 0.999951$. Then by using (22.7.4) we have $\operatorname{dn}\left(0.2, \sqrt{0.19}\right) = 0.996253$.

If needed, the corresponding values of sn and cn can be found subsequently by applying (22.10.4) and (22.7.2), followed by (22.10.5) and (22.7.3).

22.20(iv) Lattice Calculations

If either τ or $q = e^{i\pi\tau}$ is given, then we use $k = \theta_2^2(0, q)/\theta_3^2(0, q)$, $k' = \theta_4^2(0, q)/\theta_3^2(0, q)$, $K = \tfrac{1}{2}\pi\theta_3^2(0, q)$, and $K' = -i\tau K$, obtaining the values of the theta functions as in §20.14.

If k, k' are given with $k^2 + k'^2 = 1$ and $\Im k'/\Im k < 0$, then K, K' can be found from

22.20.6 $K = \dfrac{\pi}{2M(1, k')}, \quad K' = \dfrac{\pi}{2M(1, k)},$

using the arithmetic-geometric mean.

Example 1

If $k = k' = 1/\sqrt{2}$, then three iterations of (22.20.1) give $M = 0.84721\,30848$, and from (22.20.6) $K = \pi/(2M) = 1.85407\,46773$ —in agreement with the value of $\left(\Gamma\left(\tfrac{1}{4}\right)\right)^2 / (4\sqrt{\pi})$; compare (23.17.3) and (23.22.2).

Example 2

If $k' = 1 - i$, then four iterations of (22.20.1) give $K = 1.23969\,74481 + i0.56499\,30988$.

22.20(v) Inverse Functions

See Wachspress (2000).

22.20(vi) Related Functions

$\operatorname{am}(x, k)$ can be computed from its definition (22.16.1) or from its Fourier series (22.16.9). Alternatively, Sala (1989) shows how to apply the arithmetic-geometric mean to compute $\operatorname{am}(x, k)$.

Jacobi's epsilon function can be computed from its representation (22.16.30) in terms of theta functions and complete elliptic integrals; compare §20.14. Jacobi's zeta function can then be found by use of (22.16.32).

22.20(vii) Further References

For additional information on methods of computation for the Jacobi and related functions, see the introductory sections in the following books: Lawden (1989), Curtis (1964b), Milne-Thomson (1950), and Spenceley and Spenceley (1947).

22.21 Tables

Spenceley and Spenceley (1947) tabulates $\operatorname{sn}(Kx, k)$, $\operatorname{cn}(Kx, k)$, $\operatorname{dn}(Kx, k)$, $\operatorname{am}(Kx, k)$, $\mathcal{E}(Kx, k)$ for $\arcsin k = 1°(1°)89°$ and $x = 0\left(\tfrac{1}{90}\right)1$ to 12D, or 12 decimals of a radian in the case of $\operatorname{am}(Kx, k)$.

Curtis (1964b) tabulates $\operatorname{sn}(mK/n, k)$, $\operatorname{cn}(mK/n, k)$, $\operatorname{dn}(mK/n, k)$ for $n = 2(1)15$, $m = 1(1)n-1$, and q (not k) $= 0(.005)0.35$ to 20D.

Lawden (1989, pp. 280–284 and 293–297) tabulates $\operatorname{sn}(x, k)$, $\operatorname{cn}(x, k)$, $\operatorname{dn}(x, k)$, $\mathcal{E}(x, k)$, $Z(x|k)$ to 5D for $k = 0.1(.1)0.9$, $x = 0(.1)X$, where X ranges from 1.5 to 2.2.

Zhang and Jin (1996, p. 678) tabulates $\operatorname{sn}(Kx, k)$, $\operatorname{cn}(Kx, k)$, $\operatorname{dn}(Kx, k)$ for $k = \tfrac{1}{4}, \tfrac{1}{2}$ and $x = 0(.1)4$ to 7D.

For other tables prior to 1961 see Fletcher et al. (1962, pp. 500–503) and Lebedev and Fedorova (1960, pp. 221–223).

Tables of theta functions (§20.15) can also be used to compute the twelve Jacobian elliptic functions by application of the quotient formulas given in §22.2.

22.22 Software

See http://dlmf.nist.gov/22.22.

References

General References

The main references used for the mathematical properties in this chapter are Bowman (1953), Copson (1935), Lawden (1989), McKean and Moll (1999), Walker (1996), Whittaker and Watson (1927), and for physical applications Drazin and Johnson (1993), Lawden (1989), Walker (1996).

Sources

The following list gives the references or other indications of proofs that were used in constructing the various sections in this chapter. These sources supplement the references that are quoted in the text.

§22.2 Lawden (1989, §2.1), Whittaker and Watson (1927, §22.1), Walker (1996, §§5.1, 6.2), Walker (2003).

§22.3 These graphics were produced at NIST and by the authors.

§22.4 Lawden (1989, §§2.1, 2.2), Whittaker and Watson (1927, §§22.1–22.3), Walker (1996, §6.2).

§22.5 Lawden (1989, §§2.1–2.2, 2.6), Whittaker and Watson (1927, §22.3).

§22.6 Lawden (1989, §§2.4–2.6), Whittaker and Watson (1927, §§22.1, 22.4), Walker (1996, §6.2). For (22.6.6) and (22.6.7) set $u = v = z$ in (22.8.2) and use (22.6.1) repeatedly.

§22.7 Lawden (1989, §3.9), Whittaker and Watson (1927, §22.42), Walker (1996, p. 148).

§22.8 Lawden (1989, §2.4 and p. 43), Whittaker and Watson (1927, §22.2 and p. 530), Walker (1996, §6.2).

§22.9 Khare and Sukhatme (2002), Khare et al. (2003).

§22.10 Lawden (1989, §§2.5, 3.1). The expansions in powers of k' follow from those in powers of k by use of Table 22.6.1 and (4.28.8)–(4.28.10).

§22.11 Walker (1996, §5.4), Whittaker and Watson (1927, §22.6). For (22.11.13) see Deconinck and Kutz (2006, Eq. (48)). The version of this formula in Byrd and Friedman (1971, p. 307, Eq. (911.01)) is incorrect; see Tang (1969).

§22.12 For the first right-hand sides of (22.12.2)–(22.12.4) see Lawden (1989, §8.8). The second right-hand sides of (22.12.2)–(22.12.4) can be obtained from the corresponding first right-hand sides by substituting, as appropriate, the expansions $\pi \csc(\pi\zeta) = \sum_{m=-\infty}^{\infty} (-1)^m/(\zeta - m)$ or $\pi \cot(\pi\zeta) = \lim_{M\to\infty} \sum_{m=-M}^{M} 1/(\zeta - m)$; compare (4.22.5) and (4.22.3).

§22.13 Lawden (1989, §2.5), Walker (1996, §6.2), Whittaker and Watson (1927, §§22.1–22.2). (22.13.13)–(22.13.15) are obtained by differentiation of (22.13.1)–(22.13.3).

§22.14 Lawden (1989, §2.7), Whittaker and Watson (1927, §§22.5, 22.72).

§22.15 Bowman (1953, Chapter 1), Lawden (1989, §§3.1, 3.2), Whittaker and Watson (1927, §22.72).

§22.16 Lawden (1989, §§3.4–3.6), Walker (1996, §6.5), Whittaker and Watson (1927, §§22.72–22.73). The figures were produced at NIST.

§22.17 Lawden (1989, §3.9).

§22.19 Lawden (1989, §§5.1–5.2). The figure was produced at NIST.

§22.20 Walker (1996, pp. 141–143).

Chapter 23
Weierstrass Elliptic and Modular Functions

W. P. Reinhardt[1] and P. L. Walker[2]

Notation	**570**
23.1 Special Notation	570
Weierstrass Elliptic Functions	**570**
23.2 Definitions and Periodic Properties	570
23.3 Differential Equations	571
23.4 Graphics	571
23.5 Special Lattices	574
23.6 Relations to Other Functions	574
23.7 Quarter Periods	576
23.8 Trigonometric Series and Products	576
23.9 Laurent and Other Power Series	577
23.10 Addition Theorems and Other Identities	577
23.11 Integral Representations	578
23.12 Asymptotic Approximations	578
23.13 Zeros	579
23.14 Integrals	579
Modular Functions	**579**
23.15 Definitions	579
23.16 Graphics	579
23.17 Elementary Properties	580
23.18 Modular Transformations	580
23.19 Interrelations	581
Applications	**581**
23.20 Mathematical Applications	581
23.21 Physical Applications	582
Computation	**583**
23.22 Methods of Computation	583
23.23 Tables	584
23.24 Software	584
References	**584**

[1] University of Washington, Seattle, Washington.
[2] American University of Sharjah, Sharjah, United Arab Emirates.
Acknowledgments: This chapter is based in part on Abramowitz and Stegun (1964, Chapter 18) by T. H. Southard.
Copyright © 2009 National Institute of Standards and Technology. All rights reserved.

Notation

23.1 Special Notation

(For other notation see pp. xiv and 873.)

$\mathbb{L}$	lattice in $\mathbb{C}$.
ℓ, n	integers.
m	integer, except in §23.20(ii).
$z = x + iy$	complex variable, except in §§23.20(ii), 23.21(iii).
$[a,b]$ or (a,b)	closed, or open, straight-line segment joining a and b, whether or not a and b are real.
primes	derivatives with respect to the variable, except where indicated otherwise.
$K(k), K'(k)$	complete elliptic integrals (§19.2(i)).
$2\omega_1, 2\omega_3$	lattice generators ($\Im(\omega_3/\omega_1) > 0$).
ω_2	$-\omega_1 - \omega_3$.
$\tau = \omega_3/\omega_1$	lattice parameter ($\Im\tau > 0$).
$q = e^{i\pi\omega_3/\omega_1}$ $= e^{i\pi\tau}$	nome.
g_2, g_3	lattice invariants.
e_1, e_2, e_3	zeros of Weierstrass normal cubic $4z^3 - g_2 z - g_3$.
Δ	discriminant $g_2^3 - 27g_3^2$.
$n\mathbb{Z}$	set of all integer multiples of n.
S_1/S_2	set of all elements of S_1, modulo elements of S_2. Thus two elements of S_1/S_2 are equivalent if they are both in S_1 and their difference is in S_2. (For an example see §20.12(ii).)
$G \times H$	Cartesian product of groups G and H, that is, the set of all pairs of elements (g, h) with group operation $(g_1, h_1) + (g_2, h_2) = (g_1 + g_2, h_1 + h_2)$.

The main functions treated in this chapter are the Weierstrass $\wp$-function $\wp(z) = \wp(z|\mathbb{L}) = \wp(z; g_2, g_3)$; the Weierstrass zeta function $\zeta(z) = \zeta(z|\mathbb{L}) = \zeta(z; g_2, g_3)$; the Weierstrass sigma function $\sigma(z) = \sigma(z|\mathbb{L}) = \sigma(z; g_2, g_3)$; the elliptic modular function $\lambda(\tau)$; Klein's complete invariant $J(\tau)$; Dedekind's eta function $\eta(\tau)$.

Other Notations

Whittaker and Watson (1927) requires only $\Im(\omega_3/\omega_1) \neq 0$, instead of $\Im(\omega_3/\omega_1) > 0$. Abramowitz and Stegun (1964, Chapter 18) considers only rectangular and rhombic lattices (§23.5); ω_1, ω_3 are replaced by ω, ω' for the former and by ω_2, ω' for the latter. Silverman and Tate (1992) and Koblitz (1993) replace $2\omega_1$ and $2\omega_3$ by ω_1 and ω_3, respectively. Walker (1996) normalizes $2\omega_1 = 1$, $2\omega_3 = \tau$, and uses homogeneity (§23.10(iv)). McKean and Moll (1999) replaces $2\omega_1$ and $2\omega_3$ by ω_1 and ω_2, respectively.

Weierstrass Elliptic Functions

23.2 Definitions and Periodic Properties

23.2(i) Lattices

If ω_1 and ω_3 are nonzero real or complex numbers such that $\Im(\omega_3/\omega_1) > 0$, then the set of points $2m\omega_1 + 2n\omega_3$, with $m, n \in \mathbb{Z}$, constitutes a *lattice* $\mathbb{L}$ with $2\omega_1$ and $2\omega_3$ *lattice generators*.

The generators of a given lattice $\mathbb{L}$ are not unique. For example, if

23.2.1 $$\omega_1 + \omega_2 + \omega_3 = 0,$$

then $2\omega_2, 2\omega_3$ are generators, as are $2\omega_2, 2\omega_1$. In general, if

23.2.2 $$\chi_1 = a\omega_1 + b\omega_3, \quad \chi_3 = c\omega_1 + d\omega_3,$$

where a, b, c, d are integers, then $2\chi_1, 2\chi_3$ are generators of $\mathbb{L}$ iff

23.2.3 $$ad - bc = 1.$$

23.2(ii) Weierstrass Elliptic Functions

23.2.4 $$\wp(z) = \frac{1}{z^2} + \sum_{w \in \mathbb{L}\setminus\{0\}} \left(\frac{1}{(z-w^2)} - \frac{1}{w^2}\right),$$

23.2.5 $$\zeta(z) = \frac{1}{z} + \sum_{w \in \mathbb{L}\setminus\{0\}} \left(\frac{1}{z-w} + \frac{1}{w} + \frac{z}{w^2}\right),$$

23.2.6 $$\sigma(z) = z \prod_{w \in \mathbb{L}\setminus\{0\}} \left(\left(1 - \frac{z}{w}\right)\exp\left(\frac{z}{w} + \frac{z^2}{2w^2}\right)\right).$$

The double series and double product are absolutely and uniformly convergent in compact sets in $\mathbb{C}$ that do not include lattice points. Hence the order of the terms or factors is immaterial.

When $z \notin \mathbb{L}$ the functions are related by

23.2.7 $$\wp(z) = -\zeta'(z),$$

23.2.8 $$\zeta(z) = \sigma'(z)/\sigma(z).$$

$\wp(z)$ and $\zeta(z)$ are meromorphic functions with poles at the lattice points. $\wp(z)$ is even and $\zeta(z)$ is odd. The poles of $\wp(z)$ are double with residue 0; the poles of $\zeta(z)$ are simple with residue 1. The function $\sigma(z)$ is entire and odd, with simple zeros at the lattice points. When it is important to display the lattice with the functions they are denoted by $\wp(z|\mathbb{L})$, $\zeta(z|\mathbb{L})$, and $\sigma(z|\mathbb{L})$, respectively.

23.2(iii) Periodicity

If $2\omega_1$, $2\omega_3$ is any pair of generators of $\mathbb{L}$, and ω_2 is defined by (23.2.1), then

23.2.9 $$\wp(z+2\omega_j)=\wp(z), \qquad j=1,2,3.$$

Hence $\wp(z)$ is an *elliptic function*, that is, $\wp(z)$ is meromorphic and periodic on a lattice; equivalently, $\wp(z)$ is meromorphic and has two periods whose ratio is not real. We also have

23.2.10 $$\wp'(\omega_j)=0, \qquad j=1,2,3.$$

The function $\zeta(z)$ is quasi-periodic: for $j=1,2,3$,

23.2.11 $$\zeta(z+2\omega_j)=\zeta(z)+2\eta_j,$$

where

23.2.12 $$\eta_j=\zeta(\omega_j).$$

Also,

23.2.13 $$\eta_1+\eta_2+\eta_3=0,$$

23.2.14 $\eta_3\omega_2-\eta_2\omega_3=\eta_2\omega_1-\eta_1\omega_2=\eta_1\omega_3-\eta_3\omega_1=\tfrac{1}{2}\pi i.$

For $j=1,2,3$, the function $\sigma(z)$ satisfies

23.2.15 $$\sigma(z+2\omega_j)=-e^{2\eta_j(z+\omega_j)}\sigma(z),$$

23.2.16 $$\sigma'(2\omega_j)=-e^{2\eta_j\omega_j}.$$

More generally, if $j=1,2,3$, $k=1,2,3$, $j\neq k$, and $m,n\in\mathbb{Z}$, then

23.2.17
$\sigma(z+2m\omega_j+2n\omega_k)/\sigma(z)$
$=(-1)^{m+n+mn}\exp((2m\eta_j+2n\eta_k)(m\omega_j+n\omega_k+z)).$

For further quasi-periodic properties of the σ-function see Lawden (1989, §6.2).

23.3 Differential Equations

23.3(i) Invariants, Roots, and Discriminant

The *lattice invariants* are defined by

23.3.1 $$g_2=60\sum_{w\in\mathbb{L}\setminus\{0\}}w^{-4},$$

23.3.2 $$g_3=140\sum_{w\in\mathbb{L}\setminus\{0\}}w^{-6}.$$

The *lattice roots* satisfy the cubic equation

23.3.3 $$4z^3-g_2z-g_3=0,$$

and are denoted by e_1,e_2,e_3. The *discriminant* (§1.11(ii)) is given by

23.3.4 $\Delta=g_2^3-27g_3^2=16(e_2-e_3)^2(e_3-e_1)^2(e_1-e_2)^2.$

In consequence,

23.3.5 $$e_1+e_2+e_3=0,$$

23.3.6 $g_2=2(e_1^2+e_2^2+e_3^2)=-4(e_2e_3+e_3e_1+e_1e_2),$

23.3.7 $g_3=4e_1e_2e_3=\tfrac{4}{3}(e_1^3+e_2^3+e_3^3).$

Let $g_2^3\neq 27g_3^2$, or equivalently Δ be nonzero, or e_1,e_2,e_3 be distinct. Given g_2 and g_3 there is a unique lattice $\mathbb{L}$ such that (23.3.1) and (23.3.2) are satisfied. We may therefore define

23.3.8 $$\wp(z;g_2,g_3)=\wp(z|\mathbb{L}).$$

Similarly for $\zeta(z;g_2,g_3)$ and $\sigma(z;g_2,g_3)$. As functions of g_2 and g_3, $\wp(z;g_2,g_3)$ and $\zeta(z;g_2,g_3)$ are meromorphic and $\sigma(z;g_2,g_3)$ is entire.

Conversely, g_2, g_3, and the set $\{e_1,e_2,e_3\}$ are determined uniquely by the lattice $\mathbb{L}$ independently of the choice of generators. However, given any pair of generators $2\omega_1$, $2\omega_3$ of $\mathbb{L}$, and with ω_2 defined by (23.2.1), we can identify the e_j individually, via

23.3.9 $$e_j=\wp(\omega_j|\mathbb{L}), \qquad j=1,2,3.$$

In what follows, *it will be assumed that* (23.3.9) *always applies*.

23.3(ii) Differential Equations and Derivatives

23.3.10 $$\wp'^2(z)=4\wp^3(z)-g_2\wp(z)-g_3,$$

23.3.11 $$\wp'^2(z)=4(\wp(z)-e_1)(\wp(z)-e_2)(\wp(z)-e_3),$$

23.3.12 $$\wp''(z)=6\wp^2(z)-\tfrac{1}{2}g_2,$$

23.3.13 $$\wp'''(z)=12\wp(z)\wp'(z).$$

See also (23.2.7) and (23.2.8).

23.4 Graphics

23.4(i) Real Variables

See Figures 23.4.1–23.4.7 for line graphs of the Weierstrass functions $\wp(x)$, $\zeta(x)$, and $\sigma(x)$, illustrating the lemniscatic and equianharmonic cases. (The figures in this subsection may be compared with the figures in §22.3(i).)

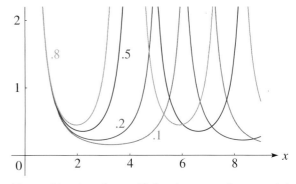

Figure 23.4.1: $\wp(x;g_2,0)$ for $0\leq x\leq 9$, $g_2=0.1, 0.2, 0.5, 0.8$. (Lemniscatic case.)

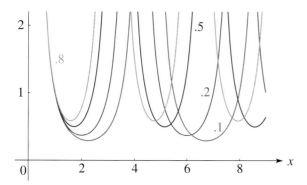

Figure 23.4.2: $\wp(x;0,g_3)$ for $0 \leq x \leq 9$, $g_3 = 0.1, 0.2, 0.5, 0.8$. (Equianharmonic case.)

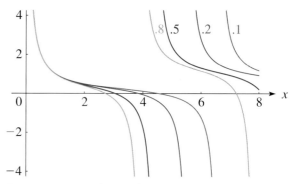

Figure 23.4.3: $\zeta(x;g_2,0)$ for $0 \leq x \leq 8$, $g_2 = 0.1, 0.2, 0.5, 0.8$. (Lemniscatic case.)

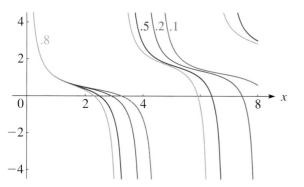

Figure 23.4.4: $\zeta(x;0,g_3)$ for $0 \leq x \leq 8$, $g_3 = 0.1, 0.2, 0.5, 0.8$. (Equianharmonic case.)

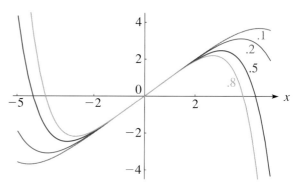

Figure 23.4.5: $\sigma(x;g_2,0)$ for $-5 \leq x \leq 5$, $g_2 = 0.1, 0.2, 0.5, 0.8$. (Lemniscatic case.)

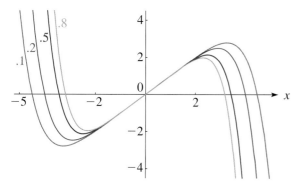

Figure 23.4.6: $\sigma(x;0,g_3)$ for $-5 \leq x \leq 5$, $g_3 = 0.1, 0.2, 0.5, 0.8$. (Equianharmonic case.)

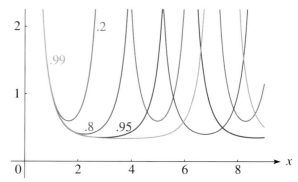

Figure 23.4.7: $\wp(x)$ with $\omega_1 = K(k)$, $\omega_3 = iK'(k)$ for $0 \leq x \leq 9$, $k^2 = 0.2, 0.8, 0.95, 0.99$. (Lemniscatic case.)

23.4(ii) Complex Variables

See Figures 23.4.8–23.4.12 for surfaces for the Weierstrass functions $\wp(z)$, $\zeta(z)$, and $\sigma(z)$. Height corresponds to the absolute value of the function and color to the phase. See also p. xiv. (The figures in this subsection may be compared with the figures in §22.3(iii).)

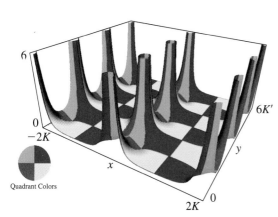

Figure 23.4.8: $\wp(x+iy)$ with $\omega_1 = K(k)$, $\omega_3 = iK'(k)$ for $-2K(k) \leq x \leq 2K(k)$, $0 \leq y \leq 6K'(k)$, $k^2 = 0.9$. (The scaling makes the lattice appear to be square.)

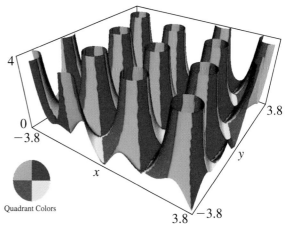

Figure 23.4.9: $\wp(x+iy;1,4i)$ for $-3.8 \leq x \leq 3.8$, $-3.8 \leq y \leq 3.8$. (The variables are unscaled and the lattice is skew.)

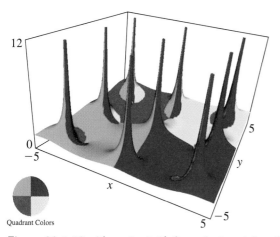

Figure 23.4.10: $\zeta(x+iy;1,0)$ for $-5 \leq x \leq 5$, $-5 \leq y \leq 5$.

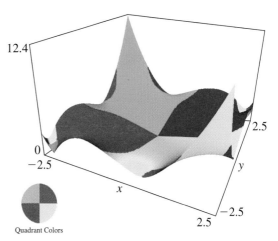

Figure 23.4.11: $\sigma(x+iy;1,i)$ for $-2.5 \leq x \leq 2.5$, $-2.5 \leq y \leq 2.5$.

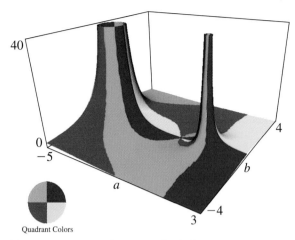

Figure 23.4.12: $\wp(3.7; a+ib, 0)$ for $-5 \le a \le 3$, $-4 \le b \le 4$. There is a double zero at $a = b = 0$ and double poles on the real axis.

23.5 Special Lattices

23.5(i) Real-Valued Functions

The Weierstrass functions take real values on the real axis iff the lattice is fixed under complex conjugation: $\mathbb{L} = \overline{\mathbb{L}}$; equivalently, when $g_2, g_3 \in \mathbb{R}$. This happens in the cases treated in the following four subsections.

23.5(ii) Rectangular Lattice

This occurs when both ω_1 and ω_3/i are real and positive. Then $\Delta > 0$ and the parallelogram with vertices at $0, 2\omega_1, 2\omega_1 + 2\omega_3, 2\omega_3$ is a rectangle.

In this case the lattice roots e_1, e_2, and e_3 are real and distinct. When they are identified as in (23.3.9)

23.5.1 $\qquad e_1 > e_2 > e_3, \quad e_1 > 0 > e_3.$

Also, e_2 and g_3 have opposite signs unless $\omega_3 = i\omega_1$, in which event both are zero.

As functions of $\Im \omega_3$, e_1 and e_2 are decreasing and e_3 is increasing.

23.5(iii) Lemniscatic Lattice

This occurs when ω_1 is real and positive and $\omega_3 = i\omega_1$. The parallelogram $0, 2\omega_1, 2\omega_1 + 2\omega_3, 2\omega_3$ is a square, and

23.5.2 $\qquad \eta_1 = i\eta_3 = \pi/(4\omega_1),$

23.5.3 $\qquad e_1 = -e_3 = \left(\Gamma\left(\tfrac{1}{4}\right)\right)^4/(32\pi\omega_1^2), \quad e_2 = 0,$

23.5.4 $\qquad g_2 = \left(\Gamma\left(\tfrac{1}{4}\right)\right)^8/(256\pi^2\omega_1^4), \quad g_3 = 0.$

Note also that in this case $\tau = i$. In consequence,

23.5.5 $\qquad k^2 = \tfrac{1}{2}, \quad K(k) = K'(k) = \left(\Gamma\left(\tfrac{1}{4}\right)\right)^2/(4\sqrt{\pi}).$

23.5(iv) Rhombic Lattice

This occurs when ω_1 is real and positive, $\Im \omega_3 > 0$, $\Re \omega_3 = \tfrac{1}{2}\omega_1$, and $\Delta < 0$. The parallelogram $0, 2\omega_1 - 2\omega_3, 2\omega_1, 2\omega_3$, is a rhombus: see Figure 23.5.1.

The lattice root e_1 is real, and $e_3 = \bar{e}_2$, with $\Im e_2 > 0$. e_1 and g_3 have the same sign unless $2\omega_3 = (1+i)\omega_1$ when both are zero: the *pseudo-lemniscatic* case. As a function of $\Im e_3$ the root e_1 is increasing. For the case $\omega_3 = e^{\pi i/3}\omega_1$ see §23.5(v).

23.5(v) Equianharmonic Lattice

This occurs when ω_1 is real and positive and $\omega_3 = e^{\pi i/3}\omega_1$. The rhombus $0, 2\omega_1 - 2\omega_3, 2\omega_1, 2\omega_3$ can be regarded as the union of two equilateral triangles: see Figure 23.5.2.

23.5.6 $\qquad \eta_1 = e^{\pi i/3}\eta_3 = \dfrac{\pi}{2\sqrt{3}\omega_1},$

and the lattice roots and invariants are given by

23.5.7 $\qquad e_1 = e^{2\pi i/3}e_3 = e^{-2\pi i/3}e_2 = \dfrac{\left(\Gamma\left(\tfrac{1}{3}\right)\right)^6}{2^{14/3}\pi^2\omega_1^2},$

23.5.8 $\qquad g_2 = 0, \quad g_3 = \dfrac{\left(\Gamma\left(\tfrac{1}{3}\right)\right)^{18}}{(4\pi\omega_1)^6}.$

Note also that in this case $\tau = e^{i\pi/3}$. In consequence,

23.5.9

$$k^2 = e^{i\pi/3}, \quad K(k) = e^{i\pi/6}K'(k) = e^{i\pi/12}\dfrac{3^{1/4}\left(\Gamma\left(\tfrac{1}{3}\right)\right)^3}{2^{7/3}\pi}.$$

23.6 Relations to Other Functions

23.6(i) Theta Functions

In this subsection $2\omega_1, 2\omega_3$ are any pair of generators of the lattice $\mathbb{L}$, and the lattice roots e_1, e_2, e_3 are given by (23.3.9).

23.6.1 $\qquad q = e^{i\pi\tau}, \quad \tau = \omega_3/\omega_1.$

23.6.2 $\qquad e_1 = \dfrac{\pi^2}{12\omega_1^2}\left(\theta_2^4(0,q) + 2\theta_4^4(0,q)\right),$

23.6.3 $\qquad e_2 = \dfrac{\pi^2}{12\omega_1^2}\left(\theta_2^4(0,q) - \theta_4^4(0,q)\right),$

23.6.4 $\qquad e_3 = -\dfrac{\pi^2}{12\omega_1^2}\left(2\theta_2^4(0,q) + \theta_4^4(0,q)\right).$

23.6.5

$$\wp(z) - e_1 = \left(\dfrac{\pi\,\theta_3(0,q)\,\theta_4(0,q)\,\theta_2(\pi z/(2\omega_1), q)}{2\omega_1\,\theta_1(\pi z/(2\omega_1), q)}\right)^2,$$

23.6.6

$$\wp(z) - e_2 = \left(\dfrac{\pi\,\theta_2(0,q)\,\theta_4(0,q)\,\theta_3(\pi z/(2\omega_1), q)}{2\omega_1\,\theta_1(\pi z/(2\omega_1), q)}\right)^2,$$

23.6.7

$$\wp(z) - e_3 = \left(\dfrac{\pi\,\theta_2(0,q)\,\theta_3(0,q)\,\theta_4(\pi z/(2\omega_1), q)}{2\omega_1\,\theta_1(\pi z/(2\omega_1), q)}\right)^2.$$

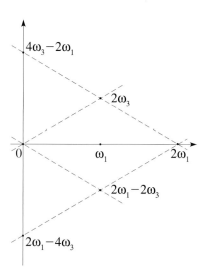

Figure 23.5.1: Rhombic lattice. $\Re(2\omega_3) = \omega_1$.

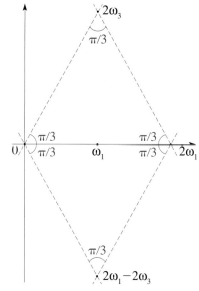

Figure 23.5.2: Equianharmonic lattice. $2\omega_3 = e^{\pi i/3} 2\omega_1$, $2\omega_1 - 2\omega_3 = e^{-\pi i/3} 2\omega_1$.

23.6.8
$$\eta_1 = -\frac{\pi^2}{12\omega_1} \frac{\theta_1'''(0,q)}{\theta_1'(0,q)}.$$

23.6.9
$$\sigma(z) = 2\omega_1 \exp\left(\frac{\eta_1 z^2}{2\omega_1}\right) \frac{\theta_1(\pi z/(2\omega_1), q)}{\pi \, \theta_1'(0,q)},$$

23.6.10
$$\sigma(\omega_1) = 2\omega_1 \frac{\exp\left(\frac{1}{2}\eta_1\omega_1\right) \theta_2(0,q)}{\pi \, \theta_1'(0,q)},$$

23.6.11
$$\sigma(\omega_2) = 2\omega_1 i \frac{\exp\left(\frac{1}{2}\eta_1\omega_1\tau^2\right) \theta_3(0,q)}{\pi q^{1/4} \, \theta_1'(0,q)},$$

23.6.12
$$\sigma(\omega_3) = -2\omega_1 \frac{\exp\left(\frac{1}{2}\eta_1\omega_1\right) \theta_4(0,q)}{\pi q^{1/4} \, \theta_1'(0,q)}.$$

With $z = \pi u/(2\omega_1)$,

23.6.13
$$\zeta(u) = \frac{\eta_1}{\omega_1} u + \frac{\pi}{2\omega_1} \frac{d}{dz} \ln \theta_1(z,q),$$

23.6.14
$$\wp(u) = \left(\frac{\pi}{2\omega_1}\right)^2 \left(\frac{\theta_1'''(0,q)}{3\theta_1'(0,q)} - \frac{d^2}{dz^2} \ln \theta_1(z,q)\right),$$

23.6.15
$$\frac{\sigma(u + \omega_j)}{\sigma(\omega_j)} = \exp\left(\eta_j u + \frac{\eta_j u^2}{2\omega_1}\right) \frac{\theta_{j+1}(z,q)}{\theta_{j+1}(0,q)}, \quad j = 1,2,3.$$

For further results for the σ-function see Lawden (1989, §6.2).

23.6(ii) Jacobian Elliptic Functions

Again, in Equations (23.6.16)–(23.6.26), $2\omega_1, 2\omega_3$ are any pair of generators of the lattice $\mathbb{L}$ and e_1, e_2, e_3 are given by (23.3.9).

23.6.16
$$k^2 = \frac{e_2 - e_3}{e_1 - e_3}, \quad k'^2 = \frac{e_1 - e_2}{e_1 - e_3},$$

23.6.17
$$K^2 = (K(k))^2 = \omega_1^2(e_1 - e_3),$$
$$K'^2 = (K(k'))^2 = \omega_3^2(e_3 - e_1).$$

23.6.18
$$e_1 = \frac{K^2}{3\omega_1^2}(1 + k'^2),$$

23.6.19
$$e_2 = \frac{K^2}{3\omega_1^2}(k^2 - k'^2),$$

23.6.20
$$e_3 = -\frac{K^2}{3\omega_1^2}(1 + k^2).$$

23.6.21
$$\wp(z) - e_1 = \frac{K^2}{\omega_1^2} \operatorname{cs}^2\left(\frac{Kz}{\omega_1}, k\right),$$

23.6.22
$$\wp(z) - e_2 = \frac{K^2}{\omega_1^2} \operatorname{ds}^2\left(\frac{Kz}{\omega_1}, k\right),$$

23.6.23
$$\wp(z) - e_3 = \frac{K^2}{\omega_1^2} \operatorname{ns}^2\left(\frac{Kz}{\omega_1}, k\right).$$

23.6.24
$$\wp(z + \omega_1) - e_1 = \left(\frac{Kk'}{\omega_1}\right)^2 \operatorname{sc}^2\left(\frac{Kz}{\omega_1}, k\right),$$

23.6.25
$$\wp(z + \omega_2) - e_2 = -\left(\frac{Kkk'}{\omega_1}\right)^2 \operatorname{sd}^2\left(\frac{Kz}{\omega_1}, k\right),$$

23.6.26
$$\wp(z + \omega_3) - e_3 = \left(\frac{Kk}{\omega_1}\right)^2 \operatorname{sn}^2\left(\frac{Kz}{\omega_1}, k\right).$$

In (23.6.27)–(23.6.29) the modulus k is given and $K = K(k)$, $K' = K(k')$ are the corresponding complete elliptic integrals (§19.2(ii)). Also, $\mathbb{L}_1, \mathbb{L}_2, \mathbb{L}_3$ are the lattices with generators $(4K, 2iK')$, $(2K - 2iK', 2K + 2iK')$, $(2K, 4iK')$, respectively.

23.6.27
$$\zeta(z|\mathbb{L}_1) - \zeta(z + 2K|\mathbb{L}_1) + \zeta(2K|\mathbb{L}_1) = \operatorname{ns}(z,k),$$

23.6.28 $\zeta(z|\mathbb{L}_2) - \zeta(z+2K|\mathbb{L}_2) + \zeta(2K|\mathbb{L}_2) = \text{ds}(z,k)$,

23.6.29
$$\zeta(z|\mathbb{L}_3) - \zeta(z+2iK'|\mathbb{L}_3) - \zeta(2iK'|\mathbb{L}_3) = \text{cs}(z,k).$$

Similar results for some of the other nine Jacobi functions can be constructed with the aid of the transformations given by Table 22.4.3, or for all nine by referring to the augmented version of Table 22.4.3 at http://dlmf.nist.gov/22.4.t3.

For representations of the Jacobi functions sn, cn, and dn as quotients of σ-functions see Lawden (1989, §§6.2, 6.3).

23.6(iii) General Elliptic Functions

For representations of general elliptic functions (§23.2(iii)) in terms of $\sigma(z)$ and $\wp(z)$ see Lawden (1989, §§8.9, 8.10), and for expansions in terms of $\zeta(z)$ see Lawden (1989, §8.11).

23.6(iv) Elliptic Integrals

Rectangular Lattice

Let z be on the perimeter of the rectangle with vertices $0, 2\omega_1, 2\omega_1 + 2\omega_3, 2\omega_3$. Then $t = \wp(z)$ is real (§§23.5(i)–23.5(ii)), and

23.6.30
$$z = \frac{1}{2}\int_t^\infty \frac{du}{\sqrt{(u-e_1)(u-e_2)(u-e_3)}},$$
$$t \geq e_1, z \in (0, \omega_1],$$

23.6.31
$$z - \omega_1 = \frac{i}{2}\int_t^{e_1} \frac{du}{\sqrt{(e_1-u)(u-e_2)(u-e_3)}},$$
$$e_2 \leq t \leq e_1, z \in [\omega_1, \omega_1+\omega_3],$$

23.6.32
$$z - \omega_3 = \frac{1}{2}\int_{e_3}^t \frac{du}{\sqrt{(e_1-u)(e_2-u)(u-e_3)}},$$
$$e_3 \leq t \leq e_2, z \in [\omega_3, \omega_1+\omega_3],$$

23.6.33
$$z = \frac{i}{2}\int_{-\infty}^t \frac{du}{\sqrt{(e_1-u)(e_2-u)(e_3-u)}},$$
$$t \leq e_3, z \in (0, \omega_3].$$

23.6.34
$$2\omega_1 = \int_{e_1}^\infty \frac{du}{\sqrt{(u-e_1)(u-e_2)(u-e_3)}}$$
$$= \int_{e_3}^{e_2} \frac{du}{\sqrt{(e_1-u)(e_2-u)(u-e_3)}},$$

23.6.35
$$2\omega_3 = i\int_{e_2}^{e_1} \frac{du}{\sqrt{(e_1-u)(u-e_2)(u-e_3)}}$$
$$= i\int_{-\infty}^{e_3} \frac{du}{\sqrt{(e_1-u)(e_2-u)(e_3-u)}}.$$

For (23.6.30)–(23.6.35) and further identities see Lawden (1989, §6.12).

See also §§19.2(i), 19.14, and Erdélyi et al. (1953b, §13.14).

For relations to symmetric elliptic integrals see §19.25(vi).

General Lattice

Let z be a point of $\mathbb{C}$ different from e_1, e_2, e_3, and define w by

23.6.36
$$w = \int_z^\infty \frac{du}{\sqrt{4u^3 - g_2 u - g_3}}$$
$$= \frac{1}{2}\int_z^\infty \frac{du}{\sqrt{(u-e_1)(u-e_2)(u-e_3)}},$$

where the integral is taken along any path from z to ∞ that does not pass through any of e_1, e_2, e_3. Then $z = \wp(w)$, where the value of w depends on the choice of path and determination of the square root; see McKean and Moll (1999, pp. 87–88 and §2.5).

23.7 Quarter Periods

23.7.1
$$\wp(\tfrac{1}{2}\omega_1) = e_1 + \sqrt{(e_1-e_3)(e_1-e_2)}$$
$$= e_1 + \omega_1^{-2}(K(k))^2 k',$$

23.7.2
$$\wp(\tfrac{1}{2}\omega_2) = e_2 - i\sqrt{(e_1-e_2)(e_2-e_3)}$$
$$= e_2 - i\omega_1^{-2}(K(k))^2 kk',$$

23.7.3
$$\wp(\tfrac{1}{2}\omega_3) = e_3 - \sqrt{(e_1-e_3)(e_2-e_3)}$$
$$= e_3 - \omega_1^{-2}(K(k))^2 k,$$

where k, k' and the square roots are real and positive when the lattice is rectangular; otherwise they are determined by continuity from the rectangular case.

23.8 Trigonometric Series and Products

23.8(i) Fourier Series

If $q = e^{i\pi\omega_3/\omega_1}$, $\Im(z/\omega_1) < 2\Im(\omega_3/\omega_1)$, and $z \notin \mathbb{L}$, then

23.8.1
$$\wp(z) + \frac{\eta_1}{\omega_1} - \frac{\pi^2}{4\omega_1^2}\csc^2\left(\frac{\pi z}{2\omega_1}\right)$$
$$= -\frac{2\pi^2}{\omega_1^2}\sum_{n=1}^\infty \frac{nq^{2n}}{1-q^{2n}}\cos\left(\frac{n\pi z}{\omega_1}\right),$$

23.8.2
$$\zeta(z) - \frac{\eta_1 z}{\omega_1} - \frac{\pi}{2\omega_1}\cot\left(\frac{\pi z}{2\omega_1}\right)$$
$$= \frac{2\pi}{\omega_1}\sum_{n=1}^\infty \frac{q^{2n}}{1-q^{2n}}\sin\left(\frac{n\pi z}{\omega_1}\right).$$

23.8(ii) Series of Cosecants and Cotangents

When $z \notin \mathbb{L}$,

23.8.3 $\quad \wp(z) = -\dfrac{\eta_1}{\omega_1} + \dfrac{\pi^2}{4\omega_1^2} \displaystyle\sum_{n=-\infty}^{\infty} \csc^2\left(\dfrac{\pi(z+2n\omega_3)}{2\omega_1}\right),$

23.8.4 $\quad \zeta(z) = \dfrac{\eta_1 z}{\omega_1} + \dfrac{\pi}{2\omega_1} \displaystyle\sum_{n=-\infty}^{\infty} \cot\left(\dfrac{\pi(z+2n\omega_3)}{2\omega_1}\right),$

where in (23.8.4) the terms in n and $-n$ are to be bracketed together (the *Eisenstein convention* or *principal value*: see Weil (1999, p. 6) or Walker (1996, p. 3)).

23.8.5 $\quad \eta_1 = \dfrac{\pi^2}{2\omega_1}\left(\dfrac{1}{6} + \displaystyle\sum_{n=1}^{\infty} \csc^2\left(\dfrac{n\pi\omega_3}{\omega_1}\right)\right),$

with similar results for η_2 and η_3 obtainable by use of (23.2.14).

23.8(iii) Infinite Products

23.8.6 $\quad \sigma(z) = \dfrac{2\omega_1}{\pi} \exp\left(\dfrac{\eta_1 z^2}{2\omega_1}\right) \sin\left(\dfrac{\pi z}{2\omega_1}\right) \displaystyle\prod_{n=1}^{\infty} \dfrac{1 - 2q^{2n}\cos(\pi z/\omega_1) + q^{4n}}{(1-q^{2n})^2},$

23.8.7 $\quad \sigma(z) = \dfrac{2\omega_1}{\pi} \exp\left(\dfrac{\eta_1 z^2}{2\omega_1}\right) \sin\left(\dfrac{\pi z}{2\omega_1}\right) \displaystyle\prod_{n=1}^{\infty} \dfrac{\sin(\pi(2n\omega_3+z)/(2\omega_1))\sin(\pi(2n\omega_3-z)/(2\omega_1))}{\sin^2(\pi n\omega_3/\omega_1)}.$

23.9 Laurent and Other Power Series

Let $z_0 (\neq 0)$ be the nearest lattice point to the origin, and define

23.9.1 $\quad c_n = (2n-1) \displaystyle\sum_{w \in \mathbb{L}\setminus\{0\}} w^{-2n}, \quad n = 2, 3, 4, \ldots.$

Then

23.9.2 $\quad \wp(z) = \dfrac{1}{z^2} + \displaystyle\sum_{n=2}^{\infty} c_n z^{2n-2}, \quad 0 < |z| < |z_0|,$

23.9.3 $\quad \zeta(z) = \dfrac{1}{z} - \displaystyle\sum_{n=2}^{\infty} \dfrac{c_n}{2n-1} z^{2n-1}, \quad 0 < |z| < |z_0|.$

Here

23.9.4 $\quad c_2 = \dfrac{1}{20}g_2, \quad c_3 = \dfrac{1}{28}g_3,$

23.9.5 $\quad c_n = \dfrac{3}{(2n+1)(n-3)} \displaystyle\sum_{m=2}^{n-2} c_m c_{n-m}, \quad n \geq 4.$

Explicit coefficients c_n in terms of c_2 and c_3 are given up to c_{19} in Abramowitz and Stegun (1964, p. 636).

For $j = 1, 2, 3$, and with e_j as in §23.3(i),

23.9.6 $\quad \wp(\omega_j + t) = e_j + (3e_j^2 - 5c_2)t^2 + (10c_2 e_j + 21c_3)t^4 + (7c_2 e_j^2 + 21c_3 e_j + 5c_2^2)t^6 + O(t^8),$

as $t \to 0$. For the next four terms see Abramowitz and Stegun (1964, (18.5.56)). Also, Abramowitz and Stegun (1964, (18.5.25)) supplies the first 22 terms in the reverted form of (23.9.2) as $1/\wp(z) \to 0$.

For $z \in \mathbb{C}$

23.9.7 $\quad \sigma(z) = \displaystyle\sum_{m,n=0}^{\infty} a_{m,n} (10c_2)^m (56c_3)^n \dfrac{z^{4m+6n+1}}{(4m+6n+1)!},$

where $a_{0,0} = 1$, $a_{m,n} = 0$ if either m or $n < 0$, and

23.9.8 $\quad a_{m,n} = 3(m+1)a_{m+1,n-1} + \tfrac{16}{3}(n+1)a_{m-2,n+1} - \tfrac{1}{3}(2m+3n-1)(4m+6n-1)a_{m-1,n}.$

For $a_{m,n}$ with $m = 0, 1, \ldots, 12$ and $n = 0, 1, \ldots, 8$, see Abramowitz and Stegun (1964, p. 637).

23.10 Addition Theorems and Other Identities

23.10(i) Addition Theorems

23.10.1 $\quad \wp(u+v) = \dfrac{1}{4}\left(\dfrac{\wp'(u) - \wp'(v)}{\wp(u) - \wp(v)}\right)^2 - \wp(u) - \wp(v),$

23.10.2 $\quad \zeta(u+v) = \zeta(u) + \zeta(v) + \dfrac{1}{2}\dfrac{\zeta''(u) - \zeta''(v)}{\zeta'(u) - \zeta'(v)},$

23.10.3 $\quad \dfrac{\sigma(u+v)\sigma(u-v)}{\sigma^2(u)\sigma^2(v)} = \wp(v) - \wp(u),$

23.10.4 $\quad \begin{aligned}&\sigma(u+v)\sigma(u-v)\sigma(x+y)\sigma(x-y)\\&+ \sigma(v+x)\sigma(v-x)\sigma(u+y)\sigma(u-y)\\&+ \sigma(x+u)\sigma(x-u)\sigma(v+y)\sigma(v-y) = 0.\end{aligned}$

For further addition-type identities for the σ-function see Lawden (1989, §6.4).

If $u + v + w = 0$, then

23.10.5 $\quad \begin{vmatrix} 1 & \wp(u) & \wp'(u) \\ 1 & \wp(v) & \wp'(v) \\ 1 & \wp(w) & \wp'(w) \end{vmatrix} = 0,$

and

23.10.6 $\quad (\zeta(u) + \zeta(v) + \zeta(w))^2 + \zeta'(u) + \zeta'(v) + \zeta'(w) = 0.$

23.10(ii) Duplication Formulas

23.10.7
$$\wp(2z) = -2\wp(z) + \frac{1}{4}\left(\frac{\wp''(z)}{\wp'(z)}\right)^2,$$

23.10.8
$$(\wp(2z)-e_1)\wp'^2(z) = \left((\wp(z)-e_1)^2 - (e_1-e_2)(e_1-e_3)\right)^2.$$
(23.10.8) continues to hold when e_1, e_2, e_3 are permuted cyclically.

23.10.9
$$\zeta(2z) = 2\zeta(z) + \frac{1}{2}\frac{\zeta'''(z)}{\zeta''(z)},$$

23.10.10
$$\sigma(2z) = -\wp'(z)\sigma^4(z).$$

23.10(iii) n-Tuple Formulas

For $n = 2, 3, \ldots,$

23.10.11
$$n^2 \wp(nz) = \sum_{j=0}^{n-1}\sum_{\ell=0}^{n-1} \wp\left(z + \frac{2j}{n}\omega_1 + \frac{2\ell}{n}\omega_3\right),$$

23.10.12
$$n\zeta(nz) = -n(n-1)(\eta_1+\eta_3) + \sum_{j=0}^{n-1}\sum_{\ell=0}^{n-1} \zeta\left(z + \frac{2j}{n}\omega_1 + \frac{2\ell}{n}\omega_3\right),$$

23.10.13
$$\sigma(nz) = A_n e^{-n(n-1)(\eta_1+\eta_3)z} \prod_{j=0}^{n-1}\prod_{\ell=0}^{n-1} \sigma\left(z + \frac{2j}{n}\omega_1 + \frac{2\ell}{n}\omega_3\right),$$

where

23.10.14
$$A_n = n\prod_{j=0}^{n-1}\prod_{\substack{\ell=0\\\ell\neq j}}^{n-1} \frac{1}{\sigma((2j\omega_1+2\ell\omega_3)/n)}.$$

Equivalently,

23.10.15
$$A_n = \left(\frac{\pi^2 G^2}{\omega_1}\right)^{n^2-1}\frac{q^{n(n-1)/2}}{i^{n-1}}\exp\left(-\frac{(n-1)\eta_1}{3\omega_1}\left((2n-1)(\omega_1^2+\omega_3^2)+3(n-1)\omega_1\omega_3\right)\right),$$

where

23.10.16
$$q = e^{\pi i \omega_3/\omega_1}, \quad G = \prod_{n=1}^{\infty}(1-q^{2n}).$$

23.10(iv) Homogeneity

For any nonzero real or complex constant c,

23.10.17
$$\wp(cz|c\mathbb{L}) = c^{-2}\wp(z|\mathbb{L}),$$

23.10.18
$$\zeta(cz|c\mathbb{L}) = c^{-1}\zeta(z|\mathbb{L}),$$

23.10.19
$$\sigma(cz|c\mathbb{L}) = c\,\sigma(z|\mathbb{L}).$$

Also, when $\mathbb{L}$ is replaced by $c\mathbb{L}$ the lattice invariants g_2 and g_3 are divided by c^4 and c^6, respectively.

For these results and further identities see Lawden (1989, §6.6) and Apostol (1990, p. 14).

23.11 Integral Representations

Let $\tau = \omega_3/\omega_1$ and

23.11.1
$$f_1(s,\tau) = \frac{\cosh^2\left(\frac{1}{2}\tau s\right)}{1 - 2e^{-s}\cosh(\tau s) + e^{-2s}},$$
$$f_2(s,\tau) = \frac{\cos^2\left(\frac{1}{2}s\right)}{1 - 2e^{i\tau s}\cos s + e^{2i\tau s}}.$$

Then

23.11.2
$$\wp(z) = \frac{1}{z^2} + 8\int_0^\infty s\left(e^{-s}\sinh^2\left(\tfrac{1}{2}zs\right)f_1(s,\tau) + e^{i\tau s}\sin^2\left(\tfrac{1}{2}zs\right)f_2(s,\tau)\right)ds,$$

and

23.11.3
$$\zeta(z) = \frac{1}{z} + \int_0^\infty \left(e^{-s}(zs-\sinh(zs))f_1(s,\tau) - e^{i\tau s}(zs-\sin(zs))f_2(s,\tau)\right)ds,$$

provided that $-1 < \Re(z+\tau) < 1$ and $|\Im z| < \Im\tau$.

23.12 Asymptotic Approximations

If $q\,(=e^{\pi i\omega_3/\omega_1}) \to 0$ with ω_1 and z fixed, then

23.12.1
$$\wp(z) = \frac{\pi^2}{4\omega_1^2}\left(-\frac{1}{3} + \csc^2\left(\frac{\pi z}{2\omega_1}\right) + 8\left(1 - \cos\left(\frac{\pi z}{\omega_1}\right)\right)q^2 + O(q^4)\right),$$

23.12.2
$$\zeta(z) = \frac{\pi^2}{4\omega_1^2}\left(\frac{z}{3} + \frac{2\omega_1}{\pi}\cot\left(\frac{\pi z}{\omega_1}\right) - 8\left(z - \frac{\omega_1}{\pi}\sin\left(\frac{\pi z}{\omega_1}\right)\right)q^2 + O(q^4)\right),$$

23.12.3
$$\sigma(z) = \frac{2\omega_1}{\pi}\exp\left(\frac{\pi^2 z^2}{24\omega_1^2}\right)\sin\left(\frac{\pi z}{2\omega_1}\right) \times \left(1 - \left(\frac{\pi^2 z^2}{\omega_1^2} - 4\sin^2\left(\frac{\pi z}{2\omega_1}\right)\right)q^2 + O(q^4)\right),$$

provided that $z \notin \mathbb{L}$ in the case of (23.12.1) and (23.12.2). Also,

23.12.4
$$\eta_1 = \frac{\pi^2}{4\omega_1}\left(\frac{1}{3} - 8q^2 + O(q^4)\right),$$

with similar results for η_2 and η_3 obtainable by use of (23.2.14).

23.13 Zeros

For information on the zeros of $\wp(z)$ see Eichler and Zagier (1982).

23.14 Integrals

23.14.1 $$\int \wp(z)\,dz = -\zeta(z),$$

23.14.2 $$\int \wp^2(z)\,dz = \tfrac{1}{6}\wp'(z) + \tfrac{1}{12}g_2 z,$$

23.14.3 $$\int \wp^3(z)\,dz = \tfrac{1}{120}\wp'''(z) - \tfrac{3}{20}g_2\zeta(z) + \tfrac{1}{10}g_3 z.$$

For further integrals see Gröbner and Hofreiter (1949, Vol. 1, pp. 161–162), Gradshteyn and Ryzhik (2000, p. 622), and Prudnikov et al. (1990, pp. 51–52).

Modular Functions

23.15 Definitions

23.15(i) General Modular Functions

In §§23.15–23.19, k and k' ($\in \mathbb{C}$) denote the Jacobi modulus and complementary modulus, respectively, and $q = e^{i\pi\tau}$ ($\Im\tau > 0$) denotes the nome; compare §§20.1 and 22.1. Thus

23.15.1 $$q = \exp\left(-\pi \frac{K'(k)}{K(k)}\right),$$

23.15.2 $$k = \frac{\theta_2^2(0,q)}{\theta_3^2(0,q)}, \quad k' = \frac{\theta_4^2(0,q)}{\theta_3^2(0,q)}.$$

Also $\mathcal{A}$ denotes a bilinear transformation on τ, given by

23.15.3 $$\mathcal{A}\tau = \frac{a\tau + b}{c\tau + d},$$

in which a, b, c, d are integers, with

23.15.4 $$ad - bc = 1.$$

The set of all bilinear transformations of this form is denoted by $\mathrm{SL}(2,\mathbb{Z})$ (Serre (1973, p. 77)).

A *modular function* $f(\tau)$ is a function of τ that is meromorphic in the half-plane $\Im\tau > 0$, and has the property that for all $\mathcal{A} \in \mathrm{SL}(2,\mathbb{Z})$, or for all $\mathcal{A}$ belonging to a subgroup of $\mathrm{SL}(2,\mathbb{Z})$,

23.15.5 $$f(\mathcal{A}\tau) = c_{\mathcal{A}}(c\tau + d)^\ell f(\tau), \qquad \Im\tau > 0,$$

where $c_{\mathcal{A}}$ is a constant depending only on $\mathcal{A}$, and ℓ (the *level*) is an integer or half an odd integer. (Some references refer to 2ℓ as the level). If, as a function of q, $f(\tau)$ is analytic at $q = 0$, then $f(\tau)$ is called a *modular form*. If, in addition, $f(\tau) \to 0$ as $q \to 0$, then $f(\tau)$ is called a *cusp form*.

23.15(ii) Functions $\lambda(\tau)$, $J(\tau)$, $\eta(\tau)$

Elliptic Modular Function

23.15.6 $$\lambda(\tau) = \frac{\theta_2^4(0,q)}{\theta_3^4(0,q)};$$

compare also (23.15.2).

Klein's Complete Invariant

23.15.7 $$J(\tau) = \frac{\left(\theta_2^8(0,q) + \theta_3^8(0,q) + \theta_4^8(0,q)\right)^3}{54\left(\theta_1'(0,q)\right)^8},$$

where (as in §20.2(i))

23.15.8 $$\theta_1'(0,q) = \partial\theta_1(z,q)/\partial z\big|_{z=0}.$$

Dedekind's Eta Function (or Dedekind Modular Function)

23.15.9 $$\eta(\tau) = \left(\tfrac{1}{2}\theta_1'(0,q)\right)^{1/3} = e^{i\pi\tau/12}\theta_3\left(\tfrac{1}{2}\pi(1+\tau)\big|3\tau\right).$$

In (23.15.9) the branch of the cube root is chosen to agree with the second equality; in particular, when τ lies on the positive imaginary axis the cube root is real and positive.

23.16 Graphics

See Figures 23.16.1–23.16.3 for the modular functions λ, J, and η. In Figures 23.16.2 and 23.16.3, height corresponds to the absolute value of the function and color to the phase. See also p. xiv.

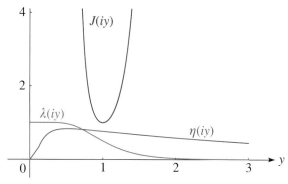

Figure 23.16.1: Modular functions $\lambda(iy)$, $J(iy)$, $\eta(iy)$ for $0 \leq y \leq 3$. See also Figure 20.3.2.

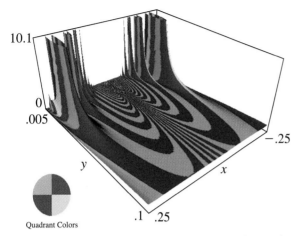

Figure 23.16.2: Elliptic modular function $\lambda(x+iy)$ for $-0.25 \le x \le 0.25$, $0.005 \le y \le 0.1$.

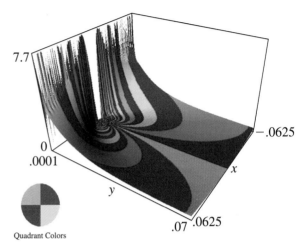

Figure 23.16.3: Dedekind's eta function $\eta(x+iy)$ for $-0.0625 \le x \le 0.0625$, $0.0001 \le y \le 0.07$.

23.17 Elementary Properties

23.17(i) Special Values

23.17.1 $\quad \lambda(i) = \tfrac{1}{2}, \quad \lambda(e^{\pi i/3}) = e^{\pi i/3},$

23.17.2 $\quad J(i) = 1, \quad J(e^{\pi i/3}) = 0,$

23.17.3
$$\eta(i) = \frac{\Gamma\!\left(\tfrac{1}{4}\right)}{2\pi^{3/4}}, \quad \eta(e^{\pi i/3}) = \frac{3^{1/8}\left(\Gamma\!\left(\tfrac{1}{3}\right)\right)^{3/2}}{2\pi} e^{\pi i/24}.$$

For further results for $J(\tau)$ see Cohen (1993, p. 376).

23.17(ii) Power and Laurent Series

When $|q| < 1$

23.17.4 $\quad \lambda(\tau) = 16q(1 - 8q + 44q^2 + \cdots),$

23.17.5
$1728 J(\tau) = q^{-2} + 744 + 1\,96884 q^2 + 214\,93760 q^4 + \cdots,$

23.17.6 $\quad \eta(\tau) = \sum_{n=-\infty}^{\infty} (-1)^n q^{(6n+1)^2/12}.$

In (23.17.5) for terms up to q^{48} see Zuckerman (1939), and for terms up to q^{100} see van Wijngaarden (1953). See also Apostol (1990, p. 22).

23.17(iii) Infinite Products

23.17.7 $\quad \lambda(\tau) = 16q \prod_{n=1}^{\infty} \left(\frac{1+q^{2n}}{1+q^{2n-1}}\right)^8,$

23.17.8 $\quad \eta(\tau) = q^{1/12} \prod_{n=1}^{\infty} (1 - q^{2n}),$

with $q^{1/12} = e^{i\pi\tau/12}$.

23.18 Modular Transformations

Elliptic Modular Function

$\lambda(\mathcal{A}\tau)$ equals

23.18.1
$$\lambda(\tau), \quad 1 - \lambda(\tau), \quad \frac{1}{\lambda(\tau)},$$
$$\frac{1}{1-\lambda(\tau)}, \quad \frac{\lambda(\tau)}{\lambda(\tau)-1}, \quad 1 - \frac{1}{\lambda(\tau)},$$

according as the elements $\begin{bmatrix} a & b \\ c & d \end{bmatrix}$ of $\mathcal{A}$ in (23.15.3) have the respective forms

23.18.2
$$\begin{bmatrix} o & e \\ e & o \end{bmatrix}, \begin{bmatrix} e & o \\ o & e \end{bmatrix}, \begin{bmatrix} o & e \\ o & o \end{bmatrix},$$
$$\begin{bmatrix} e & o \\ o & o \end{bmatrix}, \begin{bmatrix} o & o \\ e & o \end{bmatrix}, \begin{bmatrix} o & o \\ o & e \end{bmatrix}.$$

Here e and o are generic symbols for even and odd integers, respectively. In particular, if $a-1, b, c$, and $d-1$ are all even, then

23.18.3 $\quad \lambda(\mathcal{A}\tau) = \lambda(\tau),$

and $\lambda(\tau)$ is a cusp form of level zero for the corresponding subgroup of $\mathrm{SL}(2,\mathbb{Z})$.

Klein's Complete Invariant

23.18.4 $\quad J(\mathcal{A}\tau) = J(\tau).$

$J(\tau)$ is a modular form of level zero for $\mathrm{SL}(2,\mathbb{Z})$.

23.19 Interrelations

Dedekind's Eta Function

23.18.5 $\quad \eta(\mathcal{A}\tau) = \varepsilon(\mathcal{A})\left(-i(c\tau+d)\right)^{1/2} \eta(\tau),$

where the square root has its principal value and

23.18.6 $\quad \varepsilon(\mathcal{A}) = \exp\left(\pi i \left(\dfrac{a+d}{12c} + s(-d,c)\right)\right),$

23.18.7 $\quad s(d,c) = \displaystyle\sum_{\substack{r=1 \\ (r,c)=1}}^{c-1} \dfrac{r}{c}\left(\dfrac{dr}{c} - \left\lfloor\dfrac{dr}{c}\right\rfloor - \dfrac{1}{2}\right), \quad c > 0.$

Here the notation $(r,c) = 1$ means that the sum is confined to those values of r that are relatively prime to c. See §27.14(iii) and Apostol (1990, pp. 48 and 51–53). Note that $\eta(\tau)$ is of level $\frac{1}{2}$.

23.19 Interrelations

23.19.1 $\quad \lambda(\tau) = 16\left(\dfrac{\eta^2(2\tau)\,\eta(\frac{1}{2}\tau)}{\eta^3(\tau)}\right)^8,$

23.19.2 $\quad J(\tau) = \dfrac{4}{27}\dfrac{\left(1-\lambda(\tau)+\lambda^2(\tau)\right)^3}{\left(\lambda(\tau)\left(1-\lambda(\tau)\right)\right)^2},$

23.19.3 $\quad J(\tau) = \dfrac{g_2^3}{g_2^3 - 27g_3^2},$

where g_2, g_3 are the invariants of the lattice $\mathbb{L}$ with generators 1 and τ; see §23.3(i).

Also, with Δ defined as in (23.3.4),

23.19.4 $\quad \Delta = (2\pi)^{12}\,\eta^{24}(\tau).$

Applications

23.20 Mathematical Applications

23.20(i) Conformal Mappings

Rectangular Lattice

The boundary of the rectangle R, with vertices $0, \omega_1, \omega_1 + \omega_3, \omega_3$, is mapped strictly monotonically by $\wp$ onto the real line with $0 \to \infty$, $\omega_1 \to e_1$, $\omega_1 + \omega_3 \to e_2$, $\omega_3 \to e_3$, $0 \to -\infty$. There is a unique point $z_0 \in [\omega_1, \omega_1+\omega_3] \cup [\omega_1+\omega_3, \omega_3]$ such that $\wp(z_0) = 0$. The interior of R is mapped one-to-one onto the lower half-plane.

Rhombic Lattice

The two pairs of edges $[0,\omega_1] \cup [\omega_1, 2\omega_3]$ and $[2\omega_3, 2\omega_3 - \omega_1] \cup [2\omega_3 - \omega_1, 0]$ of R are each mapped strictly monotonically by $\wp$ onto the real line, with $0 \to \infty$, $\omega_1 \to e_1$, $2\omega_3 \to -\infty$; similarly for the other pair of edges. For each pair of edges there is a unique point z_0 such that $\wp(z_0) = 0$.

The interior of the rectangle with vertices $0, \omega_1, 2\omega_3, 2\omega_3 - \omega_1$ is mapped two-to-one onto the lower half-plane. The interior of the rectangle with vertices $0, \omega_1, \frac{1}{2}\omega_1 + \omega_3, \frac{1}{2}\omega_1 - \omega_3$ is mapped one-to-one onto the lower half-plane with a cut from e_3 to $\wp\left(\frac{1}{2}\omega_1 + \omega_3\right)$ ($= \wp\left(\frac{1}{2}\omega_1 - \omega_3\right)$). The cut is the image of the edge from $\frac{1}{2}\omega_1 + \omega_3$ to $\frac{1}{2}\omega_1 - \omega_3$ and is not a line segment.

For examples of conformal mappings of the function $\wp(z)$, see Abramowitz and Stegun (1964, pp. 642–648, 654–655, and 659–60).

For conformal mappings via modular functions see Apostol (1990, §2.7).

23.20(ii) Elliptic Curves

An algebraic curve that can be put either into the form

23.20.1 $\quad C: y^2 = x^3 + ax + b,$

or equivalently, on replacing x by x/z and y by y/z (projective coordinates), into the form

23.20.2 $\quad C: y^2 z = x^3 + axz^2 + bz^3,$

is an example of an *elliptic curve* (§22.18(iv)). Here a and b are real or complex constants.

Points $P = (x,y)$ on the curve can be parametrized by $x = \wp(z;g_2,g_3)$, $2y = \wp'(z;g_2,g_3)$, where $g_2 = -4a$ and $g_3 = -4b$: in this case we write $P = P(z)$. The curve C is made into an abelian group (Macdonald (1968, Chapter 5)) by defining the zero element $o = (0,1,0)$ as the point at infinity, the negative of $P = (x,y)$ by $-P = (x,-y)$, and generally $P_1 + P_2 + P_3 = 0$ on the curve iff the points P_1, P_2, P_3 are collinear. It follows from the addition formula (23.10.1) that the points $P_j = P(z_j)$, $j = 1,2,3$, have zero sum iff $z_1 + z_2 + z_3 \in \mathbb{L}$, so that addition of points on the curve C corresponds to addition of parameters z_j on the torus $\mathbb{C}/\mathbb{L}$; see McKean and Moll (1999, §§2.11, 2.14).

In terms of (x,y) the addition law can be expressed $(x,y) + o = (x,y)$, $(x,y) + (x,-y) = o$; otherwise $(x_1, y_1) + (x_2, y_2) = (x_3, y_3)$, where

23.20.3 $\quad x_3 = m^2 - x_1 - x_2, \quad y_3 = -m(x_3 - x_1) - y_1,$

and

23.20.4 $\quad m = \begin{cases} (3x_1^2 + a)/(2y_1), & P_1 = P_2, \\ (y_2 - y_1)/(x_2 - x_1), & P_1 \neq P_2. \end{cases}$

If $a, b \in \mathbb{R}$, then C intersects the plane $\mathbb{R}^2$ in a curve that is connected if $\Delta \equiv 4a^3 + 27b^2 > 0$; if $\Delta < 0$, then the intersection has two components, one of which is a closed loop. These cases correspond to rhombic and rectangular lattices, respectively. The addition law states that to find the sum of two points, take the third intersection with C of the chord joining them (or the tangent if they coincide); then its reflection in the x-axis gives the required sum. The geometric nature of this construction is illustrated in McKean and Moll (1999, §2.14), Koblitz (1993, §§6, 7), and Silverman and Tate (1992, Chapter 1, §§3, 4): each of these references makes a connection with the addition theorem (23.10.1).

If $a, b \in \mathbb{Q}$, then by rescaling we may assume $a, b \in \mathbb{Z}$. Let T denote the set of points on C that are of finite order (that is, those points P for which there exists a positive integer n with $nP = o$), and let I, K be the sets of points with integer and rational coordinates, respectively. Then $\emptyset \subseteq T \subseteq I \subseteq K \subseteq C$. Both T, K are subgroups of C, though I may not be. K always has the form $T \times \mathbb{Z}^r$ (*Mordell's Theorem*: Silverman and Tate (1992, Chapter 3, §5)); the determination of r, the rank of K, raises questions of great difficulty, many of which are still open. Both T and I are finite sets. T must have one of the forms $\mathbb{Z}/(n\mathbb{Z})$, $1 \leq n \leq 10$ or $n = 12$, or $(\mathbb{Z}/(2\mathbb{Z})) \times (\mathbb{Z}/(2n\mathbb{Z}))$, $1 \leq n \leq 4$. To determine T, we make use of the fact that if $(x, y) \in T$ then y^2 must be a divisor of Δ; hence there are only a finite number of possibilities for y. Values of x are then found as integer solutions of $x^3 + ax + b - y^2 = 0$ (in particular x must be a divisor of $b - y^2$). The resulting points are then tested for finite order as follows. Given P, calculate $2P$, $4P$, $8P$ by doubling as above. If any of these quantities is zero, then the point has finite order. If any of $2P$, $4P$, $8P$ is not an integer, then the point has infinite order. Otherwise observe any equalities between P, $2P$, $4P$, $8P$, and their negatives. The order of a point (if finite and not already determined) can have only the values 3, 5, 6, 7, 9, 10, or 12, and so can be found from $2P = -P$, $4P = -P$, $4P = -2P$, $8P = P$, $8P = -P$, $8P = -2P$, or $8P = -4P$. If none of these equalities hold, then P has infinite order.

For extensive tables of elliptic curves see Cremona (1997, pp. 84–340).

23.20(iii) Factorization

§27.16 describes the use of primality testing and factorization in cryptography. For applications of the Weierstrass function and the elliptic curve method to these problems see Bressoud (1989) and Koblitz (1999).

23.20(iv) Modular and Quintic Equations

The *modular equation* of degree p, p prime, is an algebraic equation in $\alpha = \lambda(p\tau)$ and $\beta = \lambda(\tau)$. For $p = 2, 3, 5, 7$ and with $u = \alpha^{1/4}$, $v = \beta^{1/4}$, the modular equation is as follows:

23.20.5 $$v^8(1 + u^8) = 4u^4, \qquad p = 2,$$

23.20.6 $$u^4 - v^4 + 2uv(1 - u^2v^2) = 0, \qquad p = 3,$$

23.20.7 $$u^6 - v^6 + 5u^2v^2(u^2 - v^2) + 4uv(1 - u^4v^4) = 0, \quad p = 5,$$

23.20.8 $$(1 - u^8)(1 - v^8) = (1 - uv)^8, \qquad p = 7.$$

For further information, including the application of (23.20.7) to the solution of the general quintic equation, see Borwein and Borwein (1987, Chapter 4).

23.20(v) Modular Functions and Number Theory

For applications of modular functions to number theory see §27.14(iv) and Apostol (1990). See also Silverman and Tate (1992), Serre (1973, Part 2, Chapters 6, 7), Koblitz (1993), and Cornell et al. (1997).

23.21 Physical Applications

23.21(i) Classical Dynamics

In §22.19(ii) it is noted that Jacobian elliptic functions provide a natural basis of solutions for problems in Newtonian classical dynamics with quartic potentials in canonical form $(1 - x^2)(1 - k^2 x^2)$. The Weierstrass function $\wp$ plays a similar role for cubic potentials in canonical form $g_3 + g_2 x - 4x^3$. See, for example, Lawden (1989, Chapter 7) and Whittaker (1964, Chapters 4–6).

23.21(ii) Nonlinear Evolution Equations

Airault et al. (1977) applies the function $\wp$ to an integrable classical many-body problem, and relates the solutions to nonlinear partial differential equations. For applications to soliton solutions of the Korteweg–de Vries (KdV) equation see McKean and Moll (1999, p. 91), Deconinck and Segur (2000), and Walker (1996, §8.1).

23.21(iii) Ellipsoidal Coordinates

Ellipsoidal coordinates (ξ, η, ζ) may be defined as the three roots ρ of the equation

23.21.1 $$\frac{x^2}{\rho - e_1} + \frac{y^2}{\rho - e_2} + \frac{z^2}{\rho - e_3} = 1,$$

where x,y,z are the corresponding Cartesian coordinates and e_1, e_2, e_3 are constants. The Laplacian operator ∇^2 (§1.5(ii)) is given by

23.21.2
$$(\eta - \zeta)(\zeta - \xi)(\xi - \eta)\nabla^2 = (\zeta - \eta)f(\xi)f'(\xi)\frac{\partial}{\partial \xi}$$
$$+ (\xi - \zeta)f(\eta)f'(\eta)\frac{\partial}{\partial \eta}$$
$$+ (\eta - \xi)f(\zeta)f'(\zeta)\frac{\partial}{\partial \zeta},$$

where

23.21.3 $\quad f(\rho) = 2\left((\rho - e_1)(\rho - e_2)(\rho - e_3)\right)^{1/2}$.

Another form is obtained by identifying e_1, e_2, e_3 as lattice roots (§23.3(i)), and setting

23.21.4 $\quad \xi = \wp(u), \quad \eta = \wp(v), \quad \zeta = \wp(w).$

Then

23.21.5
$$(\wp(v) - \wp(w))(\wp(w) - \wp(u))(\wp(u) - \wp(v))\nabla^2$$
$$= (\wp(w) - \wp(v))\frac{\partial^2}{\partial u^2} + (\wp(u) - \wp(w))\frac{\partial^2}{\partial v^2}$$
$$+ (\wp(v) - \wp(u))\frac{\partial^2}{\partial w^2}.$$

See also §29.18(ii).

23.21(iv) Modular Functions

Physical applications of modular functions include:

- Quantum field theory. See Witten (1987).
- Statistical mechanics. See Baxter (1982, p. 434) and Itzykson and Drouffe (1989, §9.3).
- String theory. See Green et al. (1988a, §8.2) and Polchinski (1998, §7.2).

Computation

23.22 Methods of Computation

23.22(i) Function Values

Given ω_1 and ω_3, with $\Im(\omega_3/\omega_1) > 0$, the nome q is computed from $q = e^{i\pi\omega_3/\omega_1}$. For $\wp(z)$ we apply (23.6.2) and (23.6.5), generating all needed values of the theta functions by the methods described in §20.14.

The functions $\zeta(z)$ and $\sigma(z)$ are computed in a similar manner: the former by replacing u and z in (23.6.13) by z and $\pi z/(2\omega_1)$, respectively, and also referring to (23.6.8); the latter by applying (23.6.9).

The modular functions $\lambda(\tau)$, $J(\tau)$, and $\eta(\tau)$ are also obtainable in a similar manner from their definitions in §23.15(ii).

23.22(ii) Lattice Calculations

Starting from Lattice

Suppose that the lattice $\mathbb{L}$ is given. Then a pair of generators $2\omega_1$ and $2\omega_3$ can be chosen in an almost canonical way as follows. For $2\omega_1$ choose a nonzero point of $\mathbb{L}$ of smallest absolute value. (There will be 2, 4, or 6 possible choices.) For $2\omega_3$ choose a nonzero point that is not a multiple of $2\omega_1$ and is such that $\Im\tau > 0$ and $|\tau|$ is as small as possible, where $\tau = \omega_3/\omega_1$. (There will be either 1 or 2 possible choices.) This yields a pair of generators that satisfy $\Im\tau > 0$, $|\Re\tau| \leq \frac{1}{2}$, $|\tau| > 1$. In consequence, $q = e^{i\pi\omega_3/\omega_1}$ satisfies $|q| \leq e^{-\pi\sqrt{3}/2} = 0.0658\ldots$. The corresponding values of e_1, e_2, e_3 are calculated from (23.6.2)–(23.6.4), then g_2 and g_3 are obtained from (23.3.6) and (23.3.7).

Starting from Invariants

Suppose that the invariants $g_2 = c$, $g_3 = d$, are given, for example in the differential equation (23.3.10) or via coefficients of an elliptic curve (§23.20(ii)). The determination of suitable generators $2\omega_1$ and $2\omega_3$ is the classical *inversion problem* (Whittaker and Watson (1927, §21.73), McKean and Moll (1999, §2.12); see also §20.9(i) and McKean and Moll (1999, §2.16)). This problem is solvable as follows:

(a) In the general case, given by $cd \neq 0$, we compute the roots α, β, γ, say, of the cubic equation $4t^3 - ct - d = 0$; see §1.11(iii). These roots are necessarily distinct and represent e_1, e_2, e_3 in some order.

If c and d are real, then e_1, e_2, e_3 can be identified via (23.5.1), and k^2, k'^2 obtained from (23.6.16).

If c and d are not both real, then we label α, β, γ so that the triangle with vertices α, β, γ is positively oriented and $[\alpha, \gamma]$ is its longest side (chosen arbitrarily if there is more than one). In particular, if α, β, γ are collinear, then we label them so that β is on the line segment (α, γ). In consequence, $k^2 = (\beta - \gamma)/(\alpha - \gamma)$, $k'^2 = (\alpha - \beta)/(\alpha - \gamma)$ satisfy $\Im k^2 \geq 0 \geq \Im k'^2$ (with strict inequality unless α, β, γ are collinear); also $|k^2|, |k'^2| \leq 1$.

Finally, on taking the principal square roots of k^2 and k'^2 we obtain values for k and k' that lie in the 1st and 4th quadrants, respectively, and $2\omega_1$, $2\omega_3$ are given by

23.22.1
$$2\omega_1 M(1, k') = -2i\omega_3 M(1, k)$$
$$= \frac{\pi}{3}\sqrt{\frac{c(2 + k^2 k'^2)(k'^2 - k^2)}{d(1 - k^2 k'^2)}},$$

where M denotes the arithmetic-geometric mean (see §§19.8(i) and 22.20(ii)). This process yields 2 possible pairs $(2\omega_1, 2\omega_3)$, corresponding to the 2 possible choices of the square root.

(b) If $d = 0$, then

23.22.2
$$2\omega_1 = -2i\omega_3 = \frac{\left(\Gamma\left(\tfrac{1}{4}\right)\right)^2}{2\sqrt{\pi}c^{1/4}}.$$

There are 4 possible pairs $(2\omega_1, 2\omega_3)$, corresponding to the 4 rotations of a square lattice. The lemniscatic case occurs when $c > 0$ and $\omega_1 > 0$.

(c) If $c = 0$, then

23.22.3
$$2\omega_1 = 2e^{-\pi i/3}\omega_3 = \frac{\left(\Gamma\left(\tfrac{1}{3}\right)\right)^3}{2\pi d^{1/6}}.$$

There are 6 possible pairs $(2\omega_1, 2\omega_3)$, corresponding to the 6 rotations of a lattice of equilateral triangles. The equianharmonic case occurs when $d > 0$ and $\omega_1 > 0$.

Example

Assume $c = g_2 = -4(3 - 2i)$ and $d = g_3 = 4(4 - 2i)$. Then $\alpha = -1 - 2i$, $\beta = 1$, $\gamma = 2i$; $k^2 = (7 + 6i)/17$, and $k'^2 = (10 - 6i)/17$. Working to 6 decimal places we obtain

23.22.4
$$\begin{aligned}2\omega_1 &= 0.867568 + i1.466607,\\ 2\omega_3 &= -1.223741 + i1.328694,\\ \tau &= 0.305480 + i1.015109.\end{aligned}$$

23.23 Tables

Table 18.2 in Abramowitz and Stegun (1964) gives values of $\wp(z)$, $\wp'(z)$, and $\zeta(z)$ to 7 or 8D in the rectangular and rhombic cases, normalized so that $\omega_1 = 1$ and $\omega_3 = ia$ (rectangular case), or $\omega_1 = 1$ and $\omega_3 = \tfrac{1}{2} + ia$ (rhombic case), for $a = 1.00, 1.05, 1.1, 1.2, 1.4, 2, 4$. The values are tabulated on the real and imaginary z-axes, mostly ranging from 0 to 1 or i in steps of length 0.05, and in the case of $\wp(z)$ the user may deduce values for complex z by application of the addition theorem (23.10.1).

Abramowitz and Stegun (1964) also includes other tables to assist the computation of the Weierstrass functions, for example, the generators as functions of the lattice invariants g_2 and g_3.

For earlier tables related to Weierstrass functions see Fletcher et al. (1962, pp. 503–505) and Lebedev and Fedorova (1960, pp. 223–226).

23.24 Software

See http://dlmf.nist.gov/23.24.

References

General References

The main references used in writing this chapter are Lawden (1989, Chapters 6, 7, 9), McKean and Moll (1999, Chapters 1–5), Walker (1996, Chapter 7 and §§3.4, 8.4), and Whittaker and Watson (1927, Chapter 20 and §21.7). For additional bibliographic reading see Apostol (1990, Chapters 1–6), Copson (1935, Chapters 13 and 15), Erdélyi et al. (1953b, §§13.12–13.15 and 13.24), and Koblitz (1993, Chapters 1–4).

Sources

The following list gives the references or other indications of proofs that were used in constructing the various sections of this chapter. These sources supplement the references that are quoted in the text.

§23.2 Whittaker and Watson (1927, §§20.2–20.21, 20.4–20.421), Walker (1996, §3.1), Lawden (1989, Chapter 6). For (23.2.16) differentiate (23.2.15) and use (23.2.6). For (23.2.17) use (23.2.15) and induction.

§23.3 Whittaker and Watson (1927, §§20.22, 20.32, 21.73), Lawden (1989, §6.7), Walker (1996, §3.4). (23.3.11) follows from (23.3.3), (23.3.10). For (23.3.12), (23.3.13) differentiate (23.3.10).

§23.4 These graphics were produced at NIST.

§23.5 Walker (1996, §§7.5, 8.4.2). (Some errors in §7.5 are corrected here.)

§23.6 For (23.6.2)–(23.6.7) see Walker (1996, pp. 94 and 103). For (23.6.8) and (23.6.9) see Whittaker and Watson (1927, §21.43). For (23.6.10)–(23.6.12) use (23.6.9). For (23.6.13) and (23.6.14) see Lawden (1989, §6.6). For (23.6.15) combine (20.2.6) and (23.6.9). For (23.6.16) and (23.6.17) combine (23.6.2)–(23.6.4) with (22.2.2) and (20.7.5). For (23.6.18)–(23.6.20) combine (20.9.1) and (20.9.2) with (23.6.2)–(23.6.4). For (23.6.21)–(23.6.23) combine (23.6.5)–(23.6.7) with (22.2.4)–(22.2.9). For (23.6.24)–(23.6.26) combine (23.6.21)–(23.6.23) with §22.4(iii). (23.6.27)–(23.6.29) can be verified by matching periods, poles, and residues as in Lawden (1989, §8.11).

§23.7 Lawden (1989, p. 182).

§23.8 Lawden (1989, §6.5, pp. 183–184, §8.6).

§**23.9** For (23.9.2)–(23.9.5) equate coefficients in (23.2.4), (23.2.5), and also apply (23.10.1), (23.10.2). The first two coefficients in the Maclaurin expansion (23.9.6) are given by (23.3.9), (23.2.10); the others are obtained from §23.3(ii) combined with (23.9.4). (23.9.7) follows from (23.2.8) and (23.9.3).

§**23.10** Whittaker and Watson (1927, §§20.3–20.311, 20.41), Lawden (1989, pp. 152–158, 161–162). For (23.10.7), (23.10.9), (23.10.10) let $v \to u$ in (23.10.1)–(23.10.3). For (23.10.8) see Walker (1996, p. 83). For (23.10.11) and (23.10.12) compare the poles and residues of the two sides. (23.10.13) follows by integration. For (23.10.15) combine (23.10.14), (23.8.7), and (4.21.35).

§**23.11** Dienstfrey and Huang (2006).

§**23.12** These approximations follow from the expansions given in §23.8(ii). For (23.12.4) use Lawden (1989, Eq. 6.2.7) and (20.4.8).

§**23.14** To verify these results differentiate and use (23.2.7), §23.3(ii).

§**23.15** Apostol (1990, Chapters 1, 2), Walker (1996, Chapter 7), McKean and Moll (1999, Chapters 4, 6). For (23.15.9) use (20.5.3) and (23.17.8).

§**23.16** These graphics were produced at NIST.

§**23.17** Walker (1996, §7.5). For (23.17.4)–(23.17.6) combine §23.15(ii) with the q-expansions of the theta functions obtained by setting $z = 0$ in §20.2(i). For (23.17.7), (23.17.8) combine (23.15.6), (23.15.9), and (20.5.1)–(20.5.3).

§**23.18** See Walker (1996, Chapter 7), Ahlfors (1966, pp. 271–274), and Serre (1973, Chapter 7). For (23.18.4)–(23.18.7) see Apostol (1990, pp. 17, 52).

§**23.19** Apostol (1990, Chapters 2, 3), Serre (1973, Chapter 7). (23.19.4) follows from (20.5.3) and (23.17.8).

§**23.20** McKean and Moll (1999, §2.8).

§**23.21** Jones (1964, pp. 31–33).

§**23.22** For (23.22.1) combine (23.6.2)–(23.6.4) and §23.10(iv). (23.22.2) and (23.22.3) follow from (23.5.3) and (23.5.7), respectively.

000# Chapter 24

Bernoulli and Euler Polynomials

K. Dilcher[1]

Notation — **588**
- 24.1 Special Notation — 588

Properties — **588**
- 24.2 Definitions and Generating Functions — 588
- 24.3 Graphs — 589
- 24.4 Basic Properties — 589
- 24.5 Recurrence Relations — 591
- 24.6 Explicit Formulas — 591
- 24.7 Integral Representations — 592
- 24.8 Series Expansions — 592
- 24.9 Inequalities — 593
- 24.10 Arithmetic Properties — 593
- 24.11 Asymptotic Approximations — 593
- 24.12 Zeros — 594
- 24.13 Integrals — 594
- 24.14 Sums — 595
- 24.15 Related Sequences of Numbers — 595
- 24.16 Generalizations — 596

Applications — **597**
- 24.17 Mathematical Applications — 597
- 24.18 Physical Applications — 598

Computation — **598**
- 24.19 Methods of Computation — 598
- 24.20 Tables — 598
- 24.21 Software — 598

References — **598**

---

[1]Dalhousie University, Halifax, Nova Scotia, Canada.
Acknowledgments: This chapter is based in part on Abramowitz and Stegun (1964, Chapter 23) by E. V. Haynsworth and K. Goldberg.
Copyright © 2009 National Institute of Standards and Technology. All rights reserved.

Notation

24.1 Special Notation

(For other notation see pp. xiv and 873.)

j, k, ℓ, m, n	integers, nonnegative unless stated otherwise.
t, x	real or complex variables.
p	prime.
$p \mid m$	p divides m.
(k, m)	greatest common divisor of m, n.
$(k, m) = 1$	k and m relatively prime.

Unless otherwise noted, the formulas in this chapter hold for all values of the variables x and t, and for all nonnegative integers n.

Bernoulli Numbers and Polynomials

The origin of the notation B_n, $B_n(x)$, is not clear. The present notation, as defined in §24.2(i), was used in Lucas (1891) and Nörlund (1924), and has become the prevailing notation; see Table 24.2.1. Among various older notations, the most common one is

$$B_1 = \tfrac{1}{6}, \quad B_2 = \tfrac{1}{30}, \quad B_3 = \tfrac{1}{42}, \quad B_4 = \tfrac{1}{30}, \dots.$$

It was used in Saalschütz (1893), Nielsen (1923), Schwatt (1962), and Whittaker and Watson (1927).

Euler Numbers and Polynomials

The secant series ((4.19.5)) first occurs in the work of Gregory in 1671. Its coefficients were first studied in Euler (1755); they were called Euler numbers by Raabe in 1851. The notations E_n, $E_n(x)$, as defined in §24.2(ii), were used in Lucas (1891) and Nörlund (1924).

Other historical remarks on notations can be found in Cajori (1929, pp. 42–44). Various systems of notation are summarized in Adrian (1959) and D'Ocagne (1904).

Properties

24.2 Definitions and Generating Functions

24.2(i) Bernoulli Numbers and Polynomials

24.2.1 $\quad \dfrac{t}{e^t - 1} = \sum_{n=0}^{\infty} B_n \dfrac{t^n}{n!}, \qquad |t| < 2\pi.$

24.2.2 $\quad B_{2n+1} = 0, \quad (-1)^{n+1} B_{2n} > 0, \quad n = 1, 2, \dots.$

24.2.3 $\quad \dfrac{t e^{xt}}{e^t - 1} = \sum_{n=0}^{\infty} B_n(x) \dfrac{t^n}{n!}, \qquad |t| < 2\pi.$

24.2.4 $\quad B_n = B_n(0),$

24.2.5 $\quad B_n(x) = \sum_{k=0}^{n} \binom{n}{k} B_k \, x^{n-k}.$

See also §§4.19 and 4.33.

24.2(ii) Euler Numbers and Polynomials

24.2.6 $\quad \dfrac{2 e^t}{e^{2t} + 1} = \sum_{n=0}^{\infty} E_n \dfrac{t^n}{n!}, \qquad |t| < \tfrac{1}{2}\pi,$

24.2.7 $\quad E_{2n+1} = 0, \quad (-1)^n E_{2n} > 0.$

24.2.8 $\quad \dfrac{2 e^{xt}}{e^t + 1} = \sum_{n=0}^{\infty} E_n(x) \dfrac{t^n}{n!}, \qquad |t| < \pi,$

24.2.9 $\quad E_n = 2^n E_n\!\left(\tfrac{1}{2}\right) = \text{integer},$

24.2.10 $\quad E_n(x) = \sum_{k=0}^{n} \binom{n}{k} \dfrac{E_k}{2^k} (x - \tfrac{1}{2})^{n-k}.$

See also (4.19.5).

24.2(iii) Periodic Bernoulli and Euler Functions

24.2.11 $\quad \widetilde{B}_n(x) = B_n(x), \quad \widetilde{E}_n(x) = E_n(x), \quad 0 \le x < 1,$

24.2.12 $\quad \widetilde{B}_n(x+1) = \widetilde{B}_n(x), \quad \widetilde{E}_n(x+1) = -\widetilde{E}_n(x),$
$\qquad x \in \mathbb{R}.$

24.2(iv) Tables

Table 24.2.1: Bernoulli and Euler numbers.

n	B_n	E_n
0	1	1
1	$-\frac{1}{2}$	0
2	$\frac{1}{6}$	-1
4	$-\frac{1}{30}$	5
6	$\frac{1}{42}$	-61
8	$-\frac{1}{30}$	1385
10	$\frac{5}{66}$	-50521
12	$-\frac{691}{2730}$	27 02765
14	$\frac{7}{6}$	$-1993\,60981$
16	$-\frac{3617}{510}$	$1\,93915\,12145$

Table 24.2.2: Bernoulli and Euler polynomials.

n	$B_n(x)$	$E_n(x)$
0	1	1
1	$x - \frac{1}{2}$	$x - \frac{1}{2}$
2	$x^2 - x + \frac{1}{6}$	$x^2 - x$
3	$x^3 - \frac{3}{2}x^2 + \frac{1}{2}x$	$x^3 - \frac{3}{2}x^2 + \frac{1}{4}$
4	$x^4 - 2x^3 + x^2 - \frac{1}{30}$	$x^4 - 2x^3 + x$
5	$x^5 - \frac{5}{2}x^4 + \frac{5}{3}x^3 - \frac{1}{6}x$	$x^5 - \frac{5}{2}x^4 + \frac{5}{2}x^2 - \frac{1}{2}$

For extensions of Tables 24.2.1 and 24.2.2 see http://dlmf.nist.gov/24.2.iv.

24.3 Graphs

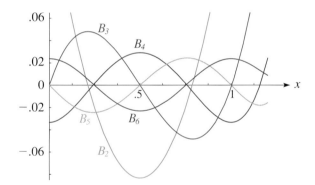

Figure 24.3.1: Bernoulli polynomials $B_n(x)$, $n = 2, 3, \ldots, 6$.

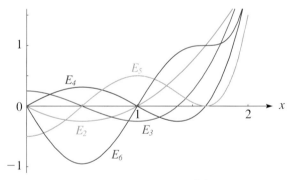

Figure 24.3.2: Euler polynomials $E_n(x)$, $n = 2, 3, \ldots, 6$.

24.4 Basic Properties

24.4(i) Difference Equations

24.4.1 $\qquad B_n(x+1) - B_n(x) = nx^{n-1},$

24.4.2 $\qquad E_n(x+1) + E_n(x) = 2x^n.$

24.4(ii) Symmetry

24.4.3 $\qquad B_n(1-x) = (-1)^n B_n(x),$

24.4.4 $\qquad E_n(1-x) = (-1)^n E_n(x).$

24.4.5 $\qquad (-1)^n B_n(-x) = B_n(x) + nx^{n-1},$

24.4.6 $\qquad (-1)^{n+1} E_n(-x) = E_n(x) - 2x^n.$

24.4(iii) Sums of Powers

24.4.7 $\qquad \displaystyle\sum_{k=1}^{m} k^n = \frac{B_{n+1}(m+1) - B_{n+1}}{n+1},$

24.4.8 $\qquad \displaystyle\sum_{k=1}^{m} (-1)^{m-k} k^n = \frac{E_n(m+1) + (-1)^m E_n(0)}{2}.$

24.4.9
$$\sum_{k=0}^{m-1}(a+dk)^n = \frac{d^n}{n+1}\left(B_{n+1}\left(m+\frac{a}{d}\right) - B_{n+1}\left(\frac{a}{d}\right)\right),$$

24.4.10
$$\sum_{k=0}^{m-1}(-1)^k(a+dk)^n = \frac{d^n}{2}\left((-1)^{m-1}E_n\left(m+\frac{a}{d}\right) + E_n\left(\frac{a}{d}\right)\right).$$

24.4.11
$$\sum_{\substack{k=1\\(k,m)=1}}^{m} k^n = \frac{1}{n+1}\sum_{j=1}^{n+1}\binom{n+1}{j} \times \left(\prod_{p|m}(1-p^{n-j})B_{n+1-j}\right)m^j.$$

24.4(iv) Finite Expansions

24.4.12 $B_n(x+h) = \sum_{k=0}^{n}\binom{n}{k}B_k(x)h^{n-k},$

24.4.13 $E_n(x+h) = \sum_{k=0}^{n}\binom{n}{k}E_k(x)h^{n-k},$

24.4.14 $E_{n-1}(x) = \frac{2}{n}\sum_{k=0}^{n}\binom{n}{k}(1-2^k)B_k x^{n-k},$

24.4.15
$$B_{2n} = \frac{2n}{2^{2n}(2^{2n}-1)}\sum_{k=0}^{n-1}\binom{2n-1}{2k}E_{2k},$$

24.4.16
$$E_{2n} = \frac{1}{2n+1} - \sum_{k=1}^{n}\binom{2n}{2k-1}\frac{2^{2k}(2^{2k-1}-1)B_{2k}}{k},$$

24.4.17
$$E_{2n} = 1 - \sum_{k=1}^{n}\binom{2n}{2k-1}\frac{2^{2k}(2^{2k}-1)B_{2k}}{2k}.$$

24.4(v) Multiplication Formulas

Raabe's Theorem

24.4.18 $B_n(mx) = m^{n-1}\sum_{k=0}^{m-1}B_n\left(x+\frac{k}{m}\right).$

Next,

24.4.19 $E_n(mx) = -\frac{2m^n}{n+1}\sum_{k=0}^{m-1}(-1)^k B_{n+1}\left(x+\frac{k}{m}\right),$ $m=2,4,6,\ldots,$

24.4.20 $E_n(mx) = m^n \sum_{k=0}^{m-1}(-1)^k E_n\left(x+\frac{k}{m}\right),\ m=1,3,5,\ldots.$

24.4.21 $B_n(x) = 2^{n-1}\left(B_n\left(\tfrac{1}{2}x\right) + B_n\left(\tfrac{1}{2}x+\tfrac{1}{2}\right)\right),$

24.4.22 $E_{n-1}(x) = \frac{2}{n}\left(B_n(x) - 2^n B_n\left(\tfrac{1}{2}x\right)\right),$

24.4.23 $E_{n-1}(x) = \frac{2^n}{n}\left(B_n\left(\tfrac{1}{2}x+\tfrac{1}{2}\right) - B_n\left(\tfrac{1}{2}x\right)\right),$

24.4.24
$$B_n(mx) = m^n B_n(x) + n\sum_{k=1}^{n}\sum_{j=0}^{k-1}(-1)^j\binom{n}{k} \times \left(\sum_{r=1}^{m-1}\frac{e^{2\pi i(k-j)r/m}}{(1-e^{2\pi ir/m})^n}\right)(j+mx)^{n-1},$$
$n=1,2,\ldots,\ m=2,3,\ldots.$

24.4(vi) Special Values

24.4.25 $B_n(0) = (-1)^n B_n(1) = B_n,$

24.4.26 $E_n(0) = -E_n(1) = -\frac{2}{n+1}(2^{n+1}-1)B_{n+1}.$

24.4.27 $B_n\left(\tfrac{1}{2}\right) = -(1-2^{1-n})B_n,$

24.4.28 $E_n\left(\tfrac{1}{2}\right) = 2^{-n}E_n.$

24.4.29 $B_{2n}\left(\tfrac{1}{3}\right) = B_{2n}\left(\tfrac{2}{3}\right) = -\tfrac{1}{2}(1-3^{1-2n})B_{2n}.$

24.4.30
$$E_{2n-1}\left(\tfrac{1}{3}\right) = -E_{2n-1}\left(\tfrac{2}{3}\right) = -\frac{(1-3^{1-2n})(2^{2n}-1)}{2n}B_{2n},$$
$n=1,2,\ldots.$

24.4.31
$$B_n\left(\tfrac{1}{4}\right) = (-1)^n B_n\left(\tfrac{3}{4}\right) = -\frac{1-2^{1-n}}{2^n}B_n - \frac{n}{4^n}E_{n-1},$$
$n=1,2,\ldots.$

24.4.32 $B_{2n}\left(\tfrac{1}{6}\right) = B_{2n}\left(\tfrac{5}{6}\right) = \tfrac{1}{2}(1-2^{1-2n})(1-3^{1-2n})B_{2n},$

24.4.33 $E_{2n}\left(\tfrac{1}{6}\right) = E_{2n}\left(\tfrac{5}{6}\right) = \frac{1+3^{-2n}}{2^{2n+1}}E_{2n}.$

24.4(vii) Derivatives

24.4.34 $\frac{d}{dx}B_n(x) = n B_{n-1}(x),\qquad n=1,2,\ldots,$

24.4.35 $\frac{d}{dx}E_n(x) = n E_{n-1}(x),\qquad n=1,2,\ldots.$

24.4(viii) Symbolic Operations

Let $P(x)$ denote any polynomial in x, and after expanding set $(B(x))^n = B_n(x)$ and $(E(x))^n = E_n(x)$. Then

24.4.36 $P(B(x)+1) - P(B(x)) = P'(x),$

24.4.37 $B_n(x+h) = (B(x)+h)^n,$

24.4.38 $P(E(x)+1) + P(E(x)) = 2P(x),$

24.4.39 $E_n(x+h) = (E(x)+h)^n.$

For these results and also connections with the umbral calculus see Gessel (2003).

24.4(ix) Relations to Other Functions

For the relation of Bernoulli numbers to the Riemann zeta function see §25.6, and to the Eulerian numbers see (26.14.11).

24.5 Recurrence Relations

24.5(i) Basic Relations

24.5.1 $$\sum_{k=0}^{n-1} \binom{n}{k} B_k(x) = nx^{n-1}, \quad n = 2, 3, \ldots,$$

24.5.2 $$\sum_{k=0}^{n} \binom{n}{k} E_k(x) + E_n(x) = 2x^n, \quad n = 1, 2, \ldots.$$

24.5.3 $$\sum_{k=0}^{n-1} \binom{n}{k} B_k = 0, \quad n = 2, 3, \ldots,$$

24.5.4 $$\sum_{k=0}^{n} \binom{2n}{2k} E_{2k} = 0, \quad n = 1, 2, \ldots,$$

24.5.5 $$\sum_{k=0}^{n} \binom{n}{k} 2^k E_{n-k} + E_n = 2.$$

24.5(ii) Other Identities

24.5.6 $$\sum_{k=2}^{n} \binom{n}{k-2} \frac{B_k}{k} = \frac{1}{(n+1)(n+2)} - B_{n+1}, \quad n = 2, 3, \ldots,$$

24.5.7 $$\sum_{k=0}^{n} \binom{n}{k} \frac{B_k}{n+2-k} = \frac{B_{n+1}}{n+1}, \quad n = 1, 2, \ldots,$$

24.5.8 $$\sum_{k=0}^{n} \frac{2^{2k} B_{2k}}{(2k)!(2n+1-2k)!} = \frac{1}{(2n)!}, \quad n = 1, 2, \ldots.$$

24.5(iii) Inversion Formulas

In each of (24.5.9) and (24.5.10) the first identity implies the second one and vice-versa.

24.5.9 $$a_n = \sum_{k=0}^{n} \binom{n}{k} \frac{b_{n-k}}{k+1}, \quad b_n = \sum_{k=0}^{n} \binom{n}{k} B_k a_{n-k}.$$

24.5.10 $$a_n = \sum_{k=0}^{\lfloor n/2 \rfloor} \binom{n}{2k} b_{n-2k},$$
$$b_n = \sum_{k=0}^{\lfloor n/2 \rfloor} \binom{n}{2k} E_{2k} a_{n-2k}.$$

24.6 Explicit Formulas

The identities in this section hold for $n = 1, 2, \ldots$. (24.6.7), (24.6.8), (24.6.10), and (24.6.12) are valid also for $n = 0$.

24.6.1 $$B_{2n} = \sum_{k=2}^{2n+1} \frac{(-1)^{k-1}}{k} \binom{2n+1}{k} \sum_{j=1}^{k-1} j^{2n},$$

24.6.2 $$B_n = \frac{1}{n+1} \sum_{k=1}^{n} \sum_{j=1}^{k} (-1)^j j^n \binom{n+1}{k-j} \bigg/ \binom{n}{k},$$

24.6.3 $$B_{2n} = \sum_{k=1}^{n} \frac{(k-1)!k!}{(2k+1)!} \sum_{j=1}^{k} (-1)^{j-1} \binom{2k}{k+j} j^{2n}.$$

24.6.4 $$E_{2n} = \sum_{k=1}^{n} \frac{1}{2^{k-1}} \sum_{j=1}^{k} (-1)^j \binom{2k}{k-j} j^{2n},$$

24.6.5 $$E_{2n} = \frac{1}{2^{n-1}} \sum_{k=0}^{n-1} (-1)^{n-k} (n-k)^{2n} \sum_{j=0}^{k} \binom{2n-2j}{k-j} 2^j,$$

24.6.6 $$E_{2n} = \sum_{k=1}^{2n} \frac{(-1)^k}{2^{k-1}} \binom{2n+1}{k+1} \sum_{j=0}^{\lfloor \frac{1}{2}k - \frac{1}{2} \rfloor} \binom{k}{j} (k-2j)^{2n}.$$

24.6.7 $$B_n(x) = \sum_{k=0}^{n} \frac{1}{k+1} \sum_{j=0}^{k} (-1)^j \binom{k}{j} (x+j)^n,$$

24.6.8 $$E_n(x) = \frac{1}{2^n} \sum_{k=1}^{n+1} \sum_{j=0}^{k-1} (-1)^j \binom{n+1}{k} (x+j)^n.$$

24.6.9 $$B_n = \sum_{k=0}^{n} \frac{1}{k+1} \sum_{j=0}^{k} (-1)^j \binom{k}{j} j^n,$$

24.6.10 $$E_n = \frac{1}{2^n} \sum_{k=1}^{n+1} \binom{n+1}{k} \sum_{j=0}^{k-1} (-1)^j (2j+1)^n.$$

24.6.11 $$B_n = \frac{n}{2^n(2^n-1)} \sum_{k=1}^{n} \sum_{j=0}^{k-1} (-1)^{j+1} \binom{n}{k} j^{n-1},$$

24.6.12 $$E_{2n} = \sum_{k=0}^{2n} \frac{1}{2^k} \sum_{j=0}^{k} (-1)^j \binom{k}{j} (1+2j)^{2n}.$$

24.7 Integral Representations

24.7(i) Bernoulli and Euler Numbers

The identities in this subsection hold for $n = 1, 2, \ldots$. (24.7.6) also holds for $n = 0$.

24.7.1
$$B_{2n} = (-1)^{n+1} \frac{4n}{1 - 2^{1-2n}} \int_0^\infty \frac{t^{2n-1}}{e^{2\pi t} + 1} \, dt$$
$$= (-1)^{n+1} \frac{2n}{1 - 2^{1-2n}} \int_0^\infty t^{2n-1} e^{-\pi t} \operatorname{sech}(\pi t) \, dt,$$

24.7.2
$$B_{2n} = (-1)^{n+1} 4n \int_0^\infty \frac{t^{2n-1}}{e^{2\pi t} - 1} \, dt$$
$$= (-1)^{n+1} 2n \int_0^\infty t^{2n-1} e^{-\pi t} \operatorname{csch}(\pi t) \, dt,$$

24.7.3
$$B_{2n} = (-1)^{n+1} \frac{\pi}{1 - 2^{1-2n}} \int_0^\infty t^{2n} \operatorname{sech}^2(\pi t) \, dt,$$

24.7.4
$$B_{2n} = (-1)^{n+1} \pi \int_0^\infty t^{2n} \operatorname{csch}^2(\pi t) \, dt,$$

24.7.5
$$B_{2n} = (-1)^n \frac{2n(2n-1)}{\pi} \int_0^\infty t^{2n-2} \ln\!\left(1 - e^{-2\pi t}\right) dt.$$

24.7.6
$$E_{2n} = (-1)^n 2^{2n+1} \int_0^\infty t^{2n} \operatorname{sech}(\pi t) \, dt.$$

24.7(ii) Bernoulli and Euler Polynomials

The following four equations hold for $0 < \Re x < 1$.

24.7.7
$$B_{2n}(x) = (-1)^{n+1} 2n$$
$$\times \int_0^\infty \frac{\cos(2\pi x) - e^{-2\pi t}}{\cosh(2\pi t) - \cos(2\pi x)} t^{2n-1} \, dt,$$
$$n = 1, 2, \ldots,$$

24.7.8
$$B_{2n+1}(x) = (-1)^{n+1} (2n+1)$$
$$\times \int_0^\infty \frac{\sin(2\pi x)}{\cosh(2\pi t) - \cos(2\pi x)} t^{2n} \, dt.$$

24.7.9
$$E_{2n}(x) = (-1)^n 4 \int_0^\infty \frac{\sin(\pi x) \cosh(\pi t)}{\cosh(2\pi t) - \cos(2\pi x)} t^{2n} \, dt,$$

24.7.10
$$E_{2n+1}(x) = (-1)^{n+1} 4$$
$$\times \int_0^\infty \frac{\cos(\pi x) \sinh(\pi t)}{\cosh(2\pi t) - \cos(2\pi x)} t^{2n+1} \, dt.$$

Mellin–Barnes Integral

24.7.11
$$B_n(x) = \frac{1}{2\pi i} \int_{-c-i\infty}^{-c+i\infty} (x+t)^n \left(\frac{\pi}{\sin(\pi t)}\right)^2 dt, \quad 0 < c < 1.$$

24.7(iii) Compendia

For further integral representations see Prudnikov et al. (1986a, §§2.3–2.6) and Gradshteyn and Ryzhik (2000, Chapters 3 and 4).

24.8 Series Expansions

24.8(i) Fourier Series

If $n = 1, 2, \ldots$ and $0 \leq x \leq 1$, then

24.8.1
$$B_{2n}(x) = (-1)^{n+1} \frac{2(2n)!}{(2\pi)^{2n}} \sum_{k=1}^\infty \frac{\cos(2\pi kx)}{k^{2n}},$$

24.8.2
$$B_{2n+1}(x) = (-1)^{n+1} \frac{2(2n+1)!}{(2\pi)^{2n+1}} \sum_{k=1}^\infty \frac{\sin(2\pi kx)}{k^{2n+1}}.$$

The second expansion holds also for $n = 0$ and $0 < x < 1$.

If $n = 1$ with $0 < x < 1$, or $n = 2, 3, \ldots$ with $0 \leq x \leq 1$, then

24.8.3
$$B_n(x) = -\frac{n!}{(2\pi i)^n} \sum_{\substack{k=-\infty \\ k \neq 0}}^\infty \frac{e^{2\pi i k x}}{k^n}.$$

If $n = 1, 2, \ldots$ and $0 \leq x \leq 1$, then

24.8.4
$$E_{2n}(x) = (-1)^n \frac{4(2n)!}{\pi^{2n+1}} \sum_{k=0}^\infty \frac{\sin((2k+1)\pi x)}{(2k+1)^{2n+1}},$$

24.8.5
$$E_{2n-1}(x) = (-1)^n \frac{4(2n-1)!}{\pi^{2n}} \sum_{k=0}^\infty \frac{\cos((2k+1)\pi x)}{(2k+1)^{2n}}.$$

24.8(ii) Other Series

24.8.6
$$B_{4n+2} = (8n+4) \sum_{k=1}^\infty \frac{k^{4n+1}}{e^{2\pi k} - 1}, \quad n = 1, 2, \ldots,$$

24.8.7
$$B_{2n} = \frac{(-1)^{n+1} 4n}{2^{2n} - 1} \sum_{k=1}^\infty \frac{k^{2n-1}}{e^{\pi k} + (-1)^{k+n}}, \quad n = 2, 3, \ldots.$$

Let $\alpha\beta = \pi^2$. Then

24.8.8
$$\frac{B_{2n}}{4n} (\alpha^n - (-\beta)^n) = \alpha^n \sum_{k=1}^\infty \frac{k^{2n-1}}{e^{2\alpha k} - 1}$$
$$- (-\beta)^n \sum_{k=1}^\infty \frac{k^{2n-1}}{e^{2\beta k} - 1},$$
$$n = 2, 3, \ldots.$$

24.8.9
$$E_{2n} = (-1)^n \sum_{k=1}^\infty \frac{k^{2n}}{\cosh\!\left(\tfrac{1}{2}\pi k\right)}$$
$$- 4 \sum_{k=0}^\infty \frac{(-1)^k (2k+1)^{2n}}{e^{2\pi(2k+1)} - 1}, \quad n = 1, 2, \ldots.$$

24.9 Inequalities

Except where otherwise noted, the inequalities in this section hold for $n = 1, 2, \ldots$.

24.9.1 $\quad |B_{2n}| > |B_{2n}(x)|, \qquad 1 > x > 0,$

24.9.2
$$(2 - 2^{1-2n})|B_{2n}| \geq |B_{2n}(x) - B_{2n}|, \qquad 1 \geq x \geq 0.$$

(24.9.3)–(24.9.5) hold for $\frac{1}{2} > x > 0$.

24.9.3 $\quad 4^{-n}|E_{2n}| > (-1)^n E_{2n}(x) > 0,$

24.9.4
$$\frac{2(2n+1)!}{(2\pi)^{2n+1}} > (-1)^{n+1} B_{2n+1}(x) > 0, \qquad n = 2, 3, \ldots,$$

24.9.5
$$\frac{4(2n-1)!}{\pi^{2n}} \frac{2^{2n} - 1}{2^{2n} - 2} > (-1)^n E_{2n-1}(x) > 0.$$

(24.9.6)–(24.9.7) hold for $n = 2, 3, \ldots$.

24.9.6 $\quad 5\sqrt{\pi n}\left(\dfrac{n}{\pi e}\right)^{2n} > (-1)^{n+1} B_{2n} > 4\sqrt{\pi n}\left(\dfrac{n}{\pi e}\right)^{2n},$

24.9.7
$$8\sqrt{\frac{n}{\pi}}\left(\frac{4n}{\pi e}\right)^{2n}\left(1 + \frac{1}{12n}\right) > (-1)^n E_{2n} > 8\sqrt{\frac{n}{\pi}}\left(\frac{4n}{\pi e}\right)^{2n}.$$

Lastly,

24.9.8
$$\frac{2(2n)!}{(2\pi)^{2n}} \frac{1}{1 - 2^{\beta - 2n}} \geq (-1)^{n+1} B_{2n} \geq \frac{2(2n)!}{(2\pi)^{2n}} \frac{1}{1 - 2^{-2n}}$$

with

24.9.9 $\quad \beta = 2 + \dfrac{\ln(1 - 6\pi^{-2})}{\ln 2} = 0.6491\ldots.$

24.9.10
$$\frac{4^{n+1}(2n)!}{\pi^{2n+1}} > (-1)^n E_{2n} > \frac{4^{n+1}(2n)!}{\pi^{2n+1}} \frac{1}{1 + 3^{-1-2n}}.$$

24.10 Arithmetic Properties

24.10(i) Von Staudt–Clausen Theorem

Here and elsewhere in §24.10 the symbol p denotes a prime number.

24.10.1 $\quad B_{2n} + \displaystyle\sum_{(p-1)|2n} \frac{1}{p} = \text{integer},$

where the summation is over all p such that $p-1$ divides $2n$. The denominator of B_{2n} is the product of all these primes p.

24.10.2 $\quad p B_{2n} \equiv p - 1 \pmod{p^{\ell+1}},$

where $n \geq 2$, and $\ell(\geq 1)$ is an arbitrary integer such that $(p-1)p^\ell \,|\, 2n$. Here and elsewhere two rational numbers are *congruent* if the modulus divides the numerator of their difference.

24.10(ii) Kummer Congruences

24.10.3 $\quad \dfrac{B_m}{m} \equiv \dfrac{B_n}{n} \pmod{p},$

where $m \equiv n \not\equiv 0 \pmod{p-1}$.

24.10.4 $\quad (1 - p^{m-1})\dfrac{B_m}{m} \equiv (1 - p^{n-1})\dfrac{B_n}{n} \pmod{p^{\ell+1}},$

valid when $m \equiv n \pmod{(p-1)p^\ell}$ and $n \not\equiv 0 \pmod{p-1}$, where $\ell(\geq 0)$ is a fixed integer.

24.10.5 $\quad E_n \equiv E_{n+p-1} \pmod{p},$

where $p(>2)$ is a prime and $n \geq 2$.

24.10.6 $\quad E_{2n} \equiv E_{2n+w} \pmod{2^\ell},$

valid for fixed integers $\ell(\geq 0)$, and for all $n(\geq 0)$ and $w(\geq 0)$ such that $2^\ell \,|\, w$.

24.10(iii) Voronoi's Congruence

Let $B_{2n} = N_{2n}/D_{2n}$, with N_{2n} and D_{2n} relatively prime and $D_{2n} > 0$. Then

24.10.7
$$(b^{2n} - 1)N_{2n} \equiv 2nb^{2n-1}D_{2n} \sum_{k=1}^{M-1} k^{2n-1} \left\lfloor \frac{kb}{M} \right\rfloor \pmod{M},$$

where $M(\geq 2)$ and b are integers, with b relatively prime to M.

For historical notes, generalizations, and applications, see Porubský (1998).

24.10(iv) Factors

With N_{2n} as in §24.10(iii)

24.10.8 $\quad N_{2n} \equiv 0 \pmod{p^\ell},$

valid for fixed integers $\ell(\geq 1)$, and for all $n(\geq 1)$ such that $2n \not\equiv 0 \pmod{p-1}$ and $p^\ell \,|\, 2n$.

24.10.9 $\quad E_{2n} \equiv \begin{cases} 0 \pmod{p^\ell} & \text{if } p \equiv 1 \pmod{4}, \\ 2 \pmod{p^\ell} & \text{if } p \equiv 3 \pmod{4}, \end{cases}$

valid for fixed integers $\ell(\geq 1)$ and for all $n(\geq 1)$ such that $(p-1)p^{\ell-1} \,|\, 2n$.

24.11 Asymptotic Approximations

As $n \to \infty$

24.11.1 $\quad (-1)^{n+1} B_{2n} \sim \dfrac{2(2n)!}{(2\pi)^{2n}},$

24.11.2 $\quad (-1)^{n+1} B_{2n} \sim 4\sqrt{\pi n}\left(\dfrac{n}{\pi e}\right)^{2n},$

24.11.3 $\quad (-1)^n E_{2n} \sim \dfrac{2^{2n+2}(2n)!}{\pi^{2n+1}},$

24.11.4 $\quad (-1)^n E_{2n} \sim 8\sqrt{\dfrac{n}{\pi}}\left(\dfrac{4n}{\pi e}\right)^{2n}.$

Also,

24.11.5
$$(-1)^{\lfloor n/2\rfloor -1}\frac{(2\pi)^n}{2(n!)}B_n(x) \to \begin{cases}\cos(2\pi x), & n \text{ even},\\ \sin(2\pi x), & n \text{ odd},\end{cases}$$

24.11.6
$$(-1)^{\lfloor (n+1)/2\rfloor}\frac{\pi^{n+1}}{4(n!)}E_n(x) \to \begin{cases}\sin(\pi x), & n \text{ even},\\ \cos(\pi x), & n \text{ odd},\end{cases}$$

uniformly for x on compact subsets of $\mathbb{C}$.

For further results see Temme (1995b) and López and Temme (1999b).

24.12 Zeros

24.12(i) Bernoulli Polynomials: Real Zeros

In the interval $0 \le x \le 1$ the only zeros of $B_{2n+1}(x)$, $n=1,2,\ldots$, are $0, \tfrac{1}{2}, 1$, and the only zeros of $B_{2n}(x) - B_{2n}$, $n=1,2,\ldots$, are $0,1$.

For the interval $\tfrac{1}{2} \le x < \infty$ denote the zeros of $B_n(x)$ by $x_j^{(n)}$, $j=1,2,\ldots$, with

24.12.1
$$\tfrac{1}{2} \le x_1^{(n)} \le x_2^{(n)} \le \cdots.$$

Then the zeros in the interval $-\infty < x \le \tfrac{1}{2}$ are $1-x_j^{(n)}$.

When $n(\ge 2)$ is even

24.12.2
$$\frac{3}{4} + \frac{1}{2^{n+2}\pi} < x_1^{(n)} < \frac{3}{4} + \frac{1}{2^{n+1}\pi},$$

24.12.3
$$x_1^{(n)} - \frac{3}{4} \sim \frac{1}{2^{n+1}\pi}, \qquad n \to \infty,$$

and as $n \to \infty$ with $m(\ge 1)$ fixed,

24.12.4
$$x_{2m-1}^{(n)} \to m - \tfrac{1}{4}, \quad x_{2m}^{(n)} \to m + \tfrac{1}{4}.$$

When n is odd $x_1^{(n)} = \tfrac{1}{2}$, $x_2^{(n)} = 1$ ($n \ge 3$), and as $n \to \infty$ with $m(\ge 1)$ fixed,

24.12.5
$$x_{2m-1}^{(n)} \to m - \tfrac{1}{2}, \quad x_{2m}^{(n)} \to m.$$

Let $R(n)$ be the total number of real zeros of $B_n(x)$. Then $R(n) = n$ when $1 \le n \le 5$, and

24.12.6
$$R(n) \sim 2n/(\pi e), \qquad n \to \infty.$$

24.12(ii) Euler Polynomials: Real Zeros

For the interval $\tfrac{1}{2} \le x < \infty$ denote the zeros of $E_n(x)$ by $y_j^{(n)}$, $j=1,2,\ldots$, with

24.12.7
$$\tfrac{1}{2} \le y_1^{(n)} \le y_2^{(n)} \le \cdots.$$

Then the zeros in the interval $-\infty < x \le \tfrac{1}{2}$ are $1-y_j^{(n)}$.

When $n(\ge 2)$ is even $y_1^{(n)} = 1$, and as $n \to \infty$ with $m(\ge 1)$ fixed,

24.12.8
$$y_m^{(n)} \to m.$$

When n is odd $y_1^{(n)} = \tfrac{1}{2}$,

24.12.9
$$\frac{3}{2} - \frac{\pi^{n+1}}{3(n!)} < y_2^{(n)} < \frac{3}{2}, \quad n=3,7,11,\ldots,$$

24.12.10
$$\frac{3}{2} < y_2^{(n)} < \frac{3}{2} + \frac{\pi^{n+1}}{3(n!)}, \quad n=5,9,13,\ldots,$$

and as $n \to \infty$ with $m(\ge 1)$ fixed,

24.12.11
$$y_{2m}^{(n)} \to m - \tfrac{1}{2}.$$

24.12(iii) Complex Zeros

For complex zeros of Bernoulli and Euler polynomials, see Delange (1987) and Dilcher (1988). A related topic is the irreducibility of Bernoulli and Euler polynomials. For details and references, see Dilcher (1987b), Kimura (1988), or Adelberg (1992).

24.12(iv) Multiple Zeros

$B_n(x)$, $n=1,2,\ldots$, has no multiple zeros. The only polynomial $E_n(x)$ with multiple zeros is $E_5(x) = (x-\tfrac{1}{2})(x^2 - x - 1)^2$.

24.13 Integrals

24.13(i) Bernoulli Polynomials

24.13.1
$$\int B_n(t)\,dt = \frac{B_{n+1}(t)}{n+1} + \text{const.},$$

24.13.2
$$\int_x^{x+1} B_n(t)\,dt = x^n, \qquad n=1,2,\ldots,$$

24.13.3
$$\int_x^{x+(1/2)} B_n(t)\,dt = \frac{E_n(2x)}{2^{n+1}},$$

24.13.4
$$\int_0^{1/2} B_n(t)\,dt = \frac{1-2^{n+1}}{2^n}\frac{B_{n+1}}{n+1},$$

24.13.5
$$\int_{1/4}^{3/4} B_n(t)\,dt = \frac{E_n}{2^{2n+1}}.$$

For $m,n = 1,2,\ldots$,

24.13.6
$$\int_0^1 B_n(t) B_m(t)\,dt = \frac{(-1)^{n-1} m! n!}{(m+n)!} B_{m+n}.$$

24.13(ii) Euler Polynomials

24.13.7
$$\int E_n(t)\,dt = \frac{E_{n+1}(t)}{n+1} + \text{const.},$$

24.13.8
$$\int_0^1 E_n(t)\,dt = -2\frac{E_{n+1}(0)}{n+1} = \frac{4(2^{n+2}-1)}{(n+1)(n+2)} B_{n+2},$$

24.13.9
$$\int_0^{1/2} E_{2n}(t)\,dt = -\frac{E_{2n+1}(0)}{2n+1} = \frac{2(2^{2n+2}-1) B_{2n+2}}{(2n+1)(2n+2)},$$

24.14 Sums

24.14(i) Quadratic Recurrence Relations

24.13.10
$$\int_0^{1/2} E_{2n-1}(t)\,dt = \frac{E_{2n}}{n 2^{2n+1}}, \quad n=1,2,\ldots.$$

For $m,n = 1,2,\ldots,$

24.13.11
$$\int_0^1 E_n(t)\,E_m(t)\,dt = (-1)^n 4 \frac{(2^{m+n+2}-1)m!n!}{(m+n+2)!} B_{m+n+2}.$$

24.13(iii) Compendia

For Laplace and inverse Laplace transforms see Prudnikov et al. (1992a, §§3.28.1–3.28.2) and Prudnikov et al. (1992b, §§3.26.1–3.26.2). For other integrals see Prudnikov et al. (1990, pp. 55–57).

24.14 Sums

24.14(i) Quadratic Recurrence Relations

24.14.1
$$\sum_{k=0}^n \binom{n}{k} B_k(x) B_{n-k}(y) = n(x+y-1) B_{n-1}(x+y) - (n-1) B_n(x+y),$$

24.14.2
$$\sum_{k=0}^n \binom{n}{k} B_k B_{n-k} = (1-n) B_n - n B_{n-1}.$$

24.14.3
$$\sum_{k=0}^n \binom{n}{k} E_k(h) E_{n-k}(x) = 2(E_{n+1}(x+h) - (x+h-1) E_n(x+h)),$$

24.14.4
$$\sum_{k=0}^n \binom{n}{k} E_k E_{n-k} = -2^{n+1} E_{n+1}(0) = -2^{n+2}(1-2^{n+2}) \frac{B_{n+2}}{n+2}.$$

24.14.5
$$\sum_{k=0}^n \binom{n}{k} E_k(h) B_{n-k}(x) = 2^n B_n\!\left(\tfrac{1}{2}(x+h)\right),$$

24.14.6
$$\sum_{k=0}^n \binom{n}{k} 2^k B_k E_{n-k} = 2(1-2^{n-1}) B_n - n E_{n-1}.$$

Let $m+n$ be even with m and n nonzero. Then

24.14.7
$$\sum_{j=0}^m \sum_{k=0}^n \binom{m}{j}\binom{n}{k} \frac{B_j B_k}{m+n-j-k+1} = (-1)^{m-1} \frac{m!n!}{(m+n)!} B_{m+n}.$$

24.14(ii) Higher-Order Recurrence Relations

In the following two identities, valid for $n \geq 2$, the sums are taken over all nonnegative integers j,k,ℓ with $j+k+\ell = n$.

24.14.8
$$\sum \frac{(2n)!}{(2j)!(2k)!(2\ell)!} B_{2j} B_{2k} B_{2\ell} = (n-1)(2n-1) B_{2n} + n(n-\tfrac{1}{2}) B_{2n-2},$$

24.14.9
$$\sum \frac{(2n)!}{(2j)!(2k)!(2\ell)!} E_{2j} E_{2k} E_{2\ell} = \tfrac{1}{2}(E_{2n} - E_{2n+2}).$$

In the next identity, valid for $n \geq 4$, the sum is taken over all positive integers j,k,ℓ,m with $j+k+\ell+m = n$.

24.14.10
$$\sum \frac{(2n)!}{(2j)!(2k)!(2\ell)!(2m)!} B_{2j} B_{2k} B_{2\ell} B_{2m} = -\binom{2n+3}{3} B_{2n} - \tfrac{4}{3} n^2(2n-1) B_{2n-2}.$$

For (24.14.11) and (24.14.12), see Al-Salam and Carlitz (1959). These identities can be regarded as higher-order recurrences. Let $\det[a_{r+s}]$ denote a *Hankel* (or *persymmetric*) *determinant*, that is, an $(n+1) \times (n+1)$ determinant with element a_{r+s} in row r and column s for $r,s = 0,1,\ldots,n$. Then

24.14.11
$$\det[B_{r+s}] = (-1)^{n(n+1)/2} \left(\prod_{k=1}^n k!\right)^6 \Big/ \left(\prod_{k=1}^{2n+1} k!\right),$$

24.14.12
$$\det[E_{r+s}] = (-1)^{n(n+1)/2} \left(\prod_{k=1}^n k!\right)^2.$$

See also Sachse (1882).

24.14(iii) Compendia

For other sums involving Bernoulli and Euler numbers and polynomials see Hansen (1975, pp. 331–347) and Prudnikov et al. (1990, pp. 383–386).

24.15 Related Sequences of Numbers

24.15(i) Genocchi Numbers

24.15.1
$$\frac{2t}{e^t+1} = \sum_{n=1}^\infty G_n \frac{t^n}{n!},$$

24.15.2
$$G_n = 2(1-2^n) B_n.$$

See Table 24.15.1.

24.15(ii) Tangent Numbers

24.15.3
$$\tan t = \sum_{n=0}^{\infty} T_n \frac{t^n}{n!},$$

24.15.4
$$T_{2n-1} = (-1)^{n-1} \frac{2^{2n}(2^{2n}-1)}{2n} B_{2n}, \quad n = 1, 2, \ldots,$$

24.15.5
$$T_{2n} = 0, \quad n = 0, 1, \ldots.$$

Table 24.15.1: Genocchi and Tangent numbers.

n	0	1	2	3	4	5	6	7	8
G_n	0	1	−1	0	1	0	−3	0	17
T_n	0	1	0	2	0	16	0	272	0

24.15(iii) Stirling Numbers

The Stirling numbers of the first kind $s(n,m)$, and the second kind $S(n,m)$, are as defined in §26.8(i).

24.15.6
$$B_n = \sum_{k=0}^{n} (-1)^k \frac{k!\, S(n,k)}{k+1},$$

24.15.7
$$B_n = \sum_{k=0}^{n} (-1)^k \binom{n+1}{k+1} S(n+k,k) \bigg/ \binom{n+k}{k},$$

24.15.8
$$\sum_{k=0}^{n} (-1)^{n+k} s(n+1, k+1) B_k = \frac{n!}{n+1}.$$

In (24.15.9) and (24.15.10) p denotes a prime. See Horata (1991).

24.15.9
$$p \frac{B_n}{n} \equiv S(p-1+n, p-1) \pmod{p^2}, \quad 1 \le n \le p-2,$$

24.15.10
$$\frac{2n-1}{4n} p^2 B_{2n} \equiv S(p+2n, p-1) \pmod{p^3},$$
$$2 \le 2n \le p-3.$$

24.15(iv) Fibonacci and Lucas Numbers

The Fibonacci numbers are defined by $u_0 = 0$, $u_1 = 1$, and $u_{n+1} = u_n + u_{n-1}$, $n \ge 1$. The Lucas numbers are defined by $v_0 = 2$, $v_1 = 1$, and $v_{n+1} = v_n + v_{n-1}$, $n \ge 1$.

24.15.11
$$\sum_{k=0}^{\lfloor n/2 \rfloor} \binom{n}{2k} \left(\frac{5}{9}\right)^k B_{2k} u_{n-2k} = \frac{n}{6} v_{n-1} + \frac{n}{3^n} v_{2n-2},$$

24.15.12
$$\sum_{k=0}^{\lfloor n/2 \rfloor} \binom{n}{2k} \left(\frac{5}{4}\right)^k E_{2k} v_{n-2k} = \frac{1}{2^{n-1}}.$$

For further information on the Fibonacci numbers see §26.11.

24.16 Generalizations

24.16(i) Higher-Order Analogs

Polynomials and Numbers of Integer Order

For $\ell = 0, 1, 2, \ldots$, *Bernoulli* and *Euler polynomials of order* ℓ are defined respectively by

24.16.1
$$\left(\frac{t}{e^t - 1}\right)^\ell e^{xt} = \sum_{n=0}^{\infty} B_n^{(\ell)}(x) \frac{t^n}{n!}, \quad |t| < 2\pi,$$

24.16.2
$$\left(\frac{2}{e^t + 1}\right)^\ell e^{xt} = \sum_{n=0}^{\infty} E_n^{(\ell)}(x) \frac{t^n}{n!}, \quad |t| < \pi.$$

When $x = 0$ they reduce to the *Bernoulli* and *Euler numbers of order* ℓ:

24.16.3
$$B_n^{(\ell)} = B_n^{(\ell)}(0), \quad E_n^{(\ell)} = E_n^{(\ell)}(0).$$

Also for $\ell = 1, 2, 3, \ldots$,

24.16.4
$$\left(\frac{\ln(1+t)}{t}\right)^\ell = \ell \sum_{n=0}^{\infty} \frac{B_n^{(\ell+n)}}{\ell+n} \frac{t^n}{n!}, \quad |t| < 1.$$

For this and other properties see Milne-Thomson (1933, pp. 126–153) or Nörlund (1924, pp. 144–162).

For extensions of $B_n^{(\ell)}(x)$ to complex values of x, n, and ℓ, and also for uniform asymptotic expansions for large x and large n, see Temme (1995b).

Bernoulli Numbers of the Second Kind

24.16.5
$$\frac{t}{\ln(1+t)} = \sum_{n=0}^{\infty} b_n t^n, \quad |t| < 1,$$

24.16.6
$$n! b_n = -\frac{1}{n-1} B_n^{(n-1)}, \quad n = 2, 3, \ldots.$$

Degenerate Bernoulli Numbers

For sufficiently small $|t|$,

24.16.7
$$\frac{t}{(1+\lambda t)^{1/\lambda} - 1} = \sum_{n=0}^{\infty} \beta_n(\lambda) \frac{t^n}{n!},$$

24.16.8
$$\beta_n(\lambda) = n! b_n \lambda^n + \sum_{k=1}^{\lfloor n/2 \rfloor} \frac{n}{2k} B_{2k}\, s(n-1, 2k-1) \lambda^{n-2k},$$
$$n = 2, 3, \ldots.$$

Here $s(n,m)$ again denotes the Stirling number of the first kind.

Nörlund Polynomials

24.16.9
$$\left(\frac{t}{e^t - 1}\right)^x = \sum_{n=0}^{\infty} B_n^{(x)} \frac{t^n}{n!}, \quad |t| < 2\pi.$$

$B_n^{(x)}$ is a polynomial in x of degree n. (This notation is consistent with (24.16.3) when $x = \ell$.)

24.16(ii) Character Analogs

Let χ be a primitive Dirichlet character mod f (see §27.8). Then f is called the *conductor* of χ. Generalized Bernoulli numbers and polynomials belonging to χ are defined by

24.16.10 $$\sum_{a=1}^{f} \frac{\chi(a)te^{at}}{e^{ft}-1} = \sum_{n=0}^{\infty} B_{n,\chi}\frac{t^n}{n!},$$

24.16.11 $$B_{n,\chi}(x) = \sum_{k=0}^{n} \binom{n}{k} B_{k,\chi} x^{n-k}.$$

Let χ_0 be the trivial character and χ_4 the unique (nontrivial) character with $f = 4$; that is, $\chi_4(1) = 1$, $\chi_4(3) = -1$, $\chi_4(2) = \chi_4(4) = 0$. Then

24.16.12 $$B_n(x) = B_{n,\chi_0}(x-1),$$

24.16.13 $$E_n(x) = -\frac{2^{1-n}}{n+1} B_{n+1,\chi_4}(2x-1).$$

For further properties see Berndt (1975a).

24.16(iii) Other Generalizations

In no particular order, other generalizations include: Bernoulli numbers and polynomials with arbitrary complex index (Butzer *et al.* (1992)); Euler numbers and polynomials with arbitrary complex index (Butzer *et al.* (1994)); q-analogs (Carlitz (1954b), Andrews and Foata (1980)); conjugate Bernoulli and Euler polynomials (Hauss (1997, 1998)); Bernoulli–Hurwitz numbers (Katz (1975)); poly-Bernoulli numbers (Kaneko (1997)); Universal Bernoulli numbers (Clarke (1989)); p-adic integer order Bernoulli numbers (Adelberg (1996)); p-adic q-Bernoulli numbers (Kim and Kim (1999)); periodic Bernoulli numbers (Berndt (1975b)); cotangent numbers (Girstmair (1990a)); Bernoulli–Carlitz numbers (Goss (1978)); Bernoulli-Padé numbers (Dilcher (2002)); Bernoulli numbers belonging to periodic functions (Urbanowicz (1988)); cyclotomic Bernoulli numbers (Girstmair (1990b)); modified Bernoulli numbers (Zagier (1998)); higher-order Bernoulli and Euler polynomials with multiple parameters (Erdélyi *et al.* (1953a), §§1.13.1, 1.14.1)).

Applications

24.17 Mathematical Applications

24.17(i) Summation

Euler–Maclaurin Summation Formula

See §2.10(i). For a generalization see Olver (1997b, p. 284).

Boole Summation Formula

Let $0 \leq h \leq 1$ and a, m, and n be integers such that $n > a$, $m > 0$, and $f^{(m)}(x)$ is absolutely integrable over $[a, n]$. Then with the notation of §24.2(iii)

24.17.1 $$\sum_{j=a}^{n-1}(-1)^j f(j+h) = \frac{1}{2}\sum_{k=0}^{m-1} \frac{E_k(h)}{k!}\left((-1)^{n-1}f^{(k)}(n) + (-1)^a f^{(k)}(a)\right) + R_m(n),$$

where

24.17.2 $$R_m(n) = \frac{1}{2(m-1)!}\int_a^n f^{(m)}(x)\,\widetilde{E}_{m-1}(h-x)\,dx.$$

Calculus of Finite Differences

See Milne-Thomson (1933), Nörlund (1924), or Jordan (1965). For a more modern perspective see Graham *et al.* (1994).

24.17(ii) Spline Functions

Euler Splines

Let $\mathcal{S}_n$ denote the class of functions that have $n-1$ continuous derivatives on $\mathbb{R}$ and are polynomials of degree at most n in each interval $(k, k+1)$, $k \in \mathbb{Z}$. The members of $\mathcal{S}_n$ are called *cardinal spline functions*. The functions

24.17.3 $$S_n(x) = \frac{\widetilde{E}_n\left(x + \tfrac{1}{2}n + \tfrac{1}{2}\right)}{\widetilde{E}_n\left(\tfrac{1}{2}n + \tfrac{1}{2}\right)}, \quad n = 0, 1, \dots,$$

are called *Euler splines of degree* n. For each n, $S_n(x)$ is the unique bounded function such that $S_n(x) \in \mathcal{S}_n$ and

24.17.4 $$S_n(k) = (-1)^k, \qquad k \in \mathbb{Z}.$$

The function $S_n(x)$ is also optimal in a certain sense; see Schoenberg (1971).

Bernoulli Monosplines

A function of the form $x^n - S(x)$, with $S(x) \in \mathcal{S}_{n-1}$ is called a *cardinal monospline of degree* n. Again with the notation of §24.2(iii) define

24.17.5 $$M_n(x) = \begin{cases} \widetilde{B}_n(x) - B_n, & n \text{ even}, \\ \widetilde{B}_n\left(x + \tfrac{1}{2}\right), & n \text{ odd}. \end{cases}$$

$M_n(x)$ is a monospline of degree n, and it follows from (24.4.25) and (24.4.27) that

24.17.6 $$M_n(k) = 0, \qquad k \in \mathbb{Z}.$$

For each $n = 1, 2, \dots$ the function $M_n(x)$ is also the unique cardinal monospline of degree n satisfying (24.17.6), provided that

24.17.7 $$M_n(x) = O(|x|^\gamma), \qquad x \to \pm\infty,$$

for some positive constant γ.

For any $n \geq 2$ the function

24.17.8 $\qquad F(x) = \widetilde{B}_n(x) - 2^{-n} B_n$

is the unique cardinal monospline of degree n having the least supremum norm $\|F\|_\infty$ on $\mathbb{R}$ (minimality property).

24.17(iii) Number Theory

Bernoulli and Euler numbers and polynomials occur in: number theory via (24.4.7), (24.4.8), and other identities involving sums of powers; the Riemann zeta function and L-series (§25.15, Apostol (1976), and Ireland and Rosen (1990)); arithmetic of cyclotomic fields and the classical theory of Fermat's last theorem (Ribenboim (1979) and Washington (1997)); p-adic analysis (Koblitz (1984, Chapter 2)).

24.18 Physical Applications

Bernoulli polynomials appear in statistical physics (Ordóñez and Driebe (1996)), in discussions of Casimir forces (Li et al. (1991)), and in a study of quark-gluon plasma (Meisinger et al. (2002)).

Euler polynomials also appear in statistical physics as well as in semi-classical approximations to quantum probability distributions (Ballentine and McRae (1998)).

Computation

24.19 Methods of Computation

24.19(i) Bernoulli and Euler Numbers and Polynomials

Equations (24.5.3) and (24.5.4) enable B_n and E_n to be computed by recurrence. For higher values of n more efficient methods are available. For example, the tangent numbers T_n can be generated by simple recurrence relations obtained from (24.15.3), then (24.15.4) is applied. A similar method can be used for the Euler numbers based on (4.19.5). For details see Knuth and Buckholtz (1967).

Another method is based on the identities

24.19.1 $\qquad N_{2n} = \dfrac{2(2n)!}{(2\pi)^{2n}} \left(\prod_{p-1 \mid 2n} p \right) \left(\prod_{p} \dfrac{p^{2n}}{p^{2n} - 1} \right),$

24.19.2 $\qquad D_{2n} = \prod_{p-1 \mid 2n} p, \quad B_{2n} = \dfrac{N_{2n}}{D_{2n}}.$

If $\widetilde{N}_{2n}$ denotes the right-hand side of (24.19.1) but with the second product taken only for $p \leq \lfloor (\pi e)^{-1} 2n \rfloor + 1$,

then $N_{2n} = \left\lceil \widetilde{N}_{2n} \right\rceil$ for $n \geq 2$. For proofs and further information see Fillebrown (1992).

For other information see Chellali (1988) and Zhang and Jin (1996, pp. 1–11). For algorithms for computing B_n, E_n, $B_n(x)$, and $E_n(x)$ see Spanier and Oldham (1987, pp. 37, 41, 171, and 179–180).

24.19(ii) Values of B_n Modulo p

For number-theoretic applications it is important to compute $B_{2n} \pmod{p}$ for $2n \leq p - 3$; in particular to find the *irregular pairs* $(2n, p)$ for which $B_{2n} \equiv 0 \pmod{p}$. We list here three methods, arranged in increasing order of efficiency.

- Tanner and Wagstaff (1987) derives a congruence $\pmod{p}$ for Bernoulli numbers in terms of sums of powers. See also §24.10(iii).

- Buhler et al. (1992) uses the expansion

24.19.3 $\qquad \dfrac{t^2}{\cosh t - 1} = -2 \sum_{n=0}^{\infty} (2n - 1) B_{2n} \dfrac{t^{2n}}{(2n)!},$

and computes inverses modulo p of the left-hand side. Multisectioning techniques are applied in implementations. See also Crandall (1996, pp. 116–120).

- A method related to "Stickelberger codes" is applied in Buhler et al. (2001); in particular, it allows for an efficient search for the irregular pairs $(2n, p)$. Discrete Fourier transforms are used in the computations. See also Crandall (1996, pp. 120–124).

24.20 Tables

Abramowitz and Stegun (1964, Chapter 23) includes exact values of $\sum_{k=1}^{m} k^n$, $m = 1(1)100$, $n = 1(1)10$; $\sum_{k=1}^{\infty} k^{-n}$, $\sum_{k=1}^{\infty} (-1)^{k-1} k^{-n}$, $\sum_{k=0}^{\infty} (2k+1)^{-n}$, $n = 1, 2, \ldots, 20\text{D}$; $\sum_{k=0}^{\infty} (-1)^k (2k+1)^{-n}$, $n = 1, 2, \ldots, 18\text{D}$.

Wagstaff (1978) gives complete prime factorizations of N_n and E_n for $n = 20(2)60$ and $n = 8(2)42$, respectively. In Wagstaff (2002) these results are extended to $n = 60(2)152$ and $n = 40(2)88$, respectively, with further complete and partial factorizations listed up to $n = 300$ and $n = 200$, respectively.

For information on tables published before 1961 see Fletcher et al. (1962, v. 1, §4) and Lebedev and Fedorova (1960, Chapters 11 and 14).

24.21 Software

See http://dlmf.nist.gov/24.21.

References

General References

The main references used in writing this chapter are Erdélyi et al. (1953a, Chapter 1), Nörlund (1924, Chapter 2), and Nörlund (1922). Introductions to the subject are contained in Dence and Dence (1999) and Rademacher (1973); see also Apostol (2008). A comprehensive bibliography on the topics of this chapter can be found in Dilcher et al. (1991).

Sources

The following list gives the references or other indications of proofs that were used in constructing the various sections of this chapter. These sources supplement the references that are quoted in the text.

§24.2 Nörlund (1924, Chapter 2), Milne-Thomson (1933, Chapter 6). Tables are from Abramowitz and Stegun (1964, pp. 809–810).

§24.3 These graphs were produced at NIST.

§24.4 Nörlund (1924, Chapter 2), Milne-Thomson (1933, Chapter 6), Howard (1996b), Slavutskiĭ (2000), Apostol (2006), Todorov (1991). For (24.4.12)–(24.4.17) use (24.2.3) and (24.2.8). For (24.4.25)–(24.4.33) use §§24.4(ii) and 24.4(v). For (24.4.34) and (24.4.35) use (24.2.3) and (24.2.8).

§24.5 Erdélyi et al. (1953a, Chapter 1), Nörlund (1924, pp. 19 and 24), Apostol (2008), Apostol (1976, p. 275), Riordan (1979, p. 114).

§24.6 Gould (1972, pp. 45–46), Horata (1989), Todorov (1978), Schwatt (1962, p. 270), Carlitz (1961b, p. 134), Todorov (1991, pp. 176–177), Jordan (1965, p. 236).

§24.7 Erdélyi et al. (1953a, Chapter 1), Ramanujan (1927, p. 7), Paris and Kaminski (2001, p. 173).

§24.8 Apostol (1976, p. 267), Erdélyi et al. (1953a, p. 42), Berndt (1975b, pp. 176–178).

§24.9 Olver (1997b, p. 283), Temme (1996a, p. 16), Lehmer (1940, p. 538), Leeming (1989), Alzer (2000). For (24.9.5) use (24.9.8), (24.4.35), (24.2.7), and (24.4.26). For (24.9.10) use (24.4.28) and (24.8.4) with $x = \frac{1}{2}$; see also Lehmer (1940, p. 538).

§24.10 Ireland and Rosen (1990, Chapter 15), Washington (1997, Chapter 5), Carlitz (1953, p. 167), Ernvall (1979, pp. 36 and 24), Slavutskiĭ (1995, 1999), Uspensky and Heaslet (1939, p. 261), Ribenboim (1979, p. 105), Girstmair (1990b), Carlitz (1954a).

§24.11 Leeming (1977), Dilcher (1987a).

§24.12 Olver (1997b, p. 283), Inkeri (1959), Leeming (1989), Delange (1991), Lehmer (1940), Dilcher (1988, p. 77), Howard (1976), Delange (1988), Dilcher (2008), Brillhart (1969).

§24.13 Nörlund (1922, p. 143), Apostol (1976, p. 276), Nörlund (1924, pp. 31 and 36). For (24.13.1) and (24.13.2) use (24.4.34) and (24.4.1).

§24.14 Nörlund (1922, pp. 135–142), Carlitz (1961a, p. 992), Dilcher (1996), Sitaramachandrarao and Davis (1986), Huang and Huang (1999).

§24.15 Dumont and Viennot (1980), Graham et al. (1994, Chapter 6), Knuth and Buckholtz (1967), Todorov (1984, pp. 310 and 343), Gould (1972, pp. 44 and 48), Kelisky (1957, pp. 32 and 34).

§24.16 Howard (1996a), Washington (1997, pp. 31–34), Dilcher (1988, pp. 8 and 9).

§24.17 Temme (1996a, pp. 17 and 18), Nörlund (1924, pp. 29–36), Schumaker (1981, pp. 152–153), Schoenberg (1973, pp. 40–41 and 101).

§24.19 For (24.19.3) use (24.2.1).

Chapter 25

Zeta and Related Functions

T. M. Apostol[1]

Notation   **602**
- 25.1 Special Notation 602

Riemann Zeta Function   **602**
- 25.2 Definition and Expansions 602
- 25.3 Graphics 603
- 25.4 Reflection Formulas 603
- 25.5 Integral Representations 604
- 25.6 Integer Arguments 605
- 25.7 Integrals 606
- 25.8 Sums 606
- 25.9 Asymptotic Approximations 606
- 25.10 Zeros 606

Related Functions   **607**
- 25.11 Hurwitz Zeta Function 607
- 25.12 Polylogarithms 610
- 25.13 Periodic Zeta Function 612
- 25.14 Lerch's Transcendent 612
- 25.15 Dirichlet L-functions 612

Applications   **613**
- 25.16 Mathematical Applications 613
- 25.17 Physical Applications 614

Computation   **614**
- 25.18 Methods of Computation 614
- 25.19 Tables 614
- 25.20 Approximations 615
- 25.21 Software 615

References   **615**

---

[1] California Institute of Technology, Pasadena, California.
Copyright © 2009 National Institute of Standards and Technology. All rights reserved.

Notation

25.1 Special Notation

(For other notation see pp. xiv and 873.)

k, m, n	nonnegative integers.
p	prime number.
x	real variable.
a	real or complex parameter.
$s = \sigma + it$	complex variable.
$z = x + iy$	complex variable.
γ	Euler's constant (§5.2(ii)).
$\psi(x)$	digamma function $\Gamma'(x)/\Gamma(x)$ except in §25.16. See §5.2(i).
$B_n, B_n(x)$	Bernoulli number and polynomial (§24.2(i)).
$\widetilde{B}_n(x)$	periodic Bernoulli function $B_n(x - \lfloor x \rfloor)$.
$m \mid n$	m divides n.
primes	on function symbols: derivatives with respect to argument.

The main function treated in this chapter is the Riemann zeta function $\zeta(s)$. This notation was introduced in Riemann (1859).

The main related functions are the Hurwitz zeta function $\zeta(s, a)$, the dilogarithm $\operatorname{Li}_2(z)$, the polylogarithm $\operatorname{Li}_s(z)$ (also known as Jonquière's function $\phi(z, s)$), Lerch's transcendent $\Phi(z, s, a)$, and the Dirichlet L-functions $L(s, \chi)$.

Riemann Zeta Function

25.2 Definition and Expansions

25.2(i) Definition

When $\Re s > 1$,

25.2.1 $$\zeta(s) = \sum_{n=1}^{\infty} \frac{1}{n^s}.$$

Elsewhere $\zeta(s)$ is defined by analytic continuation. It is a meromorphic function whose only singularity in $\mathbb{C}$ is a simple pole at $s = 1$, with residue 1.

25.2(ii) Other Infinite Series

25.2.2 $$\zeta(s) = \frac{1}{1 - 2^{-s}} \sum_{n=0}^{\infty} \frac{1}{(2n+1)^s}, \qquad \Re s > 1.$$

25.2.3 $$\zeta(s) = \frac{1}{1 - 2^{1-s}} \sum_{n=1}^{\infty} \frac{(-1)^{n-1}}{n^s}, \qquad \Re s > 0.$$

25.2.4 $$\zeta(s) = \frac{1}{s-1} + \sum_{n=0}^{\infty} \frac{(-1)^n}{n!} \gamma_n (s-1)^n, \quad \Re s > 0,$$

where

25.2.5 $$\gamma_n = \lim_{m \to \infty} \left(\sum_{k=1}^{m} \frac{(\ln k)^n}{k} - \frac{(\ln m)^{n+1}}{n+1} \right).$$

25.2.6 $$\zeta'(s) = -\sum_{n=2}^{\infty} (\ln n) n^{-s}, \qquad \Re s > 1.$$

25.2.7
$$\zeta^{(k)}(s) = (-1)^k \sum_{n=2}^{\infty} (\ln n)^k n^{-s}, \quad \Re s > 1,\ k = 1, 2, 3, \ldots.$$

For further expansions of functions similar to (25.2.1) (Dirichlet series) see §27.4. This includes, for example, $1/\zeta(s)$.

25.2(iii) Representations by the Euler–Maclaurin Formula

25.2.8 $$\zeta(s) = \sum_{k=1}^{N} \frac{1}{k^s} + \frac{N^{1-s}}{s-1} - s \int_N^\infty \frac{x - \lfloor x \rfloor}{x^{s+1}} \, dx,$$
$$\Re s > 0,\ N = 1, 2, 3, \ldots.$$

25.2.9
$$\zeta(s) = \sum_{k=1}^{N} \frac{1}{k^s} + \frac{N^{1-s}}{s-1} - \frac{1}{2} N^{-s}$$
$$+ \sum_{k=1}^{n} \binom{s+2k-2}{2k-1} \frac{B_{2k}}{2k} N^{1-s-2k}$$
$$- \binom{s+2n}{2n+1} \int_N^\infty \frac{\widetilde{B}_{2n+1}(x)}{x^{s+2n+1}} \, dx,$$
$$\Re s > -2n;\ n, N = 1, 2, 3, \ldots.$$

25.2.10
$$\zeta(s) = \frac{1}{s-1} + \frac{1}{2} + \sum_{k=1}^{n} \binom{s+2k-2}{2k-1} \frac{B_{2k}}{2k}$$
$$- \binom{s+2n}{2n+1} \int_1^\infty \frac{\widetilde{B}_{2n+1}(x)}{x^{s+2n+1}} \, dx,$$
$$\Re s > -2n,\ n = 1, 2, 3, \ldots.$$

For B_{2k} see §24.2(i), and for $\widetilde{B}_n(x)$ see §24.2(iii).

25.2(iv) Infinite Products

25.2.11 $$\zeta(s) = \prod_{p} (1 - p^{-s})^{-1}, \qquad \Re s > 1,$$

product over all primes p.

25.2.12 $$\zeta(s) = \frac{(2\pi)^s e^{-s-(\gamma s/2)}}{2(s-1)\Gamma(\tfrac{1}{2}s+1)} \prod_{\rho} \left(1 - \frac{s}{\rho}\right) e^{s/\rho},$$

product over zeros ρ of ζ with $\Re \rho > 0$ (see §25.10(i)); γ is Euler's constant (§5.2(ii)).

25.3 Graphics

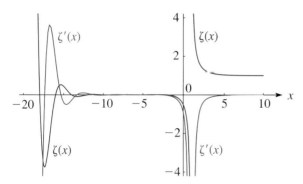

Figure 25.3.1: Riemann zeta function $\zeta(x)$ and its derivative $\zeta'(x)$, $-20 \leq x \leq 10$.

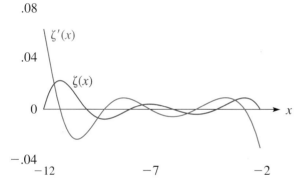

Figure 25.3.2: Riemann zeta function $\zeta(x)$ and its derivative $\zeta'(x)$, $-12 \leq x \leq -2$.

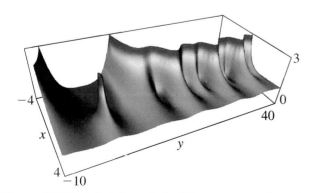

Figure 25.3.3: Modulus of the Riemann zeta function $|\zeta(x+iy)|$, $-4 \leq x \leq 4$, $-10 \leq y \leq 40$.

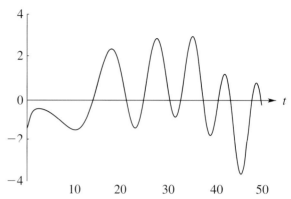

Figure 25.3.4: $Z(t)$, $0 \leq t \leq 50$. $Z(t)$ and $\zeta\!\left(\tfrac{1}{2}+it\right)$ have the same zeros. See §25.10(i).

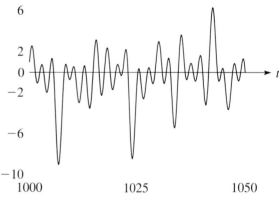

Figure 25.3.5: $Z(t)$, $1000 \leq t \leq 1050$.

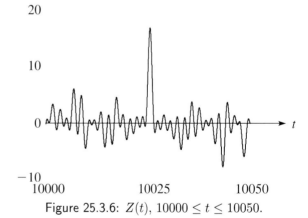

Figure 25.3.6: $Z(t)$, $10000 \leq t \leq 10050$.

25.4 Reflection Formulas

For $s \neq 0, 1$,

25.4.1 $\quad \zeta(1-s) = 2(2\pi)^{-s}\cos\!\left(\tfrac{1}{2}\pi s\right)\Gamma(s)\,\zeta(s),$

25.4.2 $\quad \zeta(s) = 2(2\pi)^{s-1}\sin\!\left(\tfrac{1}{2}\pi s\right)\Gamma(1-s)\,\zeta(1-s).$

Equivalently,

25.4.3 $\quad\quad\quad\quad \xi(s) = \xi(1-s),$

where $\xi(s)$ is *Riemann's ξ-function*, defined by:

25.4.4 $\quad \xi(s) = \tfrac{1}{2}s(s-1)\,\Gamma\!\left(\tfrac{1}{2}s\right)\pi^{-s/2}\,\zeta(s).$

For $s \neq 0, 1$ and $k = 1, 2, 3, \ldots,$

25.4.5
$$(-1)^k \zeta^{(k)}(1-s) = \frac{2}{(2\pi)^s} \sum_{m=0}^{k} \sum_{r=0}^{m} \binom{k}{m}\binom{m}{r} \left(\Re(c^{k-m})\cos\!\left(\tfrac{1}{2}\pi s\right) + \Im(c^{k-m})\sin\!\left(\tfrac{1}{2}\pi s\right)\right) \Gamma^{(r)}(s)\,\zeta^{(m-r)}(s),$$

where

25.4.6 $\quad c = -\ln(2\pi) - \tfrac{1}{2}\pi i.$

25.5 Integral Representations

25.5(i) In Terms of Elementary Functions

Throughout this subsection $s \neq 1$.

25.5.1 $\quad \zeta(s) = \dfrac{1}{\Gamma(s)}\displaystyle\int_0^{\infty} \dfrac{x^{s-1}}{e^x - 1}\,dx, \qquad \Re s > 1.$

25.5.2 $\quad \zeta(s) = \dfrac{1}{\Gamma(s+1)}\displaystyle\int_0^{\infty} \dfrac{e^x x^s}{(e^x - 1)^2}\,dx, \qquad \Re s > 1.$

25.5.3 $\quad \zeta(s) = \dfrac{1}{(1-2^{1-s})\Gamma(s)}\displaystyle\int_0^{\infty} \dfrac{x^{s-1}}{e^x + 1}\,dx, \qquad \Re s > 0.$

25.5.4 $\quad \zeta(s) = \dfrac{1}{(1-2^{1-s})\Gamma(s+1)}\displaystyle\int_0^{\infty} \dfrac{e^x x^s}{(e^x + 1)^2}\,dx, \qquad \Re s > 0.$

25.5.5 $\quad \zeta(s) = -s \displaystyle\int_0^{\infty} \dfrac{x - \lfloor x \rfloor - \tfrac{1}{2}}{x^{s+1}}\,dx, \quad -1 < \Re s < 0.$

25.5.6
$$\zeta(s) = \frac{1}{2} + \frac{1}{s-1} + \frac{1}{\Gamma(s)}\int_0^{\infty} \left(\frac{1}{e^x - 1} - \frac{1}{x} + \frac{1}{2}\right)\frac{x^{s-1}}{e^x}\,dx, \quad \Re s > -1.$$

25.5.7
$$\zeta(s) = \frac{1}{2} + \frac{1}{s-1} + \sum_{m=1}^{n} \frac{B_{2m}}{(2m)!}\frac{\Gamma(s+2m-1)}{\Gamma(s)} + \frac{1}{\Gamma(s)}\int_0^{\infty}\left(\frac{1}{e^x - 1} - \frac{1}{x} + \frac{1}{2} - \sum_{m=1}^{n}\frac{B_{2m}}{(2m)!}x^{2m-1}\right)\frac{x^{s-1}}{e^x}\,dx,$$
$$\Re s > -(2n+1),\ n = 1, 2, 3, \ldots.$$

25.5.8 $\quad \zeta(s) = \dfrac{1}{2(1-2^{-s})\Gamma(s)}\displaystyle\int_0^{\infty} \dfrac{x^{s-1}}{\sinh x}\,dx, \quad \Re s > 1.$

25.5.9 $\quad \zeta(s) = \dfrac{2^{s-1}}{\Gamma(s+1)}\displaystyle\int_0^{\infty} \dfrac{x^s}{(\sinh x)^2}\,dx, \quad \Re s > 1.$

25.5.10 $\quad \zeta(s) = \dfrac{2^{s-1}}{1-2^{1-s}}\displaystyle\int_0^{\infty} \dfrac{\cos(s \arctan x)}{(1+x^2)^{s/2}\cosh\!\left(\tfrac{1}{2}\pi x\right)}\,dx.$

25.5.11
$$\zeta(s) = \frac{1}{2} + \frac{1}{s-1} + 2\int_0^{\infty} \frac{\sin(s \arctan x)}{(1+x^2)^{s/2}(e^{2\pi x} - 1)}\,dx.$$

25.5.12 $\quad \zeta(s) = \dfrac{2^{s-1}}{s-1} - 2^s \displaystyle\int_0^{\infty} \dfrac{\sin(s \arctan x)}{(1+x^2)^{s/2}(e^{\pi x} + 1)}\,dx.$

25.5(ii) In Terms of Other Functions

25.5.13
$$\zeta(s) = \frac{\pi^{s/2}}{s(s-1)\,\Gamma\!\left(\tfrac{1}{2}s\right)} + \frac{\pi^{s/2}}{\Gamma\!\left(\tfrac{1}{2}s\right)}\int_1^{\infty}\left(x^{s/2} + x^{(1-s)/2}\right)\frac{\omega(x)}{x}\,dx, \quad s \neq 1,$$

where

25.5.14 $\quad \omega(x) = \displaystyle\sum_{n=1}^{\infty} e^{-n^2 \pi x} = \tfrac{1}{2}\left(\theta_3(0|ix) - 1\right).$

For θ_3 see §20.2(i). For similar representations involving other theta functions see Erdélyi et al. (1954a, p. 339).

In (25.5.15)–(25.5.19), $0 < \Re s < 1$, $\psi(x)$ is the digamma function, and γ is Euler's constant (§5.2). (25.5.16) is also valid for $0 < \Re s < 2$, $s \neq 1$.

25.5.15
$$\zeta(s) = \frac{1}{s-1} + \frac{\sin(\pi s)}{\pi} \times \int_0^{\infty} (\ln(1+x) - \psi(1+x))x^{-s}\,dx,$$

25.5.16
$$\zeta(s) = \frac{1}{s-1} + \frac{\sin(\pi s)}{\pi(s-1)} \times \int_0^{\infty} \left(\frac{1}{1+x} - \psi'(1+x)\right) x^{1-s}\,dx,$$

25.5.17 $\quad \zeta(1+s) = \dfrac{\sin(\pi s)}{\pi}\displaystyle\int_0^{\infty}(\gamma + \psi(1+x))\,x^{-s-1}\,dx,$

25.5.18 $\quad \zeta(1+s) = \dfrac{\sin(\pi s)}{\pi s}\displaystyle\int_0^{\infty}\psi'(1+x)x^{-s}\,dx,$

25.5.19
$$\zeta(m+s) = (-1)^{m-1}\frac{\Gamma(s)\sin(\pi s)}{\pi\,\Gamma(m+s)} \times \int_0^{\infty}\psi^{(m)}(1+x)x^{-s}\,dx,$$
$$m = 1, 2, 3, \ldots.$$

25.5(iii) Contour Integrals

25.5.20
$$\zeta(s) = \frac{\Gamma(1-s)}{2\pi i} \int_{-\infty}^{(0+)} \frac{z^{s-1}}{e^{-z}-1} dz, \quad s \neq 1, 2, \ldots,$$

where the integration contour is a loop around the negative real axis; it starts at $-\infty$, encircles the origin once in the positive direction without enclosing any of the points $z = \pm 2\pi i, \pm 4\pi i, \ldots$, and returns to $-\infty$. Equivalently,

25.5.21
$$\zeta(s) = \frac{\Gamma(1-s)}{2\pi i(1-2^{1-s})} \int_{-\infty}^{(0+)} \frac{z^{s-1}}{e^{-z}+1} dz, \quad s \neq 1, 2, \ldots.$$

The contour here is any loop that encircles the origin in the positive direction not enclosing any of the points $\pm \pi i, \pm 3\pi i, \ldots$.

25.6 Integer Arguments

25.6(i) Function Values

25.6.1
$$\zeta(0) = -\frac{1}{2}, \quad \zeta(2) = \frac{\pi^2}{6}, \quad \zeta(4) = \frac{\pi^4}{90}, \quad \zeta(6) = \frac{\pi^6}{945}.$$

25.6.2 $\quad \zeta(2n) = \dfrac{(2\pi)^{2n}}{2(2n)!} |B_{2n}|, \quad n = 1, 2, 3, \ldots.$

25.6.3 $\quad \zeta(-n) = -\dfrac{B_{n+1}}{n+1}, \quad n = 1, 2, 3, \ldots.$

25.6.4 $\quad \zeta(-2n) = 0, \quad n = 1, 2, 3, \ldots.$

25.6.5
$$\zeta(k+1) = \frac{1}{k!} \sum_{n_1=1}^{\infty} \cdots \sum_{n_k=1}^{\infty} \frac{1}{n_1 \cdots n_k(n_1 + \cdots + n_k)}, \quad k = 1, 2, 3, \ldots.$$

25.6.6
$$\zeta(2k+1) = \frac{(-1)^{k+1}(2\pi)^{2k+1}}{2(2k+1)!} \int_0^1 B_{2k+1}(t) \cot(\pi t) dt, \quad k = 1, 2, 3, \ldots.$$

25.6.7 $\quad \zeta(2) = \displaystyle\int_0^1 \int_0^1 \frac{1}{1-xy} dx\, dy.$

25.6.8 $\quad \zeta(2) = 3 \displaystyle\sum_{k=1}^{\infty} \frac{1}{k^2 \binom{2k}{k}}.$

25.6.9 $\quad \zeta(3) = \dfrac{5}{2} \displaystyle\sum_{k=1}^{\infty} \frac{(-1)^{k-1}}{k^3 \binom{2k}{k}}.$

25.6.10 $\quad \zeta(4) = \dfrac{36}{17} \displaystyle\sum_{k=1}^{\infty} \frac{1}{k^4 \binom{2k}{k}}.$

25.6(ii) Derivative Values

25.6.11 $\quad \zeta'(0) = -\frac{1}{2} \ln(2\pi).$

25.6.12 $\quad \zeta''(0) = -\frac{1}{2} (\ln(2\pi))^2 + \frac{1}{2}\gamma^2 - \frac{1}{24}\pi^2 + \gamma_1,$
where γ_1 is given by (25.2.5).

With c defined by (25.4.6) and $n = 1, 2, 3, \ldots$,

25.6.13
$$(-1)^k \zeta^{(k)}(-2n) = \frac{2(-1)^n}{(2\pi)^{2n+1}} \sum_{m=0}^{k} \sum_{r=0}^{m} \binom{k}{m}\binom{m}{r} \Im(c^{k-m})\, \Gamma^{(r)}(2n+1)\, \zeta^{(m-r)}(2n+1),$$

25.6.14
$$(-1)^k \zeta^{(k)}(1-2n) = \frac{2(-1)^n}{(2\pi)^{2n}} \sum_{m=0}^{k} \sum_{r=0}^{m} \binom{k}{m}\binom{m}{r} \Re(c^{k-m})\, \Gamma^{(r)}(2n)\, \zeta^{(m-r)}(2n),$$

25.6.15
$$\zeta'(2n) = \frac{(-1)^{n+1}(2\pi)^{2n}}{2(2n)!} \left(2n\, \zeta'(1-2n) - (\psi(2n) - \ln(2\pi)) B_{2n} \right).$$

25.6(iii) Recursion Formulas

25.6.16 $\quad \left(n + \frac{1}{2}\right) \zeta(2n) = \displaystyle\sum_{k=1}^{n-1} \zeta(2k) \zeta(2n-2k), \quad n \geq 2.$

25.6.17
$$\left(n + \tfrac{3}{4}\right) \zeta(4n+2) = \sum_{k=1}^{n} \zeta(2k) \zeta(4n+2-2k), \quad n \geq 1.$$

25.6.18
$$\left(n + \tfrac{1}{4}\right) \zeta(4n) + \tfrac{1}{2}(\zeta(2n))^2 = \sum_{k=1}^{n} \zeta(2k) \zeta(4n-2k), \quad n \geq 1.$$

25.6.19
$$\left(m + n + \tfrac{3}{2}\right) \zeta(2m+2n+2)$$
$$= \left(\sum_{k=1}^{m} + \sum_{k=1}^{n} \right) \zeta(2k) \zeta(2m+2n+2-2k),$$
$$m \geq 0, n \geq 0, m+n \geq 1.$$

25.6.20
$$\tfrac{1}{2}(2^{2n}-1)\zeta(2n) = \sum_{k=1}^{n-1}(2^{2n-2k}-1)\zeta(2n-2k)\zeta(2k),$$
$$n \geq 2.$$

For related results see Basu and Apostol (2000).

25.7 Integrals

For definite integrals of the Riemann zeta function see Prudnikov et al. (1986b, §2.4), Prudnikov et al. (1992a, §3.2), and Prudnikov et al. (1992b, §3.2).

25.8 Sums

25.8.1 $$\sum_{k=2}^{\infty}(\zeta(k)-1) = 1.$$

25.8.2 $$\sum_{k=0}^{\infty}\frac{\Gamma(s+k)}{(k+1)!}(\zeta(s+k)-1) = \Gamma(s-1), \quad s \neq 1, 0, -1, -2, \ldots.$$

25.8.3 $$\sum_{k=0}^{\infty}\frac{\Gamma(s+k)\zeta(s+k)}{k!\,\Gamma(s)2^{s+k}} = (1-2^{-s})\zeta(s), \quad s \neq 1.$$

25.8.4
$$\sum_{k=1}^{\infty}\frac{(-1)^k}{k}(\zeta(nk)-1) = \ln\left(\prod_{j=0}^{n-1}\Gamma\left(2-e^{(2j+1)\pi i/n}\right)\right),$$
$$n = 2, 3, 4, \ldots.$$

25.8.5 $$\sum_{k=2}^{\infty}\zeta(k)z^k = -\gamma z - z\psi(1-z), \quad |z|<1.$$

25.8.6 $$\sum_{k=0}^{\infty}\zeta(2k)z^{2k} = -\tfrac{1}{2}\pi z\cot(\pi z), \quad |z|<1.$$

25.8.7 $$\sum_{k=2}^{\infty}\frac{\zeta(k)}{k}z^k = -\gamma z + \ln\Gamma(1-z), \quad |z|<1.$$

25.8.8 $$\sum_{k=1}^{\infty}\frac{\zeta(2k)}{k}z^{2k} = \ln\left(\frac{\pi z}{\sin(\pi z)}\right), \quad |z|<1.$$

25.8.9 $$\sum_{k=1}^{\infty}\frac{\zeta(2k)}{(2k+1)2^{2k}} = \frac{1}{2} - \frac{1}{2}\ln 2.$$

25.8.10 $$\sum_{k=1}^{\infty}\frac{\zeta(2k)}{(2k+1)(2k+2)2^{2k}} = \frac{1}{4} - \frac{7}{4\pi^2}\zeta(3).$$

For other sums see Prudnikov et al. (1986b, pp. 648–649), Hansen (1975, pp. 355–357), Ogreid and Osland (1998), and Srivastava and Choi (2001, Chapter 3).

25.9 Asymptotic Approximations

If $x \geq 1$, $y \geq 1$, $2\pi xy = t$, and $0 \leq \sigma \leq 1$, then as $t \to \infty$ with σ fixed,

25.9.1
$$\zeta(\sigma+it) = \sum_{1\leq n\leq x}\frac{1}{n^s} + \chi(s)\sum_{1\leq n\leq y}\frac{1}{n^{1-s}} + O(x^{-\sigma}) + O\left(y^{\sigma-1}t^{\frac{1}{2}-\sigma}\right),$$

where $s = \sigma + it$ and

25.9.2 $$\chi(s) = \pi^{s-\frac{1}{2}}\,\Gamma\!\left(\tfrac{1}{2}-\tfrac{1}{2}s\right)/\Gamma\!\left(\tfrac{1}{2}s\right).$$

If $\sigma = \tfrac{1}{2}$, $x = y = \sqrt{t/(2\pi)}$, and $m = \lfloor x \rfloor$, then (25.9.1) becomes

25.9.3
$$\zeta\!\left(\tfrac{1}{2}+it\right) = \sum_{n=1}^{m}\frac{1}{n^{\frac{1}{2}+it}} + \chi\!\left(\tfrac{1}{2}+it\right)\sum_{n=1}^{m}\frac{1}{n^{\frac{1}{2}-it}} + O\!\left(t^{-1/4}\right).$$

For other asymptotic approximations see Berry and Keating (1992), Paris and Cang (1997); see also Paris and Kaminski (2001, pp. 380–389).

25.10 Zeros

25.10(i) Distribution

The product representation (25.2.11) implies $\zeta(s) \neq 0$ for $\Re s > 1$. Also, $\zeta(s) \neq 0$ for $\Re s = 1$, a property first established in Hadamard (1896) and de la Vallée Poussin (1896a,b) in the proof of the prime number theorem (25.16.3). The functional equation (25.4.1) implies $\zeta(-2n) = 0$ for $n = 1, 2, 3, \ldots$. These are called the *trivial zeros*. Except for the trivial zeros, $\zeta(s) \neq 0$ for $\Re s \leq 0$. In the region $0 < \Re s < 1$, called the *critical strip*, $\zeta(s)$ has infinitely many zeros, distributed symmetrically about the real axis and about the *critical line* $\Re s = \tfrac{1}{2}$. The *Riemann hypothesis* states that all nontrivial zeros lie on this line.

Calculations relating to the zeros on the critical line make use of the real-valued function

25.10.1 $$Z(t) = \exp(i\vartheta(t))\,\zeta\!\left(\tfrac{1}{2}+it\right),$$

where

25.10.2 $$\vartheta(t) \equiv \operatorname{ph}\Gamma\!\left(\tfrac{1}{4}+\tfrac{1}{2}it\right) - \tfrac{1}{2}t\ln\pi$$

is chosen to make $Z(t)$ real, and $\operatorname{ph}\Gamma\!\left(\tfrac{1}{4}+\tfrac{1}{2}it\right)$ assumes its principal value. Because $|Z(t)| = |\zeta\!\left(\tfrac{1}{2}+it\right)|$, $Z(t)$ vanishes at the zeros of $\zeta\!\left(\tfrac{1}{2}+it\right)$, which can be separated by observing sign changes of $Z(t)$. Because $Z(t)$ changes sign infinitely often, $\zeta\!\left(\tfrac{1}{2}+it\right)$ has infinitely many zeros with t real.

25.10(ii) Riemann–Siegel Formula

Riemann developed a method for counting the total number $N(T)$ of zeros of $\zeta(s)$ in that portion of the critical strip with $0 < t < T$. By comparing $N(T)$ with the number of sign changes of $Z(t)$ we can decide whether $\zeta(s)$ has any zeros off the line in this region. Sign changes of $Z(t)$ are determined by multiplying (25.9.3) by $\exp(i\vartheta(t))$ to obtain the *Riemann–Siegel formula*:

$$25.10.3 \quad Z(t) = 2\sum_{n=1}^{m} \frac{\cos(\vartheta(t) - t\ln n)}{n^{1/2}} + R(t),$$

where $R(t) = O(t^{-1/4})$ as $t \to \infty$.

The error term $R(t)$ can be expressed as an asymptotic series that begins

$$25.10.4 \quad R(t) = (-1)^{m-1}\left(\frac{2\pi}{t}\right)^{1/4} \frac{\cos\left(t - (2m+1)\sqrt{2\pi t} - \tfrac{1}{8}\pi\right)}{\cos(\sqrt{2\pi t})} + O\left(t^{-3/4}\right).$$

Riemann also developed a technique for determining further terms. Calculations based on the Riemann–Siegel formula reveal that the first ten billion zeros of $\zeta(s)$ in the critical strip are on the critical line (van de Lune et al. (1986)). More than one-third of all the zeros in the critical strip lie on the critical line (Levinson (1974)).

For further information on the Riemann–Siegel expansion see Berry (1995).

Related Functions

25.11 Hurwitz Zeta Function

25.11(i) Definition

The function $\zeta(s,a)$ was introduced in Hurwitz (1882) and defined by the series expansion

$$25.11.1 \quad \zeta(s,a) = \sum_{n=0}^{\infty} \frac{1}{(n+a)^s}, \quad \Re s > 1,\, a \neq 0, -1, -2, \ldots.$$

$\zeta(s,a)$ has a meromorphic continuation in the s-plane, its only singularity in $\mathbb{C}$ being a simple pole at $s = 1$ with residue 1. As a function of a, with $s\ (\neq 1)$ fixed, $\zeta(s,a)$ is analytic in the half-plane $\Re a > 0$. The Riemann zeta function is a special case:

$$25.11.2 \quad \zeta(s,1) = \zeta(s).$$

For most purposes it suffices to restrict $0 < \Re a \leq 1$ because of the following straightforward consequences of (25.11.1):

$$25.11.3 \quad \zeta(s,a) = \zeta(s,a+1) + a^{-s},$$

$$25.11.4 \quad \zeta(s,a) = \zeta(s,a+m) + \sum_{n=0}^{m-1} \frac{1}{(n+a)^s}, \quad m = 1, 2, 3, \ldots.$$

Most references treat real a with $0 < a \leq 1$.

25.11(ii) Graphics

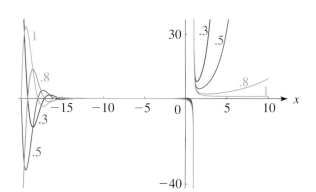

Figure 25.11.1: Hurwitz zeta function $\zeta(x,a)$, $a = 0.3$, 0.5, 0.8, 1, $-20 \leq x \leq 10$. The curves are almost indistinguishable for $-14 < x < -1$, approximately.

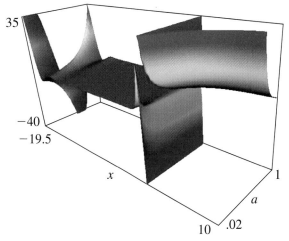

Figure 25.11.2: Hurwitz zeta function $\zeta(x,a)$, $-19.5 \leq x \leq 10$, $0.02 \leq a \leq 1$.

25.11(iii) Representations by the Euler–Maclaurin Formula

25.11.5
$$\zeta(s,a) = \sum_{n=0}^{N} \frac{1}{(n+a)^s} + \frac{(N+a)^{1-s}}{s-1} - s \int_{N}^{\infty} \frac{x - \lfloor x \rfloor}{(x+a)^{s+1}}\, dx, \quad s \neq 1,\ \Re s > 0,\ a > 0,\ N = 0, 1, 2, 3, \ldots.$$

25.11.6
$$\zeta(s,a) = \frac{1}{a^s}\left(\frac{1}{2} + \frac{a}{s-1}\right) - s(s+1)\int_{0}^{\infty} \frac{\widetilde{B}_2(x)}{(x+a)^{s+2}}\, dx, \qquad s \neq 1,\ \Re s > -1,\ a > 0.$$

25.11.7
$$\zeta(s,a) = \frac{1}{a^s} + \frac{1}{(1+a)^s}\left(\frac{1}{2} + \frac{1+a}{s-1}\right) + \sum_{k=1}^{n}\binom{s+2k-2}{2k-1}\frac{B_{2k}}{2k}\frac{1}{(1+a)^{s+2k-1}} - \binom{s+2n}{2n+1}\int_{1}^{\infty} \frac{\widetilde{B}_{2n+1}(x)}{(x+a)^{s+2n+1}}\, dx,$$
$$s \neq 1,\ a > 0,\ n = 1, 2, 3, \ldots,\ \Re s > -2n.$$

For $\widetilde{B}_n(x)$ see §24.2(iii).

25.11(iv) Series Representations

25.11.8
$$\zeta\!\left(s, \tfrac{1}{2}a\right) = \zeta\!\left(s, \tfrac{1}{2}a + \tfrac{1}{2}\right) + 2^s \sum_{n=0}^{\infty} \frac{(-1)^n}{(n+a)^s},$$
$$\Re s > 0,\ s \neq 1,\ 0 < a \leq 1.$$

25.11.9
$$\zeta(1-s, a) = \frac{2\,\Gamma(s)}{(2\pi)^s} \sum_{n=1}^{\infty} \frac{1}{n^s} \cos\!\left(\tfrac{1}{2}\pi s - 2n\pi a\right),$$
$$\Re s > 1,\ 0 < a \leq 1.$$

25.11.10
$$\zeta(s, a) = \sum_{n=0}^{\infty} \frac{\Gamma(n+s)}{n!\,\Gamma(s)} \zeta(n+s)(1-a)^n,$$
$$s \neq 1,\ |a - 1| < 1.$$

When $a = \tfrac{1}{2}$, (25.11.10) reduces to (25.8.3); compare (25.11.11).

25.11(v) Special Values

Throughout this subsection $\Re a > 0$.

25.11.11
$$\zeta\!\left(s, \tfrac{1}{2}\right) = (2^s - 1)\,\zeta(s), \qquad s \neq 1.$$

25.11.12
$$\zeta(n+1, a) = \frac{(-1)^{n+1}\psi^{(n)}(a)}{n!}, \quad n = 1, 2, 3, \ldots.$$

25.11.13
$$\zeta(0, a) = \tfrac{1}{2} - a.$$

25.11.14
$$\zeta(-n, a) = -\frac{B_{n+1}(a)}{n+1}, \quad n = 0, 1, 2, \ldots.$$

25.11.15
$$\zeta(s, ka) = k^{-s} \sum_{n=0}^{k-1} \zeta\!\left(s, a + \frac{n}{k}\right), \quad s \neq 1,\ k = 1, 2, 3, \ldots.$$

25.11.16
$$\zeta\!\left(1 - s, \frac{h}{k}\right) = \frac{2\,\Gamma(s)}{(2\pi k)^s} \sum_{r=1}^{k} \cos\!\left(\frac{\pi s}{2} - \frac{2\pi r h}{k}\right) \zeta\!\left(s, \frac{r}{k}\right),$$
$$s \neq 0, 1;\ h, k\ \text{integers},\ 1 \leq h \leq k.$$

25.11(vi) Derivatives

a-Derivative

25.11.17
$$\frac{\partial}{\partial a}\zeta(s, a) = -s\,\zeta(s+1, a), \quad s \neq 0, 1;\ \Re a > 0.$$

s-Derivatives

In (25.11.18)–(25.11.24) primes on ζ denote derivatives with respect to s. Similarly in §§25.11(viii) and 25.11(xii).

25.11.18
$$\zeta'(0, a) = \ln\Gamma(a) - \tfrac{1}{2}\ln(2\pi), \qquad a > 0.$$

25.11.19
$$\zeta'(s, a) = -\frac{\ln a}{a^s}\left(\frac{1}{2} + \frac{a}{s-1}\right) - \frac{a^{1-s}}{(s-1)^2} + s(s+1)\int_0^\infty \frac{\widetilde{B}_2(x)\ln(x+a)}{(x+a)^{s+2}}\,dx - (2s+1)\int_0^\infty \frac{\widetilde{B}_2(x)}{(x+a)^{s+2}}\,dx,$$
$$\Re s > -1,\ s \neq 1,\ a > 0.$$

25.11.20
$$(-1)^k \zeta^{(k)}(s, a) = \frac{(\ln a)^k}{a^s}\left(\frac{1}{2} + \frac{a}{s-1}\right) + k!\,a^{1-s}\sum_{r=0}^{k-1}\frac{(\ln a)^r}{r!(s-1)^{k-r+1}} - s(s+1)\int_0^\infty \frac{\widetilde{B}_2(x)(\ln(x+a))^k}{(x+a)^{s+2}}\,dx$$
$$+ k(2s+1)\int_0^\infty \frac{\widetilde{B}_2(x)(\ln(x+a))^{k-1}}{(x+a)^{s+2}}\,dx - k(k-1)\int_0^\infty \frac{\widetilde{B}_2(x)(\ln(x+a))^{k-2}}{(x+a)^{s+2}}\,dx,$$
$$\Re s > -1,\ s \neq 1,\ a > 0.$$

25.11.21
$$\zeta'\left(1-2n,\frac{h}{k}\right) = \frac{(\psi(2n)-\ln(2\pi k))\,B_{2n}(h/k)}{2n} - \frac{(\psi(2n)-\ln(2\pi))\,B_{2n}}{2nk^{2n}} + \frac{(-1)^{n+1}\pi}{(2\pi k)^{2n}}\sum_{r=1}^{k-1}\sin\left(\frac{2\pi rh}{k}\right)\psi^{(2n-1)}\left(\frac{r}{k}\right)$$
$$+ \frac{(-1)^{n+1}2\cdot(2n-1)!}{(2\pi k)^{2n}}\sum_{r=1}^{k-1}\cos\left(\frac{2\pi rh}{k}\right)\zeta'\left(2n,\frac{r}{k}\right) + \frac{\zeta'(1-2n)}{k^{2n}},$$

where h, k are integers with $1 \leq h \leq k$ and $n = 1, 2, 3, \ldots$.

25.11.22
$$\zeta'\left(1-2n,\tfrac{1}{2}\right) = -\frac{B_{2n}\ln 2}{n\cdot 4^n} - \frac{(2^{2n-1}-1)\,\zeta'(1-2n)}{2^{2n-1}}, \qquad n = 1, 2, 3, \ldots.$$

25.11.23
$$\zeta'\left(1-2n,\tfrac{1}{3}\right) = -\frac{\pi(9^n-1)\,B_{2n}}{8n\sqrt{3}(3^{2n-1}-1)} - \frac{B_{2n}\ln 3}{4n\cdot 3^{2n-1}} - \frac{(-1)^n\,\psi^{(2n-1)}\!\left(\tfrac{1}{3}\right)}{2\sqrt{3}(6\pi)^{2n-1}} - \frac{(3^{2n-1}-1)\,\zeta'(1-2n)}{2\cdot 3^{2n-1}}, \quad n=1,2,3,\ldots.$$

25.11.24
$$\sum_{r=1}^{k-1}\zeta'\left(s,\frac{r}{k}\right) = (k^s-1)\,\zeta'(s) + k^s\,\zeta(s)\ln k, \qquad s\neq 1,\ k=1,2,3,\ldots.$$

25.11(vii) Integral Representations

25.11.25
$$\zeta(s,a) = \frac{1}{\Gamma(s)}\int_0^\infty \frac{x^{s-1}e^{-ax}}{1-e^{-x}}\,dx, \qquad \Re s > 1,\ \Re a > 0.$$

25.11.26
$$\zeta(s,a) = -s\int_{-a}^\infty \frac{x-\lfloor x\rfloor - \tfrac{1}{2}}{(x+a)^{s+1}}\,dx, \qquad -1 < \Re s < 0,\ 0 < a \leq 1.$$

25.11.27
$$\zeta(s,a) = \frac{1}{2}a^{-s} + \frac{a^{1-s}}{s-1} + \frac{1}{\Gamma(s)}\int_0^\infty\left(\frac{1}{e^x-1} - \frac{1}{x} + \frac{1}{2}\right)\frac{x^{s-1}}{e^{ax}}\,dx, \quad \Re s > -1,\ s\neq 1,\ \Re a > 0.$$

25.11.28
$$\zeta(s,a) = \frac{1}{2}a^{-s} + \frac{a^{1-s}}{s-1} + \sum_{k=1}^n \frac{\Gamma(s+2k-1)}{\Gamma(s)}\frac{B_{2k}}{(2k)!}a^{-2k-s+1}$$
$$+ \frac{1}{\Gamma(s)}\int_0^\infty\left(\frac{1}{e^x-1}-\frac{1}{x}+\frac{1}{2}-\sum_{k=1}^n \frac{B_{2k}}{(2k)!}x^{2k-1}\right)x^{s-1}e^{-ax}\,dx, \quad \Re s > -(2n+1),\ s\neq 1,\ \Re a > 0.$$

25.11.29
$$\zeta(s,a) = \frac{1}{2}a^{-s} + \frac{a^{1-s}}{s-1} + 2\int_0^\infty \frac{\sin(s\arctan(x/a))}{(a^2+x^2)^{s/2}(e^{2\pi x}-1)}\,dx, \qquad s\neq 1,\ \Re a > 0.$$

25.11.30
$$\zeta(s,a) = \frac{\Gamma(1-s)}{2\pi i}\int_{-\infty}^{(0+)} \frac{e^{az}z^{s-1}}{1-e^z}\,dz, \qquad s\neq 1,\ \Re a > 0,$$

where the integration contour is a loop around the negative real axis as described for (25.5.20).

25.11(viii) Further Integral Representations

25.11.31
$$\frac{1}{\Gamma(s)}\int_0^\infty \frac{x^{s-1}e^{-ax}}{2\cosh x}\,dx = 4^{-s}\left(\zeta\!\left(s,\tfrac{1}{4}+\tfrac{1}{4}a\right) - \zeta\!\left(s,\tfrac{3}{4}+\tfrac{1}{4}a\right)\right), \qquad \Re s > 0,\ \Re a > -1.$$

25.11.32
$$\int_0^a x^n\,\psi(x)\,dx = (-1)^{n-1}\zeta'(-n) + (-1)^n h(n)\frac{B_{n+1}}{n+1} - \sum_{k=0}^n (-1)^k\binom{n}{k}h(k)\frac{B_{k+1}(a)}{k+1}a^{n-k}$$
$$+ \sum_{k=0}^n (-1)^k\binom{n}{k}\zeta'(-k,a)a^{n-k}, \qquad n=1,2,\ldots,\ \Re a > 0,$$

where

25.11.33
$$h(n) = \sum_{k=1}^n k^{-1}.$$

25.11.34
$$n\int_0^a \zeta'(1-n,x)\,dx = \zeta'(-n,a) - \zeta'(-n) + \frac{B_{n+1} - B_{n+1}(a)}{n(n+1)}, \qquad n=1,2,\ldots,\ \Re a > 0.$$

25.11(ix) Integrals

See Prudnikov et al. (1990, §2.3), Prudnikov et al. (1992a, §3.2), and Prudnikov et al. (1992b, §3.2).

25.11(x) Further Series Representations

25.11.35
$$\sum_{n=0}^{\infty} \frac{(-1)^n}{(n+a)^s}$$
$$= \frac{1}{\Gamma(s)} \int_0^{\infty} \frac{x^{s-1} e^{-ax}}{1+e^{-x}} \, dx$$
$$= 2^{-s} \left(\zeta(s, \tfrac{1}{2}a) - \zeta(s, \tfrac{1}{2}(1+a)) \right),$$
$$\Re a > 0, \Re s > 0; \text{ or } \Re a = 0, \Im a \neq 0, 0 < \Re s < 1.$$

When $a = 1$, (25.11.35) reduces to (25.2.3).

25.11.36 $\quad \displaystyle\sum_{n=1}^{\infty} \frac{\chi(n)}{n^s} = k^{-s} \sum_{r=1}^{k} \chi(r) \, \zeta\!\left(s, \frac{r}{k}\right), \quad \Re s > 1,$

where $\chi(n)$ is a Dirichlet character (mod k) (§27.8).

See also Srivastava and Choi (2001).

25.11(xi) Sums

25.11.37
$$\sum_{k=1}^{\infty} \frac{(-1)^k}{k} \zeta(nk, a) = -n \ln \Gamma(a)$$
$$+ \ln \left(\prod_{j=0}^{n-1} \Gamma\!\left(a - e^{(2j+1)\pi i/n}\right) \right),$$
$$n = 2, 3, 4, \ldots, \Re a \geq 1.$$

25.11.38
$$\sum_{k=1}^{\infty} \binom{n+k}{k} \zeta(n+k+1, a) z^k$$
$$= \frac{(-1)^n}{n!} \left(\psi^{(n)}(a) - \psi^{(n)}(a-z) \right),$$
$$n = 1, 2, 3, \ldots, \Re a > 0, |z| < |a|.$$

25.11.39 $\quad \displaystyle\sum_{k=2}^{\infty} \frac{k}{2^k} \zeta\!\left(k+1, \tfrac{3}{4}\right) = 8G,$

where G is *Catalan's constant*:

25.11.40 $\quad G = \displaystyle\sum_{n=0}^{\infty} \frac{(-1)^n}{(2n+1)^2} = 0.91596\,55941\,772\ldots.$

For further sums see Prudnikov et al. (1990, pp. 396–397) and Hansen (1975, pp. 358–360).

25.11(xii) a-Asymptotic Behavior

As $a \to 0$ with $s \, (\neq 1)$ fixed,

25.11.41 $\quad \zeta(s, a+1) = \zeta(s) - s\zeta(s+1) a + O(a^2).$

As $\beta \to \pm\infty$ with s fixed, $\Re s > 1$,

25.11.42 $\quad \zeta(s, \alpha + i\beta) \to 0,$

uniformly with respect to bounded nonnegative values of α.

As $a \to \infty$ in the sector $|\mathrm{ph}\, a| \leq \pi - \delta (< \pi)$, with $s (\neq 1)$ and δ fixed, we have the asymptotic expansion

25.11.43
$$\zeta(s,a) - \frac{a^{1-s}}{s-1} - \frac{1}{2} a^{-s} \sim \sum_{k=1}^{\infty} \frac{B_{2k}}{(2k)!} \frac{\Gamma(s+2k-1)}{\Gamma(s)} a^{1-s-2k}.$$

Similarly, as $a \to \infty$ in the sector $|\mathrm{ph}\, a| \leq \tfrac{1}{2}\pi - \delta (< \tfrac{1}{2}\pi)$,

25.11.44
$$\zeta'(-1, a) - \frac{1}{12} + \frac{1}{4} a^2 - \left(\frac{1}{12} - \frac{1}{2} a + \frac{1}{2} a^2 \right) \ln a$$
$$\sim -\sum_{k=1}^{\infty} \frac{B_{2k+2}}{(2k+2)(2k+1) 2k} a^{-2k},$$

and

25.11.45
$$\zeta'(-2, a) - \frac{1}{12} a + \frac{1}{9} a^3 - \left(\frac{1}{6} a - \frac{1}{2} a^2 + \frac{1}{3} a^3 \right) \ln a$$
$$\sim \sum_{k=1}^{\infty} \frac{2 B_{2k+2}}{(2k+2)(2k+1) 2k (2k-1)} a^{-(2k-1)}.$$

For the more general case $\zeta'(-m, a)$, $m = 1, 2, \ldots$, see Elizalde (1986).

For an exponentially-improved form of (25.11.43) see Paris (2005b).

25.12 Polylogarithms

25.12(i) Dilogarithms

The notation $\mathrm{Li}_2(z)$ was introduced in Lewin (1981) for a function discussed in Euler (1768) and called the *dilogarithm* in Hill (1828):

25.12.1 $\quad \mathrm{Li}_2(z) = \displaystyle\sum_{n=1}^{\infty} \frac{z^n}{n^2}, \qquad |z| \leq 1.$

25.12.2 $\quad \mathrm{Li}_2(z) = -\displaystyle\int_0^z t^{-1} \ln(1-t) \, dt, \quad z \in \mathbb{C} \setminus (1, \infty).$

Other notations and names for $\mathrm{Li}_2(z)$ include $S_2(z)$ (Kölbig et al. (1970)), Spence function $\mathrm{Sp}(z)$ ('t Hooft and Veltman (1979)), and $\mathrm{L}_2(z)$ (Maximon (2003)).

In the complex plane $\mathrm{Li}_2(z)$ has a branch point at $z = 1$. The principal branch has a cut along the interval $[1, \infty)$ and agrees with (25.12.1) when $|z| \leq 1$; see also §4.2(i). The remainder of the equations in this subsection apply to principal branches.

25.12.3
$$\mathrm{Li}_2(z) + \mathrm{Li}_2\!\left(\frac{z}{z-1}\right) = -\frac{1}{2} (\ln(1-z))^2, \quad z \in \mathbb{C} \setminus [1, \infty).$$

25.12.4
$$\mathrm{Li}_2(z) + \mathrm{Li}_2\!\left(\frac{1}{z}\right) = -\frac{1}{6} \pi^2 - \frac{1}{2} (\ln(-z))^2, \quad z \in \mathbb{C} \setminus [0, \infty).$$

25.12.5 $\quad \mathrm{Li}_2(z^m) = m \displaystyle\sum_{k=0}^{m-1} \mathrm{Li}_2\!\left(z e^{2\pi i k/m}\right),$
$$m = 1, 2, 3, \ldots, |z| < 1.$$

25.12.6
$$\operatorname{Li}_2(x) + \operatorname{Li}_2(1-x) = \tfrac{1}{6}\pi^2 - (\ln x)\ln(1-x), \quad 0 < x < 1.$$

When $z = e^{i\theta}$, $0 \leq \theta \leq 2\pi$, (25.12.1) becomes

25.12.7
$$\operatorname{Li}_2(e^{i\theta}) = \sum_{n=1}^{\infty} \frac{\cos(n\theta)}{n^2} + i\sum_{n=1}^{\infty} \frac{\sin(n\theta)}{n^2}.$$

The cosine series in (25.12.7) has the elementary sum

25.12.8
$$\sum_{n=1}^{\infty} \frac{\cos(n\theta)}{n^2} = \frac{\pi^2}{6} - \frac{\pi\theta}{2} + \frac{\theta^2}{4}.$$

By (25.12.2)

25.12.9
$$\sum_{n=1}^{\infty} \frac{\sin(n\theta)}{n^2} = -\int_0^{\theta} \ln\!\bigl(2\sin(\tfrac{1}{2}x)\bigr)\, dx.$$

The right-hand side is called *Clausen's integral*.

For graphics see Figures 25.12.1 and 25.12.2, and for further properties see Maximon (2003), Kirillov (1995), Lewin (1981), Nielsen (1909), and Zagier (1989).

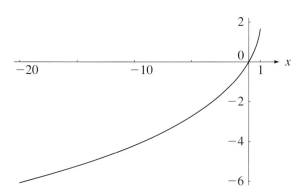

Figure 25.12.1: Dilogarithm function $\operatorname{Li}_2(x)$, $-20 \leq x < 1$.

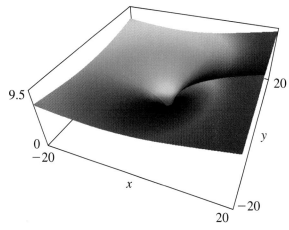

Figure 25.12.2: Absolute value of the dilogarithm function $|\operatorname{Li}_2(x+iy)|$, $-20 \leq x \leq 20$, $-20 \leq y \leq 20$. Principal value. There is a cut along the real axis from 1 to ∞.

25.12(ii) Polylogarithms

For real or complex s and z the *polylogarithm* $\operatorname{Li}_s(z)$ is defined by

25.12.10
$$\operatorname{Li}_s(z) = \sum_{n=1}^{\infty} \frac{z^n}{n^s}.$$

For each fixed complex s the series defines an analytic function of z for $|z| < 1$. The series also converges when $|z| = 1$, provided that $\Re s > 1$. For other values of z, $\operatorname{Li}_s(z)$ is defined by analytic continuation.

The notation $\phi(z,s)$ was used for $\operatorname{Li}_s(z)$ in Truesdell (1945) for a series treated in Jonquière (1889), hence the alternative name *Jonquière's function*. The special case $z = 1$ is the Riemann zeta function: $\zeta(s) = \operatorname{Li}_s(1)$.

Integral Representation

25.12.11
$$\operatorname{Li}_s(z) = \frac{z}{\Gamma(s)} \int_0^{\infty} \frac{x^{s-1}}{e^x - z}\, dx,$$

valid when $\Re s > 0$ and $|\operatorname{ph}(1-z)| < \pi$, or $\Re s > 1$ and $z = 1$. (In the latter case (25.12.11) becomes (25.5.1)).

Further properties include

25.12.12
$$\operatorname{Li}_s(z) = \Gamma(1-s)\left(\ln\frac{1}{z}\right)^{s-1} + \sum_{n=0}^{\infty} \zeta(s-n)\frac{(\ln z)^n}{n!},$$
$$s \neq 1, 2, 3, \ldots,\ |\ln z| < 2\pi,$$

and

25.12.13
$$\operatorname{Li}_s(e^{2\pi i a}) + e^{\pi i s}\operatorname{Li}_s(e^{-2\pi i a}) = \frac{(2\pi)^s e^{\pi i s/2}}{\Gamma(s)}\zeta(1-s,a),$$

valid when $\Re s > 0$, $\Im a > 0$ or $\Re s > 1$, $\Im a = 0$. When $s = 2$ and $e^{2\pi i a} = z$, (25.12.13) becomes (25.12.4).

See also Lewin (1981), Kölbig (1986), Maximon (2003), Prudnikov et al. (1990, §§1.2 and 2.5), Prudnikov et al. (1992a, §3.3), and Prudnikov et al. (1992b, §3.3).

25.12(iii) Fermi–Dirac and Bose–Einstein Integrals

The Fermi–Dirac and Bose–Einstein integrals are defined by

$$25.12.14 \quad F_s(x) = \frac{1}{\Gamma(s+1)} \int_0^\infty \frac{t^s}{e^{t-x}+1} \, dt, \quad s > -1,$$

$$25.12.15 \quad G_s(x) = \frac{1}{\Gamma(s+1)} \int_0^\infty \frac{t^s}{e^{t-x}-1} \, dt,$$
$$s > -1, \, x < 0; \text{ or } s > 0, \, x \le 0,$$

respectively. Sometimes the factor $1/\Gamma(s+1)$ is omitted. See Cloutman (1989) and Gautschi (1993).

In terms of polylogarithms

$$25.12.16 \quad F_s(x) = -\operatorname{Li}_{s+1}(-e^x), \quad G_s(x) = \operatorname{Li}_{s+1}(e^x).$$

For a uniform asymptotic approximation for $F_s(x)$ see Temme and Olde Daalhuis (1990).

25.13 Periodic Zeta Function

The notation $F(x,s)$ is used for the polylogarithm $\operatorname{Li}_s\!\left(e^{2\pi i x}\right)$ with x real:

$$25.13.1 \quad F(x,s) = \sum_{n=1}^\infty \frac{e^{2\pi i n x}}{n^s},$$

where $\Re s > 1$ if x is an integer, $\Re s > 0$ otherwise.

$F(x,s)$ is periodic in x with period 1, and equals $\zeta(s)$ when x is an integer. Also,

$$25.13.2 \quad F(x,s) = \frac{\Gamma(1-s)}{(2\pi)^{1-s}} \left(e^{\pi i (1-s)/2} \zeta(1-s,x) \right.$$
$$\left. + e^{\pi i (s-1)/2} \zeta(1-s, 1-x) \right),$$
$$0 < x < 1, \Re s > 1,$$

$$25.13.3$$
$$\zeta(1-s,x) = \frac{\Gamma(s)}{(2\pi)^s} \left(e^{-\pi i s/2} F(x,s) + e^{\pi i s/2} F(-x,s) \right),$$
$$0 < x < 1, \Re s > 0.$$

25.14 Lerch's Transcendent

25.14(i) Definition

$$25.14.1 \quad \Phi(z,s,a) = \sum_{n=0}^\infty \frac{z^n}{(a+n)^s},$$
$$a \ne 0, -1, -2, \ldots; |z| < 1; \Re s > 1, |z| = 1.$$

For other values of z, $\Phi(z,s,a)$ is defined by analytic continuation. This is the notation used in Erdélyi et al. (1953a, p. 27). Lerch (1887) used $\mathfrak{K}(a,x,s) = \Phi(e^{2\pi i x}, s, a)$.

The Hurwitz zeta function $\zeta(s,a)$ (§25.11) and the polylogarithm $\operatorname{Li}_s(z)$ (§25.12(ii)) are special cases:

$$25.14.2 \quad \zeta(s,a) = \Phi(1,s,a), \quad \Re s > 1, a \ne 0, -1, -2, \ldots,$$

$$25.14.3 \quad \operatorname{Li}_s(z) = z\,\Phi(z,s,1), \quad \Re s > 1, |z| \le 1.$$

25.14(ii) Properties

With the conditions of (25.14.1) and $m = 1, 2, 3, \ldots$,

$$25.14.4 \quad \Phi(z,s,a) = z^m \Phi(z,s,a+m) + \sum_{n=0}^{m-1} \frac{z^n}{(a+n)^s}.$$

$$25.14.5 \quad \Phi(z,s,a) = \frac{1}{\Gamma(s)} \int_0^\infty \frac{x^{s-1} e^{-ax}}{1 - ze^{-x}} \, dx,$$
$$\Re s > 0, \Re a > 0, z \in \mathbb{C}\setminus[1,\infty).$$

$$25.14.6$$
$$\Phi(z,s,a) = \frac{1}{2} a^{-s} + \int_0^\infty \frac{z^x}{(a+x)^s} \, dx$$
$$- 2 \int_0^\infty \frac{\sin(x \ln z - s \arctan(x/a))}{(a^2+x^2)^{s/2}(e^{2\pi x}-1)} \, dx,$$
$$\Re s > 0 \text{ if } |z| < 1;\ \Re s > 1 \text{ if } |z| = 1, \Re a > 0.$$

For these and further properties see Erdélyi et al. (1953a, pp. 27–31).

25.15 Dirichlet L-functions

25.15(i) Definitions and Basic Properties

The notation $L(s,\chi)$ was introduced by Dirichlet (1837) for the meromorphic continuation of the function defined by the series

$$25.15.1 \quad L(s,\chi) = \sum_{n=1}^\infty \frac{\chi(n)}{n^s}, \qquad \Re s > 1,$$

where $\chi(n)$ is a Dirichlet character $(\operatorname{mod} k)$ (§27.8). For the principal character χ_1 $(\operatorname{mod} k)$, $L(s,\chi_1)$ is analytic everywhere except for a simple pole at $s=1$ with residue $\phi(k)/k$, where $\phi(k)$ is Euler's totient function (§27.2). If $\chi \ne \chi_1$, then $L(s,\chi)$ is an entire function of s.

$$25.15.2 \quad L(s,\chi) = \prod_p \left(1 - \frac{\chi(p)}{p^s}\right)^{-1}, \qquad \Re s > 1,$$

with the product taken over all primes p, beginning with $p = 2$. This implies that $L(s,\chi) \ne 0$ if $\Re s > 1$.

Equations (25.15.3) and (25.15.4) hold for all s if $\chi \ne \chi_1$, and for all s $(\ne 1)$ if $\chi = \chi_1$:

$$25.15.3 \quad L(s,\chi) = k^{-s} \sum_{r=1}^{k-1} \chi(r) \, \zeta\!\left(s, \frac{r}{k}\right),$$

$$25.15.4 \quad L(s,\chi) = L(s,\chi_0) \prod_{p \mid k} \left(1 - \frac{\chi_0(p)}{p^s}\right),$$

where χ_0 is a primitive character $(\operatorname{mod} d)$ for some positive divisor d of k (§27.8).

When χ is a primitive character $(\operatorname{mod} k)$ the L-functions satisfy the functional equation:

$$25.15.5$$
$$L(1-s,\chi) = \frac{k^{s-1} \Gamma(s)}{(2\pi)^s} \left(e^{-\pi i s/2} + \chi(-1) e^{\pi i s/2} \right)$$
$$\times G(\chi) \, L(s, \overline{\chi}),$$

where $\overline{\chi}$ is the complex conjugate of χ, and

25.15.6 $$G(\chi) = \sum_{r=1}^{k} \chi(r) e^{2\pi i r/k}.$$

25.15(ii) Zeros

Since $L(s,\chi) \neq 0$ if $\Re s > 1$, (25.15.5) shows that for a primitive character χ the only zeros of $L(s,\chi)$ for $\Re s < 0$ (the so-called trivial zeros) are as follows:

25.15.7 $\quad L(-2n,\chi) = 0$ if $\chi(-1) = 1$, $\quad n = 0, 1, 2, \ldots,$

25.15.8
$$L(-2n-1,\chi) = 0 \text{ if } \chi(-1) = -1, \quad n = 0, 1, 2, \ldots.$$

There are also infinitely many zeros in the critical strip $0 \leq \Re s \leq 1$, located symmetrically about the critical line $\Re s = \tfrac{1}{2}$, but not necessarily symmetrically about the real axis.

25.15.9 $\quad L(1,\chi) \neq 0$ if $\chi \neq \chi_1,$

where χ_1 is the principal character (mod k). This result plays an important role in the proof of Dirichlet's theorem on primes in arithmetic progressions (§27.11). Related results are:

25.15.10 $\quad L(0,\chi) = \begin{cases} -\dfrac{1}{k} \sum\limits_{r=1}^{k} r\chi(r), & \chi \neq \chi_1, \\ 0, & \chi = \chi_1. \end{cases}$

Applications

25.16 Mathematical Applications

25.16(i) Distribution of Primes

In studying the distribution of primes $p \leq x$, Chebyshev (1851) introduced a function $\psi(x)$ (not to be confused with the digamma function used elsewhere in this chapter), given by

25.16.1 $$\psi(x) = \sum_{m=1}^{\infty} \sum_{p^m \leq x} \ln p,$$

which is related to the Riemann zeta function by

25.16.2 $$\psi(x) = x - \frac{\zeta'(0)}{\zeta(0)} - \sum_{\rho} \frac{x^{\rho}}{\rho} + o(1), \quad x \to \infty,$$

where the sum is taken over the nontrivial zeros ρ of $\zeta(s)$.

The prime number theorem (27.2.3) is equivalent to the statement

25.16.3 $\quad \psi(x) = x + o(x), \qquad x \to \infty.$

The Riemann hypothesis is equivalent to the statement

25.16.4 $\quad \psi(x) = x + O\!\left(x^{\frac{1}{2}+\epsilon}\right), \qquad x \to \infty,$

for every $\epsilon > 0$.

25.16(ii) Euler Sums

Euler sums have the form

25.16.5 $$H(s) = \sum_{n=1}^{\infty} \frac{h(n)}{n^s},$$

where $h(n)$ is given by (25.11.33).

$H(s)$ is analytic for $\Re s > 1$, and can be extended meromorphically into the half-plane $\Re s > -2k$ for every positive integer k by use of the relations

25.16.6 $\quad H(s) = -\zeta'(s) + \gamma\,\zeta(s) + \dfrac{1}{2}\zeta(s+1) + \sum\limits_{r=1}^{k} \zeta(1-2r)\,\zeta(s+2r) + \sum\limits_{n=1}^{\infty} \dfrac{1}{n^s} \int_{n}^{\infty} \dfrac{\widetilde{B}_{2k+1}(x)}{x^{2k+2}}\,dx,$

25.16.7 $\quad H(s) = \dfrac{1}{2}\zeta(s+1) + \dfrac{\zeta(s)}{s-1} - \sum\limits_{r=1}^{k} \binom{s+2r-2}{2r-1} \zeta(1-2r)\,\zeta(s+2r) - \binom{s+2k}{2k+1} \sum\limits_{n=1}^{\infty} \dfrac{1}{n} \int_{n}^{\infty} \dfrac{\widetilde{B}_{2k+1}(x)}{x^{s+2k+1}}\,dx.$

For integer $s\ (\geq 2)$, $H(s)$ can be evaluated in terms of the zeta function:

25.16.8 $\quad H(2) = 2\zeta(3), \quad H(3) = \tfrac{5}{4}\zeta(4),$

25.16.9 $\quad H(a) = \dfrac{a+2}{2}\zeta(a+1) - \dfrac{1}{2}\sum\limits_{r=1}^{a-2} \zeta(r+1)\,\zeta(a-r),$
$\qquad a = 2, 3, 4, \ldots.$

Also,

25.16.10
$$H(-2a) = \tfrac{1}{2}\zeta(1-2a) = -\frac{B_{2a}}{4a}, \quad a = 1, 2, 3, \ldots.$$

$H(s)$ has a simple pole with residue $\zeta(1-2r)$ ($= -B_{2r}/(2r)$) at each odd negative integer $s = 1 - 2r$, $r = 1, 2, 3, \ldots$.

$H(s)$ is the special case $H(s,1)$ of the function

25.16.11 $$H(s,z) = \sum_{n=1}^{\infty} \frac{1}{n^s} \sum_{m=1}^{n} \frac{1}{m^z}, \quad \Re(s+z) > 1,$$

which satisfies the reciprocity law

25.16.12 $H(s,z) + H(z,s) = \zeta(s)\zeta(z) + \zeta(s+z),$

when both $H(s,z)$ and $H(z,s)$ are finite.

For further properties of $H(s,z)$ see Apostol and Vu (1984). Related results are:

25.16.13 $$\sum_{n=1}^{\infty} \left(\frac{h(n)}{n}\right)^2 = \frac{17}{4}\zeta(4),$$

25.16.14 $$\sum_{r=1}^{\infty} \sum_{k=1}^{r} \frac{1}{rk(r+k)} = \frac{5}{4}\zeta(3),$$

25.16.15 $$\sum_{r=1}^{\infty} \sum_{k=1}^{r} \frac{1}{r^2(r+k)} = \frac{3}{4}\zeta(3).$$

For further generalizations, see Flajolet and Salvy (1998).

25.17 Physical Applications

Analogies exist between the distribution of the zeros of $\zeta(s)$ on the critical line and of semiclassical quantum eigenvalues. This relates to a suggestion of Hilbert and Pólya that the zeros are eigenvalues of some operator, and the Riemann hypothesis is true if that operator is Hermitian. See Armitage (1989), Berry and Keating (1998, 1999), Keating (1993, 1999), and Sarnak (1999).

The zeta function arises in the calculation of the partition function of ideal quantum gases (both Bose–Einstein and Fermi–Dirac cases), and it determines the critical gas temperature and density for the Bose–Einstein condensation phase transition in a dilute gas (Lifshitz and Pitaevskiĭ (1980)). Quantum field theory often encounters formally divergent sums that need to be evaluated by a process of regularization: for example, the energy of the electromagnetic vacuum in a confined space (*Casimir–Polder effect*). It has been found possible to perform such regularizations by equating the divergent sums to zeta functions and associated functions (Elizalde (1995)).

Computation

25.18 Methods of Computation

25.18(i) Function Values and Derivatives

The principal tools for computing $\zeta(s)$ are the expansion (25.2.9) for general values of s, and the Riemann–Siegel formula (25.10.3) (extended to higher terms) for $\zeta(\tfrac{1}{2} + it)$. Details are provided in Haselgrove and Miller (1960). See also Allasia and Besenghi (1989), Butzer and Hauss (1992), Kerimov (1980), and Yeremin et al. (1985). Calculations relating to derivatives of $\zeta(s)$ and/or $\zeta(s,a)$ can be found in Apostol (1985a), Choudhury (1995), Miller and Adamchik (1998), and Yeremin et al. (1988).

For the Hurwitz zeta function $\zeta(s,a)$ see Spanier and Oldham (1987, p. 653).

For dilogarithms and polylogarithms see Jacobs and Lambert (1972), Osácar et al. (1995), and Spanier and Oldham (1987, pp. 231–232).

For Fermi–Dirac and Bose–Einstein integrals see Cloutman (1989), Gautschi (1993), Mohankumar and Natarajan (1997), Natarajan and Mohankumar (1993), Paszkowski (1988, 1991), Pichon (1989), and Sagar (1991a,b).

25.18(ii) Zeros

Most numerical calculations of the Riemann zeta function are concerned with locating zeros of $\zeta(\tfrac{1}{2}+it)$ in an effort to prove or disprove the Riemann hypothesis, which states that all nontrivial zeros of $\zeta(s)$ lie on the critical line $\Re s = \tfrac{1}{2}$. Calculations to date (2008) have found no nontrivial zeros off the critical line. For recent investigations see, for example, van de Lune et al. (1986) and Odlyzko (1987). For earlier work see Haselgrove and Miller (1960).

25.19 Tables

- Abramowitz and Stegun (1964) tabulates: $\zeta(n)$, $n = 2,3,4,\ldots$, 20D (p. 811); $\mathrm{Li}_2(1-x)$, $x = 0(.01)0.5$, 9D (p. 1005); $f(\theta)$, $\theta = 15°(1°)30°(2°)90°(5°)180°$, $f(\theta) + \theta \ln \theta$, $\theta = 0(1°)15°$, 6D (p. 1006). Here $f(\theta)$ denotes Clausen's integral, given by the right-hand side of (25.12.9).

- Morris (1979) tabulates $\mathrm{Li}_2(x)$ (§25.12(i)) for $\pm x = 0.02(.02)1(.1)6$ to 30D.

- Cloutman (1989) tabulates $\Gamma(s+1)F_s(x)$, where $F_s(x)$ is the Fermi–Dirac integral (25.12.14), for $s = -\tfrac{1}{2}, \tfrac{1}{2}, \tfrac{3}{2}, \tfrac{5}{2}$, $x = -5(.05)25$, to 12S.

- Fletcher et al. (1962, §22.1) lists many sources for earlier tables of $\zeta(s)$ for both real and complex s. §22.133 gives sources for numerical values of coefficients in the Riemann–Siegel formula, §22.15 describes tables of values of $\zeta(s,a)$, and §22.17 lists tables for some Dirichlet L-functions for real characters. For tables of dilogarithms, polylogarithms, and Clausen's integral see §§22.84–22.858.

25.20 Approximations

- Cody et al. (1971) gives rational approximations for $\zeta(s)$ in the form of quotients of polynomials or quotients of Chebyshev series. The ranges covered are $0.5 \leq s \leq 5$, $5 \leq s \leq 11$, $11 \leq s \leq 25$, $25 \leq s \leq 55$. Precision is varied, with a maximum of 20S.

- Piessens and Branders (1972) gives the coefficients of the Chebyshev-series expansions of $s\zeta(s+1)$ and $\zeta(s+k)$, $k=2,3,4,5,8$, for $0 \leq s \leq 1$ (23D).

- Luke (1969b, p. 306) gives coefficients in Chebyshev-series expansions that cover $\zeta(s)$ for $0 \leq s \leq 1$ (15D), $\zeta(s+1)$ for $0 \leq s \leq 1$ (20D), and $\ln\xi(\tfrac{1}{2}+ix)$ (§25.4) for $-1 \leq x \leq 1$ (20D). For errata see Piessens and Branders (1972).

- Morris (1979) gives rational approximations for $\mathrm{Li}_2(x)$ (§25.12(i)) for $0.5 \leq x \leq 1$. Precision is varied with a maximum of 24S.

- Antia (1993) gives minimax rational approximations for $\Gamma(s+1)F_s(x)$, where $F_s(x)$ is the Fermi–Dirac integral (25.12.14), for the intervals $-\infty < x \leq 2$ and $2 \leq x < \infty$, with $s = -\tfrac{1}{2}, \tfrac{1}{2}, \tfrac{3}{2}, \tfrac{5}{2}$. For each s there are three sets of approximations, with relative maximum errors $10^{-4}, 10^{-8}, 10^{-12}$.

25.21 Software

See http://dlmf.nist.gov/25.21.

References

General References

The main references used in writing this chapter are Apostol (1976), Erdélyi et al. (1953a), and Titchmarsh (1986b). For additional bibliographic reading see Edwards (1974), Ivić (1985), Karatsuba and Voronin (1992).

Sources

The following list gives the references or other indications of proofs that were used in constructing the various sections of this chapter. These sources supplement the references that are quoted in the text.

§25.2 Apostol (1976, Chapter 12). For (25.2.2)–(25.2.7) see also Hardy (1912). For (25.2.8)–(25.2.10) see also Knopp (1948, p. 533). (25.2.9) follows from (25.2.8) by repeated integration by parts. For (25.2.11), (25.2.12) see also Titchmarsh (1986b, p. 30).

§25.3 These graphics were constructed at NIST.

§25.4 Apostol (1976, Chapter 12).

§25.5 Apostol (1976, Chapter 12), Erdélyi et al. (1953a, Chapter I). For (25.5.2) and (25.5.4) integrate (25.5.1) and (25.5.3) by parts. For (25.5.5) see Titchmarsh (1986b, p. 15). (25.5.6) comes from (25.5.1) by using the identity $e^{-x} = (1-e^{-x})/(e^x-1)$ in the integral $\Gamma(s) = \int_0^\infty e^{-x}x^{s-1}\,dx$ together with (5.5.1). (25.5.7) follows from (25.5.6) because $\Gamma(s+2m-1) = \int_0^\infty e^{-x}x^{s+2m-2}\,dx$. For (25.5.10) and (25.5.11) see Lindelöf (1905, p. 103). For (25.5.12) see Srivastava and Choi (2001, p. 12). For (25.5.13) see Titchmarsh (1986b, p. 22). For (25.5.14)–(25.5.19) see de Bruijn (1937). For (25.5.21) see Erdélyi et al. (1953a, p. 32).

§25.6 For (25.6.1)–(25.6.4) see Apostol (1976, pp. 266–268). For (25.6.5) see Mordell (1958). For (25.6.6) see Nörlund (1924, p. 66). For (25.6.7) see Apostol (1983). For (25.6.8)–(25.6.10) see van der Poorten (1980, pp. 271, 274). For (25.6.11)–(25.6.14) see Apostol (1985a). For (25.6.15) see Miller and Adamchik (1998). For (25.6.16)–(25.6.20) see Basu and Apostol (2000).

§25.8 Titchmarsh (1986b, Chapter IV), Adamchik and Srivastava (1998), Erdélyi et al. (1953a, pp. 45 and 51). For (25.8.2) see Landau (1953, p. 274). For (25.8.3) see Srivastava (1988). For (25.8.7), (25.8.8) divide by x in (25.8.5), (25.8.6) and integrate. For (25.8.9) see Srivastava and Choi (2001, p. 212). For (25.8.10) see Ewell (1990).

§25.9 Titchmarsh (1986b, Chapter XV), Berry (1995).

§25.10 Apostol (1976, Chapter 12), Titchmarsh (1986b, pp. 89 and 263).

§25.11 Apostol (1976, Chapter 12). Analytic properties of $\zeta(s,a)$ with respect to a follow from (25.11.30). For (25.11.5)–(25.11.6) see Apostol (1985a). For (25.11.7) take $N=1$ in (25.11.5) and integrate by parts. For (25.11.8)–(25.11.9) see Srivastava and Choi (2001, p. 89). For (25.11.10) use Taylor's theorem (§§1.4(vi), 1.10(i)) and (25.11.17). For (25.11.11) apply (25.2.2) and (25.11.1). For (25.11.12) see Erdélyi et al. (1953a,

p. 45). For (25.11.13) and (25.11.14) see Apostol (1976, pp. 268, 264). For (25.11.15) use (25.11.1) and analytic continuation. For (25.11.16) see Apostol (1976, p. 263). For (25.11.17) differentiate (25.11.1). For (25.11.18) see Erdélyi et al. (1953a, p. 26). For (25.11.19)–(25.11.23) see Apostol (1985a, p. 231) and Miller and Adamchik (1998). For (25.11.24) use (25.11.15) with $a = 1/k$, multiply by k^s and differentiate. For (25.11.25) see Srivastava and Choi (2001, p. 89) For (25.11.26) see Berndt (1972). For (25.11.27) and (25.11.28) argue as indicated above for (25.5.6) and (25.5.7). For (25.11.29) see Lindelöf (1905, p. 106). For (25.11.30) assume $\Re s > 1$, collapse the integration path onto the real axis, apply (25.11.25) and (5.5.3) followed by analytic continuation. For (25.11.31) use (25.11.25). For (25.11.32)–(25.11.34) see Adamchik (1998). For (25.11.35) use (25.11.25) and (25.11.8). For (25.11.36) see Apostol (1976). For (25.11.37)–(25.11.40) see Adamchik and Srivastava (1998). For (25.11.41) and (25.11.42) see Apostol (1952). For (25.11.43) see Paris (2005b). For (25.11.44) and (25.11.45) see Elizalde (1986). The graphics were constructed at NIST.

§25.12 Erdélyi et al. (1953a, pp. 27, 29), Maximon (2003). For (25.12.13) see Erdélyi et al. (1953a, p. 31) with change of notation. The graphics were constructed at NIST.

§25.13 Apostol (1976, Chapter 13).

§25.15 Apostol (1976, Chapter 12), Apostol (1985b). For (25.15.9) see Apostol (1976, pp. 142, 149).

§25.16 Apostol (1976, Chapter 13). For (25.16.2) see Apostol (2000). For (25.16.4) see Ingham (1932, p. 84). For (25.16.5)–(25.16.15) see Apostol and Vu (1984) and Basu and Apostol (2000).

Chapter 26
Combinatorial Analysis

D. M. Bressoud[1]

Notation — **618**
- 26.1 Special Notation — 618

Properties — **618**
- 26.2 Basic Definitions — 618
- 26.3 Lattice Paths: Binomial Coefficients — 619
- 26.4 Lattice Paths: Multinomial Coefficients and Set Partitions — 620
- 26.5 Lattice Paths: Catalan Numbers — 620
- 26.6 Other Lattice Path Numbers — 621
- 26.7 Set Partitions: Bell Numbers — 623
- 26.8 Set Partitions: Stirling Numbers — 624
- 26.9 Integer Partitions: Restricted Number and Part Size — 626
- 26.10 Integer Partitions: Other Restrictions — 627
- 26.11 Integer Partitions: Compositions — 628
- 26.12 Plane Partitions — 629
- 26.13 Permutations: Cycle Notation — 631
- 26.14 Permutations: Order Notation — 632
- 26.15 Permutations: Matrix Notation — 633
- 26.16 Multiset Permutations — 634
- 26.17 The Twelvefold Way — 634
- 26.18 Counting Techniques — 634

Applications — **635**
- 26.19 Mathematical Applications — 635
- 26.20 Physical Applications — 635

Computation — **635**
- 26.21 Tables — 635
- 26.22 Software — 635

References — **635**

---

[1] Macalester College, Saint Paul, Minnesota.
Copyright © 2009 National Institute of Standards and Technology. All rights reserved.

Notation

26.1 Special Notation

(For other notation see pp. xiv and 873.)

x	real variable.
h, j, k, ℓ, m, n	nonnegative integers.
λ	integer partition.
π	plane partition.
$\|A\|$	number of elements of a finite set A.
$j \mid k$	j divides k.
(h, k)	greatest common divisor of positive integers h and k.

The main functions treated in this chapter are:

$\binom{m}{n}$	binomial coefficient.
$\binom{m}{n_1, n_2, \ldots, n_k}$	multinomial coefficient.
$\left\langle {m \atop n} \right\rangle$	Eulerian number.
$\left[{m \atop n} \right]_q$	Gaussian polynomial.
$B(n)$	Bell number.
$C(n)$	Catalan number.
$p(n)$	number of partitions of n.
$p_k(n)$	number of partitions of n into at most k parts.
$pp(n)$	number of plane partitions of n.
$s(n, k)$	Stirling numbers of the first kind.
$S(n, k)$	Stirling numbers of the second kind.

Alternative Notations

Many combinatorics references use the rising and falling factorials:

$$26.1.1 \quad \begin{aligned} x^{\overline{n}} &= x(x+1)(x+2) \cdots (x+n-1), \\ x^{\underline{n}} &= x(x-1)(x-2) \cdots (x-n+1). \end{aligned}$$

Other notations for $s(n, k)$, the Stirling numbers of the first kind, include $S_n^{(k)}$ (Abramowitz and Stegun (1964, Chapter 24), Fort (1948)), S_n^k (Jordan (1939), Moser and Wyman (1958a)), $\binom{n-1}{k-1} B_{n-k}^{(n)}$ (Milne-Thomson (1933)), $(-1)^{n-k} S_1(n-1, n-k)$ (Carlitz (1960), Gould (1960)), $(-1)^{n-k} \left[{n \atop k} \right]$ (Knuth (1992), Graham et al. (1994), Rosen et al. (2000)).

Other notations for $S(n, k)$, the Stirling numbers of the second kind, include $\mathscr{S}_n^{(k)}$ (Fort (1948)), $\mathfrak{S}_n^k$ (Jordan (1939)), σ_n^k (Moser and Wyman (1958b)), $\binom{n}{k} B_{n-k}^{(-k)}$ (Milne-Thomson (1933)), $S_2(k, n-k)$ (Carlitz (1960), Gould (1960)), $\left\{ {n \atop k} \right\}$ (Knuth (1992), Graham et al. (1994), Rosen et al. (2000)), and also an unconventional symbol in Abramowitz and Stegun (1964, Chapter 24).

Properties

26.2 Basic Definitions

Permutation

A *permutation* is a one-to-one and onto function from a non-empty set to itself. If the set consists of the integers 1 through n, a permutation σ can be thought of as a rearrangement of these integers where the integer in position j is $\sigma(j)$. Thus 231 is the permutation $\sigma(1) = 2$, $\sigma(2) = 3$, $\sigma(3) = 1$.

Cycle

Given a finite set S with permutation σ, a *cycle* is an ordered equivalence class of elements of S where j is equivalent to k if there exists an $\ell = \ell(j, k)$ such that $j = \sigma^\ell(k)$, where $\sigma^1 = \sigma$ and σ^ℓ is the composition of σ with $\sigma^{\ell-1}$. It is ordered so that $\sigma(j)$ follows j. If, for example, a permutation of the integers 1 through 6 is denoted by 256413, then the cycles are $(1, 2, 5)$, $(3, 6)$, and (4). Here $\sigma(1) = 2, \sigma(2) = 5$, and $\sigma(5) = 1$. The function σ also interchanges 3 and 6, and sends 4 to itself.

Lattice Path

A *lattice path* is a directed path on the plane integer lattice $\{0, 1, 2, \ldots\} \times \{0, 1, 2, \ldots\}$. Unless otherwise specified, it consists of horizontal segments corresponding to the vector $(1, 0)$ and vertical segments corresponding to the vector $(0, 1)$. For an example see Figure 26.9.2.

A *k-dimensional lattice path* is a directed path composed of segments that connect vertices in $\{0, 1, 2, \ldots\}^k$ so that each segment increases one coordinate by exactly one unit.

Partition

A *partition of a set S* is an unordered collection of pairwise disjoint nonempty sets whose union is S. As an example, $\{1, 3, 4\}$, $\{2, 6\}$, $\{5\}$ is a partition of $\{1, 2, 3, 4, 5, 6\}$.

A *partition of a nonnegative integer n* is an unordered collection of positive integers whose sum is n. As an example, $\{1, 1, 1, 2, 4, 4\}$ is a partition of 13. The total number of partitions of n is denoted by $p(n)$. See Table 26.2.1 for $n = 0(1)50$. For the actual partitions (π) for $n = 1(1)5$ see Table 26.4.1.

The integers whose sum is n are referred to as the *parts* in the partition. The example $\{1, 1, 1, 2, 4, 4\}$ has six parts, three of which equal 1.

26.3 Lattice Paths: Binomial Coefficients

Table 26.2.1: Partitions $p(n)$.

n	$p(n)$	n	$p(n)$	n	$p(n)$
0	1	17	297	34	12310
1	1	18	385	35	14883
2	2	19	490	36	17977
3	3	20	627	37	21637
4	5	21	792	38	26015
5	7	22	1002	39	31185
6	11	23	1255	40	37338
7	15	24	1575	41	44583
8	22	25	1958	42	53174
9	30	26	2436	43	63261
10	42	27	3010	44	75175
11	56	28	3718	45	89134
12	77	29	4565	46	1 05558
13	101	30	5604	47	1 24754
14	135	31	6842	48	1 47273
15	176	32	8349	49	1 73525
16	231	33	10143	50	2 04226

26.3(i) Definitions

$\binom{m}{n}$ is the number of ways of choosing n objects from a collection of m distinct objects without regard to order. $\binom{m+n}{n}$ is the number of lattice paths from $(0,0)$ to (m,n). The number of lattice paths from $(0,0)$ to (m,n), $m \leq n$, that stay on or above the line $y = x$ is $\binom{m+n}{m} - \binom{m+n}{m-1}$.

26.3.1 $$\binom{m}{n} = \binom{m}{m-n} = \frac{m!}{(m-n)!\, n!}, \qquad m \geq n,$$

26.3.2 $$\binom{m}{n} = 0, \qquad n > m.$$

For numerical values of $\binom{m}{n}$ and $\binom{m+n}{n}$ see Tables 26.3.1 and 26.3.2.

Table 26.3.1: Binomial coefficients $\binom{m}{n}$.

m	\multicolumn{11}{c}{n}
	0	1	2	3	4	5	6	7	8	9	10
0	1										
1	1	1									
2	1	2	1								
3	1	3	3	1							
4	1	4	6	4	1						
5	1	5	10	10	5	1					
6	1	6	15	20	15	6	1				
7	1	7	21	35	35	21	7	1			
8	1	8	28	56	70	56	28	8	1		
9	1	9	36	84	126	126	84	36	9	1	
10	1	10	45	120	210	252	210	120	45	10	1

Table 26.3.2: Binomial coefficients $\binom{m+n}{m}$ for lattice paths.

m	\multicolumn{9}{c}{n}
	0	1	2	3	4	5	6	7	8
0	1	1	1	1	1	1	1	1	1
1	1	2	3	4	5	6	7	8	9
2	1	3	6	10	15	21	28	36	45
3	1	4	10	20	35	56	84	120	165
4	1	5	15	35	70	126	210	330	495
5	1	6	21	56	126	252	462	792	1287
6	1	7	28	84	210	462	924	1716	3003
7	1	8	36	120	330	792	1716	3432	6435
8	1	9	45	165	495	1287	3003	6435	12870

26.3(ii) Generating Functions

26.3.3 $$\sum_{n=0}^{m} \binom{m}{n} x^n = (1+x)^m, \quad m = 0, 1, \ldots,$$

26.3.4 $$\sum_{m=0}^{\infty} \binom{m+n}{m} x^m = \frac{1}{(1-x)^{n+1}}, \qquad |x| < 1.$$

26.3(iii) Recurrence Relations

26.3.5 $$\binom{m}{n} = \binom{m-1}{n} + \binom{m-1}{n-1}, \qquad m \geq n \geq 1,$$

26.3.6 $$\binom{m}{n} = \frac{m}{n}\binom{m-1}{n-1} = \frac{m-n+1}{n}\binom{m}{n-1}, \qquad m \geq n \geq 1,$$

26.3.7 $$\binom{m+1}{n+1} = \sum_{k=n}^{m} \binom{k}{n}, \qquad m \geq n \geq 0,$$

26.3.8 $$\binom{m}{n} = \sum_{k=0}^{n} \binom{m-n-1+k}{k}, \quad m \geq n \geq 0.$$

26.3(iv) Identities

26.3.9 $$\binom{n}{0} = \binom{n}{n} = 1,$$

26.3.10 $$\binom{m}{n} = \sum_{k=0}^{n} (-1)^{n-k}\binom{m+1}{k}, \quad m \geq n \geq 0,$$

26.3.11 $$\binom{2n}{n} = \frac{2^n (2n-1)(2n-3) \cdots 3 \cdot 1}{n!}.$$

See also §1.2(i).

26.3(v) Limiting Form

26.3.12 $$\binom{2n}{n} \sim \frac{4^n}{\sqrt{\pi n}}, \qquad n \to \infty.$$

26.4 Lattice Paths: Multinomial Coefficients and Set Partitions

26.4(i) Definitions

$\binom{n}{n_1, n_2, \ldots, n_k}$ is the number of ways of placing $n = n_1 + n_2 + \cdots + n_k$ distinct objects into k labeled boxes so that there are n_j objects in the jth box. It is also the number of k-dimensional lattice paths from $(0, 0, \ldots, 0)$ to $(n_1, n_2, \ldots, n_k)$. For $k = 0, 1$, the multinomial coefficient is defined to be 1. For $k = 2$

26.4.1 $\quad \binom{n_1 + n_2}{n_1, n_2} = \binom{n_1 + n_2}{n_1} = \binom{n_1 + n_2}{n_2},$

and in general,

26.4.2
$$\binom{n_1 + n_2 + \cdots + n_k}{n_1, n_2, \ldots, n_k} = \frac{(n_1 + n_2 + \cdots + n_k)!}{n_1! n_2! \cdots n_k!}$$
$$= \prod_{j=1}^{k-1} \binom{n_j + n_{j+1} + \cdots + n_k}{n_j}.$$

Table 26.4.1 gives numerical values of multinomials and partitions λ, M_1, M_2, M_3 for $1 \leq m \leq n \leq 5$. These are given by the following equations in which $a_1, a_2, \ldots, a_n$ are nonnegative integers such that

26.4.3 $\qquad n = a_1 + 2a_2 + \cdots + na_n,$

26.4.4 $\qquad m = a_1 + a_2 + \cdots + a_n.$

λ is a partition of n:

26.4.5 $\qquad \lambda = 1^{a_1}, 2^{a_2}, \ldots, n^{a_n}.$

M_1 is the multinominal coefficient (26.4.2):

26.4.6
$$M_1 = \binom{n}{\underbrace{1, \ldots, 1}_{a_1}, \ldots, \underbrace{n, \ldots, n}_{a_n}}$$
$$= \frac{n!}{(1!)^{a_1}(2!)^{a_2} \cdots (n!)^{a_n}}.$$

M_2 is the number of permutations of $\{1, 2, \ldots, n\}$ with a_1 cycles of length 1, a_2 cycles of length 2, $\ldots$, and a_n cycles of length n:

26.4.7 $\qquad M_2 = \dfrac{n!}{1^{a_1}(a_1!) \, 2^{a_2}(a_2!) \cdots n^{a_n}(a_n!)}.$

(The empty set is considered to have one permutation consisting of no cycles.) M_3 is the number of set partitions of $\{1, 2, \ldots, n\}$ with a_1 subsets of size 1, a_2 subsets of size 2, $\ldots$, and a_n subsets of size n:

26.4.8 $\quad M_3 = \dfrac{n!}{(1!)^{a_1}(a_1!)\,(2!)^{a_2}(a_2!) \cdots (n!)^{a_n}(a_n!)}.$

For each n all possible values of $a_1, a_2, \ldots, a_n$ are covered.

Table 26.4.1: Multinomials and partitions.

n	m	λ	M_1	M_2	M_3
1	1	1^1	1	1	1
2	1	2^1	1	1	1
2	2	1^2	2	1	1
3	1	3^1	1	2	1
3	2	$1^1, 2^1$	3	3	3
3	3	1^3	6	1	1
4	1	4^1	1	6	1
4	2	$1^1, 3^1$	4	8	4
4	2	2^2	6	3	3
4	3	$1^2, 2^1$	12	6	6
4	4	1^4	24	1	1
5	1	5^1	1	24	1
5	2	$1^1, 4^1$	5	30	5
5	2	$2^1, 3^1$	10	20	10
5	3	$1^2, 3^1$	20	20	10
5	3	$1^1, 2^2$	30	15	15
5	4	$1^3, 2^1$	60	10	10
5	5	1^5	120	1	1

26.4(ii) Generating Function

26.4.9
$$(x_1 + x_2 + \cdots + x_k)^n = \sum \binom{n}{n_1, n_2, \ldots, n_k} x_1^{n_1} x_2^{n_2} \cdots x_k^{n_k},$$
where the summation is over all nonnegative integers $n_1, n_2, \ldots, n_k$ such that $n_1 + n_2 + \cdots + n_k = n$.

26.4(iii) Recurrence Relation

26.4.10
$$\binom{n_1 + n_2 + \cdots + n_m}{n_1, n_2, \ldots, n_m}$$
$$= \sum_{k=1}^{m} \binom{n_1 + n_2 + \cdots + n_m - 1}{n_1, n_2, \ldots, n_{k-1}, n_k - 1, n_{k+1}, \ldots, n_m},$$
$$n_1, n_2, \ldots, n_m \geq 1.$$

26.5 Lattice Paths: Catalan Numbers

26.5(i) Definitions

$C(n)$ is the Catalan number. It counts the number of lattice paths from $(0, 0)$ to (n, n) that stay on or above the line $y = x$.

26.5.1
$$C(n) = \frac{1}{n+1} \binom{2n}{n} = \frac{1}{2n+1} \binom{2n+1}{n}$$
$$= \binom{2n}{n} - \binom{2n}{n-1} = \binom{2n-1}{n} - \binom{2n-1}{n+1}.$$

(Sixty-six equivalent definitions of $C(n)$ are given in Stanley (1999, pp. 219–229).)

See Table 26.5.1.

Table 26.5.1: Catalan numbers.

n	$C(n)$	n	$C(n)$	n	$C(n)$
0	1	7	429	14	26 74440
1	1	8	1430	15	96 94845
2	2	9	4862	16	353 57670
3	5	10	16796	17	1296 44790
4	14	11	58786	18	4776 38700
5	42	12	2 08012	19	17672 63190
6	132	13	7 42900	20	65641 20420

26.5(ii) Generating Function

26.5.2 $$\sum_{n=0}^{\infty} C(n) x^n = \frac{1 - \sqrt{1-4x}}{2x}, \qquad |x| < \tfrac{1}{4}.$$

26.5(iii) Recurrence Relations

26.5.3 $$C(n+1) = \sum_{k=0}^{n} C(k) \, C(n-k),$$

26.5.4 $$C(n+1) = \frac{2(2n+1)}{n+2} C(n),$$

26.5.5 $$C(n+1) = \sum_{k=0}^{\lfloor n/2 \rfloor} \binom{n}{2k} 2^{n-2k} C(k).$$

26.5(iv) Limiting Forms

26.5.6 $$C(n) \sim \frac{4^n}{\sqrt{\pi n^3}}, \qquad n \to \infty,$$

26.5.7 $$\lim_{n \to \infty} \frac{C(n+1)}{C(n)} = 4.$$

26.6 Other Lattice Path Numbers

26.6(i) Definitions

Dellanoy Number $D(m,n)$

$D(m,n)$ is the number of paths from $(0,0)$ to (m,n) that are composed of directed line segments of the form $(1,0)$, $(0,1)$, or $(1,1)$.

26.6.1 $$D(m,n) = \sum_{k=0}^{n} \binom{n}{k} \binom{m+n-k}{n} = \sum_{k=0}^{n} 2^k \binom{m}{k} \binom{n}{k}.$$

See Table 26.6.1.

Table 26.6.1: Dellanoy numbers $D(m,n)$.

m						n					
	0	1	2	3	4	5	6	7	8	9	10
0	1	1	1	1	1	1	1	1	1	1	1
1	1	3	5	7	9	11	13	15	17	19	21
2	1	5	13	25	41	61	85	113	145	181	221
3	1	7	25	63	129	231	377	575	833	1159	1561
4	1	9	41	129	321	681	1289	2241	3649	5641	8361
5	1	11	61	231	681	1683	3653	7183	13073	22363	36365
6	1	13	85	377	1289	3653	8989	19825	40081	75517	1 34245
7	1	15	113	575	2241	7183	19825	48639	1 08545	2 24143	4 33905
8	1	17	145	833	3649	13073	40081	1 08545	2 65729	5 98417	12 56465
9	1	19	181	1159	5641	22363	75517	2 24143	5 98417	14 62563	33 17445
10	1	21	221	1561	8361	36365	1 34245	4 33905	12 56465	33 17445	80 97453

Motzkin Number $M(n)$

$M(n)$ is the number of lattice paths from $(0,0)$ to (n,n) that stay on or above the line $y=x$ and are composed of directed line segments of the form $(2,0)$, $(0,2)$, or $(1,1)$.

26.6.2 $$M(n) = \sum_{k=0}^{n} \frac{(-1)^k}{n+2-k} \binom{n}{k} \binom{2n+2-2k}{n+1-k}.$$

See Table 26.6.2.

Table 26.6.2: Motzkin numbers $M(n)$.

n	$M(n)$	n	$M(n)$	n	$M(n)$	n	$M(n)$	n	$M(n)$
0	1	4	9	8	323	12	15511	16	8 53467
1	1	5	21	9	835	13	41835	17	23 56779
2	2	6	51	10	2188	14	1 13634	18	65 36382
3	4	7	127	11	5798	15	3 10572	19	181 99284

Narayana Number $N(n,k)$

$N(n,k)$ is the number of lattice paths from $(0,0)$ to (n,n) that stay on or above the line $y = x$, are composed of directed line segments of the form $(1,0)$ or $(0,1)$, and for which there are exactly k occurrences at which a segment of the form $(0,1)$ is followed by a segment of the form $(1,0)$.

26.6.3
$$N(n,k) = \frac{1}{n}\binom{n}{k}\binom{n}{k-1}.$$

See Table 26.6.3.

Table 26.6.3: Narayana numbers $N(n,k)$.

n	k									
	0	1	2	3	4	5	6	7	8	9	10
0	1									
1	0	1								
2	0	1	1							
3	0	1	3	1						
4	0	1	6	6	1					
5	0	1	10	20	10	1				
6	0	1	15	50	50	15	1			
7	0	1	21	105	175	105	21	1		
8	0	1	28	196	490	490	196	28	1	
9	0	1	36	336	1176	1764	1176	336	36	1
10	0	1	45	540	2520	5292	5292	2520	540	45	1

Schröder Number $r(n)$

$r(n)$ is the number of paths from $(0,0)$ to (n,n) that stay on or above the diagonal $y = x$ and are composed of directed line segments of the form $(1,0)$, $(0,1)$, or $(1,1)$.

26.6.4
$$r(n) = D(n,n) - D(n+1, n-1), \qquad n \geq 1.$$

See Table 26.6.4.

Table 26.6.4: Schröder numbers $r(n)$.

n	$r(n)$	n	$r(n)$	n	$r(n)$	n	$r(n)$	n	$r(n)$
0	1	4	90	8	41586	12	272 97738	16	2 09271 56706
1	2	5	394	9	2 06098	13	1420 78746	17	11 18180 26018
2	6	6	1806	10	10 37718	14	7453 87038	18	60 03188 53926
3	22	7	8558	11	52 93446	15	39376 03038	19	323 67243 17174

26.6(ii) Generating Functions

For sufficiently small $|x|$ and $|y|$,

26.6.5
$$\sum_{m,n=0}^{\infty} D(m,n) x^m y^n = \frac{1}{1-x-y-xy},$$

26.6.6
$$\sum_{n=0}^{\infty} D(n,n) x^n = \frac{1}{\sqrt{1-6x+x^2}},$$

26.6.7
$$\sum_{n=0}^{\infty} M(n) x^n = \frac{1-x-\sqrt{1-2x-3x^2}}{2x^2},$$

26.6.8
$$\sum_{n,k=1}^{\infty} N(n,k) x^n y^k = \frac{1-x-xy-\sqrt{(1-x-xy)^2-4x^2 y}}{2x},$$

26.6.9
$$\sum_{n=0}^{\infty} r(n) x^n = \frac{1-x-\sqrt{1-6x+x^2}}{2x}.$$

26.6(iii) Recurrence Relations

26.6.10
$$D(m,n) = D(m, n-1) + D(m-1, n) + D(m-1, n-1), \quad m,n \geq 1,$$

26.6.11
$$M(n) = M(n-1) + \sum_{k=2}^{n} M(k-2) M(n-k), \quad n \geq 2.$$

26.6(iv) Identities

26.6.12
$$C(n) = \sum_{k=1}^{n} N(n,k),$$

26.6.13
$$M(n) = \sum_{k=0}^{n} (-1)^k \binom{n}{k} C(n+1-k),$$

26.6.14
$$C(n) = \sum_{k=0}^{2n} (-1)^k \binom{2n}{k} M(2n-k).$$

26.7 Set Partitions: Bell Numbers

26.7(i) Definitions

$B(n)$ is the number of partitions of $\{1, 2, \ldots, n\}$. For $S(n,k)$ see §26.8(i).

26.7.1
$$B(0) = 1,$$

26.7.2
$$B(n) = \sum_{k=0}^{n} S(n,k),$$

26.7.3
$$B(n) = \sum_{k=1}^{m} \frac{k^n}{k!} \sum_{j=0}^{m-k} \frac{(-1)^j}{j!}, \quad m \geq n,$$

26.7.4
$$B(n) = e^{-1} \sum_{k=1}^{\infty} \frac{k^n}{k!} = 1 + \left\lfloor e^{-1} \sum_{k=1}^{2n} \frac{k^n}{k!} \right\rfloor.$$

See Table 26.7.1.

Table 26.7.1: Bell numbers.

n	$B(n)$	n	$B(n)$
0	1	10	1 15975
1	1	11	6 78570
2	2	12	42 13597
3	5	13	276 44437
4	15	14	1908 99322
5	52	15	13829 58545
6	203	16	1 04801 42147
7	877	17	8 28648 69804
8	4140	18	68 20768 06159
9	21147	19	583 27422 05057

26.7(ii) Generating Function

26.7.5
$$\sum_{n=0}^{\infty} B(n) \frac{x^n}{n!} = \exp(e^x - 1).$$

26.7(iii) Recurrence Relation

26.7.6
$$B(n+1) = \sum_{k=0}^{n} \binom{n}{k} B(n).$$

26.7(iv) Asymptotic Approximation

26.7.7
$$B(n) = \frac{N^n e^{N-n-1}}{(1+\ln N)^{1/2}} \left(1 + O\left(\frac{(\ln n)^{1/2}}{n^{1/2}} \right) \right), \quad n \to \infty,$$

where

26.7.8
$$N \ln N = n,$$

or, equivalently, $N = e^{\mathrm{Wm}(n)}$, with properties of the Lambert function $\mathrm{Wm}(n)$ given in §4.13. For higher approximations to $B(n)$ as $n \to \infty$ see de Bruijn (1961, pp. 104–108).

26.8 Set Partitions: Stirling Numbers

26.8(i) Definitions

$s(n,k)$ denotes the *Stirling number of the first kind*: $(-1)^{n-k}$ times the number of permutations of $\{1,2,\ldots,n\}$ with exactly k cycles. See Table 26.8.1.

26.8.1 $\qquad s(n,n) = 1, \qquad\qquad n \geq 0,$

26.8.2 $\qquad s(1,k) = \delta_{1,k},$

26.8.3
$$(-1)^{n-k} s(n,k) = \sum_{1 \leq b_1 < \cdots < b_{n-k} \leq n-1} b_1 b_2 \cdots b_{n-k},$$
$$n > k \geq 1.$$

$S(n,k)$ denotes the *Stirling number of the second kind*: the number of partitions of $\{1, 2, \ldots, n\}$ into exactly k nonempty subsets. See Table 26.8.2.

26.8.4 $\qquad S(n,n) = 1, \qquad\qquad n \geq 0,$

26.8.5 $\qquad S(n,k) = \sum 1^{c_1} 2^{c_2} \cdots k^{c_k},$

where the summation is over all nonnegative integers $c_1, c_2, \ldots, c_k$ such that $c_1 + c_2 + \cdots + c_k = n - k$.

26.8.6 $\qquad S(n,k) = \dfrac{1}{k!} \sum_{j=0}^{k} (-1)^{k-j} \binom{k}{j} j^n.$

Table 26.8.1: Stirling numbers of the first kind $s(n,k)$.

n	k										
	0	1	2	3	4	5	6	7	8	9	10
0	1										
1	0	1									
2	0	−1	1								
3	0	2	−3	1							
4	0	−6	11	−6	1						
5	0	24	−50	35	−10	1					
6	0	−120	274	−225	85	−15	1				
7	0	720	−1764	1624	−735	175	−21	1			
8	0	−5040	13068	−13132	6769	−1960	322	−28	1		
9	0	40320	−109584	118124	−67284	22449	−4536	546	−36	1	
10	0	−362880	1026576	−1172700	723680	−269325	6327	−9450	870	−45	1

Table 26.8.2: Stirling numbers of the second kind $S(n,k)$.

n	k										
	0	1	2	3	4	5	6	7	8	9	10
0	1										
1	0	1									
2	0	1	1								
3	0	1	3	1							
4	0	1	7	6	1						
5	0	1	15	25	10	1					
6	0	1	31	90	65	15	1				
7	0	1	63	301	350	140	21	1			
8	0	1	127	966	1701	1050	266	28	1		
9	0	1	255	3025	7770	6951	2646	462	36	1	
10	0	1	511	9330	34105	42525	22827	5880	750	45	1

26.8(ii) Generating Functions

26.8.7 $\qquad \displaystyle\sum_{k=0}^{n} s(n,k) x^k = (x-n+1)_n,$

where $(x)_n$ is the Pochhammer symbol: $x(x+1)\cdots(x+n-1)$.

26.8.8 $\qquad \displaystyle\sum_{n=0}^{\infty} s(n,k) \dfrac{x^n}{n!} = \dfrac{(\ln(1+x))^k}{k!}, \qquad |x| < 1,$

26.8 Set Partitions: Stirling Numbers

26.8.9 $$\sum_{n,k=0}^{\infty} s(n,k) \frac{x^n}{n!} y^k = (1+x)^y, \qquad |x| < 1.$$

26.8.10 $$\sum_{k=1}^{n} S(n,k)(x-k+1)_k = x^n,$$

26.8.11 $$\sum_{n=0}^{\infty} S(n,k) x^n = \frac{x^k}{(1-x)(1-2x)\cdots(1-kx)}, \quad |x| < 1/k,$$

26.8.12 $$\sum_{n=0}^{\infty} S(n,k) \frac{x^n}{n!} = \frac{(e^x-1)^k}{k!},$$

26.8.13 $$\sum_{n,k=0}^{\infty} S(n,k) \frac{x^n}{n!} y^k = \exp(y(e^x-1)).$$

26.8(iii) Special Values

For $n \geq 1$,

26.8.14 $s(n,0) = 0, \quad s(n,1) = (-1)^{n-1}(n-1)!,$

26.8.15 $s(n,2) = (-1)^n (n-1)! \left(1 + \frac{1}{2} + \cdots + \frac{1}{n-1}\right),$

26.8.16 $-s(n,n-1) = S(n,n-1) = \binom{n}{2},$

26.8.17 $S(n,0) = 0, \quad S(n,1) = 1, \quad S(n,2) = 2^{n-1} - 1.$

26.8(iv) Recurrence Relations

26.8.18 $s(n,k) = s(n-1,k-1) - (n-1)s(n-1,k),$

26.8.19 $\binom{k}{h} s(n,k) = \sum_{j=k-h}^{n-h} \binom{n}{j} s(n-j,h) s(j,k-h),$
$n \geq k \geq h,$

26.8.20 $s(n+1, k+1) = n! \sum_{j=k}^{n} \frac{(-1)^{n-j}}{j!} s(j,k),$

26.8.21 $s(n+k+1, k) = -\sum_{j=0}^{k} (n+j) s(n+j, j).$

26.8.22 $S(n,k) = k S(n-1,k) + S(n-1,k-1),$

26.8.23 $\binom{k}{h} S(n,k) = \sum_{j=k-h}^{n-h} \binom{n}{j} S(n-j,h) S(j,k-h),$
$n \geq k \geq h,$

26.8.24 $S(n,k) = \sum_{j=k}^{n} S(j-1,k-1) k^{n-j},$

26.8.25 $S(n+1, k+1) = \sum_{j=k}^{n} \binom{n}{j} S(j,k),$

26.8.26 $S(n+k+1, k) = \sum_{j=0}^{k} j S(n+j, j).$

26.8(v) Identities

26.8.27 $$s(n, n-k) = \sum_{j=0}^{k} (-1)^j \binom{n-1+j}{k+j} \binom{n+k}{k-j} S(k+j, j),$$

26.8.28 $$\sum_{k=1}^{n} s(n,k) = 0, \qquad n > 1,$$

26.8.29 $$\sum_{k=1}^{n} (-1)^{n-k} s(n,k) = n!,$$

26.8.30 $$\sum_{j=k}^{n} s(n+1, j+1) n^{j-k} = s(n,k).$$

26.8.31 $$\frac{1}{k!} \frac{d^k}{dx^k} f(x) = \sum_{n=k}^{\infty} \frac{s(n,k)}{n!} \Delta^n f(x),$$

when $f(x)$ is analytic for all x, and the series converges, where

26.8.32 $\Delta f(x) = f(x+1) - f(x);$

compare §3.6(i).

26.8.33 $$S(n, n-k) = \sum_{j=0}^{k} (-1)^j \binom{n-1+j}{k+j} \binom{n+k}{k-j} s(k+j, j),$$

26.8.34 $$\sum_{j=0}^{n} j^k x^j = \sum_{j=0}^{k} S(k,j) x^j \frac{d^j}{dx^j}\left(\frac{1-x^{n+1}}{1-x}\right),$$

26.8.35 $$\sum_{j=0}^{n} j^k = \sum_{j=0}^{k} j! S(k,j) \binom{n+1}{j+1},$$

26.8.36 $$\sum_{k=0}^{n} (-1)^{n-k} k! S(n,k) = 1.$$

26.8.37 $$\frac{1}{k!} \Delta^k f(x) = \sum_{n=k}^{\infty} \frac{S(n,k)}{n!} \frac{d^n}{dx^n} f(x),$$

when $f(x)$ is analytic for all x, and the series converges.

Let A and B be the $n \times n$ matrices with (j,k)th elements $s(j,k)$, and $S(j,k)$, respectively. Then

26.8.38 $A^{-1} = B.$

26.8.39 $$\sum_{j=k}^{n} s(j,k) S(n,j) = \sum_{j=k}^{n} s(n,j) S(j,k) = \delta_{n,k}.$$

26.8(vi) Relations to Bernoulli Numbers

See §24.15(iii).

26.8(vii) Asymptotic Approximations

26.8.40
$$s(n+1, k+1) \sim (-1)^{n-k} \frac{n!}{k!} (\gamma + \ln n)^k, \quad n \to \infty,$$
uniformly for $k = o(\ln n)$, where γ is Euler's constant (§5.2(ii)).

26.8.41
$$s(n+k, k) \sim \frac{(-1)^n}{2^n n!} k^{2n}, \quad k \to \infty,$$
n fixed.

26.8.42
$$S(n, k) \sim \frac{k^n}{k!}, \quad n \to \infty,$$
k fixed.

26.8.43
$$S(n+k, k) \sim \frac{k^{2n}}{2^n n!}, \quad k \to \infty,$$
uniformly for $n = o(k^{1/2})$.

For asymptotic approximations for $s(n+1, k+1)$ and $S(n, k)$ that apply uniformly for $1 \leq k \leq n$ as $n \to \infty$ see Temme (1993).

For other asymptotic approximations and also expansions see Moser and Wyman (1958a) for Stirling numbers of the first kind, and Moser and Wyman (1958b), Bleick and Wang (1974) for Stirling numbers of the second kind.

For asymptotic estimates for generalized Stirling numbers see Chelluri et al. (2000).

26.9 Integer Partitions: Restricted Number and Part Size

26.9(i) Definitions

$p_k(n)$ denotes the number of partitions of n into at most k parts. See Table 26.9.1.

26.9.1
$$p_k(n) = p(n), \quad k \geq n.$$

Unrestricted partitions are covered in §27.14.

Table 26.9.1: Partitions $p_k(n)$.

n	\multicolumn{11}{c}{k}										
	0	1	2	3	4	5	6	7	8	9	10
0	1	1	1	1	1	1	1	1	1	1	1
1	0	1	1	1	1	1	1	1	1	1	1
2	0	1	2	2	2	2	2	2	2	2	2
3	0	1	2	3	3	3	3	3	3	3	3
4	0	1	3	4	5	5	5	5	5	5	5
5	0	1	3	5	6	7	7	7	7	7	7
6	0	1	4	7	9	10	11	11	11	11	11
7	0	1	4	8	11	13	14	15	15	15	15
8	0	1	5	10	15	18	20	21	22	22	22
9	0	1	5	12	18	23	26	28	29	30	30
10	0	1	6	14	23	30	35	38	40	41	42

A useful representation for a partition is the *Ferrers graph* in which the integers in the partition are each represented by a row of dots. An example is provided in Figure 26.9.1.

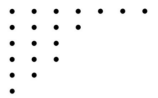

Figure 26.9.1: Ferrers graph of the partition $7 + 4 + 3 + 3 + 2 + 1$.

The *conjugate* partition is obtained by reflecting the Ferrers graph across the main diagonal or, equivalently, by representing each integer by a column of dots. The conjugate to the example in Figure 26.9.1 is $6 + 5 + 4 + 2 + 1 + 1 + 1$. Conjugation establishes a one-to-one correspondence between partitions of n into at most k parts and partitions of n into parts with largest part less than or equal to k. It follows that $p_k(n)$ also equals the number of partitions of n into parts that are less than or equal to k.

$p_k(\leq m, n)$ is the number of partitions of n into at most k parts, each less than or equal to m. It is also equal to the number of lattice paths from $(0, 0)$ to (m, k) that have exactly n vertices (h, j), $1 \leq h \leq m$, $1 \leq j \leq k$, above and to the left of the lattice path. See Figure 26.9.2.

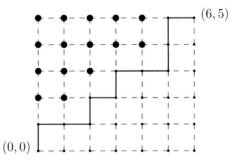

Figure 26.9.2: The partition $5 + 5 + 3 + 2$ represented as a lattice path.

Equations (26.9.2)–(26.9.3) are examples of closed forms that can be computed explicitly for any positive integer k. See Andrews (1976, p. 81).

26.9.2
$$p_0(n) = 0, \quad n > 0,$$

26.9.3
$$p_1(n) = 1, \quad p_2(n) = 1 + \lfloor n/2 \rfloor,$$
$$p_3(n) = 1 + \left\lfloor \frac{n^2 + 6n}{12} \right\rfloor.$$

26.9(ii) Generating Functions

In what follows

$$26.9.4 \qquad \begin{bmatrix} m \\ n \end{bmatrix}_q = \prod_{j=1}^{n} \frac{1-q^{m-n+j}}{1-q^j}, \qquad n \geq 0,$$

is the *Gaussian polynomial* (or *q-binomial coefficient*); compare §§17.2(i)–17.2(ii). In the present chapter $m \geq n \geq 0$ in all cases. It is also assumed everywhere that $|q| < 1$.

$$26.9.5 \qquad \sum_{n=0}^{\infty} p_k(n) q^n = \prod_{j=1}^{k} \frac{1}{1-q^j} = 1 + \sum_{m=1}^{\infty} \begin{bmatrix} k+m-1 \\ m \end{bmatrix}_q q^m,$$

$$26.9.6 \qquad \sum_{n=0}^{\infty} p_k(\leq m, n) q^n = \begin{bmatrix} m+k \\ k \end{bmatrix}_q.$$

Also, when $|xq| < 1$

$$26.9.7 \qquad \sum_{m,n=0}^{\infty} p_k(\leq m, n) x^k q^n = 1 + \sum_{k=1}^{\infty} \begin{bmatrix} m+k \\ k \end{bmatrix}_q x^k = \prod_{j=0}^{m} \frac{1}{1-xq^j}.$$

26.9(iii) Recurrence Relations

$$26.9.8 \qquad p_k(n) = p_k(n-k) + p_{k-1}(n);$$

equivalently, partitions into at most k parts either have exactly k parts, in which case we can subtract one from each part, or they have strictly fewer than k parts.

$$26.9.9 \qquad p_k(n) = \frac{1}{n} \sum_{t=1}^{n} p_k(n-t) \sum_{\substack{j \mid t \\ j \leq k}} j,$$

where the inner sum is taken over all positive divisors of t that are less than or equal to k.

26.9(iv) Limiting Form

As $n \to \infty$ with k fixed,

$$26.9.10 \qquad p_k(n) \sim \frac{n^{k-1}}{k!(k-1)!}.$$

26.10 Integer Partitions: Other Restrictions

26.10(i) Definitions

$p(\mathcal{D}, n)$ denotes the number of partitions of n into distinct parts. $p_m(\mathcal{D}, n)$ denotes the number of partitions of n into at most m distinct parts. $p(\mathcal{D}k, n)$ denotes the number of partitions of n into parts with difference at least k. $p(\mathcal{D}'3, n)$ denotes the number of partitions of n into parts with difference at least 3, except that multiples of 3 must differ by at least 6. $p(\mathcal{O}, n)$ denotes the number of partitions of n into odd parts. $p(\in S, n)$ denotes the number of partitions of n into parts taken from the set S. The set $\{n \geq 1 \mid n \equiv \pm j \pmod{k}\}$ is denoted by $A_{j,k}$. The set $\{2, 3, 4, \ldots\}$ is denoted by T. If more than one restriction applies, then the restrictions are separated by commas, for example, $p(\mathcal{D}2, \in T, n)$. See Table 26.10.1.

$$26.10.1 \qquad p(\mathcal{D}, 0) = p(\mathcal{D}k, 0) = p(\subset S, 0) = 1.$$

Table 26.10.1: Partitions restricted by difference conditions, or equivalently with parts from $A_{j,k}$.

n	$p(\mathcal{D}, n)$ and $p(\mathcal{O}, n)$	$p(\mathcal{D}2, n)$ and $p(\in A_{1,5}, n)$	$p(\mathcal{D}2, \in T, n)$ and $p(\in A_{2,5}, n)$	$p(\mathcal{D}'3, n)$ and $p(\in A_{1,6}, n)$
0	1	1	1	1
1	1	1	0	1
2	1	1	1	1
3	2	1	1	1
4	2	2	1	1
5	3	2	1	2
6	4	3	2	2
7	5	3	2	3
8	6	4	3	3
9	8	5	3	3
10	10	6	4	4
11	12	7	4	5
12	15	9	6	6
13	18	10	6	7
14	22	12	8	8
15	27	14	9	9
16	32	17	11	10
17	38	19	12	12
18	46	23	15	14
19	54	26	16	16
20	64	31	20	18

26.10(ii) Generating Functions

Throughout this subsection it is assumed that $|q| < 1$.

$$26.10.2 \qquad \sum_{n=0}^{\infty} p(\mathcal{D}, n) q^n = \prod_{j=1}^{\infty} (1+q^j) = \prod_{j=1}^{\infty} \frac{1}{1-q^{2j-1}}$$
$$= 1 + \sum_{m=1}^{\infty} \frac{q^{m(m+1)/2}}{(1-q)(1-q^2)\cdots(1-q^m)}$$
$$= 1 + \sum_{m=1}^{\infty} q^m (1+q)(1+q^2) \cdots (1+q^{m-1}),$$

where the last right-hand side is the sum over $m \geq 0$ of the generating functions for partitions into distinct parts with largest part equal to m.

$$(1-x)\sum_{m,n=0}^{\infty} p_m(\le k, \mathcal{D}, n)x^m q^n$$

26.10.3
$$= \sum_{m=0}^{k} \begin{bmatrix} k \\ m \end{bmatrix}_q q^{m(m+1)/2} x^m = \prod_{j=1}^{k}(1+x\,q^j),$$
$$|x| < 1,$$

26.10.4
$$\sum_{n=0}^{\infty} p(\mathcal{D}k, n)q^n = 1 + \sum_{m=1}^{\infty} \frac{q^{(km^2+(2-k)m)/2}}{(1-q)(1-q^2)\cdots(1-q^m)},$$

26.10.5
$$\sum_{n=0}^{\infty} p(\in S, n)q^n = \prod_{j \in S} \frac{1}{1-q^j}.$$

26.10(iii) Recurrence Relations

26.10.6
$$p(\mathcal{D}, n) = \frac{1}{n} \sum_{t=1}^{n} p(\mathcal{D}, n-t) \sum_{\substack{j \mid t \\ j \text{ odd}}} j,$$

where the inner sum is the sum of all positive odd divisors of t.

26.10.7
$$\sum (-1)^k p\left(\mathcal{D}, n - \tfrac{1}{2}(3k^2 \pm k)\right)$$
$$= \begin{cases} (-1)^r, & n = 3r^2 \pm r, \\ 0, & \text{otherwise}, \end{cases}$$

where the sum is over nonnegative integer values of k for which $n - \tfrac{1}{2}(3k^2 \pm k) \ge 0$.

26.10.8
$$\sum (-1)^k p\left(\mathcal{D}, n - (3k^2 \pm k)\right) = \begin{cases} 1, & n = \tfrac{1}{2}(r^2 \pm r), \\ 0, & \text{otherwise}, \end{cases}$$

where the sum is over nonnegative integer values of k for which $n - (3k^2 \pm k) \ge 0$.

In exact analogy with (26.9.8), we have

26.10.9 $p_m(\mathcal{D}, n) = p_m(\mathcal{D}, n-m) + p_{m-1}(\mathcal{D}, n),$

26.10.10 $p(\mathcal{D}k, n) = \sum p_m\left(n - \tfrac{1}{2}km^2 - m + \tfrac{1}{2}km\right),$

where the sum is over nonnegative integer values of m for which $n - \tfrac{1}{2}km^2 - m + \tfrac{1}{2}km \ge 0$.

26.10.11
$$p(\in S, n) = \frac{1}{n} \sum_{t=1}^{n} p(\in S, n-t) \sum_{\substack{j \mid t \\ j \in S}} j,$$

where the inner sum is the sum of all positive divisors of t that are in S.

26.10(iv) Identities

Equations (26.10.13) and (26.10.14) are the *Rogers–Ramanujan identities*. See also §17.2(vi).

26.10.12 $p(\mathcal{D}, n) = p(\mathcal{O}, n),$

26.10.13 $p(\mathcal{D}2, n) = p(\in A_{1,5}, n),$

26.10.14 $p(\mathcal{D}2, \in T, n) = p(\in A_{2,5}, n), \quad T = \{2,3,4,\ldots\},$

26.10.15 $p(\mathcal{D}'3, n) = p(\in A_{1,6}, n).$

Note that $p(\mathcal{D}'3, n) \le p(\mathcal{D}3, n)$, with strict inequality for $n \ge 9$. It is known that for $k > 3$, $p(\mathcal{D}k, n) \ge p(\in A_{1,k+3}, n)$, with strict inequality for n sufficiently large, provided that $k = 2^m - 1, m = 3, 4, 5$, or $k \ge 32$; see Yee (2004).

26.10(v) Limiting Form

26.10.16
$$p(\mathcal{D}, n) \sim \frac{e^{\pi\sqrt{n/3}}}{(768n^3)^{1/4}}, \qquad n \to \infty.$$

26.10(vi) Bessel-Function Expansion

26.10.17
$$p(\mathcal{D}, n)$$
$$= \pi \sum_{k=1}^{\infty} \frac{A_{2k-1}(n)}{(2k-1)\sqrt{24n+1}} I_1\left(\frac{\pi}{2k-1}\sqrt{\frac{24n+1}{72}}\right),$$

where $I_1(x)$ is the modified Bessel function (§10.25(ii)), and

26.10.18
$$A_k(n) = \sum_{\substack{1 < h \le k \\ (h,k)=1}} e^{\pi i f(h,k) - (2\pi i n h/k)},$$

with

26.10.19
$$f(h,k) = \sum_{j=1}^{k} \left[\!\!\left[\frac{2j-1}{2k}\right]\!\!\right] \left[\!\!\left[\frac{h(2j-1)}{k}\right]\!\!\right],$$

and

26.10.20
$$[\![x]\!] = \begin{cases} x - \lfloor x \rfloor - \tfrac{1}{2}, & x \notin \mathbb{Z}, \\ 0, & x \in \mathbb{Z}. \end{cases}$$

The quantity $A_k(n)$ is real-valued.

26.11 Integer Partitions: Compositions

A *composition* is an integer partition in which order is taken into account. For example, there are eight compositions of 4: $4, 3+1, 1+3, 2+2, 2+1+1, 1+2+1, 1+1+2,$ and $1+1+1+1$. $c(n)$ denotes the number of compositions of n, and $c_m(n)$ is the number of compositions into *exactly* m parts. $c(\in T, n)$ is the number of compositions of n with no 1's, where again $T = \{2, 3, 4, \ldots\}$. The integer 0 is considered to have one composition consisting of no parts:

26.11.1 $c(0) = c(\in T, 0) = 1.$

Also,

26.11.2 $c_m(0) = \delta_{0,m},$

26.11.3 $c_m(n) = \binom{n-1}{m-1},$

26.11.4
$$\sum_{n=0}^{\infty} c_m(n) q^n = \frac{q^m}{(1-q)^m}.$$

26.12 Plane Partitions

The *Fibonacci numbers* are determined recursively by

26.11.5
$$F_0 = 0, \quad F_1 = 1, \quad F_n = F_{n-1} + F_{n-2}, \quad n \geq 2.$$

26.11.6
$$c(\in T, n) = F_{n-1}, \quad n \geq 1.$$

Explicitly,

26.11.7
$$F_n = \frac{(1+\sqrt{5})^n - (1-\sqrt{5})^n}{2^n \sqrt{5}}.$$

Additional information on Fibonacci numbers can be found in Rosen et al. (2000, pp. 140–145).

26.12(i) Definitions

A *plane partition*, π, of a positive integer n, is a partition of n in which the parts have been arranged in a 2-dimensional array that is weakly decreasing (nonincreasing) across rows and down columns. Different configurations are counted as different plane partitions. As an example, there are six plane partitions of 3:

26.12.1
$$3, \quad 2\ 1, \quad \begin{matrix}2\\1\end{matrix}, \quad 1\ 1\ 1, \quad \begin{matrix}1\ 1\\1\end{matrix}, \quad \begin{matrix}1\\1\\1\end{matrix}.$$

An equivalent definition is that a plane partition is a finite subset of $\mathbb{N} \times \mathbb{N} \times \mathbb{N}$ with the property that if $(r,s,t) \in \pi$ and $(1,1,1) \leq (h,j,k) \leq (r,s,t)$, then (h,j,k) must be an element of π. Here $(h,j,k) \leq (r,s,t)$ means $h \leq r$, $j \leq s$, and $k \leq t$. It is useful to be able to visualize a plane partition as a pile of blocks, one block at each lattice point $(h,j,k) \in \pi$. For example, Figure 26.12.1 depicts the pile of blocks that represents the plane partition of 75 given by (26.12.2).

Figure 26.12.1: A plane partition of 75.

26.12.2
$$\begin{matrix}6&5&5&4&3&3\\6&4&3&3&1\\6&4&3&1&1\\4&2&2&1\\3&1&1\\1&1&1\end{matrix}$$

The number of plane partitions of n is denoted by $pp(n)$, with $pp(0) = 1$. See Table 26.12.1.

Table 26.12.1: Plane partitions.

n	$pp(n)$	n	$pp(n)$	n	$pp(n)$
0	1	17	18334	34	281 75955
1	1	18	29601	35	416 91046
2	3	19	47330	36	614 84961
3	6	20	75278	37	903 79784
4	13	21	1 18794	38	1324 41995
5	24	22	1 86475	39	1934 87501
6	48	23	2 90783	40	2818 46923
7	86	24	4 51194	41	4093 83981
8	160	25	6 96033	42	5930 01267
9	282	26	10 68745	43	8566 67495
10	500	27	16 32658	44	12343 63833
11	859	28	24 83234	45	17740 79109
12	1479	29	37 59612	46	25435 35902
13	2485	30	56 68963	47	36379 93036
14	4167	31	85 12309	48	51913 04973
15	6879	32	127 33429	49	73910 26522
16	11297	33	189 74973	50	1 04996 40707

We define the $r \times s \times t$ *box* $B(r,s,t)$ as

26.12.3
$$B(r,s,t) = \{(h,j,k) \mid 1 \leq h \leq r, 1 \leq j \leq s, 1 \leq k \leq t\}.$$

Then the number of plane partitions in $B(r,s,t)$ is

26.12.4
$$\prod_{(h,j,k) \in B(r,s,t)} \frac{h+j+k-1}{h+j+k-2} = \prod_{h=1}^{r}\prod_{j=1}^{s} \frac{h+j+t-1}{h+j-1}.$$

A plane partition is *symmetric* if $(h,j,k) \in \pi$ implies that $(j,h,k) \in \pi$. The number of symmetric plane partitions in $B(r,r,t)$ is

26.12.5
$$\prod_{h=1}^{r} \frac{2h+t-1}{2h-1} \prod_{1 \leq h < j \leq r} \frac{h+j+t-1}{h+j-1}.$$

A plane partition is *cyclically symmetric* if $(h,j,k) \in \pi$ implies $(j,k,h) \in \pi$. The plane partition in Figure 26.12.1 is an example of a cyclically symmetric plane partition. The number of cyclically symmetric plane partitions in $B(r,r,r)$ is

26.12.6
$$\prod_{h=1}^{r} \frac{3h-1}{3h-2} \prod_{1 \leq h < j \leq r} \frac{h+2j-1}{h+j-1},$$

or equivalently,

$$26.12.7 \qquad \prod_{h=1}^{r}\left(\frac{3h-1}{3h-2}\prod_{j=h}^{r}\frac{r+h+j-1}{2h+j-1}\right).$$

A plane partition is *totally symmetric* if it is both symmetric and cyclically symmetric. The number of totally symmetric plane partitions in $B(r,r,r)$ is

$$26.12.8 \qquad \prod_{1\leq h\leq j\leq r}\frac{h+j+r-1}{h+2j-2}.$$

The *complement* of $\pi \subseteq B(r,s,t)$ is $\pi^c = \{(h,j,k) \mid (r-h+1, s-j+1, t-k+1) \notin \pi\}$. A plane partition is *self-complementary* if it is equal to its complement. The number of self-complementary plane partitions in $B(2r,2s,2t)$ is

$$26.12.9 \qquad \left(\prod_{h=1}^{r}\prod_{j=1}^{s}\frac{h+j+t-1}{h+j-1}\right)^2;$$

in $B(2r+1,2s,2t)$ it is

$$26.12.10 \qquad \left(\prod_{h=1}^{r}\prod_{j=1}^{s}\frac{h+j+t-1}{h+j-1}\right)\left(\prod_{h=1}^{r+1}\prod_{j=1}^{s}\frac{h+j+t-1}{h+j-1}\right);$$

in $B(2r+1,2s+1,2t)$ it is

$$26.12.11 \qquad \left(\prod_{h=1}^{r+1}\prod_{j=1}^{s}\frac{h+j+t-1}{h+j-1}\right)\left(\prod_{h=1}^{r}\prod_{j=1}^{s+1}\frac{h+j+t-1}{h+j-1}\right).$$

A plane partition is *transpose complement* if it is equal to the reflection through the (x,y)-plane of its complement. The number of transpose complement plane partitions in $B(r,r,2t)$ is

$$26.12.12 \qquad \binom{t+r-1}{r-1}\prod_{1\leq h\leq j\leq r-2}\frac{h+j+2t+1}{h+j+1}.$$

The number of symmetric self-complementary plane partitions in $B(2r,2r,2t)$ is

$$26.12.13 \qquad \prod_{h=1}^{r}\prod_{j=1}^{r}\frac{h+j+t-1}{h+j-1};$$

in $B(2r+1,2r+1,2t)$ it is

$$26.12.14 \qquad \prod_{h=1}^{r}\prod_{j=1}^{r+1}\frac{h+j+t-1}{h+j-1}.$$

The number of cyclically symmetric transpose complement plane partitions in $B(2r,2r,2r)$ is

$$26.12.15 \qquad \prod_{h=0}^{r-1}\frac{(3h+1)(6h)!(2h)!}{(4h+1)!(4h)!}.$$

The number of cyclically symmetric self-complementary plane partitions in $B(2r,2r,2r)$ is

$$26.12.16 \qquad \left(\prod_{h=0}^{r-1}\frac{(3h+1)!}{(r+h)!}\right)^2.$$

The number of totally symmetric self-complementary plane partitions in $B(2r,2r,2r)$ is

$$26.12.17 \qquad \prod_{h=0}^{r-1}\frac{(3h+1)!}{(r+h)!}.$$

A *strict shifted plane partition* is an arrangement of the parts in a partition so that each row is indented one space from the previous row and there is weak decrease across rows and strict decrease down columns. An example is given by:

$$26.12.18 \qquad \begin{array}{ccccc} 6 & 6 & 6 & 4 & 3 \\ & 3 & 3 & & \\ & & 2 & & \end{array}$$

A *descending plane partition* is a strict shifted plane partition in which the number of parts in each row is strictly less than the largest part in that row and is greater than or equal to the largest part in the next row. The example of a strict shifted plane partition also satisfies the conditions of a descending plane partition. The number of descending plane partitions in $B(r,r,r)$ is

$$26.12.19 \qquad \prod_{h=0}^{r-1}\frac{(3h+1)!}{(r+h)!}.$$

26.12(ii) Generating Functions

The notation $\sum_{\pi\subseteq B(r,s,t)}$ denotes the sum over all plane partitions contained in $B(r,s,t)$, and $|\pi|$ denotes the number of elements in π.

$$26.12.20 \qquad \sum_{\pi\subseteq\mathbb{N}\times\mathbb{N}\times\mathbb{N}} q^{|\pi|} = \prod_{k=1}^{\infty}\frac{1}{(1-q^k)^k},$$

$$26.12.21 \qquad \sum_{\pi\subseteq B(r,s,t)} q^{|\pi|} = \prod_{(h,j,k)\in B(r,s,t)}\frac{1-q^{h+j+k-1}}{1-q^{h+j+k-2}}$$
$$= \prod_{h=1}^{r}\prod_{j=1}^{s}\frac{1-q^{h+j+t-1}}{1-q^{h+j-1}},$$

$$26.12.22 \qquad \sum_{\substack{\pi\subseteq B(r,r,t)\\ \pi\text{ symmetric}}} q^{|\pi|}$$
$$= \prod_{h=1}^{r}\frac{1-q^{2h+t-1}}{1-q^{2h-1}}\prod_{1\leq h<j\leq r}\frac{1-q^{2(h+j+t-1)}}{1-q^{2(h+j-1)}}.$$

$$\sum_{\substack{\pi \subseteq B(r,r,r) \\ \pi \text{ cyclically symmetric}}} q^{|\pi|}$$

26.12.23
$$= \prod_{h=1}^{r} \frac{1-q^{3h-1}}{1-q^{3h-2}} \prod_{1 \le h < j \le r} \frac{1-q^{3(h+2j-1)}}{1-q^{3(h+j-1)}}$$

$$= \prod_{h=1}^{r} \left(\frac{1-q^{3h-1}}{1-q^{3h-2}} \prod_{j=h}^{r} \frac{1-q^{3(r+h+j-1)}}{1-q^{3(2h+j-1)}} \right).$$

26.12.24
$$\sum_{\pi \subseteq B(r,r,r) \atop \pi \text{ descending plane partition}} q^{|\pi|} = \prod_{1 \le h < j \le r} \frac{1-q^{r+h+j-1}}{1-q^{2h+j-1}}.$$

26.12(iii) Recurrence Relation

26.12.25
$$pp(n) = \frac{1}{n} \sum_{j=1}^{n} pp(n-j)\sigma_2(j),$$

where $\sigma_2(j)$ is the sum of the squares of the divisors of j.

26.12(iv) Limiting Form

As $n \to \infty$

26.12.26
$$pp(n) \sim \left(\frac{\zeta(3)}{2^{11}n^{25}} \right)^{1/36} \exp\left(3 \left(\frac{\zeta(3)n^2}{4} \right)^{1/3} + \zeta'(-1) \right),$$

where ζ is the Riemann ζ-function (§25.2(i)).

26.13 Permutations: Cycle Notation

$\mathfrak{S}_n$ denotes the set of permutations of $\{1, 2, \ldots, n\}$. $\sigma \in \mathfrak{S}_n$ is a one-to-one and onto mapping from $\{1, 2, \ldots, n\}$ to itself. An explicit representation of σ can be given by the $2 \times n$ matrix:

26.13.1
$$\begin{bmatrix} 1 & 2 & 3 & \cdots & n \\ \sigma(1) & \sigma(2) & \sigma(3) & \cdots & \sigma(n) \end{bmatrix}.$$

In cycle notation, the elements in each cycle are put inside parentheses, ordered so that $\sigma(j)$ immediately follows j or, if j is the last listed element of the cycle, then $\sigma(j)$ is the first element of the cycle. The permutation

26.13.2
$$\begin{bmatrix} 1 & 2 & 3 & 4 & 5 & 6 & 7 & 8 \\ 3 & 5 & 2 & 4 & 7 & 8 & 1 & 6 \end{bmatrix}$$

is $(1,3,2,5,7)(4)(6,8)$ in cycle notation. Cycles of length one are *fixed points*. They are often dropped from the cycle notation. In consequence, (26.13.2) can also be written as $(1,3,2,5,7)(6,8)$.

An element of $\mathfrak{S}_n$ with a_1 fixed points, a_2 cycles of length $2, \ldots, a_n$ cycles of length n, where $n = a_1 + 2a_2 + \cdots + na_n$, is said to have *cycle type* $(a_1, a_2, \ldots, a_n)$. The number of elements of $\mathfrak{S}_n$ with cycle type $(a_1, a_2, \ldots, a_n)$ is given by (26.4.7).

The *Stirling cycle numbers* of the first kind, denoted by $\begin{bmatrix} n \\ k \end{bmatrix}$, count the number of permutations of $\{1, 2, \ldots, n\}$ with exactly k cycles. They are related to Stirling numbers of the first kind by

26.13.3
$$\begin{bmatrix} n \\ k \end{bmatrix} = |s(n,k)|.$$

See §26.8 for generating functions, recurrence relations, identities, and asymptotic approximations.

A *derangement* is a permutation with no fixed points. The *derangement number*, $d(n)$, is the number of elements of $\mathfrak{S}_n$ with no fixed points:

26.13.4
$$d(n) = n! \sum_{j=0}^{n} (-1)^j \frac{1}{j!} = \left\lfloor \frac{n!+e-2}{e} \right\rfloor.$$

A *transposition* is a permutation that consists of a single cycle of length two. An *adjacent transposition* is a transposition of two consecutive integers. A permutation that consists of a single cycle of length k can be written as the composition of $k-1$ two-cycles (read from right to left):

26.13.5
$$(j_1, j_2, \ldots, j_k) = (j_1, j_2)(j_2, j_3) \cdots (j_{k-2}, j_{k-1})(j_{k-1}, j_k).$$

Every permutation is a product of transpositions. A permutation with cycle type $(a_1, a_2, \ldots, a_n)$ can be written as a product of $a_2 + 2a_3 + \cdots + (n-1)a_n = n - (a_1 + a_2 + \cdots + a_n)$ transpositions, and no fewer. For the example (26.13.2), this decomposition is given by $(1,3,2,5,7)(6,8) = (1,3)(2,3)(2,5)(5,7)(6,8)$.

A permutation is *even* or *odd* according to the parity of the number of transpositions. The *sign of a permutation* is $+$ if the permutation is even, $-$ if it is odd.

Every transposition is the product of adjacent transpositions. If $j < k$, then (j,k) is a product of $2k-2j-1$ adjacent transpositions:

26.13.6
$$(j,k) = (k-1,k)(k-2,k-1) \cdots (j+1,j+2)$$
$$\times (j,j+1)(j+1,j+2) \cdots (k-1,k).$$

Every permutation is a product of adjacent transpositions. Given a permutation $\sigma \in \mathfrak{S}_n$, the *inversion number* of σ, denoted $\text{inv}(\sigma)$, is the least number of adjacent transpositions required to represent σ. Again, for the example (26.13.2) a minimal decomposition into adjacent transpositions is given by $(1,3,2,5,7)(6,8) = (2,3)(1,2)(4,5)(3,4)(2,3)(3,4)(4,5)(6,7)(5,6)(7,8) \times (6,7)$: $\text{inv}((1,3,2,5,7)(6,8)) = 11$.

26.14 Permutations: Order Notation

26.14(i) Definitions

The set $\mathfrak{S}_n$ (§26.13) can be viewed as the collection of all ordered lists of elements of $\{1, 2, \ldots, n\}$: $\{\sigma(1)\sigma(2)\cdots\sigma(n)\}$. As an example, 35247816 is an element of $\mathfrak{S}_8$. The *inversion number* is the number of pairs of elements for which the larger element precedes the smaller:

26.14.1
$$\operatorname{inv}(\sigma) = \sum_{\substack{1 \leq j < k \leq n \\ \sigma(j) > \sigma(k)}} 1.$$

Equivalently, this is the sum over $1 \leq j < n$ of the number of integers less than $\sigma(j)$ that lie in positions to the right of the jth position: $\operatorname{inv}(35247816) = 2 + 3 + 1 + 1 + 2 + 2 + 0 = 11$.

A *descent* of a permutation is a pair of adjacent elements for which the first is larger than the second. The permutation 35247816 has two descents: 52 and 81. The *major index* is the sum of all positions that mark the first element of a descent:

26.14.2
$$\operatorname{maj}(\sigma) = \sum_{\substack{1 \leq j < n \\ \sigma(j) > \sigma(j+1)}} j.$$

For example, $\operatorname{maj}(35247816) = 2 + 6 = 8$. The major index is also called the *greater index* of the permutation.

The *Eulerian number*, denoted $\left\langle {n \atop k} \right\rangle$, is the number of permutations in $\mathfrak{S}_n$ with exactly k descents. An *excedance* in $\sigma \in \mathfrak{S}_n$ is a position j for which $\sigma(j) > j$. A *weak excedance* is a position j for which $\sigma(j) \geq j$. The Eulerian number $\left\langle {n \atop k} \right\rangle$ is equal to the number of permutations in $\mathfrak{S}_n$ with exactly k excedances. It is also equal to the number of permutations in $\mathfrak{S}_n$ with exactly $k+1$ weak excedances. See Table 26.14.1.

Table 26.14.1: Eulerian numbers $\left\langle {n \atop k} \right\rangle$.

n	k									
	0	1	2	3	4	5	6	7	8	9
0	1									
1	1									
2	1	1								
3	1	4	1							
4	1	11	11	1						
5	1	26	66	26	1					
6	1	57	302	302	57	1				
7	1	120	1191	2416	1191	120	1			
8	1	247	4293	15619	15619	4293	247	1		
9	1	502	14608	88234	1 56190	88234	14608	502	1	
10	1	1013	47840	4 55192	13 10354	13 10354	4 55192	47840	1013	1

26.14(ii) Generating Functions

26.14.3
$$\sum_{\sigma \in \mathfrak{S}_n} q^{\operatorname{inv}(\sigma)} = \sum_{\sigma \in \mathfrak{S}_n} q^{\operatorname{maj}(\sigma)} = \prod_{j=1}^{n} \frac{1 - q^j}{1 - q}.$$

26.14.4
$$\sum_{n,k=0}^{\infty} \left\langle {n \atop k} \right\rangle x^k \frac{t^n}{n!} = \frac{1 - x}{\exp((x-1)t) - x}, \quad |x| < 1, |t| < 1.$$

26.14.5
$$\sum_{k=0}^{n-1} \left\langle {n \atop k} \right\rangle \binom{x+k}{n} = x^n.$$

26.14(iii) Identities

In this subsection $S(n, k)$ is again the Stirling number of the second kind (§26.8), and B_m is the mth Bernoulli number (§24.2(i)).

26.14.6
$$\left\langle {n \atop k} \right\rangle = \sum_{j=0}^{k} (-1)^j \binom{n+1}{j} (k+1-j)^n, \quad n \geq 1,$$

26.14.7
$$\left\langle {n \atop k} \right\rangle = \sum_{j=0}^{n-k} (-1)^{n-k-j} j! \binom{n-j}{k} S(n, j),$$

26.14.8
$$\left\langle {n \atop k} \right\rangle = (k+1) \left\langle {n-1 \atop k} \right\rangle + (n-k) \left\langle {n-1 \atop k-1} \right\rangle, \quad n \geq 2,$$

26.14.9
$$\left\langle {n \atop k} \right\rangle = \left\langle {n \atop n-1-k} \right\rangle, \quad n \geq 1,$$

26.14.10
$$\sum_{k=0}^{n-1} \left\langle {n \atop k} \right\rangle = n!, \quad n \geq 1.$$

26.14.11
$$B_m = \frac{m}{2^m(2^m-1)} \sum_{k=0}^{m-2} (-1)^k \left\langle {m-1 \atop k} \right\rangle, \quad m \geq 2.$$

26.14.12
$$S(n,m) = \frac{1}{m!} \sum_{k=0}^{n-1} \left\langle {n \atop k} \right\rangle \binom{k}{n-m}, \quad n \geq m, \, n \geq 1.$$

26.14(iv) Special Values

26.14.13 $\left\langle {0 \atop k} \right\rangle = \delta_{0,k},$

26.14.14 $\left\langle {n \atop 0} \right\rangle = 1,$

26.14.15 $\left\langle {n \atop 1} \right\rangle = 2^n - n - 1, \qquad n \geq 1,$

26.14.16 $\left\langle {n \atop 2} \right\rangle = 3^n - (n+1)2^n + \binom{n+1}{2}, \quad n \geq 1.$

26.15 Permutations: Matrix Notation

The set $\mathfrak{S}_n$ (§26.13) can be identified with the set of $n \times n$ matrices of 0's and 1's with exactly one 1 in each row and column. The permutation σ corresponds to the matrix in which there is a 1 at the intersection of row j with column $\sigma(j)$, and 0's in all other positions. The permutation 35247816 corresponds to the matrix

26.15.1
$$\begin{bmatrix} 0 & 0 & 1 & 0 & 0 & 0 & 0 & 0 \\ 0 & 0 & 0 & 0 & 1 & 0 & 0 & 0 \\ 0 & 1 & 0 & 0 & 0 & 0 & 0 & 0 \\ 0 & 0 & 0 & 1 & 0 & 0 & 0 & 0 \\ 0 & 0 & 0 & 0 & 0 & 0 & 1 & 0 \\ 0 & 0 & 0 & 0 & 0 & 0 & 0 & 1 \\ 1 & 0 & 0 & 0 & 0 & 0 & 0 & 0 \\ 0 & 0 & 0 & 0 & 0 & 1 & 0 & 0 \end{bmatrix}$$

The *sign of the permutation* σ is the sign of the determinant of its matrix representation. The *inversion number* of σ is a sum of products of pairs of entries in the matrix representation of σ:

26.15.2 $\quad \text{inv}(\sigma) = \sum a_{gh} a_{k\ell},$

where the sum is over $1 \leq g < k \leq n$ and $n \geq h > \ell \geq 1$.

The matrix represents the placement of n nonattacking rooks on an $n \times n$ chessboard, that is, rooks that share neither a row nor a column with any other rook. A *permutation with restricted position* specifies a subset $B \subseteq \{1,2,\ldots,n\} \times \{1,2,\ldots,n\}$. If $(j,k) \in B$, then $\sigma(j) \neq k$. The number of *derangements* of n is the number of permutations with forbidden positions $B = \{(1,1), (2,2), \ldots, (n,n)\}$.

Let $r_j(B)$ be the number of ways of placing j nonattacking rooks on the squares of B. Define $r_0(B) = 1$.

For the problem of derangements, $r_j(B) = \binom{n}{j}$. The *rook polynomial* is the generating function for $r_j(B)$:

26.15.3 $\quad R(x,B) = \sum_{j=0}^{n} r_j(B) x^j.$

If $B = B_1 \cup B_2$, where no element of B_1 is in the same row or column as any element of B_2, then

26.15.4 $\quad R(x,B) = R(x,B_1) R(x,B_2).$

For $(j,k) \in B$, $B \setminus [j,k]$ denotes B after removal of all elements of the form (j,t) or (t,k), $t = 1, 2, \ldots, n$. $B \setminus (j,k)$ denotes B with the element (j,k) removed.

26.15.5 $\quad R(x,B) = x R(x, B \setminus [j,k]) + R(x, B \setminus (j,k)).$

$N_k(B)$ is the number of permutations in $\mathfrak{S}_n$ for which exactly k of the pairs $(j, \sigma(j))$ are elements of B. $N(x,B)$ is the generating function:

26.15.6 $\quad N(x,B) = \sum_{k=0}^{n} N_k(B) x^k,$

and

26.15.7 $\quad N(x,B) = \sum_{k=0}^{n} r_k(B)(n-k)!(x-1)^k.$

The number of permutations that avoid B is

26.15.8 $\quad N_0(B) \equiv N(0, B) = \sum_{k=0}^{n} (-1)^k r_k(B)(n-k)!.$

Example 1

The *problème des ménages* asks for the number of ways of seating n married couples around a circular table with labeled seats so that no men are adjacent, no women are adjacent, and no husband and wife are adjacent. There are $2(n!)$ ways to place the wives. Let $B = \{(j,j), (j, j+1) \mid 1 \leq j < n\} \cup \{(n,n), (n,1)\}$. Then

26.15.9 $\quad r_k(B) = \frac{2n}{2n-k} \binom{2n-k}{k}.$

The solution is

26.15.10
$$2(n!) N_0(B) = 2(n!) \sum_{k=0}^{n} (-1)^k \frac{2n}{2n-k} \binom{2n-k}{k} (n-k)!.$$

Example 2

The *Ferrers board* of shape $(b_1, b_2, \ldots, b_n)$, $0 \leq b_1 \leq b_2 \leq \cdots \leq b_n$, is the set $B = \{(j,k) \mid 1 \leq j \leq n, 1 \leq k \leq b_j\}$. For this set,

26.15.11 $\quad \sum_{k=0}^{n} r_{n-k}(B)(x-k+1)_k = \prod_{j=1}^{n} (x + b_j - j + 1).$

If B is the Ferrers board of shape $(0, 1, 2, \ldots, n-1)$, then

26.15.12 $\quad \sum_{k=0}^{n} r_{n-k}(B)(x-k+1)_k = x^n,$

and therefore by (26.8.10),

26.15.13 $\quad r_{n-k}(B) = S(n,k).$

26.16 Multiset Permutations

Let $S = \{1^{a_1}, 2^{a_2}, \ldots, n^{a_n}\}$ be the multiset that has a_j copies of j, $1 \leq j \leq n$. $\mathfrak{S}_S$ denotes the set of permutations of S for all distinct orderings of the $a_1+a_2+\cdots+a_n$ integers. The number of elements in $\mathfrak{S}_S$ is the multinomial coefficient (§26.4) $\binom{a_1+a_2+\cdots+a_n}{a_1,a_2,\ldots,a_n}$. Additional information can be found in Andrews (1976, pp. 39–45).

The definitions of inversion number and major index can be extended to permutations of a multiset such as $351322453154 \in \mathfrak{S}_{\{1^2,2^2,3^3,4^2,5^3\}}$. Thus $\mathrm{inv}(351322453154) = 4+8+0+3+1+1+2+3+1+0+1 = 24$, and $\mathrm{maj}(351322453154) = 2+4+8+9+11 = 34$.

The *q-multinomial coefficient* is defined in terms of Gaussian polynomials (§26.9(ii)) by

26.16.1
$$\begin{bmatrix} a_1 + a_2 + \cdots + a_n \\ a_1, a_2, \ldots, a_n \end{bmatrix}_q = \prod_{k=1}^{n-1} \begin{bmatrix} a_k + a_{k+1} + \cdots + a_n \\ a_k \end{bmatrix}_q,$$

and again with $S = \{1^{a_1}, 2^{a_2}, \ldots, n^{a_n}\}$ we have

26.16.2
$$\sum_{\sigma \in \mathfrak{S}_S} q^{\mathrm{inv}(\sigma)} = \begin{bmatrix} a_1 + a_2 + \cdots + a_n \\ a_1, a_2, \ldots, a_n \end{bmatrix}_q,$$

26.16.3
$$\sum_{\sigma \in \mathfrak{S}_S} q^{\mathrm{maj}(\sigma)} = \begin{bmatrix} a_1 + a_2 + \cdots + a_n \\ a_1, a_2, \ldots, a_n \end{bmatrix}_q.$$

26.17 The Twelvefold Way

The *twelvefold way* gives the number of mappings f from set N of n objects to set K of k objects (putting balls from set N into boxes in set K). See Table 26.17.1. In this table $(k)_n$ is Pochhammer's symbol, and $S(n,k)$ and $p_k(n)$ are defined in §§26.8(i) and 26.9(i).

Table 26.17.1 is reproduced (in modified form) from Stanley (1997, p. 33). See also Example 3 in §26.18.

Table 26.17.1: The twelvefold way.

elements of N	elements of K	f unrestricted	f one-to-one	f onto
labeled	labeled	k^n	$(k-n+1)_n$	$k!\,S(n,k)$
unlabeled	labeled	$\binom{k+n-1}{n}$	$\binom{k}{n}$	$\binom{n-1}{n-k}$
labeled	unlabeled	$S(n,1) + S(n,2) + \cdots + S(n,k)$	$\begin{cases} 1 & n \leq k \\ 0 & n > k \end{cases}$	$S(n,k)$
unlabeled	unlabeled	$p_k(n)$	$\begin{cases} 1 & n \leq k \\ 0 & n > k \end{cases}$	$p_k(n) - p_{k-1}(n)$

26.18 Counting Techniques

Let $A_1, A_2, \ldots, A_n$ be subsets of a set S that are not necessarily disjoint. Then the number of elements in the set $S \backslash (A_1 \cup A_2 \cup \cdots \cup A_n)$ is

26.18.1
$$|S \backslash (A_1 \cup A_2 \cup \cdots \cup A_n)| = |S| + \sum_{t=1}^{n} (-1)^t \sum_{1 \leq j_1 < j_2 < \cdots < j_t \leq n} |A_{j_1} \cap A_{j_2} \cap \cdots \cap A_{j_t}|.$$

Example 1

The number of positive integers $\leq N$ that are not divisible by any of the primes $p_1, p_2, \ldots, p_n$ (§27.2(i)) is

26.18.2
$$N + \sum_{t=1}^{n} (-1)^t \sum_{1 \leq j_1 < j_2 < \cdots < j_t \leq n} \left\lfloor \frac{N}{p_{j_1} p_{j_2} \cdots p_{j_t}} \right\rfloor.$$

Applications

Example 2

With the notation of §26.15, the number of placements of n nonattacking rooks on an $n \times n$ chessboard that avoid the squares in a specified subset B is

26.18.3 $$n! + \sum_{t=1}^{n} (-1)^t r_t(B)(n-t)!.$$

Example 3

The number of ways of placing n labeled objects into k labeled boxes so that at least one object is in each box is

26.18.4 $$k^n + \sum_{t=1}^{n} (-1)^t \binom{k}{t}(k-t)^n.$$

Note that this is also one of the counting problems for which a formula is given in Table 26.17.1. Elements of N are labeled, elements of K are labeled, and f is onto.

For further examples in the use of generating functions, see Stanley (1997, 1999) and Wilf (1994). See also Pólya et al. (1983).

Applications

26.19 Mathematical Applications

Combinatorics has applications to analysis, algebra, and geometry. Examples can be found in Beckenbach (1981), Billera et al. (1996), and Lovász et al. (1995). Partitions and plane partitions have applications to representation theory (Bressoud (1999), Macdonald (1995), and Sagan (2001)) and to special functions (Andrews et al. (1999) and Gasper and Rahman (2004)).

Other areas of combinatorial analysis include graph theory, coding theory, and combinatorial designs. These have applications in operations research, probability theory, and statistics. See Graham et al. (1995) and Rosen et al. (2000).

26.20 Physical Applications

An English translation of Pólya (1937) on applications of combinatorics to chemistry has been published as Pólya and Read (1987). Other articles on this subject are de Bruijn (1981) and Rouvray (1995). The latter reference also describes chemical applications of other combinatorial techniques.

Applications of combinatorics, especially integer and plane partitions, to counting lattice structures and other problems of statistical mechanics, of which the Ising model is the principal example, can be found in Montroll (1964), Godsil et al. (1995), Baxter (1982), and Korepin et al. (1993). For an application of statistical mechanics to combinatorics, see Bressoud (1999).

Other applications to problems in engineering, crystallography, biology, and computer science can be found in Beckenbach (1981) and Graham et al. (1995).

Computation

26.21 Tables

Abramowitz and Stegun (1964, Chapter 24) tabulates binomial coefficients $\binom{m}{n}$ for m up to 50 and n up to 25; extends Table 26.4.1 to $n=10$; tabulates Stirling numbers of the first and second kinds, $s(n,k)$ and $S(n,k)$, for n up to 25 and k up to n; tabulates partitions $p(n)$ and partitions into distinct parts $p(\mathcal{D},n)$ for n up to 500.

Andrews (1976) contains tables of the number of unrestricted partitions, partitions into odd parts, partitions into parts $\not\equiv \pm 2 \pmod 5$, partitions into parts $\not\equiv \pm 1 \pmod 5$, and unrestricted plane partitions up to 100. It also contains a table of Gaussian polynomials up to $\begin{bmatrix} 12 \\ 6 \end{bmatrix}_q$.

Goldberg et al. (1976) contains tables of binomial coefficients to $n=100$ and Stirling numbers to $n=40$.

26.22 Software

See http://dlmf.nist.gov/26.22.

References

General References

Comprehensive references include Graham et al. (1995) and Rosen et al. (2000). Most of this chapter is treated in detail in Comtet (1974), Riordan (1958), and Stanley (1997, 1999).

Sources

The following list gives the references or other indications of proofs that were used in constructing the various sections of this chapter. These sources supplement the references that are quoted in the text.

§26.2 Table 26.2.1 is from Abramowitz and Stegun (1964, Table 24.5).

§26.3 Comtet (1974, pp. 8–10, 22–23, 292), Riordan (1958, pp. 4–11), and (5.11.7). Tables 26.3.1 and 26.3.2 are from Abramowitz and Stegun (1964, Table 24.1).

§26.4 Comtet (1974, pp. 28–29). Table 26.4.1 is from Abramowitz and Stegun (1964, Table 24.2).

§26.5 Comtet (1974, pp. 52–54), Riordan (1979, p. 157). For (26.5.6) and (26.5.7) use (26.3.12) and (26.5.1). Table 26.5.1 was computed by the author.

§26.6 Comtet (1974, pp. 80–81), Stanley (1999, pp. 237–241). (26.6.4) is a consequence of André's reflection principle; see Comtet (1974, pp. 22–23). For (26.6.11) use (26.6.7). Tables 26.6.1–26.6.4 were computed by the author.

§26.7 Comtet (1974, pp. 210–211). For (26.7.3) see Wilf (1994, p. 22). For (26.7.7) see de Bruijn (1961, pp. 104–108) and Olver (1997b, pp. 329–331). Table 26.7.1 was computed by the author.

§26.8 Comtet (1974, pp. 206–216), Riordan (1979, pp. 195 and 203–227), Graham *et al.* (1994, pp. 264–265). For (26.8.31) use (26.8.37) and (26.8.39). For (26.8.40)–(26.8.43) see Jordan (1939, pp. 161–174) (as corrected here). Tables 26.8.1 and 26.8.2 are from Abramowitz and Stegun (1964, Tables 24.3 and 24.4).

§26.9 Andrews (1976, Chapter 6 and pp. 1–13, 36, 47, 81). For (26.9.9) see Bressoud (1999, p. 60, Eq. (2.23)). Table 26.9.1 was computed by the author.

§26.10 Andrews (1976, pp. 5, 11–12, 16–17, 19, 36, 82, 97, 104, 116), Bressoud (1999, pp. 60, 78–79). Table 26.10.1 was computed by the author.

§26.11 Andrews (1976, Chapter 4).

§26.12 Bressoud (1999, pp. 11, 13–18, 22, 57, 197–199 (with corrections)), Andrews (1976, p. 199), Andrews (1979, p. 195). Table 26.12.1 was computed by the author.

§26.13 Cameron (1994, pp. 77, 80–84), Stanley (1997, pp. 20–21, 67).

§26.14 Andrews (1976, pp. 39–42), Graham *et al.* (1994, pp. 267–272), Riordan (1958, pp. 38–39), Stanley (1997, pp. 20–23). For (26.14.10) and (26.14.11) use (26.14.4) and (24.2.1). Table 26.14.1 was computed by the author.

§26.15 Stanley (1997, pp. 71–76), Tucker (2006, pp. 335–345).

§26.16 Andrews (1976, pp. 39–45).

§26.18 Riordan (1958, pp. 50–65).

Chapter 27
Functions of Number Theory

T. M. Apostol[1]

Notation — 638
27.1 Special Notation 638

Multiplicative Number Theory — 638
27.2 Functions 638
27.3 Multiplicative Properties 640
27.4 Euler Products and Dirichlet Series ... 640
27.5 Inversion Formulas 641
27.6 Divisor Sums 641
27.7 Lambert Series as Generating Functions . 641
27.8 Dirichlet Characters 642
27.9 Quadratic Characters 642
27.10 Periodic Number-Theoretic Functions .. 642
27.11 Asymptotic Formulas: Partial Sums ... 643
27.12 Asymptotic Formulas: Primes 644

Additive Number Theory — 644
27.13 Functions 644
27.14 Unrestricted Partitions 645

Applications — 647
27.15 Chinese Remainder Theorem 647
27.16 Cryptography 647
27.17 Other Applications 647

Computation — 648
27.18 Methods of Computation: Primes 648
27.19 Methods of Computation: Factorization . 648
27.20 Methods of Computation: Other Number-Theoretic Functions 649
27.21 Tables 649
27.22 Software 649

References — 649

---

[1]California Institute of Technology, Pasadena, California.
Acknowledgments: The author thanks Basil Gordon for comments on an earlier draft, and David Bressoud for providing §§27.12, 27.18, 27.19, and 27.22.
Copyright © 2009 National Institute of Standards and Technology. All rights reserved.

Notation

27.1 Special Notation

(For other notation see pp. xiv and 873.)

d, k, m, n	positive integers (unless otherwise indicated).
$d \mid n$	d divides n.
(m, n)	greatest common divisor of m, n. If $(m, n) = 1$, m and n are called relatively prime, or coprime.
$(d_1, \ldots, d_n)$	greatest common divisor of $d_1, \ldots, d_n$.
$\sum_{d\mid n}, \prod_{d\mid n}$	sum, product taken over divisors of n.
$\sum_{(m,n)=1}$	sum taken over m, $1 \leq m \leq n$ and m relatively prime to n.
$p, p_1, p_2, \ldots$	prime numbers (or primes): integers (> 1) with only two positive integer divisors, 1 and the number itself.
$\sum_p, \prod_p$	sum, product extended over all primes.
x, y	real numbers.
$\sum_{n \leq x}$	$\sum_{n=1}^{\lfloor x \rfloor}$.
$\log x$	natural logarithm of x, written as $\ln x$ in other chapters.
$\zeta(s)$	Riemann zeta function; see §25.2(i).
$(n\mid P)$	Jacobi symbol; see §27.9.
$(n\mid p)$	Legendre symbol; see §27.9.

Multiplicative Number Theory

27.2 Functions

27.2(i) Definitions

Functions in this section derive their properties from the *fundamental theorem of arithmetic*, which states that every integer $n > 1$ can be represented uniquely as a product of prime powers,

$$27.2.1 \qquad n = \prod_{r=1}^{\nu(n)} p_r^{a_r},$$

where $p_1, p_2, \ldots, p_{\nu(n)}$ are the distinct prime factors of n, each exponent a_r is positive, and $\nu(n)$ is the number of distinct primes dividing n. ($\nu(1)$ is defined to be 0.) Euclid's Elements (Euclid (1908, Book IX, Proposition 20)) gives an elegant proof that there are infinitely many primes. Tables of primes (§27.21) reveal great irregularity in their distribution. They tend to thin out among the large integers, but this thinning out is not completely regular. There is great interest in the function $\pi(x)$ that counts the number of primes not exceeding x. It can be expressed as a sum over all primes $p \leq x$:

$$27.2.2 \qquad \pi(x) = \sum_{p \leq x} 1.$$

Gauss and Legendre conjectured that $\pi(x)$ is asymptotic to $x/\log x$ as $x \to \infty$:

$$27.2.3 \qquad \pi(x) \sim \frac{x}{\log x}.$$

(See Gauss (1863, Band II, pp. 437–477) and Legendre (1808, p. 394).)

This result, first proved in Hadamard (1896) and de la Vallée Poussin (1896a,b), is known as the *prime number theorem*. An equivalent form states that the nth prime p_n (when the primes are listed in increasing order) is asymptotic to $n \log n$ as $n \to \infty$:

$$27.2.4 \qquad p_n \sim n \log n.$$

(See also §27.12.) Other examples of number-theoretic functions treated in this chapter are as follows.

$$27.2.5 \qquad \left\lfloor \frac{1}{n} \right\rfloor = \begin{cases} 1, & n = 1, \\ 0, & n > 1. \end{cases}$$

$$27.2.6 \qquad \phi_k(n) = \sum_{(m,n)=1} m^k,$$

the sum of the kth powers of the positive integers $m \leq n$ that are relatively prime to n.

$$27.2.7 \qquad \phi(n) = \phi_0(n).$$

This is the number of positive integers $\leq n$ that are relatively prime to n; $\phi(n)$ is *Euler's totient*.

If $(a, n) = 1$, then the *Euler–Fermat theorem* states that

$$27.2.8 \qquad a^{\phi(n)} \equiv 1 \pmod{n},$$

and if $\phi(n)$ is the smallest positive integer f such that $a^f \equiv 1 \pmod{n}$, then a is a *primitive root* mod n. The $\phi(n)$ numbers $a, a^2, \ldots, a^{\phi(n)}$ are relatively prime to n and distinct (mod n). Such a set is a *reduced residue system* modulo n.

$$27.2.9 \qquad d(n) = \sum_{d\mid n} 1$$

is the number of divisors of n and is the *divisor function*. It is the special case $k = 2$ of the function $d_k(n)$ that counts the number of ways of expressing n as the product of k factors, with the order of factors taken into account.

$$27.2.10 \qquad \sigma_\alpha(n) = \sum_{d\mid n} d^\alpha,$$

is the sum of the αth powers of the divisors of n, where the exponent α can be real or complex. Note that $\sigma_0(n) = d(n)$.

$$27.2.11 \qquad J_k(n) = \sum_{((d_1,\ldots,d_k),n)=1} 1,$$

27.2 Functions

is the number of k-tuples of integers $\leq n$ whose greatest common divisor is relatively prime to n. This is *Jordan's function*. Note that $J_1(n) = \phi(n)$.

In the following examples, $a_1, \ldots, a_{\nu(n)}$ are the exponents in the factorization of n in (27.2.1).

27.2.12 $\quad \mu(n) = \begin{cases} 1, & n = 1, \\ (-1)^{\nu(n)}, & a_1 = a_2 = \cdots = a_{\nu(n)} = 1, \\ 0, & \text{otherwise}. \end{cases}$

This is the *Möbius function*.

27.2.13 $\quad \lambda(n) = \begin{cases} 1, & n = 1, \\ (-1)^{a_1 + \cdots + a_{\nu(n)}}, & n > 1. \end{cases}$

This is *Liouville's function*.

27.2.14 $\quad \Lambda(n) = \log p, \qquad n = p^a,$

where p^a is a prime power with $a \geq 1$; otherwise $\Lambda(n) = 0$. This is *Mangoldt's function*.

27.2(ii) Tables

Table 27.2.1 lists the first 100 prime numbers p_n. Table 27.2.2 tabulates the Euler totient function $\phi(n)$, the divisor function $d(n)$ $(= \sigma_0(n))$, and the sum of the divisors $\sigma(n)$ $(= \sigma_1(n))$, for $n = 1(1)52$.

Table 27.2.1: Primes.

n	p_n	p_{n+10}	p_{n+20}	p_{n+30}	p_{n+40}	p_{n+50}	p_{n+60}	p_{n+70}	p_{n+80}	p_{n+90}
1	2	31	73	127	179	233	283	353	419	467
2	3	37	79	131	181	239	293	359	421	479
3	5	41	83	137	191	241	307	367	431	487
4	7	43	89	139	193	251	311	373	433	491
5	11	47	97	149	197	257	313	379	439	499
6	13	53	101	151	199	263	317	383	443	503
7	17	59	103	157	211	269	331	389	449	509
8	19	61	107	163	223	271	337	397	457	521
9	23	67	109	167	227	277	347	401	461	523
10	29	71	113	173	229	281	349	409	463	541

Table 27.2.2: Functions related to division.

n	$\phi(n)$	$d(n)$	$\sigma(n)$	n	$\phi(n)$	$d(n)$	$\sigma(n)$	n	$\phi(n)$	$d(n)$	$\sigma(n)$	n	$\phi(n)$	$d(n)$	$\sigma(n)$
1	1	1	1	14	6	4	24	27	18	4	40	40	16	8	90
2	1	2	3	15	8	4	24	28	12	6	56	41	40	2	42
3	2	2	4	16	8	5	31	29	28	2	30	42	12	8	96
4	2	3	7	17	16	2	18	30	8	8	72	43	42	2	44
5	4	2	6	18	6	6	39	31	30	2	32	44	20	6	84
6	2	4	12	19	18	2	20	32	16	6	63	45	24	6	78
7	6	2	8	20	8	6	42	33	20	4	48	46	22	4	72
8	4	4	15	21	12	4	32	34	16	4	54	47	46	2	48
9	6	3	13	22	10	4	36	35	24	4	48	48	16	10	124
10	4	4	18	23	22	2	24	36	12	9	91	49	42	3	57
11	10	2	12	24	8	8	60	37	36	2	38	50	20	6	93
12	4	6	28	25	20	3	31	38	18	4	60	51	32	4	72
13	12	2	14	26	12	4	42	39	24	4	56	52	24	6	98

27.3 Multiplicative Properties

Except for $\nu(n)$, $\Lambda(n)$, p_n, and $\pi(x)$, the functions in §27.2 are *multiplicative*, which means $f(1) = 1$ and

27.3.1 $$f(mn) = f(m)f(n), \quad (m,n) = 1.$$

If f is multiplicative, then the values $f(n)$ for $n > 1$ are determined by the values at the prime powers. Specifically, if n is factored as in (27.2.1), then

27.3.2 $$f(n) = \prod_{r=1}^{\nu(n)} f(p_r^{a_r}).$$

In particular,

27.3.3 $$\phi(n) = n \prod_{p|n}(1 - p^{-1}),$$

27.3.4 $$J_k(n) = n^k \prod_{p|n}(1 - p^{-k}),$$

27.3.5 $$d(n) = \prod_{r=1}^{\nu(n)}(1 + a_r),$$

27.3.6 $$\sigma_\alpha(n) = \prod_{r=1}^{\nu(n)} \frac{p_r^{\alpha(1+a_r)} - 1}{p_r^\alpha - 1}, \quad \alpha \neq 0.$$

Related multiplicative properties are

27.3.7 $$\sigma_\alpha(m)\sigma_\alpha(n) = \sum_{d|(m,n)} d^\alpha \sigma_\alpha\left(\frac{mn}{d^2}\right),$$

27.3.8 $$\phi(m)\phi(n) = \phi(mn)\phi((m,n))/(m,n).$$

A function f is *completely multiplicative* if $f(1) = 1$ and

27.3.9 $$f(mn) = f(m)f(n), \quad m,n = 1,2,\ldots.$$

Examples are $\lfloor 1/n \rfloor$ and $\lambda(n)$, and the Dirichlet characters, defined in §27.8.

If f is completely multiplicative, then (27.3.2) becomes

27.3.10 $$f(n) = \prod_{r=1}^{\nu(n)}(f(p_r))^{a_r}.$$

27.4 Euler Products and Dirichlet Series

The fundamental theorem of arithmetic is linked to analysis through the concept of the Euler product. Every multiplicative f satisfies the identity

27.4.1 $$\sum_{n=1}^{\infty} f(n) = \prod_p \left(1 + \sum_{r=1}^{\infty} f(p^r)\right),$$

if the series on the left is absolutely convergent. In this case the infinite product on the right (extended over all primes p) is also absolutely convergent and is called the *Euler product* of the series. If $f(n)$ is completely multiplicative, then each factor in the product is a geometric series and the Euler product becomes

27.4.2 $$\sum_{n=1}^{\infty} f(n) = \prod_p (1 - f(p))^{-1}.$$

Euler products are used to find series that generate many functions of multiplicative number theory. The completely multiplicative function $f(n) = n^{-s}$ gives the Euler product representation of the *Riemann zeta function* $\zeta(s)$ (§25.2(i)):

27.4.3 $$\zeta(s) = \sum_{n=1}^{\infty} n^{-s} = \prod_p (1 - p^{-s})^{-1}, \quad \Re s > 1.$$

The Riemann zeta function is the prototype of series of the form

27.4.4 $$F(s) = \sum_{n=1}^{\infty} f(n) n^{-s},$$

called *Dirichlet series* with coefficients $f(n)$. The function $F(s)$ is a *generating function*, or more precisely, a *Dirichlet generating function*, for the coefficients. The following examples have generating functions related to the zeta function:

27.4.5 $$\sum_{n=1}^{\infty} \mu(n) n^{-s} = \frac{1}{\zeta(s)}, \quad \Re s > 1,$$

27.4.6 $$\sum_{n=1}^{\infty} \phi(n) n^{-s} = \frac{\zeta(s-1)}{\zeta(s)}, \quad \Re s > 2,$$

27.4.7 $$\sum_{n=1}^{\infty} \lambda(n) n^{-s} = \frac{\zeta(2s)}{\zeta(s)}, \quad \Re s > 1,$$

27.4.8 $$\sum_{n=1}^{\infty} |\mu(n)| n^{-s} = \frac{\zeta(s)}{\zeta(2s)}, \quad \Re s > 1,$$

27.4.9 $$\sum_{n=1}^{\infty} 2^{\nu(n)} n^{-s} = \frac{(\zeta(s))^2}{\zeta(2s)}, \quad \Re s > 1,$$

27.4.10 $$\sum_{n=1}^{\infty} d_k(n) n^{-s} = (\zeta(s))^k, \quad \Re s > 1,$$

27.4.11 $$\sum_{n=1}^{\infty} \sigma_\alpha(n) n^{-s} = \zeta(s)\zeta(s-\alpha), \quad \Re s > \max(1, 1 + \Re\alpha),$$

27.4.12 $$\sum_{n=1}^{\infty} \Lambda(n) n^{-s} = -\frac{\zeta'(s)}{\zeta(s)}, \quad \Re s > 1,$$

27.4.13 $$\sum_{n=2}^{\infty} (\log n) n^{-s} = -\zeta'(s), \quad \Re s > 1.$$

In (27.4.12) and (27.4.13) $\zeta'(s)$ is the derivative of $\zeta(s)$.

27.5 Inversion Formulas

If a Dirichlet series $F(s)$ generates $f(n)$, and $G(s)$ generates $g(n)$, then the product $F(s)G(s)$ generates

27.5.1 $$h(n) = \sum_{d|n} f(d) g\left(\frac{n}{d}\right),$$

called the *Dirichlet product* (or *convolution*) of f and g. The set of all number-theoretic functions f with $f(1) \neq 0$ forms an abelian group under Dirichlet multiplication, with the function $\lfloor 1/n \rfloor$ in (27.2.5) as identity element; see Apostol (1976, p. 129). The multiplicative functions are a subgroup of this group. Generating functions yield many relations connecting number-theoretic functions. For example, the equation $\zeta(s) \cdot (1/\zeta(s)) = 1$ is equivalent to the identity

27.5.2 $$\sum_{d|n} \mu(d) = \left\lfloor \frac{1}{n} \right\rfloor,$$

which, in turn, is the basis for the *Möbius inversion formula* relating sums over divisors:

27.5.3 $$g(n) = \sum_{d|n} f(d) \iff f(n) = \sum_{d|n} g(d) \mu\left(\frac{n}{d}\right).$$

Special cases of Möbius inversion pairs are:

27.5.4 $$n = \sum_{d|n} \phi(d) \iff \phi(n) = \sum_{d|n} d\, \mu\left(\frac{n}{d}\right),$$

27.5.5 $$\log n = \sum_{d|n} \Lambda(d) \iff \Lambda(n) = \sum_{d|n} (\log d)\, \mu\left(\frac{n}{d}\right).$$

Other types of Möbius inversion formulas include:

27.5.6 $$G(x) = \sum_{n \leq x} F\left(\frac{x}{n}\right) \iff F(x) = \sum_{n \leq x} \mu(n) G\left(\frac{x}{n}\right),$$

27.5.7 $$G(x) = \sum_{m=1}^{\infty} \frac{F(mx)}{m^s} \iff F(x) = \sum_{m=1}^{\infty} \mu(m) \frac{G(mx)}{m^s},$$

27.5.8 $$g(n) = \prod_{d|n} f(d) \iff f(n) = \prod_{d|n} \left(g\left(\frac{n}{d}\right)\right)^{\mu(d)}.$$

For a general theory of Möbius inversion with applications to combinatorial theory see Rota (1964).

27.6 Divisor Sums

Sums of number-theoretic functions extended over divisors are of special interest. For example,

27.6.1 $$\sum_{d|n} \lambda(d) = \begin{cases} 1, & n \text{ is a square,} \\ 0, & \text{otherwise.} \end{cases}$$

If f is multiplicative, then

27.6.2 $$\sum_{d|n} \mu(d) f(d) = \prod_{p|n}(1 - f(p)), \qquad n > 1.$$

Generating functions, Euler products, and Möbius inversion are used to evaluate many sums extended over divisors. Examples include:

27.6.3 $$\sum_{d|n} |\mu(d)| = 2^{\nu(n)},$$

27.6.4 $$\sum_{d^2|n} \mu(d) = |\mu(n)|,$$

27.6.5 $$\sum_{d|n} \frac{|\mu(d)|}{\phi(d)} = \frac{n}{\phi(n)},$$

27.6.6 $$\sum_{d|n} \phi_k(d) \left(\frac{n}{d}\right)^k = 1^k + 2^k + \cdots + n^k,$$

27.6.7 $$\sum_{d|n} \mu(d) \left(\frac{n}{d}\right)^k = J_k(n),$$

27.6.8 $$\sum_{d|n} J_k(d) = n^k.$$

27.7 Lambert Series as Generating Functions

Lambert series have the form

27.7.1 $$\sum_{n=1}^{\infty} f(n) \frac{x^n}{1 - x^n}.$$

If $|x| < 1$, then the quotient $x^n/(1-x^n)$ is the sum of a geometric series, and when the series (27.7.1) converges absolutely it can be rearranged as a power series:

27.7.2 $$\sum_{n=1}^{\infty} f(n) \frac{x^n}{1 - x^n} = \sum_{n=1}^{\infty} \sum_{d|n} f(d) x^n.$$

Again with $|x| < 1$, special cases of (27.7.2) include:

27.7.3 $$\sum_{n=1}^{\infty} \mu(n) \frac{x^n}{1 - x^n} = x,$$

27.7.4 $$\sum_{n=1}^{\infty} \phi(n) \frac{x^n}{1 - x^n} = \frac{x}{(1 - x)^2},$$

27.7.5 $$\sum_{n=1}^{\infty} n^\alpha \frac{x^n}{1 - x^n} = \sum_{n=1}^{\infty} \sigma_\alpha(n) x^n,$$

27.7.6 $$\sum_{n=1}^{\infty} \lambda(n) \frac{x^n}{1 - x^n} = \sum_{n=1}^{\infty} x^{n^2}.$$

27.8 Dirichlet Characters

If k (> 1) is a given integer, then a function $\chi(n)$ is called a *Dirichlet character* (mod k) if it is completely multiplicative, periodic with period k, and vanishes when $(n,k) > 1$. In other words, Dirichlet characters (mod k) satisfy the four conditions:

27.8.1 $\quad \chi(1) = 1,$

27.8.2 $\quad \chi(mn) = \chi(m)\,\chi(n), \qquad m,n = 1,2,\ldots,$

27.8.3 $\quad \chi(n+k) = \chi(n), \qquad n = 1,2,\ldots,$

27.8.4 $\quad \chi(n) = 0, \qquad (n,k) > 1.$

An example is the *principal character* (mod k):

27.8.5 $\quad \chi_1(n) = \begin{cases} 1, & (n,k) = 1, \\ 0, & (n,k) > 1. \end{cases}$

For any character χ (mod k), $\chi(n) \neq 0$ if and only if $(n,k) = 1$, in which case the Euler–Fermat theorem (27.2.8) implies $(\chi(n))^{\phi(k)} = 1$. There are exactly $\phi(k)$ different characters (mod k), which can be labeled as $\chi_1, \ldots, \chi_{\phi(k)}$. If χ is a character (mod k), so is its complex conjugate $\overline{\chi}$. If $(n,k) = 1$, then the characters satisfy the *orthogonality relation*

27.8.6 $\quad \displaystyle\sum_{r=1}^{\phi(k)} \chi_r(m)\overline{\chi}_r(n) = \begin{cases} \phi(k), & m \equiv n \pmod{k}, \\ 0, & \text{otherwise}. \end{cases}$

A Dirichlet character χ (mod k) is called *primitive* (mod k) if for every proper divisor d of k (that is, a divisor $d < k$), there exists an integer $a \equiv 1 \pmod{d}$, with $(a,k) = 1$ and $\chi(a) \neq 1$. If k is prime, then every nonprincipal character χ (mod k) is primitive. A divisor d of k is called an *induced modulus* for χ if

27.8.7 $\quad \chi(a) = 1 \text{ for all } a \equiv 1 \pmod{d}, \ (a,k) = 1.$

Every Dirichlet character χ (mod k) is a product

27.8.8 $\quad \chi(n) = \chi_0(n)\,\chi_1(n),$

where χ_0 is a character (mod d) for some induced modulus d for χ, and χ_1 is the principal character (mod k). A character is *real* if all its values are real. If k is odd, then the real characters (mod k) are the principal character and the quadratic characters described in the next section.

27.9 Quadratic Characters

For an odd prime p, the *Legendre symbol* $(n|p)$ is defined as follows. If p divides n, then the value of $(n|p)$ is 0. If p does not divide n, then $(n|p)$ has the value 1 when the quadratic congruence $x^2 \equiv n \pmod{p}$ has a solution, and the value -1 when this congruence has no solution. The Legendre symbol $(n|p)$, as a function of n, is a Dirichlet character (mod p). It is sometimes written as $\left(\frac{n}{p}\right)$. Special values include:

27.9.1 $\quad (-1|p) = (-1)^{(p-1)/2},$

27.9.2 $\quad (2|p) = (-1)^{(p^2-1)/8}.$

If p,q are distinct odd primes, then the *quadratic reciprocity law* states that

27.9.3 $\quad (p|q)\,(q|p) = (-1)^{(p-1)(q-1)/4}.$

If an odd integer P has prime factorization $P = \prod_{r=1}^{\nu(n)} p_r^{a_r}$, then the *Jacobi symbol* $(n|P)$ is defined by $(n|P) = \prod_{r=1}^{\nu(n)} (n|p_r)^{a_r}$, with $(n|1) = 1$. The Jacobi symbol $(n|P)$ is a Dirichlet character (mod P). Both (27.9.1) and (27.9.2) are valid with p replaced by P; the reciprocity law (27.9.3) holds if p,q are replaced by any two relatively prime odd integers P,Q.

27.10 Periodic Number-Theoretic Functions

If k is a fixed positive integer, then a number-theoretic function f is *periodic* (mod k) if

27.10.1 $\quad f(n+k) = f(n), \qquad n = 1,2,\ldots.$

Examples are the Dirichlet characters (mod k) and the greatest common divisor (n,k) regarded as a function of n.

Every function periodic (mod k) can be expressed as a *finite Fourier series* of the form

27.10.2 $\quad \displaystyle f(n) = \sum_{m=1}^{k} g(m) e^{2\pi i m n/k},$

where $g(m)$ is also periodic (mod k), and is given by

27.10.3 $\quad \displaystyle g(m) = \frac{1}{k}\sum_{n=1}^{k} f(n) e^{-2\pi i m n/k}.$

An example is *Ramanujan's sum*:

27.10.4 $\quad \displaystyle c_k(n) = \sum_{m=1}^{k} \chi_1(m) e^{2\pi i m n/k},$

where χ_1 is the principal character (mod k). This is the sum of the nth powers of the primitive kth roots of unity. It can also be expressed in terms of the Möbius function as a divisor sum:

27.10.5 $\quad \displaystyle c_k(n) = \sum_{d | (n,k)} d\,\mu\!\left(\frac{k}{d}\right).$

More generally, if f and g are arbitrary, then the sum

27.10.6 $\quad \displaystyle s_k(n) = \sum_{d | (n,k)} f(d) g\!\left(\frac{k}{d}\right)$

is a periodic function of n (mod k) and has the finite Fourier-series expansion

$$27.10.7 \qquad s_k(n) = \sum_{m=1}^{k} a_k(m) e^{2\pi i m n/k},$$

where

$$27.10.8 \qquad a_k(m) = \sum_{d|(m,k)} g(d) f\left(\frac{k}{d}\right) \frac{d}{k}.$$

Another generalization of Ramanujan's sum is the *Gauss sum* $G(n,\chi)$ associated with a Dirichlet character χ (mod k). It is defined by the relation

$$27.10.9 \qquad G(n,\chi) = \sum_{m=1}^{k} \chi(m) e^{2\pi i m n/k}.$$

In particular, $G(n,\chi_1) = c_k(n)$.

$G(n,\chi)$ is *separable* for some n if

$$27.10.10 \qquad G(n,\chi) = \overline{\chi}(n) G(1,\chi).$$

For any Dirichlet character χ (mod k), $G(n,\chi)$ is separable for n if $(n,k) = 1$, and is separable for every n if and only if $G(n,\chi) = 0$ whenever $(n,k) > 1$. For a primitive character χ (mod k), $G(n,\chi)$ is separable for every n, and

$$27.10.11 \qquad |G(1,\chi)|^2 = k.$$

Conversely, if $G(n,\chi)$ is separable for every n, then χ is primitive (mod k).

The finite Fourier expansion of a primitive Dirichlet character χ (mod k) has the form

$$27.10.12 \qquad \chi(n) = \frac{G(1,\chi)}{k} \sum_{m=1}^{k} \overline{\chi}(m) e^{-2\pi i m n/k}.$$

27.11 Asymptotic Formulas: Partial Sums

The behavior of a number-theoretic function $f(n)$ for large n is often difficult to determine because the function values can fluctuate considerably as n increases. It is more fruitful to study partial sums and seek asymptotic formulas of the form

$$27.11.1 \qquad \sum_{n \leq x} f(n) = F(x) + O(g(x)),$$

where $F(x)$ is a known function of x, and $O(g(x))$ represents the error, a function of smaller order than $F(x)$ for all x in some prescribed range. For example, Dirichlet (1849) proves that for all $x \geq 1$,

$$27.11.2 \qquad \sum_{n \leq x} d(n) = x \log x + (2\gamma - 1)x + O(\sqrt{x}),$$

where γ is Euler's constant (§5.2(ii)). *Dirichlet's divisor problem* (unsolved in 2009) is to determine the least number θ_0 such that the error term in (27.11.2) is $O(x^\theta)$ for all $\theta > \theta_0$. Kolesnik (1969) proves that $\theta_0 \leq \frac{12}{37}$.

Equations (27.11.3)–(27.11.11) list further asymptotic formulas related to some of the functions listed in §27.2. They are valid for all $x \geq 2$. The error terms given here are not necessarily the best known.

$$27.11.3 \qquad \sum_{n \leq x} \frac{d(n)}{n} = \frac{1}{2}(\log x)^2 + 2\gamma \log x + O(1),$$

where γ again is Euler's constant.

$$27.11.4 \qquad \sum_{n \leq x} \sigma_1(n) = \frac{\pi^2}{12} x^2 + O(x \log x).$$

$$27.11.5 \qquad \sum_{n \leq x} \sigma_\alpha(n) = \frac{\zeta(\alpha+1)}{\alpha+1} x^{\alpha+1} + O(x^\beta),$$
$$\alpha > 0, \, \alpha \neq 1, \, \beta = \max(1,\alpha).$$

$$27.11.6 \qquad \sum_{n \leq x} \phi(n) = \frac{3}{\pi^2} x^2 + O(x \log x).$$

$$27.11.7 \qquad \sum_{n \leq x} \frac{\phi(n)}{n} = \frac{6}{\pi^2} x + O(\log x).$$

$$27.11.8 \qquad \sum_{p \leq x} \frac{1}{p} = \log \log x + A + O\left(\frac{1}{\log x}\right),$$

where A is a constant.

$$27.11.9 \qquad \sum_{\substack{p \leq x \\ p \equiv h \,(\mathrm{mod}\, k)}} \frac{1}{p} = \frac{1}{\phi(k)} \log \log x + B + O\left(\frac{1}{\log x}\right),$$

where $(h,k) = 1$, $k > 0$, and B is a constant depending on h and k.

$$27.11.10 \qquad \sum_{p \leq x} \frac{\log p}{p} = \log x + O(1).$$

$$27.11.11 \qquad \sum_{\substack{p \leq x \\ p \equiv h \,(\mathrm{mod}\, k)}} \frac{\log p}{p} = \frac{1}{\phi(k)} \log x + O(1),$$

where $(h,k) = 1$, $k > 0$.

Letting $x \to \infty$ in (27.11.9) or in (27.11.11) we see that there are infinitely many primes $p \equiv h$ (mod k) if h, k are coprime; this is *Dirichlet's theorem on primes in arithmetic progressions*.

$$27.11.12 \qquad \sum_{n \leq x} \mu(n) = O\left(x e^{-C\sqrt{\log x}}\right), \qquad x \to \infty,$$

for some positive constant C,

$$27.11.13 \qquad \lim_{x \to \infty} \frac{1}{x} \sum_{n \leq x} \mu(n) = 0,$$

$$27.11.14 \qquad \lim_{x \to \infty} \sum_{n \leq x} \frac{\mu(n)}{n} = 0,$$

$$27.11.15 \qquad \lim_{x \to \infty} \sum_{n \leq x} \frac{\mu(n) \log n}{n} = -1.$$

Each of (27.11.13)–(27.11.15) is equivalent to the prime number theorem (27.2.3). The *prime number theorem for arithmetic progressions*—an extension of (27.2.3) and first proved in de la Vallée Poussin (1896a,b)—states that if $(h, k) = 1$, then the number of primes $p \leq x$ with $p \equiv h \pmod{k}$ is asymptotic to $x/(\phi(k) \log x)$ as $x \to \infty$.

27.12 Asymptotic Formulas: Primes

p_n is the nth prime, beginning with $p_1 = 2$. $\pi(x)$ is the number of primes less than or equal to x.

27.12.1
$$\lim_{n \to \infty} \frac{p_n}{n \log n} = 1,$$

27.12.2
$$p_n > n \log n, \qquad n = 1, 2, \ldots.$$

27.12.3
$$\pi(x) = \lfloor x \rfloor - 1 - \sum_{p_j \leq \sqrt{x}} \left\lfloor \frac{x}{p_j} \right\rfloor$$
$$+ \sum_{r \geq 2} (-1)^r \sum_{p_{j_1} < p_{j_2} < \cdots < p_{j_r} \leq \sqrt{x}} \left\lfloor \frac{x}{p_{j_1} p_{j_2} \cdots p_{j_r}} \right\rfloor,$$
$$x \geq 1,$$

where the series terminates when the product of the first r primes exceeds x.

As $x \to \infty$

27.12.4
$$\pi(x) \sim \sum_{k=1}^{\infty} \frac{(k-1)! \, x}{(\log x)^k}.$$

Prime Number Theorem

There exists a positive constant c such that

27.12.5
$$|\pi(x) - \mathrm{li}(x)| = O\!\left(x \exp\!\left(-c\sqrt{\log x}\right)\right), \quad x \to \infty.$$

For the logarithmic integral $\mathrm{li}(x)$ see (6.2.8). The best available asymptotic error estimate (2009) appears in Korobov (1958) and Vinogradov (1958): there exists a positive constant d such that

27.12.6
$$|\pi(x) - \mathrm{li}(x)| = O\!\left(x \exp\!\left(-d(\log x)^{3/5} (\log \log x)^{-1/5}\right)\right).$$

$\pi(x) - \mathrm{li}(x)$ changes sign infinitely often as $x \to \infty$; see Littlewood (1914), Bays and Hudson (2000).

The *Riemann hypothesis* (§25.10(i)) is equivalent to the statement that for every $x \geq 2657$,

27.12.7
$$|\pi(x) - \mathrm{li}(x)| < \frac{1}{8\pi} \sqrt{x} \log x.$$

If a is relatively prime to the modulus m, then there are infinitely many primes congruent to $a \pmod{m}$.

The number of such primes not exceeding x is

27.12.8
$$\frac{x}{\phi(m)} + O\!\left(x \exp\!\left(-\lambda(\alpha)(\log x)^{1/2}\right)\right),$$
$$m \leq (\log x)^\alpha, \; \alpha > 0,$$

where $\lambda(\alpha)$ depends only on α, and $\phi(m)$ is the Euler totient function (§27.2).

A *Mersenne prime* is a prime of the form $2^p - 1$. The largest known prime (2009) is the Mersenne prime $2^{43,112,609} - 1$. For current records online, see http://dlmf.nist.gov/27.12.

A *pseudoprime test* is a test that correctly identifies most composite numbers. For example, if $2^n \not\equiv 2 \pmod{n}$, then n is composite. Descriptions and comparisons of pseudoprime tests are given in Bressoud and Wagon (2000, §§2.4, 4.2, and 8.2) and Crandall and Pomerance (2005, §§3.4–3.6).

A *Carmichael number* is a composite number n for which $b^n \equiv b \pmod{n}$ for all $b \in \mathbb{N}$. There are infinitely many Carmichael numbers.

Additive Number Theory

27.13 Functions

27.13(i) Introduction

Whereas multiplicative number theory is concerned with functions arising from prime factorization, additive number theory treats functions related to addition of integers. The basic problem is that of expressing a given positive integer n as a sum of integers from some prescribed set S whose members are primes, squares, cubes, or other special integers. Each representation of n as a sum of elements of S is called a *partition* of n, and the number $S(n)$ of such partitions is often of great interest. The subsections that follow describe problems from additive number theory. See also Apostol (1976, Chapter 14) and Apostol and Niven (1994, pp. 33–34).

27.13(ii) Goldbach Conjecture

Every even integer $n > 4$ is the sum of two odd primes. In this case, $S(n)$ is the number of solutions of the equation $n = p + q$, where p and q are odd primes. Goldbach's assertion is that $S(n) \geq 1$ for all even $n > 4$. This conjecture dates back to 1742 and was undecided in 2009, although it has been confirmed numerically up to very large numbers. Vinogradov (1937) proves that every sufficiently large odd integer is the sum of three odd primes, and Chen (1966) shows that every sufficiently large even integer is the sum of a prime and a number with no more than two prime factors.

For an online account of the current status of Goldbach's conjecture see http://dlmf.nist.gov/27.13.ii.

27.13(iii) Waring's Problem

This problem is named after Edward Waring who, in 1770, stated without proof and with limited numerical evidence, that every positive integer n is the sum of four squares, of nine cubes, of nineteen fourth powers, and so on. Waring's problem is to find, for each positive integer k, whether there is an integer m (depending only on k) such that the equation

$$27.13.1 \qquad n = x_1^k + x_2^k + \cdots + x_m^k$$

has nonnegative integer solutions for all $n \geq 1$. The smallest m that exists for a given k is denoted by $g(k)$. Similarly, $G(k)$ denotes the smallest m for which (27.13.1) has nonnegative integer solutions for all sufficiently large n.

Lagrange (1770) proves that $g(2) = 4$, and during the next 139 years the existence of $g(k)$ was shown for $k = 3, 4, 5, 6, 7, 8, 10$. Hilbert (1909) proves the existence of $g(k)$ for every k but does not determine its corresponding numerical value. The exact value of $g(k)$ is now known for every $k \leq 200{,}000$. For example, $g(3) = 9$, $g(4) = 19$, $g(5) = 37$, $g(6) = 73$, $g(7) = 143$, and $g(8) = 279$. A general formula states that

$$27.13.2 \qquad g(k) \geq 2^k + \left\lfloor \frac{3^k}{2^k} \right\rfloor - 2,$$

for all $k \geq 2$, with equality if $4 \leq k \leq 200{,}000$. If $3^k = q 2^k + r$ with $0 < r < 2^k$, then equality holds in (27.13.2) provided $r + q \leq 2^k$, a condition that is satisfied with at most a finite number of exceptions.

The existence of $G(k)$ follows from that of $g(k)$ because $G(k) \leq g(k)$, but only the values $G(2) = 4$ and $G(4) = 16$ are known exactly. Some upper bounds smaller than $g(k)$ are known. For example, $G(3) \leq 7$, $G(5) \leq 23$, $G(6) \leq 36$, $G(7) \leq 53$, and $G(8) \leq 73$. Hardy and Littlewood (1925) conjectures that $G(k) < 2k + 1$ when k is not a power of 2, and that $G(k) \leq 4k$ when k is a power of 2, but the most that is known (in 2009) is $G(k) < ck \log k$ for some constant c. A survey is given in Ellison (1971).

27.13(iv) Representation by Squares

For a given integer $k \geq 2$ the function $r_k(n)$ is defined as the number of solutions of the equation

$$27.13.3 \qquad n = x_1^2 + x_2^2 + \cdots + x_k^2,$$

where the x_j are integers, positive, negative, or zero, and the order of the summands is taken into account.

Jacobi (1829) notes that $r_2(n)$ is the coefficient of x^n in the square of the theta function $\vartheta(x)$:

$$27.13.4 \qquad \vartheta(x) = 1 + 2 \sum_{m=1}^{\infty} x^{m^2}, \qquad |x| < 1.$$

(In §20.2(i), $\vartheta(x)$ is denoted by $\theta_3(0, x)$.) Thus,

$$27.13.5 \qquad (\vartheta(x))^2 = 1 + \sum_{n=1}^{\infty} r_2(n) x^n.$$

One of Jacobi's identities implies that

$$27.13.6 \qquad (\vartheta(x))^2 = 1 + 4 \sum_{n=1}^{\infty} (\delta_1(n) - \delta_3(n)) x^n,$$

where $\delta_1(n)$ and $\delta_3(n)$ are the number of divisors of n congruent respectively to 1 and 3 (mod 4), and by equating coefficients in (27.13.5) and (27.13.6) Jacobi deduced that

$$27.13.7 \qquad r_2(n) = 4 \left(\delta_1(n) - \delta_3(n) \right).$$

Hence $r_2(5) = 8$ because both divisors, 1 and 5, are congruent to 1 (mod 4). In fact, there are four representations, given by $5 = 2^2 + 1^2 = 2^2 + (-1)^2 = (-2)^2 + 1^2 = (-2)^2 + (-1)^2$, and four more with the order of summands reversed.

By similar methods Jacobi proved that $r_4(n) = 8\sigma_1(n)$ if n is odd, whereas, if n is even, $r_4(n) = 24$ times the sum of the odd divisors of n. Mordell (1917) notes that $r_k(n)$ is the coefficient of x^n in the power-series expansion of the kth power of the series for $\vartheta(x)$. Explicit formulas for $r_k(n)$ have been obtained by similar methods for $k = 6, 8, 10$, and 12, but they are more complicated. Exact formulas for $r_k(n)$ have also been found for $k = 3, 5$, and 7, and for all even $k \leq 24$. For values of $k > 24$ the analysis of $r_k(n)$ is considerably more complicated (see Hardy (1940)). Also, Milne (1996, 2002) announce new infinite families of explicit formulas extending Jacobi's identities. For more than 8 squares, Milne's identities are not the same as those obtained earlier by Mordell and others.

27.14 Unrestricted Partitions

27.14(i) Partition Functions

A fundamental problem studies the number of ways n can be written as a sum of positive integers $\leq n$, that is, the number of solutions of

$$27.14.1 \qquad n = a_1 + a_2 + \cdots, \quad a_1 \geq a_2 \geq \cdots \geq 1.$$

The number of summands is unrestricted, repetition is allowed, and the order of the summands is not taken into account. The corresponding *unrestricted partition function* is denoted by $p(n)$, and the summands are called *parts*; see §26.9(i). For example, $p(5) = 7$ because there are exactly seven partitions of 5: $5 = 4+1 = 3+2 = 3+1+1 = 2+2+1 = 2+1+1+1 = 1+1+1+1+1$.

The number of partitions of n into at most k parts is denoted by $p_k(n)$; again see §26.9(i).

27.14(ii) Generating Functions and Recursions

Euler introduced the reciprocal of the infinite product

$$27.14.2 \qquad f(x) = \prod_{m=1}^{\infty}(1-x^m), \qquad |x| < 1,$$

as a generating function for the function $p(n)$ defined in §27.14(i):

$$27.14.3 \qquad \frac{1}{f(x)} = \sum_{n=0}^{\infty} p(n)x^n,$$

with $p(0) = 1$. Euler's *pentagonal number theorem* states that

$$27.14.4 \quad \begin{aligned} f(x) &= 1 - x - x^2 + x^5 + x^7 - x^{12} - x^{15} + \cdots \\ &= 1 + \sum_{k=1}^{\infty}(-1)^k\left(x^{\omega(k)} + x^{\omega(-k)}\right), \end{aligned}$$

where the exponents 1, 2, 5, 7, 12, 15, ... are the *pentagonal numbers*, defined by

$$27.14.5 \qquad \omega(\pm k) = (3k^2 \mp k)/2, \quad k = 1, 2, 3, \ldots.$$

Multiplying the power series for $f(x)$ with that for $1/f(x)$ and equating coefficients, we obtain the recursion formula

$$27.14.6$$
$$p(n) = \sum_{k=1}^{\infty}(-1)^{k+1}\left(p(n-\omega(k)) + p(n-\omega(-k))\right)$$
$$= p(n-1) + p(n-2) - p(n-5) - p(n-7) + \cdots,$$

where $p(k)$ is defined to be 0 if $k < 0$. Logarithmic differentiation of the generating function $1/f(x)$ leads to another recursion:

$$27.14.7 \qquad np(n) = \sum_{k=1}^{n} \sigma_1(n) p(n-k),$$

where $\sigma_1(n)$ is defined by (27.2.10) with $\alpha = 1$.

27.14(iii) Asymptotic Formulas

These recursions can be used to calculate $p(n)$, which grows very rapidly. For example, $p(10) = 42, p(100) = 1905\,69292$, and $p(200) = 397\,29990\,29388$. For large n

$$27.14.8 \qquad p(n) \sim e^{K\sqrt{n}}/(4n\sqrt{3}),$$

where $K = \pi\sqrt{2/3}$ (Hardy and Ramanujan (1918)). Rademacher (1938) derives a convergent series that also provides an asymptotic expansion for $p(n)$:

$$27.14.9$$
$$p(n) = \frac{1}{\pi\sqrt{2}} \sum_{k=1}^{\infty} \sqrt{k} A_k(n) \left[\frac{d}{dt} \frac{\sinh(K\sqrt{t}/k)}{\sqrt{t}}\right]_{t=n-(1/24)},$$

where

$$27.14.10 \qquad A_k(n) = \sum_{\substack{h=1 \\ (h,k)=1}}^{k} \exp\left(\pi i s(h,k) - 2\pi i n \frac{h}{k}\right),$$

and $s(h,k)$ is a *Dedekind sum* given by

$$27.14.11 \qquad s(h,k) = \sum_{r=1}^{k-1} \frac{r}{k}\left(\frac{hr}{k} - \left\lfloor\frac{hr}{k}\right\rfloor - \frac{1}{2}\right).$$

27.14(iv) Relation to Modular Functions

Dedekind sums occur in the transformation theory of the *Dedekind modular function* $\eta(\tau)$, defined by

$$27.14.12 \qquad \eta(\tau) = e^{\pi i \tau/12} \prod_{n=1}^{\infty}(1-e^{2\pi i n\tau}), \qquad \Im \tau > 0.$$

This is related to the function $f(x)$ in (27.14.2) by

$$27.14.13 \qquad \eta(\tau) = e^{\pi i \tau/12} f(e^{2\pi i \tau}).$$

$\eta(\tau)$ satisfies the following functional equation: if a, b, c, d are integers with $ad - bc = 1$ and $c > 0$, then

$$27.14.14 \qquad \eta\left(\frac{a\tau+b}{c\tau+d}\right) = \varepsilon(-i(c\tau+d))^{\frac{1}{2}} \eta(\tau),$$

where $\varepsilon = \exp(\pi i(((a+d)/(12c)) - s(d,c)))$ and $s(d,c)$ is given by (27.14.11).

For further properties of the function $\eta(\tau)$ see §§23.15–23.19.

27.14(v) Divisibility Properties

Ramanujan (1921) gives identities that imply divisibility properties of the partition function. For example, the Ramanujan identity

$$27.14.15 \qquad 5\frac{(f(x^5))^5}{(f(x))^6} = \sum_{n=0}^{\infty} p(5n+4)x^n$$

implies $p(5n+4) \equiv 0 \pmod 5$. Ramanujan also found that $p(7n+5) \equiv 0 \pmod 7$ and $p(11n+6) \equiv 0 \pmod{11}$ for all n. After decades of nearly fruitless searching for further congruences of this type, it was believed that no others existed, until it was shown in Ono (2000) that there are infinitely many. Ono proved that for every prime $q > 3$ there are integers a and b such that $p(an+b) \equiv 0 \pmod q$ for all n. For example, $p(1575\,25693n + 1\,11247) \equiv 0 \pmod{13}$.

27.14(vi) Ramanujan's Tau Function

The *discriminant function* $\Delta(\tau)$ is defined by

$$27.14.16 \qquad \Delta(\tau) = (2\pi)^{12}(\eta(\tau))^{24}, \qquad \Im \tau > 0,$$

and satisfies the functional equation

$$27.14.17 \qquad \Delta\left(\frac{a\tau+b}{c\tau+d}\right) = (c\tau+d)^{12}\Delta(\tau),$$

if a, b, c, d are integers with $ad - bc = 1$ and $c > 0$.

The 24th power of $\eta(\tau)$ in (27.14.12) with $e^{2\pi i \tau} = x$ is an infinite product that generates a power series in

x with integer coefficients called *Ramanujan's tau function* $\tau(n)$:

27.14.18
$$x \prod_{n=1}^{\infty} (1-x^n)^{24} = \sum_{n=1}^{\infty} \tau(n) x^n, \qquad |x| < 1.$$

The tau function is multiplicative and satisfies the more general relation:

27.14.19
$$\tau(m)\,\tau(n) = \sum_{d|(m,n)} d^{11}\,\tau\!\left(\frac{mn}{d^2}\right), \quad m,n = 1,2,\ldots.$$

Lehmer (1947) conjectures that $\tau(n)$ is never 0 and verifies this for all $n < 21\,49286\,39999$ by studying various congruences satisfied by $\tau(n)$, for example:

27.14.20 $\qquad \tau(n) \equiv \sigma_{11}(n) \pmod{691}$.

For further information on partitions and generating functions see Andrews (1976); also §§17.2–17.14, and §§26.9–26.10.

Applications

27.15 Chinese Remainder Theorem

The Chinese remainder theorem states that a system of congruences $x \equiv a_1 \pmod{m_1}, \ldots, x \equiv a_k \pmod{m_k}$, always has a solution if the moduli are relatively prime in pairs; the solution is unique $\pmod{m}$, where m is the product of the moduli.

This theorem is employed to increase efficiency in calculating with large numbers by making use of smaller numbers in most of the calculation. For example, suppose a lengthy calculation involves many 10-digit integers. Most of the calculation can be done with five-digit integers as follows. Choose four relatively prime moduli m_1, m_2, m_3, and m_4 of five digits each, for example $2^{16} - 3$, $2^{16} - 1$, $2^{16} + 1$, and $2^{16} + 3$. Their product m has 20 digits, twice the number of digits in the data. By the Chinese remainder theorem each integer in the data can be uniquely represented by its residues $\pmod{m_1}$, $\pmod{m_2}$, $\pmod{m_3}$, and $\pmod{m_4}$, respectively. Because each residue has no more than five digits, the arithmetic can be performed efficiently on these residues with respect to each of the moduli, yielding answers $a_1 \pmod{m_1}$, $a_2 \pmod{m_2}$, $a_3 \pmod{m_3}$, and $a_4 \pmod{m_4}$, where each a_j has no more than five digits. These numbers, in turn, are combined by the Chinese remainder theorem to obtain the final result $\pmod{m}$, which is correct to 20 digits.

Even though the lengthy calculation is repeated four times, once for each modulus, most of it only uses five-digit integers and is accomplished quickly without overwhelming the machine's memory. Details of a machine program describing the method together with typical numerical results can be found in Newman (1967). See also Apostol and Niven (1994, pp. 18–19).

27.16 Cryptography

Applications to cryptography rely on the disparity in computer time required to find large primes and to factor large integers.

For example, a code maker chooses two large primes p and q of about 100 decimal digits each. Procedures for finding such primes require very little computer time. The primes are kept secret but their product $n = pq$, a 200-digit number, is made public. For this reason, these are often called public key codes. Messages are coded by a method (described below) that requires only the knowledge of n. But to decode, both factors p and q must be known. With the most efficient computer techniques devised to date (2009), factoring a 200-digit number may require billions of years on a single computer. For this reason, the codes are considered unbreakable, at least with the current state of knowledge on factoring large numbers.

To code a message by this method, we replace each letter by two digits, say $A = 01$, $B = 02$, ..., $Z = 26$, and divide the message into pieces of convenient length smaller than the public value $n = pq$. Choose a prime r that does not divide either $p - 1$ or $q - 1$. Like n, the prime r is made public. To code a piece x, raise x to the power r and reduce x^r modulo n to obtain an integer y (the coded form of x) between 1 and n. Thus, $y \equiv x^r \pmod{n}$ and $1 \leq y < n$.

To decode, we must recover x from y. To do this, let s denote the reciprocal of r modulo $\phi(n)$, so that $rs = 1 + t\,\phi(n)$ for some integer t. (Here $\phi(n)$ is Euler's totient (§27.2).) By the Euler–Fermat theorem (27.2.8), $x^{\phi(n)} \equiv 1 \pmod{n}$; hence $x^{t\,\phi(n)} \equiv 1 \pmod{n}$. But $y^s \equiv x^{rs} \equiv x^{1+t\,\phi(n)} \equiv x \pmod{n}$, so y^s is the same as x modulo n. In other words, to recover x from y we simply raise y to the power s and reduce modulo n. If p and q are known, s and y^s can be determined $\pmod{n}$ by straightforward calculations that require only a few minutes of machine time. But if p and q are not known, the problem of recovering x from y seems insurmountable.

For further information see Apostol and Niven (1994, p. 24), and for other applications to cryptography see Menezes *et al.* (1997) and Schroeder (2006).

27.17 Other Applications

Reed *et al.* (1990, pp. 458–470) describes a number-theoretic approach to Fourier analysis (called the *arithmetic Fourier transform*) that uses the Möbius inversion

(27.5.7) to increase efficiency in computing coefficients of Fourier series.

Congruences are used in constructing perpetual calendars, splicing telephone cables, scheduling round-robin tournaments, devising systematic methods for storing computer files, and generating pseudorandom numbers. Rosen (2004, Chapters 5 and 10) describes many of these applications. Apostol and Zuckerman (1951) uses congruences to construct magic squares.

There are also applications of number theory in many diverse areas, including physics, biology, chemistry, communications, and art. Schroeder (2006) describes many of these applications, including the design of concert hall ceilings to scatter sound into broad lateral patterns for improved acoustic quality, precise measurements of delays of radar echoes from Venus and Mercury to confirm one of the relativistic effects predicted by Einstein's theory of general relativity, and the use of primes in creating artistic graphical designs.

Computation

27.18 Methods of Computation: Primes

An overview of methods for precise counting of the number of primes not exceeding an arbitrary integer x is given in Crandall and Pomerance (2005, §3.7). T. Oliveira e Silva has calculated $\pi(x)$ for $x = 10^{23}$, using the combinatorial methods of Lagarias *et al.* (1985) and Deléglise and Rivat (1996); see Oliveira e Silva (2006). An analytic approach using a contour integral of the Riemann zeta function (§25.2(i)) is discussed in Borwein *et al.* (2000).

The *Sieve of Eratosthenes* (Crandall and Pomerance (2005, §3.2)) generates a list of all primes below a given bound. An alternative procedure is the *binary quadratic sieve* of Atkin and Bernstein (Crandall and Pomerance (2005, p. 170)).

For small values of n, primality is proven by showing that n is not divisible by any prime not exceeding $\sqrt{n}$.

Two simple algorithms for proving primality require a knowledge of all or part of the factorization of $n-1, n+1$, or both; see Crandall and Pomerance (2005, §§4.1–4.2). These algorithms are used for testing primality of *Mersenne numbers*, $2^n - 1$, and *Fermat numbers*, $2^{2^n} + 1$.

The *APR (Adleman–Pomerance–Rumely)* algorithm for primality testing is based on Jacobi sums. It runs in time $O\big((\log n)^{c \log \log \log n}\big)$. Explanations are given in Cohen (1993, §9.1) and Crandall and Pomerance (2005, §4.4). A practical version is described in Bosma and van der Hulst (1990).

The *AKS (Agrawal–Kayal–Saxena)* algorithm is the first deterministic, polynomial-time, primality test. That is to say, it runs in time $O((\log n)^c)$ for some constant c. An explanation is given in Crandall and Pomerance (2005, §4.5).

The *ECPP (Elliptic Curve Primality Proving)* algorithm handles primes with over 20,000 digits. Explanations are given in Cohen (1993, §9.2) and Crandall and Pomerance (2005, §7.6).

27.19 Methods of Computation: Factorization

Techniques for factorization of integers fall into three general classes: *Deterministic algorithms*, *Type I probabilistic algorithms* whose expected running time depends on the size of the smallest prime factor, and *Type II probabilistic algorithms* whose expected running time depends on the size of the number to be factored.

Deterministic algorithms are slow but are guaranteed to find the factorization within a known period of time. Trial division is one example. Fermat's algorithm is another; see Bressoud (1989, §5.1).

Type I probabilistic algorithms include the *Brent–Pollard rho algorithm* (also called *Monte Carlo method*), the *Pollard $p - 1$ algorithm*, and the *Elliptic Curve Method* (ECM). Descriptions of these algorithms are given in Crandall and Pomerance (2005, §§5.2, 5.4, and 7.4). As of January 2009 the largest prime factors found by these methods are a 19-digit prime for Brent–Pollard rho, a 58-digit prime for Pollard $p - 1$, and a 67-digit prime for ECM.

Type II probabilistic algorithms for factoring n rely on finding a pseudo-random pair of integers (x, y) that satisfy $x^2 \equiv y^2 \pmod{n}$. These algorithms include the *Continued Fraction Algorithm* (CFRAC), the *Multiple Polynomial Quadratic Sieve* (MPQS), the *General Number Field Sieve* (GNFS), and the *Special Number Field Sieve* (SNFS). A description of CFRAC is given in Bressoud and Wagon (2000). Descriptions of MPQS, GNFS, and SNFS are given in Crandall and Pomerance (2005, §§6.1 and 6.2). As of January 2009 the SNFS holds the record for the largest integer that has been factored by a Type II probabilistic algorithm, a 307-digit composite integer. The SNFS can be applied only to numbers that are very close to a power of a very small base. The largest composite numbers that have been factored by other Type II probabilistic algorithms are a 63-digit integer by CFRAC, a 135-digit integer by MPQS, and a 182-digit integer by GNFS.

For further information see Crandall and Pomerance (2005) and §26.22.

For current records online, see http://dlmf.nist.gov/27.19.

27.20 Methods of Computation: Other Number-Theoretic Functions

To calculate a multiplicative function it suffices to determine its values at the prime powers and then use (27.3.2). For a completely multiplicative function we use the values at the primes together with (27.3.10). The recursion formulas (27.14.6) and (27.14.7) can be used to calculate the partition function $p(n)$. A similar recursion formula obtained by differentiating (27.14.18) can be used to calculate Ramanujan's function $\tau(n)$, and the values can be checked by the congruence (27.14.20).

For further information see Lehmer (1941, pp. 5–83) and Lehmer (1943, pp. 483–492).

27.21 Tables

Lehmer (1914) lists all primes up to 100 06721. Bressoud and Wagon (2000, pp. 103–104) supplies tables and graphs that compare $\pi(x)$, $x/\log x$, and $\text{li}(x)$. Glaisher (1940) contains four tables: Table I tabulates, for all $n \leq 10^4$: (a) the canonical factorization of n into powers of primes; (b) the Euler totient $\phi(n)$; (c) the divisor function $d(n)$; (d) the sum $\sigma(n)$ of these divisors. Table II lists all solutions n of the equation $f(n) = m$ for all $m \leq 2500$, where $f(n)$ is defined by (27.14.2). Table III lists all solutions $n \leq 10^4$ of the equation $d(n) = m$, and Table IV lists all solutions n of the equation $\sigma(n) = m$ for all $m \leq 10^4$. Table 24.7 of Abramowitz and Stegun (1964) also lists the factorizations in Glaisher's Table I(a); Table 24.6 lists $\phi(n), d(n)$, and $\sigma(n)$ for $n \leq 1000$; Table 24.8 gives examples of primitive roots of all primes ≤ 9973; Table 24.9 lists all primes that are less than 1 00000.

The partition function $p(n)$ is tabulated in Gupta (1935, 1937), Watson (1937), and Gupta et al. (1958). Tables of the Ramanujan function $\tau(n)$ are published in Lehmer (1943) and Watson (1949). Lehmer (1941) gives a comprehensive account of tables in the theory of numbers, including virtually every table published from 1918 to 1941. Those published prior to 1918 are mentioned in Dickson (1919). The bibliography in Lehmer (1941) gives references to the places in Dickson's History where the older tables are cited. Lehmer (1941) also has a section that supplies errata and corrections to all tables cited.

No sequel to Lehmer (1941) exists to date, but many tables of functions of number theory are included in Unpublished Mathematical Tables (1944).

27.22 Software

See http://dlmf.nist.gov/27.22.

References

General References

The main references used in writing this chapter are Apostol (1976, 1990), and Apostol and Niven (1994). Further information can be found in Andrews (1976), Erdélyi et al. (1955, Chapter XVII), Hardy and Wright (1979), and Niven et al. (1991).

Sources

The following list gives the references or other indications of proofs that were used in constructing the various sections of this chapter. These sources supplement the references quoted in the text.

§27.2 Apostol (1976, Chapter 2). For (27.2.11) see Erdélyi et al. (1955, p. 168). Tables 27.2.1 and 27.2.2 are from Abramowitz and Stegun (1964, Tables 24.6 and 24.9).

§27.3 Apostol (1976, Chapter 2).

§27.4 Apostol (1976, Chapter 11). For (27.4.10) see Titchmarsh (1986b, p. 4).

§27.5 Apostol (1976, Chapter 2 and p. 228). For (27.5.7) use (27.5.2) and formal substitution.

§27.6 Apostol (1976, Chapter 2).

§27.7 Apostol (1990, Chapter 1).

§27.8 Apostol (1976, Chapter 6).

§27.9 Apostol (1976, Chapter 9).

§27.10 Apostol (1976, Chapter 8).

§27.11 Apostol (1976, Chapters 3, 4). For (27.11.12), (27.11.14), and (27.11.15) see Prachar (1957, pp. 71–74).

§27.12 Crandall and Pomerance (2005, pp. 131–152), Davenport (2000), Narkiewicz (2000), Rosser (1939). For (27.12.7) see Schoenfeld (1976) and Crandall and Pomerance (2005, pp. 37, 60). For the proof that there are infinitely many Carmichael numbers see Alford et al. (1994).

§27.13 Apostol (1976, Chapter 14), Ellison (1971), Grosswald (1985, pp. 8, 32). For (27.13.4) see (20.2.3).

§27.14(ii) Apostol (1976, Chapter 14), Apostol (1990, Chapters 3–5).

Chapter 28
Mathieu Functions and Hill's Equation
G. Wolf[1]

Notation   **652**
28.1   Special Notation 652

Mathieu Functions of Integer Order   **652**
28.2   Definitions and Basic Properties 652
28.3   Graphics 655
28.4   Fourier Series 656
28.5   Second Solutions fe_n, ge_n 657
28.6   Expansions for Small q 659
28.7   Analytic Continuation of Eigenvalues . . . 661
28.8   Asymptotic Expansions for Large q 661
28.9   Zeros 663
28.10   Integral Equations 663
28.11   Expansions in Series of Mathieu Functions 664

Mathieu Functions of Noninteger Order   **664**
28.12   Definitions and Basic Properties 664
28.13   Graphics 665
28.14   Fourier Series 666
28.15   Expansions for Small q 666
28.16   Asymptotic Expansions for Large q 666
28.17   Stability as $x \to \pm\infty$ 667
28.18   Integrals and Integral Equations 667
28.19   Expansions in Series of $\text{me}_{\nu+2n}$ Functions 667

Modified Mathieu Functions   **667**
28.20   Definitions and Basic Properties 667

28.21   Graphics 669
28.22   Connection Formulas 669
28.23   Expansions in Series of Bessel Functions . 670
28.24   Expansions in Series of Cross-Products of Bessel Functions or Modified Bessel Functions 671
28.25   Asymptotic Expansions for Large $\Re z$ . . . 672
28.26   Asymptotic Approximations for Large q . 672
28.27   Addition Theorems 672
28.28   Integrals, Integral Representations, and Integral Equations 672

Hill's Equation   **674**
28.29   Definitions and Basic Properties 674
28.30   Expansions in Series of Eigenfunctions . . 676
28.31   Equations of Whittaker–Hill and Ince . . 676

Applications   **677**
28.32   Mathematical Applications 677
28.33   Physical Applications 678

Computation   **679**
28.34   Methods of Computation 679
28.35   Tables 680
28.36   Software 681

References   **681**

---

[1] Fachbereich Mathematik, University Duisburg-Essen, Essen, Germany.
  Acknowledgments: This chapter is based in part on Abramowitz and Stegun (1964, Chapter 20) by G. Blanch.
  Copyright © 2009 National Institute of Standards and Technology. All rights reserved.

Notation

28.1 Special Notation

(For other notation see pp. xiv and 873.)

m, n	integers.
x, y	real variables.
$z = x + iy$	complex variable.
ν	order of the Mathieu function or modified Mathieu function. (When ν is an integer it is often replaced by n.)
δ	arbitrary small positive number.
a, q, h	real or complex parameters of Mathieu's equation with $q = h^2$.
primes	unless indicated otherwise, derivatives with respect to the argument

The main functions treated in this chapter are the Mathieu functions

$$\mathrm{ce}_\nu(z,q),\ \mathrm{se}_\nu(z,q),\ \mathrm{fe}_n(z,q),\ \mathrm{ge}_n(z,q),\ \mathrm{me}_\nu(z,q),$$

and the modified Mathieu functions

$$\mathrm{Ce}_\nu(z,q),\quad \mathrm{Se}_\nu(z,q),\quad \mathrm{Fe}_n(z,q),\quad \mathrm{Ge}_n(z,q),$$
$$\mathrm{Me}_\nu(z,q),\quad \mathrm{M}_\nu^{(j)}(z,h),\quad \mathrm{Mc}_n^{(j)}(z,h),\quad \mathrm{Ms}_n^{(j)}(z,h),$$
$$\mathrm{Ie}_n(z,h),\quad \mathrm{Io}_n(z,h),\quad \mathrm{Ke}_n(z,h),\quad \mathrm{Ko}_n(z,h).$$

The functions $\mathrm{Mc}_n^{(j)}(z,h)$ and $\mathrm{Ms}_n^{(j)}(z,h)$ are also known as the radial Mathieu functions.

The eigenvalues of Mathieu's equation are denoted by

$$a_n(q),\quad b_n(q),\quad \lambda_\nu(q).$$

The notation for the joining factors is

$$g_{e,n}(h),\quad g_{o,n}(h),\quad f_{e,n}(h),\quad f_{o,n}(h).$$

Alternative notations for the parameters a and q are shown in Table 28.1.1.

Table 28.1.1: Notations for parameters in Mathieu's equation.

Reference	a	q
Erdélyi et al. (1955)	h	θ
Meixner and Schäfke (1954)	λ	h^2
Moon and Spencer (1971)	λ	q
Strutt (1932)	λ	h^2
Whittaker and Watson (1927)	a	$8q$

Alternative notations for the functions are as follows.

Arscott (1964b) and McLachlan (1947)

$$\mathrm{Fey}_n(z,q) = \sqrt{\tfrac{1}{2}\pi}\, g_{e,n}(h)\, \mathrm{ce}_n(0,q)\, \mathrm{Mc}_n^{(2)}(z,h),$$
$$\mathrm{Me}_n^{(1,2)}(z,q) = \sqrt{\tfrac{1}{2}\pi}\, g_{e,n}(h)\, \mathrm{ce}_n(0,q)\, \mathrm{Mc}_n^{(3,4)}(z,h),$$
$$\mathrm{Gey}_n(z,q) = \sqrt{\tfrac{1}{2}\pi}\, g_{o,n}(h)\, \mathrm{se}_n'(0,q)\, \mathrm{Ms}_n^{(2)}(z,h),$$
$$\mathrm{Ne}_n^{(1,2)}(z,q) = \sqrt{\tfrac{1}{2}\pi}\, g_{o,n}(h)\, \mathrm{se}_n'(0,q)\, \mathrm{Ms}_n^{(3,4)}(z,h).$$

Arscott (1964b) also uses $-i\mu$ for ν.

Campbell (1955)

$$\mathrm{in}_n = \mathrm{fe}_n,\qquad \mathrm{ceh}_n = \mathrm{Ce}_n,\qquad \mathrm{inh}_n = \mathrm{Fe}_n,$$
$$\mathrm{jn}_n = \mathrm{ge}_n,\qquad \mathrm{seh}_n = \mathrm{Se}_n,\qquad \mathrm{jnh}_n = \mathrm{Ge}_n.$$

Abramowitz and Stegun (1964, Chapter 20)

$$F_\nu(z) = \mathrm{Me}_\nu(z,q).$$

NBS (1967)

With $s = 4q$,

$$\mathrm{Se}_n(s,z) = \frac{\mathrm{ce}_n(z,q)}{\mathrm{ce}_n(0,q)},\quad \mathrm{So}_n(s,z) = \frac{\mathrm{se}_n(z,q)}{\mathrm{se}_n'(0,q)}.$$

Stratton et al. (1941)

With $c = 2\sqrt{q}$,

$$\mathrm{Se}_n(c,z) = \frac{\mathrm{ce}_n(z,q)}{\mathrm{ce}_n(0,q)},\quad \mathrm{So}_n(c,z) = \frac{\mathrm{se}_n(z,q)}{\mathrm{se}_n'(0,q)}.$$

Zhang and Jin (1996)

The radial functions $\mathrm{Mc}_n^{(j)}(z,h)$ and $\mathrm{Ms}_n^{(j)}(z,h)$ are denoted by $\mathrm{Mc}_n^{(j)}(z,q)$ and $\mathrm{Ms}_n^{(j)}(z,q)$, respectively.

Mathieu Functions of Integer Order

28.2 Definitions and Basic Properties

28.2(i) Mathieu's Equation

The *standard form* of Mathieu's equation with parameters (a, q) is

28.2.1 $\qquad w'' + (a - 2q\cos(2z))w = 0.$

With $\zeta = \sin^2 z$ we obtain the *algebraic form* of Mathieu's equation

28.2.2
$$\zeta(1-\zeta)w'' + \tfrac{1}{2}(1-2\zeta)w' + \tfrac{1}{4}(a - 2q(1-2\zeta))\,w = 0.$$

This equation has regular singularities at 0 and 1, both with exponents 0 and $\tfrac{1}{2}$, and an irregular singular point at ∞. With $\zeta = \cos z$ we obtain another algebraic form:

28.2.3 $\quad (1-\zeta^2)w'' - \zeta w' + \left(a + 2q - 4q\zeta^2\right)w = 0.$

28.2 Definitions and Basic Properties

28.2(ii) Basic Solutions w_{I}, w_{II}

Since (28.2.1) has no finite singularities its solutions are entire functions of z. Furthermore, a solution w with given initial constant values of w and w' at a point z_0 is an entire function of the three variables z, a, and q.

The following three transformations

28.2.4 $\quad z \to -z; \quad z \to z \pm \pi; \quad z \to z \pm \tfrac{1}{2}\pi, q \to -q;$

each leave (28.2.1) unchanged. (28.2.1) possesses a fundamental pair of solutions $w_{\text{I}}(z;a,q), w_{\text{II}}(z;a,q)$ called *basic solutions* with

28.2.5 $\quad \begin{bmatrix} w_{\text{I}}(0;a,q) & w_{\text{II}}(0;a,q) \\ w_{\text{I}}'(0;a,q) & w_{\text{II}}'(0;a,q) \end{bmatrix} = \begin{bmatrix} 1 & 0 \\ 0 & 1 \end{bmatrix}.$

$w_{\text{I}}(z;a,q)$ is even and $w_{\text{II}}(z;a,q)$ is odd. Other properties are as follows.

28.2.6 $\quad \mathscr{W}\{w_{\text{I}}, w_{\text{II}}\} = 1,$

28.2.7 $\quad \begin{aligned} w_{\text{I}}(z \pm \pi; a, q) &= w_{\text{I}}(\pi; a, q) w_{\text{I}}(z; a, q) \\ &\quad \pm w_{\text{I}}'(\pi; a, q) w_{\text{II}}(z; a, q), \end{aligned}$

28.2.8 $\quad \begin{aligned} w_{\text{II}}(z \pm \pi; a, q) &= \pm w_{\text{II}}(\pi; a, q) w_{\text{I}}(z; a, q) \\ &\quad + w_{\text{II}}'(\pi; a, q) w_{\text{II}}(z; a, q), \end{aligned}$

28.2.9 $\quad w_{\text{I}}(\pi; a, q) = w_{\text{II}}'(\pi; a, q),$

28.2.10 $\quad w_{\text{I}}(\pi; a, q) - 1 = 2 w_{\text{I}}'(\tfrac{1}{2}\pi; a, q) w_{\text{II}}(\tfrac{1}{2}\pi; a, q),$

28.2.11 $\quad w_{\text{I}}(\pi; a, q) + 1 = 2 w_{\text{I}}(\tfrac{1}{2}\pi; a, q) w_{\text{II}}'(\tfrac{1}{2}\pi; a, q),$

28.2.12 $\quad w_{\text{I}}'(\pi; a, q) = 2 w_{\text{I}}(\tfrac{1}{2}\pi; a, q) w_{\text{I}}'(\tfrac{1}{2}\pi; a, q),$

28.2.13 $\quad w_{\text{II}}(\pi; a, q) = 2 w_{\text{II}}(\tfrac{1}{2}\pi; a, q) w_{\text{II}}'(\tfrac{1}{2}\pi; a, q).$

28.2(iii) Floquet's Theorem and the Characteristic Exponents

Let ν be any real or complex constant. Then Mathieu's equation (28.2.1) has a nontrivial solution $w(z)$ such that

28.2.14 $\quad w(z+\pi) = e^{\pi i \nu} w(z),$

iff $e^{\pi i \nu}$ is an eigenvalue of the matrix

28.2.15 $\quad \begin{bmatrix} w_{\text{I}}(\pi; a, q) & w_{\text{II}}(\pi; a, q) \\ w_{\text{I}}'(\pi; a, q) & w_{\text{II}}'(\pi; a, q) \end{bmatrix}.$

Equivalently,

28.2.16 $\quad \cos(\pi \nu) = w_{\text{I}}(\pi; a, q) = w_{\text{I}}(\pi; a, -q).$

This is the *characteristic equation* of Mathieu's equation (28.2.1). $\cos(\pi \nu)$ is an entire function of a, q^2. The solutions of (28.2.16) are given by $\nu = \pi^{-1} \arccos(w_{\text{I}}(\pi; a, q))$. If the inverse cosine takes its principal value (§4.23(ii)), then $\nu = \hat{\nu}$, where $0 \leq \Re \hat{\nu} \leq 1$. The general solution of (28.2.16) is $\nu = \pm \hat{\nu} + 2n$, where $n \in \mathbb{Z}$. Either $\hat{\nu}$ or ν is called a *characteristic exponent* of (28.2.1). If $\hat{\nu} = 0$ or 1, or equivalently, $\nu = n$, then ν is a double root of the characteristic equation, otherwise it is a simple root.

28.2(iv) Floquet Solutions

A solution with the *pseudoperiodic property* (28.2.14) is called a *Floquet solution with respect to ν*. (28.2.9), (28.2.16), and (28.2.7) give for each solution $w(z)$ of (28.2.1) the connection formula

28.2.17 $\quad w(z+\pi) + w(z-\pi) = 2\cos(\pi\nu) w(z).$

Therefore a nontrivial solution $w(z)$ is either a Floquet solution with respect to ν, or $w(z+\pi) - e^{i\nu\pi} w(z)$ is a Floquet solution with respect to $-\nu$.

If $q \neq 0$, then for a given value of ν the corresponding Floquet solution is unique, except for an arbitrary constant factor (Theorem of Ince; see also 28.5(i)).

The Fourier series of a Floquet solution

28.2.18 $\quad w(z) = \sum_{n=-\infty}^{\infty} c_{2n} e^{i(\nu+2n)z}$

converges absolutely and uniformly in compact subsets of $\mathbb{C}$. The coefficients c_{2n} satisfy

28.2.19
$$q c_{2n+2} - \left(a - (\nu+2n)^2\right) c_{2n} + q c_{2n-2} = 0, \quad n \in \mathbb{Z}.$$

Conversely, a nontrivial solution c_{2n} of (28.2.19) that satisfies

28.2.20 $\quad \lim_{n \to \pm\infty} |c_{2n}|^{1/|n|} = 0$

leads to a Floquet solution.

28.2(v) Eigenvalues a_n, b_n

For given ν and q, equation (28.2.16) determines an infinite discrete set of values of a, the *eigenvalues* or *characteristic values*, of Mathieu's equation. When $\hat{\nu} = 0$ or 1, the notation for the two sets of eigenvalues corresponding to each $\hat{\nu}$ is shown in Table 28.2.1, together with the boundary conditions of the associated eigenvalue problem. In Table 28.2.1 $n = 0, 1, 2, \ldots$.

Table 28.2.1: Eigenvalues of Mathieu's equation.

$\hat{\nu}$	Boundary Conditions	Eigenvalues
0	$w'(0) = w'(\tfrac{1}{2}\pi) = 0$	$a_{2n}(q)$
1	$w'(0) = w(\tfrac{1}{2}\pi) = 0$	$a_{2n+1}(q)$
1	$w(0) = w'(\tfrac{1}{2}\pi) = 0$	$b_{2n+1}(q)$
0	$w(0) = w(\tfrac{1}{2}\pi) = 0$	$b_{2n+2}(q)$

An equivalent formulation is given by

28.2.21 $\quad \begin{aligned} w_{\text{I}}'(\tfrac{1}{2}\pi; a, q) &= 0, & a &= a_{2n}(q), \\ w_{\text{I}}(\tfrac{1}{2}\pi; a, q) &= 0, & a &= a_{2n+1}(q), \end{aligned}$

and

28.2.22 $\quad \begin{aligned} w_{\text{II}}'(\tfrac{1}{2}\pi; a, q) &= 0, & a &= b_{2n+1}(q), \\ w_{\text{II}}(\tfrac{1}{2}\pi; a, q) &= 0, & a &= b_{2n+2}(q), \end{aligned}$

where $n = 0, 1, 2, \ldots$. When $q = 0$,

28.2.23 $\qquad a_n(0) = n^2, \qquad\qquad n = 0, 1, 2, \ldots,$

28.2.24 $\qquad b_n(0) = n^2, \qquad\qquad n = 1, 2, 3, \ldots.$

Near $q = 0$, $a_n(q)$ and $b_n(q)$ can be expanded in power series in q (see §28.6(i)); elsewhere they are determined by analytic continuation (see §28.7). For nonnegative real values of q, see Figure 28.2.1.

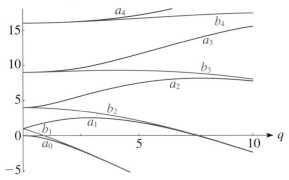

Figure 28.2.1: Eigenvalues $a_n(q)$, $b_n(q)$ of Mathieu's equation as functions of q for $0 \leq q \leq 10$, $n = 0, 1, 2, 3, 4$ (a's), $n = 1, 2, 3, 4$ (b's).

Distribution

28.2.25

for $q > 0$: $\quad a_0 < b_1 < a_1 < b_2 < a_2 < b_3 < \cdots,$

for $q < 0$: $\quad a_0 < a_1 < b_1 < b_2 < a_2 < a_3 < \cdots.$

Change of Sign of q

28.2.26 $\qquad a_{2n}(-q) = a_{2n}(q),$

28.2.27 $\qquad a_{2n+1}(-q) = b_{2n+1}(q),$

28.2.28 $\qquad b_{2n+2}(-q) = b_{2n+2}(q).$

28.2(vi) Eigenfunctions

Table 28.2.2 gives the notation for the eigenfunctions corresponding to the eigenvalues in Table 28.2.1. *Period* π means that the eigenfunction has the property $w(z + \pi) = w(z)$, whereas *antiperiod* π means that $w(z + \pi) = -w(z)$. *Even parity* means $w(-z) = w(z)$, and *odd parity* means $w(-z) = -w(z)$.

Table 28.2.2: Eigenfunctions of Mathieu's equation.

Eigenvalues	Eigenfunctions	Periodicity	Parity
$a_{2n}(q)$	$\mathrm{ce}_{2n}(z, q)$	Period π	Even
$a_{2n+1}(q)$	$\mathrm{ce}_{2n+1}(z, q)$	Antiperiod π	Even
$b_{2n+1}(q)$	$\mathrm{se}_{2n+1}(z, q)$	Antiperiod π	Odd
$b_{2n+2}(q)$	$\mathrm{se}_{2n+2}(z, q)$	Period π	Odd

When $q = 0$,

28.2.29
$$\mathrm{ce}_0(z, 0) = 1/\sqrt{2}, \quad \mathrm{ce}_n(z, 0) = \cos(nz),$$
$$\mathrm{se}_n(z, 0) = \sin(nz), \qquad n = 1, 2, 3, \ldots.$$

For simple roots q of the corresponding equations (28.2.21) and (28.2.22), the functions are made unique by the normalizations

28.2.30
$$\int_0^{2\pi} (\mathrm{ce}_n(x, q))^2 \, dx = \pi, \quad \int_0^{2\pi} (\mathrm{se}_n(x, q))^2 \, dx = \pi,$$

the ambiguity of sign being resolved by (28.2.29) when $q = 0$ and by continuity for the other values of q.

The functions are orthogonal, that is,

28.2.31 $\qquad \int_0^{2\pi} \mathrm{ce}_m(x, q) \, \mathrm{ce}_n(x, q) \, dx = 0, \qquad n \neq m,$

28.2.32 $\qquad \int_0^{2\pi} \mathrm{se}_m(x, q) \, \mathrm{se}_n(x, q) \, dx = 0, \qquad n \neq m,$

28.2.33 $\qquad \int_0^{2\pi} \mathrm{ce}_m(x, q) \, \mathrm{se}_n(x, q) \, dx = 0.$

For change of sign of q (compare (28.2.4))

28.2.34 $\quad \mathrm{ce}_{2n}(z, -q) = (-1)^n \mathrm{ce}_{2n}\left(\tfrac{1}{2}\pi - z, q\right),$

28.2.35 $\quad \mathrm{ce}_{2n+1}(z, -q) = (-1)^n \mathrm{se}_{2n+1}\left(\tfrac{1}{2}\pi - z, q\right),$

28.2.36 $\quad \mathrm{se}_{2n+1}(z, -q) = (-1)^n \mathrm{ce}_{2n+1}\left(\tfrac{1}{2}\pi - z, q\right),$

28.2.37 $\quad \mathrm{se}_{2n+2}(z, -q) = (-1)^n \mathrm{se}_{2n+2}\left(\tfrac{1}{2}\pi - z, q\right).$

For the connection with the basic solutions in §28.2(ii),

28.2.38 $\qquad \dfrac{\mathrm{ce}_n(z, q)}{\mathrm{ce}_n(0, q)} = w_{\mathrm{I}}(z; a_n(q), q), \qquad n = 0, 1, \ldots,$

28.2.39 $\qquad \dfrac{\mathrm{se}_n(z, q)}{\mathrm{se}_n'(0, q)} = w_{\mathrm{II}}(z; b_n(q), q), \qquad n = 1, 2, \ldots.$

28.3 Graphics

28.3(i) Line Graphs: Mathieu Functions with Fixed q and Variable x

Even π-Periodic Solutions

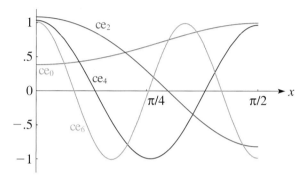

Figure 28.3.1: $\mathrm{ce}_{2n}(x,1)$ for $0 \leq x \leq \pi/2$, $n = 0, 1, 2, 3$.

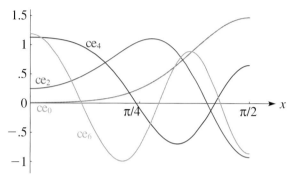

Figure 28.3.2: $\mathrm{ce}_{2n}(x,10)$ for $0 \leq x \leq \pi/2$, $n = 0, 1, 2, 3$.

Even π-Antiperiodic Solutions

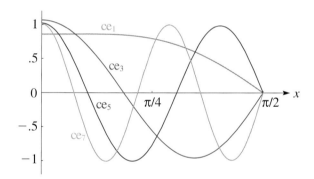

Figure 28.3.3: $\mathrm{ce}_{2n+1}(x,1)$ for $0 \leq x \leq \pi/2$, $n = 0, 1, 2, 3$.

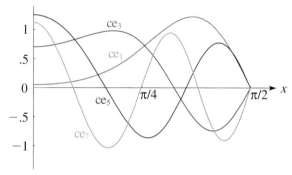

Figure 28.3.4: $\mathrm{ce}_{2n+1}(x,10)$ for $0 \leq x \leq \pi/2$, $n = 0, 1, 2, 3$.

Odd π-Antiperiodic Solutions

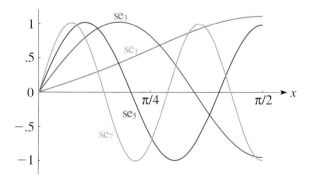

Figure 28.3.5: $\mathrm{se}_{2n+1}(x,1)$ for $0 \leq x \leq \pi/2$, $n = 0, 1, 2, 3$.

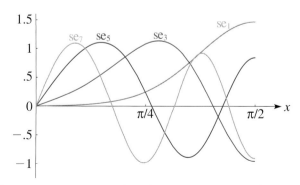

Figure 28.3.6: $\mathrm{se}_{2n+1}(x,10)$ for $0 \leq x \leq \pi/2$, $n = 0, 1, 2, 3$.

Odd π-Periodic Solutions

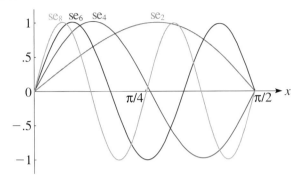

Figure 28.3.7: $\mathrm{se}_{2n}(x,1)$ for $0 \leq x \leq \pi/2$, $n=1,2,3,4$.

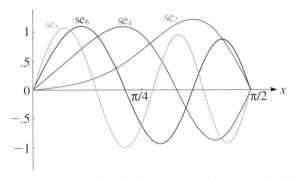

Figure 28.3.8: $\mathrm{se}_{2n}(x,10)$ for $0 \leq x \leq \pi/2$, $n=1,2,3,4$.

For further graphs see Jahnke *et al.* (1966, pp. 264–265 and 268–275).

28.3(ii) Surfaces: Mathieu Functions with Variable x and q

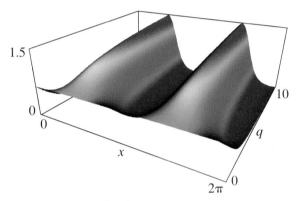

Figure 28.3.9: $\mathrm{ce}_0(x,q)$ for $0 \leq x \leq 2\pi$, $0 \leq q \leq 10$.

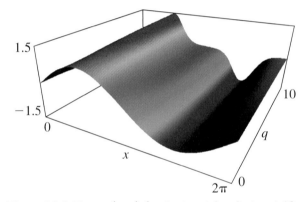

Figure 28.3.10: $\mathrm{se}_1(x,q)$ for $0 \leq x \leq 2\pi$, $0 \leq q \leq 10$.

For further graphics see http://dlmf.nist.gov/28.3.ii.

28.4 Fourier Series

28.4(i) Definitions

The Fourier series of the periodic Mathieu functions converge absolutely and uniformly on all compact sets in the z-plane. For $n = 0, 1, 2, 3, \ldots$,

28.4.1 $$\mathrm{ce}_{2n}(z,q) = \sum_{m=0}^{\infty} A_{2m}^{2n}(q) \cos 2mz,$$

28.4.2 $$\mathrm{ce}_{2n+1}(z,q) = \sum_{m=0}^{\infty} A_{2m+1}^{2n+1}(q) \cos (2m+1)z,$$

28.4.3 $$\mathrm{se}_{2n+1}(z,q) = \sum_{m=0}^{\infty} B_{2m+1}^{2n+1}(q) \sin (2m+1)z,$$

28.4.4 $$\mathrm{se}_{2n+2}(z,q) = \sum_{m=0}^{\infty} B_{2m+2}^{2n+2}(q) \sin (2m+2)z.$$

28.4(ii) Recurrence Relations

28.4.5
$$aA_0 - qA_2 = 0, \quad (a-4)A_2 - q(2A_0 + A_4) = 0,$$
$$(a-4m^2)A_{2m} - q(A_{2m-2} + A_{2m+2}) = 0,$$
$$m = 2,3,4,\ldots, \ a = a_{2n}(q), \ A_{2m} = A_{2m}^{2n}(q).$$

28.4.6
$$(a-1-q)A_1 - qA_3 = 0,$$
$$(a-(2m+1)^2)A_{2m+1} - q(A_{2m-1} + A_{2m+3}) = 0,$$
$$m = 1,2,3,\ldots, \ a = a_{2n+1}(q), \ A_{2m+1} = A_{2m+1}^{2n+1}(q).$$

28.4.7
$$(a-1+q)B_1 - qB_3 = 0,$$
$$(a-(2m+1)^2)B_{2m+1} - q(B_{2m-1} + B_{2m+3}) = 0,$$
$$m = 1,2,3,\ldots, \ a = b_{2n+1}(q), \ B_{2m+1} = B_{2m+1}^{2n+1}(q).$$

28.4.8
$$(a-4)B_2 - qB_4 = 0,$$
$$(a-4m^2)B_{2m} - q(B_{2m-2} + B_{2m+2}) = 0,$$
$$m = 2, 3, 4, \ldots, a = b_{2n+2}(q), B_{2m+2} = B_{2m+2}^{2n+2}(q).$$

28.4(iii) Normalization

28.4.9 $\quad 2\left(A_0^{2n}(q)\right)^2 + \sum_{m=1}^{\infty} \left(A_{2m}^{2n}(q)\right)^2 = 1,$

28.4.10 $\quad \sum_{m=0}^{\infty} \left(A_{2m+1}^{2n+1}(q)\right)^2 = 1,$

28.4.11 $\quad \sum_{m=0}^{\infty} \left(B_{2m+1}^{2n+1}(q)\right)^2 = 1,$

28.4.12 $\quad \sum_{m=0}^{\infty} \left(B_{2m+2}^{2n+2}(q)\right)^2 = 1.$

Ambiguities in sign are resolved by (28.4.13)–(28.4.16) when $q = 0$, and by continuity for the other values of q.

28.4(iv) Case $q = 0$

28.4.13 $\quad A_0^0(0) = 1/\sqrt{2}, \quad A_{2n}^{2n}(0) = 1, \quad n > 0,$
$\quad A_{2m}^{2n}(0) = 0, \quad n \neq m,$

28.4.14 $\quad A_{2n+1}^{2n+1}(0) = 1, \quad A_{2m+1}^{2n+1}(0) = 0, \quad n \neq m,$

28.4.15 $\quad B_{2n+1}^{2n+1}(0) = 1, \quad B_{2m+1}^{2n+1}(0) = 0, \quad n \neq m,$

28.4.16 $\quad B_{2n+2}^{2n+2}(0) = 1, \quad B_{2m+2}^{2n+2}(0) = 0, \quad n \neq m.$

28.4(v) Change of Sign of q

28.4.17 $\quad A_{2m}^{2n}(-q) = (-1)^{n-m} A_{2m}^{2n}(q),$

28.4.18 $\quad B_{2m+2}^{2n+2}(-q) = (-1)^{n-m} B_{2m+2}^{2n+2}(q),$

28.4.19 $\quad A_{2m+1}^{2n+1}(-q) = (-1)^{n-m} B_{2m+1}^{2n+1}(q),$

28.4.20 $\quad B_{2m+1}^{2n+1}(-q) = (-1)^{n-m} A_{2m+1}^{2n+1}(q).$

28.4(vi) Behavior for Small q

For fixed $s = 1, 2, 3, \ldots$ and fixed $m = 1, 2, 3, \ldots,$

28.4.21 $\quad A_{2s}^0(q) = \left(\frac{(-1)^s 2}{(s!)^2} \left(\frac{q}{4}\right)^s + O(q^{s+2})\right) A_0^0(q),$

28.4.22
$$\left.\begin{array}{l} A_{m+2s}^m(q) \\ B_{m+2s}^m(q) \end{array}\right\} = \left(\frac{(-1)^s m!}{s!(m+s)!} \left(\frac{q}{4}\right)^s + O(q^{s+1})\right) \left\{\begin{array}{l} A_m^m(q), \\ B_m^m(q), \end{array}\right.$$

28.4.23
$$\left.\begin{array}{l} A_{m-2s}^m(q) \\ B_{m-2s}^m(q) \end{array}\right\} = \left(\frac{(m-s-1)!}{s!(m-1)!} \left(\frac{q}{4}\right)^s + O(q^{s+1})\right) \left\{\begin{array}{l} A_m^m(q), \\ B_m^m(q). \end{array}\right.$$

For further terms and expansions see Meixner and Schäfke (1954, p. 122) and McLachlan (1947, §3.33).

28.4(vii) Asymptotic Forms for Large m

As $m \to \infty$, with fixed $q \, (\neq 0)$ and fixed n,

28.4.24 $\quad \dfrac{A_{2m}^{2n}(q)}{A_0^{2n}(q)} = \dfrac{(-1)^m}{(m!)^2} \left(\dfrac{q}{4}\right)^m \dfrac{\pi \left(1 + O(m^{-1})\right)}{w_{\text{II}}(\frac{1}{2}\pi; a_{2n}(q), q)},$

28.4.25 $\quad \dfrac{A_{2m+1}^{2n+1}(q)}{A_1^{2n+1}(q)} = \dfrac{(-1)^{m+1}}{\left(\left(\frac{1}{2}\right)_{m+1}\right)^2} \left(\dfrac{q}{4}\right)^{m+1} \dfrac{2 \left(1 + O(m^{-1})\right)}{w'_{\text{II}}(\frac{1}{2}\pi; a_{2n+1}(q), q)},$

28.4.26 $\quad \dfrac{B_{2m+1}^{2n+1}(q)}{B_1^{2n+1}(q)} = \dfrac{(-1)^m}{\left(\left(\frac{1}{2}\right)_{m+1}\right)^2} \left(\dfrac{q}{4}\right)^{m+1} \dfrac{2 \left(1 + O(m^{-1})\right)}{w_{\text{I}}(\frac{1}{2}\pi; b_{2n+1}(q), q)},$

28.4.27 $\quad \dfrac{B_{2m}^{2n+2}(q)}{B_2^{2n+2}(q)} = \dfrac{(-1)^m}{(m!)^2} \left(\dfrac{q}{4}\right)^m \dfrac{q\pi \left(1 + O(m^{-1})\right)}{w'_{\text{I}}(\frac{1}{2}\pi; b_{2n+2}(q), q)}.$

For the basic solutions w_{I} and w_{II} see §28.2(ii).

28.5 Second Solutions fe_n, ge_n

28.5(i) Definitions

Theorem of Ince (1922)

If a nontrivial solution of Mathieu's equation with $q \neq 0$ has period π or 2π, then any linearly independent solution cannot have either period.

Second solutions of (28.2.1) are given by

28.5.1 $\quad \mathrm{fe}_n(z, q) = C_n(q) \left(z \, \mathrm{ce}_n(z, q) + f_n(z, q)\right),$

when $a = a_n(q)$, $n = 0, 1, 2, \ldots,$ and by

28.5.2 $\quad \mathrm{ge}_n(z, q) = S_n(q) \left(z \, \mathrm{se}_n(z, q) + g_n(z, q)\right),$

when $a = b_n(q)$, $n = 1, 2, 3, \ldots.$ For $m = 0, 1, 2, \ldots,$ we have

28.5.3 $\quad \begin{array}{l} f_{2m}(z, q) \quad \pi\text{-periodic, odd,} \\ f_{2m+1}(z, q) \quad \pi\text{-antiperiodic, odd,} \end{array}$

and

28.5.4 $\quad \begin{array}{l} g_{2m+1}(z, q) \quad \pi\text{-antiperiodic, even,} \\ g_{2m+2}(z, q) \quad \pi\text{-periodic, even;} \end{array}$

compare §28.2(vi). The functions $f_n(z, q)$, $g_n(z, q)$ are unique.

The factors $C_n(q)$ and $S_n(q)$ in (28.5.1) and (28.5.2) are normalized so that

28.5.5
$$(C_n(q))^2 \int_0^{2\pi} (f_n(x, q))^2 \, dx$$
$$= (S_n(q))^2 \int_0^{2\pi} (g_n(x, q))^2 \, dx = \pi.$$

As $q \to 0$ with $n \neq 0$, $C_n(q) \to 0$, $S_n(q) \to 0$, $C_n(q) f_n(z, q) \to \sin nz$, and $S_n(q) g_n(z, q) \to \cos nz$. This determines the signs of $C_n(q)$ and $S_n(q)$. (Other normalizations for $C_n(q)$ and $S_n(q)$ can be found in the literature, but most formulas—including connection formulas—are unaffected since $\operatorname{fe}_n(z,q)/C_n(q)$ and $\operatorname{ge}_n(z,q)/S_n(q)$ are invariant.)

28.5.6
$$C_{2m}(-q) = C_{2m}(q),$$
$$C_{2m+1}(-q) = S_{2m+1}(q),$$
$$S_{2m+2}(-q) = S_{2m+2}(q).$$

For $q = 0$,

28.5.7
$$\operatorname{fe}_0(z,0) = z, \quad \operatorname{fe}_n(z,0) = \sin nz,$$
$$\operatorname{ge}_n(z,0) = \cos nz, \qquad n = 1,2,3,\ldots;$$
compare (28.2.29).

As a consequence of the factor z on the right-hand sides of (28.5.1), (28.5.2), all solutions of Mathieu's equation that are linearly independent of the periodic solutions are unbounded as $z \to \pm\infty$ on $\mathbb{R}$.

Wronskians

28.5.8 $\quad \mathscr{W}\{\operatorname{ce}_n, \operatorname{fe}_n\} = \operatorname{ce}_n(0,q)\operatorname{fe}_n'(0,q),$

28.5.9 $\quad \mathscr{W}\{\operatorname{se}_n, \operatorname{ge}_n\} = -\operatorname{se}_n'(0,q)\operatorname{ge}_n(0,q).$

See (28.22.12) for $\operatorname{fe}_n'(0,q)$ and $\operatorname{ge}_n(0,q)$.

For further information on $C_n(q)$, $S_n(q)$, and expansions of $f_n(z,q)$, $g_n(z,q)$ in Fourier series or in series of ce_n, se_n functions, see McLachlan (1947, Chapter VII) or Meixner and Schäfke (1954, §2.72).

28.5(ii) Graphics: Line Graphs of Second Solutions of Mathieu's Equation

Odd Second Solutions

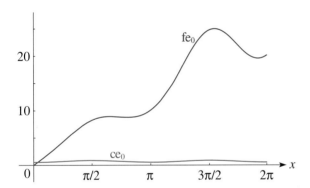

Figure 28.5.1: $\operatorname{fe}_0(x,0.5)$ for $0 \leq x \leq 2\pi$ and (for comparison) $\operatorname{ce}_0(x,0.5)$.

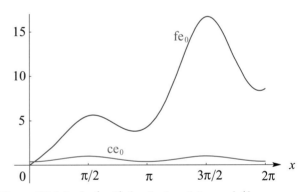

Figure 28.5.2: $\operatorname{fe}_0(x,1)$ for $0 \leq x \leq 2\pi$ and (for comparison) $\operatorname{ce}_0(x,1)$.

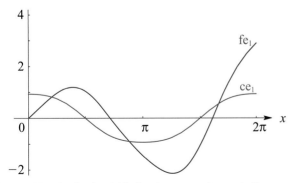

Figure 28.5.3: $\operatorname{fe}_1(x,0.5)$ for $0 \leq x \leq 2\pi$ and (for comparison) $\operatorname{ce}_1(x,0.5)$.

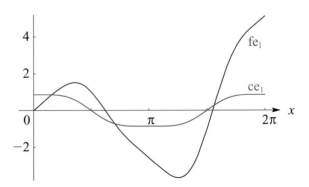

Figure 28.5.4: $\operatorname{fe}_1(x,1)$ for $0 \leq x \leq 2\pi$ and (for comparison) $\operatorname{ce}_1(x,1)$.

Even Second Solutions

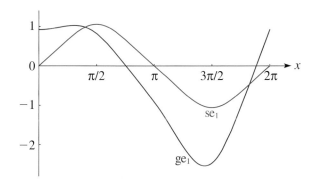

Figure 28.5.5: $ge_1(x, 0.5)$ for $0 \leq x \leq 2\pi$ and (for comparison) $se_1(x, 0.5)$.

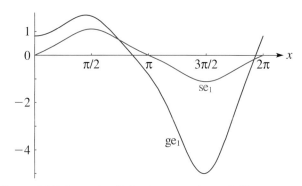

Figure 28.5.6: $ge_1(x, 1)$ for $0 \leq x \leq 2\pi$ and (for comparison) $se_1(x, 1)$.

28.6 Expansions for Small q

28.6(i) Eigenvalues

Leading terms of the power series for $a_m(q)$ and $b_m(q)$ for $m \leq 6$ are:

28.6.1 $\quad a_0(q) = -\frac{1}{2}q^2 + \frac{7}{128}q^4 - \frac{29}{2304}q^6 + \frac{68687}{188\,74368}q^8 + \cdots,$

28.6.2 $\quad a_1(q) = 1 + q - \frac{1}{8}q^2 - \frac{1}{64}q^3 - \frac{1}{1536}q^4 + \frac{11}{36864}q^5 + \frac{49}{5\,89824}q^6 + \frac{55}{94\,37184}q^7 - \frac{83}{353\,89440}q^8 + \cdots,$

28.6.3 $\quad b_1(q) = 1 - q - \frac{1}{8}q^2 + \frac{1}{64}q^3 - \frac{1}{1536}q^4 - \frac{11}{36864}q^5 + \frac{49}{5\,89824}q^6 - \frac{55}{94\,37184}q^7 - \frac{83}{353\,89440}q^8 + \cdots,$

28.6.4 $\quad a_2(q) = 4 + \frac{5}{12}q^2 - \frac{763}{13824}q^4 + \frac{10\,02401}{796\,26240}q^6 - \frac{16690\,68401}{45\,86471\,42400}q^8 + \cdots,$

28.6.5 $\quad b_2(q) = 4 - \frac{1}{12}q^2 + \frac{5}{13824}q^4 - \frac{289}{796\,26240}q^6 + \frac{21391}{45\,86471\,42400}q^8 + \cdots,$

28.6.6 $\quad a_3(q) = 9 + \frac{1}{16}q^2 + \frac{1}{64}q^3 + \frac{13}{20480}q^4 - \frac{5}{16384}q^5 - \frac{1961}{235\,92960}q^6 - \frac{609}{1048\,57600}q^7 + \cdots,$

28.6.7 $\quad b_3(q) = 9 + \frac{1}{16}q^2 - \frac{1}{64}q^3 + \frac{13}{20480}q^4 + \frac{5}{16384}q^5 - \frac{1961}{235\,92960}q^6 + \frac{609}{1048\,57600}q^7 + \cdots,$

28.6.8 $\quad a_4(q) = 16 + \frac{1}{30}q^2 + \frac{433}{8\,64000}q^4 - \frac{5701}{27216\,00000}q^6 + \cdots,$

28.6.9 $\quad b_4(q) = 16 + \frac{1}{30}q^2 - \frac{317}{8\,64000}q^4 + \frac{10049}{27216\,00000}q^6 + \cdots,$

28.6.10 $\quad a_5(q) = 25 + \frac{1}{48}q^2 + \frac{11}{7\,74144}q^4 + \frac{1}{1\,47456}q^5 + \frac{37}{8918\,13888}q^6 + \cdots,$

28.6.11 $\quad b_5(q) = 25 + \frac{1}{48}q^2 + \frac{11}{7\,74144}q^4 - \frac{1}{1\,47456}q^5 + \frac{37}{8918\,13888}q^6 + \cdots,$

28.6.12 $\quad a_6(q) = 36 + \frac{1}{70}q^2 + \frac{187}{439\,04000}q^4 + \frac{67\,43617}{9293\,59872\,00000}q^6 + \cdots,$

28.6.13 $\quad b_6(q) = 36 + \frac{1}{70}q^2 + \frac{187}{439\,04000}q^4 - \frac{58\,61633}{9293\,59872\,00000}q^6 + \cdots.$

Leading terms of the of the power series for $m = 7, 8, 9, \ldots$ are:

28.6.14 $\quad \left.\begin{array}{c} a_m(q) \\ b_m(q) \end{array}\right\} = m^2 + \frac{1}{2(m^2-1)}q^2 + \frac{5m^2+7}{32(m^2-1)^3(m^2-4)}q^4 + \frac{9m^4+58m^2+29}{64(m^2-1)^5(m^2-4)(m^2-9)}q^6 + \cdots.$

The coefficients of the power series of $a_{2n}(q)$, $b_{2n}(q)$ and also $a_{2n+1}(q)$, $b_{2n+1}(q)$ are the same until the terms in q^{2n-2} and q^{2n}, respectively. Then

28.6.15 $\quad a_m(q) - b_m(q) = \frac{2q^m}{(2^{m-1}(m-1)!)^2}\left(1 + O(q^2)\right).$

Higher coefficients in the foregoing series can be found by equating coefficients in the following continued-fraction equations:

28.6.16

$a - (2n)^2 - \cfrac{q^2}{a-(2n-2)^2-} \cfrac{q^2}{a-(2n-4)^2-} \cdots \cfrac{q^2}{a-2^2-} \cfrac{2q^2}{a} = -\cfrac{q^2}{(2n+2)^2-a-} \cfrac{q^2}{(2n+4)^2-a-} \cdots, \quad a = a_{2n}(q),$

28.6.17
$$a - (2n+1)^2 - \cfrac{q^2}{a-(2n-1)^2-} \cdots \cfrac{q^2}{a-3^2-} \cfrac{q^2}{a-1^2-q} = -\cfrac{q^2}{(2n+3)^2-a-} \cfrac{q^2}{(2n+5)^2-a-} \cdots, \quad a = a_{2n+1}(q),$$

28.6.18
$$a - (2n+1)^2 - \cfrac{q^2}{a-(2n-1)^2-} \cdots \cfrac{q^2}{a-3^2-} \cfrac{q^2}{a-1^2+q} = -\cfrac{q^2}{(2n+3)^2-a-} \cfrac{q^2}{(2n+5)^2-a-} \cdots, \quad a = b_{2n+1}(q),$$

28.6.19
$$a - (2n+2)^2 - \cfrac{q^2}{a-(2n)^2-} \cfrac{q^2}{a-(2n-2)^2-} \cdots \cfrac{q^2}{a-2^2} = -\cfrac{q^2}{(2n+4)^2-a-} \cfrac{q^2}{(2n+6)^2-a-} \cdots, \quad a = b_{2n+2}(q).$$

Numerical values of the radii of convergence $\rho_n^{(j)}$ of the power series (28.6.1)–(28.6.14) for $n = 0, 1, \ldots, 9$ are given in Table 28.6.1. Here $j = 1$ for $a_{2n}(q)$, $j = 2$ for $b_{2n+2}(q)$, and $j = 3$ for $a_{2n+1}(q)$ and $b_{2n+1}(q)$. (Table 28.6.1 is reproduced from Meixner et al. (1980, §2.4).)

Table 28.6.1: Radii of convergence for power-series expansions of eigenvalues of Mathieu's equation.

n	$\rho_n^{(1)}$	$\rho_n^{(2)}$	$\rho_n^{(3)}$
0 or 1	1.46876 86138	6.92895 47588	3.76995 74940
2	7.26814 68935	16.80308 98254	11.27098 52655
3	16.47116 58923	30.09677 28376	22.85524 71216
4	30.42738 20960	48.13638 18593	38.52292 50099
5	47.80596 57026	69.59879 32769	58.27413 84472
6	69.92930 51764	95.80595 67052	82.10894 36067
7	95.47527 27072	125.43541 1314	110.02736 9210
8	125.76627 89677	159.81025 4642	142.02943 1279
9	159.47921 26694	197.60667 8692	178.11513 940

It is conjectured that for large n, the radii increase in proportion to the square of the eigenvalue number n; see Meixner et al. (1980, §2.4). It is known that

28.6.20
$$\liminf_{n \to \infty} \frac{\rho_n^{(j)}}{n^2} \geq kk'(K(k))^2 = 2.04183\ 4\ldots,$$

where k is the unique root of the equation $2E(k) = K(k)$ in the interval $(0, 1)$, and $k' = \sqrt{1-k^2}$. For $E(k)$ and $K(k)$ see §19.2(ii).

28.6(ii) Functions ce_n and se_n

Leading terms of the power series for the normalized functions are:

28.6.21 $2^{1/2} \mathrm{ce}_0(z, q) = 1 - \tfrac{1}{2}q\cos 2z + \tfrac{1}{32}q^2(\cos 4z - 2) - \tfrac{1}{128}q^3\left(\tfrac{1}{9}\cos 6z - 11\cos 2z\right) + \cdots,$

28.6.22
$$\mathrm{ce}_1(z,q) = \cos z - \tfrac{1}{8}q\cos 3z + \tfrac{1}{128}q^2\left(\tfrac{2}{3}\cos 5z - 2\cos 3z - \cos z\right) - \tfrac{1}{1024}q^3\left(\tfrac{1}{9}\cos 7z - \tfrac{8}{9}\cos 5z - \tfrac{1}{3}\cos 3z + 2\cos z\right) + \cdots,$$

28.6.23
$$\mathrm{se}_1(z,q) = \sin z - \tfrac{1}{8}q\sin 3z + \tfrac{1}{128}q^2\left(\tfrac{2}{3}\sin 5z + 2\sin 3z - \sin z\right) - \tfrac{1}{1024}q^3\left(\tfrac{1}{9}\sin 7z + \tfrac{8}{9}\sin 5z - \tfrac{1}{3}\sin 3z - 2\sin z\right) + \cdots,$$

28.6.24 $\mathrm{ce}_2(z,q) = \cos 2z - \tfrac{1}{4}q\left(\tfrac{1}{3}\cos 4z - 1\right) + \tfrac{1}{128}q^2\left(\tfrac{1}{3}\cos 6z - \tfrac{76}{9}\cos 2z\right) + \cdots,$

28.6.25 $\mathrm{se}_2(z,q) = \sin 2z - \tfrac{1}{12}q\sin 4z + \tfrac{1}{128}q^2\left(\tfrac{1}{3}\sin 6z - \tfrac{4}{9}\sin 2z\right) + \cdots.$

For $m = 3, 4, 5, \ldots,$

28.6.26
$$\mathrm{ce}_m(z,q) = \cos mz - \frac{q}{4}\left(\frac{1}{m+1}\cos(m+2)z - \frac{1}{m-1}\cos(m-2)z\right)$$
$$+ \frac{q^2}{32}\left(\frac{1}{(m+1)(m+2)}\cos(m+4)z + \frac{1}{(m-1)(m-2)}\cos(m-4)z - \frac{2(m^2+1)}{(m^2-1)^2}\cos mz\right) + \cdots.$$

For the corresponding expansions of $\mathrm{se}_m(z, q)$ for $m = 3, 4, 5, \ldots$ change cos to sin everywhere in (28.6.26).

The radii of convergence of the series (28.6.21)–(28.6.26) are the same as the radii of the corresponding series for $a_n(q)$ and $b_n(q)$; compare Table 28.6.1 and (28.6.20).

28.7 Analytic Continuation of Eigenvalues

As functions of q, $a_n(q)$ and $b_n(q)$ can be continued analytically in the complex q-plane. The only singularities are algebraic branch points, with $a_n(q)$ and $b_n(q)$ finite at these points. The number of branch points is infinite, but countable, and there are no finite limit points. In consequence, the functions can be defined uniquely by introducing suitable cuts in the q-plane. See Meixner and Schäfke (1954, §2.22). The branch points are called the *exceptional values*, and the other points *normal values*. The normal values are simple roots of the corresponding equations (28.2.21) and (28.2.22). All real values of q are normal values. To 4D the first branch points between $a_0(q)$ and $a_2(q)$ are at $q_0 = \pm i1.4688$ with $a_0(q_0) = a_2(q_0) = 2.0886$, and between $b_2(q)$ and $b_4(q)$ they are at $q_1 = \pm i6.9289$ with $b_2(q_1) = b_4(q_1) = 11.1904$. For real q with $|q| < |q_0|$, $a_0(iq)$ and $a_2(iq)$ are real-valued, whereas for real q with $|q| > |q_0|$, $a_0(iq)$ and $a_2(iq)$ are complex conjugates. See also Mulholland and Goldstein (1929), Bouwkamp (1948), Meixner et al. (1980), Hunter and Guerrieri (1981), Hunter (1981), and Shivakumar and Xue (1999).

For a visualization of the first branch point of $a_0(i\hat{q})$ and $a_2(i\hat{q})$ see Figure 28.7.1.

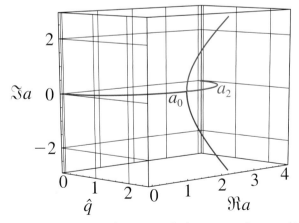

Figure 28.7.1: Branch point of the eigenvalues $a_0(i\hat{q})$ and $a_2(i\hat{q})$: $0 \leq \hat{q} \leq 2.5$.

All the $a_{2n}(q)$, $n = 0, 1, 2, \ldots$, can be regarded as belonging to a complete analytic function (in the large). Therefore $w'_{\mathrm{I}}(\tfrac{1}{2}\pi; a, q)$ is irreducible, in the sense that it cannot be decomposed into a product of entire functions that contain its zeros; see Meixner et al. (1980, p. 88). Analogous statements hold for $a_{2n+1}(q)$, $b_{2n+1}(q)$, and $b_{2n+2}(q)$, also for $n = 0, 1, 2, \ldots$. Closely connected with the preceding statements, we have

28.7.1
$$\sum_{n=0}^{\infty} \left(a_{2n}(q) - (2n)^2\right) = 0,$$

28.7.2
$$\sum_{n=0}^{\infty} \left(a_{2n+1}(q) - (2n+1)^2\right) = q,$$

28.7.3
$$\sum_{n=0}^{\infty} \left(b_{2n+1}(q) - (2n+1)^2\right) = -q,$$

28.7.4
$$\sum_{n=0}^{\infty} \left(b_{2n+2}(q) - (2n+2)^2\right) = 0.$$

28.8 Asymptotic Expansions for Large q

28.8(i) Eigenvalues

Denote $h = \sqrt{q}$ and $s = 2m + 1$. Then as $h \to +\infty$ with $m = 0, 1, 2, \ldots$,

28.8.1
$$\begin{aligned}a_m(h^2)\\b_{m+1}(h^2)\end{aligned}\bigg\} \sim -2h^2 + 2sh - \frac{1}{8}(s^2 + 1) - \frac{1}{2^7 h}(s^3 + 3s)$$
$$- \frac{1}{2^{12} h^2}(5s^4 + 34s^2 + 9)$$
$$- \frac{1}{2^{17} h^3}(33s^5 + 410s^3 + 405s)$$
$$- \frac{1}{2^{20} h^4}(63s^6 + 1260s^4 + 2943s^2 + 486)$$
$$- \frac{1}{2^{25} h^5}(527s^7 + 15617s^5 + 69001s^3 + 41607s) + \cdots.$$

For error estimates see Kurz (1979), and for graphical interpretation see Figure 28.2.1. Also,

28.8.2
$$b_{m+1}(h^2) - a_m(h^2)$$
$$= \frac{2^{4m+5}}{m!}\left(\frac{2}{\pi}\right)^{1/2} h^{m+(3/2)} e^{-4h}$$
$$\times \left(1 - \frac{6m^2 + 14m + 7}{32h} + O\left(\frac{1}{h^2}\right)\right).$$

28.8(ii) Sips' Expansions

Let $x = \tfrac{1}{2}\pi + \lambda h^{-1/4}$, where λ is a real constant such that $|\lambda| < 2^{1/4}$. Also let $\xi = 2\sqrt{h}\cos x$ and $D_m(\xi) = e^{-\xi^2/4} \mathit{He}_m(\xi)$ (§18.3). Then as $h \to +\infty$

28.8.3
$$\mathrm{ce}_m(x, h^2) = \widehat{C}_m \left(U_m(\xi) + V_m(\xi)\right),$$
$$\frac{\mathrm{se}_{m+1}(x, h^2)}{\sin x} = \widehat{S}_m \left(U_m(\xi) - V_m(\xi)\right),$$

where

28.8.4
$$U_m(\xi) \sim D_m(\xi) - \frac{1}{2^6 h}\left(D_{m+4}(\xi) - 4!\binom{m}{4}D_{m-4}(\xi)\right)$$
$$+ \frac{1}{2^{13}h^2}\left(D_{m+8}(\xi) - 2^5(m+2)D_{m+4}(\xi) + 4! \, 2^5(m-1)\binom{m}{4}D_{m-4}(\xi) + 8!\binom{m}{8}D_{m-8}(\xi)\right) + \cdots,$$

28.8.5
$$V_m(\xi) \sim \frac{1}{2^4 h}\left(-D_{m+2}(\xi) - m(m-1)D_{m-2}(\xi)\right) + \frac{1}{2^{10}h^2}\Bigg(D_{m+6}(\xi) + (m^2 - 25m - 36)D_{m+2}(\xi)$$
$$- m(m-1)(m^2 + 27m - 10)D_{m-2}(\xi) + 6!\binom{m}{6}D_{m-6}(\xi)\Bigg) + \cdots,$$

and

28.8.6
$$\widehat{C}_m \sim \left(\frac{\pi h}{2(m!)^2}\right)^{1/4}\left(1 + \frac{2m+1}{8h} + \frac{m^4 + 2m^3 + 263m^2 + 262m + 108}{2048h^2} + \cdots\right)^{-1/2},$$

28.8.7
$$\widehat{S}_m \sim \left(\frac{\pi h}{2(m!)^2}\right)^{1/4}\left(1 - \frac{2m+1}{8h} + \frac{m^4 + 2m^3 - 121m^2 - 122m - 84}{2048h^2} + \cdots\right)^{-1/2}.$$

These results are derived formally in Sips (1949, 1959, 1965). See also Meixner and Schäfke (1954, §2.84).

28.8(iii) Goldstein's Expansions

Let $x = \frac{1}{2}\pi - \mu h^{-1/4}$, where μ is a constant such that $\mu \geq 1$, and $s = 2m + 1$. Then as $h \to +\infty$

28.8.8
$$\frac{\operatorname{ce}_m(x, h^2)}{\operatorname{ce}_m(0, h^2)} = \frac{2^{m-(1/2)}}{\sigma_m}\left(W_m^+(x)(P_m(x) - Q_m(x)) + W_m^-(x)(P_m(x) + Q_m(x))\right),$$
$$\frac{\operatorname{se}_{m+1}(x, h^2)}{\operatorname{se}'_{m+1}(0, h^2)} = \frac{2^{m-(1/2)}}{\tau_{m+1}}\left(W_m^+(x)(P_m(x) - Q_m(x)) - W_m^-(x)(P_m(x) + Q_m(x))\right),$$

where

28.8.9
$$W_m^\pm(x) = \frac{e^{\pm 2h \sin x}}{(\cos x)^{m+1}}\begin{cases}\left(\cos\left(\frac{1}{2}x + \frac{1}{4}\pi\right)\right)^{2m+1}, \\ \left(\sin\left(\frac{1}{2}x + \frac{1}{4}\pi\right)\right)^{2m+1},\end{cases}$$

and

28.8.10
$$\sigma_m \sim 1 + \frac{s}{2^3 h} + \frac{4s^2 + 3}{2^7 h^2} + \frac{19s^3 + 59s}{2^{11}h^3} + \cdots, \qquad \tau_{m+1} \sim 2h - \frac{1}{4}s - \frac{2s^2 + 3}{2^6 h} - \frac{7s^3 + 47s}{2^{10}h^2} - \cdots,$$

28.8.11
$$P_m(x) \sim 1 + \frac{s}{2^3 h \cos^2 x} + \frac{1}{h^2}\left(\frac{s^4 + 86s^2 + 105}{2^{11} \cos^4 x} - \frac{s^4 + 22s^2 + 57}{2^{11}\cos^2 x}\right) + \cdots,$$

28.8.12
$$Q_m(x) \sim \frac{\sin x}{\cos^2 x}\left(\frac{1}{2^5 h}(s^2 + 3) + \frac{1}{2^9 h^2}\left(s^3 + 3s + \frac{4s^3 + 44s}{\cos^2 x}\right)\right) + \cdots.$$

28.8(iv) Uniform Approximations

Barrett's Expansions

Barrett (1981) supplies asymptotic approximations for numerically satisfactory pairs of solutions of both Mathieu's equation (28.2.1) and the modified Mathieu equation (28.20.1). The approximations apply when the parameters a and q are real and large, and are uniform with respect to various regions in the z-plane. The approximants are elementary functions, Airy functions, Bessel functions, and parabolic cylinder functions; compare §2.8. It is stated that corresponding uniform approximations can be obtained for other solutions, including the eigensolutions, of the differential equations by application of the results, but these approximations are not included.

Dunster's Approximations

Dunster (1994a) supplies uniform asymptotic approximations for numerically satisfactory pairs of solutions of Mathieu's equation (28.2.1). These approximations apply when q and a are real and $q \to \infty$. They are uniform with respect to a when $-2q \leq a \leq (2-\delta)q$, where δ is an arbitrary constant such that $0 < \delta < 4$, and also with respect to z in the semi-infinite strip given by $0 \leq \Re z \leq \pi$ and $\Im z \geq 0$.

The approximations are expressed in terms of Whittaker functions $W_{\kappa,\mu}(z)$ and $M_{\kappa,\mu}(z)$ with $\mu = \frac{1}{4}$; compare §2.8(vi). They are derived by rigorous analysis and accompanied by strict and realistic error bounds. With additional restrictions on z, uniform asymptotic approximations for solutions of (28.2.1) and (28.20.1) are also obtained in terms of elementary functions by re-expansions of the Whittaker functions; compare §2.8(ii).

Subsequently the asymptotic solutions involving either elementary or Whittaker functions are identified in terms of the Floquet solutions $\mathrm{me}_\nu(z,q)$ (§28.12(ii)) and modified Mathieu functions $\mathrm{M}_\nu^{(j)}(z,h)$ (§28.20(iii)).

For related results see Langer (1934) and Sharples (1967, 1971).

28.9 Zeros

For real q each of the functions $\mathrm{ce}_{2n}(z,q)$, $\mathrm{se}_{2n+1}(z,q)$, $\mathrm{ce}_{2n+1}(z,q)$, and $\mathrm{se}_{2n+2}(z,q)$ has exactly n zeros in $0 < z < \frac{1}{2}\pi$. They are continuous in q. For $q \to \infty$ the zeros of $\mathrm{ce}_{2n}(z,q)$ and $\mathrm{se}_{2n+1}(z,q)$ approach asymptotically the zeros of $He_{2n}(q^{1/4}(\pi - 2z))$, and the zeros of $\mathrm{ce}_{2n+1}(z,q)$ and $\mathrm{se}_{2n+2}(z,q)$ approach asymptotically the zeros of $He_{2n+1}(q^{1/4}(\pi - 2z))$. Here $He_n(z)$ denotes the Hermite polynomial of degree n (§18.3). Furthermore, for $q > 0$ $\mathrm{ce}_m(z,q)$ and $\mathrm{se}_m(z,q)$ also have purely imaginary zeros that correspond uniquely to the purely imaginary z-zeros of $J_m(2\sqrt{q}\cos z)$ (§10.21(i)), and they are asymptotically equal as $q \to 0$ and $|\Im z| \to \infty$. There are no zeros within the strip $|\Re z| < \frac{1}{2}\pi$ other than those on the real and imaginary axes.

For further details see McLachlan (1947, pp. 234–239) and Meixner and Schäfke (1954, §§2.331, 2.8, 2.81, and 2.85).

28.10 Integral Equations

28.10(i) Equations with Elementary Kernels

With the notation of §28.4 for Fourier coefficients,

28.10.1
$$\frac{2}{\pi}\int_0^{\pi/2} \cos(2h\cos z \cos t)\,\mathrm{ce}_{2n}(t,h^2)\,dt = \frac{A_0^{2n}(h^2)}{\mathrm{ce}_{2n}(\tfrac{1}{2}\pi, h^2)}\,\mathrm{ce}_{2n}(z,h^2),$$

28.10.2
$$\frac{2}{\pi}\int_0^{\pi/2} \cosh(2h\sin z \sin t)\,\mathrm{ce}_{2n}(t,h^2)\,dt = \frac{A_0^{2n}(h^2)}{\mathrm{ce}_{2n}(0, h^2)}\,\mathrm{ce}_{2n}(z,h^2),$$

28.10.3
$$\frac{2}{\pi}\int_0^{\pi/2} \sin(2h\cos z \cos t)\,\mathrm{ce}_{2n+1}(t,h^2)\,dt = -\frac{hA_1^{2n+1}(h^2)}{\mathrm{ce}'_{2n+1}(\tfrac{1}{2}\pi, h^2)}\,\mathrm{ce}_{2n+1}(z,h^2),$$

28.10.4
$$\frac{2}{\pi}\int_0^{\pi/2} \cos z \cos t \cosh(2h\sin z \sin t)\,\mathrm{ce}_{2n+1}(t,h^2)\,dt = \frac{A_1^{2n+1}(h^2)}{2\,\mathrm{ce}_{2n+1}(0, h^2)}\,\mathrm{ce}_{2n+1}(z,h^2),$$

28.10.5
$$\frac{2}{\pi}\int_0^{\pi/2} \sinh(2h\sin z \sin t)\,\mathrm{se}_{2n+1}(t,h^2)\,dt = \frac{hB_1^{2n+1}(h^2)}{\mathrm{se}'_{2n+1}(0, h^2)}\,\mathrm{se}_{2n+1}(z,h^2),$$

28.10.6
$$\frac{2}{\pi}\int_0^{\pi/2} \sin z \sin t \cos(2h\cos z \cos t)\,\mathrm{se}_{2n+1}(t,h^2)\,dt = \frac{B_1^{2n+1}(h^2)}{2\,\mathrm{se}_{2n+1}(\tfrac{1}{2}\pi, h^2)}\,\mathrm{se}_{2n+1}(z,h^2),$$

28.10.7
$$\frac{2}{\pi}\int_0^{\pi/2} \sin z \sin t \sin(2h\cos z \cos t)\,\mathrm{se}_{2n+2}(t,h^2)\,dt = -\frac{hB_2^{2n+2}(h^2)}{2\,\mathrm{se}'_{2n+2}(\tfrac{1}{2}\pi, h^2)}\,\mathrm{se}_{2n+2}(z,h^2),$$

28.10.8
$$\frac{2}{\pi}\int_0^{\pi/2} \cos z \cos t \sinh(2h\sin z \sin t)\,\mathrm{se}_{2n+2}(t,h^2)\,dt = \frac{hB_2^{2n+2}(h^2)}{2\,\mathrm{se}'_{2n+2}(0, h^2)}\,\mathrm{se}_{2n+2}(z,h^2).$$

28.10(ii) Equations with Bessel-Function Kernels

28.10.9
$$\int_0^{\pi/2} J_0\left(2\sqrt{q(\cos^2\tau - \sin^2\zeta)}\right)\mathrm{ce}_{2n}(\tau,q)\,d\tau = w_{\mathrm{II}}(\tfrac{1}{2}\pi; a_{2n}(q), q)\,\mathrm{ce}_{2n}(\zeta,q),$$

28.10.10
$$\int_0^{\pi} J_0(2\sqrt{q}(\cos\tau + \cos\zeta))\,\mathrm{ce}_n(\tau,q)\,d\tau = w_{\mathrm{II}}(\pi; a_n(q), q)\,\mathrm{ce}_n(\zeta,q).$$

28.10(iii) Further Equations

See §28.28. See also Prudnikov et al. (1990, pp. 359–368), Erdélyi et al. (1955, p. 115), and Gradshteyn and Ryzhik (2000, pp. 755–759). For relations with variable boundaries see Volkmer (1983).

28.11 Expansions in Series of Mathieu Functions

Let $f(z)$ be a 2π-periodic function that is analytic in an open doubly-infinite strip S that contains the real axis, and q be a normal value (§28.7). Then

28.11.1
$$f(z) = \alpha_0 \operatorname{ce}_0(z,q) + \sum_{n=1}^{\infty} \left(\alpha_n \operatorname{ce}_n(z,q) + \beta_n \operatorname{se}_n(z,q) \right),$$

where

28.11.2
$$\alpha_n = \frac{1}{\pi} \int_0^{2\pi} f(x) \operatorname{ce}_n(x,q) \, dx,$$
$$\beta_n = \frac{1}{\pi} \int_0^{2\pi} f(x) \operatorname{se}_n(x,q) \, dx.$$

The series (28.11.1) converges absolutely and uniformly on any compact subset of the strip S. See Meixner and Schäfke (1954, §2.28), and for expansions in the case of the exceptional values of q see Meixner et al. (1980, p. 33).

Examples

With the notation of §28.4,

28.11.3
$$1 = 2 \sum_{n=0}^{\infty} A_0^{2n}(q) \operatorname{ce}_{2n}(z,q),$$

28.11.4
$$\cos 2mz = \sum_{n=0}^{\infty} A_{2m}^{2n}(q) \operatorname{ce}_{2n}(z,q), \quad m \neq 0,$$

28.11.5
$$\cos(2m+1)z = \sum_{n=0}^{\infty} A_{2m+1}^{2n+1}(q) \operatorname{ce}_{2n+1}(z,q),$$

28.11.6
$$\sin(2m+1)z = \sum_{n=0}^{\infty} B_{2m+1}^{2n+1}(q) \operatorname{se}_{2n+1}(z,q),$$

28.11.7
$$\sin(2m+2)z = \sum_{n=0}^{\infty} B_{2m+2}^{2n+2}(q) \operatorname{se}_{2n+2}(z,q).$$

Mathieu Functions of Noninteger Order

28.12 Definitions and Basic Properties

28.12(i) Eigenvalues $\lambda_{\nu+2n}(q)$

The introduction to the eigenvalues and the functions of general order proceeds as in §§28.2(i), 28.2(ii), and 28.2(iii), except that we now restrict $\hat{\nu} \neq 0, 1$; equivalently $\nu \neq n$. In consequence, for the Floquet solutions $w(z)$ the factor $e^{\pi i \nu}$ in (28.2.14) is no longer ± 1.

For given ν (or $\cos(\nu\pi)$) and q, equation (28.2.16) determines an infinite discrete set of values of a, denoted by $\lambda_{\nu+2n}(q)$, $n = 0, \pm 1, \pm 2, \ldots$. When $q = 0$ Equation (28.2.16) has simple roots, given by

28.12.1
$$\lambda_{\nu+2n}(0) = (\nu + 2n)^2.$$

For other values of q, $\lambda_{\nu+2n}(q)$ is determined by analytic continuation. Without loss of generality, from now on we replace $\nu + 2n$ by ν.

For change of signs of ν and q,

28.12.2
$$\lambda_\nu(-q) = \lambda_\nu(q) = \lambda_{-\nu}(q).$$

As in §28.7 values of q for which (28.2.16) has simple roots λ are called *normal values* with respect to ν. For real values of ν and q all the $\lambda_\nu(q)$ are real, and q is normal. For graphical interpretation see Figure 28.13.1. To complete the definition we require

28.12.3
$$\lambda_m(q) = \begin{cases} a_m(q), & m = 0, 1, \ldots, \\ b_{-m}(q), & m = -1, -2, \ldots. \end{cases}$$

As a function of ν with fixed q ($\neq 0$), $\lambda_\nu(q)$ is discontinuous at $\nu = \pm 1, \pm 2, \ldots$. See Figure 28.13.2.

28.12(ii) Eigenfunctions $\operatorname{me}_\nu(z,q)$

Two eigenfunctions correspond to each eigenvalue $a = \lambda_\nu(q)$. The Floquet solution with respect to ν is denoted by $\operatorname{me}_\nu(z,q)$. For $q = 0$,

28.12.4
$$\operatorname{me}_\nu(z,0) = e^{i\nu z}.$$

The other eigenfunction is $\operatorname{me}_\nu(-z,q)$, a Floquet solution with respect to $-\nu$ with $a = \lambda_\nu(q)$. If q is a normal value of the corresponding equation (28.2.16), then these functions are uniquely determined as analytic functions of z and q by the normalization

28.12.5
$$\int_0^\pi \operatorname{me}_\nu(x,q) \operatorname{me}_\nu(-x,q) \, dx = \pi.$$

They have the following pseudoperiodic and orthogonality properties:

28.12.6
$$\operatorname{me}_\nu(z+\pi, q) = e^{\pi i \nu} \operatorname{me}_\nu(z,q),$$

28.12.7
$$\int_0^\pi \operatorname{me}_{\nu+2m}(x,q) \operatorname{me}_{\nu+2n}(-x,q) \, dx = 0, \quad m \neq n.$$

For changes of sign of ν, q, and z,

28.12.8
$$\operatorname{me}_{-\nu}(z,q) = \operatorname{me}_\nu(-z,q),$$

28.12.9
$$\operatorname{me}_\nu(z,-q) = e^{i\nu\pi/2} \operatorname{me}_\nu\left(z - \tfrac{1}{2}\pi, q\right),$$

28.12.10
$$\overline{\operatorname{me}_\nu(z,q)} = \operatorname{me}_{\bar\nu}(-\bar z, \bar q).$$

(28.12.10) is not valid for cuts on the real axis in the q-plane for special complex values of ν; but it remains valid for small q; compare §28.7.

28.13 Graphics

To complete the definitions of the me_ν functions we set

28.12.11
$$\text{me}_n(z,q) = \sqrt{2}\,\text{ce}_n(z,q), \qquad n=0,1,2,\ldots,$$
$$\text{me}_{-n}(z,q) = -\sqrt{2}i\,\text{se}_n(z,q), \qquad n=1,2,\ldots;$$

compare (28.12.3). However, these functions are *not* the limiting values of $\text{me}_{\pm\nu}(z,q)$ as $\nu \to n\, (\neq 0)$.

28.12(iii) Functions $\text{ce}_\nu(z,q)$, $\text{se}_\nu(z,q)$, when $\nu \notin \mathbb{Z}$

28.12.12 $\quad \text{ce}_\nu(z,q) = \tfrac{1}{2}\left(\text{me}_\nu(z,q) + \text{me}_\nu(-z,q)\right),$

28.12.13 $\quad \text{se}_\nu(z,q) = -\tfrac{1}{2}i\left(\text{me}_\nu(z,q) - \text{me}_\nu(-z,q)\right).$

These functions are real-valued for real ν, real q, and $z=x$, whereas $\text{me}_\nu(x,q)$ is complex. When $\nu = s/m$ is a rational number, but not an integer, all solutions of Mathieu's equation are periodic with period $2m\pi$.

For change of signs of ν and z,

28.12.14 $\quad \text{ce}_\nu(z,q) = \text{ce}_\nu(-z,q) = \text{ce}_{-\nu}(z,q),$

28.12.15 $\quad \text{se}_\nu(z,q) = -\text{se}_\nu(-z,q) = -\text{se}_{-\nu}(z,q).$

Again, the limiting values of $\text{ce}_\nu(z,q)$ and $\text{se}_\nu(z,q)$ as $\nu \to n\, (\neq 0)$ are *not* the functions $\text{ce}_n(z,q)$ and $\text{se}_n(z,q)$ defined in §28.2(vi). Compare e.g. Figure 28.13.3.

28.13 Graphics

28.13(i) Eigenvalues $\lambda_\nu(q)$ for General ν

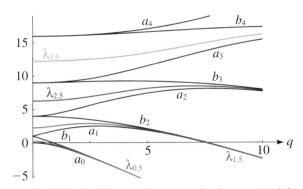

Figure 28.13.1: $\lambda_\nu(q)$ as a function of q for $\nu = 0.5(1)3.5$ and $a_n(q), b_n(q)$ for $n=0,1,2,3,4$ (a's), $n=1,2,3,4$ (b's). (Compare Figure 28.2.1.)

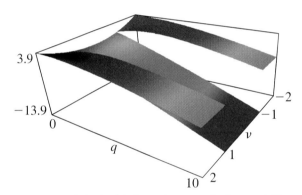

Figure 28.13.2: $\lambda_\nu(q)$ for $-2 < \nu < 2$, $0 \leq q \leq 10$.

28.13(ii) Solutions $\text{ce}_\nu(x,q)$, $\text{se}_\nu(x,q)$, and $\text{me}_\nu(x,q)$ for General ν

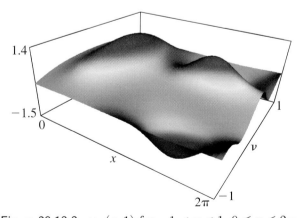

Figure 28.13.3: $\text{ce}_\nu(x,1)$ for $-1 < \nu < 1$, $0 \leq x \leq 2\pi$.

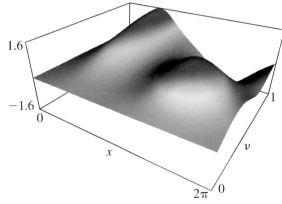

Figure 28.13.4: $\text{se}_\nu(x,1)$ for $0 < \nu < 1$, $0 \leq x \leq 2\pi$.

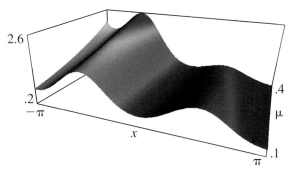

Figure 28.13.5: $\mathrm{me}_{i\mu}(x,1)$ for $0.1 \le \mu \le 0.4$, $-\pi \le x \le \pi$.

28.14 Fourier Series

The Fourier series

28.14.1 $\quad \mathrm{me}_\nu(z, q) = \sum_{m=-\infty}^{\infty} c_{2m}^\nu(q) e^{i(\nu+2m)z},$

28.14.2 $\quad \mathrm{ce}_\nu(z, q) = \sum_{m=-\infty}^{\infty} c_{2m}^\nu(q) \cos(\nu+2m)z,$

28.14.3 $\quad \mathrm{se}_\nu(z, q) = \sum_{m=-\infty}^{\infty} c_{2m}^\nu(q) \sin(\nu+2m)z,$

converge absolutely and uniformly on all compact sets in the z-plane. The coefficients satisfy

28.14.4 $\quad qc_{2m+2} - \bigl(a - (\nu+2m)^2\bigr) c_{2m} + qc_{2m-2} = 0,$
$\qquad a = \lambda_\nu(q), c_{2m} = c_{2m}^\nu(q),$

and the normalization relation

28.14.5 $\quad \sum_{m=-\infty}^{\infty} (c_{2m}^\nu(q))^2 = 1;$

compare (28.12.5). Ambiguities in sign are resolved by (28.14.9) when $q = 0$, and by continuity for other values of q.

The rate of convergence is indicated by

28.14.6 $\quad \dfrac{c_{2m}^\nu(q)}{c_{2m\mp 2}^\nu(q)} = \dfrac{-q}{4m^2}\left(1 + O\!\left(\dfrac{1}{m}\right)\right), \quad m \to \pm\infty.$

For changes of sign of ν, q, and m,

28.14.7 $\quad c_{-2m}^{-\nu}(q) = c_{2m}^\nu(q),$

28.14.8 $\quad c_{2m}^\nu(-q) = (-1)^m c_{2m}^\nu(q).$

When $q = 0$,

28.14.9 $\quad c_0^\nu(0) = 1, \quad c_{2m}^\nu(0) = 0, \qquad m \ne 0.$

When $q \to 0$ with $m\ (\ge 1)$ and ν fixed,

28.14.10 $\quad c_{2m}^\nu(q) = \left(\dfrac{(-1)^m q^m\, \Gamma(\nu+1)}{m!\, 2^{2m}\, \Gamma(\nu+m+1)} + O\!\left(q^{m+2}\right)\right) c_0^\nu(q).$

28.15 Expansions for Small q

28.15(i) Eigenvalues $\lambda_\nu(q)$

28.15.1
$$\lambda_\nu(q) = \nu^2 + \dfrac{1}{2(\nu^2-1)} q^2 + \dfrac{5\nu^2+7}{32(\nu^2-1)^3(\nu^2-4)} q^4 + \dfrac{9\nu^4+58\nu^2+29}{64(\nu^2-1)^5(\nu^2-4)(\nu^2-9)} q^6 + \cdots.$$

Higher coefficients can be found by equating powers of q in the following continued-fraction equation, with $a = \lambda_\nu(q)$:

28.15.2
$$a - \nu^2 - \dfrac{q^2}{a - (\nu+2)^2 -} \dfrac{q^2}{a - (\nu+4)^2 -} \cdots$$
$$= \dfrac{q^2}{a - (\nu-2)^2 -} \dfrac{q^2}{a - (\nu-4)^2 -} \cdots.$$

28.15(ii) Solutions $\mathrm{me}_\nu(z, q)$

28.15.3
$\mathrm{me}_\nu(z, q)$
$= e^{i\nu z} - \dfrac{q}{4}\left(\dfrac{1}{\nu+1} e^{i(\nu+2)z} - \dfrac{1}{\nu-1} e^{i(\nu-2)z}\right)$
$\quad + \dfrac{q^2}{32}\left(\dfrac{1}{(\nu+1)(\nu+2)} e^{i(\nu+4)z}\right.$
$\qquad\left. + \dfrac{1}{(\nu-1)(\nu-2)} e^{i(\nu-4)z} - \dfrac{2(\nu^2+1)}{(\nu^2-1)^2} e^{i\nu z}\right) + \cdots;$

compare §28.6(ii).

28.16 Asymptotic Expansions for Large q

Let $s = 2m+1$, $m = 0, 1, 2, \ldots$, and ν be fixed with $m < \nu < m+1$. Then as $h(=\sqrt{q}) \to +\infty$

28.16.1
$$\lambda_\nu(h^2) \sim -2h^2 + 2sh - \dfrac{1}{8}(s^2+1) - \dfrac{1}{2^7 h}(s^3+3s)$$
$$\quad - \dfrac{1}{2^{12} h^2}(5s^4+34s^2+9)$$
$$\quad - \dfrac{1}{2^{17} h^3}(33s^5+410s^3+405s)$$
$$\quad - \dfrac{1}{2^{20} h^4}(63s^6+1260s^4+2943s^2+486)$$
$$\quad - \dfrac{1}{2^{25} h^5}(527s^7+15617s^5+69001s^3+41607s)$$
$$\quad + \cdots.$$

For graphical interpretation, see Figures 28.13.1 and 28.13.2.

See also §28.8(iv).

28.17 Stability as $x \to \pm\infty$

If all solutions of (28.2.1) are bounded when $x \to \pm\infty$ along the real axis, then the corresponding pair of parameters (a,q) is called *stable*. All other pairs are *unstable*.

For example, positive real values of a with $q = 0$ comprise stable pairs, as do values of a and q that correspond to real, but noninteger, values of ν.

However, if $\Im\nu \neq 0$, then (a,q) always comprises an unstable pair. For example, as $x \to +\infty$ one of the solutions $\mathrm{me}_\nu(x,q)$ and $\mathrm{me}_\nu(-x,q)$ tends to 0 and the other is unbounded (compare Figure 28.13.5). Also, all nontrivial solutions of (28.2.1) are unbounded on $\mathbb{R}$.

For real a and q ($\neq 0$) the stable regions are the open regions indicated in color in Figure 28.17.1. The boundary of each region comprises the *characteristic curves* $a = a_n(q)$ and $a = b_n(q)$; compare Figure 28.2.1.

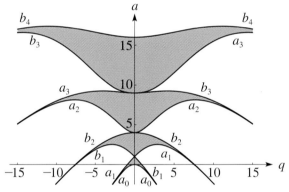

Figure 28.17.1: Stability chart for eigenvalues of Mathieu's equation (28.2.1).

28.18 Integrals and Integral Equations

See §28.28.

28.19 Expansions in Series of $\mathrm{me}_{\nu+2n}$ Functions

Let q be a normal value (§28.12(i)) with respect to ν, and $f(z)$ be a function that is analytic on a doubly-infinite open strip S that contains the real axis. Assume also

28.19.1 $$f(z+\pi) = e^{i\nu\pi}f(z).$$

Then

28.19.2 $$f(z) = \sum_{n=-\infty}^{\infty} f_n \, \mathrm{me}_{\nu+2n}(z,q),$$

where

28.19.3 $$f_n = \frac{1}{\pi}\int_0^\pi f(z)\,\mathrm{me}_{\nu+2n}(-z,q)\,dz.$$

The series (28.19.2) converges absolutely and uniformly on compact subsets within S.

Example

28.19.4 $$e^{i\nu z} = \sum_{n=-\infty}^{\infty} c_{-2n}^{\nu+2n}(q)\,\mathrm{me}_{\nu+2n}(z,q),$$

where the coefficients are as in §28.14.

Modified Mathieu Functions

28.20 Definitions and Basic Properties

28.20(i) Modified Mathieu's Equation

When z is replaced by $\pm iz$, (28.2.1) becomes the *modified Mathieu's equation*:

28.20.1 $$w'' - (a - 2q\cosh(2z))\,w = 0,$$

with its *algebraic form*

28.20.2 $$(\zeta^2-1)w'' + \zeta w' + (4q\zeta^2 - 2q - a)\,w = 0, \quad \zeta = \cosh z.$$

28.20(ii) Solutions Ce_ν, Se_ν, Me_ν, Fe_n, Ge_n

28.20.3 $\quad \mathrm{Ce}_\nu(z,q) = \mathrm{ce}_\nu(\pm iz, q), \qquad \nu \neq -1,-2,\dots,$

28.20.4 $\quad \mathrm{Se}_\nu(z,q) = \mp i\,\mathrm{se}_\nu(\pm iz, q), \qquad \nu \neq 0,-1,\dots,$

28.20.5 $\quad \mathrm{Me}_\nu(z,q) = \mathrm{me}_\nu(-iz,q),$

28.20.6 $\quad \mathrm{Fe}_n(z,q) = \mp i\,\mathrm{fe}_n(\pm iz, q), \qquad n = 0,1,\dots,$

28.20.7 $\quad \mathrm{Ge}_n(z,q) = \mathrm{ge}_n(\pm iz, q), \qquad n = 1,2,\dots.$

28.20(iii) Solutions $\mathrm{M}_\nu^{(j)}$

Assume first that ν is real, q is positive, and $a = \lambda_\nu(q)$; see §28.12(i). Write

28.20.8 $$h = \sqrt{q} \; (>0).$$

Then from §2.7(ii) it is seen that equation (28.20.2) has independent and unique solutions that are asymptotic to $\zeta^{1/2} e^{\pm 2ih\zeta}$ as $\zeta \to \infty$ in the respective sectors $|\mathrm{ph}(\mp i\zeta)| \leq \frac{3}{2}\pi - \delta$, δ being an arbitrary small positive constant. It follows that (28.20.1) has independent and unique solutions $\mathrm{M}_\nu^{(3)}(z,h)$, $\mathrm{M}_\nu^{(4)}(z,h)$ such that

28.20.9 $\quad \mathrm{M}_\nu^{(3)}(z,h) = H_\nu^{(1)}(2h\cosh z)\,(1 + O(\mathrm{sech}\,z)),$

as $\Re z \to +\infty$ with $-\pi + \delta \leq \Im z \leq 2\pi - \delta$, and

28.20.10 $\quad \mathrm{M}_\nu^{(4)}(z,h) = H_\nu^{(2)}(2h\cosh z)\,(1 + O(\mathrm{sech}\,z)),$

as $\Re z \to +\infty$ with $-2\pi + \delta \leq \Im z \leq \pi - \delta$. See §10.2(ii) for the notation. In addition, there are unique solutions $\mathrm{M}_\nu^{(1)}(z,h)$, $\mathrm{M}_\nu^{(2)}(z,h)$ that are real when z is real and have the properties

28.20.11
$$\mathrm{M}_\nu^{(1)}(z,h) = J_\nu(2h\cosh z) + e^{|\Im(2h\cosh z)|}O\left((\mathrm{sech}\,z)^{3/2}\right),$$

28.20.12
$$\mathrm{M}_\nu^{(2)}(z,h) = Y_\nu(2h\cosh z) + e^{|\Im(2h\cosh z)|} O\!\left((\operatorname{sech} z)^{3/2}\right),$$
as $\Re z \to +\infty$ with $|\Im z| \leq \pi - \delta$.

For other values of z, h, and ν the functions $\mathrm{M}_\nu^{(j)}(z,h)$, $j=1,2,3,4$, are determined by analytic continuation. Furthermore,

28.20.13 $\quad \mathrm{M}_\nu^{(3)}(z,h) = \mathrm{M}_\nu^{(1)}(z,h) + i\,\mathrm{M}_\nu^{(2)}(z,h),$

28.20.14 $\quad \mathrm{M}_\nu^{(4)}(z,h) = \mathrm{M}_\nu^{(1)}(z,h) - i\,\mathrm{M}_\nu^{(2)}(z,h).$

28.20(iv) Radial Mathieu Functions $\mathrm{Mc}_n^{(j)}$, $\mathrm{Ms}_n^{(j)}$

For $j=1,2,3,4$,

28.20.15 $\quad \mathrm{Mc}_n^{(j)}(z,h) = \mathrm{M}_n^{(j)}(z,h), \qquad n=0,1,\dots,$

28.20.16 $\quad \mathrm{Ms}_n^{(j)}(z,h) = (-1)^n\,\mathrm{M}_{-n}^{(j)}(z,h), \quad n=1,2,\dots.$

28.20(v) Solutions Ie_n, Io_n, Ke_n, Ko_n

28.20.17 $\qquad \mathrm{Ie}_n(z,h) = i^{-n}\,\mathrm{Mc}_n^{(1)}(z,ih),$

28.20.18 $\qquad \mathrm{Io}_n(z,h) = i^{-n}\,\mathrm{Ms}_n^{(1)}(z,ih),$

28.20.19
$$\mathrm{Ke}_{2m}(z,h) = (-1)^m \tfrac{1}{2}\pi i\,\mathrm{Mc}_{2m}^{(3)}(z,ih),$$
$$\mathrm{Ke}_{2m+1}(z,h) = (-1)^{m+1}\tfrac{1}{2}\pi\,\mathrm{Mc}_{2m+1}^{(3)}(z,ih),$$

28.20.20
$$\mathrm{Ko}_{2m}(z,h) = (-1)^m \tfrac{1}{2}\pi i\,\mathrm{Ms}_{2m}^{(3)}(z,ih),$$
$$\mathrm{Ko}_{2m+1}(z,h) = (-1)^{m+1}\tfrac{1}{2}\pi\,\mathrm{Ms}_{2m+1}^{(3)}(z,ih).$$

28.20(vi) Wronskians

28.20.21
$$\mathscr{W}\!\left\{\mathrm{M}_\nu^{(1)},\mathrm{M}_\nu^{(2)}\right\} = -\mathscr{W}\!\left\{\mathrm{M}_\nu^{(2)},\mathrm{M}_\nu^{(3)}\right\}$$
$$= -\mathscr{W}\!\left\{\mathrm{M}_\nu^{(2)},\mathrm{M}_\nu^{(4)}\right\} = 2/\pi,$$
$$\mathscr{W}\!\left\{\mathrm{M}_\nu^{(1)},\mathrm{M}_\nu^{(3)}\right\} = -\mathscr{W}\!\left\{\mathrm{M}_\nu^{(1)},\mathrm{M}_\nu^{(4)}\right\}$$
$$= -\tfrac{1}{2}\mathscr{W}\!\left\{\mathrm{M}_\nu^{(3)},\mathrm{M}_\nu^{(4)}\right\} = 2i/\pi.$$

28.20(vii) Shift of Variable

28.20.22 $\quad \mathrm{M}_\nu^{(j)}\!\left(z \pm \tfrac{1}{2}\pi i, h\right) = \mathrm{M}_\nu^{(j)}(z, \pm ih), \qquad \nu \notin \mathbb{Z}.$

For $n=0,1,2,\dots$,

28.20.23
$$\mathrm{Mc}_{2n}^{(j)}\!\left(z \pm \tfrac{1}{2}\pi i, h\right) = \mathrm{Mc}_{2n}^{(j)}(z, \pm ih),$$
$$\mathrm{Ms}_{2n+1}^{(j)}\!\left(z \pm \tfrac{1}{2}\pi i, h\right) = \mathrm{Mc}_{2n+1}^{(j)}(z, \pm ih),$$

28.20.24
$$\mathrm{Mc}_{2n+1}^{(j)}\!\left(z \pm \tfrac{1}{2}\pi i, h\right) = \mathrm{Ms}_{2n+1}^{(j)}(z, \pm ih),$$
$$\mathrm{Ms}_{2n+2}^{(j)}\!\left(z \pm \tfrac{1}{2}\pi i, h\right) = \mathrm{Ms}_{2n+2}^{(j)}(z, \pm ih).$$

For $s \in \mathbb{Z}$,

28.20.25
$$\mathrm{M}_\nu^{(1)}(z+s\pi i, h) = e^{is\pi\nu}\,\mathrm{M}_\nu^{(1)}(z,h),$$
$$\mathrm{M}_\nu^{(2)}(z+s\pi i, h) = e^{-is\pi\nu}\,\mathrm{M}_\nu^{(2)}(z,h)$$
$$\qquad + 2i\cot(\pi\nu)\sin(s\pi\nu)\,\mathrm{M}_\nu^{(1)}(z,h),$$
$$\mathrm{M}_\nu^{(3)}(z+s\pi i, h) = -\frac{\sin((s-1)\pi\nu)}{\sin(\pi\nu)}\,\mathrm{M}_\nu^{(3)}(z,h)$$
$$\qquad - e^{-i\pi\nu}\frac{\sin(s\pi\nu)}{\sin(\pi\nu)}\,\mathrm{M}_\nu^{(4)}(z,h),$$
$$\mathrm{M}_\nu^{(4)}(z+s\pi i, h) = e^{i\pi\nu}\frac{\sin(s\pi\nu)}{\sin(\pi\nu)}\,\mathrm{M}_\nu^{(3)}(z,h)$$
$$\qquad + \frac{\sin((s+1)\pi\nu)}{\sin(\pi\nu)}\,\mathrm{M}_\nu^{(4)}(z,h).$$

When ν is an integer the right-hand sides of (28.20.25) are replaced by the their limiting values. And for the corresponding identities for the radial functions use (28.20.15) and (28.20.16).

28.21 Graphics

Radial Mathieu Functions: Surfaces

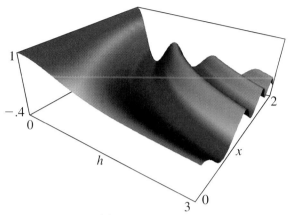

Figure 28.21.1: $\mathrm{Mc}_0^{(1)}(x, h)$ for $0 \le h \le 3$, $0 \le x \le 2$.

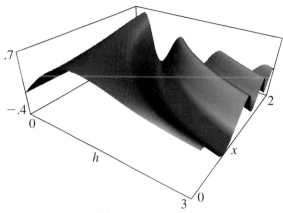

Figure 28.21.2: $\mathrm{Mc}_1^{(1)}(x, h)$ for $0 \le h \le 3$, $0 \le x \le 2$.

For further graphics see http://dlmf.nist.gov/28.21.

28.22 Connection Formulas

28.22(i) Integer ν

28.22.1
$$\mathrm{Mc}_m^{(1)}(z, h) = \sqrt{\frac{2}{\pi}} \frac{1}{g_{e,m}(h) \, \mathrm{ce}_m(0, h^2)} \, \mathrm{Ce}_m(z, h^2),$$

28.22.2
$$\mathrm{Ms}_m^{(1)}(z, h) = \sqrt{\frac{2}{\pi}} \frac{1}{g_{o,m}(h) \, \mathrm{se}'_m(0, h^2)} \, \mathrm{Se}_m(z, h^2),$$

28.22.3
$$\mathrm{Mc}_m^{(2)}(z, h) = \sqrt{\frac{2}{\pi}} \frac{1}{g_{e,m}(h) \, \mathrm{ce}_m(0, h^2)}$$
$$\times \left(-f_{e,m}(h) \, \mathrm{Ce}_m(z, h^2) + \frac{2}{\pi C_m(h^2)} \, \mathrm{Fe}_m(z, h^2) \right),$$

28.22.4
$$\mathrm{Ms}_m^{(2)}(z, h) = \sqrt{\frac{2}{\pi}} \frac{1}{g_{o,m}(h) \, \mathrm{se}'_m(0, h^2)}$$
$$\times \left(-f_{o,m}(h) \, \mathrm{Se}_m(z, h^2) - \frac{2}{\pi S_m(h^2)} \, \mathrm{Ge}_m(z, h^2) \right).$$

The *joining factors* in the above formulas are given by

28.22.5 $\quad g_{e,2m}(h) = (-1)^m \sqrt{\dfrac{2}{\pi}} \dfrac{\mathrm{ce}_{2m}\left(\frac{1}{2}\pi, h^2\right)}{A_0^{2m}(h^2)},$

28.22.6 $\quad g_{e,2m+1}(h) = (-1)^{m+1} \sqrt{\dfrac{2}{\pi}} \dfrac{\mathrm{ce}'_{2m+1}\left(\frac{1}{2}\pi, h^2\right)}{h A_1^{2m+1}(h^2)},$

28.22.7 $\quad g_{o,2m+1}(h) = (-1)^m \sqrt{\dfrac{2}{\pi}} \dfrac{\mathrm{se}_{2m+1}\left(\frac{1}{2}\pi, h^2\right)}{h B_1^{2m+1}(h^2)},$

28.22.8 $\quad g_{o,2m+2}(h) = (-1)^{m+1} \sqrt{\dfrac{2}{\pi}} \dfrac{\mathrm{se}'_{2m+2}\left(\frac{1}{2}\pi, h^2\right)}{h^2 B_2^{2m+2}(h^2)},$

28.22.9 $\quad f_{e,m}(h) = -\sqrt{\pi/2}\, g_{e,m}(h) \, \mathrm{Mc}_m^{(2)}(0, h),$

28.22.10 $\quad f_{o,m}(h) = -\sqrt{\pi/2}\, g_{o,m}(h) \, \mathrm{Ms}_m^{(2)\prime}(0, h),$

where $A_n^m(h^2)$, $B_n^m(h^2)$ are as in §28.4(i), and $C_m(h^2)$, $S_m(h^2)$ are as in §28.5(i). Furthermore,

28.22.11
$$\mathrm{Mc}_m^{(2)\prime}(0, h) = \sqrt{2/\pi}\, g_{e,m}(h),$$
$$\mathrm{Ms}_m^{(2)}(0, h) = -\sqrt{2/\pi}\, g_{o,m}(h),$$

28.22.12
$$\mathrm{fe}'_m(0, h^2) = \tfrac{1}{2}\pi C_m(h^2) \left(g_{e,m}(h)\right)^2 \mathrm{ce}_m(0, h^2),$$
$$\mathrm{ge}_m(0, h^2) = \tfrac{1}{2}\pi S_m(h^2) \left(g_{o,m}(h)\right)^2 \mathrm{se}'_m(0, h^2).$$

28.22(ii) Noninteger ν

28.22.13 $\quad \mathrm{M}_\nu^{(1)}(z, h) = \dfrac{\mathrm{M}_\nu^{(1)}(0, h)}{\mathrm{me}_\nu(0, h^2)} \, \mathrm{Me}_\nu(z, h^2).$

Here $\mathrm{me}_\nu(0, h^2)$ ($\ne 0$) is given by (28.14.1) with $z = 0$, and $\mathrm{M}_\nu^{(1)}(0, h)$ is given by (28.24.1) with $j = 1$, $z = 0$, and n chosen so that $|c_{2n}^\nu(h^2)| = \max(|c_{2\ell}^\nu(h^2)|)$, where the maximum is taken over all integers ℓ.

28.22.14
$$\mathrm{M}_\nu^{(2)}(z, h) = \cot(\nu\pi)\, \mathrm{M}_\nu^{(1)}(z, h) - \frac{1}{\sin(\nu\pi)} \mathrm{M}_{-\nu}^{(1)}(z, h).$$

See also (28.20.13) and (28.20.14).

28.23 Expansions in Series of Bessel Functions

We use the following notations:

28.23.1
$$\mathcal{C}_\mu^{(1)} = J_\mu, \quad \mathcal{C}_\mu^{(2)} = Y_\mu, \quad \mathcal{C}_\mu^{(3)} = H_\mu^{(1)}, \quad \mathcal{C}_\mu^{(4)} = H_\mu^{(2)};$$

compare §10.2(ii). For the coefficients $c_{2n}^\nu(q)$ see §28.14. For $A_n^m(q)$ and $B_n^m(q)$ see §28.4.

28.23.2
$$\mathrm{me}_\nu(0, h^2) \, \mathrm{M}_\nu^{(j)}(z, h) = \sum_{n=-\infty}^{\infty} (-1)^n c_{2n}^\nu(h^2) \mathcal{C}_{\nu+2n}^{(j)}(2h \cosh z),$$

28.23.3
$$\mathrm{me}_\nu'(0, h^2) \, \mathrm{M}_\nu^{(j)}(z, h) = i \tanh z \sum_{n=-\infty}^{\infty} (-1)^n (\nu + 2n) c_{2n}^\nu(h^2) \mathcal{C}_{\nu+2n}^{(j)}(2h \cosh z),$$

valid for all z when $j=1$, and for $\Re z > 0$ and $|\cosh z| > 1$ when $j = 2, 3, 4$.

28.23.4
$$\mathrm{me}_\nu\left(\tfrac{1}{2}\pi, h^2\right) \mathrm{M}_\nu^{(j)}(z, h) = e^{i\nu\pi/2} \sum_{n=-\infty}^{\infty} c_{2n}^\nu(h^2) \mathcal{C}_{\nu+2n}^{(j)}(2h \sinh z),$$

28.23.5
$$\mathrm{me}_\nu'\left(\tfrac{1}{2}\pi, h^2\right) \mathrm{M}_\nu^{(j)}(z, h) = i e^{i\nu\pi/2} \coth z \sum_{n=-\infty}^{\infty} (\nu + 2n) c_{2n}^\nu(h^2) \mathcal{C}_{\nu+2n}^{(j)}(2h \sinh z),$$

valid for all z when $j=1$, and for $\Re z > 0$ and $|\sinh z| > 1$ when $j = 2, 3, 4$.

In the case when ν is an integer

28.23.6
$$\mathrm{Mc}_{2m}^{(j)}(z, h) = (-1)^m \left(\mathrm{ce}_{2m}(0, h^2)\right)^{-1} \sum_{\ell=0}^{\infty} (-1)^\ell A_{2\ell}^{2m}(h^2) \mathcal{C}_{2\ell}^{(j)}(2h \cosh z),$$

28.23.7
$$\mathrm{Mc}_{2m}^{(j)}(z, h) = (-1)^m \left(\mathrm{ce}_{2m}\left(\tfrac{1}{2}\pi, h^2\right)\right)^{-1} \sum_{\ell=0}^{\infty} A_{2\ell}^{2m}(h^2) \mathcal{C}_{2\ell}^{(j)}(2h \sinh z),$$

28.23.8
$$\mathrm{Mc}_{2m+1}^{(j)}(z, h) = (-1)^m \left(\mathrm{ce}_{2m+1}(0, h^2)\right)^{-1} \sum_{\ell=0}^{\infty} (-1)^\ell A_{2\ell+1}^{2m+1}(h^2) \mathcal{C}_{2\ell+1}^{(j)}(2h \cosh z),$$

28.23.9
$$\mathrm{Mc}_{2m+1}^{(j)}(z, h) = (-1)^{m+1} \left(\mathrm{ce}_{2m+1}'\left(\tfrac{1}{2}\pi, h^2\right)\right)^{-1} \coth z \sum_{\ell=0}^{\infty} (2\ell+1) A_{2\ell+1}^{2m+1}(h^2) \mathcal{C}_{2\ell+1}^{(j)}(2h \sinh z),$$

28.23.10
$$\mathrm{Ms}_{2m+1}^{(j)}(z, h) = (-1)^m \left(\mathrm{se}_{2m+1}'(0, h^2)\right)^{-1} \tanh z \sum_{\ell=0}^{\infty} (-1)^\ell (2\ell+1) B_{2\ell+1}^{2m+1}(h^2) \mathcal{C}_{2\ell+1}^{(j)}(2h \cosh z),$$

28.23.11
$$\mathrm{Ms}_{2m+1}^{(j)}(z, h) = (-1)^m \left(\mathrm{se}_{2m+1}\left(\tfrac{1}{2}\pi, h^2\right)\right)^{-1} \sum_{\ell=0}^{\infty} B_{2\ell+1}^{2m+1}(h^2) \mathcal{C}_{2\ell+1}^{(j)}(2h \sinh z),$$

28.23.12
$$\mathrm{Ms}_{2m+2}^{(j)}(z, h) = (-1)^m \left(\mathrm{se}_{2m+2}'(0, h^2)\right)^{-1} \tanh z \sum_{\ell=0}^{\infty} (-1)^\ell (2\ell+2) B_{2\ell+2}^{2m+2}(h^2) \mathcal{C}_{2\ell+2}^{(j)}(2h \cosh z),$$

28.23.13
$$\mathrm{Ms}_{2m+2}^{(j)}(z, h) = (-1)^{m+1} \left(\mathrm{se}_{2m+2}'\left(\tfrac{1}{2}\pi, h^2\right)\right)^{-1} \coth z \sum_{\ell=0}^{\infty} (2\ell+2) B_{2\ell+2}^{2m+2}(h^2) \mathcal{C}_{2\ell+2}^{(j)}(2h \sinh z).$$

When $j=1$, each of the series (28.23.6)–(28.23.13) converges for all z. When $j = 2, 3, 4$ the series in the even-numbered equations converge for $\Re z > 0$ and $|\cosh z| > 1$, and the series in the odd-numbered equations converge for $\Re z > 0$ and $|\sinh z| > 1$.

For proofs and generalizations, see Meixner and Schäfke (1954, §§2.62 and 2.64).

28.24 Expansions in Series of Cross-Products of Bessel Functions or Modified Bessel Functions

Throughout this section $\varepsilon_0 = 2$ and $\varepsilon_s = 1$, $s = 1, 2, 3, \ldots$.

With $\mathcal{C}_\mu^{(j)}$, $c_n^\nu(q)$, $A_n^m(q)$, and $B_n^m(q)$ as in §28.23,

28.24.1 $$c_{2n}^\nu(h^2)\, \mathrm{M}_\nu^{(j)}(z,h) = \sum_{\ell=-\infty}^{\infty} (-1)^\ell c_{2\ell}^\nu(h^2)\, J_{\ell-n}(he^{-z})\, \mathcal{C}_{\nu+n+\ell}^{(j)}(he^z),$$

where $j = 1, 2, 3, 4$ and $n \in \mathbb{Z}$.

In the case when ν is an integer,

28.24.2 $$\varepsilon_s\, \mathrm{Mc}_{2m}^{(j)}(z,h) = (-1)^m \sum_{\ell=0}^{\infty} (-1)^\ell \frac{A_{2\ell}^{2m}(h^2)}{A_{2s}^{2m}(h^2)} \left(J_{\ell-s}(he^{-z})\mathcal{C}_{\ell+s}^{(j)}(he^z) + J_{\ell+s}(he^{-z})\mathcal{C}_{\ell-s}^{(j)}(he^z) \right),$$

28.24.3 $$\mathrm{Mc}_{2m+1}^{(j)}(z,h) = (-1)^m \sum_{\ell=0}^{\infty} (-1)^\ell \frac{A_{2\ell+1}^{2m+1}(h^2)}{A_{2s+1}^{2m+1}(h^2)} \left(J_{\ell-s}(he^{-z})\mathcal{C}_{\ell+s+1}^{(j)}(he^z) + J_{\ell+s+1}(he^{-z})\mathcal{C}_{\ell-s}^{(j)}(he^z) \right),$$

28.24.4 $$\mathrm{Ms}_{2m+1}^{(j)}(z,h) = (-1)^m \sum_{\ell=0}^{\infty} (-1)^\ell \frac{B_{2\ell+1}^{2m+1}(h^2)}{B_{2s+1}^{2m+1}(h^2)} \left(J_{\ell-s}(he^{-z})\mathcal{C}_{\ell+s+1}^{(j)}(he^z) - J_{\ell+s+1}(he^{-z})\mathcal{C}_{\ell-s}^{(j)}(he^z) \right),$$

28.24.5 $$\mathrm{Ms}_{2m+2}^{(j)}(z,h) = (-1)^m \sum_{\ell=0}^{\infty} (-1)^\ell \frac{B_{2\ell+2}^{2m+2}(h^2)}{B_{2s+2}^{2m+2}(h^2)} \left(J_{\ell-s}(he^{-z})\mathcal{C}_{\ell+s+2}^{(j)}(he^z) - J_{\ell+s+2}(he^{-z})\mathcal{C}_{\ell-s}^{(j)}(he^z) \right),$$

where $j = 1, 2, 3, 4$, and $s = 0, 1, 2, \ldots$.

Also, with I_n and K_n denoting the modified Bessel functions (§10.25(ii)), and again with $s = 0, 1, 2, \ldots$,

28.24.6 $$\varepsilon_s\, \mathrm{Ie}_{2m}(z,h) = (-1)^s \sum_{\ell=0}^{\infty} (-1)^\ell \frac{A_{2\ell}^{2m}(h^2)}{A_{2s}^{2m}(h^2)} \left(I_{\ell-s}(he^{-z})\, I_{\ell+s}(he^z) + I_{\ell+s}(he^{-z})\, I_{\ell-s}(he^z) \right),$$

28.24.7 $$\mathrm{Io}_{2m+2}(z,h) = (-1)^s \sum_{\ell=0}^{\infty} (-1)^\ell \frac{B_{2\ell+2}^{2m+2}(h^2)}{B_{2s+2}^{2m+2}(h^2)} \left(I_{\ell-s}(he^{-z})\, I_{\ell+s+2}(he^z) - I_{\ell+s+2}(he^{-z})\, I_{\ell-s}(he^z) \right),$$

28.24.8 $$\mathrm{Ie}_{2m+1}(z,h) = (-1)^s \sum_{\ell=0}^{\infty} (-1)^\ell \frac{B_{2\ell+1}^{2m+1}(h^2)}{B_{2s+1}^{2m+1}(h^2)} \left(I_{\ell-s}(he^{-z})\, I_{\ell+s+1}(he^z) + I_{\ell+s+1}(he^{-z})\, I_{\ell-s}(he^z) \right),$$

28.24.9 $$\mathrm{Io}_{2m+1}(z,h) = (-1)^s \sum_{\ell=0}^{\infty} (-1)^\ell \frac{A_{2\ell+1}^{2m+1}(h^2)}{A_{2s+1}^{2m+1}(h^2)} \left(I_{\ell-s}(he^{-z})\, I_{\ell+s+1}(he^z) - I_{\ell+s+1}(he^{-z})\, I_{\ell-s}(he^z) \right),$$

28.24.10 $$\varepsilon_s\, \mathrm{Ke}_{2m}(z,h) = \sum_{\ell=0}^{\infty} \frac{A_{2\ell}^{2m}(h^2)}{A_{2s}^{2m}(h^2)} \left(I_{\ell-s}(he^{-z})\, K_{\ell+s}(he^z) + I_{\ell+s}(he^{-z})\, K_{\ell-s}(he^z) \right),$$

28.24.11 $$\mathrm{Ko}_{2m+2}(z,h) = \sum_{\ell=0}^{\infty} \frac{B_{2\ell+2}^{2m+2}(h^2)}{B_{2s+2}^{2m+2}(h^2)} \left(I_{\ell-s}(he^{-z})\, K_{\ell+s+2}(he^z) - I_{\ell+s+2}(he^{-z})\, K_{\ell-s}(he^z) \right),$$

28.24.12 $$\mathrm{Ke}_{2m+1}(z,h) = \sum_{\ell=0}^{\infty} \frac{B_{2\ell+1}^{2m+1}(h^2)}{B_{2s+1}^{2m+1}(h^2)} \left(I_{\ell-s}(he^{-z})\, K_{\ell+s+1}(he^z) - I_{\ell+s+1}(he^{-z})\, K_{\ell-s}(he^z) \right),$$

28.24.13 $$\mathrm{Ko}_{2m+1}(z,h) = \sum_{\ell=0}^{\infty} \frac{A_{2\ell+1}^{2m+1}(h^2)}{A_{2s+1}^{2m+1}(h^2)} \left(I_{\ell-s}(he^{-z})\, K_{\ell+s+1}(he^z) + I_{\ell+s+1}(he^{-z})\, K_{\ell-s}(he^z) \right).$$

The expansions (28.24.1)–(28.24.13) converge absolutely and uniformly on compact sets of the z-plane.

28.25 Asymptotic Expansions for Large $\Re z$

For fixed $h(\neq 0)$ and fixed ν,

28.25.1
$$M_\nu^{(3,4)}(z,h) \sim \frac{e^{\pm i(2h\cosh z - (\frac{1}{2}\nu + \frac{1}{4})\pi)}}{(\pi h(\cosh z + 1))^{\frac{1}{2}}} \times \sum_{m=0}^{\infty} \frac{D_m^\pm}{(\mp 4ih(\cosh z + 1))^m},$$

where the coefficients are given by

28.25.2
$$D_{-1}^\pm = 0, \quad D_0^\pm = 1,$$

and

28.25.3
$$(m+1)D_{m+1}^\pm + \left(\left(m+\tfrac{1}{2}\right)^2 \pm \left(m+\tfrac{1}{4}\right)8ih + 2h^2 - a\right)D_m^\pm \pm \left(m - \tfrac{1}{2}\right)(8ihm)D_{m-1}^\pm = 0, \quad m \geq 0.$$

The upper signs correspond to $M_\nu^{(3)}(z,h)$ and the lower signs to $M_\nu^{(4)}(z,h)$. The expansion (28.25.1) is valid for $M_\nu^{(3)}(z,h)$ when

28.25.4 $\quad \Re z \to +\infty, \quad -\pi + \delta \leq \mathrm{ph}\,h + \Im z \leq 2\pi - \delta,$

and for $M_\nu^{(4)}(z,h)$ when

28.25.5 $\quad \Re z \to +\infty, \quad -2\pi + \delta \leq \mathrm{ph}\,h + \Im z \leq \pi - \delta,$

where δ again denotes an arbitrary small positive constant.

For proofs and generalizations see Meixner and Schäfke (1954, §2.63).

28.26 Asymptotic Approximations for Large q

28.26(i) Goldstein's Expansions

Denote

28.26.1
$$Mc_m^{(3)}(z,h) = \frac{e^{i\phi}}{(\pi h \cosh z)^{1/2}} \times (Fc_m(z,h) - i\,Gc_m(z,h)),$$

28.26.2
$$i\,Ms_{m+1}^{(3)}(z,h) = \frac{e^{i\phi}}{(\pi h \cosh z)^{1/2}} \times (Fs_m(z,h) - i\,Gs_m(z,h)),$$

where

28.26.3 $\quad \phi = 2h\sinh z - \left(m + \tfrac{1}{2}\right)\arctan(\sinh z).$

Then as $h \to +\infty$ with fixed z in $\Re z > 0$ and fixed $s = 2m + 1$,

28.26.4
$$Fc_m(z,h) \sim 1 + \frac{s}{8h\cosh^2 z} + \frac{1}{2^{11}h^2}\left(\frac{s^4 + 86s^2 + 105}{\cosh^4 z} - \frac{s^4 + 22s^2 + 57}{\cosh^2 z}\right) + \frac{1}{2^{14}h^3}\left(-\frac{s^5 + 14s^3 + 33s}{\cosh^2 z} - \frac{2s^5 + 124s^3 + 1122s}{\cosh^4 z} + \frac{3s^5 + 290s^3 + 1627s}{\cosh^6 z}\right) + \cdots,$$

28.26.5
$$Gc_m(z,h) \sim \frac{\sinh z}{\cosh^2 z}\left(\frac{s^2+3}{2^5 h} + \frac{1}{2^9 h^2}\left(s^3 + 3s + \frac{4s^3 + 44s}{\cosh^2 z}\right) + \frac{1}{2^{14}h^3}\left(5s^4 + 34s^2 + 9 - \frac{s^6 - 47s^4 + 667s^2 + 2835}{12\cosh^2 z} + \frac{s^6 + 505s^4 + 12139s^2 + 10395}{12\cosh^4 z}\right)\right) + \cdots.$$

The asymptotic expansions of $Fs_m(z,h)$ and $Gs_m(z,h)$ in the same circumstances are also given by the right-hand sides of (28.26.4) and (28.26.5), respectively.

For additional terms see Goldstein (1927).

28.26(ii) Uniform Approximations

See §28.8(iv). For asymptotic approximations for $M_\nu^{(3,4)}(z,h)$ see also Naylor (1984, 1987, 1989).

28.27 Addition Theorems

Addition theorems provide important connections between Mathieu functions with different parameters and in different coordinate systems. They are analogous to the addition theorems for Bessel functions (§10.23(ii)) and modified Bessel functions (§10.44(ii)). For a comprehensive treatment see Meixner et al. (1980, §2.2).

28.28 Integrals, Integral Representations, and Integral Equations

28.28(i) Equations with Elementary Kernels

Let

28.28.1 $\quad w = \cosh z \cos t \cos \alpha + \sinh z \sin t \sin \alpha.$

Then

28.28.2
$$\frac{1}{2\pi}\int_0^{2\pi} e^{2ihw}\,\mathrm{ce}_n(t,h^2)\,dt = i^n\,\mathrm{ce}_n(\alpha,h^2)\,Mc_n^{(1)}(z,h),$$

28.28.3
$$\frac{1}{2\pi}\int_0^{2\pi} e^{2ihw}\,\mathrm{se}_n(t,h^2)\,dt = i^n\,\mathrm{se}_n(\alpha,h^2)\,Ms_n^{(1)}(z,h),$$

28.28.4
$$\frac{ih}{\pi}\int_0^{2\pi}\frac{\partial w}{\partial \alpha}e^{2ihw}\,\mathrm{ce}_n(t,h^2)\,dt$$
$$= i^n\,\mathrm{ce}'_n(\alpha,h^2)\,\mathrm{Mc}_n^{(1)}(z,h),$$

28.28.5
$$\frac{ih}{\pi}\int_0^{2\pi}\frac{\partial w}{\partial \alpha}e^{2ihw}\,\mathrm{se}_n(t,h^2)\,dt$$
$$= i^n\,\mathrm{se}'_n(\alpha,h^2)\,\mathrm{Ms}_n^{(1)}(z,h).$$

In (28.28.7)–(28.28.9) the paths of integration $\mathcal{L}_j$ are given by

28.28.6
$$\mathcal{L}_1: \text{from } -\eta_1+i\infty \text{ to } 2\pi-\eta_1+i\infty,$$
$$\mathcal{L}_3: \text{from } -\eta_1+i\infty \text{ to } \eta_2-i\infty,$$
$$\mathcal{L}_4: \text{from } \eta_2-i\infty \text{ to } 2\pi-\eta_1+i\infty,$$

where η_1 and η_2 are real constants.

28.28.7
$$\frac{1}{\pi}\int_{\mathcal{L}_j} e^{2ihw}\,\mathrm{me}_\nu(t,h^2)\,dt$$
$$= e^{i\nu\pi/2}\,\mathrm{me}_\nu(\alpha,h^2)\,\mathrm{M}_\nu^{(j)}(z,h), \qquad j=3,4,$$

28.28.8
$$\frac{1}{\pi}\int_{\mathcal{L}_j} 2ih\frac{\partial w}{\partial \alpha}e^{2ihw}\,\mathrm{me}_\nu(t,h^2)\,dt$$
$$= e^{i\nu\pi/2}\,\mathrm{me}'_\nu(\alpha,h^2)\,\mathrm{M}_\nu^{(j)}(z,h), \qquad j=3,4,$$

28.28.9
$$\frac{1}{2\pi}\int_{\mathcal{L}_1} e^{2ihw}\,\mathrm{me}_\nu(t,h^2)\,dt$$
$$= e^{i\nu\pi/2}\,\mathrm{me}_\nu(\alpha,h^2)\,\mathrm{M}_\nu^{(1)}(z,h).$$

In (28.28.11)–(28.28.14)

28.28.10
$$0 < \mathrm{ph}(h(\cosh z \pm 1)) < \pi.$$

28.28.11
$$\int_0^\infty e^{2ih\cosh z\cosh t}\,\mathrm{Ce}_\nu(t,h^2)\,dt$$
$$= \tfrac{1}{2}\pi i e^{i\nu\pi}\,\mathrm{ce}_\nu(0,h^2)\,\mathrm{M}_\nu^{(3)}(z,h),$$

28.28.12
$$\int_0^\infty e^{2ih\cosh z\cosh t}\sinh z\sinh t\,\mathrm{Se}_\nu(t,h^2)\,dt$$
$$= -\frac{\pi}{4h}e^{i\nu\pi/2}\,\mathrm{se}'_\nu(0,h^2)\,\mathrm{M}_\nu^{(3)}(z,h),$$

28.28.13
$$\int_0^\infty e^{2ih\cosh z\cosh t}\sinh z\sinh t\,\mathrm{Fe}_m(t,h^2)\,dt$$
$$= -\frac{\pi}{4h}i^m\,\mathrm{fe}'_m(0,h^2)\,\mathrm{Mc}_m^{(3)}(z,h),$$

28.28.14
$$\int_0^\infty e^{2ih\cosh z\cosh t}\,\mathrm{Ge}_m(t,h^2)\,dt$$
$$= \tfrac{1}{2}\pi i^{m+1}\,\mathrm{ge}_m(0,h^2)\,\mathrm{Ms}_m^{(3)}(z,h).$$

In particular, when $h > 0$ the integrals (28.28.11), (28.28.14) converge absolutely and uniformly in the half strip $\Re z \geq 0$, $0 \leq \Im z \leq \pi$.

28.28.15
$$\int_0^\infty \cos(2h\cos y\cosh t)\,\mathrm{Ce}_{2n}(t,h^2)\,dt = (-1)^{n+1}\tfrac{1}{2}\pi\,\mathrm{Mc}_{2n}^{(2)}(0,h)\,\mathrm{ce}_{2n}(y,h^2),$$

28.28.16
$$\int_0^\infty \sin(2h\cos y\cosh t)\,\mathrm{Ce}_{2n}(t,h^2)\,dt = -\frac{\pi A_0^{2n}(h^2)}{2\,\mathrm{ce}_{2n}(\tfrac{1}{2}\pi,h^2)}\left(\mathrm{ce}_{2n}(y,h^2)\mp\frac{2}{\pi C_{2n}(h^2)}\,\mathrm{fe}_{2n}(y,h^2)\right),$$

where the upper or lower sign is taken according as $0 \leq y \leq \pi$ or $\pi \leq y \leq 2\pi$. For $A_0^{2n}(q)$ and $C_{2n}(q)$ see §§28.4 and 28.5(i).

For details and further equations see Meixner et al. (1980, §2.1.1) and Sips (1970).

28.28(ii) Integrals of Products with Bessel Functions

With the notations of §28.4 for $A_m^n(q)$ and $B_m^n(q)$, §28.14 for $c_n^\nu(q)$, and (28.23.1) for $\mathcal{C}_\mu^{(j)}$, $j=1,2,3,4$,

28.28.17
$$\frac{1}{\pi}\int_0^\pi \mathcal{C}_{\nu+2s}^{(j)}(2hR)e^{-i(\nu+2s)\phi}\,\mathrm{me}_\nu(t,h^2)\,dt$$
$$= (-1)^s c_{2s}^\nu(h^2)\,\mathrm{M}_\nu^{(j)}(z,h), \qquad s\in\mathbb{Z},$$

where $R = R(z,t)$ and $\phi = \phi(z,t)$ are analytic functions for $\Re z > 0$ and real t with

28.28.18
$$R(z,t) = \left(\tfrac{1}{2}(\cosh(2z)+\cos(2t))\right)^{1/2},$$
$$R(z,0) = \cosh z,$$

and

28.28.19
$$e^{2i\phi} = \frac{\cosh(z+it)}{\cosh(z-it)},$$
$$\phi(z,0) = 0.$$

In particular, for integer ν and $\ell = 0,1,2,\ldots$,

28.28.20
$$\frac{2}{\pi}\int_0^\pi \mathcal{C}_{2\ell}^{(j)}(2hR)\cos(2\ell\phi)\,\mathrm{ce}_{2m}(t,h^2)\,dt$$
$$= \varepsilon_\ell(-1)^{\ell+m}A_{2\ell}^{2m}(h^2)\,\mathrm{Mc}_{2m}^{(j)}(z,h),$$

where again $\varepsilon_0 = 2$ and $\varepsilon_\ell = 1$, $\ell = 1,2,3,\ldots$.

28.28.21
$$\frac{2}{\pi}\int_0^\pi \mathcal{C}_{2\ell+1}^{(j)}(2hR)\cos((2\ell+1)\phi)\,\mathrm{ce}_{2m+1}(t,h^2)\,dt$$
$$= (-1)^{\ell+m}A_{2\ell+1}^{2m+1}(h^2)\,\mathrm{Mc}_{2m+1}^{(j)}(z,h),$$

28.28.22
$$\frac{2}{\pi}\int_0^\pi \mathcal{C}_{2\ell+1}^{(j)}(2hR)\sin((2\ell+1)\phi)\,\mathrm{se}_{2m+1}(t,h^2)\,dt$$
$$= (-1)^{\ell+m}B_{2\ell+1}^{2m+1}(h^2)\,\mathrm{Ms}_{2m+1}^{(j)}(z,h),$$

28.28.23
$$\frac{2}{\pi}\int_0^\pi \mathcal{C}_{2\ell+2}^{(j)}(2hR)\sin((2\ell+2)\phi)\,\mathrm{se}_{2m+2}(t,h^2)\,dt$$
$$= (-1)^{\ell+m} B_{2\ell+2}^{2m+2}(h^2)\,\mathrm{Ms}_{2m+2}^{(j)}(z,h).$$

28.28(iii) Integrals of Products of Mathieu Functions of Noninteger Order

With the parameter h suppressed we use the notation

28.28.24
$$\mathrm{D}_0(\nu,\mu,z) = \mathrm{M}_\nu^{(3)}(z)\,\mathrm{M}_\mu^{(4)}(z) - \mathrm{M}_\nu^{(4)}(z)\,\mathrm{M}_\mu^{(3)}(z),$$
$$\mathrm{D}_1(\nu,\mu,z) = \mathrm{M}_\nu^{(3)'}(z)\,\mathrm{M}_\mu^{(4)}(z) - \mathrm{M}_\nu^{(4)'}(z)\,\mathrm{M}_\mu^{(3)}(z),$$

and assume $\nu \notin \mathbb{Z}$ and $m \in \mathbb{Z}$. Then

28.28.25
$$\frac{\sinh z}{\pi^2}\int_0^{2\pi} \frac{\cos t\,\mathrm{me}_\nu(t,h^2)\,\mathrm{me}_{-\nu-2m-1}(t,h^2)}{\sinh^2 z + \sin^2 t}\,dt$$
$$= (-1)^{m+1}ih\alpha_{\nu,m}^{(0)}\,\mathrm{D}_0(\nu,\nu+2m+1,z),$$

28.28.26
$$\frac{\cosh z}{\pi^2}\int_0^{2\pi} \frac{\sin t\,\mathrm{me}_\nu(t,h^2)\,\mathrm{me}_{-\nu-2m-1}(t,h^2)}{\sinh^2 z + \sin^2 t}\,dt$$
$$= (-1)^{m+1}ih\alpha_{\nu,m}^{(1)}\,\mathrm{D}_0(\nu,\nu+2m+1,z),$$

where

28.28.27
$$\alpha_{\nu,m}^{(0)} = \frac{1}{2\pi}\int_0^{2\pi} \cos t\,\mathrm{me}_\nu(t,h^2)\,\mathrm{me}_{-\nu-2m-1}(t,h^2)\,dt$$
$$= (-1)^m \frac{2i}{\pi} \frac{\mathrm{me}_\nu(0,h^2)\,\mathrm{me}_{-\nu-2m-1}(0,h^2)}{h\,\mathrm{D}_0(\nu,\nu+2m+1,0)},$$

28.28.28
$$\alpha_{\nu,m}^{(1)} = \frac{1}{2\pi}\int_0^{2\pi} \sin t\,\mathrm{me}_\nu(t,h^2)\,\mathrm{me}_{-\nu-2m-1}(t,h^2)\,dt$$
$$= (-1)^{m+1} \frac{2i}{\pi} \frac{\mathrm{me}'_\nu(0,h^2)\,\mathrm{me}_{-\nu-2m-1}(0,h^2)}{h\,\mathrm{D}_1(\nu,\nu+2m+1,0)}.$$

For further integrals see http://dlmf.nist.gov/28.28.iii.

28.28(iv) Integrals of Products of Mathieu Functions of Integer Order

Again with the parameter h suppressed, let

28.28.35
$$\mathrm{Ds}_0(n,m,z) = \mathrm{Ms}_n^{(3)}(z)\,\mathrm{Ms}_m^{(4)}(z) - \mathrm{Ms}_n^{(4)}(z)\,\mathrm{Ms}_m^{(3)}(z),$$
$$\mathrm{Ds}_1(n,m,z) = \mathrm{Ms}_n^{(3)'}(z)\,\mathrm{Ms}_m^{(4)}(z) - \mathrm{Ms}_n^{(4)'}(z)\,\mathrm{Ms}_m^{(3)}(z),$$
$$\mathrm{Ds}_2(n,m,z) = \mathrm{Ms}_n^{(3)'}(z)\,\mathrm{Ms}_m^{(4)'}(z) - \mathrm{Ms}_n^{(4)'}(z)\,\mathrm{Ms}_m^{(3)'}(z).$$

Then

28.28.36
$$\frac{\sinh z}{\pi^2}\int_0^{2\pi} \frac{\cos t\,\mathrm{se}_n(t,h^2)\,\mathrm{se}_m(t,h^2)}{\sinh^2 z + \sin^2 t}\,dt$$
$$= (-1)^{p+1}ih\widehat{\alpha}_{n,m}^{(s)}\,\mathrm{Ds}_0(n,m,z),$$

28.28.37
$$\frac{\cosh z}{\pi^2}\int_0^{2\pi} \frac{\sin t\,\mathrm{se}'_n(t,h^2)\,\mathrm{se}_m(t,h^2)}{\sinh^2 z + \sin^2 t}\,dt$$
$$= (-1)^{p+1}ih\widehat{\alpha}_{n,m}^{(s)}\,\mathrm{Ds}_1(n,m,z),$$

where $m - n = 2p+1$, $p \in \mathbb{Z}$; $m,n = 1,2,3,\ldots$. Also,

28.28.38
$$\widehat{\alpha}_{n,m}^{(s)} = \frac{1}{2\pi}\int_0^{2\pi} \cos t\,\mathrm{se}_n(t,h^2)\,\mathrm{se}_m(t,h^2)\,dt$$
$$= (-1)^p \frac{2}{i\pi} \frac{\mathrm{se}'_n(0,h^2)\,\mathrm{se}'_m(0,h^2)}{h\,\mathrm{Ds}_2(n,m,0)}.$$

For further integrals see http://dlmf.nist.gov/28.28.iv and Schäfke (1983).

28.28(v) Compendia

See Prudnikov et al. (1990, pp. 359–368), Gradshteyn and Ryzhik (2000, pp. 755–759), Sips (1970), and Meixner et al. (1980, §2.1.1).

Hill's Equation

28.29 Definitions and Basic Properties

28.29(i) Hill's Equation

A generalization of Mathieu's equation (28.2.1) is *Hill's equation*

28.29.1
$$w''(z) + (\lambda + Q(z))w = 0,$$

with

28.29.2
$$Q(z+\pi) = Q(z),$$

and

28.29.3
$$\int_0^\pi Q(z)\,dz = 0.$$

$Q(z)$ is either a continuous and real-valued function for $z \in \mathbb{R}$ or an analytic function of z in a doubly-infinite open strip that contains the real axis. π is the minimum period of Q.

28.29(ii) Floquet's Theorem and the Characteristic Exponent

The *basic solutions* $w_\mathrm{I}(z,\lambda)$, $w_\mathrm{II}(z,\lambda)$ are defined in the same way as in §28.2(ii) (compare (28.2.5), (28.2.6)). Then

28.29.4
$$w_\mathrm{I}(z+\pi,\lambda) = w_\mathrm{I}(\pi,\lambda)w_\mathrm{I}(z,\lambda) + w'_\mathrm{I}(\pi,\lambda)w_\mathrm{II}(z,\lambda),$$

28.29.5
$$w_\mathrm{II}(z+\pi,\lambda) = w_\mathrm{II}(\pi,\lambda)w_\mathrm{I}(z,\lambda) + w'_\mathrm{II}(\pi,\lambda)w_\mathrm{II}(z,\lambda).$$

Let ν be a real or complex constant satisfying (without loss of generality)

28.29.6
$$-1 < \Re\nu \leq 1$$

throughout this section. Then (28.29.1) has a nontrivial solution $w(z)$ with the pseudoperiodic property

28.29.7
$$w(z+\pi) = e^{\pi i\nu}w(z),$$

iff $e^{\pi i\nu}$ is an eigenvalue of the matrix

$$28.29.8 \qquad \begin{bmatrix} w_{\text{I}}(\pi,\lambda) & w_{\text{II}}(\pi,\lambda) \\ w_{\text{I}}'(\pi,\lambda) & w_{\text{II}}'(\pi,\lambda) \end{bmatrix}.$$

Equivalently,

$$28.29.9 \qquad 2\cos(\pi\nu) = w_{\text{I}}(\pi,\lambda) + w_{\text{II}}'(\pi,\lambda).$$

This is the *characteristic equation* of (28.29.1), and $\cos(\pi\nu)$ is an entire function of λ. Given λ together with the condition (28.29.6), the solutions $\pm\nu$ of (28.29.9) are the *characteristic exponents* of (28.29.1). A solution satisfying (28.29.7) is called a *Floquet solution with respect to ν* (or *Floquet solution*). It has the form

$$28.29.10 \qquad F_\nu(z) = e^{i\nu z} P_\nu(z),$$

where the function $P_\nu(z)$ is π-periodic.

If ν ($\neq 0, 1$) is a solution of (28.29.9), then $F_\nu(z)$, $F_{-\nu}(z)$ comprise a fundamental pair of solutions of Hill's equation.

If $\nu = 0$ or 1, then (28.29.1) has a nontrivial solution $P(z)$ which is periodic with period π (when $\nu = 0$) or 2π (when $\nu = 1$). Let $w(z)$ be a solution linearly independent of $P(z)$. Then

$$28.29.11 \qquad w(z+\pi) = (-1)^\nu w(z) + cP(z),$$

where c is a constant. The case $c = 0$ is equivalent to

$$28.29.12 \qquad \begin{bmatrix} w_{\text{I}}(\pi,\lambda) & w_{\text{II}}(\pi,\lambda) \\ w_{\text{I}}'(\pi,\lambda) & w_{\text{II}}'(\pi,\lambda) \end{bmatrix} = \begin{bmatrix} (-1)^\nu & 0 \\ 0 & (-1)^\nu \end{bmatrix}.$$

The solutions of period π or 2π are exceptional in the following sense. If (28.29.1) has a periodic solution with minimum period $n\pi$, $n = 3, 4, \ldots$, then all solutions are periodic with period $n\pi$.

Furthermore, for each solution $w(z)$ of (28.29.1)

$$28.29.13 \qquad w(z+\pi) + w(z-\pi) = 2\cos(\pi\nu)w(z).$$

A nontrivial solution $w(z)$ is either a Floquet solution with respect to ν, or $w(z+\pi) - e^{i\nu\pi}w(z)$ is a Floquet solution with respect to $-\nu$.

In the *symmetric case* $Q(z) = Q(-z)$, $w_{\text{I}}(z,\lambda)$ is an even solution and $w_{\text{II}}(z,\lambda)$ is an odd solution; compare §28.2(ii). (28.29.9) reduces to

$$28.29.14 \qquad \cos(\pi\nu) = w_{\text{I}}(\pi,\lambda).$$

The cases $\nu = 0$ and $\nu = 1$ split into four subcases as in (28.2.21) and (28.2.22). The π-periodic or π-antiperiodic solutions are multiples of $w_{\text{I}}(z,\lambda)$, $w_{\text{II}}(z,\lambda)$, respectively.

For details and proofs see Magnus and Winkler (1966, §1.3).

28.29(iii) Discriminant and Eigenvalues in the Real Case

$Q(x)$ is assumed to be real-valued throughout this subsection.

The function

$$28.29.15 \qquad \triangle(\lambda) = w_{\text{I}}(\pi,\lambda) + w_{\text{II}}'(\pi,\lambda)$$

is called the *discriminant* of (28.29.1). It is an entire function of λ. Its order of growth for $|\lambda| \to \infty$ is exactly $\frac{1}{2}$; see Magnus and Winkler (1966, Chapter II, pp. 19–28).

For a given ν, the characteristic equation $\triangle(\lambda) - 2\cos(\pi\nu) = 0$ has infinitely many roots λ. Conversely, for a given λ, the value of $\triangle(\lambda)$ is needed for the computation of ν. For this purpose the discriminant can be expressed as an infinite determinant involving the Fourier coefficients of $Q(x)$; see Magnus and Winkler (1966, §2.3, pp. 28–36).

To every equation (28.29.1), there belong two increasing infinite sequences of real *eigenvalues*:

$$28.29.16 \quad \lambda_n,\ n = 0, 1, 2, \ldots,\ \text{with}\ \triangle(\lambda_n) = 2,$$

$$28.29.17 \quad \mu_n,\ n = 1, 2, 3, \ldots,\ \text{with}\ \triangle(\mu_n) = -2.$$

In consequence, (28.29.1) has a solution of period π iff $\lambda = \lambda_n$, and a solution of period 2π iff $\lambda = \mu_n$. Both λ_n and $\mu_n \to \infty$ as $n \to \infty$, and interlace according to the inequalities

28.29.18
$$\lambda_0 < \mu_1 \leq \mu_2 < \lambda_1 \leq \lambda_2 < \mu_3 \leq \mu_4 < \lambda_3 \leq \lambda_4 < \cdots.$$

Assume that the second derivative of $Q(x)$ in (28.29.1) exists and is continuous. Then with

$$28.29.19 \qquad N = \frac{1}{\pi} \int_0^\pi (Q(x))^2\, dx,$$

we have for $m \to \infty$

$$28.29.20 \qquad \begin{aligned} \mu_{2m-1} - (2m-1)^2 - \frac{N}{(4m)^2} &= o(m^{-2}), \\ \mu_{2m} - (2m-1)^2 - \frac{N}{(4m)^2} &= o(m^{-2}), \end{aligned}$$

$$28.29.21 \qquad \begin{aligned} \lambda_{2m-1} - (2m)^2 - \frac{N}{(4m)^2} &= o(m^{-2}), \\ \lambda_{2m} - (2m)^2 - \frac{N}{(4m)^2} &= o(m^{-2}). \end{aligned}$$

If $Q(x)$ has k continuous derivatives, then as $m \to \infty$

$$28.29.22 \qquad \begin{aligned} \lambda_{2m} - \lambda_{2m-1} &= o(1/m^k), \\ \mu_{2m} - \mu_{2m-1} &= o(1/m^k); \end{aligned}$$

see Hochstadt (1963).

For further results, especially when $Q(z)$ is analytic in a strip, see Weinstein and Keller (1987).

28.30 Expansions in Series of Eigenfunctions

28.30(i) Real Variable

Let $\widehat{\lambda}_m$, $m = 0, 1, 2, \ldots$, be the set of characteristic values (28.29.16) and (28.29.17), arranged in their natural order (see (28.29.18)), and let $w_m(x)$, $m = 0, 1, 2, \ldots$, be the *eigenfunctions*, that is, an orthonormal set of 2π-periodic solutions; thus

28.30.1 $\quad w_m'' + (\widehat{\lambda}_m + Q(x))w_m = 0,$

28.30.2 $\quad \dfrac{1}{2\pi} \displaystyle\int_0^{2\pi} w_m(x) w_n(x)\, dx = \delta_{m,n}.$

Then every continuous 2π-periodic function $f(x)$ whose second derivative is square-integrable over the interval $[0, 2\pi]$ can be expanded in a uniformly and absolutely convergent series

28.30.3 $\quad f(x) = \displaystyle\sum_{m=0}^{\infty} f_m w_m(x),$

where

28.30.4 $\quad f_m = \dfrac{1}{2\pi} \displaystyle\int_0^{2\pi} f(x) w_m(x)\, dx.$

28.30(ii) Complex Variable

For analogous results to those of §28.19, see Schäfke (1960, 1961b), and Meixner et al. (1980, §1.1.11).

28.31 Equations of Whittaker–Hill and Ince

28.31(i) Whittaker–Hill Equation

Hill's equation with three terms

28.31.1 $\quad W'' + \left(A + B\cos(2z) - \tfrac{1}{2}(kc)^2\cos(4z)\right)W = 0$

and constant values of A, B, k, and c, is called the *Equation of Whittaker–Hill*. It has been discussed in detail by Arscott (1967) for $k^2 < 0$, and by Urwin and Arscott (1970) for $k^2 > 0$.

28.31(ii) Equation of Ince; Ince Polynomials

When $k^2 < 0$, we substitute

28.31.2 $\quad \begin{aligned}&\xi^2 = -4k^2c^2, \quad A = \eta - \tfrac{1}{8}\xi^2, \quad B = -(p+1)\xi,\\ &W(z) = w(z)\exp\!\left(-\tfrac{1}{4}\xi\cos(2z)\right),\end{aligned}$

in (28.31.1). The result is the *Equation of Ince*:

28.31.3 $\quad w'' + \xi\sin(2z)w' + (\eta - p\xi\cos(2z))w = 0.$

Formal 2π-periodic solutions can be constructed as Fourier series; compare §28.4:

28.31.4 $\quad w_{e,s}(z) = \displaystyle\sum_{\ell=0}^{\infty} A_{2\ell+s}\cos(2\ell+s)z, \quad s = 0, 1,$

28.31.5 $\quad w_{o,s}(z) = \displaystyle\sum_{\ell=0}^{\infty} B_{2\ell+s}\sin(2\ell+s)z, \quad s = 1, 2,$

where the coefficients satisfy

28.31.6 $\quad \begin{aligned}&-2\eta A_0 + (2+p)\xi A_2 = 0, \quad p\xi A_0 + (4-\eta)A_2 + \left(\tfrac{1}{2}p + 2\right)\xi A_4 = 0,\\ &\left(\tfrac{1}{2}p - \ell + 1\right)\xi A_{2\ell-2} + \left(4\ell^2 - \eta\right)A_{2\ell} + \left(\tfrac{1}{2}p + \ell + 1\right)\xi A_{2\ell+2} = 0,\end{aligned} \qquad \ell \geq 2,$

28.31.7 $\quad \begin{aligned}&\left(1 - \eta + \left(\tfrac{1}{2}p + \tfrac{1}{2}\right)\xi\right)A_1 + \left(\tfrac{1}{2}p + \tfrac{3}{2}\right)\xi A_3 = 0,\\ &\left(\tfrac{1}{2}p - \ell + \tfrac{1}{2}\right)\xi A_{2\ell-1} + \left((2\ell+1)^2 - \eta\right)A_{2\ell+1} + \left(\tfrac{1}{2}p + \ell + \tfrac{3}{2}\right)\xi A_{2\ell+3} = 0,\end{aligned} \qquad \ell \geq 1,$

28.31.8 $\quad \begin{aligned}&\left(1 - \eta - \left(\tfrac{1}{2}p + \tfrac{1}{2}\right)\xi\right)B_1 + \left(\tfrac{1}{2}p + \tfrac{3}{2}\right)\xi B_3 = 0,\\ &\left(\tfrac{1}{2}p - \ell + \tfrac{1}{2}\right)\xi B_{2\ell-1} + \left((2\ell+1)^2 - \eta\right)B_{2\ell+1} + \left(\tfrac{1}{2}p + \ell + \tfrac{3}{2}\right)\xi B_{2\ell+3} = 0,\end{aligned} \qquad \ell \geq 1,$

28.31.9 $\quad \begin{aligned}&(4-\eta)B_2 + \left(\tfrac{1}{2}p + 2\right)\xi B_4 = 0,\\ &\left(\tfrac{1}{2}p - \ell + 1\right)\xi B_{2\ell-2} + (4\ell^2 - \eta)B_{2\ell} + \left(\tfrac{1}{2}p + \ell + 1\right)\xi B_{2\ell+2} = 0,\end{aligned} \qquad \ell \geq 2.$

When p is a nonnegative integer, the parameter η can be chosen so that solutions of (28.31.3) are trigonometric polynomials, called *Ince polynomials*. They are denoted by

28.31.10 $\quad \begin{aligned}&C_{2n}^{2m}(z,\xi) \quad \text{with } p = 2n,\\ &C_{2n+1}^{2m+1}(z,\xi) \quad \text{with } p = 2n+1,\end{aligned}$

28.31.11 $\quad \begin{aligned}&S_{2n+1}^{2m+1}(z,\xi) \quad \text{with } p = 2n+1,\\ &S_{2n+2}^{2m+2}(z,\xi) \quad \text{with } p = 2n+2,\end{aligned}$

and $m = 0, 1, \ldots, n$ in all cases.

The values of η corresponding to $C_p^m(z,\xi)$, $S_p^m(z,\xi)$ are denoted by $a_p^m(\xi)$, $b_p^m(\xi)$, respectively. They are real and distinct, and can be ordered so that $C_p^m(z,\xi)$ and $S_p^m(z,\xi)$ have precisely m zeros, all simple, in $0 \leq z < \pi$. The normalization is given by

28.31.12 $\quad \dfrac{1}{\pi}\displaystyle\int_0^{2\pi}\left(C_p^m(x,\xi)\right)^2 dx = \dfrac{1}{\pi}\displaystyle\int_0^{2\pi}\left(S_p^m(x,\xi)\right)^2 dx = 1,$

APPLICATIONS

ambiguities in sign being resolved by requiring $C_p^m(x,\xi)$ and $S_p^{m\prime}(x,\xi)$ to be continuous functions of x and positive when $x = 0$.

For $\xi \to 0$, with x fixed,

28.31.13
$$C_p^0(x,\xi) \to 1/\sqrt{2}, \quad C_p^m(x,\xi) \to \cos(mx),$$
$$S_p^m(x,\xi) \to \sin(mx), \quad m \neq 0; \quad a_p^m(\xi), b_p^m(\xi) \to m^2.$$

If $p \to \infty$ and $\xi \to 0$ in such a way that $p\xi \to 2q$, then in the notation of §§28.2(v) and 28.2(vi)

28.31.14 $\quad C_p^m(x,\xi) \to \mathrm{ce}_m(x,q), \quad S_p^m(x,\xi) \to \mathrm{se}_m(x,q),$

28.31.15 $\quad a_p^m(\xi) \to a_m(q), \quad b_p^m(\xi) \to b_m(q).$

For proofs and further information, including convergence of the series (28.31.4), (28.31.5), see Arscott (1967).

28.31(iii) Paraboloidal Wave Functions

With (28.31.10) and (28.31.11),

28.31.16 $\quad hc_p^m(z,\xi) = e^{-\frac{1}{4}\xi\cos(2z)} C_p^m(z,\xi),$

28.31.17 $\quad hs_p^m(z,\xi) = e^{-\frac{1}{4}\xi\cos(2z)} S_p^m(z,\xi),$

are called *paraboloidal wave functions*. They satisfy the differential equation

28.31.18
$$w'' + \left(\eta - \tfrac{1}{8}\xi^2 - (p+1)\xi\cos(2z) + \tfrac{1}{8}\xi^2\cos(4z)\right)w = 0,$$
with $\eta = a_p^m(\xi)$, $\eta = b_p^m(\xi)$, respectively.

For change of sign of ξ,

28.31.19
$$hc_{2n}^{2m}(z,-\xi) = (-1)^m hc_{2n}^{2m}(\tfrac{1}{2}\pi - z, \xi),$$
$$hc_{2n+1}^{2m+1}(z,-\xi) = (-1)^m hs_{2n+1}^{2m+1}(\tfrac{1}{2}\pi - z, \xi),$$

and

28.31.20
$$hs_{2n+1}^{2m+1}(z,-\xi) = (-1)^m hc_{2n+1}^{2m+1}(\tfrac{1}{2}\pi - z, \xi),$$
$$hs_{2n+2}^{2m+2}(z,-\xi) = (-1)^m hs_{2n+2}^{2m+2}(\tfrac{1}{2}\pi - z, \xi).$$

For $m_1 \neq m_2$,

28.31.21
$$\int_0^{2\pi} hc_p^{m_1}(x,\xi) hc_p^{m_2}(x,\xi)\,dx$$
$$= \int_0^{2\pi} hs_p^{m_1}(x,\xi) hs_p^{m_2}(x,\xi)\,dx = 0.$$

More important are the *double orthogonality relations* for $p_1 \neq p_2$ or $m_1 \neq m_2$ or both, given by

28.31.22
$$\int_{u_0}^{u_\infty}\int_0^{2\pi} hc_{p_1}^{m_1}(u,\xi) hc_{p_1}^{m_1}(v,\xi) hc_{p_2}^{m_2}(u,\xi) hc_{p_2}^{m_2}(v,\xi)$$
$$\times (\cos(2u) - \cos(2v))\,dv\,du = 0,$$

and

28.31.23
$$\int_{u_0}^{u_\infty}\int_0^{2\pi} hs_{p_1}^{m_1}(u,\xi) hs_{p_1}^{m_1}(v,\xi) hs_{p_2}^{m_2}(u,\xi) hs_{p_2}^{m_2}(v,\xi)$$
$$\times (\cos(2u) - \cos(2v))\,dv\,du = 0,$$

and also for all p_1, p_2, m_1, m_2, given by

28.31.24
$$\int_{u_0}^{u_\infty}\int_0^{2\pi} hc_{p_1}^{m_1}(u,\xi) hc_{p_1}^{m_1}(v,\xi) hs_{p_2}^{m_2}(u,\xi) hs_{p_2}^{m_2}(v,\xi)$$
$$\times (\cos(2u) - \cos(2v))\,dv\,du = 0,$$

where $(u_0, u_\infty) = (0, i\infty)$ when $\xi > 0$, and $(u_0, u_\infty) = (\tfrac{1}{2}\pi, \tfrac{1}{2}\pi + i\infty)$ when $\xi < 0$.

For proofs and further integral equations see Urwin (1964, 1965).

Asymptotic Behavior

For $\xi > 0$, the functions $hc_p^m(z,\xi)$, $hs_p^m(z,\xi)$ behave asymptotically as multiples of $\exp(-\tfrac{1}{4}\xi\cos(2z))(\cos z)^p$ as $z \to \pm i\infty$. All other periodic solutions behave as multiples of $\exp(\tfrac{1}{4}\xi\cos(2z))(\cos z)^{-p-2}$.

For $\xi > 0$, the functions $hc_p^m(z,-\xi)$, $hs_p^m(z,-\xi)$ behave asymptotically as multiples of $\exp(\tfrac{1}{4}\xi\cos(2z))(\cos z)^{-p-2}$ as $z \to \tfrac{1}{2}\pi \pm i\infty$. All other periodic solutions behave as multiples of $\exp(-\tfrac{1}{4}\xi\cos(2z))(\cos z)^p$.

Applications

28.32 Mathematical Applications

28.32(i) Elliptical Coordinates and an Integral Relationship

If the boundary conditions in a physical problem relate to the perimeter of an ellipse, then elliptical coordinates are convenient. These are given by

28.32.1 $\quad x = c\cosh\xi\cos\eta, \quad y = c\sinh\xi\sin\eta.$

The two-dimensional wave equation

28.32.2
$$\frac{\partial^2 V}{\partial x^2} + \frac{\partial^2 V}{\partial y^2} + k^2 V = 0$$

then becomes

28.32.3 $\quad \dfrac{\partial^2 V}{\partial \xi^2} + \dfrac{\partial^2 V}{\partial \eta^2} + \dfrac{1}{2}c^2k^2(\cosh(2\xi) - \cos(2\eta))V = 0.$

The separated solutions $V(\xi,\eta) = v(\xi)w(\eta)$ can be obtained from the modified Mathieu's equation (28.20.1) for v and from Mathieu's equation (28.2.1) for w, where a is the separation constant and $q = \tfrac{1}{4}c^2k^2$.

This leads to integral equations and an integral relation between the solutions of Mathieu's equation (setting $\zeta = i\xi$, $z = \eta$ in (28.32.3)).

Let $u(\zeta)$ be a solution of Mathieu's equation (28.2.1) and $K(z,\zeta)$ be a solution of

28.32.4 $\quad \dfrac{\partial^2 K}{\partial z^2} - \dfrac{\partial^2 K}{\partial \zeta^2} = 2q\left(\cos(2z) - \cos(2\zeta)\right)K.$

Also let $\mathcal{L}$ be a curve (possibly improper) such that the quantity

28.32.5 $\quad K(z,\zeta)\dfrac{du(\zeta)}{d\zeta} - u(\zeta)\dfrac{\partial K(z,\zeta)}{\partial \zeta}$

approaches the same value when ζ tends to the endpoints of $\mathcal{L}$. Then

28.32.6 $\quad w(z) = \displaystyle\int_{\mathcal{L}} K(z,\zeta)u(\zeta)\,d\zeta$

defines a solution of Mathieu's equation, provided that (in the case of an improper curve) the integral converges with respect to z uniformly on compact subsets of $\mathbb{C}$.

Kernels K can be found, for example, by separating solutions of the wave equation in other systems of orthogonal coordinates. See Schmidt and Wolf (1979).

28.32(ii) Paraboloidal Coordinates

The general paraboloidal coordinate system is linked with Cartesian coordinates via

28.32.7
$x_1 = \tfrac{1}{2}c\left(\cosh(2\alpha) + \cos(2\beta) - \cosh(2\gamma)\right),$
$x_2 = 2c\cosh\alpha\cos\beta\sinh\gamma, \quad x_3 = 2c\sinh\alpha\sin\beta\cosh\gamma,$
where c is a parameter, $0 \le \alpha < \infty$, $-\pi < \beta \le \pi$, and $0 \le \gamma < \infty$. When the Helmholtz equation

28.32.8 $\quad \nabla^2 V + k^2 V = 0$

is separated in this system, each of the separated equations can be reduced to the Whittaker–Hill equation (28.31.1), in which A, B are *separation constants*. Two conditions are used to determine A, B. The first is the 2π-periodicity of the solutions; the second can be their asymptotic form. For further information see Arscott (1967) for $k^2 < 0$, and Urwin and Arscott (1970) for $k^2 > 0$.

28.33 Physical Applications

28.33(i) Introduction

Mathieu functions occur in practical applications in two main categories:

- Boundary-values problems arising from solution of the two-dimensional wave equation in elliptical coordinates. This yields a pair of equations of the form (28.2.1) and (28.20.1), and the appropriate solution of (28.2.1) is usually a periodic solution of integer order. See §28.33(ii).

- Initial-value problems, in which only one equation (28.2.1) or (28.20.1) is involved. See §28.33(iii).

28.33(ii) Boundary-Value Problems

Physical problems involving Mathieu functions include vibrational problems in elliptical coordinates; see (28.32.1). We shall derive solutions to the uniform, homogeneous, loss-free, and stretched elliptical ring membrane with mass ρ per unit area, and radial tension τ per unit arc length. The wave equation

28.33.1 $\quad \dfrac{\partial^2 W}{\partial x^2} + \dfrac{\partial^2 W}{\partial y^2} - \dfrac{\rho}{\tau}\dfrac{\partial^2 W}{\partial t^2} = 0,$

with $W(x,y,t) = e^{i\omega t}V(x,y)$, reduces to (28.32.2) with $k^2 = \omega^2\rho/\tau$. In elliptical coordinates (28.32.2) becomes (28.32.3). The separated solutions $V_n(\xi,\eta)$ must be 2π-periodic in η, and have the form

28.33.2
$V_n(\xi,\eta) = \left(c_n\,\mathrm{M}_n^{(1)}(\xi,\sqrt{q}) + d_n\,\mathrm{M}_n^{(2)}(\xi,\sqrt{q})\right)\mathrm{me}_n(\eta,q),$

where $q = \tfrac{1}{4}c^2 k^2$ and $a_n(q)$ or $b_n(q)$ is the separation constant; compare (28.12.11), (28.20.11), and (28.20.12). Here c_n and d_n are constants. The boundary conditions for $\xi = \xi_0$ (outer clamp) and $\xi = \xi_1$ (inner clamp) yield the following equation for q:

28.33.3 $\quad \mathrm{M}_n^{(1)}(\xi_0,\sqrt{q})\,\mathrm{M}_n^{(2)}(\xi_1,\sqrt{q}) - \mathrm{M}_n^{(1)}(\xi_1,\sqrt{q})\,\mathrm{M}_n^{(2)}(\xi_0,\sqrt{q}) = 0.$

If we denote the positive solutions q of (28.33.3) by $q_{n,m}$, then the vibration of the membrane is given by $\omega_{n,m}^2 = 4q_{n,m}\tau/(c^2\rho)$. The general solution of the problem is a superposition of the separated solutions.

For a visualization see Gutiérrez-Vega et al. (2003), and for references to other boundary-value problems see:

- McLachlan (1947, Chapters XVI–XIX) for applications of the wave equation to vibrational systems, electrical and thermal diffusion, electromagnetic wave guides, elliptical cylinders in viscous fluids, and diffraction of sound and electromagnetic waves.

- Meixner and Schäfke (1954, §§4.3, 4.4) for elliptic membranes and electromagnetic waves.

- Daymond (1955) for vibrating systems.

- Troesch and Troesch (1973) for elliptic membranes.

- Alhargan and Judah (1995), Bhattacharyya and Shafai (1988), and Shen (1981) for ring antennas.

- Alhargan and Judah (1992), Germey (1964), Ragheb et al. (1991), and Sips (1967) for electromagnetic waves.

More complete bibliographies will be found in McLachlan (1947) and Meixner and Schäfke (1954).

28.33(iii) Stability and Initial-Value Problems

If the parameters of a physical system vary periodically with time, then the question of stability arises, for example, a mathematical pendulum whose length varies as $\cos(2\omega t)$. The equation of motion is given by

28.33.4 $\quad w''(t) + (b - f\cos(2\omega t))\, w(t) = 0,$

with b, f, and ω positive constants. Substituting $z = \omega t$, $a = b/\omega^2$, and $2q = f/\omega^2$, we obtain Mathieu's standard form (28.2.1).

As ω runs from 0 to $+\infty$, with b and f fixed, the point (q,a) moves from ∞ to 0 along the ray $\mathcal{L}$ given by the part of the line $a = (2b/f)q$ that lies in the first quadrant of the (q,a)-plane. Hence from §28.17 the corresponding Mathieu equation is stable or unstable according as (q,a) is in the intersection of $\mathcal{L}$ with the colored or the uncolored open regions depicted in Figure 28.17.1. In particular, the equation is stable for all sufficiently large values of ω.

For points (q,a) that are at intersections of $\mathcal{L}$ with the characteristic curves $a = a_n(q)$ or $a = b_n(q)$, a periodic solution is possible. However, in response to a small perturbation at least one solution may become unbounded.

References for other initial-value problems include:

- McLachlan (1947, Chapter XV) for amplitude distortion in moving-coil loud-speakers, frequency modulation, dynamical systems, and vibration of stretched strings.

- Vedeler (1950) for ships rolling among waves.

- Meixner and Schäfke (1954, §§4.1, 4.2, and 4.7) for quantum mechanical problems and rotation of molecules.

- Aly et al. (1975) for scattering theory.

- Hunter and Kuriyan (1976) and Rushchitsky and Rushchitska (2000) for wave mechanics.

- Fukui and Horiguchi (1992) for quantum theory.

- Jager (1997, 1998) for relativistic oscillators.

- Torres-Vega et al. (1998) for Mathieu functions in phase space.

Computation

28.34 Methods of Computation

28.34(i) Characteristic Exponents

Methods available for computing the values of $w_\mathrm{I}(\pi; a, \pm q)$ needed in (28.2.16) include:

(a) Direct numerical integration of the differential equation (28.2.1), with initial values given by (28.2.5) (§§3.7(ii), 3.7(v)).

(b) Representations for $w_\mathrm{I}(\pi; a, \pm q)$ with limit formulas for special solutions of the recurrence relations §28.4(ii) for fixed a and q; see Schäfke (1961a).

28.34(ii) Eigenvalues

Methods for computing the eigenvalues $a_n(q)$, $b_n(q)$, and $\lambda_\nu(q)$, defined in §§28.2(v) and 28.12(i), include:

(a) Summation of the power series in §§28.6(i) and 28.15(i) when $|q|$ is small.

(b) Use of asymptotic expansions and approximations for large q (§§28.8(i), 28.16). See also Zhang and Jin (1996, pp. 482–485).

(c) Methods described in §3.7(iv) applied to the differential equation (28.2.1) with the conditions (28.2.5) and (28.2.16).

(d) Solution of the matrix eigenvalue problem for each of the five infinite matrices that correspond to the linear algebraic equations (28.4.5)–(28.4.8) and (28.14.4). See Zhang and Jin (1996, pp. 479–482) and §3.2(iv).

(e) Solution of the continued-fraction equations (28.6.16)–(28.6.19) and (28.15.2) by successive approximation. See Blanch (1966), Shirts (1993), and Meixner and Schäfke (1954, §2.87).

28.34(iii) Floquet Solutions

(a) Summation of the power series in §§28.6(ii) and 28.15(ii) when $|q|$ is small.

(b) Use of asymptotic expansions and approximations for large q (§§28.8(ii)–28.8(iv)).

Also, once the eigenvalues $a_n(q)$, $b_n(q)$, and $\lambda_\nu(q)$ have been computed the following methods are applicable:

(c) Solution of (28.2.1) by boundary-value methods; see §3.7(iii). This can be combined with §28.34(ii)(c).

(d) Solution of the systems of linear algebraic equations (28.4.5)–(28.4.8) and (28.14.4), with the conditions (28.4.9)–(28.4.12) and (28.14.5), by boundary-value methods (§3.6) to determine the Fourier coefficients. Subsequently, the Fourier series can be summed with the aid of Clenshaw's algorithm (§3.11(ii)). See Meixner and Schäfke (1954, §2.87). This procedure can be combined with §28.34(ii)(d).

28.34(iv) Modified Mathieu Functions

For the modified functions we have:

(a) Numerical summation of the expansions in series of Bessel functions (28.24.1)–(28.24.13). These series converge quite rapidly for a wide range of values of q and z.

(b) Direct numerical integration (§3.7) of the differential equation (28.20.1) for moderate values of the parameters.

(c) Use of asymptotic expansions for large z or large q. See §§28.25 and 28.26.

28.35 Tables

28.35(i) Real Variables

- Blanch and Clemm (1962) includes values of $\mathrm{Mc}_n^{(1)}(x,\sqrt{q})$ and $\mathrm{Mc}_n^{(1)'}(x,\sqrt{q})$ for $n = 0(1)15$ with $q = 0(.05)1$, $x = 0(.02)1$. Also $\mathrm{Ms}_n^{(1)}(x,\sqrt{q})$ and $\mathrm{Ms}_n^{(1)'}(x,\sqrt{q})$ for $n = 1(1)15$ with $q = 0(.05)1$, $x = 0(.02)1$. Precision is generally 7D.

- Blanch and Clemm (1965) includes values of $\mathrm{Mc}_n^{(2)}(x,\sqrt{q})$, $\mathrm{Mc}_n^{(2)'}(x,\sqrt{q})$ for $n = 0(1)7$, $x = 0(.02)1$; $n = 8(1)15$, $x = 0(.01)1$. Also $\mathrm{Ms}_n^{(2)}(x,\sqrt{q})$, $\mathrm{Ms}_n^{(2)'}(x,\sqrt{q})$ for $n = 1(1)7$, $x = 0(.02)1$; $n = 8(1)15$, $x = 0(.01)1$. In all cases $q = 0(.05)1$. Precision is generally 7D. Approximate formulas and graphs are also included.

- Blanch and Rhodes (1955) includes $Be_n(t)$, $Bo_n(t)$, $t = \frac{1}{2}\sqrt{q}$, $n = 0(1)15$; 8D. The range of t is 0 to 0.1, with step sizes ranging from 0.002 down to 0.00025. Notation: $Be_n(t) = a_n(q) + 2q - (4n+2)\sqrt{q}$, $Bo_n(t) = b_n(q) + 2q - (4n-2)\sqrt{q}$.

- Ince (1932) includes eigenvalues a_n, b_n, and Fourier coefficients for $n = 0$ or $1(1)6$, $q = 0(1)10(2)20(4)40$; 7D. Also $\mathrm{ce}_n(x,q)$, $\mathrm{se}_n(x,q)$ for $q = 0(1)10$, $x = 1(1)90$, corresponding to the eigenvalues in the tables; 5D. Notation: $a_n = be_n - 2q$, $b_n = bo_n - 2q$.

- Kirkpatrick (1960) contains tables of the modified functions $\mathrm{Ce}_n(x,q)$, $\mathrm{Se}_{n+1}(x,q)$ for $n = 0(1)5$, $q = 1(1)20$, $x = 0.1(.1)1$; 4D or 5D.

- NBS (1967) includes the eigenvalues $a_n(q)$, $b_n(q)$ for $n = 0(1)3$ with $q = 0(.2)20(.5)37(1)100$, and $n = 4(1)15$ with $q = 0(2)100$; Fourier coefficients for $\mathrm{ce}_n(x,q)$ and $\mathrm{se}_n(x,q)$ for $n = 0(1)15$, $n = 1(1)15$, respectively, and various values of q in the interval $[0,100]$; joining factors $g_{e,n}(\sqrt{q})$, $f_{e,n}(\sqrt{q})$ for $n = 0(1)15$ with $q = 0(.5 \text{ to } 10)100$ (but in a different notation). Also, eigenvalues for large values of q. Precision is generally 8D.

- Stratton et al. (1941) includes b_n, b'_n, and the corresponding Fourier coefficients for $\mathrm{Se}_n(c,x)$ and $\mathrm{So}_n(c,x)$ for $n = 0$ or $1(1)4$, $c = 0(.1 \text{ or } .2)4.5$. Precision is mostly 5S. Notation: $c = 2\sqrt{q}$, $b_n = a_n + 2q$, $b'_n = b_n + 2q$, and for $\mathrm{Se}_n(c,x)$, $\mathrm{So}_n(c,x)$ see §28.1.

- Zhang and Jin (1996, pp. 521–532) includes the eigenvalues $a_n(q)$, $b_{n+1}(q)$ for $n = 0(1)4$, $q = 0(1)50$; $n = 0(1)20$ (a's) or 19 (b's), $q = 1, 3, 5, 10, 15, 25, 50(50)200$. Fourier coefficients for $\mathrm{ce}_n(x,10)$, $\mathrm{se}_{n+1}(x,10)$, $n = 0(1)7$. Mathieu functions $\mathrm{ce}_n(x,10)$, $\mathrm{se}_{n+1}(x,10)$, and their first x-derivatives for $n = 0(1)4$, $x = 0(5°)90°$. Modified Mathieu functions $\mathrm{Mc}_n^{(j)}(x,\sqrt{10})$, $\mathrm{Ms}_{n+1}^{(j)}(x,\sqrt{10})$, and their first x-derivatives for $n = 0(1)4$, $j = 1, 2$, $x = 0(.2)4$. Precision is mostly 9S.

28.35(ii) Complex Variables

- Blanch and Clemm (1969) includes eigenvalues $a_n(q)$, $b_n(q)$ for $q = \rho e^{i\phi}$, $\rho = 0(.5)25$, $\phi = 5°(5°)90°$, $n = 0(1)15$; 4D. Also $a_n(q)$ and $b_n(q)$ for $q = i\rho$, $\rho = 0(.5)100$, $n = 0(2)14$ and $n = 2(2)16$, respectively; 8D. Double points for $n = 0(1)15$; 8D. Graphs are included.

28.35(iii) Zeros

- Blanch and Clemm (1965) includes the first and second zeros of $\mathrm{Mc}_n^{(2)}(x,\sqrt{q})$, $\mathrm{Mc}_n^{(2)'}(x,\sqrt{q})$ for $n = 0, 1$, and $\mathrm{Ms}_n^{(2)}(x,\sqrt{q})$, $\mathrm{Ms}_n^{(2)'}(x,\sqrt{q})$ for $n = 1, 2$, with $q = 0(.05)1$; 7D.

- Ince (1932) includes the first zero for ce_n, se_n for $n = 2(1)5$ or 6, $q = 0(1)10(2)40$; 4D. This reference also gives zeros of the first derivatives, together with expansions for small q.

- Zhang and Jin (1996, pp. 533–535) includes the zeros (in degrees) of $\mathrm{ce}_n(x,10)$, $\mathrm{se}_n(x,10)$ for $n = 1(1)10$, and the first 5 zeros of $\mathrm{Mc}_n^{(j)}(x,\sqrt{10})$, $\mathrm{Ms}_n^{(j)}(x,\sqrt{10})$ for $n = 0$ or $1(1)8$, $j = 1, 2$. Precision is mostly 9S.

28.35(iv) Further Tables

For other tables prior to 1961 see Fletcher et al. (1962, §2.2) and Lebedev and Fedorova (1960, Chapter 11).

28.36 Software

See http://dlmf.nist.gov/28.36.

References

General References

The main references used in writing this chapter are Arscott (1964b), McLachlan (1947), Meixner and Schäfke (1954), and Meixner et al. (1980). For §§28.29–28.30 the main source is Magnus and Winkler (1966).

Sources

The following list gives the references or other indications of proofs that were used in constructing the various sections of this chapter. These sources supplement the references that are quoted in the text.

§28.2 Arscott (1964b, Chapter II), Erdélyi et al. (1955, §§16.2, 16.4), McLachlan (1947, Chapter II), Meixner and Schäfke (1954, §2.1), Meixner et al. (1980, Chapter 2). Figure 28.2.1 was produced at NIST.

§28.3 These graphics were produced at NIST.

§28.4 Arscott (1964b, Chapter III), McLachlan (1947, Chapter III), Meixner and Schäfke (1954, §§2.25, 2.71), Wolf (2008).

§28.5 Arscott (1964b, §2.4), McLachlan (1947, Chapter VII), Meixner and Schäfke (1954, §2.7). The graphics were produced at NIST.

§28.6 McLachlan (1947, Chapter II), Meixner and Schäfke (1954, §2.2), Meixner et al. (1980, §2.4), Volkmer (1998).

§28.7 Meixner and Schäfke (1954, §§2.22, 2.25). Figure 28.7.1 was provided by the author.

§28.8 Goldstein (1927), Meixner and Schäfke (1954, §§2.33, 2.84).

§28.10 Arscott (1964b, Chapter IV), Meixner and Schäfke (1954, §2.6), Meixner et al. (1980, §2.1.2).

§28.11 Arscott (1964b, §3.9.1).

§28.12 Arscott (1964b, Chapter VI), McLachlan (1947, Chapter IV), Meixner and Schäfke (1954, §2.2).

§28.13 These graphics were produced at NIST.

§28.14 Arscott (1964b, Chapter VI), McLachlan (1947, Chapter IV), Meixner and Schäfke (1954, §2.2).

§28.15 Meixner and Schäfke (1954, §2.2).

§28.16 Meixner and Schäfke (1954, §2.2).

§28.17 Arscott (1964b, §6.2), McLachlan (1947, Chapter III), Meixner and Schäfke (1954, §2.3). Figure 28.17.1 was recomputed by the author.

§28.19 Meixner and Schäfke (1954, §2.28).

§28.20 Arscott (1964b, Chapter VI), Meixner and Schäfke (1954, §2.4).

§28.21 These graphics were produced at NIST.

§28.22 Meixner and Schäfke (1954, §§2.29, 2.65, 2.73, 2.76).

§28.23 Meixner and Schäfke (1954, §2.6).

§28.24 Meixner and Schäfke (1954, §§2.6, 2.7).

§28.26 Goldstein (1927), Meixner and Schäfke (1954, §2.84), NBS (1967, IV).

§28.28 Arscott (1964b, Chapters IV and VI), McLachlan (1947, Chapters IX and XIV), Meixner and Schäfke (1954, §2.7), Meixner et al. (1980, §2.1), Schäfke (1983). There is a sign error on p. 158 of the last reference.

§28.29 Magnus and Winkler (1966, Part I, pp. 1–43), Arscott (1964b, Chapter VII), McLachlan (1947, §6.10).

§28.30 Magnus and Winkler (1966, Part I, §2.5).

§28.31 Arscott (1967), Urwin and Arscott (1970), Urwin (1964, 1965).

§28.32 Arscott (1967, §§1.3 and 2.6), Meixner and Schäfke (1954, §1.135).

Chapter 29

Lamé Functions

H. Volkmer[1]

Notation — **684**
- 29.1 Special Notation — 684

Lamé Functions — **684**
- 29.2 Differential Equations — 684
- 29.3 Definitions and Basic Properties — 685
- 29.4 Graphics — 686
- 29.5 Special Cases and Limiting Forms — 688
- 29.6 Fourier Series — 688
- 29.7 Asymptotic Expansions — 689
- 29.8 Integral Equations — 689
- 29.9 Stability — 690
- 29.10 Lamé Functions with Imaginary Periods — 690
- 29.11 Lamé Wave Equation — 690

Lamé Polynomials — **690**
- 29.12 Definitions — 690
- 29.13 Graphics — 691
- 29.14 Orthogonality — 692
- 29.15 Fourier Series and Chebyshev Series — 692
- 29.16 Asymptotic Expansions — 693
- 29.17 Other Solutions — 693

Applications — **693**
- 29.18 Mathematical Applications — 693
- 29.19 Physical Applications — 694

Computation — **694**
- 29.20 Methods of Computation — 694
- 29.21 Tables — 694
- 29.22 Software — 695

References — **695**

[1]Department of Mathematical Sciences, University of Wisconsin–Milwaukee, Milwaukee, Wisconsin.
Copyright © 2009 National Institute of Standards and Technology. All rights reserved.

Notation

29.1 Special Notation

(For other notation see pp. xiv and 873.)

m, n, p	nonnegative integers.
x	real variable.
z	complex variable.
h, k, ν	real parameters, $0 < k < 1$, $\nu \geq -\tfrac{1}{2}$.
k'	$\sqrt{1-k^2}$, $0 < k' < 1$.
K, K'	complete elliptic integrals of the first kind with moduli k, k', respectively (see §19.2(ii)).

All derivatives are denoted by differentials, not by primes.

The main functions treated in this chapter are the eigenvalues $a_\nu^{2m}(k^2)$, $a_\nu^{2m+1}(k^2)$, $b_\nu^{2m+1}(k^2)$, $b_\nu^{2m+2}(k^2)$, the Lamé functions $Ec_\nu^{2m}(z, k^2)$, $Ec_\nu^{2m+1}(z, k^2)$, $Es_\nu^{2m+1}(z, k^2)$, $Es_\nu^{2m+2}(z, k^2)$, and the Lamé polynomials $uE_{2n}^m(z, k^2)$, $sE_{2n+1}^m(z, k^2)$, $cE_{2n+1}^m(z, k^2)$, $dE_{2n+1}^m(z, k^2)$, $scE_{2n+2}^m(z, k^2)$, $sdE_{2n+2}^m(z, k^2)$, $cdE_{2n+2}^m(z, k^2)$, $scdE_{2n+3}^m(z, k^2)$. The notation for the eigenvalues and functions is due to Erdélyi et al. (1955, §15.5.1) and that for the polynomials is due to Arscott (1964b, §9.3.2). The normalization is that of Jansen (1977, §3.1).

Other notations that have been used are as follows: Ince (1940a) interchanges $a_\nu^{2m+1}(k^2)$ with $b_\nu^{2m+1}(k^2)$. The relation to the Lamé functions $L_{c\nu}^{(m)}$, $L_{s\nu}^{(m)}$ of Jansen (1977) is given by

$$Ec_\nu^{2m}(z, k^2) = (-1)^m L_{c\nu}^{(2m)}(\psi, k'^2),$$
$$Ec_\nu^{2m+1}(z, k^2) = (-1)^m L_{s\nu}^{(2m+1)}(\psi, k'^2),$$
$$Es_\nu^{2m+1}(z, k^2) = (-1)^m L_{c\nu}^{(2m+1)}(\psi, k'^2),$$
$$Es_\nu^{2m+2}(z, k^2) = (-1)^m L_{s\nu}^{(2m+2)}(\psi, k'^2),$$

where $\psi = \operatorname{am}(z, k)$; see §22.16(i). The relation to the Lamé functions Ec_ν^m, Es_ν^m of Ince (1940b) is given by

$$Ec_\nu^{2m}(z, k^2) = c_\nu^{2m}(k^2)\operatorname{Ec}_\nu^{2m}(z, k^2),$$
$$Ec_\nu^{2m+1}(z, k^2) = c_\nu^{2m+1}(k^2)\operatorname{Es}_\nu^{2m+1}(z, k^2),$$
$$Es_\nu^{2m+1}(z, k^2) = s_\nu^{2m+1}(k^2)\operatorname{Ec}_\nu^{2m+1}(z, k^2),$$
$$Es_\nu^{2m+2}(z, k^2) = s_\nu^{2m+2}(k^2)\operatorname{Es}_\nu^{2m+2}(z, k^2),$$

where the positive factors $c_\nu^m(k^2)$ and $s_\nu^m(k^2)$ are determined by

$$(c_\nu^m(k^2))^2 = \frac{4}{\pi}\int_0^K \left(Ec_\nu^m(x, k^2)\right)^2 dx,$$
$$(s_\nu^m(k^2))^2 = \frac{4}{\pi}\int_0^K \left(Es_\nu^m(x, k^2)\right)^2 dx.$$

Lamé Functions

29.2 Differential Equations

29.2(i) Lamé's Equation

29.2.1 $$\frac{d^2w}{dz^2} + (h - \nu(\nu+1)k^2 \operatorname{sn}^2(z, k))w = 0,$$

where k and ν are real parameters such that $0 < k < 1$ and $\nu \geq -\tfrac{1}{2}$. For $\operatorname{sn}(z, k)$ see §22.2. This equation has regular singularities at the points $2pK + (2q+1)iK'$, where $p, q \in \mathbb{Z}$, and K, K' are the complete elliptic integrals of the first kind with moduli k, $k'(=(1-k^2)^{1/2})$, respectively; see §19.2(ii). In general, at each singularity each solution of (29.2.1) has a branch point (§2.7(i)). See Figure 29.2.1.

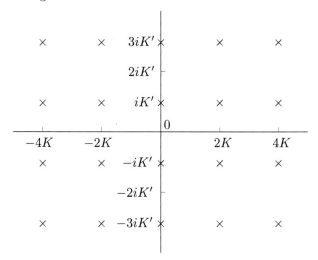

Figure 29.2.1: z-plane: singularities $\times\times\times$ of Lamé's equation.

29.2(ii) Other Forms

29.2.2
$$\frac{d^2w}{d\xi^2} + \frac{1}{2}\left(\frac{1}{\xi} + \frac{1}{\xi-1} + \frac{1}{\xi-k^{-2}}\right)\frac{dw}{d\xi} + \frac{hk^{-2} - \nu(\nu+1)\xi}{4\xi(\xi-1)(\xi-k^{-2})}w = 0,$$

where

29.2.3 $$\xi = \operatorname{sn}^2(z, k).$$

29.2.4
$$(1 - k^2 \cos^2\phi)\frac{d^2w}{d\phi^2} + k^2 \cos\phi \sin\phi \frac{dw}{d\phi} + (h - \nu(\nu+1)k^2 \cos^2\phi)w = 0,$$

where

29.2.5 $$\phi = \tfrac{1}{2}\pi - \operatorname{am}(z, k).$$

For $\operatorname{am}(z, k)$ see §22.16(i).

Next, let e_1, e_2, e_3 be any real constants that satisfy $e_1 > e_2 > e_3$ and

29.2.6 $\quad e_1 + e_2 + e_3 = 0, \quad (e_2 - e_3)/(e_1 - e_3) = k^2.$

(These constants are not unique.) Then with

29.2.7 $\quad g = (e_1 - e_3)h + \nu(\nu + 1)e_3,$

29.2.8 $\quad \eta = (e_1 - e_3)^{-1/2}(z - iK'),$

we have

29.2.9 $\quad \dfrac{d^2 w}{d\eta^2} + (g - \nu(\nu+1)\wp(\eta))w = 0,$

and

29.2.10 $\quad \dfrac{d^2 w}{d\zeta^2} + \dfrac{1}{2}\left(\dfrac{1}{\zeta - e_1} + \dfrac{1}{\zeta - e_2} + \dfrac{1}{\zeta - e_3}\right)\dfrac{dw}{d\zeta} + \dfrac{g - \nu(\nu+1)\zeta}{4(\zeta - e_1)(\zeta - e_2)(\zeta - e_3)}w = 0,$

where

29.2.11 $\quad \zeta = \wp(\eta; g_2, g_3) = \wp(\eta),$

with

29.2.12 $\quad g_2 = -4(e_2 e_3 + e_3 e_1 + e_1 e_2), \quad g_3 = 4 e_1 e_2 e_3.$

For the Weierstrass function $\wp$ see §23.2(ii).

Equation (29.2.10) is a special case of Heun's equation (31.2.1).

29.3 Definitions and Basic Properties

29.3(i) Eigenvalues

For each pair of values of ν and k there are four infinite unbounded sets of real eigenvalues h for which equation (29.2.1) has even or odd solutions with periods $2K$ or $4K$. They are denoted by $a_\nu^{2m}(k^2)$, $a_\nu^{2m+1}(k^2)$, $b_\nu^{2m+1}(k^2)$, $b_\nu^{2m+2}(k^2)$, where $m = 0, 1, 2, \ldots$; see Table 29.3.1.

Table 29.3.1: Eigenvalues of Lamé's equation.

eigenvalue h	parity	period
$a_\nu^{2m}(k^2)$	even	$2K$
$a_\nu^{2m+1}(k^2)$	odd	$4K$
$b_\nu^{2m+1}(k^2)$	even	$4K$
$b_\nu^{2m+2}(k^2)$	odd	$2K$

29.3(ii) Distribution

The eigenvalues interlace according to

29.3.1 $\quad a_\nu^m(k^2) < a_\nu^{m+1}(k^2),$

29.3.2 $\quad a_\nu^m(k^2) < b_\nu^{m+1}(k^2),$

29.3.3 $\quad b_\nu^m(k^2) < b_\nu^{m+1}(k^2),$

29.3.4 $\quad b_\nu^m(k^2) < a_\nu^{m+1}(k^2).$

The eigenvalues coalesce according to

29.3.5 $\quad a_\nu^m(k^2) = b_\nu^m(k^2), \quad \nu = 0, 1, \ldots, m-1.$

If ν is distinct from $0, 1, \ldots, m-1$, then

29.3.6 $\quad \left(a_\nu^m(k^2) - b_\nu^m(k^2)\right)\nu(\nu - 1)\cdots(\nu - m + 1) > 0.$

If ν is a nonnegative integer, then

29.3.7 $\quad a_\nu^m(k^2) + a_\nu^{\nu-m}(1 - k^2) = \nu(\nu + 1), \quad m = 0, 1, \ldots, \nu,$

29.3.8 $\quad b_\nu^m(k^2) + b_\nu^{\nu-m+1}(1 - k^2) = \nu(\nu + 1), \quad m = 1, 2, \ldots, \nu.$

For the special case $k = k' = 1/\sqrt{2}$ see Erdélyi et al. (1955, §15.5.2).

29.3(iii) Continued Fractions

The quantity

29.3.9 $\quad H = 2 a_\nu^{2m}(k^2) - \nu(\nu + 1) k^2$

satisfies the continued-fraction equation

29.3.10 $\quad \beta_p - H - \dfrac{\alpha_{p-1}\gamma_p}{\beta_{p-1} - H -}\dfrac{\alpha_{p-2}\gamma_{p-1}}{\beta_{p-2} - H-}\cdots$
$\quad = \dfrac{\alpha_p \gamma_{p+1}}{\beta_{p+1} - H-}\dfrac{\alpha_{p+1}\gamma_{p+2}}{\beta_{p+2} - H-}\cdots,$

where p is any nonnegative integer, and

29.3.11 $\quad \alpha_p = \begin{cases} (\nu - 1)(\nu + 2)k^2, & p = 0, \\ \tfrac{1}{2}(\nu - 2p - 1)(\nu + 2p + 2)k^2, & p \geq 1, \end{cases}$

29.3.12 $\quad \beta_p = 4p^2(2 - k^2),$
$\quad \gamma_p = \tfrac{1}{2}(\nu - 2p + 2)(\nu + 2p - 1)k^2.$

The continued fraction following the second negative sign on the left-hand side of (29.3.10) is finite: it equals 0 if $p = 0$, and if $p > 0$, then the last denominator is $\beta_0 - H$. If ν is a nonnegative integer and $2p \leq \nu$, then the continued fraction on the right-hand side of (29.3.10) terminates, and (29.3.10) has only the solutions (29.3.9) with $2m \leq \nu$. If ν is a nonnegative integer and $2p > \nu$, then (29.3.10) has only the solutions (29.3.9) with $2m > \nu$.

For the corresponding continued-fraction equations for $a_\nu^{2m+1}(k^2)$, $b_\nu^{2m+1}(k^2)$, and $b_\nu^{2m+2}(k^2)$ see http://dlmf.nist.gov/29.3.iii.

29.3(iv) Lamé Functions

The eigenfunctions corresponding to the eigenvalues of §29.3(i) are denoted by $Ec_\nu^{2m}(z, k^2)$, $Ec_\nu^{2m+1}(z, k^2)$, $Es_\nu^{2m+1}(z, k^2)$, $Es_\nu^{2m+2}(z, k^2)$. They are called *Lamé functions with real periods and of order ν*, or more simply, *Lamé functions*. See Table 29.3.2. In this table the nonnegative integer m corresponds to the number of zeros of each Lamé function in $(0, K)$, whereas the superscripts $2m$, $2m+1$, or $2m+2$ correspond to the number of zeros in $[0, 2K)$.

Table 29.3.2: Lamé functions.

boundary conditions	eigenvalue h	eigenfunction $w(z)$	parity of $w(z)$	parity of $w(z-K)$	period of $w(z)$
$dw/dz\|_{z=0} = dw/dz\|_{z=K} = 0$	$a_\nu^{2m}(k^2)$	$Ec_\nu^{2m}(z,k^2)$	even	even	$2K$
$w(0) = dw/dz\|_{z=K} = 0$	$a_\nu^{2m+1}(k^2)$	$Ec_\nu^{2m+1}(z,k^2)$	odd	even	$4K$
$dw/dz\|_{z=0} = w(K) = 0$	$b_\nu^{2m+1}(k^2)$	$Es_\nu^{2m+1}(z,k^2)$	even	odd	$4K$
$w(0) = w(K) = 0$	$b_\nu^{2m+2}(k^2)$	$Es_\nu^{2m+2}(z,k^2)$	odd	odd	$2K$

29.3(v) Normalization

29.3.18
$$\int_0^K \operatorname{dn}(x,k) \left(Ec_\nu^{2m}(x,k^2)\right)^2 dx = \frac{1}{4}\pi,$$
$$\int_0^K \operatorname{dn}(x,k) \left(Ec_\nu^{2m+1}(x,k^2)\right)^2 dx = \frac{1}{4}\pi,$$
$$\int_0^K \operatorname{dn}(x,k) \left(Es_\nu^{2m+1}(x,k^2)\right)^2 dx = \frac{1}{4}\pi,$$
$$\int_0^K \operatorname{dn}(x,k) \left(Es_\nu^{2m+2}(x,k^2)\right)^2 dx = \frac{1}{4}\pi.$$

For $\operatorname{dn}(z,k)$ see §22.2.

To complete the definitions, $Ec_\nu^m(K,k^2)$ is positive and $dEs_\nu^m(z,k^2)/dz\big|_{z=K}$ is negative.

29.3(vi) Orthogonality

For $m \neq p$,

29.3.19
$$\int_0^K Ec_\nu^{2m}(x,k^2) Ec_\nu^{2p}(x,k^2) dx = 0,$$
$$\int_0^K Ec_\nu^{2m+1}(x,k^2) Ec_\nu^{2p+1}(x,k^2) dx = 0,$$
$$\int_0^K Es_\nu^{2m+1}(x,k^2) Es_\nu^{2p+1}(x,k^2) dx = 0,$$
$$\int_0^K Es_\nu^{2m+2}(x,k^2) Es_\nu^{2p+2}(x,k^2) dx = 0.$$

For the values of these integrals when $m = p$ see §29.6.

29.3(vii) Power Series

For power-series expansions of the eigenvalues see Volkmer (2004b).

29.4 Graphics

29.4(i) Eigenvalues of Lamé's Equation: Line Graphs

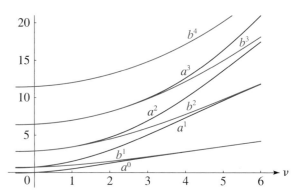

Figure 29.4.1: $a_\nu^m(0.5)$, $b_\nu^{m+1}(0.5)$ as functions of ν for $m = 0, 1, 2, 3$.

29.4 GRAPHICS

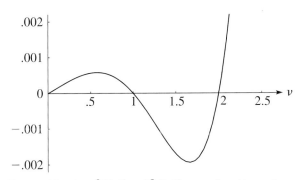

Figure 29.4.2: $a_\nu^3(0.5) - b_\nu^3(0.5)$ as a function of ν.

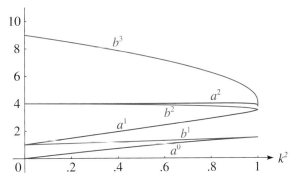

Figure 29.4.3: $a_{1.5}^m(k^2)$, $b_{1.5}^{m+1}(k^2)$ as functions of k^2 for $m = 0, 1, 2$.

For additional graphs see http://dlmf.nist.gov/29.4.i.

29.4(ii) Eigenvalues of Lamé's Equation: Surfaces

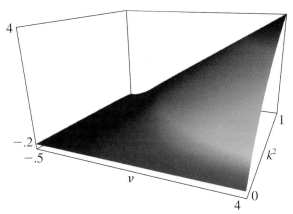

Figure 29.4.9: $a_\nu^0(k^2)$ as a function of ν and k^2.

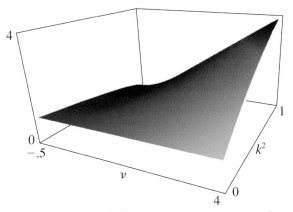

Figure 29.4.10: $b_\nu^1(k^2)$ as a function of ν and k^2.

For additional surfaces see http://dlmf.nist.gov/29.4.ii.

29.4(iii) Lamé Functions: Line Graphs

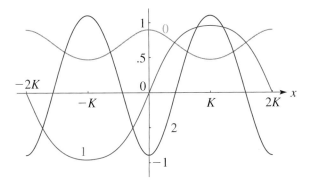

Figure 29.4.13: $Ec_{1.5}^m(x, 0.5)$ for $-2K \leq x \leq 2K$, $m = 0, 1, 2$. $K = 1.85407\ldots$.

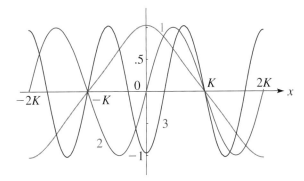

Figure 29.4.14: $Es_{1.5}^m(x, 0.5)$ for $-2K \leq x \leq 2K$, $m = 1, 2, 3$. $K = 1.85407\ldots$.

For additional graphs see http://dlmf.nist.gov/29.4.iii.

29.4(iv) Lamé Functions: Surfaces

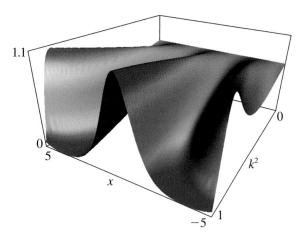

Figure 29.4.25: $Ec_{1.5}^0(x,k^2)$ as a function of x and k^2.

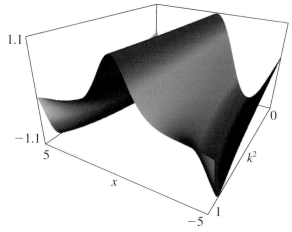

Figure 29.4.26: $Es_{1.5}^1(x,k^2)$ as a function of x and k^2.

For additional surfaces see http://dlmf.nist.gov/29.4.iv.

29.5 Special Cases and Limiting Forms

29.5.1
$$a_\nu^m(0) = b_\nu^m(0) = m^2,$$

29.5.2
$$Ec_\nu^0(z,0) = 2^{-\frac{1}{2}},$$

29.5.3
$$Ec_\nu^m(z,0) = \cos(m(\tfrac{1}{2}\pi - z)), \qquad m \geq 1,$$
$$Es_\nu^m(z,0) = \sin(m(\tfrac{1}{2}\pi - z)), \qquad m \geq 1.$$

Let $\mu = \max(\nu - m, 0)$. Then

29.5.4
$$\lim_{k \to 1-} a_\nu^m(k^2) = \lim_{k \to 1-} b_\nu^{m+1}(k^2) = \nu(\nu+1) - \mu^2,$$

29.5.5
$$\lim_{k \to 1-} \frac{Ec_\nu^m(z,k^2)}{Ec_\nu^m(0,k^2)} = \lim_{k \to 1-} \frac{Es_\nu^{m+1}(z,k^2)}{Es_\nu^{m+1}(0,k^2)}$$
$$= \frac{1}{(\cosh z)^\mu} F\left(\genfrac{}{}{0pt}{}{\tfrac{1}{2}\mu - \tfrac{1}{2}\nu, \tfrac{1}{2}\mu + \tfrac{1}{2}\nu + \tfrac{1}{2}}{\tfrac{1}{2}}; \tanh^2 z\right),$$
$$m \text{ even,}$$

29.5.6
$$\lim_{k \to 1-} \frac{Ec_\nu^m(z,k^2)}{dEc_\nu^m(z,k^2)/dz|_{z=0}}$$
$$= \lim_{k \to 1-} \frac{Es_\nu^{m+1}(z,k^2)}{dEs_\nu^{m+1}(z,k^2)/dz|_{z=0}}$$
$$= \frac{\tanh z}{(\cosh z)^\mu} F\left(\genfrac{}{}{0pt}{}{\tfrac{1}{2}\mu - \tfrac{1}{2}\nu + \tfrac{1}{2}, \tfrac{1}{2}\mu + \tfrac{1}{2}\nu + 1}{\tfrac{3}{2}}; \tanh^2 z\right),$$
$$m \text{ odd,}$$

where F is the hypergeometric function; see §15.2(i).

If $k \to 0+$ and $\nu \to \infty$ in such a way that $k^2\nu(\nu+1) = 4\theta$ (a positive constant), then

29.5.7
$$\lim Ec_\nu^m(z,k^2) = \mathrm{ce}_m(\tfrac{1}{2}\pi - z, \theta),$$
$$\lim Es_\nu^m(z,k^2) = \mathrm{se}_m(\tfrac{1}{2}\pi - z, \theta),$$

where $\mathrm{ce}_m(z,\theta)$ and $\mathrm{se}_m(z,\theta)$ are Mathieu functions; see §28.2(vi).

29.6 Fourier Series

29.6(i) Function $Ec_\nu^{2m}(z,k^2)$

With $\phi = \tfrac{1}{2}\pi - \mathrm{am}(z,k)$, as in (29.2.5), we have

29.6.1
$$Ec_\nu^{2m}(z,k^2) = \tfrac{1}{2}A_0 + \sum_{p=1}^\infty A_{2p}\cos(2p\phi).$$

Here

29.6.2
$$H = 2a_\nu^{2m}(k^2) - \nu(\nu+1)k^2,$$

29.6.3
$$(\beta_0 - H)A_0 + \alpha_0 A_2 = 0,$$

29.6.4
$$\gamma_p A_{2p-2} + (\beta_p - H)A_{2p} + \alpha_p A_{2p+2} = 0, \quad p \geq 1,$$

with α_p, β_p, and γ_p as in (29.3.11) and (29.3.12), and

29.6.5
$$\tfrac{1}{2}A_0^2 + \sum_{p=1}^\infty A_{2p}^2 = 1,$$

29.6.6
$$\tfrac{1}{2}A_0 + \sum_{p=1}^\infty A_{2p} > 0.$$

When $\nu \neq 2n$, where n is a nonnegative integer, it follows from §2.9(i) that for any value of H the system (29.6.4)–(29.6.6) has a unique recessive solution $A_0, A_2, A_4, \ldots$; furthermore

29.6.7
$$\lim_{p\to\infty} \frac{A_{2p+2}}{A_{2p}} = \frac{k^2}{(1+k')^2}, \quad \nu \neq 2n, \text{ or } \nu = 2n \text{ and } m > n.$$

In addition, if H satisfies (29.6.2), then (29.6.3) applies.

In the special case $\nu = 2n$, $m = 0, 1, \ldots, n$, there is a unique nontrivial solution with the property $A_{2p} = 0$, $p = n+1, n+2, \ldots$. This solution can be constructed from (29.6.4) by backward recursion, starting with $A_{2n+2} = 0$ and an arbitrary nonzero value of A_{2n}, followed by normalization via (29.6.5) and (29.6.6). Consequently, $Ec_\nu^{2m}(z, k^2)$ reduces to a Lamé polynomial; compare §§29.12(i) and 29.15(i).

An alternative version of the Fourier series expansion (29.6.1) is given by

29.6.8
$$Ec_\nu^{2m}(z, k^2) = \mathrm{dn}\,(z, k) \left(\tfrac{1}{2} C_0 + \sum_{p=1}^{\infty} C_{2p} \cos(2p\phi) \right).$$

Here $\mathrm{dn}\,(z, k)$ is as in §22.2, and

29.6.9
$$(\beta_0 - H)C_0 + \alpha_0 C_2 = 0,$$

29.6.10
$$\gamma_p C_{2p-2} + (\beta_p - H)C_{2p} + \alpha_p C_{2p+2} = 0, \quad p \geq 1,$$

with $\alpha_p, \beta_p,$ and γ_p now defined by

29.6.11
$$\alpha_p = \begin{cases} \nu(\nu+1)k^2, & p = 0, \\ \tfrac{1}{2}(\nu-2p)(\nu+2p+1)k^2, & p \geq 1, \end{cases}$$
$$\beta_p = 4p^2(2 - k^2),$$
$$\gamma_p = \tfrac{1}{2}(\nu-2p+1)(\nu+2p)k^2,$$

and

29.6.12
$$(1 - \tfrac{1}{2}k^2)\left(\tfrac{1}{2}C_0^2 + \sum_{p=1}^{\infty} C_{2p}^2 \right) - \tfrac{1}{2}k^2 \sum_{p=0}^{\infty} C_{2p} C_{2p+2} = 1,$$

29.6.13
$$\tfrac{1}{2}C_0 + \sum_{p=1}^{\infty} C_{2p} > 0,$$

29.6.14
$$\lim_{p\to\infty} \frac{C_{2p+2}}{C_{2p}} = \frac{k^2}{(1+k')^2},$$
$$\nu \neq 2n+1, \text{ or } \nu = 2n+1 \text{ and } m > n.$$

29.6.15
$$\tfrac{1}{2}A_0 C_0 + \sum_{p=1}^{\infty} A_{2p} C_{2p} = \frac{4}{\pi} \int_0^K \left(Ec_\nu^{2m}(x, k^2) \right)^2 dx.$$

For the corresponding expansions for $Ec_\nu^{2m+1}(z, k^2)$, $Es_\nu^{2m+1}(z, k^2)$, and $Es_\nu^{2m+2}(z, k^2)$ see http://dlmf.nist.gov/29.6.ii.

29.7 Asymptotic Expansions

29.7(i) Eigenvalues

As $\nu \to \infty$,

29.7.1
$$a_\nu^m(k^2) \sim p\kappa - \tau_0 - \tau_1 \kappa^{-1} - \tau_2 \kappa^{-2} - \cdots,$$

where

29.7.2
$$\kappa = k(\nu(\nu+1))^{1/2}, \quad p = 2m+1,$$

29.7.3
$$\tau_0 = \frac{1}{2^3}(1 + k^2)(1 + p^2),$$

29.7.4
$$\tau_1 = \frac{p}{2^6}((1+k^2)^2(p^2+3) - 4k^2(p^2+5)).$$

The same Poincaré expansion holds for $b_\nu^{m+1}(k^2)$, since

29.7.5
$$b_\nu^{m+1}(k^2) - a_\nu^m(k^2) = O\!\left(\nu^{m+\frac{3}{2}} \left(\frac{1-k}{1+k} \right)^\nu \right), \quad \nu \to \infty.$$

See also Volkmer (2004b).

For higher terms in (29.7.1) see http://dlmf.nist.gov/29.7.i.

29.7(ii) Lamé Functions

Müller (1966a,b) found three formal asymptotic expansions for a fundamental system of solutions of (29.2.1) (and (29.11.1)) as $\nu \to \infty$, one in terms of Jacobian elliptic functions and two in terms of Hermite polynomials. In Müller (1966c) it is shown how these expansions lead to asymptotic expansions for the Lamé functions $Ec_\nu^m(z, k^2)$ and $Es_\nu^m(z, k^2)$. Weinstein and Keller (1985) give asymptotics for solutions of Hill's equation (§28.29(i)) that are applicable to the Lamé equation.

29.8 Integral Equations

Let $w(z)$ be any solution of (29.2.1) of period $4K$, $w_2(z)$ be a linearly independent solution, and $\mathscr{W}\{w, w_2\}$ denote their Wronskian. Also let x be defined by

29.8.1
$$\begin{aligned} x = {} & k^2 \,\mathrm{sn}\,(z, k)\,\mathrm{sn}\,(z_1, k)\,\mathrm{sn}\,(z_2, k)\,\mathrm{sn}\,(z_3, k) \\ & - \frac{k^2}{k'^2}\,\mathrm{cn}\,(z, k)\,\mathrm{cn}\,(z_1, k)\,\mathrm{cn}\,(z_2, k)\,\mathrm{cn}\,(z_3, k) \\ & + \frac{1}{k'^2}\,\mathrm{dn}\,(z, k)\,\mathrm{dn}\,(z_1, k)\,\mathrm{dn}\,(z_2, k)\,\mathrm{dn}\,(z_3, k), \end{aligned}$$

where z, z_1, z_2, z_3 are real, and sn, cn, dn are the Jacobian elliptic functions (§22.2). Then

29.8.2
$$\mu w(z_1) w(z_2) w(z_3) = \int_{-2K}^{2K} \mathsf{P}_\nu(x) w(z) \, dz,$$

where $\mathsf{P}_\nu(x)$ is the Ferrers function of the first kind (§14.3(i)),

29.8.3
$$\mu = \frac{2\sigma\tau}{\mathscr{W}\{w, w_2\}},$$

and σ ($=\pm 1$) and τ are determined by

29.8.4
$$w(z+2K) = \sigma w(z),$$
$$w_2(z+2K) = \tau w(z) + \sigma w_2(z).$$

A special case of (29.8.2) is

29.8.5
$$Ec_\nu^{2m}(z_1, k^2) \frac{w_2(K) - w_2(-K)}{dw_2(z)/dz|_{z=0}}$$
$$= \int_{-K}^{K} \mathsf{P}_\nu(y) \, Ec_\nu^{2m}(z, k^2) \, dz,$$

where

29.8.6
$$y = \frac{1}{k'} \operatorname{dn}(z, k) \operatorname{dn}(z_1, k).$$

For results corresponding to (29.8.5) for Ec_ν^{2m+1}, Es_ν^{2m+1}, Es_ν^{2m+2} see http://dlmf.nist.gov/29.8.

For further integral equations see Arscott (1964a), Erdélyi et al. (1955, §15.5.3), Shail (1980), Sleeman (1968a), and Volkmer (1982, 1983, 1984).

29.9 Stability

The Lamé equation (29.2.1) with specified values of k, h, ν is called *stable* if all of its solutions are bounded on $\mathbb{R}$; otherwise the equation is called *unstable*. If ν is not an integer, then (29.2.1) is unstable iff $h \leq a_\nu^0(k^2)$ or h lies in one of the closed intervals with endpoints $a_\nu^m(k^2)$ and $b_\nu^m(k^2)$, $m = 1, 2, \ldots$. If ν is a nonnegative integer, then (29.2.1) is unstable iff $h \leq a_\nu^0(k^2)$ or $h \in [b_\nu^m(k^2), a_\nu^m(k^2)]$ for some $m = 1, 2, \ldots, \nu$.

29.10 Lamé Functions with Imaginary Periods

The substitutions

29.10.1 $$h = \nu(\nu+1) - h',$$

29.10.2 $$z' = i(z - K - iK'),$$

transform (29.2.1) into

29.10.3 $$\frac{d^2w}{dz'^2} + (h' - \nu(\nu+1)k'^2 \operatorname{sn}^2(z', k'))w = 0.$$

In consequence, the functions

29.10.4
$$Ec_\nu^{2m}\left(i(z-K-iK'), k'^2\right),$$
$$Ec_\nu^{2m+1}\left(i(z-K-iK'), k'^2\right),$$
$$Es_\nu^{2m+1}\left(i(z-K-iK'), k'^2\right),$$
$$Es_\nu^{2m+2}\left(i(z-K-iK'), k'^2\right),$$

are solutions of (29.2.1). The first and the fourth functions have period $2iK'$; the second and the third have period $4iK'$.

For these results and further information see Erdélyi et al. (1955, §15.5.2).

29.11 Lamé Wave Equation

The *Lamé* (or *ellipsoidal*) *wave equation* is given by

29.11.1
$$\frac{d^2w}{dz^2} + (h - \nu(\nu+1)k^2 \operatorname{sn}^2(z,k) + k^2\omega^2 \operatorname{sn}^4(z,k))w = 0,$$

in which ω is another parameter. In the case $\omega = 0$, (29.11.1) reduces to Lamé's equation (29.2.1).

For properties of the solutions of (29.11.1) see Arscott (1956, 1959), Arscott (1964b, Chapter X), Erdélyi et al. (1955, §16.14), Fedoryuk (1989), and Müller (1966a,b,c).

Lamé Polynomials

29.12 Definitions

29.12(i) Elliptic-Function Form

Throughout §§29.12–29.16 the order ν in the differential equation (29.2.1) is assumed to be a nonnegative integer.

The Lamé functions $Ec_\nu^m(z, k^2)$, $m = 0, 1, \ldots, \nu$, and $Es_\nu^m(z, k^2)$, $m = 1, 2, \ldots, \nu$, are called the *Lamé polynomials*. There are eight types of Lamé polynomials, defined as follows:

29.12.1 $$uE_{2n}^m(z, k^2) = Ec_{2n}^{2m}(z, k^2),$$

29.12.2 $$sE_{2n+1}^m(z, k^2) = Ec_{2n+1}^{2m+1}(z, k^2),$$

29.12.3 $$cE_{2n+1}^m(z, k^2) = Es_{2n+1}^{2m+1}(z, k^2),$$

29.12.4 $$dE_{2n+1}^m(z, k^2) = Ec_{2n+1}^{2m}(z, k^2),$$

29.12.5 $$scE_{2n+2}^m(z, k^2) = Es_{2n+2}^{2m+2}(z, k^2),$$

29.12.6 $$sdE_{2n+2}^m(z, k^2) = Ec_{2n+2}^{2m+1}(z, k^2),$$

29.12.7 $$cdE_{2n+2}^m(z, k^2) = Es_{2n+2}^{2m+1}(z, k^2),$$

29.12.8 $$scdE_{2n+3}^m(z, k^2) = Es_{2n+3}^{2m+2}(z, k^2),$$

where $n = 0, 1, 2, \ldots$, $m = 0, 1, 2, \ldots, n$. These functions are polynomials in $\operatorname{sn}(z,k)$, $\operatorname{cn}(z,k)$, and $\operatorname{dn}(z,k)$. In consequence they are doubly-periodic meromorphic functions of z.

The superscript m on the left-hand sides of (29.12.1)–(29.12.8) agrees with the number of z-zeros of each Lamé polynomial in the interval $(0, K)$, while $n - m$ is the number of z-zeros in the open line segment from K to $K + iK'$.

The prefixes u, s, c, d, sc, sd, cd, scd indicate the type of the polynomial form of the Lamé polynomial; compare the 3rd and 4th columns in Table 29.12.1. In the fourth column the variable z and modulus k of the Jacobian elliptic functions have been suppressed, and $P(\operatorname{sn}^2)$ denotes a polynomial of degree n in $\operatorname{sn}^2(z,k)$ (different for each type). For the determination of the coefficients of the P's see §29.15(ii).

29.13 Graphics

Table 29.12.1: Lamé polynomials.

ν	eigenvalue h	eigenfunction $w(z)$	polynomial form	real period	imag. period	parity of $w(z)$	parity of $w(z-K)$	parity of $w(z-K-iK')$
$2n$	$a_\nu^{2m}(k^2)$	$uE_\nu^m(z,k^2)$	$P(\operatorname{sn}^2)$	$2K$	$2iK'$	even	even	even
$2n+1$	$a_\nu^{2m+1}(k^2)$	$sE_\nu^m(z,k^2)$	$\operatorname{sn} P(\operatorname{sn}^2)$	$4K$	$2iK'$	odd	even	even
$2n+1$	$b_\nu^{2m+1}(k^2)$	$cE_\nu^m(z,k^2)$	$\operatorname{cn} P(\operatorname{sn}^2)$	$4K$	$4iK'$	even	odd	even
$2n+1$	$a_\nu^{2m}(k^2)$	$dE_\nu^m(z,k^2)$	$\operatorname{dn} P(\operatorname{sn}^2)$	$2K$	$4iK'$	even	even	odd
$2n+2$	$b_\nu^{2m+2}(k^2)$	$scE_\nu^m(z,k^2)$	$\operatorname{sn}\operatorname{cn} P(\operatorname{sn}^2)$	$2K$	$4iK'$	odd	odd	even
$2n+2$	$a_\nu^{2m+1}(k^2)$	$sdE_\nu^m(z,k^2)$	$\operatorname{sn}\operatorname{dn} P(\operatorname{sn}^2)$	$4K$	$4iK'$	odd	even	odd
$2n+2$	$b_\nu^{2m+1}(k^2)$	$cdE_\nu^m(z,k^2)$	$\operatorname{cn}\operatorname{dn} P(\operatorname{sn}^2)$	$4K$	$2iK'$	even	odd	odd
$2n+3$	$b_\nu^{2m+2}(k^2)$	$scdE_\nu^m(z,k^2)$	$\operatorname{sn}\operatorname{cn}\operatorname{dn} P(\operatorname{sn}^2)$	$2K$	$2iK'$	odd	odd	odd

29.12(ii) Algebraic Form

With the substitution $\xi = \operatorname{sn}^2(z,k)$ every Lamé polynomial in Table 29.12.1 can be written in the form

$$29.12.9 \qquad \xi^\rho (\xi-1)^\sigma (\xi - k^{-2})^\tau P(\xi),$$

where ρ, σ, τ are either 0 or $\tfrac{1}{2}$. The polynomial $P(\xi)$ is of degree n and has m zeros (all simple) in $(0,1)$ and $n-m$ zeros (all simple) in $(1, k^{-2})$. The functions (29.12.9) satisfy (29.2.2).

29.12(iii) Zeros

Let $\xi_1, \xi_2, \ldots, \xi_n$ denote the zeros of the polynomial P in (29.12.9) arranged according to

$$29.12.10 \qquad 0 < \xi_1 < \cdots < \xi_m < 1 < \xi_{m+1} < \cdots < \xi_n < k^{-2}.$$

Then the function

29.12.11

$$g(t_1, t_2, \ldots, t_n) = \left(\prod_{p=1}^n t_p^{\rho + \tfrac{1}{4}} |t_p - 1|^{\sigma + \tfrac{1}{4}} (k^{-2} - t_p)^{\tau + \tfrac{1}{4}} \right) \prod_{q < r} (t_r - t_q),$$

defined for $(t_1, t_2, \ldots, t_n)$ with

$$29.12.12 \qquad 0 \leq t_1 \leq \cdots \leq t_m \leq 1 \leq t_{m+1} \leq \cdots \leq t_n \leq k^{-2},$$

attains its absolute maximum iff $t_j = \xi_j$, $j = 1, 2, \ldots, n$. Moreover,

29.12.13

$$\frac{\rho + \tfrac{1}{4}}{\xi_p} + \frac{\sigma + \tfrac{1}{4}}{\xi_p - 1} + \frac{\tau + \tfrac{1}{4}}{\xi_p - k^{-2}} + \sum_{\substack{q=1\\q \neq p}}^n \frac{1}{\xi_p - \xi_q} = 0,$$

$$p = 1, 2, \ldots, n.$$

This result admits the following electrostatic interpretation: Given three point masses fixed at $t = 0$, $t = 1$, and $t = k^{-2}$ with positive charges $\rho + \tfrac{1}{4}$, $\sigma + \tfrac{1}{4}$, and $\tau + \tfrac{1}{4}$, respectively, and n movable point masses at $t_1, t_2, \ldots, t_n$ arranged according to (29.12.12) with unit positive charges, the equilibrium position is attained when $t_j = \xi_j$ for $j = 1, 2, \ldots, n$.

29.13 Graphics

29.13(i) Eigenvalues for Lamé Polynomials

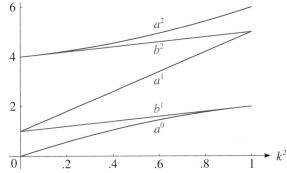

Figure 29.13.1: $a_2^m(k^2)$, $b_2^m(k^2)$ as functions of k^2 for $m = 0, 1, 2$ (a's), $m = 1, 2$ (b's).

For additional graphs see http://dlmf.nist.gov/29.13.i.

29.13(ii) Lamé Polynomials: Real Variable

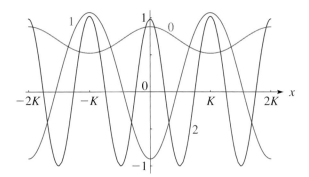

Figure 29.13.5: $uE_4^m(x, 0.1)$ for $-2K \leq x \leq 2K$, $m = 0, 1, 2$. $K = 1.61244\ldots$.

For additional graphs see http://dlmf.nist.gov/29.13.ii.

29.13(iii) Lamé Polynomials: Complex Variable

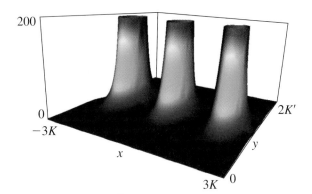

Figure 29.13.21: $|uE_4^1(x+iy, 0.1)|$ for $-3K \leq x \leq 3K$, $0 \leq y \leq 2K'$. $K = 1.61244\ldots, K' = 2.57809\ldots$.

For additional graphics see http://dlmf.nist.gov/29.13.iii.

29.14 Orthogonality

Lamé polynomials are orthogonal in two ways. First, the orthogonality relations (29.3.19) apply; see §29.12(i). Secondly, the system of functions

29.14.1 $$f_n^m(s,t) = uE_{2n}^m(s, k^2)\, uE_{2n}^m(K+it, k^2),$$
$$n = 0, 1, 2, \ldots, m = 0, 1, \ldots, n,$$

is orthogonal and complete with respect to the inner product

29.14.2 $$\langle g, h \rangle = \int_0^K \int_0^{K'} w(s,t) g(s,t) h(s,t) \, dt \, ds,$$

where

29.14.3 $$w(s,t) = \operatorname{sn}^2(K+it, k) - \operatorname{sn}^2(s, k).$$

For the corresponding results for the other seven types of Lamé polynomials see http://dlmf.nist.gov/29.14.

29.15 Fourier Series and Chebyshev Series

29.15(i) Fourier Coefficients

Polynomial $uE_{2n}^m(z, k^2)$

When $\nu = 2n$, $m = 0, 1, \ldots, n$, the Fourier series (29.6.1) terminates:

29.15.1 $$uE_{2n}^m(z, k^2) = \tfrac{1}{2} A_0 + \sum_{p=1}^{n} A_{2p} \cos(2p\phi).$$

A convenient way of constructing the coefficients, together with the eigenvalues, is as follows. Equations (29.6.4), with $p = 1, 2, \ldots, n$, (29.6.3), and $A_{2n+2} = 0$ can be cast as an algebraic eigenvalue problem in the following way. Let

29.15.2 $$\mathbf{M} = \begin{bmatrix} \beta_0 & \alpha_0 & 0 & \cdots & 0 \\ \gamma_1 & \beta_1 & \alpha_1 & \ddots & \vdots \\ 0 & \ddots & \ddots & \ddots & 0 \\ \vdots & \ddots & \gamma_{n-1} & \beta_{n-1} & \alpha_{n-1} \\ 0 & \cdots & 0 & \gamma_n & \beta_n \end{bmatrix}$$

be the tridiagonal matrix with $\alpha_p, \beta_p, \gamma_p$ as in (29.3.11), (29.3.12). Let the eigenvalues of $\mathbf{M}$ be H_p with

29.15.3 $$H_0 < H_1 < \cdots < H_n,$$

and also let

29.15.4 $$[A_0, A_2, \ldots, A_{2n}]^{\mathrm{T}}$$

be the eigenvector corresponding to H_m and normalized so that

29.15.5 $$\tfrac{1}{2} A_0^2 + \sum_{p=1}^{n} A_{2p}^2 = 1$$

and

29.15.6 $$\tfrac{1}{2} A_0 + \sum_{p=1}^{n} A_{2p} > 0.$$

Then

29.15.7 $$a_\nu^{2m}(k^2) = \tfrac{1}{2}(H_m + \nu(\nu+1)k^2),$$

and (29.15.1) applies, with ϕ again defined as in (29.2.5).

For the corresponding formulations for the other seven types of Lamé polynomials see http://dlmf.nist.gov/29.15.i.

29.15(ii) Chebyshev Series

The Chebyshev polynomial T of the first kind (§18.3) satisfies $\cos(p\phi) = T_p(\cos\phi)$. Since (29.2.5) implies that $\cos\phi = \operatorname{sn}(z,k)$, (29.15.1) can be rewritten in the form

$$\textbf{29.15.43} \quad uE_{2n}^m(z, k^2) = \tfrac{1}{2}A_0 + \sum_{p=1}^{n} A_{2p}\, T_{2p}(\operatorname{sn}(z,k)).$$

This determines the polynomial P of degree n for which $uE_{2n}^m(z, k^2) = P(\operatorname{sn}^2(z,k))$; compare Table 29.12.1. The set of coefficients of this polynomial (without normalization) can also be found directly as an eigenvector of an $(n+1) \times (n+1)$ tridiagonal matrix; see Arscott and Khabaza (1962).

For the corresponding expansions of the other seven types of Lamé polynomials see http://dlmf.nist.gov/29.15.ii.

For explicit formulas for Lamé polynomials of low degree, see Arscott (1964b, p. 205).

29.16 Asymptotic Expansions

Hargrave and Sleeman (1977) give asymptotic approximations for Lamé polynomials and their eigenvalues, including error bounds. The approximations for Lamé polynomials hold uniformly on the rectangle $0 \leq \Re z \leq K$, $0 \leq \Im z \leq K'$, when nk and nk' assume large real values. The approximating functions are exponential, trigonometric, and parabolic cylinder functions.

29.17 Other Solutions

29.17(i) Second Solution

If (29.2.1) admits a Lamé polynomial solution E, then a second linearly independent solution F is given by

$$\textbf{29.17.1} \quad F(z) = E(z) \int_{iK'}^{z} \frac{du}{(E(u))^2}.$$

For properties of these solutions see Arscott (1964b, §9.7), Erdélyi et al. (1955, §15.5.1), Shail (1980), and Sleeman (1966a).

29.17(ii) Algebraic Lamé Functions

Algebraic Lamé functions are solutions of (29.2.1) when ν is half an odd integer. They are algebraic functions of $\operatorname{sn}(z,k)$, $\operatorname{cn}(z,k)$, and $\operatorname{dn}(z,k)$, and have primitive period $8K$. See Erdélyi (1941c), Ince (1940b), and Lambe (1952).

29.17(iii) Lamé–Wangerin Functions

Lamé–Wangerin functions are solutions of (29.2.1) with the property that $(\operatorname{sn}(z,k))^{1/2} w(z)$ is bounded on the line segment from iK' to $2K + iK'$. See Erdélyi et al. (1955, §15.6).

Applications

29.18 Mathematical Applications

29.18(i) Sphero-Conal Coordinates

The wave equation

$$\textbf{29.18.1} \quad \nabla^2 u + \omega^2 u = 0,$$

when transformed to *sphero-conal coordinates* r, β, γ:

$$\textbf{29.18.2}$$
$$x = kr\operatorname{sn}(\beta,k)\operatorname{sn}(\gamma,k), \quad y = i\frac{k}{k'} r\operatorname{cn}(\beta,k)\operatorname{cn}(\gamma,k),$$
$$z = \frac{1}{k'} r\operatorname{dn}(\beta,k)\operatorname{dn}(\gamma,k),$$

with

$$\textbf{29.18.3} \quad r \geq 0, \quad \beta = K + i\beta', \quad 0 \leq \beta' \leq 2K', \quad 0 \leq \gamma \leq 4K,$$

admits solutions

$$\textbf{29.18.4} \quad u(r, \beta, \gamma) = u_1(r) u_2(\beta) u_3(\gamma),$$

where u_1, u_2, u_3 satisfy the differential equations

$$\textbf{29.18.5} \quad \frac{d}{dr}\left(r^2 \frac{du_1}{dr}\right) + (\omega^2 r^2 - \nu(\nu+1)) u_1 = 0,$$

$$\textbf{29.18.6} \quad \frac{d^2 u_2}{d\beta^2} + (h - \nu(\nu+1) k^2 \operatorname{sn}^2(\beta, k)) u_2 = 0,$$

$$\textbf{29.18.7} \quad \frac{d^2 u_3}{d\gamma^2} + (h - \nu(\nu+1) k^2 \operatorname{sn}^2(\gamma, k)) u_3 = 0,$$

with *separation constants* h and ν. (29.18.5) is the differential equation of spherical Bessel functions (§10.47(i)), and (29.18.6), (29.18.7) agree with the Lamé equation (29.2.1).

29.18(ii) Ellipsoidal Coordinates

The wave equation (29.18.1), when transformed to *ellipsoidal coordinates* α, β, γ:

$$\textbf{29.18.8}$$
$$x = k\operatorname{sn}(\alpha,k)\operatorname{sn}(\beta,k)\operatorname{sn}(\gamma,k),$$
$$y = -\frac{k}{k'}\operatorname{cn}(\alpha,k)\operatorname{cn}(\beta,k)\operatorname{cn}(\gamma,k),$$
$$z = \frac{i}{kk'}\operatorname{dn}(\alpha,k)\operatorname{dn}(\beta,k)\operatorname{dn}(\gamma,k),$$

with

$$\textbf{29.18.9}$$
$$\alpha = K + iK' - \alpha', \quad 0 \leq \alpha' < K,$$
$$\beta = K + i\beta', \quad 0 \leq \beta' \leq 2K', 0 \leq \gamma \leq 4K,$$

admits solutions

$$\textbf{29.18.10} \quad u(\alpha, \beta, \gamma) = u_1(\alpha) u_2(\beta) u_3(\gamma),$$

where u_1, u_2, u_3 each satisfy the Lamé wave equation (29.11.1).

29.18(iii) Spherical and Ellipsoidal Harmonics

See Erdélyi et al. (1955, §15.7).

29.18(iv) Other Applications

Triebel (1965) gives applications of Lamé functions to the theory of conformal mappings. Patera and Winternitz (1973) finds bases for the rotation group.

29.19 Physical Applications

29.19(i) Lamé Functions

Simply-periodic Lamé functions (ν noninteger) can be used to solve boundary-value problems for Laplace's equation in elliptical cones. For applications in antenna research see Jansen (1977). Brack et al. (2001) shows that Lamé functions occur at bifurcations in chaotic Hamiltonian systems. Bronski et al. (2001) uses Lamé functions in the theory of Bose–Einstein condensates.

29.19(ii) Lamé Polynomials

Ward (1987) computes finite-gap potentials associated with the periodic Korteweg–de Vries equation. Shail (1978) treats applications to solutions of elliptic crack and punch problems. Hargrave (1978) studies high frequency solutions of the delta wing equation. Macfadyen and Winternitz (1971) finds expansions for the two-body relativistic scattering amplitudes. Roper (1951) solves the linearized supersonic flow equations. Clarkson (1991) solves nonlinear evolution equations. Strutt (1932) describes various applications and provides an extensive list of references.

See also §29.12(iii).

Computation

29.20 Methods of Computation

29.20(i) Lamé Functions

The eigenvalues $a_\nu^m(k^2)$, $b_\nu^m(k^2)$, and the Lamé functions $\mathit{Ec}_\nu^m(z, k^2)$, $\mathit{Es}_\nu^m(z, k^2)$, can be calculated by direct numerical methods applied to the differential equation (29.2.1); see §3.7. The normalization of Lamé functions given in §29.3(v) can be carried out by quadrature (§3.5).

A second approach is to solve the continued-fraction equations typified by (29.3.10) by Newton's rule or other iterative methods; see §3.8. Initial approximations to the eigenvalues can be found, for example, from the asymptotic expansions supplied in §29.7(i). Subsequently, formulas typified by (29.6.4) can be applied to compute the coefficients of the Fourier expansions of the corresponding Lamé functions by backward recursion followed by application of formulas typified by (29.6.5) and (29.6.6) to achieve normalization; compare §3.6. (Equation (29.6.3) serves as a check.) The Fourier series may be summed using Clenshaw's algorithm; see §3.11(ii). For further information see Jansen (1977).

A third method is to approximate eigenvalues and Fourier coefficients of Lamé functions by eigenvalues and eigenvectors of finite matrices using the methods of §§3.2(vi) and 3.8(iv). These matrices are the same as those provided in §29.15(i) for the computation of Lamé polynomials with the difference that n has to be chosen sufficiently large. The approximations converge geometrically (§3.8(i)) to the eigenvalues and coefficients of Lamé functions as $n \to \infty$. The numerical computations described in Jansen (1977) are based in part upon this method.

29.20(ii) Lamé Polynomials

The eigenvalues corresponding to Lamé polynomials are computed from eigenvalues of the finite tridiagonal matrices $\mathbf{M}$ given in §29.15(i), using methods described in §3.2(vi) and Ritter (1998). The corresponding eigenvectors yield the coefficients in the finite Fourier series for Lamé polynomials. §29.15(i) includes formulas for normalizing the eigenvectors.

29.20(iii) Zeros

Zeros of Lamé polynomials can be computed by solving the system of equations (29.12.13) by employing Newton's method; see §3.8(ii). Alternatively, the zeros can be found by locating the maximum of function g in (29.12.11).

29.21 Tables

- Ince (1940a) tabulates the eigenvalues $a_\nu^m(k^2)$, $b_\nu^{m+1}(k^2)$ (with a_ν^{2m+1} and b_ν^{2m+1} interchanged) for $k^2 = 0.1, 0.5, 0.9$, $\nu = -\frac{1}{2}, 0(1)25$, and $m = 0, 1, 2, 3$. Precision is 4D.

- Arscott and Khabaza (1962) tabulates the coefficients of the polynomials P in Table 29.12.1 (normalized so that the numerically largest coefficient is unity, i.e. monic polynomials), and the corresponding eigenvalues h for $k^2 = 0.1(.1)0.9$, $n = 1(1)30$. Equations from §29.6 can be used to transform to the normalization adopted in this chapter. Precision is 6S.

29.22 Software

See http://dlmf.nist.gov/29.22.

References

General References

The main references used in writing this chapter are Arscott (1964b), Erdélyi et al. (1955), Ince (1940b), and Jansen (1977). For additional bibliographic reading see Hobson (1931), Strutt (1932), and Whittaker and Watson (1927).

Sources

The following list gives the references or other indications of proofs that were used in constructing the various sections of this chapter. These sources supplement the references that are quoted in the text.

§29.2 Erdélyi et al. (1955, §15.2).

§29.3 Erdélyi (1941b), Erdélyi et al. (1955, §15.5.1), Ince (1940b), Jansen (1977, §3.1), Magnus and Winkler (1966, §2.1), Volkmer (2004b). For (29.3.19) combine Hochstadt (1964, p. 148) and Table 29.3.2.

§29.4 The graphics were produced at NIST.

§29.5 Erdélyi et al. (1955, §15.5.4), Volkmer (2004b).

§29.6 Erdélyi et al. (1955, §15.5.1), Ince (1940b), Jansen (1977).

§29.7 Ince (1940a), Müller (1966a). For (29.7.5) see Volkmer (2004b).

§29.8 Erdélyi et al. (1955, §15.5.3), Volkmer (1982, 1983).

§29.9 Magnus and Winkler (1966, §2.1).

§29.12 Arscott (1964b, Chapter 9), Whittaker and Watson (1927, Chapter 23).

§29.13 The graphics were produced at NIST.

§29.14 Arscott (1964b, §9.4), Erdélyi et al. (1955, §15.7).

§29.15 The method for constructing the Fourier coefficients follows from §29.6. For the Chebyshev coefficients see Arscott (1964b, §9.6.2), Ince (1940a).

§29.18 Arscott (1964b, §9.8.1), Erdélyi et al. (1955, §§15.1.2, 15.1.3, 15.7), Hobson (1931, Chapter XI), Jansen (1977).

Chapter 30
Spheroidal Wave Functions

H. Volkmer[1]

Notation   **698**
- 30.1 Special Notation 698

Properties   **698**
- 30.2 Differential Equations 698
- 30.3 Eigenvalues 698
- 30.4 Functions of the First Kind 699
- 30.5 Functions of the Second Kind 700
- 30.6 Functions of Complex Argument 700
- 30.7 Graphics 700
- 30.8 Expansions in Series of Ferrers Functions 702
- 30.9 Asymptotic Approximations and Expansions 702
- 30.10 Series and Integrals 703
- 30.11 Radial Spheroidal Wave Functions . . . 703
- 30.12 Generalized and Coulomb Spheroidal Functions 704

Applications   **704**
- 30.13 Wave Equation in Prolate Spheroidal Coordinates 704
- 30.14 Wave Equation in Oblate Spheroidal Coordinates 705
- 30.15 Signal Analysis 706

Computation   **707**
- 30.16 Methods of Computation 707
- 30.17 Tables 708
- 30.18 Software 708

References   **708**

---

[1]Department of Mathematical Sciences, University of Wisconsin-Milwaukee, Milwaukee, Wisconsin.
Acknowledgments: This chapter is based in part on Abramowitz and Stegun (1964, Chapter 21) by A. N. Lowan.
Copyright © 2009 National Institute of Standards and Technology. All rights reserved.

Notation

30.1 Special Notation

(For other notation see pp. xiv and 873.)

- x real variable. Except in §§30.7(iv), 30.11(ii), 30.13, and 30.14, $-1 < x < 1$.
- γ^2 real parameter (positive, zero, or negative).
- m order, a nonnegative integer.
- n degree, an integer $n = m, m+1, m+2, \ldots$.
- k integer.
- δ arbitrary small positive constant.

The main functions treated in this chapter are the eigenvalues $\lambda_n^m(\gamma^2)$ and the spheroidal wave functions $\mathsf{Ps}_n^m(x,\gamma^2)$, $\mathsf{Qs}_n^m(x,\gamma^2)$, $\mathit{Ps}_n^m(z,\gamma^2)$, $\mathit{Qs}_n^m(z,\gamma^2)$, and $S_n^{m(j)}(z,\gamma)$, $j=1,2,3,4$. These notations are similar to those used in Arscott (1964b) and Erdélyi et al. (1955). Meixner and Schäfke (1954) use ps, qs, Ps, Qs for Ps, Qs, Ps, Qs, respectively.

Other Notations

Flammer (1957) and Abramowitz and Stegun (1964) use $\lambda_{mn}(\gamma)$ for $\lambda_n^m(\gamma^2) + \gamma^2$, $R_{mn}^{(j)}(\gamma,z)$ for $S_n^{m(j)}(z,\gamma)$, and

30.1.1
$$S_{mn}^{(1)}(\gamma,x) = d_{mn}(\gamma)\,\mathsf{Ps}_n^m(x,\gamma^2),$$
$$S_{mn}^{(2)}(\gamma,x) = d_{mn}(\gamma)\,\mathsf{Qs}_n^m(x,\gamma^2),$$

where $d_{mn}(\gamma)$ is a normalization constant determined by

30.1.2
$$S_{mn}^{(1)}(\gamma,0) = (-1)^m\,\mathsf{P}_n^m(0), \quad n-m \text{ even},$$
$$\left.\frac{d}{dx}S_{mn}^{(1)}(\gamma,x)\right|_{x=0} = (-1)^m\left.\frac{d}{dx}\mathsf{P}_n^m(x)\right|_{x=0},$$
$$n-m \text{ odd}.$$

For older notations see Abramowitz and Stegun (1964, §21.11) and Flammer (1957, pp. 14,15).

Properties

30.2 Differential Equations

30.2(i) Spheroidal Differential Equation

30.2.1
$$\frac{d}{dz}\left((1-z^2)\frac{dw}{dz}\right) + \left(\lambda + \gamma^2(1-z^2) - \frac{\mu^2}{1-z^2}\right)w = 0.$$

This equation has regular singularities at $z = \pm 1$ with exponents $\pm\frac{1}{2}\mu$ and an irregular singularity of rank 1 at $z = \infty$ (if $\gamma \neq 0$). The equation contains three real parameters λ, γ^2, and μ. In applications involving prolate spheroidal coordinates γ^2 is positive, in applications involving oblate spheroidal coordinates γ^2 is negative; see §§30.13, 30.14.

30.2(ii) Other Forms

The *Liouville normal form* of equation (30.2.1) is

30.2.2
$$\frac{d^2g}{dt^2} + \left(\lambda + \frac{1}{4} + \gamma^2 \sin^2 t - \frac{\mu^2 - \frac{1}{4}}{\sin^2 t}\right)g = 0,$$

30.2.3
$$z = \cos t, \quad w(z) = (1-z^2)^{-\frac{1}{4}} g(t).$$

With $\zeta = \gamma z$ Equation (30.2.1) changes to

30.2.4
$$(\zeta^2 - \gamma^2)\frac{d^2w}{d\zeta^2} + 2\zeta\frac{dw}{d\zeta} + \left(\zeta^2 - \lambda - \gamma^2 - \frac{\gamma^2 \mu^2}{\zeta^2 - \gamma^2}\right)w = 0.$$

30.2(iii) Special Cases

If $\gamma = 0$, Equation (30.2.1) is the associated Legendre differential equation; see (14.2.2). If $\mu^2 = \frac{1}{4}$, Equation (30.2.2) reduces to the Mathieu equation; see (28.2.1). If $\gamma = 0$, Equation (30.2.4) is satisfied by spherical Bessel functions; see (10.47.1).

30.3 Eigenvalues

30.3(i) Definition

With $\mu = m = 0, 1, 2, \ldots$, the spheroidal wave functions $\mathsf{Ps}_n^m(x,\gamma^2)$ are solutions of Equation (30.2.1) which are bounded on $(-1,1)$, or equivalently, which are of the form $(1-x^2)^{\frac{1}{2}m}g(x)$ where $g(z)$ is an entire function of z. These solutions exist only for eigenvalues $\lambda_n^m(\gamma^2)$, $n = m, m+1, m+2, \ldots$, of the parameter λ.

30.3(ii) Properties

The eigenvalues $\lambda_n^m(\gamma^2)$ are analytic functions of the real variable γ^2 and satisfy

30.3.1
$$\lambda_m^m(\gamma^2) < \lambda_{m+1}^m(\gamma^2) < \lambda_{m+2}^m(\gamma^2) < \cdots,$$

30.3.2
$$\lambda_n^m(\gamma^2) = n(n+1) - \tfrac{1}{2}\gamma^2 + O(n^{-2}), \quad n \to \infty,$$

30.3.3
$$\lambda_n^m(0) = n(n+1),$$

30.3.4
$$-1 < \frac{d\lambda_n^m(\gamma^2)}{d(\gamma^2)} < 0.$$

30.3(iii) Transcendental Equation

If p is an even nonnegative integer, then the continued-fraction equation

$$
\beta_p - \lambda - \cfrac{\alpha_{p-2}\gamma_p}{\beta_{p-2} - \lambda -} \cfrac{\alpha_{p-4}\gamma_{p-2}}{\beta_{p-4} - \lambda -} \cdots
$$

30.3.5

$$
= \cfrac{\alpha_p \gamma_{p+2}}{\beta_{p+2} - \lambda -} \cfrac{\alpha_{p+2}\gamma_{p+4}}{\beta_{p+4} - \lambda -} \cdots ,
$$

where $\alpha_k, \beta_k, \gamma_k$ are defined by

30.3.6
$$\alpha_k = -(k+1)(k+2),$$
$$\beta_k = (m+k)(m+k+1) - \gamma^2, \quad \gamma_k = \gamma^2,$$

has the solutions $\lambda = \lambda^m_{m+2j}(\gamma^2)$, $j = 0, 1, 2, \ldots$. If p is an odd positive integer, then Equation (30.3.5) has the solutions $\lambda = \lambda^m_{m+2j+1}(\gamma^2)$, $j = 0, 1, 2, \ldots$. If $p = 0$ or $p = 1$, the finite continued-fraction on the left-hand side of (30.3.5) equals 0; if $p > 1$ its last denominator is $\beta_0 - \lambda$ or $\beta_1 - \lambda$.

For a different choice of $\alpha_p, \beta_p, \gamma_p$ in (30.3.5) see http://dlmf.nist.gov/30.3.iii.

30.3(iv) Power-Series Expansion

30.3.8
$$\lambda^m_n(\gamma^2) = \sum_{k=0}^{\infty} \ell_{2k} \gamma^{2k}, \qquad |\gamma^2| < r^m_n.$$

For values of r^m_n see Meixner et al. (1980, p. 109).

30.3.9
$$\ell_0 = n(n+1), \quad 2\ell_2 = -1 - \frac{(2m-1)(2m+1)}{(2n-1)(2n+3)},$$
$$2\ell_4 = \frac{(n-m-1)(n-m)(n+m-1)(n+m)}{(2n-3)(2n-1)^3(2n+1)} - \frac{(n-m+1)(n-m+2)(n+m+1)(n+m+2)}{(2n+1)(2n+3)^3(2n+5)}.$$

For additional coefficients see http://dlmf.nist.gov/30.3.iv.

30.4 Functions of the First Kind

30.4(i) Definitions

The eigenfunctions of (30.2.1) that correspond to the eigenvalues $\lambda^m_n(\gamma^2)$ are denoted by $\mathsf{Ps}^m_n(x, \gamma^2)$, $n = m, m+1, m+2, \ldots$. They are normalized by the condition

30.4.1
$$\int_{-1}^{1} \left(\mathsf{Ps}^m_n(x, \gamma^2)\right)^2 dx = \frac{2}{2n+1} \frac{(n+m)!}{(n-m)!},$$

the sign of $\mathsf{Ps}^m_n(0, \gamma^2)$ being $(-1)^{(n+m)/2}$ when $n-m$ is even, and the sign of $d\mathsf{Ps}^m_n(x, \gamma^2)/dx\,|_{x=0}$ being $(-1)^{(n+m-1)/2}$ when $n-m$ is odd.

When $\gamma^2 > 0$ $\mathsf{Ps}^m_n(x, \gamma^2)$ is the *prolate angular spheroidal wave function*, and when $\gamma^2 < 0$ $\mathsf{Ps}^m_n(x, \gamma^2)$ is the *oblate angular spheroidal wave function*. If $\gamma = 0$, $\mathsf{Ps}^m_n(x, 0)$ reduces to the Ferrers function $\mathsf{P}^m_n(x)$:

30.4.2
$$\mathsf{Ps}^m_n(x, 0) = \mathsf{P}^m_n(x);$$

compare §14.3(i).

30.4(ii) Elementary Properties

30.4.3
$$\mathsf{Ps}^m_n(-x, \gamma^2) = (-1)^{n-m} \mathsf{Ps}^m_n(x, \gamma^2).$$

$\mathsf{Ps}^m_n(x, \gamma^2)$ has exactly $n-m$ zeros in the interval $-1 < x < 1$.

30.4(iii) Power-Series Expansion

30.4.4
$$\mathsf{Ps}^m_n(x, \gamma^2) = (1-x^2)^{\frac{1}{2}m} \sum_{k=0}^{\infty} g_k x^k, \quad -1 \leq x \leq 1,$$

where

30.4.5
$$\alpha_k g_{k+2} + (\beta_k - \lambda^m_n(\gamma^2))g_k + \gamma_k g_{k-2} = 0$$

with $\alpha_k, \beta_k, \gamma_k$ from (30.3.6), and $g_{-1} = g_{-2} = 0$, $g_k = 0$ for even k if $n-m$ is odd and $g_k = 0$ for odd k if $n-m$ is even. Normalization of the coefficients g_k is effected by application of (30.4.1).

30.4(iv) Orthogonality

30.4.6
$$\int_{-1}^{1} \mathsf{Ps}^m_k(x, \gamma^2) \mathsf{Ps}^m_n(x, \gamma^2)\, dx = \frac{2}{2n+1} \frac{(n+m)!}{(n-m)!} \delta_{k,n}.$$

If $f(x)$ is mean-square integrable on $[-1, 1]$, then formally

30.4.7
$$f(x) = \sum_{n=m}^{\infty} c_n \mathsf{Ps}^m_n(x, \gamma^2),$$

where

30.4.8
$$c_n = (n + \tfrac{1}{2}) \frac{(n-m)!}{(n+m)!} \int_{-1}^{1} f(t)\, \mathsf{Ps}^m_n(t, \gamma^2)\, dt.$$

The expansion (30.4.7) converges in the norm of $L^2(-1, 1)$, that is,

30.4.9
$$\lim_{N \to \infty} \int_{-1}^{1} \left| f(x) - \sum_{n=m}^{N} c_n \mathsf{Ps}^m_n(x, \gamma^2) \right|^2 dx = 0.$$

It is also equiconvergent with its expansion in Ferrers functions (as in (30.4.2)), that is, the difference of corresponding partial sums converges to 0 uniformly for $-1 \leq x \leq 1$.

30.5 Functions of the Second Kind

Other solutions of (30.2.1) with $\mu = m$, $\lambda = \lambda_n^m(\gamma^2)$, and $z = x$ are

30.5.1 $\qquad \mathsf{Qs}_n^m(x, \gamma^2), \quad n = m, m+1, m+2, \ldots.$

They satisfy

30.5.2 $\quad \mathsf{Qs}_n^m(-x, \gamma^2) = (-1)^{n-m+1} \mathsf{Qs}_n^m(x, \gamma^2),$

and

30.5.3 $\qquad \mathsf{Qs}_n^m(x, 0) = \mathsf{Q}_n^m(x);$

compare §14.3(i). Also,

30.5.4
$$\mathscr{W}\left\{\mathsf{Ps}_n^m(x, \gamma^2), \mathsf{Qs}_n^m(x, \gamma^2)\right\} \\ = \frac{(n+m)!}{(1-x^2)(n-m)!} A_n^m(\gamma^2) A_n^{-m}(\gamma^2) \quad (\neq 0),$$

with $A_n^{\pm m}(\gamma^2)$ as in (30.11.4).

For further properties see Meixner and Schäfke (1954) and §30.8(ii).

30.6 Functions of Complex Argument

The solutions

30.6.1 $\qquad Ps_n^m(z, \gamma^2), \quad Qs_n^m(z, \gamma^2),$

of (30.2.1) with $\mu = m$ and $\lambda = \lambda_n^m(\gamma^2)$ are real when $z \in (1, \infty)$, and their principal values (§4.2(i)) are obtained by analytic continuation to $\mathbb{C} \setminus (-\infty, 1]$.

Relations to Associated Legendre Functions

30.6.2 $\quad Ps_n^m(z, 0) = P_n^m(z), \quad Qs_n^m(z, 0) = Q_n^m(z);$

compare §14.3(ii).

Wronskian

30.6.3
$$\mathscr{W}\left\{Ps_n^m(z, \gamma^2), Qs_n^m(z, \gamma^2)\right\} \\ = \frac{(-1)^m(n+m)!}{(1-z^2)(n-m)!} A_n^m(\gamma^2) A_n^{-m}(\gamma^2),$$

with $A_n^{\pm m}(\gamma^2)$ as in (30.11.4).

Values on $(-1, 1)$

30.6.4 $\quad Ps_n^m(x \pm i0, \gamma^2) = (\mp i)^m \mathsf{Ps}_n^m(x, \gamma^2),$

30.6.5 $\quad \begin{aligned} &Qs_n^m(x \pm i0, \gamma^2) \\ &= (\mp i)^m \left(\mathsf{Qs}_n^m(x, \gamma^2) \mp \tfrac{1}{2} i \pi \mathsf{Ps}_n^m(x, \gamma^2)\right). \end{aligned}$

For further properties see Arscott (1964b).

For results for Equation (30.2.1) with complex parameters see Meixner and Schäfke (1954).

30.7 Graphics

30.7(i) Eigenvalues

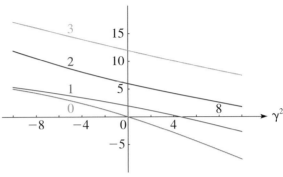

Figure 30.7.1: Eigenvalues $\lambda_n^0(\gamma^2)$, $n = 0, 1, 2, 3$, $-10 \leq \gamma^2 \leq 10$.

For additional graphs see http://dlmf.nist.gov/30.7.i.

30.7(ii) Functions of the First Kind

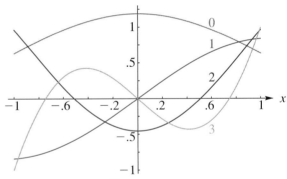

Figure 30.7.5: $\mathsf{Ps}_n^0(x, 4)$, $n = 0, 1, 2, 3$, $-1 \leq x \leq 1$.

For additional graphs see http://dlmf.nist.gov/30.7.ii.

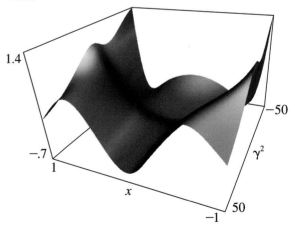

Figure 30.7.9: $\mathsf{Ps}_2^0(x, \gamma^2)$, $-1 \leq x \leq 1$, $-50 \leq \gamma^2 \leq 50$.

30.7 Graphics

For an additional surface see http://dlmf.nist.gov/30.7.ii.

30.7(iii) Functions of the Second Kind

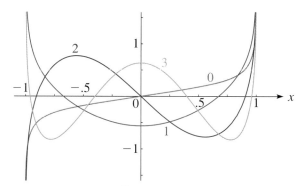

Figure 30.7.11: $\mathsf{Qs}_n^0(x,4)$, $n = 0, 1, 2, 3$, $-1 < x < 1$.

For additional graphs see http://dlmf.nist.gov/30.7.iii.

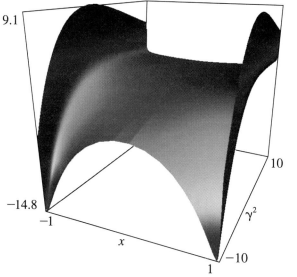

Figure 30.7.15: $\mathsf{Qs}_1^0(x, \gamma^2)$, $-1 < x < 1$, $-10 \leq \gamma^2 \leq 10$.

30.7(iv) Functions of Complex Argument

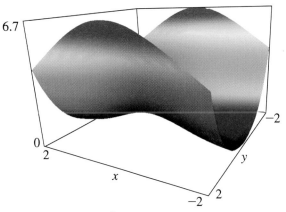

Figure 30.7.16: $|Ps_0^0(x+iy, 4)|$, $-2 \leq x \leq 2$, $-2 \leq y \leq 2$.

For additional surfaces see http://dlmf.nist.gov/30.7.iv.

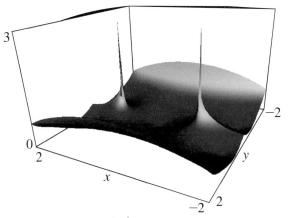

Figure 30.7.20: $|Qs_0^0(x+iy, 4)|$, $-2 \leq x \leq 2$, $-2 \leq y \leq 2$.

For an additional surface see http://dlmf.nist.gov/30.7.iv.

30.8 Expansions in Series of Ferrers Functions

30.8(i) Functions of the First Kind

30.8.1 $\quad \mathsf{Ps}_n^m(x, \gamma^2) = \sum_{k=-R}^{\infty} (-1)^k a_{n,k}^m(\gamma^2) \mathsf{P}_{n+2k}^m(x),$

where $\mathsf{P}_{n+2k}^m(x)$ is the Ferrers function of the first kind (§14.3(i)), $R = \lfloor \tfrac{1}{2}(n-m) \rfloor$, and the coefficients $a_{n,k}^m(\gamma^2)$ are given by

30.8.2
$$a_{n,k}^m(\gamma^2) = (-1)^k \left(n+2k+\tfrac{1}{2}\right) \frac{(n-m+2k)!}{(n+m+2k)!} \\ \times \int_{-1}^{1} \mathsf{Ps}_n^m(x, \gamma^2) \mathsf{P}_{n+2k}^m(x)\, dx.$$

Let

30.8.3
$$A_k = -\gamma^2 \frac{(n-m+2k-1)(n-m+2k)}{(2n+4k-3)(2n+4k-1)},$$
$$B_k = (n+2k)(n+2k+1) \\ - 2\gamma^2 \frac{(n+2k)(n+2k+1) - 1 + m^2}{(2n+4k-1)(2n+4k+3)},$$
$$C_k = -\gamma^2 \frac{(n+m+2k+1)(n+m+2k+2)}{(2n+4k+3)(2n+4k+5)}.$$

Then the set of coefficients $a_{n,k}^m(\gamma^2)$, $k = -R, -R+1, -R+2, \ldots$ is the solution of the difference equation

30.8.4 $\quad A_k f_{k-1} + \left(B_k - \lambda_n^m(\gamma^2)\right) f_k + C_k f_{k+1} = 0,$

(note that $A_{-R} = 0$) that satisfies the normalizing condition

30.8.5 $\quad \sum_{k=-R}^{\infty} a_{n,k}^m(\gamma^2) a_{n,k}^{-m}(\gamma^2) \frac{1}{2n+4k+1} = \frac{1}{2n+1},$

with

30.8.6 $\quad a_{n,k}^{-m}(\gamma^2) = \frac{(n-m)!(n+m+2k)!}{(n+m)!(n-m+2k)!} a_{n,k}^m(\gamma^2).$

Also, as $k \to \infty$,

30.8.7 $\quad \dfrac{k^2 a_{n,k}^m(\gamma^2)}{a_{n,k-1}^m(\gamma^2)} = \dfrac{\gamma^2}{16} + O\!\left(\dfrac{1}{k}\right),$

and

30.8.8 $\quad \dfrac{\lambda_n^m(\gamma^2) - B_k}{A_k} \dfrac{a_{n,k}^m(\gamma^2)}{a_{n,k-1}^m(\gamma^2)} = 1 + O\!\left(\dfrac{1}{k^4}\right).$

30.8(ii) Functions of the Second Kind

30.8.9
$$\mathsf{Qs}_n^m(x, \gamma^2) = \sum_{k=-\infty}^{-N-1} (-1)^k a'^m_{n,k}(\gamma^2) \mathsf{P}_{n+2k}^m(x) \\ + \sum_{k=-N}^{\infty} (-1)^k a_{n,k}^m(\gamma^2) \mathsf{Q}_{n+2k}^m(x),$$

where P_n^m and Q_n^m are again the Ferrers functions and $N = \lfloor \tfrac{1}{2}(n+m) \rfloor$. The coefficients $a_{n,k}^m(\gamma^2)$ satisfy (30.8.4) for all k when we set $a_{n,k}^m(\gamma^2) = 0$ for $k < -N$. For $k \geq -R$ they agree with the coefficients defined in §30.8(i). For $k = -N, -N+1, \ldots, -R-1$ they are determined from (30.8.4) by forward recursion using $a_{n,-N-1}^m(\gamma^2) = 0$. The set of coefficients $a'^m_{n,k}(\gamma^2)$, $k = -N-1, -N-2, \ldots$, is the recessive solution of (30.8.4) as $k \to -\infty$ that is normalized by

30.8.10
$$A_{-N-1} a'^m_{n,-N-2}(\gamma^2) \\ + \left(B_{-N-1} - \lambda_n^m(\gamma^2)\right) a'^m_{n,-N-1}(\gamma^2) \\ + C' a^m_{n,-N}(\gamma^2) = 0,$$

with

30.8.11 $\quad C' = \begin{cases} \dfrac{\gamma^2}{4m^2 - 1}, & n-m \text{ even,} \\ -\dfrac{\gamma^2}{(2m-1)(2m-3)}, & n-m \text{ odd.} \end{cases}$

It should be noted that if the forward recursion (30.8.4) beginning with $f_{-N-1} = 0$, $f_{-N} = 1$ leads to $f_{-R} = 0$, then $a_{n,k}^m(\gamma^2)$ is undefined for $n < -R$ and $\mathsf{Qs}_n^m(x, \gamma^2)$ does not exist.

30.9 Asymptotic Approximations and Expansions

30.9(i) Prolate Spheroidal Wave Functions

As $\gamma^2 \to +\infty$, with $q = 2(n-m) + 1$,

30.9.1 $\quad \lambda_n^m(\gamma^2) \sim -\gamma^2 + \gamma q + \beta_0 + \beta_1 \gamma^{-1} + \beta_2 \gamma^{-2} + \cdots,$

where

30.9.2
$$8\beta_0 = 8m^2 - q^2 - 5, \quad 2^6 \beta_1 = -q^3 - 11q + 32m^2 q,$$
$$2^{10} \beta_2 = -5(q^4 + 26q^2 + 21) + 384 m^2 (q^2 + 1),$$
$$2^{14} \beta_3 = -33q^5 - 1594q^3 - 5621q \\ + 128 m^2 (37 q^3 + 167 q) - 2048 m^4 q.$$

For additional coefficients see http://dlmf.nist.gov/30.9.i.

For the eigenfunctions see Meixner and Schäfke (1954, §3.251) and Müller (1963).

For uniform asymptotic expansions in terms of Airy or Bessel functions for real values of the parameters, complex values of the variable, and with explicit error bounds see Dunster (1986). See also Miles (1975).

30.9(ii) Oblate Spheroidal Wave Functions

As $\gamma^2 \to -\infty$, with $q = n+1$ if $n-m$ is even, or $q = n$ if $n-m$ is odd, we have

30.9.4 $\quad \lambda_n^m(\gamma^2) \sim 2q|\gamma| + c_0 + c_1|\gamma|^{-1} + c_2|\gamma|^{-2} + \cdots,$

where

30.9.5
$$2c_0 = -q^2 - 1 + m^2, \quad 8c_1 = -q^3 - q + m^2 q,$$
$$2^6 c_2 = -5q^4 - 10q^2 - 1 + 2m^2(3q^2+1) - m^4,$$
$$2^9 c_3 = -33q^5 - 114q^3 - 37q + 2m^2(23q^3 + 25q) - 13m^4 q.$$

For additional coefficients see http://dlmf.nist.gov/30.9.ii.

For the eigenfunctions see Meixner and Schäfke (1954, §3.252) and Müller (1962).

For uniform asymptotic expansions in terms of elementary, Airy, or Bessel functions for real values of the parameters, complex values of the variable, and with explicit error bounds see Dunster (1992, 1995). See also Jorna and Springer (1971).

30.9(iii) Other Approximations and Expansions

The asymptotic behavior of $\lambda_n^m(\gamma^2)$ and $a_{n,k}^m(\gamma^2)$ as $n \to \infty$ in descending powers of $2n+1$ is derived in Meixner (1944). The cases of large m, and of large m and large $|\gamma|$, are studied in Abramowitz (1949). The asymptotic behavior of $\mathsf{Ps}_n^m(x,\gamma^2)$ and $\mathsf{Qs}_n^m(x,\gamma^2)$ as $x \to \pm 1$ is given in Erdélyi et al. (1955, p. 151). The behavior of $\lambda_n^m(\gamma^2)$ for complex γ^2 and large $|\lambda_n^m(\gamma^2)|$ is investigated in Hunter and Guerrieri (1982).

30.10 Series and Integrals

Integrals and integral equations for $\mathsf{Ps}_n^m(x,\gamma^2)$ are given in Arscott (1964b, §8.6), Erdélyi et al. (1955, §16.13), Flammer (1957, Chapter 5), and Meixner (1951). For product formulas and convolutions see Connett et al. (1993). For an addition theorem, see Meixner and Schäfke (1954, p. 300) and King and Van Buren (1973). For expansions in products of spherical Bessel functions, see Flammer (1957, Chapter 6).

30.11 Radial Spheroidal Wave Functions

30.11(i) Definitions

Denote

30.11.1 $\quad \psi_k^{(j)}(z) = \left(\dfrac{\pi}{2z}\right)^{\frac{1}{2}} \mathcal{C}_{k+\frac{1}{2}}^{(j)}(z), \quad j = 1, 2, 3, 4,$

where

30.11.2
$$\mathcal{C}_\nu^{(1)} = J_\nu, \quad \mathcal{C}_\nu^{(2)} = Y_\nu, \quad \mathcal{C}_\nu^{(3)} = H_\nu^{(1)}, \quad \mathcal{C}_\nu^{(4)} = H_\nu^{(2)},$$

with J_ν, Y_ν, $H_\nu^{(1)}$, and $H_\nu^{(2)}$ as in §10.2(ii). Then solutions of (30.2.1) with $\mu = m$ and $\lambda = \lambda_n^m(\gamma^2)$ are given by

30.11.3
$$S_n^{m(j)}(z,\gamma) = \frac{(1-z^{-2})^{\frac{1}{2}m}}{A_n^{-m}(\gamma^2)} \sum_{2k \geq m-n} a_{n,k}^{-m}(\gamma^2) \psi_{n+2k}^{(j)}(\gamma z).$$

Here $a_{n,k}^{-m}(\gamma^2)$ is defined by (30.8.2) and (30.8.6), and

30.11.4 $\quad A_n^{\pm m}(\gamma^2) = \displaystyle\sum_{2k \geq \mp m - n} (-1)^k a_{n,k}^{\pm m}(\gamma^2) \quad (\neq 0).$

In (30.11.3) $z \neq 0$ when $j = 1$, and $|z| > 1$ when $j = 2, 3, 4$.

Connection Formulas

30.11.5
$$S_n^{m(3)}(z,\gamma) = S_n^{m(1)}(z,\gamma) + i S_n^{m(2)}(z,\gamma),$$
$$S_n^{m(4)}(z,\gamma) = S_n^{m(1)}(z,\gamma) - i S_n^{m(2)}(z,\gamma).$$

30.11(ii) Graphics

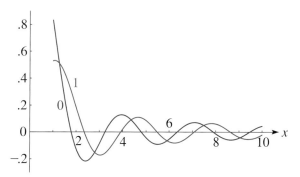

Figure 30.11.1: $S_n^{0(1)}(x,2)$, $n = 0, 1$, $1 \leq x \leq 10$.

For additional graphs see http://dlmf.nist.gov/30.11.ii.

30.11(iii) Asymptotic Behavior

For fixed γ, as $z \to \infty$ in the sector $|\operatorname{ph} z| \leq \pi - \delta \, (< \pi)$,

30.11.6
$$S_n^{m(j)}(z,\gamma) = \begin{cases} \psi_n^{(j)}(\gamma z) + O(z^{-2} e^{|\Im z|}), & j = 1, 2 \\ \psi_n^{(j)}(\gamma z)(1 + O(z^{-1})), & j = 3, 4. \end{cases}$$

For asymptotic expansions in negative powers of z see Meixner and Schäfke (1954, p. 293).

30.11(iv) Wronskian

30.11.7 $\quad \mathscr{W}\left\{S_n^{m(1)}(z,\gamma), S_n^{m(2)}(z,\gamma)\right\} = \dfrac{1}{\gamma(z^2-1)}.$

30.11(v) Connection with the Ps and Qs Functions

30.11.8 $\quad S_n^{m(1)}(z,\gamma) = K_n^m(\gamma)\, Ps_n^m(z,\gamma^2),$

30.11.9
$$S_n^{m(2)}(z,\gamma) = \frac{(n-m)!}{(n+m)!}\frac{(-1)^{m+1}\,Qs_n^m(z,\gamma^2)}{\gamma K_n^m(\gamma)A_n^m(\gamma^2)A_n^{-m}(\gamma^2)},$$

where

30.11.10
$$K_n^m(\gamma) = \frac{\sqrt{\pi}}{2}\left(\frac{\gamma}{2}\right)^m \frac{(-1)^m a_{n,\frac{1}{2}(m-n)}^{-m}(\gamma^2)}{\Gamma\!\left(\frac{3}{2}+m\right)A_n^{-m}(\gamma^2)\,\mathsf{Ps}_n^m(0,\gamma^2)},$$
$n-m$ even,

or

30.11.11
$$K_n^m(\gamma) = \frac{\sqrt{\pi}}{2}\left(\frac{\gamma}{2}\right)^{m+1}$$
$$\times \frac{(-1)^m a_{n,\frac{1}{2}(m-n+1)}^{-m}(\gamma^2)}{\Gamma\!\left(\frac{5}{2}+m\right)A_n^{-m}(\gamma^2)\left(\,d\mathsf{Ps}_n^m(z,\gamma^2)/dz\,|_{z=0}\right)},$$
$n-m$ odd.

30.11(vi) Integral Representations

When $z \in \mathbb{C}\setminus(-\infty,1]$

30.11.12
$$\begin{aligned}A_n^{-m}(\gamma^2)\,S_n^{m(1)}(z,\gamma) &= \frac{1}{2}i^{m+n}\gamma^m \frac{(n-m)!}{(n+m)!}z^m(1-z^{-2})^{\frac{1}{2}m}\\ &\quad\times \int_{-1}^{1} e^{-i\gamma z t}(1-t^2)^{\frac{1}{2}m}\,\mathsf{Ps}_n^m(t,\gamma^2)\,dt.\end{aligned}$$

For further relations see Arscott (1964b, §8.6), Connett et al. (1993), Erdélyi et al. (1955, §16.13), Meixner and Schäfke (1954), and Meixner et al. (1980, §3.1).

30.12 Generalized and Coulomb Spheroidal Functions

Generalized spheroidal wave functions and Coulomb spheroidal functions are solutions of the differential equation

30.12.1
$$\frac{d}{dz}\!\left((1-z^2)\frac{dw}{dz}\right)\\ + \left(\lambda + \alpha z + \gamma^2(1-z^2) - \frac{\mu^2}{1-z^2}\right)w = 0,$$

which reduces to (30.2.1) if $\alpha = 0$. Equation (30.12.1) appears in astrophysics and molecular physics. For the theory and computation of solutions of (30.12.1) see Falloon (2001), Judd (1975), Leaver (1986), and Komarov et al. (1976).

Another generalization is provided by the differential equation

30.12.2
$$\frac{d}{dz}\!\left((1-z^2)\frac{dw}{dz}\right) + \Big(\lambda + \gamma^2(1-z^2)\\ -\frac{\alpha(\alpha+1)}{z^2} - \frac{\mu^2}{1-z^2}\Big)w = 0,$$

which also reduces to (30.2.1) when $\alpha = 0$. See Leitner and Meixner (1960), Slepian (1964) with $\mu = 0$, and Meixner et al. (1980).

Applications

30.13 Wave Equation in Prolate Spheroidal Coordinates

30.13(i) Prolate Spheroidal Coordinates

Prolate spheroidal coordinates ξ,η,ϕ are related to Cartesian coordinates x,y,z by

30.13.1
$$\begin{aligned}x &= c\sqrt{(\xi^2-1)(1-\eta^2)}\cos\phi,\\ y &= c\sqrt{(\xi^2-1)(1-\eta^2)}\sin\phi, \quad z = c\xi\eta,\end{aligned}$$

where c is a positive constant. The (x,y,z)-space without the z-axis corresponds to

30.13.2 $\quad 1 < \xi < \infty, \quad -1 < \eta < 1, \quad 0 \le \phi < 2\pi.$

The coordinate surfaces $\xi = $ const. are prolate ellipsoids of revolution with foci at $x=y=0$, $z=\pm c$. The coordinate surfaces $\eta = $ const. are sheets of two-sheeted hyperboloids of revolution with the same foci. The focal line is given by $\xi = 1$, $-1 \le \eta \le 1$, and the rays $\pm z \ge c$, $x=y=0$ are given by $\eta = \pm 1$, $\xi \ge 1$.

30.13(ii) Metric Coefficients

30.13.3 $\quad h_\xi^2 = \left(\dfrac{\partial x}{\partial \xi}\right)^2 + \left(\dfrac{\partial y}{\partial \xi}\right)^2 + \left(\dfrac{\partial z}{\partial \xi}\right)^2 = \dfrac{c^2(\xi^2-\eta^2)}{\xi^2-1},$

30.13.4 $\quad h_\eta^2 = \left(\dfrac{\partial x}{\partial \eta}\right)^2 + \left(\dfrac{\partial y}{\partial \eta}\right)^2 + \left(\dfrac{\partial z}{\partial \eta}\right)^2 = \dfrac{c^2(\xi^2-\eta^2)}{1-\eta^2},$

30.13.5 $\quad h_\phi^2 = \left(\dfrac{\partial x}{\partial \phi}\right)^2 + \left(\dfrac{\partial y}{\partial \phi}\right)^2 + \left(\dfrac{\partial z}{\partial \phi}\right)^2$
$= c^2(\xi^2-1)(1-\eta^2).$

30.13(iii) Laplacian

30.13.6
$$\nabla^2 = \frac{1}{h_\xi h_\eta h_\phi}\left(\frac{\partial}{\partial\xi}\left(\frac{h_\eta h_\phi}{h_\xi}\frac{\partial}{\partial\xi}\right) + \frac{\partial}{\partial\eta}\left(\frac{h_\xi h_\phi}{h_\eta}\frac{\partial}{\partial\eta}\right)\right.$$
$$\left. + \frac{\partial}{\partial\phi}\left(\frac{h_\xi h_\eta}{h_\phi}\frac{\partial}{\partial\phi}\right)\right)$$
$$= \frac{1}{c^2(\xi^2-\eta^2)}\left(\frac{\partial}{\partial\xi}\left((\xi^2-1)\frac{\partial}{\partial\xi}\right)\right.$$
$$\left. + \frac{\partial}{\partial\eta}\left((1-\eta^2)\frac{\partial}{\partial\eta}\right) + \frac{\xi^2-\eta^2}{(\xi^2-1)(1-\eta^2)}\frac{\partial^2}{\partial\phi^2}\right).$$

30.13(iv) Separation of Variables

The wave equation

30.13.7 $$\nabla^2 w + \kappa^2 w = 0,$$

transformed to prolate spheroidal coordinates (ξ, η, ϕ), admits solutions

30.13.8 $$w(\xi,\eta,\phi) = w_1(\xi)w_2(\eta)w_3(\phi),$$

where w_1, w_2, w_3 satisfy the differential equations

30.13.9
$$\frac{d}{d\xi}\left((1-\xi^2)\frac{dw_1}{d\xi}\right) + \left(\lambda + \gamma^2(1-\xi^2) - \frac{\mu^2}{1-\xi^2}\right)w_1 = 0,$$

30.13.10
$$\frac{d}{d\eta}\left((1-\eta^2)\frac{dw_2}{d\eta}\right) + \left(\lambda + \gamma^2(1-\eta^2) - \frac{\mu^2}{1-\eta^2}\right)w_2 = 0,$$

30.13.11 $$\frac{d^2 w_3}{d\phi^2} + \mu^2 w_3 = 0,$$

with $\gamma^2 = \kappa^2 c^2 \geq 0$ and separation constants λ and μ^2. Equations (30.13.9) and (30.13.10) agree with (30.2.1).

In most applications the solution w has to be a single-valued function of (x, y, z), which requires $\mu = m$ (a nonnegative integer) and

30.13.12 $$w_3(\phi) = a_3 \cos(m\phi) + b_3 \sin(m\phi).$$

Moreover, w has to be bounded along the z-axis away from the focal line: this requires $w_2(\eta)$ to be bounded when $-1 < \eta < 1$. Then $\lambda = \lambda_n^m(\gamma^2)$ for some $n = m, m+1, m+2,\ldots$, and the general solution of (30.13.10) is

30.13.13 $$w_2(\eta) = a_2 \, \mathsf{Ps}_n^m(\eta,\gamma^2) + b_2 \, \mathsf{Qs}_n^m(\eta,\gamma^2).$$

The solution of (30.13.9) with $\mu = m$ is

30.13.14 $$w_1(\xi) = a_1 \, S_n^{m(1)}(\xi,\gamma) + b_1 \, S_n^{m(2)}(\xi,\gamma).$$

If $b_1 = b_2 = 0$, then the function (30.13.8) is a twice-continuously differentiable solution of (30.13.7) in the entire (x,y,z)-space. If $b_2 = 0$, then this property holds outside the focal line.

30.13(v) The Interior Dirichlet Problem for Prolate Ellipsoids

Equation (30.13.7) for $\xi \leq \xi_0$, and subject to the boundary condition $w = 0$ on the ellipsoid given by $\xi = \xi_0$, poses an eigenvalue problem with κ^2 as spectral parameter. The eigenvalues are given by $c^2\kappa^2 = \gamma^2$, where γ is determined from the condition

30.13.15 $$S_n^{m(1)}(\xi_0, \gamma) = 0.$$

The corresponding eigenfunctions are given by (30.13.8), (30.13.14), (30.13.13), (30.13.12), with $b_1 = b_2 = 0$. For the Dirichlet boundary-value problem of the region $\xi_1 \leq \xi \leq \xi_2$ between two ellipsoids, the eigenvalues are determined from

30.13.16 $$w_1(\xi_1) = w_1(\xi_2) = 0,$$

with w_1 as in (30.13.14). The corresponding eigenfunctions are given as before with $b_2 = 0$.

For further applications see Meixner and Schäfke (1954), Meixner et al. (1980) and the references cited therein; also Ong (1986), Müller et al. (1994), and Xiao et al. (2001).

30.14 Wave Equation in Oblate Spheroidal Coordinates

30.14(i) Oblate Spheroidal Coordinates

Oblate spheroidal coordinates ξ, η, ϕ are related to Cartesian coordinates x, y, z by

30.14.1
$$x = c\sqrt{(\xi^2+1)(1-\eta^2)}\cos\phi,$$
$$y = c\sqrt{(\xi^2+1)(1-\eta^2)}\sin\phi, \quad z = c\xi\eta,$$

where c is a positive constant. The (x, y, z)-space without the z-axis and the disk $z = 0$, $x^2 + y^2 \leq c^2$ corresponds to

30.14.2 $$0 < \xi < \infty, \quad -1 < \eta < 1, \quad 0 \leq \phi < 2\pi.$$

The coordinate surfaces $\xi = $ const. are oblate ellipsoids of revolution with focal circle $z = 0$, $x^2 + y^2 = c^2$. The coordinate surfaces $\eta = $ const. are halves of one-sheeted hyperboloids of revolution with the same focal circle. The disk $z = 0$, $x^2 + y^2 \leq c^2$ is given by $\xi = 0$, $-1 \leq \eta \leq 1$, and the rays $\pm z \geq 0$, $x = y = 0$ are given by $\eta = \pm 1$, $\xi \geq 0$.

30.14(ii) Metric Coefficients

30.14.3 $$h_\xi^2 = \frac{c^2(\xi^2+\eta^2)}{1+\xi^2},$$

30.14.4 $$h_\eta^2 = \frac{c^2(\xi^2+\eta^2)}{1-\eta^2},$$

30.14.5 $$h_\phi^2 = c^2(\xi^2+1)(1-\eta^2).$$

30.14(iii) Laplacian

30.14.6
$$\nabla^2 = \frac{1}{c^2(\xi^2+\eta^2)} \left(\frac{\partial}{\partial \xi}\left((\xi^2+1)\frac{\partial}{\partial \xi}\right) \right.$$
$$\left. + \frac{\partial}{\partial \eta}\left((1-\eta^2)\frac{\partial}{\partial \eta}\right) + \frac{\xi^2+\eta^2}{(\xi^2+1)(1-\eta^2)}\frac{\partial^2}{\partial \phi^2} \right).$$

30.14(iv) Separation of Variables

The wave equation (30.13.7), transformed to oblate spheroidal coordinates (ξ, η, ϕ), admits solutions of the form (30.13.8), where w_1 satisfies the differential equation

30.14.7
$$\frac{d}{d\xi}\left((1+\xi^2)\frac{dw_1}{d\xi}\right) - \left(\lambda + \gamma^2(1+\xi^2) - \frac{\mu^2}{1+\xi^2}\right)w_1 = 0,$$

and w_2, w_3 satisfy (30.13.10) and (30.13.11), respectively, with $\gamma^2 = -\kappa^2 c^2 \leq 0$ and separation constants λ and μ^2. Equation (30.14.7) can be transformed to equation (30.2.1) by the substitution $z = \pm i\xi$.

In most applications the solution w has to be a single-valued function of (x, y, z), which requires $\mu = m$ (a nonnegative integer). Moreover, the solution w has to be bounded along the z-axis: this requires $w_2(\eta)$ to be bounded when $-1 < \eta < 1$. Then $\lambda = \lambda_n^m(\gamma^2)$ for some $n = m, m+1, m+2, \ldots$, and the solution of (30.13.10) is given by (30.13.13). The solution of (30.14.7) is given by

30.14.8 $\quad w_1(\xi) = a_1 S_n^{m(1)}(i\xi, \gamma) + b_1 S_n^{m(2)}(i\xi, \gamma).$

If $b_1 = b_2 = 0$, then the function (30.13.8) is a twice-continuously differentiable solution of (30.13.7) in the entire (x, y, z)-space. If $b_2 = 0$, then this property holds outside the focal disk.

30.14(v) The Interior Dirichlet Problem for Oblate Ellipsoids

Equation (30.13.7) for $\xi \leq \xi_0$ together with the boundary condition $w = 0$ on the ellipsoid given by $\xi = \xi_0$, poses an eigenvalue problem with κ^2 as spectral parameter. The eigenvalues are given by $c^2\kappa^2 = -\gamma^2$, where γ^2 is determined from the condition

30.14.9 $\quad S_n^{m(1)}(i\xi_0, \gamma) = 0.$

The corresponding eigenfunctions are then given by (30.13.8), (30.14.8), (30.13.13), (30.13.12), with $b_1 = b_2 = 0$.

For further applications see Meixner and Schäfke (1954), Meixner et al. (1980) and the references cited therein; also Kokkorakis and Roumeliotis (1998) and Li et al. (1998).

30.15 Signal Analysis

30.15(i) Scaled Spheroidal Wave Functions

Let τ (> 0) and σ (> 0) be given. Set $\gamma = \tau\sigma$ and define

30.15.1
$$\phi_n(t) = \sqrt{\frac{2n+1}{2\tau}}\sqrt{\Lambda_n}\,\mathsf{Ps}_n^0\left(\frac{t}{\tau}, \gamma^2\right), \quad n = 0, 1, 2, \ldots,$$

30.15.2 $\quad \Lambda_n = \frac{2\gamma}{\pi}\left(K_n^0(\gamma)A_n^0(\gamma^2)\right)^2;$

see §30.11(v).

30.15(ii) Integral Equation

30.15.3 $\quad \int_{-\tau}^{\tau} \frac{\sin\sigma(t-s)}{\pi(t-s)}\phi_n(s)\,ds = \Lambda_n\phi_n(t).$

30.15(iii) Fourier Transform

30.15.4
$$\int_{-\infty}^{\infty} e^{-it\omega}\phi_n(t)\,dt = (-i)^n\sqrt{\frac{2\pi\tau}{\sigma\Lambda_n}}\,\phi_n\left(\frac{\tau}{\sigma}\omega\right)\chi_\sigma(\omega),$$

30.15.5 $\quad \int_{-\tau}^{\tau} e^{-it\omega}\phi_n(t)\,dt = (-i)^n\sqrt{\frac{2\pi\tau\Lambda_n}{\sigma}}\,\phi_n\left(\frac{\tau}{\sigma}\omega\right),$

where

30.15.6 $\quad \chi_\sigma(\omega) = \begin{cases} 1, & |\omega| \leq \sigma, \\ 0, & |\omega| > \sigma. \end{cases}$

Equations (30.15.4) and (30.15.6) show that the functions ϕ_n are σ-*bandlimited*, that is, their Fourier transform vanishes outside the interval $[-\sigma, \sigma]$.

30.15(iv) Orthogonality

30.15.7 $\quad \int_{-\tau}^{\tau} \phi_k(t)\phi_n(t)\,dt = \Lambda_n\delta_{k,n},$

30.15.8 $\quad \int_{-\infty}^{\infty} \phi_k(t)\phi_n(t)\,dt = \delta_{k,n}.$

The sequence ϕ_n, $n = 0, 1, 2, \ldots$ forms an orthonormal basis in the space of σ-bandlimited functions, and, after normalization, an orthonormal basis in $L^2(-\tau, \tau)$.

30.15(v) Extremal Properties

The maximum (or least upper bound) B of all numbers

$$30.15.9 \quad \beta = \frac{1}{2\pi} \int_{-\sigma}^{\sigma} \left| \int_{-\infty}^{\infty} e^{-it\omega} f(t)\, dt \right|^2 d\omega$$

taken over all $f \in L^2(-\infty, \infty)$ subject to

$$30.15.10 \quad \int_{-\infty}^{\infty} |f(t)|^2\, dt = 1, \quad \int_{-\tau}^{\tau} |f(t)|^2\, dt = \alpha,$$

for (fixed) $\Lambda_0 < \alpha \leq 1$, is given by

$$30.15.11 \quad \arccos \sqrt{B} + \arccos \sqrt{\alpha} = \arccos \sqrt{\Lambda_0},$$

or equivalently,

$$30.15.12 \quad B = \left(\sqrt{\Lambda_0 \alpha} + \sqrt{1 - \Lambda_0}\sqrt{1-\alpha} \right)^2.$$

The corresponding function f is given by

$$30.15.13 \quad \begin{aligned} f(t) &= a\phi_0(t)\chi_\tau(t) + b\phi_0(t)(1 - \chi_\tau(t)), \\ a &= \sqrt{\frac{\alpha}{\Lambda_0}}, \quad b = \sqrt{\frac{1-\alpha}{1-\Lambda_0}}. \end{aligned}$$

If $0 < \alpha \leq \Lambda_0$, then $B = 1$.

For further information see Frieden (1971), Lyman and Edmonson (2001), Papoulis (1977, Chapter 6), Slepian (1983), and Slepian and Pollak (1961).

Computation

30.16 Methods of Computation

30.16(i) Eigenvalues

For small $|\gamma^2|$ we can use the power-series expansion (30.3.8). Schäfke and Groh (1962) gives corresponding error bounds. If $|\gamma^2|$ is large we can use the asymptotic expansions in §30.9. Approximations to eigenvalues can be improved by using the continued-fraction equations from §30.3(iii) and §30.8; see Bouwkamp (1947) and Meixner and Schäfke (1954, §3.93).

Another method is as follows. Let $n - m$ be even. For d sufficiently large, construct the $d \times d$ tridiagonal matrix $\mathbf{A} = [A_{j,k}]$ with nonzero elements

$$30.16.1 \quad \begin{aligned} A_{j,j} &= (m + 2j - 2)(m + 2j - 1) \\ &\quad - 2\gamma^2 \frac{(m + 2j - 2)(m + 2j - 1) - 1 + m^2}{(2m + 4j - 5)(2m + 4j - 1)}, \\ A_{j,j+1} &= -\gamma^2 \frac{(2m + 2j - 1)(2m + 2j)}{(2m + 4j - 1)(2m + 4j + 1)}, \\ A_{j,j-1} &= -\gamma^2 \frac{(2j - 3)(2j - 2)}{(2m + 4j - 7)(2m + 4j - 5)}, \end{aligned}$$

and real eigenvalues $\alpha_{1,d}, \alpha_{2,d}, \ldots, \alpha_{d,d}$, arranged in ascending order of magnitude. Then

$$30.16.2 \quad \alpha_{j,d+1} \leq \alpha_{j,d},$$

and

$$30.16.3 \quad \lambda_n^m(\gamma^2) = \lim_{d \to \infty} \alpha_{p,d}, \quad p = \left\lfloor \tfrac{1}{2}(n-m) \right\rfloor + 1.$$

The eigenvalues of $\mathbf{A}$ can be computed by methods indicated in §§3.2(vi), 3.2(vii). The error satisfies

$$30.16.4 \quad \begin{aligned} &\alpha_{p,d} - \lambda_n^m(\gamma^2) \\ &= O\left(\frac{\gamma^{4d}}{4^{2d+1}((m + 2d - 1)!(m + 2d + 1)!)^2} \right), \\ &\quad d \to \infty. \end{aligned}$$

Example

For $m = 2$, $n = 4$, $\gamma^2 = 10$,

$$30.16.5 \quad \begin{aligned} \alpha_{2,2} &= 14.18833\,246, \quad \alpha_{2,3} = 13.98002\,013, \\ \alpha_{2,4} &= 13.97907\,459, \quad \alpha_{2,5} = 13.97907\,345, \\ \alpha_{2,6} &= 13.97907\,345, \end{aligned}$$

which yields $\lambda_4^2(10) = 13.97907\,345$. If $n - m$ is odd, then (30.16.1) is replaced by

$$30.16.6 \quad \begin{aligned} A_{j,j} &= (m + 2j - 1)(m + 2j) \\ &\quad - 2\gamma^2 \frac{(m + 2j - 1)(m + 2j) - 1 + m^2}{(2m + 4j - 3)(2m + 4j + 1)}, \\ A_{j,j+1} &= -\gamma^2 \frac{(2m + 2j)(2m + 2j + 1)}{(2m + 4j + 1)(2m + 4j + 3)}, \\ A_{j,j-1} &= -\gamma^2 \frac{(2j - 2)(2j - 1)}{(2m + 4j - 5)(2m + 4j - 3)}. \end{aligned}$$

30.16(ii) Spheroidal Wave Functions of the First Kind

If $|\gamma^2|$ is large, then we can use the asymptotic expansions referred to in §30.9 to approximate $\mathsf{Ps}_n^m(x, \gamma^2)$.

If $\lambda_n^m(\gamma^2)$ is known, then we can compute $\mathsf{Ps}_n^m(x, \gamma^2)$ (not normalized) by solving the differential equation (30.2.1) numerically with initial conditions $w(0) = 1$, $w'(0) = 0$ if $n - m$ is even, or $w(0) = 0$, $w'(0) = 1$ if $n - m$ is odd.

If $\lambda_n^m(\gamma^2)$ is known, then $\mathsf{Ps}_n^m(x, \gamma^2)$ can be found by summing (30.8.1). The coefficients $a_{n,r}^m(\gamma^2)$ are computed as the recessive solution of (30.8.4) (§3.6), and normalized via (30.8.5).

A fourth method, based on the expansion (30.8.1), is as follows. Let $\mathbf{A}$ be the $d \times d$ matrix given by (30.16.1) if $n - m$ is even, or by (30.16.6) if $n - m$ is odd. Form the eigenvector $[e_{1,d}, e_{2,d}, \ldots, e_{d,d}]^{\mathrm{T}}$ of $\mathbf{A}$ associated with the eigenvalue $\alpha_{p,d}$, $p = \left\lfloor \tfrac{1}{2}(n-m) \right\rfloor + 1$, normalized according to

$$30.16.7 \quad \begin{aligned} &\sum_{j=1}^{d} e_{j,d}^2 \frac{(n + m + 2j - 2p)!}{(n - m + 2j - 2p)!} \frac{1}{2n + 4j - 4p + 1} \\ &= \frac{(n+m)!}{(n-m)!} \frac{1}{2n + 1}. \end{aligned}$$

Then

30.16.8 $$a_{n,k}^m(\gamma^2) = \lim_{d\to\infty} e_{k+p,d},$$

30.16.9 $$\mathsf{Ps}_n^m(x,\gamma^2) = \lim_{d\to\infty} \sum_{j=1}^d (-1)^{j-p} e_{j,d}\, \mathsf{P}_{n+2(j-p)}^m(x).$$

For error estimates see Volkmer (2004a).

30.16(iii) Radial Spheroidal Wave Functions

The coefficients $a_{n,k}^m(\gamma^2)$ calculated in §30.16(ii) can be used to compute $S_n^{m(j)}(z,\gamma)$, $j = 1, 2, 3, 4$ from (30.11.3) as well as the connection coefficients $K_n^m(\gamma)$ from (30.11.10) and (30.11.11).

For another method see Van Buren and Boisvert (2002).

30.17 Tables

- Stratton et al. (1956) tabulates quantities closely related to $\lambda_n^m(\gamma^2)$ and $a_{n,k}^m(\gamma^2)$ for $0 \le m \le 8$, $m \le n \le 8$, $-64 \le \gamma^2 \le 64$. Precision is 7S.

- Flammer (1957) includes 18 tables of eigenvalues, expansion coefficients, spheroidal wave functions, and other related quantities. Precision varies between 4S and 10S.

- Hanish et al. (1970) gives $\lambda_n^m(\gamma^2)$ and $S_n^{m(j)}(z,\gamma)$, $j = 1, 2$, and their first derivatives, for $0 \le m \le 2$, $m \le n \le m+49$, $-1600 \le \gamma^2 \le 1600$. The range of z is given by $1 \le z \le 10$ if $\gamma^2 > 0$, or $z = -i\xi$, $0 \le \xi \le 2$ if $\gamma^2 < 0$. Precision is 18S.

- EraŠevskaja et al. (1973, 1976) gives $S^{m(j)}(iy,-ic)$, $S^{m(j)}(z,\gamma)$ and their first derivatives for $j = 1, 2$, $0.5 \le c \le 8$, $y = 0, 0.5, 1, 1.5$, $0.5 \le \gamma \le 8$, $z = 1.01, 1.1, 1.4, 1.8$. Precision is 15S.

- Van Buren et al. (1975) gives $\lambda_n^0(\gamma^2)$, $\mathsf{Ps}_n^0(x,\gamma^2)$ for $0 \le n \le 49$, $-1600 \le \gamma^2 \le 1600$, $-1 \le x \le 1$. Precision is 8S.

- Zhang and Jin (1996) includes 24 tables of eigenvalues, spheroidal wave functions and their derivatives. Precision varies between 6S and 8S.

Fletcher et al. (1962, §22.28) provides additional information on tables prior to 1961.

30.18 Software

See http://dlmf.nist.gov/30.18.

References

General References

The main references used in writing this chapter are Arscott (1964b), Erdélyi et al. (1955), Meixner and Schäfke (1954), and Meixner et al. (1980). For additional bibliographic reading see Flammer (1957), Komarov et al. (1976), and Stratton et al. (1956).

Sources

The following list gives the references or other indications of proofs that were used in constructing the various sections of this chapter. These sources supplement the references that are quoted in the text.

§30.2 Meixner and Schäfke (1954, §3.1), Arscott (1964b, §8.1).

§30.3 Meixner and Schäfke (1954, §§3.2, 3.531).

§30.4 Meixner and Schäfke (1954, §3.2), Arscott (1964b, §8.2).

§§30.5, 30.6 Meixner and Schäfke (1954, §3.6).

§30.7 These graphics were produced at NIST with the aid of Maple procedures provided by the author.

§30.8 Arscott (1964b, §§8.2, 8.5), Meixner and Schäfke (1954, §§3.542, 3.62).

§30.9 Meixner and Schäfke (1954, §§3.251, 3.252), Müller (1962, 1963).

§30.11 Arscott (1964b, §8.5), Meixner and Schäfke (1954, §§3.64–3.66, 3.84), Erdélyi et al. (1955, §16.11). Figure 30.11.1 was produced at NIST with the aid of Maple procedures provided by the author.

§30.13 Erdélyi et al. (1955, §16.1.2), Meixner and Schäfke (1954, §§1.123, 1.133, Chapter 4).

§30.14 Erdélyi et al. (1955, §16.1.3), Meixner and Schäfke (1954, §§1.124, 1.134, Chapter 4).

§30.15 Frieden (1971, pp. 321–324, §2.10), Meixner et al. (1980, p. 114), Papoulis (1977, pp. 205–210), Slepian (1983).

§30.16 Meixner and Schäfke (1954, §3.93), Volkmer (2004a), Van Buren et al. (1972).

Chapter 31

Heun Functions

B. D. Sleeman[1] and V. B. Kuznetsov[2]

Notation   **710**
- 31.1 Special Notation 710

Properties   **710**
- 31.2 Differential Equations 710
- 31.3 Basic Solutions 711
- 31.4 Solutions Analytic at Two Singularities: Heun Functions 712
- 31.5 Solutions Analytic at Three Singularities: Heun Polynomials 712
- 31.6 Path-Multiplicative Solutions 712
- 31.7 Relations to Other Functions 713
- 31.8 Solutions via Quadratures 713
- 31.9 Orthogonality 714
- 31.10 Integral Equations and Representations . 714
- 31.11 Expansions in Series of Hypergeometric Functions 716
- 31.12 Confluent Forms of Heun's Equation . . . 717
- 31.13 Asymptotic Approximations 718
- 31.14 General Fuchsian Equation 718
- 31.15 Stieltjes Polynomials 718

Applications   **719**
- 31.16 Mathematical Applications 719
- 31.17 Physical Applications 720

Computation   **720**
- 31.18 Methods of Computation 720

References   **721**

---

[1]Department of Applied Mathematics, University of Leeds, Leeds, United Kingdom.
[2]Department of Applied Mathematics, University of Leeds, Leeds, United Kingdom.
Copyright © 2009 National Institute of Standards and Technology. All rights reserved.

Notation

31.1 Special Notation

(For other notation see pp. xiv and 873.)

x, y	real variables.
z, ζ, w, W	complex variables.
j, k, ℓ, m, n	nonnegative integers.
a	complex parameter, $\|a\| \geq 1, a \neq 1$.
$q, \alpha, \beta, \gamma, \delta, \epsilon, \nu$	complex parameters.

The main functions treated in this chapter are $H\ell(a,q;\alpha,\beta,\gamma,\delta;z)$, $(s_1,s_2)Hf_m(a,q_m;\alpha,\beta,\gamma,\delta;z)$, $(s_1,s_2)Hf_m^\nu(a,q_m;\alpha,\beta,\gamma,\delta;z)$, and the polynomial $Hp_{n,m}(a,q_{n,m};-n,\beta,\gamma,\delta;z)$. These notations were introduced by Arscott in Ronveaux (1995, pp. 34–44). Sometimes the parameters are suppressed.

Properties

31.2 Differential Equations

31.2(i) Heun's Equation

31.2.1
$$\frac{d^2w}{dz^2} + \left(\frac{\gamma}{z} + \frac{\delta}{z-1} + \frac{\epsilon}{z-a}\right)\frac{dw}{dz} + \frac{\alpha\beta z - q}{z(z-1)(z-a)}w = 0, \qquad \alpha + \beta + 1 = \gamma + \delta + \epsilon.$$

This equation has regular singularities at $0, 1, a, \infty$, with corresponding exponents $\{0, 1-\gamma\}$, $\{0, 1-\delta\}$, $\{0, 1-\epsilon\}$, $\{\alpha, \beta\}$, respectively (§2.7(i)). All other homogeneous linear differential equations of the second order having four regular singularities in the extended complex plane, $\mathbb{C} \cup \{\infty\}$, can be transformed into (31.2.1).

The parameters play different roles: a is the *singularity parameter*; $\alpha, \beta, \gamma, \delta, \epsilon$ are *exponent parameters*; q is the *accessory parameter*. The total number of free parameters is six.

31.2(ii) Normal Form of Heun's Equation

31.2.2 $\quad w(z) = z^{-\gamma/2}(z-1)^{-\delta/2}(z-a)^{-\epsilon/2}W(z),$

31.2.3
$$\frac{d^2W}{dz^2} = \left(\frac{A}{z} + \frac{B}{z-1} + \frac{C}{z-a} + \frac{D}{z^2} + \frac{E}{(z-1)^2} + \frac{F}{(z-a)^2}\right)W,$$
$$A + B + C = 0,$$

31.2.4
$$A = -\frac{\gamma\delta}{2} - \frac{\gamma\epsilon}{2a} + \frac{q}{a}, \quad B = \frac{\gamma\delta}{2} - \frac{\delta\epsilon}{2(a-1)} - \frac{q-\alpha\beta}{a-1},$$
$$C = \frac{\gamma\epsilon}{2a} + \frac{\delta\epsilon}{2(a-1)} - \frac{a\alpha\beta - q}{a(a-1)}, \quad D = \tfrac{1}{2}\gamma\left(\tfrac{1}{2}\gamma - 1\right),$$
$$E = \tfrac{1}{2}\delta\left(\tfrac{1}{2}\delta - 1\right), \quad F = \tfrac{1}{2}\epsilon\left(\tfrac{1}{2}\epsilon - 1\right).$$

31.2(iii) Trigonometric Form

31.2.5 $\qquad z = \sin^2\theta,$

31.2.6
$$\frac{d^2w}{d\theta^2} + \left((2\gamma-1)\cot\theta - (2\delta-1)\tan\theta - \frac{\epsilon\sin(2\theta)}{a-\sin^2\theta}\right)\frac{dw}{d\theta} + 4\frac{\alpha\beta\sin^2\theta - q}{a-\sin^2\theta}w = 0.$$

31.2(iv) Doubly-Periodic Forms

Jacobi's Elliptic Form

With the notation of §22.2 let

31.2.7 $\qquad a = k^{-2}, \quad z = \operatorname{sn}^2(\zeta, k).$

Then (suppressing the parameter k)

31.2.8
$$\frac{d^2w}{d\zeta^2} + \left((2\gamma-1)\frac{\operatorname{cn}\zeta\operatorname{dn}\zeta}{\operatorname{sn}\zeta} - (2\delta-1)\frac{\operatorname{sn}\zeta\operatorname{dn}\zeta}{\operatorname{cn}\zeta} - (2\epsilon-1)k^2\frac{\operatorname{sn}\zeta\operatorname{cn}\zeta}{\operatorname{dn}\zeta}\right)\frac{dw}{d\zeta} + 4k^2(\alpha\beta\operatorname{sn}^2\zeta - q)w = 0.$$

Weierstrass's Form

With the notation of §§19.2(ii) and 23.2 let

31.2.9
$$k^2 = (e_2 - e_3)/(e_1 - e_3),$$
$$\zeta = iK' + \xi(e_1 - e_3)^{1/2}, \quad e_1 = \wp(\omega_1),$$
$$e_2 = \wp(\omega_2), \quad e_3 = \wp(\omega_3), \quad e_1 + e_2 + e_3 = 0,$$

where $2\omega_1$ and $2\omega_3$ with $\Im(\omega_3/\omega_1) > 0$ are generators of the lattice $\mathbb{L}$ for $\wp(z|\mathbb{L})$. Then

31.2.10
$$w(\xi) = (\wp(\xi) - e_3)^{(1-2\gamma)/4}(\wp(\xi) - e_2)^{(1-2\delta)/4} \times (\wp(\xi) - e_1)^{(1-2\epsilon)/4}W(\xi),$$

where $W(\xi)$ satisfies

31.2.11
$$d^2W/d\xi^2 + (H + b_0\,\wp(\xi) + b_1\,\wp(\xi+\omega_1) + b_2\,\wp(\xi+\omega_2) + b_3\,\wp(\xi+\omega_3))W = 0,$$

with

31.2.12
$$b_0 = 4\alpha\beta - (\gamma+\delta+\epsilon-\tfrac{1}{2})(\gamma+\delta+\epsilon-\tfrac{3}{2}),$$
$$b_1 = -(\epsilon-\tfrac{1}{2})(\epsilon-\tfrac{3}{2}), \quad b_2 = -(\delta-\tfrac{1}{2})(\delta-\tfrac{3}{2}),$$
$$b_3 = -(\gamma-\tfrac{1}{2})(\gamma-\tfrac{3}{2}),$$
$$H = e_1(\gamma+\delta-1)^2 + e_2(\gamma+\epsilon-1)^2 + e_3(\delta+\epsilon-1)^2 - 4\alpha\beta e_3 - 4q(e_2 - e_3).$$

31.2(v) Heun's Equation Automorphisms

F-Homotopic Transformations

$w(z) = z^{1-\gamma}w_1(z)$ satisfies (31.2.1) if w_1 is a solution of (31.2.1) with transformed parameters $q_1 = q + (a\delta + \epsilon)(1-\gamma)$; $\alpha_1 = \alpha + 1 - \gamma$, $\beta_1 = \beta + 1 - \gamma$, $\gamma_1 = 2 - \gamma$. Next, $w(z) = (z-1)^{1-\delta}w_2(z)$ satisfies (31.2.1) if w_2 is a solution of (31.2.1) with transformed parameters $q_2 = q + a\gamma(1-\delta)$; $\alpha_2 = \alpha + 1 - \delta$, $\beta_2 = \beta + 1 - \delta$, $\delta_2 = 2 - \delta$. Lastly, $w(z) = (z-a)^{1-\epsilon}w_3(z)$ satisfies (31.2.1) if w_3 is a solution of (31.2.1) with transformed parameters $q_3 = q + \gamma(1-\epsilon)$; $\alpha_3 = \alpha + 1 - \epsilon$, $\beta_3 = \beta + 1 - \epsilon$, $\epsilon_3 = 2 - \epsilon$. By composing these three steps, there result $2^3 = 8$ possible transformations of the dependent variable (including the identity transformation) that preserve the form of (31.2.1).

Homographic Transformations

There are $4! = 24$ homographies $\tilde{z}(z) = (Az+B)/(Cz+D)$ that take $0, 1, a, \infty$ to some permutation of $0, 1, a', \infty$, where a' may differ from a. If $\tilde{z} = \tilde{z}(z)$ is one of the $3! = 6$ homographies that map ∞ to ∞, then $w(z) = \tilde{w}(\tilde{z})$ satisfies (31.2.1) if $\tilde{w}(\tilde{z})$ is a solution of (31.2.1) with z replaced by $\tilde{z}$ and appropriately transformed parameters. For example, if $\tilde{z} = z/a$, then the parameters are $\tilde{a} = 1/a$, $\tilde{q} = q/a$; $\tilde{\delta} = \epsilon$, $\tilde{\epsilon} = \delta$. If $\tilde{z} = \tilde{z}(z)$ is one of the $4! - 3! = 18$ homographies that do not map ∞ to ∞, then an appropriate prefactor must be included on the right-hand side. For example, $w(z) = (1-z)^{-\alpha}\tilde{w}(z/(z-1))$, which arises from $\tilde{z} = z/(z-1)$, satisfies (31.2.1) if $\tilde{w}(\tilde{z})$ is a solution of (31.2.1) with z replaced by $\tilde{z}$ and transformed parameters $\tilde{a} = a/(a-1)$, $\tilde{q} = -(q - a\alpha\gamma)/(a-1)$; $\tilde{\beta} = \alpha + 1 - \delta$,

$\tilde{\delta} = \alpha + 1 - \beta$.

Composite Transformations

There are $8 \cdot 24 = 192$ automorphisms of equation (31.2.1) by compositions of F-homotopic and homographic transformations. Each is a substitution of dependent and/or independent variables that preserves the form of (31.2.1). Except for the identity automorphism, each alters the parameters.

31.3 Basic Solutions

31.3(i) Fuchs–Frobenius Solutions at $z = 0$

$H\ell(a, q; \alpha, \beta, \gamma, \delta; z)$ denotes the solution of (31.2.1) that corresponds to the exponent 0 at $z = 0$ and assumes the value 1 there. If the other exponent is not a positive integer, that is, if $\gamma \neq 0, -1, -2, \ldots$, then from §2.7(i) it follows that $H\ell(a, q; \alpha, \beta, \gamma, \delta; z)$ exists, is analytic in the disk $|z| < 1$, and has the Maclaurin expansion

31.3.1
$$H\ell(a, q; \alpha, \beta, \gamma, \delta; z) = \sum_{j=0}^{\infty} c_j z^j, \qquad |z| < 1,$$

where $c_0 = 1$,

31.3.2 $$a\gamma c_1 - qc_0 = 0,$$

31.3.3 $$R_j c_{j+1} - (Q_j + q)c_j + P_j c_{j-1} = 0, \qquad j \geq 1,$$

with

31.3.4
$$P_j = (j - 1 + \alpha)(j - 1 + \beta),$$
$$Q_j = j\left((j - 1 + \gamma)(1 + a) + a\delta + \epsilon\right),$$
$$R_j = a(j + 1)(j + \gamma).$$

Similarly, if $\gamma \neq 1, 2, 3, \ldots$, then the solution of (31.2.1) that corresponds to the exponent $1 - \gamma$ at $z = 0$ is

31.3.5 $$z^{1-\gamma} H\ell(a, (a\delta + \epsilon)(1-\gamma) + q; \alpha + 1 - \gamma, \beta + 1 - \gamma, 2 - \gamma, \delta; z).$$

When $\gamma \in \mathbb{Z}$, linearly independent solutions can be constructed as in §2.7(i). In general, one of them has a logarithmic singularity at $z = 0$.

31.3(ii) Fuchs–Frobenius Solutions at Other Singularities

With similar restrictions to those given in §31.3(i), the following results apply. Solutions of (31.2.1) corresponding to the exponents 0 and $1 - \delta$ at $z = 1$ are respectively,

31.3.6 $$H\ell(1 - a, \alpha\beta - q; \alpha, \beta, \delta, \gamma; 1 - z),$$

31.3.7 $$(1-z)^{1-\delta} H\ell(1-a, ((1-a)\gamma + \epsilon)(1-\delta) + \alpha\beta - q; \alpha + 1 - \delta, \beta + 1 - \delta, 2 - \delta, \gamma; 1 - z).$$

Solutions of (31.2.1) corresponding to the exponents 0 and $1 - \epsilon$ at $z = a$ are respectively,

31.3.8 $$H\ell\left(\frac{a}{a-1}, \frac{\alpha\beta a}{a-1} - \frac{q}{a-1}; \alpha, \beta, \epsilon, \delta; \frac{a-z}{a-1}\right),$$

31.3.9 $$\left(\frac{a-z}{a-1}\right)^{1-\epsilon} H\ell\left(\frac{a}{a-1}, \frac{(a(\delta + \gamma) - \gamma)(1-\epsilon)}{a-1} + \frac{\alpha\beta a - q}{a-1}; \alpha + 1 - \epsilon, \beta + 1 - \epsilon, 2 - \epsilon, \delta; \frac{a-z}{a-1}\right).$$

Solutions of (31.2.1) corresponding to the exponents α and β at $z = \infty$ are respectively,

31.3.10 $$z^{-\alpha} \, H\ell\left(\frac{1}{a}, \alpha(\beta - \epsilon) + \frac{\alpha}{a}(\beta - \delta) - \frac{q}{a}; \alpha, \alpha - \gamma + 1, \alpha - \beta + 1, \delta; \frac{1}{z}\right),$$

31.3.11 $$z^{-\beta} \, H\ell\left(\frac{1}{a}, \beta(\alpha - \epsilon) + \frac{\beta}{a}(\alpha - \delta) - \frac{q}{a}; \beta, \beta - \gamma + 1, \beta - \alpha + 1, \delta; \frac{1}{z}\right).$$

31.3(iii) Equivalent Expressions

Solutions (31.3.1) and (31.3.5)–(31.3.11) comprise a set of 8 local solutions of (31.2.1): 2 per singular point. Each is related to the solution (31.3.1) by one of the automorphisms of §31.2(v). There are 192 automorphisms in all, so there are $192/8 = 24$ equivalent expressions for each of the 8. For example, $H\ell(a, q; \alpha, \beta, \gamma, \delta; z)$ is equal to

31.3.12 $$H\ell(1/a, q/a; \alpha, \beta, \gamma, \alpha + \beta + 1 - \gamma - \delta; z/a),$$

which arises from the homography $\tilde{z} = z/a$, and to

31.3.13 $$(1-z)^{-\alpha} \, H\ell\left(\frac{a}{a-1}, -\frac{q - a\alpha\gamma}{a-1}; \alpha, \alpha + 1 - \delta, \gamma, \alpha + 1 - \beta; \frac{z}{z-1}\right),$$

which arises from $\tilde{z} = z/(z-1)$, and also to 21 further expressions. The full set of 192 local solutions of (31.2.1), equivalent in 8 sets of 24, resembles Kummer's set of 24 local solutions of the hypergeometric equation, which are equivalent in 4 sets of 6 solutions (§15.10(ii)); see Maier (2007).

31.4 Solutions Analytic at Two Singularities: Heun Functions

For an infinite set of discrete values q_m, $m = 0, 1, 2, \ldots$, of the accessory parameter q, the function $H\ell(a, q; \alpha, \beta, \gamma, \delta; z)$ is analytic at $z = 1$, and hence also throughout the disk $|z| < a$. To emphasize this property this set of functions is denoted by

31.4.1 $$(0,1)Hf_m(a, q_m; \alpha, \beta, \gamma, \delta; z), \quad m = 0, 1, 2, \ldots.$$

The eigenvalues q_m satisfy the continued-fraction equation

31.4.2 $$q = \frac{a\gamma P_1}{Q_1 + q -} \frac{R_1 P_2}{Q_2 + q -} \frac{R_2 P_3}{Q_3 + q -} \cdots,$$

in which P_j, Q_j, R_j are as in §31.3(i).

More generally,

31.4.3 $$(s_1, s_2)Hf_m(a, q_m; \alpha, \beta, \gamma, \delta; z), \quad m = 0, 1, 2, \ldots,$$

with $(s_1, s_2) \in \{0, 1, a, \infty\}$, denotes a set of solutions of (31.2.1), each of which is analytic at s_1 and s_2. The set q_m depends on the choice of s_1 and s_2.

The solutions (31.4.3) are called the *Heun functions*. See Ronveaux (1995, pp. 39–41).

31.5 Solutions Analytic at Three Singularities: Heun Polynomials

Let $\alpha = -n$, $n = 0, 1, 2, \ldots$, and $q_{n,m}$, $m = 0, 1, \ldots, n$, be the eigenvalues of the tridiagonal matrix

31.5.1 $$\begin{bmatrix} 0 & a\gamma & 0 & \ldots & 0 \\ P_1 & -Q_1 & R_1 & \ldots & 0 \\ 0 & P_2 & -Q_2 & & \vdots \\ \vdots & \vdots & & \ddots & R_{n-1} \\ 0 & 0 & \ldots & P_n & -Q_n \end{bmatrix},$$

where P_j, Q_j, R_j are again defined as in §31.3(i). Then

31.5.2 $$Hp_{n,m}(a, q_{n,m}; -n, \beta, \gamma, \delta; z) = H\ell(a, q_{n,m}; -n, \beta, \gamma, \delta; z)$$

is a polynomial of degree n, and hence a solution of (31.2.1) that is analytic at all three finite singularities $0, 1, a$. These solutions are the *Heun polynomials*. Some properties are included as special cases of properties given in §31.15 below.

31.6 Path-Multiplicative Solutions

A further extension of the notation (31.4.1) and (31.4.3) is given by

31.6.1 $$(s_1, s_2)Hf_m^\nu(a, q_m; \alpha, \beta, \gamma, \delta; z), \quad m = 0, 1, 2, \ldots,$$

with $(s_1, s_2) \in \{0, 1, a\}$, but with another set of $\{q_m\}$. This denotes a set of solutions of (31.2.1) with the property that if we pass around a simple closed contour in the z-plane that encircles s_1 and s_2 once in the positive sense, but not the remaining finite singularity, then the solution is multiplied by a constant factor $e^{2\nu\pi i}$. These solutions are called *path-multiplicative*. See Schmidt (1979).

31.7 Relations to Other Functions

31.7(i) Reductions to the Gauss Hypergeometric Function

31.7.1
$$\begin{aligned}{}_2F_1(\alpha,\beta;\gamma;z) &= H\ell(1,\alpha\beta;\alpha,\beta,\gamma,\delta;z) \\ &= H\ell(0,0;\alpha,\beta,\gamma,\alpha+\beta+1-\gamma;z) \\ &= H\ell(a,a\alpha\beta;\alpha,\beta,\gamma,\alpha+\beta+1-\gamma;z).\end{aligned}$$

Other reductions of $H\ell$ to a ${}_2F_1$, with at least one free parameter, exist iff the pair (a,p) takes one of a finite number of values, where $q = \alpha\beta p$. Below are three such reductions with three and two parameters. They are analogous to quadratic and cubic hypergeometric transformations (§§15.8(iii)–15.8(v)).

31.7.2
$$\begin{aligned} H\ell(2,\alpha\beta;\alpha,\beta,\gamma,\alpha+\beta-2\gamma+1;z) \\ = {}_2F_1\bigl(\tfrac{1}{2}\alpha,\tfrac{1}{2}\beta;\gamma;1-(1-z)^2\bigr),\end{aligned}$$

31.7.3
$$\begin{aligned} H\ell\bigl(4,\alpha\beta;\alpha,\beta,\tfrac{1}{2},\tfrac{2}{3}(\alpha+\beta);z\bigr) \\ = {}_2F_1\bigl(\tfrac{1}{3}\alpha,\tfrac{1}{3}\beta;\tfrac{1}{2};1-(1-z)^2(1-\tfrac{1}{4}z)\bigr),\end{aligned}$$

31.7.4
$$\begin{aligned} H\ell\bigl(\tfrac{1}{2}+i\tfrac{\sqrt{3}}{2},\alpha\beta(\tfrac{1}{2}+i\tfrac{\sqrt{3}}{6});\alpha,\beta,\tfrac{1}{3}(\alpha+\beta+1),\tfrac{1}{3}(\alpha+\beta+1);z\bigr) \\ = {}_2F_1\bigl(\tfrac{1}{3}\alpha,\tfrac{1}{3}\beta;\tfrac{1}{3}(\alpha+\beta+1);1-\bigl(1-\bigl(\tfrac{3}{2}-i\tfrac{\sqrt{3}}{2}\bigr)z\bigr)^3\bigr). \end{aligned}$$

For additional reductions, see Maier (2005). Joyce (1994) gives a reduction in which the independent variable is transformed not polynomially or rationally, but algebraically.

31.7(ii) Relations to Lamé Functions

With $z = \operatorname{sn}^2(\zeta,k)$ and

31.7.5
$$\begin{aligned} a &= k^{-2}, & q &= -\tfrac{1}{4}ah, & \alpha &= -\tfrac{1}{2}\nu, \\ \beta &= \tfrac{1}{2}(\nu+1), & \gamma &= \delta = \epsilon = \tfrac{1}{2}, \end{aligned}$$

equation (31.2.1) becomes Lamé's equation with independent variable ζ; compare (29.2.1) and (31.2.8). The solutions (31.3.1) and (31.3.5) transform into even and odd solutions of Lamé's equation, respectively. Similar specializations of formulas in §31.3(ii) yield solutions in the neighborhoods of the singularities $\zeta = K$, $K + iK'$, and iK', where K and K' are related to k as in §19.2(ii).

31.8 Solutions via Quadratures

For half-odd-integer values of the exponent parameters:

31.8.1
$$\begin{aligned} \beta-\alpha &= m_0+\tfrac{1}{2}, & \gamma &= -m_1+\tfrac{1}{2}, & \delta &= -m_2+\tfrac{1}{2}, \\ \epsilon &= -m_3+\tfrac{1}{2}, & & m_0,m_1,m_2,m_3 &= 0,1,2,\ldots,\end{aligned}$$

the Hermite–Darboux method (see Whittaker and Watson (1927, pp. 570–572)) can be applied to construct solutions of (31.2.1) expressed in quadratures, as follows. Denote $\mathbf{m} = (m_0,m_1,m_2,m_3)$ and $\lambda = -4q$. Then

31.8.2
$$\begin{aligned} w_\pm(\mathbf{m};\lambda;z) &= \sqrt{\Psi_{g,N}(\lambda,z)} \\ &\quad \times \exp\left(\pm\frac{i\nu(\lambda)}{2}\int_{z_0}^{z}\frac{t^{m_1}(t-1)^{m_2}(t-a)^{m_3}\,dt}{\Psi_{g,N}(\lambda,t)\sqrt{t(t-1)(t-a)}}\right) \end{aligned}$$

are two independent solutions of (31.2.1). Here $\Psi_{g,N}(\lambda,z)$ is a polynomial of degree g in λ and of degree $N = m_0 + m_1 + m_2 + m_3$ in z, that is a solution of the third-order differential equation satisfied by a product of any two solutions of Heun's equation. The degree g is given by

31.8.3
$$g = \tfrac{1}{2}\max\left(2\max_{0\le k\le 3}m_k, 1+N - (1+(-1)^N)\left(\tfrac{1}{2}+\min_{0\le k\le 3}m_k\right)\right).$$

The variables λ and ν are two coordinates of the associated hyperelliptic (spectral) curve $\Gamma: \nu^2 = \prod_{j=1}^{2g+1}(\lambda-\lambda_j)$. (This ν is unrelated to the ν in §31.6.) Lastly, λ_j, $j = 1,2,\ldots,2g+1$, are the zeros of the Wronskian of $w_+(\mathbf{m};\lambda;z)$ and $w_-(\mathbf{m};\lambda;z)$.

By automorphisms from §31.2(v), similar solutions also exist for $m_0,m_1,m_2,m_3 \in \mathbb{Z}$, and $\Psi_{g,N}(\lambda,z)$ may become a rational function in z. For instance,

31.8.4
$$\begin{aligned} \Psi_{1,2} &= z^2+\lambda z+a, & \nu^2 &= (\lambda+a+1)(\lambda^2-4a), \\ & & \mathbf{m} &= (1,1,0,0), \end{aligned}$$

and

31.8.5
$$\begin{aligned} \Psi_{1,-1} &= \bigl(z^3+(\lambda+3a+3)z+a\bigr)/z^3, \\ \nu^2 &= (\lambda+4a+4)\bigl((\lambda+3a+3)^2-4a\bigr), \\ \mathbf{m} &= (1,-2,0,0).\end{aligned}$$

For $\mathbf{m} = (m_0,0,0,0)$, these solutions reduce to Hermite's solutions (Whittaker and Watson (1927, §23.7)) of the Lamé equation in its algebraic form. The curve Γ reflects the finite-gap property of Equation (31.2.1) when the exponent parameters satisfy (31.8.1) for $m_j \in \mathbb{Z}$. When $\lambda = -4q$ approaches the ends of the gaps, the solution (31.8.2) becomes the corresponding Heun polynomial. For more details see Smirnov (2002).

The solutions in this section are finite-term Liouvillean solutions which can be constructed via Kovacic's algorithm; see §31.14(ii).

31.9 Orthogonality

31.9(i) Single Orthogonality

With

31.9.1 $\quad w_m(z) = (0,1)Hf_m(a, q_m; \alpha, \beta, \gamma, \delta; z),$

we have

31.9.2 $\quad \int_\zeta^{(1+,0+,1-,0-)} t^{\gamma-1}(1-t)^{\delta-1}(t-a)^{\epsilon-1}$
$\times w_m(t) w_k(t)\, dt = \delta_{m,k} \theta_m.$

Here ζ is an arbitrary point in the interval $(0,1)$. The integration path begins at $z = \zeta$, encircles $z = 1$ once in the positive sense, followed by $z = 0$ once in the positive sense, and so on, returning finally to $z = \zeta$. The integration path is called a *Pochhammer double-loop contour* (compare Figure 5.12.3). The branches of the many-valued functions are continuous on the path, and assume their principal values at the beginning.

The normalization constant θ_m is given by

31.9.3 $\quad \theta_m = (1 - e^{2\pi i \gamma})(1 - e^{2\pi i \delta}) \zeta^\gamma (1-\zeta)^\delta (\zeta - a)^\epsilon$
$\times \frac{f_0(q,\zeta)}{f_1(q,\zeta)} \frac{\partial}{\partial q} \mathscr{W}\{f_0(q,\zeta), f_1(q,\zeta)\}\Big|_{q=q_m},$

where

31.9.4
$f_0(q_m, z) = H\ell(a, q_m; \alpha, \beta, \gamma, \delta; z),$
$f_1(q_m, z) = H\ell(1 - a, \alpha\beta - q_m; \alpha, \beta, \delta, \gamma; 1 - z),$

and $\mathscr{W}$ denotes the Wronskian (§1.13(i)). The right-hand side may be evaluated at any convenient value, or limiting value, of ζ in $(0,1)$ since it is independent of ζ.

For corresponding orthogonality relations for Heun functions (§31.4) and Heun polynomials (§31.5), see Lambe and Ward (1934), Erdélyi (1944), Sleeman (1966b), and Ronveaux (1995, Part A, pp. 59–64).

31.9(ii) Double Orthogonality

Heun polynomials $w_j = Hp_{n_j, m_j}$, $j = 1, 2$, satisfy

31.9.5 $\quad \int_{\mathcal{L}_1} \int_{\mathcal{L}_2} \rho(s,t) w_1(s) w_1(t) w_2(s) w_2(t) \, ds\, dt$
$= 0, \qquad |n_1 - n_2| + |m_1 - m_2| \neq 0,$

where

31.9.6 $\quad \rho(s,t) = (s-t)(st)^{\gamma-1}((s-1)(t-1))^{\delta-1}$
$\times ((s-a)(t-a))^{\epsilon-1},$

and the integration paths $\mathcal{L}_1$, $\mathcal{L}_2$ are Pochhammer double-loop contours encircling distinct pairs of singularities $\{0,1\}$, $\{0,a\}$, $\{1,a\}$.

For further information, including normalization constants, see Sleeman (1966b). For bi-orthogonal relations for path-multiplicative solutions see Schmidt (1979, §2.2). For other generalizations see Arscott (1964b, pp. 206–207 and 241).

31.10 Integral Equations and Representations

31.10(i) Type I

If $w(z)$ is a solution of Heun's equation, then another solution $W(z)$ (possibly a multiple of $w(z)$) can be represented as

31.10.1 $\quad W(z) = \int_C \mathcal{K}(z,t) w(t) \rho(t)\, dt$

for a suitable contour C. The weight function is given by

31.10.2 $\quad \rho(t) = t^{\gamma-1}(t-1)^{\delta-1}(t-a)^{\epsilon-1},$

and the kernel $\mathcal{K}(z,t)$ is a solution of the partial differential equation

31.10.3 $\quad (\mathcal{D}_z - \mathcal{D}_t)\mathcal{K} = 0,$

where $\mathcal{D}_z$ is *Heun's operator in the variable z*:

31.10.4 $\quad \mathcal{D}_z = z(z-1)(z-a)(\partial^2/\partial z^2) + (\gamma(z-1)(z-a)$
$+ \delta z(z-a) + \epsilon z(z-1))(\partial/\partial z) + \alpha\beta z.$

The contour C must be such that

31.10.5 $\quad p(t)\left(\frac{\partial \mathcal{K}}{\partial t} w(t) - \mathcal{K}\frac{dw(t)}{dt}\right)\Big|_C = 0,$

where

31.10.6 $\quad p(t) = t^\gamma (t-1)^\delta (t-a)^\epsilon.$

Kernel Functions

Set

31.10.7 $\quad \cos\theta = \left(\frac{zt}{a}\right)^{1/2}, \quad \sin\theta\cos\phi = i\left(\frac{(z-a)(t-a)}{a(1-a)}\right)^{1/2}, \quad \sin\theta\sin\phi = \left(\frac{(z-1)(t-1)}{1-a}\right)^{1/2}.$

31.10 Integral Equations and Representations

The kernel $\mathcal{K}$ must satisfy

31.10.8 $\sin^2\theta \left(\dfrac{\partial^2 \mathcal{K}}{\partial \theta^2} + \left((1-2\gamma)\tan\theta + 2(\delta+\epsilon-\tfrac{1}{2})\cot\theta\right)\dfrac{\partial \mathcal{K}}{\partial \theta} - 4\alpha\beta\mathcal{K} \right) + \dfrac{\partial^2 \mathcal{K}}{\partial \phi^2} + \left((1-2\delta)\cot\phi - (1-2\epsilon)\tan\phi\right)\dfrac{\partial \mathcal{K}}{\partial \phi} = 0.$

The solutions of (31.10.8) are given in terms of the Riemann P-symbol (see §15.11(i)) as

31.10.9 $\mathcal{K}(\theta, \phi) = P\begin{Bmatrix} 0 & 1 & \infty & \\ 0 & \tfrac{1}{2}-\delta-\sigma & \alpha & \cos^2\theta \\ 1-\gamma & \tfrac{1}{2}-\epsilon+\sigma & \beta & \end{Bmatrix} P\begin{Bmatrix} 0 & 1 & \infty & \\ 0 & 0 & -\tfrac{1}{2}+\delta+\sigma & \cos^2\phi \\ 1-\epsilon & 1-\delta & -\tfrac{1}{2}+\epsilon-\sigma & \end{Bmatrix},$

where σ is a *separation constant*. For integral equations satisfied by the Heun polynomial $Hp_{n,m}(z)$ we have $\sigma = \tfrac{1}{2} - \delta - j$, $j = 0, 1, \ldots, n$.

For suitable choices of the branches of the P-symbols in (31.10.9) and the contour C, we can obtain both integral equations satisfied by Heun functions, as well as the integral representations of a distinct solution of Heun's equation in terms of a Heun function (polynomial, path-multiplicative solution).

Example 1

Let

31.10.10
$\mathcal{K}(z, t) = (zt-a)^{\tfrac{1}{2}-\delta-\sigma}\,{}_2F_1\left(\begin{matrix}\tfrac{1}{2}-\delta-\sigma+\alpha, \tfrac{1}{2}-\delta-\sigma+\beta \\ \gamma\end{matrix}; \dfrac{zt}{a}\right)\,{}_2F_1\left(\begin{matrix}-\tfrac{1}{2}+\delta+\sigma, -\tfrac{1}{2}+\epsilon-\sigma \\ \delta\end{matrix}; \dfrac{a(z-1)(t-1)}{(a-1)(zt-a)}\right),$

where $\Re\gamma > 0$, $\Re\delta > 0$, and C be the Pochhammer double-loop contour about 0 and 1 (as in §31.9(i)). Then the integral equation (31.10.1) is satisfied by $w(z) = w_m(z)$ and $W(z) = \kappa_m w_m(z)$, where $w_m(z) = (0,1)Hf_m(a, q_m; \alpha, \beta, \gamma, \delta; z)$ and κ_m is the corresponding eigenvalue.

Example 2

Fuchs–Frobenius solutions $W_m(z) = \tilde{\kappa}_m z^{-\alpha} H\ell(1/a, q_m; \alpha, \alpha-\gamma+1, \alpha-\beta+1, \delta; 1/z)$ are represented in terms of Heun functions $w_m(z) = (0,1)Hf_m(a, q_m; \alpha, \beta, \gamma, \delta; z)$ by (31.10.1) with $W(z) = W_m(z)$, $w(z) = w_m(z)$, and with kernel chosen from

31.10.11
$\mathcal{K}(z,t) = (zt-a)^{\tfrac{1}{2}-\delta-\sigma}(zt/a)^{-\tfrac{1}{2}+\delta+\sigma-\alpha}\,{}_2F_1\left(\begin{matrix}\tfrac{1}{2}-\delta-\sigma+\alpha, \tfrac{3}{2}-\delta-\sigma+\alpha-\gamma \\ \alpha-\beta+1\end{matrix}; \dfrac{a}{zt}\right)$
$\times P\begin{Bmatrix} 0 & 1 & \infty & \\ 0 & 0 & -\tfrac{1}{2}+\delta+\sigma & \dfrac{(z-a)(t-a)}{(1-a)(zt-a)} \\ 1-\epsilon & 1-\delta & -\tfrac{1}{2}+\epsilon-\sigma & \end{Bmatrix}.$

Here $\tilde{\kappa}_m$ is a normalization constant and C is the contour of Example 1.

31.10(ii) Type II

If $w(z)$ is a solution of Heun's equation, then another solution $W(z)$ (possibly a multiple of $w(z)$) can be represented as

31.10.12 $W(z) = \displaystyle\int_{C_1}\int_{C_2} \mathcal{K}(z;s,t) w(s) w(t) \rho(s,t)\, ds\, dt$

for suitable contours C_1, C_2. The weight function is

31.10.13 $\rho(s,t) = (s-t)(st)^{\gamma-1}\left((1-s)(1-t)\right)^{\delta-1} \times \left((1-(s/a))(1-(t/a))\right)^{\epsilon-1},$

and the kernel $\mathcal{K}(z;s,t)$ is a solution of the partial differential equation

31.10.14 $\left((t-z)\mathcal{D}_s + (z-s)\mathcal{D}_t + (s-t)\mathcal{D}_z\right)\mathcal{K} = 0,$

where $\mathcal{D}_z$ is given by (31.10.4). The contours C_1, C_2 must be chosen so that

31.10.15 $\left. p(t)\left(\dfrac{\partial \mathcal{K}}{\partial t}w(t) - \mathcal{K}\dfrac{dw(t)}{dt}\right)\right|_{C_1} = 0,$

and

31.10.16 $\left. p(s)\left(\dfrac{\partial \mathcal{K}}{\partial s}w(s) - \mathcal{K}\dfrac{dw(s)}{ds}\right)\right|_{C_2} = 0,$

where $p(t)$ is given by (31.10.6).

Kernel Functions

Set

31.10.17
$$u = \frac{(stz)^{1/2}}{a}, \quad v = \left(\frac{(s-1)(t-1)(z-1)}{1-a}\right)^{1/2},$$
$$w = i\left(\frac{(s-a)(t-a)(z-a)}{a(1-a)}\right)^{1/2}.$$

The kernel $\mathcal{K}$ must satisfy

31.10.18
$$\frac{\partial^2 \mathcal{K}}{\partial u^2} + \frac{\partial^2 \mathcal{K}}{\partial v^2} + \frac{\partial^2 \mathcal{K}}{\partial w^2} + \frac{2\gamma - 1}{u}\frac{\partial \mathcal{K}}{\partial u}$$
$$+ \frac{2\delta - 1}{v}\frac{\partial \mathcal{K}}{\partial v} + \frac{2\epsilon - 1}{w}\frac{\partial \mathcal{K}}{\partial w} = 0.$$

This equation can be solved in terms of cylinder functions $\mathscr{C}_\nu(z)$ (§10.2(ii)):

31.10.19
$$\mathcal{K}(u,v,w) = u^{1-\gamma}v^{1-\delta}w^{1-\epsilon}\,\mathscr{C}_{1-\gamma}(u\sqrt{\sigma_1})$$
$$\times \mathscr{C}_{1-\delta}(v\sqrt{\sigma_2})\,\mathscr{C}_{1-\epsilon}(iw\sqrt{\sigma_1+\sigma_2}),$$

where σ_1 and σ_2 are separation constants.

Transformation of Independent Variable

A further change of variables, to spherical coordinates,

31.10.20
$$u = r\cos\theta, \quad v = r\sin\theta\sin\phi, \quad w = r\sin\theta\cos\phi,$$

leads to the kernel equation

31.10.21
$$\frac{\partial^2 \mathcal{K}}{\partial r^2} + \frac{2(\gamma+\delta+\epsilon)-1}{r}\frac{\partial \mathcal{K}}{\partial r} + \frac{1}{r^2}\frac{\partial^2 \mathcal{K}}{\partial \theta^2}$$
$$+ \frac{(2(\delta+\epsilon)-1)\cot\theta - (2\gamma-1)\tan\theta}{r^2}\frac{\partial \mathcal{K}}{\partial \theta}$$
$$+ \frac{1}{r^2\sin^2\theta}\frac{\partial^2 \mathcal{K}}{\partial \phi^2} + \frac{(2\delta-1)\cot\phi - (2\epsilon-1)\tan\phi}{r^2\sin^2\theta}\frac{\partial \mathcal{K}}{\partial \phi} = 0.$$

This equation can be solved in terms of hypergeometric functions (§15.11(i)):

31.10.22
$$\mathcal{K}(r,\theta,\phi) = r^m \sin^{2p}\theta\, P\begin{Bmatrix} 0 & 1 & \infty & \\ 0 & 0 & a & \cos^2\theta \\ \tfrac{1}{2}(3-\gamma) & c & b & \end{Bmatrix}$$
$$\times P\begin{Bmatrix} 0 & 1 & \infty & \\ 0 & 0 & a' & \cos^2\phi \\ 1-\epsilon & 1-\delta & b' & \end{Bmatrix},$$

with

31.10.23
$$m^2 + 2(\alpha+\beta)m - \sigma_1 = 0,$$
$$p^2 + (\alpha+\beta-\gamma-\tfrac{1}{2})p - \tfrac{1}{4}\sigma_2 = 0,$$
$$a + b = 2(\alpha+\beta+p) - 1,$$
$$ab = p^2 - p(1-\alpha-\beta) - \tfrac{1}{4}\sigma_1,$$
$$c = \gamma - \tfrac{1}{2} - 2(\alpha+\beta+p),$$
$$a' + b' = \delta + \epsilon - 1, \quad a'b' = -\tfrac{1}{4}\sigma_2,$$

and σ_1 and σ_2 are separation constants.

For integral equations for special confluent Heun functions (§31.12) see Kazakov and Slavyanov (1996).

31.11 Expansions in Series of Hypergeometric Functions

31.11(i) Introduction

The formulas in this section are given in Svartholm (1939) and Erdélyi (1942a, 1944).

The series of Type I (§31.11(iii)) are useful since they represent the functions in large domains. Series of Type II (§31.11(iv)) are expansions in orthogonal polynomials, which are useful in calculations of normalization integrals for Heun functions; see Erdélyi (1944) and §31.9(i).

For other expansions see §31.16(ii).

31.11(ii) General Form

Let $w(z)$ be any Fuchs–Frobenius solution of Heun's equation. Expand

31.11.1
$$w(z) = \sum_{j=0}^{\infty} c_j P_j,$$

where (§15.11(i))

31.11.2
$$P_j = P\begin{Bmatrix} 0 & 1 & \infty & \\ 0 & 0 & \lambda+j & z \\ 1-\gamma & 1-\delta & \mu-j & \end{Bmatrix},$$

with

31.11.3 $\quad \lambda + \mu = \gamma + \delta - 1 = \alpha + \beta - \epsilon.$

The coefficients c_j satisfy the equations

31.11.4 $\quad L_0 c_0 + M_0 c_1 = 0,$

31.11.5 $\quad K_j c_{j-1} + L_j c_j + M_j c_{j+1} = 0, \quad j = 1, 2, \ldots,$

where

31.11.6
$$K_j = -\frac{(j+\alpha-\mu-1)(j+\beta-\mu-1)(j+\gamma-\mu-1)(j+\lambda-1)}{(2j+\lambda-\mu-1)(2j+\lambda-\mu-2)},$$

31.11.7
$$L_j = a(\lambda+j)(\mu-j) - q + \frac{(j+\alpha-\mu)(j+\beta-\mu)(j+\gamma-\mu)(j+\lambda)}{(2j+\lambda-\mu)(2j+\lambda-\mu+1)}$$
$$+ \frac{(j-\alpha+\lambda)(j-\beta+\lambda)(j-\gamma+\lambda)(j-\mu)}{(2j+\lambda-\mu)(2j+\lambda-\mu-1)},$$

31.11.8
$$M_j = -\frac{(j-\alpha+\lambda+1)(j-\beta+\lambda+1)(j-\gamma+\lambda+1)(j-\mu+1)}{(2j+\lambda-\mu+1)(2j+\lambda-\mu+2)}.$$

λ, μ must also satisfy the condition

31.11.9
$$M_{-1}P_{-1} = 0.$$

31.11(iii) Type I

Here

31.11.10
$$\lambda = \alpha, \quad \mu = \beta - \epsilon,$$

or

31.11.11
$$\lambda = \beta, \quad \mu = \alpha - \epsilon.$$

Then condition (31.11.9) is satisfied.

Every Fuchs–Frobenius solution of Heun's equation (31.2.1) can be represented by a series of Type I. For instance, choose (31.11.10). Then the Fuchs–Frobenius solution at ∞ belonging to the exponent α has the expansion (31.11.1) with

31.11.12
$$P_j = \frac{\Gamma(\alpha+j)\,\Gamma(1-\gamma+\alpha+j)}{\Gamma(1+\alpha-\beta+\epsilon+2j)} z^{-\alpha-j}$$
$$\times {}_2F_1\!\left(\begin{matrix}\alpha+j, 1-\gamma+\alpha+j\\1+\alpha-\beta+\epsilon+2j\end{matrix}; \frac{1}{z}\right),$$

and (31.11.1) converges outside the ellipse $\mathcal{E}$ in the z-plane with foci at 0, 1, and passing through the third finite singularity at $z = a$.

Every Heun function (§31.4) can be represented by a series of Type I convergent in the whole plane cut along a line joining the two singularities of the Heun function.

For example, consider the Heun function which is analytic at $z = a$ and has exponent α at ∞. The expansion (31.11.1) with (31.11.12) is convergent in the plane cut along the line joining the two singularities $z = 0$ and $z = 1$. In this case the accessory parameter q is a root of the continued-fraction equation

31.11.13
$$(L_0/M_0) - \frac{K_1/M_1}{L_1/M_1 -}\frac{K_2/M_2}{L_2/M_2 -} \cdots = 0.$$

The case $\alpha = -n$ for nonnegative integer n corresponds to the Heun polynomial $Hp_{n,m}(z)$.

The expansion (31.11.1) for a Heun function that is associated with any branch of (31.11.2)—other than a multiple of the right-hand side of (31.11.12)—is convergent inside the ellipse $\mathcal{E}$.

31.11(iv) Type II

Here one of the following four pairs of conditions is satisfied:

31.11.14 $\quad \lambda = \gamma + \delta - 1, \quad \mu = 0,$

31.11.15 $\quad \lambda = \gamma, \quad \mu = \delta - 1,$

31.11.16 $\quad \lambda = \delta, \quad \mu = \gamma - 1,$

31.11.17 $\quad \lambda = 1, \quad \mu = \gamma + \delta - 2.$

In each case P_j can be expressed in terms of a Jacobi polynomial (§18.3). Such series diverge for Fuchs–Frobenius solutions. For Heun functions they are convergent inside the ellipse $\mathcal{E}$. Every Heun function can be represented by a series of Type II.

31.11(v) Doubly-Infinite Series

Schmidt (1979) gives expansions of path-multiplicative solutions (§31.6) in terms of doubly-infinite series of hypergeometric functions.

31.12 Confluent Forms of Heun's Equation

Confluent forms of Heun's differential equation (31.2.1) arise when two or more of the regular singularities merge to form an irregular singularity. This is analogous to the derivation of the confluent hypergeometric equation from the hypergeometric equation in §13.2(i). There are four standard forms, as follows:

Confluent Heun Equation

31.12.1
$$\frac{d^2w}{dz^2} + \left(\frac{\gamma}{z} + \frac{\delta}{z-1} + \epsilon\right)\frac{dw}{dz} + \frac{\alpha z - q}{z(z-1)}w = 0.$$

This has regular singularities at $z = 0$ and 1, and an irregular singularity of rank 1 at $z = \infty$.

Mathieu functions (Chapter 28), spheroidal wave functions (Chapter 30), and Coulomb spheroidal functions (§30.12) are special cases of solutions of the confluent Heun equation.

Doubly-Confluent Heun Equation

31.12.2
$$\frac{d^2w}{dz^2} + \left(\frac{\delta}{z^2} + \frac{\gamma}{z} + 1\right)\frac{dw}{dz} + \frac{\alpha z - q}{z^2}w = 0.$$

This has irregular singularities at $z = 0$ and ∞, each of rank 1.

Biconfluent Heun Equation

31.12.3 $\quad \dfrac{d^2w}{dz^2} + \left(\dfrac{\gamma}{z} + \delta + z\right)\dfrac{dw}{dz} + \dfrac{\alpha z - q}{z} w = 0.$

This has a regular singularity at $z = 0$, and an irregular singularity at ∞ of rank 2.

Triconfluent Heun Equation

31.12.4 $\quad \dfrac{d^2w}{dz^2} + (\gamma + z)z\dfrac{dw}{dz} + (\alpha z - q)w = 0.$

This has one singularity, an irregular singularity of rank 3 at $z = \infty$.

For properties of the solutions of (31.12.1)–(31.12.4), including connection formulas, see Bühring (1994), Ronveaux (1995, Parts B,C,D,E), Wolf (1998), Lay and Slavyanov (1998), and Slavyanov and Lay (2000).

31.13 Asymptotic Approximations

For asymptotic approximations for the accessory parameter eigenvalues q_m, see Fedoryuk (1991) and Slavyanov (1996).

For asymptotic approximations of the solutions of Heun's equation (31.2.1) when two singularities are close together, see Lay and Slavyanov (1999).

For asymptotic approximations of the solutions of confluent forms of Heun's equation in the neighborhood of irregular singularities, see Komarov et al. (1976), Ronveaux (1995, Parts B,C,D,E), Bogush and Otchik (1997), Slavyanov and Veshev (1997), and Lay et al. (1998).

31.14 General Fuchsian Equation

31.14(i) Definitions

The general second-order *Fuchsian equation* with $N+1$ regular singularities at $z = a_j$, $j = 1, 2, \ldots, N$, and at ∞, is given by

31.14.1
$$\dfrac{d^2w}{dz^2} + \left(\sum_{j=1}^{N} \dfrac{\gamma_j}{z - a_j}\right)\dfrac{dw}{dz} + \left(\sum_{j=1}^{N} \dfrac{q_j}{z - a_j}\right)w = 0,$$
$$\sum_{j=1}^{N} q_j = 0.$$

The exponents at the finite singularities a_j are $\{0, 1 - \gamma_j\}$ and those at ∞ are $\{\alpha, \beta\}$, where

31.14.2 $\quad \alpha + \beta + 1 = \sum\limits_{j=1}^{N}\gamma_j, \quad \alpha\beta = \sum\limits_{j=1}^{N} a_j q_j.$

The three sets of parameters comprise the *singularity parameters* a_j, the *exponent parameters* α, β, γ_j, and the $N - 2$ free *accessory parameters* q_j. With $a_1 = 0$ and $a_2 = 1$ the total number of free parameters is $3N-3$. Heun's equation (31.2.1) corresponds to $N = 3$.

Normal Form

31.14.3 $\quad w(z) = \left(\prod\limits_{j=1}^{N}(z - a_j)^{-\gamma_j/2}\right) W(z),$

31.14.4
$$\dfrac{d^2 W}{dz^2} = \sum_{j=1}^{N}\left(\dfrac{\tilde{\gamma}_j}{(z - a_j)^2} + \dfrac{\tilde{q}_j}{z - a_j}\right)W, \quad \sum_{j=1}^{N}\tilde{q}_j = 0,$$

31.14.5 $\quad \tilde{q}_j = \dfrac{1}{2}\sum\limits_{\substack{k=1 \\ k \neq j}}^{N}\dfrac{\gamma_j \gamma_k}{a_j - a_k} - q_j, \quad \tilde{\gamma}_j = \dfrac{\gamma_j}{2}\left(\dfrac{\gamma_j}{2} - 1\right).$

31.14(ii) Kovacic's Algorithm

An algorithm given in Kovacic (1986) determines if a given (not necessarily Fuchsian) second-order homogeneous linear differential equation with rational coefficients has solutions expressible in finite terms (Liouvillean solutions). The algorithm returns a list of solutions if they exist.

For applications of Kovacic's algorithm in spatiotemporal dynamics see Rod and Sleeman (1995).

31.15 Stieltjes Polynomials

31.15(i) Definitions

Stieltjes polynomials are polynomial solutions of the Fuchsian equation (31.14.1). Rewrite (31.14.1) in the form

31.15.1
$$\dfrac{d^2 w}{dz^2} + \left(\sum_{j=1}^{N}\dfrac{\gamma_j}{z - a_j}\right)\dfrac{dw}{dz} + \dfrac{\Phi(z)}{\prod_{j=1}^{N}(z - a_j)} w = 0,$$

where $\Phi(z)$ is a polynomial of degree not exceeding $N - 2$. There exist at most $\binom{n+N-2}{N-2}$ polynomials $V(z)$ of degree not exceeding $N-2$ such that for $\Phi(z) = V(z)$, (31.15.1) has a polynomial solution $w = S(z)$ of degree n. The $V(z)$ are called *Van Vleck polynomials* and the corresponding $S(z)$ *Stieltjes polynomials*.

31.15(ii) Zeros

If $z_1, z_2, \ldots, z_n$ are the zeros of an nth degree Stieltjes polynomial $S(z)$, then every zero z_k is either one of the parameters a_j or a solution of the system of equations

31.15.2 $\quad \sum\limits_{j=1}^{N}\dfrac{\gamma_j/2}{z_k - a_j} + \sum\limits_{\substack{j=1 \\ j \neq k}}^{n}\dfrac{1}{z_k - z_j} = 0, \quad k = 1, 2, \ldots, n.$

If t_k is a zero of the Van Vleck polynomial $V(z)$, corresponding to an nth degree Stieltjes polynomial $S(z)$, and $z'_1, z'_2, \ldots, z'_{n-1}$ are the zeros of $S'(z)$ (the derivative of $S(z)$), then t_k is either a zero of $S'(z)$ or a solution of the equation

31.15.3 $$\sum_{j=1}^{N} \frac{\gamma_j}{t_k - a_j} + \sum_{j=1}^{n-1} \frac{1}{t_k - z'_j} = 0.$$

The system (31.15.2) determines the z_k as the points of equilibrium of n movable (interacting) particles with unit charges in a field of N particles with the charges $\gamma_j/2$ fixed at a_j. This is the *Stieltjes electrostatic interpretation*.

The zeros z_k, $k = 1, 2, \ldots, n$, of the Stieltjes polynomial $S(z)$ are the critical points of the function G, that is, points at which $\partial G/\partial \zeta_k = 0$, $k = 1, 2, \ldots, n$, where

$$G(\zeta_1, \zeta_2, \ldots, \zeta_n)$$
31.15.4 $$= \prod_{k=1}^{n} \prod_{\ell=1}^{N} (\zeta_k - a_\ell)^{\gamma_\ell/2} \prod_{j=k+1}^{n} (\zeta_k - \zeta_j).$$

If the following conditions are satisfied:

31.15.5 $$\gamma_j > 0, \quad a_j \in \mathbb{R}, \quad j = 1, 2, \ldots, N,$$
and

31.15.6 $$a_j < a_{j+1}, \quad j = 1, 2, \ldots, N-1,$$
then there are *exactly* $\binom{n+N-2}{N-2}$ polynomials $S(z)$, each of which corresponds to each of the $\binom{n+N-2}{N-2}$ ways of distributing its n zeros among $N-1$ intervals (a_j, a_{j+1}), $j = 1, 2, \ldots, N-1$. In this case the accessory parameters q_j are given by

31.15.7 $$q_j = \gamma_j \sum_{k=1}^{n} \frac{1}{z_k - a_j}, \quad j = 1, 2, \ldots, N.$$

See Marden (1966), Alam (1979), and Al-Rashed and Zaheer (1985) for further results on the location of the zeros of Stieltjes and Van Vleck polynomials.

31.15(iii) Products of Stieltjes Polynomials

If the exponent and singularity parameters satisfy (31.15.5)–(31.15.6), then for every multi-index $\mathbf{m} = (m_1, m_2, \ldots, m_{N-1})$, where each m_j is a nonnegative integer, there is a unique Stieltjes polynomial with m_j zeros in the open interval (a_j, a_{j+1}) for each $j = 1, 2, \ldots, N-1$. We denote this Stieltjes polynomial by $S_\mathbf{m}(z)$.

Let $S_\mathbf{m}(z)$ and $S_\mathbf{l}(z)$ be Stieltjes polynomials corresponding to two distinct multi-indices $\mathbf{m} = (m_1, m_2, \ldots, m_{N-1})$ and $\mathbf{l} = (\ell_1, \ell_2, \ldots, \ell_{N-1})$. The products

31.15.8 $$S_\mathbf{m}(z_1) S_\mathbf{m}(z_2) \cdots S_\mathbf{m}(z_{N-1}), \quad z_j \in (a_j, a_{j+1}),$$

31.15.9 $$S_\mathbf{l}(z_1) S_\mathbf{l}(z_2) \cdots S_\mathbf{l}(z_{N-1}), \quad z_j \in (a_j, a_{j+1}),$$
are mutually orthogonal over the set Q:

31.15.10 $$Q = (a_1, a_2) \times (a_2, a_3) \times \cdots \times (a_{N-1}, a_N),$$

with respect to the inner product

31.15.11 $$(f, g)_\rho = \int_Q f(z) \bar{g}(z) \rho(z)\, dz,$$
with weight function

31.15.12
$$\rho(z) = \left(\prod_{j=1}^{N-1} \prod_{k=1}^{N} |z_j - a_k|^{\gamma_k - 1} \right) \left(\prod_{j<k}^{N-1} (z_k - z_j) \right).$$

The normalized system of products (31.15.8) forms an orthonormal basis in the Hilbert space $L^2_\rho(Q)$. For further details and for the expansions of analytic functions in this basis see Volkmer (1999).

Applications

31.16 Mathematical Applications

31.16(i) Uniformization Problem for Heun's Equation

The main part of Smirnov (1996) consists of V. I. Smirnov's 1918 M. Sc. thesis "Inversion problem for a second-order linear differential equation with four singular points". It describes the monodromy group of Heun's equation for specific values of the accessory parameter.

31.16(ii) Heun Polynomial Products

Expansions of Heun polynomial products in terms of Jacobi polynomial (§18.3) products are derived in Kalnins and Miller (1991a,b, 1993) from the viewpoint of interrelation between two bases in a Hilbert space:

31.16.1
$$Hp_{n,m}(x) Hp_{n,m}(y)$$
$$= \sum_{j=0}^{n} A_j \sin^{2j} \theta$$
$$\times P_{n-j}^{(\gamma+\delta+2j-1,\epsilon-1)}(\cos 2\theta) P_j^{(\delta-1,\gamma-1)}(\cos 2\phi),$$
where $n = 0, 1, \ldots$, $m = 0, 1, \ldots, n$, and

31.16.2 $$x = \sin^2 \theta \cos^2 \phi, \quad y = \sin^2 \theta \sin^2 \phi.$$

The coefficients A_j satisfy the relations:

31.16.3 $$Q_0 A_0 + R_0 A_1 = 0,$$

31.16.4 $$P_j A_{j-1} + Q_j A_j + R_j A_{j+1} = 0, \quad j = 1, 2, \ldots, n,$$
where

31.16.5
$$P_j = \frac{(\epsilon - j + n)j(\beta + j - 1)(\gamma + \delta + j - 2)}{(\gamma + \delta + 2j - 3)(\gamma + \delta + 2j - 2)},$$

31.16.6
$$Q_j = -aj(j + \gamma + \delta - 1) - q + \frac{(j-n)(j+\beta)(j+\gamma)(j+\gamma+\delta-1)}{(2j+\gamma+\delta)(2j+\gamma+\delta-1)} + \frac{(j+n+\gamma+\delta-1)j(j+\delta-1)(j-\beta+\gamma+\delta-1)}{(2j+\gamma+\delta-1)(2j+\gamma+\delta-2)},$$

31.16.7
$$R_j = \frac{(n-j)(j+n+\gamma+\delta)(j+\gamma)(j+\delta)}{(\gamma+\delta+2j)(\gamma+\delta+2j+1)}.$$

By specifying either θ or ϕ in (31.16.1) and (31.16.2) we obtain expansions in terms of one variable.

31.17 Physical Applications

31.17(i) Addition of Three Quantum Spins

The problem of adding three quantum spins $\mathbf{s}$, $\mathbf{t}$, and $\mathbf{u}$ can be solved by the *method of separation of variables*, and the solution is given in terms of a product of two Heun functions. We use vector notation $[\mathbf{s}, \mathbf{t}, \mathbf{u}]$ (respective scalar (s, t, u)) for any one of the three spin operators (respective spin values).

Consider the following spectral problem on the sphere S_2: $\mathbf{x}^2 = x_s^2 + x_t^2 + x_u^2 = R^2$.

31.17.1
$$\mathbf{J}^2 \Psi(\mathbf{x}) \equiv (\mathbf{s} + \mathbf{t} + \mathbf{u})^2 \Psi(\mathbf{x}) = j(j+1)\Psi(\mathbf{x}),$$
$$H_s \Psi(\mathbf{x}) \equiv (-2\mathbf{s} \cdot \mathbf{t} - (2/a)\mathbf{s} \cdot \mathbf{u})\Psi(\mathbf{x}) = h_s \Psi(\mathbf{x}),$$

for the common eigenfunction $\Psi(\mathbf{x}) = \Psi(x_s, x_t, x_u)$, where a is the coupling parameter of interacting spins. Introduce elliptic coordinates z_1 and z_2 on S_2. Then

31.17.2
$$\frac{x_s^2}{z_k} + \frac{x_t^2}{z_k - 1} + \frac{x_u^2}{z_k - a} = 0, \qquad k = 1, 2,$$

with

31.17.3
$$x_s^2 = R^2 \frac{z_1 z_2}{a}, \quad x_t^2 = R^2 \frac{(z_1-1)(z_2-1)}{1-a},$$
$$x_u^2 = R^2 \frac{(z_1-a)(z_2-a)}{a(a-1)}.$$

The operators $\mathbf{J}^2$ and H_s admit separation of variables in z_1, z_2, leading to the following factorization of the eigenfunction $\Psi(\mathbf{x})$:

31.17.4
$$\Psi(\mathbf{x}) = (z_1 z_2)^{-s-\frac{1}{4}}((z_1-1)(z_2-1))^{-t-\frac{1}{4}} \times ((z_1-a)(z_2-a))^{-u-\frac{1}{4}} w(z_1) w(z_2),$$

where $w(z)$ satisfies Heun's equation (31.2.1) with a as in (31.17.1) and the other parameters given by

31.17.5
$$\alpha = -s-t-u-j-1, \quad \beta = j-s-t-u, \quad \gamma = -2s,$$
$$\delta = -2t, \quad \epsilon = -2u; \quad q = ah_s + 2s(at+u).$$

For more details about the method of separation of variables and relation to special functions see Olevskiĭ (1950), Kalnins et al. (1976), Miller (1977), and Kalnins (1986).

31.17(ii) Other Applications

Heun functions appear in the theory of black holes (Kerr (1963), Teukolsky (1972), Chandrasekhar (1984), Suzuki et al. (1998), Kalnins et al. (2000)), lattice systems in statistical mechanics (Joyce (1973, 1994)), dislocation theory (Lay and Slavyanov (1999)), and quantum systems (Bay et al. (1997), Tolstikhin and Matsuzawa (2001)).

For applications of Heun's equation and functions in astrophysics see Debosscher (1998) where different spectral problems for Heun's equation are also considered. More applications—including those of generalized spheroidal wave functions and confluent Heun functions in mathematical physics, astrophysics, and the two-center problem in molecular quantum mechanics—can be found in Leaver (1986) and Slavyanov and Lay (2000, Chapter 4). For application of biconfluent Heun functions in a model of an equatorially trapped Rossby wave in a shear flow in the ocean or atmosphere see Boyd and Natarov (1998).

Computation

31.18 Methods of Computation

Independent solutions of (31.2.1) can be computed in the neighborhoods of singularities from their Fuchs–Frobenius expansions (§31.3), and elsewhere by numerical integration of (31.2.1). Subsequently, the coefficients in the necessary connection formulas can be calculated numerically by matching the values of solutions and their derivatives at suitably chosen values of z; see Laĭ (1994) and Lay et al. (1998). Care needs to be taken to choose integration paths in such a way that the wanted solution is growing in magnitude along the path at least as rapidly as all other solutions (§3.7(ii)). The computation of the accessory parameter for the Heun functions is carried out via the continued-fraction equations (31.4.2) and (31.11.13) in the same way as for the Mathieu, Lamé, and spheroidal wave functions in Chapters 28–30.

References

General References

The main references used in writing this chapter are Sleeman (1966b) and Ronveaux (1995). For additional bibliographic reading see Erdélyi et al. (1955).

Sources

The following list gives the references or other indications of proofs that were used in constructing the various sections of this chapter. These sources supplement the references that are quoted in the text.

§31.2 Erdélyi et al. (1955, Chapter XV), Ronveaux (1995, Part A, Chapters 1 and 2).

§31.3 Snow (1952), Ronveaux (1995, Part A, Chapters 2 and 3).

§31.4 Erdélyi et al. (1955, Chapter XV), Arscott (1964b, Chapter IX).

§31.5 Erdélyi et al. (1955, Chapter XV), Arscott (1964b, Chapter IX).

§31.7 Ronveaux (1995, Part A, Chapter 1).

§31.9 Becker (1997).

§31.10 Lambe and Ward (1934) Erdélyi (1942b), Valent (1986), Sleeman (1969), An error in the last reference is corrected here.

§31.12 The process of confluence is discussed in Ince (1926, Chapter XX). See Decarreau et al. (1978a,b) for the classification of confluent forms.

§31.14 Ince (1926, Chapter XV).

§31.15 Marden (1966).

§31.17 Gaudin (1983), Kuznetsov (1992).

Chapter 32

Painlevé Transcendents

P. A. Clarkson[1]

Notation   **724**
- 32.1 Special Notation 724

Properties   **724**
- 32.2 Differential Equations 724
- 32.3 Graphics 726
- 32.4 Isomonodromy Problems 728
- 32.5 Integral Equations 729
- 32.6 Hamiltonian Structure 729
- 32.7 Bäcklund Transformations 730
- 32.8 Rational Solutions 732
- 32.9 Other Elementary Solutions 734
- 32.10 Special Function Solutions 735
- 32.11 Asymptotic Approximations for Real Variables 736
- 32.12 Asymptotic Approximations for Complex Variables 738

Applications   **738**
- 32.13 Reductions of Partial Differential Equations 738
- 32.14 Combinatorics 739
- 32.15 Orthogonal Polynomials 739
- 32.16 Physical 739

Computation   **739**
- 32.17 Methods of Computation 740

References   **740**

---

[1]School of Mathematics, Statistics & Actuarial Science, University of Kent, Canterbury, United Kingdom.
Copyright © 2009 National Institute of Standards and Technology. All rights reserved.

Notation

32.1 Special Notation

(For other notation see pp. xiv and 873.)

m, n integers.
x real variable.
z complex variable.
k real parameter.

Unless otherwise noted, primes indicate derivatives with respect to the argument.

The functions treated in this chapter are the solutions of the Painlevé equations P_I–P_{VI}.

Properties

32.2 Differential Equations

32.2(i) Introduction

The six Painlevé equations P_I–P_{VI} are as follows:

32.2.1 $$\frac{d^2w}{dz^2} = 6w^2 + z,$$

32.2.2 $$\frac{d^2w}{dz^2} = 2w^3 + zw + \alpha,$$

32.2.3 $$\frac{d^2w}{dz^2} = \frac{1}{w}\left(\frac{dw}{dz}\right)^2 - \frac{1}{z}\frac{dw}{dz} + \frac{\alpha w^2 + \beta}{z} + \gamma w^3 + \frac{\delta}{w},$$

32.2.4 $$\frac{d^2w}{dz^2} = \frac{1}{2w}\left(\frac{dw}{dz}\right)^2 + \frac{3}{2}w^3 + 4zw^2 + 2(z^2 - \alpha)w + \frac{\beta}{w},$$

32.2.5 $$\frac{d^2w}{dz^2} = \left(\frac{1}{2w} + \frac{1}{w-1}\right)\left(\frac{dw}{dz}\right)^2 - \frac{1}{z}\frac{dw}{dz} + \frac{(w-1)^2}{z^2}\left(\alpha w + \frac{\beta}{w}\right) + \frac{\gamma w}{z} + \frac{\delta w(w+1)}{w-1},$$

32.2.6 $$\frac{d^2w}{dz^2} = \frac{1}{2}\left(\frac{1}{w} + \frac{1}{w-1} + \frac{1}{w-z}\right)\left(\frac{dw}{dz}\right)^2 - \left(\frac{1}{z} + \frac{1}{z-1} + \frac{1}{w-z}\right)\frac{dw}{dz}$$
$$+ \frac{w(w-1)(w-z)}{z^2(z-1)^2}\left(\alpha + \frac{\beta z}{w^2} + \frac{\gamma(z-1)}{(w-1)^2} + \frac{\delta z(z-1)}{(w-z)^2}\right),$$

with α, β, γ, and δ arbitrary constants. The solutions of P_I–P_{VI} are called the *Painlevé transcendents*. The six equations are sometimes referred to as the Painlevé transcendents, but in this chapter this term will be used only for their solutions.

Let

32.2.7 $$\frac{d^2w}{dz^2} = F\left(z, w, \frac{dw}{dz}\right),$$

be a nonlinear second-order differential equation in which F is a rational function of w and dw/dz, and is *locally analytic* in z, that is, analytic except for isolated singularities in $\mathbb{C}$. In general the singularities of the solutions are *movable* in the sense that their location depends on the constants of integration associated with the initial or boundary conditions. An equation is said to have the *Painlevé property* if all its solutions are free from *movable branch points*; the solutions may have movable poles or movable isolated essential singularities (§1.10(iii)), however.

There are fifty equations with the Painlevé property. They are distinct modulo Möbius (bilinear) transformations

32.2.8 $$W(\zeta) = \frac{a(z)w + b(z)}{c(z)w + d(z)}, \quad \zeta = \phi(z),$$

in which $a(z)$, $b(z)$, $c(z)$, $d(z)$, and $\phi(z)$ are locally analytic functions. The fifty equations can be reduced to linear equations, solved in terms of elliptic functions (Chapters 22 and 23), or reduced to one of P_I–P_{VI}.

For arbitrary values of the parameters α, β, γ, and δ, the general solutions of P_I–P_{VI} are *transcendental*, that is, they cannot be expressed in closed-form elementary functions. However, for special values of the parameters, equations P_{II}–P_{VI} have special solutions in terms of elementary functions, or special functions defined elsewhere in this Handbook.

32.2(ii) Renormalizations

If $\gamma\delta \neq 0$ in P_{III}, then set $\gamma = 1$ and $\delta = -1$, without loss of generality, by rescaling w and z if necessary. If $\gamma = 0$ and $\alpha\delta \neq 0$ in P_{III}, then set $\alpha = 1$ and $\delta = -1$, without loss of generality. Lastly, if $\delta = 0$ and $\beta\gamma \neq 0$, then set $\beta = -1$ and $\gamma = 1$, without loss of generality.

If $\delta \neq 0$ in P_V, then set $\delta = -\frac{1}{2}$, without loss of generality.

32.2(iii) Alternative Forms

In P_{III}, if $w(z) = \zeta^{-1/2}u(\zeta)$ with $\zeta = z^2$, then

32.2.9 $$\frac{d^2u}{d\zeta^2} = \frac{1}{u}\left(\frac{du}{d\zeta}\right)^2 - \frac{1}{\zeta}\frac{du}{d\zeta} + \frac{u^2(\alpha + \gamma u)}{4\zeta^2} + \frac{\beta}{4\zeta} + \frac{\delta}{4u},$$

which is known as P'_{III}.

In P_{III}, if $w(z) = \exp(-iu(z))$, $\beta = -\alpha$, and $\delta = -\gamma$, then

32.2.10 $\quad \dfrac{d^2 u}{dz^2} + \dfrac{1}{z}\dfrac{du}{dz} = \dfrac{2\alpha}{z}\sin u + 2\gamma \sin(2u).$

In P_{IV}, if $w(z) = 2\sqrt{2}(u(\zeta))^2$ with $\zeta = \sqrt{2}z$ and $\alpha = 2\nu + 1$, then

32.2.11 $\quad \dfrac{d^2 u}{d\zeta^2} = 3u^5 + 2\zeta u^3 + \left(\tfrac{1}{4}\zeta^2 - \nu - \tfrac{1}{2}\right)u + \dfrac{\beta}{32u^3}.$

When $\beta = 0$ this is a nonlinear harmonic oscillator.

In P_V, if $w(z) = (\coth u(\zeta))^2$ with $\zeta = \ln z$, then

32.2.12 $\quad \begin{aligned}\dfrac{d^2 u}{d\zeta^2} &= -\dfrac{\alpha \cosh u}{2(\sinh u)^3} - \dfrac{\beta \sinh u}{2(\cosh u)^3} \\ &\quad - \tfrac{1}{4}\gamma e^\zeta \sinh(2u) - \tfrac{1}{8}\delta e^{2\zeta}\sinh(4u).\end{aligned}$

See also Okamoto (1987c), McCoy et al. (1977), Bassom et al. (1992), Bassom et al. (1995), and Takasaki (2001).

32.2(iv) Elliptic Form

P_{VI} can be written in the form

32.2.13
$$\begin{aligned}&z(1-z) I\left(\int_\infty^w \dfrac{dt}{\sqrt{t(t-1)(t-z)}}\right) \\ &= \sqrt{w(w-1)(w-z)} \\ &\quad \times \left(\alpha + \dfrac{\beta z}{w^2} + \dfrac{\gamma(z-1)}{(w-1)^2} + \left(\delta - \tfrac{1}{2}\right)\dfrac{z(z-1)}{(w-z)^2}\right),\end{aligned}$$

where

32.2.14 $\quad I = z(1-z)\dfrac{d^2}{dz^2} + (1-2z)\dfrac{d}{dz} - \dfrac{1}{4}.$

See Fuchs (1907), Painlevé (1906), Gromak et al. (2002, §42); also Manin (1998).

32.2(v) Symmetric Forms

Let

32.2.15
$$\begin{aligned}\dfrac{df_1}{dz} + f_1(f_2 - f_3) + 2\mu_1 &= 0, \\ \dfrac{df_2}{dz} + f_2(f_3 - f_1) + 2\mu_2 &= 0, \\ \dfrac{df_3}{dz} + f_3(f_1 - f_2) + 2\mu_3 &= 0,\end{aligned}$$

where μ_1, μ_2, μ_3 are constants, f_1, f_2, f_3 are functions of z, with

32.2.16 $\quad \mu_1 + \mu_2 + \mu_3 = 1,$

32.2.17 $\quad f_1(z) + f_2(z) + f_3(z) + 2z = 0.$

Then $w(z) = f_1(z)$ satisfies P_{IV} with

32.2.18 $\quad (\alpha, \beta) = (\mu_3 - \mu_2, -2\mu_1^2).$

See Noumi and Yamada (1998).

Next, let

32.2.19
$$\begin{aligned}z\dfrac{df_1}{dz} &= f_1 f_3(f_2 - f_4) + \left(\tfrac{1}{2} - \mu_3\right)f_1 + \mu_1 f_3, \\ z\dfrac{df_2}{dz} &= f_2 f_4(f_3 - f_1) + \left(\tfrac{1}{2} - \mu_4\right)f_2 + \mu_2 f_4, \\ z\dfrac{df_3}{dz} &= f_3 f_1(f_4 - f_2) + \left(\tfrac{1}{2} - \mu_1\right)f_3 + \mu_3 f_1, \\ z\dfrac{df_4}{dz} &= f_4 f_2(f_1 - f_3) + \left(\tfrac{1}{2} - \mu_2\right)f_4 + \mu_4 f_2,\end{aligned}$$

where $\mu_1, \mu_2, \mu_3, \mu_4$ are constants, f_1, f_2, f_3, f_4 are functions of z, with

32.2.20 $\quad \mu_1 + \mu_2 + \mu_3 + \mu_4 = 1,$

32.2.21 $\quad f_1(z) + f_3(z) = \sqrt{z},$

32.2.22 $\quad f_2(z) + f_4(z) = \sqrt{z}.$

Then $w(z) = 1 - (\sqrt{z}/f_1(z))$ satisfies P_V with

32.2.23 $\quad (\alpha, \beta, \gamma, \delta) = (\tfrac{1}{2}\mu_1^2, -\tfrac{1}{2}\mu_3^2, \mu_4 - \mu_2, -\tfrac{1}{2}).$

32.2(vi) Coalescence Cascade

P_I–P_V are obtained from P_{VI} by a coalescence cascade:

32.2.24
$$\begin{array}{ccccc} P_{VI} & \longrightarrow & P_V & \longrightarrow & P_{IV} \\ & & \downarrow & & \downarrow \\ & & P_{III} & \longrightarrow & P_{II} & \longrightarrow & P_I \end{array}$$

For example, if in P_{II}

32.2.25 $\quad w(z; \alpha) = \epsilon W(\zeta) + \dfrac{1}{\epsilon^5},$

32.2.26 $\quad z = \epsilon^2 \zeta - \dfrac{6}{\epsilon^{10}}, \quad \alpha = \dfrac{4}{\epsilon^{15}},$

then

32.2.27 $\quad \dfrac{d^2 W}{d\zeta^2} = 6W^2 + \zeta + \epsilon^6(2W^3 + \zeta W);$

thus in the limit as $\epsilon \to 0$, $W(\zeta)$ satisfies P_I with $z = \zeta$.

If in P_{III}

32.2.28 $\quad w(z; \alpha, \beta, \gamma, \delta) = 1 + 2\epsilon W(\zeta; a),$

32.2.29 $\quad \begin{aligned}z &= 1 + \epsilon^2 \zeta, \quad \alpha = -\tfrac{1}{2}\epsilon^{-6}, \\ \beta &= \tfrac{1}{2}\epsilon^{-6} + 2a\epsilon^{-3}, \quad \gamma = -\delta = \tfrac{1}{4}\epsilon^{-6},\end{aligned}$

then as $\epsilon \to 0$, $W(\zeta; a)$ satisfies P_{II} with $z = \zeta$, $\alpha = a$.

If in P_{IV}

32.2.30 $\quad w(z; \alpha, \beta) = 2^{2/3}\epsilon^{-1} W(\zeta; a) + \epsilon^{-3},$

32.2.31
$z = 2^{-2/3}\epsilon\zeta - \epsilon^{-3}, \quad \alpha = -2a - \tfrac{1}{2}\epsilon^{-6}, \quad \beta = -\tfrac{1}{2}\epsilon^{-12},$

then as $\epsilon \to 0$, $W(\zeta; a)$ satisfies P_{II} with $z = \zeta$, $\alpha = a$.

If in P_V

32.2.32 $\quad w(z; \alpha, \beta, \gamma, \delta) = 1 + \epsilon\zeta W(\zeta; a, b, c, d),$

32.2.33 $\quad \begin{aligned}z &= \zeta^2, \quad \alpha = \tfrac{1}{4}a\epsilon^{-1} + \tfrac{1}{8}c\epsilon^{-2}, \\ \beta &= -\tfrac{1}{8}c\epsilon^{-2}, \quad \gamma = \tfrac{1}{4}\epsilon b, \quad \delta = \tfrac{1}{8}\epsilon^2 d,\end{aligned}$

then as $\epsilon \to 0$, $W(\zeta; a, b, c, d)$ satisfies P_{III} with $z = \zeta$, $\alpha = a$, $\beta = b$, $\gamma = c$, $\delta = d$.

If in P_V

32.2.34 $\quad w(z; \alpha, \beta, \gamma, \delta) = \tfrac{1}{2}\sqrt{2}\epsilon W(\zeta; a, b),$

32.2.35 $z = 1 + \sqrt{2}\epsilon\zeta, \quad \alpha = \tfrac{1}{2}\epsilon^{-4}, \quad \beta = \tfrac{1}{4}b,$
$\gamma = -\epsilon^{-4}, \quad \delta = a\epsilon^{-2} - \tfrac{1}{2}\epsilon^{-4},$

then as $\epsilon \to 0$, $W(\zeta; a, b)$ satisfies P_{IV} with $z = \zeta$, $\alpha = a$, $\beta = b$.

Lastly, if in P_{VI}

32.2.36 $w(z; \alpha, \beta, \gamma, \delta) = W(\zeta; a, b, c, d),$

32.2.37 $z = 1 + \epsilon\zeta, \quad \gamma = c\epsilon^{-1} - d\epsilon^{-2}, \quad \delta = d\epsilon^{-2},$

then as $\epsilon \to 0$, $W(\zeta; a, b, c, d)$ satisfies P_{V} with $z = \zeta$, $\alpha = a$, $\beta = b$, $\gamma = c$, $\delta = d$.

32.3 Graphics

32.3(i) First Painlevé Equation

Plots of solutions $w_k(x)$ of P_{I} with $w_k(0) = 0$ and $w_k'(0) = k$ for various values of k, and the parabola $6w^2 + x = 0$. For analytical explanation see §32.11(i).

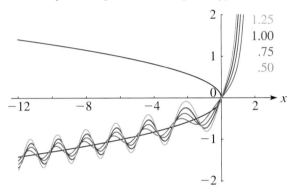

Figure 32.3.1: $w_k(x)$ for $-12 \le x \le 1.33$ and $k = 0.5$, 0.75, 1, 1.25, and the parabola $6w^2 + x = 0$, shown in black.

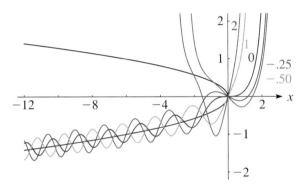

Figure 32.3.2: $w_k(x)$ for $-12 \le x \le 2.43$ and $k = -0.5$, -0.25, 0, 1, 2, and the parabola $6w^2 + x = 0$, shown in black.

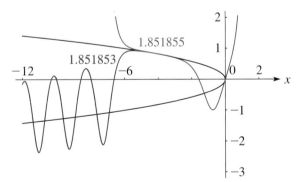

Figure 32.3.3: $w_k(x)$ for $-12 \le x \le 0.73$ and $k = 1.85185\,3, 1.85185\,5$. The two graphs are indistinguishable when x exceeds -5.2, approximately. The parabola $6w^2 + x = 0$ is shown in black.

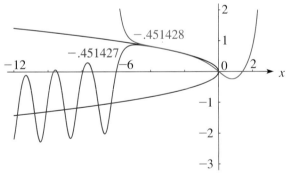

Figure 32.3.4: $w_k(x)$ for $-12 \le x \le 2.3$ and $k = -0.45142\,7, -0.45142\,8$. The two graphs are indistinguishable when x exceeds -4.8, approximately. The parabola $6w^2 + x = 0$ is shown in black.

32.3(ii) Second Painlevé Equation with $\alpha = 0$

Here $w_k(x)$ is the solution of $\mathrm{P_{II}}$ with $\alpha = 0$ and such that

32.3.1
$$w_k(x) \sim k\,\mathrm{Ai}(x), \qquad x \to +\infty;$$

compare §32.11(ii).

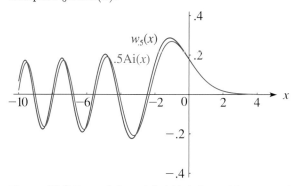

Figure 32.3.5: $w_k(x)$ and $k\,\mathrm{Ai}(x)$ for $-10 \le x \le 4$ with $k = 0.5$. The two graphs are indistinguishable when x exceeds -0.4, approximately.

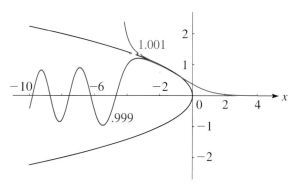

Figure 32.3.6: $w_k(x)$ for $-10 \le x \le 4$ with $k = 0.999$, 1.001. The two graphs are indistinguishable when x exceeds -2.8, approximately. The parabola $2w^2 + x = 0$ is shown in black.

32.3(iii) Fourth Painlevé Equation with $\beta = 0$

Here $u = u_k(x; \nu)$ is the solution of

32.3.2
$$\frac{d^2 u}{dx^2} = 3u^5 + 2xu^3 + \left(\tfrac{1}{4}x^2 - \nu - \tfrac{1}{2}\right)u,$$

such that

32.3.3
$$u \sim k\,U\!\left(-\nu - \tfrac{1}{2}, x\right), \qquad x \to +\infty.$$

The corresponding solution of $\mathrm{P_{IV}}$ is given by

32.3.4
$$w(x) = 2\sqrt{2}\,u_k^2(\sqrt{2}x, \nu),$$

with $\beta = 0$, $\alpha = 2\nu + 1$, and

32.3.5
$$w(x) \sim 2\sqrt{2}\,k^2\,U^2\!\left(-\nu - \tfrac{1}{2}, \sqrt{2}x\right), \qquad x \to +\infty;$$

compare (32.2.11) and §32.11(v). If we set $d^2u/dx^2 = 0$ in (32.3.2) and solve for u, then

32.3.6
$$u^2 = -\tfrac{1}{3}x \pm \tfrac{1}{6}\sqrt{x^2 + 12\nu + 6}.$$

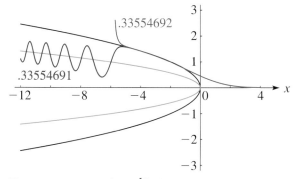

Figure 32.3.7: $u_k(x; -\tfrac{1}{2})$ for $-12 \le x \le 4$ with $k = 0.33554\,691$, $0.33554\,692$. The two graphs are indistinguishable when x exceeds -5.0, approximately. The parabolas $u^2 + \tfrac{1}{2}x = 0$, $u^2 + \tfrac{1}{6}x = 0$ are shown in black and green, respectively.

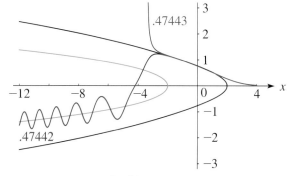

Figure 32.3.8: $u_k(x; \tfrac{1}{2})$ for $-12 \le x \le 4$ with $k = 0.47442$, 0.47443. The two graphs are indistinguishable when x exceeds -2.2, approximately. The curves $u^2 + \tfrac{1}{3}x \pm \tfrac{1}{6}\sqrt{x^2 + 12} = 0$ are shown in green and black, respectively.

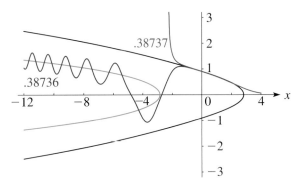

Figure 32.3.9: $u_k(x; \frac{3}{2})$ for $-12 \leq x \leq 4$ with $k = 0.38736, 0.38737$. The two graphs are indistinguishable when x exceeds -1.0, approximately. The curves $u^2 + \frac{1}{3}x \pm \frac{1}{6}\sqrt{x^2 + 24} = 0$ are shown in green and black, respectively.

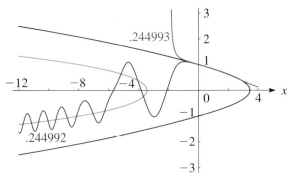

Figure 32.3.10: $u_k(x; \frac{5}{2})$ for $-12 \leq x \leq 4$ with $k = 0.244992, 0.244993$. The two graphs are indistinguishable when x exceeds -0.6, approximately. The curves $u^2 + \frac{1}{3}x \pm \frac{1}{6}\sqrt{x^2 + 36} = 0$ are shown in green and black, respectively.

32.4 Isomonodromy Problems

32.4(i) Definition

P_I–P_{VI} can be expressed as the compatibility condition of a linear system, called an *isomonodromy problem* or *Lax pair*. Suppose

32.4.1 $$\frac{\partial \mathbf{\Psi}}{\partial \lambda} = \mathbf{A}(z, \lambda)\mathbf{\Psi}, \quad \frac{\partial \mathbf{\Psi}}{\partial z} = \mathbf{B}(z, \lambda)\mathbf{\Psi},$$

is a linear system in which $\mathbf{A}$ and $\mathbf{B}$ are matrices and λ is independent of z. Then the equation

32.4.2 $$\frac{\partial^2 \mathbf{\Psi}}{\partial z \, \partial \lambda} = \frac{\partial^2 \mathbf{\Psi}}{\partial \lambda \, \partial z},$$

is satisfied provided that

32.4.3 $$\frac{\partial \mathbf{A}}{\partial z} - \frac{\partial \mathbf{B}}{\partial \lambda} + \mathbf{A}\mathbf{B} - \mathbf{B}\mathbf{A} = 0.$$

(32.4.3) is the *compatibility condition* of (32.4.1). Isomonodromy problems for Painlevé equations are not unique.

32.4(ii) First Painlevé Equation

P_I is the compatibility condition of (32.4.1) with

32.4.4 $$\mathbf{A}(z, \lambda) = (4\lambda^4 + 2w^2 + z)\begin{bmatrix} 1 & 0 \\ 0 & -1 \end{bmatrix} \\ - i(4\lambda^2 w + 2w^2 + z)\begin{bmatrix} 0 & -i \\ i & 0 \end{bmatrix} \\ - \left(2\lambda w' + \frac{1}{2\lambda}\right)\begin{bmatrix} 0 & 1 \\ 1 & 0 \end{bmatrix},$$

32.4.5 $$\mathbf{B}(z, \lambda) = \left(\lambda + \frac{w}{\lambda}\right)\begin{bmatrix} 1 & 0 \\ 0 & -1 \end{bmatrix} - \frac{iw}{\lambda}\begin{bmatrix} 0 & -i \\ i & 0 \end{bmatrix}.$$

32.4(iii) Second Painlevé Equation

P_{II} is the compatibility condition of (32.4.1) with

32.4.6 $$\mathbf{A}(z, \lambda) = -i(4\lambda^2 + 2w^2 + z)\begin{bmatrix} 1 & 0 \\ 0 & -1 \end{bmatrix} \\ - 2w'\begin{bmatrix} 0 & -i \\ i & 0 \end{bmatrix} + \left(4\lambda w - \frac{\alpha}{\lambda}\right)\begin{bmatrix} 0 & 1 \\ 1 & 0 \end{bmatrix},$$

32.4.7 $$\mathbf{B}(z, \lambda) = \begin{bmatrix} -i\lambda & w \\ w & i\lambda \end{bmatrix}.$$

See Flaschka and Newell (1980).

32.4(iv) Third Painlevé Equation

The compatibility condition of (32.4.1) with

32.4.8 $$\mathbf{A}(z, \lambda) = \begin{bmatrix} \frac{1}{4}z & 0 \\ 0 & -\frac{1}{4}z \end{bmatrix} + \begin{bmatrix} -\frac{1}{2}\theta_\infty & u_0 \\ u_1 & \frac{1}{2}\theta_\infty \end{bmatrix}\frac{1}{\lambda} \\ + \begin{bmatrix} v_0 - \frac{1}{4}z & -v_1 v_0 \\ (v_0 - \frac{1}{2}z)/v_1 & \frac{1}{4}z - v_0 \end{bmatrix}\frac{1}{\lambda^2},$$

32.4.9 $$\mathbf{B}(z, \lambda) = \begin{bmatrix} \frac{1}{4} & 0 \\ 0 & -\frac{1}{4} \end{bmatrix}\lambda + \begin{bmatrix} 0 & u_0 \\ u_1 & 0 \end{bmatrix}\frac{1}{z} \\ - \begin{bmatrix} v_0 - \frac{1}{4}z & -v_1 v_0 \\ (v_0 - \frac{1}{2}z)/v_1 & \frac{1}{4}z - v_0 \end{bmatrix}\frac{1}{z\lambda},$$

where θ_∞ is an arbitrary constant, is

32.4.10 $$zu_0' = \theta_\infty u_0 - zv_0 v_1,$$

32.4.11 $$zu_1' = -\theta_\infty u_1 - (z(2v_0 - z)/(2v_1)),$$

32.4.12 $$zv_0' = 2v_0 u_1 v_1 + v_0 + (u_0(2v_0 - z)/v_1),$$

32.4.13 $$zv_1' = 2u_0 - 2u_1 v_1^2 - \theta_\infty v_1.$$

If $w = -u_0/(v_0 v_1)$, then

32.4.14 $$zw' = (4v_0 - z)w^2 + (2\theta_\infty - 1)w + z,$$

and w satisfies $\mathrm{P_{III}}$ with

32.4.15 $\quad (\alpha, \beta, \gamma, \delta) = (2\theta_0, 2(1-\theta_\infty), 1, -1)$,

where

32.4.16 $\quad \theta_0 = \dfrac{4v_0}{z}\left(\theta_\infty\left(1-\dfrac{z}{4v_0}\right) + \dfrac{z-2v_0}{2v_0v_1}u_0 + u_1 v_1\right)$.

Note that the right-hand side of the last equation is a first integral of the system (32.4.10)–(32.4.13).

32.4(v) Other Painlevé Equations

For isomonodromy problems for $\mathrm{P_{IV}}$, $\mathrm{P_V}$, and $\mathrm{P_{VI}}$ see Jimbo and Miwa (1981).

32.5 Integral Equations

Let $K(z,\zeta)$ be the solution of

32.5.1
$$K(z,\zeta) = k\operatorname{Ai}\left(\dfrac{z+\zeta}{2}\right) + \dfrac{k^2}{4}\int_z^\infty \int_z^\infty K(z,s)\operatorname{Ai}\left(\dfrac{s+t}{2}\right)\operatorname{Ai}\left(\dfrac{t+\zeta}{2}\right)ds\,dt,$$

where k is a real constant, and $\operatorname{Ai}(z)$ is defined in §9.2. Then

32.5.2 $\quad w(z) = K(z,z)$,

satisfies $\mathrm{P_{II}}$ with $\alpha = 0$ and the boundary condition

32.5.3 $\quad w(z) \sim k\operatorname{Ai}(z), \qquad z \to +\infty$.

32.6 Hamiltonian Structure

32.6(i) Introduction

$\mathrm{P_I}$–$\mathrm{P_{VI}}$ can be written as a Hamiltonian system

32.6.1 $\quad \dfrac{dq}{dz} = \dfrac{\partial \mathrm{H}}{\partial p}, \quad \dfrac{dp}{dz} = -\dfrac{\partial \mathrm{H}}{\partial q}$,

for suitable (non-autonomous) Hamiltonian functions $\mathrm{H}(q,p,z)$.

32.6(ii) First Painlevé Equation

The Hamiltonian for $\mathrm{P_I}$ is

32.6.2 $\quad \mathrm{H_I}(q,p,z) = \tfrac{1}{2}p^2 - 2q^3 - zq$,

and so

32.6.3 $\quad q' = p$,

32.6.4 $\quad p' = 6q^2 + z$.

Then $q = w$ satisfies $\mathrm{P_I}$. The function

32.6.5 $\quad \sigma = \mathrm{H_I}(q,p,z)$,

defined by (32.6.2) satisfies

32.6.6 $\quad (\sigma'')^2 + 4(\sigma')^3 + 2z\sigma' - 2\sigma = 0$.

Conversely, if σ is a solution of (32.6.6), then

32.6.7 $\quad q = -\sigma'$,

32.6.8 $\quad p = -\sigma''$,

are solutions of (32.6.3) and (32.6.4).

32.6(iii) Second Painlevé Equation

The Hamiltonian for $\mathrm{P_{II}}$ is

32.6.9 $\quad \mathrm{H_{II}}(q,p,z) = \tfrac{1}{2}p^2 - (q^2 + \tfrac{1}{2}z)p - (\alpha + \tfrac{1}{2})q$,

and so

32.6.10 $\quad q' = p - q^2 - \tfrac{1}{2}z$,

32.6.11 $\quad p' = 2qp + \alpha + \tfrac{1}{2}$.

Then $q = w$ satisfies $\mathrm{P_{II}}$ and p satisfies

32.6.12 $\quad pp'' = \tfrac{1}{2}(p')^2 + 2p^3 - zp^2 - \tfrac{1}{2}(\alpha + \tfrac{1}{2})^2$.

The function $\sigma(z) = \mathrm{H_{II}}(q,p,z)$ defined by (32.6.9) satisfies

32.6.13 $\quad (\sigma'')^2 + 4(\sigma')^3 + 2\sigma'(z\sigma' - \sigma) = \tfrac{1}{4}(\alpha + \tfrac{1}{2})^2$.

Conversely, if $\sigma(z)$ is a solution of (32.6.13), then

32.6.14 $\quad q = (4\sigma'' + 2\alpha + 1)/(8\sigma')$,

32.6.15 $\quad p = -2\sigma'$,

are solutions of (32.6.10) and (32.6.11).

32.6(iv) Third Painlevé Equation

The Hamiltonian for $\mathrm{P_{III}}$ is

32.6.16
$$z\mathrm{H_{III}}(q,p,z) = q^2p^2 - \left(\kappa_\infty zq^2 + (2\theta_0 + 1)q - \kappa_0 z\right)p + \kappa_\infty(\theta_0 + \theta_\infty)zq,$$

and so

32.6.17 $\quad zq' = 2q^2p - \kappa_\infty zq^2 - (2\theta_0+1)q + \kappa_0 z$,

32.6.18 $\quad \begin{aligned}zp' &= -2qp^2 + 2\kappa_\infty zqp \\ &\quad + (2\theta_0 + 1)p - \kappa_\infty(\theta_0 + \theta_\infty)z.\end{aligned}$

Then $q = w$ satisfies $\mathrm{P_{III}}$ with

32.6.19 $\quad (\alpha, \beta, \gamma, \delta) = \left(-2\kappa_\infty\theta_\infty, 2\kappa_0(\theta_0+1), \kappa_\infty^2, -\kappa_0^2\right)$.

The function

32.6.20 $\quad \sigma = z\mathrm{H_{III}}(q,p,z) + pq + \theta_0^2 - \tfrac{1}{2}\kappa_0\kappa_\infty z^2$

defined by (32.6.16) satisfies

32.6.21 $\quad \begin{aligned}&(z\sigma'' - \sigma')^2 + 2\left((\sigma')^2 - \kappa_0^2\kappa_\infty^2 z^2\right)(z\sigma' - 2\sigma) \\ &\quad + 8\kappa_0\kappa_\infty\theta_0\theta_\infty z\sigma' = 4\kappa_0^2\kappa_\infty^2(\theta_0^2 + \theta_\infty^2)z^2.\end{aligned}$

Conversely, if σ is a solution of (32.6.21), then

32.6.22 $\quad q = \dfrac{\kappa_0\left(z\sigma'' - (2\theta_0 + 1)\sigma' + 2\kappa_0\kappa_\infty\theta_\infty z\right)}{\kappa_0^2\kappa_\infty^2 z^2 - (\sigma')^2}$,

32.6.23 $\quad p = (\sigma' + \kappa_0\kappa_\infty z)/(2\kappa_0)$,

are solutions of (32.6.17) and (32.6.18).

The Hamiltonian for P'_{III} (§32.2(iii)) is

32.6.24 $\quad \zeta H_{III}(q,p,\zeta) = q^2 p^2 - \left(\eta_\infty q^2 + \theta_0 q - \eta_0 \zeta\right) p + \tfrac{1}{2}\eta_\infty(\theta_0 + \theta_\infty)q,$

and so

32.6.25 $\quad \zeta q' = 2q^2 p - \eta_\infty q^2 - \theta_0 q + \eta_0 \zeta,$

32.6.26 $\quad \zeta p' = -2qp^2 + 2\eta_\infty q p + \theta_0 p - \tfrac{1}{2}\eta_\infty(\theta_0 + \theta_1).$

Then $q = u$ satisfies P'_{III} with

32.6.27 $\quad (\alpha, \beta, \gamma, \delta) = \left(-4\eta_\infty \theta_\infty, 4\eta_0(\theta_0+1), 4\eta_\infty^2, -4\eta_0^2\right).$

The function

32.6.28 $\quad \sigma = \zeta H_{III}(q,p,\zeta) + \tfrac{1}{4}\theta_0^2 - \tfrac{1}{2}\eta_0 \eta_\infty \zeta$

defined by (32.6.24) satisfies

32.6.29 $\quad \begin{aligned}&\zeta^2 (\sigma'')^2 + \left(4(\sigma')^2 - \eta_0^2 \eta_\infty^2\right)(\zeta \sigma' - \sigma) \\ & + \eta_0 \eta_\infty \theta_0 \theta_\infty \sigma' = \tfrac{1}{4}\eta_0^2 \eta_\infty^2 (\theta_0^2 + \theta_\infty^2).\end{aligned}$

Conversely, if σ is a solution of (32.6.29), then

32.6.30 $\quad q = \dfrac{\eta_0 \left(\zeta \sigma'' - 2\theta_0 \sigma' + \eta_0 \eta_\infty \theta_\infty\right)}{\eta_0^2 \eta_\infty^2 - 4(\sigma')^2},$

32.6.31 $\quad p = (2\sigma' + \eta_0 \eta_\infty \zeta)/(2\eta_0),$

are solutions of (32.6.25) and (32.6.26).

The Hamiltonian for P_{III} with $\gamma = 0$ is

32.6.32 $\quad zH_{III}(q,p,z) = q^2 p^2 + (\theta q - \kappa_0 z)p - \kappa_\infty z q,$

and so

32.6.33 $\quad z q' = 2q^2 p + \theta q - \kappa_0 z,$

32.6.34 $\quad z p' = -2qp^2 - \theta p + \kappa_\infty z.$

Then $q = w$ satisfies P_{III} with

32.6.35 $\quad (\alpha, \beta, \gamma, \delta) = \left(2\kappa_\infty, \kappa_0(\theta - 1), 0, -\kappa_0^2\right).$

The function

32.6.36 $\quad \sigma = zH_{III}(q,p,z) + pq + \tfrac{1}{4}(\theta + 1)^2$

defined by (32.6.32) satisfies

32.6.37 $\quad \begin{aligned}&(z\sigma'' - \sigma')^2 + 2(\sigma')^2(z\sigma' - 2\sigma) \\ & - 4\kappa_0 \kappa_\infty(\theta+1)\theta_\infty z\sigma' = 4\kappa_0^2 \kappa_\infty^2 z^2.\end{aligned}$

Conversely, if σ is a solution of (32.6.37), then

32.6.38 $\quad q = \kappa_0 \left(z\sigma'' - \theta \sigma' + 2\kappa_0 \kappa_\infty z\right)/(\sigma')^2,$

32.6.39 $\quad p = \sigma'/(2\kappa_0),$

are solutions of (32.6.33) and (32.6.34).

32.6(v) Other Painlevé Equations

For Hamiltonian structure for P_{IV} see Jimbo and Miwa (1981), Okamoto (1986); also Forrester and Witte (2001).

For Hamiltonian structure for P_V see Jimbo and Miwa (1981), Okamoto (1987b); also Forrester and Witte (2002).

For Hamiltonian structure for P_{VI} see Jimbo and Miwa (1981) and Okamoto (1987a); also Forrester and Witte (2004).

32.7 Bäcklund Transformations

32.7(i) Definition

With the exception of P_I, a *Bäcklund transformation* relates a Painlevé transcendent of one type either to another of the same type but with different values of the parameters, or to another type.

32.7(ii) Second Painlevé Equation

Let $w = w(z;\alpha)$ be a solution of P_{II}. Then the transformations

32.7.1 $\quad \mathcal{S}: \ w(z;-\alpha) = -w,$

and

32.7.2 $\quad \mathcal{T}^\pm: \ w(z;\alpha \pm 1) = -w - \dfrac{2\alpha \pm 1}{2w^2 \pm 2w' + z},$

furnish solutions of P_{II}, provided that $\alpha \neq \mp \tfrac{1}{2}$. P_{II} also has the special transformation

32.7.3 $\quad W(\zeta; \tfrac{1}{2}\varepsilon) = \dfrac{2^{-1/3}\varepsilon}{w(z;0)}\dfrac{d}{dz}w(z;0),$

or equivalently,

32.7.4 $\quad w^2(z;0) = 2^{-1/3}\left(W^2(\zeta; \tfrac{1}{2}\varepsilon) - \varepsilon \dfrac{d}{d\zeta}W(\zeta; \tfrac{1}{2}\varepsilon) + \tfrac{1}{2}\zeta\right),$

with $\zeta = -2^{1/3}z$ and $\varepsilon = \pm 1$, where $W(\zeta; \tfrac{1}{2}\varepsilon)$ satisfies P_{II} with $z = \zeta$, $\alpha = \tfrac{1}{2}\varepsilon$, and $w(z;0)$ satisfies P_{II} with $\alpha = 0$.

The solutions $w_\alpha = w(z;\alpha)$, $w_{\alpha \pm 1} = w(z; \alpha \pm 1)$, satisfy the nonlinear recurrence relation

32.7.5 $\quad \dfrac{\alpha + \tfrac{1}{2}}{w_{\alpha+1} + w_\alpha} + \dfrac{\alpha - \tfrac{1}{2}}{w_\alpha + w_{\alpha-1}} + 2w_\alpha^2 + z = 0.$

See Fokas *et al.* (1993).

32.7(iii) Third Painlevé Equation

Let $w_j = w(z; \alpha_j, \beta_j, \gamma_j, \delta_j)$, $j = 0, 1, 2$, be solutions of P_{III} with

32.7.6 $\quad (\alpha_1, \beta_1, \gamma_1, \delta_1) = (-\alpha_0, -\beta_0, \gamma_0, \delta_0),$

32.7.7 $\quad (\alpha_2, \beta_2, \gamma_2, \delta_2) = (-\beta_0, -\alpha_0, -\delta_0, -\gamma_0).$

Then

32.7.8 $\quad\quad\quad \mathcal{S}_1: \ w_1 = -w_0,$

32.7.9 $\quad\quad\quad \mathcal{S}_2: \ w_2 = 1/w_0\,.$

Next, let $W_j = W(z; \alpha_j, \beta_j, 1, -1)$, $j = 0, 1, 2, 3, 4$, be solutions of P_{III} with

32.7.10 $\quad \alpha_1 = \alpha_3 = \alpha_0 + 2, \quad \alpha_2 = \alpha_4 = \alpha_0 - 2,$
$\quad\quad\quad \beta_1 = \beta_2 = \beta_0 + 2, \quad \beta_3 = \beta_4 = \beta_0 - 2.$

Then

32.7.11
$$\mathcal{T}_1: \ W_1 = \frac{zW_0' + zW_0^2 - \beta W_0 - W_0 + z}{W_0(zW_0' + zW_0^2 + \alpha W_0 + W_0 + z)},$$

32.7.12
$$\mathcal{T}_2: \ W_2 = -\frac{zW_0' - zW_0^2 - \beta W_0 - W_0 + z}{W_0(zW_0' - zW_0^2 - \alpha W_0 + W_0 + z)},$$

32.7.13
$$\mathcal{T}_3: \ W_3 = -\frac{zW_0' + zW_0^2 + \beta W_0 - W_0 - z}{W_0(zW_0' + zW_0^2 + \alpha W_0 + W_0 - z)},$$

32.7.14
$$\mathcal{T}_4: \ W_4 = \frac{zW_0' - zW_0^2 + \beta W_0 - W_0 - z}{W_0(zW_0' - zW_0^2 - \alpha W_0 + W_0 - z)}.$$

See Milne et al. (1997).

If $\gamma = 0$ and $\alpha\delta \neq 0$, then set $\alpha = 1$ and $\delta = -1$, without loss of generality. Let $u_j = w(z; 1, \beta_j, 0, -1)$, $j = 0, 5, 6$, be solutions of P_{III} with

32.7.15 $\quad\quad\quad \beta_5 = \beta_0 + 2, \quad \beta_6 = \beta_0 - 2.$

Then

32.7.16 $\quad \mathcal{T}_5: \ u_5 = (zu_0' + z - (\beta_0 + 1)u_0)/u_0^2\,,$

32.7.17 $\quad \mathcal{T}_6: \ u_6 = -(zu_0' - z + (\beta_0 - 1)u_0)/u_0^2\,.$

Similar results hold for P_{III} with $\delta = 0$ and $\beta\gamma \neq 0$.

Furthermore,

32.7.18 $\quad w(z; a, b, 0, 0) = W^2(\zeta; 0, 0, a, b), \quad z = \tfrac{1}{2}\zeta^2.$

32.7(iv) Fourth Painlevé Equation

Let $w_0 = w(z; \alpha_0, \beta_0)$ and $w_j^\pm = w(z; \alpha_j^\pm, \beta_j^\pm)$, $j = 1, 2, 3, 4$, be solutions of P_{IV} with

32.7.19
$$\alpha_1^\pm = \tfrac{1}{4}\left(2 - 2\alpha_0 \pm 3\sqrt{-2\beta_0}\right),$$
$$\beta_1^\pm = -\tfrac{1}{2}\left(1 + \alpha_0 \pm \tfrac{1}{2}\sqrt{-2\beta_0}\right)^2,$$
$$\alpha_2^\pm = -\tfrac{1}{4}\left(2 + 2\alpha_0 \pm 3\sqrt{-2\beta_0}\right),$$
$$\beta_2^\pm = -\tfrac{1}{2}\left(1 - \alpha_0 \pm \tfrac{1}{2}\sqrt{-2\beta_0}\right)^2,$$
$$\alpha_3^\pm = \tfrac{3}{2} - \tfrac{1}{2}\alpha_0 \mp \tfrac{3}{4}\sqrt{-2\beta_0},$$
$$\beta_3^\pm = -\tfrac{1}{2}\left(1 - \alpha_0 \pm \tfrac{1}{2}\sqrt{-2\beta_0}\right)^2,$$
$$\alpha_4^\pm = -\tfrac{3}{2} - \tfrac{1}{2}\alpha_0 \mp \tfrac{3}{4}\sqrt{-2\beta_0},$$
$$\beta_4^\pm = -\tfrac{1}{2}\left(-1 - \alpha_0 \pm \tfrac{1}{2}\sqrt{-2\beta_0}\right)^2.$$

Then

32.7.20 $\quad \mathcal{T}_1^\pm: \ w_1^\pm = \dfrac{w_0' - w_0^2 - 2zw_0 \mp \sqrt{-2\beta_0}}{2w_0},$

32.7.21 $\quad \mathcal{T}_2^\pm: \ w_2^\pm = -\dfrac{w_0' + w_0^2 + 2zw_0 \mp \sqrt{-2\beta_0}}{2w_0},$

32.7.22 $\quad \mathcal{T}_3^\pm: \ w_3^\pm = w_0 + \dfrac{2\left(1 - \alpha_0 \mp \tfrac{1}{2}\sqrt{-2\beta_0}\right)w_0}{w_0' \pm \sqrt{-2\beta_0} + 2zw_0 + w_0^2},$

32.7.23 $\quad \mathcal{T}_4^\pm: \ w_4^\pm = w_0 + \dfrac{2\left(1 + \alpha_0 \pm \tfrac{1}{2}\sqrt{-2\beta_0}\right)w_0}{w_0' \mp \sqrt{-2\beta_0} - 2zw_0 - w_0^2},$

valid when the denominators are nonzero, and where the upper signs or the lower signs are taken throughout each transformation. See Bassom et al. (1995).

32.7(v) Fifth Painlevé Equation

Let $w_j(z_j) = w(z_j; \alpha_j, \beta_j, \gamma_j, \delta_j)$, $j = 0, 1, 2$, be solutions of P_V with

32.7.24
$z_1 = -z_0, \quad z_2 = z_0, \quad (\alpha_1, \beta_1, \gamma_1, \delta_1) = (\alpha_0, \beta_0, -\gamma_0, \delta_0),$
$(\alpha_2, \beta_2, \gamma_2, \delta_2) = (-\beta_0, -\alpha_0, -\gamma_0, \delta_0).$

Then

32.7.25 $\quad\quad\quad \mathcal{S}_1: \ w_1(z_1) = w(z_0),$

32.7.26 $\quad\quad\quad \mathcal{S}_2: \ w_2(z_2) = 1/w(z_0)\,.$

Let $W_0 = W(z; \alpha_0, \beta_0, \gamma_0, -\tfrac{1}{2})$ and $W_1 = W(z; \alpha_1, \beta_1, \gamma_1, -\tfrac{1}{2})$ be solutions of P_V, where

32.7.27
$$\alpha_1 = \tfrac{1}{8}\left(\gamma_0 + \varepsilon_1\left(1 - \varepsilon_3\sqrt{-2\beta_0} - \varepsilon_2\sqrt{2\alpha_0}\right)\right)^2,$$
$$\beta_1 = -\tfrac{1}{8}\left(\gamma_0 - \varepsilon_1\left(1 - \varepsilon_3\sqrt{-2\beta_0} - \varepsilon_2\sqrt{2\alpha_0}\right)\right)^2,$$
$$\gamma_1 = \varepsilon_1\left(\varepsilon_3\sqrt{-2\beta_0} - \varepsilon_2\sqrt{2\alpha_0}\right),$$

and $\varepsilon_j = \pm 1$, $j = 1, 2, 3$, independently. Also let

32.7.28
$$\Phi = zW_0' - \varepsilon_2\sqrt{2\alpha_0}W_0^2 + \varepsilon_3\sqrt{-2\beta_0} \\ + \left(\varepsilon_2\sqrt{2\alpha_0} - \varepsilon_3\sqrt{-2\beta_0} + \varepsilon_1 z\right)W_0,$$

and assume $\Phi \neq 0$. Then

32.7.29 $\quad \mathcal{T}_{\varepsilon_1, \varepsilon_2, \varepsilon_3}: \; W_1 = (\Phi - 2\varepsilon_1 zW_0)/\Phi$,

provided that the numerator on the right-hand side does not vanish. Again, since $\varepsilon_j = \pm 1$, $j = 1, 2, 3$, independently, there are eight distinct transformations of type $\mathcal{T}_{\varepsilon_1, \varepsilon_2, \varepsilon_3}$.

32.7(vi) Relationship Between the Third and Fifth Painlevé Equations

Let $w = w(z; \alpha, \beta, 1, -1)$ be a solution of P_{III} and

32.7.30 $\quad v = w' - \varepsilon w^2 + ((1 - \varepsilon \alpha)w/z)$,

with $\varepsilon = \pm 1$. Then

32.7.31 $\quad W(\zeta; \alpha_0, \beta_0, \gamma_0, \delta_0) = \dfrac{v-1}{v+1}, \quad z = \sqrt{2\zeta}$,

satisfies P_V with

32.7.32
$$(\alpha_0, \beta_0, \gamma_0, \delta_0) \\ = \left((\beta - \varepsilon\alpha + 2)^2/32, -(\beta + \varepsilon\alpha - 2)^2/32, -\varepsilon, 0\right).$$

32.7(vii) Sixth Painlevé Equation

Let $w_j(z_j) = w_j(z_j; \alpha_j, \beta_j, \gamma_j, \delta_j)$, $j = 0, 1, 2, 3$, be solutions of P_{VI} with

32.7.33 $\quad z_1 = 1/z_0$,

32.7.34 $\quad z_2 = 1 - z_0$,

32.7.35 $\quad z_3 = 1/z_0$,

32.7.36 $\quad (\alpha_1, \beta_1, \gamma_1, \delta_1) = (\alpha_0, \beta_0, -\delta_0 + \tfrac{1}{2}, -\gamma_0 + \tfrac{1}{2})$,

32.7.37 $\quad (\alpha_2, \beta_2, \gamma_2, \delta_2) = (\alpha_0, -\gamma_0, -\beta_0, \delta_0)$,

32.7.38 $\quad (\alpha_3, \beta_3, \gamma_3, \delta_3) = (-\beta_0, -\alpha_0, \gamma_0, \delta_0)$.

Then

32.7.39 $\quad \mathcal{S}_1: \; w_1(z_1) = w_0(z_0)/z_0$,

32.7.40 $\quad \mathcal{S}_2: \; w_2(z_2) = 1 - w_0(z_0)$,

32.7.41 $\quad \mathcal{S}_3: \; w_3(z_3) = 1/w_0(z_0)$.

The transformations $\mathcal{S}_j$, for $j = 1, 2, 3$, generate a group of order 24. See Iwasaki et al. (1991, p. 127).

Let $w(z; \alpha, \beta, \gamma, \delta)$ and $W(z; A, B, C, D)$ be solutions of P_{VI} with

32.7.42 $\quad (\alpha, \beta, \gamma, \delta) = \left(\tfrac{1}{2}(\theta_\infty - 1)^2, -\tfrac{1}{2}\theta_0^2, \tfrac{1}{2}\theta_1^2, \tfrac{1}{2}(1 - \theta_2^2)\right)$,

32.7.43
$$(A, B, C, D) = \left(\tfrac{1}{2}(\Theta_\infty - 1)^2, -\tfrac{1}{2}\Theta_0^2, \tfrac{1}{2}\Theta_1^2, \tfrac{1}{2}(1 - \Theta_2^2)\right),$$

and

32.7.44 $\quad \theta_j = \Theta_j + \tfrac{1}{2}\sigma$,

for $j = 0, 1, 2, \infty$, where

32.7.45 $\quad \sigma = \theta_0 + \theta_1 + \theta_2 + \theta_\infty - 1 = 1 - (\Theta_0 + \Theta_1 + \Theta_2 + \Theta_\infty)$.

Then

32.7.46
$$\dfrac{\sigma}{w - W} = \dfrac{z(z-1)W'}{W(W-1)(W-z)} + \dfrac{\Theta_0}{W} + \dfrac{\Theta_1}{W-1} + \dfrac{\Theta_2 - 1}{W-z} \\ = \dfrac{z(z-1)w'}{w(w-1)(w-z)} + \dfrac{\theta_0}{w} + \dfrac{\theta_1}{w-1} + \dfrac{\theta_2 - 1}{w-z}.$$

P_{VI} also has quadratic and quartic transformations. Let $w = w(z; \alpha, \beta, \gamma, \delta)$ be a solution of P_{VI}. The quadratic transformation

32.7.47 $\quad u_1(\zeta_1) = \dfrac{(1-w)(w-z)}{(1+\sqrt{z})^2 w}, \quad \zeta_1 = \left(\dfrac{1-\sqrt{z}}{1+\sqrt{z}}\right)^2$,

transforms P_{VI} with $\alpha = -\beta$ and $\gamma = \tfrac{1}{2} - \delta$ to P_{VI} with $(\alpha_1, \beta_1, \gamma_1, \delta_1) = (4\alpha, -4\gamma, 0, \tfrac{1}{2})$. The quartic transformation

32.7.48 $\quad u_2(\zeta_2) = \dfrac{(w^2 - z)^2}{4w(w-1)(w-z)}, \quad \zeta_2 = z$,

transforms P_{VI} with $\alpha = -\beta = \gamma = \tfrac{1}{2} - \delta$ to P_{VI} with $(\alpha_2, \beta_2, \gamma_2, \delta_2) = (16\alpha, 0, 0, \tfrac{1}{2})$. Also,

32.7.49 $\quad u_3(\zeta_3) = \left(\dfrac{1 - z^{1/4}}{1 + z^{1/4}}\right)^2 \left(\dfrac{\sqrt{w} + z^{1/4}}{\sqrt{w} - z^{1/4}}\right)^2$,

32.7.50 $\quad \zeta_3 = \left(\dfrac{1 - z^{1/4}}{1 + z^{1/4}}\right)^4$,

transforms P_{VI} with $\alpha = \beta = 0$ and $\gamma = \tfrac{1}{2} - \delta$ to P_{VI} with $\alpha_3 = \beta_3$ and $\gamma_3 = \tfrac{1}{2} - \delta_3$.

32.7(viii) Affine Weyl Groups

See Okamoto (1986, 1987a,b,c), Sakai (2001), Umemura (2000).

32.8 Rational Solutions

32.8(i) Introduction

P_{II}–P_{VI} possess hierarchies of rational solutions for special values of the parameters which are generated from "seed solutions" using the Bäcklund transformations and often can be expressed in the form of determinants. See Airault (1979).

32.8(ii) Second Painlevé Equation

Rational solutions of P_{II} exist for $\alpha = n(\in \mathbb{Z})$ and are generated using the seed solution $w(z;0) = 0$ and the Bäcklund transformations (32.7.1) and (32.7.2). The first four are

32.8.1 $$w(z;1) = -1/z,$$

32.8.2 $$w(z;2) = \frac{1}{z} - \frac{3z^2}{z^3 + 4},$$

32.8.3 $$w(z;3) = \frac{3z^2}{z^3 + 4} - \frac{6z^2(z^3 + 10)}{z^6 + 20z^3 - 80},$$

32.8.4
$$w(z;4) = -\frac{1}{z} + \frac{6z^2(z^3 + 10)}{z^6 + 20z^3 - 80} - \frac{9z^5(z^3 + 40)}{z^9 + 60z^6 + 11200}.$$

More generally,

32.8.5 $$w(z;n) = \frac{d}{dz}\left(\ln\left(\frac{Q_{n-1}(z)}{Q_n(z)}\right)\right),$$

where the $Q_n(z)$ are monic polynomials (coefficient of highest power of z is 1) satisfying

32.8.6
$$Q_{n+1}(z)Q_{n-1}(z) = zQ_n^2(z) + 4\left(Q_n'(z)\right)^2 - 4Q_n(z)Q_n''(z),$$

with $Q_0(z) = 1$, $Q_1(z) = z$. Thus

$$Q_2(z) = z^3 + 4,$$
$$Q_3(z) = z^6 + 20z^3 - 80,$$
$$Q_4(z) = z^{10} + 60z^7 + 11200z,$$

32.8.7
$$Q_5(z) = z^{15} + 140z^{12} + 2800z^9 + 78400z^6 - 3\,13600z^3 - 62\,72000,$$
$$Q_6(z) = z^{21} + 280z^{18} + 18480z^{15} + 6\,27200z^{12} - 172\,48000z^9 + 14488\,32000z^6 + 1\,93177\,60000z^3 - 3\,86355\,20000.$$

Next, let $p_m(z)$ be the polynomials defined by $p_m(z) = 0$ for $m < 0$, and

32.8.8 $$\sum_{m=0}^{\infty} p_m(z)\lambda^m = \exp\left(z\lambda - \tfrac{4}{3}\lambda^3\right).$$

Then for $n \geq 2$

32.8.9 $$w(z;n) = \frac{d}{dz}\left(\ln\left(\frac{\tau_{n-1}(z)}{\tau_n(z)}\right)\right),$$

where $\tau_n(z)$ is the $n \times n$ determinant

32.8.10
$$\tau_n(z) = \begin{vmatrix} p_1(z) & p_3(z) & \cdots & p_{2n-1}(z) \\ p_1'(z) & p_3'(z) & \cdots & p_{2n-1}'(z) \\ \vdots & \vdots & \ddots & \vdots \\ p_1^{(n-1)}(z) & p_3^{(n-1)}(z) & \cdots & p_{2n-1}^{(n-1)}(z) \end{vmatrix}.$$

For plots of the zeros of $Q_n(z)$ see Clarkson and Mansfield (2003).

32.8(iii) Third Painlevé Equation

Special rational solutions of P_{III} are

32.8.11 $$w(z;\mu, -\mu\kappa^2, \lambda, -\lambda\kappa^4) = \kappa,$$

32.8.12 $$w(z;0, -\mu, 0, \mu\kappa) = \kappa z,$$

32.8.13 $$w(z;2\kappa+3, -2\kappa+1, 1, -1) = \frac{z+\kappa}{z+\kappa+1},$$

with κ, λ, and μ arbitrary constants.

In the general case assume $\gamma\delta \neq 0$, so that as in §32.2(ii) we may set $\gamma = 1$ and $\delta = -1$. Then P_{III} has rational solutions iff

32.8.14 $$\alpha \pm \beta = 4n,$$

with $n \in \mathbb{Z}$. These solutions have the form

32.8.15 $$w(z) = P_m(z)/Q_m(z),$$

where $P_m(z)$ and $Q_m(z)$ are polynomials of degree m, with no common zeros.

For examples and plots see Milne et al. (1997); also Clarkson (2003a). For determinantal representations see Kajiwara and Masuda (1999).

32.8(iv) Fourth Painlevé Equation

Special rational solutions of P_{IV} are

32.8.16 $$w_1(z;\pm 2, -2) = \pm 1/z,$$

32.8.17 $$w_2(z;0,-2) = -2z,$$

32.8.18 $$w_3(z;0,-\tfrac{2}{9}) = -\tfrac{2}{3}z.$$

There are also three families of solutions of P_{IV} of the form

32.8.19 $$w_1(z;\alpha_1, \beta_1) = P_{1,n-1}(z)/Q_{1,n}(z),$$

32.8.20 $$w_2(z;\alpha_2, \beta_2) = -2z + (P_{2,n-1}(z)/Q_{2,n}(z)),$$

32.8.21 $$w_3(z;\alpha_3, \beta_3) = -\tfrac{2}{3}z + (P_{3,n-1}(z)/Q_{3,n}(z)),$$

where $P_{j,n-1}(z)$ and $Q_{j,n}(z)$ are polynomials of degrees $n-1$ and n, respectively, with no common zeros.

In general, P_{IV} has rational solutions iff either

32.8.22 $$\alpha = m, \quad \beta = -2(1+2n-m)^2,$$

or

32.8.23 $$\alpha = m, \quad \beta = -2(\tfrac{1}{3}+2n-m)^2,$$

with $m, n \in \mathbb{Z}$. The rational solutions when the parameters satisfy (32.8.22) are special cases of §32.10(iv).

For examples and plots see Bassom et al. (1995); also Clarkson (2003b). For determinantal representations see Kajiwara and Ohta (1998) and Noumi and Yamada (1999).

32.8(v) Fifth Painlevé Equation

Special rational solutions of P_V are

32.8.24 $w(z; \frac{1}{2}, -\frac{1}{2}\mu^2, \kappa(2-\mu), -\frac{1}{2}\kappa^2) = \kappa z + \mu$,

32.8.25 $w(z; \frac{1}{2}, \kappa^2\mu, 2\kappa\mu, \mu) = \kappa/(z+\kappa)$,

32.8.26 $w(z; \frac{1}{8}, -\frac{1}{8}, -\kappa\mu, \mu) = (\kappa+z)/(\kappa-z)$,

with κ and μ arbitrary constants.

In the general case assume $\delta \neq 0$, so that as in §32.2(ii) we may set $\delta = -\frac{1}{2}$. Then P_V has a rational solution iff one of the following holds with $m, n \in \mathbb{Z}$ and $\varepsilon = \pm 1$:

(a) $\alpha = \frac{1}{2}(m+\varepsilon\gamma)^2$ and $\beta = -\frac{1}{2}n^2$, where $n > 0$, $m+n$ is odd, and $\alpha \neq 0$ when $|m| < n$.

(b) $\alpha = \frac{1}{2}n^2$ and $\beta = -\frac{1}{2}(m+\varepsilon\gamma)^2$, where $n > 0$, $m+n$ is odd, and $\beta \neq 0$ when $|m| < n$.

(c) $\alpha = \frac{1}{2}a^2$, $\beta = -\frac{1}{2}(a+n)^2$, and $\gamma = m$, with $m+n$ even.

(d) $\alpha = \frac{1}{2}(b+n)^2$, $\beta = -\frac{1}{2}b^2$, and $\gamma = m$, with $m+n$ even.

(e) $\alpha = \frac{1}{8}(2m+1)^2$, $\beta = -\frac{1}{8}(2n+1)^2$, and $\gamma \notin \mathbb{Z}$.

These rational solutions have the form

32.8.27 $w(z) = \lambda z + \mu + (P_{n-1}(z)/Q_n(z))$,

where λ, μ are constants, and $P_{n-1}(z)$, $Q_n(z)$ are polynomials of degrees $n-1$ and n, respectively, with no common zeros. Cases (a) and (b) are special cases of §32.10(v).

For examples and plots see Clarkson (2005). For determinantal representations see Masuda et al. (2002). For the case $\delta = 0$ see Airault (1979) and Lukaševič (1968).

32.8(vi) Sixth Painlevé Equation

Special rational solutions of P_{VI} are

32.8.28 $w(z; \mu, -\mu\kappa^2, \frac{1}{2}, \frac{1}{2} - \mu(\kappa-1)^2) = \kappa z$,

32.8.29 $w(z; 0, 0, 2, 0) = \kappa z^2$,

32.8.30 $w(z; 0, 0, \frac{1}{2}, -\frac{3}{2}) = \kappa/z$,

32.8.31 $w(z; 0, 0, 2, -4) = \kappa/z^2$,

32.8.32
$w(z; \frac{1}{2}(\kappa+\mu)^2, -\frac{1}{2}, \frac{1}{2}(\mu-1)^2, \frac{1}{2}\kappa(2-\kappa)) = \dfrac{z}{\kappa+\mu z}$,

with κ and μ arbitrary constants.

In the general case, P_{VI} has rational solutions if

32.8.33 $a + b + c + d = 2n + 1$,

where $n \in \mathbb{Z}$, $a = \varepsilon_1\sqrt{2\alpha}$, $b = \varepsilon_2\sqrt{-2\beta}$, $c = \varepsilon_3\sqrt{2\gamma}$, and $d = \varepsilon_4\sqrt{1-2\delta}$, with $\varepsilon_j = \pm 1$, $j = 1, 2, 3, 4$, independently, and at least one of a, b, c or d is an integer. These are special cases of §32.10(vi).

32.9 Other Elementary Solutions

32.9(i) Third Painlevé Equation

Elementary nonrational solutions of P_{III} are

32.9.1 $w(z; \mu, 0, 0, -\mu\kappa^3) = \kappa z^{1/3}$,

32.9.2
$w(z; 0, -2\kappa, 0, 4\kappa\mu - \lambda^2) = z(\kappa(\ln z)^2 + \lambda \ln z + \mu)$,

32.9.3 $w(z; -\nu^2\lambda, 0, \nu^2(\lambda^2 - 4\kappa\mu), 0) = \dfrac{z^{\nu-1}}{\kappa z^{2\nu} + \lambda z^\nu + \mu}$,

with κ, λ, μ, and ν arbitrary constants.

In the case $\gamma = 0$ and $\alpha\delta \neq 0$ we assume, as in §32.2(ii), $\alpha = 1$ and $\delta = -1$. Then P_{III} has algebraic solutions iff

32.9.4 $\beta = 2n$,

with $n \in \mathbb{Z}$. These are rational solutions in $\zeta = z^{1/3}$ of the form

32.9.5 $w(z) = P_{n^2+1}(\zeta)/Q_{n^2}(\zeta)$,

where $P_{n^2+1}(\zeta)$ and $Q_{n^2}(\zeta)$ are polynomials of degrees $n^2 + 1$ and n^2, respectively, with no common zeros. For examples and plots see Clarkson (2003a) and Milne et al. (1997). Similar results hold when $\delta = 0$ and $\beta\gamma \neq 0$.

P_{III} with $\beta = \delta = 0$ has a first integral

32.9.6 $z^2(w')^2 + 2zww' = (C + 2\alpha z w + \gamma z^2 w^2)w^2$,

with C an arbitrary constant, which is solvable by quadrature. A similar result holds when $\alpha = \gamma = 0$. P_{III} with $\alpha = \beta = \gamma = \delta = 0$, has the general solution $w(z) = Cz^\mu$, with C and μ arbitrary constants.

32.9(ii) Fifth Painlevé Equation

Elementary nonrational solutions of P_V are

32.9.7 $w(z; \mu, -\frac{1}{8}, -\mu\kappa^2, 0) = 1 + \kappa z^{1/2}$,

32.9.8 $w(z; 0, 0, \mu, -\frac{1}{2}\mu^2) = \kappa \exp(\mu z)$,

with κ and μ arbitrary constants.

P_V, with $\delta = 0$, has algebraic solutions if either

32.9.9 $(\alpha, \beta, \gamma) = (\frac{1}{2}\mu^2, -\frac{1}{8}(2n-1)^2, -1)$,

or

32.9.10 $(\alpha, \beta, \gamma) = (\frac{1}{8}(2n-1)^2, -\frac{1}{2}\mu^2, 1)$,

with $n \in \mathbb{Z}$ and μ arbitrary. These are rational solutions in $\zeta = z^{1/2}$ of the form

32.9.11 $w(z) = P_{n^2-n+1}(\zeta)/Q_{n^2-n}(\zeta)$,

where $P_{n^2-n+1}(\zeta)$ and $Q_{n^2-n}(\zeta)$ are polynomials of degrees n^2-n+1 and n^2-n, respectively, with no common zeros.

P_V, with $\gamma = \delta = 0$, has a first integral

32.9.12 $z^2(w')^2 = (w-1)^2(2\alpha w^2 + Cw - 2\beta)$, with C an arbitrary constant, which is solvable by quadrature. For examples and plots see Clarkson (2005). P_V, with $\alpha = \beta = 0$ and $\gamma^2 + 2\delta = 0$, has solutions $w(z) = C \exp(\pm\sqrt{-2\delta}z)$, with C an arbitrary constant.

32.9(iii) Sixth Painlevé Equation

An elementary algebraic solution of P_{VI} is

32.9.13 $\quad w(z; \tfrac{1}{2}\kappa^2, -\tfrac{1}{2}\kappa^2, \tfrac{1}{2}\mu^2, \tfrac{1}{2}(1-\mu^2)) = z^{1/2}$,

with κ and μ arbitrary constants.

Dubrovin and Mazzocco (2000) classifies all algebraic solutions for the special case of P_{VI} with $\beta = \gamma = 0$, $\delta = \tfrac{1}{2}$. For further examples of algebraic solutions see Andreev and Kitaev (2002), Boalch (2005, 2006), Gromak et al. (2002, §48), Hitchin (2003), Masuda (2003), and Mazzocco (2001b).

32.10 Special Function Solutions

32.10(i) Introduction

For certain combinations of the parameters, P_{II}–P_{VI} have particular solutions expressible in terms of the solution of a Riccati differential equation, which can be solved in terms of special functions defined in other chapters. All solutions of P_{II}–P_{VI} that are expressible in terms of special functions satisfy a first-order equation of the form

32.10.1 $\quad (w')^n + \sum_{j=0}^{n-1} F_j(w,z)(w')^j = 0$,

where $F_j(w,z)$ is polynomial in w with coefficients that are rational functions of z.

32.10(ii) Second Painlevé Equation

P_{II} has solutions expressible in terms of Airy functions (§9.2) iff

32.10.2 $\quad \alpha = n + \tfrac{1}{2}$,

with $n \in \mathbb{Z}$. For example, if $\alpha = \tfrac{1}{2}\varepsilon$, with $\varepsilon = \pm 1$, then the Riccati equation is

32.10.3 $\quad \varepsilon w' = w^2 + \tfrac{1}{2}z$,

with solution

32.10.4 $\quad w(z; \tfrac{1}{2}\varepsilon) = -\varepsilon \phi'(z)/\phi(z)$,

where

32.10.5 $\quad \phi(z) = C_1 \operatorname{Ai}(-2^{-1/3}z) + C_2 \operatorname{Bi}(-2^{-1/3}z)$,

with C_1, C_2 arbitrary constants.

Solutions for other values of α are derived from $w(z; \pm\tfrac{1}{2})$ by application of the Bäcklund transformations (32.7.1) and (32.7.2). For example,

32.10.6 $\quad w(z; \tfrac{3}{2}) = \Phi - \dfrac{1}{2\Phi^2 + z}$,

32.10.7 $\quad w(z; \tfrac{5}{2}) = \dfrac{1}{2\Phi^2 + z} + \dfrac{2z\Phi^2 + \Phi + z^2}{4\Phi^3 + 2z\Phi - 1}$,

where $\Phi = \phi'(z)/\phi(z)$, with $\phi(z)$ given by (32.10.5).

More generally, if $n = 1, 2, 3, \ldots$, then

32.10.8 $\quad w(z; n + \tfrac{1}{2}) = \dfrac{d}{dz}\left(\ln\left(\dfrac{\tau_n(z)}{\tau_{n+1}(z)}\right)\right)$,

where $\tau_n(z)$ is the $n \times n$ determinant

32.10.9 $\quad \tau_n(z) = \begin{vmatrix} \phi(z) & \phi'(z) & \cdots & \phi^{(n-1)}(z) \\ \phi'(z) & \phi''(z) & \cdots & \phi^{(n)}(z) \\ \vdots & \vdots & \ddots & \vdots \\ \phi^{(n-1)}(z) & \phi^{(n)}(z) & \cdots & \phi^{(2n-2)}(z) \end{vmatrix}$,

and

32.10.10 $\quad w(z; -n - \tfrac{1}{2}) = -w(z; n + \tfrac{1}{2})$.

32.10(iii) Third Painlevé Equation

If $\gamma\delta \neq 0$, then as in §32.2(ii) we may set $\gamma = 1$ and $\delta = -1$. P_{III} then has solutions expressible in terms of Bessel functions (§10.2) iff

32.10.11 $\quad \varepsilon_1\alpha + \varepsilon_2\beta = 4n + 2$,

with $n \in \mathbb{Z}$, and $\varepsilon_1 = \pm 1$, $\varepsilon_2 = \pm 1$, independently. In the case $\varepsilon_1\alpha + \varepsilon_2\beta = 2$, the Riccati equation is

32.10.12 $\quad zw' = \varepsilon_1 zw^2 + (\alpha\varepsilon_1 - 1)w + \varepsilon_2 z$.

If $\alpha \neq \varepsilon_1$, then (32.10.12) has the solution

32.10.13 $\quad w(z) = -\varepsilon_1 \phi'(z)/\phi(z)$,

where

32.10.14 $\quad \phi(z) = z^\nu (C_1 J_\nu(\zeta) + C_2 Y_\nu(\zeta))$,

with $\zeta = \sqrt{\varepsilon_1\varepsilon_2}\,z$, $\nu = \tfrac{1}{2}\alpha\varepsilon_1$, and C_1, C_2 arbitrary constants.

For examples and plots see Milne et al. (1997). For determinantal representations see Forrester and Witte (2002) and Okamoto (1987c).

32.10(iv) Fourth Painlevé Equation

P_{IV} has solutions expressible in terms of parabolic cylinder functions (§12.2) iff either

32.10.15 $\quad \beta = -2(2n + 1 + \varepsilon\alpha)^2$,

or

32.10.16 $\quad \beta = -2n^2$,

with $n \in \mathbb{Z}$ and $\varepsilon = \pm 1$. In the case when $n = 0$ in (32.10.15), the Riccati equation is

32.10.17 $\quad w' = \varepsilon(w^2 + 2zw) - 2(1 + \varepsilon\alpha)$,

which has the solution

32.10.18 $\quad w(z) = -\varepsilon\phi'(z)/\phi(z)$,

where

32.10.19
$$\phi(z) = \left(C_1 U\left(a, \sqrt{2}z\right) + C_2 V\left(a, \sqrt{2}z\right)\right) \exp\left(\tfrac{1}{2}\varepsilon z^2\right),$$

with $a = \alpha + \tfrac{1}{2}\varepsilon$, and C_1, C_2 arbitrary constants. When $a + \tfrac{1}{2}$ is zero or a negative integer the U parabolic cylinder functions reduce to Hermite polynomials (§18.3) times an exponential function; thus

32.10.20
$$w(z; -m, -2(m-1)^2) = -\frac{H'_{m-1}(z)}{H_{m-1}(z)}, \quad m = 1, 2, 3, \ldots,$$

and

32.10.21
$$w(z; -m, -2(m+1)^2) = -2z + \frac{H'_m(z)}{H_m(z)}, \quad m = 0, 1, 2, \ldots.$$

If $1 + \varepsilon\alpha = 0$, then (32.10.17) has solutions

32.10.22
$$w(z) = \begin{cases} \dfrac{2\exp(z^2)}{\sqrt{\pi}\,(C - i\,\mathrm{erfc}(iz))}, & \varepsilon = 1, \\ \dfrac{2\exp(-z^2)}{\sqrt{\pi}\,(C - \mathrm{erfc}(z))}, & \varepsilon = -1, \end{cases}$$

where C is an arbitrary constant and erfc is the complementary error function (§7.2(i)).

For examples and plots see Bassom et al. (1995). For determinantal representations see Forrester and Witte (2001) and Okamoto (1986).

32.10(v) Fifth Painlevé Equation

If $\delta \neq 0$, then as in §32.2(ii) we may set $\delta = -\tfrac{1}{2}$. P_V then has solutions expressible in terms of Whittaker functions (§13.14(i)), iff

32.10.23
$$a + b + \varepsilon_3 \gamma = 2n + 1,$$

or

32.10.24
$$(a - n)(b - n) = 0,$$

where $n \in \mathbb{Z}$, $a = \varepsilon_1\sqrt{2\alpha}$, and $b = \varepsilon_2\sqrt{-2\beta}$, with $\varepsilon_j = \pm 1$, $j = 1, 2, 3$, independently. In the case when $n = 0$ in (32.10.23), the Riccati equation is

32.10.25
$$zw' = aw^2 + (b - a + \varepsilon_3 z)w - b.$$

If $a \neq 0$, then (32.10.25) has the solution

32.10.26
$$w(z) = -z\phi'(z)/(a\phi(z)),$$

where

32.10.27
$$\phi(z) = \frac{C_1 M_{\kappa,\mu}(\zeta) + C_2 W_{\kappa,\mu}(\zeta)}{\zeta^{(a-b+1)/2}} \exp\left(\tfrac{1}{2}\zeta\right),$$

with $\zeta = \varepsilon_3 z$, $\kappa = \tfrac{1}{2}(a - b + 1)$, $\mu = \tfrac{1}{2}(a + b)$, and C_1, C_2 arbitrary constants.

For determinantal representations see Forrester and Witte (2002), Masuda (2004), and Okamoto (1987b).

32.10(vi) Sixth Painlevé Equation

P_VI has solutions expressible in terms of hypergeometric functions (§15.2(i)) iff

32.10.28
$$a + b + c + d = 2n + 1,$$

where $n \in \mathbb{Z}$, $a = \varepsilon_1\sqrt{2\alpha}$, $b = \varepsilon_2\sqrt{-2\beta}$, $c = \varepsilon_3\sqrt{2\gamma}$, and $d = \varepsilon_4\sqrt{1 - 2\delta}$, with $\varepsilon_j = \pm 1$, $j = 1, 2, 3, 4$, independently. If $n = 1$, then the Riccati equation is

32.10.29
$$w' = \frac{aw^2}{z(z-1)} + \frac{(b+c)z - a - c}{z(z-1)} w - \frac{b}{z-1}.$$

If $a \neq 0$, then (32.10.29) has the solution

32.10.30
$$w(z) = \frac{\zeta - 1}{a\phi(\zeta)}\frac{d\phi}{d\zeta}, \quad \zeta = \frac{1}{1-z},$$

where

32.10.31
$$\phi(\zeta) = C_1 F(b, -a; b+c; \zeta) + C_2 \zeta^{-b+1-c}$$
$$\times F(-a-b-c+1, -c+1; 2-b-c; \zeta),$$

with C_1, C_2 arbitrary constants.

Next, let $\Lambda = \Lambda(u, z)$ be the elliptic function (§§22.15(ii), 23.2(iii)) defined by

32.10.32
$$u = \int_0^\Lambda \frac{dt}{\sqrt{t(t-1)(t-z)}},$$

where the fundamental periods $2\phi_1$ and $2\phi_2$ are linearly independent functions satisfying the hypergeometric equation

32.10.33
$$z(1-z)\frac{d^2\phi}{dz^2} + (1 - 2z)\frac{d\phi}{dz} - \tfrac{1}{4}\phi = 0.$$

Then P_VI, with $\alpha = \beta = \gamma = 0$ and $\delta = \tfrac{1}{2}$, has the general solution

32.10.34
$$w(z; 0, 0, 0, \tfrac{1}{2}) = \Lambda(C_1\phi_1 + C_2\phi_2, z),$$

with C_1, C_2 arbitrary constants. The solution (32.10.34) is an essentially transcendental function of both constants of integration since P_VI with $\alpha = \beta = \gamma = 0$ and $\delta = \tfrac{1}{2}$ does not admit an algebraic first integral of the form $P(z, w, w', C) = 0$, with C a constant.

For determinantal representations see Forrester and Witte (2004) and Masuda (2004).

32.11 Asymptotic Approximations for Real Variables

32.11(i) First Painlevé Equation

There are solutions of (32.2.1) such that

32.11.1
$$w(x) = -\sqrt{\tfrac{1}{6}|x|} + d|x|^{-1/8}\sin(\phi(x) - \theta_0)$$
$$+ o\left(|x|^{-1/8}\right), \qquad x \to -\infty,$$

where

32.11.2
$$\phi(x) = (24)^{1/4}\left(\tfrac{4}{5}|x|^{5/4} - \tfrac{5}{8}d^2 \ln|x|\right),$$

and d and θ_0 are constants.

There are also solutions of (32.2.1) such that

32.11.3 $$w(x) \sim \sqrt{\tfrac{1}{6}|x|}, \qquad x \to -\infty.$$

Next, for given initial conditions $w(0) = 0$ and $w'(0) = k$, with k real, $w(x)$ has at least one pole on the real axis. There are two special values of k, k_1 and k_2, with the properties $-0.451428 < k_1 < -0.451427$, $1.851853 < k_2 < 1.851855$, and such that:

(a) If $k < k_1$, then $w(x) > 0$ for $x_0 < x < 0$, where x_0 is the first pole on the negative real axis.

(b) If $k_1 < k < k_2$, then $w(x)$ oscillates about, and is asymptotic to, $-\sqrt{\tfrac{1}{6}|x|}$ as $x \to -\infty$.

(c) If $k_2 < k$, then $w(x)$ changes sign once, from positive to negative, as x passes from x_0 to 0.

For illustration see Figures 32.3.1 to 32.3.4, and for further information see Joshi and Kitaev (2005), Joshi and Kruskal (1992), Kapaev (1988), Kapaev and Kitaev (1993), and Kitaev (1994).

32.11(ii) Second Painlevé Equation

Consider the special case of P_{II} with $\alpha = 0$:

32.11.4 $$w'' = 2w^3 + xw,$$

with boundary condition

32.11.5 $$w(x) \to 0, \qquad x \to +\infty.$$

Any nontrivial real solution of (32.11.4) that satisfies (32.11.5) is asymptotic to $k\,\mathrm{Ai}(x)$, for some nonzero real k, where Ai denotes the Airy function (§9.2). Conversely, for any nonzero real k, there is a unique solution $w_k(x)$ of (32.11.4) that is asymptotic to $k\,\mathrm{Ai}(x)$ as $x \to +\infty$.

If $|k| < 1$, then $w_k(x)$ exists for all sufficiently large $|x|$ as $x \to -\infty$, and

32.11.6 $$w_k(x) = d|x|^{-1/4}\sin(\phi(x) - \theta_0) + o\!\left(|x|^{-1/4}\right),$$

where

32.11.7 $$\phi(x) = \tfrac{2}{3}|x|^{3/2} - \tfrac{3}{4}d^2\ln|x|,$$

and $d\,(\neq 0)$, θ_0 are real constants. Connection formulas for d and θ_0 are given by

32.11.8 $$d^2 = -\pi^{-1}\ln(1 - k^2),$$

32.11.9 $$\theta_0 = \tfrac{3}{2}d^2\ln 2 + \mathrm{ph}\,\Gamma\!\left(1 - \tfrac{1}{2}id^2\right) + \tfrac{1}{4}\pi(1 - 2\,\mathrm{sign}(k)),$$

where Γ is the gamma function (§5.2(i)), and the branch of the ph function is immaterial.

If $|k| = 1$, then

32.11.10 $$w_k(x) \sim \mathrm{sign}(k)\sqrt{\tfrac{1}{2}|x|}, \qquad x \to -\infty.$$

If $|k| > 1$, then $w_k(x)$ has a pole at a finite point $x = c_0$, dependent on k, and

32.11.11 $$w_k(x) \sim \mathrm{sign}(k)(x - c_0)^{-1}, \qquad x \to c_0+.$$

For illustration see Figures 32.3.5 and 32.3.6, and for further information see Ablowitz and Clarkson (1991), Bassom et al. (1998), Clarkson and McLeod (1988), Deift and Zhou (1995), Segur and Ablowitz (1981), and Suleĭmanov (1987). For numerical studies see Miles (1978, 1980) and Rosales (1978).

32.11(iii) Modified Second Painlevé Equation

Replacement of w by iw in (32.11.4) gives

32.11.12 $$w'' = -2w^3 + xw.$$

Any nontrivial real solution of (32.11.12) satisfies

32.11.13 $$\begin{aligned} w(x) &= d|x|^{-1/4}\sin(\phi(x) - \chi) \\ &\quad + O\!\left(|x|^{-5/4}\ln|x|\right), \end{aligned} \qquad x \to -\infty,$$

where

32.11.14 $$\phi(x) = \tfrac{2}{3}|x|^{3/2} + \tfrac{3}{4}d^2\ln|x|,$$

with $d\,(\neq 0)$ and χ arbitrary real constants.

In the case when

32.11.15 $$\chi + \tfrac{3}{2}d^2\ln 2 - \tfrac{1}{4}\pi - \mathrm{ph}\,\Gamma\!\left(\tfrac{1}{2}id^2\right) = n\pi,$$

with $n \in \mathbb{Z}$, we have

32.11.16 $$w(x) \sim k\,\mathrm{Ai}(x), \qquad x \to +\infty,$$

where k is a nonzero real constant. The connection formulas for k are

32.11.17 $$d^2 = \pi^{-1}\ln(1 + k^2), \quad \mathrm{sign}(k) = (-1)^n.$$

In the generic case

32.11.18 $$\chi + \tfrac{3}{2}d^2\ln 2 - \tfrac{1}{4}\pi - \mathrm{ph}\,\Gamma\!\left(\tfrac{1}{2}id^2\right) \neq n\pi,$$

we have

32.11.19 $$\begin{aligned} w(x) &= \sigma\sqrt{\tfrac{1}{2}x} + \sigma\rho(2x)^{-1/4}\cos(\psi(x) + \theta) \\ &\quad + O\!\left(x^{-1}\right), \end{aligned} \qquad x \to +\infty,$$

where σ, $\rho\,(>0)$, and θ are real constants, and

32.11.20 $$\psi(x) = \tfrac{2}{3}\sqrt{2}x^{3/2} - \tfrac{3}{2}\rho^2\ln x.$$

The connection formulas for σ, ρ, and θ are

32.11.21 $$\sigma = -\mathrm{sign}(\Im s),$$

32.11.22 $$\rho^2 = \pi^{-1}\ln\!\left((1 + |s|^2)/|2\Im s|\right),$$

32.11.23 $$\theta = -\tfrac{3}{4}\pi - \tfrac{7}{2}\rho^2\ln 2 + \mathrm{ph}(1 + s^2) + \mathrm{ph}\,\Gamma(i\rho^2),$$

where

32.11.24 $$\begin{aligned} s &= \left(\exp(\pi d^2) - 1\right)^{1/2} \\ &\quad \times \exp\!\left(i\!\left(\tfrac{3}{2}d^2\ln 2 - \tfrac{1}{4}\pi + \chi - \mathrm{ph}\,\Gamma\!\left(\tfrac{1}{2}id^2\right)\right)\right). \end{aligned}$$

32.11(iv) Third Painlevé Equation

For $\mathrm{P}_{\mathrm{III}}$, with $\alpha = -\beta = 2\nu\ (\in \mathbb{R})$ and $\gamma = -\delta = 1$,

32.11.25
$$w(x) - 1 \sim -\lambda\,\Gamma\left(\nu + \tfrac{1}{2}\right) 2^{-2\nu} x^{-\nu-(1/2)} e^{-2x}, \quad x \to +\infty,$$

where λ is an arbitrary constant such that $-1/\pi < \lambda < 1/\pi$, and

32.11.26
$$w(x) \sim Bx^{\sigma}, \qquad x \to 0,$$

where B and σ are arbitrary constants such that $B \neq 0$ and $|\Re\sigma| < 1$. The connection formulas relating (32.11.25) and (32.11.26) are

32.11.27
$$\sigma = (2/\pi)\arcsin(\pi\lambda),$$

32.11.28
$$B = 2^{-2\sigma} \frac{\Gamma^2\left(\tfrac{1}{2}(1-\sigma)\right)\Gamma\left(\tfrac{1}{2}(1+\sigma) + \nu\right)}{\Gamma^2\left(\tfrac{1}{2}(1+\sigma)\right)\Gamma\left(\tfrac{1}{2}(1-\sigma) + \nu\right)}.$$

See also Abdullaev (1985), Novokshënov (1985), Its and Novokshënov (1986), Kitaev (1987), Bobenko (1991), Bobenko and Its (1995), Tracy and Widom (1997), and Kitaev and Vartanian (2004).

32.11(v) Fourth Painlevé Equation

Consider P_{IV} with $\alpha = 2\nu + 1\ (\in \mathbb{R})$ and $\beta = 0$, that is,

32.11.29
$$w'' = \frac{(w')^2}{2w} + \frac{3}{2}w^3 + 4xw^2 + 2(x^2 - 2\nu - 1)w,$$

and with boundary condition

32.11.30
$$w(x) \to 0, \qquad x \to +\infty.$$

Any nontrivial solution of (32.11.29) that satisfies (32.11.30) is asymptotic to $hU^2\left(-\nu - \tfrac{1}{2}, \sqrt{2}x\right)$ as $x \to +\infty$, where $h\ (\neq 0)$ is a constant. Conversely, for any $h\ (\neq 0)$ there is a unique solution $w_h(x)$ of (32.11.29) that is asymptotic to $hU^2\left(-\nu - \tfrac{1}{2}, \sqrt{2}x\right)$ as $x \to +\infty$. Here U denotes the parabolic cylinder function (§12.2).

Now suppose $x \to -\infty$. If $0 \leq h < h^*$, where

32.11.31
$$h^* = 1 \Big/ \left(\pi^{1/2}\,\Gamma(\nu+1)\right),$$

then $w_h(x)$ has no poles on the real axis. Furthermore, if $\nu = n = 0, 1, 2, \ldots$, then

32.11.32
$$w_h(x) \sim h2^n x^{2n} \exp(-x^2), \quad x \to -\infty.$$

Alternatively, if ν is not zero or a positive integer, then

32.11.33
$$w_h(x) = -\tfrac{2}{3}x + \tfrac{4}{3}d\sqrt{3}\sin(\phi(x) - \theta_0) + O\left(x^{-1}\right),$$
$$x \to -\infty,$$

where

32.11.34
$$\phi(x) = \tfrac{1}{3}\sqrt{3}x^2 - \tfrac{4}{3}d^2\sqrt{3}\ln\left(\sqrt{2}|x|\right),$$

and $d\ (> 0)$ and θ_0 are real constants. Connection formulas for d and θ_0 are given by

32.11.35
$$d^2 = -\tfrac{1}{4}\sqrt{3}\pi^{-1}\ln\left(1 - |\mu|^2\right),$$

32.11.36
$$\theta_0 = \tfrac{1}{3}d^2\sqrt{3}\ln 3 + \tfrac{2}{3}\pi\nu + \tfrac{7}{12}\pi$$
$$+ \mathrm{ph}\,\mu + \mathrm{ph}\,\Gamma\!\left(-\tfrac{2}{3}i\sqrt{3}d^2\right),$$

where

32.11.37
$$\mu = 1 + \left(2ih\pi^{3/2}\exp(-i\pi\nu)\Big/\Gamma(-\nu)\right),$$

and the branch of the ph function is immaterial.

Next if $h = h^*$, then

32.11.38
$$w_{h^*}(x) \sim -2x, \qquad x \to -\infty,$$

and $w_{h^*}(x)$ has no poles on the real axis.

Lastly if $h > h^*$, then $w_h(x)$ has a simple pole on the real axis, whose location is dependent on h.

For illustration see Figures 32.3.7–32.3.10. In terms of the parameter k that is used in these figures $h = 2^{3/2}k^2$.

32.12 Asymptotic Approximations for Complex Variables

32.12(i) First Painlevé Equation

See Boutroux (1913), Kapaev and Kitaev (1993), Takei (1995), Costin (1999), Joshi and Kitaev (2001), Kapaev (2004), and Olde Daalhuis (2005b).

32.12(ii) Second Painlevé Equation

See Boutroux (1913), Novokshënov (1990), Kapaev (1991), Joshi and Kruskal (1992), Kitaev (1994), Its and Kapaev (2003), and Fokas et al. (2006, Chapter 7).

32.12(iii) Third Painlevé Equation

See Fokas et al. (2006, Chapter 16).

Applications

32.13 Reductions of Partial Differential Equations

32.13(i) Korteweg–de Vries and Modified Korteweg–de Vries Equations

The *modified Korteweg–de Vries* (mKdV) equation

32.13.1
$$v_t - 6v^2 v_x + v_{xxx} = 0,$$

has the scaling reduction

32.13.2
$$z = x(3t)^{-1/3}, \quad v(x,t) = (3t)^{-1/3} w(z),$$

where $w(z)$ satisfies P_{II} with α a constant of integration.

The *Korteweg–de Vries* (KdV) equation

32.13.3
$$u_t + 6uu_x + u_{xxx} = 0,$$

has the scaling reduction

32.13.4
$$z = x(3t)^{-1/3}, \quad u(x,t) = -(3t)^{-2/3}(w' + w^2),$$

where $w(z)$ satisfies P_{II}.

Equation (32.13.3) also has the similarity reduction

32.13.5 $\quad z = x + 3\lambda t^2, \quad u(x,t) = W(z) - \lambda t,$

where λ is an arbitrary constant and $W(z)$ is expressible in terms of solutions of P_I. See Fokas and Ablowitz (1982) and P. J. Olver (1993b, p. 194).

32.13(ii) Sine-Gordon Equation

The *sine-Gordon* equation

32.13.6 $\quad\quad\quad\quad u_{xt} = \sin u,$

has the scaling reduction

32.13.7 $\quad\quad z = xt, \quad u(x,t) = v(z),$

where $v(z)$ satisfies (32.2.10) with $\alpha = \frac{1}{2}$ and $\gamma = 0$. In consequence if $w = \exp(-iv)$, then $w(z)$ satisfies P_{III} with $\alpha = -\beta = \frac{1}{2}$ and $\gamma = \delta = 0$.

32.13(iii) Boussinesq Equation

The *Boussinesq* equation

32.13.8 $\quad\quad u_{tt} = u_{xx} - 6(u^2)_{xx} + u_{xxxx},$

has the traveling wave solution

32.13.9 $\quad\quad z = x - ct, \quad u(x,t) = v(z),$

where c is an arbitrary constant and $v(z)$ satisfies

32.13.10 $\quad v'' = 6v^2 + (c^2 - 1)v + Az + B,$

with A and B constants of integration. Depending whether $A = 0$ or $A \neq 0$, $v(z)$ is expressible in terms of the Weierstrass elliptic function (§23.2) or solutions of P_I, respectively.

32.14 Combinatorics

Let S_N be the group of permutations π of the numbers $1, 2, \ldots, N$ (§26.2). With $1 \leq m_1 < \cdots < m_n \leq N$, $\pi(m_1), \pi(m_2), \ldots, \pi(m_n)$ is said to be an *increasing subsequence* of π of *length* n when $\pi(m_1) < \pi(m_2) < \cdots < \pi(m_n)$. Let $\ell_N(\pi)$ be the length of the longest increasing subsequence of π. Then

32.14.1 $\quad \lim_{N \to \infty} \text{Prob}\left(\frac{\ell_N(\pi) - 2\sqrt{N}}{N^{1/6}} \leq s\right) = F(s),$

where the *distribution function* $F(s)$ is defined here by

32.14.2 $\quad F(s) = \exp\left(-\int_s^\infty (x-s)w^2(x)\,dx\right),$

and $w(x)$ satisfies P_{II} with $\alpha = 0$ and boundary conditions

32.14.3 $\quad\quad w(x) \sim \text{Ai}(x), \quad\quad\quad x \to +\infty,$

32.14.4 $\quad\quad w(x) \sim \sqrt{-\tfrac{1}{2}x}, \quad\quad x \to -\infty,$

where Ai denotes the Airy function (§9.2).

The distribution function $F(s)$ given by (32.14.2) arises in random matrix theory where it gives the limiting distribution for the normalized largest eigenvalue in the Gaussian Unitary Ensemble of $n \times n$ Hermitian matrices; see Tracy and Widom (1994).

See Forrester and Witte (2001, 2002) for other instances of Painlevé equations in random matrix theory.

32.15 Orthogonal Polynomials

Let $p_n(\xi)$, $n = 0, 1, \ldots$, be the orthonormal set of polynomials defined by

32.15.1 $\quad \int_{-\infty}^{\infty} \exp\left(-\tfrac{1}{4}\xi^4 - z\xi^2\right) p_m(\xi) p_n(\xi)\,d\xi = \delta_{m,n},$

with recurrence relation

32.15.2 $\quad a_{n+1}(z)p_{n+1}(\xi) = \xi p_n(\xi) - a_n(z)p_{n-1}(\xi),$

for $n = 1, 2, \ldots$; compare §18.2. Then $u_n(z) = (a_n(z))^2$ satisfies the nonlinear recurrence relation

32.15.3 $\quad (u_{n+1} + u_n + u_{n-1})u_n = n - 2zu_n,$

for $n = 1, 2, \ldots$, and also P_{IV} with $\alpha = -\tfrac{1}{2}n$ and $\beta = -\tfrac{1}{2}n^2$.

For this result and applications see Fokas et al. (1991): in this reference, on the right-hand side of Eq. (1.10), $(n + \gamma)^2$ should be replaced by $n + \gamma$ at its first appearance. See also Freud (1976), Brézin et al. (1978), Fokas et al. (1992), and Magnus (1995).

32.16 Physical

Statistical Physics

Statistical physics, especially classical and quantum spin models, has proved to be a major area for research problems in the modern theory of Painlevé transcendents. For a survey see McCoy (1992). See also McCoy et al. (1977), Jimbo et al. (1980), Essler et al. (1996), and Kanzieper (2002).

Integrable Continuous Dynamical Systems

See Bountis et al. (1982) and Grammaticos et al. (1991).

Other Applications

For the Ising model see Barouch et al. (1973).

For applications in 2D quantum gravity and related aspects of the enumerative topology see Di Francesco et al. (1995). For applications in string theory see Seiberg and Shih (2005).

Computation

32.17 Methods of Computation

The Painlevé equations can be integrated by Runge–Kutta methods for ordinary differential equations; see §3.7(v), Butcher (2003), and Hairer *et al.* (2000). For numerical studies of P_I see Holmes and Spence (1984) and Noonburg (1995). For numerical studies of P_{II} see Kashevarov (1998, 2004), Miles (1978, 1980), and Rosales (1978). For numerical studies of P_{IV} see Bassom *et al.* (1993).

References

General Reference

The survey article Clarkson (2006) covers all topics treated in this chapter.

Sources

The following list gives the references or other indications of proofs that were used in constructing the various sections of this chapter. These sources supplement the references that are quoted in the text.

§32.2 Adler (1994), Hille (1976, pp. 439–444), Ince (1926, Chapter XIV), Kruskal and Clarkson (1992), Iwasaki *et al.* (1991, pp. 119–126), Noumi (2004, pp. 13–23).

§32.3 The graphs were produced at NIST. See also Bassom *et al.* (1993).

§32.4 Jimbo and Miwa (1981), Fokas *et al.* (2006, Chapter 5), Its and Novokshënov (1986).

§32.5 Ablowitz and Clarkson (1991), Ablowitz and Segur (1977, 1981).

§32.6 Forrester and Witte (2002), Jimbo and Miwa (1981), Okamoto (1981, 1986, 1987c).

§32.7 Cosgrove (2006), Fokas and Ablowitz (1982), Gambier (1910), Gromak (1975, 1976, 1978, 1987), Gromak *et al.* (2002, §§25,34,39,42,47), Lukaševič (1967a, 1971), Okamoto (1987a).

§32.8 Flaschka and Newell (1980), Gromak (1987), Gromak *et al.* (2002, §§20,26,35,40), Gromak and Lukaševič (1982), Kajiwara and Ohta (1996), Kitaev *et al.* (1994), Lukaševič (1967a,b), Mazzocco (2001a), Murata (1985, 1995), Vorob'ev (1965), Yablonskiĭ (1959).

§32.9 Gromak *et al.* (2002, §§33,38), Gromak and Lukaševič (1982), Hitchin (1995), Lukaševič (1965, 1967b).

§32.10 Airault (1979), Albrecht *et al.* (1996), Flaschka and Newell (1980), Fokas and Yortsos (1981), Gambier (1910), Gromak (1978, 1987), Gromak *et al.* (2002, Chapter 6, §§35,40,44), Gromak and Lukaševič (1982), Lukaševič (1965, 1967a,b, 1968), Lukaševič and Yablonskiĭ (1967), Mansfield and Webster (1998), Okamoto (1986, 1987a), Umemura and Watanabe (1998), Watanabe (1995).

§32.11 Ablowitz and Segur (1977), Bassom *et al.* (1992), Bender and Orszag (1978, pp. 158–166), Deift and Zhou (1995), Fokas *et al.* (2006, Chapters 9, 10, 14), Hastings and McLeod (1980), Holmes and Spence (1984), Its *et al.* (1994), Its and Kapaev (1987, 1998), McCoy *et al.* (1977). For (32.11.2) see Qin and Lu (2008).

§32.13 Ablowitz and Segur (1977).

§32.14 Baik *et al.* (1999).

Chapter 33

Coulomb Functions

I. J. Thompson[1]

Notation — 742
- 33.1 Special Notation 742

Variables ρ, η — 742
- 33.2 Definitions and Basic Properties 742
- 33.3 Graphics 743
- 33.4 Recurrence Relations and Derivatives . . 744
- 33.5 Limiting Forms for Small ρ, Small $|\eta|$, or Large ℓ 744
- 33.6 Power-Series Expansions in ρ 745
- 33.7 Integral Representations 745
- 33.8 Continued Fractions 745
- 33.9 Expansions in Series of Bessel Functions . 745
- 33.10 Limiting Forms for Large ρ or Large $|\eta|$. 746
- 33.11 Asymptotic Expansions for Large ρ 747
- 33.12 Asymptotic Expansions for Large η 747
- 33.13 Complex Variable and Parameters 748

Variables r, ϵ — 748
- 33.14 Definitions and Basic Properties 748
- 33.15 Graphics 749
- 33.16 Connection Formulas 751
- 33.17 Recurrence Relations and Derivatives . . 752
- 33.18 Limiting Forms for Large ℓ 752
- 33.19 Power-Series Expansions in r 752
- 33.20 Expansions for Small $|\epsilon|$ 752
- 33.21 Asymptotic Approximations for Large $|r|$. 753

Physical Applications — 753
- 33.22 Particle Scattering and Atomic and Molecular Spectra 753

Computation — 755
- 33.23 Methods of Computation 755
- 33.24 Tables 755
- 33.25 Approximations 756
- 33.26 Software 756

References — 756

---

[1] Lawrence Livermore National Laboratory, Livermore, California.
Acknowledgments: This chapter is based in part on Abramowitz and Stegun (1964, Chapter 14) by M. Abramowitz.
Copyright © 2009 National Institute of Standards and Technology. All rights reserved.

Notation

33.1 Special Notation

(For other notation see pp. xiv and 873.)

k, ℓ	nonnegative integers.
r, x	real variables.
ρ	nonnegative real variable.
ϵ, η	real parameters.
$\psi(x)$	logarithmic derivative of $\Gamma(x)$; see §5.2(i).
$\delta(x)$	Dirac delta; see §1.17.
primes	derivatives with respect to the variable.

The main functions treated in this chapter are first the Coulomb radial functions $F_\ell(\eta, \rho)$, $G_\ell(\eta, \rho)$, $H_\ell^\pm(\eta, \rho)$ (Sommerfeld (1928)), which are used in the case of repulsive Coulomb interactions, and secondly the functions $f(\epsilon, \ell; r)$, $h(\epsilon, \ell; r)$, $s(\epsilon, \ell; r)$, $c(\epsilon, \ell; r)$ (Seaton (1982, 2002)), which are used in the case of attractive Coulomb interactions.

Alternative Notations

Curtis (1964a): $P_\ell(\epsilon, r) = (2\ell + 1)! f(\epsilon, \ell; r)/2^{\ell+1}$, $Q_\ell(\epsilon, r) = -(2\ell + 1)! h(\epsilon, \ell; r)/(2^{\ell+1} A(\epsilon, \ell))$.

Greene et al. (1979): $f^{(0)}(\epsilon, \ell; r) = f(\epsilon, \ell; r)$, $f(\epsilon, \ell; r) = s(\epsilon, \ell; r)$, $g(\epsilon, \ell; r) = c(\epsilon, \ell; r)$.

Variables ρ, η

33.2 Definitions and Basic Properties

33.2(i) Coulomb Wave Equation

33.2.1
$$\frac{d^2 w}{d\rho^2} + \left(1 - \frac{2\eta}{\rho} - \frac{\ell(\ell+1)}{\rho^2}\right) w = 0, \quad \ell = 0, 1, 2, \ldots.$$

This differential equation has a regular singularity at $\rho = 0$ with indices $\ell + 1$ and $-\ell$, and an irregular singularity of rank 1 at $\rho = \infty$ (§§2.7(i), 2.7(ii)). There are two turning points, that is, points at which $d^2w/d\rho^2 = 0$ (§2.8(i)). The outer one is given by

33.2.2 $\qquad \rho_{\text{tp}}(\eta, \ell) = \eta + (\eta^2 + \ell(\ell+1))^{1/2}.$

33.2(ii) Regular Solution $F_\ell(\eta, \rho)$

The function $F_\ell(\eta, \rho)$ is recessive (§2.7(iii)) at $\rho = 0$, and is defined by

33.2.3 $\quad F_\ell(\eta, \rho) = C_\ell(\eta) 2^{-\ell-1} (\mp i)^{\ell+1} M_{\pm i\eta, \ell+\frac{1}{2}}(\pm 2i\rho),$

or equivalently

33.2.4
$$F_\ell(\eta, \rho) = C_\ell(\eta) \rho^{\ell+1} e^{\mp i\rho} M(\ell + 1 \mp i\eta, 2\ell + 2, \pm 2i\rho),$$

where $M_{\kappa,\mu}(z)$ and $M(a, b, z)$ are defined in §§13.14(i) and 13.2(i), and

33.2.5 $\qquad C_\ell(\eta) = \dfrac{2^\ell e^{-\pi \eta/2} |\Gamma(\ell + 1 + i\eta)|}{(2\ell + 1)!}.$

The choice of ambiguous signs in (33.2.3) and (33.2.4) is immaterial, provided that either all upper signs are taken, or all lower signs are taken. This is a consequence of Kummer's transformation (§13.2(vii)).

$F_\ell(\eta, \rho)$ is a real and analytic function of ρ on the open interval $0 < \rho < \infty$, and also an analytic function of η when $-\infty < \eta < \infty$.

The *normalizing constant* $C_\ell(\eta)$ is always positive, and has the alternative form

33.2.6
$$C_\ell(\eta) = \frac{2^\ell \left((2\pi\eta/(e^{2\pi\eta} - 1)) \prod_{k=1}^{\ell}(\eta^2 + k^2)\right)^{1/2}}{(2\ell + 1)!}.$$

33.2(iii) Irregular Solutions $G_\ell(\eta, \rho)$, $H_\ell^\pm(\eta, \rho)$

The functions $H_\ell^\pm(\eta, \rho)$ are defined by

33.2.7 $\quad H_\ell^\pm(\eta, \rho) = (\mp i)^\ell e^{(\pi\eta/2) \pm i\sigma_\ell(\eta)} W_{\mp i\eta, \ell+\frac{1}{2}}(\mp 2i\rho),$

or equivalently

33.2.8
$$H_\ell^\pm(\eta, \rho) = e^{\pm i\theta_\ell(\eta, \rho)} (\mp 2i\rho)^{\ell+1 \pm i\eta} U(\ell + 1 \pm i\eta, 2\ell + 2, \mp 2i\rho),$$

where $W_{\kappa,\mu}(z)$, $U(a, b, z)$ are defined in §§13.14(i) and 13.2(i),

33.2.9 $\qquad \theta_\ell(\eta, \rho) = \rho - \eta \ln(2\rho) - \tfrac{1}{2} \ell \pi + \sigma_\ell(\eta),$

and

33.2.10 $\qquad \sigma_\ell(\eta) = \operatorname{ph} \Gamma(\ell + 1 + i\eta),$

the branch of the phase in (33.2.10) being zero when $\eta = 0$ and continuous elsewhere. $\sigma_\ell(\eta)$ is the *Coulomb phase shift*.

$H_\ell^+(\eta, \rho)$ and $H_\ell^-(\eta, \rho)$ are complex conjugates, and their real and imaginary parts are given by

33.2.11
$$H_\ell^+(\eta, \rho) = G_\ell(\eta, \rho) + i F_\ell(\eta, \rho),$$
$$H_\ell^-(\eta, \rho) = G_\ell(\eta, \rho) - i F_\ell(\eta, \rho).$$

As in the case of $F_\ell(\eta, \rho)$, the solutions $H_\ell^\pm(\eta, \rho)$ and $G_\ell(\eta, \rho)$ are analytic functions of ρ when $0 < \rho < \infty$. Also, $e^{\mp i\sigma_\ell(\eta)} H_\ell^\pm(\eta, \rho)$ are analytic functions of η when $-\infty < \eta < \infty$.

33.2(iv) Wronskians and Cross-Product

With arguments η, ρ suppressed,

33.2.12 $\qquad \mathscr{W}\{G_\ell, F_\ell\} = \mathscr{W}\{H_\ell^\pm, F_\ell\} = 1.$

33.2.13 $\qquad F_{\ell-1} G_\ell - F_\ell G_{\ell-1} = \ell/(\ell^2 + \eta^2)^{1/2}, \quad \ell \geq 1.$

33.3 Graphics

33.3(i) Line Graphs of the Coulomb Radial Functions $F_\ell(\eta,\rho)$ and $G_\ell(\eta,\rho)$

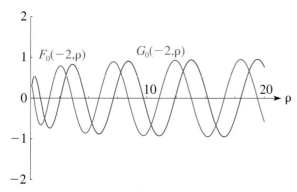

Figure 33.3.1: $F_\ell(\eta,\rho)$, $G_\ell(\eta,\rho)$ with $\ell = 0$, $\eta = -2$.

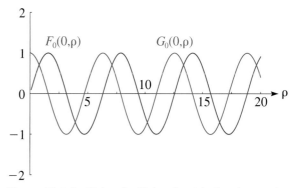

Figure 33.3.2: $F_\ell(\eta,\rho)$, $G_\ell(\eta,\rho)$ with $\ell = 0$, $\eta = 0$.

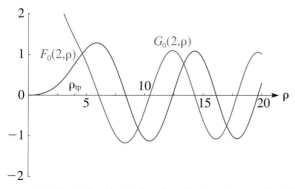

Figure 33.3.3: $F_\ell(\eta,\rho)$, $G_\ell(\eta,\rho)$ with $\ell = 0$, $\eta = 2$. The turning point is at $\rho_{\text{tp}}(2,0) = 4$.

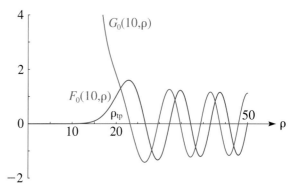

Figure 33.3.4: $F_\ell(\eta,\rho)$, $G_\ell(\eta,\rho)$ with $\ell = 0$, $\eta = 10$. The turning point is at $\rho_{\text{tp}}(10,0) = 20$.

In Figures 33.3.5 and 33.3.6

33.3.1 $$M_\ell(\eta,\rho) = (F_\ell^2(\eta,\rho) + G_\ell^2(\eta,\rho))^{1/2} = \left|H_\ell^\pm(\eta,\rho)\right|.$$

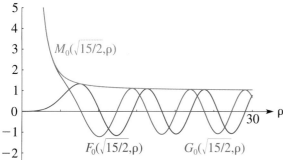

Figure 33.3.5: $F_\ell(\eta,\rho)$, $G_\ell(\eta,\rho)$, and $M_\ell(\eta,\rho)$ with $\ell = 0$, $\eta = \sqrt{15/2}$. The turning point is at $\rho_{\text{tp}}\left(\sqrt{15/2},0\right) = \sqrt{30} = 5.47\ldots$.

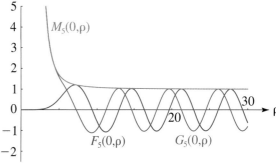

Figure 33.3.6: $F_\ell(\eta,\rho)$, $G_\ell(\eta,\rho)$, and $M_\ell(\eta,\rho)$ with $\ell = 5$, $\eta = 0$. The turning point is at $\rho_{\text{tp}}(0,5) = \sqrt{30}$ (as in Figure 33.3.5).

33.3(ii) Surfaces of the Coulomb Radial Functions $F_0(\eta,\rho)$ and $G_0(\eta,\rho)$

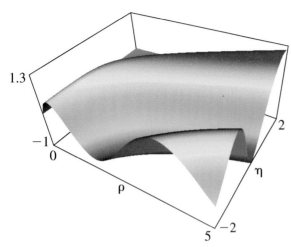

Figure 33.3.7: $F_0(\eta,\rho)$, $-2 \leq \eta \leq 2$, $0 \leq \rho \leq 5$.

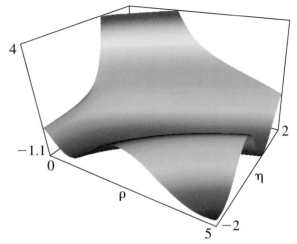

Figure 33.3.8: $G_0(\eta,\rho)$, $-2 \leq \eta \leq 2$, $0 < \rho \leq 5$.

33.4 Recurrence Relations and Derivatives

For $\ell = 1, 2, 3, \ldots$, let

33.4.1 $\quad R_\ell = \sqrt{1 + \dfrac{\eta^2}{\ell^2}}, \quad S_\ell = \dfrac{\ell}{\rho} + \dfrac{\eta}{\ell}, \quad T_\ell = S_\ell + S_{\ell+1}.$

Then, with X_ℓ denoting any of $F_\ell(\eta,\rho)$, $G_\ell(\eta,\rho)$, or $H_\ell^\pm(\eta,\rho)$,

33.4.2 $\quad R_\ell X_{\ell-1} - T_\ell X_\ell + R_{\ell+1} X_{\ell+1} = 0, \qquad \ell \geq 1,$

33.4.3 $\quad X_\ell' = R_\ell X_{\ell-1} - S_\ell X_\ell, \qquad \ell \geq 1,$

33.4.4 $\quad X_\ell' = S_{\ell+1} X_\ell - R_{\ell+1} X_{\ell+1}, \qquad \ell \geq 0.$

33.5 Limiting Forms for Small ρ, Small $|\eta|$, or Large ℓ

33.5(i) Small ρ

As $\rho \to 0$ with η fixed,

33.5.1 $\quad F_\ell(\eta,\rho) \sim C_\ell(\eta)\rho^{\ell+1}, \quad F_\ell'(\eta,\rho) \sim (\ell+1)\,C_\ell(\eta)\rho^\ell.$

33.5.2
$$G_\ell(\eta,\rho) \sim \frac{\rho^{-\ell}}{(2\ell+1)\,C_\ell(\eta)}, \qquad \ell = 0,1,2,\ldots,$$
$$G_\ell'(\eta,\rho) \sim -\frac{\ell\rho^{-\ell-1}}{(2\ell+1)\,C_\ell(\eta)}, \qquad \ell = 1,2,3,\ldots.$$

33.5(ii) $\eta = 0$

33.5.3 $\quad F_\ell(0,\rho) = \rho\,\mathsf{j}_\ell(\rho), \quad G_\ell(0,\rho) = -\rho\,\mathsf{y}_\ell(\rho).$

Equivalently,

33.5.4
$$F_\ell(0,\rho) = (\pi\rho/2)^{1/2}\,J_{\ell+\frac{1}{2}}(\rho),$$
$$G_\ell(0,\rho) = -(\pi\rho/2)^{1/2}\,Y_{\ell+\frac{1}{2}}(\rho).$$

For the functions j, y, J, Y see §§10.47(ii), 10.2(ii).

33.5.5 $\quad F_0(0,\rho) = \sin\rho, \quad G_0(0,\rho) = \cos\rho, \quad H_0^\pm(0,\rho) = e^{\pm i\rho}.$

33.5.6 $\quad C_\ell(0) = \dfrac{2^\ell \ell!}{(2\ell+1)!} = \dfrac{1}{(2\ell+1)!!}.$

33.5(iii) Small $|\eta|$

33.5.7 $\quad \sigma_0(\eta) \sim -\gamma\eta, \qquad \eta \to 0,$

where γ is Euler's constant (§5.2(ii)).

33.5(iv) Large ℓ

As $\ell \to \infty$ with η and ρ ($\neq 0$) fixed,

33.5.8
$$F_\ell(\eta,\rho) \sim C_\ell(\eta)\rho^{\ell+1}, \quad G_\ell(\eta,\rho) \sim \frac{\rho^{-\ell}}{(2\ell+1)\,C_\ell(\eta)},$$

33.5.9 $\quad C_\ell(\eta) \sim \dfrac{e^{-\pi\eta/2}}{(2\ell+1)!!} \sim e^{-\pi\eta/2}\dfrac{e^\ell}{\sqrt{2}(2\ell)^{\ell+1}}.$

33.6 Power-Series Expansions in ρ

$$F_\ell(\eta,\rho) = C_\ell(\eta) \sum_{k=\ell+1}^{\infty} A_k^\ell(\eta)\rho^k,$$
33.6.1

$$F'_\ell(\eta,\rho) = C_\ell(\eta) \sum_{k=\ell+1}^{\infty} k A_k^\ell(\eta)\rho^{k-1},$$
33.6.2

$$H_\ell^\pm(\eta,\rho) = \frac{e^{\pm i\theta_\ell(\eta,\rho)}}{(2\ell+1)!\,\Gamma(-\ell+i\eta)} \left(\sum_{k=0}^{\infty} \frac{(a)_k}{(2\ell+2)_k k!}(\mp 2i\rho)^{a+k}\left(\ln(\mp 2i\rho) + \psi(a+k) - \psi(1+k) - \psi(2\ell+2+k)\right) \right.$$
33.6.5
$$\left. - \sum_{k=1}^{2\ell+1} \frac{(2\ell+1)!(k-1)!}{(2\ell+1-k)!(1-a)_k}(\mp 2i\rho)^{a-k} \right),$$

where $A_{\ell+1}^\ell = 1$, $A_{\ell+2}^\ell = \eta/(\ell+1)$, and

33.6.3
$$(k+\ell)(k-\ell-1)A_k^\ell = 2\eta A_{k-1}^\ell - A_{k-2}^\ell,$$
$$k = \ell+3, \ell+4, \ldots,$$

or in terms of the hypergeometric function (§§15.1, 15.2(i)),

33.6.4
$$A_k^\ell(\eta) = \frac{(-i)^{k-\ell-1}}{(k-\ell-1)!}\,{}_2F_1(\ell+1-k,\ell+1-i\eta;2\ell+2;2).$$

where $a = 1 + \ell \pm i\eta$ and $\psi(x) = \Gamma'(x)/\Gamma(x)$ (§5.2(i)).

The series (33.6.1), (33.6.2), and (33.6.5) converge for all finite values of ρ. Corresponding expansions for $H_\ell^{\pm\prime}(\eta,\rho)$ can be obtained by combining (33.6.5) with (33.4.3) or (33.4.4).

33.7 Integral Representations

33.7.1
$$F_\ell(\eta,\rho) = \frac{\rho^{\ell+1} 2^\ell e^{i\rho - (\pi\eta/2)}}{|\Gamma(\ell+1+i\eta)|} \int_0^1 e^{-2i\rho t} t^{\ell+i\eta}(1-t)^{\ell-i\eta}\,dt,$$

33.7.2
$$H_\ell^-(\eta,\rho) = \frac{e^{-i\rho}\rho^{-\ell}}{(2\ell+1)!\,C_\ell(\eta)} \int_0^\infty e^{-t} t^{\ell-i\eta}(t+2i\rho)^{\ell+i\eta}\,dt,$$

33.7.3
$$H_\ell^-(\eta,\rho) = \frac{-ie^{-\pi\eta}\rho^{\ell+1}}{(2\ell+1)!\,C_\ell(\eta)} \int_0^\infty \left(\frac{\exp(-i(\rho\tanh t - 2\eta t))}{(\cosh t)^{2\ell+2}} + i(1+t^2)^\ell \exp(-\rho t + 2\eta\arctan t) \right)dt,$$

33.7.4
$$H_\ell^+(\eta,\rho) = \frac{ie^{-\pi\eta}\rho^{\ell+1}}{(2\ell+1)!\,C_\ell(\eta)} \int_{-1}^{-i\infty} e^{-i\rho t}(1-t)^{\ell-i\eta}(1+t)^{\ell+i\eta}\,dt.$$

Noninteger powers in (33.7.1)–(33.7.4) and the arctangent assume their principal values (§§4.2(i), 4.2(iv), 4.23(ii)).

33.8 Continued Fractions

With arguments η, ρ suppressed,

33.8.1
$$\frac{F'_\ell}{F_\ell} = S_{\ell+1} - \frac{R_{\ell+1}^2}{T_{\ell+1}-} \frac{R_{\ell+2}^2}{T_{\ell+2}-} \cdots.$$

For R, S, and T see (33.4.1).

33.8.2
$$\frac{H_\ell^{\pm\prime}}{H_\ell^\pm} = c \pm \frac{i}{\rho}\,\frac{ab}{2(\rho-\eta\pm i)+}\,\frac{(a+1)(b+1)}{2(\rho-\eta\pm 2i)+}\,\cdots,$$

where

33.8.3 $a = 1 + \ell \pm i\eta$, $b = -\ell \pm i\eta$, $c = \pm i(1-(\eta/\rho))$.

The continued fraction (33.8.1) converges for all finite values of ρ, and (33.8.2) converges for all $\rho \neq 0$.

If we denote $u = F'_\ell/F_\ell$ and $p + iq = H_\ell^{+\prime}/H_\ell^+$, then

33.8.4 $F_\ell = \pm(q^{-1}(u-p)^2 + q)^{-1/2}$, $\quad F'_\ell = uF_\ell$,

33.8.5 $G_\ell = q^{-1}(u-p)F_\ell$, $\quad G'_\ell = q^{-1}(up - p^2 - q^2)F_\ell$.

The ambiguous sign in (33.8.4) has to agree with that of the final denominator in (33.8.1) when the continued fraction has converged to the required precision. For proofs and further information see Barnett et al. (1974) and Barnett (1996).

33.9 Expansions in Series of Bessel Functions

33.9(i) Spherical Bessel Functions

33.9.1
$$F_\ell(\eta,\rho) = \rho \sum_{k=0}^{\infty} a_k\,\mathsf{j}_{\ell+k}(\rho),$$

where the function j is as in §10.47(ii), $a_{-1} = 0$, $a_0 = (2\ell+1)!! \, C_\ell(\eta)$, and

33.9.2
$$\frac{k(k+2\ell+1)}{2k+2\ell+1}a_k - 2\eta a_{k-1} + \frac{(k-2)(k+2\ell-1)}{2k+2\ell-3}a_{k-2} = 0, \quad k=1,2,\ldots.$$

The series (33.9.1) converges for all finite values of η and ρ.

33.9(ii) Bessel Functions and Modified Bessel Functions

In this subsection the functions J, I, and K are as in §§10.2(ii) and 10.25(ii).

With $t = 2|\eta|\rho$,

33.9.3
$$F_\ell(\eta,\rho) = C_\ell(\eta)\frac{(2\ell+1)!}{(2\eta)^{2\ell+1}}\rho^{-\ell}\sum_{k=2\ell+1}^{\infty} b_k t^{k/2} I_k(2\sqrt{t}), \quad \eta>0,$$

33.9.4
$$F_\ell(\eta,\rho) = C_\ell(\eta)\frac{(2\ell+1)!}{(2|\eta|)^{2\ell+1}}\rho^{-\ell}\sum_{k=2\ell+1}^{\infty} b_k t^{k/2} J_k(2\sqrt{t}), \quad \eta<0.$$

Here $b_{2\ell} = b_{2\ell+2} = 0$, $b_{2\ell+1} = 1$, and

33.9.5
$$4\eta^2(k-2\ell)b_{k+1} + kb_{k-1} + b_{k-2} = 0, \quad k = 2\ell+2, 2\ell+3, \ldots.$$

The series (33.9.3) and (33.9.4) converge for all finite positive values of $|\eta|$ and ρ.

Next, as $\eta \to +\infty$ with $\rho\,(>0)$ fixed,

33.9.6
$$G_\ell(\eta,\rho) \sim \frac{\rho^{-\ell}}{(\ell+\tfrac{1}{2})\lambda_\ell(\eta)\,C_\ell(\eta)}\sum_{k=2\ell+1}^{\infty}(-1)^k b_k t^{k/2} K_k(2\sqrt{t}),$$

where

33.9.7
$$\lambda_\ell(\eta) \sim \sum_{k=2\ell+1}^{\infty}(-1)^k(k-1)!\,b_k.$$

For other asymptotic expansions of $G_\ell(\eta,\rho)$ see Fröberg (1955, §8) and Humblet (1985).

33.10 Limiting Forms for Large ρ or Large $|\eta|$

33.10(i) Large ρ

As $\rho \to \infty$ with η fixed,

33.10.1
$$F_\ell(\eta,\rho) = \sin(\theta_\ell(\eta,\rho)) + o(1),$$
$$G_\ell(\eta,\rho) = \cos(\theta_\ell(\eta,\rho)) + o(1),$$

33.10.2
$$H_\ell^\pm(\eta,\rho) \sim \exp(\pm i\,\theta_\ell(\eta,\rho)),$$

where $\theta_\ell(\eta,\rho)$ is defined by (33.2.9).

33.10(ii) Large Positive η

As $\eta \to \infty$ with ρ fixed,

33.10.3
$$F_\ell(\eta,\rho) \sim \frac{(2\ell+1)!\,C_\ell(\eta)}{(2\eta)^{\ell+1}}(2\eta\rho)^{1/2}\,I_{2\ell+1}\!\left((8\eta\rho)^{1/2}\right),$$
$$G_\ell(\eta,\rho) \sim \frac{2(2\eta)^\ell}{(2\ell+1)!\,C_\ell(\eta)}(2\eta\rho)^{1/2}\,K_{2\ell+1}\!\left((8\eta\rho)^{1/2}\right).$$

In particular, for $\ell=0$,

33.10.4
$$F_0(\eta,\rho) \sim e^{-\pi\eta}(\pi\rho)^{1/2}\,I_1\!\left((8\eta\rho)^{1/2}\right),$$
$$G_0(\eta,\rho) \sim 2e^{\pi\eta}(\rho/\pi)^{1/2}\,K_1\!\left((8\eta\rho)^{1/2}\right),$$

33.10.5
$$F_0'(\eta,\rho) \sim e^{-\pi\eta}(2\pi\eta)^{1/2}\,I_0\!\left((8\eta\rho)^{1/2}\right),$$
$$G_0'(\eta,\rho) \sim -2e^{\pi\eta}(2\eta/\pi)^{1/2}\,K_0\!\left((8\eta\rho)^{1/2}\right).$$

Also,

33.10.6
$$\sigma_0(\eta) = \eta(\ln\eta - 1) + \tfrac{1}{4}\pi + o(1),$$
$$C_0(\eta) \sim (2\pi\eta)^{1/2} e^{-\pi\eta}.$$

33.10(iii) Large Negative η

As $\eta \to -\infty$ with ρ fixed,

33.10.7
$$F_\ell(\eta,\rho) = \frac{(2\ell+1)!\,C_\ell(\eta)}{(-2\eta)^{\ell+1}}\left((-2\eta\rho)^{1/2}\right.$$
$$\left. \times J_{2\ell+1}\!\left((-8\eta\rho)^{1/2}\right) + o\!\left(|\eta|^{1/4}\right)\right),$$
$$G_\ell(\eta,\rho) = -\frac{\pi(-2\eta)^\ell}{(2\ell+1)!\,C_\ell(\eta)}\left((-2\eta\rho)^{1/2}\right.$$
$$\left. \times Y_{2\ell+1}\!\left((-8\eta\rho)^{1/2}\right) + o\!\left(|\eta|^{1/4}\right)\right).$$

In particular, for $\ell=0$,

33.10.8
$$F_0(\eta,\rho) = (\pi\rho)^{1/2}\,J_1\!\left((-8\eta\rho)^{1/2}\right) + o\!\left(|\eta|^{-1/4}\right),$$
$$G_0(\eta,\rho) = -(\pi\rho)^{1/2}\,Y_1\!\left((-8\eta\rho)^{1/2}\right) + o\!\left(|\eta|^{-1/4}\right).$$

33.10.9
$$F_0'(\eta,\rho) = (-2\pi\eta)^{1/2}\,J_0\!\left((-8\eta\rho)^{1/2}\right) + o\!\left(|\eta|^{1/4}\right),$$
$$G_0'(\eta,\rho) = -(-2\pi\eta)^{1/2}\,Y_0\!\left((-8\eta\rho)^{1/2}\right) + o\!\left(|\eta|^{1/4}\right).$$

Also,

33.10.10
$$\sigma_0(\eta) = \eta(\ln(-\eta) - 1) - \tfrac{1}{4}\pi + o(1), \quad C_0(\eta) \sim (-2\pi\eta)^{1/2}.$$

33.11 Asymptotic Expansions for Large ρ

For large ρ, with ℓ and η fixed,

33.11.1 $\quad H_\ell^\pm(\eta, \rho) = e^{\pm i\theta_\ell(\eta,\rho)} \sum_{k=0}^\infty \frac{(a)_k(b)_k}{k!(\mp 2i\rho)^k},$

where $\theta_\ell(\eta, \rho)$ is defined by (33.2.9), and a and b are defined by (33.8.3).

With arguments (η, ρ) suppressed, an equivalent formulation is given by

33.11.2 $\quad F_\ell = g\cos\theta_\ell + f\sin\theta_\ell, \quad G_\ell = f\cos\theta_\ell - g\sin\theta_\ell,$

33.11.3 $\quad F_\ell' = \widehat{g}\cos\theta_\ell + \widehat{f}\sin\theta_\ell, \quad G_\ell' = \widehat{f}\cos\theta_\ell - \widehat{g}\sin\theta_\ell,$

33.11.4 $\quad H_\ell^\pm = e^{\pm i\theta_\ell}(f \pm ig),$

where

33.11.5 $\quad f \sim \sum_{k=0}^\infty f_k, \quad g \sim \sum_{k=0}^\infty g_k,$

33.11.6 $\quad \widehat{f} \sim \sum_{k=0}^\infty \widehat{f}_k, \quad \widehat{g} \sim \sum_{k=0}^\infty \widehat{g}_k,$

33.11.7 $\quad g\widehat{f} - f\widehat{g} = 1.$

Here $f_0 = 1$, $g_0 = 0$, $\widehat{f}_0 = 0$, $\widehat{g}_0 = 1 - (\eta/\rho)$, and for $k = 0, 1, 2, \ldots,$

33.11.8
$$\begin{aligned} f_{k+1} &= \lambda_k f_k - \mu_k g_k, \\ g_{k+1} &= \lambda_k g_k + \mu_k f_k, \\ \widehat{f}_{k+1} &= \lambda_k \widehat{f}_k - \mu_k \widehat{g}_k - (f_{k+1}/\rho), \\ \widehat{g}_{k+1} &= \lambda_k \widehat{g}_k + \mu_k \widehat{f}_k - (g_{k+1}/\rho), \end{aligned}$$

where

33.11.9 $\quad \lambda_k = \frac{(2k+1)\eta}{(2k+2)\rho}, \quad \mu_k = \frac{\ell(\ell+1) - k(k+1) + \eta^2}{(2k+2)\rho}.$

33.12 Asymptotic Expansions for Large η

33.12(i) Transition Region

When $\ell = 0$ and $\eta > 0$, the outer turning point is given by $\rho_{\mathrm{tp}}(\eta, 0) = 2\eta$; compare (33.2.2). Define

33.12.1 $\quad x = (2\eta - \rho)/(2\eta)^{1/3}, \quad \mu = (2\eta)^{2/3}.$

Then as $\eta \to \infty$,

33.12.2 $\quad \begin{matrix} F_0(\eta,\rho) \\ G_0(\eta,\rho) \end{matrix} \sim \pi^{1/2}(2\eta)^{1/6} \left\{ \begin{matrix} \mathrm{Ai}(x) \\ \mathrm{Bi}(x) \end{matrix} \left(1 + \frac{B_1}{\mu} + \frac{B_2}{\mu^2} + \cdots\right) + \begin{matrix} \mathrm{Ai}'(x) \\ \mathrm{Bi}'(x) \end{matrix} \left(\frac{A_1}{\mu} + \frac{A_2}{\mu^2} + \cdots\right) \right\},$

33.12.3 $\quad \begin{matrix} F_0'(\eta,\rho) \\ G_0'(\eta,\rho) \end{matrix} \sim -\pi^{1/2}(2\eta)^{-1/6} \left\{ \begin{matrix} \mathrm{Ai}(x) \\ \mathrm{Bi}(x) \end{matrix} \left(\frac{B_1' + xA_1}{\mu} + \frac{B_2' + xA_2}{\mu^2} + \cdots\right) + \begin{matrix} \mathrm{Ai}'(x) \\ \mathrm{Bi}'(x) \end{matrix} \left(\frac{B_1 + A_1'}{\mu} + \frac{B_2 + A_2'}{\mu^2} + \cdots\right) \right\},$

uniformly for bounded values of $|(\rho - 2\eta)/\eta^{1/3}|$. Here Ai and Bi are the Airy functions (§9.2), and

33.12.4 $\quad A_1 = \frac{1}{5}x^2, \quad A_2 = \frac{1}{35}(2x^3 + 6), \quad A_3 = \frac{1}{15750}(21x^7 + 370x^4 + 580x),$

33.12.5 $\quad B_1 = -\frac{1}{5}x, \quad B_2 = \frac{1}{350}(7x^5 - 30x^2), \quad B_3 = \frac{1}{15750}(264x^6 - 290x^3 - 560).$

In particular,

33.12.6 $\quad \begin{matrix} F_0(\eta, 2\eta) \\ 3^{-1/2}\,G_0(\eta, 2\eta) \end{matrix} \sim \frac{\Gamma(\frac{1}{3})\omega^{1/2}}{2\sqrt{\pi}} \left(1 \mp \frac{2}{35}\frac{\Gamma(\frac{2}{3})}{\Gamma(\frac{1}{3})}\frac{1}{\omega^4} - \frac{8}{2025}\frac{1}{\omega^6} \mp \frac{5792}{46\,06875}\frac{\Gamma(\frac{2}{3})}{\Gamma(\frac{1}{3})}\frac{1}{\omega^{10}} - \cdots \right),$

33.12.7 $\quad \begin{matrix} F_0'(\eta, 2\eta) \\ 3^{-1/2}\,G_0'(\eta, 2\eta) \end{matrix} \sim \frac{\Gamma(\frac{2}{3})}{2\sqrt{\pi}\omega^{1/2}} \left(\pm 1 + \frac{1}{15}\frac{\Gamma(\frac{1}{3})}{\Gamma(\frac{2}{3})}\frac{1}{\omega^2} \pm \frac{2}{14175}\frac{1}{\omega^6} + \frac{1436}{23\,38875}\frac{\Gamma(\frac{1}{3})}{\Gamma(\frac{2}{3})}\frac{1}{\omega^8} \pm \cdots\right),$

where $\omega = (\frac{2}{3}\eta)^{1/3}$.

For derivations and additional terms in the expansions in this subsection see Abramowitz and Rabinowitz (1954) and Fröberg (1955).

33.12(ii) Uniform Expansions

With the substitution $\rho = 2\eta z$, Equation (33.2.1) becomes

33.12.8 $\quad \frac{d^2w}{dz^2} = \left(4\eta^2 \left(\frac{1-z}{z}\right) + \frac{\ell(\ell+1)}{z^2}\right)w.$

Then, by application of the results given in §§2.8(iii) and 2.8(iv), two sets of asymptotic expansions can be constructed for $F_\ell(\eta, \rho)$ and $G_\ell(\eta, \rho)$ when $\eta \to \infty$.

The first set is in terms of Airy functions and the expansions are uniform for fixed ℓ and $\delta \le z < \infty$, where δ is an arbitrary small positive constant. They would include the results of §33.12(i) as a special case.

The second set is in terms of Bessel functions of orders $2\ell + 1$ and $2\ell + 2$, and they are uniform for fixed ℓ

and $0 \leq z \leq 1 - \delta$, where δ again denotes an arbitrary small positive constant.

Compare also §33.20(iv).

33.13 Complex Variable and Parameters

The functions $F_\ell(\eta,\rho)$, $G_\ell(\eta,\rho)$, and $H_\ell^\pm(\eta,\rho)$ may be extended to noninteger values of ℓ by generalizing $(2\ell+1)! = \Gamma(2\ell+2)$, and supplementing (33.6.5) by a formula derived from (33.2.8) with $U(a,b,z)$ expanded via (13.2.42).

These functions may also be continued analytically to complex values of ρ, η, and ℓ. The quantities $C_\ell(\eta)$, $\sigma_\ell(\eta)$, and R_ℓ, given by (33.2.6), (33.2.10), and (33.4.1), respectively, must be defined consistently so that

33.13.1
$$C_\ell(\eta) = 2^\ell e^{i\sigma_\ell(\eta)-(\pi\eta/2)} \Gamma(\ell+1-i\eta)/\Gamma(2\ell+2),$$
and

33.13.2 $\qquad R_\ell = (2\ell+1)\, C_\ell(\eta)/C_{\ell-1}(\eta).$

For further information see Dzieciol et al. (1999), Thompson and Barnett (1986), and Humblet (1984).

Variables r, ϵ

33.14 Definitions and Basic Properties

33.14(i) Coulomb Wave Equation

Another parametrization of (33.2.1) is given by

33.14.1 $\qquad \dfrac{d^2 w}{dr^2} + \left(\epsilon + \dfrac{2}{r} - \dfrac{\ell(\ell+1)}{r^2}\right) w = 0,$

where

33.14.2 $\qquad r = -\eta\rho, \quad \epsilon = 1/\eta^2.$

Again, there is a regular singularity at $r=0$ with indices $\ell+1$ and $-\ell$, and an irregular singularity of rank 1 at $r=\infty$. When $\epsilon > 0$ the outer turning point is given by

33.14.3 $\qquad r_{\rm tp}(\epsilon,\ell) = \left(\sqrt{1+\epsilon\ell(\ell+1)} - 1\right)/\epsilon;$

compare (33.2.2).

33.14(ii) Regular Solution $f(\epsilon,\ell;r)$

The function $f(\epsilon,\ell;r)$ is recessive (§2.7(iii)) at $r=0$, and is defined by

33.14.4 $\quad f(\epsilon,\ell;r) = \kappa^{\ell+1} M_{\kappa,\ell+\frac{1}{2}}(2r/\kappa)/(2\ell+1)!,$

or equivalently

33.14.5
$$f(\epsilon,\ell;r) = (2r)^{\ell+1} e^{-r/\kappa} M(\ell+1-\kappa, 2\ell+2, 2r/\kappa)/(2\ell+1)!,$$

where $M_{\kappa,\mu}(z)$ and $M(a,b,z)$ are defined in §§13.14(i) and 13.2(i), and

33.14.6 $\qquad \kappa = \begin{cases} (-\epsilon)^{-1/2}, & \epsilon<0, r>0, \\ -(-\epsilon)^{-1/2}, & \epsilon<0, r<0, \\ \pm i\epsilon^{-1/2}, & \epsilon>0. \end{cases}$

The choice of sign in the last line of (33.14.6) is immaterial: the same function $f(\epsilon,\ell;r)$ is obtained. This is a consequence of Kummer's transformation (§13.2(vii)).

$f(\epsilon,\ell;r)$ is real and an analytic function of r in the interval $-\infty < r < \infty$, and it is also an analytic function of ϵ when $-\infty < \epsilon < \infty$. This includes $\epsilon = 0$, hence $f(\epsilon,\ell;r)$ can be expanded in a convergent power series in ϵ in a neighborhood of $\epsilon = 0$ (§33.20(ii)).

33.14(iii) Irregular Solution $h(\epsilon,\ell;r)$

For nonzero values of ϵ and r the function $h(\epsilon,\ell;r)$ is defined by

33.14.7
$$h(\epsilon,\ell;r) = \frac{\Gamma(\ell+1-\kappa)}{\pi\kappa^\ell} \left(W_{\kappa,\ell+\frac{1}{2}}(2r/\kappa) \right.$$
$$\left. + (-1)^\ell S(\epsilon,r) \frac{\Gamma(\ell+1+\kappa)}{2(2\ell+1)!} M_{\kappa,\ell+\frac{1}{2}}(2r/\kappa) \right),$$

where κ is given by (33.14.6) and

33.14.8 $\quad S(\epsilon,r) = \begin{cases} 2\cos(\pi|\epsilon|^{-1/2}), & \epsilon<0, r>0, \\ 0, & \epsilon<0, r<0, \\ e^{\pi\epsilon^{-1/2}}, & \epsilon>0, r>0, \\ e^{-\pi\epsilon^{-1/2}}, & \epsilon>0, r<0. \end{cases}$

(Again, the choice of the ambiguous sign in the last line of (33.14.6) is immaterial.)

$h(\epsilon,\ell;r)$ is real and an analytic function of each of r and ϵ in the intervals $-\infty < r < \infty$ and $-\infty < \epsilon < \infty$, except when $r=0$ or $\epsilon=0$.

33.14(iv) Solutions $s(\epsilon,\ell;r)$ and $c(\epsilon,\ell;r)$

The functions $s(\epsilon,\ell;r)$ and $c(\epsilon,\ell;r)$ are defined by

33.14.9
$$s(\epsilon,\ell;r) = (B(\epsilon,\ell)/2)^{1/2} f(\epsilon,\ell;r),$$
$$c(\epsilon,\ell;r) = (2B(\epsilon,\ell))^{-1/2} h(\epsilon,\ell;r),$$

provided that $\ell < (-\epsilon)^{-1/2}$ when $\epsilon < 0$, where

33.14.10
$$B(\epsilon,\ell) = \begin{cases} A(\epsilon,\ell)\left(1-\exp(-2\pi/\epsilon^{1/2})\right)^{-1}, & \epsilon>0, \\ A(\epsilon,\ell), & \epsilon\leq 0, \end{cases}$$

and

33.14.11 $\qquad A(\epsilon,\ell) = \prod_{k=0}^{\ell}(1+\epsilon k^2).$

An alternative formula for $A(\epsilon,\ell)$ is

33.14.12 $\qquad A(\epsilon,\ell) = \dfrac{\Gamma(1+\ell+\kappa)}{\Gamma(\kappa-\ell)} \kappa^{-2\ell-1},$

the choice of sign in the last line of (33.14.6) again being immaterial.

When $\epsilon < 0$ and $\ell > (-\epsilon)^{-1/2}$ the quantity $A(\epsilon, \ell)$ may be negative, causing $s(\epsilon, \ell; r)$ and $c(\epsilon, \ell; r)$ to become imaginary.

The function $s(\epsilon, \ell; r)$ has the following properties:

33.14.13 $\quad \int_0^\infty s(\epsilon_1, \ell; r) \, s(\epsilon_2, \ell; r) \, dr = \delta(\epsilon_1 - \epsilon_2),$

where the right-hand side is the Dirac delta (§1.17). When $\epsilon = -1/n^2$, $n = \ell+1, \ell+2, \ldots$, $s(\epsilon, \ell; r)$ is $\exp(-r/n)$ times a polynomial in r, and

33.14.14 $\quad \phi_{n,\ell}(r) = (-1)^{\ell+1+n} (2/n^3)^{1/2} s(-1/n^2, \ell; r)$

satisfies

33.14.15 $\quad \int_0^\infty \phi_{n,\ell}^2(r) \, dr = 1.$

33.14(v) Wronskians

With arguments ϵ, ℓ, r suppressed,

33.14.16 $\quad \mathscr{W}\{h, f\} = 2/\pi, \quad \mathscr{W}\{c, s\} = 1/\pi.$

33.15 Graphics

33.15(i) Line Graphs of the Coulomb Functions $f(\epsilon, \ell; r)$ and $h(\epsilon, \ell; r)$

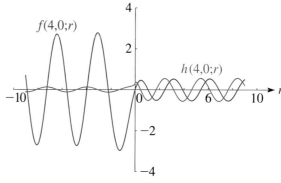

Figure 33.15.1: $f(\epsilon, \ell; r), h(\epsilon, \ell; r)$ with $\ell = 0, \epsilon = 4$.

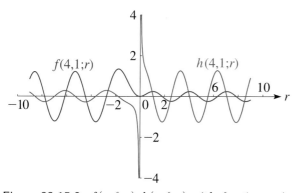

Figure 33.15.2: $f(\epsilon, \ell; r), h(\epsilon, \ell; r)$ with $\ell = 1, \epsilon = 4$.

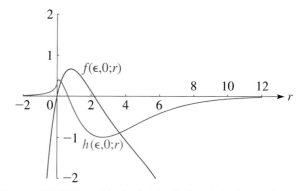

Figure 33.15.3: $f(\epsilon, \ell; r), h(\epsilon, \ell; r)$ with $\ell = 0, \epsilon = -1/\nu^2, \nu = 1.5$.

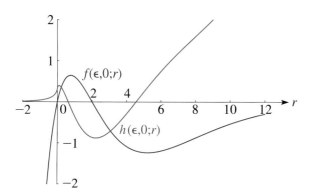

Figure 33.15.4: $f(\epsilon, \ell; r), h(\epsilon, \ell; r)$ with $\ell = 0, \epsilon = -1/\nu^2, \nu = 2$.

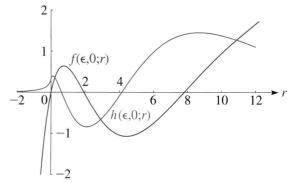

Figure 33.15.5: $f(\epsilon, \ell; r), h(\epsilon, \ell; r)$ with $\ell = 0, \epsilon = -1/\nu^2, \nu = 2.5$.

33.15(ii) Surfaces of the Coulomb Functions $f(\epsilon,\ell;r)$, $h(\epsilon,\ell;r)$, $s(\epsilon,\ell;r)$, and $c(\epsilon,\ell;r)$

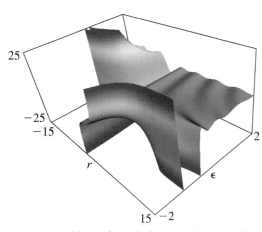

Figure 33.15.6: $f(\epsilon,\ell;r)$ with $\ell=0, -2<\epsilon<2, -15<r<15$.

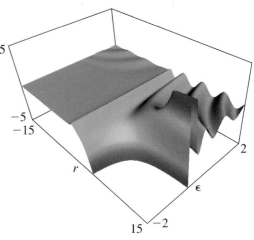

Figure 33.15.7: $h(\epsilon,\ell;r)$ with $\ell=0, -2<\epsilon<2, -15<r<15$.

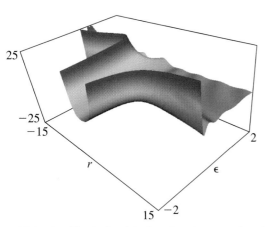

Figure 33.15.8: $f(\epsilon,\ell;r)$ with $\ell=1, -2<\epsilon<2, -15<r<15$.

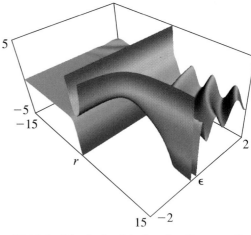

Figure 33.15.9: $h(\epsilon,\ell;r)$ with $\ell=1, -2<\epsilon<2, -15<r<15$.

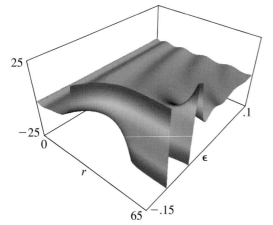

Figure 33.15.10: $s(\epsilon,\ell;r)$ with $\ell=0, -0.15<\epsilon<0.10, 0<r<65$.

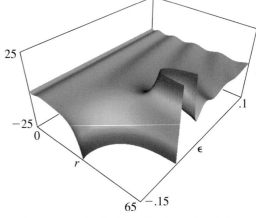

Figure 33.15.11: $c(\epsilon,\ell;r)$ with $\ell=0, -0.15<\epsilon<0.10, 0<r<65$.

33.16 Connection Formulas

33.16(i) F_ℓ and G_ℓ in Terms of f and h

33.16.1 $\quad F_\ell(\eta,\rho) = \dfrac{(2\ell+1)!\, C_\ell(\eta)}{(-2\eta)^{\ell+1}}\, f\bigl(1/\eta^2, \ell; -\eta\rho\bigr),$

33.16.2 $\quad G_\ell(\eta,\rho) = \dfrac{\pi(-2\eta)^\ell}{(2\ell+1)!\, C_\ell(\eta)}\, h\bigl(1/\eta^2, \ell; -\eta\rho\bigr),$

where $C_\ell(\eta)$ is given by (33.2.5) or (33.2.6).

33.16(ii) f and h in Terms of F_ℓ and G_ℓ when $\epsilon > 0$

When $\epsilon > 0$ denote

33.16.3 $\quad\quad\quad \tau = \epsilon^{1/2}(>0),$

and again define $A(\epsilon,\ell)$ by (33.14.11) or (33.14.12). Then for $r > 0$

33.16.4 $\quad f(\epsilon,\ell;r) = \left(\dfrac{2}{\pi\tau}\dfrac{1-e^{-2\pi/\tau}}{A(\epsilon,\ell)}\right)^{1/2} F_\ell(-1/\tau, \tau r),$

33.16.5 $\quad h(\epsilon,\ell;r) = \left(\dfrac{2}{\pi\tau}\dfrac{A(\epsilon,\ell)}{1-e^{-2\pi/\tau}}\right)^{1/2} G_\ell(-1/\tau, \tau r).$

Alternatively, for $r < 0$

33.16.6
$$f(\epsilon,\ell;r) = (-1)^{\ell+1}\left(\dfrac{2}{\pi\tau}\dfrac{e^{2\pi/\tau}-1}{A(\epsilon,\ell)}\right)^{1/2} F_\ell(1/\tau, -\tau r),$$

33.16.7
$$h(\epsilon,\ell;r) = (-1)^\ell \left(\dfrac{2}{\pi\tau}\dfrac{A(\epsilon,\ell)}{e^{2\pi/\tau}-1}\right)^{1/2} G_\ell(1/\tau, -\tau r).$$

33.16(iii) f and h in Terms of $W_{\kappa,\mu}(z)$ when $\epsilon < 0$

When $\epsilon < 0$ denote

33.16.8 $\quad\quad\quad \nu = 1/(-\epsilon)^{1/2}(>0),$

33.16.9
$$\zeta_\ell(\nu,r) = W_{\nu,\ell+\frac{1}{2}}(2r/\nu),$$
$$\xi_\ell(\nu,r) = \Re\!\left(e^{i\pi\nu} W_{-\nu,\ell+\frac{1}{2}}\bigl(e^{i\pi}2r/\nu\bigr)\right),$$

and again define $A(\epsilon,\ell)$ by (33.14.11) or (33.14.12). Then for $r > 0$

33.16.10
$$f(\epsilon,\ell;r) = (-1)^\ell \nu^{\ell+1}\left(-\dfrac{\cos(\pi\nu)\zeta_\ell(\nu,r)}{\Gamma(\ell+1+\nu)} + \dfrac{\sin(\pi\nu)\,\Gamma(\nu-\ell)\xi_\ell(\nu,r)}{\pi}\right),$$

33.16.11
$$h(\epsilon,\ell;r) = (-1)^\ell \nu^{\ell+1} A(\epsilon,\ell)\left(\dfrac{\sin(\pi\nu)\zeta_\ell(\nu,r)}{\Gamma(\ell+1+\nu)} + \dfrac{\cos(\pi\nu)\,\Gamma(\nu-\ell)\xi_\ell(\nu,r)}{\pi}\right).$$

Alternatively, for $r < 0$

33.16.12
$$f(\epsilon,\ell;r) = \dfrac{(-1)^\ell \nu^{\ell+1}}{\pi}\left(-\dfrac{\pi\xi_\ell(-\nu,r)}{\Gamma(\ell+1+\nu)} + \sin(\pi\nu)\cos(\pi\nu)\,\Gamma(\nu-\ell)\zeta_\ell(-\nu,r)\right),$$

33.16.13 $\quad h(\epsilon,\ell;r) = (-1)^\ell \nu^{\ell+1} A(\epsilon,\ell)\,\Gamma(\nu-\ell)\zeta_\ell(-\nu,r)/\pi.$

33.16(iv) s and c in Terms of F_ℓ and G_ℓ when $\epsilon > 0$

When $\epsilon > 0$, again denote τ by (33.16.3). Then for $r > 0$

33.16.14
$$s(\epsilon,\ell;r) = (\pi\tau)^{-1/2} F_\ell(-1/\tau, \tau r),$$
$$c(\epsilon,\ell;r) = (\pi\tau)^{-1/2} G_\ell(-1/\tau, \tau r).$$

Alternatively, for $r < 0$

33.16.15
$$s(\epsilon,\ell;r) = (\pi\tau)^{-1/2} F_\ell(1/\tau, -\tau r),$$
$$c(\epsilon,\ell;r) = (\pi\tau)^{-1/2} G_\ell(1/\tau, -\tau r).$$

33.16(v) s and c in Terms of $W_{\kappa,\mu}(z)$ when $\epsilon < 0$

When $\epsilon < 0$ denote ν, $\zeta_\ell(\nu,r)$, and $\xi_\ell(\nu,r)$ by (33.16.8) and (33.16.9). Also denote

33.16.16 $\quad K(\nu,\ell) = \bigl(\nu^2\, \Gamma(\nu+\ell+1)\,\Gamma(\nu-\ell)\bigr)^{-1/2}.$

Then for $r > 0$

33.16.17
$$s(\epsilon,\ell;r) = \dfrac{(-1)^\ell}{2\nu^{1/2}}\left(\dfrac{\sin(\pi\nu)}{\pi K(\nu,\ell)}\xi_\ell(\nu,r) - \cos(\pi\nu)\nu^2 K(\nu,\ell)\zeta_\ell(\nu,r)\right),$$
$$c(\epsilon,\ell;r) = \dfrac{(-1)^\ell}{2\nu^{1/2}}\left(\dfrac{\cos(\pi\nu)}{\pi K(\nu,\ell)}\xi_\ell(\nu,r) + \sin(\pi\nu)\nu^2 K(\nu,\ell)\zeta_\ell(\nu,r)\right).$$

Alternatively, for $r < 0$

33.16.18
$$s(\epsilon,\ell;r) = \dfrac{(-1)^{\ell+1}}{2^{1/2}}\left(\dfrac{\nu^{3/2}}{K(\nu,\ell)}\xi_\ell(-\nu,r) - \dfrac{\sin(\pi\nu)\cos(\pi\nu)}{\pi\nu^{1/2}}K(\nu,\ell)\zeta_\ell(-\nu,r)\right),$$
$$c(\epsilon,\ell;r) = \dfrac{(-1)^\ell}{\pi(2\nu)^{1/2}} K(\nu,\ell)\zeta_\ell(-\nu,r).$$

33.17 Recurrence Relations and Derivatives

33.17.1 $(\ell+1)r\, f(\epsilon,\ell-1;r) - (2\ell+1)\left(\ell(\ell+1) - r\right) f(\epsilon,\ell;r) + \ell\left(1 + (\ell+1)^2\epsilon\right) r\, f(\epsilon,\ell+1;r) = 0,$

33.17.2 $(\ell+1)\left(1+\ell^2\epsilon\right) r\, h(\epsilon,\ell-1;r) - (2\ell+1)\left(\ell(\ell+1) - r\right) h(\epsilon,\ell;r) + \ell r\, h(\epsilon,\ell+1;r) = 0,$

33.17.3 $(\ell+1)r\, f'(\epsilon,\ell;r) = \left((\ell+1)^2 - r\right) f(\epsilon,\ell;r) - \left(1 + (\ell+1)^2\epsilon\right) r\, f(\epsilon,\ell+1;r),$

33.17.4 $(\ell+1)r\, h'(\epsilon,\ell;r) = \left((\ell+1)^2 - r\right) h(\epsilon,\ell;r) - r\, h(\epsilon,\ell+1;r).$

33.18 Limiting Forms for Large ℓ

As $\ell \to \infty$ with ϵ and r ($\neq 0$) fixed,

33.18.1 $f(\epsilon,\ell;r) \sim \dfrac{(2r)^{\ell+1}}{(2\ell+1)!}, \quad h(\epsilon,\ell;r) \sim \dfrac{(2\ell)!}{\pi(2r)^\ell}.$

33.19 Power-Series Expansions in r

33.19.1 $f(\epsilon,\ell;r) = r^{\ell+1} \sum_{k=0}^{\infty} \alpha_k r^k,$

where

33.19.2
$\alpha_0 = 2^{\ell+1}/(2\ell+1)!, \quad \alpha_1 = -\alpha_0/(\ell+1),$
$k(k+2\ell+1)\alpha_k + 2\alpha_{k-1} + \epsilon\alpha_{k-2} = 0, \quad k = 2, 3, \ldots.$

33.19.3
$2\pi\, h(\epsilon,\ell;r) = \sum_{k=0}^{2\ell} \dfrac{(2\ell-k)!\gamma_k}{k!}(2r)^{k-\ell} - \sum_{k=0}^{\infty} \delta_k r^{k+\ell+1}$
$- A(\epsilon,\ell)\left(2\ln|2r/\kappa| + \Re\psi(\ell+1+\kappa) + \Re\psi(-\ell+\kappa)\right) f(\epsilon,\ell;r), \quad r \neq 0.$

Here κ is defined by (33.14.6), $A(\epsilon,\ell)$ is defined by (33.14.11) or (33.14.12), $\gamma_0 = 1$, $\gamma_1 = 1$, and

33.19.4 $\gamma_k - \gamma_{k-1} + \tfrac{1}{4}(k-1)(k-2\ell-2)\epsilon\gamma_{k-2} = 0, \quad k = 2, 3, \ldots.$

Also,

33.19.5
$\delta_0 = (\beta_{2\ell+1} - 2(\psi(2\ell+2) + \psi(1))A(\epsilon,\ell))\alpha_0,$
$\delta_1 = (\beta_{2\ell+2} - 2(\psi(2\ell+3) + \psi(2))A(\epsilon,\ell))\alpha_1,$

33.19.6
$k(k+2\ell+1)\delta_k + 2\delta_{k-1} + \epsilon\delta_{k-2}$
$+ 2(2k+2\ell+1)A(\epsilon,\ell)\alpha_k = 0, \quad k = 2, 3, \ldots,$

with $\beta_0 = \beta_1 = 0$, and

33.19.7 $\beta_k - \beta_{k-1} + \tfrac{1}{4}(k-1)(k-2\ell-2)\epsilon\beta_{k-2} + \tfrac{1}{2}(k-1)\epsilon\gamma_{k-2} = 0, \quad k = 2, 3, \ldots.$

The expansions (33.19.1) and (33.19.3) converge for all finite values of r, except $r = 0$ in the case of (33.19.3).

33.20 Expansions for Small $|\epsilon|$

33.20(i) Case $\epsilon = 0$

33.20.1
$f(0,\ell;r) = (2r)^{1/2} J_{2\ell+1}\left(\sqrt{8r}\right),$
$h(0,\ell;r) = -(2r)^{1/2} Y_{2\ell+1}\left(\sqrt{8r}\right), \quad r > 0,$

33.20.2
$f(0,\ell;r) = (-1)^{\ell+1}(2|r|)^{1/2} I_{2\ell+1}\left(\sqrt{8|r|}\right),$
$h(0,\ell;r) = (-1)^{\ell}(2/\pi)(2|r|)^{1/2} K_{2\ell+1}\left(\sqrt{8|r|}\right), \quad r < 0.$

For the functions J, Y, I, and K see §§10.2(ii), 10.25(ii).

33.20(ii) Power-Series in ϵ for the Regular Solution

33.20.3 $f(\epsilon,\ell;r) = \sum_{k=0}^{\infty} \epsilon^k \mathsf{F}_k(\ell;r),$

where

33.20.4 $\mathsf{F}_k(\ell;r) = \sum_{p=2k}^{3k} (2r)^{(p+1)/2} C_{k,p} J_{2\ell+1+p}\left(\sqrt{8r}\right), \quad r > 0,$

33.20.5
$\mathsf{F}_k(\ell;r) = \sum_{p=2k}^{3k} (-1)^{\ell+1+p}(2|r|)^{(p+1)/2} C_{k,p} I_{2\ell+1+p}\left(\sqrt{8|r|}\right), \quad r < 0.$

The functions J and I are as in §§10.2(ii), 10.25(ii), and the coefficients $C_{k,p}$ are given by $C_{0,0} = 1$, $C_{1,0} = 0$, and

33.20.6
$C_{k,p} = 0, \quad p < 2k \text{ or } p > 3k,$
$C_{k,p} = \left(-(2\ell+p)C_{k-1,p-2} + C_{k-1,p-3}\right)/(4p), \quad k > 0, \; 2k \leq p \leq 3k.$

The series (33.20.3) converges for all r and ϵ.

33.20(iii) Asymptotic Expansion for the Irregular Solution

As $\epsilon \to 0$ with ℓ and r fixed,

33.20.7 $\quad h(\epsilon, \ell; r) \sim -A(\epsilon, \ell) \sum_{k=0}^{\infty} \epsilon^k \mathsf{H}_k(\ell; r),$

where $A(\epsilon, \ell)$ is given by (33.14.11), (33.14.12), and

33.20.8
$$\mathsf{H}_k(\ell; r) = \sum_{p=2k}^{3k} (2r)^{(p+1)/2} C_{k,p} \, Y_{2\ell+1+p}\left(\sqrt{8r}\right), \quad r > 0,$$

33.20.9
$$\mathsf{H}_k(\ell; r) = (-1)^{\ell+1} \frac{2}{\pi} \sum_{p=2k}^{3k} (2|r|)^{(p+1)/2} C_{k,p} \, K_{2\ell+1+p}\left(\sqrt{8|r|}\right),$$
$$r < 0.$$

The functions Y and K are as in §§10.2(ii), 10.25(ii), and the coefficients $C_{k,p}$ are given by (33.20.6).

33.20(iv) Uniform Asymptotic Expansions

For a comprehensive collection of asymptotic expansions that cover $f(\epsilon, \ell; r)$ and $h(\epsilon, \ell; r)$ as $\epsilon \to 0\pm$ and are uniform in r, including unbounded values, see Curtis (1964a, §7). These expansions are in terms of elementary functions, Airy functions, and Bessel functions of orders $2\ell+1$ and $2\ell+2$.

33.21 Asymptotic Approximations for Large $|r|$

33.21(i) Limiting Forms

We indicate here how to obtain the limiting forms of $f(\epsilon, \ell; r)$, $h(\epsilon, \ell; r)$, $s(\epsilon, \ell; r)$, and $c(\epsilon, \ell; r)$ as $r \to \pm\infty$, with ϵ and ℓ fixed, in the following cases:

(a) When $r \to \pm\infty$ with $\epsilon > 0$, Equations (33.16.4)–(33.16.7) are combined with (33.10.1).

(b) When $r \to \pm\infty$ with $\epsilon < 0$, Equations (33.16.10)–(33.16.13) are combined with

33.21.1 $\quad \begin{aligned} \zeta_\ell(\nu, r) &\sim e^{-r/\nu}(2r/\nu)^\nu, \\ \xi_\ell(\nu, r) &\sim e^{r/\nu}(2r/\nu)^{-\nu}, \end{aligned} \quad r \to \infty,$

33.21.2 $\quad \begin{aligned} \zeta_\ell(-\nu, r) &\sim e^{r/\nu}(-2r/\nu)^{-\nu}, \\ \xi_\ell(-\nu, r) &\sim e^{-r/\nu}(-2r/\nu)^\nu, \end{aligned} \quad r \to -\infty.$

Corresponding approximations for $s(\epsilon, \ell; r)$ and $c(\epsilon, \ell; r)$ as $r \to \infty$ can be obtained via (33.16.17), and as $r \to -\infty$ via (33.16.18).

(c) When $r \to \pm\infty$ with $\epsilon = 0$, combine (33.20.1), (33.20.2) with §§10.7(ii), 10.30(ii).

33.21(ii) Asymptotic Expansions

For asymptotic expansions of $f(\epsilon, \ell; r)$ and $h(\epsilon, \ell; r)$ as $r \to \pm\infty$ with ϵ and ℓ fixed, see Curtis (1964a, §6).

Physical Applications

33.22 Particle Scattering and Atomic and Molecular Spectra

33.22(i) Schrödinger Equation

With e denoting here the elementary charge, the Coulomb potential between two point particles with charges $Z_1 e, Z_2 e$ and masses m_1, m_2 separated by a distance s is $V(s) = Z_1 Z_2 e^2/(4\pi\epsilon_0 s) = Z_1 Z_2 \alpha \hbar c / s$, where Z_j are atomic numbers, ϵ_0 is the electric constant, α is the fine structure constant, and $\hbar$ is the reduced Planck's constant. The reduced mass is $m = m_1 m_2/(m_1 + m_2)$, and at energy of relative motion E with relative orbital angular momentum $\ell\hbar$, the Schrödinger equation for the radial wave function $w(s)$ is given by

33.22.1
$$\left(-\frac{\hbar^2}{2m}\left(\frac{d^2}{ds^2} - \frac{\ell(\ell+1)}{s^2}\right) + \frac{Z_1 Z_2 \alpha \hbar c}{s}\right) w = E w,$$

With the substitutions

33.22.2 $\quad \mathsf{k} = (2mE/\hbar^2)^{1/2}, \quad Z = mZ_1 Z_2 \alpha c/\hbar, \quad x = s,$

(33.22.1) becomes

33.22.3 $\quad \dfrac{d^2 w}{dx^2} + \left(\mathsf{k}^2 - \dfrac{2Z}{x} - \dfrac{\ell(\ell+1)}{x^2}\right) w = 0.$

33.22(ii) Definitions of Variables

k Scaling

The k-scaled variables ρ and η of §33.2 are given by

33.22.4 $\quad \rho = s(2mE/\hbar^2)^{1/2}, \quad \eta = Z_1 Z_2 \alpha c(m/(2E))^{1/2}.$

At positive energies $E > 0$, $\rho \geq 0$, and:

Attractive potentials: $\quad Z_1 Z_2 < 0, \eta < 0.$
Zero potential $(V = 0)$: $\quad Z_1 Z_2 = 0, \eta = 0.$
Repulsive potentials: $\quad Z_1 Z_2 > 0, \eta > 0.$

Positive-energy functions correspond to processes such as Rutherford scattering and Coulomb excitation of nuclei (Alder et al. (1956)), and atomic photo-ionization and electron-ion collisions (Bethe and Salpeter (1977)).

At negative energies $E < 0$ and both ρ and η are purely imaginary. The negative-energy functions are widely used in the description of atomic and molecular spectra; see Bethe and Salpeter (1977), Seaton (1983), and Aymar et al. (1996). In these applications, the Z-scaled variables r and ϵ are more convenient.

Z Scaling

The Z-scaled variables r and ϵ of §33.14 are given by

33.22.5 $\quad r = -Z_1 Z_2 (mc\alpha/\hbar) s, \quad \epsilon = E/(Z_1^2 Z_2^2 mc^2 \alpha^2/2).$

For $Z_1 Z_2 = -1$ and $m = m_e$, the electron mass, the scaling factors in (33.22.5) reduce to the Bohr radius, $a_0 = \hbar/(m_e c \alpha)$, and to a multiple of the Rydberg constant,

$$R_\infty = m_e c \alpha^2 / (2\hbar).$$

Attractive potentials: $\quad Z_1 Z_2 < 0, r > 0.$
Zero potential ($V = 0$): $\quad Z_1 Z_2 = 0, r = 0.$
Repulsive potentials: $\quad Z_1 Z_2 > 0, r < 0.$

ik Scaling

The ik-scaled variables z and κ of §13.2 are given by

33.22.6
$$z = 2is(2mE/\hbar^2)^{1/2}, \quad \kappa = iZ_1 Z_2 \alpha c (m/(2E))^{1/2}.$$

Attractive potentials: $\quad Z_1 Z_2 < 0, \Im\kappa < 0.$
Zero potential ($V = 0$): $\quad Z_1 Z_2 = 0, \kappa = 0.$
Repulsive potentials: $\quad Z_1 Z_2 > 0, \Im\kappa > 0.$

Customary variables are (ϵ, r) in atomic physics and (η, ρ) in atomic and nuclear physics. Both variable sets may be used for attractive and repulsive potentials: the (ϵ, r) set cannot be used for a zero potential because this would imply $r = 0$ for all s, and the (η, ρ) set cannot be used for zero energy E because this would imply $\rho = 0$ always.

33.22(iii) Conversions Between Variables

33.22.7 $\quad r = -\eta\rho, \quad \epsilon = 1/\eta^2, \quad$ Z from k.
33.22.8 $\quad z = 2i\rho, \quad \kappa = i\eta, \quad$ ik from k.
33.22.9 $\quad \rho = z/(2i), \quad \eta = \kappa/i, \quad$ k from ik.
33.22.10 $\quad r = \kappa z/2, \quad \epsilon = -1/\kappa^2, \quad$ Z from ik.
33.22.11 $\quad \eta = \pm\epsilon^{-1/2}, \quad \rho = -r/\eta, \quad$ k from Z.
33.22.12 $\quad \kappa = \pm(-\epsilon)^{-1/2}, \quad z = 2r/\kappa, \quad$ ik from Z.

Resolution of the ambiguous signs in (33.22.11), (33.22.12) depends on the sign of $Z/$k in (33.22.3). See also §§33.14(ii), 33.14(iii), 33.22(i), and 33.22(ii).

33.22(iv) Klein–Gordon and Dirac Equations

The relativistic motion of spinless particles in a Coulomb field, as encountered in pionic atoms and pion-nucleon scattering (Backenstoss (1970)) is described by a Klein–Gordon equation equivalent to (33.2.1); see Barnett (1981a). The motion of a relativistic electron in a Coulomb field, which arises in the theory of the electronic structure of heavy elements (Johnson (2007)), is described by a Dirac equation. The solutions to this equation are closely related to the Coulomb functions; see Greiner et al. (1985).

33.22(v) Asymptotic Solutions

The Coulomb solutions of the Schrödinger and Klein–Gordon equations are almost always used in the external region, outside the range of any non-Coulomb forces or couplings.

For scattering problems, the interior solution is then matched to a linear combination of a pair of Coulomb functions, $F_\ell(\eta, \rho)$ and $G_\ell(\eta, \rho)$, or $f(\epsilon, \ell; r)$ and $h(\epsilon, \ell; r)$, to determine the scattering S-matrix and also the correct normalization of the interior wave solutions; see Bloch et al. (1951).

For bound-state problems only the exponentially decaying solution is required, usually taken to be the Whittaker function $W_{-\eta, \ell+\frac{1}{2}}(2\rho)$. The functions $\phi_{n,\ell}(r)$ defined by (33.14.14) are the hydrogenic bound states in attractive Coulomb potentials; their polynomial components are often called *associated Laguerre functions*; see Christy and Duck (1961) and Bethe and Salpeter (1977).

33.22(vi) Solutions Inside the Turning Point

The penetrability of repulsive Coulomb potential barriers is normally expressed in terms of the quantity $\rho/(F_\ell^2(\eta, \rho) + G_\ell^2(\eta, \rho))$ (Mott and Massey (1956, pp. 63–65)). The WKBJ approximations of §33.23(vii) may also be used to estimate the penetrability.

33.22(vii) Complex Variables and Parameters

The Coulomb functions given in this chapter are most commonly evaluated for real values of ρ, r, η, ϵ and nonnegative integer values of ℓ, but they may be continued analytically to complex arguments and order ℓ as indicated in §33.13.

Examples of applications to noninteger and/or complex variables are as follows.

- Scattering at complex energies. See for example McDonald and Nuttall (1969).

- Searches for resonances as poles of the S-matrix in the complex half-plane $\Im$k < 0. See for example Csótó and Hale (1997).

- Regge poles at complex values of ℓ. See for example Takemasa et al. (1979).

- Eigenstates using complex-rotated coordinates $r \to re^{i\theta}$, so that resonances have square-integrable eigenfunctions. See for example Halley et al. (1993).

- Solution of relativistic Coulomb equations. See for example Cooper et al. (1979) and Barnett (1981b).

- Gravitational radiation. See for example Berti and Cardoso (2006).

For further examples see Humblet (1984).

Computation

33.23 Methods of Computation

33.23(i) Methods for the Confluent Hypergeometric Functions

The methods used for computing the Coulomb functions described below are similar to those in §13.29.

33.23(ii) Series Solutions

The power-series expansions of §§33.6 and 33.19 converge for all finite values of the radii ρ and r, respectively, and may be used to compute the regular and irregular solutions. Cancellation errors increase with increases in ρ and $|r|$, and may be estimated by comparing the final sum of the series with the largest partial sum. Use of extended-precision arithmetic increases the radial range that yields accurate results, but eventually other methods must be employed, for example, the asymptotic expansions of §§33.11 and 33.21.

33.23(iii) Integration of Defining Differential Equations

When numerical values of the Coulomb functions are available for some radii, their values for other radii may be obtained by direct numerical integration of equations (33.2.1) or (33.14.1), provided that the integration is carried out in a stable direction (§3.7). Thus the regular solutions can be computed from the power-series expansions (§§33.6, 33.19) for small values of the radii and then integrated in the direction of increasing values of the radii. On the other hand, the irregular solutions of §§33.2(iii) and 33.14(iii) need to be integrated in the direction of decreasing radii beginning, for example, with values obtained from asymptotic expansions (§§33.11 and 33.21).

33.23(iv) Recurrence Relations

In a manner similar to §33.23(iii) the recurrence relations of §§33.4 or 33.17 can be used for a range of values of the integer ℓ, provided that the recurrence is carried out in a stable direction (§3.6). This implies decreasing ℓ for the regular solutions and increasing ℓ for the irregular solutions of §§33.2(iii) and 33.14(iii).

33.23(v) Continued Fractions

§33.8 supplies continued fractions for F_ℓ'/F_ℓ and $H_\ell^{\pm\prime}/H_\ell^\pm$. Combined with the Wronskians (33.2.12), the values of F_ℓ, G_ℓ, and their derivatives can be extracted. Inside the turning points, that is, when $\rho < \rho_{\mathrm{tp}}(\eta,\ell)$, there can be a loss of precision by a factor of approximately $|G_\ell|^2$.

33.23(vi) Other Numerical Methods

Curtis (1964a, §10) describes the use of series, radial integration, and other methods to generate the tables listed in §33.24.

Bardin et al. (1972) describes ten different methods for the calculation of F_ℓ and G_ℓ, valid in different regions of the (η,ρ)-plane.

Thompson and Barnett (1985, 1986) and Thompson (2004) use combinations of series, continued fractions, and Padé-accelerated asymptotic expansions (§3.11(iv)) for the analytic continuations of Coulomb functions.

Noble (2004) obtains double-precision accuracy for $W_{-\eta,\mu}(2\rho)$ for a wide range of parameters using a combination of recurrence techniques, power-series expansions, and numerical quadrature; compare (33.2.7).

33.23(vii) WKBJ Approximations

WKBJ approximations (§2.7(iii)) for $\rho > \rho_{\mathrm{tp}}(\eta,\ell)$ are presented in Hull and Breit (1959) and Seaton and Peach (1962: in Eq. (12) $(\rho-c)/c$ should be $(\rho-c)/\rho$). A set of consistent second-order WKBJ formulas is given by Burgess (1963: in Eq. (16) $3\kappa^2+2$ should be $3\kappa^2 c+2$). Seaton (1984) estimates the accuracies of these approximations.

Hull and Breit (1959) and Barnett (1981b) give WKBJ approximations for F_0 and G_0 in the region inside the turning point: $\rho < \rho_{\mathrm{tp}}(\eta,\ell)$.

33.24 Tables

- Abramowitz and Stegun (1964, Chapter 14) tabulates $F_0(\eta,\rho)$, $G_0(\eta,\rho)$, $F_0'(\eta,\rho)$, and $G_0'(\eta,\rho)$ for $\eta = 0.5(.5)20$ and $\rho = 1(1)20$, 5S; $C_0(\eta)$ for $\eta = 0(.05)3$, 6S.

- Curtis (1964a) tabulates $P_\ell(\epsilon,r)$, $Q_\ell(\epsilon,r)$ (§33.1), and related functions for $\ell = 0,1,2$ and $\epsilon = -2(.2)2$, with $x = 0(.1)4$ for $\epsilon < 0$ and $x = 0(.1)10$ for $\epsilon \geq 0$; 6D.

For earlier tables see Hull and Breit (1959) and Fletcher et al. (1962, §22.59).

33.25 Approximations

Cody and Hillstrom (1970) provides rational approximations of the phase shift $\sigma_0(\eta) = \operatorname{ph} \Gamma(1+i\eta)$ (see (33.2.10)) for the ranges $0 \leq \eta \leq 2$, $2 \leq \eta \leq 4$, and $4 \leq \eta \leq \infty$. Maximum relative errors range from 1.09×10^{-20} to 4.24×10^{-19}.

33.26 Software

See http://dlmf.nist.gov/33.26.

References

General References

The main references used in writing this chapter are Hull and Breit (1959), Thompson and Barnett (1986), and Seaton (2002). For additional bibliographic reading see also the General References in Chapter 13.

Sources

The following list gives the references or other indications of proofs that were used in constructing the various sections of this chapter. These sources supplement the references that are quoted in the text.

§33.2 Yost et al. (1936), Hull and Breit (1959, pp. 409–410).

§33.3 These graphics were produced at NIST.

§33.4 Powell (1947).

§33.5 Yost et al. (1936), Hull and Breit (1959, pp. 435–436), Wheeler (1937), Biedenharn et al. (1955). For (33.5.9) combine the second formula in (5.4.2) with (5.11.7).

§33.6 For (33.6.5) use the definition (33.2.8) with $U(a,b,z)$ expanded as in (13.2.9). For (33.6.4) use (33.2.4) with Eq. (1.12) of Buchholz (1969).

§33.7 Hull and Breit (1959, pp. 413–416). For (33.7.1) see also Lowan and Horenstein (1942), with change of variable $\xi = 1-t$ in the integral that follows Eq. (8). For (33.7.2) see also Hoisington and Breit (1938). For (33.7.3) see also Bloch et al. (1950). For (33.7.4) see also Newton (1952).

§33.9 The convergence of (33.9.1) follows from the asymptotic forms, for large k, of a_k (obtained by application of §2.9(i)) and $j_{\ell+k}(\rho)$ (obtained from (10.19.1) and (10.47.3)). For (33.9.3) see Yost et al. (1936), Abramowitz (1954), and Humblet (1985). For (33.9.4) see Curtis (1964a, §5.1). For (33.9.6) see Yost et al. (1936) and Abramowitz (1954).

§33.10 Yost et al. (1936), Fröberg (1955), Humblet (1984), Humblet (1985, Eqs. 2.10a,b and 4.7a,b). For (33.10.6) and (33.10.10) use (33.2.5), (33.2.10), and §5.11(i).

§33.11 Fröberg (1955).

§33.14 Curtis (1964a, pp. ix–xxv), Seaton (1983), Seaton (2002, Eqs. 3, 4, 7, 9, 14, 22, 47, 49, 51, 109, 113–116, 122–125, 131, and §2.3). For (33.14.11) and (33.14.12) see Humblet (1985, Eqs. 1.4a,b), Seaton (1982, Eq. 2.4.4).

§33.15 These graphics were produced at NIST.

§33.16 Seaton (2002, Eqs. 104–109, 119–121, 130, 131). (33.16.3)–(33.16.7) are generalizations of Seaton (2002, Eqs. 88, 90, 93, 95). For (33.16.14) and (33.16.15) combine (33.14.9) with (33.16.4)–(33.16.7). For (33.16.17) and (33.16.18) combine (33.14.6), (33.14.9)–(33.14.12), (33.16.10)–(33.16.13), and (33.16.16).

§33.17 Seaton (2002, Eqs. 77, 78, 82).

§33.18 Combine (33.5.8) and (33.16.1), (33.16.2). For $f(\epsilon, \ell; r)$ (33.19.1) can also be used.

§33.19 Seaton (2002, Eqs. 15–17, 31–48).

§33.20 Seaton (2002, Eqs. 58, 59, 64, 67–70, 96, 98, 100, 102 (corrected)).

§33.21 Seaton (2002, Eqs. 104, 107), or apply (13.14.21) to (33.16.9).

§33.23 Stable integration directions for the differential equations are determined by comparison of the asymptotic behavior of the solutions as the radii tend to infinity and also as the radii tend to zero (§§33.11, 33.21; §§33.6, 33.19). Stable recurrence directions for §33.4 are determined by the asymptotic form of $F_\ell(\eta,\rho)/G_\ell(\eta,\rho)$ as $\ell \to \infty$; see (33.5.8) and (33.5.9). For §33.17 see §33.18.

Chapter 34

$3j, 6j, 9j$ Symbols

L. C. Maximon[1]

Notation   **758**
34.1   Special Notation 758

Properties   **758**
34.2   Definition: $3j$ Symbol 758
34.3   Basic Properties: $3j$ Symbol 759
34.4   Definition: $6j$ Symbol 761
34.5   Basic Properties: $6j$ Symbol 762
34.6   Definition: $9j$ Symbol 763
34.7   Basic Properties: $9j$ Symbol 764
34.8   Approximations for Large Parameters . . 764
34.9   Graphical Method 765

34.10   Zeros 765
34.11   Higher-Order $3nj$ Symbols 765

Applications   **765**
34.12   Physical Applications 765

Computation   **765**
34.13   Methods of Computation 765
34.14   Tables 765
34.15   Software 766

References   **766**

[1]Center for Nuclear Studies, Department of Physics, The George Washington University, Washington, D.C.
Copyright © 2009 National Institute of Standards and Technology. All rights reserved.

Notation

34.1 Special Notation

(For other notation see pp. xiv and 873.)

$2j_1, 2j_2, 2j_3, 2l_1, 2l_2, 2l_3$ nonnegative integers.
r, s, t nonnegative integers.

The main functions treated in this chapter are the Wigner $3j, 6j, 9j$ symbols, respectively,

$$\begin{pmatrix} j_1 & j_2 & j_3 \\ m_1 & m_2 & m_3 \end{pmatrix}, \quad \begin{Bmatrix} j_1 & j_2 & j_3 \\ l_1 & l_2 & l_3 \end{Bmatrix}, \quad \begin{Bmatrix} j_{11} & j_{12} & j_{13} \\ j_{21} & j_{22} & j_{23} \\ j_{31} & j_{32} & j_{33} \end{Bmatrix}.$$

The most commonly used alternative notation for the $3j$ symbol is the Clebsch–Gordan coefficient

$$(j_1\, m_1\, j_2\, m_2 | j_1\, j_2\, j_3\, -m_3)$$
$$= (-1)^{j_1-j_2-m_3}(2j_3+1)^{\frac{1}{2}} \begin{pmatrix} j_1 & j_2 & j_3 \\ m_1 & m_2 & m_3 \end{pmatrix};$$

see Condon and Shortley (1935). For other notations see Edmonds (1974, pp. 52, 97, 104–105) and Varshalovich et al. (1988, §§8.11, 9.10, 10.10).

Properties

34.2 Definition: $3j$ Symbol

The quantities j_1, j_2, j_3 in the $3j$ symbol are called *angular momenta*. Either all of them are nonnegative integers, or one is a nonnegative integer and the other two are half-odd positive integers. They must form the sides of a triangle (possibly degenerate). They therefore satisfy the *triangle conditions*

34.2.1 $$|j_r - j_s| \leq j_t \leq j_r + j_s,$$

where r, s, t is any permutation of $1, 2, 3$. The corresponding *projective quantum numbers* m_1, m_2, m_3 are given by

34.2.2 $$m_r = -j_r, -j_r + 1, \ldots, j_r - 1, j_r, \quad r = 1, 2, 3,$$

and satisfy

34.2.3 $$m_1 + m_2 + m_3 = 0.$$

See Figure 34.2.1 for a schematic representation.

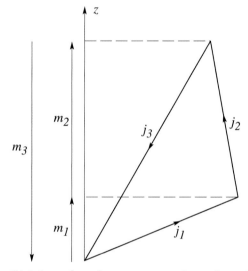

Figure 34.2.1: Angular momenta j_r and projective quantum numbers m_r, $r = 1, 2, 3$.

If either of the conditions (34.2.1) or (34.2.3) is not satisfied, then the $3j$ symbol is zero. When both conditions are satisfied the $3j$ symbol can be expressed as the finite sum

34.2.4
$$\begin{pmatrix} j_1 & j_2 & j_3 \\ m_1 & m_2 & m_3 \end{pmatrix} = (-1)^{j_1-j_2-m_3} \Delta(j_1 j_2 j_3) \left((j_1+m_1)!(j_1-m_1)!(j_2+m_2)!(j_2-m_2)!(j_3+m_3)!(j_3-m_3)!\right)^{\frac{1}{2}}$$
$$\times \sum_s \frac{(-1)^s}{s!(j_1+j_2-j_3-s)!(j_1-m_1-s)!(j_2+m_2-s)!(j_3-j_2+m_1+s)!(j_3-j_1-m_2+s)!},$$

where

34.2.5 $$\Delta(j_1 j_2 j_3) = \left(\frac{(j_1+j_2-j_3)!(j_1-j_2+j_3)!(-j_1+j_2+j_3)!}{(j_1+j_2+j_3+1)!}\right)^{\frac{1}{2}},$$

and the summation is over all nonnegative integers s such that the arguments in the factorials are nonnegative.

Equivalently,

34.2.6
$$\begin{pmatrix} j_1 & j_2 & j_3 \\ m_1 & m_2 & m_3 \end{pmatrix} = (-1)^{j_2-m_1+m_3} \frac{(j_1+j_2+m_3)!(j_2+j_3-m_1)!}{\Delta(j_1 j_2 j_3)(j_1+j_2+j_3+1)!} \left(\frac{(j_1+m_1)!(j_3-m_3)!}{(j_1-m_1)!(j_2+m_2)!(j_2-m_2)!(j_3+m_3)!}\right)^{\frac{1}{2}}$$
$$\times {}_3F_2(-j_1-j_2-j_3-1, -j_1+m_1, -j_3-m_3; -j_1-j_2-m_3, -j_2-j_3+m_1; 1),$$

where ${}_3F_2$ is defined as in §16.2.

For alternative expressions for the $3j$ symbol, written either as a finite sum or as other terminating generalized hypergeometric series $_3F_2$ of unit argument, see Varshalovich et al. (1988, §§8.21, 8.24–8.26).

34.3 Basic Properties: $3j$ Symbol

34.3(i) Special Cases

When any one of j_1, j_2, j_3 is equal to $0, \tfrac{1}{2}$, or 1, the $3j$ symbol has a simple algebraic form. Examples are provided by

34.3.1
$$\begin{pmatrix} j & j & 0 \\ m & -m & 0 \end{pmatrix} = \frac{(-1)^{j-m}}{(2j+1)^{\frac{1}{2}}},$$

34.3.2
$$\begin{pmatrix} j & j & 1 \\ m & -m & 0 \end{pmatrix} = (-1)^{j-m}\frac{2m}{(2j(2j+1)(2j+2))^{\frac{1}{2}}}, \qquad j \geq \tfrac{1}{2},$$

34.3.3
$$\begin{pmatrix} j & j & 1 \\ m & -m-1 & 1 \end{pmatrix} = (-1)^{j-m}\left(\frac{2(j-m)(j+m+1)}{2j(2j+1)(2j+2)}\right)^{\frac{1}{2}}, \qquad j \geq \tfrac{1}{2}.$$

For these and other results, and also cases in which any one of j_1, j_2, j_3 is $\tfrac{3}{2}$ or 2, see Edmonds (1974, pp. 125–127).

Next define

34.3.4
$$J = j_1 + j_2 + j_3.$$

Then assuming the triangle conditions are satisfied

34.3.5
$$\begin{pmatrix} j_1 & j_2 & j_3 \\ 0 & 0 & 0 \end{pmatrix} = \begin{cases} 0, & J \text{ odd,} \\ (-1)^{\frac{1}{2}J}\left(\frac{(J-2j_1)!(J-2j_2)!(J-2j_3)!}{(J+1)!}\right)^{\frac{1}{2}} \frac{(\tfrac{1}{2}J)!}{(\tfrac{1}{2}J-j_1)!(\tfrac{1}{2}J-j_2)!(\tfrac{1}{2}J-j_3)!}, & J \text{ even.} \end{cases}$$

Lastly,

34.3.6
$$\begin{pmatrix} j_1 & j_2 & j_1+j_2 \\ m_1 & m_2 & -m_1-m_2 \end{pmatrix} = (-1)^{j_1-j_2+m_1+m_2}\left(\frac{(2j_1)!(2j_2)!(j_1+j_2+m_1+m_2)!(j_1+j_2-m_1-m_2)!}{(2j_1+2j_2+1)!(j_1+m_1)!(j_1-m_1)!(j_2+m_2)!(j_2-m_2)!}\right)^{\frac{1}{2}},$$

34.3.7
$$\begin{pmatrix} j_1 & j_2 & j_3 \\ j_1 & -j_1-m_3 & m_3 \end{pmatrix} = (-1)^{-j_2+j_3+m_3}\left(\frac{(2j_1)!(-j_1+j_2+j_3)!(j_1+j_2+m_3)!(j_3-m_3)!}{(j_1+j_2+j_3+1)!(j_1-j_2+j_3)!(j_1+j_2-j_3)!(-j_1+j_2-m_3)!(j_3+m_3)!}\right)^{\frac{1}{2}}.$$

Again it is assumed that in (34.3.7) the triangle conditions are satisfied.

34.3(ii) Symmetry

Even permutations of columns of a $3j$ symbol leave it unchanged; odd permutations of columns produce a phase factor $(-1)^{j_1+j_2+j_3}$, for example,

34.3.8
$$\begin{pmatrix} j_1 & j_2 & j_3 \\ m_1 & m_2 & m_3 \end{pmatrix} = \begin{pmatrix} j_2 & j_3 & j_1 \\ m_2 & m_3 & m_1 \end{pmatrix} = \begin{pmatrix} j_3 & j_1 & j_2 \\ m_3 & m_1 & m_2 \end{pmatrix},$$

34.3.9
$$\begin{pmatrix} j_1 & j_2 & j_3 \\ m_1 & m_2 & m_3 \end{pmatrix} = (-1)^{j_1+j_2+j_3}\begin{pmatrix} j_2 & j_1 & j_3 \\ m_2 & m_1 & m_3 \end{pmatrix}.$$

Next,

34.3.10
$$\begin{pmatrix} j_1 & j_2 & j_3 \\ m_1 & m_2 & m_3 \end{pmatrix} = (-1)^{j_1+j_2+j_3}\begin{pmatrix} j_1 & j_2 & j_3 \\ -m_1 & -m_2 & -m_3 \end{pmatrix},$$

34.3.11
$$\begin{pmatrix} j_1 & j_2 & j_3 \\ m_1 & m_2 & m_3 \end{pmatrix} = \begin{pmatrix} j_1 & \tfrac{1}{2}(j_2+j_3+m_1) & \tfrac{1}{2}(j_2+j_3-m_1) \\ j_2-j_3 & \tfrac{1}{2}(j_3-j_2+m_1)+m_2 & \tfrac{1}{2}(j_3-j_2+m_1)+m_3 \end{pmatrix},$$

34.3.12
$$\begin{pmatrix} j_1 & j_2 & j_3 \\ m_1 & m_2 & m_3 \end{pmatrix} = \begin{pmatrix} \tfrac{1}{2}(j_1+j_2-m_3) & \tfrac{1}{2}(j_2+j_3-m_1) & \tfrac{1}{2}(j_1+j_3-m_2) \\ j_3-\tfrac{1}{2}(j_1+j_2+m_3) & j_1-\tfrac{1}{2}(j_2+j_3+m_1) & j_2-\tfrac{1}{2}(j_1+j_3+m_2) \end{pmatrix}.$$

Equations (34.3.11) and (34.3.12) are called *Regge symmetries*. Additional symmetries are obtained by applying (34.3.8)–(34.3.10) to (34.3.11)) and (34.3.12). See Srinivasa Rao and Rajeswari (1993, pp. 44–47) and references given there.

34.3(iii) Recursion Relations

In the following three equations it is assumed that the triangle conditions are satisfied by each $3j$ symbol.

34.3.13
$$((j_1+j_2+j_3+1)(-j_1+j_2+j_3))^{\frac{1}{2}} \begin{pmatrix} j_1 & j_2 & j_3 \\ m_1 & m_2 & m_3 \end{pmatrix} = ((j_2+m_2)(j_3-m_3))^{\frac{1}{2}} \begin{pmatrix} j_1 & j_2-\frac{1}{2} & j_3-\frac{1}{2} \\ m_1 & m_2-\frac{1}{2} & m_3+\frac{1}{2} \end{pmatrix}$$
$$- ((j_2-m_2)(j_3+m_3))^{\frac{1}{2}} \begin{pmatrix} j_1 & j_2-\frac{1}{2} & j_3-\frac{1}{2} \\ m_1 & m_2+\frac{1}{2} & m_3-\frac{1}{2} \end{pmatrix},$$

34.3.14
$$(j_1(j_1+1) - j_2(j_2+1) - j_3(j_3+1) - 2m_2m_3) \begin{pmatrix} j_1 & j_2 & j_3 \\ m_1 & m_2 & m_3 \end{pmatrix}$$
$$= ((j_2-m_2)(j_2+m_2+1)(j_3-m_3+1)(j_3+m_3))^{\frac{1}{2}} \begin{pmatrix} j_1 & j_2 & j_3 \\ m_1 & m_2+1 & m_3-1 \end{pmatrix}$$
$$+ ((j_2-m_2+1)(j_2+m_2)(j_3-m_3)(j_3+m_3+1))^{\frac{1}{2}} \begin{pmatrix} j_1 & j_2 & j_3 \\ m_1 & m_2-1 & m_3+1 \end{pmatrix},$$

34.3.15
$$(2j_1+1) \left((j_2(j_2+1) - j_3(j_3+1))m_1 - j_1(j_1+1)(m_3-m_2) \right) \begin{pmatrix} j_1 & j_2 & j_3 \\ m_1 & m_2 & m_3 \end{pmatrix}$$
$$= (j_1+1) \left(j_1^2 - (j_2-j_3)^2 \right)^{\frac{1}{2}} \left((j_2+j_3+1)^2 - j_1^2 \right)^{\frac{1}{2}} \left(j_1^2 - m_1^2 \right)^{\frac{1}{2}} \begin{pmatrix} j_1-1 & j_2 & j_3 \\ m_1 & m_2 & m_3 \end{pmatrix}$$
$$+ j_1 \left((j_1+1)^2 - (j_2-j_3)^2 \right)^{\frac{1}{2}} \left((j_2+j_3+1)^2 - (j_1+1)^2 \right)^{\frac{1}{2}} \left((j_1+1)^2 - m_1^2 \right)^{\frac{1}{2}} \begin{pmatrix} j_1+1 & j_2 & j_3 \\ m_1 & m_2 & m_3 \end{pmatrix}.$$

For these and other recursion relations see Varshalovich et al. (1988, §8.6). See also Micu (1968), Louck (1958), Schulten and Gordon (1975a), Srinivasa Rao and Rajeswari (1993, pp. 220–225), and Luscombe and Luban (1998).

34.3(iv) Orthogonality

34.3.16
$$\sum_{m_1 m_2} (2j_3+1) \begin{pmatrix} j_1 & j_2 & j_3 \\ m_1 & m_2 & m_3 \end{pmatrix} \begin{pmatrix} j_1 & j_2 & j_3' \\ m_1 & m_2 & m_3' \end{pmatrix} = \delta_{j_3, j_3'} \delta_{m_3, m_3'},$$

34.3.17
$$\sum_{j_3 m_3} (2j_3+1) \begin{pmatrix} j_1 & j_2 & j_3 \\ m_1 & m_2 & m_3 \end{pmatrix} \begin{pmatrix} j_1 & j_2 & j_3 \\ m_1' & m_2' & m_3 \end{pmatrix} = \delta_{m_1, m_1'} \delta_{m_2, m_2'},$$

34.3.18
$$\sum_{m_1 m_2 m_3} \begin{pmatrix} j_1 & j_2 & j_3 \\ m_1 & m_2 & m_3 \end{pmatrix} \begin{pmatrix} j_1 & j_2 & j_3 \\ m_1 & m_2 & m_3 \end{pmatrix} = 1.$$

In the summations (34.3.16)–(34.3.18) the summation variables range over all values that satisfy the conditions given in (34.2.1)–(34.2.3). *Similar conventions apply to all subsequent summations in this chapter.*

34.3(v) Generating Functions

For generating functions for the $3j$ symbol see Biedenharn and van Dam (1965, p. 245, Eq. (3.42) and p. 247, Eq. (3.55)).

34.3(vi) Sums

For sums of products of $3j$ symbols, see Varshalovich et al. (1988, pp. 259–262).

34.3(vii) Relations to Legendre Polynomials and Spherical Harmonics

For the polynomials P_l see §18.3, and for the functions $Y_{l,m}$ and $Y_{l,m}^*$ see §14.30.

34.3.19
$$P_{l_1}(\cos\theta) P_{l_2}(\cos\theta) = \sum_{l} (2l+1) \begin{pmatrix} l_1 & l_2 & l \\ 0 & 0 & 0 \end{pmatrix}^2 P_l(\cos\theta),$$

34.3.20
$$Y_{l_1,m_1}(\theta,\phi) Y_{l_2,m_2}(\theta,\phi) = \sum_{l,m} \left(\frac{(2l_1+1)(2l_2+1)(2l+1)}{4\pi} \right)^{\frac{1}{2}} \begin{pmatrix} l_1 & l_2 & l \\ m_1 & m_2 & m \end{pmatrix} Y_{l,m}^*(\theta,\phi) \begin{pmatrix} l_1 & l_2 & l \\ 0 & 0 & 0 \end{pmatrix},$$

34.3.21
$$\int_0^\pi P_{l_1}(\cos\theta) P_{l_2}(\cos\theta) P_{l_3}(\cos\theta) \sin\theta\, d\theta = 2 \begin{pmatrix} l_1 & l_2 & l_3 \\ 0 & 0 & 0 \end{pmatrix}^2,$$

34.3.22
$$\int_0^{2\pi}\int_0^\pi Y_{l_1,m_1}(\theta,\phi) Y_{l_2,m_2}(\theta,\phi) Y_{l_3,m_3}(\theta,\phi) \sin\theta\, d\theta\, d\phi$$
$$= \left(\frac{(2l_1+1)(2l_2+1)(2l_3+1)}{4\pi}\right)^{\frac{1}{2}} \begin{pmatrix} l_1 & l_2 & l_3 \\ 0 & 0 & 0 \end{pmatrix} \begin{pmatrix} l_1 & l_2 & l_3 \\ m_1 & m_2 & m_3 \end{pmatrix}.$$

Equations (34.3.19)–(34.3.22) are particular cases of more general results that relate rotation matrices to $3j$ symbols, for which see Edmonds (1974, Chapter 4). The left- and right-hand sides of (34.3.22) are known, respectively, as *Gaunt's integral* and the *Gaunt coefficient* (Gaunt (1929)).

34.4 Definition: $6j$ Symbol

The $6j$ symbol is defined by the following double sum of products of $3j$ symbols:

34.4.1
$$\begin{Bmatrix} j_1 & j_2 & j_3 \\ l_1 & l_2 & l_3 \end{Bmatrix} = \sum_{m_r m'_s} (-1)^{l_1+m'_1+l_2+m'_2+l_3+m'_3}$$
$$\times \begin{pmatrix} j_1 & j_2 & j_3 \\ m_1 & m_2 & m_3 \end{pmatrix} \begin{pmatrix} j_1 & l_2 & l_3 \\ m_1 & m'_2 & -m'_3 \end{pmatrix} \begin{pmatrix} l_1 & j_2 & l_3 \\ -m'_1 & m_2 & m'_3 \end{pmatrix} \begin{pmatrix} l_1 & l_2 & j_3 \\ m'_1 & -m'_2 & m_3 \end{pmatrix},$$

where the summation is taken over all admissible values of the m's and m''s for each of the four $3j$ symbols; compare (34.2.2) and (34.2.3).

Except in degenerate cases the combination of the triangle inequalities for the four $3j$ symbols in (34.4.1) is equivalent to the existence of a tetrahedron (possibly degenerate) with edges of lengths $j_1, j_2, j_3, l_1, l_2, l_3$; see Figure 34.4.1.

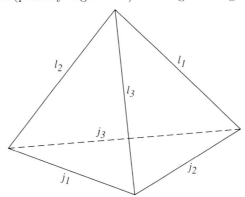

Figure 34.4.1: Tetrahedron corresponding to $6j$ symbol.

The $6j$ symbol can be expressed as the finite sum

34.4.2
$$\begin{Bmatrix} j_1 & j_2 & j_3 \\ l_1 & l_2 & l_3 \end{Bmatrix} = \sum_s \frac{(-1)^s (s+1)!}{(s-j_1-j_2-j_3)!(s-j_1-l_2-l_3)!(s-l_1-j_2-l_3)!(s-l_1-l_2-j_3)!}$$
$$\times \frac{1}{(j_1+j_2+l_1+l_2-s)!(j_2+j_3+l_2+l_3-s)!(j_3+j_1+l_3+l_1-s)!},$$

where the summation is over all nonnegative integers s such that the arguments in the factorials are nonnegative.

Equivalently,

34.4.3
$$\begin{Bmatrix} j_1 & j_2 & j_3 \\ l_1 & l_2 & l_3 \end{Bmatrix} = (-1)^{j_1+j_3+l_1+l_3} \frac{\Delta(j_1 j_2 j_3)\Delta(j_2 l_1 l_3)(j_1-j_2+l_1+l_2)!(-j_2+j_3+l_2+l_3)!(j_1+j_3+l_1+l_3+1)!}{\Delta(j_1 l_2 l_3)\Delta(j_3 l_1 l_2)(j_1-j_2+j_3)!(-j_2+l_1+l_3)!(j_1+l_2+l_3+1)!(j_3+l_1+l_2+1)!}$$
$$\times {}_4F_3\left(\begin{array}{c} -j_1+j_2-j_3, j_2-l_1-l_3, -j_1-l_2-l_3-1, -j_3-l_1-l_2-1 \\ -j_1+j_2-l_1-l_2, j_2-j_3-l_2-l_3, -j_1-j_3-l_1-l_3-1 \end{array}; 1\right),$$

where ${}_4F_3$ is defined as in §16.2.

For alternative expressions for the $6j$ symbol, written either as a finite sum or as other terminating generalized hypergeometric series ${}_4F_3$ of unit argument, see Varshalovich et al. (1988, §§9.2.1, 9.2.3).

34.5 Basic Properties: $6j$ Symbol

34.5(i) Special Cases

In the following equations it is assumed that the triangle inequalities are satisfied and that J is again defined by (34.3.4).

If any lower argument in a $6j$ symbol is 0, $\tfrac{1}{2}$, or 1, then the $6j$ symbol has a simple algebraic form. Examples are provided by:

34.5.1
$$\begin{Bmatrix} j_1 & j_2 & j_3 \\ 0 & j_3 & j_2 \end{Bmatrix} = \frac{(-1)^J}{((2j_2+1)(2j_3+1))^{\frac{1}{2}}},$$

34.5.2
$$\begin{Bmatrix} j_1 & j_2 & j_3 \\ \tfrac{1}{2} & j_3-\tfrac{1}{2} & j_2+\tfrac{1}{2} \end{Bmatrix} = (-1)^J \left(\frac{(j_1+j_3-j_2)(j_1+j_2-j_3+1)}{(2j_2+1)(2j_2+2)2j_3(2j_3+1)} \right)^{\frac{1}{2}},$$

34.5.3
$$\begin{Bmatrix} j_1 & j_2 & j_3 \\ \tfrac{1}{2} & j_3-\tfrac{1}{2} & j_2-\tfrac{1}{2} \end{Bmatrix} = (-1)^J \left(\frac{(j_2+j_3-j_1)(j_1+j_2+j_3+1)}{2j_2(2j_2+1)2j_3(2j_3+1)} \right)^{\frac{1}{2}},$$

34.5.4
$$\begin{Bmatrix} j_1 & j_2 & j_3 \\ 1 & j_3-1 & j_2-1 \end{Bmatrix} = (-1)^J \left(\frac{J(J+1)(J-2j_1)(J-2j_1-1)}{(2j_2-1)2j_2(2j_2+1)(2j_3-1)2j_3(2j_3+1)} \right)^{\frac{1}{2}},$$

34.5.5
$$\begin{Bmatrix} j_1 & j_2 & j_3 \\ 1 & j_3-1 & j_2 \end{Bmatrix} = (-1)^J \left(\frac{2(J+1)(J-2j_1)(J-2j_2)(J-2j_3+1)}{2j_2(2j_2+1)(2j_2+2)(2j_3-1)2j_3(2j_3+1)} \right)^{\frac{1}{2}},$$

34.5.6
$$\begin{Bmatrix} j_1 & j_2 & j_3 \\ 1 & j_3-1 & j_2+1 \end{Bmatrix} = (-1)^J \left(\frac{(J-2j_2-1)(J-2j_2)(J-2j_3+1)(J-2j_3+2)}{(2j_2+1)(2j_2+2)(2j_2+3)(2j_3-1)2j_3(2j_3+1)} \right)^{\frac{1}{2}},$$

34.5.7
$$\begin{Bmatrix} j_1 & j_2 & j_3 \\ 1 & j_3 & j_2 \end{Bmatrix} = (-1)^{J+1} \frac{2(j_2(j_2+1)+j_3(j_3+1)-j_1(j_1+1))}{(2j_2(2j_2+1)(2j_2+2)2j_3(2j_3+1)(2j_3+2))^{\frac{1}{2}}}.$$

34.5(ii) Symmetry

The $6j$ symbol is invariant under interchange of any two columns and also under interchange of the upper and lower arguments in each of any two columns, for example,

34.5.8
$$\begin{Bmatrix} j_1 & j_2 & j_3 \\ l_1 & l_2 & l_3 \end{Bmatrix} = \begin{Bmatrix} j_2 & j_1 & j_3 \\ l_2 & l_1 & l_3 \end{Bmatrix} = \begin{Bmatrix} j_1 & l_2 & l_3 \\ l_1 & j_2 & j_3 \end{Bmatrix}.$$

Next,

34.5.9
$$\begin{Bmatrix} j_1 & j_2 & j_3 \\ l_1 & l_2 & l_3 \end{Bmatrix} = \begin{Bmatrix} j_1 & \tfrac{1}{2}(j_2+l_2+j_3-l_3) & \tfrac{1}{2}(j_2-l_2+j_3+l_3) \\ l_1 & \tfrac{1}{2}(j_2+l_2-j_3+l_3) & \tfrac{1}{2}(-j_2+l_2+j_3+l_3) \end{Bmatrix},$$

34.5.10
$$\begin{Bmatrix} j_1 & j_2 & j_3 \\ l_1 & l_2 & l_3 \end{Bmatrix} = \begin{Bmatrix} \tfrac{1}{2}(j_2+l_2+j_3-l_3) & \tfrac{1}{2}(j_1-l_1+j_3+l_3) & \tfrac{1}{2}(j_1+l_1+j_2-l_2) \\ \tfrac{1}{2}(j_2+l_2-j_3+l_3) & \tfrac{1}{2}(-j_1+l_1+j_3+l_3) & \tfrac{1}{2}(j_1+l_1-j_2+l_2) \end{Bmatrix}.$$

Equations (34.5.9) and (34.5.10) are called *Regge symmetries*. Additional symmetries are obtained by applying (34.5.8) to (34.5.9) and (34.5.10). See Srinivasa Rao and Rajeswari (1993, pp. 102–103) and references given there.

34.5(iii) Recursion Relations

In the following equation it is assumed that the triangle conditions are satisfied.

34.5.11
$$(2j_1+1)\left((J_3+J_2-J_1)(L_3+L_2-L_1) - 2(J_3L_3 + J_2L_2 - J_1L_1)\right) \begin{Bmatrix} j_1 & j_2 & j_3 \\ l_1 & l_2 & l_3 \end{Bmatrix}$$
$$= j_1 E(j_1+1) \begin{Bmatrix} j_1+1 & j_2 & j_3 \\ l_1 & l_2 & l_3 \end{Bmatrix} + (j_1+1) E(j_1) \begin{Bmatrix} j_1-1 & j_2 & j_3 \\ l_1 & l_2 & l_3 \end{Bmatrix},$$

where

34.5.12
$$J_r = j_r(j_r+1), \quad L_r = l_r(l_r+1),$$

34.5.13
$$E(j) = \left((j^2 - (j_2-j_3)^2)((j_2+j_3+1)^2 - j^2)(j^2 - (l_2-l_3)^2)((l_2+l_3+1)^2 - j^2)\right)^{\frac{1}{2}}.$$

For further recursion relations see Varshalovich et al. (1988, §9.6) and Edmonds (1974, pp. 98–99).

34.5(iv) Orthogonality

34.5.14
$$\sum_{j_3}(2j_3+1)(2l_3+1)\begin{Bmatrix} j_1 & j_2 & j_3 \\ l_1 & l_2 & l_3 \end{Bmatrix}\begin{Bmatrix} j_1 & j_2 & j_3 \\ l_1 & l_2 & l_3' \end{Bmatrix} = \delta_{l_3,l_3'}.$$

34.5(v) Generating Functions

For generating functions for the $6j$ symbol see Biedenharn and van Dam (1965, p. 255, eq. (4.18)).

34.5(vi) Sums

34.5.15
$$\sum_{j}(-1)^{j+j'+j''}(2j+1)\begin{Bmatrix} j_1 & j_2 & j \\ j_3 & j_4 & j' \end{Bmatrix}\begin{Bmatrix} j_1 & j_2 & j \\ j_4 & j_3 & j'' \end{Bmatrix} = \begin{Bmatrix} j_1 & j_4 & j' \\ j_2 & j_3 & j'' \end{Bmatrix},$$

34.5.16
$$(-1)^{j_1+j_2+j_3+j_1'+j_2'+l_1+l_2}\begin{Bmatrix} j_1 & j_2 & j_3 \\ l_1 & l_2 & l_3 \end{Bmatrix}\begin{Bmatrix} j_1' & j_2' & j_3 \\ l_1 & l_2 & l_3' \end{Bmatrix}$$
$$= \sum_{j}(-1)^{l_3+l_3'+j}(2j+1)\begin{Bmatrix} j_1 & j_1' & j \\ j_2' & j_2 & j_3 \end{Bmatrix}\begin{Bmatrix} l_3 & l_3' & j \\ j_1' & j_1 & l_2 \end{Bmatrix}\begin{Bmatrix} l_3 & l_3' & j \\ j_2' & j_2 & l_1 \end{Bmatrix}.$$

Equations (34.5.15) and (34.5.16) are the *sum rules*. They constitute addition theorems for the $6j$ symbol.

34.5.17
$$\sum_{j}(2j+1)\begin{Bmatrix} j_1 & j_2 & j \\ j_1 & j_2 & j' \end{Bmatrix} = (-1)^{2(j_1+j_2)},$$

34.5.18
$$\sum_{j}(-1)^{j_1+j_2+j}(2j+1)\begin{Bmatrix} j_1 & j_2 & j \\ j_2 & j_1 & j' \end{Bmatrix} = \sqrt{(2j_1+1)(2j_2+1)}\,\delta_{j',0},$$

34.5.19
$$\sum_{l}\begin{Bmatrix} j_1 & j_2 & l \\ j_2 & j_1 & j \end{Bmatrix} = 0, \qquad 2\mu-j \text{ odd}, \mu=\min(j_1,j_2),$$

34.5.20
$$\sum_{l}(-1)^{l+j}\begin{Bmatrix} j_1 & j_2 & l \\ j_1 & j_2 & j \end{Bmatrix} = \frac{(-1)^{2\mu}}{2j+1}, \qquad \mu=\min(j_1,j_2),$$

34.5.21
$$\sum_{l}(-1)^{l+j+j_1+j_2}\begin{Bmatrix} j_1 & j_2 & l \\ j_2 & j_1 & j \end{Bmatrix} = \frac{1}{2j+1}\left(\frac{(2j_1-j)!(2j_2+j+1)!}{(2j_2-j)!(2j_1+j+1)!}\right)^{\frac{1}{2}}, \qquad j_2\leq j_1,$$

34.5.22
$$\sum_{l}(-1)^{l+j+j_1+j_2}\frac{1}{l(l+1)}\begin{Bmatrix} j_1 & j_2 & l \\ j_2 & j_1 & j \end{Bmatrix} = \frac{1}{j_1(j_1+1)-j_2(j_2+1)}\left(\frac{(2j_1-j)!(2j_2+j+1)!}{(2j_2-j)!(2j_1+j+1)!}\right)^{\frac{1}{2}}, \qquad j_2<j_1.$$

34.5.23
$$\begin{pmatrix} j_1 & j_2 & j_3 \\ m_1 & m_2 & m_3 \end{pmatrix}\begin{Bmatrix} j_1 & j_2 & j_3 \\ l_1 & l_2 & l_3 \end{Bmatrix}$$
$$= \sum_{m_1'm_2'm_3'}(-1)^{l_1+l_2+l_3+m_1'+m_2'+m_3'}\begin{pmatrix} j_1 & l_2 & l_3 \\ m_1 & m_2' & -m_3' \end{pmatrix}\begin{pmatrix} l_1 & j_2 & l_3 \\ -m_1' & m_2 & m_3' \end{pmatrix}\begin{pmatrix} l_1 & l_2 & j_3 \\ m_1' & -m_2' & m_3 \end{pmatrix}.$$

Equation (34.5.23) can be regarded as an alternative definition of the $6j$ symbol.

For other sums see Ginocchio (1991).

34.6 Definition: $9j$ Symbol

The $9j$ symbol may be defined either in terms of $3j$ symbols or equivalently in terms of $6j$ symbols:

34.6.1
$$\begin{Bmatrix} j_{11} & j_{12} & j_{13} \\ j_{21} & j_{22} & j_{23} \\ j_{31} & j_{32} & j_{33} \end{Bmatrix} = \sum_{\text{all } m_{rs}}\begin{pmatrix} j_{11} & j_{12} & j_{13} \\ m_{11} & m_{12} & m_{13} \end{pmatrix}\begin{pmatrix} j_{21} & j_{22} & j_{23} \\ m_{21} & m_{22} & m_{23} \end{pmatrix}\begin{pmatrix} j_{31} & j_{32} & j_{33} \\ m_{31} & m_{32} & m_{33} \end{pmatrix}$$
$$\times \begin{pmatrix} j_{11} & j_{21} & j_{31} \\ m_{11} & m_{21} & m_{31} \end{pmatrix}\begin{pmatrix} j_{12} & j_{22} & j_{32} \\ m_{12} & m_{22} & m_{32} \end{pmatrix}\begin{pmatrix} j_{13} & j_{23} & j_{33} \\ m_{13} & m_{23} & m_{33} \end{pmatrix},$$

34.6.2
$$\begin{Bmatrix} j_{11} & j_{12} & j_{13} \\ j_{21} & j_{22} & j_{23} \\ j_{31} & j_{32} & j_{33} \end{Bmatrix} = \sum_{j}(-1)^{2j}(2j+1)\begin{Bmatrix} j_{11} & j_{21} & j_{31} \\ j_{32} & j_{33} & j \end{Bmatrix}\begin{Bmatrix} j_{12} & j_{22} & j_{32} \\ j_{21} & j & j_{23} \end{Bmatrix}\begin{Bmatrix} j_{13} & j_{23} & j_{33} \\ j & j_{11} & j_{12} \end{Bmatrix}.$$

The $9j$ symbol may also be written as a finite triple sum equivalent to a terminating generalized hypergeometric series of three variables with unit arguments. See Srinivasa Rao and Rajeswari (1993, pp. 7 and 125–132) and Rosengren (1999).

34.7 Basic Properties: $9j$ Symbol

34.7(i) Special Case

34.7.1
$$\begin{Bmatrix} j_{11} & j_{12} & j_{13} \\ j_{21} & j_{22} & j_{13} \\ j_{31} & j_{31} & 0 \end{Bmatrix} = \frac{(-1)^{j_{12}+j_{21}+j_{13}+j_{31}}}{((2j_{13}+1)(2j_{31}+1))^{\frac{1}{2}}} \begin{Bmatrix} j_{11} & j_{12} & j_{13} \\ j_{22} & j_{21} & j_{31} \end{Bmatrix}.$$

34.7(ii) Symmetry

The $9j$ symbol has symmetry properties with respect to permutation of columns, permutation of rows, and transposition of rows and columns; these relate 72 independent $9j$ symbols. Even (cyclic) permutations of either columns or rows, as well as transpositions, leave the $9j$ symbol unchanged. Odd permutations of columns or rows introduce a phase factor $(-1)^R$, where R is the sum of all arguments of the $9j$ symbol.

For further symmetry properties of the $9j$ symbol see Edmonds (1974, pp. 102–103) and Varshalovich *et al.* (1988, §10.4.1).

34.7(iii) Recursion Relations

For recursion relations see Varshalovich *et al.* (1988, §10.5).

34.7(iv) Orthogonality

34.7.2
$$\sum_{j_{12}\,j_{34}} (2j_{12}+1)(2j_{34}+1)(2j_{13}+1)(2j_{24}+1) \begin{Bmatrix} j_1 & j_2 & j_{12} \\ j_3 & j_4 & j_{34} \\ j_{13} & j_{24} & j \end{Bmatrix} \begin{Bmatrix} j_1 & j_2 & j_{12} \\ j_3 & j_4 & j_{34} \\ j'_{13} & j'_{24} & j \end{Bmatrix} = \delta_{j_{13},j'_{13}} \delta_{j_{24},j'_{24}}.$$

34.7(v) Generating Functions

For generating functions for the $9j$ symbol see Biedenharn and van Dam (1965, p. 258, eq. (4.37)).

34.7(vi) Sums

34.7.3
$$\sum_{j_{13}\,j_{24}} (-1)^{2j_2+j_{24}+j_{23}-j_{34}} (2j_{13}+1)(2j_{24}+1) \begin{Bmatrix} j_1 & j_2 & j_{12} \\ j_3 & j_4 & j_{34} \\ j_{13} & j_{24} & j \end{Bmatrix} \begin{Bmatrix} j_1 & j_3 & j_{13} \\ j_4 & j_2 & j_{24} \\ j_{14} & j_{23} & j \end{Bmatrix} = \begin{Bmatrix} j_1 & j_2 & j_{12} \\ j_4 & j_3 & j_{34} \\ j_{14} & j_{23} & j \end{Bmatrix}.$$

This equation is the *sum rule*. It constitutes an addition theorem for the $9j$ symbol.

34.7.4
$$\begin{pmatrix} j_{13} & j_{23} & j_{33} \\ m_{13} & m_{23} & m_{33} \end{pmatrix} \begin{Bmatrix} j_{11} & j_{12} & j_{13} \\ j_{21} & j_{22} & j_{23} \\ j_{31} & j_{32} & j_{33} \end{Bmatrix} = \sum_{m_{r1},m_{r2},r=1,2,3} \begin{pmatrix} j_{11} & j_{12} & j_{13} \\ m_{11} & m_{12} & m_{13} \end{pmatrix} \begin{pmatrix} j_{21} & j_{22} & j_{23} \\ m_{21} & m_{22} & m_{23} \end{pmatrix}$$
$$\times \begin{pmatrix} j_{31} & j_{32} & j_{33} \\ m_{13} & m_{23} & m_{33} \end{pmatrix} \begin{pmatrix} j_{11} & j_{21} & j_{31} \\ m_{11} & m_{21} & m_{31} \end{pmatrix} \begin{pmatrix} j_{12} & j_{22} & j_{32} \\ m_{12} & m_{22} & m_{32} \end{pmatrix}.$$

34.7.5
$$\sum_{j'} (2j'+1) \begin{Bmatrix} j_{11} & j_{12} & j' \\ j_{21} & j_{22} & j_{23} \\ j_{31} & j_{32} & j_{33} \end{Bmatrix} \begin{Bmatrix} j_{11} & j_{12} & j' \\ j_{23} & j_{33} & j \end{Bmatrix} = (-1)^{2j} \begin{Bmatrix} j_{21} & j_{22} & j_{23} \\ j_{12} & j & j_{32} \end{Bmatrix} \begin{Bmatrix} j_{31} & j_{32} & j_{33} \\ j & j_{11} & j_{21} \end{Bmatrix}.$$

34.8 Approximations for Large Parameters

For large values of the parameters in the $3j$, $6j$, and $9j$ symbols, different asymptotic forms are obtained depending on which parameters are large. For example,

34.8.1
$$\begin{Bmatrix} j_1 & j_2 & j_3 \\ j_2 & j_1 & l_3 \end{Bmatrix} = (-1)^{j_1+j_2+j_3+l_3} \left(\frac{4}{\pi(2j_1+1)(2j_2+1)(2l_3+1)\sin\theta} \right)^{\frac{1}{2}} \left(\cos\left((l_3+\tfrac{1}{2})\theta - \tfrac{1}{4}\pi\right) + o(1) \right),$$
$$j_1, j_2, j_3 \gg l_3 \gg 1,$$

where

34.8.2 $\quad \cos\theta = \dfrac{j_1(j_1+1)+j_2(j_2+1)-j_3(j_3+1)}{2\sqrt{j_1(j_1+1)j_2(j_2+1)}},$

and the symbol $o(1)$ denotes a quantity that tends to zero as the parameters tend to infinity, as in §2.1(i).

Semiclassical (WKBJ) approximations in terms of trigonometric or exponential functions are given in Varshalovich et al. (1988, §§8.9, 9.9, 10.7). Uniform approximations in terms of Airy functions for the $3j$ and $6j$ symbols are given in Schulten and Gordon (1975b). For approximations for the $3j$, $6j$, and $9j$ symbols with error bounds see Flude (1998), Chen et al. (1999), and Watson (1999): these references also cite earlier work.

34.9 Graphical Method

The graphical method establishes a one-to-one correspondence between an analytic expression and a diagram by assigning a graphical symbol to each function and operation of the analytic expression. Thus, any analytic expression in the theory, for example equations (34.3.16), (34.4.1), (34.5.15), and (34.7.3), may be represented by a diagram; conversely, any diagram represents an analytic equation. For an account of this method see Brink and Satchler (1993, Chapter VII). For specific examples of the graphical method of representing sums involving the $3j, 6j$, and $9j$ symbols, see Varshalovich et al. (1988, Chapters 11, 12) and Lehman and O'Connell (1973, §3.3).

34.10 Zeros

In a $3j$ symbol, if the three angular momenta j_1, j_2, j_3 do not satisfy the triangle conditions (34.2.1), or if the projective quantum numbers do not satisfy (34.2.3), then the $3j$ symbol is zero. Similarly the $6j$ symbol (34.4.1) vanishes when the triangle conditions are not satisfied by any of the four $3j$ symbols in the summation. Such zeros are called *trivial zeros*. However, the $3j$ and $6j$ symbols may vanish for certain combinations of the angular momenta and projective quantum numbers even when the triangle conditions are fulfilled. Such zeros are called *nontrivial zeros*.

For further information, including examples of nontrivial zeros and extensions to $9j$ symbols, see Srinivasa Rao and Rajeswari (1993, pp. 133–215, 294–295, 299–310).

34.11 Higher-Order $3nj$ Symbols

For information on $12j, 15j,...,$ symbols, see Varshalovich et al. (1988, §10.12) and Yutsis et al. (1962, pp. 62–65 and 122–153).

Applications

34.12 Physical Applications

The angular momentum coupling coefficients ($3j$, $6j$, and $9j$ symbols) are essential in the fields of nuclear, atomic, and molecular physics. For applications in nuclear structure, see de Shalit and Talmi (1963); in atomic spectroscopy, see Biedenharn and van Dam (1965, pp. 134–200), Judd (1998), Sobelman (1992, Chapter 4), Shore and Menzel (1968, pp. 268–303), and Wigner (1959); in molecular spectroscopy and chemical reactions, see Burshtein and Temkin (1994, Chapter 5), and Judd (1975). $3j, 6j,$ and $9j$ symbols are also found in multipole expansions of solutions of the Laplace and Helmholtz equations; see Carlson and Rushbrooke (1950) and Judd (1976).

Computation

34.13 Methods of Computation

Methods of computation for $3j$ and $6j$ symbols include recursion relations, see Schulten and Gordon (1975a), Luscombe and Luban (1998), and Edmonds (1974, pp. 42–45, 48–51, 97–99); summation of single-sum expressions for these symbols, see Varshalovich et al. (1988, §§8.2.6, 9.2.1) and Fang and Shriner (1992); evaluation of the generalized hypergeometric functions of unit argument that represent these symbols, see Srinivasa Rao and Venkatesh (1978) and Srinivasa Rao (1981).

For $9j$ symbols, methods include evaluation of the single-sum series (34.6.2), see Fang and Shriner (1992); evaluation of triple-sum series, see Varshalovich et al. (1988, §10.2.1) and Srinivasa Rao et al. (1989). A review of methods of computation is given in Srinivasa Rao and Rajeswari (1993, Chapter VII, pp. 235–265). See also Roothaan and Lai (1997) and references given there.

34.14 Tables

Tables of exact values of the squares of the $3j$ and $6j$ symbols in which all parameters are ≤ 8 are given in Rotenberg et al. (1959), together with a bibliography of earlier tables of $3j, 6j$, and $9j$ symbols on pp. 33–36.

Tables of $3j$ and $6j$ symbols in which all parameters are $\leq 17/2$ are given in Appel (1968) to 6D. Some selected $9j$ symbols are also given. Other tabulations for $3j$ symbols are listed on pp. 11-12; for $6j$ symbols on pp. 16-17; for $9j$ symbols on p. 21.

Biedenharn and Louck (1981) give tables of algebraic expressions for Clebsch–Gordan coefficients and $6j$ symbols, together with a bibliography of tables produced prior to 1975. In Varshalovich *et al.* (1988) algebraic expressions for the Clebsch–Gordan coefficients with all parameters ≤ 5 and numerical values for all parameters ≤ 3 are given on pp. 270–289; similar tables for the $6j$ symbols are given on pp. 310–332, and for the $9j$ symbols on pp. 359, 360, 372–411. Earlier tables are listed on p. 513.

34.15 Software

See http://dlmf.nist.gov/34.15.

References

General References

The main references used in writing this chapter are Edmonds (1974), Varshalovich *et al.* (1988), and de Shalit and Talmi (1963).

Sources

The following list gives the references or other indications of proofs that were used in constructing the various sections of this chapter. These sources supplement the references that are quoted in the text.

§34.2 Edmonds (1974, pp. 44–45).

§34.3 Edmonds (1974, pp. 46–50, 63), de Shalit and Talmi (1963, pp. 515, 519), Thompson (1994, p. 288).

§34.4 Varshalovich *et al.* (1988, §9.2.4), de Shalit and Talmi (1963, p. 131).

§34.5 Edmonds (1974, pp. 94–98, 130–132), de Shalit and Talmi (1963, pp. 517–518, 520), Varshalovich *et al.* (1988, §9.8), Dunlap and Judd (1975).

§34.6 Edmonds (1974, p. 101), de Shalit and Talmi (1963, p. 516).

§34.7 Edmonds (1974, pp. 103–106), de Shalit and Talmi (1963, pp. 127, 517–518).

§34.8 Watson (1999), Chen *et al.* (1999).

Chapter 35
Functions of Matrix Argument
D. St. P. Richards[1]

Notation — **768**
- 35.1 Special Notation 768

Properties — **768**
- 35.2 Laplace Transform 768
- 35.3 Multivariate Gamma and Beta Functions — 768
- 35.4 Partitions and Zonal Polynomials 769
- 35.5 Bessel Functions of Matrix Argument . . 769
- 35.6 Confluent Hypergeometric Functions of Matrix Argument 770
- 35.7 Gaussian Hypergeometric Function of Matrix Argument 771
- 35.8 Generalized Hypergeometric Functions of Matrix Argument 772

Applications — **773**
- 35.9 Applications 773

Computation — **773**
- 35.10 Methods of Computation 773
- 35.11 Tables 773
- 35.12 Software 773

References — **773**

---

[1] Department of Statistics, Pennsylvania State University.
Acknowledgments: With deep gratitude to Ingram Olkin for advice and support regarding the final version of this material.
Copyright © 2009 National Institute of Standards and Technology. All rights reserved.

Notation

35.1 Special Notation

(For other notation see pp. xiv and 873.)

All matrices are of order $m \times m$, unless specified otherwise. All fractional or complex powers are principal values.

a, b	complex variables.
j, k	nonnegative integers.
m	positive integer.
$[a]_\kappa$	partitional shifted factorial (§35.4(i)).
$\mathbf{0}$	zero matrix.
$\mathbf{I}$	identity matrix.
$\mathcal{S}$	space of all real symmetric matrices.
$\mathbf{S}, \mathbf{T}, \mathbf{X}$	real symmetric matrices.
$\operatorname{tr} \mathbf{X}$	trace of $\mathbf{X}$.
$\operatorname{etr}(\mathbf{X})$	$\exp(\operatorname{tr} \mathbf{X})$.
$|\mathbf{X}|$	determinant of $\mathbf{X}$ (except when $m = 1$ where it means either determinant or absolute value, depending on the context).
$|(\mathbf{X})_j|$	jth principal minor of $\mathbf{X}$.
$x_{j,k}$	(j, k)th element of $\mathbf{X}$.
$d\mathbf{X}$	$\prod_{1 \le j \le k \le m} dx_{j,k}$.
$\mathbf{\Omega}$	space of positive-definite real symmetric matrices.
$t_1, \ldots, t_m$	eigenvalues of $\mathbf{T}$.
$\|\mathbf{T}\|$	spectral norm of $\mathbf{T}$.
$\mathbf{X} > \mathbf{T}$	$\mathbf{X} - \mathbf{T}$ is positive definite.
$\mathbf{Z}$	complex symmetric matrix.
$\mathbf{U}, \mathbf{V}$	real and complex parts of $\mathbf{Z}$.
$f(\mathbf{X})$	complex-valued function with $\mathbf{X} \in \mathbf{\Omega}$.
$\mathbf{O}(m)$	space of orthogonal matrices.
$\mathbf{H}$	orthogonal matrix.
$d\mathbf{H}$	normalized Haar measure on $\mathbf{O}(m)$.
$Z_\kappa(\mathbf{T})$	zonal polynomials.

The main functions treated in this chapter are the multivariate gamma and beta functions, respectively $\Gamma_m(a)$ and $\mathrm{B}_m(a,b)$, and the special functions of matrix argument: Bessel (of the first kind) $A_\nu(\mathbf{T})$ and (of the second kind) $B_\nu(\mathbf{T})$; confluent hypergeometric (of the first kind) ${}_1F_1(a;b;\mathbf{T})$ or ${}_1F_1\!\left(\begin{matrix}a\\b\end{matrix};\mathbf{T}\right)$ and (of the second kind) $\Psi(a;b;\mathbf{T})$; Gaussian hypergeometric ${}_2F_1(a_1,a_2;b;\mathbf{T})$ or ${}_2F_1\!\left(\begin{matrix}a_1,a_2\\b\end{matrix};\mathbf{T}\right)$; generalized hypergeometric ${}_pF_q(a_1,\ldots,a_p;b_1,\ldots,b_q;\mathbf{T})$ or ${}_pF_q\!\left(\begin{matrix}a_1,\ldots,a_p\\b_1,\ldots,b_q\end{matrix};\mathbf{T}\right)$.

An alternative notation for the multivariate gamma function is $\Pi_m(a) = \Gamma_m\!\left(a + \tfrac{1}{2}(m+1)\right)$ (Herz (1955, p. 480)). Related notations for the Bessel functions are $\mathcal{J}_{\nu+\frac{1}{2}(m+1)}(\mathbf{T}) = A_\nu(\mathbf{T})/A_\nu(\mathbf{0})$ (Faraut and Korányi (1994, pp. 320–329)), $K_m(0,\ldots,0,\nu|\mathbf{S},\mathbf{T}) = |\mathbf{T}|^\nu B_\nu(\mathbf{ST})$ (Terras (1988, pp. 49–64)), and $\mathcal{K}_\nu(\mathbf{T}) = |\mathbf{T}|^\nu B_\nu(\mathbf{ST})$ (Faraut and Korányi (1994, pp. 357–358)).

Properties

35.2 Laplace Transform

Definition

For any complex symmetric matrix $\mathbf{Z}$,

35.2.1 $$g(\mathbf{Z}) = \int_{\mathbf{\Omega}} \operatorname{etr}(-\mathbf{Z}\mathbf{X}) f(\mathbf{X}) \, d\mathbf{X},$$

where the integration variable $\mathbf{X}$ ranges over the space $\mathbf{\Omega}$.

Suppose there exists a constant $\mathbf{X}_0 \in \mathbf{\Omega}$ such that $|f(\mathbf{X})| < \operatorname{etr}(-\mathbf{X}_0 \mathbf{X})$ for all $\mathbf{X} \in \mathbf{\Omega}$. Then (35.2.1) converges absolutely on the region $\Re(\mathbf{Z}) > \mathbf{X}_0$, and $g(\mathbf{Z})$ is a complex analytic function of all elements $z_{j,k}$ of $\mathbf{Z}$.

Inversion Formula

Assume that $\int_{\mathcal{S}} |g(\mathbf{Z})| \, d\mathbf{V}$ converges, and also that $\lim_{\mathbf{U} \to \infty} \int_{\mathcal{S}} |g(\mathbf{Z})| \, d\mathbf{V} = 0$. Then

35.2.2 $$f(\mathbf{X}) = \frac{1}{(2\pi i)^{m(m+1)/2}} \int \operatorname{etr}(\mathbf{Z}\mathbf{X}) g(\mathbf{Z}) \, d\mathbf{Z},$$

where the integral is taken over all $\mathbf{Z} = \mathbf{U} + i\mathbf{V}$ such that $\mathbf{U} > \mathbf{X}_0$ and $\mathbf{V}$ ranges over $\mathcal{S}$.

Convolution Theorem

If g_j is the Laplace transform of f_j, $j = 1, 2$, then $g_1 g_2$ is the Laplace transform of the convolution $f_1 * f_2$, where

35.2.3 $$f_1 * f_2(\mathbf{T}) = \int_{0 < \mathbf{X} < \mathbf{T}} f_1(\mathbf{T} - \mathbf{X}) f_2(\mathbf{X}) \, d\mathbf{X}.$$

35.3 Multivariate Gamma and Beta Functions

35.3(i) Definitions

35.3.1 $$\Gamma_m(a) = \int_{\mathbf{\Omega}} \operatorname{etr}(-\mathbf{X}) |\mathbf{X}|^{a - \frac{1}{2}(m+1)} \, d\mathbf{X},$$
$$\Re(a) > \tfrac{1}{2}(m-1).$$

35.3.2
$$\Gamma_m(s_1, \ldots, s_m)$$
$$= \int_{\mathbf{\Omega}} \operatorname{etr}(-\mathbf{X}) |\mathbf{X}|^{s_m - \frac{1}{2}(m+1)} \prod_{j=1}^{m-1} |(\mathbf{X})_j|^{s_j - s_{j+1}} \, d\mathbf{X},$$
$$s_j \in \mathbb{C}, \ \Re(s_j) > \tfrac{1}{2}(j-1), \ j = 1, \ldots, m.$$

35.3.3
$$\mathrm{B}_m(a, b) = \int_{0 < \mathbf{X} < \mathbf{I}} |\mathbf{X}|^{a - \frac{1}{2}(m+1)} |\mathbf{I} - \mathbf{X}|^{b - \frac{1}{2}(m+1)} \, d\mathbf{X},$$
$$\Re(a), \Re(b) > \tfrac{1}{2}(m-1).$$

35.3(ii) Properties

35.3.4 $\quad \Gamma_m(a) = \pi^{m(m-1)/4} \prod_{j=1}^{m} \Gamma\left(a - \tfrac{1}{2}(j-1)\right).$

35.3.5
$$\Gamma_m(s_1,\ldots,s_m) = \pi^{m(m-1)/4} \prod_{j=1}^{m} \Gamma\left(s_j - \tfrac{1}{2}(j-1)\right).$$

35.3.6 $\quad \Gamma_m(a,\ldots,a) = \Gamma_m(a).$

35.3.7 $\quad \mathrm{B}_m(a,b) = \dfrac{\Gamma_m(a)\,\Gamma_m(b)}{\Gamma_m(a+b)}.$

35.3.8
$$\mathrm{B}_m(a,b) = \int_{\boldsymbol{\Omega}} |\mathbf{X}|^{a-\frac{1}{2}(m+1)} |\mathbf{I}+\mathbf{X}|^{-(a+b)}\, d\mathbf{X},$$
$$\Re(a), \Re(b) > \tfrac{1}{2}(m-1).$$

35.4 Partitions and Zonal Polynomials

35.4(i) Definitions

A *partition* $\kappa = (k_1,\ldots,k_m)$ is a vector of nonnegative integers, listed in nonincreasing order. Also, $|\kappa|$ denotes $k_1+\cdots+k_m$, the *weight* of κ; $\ell(\kappa)$ denotes the number of nonzero k_j; $a+\kappa$ denotes the vector $(a+k_1,\ldots,a+k_m)$.

The *partitional shifted factorial* is given by

35.4.1 $\quad [a]_\kappa = \dfrac{\Gamma_m(a+\kappa)}{\Gamma_m(a)} = \prod_{j=1}^{m}\left(a - \tfrac{1}{2}(j-1)\right)_{k_j},$

where $(a)_k = a(a+1)\cdots(a+k-1)$.

For any partition κ, the *zonal polynomial* $Z_\kappa : \boldsymbol{\mathcal{S}} \to \mathbb{R}$ is defined by the properties

35.4.2
$$Z_\kappa(\mathbf{I}) = |\kappa|!\, 2^{2|\kappa|}\, [m/2]_\kappa \dfrac{\prod_{1\le j<l\le \ell(\kappa)}(2k_j - 2k_l - j + l)}{\prod_{j=1}^{\ell(\kappa)}(2k_j + \ell(\kappa) - j)!}$$

and

35.4.3
$$Z_\kappa(\mathbf{T}) = Z_\kappa(\mathbf{I})\,|\mathbf{T}|^{k_m} \int_{\mathbf{O}(m)} \prod_{j=1}^{m-1} |(\mathbf{H}\mathbf{T}\mathbf{H}^{-1})_j|^{k_j - k_{j+1}}\, d\mathbf{H},$$
$$\mathbf{T} \in \boldsymbol{\mathcal{S}}.$$

See Muirhead (1982, pp. 68–72) for the definition and properties of the *Haar measure* $d\mathbf{H}$. See Hua (1963, p. 30), Constantine (1963), James (1964), and Macdonald (1995, pp. 425–431) for further information on (35.4.2) and (35.4.3). Alternative notations for the zonal polynomials are $C_\kappa(\mathbf{T})$ (Muirhead (1982, pp. 227–239)), $\mathcal{Y}_\kappa(\mathbf{T})$ (Takemura (1984, p. 22)), and $\Phi_\kappa(\mathbf{T})$ (Faraut and Korányi (1994, pp. 228–236)).

35.4(ii) Properties

Normalization

35.4.4 $\quad Z_\kappa(\mathbf{0}) = \begin{cases} 1, & \kappa = (0,\ldots,0), \\ 0, & \kappa \ne (0,\ldots,0). \end{cases}$

Orthogonal Invariance

35.4.5 $\quad Z_\kappa(\mathbf{H}\mathbf{T}\mathbf{H}^{-1}) = Z_\kappa(\mathbf{T}), \qquad \mathbf{H} \in \mathbf{O}(m).$

Therefore $Z_\kappa(\mathbf{T})$ is a symmetric polynomial in the eigenvalues of $\mathbf{T}$.

Summation

For $k = 0,1,2,\ldots$,

35.4.6 $\quad \displaystyle\sum_{|\kappa|=k} Z_\kappa(\mathbf{T}) = (\mathrm{tr}\,\mathbf{T})^k.$

Mean-Value

35.4.7 $\quad \displaystyle\int_{\mathbf{O}(m)} Z_\kappa(\mathbf{S}\mathbf{H}\mathbf{T}\mathbf{H}^{-1})\, d\mathbf{H} = \dfrac{Z_\kappa(\mathbf{S})\,Z_\kappa(\mathbf{T})}{Z_\kappa(\mathbf{I})}.$

Laplace and Beta Integrals

For $\mathbf{T} \in \boldsymbol{\Omega}$ and $\Re(a), \Re(b) > \tfrac{1}{2}(m-1)$,

35.4.8
$$\int_{\boldsymbol{\Omega}} \mathrm{etr}(-\mathbf{T}\mathbf{X})\,|\mathbf{X}|^{a-\frac{1}{2}(m+1)}\,Z_\kappa(\mathbf{X})\, d\mathbf{X}$$
$$= \Gamma_m(a+\kappa)\,|\mathbf{T}|^{-a}\,Z_\kappa(\mathbf{T}^{-1}),$$

35.4.9
$$\int_{0<\mathbf{X}<\mathbf{I}} |\mathbf{X}|^{a-\frac{1}{2}(m+1)}\,|\mathbf{I}-\mathbf{X}|^{b-\frac{1}{2}(m+1)}\,Z_\kappa(\mathbf{T}\mathbf{X})\, d\mathbf{X}$$
$$= \dfrac{[a]_\kappa}{[a+b]_\kappa}\,\mathrm{B}_m(a,b)\,Z_\kappa(\mathbf{T}).$$

35.5 Bessel Functions of Matrix Argument

35.5(i) Definitions

35.5.1 $\quad A_\nu(\mathbf{0}) = \dfrac{1}{\Gamma_m\!\left(\nu + \tfrac{1}{2}(m+1)\right)}, \qquad \nu \in \mathbb{C}.$

35.5.2
$$A_\nu(\mathbf{T}) = A_\nu(\mathbf{0}) \sum_{k=0}^{\infty} \dfrac{(-1)^k}{k!} \sum_{|\kappa|=k} \dfrac{1}{\left[\nu + \tfrac{1}{2}(m+1)\right]_\kappa} Z_\kappa(\mathbf{T}),$$
$$\nu \in \mathbb{C},\ \mathbf{T} \in \boldsymbol{\mathcal{S}}.$$

35.5.3
$$B_\nu(\mathbf{T}) = \int_{\boldsymbol{\Omega}} \mathrm{etr}\!\left(-(\mathbf{T}\mathbf{X}+\mathbf{X}^{-1})\right)|\mathbf{X}|^{\nu-\frac{1}{2}(m+1)}\, d\mathbf{X},$$
$$\nu \in \mathbb{C},\ \mathbf{T} \in \boldsymbol{\Omega}.$$

35.5(ii) Properties

35.5.4
$$\int_{\boldsymbol{\Omega}} \mathrm{etr}(-\mathbf{T}\mathbf{X})|\mathbf{X}|^\nu\,A_\nu(\mathbf{S}\mathbf{X})\, d\mathbf{X}$$
$$= \mathrm{etr}(-\mathbf{S}\mathbf{T}^{-1})|\mathbf{T}|^{-\nu-\frac{1}{2}(m+1)},$$
$$\mathbf{S} \in \boldsymbol{\mathcal{S}},\ \mathbf{T} \in \boldsymbol{\Omega};\ \Re(\nu) > -1.$$

$$35.5.5 \quad \int_{0<\mathbf{X}<\mathbf{T}} A_{\nu_1}(\mathbf{S}_1\mathbf{X})|\mathbf{X}|^{\nu_1} A_{\nu_2}(\mathbf{S}_2(\mathbf{T}-\mathbf{X}))|\mathbf{T}-\mathbf{X}|^{\nu_2} d\mathbf{X} = |\mathbf{T}|^{\nu_1+\nu_2+\frac{1}{2}(m+1)} A_{\nu_1+\nu_2+\frac{1}{2}(m+1)}((\mathbf{S}_1+\mathbf{S}_2)\mathbf{T}),$$
$$\nu_j \in \mathbb{C},\ \Re(\nu_j) > -1,\ j=1,2;\ \mathbf{S}_1,\mathbf{S}_2 \in \mathcal{S};\ \mathbf{T} \in \Omega.$$

$$35.5.6 \quad B_\nu(\mathbf{T}) = |\mathbf{T}|^{-\nu} B_{-\nu}(\mathbf{T}), \qquad \nu \in \mathbb{C},\ \mathbf{T} \in \Omega.$$

$$35.5.7 \quad \int_\Omega A_{\nu_1}(\mathbf{T}\mathbf{X})\, B_{-\nu_2}(\mathbf{S}\mathbf{X})|\mathbf{X}|^{\nu_1} d\mathbf{X} = \frac{1}{A_{\nu_1+\nu_2}(\mathbf{0})}|\mathbf{S}|^{\nu_2}|\mathbf{T}+\mathbf{S}|^{-(\nu_1+\nu_2+\frac{1}{2}(m+1))}, \quad \Re(\nu_1+\nu_2) > -1;\ \mathbf{S},\mathbf{T} \in \Omega.$$

$$35.5.8 \quad \int_{\mathbf{O}(m)} \operatorname{etr}(\mathbf{S}\mathbf{H})\, d\mathbf{H} = \frac{A_{-1/2}\!\left(-\tfrac{1}{4}\mathbf{S}\mathbf{S}^{\mathrm{T}}\right)}{A_{-1/2}(\mathbf{0})}, \qquad \mathbf{S}\ \text{arbitrary.}$$

35.5(iii) Asymptotic Approximations

For asymptotic approximations for Bessel functions of matrix argument, see Herz (1955) and Butler and Wood (2003).

35.6 Confluent Hypergeometric Functions of Matrix Argument

35.6(i) Definitions

$$35.6.1 \quad {}_1F_1\!\left(\begin{matrix}a\\b\end{matrix};\mathbf{T}\right) = \sum_{k=0}^{\infty} \frac{1}{k!} \sum_{|\kappa|=k} \frac{[a]_\kappa}{[b]_\kappa} Z_\kappa(\mathbf{T}).$$

$$35.6.2 \quad \Psi(a;b;\mathbf{T}) = \frac{1}{\Gamma_m(a)} \int_\Omega \operatorname{etr}(-\mathbf{T}\mathbf{X})|\mathbf{X}|^{a-\frac{1}{2}(m+1)}|\mathbf{I}+\mathbf{X}|^{b-a-\frac{1}{2}(m+1)} d\mathbf{X},\quad \Re(a) > \tfrac{1}{2}(m-1),\ \mathbf{T} \in \Omega.$$

Laguerre Form

$$35.6.3 \quad L_\nu^{(\gamma)}(\mathbf{T}) = \frac{\Gamma_m\!\left(\gamma+\nu+\tfrac{1}{2}(m+1)\right)}{\Gamma_m\!\left(\gamma+\tfrac{1}{2}(m+1)\right)}\, {}_1F_1\!\left(\begin{matrix}-\nu\\ \gamma+\tfrac{1}{2}(m+1)\end{matrix};\mathbf{T}\right),\qquad \Re(\gamma),\Re(\gamma+\nu) > -1.$$

35.6(ii) Properties

$$35.6.4 \quad {}_1F_1\!\left(\begin{matrix}a\\b\end{matrix};\mathbf{T}\right) = \frac{1}{\mathrm{B}_m(a,b-a)} \int_{0<\mathbf{X}<\mathbf{I}} \operatorname{etr}(\mathbf{T}\mathbf{X})|\mathbf{X}|^{a-\frac{1}{2}(m+1)}|\mathbf{I}-\mathbf{X}|^{b-a-\frac{1}{2}(m+1)} d\mathbf{X},\quad \Re(a),\Re(b-a) > \tfrac{1}{2}(m-1).$$

$$35.6.5 \quad \int_\Omega \operatorname{etr}(-\mathbf{T}\mathbf{X})|\mathbf{X}|^{b-\frac{1}{2}(m+1)}\, {}_1F_1\!\left(\begin{matrix}a\\b\end{matrix};\mathbf{S}\mathbf{X}\right) d\mathbf{X} = \Gamma_m(b)|\mathbf{I}-\mathbf{S}\mathbf{T}^{-1}|^{-a}|\mathbf{T}|^{-b},\quad \mathbf{T}>\mathbf{S},\ \Re(b) > \tfrac{1}{2}(m-1).$$

$$35.6.6 \quad \mathrm{B}_m(b_1,b_2)|\mathbf{T}|^{b_1+b_2-\frac{1}{2}(m+1)}\, {}_1F_1\!\left(\begin{matrix}a_1+a_2\\b_1+b_2\end{matrix};\mathbf{T}\right)$$
$$= \int_{0<\mathbf{X}<\mathbf{T}} |\mathbf{X}|^{b_1-\frac{1}{2}(m+1)}\, {}_1F_1\!\left(\begin{matrix}a_1\\b_1\end{matrix};\mathbf{X}\right) |\mathbf{T}-\mathbf{X}|^{b_2-\frac{1}{2}(m+1)}\, {}_1F_1\!\left(\begin{matrix}a_2\\b_2\end{matrix};\mathbf{T}-\mathbf{X}\right) d\mathbf{X},\quad \Re(b_1),\Re(b_2) > \tfrac{1}{2}(m-1).$$

$$35.6.7 \quad {}_1F_1\!\left(\begin{matrix}a\\b\end{matrix};\mathbf{T}\right) = \operatorname{etr}(\mathbf{T})\, {}_1F_1\!\left(\begin{matrix}b-a\\b\end{matrix};-\mathbf{T}\right).$$

$$35.6.8 \quad \int_\Omega |\mathbf{T}|^{c-\frac{1}{2}(m+1)}\, \Psi(a;b;\mathbf{T})\, d\mathbf{T} = \frac{\Gamma_m(c)\,\Gamma_m(a-c)\,\Gamma_m\!\left(c-b+\tfrac{1}{2}(m+1)\right)}{\Gamma_m(a)\,\Gamma_m\!\left(a-b+\tfrac{1}{2}(m+1)\right)},$$
$$\Re(a) > \Re(c) + \tfrac{1}{2}(m-1) > m-1,\ \Re(c-b) > -1.$$

35.6(iii) Relations to Bessel Functions of Matrix Argument

$$35.6.9 \quad \lim_{a\to\infty} {}_1F_1\!\left(\begin{matrix}a\\ \nu+\tfrac{1}{2}(m+1)\end{matrix};-a^{-1}\mathbf{T}\right) = \frac{A_\nu(\mathbf{T})}{A_\nu(\mathbf{0})}.$$

$$35.6.10 \quad \lim_{a\to\infty} \Gamma_m(a)\,\Psi\!\left(a+\nu;\nu+\tfrac{1}{2}(m+1);a^{-1}\mathbf{T}\right) = B_\nu(\mathbf{T}).$$

35.6(iv) Asymptotic Approximations

For asymptotic approximations for confluent hypergeometric functions of matrix argument, see Herz (1955) and Butler and Wood (2002).

35.7 Gaussian Hypergeometric Function of Matrix Argument

35.7(i) Definition

35.7.1
$$ {}_2F_1\left(\begin{matrix}a,b\\c\end{matrix};\mathbf{T}\right) = \sum_{k=0}^{\infty}\frac{1}{k!}\sum_{|\kappa|=k}\frac{[a]_\kappa[b]_\kappa}{[c]_\kappa}Z_\kappa(\mathbf{T}), \quad -c+\tfrac{1}{2}(j+1)\notin\mathbb{N},\ 1\le j\le m;\ \|\mathbf{T}\|<1. $$

Jacobi Form

35.7.2
$$ P_\nu^{(\gamma,\delta)}(\mathbf{T}) = \frac{\Gamma_m\!\left(\gamma+\nu+\tfrac{1}{2}(m+1)\right)}{\Gamma_m\!\left(\gamma+\tfrac{1}{2}(m+1)\right)}\,{}_2F_1\!\left(\begin{matrix}-\nu,\gamma+\delta+\nu+\tfrac{1}{2}(m+1)\\ \gamma+\tfrac{1}{2}(m+1)\end{matrix};\mathbf{T}\right),\quad \mathbf{0}<\mathbf{T}<\mathbf{I};\ \gamma,\delta,\nu\in\mathbb{C};\ \Re(\gamma)>-1. $$

35.7(ii) Basic Properties

Case $m=2$

35.7.3
$$ {}_2F_1\!\left(\begin{matrix}a,b\\c\end{matrix};\begin{pmatrix}t_1&0\\0&t_2\end{pmatrix}\right) = \sum_{k=0}^{\infty}\frac{(a)_k(c-a)_k(b)_k(c-b)_k}{k!\,(c)_{2k}\left(c-\tfrac{1}{2}\right)_k}(t_1t_2)^k\,{}_2F_1\!\left(\begin{matrix}a+k,b+k\\c+2k\end{matrix};t_1+t_2-t_1t_2\right). $$

Confluent Form

35.7.4
$$ \lim_{c\to\infty}{}_2F_1\!\left(\begin{matrix}a,b\\c\end{matrix};\mathbf{I}-c\mathbf{T}^{-1}\right)=|\mathbf{T}|^b\,\Psi(b;b-a+\tfrac{1}{2}(m+1);\mathbf{T}). $$

Integral Representation

35.7.5
$$ {}_2F_1\!\left(\begin{matrix}a,b\\c\end{matrix};\mathbf{T}\right)=\frac{1}{\mathrm{B}_m(a,c-a)}\int_{\mathbf{0}<\mathbf{X}<\mathbf{I}}|\mathbf{X}|^{a-\tfrac{1}{2}(m+1)}|\mathbf{I}-\mathbf{X}|^{c-a-\tfrac{1}{2}(m+1)}|\mathbf{I}-\mathbf{TX}|^{-b}\,d\mathbf{X}, $$
$$ \Re(a),\Re(c-a)>\tfrac{1}{2}(m-1),\ \mathbf{0}<\mathbf{T}<\mathbf{I}. $$

Transformations of Parameters

35.7.6
$$ {}_2F_1\!\left(\begin{matrix}a,b\\c\end{matrix};\mathbf{T}\right)=|\mathbf{I}-\mathbf{T}|^{c-a-b}\,{}_2F_1\!\left(\begin{matrix}c-a,c-b\\c\end{matrix};\mathbf{T}\right)=|\mathbf{I}-\mathbf{T}|^{-a}\,{}_2F_1\!\left(\begin{matrix}a,c-b\\c\end{matrix};-\mathbf{T}(\mathbf{I}-\mathbf{T})^{-1}\right) $$
$$ =|\mathbf{I}-\mathbf{T}|^{-b}\,{}_2F_1\!\left(\begin{matrix}c-a,b\\c\end{matrix};-\mathbf{T}(\mathbf{I}-\mathbf{T})^{-1}\right). $$

Gauss Formula

35.7.7
$$ {}_2F_1\!\left(\begin{matrix}a,b\\c\end{matrix};\mathbf{I}\right)=\frac{\Gamma_m(c)\,\Gamma_m(c-a-b)}{\Gamma_m(c-a)\,\Gamma_m(c-b)},\qquad \Re(c),\Re(c-a-b)>\tfrac{1}{2}(m-1). $$

Reflection Formula

35.7.8
$$ {}_2F_1\!\left(\begin{matrix}a,b\\c\end{matrix};\mathbf{T}\right)=\frac{\Gamma_m(c)\,\Gamma_m(c-a-b)}{\Gamma_m(c-a)\,\Gamma_m(c-b)}\,{}_2F_1\!\left(\begin{matrix}a,b\\a+b-c+\tfrac{1}{2}(m+1)\end{matrix};\mathbf{I}-\mathbf{T}\right),\quad \Re(c),\Re(c-a-b)>\tfrac{1}{2}(m-1). $$

35.7(iii) Partial Differential Equations

Let $f:\Omega\to\mathbb{C}$ (a) be *orthogonally invariant*, so that $f(\mathbf{T})$ is a symmetric function of $t_1,\ldots,t_m$, the eigenvalues of the matrix argument $\mathbf{T}\in\Omega$; (b) be analytic in $t_1,\ldots,t_m$ in a neighborhood of $\mathbf{T}=\mathbf{0}$; (c) satisfy $f(\mathbf{0})=1$. Subject to the conditions (a)–(c), the function $f(\mathbf{T})={}_2F_1(a,b;c;\mathbf{T})$ is the unique solution of each partial differential equation

35.7.9
$$ t_j(1-t_j)\frac{\partial^2 F}{\partial t_j{}^2}-\frac{1}{2}\sum_{\substack{k=1\\k\ne j}}^{m}\frac{t_k(1-t_k)}{t_j-t_k}\frac{\partial F}{\partial t_k}+\left(c-\tfrac{1}{2}(m-1)-\big(a+b-\tfrac{1}{2}(m-3)\big)t_j+\frac{1}{2}\sum_{\substack{k=1\\k\ne j}}^{m}\frac{t_j(1-t_j)}{t_j-t_k}\right)\frac{\partial F}{\partial t_j}=abF, $$

for $j=1,\ldots,m$.

Systems of partial differential equations for the $_0F_1$ (defined in §35.8) and $_1F_1$ functions of matrix argument can be obtained by applying (35.8.9) and (35.8.10) to (35.7.9).

35.7(iv) Asymptotic Approximations

Butler and Wood (2002) applies Laplace's method (§2.3(iii)) to (35.7.5) to derive uniform asymptotic approximations for the functions

35.7.10
$$_2F_1\left(\begin{matrix}\alpha a, \alpha b\\ \alpha c\end{matrix}; \mathbf{T}\right)$$

and

35.7.11
$$_2F_1\left(\begin{matrix}a, b\\ c\end{matrix}; \mathbf{I}-\alpha^{-1}\mathbf{T}\right)$$

as $\alpha \to \infty$. These approximations are in terms of elementary functions.

For other asymptotic approximations for Gaussian hypergeometric functions of matrix argument, see Herz (1955), Muirhead (1982, pp. 264–281, 290, 472, 563), and Butler and Wood (2002).

35.8 Generalized Hypergeometric Functions of Matrix Argument

35.8(i) Definition

Let p and q be nonnegative integers; $a_1, \ldots, a_p \in \mathbb{C}$; $b_1, \ldots, b_q \in \mathbb{C}$; $-b_j + \frac{1}{2}(k+1) \notin \mathbb{N}$, $1 \le j \le q$, $1 \le k \le m$. The generalized hypergeometric function $_pF_q$ with matrix argument $\mathbf{T} \in \mathcal{S}$, numerator parameters $a_1, \ldots, a_p$, and denominator parameters $b_1, \ldots, b_q$ is

35.8.1
$$_pF_q\left(\begin{matrix}a_1,\ldots,a_p\\ b_1,\ldots,b_q\end{matrix}; \mathbf{T}\right) = \sum_{k=0}^{\infty}\frac{1}{k!}\sum_{|\kappa|=k}\frac{[a_1]_\kappa\cdots[a_p]_\kappa}{[b_1]_\kappa\cdots[b_q]_\kappa}Z_\kappa(\mathbf{T}).$$

Convergence Properties

If $-a_j + \frac{1}{2}(k+1) \in \mathbb{N}$ for some j, k satisfying $1 \le j \le p$, $1 \le k \le m$, then the series expansion (35.8.1) terminates.

If $p \le q$, then (35.8.1) converges for all $\mathbf{T}$.

If $p = q+1$, then (35.8.1) converges absolutely for $\|\mathbf{T}\| < 1$ and diverges for $\|\mathbf{T}\| > 1$.

If $p > q+1$, then (35.8.1) diverges unless it terminates.

35.8(ii) Relations to Other Functions

35.8.2
$$_0F_0\left(\begin{matrix}-\\ -\end{matrix}; \mathbf{T}\right) = \operatorname{etr}(\mathbf{T}), \qquad \mathbf{T}\in\mathcal{S}.$$

35.8.3
$$_2F_1\left(\begin{matrix}a,b\\ b\end{matrix}; \mathbf{T}\right) = {}_1F_0\left(\begin{matrix}a\\ -\end{matrix}; \mathbf{T}\right) = |\mathbf{I}-\mathbf{T}|^{-a}, \quad 0 < \mathbf{T} < \mathbf{I}.$$

35.8.4
$$A_\nu(\mathbf{T}) = \frac{1}{\Gamma_m(\nu+\frac{1}{2}(m+1))}\,_0F_1\left(\begin{matrix}-\\ \nu+\frac{1}{2}(m+1)\end{matrix}; -\mathbf{T}\right),$$
$$\mathbf{T}\in\mathcal{S}.$$

35.8(iii) $_3F_2$ Case

Kummer Transformation

Let $c = b_1 + b_2 - a_1 - a_2 - a_3$. Then

35.8.5
$$_3F_2\left(\begin{matrix}a_1,a_2,a_3\\ b_1,b_2\end{matrix}; \mathbf{I}\right) = \frac{\Gamma_m(b_2)\,\Gamma_m(c)}{\Gamma_m(b_2-a_3)\,\Gamma_m(c+a_3)}$$
$$\times\,_3F_2\left(\begin{matrix}b_1-a_1, b_1-a_2, a_3\\ b_1, c+a_3\end{matrix}; \mathbf{I}\right),$$
$$\Re(b_2), \Re(c) > \tfrac{1}{2}(m-1).$$

Pfaff–Saalschutz Formula

Let $a_1+a_2+a_3+\frac{1}{2}(m+1) = b_1+b_2$; one of the a_j be a negative integer; $\Re(b_1-a_1), \Re(b_1-a_2), \Re(b_1-a_3), \Re(b_1-a_1-a_2-a_3) > \frac{1}{2}(m-1)$. Then

35.8.6
$$_3F_2\left(\begin{matrix}a_1,a_2,a_3\\ b_1,b_2\end{matrix}; \mathbf{I}\right)$$
$$= \frac{\Gamma_m(b_1-a_1)\,\Gamma_m(b_1-a_2)}{\Gamma_m(b_1)\,\Gamma_m(b_1-a_1-a_2)}$$
$$\times \frac{\Gamma_m(b_1-a_3)\,\Gamma_m(b_1-a_1-a_2-a_3)}{\Gamma_m(b_1-a_1-a_3)\,\Gamma_m(b_1-a_2-a_3)}.$$

Thomae Transformation

Again, let $c = b_1 + b_2 - a_1 - a_2 - a_3$. Then

35.8.7
$$_3F_2\left(\begin{matrix}a_1,a_2,a_3\\ b_1,b_2\end{matrix}; \mathbf{I}\right) = \frac{\Gamma_m(b_1)\,\Gamma_m(b_2)\,\Gamma(c)}{\Gamma_m(a_1)\,\Gamma_m(c+a_2)\,\Gamma(c+a_3)}$$
$$\times\,_3F_2\left(\begin{matrix}b_1-a_1, b_2-a_2, c\\ c+a_2, c+a_3\end{matrix}; \mathbf{I}\right),$$
$$\Re(b_1), \Re(b_2), \Re(c) > \tfrac{1}{2}(m-1).$$

35.8(iv) General Properties

Value at $\mathbf{T} = \mathbf{0}$

35.8.8
$$_pF_q\left(\begin{matrix}a_1,\ldots,a_p\\ b_1,\ldots,b_q\end{matrix}; \mathbf{0}\right) = 1.$$

Confluence

35.8.9
$$\lim_{\gamma\to\infty}\,_{p+1}F_q\left(\begin{matrix}a_1,\ldots,a_p,\gamma\\ b_1,\ldots,b_q\end{matrix}; \gamma^{-1}\mathbf{T}\right)$$
$$= {}_pF_q\left(\begin{matrix}a_1,\ldots,a_p\\ b_1,\ldots,b_q\end{matrix}; \mathbf{T}\right),$$

35.8.10
$$\lim_{\gamma\to\infty}\,_pF_{q+1}\left(\begin{matrix}a_1,\ldots,a_p\\ b_1,\ldots,b_q,\gamma\end{matrix}; \gamma\mathbf{T}\right) = {}_pF_q\left(\begin{matrix}a_1,\ldots,a_p\\ b_1,\ldots,b_q\end{matrix}; \mathbf{T}\right).$$

Invariance

35.8.11
$$_pF_q\left(\begin{matrix}a_1,\ldots,a_p\\b_1,\ldots,b_q\end{matrix};\mathbf{HTH}^{-1}\right) = {}_pF_q\left(\begin{matrix}a_1,\ldots,a_p\\b_1,\ldots,b_q\end{matrix};\mathbf{T}\right),$$
$$\mathbf{H}\in\mathbf{O}(m).$$

Laplace Transform

35.8.12
$$\int_\Omega \text{etr}(-\mathbf{TX})|\mathbf{X}|^{\gamma-\frac{1}{2}(m+1)} {}_pF_q\left(\begin{matrix}a_1,\ldots,a_p\\b_1,\ldots,b_q\end{matrix};-\mathbf{X}\right)d\mathbf{X}$$
$$= \Gamma_m(\gamma)|\mathbf{T}|^{-\gamma} {}_{p+1}F_q\left(\begin{matrix}a_1,\ldots,a_p,\gamma\\b_1,\ldots,b_q\end{matrix};-\mathbf{T}^{-1}\right),$$
$$\Re(\gamma) > \tfrac{1}{2}(m-1).$$

Euler Integral

35.8.13
$$\int_{0<\mathbf{X}<\mathbf{I}} |\mathbf{X}|^{a_1-\frac{1}{2}(m+1)}|\mathbf{I}-\mathbf{X}|^{b_1-a_1-\frac{1}{2}(m+1)}$$
$$\times {}_pF_q\left(\begin{matrix}a_2,\ldots,a_{p+1}\\b_2,\ldots,b_{q+1}\end{matrix};\mathbf{TX}\right)d\mathbf{X}$$
$$= \frac{1}{\mathrm{B}_m(b_1-a_1,a_1)} {}_{p+1}F_{q+1}\left(\begin{matrix}a_1,\ldots,a_{p+1}\\b_1,\ldots,b_{q+1}\end{matrix};\mathbf{T}\right),$$
$$\Re(b_1-a_1),\Re(a_1) > \tfrac{1}{2}(m-1).$$

35.8(v) Mellin–Barnes Integrals

Multidimensional Mellin–Barnes integrals are established in Ding et al. (1996) for the functions $_pF_q$ and $_{p+1}F_p$ of matrix argument. A similar result for the $_0F_1$ function of matrix argument is given in Faraut and Korányi (1994, p. 346). These multidimensional integrals reduce to the classical Mellin–Barnes integrals (§5.19(ii)) in the special case $m=1$.

See also Faraut and Korányi (1994, pp. 318–340).

Applications

35.9 Applications

In multivariate statistical analysis based on the multivariate normal distribution, the probability density functions of many random matrices are expressible in terms of generalized hypergeometric functions of matrix argument $_pF_q$, with $p\leq 2$ and $q\leq 1$. See James (1964), Muirhead (1982), Takemura (1984), Farrell (1985), and Chikuse (2003) for extensive treatments.

For other statistical applications of $_pF_q$ functions of matrix argument see Perlman and Olkin (1980), Groeneboom and Truax (2000), Bhaumik and Sarkar (2002), Richards (2004) (monotonicity of power functions of multivariate statistical test criteria), Bingham et al. (1992) (Procrustes analysis), and Phillips (1986) (exact distributions of statistical test criteria). These references all use results related to the integral formulas (35.4.7) and (35.5.8).

For applications of the integral representation (35.5.3) see McFarland and Richards (2001, 2002) (statistical estimation of misclassification probabilities for discriminating between multivariate normal populations). The asymptotic approximations of §35.7(iv) are applied in numerous statistical contexts in Butler and Wood (2002).

In chemistry, Wei and Eichinger (1993) expresses the probability density functions of macromolecules in terms of generalized hypergeometric functions of matrix argument, and develop asymptotic approximations for these density functions.

In the nascent area of applications of zonal polynomials to the limiting probability distributions of symmetric random matrices, one of the most comprehensive accounts is Rains (1998).

Computation

35.10 Methods of Computation

For small values of $\|\mathbf{T}\|$ the zonal polynomial expansion given by (35.8.1) can be summed numerically. For large $\|\mathbf{T}\|$ the asymptotic approximations referred to in §35.7(iv) are available.

Other methods include numerical quadrature applied to double and multiple integral representations. See Yan (1992) for the $_1F_1$ and $_2F_1$ functions of matrix argument in the case $m=2$, and Bingham et al. (1992) for Monte Carlo simulation on $\mathbf{O}(m)$ applied to a generalization of the integral (35.5.8).

Koev and Edelman (2006) utilizes combinatorial identities for the zonal polynomials to develop computational algorithms for approximating the series expansion (35.8.1). These algorithms are extremely efficient, converge rapidly even for large values of m, and have complexity linear in m.

35.11 Tables

Tables of zonal polynomials are given in James (1964) for $|\kappa|\leq 6$, Parkhurst and James (1974) for $|\kappa|\leq 12$, and Muirhead (1982, p. 238) for $|\kappa|\leq 5$. Each table expresses the zonal polynomials as linear combinations of monomial symmetric functions.

35.12 Software

See http://dlmf.nist.gov/35.12.

References

General References

The main references used in writing this chapter are Herz (1955), James (1964), Muirhead (1982), Gross and Richards (1987), and Richards (1992). For additional bibliographic reading see Vilenkin and Klimyk (1992) and Faraut and Korányi (1994).

Sources

The following list gives the references or other indications of proofs that were used in constructing the various sections in this chapter. These sources supplement the references that are quoted in the text.

§35.2 Gårding (1947), Herz (1955, p. 479), Muirhead (1982, p. 252). See also Siegel (1935), Bochner and Martin (1948, pp. 90–92, 113–132).

§35.3 Wishart (1928), Ingham (1933), Gindikin (1964), Gårding (1947), Herz (1955), Olkin (1959).

§35.4 James (1964), Muirhead (1982, Chapter 7), Macdonald (1995, p. 425). See also Constantine (1963), Maass (1971, pp. 64–71), Macdonald (1995, pp. 388–439).

§35.5 Herz (1955). See also Bochner (1952), Gross and Kunze (1976), Terras (1988, pp. 49–63), Butler and Wood (2003).

§35.6 Koecher (1954), Muirhead (1978), Muirhead (1982, pp. 264–266, 472–473), Herz (1955). For (35.6.6) apply (35.2.3) and (35.6.5). See also Shimura (1982).

§35.7 Herz (1955), Muirhead (1982, pp. 264–281, 290, 472), Faraut and Korányi (1994, pp. 337–340). For (35.7.8) see Zheng (1997). See also Macdonald (1990), Ding et al. (1996), Koornwinder and Sprinkhuizen-Kuyper (1978), Gross and Richards (1991).

§35.8 Gross and Richards (1987, 1991), Faraut and Korányi (1994, pp. 318–340), Herz (1955), Muirhead (1982, pp. 259–262), Macdonald (1990), James (1964), Ding et al. (1996).

Chapter 36

Integrals with Coalescing Saddles

M. V. Berry[1] and C. J. Howls[2]

Notation   **776**
- 36.1 Special Notation 776

Properties   **776**
- 36.2 Catastrophes and Canonical Integrals . . 776
- 36.3 Visualizations of Canonical Integrals . . . 778
- 36.4 Bifurcation Sets 781
- 36.5 Stokes Sets 782
- 36.6 Scaling Relations 785
- 36.7 Zeros 785
- 36.8 Convergent Series Expansions 787
- 36.9 Integral Identities 787
- 36.10 Differential Equations 788
- 36.11 Leading-Order Asymptotics 789

Applications   **789**
- 36.12 Uniform Approximation of Integrals . . . 789
- 36.13 Kelvin's Ship-Wave Pattern 790
- 36.14 Other Physical Applications 791

Computation   **792**
- 36.15 Methods of Computation 792

References   **792**

[1] H H Wills Physics Laboratory, Bristol, United Kingdom.
[2] School of Mathematics, University of Southampton, Southampton, United Kingdom.
Copyright © 2009 National Institute of Standards and Technology. All rights reserved.

Notation

36.1 Special Notation

(For other notation see pp. xiv and 873.)

l, m, n integers.
k, t, s real or complex variables.
K codimension.
$\mathbf{x}$ $\{x_1, x_2, \ldots, x_K\}$, where $x_1, x_2, \ldots, x_K$ are real parameters; also $x_1 = x$, $x_2 = y$, $x_3 = z$ when $K \leq 3$.
Ai, Bi   Airy functions (§9.2).
$*$ complex conjugate.

The main functions covered in this chapter are cuspoid catastrophes $\Phi_K(t; \mathbf{x})$; umbilic catastrophes with codimension three $\Phi^{(E)}(s, t; \mathbf{x})$, $\Phi^{(H)}(s, t; \mathbf{x})$; canonical integrals $\Psi_K(\mathbf{x})$, $\Psi^{(E)}(\mathbf{x})$, $\Psi^{(H)}(\mathbf{x})$; diffraction catastrophes $\Psi_K(\mathbf{x}; k)$, $\Psi^{(E)}(\mathbf{x}; k)$, $\Psi^{(H)}(\mathbf{x}; k)$ generated by the catastrophes. (There is no standard nomenclature for these functions.)

Properties

36.2 Catastrophes and Canonical Integrals

36.2(i) Definitions

Normal Forms Associated with Canonical Integrals: Cuspoid Catastrophe with Codimension K

36.2.1 $$\Phi_K(t; \mathbf{x}) = t^{K+2} + \sum_{m=1}^{K} x_m t^m.$$

Special cases: $K = 1$, *fold catastrophe*; $K = 2$, *cusp catastrophe*; $K = 3$, *swallowtail catastrophe*.

Normal Forms for Umbilic Catastrophes with Codimension $K = 3$

36.2.2 $$\Phi^{(E)}(s, t; \mathbf{x}) = s^3 - 3st^2 + z(s^2 + t^2) + yt + xs,$$
$$\mathbf{x} = \{x, y, z\},$$

(elliptic umbilic).

36.2.3 $$\Phi^{(H)}(s, t; \mathbf{x}) = s^3 + t^3 + zst + yt + xs,$$
$$\mathbf{x} = \{x, y, z\},$$

(hyperbolic umbilic).

Canonical Integrals

36.2.4 $$\Psi_K(\mathbf{x}) = \int_{-\infty}^{\infty} \exp(i\, \Phi_K(t; \mathbf{x}))\, dt.$$

36.2.5 $$\Psi^{(U)}(\mathbf{x}) = \int_{-\infty}^{\infty} \int_{-\infty}^{\infty} \exp\!\left(i\, \Phi^{(U)}(s, t; \mathbf{x})\right) ds\, dt, \qquad U = E, H.$$

36.2.6 $$\Psi^{(E)}(\mathbf{x}) = 2\sqrt{\pi/3}\, \exp\!\left(i\left(\tfrac{4}{27} z^3 + \tfrac{1}{3} xz - \tfrac{1}{4}\pi\right)\right) \int_{\infty \exp(-7\pi i/12)}^{\infty \exp(\pi i/12)} \exp\!\left(i\left(u^6 + 2zu^4 + (z^2 + x)u^2 + \frac{y^2}{12u^2}\right)\right) du,$$

with the contour passing to the lower right of $u = 0$.

36.2.7 $$\Psi^{(E)}(\mathbf{x}) = \frac{4\pi}{3^{1/3}} \exp\!\left(i\left(\tfrac{2}{27} z^3 - \tfrac{1}{3} xz\right)\right) \left(\exp\!\left(-i\tfrac{\pi}{6}\right) F_+(\mathbf{x}) + \exp\!\left(i\tfrac{\pi}{6}\right) F_-(\mathbf{x})\right),$$
$$F_\pm(\mathbf{x}) = \int_0^\infty \cos\!\left(ry \exp\!\left(\pm i \tfrac{\pi}{6}\right)\right) \exp\!\left(2ir^2 z \exp\!\left(\pm i \tfrac{\pi}{3}\right)\right) \operatorname{Ai}\!\left(3^{2/3} r^2 + 3^{-1/3} \exp\!\left(\mp i \tfrac{\pi}{3}\right) \left(\tfrac{1}{3} z^2 - x\right)\right) dr.$$

36.2.8 $$\Psi^{(H)}(\mathbf{x}) = 4\sqrt{\pi/6}\, \exp\!\left(i\left(\tfrac{1}{27} z^3 + \tfrac{1}{6} z(y + x) + \tfrac{1}{4}\pi\right)\right)$$
$$\times \int_{\infty \exp(5\pi i/12)}^{\infty \exp(\pi i/12)} \exp\!\left(i\left(2u^6 + 2zu^4 + \left(\tfrac{1}{2} z^2 + x + y\right)u^2 - \frac{(y-x)^2}{24u^2}\right)\right) du,$$

with the contour passing to the upper right of $u = 0$.

36.2.9 $$\Psi^{(H)}(\mathbf{x}) = \frac{2\pi}{3^{1/3}} \int_{\infty \exp(5\pi i/6)}^{\infty \exp(\pi i/6)} \exp\!\left(i(s^3 + xs)\right) \operatorname{Ai}\!\left(\frac{zs + y}{3^{1/3}}\right) ds.$$

Diffraction Catastrophes

36.2.10
$$\Psi_K(\mathbf{x}; k) = \sqrt{k} \int_{-\infty}^{\infty} \exp(ik\,\Phi_K(t; \mathbf{x}))\,dt, \quad k > 0.$$

36.2.11
$$\Psi^{(\mathrm{U})}(\mathbf{x}; k) = k \int_{-\infty}^{\infty} \int_{-\infty}^{\infty} \exp\left(ik\,\Phi^{(\mathrm{U})}(s, t; \mathbf{x})\right) ds\,dt,$$
$$\mathrm{U} = \mathrm{E}, \mathrm{H}, \quad k > 0.$$

For more extensive lists of normal forms of catastrophes (umbilic and beyond) involving two variables ("corank two") see Arnol'd (1972, 1974, 1975).

36.2(ii) Special Cases

36.2.12
$$\Psi_0 = \sqrt{\pi} \exp\left(i\frac{\pi}{4}\right).$$

Ψ_1 is related to the Airy function (§9.2):

36.2.13
$$\Psi_1(x) = \frac{2\pi}{3^{1/3}} \operatorname{Ai}\left(\frac{x}{3^{1/3}}\right).$$

Ψ_2 is the Pearcey integral (Pearcey (1946)):

36.2.14
$$\Psi_2(\mathbf{x}) = P(x_2, x_1) = \int_{-\infty}^{\infty} \exp\left(i(t^4 + x_2 t^2 + x_1 t)\right) dt.$$

(Other notations also appear in the literature.)

36.2.15
$$\Psi_K(\mathbf{0}) = \frac{2}{K+2} \Gamma\left(\frac{1}{K+2}\right) \begin{cases} \exp\left(i\dfrac{\pi}{2(K+2)}\right), & K \text{ even}, \\ \cos\left(\dfrac{\pi}{2(K+2)}\right), & K \text{ odd}. \end{cases}$$

36.2.16
$$\Psi_1(\mathbf{0}) = 1.54669, \quad \Psi_2(\mathbf{0}) = 1.67481 + i\,0.69373$$
$$\Psi_3(\mathbf{0}) = 1.74646, \quad \Psi_4(\mathbf{0}) = 1.79222 + i\,0.48022.$$

36.2.17
$$\frac{\partial^p}{\partial x_1{}^p} \Psi_K(\mathbf{0}) = \frac{2}{K+2} \Gamma\left(\frac{p+1}{K+2}\right) \cos\left(\frac{\pi}{2}\left(\frac{p+1}{K+2} + p\right)\right), \quad K \text{ odd},$$

$$\frac{\partial^{2q+1}}{\partial x_1{}^{2q+1}} \Psi_K(\mathbf{0}) = 0, \quad K \text{ even},$$

$$\frac{\partial^{2q}}{\partial x_1{}^{2q}} \Psi_K(\mathbf{0}) = \frac{2}{K+2} \Gamma\left(\frac{2q+1}{K+2}\right) \exp\left(i\frac{\pi}{2}\left(\frac{2q+1}{K+2} + 2q\right)\right), \quad K \text{ even}.$$

36.2.18
$$\Psi^{(\mathrm{E})}(\mathbf{0}) = \tfrac{1}{3}\sqrt{\pi}\,\Gamma\!\left(\tfrac{1}{6}\right) = 3.28868,$$
$$\Psi^{(\mathrm{H})}(0) = \tfrac{1}{3}\Gamma^2\!\left(\tfrac{1}{3}\right) = 2.39224.$$

36.2.19
$$\Psi_2(0, y) = \frac{\pi}{2}\sqrt{\frac{|y|}{2}} \exp\!\left(-i\frac{y^2}{8}\right) \left(\exp\!\left(i\frac{\pi}{8}\right) J_{-1/4}\!\left(\frac{y^2}{8}\right) - \operatorname{sign}(y)\exp\!\left(-i\frac{\pi}{8}\right) J_{1/4}\!\left(\frac{y^2}{8}\right)\right).$$

For the Bessel function J see §10.2(ii).

36.2.20
$$\Psi^{(\mathrm{E})}(x, y, 0) = 2\pi^2 (\tfrac{2}{3})^{2/3} \Re\!\left(\operatorname{Ai}\!\left(\frac{x+iy}{12^{1/3}}\right) \operatorname{Bi}\!\left(\frac{x-iy}{12^{1/3}}\right)\right),$$

36.2.21
$$\Psi^{(\mathrm{H})}(x, y, 0) = \frac{4\pi^2}{3^{2/3}} \operatorname{Ai}\!\left(\frac{x}{3^{1/3}}\right) \operatorname{Ai}\!\left(\frac{y}{3^{1/3}}\right).$$

36.2(iii) Symmetries

36.2.22
$$\Psi_{2K}(\mathbf{x}') = \Psi_{2K}(\mathbf{x}), \quad x'_{2m+1} = -x_{2m+1},\; x'_{2m} = x_{2m}.$$

36.2.23
$$\Psi_{2K+1}(\mathbf{x}') = \Psi^*_{2K+1}(\mathbf{x}), \quad x'_{2m+1} = x_{2m+1},\; x'_{2m} = -x_{2m}.$$

36.2.24
$$\Psi^{(\mathrm{U})}(x, y, z) = \Psi^{*(\mathrm{U})}(x, y, -z), \quad \mathrm{U} = \mathrm{E}, \mathrm{H}.$$

36.2.25
$$\Psi^{(\mathrm{E})}(x, -y, z) = \Psi^{(\mathrm{E})}(x, y, z).$$

36.2.26
$$\Psi^{(\mathrm{E})}\!\left(-\tfrac{1}{2}x \mp \tfrac{\sqrt{3}}{2}y, \pm\tfrac{\sqrt{3}}{2}x - \tfrac{1}{2}y, z\right) = \Psi^{(\mathrm{E})}(x, y, z),$$

(rotation by $\pm\tfrac{2}{3}\pi$ in x, y plane).

36.2.27
$$\Psi^{(\mathrm{H})}(x, y, z) = \Psi^{(\mathrm{H})}(y, x, z).$$

36.3 Visualizations of Canonical Integrals

36.3(i) Canonical Integrals: Modulus

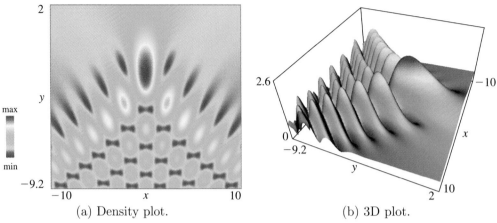

(a) Density plot.   (b) 3D plot.

Figure 36.3.1: Modulus of Pearcey integral $|\Psi_2(x,y)|$.

For additional figures see http://dlmf.nist.gov/36.3.i.

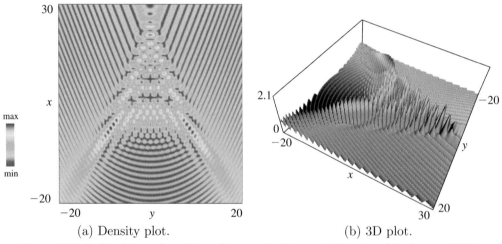

(a) Density plot.   (b) 3D plot.

Figure 36.3.5: Modulus of swallowtail canonical integral function $|\Psi_3(x,y,-7.5)|$.

For additional figures see http://dlmf.nist.gov/36.3.i.

36.3 Visualizations of Canonical Integrals

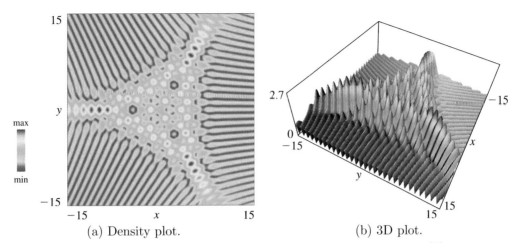

(a) Density plot. (b) 3D plot.

Figure 36.3.8: Modulus of elliptic umbilic canonical integral function $|\Psi^{(E)}(x,y,4)|$.

For additional figures see http://dlmf.nist.gov/36.3.i.

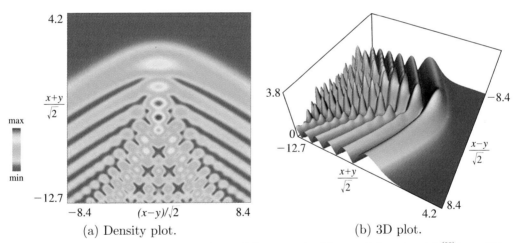

(a) Density plot. (b) 3D plot.

Figure 36.3.12: Modulus of hyperbolic umbilic canonical integral function $|\Psi^{(H)}(x,y,3)|$.

36.3(ii) Canonical Integrals: Phase

In Figure 36.3.13(a) points of confluence of phase contours are zeros of $\Psi_2(x,y)$; similarly for other contour plots in this subsection. In Figure 36.3.13(b) points of confluence of all colors are zeros of $\Psi_2(x,y)$; similarly for other density plots in this subsection.

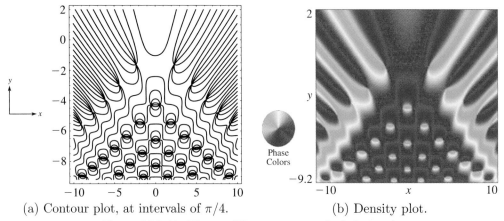

(a) Contour plot, at intervals of $\pi/4$.   (b) Density plot.

Figure 36.3.13: Phase of Pearcey integral ph $\Psi_2(x,y)$.

For additional figures see http://dlmf.nist.gov/36.3.ii.

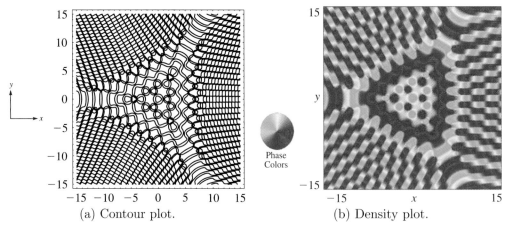

(a) Contour plot.   (b) Density plot.

Figure 36.3.17: Phase of elliptic umbilic canonical integral ph $\Psi^{(E)}(x,y,4)$.

For additional figures see http://dlmf.nist.gov/36.3.ii.

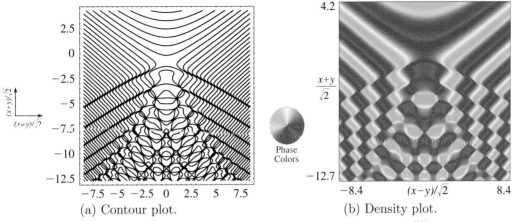

(a) Contour plot. (b) Density plot.

Figure 36.3.21: Phase of hyperbolic umbilic canonical integral ph $\Psi^{(H)}(x,y,3)$.

36.4 Bifurcation Sets

36.4(i) Formulas

Critical Points for Cuspoids

These are real solutions $t_j(\mathbf{x})$, $1 \leq j \leq j_{\max}(\mathbf{x}) \leq K+1$, of

36.4.1 $$\frac{\partial}{\partial t}\Phi_K(t_j(\mathbf{x});\mathbf{x}) = 0.$$

Critical Points for Umbilics

These are real solutions $\{s_j(\mathbf{x}),t_j(\mathbf{x})\}$, $1 \leq j \leq j_{\max}(\mathbf{x}) \leq 4$, of

36.4.2 $$\frac{\partial}{\partial s}\Phi^{(U)}(s_j(\mathbf{x}),t_j(\mathbf{x});\mathbf{x}) = 0,$$
$$\frac{\partial}{\partial t}\Phi^{(U)}(s_j(\mathbf{x}),t_j(\mathbf{x});\mathbf{x}) = 0.$$

Bifurcation (Catastrophe) Set for Cuspoids

This is the codimension-one surface in $\mathbf{x}$ space where critical points coalesce, satisfying (36.4.1) and

36.4.3 $$\frac{\partial^2}{\partial t^2}\Phi_K(t;\mathbf{x}) = 0.$$

Bifurcation (Catastrophe) Set for Umbilics

This is the codimension-one surface in $\mathbf{x}$ space where critical points coalesce, satisfying (36.4.2) and

36.4.4 $$\frac{\partial^2}{\partial s^2}\Phi^{(U)}(s,t;\mathbf{x})\frac{\partial^2}{\partial t^2}\Phi^{(U)}(s,t;\mathbf{x}) - \left(\frac{\partial^2}{\partial s\,\partial t}\Phi^{(U)}(s,t;\mathbf{x})\right)^2 = 0.$$

Special Cases

$K=1$, fold bifurcation set:

36.4.5 $$x = 0.$$

$K=2$, cusp bifurcation set:

36.4.6 $$27x^2 = -8y^3.$$

$K=3$, swallowtail bifurcation set:

36.4.7 $$x = 3t^2(z+5t^2), \quad y = -t(3z+10t^2), \quad -\infty < t < \infty.$$

Swallowtail self-intersection line:

36.4.8 $$y=0, \quad z \leq 0, \quad x = \tfrac{9}{20}z^2.$$

Swallowtail cusp lines (ribs):

36.4.9 $$z \leq 0, \quad x = -\tfrac{3}{20}z^2, \quad 10y^2 = -4z^3.$$

Elliptic umbilic bifurcation set (codimension three): for fixed z, the section of the bifurcation set is a three-cusped astroid

36.4.10 $$x = \tfrac{1}{3}z^2(-\cos(2\phi)-2\cos\phi),$$
$$y = \tfrac{1}{3}z^2(\sin(2\phi)-2\sin\phi), \qquad 0 \leq \phi \leq 2\pi.$$

Elliptic umbilic cusp lines (ribs):

36.4.11 $$x+iy = -z^2\exp\!\left(\tfrac{2}{3}i\pi m\right), \quad m=0,1,2.$$

Hyperbolic umbilic bifurcation set (codimension three):

36.4.12
$$x = -\tfrac{1}{12}z^2(\exp(2\tau)\pm 2\exp(-\tau)),$$
$$y = -\tfrac{1}{12}z^2(\exp(-2\tau)\pm 2\exp(\tau)), \quad -\infty \leq \tau < \infty.$$

The $+$ sign labels the cusped sheet; the $-$ sign labels the sheet that is smooth for $z \neq 0$ (see Figure 36.4.4).

Hyperbolic umbilic cusp line (rib):

36.4.13 $$x = y = -\tfrac{1}{4}z^2.$$

For derivations of the results in this subsection see Poston and Stewart (1978, Chapter 9).

36.4(ii) Visualizations

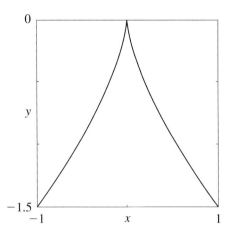

Figure 36.4.1: Bifurcation set of cusp catastrophe.

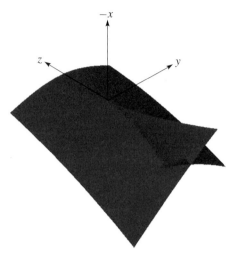

Figure 36.4.2: Bifurcation set of swallowtail catastrophe.

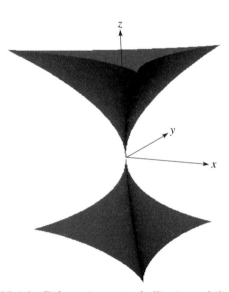

Figure 36.4.3: Bifurcation set of elliptic umbilic catastrophe.

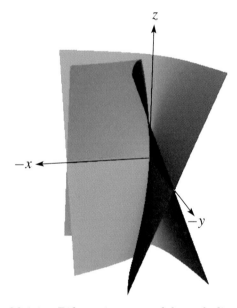

Figure 36.4.4: Bifurcation set of hyperbolic umbilic catastrophe.

36.5 Stokes Sets

36.5(i) Definitions

Stokes sets are surfaces (codimension one) in $\mathbf{x}$ space, across which $\Psi_K(\mathbf{x};k)$ or $\Psi^{(U)}(\mathbf{x};k)$ acquires an exponentially-small asymptotic contribution (in k), associated with a complex critical point of Φ_K or $\Phi^{(U)}$. The Stokes sets are defined by the exponential dominance condition:

36.5.1
$$\Re(\Phi_K(t_j(\mathbf{x});\mathbf{x}) - \Phi_K(t_\mu(\mathbf{x});\mathbf{x})) = 0,$$
$$\Re\Big(\Phi^{(U)}(s_j(\mathbf{x}),t_j(\mathbf{x});\mathbf{x}) - \Phi^{(U)}(s_\mu(\mathbf{x}),t_\mu(\mathbf{x});\mathbf{x})\Big) = 0,$$

where j denotes a real critical point (36.4.1) or (36.4.2), and μ denotes a critical point with complex t or s,t, connected with j by a steepest-descent path (that is, a path where $\Re\Phi =$ constant) in complex t or (s,t) space.

In the following subsections, only Stokes sets involving at least one real saddle are included unless stated otherwise.

36.5(ii) Cuspoids

$K = 1$. Airy Function

The Stokes set consists of the rays $\operatorname{ph} x = \pm 2\pi/3$ in the complex x-plane.

$K = 2$. Cusp

The Stokes set is itself a cusped curve, connected to the cusp of the bifurcation set:

36.5.2 $\quad y^3 = \frac{27}{4}\left(\sqrt{27} - 5\right)x^2 = 1.32403x^2.$

$K = 3$. Swallowtail

The Stokes set takes different forms for $z = 0$, $z < 0$, and $z > 0$.

For $z = 0$, the set consists of the two curves

36.5.3 $\quad x = B_{\pm}|y|^{4/3}, \quad B_{\pm} = 10^{-1/3}\left(2x_{\pm}^{4/3} - \frac{1}{2}x_{\pm}^{-2/3}\right),$

where $x_{\pm}$ are the two smallest positive roots of the equation

36.5.4 $\quad 80x^5 - 40x^4 - 55x^3 + 5x^2 + 20x - 1 = 0,$

and

36.5.5 $\quad B_{-} = -1.69916, \quad B_{+} = 0.33912.$

For $z \neq 0$, the Stokes set is expressed in terms of scaled coordinates

36.5.6 $\quad X = x/z^2, \quad Y = y/|z|^{3/2},$

by

36.5.7 $\quad X = \frac{9}{20} + 20u^4 - \frac{Y^2}{20u^2} + 6u^2 \operatorname{sign}(z),$

where u satisfies the equation

36.5.8 $\quad 16u^5 - \frac{Y^2}{10u} + 4u^3 \operatorname{sign}(z) - \frac{3}{10}|Y|\operatorname{sign}(z) + 4t^5 + 2t^3 \operatorname{sign}(z) + |Y|t^2 = 0,$

in which

36.5.9 $\quad t = -u + \left(\frac{|Y|}{10u} - u^2 - \frac{3}{10}\operatorname{sign}(z)\right)^{1/2}.$

For $z < 0$, there are two solutions u, provided that $|Y| > (\frac{2}{5})^{1/2}$. They generate a pair of cusp-edged sheets connected to the cusped sheets of the swallowtail bifurcation set (§36.4).

For $z > 0$ the Stokes set has two sheets. The first sheet corresponds to $x < 0$ and is generated as a solution of Equations (36.5.6)–(36.5.9). The second sheet corresponds to $x > 0$ and it intersects the bifurcation set (§36.4) smoothly along the line generated by $X = X_1 = 6.95643$, $|Y| = |Y_1| = 6.81337$. For $|Y| > Y_1$ the second sheet is generated by a second solution of (36.5.6)–(36.5.9), and for $|Y| < Y_1$ it is generated by the roots of the polynomial equation

36.5.10 $\quad 160u^6 + 40u^4 = Y^2.$

36.5(iii) Umbilics

Elliptic Umbilic Stokes Set (Codimension three)

This consists of three separate cusp-edged sheets connected to the cusp-edged sheets of the bifurcation set, and related by rotation about the z-axis by $2\pi/3$. One of the sheets is symmetrical under reflection in the plane $y = 0$, and is given by

36.5.11 $\quad \dfrac{x}{z^2} = -1 - 12u^2 + 8u - \left|\dfrac{y}{z^2}\right|\dfrac{\frac{1}{3} - u}{\left(u\left(\frac{2}{3} - u\right)\right)^{1/2}}.$

Here u is the root of the equation

36.5.12

$$8u^3 - 4u^2 - \left|\frac{y}{3z^2}\right|\left(\frac{u}{\frac{2}{3} - u}\right)^{1/2} = \frac{y^2}{6wz^4} - 2w^3 - 2w^2,$$

with

36.5.13 $\quad w = u - \dfrac{2}{3} + \left(\left(\dfrac{2}{3} - u\right)^2 + \left|\dfrac{y}{6z^2}\right|\left(\dfrac{\frac{2}{3}-u}{u}\right)^{1/2}\right)^{1/2},$

and such that

36.5.14 $\quad 0 < u < \frac{1}{6}.$

Hyperbolic Umbilic Stokes Set (Codimension three)

This consists of a cusp-edged sheet connected to the cusp-edged sheet of the bifurcation set and intersecting the smooth sheet of the bifurcation set. With coordinates

36.5.15 $\quad X = (x - y)/z^2, \quad Y = \frac{1}{2} + ((x+y)/z^2),$

the intersection lines with the bifurcation set are generated by $|X| = X_2 = 0.45148$, $Y = Y_2 = 0.59693$. Define

36.5.16

$$Y(u,X) = 8u - 24u^2 + X\frac{u - \frac{1}{6}}{\left(u\left(u - \frac{1}{3}\right)\right)^{1/2}},$$

$$f(u,X) = 16u^3 - 4u^2 - \frac{1}{6}|X|\left(\frac{u}{u - \frac{1}{3}}\right)^{1/2}.$$

When $|X| > X_2$ the Stokes set $Y_S(X)$ is given by

36.5.17 $\quad Y_S(X) = Y(u, |X|),$

where u is the root of the equation

36.5.18 $\quad f(u, X) = f(-u + \frac{1}{3}, X),$

such that $u > \frac{1}{3}$. This part of the Stokes set connects two complex saddles.

Alternatively, when $|X| < X_2$

36.5.19 $\quad Y_S(X) = Y(-u, -|X|),$

where u is the positive root of the equation

36.5.20 $\quad f(-u, X) = \dfrac{X^2}{12w} + 4w^3 - 2w^2,$

in which

36.5.21 $\quad w = (\frac{1}{3} + u)\left(1 - \left(1 - \dfrac{|X|}{12u^{1/2}(\frac{1}{3}+u)^{3/2}}\right)^{1/2}\right).$

36.5(iv) Visualizations

In Figures 36.5.1–36.5.6 the plane is divided into regions by the dashed curves (Stokes sets) and the continuous curves (bifurcation sets). Red and blue numbers in each region correspond, respectively, to the numbers of real and complex critical points that contribute to the asymptotics of the canonical integral away from the bifurcation sets. In Figure 36.5.4 the part of the Stokes surface inside the bifurcation set connects two complex saddles. The distribution of real and complex critical points in Figures 36.5.5 and 36.5.6 follows from consistency with Figure 36.5.1 and the fact that there are four real saddles in the inner regions.

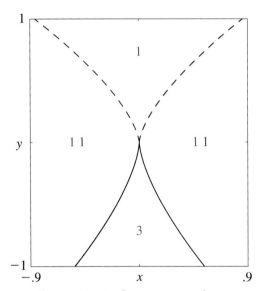

Figure 36.5.1: Cusp catastrophe.

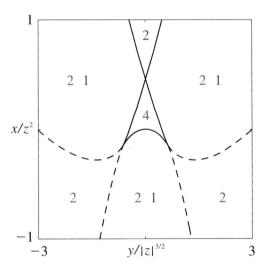

Figure 36.5.2: Swallowtail catastrophe with $z < 0$.

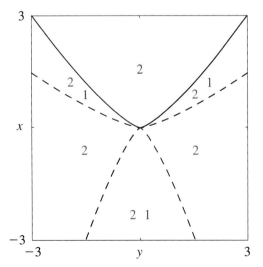

Figure 36.5.3: Swallowtail catastrophe with $z = 0$.

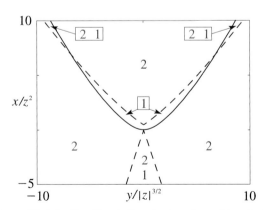

Figure 36.5.4: Swallowtail catastrophe with $z > 0$.

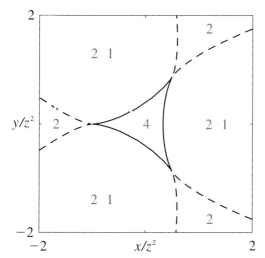

Figure 36.5.5: Elliptic umbilic catastrophe with $z = $ constant.

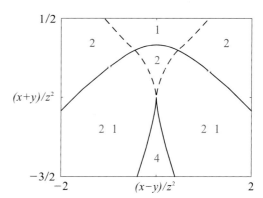

Figure 36.5.6: Hyperbolic umbilic catastrophe with $z = $ constant.

For additional figures see http://dlmf.nist.gov/36.5.iv.

36.6 Scaling Relations

Diffraction Catastrophe Scaling

36.6.1
$$\Psi_K(\mathbf{x}; k) = k^{\beta_K} \Psi_K(\mathbf{y}(k)),$$
$$\Psi^{(\mathrm{U})}(\mathbf{x}; k) = k^{\beta^{(\mathrm{U})}} \Psi^{(\mathrm{U})}\!\left(\mathbf{y}^{(\mathrm{U})}(k)\right),$$

where

36.6.2
cuspoids: $\mathbf{y}(k) = (x_1 k^{\gamma_{1K}}, x_2 k^{\gamma_{2K}}, \ldots, x_K k^{\gamma_{KK}})$,
umbilics: $\mathbf{y}^{(\mathrm{U})}(k) = \left(x k^{2/3}, y k^{2/3}, z k^{1/3}\right)$.

Indices for k-Scaling of Magnitude of Ψ_K or $\Psi^{(\mathrm{U})}$ (Singularity Index)

36.6.3 cuspoids: $\beta_K = \dfrac{K}{2(K+2)}$, umbilics: $\beta^{(\mathrm{U})} = \dfrac{1}{3}$.

Indices for k-Scaling of Coordinates x_m

36.6.4
cuspoids: $\gamma_{mK} = 1 - \dfrac{m}{K+2}$,
umbilics: $\gamma_x^{(\mathrm{U})} = \tfrac{2}{3}$, $\gamma_y^{(\mathrm{U})} = \tfrac{2}{3}$, $\gamma_z^{(\mathrm{U})} = \tfrac{1}{3}$.

Indices for k-Scaling of x Hypervolume

36.6.5
cuspoids: $\gamma_K = \sum_{m=1}^{K} \gamma_{mK} = \dfrac{K(K+3)}{2(K+2)}$,
umbilics: $\gamma^{(\mathrm{U})} = \sum_{m=1}^{3} \gamma_m^{(\mathrm{U})} = \tfrac{5}{3}$.

Table 36.6.1: Special cases of scaling exponents for cuspoids.

singularity	K	β_K	γ_{1K}	γ_{2K}	γ_{3K}	γ_K
fold	1	$\tfrac{1}{6}$	$\tfrac{2}{3}$	—	—	$\tfrac{2}{3}$
cusp	2	$\tfrac{1}{4}$	$\tfrac{3}{4}$	$\tfrac{1}{2}$	—	$\tfrac{5}{4}$
swallowtail	3	$\tfrac{3}{10}$	$\tfrac{4}{5}$	$\tfrac{3}{5}$	$\tfrac{2}{5}$	$\tfrac{9}{5}$

For the results in this section and more extensive lists of exponents see Berry (1977) and Varčenko (1976).

36.7 Zeros

36.7(i) Fold Canonical Integral

This is the Airy function Ai (§9.2).

36.7(ii) Cusp Canonical Integral

This is (36.2.4) and (36.2.1) with $K = 2$.

The zeros in Table 36.7.1 are points in the $\mathbf{x} = (x, y)$ plane, where $\operatorname{ph} \Psi_2(\mathbf{x})$ is undetermined. All zeros have $y < 0$, and fall into two classes. Inside the cusp, that is, for $x^2 < 8|y|^3/27$, the zeros form pairs lying in curved rows. Close to the y-axis the approximate location of these zeros is given by

36.7.1
$$y_m = -\sqrt{2\pi(2m+1)}, \qquad m = 1, 2, 3, \ldots,$$
$$x^{\pm}_{m,n} = \sqrt{\dfrac{2}{-y_m}}\left(2n + \tfrac{1}{2} + (-1)^m \tfrac{1}{2} \pm \tfrac{1}{4}\right)\pi,$$
$$m = 1, 2, 3, \ldots, \; n = 0, \pm 1, \pm 2, \ldots.$$

Table 36.7.1: Zeros of cusp diffraction catastrophe to 5D.

Zeros $\left\{\begin{smallmatrix}x\\y\end{smallmatrix}\right\}$ inside, and zeros $\left[\begin{smallmatrix}x\\y\end{smallmatrix}\right]$ outside, the cusp $x^2 = \frac{8}{27}|y|^3$.

$\left\{\begin{matrix}\pm 0.52768\\-4.37804\end{matrix}\right\}$	$\left[\begin{matrix}\pm 2.35218\\-1.74360\end{matrix}\right]$				
$\left\{\begin{matrix}\pm 1.41101\\-5.55470\end{matrix}\right\}$	$\left\{\begin{matrix}\pm 2.36094\\-5.52321\end{matrix}\right\}$	$\left[\begin{matrix}\pm 4.42707\\-3.05791\end{matrix}\right]$			
$\left\{\begin{matrix}\pm 0.43039\\-6.64285\end{matrix}\right\}$	$\left\{\begin{matrix}\pm 3.06389\\-6.44624\end{matrix}\right\}$	$\left\{\begin{matrix}\pm 3.95806\\-6.40312\end{matrix}\right\}$	$\left[\begin{matrix}\pm 6.16185\\-4.03551\end{matrix}\right]$		
$\left\{\begin{matrix}\pm 1.21605\\-7.49906\end{matrix}\right\}$	$\left\{\begin{matrix}\pm 2.02922\\-7.48629\end{matrix}\right\}$	$\left\{\begin{matrix}\pm 4.56537\\-7.19629\end{matrix}\right\}$	$\left\{\begin{matrix}\pm 5.42206\\-7.14718\end{matrix}\right\}$	$\left[\begin{matrix}\pm 7.72352\\-4.84817\end{matrix}\right]$	
$\left\{\begin{matrix}\pm 0.38488\\-8.31916\end{matrix}\right\}$	$\left\{\begin{matrix}\pm 2.71193\\-8.22315\end{matrix}\right\}$	$\left\{\begin{matrix}\pm 3.49286\\-8.20326\end{matrix}\right\}$	$\left\{\begin{matrix}\pm 5.96669\\-7.85723\end{matrix}\right\}$	$\left\{\begin{matrix}\pm 6.79538\\-7.80456\end{matrix}\right\}$	$\left[\begin{matrix}\pm 9.17308\\-5.55831\end{matrix}\right]$

More general asymptotic formulas are given in Kaminski and Paris (1999). Just outside the cusp, that is, for $x^2 > 8|y|^3/27$, there is a single row of zeros on each side. With $n = 0, 1, 2, \ldots$, they are located approximately at

36.7.2
$$x_n = \pm \left(\frac{8}{27}\right)^{1/2} |y_n|^{3/2}(1 + \xi_n),$$
$$y_n = -\left(\frac{3\pi(8n+5)}{9+8\xi_n}\right)^{1/2},$$

where ξ_n is the real solution of

36.7.3
$$\frac{3\pi(8n+5)}{9+8\xi_n}\xi_n^{3/2} = \frac{27}{16}\left(\frac{3}{2}\right)^{1/2}\left(\ln\left(\frac{1}{\xi_n}\right) + 3\ln\left(\frac{3}{2}\right)\right).$$

For a more extensive asymptotic analysis and further tabulations, see Kaminski and Paris (1999).

36.7(iii) Elliptic Umbilic Canonical Integral

This is (36.2.5) with (36.2.2). The zeros are lines in $\mathbf{x} = (x, y, z)$ space where $\mathrm{ph}\,\Psi^{(\mathrm{E})}(\mathbf{x})$ is undetermined. Deep inside the bifurcation set, that is, inside the three-cusped astroid (36.4.10) and close to the part of the z-axis that is far from the origin, the zero contours form an array of rings close to the planes

36.7.4
$$z_n = \pm 3(\tfrac{1}{4}\pi(2n - \tfrac{1}{2}))^{1/3}$$
$$= 3.48734(n - \tfrac{1}{4})^{1/3}, \qquad n = 1, 2, 3, \ldots.$$

Near $z = z_n$, and for small x and y, the modulus $|\Psi^{(\mathrm{E})}(\mathbf{x})|$ has the symmetry of a lattice with a rhombohedral unit cell that has a mirror plane and an inverse threefold axis whose z and x repeat distances are given by

36.7.5
$$\Delta z = \frac{9\pi}{2z_n^2}, \qquad \Delta x = \frac{6\pi}{z_n}.$$

The zeros are approximated by solutions of the equation

36.7.6
$$\exp\left(-2\pi i\left(\frac{z - z_n}{\Delta z} + \frac{2x}{\Delta x}\right)\right)$$
$$\times \left(2\exp\left(\frac{-6\pi i x}{\Delta x}\right)\cos\left(\frac{2\sqrt{3}\pi y}{\Delta x}\right) + 1\right)$$
$$= \sqrt{3}.$$

The rings are almost circular (radii close to $(\Delta x)/9$ and varying by less than 1%), and almost flat (deviating from the planes z_n by at most $(\Delta z)/36$). Away from the z-axis and approaching the cusp lines (ribs) (36.4.11), the lattice becomes distorted and the rings are deformed, eventually joining to form "hairpins" whose arms become the pairs of zeros (36.7.1) of the cusp canonical integral. In the symmetry planes (e.g., $y = 0$), the number of rings in the mth row, measured from the origin and before the transition to hairpins, is given by

36.7.7
$$n_{\max}(m) = \left\lfloor \tfrac{256}{13}m - \tfrac{269}{52}\right\rfloor.$$

Outside the bifurcation set (36.4.10), each rib is flanked by a series of zero lines in the form of curly "antelope horns" related to the "outside" zeros (36.7.2) of the cusp canonical integral. There are also three sets of zero lines in the plane $z = 0$ related by $2\pi/3$ rotation; these are zeros of (36.2.20), whose asymptotic form in polar coordinates ($x = r\cos\theta$, $y = r\sin\theta$) is given by

36.7.8
$$r = 3\left(\frac{(2n-1)\pi}{4|\sin(\tfrac{3}{2}\theta)|}\right)^{2/3}(1 + O(n^{-1})), \quad n \to \infty.$$

36.7(iv) Swallowtail and Hyperbolic Umbilic Canonical Integrals

The zeros of these functions are curves in $\mathbf{x} = (x, y, z)$ space; see Nye (2007) for Φ_3 and Nye (2006) for $\Phi^{(\mathrm{H})}$.

36.8 Convergent Series Expansions

36.8.1
$$\Psi_K(\mathbf{x}) = \frac{2}{K+2} \sum_{n=0}^{\infty} \exp\left(i\frac{\pi(2n+1)}{2(K+2)}\right) \Gamma\left(\frac{2n+1}{K+2}\right) a_{2n}(\mathbf{x}), \qquad K \text{ even,}$$

$$\Psi_K(\mathbf{x}) = \frac{2}{K+2} \sum_{n=0}^{\infty} i^n \cos\left(\frac{\pi(n(K+1)-1)}{2(K+2)}\right) \Gamma\left(\frac{n+1}{K+2}\right) a_n(\mathbf{x}), \qquad K \text{ odd,}$$

where

36.8.2
$$a_0(\mathbf{x}) = 1, \quad a_{n+1}(\mathbf{x}) = \frac{i}{n+1} \sum_{p=0}^{\min(n, K-1)} (p+1) x_{p+1} a_{n-p}(\mathbf{x}), \qquad n = 0, 1, 2, \ldots.$$

For multinomial power series for $\Psi_K(\mathbf{x})$, see Connor and Curtis (1982).

36.8.3
$$\frac{3^{2/3}}{4\pi^2} \Psi^{(\mathrm{H})}\left(3^{1/3}\mathbf{x}\right) = \operatorname{Ai}(x)\operatorname{Ai}(y) \sum_{n=0}^{\infty}(-3^{-1/3}iz)^n \frac{c_n(x)c_n(y)}{n!} + \operatorname{Ai}(x)\operatorname{Ai}'(y)\sum_{n=2}^{\infty}(-3^{-1/3}iz)^n \frac{c_n(x)d_n(y)}{n!}$$
$$+ \operatorname{Ai}'(x)\operatorname{Ai}(y) \sum_{n=2}^{\infty}(-3^{-1/3}iz)^n \frac{d_n(x)c_n(y)}{n!} + \operatorname{Ai}'(x)\operatorname{Ai}'(y)\sum_{n=1}^{\infty}(-3^{-1/3}iz)^n \frac{d_n(x)d_n(y)}{n!},$$

and

36.8.4
$$\Psi^{(\mathrm{E})}(\mathbf{x}) = 2\pi^2 \left(\frac{2}{3}\right)^{2/3} \sum_{n=0}^{\infty} \frac{\left(-i(2/3)^{2/3}z\right)^n}{n!} \Re\left(f_n\left(\frac{x+iy}{12^{1/3}}, \frac{x-iy}{12^{1/3}}\right)\right),$$

where

36.8.5
$$f_n(\zeta, \zeta^*)$$
$$= c_n(\zeta)c_n(\zeta^*)\operatorname{Ai}(\zeta)\operatorname{Bi}(\zeta^*) + c_n(\zeta)d_n(\zeta^*)\operatorname{Ai}(\zeta)\operatorname{Bi}'(\zeta^*) + d_n(\zeta)c_n(\zeta^*)\operatorname{Ai}'(\zeta)\operatorname{Bi}(\zeta^*) + d_n(\zeta)d_n(\zeta^*)\operatorname{Ai}'(\zeta)\operatorname{Bi}'(\zeta^*),$$

with asterisks denoting complex conjugates, and

36.8.6
$$c_0(t) = 1, \quad d_0(t) = 0, \quad c_{n+1}(t) = c_n'(t) + td_n(t), \quad d_{n+1}(t) = c_n(t) + d_n'(t).$$

36.9 Integral Identities

36.9.1
$$|\Psi_1(x)|^2 = 2^{5/3} \int_0^{\infty} \Psi_1\left(2^{2/3}(3u^2 + x)\right) du;$$

equivalently,

36.9.2
$$(\operatorname{Ai}(x))^2 = \frac{2^{2/3}}{\pi} \int_0^{\infty} \operatorname{Ai}\left(2^{2/3}(u^2 + x)\right) du.$$

36.9.3
$$|\Psi_1(x)|^2 = \sqrt{\frac{8\pi}{3}} \int_0^{\infty} u^{-1/2} \cos\left(2u(x + u^2) + \tfrac{1}{4}\pi\right) du.$$

36.9.4
$$|\Psi_2(x, y)|^2 = \int_0^{\infty} \left(\Psi_1\left(\frac{4u^3 + 2uy + x}{u^{1/3}}\right) + \Psi_1\left(\frac{4u^3 + 2uy - x}{u^{1/3}}\right)\right) \frac{du}{u^{1/3}}.$$

36.9.5
$$|\Psi_2(x, y)|^2 = 2 \int_0^{\infty} \cos(2xu) \Psi_1\left(2u^{2/3}(y + 2u^2)\right) \frac{du}{u^{1/3}}.$$

36.9.6
$$|\Psi_3(x, y, z)|^2 = 2^{4/5} \int_{-\infty}^{\infty} \Psi_3\left(2^{4/5}(x + 2uy + 3u^2 z + 5u^4), 0, 2^{2/5}(z + 10u^2)\right) du.$$

36.9.7
$$|\Psi_3(x, y, z)|^2 = \frac{2^{7/4}}{5^{1/4}} \int_0^{\infty} \Re\left(e^{2iu(u^4 + zu^2 + x)} \Psi_2\left(\frac{2^{7/4}}{5^{1/4}} y u^{3/4}, \sqrt{\frac{2u}{5}}(3z + 10u^2)\right)\right) \frac{du}{u^{1/4}}.$$

36.9.8 $\quad \left|\Psi^{(\mathrm{H})}(x,y,z)\right|^2 = 8\pi^2 \left(\frac{2}{9}\right)^{1/3} \int_{-\infty}^{\infty}\int_{-\infty}^{\infty} \operatorname{Ai}\left(\left(\frac{4}{3}\right)^{1/3}(x+zv+3u^2)\right)\operatorname{Ai}\left(\left(\frac{4}{3}\right)^{1/3}(y+zu+3v^2)\right) du\, dv.$

36.9.9 $\quad \left|\Psi^{(\mathrm{E})}(x,y,z)\right|^2 = \frac{8\pi^2}{3^{2/3}} \int_0^{\infty}\int_0^{2\pi} \Re\left(\operatorname{Ai}\left(\frac{1}{3^{1/3}}\left(x+iy+2zu\exp(i\theta)+3u^2\exp(-2i\theta)\right)\right)\right.$
$\qquad\qquad \times \left.\operatorname{Bi}\left(\frac{1}{3^{1/3}}\left(x-iy+2zu\exp(-i\theta)+3u^2\exp(2i\theta)\right)\right)\right) u\, du\, d\theta.$

For these results and also integrals over doubly-infinite intervals see Berry and Wright (1980). This reference also provides a physical interpretation in terms of Lagrangian manifolds and Wigner functions in phase space.

36.10 Differential Equations

36.10(i) Equations for $\Psi_K(\mathbf{x})$

In terms of the normal form (36.2.1) the $\Psi_K(\mathbf{x})$ satisfy the operator equation

36.10.1 $\quad \Phi'_K\left(-i\frac{\partial}{\partial x_1};\mathbf{x}\right)\Psi_K(\mathbf{x}) = 0,$

or explicitly,

36.10.2
$$\frac{\partial^{K+1}\Psi_K(\mathbf{x})}{\partial x_1^{K+1}} + \sum_{m=1}^{K}(-i)^{m-K-2}\left(\frac{mx_m}{K+2}\right)\frac{\partial^{m-1}\Psi_K(\mathbf{x})}{\partial x_1^{m-1}} = 0.$$

Special Cases

$K=1$, fold: (36.10.1) becomes Airy's equation (§9.2(i))

36.10.3 $\quad \dfrac{\partial^2 \Psi_1}{\partial x^2} - \dfrac{x}{3}\Psi_1 = 0.$

$K=2$, cusp:

36.10.4 $\quad \dfrac{\partial^3 \Psi_2}{\partial x^3} - \dfrac{1}{2}y\dfrac{\partial \Psi_2}{\partial x} - \dfrac{i}{4}x\,\Psi_2 = 0.$

$K=3$, swallowtail:

36.10.5 $\quad \dfrac{\partial^4 \Psi_3}{\partial x^4} - \dfrac{3}{5}z\dfrac{\partial^2 \Psi_3}{\partial x^2} - \dfrac{2i}{5}y\dfrac{\partial \Psi_3}{\partial x} + \dfrac{1}{5}x\,\Psi_3 = 0.$

36.10(ii) Partial Derivatives with Respect to the x_n

36.10.6 $\quad \dfrac{\partial^{ln}\Psi_K}{\partial x_m^{ln}} = i^{n(l-m)}\dfrac{\partial^{mn}\Psi_K}{\partial x_l^{mn}}, \quad 1\leq m\leq K, 1\leq l\leq K.$

Special Cases

$K=1$, fold: (36.10.6) is an identity.
$K=2$, cusp:

36.10.7 $\quad \dfrac{\partial^{2n}\Psi_2}{\partial x^{2n}} = i^n\dfrac{\partial^n \Psi_2}{\partial y^n}.$

$K=3$, swallowtail:

36.10.8 $\quad \dfrac{\partial^{2n}\Psi_3}{\partial x^{2n}} = i^n\dfrac{\partial^n \Psi_3}{\partial y^n},$

36.10.9 $\quad \dfrac{\partial^{3n}\Psi_3}{\partial x^{3n}} = (-1)^n\dfrac{\partial^n \Psi_3}{\partial z^n},$

36.10.10 $\quad \dfrac{\partial^{3n}\Psi_3}{\partial y^{3n}} = i^n\dfrac{\partial^{2n}\Psi_3}{\partial z^{2n}}.$

36.10(iii) Operator Equations

In terms of the normal forms (36.2.2) and (36.2.3), the $\Psi^{(\mathrm{U})}(\mathbf{x})$ satisfy the following operator equations

36.10.11
$$\Phi_s^{(\mathrm{U})}\left(-i\frac{\partial}{\partial x},-i\frac{\partial}{\partial y};\mathbf{x}\right)\Psi^{(\mathrm{U})}(\mathbf{x}) = 0,$$
$$\Phi_t^{(\mathrm{U})}\left(-i\frac{\partial}{\partial x},-i\frac{\partial}{\partial y};\mathbf{x}\right)\Psi^{(\mathrm{U})}(\mathbf{x}) = 0,$$

where

36.10.12
$$\Phi_s^{(\mathrm{U})}(s,t;\mathbf{x}) = \frac{\partial}{\partial s}\Phi^{(\mathrm{U})}(s,t;\mathbf{x}),$$
$$\Phi_t^{(\mathrm{U})}(s,t;\mathbf{x}) = \frac{\partial}{\partial t}\Phi^{(\mathrm{U})}(s,t;\mathbf{x}).$$

Explicitly,

36.10.13 $\quad 6\dfrac{\partial^2 \Psi^{(\mathrm{E})}}{\partial x\,\partial y} - 2iz\dfrac{\partial \Psi^{(\mathrm{E})}}{\partial y} + y\,\Psi^{(\mathrm{E})} = 0,$

36.10.14
$$3\left(\frac{\partial^2 \Psi^{(\mathrm{E})}}{\partial x^2} - \frac{\partial^2 \Psi^{(\mathrm{E})}}{\partial y^2}\right) + 2iz\frac{\partial \Psi^{(\mathrm{H})}}{\partial x} - x\,\Psi^{(\mathrm{E})} = 0.$$

36.10.15 $\quad 3\dfrac{\partial^2 \Psi^{(\mathrm{H})}}{\partial x^2} + iz\dfrac{\partial \Psi^{(\mathrm{H})}}{\partial y} - x\,\Psi^{(\mathrm{H})} = 0,$

36.10.16 $\quad 3\dfrac{\partial^2 \Psi^{(\mathrm{H})}}{\partial y^2} + iz\dfrac{\partial \Psi^{(\mathrm{H})}}{\partial x} - y\,\Psi^{(\mathrm{H})} = 0.$

36.10(iv) Partial z-Derivatives

36.10.17 $\quad i\dfrac{\partial \Psi^{(\mathrm{E})}}{\partial z} = \dfrac{\partial^2 \Psi^{(\mathrm{E})}}{\partial x^2} + \dfrac{\partial^2 \Psi^{(\mathrm{E})}}{\partial y^2},$

36.10.18 $\quad i\dfrac{\partial \Psi^{(\mathrm{H})}}{\partial z} = \dfrac{\partial^2 \Psi^{(\mathrm{H})}}{\partial x\,\partial y}.$

Equation (36.10.17) is the *paraxial wave equation*.

36.11 Leading-Order Asymptotics

With real critical points (36.4.1) ordered so that

36.11.1
$$t_1(\mathbf{x}) < t_2(\mathbf{x}) < \cdots < t_{j_{\max}}(\mathbf{x}),$$

and far from the bifurcation set, the cuspoid canonical integrals are approximated by

36.11.2
$$\Psi_K(\mathbf{x}) = \sqrt{2\pi} \sum_{j=1}^{j_{\max}(\mathbf{x})} \exp\bigl(i\bigl(\Phi_K(t_j(\mathbf{x});\mathbf{x}) + \tfrac{1}{4}\pi(-1)^{j+K+1}\bigr)\bigr) \left|\frac{\partial^2 \Phi_K(t_j(\mathbf{x});\mathbf{x})}{\partial t^2}\right|^{-1/2} (1+o(1)).$$

Asymptotics along Symmetry Lines

36.11.3
$$\Psi_2(0,y) = \begin{cases} \sqrt{\pi/y}\,\bigl(\exp(\tfrac{1}{4}i\pi) + o(1)\bigr), & y \to +\infty, \\ \sqrt{\pi/|y|}\,\exp(-\tfrac{1}{4}i\pi)\bigl(1 + i\sqrt{2}\exp(-\tfrac{1}{4}iy^2)\bigr) + o(1), & y \to -\infty. \end{cases}$$

36.11.4
$$\Psi_3(x,0,0) = \frac{\sqrt{2\pi}}{(5|x|^3)^{1/8}} \begin{cases} \exp\bigl(-2\sqrt{2}(x/5)^{5/4}\bigr)\bigl(\cos(2\sqrt{2}(x/5)^{5/4} - \tfrac{1}{8}\pi) + o(1)\bigr), & x \to +\infty, \\ \cos\bigl(4(|x|/5)^{5/4} - \tfrac{1}{4}\pi\bigr) + o(1), & x \to -\infty. \end{cases}$$

36.11.5
$$\Psi_3(0,y,0) = \Psi_3^*(0,-y,0) = \exp(\tfrac{1}{4}i\pi)\sqrt{\pi/y}\,\bigl(1 - (i/\sqrt{3})\exp\bigl(\tfrac{3}{2}i(2y/5)^{5/3}\bigr) + o(1)\bigr), \qquad y \to +\infty.$$

36.11.6
$$\Psi_3(0,0,z) = \frac{\Gamma(\tfrac{1}{3})}{|z|^{1/3}\sqrt{3}} + \begin{cases} o(1), & z \to +\infty, \\ \dfrac{2\sqrt{\pi}5^{1/4}}{(3|z|)^{3/4}}\left(\cos\left(\dfrac{2}{3}\left(\dfrac{3|z|}{5}\right)^{5/2} - \dfrac{1}{4}\pi\right) + o(1)\right), & z \to -\infty. \end{cases}$$

36.11.7
$$\Psi^{(E)}(0,0,z) = \frac{\pi}{z}\left(i + \sqrt{3}\exp\left(\frac{4}{27}iz^3\right) + o(1)\right), \qquad z \to \pm\infty,$$

36.11.8
$$\Psi^{(H)}(0,0,z) = \frac{2\pi}{z}\left(1 - \frac{i}{\sqrt{3}}\exp\left(\frac{1}{27}iz^3\right) + o(1)\right), \qquad z \to \pm\infty.$$

Applications

36.12 Uniform Approximation of Integrals

36.12(i) General Theory for Cuspoids

The canonical integrals (36.2.4) provide a basis for uniform asymptotic approximations of oscillatory integrals. In the cuspoid case (one integration variable)

36.12.1
$$I(\mathbf{y},k) = \int_{-\infty}^{\infty} \exp(ikf(u;\mathbf{y}))g(u,\mathbf{y})\,du,$$

where k is a large real parameter and $\mathbf{y} = \{y_1, y_2, \dots\}$ is a set of additional (nonasymptotic) parameters. As $\mathbf{y}$ varies as many as $K+1$ (real or complex) critical points of the smooth phase function f can coalesce in clusters of two or more. The function g has a smooth amplitude. Also, f is real analytic, and $\partial^{K+2}f/\partial u^{K+2} > 0$ for all $\mathbf{y}$ such that all $K+1$ critical points coincide. If $\partial^{K+2}f/\partial u^{K+2} < 0$, then we may evaluate the complex conjugate of I for real values of $\mathbf{y}$ and g, and obtain I by conjugation and analytic continuation. The critical points $u_j(\mathbf{y})$, $1 \le j \le K+1$, are defined by

36.12.2
$$\frac{\partial}{\partial u}f(u_j(\mathbf{y});\mathbf{y}) = 0.$$

The leading-order uniform asymptotic approximation is given by

36.12.3
$$I(\mathbf{y},k) = \frac{\exp(ikA(\mathbf{y}))}{k^{1/(K+2)}} \sum_{m=0}^{K} \frac{a_m(\mathbf{y})}{k^{m/(K+2)}} \left(\delta_{m,0} - (1-\delta_{m,0})\,i\frac{\partial}{\partial z_m}\right) \Psi_K(\mathbf{z}(\mathbf{y});k)\left(1 + O\left(\frac{1}{k}\right)\right),$$

where $A(\mathbf{y})$, $\mathbf{z}(\mathbf{y},k)$, $a_m(\mathbf{y})$ are as follows. Define a mapping $u(t;\mathbf{y})$ by relating $f(u;\mathbf{y})$ to the normal form (36.2.1) of $\Phi_K(t;\mathbf{x})$ in the following way:

36.12.4 $\quad f(u(t,\mathbf{y});\mathbf{y}) = A(\mathbf{y}) + \Phi_K(t;\mathbf{x}(\mathbf{y}))$,

with the $K+1$ functions $A(\mathbf{y})$ and $\mathbf{x}(\mathbf{y})$ determined by correspondence of the $K+1$ critical points of f and Φ_K. Then

36.12.5 $\quad f(u_j(\mathbf{y});\mathbf{y}) = A(\mathbf{y}) + \Phi_K(t_j(\mathbf{x}(\mathbf{y}));\mathbf{x}(\mathbf{y}))$,

where $t_j(\mathbf{x})$, $1 \leq j \leq K+1$, are the critical points of Φ_K, that is, the solutions (real and complex) of (36.4.1). Correspondence between the $u_j(\mathbf{y})$ and the $t_j(\mathbf{x})$ is established by the order of critical points along the real axis when $\mathbf{y}$ and $\mathbf{x}$ are such that these critical points are all real, and by continuation when some or all of the critical points are complex. The branch for $\mathbf{x}(\mathbf{y})$ is such that $\mathbf{x}$ is real when $\mathbf{y}$ is real. In consequence,

36.12.6 $\quad A(\mathbf{y}) = f(u(0,\mathbf{y});\mathbf{y})$,

36.12.7 $\quad \begin{aligned}\mathbf{z}(\mathbf{y};k) &= \{z_1(\mathbf{y};k), z_2(\mathbf{y};k), \ldots, z_K(\mathbf{y};k)\}, \\ z_m(\mathbf{y};k) &= x_m(\mathbf{y})k^{1-(m/(K+2))},\end{aligned}$

36.12.8

$$a_m(\mathbf{y}) = \sum_{n=1}^{K+1} \frac{P_{mn}(\mathbf{y})G_n(\mathbf{y})}{(t_n(\mathbf{x}(\mathbf{y})))^{m+1} \prod_{\substack{l=1\\l\neq n}}^{K+1} (t_n(\mathbf{x}(\mathbf{y})) - t_l(\mathbf{x}(\mathbf{y})))},$$

where

36.12.9

$$P_{mn}(\mathbf{y}) = (t_n(\mathbf{x}(\mathbf{y})))^{K+1} + \sum_{l=m+2}^{K} \frac{l}{K+2} x_l(\mathbf{y})(t_n(\mathbf{x}(\mathbf{y})))^{l-1},$$

and

36.12.10

$$G_n(\mathbf{y}) = g(t_n(\mathbf{y}),\mathbf{y})\sqrt{\frac{\partial^2 \Phi_K(t_n(\mathbf{x}(\mathbf{y}));\mathbf{x}(\mathbf{y}))/\partial t^2}{\partial^2 f(u_n(\mathbf{y}))/\partial u^2}}.$$

In (36.12.10), both second derivatives vanish when critical points coalesce, but their ratio remains finite. The square roots are real and positive when $\mathbf{y}$ is such that all the critical points are real, and are defined by analytic continuation elsewhere. The quantities $a_m(\mathbf{y})$ are real for real $\mathbf{y}$ when g is real analytic.

This technique can be applied to generate a hierarchy of approximations for the diffraction catastrophes $\Psi_K(\mathbf{x};k)$ in (36.2.10) away from $\mathbf{x} = 0$, in terms of canonical integrals $\Psi_J(\xi(\mathbf{x};k))$ for $J < K$. For example, the diffraction catastrophe $\Psi_2(x,y;k)$ defined by (36.2.10), and corresponding to the Pearcey integral (36.2.14), can be approximated by the Airy function $\Psi_1(\xi(x,y;k))$ when k is large, provided that x and y are not small. For details of this example, see Paris (1991).

For further information see Berry and Howls (1993).

36.12(ii) Special Case

For $K = 1$, with a single parameter y, let the two critical points of $f(u;y)$ be denoted by $u_\pm(y)$, with $u_+ > u_-$ for those values of y for which these critical points are real. Then

36.12.11

$$I(y,k) = \frac{\Delta^{1/4}\pi\sqrt{2}}{k^{1/3}}\exp\left(ik\widetilde{f}\right)\left(\left(\frac{g_+}{\sqrt{f_+''}} + \frac{g_-}{\sqrt{-f_-''}}\right)\mathrm{Ai}\!\left(-k^{2/3}\Delta\right)\left(1 + O\!\left(\frac{1}{k}\right)\right) - i\left(\frac{g_+}{\sqrt{f_+''}} - \frac{g_-}{\sqrt{-f_-''}}\right)\frac{\mathrm{Ai}'\!\left(-k^{2/3}\Delta\right)}{k^{1/3}\Delta^{1/2}}\left(1 + O\!\left(\frac{1}{k}\right)\right)\right),$$

where

36.12.12 $\quad\begin{aligned}\widetilde{f} &= \tfrac{1}{2}(f(u_+(y),y) + f(u_-(y),y)), \\ g_\pm &= g(u_\pm(y),y), \quad f_\pm'' = \frac{\partial^2}{\partial u^2}f(u_\pm(y),y), \\ \Delta &= \left(\tfrac{3}{4}(f(u_-(y),y) - f(u_+(y),y))\right)^{2/3}.\end{aligned}$

For Ai and Ai$'$ see §9.2. Branches are chosen so that Δ is real and positive if the critical points are real, or real and negative if they are complex. The coefficients of Ai and Ai$'$ are real if y is real and g is real analytic. Also, $\Delta^{1/4}/\sqrt{f_+''}$ and $\Delta^{1/4}/\sqrt{-f_-''}$ are chosen to be positive real when y is such that both critical points are real, and by analytic continuation otherwise.

36.12(iii) Additional References

For further information concerning integrals with several coalescing saddle points see Arnol'd et al. (1988), Berry and Howls (1993, 1994), Bleistein (1967), Duistermaat (1974), Ludwig (1966), Olde Daalhuis (2000), and Ursell (1972, 1980).

36.13 Kelvin's Ship-Wave Pattern

A ship moving with constant speed V on deep water generates a surface gravity wave. In a reference frame where the ship is at rest we use polar coordinates r and ϕ with $\phi = 0$ in the direction of the velocity of the water relative to the ship. Then with g denoting the acceleration due to gravity, the wave height is approximately given by

36.13.1 $$z(\phi, \rho) = \int_{-\pi/2}^{\pi/2} \cos\left(\rho \frac{\cos(\theta + \phi)}{\cos^2 \theta}\right) d\theta,$$

where

36.13.2 $$\rho = gr/V^2.$$

The integral is of the form of the real part of (36.12.1) with $y = \phi$, $u = \theta$, $g = 1$, $k = \rho$, and

36.13.3 $$f(\theta, \phi) = -\frac{\cos(\theta + \phi)}{\cos^2 \theta}.$$

When $\rho > 1$, that is, everywhere except close to the ship, the integrand oscillates rapidly. There are two stationary points, given by

36.13.4 $$\begin{aligned}\theta_+(\phi) &= \tfrac{1}{2}(\arcsin(3 \sin \phi) - \phi),\\ \theta_-(\phi) &= \tfrac{1}{2}(\pi - \phi - \arcsin(3 \sin \phi)).\end{aligned}$$

These coalesce when

36.13.5 $$|\phi| = \phi_c = \arcsin\left(\tfrac{1}{3}\right) = 19°.47122.$$

This is the angle of the familiar V-shaped wake. The wake is a caustic of the "rays" defined by the dispersion relation ("Hamiltonian") giving the frequency ω as a function of wavevector $\mathbf{k}$:

36.13.6 $$\omega(\mathbf{k}) = \sqrt{gk} + \mathbf{V} \cdot \mathbf{k}.$$

Here $k = |\mathbf{k}|$, and $\mathbf{V}$ is the ship velocity (so that $V = |\mathbf{V}|$).

The disturbance $z(\rho, \phi)$ can be approximated by the method of uniform asymptotic approximation for the case of two coalescing stationary points (36.12.11), using the fact that $\theta_\pm(\phi)$ are real for $|\phi| < \phi_c$ and complex for $|\phi| > \phi_c$. (See also §2.4(v).) Then with the definitions (36.12.12), and the real functions

36.13.7 $$\begin{aligned}u(\phi) &= \sqrt{\frac{\Delta^{1/2}(\phi)}{2}}\left(\frac{1}{\sqrt{f''_+(\phi)}} + \frac{1}{\sqrt{-f''_-(\phi)}}\right),\\ v(\phi) &= \sqrt{\frac{1}{2\Delta^{1/2}(\phi)}}\left(\frac{1}{\sqrt{f''_+(\phi)}} - \frac{1}{\sqrt{-f''_-(\phi)}}\right),\end{aligned}$$

the disturbance is

36.13.8
$$\begin{aligned}z(\rho, \phi) = 2\pi \Big(&\rho^{-1/3} u(\phi) \cos\left(\rho \widetilde{f}(\phi)\right) \mathrm{Ai}\left(-\rho^{2/3} \Delta(\phi)\right)\\ &\times (1 + O(1/\rho))\\ +\, &\rho^{-2/3} v(\phi) \sin\left(\rho \widetilde{f}(\phi)\right) \mathrm{Ai}'\left(-\rho^{2/3} \Delta(\phi)\right)\\ &\times (1 + O(1/\rho))\Big), \qquad \rho \to \infty.\end{aligned}$$

See Figure 36.13.1.

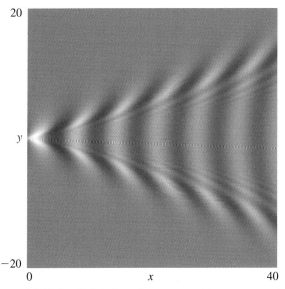

Figure 36.13.1: Kelvin's ship wave pattern, computed from the uniform asymptotic approximation (36.13.8), as a function of $x = \rho \cos \phi$, $y = \rho \sin \phi$.

For further information see Lord Kelvin (1891, 1905) and Ursell (1960, 1994).

36.14 Other Physical Applications

36.14(i) Caustics

The physical manifestations of bifurcation sets are caustics. These are the structurally stable focal singularities (envelopes) of families of rays, on which the intensities of the geometrical (ray) theory diverge. Diffraction catastrophes describe the (linear) wave amplitudes that smooth the geometrical caustic singularities and decorate them with interference patterns. See Berry (1969, 1976, 1980, 1981), Kravtsov (1964, 1988), and Ludwig (1966).

36.14(ii) Optics

Diffraction catastrophes describe the connection between ray optics and wave optics. Applications include twinkling starlight, focusing of sunlight by rippling water (e.g., swimming-pool patterns), and water-droplet "lenses" (e.g., rainbows). See Adler et al. (1997), Berry and Upstill (1980), Marston (1992, 1999), Nye (1999), Walker (1983, 1988, 1989).

36.14(iii) Quantum Mechanics

Diffraction catastrophes describe the "semiclassical" connections between classical orbits and quantum wavefunctions, for integrable (non-chaotic) systems. Applications include scattering of elementary particles, atoms and molecules from particles and surfaces, and chemical reactions. See Berry (1966, 1975), Connor (1974, 1976), Connor and Farrelly (1981), Trinkaus and Drepper (1977), and Uzer et al. (1983).

36.14(iv) Acoustics

Applications include the reflection of ultrasound pulses, and acoustical waveguides. See Chapman (1999), Frederickson and Marston (1992, 1994), and Kravtsov (1968).

Computation

36.15 Methods of Computation

36.15(i) Convergent Series

Close to the origin $\mathbf{x} = 0$ of parameter space, the series in §36.8 can be used.

36.15(ii) Asymptotics

Far from the bifurcation set, the leading-order asymptotic formulas of §36.11 reproduce accurately the form of the function, including the geometry of the zeros described in §36.7. Close to the bifurcation set but far from $\mathbf{x} = 0$, the uniform asymptotic approximations of §36.12 can be used.

36.15(iii) Integration along Deformed Contour

Direct numerical evaluation can be carried out along a contour that runs along the segment of the real t-axis containing all real critical points of Φ and is deformed outside this range so as to reach infinity along the asymptotic valleys of $\exp(i\Phi)$. (For the umbilics, representations as one-dimensional integrals (§36.2) are used.) For details, see Connor and Curtis (1982) and Kirk et al. (2000). There is considerable freedom in the choice of deformations.

36.15(iv) Integration along Finite Contour

This can be carried out by direct numerical evaluation of canonical integrals along a finite segment of the real axis including all real critical points of Φ, with contributions from the contour outside this range approximated by the first terms of an asymptotic series associated with the endpoints. See Berry et al. (1979).

36.15(v) Differential Equations

For numerical solution of partial differential equations satisfied by the canonical integrals see Connor et al. (1983).

References

General References

There is no single source covering the material in this chapter. An overview of some of the mathematical analysis is given in Arnol'd (1975, 1986). Many physical applications can be found in Poston and Stewart (1978). For applications to wave physics, especially optics, see Berry and Upstill (1980).

Sources

The following list gives the references or other indications of proofs that were used in constructing the various sections of this chapter. These sources supplement the references that are quoted in the text.

§36.2 The convergence of the oscillatory integrals (36.2.4)–(36.2.11) can be confirmed by rotating the integration paths in the complex plane. For (36.2.6) see Berry et al. (1979). For (36.2.7) shift the s variable in (36.2.5) (with (36.2.2)) to remove the quadratic term, integrate, and then deform the contour of the remaining t integration. For (36.2.8) see Berry and Howls (1990). For (36.2.9) integrate (36.2.5) (with (36.2.3)) with respect to t. For (36.2.12) and (36.2.13) use (4.10.11) and (9.5.4), respectively. For (36.2.15) and (36.2.17) use (5.9.1). For (36.2.18) combine (36.2.6), (36.2.8), and (5.9.1) For (36.2.19) use (12.5.1) and (12.14.13). For (36.2.20) see Trinkaus and Drepper (1977). For (36.2.21) use (36.2.9). Eqs. (36.2.22)–(36.2.27) follow from the definitions given in §36.2(i).

§§36.3, 36.4 The graphics were generated by the authors.

§36.5 Wright (1980) and Berry and Howls (1990). The common strategy employed in deriving the formulas in this section involves using the critical-point condition (36.4.1) to reduce the order of the catastrophe polynomials in (36.2.1), then solving (36.5.1) for the imaginary part of the complex critical point in terms of the value of the real critical point, which is itself determined by (36.4.1) and then used to generate the Stokes sets parametrically. For (36.5.11)–(36.5.21) we also use the exponents in the representations (36.2.6) and (36.2.8). The graphics were generated by the authors. For Figures 36.5.2–36.5.6, Eqs. (36.5.11)–(36.5.21) were used in parametric form $x = x(y)$, and checked against the numerical computations in Berry and Howls (1990) (which were based directly on the definitions given in §36.5(i)).

§36.7 Berry *et al.* (1979). (36.7.2) and (36.7.3) may be derived by setting to zero the stationary-phase approximation (§2.3(iv)) of the Pearcey integral $\Psi_2(x,y)$ just outside the caustic; this involves one real saddle and one complex saddle. Table 36.7.1 was computed by the authors.

§36.8 Connor (1973) and Connor *et al.* (1983). For (36.8.1), in the integral (36.2.4) retain the highest power of t in (36.2.1) in the exponent, expand the rest of the exponential as a power series in t, and evaluate the resulting integrals in terms of gamma functions. For (36.8.3), in the integral (36.2.5) with the polynomial (36.2.3) expand the z-dependent part of the exponential in powers of z, and then repeatedly use the differential equation (9.2.1) to express higher derivatives of the Airy function in terms of Ai and Ai'. For (36.8.4), in the integral (36.2.5) with the polynomial (36.2.2) expand the z-dependent part of the exponential in powers of z, and then repeatedly use (9.2.1), and (36.2.20).

§36.10 For (36.10.1) to (36.10.10) see Connor *et al.* (1983). (36.10.11) to (36.10.18) are derived by repeated differentiations with respect to x, y, or z, in combinations that generate exact derivatives of the exponents in (36.2.5).

§36.11 The formulas in this section are derived by the method of stationary phase, applied to the real critical points of the integral representations in §36.2. See §2.3(iv) and also Berry and Howls (1991). For (36.11.4) the integral is exponentially small when $x > 0$ and the dominant contribution is from a critical point off the real axis.

§36.12 Berry and Howls (1993).

§36.13 The figure was generated by the authors.

Bibliography

A. S. Abdullaev (1985). *Asymptotics of solutions of the generalized sine-Gordon equation, the third Painlevé equation and the d'Alembert equation.* Dokl. Akad. Nauk SSSR **280**(2), pp. 265–268. English translation: Soviet Math. Dokl. **31**(1985), no. 1, pp. 45–47.

M. J. Ablowitz and P. A. Clarkson (1991). *Solitons, Nonlinear Evolution Equations and Inverse Scattering*, Volume 149 of *London Mathematical Society Lecture Note Series*. Cambridge: Cambridge University Press.

M. J. Ablowitz and H. Segur (1977). *Exact linearization of a Painlevé transcendent.* Phys. Rev. Lett. **38**(20), pp. 1103–1106.

M. J. Ablowitz and H. Segur (1981). *Solitons and the Inverse Scattering Transform*, Volume 4 of *SIAM Studies in Applied Mathematics*. Philadelphia, PA: Society for Industrial and Applied Mathematics (SIAM).

A. Abramov (1960). *Tables of $\ln\Gamma(z)$ for Complex Argument.* New York: Pergamon Press.

M. Abramowitz (1949). *Asymptotic expansions of spheroidal wave functions.* J. Math. Phys. Mass. Inst. Tech. **28**, pp. 195–199.

M. Abramowitz (1954). *Regular and irregular Coulomb wave functions expressed in terms of Bessel-Clifford functions.* J. Math. Physics **33**, pp. 111–116.

M. Abramowitz and P. Rabinowitz (1954). *Evaluation of Coulomb wave functions along the transition line.* Physical Rev. (2) **96**, pp. 77–79.

M. Abramowitz and I. A. Stegun (Eds.) (1964). *Handbook of Mathematical Functions with Formulas, Graphs, and Mathematical Tables.* Number 55 in National Bureau of Standards Applied Mathematics Series. U.S. Government Printing Office, Washington, D.C. Corrections appeared in later printings up to the 10th Printing, December, 1972. Reproductions by other publishers, in whole or in part, have been available since 1965.

J.-J. Achenbach (1986). *Numerik: Implementierung von Zylinderfunktionen.* Elektrotechnik. Braunschweig-Wiesbaden: Friedr. Vieweg & Sohn.

F. S. Acton (1974). *Recurrence relations for the Fresnel integral $\int_0^\infty \frac{\exp(-ct)\,dt}{\sqrt{t}(1+t^2)}$ and similar integrals.* Comm. ACM **17**(8), pp. 480–481.

V. S. Adamchik (1998). *Polygamma functions of negative order.* J. Comput. Appl. Math. **100**(2), pp. 191–199.

V. S. Adamchik and H. M. Srivastava (1998). *Some series of the zeta and related functions.* Analysis (Munich) **18**(2), pp. 131–144.

A. Adelberg (1992). *On the degrees of irreducible factors of higher order Bernoulli polynomials.* Acta Arith. **62**(4), pp. 329–342.

A. Adelberg (1996). *Congruences of p-adic integer order Bernoulli numbers.* J. Number Theory **59**(2), pp. 374–388.

C. Adiga, B. C. Berndt, S. Bhargava, and G. N. Watson (1985). *Chapter 16 of Ramanujan's second notebook: Theta-functions and q-series.* Mem. Amer. Math. Soc. **53**(315), pp. v+85.

C. L. Adler, J. A. Lock, B. R. Stone, and C. J. Garcia (1997). *High-order interior caustics produced in scattering of a diagonally incident plane wave by a circular cylinder.* J. Opt. Soc. Amer. A **14**(6), pp. 1305–1315.

V. È. Adler (1994). *Nonlinear chains and Painlevé equations.* Phys. D **73**(4), pp. 335–351.

P. Adrian (1959). *Die Bezeichnungsweise der Bernoullischen Zahlen.* Mitt. Verein. Schweiz. Versicherungsmath. **59**, pp. 199–206.

M. M. Agrest, S. I. Labakhua, M. M. Rikenglaz, and T. S. Chachibaya (1982). *Tablitsy funktsii Struve i integralov ot nikh.* Moscow: "Nauka". Translated title: *Tables of Struve Functions and their Integrals*.

M. M. Agrest and M. S. Maksimov (1971). *Theory of Incomplete Cylindrical Functions and Their Applications.* Berlin: Springer-Verlag. Translated from the Russian by H. E. Fettis, J. W. Goresh and D. A. Lee, Die Grundlehren der mathematischen Wissenschaften, Band 160.

L. V. Ahlfors (1966). *Complex Analysis: An Introduction of the Theory of Analytic Functions of One Complex Variable* (2nd ed.). New York: McGraw-Hill Book Co. A third edition was published in 1978 by McGraw-Hill, New York.

A. R. Ahmadi and S. E. Widnall (1985). *Unsteady lifting-line theory as a singular-perturbation problem.* J. Fluid Mech **153**, pp. 59–81.

S. Ahmed and M. E. Muldoon (1980). *On the zeros of confluent hypergeometric functions. III. Characterization by means of nonlinear equations.* Lett. Nuovo Cimento (2) **29**(11), pp. 353–358.

S.-H. Ahn, H.-W. Lee, and H. M. Lee (2001). *Lyα line formation in starbursting galaxies. I. Moderately thick, dustless, and static H I media*. Astrophysical J. **554**, pp. 604–614.

H. Airault (1979). *Rational solutions of Painlevé equations*. Stud. Appl. Math. **61**(1), pp. 31–53.

H. Airault, H. P. McKean, and J. Moser (1977). *Rational and elliptic solutions of the Korteweg-de Vries equation and a related many-body problem*. Comm. Pure Appl. Math. **30**(1), pp. 95–148.

G. B. Airy (1838). *On the intensity of light in the neighbourhood of a caustic*. Trans. Camb. Phil. Soc. **6**, pp. 379–402.

G. B. Airy (1849). *Supplement to a paper "On the intensity of light in the neighbourhood of a caustic"*. Trans. Camb. Phil. Soc. **8**, pp. 595–599.

N. I. Akhiezer (1988). *Lectures on Integral Transforms*, Volume 70 of *Translations of Mathematical Monographs*. Providence, RI: American Mathematical Society. Translated from the Russian by H. H. McFaden.

N. I. Akhiezer (1990). *Elements of the Theory of Elliptic Functions*, Volume 79 of *Translations of Mathematical Monographs*. Providence, RI: American Mathematical Society. Translated from the second Russian edition by H. H. McFaden.

A. M. Al-Rashed and N. Zaheer (1985). *Zeros of Stieltjes and Van Vleck polynomials and applications*. J. Math. Anal. Appl. **110**(2), pp. 327–339.

W. A. Al-Salam and L. Carlitz (1959). *Some determinants of Bernoulli, Euler and related numbers*. Portugal. Math. **18**, pp. 91–99.

W. A. Al-Salam and L. Carlitz (1965). *Some orthogonal q-polynomials*. Math. Nachr. **30**, pp. 47–61.

W. A. Al-Salam and M. E. H. Ismail (1994). *A q-beta integral on the unit circle and some biorthogonal rational functions*. Proc. Amer. Math. Soc. **121**(2), pp. 553–561.

M. Alam (1979). *Zeros of Stieltjes and Van Vleck polynomials*. Trans. Amer. Math. Soc. **252**, pp. 197–204.

D. W. Albrecht, E. L. Mansfield, and A. E. Milne (1996). *Algorithms for special integrals of ordinary differential equations*. J. Phys. A **29**(5), pp. 973–991.

J. R. Albright (1977). *Integrals of products of Airy functions*. J. Phys. A **10**(4), pp. 485–490.

J. R. Albright and E. P. Gavathas (1986). *Integrals involving Airy functions*. J. Phys. A **19**(13), pp. 2663–2665.

K. Alder, A. Bohr, T. Huus, B. Mottelson, and A. Winther (1956). *Study of nuclear structure by electromagnetic excitation with accelerated ions*. Rev. Mod. Phys. **28**, pp. 432–542.

G. Alefeld and J. Herzberger (1983). *Introduction to Interval Computations*. Computer Science and Applied Mathematics. New York: Academic Press Inc. Translated from the German by Jon Rokne.

W. R. Alford, A. Granville, and C. Pomerance (1994). *There are infinitely many Carmichael numbers*. Ann. of Math. (2) **139**(3), pp. 703–722.

F. Alhargan and S. Judah (1992). *Frequency response characteristics of the multiport planar elliptic patch*. IEEE Trans. Microwave Theory Tech. **40**(8), pp. 1726–1730.

F. Alhargan and S. Judah (1995). *A general mode theory for the elliptic disk microstrip antenna*. IEEE Trans. Antennas and Propagation **43**(6), pp. 560–568.

G. Allasia and R. Besenghi (1987a). *Numerical calculation of incomplete gamma functions by the trapezoidal rule*. Numer. Math. **50**(4), pp. 419–428.

G. Allasia and R. Besenghi (1987b). *Numerical computation of Tricomi's psi function by the trapezoidal rule*. Computing **39**(3), pp. 271–279.

G. Allasia and R. Besenghi (1989). *Numerical Calculation of the Riemann Zeta Function and Generalizations by Means of the Trapezoidal Rule*. In C. Brezinski (Ed.), *Numerical and Applied Mathematics, Part II (Paris, 1988)*, Volume 1 of *IMACS Ann. Comput. Appl. Math.*, pp. 467–472. Basel: Baltzer.

G. Allasia and R. Besenghi (1991). *Numerical evaluation of the Kummer function with complex argument by the trapezoidal rule*. Rend. Sem. Mat. Univ. Politec. Torino **49**(3), pp. 315–327.

G. Almkvist and B. Berndt (1988). *Gauss, Landen, Ramanujan, the arithmetic-geometric mean, ellipses, π, and the Ladies Diary*. Amer. Math. Monthly **95**(7), pp. 585–608.

Z. Altaç (1996). *Integrals involving Bickley and Bessel functions in radiative transfer, and generalized exponential integral functions*. J. Heat Transfer **118**(3), pp. 789–792.

H. H. Aly, H. J. W. Müller-Kirsten, and N. Vahedi-Faridi (1975). *Scattering by singular potentials with a perturbation – Theoretical introduction to Mathieu functions*. J. Mathematical Phys. **16**, pp. 961–970. Erratum: same journal **17** (1975), no. 7, p. 1361.

H. Alzer (1997a). *A harmonic mean inequality for the gamma function*. J. Comput. Appl. Math. **87**(2), pp. 195–198. For corrigendum see same journal v. 90 (1998), no. 2, p. 265.

H. Alzer (1997b). *On some inequalities for the incomplete gamma function*. Math. Comp. **66**(218), pp. 771–778.

H. Alzer (2000). *Sharp bounds for the Bernoulli numbers*. Arch. Math. (Basel) **74**(3), pp. 207–211.

H. Alzer and S.-L. Qiu (2004). *Monotonicity theorems and inequalities for the complete elliptic integrals.* J. Comput. Appl. Math. **172**(2), pp. 289–312.

D. E. Amos (1974). *Computation of modified Bessel functions and their ratios.* Math. Comp. **28**(125), pp. 239–251.

D. E. Amos (1980). *Computation of exponential integrals.* ACM Trans. Math. Software **6**(3), pp. 365–377.

D. E. Amos (1983). *Uniform asymptotic expansions for exponential integrals $E_n(x)$ and Bickley functions $\text{Ki}_n(x)$.* ACM Trans. Math. Software **9**(4), pp. 467–479.

D. E. Amos (1985). *A subroutine package for Bessel functions of a complex argument and nonnegative order.* Technical Report SAND85-1018, Sandia National Laboratories, Albuquerque, NM.

D. E. Amos (1989). *Repeated integrals and derivatives of K Bessel functions.* SIAM J. Math. Anal. **20**(1), pp. 169–175.

G. D. Anderson, S.-L. Qiu, M. K. Vamanamurthy, and M. Vuorinen (2000). *Generalized elliptic integrals and modular equations.* Pacific J. Math. **192**(1), pp. 1–37.

G. D. Anderson and M. K. Vamanamurthy (1985). *Inequalities for elliptic integrals.* Publ. Inst. Math. (Beograd) (N.S.) **37(51)**, pp. 61–63.

G. D. Anderson, M. K. Vamanamurthy, and M. Vuorinen (1990). *Functional inequalities for complete elliptic integrals and their ratios.* SIAM J. Math. Anal. **21**(2), pp. 536–549.

G. D. Anderson, M. K. Vamanamurthy, and M. Vuorinen (1992a). *Functional inequalities for hypergeometric functions and complete elliptic integrals.* SIAM J. Math. Anal. **23**(2), pp. 512–524.

G. D. Anderson, M. K. Vamanamurthy, and M. Vuorinen (1992b). *Hypergeometric Functions and Elliptic Integrals.* In H. M. Srivastava and S. Owa (Eds.), *Current Topics in Analytic Function Theory*, pp. 48–85. River Edge, NJ: World Sci. Publishing.

G. D. Anderson, M. K. Vamanamurthy, and M. K. Vuorinen (1997). *Conformal Invariants, Inequalities, and Quasiconformal Maps.* New York: John Wiley & Sons Inc. With PC diskette, a Wiley-Interscience Publication.

F. V. Andreev and A. V. Kitaev (2002). *Transformations $RS_4^2(3)$ of the ranks ≤ 4 and algebraic solutions of the sixth Painlevé equation.* Comm. Math. Phys. **228**(1), pp. 151–176.

G. E. Andrews (1966a). *On basic hypergeometric series, mock theta functions, and partitions. II.* Quart. J. Math. Oxford Ser. (2) **17**, pp. 132–143.

G. E. Andrews (1966b). *q-identities of Auluck, Carlitz, and Rogers.* Duke Math. J. **33**(3), pp. 575–581.

G. E. Andrews (1972). *Summations and transformations for basic Appell series.* J. London Math. Soc. (2) **4**, pp. 618–622.

G. E. Andrews (1974). *Applications of basic hypergeometric functions.* SIAM Rev. **16**(4), pp. 441–484.

G. E. Andrews (1976). *The Theory of Partitions*, Volume 2 of *Encyclopedia of Mathematics and its Applications*. Reading, MA-London-Amsterdam: Addison-Wesley Publishing Co. Reprinted by Cambridge University Press, Cambridge, 1998.

G. E. Andrews (1979). *Plane partitions. III. The weak Macdonald conjecture.* Invent. Math. **53**(3), pp. 193–225.

G. E. Andrews (1984). *Multiple series Rogers-Ramanujan type identities.* Pacific J. Math. **114**(2), pp. 267–283.

G. E. Andrews (1986). *q-Series: Their Development and Application in Analysis, Number Theory, Combinatorics, Physics, and Computer Algebra*, Volume 66 of *CBMS Regional Conference Series in Mathematics*. Providence, RI: Amer. Math. Soc.

G. E. Andrews (1996). *Pfaff's method II: Diverse applications.* J. Comput. Appl. Math. **68**(1-2), pp. 15–23.

G. E. Andrews (2000). *Umbral calculus, Bailey chains, and pentagonal number theorems.* J. Combin. Theory Ser. A **91**(1-2), pp. 464–475. In memory of Gian-Carlo Rota.

G. E. Andrews (2001). *Bailey's Transform, Lemma, Chains and Tree.* In J. Bustoz, M. E. H. Ismail, and S. K. Suslov (Eds.), *Special Functions 2000: Current Perspective and Future Directions (Tempe, AZ)*, Volume 30 of *NATO Sci. Ser. II Math. Phys. Chem.*, pp. 1–22. Dordrecht: Kluwer Acad. Publ.

G. E. Andrews, R. Askey, and R. Roy (1999). *Special Functions*, Volume 71 of *Encyclopedia of Mathematics and its Applications*. Cambridge: Cambridge University Press.

G. E. Andrews, R. A. Askey, B. C. Berndt, and R. A. Rankin (Eds.) (1988). *Ramanujan Revisited*, Boston, MA. Academic Press Inc.

G. E. Andrews and A. Berkovich (1998). *A trinomial analogue of Bailey's lemma and $N = 2$ superconformal invariance.* Comm. Math. Phys. **192**(2), pp. 245–260.

G. E. Andrews and D. Foata (1980). *Congruences for the q-secant numbers.* European J. Combin. **1**(4), pp. 283–287.

G. E. Andrews, I. P. Goulden, and D. M. Jackson (1986). *Shanks' convergence acceleration transform, Padé approximants and partitions.* J. Combin. Theory Ser. A **43**(1), pp. 70–84.

H. M. Antia (1993). *Rational function approximations for Fermi-Dirac integrals.* The Astrophysical Journal Supplement Series **84**, pp. 101–108. Double-precision Fortran, maximum precision 8D.

M. A. Anuta, D. W. Lozier, and P. R. Turner (1996). *The MasPar MP-1 as a computer arithmetic laboratory.* J. Res. Nat. Inst. Stand. Technol. **101**(2), pp. 165–174.

K. Aomoto (1987). *Special value of the hypergeometric function $_3F_2$ and connection formulae among asymptotic expansions.* J. Indian Math. Soc. (N.S.) **51**, pp. 161–221.

A. Apelblat (1983). *Table of Definite and Infinite Integrals*, Volume 13 of *Physical Sciences Data*. Amsterdam: Elsevier Scientific Publishing Co. Table errata: Math. Comp. v. 65 (1996), no. 215, p. 1386.

A. Apelblat (1989). *Derivatives and integrals with respect to the order of the Struve functions $\mathbf{H}_\nu(x)$ and $\mathbf{L}_\nu(x)$.* J. Math. Anal. Appl. **137**(1), pp. 17–36.

A. Apelblat (1991). *Integral representation of Kelvin functions and their derivatives with respect to the order.* Z. Angew. Math. Phys. **42**(5), pp. 708–714.

T. M. Apostol (1952). *Theorems on generalized Dedekind sums.* Pacific J. Math. **2**(1), pp. 1–9.

T. M. Apostol (1976). *Introduction to Analytic Number Theory.* New York: Springer-Verlag. Undergraduate Texts in Mathematics.

T. M. Apostol (1983). *A proof that Euler missed: Evaluating $\zeta(2)$ the easy way.* Math. Intelligencer **5**(3), pp. 59–60.

T. M. Apostol (1985a). *Formulas for higher derivatives of the Riemann zeta function.* Math. Comp. **44**(169), pp. 223–232.

T. M. Apostol (1985b). *Note on the trivial zeros of Dirichlet L-functions.* Proc. Amer. Math. Soc. **94**(1), pp. 29–30.

T. M. Apostol (1990). *Modular Functions and Dirichlet Series in Number Theory* (2nd ed.), Volume 41 of *Graduate Texts in Mathematics*. New York: Springer-Verlag.

T. M. Apostol (2000). *A Centennial History of the Prime Number Theorem.* In *Number Theory*, Trends Math., pp. 1–14. Basel: Birkhäuser.

T. M. Apostol (2006). *Bernoulli's power-sum formulas revisited.* Math. Gaz. **90**(518), pp. 276–279.

T. M. Apostol (2008). *A primer on Bernoulli numbers and polynomials.* Math. Mag. **81**(3), pp. 178–190.

T. M. Apostol and I. Niven (1994). *Number Theory.* In *The New Encyclopaedia Britannica* (15th ed.), Volume 25, pp. 14–37.

T. M. Apostol and T. H. Vu (1984). *Dirichlet series related to the Riemann zeta function.* J. Number Theory **19**(1), pp. 85–102.

T. M. Apostol and H. S. Zuckerman (1951). *On magic squares constructed by the uniform step method.* Proc. Amer. Math. Soc. **2**(4), pp. 557–565.

H. Appel (1968). *Numerical Tables for Angular Correlation Computations in α-, β- and γ-Spectroscopy: 3j-, 6j-, 9j-Symbols, F- and Γ-Coefficients.* Landolt-Börnstein Numerical Data and Functional Relationships in Science and Technology. Springer-Verlag.

G. B. Arfken and H. J. Weber (2005). *Mathematical Methods for Physicists* (6th ed.). Oxford: Elsevier.

J. V. Armitage (1989). *The Riemann Hypothesis and the Hamiltonian of a Quantum Mechanical System.* In M. M. Dodson and J. A. G. Vickers (Eds.), *Number Theory and Dynamical Systems (York, 1987)*, Volume 134 of *London Math. Soc. Lecture Note Ser.*, pp. 153–172. Cambridge: Cambridge University Press.

B. H. Armstrong (1967). *Spectrum line profiles: The Voigt function.* J. Quant. Spectrosc. Radiat. Transfer **7**, pp. 61–88.

V. I. Arnol'd (1972). *Normal forms of functions near degenerate critical points, the Weyl groups A_k, D_k, E_k and Lagrangian singularities.* Funkcional. Anal. i Priložen. **6**(4), pp. 3–25. In Russian. English translation: Functional Anal. Appl., **6**(1973), pp. 254–272.

V. I. Arnol'd (1974). *Normal forms of functions in the neighborhood of degenerate critical points.* Uspehi Mat. Nauk **29**(2(176)), pp. 11–49. Collection of articles dedicated to the memory of Ivan Georgievič Petrovskiĭ (1901–1973), I. In Russian. English translation: Russian Math. Surveys, **29**(1974), no. 2, pp. 10–50.

V. I. Arnol'd (1975). *Critical points of smooth functions, and their normal forms.* Uspehi Mat. Nauk **30**(5(185)), pp. 3–65. In Russian. English translation: Russian Math. Surveys, **30**(1975), no. 5, pp. 1–75.

V. I. Arnol'd (1986). *Catastrophe Theory* (2nd ed.). Berlin: Springer-Verlag. Translated from the Russian by G. S. Wassermann, based on a translation by R. K. Thomas. See also Arnol'd (1992).

V. I. Arnol'd (1992). *Catastrophe Theory* (3rd ed.). Berlin: Springer-Verlag. Translated from the Russian by G. S. Wassermann, Based on a translation by R. K. Thomas.

V. I. Arnol'd (1997). *Mathematical Methods of Classical Mechanics*, Volume 60 of *Graduate Texts in Mathematics*. New York: Springer-Verlag. Translated from the 1974 Russian original by K. Vogtmann and A. Weinstein, Corrected reprint of the second (1989) edition.

V. I. Arnol'd, S. M. Guseĭn-Zade, and A. N. Varchenko (1988). *Singularities of Differentiable Maps. Vol. II.* Boston-Berlin: Birkhäuser. Monodromy and asymptotics of integrals, Translated from the Russian by Hugh Porteous, Translation revised by the authors and James Montaldi.

F. M. Arscott (1956). *Perturbation solutions of the ellipsoidal wave equation.* Quart. J. Math. Oxford Ser. (2) **7**, pp. 161–174.

F. M. Arscott (1959). *A new treatment of the ellipsoidal wave equation.* Proc. London Math. Soc. (3) **9**, pp. 21–50.

F. M. Arscott (1964a). *Integral equations and relations for Lamé functions.* Quart. J. Math. Oxford Ser. (2) **15**, pp. 103–115.

F. M. Arscott (1964b). *Periodic Differential Equations. An Introduction to Mathieu, Lamé, and Allied Functions.* International Series of Monographs in Pure and Applied Mathematics, Vol. 66. New York: Pergamon Press, The Macmillan Co.

F. M. Arscott (1967). *The Whittaker-Hill equation and the wave equation in paraboloidal co-ordinates.* Proc. Roy. Soc. Edinburgh Sect. A **67**, pp. 265–276.

F. M. Arscott and I. M. Khabaza (1962). *Tables of Lamé Polynomials.* New York: Pergamon Press, The Macmillan Co.

G. Arutyunov and M. Staudacher (2004). *Matching higher conserved charges for strings and spins.* J. High Energy Phys. (electronic only, article id. JHEP03(2004)004, 32 pp.).

U. M. Ascher, R. M. M. Mattheij, and R. D. Russell (1995). *Numerical Solution of Boundary Value Problems for Ordinary Differential Equations*, Volume 13 of *Classics in Applied Mathematics*. Philadelphia, PA: Society for Industrial and Applied Mathematics (SIAM). Corrected reprint of the 1988 original.

U. M. Ascher and L. R. Petzold (1998). *Computer Methods for Ordinary Differential Equations and Differential-Algebraic Equations.* Philadelphia, PA: Society for Industrial and Applied Mathematics (SIAM).

R. Askey (1974). *Jacobi polynomials. I. New proofs of Koornwinder's Laplace type integral representation and Bateman's bilinear sum.* SIAM J. Math. Anal. **5**, pp. 119–124.

R. Askey (1975). *Orthogonal Polynomials and Special Functions*, Volume 21 of *CBMS-NSF Regional Conference Series in Applied Mathematics*. Philadelphia, PA: Society for Industrial and Applied Mathematics.

R. Askey (1980). *Some basic hypergeometric extensions of integrals of Selberg and Andrews.* SIAM J. Math. Anal. **11**(6), pp. 938–951.

R. Askey (1982). *Commentary on the Paper "Beiträge zur Theorie der Toeplitzschen Form".* In *Gábor Szegő, Collected Papers. Vol. 1*, Contemporary Mathematicians, pp. 303–305. MA: Birkhäuser Boston.

R. Askey (1985). *Continuous Hahn polynomials.* J. Phys. A **18**(16), pp. L1017–L1019.

R. Askey (1989). *Continuous q-Hermite Polynomials when $q > 1$.* In *q-series and Partitions (Minneapolis, MN, 1988)*, Volume 18 of *IMA Vol. Math. Appl.*, pp. 151–158. New York: Springer.

R. Askey (1990). *Graphs as an Aid to Understanding Special Functions.* In R. Wong (Ed.), *Asymptotic and Computational Analysis*, Volume 124 of *Lecture Notes in Pure and Appl. Math.*, pp. 3–33. New York: Dekker.

R. Askey and J. Fitch (1969). *Integral representations for Jacobi polynomials and some applications.* J. Math. Anal. Appl. **26**(2), pp. 411–437.

R. Askey and G. Gasper (1976). *Positive Jacobi polynomial sums. II.* Amer. J. Math. **98**(3), pp. 709–737.

R. Askey and M. E. H. Ismail (1984). *Recurrence relations, continued fractions, and orthogonal polynomials.* Mem. Amer. Math. Soc. **49**(300), pp. iv+108.

R. Askey, T. H. Koornwinder, and M. Rahman (1986). *An integral of products of ultraspherical functions and a q-extension.* J. London Math. Soc. (2) **33**(1), pp. 133–148.

R. Askey and B. Razban (1972). *An integral for Jacobi polynomials.* Simon Stevin **46**, pp. 165–169.

R. Askey and J. Wilson (1985). *Some basic hypergeometric orthogonal polynomials that generalize Jacobi polynomials.* Mem. Amer. Math. Soc. **54**(319), pp. iv+55.

D. Atkinson and P. W. Johnson (1988). *Chiral-symmetry breaking in QCD. I. The infrared domain.* Phys. Rev. D (3) **37**(8), pp. 2290–2295.

K. E. Atkinson (1989). *An Introduction to Numerical Analysis* (2nd ed.). New York: John Wiley & Sons Inc.

M. Audin (1999). *Spinning Tops: A Course on Integrable Systems*, Volume 51 of *Cambridge Studies in Advanced Mathematics*. Cambridge: Cambridge University Press.

M. Aymar, C. H. Greene, and E. Luc-Koenig (1996). *Multichannel Rydberg spectroscopy of complex atoms.* Reviews of Modern Physics **68**, pp. 1015–1123.

A. W. Babister (1967). *Transcendental Functions Satisfying Nonhomogeneous Linear Differential Equations.* New York: The Macmillan Co.

L. V. Babushkina, M. K. Kerimov, and A. I. Nikitin (1997). *New tables of Bessel functions of complex argument.* Comput. Math. Math. Phys. **37**(12), pp. 1480–1482. Russian original in Zh. Vychisl. Mat. i Mat. Fiz. **37**(1997), no. 12, 1526–1528.

G. Backenstoss (1970). *Pionic atoms.* Annual Review of Nuclear and Particle Science **20**, pp. 467–508.

J. Baik, P. Deift, and K. Johansson (1999). *On the distribution of the length of the longest increasing subsequence of random permutations.* J. Amer. Math. Soc. **12**(4), pp. 1119–1178.

W. N. Bailey (1928). *Products of generalized hypergeometric series.* Proc. London Math. Soc. (2) **28**(2), pp. 242–254.

W. N. Bailey (1929). *Transformations of generalized hypergeometric series.* Proc. London Math. Soc. (2) **29**(2), pp. 495–502.

W. N. Bailey (1935). *Generalized Hypergeometric Series.* Cambridge: Cambridge University Press. Reissued by Hafner, New York, 1972.

W. N. Bailey (1964). *Generalized Hypergeometric Series.* New York: Stechert-Hafner, Inc.

G. A. Baker, Jr. and P. Graves-Morris (1996). *Padé Approximants* (2nd ed.), Volume 59 of *Encyclopedia of Mathematics and its Applications.* Cambridge: Cambridge University Press.

H. F. Baker (1995). *Abelian Functions: Abel's Theorem and the Allied Theory of Theta Functions.* Cambridge: Cambridge University Press. Reprint of the 1897 original, with a foreword by Igor Krichever.

P. Baldwin (1985). *Zeros of generalized Airy functions.* Mathematika **32**(1), pp. 104–117.

P. Baldwin (1991). *Coefficient functions for an inhomogeneous turning-point problem.* Mathematika **38**(2), pp. 217–238.

J. S. Ball (2000). *Automatic computation of zeros of Bessel functions and other special functions.* SIAM J. Sci. Comput. **21**(4), pp. 1458–1464.

L. E. Ballentine and S. M. McRae (1998). *Moment equations for probability distributions in classical and quantum mechanics.* Phys. Rev. A **58**(3), pp. 1799–1809.

C. B. Balogh (1967). *Asymptotic expansions of the modified Bessel function of the third kind of imaginary order.* SIAM J. Appl. Math. **15**, pp. 1315–1323.

E. Bannai (1990). *Orthogonal Polynomials in Coding Theory and Algebraic Combinatorics.* In *Orthogonal Polynomials (Columbus, OH, 1989)*, Volume 294 of *NATO Adv. Sci. Inst. Ser. C Math. Phys. Sci.*, pp. 25–53. Dordrecht: Kluwer Acad. Publ.

R. Barakat (1961). *Evaluation of the incomplete gamma function of imaginary argument by Chebyshev polynomials.* Math. Comp. **15**(73), pp. 7–11.

R. Barakat and E. Parshall (1996). *Numerical evaluation of the zero-order Hankel transform using Filon quadrature philosophy.* Appl. Math. Lett. **9**(5), pp. 21–26.

P. Baratella and L. Gatteschi (1988). *The Bounds for the Error Term of an Asymptotic Approximation of Jacobi Polynomials.* In *Orthogonal Polynomials and Their Applications (Segovia, 1986)*, Volume 1329 of *Lecture Notes in Math.*, pp. 203–221. Berlin: Springer.

M. N. Barber and B. W. Ninham (1970). *Random and Restricted Walks: Theory and Applications.* New York: Gordon and Breach.

C. Bardin, Y. Dandeu, L. Gauthier, J. Guillermin, T. Lena, J. M. Pernet, H. H. Wolter, and T. Tamura (1972). *Coulomb functions in entire (η, ρ)-plane.* Comput. Phys. Comm. **3**(2), pp. 73–87.

R. W. Barnard, K. Pearce, and K. C. Richards (2000). *A monotonicity property involving $_3F_2$ and comparisons of the classical approximations of elliptical arc length.* SIAM J. Math. Anal. **32**(2), pp. 403–419.

R. W. Barnard, K. Pearce, and L. Schovanec (2001). *Inequalities for the perimeter of an ellipse.* J. Math. Anal. Appl. **260**(2), pp. 295–306.

A. R. Barnett (1981a). *An algorithm for regular and irregular Coulomb and Bessel functions of real order to machine accuracy.* Comput. Phys. Comm. **21**(3), pp. 297–314.

A. R. Barnett (1981b). *KLEIN: Coulomb functions for real λ and positive energy to high accuracy.* Comput. Phys. Comm. **24**(2), pp. 141–159. Double-precision Fortran code for positive energies.

A. R. Barnett (1996). *The Calculation of Spherical Bessel Functions and Coulomb Functions.* In K. Bartschat and J. Hinze (Eds.), *Computational Atomic Physics: Electron and Positron Collisions with Atoms and Ions*, pp. 181–202. Berlin: Springer. Includes program disk. Double-precision Fortran code for positive energies. Maximum accuracy: 15D.

A. R. Barnett, D. H. Feng, J. W. Steed, and L. J. B. Goldfarb (1974). *Coulomb wave functions for all real η and ρ.* Comput. Phys. Comm. **8**(5), pp. 377–395. Single-precision Fortran code for positive energies. For modification see same journal, **11**(1976), p. 141.

E. Barouch, B. M. McCoy, and T. T. Wu (1973). *Zero-field susceptibility of the two-dimensional Ising model near T_c.* Phys. Rev. Lett. **31**, pp. 1409–1411.

G. E. Barr (1968). *A note on integrals involving parabolic cylinder functions.* SIAM J. Appl. Math. **16**(1), pp. 71–74.

R. F. Barrett (1964). *Tables of modified Struve functions of orders zero and unity.* Part of the Unpublished Mathematical Tables. Reviewed in Math. Comp., v. 18 (1964), no. 86, p. 332.

W. Barrett (1981). *Mathieu functions of general order: Connection formulae, base functions and asymptotic formulae. I–V.* Philos. Trans. Roy. Soc. London Ser. A **301**, pp. 75–162.

P. Barrucand and D. Dickinson (1968). *On the Associated Legendre Polynomials.* In *Orthogonal Expansions and their Continuous Analogues (Proc. Conf., Edwardsville, Ill., 1967)*, pp. 43–50. Carbondale, Ill.: Southern Illinois Univ. Press.

D. A. Barry, P. J. Culligan-Hensley, and S. J. Barry (1995). *Real values of the W-function.* ACM Trans. Math. Software **21**(2), pp. 161–171.

W. Bartky (1938). *Numerical calculation of a generalized complete elliptic integral.* Rev. Mod. Phys. **10**, pp. 264–269.

A. O. Barut and L. Girardello (1971). *New "coherent" states associated with non-compact groups.* Comm. Math. Phys. **21**(1), pp. 41–55.

A. P. Bassom, P. A. Clarkson, and A. C. Hicks (1993). *Numerical studies of the fourth Painlevé equation.* IMA J. Appl. Math. **50**(2), pp. 167–193.

A. P. Bassom, P. A. Clarkson, and A. C. Hicks (1995). *Bäcklund transformations and solution hierarchies for the fourth Painlevé equation.* Stud. Appl. Math. **95**(1), pp. 1–71.

A. P. Bassom, P. A. Clarkson, A. C. Hicks, and J. B. McLeod (1992). *Integral equations and exact solutions for the fourth Painlevé equation.* Proc. Roy. Soc. London Ser. A **437**, pp. 1–24.

A. P. Bassom, P. A. Clarkson, C. K. Law, and J. B. McLeod (1998). *Application of uniform asymptotics to the second Painlevé transcendent.* Arch. Rational Mech. Anal. **143**(3), pp. 241–271.

A. Basu and T. M. Apostol (2000). *A new method for investigating Euler sums.* Ramanujan J. **4**(4), pp. 397–419.

P. M. Batchelder (1967). *An Introduction to Linear Difference Equations.* New York: Dover Publications Inc. Unaltered republication of the original 1927 Harvard University edition.

H. Bateman (1905). *A generalisation of the Legendre polynomial.* Proc. London Math. Soc. (2) **3**(3), pp. 111–123. Reviewed in JFM 36.0512.01.

H. Bateman and R. C. Archibald (1944). *A guide to tables of Bessel functions.* Mathematical Tables and Other Aids to Computation (now Mathematics of Computation) **1**(7), pp. 205–308.

F. L. Bauer, H. Rutishauser, and E. Stiefel (1963). *New Aspects in Numerical Quadrature.* In *Proc. Sympos. Appl. Math., Vol. XV*, pp. 199–218. Providence, RI: American Mathematical Society.

G. Baxter (1961). *Polynomials defined by a difference system.* J. Math. Anal. Appl. **2**(2), pp. 223–263.

R. J. Baxter (1981). *Rogers-Ramanujan identities in the hard hexagon model.* J. Statist. Phys. **26**(3), pp. 427–452.

R. J. Baxter (1982). *Exactly Solved Models in Statistical Mechanics.* London-New York: Academic Press Inc. Reprinted in 1989.

K. Bay, W. Lay, and A. Akopyan (1997). *Avoided crossings of the quartic oscillator.* J. Phys. A **30**(9), pp. 3057–3067.

C. Bays and R. H. Hudson (2000). *A new bound for the smallest x with $\pi(x) > \mathrm{li}(x)$.* Math. Comp. **69**(231), pp. 1285–1296.

L. P. Bayvel and A. R. Jones (1981). *Electromagnetic Scattering and its Applications.* London: Applied Science Publishers.

E. F. Beckenbach (Ed.) (1981). *Applied Combinatorial Mathematics*, Malabar, FL: Robert E. Krieger.

P. A. Becker (1997). *Normalization integrals of orthogonal Heun functions.* J. Math. Phys. **38**(7), pp. 3692–3699.

R. Becker and F. Sauter (1964). *Electromagnetic Fields and Interactions*, Volume I. New York: Blaisdell. Replaces Abraham & Becker, Classical Theory of Electricity and Magnetism.

P. Beckmann and A. Spizzichino (1963). *The Scattering of Electromagnetic Waves from Rough Surfaces.* New York: Pergamon Press. Reprinted in 1987 by Artech House Publishers, Norwood, Massachusetts.

V. M. Belîakov, R. I. Kravtŝova, and M. G. Rappoport (1962). *Tablitŝy Elliptičeskikh Integralov. Tom I.* Mathematical tables of the Computing Center of the Academy of Sciences of the USSR]. Moscow: Izdat. Akad. Nauk SSSR. Reviewed in Math. Comp. v. 18(1964), pp. 676–677.

R. Bellman (1961). *A Brief Introduction to Theta Functions.* Athena Series: Selected Topics in Mathematics. New York: Holt, Rinehart and Winston.

E. D. Belokolos, A. I. Bobenko, V. Z. Enol'skii, A. R. Its, and V. B. Matveev (1994). *Algebro-geometric Approach to Nonlinear Integrable Problems.* Springer Series in Nonlinear Dynamics. Berlin: Springer-Verlag.

S. L. Belousov (1962). *Tables of Normalized Associated Legendre Polynomials.* Oxford-New York: Pergamon Press, The Macmillan Co.

C. M. Bender and S. A. Orszag (1978). *Advanced Mathematical Methods for Scientists and Engineers.* New York: McGraw-Hill Book Co. Reprinted by Springer-Verlag, New York, 1999.

C. M. Bender and T. T. Wu (1973). *Anharmonic oscillator. II. A study of perturbation theory in large order.* Phys. Rev. D **7**, pp. 1620–1636.

E. A. Bender (1974). *Asymptotic methods in enumeration.* SIAM Rev. **16**(4), pp. 485–515.

P. Berglund, P. Candelas, X. de la Ossa, and et al. (1994). *Periods for Calabi-Yau and Landau-Ginzburg vacua.* Nuclear Phys. B **419**(2), pp. 352–403.

A. Berkovich and B. M. McCoy (1998). *Rogers-Ramanujan Identities: A Century of Progress from Mathematics to Physics.* In *Proceedings of the International Congress of Mathematicians, Vol. III (Berlin, 1998)*, pp. 163–172.

G. D. Bernard and A. Ishimaru (1962). *Tables of the Anger and Lommel-Weber Functions*. Technical Report 53 and AFCRL 796, University Washington Press, Seattle. Reviewed in Math. Comp., v. 17 (1963), pp.315–317.

B. C. Berndt (1972). *On the Hurwitz zeta-function*. Rocky Mountain J. Math. **2**(1), pp. 151–157.

B. C. Berndt (1975a). *Character analogues of the Poisson and Euler-MacLaurin summation formulas with applications*. J. Number Theory **7**(4), pp. 413–445.

B. C. Berndt (1975b). *Periodic Bernoulli numbers, summation formulas and applications*. In *Theory and Application of Special Functions (Proc. Advanced Sem., Math. Res. Center, Univ. Wisconsin, Madison, Wis., 1975)*, pp. 143–189. New York: Academic Press.

B. C. Berndt (1989). *Ramanujan's Notebooks. Part II*. New York: Springer-Verlag.

B. C. Berndt (1991). *Ramanujan's Notebooks. Part III*. Berlin-New York: Springer-Verlag.

B. C. Berndt, S. Bhargava, and F. G. Garvan (1995). *Ramanujan's theories of elliptic functions to alternative bases*. Trans. Amer. Math. Soc. **347**(11), pp. 4163–4244.

B. C. Berndt and R. J. Evans (1984). *Chapter 13 of Ramanujan's second notebook: Integrals and asymptotic expansions*. Expo. Math. **2**(4), pp. 289–347.

M. V. Berry (1966). *Uniform approximation for potential scattering involving a rainbow*. Proc. Phys. Soc. **89**(3), pp. 479–490.

M. V. Berry (1969). *Uniform approximation: A new concept in wave theory*. Science Progress (Oxford) **57**, pp. 43–64.

M. V. Berry (1975). *Cusped rainbows and incoherence effects in the rippling-mirror model for particle scattering from surfaces*. J. Phys. A **8**(4), pp. 566–584.

M. V. Berry (1976). *Waves and Thom's theorem*. Advances in Physics **25**(1), pp. 1–26.

M. V. Berry (1977). *Focusing and twinkling: Critical exponents from catastrophes in non-Gaussian random short waves*. J. Phys. A **10**(12), pp. 2061–2081.

M. V. Berry (1980). *Some Geometric Aspects of Wave Motion: Wavefront Dislocations, Diffraction Catastrophes, Diffractals*. In *Geometry of the Laplace Operator (Proc. Sympos. Pure Math., Univ. Hawaii, Honolulu, Hawaii, 1979)*, Volume 36, pp. 13–28. Providence, R.I.: Amer. Math. Soc.

M. V. Berry (1981). *Singularities in Waves and Rays*. In R. Balian, M. Kléman, and J.-P. Poirier (Eds.), *Les Houches Lecture Series Session XXXV*, Volume 35, pp. 453–543. Amsterdam: North-Holland.

M. V. Berry (1989). *Uniform asymptotic smoothing of Stokes's discontinuities*. Proc. Roy. Soc. London Ser. A **422**, pp. 7–21.

M. V. Berry (1991). *Infinitely many Stokes smoothings in the gamma function*. Proc. Roy. Soc. London Ser. A **434**, pp. 465–472.

M. V. Berry (1995). *The Riemann-Siegel expansion for the zeta function: High orders and remainders*. Proc. Roy. Soc. London Ser. A **450**, pp. 439–462.

M. V. Berry and C. J. Howls (1990). *Stokes surfaces of diffraction catastrophes with codimension three*. Nonlinearity **3**(2), pp. 281–291.

M. V. Berry and C. J. Howls (1991). *Hyperasymptotics for integrals with saddles*. Proc. Roy. Soc. London Ser. A **434**, pp. 657–675.

M. V. Berry and C. J. Howls (1993). *Unfolding the high orders of asymptotic expansions with coalescing saddles: Singularity theory, crossover and duality*. Proc. Roy. Soc. London Ser. A **443**, pp. 107–126.

M. V. Berry and C. J. Howls (1994). *Overlapping Stokes smoothings: Survival of the error function and canonical catastrophe integrals*. Proc. Roy. Soc. London Ser. A **444**, pp. 201–216.

M. V. Berry and J. P. Keating (1992). *A new asymptotic representation for $\zeta(\frac{1}{2}+it)$ and quantum spectral determinants*. Proc. Roy. Soc. London Ser. A **437**, pp. 151–173.

M. V. Berry and J. P. Keating (1998). *H = xp and the Riemann Zeros*. In I. V. Lerner, J. P. Keating, and D. E. Khmelnitskii (Eds.), *Supersymmetry and Trace Formulae: Chaos and Disorder*, pp. 355–367. New York: Plenum.

M. V. Berry and J. P. Keating (1999). *The Riemann zeros and eigenvalue asymptotics*. SIAM Rev. **41**(2), pp. 236–266.

M. V. Berry, J. F. Nye, and F. J. Wright (1979). *The elliptic umbilic diffraction catastrophe*. Phil. Trans. Roy. Soc. Ser. A **291**(1382), pp. 453–484.

M. V. Berry and C. Upstill (1980). *Catastrophe optics: Morphologies of caustics and their diffraction patterns*. In E. Wolf (Ed.), *Progress in Optics*, Volume 18, pp. 257–346. Amsterdam: North-Holland.

M. V. Berry and F. J. Wright (1980). *Phase-space projection identities for diffraction catastrophes*. J. Phys. A **13**(1), pp. 149–160.

E. Berti and V. Cardoso (2006). *Quasinormal ringing of Kerr black holes: The excitation factors*. Phys. Rev. D **74**(104020), pp. 1–27.

H. A. Bethe and E. E. Salpeter (1977). *Quantum Mechanics of One- and Two-electron Atoms* (Rosetta ed.). New York: Plenum Publishing Corp. Reprint (with corrections) of the original edition published in 1957 by Springer and Academic Press.

F. Bethuel (1998). *Vortices in Ginzburg-Landau Equations.* In *Proceedings of the International Congress of Mathematicians, Vol. III (Berlin, 1998)*, pp. 11–19.

V. Bezvoda, R. Farzan, K. Segeth, and G. Takó (1986). *On numerical evaluation of integrals involving Bessel functions.* Apl. Mat. **31**(5), pp. 396–410.

A. Bhattacharjie and E. C. G. Sudarshan (1962). *A class of solvable potentials.* Nuovo Cimento (10) **25**, pp. 864–879.

A. Bhattacharyya and L. Shafai (1988). *Theoretical and experimental investigation of the elliptical annual ring antenna.* IEEE Trans. Antennas and Propagation **36**(11), pp. 1526–1530.

D. K. Bhaumik and S. K. Sarkar (2002). *On the power function of the likelihood ratio test for MANOVA.* J. Multivariate Anal. **82**(2), pp. 416–421.

W. G. Bickley (1935). *Some solutions of the problem of forced convection.* Philos. Mag. Series 7 **20**, pp. 322–343.

W. G. Bickley, L. J. Comrie, J. C. P. Miller, D. H. Sadler, and A. J. Thompson (1952). *Bessel Functions. Part II: Functions of Positive Integer Order.* British Association for the Advancement of Science, Mathematical Tables, Volume 10. Cambridge: Cambridge University Press. Reprinted, with corrections, in 1960.

W. G. Bickley and J. Nayler (1935). *A short table of the functions* $\text{Ki}_n(x)$, *from* $n=1$ *to* $n=16$. Phil. Mag. Series 7 **20**, pp. 343–347.

L. C. Biedenharn, R. L. Gluckstern, M. H. Hull, Jr., and G. Breit (1955). *Coulomb functions for large charges and small velocities.* Phys. Rev. (2) **97**(2), pp. 542–554.

L. C. Biedenharn and J. D. Louck (1981). *Angular Momentum in Quantum Physics: Theory and Application*, Volume 8 of *Encyclopedia of Mathematics and its Applications*. Reading, MA.: Addison-Wesley Publishing Co.

L. C. Biedenharn and H. van Dam (Eds.) (1965). *Quantum Theory of Angular Momentum. A Collection of Reprints and Original Papers.* New York: Academic Press.

D. Bierens de Haan (1867). *Nouvelles Tables d'Intégrales Définies.* Leide: P. Engels. Corrected edition printed by Hafner Press, 1965.

D. Bierens de Haan (1939). *Nouvelles Tables d'Intégrales Définies.* New York: G.E. Stechert & Co.

L. J. Billera, C. Greene, R. Simion, and R. P. Stanley (Eds.) (1996). *Formal Power Series and Algebraic Combinatorics*, Volume 24 of *DIMACS Series in Discrete Mathematics and Theoretical Computer Science*. Providence, RI: American Mathematical Society.

C. Bingham, T. Chang, and D. Richards (1992). *Approximating the matrix Fisher and Bingham distributions: Applications to spherical regression and Procrustes analysis.* J. Multivariate Anal. **41**(2), pp. 314–337.

Å. Björck (1996). *Numerical Methods for Least Squares Problems.* Philadelphia, PA: Society for Industrial and Applied Mathematics (SIAM).

J. M. Blair, C. A. Edwards, and J. H. Johnson (1976). *Rational Chebyshev approximations for the inverse of the error function.* Math. Comp. **30**(136), pp. 827–830.

J. M. Blair, C. A. Edwards, and J. H. Johnson (1978). *Rational Chebyshev approximations for the Bickley functions* $Ki_n(x)$. Math. Comp. **32**(143), pp. 876–886.

G. Blanch (1964). *Numerical evaluation of continued fractions.* SIAM Rev. **6**(4), pp. 383–421.

G. Blanch (1966). *Numerical aspects of Mathieu eigenvalues.* Rend. Circ. Mat. Palermo (2) **15**, pp. 51–97.

G. Blanch and D. S. Clemm (1962). *Tables Relating to the Radial Mathieu Functions. Vol. 1: Functions of the First Kind.* Washington, D.C.: U.S. Government Printing Office.

G. Blanch and D. S. Clemm (1965). *Tables Relating to the Radial Mathieu Functions. Vol. 2: Functions of the Second Kind.* Washington, D.C.: U.S. Government Printing Office.

G. Blanch and D. S. Clemm (1969). *Mathieu's Equation for Complex Parameters. Tables of Characteristic Values.* Washington, D.C.: U.S. Government Printing Office.

G. Blanch and I. Rhodes (1955). *Table of characteristic values of Mathieu's equation for large values of the parameter.* J. Washington Acad. Sci. **45**(6), pp. 166–196.

W. E. Bleick and P. C. C. Wang (1974). *Asymptotics of Stirling numbers of the second kind.* Proc. Amer. Math. Soc. **42**(2), pp. 575–580. Erratum: same journal v. 48 (1975), p. 518.

N. Bleistein (1966). *Uniform asymptotic expansions of integrals with stationary point near algebraic singularity.* Comm. Pure Appl. Math. **19**, pp. 353–370.

N. Bleistein (1967). *Uniform asymptotic expansions of integrals with many nearby stationary points and algebraic singularities.* J. Math. Mech. **17**, pp. 533–559.

N. Bleistein and R. A. Handelsman (1975). *Asymptotic Expansions of Integrals.* New York: Holt, Rinehart, and Winston. Reprinted with corrections by Dover Publications Inc., New York, 1986.

I. Bloch, M. H. Hull, Jr., A. A. Broyles, W. G. Bouricius, B. E. Freeman, and G. Breit (1950). *Methods of calculation of radial wave functions and new tables of Coulomb functions.* Physical Rev. (2) **80**, pp. 553–560.

I. Bloch, M. H. Hull, Jr., A. A. Broyles, W. G. Bouricius, B. E. Freeman, and G. Breit (1951). *Coulomb functions for reactions of protons and alpha-particles with the lighter nuclei.* Rev. Modern Physics **23**(2), pp. 147–182.

R. Bo and R. Wong (1994). *Uniform asymptotic expansion of Charlier polynomials.* Methods Appl. Anal. **1**(3), pp. 294–313.

R. Bo and R. Wong (1996). *Asymptotic behavior of the Pollaczek polynomials and their zeros.* Stud. Appl. Math. **96**, pp. 307–338.

R. Bo and R. Wong (1999). *A uniform asymptotic formula for orthogonal polynomials associated with* $\exp(-x^4)$. J. Approx. Theory **98**, pp. 146–166.

P. Boalch (2005). *From Klein to Painlevé via Fourier, Laplace and Jimbo.* Proc. London Math. Soc. (3) **90**(1), pp. 167–208.

P. Boalch (2006). *The fifty-two icosahedral solutions to Painlevé VI.* J. Reine Angew. Math. **596**, pp. 183–214.

A. Bobenko and A. Its (1995). *The Painlevé III equation and the Iwasawa decomposition.* Manuscripta Math. **87**(3), pp. 369–377.

A. I. Bobenko (1991). *Constant mean curvature surfaces and integrable equations.* Uspekhi Mat. Nauk **46**(4(280)), pp. 3–42, 192. In Russian. English translation: Russian Math. Surveys, **46**(1991), no. 4, pp. 1–45.

S. Bochner (1952). *Bessel functions and modular relations of higher type and hyperbolic differential equations.* Comm. Sém. Math. Univ. Lund [Medd. Lunds Univ. Mat. Sem.] **1952**(Tome Supplementaire), pp. 12–20.

S. Bochner and W. T. Martin (1948). *Several Complex Variables*, Volume 10 of *Princeton Mathematical Series*. Princeton, N. J.: Princeton University Press.

A. A. Bogush and V. S. Otchik (1997). *Problem of two Coulomb centres at large intercentre separation: Asymptotic expansions from analytical solutions of the Heun equation.* J. Phys. A **30**(2), pp. 559–571.

M. Born and E. Wolf (1999). *Principles of Optics: Electromagnetic Theory of Propagation, Interference and Diffraction of Light* (7th ed.). Cambridge: Cambridge University Press.

J. M. Borwein and P. B. Borwein (1987). *Pi and the AGM, A Study in Analytic Number Theory and Computational Complexity.* Canadian Mathematical Society Series of Monographs and Advanced Texts. New York: John Wiley & Sons Inc.

J. M. Borwein and P. B. Borwein (1991). *A cubic counterpart of Jacobi's identity and the AGM.* Trans. Amer. Math. Soc. **323**(2), pp. 691–701.

J. M. Borwein, D. M. Bradley, and R. E. Crandall (2000). *Computational strategies for the Riemann zeta function.* J. Comput. Appl. Math. **121**(1-2), pp. 247–296.

J. M. Borwein and R. M. Corless (1999). *Emerging tools for experimental mathematics.* Amer. Math. Monthly **106**(10), pp. 889–909.

J. M. Borwein and I. J. Zucker (1992). *Fast evaluation of the gamma function for small rational fractions using complete elliptic integrals of the first kind.* IMA J. Numer. Anal. **12**(4), pp. 519–526.

D. L. Bosley (1996). *A technique for the numerical verification of asymptotic expansions.* SIAM Rev. **38**(1), pp. 128–135.

W. Bosma and M.-P. van der Hulst (1990). *Faster Primality Testing.* In J.-J. Quisquater and J. Vandewalle (Eds.), *Advances in Cryptology—EUROCRYPT '89 Proceedings*, Volume 434 of *Lecture Notes in Computer Science*, New York, pp. 652–656. Springer-Verlag.

T. Bountis, H. Segur, and F. Vivaldi (1982). *Integrable Hamiltonian systems and the Painlevé property.* Phys. Rev. A (3) **25**(3), pp. 1257–1264.

P. Boutroux (1913). *Recherches sur les transcendantes de M. Painlevé et l'étude asymptotique des équations différentielles du second ordre.* Ann. Sci. École Norm. Sup. (3) **30**, pp. 255–375.

C. J. Bouwkamp (1947). *On spheroidal wave functions of order zero.* J. Math. Phys. Mass. Inst. Tech. **26**, pp. 79–92.

C. J. Bouwkamp (1948). *A note on Mathieu functions.* Proc. Nederl. Akad. Wetensch. **51**(7), pp. 891–893=Indagationes Math. **10**, 319–321 (1948).

F. Bowman (1953). *Introduction to Elliptic Functions with Applications.* London: English Universities Press, Ltd. Reprinted by Dover Publications Inc., New York, 1961.

F. Bowman (1958). *Introduction to Bessel Functions.* New York: Dover Publications Inc. Unaltered republication of the 1938 edition, published by Longmans, Green and Co., London-New York.

J. P. Boyd (1998). *Weakly Nonlocal Solitary Waves and Beyond-All-Orders Asymptotics*, Volume 442 of *Mathematics and its Applications*. Boston-Dordrecht: Kluwer Academic Publishers.

J. P. Boyd and A. Natarov (1998). *A Sturm-Liouville eigenproblem of the fourth kind: A critical latitude with equatorial trapping.* Stud. Appl. Math. **101**(4), pp. 433–455.

W. G. C. Boyd (1973). *The asymptotic analysis of canonical problems in high-frequency scattering theory. II. The circular and parabolic cylinders.* Proc. Cambridge Philos. Soc. **74**, pp. 313–332.

W. G. C. Boyd (1987). *Asymptotic expansions for the coefficient functions that arise in turning-point problems.* Proc. Roy. Soc. London Ser. A **410**, pp. 35–60.

W. G. C. Boyd (1990a). *Asymptotic Expansions for the Coefficient Functions Associated with Linear Second-order Differential Equations: The Simple Pole Case.* In R. Wong (Ed.), *Asymptotic and Computational Analysis (Winnipeg, MB, 1989)*, Volume 124 of *Lecture Notes in Pure and Applied Mathematics*, pp. 53–73. New York: Dekker.

W. G. C. Boyd (1990b). *Stieltjes transforms and the Stokes phenomenon.* Proc. Roy. Soc. London Ser. A **429**, pp. 227–246.

W. G. C. Boyd (1993). *Error bounds for the method of steepest descents.* Proc. Roy. Soc. London Ser. A **440**, pp. 493–518.

W. G. C. Boyd (1994). *Gamma function asymptotics by an extension of the method of steepest descents.* Proc. Roy. Soc. London Ser. A **447**, pp. 609–630.

W. G. C. Boyd (1995). *Approximations for the late coefficients in asymptotic expansions arising in the method of steepest descents.* Methods Appl. Anal. **2**(4), pp. 475–489.

W. G. C. Boyd and T. M. Dunster (1986). *Uniform asymptotic solutions of a class of second-order linear differential equations having a turning point and a regular singularity, with an application to Legendre functions.* SIAM J. Math. Anal. **17**(2), pp. 422–450.

T. H. Boyer (1969). *Concerning the zeros of some functions related to Bessel functions.* J. Mathematical Phys. **10**(9), pp. 1729–1744.

B. L. J. Braaksma and B. Meulenbeld (1967). *Integral transforms with generalized Legendre functions as kernels.* Compositio Math. **18**, pp. 235–287.

M. Brack, M. Mehta, and K. Tanaka (2001). *Occurrence of periodic Lamé functions at bifurcations in chaotic Hamiltonian systems.* J. Phys. A **34**(40), pp. 8199–8220.

N. Brazel, F. Lawless, and A. Wood (1992). *Exponential asymptotics for an eigenvalue of a problem involving parabolic cylinder functions.* Proc. Amer. Math. Soc. **114**(4), pp. 1025–1032.

R. P. Brent (1976). *Fast multiple-precision evaluation of elementary functions.* J. Assoc. Comput. Mach. **23**(2), pp. 242–251.

D. Bressoud and S. Wagon (2000). *A Course in Computational Number Theory.* Emeryville, CA: Key College Publishing. With 1 CD-ROM (Windows, Macintosh and UNIX) containing a collection of Mathematica programs for computations in number theory.

D. M. Bressoud (1989). *Factorization and Primality Testing.* New York: Springer-Verlag.

D. M. Bressoud (1999). *Proofs and Confirmations: The Story of the Alternating Sign Matrix Conjecture.* Cambridge: Cambridge University Press.

E. Brézin, C. Itzykson, G. Parisi, and J. B. Zuber (1978). *Planar diagrams.* Comm. Math. Phys. **59**(1), pp. 35–51.

C. Brezinski (1980). *Padé-type Approximation and General Orthogonal Polynomials*, Volume 50 of *International Series of Numerical Mathematics*. Basel: Birkhäuser Verlag.

C. Brezinski (1999). *Error estimates for the solution of linear systems.* SIAM J. Sci. Comput. **21**(2), pp. 764–781.

C. Brezinski and M. Redivo Zaglia (1991). *Extrapolation Methods. Theory and Practice*, Volume 2 of *Studies in Computational Mathematics*. Amsterdam: North-Holland Publishing Co. With 1 IBM-PC floppy disk (5.25 inch).

E. Brieskorn and H. Knörrer (1986). *Plane Algebraic Curves.* Basel: Birkhäuser Verlag. Translated from the German by John Stillwell.

J. Brillhart (1969). *On the Euler and Bernoulli polynomials.* J. Reine Angew. Math. **234**, pp. 45–64.

D. M. Brink and G. R. Satchler (1993). *Angular Momentum* (3rd ed.). Oxford: Oxford University Press.

British Association for the Advancement of Science (1937). *Bessel Functions. Part I: Functions of Orders Zero and Unity.* Mathematical Tables, Volume 6. Cambridge: Cambridge University Press.

J. C. Bronski, L. D. Carr, B. Deconinck, J. N. Kutz, and K. Promislow (2001). *Stability of repulsive Bose-Einstein condensates in a periodic potential.* Phys. Rev. E (3) **63**(036612), pp. 1–11.

J. Brüning (1984). *On the asymptotic expansion of some integrals.* Arch. Math. (Basel) **42**(3), pp. 253–259.

Yu. A. Brychkov and K. O. Geddes (2005). *On the derivatives of the Bessel and Struve functions with respect to the order.* Integral Transforms Spec. Funct. **16**(3), pp. 187–198.

H. Buchholz (1969). *The Confluent Hypergeometric Function with Special Emphasis on Its Applications.* New York: Springer-Verlag. Translated from the German by H. Lichtbau and K. Wetzel.

J. D. Buckholtz (1963). *Concerning an approximation of Copson.* Proc. Amer. Math. Soc. **14**(4), pp. 564–568.

J. Buhler, R. Crandall, R. Ernvall, T. Metsänkylä, and M. A. Shokrollahi (2001). *Irregular primes and cyclotomic invariants to 12 million.* J. Symbolic Comput. **31**(1-2), pp. 89–96. Computational algebra and number theory (Milwaukee, WI, 1996).

J. P. Buhler, R. E. Crandall, and R. W. Sompolski (1992). *Irregular primes to one million.* Math. Comp. **59**(200), pp. 717–722.

W. Bühring (1987a). *An analytic continuation of the hypergeometric series.* SIAM J. Math. Anal. **18**(3), pp. 884–889.

W. Bühring (1987b). *The behavior at unit argument of the hypergeometric function $_3F_2$*. SIAM J. Math. Anal. **18**(5), pp. 1227–1234.

W. Bühring (1988). *An analytic continuation formula for the generalized hypergeometric function*. SIAM J. Math. Anal. **19**(5), pp. 1249–1251.

W. Bühring (1992). *Generalized hypergeometric functions at unit argument*. Proc. Amer. Math. Soc. **114**(1), pp. 145–153.

W. Bühring (1994). *The double confluent Heun equation: Characteristic exponent and connection formulae*. Methods Appl. Anal. **1**(3), pp. 348–370.

R. Bulirsch (1965a). *Numerical calculation of elliptic integrals and elliptic functions*. Numer. Math. **7**(1), pp. 78–90. Algol 60 programs are included for real and complex elliptic integrals, and for Jacobian elliptic functions of real argument.

R. Bulirsch (1965b). *Numerical calculation of elliptic integrals and elliptic functions. II*. Numer. Math. **7**(4), pp. 353–354. An improved Algol 60 program for the complete elliptic integral of the third kind is included, but superseded by a program in Bulirsch (1969b).

R. Bulirsch (1967). *Numerical calculation of the sine, cosine and Fresnel integrals*. Numer. Math. **9**(5), pp. 380–385. Algol 60 procedures for sine, cosine, and 2 Fresnel integrals are included.

R. Bulirsch (1969a). *An extension of the Bartky-transformation to incomplete elliptic integrals of the third kind*. Numer. Math. **13**(3), pp. 266–284.

R. Bulirsch (1969b). *Numerical calculation of elliptic integrals and elliptic functions. III*. Numer. Math. **13**(4), pp. 305–315. Algol 60 programs for the incomplete elliptic integral of the third kind, and for a general complete elliptic integral, are included.

R. Bulirsch and H. Rutishauser (1968). *Interpolation und genäherte Quadratur*. In R. Sauer and I. Szabó (Eds.), *Mathematische Hilfsmittel des Ingenieurs. Teil III*, Volume 141 of *Die Grundlehren der mathematischen Wissenschaften in Einzeldarstellungen*, pp. 232–319. Berlin: Springer-Verlag.

R. Bulirsch and J. Stoer (1968). *II. Darstellung von Funktionen in Rechenautomaten*. In R. Sauer and I. Szabó (Eds.), *Mathematische Hilfsmittel des Ingenieurs. Teil III*. Berlin-New York: Springer-Verlag.

P. S. Bullen (1998). *A Dictionary of Inequalities*, Volume 97 of *Pitman Monographs and Surveys in Pure and Applied Mathematics*. Harlow: Longman.

J. L. Burchnall and T. W. Chaundy (1940). *Expansions of Appell's double hypergeometric functions*. Quart. J. Math., Oxford Ser. **11**, pp. 249–270.

J. L. Burchnall and T. W. Chaundy (1941). *Expansions of Appell's double hypergeometric functions. II*. Quart. J. Math., Oxford Ser. **12**, pp. 112–128.

J. L. Burchnall and T. W. Chaundy (1948). *The hypergeometric identities of Cayley, Orr, and Bailey*. Proc. London Math. Soc. (2) **50**, pp. 56–74.

A. Burgess (1963). *The determination of phases and amplitudes of wave functions*. Proc. Phys. Soc. **81**(3), pp. 442–452.

T. W. Burkhardt and T. Xue (1991). *Density profiles in confined critical systems and conformal invariance*. Phys. Rev. Lett. **66**(7), pp. 895–898.

W. S. Burnside and A. W. Panton (1960). *The Theory of Equations: With an Introduction to the Theory of Binary Algebraic Forms*. New York: Dover Publications. Two volumes. A reproduction of the seventh edition [vol. 1, Longmans, Green, London 1912; vol. 2, Longmans, Green, London 1928].

A. I. Burshtein and S. I. Temkin (1994). *Spectroscopy of Molecular Rotation in Gases and Liquids*. Cambridge: Cambridge University Press.

N. M. Burunova (1960). *A Guide to Mathematical Tables: Supplement No. 1*. New York: Pergamon Press. Supplement to "A Guide to Mathematical Tables" by A. V. Lebedev and R. M. Fedorova.

T. Busch, B.-G. Englert, K. Rzążewski, and M. Wilkens (1998). *Two cold atoms in a harmonic trap*. Found. Phys. **28**(4), pp. 549–559.

P. J. Bushell (1987). *On a generalization of Barton's integral and related integrals of complete elliptic integrals*. Math. Proc. Cambridge Philos. Soc. **101**(1), pp. 1–5.

J. C. Butcher (1987). *The Numerical Analysis of Ordinary Differential Equations. Runge-Kutta and General Linear Methods*. Chichester: John Wiley & Sons Ltd.

J. C. Butcher (2003). *Numerical Methods for Ordinary Differential Equations*. Chichester: John Wiley & Sons Ltd.

R. W. Butler and A. T. A. Wood (2002). *Laplace approximations for hypergeometric functions with matrix argument*. Ann. Statist. **30**(4), pp. 1155–1177.

R. W. Butler and A. T. A. Wood (2003). *Laplace approximation for Bessel functions of matrix argument*. J. Comput. Appl. Math. **155**(2), pp. 359–382.

P. L. Butzer, S. Flocke, and M. Hauss (1994). *Euler functions $E_\alpha(z)$ with complex α and applications*. In G. Anastassiou and S. T. Rachev (Eds.), *Approximation, probability, and related fields (Santa Barbara, CA, 1993)*, pp. 127–150. New York: Plenum Press.

P. L. Butzer and M. Hauss (1992). *Riemann zeta function: Rapidly converging series and integral representations*. Appl. Math. Lett. **5**(2), pp. 83–88.

P. L. Butzer, M. Hauss, and M. Leclerc (1992). *Bernoulli numbers and polynomials of arbitrary complex indices.* Appl. Math. Lett. **5**(6), pp. 83–88.

J. G. Byatt-Smith (2000). *The Borel transform and its use in the summation of asymptotic expansions.* Stud. Appl. Math. **105**(2), pp. 83–113.

W. E. Byerly (1888). *Elements of the Integral Calculus* (2nd ed.). Boston: Ginn & Co. Reprinted by G. E. Stechert, New York, 1926.

P. F. Byrd and M. D. Friedman (1971). *Handbook of Elliptic Integrals for Engineers and Scientists* (2nd ed.). Die Grundlehren der mathematischen Wissenschaften, Band 67. New York: Springer-Verlag. Table errata: Math. Comp. v. 66 (1997), no. 220, p. 1767, v. 36 (1981), no. 153, p. 317,319, v. 26 (1972), no. 118, p. 597.

L. G. Cabral-Rosetti and M. A. Sanchis-Lozano (2000). *Generalized hypergeometric functions and the evaluation of scalar one-loop integrals in Feynman diagrams.* J. Comput. Appl. Math. **115**(1-2), pp. 93–99.

F. Cajori (1929). *A History of Mathematical Notations, Volume II.* Chicago: Open Court Publishing Company. Reprinted by Cosimo in 2007.

F. Calogero (1978). *Asymptotic behaviour of the zeros of the (generalized) Laguerre polynomial $L_n^\alpha(x)$ as the index $\alpha \to \infty$ and limiting formula relating Laguerre polynomials of large index and large argument to Hermite polynomials.* Lett. Nuovo Cimento (2) **23**(3), pp. 101–102.

J. Camacho, R. Guimerà, and L. A. N. Amaral (2002). *Analytical solution of a model for complex food webs.* Phys. Rev. E **65**(3), pp. (030901–1)–(030901–4).

P. J. Cameron (1994). *Combinatorics: Topics, Techniques, Algorithms.* Cambridge: Cambridge University Press.

J. B. Campbell (1980). *On Temme's algorithm for the modified Bessel function of the third kind.* ACM Trans. Math. Software **6**(4), pp. 581–586.

R. Campbell (1955). *Théorie Générale de L'Équation de Mathieu et de quelques autres Équations différentielles de la mécanique.* Paris: Masson et Cie.

R. G. Campos (1995). *A quadrature formula for the Hankel transform.* Numer. Algorithms **9**(2), pp. 343–354.

S. M. Candel (1981). *An algorithm for the Fourier-Bessel transform.* Comput. Phys. Comm. **23**(4), pp. 343–353.

L. Carlitz (1953). *Some congruences for the Bernoulli numbers.* Amer. J. Math. **75**(1), pp. 163–172.

L. Carlitz (1954a). *A note on Euler numbers and polynomials.* Nagoya Math. J. **7**, pp. 35–43.

L. Carlitz (1954b). *q-Bernoulli and Eulerian numbers.* Trans. Amer. Math. Soc. **76**(2), pp. 332–350.

L. Carlitz (1958). *Expansions of q-Bernoulli numbers.* Duke Math. J. **25**(2), pp. 355–364.

L. Carlitz (1960). *Note on Nörlund's polynomial $B_n^{(z)}$.* Proc. Amer. Math. Soc. **11**(3), pp. 452–455.

L. Carlitz (1961a). *A recurrence formula for $\zeta(2n)$.* Proc. Amer. Math. Soc. **12**(6), pp. 991–992.

L. Carlitz (1961b). *The Staudt-Clausen theorem.* Math. Mag. **34**, pp. 131–146.

L. Carlitz (1963). *The inverse of the error function.* Pacific J. Math. **13**(2), pp. 459–470.

R. D. Carlitz (1972, June). *Hadronic matter at high density.* Phys. Rev. D **5**(12), pp. 3231–3242.

B. C. Carlson (1961a). *Ellipsoidal distributions of charge or mass.* J. Mathematical Phys. **2**, pp. 441–450.

B. C. Carlson (1961b). *Some series and bounds for incomplete elliptic integrals.* J. Math. and Phys. **40**, pp. 125–134.

B. C. Carlson (1963). *Lauricella's hypergeometric function F_D.* J. Math. Anal. Appl. **7**(3), pp. 452–470.

B. C. Carlson (1964). *Normal elliptic integrals of the first and second kinds.* Duke Math. J. **31**(3), pp. 405–419.

B. C. Carlson (1965). *On computing elliptic integrals and functions.* J. Math. and Phys. **44**, pp. 36–51.

B. C. Carlson (1966). *Some inequalities for hypergeometric functions.* Proc. Amer. Math. Soc. **17**(1), pp. 32–39.

B. C. Carlson (1970). *Inequalities for a symmetric elliptic integral.* Proc. Amer. Math. Soc. **25**(3), pp. 698–703.

B. C. Carlson (1971). *New proof of the addition theorem for Gegenbauer polynomials.* SIAM J. Math. Anal. **2**, pp. 347–351.

B. C. Carlson (1972a). *An algorithm for computing logarithms and arctangents.* Math. Comp. **26**(118), pp. 543–549.

B. C. Carlson (1972b). *Intégrandes à deux formes quadratiques.* C. R. Acad. Sci. Paris Sér. A–B **274** (15 May, 1972, Sér. A), pp. 1458–1461.

B. C. Carlson (1976). *Quadratic transformations of Appell functions.* SIAM J. Math. Anal. **7**(2), pp. 291–304.

B. C. Carlson (1977a). *Elliptic integrals of the first kind.* SIAM J. Math. Anal. **8**(2), pp. 231–242.

B. C. Carlson (1977b). *Special Functions of Applied Mathematics.* New York: Academic Press.

B. C. Carlson (1978). *Short proofs of three theorems on elliptic integrals.* SIAM J. Math. Anal. **9**(3), pp. 524–528.

B. C. Carlson (1979). *Computing elliptic integrals by duplication.* Numer. Math. **33**(1), pp. 1–16.

B. C. Carlson (1985). *The hypergeometric function and the R-function near their branch points.* Rend. Sem. Mat. Univ. Politec. Torino (Fascicolo Speciale), pp. 63–89.

B. C. Carlson (1987). *A table of elliptic integrals of the second kind.* Math. Comp. **49**(180), pp. 595–606, S13–S17.

B. C. Carlson (1988). *A table of elliptic integrals of the third kind.* Math. Comp. **51**(183), pp. 267–280, S1–S5.

B. C. Carlson (1990). *Landen Transformations of Integrals.* In R. Wong (Ed.), *Asymptotic and Computational Analysis (Winnipeg, MB, 1989)*, Volume 124 of *Lecture Notes in Pure and Appl. Math.*, pp. 75–94. New York-Basel: Marcel Dekker.

B. C. Carlson (1995). *Numerical computation of real or complex elliptic integrals.* Numer. Algorithms **10**(1-2), pp. 13–26.

B. C. Carlson (1998). *Elliptic Integrals: Symmetry and Symbolic Integration.* In *Tricomi's Ideas and Contemporary Applied Mathematics (Rome/Turin, 1997)*, Volume 147 of *Atti dei Convegni Lincei*, pp. 161–181. Rome: Accad. Naz. Lincei.

B. C. Carlson (1999). *Toward symbolic integration of elliptic integrals.* J. Symbolic Comput. **28**(6), pp. 739–753.

B. C. Carlson (2002). *Three improvements in reduction and computation of elliptic integrals.* J. Res. Nat. Inst. Standards Tech. **107**(5), pp. 413–418.

B. C. Carlson (2004). *Symmetry in c, d, n of Jacobian elliptic functions.* J. Math. Anal. Appl. **299**(1), pp. 242–253.

B. C. Carlson (2005). *Jacobian elliptic functions as inverses of an integral.* J. Comput. Appl. Math. **174**(2), pp. 355–359.

B. C. Carlson (2006a). *Some reformulated properties of Jacobian elliptic functions.* J. Math. Anal. Appl. **323**(1), pp. 522–529.

B. C. Carlson (2006b). *Table of integrals of squared Jacobian elliptic functions and reductions of related hypergeometric R-functions.* Math. Comp. **75**(255), pp. 1309–1318.

B. C. Carlson (2008). *Power series for inverse Jacobian elliptic functions.* Math. Comp. **77**(263), pp. 1615–1621.

B. C. Carlson and J. FitzSimons (2000). *Reduction theorems for elliptic integrands with the square root of two quadratic factors.* J. Comput. Appl. Math. **118**(1-2), pp. 71–85.

B. C. Carlson and J. L. Gustafson (1985). *Asymptotic expansion of the first elliptic integral.* SIAM J. Math. Anal. **16**(5), pp. 1072–1092.

B. C. Carlson and J. L. Gustafson (1994). *Asymptotic approximations for symmetric elliptic integrals.* SIAM J. Math. Anal. **25**(2), pp. 288–303.

B. C. Carlson and G. S. Rushbrooke (1950). *On the expansion of a Coulomb potential in spherical harmonics.* Proc. Cambridge Philos. Soc. **46**, pp. 626–633.

L. D. Carr, C. W. Clark, and W. P. Reinhardt (2000). *Stationary solutions of the one-dimensional nonlinear Schrödinger equation. I. Case of repulsive nonlinearity.* Phys. Rev. A **62**(063610), pp. 1–10.

H. S. Carslaw (1930). *Introduction to the Theory of Fourier's Series and Integrals* (3rd ed.). London: Macmillan. Reprinted by Dover, New York, 1950.

H. S. Carslaw and J. C. Jaeger (1959). *Conduction of Heat in Solids* (2nd ed.). Oxford: Clarendon Press.

J. R. Cash and R. V. M. Zahar (1994). *A Unified Approach to Recurrence Algorithms.* In R. V. M. Zahar (Ed.), *Approximation and Computation (West Lafayette, IN, 1993)*, Volume 119 of *International Series of Computational Mathematics*, pp. 97–120. Boston, MA: Birkhäuser Boston.

A. Cayley (1895). *An Elementary Treatise on Elliptic Functions.* London: George Bell and Sons. Republished by Dover Publications, Inc,, New York, 1961. Table errataum: Math. Comp. v. 29 (1975), no. 130, p. 670.

R. Cazenave (1969). *Intégrales et Fonctions Elliptiques en Vue des Applications.* Préface de Henri Villat. Publications Scientifiques et Techniques du Ministère de l'Air, No. 452. Paris: Centre de Documentation de l'Armement.

H. H. Chan (1998). *On Ramanujan's cubic transformation formula for $_2F_1\left(\frac{1}{3}, \frac{2}{3}; 1; z\right)$.* Math. Proc. Cambridge Philos. Soc. **124**(2), pp. 193–204.

S. Chandrasekhar (1984). *The Mathematical Theory of Black Holes.* In *General Relativity and Gravitation (Padova, 1983)*, pp. 5–26. Dordrecht: Reidel.

F. Chapeau-Blondeau and A. Monir (2002). *Numerical evaluation of the Lambert W function and application to generation of generalized Gaussian noise with exponent 1/2.* IEEE Trans. Signal Process. **50**(9), pp. 2160–2165.

C. J. Chapman (1999). *Caustics in cylindrical ducts.* Proc. Roy. Soc. London Ser. A **455**, pp. 2529–2548.

B. W. Char (1980). *On Stieltjes' continued fraction for the gamma function.* Math. Comp. **34**(150), pp. 547–551.

M. A. Chaudhry, N. M. Temme, and E. J. M. Veling (1996). *Asymptotics and closed form of a generalized incomplete gamma function.* J. Comput. Appl. Math. **67**(2), pp. 371–379.

M. A. Chaudhry and S. M. Zubair (1994). *Generalized incomplete gamma functions with applications.* J. Comput. Appl. Math. **55**(1), pp. 99–124.

M. A. Chaudhry and S. M. Zubair (2001). *On a Class of Incomplete Gamma Functions with Applications.* Boca Raton, FL: Chapman & Hall/CRC.

T. W. Chaundy (1969). *Elementary Differential Equations.* Oxford: Clarendon Press.

P. L. Chebyshev (1851). *Sur la fonction qui détermine la totalité des nombres premiers inférieurs à une limite donnée*. Mem. Ac. Sc. St. Pétersbourg **6**, pp. 141–157.

M. Chellali (1988). *Accélération de calcul de nombres de Bernoulli*. J. Number Theory **28**(3), pp. 347–362.

R. Chelluri, L. B. Richmond, and N. M. Temme (2000). *Asymptotic estimates for generalized Stirling numbers*. Analysis (Munich) **20**(1), pp. 1–13.

J.-r. Chen (1966). *On the representation of a large even integer as the sum of a prime and the product of at most two primes*. Kexue Tongbao (Foreign Lang. Ed.) **17**, pp. 385–386.

L.-C. Chen, M. E. H. Ismail, and P. Simeonov (1999). *Asymptotics of Racah coefficients and polynomials*. J. Phys. A **32**(3), pp. 537–553.

Y. Chen and M. E. H. Ismail (1998). *Asymptotics of the largest zeros of some orthogonal polynomials*. J. Phys. A **31**(25), pp. 5525–5544.

E. W. Cheney (1982). *Introduction to Approximation Theory* (2nd ed.). New York: Chelsea Publishing Co. Reprinted by AMS Chelsea Publishing, Providence, RI, 1998.

I. Cherednik (1995). *Macdonald's evaluation conjectures and difference Fourier transform*. Invent. Math. **122**(1), pp. 119–145. Erratum: see same journal **125**(1996), no. 2, p. 391.

T. M. Cherry (1948). *Expansions in terms of parabolic cylinder functions*. Proc. Edinburgh Math. Soc. (2) **8**, pp. 50–65.

C. Chester, B. Friedman, and F. Ursell (1957). *An extension of the method of steepest descents*. Proc. Cambridge Philos. Soc. **53**, pp. 599–611.

C. Chiccoli, S. Lorenzutta, and G. Maino (1987). *A numerical method for generalized exponential integrals*. Comput. Math. Appl. **14**(4), pp. 261–268.

C. Chiccoli, S. Lorenzutta, and G. Maino (1988). *On the evaluation of generalized exponential integrals $E_v(x)$*. J. Comput. Phys. **78**(2), pp. 278–287.

C. Chiccoli, S. Lorenzutta, and G. Maino (1990). *An algorithm for exponential integrals of real order*. Computing **45**(3), pp. 269–276. Includes Fortran code.

L. Chihara (1987). *On the zeros of the Askey-Wilson polynomials, with applications to coding theory*. SIAM J. Math. Anal. **18**(1), pp. 191–207.

T. S. Chihara (1978). *An Introduction to Orthogonal Polynomials*, Volume 13 of *Mathematics and its Applications*. New York: Gordon and Breach Science Publishers.

T. S. Chihara and M. E. H. Ismail (1993). *Extremal measures for a system of orthogonal polynomials*. Constr. Approx. **9**, pp. 111–119.

Y. Chikuse (2003). *Statistics on Special Manifolds*, Volume 174 of *Lecture Notes in Statistics*. New York: Springer-Verlag.

R. C. Y. Chin and G. W. Hedstrom (1978). *A dispersion analysis for difference schemes: Tables of generalized Airy functions*. Math. Comp. **32**(144), pp. 1163–1170.

B. K. Choudhury (1995). *The Riemann zeta-function and its derivatives*. Proc. Roy. Soc. London Ser. A **450**, pp. 477–499.

Y. Chow, L. Gatteschi, and R. Wong (1994). *A Bernstein-type inequality for the Jacobi polynomial*. Proc. Amer. Math. Soc. **121**(3), pp. 703–709.

N. B. Christensen (1990). *Optimized fast Hankel transform filters*. Geophysical Prospecting **38**(5), pp. 545–568.

J. S. Christiansen and M. E. H. Ismail (2006). *A moment problem and a family of integral evaluations*. Trans. Amer. Math. Soc. **358**(9), pp. 4071–4097.

R. F. Christy and I. Duck (1961). *γ rays from an extranuclear direct capture process*. Nuclear Physics **24**(1), pp. 89–101.

G. Chrystal (1959). *Algebra: An Elementary Textbook for the Higher Classes of Secondary Schools and for Colleges* (6th ed.), Volume 1. New York: Chelsea Publishing Co. A seventh edition (reprinting of the sixth edition) was published by AMS Chelsea Publishing Series, American Mathematical Society, 1964.

D. V. Chudnovsky and G. V. Chudnovsky (1988). *Approximations and Complex Multiplication According to Ramanujan*. In G. E. Andrews, R. A. Askey, B. C. Bernd, K. G. Ramanathan, and R. A. Rankin (Eds.), *Ramanujan Revisited (Urbana-Champaign, Ill., 1987)*, pp. 375–472. Boston, MA: Academic Press.

C. K. Chui (1988). *Multivariate Splines*, Volume 54 of *CBMS-NSF Regional Conference Series in Applied Mathematics*. Philadelphia, PA: Society for Industrial and Applied Mathematics (SIAM). With an appendix by Harvey Diamond.

A. Ciarkowski (1989). *Uniform asymptotic expansion of an integral with a saddle point, a pole and a branch point*. Proc. Roy. Soc. London Ser. A **426**, pp. 273–286.

R. Cicchetti and A. Faraone (2004). *Incomplete Hankel and modified Bessel functions: A class of special functions for electromagnetics*. IEEE Trans. Antennas and Propagation **52**(12), pp. 3373–3389.

G. M. Cicuta and E. Montaldi (1975). *Remarks on the full asymptotic expansion of Feynman parametrized integrals*. Lett. Nuovo Cimento (2) **13**(8), pp. 310–312.

C. W. Clark (1979). *Coulomb phase shift*. American Journal of Physics **47**(8), pp. 683–684.

F. Clarke (1989). *The universal von Staudt theorems*. Trans. Amer. Math. Soc. **315**(2), pp. 591–603.

P. A. Clarkson (1991). *Nonclassical Symmetry Reductions and Exact Solutions for Physically Significant Nonlinear Evolution Equations*. In W. Rozmus and J. A. Tuszynski (Eds.), *Nonlinear and Chaotic Phenomena in Plasmas, Solids and Fluids (Edmonton, AB, 1990)*, pp. 72–79. Singapore: World Scientific Publishing.

P. A. Clarkson (2003a). *The third Painlevé equation and associated special polynomials*. J. Phys. A **36**(36), pp. 9507–9532.

P. A. Clarkson (2003b). *The fourth Painlevé equation and associated special polynomials*. J. Math. Phys. **44**(11), pp. 5350–5374.

P. A. Clarkson (2005). *Special polynomials associated with rational solutions of the fifth Painlevé equation*. J. Comput. Appl. Math. **178**(1-2), pp. 111–129.

P. A. Clarkson (2006). *Painlevé Equations—Nonlinear Special Functions: Computation and Application*. In F. Marcellàn and W. van Assche (Eds.), *Orthogonal Polynomials and Special Functions*, Volume 1883 of *Lecture Notes in Math.*, pp. 331–411. Berlin: Springer.

P. A. Clarkson and E. L. Mansfield (2003). *The second Painlevé equation, its hierarchy and associated special polynomials*. Nonlinearity **16**(3), pp. R1–R26.

P. A. Clarkson and J. B. McLeod (1988). *A connection formula for the second Painlevé transcendent*. Arch. Rational Mech. Anal. **103**(2), pp. 97–138.

T. Clausen (1828). *Über die Fälle, wenn die Reihe von der Form $y = 1 + \frac{\alpha}{1} \cdot \frac{\beta}{\gamma} x + \frac{\alpha \cdot \alpha+1}{1 \cdot 2} \cdot \frac{\beta \cdot \beta+1}{\gamma \cdot \gamma+1} x^2 +$ etc. ein Quadrat von der Form $z = 1 + \frac{\alpha'}{1} \cdot \frac{\beta'}{\gamma'} \cdot \frac{\delta'}{\epsilon'} x + \frac{\alpha' \cdot \alpha'+1}{1 \cdot 2} \cdot \frac{\beta' \cdot \beta'+1}{\gamma' \cdot \gamma'+1} \cdot \frac{\delta' \cdot \delta'+1}{\epsilon' \cdot \epsilon'+1} x^2 +$ etc. hat*. J. Reine Angew. Math. **3**, pp. 89–91.

W. W. Clendenin (1966). *A method for numerical calculation of Fourier integrals*. Numer. Math. **8**(5), pp. 422–436.

C. W. Clenshaw (1955). *A note on the summation of Chebyshev series*. Math. Tables Aids Comput. **9**(51), pp. 118–120.

C. W. Clenshaw (1957). *The numerical solution of linear differential equations in Chebyshev series*. Proc. Cambridge Philos. Soc. **53**(1), pp. 134–149.

C. W. Clenshaw (1962). *Chebyshev Series for Mathematical Functions*. National Physical Laboratory Mathematical Tables, Vol. 5. Department of Scientific and Industrial Research. London: Her Majesty's Stationery Office.

C. W. Clenshaw and A. R. Curtis (1960). *A method for numerical integration on an automatic copmputer*. Numer. Math. **2**(4), pp. 197–205.

C. W. Clenshaw, D. W. Lozier, F. W. J. Olver, and P. R. Turner (1986). *Generalized exponential and logarithmic functions*. Comput. Math. Appl. Part B **12**(5-6), pp. 1091–1101.

C. W. Clenshaw and F. W. J. Olver (1984). *Beyond floating point*. J. Assoc. Comput. Mach. **31**(2), pp. 319–328.

C. W. Clenshaw, F. W. J. Olver, and P. R. Turner (1989). *Level-Index Arithmetic: An Introductory Survey*. In P. R. Turner (Ed.), *Numerical Analysis and Parallel Processing (Lancaster, 1987)*, Volume 1397 of *Lecture Notes in Math.*, pp. 95–168. Berlin-Heidelberg: Springer.

L. D. Cloutman (1989). *Numerical evaluation of the Fermi-Dirac integrals*. The Astrophysical Journal Supplement Series **71**, pp. 677–699. Double-precision Fortran, maximum precision 12D.

J. A. Cochran (1963). *Further formulas for calculating approximate values of the zeros of certain combinations of Bessel functions*. IEEE Trans. Microwave Theory Tech. **11**(6), pp. 546–547.

J. A. Cochran (1964). *Remarks on the zeros of cross-product Bessel functions*. J. Soc. Indust. Appl. Math. **12**(3), pp. 580–587.

J. A. Cochran (1965). *The zeros of Hankel functions as functions of their order*. Numer. Math. **7**(3), pp. 238–250.

J. A. Cochran (1966a). *The analyticity of cross-product Bessel function zeros*. Proc. Cambridge Philos. Soc. **62**, pp. 215–226.

J. A. Cochran (1966b). *The asymptotic nature of zeros of cross-product Bessel functions*. Quart. J. Mech. Appl. Math. **19**(4), pp. 511–522.

J. A. Cochran and J. N. Hoffspiegel (1970). *Numerical techniques for finding ν-zeros of Hankel functions*. Math. Comp. **24**(110), pp. 413–422.

W. J. Cody (1965a). *Chebyshev approximations for the complete elliptic integrals K and E*. Math. Comp. **19**(89), pp. 105–112.

W. J. Cody (1965b). *Chebyshev polynomial expansions of complete elliptic integrals*. Math. Comp. **19**(90), pp. 249–259.

W. J. Cody (1968). *Chebyshev approximations for the Fresnel integrals*. Math. Comp. **22**(102), pp. 450–453. with Microfiche Supplement.

W. J. Cody (1969). *Rational Chebyshev approximations for the error function*. Math. Comp. **23**(107), pp. 631–637.

W. J. Cody (1970). *A survey of practical rational and polynomial approximation of functions*. SIAM Rev. **12**(3), pp. 400–423.

W. J. Cody and K. E. Hillstrom (1967). *Chebyshev approximations for the natural logarithm of the gamma function*. Math. Comp. **21**(98), pp. 198–203.

W. J. Cody and K. E. Hillstrom (1970). *Chebyshev approximations for the Coulomb phase shift.* Math. Comp. **24**(111), pp. 671–677. Corrigendum: same journal, **26**(1972), p. 1031.

W. J. Cody, K. E. Hillstrom, and H. C. Thacher, Jr. (1971). *Chebyshev approximations for the Riemann zeta function.* Math. Comp. **25**(115), pp. 537–547.

W. J. Cody, K. A. Paciorek, and H. C. Thacher, Jr. (1970). *Chebyshev approximations for Dawson's integral.* Math. Comp. **24**(109), pp. 171–178.

W. J. Cody, A. J. Strecok, and H. C. Thacher, Jr. (1973). *Chebyshev approximations for the psi function.* Math. Comp. **27**(121), pp. 123–127.

W. J. Cody and H. C. Thacher, Jr. (1968). *Rational Chebyshev approximations for the exponential integral $E_1(x)$.* Math. Comp. **22**(103), pp. 641–649.

W. J. Cody and H. C. Thacher, Jr. (1969). *Chebyshev approximations for the exponential integral $\mathrm{Ei}(x)$.* Math. Comp. **23**(106), pp. 289–303.

W. J. Cody and W. Waite (1980). *Software Manual for the Elementary Functions.* Englewood Cliffs: Prentice-Hall.

H. Cohen (1993). *A Course in Computational Algebraic Number Theory.* Berlin-New York: Springer-Verlag.

J. P. Coleman (1987). *Polynomial approximations in the complex plane.* J. Comput. Appl. Math. **18**(2), pp. 193–211.

J. P. Coleman and A. J. Monaghan (1983). *Chebyshev expansions for the Bessel function $J_n(z)$ in the complex plane.* Math. Comp. **40**(161), pp. 343–366.

L. Collatz (1960). *The Numerical Treatment of Differential Equations* (3rd ed.), Volume 60 of *Die Grundlehren der Mathematischen Wissenschaften.* Berlin: Springer. Translated by P. G. Williams from a supplemented version of the 2d German edition.

D. Colton and R. Kress (1998). *Inverse Acoustic and Electromagnetic Scattering Theory* (2nd ed.), Volume 93 of *Applied Mathematical Sciences.* Berlin: Springer-Verlag.

L. Comtet (1974). *Advanced Combinatorics: The Art of Finite and Infinite Expansions* (enlarged ed.). Dordrecht: D. Reidel Publishing Co. Revised and enlarged English translation by D. Reidel of Analyse Combinatoire, Tomes I, II, Presses Universitaires de France, Paris, 1970.

S. Conde and S. L. Kalla (1979). *The ν-zeros of $J_{-\nu}(x)$.* Math. Comp. **33**(145), pp. 423–426.

S. Conde and S. L. Kalla (1981). *On zeros of the hypergeometric function.* Serdica **7**(3), pp. 243–249.

E. U. Condon and G. H. Shortley (1935). *The Theory of Atomic Spectra.* Cambridge: Cambridge University Press. Reprinted with corrections in 1991. Transferred to digital reprinting, 1999.

W. C. Connett, C. Markett, and A. L. Schwartz (1993). *Product formulas and convolutions for angular and radial spheroidal wave functions.* Trans. Amer. Math. Soc. **338**(2), pp. 695–710.

J. N. L. Connor (1973). *Evaluation of multidimensional canonical integrals in semiclassical collision theory.* Molecular Phys. **26**(6), pp. 1371–1377.

J. N. L. Connor (1974). *Semiclassical theory of molecular collisions: Many nearly coincident classical trajectories.* Molecular Phys. **27**(4), pp. 853–866.

J. N. L. Connor (1976). *Catastrophes and molecular collisions.* Molecular Phys. **31**(1), pp. 33–55.

J. N. L. Connor and P. R. Curtis (1982). *A method for the numerical evaluation of the oscillatory integrals associated with the cuspoid catastrophes: Application to Pearcey's integral and its derivatives.* J. Phys. A **15**(4), pp. 1179–1190.

J. N. L. Connor, P. R. Curtis, and D. Farrelly (1983). *A differential equation method for the numerical evaluation of the Airy, Pearcey and swallowtail canonical integrals and their derivatives.* Molecular Phys. **48**(6), pp. 1305–1330.

J. N. L. Connor and D. Farrelly (1981). *Molecular collisions and cusp catastrophes: Three methods for the calculation of Pearcey's integral and its derivatives.* Chem. Phys. Lett. **81**(2), pp. 306–310.

J. N. L. Connor and D. C. Mackay (1979). *Calculation of angular distributions in complex angular momentum theories of elastic scattering.* Molecular Physics **37**(6), pp. 1703–1712.

A. G. Constantine (1963). *Some non-central distribution problems in multivariate analysis.* Ann. Math. Statist. **34**(4), pp. 1270–1285.

E. D. Constantinides and R. J. Marhefka (1993). *Efficient and accurate computation of the incomplete Airy functions.* Radio Science **28**(4), pp. 441–457.

J. W. Cooley and J. W. Tukey (1965). *An algorithm for the machine calculation of complex Fourier series.* Math. Comp. **19**(90), pp. 297–301.

R. Cools (2003). *An encyclopaedia of cubature formulas.* J. Complexity **19**(3), pp. 445–453. Numerical integration and its complexity (Oberwolfach, 2001).

F. Cooper, A. Khare, and A. Saxena (2006). *Exact elliptic compactons in generalized Korteweg-de Vries equations.* Complexity **11**(6), pp. 30–34.

M. D. Cooper, R. H. Jeppesen, and M. B. Johnson (1979). *Coulomb effects in the Klein-Gordon equation for pions*. Phys. Rev. C **20**(2), pp. 696–704.

R. B. Cooper (1981). *Introduction to Queueing Theory* (2nd ed.). New York: North-Holland Publishing Co.

E. T. Copson (1933). *An approximation connected with e^{-x}*. Proc. Edinburgh Math. Soc. (2) **3**, pp. 201–206.

E. T. Copson (1935). *An Introduction to the Theory of Functions of a Complex Variable*. Oxford: Oxford University Press.

E. T. Copson (1963). *On the asymptotic expansion of Airy's integral*. Proc. Glasgow Math. Assoc. **6**, pp. 113–115.

E. T. Copson (1965). *Asymptotic Expansions*. Number 55 in Cambridge Tracts in Mathematics and Mathematical Physics. New York: Cambridge University Press.

R. M. Corless, G. H. Gonnet, D. E. G. Hare, D. J. Jeffrey, and D. E. Knuth (1996). *On the Lambert W function*. Adv. Comput. Math. **5**(4), pp. 329–359.

R. M. Corless, D. J. Jeffrey, and H. Rasmussen (1992). *Numerical evaluation of Airy functions with complex arguments*. J. Comput. Phys. **99**(1), pp. 106–114.

G. Cornell, J. H. Silverman, and G. Stevens (Eds.) (1997). *Modular Forms and Fermat's Last Theorem*. New York: Springer-Verlag.

H. Cornille and A. Martin (1972). *Constraints on the phase of scattering amplitudes due to positivity*. Nuclear Phys. B **49**, pp. 413–440.

H. Cornille and A. Martin (1974). *Constraints on the phases of helicity amplitudes due to positivity*. Nuclear Phys. B **77**, pp. 141–162.

P. Cornille (1972). *Computation of Hankel transforms*. SIAM Rev. **14**(2), pp. 278–285.

M. S. Corrington (1961). *Applications of the complex exponential integral*. Math. Comp. **15**(73), pp. 1–6.

C. M. Cosgrove (2006). *Chazy's second-degree Painlevé equations*. J. Phys. A **39**(39), pp. 11955–11971.

O. Costin (1999). *Correlation between pole location and asymptotic behavior for Painlevé I solutions*. Comm. Pure Appl. Math. **52**(4), pp. 461–478.

D. A. Cox (1984). *The arithmetic-geometric mean of Gauss*. Enseign. Math. (2) **30**(3-4), pp. 275–330.

D. A. Cox (1985). *Gauss and the arithmetic-geometric mean*. Notices Amer. Math. Soc. **32**(2), pp. 147–151.

R. Crandall and C. Pomerance (2005). *Prime Numbers: A Computational Perspective* (2nd ed.). New York: Springer-Verlag.

R. E. Crandall (1996). *Topics in Advanced Scientific Computation*. New York: TELOS/Springer-Verlag.

J. E. Cremona (1997). *Algorithms for Modular Elliptic Curves* (2nd ed.). Cambridge: Cambridge University Press.

J. Crisóstomo, S. Lepe, and J. Saavedra (2004). *Quasinormal modes of the extremal BTZ black hole*. Classical Quantum Gravity **21**(12), pp. 2801–2809.

D. C. Cronemeyer (1991). *Demagnetization factors for general ellipsoids*. J. Appl. Phys. **70**(6), pp. 2911–2914. For an erratum concerning availability of tables, see J. Appl. Phys 70(1991)7660.

B. Crstici and G. Tudor (1975). *Compléments au traité de D. S. Mitrinović. VII. Sur une inégalité de D. S. Mitrinović*. Univ. Beograd. Publ. Elektrotehn. Fak. Ser. Mat. Fiz. (498-541), pp. 153–154.

A. Cruz, J. Esparza, and J. Sesma (1991). *Zeros of the Hankel function of real order out of the principal Riemann sheet*. J. Comput. Appl. Math. **37**(1-3), pp. 89–99.

A. Cruz and J. Sesma (1982). *Zeros of the Hankel function of real order and of its derivative*. Math. Comp. **39**(160), pp. 639–645.

A. Csótó and G. M. Hale (1997). *S-matrix and R-matrix determination of the low-energy ^{5}He and ^{5}Li resonance parameters*. Phys. Rev. C **55**(1), pp. 536–539.

A. R. Curtis (1964a). *Coulomb Wave Functions*, Volume 11 of *Roy. Soc. Math. Tables*. Cambridge: Cambridge University Press.

A. R. Curtis (1964b). *Tables of Jacobian Elliptic Functions Whose Arguments are Rational Fractions of the Quarter Period*, Volume 7 of *National Physical Laboratory Mathematical Tables*. London: Her Majesty's Stationery Office.

A. Cuyt, V. B. Petersen, B. Verdonk, H. Waadeland, and W. B. Jones (2008). *Handbook of Continued Fractions for Special Functions*. New York: Springer.

D. Cvijović and J. Klinowski (1994). *On the integration of incomplete elliptic integrals*. Proc. Roy. Soc. London Ser. A **444**, pp. 525–532.

D. Cvijović and J. Klinowski (1999). *Integrals involving complete elliptic integrals*. J. Comput. Appl. Math. **106**(1), pp. 169–175.

G. M. D'Ariano, C. Macchiavello, and M. G. A. Paris (1994). *Detection of the density matrix through optical homodyne tomography without filtered back projection*. Phys. Rev. A **50**(5), pp. 4298–4302.

H. Davenport (2000). *Multiplicative Number Theory* (3rd ed.), Volume 74 of *Graduate Texts in Mathematics*. New York: Springer-Verlag. Revised and with a preface by Hugh L. Montgomery.

B. Davies (1973). *Complex zeros of linear combinations of spherical Bessel functions and their derivatives.* SIAM J. Math. Anal. **4**(1), pp. 128–133.

B. Davies (1984). *Integral Transforms and their Applications* (2nd ed.), Volume 25 of *Applied Mathematical Sciences.* New York: Springer-Verlag.

H. F. Davis and A. D. Snider (1987). *Introduction to Vector Analysis* (5th ed.). Boston, MA: Allyn and Bacon Inc. Seventh edition printed by WCB/McGraw-Hill, 1995.

H. T. Davis (1933). *Tables of Higher Mathematical Functions I.* Bloomington, Indiana: Principia Press.

P. J. Davis (1975). *Interpolation and Approximation.* New York: Dover Publications Inc. Republication of the original 1963 edition (Blaisdell Publishing Co.), with minor corrections and a new preface and bibliography.

P. J. Davis and P. Rabinowitz (1984). *Methods of Numerical Integration* (2nd ed.). Computer Science and Applied Mathematics. Orlando, FL: Academic Press Inc.

S. D. Daymond (1955). *The principal frequencies of vibrating systems with elliptic boundaries.* Quart. J. Mech. Appl. Math. **8**(3), pp. 361–372.

C. de Boor (2001). *A Practical Guide to Splines* (Revised ed.), Volume 27 of *Applied Mathematical Sciences.* New York: Springer-Verlag.

L. de Branges (1985). *A proof of the Bieberbach conjecture.* Acta Math. **154**(1-2), pp. 137–152.

N. G. de Bruijn (1937). *Integralen voor de ζ-functie van Riemann.* Mathematica (Zutphen) **B5**, pp. 170–180. In Dutch.

N. G. de Bruijn (1961). *Asymptotic Methods in Analysis* (2nd ed.). Bibliotheca Mathematica, Vol. IV. Amsterdam: North-Holland Publishing Co.

N. G. de Bruijn (1981). *Pólya's Theory of Counting.* In E. F. Beckenbach (Ed.), *Applied Combinatorial Mathematics*, pp. 144–184. Malabar, FL: Robert E. Krieger.

M. G. de Bruin, E. B. Saff, and R. S. Varga (1981a). *On the zeros of generalized Bessel polynomials. I.* Nederl. Akad. Wetensch. Indag. Math. **84**(1), pp. 1–13.

M. G. de Bruin, E. B. Saff, and R. S. Varga (1981b). *On the zeros of generalized Bessel polynomials. II.* Nederl. Akad. Wetensch. Indag. Math. **84**(1), pp. 14–25.

C.-J. de la Vallée Poussin (1896a). *Recherches analytiques sur la théorie des nombres premiers. Première partie. La fonction $\zeta(s)$ de Riemann et les nombres premiers en général, suivi d'un Appendice sur des réflexions applicables à une formule donnée par Riemann.* Ann. Soc. Sci. Bruxelles **20**, pp. 183–256. Reprinted in Collected works/Oeuvres scientifiques, Vol. I, pp. 223–296 Académie Royale de Belgique, Brussels, 2000.

C.-J. de la Vallée Poussin (1896b). *Recherches analytiques sur la théorie des nombres premiers. Deuxième partie. Les fonctions de Dirichlet et les nombres premiers de la forme linéaire $Mx+N$.* Ann. Soc. Sci. Bruxelles **20**, pp. 281–397. Reprinted in Collected works/Oeuvres scientifiques, Vol. I, pp. 309–425, Académie Royale de Belgique, Brussels, 2000.

A. de Shalit and I. Talmi (1963). *Nuclear Shell Theory.* Pure and Applied Physics, Vol. 14. New York: Academic Press. Reprinted by Dover Publications Inc., New York, 2004.

P. Dean (1966). *The constrained quantum mechanical harmonic oscillator.* Proc. Cambridge Philos. Soc. **62**, pp. 277–286.

S. R. Deans (1983). *The Radon Transform and Some of Its Applications.* A Wiley-Interscience Publication. New York: John Wiley & Sons Inc. Reprinted by Dover in 2007.

A. Debosscher (1998). *Unification of one-dimensional Fokker-Planck equations beyond hypergeometrics: Factorizer solution method and eigenvalue schemes.* Phys. Rev. E (3) **57**(1), pp. 252–275.

A. Decarreau, M.-C. Dumont-Lepage, P. Maroni, A. Robert, and A. Ronveaux (1978a). *Formes canoniques des équations confluentes de l'équation de Heun.* Ann. Soc. Sci. Bruxelles Sér. I **92**(1-2), pp. 53–78.

A. Decarreau, P. Maroni, and A. Robert (1978b). *Sur les équations confluentes de l'équation de Heun.* Ann. Soc. Sci. Bruxelles Sér. I **92**(3), pp. 151–189.

B. Deconinck, M. Heil, A. Bobenko, M. van Hoeij, and M. Schmies (2004). *Computing Riemann theta functions.* Math. Comp. **73**(247), pp. 1417–1442.

B. Deconinck and J. N. Kutz (2006). *Computing spectra of linear operators using the Floquet-Fourier-Hill method.* J. Comput. Phys. **219**(1), pp. 296–321.

B. Deconinck and H. Segur (1998). *The KP equation with quasiperiodic initial data.* Phys. D **123**(1-4), pp. 123–152. Nonlinear waves and solitons in physical systems (Los Alamos, NM, 1997).

B. Deconinck and H. Segur (2000). *Pole dynamics for elliptic solutions of the Korteweg-de Vries equation.* Math. Phys. Anal. Geom. **3**(1), pp. 49–74.

B. Deconinck and M. van Hoeij (2001). *Computing Riemann matrices of algebraic curves.* Phys. D **152/153**, pp. 28–46.

P. A. Deift (1998). *Orthogonal Polynomials and Random Matrices: A Riemann-Hilbert Approach*, Volume 3 of *Courant Lecture Notes in Mathematics.* New York: New York University Courant Institute of Mathematical Sciences.

P. A. Deift and X. Zhou (1995). *Asymptotics for the Painlevé II equation.* Comm. Pure Appl. Math. **48**(3), pp. 277–337.

L. Dekar, L. Chetouani, and T. F. Hammann (1999). *Wave function for smooth potential and mass step*. Phys. Rev. A **59**(1), pp. 107–112.

K. Dekker and J. G. Verwer (1984). *Stability of Runge-Kutta Methods for Stiff Nonlinear Differential Equations*, Volume 2 of *CWI Monographs*. Amsterdam: North-Holland Publishing Co.

H. Delange (1987). *Sur les zéros imaginaires des polynômes de Bernoulli*. C. R. Acad. Sci. Paris Sér. I Math. **304**(6), pp. 147–150.

H. Delange (1988). *On the real roots of Euler polynomials*. Monatsh. Math. **106**(2), pp. 115–138.

H. Delange (1991). *Sur les zéros réels des polynômes de Bernoulli*. Ann. Inst. Fourier (Grenoble) **41**(2), pp. 267–309.

M. Deléglise and J. Rivat (1996). *Computing $\pi(x)$: The Meissel, Lehmer, Lagarias, Miller, Odlyzko method*. Math. Comp. **65**(213), pp. 235–245.

G. Delic (1979). *Chebyshev series for the spherical Bessel function $j_l(r)$*. Comput. Phys. Comm. **18**(1), pp. 73–86.

P. Deligne, P. Etingof, D. S. Freed, D. Kazhdan, J. W. Morgan, and D. R. Morrison (Eds.) (1999). *Quantum Fields and Strings: A Course for Mathematicians. Vol. 1, 2*. Providence, RI: American Mathematical Society. Material from the Special Year on Quantum Field Theory held at the Institute for Advanced Study, Princeton, NJ, 1996–1997.

J. B. Dence and T. P. Dence (1999). *Elements of the Theory of Numbers*. San Diego, CA: Harcourt/Academic Press.

R. L. Devaney (1986). *An Introduction to Chaotic Dynamical Systems*. Menlo Park, CA: The Benjamin/Cummings Publishing Co. Inc.

J. Dexter and E. Agol (2009). *A fast new public code for computing photon orbits in a Kerr spacetime*. The Astrophysical Journal **696**, pp. 1616–1629.

S. C. Dhar (1940). *Note on the addition theorem of parabolic cylinder functions*. J. Indian Math. Soc. (N. S.) **4**, pp. 29–30.

P. Di Francesco, P. Ginsparg, and J. Zinn-Justin (1995). *2D gravity and random matrices*. Phys. Rep. **254**(1-2), pp. 1–133.

L. E. Dickson (1919). *History of the Theory of Numbers (3 volumes)*. Washington, DC: Carnegie Institution of Washington. Reprinted by Chelsea Publishing Co., New York, 1966.

A. R. DiDonato (1978). *An approximation for $\int_\chi^\infty e^{-t^2/2} t^p dt$, $\chi > 0$, p real*. Math. Comp. **32**(141), pp. 271–275.

A. R. DiDonato and A. H. Morris (1986). *Computation of the incomplete gamma function ratios and their inverses*. ACM Trans. Math. Software **12**(4), pp. 377–393.

P. Dienes (1931). *The Taylor Series*. Oxford: Oxford University Press. Reprinted by Dover Publications Inc., New York, 1957.

A. Dienstfrey and J. Huang (2006). *Integral representations for elliptic functions*. J. Math. Anal. Appl. **316**(1), pp. 142–160.

K. Dilcher (1987a). *Asymptotic behaviour of Bernoulli, Euler, and generalized Bernoulli polynomials*. J. Approx. Theory **49**(4), pp. 321–330.

K. Dilcher (1987b). *Irreducibility of certain generalized Bernoulli polynomials belonging to quadratic residue class characters*. J. Number Theory **25**(1), pp. 72–80.

K. Dilcher (1988). *Zeros of Bernoulli, generalized Bernoulli and Euler polynomials*. Mem. Amer. Math. Soc. **73**(386), pp. iv+94.

K. Dilcher (1996). *Sums of products of Bernoulli numbers*. J. Number Theory **60**(1), pp. 23–41.

K. Dilcher (2002). *Bernoulli Numbers and Confluent Hypergeometric Functions*. In *Number Theory for the Millennium, I (Urbana, IL, 2000)*, pp. 343–363. Natick, MA: A. K. Peters.

K. Dilcher (2008). *On multiple zeros of Bernoulli polynomials*. Acta Arith. **134**(2), pp. 149–155.

K. Dilcher, L. Skula, and I. Sh. Slavutskiĭ (1991). *Bernoulli Numbers. Bibliography (1713–1990)*, Volume 87 of *Queen's Papers in Pure and Applied Mathematics*. Kingston, ON: Queen's University.

A. M. Din (1981). *A simple sum formula for Clebsch-Gordan coefficients*. Lett. Math. Phys. **5**(3), pp. 207–211.

D. Ding (2000). *A simplified algorithm for the second-order sound fields*. J. Acoust. Soc. Amer. **108**(6), pp. 2759–2764.

H. Ding, K. I. Gross, and D. St. P. Richards (1996). *Ramanujan's master theorem for symmetric cones*. Pacific J. Math. **175**(2), pp. 447–490.

R. B. Dingle (1973). *Asymptotic Expansions: Their Derivation and Interpretation*. London-New York: Academic Press.

P. G. L. Dirichlet (1837). *Beweis des Satzes, dass jede unbegrenzte arithmetische Progression, deren erstes Glied und Differenz ganze Zahlen ohne gemeinschaftlichen Factor sind, unendlich viele Primzahlen enthält*. Abhandlungen der Königlich Preussischen Akademie der Wissenschaften von 1837, pp. 45–81. Reprinted in G. Lejeune Dirichlet's Werke, Band I, pp. 313–342, Reimer, Berlin, 1889 and with corrections in G. Lejeune Dirichlet's Werke, AMS Chelsea Publishing Series, Providence, RI, 1969.

P. G. L. Dirichlet (1849). *Über die Bestimmung der mittleren Werthe in der Zahlentheorie*. Abhandlungen der Königlich Preussischen Akademie der Wissenschaften von 1849, pp. 69–83. Reprinted in G. Lejeune Dirichlet's Werke, Band II, pp. 49–66, Reimer, Berlin, 1897 and with corrections in G. Lejeune Dirichlet's Werke, AMS Chelsea Publishing Series, Providence, RI, 1969.

A. L. Dixon and W. L. Ferrar (1930). *Infinite integrals in the theory of Bessel functions*. Quart. J. Math., Oxford Ser. **1**(1), pp. 122–145.

M. D'Ocagne (1904). *Sur une classe de nombres rationnels réductibles aux nombres de Bernoulli*. Bull. Sci. Math. (2) **28**, pp. 29–32.

G. Doetsch (1955). *Handbuch der Laplace-Transformation. Bd. II. Anwendungen der Laplace-Transformation. 1. Abteilung*. Basel und Stuttgart: Birkhäuser Verlag.

G. C. Donovan, J. S. Geronimo, and D. P. Hardin (1999). *Orthogonal polynomials and the construction of piecewise polynomial smooth wavelets*. SIAM J. Math. Anal. **30**(5), pp. 1029–1056.

B. Döring (1966). *Complex zeros of cylinder functions*. Math. Comp. **20**(94), pp. 215–222. For a numerical error in one of the zeros see Amos (1985).

B. Döring (1971). *Über die Doppelnullstellen der Ableitung der Besselfunktion*. Angewandte Informatik **13**, pp. 402–406.

J. Dougall (1907). *On Vandermonde's theorem, and some more general expansions*. Proc. Edinburgh Math. Soc. **25**, pp. 114–132.

P. G. Drazin and R. S. Johnson (1993). *Solitons: An Introduction*. Cambridge Texts in Applied Mathematics. Cambridge: Cambridge University Press.

P. G. Drazin and W. H. Reid (1981). *Hydrodynamic Stability*. Cambridge: Cambridge University Press. Cambridge Monographs on Mechanics and Applied Mathematics.

B. Dubrovin and M. Mazzocco (2000). *Monodromy of certain Painlevé-VI transcendents and reflection groups*. Invent. Math. **141**(1), pp. 55–147.

B. A. Dubrovin (1981). *Theta functions and non-linear equations*. Uspekhi Mat. Nauk **36**(2(218)), pp. 11–80. With an appendix by I. M. Krichever. In Russian. English translation: Russian Math. Surveys **36**(1981), no. 2, pp. 11–92 (1982).

J. J. Duistermaat (1974). *Oscillatory integrals, Lagrange immersions and unfolding of singularities*. Comm. Pure Appl. Math. **27**, pp. 207–281.

D. S. Dummit and R. M. Foote (1999). *Abstract Algebra* (2nd ed.). Englewood Cliffs, NJ: Prentice Hall Inc.

D. Dumont and G. Viennot (1980). *A combinatorial interpretation of the Seidel generation of Genocchi numbers*. Ann. Discrete Math. **6**, pp. 77–87. Combinatorial mathematics, optimal designs and their applications (Proc. Sympos. Combin. Math. and Optimal Design, Colorado State Univ., Fort Collins, Colo., 1978).

C. F. Dunkl and Y. Xu (2001). *Orthogonal Polynomials of Several Variables*, Volume 81 of *Encyclopedia of Mathematics and its Applications*. Cambridge: Cambridge University Press.

B. I. Dunlap and B. R. Judd (1975). *Novel identities for simple n-j symbols*. J. Mathematical Phys. **16**, pp. 318–319.

G. V. Dunne and K. Rao (2000). *Lamé instantons*. J. High Energy Phys. (electronic only, article id. JHEP01(2000)019, 8 pp.).

T. M. Dunster (1986). *Uniform asymptotic expansions for prolate spheroidal functions with large parameters*. SIAM J. Math. Anal. **17**(6), pp. 1495–1524.

T. M. Dunster (1989). *Uniform asymptotic expansions for Whittaker's confluent hypergeometric functions*. SIAM J. Math. Anal. **20**(3), pp. 744–760. (This reference has several typographical errors.).

T. M. Dunster (1990a). *Bessel functions of purely imaginary order, with an application to second-order linear differential equations having a large parameter*. SIAM J. Math. Anal. **21**(4), pp. 995–1018. Errata: In eq. (2.8) replace $(x^2/4)^2$ by $(x^2/4)^s$. In the second line of eq. (4.7) insert an external factor $e^{-2\pi ij/3}$ and change the upper limit of the sum to $n-1$.

T. M. Dunster (1990b). *Uniform asymptotic solutions of second-order linear differential equations having a double pole with complex exponent and a coalescing turning point*. SIAM J. Math. Anal. **21**(6), pp. 1594–1618.

T. M. Dunster (1991). *Conical functions with one or both parameters large*. Proc. Roy. Soc. Edinburgh Sect. A **119**(3-4), pp. 311–327.

T. M. Dunster (1992). *Uniform asymptotic expansions for oblate spheroidal functions I: Positive separation parameter λ*. Proc. Roy. Soc. Edinburgh Sect. A **121**(3-4), pp. 303–320.

T. M. Dunster (1994a). *Uniform asymptotic approximation of Mathieu functions*. Methods Appl. Anal. **1**(2), pp. 143–168.

T. M. Dunster (1994b). *Uniform asymptotic solutions of second-order linear differential equations having a simple pole and a coalescing turning point in the complex plane*. SIAM J. Math. Anal. **25**(2), pp. 322–353.

T. M. Dunster (1995). *Uniform asymptotic expansions for oblate spheroidal functions II: Negative separation parameter λ*. Proc. Roy. Soc. Edinburgh Sect. A **125**(4), pp. 719–737.

T. M. Dunster (1996a). *Asymptotic solutions of second-order linear differential equations having almost coalescent turning points, with an application to the incomplete gamma function.* Proc. Roy. Soc. London Ser. A **452**, pp. 1331–1349.

T. M. Dunster (1996b). *Asymptotics of the generalized exponential integral, and error bounds in the uniform asymptotic smoothing of its Stokes discontinuities.* Proc. Roy. Soc. London Ser. A **452**, pp. 1351–1367.

T. M. Dunster (1996c). *Error bounds for exponentially improved asymptotic solutions of ordinary differential equations having irregular singularities of rank one.* Methods Appl. Anal. **3**(1), pp. 109–134.

T. M. Dunster (1997). *Error analysis in a uniform asymptotic expansion for the generalised exponential integral.* J. Comput. Appl. Math. **80**(1), pp. 127–161.

T. M. Dunster (1999). *Asymptotic approximations for the Jacobi and ultraspherical polynomials, and related functions.* Methods Appl. Anal. **6**(3), pp. 21–56.

T. M. Dunster (2001a). *Convergent expansions for solutions of linear ordinary differential equations having a simple turning point, with an application to Bessel functions.* Stud. Appl. Math. **107**(3), pp. 293–323.

T. M. Dunster (2001b). *Uniform asymptotic expansions for Charlier polynomials.* J. Approx. Theory **112**(1), pp. 93–133.

T. M. Dunster (2001c). *Uniform asymptotic expansions for the reverse generalized Bessel polynomials, and related functions.* SIAM J. Math. Anal. **32**(5), pp. 987–1013.

T. M. Dunster (2003a). *Uniform asymptotic approximations for the Whittaker functions $M_{\kappa,i\mu}(z)$ and $W_{\kappa,i\mu}(z)$.* Anal. Appl. (Singap.) **1**(2), pp. 199–212.

T. M. Dunster (2003b). *Uniform asymptotic expansions for associated Legendre functions of large order.* Proc. Roy. Soc. Edinburgh Sect. A **133**(4), pp. 807–827.

T. M. Dunster (2004). *Convergent expansions for solutions of linear ordinary differential equations having a simple pole, with an application to associated Legendre functions.* Stud. Appl. Math. **113**(3), pp. 245–270.

T. M. Dunster (2006). *Uniform asymptotic approximations for incomplete Riemann zeta functions.* J. Comput. Appl. Math. **190**(1-2), pp. 339–353.

T. M. Dunster, D. A. Lutz, and R. Schäfke (1993). *Convergent Liouville-Green expansions for second-order linear differential equations, with an application to Bessel functions.* Proc. Roy. Soc. London Ser. A **440**, pp. 37–54.

T. M. Dunster, R. B. Paris, and S. Cang (1998). *On the high-order coefficients in the uniform asymptotic expansion for the incomplete gamma function.* Methods Appl. Anal. **5**(3), pp. 223–247.

A. J. Durán (1993). *Functions with given moments and weight functions for orthogonal polynomials.* Rocky Mountain J. Math. **23**, pp. 87–104.

A. J. Durán and F. A. Grünbaum (2005). *A survey on orthogonal matrix polynomials satisfying second order differential equations.* J. Comput. Appl. Math. **178**(1-2), pp. 169–190.

L. Durand (1975). *Nicholson-type Integrals for Products of Gegenbauer Functions and Related Topics.* In R. A. Askey (Ed.), *Theory and Application of Special Functions (Proc. Advanced Sem., Math. Res. Center, Univ. Wisconsin, Madison, Wis., 1975)*, pp. 353–374. Math. Res. Center, Univ. Wisconsin, Publ. No. 35. New York: Academic Press.

L. Durand (1978). *Product formulas and Nicholson-type integrals for Jacobi functions. I. Summary of results.* SIAM J. Math. Anal. **9**(1), pp. 76–86.

P. L. Duren (1991). *The Legendre Relation for Elliptic Integrals.* In J. H. Ewing and F. W. Gehring (Eds.), *Paul Halmos: Celebrating 50 Years of Mathematics*, pp. 305–315. New York: Springer-Verlag.

J. Dutka (1981). *The incomplete beta function—a historical profile.* Arch. Hist. Exact Sci. **24**(1), pp. 11–29.

A. Dzieciol, S. Yngve, and P. O. Fröman (1999). *Coulomb wave functions with complex values of the variable and the parameters.* J. Math. Phys. **40**(12), pp. 6145–6166.

J. Écalle (1981a). *Les fonctions résurgentes. Tome I.* Orsay: Université de Paris-Sud Département de Mathématique.

J. Écalle (1981b). *Les fonctions résurgentes. Tome II.* Orsay: Université de Paris-Sud Département de Mathématique.

C. Eckart (1930). *The penetration of a potential barrier by electrons.* Phys. Rev. **35**(11), pp. 1303–1309.

A. R. Edmonds (1974). *Angular Momentum in Quantum Mechanics* (3rd printing, with corrections, 2nd ed.). Princeton, NJ: Princeton University Press. Reprinted in 1996.

H. M. Edwards (1974). *Riemann's Zeta Function.* New York-London: Academic Press,. Pure and Applied Mathematics, Vol. 58. Reprinted by Dover Publications Inc., Mineola, NY, 2001.

J. Edwards (1954). *A Treatise on the Integral Calculus*, Volume 1-2. New York: Chelsea Publishing Co. Originally published by Macmillan, 1921-1930.

M. Edwards, D. A. Griggs, P. L. Holman, C. W. Clark, S. L. Rolston, and W. D. Phillips (1999). *Properties of a Raman atom-laser output coupler.* J. Phys. B **32**(12), pp. 2935–2950.

G. P. Egorychev (1984). *Integral Representation and the Computation of Combinatorial Sums*, Volume 59 of *Translations of Mathematical Monographs*. Providence, RI: American Mathematical Society. Translated from the Russian by H. H. McFadden, Translation edited by Lev J. Leifman.

U. T. Ehrenmark (1995). *The numerical inversion of two classes of Kontorovich-Lebedev transform by direct quadrature*. J. Comput. Appl. Math. **61**(1), pp. 43–72.

M. Eichler and D. Zagier (1982). *On the zeros of the Weierstrass $\wp$-function*. Math. Ann. **258**(4), pp. 399–407.

Á. Elbert (2001). *Some recent results on the zeros of Bessel functions and orthogonal polynomials*. J. Comput. Appl. Math. **133**(1-2), pp. 65–83.

Á. Elbert and A. Laforgia (1994). *Interlacing properties of the zeros of Bessel functions*. Atti Sem. Mat. Fis. Univ. Modena **XLII**(2), pp. 525–529.

Á. Elbert and A. Laforgia (1997). *An upper bound for the zeros of the derivative of Bessel functions*. Rend. Circ. Mat. Palermo (2) **46**(1), pp. 123–130.

Á. Elbert and A. Laforgia (2000). *Further results on McMahon's asymptotic approximations*. J. Phys. A **33**(36), pp. 6333–6341.

E. Elizalde (1986). *An asymptotic expansion for the first derivative of the generalized Riemann zeta function*. Math. Comp. **47**(175), pp. 347–350.

E. Elizalde (1995). *Ten Physical Applications of Spectral Zeta Functions*, Volume 35 of *Lecture Notes in Physics. New Series m: Monographs*. Berlin: Springer-Verlag.

D. Elliott (1971). *Uniform asymptotic expansions of the Jacobi polynomials and an associated function*. Math. Comp. **25**(114), pp. 309–315.

D. Elliott (1998). *The Euler-Maclaurin formula revisited*. J. Austral. Math. Soc. Ser. B **40**(E), pp. E27–E76 (electronic).

E. B. Elliott (1903). *A formula including Legendre's $EK' + KE' - KK' = \frac{1}{2}\pi$*. Messenger of Math. **33**, pp. 31–32. Errata on p. 45 of the same volume.

W. J. Ellison (1971). *Waring's problem*. Amer. Math. Monthly **78**(1), pp. 10–36.

S. P. Eraševskaja, E. A. Ivanov, A. A. Pal'cev, and N. D. Sokolova (1973). *Tablicy sferoidal'nyh volnovyh funkcii i ih pervyh proizvodnyh. Tom I*. Minsk: Izdat. "Nauka i Tehnika". Translated title: *Tables of Spheroidal Wave Functions and Their First Derivatives*.

S. P. Eraševskaja, E. A. Ivanov, A. A. Pal'cev, and N. D. Sokolova (1976). *Tablicy sferoidal'nyh volnovyh funkcii iih pervyh proizvodnyh. Tom II*. Minsk: Izdat. "Nauka i Tehnika". Edited by V. I. Krylov. Translated title: *Tables of Spheroidal Wave Functions and Their First Derivatives*.

N. Ercolani and A. Sinha (1989). *Monopoles and Baker functions*. Comm. Math. Phys. **125**(3), pp. 385–416.

A. Erdélyi (1941a). *Generating functions of certain continuous orthogonal systems*. Proc. Roy. Soc. Edinburgh. Sect. A. **61**, pp. 61–70.

A. Erdélyi (1941b). *On Lamé functions*. Philos. Mag. (7) **31**, pp. 123–130.

A. Erdélyi (1941c). *On algebraic Lamé functions*. Philos. Mag. (7) **32**, pp. 348–350.

A. Erdélyi (1942a). *The Fuchsian equation of second order with four singularities*. Duke Math. J. **9**(1), pp. 48–58.

A. Erdélyi (1942b). *Integral equations for Heun functions*. Quart. J. Math., Oxford Ser. **13**, pp. 107–112.

A. Erdélyi (1944). *Certain expansions of solutions of the Heun equation*. Quart. J. Math., Oxford Ser. **15**, pp. 62–69.

A. Erdélyi (1956). *Asymptotic Expansions*. New York: Dover Publications Inc.

A. Erdélyi, W. Magnus, F. Oberhettinger, and F. G. Tricomi (1953a). *Higher Transcendental Functions. Vol. I*. New York-Toronto-London: McGraw-Hill Book Company, Inc. Reprinted by Robert E. Krieger Publishing Co. Inc., 1981. Table errata: Math. Comp. v. 65 (1996), no. 215, p. 1385, v. 41 (1983), no. 164, p. 778, v. 30 (1976), no. 135, p. 675, v. 25 (1971), no. 115, p. 635, v. 25 (1971), no. 113, p. 199, v. 24 (1970), no. 112, p. 999, v. 24 (1970), no. 110, p. 504.

A. Erdélyi, W. Magnus, F. Oberhettinger, and F. G. Tricomi (1953b). *Higher Transcendental Functions. Vol. II*. New York-Toronto-London: McGraw-Hill Book Company, Inc. Reprinted by Robert E. Krieger Publishing Co. Inc., 1981. Table errata: Math. Comp. v. 41 (1983), no. 164, p. 778, v. 36 (1981), no. 153, p. 315, v. 30 (1976), no. 135, p. 675, v. 29 (1975), no. 130, p. 670, v. 26 (1972), no. 118, p. 598, v. 25 (1971), no. 113, p. 199, v. 24 (1970), no. 109, p. 239.

A. Erdélyi, W. Magnus, F. Oberhettinger, and F. G. Tricomi (1954a). *Tables of Integral Transforms. Vol. I*. New York-Toronto-London: McGraw-Hill Book Company, Inc. Table errata: Math. Comp. v. 66 (1997), no. 220, p. 1766–1767, v. 65 (1996), no. 215, p. 1384, v. 50 (1988), no. 182, p. 653, v. 41 (1983), no. 164, p. 778–779, v. 27 (1973), no. 122, p. 451, v. 26 (1972), no. 118, p. 599, v. 25 (1971), no. 113, p. 199, v. 24 (1970), no. 109, p. 239-240.

A. Erdélyi, W. Magnus, F. Oberhettinger, and F. G. Tricomi (1954b). *Tables of Integral Transforms. Vol. II*. New York-Toronto-London: McGraw-Hill Book Company, Inc. Table errata: Math. Comp. v. 65 (1996), no. 215, p. 1385, v. 41 (1983), no. 164, pp. 779–780, v. 31 (1977), no. 138, p. 614, v. 31 (1977), no. 137, pp. 328–329, v. 26 (1972), no. 118, p. 599, v. 25 (1971), no. 113, p. 199, v. 23 (1969), no. 106, p. 468.

A. Erdélyi, W. Magnus, F. Oberhettinger, and F. G. Tricomi (1955). *Higher Transcendental Functions. Vol. III*. New York-Toronto-London: McGraw-Hill Book Company, Inc. Reprinted by Robert E. Krieger Publishing Co. Inc., 1981. Table errata: Math. Comp. v. 41 (1983), no. 164, p. 778.

R. Ernvall (1979). *Generalized Bernoulli numbers, generalized irregular primes, and class number*. Ann. Univ. Turku. Ser. A I **178**, pp. 1–72.

F. H. L. Essler, H. Frahm, A. R. Its, and V. E. Korepin (1996). *Painlevé transcendent describes quantum correlation function of the XXZ antiferromagnet away from the free-fermion point*. J. Phys. A **29**(17), pp. 5619–5626.

T. Estermann (1959). *On the representations of a number as a sum of three squares*. Proc. London Math. Soc. (3) **9**, pp. 575–594.

Euclid (1908). *The Thirteen Books of Euclid's Elements*. Cambridge: Cambridge University Press. Translated from the text of Heiberg, with introduction and commentary by Thomas L. Heath. Reprint of the Cambridge Univeristy second edition by Dover Publications, Inc., New York, 1956.

L. Euler (1768). *Institutiones Calculi Integralis*, Volume 11 of *Opera Omnia (1)*, pp. 110–113. Leipzig-Berlin: B. G. Teubner.

G. A. Evans and J. R. Webster (1999). *A comparison of some methods for the evaluation of highly oscillatory integrals*. J. Comput. Appl. Math. **112**(1-2), pp. 55–69.

W. D. Evans, W. N. Everitt, K. H. Kwon, and L. L. Littlejohn (1993). *Real orthogonalizing weights for Bessel polynomials*. J. Comput. Appl. Math. **49**(1-3), pp. 51–57.

W. N. Everitt and D. S. Jones (1977). *On an integral inequality*. Proc. Roy. Soc. London Ser. A **357**, pp. 271–288.

J. A. Ewell (1990). *A new series representation for $\zeta(3)$*. Amer. Math. Monthly **97**(3), pp. 219–220.

H. Exton (1983). *The asymptotic behaviour of the inhomogeneous Airy function* $\mathrm{Hi}(z)$. Math. Chronicle **12**, pp. 99–104.

B. R. Fabijonas, D. W. Lozier, and F. W. J. Olver (2004). *Computation of complex Airy functions and their zeros using asymptotics and the differential equation*. ACM Trans. Math. Software **30**(4), pp. 471–490.

B. R. Fabijonas and F. W. J. Olver (1999). *On the reversion of an asymptotic expansion and the zeros of the Airy functions*. SIAM Rev. **41**(4), pp. 762–773.

V. N. Faddeeva and N. M. Terent'ev (1954). *Tablicy značeniĭ funkcii $w(z) = e^{-z^2}(1 + \frac{2i}{\sqrt{\pi}} \int_0^z e^{t^2} dt)$ ot kompleksnogo argumenta*. Moscow: Gosudarstv. Izdat. Tehn.-Teor. Lit. In Russian. English translation, see Faddeyeva and Terent'ev (1961).

V. N. Faddeyeva and N. M. Terent'ev (1961). *Tables of Values of the Function $w(z) = e^{-z^2}(1 + 2i\pi^{-1/2}\int_0^z e^{t^2} dt)$ for Complex Argument*. Edited by V. A. Fok; translated from the Russian by D. G. Fry. Mathematical Tables Series, Vol. 11. Oxford: Pergamon Press.

M. Faierman (1992). *Generalized parabolic cylinder functions*. Asymptotic Anal. **5**(6), pp. 517–531.

P. Falloon (2001). *Theory and Computation of Spheroidal Harmonics with General Arguments*. Master's thesis, The University of Western Australia, Department of Physics. Contains Mathematica package for spheroidal wave functions of complex arguments with complex parameters.

D. F. Fang and J. F. Shriner, Jr. (1992). *A computer program for the calculation of angular-momentum coupling coefficients*. Comput. Phys. Comm. **70**(1), pp. 147–153. Fortran for exact calculation using integers.

J. Faraut and A. Korányi (1994). *Analysis on Symmetric Cones*. Oxford Mathematical Monographs. Oxford-New York: The Clarendon Press, Oxford University Press.

R. H. Farrell (1985). *Multivariate Calculation. Use of the Continuous Groups*. Springer Series in Statistics. New York: Springer-Verlag.

J. D. Fay (1973). *Theta Functions on Riemann Surfaces*. Berlin: Springer-Verlag. Lecture Notes in Mathematics, Vol. 352.

M. V. Fedoryuk (1989). *The Lamé wave equation*. Uspekhi Mat. Nauk **44**(1(265)), pp. 123–144, 248. In Russian. English translation: Russian Math. Surveys, **44**(1989), no. 1, pp. 153–180.

M. V. Fedoryuk (1991). *Asymptotics of the spectrum of the Heun equation and of Heun functions*. Izv. Akad. Nauk SSSR Ser. Mat. **55**(3), pp. 631–646. In Russian. English translation: Math. USSR-Izv., **38**(1992), no. 3, pp. 621–635 (1992).

S. Fempl (1960). *Sur certaines sommes des intégral-cosinus*. Bull. Soc. Math. Phys. Serbie **12**, pp. 13–20.

C. Ferreira and J. L. López (2001). *An asymptotic expansion of the double gamma function*. J. Approx. Theory **111**(2), pp. 298–314.

C. Ferreira, J. L. López, and E. Pérez Sinusía (2005). *Incomplete gamma functions for large values of their variables*. Adv. in Appl. Math. **34**(3), pp. 467–485.

H. E. Fettis (1965). *Calculation of elliptic integrals of the third kind by means of Gauss' transformation*. Math. Comp. **19**(89), pp. 97–104.

H. E. Fettis (1970). *On the reciprocal modulus relation for elliptic integrals*. SIAM J. Math. Anal. **1**(4), pp. 524–526.

H. E. Fettis (1976). *Complex roots of $\sin z = az$, $\cos z = az$, and $\cosh z = az$*. Math. Comp. **30**(135), pp. 541–545.

H. E. Fettis and J. C. Caslin (1964, December). *Tables of Elliptic Integrals of the First, Second, and Third Kind.* Technical Report ARL 64-232, Aerospace Research Laboratories, Wright-Patterson Air Force Base, Ohio. Reviewed in Math. Comp. v. 1919(1965)509. Table erratum: Math. Comp. v. 20 (1966), no. 96, pp. 639-640.

H. E. Fettis and J. C. Caslin (1969, November). *A Table of the Complete Elliptic Integral of the First Kind for Complex Values of the Modulus. Part I.* Technical Report ARL 69-0172, Aerospace Research Laboratories, Office of Aerospace Research, Wright-Patterson Air Force Base, Ohio. Table erratum: Math. Comp. v. 36 (1981), no. 153, p. 318. Part II of this report, with the same date but numbered ARL 69-0173, again puts $k = R\exp(i\theta)$ but with R as parameter and θ as variable instead of the other way around. Part III, dated May 1970 and numbered ARL 70-0081, contains tables of auxiliary functions to help interpolation.

H. E. Fettis and J. C. Caslin (1973). *Table errata; Complex zeros of Fresnel integrals.* Math. Comp. **27**(121), p. 219. Errata in Abramowitz and Stegun (1964).

H. E. Fettis, J. C. Caslin, and K. R. Cramer (1973). *Complex zeros of the error function and of the complementary error function.* Math. Comp. **27**(122), pp. 401–407.

J. L. Fields (1965). *Asymptotic expansions of a class of hypergeometric polynomials with respect to the order. III.* J. Math. Anal. Appl. **12**(3), pp. 593–601.

J. L. Fields (1973). *Uniform asymptotic expansions of certain classes of Meijer G-functions for a large parameter.* SIAM J. Math. Anal. **4**(3), pp. 482–507.

J. L. Fields (1983). *Uniform asymptotic expansions of a class of Meijer G-functions for a large parameter.* SIAM J. Math. Anal. **14**(6), pp. 1204–1253.

J. L. Fields and Y. L. Luke (1963a). *Asymptotic expansions of a class of hypergeometric polynomials with respect to the order.* J. Math. Anal. Appl. **6**(3), pp. 394–403.

J. L. Fields and Y. L. Luke (1963b). *Asymptotic expansions of a class of hypergeometric polynomials with respect to the order. II.* J. Math. Anal. Appl. **7**(3), pp. 440–451.

J. L. Fields and J. Wimp (1961). *Expansions of hypergeometric functions in hypergeometric functions.* Math. Comp. **15**(76), pp. 390–395.

S. Fillebrown (1992). *Faster computation of Bernoulli numbers.* J. Algorithms **13**(3), pp. 431–445.

S. R. Finch (2003). *Mathematical Constants*, Volume 94 of *Encyclopedia of Mathematics and its Applications*. Cambridge: Cambridge University Press.

N. J. Fine (1988). *Basic Hypergeometric Series and Applications*, Volume 27 of *Mathematical Surveys and Monographs*. Providence, RI: American Mathematical Society. With a foreword by George E. Andrews.

G. D. Finn and D. Mugglestone (1965). *Tables of the line broadening function $H(a, v)$.* Monthly Notices Roy. Astronom. Soc. **129**, pp. 221–235.

P. Flajolet and A. Odlyzko (1990). *Singularity analysis of generating functions.* SIAM J. Discrete Math. **3**(2), pp. 216–240.

P. Flajolet and B. Salvy (1998). *Euler sums and contour integral representations.* Experiment. Math. **7**(1), pp. 15–35.

C. Flammer (1957). *Spheroidal Wave Functions.* Stanford, CA: Stanford University Press.

H. Flaschka and A. C. Newell (1980). *Monodromy- and spectrum-preserving deformations. I.* Comm. Math. Phys. **76**(1), pp. 65–116.

A. Fletcher (1948). *Guide to tables of elliptic functions.* Math. Tables and Other Aids to Computation **3**(24), pp. 229–281.

A. Fletcher, J. C. P. Miller, L. Rosenhead, and L. J. Comrie (1962). *An Index of Mathematical Tables. Vols. I, II* (2nd ed.). Reading, MA: Published for Scientific Computing Service Ltd., London, by Addison-Wesley Publishing Co., Inc. Table errata: Math. Comp. v.27 (1973), p. 1009.

N. Fleury and A. Turbiner (1994). *Polynomial relations in the Heisenberg algebra.* J. Math. Phys. **35**(11), pp. 6144–6149.

J. P. M. Flude (1998). *The Edmonds asymptotic formulas for the $3j$ and $6j$ symbols.* J. Math. Phys. **39**(7), pp. 3906–3915.

V. Fock (1945). *Diffraction of radio waves around the earth's surface.* Acad. Sci. USSR. J. Phys. **9**, pp. 255–266.

V. A. Fock (1965). *Electromagnetic Diffraction and Propagation Problems.* International Series of Monographs on Electromagnetic Waves, Vol. 1. Oxford: Pergamon Press.

A. S. Fokas and M. J. Ablowitz (1982). *On a unified approach to transformations and elementary solutions of Painlevé equations.* J. Math. Phys. **23**(11), pp. 2033–2042.

A. S. Fokas, B. Grammaticos, and A. Ramani (1993). *From continuous to discrete Painlevé equations.* J. Math. Anal. Appl. **180**(2), pp. 342–360.

A. S. Fokas, A. R. Its, A. A. Kapaev, and V. Y. Novokshenov (2006). *Painlevé Transcendents: The Riemann-Hilbert Approach*, Volume 128 of *Mathematical Surveys and Monographs*. Providence, RI: American Mathematical Society.

A. S. Fokas, A. R. Its, and A. V. Kitaev (1991). *Discrete Painlevé equations and their appearance in quantum gravity.* Comm. Math. Phys. **142**(2), pp. 313–344.

A. S. Fokas, A. R. Its, and X. Zhou (1992). *Continuous and Discrete Painlevé Equations*. In D. Levi and P. Winternitz (Eds.), *Painlevé Transcendents: Their Asymptotics and Physical Applications*, Volume 278 of *NATO Adv. Sci. Inst. Ser. B Phys.*, pp. 33–47. New York: Plenum. Proc. NATO Adv. Res. Workshop, Sainte-Adèle, Canada, 1990.

A. S. Fokas and Y. C. Yortsos (1981). *The transformation properties of the sixth Painlevé equation and one-parameter families of solutions*. Lett. Nuovo Cimento (2) **30**(17), pp. 539–544.

K. W. Ford and J. A. Wheeler (1959a). *Semiclassical description of scattering*. Ann. Physics **7**(3), pp. 259–286.

K. W. Ford and J. A. Wheeler (1959b). *Application of semiclassical scattering analysis*. Ann. Physics **7**(3), pp. 287–322.

W. B. Ford (1960). *Studies on Divergent Series and Summability & The Asymptotic Developments of Functions Defined by Maclaurin Series*. New York: Chelsea Publishing Co. Reprint with corrections of two books published originally in 1916 and 1936.

P. J. Forrester and N. S. Witte (2001). *Application of the τ-function theory of Painlevé equations to random matrices: PIV, PII and the GUE*. Comm. Math. Phys. **219**(2), pp. 357–398.

P. J. Forrester and N. S. Witte (2002). *Application of the τ-function theory of Painlevé equations to random matrices: P_V, P_{III}, the LUE, JUE, and CUE*. Comm. Pure Appl. Math. **55**(6), pp. 679–727.

P. J. Forrester and N. S. Witte (2004). *Application of the τ-function theory of Painlevé equations to random matrices: P_{VI}, the JUE, CyUE, cJUE and scaled limits*. Nagoya Math. J. **174**, pp. 29–114.

R. C. Forrey (1997). *Computing the hypergeometric function*. J. Comput. Phys. **137**(1), pp. 79–100. Double-precision Fortran.

T. Fort (1948). *Finite Differences and Difference Equations in the Real Domain*. Oxford: Clarendon Press.

L. Fox (1960). *Tables of Weber Parabolic Cylinder Functions and Other Functions for Large Arguments*. National Physical Laboratory Mathematical Tables, Vol. 4. Department of Scientific and Industrial Research. London: Her Majesty's Stationery Office.

L. Fox and I. B. Parker (1968). *Chebyshev Polynomials in Numerical Analysis*. London: Oxford University Press. Reprinted in 1972 with corrections.

C. H. Franke (1965). *Numerical evaluation of the elliptic integral of the third kind*. Math. Comp. **19**(91), pp. 494–496.

C. K. Frederickson and P. L. Marston (1992). *Transverse cusp diffraction catastrophes produced by the reflection of ultrasonic tone bursts from a curved surface in water*. J. Acoust. Soc. Amer. **92**(5), pp. 2869–2877.

C. K. Frederickson and P. L. Marston (1994). *Travel time surface of a transverse cusp caustic produced by reflection of acoustical transients from a curved metal surface*. J. Acoust. Soc. Amer. **95**(2), pp. 650–660.

C. L. Frenzen (1990). *Error bounds for a uniform asymptotic expansion of the Legendre function $Q_n^{-m}(\cosh z)$*. SIAM J. Math. Anal. **21**(2), pp. 523–535.

C. L. Frenzen and R. Wong (1985). *A note on asymptotic evaluation of some Hankel transforms*. Math. Comp. **45**(172), pp. 537–548.

C. L. Frenzen and R. Wong (1986). *Asymptotic expansions of the Lebesgue constants for Jacobi series*. Pacific J. Math. **122**(2), pp. 391–415.

C. L. Frenzen and R. Wong (1988). *Uniform asymptotic expansions of Laguerre polynomials*. SIAM J. Math. Anal. **19**(5), pp. 1232–1248.

A. Fresnel (1818). *Mémoire sur la diffraction de la lumière*. Mém. de l'Académie des Sciences, pp. 247–382. Œvres Complètes d'Augustin Fresnel, Paris, 1866.

G. Freud (1976). *On the coefficients in the recursion formulae of orthogonal polynomials*. Proc. Roy. Irish Acad. Sect. A **76**(1), pp. 1–6.

B. D. Fried and S. D. Conte (1961). *The Plasma Dispersion Function: The Hilbert Transform of the Gaussian*. London-New York: Academic Press. Erratum: Math. Comp. v. 26 (1972), no. 119, p. 814.

B. R. Frieden (1971). *Evaluation, design and extrapolation methods for optical signals, based on use of the prolate functions*. In E. Wolf (Ed.), *Progress in Optics*, Volume 9, pp. 311–407. Amsterdam: North-Holland.

F. G. Friedlander (1958). *Sound Pulses*. Cambridge-New York: Cambridge University Press.

C.-E. Fröberg (1955). *Numerical treatment of Coulomb wave functions*. Rev. Mod. Phys. **27**(4), pp. 399–411.

R. Fuchs (1907). *Über lineare homogene Differentialgleichungen zweiter Ordnung mit drei im Endlichen gelegenen wesentlich singulären Stellen*. Math. Ann. **63**(3), pp. 301–321.

Y. Fukui and T. Horiguchi (1992). *Characteristic values of the integral equation satisfied by the Mathieu functions and its application to a system with chirality-pair interaction on a one-dimensional lattice*. Phys. A **190**(3-4), pp. 346–362.

Y. V. Fyodorov (2005). *Introduction to the Random Matrix Theory: Gaussian Unitary Ensemble and Beyond.* In *Recent Perspectives in Random Matrix Theory and Number Theory*, Volume 322 of *London Math. Soc. Lecture Note Ser.*, pp. 31–78. Cambridge: Cambridge Univ. Press.

B. Gabutti (1979). *On high precision methods for computing integrals involving Bessel functions.* Math. Comp. **33**(147), pp. 1049–1057.

B. Gabutti (1980). *On the generalization of a method for computing Bessel function integrals.* J. Comput. Appl. Math. **6**(2), pp. 167–168.

B. Gabutti and B. Minetti (1981). *A new application of the discrete Laguerre polynomials in the numerical evaluation of the Hankel transform of a strongly decreasing even function.* J. Comput. Phys. **42**(2), pp. 277–287.

B. Gambier (1910). *Sur les équations différentielles du second ordre et du premier degré dont l'intégrale générale est a points critiques fixes.* Acta Math. **33**(1), pp. 1–55.

L. Gårding (1947). *The solution of Cauchy's problem for two totally hyperbolic linear differential equations by means of Riesz integrals.* Ann. of Math. (2) **48**(4), pp. 785–826.

I. Gargantini and P. Henrici (1967). *A continued fraction algorithm for the computation of higher transcendental functions in the complex plane.* Math. Comp. **21**(97), pp. 18–29.

G. Gasper (1972). *An inequality of Turán type for Jacobi polynomials.* Proc. Amer. Math. Soc. **32**, pp. 435–439.

G. Gasper (1975). *Formulas of the Dirichlet-Mehler Type.* In B. Ross (Ed.), *Fractional Calculus and its Applications*, Volume 457 of *Lecture Notes in Math.*, pp. 207–215. Berlin: Springer.

G. Gasper (1981). *Orthogonality of certain functions with respect to complex valued weights.* Canad. J. Math. **33**(5), pp. 1261–1270.

G. Gasper and M. Rahman (2004). *Basic Hypergeometric Series* (Second ed.), Volume 96 of *Encyclopedia of Mathematics and its Applications*. Cambridge: Cambridge University Press. With a foreword by Richard Askey.

L. Gatteschi (1987). *New inequalities for the zeros of Jacobi polynomials.* SIAM J. Math. Anal. **18**(6), pp. 1549–1562.

L. Gatteschi (1990). *New inequalities for the zeros of confluent hypergeometric functions.* In *Asymptotic and computational analysis (Winnipeg, MB, 1989)*, pp. 175–192. New York: Dekker.

L. Gatteschi (2002). *Asymptotics and bounds for the zeros of Laguerre polynomials: A survey.* J. Comput. Appl. Math. **144**(1-2), pp. 7–27.

M. Gaudin (1983). *La fonction d'onde de Bethe.* Paris: Masson.

J. A. Gaunt (1929). *The triplets of helium.* Philos. Trans. Roy. Soc. London Ser. A **228**, pp. 151–196.

C. F. Gauss (1863). *Werke. Band II*, pp. 436–447. Königlichen Gesellschaft der Wissenschaften zu Göttingen. Reprinted by Georg Olms Verlag, Hildesheim, 1973.

W. Gautschi (1959a). *Exponential integral $\int_1^\infty e^{-xt} t^{-n} dt$ for large values of n.* J. Res. Nat. Bur. Standards **62**, pp. 123–125.

W. Gautschi (1959b). *Some elementary inequalities relating to the gamma and incomplete gamma function.* J. Math. Phys. **38**(1), pp. 77–81.

W. Gautschi (1961). *Recursive computation of the repeated integrals of the error function.* Math. Comp. **15**(75), pp. 227–232.

W. Gautschi (1967). *Computational aspects of three-term recurrence relations.* SIAM Rev. **9**(1), pp. 24–82.

W. Gautschi (1970). *Efficient computation of the complex error function.* SIAM J. Numer. Anal. **7**(1), pp. 187–198.

W. Gautschi (1974). *A harmonic mean inequality for the gamma function.* SIAM J. Math. Anal. **5**(2), pp. 278–281.

W. Gautschi (1975). *Computational Methods in Special Functions – A Survey.* In R. A. Askey (Ed.), *Theory and Application of Special Functions (Proc. Advanced Sem., Math. Res. Center, Univ. Wisconsin, Madison, Wis., 1975)*, pp. 1–98. Math. Res. Center, Univ. Wisconsin Publ., No. 35. New York: Academic Press.

W. Gautschi (1977a). *Evaluation of the repeated integrals of the coerror function.* ACM Trans. Math. Software **3**, pp. 240–252. For software see Gautschi (1977b).

W. Gautschi (1977b). *Algorithm 521: Repeated integrals of the coerror function.* ACM Trans. Math. Software **3**, pp. 301–302. Single-precision Fortran, maximum accuracy 14D.

W. Gautschi (1979a). *A computational procedure for incomplete gamma functions.* ACM Trans. Math. Software **5**(4), pp. 466–481.

W. Gautschi (1979b). *Un procedimento di calcolo per le funzioni gamma incomplete.* Rend. Sem. Mat. Univ. Politec. Torino **37**(1), pp. 1–9.

W. Gautschi (1983). *How and how not to check Gaussian quadrature formulae.* BIT **23**(2), pp. 209–216.

W. Gautschi (1984). *Questions of Numerical Condition Related to Polynomials.* In G. H. Golub (Ed.), *Studies in Numerical Analysis*, pp. 140–177. Washington, DC: Mathematical Association of America.

W. Gautschi (1993). *On the computation of generalized Fermi-Dirac and Bose-Einstein integrals.* Comput. Phys. Comm. **74**(2), pp. 233–238.

W. Gautschi (1994). *Algorithm 726: ORTHPOL — a package of routines for generating orthogonal polynomials and Gauss-type quadrature rules*. ACM Trans. Math. Software **20**(1), pp. 21–62. For remark see same journal **24**(1998), p. 355.

W. Gautschi (1996). *Orthogonal Polynomials: Applications and Computation*. In A. Iserles (Ed.), *Acta Numerica, 1996*, Volume 5 of *Acta Numerica*, pp. 45–119. Cambridge: Cambridge Univ. Press.

W. Gautschi (1997a). *The Computation of Special Functions by Linear Difference Equations*. In S. Elaydi, I. Győri, and G. Ladas (Eds.), *Advances in Difference Equations (Veszprém, 1995)*, pp. 213–243. Amsterdam: Gordon and Breach.

W. Gautschi (1997b). *Numerical Analysis. An Introduction*. Boston, MA: Birkhäuser Boston Inc.

W. Gautschi (1998). *The incomplete gamma functions since Tricomi*. In *Tricomi's Ideas and Contemporary Applied Mathematics (Rome/Turin, 1997)*, Volume 147 of *Atti Convegni Lincei*, pp. 203–237. Rome: Accad. Naz. Lincei.

W. Gautschi (1999). *A note on the recursive calculation of incomplete gamma functions*. ACM Trans. Math. Software **25**(1), pp. 101–107.

W. Gautschi (2002a). *Computation of Bessel and Airy functions and of related Gaussian quadrature formulae*. BIT **42**(1), pp. 110–118.

W. Gautschi (2002b). *Gauss quadrature approximations to hypergeometric and confluent hypergeometric functions*. J. Comput. Appl. Math. **139**(1), pp. 173–187.

W. Gautschi (2004). *Orthogonal Polynomials: Computation and Approximation*. Numerical Mathematics and Scientific Computation. New York: Oxford University Press.

W. Gautschi and J. Slavik (1978). *On the computation of modified Bessel function ratios*. Math. Comp. **32**(143), pp. 865–875.

M. Gavrila (1967). *Elastic scattering of photons by a hydrogen atom*. Phys. Rev. **163**(1), pp. 147–155.

I. M. Gel'fand and G. E. Shilov (1964). *Generalized Functions. Vol. 1: Properties and Operations*. New York: Academic Press. Translated from the Russian by Eugene Saletan. A paperbound reprinting by the same publisher appeared in 1977.

M. Geller and E. W. Ng (1969). *A table of integrals of the exponential integral*. J. Res. Nat. Bur. Standards Sect. B **73B**, pp. 191–210.

M. Geller and E. W. Ng (1971). *A table of integrals of the error function. II. Additions and corrections*. J. Res. Nat. Bur. Standards Sect. B **75B**, pp. 149–163. See for the first part same journal, Sect. B 73, 1-20 (1969).

K. Germey (1964). *Die Beugung einer ebenen elektromanetischen Welle an zwei parallelen unendlich langen ideal leitenden Zylindern von elliptischem Querschnitt*. Ann. Physik (7) **468**, pp. 237–251.

A. Gervois and H. Navelet (1984). *Some integrals involving three Bessel functions when their arguments satisfy the triangle inequalities*. J. Math. Phys. **25**(11), pp. 3350–3356.

A. Gervois and H. Navelet (1985a). *Integrals of three Bessel functions and Legendre functions. I*. J. Math. Phys. **26**(4), pp. 633–644.

A. Gervois and H. Navelet (1985b). *Integrals of three Bessel functions and Legendre functions. II*. J. Math. Phys. **26**(4), pp. 645–655.

A. Gervois and H. Navelet (1986a). *Some integrals involving three modified Bessel functions. I*. J. Math. Phys. **27**(3), pp. 682–687.

A. Gervois and H. Navelet (1986b). *Some integrals involving three modified Bessel functions. II*. J. Math. Phys. **27**(3), pp. 688–695.

I. M. Gessel (2003). *Applications of the classical umbral calculus*. Algebra Universalis **49**(4), pp. 397–434.

P. Gianni, M. Seppälä, R. Silhol, and B. Trager (1998). *Riemann surfaces, plane algebraic curves and their period matrices*. J. Symbolic Comput. **26**(6), pp. 789–803.

A. G. Gibbs (1973). *Problem 72-21, Laplace transforms of Airy functions*. SIAM Rev. **15**(4), pp. 796–798. Problem proposed by P. Smith.

A. Gil and J. Segura (2000). *Evaluation of toroidal harmonics*. Comput. Phys. Comm. **124**(1), pp. 104–122.

A. Gil and J. Segura (2003). *Computing the zeros and turning points of solutions of second order homogeneous linear ODEs*. SIAM J. Numer. Anal. **41**(3), pp. 827–855.

A. Gil, J. Segura, and N. M. Temme (2000). *Computing toroidal functions for wide ranges of the parameters*. J. Comput. Phys. **161**(1), pp. 204–217.

A. Gil, J. Segura, and N. M. Temme (2001). *On nonoscillating integrals for computing inhomogeneous Airy functions*. Math. Comp. **70**(235), pp. 1183–1194.

A. Gil, J. Segura, and N. M. Temme (2002a). *Computing complex Airy functions by numerical quadrature*. Numer. Algorithms **30**(1), pp. 11–23.

A. Gil, J. Segura, and N. M. Temme (2002b). *Evaluation of the modified Bessel function of the third kind of imaginary orders*. J. Comput. Phys. **175**(2), pp. 398–411.

A. Gil, J. Segura, and N. M. Temme (2003a). *Computation of the modified Bessel function of the third kind of imaginary orders: Uniform Airy-type asymptotic expansion*. J. Comput. Appl. Math. **153**(1-2), pp. 225–234.

A. Gil, J. Segura, and N. M. Temme (2003b). *Computing special functions by using quadrature rules*. Numer. Algorithms **33**(1-4), pp. 265–275. International Conference on Numerical Algorithms, Vol. I (Marrakesh, 2001).

A. Gil, J. Segura, and N. M. Temme (2003c). *On the zeros of the Scorer functions*. J. Approx. Theory **120**(2), pp. 253–266.

A. Gil, J. Segura, and N. M. Temme (2004a). *Computing solutions of the modified Bessel differential equation for imaginary orders and positive arguments*. ACM Trans. Math. Software **30**(2), pp. 145–158.

A. Gil, J. Segura, and N. M. Temme (2004b). *Integral representations for computing real parabolic cylinder functions*. Numer. Math. **98**(1), pp. 105–134.

A. Gil, J. Segura, and N. M. Temme (2006a). *The ABC of hyper recursions*. J. Comput. Appl. Math. **190**(1-2), pp. 270–286.

A. Gil, J. Segura, and N. M. Temme (2006b). *Computing the real parabolic cylinder functions $U(a,x)$, $V(a,x)$*. ACM Trans. Math. Software **32**(1), pp. 70–101.

A. Gil, J. Segura, and N. M. Temme (2006c). *Algorithm 850: Real parabolic cylinder functions $U(a,x)$, $V(a,x)$*. ACM Trans. Math. Software **32**(1), pp. 102–112.

A. Gil, J. Segura, and N. M. Temme (2007a). *Numerical Methods for Special Functions*. Philadelphia, PA: Society for Industrial and Applied Mathematics (SIAM).

A. Gil, J. Segura, and N. M. Temme (2007b). *Numerically satisfactory solutions of hypergeometric recursions*. Math. Comp. **76**(259), pp. 1449–1468.

S. G. Gindikin (1964). *Analysis in homogeneous domains*. Uspehi Mat. Nauk **19**(4 (118)), pp. 3–92. In Russian. English translation: Russian Math. Surveys, **19**(1964), no. 4, pp. 1–89.

J. N. Ginocchio (1991). *A new identity for some six-j symbols*. J. Math. Phys. **32**(6), pp. 1430–1432.

K. Girstmair (1990a). *Dirichlet convolution of cotangent numbers and relative class number formulas*. Monatsh. Math. **110**(3-4), pp. 231–256.

K. Girstmair (1990b). *A theorem on the numerators of the Bernoulli numbers*. Amer. Math. Monthly **97**(2), pp. 136–138.

J. W. L. Glaisher (1940). *Number-Divisor Tables*. British Association Mathematical Tables, Vol. VIII. Cambridge, England: Cambridge University Press.

M. L. Glasser (1976). *Definite integrals of the complete elliptic integral K*. J. Res. Nat. Bur. Standards Sect. B **80B**(2), pp. 313–323.

M. L. Glasser (1979). *A method for evaluating certain Bessel integrals*. Z. Angew. Math. Phys. **30**(4), pp. 722–723.

C. D. Godsil, M. Grötschel, and D. J. A. Welsh (1995). *Combinatorics in Statistical Physics*. In R. L. Graham, M. Grötschel, and L. Lovász (Eds.), *Handbook of Combinatorics, Vol. 2*, pp. 1925–1954. Amsterdam: Elsevier.

W. M. Y. Goh (1998). *Plancherel-Rotach asymptotics for the Charlier polynomials*. Constr. Approx. **14**(2), pp. 151–168.

D. Goldberg (1991). *What every computer scientist should know about floating-point arithmetic*. ACM Computing Surveys **23**(1), pp. 5–48.

K. Goldberg, F. T. Leighton, M. Newman, and S. L. Zuckerman (1976). *Tables of binomial coefficients and Stirling numbers*. J. Res. Nat. Bur. Standards Sect. B **80B**(1), pp. 99–171.

S. Goldstein (1927). *Mathieu functions*. Trans. Camb. Philos. Soc. **23**, pp. 303–336.

G. H. Golub and C. F. Van Loan (1996). *Matrix Computations* (3rd ed.). Baltimore, MD: Johns Hopkins University Press.

G. H. Golub and J. H. Welsch (1969). *Calculation of Gauss quadrature rules*. Math. Comp. **23**(106), pp. 221–230. Loose microfiche suppl. A1–A10.

V. V. Golubev (1960). *Lectures on Integration of the Equations of Motion of a Rigid Body About a Fixed Point*. Translated from the Russian by J. Shorr-Kon. Washington, D. C.: Office of Technical Services, U. S. Department of Commerce. Published for the National Science Foundation by the Israel Program for Scientific Translations, 1960.

E. T. Goodwin (1949a). *The evaluation of integrals of the form $\int_{-\infty}^{\infty} f(x)e^{-x^2}dx$*. Proc. Cambridge Philos. Soc. **45**(2), pp. 241–245.

E. T. Goodwin (1949b). *Recurrence relations for cross-products of Bessel functions*. Quart. J. Mech. Appl. Math. **2**(1), pp. 72–74.

E. T. Goodwin and J. Staton (1948). *Table of $\int_0^\infty \frac{e^{-u^2}}{u+x} du$*. Quart. J. Mech. Appl. Math. **1**(1), pp. 319–326.

R. G. Gordon (1969). *New method for constructing wavefunctions for bound states and scattering*. J. Chem. Phys. **51**, pp. 14–25.

R. G. Gordon (1970). *Constructing wavefunctions for nonlocal potentials*. J. Chem. Phys. **52**, pp. 6211–6217.

D. Goss (1978). *Von Staudt for $\mathbf{F}_q[T]$*. Duke Math. J. **45**(4), pp. 885–910.

D. Gottlieb and S. A. Orszag (1977). *Numerical Analysis of Spectral Methods: Theory and Applications*. Philadelphia, PA: Society for Industrial and Applied Mathematics. CBMS-NSF Regional Conference Series in Applied Mathematics, No. 26.

H. P. W. Gottlieb (1985). *On the exceptional zeros of cross-products of derivatives of spherical Bessel functions.* Z. Angew. Math. Phys. **36**(3), pp. 491–494.

H. W. Gould (1960). *Stirling number representation problems.* Proc. Amer. Math. Soc. **11**(3), pp. 447–451.

H. W. Gould (1972). *Explicit formulas for Bernoulli numbers.* Amer. Math. Monthly **79**, pp. 44–51.

É. Goursat (1881). *Sur l'équation différentielle linéaire, qui admet pour intégrale la série hypergéométrique.* Ann. Sci. École Norm. Sup. (2) **10**, pp. 3–142.

É. Goursat (1883). *Mémoire sur les fonctions hypergéométriques d'ordre supérieur.* Ann. Sci. École Norm. Sup. (2) **12**, pp. 261–286, 395–430.

J. Grad and E. Zakrajšek (1973). *Method for evaluation of zeros of Bessel functions.* J. Inst. Math. Appl. **11**, pp. 57–72.

I. S. Gradshteyn and I. M. Ryzhik (2000). *Table of Integrals, Series, and Products* (6th ed.). San Diego, CA: Academic Press Inc. Translated from the Russian. Translation edited and with a preface by Alan Jeffrey and Daniel Zwillinger. For a review of the fourth edition (1980), with errata list, see Math. Comp. v. 36 (1981), pp. 310–314. For a review of the fifth edition (1994) see Math. Comp. v. 64 (1995), pp. 439–441.

R. L. Graham, M. Grötschel, and L. Lovász (Eds.) (1995). *Handbook of Combinatorics. Vols. 1, 2.* Amsterdam: Elsevier Science B.V.

R. L. Graham, D. E. Knuth, and O. Patashnik (1994). *Concrete Mathematics: A Foundation for Computer Science* (2nd ed.). Reading, MA: Addison-Wesley Publishing Company.

B. Grammaticos, A. Ramani, and V. Papageorgiou (1991). *Do integrable mappings have the Painlevé property?* Phys. Rev. Lett. **67**(14), pp. 1825–1828.

T. V. Gramtcheff (1981). *An application of Airy functions to the Tricomi problem.* Math. Nachr. **102**(1), pp. 169–181.

A. Gray, G. B. Mathews, and T. M. MacRobert (1922). *A Treatise on Bessel Functions and their Applications to Physics* (2nd ed.). London: Macmillan and Co. Table errata: Math. Comp. v. 24 (1970), p. 240, v. 31 (1978), pp. 806 and 1046, and v. 32 (1978), pp. 318. Reprinted by Dover Publications Inc., New York, 1966.

J. J. Gray (2000). *Linear Differential Equations and Group Theory from Riemann to Poincaré* (2nd ed.). Boston, MA: Birkhäuser Boston Inc.

N. Gray (2002). *Automatic reduction of elliptic integrals using Carlson's relations.* Math. Comp. **71**(237), pp. 311–318.

M. B. Green, J. H. Schwarz, and E. Witten (1988a). *Superstring Theory: Introduction, Vol. 1* (2nd ed.). Cambridge Monographs on Mathematical Physics. Cambridge: Cambridge University Press.

M. B. Green, J. H. Schwarz, and E. Witten (1988b). *Superstring Theory: Loop Amplitudes, Anomalies and Phenomenolgy, Vol. 2.* (2nd ed.). Cambridge Monographs on Mathematical Physics. Cambridge: Cambridge University Press.

C. H. Greene, U. Fano, and G. Strinati (1979). *General form of the quantum-defect theory.* Phys. Rev. A **19**(4), pp. 1485–1509.

D. H. Greene and D. E. Knuth (1982). *Mathematics for the Analysis of Algorithms*, Volume 1 of *Progress in Computer Science*. Boston, MA: Birkhäuser Boston.

A. G. Greenhill (1892). *The Applications of Elliptic Functions.* London: MacMillan. Reprinted by Dover Publications Inc., New York, 1959.

A. G. Greenhill (1959). *The Applications of Elliptic Functions.* New York: Dover Publications Inc. Reprint of Macmillan, London edition, 1892.

W. Greiner, B. Müller, and J. Rafelski (1985). *Quantum Electrodynamics of Strong Fields: With an Introduction into Modern Relativistic Quantum Mechanics.* Texts and Monographs in Physics. Springer.

W. Gröbner and N. Hofreiter (1949). *Integraltafel. Erster Teil. Unbestimmte Integrale.* Vienna: Springer-Verlag. Reprintings with corrections appeared in 1961 and 1965.

W. Gröbner and N. Hofreiter (1950). *Integraltafel. Zweiter Teil. Bestimmte Integrale.* Vienna and Innsbruck: Springer-Verlag. Reprintings with corrections appeared in 1958 and 1961.

P. Groeneboom and D. R. Truax (2000). *A monotonicity property of the power function of multivariate tests.* Indag. Math. (N.S.) **11**(2), pp. 209–218.

V. I. Gromak (1975). *Theory of Painlevé's equations.* Differ. Uravn. **11**(11), pp. 373–376. In Russian. English translation: Differential Equations **11**(11), pp. 285–287.

V. I. Gromak (1976). *The solutions of Painlevé's fifth equation.* Differ. Uravn. **12**(4), pp. 740–742. In Russian. English translation: Differential Equations **12**(4), pp. 519–521 (1977).

V. I. Gromak (1978). *One-parameter systems of solutions of Painlevé equations.* Differ. Uravn. **14**(12), pp. 2131–2135. In Russian. English translation: Differential Equations **14**(12), pp. 1510–1513 (1979).

V. I. Gromak (1987). *Theory of the fourth Painlevé equation.* Differ. Uravn. **23**(5), pp. 760–768, 914. In Russian. English translation: Differential Equations **23**(5), pp. 506–513.

V. I. Gromak, I. Laine, and S. Shimomura (2002). *Painlevé Differential Equations in the Complex Plane*, Volume 28 of *Studies in Mathematics*. Berlin-New York: Walter de Gruyter & Co.

V. I. Gromak and N. A. Lukaševič (1982). *Special classes of solutions of Painlevé equations*. Differ. Uravn. **18**(3), pp. 419–429. In Russian. English translation: Differential Equations **18**(3), pp. 317–326.

K. I. Gross and R. A. Kunze (1976). *Bessel functions and representation theory. I.* J. Functional Analysis **22**(2), pp. 73–105.

K. I. Gross and D. St. P. Richards (1987). *Special functions of matrix argument. I. Algebraic induction, zonal polynomials, and hypergeometric functions*. Trans. Amer. Math. Soc. **301**(2), pp. 781–811.

K. I. Gross and D. St. P. Richards (1991). *Hypergeometric functions on complex matrix space*. Bull. Amer. Math. Soc. (N.S.) **24**(2), pp. 349–355.

E. Grosswald (1978). *Bessel Polynomials*, Volume 698 of *Lecture Notes in Mathematics*. Berlin-New York: Springer.

E. Grosswald (1985). *Representations of Integers as Sums of Squares*. New York: Springer-Verlag.

F. W. Grover (1946). *Inductance Calculations*. New York: Van Nostrand. Reprinted by Dover, New York, 1962 and 2004.

B.-Y. Guo (1998). *Spectral Methods and Their Applications*. River Edge, NJ-Singapore: World Scientific Publishing Co. Inc.

B. N. Gupta (1970). *On Mill's ratio*. Proc. Cambridge Philos. Soc. **67**, pp. 363–364.

D. P. Gupta and M. E. Muldoon (2000). *Riccati equations and convolution formulae for functions of Rayleigh type*. J. Phys. A **33**(7), pp. 1363–1368.

H. Gupta (1935). *A table of partitions*. Proc. London Math. Soc. (2) **39**, pp. 142–149.

H. Gupta (1937). *A table of partitions*. Proc. London Math. Soc. (2) **42**, pp. 546–549.

H. Gupta, C. E. Gwyther, and J. C. P. Miller (1958). *Tables of Partitions*, Volume 4 of *Royal Society Math. Tables*. Cambridge University Press.

R. A. Gustafson (1987). *Multilateral summation theorems for ordinary and basic hypergeometric series in* U(n). SIAM J. Math. Anal. **18**(6), pp. 1576–1596.

A. Guthmann (1991). *Asymptotische Entwicklungen für unvollständige Gammafunktionen*. Forum Math. **3**(2), pp. 105–141. Translated title: *Asymptotic expansions for incomplete gamma functions*.

J. C. Gutiérrez-Vega, R. M. Rodríguez-Dagnino, M. A. Meneses-Nava, and S. Chávez-Cerda (2003). *Mathieu functions, a visual approach*. Amer. J. Phys. **71**(3), pp. 233–242.

A. J. Guttmann and T. Prellberg (1993). *Staircase polygons, elliptic integrals, Heun functions, and lattice Green functions*. Phys. Rev. E **47**(4), pp. R2233–R2236.

L. Habsieger (1986). *La q-conjecture de Macdonald-Morris pour G_2*. C. R. Acad. Sci. Paris Sér. I Math. **303**(6), pp. 211–213.

L. Habsieger (1988). *Une q-intégrale de Selberg et Askey*. SIAM J. Math. Anal. **19**(6), pp. 1475–1489.

J. Hadamard (1896). *Sur la distribution des zéros de la fonction $\zeta(s)$ et ses conséquences arithmétiques*. Bull. Soc. Math. France **24**, pp. 199–220. Reprinted in Oeuvres de Jacques Hadamard. Tomes I, pp. 189–210, Éditions du Centre National de la Recherche Scientifique, Paris, 1968.

P. I. Hadži (1968). *Computation of certain integrals that contain a probability function*. Bul. Akad. Štiince RSS Moldoven **1968**(2), pp. 81–104. (errata insert).

P. I. Hadži (1969). *Certain integrals that contain a probability function and degenerate hypergeometric functions*. Bul. Akad. Štiince RSS Moldoven **1969**(2), pp. 40–47.

P. I. Hadži (1970). *Some integrals that contain a probability function and hypergeometric functions*. Bul. Akad. Štiince RSS Moldoven **1970**(1), pp. 49–62.

P. I. Hadži (1972). *Certain sums that contain cylindrical functions*. Bul. Akad. Štiince RSS Moldoven. **1972**(3), pp. 75–77, 94.

P. I. Hadži (1973). *The Laplace transform for expressions that contain a probability function*. Bul. Akad. Štiince RSS Moldoven. **1973**(2), pp. 78–80, 93.

P. I. Hadži (1975a). *Certain integrals that contain a probability function*. Bul. Akad. Štiince RSS Moldoven. **1975**(2), pp. 86–88, 95.

P. I. Hadži (1975b). *Integrals containing the Fresnel functions $S(x)$ and $C(x)$*. Bul. Akad. Štiince RSS Moldoven. **1975**(3), pp. 48–60, 93.

P. I. Hadži (1976a). *Expansions for the probability function in series of Čebyšev polynomials and Bessel functions*. Bul. Akad. Štiince RSS Moldoven. **1976**(1), pp. 77–80, 96.

P. I. Hadži (1976b). *Integrals that contain a probability function of complicated arguments*. Bul. Akad. Štiince RSS Moldoven. **1976**(1), pp. 80–84, 96.

P. I. Hadži (1978). *Sums with cylindrical functions that reduce to the probability function and to related functions*. Bul. Akad. Shtiintse RSS Moldoven. **1978**(3), pp. 80–84, 95.

E. Hahn (1980). *Asymptotik bei Jacobi-Polynomen und Jacobi-Funktionen*. Math. Z. **171**(3), pp. 201–226.

W. Hahn (1949). *Über Orthogonalpolynome, die q-Differenzengleichungen genügen*. Math. Nachr. **2**, pp. 4–34.

E. Hairer, S. P. Nørsett, and G. Wanner (1993). *Solving Ordinary Differential Equations. I. Nonstiff Problems* (2nd ed.), Volume 8 of *Springer Series in Computational Mathematics*. Berlin: Springer-Verlag. A reprinting with corrections appeared in 2000.

E. Hairer, S. P. Nørsett, and G. Wanner (2000). *Solving Ordinary Differential Equations. I. Nonstiff Problems* (2nd ed.). Berlin: Springer-Verlag.

E. Hairer and G. Wanner (1996). *Solving Ordinary Differential Equations. II. Stiff and Differential-Algebraic Problems* (2nd ed.), Volume 14 of *Springer Series in Computational Mathematics*. Berlin: Springer-Verlag.

M. H. Halley, D. Delande, and K. T. Taylor (1993). *The combination of R-matrix and complex coordinate methods: Application to the diamagnetic Rydberg spectra of Ba and Sr*. J. Phys. B **26**(12), pp. 1775–1790.

A. J. S. Hamilton (2001). *Formulae for growth factors in expanding universes containing matter and a cosmological constant*. Monthly Notices Roy. Astronom. Soc. **322**(2), pp. 419–425.

J. Hammack, D. McCallister, N. Scheffner, and H. Segur (1995). *Two-dimensional periodic waves in shallow water. II. Asymmetric waves*. J. Fluid Mech. **285**, pp. 95–122.

J. Hammack, N. Scheffner, and H. Segur (1989). *Two-dimensional periodic waves in shallow water*. J. Fluid Mech. **209**, pp. 567–589.

H. Hancock (1958). *Elliptic Integrals*. New York: Dover Publications Inc. Unaltered reprint of original edition published by Wiley, New York, 1917.

R. A. Handelsman and J. S. Lew (1970). *Asymptotic expansion of Laplace transforms near the origin*. SIAM J. Math. Anal. **1**(1), pp. 118–130.

R. A. Handelsman and J. S. Lew (1971). *Asymptotic expansion of a class of integral transforms with algebraically dominated kernels*. J. Math. Anal. Appl. **35**(2), pp. 405–433.

S. Hanish, R. V. Baier, A. L. Van Buren, and B. J. King (1970). *Tables of Radial Spheroidal Wave Functions, Vols. 1-3, Prolate, $m = 0, 1, 2$; Vols. 4-6, Oblate, $m = 0, 1, 2$*. Technical report, Naval Research Laboratory, Washington, D.C. NRL Reports 7088-7093.

E. R. Hansen (1975). *A Table of Series and Products*. Englewood Cliffs, NJ: Prentice-Hall. Table erratum: Math. Comp. v.47 (1986), no. 176, p. 767.

E. W. Hansen (1985). *Fast Hankel transform algorithm*. IEEE Trans. Acoust. Speech Signal Process. **32**(3), pp. 666–671.

J. Happel and H. Brenner (1973). *Low Reynolds Number Hydrodynamics with Special Applications to Particulate Media* (2nd ed.). Leyden: Noordhoff International Publishing. The first edition was published in 1965 by Prentice-Hall. The 2nd edition was reissued in 1983 by Springer, and in 1986 by Martinus Nijhoff Publishers (Kluwer Academic Publishers Group).

G. H. Hardy (1912). *Note on Dr. Vacca's series for γ*. Quart. J. Math. **43**, pp. 215–216.

G. H. Hardy (1940). *Ramanujan. Twelve Lectures on Subjects Suggested by His Life and Work*. Cambridge, England: Cambridge University Press. Reprinted by Chelsea Publishing Company, New York, 1959.

G. H. Hardy (1949). *Divergent Series*. Oxford: Clarendon Press. Reprinted by the American Mathematical Society in 2000.

G. H. Hardy (1952). *A Course of Pure Mathematics* (10th ed.). Cambridge University Press. Numerous reprintings exist, including the Centenary Edition with Foreword by T. W. Körner (Cambridge Univerity Press, 2008).

G. H. Hardy and J. E. Littlewood (1925). *Some problems of "Partitio Numerorum" (VI): Further researches in Waring's Problem*. Math. Z. **23**, pp. 1–37. Reprinted in Collected Papers of G. H. Hardy (Including Joint papers with J. E. Littlewood and others), Vol. I, pp. 469–505, Clarendon Press, Oxford, 1966.

G. H. Hardy, J. E. Littlewood, and G. Pólya (1967). *Inequalities* (2nd ed.). Cambridge Mathematical Library. Cambridge: Cambridge University Press. Reprint of the 1952 edition and reprinted in 1988.

G. H. Hardy and S. Ramanujan (1918). *Asymptotic formulae in combinatory analysis*. Proc. London Math. Soc. (2) **17**, pp. 75–115. Reprinted in Ramanujan (1962, pp. 276–309).

G. H. Hardy and E. M. Wright (1979). *An Introduction to the Theory of Numbers* (5th ed.). New York-Oxford: The Clarendon Press Oxford University Press.

B. A. Hargrave (1978). *High frequency solutions of the delta wing equations*. Proc. Roy. Soc. Edinburgh Sect. A **81**(3-4), pp. 299–316.

B. A. Hargrave and B. D. Sleeman (1977). *Lamé polynomials of large order*. SIAM J. Math. Anal. **8**(5), pp. 800–842.

F. E. Harris (2000). *Spherical Bessel expansions of sine, cosine, and exponential integrals*. Appl. Numer. Math. **34**(1), pp. 95–98.

F. E. Harris (2002). *Analytic evaluation of two-center STO electron repulsion integrals via ellipsoidal expansion*. Internat. J. Quantum Chem. **88**(6), pp. 701–734.

J. F. Hart, E. W. Cheney, C. L. Lawson, H. J. Maehly, C. K. Mesztenyi, J. R. Rice, H. G. Thacher, Jr., and C. Witzgall (1968). *Computer Approximations*. SIAM Ser. in Appl. Math. New York: John Wiley & Sons Inc. Reprinted by Krieger Publishing Company, 1978.

D. R. Hartree (1936). *Some properties and applications of the repeated integrals of the error function.* Proc. Manchester Lit. Philos. Soc. **80**, pp. 85–102.

Harvard (1945). Harvard University. *Tables of the Modified Hankel Functions of Order One-Third and of their Derivatives.* Cambridge, MA: Harvard University Press.

A. Hasegawa (1989). *Optical Solitons in Fibers.* Berlin, Germany: Springer-Verlag.

C. B. Haselgrove and J. C. P. Miller (1960). *Tables of the Riemann Zeta Function.* Royal Society Mathematical Tables, Vol. 6. New York: Cambridge University Press.

C. Hastings, Jr. (1955). *Approximations for Digital Computers.* Princeton, N. J.: Princeton University Press. Assisted by Jeanne T. Hayward and James P. Wong, Jr.

S. P. Hastings and J. B. McLeod (1980). *A boundary value problem associated with the second Painlevé transcendent and the Korteweg-de Vries equation.* Arch. Rational Mech. Anal. **73**(1), pp. 31–51.

M. Hauss (1997). *An Euler-Maclaurin-type formula involving conjugate Bernoulli polynomials and an application to $\zeta(2m+1)$.* Commun. Appl. Anal. **1**(1), pp. 15–32.

M. Hauss (1998). *A Boole-type Formula involving Conjugate Euler Polynomials.* In P. Butzer, H. T. Jongen, and W. Oberschelp (Eds.), *Charlemagne and his Heritage. 1200 Years of Civilization and Science in Europe, Vol. 2 (Aachen, 1995)*, pp. 361–375. Turnhout: Brepols Publishers.

B. Hayes (2009). *The higher arithmetic.* American Scientist **97**, pp. 364–368.

V. B. Headley and V. K. Barwell (1975). *On the distribution of the zeros of generalized Airy functions.* Math. Comp. **29**(131), pp. 863–877.

G. J. Heckman (1991). *An elementary approach to the hypergeometric shift operators of Opdam.* Invent. Math. **103**(2), pp. 341–350.

M. Heil (1995). *Numerical Tools for the Study of Finite Gap Solutions of Integrable Systems.* Ph. D. thesis, Technischen Universität Berlin. (All relevant material, plus corrections, are included in Deconinck et al. (2004).).

R. S. Heller (1976). *25D Table of the First One Hundred Values of $j_{0,s}, J_1(j_{0,s}), j_{1,s}, J_0(j_{1,s}) = J_0(j'_{0,s+1}), j'_{1,s}, J_1(j'_{1,s})$.* Technical report, Department of Physics, Worcester Polytechnic Institute, Worcester, MA. Ms. of six pages deposited in the UMT file.

E. Hendriksen and H. van Rossum (1986). *Orthogonal Laurent polynomials.* Nederl. Akad. Wetensch. Indag. Math. **48**(1), pp. 17–36.

P. Henrici (1974). *Applied and Computational Complex Analysis. Vol. 1: Power Series—Integration—Conformal Mapping—Location of Zeros.* Pure and Applied Mathematics. New York: Wiley-Interscience [John Wiley & Sons]. Reprinted in 1988.

P. Henrici (1977). *Applied and Computational Complex Analysis. Vol. 2: Special Functions—Integral Transforms—Asymptotics—Continued Fractions.* New York: Wiley-Interscience [John Wiley & Sons]. Reprinted in 1991.

P. Henrici (1986). *Applied and Computational Complex Analysis. Vol. 3: Discrete Fourier Analysis—Cauchy Integrals—Construction of Conformal Maps—Univalent Functions.* Pure and Applied Mathematics. New York: Wiley-Interscience [John Wiley & Sons Inc.]. Reprinted in 1993.

D. R. Herrick and S. O'Connor (1998). *Inverse virial symmetry of diatomic potential curves.* J. Chem. Phys. **109**(1), pp. 11–19.

C. S. Herz (1955). *Bessel functions of matrix argument.* Ann. of Math. (2) **61**(3), pp. 474–523.

H. W. Hethcote (1970). *Error bounds for asymptotic approximations of zeros of Hankel functions occurring in diffraction problems.* J. Mathematical Phys. **11**(8), pp. 2501–2504.

D. Hilbert (1909). *Beweis für die Darstellbarkeit der ganzen Zahlen durch eine feste Anzahl n^{ter} Potenzen (Waringsches Problem).* Nachrichten von der Gesellschaft der Wissenschaften zu Göttingen, Mathematisch-Physikalische Klasse, pp. 17–36. Also in Math. Annalen (1909), v. 67, pp. 281–300.

F. B. Hildebrand (1974). *Introduction to Numerical Analysis* (2nd ed.). New York: McGraw-Hill Book Co. Reprinted by Dover Publications Inc., New York, 1987.

C. J. Hill (1828). *Über die Integration logarithmisch-rationaler Differentiale.* J. Reine Angew. Math. **3**, pp. 101–159.

E. Hille (1929). *Note on some hypergeometric series of higher order.* J. London Math. Soc. **4**, pp. 50–54.

E. Hille (1976). *Ordinary Differential Equations in the Complex Domain.* Pure and Applied Mathematics. New York: Wiley-Interscience [John Wiley & Sons]. Reprinted by Dover Publications Inc., New York, 1997.

P. Hillion (1997). *Diffraction and Weber functions.* SIAM J. Appl. Math. **57**(6), pp. 1702–1715.

M. H. Hirata (1975). *Flow near the bow of a steadily turning ship.* J. Fluid Mech. **71**(2), pp. 283–291.

N. J. Hitchin (1995). *Poncelet Polygons and the Painlevé Equations.* In Ramanan (Ed.), *Geometry and Analysis (Bombay, 1992)*, pp. 151–185. Bombay: Tata Inst. Fund. Res.

N. J. Hitchin (2003). *A lecture on the octahedron*. Bull. London Math. Soc. **35**(5), pp. 577–600.

E. W. Hobson (1928). *A Treatise on Plane and Advanced Trigonometry* (7th ed.). Cambridge University Press. Reprinted by Dover Publications Inc., New York, 2004.

E. W. Hobson (1931). *The Theory of Spherical and Ellipsoidal Harmonics*. London-New York: Cambridge University Press. Reprinted by Chelsea Publishing Company, New York, 1955 and 1965.

H. Hochstadt (1963). *Estimates of the stability intervals for Hill's equation*. Proc. Amer. Math. Soc. **14**(6), pp. 930–932.

H. Hochstadt (1964). *Differential Equations: A Modern Approach*. New York: Holt, Rinehart and Winston. Reprinted by Dover in 1975.

H. Hochstadt (1971). *The Functions of Mathematical Physics*. New York-London-Sydney: Wiley-Interscience [John Wiley & Sons, Inc.]. Pure and Applied Mathematics, Vol. XXIII. Reprinted with corrections by Dover Publications Inc., New York, 1986.

L. E. Hoisington and G. Breit (1938). *Calculation of Coulomb wave functions for high energies*. Phys. Rev. **54**(8), pp. 627–628.

P. Holmes and D. Spence (1984). *On a Painlevé-type boundary-value problem*. Quart. J. Mech. Appl. Math. **37**(4), pp. 525–538.

E. Hopf (1934). *Mathematical Problems of Radiative Equilibrium*. Cambridge Tracts in Mathematics and Mathematical Physics No. 31. Cambridge: Cambridge University Press. Reprinted by Stechert-Hafner, Inc., New York, 1964.

K. Horata (1989). *An explicit formula for Bernoulli numbers*. Rep. Fac. Sci. Technol. Meijo Univ. **29**, pp. 1–6.

K. Horata (1991). *On congruences involving Bernoulli numbers and irregular primes. II*. Rep. Fac. Sci. Technol. Meijo Univ. **31**, pp. 1–8.

F. T. Howard (1976). *Roots of the Euler polynomials*. Pacific J. Math. **64**(1), pp. 181–191.

F. T. Howard (1996a). *Explicit formulas for degenerate Bernoulli numbers*. Discrete Math. **162**(1-3), pp. 175–185.

F. T. Howard (1996b). *Sums of powers of integers via generating functions*. Fibonacci Quart. **34**(3), pp. 244–256.

C. J. Howls (1992). *Hyperasymptotics for integrals with finite endpoints*. Proc. Roy. Soc. London Ser. A **439**, pp. 373–396.

C. J. Howls, P. J. Langman, and A. B. Olde Daalhuis (2004). *On the higher-order Stokes phenomenon*. Proc. Roy. Soc. London Ser. A **460**, pp. 2285–2303.

C. J. Howls and A. B. Olde Daalhuis (1999). *On the resurgence properties of the uniform asymptotic expansion of Bessel functions of large order*. Proc. Roy. Soc. London Ser. A **455**, pp. 3917–3930.

M. Hoyles, S. Kuyucak, and S.-H. Chung (1998). *Solutions of Poisson's equation in channel-like geometries*. Comput. Phys. Comm. **115**(1), pp. 45–68.

L. K. Hua (1963). *Harmonic Analysis of Functions of Several Complex Variables in the Classical Domains*, Volume 6 of *Translations of Mathematical Monographs*. Providence, RI: American Mathematical Society. Reprinted in 1979.

I.-C. Huang and S.-Y. Huang (1999). *Bernoulli numbers and polynomials via residues*. J. Number Theory **76**(2), pp. 178–193.

J. H. Hubbard and B. B. Hubbard (2002). *Vector Calculus, Linear Algebra, and Differential Forms: A Unified Approach* (2nd ed.). Upper Saddle River, NJ: Prentice Hall Inc.

M. H. Hull, Jr. and G. Breit (1959). *Coulomb Wave Functions*. In S. Flügge (Ed.), *Handbuch der Physik, Bd. 41/1*, pp. 408–465. Berlin: Springer.

P. Humbert (1920). *Sur les fonctions hypercylindriques*. C. R. Acad. Sci. Paris Sér. I Math. **171**, pp. 490–492.

J. Humblet (1984). *Analytical structure and properties of Coulomb wave functions for real and complex energies*. Ann. Physics **155**(2), pp. 461–493.

J. Humblet (1985). *Bessel function expansions of Coulomb wave functions*. J. Math. Phys. **26**(4), pp. 656–659.

C. Hunter (1981). *Two Parametric Eigenvalue Problems of Differential Equations*. In *Spectral Theory of Differential Operators (Birmingham, AL, 1981)*, Volume 55 of *North-Holland Math. Stud.*, pp. 233–241. Amsterdam: North-Holland.

C. Hunter and B. Guerrieri (1981). *The eigenvalues of Mathieu's equation and their branch points*. Stud. Appl. Math. **64**(2), pp. 113–141.

C. Hunter and B. Guerrieri (1982). *The eigenvalues of the angular spheroidal wave equation*. Stud. Appl. Math. **66**(3), pp. 217–240.

G. Hunter and M. Kuriyan (1976). *Asymptotic expansions of Mathieu functions in wave mechanics*. J. Comput. Phys. **21**(3), pp. 319–325.

A. Hurwitz (1882). *Einige Eigenschaften der Dirichletschen Functionen $F(s) = \sum(\frac{D}{n}) \cdot \frac{1}{n}$, die bei der Bestimmung der Klassenanzahlen binärer quadratischer Formen auftreten*. Zeitschrift für Math. u. Physik **27**, pp. 86–101.

IEEE (1985). *IEEE Standard for Binary Floating-Point Arithmetic*. The Institute of Electrical and Electronics Engineers. ANSI/IEEE Std 754-1985. Reprinted in *ACM SIGPLAN Notices* v. 22, no. 2 (1987), pp. 9–25.

J.-I. Igusa (1972). *Theta Functions.* New York: Springer-Verlag. Die Grundlehren der mathematischen Wissenschaften, Band 194.

Y. Ikebe (1975). *The zeros of regular Coulomb wave functions and of their derivatives.* Math. Comp. **29**, pp. 878–887.

Y. Ikebe, Y. Kikuchi, and I. Fujishiro (1991). *Computing zeros and orders of Bessel functions.* J. Comput. Appl. Math. **38**(1-3), pp. 169–184.

Y. Ikebe, Y. Kikuchi, I. Fujishiro, N. Asai, K. Takanashi, and M. Harada (1993). *The eigenvalue problem for infinite compact complex symmetric matrices with application to the numerical computation of complex zeros of $J_0(z) - iJ_1(z)$ and of Bessel functions $J_m(z)$ of any real order m.* Linear Algebra Appl. **194**, pp. 35–70.

M. Ikonomou, P. Köhler, and A. F. Jacob (1995). *Computation of integrals over the half-line involving products of Bessel functions, with application to microwave transmission lines.* Z. Angew. Math. Mech. **75**(12), pp. 917–926.

E. L. Ince (1926). *Ordinary Differential Equations.* London: Longmans, Green and Co. Reprinted by Dover Publications, New York, 1944, 1956.

E. L. Ince (1932). *Tables of the elliptic cylinder functions.* Proc. Roy. Soc. Edinburgh Sect. A **52**, pp. 355–433.

E. L. Ince (1940a). *The periodic Lamé functions.* Proc. Roy. Soc. Edinburgh **60**, pp. 47–63.

E. L. Ince (1940b). *Further investigations into the periodic Lamé functions.* Proc. Roy. Soc. Edinburgh **60**, pp. 83–99.

A. E. Ingham (1932). *The Distribution of Prime Numbers.* Cambridge Tracts in Mathematics and Mathematical Physics, No. 30. Cambridge: Cambridge University Press. Reissued with a foreword by R. C. Vaughan in the Cambridge Mathematical Library series, 1990.

A. E. Ingham (1933). *An integral which occurs in statistics.* Proceedings of the Cambridge Philosophical Society **29**, pp. 271–276.

K. Inkeri (1959). *The real roots of Bernoulli polynomials.* Ann. Univ. Turku. Ser. A I **37**, pp. 1–20.

K. Ireland and M. Rosen (1990). *A Classical Introduction to Modern Number Theory* (2nd ed.). New York: Springer-Verlag.

A. Iserles (1996). *A First Course in the Numerical Analysis of Differential Equations.* Cambridge Texts in Applied Mathematics, No. 15. Cambridge: Cambridge University Press.

A. Iserles, P. E. Koch, S. P. Nørsett, and J. M. Sanz-Serna (1991). *On polynomials orthogonal with respect to certain Sobolev inner products.* J. Approx. Theory **65**(2), pp. 151–175.

A. Iserles, S. P. Nørsett, and S. Olver (2006). *Highly Oscillatory Quadrature: The Story So Far.* In A. Bermudez de Castro et al. (Eds.), *Numerical Mathematics and Advanced Applications*, pp. 97–118. Berlin: Springer-Verlag. Proceedings of ENuMath, Santiago de Compostela (2005).

M. E. H. Ismail (1986). *Asymptotics of the Askey-Wilson and q-Jacobi polynomials.* SIAM J. Math. Anal. **17**(6), pp. 1475–1482.

M. E. H. Ismail (2000a). *An electrostatics model for zeros of general orthogonal polynomials.* Pacific J. Math. **193**(2), pp. 355–369.

M. E. H. Ismail (2000b). *More on electrostatic models for zeros of orthogonal polynomials.* Numer. Funct. Anal. Optim. **21**(1-2), pp. 191–204.

M. E. H. Ismail (2005). *Classical and Quantum Orthogonal Polynomials in One Variable*, Volume 98 of *Encyclopedia of Mathematics and its Applications.* Cambridge: Cambridge University Press.

M. E. H. Ismail, J. Letessier, G. Valent, and J. Wimp (1990). *Two families of associated Wilson polynomials.* Canad. J. Math. **42**(4), pp. 659–695.

M. E. H. Ismail and X. Li (1992). *Bound on the extreme zeros of orthogonal polynomials.* Proc. Amer. Math. Soc. **115**(1), pp. 131–140.

M. E. H. Ismail and D. R. Masson (1991). *Two families of orthogonal polynomials related to Jacobi polynomials.* Rocky Mountain J. Math. **21**(1), pp. 359–375.

M. E. H. Ismail and D. R. Masson (1994). *q-Hermite polynomials, biorthogonal rational functions, and q-beta integrals.* Trans. Amer. Math. Soc. **346**(1), pp. 63–116.

M. E. H. Ismail and M. E. Muldoon (1995). *Bounds for the small real and purely imaginary zeros of Bessel and related functions.* Methods Appl. Anal. **2**(1), pp. 1–21.

A. R. Its, A. S. Fokas, and A. A. Kapaev (1994). *On the asymptotic analysis of the Painlevé equations via the isomonodromy method.* Nonlinearity **7**(5), pp. 1291–1325.

A. R. Its and A. A. Kapaev (1987). *The method of isomonodromic deformations and relation formulas for the second Painlevé transcendent.* Izv. Akad. Nauk SSSR Ser. Mat. **51**(4), pp. 878–892, 912. In Russian. English translation: Math. USSR-Izv. **31**(1988), no. 1, pp. 193–207.

A. R. Its and A. A. Kapaev (1998). *Connection formulae for the fourth Painlevé transcendent; Clarkson-McLeod solution.* J. Phys. A **31**(17), pp. 4073–4113.

A. R. Its and A. A. Kapaev (2003). *Quasi-linear Stokes phenomenon for the second Painlevé transcendent.* Nonlinearity **16**(1), pp. 363–386.

A. R. Its and V. Y. Novokshënov (1986). *The Isomonodromic Deformation Method in the Theory of Painlevé Equations*, Volume 1191 of *Lecture Notes in Mathematics*. Berlin: Springer-Verlag.

C. Itzykson and J.-M. Drouffe (1989). *Statistical Field Theory: Strong Coupling, Monte Carlo Methods, Conformal Field Theory, and Random Systems*, Volume 2. Cambridge: Cambridge University Press.

C. Itzykson and J. B. Zuber (1980). *Quantum Field Theory*. International Series in Pure and Applied Physics. New York: McGraw-Hill International Book Co. Reprinted by Dover, 2006.

A. Ivić (1985). *The Riemann Zeta-Function*. A Wiley-Interscience Publication. New York: John Wiley & Sons Inc. Reprinted by Dover, 2003.

K. Iwasaki, H. Kimura, S. Shimomura, and M. Yoshida (1991). *From Gauss to Painlevé: A Modern Theory of Special Functions*, Volume 16 of *Aspects of Mathematics E*. Braunschweig, Germany: Friedr. Vieweg & Sohn.

J. D. Jackson (1999). *Classical Electrodynamics* (3rd ed.). New York: John Wiley & Sons Inc.

C. G. J. Jacobi (1829). *Fundamenta Nova Theoriae Functionum Ellipticarum*. Regiomonti, Sumptibus fratrum Bornträger. Reprinted in Werke, Vol. I, pp. 49–239, Herausgegeben von C. W. Borchardt, Berlin, Verlag von G. Reimer, 1881.

D. Jacobs and F. Lambert (1972). *On the numerical calculation of polylogarithms*. Nordisk Tidskr. Informationsbehandling (BIT) **12**(4), pp. 581–585.

L. Jacobsen, W. B. Jones, and H. Waadeland (1986). *Further results on the computation of incomplete gamma functions*. In W. J. Thron (Ed.), *Analytic Theory of Continued Fractions, II (Pitlochry/Aviemore, 1985)*, Lecture Notes in Math. 1199, pp. 67–89. Berlin: Springer-Verlag.

L. Jager (1997). *Fonctions de Mathieu et polynômes de Klein-Gordon*. C. R. Acad. Sci. Paris Sér. I Math. **325**(7), pp. 713–716. Translated title: *Mathieu functions and Klein-Gordon polynomials*.

L. Jager (1998). *Fonctions de Mathieu et fonctions propres de l'oscillateur relativiste*. Ann. Fac. Sci. Toulouse Math. (6) **7**(3), pp. 465–495. Translated title: *Mathieu functions and eigenfunctions of the relativistic oscillator*.

D. L. Jagerman (1974). *Some properties of the Erlang loss function*. Bell System Tech. J. **53**, pp. 525–551.

E. Jahnke and F. Emde (1945). *Tables of Functions with Formulae and Curves* (4th ed.). New York: Dover Publications. Contains corrections and enlargements of earlier editions. Originally published by B. G. Teubner, Leipzig, in 1933. In German and English.

E. Jahnke, F. Emde, and F. Lösch (1966). *Tafeln höherer Funktionen (Tables of Higher Functions)* (7th ed.). Stuttgart: B. G. Teubner. Reprint of sixth edition (McGraw-Hill, New York, 1960) with corrections and updated bibliography. The sixth edition is a completely revised and enlarged version of Jahnke and Emde (1945). In German and English.

A. T. James (1964). *Distributions of matrix variates and latent roots derived from normal samples*. Ann. Math. Statist. **35**(2), pp. 475–501.

J. K. M. Jansen (1977). *Simple-periodic and Non-periodic Lamé Functions*. Mathematical Centre Tracts, No. 72. Amsterdam: Mathematisch Centrum.

S. Janson, D. E. Knuth, T. Łuczak, and B. Pittel (1993). *The birth of the giant component*. Random Structures Algorithms **4**(3), pp. 231–358. With an introduction by the editors.

H. Jeffreys (1928). *The effect on Love waves of heterogeneity in the lower layer*. Monthly Notices Roy. Astronom. Soc. Geophysical Supplement **2**, pp. 101–111.

H. Jeffreys and B. S. Jeffreys (1956). *Methods of Mathematical Physics* (3rd ed.). Cambridge: Cambridge University Press. Reprinted in 1999.

A. J. Jerri (1982). *A note on sampling expansion for a transform with parabolic cylinder kernel*. Inform. Sci. **26**(2), pp. 155–158.

M. Jimbo and T. Miwa (1981). *Monodromy preserving deformation of linear ordinary differential equations with rational coefficients. II*. Phys. D **2**(3), pp. 407–448.

M. Jimbo, T. Miwa, Y. Môri, and M. Sato (1980). *Density matrix of an impenetrable Bose gas and the fifth Painlevé transcendent*. Phys. D **1**(1), pp. 80–158.

X.-S. Jin and R. Wong (1998). *Uniform asymptotic expansions for Meixner polynomials*. Constr. Approx. **14**(1), pp. 113–150.

X.-S. Jin and R. Wong (1999). *Asymptotic formulas for the zeros of the Meixner polynomials*. J. Approx. Theory **96**(2), pp. 281–300.

H. K. Johansen and K. Sørensen (1979). *Fast Hankel transforms*. Geophysical Prospecting **27**(4), pp. 876–901.

J. H. Johnson and J. M. Blair (1973). *REMES2 — a Fortran program to calculate rational minimax approximations to a given function*. Technical Report AECL-4210, Atomic Energy of Canada Limited. Chalk River Nuclear Laboratories, Chalk River, Ontario.

N. L. Johnson, S. Kotz, and N. Balakrishnan (1994). *Continuous Univariate Distributions* (2nd ed.), Volume I. New York: John Wiley & Sons Inc.

N. L. Johnson, S. Kotz, and N. Balakrishnan (1995). *Continuous Univariate Distributions* (2nd ed.), Volume II. New York: John Wiley & Sons Inc.

W. R. Johnson (2007). *Atomic Structure Theory: Lectures on Atomic Physics*. Berlin and Heidelberg: Springer.

D. S. Jones (1964). *The Theory of Electromagnetism*. International Series of Monographs on Pure and Applied Mathematics, Vol. 47. A Pergamon Press Book. New York: The Macmillan Co.

D. S. Jones (1972). *Asymptotic behavior of integrals*. SIAM Rev. **14**(2), pp. 286–317.

D. S. Jones (1986). *Acoustic and Electromagnetic Waves*. Oxford Science Publications. New York: The Clarendon Press Oxford University Press.

D. S. Jones (1997). *Introduction to Asymptotics: A Treatment Using Nonstandard Analysis*. River Edge, NJ: World Scientific Publishing Co. Inc.

D. S. Jones (2001). *Asymptotics of the hypergeometric function*. Math. Methods Appl. Sci. **24**(6), pp. 369–389.

D. S. Jones (2006). *Parabolic cylinder functions of large order*. J. Comput. Appl. Math. **190**(1-2), pp. 453–469.

W. B. Jones and W. J. Thron (1974). *Numerical stability in evaluating continued fractions*. Math. Comp. **28**(127), pp. 795–810.

W. B. Jones and W. J. Thron (1980). *Continued Fractions: Analytic Theory and Applications*, Volume 11 of *Encyclopedia of Mathematics and its Applications*. Reading, MA: Addison-Wesley Publishing Co.

W. B. Jones and W. J. Thron (1985). *On the computation of incomplete gamma functions in the complex domain*. J. Comput. Appl. Math. **12/13**, pp. 401–417.

A. Jonquière (1889). *Note sur la série $\sum_{n=1}^{\infty} x^n/n^s$*. Bull. Soc. Math. France **17**, pp. 142–152.

C. Jordan (1939). *Calculus of Finite Differences*. Budapest: Hungarian Agent Eggenberger Book-Shop. Third edition published in 1965 by Chelsea Publishing Co., New York, and American Mathematical Society, AMS Chelsea Book Series, Providence, RI.

C. Jordan (1965). *Calculus of Finite Differences* (3rd ed.). Providence, RI: AMS Chelsea.

S. Jorna and C. Springer (1971). *Derivation of Green-type, transitional and uniform asymptotic expansions from differential equations. V. Angular oblate spheroidal wavefunctions $\overline{ps}_n^r(\eta,h)$ and $\overline{qs}_n^r(\eta,h)$ for large h*. Proc. Roy. Soc. London Ser. A **321**, pp. 545–555.

N. Joshi and A. V. Kitaev (2001). *On Boutroux's tritronquée solutions of the first Painlevé equation*. Stud. Appl. Math. **107**(3), pp. 253–291.

N. Joshi and A. V. Kitaev (2005). *The Dirichlet boundary value problem for real solutions of the first Painlevé equation on segments in non-positive semi-axis*. J. Reine Angew. Math. **583**, pp. 29–86.

N. Joshi and M. D. Kruskal (1992). *The Painlevé connection problem: An asymptotic approach. I*. Stud. Appl. Math. **86**(4), pp. 315–376.

G. S. Joyce (1973). *On the simple cubic lattice Green function*. Philos. Trans. Roy. Soc. London Ser. A **273**, pp. 583–610.

G. S. Joyce (1994). *On the cubic lattice Green functions*. Proc. Roy. Soc. London Ser. A **445**, pp. 463–477.

B. R. Judd (1975). *Angular Momentum Theory for Diatomic Molecules*. New York: Academic Press.

B. R. Judd (1976). *Modifications of Coulombic interactions by polarizable atoms*. Math. Proc. Cambridge Philos. Soc. **80**(3), pp. 535–539.

B. R. Judd (1998). *Operator Techniques in Atomic Spectroscopy*. Princeton, NJ: Princeton University Press.

G. Julia (1918). *Memoire sur l'itération des fonctions rationnelles*. J. Math. Pures Appl. **8**(1), pp. 47–245.

V. Kac and P. Cheung (2002). *Quantum Calculus*. Universitext. New York: Springer-Verlag.

K. W. J. Kadell (1988). *A proof of Askey's conjectured q-analogue of Selberg's integral and a conjecture of Morris*. SIAM J. Math. Anal. **19**(4), pp. 969–986.

K. W. J. Kadell (1994). *A proof of the q-Macdonald-Morris conjecture for BC_n*. Mem. Amer. Math. Soc. **108**(516), pp. vi+80.

W. Kahan (1987). *Branch Cuts for Complex Elementary Functions or Much Ado About Nothing's Sign Bit*. In A. Iserles and M. J. D. Powell (Eds.), *The State of the Art in Numerical Analysis (Birmingham, 1986)*, Volume 9 of *Inst. Math. Appl. Conf. Ser. New Ser.*, pp. 165–211. New York: Oxford Univ. Press.

K. Kajiwara and T. Masuda (1999). *On the Umemura polynomials for the Painlevé III equation*. Phys. Lett. A **260**(6), pp. 462–467.

K. Kajiwara and Y. Ohta (1996). *Determinant structure of the rational solutions for the Painlevé II equation*. J. Math. Phys. **37**(9), pp. 4693–4704.

K. Kajiwara and Y. Ohta (1998). *Determinant structure of the rational solutions for the Painlevé IV equation*. J. Phys. A **31**(10), pp. 2431–2446.

A. Kalähne (1907). *Über die Wurzeln einiger Zylinderfunktionen und gewisser aus ihnen gebildeter Gleichungen*. Zeitschrift für Mathematik und Physik **54**, pp. 55–86.

S. L. Kalla (1992). *On the evaluation of the Gauss hypergeometric function*. C. R. Acad. Bulgare Sci. **45**(6), pp. 35–36.

E. G. Kalnins (1986). *Separation of Variables for Riemannian Spaces of Constant Curvature*. Harlow: Longman Scientific & Technical.

E. G. Kalnins and W. Miller, Jr. (1991a). *Hypergeometric expansions of Heun polynomials*. SIAM J. Math. Anal. **22**(5), pp. 1450–1459.

E. G. Kalnins and W. Miller, Jr. (1991b). *Addendum: "Hypergeometric expansions of Heun polynomials"*. SIAM J. Math. Anal. **22**(6), p. 1803.

E. G. Kalnins and W. Miller, Jr. (1993). *Orthogonal Polynomials on n-spheres: Gegenbauer, Jacobi and Heun*. In *Topics in Polynomials of One and Several Variables and their Applications*, pp. 299–322. River Edge, NJ: World Sci. Publishing.

E. G. Kalnins, W. Miller, Jr., G. F. Torres del Castillo, and G. C. Williams (2000). *Special Functions and Perturbations of Black Holes*. In *Special Functions (Hong Kong, 1999)*, pp. 140–151. River Edge, NJ: World Sci. Publishing.

E. G. Kalnins, W. Miller, Jr., and P. Winternitz (1976). *The group O(4), separation of variables and the hydrogen atom*. SIAM J. Appl. Math. **30**(4), pp. 630–664.

J. Kamimoto (1998). *On an integral of Hardy and Littlewood*. Kyushu J. Math. **52**(1), pp. 249–263.

D. Kaminski and R. B. Paris (1999). *On the zeroes of the Pearcey integral*. J. Comput. Appl. Math. **107**(1), pp. 31–52.

E. Kamke (1977). *Differentialgleichungen: Lösungsmethoden und Lösungen. Teil I*. Stuttgart: B. G. Teubner.

M. Kaneko (1997). *Poly-Bernoulli numbers*. J. Théor. Nombres Bordeaux **9**(1), pp. 221–228.

E. Kanzieper (2002). *Replica field theories, Painlevé transcendents, and exact correlation functions*. Phys. Rev. Lett. **89**(25), pp. (250201-1)–(250201-4).

A. A. Kapaev (1988). *Asymptotic behavior of the solutions of the Painlevé equation of the first kind*. Differ. Uravn. **24**(10), pp. 1684–1695. In Russian. English translation: Differential Equations **24**(1988), no. 10, pp. 1107–1115 (1989).

A. A. Kapaev (1991). *Essential singularity of the Painlevé function of the second kind and the nonlinear Stokes phenomenon*. Zap. Nauchn. Sem. Leningrad. Otdel. Mat. Inst. Steklov. (LOMI) **187**, pp. 139–170. English translation: J. Math. Sci. **73**(1995), no. 4, pp. 500–517.

A. A. Kapaev (2004). *Quasi-linear Stokes phenomenon for the Painlevé first equation*. J. Phys. A **37**(46), pp. 11149–11167.

A. A. Kapaev and A. V. Kitaev (1993). *Connection formulae for the first Painlevé transcendent in the complex domain*. Lett. Math. Phys. **27**(4), pp. 243–252.

N. S. Kapany and J. J. Burke (1972). *Optical Waveguides*. Quantum Electronics - Principles and Applications. New York: Academic Press.

P. L. Kapitsa (1951a). *The computation of the sums of negative even powers of roots of Bessel functions*. Doklady Akad. Nauk SSSR (N.S.) **77**, pp. 561–564.

P. L. Kapitsa (1951b). *Heat conduction and diffusion in a fluid medium with a periodic flow. I. Determination of the wave transfer coefficient in a tube, slot, and canal*. Akad. Nauk SSSR. Žurnal Eksper. Teoret. Fiz. **21**, pp. 964–978.

E. L. Kaplan (1948). *Auxiliary table for the incomplete elliptic integrals*. J. Math. Physics **27**, pp. 11–36.

A. A. Karatsuba and S. M. Voronin (1992). *The Riemann Zeta-Function*, Volume 5 of *de Gruyter Expositions in Mathematics*. Berlin: Walter de Gruyter & Co. Translated from the Russian by Neal Koblitz.

S. Karlin and J. L. McGregor (1961). *The Hahn polynomials, formulas and an application*. Scripta Math. **26**, pp. 33–46.

D. Karp, A. Savenkova, and S. M. Sitnik (2007). *Series expansions for the third incomplete elliptic integral via partial fraction decompositions*. J. Comput. Appl. Math. **207**(2), pp. 331–337.

D. Karp and S. M. Sitnik (2007). *Asymptotic approximations for the first incomplete elliptic integral near logarithmic singularity*. J. Comput. Appl. Math. **205**(1), pp. 186–206.

K. A. Karpov and È. A. Čistova (1964). *Tablitsy funktsii Vebera. Tom II*. Moscow: Vičisl. Centr Akad. Nauk SSSR.

K. A. Karpov and È. A. Čistova (1968). *Tablitsy funktsii Vebera. Tom III*. Moscow: Vyčisl. Centr Akad. Nauk SSSR.

A. V. Kashevarov (1998). *The second Painlevé equation in electric probe theory. Some numerical solutions*. Zh. Vychisl. Mat. Mat. Fiz. **38**(6), pp. 992–1000. In Russian. English translation: Comput. Math. Math. Phys. **38**(1998), no. 6, pp. 950–958.

A. V. Kashevarov (2004). *The second Painlevé equation in the electrostatic probe theory: Numerical solutions for the partial absorption of charged particles by the surface*. Technical Physics **49**(1), pp. 1–7.

C. Kassel (1995). *Quantum Groups*, Volume 155 of *Graduate Texts in Mathematics*. New York: Springer-Verlag.

N. M. Katz (1975). *The congruences of Clausen-von Staudt and Kummer for Bernoulli-Hurwitz numbers*. Math. Ann. **216**(1), pp. 1–4.

E. H. Kaufman, Jr. and T. D. Lenker (1986). *Linear convergence and the bisection algorithm*. Amer. Math. Monthly **93**(1), pp. 48–51.

A. Ya. Kazakov and S. Yu. Slavyanov (1996). *Integral equations for special functions of Heun class*. Methods Appl. Anal. **3**(4), pp. 447–456.

N. D. Kazarinoff (1988). *Special functions and the Bieberbach conjecture.* Amer. Math. Monthly **95**(8), pp. 689–696.

J. Keating (1993). *The Riemann Zeta-Function and Quantum Chaology.* In *Quantum Chaos (Varenna, 1991)*, Proc. Internat. School of Phys. Enrico Fermi, CXIX, pp. 145–185. Amsterdam: North-Holland.

J. P. Keating (1999). *Periodic Orbits, Spectral Statistics, and the Riemann Zeros.* In J. P. Keating, D. E. Khmelnitskii, and I. V. Lerner (Eds.), *Supersymmetry and Trace Formulae: Chaos and Disorder*, pp. 1–15. New York: Springer-Verlag.

R. P. Kelisky (1957). *On formulas involving both the Bernoulli and Fibonacci numbers.* Scripta Math. **23**, pp. 27–35.

M. K. Kerimov (1980). *Methods of computing the Riemann zeta-function and some generalizations of it.* USSR Comput. Math. and Math. Phys. **20**(6), pp. 212–230.

M. K. Kerimov (1999). *The Rayleigh function: Theory and computational methods.* Zh. Vychisl. Mat. Mat. Fiz. **39**(12), pp. 1962–2006. Translation in Comput. Math. Math. Phys. 39 (1999), No. 12, pp. 1883–1925.

M. K. Kerimov and S. L. Skorokhodov (1984a). *Calculation of modified Bessel functions in a complex domain.* Zh. Vychisl. Mat. i Mat. Fiz. **24**(5), pp. 650–664. English translation in U.S.S.R. Computational Math. and Math. Phys. **24**(3), pp. 15–24.

M. K. Kerimov and S. L. Skorokhodov (1984b). *Calculation of the complex zeros of the modified Bessel function of the second kind and its derivatives.* Zh. Vychisl. Mat. i Mat. Fiz. **24**(8), pp. 1150–1163. English translation in U.S.S.R. Computational Math. and Math. Phys. **24**(4), pp. 115–123.

M. K. Kerimov and S. L. Skorokhodov (1984c). *Evaluation of complex zeros of Bessel functions $J_\nu(z)$ and $I_\nu(z)$ and their derivatives.* Zh. Vychisl. Mat. i Mat. Fiz. **24**(10), pp. 1497–1513. English translation in U.S.S.R. Computational Math. and Math. Phys. **24**(5), pp. 131–141.

M. K. Kerimov and S. L. Skorokhodov (1985a). *Calculation of the complex zeros of a Bessel function of the second kind and its derivatives.* Zh. Vychisl. Mat. i Mat. Fiz. **25**(10), pp. 1457–1473, 1581. In Russian. English translation: USSR Comput. Math. Math. Phys., **25**(1985), no. 5, pp. 117–128.

M. K. Kerimov and S. L. Skorokhodov (1985b). *Calculation of the complex zeros of Hankel functions and their derivatives.* Zh. Vychisl. Mat. i Mat. Fiz. **25**(11), pp. 1628–1643, 1741. English translation in U.S.S.R. Computational Math. and Math. Phys. **25**(6), pp. 26–36.

M. K. Kerimov and S. L. Skorokhodov (1985c). *Calculation of the multiple zeros of the derivatives of the cylindrical Bessel functions $J_\nu(z)$ and $Y_\nu(z)$.* Zh. Vychisl. Mat. i Mat. Fiz. **25**(12), pp. 1749–1760, 1918. English translation in U.S.S.R. Computational Math. and Math. Phys. **25**(6), pp. 101–107.

M. K. Kerimov and S. L. Skorokhodov (1986). *On multiple zeros of derivatives of Bessel's cylindrical functions.* Dokl. Akad. Nauk SSSR **288**(2), pp. 285–288. In Russian. English translation: Soviet Math. Dokl. **33**(3), 650–653, (1986).

M. K. Kerimov and S. L. Skorokhodov (1987). *On the calculation of the multiple complex roots of the derivatives of cylindrical Bessel functions.* Zh. Vychisl. Mat. i Mat. Fiz. **27**(11), pp. 1628–1639, 1758. English translation in U.S.S.R. Computational Math. and Math. Phys. **27**(6), pp. 18–25.

M. K. Kerimov and S. L. Skorokhodov (1988). *Multiple complex zeros of derivatives of the cylindrical Bessel functions.* Dokl. Akad. Nauk SSSR **299**(3), pp. 614–618. In Russian. English translation: Soviet Phys. Dokl. **33**(3),196–198, (1988).

M. Kerker (1969). *The Scattering of Light and Other Electromagnetic Radiation.* New York: Academic Press.

A. D. Kerr (1978). *An indirect method for evaluating certain infinite integrals.* Z. Angew. Math. Phys. **29**(3), pp. 380–386.

R. P. Kerr (1963). *Gravitational field of a spinning mass as an example of algebraically special metrics.* Phys. Rev. Lett. **11**(5), pp. 237–238.

D. Kershaw (1983). *Some extensions of W. Gautschi's inequalities for the gamma function.* Math. Comp. **41**(164), pp. 607–611.

S. Kesavan and A. S. Vasudevamurthy (1985). *On some boundary element methods for the heat equation.* Numer. Math. **46**(1), pp. 101–120.

S. H. Khamis (1965). *Tables of the Incomplete Gamma Function Ratio: The Chi-square Integral, the Poisson Distribution.* Darmstadt: Justus von Liebig Verlag. In cooperation with W. Rudert.

A. Khare, A. Lakshminarayan, and U. Sukhatme (2003). *Cyclic identities for Jacobi elliptic and related functions.* J. Math. Phys. **44**(4), pp. 1822–1841.

A. Khare and U. Sukhatme (2002). *Cyclic identities involving Jacobi elliptic functions.* J. Math. Phys. **43**(7), pp. 3798–3806.

A. Khare and U. Sukhatme (2004). *Connecting Jacobi elliptic functions with different modulus parameters.* Pramana **63**(5), pp. 921–936.

A. I. Kheyfits (2004). *Closed-form representations of the Lambert W function.* Fract. Calc. Appl. Anal. **7**(2), pp. 177–190.

H. Ki and Y.-O. Kim (2000). *On the zeros of some generalized hypergeometric functions*. J. Math. Anal. Appl. **243**(2), pp. 249–260.

S. Kida (1981). *A vortex filament moving without change of form*. J. Fluid Mech. **112**, pp. 397–409.

S. K. Kim (1972). *The asymptotic expansion of a hypergeometric function $_2F_2(1, \alpha; \rho_1, \rho_2; z)$*. Math. Comp. **26**(120), p. 963.

T. Kim and H. S. Kim (1999). *Remark on p-adic q-Bernoulli numbers*. Adv. Stud. Contemp. Math. (Pusan) **1**, pp. 127–136. Algebraic number theory (Hapcheon/Saga, 1996).

N. Kimura (1988). *On the degree of an irreducible factor of the Bernoulli polynomials*. Acta Arith. **50**(3), pp. 243–249.

B. J. King and A. L. Van Buren (1973). *A general addition theorem for spheroidal wave functions*. SIAM J. Math. Anal. **4**(1), pp. 149–160.

I. Y. Kireyeva and K. A. Karpov (1961). *Tables of Weber functions*. Vol. I, Volume 15 of *Mathematical Tables Series*. London-New York: Pergamon Press. Translated by Prasenjit Basu.

A. N. Kirillov (1995). *Dilogarithm identities*. Progr. Theoret. Phys. Suppl. (118), pp. 61–142.

N. P. Kirk, J. N. L. Connor, and C. A. Hobbs (2000). *An adaptive contour code for the numerical evaluation of the oscillatory cuspoid canonical integrals and their derivatives*. Computer Physics Comm. **132**(1-2), pp. 142–165.

E. T. Kirkpatrick (1960). *Tables of values of the modified Mathieu functions*. Math. Comp. **14**, pp. 118–129.

A. V. Kitaev (1987). *The method of isomonodromic deformations and the asymptotics of the solutions of the "complete" third Painlevé equation*. Mat. Sb. (N.S.) **134(176)**(3), pp. 421–444, 448. In Russian. English translation: Math. USSR-Sb. **62**(1989), no. 2, pp. 421–444.

A. V. Kitaev (1994). *Elliptic asymptotics of the first and second Painlevé transcendents*. Uspekhi Mat. Nauk **49**(1(295)), pp. 77–140. In Russian. English translation: Russian Math. Surveys **49**(1994), no. 1, pp. 81–150.

A. V. Kitaev, C. K. Law, and J. B. McLeod (1994). *Rational solutions of the fifth Painlevé equation*. Differential Integral Equations **7**(3-4), pp. 967–1000.

A. V. Kitaev and A. H. Vartanian (2004). *Connection formulae for asymptotics of solutions of the degenerate third Painlevé equation. I*. Inverse Problems **20**(4), pp. 1165–1206.

Y. Kivshar and B. Luther-Davies (1998). *Dark optical solitons: Physics and applications*. Physics Reports **298**(2-3), pp. 81–197.

F. Klein (1894). *Vorlesungen über die hypergeometrische Funktion*. Göttingen. Reprint by Springer-Verlag, Berlin, 1981, of the 1933 original.

A. Kneser (1927). *Neue Untersuchungen einer Reihe aus der Theorie der elliptischen Funktionen*. Journal für die Reine und Angewandte Mathematik **158**, pp. 209–218.

H. Kneser (1950). *Reelle analytische Lösungen der Gleichung $\varphi(\varphi(x)) = e^x$ und verwandter Funktionalgleichungen*. J. Reine Angew. Math. **187**, pp. 56–67.

K. Knopp (1948). *Theory and Application of Infinite Series* (2nd ed.). New York: Hafner. R. C. Young, translator.

K. Knopp (1964). *Theorie und Anwendung der unendlichen Reihen* (4th ed.). Die Grundlehren der mathematischen Wissenschaften, Band 2. Berlin-Heidelberg: Springer-Verlag. A translation by R. C. Young appeared in 1928, Theory and Application of Infinite Series, Blackie, 571pp.

U. J. Knottnerus (1960). *Approximation Formulae for Generalized Hypergeometric Functions for Large Values of the Parameters*. Groningen: J. B. Wolters.

D. E. Knuth (1968). *The Art of Computer Programming. Vol. 1: Fundamental Algorithms* (1st ed.). Reading, MA-London-Don Mills, Ont: Addison-Wesley Publishing Co. Second printing of second edition by Addison-Wesley Publishing Co., Reading, MA-London-Amsterdam, 1975.

D. E. Knuth (1986). *METAFONT: The Program*, Volume D of *Computers and Typesetting*. Reading, MA: Addison-Wesley.

D. E. Knuth (1992). *Two notes on notation*. Amer. Math. Monthly **99**(5), pp. 403–422.

D. E. Knuth and T. J. Buckholtz (1967). *Computation of tangent, Euler, and Bernoulli numbers*. Math. Comp. **21**(100), pp. 663–688.

N. Koblitz (1984). *p-adic Numbers, p-adic Analysis, and Zeta-Functions* (2nd ed.), Volume 58 of *Graduate Texts in Mathematics*. New York: Springer-Verlag.

N. Koblitz (1993). *Introduction to Elliptic Curves and Modular Forms* (2nd ed.), Volume 97 of *Graduate Texts in Mathematics*. New York: Springer-Verlag.

N. Koblitz (1999). *Algebraic Aspects of Cryptography*. Berlin: Springer-Verlag. Second printing with corrections.

M. Koecher (1954). *Zur Theorie der Modulformen n-ten Grades. I*. Math. Z. **59**, pp. 399–416.

J. Koekoek, R. Koekoek, and H. Bavinck (1998). *On differential equations for Sobolev-type Laguerre polynomials*. Trans. Amer. Math. Soc. **350**(1), pp. 347–393.

R. Koekoek and R. F. Swarttouw (1998). *The Askey-scheme of hypergeometric orthogonal polynomials and its q-analogue*. Technical Report 98-17, Delft University of Technology, Faculty of Information Technology and Systems, Department of Technical Mathematics and Informatics. An online version exists.

P. Koev and A. Edelman (2006). *The efficient evaluation of the hypergeometric function of a matrix argument*. Math. Comp. **75**(254), pp. 833–846.

D. A. Kofke (2004). *Comment on "The incomplete beta function law for parallel tempering sampling of classical canonical systems" [J. Chem. Phys. **120**, 4119 (2004)]*. J. Chem. Phys. **121**(2), p. 1167.

S. Koizumi (1976). *Theta relations and projective normality of Abelian varieties*. Amer. J. Math. **98**(4), pp. 865–889.

G. C. Kokkorakis and J. A. Roumeliotis (1998). *Electromagnetic eigenfrequencies in a spheroidal cavity (calculation by spheroidal eigenvectors)*. J. Electromagn. Waves Appl. **12**(12), pp. 1601–1624.

C. G. Kokologiannaki, P. D. Siafarikas, and C. B. Kouris (1992). *On the complex zeros of $H_\mu(z)$, $J'_\mu(z)$, $J''_\mu(z)$ for real or complex order*. J. Comput. Appl. Math. **40**(3), pp. 337–344.

K. S. Kölbig (1970). *Complex zeros of an incomplete Riemann zeta function and of the incomplete gamma function*. Math. Comp. **24**(111), pp. 679–696.

K. S. Kölbig (1972a). *Complex zeros of two incomplete Riemann zeta functions*. Math. Comp. **26**(118), pp. 551–565.

K. S. Kölbig (1972b). *On the zeros of the incomplete gamma function*. Math. Comp. **26**(119), pp. 751–755.

K. S. Kölbig (1981). *A program for computing the conical functions of the first kind $P^m_{-\frac{1}{2}+i\tau}(x)$ for $m=0$ and $m=1$*. Comput. Phys. Comm. **23**(1), pp. 51–61.

K. S. Kölbig (1986). *Nielsen's generalized polylogarithms*. SIAM J. Math. Anal. **17**(5), pp. 1232–1258.

K. S. Kölbig, J. A. Mignaco, and E. Remiddi (1970). *On Nielsen's generalized polylogarithms and their numerical calculation*. Nordisk Tidskr. Informationsbehandling (BIT) **10**, pp. 38–73.

G. A. Kolesnik (1969). *An improvement of the remainder term in the divisor problem*. Mat. Zametki **6**, pp. 545–554. In Russian. English translation: Math. Notes **6**(1969), no. 5, pp. 784–791.

I. V. Komarov, L. I. Ponomarev, and S. Yu. Slavyanov (1976). *Sferoidalnye i kulonovskie sferoidalnye funktsii*. Moscow: Izdat. "Nauka". Translated title: *Spheroidal and Coulomb Spheroidal Functions*.

E. J. Konopinski (1981). *Electromagnetic Fields and Relativistic Particles*. International Series in Pure and Applied Physics. New York: McGraw-Hill Book Co.

T. H. Koornwinder (1974). *Jacobi polynomials. II. An analytic proof of the product formula*. SIAM J. Math. Anal. **5**, pp. 125–137.

T. H. Koornwinder (1975a). *Jacobi polynomials. III. An analytic proof of the addition formula*. SIAM. J. Math. Anal. **6**, pp. 533–543.

T. H. Koornwinder (1975b). *A new proof of a Paley-Wiener type theorem for the Jacobi transform*. Ark. Mat. **13**, pp. 145–159.

T. H. Koornwinder (1975c). *Two-variable Analogues of the Classical Orthogonal Polynomials*. In R. A. Askey (Ed.), *Theory and Application of Special Functions*, pp. 435–495. New York: Academic Press.

T. H. Koornwinder (1977). *The addition formula for Laguerre polynomials*. SIAM J. Math. Anal. **8**(3), pp. 535–540.

T. H. Koornwinder (1984a). *Jacobi Functions and Analysis on Noncompact Semisimple Lie Groups*. In *Special Functions: Group Theoretical Aspects and Applications*, pp. 1–85. Dordrecht: Reidel. Reprinted by Kluwer Academic Publishers, Boston, 2002.

T. H. Koornwinder (1984b). *Orthogonal polynomials with weight function $(1-x)^\alpha(1+x)^\beta + M\delta(x+1) + N\delta(x-1)$*. Canad. Math. Bull. **27**(2), pp. 205–214.

T. H. Koornwinder (1989). *Meixner-Pollaczek polynomials and the Heisenberg algebra*. J. Math. Phys. **30**(4), pp. 767–769.

T. H. Koornwinder (1992). *Askey-Wilson Polynomials for Root Systems of Type BC*. In *Hypergeometric Functions on Domains of Positivity, Jack Polynomials, and Applications (Tampa, FL, 1991)*, Volume 138 of *Contemp. Math.*, pp. 189–204. Providence, RI: Amer. Math. Soc.

T. H. Koornwinder (2006). *Lowering and Raising Operators for Some Special Orthogonal Polynomials*. In *Jack, Hall-Littlewood and Macdonald Polynomials*, Volume 417 of *Contemp. Math.*, pp. 227–238. Providence, RI: Amer. Math. Soc.

T. H. Koornwinder and I. Sprinkhuizen-Kuyper (1978). *Hypergeometric functions of 2×2 matrix argument are expressible in terms of Appel's functions F_4*. Proc. Amer. Math. Soc. **70**(1), pp. 39–42.

B. G. Korenev (2002). *Bessel Functions and their Applications*, Volume 8 of *Analytical Methods and Special Functions*. London-New York: Taylor & Francis Ltd. Translated from the Russian by E. V. Pankratiev.

V. E. Korepin, N. M. Bogoliubov, and A. G. Izergin (1993). *Quantum Inverse Scattering Method and Correlation Functions*. Cambridge: Cambridge University Press.

T. W. Körner (1989). *Fourier Analysis* (2nd ed.). Cambridge: Cambridge University Press.

N. M. Korobov (1958). *Estimates of trigonometric sums and their applications*. Uspehi Mat. Nauk **13**(4 (82)), pp. 185–192.

V. Kourganoff (1952). *Basic Methods in Transfer Problems. Radiative Equilibrium and Neutron Diffusion*. Oxford: Oxford University Press. With the collaboration of I. W. Busbridge.

J. J. Kovacic (1986). *An algorithm for solving second order linear homogeneous differential equations*. J. Symbolic Comput. **2**(1), pp. 3–43.

S. Kowalevski (1889). *Sur le problème de la rotation d'un corps solide autour d'un point fixe*. Acta Math. **12**(1), pp. 177–232.

Y. A. Kravtsov (1964). *Asymptotic solution of Maxwell's equations near caustics*. Izv. Vuz. Radiofiz. **7**, pp. 1049–1056.

Y. A. Kravtsov (1968). *Two new asymptotic methods in the theory of wave propagation in inhomogeneous media*. Sov. Phys. Acoust. **14**, pp. 1–17.

Y. A. Kravtsov (1988). *Rays and caustics as physical objects*. In E. Wolf (Ed.), *Progress in Optics*, Volume 26, pp. 227–348. Amsterdam: North-Holland.

R. Kress and E. Martensen (1970). *Anwendung der Rechteckregel auf die reelle Hilberttransformation mit unendlichem Intervall*. Z. Angew. Math. Mech. **50**(1-4), pp. T61–T64.

E. Kreyszig (1957). *On the zeros of the Fresnel integrals*. Canad. J. Math. **9**, pp. 118–131.

I. M. Krichever (1976). *An algebraic-geometrical construction of the Zakharov-Shabat equations and their periodic solutions*. Sov. Math. Doklady **17**, pp. 394–397.

I. M. Krichever and S. P. Novikov (1989). *Algebras of Virasoro type, the energy-momentum tensor, and operator expansions on Riemann surfaces*. Funktsional. Anal. i Prilozhen. **23**(1), pp. 24–40. In Russian. English translation: Funct. Anal. Appl. **23**(1989), no. 1, pp. 19–33.

S. G. Krivoshlykov (1994). *Quantum-Theoretical Formalism for Inhomogeneous Graded-Index Waveguides*. Berlin-New York: Akademie Verlag.

E. D. Krupnikov and K. S. Kölbig (1997). *Some special cases of the generalized hypergeometric function $_{q+1}F_q$*. J. Comput. Appl. Math. **78**(1), pp. 79–95.

M. D. Kruskal (1974). *The Korteweg-de Vries Equation and Related Evolution Equations*. In A. C. Newell (Ed.), *Nonlinear Wave Motion (Proc. AMS-SIAM Summer Sem., Clarkson Coll. Tech., Potsdam, N.Y., 1972)*, Lectures in Appl. Math., Vol. 15, pp. 61–83. Providence, RI: Amer. Math. Soc.

M. D. Kruskal and P. A. Clarkson (1992). *The Painlevé-Kowalevski and poly-Painlevé tests for integrability*. Stud. Appl. Math. **86**(2), pp. 87–165.

V. I. Krylov and N. S. Skoblya (1985). *A Handbook of Methods of Approximate Fourier Transformation and Inversion of the Laplace Transformation*. Moscow: Mir. Translated from the Russian by George Yankovsky. Reprint of the 1977 translation.

M. Kurz (1979). *Fehlerabschätzungen zu asymptotischen Entwicklungen der Eigenwerte und Eigenlösungen der Mathieuschen Differentialgleichung*. Ph. D. thesis, Universität Duisburg-Essen, Essen, D 45177.

V. B. Kuznetsov (1992). *Equivalence of two graphical calculi*. J. Phys. A **25**(22), pp. 6005–6026.

V. B. Kuznetsov and S. Sahi (Eds.) (2006). *Jack, Hall-Littlewood and Macdonald Polynomials*, Volume 417 of *Contemporary Mathematics*, Providence, RI. American Mathematical Society. Proceedings of the workshop held in Edinburgh, September 23–26, 2003.

K. H. Kwon, L. L. Littlejohn, and G. J. Yoon (2006). *Construction of differential operators having Bochner-Krall orthogonal polynomials as eigenfunctions*. J. Math. Anal. Appl. **324**(1), pp. 285–303.

G. Labahn and M. Mutrie (1997). *Reduction of Elliptic Integrals to Legendre Normal Form*. Technical Report 97-21, Department of Computer Science, University of Waterloo, Waterloo, Ontario.

A. Laforgia (1979). *On the Zeros of the Derivative of Bessel Functions of Second Kind*, Volume 179 of *Pubblicazioni Serie III [Publication Series III]*. Rome: Istituto per le Applicazioni del Calcolo "Mauro Picone" (IAC).

A. Laforgia (1984). *Further inequalities for the gamma function*. Math. Comp. **42**(166), pp. 597–600.

A. Laforgia (1986). *Inequalities for Bessel functions*. J. Comput. Appl. Math. **15**(1), pp. 75–81.

A. Laforgia (1991). *Bounds for modified Bessel functions*. J. Comput. Appl. Math. **34**(3), pp. 263–267.

A. Laforgia and M. E. Muldoon (1983). *Inequalities and approximations for zeros of Bessel functions of small order*. SIAM J. Math. Anal. **14**(2), pp. 383–388.

A. Laforgia and M. E. Muldoon (1988). *Monotonicity properties of zeros of generalized Airy functions*. Z. Angew. Math. Phys. **39**(2), pp. 267–271.

A. Laforgia and S. Sismondi (1988). *Monotonicity results and inequalities for the gamma and error functions*. J. Comput. Appl. Math. **23**(1), pp. 25–33.

J. C. Lagarias, V. S. Miller, and A. M. Odlyzko (1985). *Computing $\pi(x)$: The Meissel-Lehmer method*. Math. Comp. **44**(170), pp. 537–560.

J. Lagrange (1770). *Démonstration d'un Théoréme d'Arithmétique*. Nouveau Mém. Acad. Roy. Sci. Berlin, pp. 123–133. Reprinted in Oeuvres, Vol. 3, pp. 189–201, Gauthier-Villars, Paris, 1867.

V. Laĭ (1994). *The two-point connection problem for differential equations of the Heun class.* Teoret. Mat. Fiz. **101**(3), pp. 360–368. In Russian. English translation: Theoret. and Math. Phys. **101**(1994), no. 3, pp. 1413–1418 (1995).

H. Lamb (1932). *Hydrodynamics* (6th ed.). Cambridge: Cambridge University Press. Reprinted in 1993.

C. G. Lambe (1952). *Lamé-Wangerin functions.* Quart. J. Math., Oxford Ser. (2) **3**, pp. 107–114.

C. G. Lambe and D. R. Ward (1934). *Some differential equations and associated integral equations.* Quart. J. Math. (Oxford) **5**, pp. 81–97.

E. Landau (1953). *Handbuch der Lehre von der Verteilung der Primzahlen. 2 Bände.* New York: Chelsea Publishing Co. 2nd edition, with an appendix by Paul T. Bateman.

L. D. Landau and E. M. Lifshitz (1962). *The Classical Theory of Fields.* Oxford: Pergamon Press. Revised second edition. Course of Theoretical Physics, Vol. 2. Translated from the Russian by Morton Hamermesh.

L. D. Landau and E. M. Lifshitz (1965). *Quantum Mechanics: Non-relativistic Theory.* Oxford: Pergamon Press Ltd. Revised second edition. Course of Theoretical Physics, Vol. 3. Translated from the Russian by J. B. Sykes and J. S. Bell.

L. D. Landau and E. M. Lifshitz (1987). *Fluid Mechanics* (2nd ed.). London: Pergamon Press. Course of Theoretical Physics, Vol. 6. Translated from the Russian by J. B. Sykes and W. H. Reid.

L. J. Landau (1999). *Ratios of Bessel functions and roots of* $\alpha J_\nu(x) + x J'_\nu(x) = 0$. J. Math. Anal. Appl. **240**(1), pp. 174–204.

L. J. Landau (2000). *Bessel functions: Monotonicity and bounds.* J. London Math. Soc. (2) **61**(1), pp. 197–215.

S. Lang (1987). *Elliptic Functions* (2nd ed.), Volume 112 of *Graduate Texts in Mathematics*. New York: Springer-Verlag. With an appendix by J. Tate.

R. E. Langer (1934). *The solutions of the Mathieu equation with a complex variable and at least one parameter large.* Trans. Amer. Math. Soc. **36**(3), pp. 637–695.

B. J. Laurenzi (1993). *Moment integrals of powers of Airy functions.* Z. Angew. Math. Phys. **44**(5), pp. 891–908.

H. A. Lauwerier (1974). *Asymptotic Analysis.* Number 54 in Mathematical Centre Tracts. Amsterdam: Mathematisch Centrum.

D. F. Lawden (1989). *Elliptic Functions and Applications*, Volume 80 of *Applied Mathematical Sciences*. New York: Springer-Verlag.

W. Lay, K. Bay, and S. Yu. Slavyanov (1998). *Asymptotic and numeric study of eigenvalues of the double confluent Heun equation.* J. Phys. A **31**(42), pp. 8521–8531.

W. Lay and S. Yu. Slavyanov (1998). *The central two-point connection problem for the Heun class of ODEs.* J. Phys. A **31**(18), pp. 4249–4261.

W. Lay and S. Yu. Slavyanov (1999). *Heun's equation with nearby singularities.* Proc. Roy. Soc. London Ser. A **455**, pp. 4347–4361.

D. Le (1985). *An efficient derivative-free method for solving nonlinear equations.* ACM Trans. Math. Software **11**(3), pp. 250–262. Corrigendum, *ibid.* **15**(3) (1989), 287.

E. W. Leaver (1986). *Solutions to a generalized spheroidal wave equation: Teukolsky's equations in general relativity, and the two-center problem in molecular quantum mechanics.* J. Math. Phys. **27**(5), pp. 1238–1265.

A. V. Lebedev and R. M. Fedorova (1960). *A Guide to Mathematical Tables.* Amsterdam-New York-Oxford: Pergamon Press. See also Burunova (1960).

N. N. Lebedev (1965). *Special Functions and Their Applications.* Englewood Cliffs, N.J.: Prentice-Hall Inc. R. A. Silverman, translator and editor; reprinted by Dover, New York, 1972.

N. N. Lebedev, I. P. Skalskaya, and Y. S. Uflyand (1965). *Problems of Mathematical Physics.* Revised, enlarged and corrected English edition; translated and edited by Richard A. Silverman. With a supplement by Edward L. Reiss. Englewood Cliffs, NJ: Prentice-Hall Inc. Republished as *Worked Problems in Applied Mathematics* by Dover Publications, New York, 1979.

J. LeCaine (1945). *A table of integrals involving the functions* $E_n(x)$. National Research Council of Canada, Division of Atomic Energy, Document no. **MT-131** (NRC 1553).

D. K. Lee (1990). *Application of theta functions for numerical evaluation of complete elliptic integrals of the first and second kinds.* Comput. Phys. Comm. **60**(3), pp. 319–327.

S.-Y. Lee (1980). *The inhomogeneous Airy functions,* $\mathrm{Gi}(z)$ *and* $\mathrm{Hi}(z)$. J. Chem. Phys. **72**(1), pp. 332–336.

D. J. Leeming (1977). *An asymptotic estimate for the Bernoulli and Euler numbers.* Canad. Math. Bull. **20**(1), pp. 109–111.

D. J. Leeming (1989). *The real zeros of the Bernoulli polynomials.* J. Approx. Theory **58**(2), pp. 124–150.

A. M. Legendre (1808). *Essai sur la Théorie des Nombres* (2nd ed.). Paris: Courcier.

A. M. Legendre (1825–1832). *Traité des fonctions elliptiques et des intégrales Eulériennes.* Paris: Huzard-Courcier. vol.1, 1825; vol.2, 1826; three supplements, 1828–1832.

D. R. Lehman and J. S. O'Connell (1973). *Graphical Recoupling of Angular Momenta.* Technical report, National Bureau of Standards, Washington, D.C. Monograph 136.

D. R. Lehman, W. C. Parke, and L. C. Maximon (1981). *Numerical evaluation of integrals containing a spherical Bessel function by product integration.* J. Math. Phys. **22**(7), pp. 1399–1413.

D. H. Lehmer (1940). *On the maxima and minima of Bernoulli polynomials.* Amer. Math. Monthly **47**(8), pp. 533–538.

D. H. Lehmer (1941). *Guide to Tables in the Theory of Numbers.* Bulletin of the National Research Council, No. 105. Washington, D. C.: National Research Council. Reprinted in 1961.

D. H. Lehmer (1943). *Ramanujan's function $\tau(n)$.* Duke Math. J. **10**(3), pp. 483–492.

D. H. Lehmer (1947). *The vanishing of Ramanujan's function $\tau(n)$.* Duke Math. J. **14**(2), pp. 429–433.

D. N. Lehmer (1914). *List of Prime Numbers from 1 to 10,006,721.* Publ. No. 165. Washington, DC: Carnegie Institution of Washington.

J. Lehner (1941). *A partition function connected with the modulus five.* Duke Math. J. **8**(4), pp. 631–655.

O. Lehto and K. I. Virtanen (1973). *Quasiconformal Mappings in the Plane* (2nd ed.). New York: Springer-Verlag. Translated from the German by K. W. Lucas, Die Grundlehren der mathematischen Wissenschaften, Band 126.

A. Leitner and J. Meixner (1960). *Eine Verallgemeinerung der Sphäroidfunktionen.* Arch. Math. **11**, pp. 29–39.

D. A. Leonard (1982). *Orthogonal polynomials, duality and association schemes.* SIAM J. Math. Anal. **13**(4), pp. 656–663.

N. L. Lepe (1985). *Functions on a parabolic cylinder with a negative integer index.* Differ. Uravn. **21**(11), pp. 2001–2003, 2024. In Russian.

J. Lepowsky and S. Milne (1978). *Lie algebraic approaches to classical partition identities.* Adv. in Math. **29**(1), pp. 15–59.

J. Lepowsky and R. L. Wilson (1982). *A Lie theoretic interpretation and proof of the Rogers-Ramanujan identities.* Adv. in Math. **45**(1), pp. 21–72.

M. Lerch (1887). *Note sur la fonction $\mathfrak{K}(w,x,s) = \sum_{k=0}^{\infty} \frac{e^{2k\pi ix}}{(w+k)^s}$.* Acta Math. **11**(1-4), pp. 19–24.

M. Lerch (1903). *Zur Theorie der Gaußschen Summen.* Math. Ann. **57**(4), pp. 554–567.

J. Letessier (1995). *Co-recursive associated Jacobi polynomials.* J. Comput. Appl. Math. **57**(1-2), pp. 203–213.

J. Letessier, G. Valent, and J. Wimp (1994). *Some Differential Equations Satisfied by Hypergeometric Functions.* In *Approximation and Computation (West Lafayette, IN, 1993)*, Volume 119 of *Internat. Ser. Numer. Math.*, pp. 371–381. Boston, MA: Birkhäuser Boston.

C. Leubner and H. Ritsch (1986). *A note on the uniform asymptotic expansion of integrals with coalescing endpoint and saddle points.* J. Phys. A **19**(3), pp. 329–335.

K. V. Leung and S. S. Ghaderpanah (1979). *An application of the finite element approximation method to find the complex zeros of the modified Bessel function $K_n(z)$.* Math. Comp. **33**(148), pp. 1299–1306.

L. Levey and L. B. Felsen (1969). *On incomplete Airy functions and their application to diffraction problems.* Radio Sci. **4**(10), pp. 959–969.

E. Levin and D. S. Lubinsky (2001). *Orthogonal Polynomials for Exponential Weights.* CMS Books in Mathematics/Ouvrages de Mathématiques de la SMC, 4. New York: Springer-Verlag.

H. Levine and J. Schwinger (1948). *On the theory of diffraction by an aperture in an infinite plane screen. I.* Phys. Rev. **74**(8), pp. 958–974.

N. Levinson (1974). *More than one third of zeros of Riemann's zeta-function are on $\sigma = \frac{1}{2}$.* Advances in Math. **13**(4), pp. 383–436.

N. Levinson and R. M. Redheffer (1970). *Complex Variables.* San Francisco, CA: Holden-Day Inc.

J. S. Lew (1994). *On the Darling-Mandelbrot probability density and the zeros of some incomplete gamma functions.* Constr. Approx. **10**(1), pp. 15–30.

S. Lewanowicz (1985). *Recurrence relations for hypergeometric functions of unit argument.* Math. Comp. **45**(172), pp. 521–535.

S. Lewanowicz (1987). *Corrigenda: "Recurrence relations for hypergeometric functions of unit argument" [Math. Comp. **45** (1985), no. 172, 521–535; MR 86m:33004].* Math. Comp. **48**(178), p. 853.

S. Lewanowicz (1991). *Evaluation of Bessel function integrals with algebraic singularities.* J. Comput. Appl. Math. **37**(1-3), pp. 101–112.

L. Lewin (1981). *Polylogarithms and Associated Functions.* New York: North-Holland Publishing Co. With a foreword by A. J. Van der Poorten.

J. T. Lewis and M. E. Muldoon (1977). *Monotonicity and convexity properties of zeros of Bessel functions.* SIAM J. Math. Anal. **8**(1), pp. 171–178.

L.-W. Li, T. S. Yeo, P. S. Kooi, and M. S. Leong (1998). *Microwave specific attenuation by oblate spheroidal raindrops: An exact analysis of TCS's in terms of spheroidal wave functions.* J. Electromagn. Waves Appl. **12**(6), pp. 709–711.

X. Li, X. Shi, and J. Zhang (1991). *Generalized Riemann ζ-function regularization and Casimir energy for a piecewise uniform string*. Phys. Rev. D **44**(2), pp. 560–562.

X. Li and R. Wong (1994). *Error bounds for asymptotic expansions of Laplace convolutions*. SIAM J. Math. Anal. **25**(6), pp. 1537–1553.

X. Li and R. Wong (2000). *A uniform asymptotic expansion for Krawtchouk polynomials*. J. Approx. Theory **106**(1), pp. 155–184.

X. Li and R. Wong (2001). *On the asymptotics of the Meixner-Pollaczek polynomials and their zeros*. Constr. Approx. **17**(1), pp. 59–90.

Y. A. Li and P. J. Olver (2000). *Well-posedness and blow-up solutions for an integrable nonlinearly dispersive model wave equation*. J. Differential Equations **162**(1), pp. 27–63.

Y. T. Li and R. Wong (2008). *Integral and series representations of the Dirac delta function*. Commun. Pure Appl. Anal. **7**(2), pp. 229–247.

E. M. Lifshitz and L. P. Pitaevskiĭ (1980). *Statistical Physics, Part 2: Theory of the Condensed State*. Oxford: Pergamon Press. Course of Theoretical Physics, Vol. 9. Translated from the Russian by J. B. Sykes and M. J. Kearsley.

M. J. Lighthill (1958). *An Introduction to Fourier Analysis and Generalised Functions*. Cambridge Monographs on Mechanics and Applied Mathematics. New York: Cambridge University Press.

E. Lindelöf (1905). *Le Calcul des Résidus et ses Applications à la Théorie des Fonctions*. Paris: Gauthier-Villars. Reprinted by Éditions Jacques Gabay, Sceaux, 1989.

P. Linz and T. E. Kropp (1973). *A note on the computation of integrals involving products of trigonometric and Bessel functions*. Math. Comp. **27**(124), pp. 871–872.

J. E. Littlewood (1914). *Sur la distribution des nombres premiers*. Comptes Rendus de l'Academie des Sciences, Paris **158**, pp. 1869–1872.

I. M. Longman (1956). *Note on a method for computing infinite integrals of oscillatory functions*. Proc. Cambridge Philos. Soc. **52**(4), pp. 764–768.

J. L. López (1999). *Asymptotic expansions of the Whittaker functions for large order parameter*. Methods Appl. Anal. **6**(2), pp. 249–256. Dedicated to Richard A. Askey on the occasion of his 65th birthday, Part II.

J. L. López (2000). *Asymptotic expansions of symmetric standard elliptic integrals*. SIAM J. Math. Anal. **31**(4), pp. 754–775.

J. L. López (2001). *Uniform asymptotic expansions of symmetric elliptic integrals*. Constr. Approx. **17**(4), pp. 535–559.

J. L. López and N. M. Temme (1999a). *Approximation of orthogonal polynomials in terms of Hermite polynomials*. Methods Appl. Anal. **6**(2), pp. 131–146.

J. L. López and N. M. Temme (1999b). *Hermite polynomials in asymptotic representations of generalized Bernoulli, Euler, Bessel, and Buchholz polynomials*. J. Math. Anal. Appl. **239**(2), pp. 457–477.

L. Lorch (1984). *Inequalities for ultraspherical polynomials and the gamma function*. J. Approx. Theory **40**(2), pp. 115–120.

L. Lorch (1990). *Monotonicity in terms of order of the zeros of the derivatives of Bessel functions*. Proc. Amer. Math. Soc. **108**(2), pp. 387–389.

L. Lorch (1992). *On Bessel functions of equal order and argument*. Rend. Sem. Mat. Univ. Politec. Torino **50**(2), pp. 209–216 (1993).

L. Lorch (1993). *Some inequalities for the first positive zeros of Bessel functions*. SIAM J. Math. Anal. **24**(3), pp. 814–823.

L. Lorch (1995). *The zeros of the third derivative of Bessel functions of order less than one*. Methods Appl. Anal. **2**(2), pp. 147–159.

L. Lorch (2002). *Comparison of a pair of upper bounds for a ratio of gamma functions*. Math. Balkanica (N.S.) **16**(1-4), pp. 195–202.

L. Lorch, M. E. Muldoon, and P. Szego (1970). *Higher monotonicity properties of certain Sturm-Liouville functions. III*. Canad. J. Math. **22**, pp. 1238–1265.

L. Lorch, M. E. Muldoon, and P. Szego (1972). *Higher monotonicity properties of certain Sturm-Liouville functions. IV*. Canad. J. Math. **24**, pp. 349–368.

L. Lorch and P. Szego (1963). *Higher monotonicity properties of certain Sturm-Liouville functions*. Acta Math. **109**, pp. 55–73.

L. Lorch and P. Szego (1964). *Monotonicity of the differences of zeros of Bessel functions as a function of order*. Proc. Amer. Math. Soc. **15**(1), pp. 91–96.

L. Lorch and P. Szego (1990). *On the points of inflection of Bessel functions of positive order. I*. Canad. J. Math. **42**(5), pp. 933–948. Corrigenda see same journal, **42**(6), p. 1132.

L. Lorch and P. Szego (1995). *Monotonicity of the zeros of the third derivative of Bessel functions*. Methods Appl. Anal. **2**(1), pp. 103–111.

Lord Kelvin (1891). *Popular Lectures and Addresses*, Volume 3, pp. 481–488. London: Macmillan.

Lord Kelvin (1905). *Deep water ship-waves*. Phil. Mag. **9**, pp. 733–757.

Lord Rayleigh (1945). *The Theory of Sound* (2nd ed.). New York: Dover Publications. Photographic reproduction of the 1929 reprint of the edition of 1894–1896, published by Macmillan, London.

H. A. Lorentz, A. Einstein, H. Minkowski, and H. Weyl (1923). *The Principle of Relativity: A Collection of Original Memoirs on the Special and General Theory of Relativity*. London: Methuen and Co., Ltd. Translated from Das Relativitätsprinzip by W. Perrett and G. B. Jeffery. Reprinted by Dover Publications Inc., New York, 1952.

L. Lorentzen and H. Waadeland (1992). *Continued Fractions with Applications*. Studies in Computational Mathematics. Amsterdam: North-Holland Publishing Co.

J. D. Louck (1958). *New recursion relation for the Clebsch-Gordan coefficients*. Phys. Rev. (2) **110**(4), pp. 815–816.

L. Lovász, L. Pyber, D. J. A. Welsh, and G. M. Ziegler (1995). *Combinatorics in Pure Mathematics*. In R. L. Graham, M. Grötschel, and L. Lovász (Eds.), *Handbook of Combinatorics, Vol. 2*, pp. 2039–2082. Amsterdam: Elsevier.

E. R. Love (1970). *Changing the order of integration*. J. Austral. Math. Soc. **11**, pp. 421–432. An addendum appears in Vol. 14, pp. 383–384, of the same journal.

E. R. Love (1972a). *Addendum to: "Changing the order of integration"*. J. Austral. Math. Soc. **14**, pp. 383–384.

E. R. Love (1972b). *Two index laws for fractional integrals and derivatives*. J. Austral. Math. Soc. **14**, pp. 385–410.

A. N. Lowan and W. Horenstein (1942). *On the function $H(m,a,x) = \exp(-ix)F(m+1-ia, 2m+2; 2ix)$*. J. Math. Phys. Mass. Inst. Tech. **21**, pp. 264–283.

T. A. Lowdon (1970). *Integral representation of the Hankel function in terms of parabolic cylinder functions*. Quart. J. Mech. Appl. Math. **23**(3), pp. 315–327.

D. W. Lozier (1980). *Numerical Solution of Linear Difference Equations*. NBSIR 80-1976, National Bureau of Standards, Gaithersburg, MD 20899. (Available from National Technical Information Service, Springfield, VA 22161.).

D. W. Lozier (1993). *An underflow-induced graphics failure solved by SLI arithmetic*. In E. E. Swartzlander, Jr., M. J. Irwin, and G. A. Jullien (Eds.), *IEEE Symposium on Computer Arithmetic*, Washington, D.C., pp. 10–17. IEEE Computer Society Press.

D. W. Lozier and F. W. J. Olver (1993). *Airy and Bessel Functions by Parallel Integration of ODEs*. In R. F. Sincovec, D. E. Keyes, M. R. Leuze, L. R. Petzold, and D. A. Reed (Eds.), *Proceedings of the Sixth SIAM Conference on Parallel Processing for Scientific Computing*, Philadelphia, PA, pp. 530–538. Society for Industrial and Applied Mathematics (SIAM).

D. W. Lozier and F. W. J. Olver (1994). *Numerical Evaluation of Special Functions*. In *Mathematics of Computation 1943–1993: A Half-Century of Computational Mathematics (Vancouver, BC, 1993)*, Volume 48 of *Proc. Sympos. Appl. Math.*, pp. 79–125. Providence, RI: Amer. Math. Soc. An updated version exists online.

É. Lucas (1891). *Théorie des nombres. Tome I: Le calcul des nombres entiers, le calcul des nombres rationnels, la divisibilité arithmétique*. Paris: Gauthier-Villars.

S. K. Lucas (1995). *Evaluating infinite integrals involving products of Bessel functions of arbitrary order*. J. Comput. Appl. Math. **64**(3), pp. 269–282.

S. K. Lucas and H. A. Stone (1995). *Evaluating infinite integrals involving Bessel functions of arbitrary order*. J. Comput. Appl. Math. **64**(3), pp. 217–231.

D. Ludwig (1966). *Uniform asymptotic expansions at a caustic*. Comm. Pure Appl. Math. **19**, pp. 215–250.

N. A. Lukaševič (1965). *Elementary solutions of certain Painlevé equations*. Differ. Uravn. **1**(3), pp. 731–735. In Russian. English translation: Differential Equations **1**(3), pp. 561–564.

N. A. Lukaševič (1967a). *Theory of the fourth Painlevé equation*. Differ. Uravn. **3**(5), pp. 771–780. In Russian. English translation: Differential Equations **3**(5), pp. 395–399.

N. A. Lukaševič (1967b). *On the theory of Painlevé's third equation*. Differ. Uravn. **3**(11), pp. 1913–1923. In Russian. English translation: Differential Equations **3**(11), pp. 994–999.

N. A. Lukaševič (1968). *Solutions of the fifth Painlevé equation*. Differ. Uravn. **4**(8), pp. 1413–1420. In Russian. English translation: Differential Equations **4**(8), pp. 732–735.

N. A. Lukaševič (1971). *The second Painlevé equation*. Differ. Uravn. **7**(6), pp. 1124–1125. In Russian. English translation: Differential Equations **7**(6), pp. 853–854.

N. A. Lukaševič and A. I. Yablonskiĭ (1967). *On a set of solutions of the sixth Painlevé equation*. Differ. Uravn. **3**(3), pp. 520–523. In Russian. English translation: Differential Equations **3**(3), pp. 264–266.

Y. L. Luke (1959). *Expansion of the confluent hypergeometric function in series of Bessel functions*. Math. Tables Aids Comput. **13**(68), pp. 261–271.

Y. L. Luke (1962). *Integrals of Bessel Functions*. New York: McGraw-Hill Book Co., Inc.

Y. L. Luke (1968). *Approximations for elliptic integrals*. Math. Comp. **22**(103), pp. 627–634.

Y. L. Luke (1969a). *The Special Functions and their Approximations, Vol. 1*. New York: Academic Press.

Y. L. Luke (1969b). *The Special Functions and their Approximations. Vol. 2*. New York: Academic Press.

Y. L. Luke (1970). *Further approximations for elliptic integrals*. Math. Comp. **24**(109), pp. 191–198.

Y. L. Luke (1971a). *Miniaturized tables of Bessel functions*. Math. Comp. **25**(114), pp. 323–330.

Y. L. Luke (1971b). *Miniaturized tables of Bessel functions. II*. Math. Comp. **25**(116), pp. 789–795 and D14–E13. For Corrigendum see same journal **26**(120), pp. A1–A7.

Y. L. Luke (1972). *Miniaturized tables of Bessel functions. III*. Math. Comp. **26**(117), pp. 237–240 and A14–B5.

Y. L. Luke (1975). *Mathematical Functions and their Approximations*. New York: Academic Press Inc.

Y. L. Luke (1977a). *Algorithms for rational approximations for a confluent hypergeometric function*. Utilitas Math. **11**, pp. 123–151.

Y. L. Luke (1977b). *Algorithms for the Computation of Mathematical Functions*. New York: Academic Press.

Y. L. Luke and J. Wimp (1963). *Jacobi polynomial expansions of a generalized hypergeometric function over a semi-infinite ray*. Math. Comp. **17**(84), pp. 395–404.

J. Lund (1985). *Bessel transforms and rational extrapolation*. Numer. Math. **47**(1), pp. 1–14.

J. H. Luscombe and M. Luban (1998). *Simplified recursive algorithm for Wigner $3j$ and $6j$ symbols*. Phys. Rev. E **57**(6), pp. 7274–7277.

W. Luther (1995). *Highly accurate tables for elementary functions*. BIT **35**(3), pp. 352–360.

R. J. Lyman and W. W. Edmonson (2001). *Linear prediction of bandlimited processes with flat spectral densities*. IEEE Trans. Signal Process. **49**(7), pp. 1564–1569.

J. N. Lyness (1971). *Adjusted forms of the Fourier coefficient asymptotic expansion and applications in numerical quadrature*. Math. Comp. **25**(113), pp. 87–104.

J. N. Lyness (1985). *Integrating some infinite oscillating tails*. J. Comput. Appl. Math. **12/13**, pp. 109–117.

H. Maass (1971). *Siegel's modular forms and Dirichlet series*, Volume 216 of *Lecture Notes in Mathematics*. Berlin: Springer-Verlag.

D. A. MacDonald (1989). *The roots of $J_0(z) - iJ_1(z) = 0$*. Quart. Appl. Math. **47**(2), pp. 375–378.

D. A. MacDonald (1997). *On the computation of zeroes of $J_n(z) - iJ_{n+1}(z) = 0$*. Quart. Appl. Math. **55**(4), pp. 623–633.

I. D. Macdonald (1968). *The Theory of Groups*. Oxford: Clarendon Press. Reprinted in 1988 by Krieger Publishing Co., Malabar, FL.

I. G. Macdonald (1972). *Affine root systems and Dedekind's η-function*. Invent. Math. **15**(2), pp. 91–143.

I. G. Macdonald (1982). *Some conjectures for root systems*. SIAM J. Math. Anal. **13**(6), pp. 988–1007.

I. G. Macdonald (1990). *Hypergeometric Functions*. Unpublished Lecture Notes, University of London. An online version exists.

I. G. Macdonald (1995). *Symmetric Functions and Hall Polynomials* (2nd ed.). New York-Oxford: The Clarendon Press, Oxford University Press. With contributions by A. Zelevinsky. Paperback version published by Oxford University Press, July 1999.

I. G. Macdonald (1998). *Symmetric Functions and Orthogonal Polynomials*, Volume 12 of *University Lecture Series*. Providence, RI: American Mathematical Society.

I. G. Macdonald (2000). *Orthogonal polynomials associated with root systems*. Sém. Lothar. Combin. **45**, pp. Art. B45a, 40 pp. (electronic).

I. G. Macdonald (2003). *Affine Hecke Algebras and Orthogonal Polynomials*, Volume 157 of *Cambridge Tracts in Mathematics*. Cambridge: Cambridge University Press.

R. L. Mace and M. A. Hellberg (1995). *A dispersion function for plasmas containing superthermal particles*. Physics of Plasmas **2**(6), pp. 2098–2109.

N. W. Macfadyen and P. Winternitz (1971). *Crossing symmetric expansions of physical scattering amplitudes: The $O(2,1)$ group and Lamé functions*. J. Mathematical Phys. **12**, pp. 281–293.

A. J. MacLeod (1993). *Chebyshev expansions for modified Struve and related functions*. Math. Comp. **60**(202), pp. 735–747.

A. J. MacLeod (1994). *Computation of inhomogeneous Airy functions*. J. Comput. Appl. Math. **53**(1), pp. 109–116.

A. J. MacLeod (1996). *Rational approximations, software and test methods for sine and cosine integrals*. Numer. Algorithms **12**(3-4), pp. 259–272. Includes Fortran code, maximum accuracy 20S.

A. J. MacLeod (2002a). *Asymptotic expansions for the zeros of certain special functions*. J. Comput. Appl. Math. **145**(2), pp. 261–267.

A. J. MacLeod (2002b). *The efficient computation of some generalised exponential integrals*. J. Comput. Appl. Math. **148**(2), pp. 363–374.

T. M. MacRobert (1967). *Spherical Harmonics. An Elementary Treatise on Harmonic Functions with Applications* (3rd ed.), Volume 98 of *International Series of Monographs in Pure and Applied Mathematics*. Oxford: Pergamon Press.

A. P. Magnus (1995). *Painlevé-type differential equations for the recurrence coefficients of semi-classical orthogonal polynomials*. J. Comput. Appl. Math. **57**(1-2), pp. 215–237.

W. Magnus (1941). *Zur Theorie des zylindrischparabolischen Spiegels*. Z. Physik **118**, pp. 343–356.

W. Magnus, F. Oberhettinger, and R. P. Soni (1966). *Formulas and Theorems for the Special Functions of Mathematical Physics* (3rd ed.). New York-Berlin: Springer-Verlag. Table errata: Math. Comp. v. 23 (1969), p. 471, v. 24 (1970), pp. 240 and 505, v. 25 (1971), no. 113, p. 201, v. 29 (1975), no. 130, p. 672, v. 30 (1976), no. 135, pp. 677–678, v. 32 (1978), no. 141, pp. 319–320, v. 36 (1981), no. 153, pp. 315–317, and v. 41 (1983), no. 164, pp. 775–776.

W. Magnus and S. Winkler (1966). *Hill's Equation*. Interscience Tracts in Pure and Applied Mathematics, No. 20. New York-London-Sydney: Interscience Publishers John Wiley & Sons. Reprinted by Dover Publications, Inc., New York, 1979.

K. Mahler (1930). *Über die Nullstellen der unvollständigen Gammafunktionen*. Rend. del Circ. Matem. Palermo **54**, pp. 1–41.

R. S. Maier (2005). *On reducing the Heun equation to the hypergeometric equation*. J. Differential Equations **213**(1), pp. 171–203.

R. S. Maier (2007). *The 192 solutions of the Heun equation*. Math. Comp. **76**(258), pp. 811–843.

H. Majima, K. Matsumoto, and N. Takayama (2000). *Quadratic relations for confluent hypergeometric functions*. Tohoku Math. J. (2) **52**(4), pp. 489–513.

S. Makinouchi (1966). *Zeros of Bessel functions $J_\nu(x)$ and $Y_\nu(x)$ accurate to twenty-nine significant digits*. Technology Reports of the Osaka University **16**(685), pp. 1–44.

Yu. I. Manin (1998). *Sixth Painlevé Equation, Universal Elliptic Curve, and Mirror of $\mathbf{P}^2$*. In A. Khovanskii, A. Varchenko, and V. Vassiliev (Eds.), *Geometry of Differential Equations*, Volume 186 of *Amer. Math. Soc. Transl. Ser. 2*, pp. 131–151. Providence, RI: Amer. Math. Soc.

E. L. Mansfield and H. N. Webster (1998). *On one-parameter families of Painlevé III*. Stud. Appl. Math. **101**(3), pp. 321–341.

F. Marcellán, M. Alfaro, and M. L. Rezola (1993). *Orthogonal polynomials on Sobolev spaces: Old and new directions*. J. Comput. Appl. Math. **48**(1-2), pp. 113–131.

M. Marden (1966). *Geometry of Polynomials* (2nd ed.). Providence, RI: American Mathematical Society.

O. I. Marichev (1983). *Handbook of Integral Transforms of Higher Transcendental Functions: Theory and Algorithmic Tables*. Chichester/New York: Ellis Horwood Ltd./John Wiley & Sons, Inc. Edited by F. D. Gakhov, Translated from the Russian by L. W. Longdon.

O. I. Marichev (1984). *On the Representation of Meijer's G-Function in the Vicinity of Singular Unity*. In *Complex Analysis and Applications '81 (Varna, 1981)*, pp. 383–398. Sofia: Publ. House Bulgar. Acad. Sci.

S. M. Markov (1981). *On the interval computation of elementary functions*. C. R. Acad. Bulgare Sci. **34**(3), pp. 319–322.

A. I. Markushevich (1983). *The Theory of Analytic Functions: A Brief Course*. Moscow: "Mir". Translated from the Russian by Eugene Yankovsky.

A. I. Markushevich (1985). *Theory of Functions of a Complex Variable. Vols. I, II, III*. New York: Chelsea Publishing Co. Translated and edited by Richard A. Silverman.

A. I. Markushevich (1992). *Introduction to the Classical Theory of Abelian Functions*. Providence, RI: American Mathematical Society. Translated from the 1979 Russian original by G. Bluher.

P. Maroni (1995). *An integral representation for the Bessel form*. J. Comput. Appl. Math. **57**(1-2), pp. 251–260.

J. E. Marsden and A. J. Tromba (1996). *Vector Calculus* (4th ed.). New York: W. H. Freeman & Company. Fifth edition printed in 2003.

P. L. Marston (1992). A. D. Pierce and R. N. Thurston (Eds.). *Geometrical and Catastrophe Optics Methods in Scattering*, Volume 21, pp. 1–234. New York: Academic Press.

P. L. Marston (1999). *Catastrophe optics of spheroidal drops and generalized rainbows*. J. Quantit. Spec. and Rad. Trans. **63**, pp. 341–351.

B. Martić (1978). *Note sur certaines inégalités d'intégrales*. Akad. Nauka Umjet. Bosne Hercegov. Rad. Odjelj. Prirod. Mat. Nauka **61**(17), pp. 165–168.

P. Martín, R. Pérez, and A. L. Guerrero (1992). *Two-point quasi-fractional approximations to the Airy function Ai(x)*. J. Comput. Phys. **99**(2), pp. 337–340. There is an error in Eq. (5): the second term in parentheses needs to be multiplied by x.

J. Martinek, H. P. Thielman, and E. C. Huebschman (1966). *On the zeros of cross-product Bessel functions*. J. Math. Mech. **16**, pp. 447–452.

J. C. Mason and D. C. Handscomb (2003). *Chebyshev Polynomials*. Boca Raton, FL: Chapman & Hall/CRC.

D. R. Masson (1991). *Associated Wilson polynomials*. Constr. Approx. **7**(4), pp. 521–534.

T. Masuda (2003). *On a class of algebraic solutions to the Painlevé VI equation, its determinant formula and coalescence cascade*. Funkcial. Ekvac. **46**(1), pp. 121–171.

T. Masuda (2004). *Classical transcendental solutions of the Painlevé equations and their degeneration*. Tohoku Math. J. (2) **56**(4), pp. 467–490.

T. Masuda, Y. Ohta, and K. Kajiwara (2002). *A determinant formula for a class of rational solutions of Painlevé V equation.* Nagoya Math. J. **168**, pp. 1–25.

A. M. Mathai (1993). *A Handbook of Generalized Special Functions for Statistical and Physical Sciences.* Oxford Science Publications. New York: The Clarendon Press Oxford University Press.

F. Matta and A. Reichel (1971). *Uniform computation of the error function and other related functions.* Math. Comp. **25**(114), pp. 339–344.

D. W. Matula and P. Kornerup (1980). *Foundations of Finite Precision Rational Arithmetic.* In G. Alefeld and R. D. Grigorieff (Eds.), *Fundamentals of Numerical Computation (Computer-oriented Numerical Analysis)*, Volume 2 of *Comput. Suppl.*, Vienna, pp. 85–111. Springer.

L. C. Maximon (1955). *On the evaluation of indefinite integrals involving the special functions: Application of method.* Quart. Appl. Math. **13**, pp. 84–93.

L. C. Maximon (1991). *On the evaluation of the integral over the product of two spherical Bessel functions.* J. Math. Phys. **32**(3), pp. 642–648.

L. C. Maximon (2003). *The dilogarithm function for complex argument.* Proc. Roy. Soc. London Ser. A **459**, pp. 2807–2819.

M. Mazzocco (2001a). *Rational solutions of the Painlevé VI equation.* J. Phys. A **34**(11), pp. 2281–2294.

M. Mazzocco (2001b). *Picard and Chazy solutions to the Painlevé VI equation.* Math. Ann. **321**(1), pp. 157–195.

R. C. McCann (1977). *Inequalities for the zeros of Bessel functions.* SIAM J. Math. Anal. **8**(1), pp. 166–170.

J. P. McClure and R. Wong (1978). *Explicit error terms for asymptotic expansions of Stieltjes transforms.* J. Inst. Math. Appl. **22**(2), pp. 129–145.

J. P. McClure and R. Wong (1979). *Exact remainders for asymptotic expansions of fractional integrals.* J. Inst. Math. Appl. **24**(2), pp. 139–147.

J. P. McClure and R. Wong (1987). *Asymptotic expansion of a multiple integral.* SIAM J. Math. Anal. **18**(6), pp. 1630–1637.

B. M. McCoy (1992). *Spin Systems, Statistical Mechanics and Painlevé Functions.* In D. Levi and P. Winternitz (Eds.), *Painlevé Transcendents: Their Asymptotics and Physical Applications*, Volume 278 of *NATO Adv. Sci. Inst. Ser. B Phys.*, pp. 377–391. New York: Plenum. Proc. NATO Adv. Res. Workshop, Sainte-Adèle, Canada, 1990.

B. M. McCoy, C. A. Tracy, and T. T. Wu (1977). *Painlevé functions of the third kind.* J. Mathematical Phys. **18**(5), pp. 1058–1092.

F. A. McDonald and J. Nuttall (1969). *Complex-energy method for elastic e-H scattering above the ionization threshold.* Phys. Rev. Lett. **23**(7), pp. 361–363.

J. N. McDonald and N. A. Weiss (1999). *A Course in Real Analysis.* San Diego, CA: Academic Press Inc.

H. R. McFarland, III and D. St. P. Richards (2001). *Exact misclassification probabilities for plug-in normal quadratic discriminant functions. I. The equal-means case.* J. Multivariate Anal. **77**(1), pp. 21–53.

H. R. McFarland, III and D. St. P. Richards (2002). *Exact misclassification probabilities for plug-in normal quadratic discriminant functions. II. The heterogeneous case.* J. Multivariate Anal. **82**(2), pp. 299–330.

H. McKean and V. Moll (1999). *Elliptic Curves.* Cambridge: Cambridge University Press. Reprint with corrections of original edition, published in 1997 by Cambridge University Press.

N. M. McLachlan and A. L. Meyers (1936). *The ster and stei functions.* Phil. Mag. Series 7 **21**(140), pp. 425–436.

N. W. McLachlan (1934). *Loud Speakers: Theory, Performance, Testing and Design.* New York: Oxford University Press. Reprinted, with corrections, by Dover Publications Inc., 1960.

N. W. McLachlan (1947). *Theory and Application of Mathieu Functions.* Oxford: Clarendon Press. Corrected republication by Dover Publications, Inc., New York, 1964.

N. W. McLachlan (1961). *Bessel Functions for Engineers* (2nd ed.). Oxford: Clarendon Press. The first edition was published in 1934 by Clarendon Press. The second edition was published originally in 1955.

J. McMahon (1894/95). *On the roots of the Bessel and certain related functions.* Ann. of Math. **9**(1-6), pp. 23–30.

J. M. McNamee (2007). *Numerical Methods for Roots of Polynomials. Part I*, Volume 14 of *Studies in Computational Mathematics*. Amsterdam: Elsevier.

F. Mechel (1966). *Calculation of the modified Bessel functions of the second kind with complex argument.* Math. Comp. **20**(95), pp. 407–412.

V. Meden and K. Schönhammer (1992). *Spectral functions for the Tomonaga-Luttinger model.* Phys. Rev. B **46**(24), pp. 15753–15760.

D. S. Meek and D. J. Walton (1992). *Clothoid spline transition spirals.* Math. Comp. **59**(199), pp. 117–133.

R. Mehrem, J. T. Londergan, and M. H. Macfarlane (1991). *Analytic expressions for integrals of products of spherical Bessel functions.* J. Phys. A **24**(7), pp. 1435–1453.

M. L. Mehta (2004). *Random Matrices* (3rd ed.), Volume 142 of *Pure and Applied Mathematics (Amsterdam)*. Amsterdam: Elsevier/Academic Press.

C. S. Meijer (1946). *On the G-function. VII, VIII*. Nederl. Akad. Wetensch., Proc. **49**, pp. 1063–1072, 1165–1175 = Indagationes Math. 8, 661–670, 713–723 (1946).

J. W. Meijer and N. H. G. Baken (1987). *The exponential integral distribution*. Statist. Probab. Lett. **5**(3), pp. 209–211.

G. Meinardus (1967). *Approximation of Functions: Theory and Numerical Methods*, Volume 13 of *Springer Tracts in Natural Philosophy*. New York: Springer-Verlag. Expanded translation from the German edition. Translated by Larry L. Schumaker.

P. N. Meisinger, T. R. Miller, and M. C. Ogilvie (2002). *Phenomenological equations of state for the quark-gluon plasma*. Phys. Rev. D **65**(3), pp. (034009-1)–(034009-10).

J. Meixner (1944). *Die Laméschen Wellenfunktionen des Drehellipsoids*. Forschungsbericht no. 1952, ZWB. English translation: *Lamé wave functions of the ellipsoid of revolution* appeared as NACA Technical Memorandum No. 1224, 1949.

J. Meixner (1951). *Klassifikation, Bezeichnung und Eigenschaften der Sphäroidfunktionen*. Math. Nachr. **5**, pp. 1–18.

J. Meixner and F. W. Schäfke (1954). *Mathieusche Funktionen und Sphäroidfunktionen mit Anwendungen auf physikalische und technische Probleme*. Die Grundlehren der mathematischen Wissenschaften in Einzeldarstellungen mit besonderer Berücksichtigung der Anwendungsgebiete, Band LXXI. Berlin: Springer-Verlag.

J. Meixner, F. W. Schäfke, and G. Wolf (1980). *Mathieu Functions and Spheroidal Functions and Their Mathematical Foundations: Further Studies*, Volume 837 of *Lecture Notes in Mathematics*. Berlin-New York: Springer-Verlag.

A. J. Menezes, P. C. van Oorschot, and S. A. Vanstone (1997). *Handbook of Applied Cryptography*. Boca Raton, FL: CRC Press. With a foreword by Ronald L. Rivest.

A. McD. Mercer (1992). *The zeros of $az^2 J''_\nu(z) + bz J'_\nu(z) + c J_\nu(z)$ as functions of order*. Internat. J. Math. Math. Sci. **15**(2), pp. 319–322.

X. Merrheim (1994). *The computation of elementary functions in radix 2^p*. Computing **53**(3-4), pp. 219–232. International Symposium on Scientific Computing, Computer Arithmetic and Validated Numerics (Vienna, 1993).

A. Messiah (1961). *Quantum Mechanics. Vol. I*. Amsterdam: North-Holland Publishing Co. Translated from the French by G. M. Temmer.

R. Metzler, J. Klafter, and J. Jortner (1999). *Hierarchies and logarithmic oscillations in the temporal relaxation patterns of proteins and other complex systems*. Proc. Nat. Acad. Sci. U.S.A. **96**(20), pp. 11085–11089.

A. Michaeli (1996). *Asymptotic analysis of edge-excited currents on a convex face of a perfectly conducting wedge under overlapping penumbra region conditions*. IEEE Trans. Antennas and Propagation **44**(1), pp. 97–101.

C. Micu and E. Papp (2005). *Applying q-Laguerre polynomials to the derivation of q-deformed energies of oscillator and Coulomb systems*. Romanian Reports in Physics **57**(1), pp. 25–34.

M. Micu (1968). *Recursion relations for the 3-j symbols*. Nuclear Physics A **113**(1), pp. 215–220.

P. Midy (1975). *An improved calculation of the general elliptic integral of the second kind in the neighbourhood of $x = 0$*. Numer. Math. **25**(1), pp. 99–101.

G. J. Miel (1981). *Evaluation of complex logarithms and related functions*. SIAM J. Numer. Anal. **18**(4), pp. 744–750.

J. W. Miles (1975). *Asymptotic approximations for prolate spheroidal wave functions*. Studies in Appl. Math. **54**(4), pp. 315–349.

J. W. Miles (1978). *On the second Painlevé transcendent*. Proc. Roy. Soc. London Ser. A **361**, pp. 277–291.

J. W. Miles (1980). *The Second Painlevé Transcendent: A Nonlinear Airy Function*. In *Mechanics Today*, Volume 5, pp. 297–313. Oxford: Pergamon.

M. S. Milgram (1985). *The generalized integro-exponential function*. Math. Comp. **44**(170), pp. 443–458.

A. R. Miller (1997). *A class of generalized hypergeometric summations*. J. Comput. Appl. Math. **87**(1), pp. 79–85.

A. R. Miller (2003). *On a Kummer-type transformation for the generalized hypergeometric function $_2F_2$*. J. Comput. Appl. Math. **157**(2), pp. 507–509.

G. F. Miller (1960). *Tables of Generalized Exponential Integrals*. NPL Mathematical Tables, Vol. III. London: Her Majesty's Stationery Office.

G. F. Miller (1966). *On the convergence of the Chebyshev series for functions possessing a singularity in the range of representation*. SIAM J. Numer. Anal. **3**(3), pp. 390–409.

J. Miller and V. S. Adamchik (1998). *Derivatives of the Hurwitz zeta function for rational arguments*. J. Comput. Appl. Math. **100**(2), pp. 201–206.

J. C. P. Miller (1946). *The Airy Integral, Giving Tables of Solutions of the Differential Equation $y'' = xy$*. British Association for the Advancement of Science, Mathematical Tables Part-Vol. B. Cambridge: Cambridge University Press.

J. C. P. Miller (1950). *On the choice of standard solutions for a homogeneous linear differential equation of the second order*. Quart. J. Mech. Appl. Math. **3**(2), pp. 225–235.

J. C. P. Miller (1952). *On the choice of standard solutions to Weber's equation.* Proc. Cambridge Philos. Soc. **48**, pp. 428–435.

J. C. P. Miller (Ed.) (1955). *Tables of Weber Parabolic Cylinder Functions.* London: Her Majesty's Stationery Office.

W. Miller, Jr. (1974). *Lie theory and separation of variables. I: Parabolic cylinder coordinates.* SIAM J. Math. Anal. **5**(4), pp. 626–643.

W. Miller, Jr. (1977). *Symmetry and Separation of Variables.* Reading, MA-London-Amsterdam: Addison-Wesley Publishing Co. Encyclopedia of Mathematics and its Applications, Vol. 4.

J. P. Mills (1926). *Table of the ratio: Area to bounding ordinate, for any portion of normal curve.* Biometrika **18**, pp. 395–400.

A. E. Milne, P. A. Clarkson, and A. P. Bassom (1997). *Bäcklund transformations and solution hierarchies for the third Painlevé equation.* Stud. Appl. Math. **98**(2), pp. 139–194.

S. C. Milne (1985a). *A q-analog of the $_5F_4(;;1)$ summation theorem for hypergeometric series well-poised in $SU(n)$.* Adv. in Math. **57**(1), pp. 14–33.

S. C. Milne (1985b). *An elementary proof of the Macdonald identities for $A_l^{(1)}$.* Adv. in Math. **57**(1), pp. 34–70.

S. C. Milne (1985c). *A new symmetry related to $SU(n)$ for classical basic hypergeometric series.* Adv. in Math. **57**(1), pp. 71–90.

S. C. Milne (1985d). *A q-analog of hypergeometric series well-poised in $SU(n)$ and invariant G-functions.* Adv. in Math. **58**(1), pp. 1–60.

S. C. Milne (1988). *A q-analog of the Gauss summation theorem for hypergeometric series in $U(n)$.* Adv. in Math. **72**(1), pp. 59–131.

S. C. Milne (1994). *A q-analog of a Whipple's transformation for hypergeometric series in $U(n)$.* Adv. Math. **108**(1), pp. 1–76.

S. C. Milne (1996). *New infinite families of exact sums of squares formulas, Jacobi elliptic functions, and Ramanujan's tau function.* Proc. Nat. Acad. Sci. U.S.A. **93**(26), pp. 15004–15008.

S. C. Milne (1997). *Balanced $_3\Theta_2$ summation theorems for $U(n)$ basic hypergeometric series.* Adv. Math. **131**(1), pp. 93–187.

S. C. Milne (2002). *Infinite families of exact sums of squares formulas, Jacobi elliptic functions, continued fractions, and Schur functions.* Ramanujan J. **6**(1), pp. 7–149.

S. C. Milne and G. M. Lilly (1992). *The A_l and C_l Bailey transform and lemma.* Bull. Amer. Math. Soc. (N.S.) **26**(2), pp. 258–263.

L. M. Milne-Thomson (1933). *The Calculus of Finite Differences.* London: Macmillan and Co. Ltd. Reprinted in 1951 by Macmillan, and in 1981 by Chelsea Publishing Co., New York, and American Mathematical Society, AMS Chelsea Book Series, Providence, RI.

L. M. Milne-Thomson (1950). *Jacobian Elliptic Function Tables.* New York: Dover Publications Inc.

A. C. G. Mitchell and M. W. Zemansky (1961). *Resonance Radiation and Excited Atoms* (2nd ed.). Cambridge, England: Cambridge Univerity Press. Reprinted by Cambridge University Press in 1971.

D. S. Mitrinović (1964). *Elementary Inequalities.* Groningen: P. Noordhoff Ltd. In cooperation with E. S. Barnes, D. C. B. Marsh, J. R. M. Radok. Tutorial Text, No. 1.

D. S. Mitrinović (1970). *Analytic Inequalities.* New York: Springer-Verlag. Addenda: Univ. Beograd. Publ. Elektrotehn. Fak. Ser. Mat. Fiz., no. 634–677 (1979), pp. 3–24 and no. 577-598 (1977), pp. 3–10.

D. S. Moak (1981). *The q-analogue of the Laguerre polynomials.* J. Math. Anal. Appl. **81**(1), pp. 20–47.

D. S. Moak (1984). *The q-analogue of Stirling's formula.* Rocky Mountain J. Math. **14**(2), pp. 403–413.

S. Moch, P. Uwer, and S. Weinzierl (2002). *Nested sums, expansion of transcendental functions, and multiscale multiloop integrals.* J. Math. Phys. **43**(6), pp. 3363–3386.

V. P. Modenov and A. V. Filonov (1986). *Calculation of zeros of cylindrical functions and their derivatives.* Vestnik Moskov. Univ. Ser. XV Vychisl. Mat. Kibernet. (2), pp. 63–64, 71. In Russian.

N. Mohankumar and A. Natarajan (1997). *The accurate evaluation of a particular Fermi-Dirac integral.* Comput. Phys. Comm. **101**(1-2), pp. 47–53.

E. W. Montroll (1964). *Lattice Statistics.* In E. F. Beckenbach (Ed.), *Applied Combinatorial Mathematics*, University of California Engineering and Physical Sciences Extension Series, pp. 96–143. New York-London-Sydney: John Wiley and Sons, Inc. Reprinted by Robert E. Krieger Publishing Co. in 1981.

P. Moon and D. E. Spencer (1971). *Field Theory Handbook. Including Coordinate Systems, Differential Equations and Their Solutions* (2nd ed.). Berlin: Springer-Verlag. Reprinted in 1988.

R. E. Moore (1979). *Methods and Applications of Interval Analysis*, Volume 2 of *SIAM Studies in Applied Mathematics*. Philadelphia, PA: Society for Industrial and Applied Mathematics (SIAM).

L. J. Mordell (1917). *On the representation of numbers as a sum of $2r$ squares.* Quarterly Journal of Math. **48**, pp. 93–104.

L. J. Mordell (1958). *On the evaluation of some multiple series.* J. London Math. Soc. (2) **33**, pp. 368–371.

G. W. Morgenthaler and H. Reismann (1963). *Zeros of first derivatives of Bessel functions of the first kind, $J'_n(x)$, $21 \leq n \leq 51$, $0 \leq x \leq 100$.* J. Res. Nat. Bur. Standards Sect. B **67B**(3), pp. 181–183.

T. Morita (1978). *Calculation of the complete elliptic integrals with complex modulus.* Numer. Math. **29**(2), pp. 233–236. An Algol procedure for calculating the complete elliptic integrals of the first and the second kind with complex modulus k is included.

R. Morris (1979). *The dilogarithm function of a real argument.* Math. Comp. **33**(146), pp. 778–787.

P. M. Morse and H. Feshbach (1953a). *Methods of Theoretical Physics*, Volume 1. New York: McGraw-Hill Book Co.

P. M. Morse and H. Feshbach (1953b). *Methods of Theoretical Physics*, Volume 2. New York: McGraw-Hill Book Co.

L. Moser and M. Wyman (1958a). *Asymptotic development of the Stirling numbers of the first kind.* J. London Math. Soc. **33**, pp. 133–146.

L. Moser and M. Wyman (1958b). *Stirling numbers of the second kind.* Duke Math. J. **25**(1), pp. 29–43.

S. L. B. Moshier (1989). *Methods and Programs for Mathematical Functions.* Chichester: Ellis Horwood Ltd. Includes diskette.

N. F. Mott and H. S. W. Massey (1956). *Theory of Atomic Collisions* (3rd ed.). Oxford: Oxford Univ. Press. Two-volume paperback version reprinted in 1987.

R. J. Muirhead (1978). *Latent roots and matrix variates: A review of some asymptotic results.* Ann. Statist. **6**(1), pp. 5–33.

R. J. Muirhead (1982). *Aspects of Multivariate Statistical Theory.* New York: John Wiley & Sons Inc.

M. E. Muldoon (1970). *Singular integrals whose kernels involve certain Sturm-Liouville functions. I.* J. Math. Mech. **19**(10), pp. 855–873.

M. E. Muldoon (1977). *Higher monotonicity properties of certain Sturm-Liouville functions. V.* Proc. Roy. Soc. Edinburgh Sect. A **77**(1-2), pp. 23–37.

M. E. Muldoon (1979). *On the zeros of a cross-product of Bessel functions of different orders.* Z. Angew. Math. Mech. **59**(6), pp. 272–273.

M. E. Muldoon (1981). *The variation with respect to order of zeros of Bessel functions.* Rend. Sem. Mat. Univ. Politec. Torino **39**(2), pp. 15–25.

H. P. Mulholland and S. Goldstein (1929). *The characteristic numbers of the Mathieu equation with purely imaginary parameter.* Phil. Mag. Series 7 **8**(53), pp. 834–840.

D. Müller, B. G. Kelly, and J. J. O'Brien (1994). *Spheroidal eigenfunctions of the tidal equation.* Phys. Rev. Lett. **73**(11), pp. 1557–1560.

H. J. W. Müller (1962). *Asymptotic expansions of oblate spheroidal wave functions and their characteristic numbers.* J. Reine Angew. Math. **211**, pp. 33–47.

H. J. W. Müller (1963). *Asymptotic expansions of prolate spheroidal wave functions and their characteristic numbers.* J. Reine Angew. Math. **212**, pp. 26–48.

H. J. W. Müller (1966a). *Asymptotic expansions of ellipsoidal wave functions and their characteristic numbers.* Math. Nachr. **31**, pp. 89–101.

H. J. W. Müller (1966b). *Asymptotic expansions of ellipsoidal wave functions in terms of Hermite functions.* Math. Nachr. **32**, pp. 49–62.

H. J. W. Müller (1966c). *On asymptotic expansions of ellipsoidal wave functions.* Math. Nachr. **32**, pp. 157–172.

J.-M. Muller (1997). *Elementary Functions: Algorithms and Implementation.* Boston, MA: Birkhäuser Boston Inc. A 2nd edition was published in 2006.

K. H. Müller (1988). *Elastodynamics in parabolic cylinders.* Z. Angew. Math. Phys. **39**(5), pp. 748–752.

D. Mumford (1983). *Tata Lectures on Theta. I.* Boston, MA: Birkhäuser Boston Inc.

D. Mumford (1984). *Tata Lectures on Theta. II.* Boston, MA: Birkhäuser Boston Inc. Jacobian theta functions and differential equations.

Y. Murata (1985). *Rational solutions of the second and the fourth Painlevé equations.* Funkcial. Ekvac. **28**(1), pp. 1–32.

Y. Murata (1995). *Classical solutions of the third Painlevé equation.* Nagoya Math. J. **139**, pp. 37–65.

L. A. Muraveĭ (1976). *Zeros of the function $\operatorname{Ai}'(z) - \sigma \operatorname{Ai}(z)$.* Differential Equations **11**, pp. 797–811.

B. T. M. Murphy and A. D. Wood (1997). *Hyperasymptotic solutions of second-order ordinary differential equations with a singularity of arbitrary integer rank.* Methods Appl. Anal. **4**(3), pp. 250–260.

J. Murzewski and A. Sowa (1972). *Tables of the functions of the parabolic cylinder for negative integer parameters.* Zastos. Mat. **13**, pp. 261–273.

A. Nakamura (1996). *Toda equation and its solutions in special functions.* J. Phys. Soc. Japan **65**(6), pp. 1589–1597.

W. Narkiewicz (2000). *The Development of Prime Number Theory: From Euclid to Hardy and Littlewood.* Berlin: Springer-Verlag.

A. Natarajan and N. Mohankumar (1993). *On the numerical evaluation of the generalised Fermi-Dirac integrals.* Comput. Phys. Comm. **76**(1), pp. 48–50.

National Bureau of Standards (1944). *Tables of Lagrangian Interpolation Coefficients.* New York: Columbia University Press. Technical Director: Arnold N. Lowan.

National Physical Laboratory (1961). *Modern Computing Methods* (2nd ed.). Notes on Applied Science, No. 16. London: Her Majesty's Stationery Office.

D. Naylor (1984). *On simplified asymptotic formulas for a class of Mathieu functions.* SIAM J. Math. Anal. **15**(6), pp. 1205–1213.

D. Naylor (1987). *On a simplified asymptotic formula for the Mathieu function of the third kind.* SIAM J. Math. Anal. **18**(6), pp. 1616–1629.

D. Naylor (1989). *On an integral transform involving a class of Mathieu functions.* SIAM J. Math. Anal. **20**(6), pp. 1500–1513.

D. Naylor (1990). *On an asymptotic expansion of the Kontorovich-Lebedev transform.* Applicable Anal. **39**(4), pp. 249–263.

D. Naylor (1996). *On an asymptotic expansion of the Kontorovich-Lebedev transform.* Methods Appl. Anal. **3**(1), pp. 98–108.

NBS (1958). National Bureau of Standards. *Integrals of Airy Functions.* Number 52 in National Bureau of Standards Applied Mathematics Series. Washington, D.C.: U.S. Government Printing Office.

NBS (1967). National Bureau of Standards. *Tables Relating to Mathieu Functions: Characteristic Values, Coefficients, and Joining Factors* (2nd ed.). Number 59 in National Bureau of Standards Applied Mathematics Series. Washington, D.C.: U.S. Government Printing Office.

J. Negro, L. M. Nieto, and O. Rosas-Ortiz (2000). *Confluent hypergeometric equations and related solvable potentials in quantum mechanics.* J. Math. Phys. **41**(12), pp. 7964–7996.

W. J. Nellis and B. C. Carlson (1966). *Reduction and evaluation of elliptic integrals.* Math. Comp. **20**(94), pp. 223–231.

G. Németh (1992). *Mathematical Approximation of Special Functions.* Commack, NY: Nova Science Publishers Inc. Ten papers on Chebyshev expansions.

J. J. Nestor (1984). *Uniform Asymptotic Approximations of Solutions of Second-order Linear Differential Equations, with a Coalescing Simple Turning Point and Simple Pole.* Ph. D. thesis, University of Maryland, College Park, MD.

E. Neuman (1969). *On the calculation of elliptic integrals of the second and third kinds.* Zastos. Mat. **11**, pp. 91–94.

E. Neuman (2003). *Bounds for symmetric elliptic integrals.* J. Approx. Theory **122**(2), pp. 249–259.

E. Neuman (2004). *Inequalities involving Bessel functions of the first kind.* JIPAM. J. Inequal. Pure Appl. Math. **5**(4), pp. Article 94, 4 pp. (electronic).

P. Nevai (1986). *Géza Freud, orthogonal polynomials and Christoffel functions. A case study.* J. Approx. Theory **48**(1), pp. 3–167.

E. H. Neville (1951). *Jacobian Elliptic Functions* (2nd ed.). Oxford: Clarendon Press.

J. N. Newman (1984). *Approximations for the Bessel and Struve functions.* Math. Comp. **43**(168), pp. 551–556.

M. Newman (1967). *Solving equations exactly.* J. Res. Nat. Bur. Standards Sect. B **71B**, pp. 171–179.

T. D. Newton (1952). *Coulomb Functions for Large Values of the Parameter η.* Technical report, Atomic Energy of Canada Limited, Chalk River, Ontario. Rep. CRT-526.

E. W. Ng and M. Geller (1969). *A table of integrals of the error functions.* J. Res. Nat. Bur. Standards Sect B. **73B**, pp. 1–20. See for additions and corrections same journal, Sect. B 75, 149–163 (1975).

N. Nielsen (1906a). *Handbuch der Theorie der Gammafunktion.* Leipzig: B. G. Teubner. For a reprint with corrections see Nielsen (1965).

N. Nielsen (1906b). *Theorie des Integrallogarithmus und verwandter Transzendenten.* Leipzig: B. G. Teubner. For a reprint with corrections see Nielsen (1965).

N. Nielsen (1909). *Der Eulersche Dilogarithmus und seine Verallgemeinerungen.* Nova Acta Leopoldina **90**, pp. 123–212.

N. Nielsen (1923). *Traité Élémentaire des Nombres de Bernoulli.* Paris: Gauthier-Villars.

N. Nielsen (1965). *Die Gammafunktion. Band I. Handbuch der Theorie der Gammafunktion. Band II. Theorie des Integrallogarithmus und verwandter Transzendenten.* New York: Chelsea Publishing Co. Both books are textually unaltered except for correction of errata.

Y. Nievergelt (1995). *Bisection hardly ever converges linearly.* Numer. Math. **70**(1), pp. 111–118.

A. F. Nikiforov and V. B. Uvarov (1988). *Special Functions of Mathematical Physics: A Unified Introduction with Applications.* Basel: Birkhäuser Verlag. Translated from the Russian and with a preface by Ralph P. Boas, with a foreword by A. A. Samarskiĭ.

I. Niven, H. S. Zuckerman, and H. L. Montgomery (1991). *An Introduction to the Theory of Numbers* (5th ed.). New York: John Wiley & Sons Inc.

C. J. Noble (2004). *Evaluation of negative energy Coulomb (Whittaker) functions.* Comput. Phys. Comm. **159**(1), pp. 55–62. Double-precision Fortran-90 code.

V. A. Noonburg (1995). *A separating surface for the Painlevé differential equation $x'' = x^2 - t$.* J. Math. Anal. Appl. **193**(3), pp. 817–831.

N. E. Nörlund (1922). *Mémoire sur les polynomes de Bernoulli.* Acta Math. **43**, pp. 121–196.

N. E. Nörlund (1924). *Vorlesungen über Differenzenrechnung.* Berlin: Springer-Verlag. Reprinted 1954 by Chelsea Publishing Company, New York.

N. E. Nörlund (1955). *Hypergeometric functions.* Acta Math. **94**, pp. 289–349.

L. N. Nosova and S. A. Tumarkin (1965). *Tables of Generalized Airy Functions for the Asymptotic Solution of the Differential Equations $\epsilon(py')' + (q + \epsilon r)y = f$.* Oxford: Pergamon Press. D. E. Brown, translator.

M. Noumi (2004). *Painlevé Equations through Symmetry*, Volume 223 of *Translations of Mathematical Monographs.* Providence, RI: American Mathematical Society. Translated from the 2000 Japanese original by the author.

M. Noumi and Y. Yamada (1998). *Affine Weyl groups, discrete dynamical systems and Painlevé equations.* Comm. Math. Phys. **199**(2), pp. 281–295.

M. Noumi and Y. Yamada (1999). *Symmetries in the fourth Painlevé equation and Okamoto polynomials.* Nagoya Math. J. **153**, pp. 53–86.

V. Y. Novokshënov (1985). *The asymptotic behavior of the general real solution of the third Painlevé equation.* Dokl. Akad. Nauk SSSR **283**(5), pp. 1161–1165. In Russian. English translation: Sov. Math., Dokl. **30**(1988), no. 8, pp. 666–668.

V. Y. Novokshënov (1990). *The Boutroux ansatz for the second Painlevé equation in the complex domain.* Izv. Akad. Nauk SSSR Ser. Mat. **54**(6), pp. 1229–1251. In Russian. English translation: Math. USSR-Izv. **37**(1991), no. 3, pp. 587–609.

H. M. Nussenzveig (1965). *High-frequency scattering by an impenetrable sphere.* Ann. Physics **34**(1), pp. 23–95.

J. F. Nye (1999). *Natural Focusing and Fine Structure of Light: Caustics and Wave Dislocations.* Bristol: Institute of Physics Publishing.

J. F. Nye (2006). *Dislocation lines in the hyperbolic umbilic diffraction catastrophe.* Proc. Roy. Soc. Lond. Ser. A **462**, pp. 2299–2313.

J. F. Nye (2007). *Dislocation lines in the swallowtail diffraction catastrophe.* Proc. Roy. Soc. Lond. Ser. A **463**, pp. 343–355.

F. Oberhettinger (1972). *Tables of Bessel Transforms.* Berlin-New York: Springer-Verlag. Table errata: Math. Comp. v. 65 (1996), no. 215, p. 1386.

F. Oberhettinger (1973). *Fourier Expansions. A Collection of Formulas.* New York-London: Academic Press.

F. Oberhettinger (1974). *Tables of Mellin Transforms.* Berlin-New York: Springer-Verlag.

F. Oberhettinger (1990). *Tables of Fourier Transforms and Fourier Transforms of Distributions.* Berlin: Springer-Verlag.

F. Oberhettinger and L. Badii (1973). *Tables of Laplace Transforms.* Berlin-New York: Springer-Verlag. Table errata: Math. Comp. v. 50 (1988), no. 182, pp. 653–654.

F. Oberhettinger and T. P. Higgins (1961). *Tables of Lebedev, Mehler and Generalized Mehler Transforms.* Mathematical Note 246, Boeing Scientific Research Lab, Seattle.

A. M. Odlyzko (1987). *On the distribution of spacings between zeros of the zeta function.* Math. Comp. **48**(177), pp. 273–308.

A. M. Odlyzko (1995). *Asymptotic Enumeration Methods.* In L. Lovász, R. L. Graham, and M. Grötschel (Eds.), *Handbook of Combinatorics, Vol. 2*, pp. 1063–1229. Amsterdam: Elsevier.

O. M. Ogreid and P. Osland (1998). *Summing one- and two-dimensional series related to the Euler series.* J. Comput. Appl. Math. **98**(2), pp. 245–271.

K. Okamoto (1981). *On the τ-function of the Painlevé equations.* Phys. D **2**(3), pp. 525–535.

K. Okamoto (1986). *Studies on the Painlevé equations. III. Second and fourth Painlevé equations, P_{II} and P_{IV}.* Math. Ann. **275**(2), pp. 221–255.

K. Okamoto (1987a). *Studies on the Painlevé equations. I. Sixth Painlevé equation P_{VI}.* Ann. Mat. Pura Appl. (4) **146**, pp. 337–381.

K. Okamoto (1987b). *Studies on the Painlevé equations. II. Fifth Painlevé equation P_V.* Japan. J. Math. (N.S.) **13**(1), pp. 47–76.

K. Okamoto (1987c). *Studies on the Painlevé equations. IV. Third Painlevé equation P_{III}.* Funkcial. Ekvac. **30**(2-3), pp. 305–332.

S. Okui (1974). *Complete elliptic integrals resulting from infinite integrals of Bessel functions.* J. Res. Nat. Bur. Standards Sect. B **78B**(3), pp. 113–135.

S. Okui (1975). *Complete elliptic integrals resulting from infinite integrals of Bessel functions. II.* J. Res. Nat. Bur. Standards Sect. B **79B**(3-4), pp. 137–170.

A. B. Olde Daalhuis (1994). *Asymptotic expansions for q-gamma, q-exponential, and q-Bessel functions.* J. Math. Anal. Appl. **186**(3), pp. 896–913.

A. B. Olde Daalhuis (1995). *Hyperasymptotic solutions of second-order linear differential equations. II.* Methods Appl. Anal. **2**(2), pp. 198–211.

A. B. Olde Daalhuis (1996). *Hyperterminants. I.* J. Comput. Appl. Math. **76**(1-2), pp. 255–264.

A. B. Olde Daalhuis (1998a). *Hyperasymptotic solutions of higher order linear differential equations with a singularity of rank one.* Proc. Roy. Soc. London Ser. A **454**, pp. 1–29.

A. B. Olde Daalhuis (1998b). *Hyperterminants. II.* J. Comput. Appl. Math. **89**(1), pp. 87–95

A. B. Olde Daalhuis (1998c). *On the resurgence properties of the uniform asymptotic expansion of the incomplete gamma function.* Methods Appl. Anal. **5**(4), pp. 425–438.

A. B. Olde Daalhuis (2000). *On the asymptotics for late coefficients in uniform asymptotic expansions of integrals with coalescing saddles.* Methods Appl. Anal. **7**(4), pp. 727–745.

A. B. Olde Daalhuis (2003a). *Uniform asymptotic expansions for hypergeometric functions with large parameters. I.* Analysis and Applications (Singapore) **1**(1), pp. 111–120.

A. B. Olde Daalhuis (2003b). *Uniform asymptotic expansions for hypergeometric functions with large parameters. II.* Analysis and Applications (Singapore) **1**(1), pp. 121–128.

A. B. Olde Daalhuis (2004a). *Inverse factorial-series solutions of difference equations.* Proc. Edinb. Math. Soc. (2) **47**(2), pp. 421–448.

A. B. Olde Daalhuis (2004b). *On higher-order Stokes phenomena of an inhomogeneous linear ordinary differential equation.* J. Comput. Appl. Math. **169**(1), pp. 235–246.

A. B. Olde Daalhuis (2005a). *Hyperasymptotics for nonlinear ODEs. I. A Riccati equation.* Proc. R. Soc. Lond. Ser. A Math. Phys. Eng. Sci. **461**(2060), pp. 2503–2520.

A. B. Olde Daalhuis (2005b). *Hyperasymptotics for nonlinear ODEs. II. The first Painlevé equation and a second-order Riccati equation.* Proc. R. Soc. Lond. Ser. A Math. Phys. Eng. Sci. **461**(2062), pp. 3005–3021.

A. B. Olde Daalhuis (2010). *Uniform asymptotic expansions for hypergeometric functions with large parameters. III.* Analysis and Applications (Singapore). In press.

A. B. Olde Daalhuis and F. W. J. Olver (1994). *Exponentially improved asymptotic solutions of ordinary differential equations. II Irregular singularities of rank one.* Proc. Roy. Soc. London Ser. A **445**, pp. 39–56.

A. B. Olde Daalhuis and F. W. J. Olver (1995a). *Hyperasymptotic solutions of second-order linear differential equations. I.* Methods Appl. Anal. **2**(2), pp. 173–197.

A. B. Olde Daalhuis and F. W. J. Olver (1995b). *On the calculation of Stokes multipliers for linear differential equations of the second order.* Methods Appl. Anal. **2**(3), pp. 348–367.

A. B. Olde Daalhuis and F. W. J. Olver (1998). *On the asymptotic and numerical solution of linear ordinary differential equations.* SIAM Rev. **40**(3), pp. 463–495.

A. B. Olde Daalhuis and N. M. Temme (1994). *Uniform Airy-type expansions of integrals.* SIAM J. Math. Anal. **25**(2), pp. 304–321.

M. N. Olevskiĭ (1950). *Triorthogonal systems in spaces of constant curvature in which the equation $\Delta_2 u + \lambda u = 0$ allows a complete separation of variables.* Mat. Sbornik N.S. **27(69)**(3), pp. 379–426. In Russian.

T. Oliveira e Silva (2006). *Computing $\pi(x)$: The combinatorial method.* Revista do DETUA **4**(6), pp. 759–768.

J. Oliver (1977). *An error analysis of the modified Clenshaw method for evaluating Chebyshev and Fourier series.* J. Inst. Math. Appl. **20**(3), pp. 379–391.

I. Olkin (1959). *A class of integral identities with matrix argument.* Duke Math. J. **26**(2), pp. 207–213.

F. W. J. Olver (1950). *A new method for the evaluation of zeros of Bessel functions and of other solutions of second-order differential equations.* Proc. Cambridge Philos. Soc. **46**(4), pp. 570–580.

F. W. J. Olver (1951). *A further method for the evaluation of zeros of Bessel functions and some new asymptotic expansions for zeros of functions of large order.* Proc. Cambridge Philos. Soc. **47**, pp. 699–712.

F. W. J. Olver (1952). *Some new asymptotic expansions for Bessel functions of large orders.* Proc. Cambridge Philos. Soc. **48**(3), pp. 414–427.

F. W. J. Olver (1954). *The asymptotic expansion of Bessel functions of large order.* Philos. Trans. Roy. Soc. London. Ser. A. **247**, pp. 328–368.

F. W. J. Olver (1959). *Uniform asymptotic expansions for Weber parabolic cylinder functions of large orders.* J. Res. Nat. Bur. Standards Sect. B **63B**, pp. 131–169.

F. W. J. Olver (Ed.) (1960). *Bessel Functions. Part III: Zeros and Associated Values.* Royal Society Mathematical Tables, Volume 7. Cambridge-New York: Cambridge University Press.

F. W. J. Olver (1962). *Tables for Bessel Functions of Moderate or Large Orders.* National Physical Laboratory Mathematical Tables, Vol. 6. Department of Scientific and Industrial Research. London: Her Majesty's Stationery Office.

F. W. J. Olver (1964a). *Error analysis of Miller's recurrence algorithm.* Math. Comp. **18**(85), pp. 65–74.

F. W. J. Olver (1964b). *Error bounds for asymptotic expansions in turning-point problems.* J. Soc. Indust. Appl. Math. **12**(1), pp. 200–214.

F. W. J. Olver (1965). *On the asymptotic solution of second-order differential equations having an irregular singularity of rank one, with an application to Whittaker functions.* J. Soc. Indust. Appl. Math. Ser. B Numer. Anal. **2**(2), pp. 225–243.

F. W. J. Olver (1967a). *Numerical solution of second-order linear difference equations.* J. Res. Nat. Bur. Standards Sect. B **71B**, pp. 111–129.

F. W. J. Olver (1967b). *Bounds for the solutions of second-order linear difference equations.* J. Res. Nat. Bur. Standards Sect. B **71B**(4), pp. 161–166.

F. W. J. Olver (1970). *A paradox in asymptotics.* SIAM J. Math. Anal. **1**(4), pp. 533–534.

F. W. J. Olver (1974). *Error bounds for stationary phase approximations.* SIAM J. Math. Anal. **5**(1), pp. 19–29.

F. W. J. Olver (1975a). *Second-order linear differential equations with two turning points.* Philos. Trans. Roy. Soc. London Ser. A **278**, pp. 137–174.

F. W. J. Olver (1975b). *Legendre functions with both parameters large.* Philos. Trans. Roy. Soc. London Ser. A **278**, pp. 175–185.

F. W. J. Olver (1976). *Improved error bounds for second-order differential equations with two turning points.* J. Res. Nat. Bur. Standards Sect. B **80B**(4), pp. 437–440.

F. W. J. Olver (1977a). *Connection formulas for second-order differential equations with multiple turning points.* SIAM J. Math. Anal. **8**(1), pp. 127–154.

F. W. J. Olver (1977b). *Connection formulas for second-order differential equations having an arbitrary number of turning points of arbitrary multiplicities.* SIAM J. Math. Anal. **8**(4), pp. 673–700.

F. W. J. Olver (1977c). *Second-order differential equations with fractional transition points.* Trans. Amer. Math. Soc. **226**, pp. 227–241.

F. W. J. Olver (1978). *General connection formulae for Liouville-Green approximations in the complex plane.* Philos. Trans. Roy. Soc. London Ser. A **289**, pp. 501–548.

F. W. J. Olver (1980a). *Asymptotic approximations and error bounds.* SIAM Rev. **22**(2), pp. 188–203.

F. W. J. Olver (1980b). *Whittaker functions with both parameters large: Uniform approximations in terms of parabolic cylinder functions.* Proc. Roy. Soc. Edinburgh Sect. A **86**(3-4), pp. 213–234.

F. W. J. Olver (1983). *Error Analysis of Complex Arithmetic.* In H. Werner, L. Wuytack, E. Ng, and H. J. Bünger (Eds.), *Computational Aspects of Complex Analysis (Braunlage, 1982)*, Volume 102 of *NATO Adv. Sci. Inst. Ser. C: Math. Phys. Sci.*, pp. 279–292. Dordrecht-Boston, MA: D. Reidel.

F. W. J. Olver (1991a). *Uniform, exponentially improved, asymptotic expansions for the generalized exponential integral.* SIAM J. Math. Anal. **22**(5), pp. 1460–1474.

F. W. J. Olver (1991b). *Uniform, exponentially improved, asymptotic expansions for the confluent hypergeometric function and other integral transforms.* SIAM J. Math. Anal. **22**(5), pp. 1475–1489.

F. W. J. Olver (1993a). *Exponentially-improved asymptotic solutions of ordinary differential equations I: The confluent hypergeometric function.* SIAM J. Math. Anal. **24**(3), pp. 756–767.

F. W. J. Olver (1994a). *Asymptotic expansions of the coefficients in asymptotic series solutions of linear differential equations.* Methods Appl. Anal. **1**(1), pp. 1–13.

F. W. J. Olver (1994b). *The Generalized Exponential Integral.* In R. V. M. Zahar (Ed.), *Approximation and Computation (West Lafayette, IN, 1993)*, Volume 119 of *International Series of Numerical Mathematics*, pp. 497–510. Boston, MA: Birkhäuser Boston.

F. W. J. Olver (1995). *On an asymptotic expansion of a ratio of gamma functions.* Proc. Roy. Irish Acad. Sect. A **95**(1), pp. 5–9.

F. W. J. Olver (1997a). *Asymptotic solutions of linear ordinary differential equations at an irregular singularity of rank unity.* Methods Appl. Anal. **4**(4), pp. 375–403.

F. W. J. Olver (1997b). *Asymptotics and Special Functions.* Wellesley, MA: A. K. Peters. Reprint, with corrections, of original Academic Press edition, 1974.

F. W. J. Olver (1999). *On the uniqueness of asymptotic solutions of linear differential equations.* Methods Appl. Anal. **6**(2), pp. 165–174.

F. W. J. Olver and J. M. Smith (1983). *Associated Legendre functions on the cut.* J. Comput. Phys. **51**(3), pp. 502–518.

F. W. J. Olver and D. J. Sookne (1972). *Note on backward recurrence algorithms.* Math. Comp. **26**(120), pp. 941–947.

F. W. J. Olver and F. Stenger (1965). *Error bounds for asymptotic solutions of second-order differential equations having an irregular singularity of arbitrary rank.* J. Soc. Indust. Appl. Math. Ser. B Numer. Anal. **2**(2), pp. 244–249.

P. J. Olver (1993b). *Applications of Lie Groups to Differential Equations* (2nd ed.), Volume 107 of *Graduate Texts in Mathematics*. New York: Springer-Verlag.

M. K. Ong (1986). *A closed form solution of the s-wave Bethe-Goldstone equation with an infinite repulsive core.* J. Math. Phys. **27**(4), pp. 1154–1158.

K. Ono (2000). *Distribution of the partition function modulo m.* Ann. of Math. (2) **151**(1), pp. 293–307.

G. E. Ordóñez and D. J. Driebe (1996). *Spectral decomposition of tent maps using symmetry considerations.* J. Statist. Phys. **84**(1-2), pp. 269–276.

J. M. Ortega and W. C. Rheinboldt (1970). *Iterative Solution of Nonlinear Equations in Several Variables.* New York: Academic Press. Reprinted in 2000 by the Society for Industrial and Applied Mathematics (SIAM), Philadelphia, Pennsylvania.

C. Osácar, J. Palacián, and M. Palacios (1995). *Numerical evaluation of the dilogarithm of complex argument.* Celestial Mech. Dynam. Astronom. **62**(1), pp. 93–98.

A. M. Ostrowski (1973). *Solution of Equations in Euclidean and Banach Spaces*, Volume 9 of *Pure and Applied Mathematics*. New York-London: Academic Press. Third edition of *Solution of equations and systems of equations*.

R. H. Ott (1985). *Scattering by a parabolic cylinder—a uniform asymptotic expansion.* J. Math. Phys. **26**(4), pp. 854–860.

M. L. Overton (2001). *Numerical Computing with IEEE Floating Point Arithmetic.* Philadelphia, PA: Society for Industrial and Applied Mathematics (SIAM). Including one theorem, one rule of thumb, and one hundred and one exercises.

V. I. Pagurova (1961). *Tables of the Exponential Integral $E_\nu(x) = \int_1^\infty e^{-xu} u^{-\nu} du$.* New York: Pergamon Press.

V. I. Pagurova (1963). *Tablitsy nepolnoi gamma-funktsii.* Moscow: Vyčisl. Centr Akad. Nauk SSSR. In Russian. Translated title: *Tables of the Incomplete Gamma Function*.

V. I. Pagurova (1965). *An asymptotic formula for the incomplete gamma function.* Ž. Vyčisl. Mat. i Mat. Fiz. **5**, pp. 118–121. English translation in U.S.S.R. Computational Math. and Math. Phys. **5**(1), pp. 162–166.

P. Painlevé (1906). *Sur les équations différentielles du second ordre à points critiques fixès.* C.R. Acad. Sc. Paris **143**, pp. 1111–1117.

E. Pairman (1919). *Tables of Digamma and Trigamma Functions.* In K. Pearson (Ed.), *Tracts for Computers, No. 1*. Cambridge Univ. Press.

B. V. Pal'tsev (1999). *On two-sided estimates, uniform with respect to the real argument and index, for modified Bessel functions.* Mat. Zametki **65**(5), pp. 681–692. English translation in Math. Notes 65 (1999) pp. 571-581.

D. J. Panow (1955). *Formelsammlung zur numerischen Behandlung partieller Differentialgleichungen nach dem Differenzenverfahren.* Berlin: Akademie-Verlag.

A. Papoulis (1977). *Signal Analysis.* New York: McGraw-Hill.

R. B. Paris (1984). *An inequality for the Bessel function $J_\nu(\nu x)$.* SIAM J. Math. Anal. **15**(1), pp. 203–205.

R. B. Paris (1991). *The asymptotic behaviour of Pearcey's integral for complex variables.* Proc. Roy. Soc. London Ser. A **432**, pp. 391–426.

R. B. Paris (1992a). *Smoothing of the Stokes phenomenon for high-order differential equations.* Proc. Roy. Soc. London Ser. A **436**, pp. 165–186.

R. B. Paris (1992b). *Smoothing of the Stokes phenomenon using Mellin-Barnes integrals.* J. Comput. Appl. Math. **41**(1-2), pp. 117–133.

R. B. Paris (2001a). *On the use of Hadamard expansions in hyperasymptotic evaluation. I. Real variables.* Proc. Roy. Soc. London Ser. A **457**(2016), pp. 2835–2853.

R. B. Paris (2001b). *On the use of Hadamard expansions in hyperasymptotic evaluation. II. Complex variables.* Proc. Roy. Soc. London Ser. A **457**, pp. 2855–2869.

R. B. Paris (2002a). *Error bounds for the uniform asymptotic expansion of the incomplete gamma function.* J. Comput. Appl. Math. **147**(1), pp. 215–231.

R. B. Paris (2002b). *A uniform asymptotic expansion for the incomplete gamma function.* J. Comput. Appl. Math. **148**(2), pp. 323–339.

R. B. Paris (2002c). *Exponential asymptotics of the Mittag-Leffler function.* Proc. Roy. Soc. London Ser. A **458**, pp. 3041–3052.

R. B. Paris (2003). *The asymptotic expansion of a generalised incomplete gamma function.* J. Comput. Appl. Math. **151**(2), pp. 297–306.

R. B. Paris (2004). *Exactification of the method of steepest descents: The Bessel functions of large order and argument.* Proc. Roy. Soc. London Ser. A **460**, pp. 2737–2759.

R. B. Paris (2005a). *A Kummer-type transformation for a $_2F_2$ hypergeometric function.* J. Comput. Appl. Math. **173**(2), pp. 379–382.

R. B. Paris (2005b). *The Stokes phenomenon associated with the Hurwitz zeta function $\zeta(s,a)$.* Proc. Roy. Soc. London Ser. A **461**, pp. 297–304.

R. B. Paris and S. Cang (1997). *An asymptotic representation for $\zeta(\tfrac{1}{2}+it)$.* Methods Appl. Anal. **4**(4), pp. 449–470.

R. B. Paris and D. Kaminski (2001). *Asymptotics and Mellin-Barnes Integrals.* Cambridge: Cambridge University Press.

R. B. Paris and W. N.-C. Sy (1983). *Influence of equilibrium shear flow along the magnetic field on the resistive tearing instability.* Phys. Fluids **26**(10), pp. 2966–2975.

R. B. Paris and A. D. Wood (1995). *Stokes phenomenon demystified.* Bull. Inst. Math. Appl. **31**(1-2), pp. 21–28.

G. Parisi (1988). *Statistical Field Theory.* Reading, MA: Addison-Wesley.

A. M. Parkhurst and A. T. James (1974). *Zonal Polynomials of Order* 1 *Through* 12. In H. L. Harter and D. B. Owen (Eds.), *Selected Tables in Mathematical Statistics*, Volume 2, pp. 199–388. Providence, RI: Amer. Math. Soc.

J. B. Parkinson (1969). *Optical properties of layer antiferromagnets with* K_2NiF_4 *structure*. J. Phys. C: Solid State Physics **2**(11), pp. 2012–2021.

R. Parnes (1972). *Complex zeros of the modified Bessel function* $K_n(Z)$. Math. Comp. **26**(120), pp. 949–953.

P. I. Pastro (1985). *Orthogonal polynomials and some q-beta integrals of Ramanujan*. J. Math. Anal. Appl. **112**(2), pp. 517–540.

S. Paszkowski (1988). *Evaluation of Fermi-Dirac Integral*. In A. Cuyt (Ed.), *Nonlinear Numerical Methods and Rational Approximation (Wilrijk, 1987)*, Volume 43 of *Mathematics and Its Applications*, pp. 435–444. Dordrecht: Reidel.

S. Paszkowski (1991). *Evaluation of the Fermi-Dirac integral of half-integer order*. Zastos. Mat. **21**(2), pp. 289–301.

J. K. Patel and C. B. Read (1982). *Handbook of the Normal Distribution*, Volume 40 of *Statistics: Textbooks and Monographs*. New York: Marcel Dekker Inc. A revised and expanded 2nd edition was published in 1996.

J. Patera and P. Winternitz (1973). *A new basis for the representation of the rotation group. Lamé and Heun polynomials*. J. Mathematical Phys. **14**(8), pp. 1130–1139.

A. R. Paterson (1983). *A First Course in Fluid Dynamics*. Cambridge: Cambridge University Press.

F. A. Paxton and J. E. Rollin (1959, June). *Tables of the Incomplete Elliptic Integrals of the First and Third Kind*. Technical report, Curtiss-Wright Corp., Research Division, Quehanna, PA. Reviewed in Math. Comp. 14(1960)209-210.

T. Pearcey (1946). *The structure of an electromagnetic field in the neighbourhood of a cusp of a caustic*. Philos. Mag. (7) **37**, pp. 311–317.

K. Pearson (Ed.) (1965). *Tables of the Incomplete* Γ-*function*. Cambridge: Biometrika Office, Cambridge University Press.

K. Pearson (Ed.) (1968). *Tables of the Incomplete Beta-function* (2nd ed.). Cambridge: Published for the Biometrika Trustees at the Cambridge University Press.

T. G. Pedersen (2003). *Variational approach to excitons in carbon nanotubes*. Phys. Rev. B **67**(7), pp. (073401–1)–(073401–4).

M. D. Perlman and I. Olkin (1980). *Unbiasedness of invariant tests for MANOVA and other multivariate problems*. Ann. Statist. **8**(6), pp. 1326–1341.

G. Petiau (1955). *La Théorie des Fonctions de Bessel Exposée en vue de ses Applications à la Physique Mathématique*. Paris: Centre National de la Recherche Scientifique.

M. S. Petković and L. D. Petković (1998). *Complex Interval Arithmetic and its Applications*, Volume 105 of *Mathematical Research*. Berlin: Wiley-VCH Verlag Berlin GmbH.

M. Petkovšek, H. S. Wilf, and D. Zeilberger (1996). $A = B$. Wellesley, MA: A K Peters Ltd. With a separately available computer disk.

E. Petropoulou (2000). *Bounds for ratios of modified Bessel functions*. Integral Transform. Spec. Funct. **9**(4), pp. 293–298.

P. C. B. Phillips (1986). *The exact distribution of the Wald statistic*. Econometrica **54**(4), pp. 881–895.

B. Pichon (1989). *Numerical calculation of the generalized Fermi-Dirac integrals*. Comput. Phys. Comm. **55**(2), pp. 127–136.

R. Piessens (1984). *Chebyshev series approximations for the zeros of the Bessel functions*. J. Comput. Phys. **53**(1), pp. 188–192.

R. Piessens (1990). *On the computation of zeros and turning points of Bessel functions*. Bull. Soc. Math. Grèce (N.S.) **31**, pp. 117–122.

R. Piessens and S. Ahmed (1986). *Approximation for the turning points of Bessel functions*. J. Comput. Phys. **64**(1), pp. 253–257.

R. Piessens and M. Branders (1972). *Chebyshev polynomial expansions of the Riemann zeta function*. Math. Comp. **26**(120), pp. G1–G5.

R. Piessens and M. Branders (1983). *Modified Clenshaw-Curtis method for the computation of Bessel function integrals*. BIT **23**(3), pp. 370–381.

R. Piessens and M. Branders (1985). *A survey of numerical methods for the computation of Bessel function integrals*. Rend. Sem. Mat. Univ. Politec. Torino (Special Issue), pp. 249–265. International conference on special functions: theory and computation (Turin, 1984).

A. Pinkus and S. Zafrany (1997). *Fourier Series and Integral Transforms*. Cambridge: Cambridge University Press.

G. Pittaluga and L. Sacripante (1991). *Inequalities for the zeros of the Airy functions*. SIAM J. Math. Anal. **22**(1), pp. 260–267.

S. Pokorski (1987). *Gauge Field Theories*. Cambridge Monographs on Mathematical Physics. Cambridge: Cambridge University Press.

J. Polchinski (1998). *String Theory: An Introduction to the Bosonic String, Vol. I*. Cambridge Monographs on Mathematical Physics. Cambridge: Cambridge University Press.

G. Pólya (1937). *Kombinatorische Anzahlbestimmungen für Gruppen, Graphen und chemische Verbindungen*. Acta Mathematica **68**, pp. 145–254.

G. Pólya and R. C. Read (1987). *Combinatorial Enumeration of Groups, Graphs, and Chemical Compounds*. New York: Springer-Verlag. Pólya's contribution translated from the German by Dorothee Aeppli.

G. Pólya, R. E. Tarjan, and D. R. Woods (1983). *Notes on Introductory Combinatorics*, Volume 4 of *Progress in Computer Science*. Boston, MA: Birkhäuser Boston Inc.

A. Poquérusse and S. Alexiou (1999). *Fast analytic formulas for the modified Bessel functions of imaginary order for spectral line broadening calculations*. J. Quantit. Spec. and Rad. Trans. **62**(4), pp. 389–395.

S. Porubský (1998). *Voronoi type congruences for Bernoulli numbers*. In P. Engel and H. Syta (Eds.), *Voronoï's Impact on Modern Science. Book I*. Kyiv: Institute of Mathematics of the National Academy of Sciences of Ukraine.

T. Poston and I. Stewart (1978). *Catastrophe Theory and its Applications*. London: Pitman. Reprinted by Dover in 1996.

J. L. Powell (1947). *Recurrence formulas for Coulomb wave functions*. Physical Rev. (2) **72**(7), pp. 626–627.

M. J. D. Powell (1967). *On the maximum errors of polynomial approximations defined by interpolation and by least squares criteria*. Comput. J. **9**(4), pp. 404–407.

K. Prachar (1957). *Primzahlverteilung*, Volume 91 of *Die Grundlehren der mathematischen Wissenschaften*. Berlin-Göttingen-Heidelberg: Springer-Verlag. Reprinted in 1978.

S. Pratt (2007). *Comoving coordinate system for relativistic hydrodynamics*. Phy. Rev. C **75**, pp. (024907–1)–(024907–10).

T. Prellberg and A. L. Owczarek (1995). *Stacking models of vesicles and compact clusters*. J. Statist. Phys. **80**(3–4), pp. 755–779.

P. J. Prince (1975). *Algorithm 498: Airy functions using Chebyshev series approximations*. ACM Trans. Math. Software **1**(4), pp. 372–379.

M. H. Protter and C. B. Morrey, Jr. (1991). *A First Course in Real Analysis* (2nd ed.). Undergraduate Texts in Mathematics. New York: Springer-Verlag.

A. P. Prudnikov, Yu. A. Brychkov, and O. I. Marichev (1986a). *Integrals and Series: Elementary Functions, Vol. 1*. New York: Gordon & Breach Science Publishers. Translated from the Russian and with a preface by N. M. Queen. Table errata: Math. Comp. v. 66 (1997), no. 220, pp. 1765–1766, Math. Comp. v. 65 (1996), no. 215, pp. 1380–1381.

A. P. Prudnikov, Yu. A. Brychkov, and O. I. Marichev (1986b). *Integrals and Series: Special Functions, Vol. 2*. New York: Gordon & Breach Science Publishers. A second edition was published in 1988. Table errata: Math. Comp. v. 65 (1996), no. 215, pp. 1382–1383.

A. P. Prudnikov, Yu. A. Brychkov, and O. I. Marichev (1990). *Integrals and Series: More Special Functions, Vol. 3*. New York: Gordon and Breach Science Publishers. Translated from the Russian by G. G. Gould. Table erratum Math. Comp. v. 65 (1996), no. 215, p. 1384.

A. P. Prudnikov, Yu. A. Brychkov, and O. I. Marichev (1992a). *Integrals and Series: Direct Laplace Transforms, Vol. 4*. New York: Gordon and Breach Science Publishers. Table erratum: Math. Comp. v. 66 (1997), no. 220, p. 1766.

A. P. Prudnikov, Yu. A. Brychkov, and O. I. Marichev (1992b). *Integrals and Series: Inverse Laplace Transforms, Vol. 5*. New York: Gordon and Breach Science Publishers.

J. D. Pryce (1993). *Numerical Solution of Sturm-Liouville Problems*. Monographs on Numerical Analysis. New York: The Clarendon Press, Oxford University Press.

M. Puoskari (1988). *A method for computing Bessel function integrals*. J. Comput. Phys. **75**(2), pp. 334–344.

F. Qi and J.-Q. Mei (1999). *Some inequalities of the incomplete gamma and related functions*. Z. Anal. Anwendungen **18**(3), pp. 793–799.

H.-z. Qin and Y.-m. Lu (2008). *A note on an open problem about the first Painlevé equation*. Acta Math. Appl. Sin. Engl. Ser. **24**(2), pp. 203–210.

S.-L. Qiu and J.-M. Shen (1997). *On two problems concerning means*. J. Hangzhou Inst. Elec. Engrg. **17**, pp. 1–7. In Chinese.

S.-L. Qiu and M. K. Vamanamurthy (1996). *Sharp estimates for complete elliptic integrals*. SIAM J. Math. Anal. **27**(3), pp. 823–834.

W.-Y. Qiu and R. Wong (2000). *Uniform asymptotic expansions of a double integral: Coalescence of two stationary points*. Proc. Roy. Soc. London Ser. A **456**, pp. 407–431.

W.-Y. Qiu and R. Wong (2004). *Asymptotic expansion of the Krawtchouk polynomials and their zeros*. Comput. Methods Funct. Theory **4**(1), pp. 189–226.

C. K. Qu and R. Wong (1999). *"Best possible" upper and lower bounds for the zeros of the Bessel function $J_\nu(x)$*. Trans. Amer. Math. Soc. **351**(7), pp. 2833–2859.

H. Rademacher (1938). *On the partition function $p(n)$*. Proc. London Math. Soc. (2) **43**(4), pp. 241–254.

H. Rademacher (1973). *Topics in Analytic Number Theory*. New York: Springer-Verlag. Edited by E. Grosswald, J. Lehner and M. Newman, Die Grundlehren der mathematischen Wissenschaften, Band 169.

H. A. Ragheb, L. Shafai, and M. Hamid (1991). *Plane wave scattering by a conducting elliptic cylinder coated by a nonconfocal dielectric*. IEEE Trans. Antennas and Propagation **39**(2), pp. 218–223.

M. Rahman (2001). *The Associated Classical Orthogonal Polynomials*. In *Special Functions 2000: Current Perspective and Future Directions (Tempe, AZ)*, Volume 30 of *NATO Sci. Ser. II Math. Phys. Chem.*, pp. 255–279. Dordrecht: Kluwer Acad. Publ.

E. M. Rains (1998). *Normal limit theorems for symmetric random matrices*. Probab. Theory Related Fields **112**(3), pp. 411–423.

E. D. Rainville (1960). *Special Functions*. New York: The Macmillan Co.

A. Ralston (1965). *Rational Chebyshev approximation by Remes' algorithms*. Numer. Math. **7**(4), pp. 322–330.

S. Ramanujan (1921). *Congruence properties of partitions*. Math. Z. **9**(1-2), pp. 147–153.

S. Ramanujan (1927). *Some properties of Bernoulli's numbers (J. Indian Math. Soc. 3 (1911), 219–234.)*. In *Collected Papers*. Cambridge University Press.

S. Ramanujan (1962). *Collected Papers of Srinivasa Ramanujan*. New York: Chelsea Publishing Co. Reprinted with commentary by Bruce Berndt, AMS Chelsea Publishing Series, American Mathematical Society (2000).

Ju. M. Rappoport (1979). *Tablitsy modifitsirovannykh funktsii Besselya $K_{\frac{1}{2}+i\beta}(x)$*. Moscow: "Nauka". In Russian. Translated title: *Tables of Modified Bessel Functions $K_{\frac{1}{2}+i\beta}(x)$*.

H. E. Rauch and A. Lebowitz (1973). *Elliptic Functions, Theta Functions, and Riemann Surfaces*. Baltimore, MD: The Williams & Wilkins Co.

J. Raynal (1979). *On the definition and properties of generalized 6-j symbols*. J. Math. Phys. **20**(12), pp. 2398–2415.

M. Razaz and J. L. Schonfelder (1980). *High precision Chebyshev expansions for Airy functions and their derivatives*. Technical report, University of Birmingham Computer Centre.

M. Razaz and J. L. Schonfelder (1981). *Remark on Algorithm 498: Airy functions using Chebyshev series approximations*. ACM Trans. Math. Software **7**(3), pp. 404–405.

I. S. Reed, D. W. Tufts, X. Yu, T. K. Truong, M. T. Shih, and X. Yin (1990). *Fourier analysis and signal processing by use of the Möbius inversion formula*. IEEE Trans. Acoustics, Speech, Signal Processing **38**, pp. 458–470.

W. H. Reid (1972). *Composite approximations to the solutions of the Orr-Sommerfeld equation*. Studies in Appl. Math. **51**, pp. 341–368.

W. H. Reid (1974a). *Uniform asymptotic approximations to the solutions of the Orr-Sommerfeld equation. I. Plane Couette flow*. Studies in Appl. Math. **53**, pp. 91–110.

W. H. Reid (1974b). *Uniform asymptotic approximations to the solutions of the Orr-Sommerfeld equation. II. The general theory*. Studies in Appl. Math. **53**, pp. 217–224.

W. H. Reid (1995). *Integral representations for products of Airy functions*. Z. Angew. Math. Phys. **46**(2), pp. 159–170.

W. H. Reid (1997a). *Integral representations for products of Airy functions. II. Cubic products*. Z. Angew. Math. Phys. **48**(4), pp. 646–655.

W. H. Reid (1997b). *Integral representations for products of Airy functions. III. Quartic products*. Z. Angew. Math. Phys. **48**(4), pp. 656–664.

K.-D. Reinsch and W. Raab (2000). *Elliptic Integrals of the First and Second Kind – Comparison of Bulirsch's and Carlson's Algorithms for Numerical Calculation*. In C. Dunkl, M. Ismail, and R. Wong (Eds.), *Special Functions (Hong Kong, 1999)*, pp. 293–308. River Edge, NJ-Singapore 912805: World Sci. Publishing.

F. E. Relton (1965). *Applied Bessel Functions*. New York: Dover Publications Inc. Reprint of original Blackie & Son Limited, London, 1946.

G. F. Remenets (1973). *Computation of Hankel (Bessel) functions of complex index and argument by numerical integration of a Schläfli contour integral*. Ž. Vyčisl. Mat. i Mat. Fiz. **13**, pp. 1415–1424, 1636. English translation in U.S.S.R. Computational Math. and Math. Phys. **13**(6), pp. 58–67.

E. Y. Remez (1957). *General Computation Methods of Chebyshev Approximation. The Problems with Linear Real Parameters*. Kiev: Publishing House of the Academy of Science of the Ukrainian SSR. English translation, AEC-TR-4491, United States Atomic Energy Commission.

P. Ribenboim (1979). *13 Lectures on Fermat's Last Theorem*. New York: Springer-Verlag.

S. O. Rice (1954). *Diffraction of plane radio waves by a parabolic cylinder. Calculation of shadows behind hills*. Bell System Tech. J. **33**, pp. 417–504.

D. St. P. Richards (Ed.) (1992). *Hypergeometric Functions on Domains of Positivity, Jack Polynomials, and Applications*, Volume 138 of *Contemporary Mathematics*, Providence, RI. American Mathematical Society.

D. St. P. Richards (2004). *Total positivity properties of generalized hypergeometric functions of matrix argument*. J. Statist. Phys. **116**(1-4), pp. 907–922.

È. Y. Riekstyņs̆ (1991). *Asymptotics and Bounds of the Roots of Equations (Russian)*. Riga: Zinatne.

B. Riemann (1851). *Grundlagen für eine allgemeine Theorie der Functionen einer veränderlichen complexen Grösse.* Inauguraldissertation, Göttingen.

B. Riemann (1859). *Über die Anzahl der Primzahlen unter einer gegebenen Grösse.* Monats. Berlin Akad. November 1859, pp. 671–680. Reprinted in Gesammelte Mathematische Werke, pp. 145–155, published by Dover Publications, New York, N.Y. 1953.

B. Riemann (1899). *Elliptische Functionen.* Leipzig: Teubner. Based on notes of lectures by B. Riemann edited and published by H. Stahl. Reprinted by UMI Books on Demand (Ann Arbor, MI, USA).

J. Riordan (1958). *An Introduction to Combinatorial Analysis.* New York: John Wiley & Sons Inc. Reprinted by Princeton University Press, Princeton, NJ, 1980 and reissued by Dover Publications Inc., New York, 2002.

J. Riordan (1979). *Combinatorial Identities.* Huntington, NY: Robert E. Krieger Publishing Co. Reprint of the 1968 original with corrections.

S. Ritter (1998). *On the computation of Lamé functions, of eigenvalues and eigenfunctions of some potential operators.* Z. Angew. Math. Mech. **78**(1), pp. 66–72.

T. J. Rivlin (1969). *An Introduction to the Approximation of Functions.* Waltham, MA-Toronto-London: Blaisdell Publishing Co. (Ginn and Co.). Corrected reprint by Dover Publications, Inc., New York, 1981.

L. Robin (1957). *Fonctions sphériques de Legendre et fonctions sphéroïdales. Tome I.* Paris: Gauthier-Villars. Préface de H. Villat.

L. Robin (1958). *Fonctions sphériques de Legendre et fonctions sphéroïdales. Tome II.* Paris: Gauthier-Villars.

L. Robin (1959). *Fonctions sphériques de Legendre et fonctions sphéroïdales. Tome III.* Collection Technique et Scientifique du C. N. E. T. Gauthier-Villars, Paris.

H. P. Robinson (1972). *Roots of $\tan x = x$.* 10 typewritten pages deposited in the UMT file at Lawrence Berkeley Laboratory, University of California, Berkeley, CA.

M. Robnik (1980). *An extremum property of the n-dimensional sphere.* J. Phys. A **13**(10), pp. L349–L351.

D. L. Rod and B. D. Sleeman (1995). *Complexity in spatio-temporal dynamics.* Proc. Roy. Soc. Edinburgh Sect. A **125**(5), pp. 959–974.

M. D. Rogers (2005). *Partial fractions expansions and identities for products of Bessel functions.* J. Math. Phys. **46**(4), pp. 043509–1–043509–18.

A. Ronveaux (Ed.) (1995). *Heun's Differential Equations.* New York: The Clarendon Press Oxford University Press.

C. C. J. Roothaan and S.-T. Lai (1997). *Calculation of $3n$-j symbols by Labarthe's method.* International Journal of Quantum Chemistry **63**(1), pp. 57–64.

G. M. Roper (1951). *Some Applications of the Lamé Function Solutions of the Linearised Supersonic Flow Equations.* Technical Reports and Memoranda 2865, Aeronautical Research Council (Great Britain). Reprinted by Her Majesty's Stationery Office, London, 1962.

R. R. Rosales (1978). *The similarity solution for the Korteweg-de Vries equation and the related Painlevé transcendent.* Proc. Roy. Soc. London Ser. A **361**, pp. 265–275.

K. H. Rosen (2004). *Elementary Number Theory and its Applications* (5th ed.). Reading, MA: Addison-Wesley.

K. H. Rosen, J. G. Michaels, J. L. Gross, J. W. Grossman, and D. R. Shier (Eds.) (2000). *Handbook of Discrete and Combinatorial Mathematics.* Boca Raton, FL: CRC Press.

P. A. Rosenberg and L. P. McNamee (1976). *Precision controlled trigonometric algorithms.* Appl. Math. Comput. **2**(4), pp. 335–352.

H. Rosengren (1999). *Another proof of the triple sum formula for Wigner $9j$-symbols.* J. Math. Phys. **40**(12), pp. 6689–6691.

H. Rosengren (2004). *Elliptic hypergeometric series on root systems.* Adv. Math. **181**(2), pp. 417–447.

J. B. Rosser (1939). *The n-th prime is greater than $n \log n$.* Proceedings of the London Mathematical Society **45**, pp. 21–44.

G.-C. Rota (1964). *On the foundations of combinatorial theory. I. Theory of Möbius functions.* Z. Wahrscheinlichkeitstheorie und Verw. Gebiete **2**, pp. 340–368.

M. Rotenberg, R. Bivins, N. Metropolis, and J. K. Wooten, Jr. (1959). *The 3-j and 6-j Symbols.* Cambridge, MA: The Technology Press, MIT.

M. Rothman (1954a). *The problem of an infinite plate under an inclined loading, with tables of the integrals of $\mathrm{Ai}(\pm x)$ and $\mathrm{Bi}(\pm x)$.* Quart. J. Mech. Appl. Math. **7**(1), pp. 1–7.

M. Rothman (1954b). *Tables of the integrals and differential coefficients of $\mathrm{Gi}(+x)$ and $\mathrm{Hi}(-x)$.* Quart. J. Mech. Appl. Math. **7**(3), pp. 379–384.

K. Rottbrand (2000). *Finite-sum rules for Macdonald's functions and Hankel's symbols.* Integral Transform. Spec. Funct. **10**(2), pp. 115–124.

D. H. Rouvray (1995). *Combinatorics in Chemistry.* In R. L. Graham, M. Grötschel, and L. Lovász (Eds.), *Handbook of Combinatorics, Vol. 2*, pp. 1955–1981. Amsterdam: Elsevier.

W. Rudin (1973). *Functional Analysis.* New York: McGraw-Hill Book Co. McGraw-Hill Series in Higher Mathematics.

W. Rudin (1976). *Principles of Mathematical Analysis* (3rd ed.). New York: McGraw-Hill Book Co. International Series in Pure and Applied Mathematics.

H.-J. Runckel (1971). *On the zeros of the hypergeometric function*. Math. Ann. **191**(1), pp. 53–58.

J. Rushchitsky and S. Rushchitska (2000). *On Simple Waves with Profiles in the form of some Special Functions—Chebyshev-Hermite, Mathieu, Whittaker—in Two-phase Media*. In *Differential Operators and Related Topics, Vol. I (Odessa, 1997)*, Volume 117 of *Operator Theory: Advances and Applications*, pp. 313–322. Basel-Berlin-Boston: Birkhäuser.

A. Russell (1909). *The effective resistance and inductance of a concentric main, and methods of computing the ber and bei and allied functions*. Philos. Mag. (6) **17**, pp. 524–552.

H. Rutishauser (1957). *Der Quotienten-Differenzen-Algorithmus*. Mitteilungen aus dem Institut für Angewandte Mathematik an der Eidgenössischen Technischen Hochschule in Zürich, No. 7. Basel/Stuttgart: Birkhäuser.

L. Saalschütz (1893). *Vorlesungen über die Bernoullischen Zahlen, ihren Zusammenhang mit den Secanten-Coefficienten und ihre wichtigeren Anwendungen*. Berlin: Springer-Verlag.

A. Sachse (1882). *Über die Darstellung der Bernoullischen und Eulerschen Zahlen durch Determinanten*. Archiv für Mathematik und Physik **68**, pp. 427–432.

B. E. Sagan (2001). *The Symmetric Group: Representations, Combinatorial Algorithms, and Symmetric Functions* (2nd ed.), Volume 203 of *Graduate Texts in Mathematics*. New York: Springer-Verlag. First edition published by Wadsworth & Brooks/Cole Advanced Books & Software, 1991.

R. P. Sagar (1991a). *A Gaussian quadrature for the calculation of generalized Fermi-Dirac integrals*. Comput. Phys. Comm. **66**(2-3), pp. 271–275.

R. P. Sagar (1991b). *On the evaluation of the Fermi-Dirac integrals*. Astrophys. J. **376**(1, part 1), pp. 364–366.

H. Sakai (2001). *Rational surfaces associated with affine root systems and geometry of the Painlevé equations*. Comm. Math. Phys. **220**(1), pp. 165–229.

K. L. Sala (1989). *Transformations of the Jacobian amplitude function and its calculation via the arithmetic-geometric mean*. SIAM J. Math. Anal. **20**(6), pp. 1514–1528.

L. Z. Salchev and V. B. Popov (1976). *A property of the zeros of cross-product Bessel functions of different orders*. Z. Angew. Math. Mech. **56**(2), pp. 120–121.

H. E. Salzer (1955). *Orthogonal polynomials arising in the numerical evaluation of inverse Laplace transforms*. Math. Tables Aids Comput. **9**(52), pp. 164–177.

P. Sarnak (1999). *Quantum Chaos, Symmetry and Zeta Functions. Lecture I, Quantum Chaos*. In R. Bott (Ed.), *Current Developments in Mathematics, 1997 (Cambridge, MA)*, pp. 127–144. Boston, MA: International Press.

T. Schäfer and E. V. Shuryak (1998). *Instantons in QCD*. Rev. Modern Phys. **70**(2), pp. 323–425.

F. W. Schäfke (1960). *Reihenentwicklungen analytischer Funktionen nach Biorthogonalsystemen spezieller Funktionen. I*. Math. Z. **74**, pp. 436–470.

F. W. Schäfke (1961a). *Ein Verfahren zur Berechnung des charakteristischen Exponenten der Mathieuschen Differentialgleichung I*. Numer. Math. **3**(1), pp. 30–38.

F. W. Schäfke (1961b). *Reihenentwicklungen analytischer Funktionen nach Biorthogonalsystemen spezieller Funktionen. II*. Math. Z. **75**, pp. 154–191.

F. W. Schäfke (1983). *Über einige Integrale mit Produkten von Mathieu-Funktionen*. Arch. Math. (Basel) **41**(2), pp. 152–162.

F. W. Schäfke and A. Finsterer (1990). *On Lindelöf's error bound for Stirling's series*. J. Reine Angew. Math. **404**, pp. 135–139.

F. W. Schäfke and H. Groh (1962). *Zur Berechnung der Eigenwerte der Sphäroiddifferentialgleichung*. Numer. Math. **4**, pp. 310–312.

C. W. Schelin (1983). *Calculator function approximation*. Amer. Math. Monthly **90**(5), pp. 317–325.

J. L. Schiff (1999). *The Laplace Transform: Theory and Applications*. Undergraduate Texts in Mathematics. New York: Springer-Verlag.

T. Schmelzer and L. N. Trefethen (2007). *Computing the gamma function using contour integrals and rational approximations*. SIAM J. Numer. Anal. **45**(2), pp. 558–571.

D. Schmidt (1979). *Die Lösung der linearen Differentialgleichung 2. Ordnung um zwei einfache Singularitäten durch Reihen nach hypergeometrischen Funktionen*. J. Reine Angew. Math. **309**, pp. 127–148.

D. Schmidt and G. Wolf (1979). *A method of generating integral relations by the simultaneous separability of generalized Schrödinger equations*. SIAM J. Math. Anal. **10**(4), pp. 823–838.

I. J. Schoenberg (1971). *Norm inequalities for a certain class of C^∞ functions*. Israel J. Math. **10**, pp. 364–372.

I. J. Schoenberg (1973). *Cardinal Spline Interpolation*. Philadelphia, PA: Society for Industrial and Applied Mathematics. Conference Board of the Mathematical Sciences Regional Conference Series in Applied Mathematics, No. 12.

L. Schoenfeld (1976). *Sharper bounds for the Chebyshev functions $\theta(x)$ and $\psi(x)$. II.* Math. Comp. **30**(134), pp. 337–360. For corrigendum see same journal v. 30 (1976), no. 136, p. 900.

J. L. Schonfelder (1978). *Chebyshev expansions for the error and related functions.* Math. Comp. **32**(144), pp. 1232–1240.

J. L. Schonfelder (1980). *Very high accuracy Chebyshev expansions for the basic trigonometric functions.* Math. Comp. **34**(149), pp. 237–244.

F. Schottky (1903). *Über die Moduln der Thetafunctionen.* Acta Math. **27**(1), pp. 235–288.

M. R. Schroeder (2006). *Number Theory in Science and Communication: With Applications in Cryptography, Physics, Digital Information, Computing, and Self-Similarity* (4th ed.). Berlin: Springer-Verlag. Volume 7 in Springer Series in Information Sciences.

L. S. Schulman (1981). *Techniques and Applications of Path Integration.* New York: John Wiley & Sons Inc. Reprinted by Dover in 2005.

K. Schulten and R. G. Gordon (1975a). *Exact recursive evaluation of 3j- and 6j-coefficients for quantum-mechanical coupling of angular momenta.* J. Mathematical Phys. **16**(10), pp. 1961–1970.

K. Schulten and R. G. Gordon (1975b). *Semiclassical approximations to 3j- and 6j-coefficients for quantum-mechanical coupling of angular momenta.* J. Mathematical Phys. **16**(10), pp. 1971–1988.

Z. Schulten, D. G. M. Anderson, and R. G. Gordon (1979). *An algorithm for the evaluation of the complex Airy functions.* J. Comput. Phys. **31**(1), pp. 60–75.

L. L. Schumaker (1981). *Spline Functions: Basic Theory.* New York: John Wiley & Sons Inc. Pure and Applied Mathematics, A Wiley-Interscience Publication. Reprinted by Robert E. Krieger Publishing Co. Inc., 1993.

R. Schürer (2004). *Adaptive Quasi-Monte Carlo Integration Based on MISER and VEGAS.* In *Monte Carlo and Quasi-Monte Carlo Methods 2002*, pp. 393–406. Berlin: Springer.

I. J. Schwatt (1962). *An Introduction to the Operations with Series* (2nd ed.). New York: Chelsea Publishing Co. This is a reprint, with corrections, of the first edition [University of Pennsylvania Press, Philadelphia, Pa., 1924].

R. S. Scorer (1950). *Numerical evaluation of integrals of the form $I = \int_{x_1}^{x_2} f(x) e^{i\phi(x)} dx$ and the tabulation of the function $\text{Gi}(z) = (1/\pi) \int_0^\infty \sin(uz + \frac{1}{3}u^3) du$.* Quart. J. Mech. Appl. Math. **3**(1), pp. 107–112.

J. B. Seaborn (1991). *Hypergeometric Functions and Their Applications*, Volume 8 of *Texts in Applied Mathematics*. New York: Springer-Verlag.

M. J. Seaton (1982). *Coulomb functions analytic in the energy.* Comput. Phys. Comm. **25**(1), pp. 87–95. Double-precision Fortran code.

M. J. Seaton (1983). *Quantum defect theory.* Rep. Prog. Phys. **46**(2), pp. 167–257.

M. J. Seaton (1984). *The accuracy of iterated JWBK approximations for Coulomb radial functions.* Comput. Phys. Comm. **32**(2), pp. 115–119.

M. J. Seaton (2002). *Coulomb functions for attractive and repulsive potentials and for positive and negative energies.* Comput. Phys. Comm. **146**(2), pp. 225–249.

M. J. Seaton and G. Peach (1962). *The determination of phases of wave functions.* Proc. Phys. Soc. **79**(6), pp. 1296–1297.

J. D. Secada (1999). *Numerical evaluation of the Hankel transform.* Comput. Phys. Comm. **116**(2-3), pp. 278–294.

H. Segur and M. J. Ablowitz (1981). *Asymptotic solutions of nonlinear evolution equations and a Painlevé transcendent.* Phys. D **3**(1-2), pp. 165–184.

J. Segura (1998). *A global Newton method for the zeros of cylinder functions.* Numer. Algorithms **18**(3-4), pp. 259–276.

J. Segura (2001). *Bounds on differences of adjacent zeros of Bessel functions and iterative relations between consecutive zeros.* Math. Comp. **70**(235), pp. 1205–1220.

J. Segura (2002). *The zeros of special functions from a fixed point method.* SIAM J. Numer. Anal. **40**(1), pp. 114–133.

J. Segura and A. Gil (1999). *Evaluation of associated Legendre functions off the cut and parabolic cylinder functions.* Electron. Trans. Numer. Anal. **9**, pp. 137–146. Orthogonal polynomials: numerical and symbolic algorithms (Leganés, 1998).

N. Seiberg and D. Shih (2005). *Flux vacua and branes of the minimal superstring.* J. High Energy Phys. Electronic. E-print number: hep-th/0412315. 38 pp.

R. G. Selfridge and J. E. Maxfield (1958). *A Table of the Incomplete Elliptic Integral of the Third Kind.* New York: Dover Publications Inc.

J.-P. Serre (1973). *A Course in Arithmetic*, Volume 7 of *Graduate Texts in Mathematics*. New York: Springer-Verlag. Translated from the French.

R. Shail (1978). *Lamé polynomial solutions to some elliptic crack and punch problems.* Internat. J. Engrg. Sci. **16**(8), pp. 551–563.

R. Shail (1980). *On integral representations for Lamé and other special functions.* SIAM J. Math. Anal. **11**(4), pp. 702–723.

H. Shanker (1939). *On the expansion of the parabolic cylinder function in a series of the product of two parabolic cylinder functions.* J. Indian Math. Soc. (N. S.) **3**, pp. 226–230.

H. Shanker (1940a). *On integral representation of Weber's parabolic cylinder function and its expansion into an infinite series.* J. Indian Math. Soc. (N. S.) **4**, pp. 34–38.

H. Shanker (1940b). *On certain integrals and expansions involving Weber's parabolic cylinder functions.* J. Indian Math. Soc. (N. S.) **4**, pp. 158–166.

H. Shanker (1940c). *On the expansion of the product of two parabolic cylinder functions of non integral order.* Proc. Benares Math. Soc. (N. S.) **2**, pp. 61–68.

D. Shanks (1955). *Non-linear transformations of divergent and slowly convergent sequences.* J. Math. Phys. **34**, pp. 1–42.

G. Shanmugam (1978). *Parabolic Cylinder Functions and their Application in Symmetric Two-centre Shell Model.* In *Proceedings of the Conference on Mathematical Analysis and its Applications (Inst. Engrs., Mysore, 1977)*, Volume 91 of *Matscience Rep.*, Aarhus, pp. P81–P89. Aarhus Univ.

J. Shao and P. Hänggi (1998). *Decoherent dynamics of a two-level system coupled to a sea of spins.* Phys. Rev. Lett. **81**(26), pp. 5710–5713.

O. A. Sharafeddin, H. F. Bowen, D. J. Kouri, and D. K. Hoffman (1992). *Numerical evaluation of spherical Bessel transforms via fast Fourier transforms.* J. Comput. Phys. **100**(2), pp. 294–296.

A. Sharples (1967). *Uniform asymptotic forms of modified Mathieu functions.* Quart. J. Mech. Appl. Math. **20**(3), pp. 365–380.

A. Sharples (1971). *Uniform asymptotic expansions of modified Mathieu functions.* J. Reine Angew. Math. **247**, pp. 1–17.

I. Shavitt (1963). *The Gaussian Function in Calculations of Statistical Mechanics and Quantum Mechanics.* In B. Alder, S. Fernbach, and M. Rotenberg (Eds.), *Methods in Computational Physics: Advances in Research and Applications*, Volume 2, pp. 1–45. New York: Academic Press.

I. Shavitt and M. Karplus (1965). *Gaussian-transform method for molecular integrals. I. Formulation for energy integrals.* J. Chem. Phys. **43**(2), pp. 398–414.

D. C. Shaw (1985). *Perturbational results for diffraction of water-waves by nearly-vertical barriers.* IMA J. Appl. Math. **34**(1), pp. 99–117.

N. T. Shawagfeh (1992). *The Laplace transforms of products of Airy functions.* Dirāsāt Ser. B Pure Appl. Sci. **19**(2), pp. 7–11.

L. Shen (1981). *The elliptical microstrip antenna with circular polarization.* IEEE Trans. Antennas and Propagation **29**(1), pp. 90–94.

L.-C. Shen (1998). *On an identity of Ramanujan based on the hypergeometric series $_2F_1\left(\frac{1}{3},\frac{2}{3};\frac{1}{2};x\right)$.* J. Number Theory **69**(2), pp. 125–134.

M. M. Shepherd and J. G. Laframboise (1981). *Chebyshev approximation of $(1+2x)\exp(x^2)\operatorname{erfc} x$ in $0 \le x < \infty$.* Math. Comp. **36**(153), pp. 249–253.

M. E. Sherry (1959). *The zeros and maxima of the Airy function and its first derivative to 25 significant figures.* Report afcrc-tr-59-135, astia document no. ad214568, Air Research and Development Command, U.S. Air Force, Bedford, MA. Available from U.S. Department of Commerce, Office of Technical Services, Washington, D.C.

G. Shimura (1982). *Confluent hypergeometric functions on tube domains.* Math. Ann. **260**(3), pp. 269–302.

T. Shiota (1986). *Characterization of Jacobian varieties in terms of soliton equations.* Invent. Math. **83**(2), pp. 333–382.

R. B. Shirts (1993). *The computation of eigenvalues and solutions of Mathieu's differential equation for noninteger order.* ACM Trans. Math. Software **19**(3), pp. 377–390.

P. N. Shivakumar and R. Wong (1988). *Error bounds for a uniform asymptotic expansion of the Legendre function $P_n^{-m}(\cosh z)$.* Quart. Appl. Math. **46**(3), pp. 473–488.

P. N. Shivakumar and J. Xue (1999). *On the double points of a Mathieu equation.* J. Comput. Appl. Math. **107**(1), pp. 111–125.

B. W. Shore and D. H. Menzel (1968). *Principles of Atomic Spectra.* New York: John Wiley & Sons Ltd.

A. Sidi (1997). *Computation of infinite integrals involving Bessel functions of arbitrary order by the $\overline{D}$-transformation.* J. Comput. Appl. Math. **78**(1), pp. 125–130.

A. Sidi (2003). *Practical Extrapolation Methods: Theory and Applications*, Volume 10 of *Cambridge Monographs on Applied and Computational Mathematics*. Cambridge: Cambridge University Press.

A. Sidi (2004). *Euler-Maclaurin expansions for integrals with endpoint singularities: A new perspective.* Numer. Math. **98**(2), pp. 371–387.

C. L. Siegel (1935). *Über die analytische Theorie der quadratischen Formen.* Ann. of Math. (2) **36**(3), pp. 527–606.

C. L. Siegel (1971). *Topics in Complex Function Theory. Vol. II: Automorphic Functions and Abelian Integrals.* Interscience Tracts in Pure and Applied Mathematics, No. 25. New York: Wiley-Interscience [John Wiley & Sons

Inc.]. Translated from the original German by A. Shenitzer and M. Tretkoff. Reprinted by John Wiley & Sons, New York, 1988.

C. L. Siegel (1973). *Topics in Complex Function Theory. Vol. III: Abelian Functions and Modular Functions of Several Variables*. Interscience Tracts in Pure and Applied Mathematics, No. 25. New York-London-Sydney: Wiley-Interscience, [John Wiley & Sons, Inc]. Translated from the original German by E. Gottschling and M. Tretkoff. Reprinted by John Wiley & Sons, New York, 1989.

C. L. Siegel (1988). *Topics in Complex Function Theory. Vol. I: Elliptic Functions and Uniformization Theory*. Wiley Classics Library. New York: John Wiley & Sons Inc. Translated from the German by A. Shenitzer and D. Solitar. Reprint of the 1969 edition, a Wiley-Interscience Publication.

K. M. Siegel (1953). *An inequality involving Bessel functions of argument nearly equal to their order*. Proc. Amer. Math. Soc. **4**(6), pp. 858–859.

K. M. Siegel and F. B. Sleator (1954). *Inequalities involving cylindrical functions of nearly equal argument and order*. Proc. Amer. Math. Soc. **5**(3), pp. 337–344.

C. E. Siewert and E. E. Burniston (1973). *Exact analytical solutions of $ze^z = a$*. J. Math. Anal. Appl. **43**(3), pp. 626–632.

J. H. Silverman and J. Tate (1992). *Rational Points on Elliptic Curves*. Undergraduate Texts in Mathematics. New York: Springer-Verlag.

R. A. Silverman (1967). *Introductory Complex Analysis*. Englewood Cliffs, N.J.: Prentice-Hall, Inc. Reprinted by Dover Publications Inc. 1972.

G. F. Simmons (1972). *Differential Equations with Applications and Historical Notes*. New York: McGraw-Hill Book Co. International Series in Pure and Applied Mathematics.

B. Simon (1982). *Large orders and summability of eigenvalue perturbation theory: A mathematical overview*. Int. J. Quantum Chem. **21**, pp. 3–25.

B. Simon (2005a). *Orthogonal Polynomials on the Unit Circle. Part 1: Classical Theory*, Volume 54 of *American Mathematical Society Colloquium Publications*. Providence, RI: American Mathematical Society.

B. Simon (2005b). *Orthogonal Polynomials on the Unit Circle. Part 2: Spectral Theory*, Volume 54 of *American Mathematical Society Colloquium Publications*. Providence, RI: American Mathematical Society.

R. Sips (1949). *Représentation asymptotique des fonctions de Mathieu et des fonctions d'onde sphéroidales*. Trans. Amer. Math. Soc. **66**(1), pp. 93–134.

R. Sips (1959). *Représentation asymptotique des fonctions de Mathieu et des fonctions sphéroidales. II*. Trans. Amer. Math. Soc. **90**(2), pp. 340–368.

R. Sips (1965). *Représentation asymptotique de la solution générale de l'équation de Mathieu-Hill*. Acad. Roy. Belg. Bull. Cl. Sci. (5) **51**(11), pp. 1415–1446.

R. Sips (1967). *Répartition du courant alternatif dans un conducteur cylindrique de section elliptique*. Acad. Roy. Belg. Bull. Cl. Sci. (5) **53**(8), pp. 861–878.

R. Sips (1970). *Quelques intégrales définies discontinues contenant des fonctions de Mathieu*. Acad. Roy. Belg. Bull. Cl. Sci. (5) **56**(5), pp. 475–491.

R. Sitaramachandrarao and B. Davis (1986). *Some identities involving the Riemann zeta function. II*. Indian J. Pure Appl. Math. **17**(10), pp. 1175–1186.

S. L. Skorokhodov (1985). *On the calculation of complex zeros of the modified Bessel function of the second kind*. Dokl. Akad. Nauk SSSR **280**(2), pp. 296–299. English translation in Soviet Math. Dokl. **31**(1), pp. 78–81.

H. Skovgaard (1954). *On inequalities of the Turán type*. Math. Scand. **2**, pp. 65–73.

H. Skovgaard (1966). *Uniform Asymptotic Expansions of Confluent Hypergeometric Functions and Whittaker Functions*, Volume 1965 of *Doctoral dissertation, University of Copenhagen*. Copenhagen: Jul. Gjellerups Forlag.

J. C. Slater (1942). *Microwave Transmission*. New York: McGraw-Hill Book Co. Reprinted by Dover Publications Inc., New York, 1960.

L. J. Slater (1960). *Confluent Hypergeometric Functions*. Cambridge-New York: Cambridge University Press. Table errata: Math. Comp. v. 30 (1976), no. 135, 677–678.

L. J. Slater (1966). *Generalized Hypergeometric Functions*. Cambridge: Cambridge University Press.

D. V. Slavić (1974). *Complements to asymptotic development of sine cosine integrals, and auxiliary functions*. Univ. Beograd. Publ. Elecktrotehn. Fak., Ser. Mat. Fiz. **461–497**, pp. 185–191.

I. Sh. Slavutskiĭ (1995). *Staudt and arithmetical properties of Bernoulli numbers*. Historia Sci. (2) **5**(1), pp. 69–74.

I. Sh. Slavutskiĭ (1999). *About von Staudt congruences for Bernoulli numbers*. Comment. Math. Univ. St. Paul. **48**(2), pp. 137–144.

I. Sh. Slavutskiĭ (2000). *On the generalized Bernoulli numbers that belong to unequal characters*. Rev. Mat. Iberoamericana **16**(3), pp. 459–475.

S. Yu. Slavyanov (1996). *Asymptotic Solutions of the One-dimensional Schrödinger Equation*. Providence, RI: American Mathematical Society. Translated from the 1990 Russian original by Vadim Khidekel.

S. Yu. Slavyanov and W. Lay (2000). *Special Functions: A Unified Theory Based on Singularities*. Oxford Mathematical Monographs. Oxford: Oxford University Press.

S. Yu. Slavyanov and N. A. Veshev (1997). *Structure of avoided crossings for eigenvalues related to equations of Heun's class*. J. Phys. A **30**(2), pp. 673–687.

B. D. Sleeman (1966a). *The expansion of Lamé functions into series of associated Legendre functions of the second kind*. Proc. Cambridge Philos. Soc. **62**, pp. 441–452.

B. D. Sleeman (1966b). *Some Boundary Value Problems Associated with the Heun Equation*. Ph. D. thesis, London University. Available from the library, University of Surrey, UK.

B. D. Sleeman (1968a). *Integral equations and relations for Lamé functions and ellipsoidal wave functions*. Proc. Cambridge Philos. Soc. **64**, pp. 113–126.

B. D. Sleeman (1968b). *On parabolic cylinder functions*. J. Inst. Math. Appl. **4**(1), pp. 106–112.

B. D. Sleeman (1969). *Non-linear integral equations for Heun functions*. Proc. Edinburgh Math. Soc. (2) **16**, pp. 281–289.

D. Slepian (1964). *Prolate spheroidal wave functions, Fourier analysis and uncertainity. IV. Extensions to many dimensions; generalized prolate spheroidal functions*. Bell System Tech. J. **43**, pp. 3009–3057.

D. Slepian (1983). *Some comments on Fourier analysis, uncertainty and modeling*. SIAM Rev. **25**(3), pp. 379–393.

D. Slepian and H. O. Pollak (1961). *Prolate spheroidal wave functions, Fourier analysis and uncertainty. I*. Bell System Tech. J. **40**, pp. 43–63.

N. J. A. Sloane (2003). *The On-Line Encyclopedia of Integer Sequences*. Notices Amer. Math. Soc. **50**(8), pp. 912–915.

W. M. Smart (1962). *Text-book on Spherical Astronomy*. Fifth edition. Cambridge: Cambridge University Press.

A. D. Smirnov (1960). *Tables of Airy Functions and Special Confluent Hypergeometric Functions*. New York: Pergamon Press. Translated from the Russian by D. G. Fry.

A. O. Smirnov (2002). *Elliptic Solitons and Heun's Equation*. In V. B. Kuznetsov (Ed.), *The Kowalevski Property (Leeds, UK, 2000)*, Volume 32 of *CRM Proc. Lecture Notes*, pp. 287–306. Providence, RI: Amer. Math. Soc.

V. I. Smirnov (1996). *Izbrannye Trudy. Analiticheskaya teoriya obyknovennykh differentsialnykh uravnenii*. St. Petersburg: Izdatel' stvo Sankt-Peterburgskogo Universiteta. In Russian. Translated title: *Selected Works. Analytic Theory of Ordinary Differential Equations*.

D. M. Smith (1989). *Efficient multiple-precision evaluation of elementary functions*. Math. Comp. **52**(185), pp. 131–134.

D. R. Smith (1986). *Liouville-Green approximations via the Riccati transformation*. J. Math. Anal. Appl. **116**(1), pp. 147–165.

D. R. Smith (1990). *A Riccati approach to the Airy equation*. In R. Wong (Ed.), *Asymptotic and computational analysis (Winnipeg, MB, 1989)*, pp. 403–415. New York: Marcel Dekker.

F. C. Smith (1939a). *On the logarithmic solutions of the generalized hypergeometric equation when $p = q+1$*. Bull. Amer. Math. Soc. **45**(8), pp. 629–636.

F. C. Smith (1939b). *Relations among the fundamental solutions of the generalized hypergeometric equation when $p = q + 1$. II. Logarithmic cases*. Bull. Amer. Math. Soc. **45**(12), pp. 927–935.

G. S. Smith (1997). *An Introduction to Classical Electromagnetic Radiation*. Cambridge-New York: Cambridge University Press.

I. N. Sneddon (1966). *Mixed Boundary Value Problems in Potential Theory*. Amsterdam: North-Holland Publishing Co.

I. N. Sneddon (1972). *The Use of Integral Transforms*. New York: McGraw-Hill.

C. Snow (1952). *Hypergeometric and Legendre Functions with Applications to Integral Equations of Potential Theory*. National Bureau of Standards Applied Mathematics Series, No. 19. Washington, D. C.: U. S. Government Printing Office.

W. V. Snyder (1993). *Algorithm 723: Fresnel integrals*. ACM Trans. Math. Software **19**(4), pp. 452–456. Single- and double-precision Fortran, maximum accuracy 16D. For remarks see same journal v. 22 (1996), p. 498-500 and v. 26 (2000), p. 617.

I. I. Sobelman (1992). *Atomic Spectra and Radiative Transitions* (2nd ed.). Berlin: Springer-Verlag. A second printing with corrections was published in 1996.

A. Sommerfeld (1928). *Atombau und Spektrallinien*. Braunschweig: Vieweg. English translation: *Wave Mechanics* published by Methuen, London, 1930.

K. Soni (1980). *Exact error terms in the asymptotic expansion of a class of integral transforms. I. Oscillatory kernels*. SIAM J. Math. Anal. **11**(5), pp. 828–841.

D. Sornette (1998). *Multiplicative processes and power laws*. Phys. Rev. E **57**(4), pp. 4811–4813.

J. Spanier and K. B. Oldham (1987). *An Atlas of Functions*. Washington: Hemisphere Pub. Corp.

G. W. Spenceley and R. M. Spenceley (1947). *Smithsonian Elliptic Functions Tables*. Smithsonian Miscellaneous Collections, v. 109 (Publication 3863). Washington, D. C.: The Smithsonian Institution.

R. Spigler (1980). *Some results on the zeros of cylindrical functions and of their derivatives*. Rend. Sem. Mat. Univ. Politec. Torino **38**(1), pp. 67–85.

R. Spigler and M. Vianello (1992). *Liouville-Green approximations for a class of linear oscillatory difference equations of the second order*. J. Comput. Appl. Math. **41**(1-2), pp. 105–116.

R. Spigler and M. Vianello (1997). *A Survey on the Liouville-Green (WKB) Approximation for Linear Difference Equations of the Second Order*. In S. Elaydi, I. Győri, and G. Ladas (Eds.), *Advances in Difference Equations (Veszprém, 1995)*, pp. 567–577. Amsterdam: Gordon and Breach.

R. Spigler, M. Vianello, and F. Locatelli (1999). *Liouville-Green-Olver approximations for complex difference equations*. J. Approx. Theory **96**(2), pp. 301–322.

R. Spira (1971). *Calculation of the gamma function by Stirling's formula*. Math. Comp. **25**(114), pp. 317–322.

V. P. Spiridonov (2002). *An elliptic incarnation of the Bailey chain*. Int. Math. Res. Not. (37), pp. 1945–1977.

G. Springer (1957). *Introduction to Riemann Surfaces*. Reading, Massachusetts: Addison-Wesley Publishing Company. A second edition was published in 1981 (Chelsea Publishing Co., New York, and AMS Chelsea Book Series, American Mathematical Society, Providence, Rhode Island).

K. Srinivasa Rao (1981). *Computation of angular momentum coefficients using sets of generalized hypergeometric functions*. Comput. Phys. Comm. **22**(2-3), pp. 297–302.

K. Srinivasa Rao and V. Rajeswari (1993). *Quantum Theory of Angular Momentum: Selected Topics*. Berlin: Springer-Verlag. Fortran programs comparing methods for the calculation of Clebsch–Gordan coefficients, and of $6j$ and $9j$ symbols, are given on pp. 266–292.

K. Srinivasa Rao, V. Rajeswari, and C. B. Chiu (1989). *A new Fortran program for the 9-j angular momentum coefficient*. Comput. Phys. Comm. **56**(2), pp. 231–248.

K. Srinivasa Rao and K. Venkatesh (1978). *New Fortran programs for angular momentum coefficients*. Comput. Phys. Comm. **15**(3-4), pp. 227–235.

H. M. Srivastava (1988). *Sums of certain series of the Riemann zeta function*. J. Math. Anal. Appl. **134**(1), pp. 129–140.

H. M. Srivastava and J. Choi (2001). *Series Associated with the Zeta and Related Functions*. Dordrecht: Kluwer Academic Publishers.

F. D. Stacey (1977). *Physics of the Earth* (2nd ed.). New York: John Wiley & Sons, Inc. Fourth edition published in 2008 by Cambridge University Press.

A. Stankiewicz (1968). *Tables of the integro-exponential functions*. Acta Astronom. **18**, pp. 289–311.

R. P. Stanley (1989). *Some combinatorial properties of Jack symmetric functions*. Adv. Math. **77**(1), pp. 76–115.

R. P. Stanley (1997). *Enumerative Combinatorics. Vol. 1*. Cambridge: Cambridge University Press. Corrected reprint of the 1986 original.

R. P. Stanley (1999). *Enumerative Combinatorics. Vol. 2*. Cambridge: Cambridge University Press.

I. A. Stegun and R. Zucker (1974). *Automatic computing methods for special functions. II. The exponential integral $E_n(x)$*. J. Res. Nat. Bur. Standards Sect. B **78B**, pp. 199–216. Double-precision Fortran, maximum accuracy 16S.

E. M. Stein and R. Shakarchi (2003). *Fourier Analysis: An Introduction*, Volume 1 of *Princeton Lectures in Analysis*. Oxford-Princeton, NJ: Princeton University Press.

J. Steinig (1970). *The real zeros of Struve's function*. SIAM J. Math. Anal. **1**(3), pp. 365–375.

J. Steinig (1972). *The sign of Lommel's function*. Trans. Amer. Math. Soc. **163**, pp. 123–129.

F. Stenger (1966a). *Error bounds for asymptotic solutions of differential equations. I. The distinct eigenvalue case*. J. Res. Nat. Bur. Standards Sect. B **70B**, pp. 167–186.

F. Stenger (1966b). *Error bounds for asymptotic solutions of differential equations. II. The general case*. J. Res. Nat. Bur. Standards Sect. B **70B**, pp. 187–210.

F. Stenger (1993). *Numerical Methods Based on Sinc and Analytic Functions*, Volume 20 of *Springer Series in Computational Mathematics*. New York: Springer-Verlag.

G. W. Stewart (2001). *Matrix Algorithms. Vol. 2: Eigensystems*. Philadelphia, PA: Society for Industrial and Applied Mathematics (SIAM).

A. N. Stokes (1980). *A stable quotient-difference algorithm*. Math. Comp. **34**(150), pp. 515–519.

J. A. Stratton, P. M. Morse, L. J. Chu, and R. A. Hutner (1941). *Elliptic Cylinder and Spheroidal Wave Functions, Including Tables of Separation Constants and Coefficients*. New York: John Wiley and Sons, Inc.

J. A. Stratton, P. M. Morse, L. J. Chu, J. D. C. Little, and F. J. Corbató (1956). *Spheroidal Wave Functions: Including Tables of Separation Constants and Coefficients*. New York: Technology Press of M. I. T. and John Wiley & Sons, Inc.

A. Strecok (1968). *On the calculation of the inverse of the error function*. Math. Comp. **22**(101), pp. 144–158.

R. S. Strichartz (1994). *A Guide to Distribution Theory and Fourier Transforms*. Studies in Advanced Mathematics. Boca Raton, FL: CRC Press.

A. H. Stroud (1971). *Approximate Calculation of Multiple Integrals*. Englewood Cliffs, N.J.: Prentice-Hall Inc.

A. H. Stroud and D. Secrest (1966). *Gaussian Quadrature Formulas*. Englewood Cliffs, NJ: Prentice-Hall Inc.

M. J. O. Strutt (1932). *Lamésche, Mathieusche, und verwandte Funktionen in Physik und Technik*. Berlin: Springer. Reprinted by J. W. Edwards, Ann Arbor, MI, 1944.

B. I. Suleĭmanov (1987). *The relation between asymptotic properties of the second Painlevé equation in different directions towards infinity*. Differ. Uravn. **23**(5), pp. 834–842. In Russian. English translation: Differential Equations **23**(1987), no. 5, pp. 569–576.

W. F. Sun (1996). *Uniform asymptotic expansions of Hermite polynomials*. M. Phil. thesis, City University of Hong Kong.

S. K. Suslov (2003). *An Introduction to Basic Fourier Series*, Volume 9 of *Developments in Mathematics*. Dordrecht: Kluwer Academic Publishers.

H. Suzuki, E. Takasugi, and H. Umetsu (1998). *Perturbations of Kerr-de Sitter black holes and Heun's equations*. Progr. Theoret. Phys. **100**(3), pp. 491–505.

N. Svartholm (1939). *Die Lösung der Fuchsschen Differentialgleichung zweiter Ordnung durch hypergeometrische Polynome*. Math. Ann. **116**(1), pp. 413–421.

C. A. Swanson and V. B. Headley (1967). *An extension of Airy's equation*. SIAM J. Appl. Math. **15**(6), pp. 1400–1412.

C. E. Synolakis (1988). *On the roots of $f(z) = J_0(z) - i\, J_1(z)$*. Quart. Appl. Math. **46**(1), pp. 105–107.

O. Szász (1950). *On the relative extrema of ultraspherical polynomials*. Boll. Un. Mat. Ital. (3) **5**, pp. 125–127.

O. Szász (1951). *On the relative extrema of the Hermite orthogonal functions*. J. Indian Math. Soc. (N.S.) **15**, pp. 129–134.

G. Szegö (1933). *Asymptotische Entwicklungen der Jacobischen Polynome*. Schr. der König. Gelehr. Gesell. Naturwiss. Kl. **10**, pp. 33–112.

G. Szegö (1948). *On an inequality of P. Turán concerning Legendre polynomials*. Bull. Amer. Math. Soc. **54**, pp. 401–405.

G. Szegö (1967). *Orthogonal Polynomials* (3rd ed.). New York: American Mathematical Society. American Mathematical Society Colloquium Publications, Vol. 23.

G. Szegö (1975). *Orthogonal Polynomials* (4th ed.), Volume XXIII of *Colloquium Publications*. Providence, RI: American Mathematical Society.

G. 't Hooft and M. Veltman (1979). *Scalar one-loop integrals*. Nuclear Phys. B **153**(3-4), pp. 365–401.

K. Takasaki (2001). *Painlevé-Calogero correspondence revisited*. J. Math. Phys. **42**(3), pp. 1443–1473.

Y. Takei (1995). *On the connection formula for the first Painlevé equation—from the viewpoint of the exact WKB analysis*. Sūrikaisekikenkyūsho Kōkyūroku (931), pp. 70–99. Painlevé functions and asymptotic analysis (Japanese) (Kyoto, 1995).

T. Takemasa, T. Tamura, and H. H. Wolter (1979). *Coulomb functions with complex angular momenta*. Comput. Phys. Comm. **17**(4), pp. 351–355. Single-precision Fortran code.

A. Takemura (1984). *Zonal Polynomials*. Institute of Mathematical Statistics Lecture Notes—Monograph Series, 4. Hayward, CA: Institute of Mathematical Statistics.

I. C. Tang (1969). *Some definite integrals and Fourier series for Jacobian elliptic functions*. Z. Angew. Math. Mech. **49**, pp. 95–96.

J. W. Tanner and S. S. Wagstaff, Jr. (1987). *New congruences for the Bernoulli numbers*. Math. Comp. **48**(177), pp. 341–350.

J. G. Taylor (1978). *Error bounds for the Liouville-Green approximation to initial-value problems*. Z. Angew. Math. Mech. **58**(12), pp. 529–537.

J. G. Taylor (1982). *Improved error bounds for the Liouville-Green (or WKB) approximation*. J. Math. Anal. Appl. **85**(1), pp. 79–89.

N. M. Temme (1975). *On the numerical evaluation of the modified Bessel function of the third kind*. J. Comput. Phys. **19**(3), pp. 324–337. Algol program.

N. M. Temme (1978). *Uniform asymptotic expansions of confluent hypergeometric functions*. J. Inst. Math. Appl. **22**(2), pp. 215–223.

N. M. Temme (1979a). *An algorithm with ALGOL 60 program for the computation of the zeros of ordinary Bessel functions and those of their derivatives*. J. Comput. Phys. **32**(2), pp. 270–279.

N. M. Temme (1979b). *The asymptotic expansion of the incomplete gamma functions*. SIAM J. Math. Anal. **10**(4), pp. 757–766.

N. M. Temme (1983). *The numerical computation of the confluent hypergeometric function $U(a,b,z)$*. Numer. Math. **41**(1), pp. 63–82. Algol 60 variable-precision procedures are included.

N. M. Temme (1985). *Laplace type integrals: Transformation to standard form and uniform asymptotic expansions*. Quart. Appl. Math. **43**(1), pp. 103–123.

N. M. Temme (1986). *Laguerre polynomials: Asymptotics for large degree*. Technical Report AM-R8610, CWI, Amsterdam, The Netherlands.

N. M. Temme (1987). *On the computation of the incomplete gamma functions for large values of the parameters*. In *Algorithms for approximation (Shrivenham, 1985)*, Volume 10 of *Inst. Math. Appl. Conf. Ser. New Ser.*, pp. 479–489. New York: Oxford Univ. Press.

N. M. Temme (1990a). *Asymptotic estimates for Laguerre polynomials*. Z. Angew. Math. Phys. **41**(1), pp. 114–126.

N. M. Temme (1990b). *Uniform asymptotic expansions of a class of integrals in terms of modified Bessel functions, with application to confluent hypergeometric functions*. SIAM J. Math. Anal. **21**(1), pp. 241–261.

N. M. Temme (1992a). *Asymptotic inversion of incomplete gamma functions*. Math. Comp. **58**(198), pp. 755–764.

N. M. Temme (1992b). *Asymptotic inversion of the incomplete beta function*. J. Comput. Appl. Math. **41**(1-2), pp. 145–157. Asymptotic methods in analysis and combinatorics.

N. M. Temme (1993). *Asymptotic estimates of Stirling numbers*. Stud. Appl. Math. **89**(3), pp. 233–243.

N. M. Temme (1994a). *Computational aspects of incomplete gamma functions with large complex parameters*. In R. V. M. Zahar (Ed.), *Approximation and Computation. A Festschrift in Honor of Walter Gautschi.*, Volume 119 of *International Series of Numerical Mathematics*, pp. 551–562. Boston, MA: Birkhäuser Boston.

N. M. Temme (1994b). *Steepest descent paths for integrals defining the modified Bessel functions of imaginary order*. Methods Appl. Anal. **1**(1), pp. 14–24.

N. M. Temme (1995a). *Asymptotics of zeros of incomplete gamma functions*. Ann. Numer. Math. **2**(1-4), pp. 415–423. Special functions (Torino, 1993).

N. M. Temme (1995b). *Bernoulli polynomials old and new: Generalizations and asymptotics*. CWI Quarterly **8**(1), pp. 47–66.

N. M. Temme (1995c). *Uniform asymptotic expansions of integrals: A selection of problems*. J. Comput. Appl. Math. **65**(1-3), pp. 395–417.

N. M. Temme (1996a). *Special Functions: An Introduction to the Classical Functions of Mathematical Physics*. New York: John Wiley & Sons Inc.

N. M. Temme (1996b). *Uniform asymptotics for the incomplete gamma functions starting from negative values of the parameters*. Methods Appl. Anal. **3**(3), pp. 335–344.

N. M. Temme (1997). *Numerical algorithms for uniform Airy-type asymptotic expansions*. Numer. Algorithms **15**(2), pp. 207–225.

N. M. Temme (2000). *Numerical and asymptotic aspects of parabolic cylinder functions*. J. Comput. Appl. Math. **121**(1-2), pp. 221–246. Numerical analysis in the 20th century, Vol. I, Approximation theory.

N. M. Temme (2003). *Large parameter cases of the Gauss hypergeometric function*. J. Comput. Appl. Math. **153**(1-2), pp. 441–462.

N. M. Temme and J. L. López (2001). *The Askey scheme for hypergeometric orthogonal polynomials viewed from asymptotic analysis*. J. Comput. Appl. Math. **133**(1-2), pp. 623–633.

N. M. Temme and A. B. Olde Daalhuis (1990). *Uniform asymptotic approximation of Fermi-Dirac integrals*. J. Comput. Appl. Math. **31**(3), pp. 383–387.

A. Terras (1988). *Harmonic Analysis on Symmetric Spaces and Applications. II*. Berlin: Springer-Verlag.

A. Terras (1999). *Fourier Analysis on Finite Groups and Applications*, Volume 43 of *London Mathematical Society Student Texts*. Cambridge: Cambridge University Press.

S. A. Teukolsky (1972). *Rotating black holes: Separable wave equations for gravitational and electromagnetic perturbations*. Phys. Rev. Lett. **29**(16), pp. 1114–1118.

I. J. Thompson (2004). *Erratum to "COULCC: A continued-fraction algorithm for Coulomb functions of complex order with complex arguments"*. Comput. Phys. Comm. **159**(3), pp. 241–242.

I. J. Thompson and A. R. Barnett (1985). *COULCC: A continued-fraction algorithm for Coulomb functions of complex order with complex arguments*. Comput. Phys. Comm. **36**(4), pp. 363–372. Double-precision Fortran, minimum accuracy: 14D. See also Thompson (2004).

I. J. Thompson and A. R. Barnett (1986). *Coulomb and Bessel functions of complex arguments and order*. J. Comput. Phys. **64**(2), pp. 490–509.

W. J. Thompson (1994). *Angular Momentum: An Illustrated Guide to Rotational Symmetries for Physical Systems*. A Wiley-Interscience Publication. New York: John Wiley & Sons Inc. With 1 Macintosh floppy disk (3.5 inch; DD). Programs for the calculation of $3j, 6j, 9j$ symbols are included.

W. J. Thompson (1997). *Atlas for Computing Mathematical Functions: An Illustrated Guide for Practitioners*. New York: John Wiley & Sons Inc. With CD-ROM containing a large collection of mathematical function software written in Fortran 90 and Mathematica (an edition with the same software in C and Mathematica exists also). The functions are computed for real variables only. Maximum accuracy 12D.

E. C. Titchmarsh (1962). *The Theory of Functions* (2nd ed.). Oxford: Oxford University Press.

E. C. Titchmarsh (1986a). *Introduction to the Theory of Fourier Integrals* (Third ed.). New York: Chelsea Publishing Co. Original edition was published in 1937 by Oxford University Press. Expanded bibliography and some improvements were made for the second and third editions.

E. C. Titchmarsh (1986b). *The Theory of the Riemann Zeta-Function* (2nd ed.). New York-Oxford: The Clarendon Press Oxford University Press. Edited and with a preface by D. R. Heath-Brown.

J. Todd (1954). *Evaluation of the exponential integral for large complex arguments*. J. Research Nat. Bur. Standards **52**, pp. 313–317.

J. Todd (1975). *The lemniscate constants*. Comm. ACM **18**(1), pp. 14–19. Collection of articles honoring Alston S. Householder. For corrigendum see same journal v. 18 (1975), no. 8, p. 462.

P. G. Todorov (1978). *Une nouvelle représentation explicite des nombres d'Euler*. C. R. Acad. Sci. Paris Sér. A-B **286**(19), pp. A807–A809.

P. G. Todorov (1984). *On the theory of the Bernoulli polynomials and numbers*. J. Math. Anal. Appl. **104**(2), pp. 309–350.

P. G. Todorov (1991). *Explicit formulas for the Bernoulli and Euler polynomials and numbers*. Abh. Math. Sem. Univ. Hamburg **61**, pp. 175–180.

O. I. Tolstikhin and M. Matsuzawa (2001). *Hyperspherical elliptic harmonics and their relation to the Heun equation*. Phys. Rev. A **63**(032510), pp. 1–8.

G. P. Tolstov (1962). *Fourier Series*. Englewood Cliffs, N.J.: Prentice-Hall Inc. Translated from the Russian by Richard A. Silverman. Second English edition published by Dover Publications Inc., New York, 1976.

R. F. Tooper and J. Mark (1968). *Simplified calculation of* $\text{Ei}(x)$ *for positive arguments, and a short table of* $\text{Shi}(x)$. Math. Comp. **22**(102), pp. 448–449.

G. Torres-Vega, J. D. Morales-Guzmán, and A. Zúñiga-Segundo (1998). *Special functions in phase space: Mathieu functions*. J. Phys. A **31**(31), pp. 6725–6739.

C. A. Tracy and H. Widom (1994). *Level-spacing distributions and the Airy kernel*. Comm. Math. Phys. **159**(1), pp. 151–174.

C. A. Tracy and H. Widom (1997). *On exact solutions to the cylindrical Poisson-Boltzmann equation with applications to polyelectrolytes*. Phys. A **244**(1-4), pp. 402–413.

J. F. Traub (1964). *Iterative Methods for the Solution of Equations*. Prentice-Hall Series in Automatic Computation. Englewood Cliffs, NJ: Prentice-Hall Inc.

L. N. Trefethen (2008). *Is Gauss quadrature better than Clenshaw-Curtis?* SIAM Rev. **50**(1), pp. 67–87.

L. N. Trefethen and D. Bau, III (1997). *Numerical Linear Algebra*. Philadelphia, PA: Society for Industrial and Applied Mathematics (SIAM).

C. L. Tretkoff and M. D. Tretkoff (1984). *Combinatorial Group Theory, Riemann Surfaces and Differential Equations*. In *Contributions to Group Theory*, Volume 33 of *Contemp. Math.*, pp. 467–519. Providence, RI: Amer. Math. Soc.

M. J. Tretter and G. W. Walster (1980). *Further comments on the computation of modified Bessel function ratios*. Math. Comp. **35**(151), pp. 937–939.

F. G. Tricomi (1947). *Sugli zeri delle funzioni di cui si conosce una rappresentazione asintotica*. Ann. Mat. Pura Appl. (4) **26**, pp. 283–300.

F. G. Tricomi (1949). *Sul comportamento asintotico dell'n-esimo polinomio di Laguerre nell'intorno dell'ascissa 4n*. Comment. Math. Helv. **22**, pp. 150–167.

F. G. Tricomi (1950a). *Über die Abzählung der Nullstellen der konfluenten hypergeometrischen Funktionen*. Math. Z. **52**, pp. 669–675.

F. G. Tricomi (1950b). *Asymptotische Eigenschaften der unvollständigen Gammafunktion*. Math. Z. **53**, pp. 136–148.

F. G. Tricomi (1951). *Funzioni Ellittiche* (2nd ed.). Bologna: Nicola Zanichelli Editore. German edition (M. Krafft, Ed.) Akad. Verlag., Leipzig, 1948.

F. G. Tricomi (1954). *Funzioni ipergeometriche confluenti*. Roma: Edizioni Cremonese.

H. Triebel (1965). *Über die Lamésche Differentialgleichung*. Math. Nachr. **30**, pp. 137–154.

H. Trinkaus and F. Drepper (1977). *On the analysis of diffraction catastrophes*. J. Phys. A **10**, pp. L11–L16.

B. A. Troesch and H. R. Troesch (1973). *Eigenfrequencies of an elliptic membrane*. Math. Comp. **27**(124), pp. 755–765.

C. Truesdell (1945). *On a function which occurs in the theory of the structure of polymers*. Ann. of Math. (2) **46**, pp. 144–157.

C. Truesdell (1948). *An Essay Toward a Unified Theory of Special Functions*. Annals of Mathematics Studies, no. 18. Princeton, N. J.: Princeton University Press.

P.-H. Tseng and T.-C. Lee (1998). *Numerical evaluation of exponential integral: Theis well function approximation*. Journal of Hydrology **205**(1-2), pp. 38–51.

E. O. Tuck (1964). *Some methods for flows past blunt slender bodies*. J. Fluid Mech. **18**, pp. 619–635.

A. Tucker (2006). *Applied Combinatorics* (5th ed.). New York: John Wiley and Sons.

S. A. Tumarkin (1959). *Asymptotic solution of a linear nonhomogeneous second order differential equation with a transition point and its application to the computations of toroidal shells and propeller blades*. J. Appl. Math. Mech. **23**, pp. 1549–1565.

A. A. Tuẑilin (1971). *Theory of the Fresnel integral.* USSR Comput. Math. and Math. Phys. **9**(4), pp. 271–279. Translation of the paper in Zh. Vychisl. Mat. Mat. Fiz., Vol. 9, No. 4, 938-944, 1969.

H. Umemura (2000). *On the transformation group of the second Painlevé equation.* Nagoya Math. J. **157**, pp. 15–46.

H. Umemura and H. Watanabe (1998). *Solutions of the third Painlevé equation. I.* Nagoya Math. J. **151**, pp. 1–24.

Unpublished Mathematical Tables (1944). *Mathematics of Computation Unpublished Mathematical Tables Collection.* Archives of American Mathematics, Center for American History, The University of Texas at Austin. Covers 1944–1994. An online guide is arranged chronologically by date of issuance within *Mathematics of Computation*.

J. Urbanowicz (1988). *On the equation $f(1)1^k + f(2)2^k + \cdots + f(x)x^k + R(x) = By^2$.* Acta Arith. **51**(4), pp. 349–368.

F. Ursell (1960). *On Kelvin's ship-wave pattern.* J. Fluid Mech. **8**(3), pp. 418–431.

F. Ursell (1972). *Integrals with a large parameter. Several nearly coincident saddle-points.* Proc. Cambridge Philos. Soc. **72**, pp. 49–65.

F. Ursell (1980). *Integrals with a large parameter: A double complex integral with four nearly coincident saddle-points.* Math. Proc. Cambridge Philos. Soc. **87**(2), pp. 249–273.

F. Ursell (1984). *Integrals with a large parameter: Legendre functions of large degree and fixed order.* Math. Proc. Cambridge Philos. Soc. **95**(2), pp. 367–380.

F. Ursell (1994). *Ship Hydrodynamics, Water Waves and Asymptotics*, Volume 2 of *Collected works of F. Ursell, 1946-1992*. Singapore: World Scientific.

K. M. Urwin (1964). *Integral equations for paraboloidal wave functions. I.* Quart. J. Math. Oxford Ser. (2) **15**, pp. 309–315.

K. M. Urwin (1965). *Integral equations for the paraboloidal wave functions. II.* Quart. J. Math. Oxford Ser. (2) **16**, pp. 257–262.

K. M. Urwin and F. M. Arscott (1970). *Theory of the Whittaker-Hill equation.* Proc. Roy. Soc. Edinburgh Sect. A **69**, pp. 28–44.

J. V. Uspensky and M. A. Heaslet (1939). *Elementary Number Theory.* New York: McGraw-Hill Book Company, Inc.

T. Uzer, J. T. Muckerman, and M. S. Child (1983). *Collisions and umbilic catastrophes. The hyperbolic umbilic canonical diffraction integral.* Molecular Phys. **50**(6), pp. 1215–1230.

G. Valent (1986). *An integral transform involving Heun functions and a related eigenvalue problem.* SIAM J. Math. Anal. **17**(3), pp. 688–703.

O. Vallée and M. Soares (2004). *Airy Functions and Applications to Physics.* London-Singapore: Imperial College Press. Distributed by World Scientific, Singapore.

A. L. Van Buren, R. V. Baier, S. Hanish, and B. J. King (1972). *Calculation of spheroidal wave functions.* J. Acoust. Soc. Amer. **51**, pp. 414–416.

A. L. Van Buren and J. E. Boisvert (2002). *Accurate calculation of prolate spheroidal radial functions of the first kind and their first derivatives.* Quart. Appl. Math. **60**(3), pp. 589–599.

A. L. Van Buren, B. J. King, R. V. Baier, and S. Hanish (1975). *Tables of Angular Spheroidal Wave Functions, Vol. 1, Prolate, $m = 0$; Vol. 2, Oblate, $m=0$.* Washington, D.C.: Naval Res. Lab. Reports.

H. C. van de Hulst (1957). *Light Scattering by Small Particles.* New York: John Wiley and Sons. Inc.

H. C. van de Hulst (1980). *Multiple Light Scattering*, Volume 1. New York: Academic Press.

J. van de Lune, H. J. J. te Riele, and D. T. Winter (1986). *On the zeros of the Riemann zeta function in the critical strip. IV.* Math. Comp. **46**(174), pp. 667–681.

H. Van de Vel (1969). *On the series expansion method for computing incomplete elliptic integrals of the first and second kinds.* Math. Comp. **23**(105), pp. 61–69.

C. G. van der Laan and N. M. Temme (1984). *Calculation of Special Functions: The Gamma Function, the Exponential Integrals and Error-Like Functions*, Volume 10 of *CWI Tract*. Amsterdam: Stichting Mathematisch Centrum, Centrum voor Wiskunde en Informatica.

A. J. van der Poorten (1980). *Some Wonderful Formulas ... an Introduction to Polylogarithms.* In R. Ribenboim (Ed.), *Proceedings of the Queen's Number Theory Conference, 1979 (Kingston, Ont., 1979)*, Volume 54 of *Queen's Papers in Pure and Appl. Math.*, Kingston, Ont., pp. 269–286. Queen's University.

B. L. van der Waerden (1951). *On the method of saddle points.* Appl. Sci. Research B. **2**, pp. 33–45.

J. F. Van Diejen and V. P. Spiridonov (2001). *Modular hypergeometric residue sums of elliptic Selberg integrals.* Lett. Math. Phys. **58**(3), pp. 223–238.

C. Van Loan (1992). *Computational Frameworks for the Fast Fourier Transform*, Volume 10 of *Frontiers in Applied Mathematics*. Philadelphia, PA: Society for Industrial and Applied Mathematics (SIAM).

B. Ph. van Milligen and A. López Fraguas (1994). *Expansion of vacuum magnetic fields in toroidal harmonics.* Comput. Phys. Comm. **81**(1-2), pp. 74–90.

A. van Wijngaarden (1953). *On the coefficients of the modular invariant $J(\tau)$*. Nederl. Akad. Wetensch. Proc. Ser. A. **56** = Indagationes Math. **15 56**, pp. 389–400.

A. N. Varčenko (1976). *Newton polyhedra and estimates of oscillatory integrals*. Funkcional. Anal. i Priložen. **10**(3), pp. 13–38. English translation in Functional Anal. Appl. 18 (1976), pp. 175-196.

R. S. Varma (1941). *An infinite series of Weber's parabolic cylinder functions*. Proc. Benares Math. Soc. (N.S.) **3**, p. 37.

D. A. Varshalovich, A. N. Moskalev, and V. K. Khersonskiĭ (1988). *Quantum Theory of Angular Momentum*. Singapore: World Scientific Publishing Co. Inc.

A. N. Vavreck and W. Thompson, Jr. (1984). *Some novel infinite series of spherical Bessel functions*. Quart. Appl. Math. **42**(3), pp. 321–324.

G. Vedeler (1950). *A Mathieu equation for ships rolling among waves. I, II*. Norske Vid. Selsk. Forh., Trondheim **22**(25–26), pp. 113–123.

R. Vein and P. Dale (1999). *Determinants and Their Applications in Mathematical Physics*, Volume 134 of *Applied Mathematical Sciences*. New York: Springer-Verlag.

G. Veneziano (1968). *Construction of a crossing-symmetric, Regge-behaved amplitude for linearly rising trajectories*. Il Nuovo Cimento A **57**(1), pp. 190–197.

P. Verbeeck (1970). *Rational approximations for exponential integrals $E_n(x)$*. Acad. Roy. Belg. Bull. Cl. Sci. (5) **56**, pp. 1064–1072.

R. Vidūnas (2005). *Transformations of some Gauss hypergeometric functions*. J. Comput. Appl. Math. **178**(1-2), pp. 473–487.

R. Vidūnas and N. M. Temme (2002). *Symbolic evaluation of coefficients in Airy-type asymptotic expansions*. J. Math. Anal. Appl. **269**(1), pp. 317–331.

L. Vietoris (1983). *Dritter Beweis der die unvollständige Gammafunktion betreffenden Lochsschen Ungleichungen*. Österreich. Akad. Wiss. Math.-Natur. Kl. Sitzungsber. II **192**(1-3), pp. 83–91.

N. Ja. Vilenkin (1968). *Special Functions and the Theory of Group Representations*. Providence, RI: American Mathematical Society.

N. Ja. Vilenkin and A. U. Klimyk (1991). *Representation of Lie Groups and Special Functions. Volume 1: Simplest Lie Groups, Special Functions and Integral Transforms*, Volume 72 of *Mathematics and its Applications (Soviet Series)*. Dordrecht: Kluwer Academic Publishers Group. Translated from the Russian by V. A. Groza and A. A. Groza.

N. Ja. Vilenkin and A. U. Klimyk (1992). *Representation of Lie Groups and Special Functions. Volume 3: Classical and Quantum Groups and Special Functions*, Volume 75 of *Mathematics and its Applications (Soviet Series)*. Dordrecht: Kluwer Academic Publishers Group. Translated from the Russian by V. A. Groza and A. A. Groza.

N. Ja. Vilenkin and A. U. Klimyk (1993). *Representation of Lie Groups and Special Functions. Volume 2: Class I Representations, Special Functions, and Integral Transforms*, Volume 74 of *Mathematics and its Applications (Soviet Series)*. Dordrecht: Kluwer Academic Publishers Group. Translated from the Russian by V. A. Groza and A. A. Groza.

I. M. Vinogradov (1937). *Representation of an odd number as a sum of three primes (Russian)*. Dokl. Akad. Nauk SSSR **15**, pp. 169–172.

I. M. Vinogradov (1958). *A new estimate of the function $\zeta(1+it)$*. Izv. Akad. Nauk SSSR. Ser. Mat. **22**, pp. 161–164.

N. Virchenko and I. Fedotova (2001). *Generalized Associated Legendre Functions and their Applications*. Singapore: World Scientific Publishing Co. Inc.

H. Volkmer (1982). *Integral relations for Lamé functions*. SIAM J. Math. Anal. **13**(6), pp. 978–987.

H. Volkmer (1983). *Integralgleichungen für periodische Lösungen Hill'scher Differentialgleichungen*. Analysis **3**(1-4), pp. 189–203.

H. Volkmer (1984). *Integral representations for products of Lamé functions by use of fundamental solutions*. SIAM J. Math. Anal. **15**(3), pp. 559–569.

H. Volkmer (1998). *On the growth of convergence radii for the eigenvalues of the Mathieu equation*. Math. Nachr. **192**, pp. 239–253.

H. Volkmer (1999). *Expansions in products of Heine-Stieltjes polynomials*. Constr. Approx. **15**(4), pp. 467–480.

H. Volkmer (2004a). *Error estimates for Rayleigh-Ritz approximations of eigenvalues and eigenfunctions of the Mathieu and spheroidal wave equation*. Constr. Approx. **20**(1), pp. 39–54.

H. Volkmer (2004b). *Four remarks on eigenvalues of Lamé's equation*. Anal. Appl. (Singap.) **2**(2), pp. 161–175.

A. P. Vorob'ev (1965). *On the rational solutions of the second Painlevé equation*. Differ. Uravn. **1**(1), pp. 79–81. In Russian. English translation: Differential Equations **1**(1), pp. 58–59.

M. N. Vrahatis, T. N. Grapsa, O. Ragos, and F. A. Zafiropoulos (1997a). *On the localization and computation of zeros of Bessel functions*. Z. Angew. Math. Mech. **77**(6), pp. 467–475.

M. N. Vrahatis, O. Ragos, T. Skiniotis, F. A. Zafiropoulos, and T. N. Grapsa (1997b). *The topological degree theory for the localization and computation of complex zeros of Bessel functions.* Numer. Funct. Anal. Optim. **18**(1-2), pp. 227–234.

E. L. Wachspress (2000). *Evaluating elliptic functions and their inverses.* Comput. Math. Appl. **39**(3-4), pp. 131–136.

E. Wagner (1986). *Asymptotische Darstellungen der hypergeometrischen Funktion für große Parameter unterschiedlicher Größenordnung.* Z. Anal. Anwendungen **5**(3), pp. 265–276.

E. Wagner (1988). *Asymptotische Entwicklungen der hypergeometrischen Funktion $F(a,b,c,z)$ für $|c| \to \infty$ und konstante Werte a, b und z.* Demonstratio Math. **21**(2), pp. 441–458.

E. Wagner (1990). *Asymptotische Entwicklungen der Gaußschen hypergeometrischen Funktion für unbeschränkte Parameter.* Z. Anal. Anwendungen **9**(4), pp. 351–360.

S. S. Wagstaff, Jr. (1978). *The irregular primes to 125000.* Math. Comp. **32**(142), pp. 583–591.

S. S. Wagstaff, Jr. (2002). *Prime Divisors of the Bernoulli and Euler Numbers.* In *Number Theory for the Millennium, III (Urbana, IL, 2000)*, pp. 357–374. Natick, MA: A. K. Peters.

J. Waldvogel (2006). *Fast construction of the Fejér and Clenshaw-Curtis quadrature rules.* BIT **46**(1), pp. 195–202.

J. Walker (1983). *Caustics: Mathematical curves generated by light shined through rippled plastic.* Scientific American **249**, pp. 146–153.

J. Walker (1988). *Shadows cast on the bottom of a pool are not like other shadows. Why?* Scientific American **259**, pp. 86–89.

J. Walker (1989). *A drop of water becomes a gateway into the world of catastrophe optics.* Scientific American **261**, pp. 120–123.

P. L. Walker (1991). *Infinitely differentiable generalized logarithmic and exponential functions.* Math. Comp. **57**(196), pp. 723–733.

P. L. Walker (1996). *Elliptic Functions. A Constructive Approach.* Chichester-New York: John Wiley & Sons Ltd.

P. L. Walker (2003). *The analyticity of Jacobian functions with respect to the parameter k.* Proc. Roy. Soc. London Ser A **459**, pp. 2569–2574.

P. L. Walker (2007). *The zeros of Euler's psi function and its derivatives.* J. Math. Anal. Appl. **332**(1), pp. 607–616.

H. S. Wall (1948). *Analytic Theory of Continued Fractions.* New York: D. Van Nostrand Company, Inc. Reprinted by Chelsea Publishing, New York, 1973 and AMS Chelsea Publishing, Providence, RI, 2000.

Z. Wang and R. Wong (2002). *Uniform asymptotic expansion of $J_\nu(\nu a)$ via a difference equation.* Numer. Math. **91**(1), pp. 147–193.

Z. Wang and R. Wong (2003). *Asymptotic expansions for second-order linear difference equations with a turning point.* Numer. Math. **94**(1), pp. 147–194.

Z. Wang and R. Wong (2005). *Linear difference equations with transition points.* Math. Comp. **74**(250), pp. 629–653.

Z. Wang and R. Wong (2006). *Uniform asymptotics of the Stieltjes-Wigert polynomials via the Riemann-Hilbert approach.* J. Math. Pures Appl. (9) **85**(5), pp. 698–718.

Z. X. Wang and D. R. Guo (1989). *Special Functions.* Singapore: World Scientific Publishing Co. Inc. Translated from the Chinese by Guo and X. J. Xia.

R. S. Ward (1987). *The Nahm equations, finite-gap potentials and Lamé functions.* J. Phys. A **20**(10), pp. 2679–2683.

S. O. Warnaar (1998). *A note on the trinomial analogue of Bailey's lemma.* J. Combin. Theory Ser. A **81**(1), pp. 114–118.

L. C. Washington (1997). *Introduction to Cyclotomic Fields* (2nd ed.). New York: Springer-Verlag.

W. Wasow (1965). *Asymptotic Expansions for Ordinary Differential Equations.* New York-London-Sydney: Interscience Publishers John Wiley & Sons, Inc. Reprinted by Krieger, New York, 1976 and Dover, New York, 1987.

W. Wasow (1985). *Linear Turning Point Theory.* Applied Mathematical Sciences No. 54. New York: Springer-Verlag.

H. Watanabe (1995). *Solutions of the fifth Painlevé equation. I.* Hokkaido Math. J. **24**(2), pp. 231–267.

B. M. Watrasiewicz (1967). *Some useful integrals of $\mathrm{Si}(x)$, $\mathrm{Ci}(x)$ and related integrals.* Optica Acta **14**(3), pp. 317–322.

G. N. Watson (1910). *The cubic transformation of the hypergeometric function.* Quart. J. Pure and Applied Math. **41**, pp. 70–79.

G. N. Watson (1935a). *Generating functions of class-numbers.* Compositio Math. **1**, pp. 39–68.

G. N. Watson (1935b). *The surface of an ellipsoid.* Quart. J. Math., Oxford Ser. **6**, pp. 280–287.

G. N. Watson (1937). *Two tables of partitions.* Proc. London Math. Soc. (2) **42**, pp. 550–556.

G. N. Watson (1944). *A Treatise on the Theory of Bessel Functions* (2nd ed.). Cambridge, England: Cambridge University Press. Reprinted in 1995.

G. N. Watson (1949). *A table of Ramanujan's function $\tau(n)$.* Proc. London Math. Soc. (2) **51**, pp. 1–13.

J. K. G. Watson (1999). *Asymptotic approximations for certain 6-j and 9-j symbols.* J. Phys. A **32**(39), pp. 6901–6902.

J. V. Wehausen and E. V. Laitone (1960). *Surface Waves.* In *Handbuch der Physik, Vol. 9, Part 3*, pp. 446–778. Berlin: Springer-Verlag.

G. Wei and B. E. Eichinger (1993). *Asymptotic expansions of some matrix argument hypergeometric functions, with applications to macromolecules.* Ann. Inst. Statist. Math. **45**(3), pp. 467–475.

A. Weil (1999). *Elliptic Functions According to Eisenstein and Kronecker*. Classics in Mathematics. Berlin: Springer-Verlag. Reprint of the 1976 original.

M. I. Weinstein and J. B. Keller (1985). *Hill's equation with a large potential.* SIAM J. Appl. Math. **45**(2), pp. 200–214.

M. I. Weinstein and J. B. Keller (1987). *Asymptotic behavior of stability regions for Hill's equation.* SIAM J. Appl. Math. **47**(5), pp. 941–958.

G. Weiss (1965). *Harmonic Analysis.* In I. I. Hirschman, Jr. (Ed.), *Studies in Real and Complex Analysis*, Studies in Mathematics, Vol. 3, pp. 124–178. Buffalo, NY: The Mathematical Association of America.

E. J. Weniger (1989). *Nonlinear sequence transformations for the acceleration of convergence and the summation of divergent series.* Computer Physics Reports **10**(5-6), pp. 189–371.

E. J. Weniger (1996). *Computation of the Whittaker function of the second kind by summing its divergent asymptotic series with the help of nonlinear sequence transformations.* Computers in Physics **10**(5), pp. 496–503.

E. J. Weniger (2003). *A rational approximant for the digamma function.* Numer. Algorithms **33**(1-4), pp. 499–507. International Conference on Numerical Algorithms, Vol. I (Marrakesh, 2001).

E. J. Weniger (2007). *Asymptotic Approximations to Truncation Errors of Series Representations for Special Functions.* In A. Iske and J. Levesley (Eds.), *Algorithms for Approximation*, pp. 331–348. Berlin-Heidelberg-New York: Springer-Verlag.

E. J. Weniger and J. Čížek (1990). *Rational approximations for the modified Bessel function of the second kind.* Comput. Phys. Comm. **59**(3), pp. 471–493.

H. Werner, J. Stoer, and W. Bommas (1967). *Rational Chebyshev approximation.* Numer. Math. **10**(4), pp. 289–306.

J. A. Wheeler (1937). *Wave functions for large arguments by the amplitude-phase method.* Phys. Rev. **52**, pp. 1123–1127.

A. D. Wheelon (1968). *Tables of Summable Series and Integrals Involving Bessel Functions*. San Francisco, CA: Holden-Day.

F. J. W. Whipple (1927). *Some transformations of generalized hypergeometric series.* Proc. London Math. Soc. (2) **26**(2), pp. 257–272.

C. S. Whitehead (1911). *On a generalization of the functions ber x, bei x, ker x, kei x.* Quart. J. Pure Appl. Math. **42**, pp. 316–342.

G. B. Whitham (1974). *Linear and Nonlinear Waves*. New York: John Wiley & Sons. Reprinted in 1999.

E. T. Whittaker (1902). *On the functions associated with the parabolic cylinder in harmonic analysis.* Proc. London Math. Soc. **35**, pp. 417–427.

E. T. Whittaker (1964). *A Treatise on the Analytical Dynamics of Particles and Rigid Bodies* (4th ed.). Cambridge: Cambridge University Press. Reprinted in 1988.

E. T. Whittaker and G. N. Watson (1927). *A Course of Modern Analysis* (4th ed.). Cambridge University Press. Reprinted in 1996. Table errata: Math. Comp. v. 36 (1981), no. 153, p. 319.

D. V. Widder (1941). *The Laplace Transform*. Princeton Mathematical Series, v. 6. Princeton, NJ: Princeton University Press.

D. V. Widder (1979). *The Airy transform.* Amer. Math. Monthly **86**(4), pp. 271–277.

R. L. Wiegel (1960). *A presentation of cnoidal wave theory for practical application.* J. Fluid Mech. **7**(2), pp. 273–286.

E. P. Wigner (1959). *Group Theory and its Application to the Quantum Mechanics of Atomic Spectra*. Pure and Applied Physics. Vol. 5. New York: Academic Press. Expanded and improved ed. Translated from the German by J. J. Griffin.

H. S. Wilf (1994). *generatingfunctionology* (2nd ed.). Boston, MA: Academic Press Inc. A 3rd edition was published in 2006 by A.K. Peters, Ltd., Wellesley, MA.

H. S. Wilf and D. Zeilberger (1992a). *An algorithmic proof theory for hypergeometric (ordinary and "q") multisum/integral identities.* Invent. Math. **108**, pp. 575–633.

H. S. Wilf and D. Zeilberger (1992b). *Rational function certification of multisum/integral/"q" identities.* Bull. Amer. Math. Soc. (N.S.) **27**(1), pp. 148–153.

J. H. Wilkinson (1988). *The Algebraic Eigenvalue Problem*. Monographs on Numerical Analysis. Oxford Science Publications. Oxford: The Clarendon Press, Oxford University Press. Paperback reprint of original edition, published in 1965 by Clarendon Press, Oxford.

C. A. Wills, J. M. Blair, and P. L. Ragde (1982). *Rational Chebyshev approximations for the Bessel functions $J_0(x)$, $J_1(x)$, $Y_0(x)$, $Y_1(x)$*. Math. Comp. **39**(160), pp. 617–623. Tables on microfiche.

J. A. Wilson (1978). *Hypergeometric Series, Recurrence Relations and Some New Orthogonal Polynomials*. Ph. D. thesis, University of Wisconsin, Madison, WI.

J. A. Wilson (1980). *Some hypergeometric orthogonal polynomials*. SIAM J. Math. Anal. **11**(4), pp. 690–701.

J. A. Wilson (1991). *Asymptotics for the $_4F_3$ polynomials*. J. Approx. Theory **66**(1), pp. 58–71.

J. Wimp (1964). *A class of integral transforms*. Proc. Edinburgh Math. Soc. (2) **14**, pp. 33–40.

J. Wimp (1965). *On the zeros of a confluent hypergeometric function*. Proc. Amer. Math. Soc. **16**(2), pp. 281–283.

J. Wimp (1968). *Recursion formulae for hypergeometric functions*. Math. Comp. **22**(102), pp. 363–373.

J. Wimp (1981). *Sequence Transformations and their Applications*, Volume 154 of *Mathematics in Science and Engineering*. New York: Academic Press Inc.

J. Wimp (1984). *Computation with Recurrence Relations*. Boston, MA: Pitman.

J. Wimp (1985). *Some explicit Padé approximants for the function Φ'/Φ and a related quadrature formula involving Bessel functions*. SIAM J. Math. Anal. **16**(4), pp. 887–895.

J. Wimp (1987). *Explicit formulas for the associated Jacobi polynomials and some applications*. Canad. J. Math. **39**(4), pp. 983–1000.

J. Wishart (1928). *The generalised product moment distribution in samples from a normal multivariate population*. Biometrika **20A**, pp. 32–52.

E. Witten (1987). *Elliptic genera and quantum field theory*. Comm. Math. Phys. **109**(4), pp. 525–536.

G. Wolf (1998). *On the central connection problem for the double confluent Heun equation*. Math. Nachr. **195**, pp. 267–276.

G. Wolf (2008). *On the asymptotic behavior of the Fourier coefficients of Mathieu functions*. J. Res. Nat. Inst. Standards Tech. **113**(1), pp. 11–15.

R. Wong (1973a). *An asymptotic expansion of $W_{k,m}(z)$ with large variable and parameters*. Math. Comp. **27**(122), pp. 429–436.

R. Wong (1973b). *On uniform asymptotic expansion of definite integrals*. J. Approximation Theory **7**(1), pp. 76–86.

R. Wong (1976). *Error bounds for asymptotic expansions of Hankel transforms*. SIAM J. Math. Anal. **7**(6), pp. 799–808.

R. Wong (1977). *Asymptotic expansions of Hankel transforms of functions with logarithmic singularities*. Comput. Math. Appl. **3**(4), pp. 271–286.

R. Wong (1979). *Explicit error terms for asymptotic expansions of Mellin convolutions*. J. Math. Anal. Appl. **72**(2), pp. 740–756.

R. Wong (1981). *Asymptotic expansions of the Kontorovich-Lebedev transform*. Appl. Anal. **12**(3), pp. 161–172.

R. Wong (1982). *Quadrature formulas for oscillatory integral transforms*. Numer. Math. **39**(3), pp. 351–360.

R. Wong (1983). *Applications of some recent results in asymptotic expansions*. Congr. Numer. **37**, pp. 145–182.

R. Wong (1989). *Asymptotic Approximations of Integrals*. Boston-New York: Academic Press Inc. Reprinted with corrections by SIAM, Philadelphia, PA, 2001.

R. Wong (1995). *Error bounds for asymptotic approximations of special functions*. Ann. Numer. Math. **2**(1-4), pp. 181–197. Special functions (Torino, 1993).

R. Wong and T. Lang (1990). *Asymptotic behaviour of the inflection points of Bessel functions*. Proc. Roy. Soc. London Ser. A **431**, pp. 509–518.

R. Wong and T. Lang (1991). *On the points of inflection of Bessel functions of positive order. II*. Canad. J. Math. **43**(3), pp. 628–651.

R. Wong and H. Li (1992a). *Asymptotic expansions for second-order linear difference equations*. J. Comput. Appl. Math. **41**(1-2), pp. 65–94.

R. Wong and H. Li (1992b). *Asymptotic expansions for second-order linear difference equations. II*. Stud. Appl. Math. **87**(4), pp. 289–324.

R. Wong and J. F. Lin (1978). *Asymptotic expansions of Fourier transforms of functions with logarithmic singularities*. J. Math. Anal. Appl. **64**(1), pp. 173–180.

R. Wong and M. Wyman (1974). *The method of Darboux*. J. Approximation Theory **10**(2), pp. 159–171.

R. Wong and H. Y. Zhang (2007). *Asymptotic solutions of a fourth order differential equation*. Stud. Appl. Math. **118**(2), pp. 133–152.

R. Wong and J.-M. Zhang (1994a). *Asymptotic monotonicity of the relative extrema of Jacobi polynomials*. Canad. J. Math. **46**(6), pp. 1318–1337.

R. Wong and J.-M. Zhang (1994b). *On the relative extrema of the Jacobi polynomials $P_n^{(0,-1)}(x)$*. SIAM J. Math. Anal. **25**(2), pp. 776–811.

R. Wong and J.-M. Zhang (1997). *Asymptotic expansions of the generalized Bessel polynomials.* J. Comput. Appl. Math. **85**(1), pp. 87–112.

R. Wong and Y.-Q. Zhao (1999a). *Smoothing of Stokes's discontinuity for the generalized Bessel function.* Proc. Roy. Soc. London Ser. A **455**, pp. 1381–1400.

R. Wong and Y.-Q. Zhao (1999b). *Smoothing of Stokes's discontinuity for the generalized Bessel function. II.* Proc. Roy. Soc. London Ser. A **455**, pp. 3065–3084.

R. Wong and Y.-Q. Zhao (2002a). *Exponential asymptotics of the Mittag-Leffler function.* Constr. Approx. **18**(3), pp. 355–385.

R. Wong and Y.-Q. Zhao (2002b). *Gevrey asymptotics and Stieltjes transforms of algebraically decaying functions.* Proc. Roy. Soc. London Ser. A **458**, pp. 625–644.

R. Wong and Y.-Q. Zhao (2003). *Estimates for the error term in a uniform asymptotic expansion of the Jacobi polynomials.* Anal. Appl. (Singap.) **1**(2), pp. 213–241.

R. Wong and Y.-Q. Zhao (2004). *Uniform asymptotic expansion of the Jacobi polynomials in a complex domain.* Proc. Roy. Soc. London Ser. A **460**, pp. 2569–2586.

R. Wong and Y.-Q. Zhao (2005). *On a uniform treatment of Darboux's method.* Constr. Approx. **21**(2), pp. 225–255.

P. M. Woodward and A. M. Woodward (1946). *Four-figure tables of the Airy function in the complex plane.* Philos. Mag. (7) **37**, pp. 236–261.

J. W. Wrench, Jr. (1968). *Concerning two series for the gamma function.* Math. Comp. **22**(103), pp. 617–626.

E. M. Wright (1935). *The asymptotic expansion of the generalized Bessel function.* Proc. London Math. Soc. (2) **38**, pp. 257–270.

E. M. Wright (1940a). *The asymptotic expansion of the generalized hypergeometric function.* Proc. London Math. Soc. (2) **46**, pp. 389–408.

E. M. Wright (1940b). *The generalized Bessel function of order greater than one.* Quart. J. Math., Oxford Ser. **11**, pp. 36–48.

F. J. Wright (1980). *The Stokes set of the cusp diffraction catastrophe.* J. Phys. A **13**(9), pp. 2913–2928.

C. Y. Wu (1982). *A series of inequalities for Mills's ratio.* Acta Math. Sinica **25**(6), pp. 660–670.

P. Wynn (1966). *Upon systems of recursions which obtain among the quotients of the Padé table.* Numer. Math. **8**(3), pp. 264–269.

H. Xiao, V. Rokhlin, and N. Yarvin (2001). *Prolate spheroidal wavefunctions, quadrature and interpolation.* Inverse Problems **17**(4), pp. 805–838. Special issue to celebrate Pierre Sabatier's 65th birthday (Montpellier, 2000).

A. I. Yablonskiĭ (1959). *On rational solutions of the second Painlevé equation.* Vesti Akad. Navuk. BSSR Ser. Fiz. Tkh. Nauk. **3**, pp. 30–35.

G. D. Yakovleva (1969). *Tables of Airy Functions and Their Derivatives.* Moscow: Izdat. Nauka.

Z. M. Yan (1992). *Generalized Hypergeometric Functions and Laguerre Polynomials in Two Variables.* In *Hypergeometric Functions on Domains of Positivity, Jack Polynomials, and Applications (Tampa, FL, 1991)*, Volume 138 of *Contemporary Mathematics*, pp. 239–259. Providence, RI: Amer. Math. Soc.

A. J. Yee (2004). *Partitions with difference conditions and Alder's conjecture.* Proc. Natl. Acad. Sci. USA **101**(47), pp. 16417–16418.

A. Yu. Yeremin, I. E. Kaporin, and M. K. Kerimov (1985). *The calculation of the Riemann zeta function in the complex domain.* USSR Comput. Math. and Math. Phys. **25**(2), pp. 111–119.

A. Yu. Yeremin, I. E. Kaporin, and M. K. Kerimov (1988). *Computation of the derivatives of the Riemann zeta-function in the complex domain.* USSR Comput. Math. and Math. Phys. **28**(4), pp. 115–124. Includes Fortran programs.

F. L. Yost, J. A. Wheeler, and G. Breit (1936). *Coulomb wave functions in repulsive fields.* Phys. Rev. **49**(2), pp. 174–189.

A. Young and A. Kirk (1964). *Bessel Functions. Part IV: Kelvin Functions.* Royal Society Mathematical Tables, Volume 10. Cambridge-New York: Cambridge University Press.

D. M. Young and R. T. Gregory (1988). *A Survey of Numerical Mathematics. Vol. II.* New York: Dover Publications Inc. Corrected reprint of the 1973 original, published by Addison-Wesley.

A. P. Yutsis, I. B. Levinson, and V. V. Vanagas (1962). *Mathematical Apparatus of the Theory of Angular Momentum.* Jerusalem: Israel Program for Scientific Translations for National Science Foundation and the National Aeronautics and Space Administration. Translated from the Russian by A. Sen and R. N. Sen.

F. A. Zafiropoulos, T. N. Grapsa, O. Ragos, and M. N. Vrahatis (1996). *On the Computation of Zeros of Bessel and Bessel-related Functions.* In D. Bainov (Ed.), *Proceedings of the Sixth International Colloquium on Differential Equations (Plovdiv, Bulgaria, 1995)*, Utrecht, pp. 409–416. VSP.

D. Zagier (1989). *The Dilogarithm Function in Geometry and Number Theory.* In R. Askey et al. (Eds.), *Number Theory and Related Topics (Bombay, 1988)*, Volume 12 of *Tata Inst. Fund. Res. Stud. Math.*, pp. 231–249. Bombay/Oxford: Tata Inst. Fund. Res./Oxford University Press.

D. Zagier (1998). *A modified Bernoulli number*. Nieuw Arch. Wisk. (4) **16**(1-2), pp. 63–72.

R. Zanovello (1975). *Sul calcolo numerico della funzione di Struve $\mathbf{H}_\nu(z)$*. Rend. Sem. Mat. Univ. e Politec. Torino **32**, pp. 251–269.

R. Zanovello (1977). *Integrali di funzioni di Anger, Weber ed Airy-Hardy*. Rend. Sem. Mat. Univ. Padova **58**, pp. 275–285.

R. Zanovello (1978). *Su un integrale definito del prodotto di due funzioni di Struve*. Atti Accad. Sci. Torino Cl. Sci. Fis. Mat. Natur. **112**(1-2), pp. 63–81.

R. Zanovello (1995). *Numerical analysis of Struve functions with applications to other special functions*. Ann. Numer. Math. **2**(1-4), pp. 199–208.

A. Zarzo, J. S. Dehesa, and R. J. Yañez (1995). *Distribution of zeros of Gauss and Kummer hypergeometric functions. A semiclassical approach*. Ann. Numer. Math. **2**(1-4), pp. 457–472.

D. Zeilberger and D. M. Bressoud (1985). *A proof of Andrews' q-Dyson conjecture*. Discrete Math. **54**(2), pp. 201–224.

J. Zhang (1996). *A note on the τ-method approximations for the Bessel functions $Y_0(z)$ and $Y_1(z)$*. Comput. Math. Appl. **31**(9), pp. 63–70.

J. Zhang and J. A. Belward (1997). *Chebyshev series approximations for the Bessel function $Y_n(z)$ of complex argument*. Appl. Math. Comput. **88**(2-3), pp. 275–286.

J. M. Zhang, X. C. Li, and C. K. Qu (1996). *Error bounds for asymptotic solutions of second-order linear difference equations*. J. Comput. Appl. Math. **71**(2), pp. 191–212.

S. Zhang and J. Jin (1996). *Computation of Special Functions*. New York: John Wiley & Sons Inc. Includes diskette containing a large collection of mathematical function software written in Fortran. Implementation in double precision. Maximum accuracy 16S.

Q. Zheng (1997). *Generalized Watson Transforms and Applications to Group Representations*. Ph. D. thesis, University of Vermont, Burlington,VT.

Ya. M. Zhileĭkin and A. B. Kukarkin (1995). *A fast Fourier-Bessel transform algorithm*. Zh. Vychisl. Mat. i Mat. Fiz. **35**(7), pp. 1128–1133. English translation in Comput. Math. Math. Phys. **35**(7), pp. 901–905.

D. G. Zill and B. C. Carlson (1970). *Symmetric elliptic integrals of the third kind*. Math. Comp. **24**(109), pp. 199–214.

A. Ziv (1991). *Fast evaluation of elementary mathematical functions with correctly rounded last bit*. ACM Trans. Math. Software **17**(3), pp. 410–423.

I. J. Zucker (1979). *The summation of series of hyperbolic functions*. SIAM J. Math. Anal. **10**(1), pp. 192–206.

H. S. Zuckerman (1939). *The computation of the smaller coefficients of $J(\tau)$*. Bull. Amer. Math. Soc. **45**(12), pp. 917–919.

M. I. Žurina and L. N. Karmazina (1963). *Tablitsy funktsii Lezhandra $P^1_{-1/2+i\tau}(x)$*. Vyčisl. Centr Akad. Nauk SSSR, Moscow. Translated title: Tables of the Legendre functions $P^1_{-1/2+i\tau}(x)$.

M. I. Žurina and L. N. Karmazina (1964). *Tables of the Legendre functions $P_{-1/2+i\tau}(x)$. Part I*. Translated by D. E. Brown. Mathematical Tables Series, Vol. 22. Oxford: Pergamon Press. Translation of Žurina and Karmazina, *Tablitsy funktsii Lezhandra $P_{-1/2+i\tau}(x)$. Tom I*, Izdat. Akad. Nauk SSSR, Moscow, 1960.

M. I. Žurina and L. N. Karmazina (1965). *Tables of the Legendre functions $P_{-1/2+i\tau}(x)$. Part II*. Translated by Prasenjit Basu. Mathematical Tables Series, Vol. 38. A Pergamon Press Book. New York: The Macmillan Co. Translation of Žurina and Karmazina, *Tablitsy funktsii Lezhandra $P_{-1/2+i\tau}(x)$. Tom II*, Izdat. Akad. Nauk SSSR, Moscow, 1962.

M. I. Žurina and L. N. Karmazina (1966). *Tables and formulae for the spherical functions $P^m_{-1/2+i\tau}(z)$*. Translated by E. L. Albasiny. Oxford: Pergamon Press. Translation of Žurina and Karmazina, *Tablitsy i formuly dlya sfericheskikh funktsii $P^m_{-1/2+i\tau}(z)$*, Vyčisl. Centr Akad Nauk SSSR, Moscow. 1962.

M. I. Žurina and L. N. Karmazina (1967). *Tablitsy modifitsirovannykh funktsii Besselya s mnimym indeksom $K_{i\tau}(x)$*. Vyčisl. Centr Akad. Nauk SSSR, Moscow. Translated title: Tables of the Modified Bessel Functions with Imaginary Index $K_{i\tau}(x)$.

M. I. Žurina and L. N. Osipova (1964). *Tablitsy vyrozhdennoi gipergeometricheskoi funktsii*. Moscow: Vyčisl. Centr Akad. Nauk SSSR. Translated title: Tables of the confluent hypergeometric function.

Notations

!

$n!_q$: q-factorial 145

$\mathbf{a} \cdot \mathbf{b}$: vector dot (or scalar) product 9

∗

$f * g$: convolution for Fourier transforms 27

$f * g$: convolution for Laplace transforms 28

$f * g$: convolution for Mellin transforms 29

$f * g$: convolution product 53

×

$G \times H$: Cartesian product of groups G and H ... 570

×

$\mathbf{a} \times \mathbf{b}$: vector cross product 9

/

S_1/S_2: set of all elements of S_1 modulo elements of S_2 538

∼

asymptotic equality 42

∇

del operator 10

∇²

Laplacian 7

Laplacian for cylindrical coordinates 7

Laplacian for polar coordinates 7

Laplacian for spherical coordinates 8

∇f

gradient of differentiable scalar function f 10

∇ × F

curl of vector-valued function $\mathbf{F}$ 10

∇ · F

divergence of vector-valued function $\mathbf{F}$ 10

$\int_a^b$

Cauchy principal value 6

$\int_a^{(b+)}$

loop integral in $\mathbb{C}$: path begins at a, encircles b once in the positive sense, and returns to a 139

$\int_P^{(1+,0+,1-,0-)}$

Pochhammer's loop integral 142

$\int \cdots d_q x$

q-integral 422

$\overline{z}$

complex conjugate 15

$|z|$

modulus (or absolute value) 15

$\|\mathbf{a}\|$

magnitude of vector 9

$\|\mathbf{A}\|_p$

p-norm of a matrix 74

$\|\mathbf{x}\|_2$

Euclidean norm of a vector 74

$\|\mathbf{x}\|_p$

p-norm of a vector 74

$\|\mathbf{x}\|_\infty$

infinity (or maximum) norm of a vector 74

$f(c+)$

limit on right (or from above) 4

$f(c-)$

limit on left (or from below) 4

$f^{[n]}(z)$

nth q-derivative 421

$x^{\underline{n}}$

falling factorial 618

$x^{\overline{n}}$

rising factorial 618

$b_0 + \frac{a_1}{b_1+} \frac{a_2}{b_2+} \cdots$

continued fraction 24

$(n|P)$

Jacobi symbol 642

$(n|p)$

Legendre symbol 642

$(a;q)_n$

q-factorial (or q-shifted factorial) 145, 420

$(a;q)_\nu$

q-shifted factorial (generalized) 420

$(a;q)_\infty$

q-shifted factorial 420

$(a_1, a_2, \ldots, a_r; q)_n$

multiple q-shifted factorial 420

$(a_1, a_2, \ldots, a_r; q)_\infty$

multiple q-shifted factorial 420

$(j_1\ m_1\ j_2\ m_2|j_1\ j_2\ j_3\ -m_3)$
    Clebsch–Gordan coefficient 758
$\binom{m}{n}$
    binomial coefficient 2, 619
$\binom{n_1+n_2+\cdots+n_k}{n_1,n_2,\ldots,n_k}$
    multinomial coefficient 620
$\begin{pmatrix} j_1 & j_2 & j_3 \\ m_1 & m_2 & m_3 \end{pmatrix}$
    $3j$ symbol 758
$\langle \Lambda, \phi \rangle$
    distribution 35
$\langle f, \phi \rangle$
    tempered distribution 52
$\langle \delta, \phi \rangle$
    Dirac delta distribution 36
$\left\langle {n \atop k} \right\rangle$
    Eulerian number 632
$[z_0, z_1, \ldots, z_n]$
    divided difference 76
$[a]_\kappa$
    partitional shifted factorial 769
$[p/q]_f$
    Padé approximant 98
$\left[{n \atop k} \right]$
    Stirling cycle number 631
$\left[{n \atop m} \right]_q$
    q-binomial coefficient (or Gaussian polynomial)
    .... 421, 627
$\left[{a_1+a_2+\cdots+a_n \atop a_1,a_2,\ldots,a_n} \right]_q$
    q-multinomial coefficient 634
$\{\ldots\}$
    sequence, asymptotic sequence (or scale), or enumerable set 43
$\{z, \zeta\}$
    Schwarzian derivative 27
$\begin{Bmatrix} j_1 & j_2 & j_3 \\ l_1 & l_2 & l_3 \end{Bmatrix}$
    $6j$ symbol 761
$\begin{Bmatrix} j_{11} & j_{12} & j_{13} \\ j_{21} & j_{22} & j_{23} \\ j_{31} & j_{32} & j_{33} \end{Bmatrix}$
    $9j$ symbol 763
A
    Glaisher's constant 144
$\mathbf{A}_\nu(z)$
    Anger–Weber function 295

$A_\nu(\mathbf{T})$
    Bessel function of matrix argument (first kind) .. 769
$A_n(z)$
    generalized Airy function 206
$A_k(z, p)$
    generalized Airy function 207
$A_{m,s}(q)$
    q-Euler number 422
$a_{m,s}(q)$
    q-Stirling number 422
$\text{Ai}(z)$
    Airy function 194
$\text{am}(x, k)$
    Jacobi's amplitude function 561
$\text{arccd}(x, k)$
    inverse Jacobian elliptic function 561
$\text{arccn}(x, k)$
    inverse Jacobian elliptic function 561
$\text{Arccos } z$
    general arccosine function 118
$\arccos z$
    arccosine function 119
$\text{Arccosh } z$
    general inverse hyperbolic cosine function 127
$\text{arccosh } z$
    inverse hyperbolic cosine function 127
$\text{Arccot } z$
    general arccotangent function 118
$\text{arccot } z$
    arccotangent function 119
$\text{Arccoth } z$
    general inverse hyperbolic cotangent function 127
$\text{arccoth } z$
    inverse hyperbolic cotangent function 127
$\text{arccs}(x, k)$
    inverse Jacobian elliptic function 561
$\text{Arccsc } z$
    general arccosecant function 118
$\text{arccsc } z$
    arccosecant function 119
$\text{Arccsch } z$
    general inverse hyperbolic cosecant function 127
$\text{arccsch } z$
    inverse hyperbolic cosecant function 127
$\text{arcdc}(x, k)$
    inverse Jacobian elliptic function 561

$\operatorname{arcdn}(x,k)$
    inverse Jacobian elliptic function 561
$\operatorname{arcds}(x,k)$
    inverse Jacobian elliptic function 561
$\operatorname{arcnc}(x,k)$
    inverse Jacobian elliptic function 561
$\operatorname{arcnd}(x,k)$
    inverse Jacobian elliptic function 561
$\operatorname{arcns}(x,k)$
    inverse Jacobian elliptic function 561
$\operatorname{arcsc}(x,k)$
    inverse Jacobian elliptic function 561
$\operatorname{arcsd}(x,k)$
    inverse Jacobian elliptic function 561
$\operatorname{Arcsec} z$
    general arcsecant function 118
$\operatorname{arcsec} z$
    arcsecant function 119
$\operatorname{Arcsech} z$
    general inverse hyperbolic secant function 127
$\operatorname{arcsech} z$
    inverse hyperbolic secant function 127
$\operatorname{Arcsin} z$
    general arcsine function 118
$\operatorname{arcsin} z$
    arcsine function 119
$\operatorname{Arcsinh} z$
    general inverse hyperbolic sine function 127
$\operatorname{arcsinh} z$
    inverse hyperbolic sine function 127
$\operatorname{arcsn}(x,k)$
    inverse Jacobian elliptic function 561
$\operatorname{Arctan} z$
    general arctangent function 118
$\operatorname{arctan} z$
    arctangent function 119
$\operatorname{Arctanh} z$
    general inverse hyperbolic tangent function 127
$\operatorname{arctanh} z$
    inverse hyperbolic tangent function 127
B_n
    Bernoulli numbers 588
$B_n^{(\ell)}$
    generalized Bernoulli numbers 596
$B_n^{(x)}$
    Nörlund polynomials 596

$B(n)$
    Bell number 623
$B_n(x)$
    Bernoulli polynomials 588
$B_\nu(\mathbf{T})$
    Bessel function of matrix argument (second kind)
    769
$B_n(z)$
    generalized Airy function 206
$\widetilde{B}_n(x)$
    periodic Bernoulli functions 588
$B_k(z,p)$
    generalized Airy function 207
$B_n^{(\ell)}(x)$
    generalized Bernoulli polynomials 596
$\mathrm{B}(a,b)$
    beta function 142
$\mathrm{B}_m(a,b)$
    multivariate beta function 768
$\mathrm{B}_x(a,b)$
    incomplete beta function 183
$\mathrm{B}_q(a,b)$
    q-beta function 145
$\operatorname{bei}_\nu(x)$
    Kelvin function 267
$\operatorname{ber}_\nu(x)$
    Kelvin function 267
$\beta_n(x,q)$
    q-Bernoulli polynomial 422
$\operatorname{Bi}(z)$
    Airy function 194
$C(n)$
    Catalan number 620
$C(I)$ or $C(a,b)$
    continuous on an interval I or (a,b) 4
$C^n(I)$ or $C^n(a,b)$
    continuously differentiable n times on an interval I or (a,b) 5
$C^\infty(I)$ or $C^\infty(a,b)$
    infinitely differentiable on an interval I or (a,b) ... 5
$\chi(n)$
    Dirichlet character 642
$C(z)$
    Fresnel integral 160
$\chi(n)$
    ratio of gamma functions 198

$c(n)$
    number of compositions of n 628

$\mathscr{C}_\nu(z)$
    cylinder function 218

$C_\ell(\eta)$
    normalizing constant for Coulomb radial functions
    742

$c_m(n)$
    number of compositions of n into exactly m parts
    628

$c_k(n)$
    Ramanujan's sum 642

$C_\alpha^{(\lambda)}(z)$
    Gegenbauer function 394

$C_n^{(\lambda)}(x)$
    ultraspherical (or Gegenbauer) polynomial 439

$c(\text{condition}, n)$
    restricted number of compositions of n 628

$C_n(x, a)$
    Charlier polynomial 462

$C_n^m(z, \xi)$
    Ince polynomials 676

$c(\epsilon, \ell; r)$
    irregular Coulomb function 748

$C(f, h)(x)$
    cardinal function 77

$C_n(x; \beta \mid q)$
    continuous q-ultraspherical polynomial 473

$\text{cd}\,(z, k)$
    Jacobian elliptic function 550

$cdE_{2n+2}^m(z, k^2)$
    Lamé polynomial 690

$\text{Ce}_\nu(z, q)$
    modified Mathieu function 667

$\text{ce}_\nu(z, q)$
    Mathieu function of noninteger order 665

$\text{ce}_n(z, q)$
    Mathieu function 654

$cE_{2n+1}^m(z, k^2)$
    Lamé polynomial 690

$\text{cel}(k_c, p, a, b)$
    Bulirsch's complete elliptic integral 487

$\text{Chi}(z)$
    hyperbolic cosine integral 150

$\text{Ci}(z)$
    cosine integral 150

$\text{Ci}(a, z)$
    generalized cosine integral 188

$\text{ci}(a, z)$
    generalized cosine integral 188

$\text{Cin}(z)$
    cosine integral 150

$\text{cn}\,(z, k)$
    Jacobian elliptic function 550

$\cos z$
    cosine function 112

$\text{Cos}_q(x)$
    q-cosine function 422

$\cos_q(x)$
    q-cosine function 422

$\cosh z$
    hyperbolic cosine function 123

$\cot z$
    cotangent function 112

$\coth z$
    hyperbolic cotangent function 123

$\text{cs}\,(z, k)$
    Jacobian elliptic function 550

$\csc z$
    cosecant function 112

$\text{csch}\,z$
    hyperbolic cosecant function 123

curl
    of vector-valued function 10

$\mathcal{D}(I)$
    test function space 35

$D(k)$
    complete elliptic integral of Legendre's type 487

$d(n)$
    divisor function 638

$d(n)$
    derangement number 631

$\mathcal{D}_q$
    q-differential operator 421

$D_\nu(z)$
    parabolic cylinder function 304

$d_k(n)$
    divisor function 638

$d_q x$
    q-differential 422

D^α
    fractional derivative 35

NOTATIONS

$D(m,n)$
  Dellanoy number 621
$D(\phi,k)$
  incomplete elliptic integral of Legendre's type ... 486
$\mathrm{D}_j(\nu,\mu,z)$
  cross-products of modified Mathieu functions and their derivatives 674
$\mathrm{dc}(z,k)$
  Jacobian elliptic function 550
$\frac{\partial(f,g)}{\partial(x,y)}$
  Jacobian 9
$dE_{2n+1}^m(z,k^2)$
  Lamé polynomial 690
$\Delta(\tau)$
  discriminant function 646
$\delta(x-a)$
  Dirac delta (or Dirac delta function) 37
div
  divergence of vector-valued function 10
$\mathrm{dn}(z,k)$
  Jacobian elliptic function 550
$\mathrm{ds}(z,k)$
  Jacobian elliptic function 550
$\mathrm{Ds}_j(n,m,z)$
  cross-products of radial Mathieu functions and their derivatives 674
e
  base of exponential function 105
E_n
  Euler numbers 588
$E_n^{(\ell)}$
  generalized Euler numbers 596
$E(k)$
  Legendre's complete elliptic integral of the second kind 487
$\eta(\tau)$
  Dedekind's eta function (or Dedekind modular function) 579, 646
$E_s(\mathbf{z})$
  elementary symmetric function 501
$E_n(x)$
  Euler polynomials 588
$E_1(z)$
  exponential integral 150
$E_p(z)$
  generalized exponential integral 185

$E_{a,b}(z)$
  Mittag-Leffler function 261
$E_q(x)$
  q-exponential function 422
$\mathbf{E}_\nu(z)$
  Weber function 295
$e_q(x)$
  q-exponential function 422
$\widetilde{E}_n(x)$
  periodic Euler functions 588
$E_n^{(\ell)}(x)$
  generalized Euler polynomials 596
$E(\phi,k)$
  Legendre's incomplete elliptic integral of the second kind 486
$Ec_\nu^m(z,k^2)$
  Lamé function 685
$\mathrm{Ei}(x)$
  exponential integral 150
$\mathrm{Ein}(z)$
  complementary exponential integral 150
$\mathrm{el1}(x,k_c)$
  Bulirsch's incomplete elliptic integral of the first kind 487
$\mathrm{el2}(x,k_c,a,b)$
  Bulirsch's incomplete elliptic integral of the second kind 487
$\mathrm{el3}(x,k_c,p)$
  Bulirsch's incomplete elliptic integral of the third kind 487
$\mathrm{env}\,\mathrm{Ai}(x)$
  envelope of Airy function 59
$\mathrm{env}\,\mathrm{Bi}(x)$
  envelope of Airy function 59
$\mathrm{env}\,J_\nu(x)$
  envelope of Bessel function 61
$\mathrm{env}\,Y_\nu(x)$
  envelope of Bessel function 61
$\mathrm{env}\,U(-c,x)$
  envelope of parabolic cylinder function 367
$\mathrm{env}\,\overline{U}(-c,x)$
  envelope of parabolic cylinder function 367
$\epsilon_{jk\ell}$
  Levi-Civita symbol 10
$\mathcal{E}(x,k)$
  Jacobi's epsilon function 562

erf z
    error function 160

erfc z
    complementary error function 160

$Es_\nu^m(z, k^2)$
    Lamé function 685

exp z
    exponential function 105

F_n
    Fibonacci number 629

F_D
    Lauricella's multivariate hypergeometric function
    497

$F(z)$
    Dawson's integral 160

$\mathcal{F}(z)$
    Fresnel integral 160

$F_s(x)$
    Fermi–Dirac integral 612

$F_c(x)$
    Fourier cosine transform 27

$F_s(x)$
    Fourier sine transform 27

$F_p(z)$
    terminant function 189

$f_{e,m}(h)$
    joining factor for radial Mathieu functions 669

$f_{o,m}(h)$
    joining factor for radial Mathieu functions 669

$F(x)$
    Fourier transform 27

$F(\phi, k)$
    Legendre's incomplete elliptic integral of the first kind
    486

$F(x, s)$
    periodic zeta function 612

$F_\ell(\eta, \rho)$
    regular Coulomb radial function 742

$F\binom{a,b}{c}; z\bigr)$
    hypergeometric function 384

$F(a, b; c; z)$
    hypergeometric function 384

$\mathbf{F}\binom{a,b}{c}; z\bigr)$
    Olver's hypergeometric function 384

$\mathbf{F}(a, b; c; z)$
    Olver's hypergeometric function 384

$f(\epsilon, \ell; r)$
    regular Coulomb function 748

${}_1F_1\binom{a}{b}; \mathbf{T}\bigr)$
    confluent hypergeometric function of matrix argument (first kind) 770

${}_1F_1(a; b; \mathbf{T})$
    confluent hypergeometric function of matrix argument (first kind) 768, 770

${}_2F_1\binom{a,b}{c}; \mathbf{T}\bigr)$
    hypergeometric function of matrix argument 771

${}_2F_1(a, b; c; \mathbf{T})$
    hypergeometric function of matrix argument
    .... 768, 771

${}_2F_1(a, b; c; z)$
    hypergeometric function 384

${}_pF_q\binom{\mathbf{a}}{\mathbf{b}}; z\bigr)$
    generalized hypergeometric function 404, 408

${}_pF_q\binom{a_1,\ldots,a_p}{b_1,\ldots,b_q}; z\bigr)$
    generalized hypergeometric function 404, 408

${}_pF_q\binom{a_1,a_2,\ldots,a_p}{b_1,b_2,\ldots,b_q}; \mathbf{T}\bigr)$
    generalized hypergeometric function of matrix argument 772

${}_pF_q(\mathbf{a}; \mathbf{b}; z)$
    generalized hypergeometric function 404, 408

${}_pF_q(a_1, \ldots, a_p; b_1, \ldots, b_q; z)$
    generalized hypergeometric function 404, 408

${}_pF_q(a_1, a_2, \ldots, a_p; b_1, b_2, \ldots, b_q; \mathbf{T})$
    generalized hypergeometric function of matrix argument 768, 772

${}_2\mathbf{F}_1(a, b; c; z)$
    Olver's hypergeometric function 384

${}_p\mathbf{F}_q\binom{\mathbf{a}}{\mathbf{b}}; z\bigr)$
    scaled (or Olver's) generalized hypergeometric function 405

$F_1(\alpha; \beta, \beta'; \gamma; x, y)$
    Appell function 413

$F_2(\alpha; \beta, \beta'; \gamma, \gamma'; x, y)$
    Appell function 413

$F_3(\alpha, \alpha'; \beta, \beta'; \gamma; x, y)$
    Appell function 413

$F_4(\alpha; \beta; \gamma, \gamma'; x, y)$
    Appell function 413

$\mathrm{Fe}_n(z, q)$
    modified Mathieu function 667

$\mathrm{fe}_n(z, q)$
    second solution, Mathieu's equation 657

G_n
    Genocchi numbers 595
$G(z)$
    Barnes' G-function (or double gamma function) .. 144
$G(z)$
    Goodwin–Staton integral 160
$G(k)$
    Waring's function 645
$g(k)$
    Waring's function 645
$G_s(x)$
    Bose–Einstein integral 612
$G_p(z)$
    product of gamma and incomplete gamma functions
    .... 199, 230
$g_{e,m}(h)$
    joining factor for radial Mathieu functions 669
$g_{o,m}(h)$
    joining factor for radial Mathieu functions 669
$G(n, \chi)$
    Gauss sum 643
$G_\ell(\eta, \rho)$
    irregular Coulomb radial function 742
$G_{p,q}^{m,n}\left(z; {a_1,\ldots,a_p \atop b_1,\ldots,b_q}\right)$
    Meijer G-function 415
$G_{p,q}^{m,n}(z; \mathbf{a}; \mathbf{b})$
    Meijer G-function 415
γ
    Euler's constant 136
$\Gamma(z)$
    gamma function 136
$\Gamma_m(a)$
    multivariate gamma function 768
$\Gamma_q(z)$
    q-gamma function 145
$\Gamma(a, z)$
    incomplete gamma function 174
$\gamma(a, z)$
    incomplete gamma function 174
$\gamma^*(a, z)$
    incomplete gamma function 174
gd x
    Gudermannian function 121
$\mathrm{gd}^{-1}(x)$
    inverse Gudermannian function 121

$\mathrm{Ge}_n(z, q)$
    modified Mathieu function 667
$\mathrm{ge}_n(z, q)$
    second solution, Mathieu's equation 657
$\mathrm{Gi}(z)$
    Scorer function (inhomogeneous Airy function) .. 204
grad
    gradient of differentiable scalar function 10
$H(s)$
    Euler sums 613
$H(x)$
    Heaviside function 36
$H_n(x)$
    Hermite polynomial 439
$\mathbf{H}_\nu(z)$
    Struve function 288
$\mathrm{He}_n(x)$
    Hermite polynomial 439
$H_\nu^{(1)}(z)$
    Bessel function of the third kind (or Hankel function)
    217
$H_\nu^{(2)}(z)$
    Bessel function of the third kind (or Hankel function)
    217
$\mathrm{h}_n^{(1)}(z)$
    spherical Bessel function of the third kind 262
$\mathrm{h}_n^{(2)}(z)$
    spherical Bessel function of the third kind 262
$H(s, z)$
    generalized Euler sums 614
$\mathcal{H}(f; x)$
    Hilbert transform 29
$H(a, u)$
    line-broadening function 167
$H_n(x \mid q)$
    continuous q-Hermite polynomial 473
$h_n(x \mid q)$
    continuous q^{-1}-Hermite polynomial 473
$h_n(x; q)$
    discrete q-Hermite I polynomial 471
$\tilde{h}_n(x; q)$
    discrete q-Hermite II polynomial 472
$H_\ell^\pm(\eta, \rho)$
    irregular Coulomb radial functions 742
$h(\epsilon, \ell; r)$
    irregular Coulomb function 748

$_pH_q\left(\begin{smallmatrix}a_1,\ldots,a_p\\b_1,\ldots,b_q\end{smallmatrix};z\right)$
  bilateral hypergeometric function 408
$hc_p^m(z,\xi)$
  paraboloidal wave function 677
$(s_1,s_2)Hf_m(a,q_m;\alpha,\beta,\gamma,\delta;z)$
  Heun functions 712
$(s_1,s_2)Hf_m^\nu(a,q_m;\alpha,\beta,\gamma,\delta;z)$
  path-multiplicative solutions of Heun's equation .. 712
$Hh_n(z)$
  probability function 167, 308
$\mathrm{Hi}(z)$
  Scorer function (inhomogeneous Airy function) .. 204
$H\ell(a,q;\alpha,\beta,\gamma,\delta;z)$
  Heun functions 711
$Hp_{n,m}(a,q_{n,m};-n,\beta,\gamma,\delta;z)$
  Heun polynomials 712
$hs_p^m(z,\xi)$
  paraboloidal wave function 677
I^α
  fractional integral 35, 53
$I(\mathbf{m})$
  general elliptic integral 512
$I_\nu(z)$
  modified Bessel function 249
$\widetilde{I}_\nu(x)$
  modified Bessel function of imaginary order 261
$\mathrm{i}_n^{(1)}(z)$
  modified spherical Bessel function 262
$\mathrm{i}_n^{(2)}(z)$
  modified spherical Bessel function 262
$I_x(a,b)$
  incomplete beta function 183
$\mathrm{idem}(\chi_1;\chi_2,\ldots,\chi_n)$
  idem function 420
$\mathrm{Ie}_n(z,h)$
  modified Mathieu function 668
$\mathrm{i}^n\mathrm{erfc}(z)$
  repeated integrals of the complementary error function 167
inv
  inversion number 632
inverf x
  inverse error function 166
inverfc x
  inverse complementary error function 166

$\mathrm{Io}_n(z,h)$
  modified Mathieu function 668
$j_{\nu,m}$
  zeros of the Bessel function $J_\nu(x)$ 235
$j'_{\nu,m}$
  zeros of the Bessel function derivative $J'_\nu(x)$ 235
$J(\tau)$
  Klein's complete invariant 579
$\mathbf{J}_\nu(z)$
  Anger function 295
$J_\nu(z)$
  Bessel function of the first kind 217
$\widetilde{J}_\nu(x)$
  Bessel function of imaginary order 248
$J_k(n)$
  Jordan's function 638
$\mathrm{j}_n(z)$
  spherical Bessel function of the first kind 262
$K(k)$
  Legendre's complete elliptic integral of the first kind
  487
$\kappa(\lambda)$
  condition number 75
$K_\nu(z)$
  modified Bessel function 249
$\widetilde{K}_\nu(x)$
  modified Bessel function of imaginary order 261
$\mathbf{K}_\nu(z)$
  Struve function 288
$\mathrm{k}_n(z)$
  modified spherical Bessel function 262
$K_n(x;p,N)$
  Krawtchouk polynomial 462
$\mathrm{Ke}_n(z,h)$
  modified Mathieu function 668
$\mathrm{kei}_\nu(x)$
  Kelvin function 268
$\mathrm{ker}_\nu(x)$
  Kelvin function 268
$\mathrm{Ki}_\alpha(x)$
  Bickley function 259
$\mathrm{Ko}_n(z,h)$
  modified Mathieu function 668
$\mathbb{L}$
  lattice in $\mathbb{C}$ 570

L_n
    Lebesgue constant 13
$L_n(x)$
    Laguerre polynomial 436, 439
$\mathbf{L}_\nu(z)$
    modified Struve function....................... 288
$L_n^{(\alpha)}(x)$
    Laguerre (or generalized Laguerre) polynomial .. 439
$L(s,\chi)$
    Dirichlet L-function............................ 612
$\mathscr{L}(f;s)$
    Laplace transform............................... 28
$L_n^{(\alpha)}(x;q)$
    q-Laguerre polynomial 471
$\lambda(\tau)$
    elliptic modular function 579
$\Lambda(n)$
    Mangoldt's function............................ 639
$\lambda(n)$
    Liouville's function............................. 639
$\mathrm{li}(x)$
    logarithmic integral 150
$\mathrm{Li}_2(z)$
    dilogarithm 610
$\mathrm{Li}_s(z)$
    polylogarithm................................. 611
$\mathrm{Ln}\,z$
    general logarithm function 104
$\ln z$
    principal branch of logarithm function 104
$\log x$
    logarithm to base e (Chapter 27 only)........... 105
$\log_{10} z$
    common logarithm............................ 105
$\log_a z$
    logarithm to general base a 105
$\mathsf{M}(x)$
    Mills' ratio....................................163
$M(n)$
    Motzkin number 621
$\mathbf{M}_\nu(z)$
    modified Struve function....................... 288
$M_{\kappa,\mu}(z)$
    Whittaker confluent hypergeometric function.... 334
$M(a,g)$
    arithmetic-geometric mean 492

$\mathscr{M}(f;s)$
    Mellin transform 29
$\mathrm{M}_\nu^{(j)}(z,h)$
    modified Mathieu function 667
$M(a,b,z)$
    Kummer confluent hypergeometric function 322
$\mathbf{M}(a,b,z)$
    Olver's confluent hypergeometric function 322
$M_n(x;\beta,c)$
    Meixner polynomial........................... 462
maj
    major index................................... 632
$\mathrm{Mc}_n^{(j)}(z,h)$
    radial Mathieu function 668
$\mathrm{Me}_\nu(z,q)$
    modified Mathieu function 667
$\mathrm{me}_\nu(z,q)$
    Mathieu function of noninteger order........... 664
$\mathrm{me}_n(z,q)$
    Mathieu function 665
$\mathrm{Ms}_n^{(j)}(z,h)$
    radial Mathieu function 668
$\mu(n)$
    Möbius function 639
$\mathcal{N}$
    winding number................................ 16
$N(n,k)$
    Narayana number............................. 622
$\mathrm{nc}(z,k)$
    Jacobian elliptic function 550
$\mathrm{nd}(z,k)$
    Jacobian elliptic function 550
$\mathrm{ns}(z,k)$
    Jacobian elliptic function 550
$\nu(n)$
    number of distinct primes dividing n 638
$O(x)$
    order not exceeding 42
$o(x)$
    order less than 42
$O_n(x)$
    Neumann's polynomial........................ 247
$\mathrm{P_I}, \mathrm{P_{II}}, \mathrm{P_{III}}, \mathrm{P'_{III}}, \mathrm{P_{IV}}, \mathrm{P_V}, \mathrm{P_{VI}}$
    Painlevé transcendents........................ 724
$p(\mathrm{condition}, n)$
    restricted number of partitions of n 627

$P\left\{\begin{matrix} \alpha & \beta & \gamma \\ a_1 & b_1 & c_1 & z \\ a_2 & b_2 & c_2 \end{matrix}\right\}$

Riemann's P-symbol for solutions of the generalized hypergeometric differential equation 396

$\wp(z)$ $(=\wp(z|\mathbb{L}) = \wp(z;g_2,g_3))$
Weierstrass $\wp$-function 570

$p(n)$
total number of partitions of n 618

$\mathsf{P}_\nu(x)$: $\mathsf{P}_\nu^\mu(x)$ with $\mu = 0$ 352, 353

$P_n(x)$
Legendre polynomial 439

$P_\nu(z)$: $P_\nu^\mu(z)$ with $\mu = 0$ 352, 353, 375

$p_k(n)$
number of partitions of n into at most k parts ... 626

$\mathsf{P}_\nu^\mu(x)$
Ferrers function of the first kind 353

$P_\nu^\mu(z)$
associated Legendre function of the first kind
.... 353, 375

$P_n^*(x)$
shifted Legendre polynomial 439

$P_n^{(\alpha,\beta)}(x)$
Jacobi polynomial 439

$\mathsf{P}_{-\frac{1}{2}+i\tau}^{-\mu}(x)$
conical function 372

$P(a,z)$
normalized incomplete gamma function 174

$P_n(x;c)$
associated Legendre polynomial 474

$p_k(\leq m, n)$
number of partitions of n into at most k parts, each less than or equal to m 626

$p_k(\mathcal{D}, n)$
number of partitions of n into at most k distinct parts
.... 627

$P_n^{(\alpha,\beta)}(x;c)$
associated Jacobi polynomial 474

$P_n^{(\lambda)}(x;\phi)$
Meixner–Pollaczek polynomial 462

$P_{m,n}^{\alpha,\beta,\gamma}(x,y)$
triangle polynomial 478

$P_n^{(\lambda)}(x;a,b)$
Pollaczek polynomial 476

$p_n(x;a,b;q)$
little q-Jacobi polynomial 471

$P_n^{(\alpha,\beta)}(x;c,d;q)$
big q-Jacobi polynomial 471

$p_n(x;a,b,\overline{a},\overline{b})$
continuous Hahn polynomial 462

$P_n(x;a,b,c;q)$
big q-Jacobi polynomial 471

$p_n(x;a,b,c,d\,|\,q)$
Askey–Wilson polynomial 472

ph
phase 15

$\phi(n)$
Euler's totient 638

$\Phi_1(t;\mathbf{x})$
fold catastrophe 776

$\Phi_2(t;\mathbf{x})$
cusp catastrophe 776

$\Phi_3(t;\mathbf{x})$
swallowtail catastrophe 776

$\Phi_K(t;\mathbf{x})$
cuspoid catastrophe 776

$\phi_k(n)$
sum of powers of integers relatively prime to n .. 638

$\phi_\lambda^{(\alpha,\beta)}(t)$
Jacobi function 394

$\Phi^{(\mathrm{E})}(s,t;\mathbf{x})$
elliptic umbilic catastrophe 776

$\Phi^{(\mathrm{H})}(s,t;\mathbf{x})$
hyperbolic umbilic catastrophe 776

$\phi(\rho,\beta;z)$
generalized Bessel function 261

$\Phi(z,s,a)$
Lerch's transcendent 612

$\Phi^{(1)}(a;b,b';c;x,y)$
first q-Appell function 423

$\Phi^{(2)}(a;b,b';c,c';x,y)$
second q-Appell function 423

$\Phi^{(3)}(a,a';b,b';c;x,y)$
third q-Appell function 423

$\Phi^{(4)}(a;b;c,c';x,y)$
fourth q-Appell function 423

$_{r+1}\phi_s\left(\begin{smallmatrix}a_0,a_1,\ldots,a_r\\b_1,b_2,\ldots,b_s\end{smallmatrix};q,z\right)$
basic hypergeometric (or q-hypergeometric) function
.... 423

$_{r+1}\phi_s(a_0,a_1,\ldots,a_r;b_1,b_2,\ldots,b_s;q,z)$
basic hypergeometric (or q-hypergeometric) function
.... 423

NOTATIONS 883

π
 set of plane partitions 629
$\pi(x)$
 number of primes not exceeding x 638
$\Pi(\alpha^2, k)$
 Legendre's complete elliptic integral of the third kind
 487
$\Pi(\phi, \alpha^2, k)$
 Legendre's incomplete elliptic integral of the third kind 487
$pp(n)$
 number of plane partitions of n 629
$\mathrm{pq}(z, k)$
 generic Jacobian elliptic function 550
$Ps_n^m(z, \gamma^2)$
 spheroidal wave function of complex argument .. 700
$\mathsf{Ps}_n^m(x, \gamma^2)$
 spheroidal wave function of the first kind 699
$\psi(x)$
 Chebyshev ψ-function 613
$\psi(z)$
 psi (or digamma) function 136
$\Psi_2(\mathbf{x})$
 Pearcey integral 777
$\Psi_K(\mathbf{x})$
 canonical integral 776
$\psi^{(n)}(z)$
 polygamma functions 144
$\Psi^{(E)}(\mathbf{x})$
 canonical integral 776
$\Psi^{(H)}(\mathbf{x})$
 canonical integral 776
$\Psi_3(\mathbf{x}; k)$
 swallowtail canonical integral function 777, 778
$\Psi_K(\mathbf{x}; k)$
 diffraction catastrophe 777
$\Psi^{(E)}(\mathbf{x}; k)$
 elliptic umbilic canonical integral function
 .... 777, 779, 780
$\Psi^{(H)}(\mathbf{x}; k)$
 hyperbolic umbilic canonical integral function
 .... 777, 779, 781
$\Psi(a; b; \mathbf{T})$
 confluent hypergeometric function of matrix argument (second kind) 768, 770
${}_r\psi_s\binom{a_1, a_2, \ldots, a_r}{b_1, b_2, \ldots, b_s}; q, z\bigr)$
 bilateral basic hypergeometric (or bilateral q-hypergeometric) function 423
${}_r\psi_s(a_1, a_2, \ldots, a_r; b_1, b_2, \ldots, b_s; q, z)$
 bilateral basic hypergeometric (or bilateral q-hypergeometric) function 423

$\mathsf{Q}_\nu(x)$: $\mathsf{Q}_\nu^\mu(x)$ with $\mu = 0$ 352, 353
$Q_n(x; \alpha, \beta, N)$
 Hahn polynomial 462
$Q_\nu(z)$: $Q_\nu^\mu(z)$ with $\mu = 0$ 352, 354, 375
$\mathsf{Q}_\nu^\mu(x)$
 Ferrers function of the second kind 353
$Q_\nu^\mu(z)$
 associated Legendre function of the second kind
 .... 354, 375
$\boldsymbol{Q}_\nu^\mu(z)$
 Olver's associated Legendre function 354, 375
$\widehat{\mathsf{Q}}^{-\mu}_{-\frac{1}{2}+i\tau}(x)$
 conical function 372
$Q(a, z)$
 normalized incomplete gamma function 174
$Q_n(x; a, b \,|\, q)$
 Al-Salam–Chihara polynomial 473
$Q_n(x; a, b \,|\, q^{-1})$
 q^{-1}-Al-Salam–Chihara polynomial 473
$Q_n(x; \alpha, \beta, N; q)$
 q-Hahn polynomial 470
$Qs_n^m(z, \gamma^2)$
 spheroidal wave function of complex argument .. 700
$\mathsf{Qs}_n^m(x, \gamma^2)$
 spheroidal wave function of the second kind 700

$r(n)$
 Schröder number 622
$r_{\mathrm{tp}}(\epsilon, \ell)$
 outer turning point for Coulomb functions 748
$R_{m,n}^{(\alpha)}(z)$
 disk polynomial 477
$R_{-a}(b_1, b_2, \ldots, b_n; z_1, z_2, \ldots, z_n)$
 multivariate hypergeometric function 498
$R_{-a}(\mathbf{b}; \mathbf{z})$
 multivariate hypergeometric function 498
$R_n(x; \gamma, \delta, N)$
 dual Hahn polynomial 467
$R_n(x; \alpha, \beta, \gamma, \delta)$
 Racah polynomial 467
$R_n(x; \alpha, \beta, \gamma, \delta \,|\, q)$
 q-Racah polynomial 474

$R_C(x,y)$
    Carlson's elliptic integral with two variables 487

$R_D(x,y,z)$
    elliptic integral symmetric in only two variables .. 498

$R_F(x,y,z)$
    symmetric elliptic integral of first kind 497

$R_G(x,y,z)$
    symmetric elliptic integral of second kind 498

$\rho_{\text{tp}}(\eta,\ell)$
    outer turning point for Coulomb radial functions
    742

$R_J(x,y,z,p)$
    symmetric elliptic integral of third kind 497

$\mathfrak{S}_n$
    set of permutations of $\{1,2,\ldots,n\}$ 631

$S(z)$
    Fresnel integral 160

$S_{\mu,\nu}(z)$
    Lommel function 295

$s_{\mu,\nu}(z)$
    Lommel function 294

$S_n^{m(j)}(z,\gamma)$
    radial spheroidal wave function 703

$\mathcal{S}(f;s)$
    Stieltjes transform 29

$S(n,k)$
    Stirling number of the second kind 624

$s(n,k)$
    Stirling number of the first kind 624

$S_n(x;q)$
    Stieltjes–Wigert polynomial 471

$S_n^m(z,\xi)$
    Ince polynomials 676

$s(\epsilon,\ell;r)$
    regular Coulomb function 748

$S(k,h)(x)$
    Sinc function 77

$S_n(x;a,b,c)$
    continuous dual Hahn polynomial 467

$\text{sc}(z,k)$
    Jacobian elliptic function 550

$scdE_{2n+3}^m(z,k^2)$
    Lamé polynomial 690

$scE_{2n+2}^m(z,k^2)$
    Lamé polynomial 690

$\text{sd}(z,k)$
    Jacobian elliptic function 550

$sdE_{2n+2}^m(z,k^2)$
    Lamé polynomial 690

$\text{Se}_\nu(z,q)$
    modified Mathieu function 667

$\text{se}_\nu(z,q)$
    Mathieu function of noninteger order 665

$\text{se}_n(z,q)$
    Mathieu function 654

$sE_{2n+1}^m(z,k^2)$
    Lamé polynomial 690

$\sec z$
    secant function 112

$\text{sech}\, z$
    hyperbolic secant function 123

$\text{Shi}(z)$
    hyperbolic sine integral 150

$\text{Si}(z)$
    sine integral 150

$\text{si}(z)$
    sine integral 150

$\text{Si}(a,z)$
    generalized sine integral 188

$\text{si}(a,z)$
    generalized sine integral 188

$\sigma_n(\nu)$
    Rayleigh function 240

$\sigma_\ell(\eta)$
    Coulomb phase shift 742

$\sigma_\alpha(n)$
    sum of powers of divisors of n 638

$\sigma(z)\ (=\sigma(z|\mathbb{L})=\sigma(z;g_2,g_3))$
    Weierstrass sigma function 570

$\sin z$
    sine function 112

$\text{Sin}_q(x)$
    q-sine function 422

$\sin_q(x)$
    q-sine function 422

$\sinh z$
    hyperbolic sine function 123

$\text{sn}(z,k)$
    Jacobian elliptic function 550

T_n
    tangent numbers 596

$T_n(x)$
  Chebyshev polynomial of the first kind 439
$T_n^*(x)$
  shifted Chebyshev polynomial of the first kind .. 439
$\tan z$
  tangent function 112
$\tanh z$
  hyperbolic tangent function 123
$\tau(n)$
  Ramanujan's tau function 647
$\theta_j(z|\tau)$
  theta function 524
$\theta_j(z,q)$
  theta function 524
$\theta(\mathbf{z}|\mathbf{\Omega})$
  Riemann theta function 538
$\hat{\theta}(\mathbf{z}|\mathbf{\Omega})$
  scaled Riemann theta function 538
$\theta{\genfrac{[}{]}{0pt}{}{\alpha}{\beta}}(\mathbf{z}|\mathbf{\Omega})$
  Riemann theta function with characteristics 539
$U_n(x)$
  Chebyshev polynomial of the second kind 439
$U_m(t)$
  generalized Airy function 207
$U_n^*(x)$
  shifted Chebyshev polynomial of the second kind
  439
$U(a,z)$
  parabolic cylinder function 304
$\mathsf{U}(x,t)$
  Voigt function 167
$\overline{U}(a,x)$
  parabolic cylinder function 305
$U(a,b,z)$
  Kummer confluent hypergeometric function 322
$uE_{2n}^m(z,k^2)$
  Lamé polynomial 690
$\mathcal{V}_{a,b}(f)$
  total variation 6
$V_n(x)$
  Chebyshev polynomial of the third kind 439
$V_m(t)$
  generalized Airy function 207
$\overline{V}_m(t)$
  generalized Airy function 207

$V(a,z)$
  parabolic cylinder function 304
$\mathsf{V}(x,t)$
  Voigt function 167
$\mathscr{W}$
  Wronskian 26
$W(x)$
  Lambert W-function 111
$w(z)$
  complementary error function 160
$W_n(x)$
  Chebyshev polynomial of the fourth kind 439
$\mathrm{Wp}(x)$
  principal branch of Lambert W-function 111
$\mathrm{Wm}(x)$
  nonprincipal branch of Lambert W-function 111
$W_{\kappa,\mu}(z)$
  Whittaker confluent hypergeometric function 334
$W(a,x)$
  parabolic cylinder function 314
$w_\mathrm{I}(z,\lambda)$
  basic solution, Hill's equation 674
$w_\mathrm{II}(z,\lambda)$
  basic solution, Hill's equation 674
$w_\mathrm{I}(z;a,q)$
  basic solution, Mathieu's equation 653
$w_\mathrm{II}(z;a,q)$
  basic solution, Mathieu's equation 653
$W_n(x;a,b,c,d)$
  Wilson polynomial 467
$\xi(s)$
  Riemann's ξ-function 604
$y_{\nu,m}$
  zeros of the Bessel function $Y_\nu(x)$ 235
$y'_{\nu,m}$
  zeros of the Bessel function derivative $Y'_\nu(x)$ 235
$Y_\nu(z)$
  Bessel function of the second kind 217
$\widetilde{Y}_\nu(x)$
  Bessel function of imaginary order 248
$\mathsf{y}_n(z)$
  spherical Bessel function of the second kind 262
$y_n(x;a)$
  Bessel polynomial 476
$Y_{l,m}(\theta,\phi)$
  spherical harmonic 378

$Y_l^m(\theta,\phi)$
  surface harmonic of the first kind 378

$\mathscr{Z}_\nu(z)$
  modified cylinder function 249

$Z_\kappa(\mathbf{T})$
  zonal polynomial 769

z^a
  power function 105

$Z(x|k)$
  Jacobi's zeta function 562

$\zeta(s)$
  Riemann zeta function 602

$\zeta_x(s)$
  incomplete Riemann zeta function 189

$\zeta(s,a)$
  Hurwitz zeta function 607

$\zeta(z)\ (=\zeta(z|\mathbb{L})=\zeta(z;g_2,g_3))$
  Weierstrass zeta function 570

Index

Abel means 33
Abel summability 33, 34
Abel–Plana formula 63
Abelian functions 545
absolute error 73
acceleration of convergence
    definition 93
    for sequences 93–94
    for series 93–94
    limit-preserving 93
accumulation point 15
acoustics
    canonical integrals 792
additive number theory 644–647
    Dedekind modular function 646
    Dedekind sum 646
    discriminant function 646
    Euler's pentagonal number theorem 646
    Goldbach conjecture 644
    Jacobi's identities 645
    notation 638
    partition function 644
        unrestricted 645
    Ramanujan's identity 646
    Ramanujan's tau function 646
    representation by squares 645
    Waring's problem 645
aerodynamics
    Struve functions 298
affine Weyl groups
    Painlevé equations 732
Airy functions **194**
    analytic properties 194
    applications
        mathematical 208
        physical 209
        ship waves 790
    approximations
        expansions in Chebyshev series 211
        in terms of elementary functions 211
        in the complex plane 212
    asymptotic expansions 198–199
        error bounds 199
        exponentially-improved 199
    computation 209–210
    connection formulas 194
    definitions 194
    differential equation 194
        for products 203
        initial values 194
        numerically satisfactory solutions 194
        Riccati form 194
    Dirac delta 38
    envelope functions 59
    generalized .... *see* generalized Airy functions.
    graphics 195
    incomplete 208
    integral identities 787
    integral representations 196, 203
    integrals
        approximations 211, 212
        asymptotic approximations 202
        definite 202
        indefinite 202
        of products 204
        repeated 203
        tables 211
    Laplace transforms 203
    Maclaurin series 196
    Mellin transform 203
    modulus and phase
        asymptotic expansions 200
        definitions 199
        graphs 195
        identities 200
        monotonicity 200
        relation to Bessel functions 199
        relation to zeros 200
    notation 194
    products
        differential equation 203
        integral representations 203
        integrals 204
        Wronskian 203
    relation to umbilics 777
    relations to other functions
        Bessel functions 196–197
        confluent hypergeometric functions .. 197, 328, 338
        Hankel functions 196–197
        modified Bessel functions 196–197
    Stieltjes transforms 203
    tables
        complex variables 210
        integrals 211
        real variables 210
        zeros 201–202, 211

Wronskians 194
zeros
    asymptotic expansions 201
    computation 76, 210
    differentiation 200
    relation to modulus and phase 200
    tables 201–202, 211
Airy transform 203
Airy's equation
    .... see Airy functions, differential equation.
Aitken's Δ^2-process
    for sequences 93
    iterated 93
Al-Salam–Chihara polynomials 473
algebraic curves
    Riemann surface 543, 544, 546
algebraic equations
    parametrization via Jacobian elliptic functions .. 563
        spherical trigonometry 564
        uniformization 564
algebraic Lamé functions 693
alternant
    determinant 3
amplitude (am) function **561**
    applications 564
    approximations
        small k, k' 562
        small x 561
    computation 567
    definition 561
    Fourier series 562
    integral representation 561
    quasi-periodicity 561
    relation to elliptic integrals 562
    relation to Gudermannian function 562
    special values 561
    tables 567
analytic continuation 19
    by reflection 19
analytic function 16
    at infinity 17
    in a domain 16
    singularities 19
    zeros 19
Anger function .... see Anger–Weber functions.
Anger–Weber functions **295**
    analytic properties 295
    asymptotic expansions
        large argument 297
        large order 298
    computation 87, 299
    definitions 295
    derivatives 297
    differential equation 295
    graphics 296
    incomplete 300
    integral representations 295
    integrals 297
    interrelations 296
    Maclaurin series 296
    notation 288
    order 288
    recurrence relations 297
    relations to other functions
        Fresnel integrals 297
        Lommel functions 296
        Struve functions 297
    series expansions
        power series 296
        products of Bessel functions 297
    special values 297
    sums 297
    tables 299
angle between arcs 17
angular momenta 758
angular momentum
    generalized hypergeometric functions 418
angular momentum coupling coefficients
    ..... see $3j$ symbols, $6j$ symbols, and $9j$ symbols.
angular momentum operator
    spherical coordinates 379
annulus 19
antenna research
    Lamé functions 694
Appell functions **412**
    analytic continuation 414
    applications
        physical 417
    computation 418
    definition 412–413
    integral representations 414
    integrals 414
        inverse Laplace transform 414
    notation 412
    partial differential equations 413
    relation to Legendre's elliptic integrals 490
    relation to symmetric elliptic integrals 509
    relations to hypergeometric functions 414
    transformations of variables 414–415
        quadratic 415
        reduction formulas 414
approximation techniques
    Chebyshev-series expansions 97
    least squares 99–100
    minimax polynomials 96
    minimax rational functions 97
    Padé 98–99
    splines 100

arc length
    Jacobian elliptic functions....................563
arc(s)...16
    angle between.................................17
area of triangle..................................246
argument principle..............see phase principle.
arithmetic Fourier transform.....................647
arithmetic mean................................3, 13
arithmetic progression..............................2
arithmetic-geometric mean.....................**492**
    hypergeometric function......................400
    integral representations.....................492
    Jacobian elliptic functions..................566
    Legendre's elliptic integrals............492–493
    symmetric elliptic integrals.................505
arithmetics
    complex.......................................73
    exact rational................................72
    floating-point................................72
    interval......................................72
    level-index...................................73
Askey polynomials.................................475
Askey scheme for orthogonal polynomials...........464
Askey–Gasper inequality
    Jacobi polynomials...........................478
Askey–Wilson class orthogonal polynomials...472–474
    as eigenfunctions of a q-difference operator......472
    asymptotic approximations....................474
    interrelations with other orthogonal polynomials
   464
    orthogonality properties.....................472
    representation as q-hypergeometric functions
   472–474
Askey–Wilson polynomials..........................472
    asymptotic approximations....................474
    relation to q-hypergeometric functions......472–474
associated Anger–Weber function
    ....see Anger–Weber functions.
associated Laguerre functions.....................754
associated Legendre equation................352, 375
    exponent pairs...............................352
    numerically satisfactory solutions..........352, 375
    singularities................................352
    standard solutions..................352, 354, 375
associated Legendre functions.................**352**
    ....see also Ferrers functions.
    addition theorems.......................370, 377
    analytic continuation........................376
    analytic properties..........................375
    applications.............................378–379
    asymptotic approximations
    ....see uniform asymptotic approximations.
    behavior at singularities...............361, 375
    computation..................................379

connection formulas........................362, 375
continued fractions..............................364
cross-products...................................353
definitions..............................353–354, 375
degree...352
derivatives......................................362
    with respect to degree or order..............363
differential equation
    ....see associated Legendre equation.
expansions in series of..........................370
generalized......................................378
generating functions.......................361, 375
graphics.........................357–359, 375–376
Heine's formula..................................377
hypergeometric representations........353–354, 375
integer degree and order...............360–361, 375
integer order..............................360, 375
integral representations...................363, 377
integrals
    definite.....................................369
    Laplace transforms...........................370
    Mellin transforms............................370
    products.....................................369
notation...352
of the first kind................................353
of the second kind...............................354
Olver's....................................354, 375
order..352
orthogonality....................................369
principal values (or branches)...................375
recurrence relations.......................362, 375
relations to other functions
    elliptic integrals...........................360
    Gegenbauer function..........................355
    hypergeometric function............353, 354, 394
    Jacobi function..............................355
    Legendre polynomials.........................360
Rodrigues-type formulas..........................360
special values.............................359, 360
sums...................................370–371, 377
tables...380
uniform asymptotic approximations
    large degree.........................366–368, 377
    large order..........................365–366, 377
values on the cut................................376
Whipple's formula................................362
Wronskians.............................352–353, 375
zeros......................................368, 377
associated orthogonal polynomials.................474
    corecursive..................................474
    Jacobi.......................................474
    Legendre.....................................474
astrophysics
    error functions and Voigt functions..........169

Heun functions and Heun's equation 720
asymptotic and order symbols 42
    definition 42
    differentiation 42
    integration 42
asymptotic approximations and expansions .... *see also* asymptotic approximations of integrals, asymptotic approximations of sums and sequences, asymptotic solutions of difference equations, asymptotic solutions of differential equations, *and* asymptotic solutions of transcendental equations.
    algebraic operations 42
    cases of failure 52, 66
    differentiation 42
    double asymptotic properties
        Bessel functions 258
        Hankel functions 258
        Kelvin functions 273
        modified Bessel functions 257
        parabolic cylinder functions 311
    exponentially-improved expansions 67–69
    generalized 43
    hyperasymptotic expansions 68
    improved accuracy via numerical transformations
        69
    integration 42
    logarithms of 42
    null 42
    numerical use of 66, 69
    Poincaré type 42
    powers of 42
    re-expansion of remainder terms 67–69
    reversion of 43
    Stokes phenomenon 67
    substitution of 42
    uniform 43
    uniqueness 42
    via connection formulas 66
asymptotic approximations of integrals 43–55
    Bleistein's method 45
    Chester–Friedman–Ursell method 48
    coalescing critical points 48
    coalescing peak and endpoint 45
    coalescing saddle points 48
    distributional methods 51–55
    Fourier integrals 44
    Haar's method 46
    integration by parts 43
    inverse Laplace transforms 46–47
    Laplace transforms 43
    Laplace's method 44–45, 47
    Mellin transform methods 48
        extensions 49–51
    method of stationary phase 45
        extensions 45
    method of steepest descents 47
    multidimensional integrals 51
    Stieltjes transforms 52–53
        generalized 53
    Watson's lemma 44, 46
        generalized 44
asymptotic approximations of sums and sequences
    63–66
    Abel–Plana formula 63
    Darboux's method 65–66
    entire functions 64
    Euler–Maclaurin formula 63
    summation by parts 63
asymptotic scale or sequence 43
asymptotic solutions of difference equations 61–63
    characteristic equation 62
    coincident characteristic values 62
    Liouville–Green (or WKBJ) type approximations
        62
    transition points 63
    turning points 63
    with a parameter 62–63
asymptotic solutions of differential equations 55–61
    characteristic equation 56
    coincident characteristic values 57
    error-control function 57
    Fabry's transformation 57
    irregular singularities of rank 1 56
    Liouville–Green approximation theorem 57
    Liouville–Green (or WKBJ) approximations 57
    numerically satisfactory solutions 58
    resurgence 57, 68
    with a parameter 58–61
        classification of cases 58
        coalescing transition points 61
        connection formulas across transition points 61
        in terms of Airy functions 59
        in terms of Bessel functions of fixed order ... 60–61
        in terms of Bessel functions of variable order ... 61
        in terms of elementary functions 59
        Liouville transformation 58
        transition points 58
        turning points 58
asymptotic solutions of transcendental equations 43
    Lagrange's formula 43
atomic photo-ionization
    Coulomb functions 753
atomic physics
    Coulomb functions 754
    error functions 169
atomic spectra
    Coulomb functions 753
atomic spectroscopy

$3j, 6j, 9j$ symbols..............................765
attractive potentials
    Coulomb functions.........................753, 754
auxiliary functions for Fresnel integrals
    approximations................................170
    asymptotic expansions........................164
    computation..................................169
    definitions...................................160
    derivatives...................................164
    integral representations......................163
        Mellin–Barnes integrals....................163
    symmetry.....................................162
auxiliary functions for sine and cosine integrals
    analytic continuation..........................151
    approximations................................156
    asymptotic expansions........................153
        exponentially-improved....................154
    Chebyshev-series expansions..................157
    computation..................................156
    definition....................................150
    integral representations......................152
    principal values..............................151
    relation to confluent hypergeometric functions...153
    tables..156
axially symmetric potential theory 501

Bäcklund transformations
    classical orthogonal polynomials.................478
    Painlevé transcendents....................730–732
backward recursion...............................85
Bailey's $_2F_1(-1)$ sum
    q-analog....................................426
Bailey's $_4F_3(1)$ sum
    q-analogs (first and second)..................427
Bailey's $_2\psi_2$ transformations
    bilateral q-hypergeometric function..............429
Bailey's bilateral summations
    bilateral q-hypergeometric function..............427
bandlimited functions 706
Barnes' beta integral 143
Barnes' G-function
    asymptotic expansion..........................144
    definition....................................144
    infinite product...............................144
    integral representation........................144
    recurrence relation............................144
Barnes' integral
    Ferrers functions.............................369
Bartky's transformation
    Bulirsch's elliptic integrals.....................487
    symmetric elliptic integrals....................504
basic elliptic integrals 512
basic hypergeometric functions...*see* bilateral q-hypergeometric function *and* q-hypergeometric function.
Basset's integral

    modified Bessel functions 253
Bell numbers
    asymptotic approximations.....................623
    definition....................................623
    generating function...........................623
    recurrence relation............................623
    table..623
Bernoulli monosplines.............................597
Bernoulli numbers..............................**588**
    arithmetic properties..........................593
    asymptotic approximations.....................593
    computation..................................598
    definition....................................588
    degenerate...................................596
    explicit formulas.............................591
    factors.......................................593
    finite expansions.............................590
    generalizations...........................596, 597
    generating function...........................588
    identities....................................591
    inequalities..................................593
    integral representations........................592
    inversion formulas............................591
    irregular pairs................................598
    Kummer congruences..........................593
    notation.....................................588
    of the second kind............................596
    recurrence relations
        linear......................................591
        quadratic and higher order..................595
    relations to
        Eulerian numbers...........................591
        Genocchi numbers..........................595
        Stirling numbers............................596
        tangent numbers............................596
    sums..595
    tables...................................589, 598
Bernoulli polynomials..........................**588**
    applications
        mathematical...........................597–598
        physical...................................598
    asymptotic approximations.....................593
    computation..................................598
    definitions...................................588
    derivative....................................590
    difference equation............................589
    explicit formulas.............................591
    finite expansions.............................590
    generalized..............................596, 597
    generating function...........................588
    graphs.......................................589
    inequalities..................................593
    infinite series expansions
        Fourier....................................592

other 592
integral representations 592
integrals 594
    compendia 595
Laplace transforms 595
multiplication formulas 590
notation 588
recurrence relations
    linear 591
    quadratic 595
relation to Eulerian numbers 591
relation to Riemann zeta function 591
representation as sums of powers 589
special values 590
sums 595
symbolic operations 590
symmetry 589
tables 589
zeros
    complex 594
    multiple 594
    real 594
Bernoulli's lemniscate 515
Bernstein–Szegö polynomials 474
Bessel functions **217**
.... *see also* cylinder functions, Hankel functions, Kelvin functions, modified Bessel functions, *and* spherical Bessel functions.
    addition theorems 246
    analytic continuation 226
    applications
        asymptotic solutions of differential equations
        274–275
        electromagnetic scattering 275
        Helmholtz equation 275
        Laplace's equation 275
        oscillation of chains 275
        oscillation of plates 276
        wave equation 275
    approximations 281
    asymptotic expansions for large argument .. 228–230
        error bounds 229–230
        exponentially-improved 230
    asymptotic expansions for large order 231–235
        asymptotic forms 231
        Debye's expansions 231–232
        double asymptotic properties 235, 258
        resurgence properties of coefficients . 233
        transition region 232
        uniform 232–235
    branch conventions 218
    computation 276–277
    computation by quadrature 83
    computation by recursion 87

connection formulas 222
contiguous 235
continued fractions 226
cross-products 222, 223
    zeros 238
definite integrals 203
definitions 217–218
derivatives
    asymptotic expansions for large argument .. 229
    asymptotic expansions for large order 231–232
    explicit forms 222
    uniform asymptotic expansions for large order
    232
    with respect to order 227–228
    zeros .... *see* zeros of Bessel functions.
differential equations 217, 226
    .... *see also* Bessel's equation.
Dirac delta 38
envelope functions 61
expansions in partial fractions 247
expansions in series of 247–248
Fourier–Bessel expansion 248
generalized 261
generating functions 226
graphics 218–222
incomplete 261
inequalities 227
infinite integrals 448
infinite products 235
integral representations
    along the real line 223–224
    compendia 226
    contour integrals 224–225
    Mellin–Barnes type 225
    products 225
integrals ..... *see also* integrals of Bessel and Hankel functions *and* Hankel transforms.
    approximations 281
    computation 277
    tables 279, 280
limiting forms 223
minimax rational approximation 98
modulus and phase functions
    asymptotic expansions for large argument .. 231
    basic properties 230
    definitions 230
    graphics 218
    relation to zeros 235
monotonicity 227
multiplication theorem 246
notation 217
of imaginary argument
    .... *see* modified Bessel functions.
of imaginary order

applications 248
definitions 248
graphs 221–222
limiting forms 248
numerically satisfactory pairs 248
uniform asymptotic expansions for large order
.... 248
zeros 248
of matrix argument **769**
applications 773
asymptotic approximations 770
definitions 769
notation 768
of the first and second kinds 768
properties 769
relations to confluent hypergeometric functions of
matrix argument 770
of the first, second, and third kinds 217–218
orthogonality 243, 244
power series 223
principal branches (or values) 217–218
recurrence relations 222–223
relations to other functions
Airy functions 196–197
confluent hypergeometric functions 228
elementary functions 228
generalized Airy functions 206
generalized hypergeometric functions 228
parabolic cylinder functions 228, 315
sums 246–248
addition theorems 246–247
compendia 248
expansions in series of Bessel functions ... 247–248
multiplication theorem 246
tables 278–279
Wronskians 222
zeros .... see zeros of Bessel functions.
Bessel polynomials 264, 476
asymptotic expansions 476
definition 476
differential equations 476
generalized 476
orthogonality properties 476
recurrence relations 476
relations to other functions
complex orthogonal polynomials 83
confluent hypergeometric functions 476
generalized hypergeometric functions 476
Jacobi polynomials 476
Bessel transform .... see Hankel transform.
Bessel's equation 217
inhomogeneous forms 288, 294, 295
numerically satisfactory solutions 218
singularities 217

standard solutions 217–218
Bessel's inequality
Fourier series 13
Bessel's integral
Bessel functions 223
best uniform polynomial approximation 96
best uniform rational approximation 97
beta distribution
incomplete beta functions 189
beta function **142**
.... see also incomplete beta functions.
applications
physical 145–146
definition 142
integral representations 142
multidimensional 143
integrals 143
multivariate .... see multivariate beta function.
beta integrals 142–143
Bickley function 259
applications 276
approximations 281
biconfluent Heun equation 718
application to Rossby waves 720
Bieberbach conjecture 417, 479
Jacobi polynomials 479
bifurcation sets 781
visualizations 782
big q-Jacobi polynomials 471
bilateral basic hypergeometric function
.... see bilateral q-hypergeometric function.
bilateral hypergeometric function 408
bilateral q-hypergeometric function
Bailey's $_2\psi_2$ transformations 429
Bailey's bilateral summations 427
computation 432
definition 423
notation 420
Ramanujan's $_1\psi_1$ summation 427
special cases 427–428
transformations 432
bilateral series 408
bilinear transformation 17
cross ratio 17
$SL(2,\mathbb{Z})$ 579
binary number system 72
binary quadratic sieve
number theory 648
Binet's formula
gamma function 140
binomial coefficients
definitions 619
generating functions 619
identities 619

limiting form 619
recurrence relations 619
relation to lattice paths 619
tables 619, 635
binomial expansion 108
binomial theorem 2
binomials 2
black holes
   Heun functions 720
Bohr radius
   Coulomb functions 754
Bohr–Mollerup theorem
   gamma function 138
   q-gamma function 145
Boole summation formula 597
Borel summability 33
Borel transform theory
   applications to asymptotic expansions 68
Bose–Einstein condensates
   Lamé functions 694
Bose–Einstein integrals
   computation 614
   definition 611
   relation to polylogarithms 612
Bose–Einstein phase transition 614
bound-state problems
   hydrogenic 754
   Whittaker functions 754
boundary points 11, 15
boundary-value methods or problems
   difference equations 86, 87
   ordinary differential equations 88
   parabolic cylinder functions 317
bounded variation 6
Boussinesq equation
   Painlevé transcendents 739
box
   plane partitions 629
branch
   of multivalued function 20, 104
      construction 20
      example 20
branch cut 104
branch point 20
   movable 724
Bromwich integral 83
Bulirsch's elliptic integrals **487**
   computation 518
   first, second, and third kinds 486
   notation 486
   relation to symmetric elliptic integrals 508
calculus
   complex variable 14–18
   one variable 4–7

   two or more variables 7–9
calculus of finite differences 597
canonical integrals **776**
   applications
      acoustics 792
      caustics 791
      integrals with coalescing critical points . . . 789–790
      optics 791
      quantum mechanics 791
   asymptotic approximations 789–790
   computation 792
   convergent series 787
   definitions 776
   differential equations 788
   integral identities 787–788
   notation 776
   relations to other functions
      Airy function 777
      Pearcey integral 777
   special cases 777
   symmetries 777
   visualizations of modulus 778–779
   visualizations of phase 780–781
   zeros 785–787
cardinal function 77
cardinal monosplines 597–598
cardinal spline functions 597
Carmichael numbers
   number theory 644
Casimir forces
   Bernoulli polynomials 598
Casimir–Polder effect
   Riemann zeta function 614
Catalan numbers
   definitions 620
   generating function 621
   identities 623
   limiting forms 621
   recurrence relations 621
   relation to lattice paths 620
   table 621
Catalan's constant
   Riemann zeta function 610
Cauchy determinant 4
Cauchy principal values
   integrals 6
Cauchy's integral formula 16
   for derivatives 16
Cauchy's theorem 16
Cauchy–Riemann equations 16
Cauchy–Schwarz inequalities for sums and integrals
   . . . . 12, 13
caustics
   Airy functions 209

canonical integrals 791
Cayley's identity for Schwarzian derivatives 27
central differences in imaginary direction 436
Cesàro means 33
Cesàro summability 33, 34
chain rule
    for derivatives 5, 7
characteristic equation
    difference equations 62
    differential equations 56
characteristics
    Riemann theta functions 539
characters
    number theory
        Dirichlet 642
            induced modulus 642
            orthogonality relation 642
            primitive 642
            principal 642
            quadratic Jacobi symbol 642
            quadratic Legendre symbol 642
            real 642
Charlier polynomials
            see Hahn class orthogonal polynomials.
Chebyshev ψ-function 613
Chebyshev polynomials **438**
    ..*see also* Chebyshev-series expansions *and* classical orthogonal polynomials.
    applications
        approximation theory 478
        solutions of differential equations 478
    computation 479
    continued fractions 450
    definition 439
    derivatives 447
    differential equations 445
    dilated 437
    expansions in series of 96, 459, 461
    explicit representations 442–443
    generating functions 449
    graphs 440
    inequalities 450
    integral representations 448
    integrals 458
    interrelations with other classical orthogonal polynomials 444–445
    leading coefficients 439
    linearization formula 460
    local maxima and minima 451
    normalization 439
    notation 436
    of the first, second, third, and fourth kinds 439
    orthogonality properties
        with respect to integration 96, 439
        with respect to summation 97, 440
    recurrence relations 96, 446
    relations to other functions
        hypergeometric function 394
        Jacobi polynomials 444
        trigonometric functions 442
    Rodrigues formula 442
    scaled 478
    shifted 437, 439
    special values 444
    symmetry 444
    tables 480
        of coefficients 440
    upper bounds 451
    weight functions 439
    zeros 438, 440
Chebyshev-series expansions
    complex variables 97
    computation of coefficients 97
    relation to minimax polynomials 97
    summation 97
chemical reactions
    $3j, 6j, 9j$ symbols 765
chi-square distribution function
    incomplete gamma functions 189
Chinese remainder theorem
    number theory 647
Christoffel coefficients (or numbers) .... *see* Gauss quadrature, Christoffel coefficients (or numbers)
Christoffel–Darboux formula
    classical orthogonal polynomials 438
    confluent form 438
Chu–Vandermonde identity
    hypergeometric function 387
circular trigonometric functions
                see trigonometric functions.
classical dynamics
    Jacobian elliptic functions 566
    Weierstrass elliptic functions 582
classical orthogonal polynomials **438**
    addition theorems 459
    applications
        approximation theory 478
        Bieberbach conjecture 479
        integrable systems 478
        numerical solution of differential equations ... 478
        physical 479
        quadrature 478
        quantum mechanics 479
        Radon transform 479
        random matrix theory 479
        Riemann–Hilbert problems 479
    asymptotic approximations 451–454
    computation 479

connection formulas............................460
contiguous relations...........................446
continued fractions............................450
definitions................................438–439
derivatives................................446–447
differential equations.........................445
expansions in series of....................459–461
explicit representations...................442–443
Fourier transforms.........................456–457
generating functions...........................449
in two or more variables.......................477
inequalities
    local maxima and minima...............450–451
    Turan-type................................450
    upper bounds..............................450
integral representations...................447–448
    for products..............................455
integrals..................................455–459
    compendia.................................459
interrelations
    limiting forms............................445
    linear....................................444
    quadratic.................................445
    with other orthogonal polynomials.........464
Laplace transforms.............................457
leading coefficients...........................439
limiting forms
    Mehler–Heine type formulas................449
linearization formulas.........................460
local maxima and minima....................450–451
Mellin transforms..............................457
multiplication theorems........................460
normalization..................................439
notations......................................436
orthogonality properties..................439, 443
parameter constraints.....................439, 443
Poisson kernels................................461
recurrence relations...........................446
relations to other functions
    confluent hypergeometric functions........442
    generalized hypergeometric functions......442
    hypergeometric function..........393–394, 442
sums.......................................459–461
    Bateman-type..............................461
    compendia.................................461
tables...480
    of coefficients...........................440
upper bounds...................................450
weight functions...............................439
zeros
    asymptotic approximations.............454–455
    distribution..............................438
    inequalities..............................454
classical theta functions...........*see* theta functions.

Clausen's integral.............................611
    tables....................................614
Clebsch–Gordan coefficients............*see* $3j$ symbols.
    relation to generalized hypergeometric functions
   418
Clenshaw's algorithm
    Chebyshev series...........................97
    classical orthogonal polynomials..........480
Clenshaw–Curtis quadrature formula...........79, 82
    comparison with Gauss quadrature...........80
closed point set............................11, 15
closure
    of interval.................................6
    of point sets in complex plane.............15
coalescing saddle points...................789–790
coaxial circles
    symmetric elliptic integrals..............516
coding theory
    combinatorics.............................635
    Krawtchouk and q-Racah polynomials..........479
cofactor............................*see* determinants.
coherent states
    generalized
        confluent hypergeometric functions....346
cols...............................*see* saddle points.
combinatorial design...........................635
combinatorics..................................**618**
    applications..............................635
        mathematical..........................635
        physical..............................635
    generalized hypergeometric functions......417
    hypergeometric identities.................400
    Painlevé transcendents....................739
compact set.....................................18
complementary error function......*see* error functions.
complementary exponential integral
    ....*see* exponential integrals.
completely multiplicative functions............640
complex numbers
    arithmetic operations......................15
    complex conjugates.........................15
    DeMoivre's theorem.........................15
    imaginary part.............................14
    modulus....................................15
    phase......................................15
    polar representation.......................14
    powers.....................................15
    real part..................................14
    triangle inequality........................15
complex physical systems
    incomplete gamma functions................189
complex tori
    theta functions...........................533
computer arithmetic

generalized exponentials and logarithms.........131
computer-aided design
    Cornu's spiral...................................169
conductor
    generalized Bernoulli polynomials..............597
confluent Heun equation..........................717
    applications.....................................720
    properties of solutions..........................718
    special cases....................................717
confluent hypergeometric functions .. *see also* Kummer functions *and* Whittaker functions.
    of matrix argument...........................**770**
        asymptotic approximations................... 771
        computation.................................773
        definition...................................770
        first kind................................... 768
        Laguerre form...............................770
        notation.....................................768
        properties 770
        relations to Bessel functions of matrix argument
       770
        second kind.................................768
    relations to other functions
        Airy functions.............................. 197
        Bessel and Hankel functions 228
        classical orthogonal polynomials.............. 442
        Coulomb functions.....................742, 748
        error functions...............................164
        exponential integrals.........................153
        generalized Bessel polynomials................476
        generalized exponential integral...............186
        Hahn class orthogonal polynomials...........466
        modified Bessel functions 255
        parabolic cylinder functions..............308, 315
        repeated integrals of error functions...........167
        sine and cosine integrals 153
conformal mapping 16–17
    generalized hypergeometric functions............417
    hypergeometric function 399
    Jacobian elliptic functions......................564
    modular functions.............................581
    symmetric elliptic integrals.....................515
    Weierstrass elliptic functions...................581
congruence of rational numbers...................593
conical functions................................**372**
    applications....................................379
    asymptotic approximations
        large degree 374
        large order..................................374
        behavior at singularities......................373
        connection formulas.........................372
        definitions 372
        degree 352
        differential equation.........................372
    generalized Mehler–Fock transformation.........373
    graphics 373
    integral representation 373
    integrals with respect to degree 375
    notation 352, 372
    order 352
    tables..380
    trigonometric expansion........................373
    Wronskians 372
    zeros...375
connected point set...............................15
constants
    roots of.. 23
continued fractions............................24–25
    applications....................................25
    approximants 24
    canonical denominator (or numerator)...........24
    contraction....................................25
    convergence...................................25
    convergents 24
        existence of..................................25
    determinant formula 24
    equivalent......................................24
    even part 25
    extension 25
    fractional transformations......................25
    J-fraction 95
    Jacobi fraction 95
        associated 95
    notation 24
    numerical evaluation
        backward recurrence..........................95
        forward recurrence...........................95
        forward series recurrence.....................96
    odd part 25
    Pringsheim's theorem 25
    quotient-difference algorithm....................95
    recurrence relations 24
    relation to power series.....................94, 95
    S-fraction......................................95
    series 24
    Stieltjes fraction 95
    Van Vleck's theorem 25
continuous dual Hahn polynomials
    .... *see* Wilson class orthogonal polynomials.
continuous dynamical systems and mappings
    Painlevé transcendents.........................739
continuous function
    at a point................................4, 7, 15
    notation..4
    of two variables.............................7, 15
    on a point set..................................7
    on a region...................................15
    on an interval 4

on the left (or right) 4
piecewise 4, 7
removable discontinuity 4
sectionally 4
simple discontinuity 4
continuous Hahn polynomials
.... see Hahn class orthogonal polynomials.
continuous q-Hermite polynomials 473
continuous q^{-1}-Hermite polynomials 473
asymptotic approximations to zeros 474
continuous q-ultraspherical polynomials 473
contour 16
simple 16
simple closed 16
convergence
acceleration .... see acceleration of convergence.
cubic 90
geometric 90
linear 90
local 90
of the pth order 90
quadratic 90
convex functions 7
coordinate systems
cylindrical 7
ellipsoidal 582, 693
elliptic 720
elliptical 677–678
oblate spheroidal 705
parabolic cylinder 317
paraboloid of revolution 317
paraboloidal 346, 678
polar 7
projective 581
prolate spheroidal 704
spherical (or spherical polar) 8
sphero-conal 693
toroidal 371, 379
Cornu's spiral **168**
applications 169
connection with Fresnel integrals 168
cosecant function .... see trigonometric functions.
cosine function .... see trigonometric functions.
cosine integrals **150**
analytic continuation 151
applications 155
approximations 156
asymptotic expansions 153
exponentially-improved 154
auxiliary functions ... see auxiliary functions for sine
and cosine integrals.
Chebyshev-series expansions 156–157
computation 155
definition 150

expansion in spherical Bessel functions 153
generalized 188–189
graphics 151
hyperbolic analog 150
analytic continuation 151
integral representations 152
integrals 154
Laplace transform 154
notation 150
power series 151
principal value 150
relations to exponential integrals 151
sums 154
tables 156
value at infinity 150
zeros 154
asymptotic expansion 154
computation 156
cosmology
confluent hypergeometric functions 346
incomplete beta functions 189
cotangent function .... see trigonometric functions.
Coulomb excitation of nuclei 753
Coulomb field 754
Coulomb functions
Dirac delta 38
Coulomb functions: variables ρ, η **742**
analytic properties 742
applications 753–755
asymptotic expansions
large η 747
large ρ 747
uniform expansions 747–748
case $\eta = 0$ 744
complex variable and parameters 748, 754
computation 755
continued fractions 745
conversions between variables and parameters ... 754
cross-product 742
definitions 742
derivatives 744
expansions in Airy functions 747
expansions in Bessel functions 746
expansions in modified Bessel functions 746
expansions in spherical Bessel functions 745
functions $F_\ell(\eta,\rho), G_\ell(\eta,\rho), H_\ell^\pm(\eta,\rho)$ 742
graphics 743–744
integral representations 745
limiting forms
large ℓ 744
large $|\eta|$ 746
large ρ 746, 747
small $|\eta|$ 744
small ρ 744

INDEX 899

normalizing constant 742
phase shift (or phase) 742, 756
power-series expansions in ρ 745
recurrence relations 744
relations to other functions
    confluent hypergeometric functions 742
    Coulomb functions with variables r, ϵ 751
    Whittaker functions 742
scaling of variables and parameters 753, 754
tables 755
transition region 747
WKBJ approximations 755
Wronskians 742
Coulomb functions: variables r, ϵ **748**
analytic properties 748
applications 753–755
asymptotic approximations and expansions for large $|r|$ 753
asymptotic expansions as $\epsilon \to 0$ 753
    uniform 753
case $\epsilon = 0$ 752
complex variables and parameters 754
computation 755
conversions between variables and parameters ... 754
definitions 748
derivatives 752
expansions in Airy functions 753
expansions in Bessel functions 752, 753
expansions in modified Bessel functions 752, 753
functions $f(\epsilon, \ell; r), h(\epsilon, \ell; r)$ 748
functions $s(\epsilon, \ell; r), c(\epsilon, \ell; r)$ 748
graphics 749–750
integral representations for Dirac delta 749
limiting forms for large ℓ 752
power-series expansions in ϵ 752
power-series expansions in r 752
recurrence relations 752
relations to other functions
    confluent hypergeometric functions 748
    Coulomb functions with variables ρ, η 751
    Whittaker functions 748, 751
scaling of variables and parameters 753, 754
tables 755
Wronskians 749
Coulomb phase shift 145, 742, 755, 756
Coulomb potential barriers 754
Coulomb potentials 753–754
q-hypergeometric function 432
Coulomb radial functions
    .... see Coulomb functions: variables ρ, η.
Coulomb spheroidal functions 704
as confluent Heun functions 717
Coulomb wave equation
irregular solutions 742, 748

regular solutions 742, 748
singularities 742, 748
turning points 742, 748, 754
Coulomb wave functions .. see Coulomb functions: variables ρ, η and Coulomb functions: variables r, ϵ.
counting techniques 634
critical phenomena
elliptic integrals 517
hypergeometric function 400
critical points **781**
coalescing 789–790
cross ratio 17
cryptography 647
Weierstrass elliptic functions 582
cubature
for disks and squares 84–85
cubic equation 23
resolvent 23
cubic equations
solutions as trigonometric and hyperbolic functions
.... 131
curve
piecewise differentiable 11
simple closed 11
cusp bifurcation set
formula 781
picture 782
cusp canonical integral **776**, 785
zeros 785
    table 786
cusp catastrophe **776**, 784
cuspoids
normal forms 776
cut 20
domain 20
neighborhood 20
cycle 618
Riemann surface 543
cyclic identities
Jacobian elliptic functions 558
cyclotomic fields
Bernoulli and Euler polynomials 598
cylinder functions
addition theorems 246
definition 218
derivatives 222
differential equations 217, 226
integrals 240–241
multiplication theorem 246
recurrence relations 222
zeros .... see zeros of cylinder functions.
cylindrical coordinates 7
cylindrical polar coordinates
.... see cylindrical coordinates.

Darboux's method
　asymptotic approximations of sums and sequences
　　...65–66
Dawson's integral...............................**160**
　applications...................................169
　approximations................................170
　computation...................................169
　definition.......................................160
　generalized....................................166
　graphics.......................................161
　integral representation........................162
　notation.......................................160
　relation to error functions....................162
　relation to parabolic cylinder functions.........308
　tables..169
de Branges–Wilson beta integral 143
De Moivre's theorem
　trigonometric functions........................118
Dedekind modular function......................646
　functional equation 646
Dedekind sums
　number theory.................................646
Dedekind's eta function .... see modular functions.
Dedekind's modular function...see modular functions.
del operator.....................................10
Dellanoy numbers
　definition......................................621
　generating functions...........................623
　recurrence relation.............................623
　relation to lattice paths.......................621
　table...621
delta sequence...................................37
delta wing equation
　Lamé polynomials.............................694
derivatives
　chain rule....................................5, 7
　definition....................................5, 7
　distributional..................................36
　Faà di Bruno's formula...........................5
　Jacobian..9
　L'Hôpital's rule.................................5
　left-hand.......................................14
　Leibniz's formula................................5
　mean value theorem.............................5
　notation.....................................5, 7
　of distribution.................................35
　partial..7
　right-hand.....................................14
Descartes' rule of signs (for polynomials)..........22
determinants
　alternants......................................3
　Cauchy...4
　circulant.......................................4
　cofactor..3

　definition..3
　Hadamard's inequality 3
　Hankel.....................................93, 595
　inequalities.....................................3
　infinite
　　convergence...................................4
　　Hill's type 4
　Krattenthaler's formula 4
　minor..3
　notation..3
　persymmetric..................................595
　properties......................................3
　Vandermonde..................................3
diatomic molecules
　hypergeometric function 400
difference equations
　asymptotic solutions
　　..see asymptotic solutions of difference equations.
　distinguished solutions 85
　minimal solutions 85
　numerical solution...........................85–88
　　backward recursion method 85, 86
　　boundary-value methods 86, 87
　　homogeneous equations....................85–86
　　inhomogeneous equations 86
　　normalizing factor 86
　　stability 85
　recessive solutions..............................85
difference operators 436
　backward.....................................436
　central in imaginary direction..................436
　forward.......................................436
differentiable functions.........................5, 15
differential equations
　asymptotic solutions......see asymptotic solutions of
　　differential equations.
　change of variables
　　elimination of first derivative..................26
　　Liouville transformation 26
　　point at infinity..............................26
　classification of singularities 56, 409
　closed-form solutions...........................27
　dominant solutions.............................57
　Fuchs–Frobenius theory 55
　homogeneous................................26, 88
　indices differing by an integer 56
　indicial equation 56
　inhomogeneous..............................26, 88
　　solution by variation of parameters............26
　irregular singularity............................56
　nonhomogeneous................see inhomogeneous.
　numerical solution
　　boundary-value problems 88
　　eigenfunctions 89

eigenvalues 89
   initial-value problems 88
   Runge–Kutta method 89–90
   stability 88
   Sturm–Liouville eigenvalue problems 89
   Taylor-series methods 88–89
numerically satisfactory solutions 58
of arbitrary order 409
ordinary point 55, 409
rank of singularity 56
recessive solutions 57
regular singularity 56
Schwarzian derivative 27
solutions
   existence 25
   fundamental pair 26
   in series of Chebyshev polynomials 478, 480
   in series of classical orthogonal polynomials . . . 479
   linearly independent 26
   products 27
   Wronskian 26
subdominant solutions . . . . *see* recessive solutions.
with a parameter 26
differentiation
   Cauchy–Riemann equations 16
   numerical
      analytic functions 77
      Lagrange's formula for equally-spaced nodes . . . 77
      partial derivatives 78
   of integrals 8, 21
   partial 7
diffraction catastrophes **777**, 789
   notation 776
   scaling laws 785
diffraction of light
   Fresnel integrals and Cornu's spiral 161, 169
diffraction problems
   Mathieu functions 678
diffusion equations
   theta functions 533
diffusion problems
   Mathieu functions 678
digamma function . . . . *see* psi function.
dilogarithms
   analytic properties 610
   approximations 615
   computation 614
   definition 610
   graphics 611
   principal branch (or value) 610
   tables 614
Dirac delta 37–38
   delta sequences 37–38
   integral representations

Airy functions 38
Bessel functions 38
Coulomb functions 38
Fourier 37–38
spherical Bessel functions 38
   mathematical definitions 38
   series representations
      Fourier 38
      Hermite polynomials 38
      Laguerre polynomials 38
      Legendre polynomials 38
      spherical harmonics 38
Dirac delta distribution 53
Dirac delta function . . . . *see* Dirac delta.
Dirac equation
   Coulomb functions 754
Dirichlet *L*-functions
   analytic properties 612
   definition 612
   functional equation 612
   infinite products 612
   tables 614
   zeros 613
Dirichlet characters 642
   Gauss sum 643
Dirichlet problem
   with toroidal symmetry 379
Dirichlet product (or convolution) 641
Dirichlet series 602, 640
   generating function 640
Dirichlet's divisor problem
   number theory 643
Dirichlet's theorem
   prime numbers in arithmetic progression 643
discontinuity 4
discrete Fourier transform 99
discrete q-Hermite I and II polynomials 471
discriminant
   of a polynomial 22
discriminant function
   number theory 646
      functional equation 646
disk
   around infinity 16
   open 15
disk polynomials 477
dislocation theory
   Heun functions 720
distribution function
   Painlevé transcendents 739
distribution functions
   connection with
      incomplete beta functions 189
      incomplete gamma functions 189

distributional derivative 36
distributions 35–37
    convergence 35, 36
    convolutions 54
    derivatives 35
    Dirac delta...................................... 36
    distributional derivative 36
    Fourier transforms 37
    Heaviside function 36
    linear functionals............................... 35
    of derivatives................................... 52
    regular.. 35
    regularization 55
    several variables 36–37
    singular... 35
    support 35
    tempered .... see tempered distributions., 52
    test function space 35
    test functions 35
        convergence.................................. 35
divergence theorem
    .. see Gauss's theorem for vector-valued functions.
divergent integrals................................. 51
    regularization 55
divided differences
    definition 76
    integral representation 76
divisor function
    number theory 638
Dixon's $_3F_2(1)$ sum
    q-analog 426
Dixon's sum
    F. H. Jackson's q-analog 426
domain... 15
    closed... 15
    cut.. 20
    simply-connected................................ 25
dominated convergence theorem
    infinite series.................................. 18
double gamma function .... see Barnes' G-function.
double integrals 8–9
    change of order of integration 9
    change of variables 9
    infinite... 9
double sequence.................................... 18
    convergence..................................... 18
double series...................................... 18
    convergence..................................... 18
doubly-confluent Heun equation 717
Dougall's $_7F_6(1)$ sum
    F. H. Jackson's q-analog 427
Dougall's bilateral sum 387
Dougall's expansion
    associated Legendre functions 371

dual Hahn polynomials
    .... see Wilson class orthogonal polynomials.
Duffing's equation
    Jacobian elliptic functions......................565
dynamical systems
    Mathieu functions...............................679
    Painlevé transcendents..........................739
Dyson's integral
    gamma function 144
ecological systems
    incomplete gamma functions 189
Einstein summation convention for vectors..........10
Eisenstein convention 577
Eisenstein series
    Jacobian elliptic functions......................559
electric particle field
    Stieltjes electrostatic interpretation 719
electromagnetic scattering
    Bessel functions and spherical Bessel functions .. 275
electromagnetic theory
    sine and cosine integrals 155
electromagnetic waves
    Mathieu functions...............................678
electron-ion collisions
    Coulomb functions..............................753
electronic structure of heavy elements
    Coulomb functions..............................754
electrostatics
    Jacobian elliptic functions......................566
    zeros of classical orthogonal polynomials 479
elementary functions............. see exponential function, hyperbolic functions, inverse hyperbolic functions, inverse trigonometric functions, Lambert W-function, logarithm function, power function, and trigonometric functions.
    relation to R_C-function 495
elementary particle physics
    conical functions...............................379
ellipse
    elliptic integrals................................514
ellipse arc length
    Jacobian elliptic functions......................563
ellipsoid
    capacity 516
    depolarization.............................497, 516
    potential....................................... 516
    self-energy..................................... 516
    surface area.................................... 515
    triaxial 515
ellipsoidal coordinates............................693
    Weierstrass elliptic functions...................582
ellipsoidal harmonics
    Lamé polynomials 694
ellipsoidal wave equation..... see Lamé wave equation.

elliptic coordinates 720
elliptic crack and punch problems
    Lamé polynomials 694
elliptic curves 564
    addition law 581
    definition 581
    Jacobi–Abel addition theorem................. 564
    Jacobian normal form 564
    Mordell's theorem............................. 582
    tables 582
    Weierstrass elliptic functions 581
elliptic functions ... *see also* Jacobian elliptic functions *and* Weierstrass elliptic functions.
    general....................................... 571
        representation as Weierstrass elliptic functions 576
    Weierstrass .... *see* Weierstrass elliptic functions.
elliptic integrals .. *see* basic elliptic integrals, Bulirsch's elliptic integrals, general elliptic integrals, generalizations of elliptic integrals, Legendre's elliptic integrals, *and* symmetric elliptic integrals.
    complete
        quasiconformal mapping 399
    relations to other functions
        associated Legendre functions 360
        Ferrers functions 360
        Weierstrass elliptic functions 576
elliptic modular function .... *see* modular functions.
elliptic umbilic bifurcation set
    formula 781
    picture....................................... 782
elliptic umbilic canonical integral **776**
    asymptotic approximations 789–790
    convergent series 787
    differential equations 788
    formulas for Stokes set 783
    integral identity 788
    pictures of modulus 779
    pictures of phase 780
    scaling laws 785
    zeros... 786
elliptic umbilic catastrophe................. **776**, 785
elliptical coordinates
    Mathieu functions 677–678
entire functions 16
    asymptotic expansions 64
    Liouville's theorem 16
enumerative topology
    Painlevé transcendents 739
epsilon function .... *see* Jacobi's epsilon function.
equation of Ince .. *see* Hill's equation, equation of Ince.
equiconvergent 699
Erlang loss function
    incomplete gamma functions 189

error-control function
    differential equations 57
error functions **160**
    applications
        asymptotic approximation of integrals 168
        physics 169
        statistics 169
        Stokes phenomenon 168
    approximations 170
    asymptotic expansions 164
        exponentially-improved 164
    computation 83–84, 169
    continued fractions 163
    definitions 160
    derivatives 163
    expansions in spherical Bessel functions 162
    generalized 166
    graphics 160, 161
    inequalities 163
    integral representations 162–163
    integrals
        Fourier transform 166
        Laplace transforms 166
    interrelations 162
    inverse functions 166
        asymptotic expansions 166
        computation 169
        power-series expansions 166
    notation 160
    power-series expansions 162
    relations to other functions
        confluent hypergeometric functions .. 164, 328, 338
        Dawson's integral 162
        Fresnel integrals 162
        generalized exponential integrals 164
        incomplete gamma functions 164
        parabolic cylinder functions 308
        probability functions 160
        Voigt functions 167
    repeated integrals of .... *see* repeated integrals of the complementary error function.
    sums 166
    tables 169, 170
    values at infinity 160
    zeros 165
        asymptotic expansions 165
        tables 165, 170
error measures
    absolute error 73
    complex arithmetic 73
    mollified error 73
    relative error 73
    relative precision 73
error term 43

essential singularity 19
.... see also isolated essential singularity.
eta function .... see Dedekind's eta function.
Euler–Maclaurin formula 63
    extensions 63
    generalization 597
Euler numbers **588**
    arithmetic properties 593
    asymptotic approximations 593
    computation 598
    definition 588
    explicit formulas 591
    factors 593
    finite expansions 590
    generalizations 596, 597
    generating function 588
    identities 591
    inequalities 593
    integral representation 592
    inversion formulas 591
    Kummer congruences 593
    notation 588
    recurrence relations
        linear 591
        quadratic and higher order 595
    sums 595
    tables 589, 598
Euler polynomials **588**
    applications
        mathematical 597–598
        physical 598
    asymptotic approximations 593
    computation 598
    definition 588
    derivative 590
    difference equation 589
    explicit formulas 591
    finite expansions 590
    generalized 596, 597
    generating function 588
    graphs 589
    inequalities 593
    infinite series expansions
        Fourier 592
        other 592
    integral representations 592
    integrals 594
        compendia 595
    Laplace transforms 595
    multiplication formulas 590
    notation 588
    recurrence relations
        linear 591
        quadratic 595

representations as sums of powers 589
special values 590
sums 595
symbolic operations 590
symmetry 589
tables 589
zeros
    complex 594
    multiple 594
    real 594
Euler product
    number theory 640
Euler splines 597
Euler sums
    Riemann zeta function 613
        reciprocity law 614
Euler's beta integral 142
Euler's constant 136
    integral representations 140
Euler's homogeneity relation
    symmetric elliptic integrals 501
Euler's integral
    gamma function 136
Euler's pentagonal number theorem
    number theory 646
Euler's sums
    q-hypergeometric function 423, 424
Euler's totient
    number theory 638
Euler's transformation
    applied to asymptotic expansions 69
    of series 93
Euler–Fermat theorem
    number theory 638, 647
Euler–Poisson differential equations 501
Euler–Poisson–Darboux equation
    symmetric elliptic integrals 501
Euler–Tricomi equation
    Airy functions 209
Eulerian numbers
    definition 632
    generating functions 632
    identities 632
    notation 632
    relation to Bernoulli numbers 591
    relation to permutations 632
    special values 633
    table 632
evolution equations
    Lamé polynomials 694
exact rational arithmetic 72
exponential function **105**
    analytic properties 105
    approximations 132

INDEX 905

Chebyshev-series expansions 132
computation 131
conformal maps 106
continued fractions 110
definition 105
derivatives 109
differential equation 109
generalized 111
graphics
    complex argument 107
    real argument 106
identities 109
inequalities 108
integrals 110
limits 107
notation 104
periodicity 105
power series 105
special values 107
sums 110
tables 132
zeros 105
exponential growth 28
exponential integrals **150**
  analytic continuation 151
  applications 155
  approximations 156
  asymptotic expansions 153
    exponentially-improved 153
    re-expansion of remainder term 153
  Chebyshev-series expansions 156, 157
  computation 155
  continued fraction 153
  definition 150
  expansion in inverse factorials 153
  expansions in modified spherical Bessel functions
    153
  generalized 185
  graphics 151
  inequalities 152
  integral representations 152
  integrals 154
  interrelations 150, 151
  Laplace transform 154
  notation 150
  power series 151
  principal value 150
  relations to other functions
    confluent hypergeometric functions 153
    incomplete gamma function 153
    logarithmic integral 150
    sine and cosine integrals 151
  small argument 51
  tables 156

zeros 154
extended complex plane 16
Faà di Bruno's formula
  for derivatives 5
Fabry's transformation
  differential equations 57
factorials (rising or falling) 618
factorization
  of integers 648
    via Weierstrass elliptic functions 582
Faddeeva function 169
fast Fourier transform 100
Fay's trisecant identity
  Riemann theta functions with characteristics 544
    generalizations 544
Fejér kernel
  Fourier integral 34
  Fourier series 33
Fermat numbers
  number theory 648
Fermat's last theorem
  Bernoulli and Euler numbers and polynomials ... 598
Fermi–Dirac integrals
  approximations 615
  computation 614
  definition 611
  relation to polylogarithms 612
  tables 614
  uniform asymptotic approximation 612
Ferrers board 633
Ferrers function
  of the first kind
    integral equation for Lamé functions 689
Ferrers functions **352**
  addition theorems 370
  analytic continuation 376
  applications
    spherical harmonics 378–379
    spheroidal harmonics 378
  asymptotic approximations
    .... see uniform asymptotic approximations.
  behavior at singularities 361–362
  computation 379
  connection formulas 362
  cross-products 352
  definitions 353–354
  degree 352
  derivatives 362
    with respect to degree or order 363
  differential equation
    .... see associated Legendre equation.
  generating functions 361
  graphics 355–357
  integer degree and order 360–361

integer order 360
integral representations 363
integrals
    definite 369
    indefinite 368
    Laplace transforms 370
    Mellin transforms 370
    orthogonality properties 369
notation 352
of the first kind 353
of the second kind 353
order 352
orthogonality 369
recurrence relations 362
reflection formulas 361
relations to other functions
    elliptic integrals 360
    hypergeometric function 353, 354, 394
    Legendre polynomials 360
    ultraspherical polynomials 448
Rodrigues-type formulas 360
special values 359–360
sums 370–371
tables 380
trigonometric expansions 364
uniform asymptotic approximations
    large degree 366–368
    large order 365–366
Wronskians 352–353
zeros 368
Ferrers graph 626
Feynman diagrams
    Appell functions 417
Feynman path integrals
    theta functions 534
Fibonacci numbers 596, 629
fine structure constant
    Coulomb functions 753
finite Fourier series
    number theory 643
fixed point 90
floating-point arithmetic
    bits
        format width 72
        significant 72
    double precision 72
    exponent 72
    fractional part 72
    IEEE standard 72
    machine epsilon 72
    machine number 72
    machine precision 72
    overflow 72
    rounding

        by chopping 72
        down 72
        symmetric 72
        to nearest machine number 72
    significand 72
    single precision 72
    underflow 72
Floquet solutions
    Hill's equation 675
    Mathieu's equation 653
Floquet's theorem
    Hill's equation 674
    Mathieu's equation 653
fluid dynamics
    elliptic integrals 517
    Legendre polynomials 479
    Riemann theta functions 545
    Struve functions 298
fold canonical integral **776**, 785
    bifurcation set 781
    differential equation 788
    integral identity 787
    relation to Airy function 777
    zeros 785
fold catastrophe **776**, 785
Fourier cosine and sine transforms
    definition 27
    generalized 400
    inversion 28
    Parseval's formula 28
    tables 31, 32
Fourier integral
    asymptotic expansions 44, 45
    Dirac delta 38
    Fejér kernel 34
    Poisson kernel 34
    summability 34
Fourier series 13–14
    Bessel's inequality 13
    coefficients 13
    compendia 14
    convergence 14
    definition 13
    differentiation 14
    Dirac delta 38
    Fejér kernel 33
    finite
        number theory 643
    integration 14
    Parseval's formula 14
    Poisson kernel 34
    Poisson's summation formula 14
    properties 13
    summability 33–34

uniqueness 13
Fourier transform 27–28
    convergence 27
    convolution 27
    definitions 27
    discrete 99
    distributions 37
    fast 100
    group
        hypergeometric function 400
    inversion 27
    Parseval's formula 27
    tables 30, 32
    tempered distributions 37
    uniqueness 27
Fourier–Bessel expansion
    Bessel functions 248
        computation 278
Fourier-series expansions
    nonuniformity of convergence 155
    piecewise continuous functions 155
fractals 92
fractional derivatives 35
fractional integrals 35
    asymptotic expansions 53–55
    definition 53
fractional linear transformation
     . . . . *see* bilinear transformation.
Fresnel integrals **160**
    applications
        Cornu's spiral 168
        interference patterns 161
        physics and astronomy 169
        probability theory 169
        statistics 169
    approximations 170
    asymptotic expansions 164
        exponentially-improved 164
    auxiliary functions
       . . . . . . *see* auxiliary functions for Fresnel integrals.
    computation 169
    definition 160
    expansions in spherical Bessel functions 162
    graphics 161
    integrals
        Laplace transforms 166
    interrelations 162
    notation 160
    power-series expansions 162
    relations to other functions
        Anger–Weber functions 297
        auxiliary functions 160, 162
        confluent hypergeometric functions 164
        error functions 162

    generalized hypergeometric functions 164
    symmetry 161
    tables 169
    values at infinity 160
    zeros 165
        asymptotic expansions 165
        tables 165
Freud weight function 475
Frobenius' identity
    Riemann theta functions with characteristics 544
Fuchsian equation
    classification of parameters 718
    definitions 718
    normal form 718
    polynomial solutions 718
    relation to Heun's equation 718
functions
    analytic . . . . *see* analytic function.
    analytically continued 19
    continuous . . . . *see* continuous function.
    continuously differentiable 5, 7
    convex 7
    decreasing 4
    defined by contour integrals 21
    differentiable 5
    entire . . . . *see* entire functions.
    harmonic 16
    holomorphic . . . . *see* analytic function.
    increasing 4
    inverse 21
    limits . . . . *see* limits of functions.
    many-valued . . . . *see* multivalued function.
    meromorphic 19
    monotonic 4
    multivalued . . . . *see* multivalued function.
    nondecreasing 4
    nonincreasing 4
    of a complex variable 18–22
    of bounded variation 6
    of compact support 35
    of matrix argument
       . . . . *see* functions of matrix argument.
    strictly decreasing 4
    strictly increasing 4
    strictly monotonic 4
    support of 35
    vector-valued 10–11
functions of matrix argument **768**
    Laplace transform 768
    orthogonal invariance 771
fundamental theorem of arithmetic 638
fundamental theorem of calculus 6
gamma distribution
    incomplete gamma functions 189

gamma function **136**
.... see also incomplete gamma functions.
    analytic properties 136
    applications
        mathematical 145
        physical 145–146
    approximations
        Chebyshev series 147
        complex variables 147
        rational 146, 147
    asymptotic expansions 140–142
        error bounds 141
        exponentially-improved 141
        for ratios 141
    Bohr-Mollerup theorem 138
    computation 146
    continued fraction 140
    definition 136
    duplication formula 138
    Euler's integral 136
    extrema
        asymptotic approximation 138
        table of 138
    Gauss's multiplication formula 138
    graphics 136–137
    inequalities 138
    infinite products 139
    integral representations 139–140, 143–144, 188
        for derivatives 140
        multidimensional 143–144
    logarithm
        continued fraction 140
        convexity 136, 138
        graphics 136
        integral representations 140
        Taylor series 139
    maxima and minima 138
    multiplication formulas 138
    multivariate .... see multivariate gamma function.
    notation 136
    reciprocal
        analytic properties 136
        graphics 136, 137
        Maclaurin series 139
        zeros 136
    recurrence relation 138
    reflection formula 138
    relations to hypergeometric function 387
    scaled 185
    special values 137
    tables 146
Gaunt coefficient
    $3j$ symbol 761
Gaunt's integral

    $3j$ symbol 761
Gauss quadrature 80–83
    Christoffel coefficients (or numbers) 80
    comparison with Clenshaw–Curtis formula 80
    eigenvalue/eigenvector characterization 82
    for contour integrals 83
    Gauss–Chebyshev formula 80
    Gauss–Hermite formula 81
    Gauss–Jacobi formula 80
    Gauss–Laguerre formula 80–81
        generalized 80
    Gauss–Legendre formula 80
    logarithmic weight function 81–82
    nodes 80
        tables 80–83
    remainder terms 80
    weight functions 80
        tables 80–83
Gauss series
    hypergeometric function 384
        convergence 384
Gauss sums
    number theory
        Dirichlet character 643
        separable 643
    theta functions 532
Gauss's $_2F_1(-1)$ sum
    q-analog 426
Gauss's theorem for vector-valued functions 12
Gauss–Christoffel quadrature ... see Gauss quadrature.
Gaussian
    nonperiodic 533
Gaussian elimination 73–74
    back substitution 73
    forward elimination 73
    iterative refinement 74
    multipliers 73
    partial pivoting 73
    pivot (or pivot element) 73
    residual vector 74
    triangular decomposition 73
    tridiagonal systems 74
Gaussian hypergeometric function
    .... see also hypergeometric function.
    of matrix argument **771**
        applications 773
        asymptotic approximations 772
        basic properties 771
        case $m = 2$ 771
        computation 773
        confluent form 771
        definition 771
        Gauss formula 771
        integral representation 771

INDEX 909

    Jacobi form 771
    notation 768
        partial differential equations 771–772
    reflection formula 771
    transformations of parameters 771
Gaussian noise
    Lambert W-function 131
Gaussian polynomials
    definition 627
    tables 635
Gaussian probability functions 160
Gaussian unitary ensemble
    Painlevé transcendents 739
Gegenbauer function
    definition 394
    relation to associated Legendre functions 355
    relation to hypergeometric function 394
Gegenbauer polynomials
    . . . . *see* ultraspherical polynomials *and also* classical orthogonal polynomials.
Gegenbauer's addition theorem
    Bessel functions 247
    modified Bessel functions 261
general elliptic integrals **486**
    reduction to basic elliptic integrals 512
    reduction to Legendre's elliptic integrals 496–497
    reduction to symmetric elliptic integrals 512–514
general orthogonal polynomials **437**
    computation 479
    difference operators 437
    monic 438
    on finite point sets 437
    on intervals 437
    orthonormal 438
    recurrence relations 438
    sums of products 438
    weight functions 437
    x-difference operators 437
    zeros 438
generalizations of elliptic integrals 516
generalized Airy functions
    from differential equation 206–207
        asymptotic approximations 206
        definitions 206, 207
        relation to Bessel functions 206
        relation to modified Bessel functions 206
        tables 211
    from integral representations 207–208
        connection formulas 208
        definitions 207
        difference equation 208
        differential equation 208
generalized exponential integral **185**
    analytic continuation 187

applications
    mathematical 189
    physical 189
approximations 191
asymptotic expansions
    exponentially-improved 187
    large parameter 187
    large variable 187
Chebyshev-series expansions 191
computation 190
continued fraction 187
definition 185
derivatives 186
further generalizations 187
graphics 185–186
inequalities 187
integral representations 185
    Mellin–Barnes type 185
integrals 187
notation 174
of large argument 66
principal values 185
recurrence relation 186
relations to other functions
    confluent hypergeometric functions 186
    error functions 164
    incomplete gamma functions 185
series expansions 186
special values 186
tables 190
generalized exponentials 111
generalized functions
    distributions 55
generalized hypergeometric differential equation . . . 409
    confluence of singularities 410
    connection formula 410
    fundamental solutions 409
    singularities 409
generalized hypergeometric function $_0F_2$
    definition 404, 408
    of large argument 64
generalized hypergeometric functions **404**
    analytic continuation 408
    analytic properties 404, 405, 408
    applications
        mathematical 417
        physical 417
    approximations 418
    argument unity 405
    as functions of parameters 405
    asymptotic expansions
        formal series 411
        large parameters 412
        large variable 411

small variable	408
balanced	405
bilateral series	408
Dougall's bilateral sum	408
computation	418
contiguous balanced series	407
contiguous functions	405
contiguous relations	407
continued fractions	407
definitions	404, 408
derivatives	405
differential equation	*see* generalized hypergeometric differential equation.
Dixon's well-poised sum	406
Dougall's bilateral sum	408
Dougall's very well-poised sum	406
Džrbasjan's sum	406
expansions in series of	410
extensions of Kummer's relations	407
identities	407
integral representations	408
integrals	408
inverse Laplace transform	408
Laplace transform	408
k-balanced	405
Kummer-type transformations	407, 409
monodromy	417
notation	404
of matrix argument	**772**
applications	773
computation	773
confluence	772
convergence properties	772
definition	772
Euler integral	773
expansion in zonal polynomials	772
general properties	772
invariance	773
Kummer transformation	772
Laplace transform	773
Mellin–Barnes integrals	773
notation	768
Pfaff–Saalschutz formula	772
relations to other functions	772
Thomae transformation	772
$_3F_2$ case	772
value at $\mathbf{T} = \mathbf{0}$	772
Pfaff–Saalschütz balanced sum	406
polynomial cases	404
principal branch (value)	404
products	412
recurrence relations	407
relations to other functions	
associated Jacobi polynomials	474
Bessel functions	228
classical orthogonal polynomials	442
Fresnel integrals	164
generalized Bessel polynomials	476
Hahn class orthogonal polynomials	463
Kummer functions	328
Meijer G-function	416
modified Bessel functions	255
orthogonal polynomials *and* other functions	409
$3j, 6j, 9j$ symbols	407, 418
Wilson class orthogonal polynomials	468
Rogers–Dougall very well-poised sum	406
Saalschützian	405
terminating	404
transformation of variable	408
cubic	409
quadratic	408
very well-poised	405
Watson's sum	406
well-poised	405
Whipple's sum	406
Whipple's transformation	407
with two variables	412–415
zeros	410
generalized hypergeometric series	404
generalized integrals	
asymptotic expansions	52
generalized logarithms	73, 111
applications	131
generalized precision	73
generalized sine and cosine integrals	**188**
analytic properties	188
asymptotic expansions for large variable	189
auxiliary functions	189
asymptotic expansions for large variable	189
integral representations	189
computation	190
definitions	
general values	188
principal values	188
expansions in series of spherical Bessel functions	188
integral representations	188
interrelations	188
notation	174
power-series expansions	188
relation to sine and cosine integrals	188
special values	188
Genocchi numbers	595
table	596
genus	
Riemann surface	543
geometric mean	3, 13
geometric progression (or series)	2
geophysics	

spherical harmonics 379
Gibbs phenomenon
   sine integral 154
Glaisher's constant 63, 144
Glaisher's notation
   Jacobian elliptic functions 550
Goldbach conjecture
   number theory 644
Goodwin–Staton integral
   asymptotic expansion 164
   computation 169
   definition 160
   relations to Dawson's integral and exponential integral 162
Graf's addition theorem
   Bessel functions 247
   modified Bessel functions 261
Gram–Schmidt procedure
   for least squares approximation 99
graph theory
   combinatorics 635
gravitational radiation
   Coulomb functions 755
Green's theorem for vector-valued functions
   three dimensions 12
   two dimensions 11
group representations
   orthogonal polynomials 479
group theory
   hypergeometric function 400
Gudermannian function 121
   inverse 121
   relation to R_C-function 495
   relation to amplitude (am) function 561
   tables 132

Haar measure 769
Hadamard's inequality for determinants 3
Hahn class orthogonal polynomials 462–467
   asymptotic approximations 466–467
   computation 479
   definitions 462
   difference equations on variable 465
   differences 465
   dualities 463
   generating functions 466
   interrelations with other orthogonal polynomials
      .... 463, 464
   leading coefficients 462
   limit relations 463
   normalizations 462
   notation 436
   orthogonality properties 462
   recurrence relations 464

   relations to confluent hypergeometric functions and
      generalized hypergeometric functions 328, 463
   relations to hypergeometric function 394, 463
   Rodrigues formula 462
   special cases 463
   weight functions 462
Hahn polynomials
   .... see Hahn class orthogonal polynomials.
Hamiltonian systems
   chaotic
      Lamé functions 694
handle
   Riemann surface 543
Hankel functions **217**
   addition theorems 246–247
   analytic continuation 226
   approximations 281
   asymptotic expansions for large argument .. 229–230
      error bounds 229–230
      exponentially-improved 230
   asymptotic expansions for large order 231–235
      asymptotic forms 231
      double asymptotic properties 235, 258
      transition region 232
      uniform 232–235
   branch conventions 218
   computation 276–277
   connection formulas 222
   cross-product 222
   definitions 217
   derivatives 222
      asymptotic expansions for large argument 229
      asymptotic expansions for large order 231–232
      uniform asymptotic expansions for large order
         233
      zeros 238
   differential equations 217
   .... see also Bessel's equation.
   graphics 220–221
   incomplete 262
   integral representations
      along real line 224
      compendia 226
      contour integrals 224
   integrals
      ...... see integrals of Bessel and Hankel functions.
   limiting forms 217, 223
   multiplication theorem 246
   notation 217
   power series 223
   principal branches (or values) 217
   recurrence relations 222
   relations to other functions
      Airy functions 196–197

confluent hypergeometric functions 228
    elementary functions 228
  Wronskians 222
  zeros 238
    computation 277
    tables 279
    with respect to order (ν-zeros) 240
Hankel transform 246
  computation 277
Hankel's expansions
  for Bessel and Hankel functions 228–229
  for modified Bessel functions 255
Hankel's integrals
  Bessel functions and Hankel functions 226
Hankel's inversion theorem
  Bessel functions 246
Hankel's loop integral
  gamma function 139
harmonic analysis
  hypergeometric function 399
harmonic functions 16
  maximum modulus 20
  mean value property 16
  Poisson integral 16
harmonic mean 3, 13
harmonic oscillators
  Hermite polynomials 479
  q-hypergeometric function 432
harmonic trapping potentials
  parabolic cylinder functions 317
heat conduction in liquids
  Rayleigh function 276
heat theory
  conical functions 379
Heaviside function 36, 54
  derivative 36
Heine's formula
  associated Legendre functions 377
Heine's integral
  Legendre functions 364
Helmholtz equation
  $3j, 6j, 9j$ symbols 765
  associated Legendre functions 379
  Bessel functions and modified Bessel functions . . 275
  parabolic cylinder functions 317
  paraboloidal coordinates 678
Hermite polynomials **438**
  . . . . *see also* classical orthogonal polynomials.
  addition theorem 460
  applications
    integrable systems 478
    random matrix theory 479
    Schrödinger equation 479
  asymptotic approximations 453

computation 479
continued fractions 450
definitions 439
derivatives 447
differential equations 445
Dirac delta 38
expansions in series of 459–461
explicit representations 442–443
Fourier transforms 457
generating functions 450
graphs 441
inequalities 450, 451
  Turan-type 450
integral representations 447, 448
integrals 455, 457–459
  indefinite 455
  Nicholson-type 455
interrelations with other orthogonal polynomials
  . . . . 444–445, 463, 464
Laplace transform 457
leading coefficients 439
limiting forms as trigonometric functions 449
linearization formulas 461
local maxima and minima 451
Mellin transform 458
monic 81, 441
multiplication theorem 460
normalizations 439
notation 436
orthogonality properties 439
Poisson kernels 461
recurrence relations 446
relations to other functions
  confluent hypergeometric functions . . 328, 338, 449
  derivatives of the error function 163
  generalized hypergeometric functions 443
  parabolic cylinder functions 308
  repeated integrals of the complementary error function 167
Rodrigues formula 442
special values 444
symmetry 444
tables 480
  of coefficients 440
  of zeros 81
upper bounds 450
zeros 438, 455
  asymptotic behavior 455
  tables 81
Hermite–Darboux method
  Heun functions 713
Hermitian matrices
  Gaussian unitary ensemble
    limiting distribution of eigenvalues 739

INDEX 913

Heun equation . . . . *see* Heun's equation.
Heun functions **710**
   applications
      mathematical 719–720
      physical 720
   asymptotic approximations. 718
   computation. 720
   definition 712
   differential equation. . . . . *see* Heun's equation.
   expansions in series of
      hypergeometric functions. 716–717
      orthogonal polynomials 717
   integral equations and representations 714–716
   notation 710
   orthogonality
      double. 714
      single. 714
   relations to hypergeometric function 713
   relations to Lamé functions 713
Heun polynomials **712**
   applications. 719
   definitions 712
   integral equations and representations. 715
   notation 712
   orthogonality 714
   products. 719
Heun's equation **710**
   accessory parameter. 710
      asymptotic approximations 718
   applications. 718
      mathematical 719–720
      physical 720
   asymptotic approximations
      eigenvalues of accessory parameters 718
      solutions near irregular singularities. 718
      solutions of confluent forms. 718
      solutions with coalescing singularities 718
   automorphisms. 711–713
      composite 711
      F-homotopic transformations 711
      homographic transformations 711
   basic solutions
      equivalent expressions 712
      Fuchs–Frobenius 711
   biconfluent. 718
   classification of parameters 710
   computation of solutions 720
   confluent forms. 717–718
      asymptotic approximations 718
      integral equations. 716
      special cases 717
   doubly-confluent 717
   doubly-periodic forms
      Jacobi's elliptic 710

      Weierstrass's 710
   eigenvalues of accessory parameter 712
   expansions of solutions in series of
      hypergeometric functions. 716–717
      orthogonal polynomials 717
   exponent parameters. 710
   integral equations 714–716
   integral representation of solutions. 714–716
      kernel functions. 714, 716
      separation constant 714
   inversion problem 719
   Jacobi's elliptic form 710
   Liouvillean solutions 713
   monodromy group 719
   normal form 710
   parameters
      classification. 710
   path-multiplicative solutions 712
      biorthogonality 714
      expansions in series of hypergeometric functions
      717
   relation to Fuchsian equation 718
   relation to Lamé's equation 685
   separation of variables 720
   singularities. 710
   singularity parameter 710
   solutions analytic at three singularities
   . . . . *see* Heun polynomials.
   solutions analytic at two singularities
   . . . . *see* Heun functions.
   solutions via quadratures. 713
   triconfluent 718
   trigonometric form. 710
   uniformization problem 719
   Weierstrass's form 710
Heun's operator 714
hexadecimal number system 72
high-frequency scattering
   parabolic cylinder functions 317
higher-order $3nj$ symbols. 765
highway design
   Cornu's spiral. 169
Hilbert space
   interrelation between bases
      Heun polynomial products 719–720
   $L_\rho^2(Q)$ orthonormal basis 719
Hilbert transform
   computation 84
   definition 29
   Fourier transform of. 29
   inequalities. 29
   inversion 29
Hill's equation **674**
   . . . . *see also* Whittaker–Hill equation.

antiperiodic solutions 675
basic solutions 674
characteristic equation 675
characteristic exponents........................675
definition 674
discriminant 675
eigenfunctions 676
eigenvalues......................................675
equation of Ince.................................676
    Fourier-series solutions......................676
    polynomial solutions .... see Ince polynomials.
expansions in series of eigenfunctions............676
Floquet solutions 675
Floquet's theorem...............................674
periodic solutions 675
pseudoperiodic solutions 674
real case 675
separation constants 677, 678
symmetric case 675
Hölder's inequalities for sums and integrals 12, 13
holomorphic function............ see analytic function.
homogeneous harmonic polynomials 379
homographic transformation
    .... see bilinear transformation.
Horner's scheme for polynomials...................22
    extended..22
Hurwitz criterion for stable polynomials 23
Hurwitz system
    Riemann surface...............................546
Hurwitz zeta function............................**607**
    analytic properties.............................607
    asymptotic expansions for large parameter 610
    computation....................................614
    definition......................................607
    derivatives.....................................608
        asymptotic expansions for large parameter....610
    graphics...................................607–608
    integral representations 609
    integrals.......................................610
    relations to other functions
        Lerch's transcendent........................612
        periodic zeta function........................612
        polylogarithms..............................611
        Riemann zeta function......................607
    representations by Euler–Maclaurin formula.....608
    series representations......................608, 610
    special values..................................608
    sums..610
    tables...614
hydrodynamics
    Jacobian elliptic functions......................566
hyperasymptotic expansions 68
hyperbola
    elliptic integrals................................514

hyperbolic cosecant function.. see hyperbolic functions.
hyperbolic cosine function.... see hyperbolic functions.
hyperbolic cotangent function
    .... see hyperbolic functions.
hyperbolic functions **123**
    addition formulas 125
    analytic properties..............................123
    approximations.................................132
    computation 132
    conformal maps................................124
    continued fractions.............................129
    definitions 123
    derivatives.....................................125
    differential equations...........................125
    elementary properties..........................124
    graphics
        complex argument......................124
        real argument........................123–124
    identities.......................................125
    inequalities 125
    infinite products 126
    integrals
        definite....................................130
        indefinite..................................130
    inverse............ see inverse hyperbolic functions.
    Laurent series..................................125
    limits..125
    Maclaurin series 125
    moduli..126
    multiples of argument..........................126
    notation 104
    partial fractions................................126
    periodicity.....................................123
    poles..123
    real and imaginary parts........................126
    relations to trigonometric functions 123
    special values 125
    squares and products 126
    sums..130
    tables...132
    zeros..123
hyperbolic secant function ... see hyperbolic functions.
hyperbolic sine function ...... see hyperbolic functions.
hyperbolic tangent function.. see hyperbolic functions.
hyperbolic trigonometric functions
    .... see hyperbolic functions.
hyperbolic umbilic bifurcation set
    formula..781
    picture..782
hyperbolic umbilic canonical integral..............**776**
    asymptotic approximations.................789–790
    convergent series...............................787
    differential equations...........................788
    formulas for Stokes set 783

integral identity.................................787
pictures of modulus............................779
pictures of phase..............................781
scaling laws...................................785
zeros..787
hyperbolic umbilic catastrophe...............**776**, 785
hyperelliptic functions...........................566
hyperelliptic integrals...........................498
hypergeometric differential equation..............**394**
equivalent equation for contiguous functions.....388
fundamental solutions.....................394–395
Kummer's solutions............................395
singularities..................................395
hypergeometric equation
.... see hypergeometric differential equation.
hypergeometric function........................**384**
.... see also Gaussian hypergeometric function.
analytic properties............................384
applications
mathematical...............................399
physical................................... 400
asymptotic approximations
large a (or b) and c.......................397, 398
large a and b............................397, 398
large a or b...................................398
large a, b, and c..............................398
large c.....................................396–398
large variable..............................396
branch points..................................384
computation...................................400
contiguous.....................................388
continued fractions............................389
definition......................................384
derivatives................................387–388
Fourier transforms.............................398
graphics...................................385–386
Hankel transforms............................. 398
integral representations....................388–389
Mellin–Barnes type.....................388–389
integrals...................326, 327, 337, 398–399
compendia..............................398–399
Laplace transforms............................398
Maclaurin series..............................384
Mellin transform..............................398
multivariate...................................498
notation.......................................384
Olver's..384
polynomial cases..............................385
principal value (or branch)....................384
products
series expansions..........................399
recurrence relations...........................388
relations to other functions
associated Legendre functions.......353, 354, 394

classical orthogonal polynomials.............. 442
elementary functions........................386
Ferrers functions....................353, 354, 394
gamma function............................. 387
Gegenbauer function........................394
Hahn class orthogonal polynomials...........463
Heun functions............................. 713
incomplete beta functions....................183
Jacobi function............................. 394
orthogonal polynomials..................393–394
Painlevé transcendents......................399
Pollaczek polynomials 476
psi function.................................387
symmetric elliptic integrals.................. 509
Szegő–Askey polynomials....................475
Wilson class orthogonal polynomials......... 469
singularities......................................384
special cases
argument ± 1...............................387
argument a fraction..........................387
arguments $e^{\pm i\pi/3}$..........................387, 400
elementary functions....................386–387
sums..399
compendia...................................399
transformation of variable
cubic.......................................393
linear..............................390–391, 400
quadratic.............................391–393
with two variables............. see Appell functions.
Wronskians....................................395
zeros..398
hypergeometric functions of matrix argument
.... see confluent hypergeometric functions of matrix argument, Gaussian hypergeometric functions of matrix argument, and generalized hypergeometric functions of matrix argument.
hypergeometric R-function......................**498**
derivative....................................500
differential equation..........................501
elliptic cases 498
integral representations 498
implicit function theorem..........................7
Ince polynomials................................676
normalization.................................676
zeros..676
Ince's equation... see Hill's equation, equation of Ince.
Ince's theorem................... see Theorem of Ince.
incomplete Airy functions........................208
incomplete beta functions.......................**183**
applications
physical.................................... 189
statistical...................................189
asymptotic expansions for large parameters
general case 185

inverse function............................185
    symmetric case............................184
basic properties.............................183
continued fraction...........................184
historical profile............................183
integral representation......................183
inverse function............................185
notation....................................174
recurrence relations.........................183
relation to hypergeometric function..........183
sums..184
tables......................................190
incomplete gamma functions....................**174**
    analytic continuation.........................174
    applications
        mathematical.............................189
        physical.................................189
        statistical...............................189
    approximations.............................191
    asymptotic approximations and expansions
        exponentially-improved................179, 181
        for inverse function.......................182
        large variable and/or large parameter
        ....179–180, 182
        uniform for large parameter..........181, 182
    basic properties.............................174
    Chebyshev-series expansions..................191
    computation................................190
    continued fraction..........................179
    definitions
        general values............................174
        principal values..........................174
    derivatives................................178
    differential equations.......................174
    expansions in series of
        Bessel functions..........................178
        Laguerre polynomials.....................178
        modified spherical Bessel functions..........178
    generalizations.............................183
    graphics
        complex argument........................176
        real variables.........................175–176
    inequalities................................179
    integral representations
        along real line...........................177
        compendia..............................178
        contour integrals.........................177
        Mellin–Barnes type.......................177
    integrals..................................182
    monotonicity properties.....................176
    normalized................................174
    notation...................................174
    of imaginary argument.....................177
    Padé approximant..........................179

power-series expansions......................178
principal values..............................174
recurrence relations..........................178
relations to other functions
    confluent hypergeometric functions..177, 328, 338
    error functions.............................164
    exponential integrals.......................153
    generalized exponential integral..............185
    incomplete Riemann zeta function...........189
special values................................176
sums..183
tables.......................................190
zeros.......................................182
incomplete Riemann zeta function..............189
    asymptotic approximations...................189
    expansions in series of incomplete gamma functions
   189
    zeros......................................189
inductance
    symmetric elliptic integrals...................516
inequalities
    means.......................................13
    sums and integrals
        Cauchy–Schwarz.......................12, 13
        Hölder's.................................12, 13
        Jensen's....................................13
        Minkowski's............................12, 13
infinite partial fractions........................22
    Mittag-Leffler's expansion.....................22
infinite products
    convergence.................................21
        absolute..................................21
        uniform..................................21
    M-test for uniform convergence..................21
    relation to infinite partial fractions.............22
    Weierstrass product..........................22
infinite sequences
    convergence.................................17
        absolute..................................17
        pointwise................................17
        uniform..................................17
    double......................................18
        convergence..............................18
        relation to infinite double series..............18
infinite series
    ....*see also* power series.
    convergence.................................17
        absolute..................................17
        pointwise................................17
        uniform..................................17
        Weierstrass M-test..........................17
    divergent...................................17
    dominated convergence theorem................18
    double......................................18

doubly-infinite 17
summability methods 33–34
term-by-term integration 18
inhomogeneous Airy functions . . . . *see* Scorer functions.
initial-value problems
Mathieu functions 679
integrable differential equations
Riemann theta functions 545–546
integrable equations
. . . . *see* integrable differential equations.
integral equations
Painlevé transcendents 729
integral transforms **27**
. . . . *see also* Fourier cosine and sine
transforms, Fourier transform, Jacobi transform,
Hankel (or Bessel) transform, Hilbert transform,
Kontorovich–Lebedev transform, Laplace transform, Mellin transform, spherical Bessel transform,
and Stieltjes transform.
compendia 32
in terms of parabolic cylinder functions 317
in terms of Whittaker functions 344
integrals
asymptotic approximations
. . . . *see* asymptotic approximations of integrals.
Cauchy principal values 6
change of variables 6
convergence 6
absolute 6
uniform 8, 21
convolution product 53
definite 5
differentiation 6, 8
double . . . . *see* double integrals.
fundamental theorem of calculus 6
generalized 52
indefinite 5
infinite 6, 9, 16
Jensen's inequality 13
line 11
mean value theorems
first 6
second 6
multiple 8
over parametrized surface 12
path 11
repeated 6
square-integrable 6
summability methods 34–35
tables 5
with coalescing saddle points 789–790
integrals of Bessel and Hankel functions
compendia 246
convolutions 242

fractional 243
Hankel (or Bessel) transform 246
indefinite 240–241
orthogonal properties 243, 244
over finite intervals 241–243
over infinite intervals 243–246, 326
products 241–246
triple 245
trigonometric arguments 241
integrals of modified Bessel functions
compendia 261
computation 277
fractional 259
indefinite 258
Kontorovich–Lebedev transform 260
over finite intervals 258
over infinite intervals 205, 258–260, 326, 337
products 260
tables 279
integration 5, 16
by parts 5
numerical . . . . *see* cubature, Gauss quadrature,
Monte-Carlo methods, *and* quadrature.
term by term 18
interaction potentials
hypergeometric function 400
interior Dirichlet problem
for oblate spheroids 706
for prolate spheroids 705
interior points 15
interpolation 75–77, 91
. . . . *see also* Lagrange interpolation.
based on Chebyshev points 77
based on Sinc functions 77
bivariate 77
convergence properties 77
Hermite 77
inverse 76
inverse linear 91
linear 76
rational 77
spline 77
trigonometric 77
interval
closure 6
interval arithmetic 72
inverse function 21
Lagrange inversion theorem 21
extended 21
inverse Gudermannian function 121
relation to Legendre's elliptic integrals 491
relation to R_C-function 491
inverse hyperbolic functions **127**
addition formulas 129

918 INDEX

analytic properties............................127
approximations..............................132
branch cuts..................................127
branch points................................127
Chebyshev-series expansions...................132
computation..................................132
conformal maps..............................124
continued fractions..........................129
definitions..................................127
derivatives..................................129
fundamental property........................128
general values...............................127
graphics
   complex argument..........................124
   real argument.........................123–124
integrals....................................129
interrelations...............................128
logarithmic forms............................128
notation.....................................104
power series.................................129
principal values.............................127
reflection formulas..........................128
tables.......................................132
values on the cuts...........................128
inverse incomplete beta function..................185
inverse incomplete gamma function................182
inverse Jacobian elliptic functions................**561**
applications.................................563
as Legendre's elliptic integrals...............561
as symmetric elliptic integrals...............561
computation..................................567
definitions..................................561
equivalent forms.............................561
normal forms.................................561
notation.....................................561
power-series expansions......................561
principal values.............................561
inverse Laplace transforms
asymptotic expansions.....................46–47
inverse trigonometric functions..................**118**
addition formulas............................121
analytic properties..........................119
approximations..............................132
branch cuts..................................119
branch points................................119
Chebyshev-series expansions...................132
computation..................................132
conformal maps..............................113
continued fractions..........................121
definitions..................................118
derivatives..................................121
fundamental property........................120
general values...............................118
graphics

complex argument.......................113–115
   real argument.............................112
integrals....................................122
interrelations...............................120
logarithmic forms........................119–120
notation.....................................104
power series.................................121
principal values.............................119
real and imaginary parts.....................120
reflection formulas..........................119
special values...............................120
sums...123
tables.......................................132
values on the cuts.......................119–120
Ising model
   Appell functions..............................417
   combinatorics................................635
   generalized hypergeometric functions...........417
   Painlevé transcendents........................739
isolated essential singularity.......................19
.... *see also* essential singularity.
   movable......................................724
isolated singularity...............................19
iterative methods
   Bairstow's method (for zeros of polynomials).....91
   bisection method..............................91
   convergence
      cubic.......................................90
      geometric..................................90
      linear......................................90
      local.......................................90
      logarithmic................................94
      of the pth order.........................90
      quadratic..................................90
   eigenvalue methods............................91
   fixed-point methods...........................92
   Halley's rule.................................92
   Newton's rule (or method).....................90
   regula falsi..................................91
   secant method.................................91
   Steffensen's method...........................91

Jacobi fraction (J-fraction).........................95
Jacobi function
   applications..................................399
   definition....................................394
   relations to other functions
      associated Legendre functions..............355
      conical functions..........................379
      hypergeometric function....................394
Jacobi polynomials..............................**438**
.... *see also* classical orthogonal polynomials.
   applications
      Bieberbach conjecture......................479
      associated.................................474

asymptotic approximations 451–452
Bateman-type sums 461
computation 479
definition 439
derivatives 446
differential equations 445
expansions in series of 459–461
Fourier transform 156
generating functions 449
graphs 440
inequalities 450, 451
Szegö–Szász 451
Turan-type 450
integral representations 447, 448
integrals 455–457, 459
fractional 456
indefinite 455
interrelations with other orthogonal polynomials
.... 444–445, 463, 464
Laplace transform 457
leading coefficients 439
limiting form
as Bessel functions 449
as Bessel polynomials 476
limits to monomials 444
local maxima and minima 450–451
Mellin transform 457
monic 80
normalization 439
notation 436
orthogonality properties 439
parameter constraint 439, 443
recurrence relations 446
relations to other functions
hypergeometric function 393, 442
orthogonal polynomials on the triangle 478
Rodrigues formula 442
shifted 437
special values 444
symmetry 444
tables of coefficients 440
upper bounds 450
weight function 439
zeros 438, 454
asymptotic approximations 454
Jacobi symbol
number theory 642
Jacobi transform 379, 394
inversion 394
Jacobi's amplitude function
.... *see* amplitude (am) function.
Jacobi's epsilon function **562**
applications 563
computation 567

definition 562
graphs 563
integral representations 562
quasi-addition formula 562
quasi-periodicity 562
relation to Legendre's elliptic integrals 562
relation to theta functions 562
tables 567
Jacobi's identities
number theory 645
Jacobi's imaginary transformation 556
Jacobi's inversion problem for elliptic functions 532
Jacobi's nome
power-series expansion 490
Jacobi's theta functions .... *see* theta functions.
Jacobi's triple product 529
q-version 427
Jacobi's zeta function **562**
computation 567
definition 562
graphs 563
quasi-addition formula 562
tables 567
Jacobi–Abel addition theorem
Jacobian elliptic functions 564
Jacobi-type polynomials 477
Jacobian 9
Jacobian elliptic functions **550**
addition theorems 557
analytic properties 550, 563
applications
mathematical 563–564
physical 564–566
change of modulus 563
computation 566–567
congruent points 553
coperiodic 553
copolar 553
cyclic identities
notation 558
points 558
rank 558
simultaneously permuted 558
definitions 550
derivatives 560
differential equations
first-order 560
second-order 560
double argument 556
Eisenstein series 559
elementary identities 556
equianharmonic case 555
expansions in doubly-infinite partial fractions ... 559
Fourier series 559

for squares....................559
fundamental unit cell..........554
Glaisher's notation............550, 554
graphical interpretation via Glaisher's notation..554
graphics
    complex modulus.......552–553
    complex variable.......552
    real variable..........550–552
half argument..................556
hyperbolic series for squares..559
integrals
    definite..............561
    indefinite............560
    of squares............562
interrelations................508
inverse....... see inverse Jacobian elliptic functions.
Jacobi's imaginary transformation..556
Landen transformations
    ascending.............557, 563, 566
    descending............556, 563, 566
    generalized...........557
    theta functions.......531
lattice.......................554
    computation...........566
lemniscatic case..............555
limiting forms as $k \to 0$ or $k \to 1$..555
Maclaurin series
    in k, k'...........559
    in z...............558
modulus......................**550**
    change of............563
    complex..............552, 553, 563
    limiting values......555
    outside the interval $[0, 1]$..563
    purely imaginary.....563
    real.................563
nome........................550
notation....................550
periods.....................550, 553–554
poles.......................553–554
poristic polygon constructions..557
principal...................550
relations to other functions
    symmetric elliptic integrals..508
    theta functions.......550
    Weierstrass elliptic functions..575
rotation of argument........556
special values of the variable..554–555, 557
subsidiary..................550
sums of squares.............556
tables......................567
translation of variable.....554
trigonometric series expansions..559
zeros.......................553

Jensen's inequality for integrals..13
Jonquière's function............ see polylogarithms.
Jordan curve theorem...........16
Jordan's function
    number theory........638
Jordan's inequality
    sine function........116
Julia sets....................92
Kadomtsev–Petviashvili equation
    Riemann theta functions..545
Kapteyn's inequality
    Bessel functions.....227
Kelvin functions..............**267**
    applications.........276
    approximations.......281
    asymptotic expansions for large argument...271
        cross-products and sums of squares..271
        exponentially-small contributions..271
    asymptotic expansions for large order... see uniform asymptotic expansions for large order.
    computation..........276–277
    cross-products.......269
    definitions..........267
    derivatives..........269
        with respect to order..269
    differential equations..268
    expansions in series of Bessel functions..270
    graphs...............268
    integral representations..269
    integrals
        compendia..........274
        definite...........274
        indefinite.........274
        Laplace transforms..274
    modulus and phase functions
        asymptotic expansions for large argument..272
        definitions........272
        properties.........272
    notation.............217
    orders $\pm\frac{1}{2}$..268
    power series.........269–270
        compendia..........270
        cross-products and sums of squares..270
    recurrence relations..269
    reflection formulas for arguments and orders..268
    uniform asymptotic expansions for large order..273
        double asymptotic property..273
        exponentially-small contributions..273
    zeros
        asymptotic approximations for large zeros..273
        computation........277
        tables.............281
Kelvin's ship-wave pattern....790–791
kernel equations

Heun's equation 715, 716
kernel functions
    Heun's equation 715, 716
Klein's complete invariant ...... see modular functions.
Klein–Gordon equation
    Coulomb functions 754
Kontorovich–Lebedev transform
    modified Bessel functions 260
        computation 278
Korteweg–de Vries equation
    Airy functions 209
    Jacobian elliptic functions 565
    Lamé polynomials 694
    Painlevé transcendents 738
    Riemann theta functions 545
    Weierstrass elliptic functions 582
Kovacic's algorithm 713, 718
KP equation .... see Kadomtsev–Petviashvili equation.
Krattenthaler's formula for determinants 4
Krawtchouk polynomials
    .... see also Hahn class orthogonal polynomials.
    applications
        coding theory 479
    relation to hypergeometric function 394
Kummer congruences
    Bernoulli and Euler numbers 593
Kummer functions **322**
    .... see also confluent hypergeometric functions.
    addition theorems 333
    analytic continuation 323
    analytical properties 322
    applications
        physical 346
    approximations 347
    asymptotic approximations for large parameters
        large a 330–331
        large b 330
        uniform 330–331
    asymptotic expansions for large argument .. 328–329
        error bounds 329
        exponentially-improved 329
        hyperasymptotic 329
    Chebyshev-series expansions 347
    computation 346–347
    connection formulas 325
    continued fractions 327
    definitions 322
    derivatives 325–326
    differential equation .... see Kummer's equation
    integer parameters 322–323
    integral representations
        along the real line 326
        contour integrals 326–327
        Mellin–Barnes type 327
    integrals
        along the real line 326
        compendia 333
        Fourier transforms 332
        Hankel transforms 332–333
        indefinite 332
        Laplace transforms 332
        Mellin transforms 332
    interrelations 322, 325
    Kummer's transformations 325
    limiting forms
        as $z \to 0$ 323
        as $z \to \infty$ 323
    Maclaurin series 322
    multiplication theorems 334
    notation 322
    polynomial cases 322, 323
    principal branches (or values) 322
    products 333
    recurrence relations 325
    relations to other functions
        Airy functions 328
        elementary functions 327
        error functions 328
        generalized hypergeometric functions 328
        incomplete gamma functions 328
        modified Bessel functions 328
        orthogonal polynomials 328
        parabolic cylinder functions 328
        Whittaker functions 334
    series expansions
        addition theorems 333
        in modified Bessel functions 333
        Maclaurin 322
        multiplication theorems 334
    tables 347
    Wronskians 324
    zeros
        asymptotic approximations 331
        distribution 331
        inequalities 331
        number of 331
Kummer's equation 322
    equivalent form 325
    fundamental solutions 323–324
    numerically satisfactory solutions 323–324
    relation to hypergeometric differential equation .. 322
    relation to Whittaker's equation 334
    standard solutions 322
Kummer's transformations
    for $_3F_2$ hypergeometric functions of matrix argument
    772
    for confluent hypergeometric functions 325
L'Hôpital's rule for derivatives 5

Lagrange interpolation 75–76
   abscissas 75
   coefficients 75
   equally-spaced nodes 75–76
   error term 75
   formula 75
   Newton's interpolation formula 76
   nodal polynomials 75
   nodes 75
   polynomial 75
   remainder terms 75–76
   via divided differences 76
Lagrange inversion theorem 21
   extended 21
Lagrange's formula for reversion of series 43
Laguerre functions
   associated 754
Laguerre polynomials **438**
          see also classical orthogonal polynomials.
   addition theorem 460
   applications
      Schrödinger equation 479
   asymptotic approximations 452–453
   computation 479
   continued fraction 450
   derivatives 447
   differential equations 445
   Dirac delta 38
   expansions in series of 459, 460
   explicit representations 442–443
   Fourier transforms 457
   generalized 436
   generating functions 449
   graphics 441
   inequalities 450, 451
      Turan-type 450
   integral representations 447, 448
   integrals 455–457
      fractional 456
      indefinite 455
   interrelations with other orthogonal polynomials
          .... 445, 463, 464
   Laplace transform 457
   leading coefficients 439
   limiting form as a Bessel function 449
   limits to monomials 444
   local maxima and minima 451
   Mellin transform 458
   monic 80
   multiplication theorem 460
   normalization 439
   notation 436
   orthogonality properties 439
   parameter constraint 439, 443

Poisson kernels 461
recurrence relations 446
relation to confluent hypergeometric functions
          .... 328, 338, 443, 448
Rodrigues formula 442
tables 480
   of coefficients 440
   of zeros 81
tables of zeros 81
upper bounds 450
value at $z = 0$ 443
weight function 439
zeros 438, 454
   asymptotic behavior 454
   tables 81
Lamé functions **684**
   algebraic 693
   applications
      conformal mapping 694
      physical 694
      rotation group 694
      sphero-conal coordinates 693
   asymptotic expansions 689
   computation 694
   definition 685
   differential equation .... *see* Lamé's equation.
   eigenvalues
      asymptotic expansions 689
      coalescence 685
      computation 694
      continued-fraction equation 685
      definition 685
      distribution 685
      graphics 686–687
      interlacing 685
      limiting forms 688
      notation 684
      parity 685
      periods 685
      power-series expansions 686
      special cases 688
      tables 694
   Fourier series 688–689
   graphics 687–688
   integral equations 689
   limiting forms 688
   normalization 686
   notation 684
   order 685
   orthogonality 686
   parity 686
   period 686
   relations to Heun functions 713
   relations to Lamé polynomials 689, 690

special cases 688
with imaginary periods 690
with real periods 685
zeros 685
Lamé polynomials **690**
   algebraic form 691
   applications
      ellipsoidal harmonics 694
      physical 691, 694
      spherical harmonics 694
   asymptotic expansions 693
   Chebyshev series 693
   coefficients 694
   computation 694
   definition 690
   eigenvalues
      asymptotic expansions 693
      computation 694
      graphics 691
   elliptic-function form 690
   explicit formulas 693
   Fourier series 692
   graphics 692
   notation 684, 690, 691
   orthogonality 692
   periodicity 690
   relation to Lamé functions 689, 690
   tables 694
   zeros 690
      computation 694
      electrostatic interpretation 691
Lamé wave equation 690
Lamé's equation **684**
   algebraic form 684
   eigenfunctions 686
   eigenvalues .... *see* Lamé functions, eigenvalues.
   Jacobian elliptic-function form 684
   other forms 684–685
   relation to Heun's equation 685
   second solution 693
   singularities 684
   stability 690
   trigonometric form 684
   Weierstrass elliptic-function form 685
Lamé–Wangerin functions 693
Lambert W-function **111**
   applications 131
   asymptotic expansions 111
   computation 132
   definition 111
   graphs 111
   integral representations 111
   notation 111
   principal branch 111

other branches 111
properties 111
Lambert series
   number theory 641
Lanczos tridiagonalization of a symmetric matrix ... 75
Lanczos vectors 75
Landen transformations
   Jacobian elliptic functions 556, 557, 563, 566
   theta functions 531
Laplace equation
   $3j, 6j, 9j$ symbols 765
Laplace transform
   analyticity 28
   asymptotic expansions for large parameters
      .... 43, 44, 46
   asymptotic expansions for small parameters 51
   convergence 28
   convolution 28
   definition 28
   derivatives 28
   differentiation 28
   for functions of matrix argument 768
      analytic properties 768
      convolution theorem 768
      definition 768
      inversion formula 768
   integration 28
   inversion 28
   notation 28
   numerical inversion 83–84, 99
   of periodic functions 28
   tables 32
   translation 28
   uniqueness 29
Laplace's equation
   Bessel functions 275
   for elliptical cones 694
   spherical coordinates 379
   symmetric elliptic integrals 501
   toroidal coordinates 379
Laplace's method for asymptotic expansions of integrals
   .... 44, 47
Laplacian 7
   cylindrical coordinates 7
   ellipsoidal coordinates 583
   numerical approximations 78
   oblate spheroidal coordinates 706
   parabolic cylinder coordinates 317
   polar coordinates 7
   prolate spheroidal coordinates 705
   spherical coordinates 8
lattice
   for elliptic functions
      .... *see* Weierstrass elliptic functions, lattice.

lattice models of critical phenomena
    elliptic integrals..................................517
lattice parameter
    theta functions..................................524
lattice paths.............................618–623
    definition..618
    k-dimensional..................................618
lattice walks
    Appell functions...............................417
    generalized hypergeometric functions............417
Laurent series.......................................19
    asymptotic approximations for coefficients........65
Lauricella's function
    relation to symmetric elliptic integrals...........509
Lax pairs
    classical orthogonal polynomials.................478
    Painlevé transcendents.........................728
layered materials
    elliptic integrals................................517
least squares approximations..................99–100
    conditioning.....................................99
    normal equations................................99
    orthogonal functions with respect to weighted summation..99
Lebesgue constants...........................13, 97
    asymptotic behavior.............................13
Legendre functions...............................352
    ..*see also* associated Legendre functions *and* Ferrers functions.
    complex degree.................................379
Legendre functions on the cut ... *see* Ferrers functions.
Legendre polynomials...........................**438**
    ....*see also* classical orthogonal polynomials.
    addition theorem...............................459
    applications
        Schrödinger equation......................479
        associated.................................474
    asymptotic approximations.....................452
    computation...................................479
    continued fraction.............................450
    definition......................................439
    differential equation...........................445
    Dirac delta......................................38
    expansions in series of....................459, 461
    explicit representations....................442–443
    Fourier transforms.............................456
    generating functions...........................449
    graphs...441
    inequalities....................................450
        Turan-type................................450
    integral representations....................447, 448
        for products...............................455
    integrals..................................455, 458
        Nicholson-type.............................455
    interrelations with other orthogonal polynomials
   444
    large degree....................................65
    leading coefficients.............................439
    Mellin transforms..............................458
    monic...80
    normalization..................................439
    notation.......................................436
    orthogonality properties.......................439
    recurrence relations............................446
    relations to other functions
        associated Legendre functions.................360
        Ferrers functions...........................360
        hypergeometric function.....................394
        $3j$ symbols............................760–761
    Rodrigues formula.............................442
    shifted....................................436, 439
    special values..................................444
    symmetry......................................444
    tables...480
        of coefficients..............................440
        of zeros....................................80
    value at argument zero........................285
    weight function................................439
    zeros......................................438, 454
        tables......................................80
Legendre symbol
    prime numbers.................................642
Legendre's elliptic integrals......................**486**
    addition theorem..............................495
    applications
        mathematical..........................514–516
        physical...................................517
    approximations (except asymptotic).............519
    arithmetic-geometric mean.....................492
    asymptotic approximations.....................495
    change of amplitude...........................492
    change of modulus............................492
    change of parameter..........................492
    circular cases.............................487, 492
    complete......................................487
    computation..............................517–518
    connection formulas...........................491
    derivatives....................................490
    differential equations..........................490
    duplication formulas..........................495
    first, second, and third kinds...................486
    Gauss transformation..........................493
    graphics..................................488–489
    hyperbolic cases..........................487, 492
    imaginary-argument transformations............492
    imaginary-modulus transformations............492
    incomplete....................................486
    inequalities

complete integrals 494
    incomplete integrals 494
integration
    with respect to amplitude 496
    with respect to modulus 496
Landen transformations
    ascending 493
    descending 493
Laplace transforms 496
limiting values 491
notation 486
power-series expansions 490
quadratic transformations 492
reciprocal-modulus transformation 492
reduction of general elliptic integrals 496–497
relations to other functions
    am function 562
    Appell functions 490
    inverse Gudermannian function 491
    inverse Jacobian elliptic functions 561
    Jacobi's epsilon function 562
    Jacobi's zeta function 562
    Jacobian elliptic functions 494
    symmetric elliptic integrals 507, 508
    theta functions 494
    Weierstrass elliptic functions 494
special cases 491
tables 518–519
Legendre's equation 352
    standard solutions 352
Legendre's relation
    Legendre's elliptic integrals 492
Legendre's relation for the hypergeometric function
    generalized 399
Leibniz's formula for derivatives 5
lemniscate arc length 563
lemniscate constants 502, 503
lengths of plane curves
    Bernoulli's lemniscate 515
    ellipse 514
    hyperbola 514
Lerch's transcendent
    definition 612
    properties 612
    relation to Hurwitz zeta function 612
    relation to polylogarithms 612
level-index arithmetic 73
Levi-Civita symbol for vectors 10
Levin's transformations
    application to asymptotic expansions 69
    for sequences 94
Lie algebras
    q-series 432
light absorption

Voigt functions 169
limit points (or limiting points) 15
limits of functions
    of a complex variable 15
    of one variable 4
    of two complex variables 15
    of two variables 7
line broadening function 167
linear algebra 73–75
    .... see also Gaussian elimination.
    condition numbers 74, 75
    conditioning of linear systems 74
    error bounds 74
        a posteriori 74
    norms
        Euclidean 74
        of arbitrary order 74
        of matrices 74
        of vectors 74
linear functional 35
linear transformation 17
Liouville transformation for differential equations
    .... 26, 58
Liouville's function
    number theory 639
Liouville's theorem for entire functions 16
Liouville–Green (or WKBJ) approximation 57
    for difference equations 62
little q-Jacobi polynomials 471
local maxima and minima 450
locally analytic 724
locally integrable 48
logarithm function **104**
    analytic properties 104
    approximations 132
    branch cut 104
    Briggs 105
    Chebyshev-series expansions 132
    common 105
    computation 131
    conformal maps 106
    continued fractions 109
    definition 104
    derivatives 108
    differential equations 109
    general base 105
    general value 104
    generalized 111
    graphics
        complex argument 107
        real argument 106
    hyperbolic 105
    identities 109
    inequalities 108

integrals 110
limits 107
Napierian 105
natural 105
notation 104
power series 108
principal value 104
real and imaginary parts 104
special values 107
sums 110
tables 132
values on the cut 104
zeros 104
logarithmic integral **150**
asymptotic expansion 153
definition 150
graph 155
notation 150
number-theoretic significance 155
relation to exponential integrals 150
Lommel functions **294**
asymptotic expansions for large argument 295
computation 299
definitions 294–295
differential equation 294
integral representations 295
integrals 295
notation 288
power series 294
reflection formulas 295
relation to Anger–Weber functions ... 296
series expansions
    Bessel functions 295
    power series 294
Lucas numbers 596
M-test for uniform convergence
infinite products 21
infinite series 17
magic squares
number theory 648
magnetic monopoles
Riemann theta functions 545
Mangoldt's function
number theory 639
many-body systems
confluent hypergeometric functions 346
many-valued function .... *see* multivalued function.
mathematical constants 100
Mathieu functions **652**, **664**
...... *see also* Mathieu's equation, modified Mathieu functions, *and* radial Mathieu functions.
analytic properties 653, 661, 665
antiperiodicity 654
applications

mathematical 677–678
physical 678–679
asymptotic expansions for large q
.. *see also* uniform asymptotic approximations for large parameters.
Goldstein's 662
Sips' 661
computation 679–680
connection formulas 665
definitions 664
differential equation 652
expansions in series of 664, 667
Fourier coefficients
asymptotic forms for small q 657, 666
asymptotic forms of higher coefficients 657
normalization 657, 666
recurrence relations 656, 666
reflection properties in q 657
tables 680
values at $q = 0$ 657
Fourier series 653, 656, 666
graphics 655–656, 665
integral equations
compendia 663
variable boundaries 663
with Bessel-function kernels 663
with elementary kernels 663, 672
integral representations 672
compendia 674
integrals
compendia 674
of products 674
of products with Bessel functions 673–674
irreducibility 661
limiting forms as order tends to integers 665
normalization 654, 664
notation 652
of integer order **654**
of noninteger order **664**
orthogonality 654, 664
parity 654
periodicity 654, 664
power series in q 660, 666
pseudoperiodicity 653, 664
reflection properties in ν 664
reflection properties in q 654, 664, 665
reflection properties in z 664
relations to other functions
basic solutions of Mathieu's equation 654
confluent Heun functions 717
modified Mathieu functions 667
tables 680
uniform asymptotic approximations for large parameters

Barrett's	662
Dunster's	662–663
values at $q=0$	654
Wronskians	658
zeros	663
tables	680

Mathieu's equation **652**
 algebraic form 652
 basic solutions 653
  relation to eigenfunctions 654
 characteristic equation 653
 characteristic exponents 653
  computation 679
 eigenfunctions .... *see* Mathieu functions.
 eigenvalues (or characteristic values) 653
  analytic continuation 661
  analytic properties 661
  asymptotic expansions for large q 661, 666
  branch points 661
  characteristic curves 667
  computation 679
  continued-fraction equations 659, 666
  distribution 654, 664
  exceptional values 661
  graphics 654, 665
  normal values 661, 664
  notation 652, 653, 664
  power-series expansions in q 659–660, 666
  reflection properties in ν 664
  reflection properties in q 654, 664
  tables 680
 Floquet solutions 653
  computation 679
  Fourier-series expansions 653
  uniqueness 653
 Floquet's theorem 653
 parameters
  definition 652
  stability chart 667
  stable pairs 667
  stable regions 667
  unstable pairs 667
 second solutions
  antiperiodicity 657
  definitions 657
  expansions in Mathieu functions 658
  Fourier series 658
  graphics 658
  normalization 657
  notation 652
  periodicity 657
  reflection properties in q 658
  values at $q=0$ 658
 singularities 652
 standard form 652
 Theorem of Ince 653, 657
 transformations 653
matrix
 .... *see also* linear algebra.
 augmented 73
 characteristic polynomial 74
 condition number 74
 eigenvalues 74–75
  characteristic polynomial 74
  computation 75
  condition numbers 75
  conditioning 75
  multiplicity 74
 eigenvectors
  left 74
  normalized 74
  right 74
 equivalent 542
 factorization 73
 Jacobi 82
 nondefective 74
 norms 74
 Riemann 538
 symmetric
  tridiagonalization 75
 symplectic 541
 triangular decomposition 73
 tridiagonal 74
maximum 7
 local 5, 8
maximum-modulus principle
 analytic functions 20
 harmonic functions 20
 Schwarz's lemma 20
McKean and Moll's theta functions 524
McMahon's asymptotic expansions
 zeros of Bessel functions 236
  error bounds 236
mean value property for harmonic functions 16
mean value theorems
 differentiable functions 5
 integrals 6
means .... *see* Abel
 means, arithmetic mean, Cesàro means, geometric
 mean, harmonic mean, *and* weighted means.
measure 437
 theory 437
Mehler functions .... *see* conical functions.
Mehler–Dirichlet formula
 Ferrers functions 363
Mehler–Fock transformation 373, 379
 generalized 373, 379
Mehler–Sonine integrals

Bessel and Hankel functions.....................224
Meijer G-function................................**415**
    approximations................................. 418
    asymptotic expansions.......................... 417
    differential equation........................... 417
    identities......................................416
    integral representations 415
    integrals.......................................416
    notation 415
    relation to generalized hypergeometric function
       415–416
    special cases 416
    sums..416
Meixner polynomials
    ....*see* Hahn class orthogonal polynomials.
    relation to hypergeometric function 394
Meixner–Pollaczek polynomials
    ....*see* Hahn class orthogonal polynomials.
    relation to hypergeometric function 394
Mellin transform
    analytic properties............................. 48
    analyticity.................................... 29
    convergence.................................... 29
    convolution 29
    convolution integrals 49
    definition..................................29, 48
    inversion 29, 48
    notation 29
    Parseval-type formulas 29
    tables..32
Mellin–Barnes integrals 145
meromorphic function 19
Mersenne numbers
    number theory.................................. 648
Mersenne prime
    number theory.................................. 644
method of stationary phase
    asymptotic approximations of integrals........... 45
metric coefficients
    for oblate spheroidal coordinates 705
    for prolate spheroidal coordinates 704
Mill's ratio for complementary error function 163
    inequalities 163
Miller's algorithm
    difference equations.........................85–87
minimax polynomial approximations................96
    computation of coefficients 96
minimax rational approximations..................97
    computation of coefficients 98
    type 97
    weight function 97
minimum..7
    local..5, 8
Minkowski's inequalities for sums and series.....12, 13

minor............................*see* determinants.
Mittag-Leffler function 261
Mittag-Leffler's expansion
    infinite partial fractions......................... 22
Möbius transformation .... *see* bilinear transformation.
Möbius function
    number theory.................................. 639
Möbius inversion formulas
    number theory.................................. 641
modified Bessel functions **248**
    addition theorems..............................260
    analytic continuation...........................253
    applications
        asymptotic solutions of differential equations..274
        wave equation 275
    approximations 281
    asymptotic expansions for large argument .. 255–256
        error bounds...........................255, 256
        exponentially-improved 256
        for derivatives with respect to order...........255
        for products 255
    asymptotic expansions for large order.......256–258
        asymptotic forms 256
        double asymptotic properties............. 257–258
        in inverse factorial series.....................257
        uniform 256–257
    branch conventions.............................249
    computation................................276–277
    connection formulas............................251
    continued fractions.............................253
    cross-products 251
    definitions 248
    derivatives
        asymptotic expansions for large argument.....255
        explicit forms...............................252
        uniform asymptotic expansions for large order
           256–257
    derivatives with respect to order 254
        asymptotic expansion for large argument......255
    differential equations 248, 254
    ....*see also* modified Bessel's equation.
    generating function 254
    graphics 249
    hyperasymptotic expansions....................276
    incomplete.................................... 262
    inequalities 254
    integral representations
        along real line........................... 252–253
        compendia..................................253
        contour integrals............................253
        Mellin–Barnes type 253
        products....................................253
    integrals .. *see* integrals of modified Bessel functions.
    limiting forms.................................252

monotonicity 254
multiplication theorem 260
notation 217
of imaginary order
    approximations 281
    computation 278
    definitions 261
    graphs 250, 251
    limiting forms 261
    numerically satisfactory pairs 261
    tables 280
    uniform asymptotic expansions for large order
    261
    zeros 261
power series 252
principal branches (or values) 249
recurrence relations 251
relations to other functions
    Airy functions 197
    confluent hypergeometric functions .. 255, 328, 338
    elementary functions 254
    generalized Airy functions 206
    generalized hypergeometric functions 255
    parabolic cylinder functions 255, 308
sums
    addition theorems 260
    compendia 261
    expansions in series of 261
    multiplication theorem 260
tables 279
Wronskians 251
zeros 258
    computation 277
    tables 280
modified Bessel's equation 248
    inhomogeneous forms 288, 295
    numerically satisfactory solutions 249
    singularities 248
    standard solutions 249
modified Korteweg–de Vries equation
    Painlevé transcendents 738
modified Mathieu functions **667**
    .... *see also* radial Mathieu functions.
    addition theorems 672
    analytic continuation 668
    applications
        mathematical 677
        physical 678–679
    asymptotic approximations
     .. *see also* uniform asymptotic approximations for large parameters.
        for large $\Re z$ 667, 672
        for large q 672
    computation 680

connection formulas 667, 669
definitions 667
differential equation 667
expansions in series of
    Bessel functions 670
    cross-products of Bessel functions and modified Bessel functions 671
graphics 669
integral representations 672–674
    compendia 674
    of cross-products 674
integrals 672
    compendia 674
joining factors 652, 669
    tables 680
notation 652
relation to Mathieu functions 667
shift of variable 668
tables 680
uniform asymptotic approximations for large parameters 662, 672
Wronskians 668
zeros
    tables 680
modified Mathieu's equation **667**
    algebraic form 667
modified spherical Bessel functions
    .... *see* spherical Bessel functions.
modified Struve functions .... *see* Struve functions and modified Struve functions.
modified Struve's equation ... *see* Struve functions and modified Struve functions, differential equations.
modular equations
    modular functions 582
modular functions **579**
    analytic properties 579
    applications
        mathematical 581–582
        physical 582–583
    computation 583
    cusp form 579
    definitions 579
    elementary properties 580
    general 579
    graphics 579
    infinite products 580
    interrelations 581
    Laurent series 580
    level 579
    modular form 579
    modular transformations 580
    notation 570, 579
    power series 580
    relations to theta functions 525, 532, 579

special values 580
modular theorems
 generalized elliptic integrals 516
molecular spectra
 Coulomb functions 753
molecular spectroscopy
 $3j, 6j, 9j$ symbols 765
mollified error 73
moment functionals 476
monic polynomial 22, 80
monodromy groups
 Heun functions 719
 hypergeometric function 400
monosplines
 Bernoulli 597
 cardinal 597
monotonicity 4
Monte Carlo sampling 189
Monte-Carlo methods
 for multidimensional integrals 84
Mordell's theorem 582
 elliptic curves 581
Motzkin numbers
 definition 621
 generating function 623
 identities 623
 recurrence relation 623
 relation to lattice paths 621
 table 622
multidimensional theta functions
 .... *see* Riemann theta functions *and* Riemann theta functions with characteristics.
multinomial coefficients
 definitions 620
 generating function 620
 recurrence relation 620
 relation to lattice paths 620
 table 620
multiple orthogonal polynomials 477
multiplicative functions 640
multiplicative number theory 638–644
 completely multiplicative functions 640
 Dirichlet series 640
 Euler product 640
 fundamental theorem of arithmetic 638
 multiplicative functions 640
 notation 638
 primitive roots 638
multivalued function 20
 branch 20, 104
 branch cut 104
 principal value 104
  closed definition 104
multivariate beta function

definition 768
notation 768
properties 769
multivariate gamma function
 definition 768
 notation 768
 properties 769
multivariate hypergeometric function 498
mutual inductance of coaxial circles
 elliptic integrals 516
n-dimensional sphere
 gamma function 145
Nörlund polynomials 596
nanotubes
 Struve functions 298
Narayana numbers
 definition 622
 generating function 623
 identity 623
 relation to lattice paths 622
 table 622
negative definite
 Taylor series 8
neighborhood 7, 15
 cut 20
 of infinity 16
 punctured 19
Neumann's addition theorem
 Bessel functions 246
 modified Bessel functions 260
Neumann's expansion
 Bessel functions 247
Neumann's integral
 Bessel functions 224
 Legendre functions 364
Neumann's polynomial
 Bessel functions 247
Neumann-type expansions
 modified Bessel functions 261
Neville's theta functions 524
 relations to Jacobian elliptic functions 550
Newton's interpolation formula 76
Newton's rule (or method) 90
 convergence 90
Nicholson's integral
 Bessel functions 225
Nicholson-type integral
 parabolic cylinder functions 313
$9j$ symbols **763**
 addition theorem 764
 applications 765
 approximations for large parameters 764
 computation 765
 definition 763

generating functions 764
graphical method 765
notation 758
orthogonality 764
recursion relations 764
representation as
  finite sum of $6j$ symbols 763
  finite sum of $3j$ symbols 763
  generalized hypergeometric functions..........764
special case 764
sum rule 764
summation convention 760
sums.. 764
symmetry....................................... 764
tables.. 765
zeros... 765
nodal polynomials................................ 75
nodes..79, 80
nome
  Jacobi's..490
  Jacobian elliptic functions......................550
  theta functions.................................524
  Weierstrass elliptic functions...................570
nonlinear equations
  fixed points 90
  numerical solutions
    iterative methods..........................90–92
  systems.. 92
nonlinear evolution equations
  Weierstrass elliptic functions....................582
nonlinear harmonic oscillator
  Painlevé equations 725
nonlinear ordinary differential equations
  Jacobian elliptic functions......................565
nonlinear partial differential equations
  Jacobian elliptic functions......................565
normal probability functions 160
Novikov's conjecture
  Riemann theta functions 546
nuclear physics
  Coulomb functions............................. 754
nuclear structure
  $3j, 6j, 9j$ symbols................................765
number theory .... see also additive number theory,
  multiplicative number theory, and prime numbers.
  Bernoulli and Euler numbers and polynomials...598
  generalized hypergeometric functions............417
  Jacobian elliptic functions......................564
  modular functions..............................582
  theta functions.................................533
  Weierstrass elliptic functions....................582
number-theoretic functions....................638–643
  completely multiplicative......................640
  computation 649

Dirichlet character 642
  induced modulus 642
  Legendre symbol.............................. 642
  primitive 642
  principal...................................... 642
Dirichlet divisor problem.......................643
divisor function 638
divisor sums 641
inversion formulas..............................641
Lambert series 641
Möbius inversion......................... 641, 647
  pairs 641
multiplicative 640
orthogonality 642
periodic..642
Ramanujan's sum............................. 642
tables..649
numerical differentiation
  .... see differentiation, numerical.
oblate spheroidal coordinates.....................705
  Laplacian 706
  metric coefficients............................705
Olver's algorithm
  difference equations.........................86–87
Olver's associated Legendre function..........354, 375
Olver's confluent hypergeometric function 322
Olver's hypergeometric function 353, 384
OP's..................... see orthogonal polynomials.
open disks around infinity 16
open point set 11, 15
optical diffraction
  Struve functions 298
optics
  canonical integrals 791
Orr–Sommerfeld equation
  Airy functions 209
orthogonal matrix polynomials 477
orthogonal polynomials
  complex......................................83
  Painlevé transcendents........................ 739
  relations to confluent hypergeometric functions
    ....328, 338
  relations to hypergeometric function.........393–394
orthogonal polynomials associated with root systems
  478
orthogonal polynomials on the triangle 478
orthogonal polynomials on the unit circle
  ...... see polynomials orthogonal on the unit circle.
orthogonal polynomials with Freud weights........ 475
oscillations of chains
  Bessel functions...............................275
oscillations of plates
  Bessel functions...............................276
℘-function........... see Weierstrass elliptic functions.

packing analysis
   incomplete beta functions 189
Padé approximations 98–99
   computation of coefficients 98
   convergence 98
   Padé table 98
Painlevé equations **724**
   . . . . *see also* Painlevé transcendents.
   affine Weyl groups 732
   alternative forms 724
   Bäcklund transformations 730–732
   coalescence cascade 725
   compatibility conditions 728–729
   elementary solutions 732–735
   elliptic form 725
   graphs of solutions 726–728
   Hamiltonian structure 729–730
   interrelations 730–732
   isomonodromy problems 728
      compatibility condition 728
      Lax pair 728
   rational solutions 732–734
   renormalizations 724
   special function solutions 735–736
      Airy functions 735
      Bessel functions 735
      Hermite polynomials 735
      hypergeometric function 399, 736
      parabolic cylinder functions 735
      Whittaker functions 736
   symmetric forms 725
Painlevé property 724
   applications 739
Painlevé transcendents **724**
   . . . . *see also* Painlevé equations.
   applications 738
      Boussinesq equation 739
      combinatorics 739
      enumerative topology 739
      integrable continuous dynamical systems 739
      integral equations 729
      Ising model 739
      Korteweg–de Vries equation 738
      modified Korteweg–de Vries equation 738
      orthogonal polynomials 739
      partial differential equations 738–739
      quantum gravity 739
      sine-Gordon equation 739
      statistical physics 739
      string theory 739
   asymptotic approximations 736–738
      complex variables 738
      real variables 736–738
   Bäcklund transformations 730–732

computation 740
differential equations for 724
graphs 726–728
Hamiltonians 729–730
Lax pair 728
notation 724, 730–732
parabolic cylinder functions **304**, **314**
   addition theorems 313
   applications
      mathematical 317
      physical 317
   approximations 318
   asymptotic expansions for large parameter . . *see* uniform asymptotic expansions for large parameter.
   asymptotic expansions for large variable 309, 315
      exponentially-improved 309, 317
   computation 318
   connection formulas 304, 315
   continued fraction 308
   definitions 304, 305, 314
   derivatives 309
   differential equations 304
      numerically satisfactory solutions 304, 314
      standard solutions 304
   envelope functions 367
   expansions in Chebyshev series 318
   generalized 317
   graphics
      complex variables 306
      real variables 305–306, 314
   Hermite polynomial case 304, 308
   integral representations
      along the real line 307, 315
      compendia 308
      contour integrals 307
      Mellin–Barnes type 308
   integral transforms 317
   integrals 313
      asymptotic methods 317
      compendia 313
      Nicholson-type 313
   modulus and phase functions 305, 316
   notation 304
   orthogonality 317
   power-series expansions 307, 315
   recurrence relations 309
   reflection formulas 304
   relations to other functions
      Bessel functions 228, 315
      confluent hypergeometric functions
      . . . . 308, 315, 328, 338
      error and related functions 308
      Hermite polynomials 308
      modified Bessel functions 255, 308

probability functions 308
  repeated integrals of the complementary error function 167
  sums 313
  tables 318
  uniform asymptotic expansions for large parameter 309–312, 315–316
    double asymptotic property 311
    in terms of Airy functions 311–312, 316
    in terms of elementary functions 310–311, 316
    modified expansions in terms of Airy functions 312
    modified expansions in terms of elementary functions 311
  values at $z=0$ 304, 314
  Wronskians 304, 314
  zeros
    asymptotic expansions for large parameter 313
    asymptotic expansions for large variable .. 312, 317
    distribution 312
paraboloidal coordinates
  wave equation 346
  Whittaker–Hill equation 678
paraboloidal wave functions 677
  asymptotic behavior for large variable 677
  orthogonality properties 677
  reflection properties 677
parallelepiped
  volume 10
parallelogram
  area 10
parametrization of algebraic equations
  Jacobian elliptic functions 563
parametrized surfaces
  area 11
  integral over 12
  of revolution 12
  orientation 12
  smooth 11
  sphere 11
  tangent vector 11
paraxial wave equation 788
Parseval's formula
  Fourier cosine and sine transforms 28
  Fourier series 14
  Fourier transform 27
Parseval-type formulas
  Mellin transform 29, 49
partial derivative 7
partial differential equations
  nonlinear
    Weierstrass elliptic functions 582
    Painlevé transcendents 738
  spectral methods 479

partial differentiation 7
partial fractions 2
  .... *see also* infinite partial fractions.
particle scattering
  Coulomb functions 753
partition .... *see* partition function.
partition function
  asymptotic expansion 646
  calculation 646
  divisibility 646
  generating function 646
  hadronic matter 146
  parts 645
  Ramanujan congruences 646
  unrestricted 645
partitional shifted factorial 769
partitions 618–620, 624–631, 769
  applications 635
  compositions 628
  conjugate 626
  definition 618
  of a set 618–620, 624–626
  of integers 618, 626–628
  parts 618
  plane .... *see* plane partitions.
  restricted .... *see* restricted integer partitions.
  tables 619, 629, 635
  weight of 769
path
  integrals of vector-valued functions 11
  length 11
PCFs .... *see* parabolic cylinder functions.
Pearcey integral **777**
  asymptotic approximations 789–790
  convergent series 787
  definition 777
  differential equation 788
  formula for Stokes set 783
  integral identities 787
  picture of Stokes set 784
  pictures of modulus 778
  pictures of phase 780
  scaling laws 785
  zeros 785
    table 786
pendulum
  amplitude (am) function 564
  Jacobian elliptic functions 564
  Mathieu functions 679
pentagonal numbers
  number theory 646
periodic Bernoulli functions 588
periodic Euler functions 588
periodic zeta function

relation to Hurwitz zeta function................612
relation to polylogarithms.......................612
permutations........................... 618, 631–634
   adjacent transposition...........................631
   cycle notation..................................631
   definition......................................618
   derangement....................................631
   derangement number.............................631
   descent..632
   even or odd....................................631
   excedance......................................632
      weak..632
   fixed points....................................631
   generating function............................632
   greater index..................................632
   identities.....................................632
   inversion numbers..........................631–634
   major index...............................632, 634
   matrix notation................................633
   multiset.......................................634
   order notation.................................632
   restricted position............................633
   sign.....................................631, 633
   special values.................................633
   transpositions.................................631
   twelvefold way.................................634
Pfaff–Saalschutz formula
   $_3F_2$ functions of matrix argument..............772
phase principle............................. 20, 92
photon scattering
   hypergeometric function........................400
pi
   computation to high precision via elliptic integrals
  516
Picard's theorem.................................19
Picard–Fuchs equations
   generalized hypergeometric functions...........417
piecewise continuous...............................4
piecewise differentiable curve.....................11
pion-nucleon scattering
   Coulomb functions..............................754
pionic atoms
   Coulomb functions..............................754
plane algebraic curves............. see algebraic curves.
plane curves
   elliptic integrals.........................514–515
   Jacobian elliptic functions.....................563
plane partitions
   applications...................................635
   complementary..................................630
   definitions....................................629
   descending.....................................630
   generating functions...........................630
   limiting form..................................631
   recurrence relation............................631
   strict shifted.................................630
   symmetric......................................629
   table..629
plane polar coordinates.......... see polar coordinates.
plasma dispersion function.......................169
plasma waves
   error functions................................169
plasmas
   hypergeometric function........................400
Pochhammer double-loop contour........326, 389, 714
Pochhammer's integral
   beta function..................................142
   Heun functions.................................714
Pochhammer's symbol..............................136
point sets in complex plane
   closed..15
   closure..15
   compact..18
   connected......................................15
   domain...15
      exterior....................................16
      interior....................................16
   open...15
   region...15
points in complex plane
   accumulation...................................15
   at infinity....................................16
   boundary.......................................15
   interior.......................................15
   limit (or limiting)............................15
Poisson identity
   discrete analog................................532
      Gauss sum..................................532
Poisson integral..............................16, 34
   conjugate......................................34
   harmonic functions.............................16
Poisson kernel....................................33
   Fourier integral...............................34
   Fourier series.................................33
Poisson's equation
   in channel-like geometries.....................379
Poisson's integral
   Bessel functions...............................224
Poisson's summation formula
   Fourier series.................................14
polar coordinates..................................7
polar representation
   complex numbers................................14
pole..19
   movable.......................................724
   multiplicity...................................19
   order..19
Pollaczek polynomials............................476

INDEX 935

definition 476
    expansions in series of 477
    orthogonality properties 477
    relation to hypergeometric function 476
    relations to other orthogonal polynomials 477
polygamma functions **144**
    computation 146
    continued fractions 144
    definition 144
    properties 144
    special values 144
    sums 144
    tables 146
polylogarithms **611**
    analytic properties 611
    computation 614
    definitions 611
    integral representations 611
    relations to other functions
        Fermi–Dirac integrals 612
        Lerch's transcendent 612
        periodic zeta function 612
        Riemann zeta function 611
    series expansions 611
    tables 614
polynomials
    characteristic 74
    deflation 91
    discriminant 22
    monic 22, 80
    nodal 75
    stable .... *see* stable polynomials.
    Wilkinson's 92
    zeros .... *see* zeros of polynomials.
    zonal .... *see* zonal polynomials.
polynomials orthogonal on the unit circle 475–476
    biorthogonal 476
    connection with orthogonal polynomials on the line
        475
    definition 475
    recurrence relations 475
    special cases 475
population biology
    incomplete gamma functions 189
poristic polygon constructions of Poncelet
    Jacobian elliptic functions 557
positive definite
    Taylor series 8
potential theory
    conical functions 379
    symmetric elliptic integrals 501, 516
power function **105**
    analytic properties 105
    branch cut 104

definition 105
    derivatives 109
    general bases 105
    general value 105
    identities 109
    limits 107
    modulus 105
    notation 105
    phase 105
    principal value 105
    special values 107
power series
    addition 17
    convergence 17
        circle of 17
        radius of 17
    differentiation 18
    multiplication 17
    of logarithms 18
    of powers 18
    of reciprocals 18
    subtraction 17
primality testing
    Weierstrass elliptic functions 582
prime number theorem 638, 643, 644
    equivalent statement 613
prime numbers
    applications 647
    asymptotic formula 644
    computation 648–649
    counting 648
    cryptography 647
    distribution 613, 638
        asymptotic estimate 638
    Euler–Fermat theorem 638
    in arithmetic progressions
        Dirichlet's theorem 613, 643
    Jacobi symbol 642
    largest known 644
    Legendre symbol 642
    Mersenne prime 644, 648
    prime number theorem 638, 643, 644
    quadratic reciprocity law 642
    relation to logarithmic integral 155
    tables 639, 649
primes .... *see* prime numbers.
primitive Dirichlet characters
    relation to generalized Bernoulli polynomials 597
principal branches .... *see* principal values.
principal values 104
    .... *see also* Cauchy principal values.
    closed definition 104
principle of the argument .... *see* phase principle.
Pringsheim's theorem for continued fractions 25

probability distribution
   symmetric elliptic integrals......................515
probability functions....................160, 167, 308
   Gaussian..160
   normal..160
   relations to other functions
      error functions..............................160
         parabolic cylinder functions..................308
         repeated integrals of the complementary error function...167, 308
problème des ménages.............................633
projective coordinates.............................581
projective quantum numbers
   $3j$ symbols.......................................758
prolate spheroidal coordinates.....................704
   Laplacian.......................................705
   metric coefficients..............................704
Prym's functions..................................174
pseudoperiodic solutions
   of Hill's equation...............................674
   of Mathieu's equation.....................653, 664
pseudoprime test..................................644
pseudorandom numbers...........................648
psi function.......................................**136**
   analytic properties.............................136
   applications
      mathematical...............................145
   approximations
      Chebyshev series...........................147
      complex variable............................147
      rational..................................146, 147
   asymptotic expansion..........................140
   computation...................................146
   continued fractions.............................140
   definition......................................136
   expansions in partial fractions...................139
   graphics....................................136, 137
   inequalities....................................138
   integral representations........................140
   multiplication formula.........................138
   notation.......................................136
   recurrence relation.............................138
   reflection formula..............................138
   relation to hypergeometric function.............387
   special values..................................137
   tables...146
   Taylor series...................................139
   zeros..136
      asymptotic approximation..................138
      table of...................................138
public key codes..................................647
punctured neighborhood............................19

q-beta function...................................145
q-factorials......................................145
q-gamma function................................145
q-Appell functions...............................423
   transformations...............................430
q-Bernoulli polynomials..........................422
q-binomial coefficient.......................421, 627
q-binomial series.................................423
q-binomial theorem.........................421, 424
q-calculus..................................420–422
q-cosine function................................422
q-derivatives....................................421
q-differential equations..........................425
q-Dyson conjecture..............................431
q-elementary functions.....................422, 432
q-Euler numbers.................................422
q-exponential function...........................422
   applications...................................432
q-hypergeometric function......................**420**
   Andrews–Askey sum.....................424, 426
   Andrews' q-Dyson conjecture..................431
   applications
      mathematical...............................432
      physical...................................432
   Bailey chain...................................430
   Bailey lemma
      strong....................................430
      weak.....................................430
   Bailey pairs....................................430
   Bailey transform...............................430
   Bailey's $_2F_1(-1)$ sum
      q-analog..................................426
   Bailey's $_4F_3(1)$ sum
      q-analogs (first and second).................427
   Bailey's transformation of very-well-poised $_8\phi_7$..429
   Bailey–Daum q-Kummer sum..................424
   balanced series................................423
   bibasic sums and series........................429
   bilateral.... see bilateral q-hypergeometric function.
   Cauchy's sum..................................424
   Chu–Vandermonde sums (first and second)
      q-analogs..................................424
   computation...................................432
   constant term identities........................431
   contiguous relations (Heine's)..................425
   continued fractions.............................426
   definition......................................423
   differential equations...........................425
   Dixon's $_3F_2(1)$ sum
      q-analog..................................426
   Dixon's sum
      F. H. Jackson's q-analog....................426
   Dougall's $_7F_6(1)$ sum
      F. H. Jackson's q-analog....................427
   Euler's sums (first, second, third).........423, 424
   F. H. Jackson's transformations................428

Fine's transformations (first, second, third) 424
Gauss's $_2F_1(-1)$ sum
    q-analog 426
    generalizations 432
Heine's transformations (first, second, third) 424
idem function 420, 429
integral representations 426
integrals 431
k-balanced series 423
mixed base Heine-type transformations 429
nearly-poised 423
notation 420
q-Pfaff–Saalschütz sum 426
q-Saalschütz sum
    nonterminating form 426
q-Sheppard identity 428
quintuple product identity 427
Ramanujan's integrals 431
relations to other functions
    Askey–Wilson class orthogonal polynomials
    472–474
        q-Hahn class orthogonal polynomials 470–472
Rogers–Fine identity 424
Saalschützian series 423
Sears' balanced $_4\phi_3$ transformation 428
special cases 426
three-term $_2\phi_1$ transformation 425
transformations 428
Vandermonde sum
    nonterminating q-version 425
very-well-poised 423
well-poised 423
Zeilberger–Bressoud theorem 431
q-Hahn class orthogonal polynomials 470–472
    as eigenvalues of q-difference operator 470
    asymptotic approximations 474
    orthogonality properties 470
    relation to q-hypergeometric function 470–472
q-Hahn polynomials 470
q-hypergeometric orthogonal polynomials 470
q-integrals 422
q-Laguerre polynomials 471
    applications 432
    asymptotic approximations to zeros 474
q-Leibniz rule 421
q-multinomial coefficient 634
q-Pochhammer symbol 436
q-product 436
q-Racah polynomials 474
    applications
        coding theory 479
    relation to q-hypergeometric function 474
q-series
    classification 423

q-sine function 422
q-Stirling numbers 422
q^{-1}-Al-Salam–Chihara polynomials 473
quadratic characters
    number theory 642
quadratic equations 23
quadratic reciprocity law
    number theory 642
quadrature 78–84
    contour integrals 83–84
    interpolatory rules (or formulas)
                       see also Gauss quadrature.
        Clenshaw–Curtis 79
        closed 79
        error term 79
        Fejér's 79
        midpoint 79
        Newton–Cotes 79
        nodes 79
        open 79
        weight function 79
    oscillatory integrals
        Clenshaw–Curtis formula (extended) 82
        Filon's rule 82
        Longman's method 82
        multidimensional 82
    Romberg integration 79
    Simpson's rule
        composite 78, 79
        elementary 79
    steepest-descent paths 83–84
    trapezoidal rule
        composite 78, 79, 84
        elementary 78, 79
        improved 78
    via classical orthogonal polynomials 478
quantum chemistry
    generalized exponential integral 190
    incomplete gamma functions 189
quantum chromo-dynamics
    hypergeometric function 400
quantum field theory
    modular functions 582
    Riemann zeta function 614
quantum gravity
    Painlevé transcendents 739
quantum groups
    q-series 432
quantum mechanics
    associated Legendre functions 379
    canonical integrals 791
    classical orthogonal polynomials 479
    Heun functions 720
    Mathieu functions 679

nonrelativistic
    gamma function 145
    parabolic cylinder functions 317
    Struve functions 298
    Whittaker functions 346
quantum probability distributions
    Euler polynomials 598
quantum scattering
    hypergeometric function 400
quantum spin models
    Painlevé transcendents 739
quantum spins
    Heun's equation 720
quantum systems
    Heun's equation 720
quantum wave-packets
    theta functions 534
quark-gluon plasma
    Bernoulli polynomials 598
quartic equations 23
quartic oscillator
    Jacobian elliptic functions 565
quasiconformal mapping
    complete elliptic integrals 399
    hypergeometric function 399
queueing theory
    incomplete gamma functions 189
quintic equations
    modular functions 582
quotient-difference algorithm 95
    rhombus rule 95
    stability 95
quotient-difference scheme 95

Raabe's theorem
    Bernoulli polynomials 590
Racah polynomials 407
    .... see Wilson class orthogonal polynomials.
radial Mathieu functions **668**
    .... see also modified Mathieu functions.
    definitions 668
    expansions in series of Bessel functions 670
    expansions in series of cross-products of Bessel functions and modified Bessel functions 671–672
    graphics 669
    integral representations 672–674
        compendia 674
        of cross-products 674
    joining factors 652, 669
    notation 652
    relation to modified Mathieu functions 668
    shift of variable 668
radial spheroidal wave functions **703**
    applications 706
    asymptotic behavior for large variable 703

computation 708
connection formulas 703
connection with spheroidal wave functions 704
definitions 703
graphics 703
integral representation 704
tables 708
Wronskian 703
radiative equilibrium
    generalized exponential integral 190
Radon transform
    classical orthogonal polynomials 479
railroad track design
    Cornu's spiral 169
rainbow
    Airy functions 209
Ramanujan's $_1\psi_1$ summation
    bilateral q-hypergeometric function 427
Ramanujan's beta integral 143
Ramanujan's cubic transformation
    hypergeometric function 393
Ramanujan's partition identity
    number theory 646
Ramanujan's sum
    number theory 643
Ramanujan's tau function
    number theory 646–647
random graphs
    generalized hypergeometric functions 417
random matrix theory
    Hermite polynomials 479
    Painlevé transcendents 739
random walks 417
rational arithmetics 72
    exact 72
rational functions
    summation 145
Rayleigh function 240
    applications 276
R_C-function **487**
    asymptotic approximations 496
    limiting values 491
    relation to elementary functions 495
    relation to Gudermannian function 495
    relation to inverse Gudermannian function 491
    special values 491
reduced Planck's constant 379, 479, 753
reduced residue system
    number theory 638
reductions of partial differential equations
    Painlevé transcendents 738
Regge poles
    Coulomb functions 754
Regge symmetries

$6j$ symbols	762
$3j$ symbols	759
region	15
regularization	
distributional methods	55
relative error	73
relative precision	73
relativistic Coulomb equations	754
relaxation times for proteins	
incomplete gamma functions	189
Remez's second algorithm	
minimax rational approximations	98
removable singularity	19
repeated integrals of the complementary error function	**167**
applications	169
asymptotic expansions	167
computation	169
continued fractions	167
definition	167
derivatives	167
differential equation	167
graphics	167
power-series expansion	167
recurrence relations	167
relations to other functions	
confluent hypergeometric functions	167
Hermite polynomials	167
parabolic cylinder functions	167, 308
probability functions	167, 308
scaled	167
tables	169
representation theory	
partitions	635
repulsive potentials	
Coulomb functions	753, 754
residue	19
theorem	19
resistive MHD instability theory	
Struve functions	298
resolvent cubic equation	23
resonances	
Coulomb functions	754
restricted integer partitions	
Bessel-function expansion	628
conjugate	626
generating functions	627
identities	628
limiting form	627, 628
notation	626, 627
recurrence relations	627, 628
relation to lattice paths	626
tables	626, 627
resurgence	
asymptotic solutions of differential equations	57
reversion of series	43
Riccati–Bessel functions	240
zeros	240
Riemann hypothesis	606
equivalent statements	613, 614, 644
Riemann identity	
Riemann theta functions	542
Riemann theta functions with characteristics	542
Riemann matrix	538
computation	546
Riemann surface	**543**
connection with Riemann theta functions	543, 546
cycles	543
definition	543
genus	543
handle	543
holomorphic differentials	543
hyperelliptic	544
intersection indices	538, 543
prime form	544
representation via Hurwitz system	546
representation via plane algebraic curve	546
representation via Schottky group	546
Riemann theta functions	**538**
analytic properties	538
applications	543–546
components	538
computation	546
definition	538
dimension	538
genus	538
graphics	539–541
modular group	542
modular transformations	541–542
notation	538
period lattice	539
products	542
quasi-periodicity	539
relation to classical theta functions	539
Riemann identity	542
scaled	538, 546
symmetry	539
Riemann theta functions with characteristics	**539**
addition formulas	543
applications	
Abelian functions	545
characteristics	539
half-period	539
modular transformations	542
notation	538
quasi-periodicity	539
Riemann identity	542
symmetry	539

Riemann zeta function............................**602**
    analytic properties.............................602
    applications
        mathematical................................613
        physical......................................614
    approximations.................................615
        asymptotic..................................606
    Chebyshev series...............................615
    computation....................................614
    connection with incomplete gamma functions....189
    critical line....................................606
    critical strip...................................606
    definition......................................602
    derivatives.....................................602
        integer arguments...........................605
        series expansions............................602
    Euler-product representation....................640
    graphics..603
    incomplete.....................................189
    infinite products...............................602
    integer argument...............................605
    integral representations
        along the real line...........................604
        contour integrals............................605
    integrals..606
    notation..602
    recursion formulas.............................605
    reflection formulas.............................603
    relations to other functions
        Bernoulli and Euler numbers and polynomials
        ....598, 605
        Hurwitz zeta function.......................607
        polylogarithms..............................611
    representations by Euler–Maclaurin formula.....602
    series expansions..............................602
    sums..606
    tables...614
    zeros
        computation.................................614
        counting................................607, 614
        distribution..................................606
        on critical line or strip..................606, 614
        relation to quantum eigenvalues..............614
        Riemann hypothesis.........................606
        trivial..606
Riemann's ξ-function............................603
    approximations.................................615
Riemann's differential equation
    general form...................................396
    reduction to hypergeometric differential equation
   396
    singularities...................................396
    solutions
        P-symbol notation..........................396
    transformations................................396
Riemann–Hilbert problems
    classical orthogonal polynomials.................479
Riemann–Lebesgue lemma..........................14
Riemann–Siegel formula...........................607
    coefficients.....................................614
Riemann's P-symbol..............................396
ring functions...................*see* toroidal functions.
Ritt's theorem
    differentiation of asymptotic approximations.....42
robot trajectory planning
    Cornu's spiral..................................169
Rodrigues formulas
    classical orthogonal polynomials.................442
    Hahn class orthogonal polynomials..............462
Rogers polynomials
    ....*see* continuous q-ultraspherical polynomials.
Rogers–Ramanujan identities..................422, 430
    constant term..................................431
    partitions......................................628
Rogers–Szegö polynomials..........................475
rolling of ships
    Mathieu functions..............................679
rook polynomial...................................633
roots
    of equations.....................................90
Rossby waves
    biconfluent Heun functions.....................720
rotation matrices
    relation to $3j$ symbols..........................761
Rouché's theorem..............................20, 92
round-robin tournaments..........................648
Runge–Kutta methods
    ordinary differential equations................89–90
Rutherford scattering
    Coulomb functions.............................753
    gamma function................................145
Rydberg constant
    Coulomb functions.............................754

S-matrix scattering
    Coulomb functions.............................754
saddle points......................................47
    coalescing...............................48, 789–790
sampling expansions
    parabolic cylinder functions....................317
scaled gamma function.............................185
scaled Riemann theta functions
    computation...................................546
    definition......................................538
scaled spheroidal wave functions..............706–707
    bandlimited....................................706
    extremal properties.............................707
    Fourier transform...............................706
    integral equation................................706

INDEX 941

orthogonality 706
scaling laws
    for diffraction catastrophes 785
scattering problems
    associated Legendre functions 379
    Coulomb functions 753–755
scattering theory
    Mathieu functions.............................. 679
Schläfli's integrals
    Bessel functions 224, 225
Schläfli–Sommerfeld integrals
    Bessel and Hankel functions 224
Schläfli-type integrals
    Kelvin functions 269
Schottky group
    Riemann surface 546
Schottky problem
    Riemann surface 545
Schröder numbers
    definition 622
    generating function 623
    relation to lattice paths 622
    table 622
Schrödinger equation
    Airy functions 209
    Coulomb functions 753–755
    nonlinear
        Jacobian elliptic functions 565
        Riemann theta functions 545
    q-deformed quantum mechanical 432
    solutions in terms of classical orthogonal polynomials
        479
    theta functions 534
Schwarz reflection principle 19
Schwarz's lemma 20
Schwarzian derivative 27
Scorer functions **204**
    analytic properties 204
    applications 209
    approximations
        expansions in Chebyshev series 212
    asymptotic expansions 205
    computation 210
    computation by quadrature 84
    connection formulas 205
    definition 204
    differential equation 204
        initial values 204
        numerically satisfactory solutions 204
        standard solutions 204
    graphs 204
    integral representations 204–205
    integrals
        asymptotic expansions 206

tables 211
    Maclaurin series 205
    notation 194
    tables 211
    zeros 206
        computation 210
        tables 211
secant function .... *see* trigonometric functions.
sectorial harmonics 378
Selberg integrals
    generalized elliptic integrals 516
Selberg-type integrals
    gamma function 143
separable Gauss sum
    number theory 643
Shanks' transformation
    for sequences 93
ship wave 790–791
sieve of Eratosthenes
    prime numbers 648
sigma function .... *see* Weierstrass elliptic functions.
signal analysis
    spheroidal wave functions 706–707
simple closed contour 16
simple closed curve 11
simple discontinuity 4
simple zero 19
simply-connected domain 25
Sinc function 77
sine function .... *see* trigonometric functions.
sine integrals **150**
    applications
        Gibbs phenomenon 154
        physical 155
    approximations 156
    asymptotic expansions 153
        exponentially-improved 154
    auxiliary functions ... *see* auxiliary functions for sine and cosine integrals.
    Chebyshev-series expansions 156–157
    computation 155
    definition 150
    expansion in spherical Bessel functions 153
    generalized 188–189
    graphics 151
    hyperbolic analog 150
    integral representations 152
    integrals 154
    Laplace transform 154
    maxima and minima 155
    notation 150
    power series 151
    relations to exponential integrals 151
    sums 154

tables..156
value at infinity.............................150
zeros...154
    asymptotic expansion.......................154
    computation.................................156
sine-Gordon equation
    Jacobian elliptic functions......................565
    Painlevé transcendents.......................739
singularities
    movable..724
singularity
    branch point.....................................20
    essential..19
    isolated...19
    isolated essential................................19
    pole...19
    removable....................................4, 19
$6j$ symbols..**761**
    addition theorem..............................763
    applications...................................765
    approximations for large parameters............764
    computation...................................765
    definition.....................................761
        alternative.................................763
    generating functions 763
    graphical method 765
    notation......................................758
    orthogonality..................................763
    recursion relations 762
    Regge symmetries.............................762
    representation as
        finite sum of algebraic quantities.............762
        finite sum of $3j$ symbols....................761
        generalized hypergeometric functions..........761
    special cases..................................762
    sum rules.....................................763
    summation convention.........................760
    sums..763
    symmetry.....................................762
    tables...765
    zeros..765
$SL(2,\mathbb{Z})$ bilinear transformation....................579
Sobolev polynomials...................................477
soliton theory
    classical orthogonal polynomials.................478
solitons
    Jacobian elliptic functions......................565
    Weierstrass elliptic functions...................582
spatio-temporal dynamics
    Heun functions................................718
spectral problems
    Heun's equation...............................720
    separation of variables 720
spherical Bessel functions...............................**262**

addition theorems.............................267
analytic properties............................262
applications
    electromagnetic scattering....................276
    Helmholtz equation..........................276
    wave equation...............................276
approximations................................281
asymptotic approximations for large order........*see*
    uniform asymptotic expansions for large order
computation..............................276–277
continued fractions............................266
cross-products................................265
definitions....................................262
derivatives...................................265
    zeros....................................266, 280
differential equations..........................262
    numerically satisfactory solutions.............262
    singularities.................................262
    standard solutions...........................262
Dirac delta....................................38
duplication formulas...........................267
explicit formulas
    modified functions...........................264
    sums or differences of squares................264
    unmodified functions 264
generating functions 266
graphs..262
integral representations........................266
integrals......................................267
    computation................................278
interrelations.................................262
limiting forms................................265
modified.....................................262
notation......................................217
of the first, second, and third kinds 262
power series..................................265
Rayleigh's formulas 264
recurrence relations 265
reflection formulas 262
sums...267
    addition theorems...........................267
    compendia..................................267
    duplication formulas.........................267
tables..280
uniform asymptotic expansions for large order...266
Wronskians...................................265
zeros...266
spherical Bessel transform........................278
    computation..................................278
spherical coordinates 8
spherical harmonics...............................**378**
    addition theorem..............................379
    applications...................................379
    basic properties...........................378–379

definitions 378
Dirac delta 38
distributional completeness 379
Lamé polynomials 694
relation to $3j$ symbols 760
sums 379
zonal 479
spherical polar coordinates .. *see* spherical coordinates.
spherical triangles
    solution of 131
spherical trigonometry
    Jacobian elliptic functions 564
sphero-conal coordinates 693
spheroidal coordinates .... *see* oblate spheroidal coordinates *and* prolate spheroidal coordinates.
spheroidal differential equation **698**
    eigenvalues 698–699
        asymptotic behavior 702–703
        computation 707
        continued-fraction equation 699
        graphics 700
        power-series expansion 699
        tables 708
    Liouville normal form 698
    singularities 698
    special cases 698
    with complex parameter 700
spheroidal harmonics
    oblate 378
    prolate 378
spheroidal wave functions **698**
    addition theorem 703
    applications
        signal analysis 706–707
        wave equation 704–706
    approximations 703
    as confluent Heun functions 717
    asymptotic behavior
        as $x \to \pm 1$ 703
        for large $|\gamma^2|$ 702–703
    computation 707–708
    convolutions 703
    Coulomb 704
    definitions 699, 700
    differential equation 698
    eigenvalues 698
    elementary properties 699
    expansions in series of Ferrers functions 702
        asymptotic behavior of coefficients 702
        tables of coefficients 708
    expansions in series of spherical Bessel functions .. 703
    Fourier transform 706
    generalized 704
    graphics 700–701

integral equations 703, 706
integrals 703
notation 698
oblate angular 699
of complex argument 700
of the first kind 699
of the second kind 700
orthogonality 699
other notations 698
power-series expansions 699
products 703
prolate angular 699
radial 703
scaled 706
tables 708
with complex parameters 700
zeros 699
spline functions
    Bernoulli monosplines 597
    cardinal monosplines 597
    cardinal splines 597
    Euler splines 597
splines
    Bézier curves 100
    definitions 100
square-integrable function 6
stability problems
    Mathieu functions 679
stable polynomials 23
    Hurwitz criterion 23
statistical analysis
    multivariate
        functions of matrix argument 773
statistical applications
    functions of matrix argument 773
statistical mechanics
    application to combinatorics 635
    Heun functions 720
    incomplete beta functions 189
    Jacobian elliptic functions 564
    modular functions 582
    q-hypergeometric function 432
    solvable models 146
    theta functions 533
statistical physics
    Bernoulli and Euler polynomials 598
    Painlevé transcendents 739
Steed's algorithm
    for continued fractions 96
steepest-descent paths
    numerical integration 83–84
Stickelberger codes
    Bernoulli numbers 598
Stieltjes fraction (S-fraction) 95

Stieltjes polynomials
    definition 718
    orthogonality 719
    products 719
    zeros 718
        electrostatic interpretation 719
Stieltjes transform
    analyticity 30
    asymptotic expansions 52–53
    convergence 29
    definition 29, 52
    derivatives 30
    generalized 53
    inversion 30
    representation as double Laplace transform 30
Stieltjes–Wigert polynomials 471
    asymptotic approximations 474
Stirling cycle numbers 631
Stirling numbers (first and second kinds)
    asymptotic approximations 626
    definitions 624
    generalized 626
    generating functions 624
    identities 625
    notations 618
    recurrence relations 625
    relations to Bernoulli numbers 596
    special values 625
    tables 624, 635
Stirling's formula 141
Stirling's series 141
Stokes line 68
Stokes multipliers 57
Stokes phenomenon 67
    complementary error function 164
    incomplete gamma functions 189
    smoothing of 67
Stokes sets 782–785
    cuspoids 783
    definitions 782
    umbilics 783
    visualizations 784–785
Stokes' theorem for vector-valued functions 12
string theory
    beta function 146
    elliptic integrals 517
    modular functions 582
    Painlevé transcendents 739
    Riemann theta functions 545
    theta functions 533
Struve functions .... *see* Struve functions and modified Struve functions.
Struve functions and modified Struve functions ... **288**
    analytic continuation 291

applications
    physical 298
    approximations 300
    argument $xe^{\pm 3\pi i/4}$ 294
    asymptotic expansions
        generalized 293
        large argument 293
        large order 293
        remainder terms 293
    computation 299
    definitions 288
    derivatives 292
        with respect to order 292
    differential equations 288
        numerically satisfactory solutions 288
        particular solutions 288
    graphics 289–291
    half-integer orders 291
    incomplete 300
    inequalities 291
    integral representations
        along real line 292
        compendia 293
        contour integrals 292
        Mellin–Barnes type 293
    integrals
        compendia 294
        definite 294
        indefinite 293–294
        Laplace transforms 294
        products 294
        tables 299
        with respect to order 294
    Kelvin-function analogs 294
    notation 288
    order 288
    power series 288
    principal values 288
    recurrence relations 292
    relations to Anger–Weber functions 297
    series expansions
        Bessel functions 292
        Chebyshev 300
        power series 288
    sums 294
    tables 299
    zeros 292
Struve's equation ... *see* Struve functions and modified Struve functions, differential equations.
Sturm–Liouville eigenvalue problems
    ordinary differential equations 89
summability methods for integrals
    Abel 34
    Cesàro 34

Index

Fourier integrals
    conjugate Poisson integral 34
    Fejér kernel 34
    Poisson integral 34
    Poisson kernel 34
fractional derivatives 35
fractional integrals 35
summability methods for series
    Abel 33
    Borel 33
    Cesàro 33
    general 33
    convergence 33
Fourier series
    Abel means 33
    Cesàro means 33
    Fejér kernel 33
    Poisson kernel 33
    regular 33
    Tauberian theorems 35
summation by parts 63
summation formulas
    Boole 597
    Euler–Maclaurin 597
sums of powers
    as Bernoulli or Euler polynomials 589
    tables 598
supersonic flow
    Lamé polynomials 694
support
    of a function 35
surface .... *see* parametrized surfaces.
surface harmonics of the first kind 378
surface-wave problems
    Struve functions 298
swallowtail bifurcation set
    formula 781
    picture 782
swallowtail canonical integral **776**
    asymptotic approximations 789–790
    convergent series 787
    differential equations 788
    formulas for Stokes set 783
    integral identities 787
    picture of Stokes set 784
    pictures of modulus 778
    scaling laws 785
    zeros 787
swallowtail catastrophe **776**, 784
symmetric elliptic integrals **497**
    addition theorems 509–510
    advantages of symmetry 497
    applications
        mathematical 514–516
        physical 516–517
        statistical 515
    arithmetic-geometric mean 505
    asymptotic approximations and expansions
        .... 53, 510–511
    Bartky's transformation 504
    change of parameter of R_J 504
    circular cases 502–504
    complete 486
    computation 517–519
    connection formulas 503
    degree 498
    derivatives 500
    differential equations 501
    duplication formulas 510
    elliptic cases of $R_{-a}(\mathbf{b};\mathbf{z})$ 498
    first, second, and third kinds 486
    Gauss transformations 497, 505
    general lemniscatic case 502, 503
    graphics 499–500
    hyperbolic cases 502–504
    inequalities
        complete integrals 506–507
        incomplete integrals 507
    integral representations 506
    integrals of 511
    Landen transformations 497, 505
    notation 486
    permutation symmetry 497, 498
    power-series expansions 501–502
    reduction of general elliptic integrals . 512–514
    relations to other functions
        Appell functions 509
        Bulirsch's elliptic integrals 508
        hypergeometric function 509
        Jacobian elliptic functions 508
        Lauricella's function 509
        Legendre's elliptic integrals 507, 508
        theta functions 508
        Weierstrass elliptic functions 509
    special cases 502–503
    tables 519
    transformations replaced by symmetry
        .... 497, 505, 508
symmetries
    of canonical integrals 777
Szegő–Askey polynomials 475
Szegő–Szász inequality
    Jacobi polynomials 451
tangent function .... *see* trigonometric functions.
tangent numbers 596
    tables 596
Taylor series 18
    asymptotic approximations for coefficients .. 65

Taylor's theorem
    one variable 6, 18
    two variables 8
tempered distributions 36, 52
    convergence 36
    Fourier transform 37
term-by-term integration 18
terminant function 68
    incomplete gamma functions 189
tesseral harmonics 378
test functions
    distributions 35
Theorem of Ince
    Mathieu's equation 653, 657
theta functions **524**
    addition formulas 531
    applications
        mathematical 533
        physical 533
    computation 534
    derivatives 529–530
        of ratios 531
    discrete analog 532
    double products 530
    duplication formula 531
    Fourier series 524
    fundamental parallelogram 524
    generalizations 532
    graphics
        complex variables 527–529
        real variables 525–527
    infinite products 529–530
    integrals 532
    Jacobi's identity 529
    Jacobi's inversion formula 532, 533
    Jacobi's original notation 524
    Jacobi's triple product 529
    Landen transformation 531
    Laplace transform with respect to lattice parameter
        532
    lattice parameter 524
        transformation of 531
    lattice points 524
    limit forms as $\Im \tau \to 0+$ 534
    McKean and Moll's 524
    Mellin transform with respect to lattice parameter
        532
    modular transformations 531
    multidimensional .... see Chapter 21.
    Neville's 524, 550
    nome 524
        rectangular case 524
        transformation of 531
    notation 524
    periodicity 524
    power series 530
    quasi-periodicity 524
    Ramanujan's 533
    Ramanujan's change of base 533
    rectangular case 524
    relations to other functions
        Dedekind's eta function 525
        elliptic integrals 532
        elliptic modular function 532
        Jacobi's epsilon function 562
        Jacobian elliptic functions 532, 550
        modular functions 579
        Riemann zeta function 532
        symmetric elliptic integrals 508
        Weierstrass elliptic functions 532, 574
    Riemann 538
    Riemann with characteristics 539
    sums of squares 530
    tables 534
    translation by half-periods 525
    values at $z = 0$ 529
    Watson's expansions 531
    Watson's identities 531
    with characteristics 533
    zeros 525
Thomae transformation
    $_3F_2$ functions of matrix argument 772
$3j, 6j, 9j$ symbols
    relation to generalized hypergeometric functions
        .... 407, 418
$3j$ symbols **758**
    angular momenta 758
    applications 765
    approximations for large parameters 764
    computation 765
    definition 758
    Gaunt coefficient 761
    Gaunt's integral 761
    generating functions 760
    graphical method 765
    notation 758
    orthogonality 760
    projective quantum numbers 758
    recursion relations 760
    Regge symmetries 759
    relations to other functions
        Legendre functions 760
        rotation matrices 761
        spherical harmonics 760
    representation as
        finite sum of algebraic quantities 758
        generalized hypergeometric functions 758
    special cases 759

summation convention 760
sums 760
symmetry 759
tables 765
triangle conditions 758
zeros 765
Toda equation
    Hermite polynomials 478
tomography
    confluent hypergeometric functions 346
tops
    Jacobian elliptic, or hyperelliptic, integrals 566
toroidal coordinates 371, 379
toroidal functions **371**
    applications 379
    definitions 371
    hypergeometric representations 371
    integral representations 371
    sums 372
    Whipple's formula 372
torus
    complex 533
transcendental equations
    asymptotic solutions 43
transcendental functions 724
transition points 58, 63
transport equilibrium
    generalized exponential integral 190
triangle conditions
    $3j$ symbols 758
triangle inequality 15
triangles
    solution of 130
triangular matrices
    confluent hypergeometric functions 345
triconfluent Heun equation 718
trigonometric functions **112**
    addition formulas 117
    analytic properties 112
    applications
        cubic equations 131
        solution of triangles and spherical triangles ... 130
    approximations 132
    Chebyshev-series expansions 132
    computation 131
    conformal maps 113
    continued fractions 121
    definitions 112
    derivatives 117
    differential equations 117
    elementary properties 115–116
    graphics
        complex argument 113–115
        real argument 112

    identities 117
    inequalities 116
    infinite products 118
    integrals
        definite 122
        indefinite 122
    inverse .... *see* inverse trigonometric functions.
    Laurent series 116
    limits 116
    Maclaurin series 116
    moduli 118
    multiples of argument 118
    notation 104
    orthogonality 122
    partial fractions 118
    periodicity 112
    poles 123
    real and imaginary parts 118
    relations to hyperbolic functions 123
    special values 116
    squares and products 117
    sums 123
    tables 132
    zeros 112
triple integrals 9
truncated exponential series 180
turning points 58, 63
    fractional or multiple 61
two-body relativistic scattering
    Lamé polynomials 694

ultraspherical polynomials **438**
    .... *see also* classical orthogonal polynomials.
    addition theorem 459
    applications
        zonal spherical harmonics 479
    asymptotic approximations 452
    case $\lambda = 0$ 437
    computation 479
    definition 439
    derivatives 446
    differential equation 445
    expansions in series of 460, 461
    Fourier transforms 456
    generating functions 449
    inequalities 450
    integral representations 447, 448
        for products 455
    integrals 456
    interrelations with other orthogonal polynomials
        .... 444–445, 448
    leading coefficients 439
    limits to monomials 444
    linearization formula 460
    Mellin transform 458

normalization 439
notation 436, 437
orthogonality property 439
parameter constraint 439, 443
recurrence relations 446
relations to other functions
    Ferrers functions 448
    hypergeometric function 393, 442
Rodrigues formula 442
special values 444
symmetry 444
tables of coefficients 440
upper bound 450
weight function 439
zeros 438, 454
umbilics
    normal forms 776
umbral calculus
    Bernoulli and Euler polynomials 590
uniformization
    algebraic equations via Jacobian elliptic functions
    564
unity
    roots of 23
vacuum magnetic fields
    toroidal functions 379
validated computing 72
Van Vleck polynomials
    definition 718
    zeros 718
Van Vleck's theorem for continued fractions 25
Vandermondian 3
variation of parameters
    inhomogeneous differential equations 26
variation of real or complex functions 6
    bounded 6
    total 6
variational operator 44
vector
    equivalent 542
    norms 74
vector-valued functions 9–12
    .... see also parametrized surfaces.
    curl 10
    del operator 10
    divergence 10
    divergence (or Gauss's) theorem 12
    gradient 10
    Green's theorem
        three dimensions 12
        two dimensions 11
    line integral 11
    path integral 11
    reparametrization of integration paths

    orientation-preserving 11
    orientation-reversing 11
    Stokes' theorem 12
vectors 9
    .... see also vector-valued functions.
angle 9
cross product 9
    right-hand rule 10
dot product 9
Einstein summation convention 10
Levi-Civita symbol 10
magnitude 9
notations 9, 10
right-hand rule for cross products 10
scalar product .... see dot product.
unit 9
vector product .... see cross product.
vibrational problems
    Mathieu functions 678
Voigt functions
    applications 169
    computation 169
    definition 167
    graphs 168
    properties 168
    relation to line broadening function 167
    tables 169
von Staudt–Clausen theorem
    Bernoulli numbers 593
Voronoi's congruence
    Bernoulli numbers 593
Waring's problem
    number theory 645
water waves
    Kelvin's ship-wave pattern 790–791
    Riemann theta functions 545
    Struve functions 298
Watson integrals
    Appell functions 417
    generalized hypergeometric functions 417
Watson's $_3F_2$ sum
    Andrews' terminating q-analog 427
    Gasper–Rahman q-analog 426
Watson's expansions
    theta functions 531
Watson's identities
    theta functions 531
Watson's lemma
    asymptotic expansions of integrals 44, 46
Watson's sum
    generalized hypergeometric functions 406
wave acoustics
    generalized exponential integral 190
wave equation

.... *see also* water waves.
 Bessel functions and modified Bessel functions .. 276
 confluent hypergeometric functions 346
 ellipsoidal coordinates 693
 Mathieu functions 678
 oblate spheroidal coordinates 705–706
 paraboloidal coordinates 346
 prolate spheroidal coordinates 704–705
 separation constants 693
 spherical Bessel functions 276
 sphero-conal coordinates 693
 symmetric elliptic integrals 501
wave functions
 paraboloidal 677
waveguides 275
Weber function .... *see* Anger–Weber functions.
Weber parabolic cylinder functions
 .... *see* parabolic cylinder functions.
Weber's function
 .... *see* Bessel functions of the second kind.
Weber–Schafheitlin discontinuous integrals
 Bessel functions 244
Weierstrass M-test
 .... *see* M-test for uniform convergence.
Weierstrass elliptic functions **570**
 addition theorems 577
 analytic properties 570
 applications
  mathematical 581
  physical 582–583
 asymptotic approximations 578
 computation 583
 definitions 570
 derivatives 571
 differential equations 571
 discriminant 571
 duplication formulas 578
 equianharmonic case 571–572, 574
 Fourier series 576
 graphics
  complex variables 573–574
  real variables 571–572
 homogeneity 578
 infinite products 577
 integral representations 578
 integrals 579
 lattice 570
  computation 583
  equianharmonic 574
  generators 570
  invariants 571
  lemniscatic 571–572, 574
  notation 570
  points 570
  pseudo-lemniscatic 574
  rectangular 574
  rhombic 574
  roots 571
 Laurent series 577
 lemniscatic case 571–572, 574
 n-tuple formulas 578
 notation 570
 periodicity 571
 poles 570
 power series 577
 principal value 577
 pseudo-lemniscatic case 574
 quarter periods 576
 quasi-periodicity 571
 relations to other functions
  elliptic integrals 576
  general elliptic functions 576
  Jacobian elliptic functions 575
  symmetric elliptic integrals 509
  theta functions 574
 rhombic case 574
 series of cosecants or cotangents 577
 tables 584
 zeros 570, 579
Weierstrass $\wp$-function
 .... *see* Weierstrass elliptic functions.
Weierstrass product 22
Weierstrass sigma function
 .... *see* Weierstrass elliptic functions.
Weierstrass zeta function
 .... *see* Weierstrass elliptic functions.
weight functions
 cubature 84–85
 definition 79, 437
 Freud 475
 least squares approximations 99
 logarithmic 81–82
 minimax rational approximations 97
 quadrature 79–80
weighted means 3
Weniger's transformation
 for sequences 94
Whipple's $_3F_2$ sum
 Gasper–Rahman q-analog 427
Whipple's formula
 associated Legendre functions 362
 toroidal functions 372
Whipple's sum
 generalized hypergeometric functions 406
Whipple's theorem
 Watson's q-analog 429
Whipple's transformation
 generalized hypergeometric functions 407

Whittaker functions **334**
        see also confluent hypergeometric functions.
    addition theorems 345
    analytic continuation 334
    analytical properties 334
    applications
        Coulomb functions 346
        groups of triangular matrices 345
        physical 346, 754
        uniform asymptotic solutions of differential equations 345
    asymptotic approximations for large parameters
        imaginary κ and/or μ 340
        large κ 341–342
        large μ 339–341
        uniform 339–342
    asymptotic expansions for large argument 339
        error bounds 339
        exponentially-improved 339
    computation 346
    connection formulas 335
    continued fractions 338
    definitions 334
    derivatives 336
    differential equation .... *see* Whittaker's equation.
    expansions in series of 344
    integral representations
        along the real line 337
        contour integrals 337
        Mellin–Barnes type 337
    integral transforms in terms of 344
    integrals 337
        compendia 344
        Fourier transforms 343
        Hankel transforms 343–344
        Laplace transforms 343
        Mellin transforms 343
    interrelations 335
    large argument 69
    limiting forms
        as $z \to 0$ 335
        as $z \to \infty$ 335
    multiplication theorems 345
    notation 322
    power series 334
    principal branches (or values) 334
    products 345
    recurrence relations 336
    relations to other functions
        Airy functions 338
        Coulomb functions 742, 748, 751
        elementary functions 338
        error functions 338
        incomplete gamma functions 338
        Kummer functions 334
        modified Bessel functions 338
        orthogonal polynomials 338
        parabolic cylinder functions 338
    series expansions 344–345
        addition theorems 345
        in Bessel functions or modified Bessel functions 344
        multiplication theorems 345
        power 334
    Wronskians 335
    zeros
        asymptotic approximations 343
        distribution 342
        inequalities 343
        number of 343
Whittaker's equation 334
    fundamental solutions 335
    numerically satisfactory solutions 335
    relation to Kummer's equation 334
    standard solutions 334
Whittaker–Hill equation 676
    applications 678
    separation constants 678
Wigner $3j, 6j, 9j$ symbols
       see $3j$ symbols, $6j$ symbols, and $9j$ symbols.
Wilf–Zeilberger algorithm
    applied to generalized hypergeometric functions .. 407
Wilkinson's polynomial 92
Wilson class orthogonal polynomials 467–470
    asymptotic approximations 470
    definitions 467
    differences 469
    dualities 463
    generating functions 469
    interrelations with other orthogonal polynomials
        .... 464, 468–469
    leading coefficients 468
    normalizations 467–468
    notation 436
    orthogonality properties 467
    relation to generalized hypergeometric functions
        468–469
    transformations of variable 467
    weight functions 467–468
Wilson polynomials
        see Wilson class orthogonal polynomials.
winding number
    of closed contour 16
WKB or WKBJ approximation
    see Liouville–Green (or WKBJ) approximation.
Wronskian
    differential equations 26
Wynn's cross rule

for Padé approximations 98
Wynn's epsilon algorithm
    for sequences. 93

zero potential
    Coulomb functions 753, 754
zeros of analytic functions
    computation. 90–92
    conditioning 92
    multiplicity. 19, 90
    simple 90
zeros of Bessel functions (including derivatives)
    analytic properties 235
    approximations 281
    asymptotic expansions for large order
        uniform. 237
    asymptotic expansions for large zeros 236
        error bounds. 236
    bounds 236
    common 235
    complex. 235, 238
    computation 277
    distribution. 235, 238–240
    double 235
    interlacing 235
    monotonicity. 236
    notation 235
    of cross-products. 238
        asymptotic expansions 238
    purely imaginary. 235, 236
    relation to inverse phase functions. 235
    tables. 132, 278
    with respect to order (ν-zeros) 240
zeros of cylinder functions (including derivatives)
    235–237
    analytic properties 235
    asymptotic expansions for large order
        uniform. 236

    asymptotic expansions for large zeros 236
    forward differences. 235
    interlacing 235
    monotonicity. 236
    relation to inverse phase functions. 235
zeros of polynomials
    . . . . *see also* stable polynomials.
    computation. 91–92
    conditioning 92
    degrees two, three, four 23
    Descartes' rule of signs 22
    discriminant 22
    distribution 22
    division algorithm. 22
    elementary properties 22
    elementary symmetric functions 22
    explicit formulas 91
    Horner's scheme. 22
        extended. 22
    resolvent cubic 23
    roots of constants 23
    roots of unity 23
zeta function . . . . *see* Hurwitz zeta function,
    Jacobi's zeta function, periodic zeta function, Riemann zeta function, *and* Weierstrass zeta function.
zonal polynomials. **769**
    applications. 773
    beta integral 769
    definition 769
    Laplace integral 769
    mean-value 769
    normalization 769
    notation 769
    orthogonality 769
    summation. 769
    tables. 773
zonal spherical harmonics
    ultraspherical polynomials 479